Physics

for Scientists and Engineers with Modern Physics
Hybrid Edition

Raymond A. Serway

Emeritus, James Madison University

John W. Jewett, Jr.

Emeritus, California State Polytechnic University, Pomona

With contributions from Vahé Peroomian,
University of California at Los Angeles

About the Cover

The cover shows a view inside the new railway departures concourse opened in March 2012 at the Kings Cross Station in London. The wall of the older structure (completed in 1852) is visible at the left. The sweeping shell-like roof is claimed by the architect to be the largest single-span station structure in Europe. Many principles of physics are required to design and construct such an open semicircular roof with a radius of 74 meters and containing over 2 000 triangular panels. Other principles of physics are necessary to develop the lighting design, optimize the acoustics, and integrate the new structure with existing infrastructure, historic buildings, and railway platforms.

© Ashley Cooper/Corbis

BROOKS/COLE
CENGAGE Learning

Australia • Brazil • Japan • Korea • Mexico • Singapore • Spain • United Kingdom • United States

BROOKS/COLE
CENGAGE Learning

Physics for Scientists and Engineers with Modern Physics, Ninth Edition
HYBRID Edition
Raymond A. Serway and John W. Jewett, Jr.

Publisher, Physical Sciences: Mary Finch

Publisher, Physics and Astronomy:
    Charlie Hartford

Development Editor: Ed Dodd

Assistant Editor: Brandi Kirksey

Editorial Assistant: Brendan Killion

Media Editor: Rebecca Berardy Schwartz

Brand Manager: Nicole Hamm

Marketing Communications Manager: Linda Yip

Senior Marketing Development Manager:
    Tom Ziolkowski

Content Project Manager: Alison Eigel Zade

Senior Art Director: Cate Barr

Manufacturing Planner: Sandee Milewski

Rights Acquisition Specialist:
    Shalice Shah-Caldwell

Production Service: Lachina Publishing Services

Text and Cover Designer: Roy Neuhaus

Cover Image: The new Kings Cross railway station, London, UK

Cover Image Credit: © Ashley Cooper/Corbis

Compositor: Lachina Publishing Services

For product information and technology assistance, contact us at
Cengage Learning Customer & Sales Support, 1-800-354-9706

For permission to use material from this text or product,
submit all requests online at **www.cengage.com/permissions**
Further permissions questions can be emailed to
permissionrequest@cengage.com

Library of Congress Control Number: 2012947242

ISBN-13: 978-1-133-95398-2

ISBN-10: 1-133-95398-0

Brooks/Cole
20 Channel Center Street
Boston, MA 02210
USA

Cengage Learning is a leading provider of customized learning solutions with office locations around the globe, including Singapore, the United Kingdom, Australia, Mexico, Brazil, and Japan. Locate your local office at **www.cengage.com/global**.

Cengage Learning products are represented in Canada by Nelson Education, Ltd.

For your course and learning solutions, visit **www.cengage.com**.

Purchase any of our products at your local college store or at our preferred online store **www.CengageBrain.com**.

Instructors: Please visit **login.cengage.com** and log in to access instructor-specific resources.

We dedicate this book to our wives,
Elizabeth and Lisa, and all our children and
grandchildren for their loving understanding
when we spent time on writing
instead of being with them.

Printed in the United States of America
1 2 3 4 5 6 7 17 16 15 14 13

Brief Contents

Contents

PART 2
Oscillations and Mechanical Waves 351

PART 3
Thermodynamics 443

PART 4
Electricity and Magnetism 539

PART 5
Light and Optics 823

About the Authors

Raymond A. Serway received his doctorate at Illinois Institute of Technology and is Professor Emeritus at James Madison University. In 2011, he was awarded with an honorary doctorate degree from his alma mater, Utica College. He received the 1990 Madison Scholar Award at James Madison University, where he taught for 17 years. Dr. Serway began his teaching career at Clarkson University, where he conducted research and taught from 1967 to 1980. He was the recipient of the Distinguished Teaching Award at Clarkson University in 1977 and the Alumni Achievement Award from Utica College in 1985. As Guest Scientist at the IBM Research Laboratory in Zurich, Switzerland, he worked with K. Alex Müller, 1987 Nobel Prize recipient. Dr. Serway also was a visiting scientist at Argonne National Laboratory, where he collaborated with his mentor and friend, the late Dr. Sam Marshall. Dr. Serway is the coauthor of *College Physics,* Ninth Edition; *Principles of Physics,* Fifth Edition; *Essentials of College Physics; Modern Physics,* Third Edition; and the high school textbook *Physics,* published by Holt McDougal. In addition, Dr. Serway has published more than 40 research papers in the field of condensed matter physics and has given more than 60 presentations at professional meetings. Dr. Serway and his wife, Elizabeth, enjoy traveling, playing golf, fishing, gardening, singing in the church choir, and especially spending quality time with their four children, ten grandchildren, and a recent great grandson.

John W. Jewett, Jr. earned his undergraduate degree in physics at Drexel University and his doctorate at Ohio State University, specializing in optical and magnetic properties of condensed matter. Dr. Jewett began his academic career at Richard Stockton College of New Jersey, where he taught from 1974 to 1984. He is currently Emeritus Professor of Physics at California State Polytechnic University, Pomona. Through his teaching career, Dr. Jewett has been active in promoting effective physics education. In addition to receiving four National Science Foundation grants in physics education, he helped found and direct the Southern California Area Modern Physics Institute (SCAMPI) and Science IMPACT (Institute for Modern Pedagogy and Creative Teaching). Dr. Jewett's honors include the Stockton Merit Award at Richard Stockton College in 1980, selection as Outstanding Professor at California State Polytechnic University for 1991–1992, and the Excellence in Undergraduate Physics Teaching Award from the American Association of Physics Teachers (AAPT) in 1998. In 2010, he received an Alumni Lifetime Achievement Award from Drexel University in recognition of his contributions in physics education. He has given more than 100 presentations both domestically and abroad, including multiple presentations at national meetings of the AAPT. He has also published 25 research papers in condensed matter physics and physics education research. Dr. Jewett is the author of *The World of Physics: Mysteries, Magic, and Myth,* which provides many connections between physics and everyday experiences. In addition to his work as the coauthor for *Physics for Scientists and Engineers,* he is also the coauthor on *Principles of Physics,* Fifth Edition, as well as *Global Issues,* a four-volume set of instruction manuals in integrated science for high school. Dr. Jewett enjoys playing keyboard with his all-physicist band, traveling, underwater photography, learning foreign languages, and collecting antique quack medical devices that can be used as demonstration apparatus in physics lectures. Most importantly, he relishes spending time with his wife, Lisa, and their children and grandchildren.

Preface

About the Book

Why a hybrid text? Many traditional lecture-based courses are evolving into courses for which all homework and tests are delivered online. In addition, with the rapid growth of distance learning courses, there is an even greater need for course materials that blend traditional print resources with rich media-based tools. A hybrid text is designed to address the needs of these courses through the integration of both print and online components. For this hybrid edition, the end-of-chapter problems have been removed from the text and are available exclusively online in Enhanced WebAssign, an easy-to-use online homework system. In Enhanced WebAssign there are additional resources to help students master the course content: PreLecture Explorations, problems with targeted feedback, Analysis Model tutorials, Master It tutorials, and Watch It tutorials; all can help guide students to success in solving physics problems!

In writing this Ninth Edition of *Physics for Scientists and Engineers,* we continue our ongoing efforts to improve the clarity of presentation and include new pedagogical features that help support the learning and teaching processes. Drawing on positive feedback from users of the Eighth Edition, data gathered from both professors and students who use Enhanced WebAssign, as well as reviewers' suggestions, we have refined the text to better meet the needs of students and teachers.

This textbook is intended for a course in introductory physics for students majoring in science or engineering. The entire contents of the book in its extended version could be covered in a three-semester course, but it is possible to use the material in shorter sequences with the omission of selected chapters and sections. The mathematical background of the student taking this course should ideally include one semester of calculus. If that is not possible, the student should be enrolled in a concurrent course in introductory calculus.

Content

The material in this book covers fundamental topics in classical physics and provides an introduction to modern physics. The book is divided into six parts. Part 1 (Chapters 1 to 14) deals with the fundamentals of Newtonian mechanics and the physics of fluids; Part 2 (Chapters 15 to 18) covers oscillations, mechanical waves, and sound; Part 3 (Chapters 19 to 22) addresses heat and thermodynamics; Part 4 (Chapters 23 to 34) treats electricity and magnetism; Part 5 (Chapters 35 to 38) covers light and optics; and Part 6 (Chapters 39 to 46) deals with relativity and modern physics.

Objectives

This introductory physics textbook has three main objectives: to provide the student with a clear and logical presentation of the basic concepts and principles of physics, to strengthen an understanding of the concepts and principles through a broad range of interesting real-world applications, and to develop strong problem-solving skills through an effectively organized approach. To meet these objectives, we emphasize well-organized physical arguments and a focused problem-solving strategy. At the same time, we attempt to motivate the student through practical examples that demonstrate the role of physics in other disciplines, including engineering, chemistry, and medicine.

Changes in the Ninth Edition

A large number of changes and improvements were made for the Ninth Edition of this text. Some of the new features are based on our experiences and on current trends in science education. Other changes were incorporated in response to comments and suggestions offered by users of the Eighth Edition and by reviewers of the manuscript. The features listed here represent the major changes in the Ninth Edition.

Enhanced Integration of the Analysis Model Approach to Problem Solving. Students are faced with hundreds of problems during their physics courses. A relatively small number of fundamental principles form the basis of these problems. When faced with a new problem, a physicist forms a *model* of the problem that can be solved in a simple way by identifying the fundamental principle that is applicable in the problem. For example, many problems involve conservation of energy, Newton's second law, or kinematic equations. Because the physicist has studied these principles and their applications extensively, he or she can apply this knowledge as a model for solving a new problem. Although it would be ideal for students to follow this same process, most students have difficulty becoming familiar with the entire palette of fundamental principles that are available. It is easier for students to identify a *situation* rather than a fundamental principle.

The *Analysis Model approach* we focus on in this revision lays out a standard set of situations that appear in most physics problems. These situations are based on an entity in one of four simplification models: particle, system, rigid object, and wave. Once the simplification model is identified, the student thinks about what the entity is doing or how it interacts with its environment. This leads the student to identify a particular Analysis Model for the problem. For example, if an object is falling, the object is recognized as a particle experiencing an acceleration due to gravity that is constant. The student has learned that the Analysis Model of a *particle under constant acceleration* describes this situation. Furthermore, this model has a small number of equations associated with it for use in starting problems, the kinematic equations presented in Chapter 2. Therefore, an understanding of the situation has led to an Analysis Model, which then identifies a very small number of equations to start the problem, rather than the myriad equations that students see in the text. In this way, the use of Analysis Models leads the student to identify the fundamental principle. As the student gains more experience, he or she will lean less on the Analysis Model approach and begin to identify fundamental principles directly.

To better integrate the Analysis Model approach for this edition, **Analysis Model descriptive boxes** have been added at the end of any section that introduces a new Analysis Model. This feature recaps the Analysis Model introduced in the section and provides examples of the types of problems that a student could solve using the Analysis Model. These boxes function as a "refresher" before students see the Analysis Models in use in the worked examples for a given section.

Worked examples in the text that utilize Analysis Models are now designated with an **AM** icon for ease of reference. The solutions of these examples integrate the Analysis Model approach to problem solving. The approach is further reinforced in the end-of-chapter summary under the heading *Analysis Models for Problem Solving,* and through the new **Analysis Model Tutorials** that are based on selected end-of-chapter problems and appear in Enhanced WebAssign.

Analysis Model Tutorials. John Jewett developed 165 tutorials (available in Enhanced WebAssign) that strengthen students' problem-solving skills by guiding them through the steps in the problem-solving process. Important first steps include making predictions and focusing on physics concepts before solving the problem quantitatively. A critical component of these tutorials is the selection of an appropriate Analysis Model to describe what is going on in the problem. This step allows students to make the important link between the situation in the problem and the mathematical representation of the situation. Analysis Model tutorials include meaningful feedback at each step to help students practice the problem-solving process and improve their skills. In addition, the feedback addresses student misconceptions and helps them to catch algebraic and other mathematical errors. Solutions are carried out symbolically as long as possible, with numerical values substituted at the end. This feature helps students understand the effects of changing the values of each variable in the problem, avoids unnecessary repetitive substitution of the same numbers, and eliminates round-off errors. Feedback at the end of the tutorial encourages students to compare the final answer with their original predictions.

Annotated Instructor's Edition. New for this edition, the Annotated Instructor's Edition provides instructors with teaching tips and other notes on how to utilize the textbook in the classroom, via cyan annotations.

PreLecture Explorations. The Active Figure questions in Enhanced WebAssign from the Eighth Edition have been completely revised. The simulations have been updated, with additional parameters to enhance investigation of a physical phenomenon. Students can make predictions, change the parameters, and then observe the results. Each new PreLecture Exploration comes with conceptual and analytical questions that guide students to a deeper understanding and help promote a robust physical intuition.

New Master Its Added in Enhanced WebAssign. Approximately 50 new Master Its in Enhanced WebAssign have been added for this edition.

Chapter-by-Chapter Changes

The list below highlights some of the major changes for the Ninth Edition.

Chapter 1

- Two new Master Its were added in Enhanced WebAssign.
- Three new Analysis Model Tutorials were added for this chapter in Enhanced WebAssign.

Chapter 2

- A new introduction to the concept of Analysis Models has been included in Section 2.3.
- Three Analysis Model descriptive boxes have been added, in Sections 2.3 and 2.6.
- Several textual sections have been revised to make more explicit references to analysis models.
- Three new Master Its were added in Enhanced WebAssign.
- Five new Analysis Model Tutorials were added for this chapter in Enhanced WebAssign.

Chapter 3

- Three new Analysis Model Tutorials were added for this chapter in Enhanced WebAssign.

Chapter 4

- An Analysis Model descriptive box has been added, in Section 4.6.
- Several textual sections have been revised to make more explicit references to analysis models.
- Three new Master Its were added in Enhanced WebAssign.
- Five new Analysis Model Tutorials were added for this chapter in Enhanced WebAssign.

Chapter 5

- Two Analysis Model descriptive boxes have been added, in Section 5.7.
- Several examples have been modified so that numerical values are put in only at the end of the solution.
- Several textual sections have been revised to make more explicit references to analysis models.
- Four new Master Its were added in Enhanced WebAssign.
- Four new Analysis Model Tutorials were added for this chapter in Enhanced WebAssign.

Chapter 6

- An Analysis Model descriptive box has been added, in Section 6.1.
- Several examples have been modified so that numerical values are put in only at the end of the solution.
- Four new Analysis Model Tutorials were added for this chapter in Enhanced WebAssign.

Chapter 7

- The notation for work done on a system externally and internally within a system has been clarified.

- The equations and discussions in several sections have been modified to more clearly show the comparisons of similar potential energy equations among different situations.
- One new Master It was added in Enhanced WebAssign.
- Four new Analysis Model Tutorials were added for this chapter in Enhanced WebAssign.

Chapter 8

- Two Analysis Model descriptive boxes have been added, in Sections 8.1 and 8.2.
- The problem-solving strategy in Section 8.2 has been reworded to account for a more general application to both isolated and nonisolated systems.
- As a result of a suggestion from a PER team at University of Washington and Pennsylvania State University, Example 8.1 has been rewritten to demonstrate to students the effect of choosing different systems on the development of the solution.
- All examples in the chapter have been rewritten to begin with Equation 8.2 directly rather than beginning with the format $E_i = E_f$.
- Several examples have been modified so that numerical values are put in only at the end of the solution.
- The problem-solving strategy in Section 8.4 has been deleted and the text material revised to incorporate these ideas on handling energy changes when nonconservative forces act.
- Several textual sections have been revised to make more explicit references to analysis models.
- One new Master It was added in Enhanced WebAssign.
- Four new Analysis Model Tutorials were added for this chapter in Enhanced WebAssign.

Chapter 9

- Two Analysis Model descriptive boxes have been added, in Section 9.3.
- Several examples have been modified so that numerical values are put in only at the end of the solution.
- Five new Master Its were added in Enhanced WebAssign.
- Four new Analysis Model Tutorials were added for this chapter in Enhanced WebAssign.

Chapter 10

- The order of four sections (10.4–10.7) has been modified so as to introduce moment of inertia through torque (rather than energy) and to place the two sections on energy together. The sections have been revised accordingly to account for the revised development of concepts. This revision makes the order of approach similar to the order of approach students have already seen in translational motion.

- New introductory paragraphs have been added to several sections to show how the development of our analysis of rotational motion parallels that followed earlier for translational motion.
- Two Analysis Model descriptive boxes have been added, in Sections 10.2 and 10.5.
- Several textual sections have been revised to make more explicit references to analysis models.
- Two new Master Its were added in Enhanced WebAssign.
- Four new Analysis Model Tutorials were added for this chapter in Enhanced WebAssign.

Chapter 11

- Two Analysis Model descriptive boxes have been added, in Sections 11.2 and 11.4.
- Angular momentum conservation equations have been revised so as to be presented as $\Delta L = (0$ or $\tau dt)$ in order to be consistent with the approach in Chapter 8 for energy conservation and Chapter 9 for linear momentum conservation.
- Four new Analysis Model Tutorials were added for this chapter in Enhanced WebAssign.

Chapter 12

- One Analysis Model descriptive box has been added, in Section 12.1.
- Several examples have been modified so that numerical values are put in only at the end of the solution.
- Four new Analysis Model Tutorials were added for this chapter in Enhanced WebAssign.

Chapter 13

- Sections 13.3 and 13.4 have been interchanged to provide a better flow of concepts.
- A *new* analysis model has been introduced: *Particle in a Field (Gravitational)*. This model is introduced because it represents a physical situation that occurs often. In addition, the model is introduced to anticipate the importance of versions of this model later in electricity and magnetism, where it is even more critical. An Analysis Model descriptive box has been added in Section 13.3. In addition, a new summary flash card has been added at the end of the chapter, and textual material has been revised to make reference to the new model.
- The description of the historical goals of the Cavendish experiment in 1798 has been revised to be more consistent with Cavendish's original intent and the knowledge available at the time of the experiment.
- Newly discovered Kuiper belt objects have been added, in Section 13.4.
- Textual material has been modified to make a stronger tie-in to Analysis Models, especially in the energy sections 13.5 and 13.6.
- All conservation equations have been revised so as to be presented with the change in the system on the left and the transfer across the boundary of the system on the right, in order to be consistent with the approach in earlier chapters for energy conservation, linear momentum conservation, and angular momentum conservation.
- Four new Analysis Model Tutorials were added for this chapter in Enhanced WebAssign.

Chapter 14

- Several textual sections have been revised to make more explicit references to Analysis Models.
- Several examples have been modified so that numerical values are put in only at the end of the solution.
- One new Master It was added in Enhanced WebAssign.
- Four new Analysis Model Tutorials were added for this chapter in Enhanced WebAssign.

Chapter 15

- An Analysis Model descriptive box has been added, in Section 15.2.
- Several textual sections have been revised to make more explicit references to Analysis Models.
- Four new Master Its were added in Enhanced WebAssign.
- Four new Analysis Model Tutorials were added for this chapter in Enhanced WebAssign.

Chapter 16

- A new Analysis Model descriptive box has been added, in Section 16.2.
- Section 16.3, on the derivation of the speed of a wave on a string, has been completely rewritten to improve the logical development.
- Four new Analysis Model Tutorials were added for this chapter in Enhanced WebAssign.

Chapter 17

- One new Master It was added in Enhanced WebAssign.
- Four new Analysis Model Tutorials were added for this chapter in Enhanced WebAssign.

Chapter 18

- Two Analysis Model descriptive boxes have been added, in Sections 18.1 and 18.3.
- Two new Master Its were added in Enhanced WebAssign.
- Four new Analysis Model Tutorials were added for this chapter in Enhanced WebAssign.

Chapter 19

- Several examples have been modified so that numerical values are put in only at the end of the solution.
- One new Master It was added in Enhanced WebAssign.
- Four new Analysis Model Tutorials were added for this chapter in Enhanced WebAssign.

Chapter 20

- Section 20.3 was revised to emphasize the focus on *systems*.
- Five new Master Its were added in Enhanced WebAssign.
- Four new Analysis Model Tutorials were added for this chapter in Enhanced WebAssign.

Chapter 21

- A new introduction to Section 21.1 sets up the notion of *structural models* to be used in this chapter and future chapters for describing systems that are too large or too small to observe directly.
- Fifteen new equations have been numbered, and all equations in the chapter have been renumbered. This new program of equation numbers allows easier and more efficient referencing to equations in the development of kinetic theory.
- The order of Sections 21.3 and 21.4 has been reversed to provide a more continuous discussion of specific heats of gases.
- One new Master It was added in Enhanced WebAssign.
- Four new Analysis Model Tutorials were added for this chapter in Enhanced WebAssign.

Chapter 22

- In Section 22.4, the discussion of Carnot's theorem has been rewritten and expanded, with a new figure added that is connected to the proof of the theorem.
- The material in Sections 22.6, 22.7, and 22.8 has been completely reorganized, reordered, and rewritten. The notion of entropy as a measure of disorder has been removed in favor of more contemporary ideas from the physics education literature on entropy and its relationship to notions such as uncertainty, missing information, and energy spreading.
- Two new Pitfall Preventions have been added in Section 22.6 to help students with their understanding of entropy.
- There is a newly added argument for the equivalence of the entropy statement of the second law and the Clausius and Kelvin–Planck statements in Section 22.8.
- Two new summary flashcards have been added relating to the revised entropy discussion.
- Three new Master Its were added in Enhanced WebAssign.
- Four new Analysis Model Tutorials were added for this chapter in Enhanced WebAssign.

Chapter 23

- A *new* analysis model has been introduced: *Particle in a Field (Electrical)*. This model follows on the introduction of the Particle in a Field (Gravitational) model introduced in Chapter 13. An Analysis Model descriptive box has been added, in Section 23.4. In addition, a new summary flash card has been added at the end of the chapter, and textual material has been revised to make reference to the new model.
- A new What If? has been added to Example 23.9 in order to make a connection to infinite planes of charge, to be further studied in later chapters.
- Several textual sections and worked examples have been revised to make more explicit references to analysis models.
- One new Master It was added in Enhanced WebAssign.
- Four new Analysis Model Tutorials were added for this chapter in Enhanced WebAssign.

Chapter 24

- Section 24.1 has been significantly revised to clarify the geometry of area elements through which electric field lines pass to generate an electric flux.
- Two new figures have been added to Example 24.5 to further explore the electric fields due to single and paired infinite planes of charge.
- Two new Master Its were added in Enhanced WebAssign.
- Four new Analysis Model Tutorials were added for this chapter in Enhanced WebAssign.

Chapter 25

- Sections 25.1 and 25.2 have been significantly revised to make connections to the new particle in a field analysis models introduced in Chapters 13 and 23.
- Example 25.4 has been moved so as to appear after the Problem-Solving Strategy in Section 25.5, allowing students to compare electric fields due to a small number of charges and a continuous charge distribution.
- Two new Master Its were added in Enhanced WebAssign.
- Four new Analysis Model Tutorials were added for this chapter in Enhanced WebAssign.

Chapter 26

- The discussion of series and parallel capacitors in Section 26.3 has been revised for clarity.
- The discussion of potential energy associated with an electric dipole in an electric field in Section 26.6 has been revised for clarity.
- Four new Analysis Model Tutorials were added for this chapter in Enhanced WebAssign.

Chapter 27

- The discussion of the Drude model for electrical conduction in Section 27.3 has been revised to follow the outline of structural models introduced in Chapter 21.
- Several textual sections have been revised to make more explicit references to analysis models.
- Five new Master Its were added in Enhanced WebAssign.
- Four new Analysis Model Tutorials were added for this chapter in Enhanced WebAssign.

Chapter 28

- The discussion of series and parallel resistors in Section 28.2 has been revised for clarity.
- Time-varying charge, current, and voltage have been represented with lowercase letters for clarity in distinguishing them from constant values.
- Five new Master Its were added in Enhanced WebAssign.
- Two new Analysis Model Tutorials were added for this chapter in Enhanced WebAssign.

Chapter 29

- A *new* analysis model has been introduced: *Particle in a Field (Magnetic)*. This model follows on the introduction of the Particle in a Field (Gravitational) model introduced in Chapter 13 and the Particle in a Field (Electrical) model in Chapter 23. An Analysis Model descriptive box has been added, in Section 29.1. In addition, a new summary flash card has been added at the end of the chapter, and textual material has been revised to make reference to the new model.
- One new Master It was added in Enhanced WebAssign.
- Six new Analysis Model Tutorials were added for this chapter in Enhanced WebAssign.

Chapter 30

- Several textual sections have been revised to make more explicit references to analysis models.
- One new Master It was added in Enhanced WebAssign.
- Four new Analysis Model Tutorials were added for this chapter in Enhanced WebAssign.

Chapter 31

- Several textual sections have been revised to make more explicit references to analysis models.
- One new Master It was added in Enhanced WebAssign.
- Four new Analysis Model Tutorials were added for this chapter in Enhanced WebAssign.

Chapter 32

- Several textual sections have been revised to make more explicit references to analysis models.
- Time-varying charge, current, and voltage have been represented with lowercase letters for clarity in distinguishing them from constant values.
- Two new Master Its were added in Enhanced WebAssign.
- Three new Analysis Model Tutorials were added for this chapter in Enhanced WebAssign.

Chapter 33

- Phasor colors have been revised in many figures to improve clarity of presentation.
- Three new Analysis Model Tutorials were added for this chapter in Enhanced WebAssign.

Chapter 34

- Several textual sections have been revised to make more explicit references to analysis models.

- The status of spacecraft related to solar sailing has been updated in Section 34.5.
- Six new Analysis Model Tutorials were added for this chapter in Enhanced WebAssign.

Chapter 35

- Two new Analysis Model descriptive boxes have been added, in Sections 35.4 and 35.5.
- Several textual sections and worked examples have been revised to make more explicit references to analysis models.
- Five new Master Its were added in Enhanced WebAssign.
- Four new Analysis Model Tutorials were added for this chapter in Enhanced WebAssign.

Chapter 36

- The discussion of the Keck Telescope in Section 36.10 has been updated, and a new figure from the Keck has been included, representing the first-ever direct optical image of a solar system beyond ours.
- Five new Master Its were added in Enhanced WebAssign.
- Three new Analysis Model Tutorials were added for this chapter in Enhanced WebAssign.

Chapter 37

- An Analysis Model descriptive box has been added, in Section 37.2.
- The discussion of the Laser Interferometer Gravitational-Wave Observatory (LIGO) in Section 37.6 has been updated.
- Three new Master Its were added in Enhanced WebAssign.
- Four new Analysis Model Tutorials were added for this chapter in Enhanced WebAssign.

Chapter 38

- Four new Master Its were added in Enhanced WebAssign.
- Three new Analysis Model Tutorials were added for this chapter in Enhanced WebAssign.

Chapter 39

- Several textual sections have been revised to make more explicit references to analysis models.
- Sections 39.8 and 39.9 from the Eighth Edition have been combined into one section.
- Five new Master Its were added in Enhanced WebAssign.
- Four new Analysis Model Tutorials were added for this chapter in Enhanced WebAssign.

Chapter 40

- The discussion of the Planck model for blackbody radiation in Section 40.1 has been revised to follow the outline of structural models introduced in Chapter 21.
- The discussion of the Einstein model for the photoelectric effect in Section 40.2 has been revised to

follow the outline of structural models introduced in Chapter 21.

- Several textual sections have been revised to make more explicit references to analysis models.
- Two new Master Its were added in Enhanced WebAssign.
- Two new Analysis Model Tutorials were added for this chapter in Enhanced WebAssign.

Chapter 41

- An Analysis Model descriptive box has been added, in Section 41.2.
- One new Analysis Model Tutorial was added for this chapter in Enhanced WebAssign.

Chapter 42

- The discussion of the Bohr model for the hydrogen atom in Section 42.3 has been revised to follow the outline of structural models introduced in Chapter 21.
- In Section 42.7, the tendency for atomic systems to drop to their lowest energy levels is related to the new discussion of the second law of thermodynamics appearing in Chapter 22.
- The discussion of the applications of lasers in Section 42.10 has been updated to include laser diodes, carbon dioxide lasers, and excimer lasers.
- Several textual sections have been revised to make more explicit references to analysis models.
- Five new Master Its were added in Enhanced WebAssign.
- Three new Analysis Model Tutorials were added for this chapter in Enhanced WebAssign.

Chapter 43

- A new discussion of the contribution of carbon dioxide molecules in the atmosphere to global warming has been added to Section 43.2. A new figure has been added, showing the increasing concentration of carbon dioxide in the past decades.
- A new discussion of graphene (Nobel Prize in Physics, 2010) and its properties has been added to Section 43.4.
- The discussion of worldwide photovoltaic power plants in Section 43.7 has been updated.
- The discussion of transistor density on microchips in Section 43.7 has been updated.
- Several textual sections and worked examples have been revised to make more explicit references to analysis models.
- One new Analysis Model Tutorial was added for this chapter in Enhanced WebAssign.

Chapter 44

- Data for the helium-4 atom were added to Table 44.1.
- Several textual sections have been revised to make more explicit references to analysis models.
- Three new Master Its were added in Enhanced WebAssign.
- Two new Analysis Model Tutorials were added for this chapter in Enhanced WebAssign.

Chapter 45

- Discussion of the March 2011 nuclear disaster after the earthquake and tsunami in Japan was added to Section 45.3.
- The discussion of the International Thermonuclear Experimental Reactor (ITER) in Section 45.4 has been updated.
- The discussion of the National Ignition Facility (NIF) in Section 45.4 has been updated.
- The discussion of radiation dosage in Section 45.5 has been cast in terms of SI units grays and sieverts.
- Section 45.6 from the Eighth Edition has been deleted.
- Four new Master Its were added in Enhanced WebAssign.
- One new Analysis Model Tutorial was added for this chapter in Enhanced WebAssign.

Chapter 46

- A discussion of the ALICE (A Large Ion Collider Experiment) project searching for a quark–gluon plasma at the Large Hadron Collider (LHC) has been added to Section 46.9.
- A discussion of the July 2012 announcement of the discovery of a Higgs-like particle from the ATLAS (A Toroidal LHC Apparatus) and CMS (Compact Muon Solenoid) projects at the Large Hadron Collider (LHC) has been added to Section 46.10.
- A discussion of closures of colliders due to the beginning of operations at the Large Hadron Collider (LHC) has been added to Section 46.10.
- A discussion of recent missions and the new Planck mission to study the cosmic background radiation has been added to Section 46.11.
- Several textual sections have been revised to make more explicit references to analysis models.
- One new Master It was added in Enhanced WebAssign.
- One new Analysis Model Tutorial was added for this chapter in Enhanced WebAssign.

Text Features

Most instructors believe that the textbook selected for a course should be the student's primary guide for understanding and learning the subject matter. Furthermore, the textbook should be easily accessible and should be styled and written to facilitate instruction and learning. With these points in mind, we have included many pedagogical features, listed below, that are intended to enhance its usefulness to both students and instructors.

Problem Solving and Conceptual Understanding

General Problem-Solving Strategy. A general strategy outlined at the end of Chapter 2 (pages 41–42) provides students with a structured process for solving problems. In all remaining chapters, the strategy is employed explicitly in every example so that students learn how it is applied. Students are encouraged to follow this strategy when working end-of-chapter problems.

Worked Examples. All in-text worked examples are presented in a two-column format to better reinforce physical concepts. The left column shows textual information that describes the steps for solving the problem. The right column shows the mathematical manipulations and results of taking these steps. This layout facilitates matching the concept with its mathematical execution and helps students organize their work. The examples closely follow the General Problem-Solving Strategy introduced in Chapter 2 to reinforce effective problem-solving habits. All worked examples in the text may be assigned for homework in Enhanced WebAssign. A sample of a worked example can be found on the next page.

Examples consist of two types. The first (and most common) example type presents a problem and numerical answer. The second type of example is conceptual in nature. To accommodate increased emphasis on understanding physical concepts, the many conceptual examples are labeled as such and are designed to help students focus on the physical situation in the problem. Worked examples in the text that utilize Analysis Models are now designated with an **AM** icon for ease of reference, and the solutions of these examples now more thoroughly integrate the Analysis Model approach to problem solving.

Based on reviewer feedback from the Eighth Edition, we have made careful revisions to the worked examples so that the solutions are presented symbolically as far as possible, with numerical values substituted at the end. This approach will help students think symbolically when they solve problems instead of unnecessarily inserting numbers into intermediate equations.

What If? Approximately one-third of the worked examples in the text contain a What If? feature. At the completion of the example solution, a What If? question offers a variation on the situation posed in the text of the example. This feature encourages students to think about the results of the example, and it also assists in conceptual understanding of the principles. What If? questions also prepare students to encounter novel problems that may be included on exams. Some of the end-of-chapter problems also include this feature.

Quick Quizzes. Students are provided an opportunity to test their understanding of the physical concepts presented through Quick Quizzes. The questions require students to make decisions on the basis of sound reasoning, and some of the questions have been written to help students overcome common misconceptions. Quick Quizzes have been cast in an objective format, including multiple-choice, true–false, and ranking. Answers to all Quick Quiz questions are found at the end of the text. Many instructors choose to use such questions in a "peer instruction" teaching style or with the use of personal response system "clickers," but they can be used in standard quiz format as well. An example of a Quick Quiz follows below.

Quick Quiz 7.5 A dart is inserted into a spring-loaded dart gun by pushing the spring in by a distance x. For the next loading, the spring is compressed a distance $2x$. How much faster does the second dart leave the gun compared with the first? **(a)** four times as fast **(b)** two times as fast **(c)** the same **(d)** half as fast **(e)** one-fourth as fast

Example 3.2   **A Vacation Trip**

A car travels 20.0 km due north and then 35.0 km in a direction 60.0° west of north as shown in Figure 3.11a. Find the magnitude and direction of the car's resultant displacement.

Each solution has been written to closely follow the General Problem-Solving Strategy as outlined on pages 41–42 in Chapter 2, so as to reinforce good problem-solving habits.

SOLUTION

Conceptualize The vectors $\vec{A}$ and $\vec{B}$ drawn in Figure 3.11a help us conceptualize the problem. The resultant vector $\vec{R}$ has also been drawn. We expect its magnitude to be a few tens of kilometers. The angle β that the resultant vector makes with the y axis is expected to be less than 60°, the angle that vector $\vec{B}$ makes with the y axis.

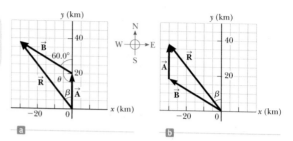

Figure 3.11 (Example 3.2) (a) Graphical method for finding the resultant displacement vector $\vec{R} = \vec{A} + \vec{B}$. (b) Adding the vectors in reverse order $(\vec{B} + \vec{A})$ gives the same result for $\vec{R}$.

Categorize We can categorize this example as a simple analysis problem in vector addition. The displacement $\vec{R}$ is the resultant when the two individual displacements $\vec{A}$ and $\vec{B}$ are added. We can further categorize it as a problem about the analysis of triangles, so we appeal to our expertise in geometry and trigonometry.

Analyze In this example, we show two ways to analyze the problem of finding the resultant of two vectors. The first way is to solve the problem geometrically, using graph paper and a protractor to measure the magnitude of $\vec{R}$ and its direction in Figure 3.11a. (In fact, even when you know you are going to be carrying out a calculation, you should sketch the vectors to check your results.) With an ordinary ruler and protractor, a large diagram typically gives answers to two-digit but not to three-digit precision. Try using these tools on $\vec{R}$ in Figure 3.11a and compare to the trigonometric analysis below!

The second way to solve the problem is to analyze it using algebra and trigonometry. The magnitude of $\vec{R}$ can be obtained from the law of cosines as applied to the triangle in Figure 3.11a (see Appendix B.4).

Each step of the solution is detailed in a two-column format. The left column provides an explanation for each mathematical step in the right column, to better reinforce the physical concepts.

Use $R^2 = A^2 + B^2 - 2AB\cos\theta$ from the law of cosines to find R:

$$R = \sqrt{A^2 + B^2 - 2AB\cos\theta}$$

Substitute numerical values, noting that $\theta = 180° - 60° = 120°$:

$$R = \sqrt{(20.0\text{ km})^2 + (35.0\text{ km})^2 - 2(20.0\text{ km})(35.0\text{ km})\cos 120°}$$
$$= \boxed{48.2\text{ km}}$$

Use the law of sines (Appendix B.4) to find the direction of $\vec{R}$ measured from the northerly direction:

$$\frac{\sin\beta}{B} = \frac{\sin\theta}{R}$$

$$\sin\beta = \frac{B}{R}\sin\theta = \frac{35.0\text{ km}}{48.2\text{ km}}\sin 120° = 0.629$$

$$\beta = \boxed{38.9°}$$

The resultant displacement of the car is 48.2 km in a direction 38.9° west of north.

Finalize Does the angle β that we calculated agree with an estimate made by looking at Figure 3.11a or with an actual angle measured from the diagram using the graphical method? Is it reasonable that the magnitude of $\vec{R}$ is larger than that of both $\vec{A}$ and $\vec{B}$? Are the units of $\vec{R}$ correct?

Although the head to tail method of adding vectors works well, it suffers from two disadvantages. First, some people find using the laws of cosines and sines to be awkward. Second, a triangle only results if you are adding two vectors. If you are adding three or more vectors, the resulting geometric shape is usually not a triangle. In Section 3.4, we explore a new method of adding vectors that will address both of these disadvantages.

WHAT IF?   Suppose the trip were taken with the two vectors in reverse order: 35.0 km at 60.0° west of north first and then 20.0 km due north. How would the magnitude and the direction of the resultant vector change?

Answer They would not change. The commutative law for vector addition tells us that the order of vectors in an addition is irrelevant. Graphically, Figure 3.11b shows that the vectors added in the reverse order give us the same resultant vector.

What If? statements appear in about one-third of the worked examples and offer a variation on the situation posed in the text of the example. For instance, this feature might explore the effects of changing the conditions of the situation, determine what happens when a quantity is taken to a particular limiting value, or question whether additional information can be determined about the problem situation. This feature encourages students to think about the results of the example and assists in conceptual understanding of the principles.

Pitfall Preventions. More than two hundred Pitfall Preventions (such as the one to the left) are provided to help students avoid common mistakes and misunderstandings. These features, which are placed in the margins of the text, address both common student misconceptions and situations in which students often follow unproductive paths.

Summaries. Each chapter contains a summary that reviews the important concepts and equations discussed in that chapter. The summary is divided into three sections: Definitions, Concepts and Principles, and Analysis Models for Problem Solving. In each section, flash card–type boxes focus on each separate definition, concept, principle, or analysis model.

Questions and Problems Sets. For the Ninth Edition, the authors reviewed each question and problem and incorporated revisions designed to improve both readability and assignability. More than 10% of the problems that appear in Enhanced Web-Assign are new to this edition.

Questions. The Questions section is divided into two sections: *Objective Questions* and *Conceptual Questions*. The instructor may select items to assign as homework or use in the classroom, possibly with "peer instruction" methods and possibly with personal response systems. More than 900 Objective and Conceptual Questions are included in this edition. Answers for selected questions are included in the *Student Solutions Manual/Study Guide,* and answers for all questions are found in the *Instructor's Solutions Manual.*

Objective Questions are multiple-choice, true–false, ranking, or other multiple guess–type questions. Some require calculations designed to facilitate students' familiarity with the equations, the variables used, the concepts the variables represent, and the relationships between the concepts. Others are more conceptual in nature and are designed to encourage conceptual thinking. Objective Questions are also written with the personal response system user in mind, and most of the questions could easily be used in these systems.

Conceptual Questions are more traditional short-answer and essay-type questions that require students to think conceptually about a physical situation.

Problems. An extensive set of problems is available in Enhanced WebAssign; in all, more than 3 700 problems. Full solutions for approximately 20% of the problems are included in the *Student Solutions Manual/Study Guide,* and solutions for all problems are found in the *Instructor's Solutions Manual.*

There are several kinds of problems featured in this edition:

Quantitative/Conceptual problems contain parts that ask students to think both quantitatively and conceptually. An example of a Quantitative/Conceptual problem appears here:

59. A horizontal spring attached to a wall has a force constant of $k = 850$ N/m. A block of mass $m = 1.00$ kg is attached to the spring and rests on a frictionless, horizontal surface as in Figure P8.59. (a) The block is pulled to a position $x_i = 6.00$ cm from equilibrium and released. Find the elastic potential energy stored in the spring when the block is 6.00 cm from equilibrium and when the block passes through equilibrium. (b) Find the speed of the block as it passes through the equilibrium point. (c) What is the speed of the block when it is at a position $x_i/2 = 3.00$ cm? (d) Why isn't the answer to part (c) half the answer to part (b)?

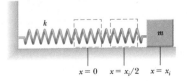

$x = 0$ $x = x_i/2$ $x = x_i$

Figure P8.59

> Parts (a)–(c) of the problem ask
> for quantitative calculations.

> Part (d) asks a conceptual
> question about the situation.

Symbolic problems ask students to solve a problem using only symbolic manipulation. Reviewers of the Eighth Edition (as well as the majority of respondents to a large survey) asked specifically for an increase in the number of symbolic problems found in the text because it better reflects the way instructors want their students to think when solving physics problems. An example of a Symbolic problem appears here:

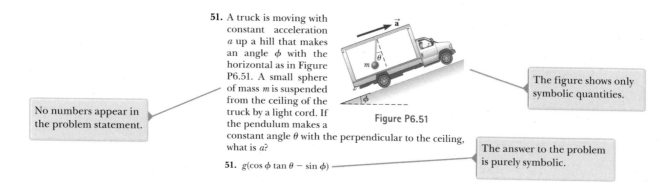

No numbers appear in the problem statement.

51. A truck is moving with constant acceleration a up a hill that makes an angle ϕ with the horizontal as in Figure P6.51. A small sphere of mass m is suspended from the ceiling of the truck by a light cord. If the pendulum makes a constant angle θ with the perpendicular to the ceiling, what is a?

Figure P6.51

The figure shows only symbolic quantities.

51. $g(\cos \phi \tan \theta - \sin \phi)$

The answer to the problem is purely symbolic.

Guided Problems help students break problems into steps. A physics problem typically asks for one physical quantity in a given context. Often, however, several concepts must be used and a number of calculations are required to obtain that final answer. Many students are not accustomed to this level of complexity and often don't know where to start. A Guided Problem breaks a standard problem into smaller steps, enabling students to grasp all the concepts and strategies required to arrive at a correct solution. Unlike standard physics problems, guidance is often built into the problem statement. Guided Problems are reminiscent of how a student might interact with a professor in an office visit. These problems (there is one in every chapter of the text) help train students to break down complex problems into a series of simpler problems, an essential problem-solving skill. An example of a Guided Problem appears here:

38. A uniform beam resting on two pivots has a length $L = 6.00$ m and mass $M = 90.0$ kg. The pivot under the left end exerts a normal force n_1 on the beam, and the second pivot located a distance $\ell = 4.00$ m from the left end exerts a normal force n_2. A woman of mass $m = 55.0$ kg steps onto the left end of the beam and begins walking to the right as in Figure P12.38. The goal is to find the woman's position when the beam begins to tip. (a) What is the appropriate analysis model for the beam before it begins to tip? (b) Sketch a force diagram for the beam, labeling the gravitational and normal forces acting on the beam and placing the woman a distance x to the right of the first pivot, which is the origin. (c) Where is the woman when the normal force n_1 is the greatest? (d) What is n_1 when the beam is about to tip? (e) Use Equation 12.1 to find the value of n_2 when the beam is about to tip. (f) Using the result of part (d) and Equation 12.2, with torques computed around the second pivot, find the woman's position x when the beam is about to tip. (g) Check the answer to part (e) by computing torques around the first pivot point.

The goal of the problem is identified.

Analysis begins by identifying the appropriate analysis model.

Students are provided with suggestions for steps to solve the problem.

The calculation associated with the goal is requested.

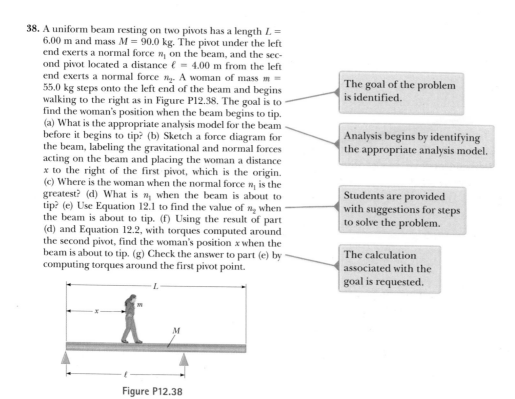

Figure P12.38

Impossibility problems. Physics education research has focused heavily on the problem-solving skills of students. Although most problems in this text are structured in the form of providing data and asking for a result of computation, two problems in each chapter, on average, are structured as impossibility problems. They begin with the phrase *Why is the following situation impossible?* That is followed by the description of a situation. The striking aspect of these problems is that no question is asked of the students, other than that in the initial italics. The student must determine what questions need to be asked and what calculations need to be performed. Based on the results of these calculations, the student must determine why the situation described is not possible. This determination may require information from personal experience, common sense, Internet or print research, measurement, mathematical skills, knowledge of human norms, or scientific thinking.

These problems can be assigned to build critical thinking skills in students. They are also fun, having the aspect of physics "mysteries" to be solved by students individually or in groups. An example of an impossibility problem appears here:

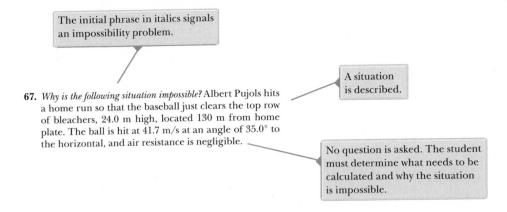

The initial phrase in italics signals an impossibility problem.

67. *Why is the following situation impossible?* Albert Pujols hits a home run so that the baseball just clears the top row of bleachers, 24.0 m high, located 130 m from home plate. The ball is hit at 41.7 m/s at an angle of 35.0° to the horizontal, and air resistance is negligible.

A situation is described.

No question is asked. The student must determine what needs to be calculated and why the situation is impossible.

Paired problems. These problems are otherwise identical, one asking for a numerical solution and one asking for a symbolic derivation. There are now three pairs of these problems in most chapters in Enhanced WebAssign.

Biomedical problems. These problems highlight the relevance of physics principles to those students taking this course who are majoring in one of the life sciences.

Review problems. Many chapters in Enhanced WebAssign include review problems requiring the student to combine concepts covered in the chapter with those discussed in previous chapters. These problems (marked **Review**) reflect the cohesive nature of the principles in the text and verify that physics is not a scattered set of ideas. When facing a real-world issue such as global warming or nuclear weapons, it may be necessary to call on ideas in physics from several parts of a textbook such as this one.

"Fermi problems." One or more problems in most chapters in Enhanced WebAssign ask the student to reason in order-of-magnitude terms.

Design problems. Several chapters in Enhanced WebAssign contain problems that ask the student to determine design parameters for a practical device so that it can function as required.

Calculus-based problems. Every chapter in Enhanced WebAssign contains at least one problem applying ideas and methods from differential calculus and one problem using integral calculus.

Integration with Enhanced WebAssign. The textbook's tight integration with Enhanced WebAssign content facilitates an online learning environment that helps students improve their problem-solving skills and gives them a variety of tools to meet their individual learning styles. Extensive user data gathered by WebAssign were used to ensure that the problems most often assigned were retained for this new edition. New Analysis Model tutorials in Enhanced WebAssign added for this edition have already been discussed (see page x). Master It tutorials help students solve problems by having them work through a stepped-out solution. In addition, Watch It solution videos in Enhanced WebAssign explain fundamental problem-solving strategies to help students step through the problem.

Artwork. Every piece of artwork in the Ninth Edition is in a modern style that helps express the physics principles at work in a clear and precise fashion. *Focus pointers* are included with many figures in the text; these either point out important aspects of a figure or guide students through a process illustrated by the artwork or photo. This format helps those students who are more visual learners. An example of a figure with a focus pointer appears below.

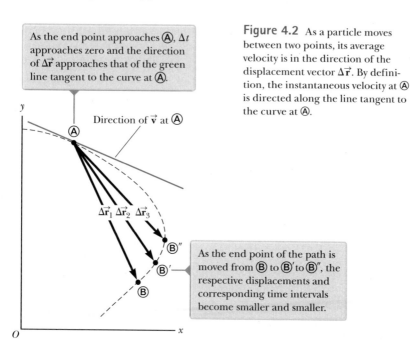

As the end point approaches Ⓐ, Δt approaches zero and the direction of $\Delta \vec{r}$ approaches that of the green line tangent to the curve at Ⓐ.

As the end point of the path is moved from Ⓑ to Ⓑ′ to Ⓑ″, the respective displacements and corresponding time intervals become smaller and smaller.

Figure 4.2 As a particle moves between two points, its average velocity is in the direction of the displacement vector $\Delta \vec{r}$. By definition, the instantaneous velocity at Ⓐ is directed along the line tangent to the curve at Ⓐ.

Math Appendix. The math appendix (Appendix B), a valuable tool for students, shows the math tools in a physics context. This resource is ideal for students who need a quick review on topics such as algebra, trigonometry, and calculus.

Helpful Features

Style. To facilitate rapid comprehension, we have written the book in a clear, logical, and engaging style. We have chosen a writing style that is somewhat informal and relaxed so that students will find the text appealing and enjoyable to read. New terms are carefully defined, and we have avoided the use of jargon.

Important Definitions and Equations. Most important definitions are set in **bold-face** or are highlighted with a background screen for added emphasis and ease of review. Similarly, important equations are also highlighted with a background screen to facilitate location.

Marginal Notes. Comments and notes appearing in the margin with a ▶ icon can be used to locate important statements, equations, and concepts in the text.

Pedagogical Use of Color. Readers should consult the **pedagogical color chart** (inside the front cover) for a listing of the color-coded symbols used in the text diagrams. This system is followed consistently throughout the text.

Mathematical Level. We have introduced calculus gradually, keeping in mind that students often take introductory courses in calculus and physics concurrently. Most steps are shown when basic equations are developed, and reference is often made to mathematical appendices near the end of the textbook. Although vectors are discussed in detail in Chapter 3, vector products are introduced later in the text, where they are needed in physical applications. The dot product is introduced in Chapter 7, which addresses energy of a system; the cross product is introduced in Chapter 11, which deals with angular momentum.

Significant Figures. In both worked examples and problems, significant figures have been handled with care. Most numerical examples are worked to either two or three significant figures, depending on the precision of the data provided. Problems regularly state data and answers to three-digit precision. When carrying out estimation calculations, we shall typically work with a single significant figure. (More discussion of significant figures can be found in Chapter 1, pages 11–13.)

Units. The international system of units (SI) is used throughout the text. The U.S. customary system of units is used only to a limited extent in the chapters on mechanics and thermodynamics.

Appendices and Endpapers. Several appendices are provided near the end of the textbook. Most of the appendix material represents a review of mathematical concepts and techniques used in the text, including scientific notation, algebra, geometry, trigonometry, differential calculus, and integral calculus. Reference to these appendices is made throughout the text. Most mathematical review sections in the appendices include worked examples and exercises with answers. In addition to the mathematical reviews, the appendices contain tables of physical data, conversion factors, and the SI units of physical quantities as well as a periodic table of the elements. Other useful information—fundamental constants and physical data, planetary data, a list of standard prefixes, mathematical symbols, the Greek alphabet, and standard abbreviations of units of measure—appears on the endpapers.

CengageCompose Options for *Physics for Scientists and Engineers*

Would you like to easily create your own personalized text, selecting the elements that meet your specific learning objectives?

CengageCompose puts the power of the vast Cengage Learning library of learning content at your fingertips to create exactly the text you need. The all-new, Web-based CengageCompose site lets you quickly scan content and review materials to pick what you need for your text. Site tools let you easily assemble the modular learning units into the order you want and immediately provide you with an online copy for review. Add enrichment content like case studies, exercises, and lab materials to further build your ideal learning materials. Even choose from hundreds of vivid, art-rich, customizable, full-color covers.

Cengage Learning offers the fastest and easiest way to create unique customized learning materials delivered the way you want. For more information about custom publishing options, visit **www.cengage.com/custom** or contact your local Cengage Learning representative.

Course Solutions That Fit Your Teaching Goals and Your Students' Learning Needs

Recent advances in educational technology have made homework management systems and audience response systems powerful and affordable tools to enhance the way you teach your course. Whether you offer a more traditional text-based course, are interested in using or are currently using an online homework management system such as Enhanced WebAssign, or are ready to turn your lecture into an interactive learning environment with JoinIn, you can be confident that the text's proven content provides the foundation for each and every component of our technology and ancillary package.

Homework Management Systems

Enhanced WebAssign for *Physics for Scientists and Engineers,* **Ninth Edition.** Exclusively from Cengage Learning, Enhanced WebAssign offers an extensive online program for physics to encourage the practice that's so critical for concept mastery. The meticulously crafted pedagogy and exercises in our proven texts become even more effective in Enhanced WebAssign. Enhanced WebAssign includes the Cengage YouBook, a highly customizable, interactive eBook. WebAssign includes:

- **All of the quantitative problems from the Ninth Edition.**
- **Selected problems enhanced with targeted feedback.** An example of targeted feedback appears below:

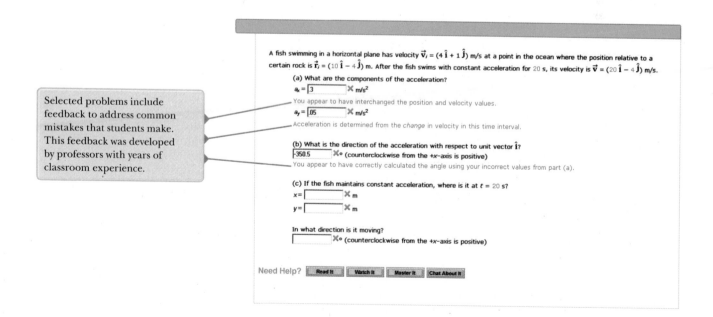

Selected problems include feedback to address common mistakes that students make. This feedback was developed by professors with years of classroom experience.

- **Master It tutorials,** to help students work through the problem one step at a time. An example of a Master It tutorial appears on page xxiv:

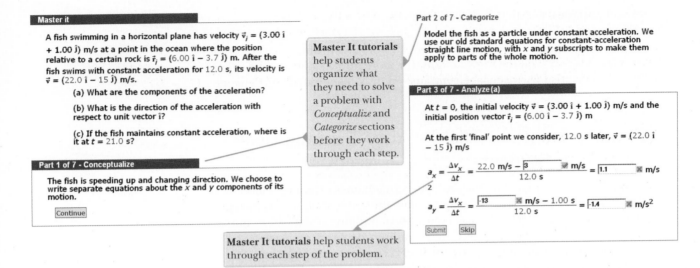

Master it

A fish swimming in a horizontal plane has velocity $\vec{v}_i = (3.00\,\hat{i} + 1.00\,\hat{j})$ m/s at a point in the ocean where the position relative to a certain rock is $\vec{r}_i = (6.00\,\hat{i} - 3.7\,\hat{j})$ m. After the fish swims with constant acceleration for 12.0 s, its velocity is $\vec{v} = (22.0\,\hat{i} - 15\,\hat{j})$ m/s.

(a) What are the components of the acceleration?

(b) What is the direction of the acceleration with respect to unit vector $\hat{i}$?

(c) If the fish maintains constant acceleration, where is it at $t = 21.0$ s?

Master It tutorials help students organize what they need to solve a problem with *Conceptualize* and *Categorize* sections before they work through each step.

Part 1 of 7 - Conceptualize

The fish is speeding up and changing direction. We choose to write separate equations about the x and y components of its motion.

Continue

Master It tutorials help students work through each step of the problem.

Part 2 of 7 - Categorize

Model the fish as a particle under constant acceleration. We use our old standard equations for constant-acceleration straight line motion, with x and y subscripts to make them apply to parts of the whole motion.

Part 3 of 7 - Analyze (a)

At $t = 0$, the initial velocity $\vec{v} = (3.00\,\hat{i} + 1.00\,\hat{j})$ m/s and the initial position vector $\vec{r}_i = (6.00\,\hat{i} - 3.7\,\hat{j})$ m

At the first 'final' point we consider, 12.0 s later, $\vec{v} = (22.0\,\hat{i} - 15\,\hat{j})$ m/s

$$a_x = \frac{\Delta v_x}{\Delta t} = \frac{22.0 \text{ m/s} - \boxed{3}}{12.0 \text{ s}} \boxed{\checkmark}\text{ m/s} = \boxed{1.1} \boxed{\times}\text{ m/s}$$

$$a_y = \frac{\Delta v_x}{\Delta t} = \frac{\boxed{-13} \boxed{\times}\text{ m/s} - 1.00 \text{ s}}{12.0 \text{ s}} = \boxed{-1.4} \boxed{\times}\text{ m/s}^2$$

Submit   Skip

- **Watch It solution videos** that explain fundamental problem-solving strategies, to help students step through the problem. In addition, instructors can choose to include video hints of problem-solving strategies. A screen shot from a Watch It solution video appears below:

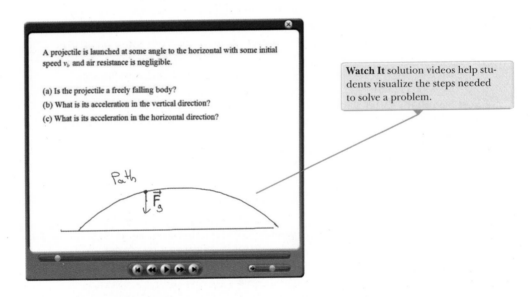

A projectile is launched at some angle to the horizontal with some initial speed v, and air resistance is negligible.

(a) Is the projectile a freely falling body?
(b) What is its acceleration in the vertical direction?
(c) What is its acceleration in the horizontal direction?

Watch It solution videos help students visualize the steps needed to solve a problem.

- **Concept Checks**
- **PhET simulations**
- **Most worked examples,** enhanced with hints and feedback, to help strengthen students' problem-solving skills
- **Every Quick Quiz,** giving your students ample opportunity to test their conceptual understanding
- **PreLecture Explorations.** The Active Figure questions in Enhanced Web-Assign have been completely revised. The simulations have been updated, with additional parameters to enhance investigation of a physical phenomenon. Students can make predictions, change the parameters, and then observe the results. Each new PreLecture Exploration comes with conceptual and analytical questions, which guide students to a deeper understanding and help promote a robust physical intuition.
- **Analysis Model tutorials.** John Jewett has developed 165 tutorials in Enhanced WebAssign that strengthen students' problem-solving skills by guiding them through the steps in the problem-solving process.

Important first steps include making predictions and focusing strategy on physics concepts before starting to solve the problem quantitatively. A critical component of these tutorials is the selection of an appropriate Analysis Model to describe what is going on in the problem. This step allows students to make the important link between the situation in the problem and the mathematical representation of the situation. Analysis Model tutorials include meaningful feedback at each step to help students practice the problem-solving process and improve their skills. In addition, the feedback addresses student misconceptions and helps them to catch algebraic and other mathematical errors. Solutions are carried out symbolically as long as possible, with numerical values substituted at the end. This feature helps students to understand the effects of changing the values of each variable in the problem, avoids unnecessary repetitive substitution of the same numbers, and eliminates round-off errors. Feedback at the end of the tutorial encourages students to think about how the final answer compares to their original predictions.

- **Personalized Study Plan.** The Personal Study Plan in Enhanced WebAssign provides chapter and section assessments that show students what material they know and what areas require more work. For items that they answer incorrectly, students can click on links to related study resources such as videos, tutorials, or reading materials. Color-coded progress indicators let them see how well they are doing on different topics. You decide what chapters and sections to include—and whether to include the plan as part of the final grade or as a study guide with no scoring involved.

- **The Cengage YouBook.** WebAssign has a customizable and interactive eBook, the **Cengage YouBook,** that lets you tailor the textbook to fit your course and connect with your students. You can remove and rearrange chapters in the table of contents and tailor assigned readings that match your syllabus exactly. Powerful editing tools let you change as much as you'd like—or leave it just like it is. You can highlight key passages or add sticky notes to pages to comment on a concept in the reading, and then share any of these individual notes and highlights with your students, or keep them personal. You can also edit narrative content in the textbook by adding a text box or striking out text. With a handy link tool, you can drop in an icon at any point in the eBook that lets you link to your own lecture notes, audio summaries, video lectures, or other files on a personal Web site or anywhere on the Web. A simple YouTube widget lets you easily find and embed videos from YouTube directly into eBook pages. The Cengage YouBook helps students go beyond just reading the textbook. Students can also highlight the text, add their own notes, and bookmark the text. Animations play right on the page at the point of learning so that they're not speed bumps to reading but true enhancements. Please visit **www.webassign.net/brookscole** to view an interactive demonstration of Enhanced WebAssign.

- Offered exclusively in WebAssign, **Quick Prep** for physics is algebra and trigonometry math remediation within the context of physics applications and principles. Quick Prep helps students succeed by using narratives illustrated throughout with video examples. The Master It tutorial problems allow students to assess and retune their understanding of the material. The Practice Problems that go along with each tutorial allow both the student and the instructor to test the student's understanding of the material.

  Quick Prep includes the following features:

  - 67 interactive tutorials
  - 67 additional practice problems
  - A thorough overview of each topic, including video examples
  - Can be taken before the semester begins or during the first few weeks of the course
  - Can also be assigned alongside each chapter for "just in time" remediation

- Topics include units, scientific notation, and significant figures; the motion of objects along a line; functions; approximation and graphing; probability and error; vectors, displacement, and velocity; spheres; force and vector projections.

MindTap™: The Personal Learning Experience

MindTap for Serway and Jewett *Physics for Scientists and Engineers* is a personalized, fully online digital learning platform of authoritative textbook content, assignments, and services that engages your students with interactivity while also offering you choice in the configuration of coursework and enhancement of the curriculum via complimentary Web-apps known as MindApps. MindApps range from ReadSpeaker (which reads the text out loud to students), to Kaltura (allowing you to insert inline video and audio into your curriculum), to ConnectYard (allowing you to create digital "yards" through social media—all without "friending" your students). MindTap is well beyond an eBook, a homework solution or digital supplement, a resource center Web site, a course delivery platform, or a Learning Management System. It is the first in a new category—the Personal Learning Experience.

CengageBrain.com

On **CengageBrain.com** students will be able to save up to 60% on their course materials through our full spectrum of options. Students will have the option to rent their textbooks, purchase print textbooks, e-textbooks, or individual e-chapters and audio books all for substantial savings over average retail prices. **CengageBrain.com** also includes access to Cengage Learning's broad range of homework and study tools and features a selection of free content.

Lecture Presentation Resources

PowerLecture with ExamView® and JoinIn for *Physics for Scientists and Engineers*, Ninth Edition. Bringing physics principles and concepts to life in your lectures has never been easier! The full-featured, two-volume **PowerLecture** Instructor's Resource DVD-ROM (Volume 1: Chapters 1–22; Volume 2: Chapters 23–46) provides everything you need for *Physics for Scientists and Engineers,* Ninth Edition. Key content includes the *Instructor's Solutions Manual,* art and images from the text, pre-made chapter-specific PowerPoint lectures, ExamView test generator software with pre-loaded test questions, JoinIn response-system "clickers," Active Figures animations, and a physics movie library.

JoinIn. Assessing to Learn in the Classroom questions developed at the University of Massachusetts Amherst. This collection of 250 advanced conceptual questions has been tested in the classroom for more than ten years and takes peer learning to a new level. JoinIn helps you turn your lectures into an interactive learning environment that promotes conceptual understanding. Available exclusively for higher education from our partnership with Turning Technologies, JoinIn™ is the easiest way to turn your lecture hall into a personal, fully interactive experience for your students!

Assessment and Course Preparation Resources

A number of resources listed below will assist with your assessment and preparation processes.

Instructor's Solutions Manual by Vahé Peroomian (University of California at Los Angeles). Thoroughly revised for this edition, the *Instructor's Solutions Manual* contains complete worked solutions to all the problems from the Ninth Edition, as well as answers to the even-numbered problems and all the questions. The solutions to problems new to the Ninth Edition are marked for easy identification. Volume 1 contains Chapters 1 through 22; Volume 2 contains Chapters 23 through 46. Electronic files of the *Instructor's Solutions Manual* are available on the PowerLecture™ DVD-ROM.

Test Bank by Ed Oberhofer (University of North Carolina at Charlotte and Lake Sumter Community College). The test bank is available on the two-volume PowerLecture™ DVD-ROM via the ExamView® test software. This two-volume test bank contains approximately 2 000 multiple-choice questions. Instructors may print and duplicate pages for distribution to students. Volume 1 contains Chapters 1 through 22, and Volume 2 contains Chapters 23 through 46. WebCT and Blackboard versions of the test bank are available on the instructor's companion site at **www.CengageBrain.com.**

Instructor's Companion Web Site. Consult the instructor's site by pointing your browser to **www.CengageBrain.com** for a problem correlation guide, PowerPoint lectures, and JoinIn audience response content. Instructors adopting the Ninth Edition of *Physics for Scientists and Engineers* may download these materials after securing the appropriate password from their local sales representative.

Supporting Materials for the Instructor

Supporting instructor materials are available to qualified adopters. Please consult your local Cengage Learning, Brooks/Cole representative for details. Visit **www.CengageBrain.com** to

- request a desk copy
- locate your local representative
- download electronic files of select support materials

Student Resources

Visit the *Physics for Scientists and Engineers* Web site at **www.cengagebrain.com/ shop/ISBN/9781133954156** to see samples of select student supplements. Go to **CengageBrain.com** to purchase and access this product at Cengage Learning's preferred online store.

CENGAGE brain.com

Student Solutions Manual/Study Guide by John R. Gordon, Vahé Peroomian, Raymond A. Serway, and John W. Jewett, Jr. This two-volume manual features detailed solutions to 20% of the problems from the Ninth Edition. The manual also features a list of important equations, concepts, and notes from key sections of the text in addition to answers to selected end-of-chapter questions. Volume 1 contains Chapters 1 through 22; and Volume 2 contains Chapters 23 through 46.

Physics Laboratory Manual, Third Edition by David Loyd (Angelo State University) supplements the learning of basic physical principles while introducing laboratory procedures and equipment. Each chapter includes a prelaboratory assignment, objectives, an equipment list, the theory behind the experiment, experimental procedures, graphing exercises, and questions. A laboratory report form is included with each experiment so that the student can record data, calculations, and experimental results. Students are encouraged to apply statistical analysis to their data. A complete *Instructor's Manual* is also available to facilitate use of this lab manual.

Physics Laboratory Experiments, Seventh Edition by Jerry D. Wilson (Lander College) and Cecilia A. Hernández (American River College). This market-leading manual for the first-year physics laboratory course offers a wide range of class-tested experiments designed specifically for use in small to midsize lab programs. A series of integrated experiments emphasizes the use of computerized instrumentation and includes a set of "computer-assisted experiments" to allow students and instructors to gain experience with modern equipment. This option also enables instructors to determine the appropriate balance between traditional and computer-based experiments for their courses. By analyzing data through two different methods, students gain a greater understanding of the concepts behind the experiments. The Seventh Edition is updated with the latest information and techniques involving state-of-the-art equipment and a new Guided Learning feature addresses the growing interest in guided-inquiry pedagogy. Fourteen additional experiments are also available through custom printing.

 ## Teaching Options

The topics in this textbook are presented in the following sequence: classical mechanics, oscillations and mechanical waves, and heat and thermodynamics, followed by electricity and magnetism, electromagnetic waves, optics, relativity, and modern physics. This presentation represents a traditional sequence, with the subject of mechanical waves being presented before electricity and magnetism. Some instructors may prefer to discuss both mechanical and electromagnetic waves together after completing electricity and magnetism. In this case, Chapters 16 through 18 could be covered along with Chapter 34. The chapter on relativity is placed near the end of the text because this topic often is treated as an introduction to the era of "modern physics." If time permits, instructors may choose to cover Chapter 39 after completing Chapter 13 as a conclusion to the material on Newtonian mechanics. For those instructors teaching a two-semester sequence, some sections and chapters could be deleted without any loss of continuity. The following sections can be considered optional for this purpose:

2.8	Kinematic Equations Derived from Calculus	**31.6**	Eddy Currents
4.6	Relative Velocity and Relative Acceleration	**33.9**	Rectifiers and Filters
6.3	Motion in Accelerated Frames	**34.6**	Production of Electromagnetic Waves by an Antenna
6.4	Motion in the Presence of Resistive Forces	**36.5**	Lens Aberrations
7.9	Energy Diagrams and Equilibrium of a System	**36.6**	The Camera
9.9	Rocket Propulsion	**36.7**	The Eye
11.5	The Motion of Gyroscopes and Tops	**36.8**	The Simple Magnifier
14.7	Other Applications of Fluid Dynamics	**36.9**	The Compound Microscope
15.6	Damped Oscillations	**36.10**	The Telescope
15.7	Forced Oscillations	**38.5**	Diffraction of X-Rays by Crystals
18.6	Standing Waves in Rods and Membranes	**39.9**	The General Theory of Relativity
18.8	Nonsinusoidal Wave Patterns	**41.6**	Applications of Tunneling
25.7	The Millikan Oil-Drop Experiment	**42.9**	Spontaneous and Stimulated Transitions
25.8	Applications of Electrostatics	**42.10**	Lasers
26.7	An Atomic Description of Dielectrics	**43.7**	Semiconductor Devices
27.5	Superconductors	**43.8**	Superconductivity
28.5	Household Wiring and Electrical Safety	**44.8**	Nuclear Magnetic Resonance and Magnetic Resonance Imaging
29.3	Applications Involving Charged Particles Moving in a Magnetic Field	**45.5**	Radiation Damage
29.6	The Hall Effect	**45.6**	Uses of Radiation
30.6	Magnetism in Matter		

 ## Acknowledgments

This Ninth Edition of *Physics for Scientists and Engineers* was prepared with the guidance and assistance of many professors who reviewed selections of the manuscript, the prerevision text, or both. We wish to acknowledge the following scholars and express our sincere appreciation for their suggestions, criticisms, and encouragement:

Benjamin C. Bromley, *University of Utah;* Elena Flitsiyan, *University of Central Florida;* Yuankun Lin, *University of North Texas;* Allen Mincer, *New York University;* Yibin Pan, *University of Wisconsin–Madison;* N. M. Ravindra, *New Jersey Institute of Technology;* Masao Sako, *University of Pennsylvania;* Charles Stone, *Colorado School of Mines;* Robert Weidman, *Michigan Technological University;* Michael Winokur, *University of Wisconsin–Madison*

Prior to our work on this revision, we conducted a survey of professors; their feedback and suggestions helped shape this revision, and so we would like to thank the survey participants:

Elise Adamson, *Wayland Baptist University;* Saul Adelman, *The Citadel;* Yiyan Bai, *Houston Community College;* Philip Blanco, *Grossmont College;* Ken Bolland, *Ohio State University;* Michael Butros, *Victor Valley College;* Brian Carter, *Grossmont College;* Jennifer Cash, *South Carolina State University;* Soumitra Chattopadhyay, *Georgia Highlands College;* John Cooper, *Brazosport College;* Gregory Dolise, *Harrisburg Area Community College;* Mike Durren, *Lake Michigan College;* Tim Farris, *Volunteer State Community College;* Mirela Fetea, *University of Richmond;* Susan Foreman, *Danville Area Community College;* Richard Gottfried, *Frederick Community College;* Christopher Gould, *University of Southern California;* Benjamin Grinstein, *University of California, San Diego;* Wayne Guinn, *Lon Morris College;* Joshua Guttman, *Bergen Community College;* Carlos Handy, *Texas Southern University;* David Heskett, *University of Rhode Island;* Ed Hungerford, *University of Houston;* Matthew Hyre, *Northwestern College;* Charles Johnson, *South Georgia College;* Lynne Lawson, *Providence College;* Byron Leles, *Northeast Alabama Community College;* Rizwan Mahmood, *Slippery Rock University;* Virginia Makepeace, *Kankakee Community College;* David Marasco, *Foothill College;* Richard McCorkle, *University of Rhode Island;* Brian Moudry, *Davis & Elkins College;* Charles Nickles, *University of Massachusetts Dartmouth;* Terrence O'Neill, *Riverside Community College;* Grant O'Rielly, *University of Massachusetts Dartmouth;* Michael Ottinger, *Missouri Western State University;* Michael Panunto, *Butte College;* Eugenia Peterson, *Richard J. Daley College;* Robert Pompi, *Binghamton University, State University of New York;* Ralph Popp, *Mercer County Community College;* Craig Rabatin, *West Virginia University at Parkersburg;* Marilyn Rands, *Lawrence Technological University;* Christina Reeves-Shull, *Cedar Valley College;* John Rollino, *Rutgers University, Newark;* Rich Schelp, *Erskine College;* Mark Semon, *Bates College;* Walther Spjeldvik, *Weber State University;* Mark Spraker, *North Georgia College and State University;* Julie Talbot, *University of West Georgia;* James Tressel, *Massasoit Community College;* Bruce Unger, *Wenatchee Valley College;* Joan Vogtman, *Potomac State College*

This title was carefully checked for accuracy by Grant Hart, *Brigham Young University;* James E. Rutledge, *University of California at Irvine;* and Som Tyagi, *Drexel University.* We thank them for their diligent efforts under schedule pressure.

Belal Abas, Zinoviy Akkerman, Eric Boyd, Hal Falk, Melanie Martin, Steve McCauley, and Glenn Stracher made corrections to problems taken from previous editions. Harvey Leff provided invaluable guidance on the restructuring of the discussion of entropy in Chapter 22. We are grateful to authors John R. Gordon and Vahé Peroomian for preparing the *Student Solutions Manual/Study Guide* and to Vahé Peroomian for preparing an excellent *Instructor's Solutions Manual.* Susan English carefully edited and improved the test bank. Linnea Cookson provided an excellent accuracy check of the Analysis Model tutorials.

Special thanks and recognition go to the professional staff at the Brooks/Cole Publishing Company—in particular, Charles Hartford, Ed Dodd, Stephanie Van-Camp, Rebecca Berardy Schwartz, Tom Ziolkowski, Alison Eigel Zade, Cate Barr, and Brendan Killion (who managed the ancillary program)—for their fine work during the development, production, and promotion of this textbook. We recognize the skilled production service and excellent artwork provided by the staff at Lachina Publishing Services and the dedicated photo research efforts of Christopher Arena at the Bill Smith Group.

Finally, we are deeply indebted to our wives, children, and grandchildren for their love, support, and long-term sacrifices.

Raymond A. Serway
St. Petersburg, Florida

John W. Jewett, Jr.
Anaheim, California

To the Student

It is appropriate to offer some words of advice that should be of benefit to you, the student. Before doing so, we assume you have read the Preface, which describes the various features of the text and support materials that will help you through the course.

 ## Getting the Most Out of the Hybrid Edition of *Physics for Scientists and Engineers,* Ninth Edition

Thank you for purchasing the Hybrid Edition of *Physics for Scientists and Engineers with Modern Physics,* Ninth Edition. The Hybrid Edition is an integrated product, designed specifically to be used with online materials that are essential to your success in your Physics course!

This trimmer version of the book doesn't include end-of-chapter problems. Problems are available online instead, in Enhanced WebAssign, and assignments of these problems may be made by your instructor. In Enhanced WebAssign you will also find additional resources to help you succeed in your course, such as problems with targeted feedback, Analysis Model tutorials, Master It tutorials, and Watch It tutorials; all of which will guide you to success in solving physics problems!

To get started and access the assignments set by your instructor, use the access code included with this text and your instructor's course key to log in at **www.webassign.net.**

 ## How to Study

Instructors are often asked, "How should I study physics and prepare for examinations?" There is no simple answer to this question, but we can offer some suggestions based on our own experiences in learning and teaching over the years.

First and foremost, maintain a positive attitude toward the subject matter, keeping in mind that physics is the most fundamental of all natural sciences. Other science courses that follow will use the same physical principles, so it is important that you understand and are able to apply the various concepts and theories discussed in the text.

 ## Concepts and Principles

It is essential that you understand the basic concepts and principles before attempting to solve assigned problems. You can best accomplish this goal by carefully reading the textbook before you attend your lecture on the covered material. When reading the text, you should jot down those points that are not clear to you. Also be sure to make a diligent attempt at answering the questions in the Quick Quizzes as you come to them in your reading. We have worked hard to prepare questions that help you judge for yourself how well you understand the material. Study the **What If?** features that appear in many of the worked examples carefully. They will help you extend your understanding beyond the simple act of arriving at a numerical result. The Pitfall Preventions will also help guide you away from common misunderstandings about physics. During class, take careful notes and ask questions about those ideas that are unclear to you. Keep in mind that few people are able to absorb the full meaning of scientific material after only one reading; several readings of the text and your notes may be necessary. Your lectures and laboratory work supplement the textbook and should clarify some of the more difficult material. You should minimize your memorization of material. Successful memorization of passages from the text, equations, and derivations does not necessarily indicate that you understand the material. Your understanding of the material will be enhanced through a combination of efficient study habits, discussions with other students and with instructors, and your ability to solve the problems presented in Enhanced WebAssign. Ask questions whenever you believe that clarification of a concept is necessary.

Study Schedule

It is important that you set up a regular study schedule, preferably a daily one. Make sure that you read the syllabus for the course and adhere to the schedule set by your instructor. The lectures will make much more sense if you read the corresponding text material *before* attending them. As a general rule, you should devote about two hours of study time for each hour you are in class. If you are having trouble with the course, seek the advice of the instructor or other students who have taken the course. You may find it necessary to seek further instruction from experienced students. Very often, instructors offer review sessions in addition to regular class periods. Avoid the practice of delaying study until a day or two before an exam. More often than not, this approach has disastrous results. Rather than undertake an all-night study session before a test, briefly review the basic concepts and equations, and then get a good night's rest. If you believe that you need additional help in understanding the concepts, in preparing for exams, or in problem solving, we suggest that you acquire a copy of the *Student Solutions Manual/Study Guide* that accompanies this textbook.

Visit the *Physics for Scientists and Engineers* Web site at **www.cengagebrain.com/shop/ISBN/9781133954156** to see samples of select student supplements. You can purchase any Cengage Learning product at your local college store or at our preferred online store **CengageBrain.com.**

 ## Use the Features

You should make full use of the various features of the text discussed in the Preface. For example, marginal notes are useful for locating and describing important equations and concepts, and **boldface** indicates important definitions. Many useful tables are contained in the appendices, but most are incorporated in the text where they are most often referenced. Appendix B is a convenient review of mathematical tools used in the text.

Answers to Quick Quizzes are given at the end of the textbook, and solutions to selected questions and problems are provided in the *Student Solutions Manual/Study Guide.* The table of contents provides an overview of the entire text, and the index enables you to locate specific material quickly. Footnotes are sometimes used to supplement the text or to cite other references on the subject discussed.

After reading a chapter, you should be able to define any new quantities introduced in that chapter and discuss the principles and assumptions that were used to arrive at certain key relations. The chapter summaries and the review sections of the *Student Solutions Manual/Study Guide* should help you in this regard. In some cases, you may find it necessary to refer to the textbook's index to locate certain topics. You should be able to associate with each physical quantity the correct symbol used to represent that quantity and the unit in which the quantity is specified. Furthermore, you should be able to express each important equation in concise and accurate prose.

 ## Problem Solving

R. P. Feynman, Nobel laureate in physics, once said, "You do not know anything until you have practiced." In keeping with this statement, we strongly advise you to develop the skills necessary to solve a wide range of problems. Your ability to solve problems will be one of the main tests of your knowledge of physics; therefore, you should try to solve as many problems as possible. It is essential that you understand basic concepts and principles before attempting to solve problems. It is good practice to try to find alternate solutions to the same problem. For example, you can solve problems in mechanics using Newton's laws, but very often an alternative method that draws on energy considerations is more direct. You should not deceive yourself into thinking that you understand a problem merely because you have seen it solved in class. You must be able to solve the problem and similar problems on your own.

The approach to solving problems should be carefully planned. A systematic plan is especially important when a problem involves several concepts. First, read the problem several times until you are confident you understand what is being asked. Look for any key words that will help you interpret the problem and perhaps allow you to make certain assumptions. Your ability to interpret a question properly is an integral part of problem solving. Second, you should acquire the habit of writing down the information given in a problem and those quantities that need to be found; for example, you might construct a table listing both the quantities given and the quantities to be found. This procedure is sometimes used in the worked examples of the textbook. Finally, after you have decided on the method you believe is appropriate for a given problem, proceed with your solution. The General Problem-Solving Strategy will guide you through complex problems. If you follow the steps of this procedure (*Conceptualize, Categorize, Analyze, Finalize*), you will find it easier to come up with a solution and gain more from your efforts. This strategy, located at the end of

Chapter 2 (pages 41–42), is used in all worked examples in the remaining chapters so that you can learn how to apply it. Specific problem-solving strategies for certain types of situations are included in the text and appear with a special heading. These specific strategies follow the outline of the General Problem-Solving Strategy.

Often, students fail to recognize the limitations of certain equations or physical laws in a particular situation. It is very important that you understand and remember the assumptions that underlie a particular theory or formalism. For example, certain equations in kinematics apply only to a particle moving with constant acceleration. These equations are not valid for describing motion whose acceleration is not constant, such as the motion of an object connected to a spring or the motion of an object through a fluid. Study the Analysis Models for Problem Solving in the chapter summaries carefully so that you know how each model can be applied to a specific situation. The analysis models provide you with a logical structure for solving problems and help you develop your thinking skills to become more like those of a physicist. Use the analysis model approach to save you hours of looking for the correct equation and to make you a faster and more efficient problem solver.

 # Experiments

Physics is a science based on experimental observations. Therefore, we recommend that you try to supplement the text by performing various types of "hands-on" experiments either at home or in the laboratory. These experiments can be used to test ideas and models discussed in class or in the textbook. For example, the common Slinky toy is excellent for studying traveling waves, a ball swinging on the end of a long string can be used to investigate pendulum motion, various masses attached to the end of a vertical spring or rubber band can be used to determine its elastic nature, an old pair of polarized sunglasses and some discarded lenses and a magnifying glass are the components of various experiments in optics, and an approximate measure of the free-fall acceleration can be determined simply by measuring with a stopwatch the time interval required for a ball to drop from a known height. The list of such experiments is endless. When physical models are not available, be imaginative and try to develop models of your own.

 # New Media

If available, we strongly encourage you to use the **Enhanced WebAssign** product that is available with this textbook. It is far easier to understand physics if you see it in action, and the materials available in Enhanced WebAssign will enable you to become a part of that action.

It is our sincere hope that you will find physics an exciting and enjoyable experience and that you will benefit from this experience, regardless of your chosen profession. Welcome to the exciting world of physics!

The scientist does not study nature because it is useful; he studies it because he delights in it, and he delights in it because it is beautiful. If nature were not beautiful, it would not be worth knowing, and if nature were not worth knowing, life would not be worth living.

—Henri Poincaré

Mechanics

The Honda FCX Clarity, a fuel-cell-powered automobile available to the public, albeit in limited quantities. A fuel cell converts hydrogen fuel into electricity to drive the motor attached to the wheels of the car. Automobiles, whether powered by fuel cells, gasoline engines, or batteries, use many of the concepts and principles of mechanics that we will study in this first part of the book. Quantities that we can use to describe the operation of vehicles include position, velocity, acceleration, force, energy, and momentum. *(PRNewsFoto/American Honda)*

Physics, the most fundamental physical science, is concerned with the fundamental principles of the Universe. It is the foundation upon which the other sciences—astronomy, biology, chemistry, and geology—are based. It is also the basis of a large number of engineering applications. The beauty of physics lies in the simplicity of its fundamental principles and in the manner in which just a small number of concepts and models can alter and expand our view of the world around us.

The study of physics can be divided into six main areas:

1. *classical mechanics,* concerning the motion of objects that are large relative to atoms and move at speeds much slower than the speed of light
2. *relativity,* a theory describing objects moving at any speed, even speeds approaching the speed of light
3. *thermodynamics,* dealing with heat, work, temperature, and the statistical behavior of systems with large numbers of particles
4. *electromagnetism,* concerning electricity, magnetism, and electromagnetic fields
5. *optics,* the study of the behavior of light and its interaction with materials
6. *quantum mechanics,* a collection of theories connecting the behavior of matter at the submicroscopic level to macroscopic observations

The disciplines of mechanics and electromagnetism are basic to all other branches of classical physics (developed before 1900) and modern physics (c. 1900–present). The first part of this textbook deals with classical mechanics, sometimes referred to as *Newtonian mechanics* or simply *mechanics.* Many principles and models used to understand mechanical systems retain their importance in the theories of other areas of physics and can later be used to describe many natural phenomena. Therefore, classical mechanics is of vital importance to students from all disciplines. ∎

Physics and Measurement

Stonehenge, in southern England, was built thousands of years ago. Various theories have been proposed about its function, including a burial ground, a healing site, and a place for ancestor worship. One of the more intriguing theories suggests that Stonehenge was an observatory, allowing measurements of some of the quantities discussed in this chapter, such as position of objects in space and time intervals between repeating celestial events. *(Stephen Inglis/Shutterstock.com)*

WebAssign Interactive content from this and other chapters may be assigned online in Enhanced WebAssign.

Like all other sciences, physics is based on experimental observations and quantitative measurements. The main objectives of physics are to identify a limited number of fundamental laws that govern natural phenomena and use them to develop theories that can predict the results of future experiments. The fundamental laws used in developing theories are expressed in the language of mathematics, the tool that provides a bridge between theory and experiment.

When there is a discrepancy between the prediction of a theory and experimental results, new or modified theories must be formulated to remove the discrepancy. Many times a theory is satisfactory only under limited conditions; a more general theory might be satisfactory without such limitations. For example, the laws of motion discovered by Isaac Newton (1642–1727) accurately describe the motion of objects moving at normal speeds but do not apply to objects moving at speeds comparable to the speed of light. In contrast, the special theory of relativity developed later by Albert Einstein (1879–1955) gives the same results as Newton's laws at low speeds but also correctly describes the motion of objects at speeds approaching the speed of light. Hence, Einstein's special theory of relativity is a more general theory of motion than that formed from Newton's laws.

Classical physics includes the principles of classical mechanics, thermodynamics, optics, and electromagnetism developed before 1900. Important contributions to classical physics were provided by Newton, who was also one of the originators of calculus as a mathematical tool. Major developments in mechanics continued in the 18th century, but the fields of thermodynamics and electromagnetism were not developed until the latter part of the 19th century, principally because before that time the apparatus for controlled experiments in these disciplines was either too crude or unavailable.

A major revolution in physics, usually referred to as *modern physics,* began near the end of the 19th century. Modern physics developed mainly because many physical phenomena could not be explained by classical physics. The two most important developments in this modern era were the theories of relativity and quantum mechanics. Einstein's special theory of relativity not only correctly describes the motion of objects moving at speeds comparable to the speed of light; it also completely modifies the traditional concepts of space, time, and energy. The theory also shows that the speed of light is the upper limit of the speed of an object and that mass and energy are related. Quantum mechanics was formulated by a number of distinguished scientists to provide descriptions of physical phenomena at the atomic level. Many practical devices have been developed using the principles of quantum mechanics.

Scientists continually work at improving our understanding of fundamental laws. Numerous technological advances in recent times are the result of the efforts of many scientists, engineers, and technicians, such as unmanned planetary explorations, a variety of developments and potential applications in nanotechnology, microcircuitry and high-speed computers, sophisticated imaging techniques used in scientific research and medicine, and several remarkable results in genetic engineering. The effects of such developments and discoveries on our society have indeed been great, and it is very likely that future discoveries and developments will be exciting, challenging, and of great benefit to humanity.

1.1 Standards of Length, Mass, and Time

To describe natural phenomena, we must make measurements of various aspects of nature. Each measurement is associated with a physical quantity, such as the length of an object. The laws of physics are expressed as mathematical relationships among physical quantities that we will introduce and discuss throughout the book. In mechanics, the three fundamental quantities are length, mass, and time. All other quantities in mechanics can be expressed in terms of these three.

If we are to report the results of a measurement to someone who wishes to reproduce this measurement, a *standard* must be defined. It would be meaningless if a visitor from another planet were to talk to us about a length of 8 "glitches" if we do not know the meaning of the unit glitch. On the other hand, if someone familiar with our system of measurement reports that a wall is 2 meters high and our unit of length is defined to be 1 meter, we know that the height of the wall is twice our basic length unit. Whatever is chosen as a standard must be readily accessible and must possess some property that can be measured reliably. Measurement standards used by different people in different places—throughout the Universe—must yield the same result. In addition, standards used for measurements must not change with time.

In 1960, an international committee established a set of standards for the fundamental quantities of science. It is called the **SI** (Système International), and its fundamental units of length, mass, and time are the *meter, kilogram,* and *second,* respectively. Other standards for SI fundamental units established by the committee are those for temperature (the *kelvin*), electric current (the *ampere*), luminous intensity (the *candela*), and the amount of substance (the *mole*).

Length

We can identify **length** as the distance between two points in space. In 1120, the king of England decreed that the standard of length in his country would be named the *yard* and would be precisely equal to the distance from the tip of his nose to the end of his outstretched arm. Similarly, the original standard for the foot adopted by the French was the length of the royal foot of King Louis XIV. Neither of these standards is constant in time; when a new king took the throne, length measurements changed! The French standard prevailed until 1799, when the legal standard of length in France became the **meter** (m), defined as one ten-millionth of the distance from the equator to the North Pole along one particular longitudinal line that passes through Paris. Notice that this value is an Earth-based standard that does not satisfy the requirement that it can be used throughout the Universe.

As recently as 1960, the length of the meter was defined as the distance between two lines on a specific platinum–iridium bar stored under controlled conditions in France. Current requirements of science and technology, however, necessitate more accuracy than that with which the separation between the lines on the bar can be determined. In the 1960s and 1970s, the meter was defined as 1 650 763.73 wavelengths[1] of orange-red light emitted from a krypton-86 lamp. In October 1983, however, the meter was redefined as **the distance traveled by light in vacuum during a time of 1/299 792 458 second.** In effect, this latest definition establishes that the speed of light in vacuum is precisely 299 792 458 meters per second. This definition of the meter is valid throughout the Universe based on our assumption that light is the same everywhere.

Table 1.1 lists approximate values of some measured lengths. You should study this table as well as the next two tables and begin to generate an intuition for what is meant by, for example, a length of 20 centimeters, a mass of 100 kilograms, or a time interval of 3.2×10^7 seconds.

Mass

The SI fundamental unit of **mass,** the **kilogram** (kg), is defined as **the mass of a specific platinum–iridium alloy cylinder kept at the International Bureau of Weights and Measures at Sèvres, France.** This mass standard was established in 1887 and

Table 1.1 **Approximate Values of Some Measured Lengths**	
	Length (m)
Distance from the Earth to the most remote known quasar	1.4×10^{26}
Distance from the Earth to the most remote normal galaxies	9×10^{25}
Distance from the Earth to the nearest large galaxy (Andromeda)	2×10^{22}
Distance from the Sun to the nearest star (Proxima Centauri)	4×10^{16}
One light-year	9.46×10^{15}
Mean orbit radius of the Earth about the Sun	1.50×10^{11}
Mean distance from the Earth to the Moon	3.84×10^{8}
Distance from the equator to the North Pole	1.00×10^{7}
Mean radius of the Earth	6.37×10^{6}
Typical altitude (above the surface) of a satellite orbiting the Earth	2×10^{5}
Length of a football field	9.1×10^{1}
Length of a housefly	5×10^{-3}
Size of smallest dust particles	$\sim 10^{-4}$
Size of cells of most living organisms	$\sim 10^{-5}$
Diameter of a hydrogen atom	$\sim 10^{-10}$
Diameter of an atomic nucleus	$\sim 10^{-14}$
Diameter of a proton	$\sim 10^{-15}$

---

[1]We will use the standard international notation for numbers with more than three digits, in which groups of three digits are separated by spaces rather than commas. Therefore, 10 000 is the same as the common American notation of 10,000. Similarly, $\pi = 3.14159265$ is written as 3.141 592 65.

Table 1.2

Approximate Masses of Various Objects

	Mass (kg)
Observable Universe	$\sim 10^{52}$
Milky Way galaxy	$\sim 10^{42}$
Sun	1.99×10^{30}
Earth	5.98×10^{24}
Moon	7.36×10^{22}
Shark	$\sim 10^{3}$
Human	$\sim 10^{2}$
Frog	$\sim 10^{-1}$
Mosquito	$\sim 10^{-5}$
Bacterium	$\sim 1 \times 10^{-15}$
Hydrogen atom	1.67×10^{-27}
Electron	9.11×10^{-31}

Table 1.3 Approximate Values of Some Time Intervals

	Time Interval (s)
Age of the Universe	4×10^{17}
Age of the Earth	1.3×10^{17}
Average age of a college student	6.3×10^{8}
One year	3.2×10^{7}
One day	8.6×10^{4}
One class period	3.0×10^{3}
Time interval between normal heartbeats	8×10^{-1}
Period of audible sound waves	$\sim 10^{-3}$
Period of typical radio waves	$\sim 10^{-6}$
Period of vibration of an atom in a solid	$\sim 10^{-13}$
Period of visible light waves	$\sim 10^{-15}$
Duration of a nuclear collision	$\sim 10^{-22}$
Time interval for light to cross a proton	$\sim 10^{-24}$

a

b

AP Photo/Focke Strangmann

Figure 1.1 (a) The National Standard Kilogram No. 20, an accurate copy of the International Standard Kilogram kept at Sèvres, France, is housed under a double bell jar in a vault at the National Institute of Standards and Technology. (b) A cesium fountain atomic clock. The clock will neither gain nor lose a second in 20 million years.

has not been changed since that time because platinum–iridium is an unusually stable alloy. A duplicate of the Sèvres cylinder is kept at the National Institute of Standards and Technology (NIST) in Gaithersburg, Maryland (Fig. 1.1a). Table 1.2 lists approximate values of the masses of various objects.

Time

Before 1967, the standard of **time** was defined in terms of the *mean solar day*. (A solar day is the time interval between successive appearances of the Sun at the highest point it reaches in the sky each day.) The fundamental unit of a **second** (s) was defined as $\left(\frac{1}{60}\right)\left(\frac{1}{60}\right)\left(\frac{1}{24}\right)$ of a mean solar day. This definition is based on the rotation of one planet, the Earth. Therefore, this motion does not provide a time standard that is universal.

In 1967, the second was redefined to take advantage of the high precision attainable in a device known as an *atomic clock* (Fig. 1.1b), which measures vibrations of cesium atoms. One second is now defined as **9 192 631 770 times the period of vibration of radiation from the cesium-133 atom.**[2] Approximate values of time intervals are presented in Table 1.3.

In addition to SI, another system of units, the *U.S. customary system,* is still used in the United States despite acceptance of SI by the rest of the world. In this system, the units of length, mass, and time are the foot (ft), slug, and second, respectively. In this book, we shall use SI units because they are almost universally accepted in science and industry. We shall make some limited use of U.S. customary units in the study of classical mechanics.

In addition to the fundamental SI units of meter, kilogram, and second, we can also use other units, such as millimeters and nanoseconds, where the prefixes *milli-* and *nano-* denote multipliers of the basic units based on various powers of ten. Prefixes for the various powers of ten and their abbreviations are listed in Table 1.4 (page 6). For example, 10^{-3} m is equivalent to 1 millimeter (mm), and 10^{3} m corresponds to 1 kilometer (km). Likewise, 1 kilogram (kg) is 10^{3} grams (g), and 1 mega volt (MV) is 10^{6} volts (V).

The variables length, time, and mass are examples of *fundamental quantities.* Most other variables are *derived quantities,* those that can be expressed as a mathematical combination of fundamental quantities. Common examples are *area* (a product of two lengths) and *speed* (a ratio of a length to a time interval).

---

[2]Period is defined as the time interval needed for one complete vibration.

Table 1.4 Prefixes for Powers of Ten

Power	Prefix	Abbreviation	Power	Prefix	Abbreviation
10^{-24}	yocto	y	10^{3}	kilo	k
10^{-21}	zepto	z	10^{6}	mega	M
10^{-18}	atto	a	10^{9}	giga	G
10^{-15}	femto	f	10^{12}	tera	T
10^{-12}	pico	p	10^{15}	peta	P
10^{-9}	nano	n	10^{18}	exa	E
10^{-6}	micro	μ	10^{21}	zetta	Z
10^{-3}	milli	m	10^{24}	yotta	Y
10^{-2}	centi	c			
10^{-1}	deci	d			

A table of the letters in the Greek alphabet is provided on the back endpaper of this book. ▶

Another example of a derived quantity is **density.** The density ρ (Greek letter rho) of any substance is defined as its *mass per unit volume:*

$$\rho \equiv \frac{m}{V} \quad (1.1)$$

In terms of fundamental quantities, density is a ratio of a mass to a product of three lengths. Aluminum, for example, has a density of 2.70×10^3 kg/m^3, and iron has a density of 7.86×10^3 kg/m^3. An extreme difference in density can be imagined by thinking about holding a 10-centimeter (cm) cube of Styrofoam in one hand and a 10-cm cube of lead in the other. See Table 14.1 in Chapter 14 for densities of several materials.

Quick Quiz 1.1 In a machine shop, two cams are produced, one of aluminum and one of iron. Both cams have the same mass. Which cam is larger? **(a)** The aluminum cam is larger. **(b)** The iron cam is larger. **(c)** Both cams have the same size.

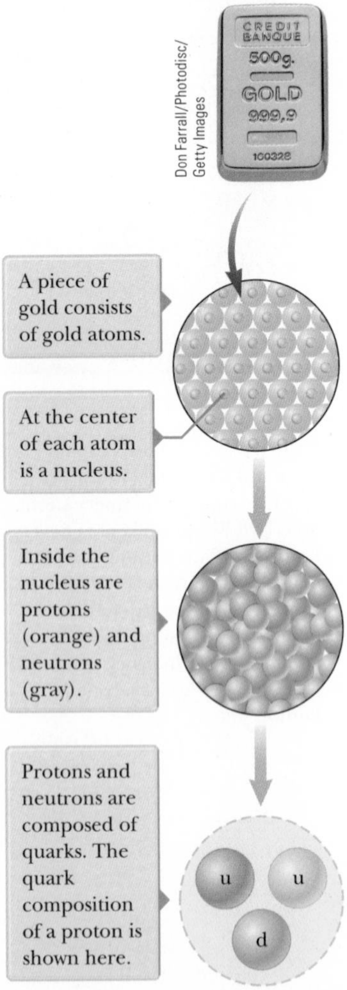
CREDIT BANQUE
500g.
GOLD
999,9
100328
Don Farrall/Photodisc/Getty Images
A piece of gold consists of gold atoms.
At the center of each atom is a nucleus.
Inside the nucleus are protons (orange) and neutrons (gray).
Protons and neutrons are composed of quarks. The quark composition of a proton is shown here.
u
u
d

Figure 1.2 Levels of organization in matter.

1.2 Matter and Model Building

If physicists cannot interact with some phenomenon directly, they often imagine a **model** for a physical system that is related to the phenomenon. For example, we cannot interact directly with atoms because they are too small. Therefore, we build a mental model of an atom based on a system of a nucleus and one or more electrons outside the nucleus. Once we have identified the physical components of the model, we make predictions about its behavior based on the interactions among the components of the system or the interaction between the system and the environment outside the system.

As an example, consider the behavior of *matter.* A sample of solid gold is shown at the top of Figure 1.2. Is this sample nothing but wall-to-wall gold, with no empty space? If the sample is cut in half, the two pieces still retain their chemical identity as solid gold. What if the pieces are cut again and again, indefinitely? Will the smaller and smaller pieces always be gold? Such questions can be traced to early Greek philosophers. Two of them—Leucippus and his student Democritus—could not accept the idea that such cuttings could go on forever. They developed a model for matter by speculating that the process ultimately must end when it produces a particle that can no longer be cut. In Greek, *atomos* means "not sliceable." From this Greek term comes our English word *atom.*

The Greek model of the structure of matter was that all ordinary matter consists of atoms, as suggested in the middle of Figure 1.2. Beyond that, no additional structure was specified in the model; atoms acted as small particles that interacted with one another, but internal structure of the atom was not a part of the model.

In 1897, J. J. Thomson identified the electron as a charged particle and as a constituent of the atom. This led to the first atomic model that contained internal structure. We shall discuss this model in Chapter 42.

Following the discovery of the nucleus in 1911, an atomic model was developed in which each atom is made up of electrons surrounding a central nucleus. A nucleus of gold is shown in Figure 1.2. This model leads, however, to a new question: Does the nucleus have structure? That is, is the nucleus a single particle or a collection of particles? By the early 1930s, a model evolved that described two basic entities in the nucleus: protons and neutrons. The proton carries a positive electric charge, and a specific chemical element is identified by the number of protons in its nucleus. This number is called the **atomic number** of the element. For instance, the nucleus of a hydrogen atom contains one proton (so the atomic number of hydrogen is 1), the nucleus of a helium atom contains two protons (atomic number 2), and the nucleus of a uranium atom contains 92 protons (atomic number 92). In addition to atomic number, a second number—**mass number,** defined as the number of protons plus neutrons in a nucleus—characterizes atoms. The atomic number of a specific element never varies (i.e., the number of protons does not vary), but the mass number can vary (i.e., the number of neutrons varies).

Is that, however, where the process of breaking down stops? Protons, neutrons, and a host of other exotic particles are now known to be composed of six different varieties of particles called **quarks,** which have been given the names of *up, down, strange, charmed, bottom,* and *top.* The up, charmed, and top quarks have electric charges of $+\frac{2}{3}$ that of the proton, whereas the down, strange, and bottom quarks have charges of $-\frac{1}{3}$ that of the proton. The proton consists of two up quarks and one down quark as shown at the bottom of Figure 1.2 and labeled u and d. This structure predicts the correct charge for the proton. Likewise, the neutron consists of two down quarks and one up quark, giving a net charge of zero.

You should develop a process of building models as you study physics. In this study, you will be challenged with many mathematical problems to solve. One of the most important problem-solving techniques is to build a model for the problem: identify a system of physical components for the problem and make predictions of the behavior of the system based on the interactions among its components or the interaction between the system and its surrounding environment.

1.3 Dimensional Analysis

In physics, the word *dimension* denotes the physical nature of a quantity. The distance between two points, for example, can be measured in feet, meters, or furlongs, which are all different ways of expressing the dimension of length.

The symbols we use in this book to specify the dimensions of length, mass, and time are L, M, and T, respectively.[3] We shall often use brackets [ ] to denote the dimensions of a physical quantity. For example, the symbol we use for speed in this book is v, and in our notation, the dimensions of speed are written $[v] = L/T$. As another example, the dimensions of area A are $[A] = L^2$. The dimensions and units of area, volume, speed, and acceleration are listed in Table 1.5. The dimensions of other quantities, such as force and energy, will be described as they are introduced in the text.

Table 1.5   **Dimensions and Units of Four Derived Quantities**

Quantity	Area (A)	Volume (V)	Speed (v)	Acceleration (a)
Dimensions	L^2	L^3	L/T	L/T^2
SI units	m^2	m^3	m/s	m/s^2
U.S. customary units	ft^2	ft^3	ft/s	ft/s^2

---

[3]The *dimensions* of a quantity will be symbolized by a capitalized, nonitalic letter such as L or T. The *algebraic symbol* for the quantity itself will be an italicized letter such as L for the length of an object or t for time.

In many situations, you may have to check a specific equation to see if it matches your expectations. A useful procedure for doing that, called **dimensional analysis,** can be used because dimensions can be treated as algebraic quantities. For example, quantities can be added or subtracted only if they have the same dimensions. Furthermore, the terms on both sides of an equation must have the same dimensions. By following these simple rules, you can use dimensional analysis to determine whether an expression has the correct form. Any relationship can be correct only if the dimensions on both sides of the equation are the same.

To illustrate this procedure, suppose you are interested in an equation for the position x of a car at a time t if the car starts from rest at $x = 0$ and moves with constant acceleration a. The correct expression for this situation is $x = \frac{1}{2}at^2$ as we show in Chapter 2. The quantity x on the left side has the dimension of length. For the equation to be dimensionally correct, the quantity on the right side must also have the dimension of length. We can perform a dimensional check by substituting the dimensions for acceleration, L/T^2 (Table 1.5), and time, T, into the equation. That is, the dimensional form of the equation $x = \frac{1}{2}at^2$ is

$$L = \frac{L}{T^2} \cdot T^2 = L$$

The dimensions of time cancel as shown, leaving the dimension of length on the right-hand side to match that on the left.

A more general procedure using dimensional analysis is to set up an expression of the form

$$x \propto a^n t^m$$

where n and m are exponents that must be determined and the symbol $\propto$ indicates a proportionality. This relationship is correct only if the dimensions of both sides are the same. Because the dimension of the left side is length, the dimension of the right side must also be length. That is,

$$[a^n t^m] = L = L^1 T^0$$

Because the dimensions of acceleration are L/T^2 and the dimension of time is T, we have

$$(L/T^2)^n T^m = L^1 T^0 \quad \rightarrow \quad (L^n T^{m-2n}) = L^1 T^0$$

The exponents of L and T must be the same on both sides of the equation. From the exponents of L, we see immediately that $n = 1$. From the exponents of T, we see that $m - 2n = 0$, which, once we substitute for n, gives us $m = 2$. Returning to our original expression $x \propto a^n t^m$, we conclude that $x \propto at^2$.

> **Q**uick Quiz 1.2 True or False: Dimensional analysis can give you the numerical value of constants of proportionality that may appear in an algebraic expression.

Example 1.1   Analysis of an Equation

Show that the expression $v = at$, where v represents speed, a acceleration, and t an instant of time, is dimensionally correct.

SOLUTION

Identify the dimensions of v from Table 1.5: $\qquad\qquad [v] = \dfrac{L}{T}$

▶ **1.1** continued

Identify the dimensions of a from Table 1.5 and multiply by the dimensions of t:

$$[at] = \frac{L}{T^2} T = \frac{L}{T}$$

Therefore, $v = at$ is dimensionally correct because we have the same dimensions on both sides. (If the expression were given as $v = at^2$, it would be dimensionally *incorrect*. Try it and see!)

Example 1.2   **Analysis of a Power Law**

Suppose we are told that the acceleration a of a particle moving with uniform speed v in a circle of radius r is proportional to some power of r, say r^n, and some power of v, say v^m. Determine the values of n and m and write the simplest form of an equation for the acceleration.

SOLUTION

Write an expression for a with a dimensionless constant of proportionality k:

$$a = kr^n v^m$$

Substitute the dimensions of a, r, and v:

$$\frac{L}{T^2} = L^n \left(\frac{L}{T}\right)^m = \frac{L^{n+m}}{T^m}$$

Equate the exponents of L and T so that the dimensional equation is balanced:

$$n + m = 1 \text{ and } m = 2$$

Solve the two equations for n:

$$n = -1$$

Write the acceleration expression:

$$a = kr^{-1} v^2 = k\frac{v^2}{r}$$

In Section 4.4 on uniform circular motion, we show that $k = 1$ if a consistent set of units is used. The constant k would not equal 1 if, for example, v were in km/h and you wanted a in m/s^2 .

1.4 Conversion of Units

Sometimes it is necessary to convert units from one measurement system to another or convert within a system (for example, from kilometers to meters). Conversion factors between SI and U.S. customary units of length are as follows:

1 mile	= 1 609 m = 1.609 km	1 ft	= 0.304 8 m = 30.48 cm
1 m	= 39.37 in. = 3.281 ft	1 in.	= 0.025 4 m = 2.54 cm (exactly)

A more complete list of conversion factors can be found in Appendix A.

Like dimensions, units can be treated as algebraic quantities that can cancel each other. For example, suppose we wish to convert 15.0 in. to centimeters. Because 1 in. is defined as exactly 2.54 cm, we find that

$$15.0 \text{ in.} = (15.0 \text{ in.})\left(\frac{2.54 \text{ cm}}{1 \text{ in.}}\right) = 38.1 \text{ cm}$$

where the ratio in parentheses is equal to 1. We express 1 as 2.54 cm/1 in. (rather than 1 in./2.54 cm) so that the unit "inch" in the denominator cancels with the unit in the original quantity. The remaining unit is the centimeter, our desired result.

Pitfall Prevention 1.3
Always Include Units When performing calculations with numerical values, include the units for every quantity and carry the units through the entire calculation. Avoid the temptation to drop the units early and then attach the expected units once you have an answer. By including the units in every step, you can detect errors if the units for the answer turn out to be incorrect.

Quick Quiz 1.3 The distance between two cities is 100 mi. What is the number of kilometers between the two cities? **(a)** smaller than 100 **(b)** larger than 100 **(c)** equal to 100

Example 1.3   **Is He Speeding?**

On an interstate highway in a rural region of Wyoming, a car is traveling at a speed of 38.0 m/s. Is the driver exceeding the speed limit of 75.0 mi/h?

SOLUTION

Convert meters in the speed to miles:

$$(38.0 \text{ m/s}) \left(\frac{1 \text{ mi}}{1\,609 \text{ m}} \right) = 2.36 \times 10^{-2} \text{ mi/s}$$

Convert seconds to hours:

$$(2.36 \times 10^{-2} \text{ mi/s}) \left(\frac{60 \text{ s}}{1 \text{ min}} \right) \left(\frac{60 \text{ min}}{1 \text{ h}} \right) = 85.0 \text{ mi/h}$$

The driver is indeed exceeding the speed limit and should slow down.

WHAT IF?   What if the driver were from outside the United States and is familiar with speeds measured in kilometers per hour? What is the speed of the car in km/h?

Answer We can convert our final answer to the appropriate units:

$$(85.0 \text{ mi/h}) \left(\frac{1.609 \text{ km}}{1 \text{ mi}} \right) = 137 \text{ km/h}$$

Figure 1.3 shows an automobile speedometer displaying speeds in both mi/h and km/h. Can you check the conversion we just performed using this photograph?

Figure 1.3 The speedometer of a vehicle that shows speeds in both miles per hour and kilometers per hour.

1.5 Estimates and Order-of-Magnitude Calculations

Suppose someone asks you the number of bits of data on a typical musical compact disc. In response, it is not generally expected that you would provide the exact number but rather an estimate, which may be expressed in scientific notation. The estimate may be made even more approximate by expressing it as an *order of magnitude*, which is a power of ten determined as follows:

1. Express the number in scientific notation, with the multiplier of the power of ten between 1 and 10 and a unit.
2. If the multiplier is less than 3.162 (the square root of 10), the order of magnitude of the number is the power of 10 in the scientific notation. If the multiplier is greater than 3.162, the order of magnitude is one larger than the power of 10 in the scientific notation.

We use the symbol ~ for "is on the order of." Use the procedure above to verify the orders of magnitude for the following lengths:

$$0.008\,6 \text{ m} \sim 10^{-2} \text{ m} \qquad 0.002\,1 \text{ m} \sim 10^{-3} \text{ m} \qquad 720 \text{ m} \sim 10^{3} \text{ m}$$

Usually, when an order-of-magnitude estimate is made, the results are reliable to within about a factor of 10. If a quantity increases in value by three orders of magnitude, its value increases by a factor of about $10^3 = 1\ 000$.

Inaccuracies caused by guessing too low for one number are often canceled by other guesses that are too high. You will find that with practice your guesstimates become better and better. Estimation problems can be fun to work because you freely drop digits, venture reasonable approximations for unknown numbers, make simplifying assumptions, and turn the question around into something you can answer in your head or with minimal mathematical manipulation on paper. Because of the simplicity of these types of calculations, they can be performed on a *small* scrap of paper and are often called "back-of-the-envelope calculations."

Example 1.4   **Breaths in a Lifetime**

Estimate the number of breaths taken during an average human lifetime.

SOLUTION

We start by guessing that the typical human lifetime is about 70 years. Think about the average number of breaths that a person takes in 1 min. This number varies depending on whether the person is exercising, sleeping, angry, serene, and so forth. To the nearest order of magnitude, we shall choose 10 breaths per minute as our estimate. (This estimate is certainly closer to the true average value than an estimate of 1 breath per minute or 100 breaths per minute.)

Find the approximate number of minutes in a year:

$$1\text{ yr}\left(\frac{400\text{ days}}{1\text{ yr}}\right)\left(\frac{25\text{ h}}{1\text{ day}}\right)\left(\frac{60\text{ min}}{1\text{ h}}\right) = 6\times10^5\text{ min}$$

Find the approximate number of minutes in a 70-year lifetime:

$$\text{number of minutes} = (70\text{ yr})(6\times10^5\text{ min/yr})$$
$$= 4\times10^7\text{ min}$$

Find the approximate number of breaths in a lifetime:

$$\text{number of breaths} = (10\text{ breaths/min})(4\times10^7\text{ min})$$
$$= 4\times10^8\text{ breaths}$$

Therefore, a person takes on the order of 10^9 breaths in a lifetime. Notice how much simpler it is in the first calculation above to multiply 400×25 than it is to work with the more accurate 365×24.

WHAT IF? What if the average lifetime were estimated as 80 years instead of 70? Would that change our final estimate?

Answer We could claim that $(80\text{ yr})(6\times10^5\text{ min/yr}) = 5\times10^7\text{ min}$, so our final estimate should be 5×10^8 breaths. This answer is still on the order of 10^9 breaths, so an order-of-magnitude estimate would be unchanged.

1.6 Significant Figures

When certain quantities are measured, the measured values are known only to within the limits of the experimental uncertainty. The value of this uncertainty can depend on various factors, such as the quality of the apparatus, the skill of the experimenter, and the number of measurements performed. The number of **significant figures** in a measurement can be used to express something about the uncertainty. The number of significant figures is related to the number of numerical digits used to express the measurement, as we discuss below.

As an example of significant figures, suppose we are asked to measure the radius of a compact disc using a meterstick as a measuring instrument. Let us assume the accuracy to which we can measure the radius of the disc is ±0.1 cm. Because of the uncertainty of ±0.1 cm, if the radius is measured to be 6.0 cm, we can claim only that its radius lies somewhere between 5.9 cm and 6.1 cm. In this case, we say that the measured value of 6.0 cm has two significant figures. Note that *the significant figures include the first estimated digit.* Therefore, we could write the radius as (6.0 ± 0.1) cm.

Zeros may or may not be significant figures. Those used to position the decimal point in such numbers as 0.03 and 0.007 5 are not significant. Therefore, there are one and two significant figures, respectively, in these two values. When the zeros come after other digits, however, there is the possibility of misinterpretation. For example, suppose the mass of an object is given as 1 500 g. This value is ambiguous because we do not know whether the last two zeros are being used to locate the decimal point or whether they represent significant figures in the measurement. To remove this ambiguity, it is common to use scientific notation to indicate the number of significant figures. In this case, we would express the mass as 1.5×10^3 g if there are two significant figures in the measured value, 1.50×10^3 g if there are three significant figures, and 1.500×10^3 g if there are four. The same rule holds for numbers less than 1, so 2.3×10^{-4} has two significant figures (and therefore could be written 0.000 23) and 2.30×10^{-4} has three significant figures (also written as 0.000 230).

In problem solving, we often combine quantities mathematically through multiplication, division, addition, subtraction, and so forth. When doing so, you must make sure that the result has the appropriate number of significant figures. A good rule of thumb to use in determining the number of significant figures that can be claimed in a multiplication or a division is as follows:

> When multiplying several quantities, the number of significant figures in the final answer is the same as the number of significant figures in the quantity having the smallest number of significant figures. The same rule applies to division.

Let's apply this rule to find the area of the compact disc whose radius we measured above. Using the equation for the area of a circle,

$$A = \pi r^2 = \pi(6.0 \text{ cm})^2 = 1.1 \times 10^2 \text{ cm}^2$$

If you perform this calculation on your calculator, you will likely see 113.097 335 5. It should be clear that you don't want to keep all of these digits, but you might be tempted to report the result as 113 cm². This result is not justified because it has three significant figures, whereas the radius only has two. Therefore, we must report the result with only two significant figures as shown above.

For addition and subtraction, you must consider the number of decimal places when you are determining how many significant figures to report:

> When numbers are added or subtracted, the number of decimal places in the result should equal the smallest number of decimal places of any term in the sum or difference.

As an example of this rule, consider the sum

$$23.2 + 5.174 = 28.4$$

Notice that we do not report the answer as 28.374 because the lowest number of decimal places is one, for 23.2. Therefore, our answer must have only one decimal place.

The rule for addition and subtraction can often result in answers that have a different number of significant figures than the quantities with which you start. For example, consider these operations that satisfy the rule:

$$1.000 \ 1 + 0.000 \ 3 = 1.000 \ 4$$

$$1.002 - 0.998 = 0.004$$

In the first example, the result has five significant figures even though one of the terms, 0.000 3, has only one significant figure. Similarly, in the second calculation, the result has only one significant figure even though the numbers being subtracted have four and three, respectively.

In this book, most of the numerical examples and end-of-chapter problems will yield answers having three significant figures. When carrying out estimation calculations, we shall typically work with a single significant figure.

◀ **Significant figure guidelines used in this book**

If the number of significant figures in the result of a calculation must be reduced, there is a general rule for rounding numbers: the last digit retained is increased by 1 if the last digit dropped is greater than 5. (For example, 1.346 becomes 1.35.) If the last digit dropped is less than 5, the last digit retained remains as it is. (For example, 1.343 becomes 1.34.) If the last digit dropped is equal to 5, the remaining digit should be rounded to the nearest even number. (This rule helps avoid accumulation of errors in long arithmetic processes.)

A technique for avoiding error accumulation is to delay the rounding of numbers in a long calculation until you have the final result. Wait until you are ready to copy the final answer from your calculator before rounding to the correct number of significant figures. In this book, we display numerical values rounded off to two or three significant figures. This occasionally makes some mathematical manipulations look odd or incorrect. For instance, looking ahead to Example 3.5 on page 59, you will see the operation $-17.7 \text{ km} + 34.6 \text{ km} = 17.0 \text{ km}$. This looks like an incorrect subtraction, but that is only because we have rounded the numbers 17.7 km and 34.6 km for display. If all digits in these two intermediate numbers are retained and the rounding is only performed on the final number, the correct three-digit result of 17.0 km is obtained.

> **Pitfall Prevention 1.5**
>
> **Symbolic Solutions** When solving problems, it is very useful to perform the solution completely in algebraic form and wait until the very end to enter numerical values into the final symbolic expression. This method will save many calculator keystrokes, especially if some quantities cancel so that you never have to enter their values into your calculator! In addition, you will only need to round once, on the final result.

Example 1.5   Installing a Carpet

A carpet is to be installed in a rectangular room whose length is measured to be 12.71 m and whose width is measured to be 3.46 m. Find the area of the room.

SOLUTION

If you multiply 12.71 m by 3.46 m on your calculator, you will see an answer of 43.976 6 m². How many of these numbers should you claim? Our rule of thumb for multiplication tells us that you can claim only the number of significant figures in your answer as are present in the measured quantity having the lowest number of significant figures. In this example, the lowest number of significant figures is three in 3.46 m, so we should express our final answer as 44.0 m².

Summary

Definitions

▌ The three fundamental physical quantities of mechanics are **length, mass,** and **time,** which in the SI system have the units **meter** (m), **kilogram** (kg), and **second** (s), respectively. These fundamental quantities cannot be defined in terms of more basic quantities.

▌ The **density** of a substance is defined as its *mass per unit volume:*

$$\rho \equiv \frac{m}{V} \tag{1.1}$$

continued

Concepts and Principles

The method of **dimensional analysis** is very powerful in solving physics problems. Dimensions can be treated as algebraic quantities. By making estimates and performing order-of-magnitude calculations, you should be able to approximate the answer to a problem when there is not enough information available to specify an exact solution completely.

When you compute a result from several measured numbers, each of which has a certain accuracy, you should give the result with the correct number of **significant figures.**

When **multiplying** several quantities, the number of significant figures in the final answer is the same as the number of significant figures in the quantity having the smallest number of significant figures. The same rule applies to **division.**

When numbers are **added** or **subtracted,** the number of decimal places in the result should equal the smallest number of decimal places of any term in the sum or difference.

Objective Questions   1. denotes answer available in *Student Solutions Manual/Study Guide*

1. One student uses a meterstick to measure the thickness of a textbook and obtains 4.3 cm ± 0.1 cm. Other students measure the thickness with vernier calipers and obtain four different measurements: (a) 4.32 cm ± 0.01 cm, (b) 4.31 cm ± 0.01 cm, (c) 4.24 cm ± 0.01 cm, and (d) 4.43 cm ± 0.01 cm. Which of these four measurements, if any, agree with that obtained by the first student?

2. A house is advertised as having 1 420 square feet under its roof. What is its area in square meters? (a) 4 660 m^2 (b) 432 m^2 (c) 158 m^2 (d) 132 m^2 (e) 40.2 m^2

3. Answer each question yes or no. Must two quantities have the same dimensions (a) if you are adding them? (b) If you are multiplying them? (c) If you are subtracting them? (d) If you are dividing them? (e) If you are equating them?

4. The price of gasoline at a particular station is 1.5 euros per liter. An American student can use 33 euros to buy gasoline. Knowing that 4 quarts make a gallon and that 1 liter is close to 1 quart, she quickly reasons that she can buy how many gallons of gasoline? (a) less than 1 gallon (b) about 5 gallons (c) about 8 gallons (d) more than 10 gallons

5. Rank the following five quantities in order from the largest to the smallest. If two of the quantities are equal,

give them equal rank in your list. (a) 0.032 kg (b) 15 g (c) 2.7 × 10^5 mg (d) 4.1 × 10^{-8} Gg (e) 2.7 × 10^8 μg

6. What is the sum of the measured values 21.4 s + 15 s + 17.17 s + 4.00 3 s? (a) 57.573 s (b) 57.57 s (c) 57.6 s (d) 58 s (e) 60 s

7. Which of the following is the best estimate for the mass of all the people living on the Earth? (a) 2 × 10^8 kg (b) 1 × 10^9 kg (c) 2 × 10^{10} kg (d) 3 × 10^{11} kg (e) 4 × 10^{12} kg

8. (a) If an equation is dimensionally correct, does that mean that the equation must be true? (b) If an equation is not dimensionally correct, does that mean that the equation cannot be true?

9. Newton's second law of motion (Chapter 5) says that the mass of an object times its acceleration is equal to the net force on the object. Which of the following gives the correct units for force? (a) kg · m/s^2 (b) kg · m^2/s^2 (c) kg/m · s^2 (d) kg · m^2/s (e) none of those answers

10. A calculator displays a result as 1.365 248 0 × 10^7 kg. The estimated uncertainty in the result is ±2%. How many digits should be included as significant when the result is written down? (a) zero (b) one (c) two (d) three (e) four

Conceptual Questions   1. denotes answer available in *Student Solutions Manual/Study Guide*

1. Suppose the three fundamental standards of the metric system were length, *density*, and time rather than length, *mass*, and time. The standard of density in this system is to be defined as that of water. What considerations about water would you need to address to make sure that the standard of density is as accurate as possible?

2. Why is the metric system of units considered superior to most other systems of units?

3. What natural phenomena could serve as alternative time standards?

4. Express the following quantities using the prefixes given in Table 1.4. (a) 3 × 10^{-4} m (b) 5 × 10^{-5} s (c) 72 × 10^2 g

Problems available in WebAssign Access end-of-chapter problems online at www.webassign.net

Section 1.1 Standards of Length, Mass, and Time
Problems 1–6

Section 1.2 Matter and Model Building
Problems 7–8

Section 1.3 Dimensional Analysis
Problems 9–14

Section 1.4 Conversion of Units
Problems 15–30

Section 1.5 Estimates and Order-of-Magnitude Calculations
Problems 31–34

Section 1.6 Significant Figures
Problems 35–53

Additional Problems
Problems 54–71

Challenge Problems
Problems 72–73

Solutions to the following Problems are available in the *Student Solutions Manual/Study Guide*:
1.9, 1.15, 1.17, 1.25, 1.27, 1.29, 1.31, 1.33, 1.36, 1.47, 1.57, 1.59, 1.63, and 1.69.

List of Enhanced Problems

Problem Number	Targeted Feedback in Enhanced WebAssign	Analysis Model Tutorial in Enhanced WebAssign	Master It in Enhanced WebAssign	Watch It in Enhanced WebAssign
1.2	✓			✓
1.5	✓			✓
1.10	✓			✓
1.12	✓			✓
1.15	✓		✓	
1.18	✓			✓
1.20	✓			✓
1.22	✓			✓
1.24	✓		✓	
1.25			✓	
1.27	✓		✓	
1.28	✓			✓
1.29	✓		✓	
1.36	✓			✓
1.38	✓			✓
1.45	✓		✓	
1.47			✓	
1.59	✓	✓	✓	
1.63		✓	✓	
1.69			✓	
1.71		✓		

Motion in One Dimension

As a first step in studying classical mechanics, we describe the motion of an object while ignoring the interactions with external agents that might be affecting or modifying that motion. This portion of classical mechanics is called *kinematics*. (The word *kinematics* has the same root as *cinema*.) In this chapter, we consider only motion in one dimension, that is, motion of an object along a straight line.

From everyday experience, we recognize that motion of an object represents a continuous change in the object's position. In physics, we can categorize motion into three types: translational, rotational, and vibrational. A car traveling on a highway is an example of translational motion, the Earth's spin on its axis is an example of rotational motion, and the back-and-forth movement of a pendulum is an example of vibrational motion. In this and the next few chapters, we are concerned only with translational motion. (Later in the book we shall discuss rotational and vibrational motions.)

In our study of translational motion, we use what is called the **particle model** and describe the moving object as a *particle* regardless of its size. Remember our discussion of making models for physical situations in Section 1.2. In general, **a particle is a point-like object, that is, an object that has mass but is of infinitesimal size.** For example, if we wish to describe the motion of the Earth around the Sun, we can treat the Earth as a particle and

In drag racing, a driver wants as large an acceleration as possible. In a distance of one-quarter mile, a vehicle reaches speeds of more than 320 mi/h, covering the entire distance in under 5 s. *(George Lepp/Stone/Getty Images)*

17

obtain reasonably accurate data about its orbit. This approximation is justified because the radius of the Earth's orbit is large compared with the dimensions of the Earth and the Sun. As an example on a much smaller scale, it is possible to explain the pressure exerted by a gas on the walls of a container by treating the gas molecules as particles, without regard for the internal structure of the molecules.

2.1 Position, Velocity, and Speed

Position ▶ A particle's **position** x is the location of the particle with respect to a chosen reference point that we can consider to be the origin of a coordinate system. The motion of a particle is completely known if the particle's position in space is known at all times.

Consider a car moving back and forth along the x axis as in Figure 2.1a. When we begin collecting position data, the car is 30 m to the right of the reference position $x = 0$. We will use the particle model by identifying some point on the car, perhaps the front door handle, as a particle representing the entire car.

We start our clock, and once every 10 s we note the car's position. As you can see from Table 2.1, the car moves to the right (which we have defined as the positive direction) during the first 10 s of motion, from position Ⓐ to position Ⓑ. After Ⓑ, the position values begin to decrease, suggesting the car is backing up from position Ⓑ through position Ⓕ. In fact, at Ⓓ, 30 s after we start measuring, the car is at the origin of coordinates (see Fig. 2.1a). It continues moving to the left and is more than 50 m to the left of $x = 0$ when we stop recording information after our sixth data point. A graphical representation of this information is presented in Figure 2.1b. Such a plot is called a *position–time graph*.

Notice the *alternative representations* of information that we have used for the motion of the car. Figure 2.1a is a *pictorial representation*, whereas Figure 2.1b is a *graphical representation*. Table 2.1 is a *tabular representation* of the same information. Using an alternative representation is often an excellent strategy for understanding the situation in a given problem. The ultimate goal in many problems is a *math-*

Table 2.1	Position of the Car at Various Times	
Position	***t* (s)**	***x* (m)**
Ⓐ	0	30
Ⓑ	10	52
Ⓒ	20	38
Ⓓ	30	0
Ⓔ	40	−37
Ⓕ	50	−53

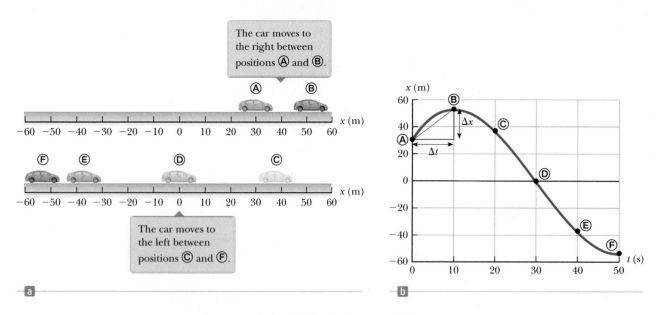

Figure 2.1 A car moves back and forth along a straight line. Because we are interested only in the car's translational motion, we can model it as a particle. Several representations of the information about the motion of the car can be used. Table 2.1 is a tabular representation of the information. (a) A pictorial representation of the motion of the car. (b) A graphical representation (position–time graph) of the motion of the car.

ematical representation, which can be analyzed to solve for some requested piece of information.

Given the data in Table 2.1, we can easily determine the change in position of the car for various time intervals. The **displacement** Δx of a particle is defined as its change in position in some time interval. As the particle moves from an initial position x_i to a final position x_f, its displacement is given by

$$\Delta x \equiv x_f - x_i \tag{2.1}$$

◀ Displacement

We use the capital Greek letter delta (Δ) to denote the *change* in a quantity. From this definition, we see that Δx is positive if x_f is greater than x_i and negative if x_f is less than x_i.

It is very important to recognize the difference between displacement and distance traveled. **Distance** is the length of a path followed by a particle. Consider, for example, the basketball players in Figure 2.2. If a player runs from his own team's basket down the court to the other team's basket and then returns to his own basket, the *displacement* of the player during this time interval is zero because he ended up at the same point as he started: $x_f = x_i$, so $\Delta x = 0$. During this time interval, however, he moved through a *distance* of twice the length of the basketball court. Distance is always represented as a positive number, whereas displacement can be either positive or negative.

Displacement is an example of a vector quantity. Many other physical quantities, including position, velocity, and acceleration, also are vectors. In general, a **vector quantity** requires the specification of both direction and magnitude. By contrast, a **scalar quantity** has a numerical value and no direction. In this chapter, we use positive (+) and negative (−) signs to indicate vector direction. For example, for horizontal motion let us arbitrarily specify to the right as being the positive direction. It follows that any object always moving to the right undergoes a positive displacement $\Delta x > 0$, and any object moving to the left undergoes a negative displacement so that $\Delta x < 0$. We shall treat vector quantities in greater detail in Chapter 3.

One very important point has not yet been mentioned. Notice that the data in Table 2.1 result only in the six data points in the graph in Figure 2.1b. Therefore, the motion of the particle is not completely known because we don't know its position at *all* times. The smooth curve drawn through the six points in the graph is only a *possibility* of the actual motion of the car. We only have information about six instants of time; we have no idea what happened between the data points. The smooth curve is a *guess* as to what happened, but keep in mind that it is *only* a guess. If the smooth curve does represent the actual motion of the car, the graph contains complete information about the entire 50-s interval during which we watch the car move.

It is much easier to see changes in position from the graph than from a verbal description or even a table of numbers. For example, it is clear that the car covers more ground during the middle of the 50-s interval than at the end. Between positions Ⓒ and Ⓓ, the car travels almost 40 m, but during the last 10 s, between positions Ⓔ and Ⓕ, it moves less than half that far. A common way of comparing these different motions is to divide the displacement Δx that occurs between two clock readings by the value of that particular time interval Δt. The result turns out to be a very useful ratio, one that we shall use many times. This ratio has been given a special name: the *average velocity*. The **average velocity** $v_{x,\text{avg}}$ of a particle is defined as the particle's displacement Δx divided by the time interval Δt during which that displacement occurs:

$$v_{x,\text{avg}} \equiv \frac{\Delta x}{\Delta t} \tag{2.2}$$

◀ Average velocity

where the subscript x indicates motion along the x axis. From this definition we see that average velocity has dimensions of length divided by time (L/T), or meters per second in SI units.

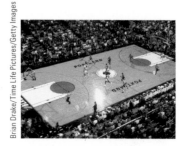

Figure 2.2 On this basketball court, players run back and forth for the entire game. The distance that the players run over the duration of the game is nonzero. The displacement of the players over the duration of the game is approximately zero because they keep returning to the same point over and over again.

The average velocity of a particle moving in one dimension can be positive or negative, depending on the sign of the displacement. (The time interval Δt is always positive.) If the coordinate of the particle increases in time (that is, if $x_f > x_i$), Δx is positive and $v_{x,avg} = \Delta x/\Delta t$ is positive. This case corresponds to a particle moving in the positive x direction, that is, toward larger values of x. If the coordinate decreases in time (that is, if $x_f < x_i$), Δx is negative and hence $v_{x,avg}$ is negative. This case corresponds to a particle moving in the negative x direction.

We can interpret average velocity geometrically by drawing a straight line between any two points on the position–time graph in Figure 2.1b. This line forms the hypotenuse of a right triangle of height Δx and base Δt. The slope of this line is the ratio $\Delta x/\Delta t$, which is what we have defined as average velocity in Equation 2.2. For example, the line between positions Ⓐ and Ⓑ in Figure 2.1b has a slope equal to the average velocity of the car between those two times, (52 m − 30 m)/(10 s − 0) = 2.2 m/s.

In everyday usage, the terms *speed* and *velocity* are interchangeable. In physics, however, there is a clear distinction between these two quantities. Consider a marathon runner who runs a distance d of more than 40 km and yet ends up at her starting point. Her total displacement is zero, so her average velocity is zero! Nonetheless, we need to be able to quantify how fast she was running. A slightly different ratio accomplishes that for us. The **average speed** v_{avg} of a particle, a scalar quantity, is defined as the total distance d traveled divided by the total time interval required to travel that distance:

Average speed ▶

$$v_{avg} \equiv \frac{d}{\Delta t} \tag{2.3}$$

The SI unit of average speed is the same as the unit of average velocity: meters per second. Unlike average velocity, however, average speed has no direction and is always expressed as a positive number. Notice the clear distinction between the definitions of average velocity and average speed: average velocity (Eq. 2.2) is the *displacement* divided by the time interval, whereas average speed (Eq. 2.3) is the *distance* divided by the time interval.

Knowledge of the average velocity or average speed of a particle does not provide information about the details of the trip. For example, suppose it takes you 45.0 s to travel 100 m down a long, straight hallway toward your departure gate at an airport. At the 100-m mark, you realize you missed the restroom, and you return back 25.0 m along the same hallway, taking 10.0 s to make the return trip. The magnitude of your average *velocity* is +75.0 m/55.0 s = +1.36 m/s. The average *speed* for your trip is 125 m/55.0 s = 2.27 m/s. You may have traveled at various speeds during the walk and, of course, you changed direction. Neither average velocity nor average speed provides information about these details.

> **Pitfall Prevention 2.1**
> **Average Speed and Average Velocity** The magnitude of the average velocity is *not* the average speed. For example, consider the marathon runner discussed before Equation 2.3. The magnitude of her average velocity is zero, but her average speed is clearly not zero.

Ⓠuick Quiz 2.1 Under which of the following conditions is the magnitude of the average velocity of a particle moving in one dimension smaller than the average speed over some time interval? **(a)** A particle moves in the $+x$ direction without reversing. **(b)** A particle moves in the $-x$ direction without reversing. **(c)** A particle moves in the $+x$ direction and then reverses the direction of its motion. **(d)** There are no conditions for which this is true.

Example 2.1   **Calculating the Average Velocity and Speed**

Find the displacement, average velocity, and average speed of the car in Figure 2.1a between positions Ⓐ and Ⓕ.

▶ **2.1** continued

SOLUTION

Consult Figure 2.1 to form a mental image of the car and its motion. We model the car as a particle. From the position–time graph given in Figure 2.1b, notice that $x_{\text{Ⓐ}} = 30$ m at $t_{\text{Ⓐ}} = 0$ s and that $x_{\text{Ⓕ}} = -53$ m at $t_{\text{Ⓕ}} = 50$ s.

Use Equation 2.1 to find the displacement of the car: $\Delta x = x_{\text{Ⓕ}} - x_{\text{Ⓐ}} = -53$ m $- 30$ m $= \boxed{-83 \text{ m}}$

This result means that the car ends up 83 m in the negative direction (to the left, in this case) from where it started. This number has the correct units and is of the same order of magnitude as the supplied data. A quick look at Figure 2.1a indicates that it is the correct answer.

Use Equation 2.2 to find the car's average velocity: $v_{x,\text{avg}} = \dfrac{x_{\text{Ⓕ}} - x_{\text{Ⓐ}}}{t_{\text{Ⓕ}} - t_{\text{Ⓐ}}}$

$$= \frac{-53 \text{ m} - 30 \text{ m}}{50 \text{ s} - 0 \text{ s}} = \frac{-83 \text{ m}}{50 \text{ s}} = \boxed{-1.7 \text{ m/s}}$$

We cannot unambiguously find the average speed of the car from the data in Table 2.1 because we do not have information about the positions of the car between the data points. If we adopt the assumption that the details of the car's position are described by the curve in Figure 2.1b, the distance traveled is 22 m (from Ⓐ to Ⓑ) plus 105 m (from Ⓑ to Ⓕ), for a total of 127 m.

Use Equation 2.3 to find the car's average speed: $v_{\text{avg}} = \dfrac{127 \text{ m}}{50 \text{ s}} = \boxed{2.5 \text{ m/s}}$

Notice that the average speed is positive, as it must be. Suppose the red-brown curve in Figure 2.1b were different so that between 0 s and 10 s it went from Ⓐ up to 100 m and then came back down to Ⓑ. The average speed of the car would change because the distance is different, but the average velocity would not change.

2.2 Instantaneous Velocity and Speed

Often we need to know the velocity of a particle at a particular instant in time t rather than the average velocity over a finite time interval Δt. In other words, you would like to be able to specify your velocity just as precisely as you can specify your position by noting what is happening at a specific clock reading, that is, at some specific instant. What does it mean to talk about how quickly something is moving if we "freeze time" and talk only about an individual instant? In the late 1600s, with the invention of calculus, scientists began to understand how to describe an object's motion at any moment in time.

To see how that is done, consider Figure 2.3a (page 22), which is a reproduction of the graph in Figure 2.1b. What is the particle's velocity at $t = 0$? We have already discussed the average velocity for the interval during which the car moved from position Ⓐ to position Ⓑ (given by the slope of the blue line) and for the interval during which it moved from Ⓐ to Ⓕ (represented by the slope of the longer blue line and calculated in Example 2.1). The car starts out by moving to the right, which we defined to be the positive direction. Therefore, being positive, the value of the average velocity during the interval from Ⓐ to Ⓑ is more representative of the initial velocity than is the value of the average velocity during the interval from Ⓐ to Ⓕ, which we determined to be negative in Example 2.1. Now let us focus on the short blue line and slide point Ⓑ to the left along the curve, toward point Ⓐ, as in Figure 2.3b. The line between the points becomes steeper and steeper, and as the two points become extremely close together, the line becomes a tangent line to the curve, indicated by the green line in Figure 2.3b. The slope of this tangent line

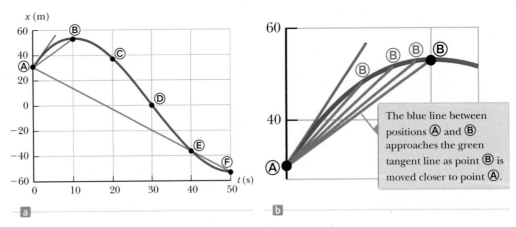

Figure 2.3 (a) Graph representing the motion of the car in Figure 2.1. (b) An enlargement of the upper-left-hand corner of the graph.

Pitfall Prevention 2.2

Slopes of Graphs In any graph of physical data, the *slope* represents the ratio of the change in the quantity represented on the vertical axis to the change in the quantity represented on the horizontal axis. Remember that *a slope has units* (unless both axes have the same units). The units of slope in Figures 2.1b and 2.3 are meters per second, the units of velocity.

▶ **Instantaneous velocity**

Pitfall Prevention 2.3

Instantaneous Speed and Instantaneous Velocity In Pitfall Prevention 2.1, we argued that the magnitude of the average velocity is not the average speed. The magnitude of the instantaneous velocity, however, *is* the instantaneous speed. In an infinitesimal time interval, the magnitude of the displacement is equal to the distance traveled by the particle.

represents the velocity of the car at point Ⓐ. What we have done is determine the *instantaneous velocity* at that moment. In other words, the **instantaneous velocity** v_x equals the limiting value of the ratio $\Delta x / \Delta t$ as Δt approaches zero:[1]

$$v_x \equiv \lim_{\Delta t \to 0} \frac{\Delta x}{\Delta t} \qquad (2.4)$$

In calculus notation, this limit is called the *derivative* of x with respect to t, written dx/dt:

$$v_x \equiv \lim_{\Delta t \to 0} \frac{\Delta x}{\Delta t} = \frac{dx}{dt} \qquad (2.5)$$

The instantaneous velocity can be positive, negative, or zero. When the slope of the position–time graph is positive, such as at any time during the first 10 s in Figure 2.3, v_x is positive and the car is moving toward larger values of x. After point Ⓑ, v_x is negative because the slope is negative and the car is moving toward smaller values of x. At point Ⓑ, the slope and the instantaneous velocity are zero and the car is momentarily at rest.

From here on, we use the word *velocity* to designate instantaneous velocity. When we are interested in *average velocity*, we shall always use the adjective *average*.

The **instantaneous speed** of a particle is defined as the magnitude of its instantaneous velocity. As with average speed, instantaneous speed has no direction associated with it. For example, if one particle has an instantaneous velocity of +25 m/s along a given line and another particle has an instantaneous velocity of −25 m/s along the same line, both have a speed[2] of 25 m/s.

Ⓠuick Quiz 2.2 Are members of the highway patrol more interested in **(a)** your average speed or **(b)** your instantaneous speed as you drive?

Conceptual Example 2.2   **The Velocity of Different Objects**

Consider the following one-dimensional motions: **(A)** a ball thrown directly upward rises to a highest point and falls back into the thrower's hand; **(B)** a race car starts from rest and speeds up to 100 m/s; and **(C)** a spacecraft drifts through space at constant velocity. Are there any points in the motion of these objects at which the instantaneous velocity has the same value as the average velocity over the entire motion? If so, identify the point(s)?

---

[1] Notice that the displacement Δx also approaches zero as Δt approaches zero, so the ratio looks like 0/0. While this ratio may appear to be difficult to evaluate, the ratio does have a specific value. As Δx and Δt become smaller and smaller, the ratio $\Delta x / \Delta t$ approaches a value equal to the slope of the line tangent to the *x*-versus-*t* curve.

[2] As with velocity, we drop the adjective for instantaneous speed. *Speed* means "instantaneous speed."

▶ **2.2** c o n t i n u e d

SOLUTION

(A) The average velocity for the thrown ball is zero because the ball returns to the starting point; therefore, its displacement is zero. There is one point at which the instantaneous velocity is zero: at the top of the motion.

(B) The car's average velocity cannot be evaluated unambiguously with the information given, but it must have some value between 0 and 100 m/s. Because the car will have every instantaneous velocity between 0 and 100 m/s at some time during the interval, there must be some instant at which the instantaneous velocity is equal to the average velocity over the entire motion.

(C) Because the spacecraft's instantaneous velocity is constant, its instantaneous velocity at *any* time and its average velocity over *any* time interval are the same.

| Example 2.3 | **Average and Instantaneous Velocity** |

A particle moves along the *x* axis. Its position varies with time according to the expression $x = -4t + 2t^2$, where *x* is in meters and *t* is in seconds.[3] The position–time graph for this motion is shown in Figure 2.4a. Because the position of the particle is given by a mathematical function, the motion of the particle is completely known, unlike that of the car in Figure 2.1. Notice that the particle moves in the negative *x* direction for the first second of motion, is momentarily at rest at the moment $t = 1$ s, and moves in the positive *x* direction at times $t > 1$ s.

(A) Determine the displacement of the particle in the time intervals $t = 0$ to $t = 1$ s and $t = 1$ s to $t = 3$ s.

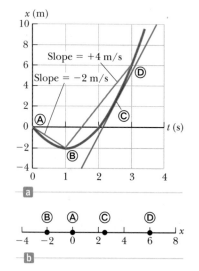

Figure 2.4 (Example 2.3) (a) Position–time graph for a particle having an *x* coordinate that varies in time according to the expression $x = -4t + 2t^2$. (b) The particle moves in one dimension along the *x* axis.

SOLUTION

From the graph in Figure 2.4a, form a mental representation of the particle's motion. Keep in mind that the particle does not move in a curved path in space such as that shown by the red-brown curve in the graphical representation. The particle moves only along the *x* axis in one dimension as shown in Figure 2.4b. At $t = 0$, is it moving to the right or to the left?

During the first time interval, the slope is negative and hence the average velocity is negative. Therefore, we know that the displacement between Ⓐ and Ⓑ must be a negative number having units of meters. Similarly, we expect the displacement between Ⓑ and Ⓓ to be positive.

In the first time interval, set $t_i = t_Ⓐ = 0$ and $t_f = t_Ⓑ = 1$ s and use Equation 2.1 to find the displacement:

$$\Delta x_{Ⓐ \to Ⓑ} = x_f - x_i = x_Ⓑ - x_Ⓐ$$

$$= [-4(1) + 2(1)^2] - [-4(0) + 2(0)^2] = \boxed{-2 \text{ m}}$$

For the second time interval ($t = 1$ s to $t = 3$ s), set $t_i = t_Ⓑ = 1$ s and $t_f = t_Ⓓ = 3$ s:

$$\Delta x_{Ⓑ \to Ⓓ} = x_f - x_i = x_Ⓓ - x_Ⓑ$$

$$= [-4(3) + 2(3)^2] - [-4(1) + 2(1)^2] = \boxed{+8 \text{ m}}$$

These displacements can also be read directly from the position–time graph.

(B) Calculate the average velocity during these two time intervals.

continued

---

[3]Simply to make it easier to read, we write the expression as $x = -4t + 2t^2$ rather than as $x = (-4.00 \text{ m/s})t + (2.00 \text{ m/s}^2)t^{2.00}$. When an equation summarizes measurements, consider its coefficients and exponents to have as many significant figures as other data quoted in a problem. Consider its coefficients to have the units required for dimensional consistency. When we start our clocks at $t = 0$, we usually do not mean to limit the precision to a single digit. Consider any zero value in this book to have as many significant figures as you need.

▶ **2.3** continued

SOLUTION

In the first time interval, use Equation 2.2 with $\Delta t =$
$t_f - t_i = t_⑧ - t_④ = 1$ s:

$$v_{x,\text{avg}\,(④\,\to\,⑧)} = \frac{\Delta x_{④\,\to\,⑧}}{\Delta t} = \frac{-2 \text{ m}}{1 \text{ s}} = \boxed{-2 \text{ m/s}}$$

In the second time interval, $\Delta t = 2$ s:

$$v_{x,\text{avg}\,(⑧\,\to\,Ⓓ)} = \frac{\Delta x_{⑧\,\to\,Ⓓ}}{\Delta t} = \frac{8 \text{ m}}{2 \text{ s}} = \boxed{+4 \text{ m/s}}$$

These values are the same as the slopes of the blue lines joining these points in Figure 2.4a.

(C) Find the instantaneous velocity of the particle at $t = 2.5$ s.

SOLUTION

Measure the slope of the green line at $t = 2.5$ s (point Ⓒ) in Figure 2.4a:

$$v_x = \frac{10 \text{ m} - (-4 \text{ m})}{3.8 \text{ s} - 1.5 \text{ s}} = \boxed{+6 \text{ m/s}}$$

Notice that this instantaneous velocity is on the same order of magnitude as our previous results, that is, a few meters per second. Is that what you would have expected?

2.3 Analysis Model: Particle Under Constant Velocity

◀ Analysis model

In Section 1.2 we discussed the importance of making models. A particularly important model used in the solution to physics problems is an *analysis model*. An **analysis model** is a common situation that occurs time and again when solving physics problems. Because it represents a common situation, it also represents a common type of problem that we have solved before. When you identify an analysis model in a new problem, the solution to the new problem can be modeled after that of the previously-solved problem. Analysis models help us to recognize those common situations and guide us toward a solution to the problem. The form that an analysis model takes is a description of either (1) the behavior of some physical entity or (2) the interaction between that entity and the environment. When you encounter a new problem, you should identify the fundamental details of the problem and attempt to recognize which of the situations you have already seen that might be used as a model for the new problem. For example, suppose an automobile is moving along a straight freeway at a constant speed. Is it important that it is an automobile? Is it important that it is a freeway? If the answers to both questions are no, but the car moves in a straight line at constant speed, we model the automobile as a *particle under constant velocity*, which we will discuss in this section. Once the problem has been modeled, it is no longer about an automobile. It is about a particle undergoing a certain type of motion, a motion that we have studied before.

This method is somewhat similar to the common practice in the legal profession of finding "legal precedents." If a previously resolved case can be found that is very similar legally to the current one, it is used as a model and an argument is made in court to link them logically. The finding in the previous case can then be used to sway the finding in the current case. We will do something similar in physics. For a given problem, we search for a "physics precedent," a model with which we are already familiar and that can be applied to the current problem.

All of the analysis models that we will develop are based on four fundamental simplification models. The first of the four is the particle model discussed in the introduction to this chapter. We will look at a particle under various behaviors and environmental interactions. Further analysis models are introduced in later chapters based on simplification models of a *system*, a *rigid object*, and a *wave*. Once we have introduced these analysis models, we shall see that they appear again and again in different problem situations.

When solving a problem, you should avoid browsing through the chapter looking for an equation that contains the unknown variable that is requested in the problem. In many cases, the equation you find may have nothing to do with the problem you are attempting to solve. It is *much* better to take this first step: **Identify the analysis model that is appropriate for the problem.** To do so, think carefully about what is going on in the problem and match it to a situation you have seen before. Once the analysis model is identified, there are a small number of equations from which to choose that are appropriate for that model, sometimes only one equation. Therefore, **the model tells you which equation(s) to use for the mathematical representation.**

Let us use Equation 2.2 to build our first analysis model for solving problems. We imagine a particle moving with a constant velocity. The model of a **particle under constant velocity** can be applied in *any* situation in which an entity that can be modeled as a particle is moving with constant velocity. This situation occurs frequently, so this model is important.

If the velocity of a particle is constant, its instantaneous velocity at any instant during a time interval is the same as the average velocity over the interval. That is, $v_x = v_{x,\text{avg}}$. Therefore, Equation 2.2 gives us an equation to be used in the mathematical representation of this situation:

$$v_x = \frac{\Delta x}{\Delta t} \qquad (2.6)$$

Remembering that $\Delta x = x_f - x_i$, we see that $v_x = (x_f - x_i)/\Delta t$, or

$$x_f = x_i + v_x \Delta t$$

This equation tells us that the position of the particle is given by the sum of its original position x_i at time $t = 0$ plus the displacement $v_x \Delta t$ that occurs during the time interval Δt. In practice, we usually choose the time at the beginning of the interval to be $t_i = 0$ and the time at the end of the interval to be $t_f = t$, so our equation becomes

$$x_f = x_i + v_x t \quad \text{(for constant } v_x) \qquad (2.7)$$

Equations 2.6 and 2.7 are the primary equations used in the model of a particle under constant velocity. Whenever you have identified the analysis model in a problem to be the particle under constant velocity, you can immediately turn to these equations.

Figure 2.5 is a graphical representation of the particle under constant velocity. On this position–time graph, the slope of the line representing the motion is constant and equal to the magnitude of the velocity. Equation 2.7, which is the equation of a straight line, is the mathematical representation of the particle under constant velocity model. The slope of the straight line is v_x and the y intercept is x_i in both representations.

Example 2.4 below shows an application of the particle under constant velocity model. Notice the analysis model icon **AM**, which will be used to identify examples in which analysis models are employed in the solution. Because of the widespread benefits of using the analysis model approach, you will notice that a large number of the examples in the book will carry such an icon.

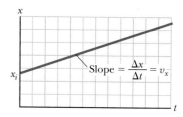

Figure 2.5 Position–time graph for a particle under constant velocity. The value of the constant velocity is the slope of the line.

◀ **Position as a function of time for the particle under constant velocity model**

| Example 2.4 | **Modeling a Runner as a Particle** AM |

A kinesiologist is studying the biomechanics of the human body. (*Kinesiology* is the study of the movement of the human body. Notice the connection to the word *kinematics*.) She determines the velocity of an experimental subject while he runs along a straight line at a constant rate. The kinesiologist starts the stopwatch at the moment the runner passes a given point and stops it after the runner has passed another point 20 m away. The time interval indicated on the stopwatch is 4.0 s.

(A) What is the runner's velocity?

continued

▶ **2.4** continued

SOLUTION

We model the moving runner as a particle because the size of the runner and the movement of arms and legs are unnecessary details. Because the problem states that the subject runs at a constant rate, we can model him as a *particle under constant velocity*.

Having identified the model, we can use Equation 2.6 to find the constant velocity of the runner:

$$v_x = \frac{\Delta x}{\Delta t} = \frac{x_f - x_i}{\Delta t} = \frac{20 \text{ m} - 0}{4.0 \text{ s}} = \boxed{5.0 \text{ m/s}}$$

(B) If the runner continues his motion after the stopwatch is stopped, what is his position after 10 s have passed?

SOLUTION

Use Equation 2.7 and the velocity found in part (A) to find the position of the particle at time $t = 10$ s:

$$x_f = x_i + v_x t = 0 + (5.0 \text{ m/s})(10 \text{ s}) = \boxed{50 \text{ m}}$$

Is the result for part (A) a reasonable speed for a human? How does it compare to world-record speeds in 100-m and 200-m sprints? Notice the value in part (B) is more than twice that of the 20-m position at which the stopwatch was stopped. Is this value consistent with the time of 10 s being more than twice the time of 4.0 s?

The mathematical manipulations for the particle under constant velocity stem from Equation 2.6 and its descendent, Equation 2.7. These equations can be used to solve for any variable in the equations that happens to be unknown if the other variables are known. For example, in part (B) of Example 2.4, we find the position when the velocity and the time are known. Similarly, if we know the velocity and the final position, we could use Equation 2.7 to find the time at which the runner is at this position.

A particle under constant velocity moves with a constant speed along a straight line. Now consider a particle moving with a constant speed through a distance d along a curved path. This situation can be represented with the model of a **particle under constant speed.** The primary equation for this model is Equation 2.3, with the average speed v_{avg} replaced by the constant speed v:

$$v = \frac{d}{\Delta t} \tag{2.8}$$

As an example, imagine a particle moving at a constant speed in a circular path. If the speed is 5.00 m/s and the radius of the path is 10.0 m, we can calculate the time interval required to complete one trip around the circle:

$$v = \frac{d}{\Delta t} \quad \rightarrow \quad \Delta t = \frac{d}{v} = \frac{2\pi r}{v} = \frac{2\pi(10.0 \text{ m})}{5.00 \text{ m/s}} = 12.6 \text{ s}$$

| Analysis Model | **Particle Under Constant Velocity** |

Imagine a moving object that can be modeled as a particle. If it moves at a constant speed through a displacement Δx in a straight line in a time interval Δt, its constant velocity is

$$v_x = \frac{\Delta x}{\Delta t} \tag{2.6}$$

The position of the particle as a function of time is given by

$$x_f = x_i + v_x t \tag{2.7}$$

Examples:

- a meteoroid traveling through gravity-free space
- a car traveling at a constant speed on a straight highway
- a runner traveling at constant speed on a perfectly straight path
- an object moving at terminal speed through a viscous medium (Chapter 6)

Analysis Model **Particle Under Constant Speed**

Imagine a moving object that can be modeled as a particle. If it moves at a constant speed through a distance d along a straight line or a curved path in a time interval Δt, its constant speed is

$$v = \frac{d}{\Delta t} \qquad (2.8)$$

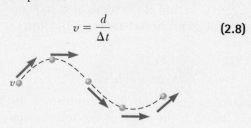

Examples:

- a planet traveling around a perfectly circular orbit
- a car traveling at a constant speed on a curved racetrack
- a runner traveling at constant speed on a curved path
- a charged particle moving through a uniform magnetic field (Chapter 29)

2.4 Acceleration

In Example 2.3, we worked with a common situation in which the velocity of a particle changes while the particle is moving. When the velocity of a particle changes with time, the particle is said to be *accelerating*. For example, the magnitude of a car's velocity increases when you step on the gas and decreases when you apply the brakes. Let us see how to quantify acceleration.

Suppose an object that can be modeled as a particle moving along the x axis has an initial velocity v_{xi} at time t_i at position Ⓐ and a final velocity v_{xf} at time t_f at position Ⓑ as in Figure 2.6a. The red-brown curve in Figure 2.6b shows how the velocity varies with time. The **average acceleration** $a_{x,\text{avg}}$ of the particle is defined as the *change* in velocity Δv_x divided by the time interval Δt during which that change occurs:

$$a_{x,\text{avg}} \equiv \frac{\Delta v_x}{\Delta t} = \frac{v_{xf} - v_{xi}}{t_f - t_i} \qquad (2.9)$$

◀ Average acceleration

As with velocity, when the motion being analyzed is one dimensional, we can use positive and negative signs to indicate the direction of the acceleration. Because the dimensions of velocity are L/T and the dimension of time is T, acceleration has dimensions of length divided by time squared, or L/T². The SI unit of acceleration is meters per second squared (m/s²). It might be easier to interpret these units if you think of them as meters per second per second. For example, suppose an object has an acceleration of +2 m/s². You can interpret this value by forming a mental image of the object having a velocity that is along a straight line and is increasing by 2 m/s during every time interval of 1 s. If the object starts from rest,

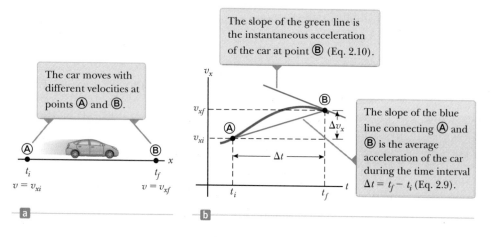

The car moves with different velocities at points Ⓐ and Ⓑ.

The slope of the green line is the instantaneous acceleration of the car at point Ⓑ (Eq. 2.10).

The slope of the blue line connecting Ⓐ and Ⓑ is the average acceleration of the car during the time interval $\Delta t = t_f - t_i$ (Eq. 2.9).

Figure 2.6 (a) A car, modeled as a particle, moving along the x axis from Ⓐ to Ⓑ, has velocity v_{xi} at $t = t_i$ and velocity v_{xf} at $t = t_f$. (b) Velocity–time graph (red-brown) for the particle moving in a straight line.

you should be able to picture it moving at a velocity of +2 m/s after 1 s, at +4 m/s after 2 s, and so on.

In some situations, the value of the average acceleration may be different over different time intervals. It is therefore useful to define the **instantaneous acceleration** as the limit of the average acceleration as Δt approaches zero. This concept is analogous to the definition of instantaneous velocity discussed in Section 2.2. If we imagine that point Ⓐ is brought closer and closer to point Ⓑ in Figure 2.6a and we take the limit of $\Delta v_x/\Delta t$ as Δt approaches zero, we obtain the instantaneous acceleration at point Ⓑ:

Instantaneous acceleration ▶

$$a_x \equiv \lim_{\Delta t \to 0} \frac{\Delta v_x}{\Delta t} = \frac{dv_x}{dt} \qquad (2.10)$$

That is, the instantaneous acceleration equals the derivative of the velocity with respect to time, which by definition is the slope of the velocity–time graph. The slope of the green line in Figure 2.6b is equal to the instantaneous acceleration at point Ⓑ. Notice that Figure 2.6b is a *velocity–time* graph, not a *position–time* graph like Figures 2.1b, 2.3, 2.4, and 2.5. Therefore, we see that just as the velocity of a moving particle is the slope at a point on the particle's x–t graph, the acceleration of a particle is the slope at a point on the particle's v_x–t graph. One can interpret the derivative of the velocity with respect to time as the time rate of change of velocity. If a_x is positive, the acceleration is in the positive x direction; if a_x is negative, the acceleration is in the negative x direction.

Figure 2.7 illustrates how an acceleration–time graph is related to a velocity–time graph. The acceleration at any time is the slope of the velocity–time graph at that time. Positive values of acceleration correspond to those points in Figure 2.7a where the velocity is increasing in the positive x direction. The acceleration reaches a maximum at time $t_Ⓐ$, when the slope of the velocity–time graph is a maximum. The acceleration then goes to zero at time $t_Ⓑ$, when the velocity is a maximum (that is, when the slope of the v_x–t graph is zero). The acceleration is negative when the velocity is decreasing in the positive x direction, and it reaches its most negative value at time $t_Ⓒ$.

Ⓠuick Quiz 2.3 Make a velocity–time graph for the car in Figure 2.1a. Suppose the speed limit for the road on which the car is driving is 30 km/h. True or False? The car exceeds the speed limit at some time within the time interval $0 - 50$ s.

For the case of motion in a straight line, the direction of the velocity of an object and the direction of its acceleration are related as follows. When the object's velocity and acceleration are in the same direction, the object is speeding up. On the other hand, when the object's velocity and acceleration are in opposite directions, the object is slowing down.

To help with this discussion of the signs of velocity and acceleration, we can relate the acceleration of an object to the total *force* exerted on the object. In Chapter 5, we formally establish that **the force on an object is proportional to the acceleration of the object:**

$$F_x \propto a_x \qquad (2.11)$$

This proportionality indicates that acceleration is caused by force. Furthermore, force and acceleration are both vectors, and the vectors are in the same direction. Therefore, let us think about the signs of velocity and acceleration by imagining a force applied to an object and causing it to accelerate. Let us assume the velocity and acceleration are in the same direction. This situation corresponds to an object that experiences a force acting in the same direction as its velocity. In this case, the object speeds up! Now suppose the velocity and acceleration are in opposite directions. In this situation, the object moves in some direction and experiences a force acting in the opposite direction. Therefore, the object slows

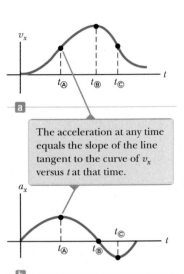

The acceleration at any time equals the slope of the line tangent to the curve of v_x versus t at that time.

Figure 2.7 (a) The velocity–time graph for a particle moving along the x axis. (b) The instantaneous acceleration can be obtained from the velocity–time graph.

down! It is very useful to equate the direction of the acceleration to the direction of a force because it is easier from our everyday experience to think about what effect a force will have on an object than to think only in terms of the direction of the acceleration.

> **Q**uick Quiz 2.4 If a car is traveling eastward and slowing down, what is the direction of the force on the car that causes it to slow down? **(a)** eastward **(b)** westward **(c)** neither eastward nor westward

From now on, we shall use the term *acceleration* to mean instantaneous acceleration. When we mean average acceleration, we shall always use the adjective *average*. Because $v_x = dx/dt$, the acceleration can also be written as

$$a_x = \frac{dv_x}{dt} = \frac{d}{dt}\left(\frac{dx}{dt}\right) = \frac{d^2x}{dt^2} \qquad (2.12)$$

That is, in one-dimensional motion, the acceleration equals the *second derivative* of x with respect to time.

Conceptual Example 2.5   Graphical Relationships Between x, v_x, and a_x

The position of an object moving along the x axis varies with time as in Figure 2.8a. Graph the velocity versus time and the acceleration versus time for the object.

SOLUTION

The velocity at any instant is the slope of the tangent to the x–t graph at that instant. Between $t = 0$ and $t = t_{Ⓐ}$, the slope of the x–t graph increases uniformly, so the velocity increases linearly as shown in Figure 2.8b. Between $t_{Ⓐ}$ and $t_{Ⓑ}$, the slope of the x–t graph is constant, so the velocity remains constant. Between $t_{Ⓑ}$ and $t_{Ⓓ}$, the slope of the x–t graph decreases, so the value of the velocity in the v_x–t graph decreases. At $t_{Ⓓ}$, the slope of the x–t graph is zero, so the velocity is zero at that instant. Between $t_{Ⓓ}$ and $t_{Ⓔ}$, the slope of the x–t graph and therefore the velocity are negative and decrease uniformly in this interval. In the interval $t_{Ⓔ}$ to $t_{Ⓕ}$, the slope of the x–t graph is still negative, and at $t_{Ⓕ}$ it goes to zero. Finally, after $t_{Ⓕ}$, the slope of the x–t graph is zero, meaning that the object is at rest for $t > t_{Ⓕ}$.

The acceleration at any instant is the slope of the tangent to the v_x–t graph at that instant. The graph of acceleration versus time for this object is shown in Figure 2.8c. The acceleration is constant and positive between 0 and $t_{Ⓐ}$, where the slope of the v_x–t graph is positive. It is zero between $t_{Ⓐ}$ and $t_{Ⓑ}$ and for $t > t_{Ⓕ}$ because the slope of the v_x–t graph is zero at these times. It is negative between $t_{Ⓑ}$ and $t_{Ⓔ}$ because the slope of the v_x–t graph is negative during this interval. Between $t_{Ⓔ}$ and $t_{Ⓕ}$, the acceleration is positive like it is between 0 and $t_{Ⓐ}$, but higher in value because the slope of the v_x–t graph is steeper.

Notice that the sudden changes in acceleration shown in Figure 2.8c are unphysical. Such instantaneous changes cannot occur in reality.

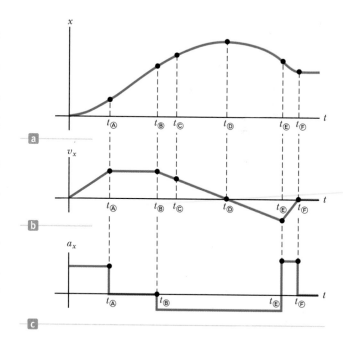

Figure 2.8 (Conceptual Example 2.5) (a) Position–time graph for an object moving along the x axis. (b) The velocity–time graph for the object is obtained by measuring the slope of the position–time graph at each instant. (c) The acceleration–time graph for the object is obtained by measuring the slope of the velocity–time graph at each instant.

Example 2.6 Average and Instantaneous Acceleration

The velocity of a particle moving along the x axis varies according to the expression $v_x = 40 - 5t^2$, where v_x is in meters per second and t is in seconds.

(A) Find the average acceleration in the time interval $t = 0$ to $t = 2.0$ s.

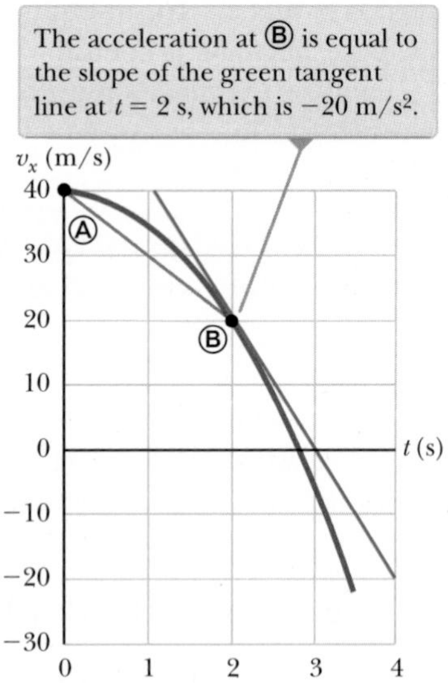

The acceleration at Ⓑ is equal to the slope of the green tangent line at $t = 2$ s, which is -20 m/s².

Figure 2.9 (Example 2.6) The velocity–time graph for a particle moving along the x axis according to the expression $v_x = 40 - 5t^2$.

SOLUTION

Think about what the particle is doing from the mathematical representation. Is it moving at $t = 0$? In which direction? Does it speed up or slow down? Figure 2.9 is a v_x–t graph that was created from the velocity versus time expression given in the problem statement. Because the slope of the entire v_x–t curve is negative, we expect the acceleration to be negative.

Find the velocities at $t_i = t_Ⓐ = 0$ and $t_f = t_Ⓑ = 2.0$ s by substituting these values of t into the expression for the velocity:

$$v_{xⒶ} = 40 - 5t_Ⓐ^2 = 40 - 5(0)^2 = +40 \text{ m/s}$$

$$v_{xⒷ} = 40 - 5t_Ⓑ^2 = 40 - 5(2.0)^2 = +20 \text{ m/s}$$

Find the average acceleration in the specified time interval $\Delta t = t_Ⓑ - t_Ⓐ = 2.0$ s:

$$a_{x,\text{avg}} = \frac{v_{xf} - v_{xi}}{t_f - t_i} = \frac{v_{xⒷ} - v_{xⒶ}}{t_Ⓑ - t_Ⓐ} = \frac{20 \text{ m/s} - 40 \text{ m/s}}{2.0 \text{ s} - 0 \text{ s}}$$

$$= \boxed{-10 \text{ m/s}^2}$$

The negative sign is consistent with our expectations: the average acceleration, represented by the slope of the blue line joining the initial and final points on the velocity–time graph, is negative.

(B) Determine the acceleration at $t = 2.0$ s.

SOLUTION

Knowing that the initial velocity at any time t is $v_{xi} = 40 - 5t^2$, find the velocity at any later time $t + \Delta t$:

$$v_{xf} = 40 - 5(t + \Delta t)^2 = 40 - 5t^2 - 10t\,\Delta t - 5(\Delta t)^2$$

Find the change in velocity over the time interval Δt:

$$\Delta v_x = v_{xf} - v_{xi} = -10t\,\Delta t - 5(\Delta t)^2$$

To find the acceleration at any time t, divide this expression by Δt and take the limit of the result as Δt approaches zero:

$$a_x = \lim_{\Delta t \to 0} \frac{\Delta v_x}{\Delta t} = \lim_{\Delta t \to 0}(-10t - 5\,\Delta t) = -10t$$

Substitute $t = 2.0$ s:

$$a_x = (-10)(2.0) \text{ m/s}^2 = \boxed{-20 \text{ m/s}^2}$$

Because the velocity of the particle is positive and the acceleration is negative at this instant, the particle is slowing down.

Notice that the answers to parts (A) and (B) are different. The average acceleration in part (A) is the slope of the blue line in Figure 2.9 connecting points Ⓐ and Ⓑ. The instantaneous acceleration in part (B) is the slope of the green line tangent to the curve at point Ⓑ. Notice also that the acceleration is *not* constant in this example. Situations involving constant acceleration are treated in Section 2.6.

So far, we have evaluated the derivatives of a function by starting with the definition of the function and then taking the limit of a specific ratio. If you are familiar with calculus, you should recognize that there are specific rules for taking derivatives. These rules, which are listed in Appendix B.6, enable us to evaluate derivatives quickly. For instance, one rule tells us that the derivative of any constant is zero. As another example, suppose x is proportional to some power of t such as in the expression

$$x = At^n$$

where A and n are constants. (This expression is a very common functional form.) The derivative of x with respect to t is

$$\frac{dx}{dt} = nAt^{n-1}$$

Applying this rule to Example 2.6, in which $v_x = 40 - 5t^2$, we quickly find that the acceleration is $a_x = dv_x/dt = -10t$, as we found in part (B) of the example.

2.5 Motion Diagrams

The concepts of velocity and acceleration are often confused with each other, but in fact they are quite different quantities. In forming a mental representation of a moving object, a pictorial representation called a *motion diagram* is sometimes useful to describe the velocity and acceleration while an object is in motion.

A motion diagram can be formed by imagining a *stroboscopic* photograph of a moving object, which shows several images of the object taken as the strobe light flashes at a constant rate. Figure 2.1a is a motion diagram for the car studied in Section 2.1. Figure 2.10 represents three sets of strobe photographs of cars moving along a straight roadway in a single direction, from left to right. The time intervals between flashes of the stroboscope are equal in each part of the diagram. So as to not confuse the two vector quantities, we use red arrows for velocity and purple arrows for acceleration in Figure 2.10. The arrows are shown at several instants during the motion of the object. Let us describe the motion of the car in each diagram.

In Figure 2.10a, the images of the car are equally spaced, showing us that the car moves through the same displacement in each time interval. This equal spacing is consistent with the car moving with *constant positive velocity* and *zero acceleration*. We could model the car as a particle and describe it with the particle under constant velocity model.

In Figure 2.10b, the images become farther apart as time progresses. In this case, the velocity arrow increases in length with time because the car's displacement between adjacent positions increases in time. These features suggest the car is moving with a *positive velocity* and a *positive acceleration*. The velocity and acceleration are in the same direction. In terms of our earlier force discussion, imagine a force pulling on the car in the same direction it is moving: it speeds up.

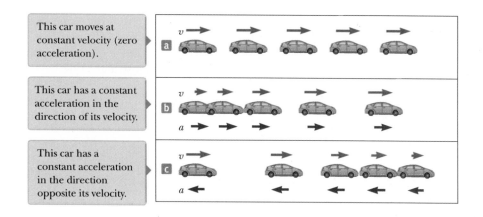

This car moves at constant velocity (zero acceleration).

This car has a constant acceleration in the direction of its velocity.

This car has a constant acceleration in the direction opposite its velocity.

Figure 2.10 Motion diagrams of a car moving along a straight roadway in a single direction. The velocity at each instant is indicated by a red arrow, and the constant acceleration is indicated by a purple arrow.

In Figure 2.10c, we can tell that the car slows as it moves to the right because its displacement between adjacent images decreases with time. This case suggests the car moves to the right with a negative acceleration. The length of the velocity arrow decreases in time and eventually reaches zero. From this diagram, we see that the acceleration and velocity arrows are *not* in the same direction. The car is moving with a *positive velocity*, but with a *negative acceleration*. (This type of motion is exhibited by a car that skids to a stop after its brakes are applied.) The velocity and acceleration are in opposite directions. In terms of our earlier force discussion, imagine a force pulling on the car opposite to the direction it is moving: it slows down.

Each purple acceleration arrow in parts (b) and (c) of Figure 2.10 is the same length. Therefore, these diagrams represent motion of a *particle under constant acceleration*. This important analysis model will be discussed in the next section.

Quick Quiz 2.5 Which one of the following statements is true? **(a)** If a car is traveling eastward, its acceleration must be eastward. **(b)** If a car is slowing down, its acceleration must be negative. **(c)** A particle with constant acceleration can never stop and stay stopped.

2.6 Analysis Model: Particle Under Constant Acceleration

If the acceleration of a particle varies in time, its motion can be complex and difficult to analyze. A very common and simple type of one-dimensional motion, however, is that in which the acceleration is constant. In such a case, the average acceleration $a_{x,\text{avg}}$ over any time interval is numerically equal to the instantaneous acceleration a_x at any instant within the interval, and the velocity changes at the same rate throughout the motion. This situation occurs often enough that we identify it as an analysis model: the **particle under constant acceleration.** In the discussion that follows, we generate several equations that describe the motion of a particle for this model.

If we replace $a_{x,\text{avg}}$ by a_x in Equation 2.9 and take $t_i = 0$ and t_f to be any later time t, we find that

$$a_x = \frac{v_{xf} - v_{xi}}{t - 0}$$

or

$$v_{xf} = v_{xi} + a_x t \quad \text{(for constant } a_x\text{)} \tag{2.13}$$

This powerful expression enables us to determine an object's velocity at *any* time t if we know the object's initial velocity v_{xi} and its (constant) acceleration a_x. A velocity–time graph for this constant-acceleration motion is shown in Figure 2.11b. The graph is a straight line, the slope of which is the acceleration a_x; the (constant) slope is consistent with $a_x = dv_x/dt$ being a constant. Notice that the slope is positive, which indicates a positive acceleration. If the acceleration were negative, the slope of the line in Figure 2.11b would be negative. When the acceleration is constant, the graph of acceleration versus time (Fig. 2.11c) is a straight line having a slope of zero.

Because velocity at constant acceleration varies linearly in time according to Equation 2.13, we can express the average velocity in any time interval as the arithmetic mean of the initial velocity v_{xi} and the final velocity v_{xf}:

$$v_{x,\text{avg}} = \frac{v_{xi} + v_{xf}}{2} \quad \text{(for constant } a_x\text{)} \tag{2.14}$$

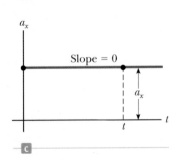

Figure 2.11 A particle under constant acceleration a_x moving along the *x* axis: (a) the position–time graph, (b) the velocity–time graph, and (c) the acceleration–time graph.

Notice that this expression for average velocity applies *only* in situations in which the acceleration is constant.

We can now use Equations 2.1, 2.2, and 2.14 to obtain the position of an object as a function of time. Recalling that Δx in Equation 2.2 represents $x_f - x_i$ and recognizing that $\Delta t = t_f - t_i = t - 0 = t$, we find that

$$x_f - x_i = v_{x,\text{avg}}\, t = \tfrac{1}{2}(v_{xi} + v_{xf})t$$

$$x_f = x_i + \tfrac{1}{2}(v_{xi} + v_{xf})t \quad \text{(for constant } a_x) \tag{2.15}$$

◀ Position as a function of velocity and time for the particle under constant acceleration model

This equation provides the final position of the particle at time t in terms of the initial and final velocities.

We can obtain another useful expression for the position of a particle under constant acceleration by substituting Equation 2.13 into Equation 2.15:

$$x_f = x_i + \tfrac{1}{2}[v_{xi} + (v_{xi} + a_x t)]t$$

$$x_f = x_i + v_{xi}t + \tfrac{1}{2}a_x t^2 \quad \text{(for constant } a_x) \tag{2.16}$$

◀ Position as a function of time for the particle under constant acceleration model

This equation provides the final position of the particle at time t in terms of the initial position, the initial velocity, and the constant acceleration.

The position–time graph for motion at constant (positive) acceleration shown in Figure 2.11a is obtained from Equation 2.16. Notice that the curve is a parabola. The slope of the tangent line to this curve at $t = 0$ equals the initial velocity v_{xi}, and the slope of the tangent line at any later time t equals the velocity v_{xf} at that time.

Finally, we can obtain an expression for the final velocity that does not contain time as a variable by substituting the value of t from Equation 2.13 into Equation 2.15:

$$x_f = x_i + \tfrac{1}{2}(v_{xi} + v_{xf})\left(\frac{v_{xf} - v_{xi}}{a_x}\right) = x_i + \frac{v_{xf}^2 - v_{xi}^2}{2a_x}$$

$$v_{xf}^2 = v_{xi}^2 + 2a_x(x_f - x_i) \quad \text{(for constant } a_x) \tag{2.17}$$

◀ Velocity as a function of position for the particle under constant acceleration model

This equation provides the final velocity in terms of the initial velocity, the constant acceleration, and the position of the particle.

For motion at *zero* acceleration, we see from Equations 2.13 and 2.16 that

$$\left.\begin{array}{l} v_{xf} = v_{xi} = v_x \\ x_f = x_i + v_x t \end{array}\right\} \quad \text{when } a_x = 0$$

That is, when the acceleration of a particle is zero, its velocity is constant and its position changes linearly with time. In terms of models, when the acceleration of a particle is zero, the particle under constant acceleration model reduces to the particle under constant velocity model (Section 2.3).

Equations 2.13 through 2.17 are **kinematic equations** that may be used to solve any problem involving a particle under constant acceleration in one dimension. These equations are listed together for convenience on page 34. The choice of which equation you use in a given situation depends on what you know beforehand. Sometimes it is necessary to use two of these equations to solve for two unknowns. You should recognize that the quantities that vary during the motion are position x_f, velocity v_{xf}, and time t.

You will gain a great deal of experience in the use of these equations by solving a number of exercises and problems. Many times you will discover that more than one method can be used to obtain a solution. Remember that these equations of kinematics *cannot* be used in a situation in which the acceleration varies with time. They can be used only when the acceleration is constant.

Ⓠuick Quiz 2.6 In Figure 2.12, match each v_x–t graph on the top with the a_x–t graph on the bottom that best describes the motion.

Figure 2.12 (Quick Quiz 2.6) Parts (a), (b), and (c) are v_x–t graphs of objects in one-dimensional motion. The possible accelerations of each object as a function of time are shown in scrambled order in (d), (e), and (f).

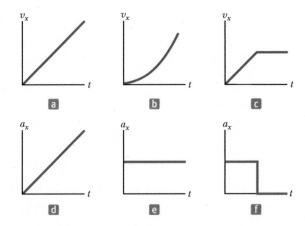

Analysis Model　**Particle Under Constant Acceleration**

Imagine a moving object that can be modeled as a particle. If it begins from position x_i and initial velocity v_{xi} and moves in a straight line with a constant acceleration a_x, its subsequent position and velocity are described by the following kinematic equations:

$$v_{xf} = v_{xi} + a_x t \tag{2.13}$$

$$v_{x,\text{avg}} = \frac{v_{xi} + v_{xf}}{2} \tag{2.14}$$

$$x_f = x_i + \tfrac{1}{2}(v_{xi} + v_{xf})t \tag{2.15}$$

$$x_f = x_i + v_{xi}t + \tfrac{1}{2}a_x t^2 \tag{2.16}$$

$$v_{xf}^2 = v_{xi}^2 + 2a_x(x_f - x_i) \tag{2.17}$$

Examples

- a car accelerating at a constant rate along a straight freeway
- a dropped object in the absence of air resistance (Section 2.7)
- an object on which a constant net force acts (Chapter 5)
- a charged particle in a uniform electric field (Chapter 23)

Example 2.7　**Carrier Landing**　AM

A jet lands on an aircraft carrier at a speed of 140 mi/h ($\approx$ 63 m/s).

(A) What is its acceleration (assumed constant) if it stops in 2.0 s due to an arresting cable that snags the jet and brings it to a stop?

SOLUTION

You might have seen movies or television shows in which a jet lands on an aircraft carrier and is brought to rest surprisingly fast by an arresting cable. A careful reading of the problem reveals that in addition to being given the initial speed of 63 m/s, we also know that the final speed is zero. Because the acceleration of the jet is assumed constant, we model it as a *particle under constant acceleration*. We define our x axis as the direction of motion of the jet. Notice that we have no information about the change in position of the jet while it is slowing down.

▶ **2.7** continued

Equation 2.13 is the only equation in the particle under constant acceleration model that does not involve position, so we use it to find the acceleration of the jet, modeled as a particle:

$$a_x = \frac{v_{xf} - v_{xi}}{t} \approx \frac{0 - 63 \text{ m/s}}{2.0 \text{ s}}$$
$$= -32 \text{ m/s}^2$$

(B) If the jet touches down at position $x_i = 0$, what is its final position?

SOLUTION

Use Equation 2.15 to solve for the final position:

$$x_f = x_i + \tfrac{1}{2}(v_{xi} + v_{xf})t = 0 + \tfrac{1}{2}(63 \text{ m/s} + 0)(2.0 \text{ s}) = \boxed{63 \text{ m}}$$

Given the size of aircraft carriers, a length of 63 m seems reasonable for stopping the jet. The idea of using arresting cables to slow down landing aircraft and enable them to land safely on ships originated at about the time of World War I. The cables are still a vital part of the operation of modern aircraft carriers.

WHAT IF? Suppose the jet lands on the deck of the aircraft carrier with a speed faster than 63 m/s but has the same acceleration due to the cable as that calculated in part (A). How will that change the answer to part (B)?

Answer If the jet is traveling faster at the beginning, it will stop farther away from its starting point, so the answer to part (B) should be larger. Mathematically, we see in Equation 2.15 that if v_{xi} is larger, then x_f will be larger.

Example 2.8 **Watch Out for the Speed Limit!** AM

A car traveling at a constant speed of 45.0 m/s passes a trooper on a motorcycle hidden behind a billboard. One second after the speeding car passes the billboard, the trooper sets out from the billboard to catch the car, accelerating at a constant rate of 3.00 m/s². How long does it take the trooper to overtake the car?

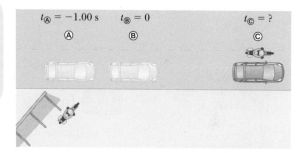

Figure 2.13 (Example 2.8) A speeding car passes a hidden trooper.

SOLUTION

A pictorial representation (Fig. 2.13) helps clarify the sequence of events. The car is modeled as a *particle under constant velocity*, and the trooper is modeled as a *particle under constant acceleration*.

First, we write expressions for the position of each vehicle as a function of time. It is convenient to choose the position of the billboard as the origin and to set $t_{\text{®}} = 0$ as the time the trooper begins moving. At that instant, the car has already traveled a distance of 45.0 m from the billboard because it has traveled at a constant speed of $v_x = 45.0$ m/s for 1 s. Therefore, the initial position of the speeding car is $x_{\text{®}} = 45.0$ m.

Using the particle under constant velocity model, apply Equation 2.7 to give the car's position at any time t:

$$x_{\text{car}} = x_{\text{®}} + v_{x\,\text{car}}t$$

A quick check shows that at $t = 0$, this expression gives the car's correct initial position when the trooper begins to move: $x_{\text{car}} = x_{\text{®}} = 45.0$ m.

The trooper starts from rest at $t_{\text{®}} = 0$ and accelerates at $a_x = 3.00$ m/s² away from the origin. Use Equation 2.16 to give her position at any time t:

$$x_f = x_i + v_{xi}t + \tfrac{1}{2}a_x t^2$$
$$x_{\text{trooper}} = 0 + (0)t + \tfrac{1}{2}a_x t^2 = \tfrac{1}{2}a_x t^2$$

Set the positions of the car and trooper equal to represent the trooper overtaking the car at position ©:

$$x_{\text{trooper}} = x_{\text{car}}$$
$$\tfrac{1}{2}a_x t^2 = x_{\text{®}} + v_{x\,\text{car}}t$$

continued

▶ **2.8** c o n t i n u e d

Rearrange to give a quadratic equation:

$$\tfrac{1}{2}a_x t^2 - v_{x\,car} t - x_{\circledR} = 0$$

Solve the quadratic equation for the time at which the trooper catches the car (for help in solving quadratic equations, see Appendix B.2.):

$$t = \frac{v_{x\,car} \pm \sqrt{v_{x\,car}^2 + 2a_x x_{\circledR}}}{a_x}$$

$$(1) \quad t = \frac{v_{x\,car}}{a_x} \pm \sqrt{\frac{v_{x\,car}^2}{a_x^2} + \frac{2x_{\circledR}}{a_x}}$$

Evaluate the solution, choosing the positive root because that is the only choice consistent with a time $t > 0$:

$$t = \frac{45.0\ \text{m/s}}{3.00\ \text{m/s}^2} + \sqrt{\frac{(45.0\ \text{m/s})^2}{(3.00\ \text{m/s}^2)^2} + \frac{2(45.0\ \text{m})}{3.00\ \text{m/s}^2}} = \boxed{31.0\ \text{s}}$$

Why didn't we choose $t = 0$ as the time at which the car passes the trooper? If we did so, we would not be able to use the particle under constant acceleration model for the trooper. Her acceleration would be zero for the first second and then 3.00 m/s^2 for the remaining time. By defining the time $t = 0$ as when the trooper begins moving, we can use the particle under constant acceleration model for her movement for all positive times.

WHAT IF? What if the trooper had a more powerful motorcycle with a larger acceleration? How would that change the time at which the trooper catches the car?

Answer If the motorcycle has a larger acceleration, the trooper should catch up to the car sooner, so the answer for the time should be less than 31 s. Because all terms on the right side of Equation (1) have the acceleration a_x in the denominator, we see symbolically that increasing the acceleration will decrease the time at which the trooper catches the car.

Georgios Kollidas/Shutterstock.com

Galileo Galilei
Italian physicist and astronomer
(1564–1642)
Galileo formulated the laws that govern the motion of objects in free fall and made many other significant discoveries in physics and astronomy. Galileo publicly defended Nicolaus Copernicus's assertion that the Sun is at the center of the Universe (the heliocentric system). He published *Dialogue Concerning Two New World Systems* to support the Copernican model, a view that the Catholic Church declared to be heretical.

2.7 Freely Falling Objects

It is well known that, in the absence of air resistance, all objects dropped near the Earth's surface fall toward the Earth with the same constant acceleration under the influence of the Earth's gravity. It was not until about 1600 that this conclusion was accepted. Before that time, the teachings of the Greek philosopher Aristotle (384–322 BC) had held that heavier objects fall faster than lighter ones.

The Italian Galileo Galilei (1564–1642) originated our present-day ideas concerning falling objects. There is a legend that he demonstrated the behavior of falling objects by observing that two different weights dropped simultaneously from the Leaning Tower of Pisa hit the ground at approximately the same time. Although there is some doubt that he carried out this particular experiment, it is well established that Galileo performed many experiments on objects moving on inclined planes. In his experiments, he rolled balls down a slight incline and measured the distances they covered in successive time intervals. The purpose of the incline was to reduce the acceleration, which made it possible for him to make accurate measurements of the time intervals. By gradually increasing the slope of the incline, he was finally able to draw conclusions about freely falling objects because a freely falling ball is equivalent to a ball moving down a vertical incline.

You might want to try the following experiment. Simultaneously drop a coin and a crumpled-up piece of paper from the same height. If the effects of air resistance are negligible, both will have the same motion and will hit the floor at the same time. In the idealized case, in which air resistance is absent, such motion is referred to as *free-fall* motion. If this same experiment could be conducted in a vacuum, in which air resistance is truly negligible, the paper and the coin would fall with the same acceleration even when the paper is not crumpled. On August 2, 1971, astronaut David Scott conducted such a demonstration on the Moon. He simultaneously released a hammer and a feather, and the two objects fell together to the lunar surface. This simple demonstration surely would have pleased Galileo!

When we use the expression *freely falling object*, we do not necessarily refer to an object dropped from rest. A freely falling object is any object moving freely under the influence of gravity alone, regardless of its initial motion. Objects thrown upward or downward and those released from rest are all falling freely once they are released. Any freely falling object experiences an acceleration directed *downward*, regardless of its initial motion.

We shall denote the magnitude of the *free-fall acceleration*, also called the *acceleration due to gravity*, by the symbol g. The value of g decreases with increasing altitude above the Earth's surface. Furthermore, slight variations in g occur with changes in latitude. At the Earth's surface, the value of g is approximately 9.80 m/s². Unless stated otherwise, we shall use this value for g when performing calculations. For making quick estimates, use $g = 10$ m/s².

If we neglect air resistance and assume the free-fall acceleration does not vary with altitude over short vertical distances, the motion of a freely falling object moving vertically is equivalent to the motion of a particle under constant acceleration in one dimension. Therefore, the equations developed in Section 2.6 for the particle under constant acceleration model can be applied. The only modification for freely falling objects that we need to make in these equations is to note that the motion is in the vertical direction (the y direction) rather than in the horizontal direction (x) and that the acceleration is downward and has a magnitude of 9.80 m/s². Therefore, we choose $a_y = -g = -9.80$ m/s², where the negative sign means that the acceleration of a freely falling object is downward. In Chapter 13, we shall study how to deal with variations in g with altitude.

Quick **Quiz** 2.7 Consider the following choices: (a) increases, (b) decreases, (c) increases and then decreases, (d) decreases and then increases, (e) remains the same. From these choices, select what happens to **(i)** the acceleration and **(ii)** the speed of a ball after it is thrown upward into the air.

Pitfall Prevention 2.6

g and g Be sure not to confuse the italic symbol g for free-fall acceleration with the nonitalic symbol g used as the abbreviation for the unit gram.

Pitfall Prevention 2.7

The Sign of g Keep in mind that g is a *positive number*. It is tempting to substitute -9.80 m/s² for g, but resist the temptation. Downward gravitational acceleration is indicated explicitly by stating the acceleration as $a_y = -g$.

Pitfall Prevention 2.8

Acceleration at the Top of the Motion A common misconception is that the acceleration of a projectile at the top of its trajectory is zero. Although the velocity at the top of the motion of an object thrown upward momentarily goes to zero, *the acceleration is still that due to gravity* at this point. If the velocity and acceleration were both zero, the projectile would stay at the top.

Conceptual Example 2.9   **The Daring Skydivers**

A skydiver jumps out of a hovering helicopter. A few seconds later, another skydiver jumps out, and they both fall along the same vertical line. Ignore air resistance so that both skydivers fall with the same acceleration. Does the difference in their speeds stay the same throughout the fall? Does the vertical distance between them stay the same throughout the fall?

SOLUTION

At any given instant, the speeds of the skydivers are different because one had a head start. In any time interval Δt after this instant, however, the two skydivers increase their speeds by the same amount because they have the same acceleration. Therefore, the difference in their speeds remains the same throughout the fall.

The first jumper always has a greater speed than the second. Therefore, in a given time interval, the first skydiver covers a greater distance than the second. Consequently, the separation distance between them increases.

Example 2.10 **Not a Bad Throw for a Rookie!** AM

A stone thrown from the top of a building is given an initial velocity of 20.0 m/s straight upward. The stone is launched 50.0 m above the ground, and the stone just misses the edge of the roof on its way down as shown in Figure 2.14.

(A) Using $t_Ⓐ = 0$ as the time the stone leaves the thrower's hand at position Ⓐ, determine the time at which the stone reaches its maximum height.

SOLUTION

You most likely have experience with dropping objects or throwing them upward and watching them fall, so this problem should describe a familiar experience. To simulate this situation, toss a small object upward and notice the time interval required for it to fall to the floor. Now imagine throwing that object upward from the roof of a building. Because the stone is in free fall, it is modeled as a *particle under constant acceleration* due to gravity.

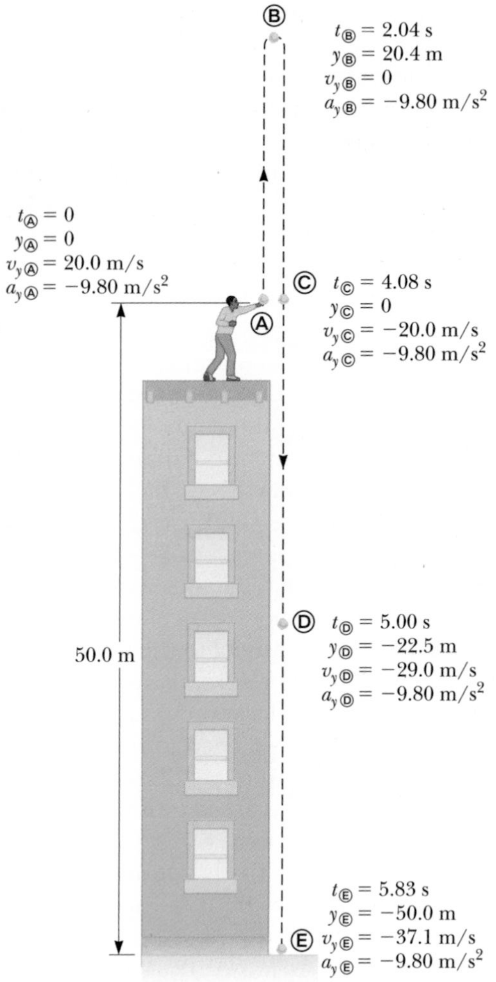

Figure 2.14 (Example 2.10) Position, velocity, and acceleration values at various times for a freely falling stone thrown initially upward with a velocity $v_{yi} = 20.0$ m/s. Many of the quantities in the labels for points in the motion of the stone are calculated in the example. Can you verify the other values that are not?

$t_Ⓑ = 2.04$ s
$y_Ⓑ = 20.4$ m
$v_{yⒷ} = 0$
$a_{yⒷ} = -9.80$ m/s^2

$t_Ⓐ = 0$
$y_Ⓐ = 0$
$v_{yⒶ} = 20.0$ m/s
$a_{yⒶ} = -9.80$ m/s^2

$t_Ⓒ = 4.08$ s
$y_Ⓒ = 0$
$v_{yⒸ} = -20.0$ m/s
$a_{yⒸ} = -9.80$ m/s^2

$t_Ⓓ = 5.00$ s
$y_Ⓓ = -22.5$ m
$v_{yⒹ} = -29.0$ m/s
$a_{yⒹ} = -9.80$ m/s^2

50.0 m

$t_Ⓔ = 5.83$ s
$y_Ⓔ = -50.0$ m
$v_{yⒺ} = -37.1$ m/s
$a_{yⒺ} = -9.80$ m/s^2

Recognize that the initial velocity is positive because the stone is launched upward. The velocity will change sign after the stone reaches its highest point, but the acceleration of the stone will *always* be downward so that it will always have a negative value. Choose an initial point just after the stone leaves the person's hand and a final point at the top of its flight.

Use Equation 2.13 to calculate the time at which the stone reaches its maximum height:

$$v_{yf} = v_{yi} + a_y t \rightarrow t = \frac{v_{yf} - v_{yi}}{a_y}$$

Substitute numerical values:

$$t = t_Ⓑ = \frac{0 - 20.0 \text{ m/s}}{-9.80 \text{ m/s}^2} = \boxed{2.04 \text{ s}}$$

(B) Find the maximum height of the stone.

SOLUTION

As in part (A), choose the initial and final points at the beginning and the end of the upward flight.

Set $y_Ⓐ = 0$ and substitute the time from part (A) into Equation 2.16 to find the maximum height:

$$y_{max} = y_Ⓑ = y_Ⓐ + v_{xⒶ} t + \tfrac{1}{2} a_y t^2$$

$$y_Ⓑ = 0 + (20.0 \text{ m/s})(2.04 \text{ s}) + \tfrac{1}{2}(-9.80 \text{ m/s}^2)(2.04 \text{ s})^2 = \boxed{20.4 \text{ m}}$$

(C) Determine the velocity of the stone when it returns to the height from which it was thrown.

SOLUTION

Choose the initial point where the stone is launched and the final point when it passes this position coming down.

Substitute known values into Equation 2.17:

$$v_{yⒸ}{}^2 = v_{yⒶ}{}^2 + 2a_y(y_Ⓒ - y_Ⓐ)$$

$$v_{yⒸ}{}^2 = (20.0 \text{ m/s})^2 + 2(-9.80 \text{ m/s}^2)(0 - 0) = 400 \text{ m}^2/\text{s}^2$$

$$v_{yⒸ} = \boxed{-20.0 \text{ m/s}}$$

▶ **2.10** continued

When taking the square root, we could choose either a positive or a negative root. We choose the negative root because we know that the stone is moving downward at point Ⓒ. The velocity of the stone when it arrives back at its original height is equal in magnitude to its initial velocity but is opposite in direction.

(D) Find the velocity and position of the stone at $t = 5.00$ s.

SOLUTION

Choose the initial point just after the throw and the final point 5.00 s later.

Calculate the velocity at Ⓓ from Equation 2.13:

$$v_{y\text{Ⓓ}} = v_{y\text{Ⓐ}} + a_y t = 20.0 \text{ m/s} + (-9.80 \text{ m/s}^2)(5.00 \text{ s}) = \boxed{-29.0 \text{ m/s}}$$

Use Equation 2.16 to find the position of the stone at $t_\text{Ⓓ} = 5.00$ s:

$$y_\text{Ⓓ} = y_\text{Ⓐ} + v_{y\text{Ⓐ}} t + \tfrac{1}{2} a_y t^2$$
$$= 0 + (20.0 \text{ m/s})(5.00 \text{ s}) + \tfrac{1}{2}(-9.80 \text{ m/s}^2)(5.00 \text{ s})^2$$
$$= \boxed{-22.5 \text{ m}}$$

The choice of the time defined as $t = 0$ is arbitrary and up to you to select as the problem solver. As an example of this arbitrariness, choose $t = 0$ as the time at which the stone is at the highest point in its motion. Then solve parts (C) and (D) again using this new initial instant and notice that your answers are the same as those above.

WHAT IF? What if the throw were from 30.0 m above the ground instead of 50.0 m? Which answers in parts (A) to (D) would change?

Answer None of the answers would change. All the motion takes place in the air during the first 5.00 s. (Notice that even for a throw from 30.0 m, the stone is above the ground at $t = 5.00$ s.) Therefore, the height of the throw is not an issue. Mathematically, if we look back over our calculations, we see that we never entered the height of the throw into any equation.

2.8 Kinematic Equations Derived from Calculus

This section assumes the reader is familiar with the techniques of integral calculus. If you have not yet studied integration in your calculus course, you should skip this section or cover it after you become familiar with integration.

The velocity of a particle moving in a straight line can be obtained if its position as a function of time is known. Mathematically, the velocity equals the derivative of the position with respect to time. It is also possible to find the position of a particle if its velocity is known as a function of time. In calculus, the procedure used to perform this task is referred to either as *integration* or as finding the *antiderivative*. Graphically, it is equivalent to finding the area under a curve.

Suppose the v_x–t graph for a particle moving along the x axis is as shown in Figure 2.15 on page 40. Let us divide the time interval $t_f - t_i$ into many small intervals, each of duration Δt_n. From the definition of average velocity, we see that the displacement of the particle during any small interval, such as the one shaded in Figure 2.15, is given by $\Delta x_n = v_{xn,\text{avg}} \, \Delta t_n$, where $v_{xn,\text{avg}}$ is the average velocity in that interval. Therefore, the displacement during this small interval is simply the area of the shaded rectangle in Figure 2.15. The total displacement for the interval $t_f - t_i$ is the sum of the areas of all the rectangles from t_i to t_f:

$$\Delta x = \sum_n v_{xn,\text{avg}} \, \Delta t_n$$

where the symbol Σ (uppercase Greek sigma) signifies a sum over all terms, that is, over all values of n. Now, as the intervals are made smaller and smaller, the number of terms in the sum increases and the sum approaches a value equal to the area

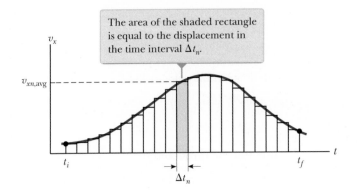

The area of the shaded rectangle is equal to the displacement in the time interval Δt_n.

under the curve in the velocity–time graph. Therefore, in the limit $n \to \infty$, or $\Delta t_n \to 0$, the displacement is

$$\Delta x = \lim_{\Delta t_n \to 0} \sum_n v_{xn,\text{avg}} \, \Delta t_n \tag{2.18}$$

If we know the v_x–t graph for motion along a straight line, we can obtain the displacement during any time interval by measuring the area under the curve corresponding to that time interval.

The limit of the sum shown in Equation 2.18 is called a **definite integral** and is written

Definite integral ▶

$$\lim_{\Delta t_n \to 0} \sum_n v_{xn,\text{avg}} \, \Delta t_n = \int_{t_i}^{t_f} v_x(t) \, dt \tag{2.19}$$

where $v_x(t)$ denotes the velocity at any time t. If the explicit functional form of $v_x(t)$ is known and the limits are given, the integral can be evaluated. Sometimes the v_x–t graph for a moving particle has a shape much simpler than that shown in Figure 2.15. For example, suppose an object is described with the particle under constant velocity model. In this case, the v_x–t graph is a horizontal line as in Figure 2.16 and the displacement of the particle during the time interval Δt is simply the area of the shaded rectangle:

$$\Delta x = v_{xi} \, \Delta t \quad (\text{when } v_x = v_{xi} = \text{constant})$$

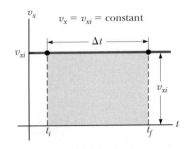

Figure 2.16 The velocity–time curve for a particle moving with constant velocity v_{xi}. The displacement of the particle during the time interval $t_f - t_i$ is equal to the area of the shaded rectangle.

Kinematic Equations

We now use the defining equations for acceleration and velocity to derive two of our kinematic equations, Equations 2.13 and 2.16.

The defining equation for acceleration (Eq. 2.10),

$$a_x = \frac{dv_x}{dt}$$

may be written as $dv_x = a_x \, dt$ or, in terms of an integral (or antiderivative), as

$$v_{xf} - v_{xi} = \int_0^t a_x \, dt$$

For the special case in which the acceleration is constant, a_x can be removed from the integral to give

$$v_{xf} - v_{xi} = a_x \int_0^t dt = a_x(t - 0) = a_x t \tag{2.20}$$

which is Equation 2.13 in the particle under constant acceleration model.

Now let us consider the defining equation for velocity (Eq. 2.5):

$$v_x = \frac{dx}{dt}$$

We can write this equation as $dx = v_x\,dt$ or in integral form as

$$x_f - x_i = \int_0^t v_x\,dt$$

Because $v_x = v_{xf} = v_{xi} + a_x t$, this expression becomes

$$x_f - x_i = \int_0^t (v_{xi} + a_x t)\,dt = \int_0^t v_{xi}\,dt + a_x \int_0^t t\,dt = v_{xi}(t - 0) + a_x\left(\frac{t^2}{2} - 0\right)$$

$$x_f - x_i = v_{xi}t + \tfrac{1}{2}a_x t^2$$

which is Equation 2.16 in the particle under constant acceleration model.

Besides what you might expect to learn about physics concepts, a very valuable skill you should hope to take away from your physics course is the ability to solve complicated problems. The way physicists approach complex situations and break them into manageable pieces is extremely useful. The following is a general problem-solving strategy to guide you through the steps. To help you remember the steps of the strategy, they are *Conceptualize*, *Categorize*, *Analyze*, and *Finalize*.

GENERAL PROBLEM-SOLVING STRATEGY

Conceptualize

- The first things to do when approaching a problem are to *think about* and *understand* the situation. Study carefully any representations of the information (for example, diagrams, graphs, tables, or photographs) that accompany the problem. Imagine a movie, running in your mind, of what happens in the problem.

- If a pictorial representation is not provided, you should almost always make a quick drawing of the situation. Indicate any known values, perhaps in a table or directly on your sketch.

- Now focus on what algebraic or numerical information is given in the problem. Carefully read the problem statement, looking for key phrases such as "starts from rest" ($v_i = 0$), "stops" ($v_f = 0$), or "falls freely" ($a_y = -g = -9.80 \text{ m/s}^2$).

- Now focus on the expected result of solving the problem. Exactly what is the question asking? Will the final result be numerical or algebraic? Do you know what units to expect?

- Don't forget to incorporate information from your own experiences and common sense. What should a reasonable answer look like? For example, you wouldn't expect to calculate the speed of an automobile to be 5×10^6 m/s.

Categorize

- Once you have a good idea of what the problem is about, you need to *simplify* the problem. Remove the details that are not important to the solution. For example, model a moving object as a particle. If appropriate, ignore air resistance or friction between a sliding object and a surface.

- Once the problem is simplified, it is important to *categorize* the problem. Is it a simple *substitution problem* such that numbers can be substituted into a simple equation or a definition? If so, the problem is likely to be finished when this substitution is done. If not, you face what we call an *analysis problem:* the situation must be analyzed more deeply to generate an appropriate equation and reach a solution.

- If it is an analysis problem, it needs to be categorized further. Have you seen this type of problem before? Does it fall into the growing list of types of problems that you have solved previously? If so, identify any analysis model(s) appropriate for the problem to prepare for the Analyze step below. We saw the first three analysis models in this chapter: the particle under constant velocity, the particle under constant speed, and the particle under constant acceleration. Being able to classify a problem with an analysis model can make it much easier to lay out a plan to solve it. For example, if your simplification shows that the problem can be treated as a particle under constant acceleration and you have already solved such a problem (such as the examples in Section 2.6), the solution to the present problem follows a similar pattern.

continued

Analyze

- Now you must analyze the problem and strive for a mathematical solution. Because you have already categorized the problem and identified an analysis model, it should not be too difficult to select relevant equations that apply to the type of situation in the problem. For example, if the problem involves a particle under constant acceleration, Equations 2.13 to 2.17 are relevant.

- Use algebra (and calculus, if necessary) to solve symbolically for the unknown variable in terms of what is given. Finally, substitute in the appropriate numbers, calculate the result, and round it to the proper number of significant figures.

Finalize

- Examine your numerical answer. Does it have the correct units? Does it meet your expectations from your conceptualization of the problem? What about the algebraic form of the result? Does it make sense? Examine the variables in the problem to see whether the answer would change in a physically meaningful way if the variables were drastically increased or decreased or even became zero. Looking at limiting cases to see whether they yield expected values is a very useful way to make sure that you are obtaining reasonable results.

- Think about how this problem compared with others you have solved. How was it similar? In what critical ways did it differ? Why was this problem assigned? Can you figure out what you have learned by doing it? If it is a new category of problem, be sure you understand it so that you can use it as a model for solving similar problems in the future.

When solving complex problems, you may need to identify a series of subproblems and apply the problem-solving strategy to each. For simple problems, you probably don't need this strategy. When you are trying to solve a problem and you don't know what to do next, however, remember the steps in the strategy and use them as a guide.

For practice, it would be useful for you to revisit the worked examples in this chapter and identify the *Conceptualize*, *Categorize*, *Analyze*, and *Finalize* steps. In the rest of this book, we will label these steps explicitly in the worked examples. Many chapters in this book include a section labeled Problem-Solving Strategy that should help you through the rough spots. These sections are organized according to the General Problem-Solving Strategy outlined above and are tailored to the specific types of problems addressed in that chapter.

To clarify how this Strategy works, we repeat Example 2.7 below with the particular steps of the Strategy identified.

> When you **Conceptualize** a problem, try to understand the situation that is presented in the problem statement. Study carefully any representations of the information (for example, diagrams, graphs, tables, or photographs) that accompany the problem. Imagine a movie, running in your mind, of what happens in the problem.

> Simplify the problem. Remove the details that are not important to the solution. Then **Categorize** the problem. Is it a simple substitution problem such that numbers can be substituted into a simple equation or a definition? If not, you face an analysis problem. In this case, identify the appropriate analysis model.

Example 2.7   Carrier Landing   AM

A jet lands on an aircraft carrier at a speed of 140 mi/h ($\approx$ 63 m/s).

(A) What is its acceleration (assumed constant) if it stops in 2.0 s due to an arresting cable that snags the jet and brings it to a stop?

SOLUTION

Conceptualize
You might have seen movies or television shows in which a jet lands on an aircraft carrier and is brought to rest surprisingly fast by an arresting cable. A careful reading of the problem reveals that in addition to being given the initial speed of 63 m/s, we also know that the final speed is zero.

Categorize
Because the acceleration of the jet is assumed constant, we model it as a *particle under constant acceleration*.

▶ **2.7** continued

....

Analyze

We define our x axis as the direction of motion of the jet. Notice that we have no information about the change in position of the jet while it is slowing down.

Equation 2.13 is the only equation in the particle under constant acceleration model that does not involve position, so we use it to find the acceleration of the jet, modeled as a particle:

$$a_x = \frac{v_{xf} - v_{xi}}{t} = \frac{0 - 63 \text{ m/s}}{2.0 \text{ s}}$$

$$= -32 \text{ m/s}^2$$

(B) If the jet touches down at position $x_i = 0$, what is its final position?

SOLUTION

Use Equation 2.15 to solve for the final position:

$$x_f = x_i + \tfrac{1}{2}(v_{xi} + v_{xf})t = 0 + \tfrac{1}{2}(63 \text{ m/s} + 0)(2.0 \text{ s}) = 63 \text{ m}$$

Finalize

Given the size of aircraft carriers, a length of 63 m seems reasonable for stopping the jet. The idea of using arresting cables to slow down landing aircraft and enable them to land safely on ships originated at about the time of World War I. The cables are still a vital part of the operation of modern aircraft carriers.

WHAT IF? Suppose the jet lands on the deck of the aircraft carrier with a speed higher than 63 m/s but has the same acceleration due to the cable as that calculated in part (A). How will that change the answer to part (B)?

Answer If the jet is traveling faster at the beginning, it will stop farther away from its starting point, so the answer to part (B) should be larger. Mathematically, we see in Equation 2.15 that if v_{xi} is larger, x_f will be larger.

Now **Analyze** the problem. Select relevant equations from the analysis model. Solve symbolically for the unknown variable in terms of what is given. Substitute in the appropriate numbers, calculate the result, and round it to the proper number of significant figures.

Finalize the problem. Examine the numerical answer. Does it have the correct units? Does it meet your expectations from your conceptualization of the problem? Does the answer make sense? What about the algebraic form of the result? Examine the variables in the problem to see whether the answer would change in a physically meaningful way if the variables were drastically increased or decreased or even became zero.

What If? questions will appear in many examples in the text, and offer a variation on the situation just explored. This feature encourages you to think about the results of the example and assists in conceptual understanding of the principles.

Summary

Definitions

When a particle moves along the x axis from some initial position x_i to some final position x_f, its **displacement** is

$$\Delta x \equiv x_f - x_i \qquad \text{(2.1)}$$

The **average velocity** of a particle during some time interval is the displacement Δx divided by the time interval Δt during which that displacement occurs:

$$v_{x,\text{avg}} \equiv \frac{\Delta x}{\Delta t} \qquad \text{(2.2)}$$

The **average speed** of a particle is equal to the ratio of the total distance it travels to the total time interval during which it travels that distance:

$$v_{\text{avg}} \equiv \frac{d}{\Delta t} \qquad \text{(2.3)}$$

continued

The **instantaneous velocity** of a particle is defined as the limit of the ratio $\Delta x/\Delta t$ as Δt approaches zero. By definition, this limit equals the derivative of x with respect to t, or the time rate of change of the position:

$$v_x \equiv \lim_{\Delta t \to 0} \frac{\Delta x}{\Delta t} = \frac{dx}{dt} \qquad \textbf{(2.5)}$$

The **instantaneous speed** of a particle is equal to the magnitude of its instantaneous velocity.

The **average acceleration** of a particle is defined as the ratio of the change in its velocity Δv_x divided by the time interval Δt during which that change occurs:

$$a_{x,avg} \equiv \frac{\Delta v_x}{\Delta t} = \frac{v_{xf} - v_{xi}}{t_f - t_i} \qquad \textbf{(2.9)}$$

The **instantaneous acceleration** is equal to the limit of the ratio $\Delta v_x/\Delta t$ as Δt approaches 0. By definition, this limit equals the derivative of v_x with respect to t, or the time rate of change of the velocity:

$$a_x \equiv \lim_{\Delta t \to 0} \frac{\Delta v_x}{\Delta t} = \frac{dv_x}{dt} \qquad \textbf{(2.10)}$$

Concepts and Principles

When an object's velocity and acceleration are in the same direction, the object is speeding up. On the other hand, when the object's velocity and acceleration are in opposite directions, the object is slowing down. Remembering that $F_x \propto a_x$ is a useful way to identify the direction of the acceleration by associating it with a force.

An object falling freely in the presence of the Earth's gravity experiences free-fall acceleration directed toward the center of the Earth. If air resistance is neglected, if the motion occurs near the surface of the Earth, and if the range of the motion is small compared with the Earth's radius, the free-fall acceleration $a_y = -g$ is constant over the range of motion, where g is equal to 9.80 m/s².

Complicated problems are best approached in an organized manner. Recall and apply the *Conceptualize, Categorize, Analyze,* and *Finalize* steps of the **General Problem-Solving Strategy** when you need them.

An important aid to problem solving is the use of **analysis models.** Analysis models are situations that we have seen in previous problems. Each analysis model has one or more equations associated with it. When solving a new problem, identify the analysis model that corresponds to the problem. The model will tell you which equations to use. The first three analysis models introduced in this chapter are summarized below.

Analysis Models for Problem-Solving

Particle Under Constant Velocity. If a particle moves in a straight line with a constant speed v_x, its constant velocity is given by

$$v_x = \frac{\Delta x}{\Delta t} \qquad \textbf{(2.6)}$$

and its position is given by

$$x_f = x_i + v_x t \qquad \textbf{(2.7)}$$

Particle Under Constant Speed. If a particle moves a distance d along a curved or straight path with a constant speed, its constant speed is given by

$$v = \frac{d}{\Delta t} \qquad \textbf{(2.8)}$$

Particle Under Constant Acceleration. If a particle moves in a straight line with a constant acceleration a_x, its motion is described by the kinematic equations:

$$v_{xf} = v_{xi} + a_x t \qquad \textbf{(2.13)}$$

$$v_{x,avg} = \frac{v_{xi} + v_{xf}}{2} \qquad \textbf{(2.14)}$$

$$x_f = x_i + \tfrac{1}{2}(v_{xi} + v_{xf})t \qquad \textbf{(2.15)}$$

$$x_f = x_i + v_{xi}t + \tfrac{1}{2}a_x t^2 \qquad \textbf{(2.16)}$$

$$v_{xf}^2 = v_{xi}^2 + 2a_x(x_f - x_i) \qquad \textbf{(2.17)}$$

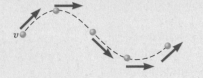

1. One drop of oil falls straight down onto the road from the engine of a moving car every 5 s. Figure OQ2.1 shows the pattern of the drops left behind on the pavement. What is the average speed of the car over this section of its motion? (a) 20 m/s (b) 24 m/s (c) 30 m/s (d) 100 m/s (e) 120 m/s

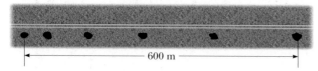

Figure OQ2.1

2. A racing car starts from rest at $t = 0$ and reaches a final speed v at time t. If the acceleration of the car is constant during this time, which of the following statements are true? (a) The car travels a distance vt. (b) The average speed of the car is $v/2$. (c) The magnitude of the acceleration of the car is v/t. (d) The velocity of the car remains constant. (e) None of statements (a) through (d) is true.

3. A juggler throws a bowling pin straight up in the air. After the pin leaves his hand and while it is in the air, which statement is true? (a) The velocity of the pin is always in the same direction as its acceleration. (b) The velocity of the pin is never in the same direction as its acceleration. (c) The acceleration of the pin is zero. (d) The velocity of the pin is opposite its acceleration on the way up. (e) The velocity of the pin is in the same direction as its acceleration on the way up.

4. When applying the equations of kinematics for an object moving in one dimension, which of the following statements *must* be true? (a) The velocity of the object must remain constant. (b) The acceleration of the object must remain constant. (c) The velocity of the object must increase with time. (d) The position of the object must increase with time. (e) The velocity of the object must always be in the same direction as its acceleration.

5. A cannon shell is fired straight up from the ground at an initial speed of 225 m/s. After how much time is the shell at a height of 6.20×10^2 m above the ground and moving downward? (a) 2.96 s (b) 17.3 s (c) 25.4 s (d) 33.6 s (e) 43.0 s

6. An arrow is shot straight up in the air at an initial speed of 15.0 m/s. After how much time is the arrow moving downward at a speed of 8.00 m/s? (a) 0.714 s (b) 1.24 s (c) 1.87 s (d) 2.35 s (e) 3.22 s

7. When the pilot reverses the propeller in a boat moving north, the boat moves with an acceleration directed south. Assume the acceleration of the boat remains constant in magnitude and direction. What happens to the boat? (a) It eventually stops and remains stopped. (b) It eventually stops and then speeds up in the forward direction. (c) It eventually stops and then speeds up in the reverse direction. (d) It never stops but loses speed more and more slowly forever. (e) It never stops but continues to speed up in the forward direction.

8. A rock is thrown downward from the top of a 40.0-m-tall tower with an initial speed of 12 m/s. Assuming negligible air resistance, what is the speed of the rock just before hitting the ground? (a) 28 m/s (b) 30 m/s (c) 56 m/s (d) 784 m/s (e) More information is needed.

9. A skateboarder starts from rest and moves down a hill with constant acceleration in a straight line, traveling for 6 s. In a second trial, he starts from rest and moves along the same straight line with the same acceleration for only 2 s. How does his displacement from his starting point in this second trial compare with that from the first trial? (a) one-third as large (b) three times larger (c) one-ninth as large (d) nine times larger (e) $1/\sqrt{3}$ times as large

10. On another planet, a marble is released from rest at the top of a high cliff. It falls 4.00 m in the first 1 s of its motion. Through what additional distance does it fall in the next 1 s? (a) 4.00 m (b) 8.00 m (c) 12.0 m (d) 16.0 m (e) 20.0 m

11. As an object moves along the x axis, many measurements are made of its position, enough to generate a smooth, accurate graph of x versus t. Which of the following quantities for the object *cannot* be obtained from this graph *alone*? (a) the velocity at any instant (b) the acceleration at any instant (c) the displacement during some time interval (d) the average velocity during some time interval (e) the speed at any instant

12. A pebble is dropped from rest from the top of a tall cliff and falls 4.9 m after 1.0 s has elapsed. How much farther does it drop in the next 2.0 s? (a) 9.8 m (b) 19.6 m (c) 39 m (d) 44 m (e) none of the above

13. A student at the top of a building of height h throws one ball upward with a speed of v_i and then throws a second ball downward with the same initial speed v_i. Just before it reaches the ground, is the final speed of the ball thrown upward (a) larger, (b) smaller, or (c) the same in magnitude, compared with the final speed of the ball thrown downward?

14. You drop a ball from a window located on an upper floor of a building. It strikes the ground with speed v. You now repeat the drop, but your friend down on the ground throws another ball upward at the same speed v, releasing her ball at the same moment that you drop yours from the window. At some location, the balls pass each other. Is this location (a) *at* the halfway point between window and ground, (b) *above* this point, or (c) *below* this point?

15. A pebble is released from rest at a certain height and falls freely, reaching an impact speed of 4 m/s at the floor. Next, the pebble is thrown down with an initial speed of 3 m/s from the same height. What is its speed at the floor? (a) 4 m/s (b) 5 m/s (c) 6 m/s (d) 7 m/s (e) 8 m/s

16. A ball is thrown straight up in the air. For which situation are both the instantaneous velocity and the acceleration zero? (a) on the way up (b) at the top of its flight path (c) on the way down (d) halfway up and halfway down (e) none of the above

17. A hard rubber ball, not affected by air resistance in its motion, is tossed upward from shoulder height, falls to the sidewalk, rebounds to a smaller maximum height, and is caught on its way down again. This motion is represented in Figure OQ2.17, where

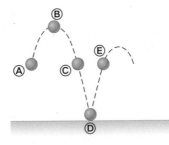

Figure OQ2.17

the successive positions of the ball Ⓐ through Ⓔ are not equally spaced in time. At point Ⓓ the center of the ball is at its lowest point in the motion. The motion of the ball is along a straight, vertical line, but the diagram shows successive positions offset to the right to avoid overlapping. Choose the positive y direction to be upward. (a) Rank the situations Ⓐ through Ⓔ according to the speed of the ball $|v_y|$ at each point, with the largest speed first. (b) Rank the same situations according to the acceleration a_y of the ball at each point. (In both rankings, remember that zero is greater than a negative value. If two values are equal, show that they are equal in your ranking.)

18. Each of the strobe photographs (a), (b), and (c) in Figure OQ2.18 was taken of a single disk moving toward the right, which we take as the positive direction. Within each photograph, the time interval between images is constant. **(i)** Which photograph shows motion with zero acceleration? **(ii)** Which photograph shows motion with positive acceleration? **(iii)** Which photograph shows motion with negative acceleration?

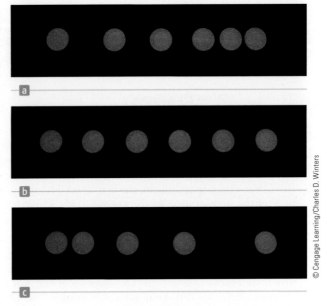

Figure OQ2.18

Conceptual Questions

1. denotes answer available in *Student Solutions Manual/Study Guide*

1. If the average velocity of an object is zero in some time interval, what can you say about the displacement of the object for that interval?

2. Try the following experiment away from traffic where you can do it safely. With the car you are driving moving slowly on a straight, level road, shift the transmission into neutral and let the car coast. At the moment the car comes to a complete stop, step hard on the brake and notice what you feel. Now repeat the same experiment on a fairly gentle, uphill slope. Explain the difference in what a person riding in the car feels in the two cases. (Brian Popp suggested the idea for this question.)

3. If a car is traveling eastward, can its acceleration be westward? Explain.

4. If the velocity of a particle is zero, can the particle's acceleration be zero? Explain.

5. If the velocity of a particle is nonzero, can the particle's acceleration be zero? Explain.

6. You throw a ball vertically upward so that it leaves the ground with velocity +5.00 m/s. (a) What is its velocity when it reaches its maximum altitude? (b) What is its acceleration at this point? (c) What is the velocity with which it returns to ground level? (d) What is its acceleration at this point?

7. (a) Can the equations of kinematics (Eqs. 2.13–2.17) be used in a situation in which the acceleration varies in time? (b) Can they be used when the acceleration is zero?

8. (a) Can the velocity of an object at an instant of time be greater in magnitude than the average velocity over a time interval containing the instant? (b) Can it be less?

9. Two cars are moving in the same direction in parallel lanes along a highway. At some instant, the velocity of car A exceeds the velocity of car B. Does that mean that the acceleration of car A is greater than that of car B? Explain.

Section 2.1 Position, Velocity, and Speed
Problems 1–5

Section 2.2 Instantaneous Velocity and Speed
Problems 6–9

Section 2.3 Analysis Model: Particle Under Constant Velocity
Problems 10–13

Section 2.4 Acceleration
Problems 14–21

Section 2.5 Motion Diagrams
Problems 22–23

Section 2.6 Analysis Model: Particle Under Constant Acceleration
Problems 24–44

Section 2.7 Freely Falling Objects
Problems 45–56

Section 2.8 Kinematic Equations Derived from Calculus
Problems 57–59

Additional Problems
Problems 60–80

Challenge Problems
Problems 81–85

Solutions to the following Problems are available in the *Student Solutions Manual/Study Guide:*
2.1, 2.3, 2.7, 2.14, 2.21, 2.29, 2.30, 2.31, 2.35, 2.48, 2.53, 2.55, 2.57, 2.63, 2.68, 2.73, 2.75, and 2.83

List of Enhanced Problems

Problem Number	Targeted Feedback in Enhanced WebAssign	Analysis Model Tutorial in Enhanced WebAssign	Master It in Enhanced WebAssign	Watch It in Enhanced WebAssign
2.1	✓			✓
2.3	✓		✓	
2.4	✓			✓
2.9	✓			✓
2.12		✓		
2.13	✓		✓	
2.14	✓			✓
2.19	✓			✓
2.20	✓			✓
2.21	✓		✓	
2.28	✓			✓
2.29			✓	
2.30	✓		✓	
2.31			✓	
2.35	✓	✓		✓
2.37		✓		
2.38	✓			✓
2.44	✓		✓	
2.48	✓			✓
2.51	✓			✓
2.52	✓		✓	
2.53			✓	
2.55		✓		
2.63			✓	
2.73		✓	✓	

Vectors

In our study of physics, we often need to work with physical quantities that have both numerical and directional properties. As noted in Section 2.1, quantities of this nature are vector quantities. This chapter is primarily concerned with general properties of vector quantities. We discuss the addition and subtraction of vector quantities, together with some common applications to physical situations.

Vector quantities are used throughout this text. Therefore, it is imperative that you master the techniques discussed in this chapter.

A signpost in Saint Petersburg, Florida, shows the distance and direction to several cities. Quantities that are defined by both a magnitude and a direction are called vector quantities.
(Raymond A. Serway)

3.1 Coordinate Systems

Many aspects of physics involve a description of a location in space. In Chapter 2, for example, we saw that the mathematical description of an object's motion requires a method for describing the object's position at various times. In two dimensions, this description is accomplished with the use of the Cartesian coordinate system, in which perpendicular axes intersect at a point defined as the origin O (Fig. 3.1). Cartesian coordinates are also called *rectangular coordinates*.

Sometimes it is more convenient to represent a point in a plane by its *plane polar coordinates* (r, θ) as shown in Figure 3.2a (page 50). In this *polar coordinate system*, r is the distance from the origin to the point having Cartesian coordinates (x, y) and θ is the angle between a fixed axis and a line drawn from the origin to the point. The fixed axis is often the positive x axis, and θ is usually measured counterclockwise

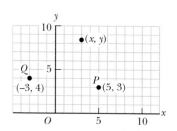

Figure 3.1 Designation of points in a Cartesian coordinate system. Every point is labeled with coordinates (x, y).

49

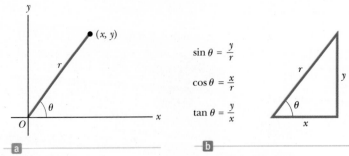

Figure 3.2 (a) The plane polar coordinates of a point are represented by the distance r and the angle θ, where θ is measured counterclockwise from the positive x axis. (b) The right triangle used to relate (x, y) to (r, θ).

from it. From the right triangle in Figure 3.2b, we find that $\sin \theta = y/r$ and that $\cos \theta = x/r$. (A review of trigonometric functions is given in Appendix B.4.) Therefore, starting with the plane polar coordinates of any point, we can obtain the Cartesian coordinates by using the equations

◄ **Cartesian coordinates in terms of polar coordinates**

$$x = r \cos \theta \tag{3.1}$$

$$y = r \sin \theta \tag{3.2}$$

Furthermore, if we know the Cartesian coordinates, the definitions of trigonometry tell us that

◄ **Polar coordinates in terms of Cartesian coordinates**

$$\tan \theta = \frac{y}{x} \tag{3.3}$$

$$r = \sqrt{x^2 + y^2} \tag{3.4}$$

Equation 3.4 is the familiar Pythagorean theorem.

These four expressions relating the coordinates (x, y) to the coordinates (r, θ) apply only when θ is defined as shown in Figure 3.2a—in other words, when positive θ is an angle measured counterclockwise from the positive x axis. (Some scientific calculators perform conversions between Cartesian and polar coordinates based on these standard conventions.) If the reference axis for the polar angle θ is chosen to be one other than the positive x axis or if the sense of increasing θ is chosen differently, the expressions relating the two sets of coordinates will change.

Example 3.1 | Polar Coordinates

The Cartesian coordinates of a point in the xy plane are $(x, y) = (-3.50, -2.50)$ m as shown in Figure 3.3. Find the polar coordinates of this point.

SOLUTION

Conceptualize The drawing in Figure 3.3 helps us conceptualize the problem. We wish to find r and θ. We expect r to be a few meters and θ to be larger than 180°.

Categorize Based on the statement of the problem and the Conceptualize step, we recognize that we are simply converting from Cartesian coordinates to polar coordinates. We therefore categorize this example as a substitution problem. Substitution problems generally do not have an extensive Analyze step other than the substitution of numbers into a given equation. Similarly, the Finalize step

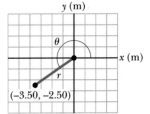

Figure 3.3 (Example 3.1) Finding polar coordinates when Cartesian coordinates are given.

▶ **3.1** continued

consists primarily of checking the units and making sure that the answer is reasonable and consistent with our expectations. Therefore, for substitution problems, we will not label Analyze or Finalize steps.

Use Equation 3.4 to find r:

$$r = \sqrt{x^2 + y^2} = \sqrt{(-3.50 \text{ m})^2 + (-2.50 \text{ m})^2} = \boxed{4.30 \text{ m}}$$

Use Equation 3.3 to find θ:

$$\tan \theta = \frac{y}{x} = \frac{-2.50 \text{ m}}{-3.50 \text{ m}} = 0.714$$

$$\theta = \boxed{216°}$$

Notice that you must use the signs of x and y to find that the point lies in the third quadrant of the coordinate system. That is, $\theta = 216°$, not 35.5°, whose tangent is also 0.714. Both answers agree with our expectations in the Conceptualize step.

3.2 Vector and Scalar Quantities

We now formally describe the difference between scalar quantities and vector quantities. When you want to know the temperature outside so that you will know how to dress, the only information you need is a number and the unit "degrees C" or "degrees F." Temperature is therefore an example of a *scalar quantity:*

> A **scalar quantity** is completely specified by a single value with an appropriate unit and has no direction.

Other examples of scalar quantities are volume, mass, speed, time, and time intervals. Some scalars are always positive, such as mass and speed. Others, such as temperature, can have either positive or negative values. The rules of ordinary arithmetic are used to manipulate scalar quantities.

If you are preparing to pilot a small plane and need to know the wind velocity, you must know both the speed of the wind and its direction. Because direction is important for its complete specification, velocity is a *vector quantity:*

> A **vector quantity** is completely specified by a number with an appropriate unit (the *magnitude* of the vector) plus a direction.

Another example of a vector quantity is displacement, as you know from Chapter 2. Suppose a particle moves from some point Ⓐ to some point Ⓑ along a straight path as shown in Figure 3.4. We represent this displacement by drawing an arrow from Ⓐ to Ⓑ, with the tip of the arrow pointing away from the starting point. The direction of the arrowhead represents the direction of the displacement, and the length of the arrow represents the magnitude of the displacement. If the particle travels along some other path from Ⓐ to Ⓑ such as shown by the broken line in Figure 3.4, its displacement is still the arrow drawn from Ⓐ to Ⓑ. Displacement depends only on the initial and final positions, so the displacement vector is independent of the path taken by the particle between these two points.

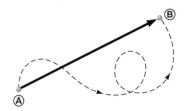

Figure 3.4 As a particle moves from Ⓐ to Ⓑ along an arbitrary path represented by the broken line, its displacement is a vector quantity shown by the arrow drawn from Ⓐ to Ⓑ.

In this text, we use a boldface letter with an arrow over the letter, such as $\vec{A}$, to represent a vector. Another common notation for vectors with which you should be familiar is a simple boldface character: **A**. The magnitude of the vector $\vec{A}$ is written either A or $|\vec{A}|$. The magnitude of a vector has physical units, such as meters for displacement or meters per second for velocity. The magnitude of a vector is *always* a positive number.

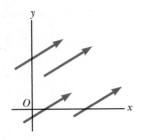

Figure 3.5 These four vectors are equal because they have equal lengths and point in the same direction.

Commutative law of addition ▶

⓪uick Quiz 3.1 Which of the following are vector quantities and which are scalar quantities? **(a)** your age **(b)** acceleration **(c)** velocity **(d)** speed **(e)** mass

3.3 Some Properties of Vectors

In this section, we shall investigate general properties of vectors representing physical quantities. We also discuss how to add and subtract vectors using both algebraic and geometric methods.

Equality of Two Vectors

For many purposes, two vectors $\vec{A}$ and $\vec{B}$ may be defined to be equal if they have the same magnitude and if they point in the same direction. That is, $\vec{A} = \vec{B}$ only if $A = B$ and if $\vec{A}$ and $\vec{B}$ point in the same direction along parallel lines. For example, all the vectors in Figure 3.5 are equal even though they have different starting points. This property allows us to move a vector to a position parallel to itself in a diagram without affecting the vector.

Adding Vectors

The rules for adding vectors are conveniently described by a graphical method. To add vector $\vec{B}$ to vector $\vec{A}$, first draw vector $\vec{A}$ on graph paper, with its magnitude represented by a convenient length scale, and then draw vector $\vec{B}$ to the same scale, with its tail starting from the tip of $\vec{A}$, as shown in Figure 3.6. The **resultant vector** $\vec{R} = \vec{A} + \vec{B}$ is the vector drawn from the tail of $\vec{A}$ to the tip of $\vec{B}$.

A geometric construction can also be used to add more than two vectors as shown in Figure 3.7 for the case of four vectors. The resultant vector $\vec{R} = \vec{A} + \vec{B} + \vec{C} + \vec{D}$ is the vector that completes the polygon. In other words, $\vec{R}$ is the vector drawn from the tail of the first vector to the tip of the last vector. This technique for adding vectors is often called the "head to tail method."

When two vectors are added, the sum is independent of the order of the addition. (This fact may seem trivial, but as you will see in Chapter 11, the order is important when vectors are multiplied. Procedures for multiplying vectors are discussed in Chapters 7 and 11.) This property, which can be seen from the geometric construction in Figure 3.8, is known as the **commutative law of addition:**

$$\vec{A} + \vec{B} = \vec{B} + \vec{A} \tag{3.5}$$

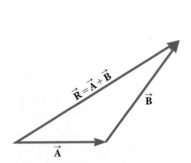

Figure 3.6 When vector $\vec{B}$ is added to vector $\vec{A}$, the resultant $\vec{R}$ is the vector that runs from the tail of $\vec{A}$ to the tip of $\vec{B}$.

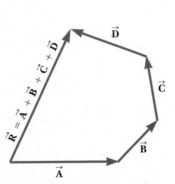

Figure 3.7 Geometric construction for summing four vectors. The resultant vector $\vec{R}$ is by definition the one that completes the polygon.

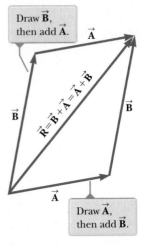

Draw $\vec{B}$, then add $\vec{A}$.

Draw $\vec{A}$, then add $\vec{B}$.

Figure 3.8 This construction shows that $\vec{A} + \vec{B} = \vec{B} + \vec{A}$ or, in other words, that vector addition is commutative.

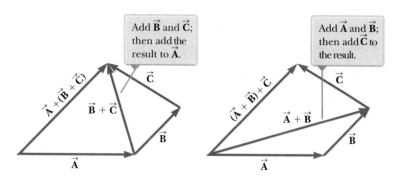

Add $\vec{B}$ and $\vec{C}$; then add the result to $\vec{A}$.

Add $\vec{A}$ and $\vec{B}$; then add $\vec{C}$ to the result.

Figure 3.9 Geometric constructions for verifying the associative law of addition.

When three or more vectors are added, their sum is independent of the way in which the individual vectors are grouped together. A geometric proof of this rule for three vectors is given in Figure 3.9. This property is called the **associative law of addition:**

$$\vec{A} + (\vec{B} + \vec{C}) = (\vec{A} + \vec{B}) + \vec{C} \qquad (3.6)$$

◀ Associative law of addition

In summary, **a vector quantity has both magnitude and direction and also obeys the laws of vector addition** as described in Figures 3.6 to 3.9. When two or more vectors are added together, they must all have the same units and they must all be the same type of quantity. It would be meaningless to add a velocity vector (for example, 60 km/h to the east) to a displacement vector (for example, 200 km to the north) because these vectors represent different physical quantities. The same rule also applies to scalars. For example, it would be meaningless to add time intervals to temperatures.

Negative of a Vector

The negative of the vector $\vec{A}$ is defined as the vector that when added to $\vec{A}$ gives zero for the vector sum. That is, $\vec{A} + (-\vec{A}) = 0$. The vectors $\vec{A}$ and $-\vec{A}$ have the same magnitude but point in opposite directions.

Subtracting Vectors

The operation of vector subtraction makes use of the definition of the negative of a vector. We define the operation $\vec{A} - \vec{B}$ as vector $-\vec{B}$ added to vector $\vec{A}$:

$$\vec{A} - \vec{B} = \vec{A} + (-\vec{B}) \qquad (3.7)$$

The geometric construction for subtracting two vectors in this way is illustrated in Figure 3.10a.

Another way of looking at vector subtraction is to notice that the difference $\vec{A} - \vec{B}$ between two vectors $\vec{A}$ and $\vec{B}$ is what you have to add to the second vector

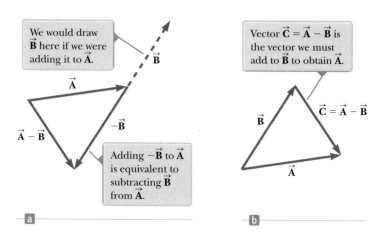

We would draw $\vec{B}$ here if we were adding it to $\vec{A}$.

Adding $-\vec{B}$ to $\vec{A}$ is equivalent to subtracting $\vec{B}$ from $\vec{A}$.

Vector $\vec{C} = \vec{A} - \vec{B}$ is the vector we must add to $\vec{B}$ to obtain $\vec{A}$.

Figure 3.10 (a) Subtracting vector $\vec{B}$ from vector $\vec{A}$. The vector $-\vec{B}$ is equal in magnitude to vector $\vec{B}$ and points in the opposite direction. (b) A second way of looking at vector subtraction.

to obtain the first. In this case, as Figure 3.10b shows, the vector $\vec{A} - \vec{B}$ points from the tip of the second vector to the tip of the first.

Multiplying a Vector by a Scalar

If vector $\vec{A}$ is multiplied by a positive scalar quantity m, the product $m\vec{A}$ is a vector that has the same direction as $\vec{A}$ and magnitude mA. If vector $\vec{A}$ is multiplied by a negative scalar quantity $-m$, the product $-m\vec{A}$ is directed opposite $\vec{A}$. For example, the vector $5\vec{A}$ is five times as long as $\vec{A}$ and points in the same direction as $\vec{A}$; the vector $-\frac{1}{3}\vec{A}$ is one-third the length of $\vec{A}$ and points in the direction opposite $\vec{A}$.

Quick Quiz 3.2 The magnitudes of two vectors $\vec{A}$ and $\vec{B}$ are $A = 12$ units and $B = 8$ units. Which pair of numbers represents the *largest* and *smallest* possible values for the magnitude of the resultant vector $\vec{R} = \vec{A} + \vec{B}$? **(a)** 14.4 units, 4 units **(b)** 12 units, 8 units **(c)** 20 units, 4 units **(d)** none of these answers

Quick Quiz 3.3 If vector $\vec{B}$ is added to vector $\vec{A}$, which *two* of the following choices must be true for the resultant vector to be equal to zero? **(a)** $\vec{A}$ and $\vec{B}$ are parallel and in the same direction. **(b)** $\vec{A}$ and $\vec{B}$ are parallel and in opposite directions. **(c)** $\vec{A}$ and $\vec{B}$ have the same magnitude. **(d)** $\vec{A}$ and $\vec{B}$ are perpendicular.

Example 3.2   A Vacation Trip

A car travels 20.0 km due north and then 35.0 km in a direction 60.0° west of north as shown in Figure 3.11a. Find the magnitude and direction of the car's resultant displacement.

SOLUTION

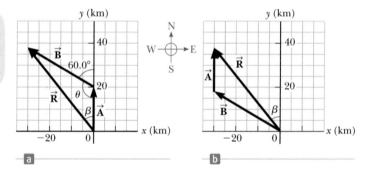

Figure 3.11 (Example 3.2) (a) Graphical method for finding the resultant displacement vector $\vec{R} = \vec{A} + \vec{B}$. (b) Adding the vectors in reverse order $(\vec{B} + \vec{A})$ gives the same result for $\vec{R}$.

Conceptualize The vectors $\vec{A}$ and $\vec{B}$ drawn in Figure 3.11a help us conceptualize the problem. The resultant vector $\vec{R}$ has also been drawn. We expect its magnitude to be a few tens of kilometers. The angle β that the resultant vector makes with the y axis is expected to be less than 60°, the angle that vector $\vec{B}$ makes with the y axis.

Categorize We can categorize this example as a simple analysis problem in vector addition. The displacement $\vec{R}$ is the resultant when the two individual displacements $\vec{A}$ and $\vec{B}$ are added. We can further categorize it as a problem about the analysis of triangles, so we appeal to our expertise in geometry and trigonometry.

Analyze In this example, we show two ways to analyze the problem of finding the resultant of two vectors. The first way is to solve the problem geometrically, using graph paper and a protractor to measure the magnitude of $\vec{R}$ and its direction in Figure 3.11a. (In fact, even when you know you are going to be carrying out a calculation, you should sketch the vectors to check your results.) With an ordinary ruler and protractor, a large diagram typically gives answers to two-digit but not to three-digit precision. Try using these tools on $\vec{R}$ in Figure 3.11a and compare to the trigonometric analysis below!

The second way to solve the problem is to analyze it using algebra and trigonometry. The magnitude of $\vec{R}$ can be obtained from the law of cosines as applied to the triangle in Figure 3.11a (see Appendix B.4).

Use $R^2 = A^2 + B^2 - 2AB \cos \theta$ from the law of cosines to find R:

$$R = \sqrt{A^2 + B^2 - 2AB \cos \theta}$$

Substitute numerical values, noting that $\theta = 180° - 60° = 120°$:

$$R = \sqrt{(20.0 \text{ km})^2 + (35.0 \text{ km})^2 - 2(20.0 \text{ km})(35.0 \text{ km}) \cos 120°}$$
$$= \boxed{48.2 \text{ km}}$$

▶ **3.2** continued

Use the law of sines (Appendix B.4) to find the direction of $\vec{\mathbf{R}}$ measured from the northerly direction:

$$\frac{\sin \beta}{B} = \frac{\sin \theta}{R}$$

$$\sin \beta = \frac{B}{R} \sin \theta = \frac{35.0 \text{ km}}{48.2 \text{ km}} \sin 120° = 0.629$$

$$\beta = \boxed{38.9°}$$

The resultant displacement of the car is 48.2 km in a direction 38.9° west of north.

Finalize Does the angle β that we calculated agree with an estimate made by looking at Figure 3.11a or with an actual angle measured from the diagram using the graphical method? Is it reasonable that the magnitude of $\vec{\mathbf{R}}$ is larger than that of both $\vec{\mathbf{A}}$ and $\vec{\mathbf{B}}$? Are the units of $\vec{\mathbf{R}}$ correct?

Although the head to tail method of adding vectors works well, it suffers from two disadvantages. First, some people find using the laws of cosines and sines to be awkward. Second, a triangle only results if you are adding two vectors. If you are adding three or more vectors, the resulting geometric shape is usually not a triangle. In Section 3.4, we explore a new method of adding vectors that will address both of these disadvantages.

WHAT IF? Suppose the trip were taken with the two vectors in reverse order: 35.0 km at 60.0° west of north first and then 20.0 km due north. How would the magnitude and the direction of the resultant vector change?

Answer They would not change. The commutative law for vector addition tells us that the order of vectors in an addition is irrelevant. Graphically, Figure 3.11b shows that the vectors added in the reverse order give us the same resultant vector.

3.4 Components of a Vector and Unit Vectors

The graphical method of adding vectors is not recommended whenever high accuracy is required or in three-dimensional problems. In this section, we describe a method of adding vectors that makes use of the projections of vectors along coordinate axes. These projections are called the **components** of the vector or its **rectangular components.** Any vector can be completely described by its components.

Consider a vector $\vec{\mathbf{A}}$ lying in the xy plane and making an arbitrary angle θ with the positive x axis as shown in Figure 3.12a. This vector can be expressed as the sum of two other *component vectors* $\vec{\mathbf{A}}_x$, which is parallel to the x axis, and $\vec{\mathbf{A}}_y$, which is parallel to the y axis. From Figure 3.12b, we see that the three vectors form a right triangle and that $\vec{\mathbf{A}} = \vec{\mathbf{A}}_x + \vec{\mathbf{A}}_y$. We shall often refer to the "components of a vector $\vec{\mathbf{A}}$," written A_x and A_y (without the boldface notation). The component A_x represents the projection of $\vec{\mathbf{A}}$ along the x axis, and the component A_y represents the projection of $\vec{\mathbf{A}}$ along the y axis. These components can be positive or negative. The component A_x is positive if the component vector $\vec{\mathbf{A}}_x$ points in the positive x direction and is negative if $\vec{\mathbf{A}}_x$ points in the negative x direction. A similar statement is made for the component A_y.

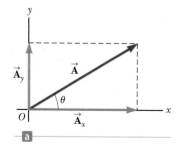

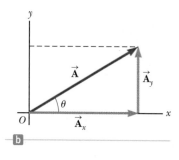

Figure 3.12 (a) A vector $\vec{\mathbf{A}}$ lying in the xy plane can be represented by its component vectors $\vec{\mathbf{A}}_x$ and $\vec{\mathbf{A}}_y$. (b) The y component vector $\vec{\mathbf{A}}_y$ can be moved to the right so that it adds to $\vec{\mathbf{A}}_x$. The vector sum of the component vectors is $\vec{\mathbf{A}}$. These three vectors form a right triangle.

From Figure 3.12 and the definition of sine and cosine, we see that $\cos \theta = A_x/A$ and that $\sin \theta = A_y/A$. Hence, the components of $\vec{A}$ are

$$A_x = A \cos \theta \tag{3.8}$$

$$A_y = A \sin \theta \tag{3.9}$$

The magnitudes of these components are the lengths of the two sides of a right triangle with a hypotenuse of length *A*. Therefore, the magnitude and direction of $\vec{A}$ are related to its components through the expressions

$$A = \sqrt{A_x^2 + A_y^2} \tag{3.10}$$

$$\theta = \tan^{-1}\left(\frac{A_y}{A_x}\right) \tag{3.11}$$

Notice that the signs of the components A_x and A_y depend on the angle θ. For example, if θ = 120°, A_x is negative and A_y is positive. If θ = 225°, both A_x and A_y are negative. Figure 3.13 summarizes the signs of the components when $\vec{A}$ lies in the various quadrants.

When solving problems, you can specify a vector $\vec{A}$ either with its components A_x and A_y or with its magnitude and direction *A* and θ.

Suppose you are working a physics problem that requires resolving a vector into its components. In many applications, it is convenient to express the components in a coordinate system having axes that are not horizontal and vertical but that are still perpendicular to each other. For example, we will consider the motion of objects sliding down inclined planes. For these examples, it is often convenient to orient the *x* axis parallel to the plane and the *y* axis perpendicular to the plane.

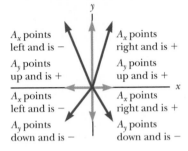

Figure 3.13 The signs of the components of a vector $\vec{A}$ depend on the quadrant in which the vector is located.

Quick Quiz 3.4 Choose the correct response to make the sentence true: A component of a vector is **(a)** always, **(b)** never, or **(c)** sometimes larger than the magnitude of the vector.

Unit Vectors

Vector quantities often are expressed in terms of unit vectors. A **unit vector** is a dimensionless vector having a magnitude of exactly 1. Unit vectors are used to specify a given direction and have no other physical significance. They are used solely as a bookkeeping convenience in describing a direction in space. We shall use the symbols $\hat{i}$, $\hat{j}$, and $\hat{k}$ to represent unit vectors pointing in the positive *x*, *y*, and *z* directions, respectively. (The "hats," or circumflexes, on the symbols are a standard notation for unit vectors.) The unit vectors $\hat{i}$, $\hat{j}$, and $\hat{k}$ form a set of mutually perpendicular vectors in a right-handed coordinate system as shown in Figure 3.14a. The magnitude of each unit vector equals 1; that is, $|\hat{i}| = |\hat{j}| = |\hat{k}| = 1$.

Consider a vector $\vec{A}$ lying in the *xy* plane as shown in Figure 3.14b. The product of the component A_x and the unit vector $\hat{i}$ is the component vector $\vec{A}_x = A_x\hat{i}$,

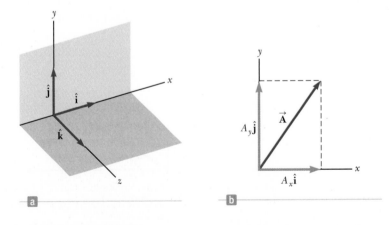

Figure 3.14 (a) The unit vectors $\hat{i}$, $\hat{j}$, and $\hat{k}$ are directed along the *x*, *y*, and *z* axes, respectively. (b) Vector $\vec{A} = A_x\hat{i} + A_y\hat{j}$ lying in the *xy* plane has components A_x and A_y.

which lies on the x axis and has magnitude $|A_x|$. Likewise, $\vec{\mathbf{A}}_y = A_y\,\vec{\mathbf{j}}$ is the component vector of magnitude $|A_y|$ lying on the y axis. Therefore, the unit-vector notation for the vector $\vec{\mathbf{A}}$ is

$$\vec{\mathbf{A}} = A_x\hat{\mathbf{i}} + A_y\hat{\mathbf{j}} \qquad (3.12)$$

For example, consider a point lying in the xy plane and having Cartesian coordinates (x, y) as in Figure 3.15. The point can be specified by the **position vector $\vec{\mathbf{r}}$**, which in unit-vector form is given by

$$\vec{\mathbf{r}} = x\hat{\mathbf{i}} + y\hat{\mathbf{j}} \qquad (3.13)$$

This notation tells us that the components of $\vec{\mathbf{r}}$ are the coordinates x and y.

Now let us see how to use components to add vectors when the graphical method is not sufficiently accurate. Suppose we wish to add vector $\vec{\mathbf{B}}$ to vector $\vec{\mathbf{A}}$ in Equation 3.12, where vector $\vec{\mathbf{B}}$ has components B_x and B_y. Because of the bookkeeping convenience of the unit vectors, all we do is add the x and y components separately. The resultant vector $\vec{\mathbf{R}} = \vec{\mathbf{A}} + \vec{\mathbf{B}}$ is

$$\vec{\mathbf{R}} = (A_x\hat{\mathbf{i}} + A_y\hat{\mathbf{j}}) + (B_x\hat{\mathbf{i}} + B_y\hat{\mathbf{j}})$$

or

$$\vec{\mathbf{R}} = (A_x + B_x)\hat{\mathbf{i}} + (A_y + B_y)\hat{\mathbf{j}} \qquad (3.14)$$

Because $\vec{\mathbf{R}} = R_x\hat{\mathbf{i}} + R_y\hat{\mathbf{j}}$, we see that the components of the resultant vector are

$$R_x = A_x + B_x$$
$$R_y = A_y + B_y \qquad (3.15)$$

Therefore, we see that in the component method of adding vectors, we add all the x components together to find the x component of the resultant vector and use the same process for the y components. We can check this addition by components with a geometric construction as shown in Figure 3.16.

The magnitude of $\vec{\mathbf{R}}$ and the angle it makes with the x axis are obtained from its components using the relationships

$$R = \sqrt{R_x^2 + R_y^2} = \sqrt{(A_x + B_x)^2 + (A_y + B_y)^2} \qquad (3.16)$$

$$\tan\theta = \frac{R_y}{R_x} = \frac{A_y + B_y}{A_x + B_x} \qquad (3.17)$$

At times, we need to consider situations involving motion in three component directions. The extension of our methods to three-dimensional vectors is straightforward. If $\vec{\mathbf{A}}$ and $\vec{\mathbf{B}}$ both have x, y, and z components, they can be expressed in the form

$$\vec{\mathbf{A}} = A_x\hat{\mathbf{i}} + A_y\hat{\mathbf{j}} + A_z\hat{\mathbf{k}} \qquad (3.18)$$

$$\vec{\mathbf{B}} = B_x\hat{\mathbf{i}} + B_y\hat{\mathbf{j}} + B_z\hat{\mathbf{k}} \qquad (3.19)$$

The sum of $\vec{\mathbf{A}}$ and $\vec{\mathbf{B}}$ is

$$\vec{\mathbf{R}} = (A_x + B_x)\hat{\mathbf{i}} + (A_y + B_y)\hat{\mathbf{j}} + (A_z + B_z)\hat{\mathbf{k}} \qquad (3.20)$$

Notice that Equation 3.20 differs from Equation 3.14: in Equation 3.20, the resultant vector also has a z component $R_z = A_z + B_z$. If a vector $\vec{\mathbf{R}}$ has x, y, and z components, the magnitude of the vector is $R = \sqrt{R_x^2 + R_y^2 + R_z^2}$. The angle θ_x that $\vec{\mathbf{R}}$ makes with the x axis is found from the expression $\cos\theta_x = R_x/R$, with similar expressions for the angles with respect to the y and z axes.

The extension of our method to adding more than two vectors is also straightforward. For example, $\vec{\mathbf{A}} + \vec{\mathbf{B}} + \vec{\mathbf{C}} = (A_x + B_x + C_x)\hat{\mathbf{i}} + (A_y + B_y + C_y)\hat{\mathbf{j}} + (A_z + B_z + C_z)\hat{\mathbf{k}}$. We have described adding displacement vectors in this section because these types of vectors are easy to visualize. We can also add other types of

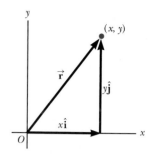

Figure 3.15 The point whose Cartesian coordinates are (x, y) can be represented by the position vector $\vec{\mathbf{r}} = x\hat{\mathbf{i}} + y\hat{\mathbf{j}}$.

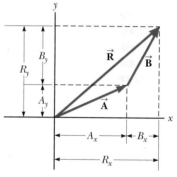

Figure 3.16 This geometric construction for the sum of two vectors shows the relationship between the components of the resultant $\vec{\mathbf{R}}$ and the components of the individual vectors.

Pitfall Prevention 3.3

Tangents on Calculators Equation 3.17 involves the calculation of an angle by means of a tangent function. Generally, the inverse tangent function on calculators provides an angle between $-90°$ and $+90°$. As a consequence, if the vector you are studying lies in the second or third quadrant, the angle measured from the positive x axis will be the angle your calculator returns plus 180°.

vectors, such as velocity, force, and electric field vectors, which we will do in later chapters.

Quick Quiz 3.5 For which of the following vectors is the magnitude of the vector equal to one of the components of the vector? (a) $\vec{\mathbf{A}} = 2\hat{\mathbf{i}} + 5\hat{\mathbf{j}}$ (b) $\vec{\mathbf{B}} = -3\hat{\mathbf{j}}$ (c) $\vec{\mathbf{C}} = +5\hat{\mathbf{k}}$

Example 3.3   **The Sum of Two Vectors**

Find the sum of two displacement vectors $\vec{\mathbf{A}}$ and $\vec{\mathbf{B}}$ lying in the xy plane and given by

$$\vec{\mathbf{A}} = (2.0\hat{\mathbf{i}} + 2.0\hat{\mathbf{j}}) \text{ m} \quad \text{and} \quad \vec{\mathbf{B}} = (2.0\hat{\mathbf{i}} - 4.0\hat{\mathbf{j}}) \text{ m}$$

SOLUTION

Conceptualize You can conceptualize the situation by drawing the vectors on graph paper. Draw an approximation of the expected resultant vector.

Categorize We categorize this example as a simple substitution problem. Comparing this expression for $\vec{\mathbf{A}}$ with the general expression $\vec{\mathbf{A}} = A_x\hat{\mathbf{i}} + A_y\hat{\mathbf{j}} + A_z\hat{\mathbf{k}}$, we see that $A_x = 2.0$ m, $A_y = 2.0$ m, and $A_z = 0$. Likewise, $B_x = 2.0$ m, $B_y = -4.0$ m, and $B_z = 0$. We can use a two-dimensional approach because there are no z components.

Use Equation 3.14 to obtain the resultant vector $\vec{\mathbf{R}}$:

$$\vec{\mathbf{R}} = \vec{\mathbf{A}} + \vec{\mathbf{B}} = (2.0 + 2.0)\hat{\mathbf{i}} \text{ m} + (2.0 - 4.0)\hat{\mathbf{j}} \text{ m}$$

Evaluate the components of $\vec{\mathbf{R}}$:

$$R_x = 4.0 \text{ m} \qquad R_y = -2.0 \text{ m}$$

Use Equation 3.16 to find the magnitude of $\vec{\mathbf{R}}$:

$$R = \sqrt{R_x^2 + R_y^2} = \sqrt{(4.0 \text{ m})^2 + (-2.0 \text{ m})^2} = \sqrt{20} \text{ m} = \boxed{4.5 \text{ m}}$$

Find the direction of $\vec{\mathbf{R}}$ from Equation 3.17:

$$\tan \theta = \frac{R_y}{R_x} = \frac{-2.0 \text{ m}}{4.0 \text{ m}} = -0.50$$

Your calculator likely gives the answer $-27°$ for $\theta = \tan^{-1}(-0.50)$. This answer is correct if we interpret it to mean 27° clockwise from the x axis. Our standard form has been to quote the angles measured counterclockwise from the $+x$ axis, and that angle for this vector is $\theta = \boxed{333°}$.

Example 3.4   **The Resultant Displacement**

A particle undergoes three consecutive displacements: $\Delta\vec{\mathbf{r}}_1 = (15\hat{\mathbf{i}} + 30\hat{\mathbf{j}} + 12\hat{\mathbf{k}})$ cm, $\Delta\vec{\mathbf{r}}_2 = (23\hat{\mathbf{i}} - 14\hat{\mathbf{j}} - 5.0\hat{\mathbf{k}})$ cm, and $\Delta\vec{\mathbf{r}}_3 = (-13\hat{\mathbf{i}} + 15\hat{\mathbf{j}})$ cm. Find unit-vector notation for the resultant displacement and its magnitude.

SOLUTION

Conceptualize Although x is sufficient to locate a point in one dimension, we need a vector $\vec{\mathbf{r}}$ to locate a point in two or three dimensions. The notation $\Delta\vec{\mathbf{r}}$ is a generalization of the one-dimensional displacement Δx in Equation 2.1. Three-dimensional displacements are more difficult to conceptualize than those in two dimensions because they cannot be drawn on paper like the latter.

For this problem, let us imagine that you start with your pencil at the origin of a piece of graph paper on which you have drawn x and y axes. Move your pencil 15 cm to the right along the x axis, then 30 cm upward along the y axis, and then 12 cm *perpendicularly toward you away*

from the graph paper. This procedure provides the displacement described by $\Delta\vec{\mathbf{r}}_1$. From this point, move your pencil 23 cm to the right parallel to the x axis, then 14 cm parallel to the graph paper in the $-y$ direction, and then 5.0 cm perpendicularly away from you toward the graph paper. You are now at the displacement from the origin described by $\Delta\vec{\mathbf{r}}_1 + \Delta\vec{\mathbf{r}}_2$. From this point, move your pencil 13 cm to the left in the $-x$ direction, and (finally!) 15 cm parallel to the graph paper along the y axis. Your final position is at a displacement $\Delta\vec{\mathbf{r}}_1 + \Delta\vec{\mathbf{r}}_2 + \Delta\vec{\mathbf{r}}_3$ from the origin.

▶ **3.4** continued

Categorize Despite the difficulty in conceptualizing in three dimensions, we can categorize this problem as a substitution problem because of the careful bookkeeping methods that we have developed for vectors. The mathematical manipulation keeps track of this motion along the three perpendicular axes in an organized, compact way, as we see below.

To find the resultant displacement, add the three vectors:

$$\Delta\vec{\mathbf{r}} = \Delta\vec{\mathbf{r}}_1 + \Delta\vec{\mathbf{r}}_2 + \Delta\vec{\mathbf{r}}_3$$

$$= (15 + 23 - 13)\hat{\mathbf{i}}\text{ cm} + (30 - 14 + 15)\hat{\mathbf{j}}\text{ cm} + (12 - 5.0 + 0)\hat{\mathbf{k}}\text{ cm}$$

$$= \boxed{(25\hat{\mathbf{i}} + 31\hat{\mathbf{j}} + 7.0\hat{\mathbf{k}})\text{ cm}}$$

Find the magnitude of the resultant vector:

$$R = \sqrt{R_x^2 + R_y^2 + R_z^2}$$

$$= \sqrt{(25\text{ cm})^2 + (31\text{ cm})^2 + (7.0\text{ cm})^2} = \boxed{40\text{ cm}}$$

Example 3.5 **Taking a Hike**

A hiker begins a trip by first walking 25.0 km southeast from her car. She stops and sets up her tent for the night. On the second day, she walks 40.0 km in a direction 60.0° north of east, at which point she discovers a forest ranger's tower.

(A) Determine the components of the hiker's displacement for each day.

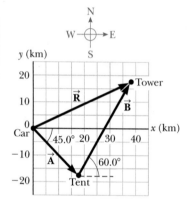

SOLUTION

Conceptualize We conceptualize the problem by drawing a sketch as in Figure 3.17. If we denote the displacement vectors on the first and second days by $\vec{\mathbf{A}}$ and $\vec{\mathbf{B}}$, respectively, and use the car as the origin of coordinates, we obtain the vectors shown in Figure 3.17. The sketch allows us to estimate the resultant vector as shown.

Categorize Having drawn the resultant $\vec{\mathbf{R}}$, we can now categorize this problem as one we've solved before: an addition of two vectors. You should now have a hint of the power of categorization in that many new problems are very similar to problems we have already solved if we are careful to conceptualize them. Once

Figure 3.17 (Example 3.5) The total displacement of the hiker is the vector $\vec{\mathbf{R}} = \vec{\mathbf{A}} + \vec{\mathbf{B}}$.

we have drawn the displacement vectors and categorized the problem, this problem is no longer about a hiker, a walk, a car, a tent, or a tower. It is a problem about vector addition, one that we have already solved.

....

Analyze Displacement $\vec{\mathbf{A}}$ has a magnitude of 25.0 km and is directed 45.0° below the positive x axis.

Find the components of $\vec{\mathbf{A}}$ using Equations 3.8 and 3.9:

$$A_x = A\cos(-45.0°) = (25.0\text{ km})(0.707) = \boxed{17.7\text{ km}}$$

$$A_y = A\sin(-45.0°) = (25.0\text{ km})(-0.707) = \boxed{-17.7\text{ km}}$$

The negative value of A_y indicates that the hiker walks in the negative y direction on the first day. The signs of A_x and A_y also are evident from Figure 3.17.

Find the components of $\vec{\mathbf{B}}$ using Equations 3.8 and 3.9:

$$B_x = B\cos 60.0° = (40.0\text{ km})(0.500) = \boxed{20.0\text{ km}}$$

$$B_y = B\sin 60.0° = (40.0\text{ km})(0.866) = \boxed{34.6\text{ km}}$$

(B) Determine the components of the hiker's resultant displacement $\vec{\mathbf{R}}$ for the trip. Find an expression for $\vec{\mathbf{R}}$ in terms of unit vectors.

SOLUTION

Use Equation 3.15 to find the components of the resultant displacement $\vec{\mathbf{R}} = \vec{\mathbf{A}} + \vec{\mathbf{B}}$:

$$R_x = A_x + B_x = 17.7\text{ km} + 20.0\text{ km} = \boxed{37.7\text{ km}}$$

$$R_y = A_y + B_y = -17.7\text{ km} + 34.6\text{ km} = \boxed{17.0\text{ km}}$$

continued

▶ **3.5** continued

Write the total displacement in unit-vector form:

$$\vec{R} = (37.7\hat{i} + 17.0\hat{j}) \text{ km}$$

Finalize Looking at the graphical representation in Figure 3.17, we estimate the position of the tower to be about (38 km, 17 km), which is consistent with the components of $\vec{R}$ in our result for the final position of the hiker. Also, both components of $\vec{R}$ are positive, putting the final position in the first quadrant of the coordinate system, which is also consistent with Figure 3.17.

WHAT IF?   After reaching the tower, the hiker wishes to return to her car along a single straight line. What are the components of the vector representing this hike? What should the direction of the hike be?

Answer The desired vector $\vec{R}_{car}$ is the negative of vector $\vec{R}$:

$$\vec{R}_{car} = -\vec{R} = (-37.7\hat{i} - 17.0\hat{j}) \text{ km}$$

The direction is found by calculating the angle that the vector makes with the x axis:

$$\tan \theta = \frac{R_{car,y}}{R_{car,x}} = \frac{-17.0 \text{ km}}{-37.7 \text{ km}} = 0.450$$

which gives an angle of $\theta = 204.2°$, or $24.2°$ south of west.

Summary

Definitions

Scalar quantities are those that have only a numerical value and no associated direction.

Vector quantities have both magnitude and direction and obey the laws of vector addition. The magnitude of a vector is *always* a positive number.

Concepts and Principles

When two or more vectors are added together, they must all have the same units and they all must be the same type of quantity. We can add two vectors $\vec{A}$ and $\vec{B}$ graphically. In this method (Fig. 3.6), the resultant vector $\vec{R} = \vec{A} + \vec{B}$ runs from the tail of $\vec{A}$ to the tip of $\vec{B}$.

A second method of adding vectors involves **components** of the vectors. The x component A_x of the vector $\vec{A}$ is equal to the projection of $\vec{A}$ along the x axis of a coordinate system, where $A_x = A \cos \theta$. The y component A_y of $\vec{A}$ is the projection of $\vec{A}$ along the y axis, where $A_y = A \sin \theta$.

If a vector $\vec{A}$ has an x component A_x and a y component A_y, the vector can be expressed in unit-vector form as $\vec{A} = A_x\hat{i} + A_y\hat{j}$. In this notation, $\hat{i}$ is a unit vector pointing in the positive x direction and $\hat{j}$ is a unit vector pointing in the positive y direction. Because $\hat{i}$ and $\hat{j}$ are unit vectors, $|\hat{i}| = |\hat{j}| = 1$.

We can find the resultant of two or more vectors by resolving all vectors into their x and y components, adding their resultant x and y components, and then using the Pythagorean theorem to find the magnitude of the resultant vector. We can find the angle that the resultant vector makes with respect to the x axis by using a suitable trigonometric function.

1. What is the magnitude of the vector $(10\hat{\mathbf{i}} - 10\hat{\mathbf{k}})$ m/s?
   (a) 0 (b) 10 m/s (c) −10 m/s (d) 10 (e) 14.1 m/s

2. A vector lying in the xy plane has components of opposite sign. The vector must lie in which quadrant? (a) the first quadrant (b) the second quadrant (c) the third quadrant (d) the fourth quadrant (e) either the second or the fourth quadrant

3. Figure OQ3.3 shows two vectors $\vec{\mathbf{D}}_1$ and $\vec{\mathbf{D}}_2$. Which of the possibilities (a) through (d) is the vector $\vec{\mathbf{D}}_2 - 2\vec{\mathbf{D}}_1$, or (e) is it none of them?

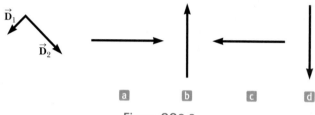

Figure OQ3.3

4. The cutting tool on a lathe is given two displacements, one of magnitude 4 cm and one of magnitude 3 cm, in each one of five situations (a) through (e) diagrammed in Figure OQ3.4. Rank these situations according to the magnitude of the total displacement of the tool, putting the situation with the greatest resultant magnitude first. If the total displacement is the same size in two situations, give those letters equal ranks.

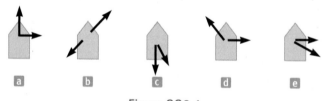

Figure OQ3.4

5. The magnitude of vector $\vec{\mathbf{A}}$ is 8 km, and the magnitude of $\vec{\mathbf{B}}$ is 6 km. Which of the following are possible values for the magnitude of $\vec{\mathbf{A}} + \vec{\mathbf{B}}$? Choose all possible answers. (a) 10 km (b) 8 km (c) 2 km (d) 0 (e) −2 km

6. Let vector $\vec{\mathbf{A}}$ point from the origin into the second quadrant of the xy plane and vector $\vec{\mathbf{B}}$ point from the origin into the fourth quadrant. The vector $\vec{\mathbf{B}} - \vec{\mathbf{A}}$ must be in which quadrant, (a) the first, (b) the second, (c) the third, or (d) the fourth, or (e) is more than one answer possible?

7. Yes or no: Is each of the following quantities a vector? (a) force (b) temperature (c) the volume of water in a can (d) the ratings of a TV show (e) the height of a building (f) the velocity of a sports car (g) the age of the Universe

8. What is the y component of the vector $(3\hat{\mathbf{i}} - 8\hat{\mathbf{k}})$ m/s?
   (a) 3 m/s (b) −8 m/s (c) 0 (d) 8 m/s (e) none of those answers

9. What is the x component of the vector shown in Figure OQ3.9? (a) 3 cm (b) 6 cm (c) −4 cm (d) −6 cm (e) none of those answers

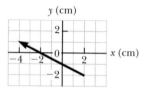

Figure OQ3.9 Objective Questions 9 and 10.

10. What is the y component of the vector shown in Figure OQ3.9? (a) 3 cm (b) 6 cm (c) −4 cm (d) −6 cm (e) none of those answers

11. Vector $\vec{\mathbf{A}}$ lies in the xy plane. Both of its components will be negative if it points from the origin into which quadrant? (a) the first quadrant (b) the second quadrant (c) the third quadrant (d) the fourth quadrant (e) the second or fourth quadrants

12. A submarine dives from the water surface at an angle of 30° below the horizontal, following a straight path 50 m long. How far is the submarine then below the water surface? (a) 50 m (b) (50 m)/sin 30° (c) (50 m) sin 30° (d) (50 m) cos 30° (e) none of those answers

13. A vector points from the origin into the second quadrant of the xy plane. What can you conclude about its components? (a) Both components are positive. (b) The x component is positive, and the y component is negative. (c) The x component is negative, and the y component is positive. (d) Both components are negative. (e) More than one answer is possible.

1. Is it possible to add a vector quantity to a scalar quantity? Explain.

2. Can the magnitude of a vector have a negative value? Explain.

3. A book is moved once around the perimeter of a tabletop with the dimensions 1.0 m by 2.0 m. The book ends up at its initial position. (a) What is its displacement? (b) What is the distance traveled?

4. If the component of vector $\vec{\mathbf{A}}$ along the direction of vector $\vec{\mathbf{B}}$ is zero, what can you conclude about the two vectors?

5. On a certain calculator, the inverse tangent function returns a value between −90° and +90°. In what cases will this value correctly state the direction of a vector in the xy plane, by giving its angle measured counterclockwise from the positive x axis? In what cases will it be incorrect?

Problems available in ENHANCED **WebAssign** Access end-of-chapter problems online at www.webassign.net

Section 3.1 Coordinate Systems
Problems 1–6

Section 3.2 Vector and Scalar Quantities
Section 3.3 Some Properties of Vectors
Problems 7–14

Section 3.4 Components of a Vector and Unit Vectors
Problems 15–47

Additional Problems
Problems 48–66

Challenge Problems
Problem 67

Solutions to the following Problems are available in the *Student Solutions Manual/Study Guide:*
3.1, 3.7, 3.9, 3.11, 3.13, 3.15, 3.19, 3.23, 3.25, 3.33, 3.35, 3.38, 3.39, 3.48, and 3.51

List of Enhanced Problems

Problem Number	Targeted Feedback in Enhanced WebAssign	Analysis Model Tutorial in Enhanced WebAssign	Master It in Enhanced WebAssign	Watch It in Enhanced WebAssign
3.1	✓			✓
3.4	✓			✓
3.7	✓			✓
3.11			✓	
3.15	✓			✓
3.19			✓	
3.23	✓		✓	
3.25			✓	
3.26	✓			✓
3.29	✓			✓
3.31	✓			✓
3.32	✓			✓
3.33			✓	
3.35			✓	
3.36	✓			✓
3.39			✓	
3.43		✓		
3.45		✓		
3.48				✓
3.51	✓		✓	
3.53		✓		
3.63	✓			✓

Motion in Two Dimensions

Fireworks erupt from the Sydney Harbour Bridge in New South Wales, Australia. Notice the parabolic paths of embers projected into the air. All projectiles follow a parabolic path in the absence of air resistance. *(Graham Monro/ Photolibrary/Jupiter Images)*

In this chapter, we explore the kinematics of a particle moving in two dimensions. Knowing the basics of two-dimensional motion will allow us—in future chapters—to examine a variety of situations, ranging from the motion of satellites in orbit to the motion of electrons in a uniform electric field. We begin by studying in greater detail the vector nature of position, velocity, and acceleration. We then treat projectile motion and uniform circular motion as special cases of motion in two dimensions. We also discuss the concept of relative motion, which shows why observers in different frames of reference may measure different positions and velocities for a given particle.

4.1 The Position, Velocity, and Acceleration Vectors

In Chapter 2, we found that the motion of a particle along a straight line such as the x axis is completely known if its position is known as a function of time. Let us now extend this idea to two-dimensional motion of a particle in the xy plane. We begin by describing the position of the particle. In one dimension, a single numerical value describes a particle's position, but in two dimensions, we indicate its position by its **position vector $\vec{r}$**, drawn from the origin of some coordinate system to the location of the particle in the xy plane as in Figure 4.1. At time t_i, the particle is at point Ⓐ, described by position vector $\vec{r}_i$. At some later time t_f, it is at point Ⓑ, described by position vector $\vec{r}_f$. The path followed by the particle from

Ⓐ to Ⓑ is not necessarily a straight line. As the particle moves from Ⓐ to Ⓑ in the time interval $\Delta t = t_f - t_i$, its position vector changes from $\vec{\mathbf{r}}_i$ to $\vec{\mathbf{r}}_f$. As we learned in Chapter 2, displacement is a vector, and the displacement of the particle is the difference between its final position and its initial position. We now define the **displacement vector** $\Delta\vec{\mathbf{r}}$ for a particle such as the one in Figure 4.1 as being the difference between its final position vector and its initial position vector:

$$\Delta\vec{\mathbf{r}} \equiv \vec{\mathbf{r}}_f - \vec{\mathbf{r}}_i \qquad (4.1)$$

◀ Displacement vector

The direction of $\Delta\vec{\mathbf{r}}$ is indicated in Figure 4.1. As we see from the figure, the magnitude of $\Delta\vec{\mathbf{r}}$ is *less* than the distance traveled along the curved path followed by the particle.

As we saw in Chapter 2, it is often useful to quantify motion by looking at the displacement divided by the time interval during which that displacement occurs, which gives the rate of change of position. Two-dimensional (or three-dimensional) kinematics is similar to one-dimensional kinematics, but we must now use full vector notation rather than positive and negative signs to indicate the direction of motion.

We define the **average velocity** $\vec{\mathbf{v}}_{avg}$ of a particle during the time interval Δt as the displacement of the particle divided by the time interval:

$$\vec{\mathbf{v}}_{avg} \equiv \frac{\Delta\vec{\mathbf{r}}}{\Delta t} \qquad (4.2)$$

◀ Average velocity

Multiplying or dividing a vector quantity by a positive scalar quantity such as Δt changes only the magnitude of the vector, not its direction. Because displacement is a vector quantity and the time interval is a positive scalar quantity, we conclude that the average velocity is a vector quantity directed along $\Delta\vec{\mathbf{r}}$. Compare Equation 4.2 with its one-dimensional counterpart, Equation 2.2.

The average velocity between points is *independent of the path* taken. That is because average velocity is proportional to displacement, which depends only on the initial and final position vectors and not on the path taken. As with one-dimensional motion, we conclude that if a particle starts its motion at some point and returns to this point via any path, its average velocity is zero for this trip because its displacement is zero. Consider again our basketball players on the court in Figure 2.2 (page 19). We previously considered only their one-dimensional motion back and forth between the baskets. In reality, however, they move over a two-dimensional surface, running back and forth between the baskets as well as left and right across the width of the court. Starting from one basket, a given player may follow a very complicated two-dimensional path. Upon returning to the original basket, however, a player's average velocity is zero because the player's displacement for the whole trip is zero.

Consider again the motion of a particle between two points in the xy plane as shown in Figure 4.2 (page 66). The dashed curve shows the path of the particle. As the time interval over which we observe the motion becomes smaller and smaller—that is, as Ⓑ is moved to Ⓑ′ and then to Ⓑ″ and so on—the direction of the displacement approaches that of the line tangent to the path at Ⓐ. The **instantaneous velocity** $\vec{\mathbf{v}}$ is defined as the limit of the average velocity $\Delta\vec{\mathbf{r}}/\Delta t$ as Δt approaches zero:

$$\vec{\mathbf{v}} \equiv \lim_{\Delta t \to 0} \frac{\Delta\vec{\mathbf{r}}}{\Delta t} = \frac{d\vec{\mathbf{r}}}{dt} \qquad (4.3)$$

◀ Instantaneous velocity

That is, the instantaneous velocity equals the derivative of the position vector with respect to time. The direction of the instantaneous velocity vector at any point in a particle's path is along a line tangent to the path at that point and in the direction of motion. Compare Equation 4.3 with the corresponding one-dimensional version, Equation 2.5.

The magnitude of the instantaneous velocity vector $v = |\vec{\mathbf{v}}|$ of a particle is called the *speed* of the particle, which is a scalar quantity.

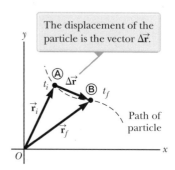

Figure 4.1 A particle moving in the xy plane is located with the position vector $\vec{\mathbf{r}}$ drawn from the origin to the particle. The displacement of the particle as it moves from Ⓐ to Ⓑ in the time interval $\Delta t = t_f - t_i$ is equal to the vector $\Delta\vec{\mathbf{r}} = \vec{\mathbf{r}}_f - \vec{\mathbf{r}}_i$.

Figure 4.2 As a particle moves between two points, its average velocity is in the direction of the displacement vector $\Delta\vec{r}$. By definition, the instantaneous velocity at Ⓐ is directed along the line tangent to the curve at Ⓐ.

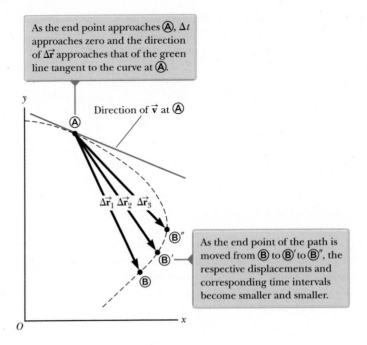

As the end point approaches Ⓐ, Δt approaches zero and the direction of $\Delta\vec{r}$ approaches that of the green line tangent to the curve at Ⓐ.

Direction of $\vec{v}$ at Ⓐ

$\Delta\vec{r}_1 \ \Delta\vec{r}_2 \ \Delta\vec{r}_3$

As the end point of the path is moved from Ⓑ to Ⓑ′ to Ⓑ″, the respective displacements and corresponding time intervals become smaller and smaller.

As a particle moves from one point to another along some path, its instantaneous velocity vector changes from $\vec{v}_i$ at time t_i to $\vec{v}_f$ at time t_f. Knowing the velocity at these points allows us to determine the average acceleration of the particle. The **average acceleration** $\vec{a}_{avg}$ of a particle is defined as the change in its instantaneous velocity vector $\Delta\vec{v}$ divided by the time interval Δt during which that change occurs:

◀ **Average acceleration**

$$\vec{a}_{avg} \equiv \frac{\Delta\vec{v}}{\Delta t} = \frac{\vec{v}_f - \vec{v}_i}{t_f - t_i} \tag{4.4}$$

Because $\vec{a}_{avg}$ is the ratio of a vector quantity $\Delta\vec{v}$ and a positive scalar quantity Δt, we conclude that average acceleration is a vector quantity directed along $\Delta\vec{v}$. As indicated in Figure 4.3, the direction of $\Delta\vec{v}$ is found by adding the vector $-\vec{v}_i$ (the negative of $\vec{v}_i$) to the vector $\vec{v}_f$ because, by definition, $\Delta\vec{v} = \vec{v}_f - \vec{v}_i$. Compare Equation 4.4 with Equation 2.9.

When the average acceleration of a particle changes during different time intervals, it is useful to define its instantaneous acceleration. The **instantaneous acceleration** $\vec{a}$ is defined as the limiting value of the ratio $\Delta\vec{v}/\Delta t$ as Δt approaches zero:

◀ **Instantaneous acceleration**

$$\vec{a} \equiv \lim_{\Delta t \to 0} \frac{\Delta\vec{v}}{\Delta t} = \frac{d\vec{v}}{dt} \tag{4.5}$$

In other words, the instantaneous acceleration equals the derivative of the velocity vector with respect to time. Compare Equation 4.5 with Equation 2.10.

Various changes can occur when a particle accelerates. First, the magnitude of the velocity vector (the speed) may change with time as in straight-line (one-

Pitfall Prevention 4.1

Vector Addition Although the vector addition discussed in Chapter 3 involves *displacement* vectors, vector addition can be applied to *any* type of vector quantity. Figure 4.3, for example, shows the addition of *velocity* vectors using the graphical approach.

Figure 4.3 A particle moves from position Ⓐ to position Ⓑ. Its velocity vector changes from $\vec{v}_i$ to $\vec{v}_f$. The vector diagrams at the upper right show two ways of determining the vector $\Delta\vec{v}$ from the initial and final velocities.

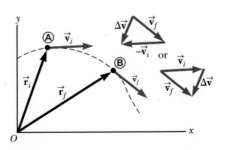

dimensional) motion. Second, the direction of the velocity vector may change with time even if its magnitude (speed) remains constant as in two-dimensional motion along a curved path. Finally, both the magnitude and the direction of the velocity vector may change simultaneously.

Quick **Quiz** 4.1 Consider the following controls in an automobile in motion: gas pedal, brake, steering wheel. What are the controls in this list that cause an acceleration of the car? **(a)** all three controls **(b)** the gas pedal and the brake **(c)** only the brake **(d)** only the gas pedal **(e)** only the steering wheel

4.2 Two-Dimensional Motion with Constant Acceleration

In Section 2.5, we investigated one-dimensional motion of a particle under constant acceleration and developed the particle under constant acceleration model. Let us now consider two-dimensional motion during which the acceleration of a particle remains constant in both magnitude and direction. As we shall see, this approach is useful for analyzing some common types of motion.

Before embarking on this investigation, we need to emphasize an important point regarding two-dimensional motion. Imagine an air hockey puck moving in a straight line along a perfectly level, friction-free surface of an air hockey table. Figure 4.4a shows a motion diagram from an overhead point of view of this puck. Recall that in Section 2.4 we related the acceleration of an object to a force on the object. Because there are no forces on the puck in the horizontal plane, it moves with constant velocity in the x direction. Now suppose you blow a puff of air on the puck as it passes your position, with the force from your puff of air *exactly* in the y direction. Because the force from this puff of air has no component in the x direction, it causes no acceleration in the x direction. It only causes a momentary acceleration in the y direction, causing the puck to have a constant y component of velocity once the force from the puff of air is removed. After your puff of air on the puck, its velocity component in the x direction is unchanged as shown in Figure 4.4b. The generalization of this simple experiment is that **motion in two dimensions can be modeled as two *independent* motions in each of the two perpendicular directions associated with the x and y axes. That is, any influence in the y direction does not affect the motion in the x direction and vice versa.**

The position vector for a particle moving in the xy plane can be written

$$\vec{\mathbf{r}} = x\hat{\mathbf{i}} + y\hat{\mathbf{j}} \qquad (4.6)$$

where x, y, and $\vec{\mathbf{r}}$ change with time as the particle moves while the unit vectors $\hat{\mathbf{i}}$ and $\hat{\mathbf{j}}$ remain constant. If the position vector is known, the velocity of the particle can be obtained from Equations 4.3 and 4.6, which give

$$\vec{\mathbf{v}} = \frac{d\vec{\mathbf{r}}}{dt} = \frac{dx}{dt}\hat{\mathbf{i}} + \frac{dy}{dt}\hat{\mathbf{j}} = v_x\hat{\mathbf{i}} + v_y\hat{\mathbf{j}} \qquad (4.7)$$

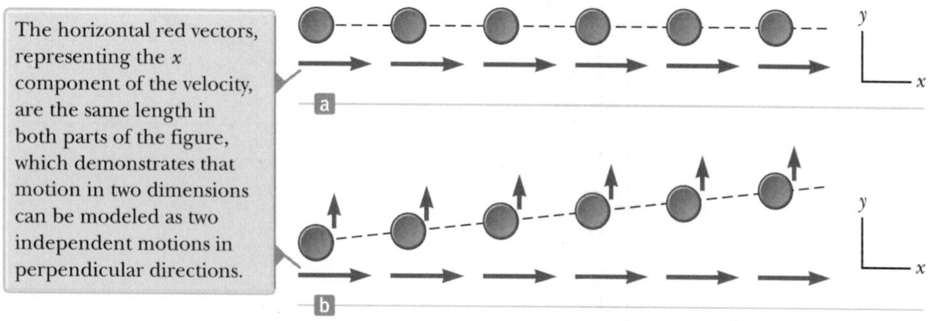

The horizontal red vectors, representing the x component of the velocity, are the same length in both parts of the figure, which demonstrates that motion in two dimensions can be modeled as two independent motions in perpendicular directions.

Figure 4.4 (a) A puck moves across a horizontal air hockey table at constant velocity in the x direction. (b) After a puff of air in the y direction is applied to the puck, the puck has gained a y component of velocity, but the x component is unaffected by the force in the perpendicular direction.

Because the acceleration $\vec{\mathbf{a}}$ of the particle is assumed constant in this discussion, its components a_x and a_y also are constants. Therefore, we can model the particle as a particle under constant acceleration independently in each of the two directions and apply the equations of kinematics separately to the x and y components of the velocity vector. Substituting, from Equation 2.13, $v_{xf} = v_{xi} + a_x t$ and $v_{yf} = v_{yi} + a_y t$ into Equation 4.7 to determine the final velocity at any time t, we obtain

$$\vec{\mathbf{v}}_f = (v_{xi} + a_x t)\hat{\mathbf{i}} + (v_{yi} + a_y t)\hat{\mathbf{j}} = (v_{xi}\hat{\mathbf{i}} + v_{yi}\hat{\mathbf{j}}) + (a_x\hat{\mathbf{i}} + a_y\hat{\mathbf{j}})t$$

Velocity vector as ▶
a function of time for a
particle under constant
acceleration in two
dimensions

$$\vec{\mathbf{v}}_f = \vec{\mathbf{v}}_i + \vec{\mathbf{a}}t \qquad (4.8)$$

This result states that the velocity of a particle at some time t equals the vector sum of its initial velocity $\vec{\mathbf{v}}_i$ at time $t = 0$ and the additional velocity $\vec{\mathbf{a}}t$ acquired at time t as a result of constant acceleration. Equation 4.8 is the vector version of Equation 2.13.

Similarly, from Equation 2.16 we know that the x and y coordinates of a particle under constant acceleration are

$$x_f = x_i + v_{xi}t + \tfrac{1}{2}a_x t^2 \qquad y_f = y_i + v_{yi}t + \tfrac{1}{2}a_y t^2$$

Substituting these expressions into Equation 4.6 (and labeling the final position vector $\vec{\mathbf{r}}_f$) gives

$$\vec{\mathbf{r}}_f = (x_i + v_{xi}t + \tfrac{1}{2}a_x t^2)\hat{\mathbf{i}} + (y_i + v_{yi}t + \tfrac{1}{2}a_y t^2)\hat{\mathbf{j}}$$
$$= (x_i\hat{\mathbf{i}} + y_i\hat{\mathbf{j}}) + (v_{xi}\hat{\mathbf{i}} + v_{yi}\hat{\mathbf{j}})t + \tfrac{1}{2}(a_x\hat{\mathbf{i}} + a_y\hat{\mathbf{j}})t^2$$

Position vector as ▶
a function of time for a
particle under constant
acceleration in two
dimensions

$$\vec{\mathbf{r}}_f = \vec{\mathbf{r}}_i + \vec{\mathbf{v}}_i t + \tfrac{1}{2}\vec{\mathbf{a}}t^2 \qquad (4.9)$$

which is the vector version of Equation 2.16. Equation 4.9 tells us that the position vector $\vec{\mathbf{r}}_f$ of a particle is the vector sum of the original position $\vec{\mathbf{r}}_i$, a displacement $\vec{\mathbf{v}}_i t$ arising from the initial velocity of the particle, and a displacement $\tfrac{1}{2}\vec{\mathbf{a}}t^2$ resulting from the constant acceleration of the particle.

We can consider Equations 4.8 and 4.9 to be the mathematical representation of a two-dimensional version of the particle under constant acceleration model. Graphical representations of Equations 4.8 and 4.9 are shown in Figure 4.5. The components of the position and velocity vectors are also illustrated in the figure. Notice from Figure 4.5a that $\vec{\mathbf{v}}_f$ is generally not along the direction of either $\vec{\mathbf{v}}_i$ or $\vec{\mathbf{a}}$ because the relationship between these quantities is a vector expression. For the same reason, from Figure 4.5b we see that $\vec{\mathbf{r}}_f$ is generally not along the direction of $\vec{\mathbf{r}}_i$, $\vec{\mathbf{v}}_i$, or $\vec{\mathbf{a}}$. Finally, notice that $\vec{\mathbf{v}}_f$ and $\vec{\mathbf{r}}_f$ are generally not in the same direction.

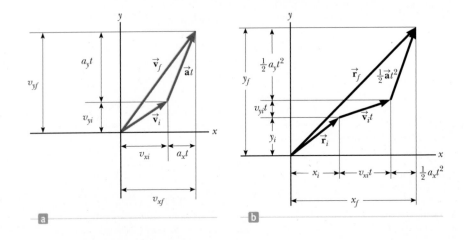

Figure 4.5 Vector representations and components of (a) the velocity and (b) the position of a particle under constant acceleration in two dimensions.

Example 4.1 Motion in a Plane AM

A particle moves in the xy plane, starting from the origin at $t = 0$ with an initial velocity having an x component of 20 m/s and a y component of -15 m/s. The particle experiences an acceleration in the x direction, given by $a_x = 4.0$ m/s^2.

(A) Determine the total velocity vector at any time.

SOLUTION

Conceptualize The components of the initial velocity tell us that the particle starts by moving toward the right and downward. The x component of velocity starts at 20 m/s and increases by 4.0 m/s every second. The y component of velocity never changes from its initial value of -15 m/s. We sketch a motion diagram of the situation in Figure 4.6. Because the particle is accelerating in the $+x$ direction, its velocity component in this direction increases and the path curves as shown in the diagram. Notice that the spacing between successive images increases as time goes on because the speed is increasing. The placement of the acceleration and velocity vectors in Figure 4.6 helps us further conceptualize the situation.

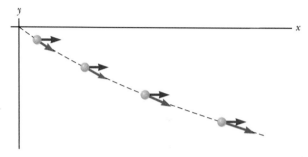

Figure 4.6 (Example 4.1) Motion diagram for the particle.

Categorize Because the initial velocity has components in both the x and y directions, we categorize this problem as one involving a particle moving in two dimensions. Because the particle only has an x component of acceleration, we model it as a *particle under constant acceleration* in the x direction and a *particle under constant velocity* in the y direction.

Analyze To begin the mathematical analysis, we set $v_{xi} = 20$ m/s, $v_{yi} = -15$ m/s, $a_x = 4.0$ m/s^2, and $a_y = 0$.

Use Equation 4.8 for the velocity vector:

$$\vec{\mathbf{v}}_f = \vec{\mathbf{v}}_i + \vec{\mathbf{a}}t = (v_{xi} + a_x t)\hat{\mathbf{i}} + (v_{yi} + a_y t)\hat{\mathbf{j}}$$

Substitute numerical values with the velocity in meters per second and the time in seconds:

$$\vec{\mathbf{v}}_f = [20 + (4.0)t]\hat{\mathbf{i}} + [-15 + (0)t]\hat{\mathbf{j}}$$

$$(1) \quad \vec{\mathbf{v}}_f = [(20 + 4.0t)\hat{\mathbf{i}} - 15\hat{\mathbf{j}}]$$

Finalize Notice that the x component of velocity increases in time while the y component remains constant; this result is consistent with our prediction.

(B) Calculate the velocity and speed of the particle at $t = 5.0$ s and the angle the velocity vector makes with the x axis.

SOLUTION

Analyze

Evaluate the result from Equation (1) at $t = 5.0$ s:

$$\vec{\mathbf{v}}_f = [(20 + 4.0(5.0))\hat{\mathbf{i}} - 15\hat{\mathbf{j}}] = (40\hat{\mathbf{i}} - 15\hat{\mathbf{j}}) \text{ m/s}$$

Determine the angle θ that $\vec{\mathbf{v}}_f$ makes with the x axis at $t = 5.0$ s:

$$\theta = \tan^{-1}\left(\frac{v_{yf}}{v_{xf}}\right) = \tan^{-1}\left(\frac{-15 \text{ m/s}}{40 \text{ m/s}}\right) = -21°$$

Evaluate the speed of the particle as the magnitude of $\vec{\mathbf{v}}_f$:

$$v_f = |\vec{\mathbf{v}}_f| = \sqrt{v_{xf}^2 + v_{yf}^2} = \sqrt{(40)^2 + (-15)^2} \text{ m/s} = 43 \text{ m/s}$$

Finalize The negative sign for the angle θ indicates that the velocity vector is directed at an angle of 21° below the positive x axis. Notice that if we calculate v_i from the x and y components of $\vec{\mathbf{v}}_i$, we find that $v_f > v_i$. Is that consistent with our prediction?

(C) Determine the x and y coordinates of the particle at any time t and its position vector at this time.

continued

▶ **4.1** continued

SOLUTION

Analyze

Use the components of Equation 4.9 with $x_i = y_i = 0$ at $t = 0$ and with x and y in meters and t in seconds:

$$x_f = v_{xi}t + \tfrac{1}{2}a_x t^2 = \boxed{20t + 2.0t^2}$$

$$y_f = v_{yi}t = \boxed{-15t}$$

Express the position vector of the particle at any time t:

$$\vec{\mathbf{r}}_f = x_f\hat{\mathbf{i}} + y_f\hat{\mathbf{j}} = \boxed{(20t + 2.0t^2)\hat{\mathbf{i}} - 15t\hat{\mathbf{j}}}$$

Finalize Let us now consider a limiting case for very large values of t.

WHAT IF?   What if we wait a very long time and then observe the motion of the particle? How would we describe the motion of the particle for large values of the time?

Answer Looking at Figure 4.6, we see the path of the particle curving toward the x axis. There is no reason to assume this tendency will change, which suggests that the path will become more and more parallel to the x axis as time grows large. Mathematically, Equation (1) shows that the y component of the velocity remains constant while the x component grows linearly with t. Therefore, when t is very large, the x component of the velocity will be much larger than the y component, suggesting that the velocity vector becomes more and more parallel to the x axis. The magnitudes of both x_f and y_f continue to grow with time, although x_f grows much faster.

Pitfall Prevention 4.2

Acceleration at the Highest Point
As discussed in Pitfall Prevention 2.8, many people claim that the acceleration of a projectile at the topmost point of its trajectory is zero. This mistake arises from confusion between zero vertical velocity and zero acceleration. If the projectile were to experience zero acceleration at the highest point, its velocity at that point would not change; rather, the projectile would move horizontally at constant speed from then on! That does not happen, however, because the acceleration is *not* zero anywhere along the trajectory.

Lester Lefkowitz/Taxi/Getty Images

A welder cuts holes through a heavy metal construction beam with a hot torch. The sparks generated in the process follow parabolic paths.

4.3 Projectile Motion

Anyone who has observed a baseball in motion has observed projectile motion. The ball moves in a curved path and returns to the ground. **Projectile motion** of an object is simple to analyze if we make two assumptions: (1) the free-fall acceleration is constant over the range of motion and is directed downward,[1] and (2) the effect of air resistance is negligible.[2] With these assumptions, we find that the path of a projectile, which we call its *trajectory,* is *always* a parabola as shown in Figure 4.7. **We use these assumptions throughout this chapter.**

The expression for the position vector of the projectile as a function of time follows directly from Equation 4.9, with its acceleration being that due to gravity, $\vec{\mathbf{a}} = \vec{\mathbf{g}}$:

$$\vec{\mathbf{r}}_f = \vec{\mathbf{r}}_i + \vec{\mathbf{v}}_i t + \tfrac{1}{2}\vec{\mathbf{g}}t^2 \tag{4.10}$$

where the initial x and y components of the velocity of the projectile are

$$v_{xi} = v_i \cos\theta_i \qquad v_{yi} = v_i \sin\theta_i \tag{4.11}$$

The expression in Equation 4.10 is plotted in Figure 4.8 for a projectile launched from the origin, so that $\vec{\mathbf{r}}_i = 0$. The final position of a particle can be considered to be the superposition of its initial position $\vec{\mathbf{r}}_i$; the term $\vec{\mathbf{v}}_i t$, which is its displacement if no acceleration were present; and the term $\tfrac{1}{2}\vec{\mathbf{g}}t^2$ that arises from its acceleration due to gravity. In other words, if there were no gravitational acceleration, the particle would continue to move along a straight path in the direction of $\vec{\mathbf{v}}_i$. Therefore, the vertical distance $\tfrac{1}{2}\vec{\mathbf{g}}t^2$ through which the particle "falls" off the straight-line path is the same distance that an object dropped from rest would fall during the same time interval.

---

[1]This assumption is reasonable as long as the range of motion is small compared with the radius of the Earth (6.4×10^6 m). In effect, this assumption is equivalent to assuming the Earth is flat over the range of motion considered.

[2]This assumption is often *not* justified, especially at high velocities. In addition, any spin imparted to a projectile, such as that applied when a pitcher throws a curve ball, can give rise to some very interesting effects associated with aerodynamic forces, which will be discussed in Chapter 14.

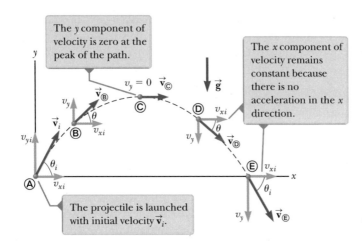

The y component of velocity is zero at the peak of the path.

The x component of velocity remains constant because there is no acceleration in the x direction.

The projectile is launched with initial velocity $\vec{\mathbf{v}}_i$.

Figure 4.7 The parabolic path of a projectile that leaves the origin with a velocity $\vec{\mathbf{v}}_i$. The velocity vector $\vec{\mathbf{v}}$ changes with time in both magnitude and direction. This change is the result of acceleration $\vec{\mathbf{a}} = \vec{\mathbf{g}}$ in the negative y direction.

In Section 4.2, we stated that two-dimensional motion with constant acceleration can be analyzed as a combination of two independent motions in the x and y directions, with accelerations a_x and a_y. Projectile motion can also be handled in this way, with acceleration $a_x = 0$ in the x direction and a constant acceleration $a_y = -g$ in the y direction. Therefore, when solving projectile motion problems, use two analysis models: (1) the particle under constant velocity in the horizontal direction (Eq. 2.7):

$$x_f = x_i + v_{xi}t$$

and (2) the particle under constant acceleration in the vertical direction (Eqs. 2.13–2.17 with x changed to y and $a_y = -g$):

$$v_{yf} = v_{yi} - gt$$

$$v_{y,\text{avg}} = \frac{v_{yi} + v_{yf}}{2}$$

$$y_f = y_i + \tfrac{1}{2}(v_{yi} + v_{yf})t$$

$$y_f = y_i + v_{yi}t - \tfrac{1}{2}gt^2$$

$$v_{yf}^2 = v_{yi}^2 - 2g(y_f - y_i)$$

The horizontal and vertical components of a projectile's motion are completely independent of each other and can be handled separately, with time t as the common variable for both components.

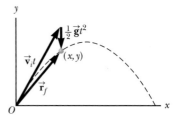

Figure 4.8 The position vector $\vec{\mathbf{r}}_f$ of a projectile launched from the origin whose initial velocity at the origin is $\vec{\mathbf{v}}_i$. The vector $\vec{\mathbf{v}}_it$ would be the displacement of the projectile if gravity were absent, and the vector $\tfrac{1}{2}\vec{\mathbf{g}}t^2$ is its vertical displacement from a straight-line path due to its downward gravitational acceleration.

Quick Quiz 4.2 **(i)** As a projectile thrown upward moves in its parabolic path (such as in Fig. 4.8), at what point along its path are the velocity and acceleration vectors for the projectile perpendicular to each other? (a) nowhere (b) the highest point (c) the launch point **(ii)** From the same choices, at what point are the velocity and acceleration vectors for the projectile parallel to each other?

Horizontal Range and Maximum Height of a Projectile

Before embarking on some examples, let us consider a special case of projectile motion that occurs often. Assume a projectile is launched from the origin at $t_i = 0$ with a positive v_{yi} component as shown in Figure 4.9 and returns to *the same horizontal level*. This situation is common in sports, where baseballs, footballs, and golf balls often land at the same level from which they were launched.

Two points in this motion are especially interesting to analyze: the peak point Ⓐ, which has Cartesian coordinates $(R/2, h)$, and the point Ⓑ, which has coordinates $(R, 0)$. The distance R is called the *horizontal range* of the projectile, and the distance h is its *maximum height*. Let us find h and R mathematically in terms of v_i, θ_i, and g.

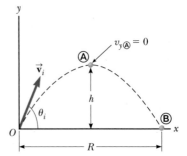

Figure 4.9 A projectile launched over a flat surface from the origin at $t_i = 0$ with an initial velocity $\vec{\mathbf{v}}_i$. The maximum height of the projectile is h, and the horizontal range is R. At Ⓐ, the peak of the trajectory, the particle has coordinates $(R/2, h)$.

We can determine h by noting that at the peak $v_{y\text{\textcircled{A}}} = 0$. Therefore, from the particle under constant acceleration model, we can use the y direction version of Equation 2.13 to determine the time $t_{\text{\textcircled{A}}}$ at which the projectile reaches the peak:

$$v_{yf} = v_{yi} - gt \quad \rightarrow \quad 0 = v_i \sin \theta_i - gt_{\text{\textcircled{A}}}$$

$$t_{\text{\textcircled{A}}} = \frac{v_i \sin \theta_i}{g}$$

Substituting this expression for $t_{\text{\textcircled{A}}}$ into the y direction version of Equation 2.16 and replacing $y_f = y_{\text{\textcircled{A}}}$ with h, we obtain an expression for h in terms of the magnitude and direction of the initial velocity vector:

$$y_f = y_i + v_{yi}t - \tfrac{1}{2}gt^2 \quad \rightarrow \quad h = (v_i \sin \theta_i)\frac{v_i \sin \theta_i}{g} - \tfrac{1}{2}g\left(\frac{v_i \sin \theta_i}{g}\right)^2$$

$$h = \frac{v_i^2 \sin^2 \theta_i}{2g} \tag{4.12}$$

The range R is the horizontal position of the projectile at a time that is twice the time at which it reaches its peak, that is, at time $t_{\text{\textcircled{B}}} = 2t_{\text{\textcircled{A}}}$. Using the particle under constant velocity model, noting that $v_{xi} = v_{x\text{\textcircled{B}}} = v_i \cos \theta_i$, and setting $x_{\text{\textcircled{B}}} = R$ at $t = 2t_{\text{\textcircled{A}}}$, we find that

$$x_f = x_i + v_{xi}t \quad \rightarrow \quad R = v_{xi}t_{\text{\textcircled{B}}} = (v_i \cos \theta_i)2t_{\text{\textcircled{A}}}$$

$$= (v_i \cos \theta_i)\frac{2v_i \sin \theta_i}{g} = \frac{2v_i^2 \sin \theta_i \cos \theta_i}{g}$$

Using the identity $\sin 2\theta = 2 \sin \theta \cos \theta$ (see Appendix B.4), we can write R in the more compact form

$$R = \frac{v_i^2 \sin 2\theta_i}{g} \tag{4.13}$$

The maximum value of R from Equation 4.13 is $R_{\max} = v_i^2/g$. This result makes sense because the maximum value of $\sin 2\theta_i$ is 1, which occurs when $2\theta_i = 90°$. Therefore, R is a maximum when $\theta_i = 45°$.

Figure 4.10 illustrates various trajectories for a projectile having a given initial speed but launched at different angles. As you can see, the range is a maximum for $\theta_i = 45°$. In addition, for any θ_i other than $45°$, a point having Cartesian coordinates $(R, 0)$ can be reached by using either one of two complementary values of θ_i, such as $75°$ and $15°$. Of course, the maximum height and time of flight for one of these values of θ_i are different from the maximum height and time of flight for the complementary value.

Quick Quiz 4.3 Rank the launch angles for the five paths in Figure 4.10 with respect to time of flight from the shortest time of flight to the longest.

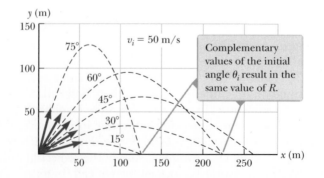

Figure 4.10 A projectile launched over a flat surface from the origin with an initial speed of 50 m/s at various angles of projection.

Problem-Solving Strategy   Projectile Motion

We suggest you use the following approach when solving projectile motion problems.

1. Conceptualize. Think about what is going on physically in the problem. Establish the mental representation by imagining the projectile moving along its trajectory.

2. Categorize. Confirm that the problem involves a particle in free fall and that air resistance is neglected. Select a coordinate system with x in the horizontal direction and y in the vertical direction. Use the particle under constant velocity model for the x component of the motion. Use the particle under constant acceleration model for the y direction. In the special case of the projectile returning to the same level from which it was launched, use Equations 4.12 and 4.13.

3. Analyze. If the initial velocity vector is given, resolve it into x and y components. Select the appropriate equation(s) from the particle under constant acceleration model for the vertical motion and use these along with Equation 2.7 for the horizontal motion to solve for the unknown(s).

4. Finalize. Once you have determined your result, check to see if your answers are consistent with the mental and pictorial representations and your results are realistic.

Example 4.2   The Long Jump

A long jumper (Fig. 4.11) leaves the ground at an angle of 20.0° above the horizontal and at a speed of 11.0 m/s.

(A) How far does he jump in the horizontal direction?

SOLUTION

Conceptualize The arms and legs of a long jumper move in a complicated way, but we will ignore this motion. We conceptualize the motion of the long jumper as equivalent to that of a simple projectile.

Categorize We categorize this example as a projectile motion problem. Because the initial speed and launch angle are given and because the final height is the same as the initial height, we further categorize this problem as satisfying the conditions for which Equations 4.12 and 4.13 can be used. This approach is the most direct way to analyze this problem, although the general methods that have been described will always give the correct answer.

Figure 4.11 (Example 4.2) Romain Barras of France competes in the men's decathlon long jump at the 2008 Beijing Olympic Games.

Sipa via AP Images

Analyze

Use Equation 4.13 to find the range of the jumper:

$$R = \frac{v_i^2 \sin 2\theta_i}{g} = \frac{(11.0 \text{ m/s})^2 \sin 2(20.0°)}{9.80 \text{ m/s}^2} = \boxed{7.94 \text{ m}}$$

(B) What is the maximum height reached?

SOLUTION

Analyze

Find the maximum height reached by using Equation 4.12:

$$h = \frac{v_i^2 \sin^2 \theta_i}{2g} = \frac{(11.0 \text{ m/s})^2 (\sin 20.0°)^2}{2(9.80 \text{ m/s}^2)} = \boxed{0.722 \text{ m}}$$

Finalize Find the answers to parts (A) and (B) using the general method. The results should agree. Treating the long jumper as a particle is an oversimplification. Nevertheless, the values obtained are consistent with experience in sports. We can model a complicated system such as a long jumper as a particle and still obtain reasonable results.

Example 4.3   **A Bull's-Eye Every Time**   AM

In a popular lecture demonstration, a projectile is fired at a target in such a way that the projectile leaves the gun at the same time the target is dropped from rest. Show that if the gun is initially aimed at the stationary target, the projectile hits the falling target as shown in Figure 4.12a.

SOLUTION

Conceptualize We conceptualize the problem by studying Figure 4.12a. Notice that the problem does not ask for numerical values. The expected result must involve an algebraic argument.

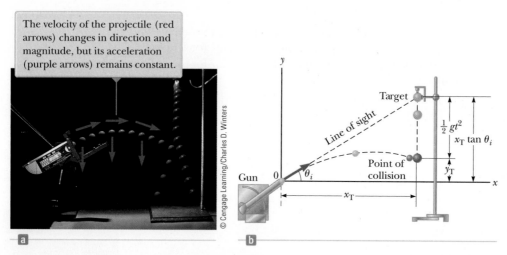

The velocity of the projectile (red arrows) changes in direction and magnitude, but its acceleration (purple arrows) remains constant.

© Cengage Learning/Charles D. Winters

Figure 4.12 (Example 4.3) (a) Multiflash photograph of the projectile–target demonstration. If the gun is aimed directly at the target and is fired at the same instant the target begins to fall, the projectile will hit the target. (b) Schematic diagram of the projectile–target demonstration.

Categorize Because both objects are subject only to gravity, we categorize this problem as one involving two objects in free fall, the target moving in one dimension and the projectile moving in two. The target T is modeled as a *particle under constant acceleration* in one dimension. The projectile P is modeled as a *particle under constant acceleration* in the *y* direction and a *particle under constant velocity* in the *x* direction.

Analyze Figure 4.12b shows that the initial *y* coordinate $y_{i\text{T}}$ of the target is $x_\text{T} \tan \theta_i$ and its initial velocity is zero. It falls with acceleration $a_y = -g$.

Write an expression for the *y* coordinate of the target at any moment after release, noting that its initial velocity is zero:

$$(1) \quad y_\text{T} = y_{i\text{T}} + (0)t - \tfrac{1}{2}gt^2 = x_\text{T} \tan \theta_i - \tfrac{1}{2}gt^2$$

Write an expression for the *y* coordinate of the projectile at any moment:

$$(2) \quad y_\text{P} = y_{i\text{P}} + v_{yi\text{P}}t - \tfrac{1}{2}gt^2 = 0 + (v_{i\text{P}} \sin\theta_i)t - \tfrac{1}{2}gt^2 = (v_{i\text{P}} \sin\theta_i)t - \tfrac{1}{2}gt^2$$

Write an expression for the *x* coordinate of the projectile at any moment:

$$x_\text{P} = x_{i\text{P}} + v_{xi\text{P}}t = 0 + (v_{i\text{P}} \cos\theta_i)t = (v_{i\text{P}} \cos\theta_i)t$$

Solve this expression for time as a function of the horizontal position of the projectile:

$$t = \frac{x_\text{P}}{v_{i\text{P}} \cos\theta_i}$$

Substitute this expression into Equation (2):

$$(3) \quad y_\text{P} = (v_{i\text{P}} \sin\theta_i)\left(\frac{x_\text{P}}{v_{i\text{P}} \cos\theta_i}\right) - \tfrac{1}{2}gt^2 = x_\text{P} \tan\theta_i - \tfrac{1}{2}gt^2$$

Finalize Compare Equations (1) and (3). We see that when the *x* coordinates of the projectile and target are the same—that is, when $x_\text{T} = x_\text{P}$—their *y* coordinates given by Equations (1) and (3) are the same and a collision results.

Example 4.4   That's Quite an Arm!   AM

A stone is thrown from the top of a building upward at an angle of 30.0° to the horizontal with an initial speed of 20.0 m/s as shown in Figure 4.13. The height from which the stone is thrown is 45.0 m above the ground.

(A) How long does it take the stone to reach the ground?

SOLUTION

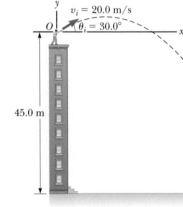

Conceptualize Study Figure 4.13, in which we have indicated the trajectory and various parameters of the motion of the stone.

Categorize We categorize this problem as a projectile motion problem. The stone is modeled as a *particle under constant acceleration* in the y direction and a *particle under constant velocity* in the x direction.

Analyze We have the information $x_i = y_i = 0$, $y_f = -45.0$ m, $a_y = -g$, and $v_i = 20.0$ m/s (the numerical value of y_f is negative because we have chosen the point of the throw as the origin).

Figure 4.13
(Example 4.4) A stone is thrown from the top of a building.

Find the initial x and y components of the stone's velocity:

$$v_{xi} = v_i \cos \theta_i = (20.0 \text{ m/s}) \cos 30.0° = 17.3 \text{ m/s}$$

$$v_{yi} = v_i \sin \theta_i = (20.0 \text{ m/s}) \sin 30.0° = 10.0 \text{ m/s}$$

Express the vertical position of the stone from the particle under constant acceleration model:

$$y_f = y_i + v_{yi}t - \tfrac{1}{2}gt^2$$

Substitute numerical values:

$$-45.0 \text{ m} = 0 + (10.0 \text{ m/s})t + \tfrac{1}{2}(-9.80 \text{ m/s}^2)t^2$$

Solve the quadratic equation for t:

$$t = 4.22 \text{ s}$$

(B) What is the speed of the stone just before it strikes the ground?

SOLUTION

Analyze Use the velocity equation in the particle under constant acceleration model to obtain the y component of the velocity of the stone just before it strikes the ground:

$$v_{yf} = v_{yi} - gt$$

Substitute numerical values, using $t = 4.22$ s:

$$v_{yf} = 10.0 \text{ m/s} + (-9.80 \text{ m/s}^2)(4.22 \text{ s}) = -31.3 \text{ m/s}$$

Use this component with the horizontal component $v_{xf} = v_{xi} = 17.3$ m/s to find the speed of the stone at $t = 4.22$ s:

$$v_f = \sqrt{v_{xf}^2 + v_{yf}^2} = \sqrt{(17.3 \text{ m/s})^2 + (-31.3 \text{ m/s})^2} = 35.8 \text{ m/s}$$

Finalize Is it reasonable that the y component of the final velocity is negative? Is it reasonable that the final speed is larger than the initial speed of 20.0 m/s?

WHAT IF?   What if a horizontal wind is blowing in the same direction as the stone is thrown and it causes the stone to have a horizontal acceleration component $a_x = 0.500$ m/s²? Which part of this example, (A) or (B), will have a different answer?

Answer Recall that the motions in the x and y directions are independent. Therefore, the horizontal wind cannot affect the vertical motion. The vertical motion determines the time of the projectile in the air, so the answer to part (A) does not change. The wind causes the horizontal velocity component to increase with time, so the final speed will be larger in part (B). Taking $a_x = 0.500$ m/s², we find $v_{xf} = 19.4$ m/s and $v_f = 36.9$ m/s.

Example 4.5	The End of the Ski Jump AM

A ski jumper leaves the ski track moving in the horizontal direction with a speed of 25.0 m/s as shown in Figure 4.14. The landing incline below her falls off with a slope of 35.0°. Where does she land on the incline?

SOLUTION

Conceptualize We can conceptualize this problem based on memories of observing winter Olympic ski competitions. We estimate the skier to be airborne for perhaps 4 s and to travel a distance of about 100 m horizontally. We should expect the value of d, the distance traveled along the incline, to be of the same order of magnitude.

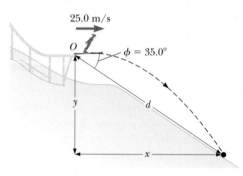

Categorize We categorize the problem as one of a particle in projectile motion. As with other projectile motion problems, we use the *particle under constant velocity* model for the horizontal motion and the *particle under constant acceleration* model for the vertical motion.

Figure 4.14 (Example 4.5) A ski jumper leaves the track moving in a horizontal direction.

Analyze It is convenient to select the beginning of the jump as the origin. The initial velocity components are $v_{xi} = 25.0$ m/s and $v_{yi} = 0$. From the right triangle in Figure 4.14, we see that the jumper's x and y coordinates at the landing point are given by $x_f = d \cos \phi$ and $y_f = -d \sin \phi$.

Express the coordinates of the jumper as a function of time, using the particle under constant velocity model for x and the position equation from the particle under constant acceleration model for y:

(1) $x_f = v_{xi} t$

(2) $y_f = v_{yi} t - \frac{1}{2} g t^2$

(3) $d \cos \phi = v_{xi} t$

(4) $-d \sin \phi = -\frac{1}{2} g t^2$

Solve Equation (3) for t and substitute the result into Equation (4):

$$-d \sin \phi = -\frac{1}{2} g \left(\frac{d \cos \phi}{v_{xi}} \right)^2$$

Solve for d and substitute numerical values:

$$d = \frac{2 v_{xi}^2 \sin \phi}{g \cos^2 \phi} = \frac{2(25.0 \text{ m/s})^2 \sin 35.0°}{(9.80 \text{ m/s}^2) \cos^2 35.0°} = 109 \text{ m}$$

Evaluate the x and y coordinates of the point at which the skier lands:

$x_f = d \cos \phi = (109 \text{ m}) \cos 35.0° = \boxed{89.3 \text{ m}}$

$y_f = -d \sin \phi = -(109 \text{ m}) \sin 35.0° = \boxed{-62.5 \text{ m}}$

Finalize Let us compare these results with our expectations. We expected the horizontal distance to be on the order of 100 m, and our result of 89.3 m is indeed on this order of magnitude. It might be useful to calculate the time interval that the jumper is in the air and compare it with our estimate of about 4 s.

WHAT IF? Suppose everything in this example is the same except the ski jump is curved so that the jumper is projected upward at an angle from the end of the track. Is this design better in terms of maximizing the length of the jump?

Answer If the initial velocity has an upward component, the skier will be in the air longer and should therefore travel farther. Tilting the initial velocity vector upward, however, will reduce the horizontal component of the initial velocity. Therefore, angling the end of the ski track upward at a *large* angle may actually *reduce* the distance. Consider the extreme case: the skier is projected at 90° to the horizontal and simply goes up and comes back down at the end of the ski track! This argument suggests that there must be an optimal angle between 0° and 90° that represents a balance between making the flight time longer and the horizontal velocity component smaller.

Let us find this optimal angle mathematically. We modify Equations (1) through (4) in the following way, assuming the skier is projected at an angle θ with respect to the horizontal over a landing incline sloped with an arbitrary angle ϕ:

(1) and (3) $\rightarrow$ $x_f = (v_i \cos \theta) t = d \cos \phi$

(2) and (4) $\rightarrow$ $y_f = (v_i \sin \theta) t - \frac{1}{2} g t^2 = -d \sin \phi$

▶ **4.5** continued

By eliminating the time t between these equations and using differentiation to maximize d in terms of θ, we arrive (after several steps; see Problem 88 in Enhanced WebAssign) at the following equation for the angle θ that gives the maximum value of d:

$$\theta = 45° - \frac{\phi}{2}$$

For the slope angle in Figure 4.14, $\phi = 35.0°$; this equation results in an optimal launch angle of $\theta = 27.5°$. For a slope angle of $\phi = 0°$, which represents a horizontal plane, this equation gives an optimal launch angle of $\theta = 45°$, as we would expect (see Figure 4.10).

4.4 Analysis Model: Particle in Uniform Circular Motion

Figure 4.15a shows a car moving in a circular path; we describe this motion by calling it **circular motion.** If the car is moving on this path with *constant speed v*, we call it **uniform circular motion.** Because it occurs so often, this type of motion is recognized as an analysis model called the **particle in uniform circular motion.** We discuss this model in this section.

It is often surprising to students to find that even though an object moves at a constant speed in a circular path, *it still has an acceleration.* To see why, consider the defining equation for acceleration, $\vec{a} = d\vec{v}/dt$ (Eq. 4.5). Notice that the acceleration depends on the change in the *velocity.* Because velocity is a vector quantity, an acceleration can occur in two ways as mentioned in Section 4.1: by a change in the *magnitude* of the velocity and by a change in the *direction* of the velocity. The latter situation occurs for an object moving with constant speed in a circular path. The constant-magnitude velocity vector is always tangent to the path of the object and perpendicular to the radius of the circular path. Therefore, the direction of the velocity vector is always changing.

Let us first argue that the acceleration vector in uniform circular motion is always perpendicular to the path and always points toward the center of the circle. If that were not true, there would be a component of the acceleration parallel to the path and therefore parallel to the velocity vector. Such an acceleration component would lead to a change in the speed of the particle along the path. This situation, however, is inconsistent with our setup of the situation: the particle moves with constant speed along the path. Therefore, for *uniform* circular motion, the acceleration vector can only have a component perpendicular to the path, which is toward the center of the circle.

Let us now find the magnitude of the acceleration of the particle. Consider the diagram of the position and velocity vectors in Figure 4.15b. The figure also shows the vector representing the change in position $\Delta\vec{r}$ for an arbitrary time interval. The particle follows a circular path of radius r, part of which is shown by the dashed

Pitfall Prevention 4.4
Acceleration of a Particle in Uniform Circular Motion
Remember that acceleration in physics is defined as a change in the *velocity*, not a change in the *speed* (contrary to the everyday interpretation). In circular motion, the velocity vector is always changing in direction, so there is indeed an acceleration.

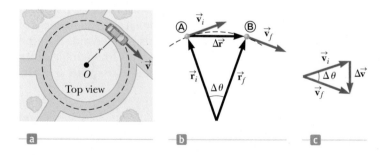

Figure 4.15 (a) A car moving along a circular path at constant speed experiences uniform circular motion. (b) As a particle moves along a portion of a circular path from Ⓐ to Ⓑ, its velocity vector changes from $\vec{v}_i$ to $\vec{v}_f$. (c) The construction for determining the direction of the change in velocity $\Delta\vec{v}$, which is toward the center of the circle for small $\Delta\vec{r}$.

curve. The particle is at Ⓐ at time t_i, and its velocity at that time is $\vec{\mathbf{v}}_i$; it is at Ⓑ at some later time t_f, and its velocity at that time is $\vec{\mathbf{v}}_f$. Let us also assume $\vec{\mathbf{v}}_i$ and $\vec{\mathbf{v}}_f$ differ only in direction; their magnitudes are the same (that is, $v_i = v_f = v$ because it is *uniform* circular motion).

In Figure 4.15c, the velocity vectors in Figure 4.15b have been redrawn tail to tail. The vector $\Delta\vec{\mathbf{v}}$ connects the tips of the vectors, representing the vector addition $\vec{\mathbf{v}}_f = \vec{\mathbf{v}}_i + \Delta\vec{\mathbf{v}}$. In both Figures 4.15b and 4.15c, we can identify triangles that help us analyze the motion. The angle $\Delta\theta$ between the two position vectors in Figure 4.15b is the same as the angle between the velocity vectors in Figure 4.15c because the velocity vector $\vec{\mathbf{v}}$ is always perpendicular to the position vector $\vec{\mathbf{r}}$. Therefore, the two triangles are *similar*. (Two triangles are similar if the angle between any two sides is the same for both triangles and if the ratio of the lengths of these sides is the same.) We can now write a relationship between the lengths of the sides for the two triangles in Figures 4.15b and 4.15c:

$$\frac{|\Delta\vec{\mathbf{v}}|}{v} = \frac{|\Delta\vec{\mathbf{r}}|}{r}$$

where $v = v_i = v_f$ and $r = r_i = r_f$. This equation can be solved for $|\Delta\vec{\mathbf{v}}|$, and the expression obtained can be substituted into Equation 4.4, $\vec{\mathbf{a}}_{avg} = \Delta\vec{\mathbf{v}}/\Delta t$, to give the magnitude of the average acceleration over the time interval for the particle to move from Ⓐ to Ⓑ:

$$|\vec{\mathbf{a}}_{avg}| = \frac{|\Delta\vec{\mathbf{v}}|}{|\Delta t|} = \frac{v|\Delta\vec{\mathbf{r}}|}{r\Delta t}$$

Now imagine that points Ⓐ and Ⓑ in Figure 4.15b become extremely close together. As Ⓐ and Ⓑ approach each other, Δt approaches zero, $|\Delta\vec{\mathbf{r}}|$ approaches the distance traveled by the particle along the circular path, and the ratio $|\Delta\vec{\mathbf{r}}|/\Delta t$ approaches the speed v. In addition, the average acceleration becomes the instantaneous acceleration at point Ⓐ. Hence, in the limit $\Delta t \rightarrow 0$, the magnitude of the acceleration is

Centripetal acceleration ▶
for a particle in uniform
circular motion

$$a_c = \frac{v^2}{r} \tag{4.14}$$

An acceleration of this nature is called a **centripetal acceleration** (*centripetal* means *center-seeking*). The subscript on the acceleration symbol reminds us that the acceleration is centripetal.

In many situations, it is convenient to describe the motion of a particle moving with constant speed in a circle of radius r in terms of the **period** T, which is defined as the time interval required for one complete revolution of the particle. In the time interval T, the particle moves a distance of $2\pi r$, which is equal to the circumference of the particle's circular path. Therefore, because its speed is equal to the circumference of the circular path divided by the period, or $v = 2\pi r/T$, it follows that

Period of circular motion ▶
for a particle in uniform
circular motion

$$T = \frac{2\pi r}{v} \tag{4.15}$$

The period of a particle in uniform circular motion is a measure of the number of seconds for one revolution of the particle around the circle. The inverse of the period is the *rotation rate* and is measured in revolutions per second. Because one full revolution of the particle around the circle corresponds to an angle of 2π radians, the product of 2π and the rotation rate gives the **angular speed** ω of the particle, measured in radians/s or s^{-1}:

$$\omega = \frac{2\pi}{T} \tag{4.16}$$

Combining this equation with Equation 4.15, we find a relationship between angular speed and the translational speed with which the particle travels in the circular path:

$$\omega = 2\pi\left(\frac{v}{2\pi r}\right) = \frac{v}{r} \quad \rightarrow \quad v = r\omega \qquad (4.17)$$

Equation 4.17 demonstrates that, for a fixed angular speed, the translational speed becomes larger as the radial position becomes larger. Therefore, for example, if a merry-go-round rotates at a fixed angular speed ω, a rider at an outer position at large r will be traveling through space faster than a rider at an inner position at smaller r. We will investigate Equations 4.16 and 4.17 more deeply in Chapter 10.

We can express the centripetal acceleration of a particle in uniform circular motion in terms of angular speed by combining Equations 4.14 and 4.17:

$$a_c = \frac{(r\omega)^2}{r}$$

$$a_c = r\omega^2 \qquad (4.18)$$

Equations 4.14–4.18 are to be used when the particle in uniform circular motion model is identified as appropriate for a given situation.

> **Pitfall Prevention 4.5**
> **Centripetal Acceleration Is Not Constant** We derived the magnitude of the centripetal acceleration vector and found it to be constant for uniform circular motion, but *the centripetal acceleration vector is not constant*. It always points toward the center of the circle, but it continuously changes direction as the object moves around the circular path.

Quick Quiz 4.4 A particle moves in a circular path of radius r with speed v. It then increases its speed to $2v$ while traveling along the same circular path. **(i)** The centripetal acceleration of the particle has changed by what factor? Choose one: (a) 0.25 (b) 0.5 (c) 2 (d) 4 (e) impossible to determine **(ii)** From the same choices, by what factor has the period of the particle changed?

Analysis Model   **Particle in Uniform Circular Motion**

Imagine a moving object that can be modeled as a particle. If it moves in a circular path of radius r at a constant speed v, the magnitude of its centripetal acceleration is

$$a_c = \frac{v^2}{r} \qquad (4.14)$$

and the **period** of the particle's motion is given by

$$T = \frac{2\pi r}{v} \qquad (4.15)$$

The **angular speed** of the particle is

$$\omega = \frac{2\pi}{T} \qquad (4.16)$$

Examples:

- a rock twirled in a circle on a string of constant length
- a planet traveling around a perfectly circular orbit (Chapter 13)
- a charged particle moving in a uniform magnetic field (Chapter 29)
- an electron in orbit around a nucleus in the Bohr model of the hydrogen atom (Chapter 42)

Example 4.6   **The Centripetal Acceleration of the Earth**   **AM**

(A) What is the centripetal acceleration of the Earth as it moves in its orbit around the Sun?

SOLUTION

Conceptualize Think about a mental image of the Earth in a circular orbit around the Sun. We will model the Earth as a particle and approximate the Earth's orbit as circular (it's actually slightly elliptical, as we discuss in Chapter 13).

Categorize The Conceptualize step allows us to categorize this problem as one of a *particle in uniform circular motion*.

Analyze We do not know the orbital speed of the Earth to substitute into Equation 4.14. With the help of Equation 4.15, however, we can recast Equation 4.14 in terms of the period of the Earth's orbit, which we know is one year, and the radius of the Earth's orbit around the Sun, which is 1.496×10^{11} m.

continued

▶ **4.6** continued

Combine Equations 4.14 and 4.15:

$$a_c = \frac{v^2}{r} = \frac{\left(\frac{2\pi r}{T}\right)^2}{r} = \frac{4\pi^2 r}{T^2}$$

Substitute numerical values:

$$a_c = \frac{4\pi^2 (1.496 \times 10^{11}\ \mathrm{m})}{(1\ \mathrm{yr})^2} \left(\frac{1\ \mathrm{yr}}{3.156 \times 10^7\ \mathrm{s}}\right)^2 = 5.93 \times 10^{-3}\ \mathrm{m/s^2}$$

(B) What is the angular speed of the Earth in its orbit around the Sun?

SOLUTION

Analyze

Substitute numerical values into Equation 4.16:

$$\omega = \frac{2\pi}{1\ \mathrm{yr}} \left(\frac{1\ \mathrm{yr}}{3.156 \times 10^7\ \mathrm{s}}\right) = 1.99 \times 10^{-7}\ \mathrm{s^{-1}}$$

Finalize The acceleration in part (A) is much smaller than the free-fall acceleration on the surface of the Earth. An important technique we learned here is replacing the speed v in Equation 4.14 in terms of the period T of the motion. In many problems, it is more likely that T is known rather than v. In part (B), we see that the angular speed of the Earth is very small, which is to be expected because the Earth takes an entire year to go around the circular path once.

4.5 Tangential and Radial Acceleration

Let us consider a more general motion than that presented in Section 4.4. A particle moves to the right along a curved path, and its velocity changes *both* in direction and in magnitude as described in Figure 4.16. In this situation, the velocity vector is always tangent to the path; the acceleration vector $\vec{a}$, however, is at some angle to the path. At each of three points Ⓐ, Ⓑ, and Ⓒ in Figure 4.16, the dashed blue circles represent the curvature of the actual path at each point. The radius of each circle is equal to the path's radius of curvature at each point.

As the particle moves along the curved path in Figure 4.16, the direction of the total acceleration vector $\vec{a}$ changes from point to point. At any instant, this vector can be resolved into two components based on an origin at the center of the dashed circle corresponding to that instant: a radial component a_r along the radius of the circle and a tangential component a_t perpendicular to this radius. The *total* acceleration vector $\vec{a}$ can be written as the vector sum of the component vectors:

◀ Total acceleration

$$\vec{a} = \vec{a}_r + \vec{a}_t \qquad (4.19)$$

The tangential acceleration component causes a change in the speed v of the particle. This component is parallel to the instantaneous velocity, and its magnitude is given by

◀ Tangential acceleration

$$a_t = \left|\frac{dv}{dt}\right| \qquad (4.20)$$

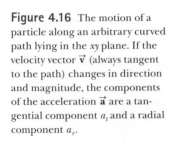

Figure 4.16 The motion of a particle along an arbitrary curved path lying in the *xy* plane. If the velocity vector $\vec{v}$ (always tangent to the path) changes in direction and magnitude, the components of the acceleration $\vec{a}$ are a tangential component a_t and a radial component a_r.

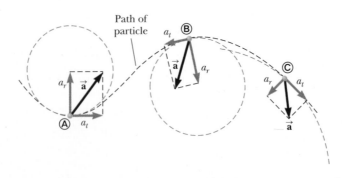

The radial acceleration component arises from a change in direction of the velocity vector and is given by

$$a_r = -a_c = -\frac{v^2}{r}$$

(4.21) ◀ **Radial acceleration**

where r is the radius of curvature of the path at the point in question. We recognize the magnitude of the radial component of the acceleration as the centripetal acceleration discussed in Section 4.4 with regard to the particle in uniform circular motion model. Even in situations in which a particle moves along a curved path with a varying speed, however, Equation 4.14 can be used for the centripetal acceleration. In this situation, the equation gives the *instantaneous* centripetal acceleration at any time. The negative sign in Equation 4.21 indicates that the direction of the centripetal acceleration is toward the center of the circle representing the radius of curvature. The direction is opposite that of the radial unit vector $\hat{\mathbf{r}}$, which always points away from the origin at the center of the circle.

Because $\vec{\mathbf{a}}_r$ and $\vec{\mathbf{a}}_t$ are perpendicular component vectors of $\vec{\mathbf{a}}$, it follows that the magnitude of $\vec{\mathbf{a}}$ is $a = \sqrt{a_r^2 + a_t^2}$. At a given speed, a_r is large when the radius of curvature is small (as at points Ⓐ and Ⓑ in Fig. 4.16) and small when r is large (as at point Ⓒ). The direction of $\vec{\mathbf{a}}_t$ is either in the same direction as $\vec{\mathbf{v}}$ (if v is increasing) or opposite $\vec{\mathbf{v}}$ (if v is decreasing, as at point Ⓑ).

In uniform circular motion, where v is constant, $a_t = 0$ and the acceleration is always completely radial as described in Section 4.4. In other words, uniform circular motion is a special case of motion along a general curved path. Furthermore, if the direction of $\vec{\mathbf{v}}$ does not change, there is no radial acceleration and the motion is one dimensional (in this case, $a_r = 0$, but a_t may not be zero).

Quick Quiz 4.5 A particle moves along a path, and its speed increases with time. **(i)** In which of the following cases are its acceleration and velocity vectors parallel? (a) when the path is circular (b) when the path is straight (c) when the path is a parabola (d) never **(ii)** From the same choices, in which case are its acceleration and velocity vectors perpendicular everywhere along the path?

Example 4.7 **Over the Rise**

A car leaves a stop sign and exhibits a constant acceleration of 0.300 m/s² parallel to the roadway. The car passes over a rise in the roadway such that the top of the rise is shaped like an arc of a circle of radius 500 m. At the moment the car is at the top of the rise, its velocity vector is horizontal and has a magnitude of 6.00 m/s. What are the magnitude and direction of the total acceleration vector for the car at this instant?

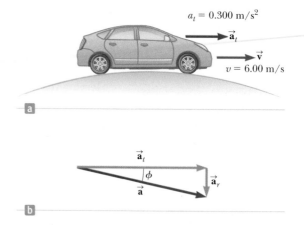

SOLUTION

Conceptualize Conceptualize the situation using Figure 4.17a and any experiences you have had in driving over rises on a roadway.

Categorize Because the accelerating car is moving along a curved path, we categorize this problem as one involving a particle experiencing both tangential and radial acceleration. We recognize that it is a relatively simple substitution problem.

Figure 4.17 (Example 4.7) (a) A car passes over a rise that is shaped like an arc of a circle. (b) The total acceleration vector $\vec{\mathbf{a}}$ is the sum of the tangential and radial acceleration vectors $\vec{\mathbf{a}}_t$ and $\vec{\mathbf{a}}_r$.

The tangential acceleration vector has magnitude 0.300 m/s² and is horizontal. The radial acceleration is given by Equation 4.21, with $v = 6.00$ m/s and $r = 500$ m. The radial acceleration vector is directed straight downward.

continued

▶ **4.7** continued

Evaluate the radial acceleration:

$$a_r = -\frac{v^2}{r} = -\frac{(6.00 \text{ m/s})^2}{500 \text{ m}} = -0.072\ 0 \text{ m/s}^2$$

Find the magnitude of $\vec{a}$:

$$\sqrt{a_r^2 + a_t^2} = \sqrt{(-0.072\ 0 \text{ m/s}^2)^2 + (0.300 \text{ m/s}^2)^2}$$
$$= \boxed{0.309 \text{ m/s}^2}$$

Find the angle ϕ (see Fig. 4.17b) between $\vec{a}$ and the horizontal:

$$\phi = \tan^{-1}\frac{a_r}{a_t} = \tan^{-1}\left(\frac{-0.072\ 0 \text{ m/s}^2}{0.300 \text{ m/s}^2}\right) = \boxed{-13.5°}$$

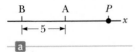

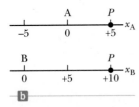

Figure 4.18 Different observers make different measurements. (a) Observer A is located 5 units to the right of Observer B. Both observers measure the position of a particle at P. (b) If both observers see themselves at the origin of their own coordinate system, they disagree on the value of the position of the particle at P.

The woman standing on the beltway sees the man moving with a slower speed than does the woman observing the man from the stationary floor.

Figure 4.19 Two observers measure the speed of a man walking on a moving beltway.

4.6 Relative Velocity and Relative Acceleration

In this section, we describe how observations made by different observers in different frames of reference are related to one another. A frame of reference can be described by a Cartesian coordinate system for which an observer is at rest with respect to the origin.

Let us conceptualize a sample situation in which there will be different observations for different observers. Consider the two observers A and B along the number line in Figure 4.18a. Observer A is located 5 units to the right of observer B. Both observers measure the position of point P, which is located 5 units to the right of observer A. Suppose each observer decides that he is located at the origin of an x axis as in Figure 4.18b. Notice that the two observers disagree on the value of the position of point P. Observer A claims point P is located at a position with a value of $x_A = +5$, whereas observer B claims it is located at a position with a value of $x_B = +10$. Both observers are correct, even though they make different measurements. Their measurements differ because they are making the measurement from different frames of reference.

Imagine now that observer B in Figure 4.18b is moving to the right along the x_B axis. Now the two measurements are even more different. Observer A claims point P remains at rest at a position with a value of +5, whereas observer B claims the position of P continuously changes with time, even passing him and moving behind him! Again, both observers are correct, with the difference in their measurements arising from their different frames of reference.

We explore this phenomenon further by considering two observers watching a man walking on a moving beltway at an airport in Figure 4.19. The woman standing on the moving beltway sees the man moving at a normal walking speed. The woman observing from the stationary floor sees the man moving with a higher speed because the beltway speed combines with his walking speed. Both observers look at the same man and arrive at different values for his speed. Both are correct; the difference in their measurements results from the relative velocity of their frames of reference.

In a more general situation, consider a particle located at point P in Figure 4.20. Imagine that the motion of this particle is being described by two observers, observer A in a reference frame S_A fixed relative to the Earth and a second observer B in a reference frame S_B moving to the right relative to S_A (and therefore relative to the Earth) with a constant velocity $\vec{v}_{BA}$. In this discussion of relative velocity, we use a double-subscript notation; the first subscript represents what is being observed, and the second represents who is doing the observing. Therefore, the notation $\vec{v}_{BA}$ means the velocity of observer B (and the attached frame S_B) as measured by observer A. With this notation, observer B measures A to be moving to the left with a velocity $\vec{v}_{AB} = -\vec{v}_{BA}$. For purposes of this discussion, let us place each observer at her or his respective origin.

We define the time $t = 0$ as the instant at which the origins of the two reference frames coincide in space. Therefore, at time t, the origins of the reference frames will be separated by a distance $v_{BA}t$. We label the position P of the particle relative to observer A with the position vector $\vec{\mathbf{r}}_{PA}$ and that relative to observer B with the position vector $\vec{\mathbf{r}}_{PB}$, both at time t. From Figure 4.20, we see that the vectors $\vec{\mathbf{r}}_{PA}$ and $\vec{\mathbf{r}}_{PB}$ are related to each other through the expression

$$\vec{\mathbf{r}}_{PA} = \vec{\mathbf{r}}_{PB} + \vec{\mathbf{v}}_{BA}t \tag{4.22}$$

By differentiating Equation 4.22 with respect to time, noting that $\vec{\mathbf{v}}_{BA}$ is constant, we obtain

$$\frac{d\vec{\mathbf{r}}_{PA}}{dt} = \frac{d\vec{\mathbf{r}}_{PB}}{dt} + \vec{\mathbf{v}}_{BA}$$

$$\vec{\mathbf{u}}_{PA} = \vec{\mathbf{u}}_{PB} + \vec{\mathbf{v}}_{BA} \tag{4.23}$$

◀ **Galilean velocity transformation**

where $\vec{\mathbf{u}}_{PA}$ is the velocity of the particle at P measured by observer A and $\vec{\mathbf{u}}_{PB}$ is its velocity measured by B. (We use the symbol $\vec{\mathbf{u}}$ for particle velocity rather than $\vec{\mathbf{v}}$, which we have already used for the relative velocity of two reference frames.) Equations 4.22 and 4.23 are known as **Galilean transformation equations.** They relate the position and velocity of a particle as measured by observers in relative motion. Notice the pattern of the subscripts in Equation 4.23. When relative velocities are added, the inner subscripts (B) are the same and the outer ones (P, A) match the subscripts on the velocity on the left of the equation.

Although observers in two frames measure different velocities for the particle, they measure the *same acceleration* when $\vec{\mathbf{v}}_{BA}$ is constant. We can verify that by taking the time derivative of Equation 4.23:

$$\frac{d\vec{\mathbf{u}}_{PA}}{dt} = \frac{d\vec{\mathbf{u}}_{PB}}{dt} + \frac{d\vec{\mathbf{v}}_{BA}}{dt}$$

Because $\vec{\mathbf{v}}_{BA}$ is constant, $d\vec{\mathbf{v}}_{BA}/dt = 0$. Therefore, we conclude that $\vec{\mathbf{a}}_{PA} = \vec{\mathbf{a}}_{PB}$ because $\vec{\mathbf{a}}_{PA} = d\vec{\mathbf{u}}_{PA}/dt$ and $\vec{\mathbf{a}}_{PB} = d\vec{\mathbf{u}}_{PB}/dt$. That is, the acceleration of the particle measured by an observer in one frame of reference is the same as that measured by any other observer moving with constant velocity relative to the first frame.

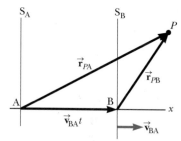

Figure 4.20 A particle located at P is described by two observers, one in the fixed frame of reference S_A and the other in the frame S_B, which moves to the right with a constant velocity $\vec{\mathbf{v}}_{BA}$. The vector $\vec{\mathbf{r}}_{PA}$ is the particle's position vector relative to S_A, and $\vec{\mathbf{r}}_{PB}$ is its position vector relative to S_B.

Example 4.8 **A Boat Crossing a River**

A boat crossing a wide river moves with a speed of 10.0 km/h relative to the water. The water in the river has a uniform speed of 5.00 km/h due east relative to the Earth.

(A) If the boat heads due north, determine the velocity of the boat relative to an observer standing on either bank.

SOLUTION

Conceptualize Imagine moving in a boat across a river while the current pushes you down the river. You will not be able to move directly across the river, but will end up downstream as suggested in Figure 4.21a.

Categorize Because of the combined velocities of you relative to the river and the river relative to the Earth, we can categorize this problem as one involving relative velocities.

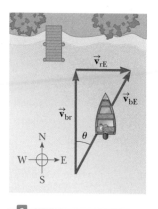
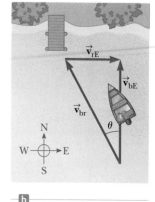

a b

Figure 4.21 (Example 4.8) (a) A boat aims directly across a river and ends up downstream. (b) To move directly across the river, the boat must aim upstream.

Analyze We know $\vec{\mathbf{v}}_{br}$, the velocity of the *boat* relative to the *river*, and $\vec{\mathbf{v}}_{rE}$, the velocity of the *river* relative to the *Earth*. What we must find is $\vec{\mathbf{v}}_{bE}$, the velocity of the *boat* relative to the *Earth*. The relationship between these three quantities is $\vec{\mathbf{v}}_{bE} = \vec{\mathbf{v}}_{br} + \vec{\mathbf{v}}_{rE}$. The terms in the equation must be manipulated as vector quantities; the vectors are shown in Figure 4.21a. The quantity $\vec{\mathbf{v}}_{br}$ is due north; $\vec{\mathbf{v}}_{rE}$ is due east; and the vector sum of the two, $\vec{\mathbf{v}}_{bE}$, is at an angle θ as defined in Figure 4.21a.

continued

▶ **4.8** continued

Find the speed v_{bE} of the boat relative to the Earth using the Pythagorean theorem:

$$v_{bE} = \sqrt{v_{br}^2 + v_{rE}^2} = \sqrt{(10.0\ \text{km/h})^2 + (5.00\ \text{km/h})^2}$$
$$= 11.2\ \text{km/h}$$

Find the direction of $\vec{\mathbf{v}}_{bE}$:

$$\theta = \tan^{-1}\left(\frac{v_{rE}}{v_{br}}\right) = \tan^{-1}\left(\frac{5.00}{10.0}\right) = \boxed{26.6°}$$

Finalize The boat is moving at a speed of 11.2 km/h in the direction 26.6° east of north relative to the Earth. Notice that the speed of 11.2 km/h is faster than your boat speed of 10.0 km/h. The current velocity adds to yours to give you a higher speed. Notice in Figure 4.21a that your resultant velocity is at an angle to the direction straight across the river, so you will end up downstream, as we predicted.

(B) If the boat travels with the same speed of 10.0 km/h relative to the river and is to travel due north as shown in Figure 4.21b, what should its heading be?

SOLUTION

Conceptualize/Categorize This question is an extension of part (A), so we have already conceptualized and categorized the problem. In this case, however, we must aim the boat upstream so as to go straight across the river.

Analyze The analysis now involves the new triangle shown in Figure 4.21b. As in part (A), we know $\vec{\mathbf{v}}_{rE}$ and the magnitude of the vector $\vec{\mathbf{v}}_{br}$, and we want $\vec{\mathbf{v}}_{bE}$ to be directed across the river. Notice the difference between the triangle in Figure 4.21a and the one in Figure 4.21b: the hypotenuse in Figure 4.21b is no longer $\vec{\mathbf{v}}_{bE}$.

Use the Pythagorean theorem to find v_{bE}:

$$v_{bE} = \sqrt{v_{br}^2 - v_{rE}^2} = \sqrt{(10.0\ \text{km/h})^2 - (5.00\ \text{km/h})^2} = 8.66\ \text{km/h}$$

Find the direction in which the boat is heading:

$$\theta = \tan^{-1}\left(\frac{v_{rE}}{v_{bE}}\right) = \tan^{-1}\left(\frac{5.00}{8.66}\right) = \boxed{30.0°}$$

Finalize The boat must head upstream so as to travel directly northward across the river. For the given situation, the boat must steer a course 30.0° west of north. For faster currents, the boat must be aimed upstream at larger angles.

WHAT IF? Imagine that the two boats in parts (A) and (B) are racing across the river. Which boat arrives at the opposite bank first?

Answer In part (A), the velocity of 10 km/h is aimed directly across the river. In part (B), the velocity that is directed across the river has a magnitude of only 8.66 km/h. Therefore, the boat in part (A) has a larger velocity component directly across the river and arrives first.

Summary

Definitions

The **displacement vector** $\Delta\vec{\mathbf{r}}$ for a particle is the difference between its final position vector and its initial position vector:

$$\Delta\vec{\mathbf{r}} \equiv \vec{\mathbf{r}}_f - \vec{\mathbf{r}}_i \tag{4.1}$$

The **average velocity** of a particle during the time interval Δt is defined as the displacement of the particle divided by the time interval:

$$\vec{\mathbf{v}}_{avg} \equiv \frac{\Delta\vec{\mathbf{r}}}{\Delta t} \tag{4.2}$$

The **instantaneous velocity** of a particle is defined as the limit of the average velocity as Δt approaches zero:

$$\vec{\mathbf{v}} \equiv \lim_{\Delta t \to 0} \frac{\Delta\vec{\mathbf{r}}}{\Delta t} = \frac{d\vec{\mathbf{r}}}{dt} \tag{4.3}$$

The **average acceleration** of a particle is defined as the change in its instantaneous velocity vector divided by the time interval Δt during which that change occurs:

$$\vec{a}_{avg} \equiv \frac{\Delta \vec{v}}{\Delta t} = \frac{\vec{v}_f - \vec{v}_i}{t_f - t_i} \qquad (4.4)$$

The **instantaneous acceleration** of a particle is defined as the limiting value of the average acceleration as Δt approaches zero:

$$\vec{a} \equiv \lim_{\Delta t \to 0} \frac{\Delta \vec{v}}{\Delta t} = \frac{d\vec{v}}{dt} \qquad (4.5)$$

Projectile motion is one type of two-dimensional motion, exhibited by an object launched into the air near the Earth's surface and experiencing free fall. This common motion can be analyzed by applying the particle under constant velocity model to the motion of the projectile in the x direction and the particle under constant acceleration model ($a_y = -g$) in the y direction.

A particle moving in a circular path with constant speed is exhibiting **uniform circular motion.**

Concepts and Principles

If a particle moves with *constant* acceleration $\vec{a}$ and has velocity $\vec{v}_i$ and position $\vec{r}_i$ at $t = 0$, its velocity and position vectors at some later time t are

$$\vec{v}_f = \vec{v}_i + \vec{a}t \qquad (4.8)$$

$$\vec{r}_f = \vec{r}_i + \vec{v}_i t + \tfrac{1}{2}\vec{a}t^2 \qquad (4.9)$$

For two-dimensional motion in the xy plane under constant acceleration, ea–ch of these vector expressions is equivalent to two component expressions: one for the motion in the x direction and one for the motion in the y direction.

It is useful to think of projectile motion in terms of a combination of two analysis models: (1) the particle under constant velocity model in the x direction and (2) the particle under constant acceleration model in the vertical direction with a constant downward acceleration of magnitude $g = 9.80$ m/s².

If a particle moves along a curved path in such a way that both the magnitude and the direction of $\vec{v}$ change in time, the particle has an acceleration vector that can be described by two component vectors: (1) a radial component vector $\vec{a}_r$ that causes the change in direction of $\vec{v}$ and (2) a tangential component vector $\vec{a}_t$ that causes the change in magnitude of $\vec{v}$. The magnitude of $\vec{a}_r$ is v^2/r, and the magnitude of $\vec{a}_t$ is $|dv/dt|$.

A particle in uniform circular motion undergoes a radial acceleration $\vec{a}_r$ because the direction of $\vec{v}$ changes in time. This acceleration is called **centripetal acceleration,** and its direction is always toward the center of the circle.

The velocity $\vec{u}_{PA}$ of a particle measured in a fixed frame of reference S_A can be related to the velocity $\vec{u}_{PB}$ of the same particle measured in a moving frame of reference S_B by

$$\vec{u}_{PA} = \vec{u}_{PB} + \vec{v}_{BA} \qquad (4.23)$$

where $\vec{v}_{BA}$ is the velocity of S_B relative to S_A.

Analysis Model for Problem Solving

Particle in Uniform Circular Motion If a particle moves in a circular path of radius r with a constant speed v, the magnitude of its centripetal acceleration is given by

$$a_c = \frac{v^2}{r} \qquad (4.14)$$

and the **period** of the particle's motion is given by

$$T = \frac{2\pi r}{v} \qquad (4.15)$$

The **angular speed** of the particle is

$$\omega = \frac{2\pi}{T} \qquad (4.16)$$

1. Figure OQ4.1 shows a bird's-eye view of a car going around a highway curve. As the car moves from point 1 to point 2, its speed doubles. Which of the vectors (a) through (e) shows the direction of the car's average acceleration between these two points?

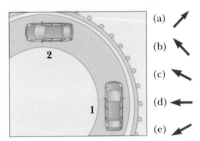

(a)
(b)
(c)
(d)
(e)

Figure OQ4.1

2. Entering his dorm room, a student tosses his book bag to the right and upward at an angle of 45° with the horizontal (Fig. OQ4.2). Air resistance does not affect the bag. The bag moves through point Ⓐ immediately after it leaves the student's hand, through point Ⓑ at the top of its flight, and through point Ⓒ immediately before it lands on the top bunk bed. **(i)** Rank the following horizontal and vertical velocity components from the largest to the smallest. (a) $v_{Ⓐx}$ (b) $v_{Ⓐy}$ (c) $v_{Ⓑx}$ (d) $v_{Ⓑy}$ (e) $v_{Ⓒy}$. Note that zero is larger than a negative number. If two quantities are equal, show them as equal in your list. If any quantity is equal to zero, show that fact in your list. **(ii)** Similarly, rank the following acceleration components. (a) $a_{Ⓐx}$ (b) $a_{Ⓐy}$ (c) $a_{Ⓑx}$ (d) $a_{Ⓑy}$ (e) $a_{Ⓒy}$.

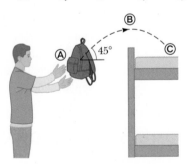

Figure OQ4.2

3. A student throws a heavy red ball horizontally from a balcony of a tall building with an initial speed v_i. At the same time, a second student drops a lighter blue ball from the balcony. Neglecting air resistance, which statement is true? (a) The blue ball reaches the ground first. (b) The balls reach the ground at the same instant. (c) The red ball reaches the ground first. (d) Both balls hit the ground with the same speed. (e) None of statements (a) through (d) is true.

4. A projectile is launched on the Earth with a certain initial velocity and moves without air resistance. Another projectile is launched with the same initial velocity on the Moon, where the acceleration due to gravity is one-sixth as large. How does the maximum altitude of the projectile on the Moon compare with that of the projectile on the Earth? (a) It is one-sixth as large. (b) It is the same. (c) It is $\sqrt{6}$ times larger. (d) It is 6 times larger. (e) It is 36 times larger.

5. Does a car moving around a circular track with constant speed have (a) zero acceleration, (b) an acceleration in the direction of its velocity, (c) an acceleration directed away from the center of its path, (d) an acceleration directed toward the center of its path, or (e) an acceleration with a direction that cannot be determined from the given information?

6. An astronaut hits a golf ball on the Moon. Which of the following quantities, if any, remain constant as a ball travels through the vacuum there? (a) speed (b) acceleration (c) horizontal component of velocity (d) vertical component of velocity (e) velocity

7. A projectile is launched on the Earth with a certain initial velocity and moves without air resistance. Another projectile is launched with the same initial velocity on the Moon, where the acceleration due to gravity is one-sixth as large. How does the range of the projectile on the Moon compare with that of the projectile on the Earth? (a) It is one-sixth as large. (b) It is the same. (c) It is $\sqrt{6}$ times larger. (d) It is 6 times larger. (e) It is 36 times larger.

8. A girl, moving at 8 m/s on in-line skates, is overtaking a boy moving at 5 m/s as they both skate on a straight path. The boy tosses a ball backward toward the girl, giving it speed 12 m/s relative to him. What is the speed of the ball relative to the girl, who catches it? (a) $(8 + 5 + 12)$ m/s (b) $(8 - 5 - 12)$ m/s (c) $(8 + 5 - 12)$ m/s (d) $(8 - 5 + 12)$ m/s (e) $(-8 + 5 + 12)$ m/s

9. A sailor drops a wrench from the top of a sailboat's vertical mast while the boat is moving rapidly and steadily straight forward. Where will the wrench hit the deck? (a) ahead of the base of the mast (b) at the base of the mast (c) behind the base of the mast (d) on the windward side of the base of the mast (e) None of the choices (a) through (d) is true.

10. A baseball is thrown from the outfield toward the catcher. When the ball reaches its highest point, which statement is true? (a) Its velocity and its acceleration are both zero. (b) Its velocity is not zero, but its acceleration is zero. (c) Its velocity is perpendicular to its acceleration. (d) Its acceleration depends on the angle at which the ball was thrown. (e) None of statements (a) through (d) is true.

11. A set of keys on the end of a string is swung steadily in a horizontal circle. In one trial, it moves at speed v in a circle of radius r. In a second trial, it moves at a

higher speed $4v$ in a circle of radius $4r$. In the second trial, how does the period of its motion compare with its period in the first trial? (a) It is the same as in the first trial. (b) It is 4 times larger. (c) It is one-fourth as large. (d) It is 16 times larger. (e) It is one-sixteenth as large.

12. A rubber stopper on the end of a string is swung steadily in a horizontal circle. In one trial, it moves at speed v in a circle of radius r. In a second trial, it moves at a higher speed $3v$ in a circle of radius $3r$. In this second trial, is its acceleration (a) the same as in the first trial, (b) three times larger, (c) one-third as large, (d) nine times larger, or (e) one-ninth as large?

13. In which of the following situations is the moving object appropriately modeled as a projectile? Choose all correct answers. (a) A shoe is tossed in an arbitrary direction. (b) A jet airplane crosses the sky with its engines thrusting the plane forward. (c) A rocket leaves the launch pad. (d) A rocket moves through the sky, at much less than the speed of sound, after its fuel has been used up. (e) A diver throws a stone under water.

14. A certain light truck can go around a curve having a radius of 150 m with a maximum speed of 32.0 m/s. To have the same acceleration, at what maximum speed can it go around a curve having a radius of 75.0 m? (a) 64 m/s (b) 45 m/s (c) 32 m/s (d) 23 m/s (e) 16 m/s

Conceptual Questions   **1.** denotes answer available in *Student Solutions Manual/Study Guide*

1. A spacecraft drifts through space at a constant velocity. Suddenly, a gas leak in the side of the spacecraft gives it a constant acceleration in a direction perpendicular to the initial velocity. The orientation of the spacecraft does not change, so the acceleration remains perpendicular to the original direction of the velocity. What is the shape of the path followed by the spacecraft in this situation?

2. An ice skater is executing a figure eight, consisting of two identically shaped, tangent circular paths. Throughout the first loop she increases her speed uniformly, and during the second loop she moves at a constant speed. Draw a motion diagram showing her velocity and acceleration vectors at several points along the path of motion.

3. If you know the position vectors of a particle at two points along its path and also know the time interval during which it moved from one point to the other, can you determine the particle's instantaneous velocity? Its average velocity? Explain.

4. Describe how a driver can steer a car traveling at constant speed so that (a) the acceleration is zero or (b) the magnitude of the acceleration remains constant.

5. A projectile is launched at some angle to the horizontal with some initial speed v_i, and air resistance is negligible. (a) Is the projectile a freely falling body? (b) What is its acceleration in the vertical direction? (c) What is its acceleration in the horizontal direction?

6. Construct motion diagrams showing the velocity and acceleration of a projectile at several points along its path, assuming (a) the projectile is launched horizontally and (b) the projectile is launched at an angle θ with the horizontal.

7. Explain whether or not the following particles have an acceleration: (a) a particle moving in a straight line with constant speed and (b) a particle moving around a curve with constant speed.

Problems available in Access end-of-chapter problems online at **www.webassign.net**

Section 4.1 The Position, Velocity, and Acceleration Vectors
Problems 1–5

Section 4.2 Two-Dimensional Motion with Constant Acceleration
Problems 6–10

Section 4.3 Projectile Motion
Problems 11–32

Section 4.4 Analysis Model: Particle in Uniform Circular Motion
Problems 33–39

Section 4.5 Tangential and Radial Acceleration
Problems 40–43

Section 4.6 Relative Velocity and Relative Acceleration
Problems 44–54

Additional Problems
Problems 55–80

Challenge Problems
Problem 81–89

Solutions to the following Problems are available in the *Student Solutions Manual/ Study Guide:*
4.1, 4.6, 4.9, 4.13, 4.23, 4.27, 4.33, 4.40, 4.41, 4.45, 4.50, 4.53, 4.67, 4.71, 4.75, 4.77, and 4.81

List of Enhanced Problems

Problem Number	Targeted Feedback in Enhanced WebAssign	Analysis Model Tutorial in Enhanced WebAssign	Master It in Enhanced WebAssign	Watch It in Enhanced WebAssign
4.6	✓			✓
4.7	✓			✓
4.9	✓	✓	✓	
4.13	✓	✓	✓	
4.16	✓			✓
4.20	✓			✓
4.23	✓	✓	✓	
4.27	✓			✓
4.32	✓		✓	
4.37		✓		
4.40	✓			✓
4.41			✓	
4.48			✓	
4.50			✓	
4.53		✓	✓	
4.64	✓		✓	
4.71			✓	
4.77			✓	

The Laws of Motion

In Chapters 2 and 4, we *described* the motion of an object in terms of its position, velocity, and acceleration without considering what might *influence* that motion. Now we consider that influence: Why does the motion of an object change? What might cause one object to remain at rest and another object to accelerate? Why is it generally easier to move a small object than a large object? The two main factors we need to consider are the *forces* acting on an object and the *mass* of the object. In this chapter, we begin our study of *dynamics* by discussing the three basic laws of motion, which deal with forces and masses and were formulated more than three centuries ago by Isaac Newton.

A person sculls on a calm waterway. The water exerts forces on the oars to accelerate the boat. *(Tetra Images/ Getty Images)*

5.1 The Concept of Force

Everyone has a basic understanding of the concept of force from everyday experience. When you push your empty dinner plate away, you exert a force on it. Similarly, you exert a force on a ball when you throw or kick it. In these examples, the word *force* refers to an interaction with an object by means of muscular activity and some change in the object's velocity. Forces do not always cause motion, however. For example, when you are sitting, a gravitational force acts on your body and yet you remain stationary. As a second example, you can push (in other words, exert a force) on a large boulder and not be able to move it.

What force (if any) causes the Moon to orbit the Earth? Newton answered this and related questions by stating that forces are what cause any change in the velocity of an object. The Moon's velocity changes in direction as it moves in a nearly circular

Figure 5.1 Some examples of applied forces. In each case, a force is exerted on the object within the boxed area. Some agent in the environment external to the boxed area exerts a force on the object.

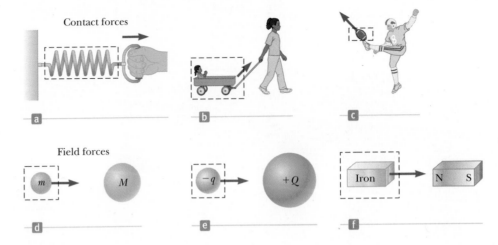

orbit around the Earth. This change in velocity is caused by the gravitational force exerted by the Earth on the Moon.

When a coiled spring is pulled, as in Figure 5.1a, the spring stretches. When a stationary cart is pulled, as in Figure 5.1b, the cart moves. When a football is kicked, as in Figure 5.1c, it is both deformed and set in motion. These situations are all examples of a class of forces called *contact forces*. That is, they involve physical contact between two objects. Other examples of contact forces are the force exerted by gas molecules on the walls of a container and the force exerted by your feet on the floor.

Another class of forces, known as *field forces*, does not involve physical contact between two objects. These forces act through empty space. The gravitational force of attraction between two objects with mass, illustrated in Figure 5.1d, is an example of this class of force. The gravitational force keeps objects bound to the Earth and the planets in orbit around the Sun. Another common field force is the electric force that one electric charge exerts on another (Fig. 5.1e), such as the attractive electric force between an electron and a proton that form a hydrogen atom. A third example of a field force is the force a bar magnet exerts on a piece of iron (Fig. 5.1f).

The distinction between contact forces and field forces is not as sharp as you may have been led to believe by the previous discussion. When examined at the atomic level, all the forces we classify as contact forces turn out to be caused by electric (field) forces of the type illustrated in Figure 5.1e. Nevertheless, in developing models for macroscopic phenomena, it is convenient to use both classifications of forces. The only known *fundamental* forces in nature are all field forces: (1) *gravitational forces* between objects, (2) *electromagnetic forces* between electric charges, (3) *strong forces* between subatomic particles, and (4) *weak forces* that arise in certain radioactive decay processes. In classical physics, we are concerned only with gravitational and electromagnetic forces. We will discuss strong and weak forces in Chapter 46.

The Vector Nature of Force

It is possible to use the deformation of a spring to measure force. Suppose a vertical force is applied to a spring scale that has a fixed upper end as shown in Figure 5.2a. The spring elongates when the force is applied, and a pointer on the scale reads the extension of the spring. We can calibrate the spring by defining a reference force $\vec{F}_1$ as the force that produces a pointer reading of 1.00 cm. If we now apply a different downward force $\vec{F}_2$ whose magnitude is twice that of the reference force $\vec{F}_1$ as seen in Figure 5.2b, the pointer moves to 2.00 cm. Figure 5.2c shows that the combined effect of the two collinear forces is the sum of the effects of the individual forces.

Now suppose the two forces are applied simultaneously with $\vec{F}_1$ downward and $\vec{F}_2$ horizontal as illustrated in Figure 5.2d. In this case, the pointer reads 2.24 cm. The single force $\vec{F}$ that would produce this same reading is the sum of the two vectors $\vec{F}_1$ and $\vec{F}_2$ as described in Figure 5.2d. That is, $|\vec{F}_1| = \sqrt{F_1^2 + F_2^2} = 2.24$ units,

Isaac Newton
English physicist and mathematician
(1642–1727)
Isaac Newton was one of the most brilliant scientists in history. Before the age of 30, he formulated the basic concepts and laws of mechanics, discovered the law of universal gravitation, and invented the mathematical methods of calculus. As a consequence of his theories, Newton was able to explain the motions of the planets, the ebb and flow of the tides, and many special features of the motions of the Moon and the Earth. He also interpreted many fundamental observations concerning the nature of light. His contributions to physical theories dominated scientific thought for two centuries and remain important today.

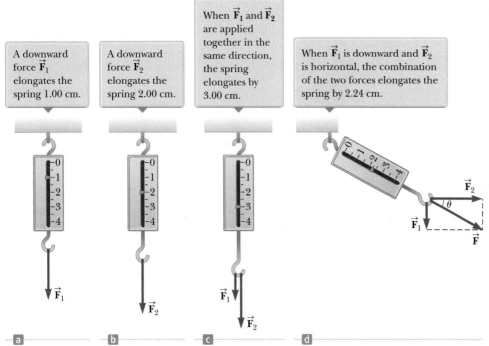

A downward force $\vec{F}_1$ elongates the spring 1.00 cm.

A downward force $\vec{F}_2$ elongates the spring 2.00 cm.

When $\vec{F}_1$ and $\vec{F}_2$ are applied together in the same direction, the spring elongates by 3.00 cm.

When $\vec{F}_1$ is downward and $\vec{F}_2$ is horizontal, the combination of the two forces elongates the spring by 2.24 cm.

a   b   c   d

Figure 5.2 The vector nature of a force is tested with a spring scale.

and its direction is $\theta = \tan^{-1}(-0.500) = -26.6°$. Because forces have been experimentally verified to behave as vectors, you *must* use the rules of vector addition to obtain the net force on an object.

5.2 Newton's First Law and Inertial Frames

We begin our study of forces by imagining some physical situations involving a puck on a perfectly level air hockey table (Fig. 5.3). You expect that the puck will remain stationary when it is placed gently at rest on the table. Now imagine your air hockey table is located on a train moving with constant velocity along a perfectly smooth track. If the puck is placed on the table, the puck again remains where it is placed. If the train were to accelerate, however, the puck would start moving along the table opposite the direction of the train's acceleration, just as a set of papers on your dashboard falls onto the floor of your car when you step on the accelerator.

As we saw in Section 4.6, a moving object can be observed from any number of reference frames. **Newton's first law of motion,** sometimes called the *law of inertia,* defines a special set of reference frames called *inertial frames.* This law can be stated as follows:

> If an object does not interact with other objects, it is possible to identify a reference frame in which the object has zero acceleration.

Such a reference frame is called an **inertial frame of reference.** When the puck is on the air hockey table located on the ground, you are observing it from an inertial reference frame; there are no horizontal interactions of the puck with any other objects, and you observe it to have zero acceleration in that direction. When you are on the train moving at constant velocity, you are also observing the puck from an inertial reference frame. Any reference frame that moves with constant velocity relative to an inertial frame is itself an inertial frame. When you and the train accelerate, however, you are observing the puck from a **noninertial reference frame** because the train is accelerating relative to the inertial reference frame of the Earth's surface. While the puck appears to be accelerating according to your observations, a reference frame can be identified in which the puck has zero acceleration.

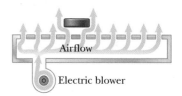

Airflow

Electric blower

Figure 5.3 On an air hockey table, air blown through holes in the surface allows the puck to move almost without friction. If the table is not accelerating, a puck placed on the table will remain at rest.

◄ **Newton's first law**

◄ **Inertial frame of reference**

For example, an observer standing outside the train on the ground sees the puck sliding relative to the table but always moving with the same velocity with respect to the ground as the train had before it started to accelerate (because there is almost no friction to "tie" the puck and the train together). Therefore, Newton's first law is still satisfied even though your observations as a rider on the train show an apparent acceleration relative to you.

A reference frame that moves with constant velocity relative to the distant stars is the best approximation of an inertial frame, and for our purposes we can consider the Earth as being such a frame. The Earth is not really an inertial frame because of its orbital motion around the Sun and its rotational motion about its own axis, both of which involve centripetal accelerations. These accelerations are small compared with g, however, and can often be neglected. For this reason, we model the Earth as an inertial frame, along with any other frame attached to it.

Let us assume we are observing an object from an inertial reference frame. (We will return to observations made in noninertial reference frames in Section 6.3.) Before about 1600, scientists believed that the natural state of matter was the state of rest. Observations showed that moving objects eventually stopped moving. Galileo was the first to take a different approach to motion and the natural state of matter. He devised thought experiments and concluded that it is not the nature of an object to stop once set in motion: rather, it is its nature to *resist changes in its motion.* In his words, "Any velocity once imparted to a moving body will be rigidly maintained as long as the external causes of retardation are removed." For example, a spacecraft drifting through empty space with its engine turned off will keep moving forever. It would *not* seek a "natural state" of rest.

Given our discussion of observations made from inertial reference frames, we can pose a more practical statement of Newton's first law of motion:

◀ Another statement of
Newton's first law

> In the absence of external forces and when viewed from an inertial reference frame, an object at rest remains at rest and an object in motion continues in motion with a constant velocity (that is, with a constant speed in a straight line).

In other words, **when no force acts on an object, the acceleration of the object is zero.** From the first law, we conclude that any *isolated object* (one that does not interact with its environment) is either at rest or moving with constant velocity. The tendency of an object to resist any attempt to change its velocity is called **inertia.** Given the statement of the first law above, we can conclude that an object that is accelerating must be experiencing a force. In turn, from the first law, we can define **force** as **that which causes a change in motion of an object.**

◀ Definition of force

Quick Quiz 5.1 Which of the following statements is correct? **(a)** It is possible for an object to have motion in the absence of forces on the object. **(b)** It is possible to have forces on an object in the absence of motion of the object. **(c)** Neither statement **(a)** nor statement **(b)** is correct. **(d)** Both statements **(a)** and **(b)** are correct.

5.3 Mass

Imagine playing catch with either a basketball or a bowling ball. Which ball is more likely to keep moving when you try to catch it? Which ball requires more effort to throw it? The bowling ball requires more effort. In the language of physics, we say that the bowling ball is more resistant to changes in its velocity than the basketball. How can we quantify this concept?

◀ Definition of mass

Mass is that property of an object that specifies how much resistance an object exhibits to changes in its velocity, and as we learned in Section 1.1, the SI unit of mass is the kilogram. Experiments show that the greater the mass of an object, the less that object accelerates under the action of a given applied force.

To describe mass quantitatively, we conduct experiments in which we compare the accelerations a given force produces on different objects. Suppose a force act-

Pitfall Prevention 5.1

Newton's First Law Newton's first law does *not* say what happens for an object with *zero net force,* that is, multiple forces that cancel; it says what happens *in the absence of external forces.* This subtle but important difference allows us to define force as that which causes a change in the motion. The description of an object under the effect of forces that balance is covered by Newton's second law.

Pitfall Prevention 5.2

Force Is the Cause of Changes in Motion An object can have motion in the absence of forces as described in Newton's first law. Therefore, don't interpret force as the cause of *motion.* Force is the cause of *changes in motion.*

ing on an object of mass m_1 produces a change in motion of the object that we can quantify with the object's acceleration $\vec{\mathbf{a}}_1$, and the *same force* acting on an object of mass m_2 produces an acceleration $\vec{\mathbf{a}}_2$. The ratio of the two masses is defined as the *inverse* ratio of the magnitudes of the accelerations produced by the force:

$$\frac{m_1}{m_2} \equiv \frac{a_2}{a_1}$$ (5.1)

For example, if a given force acting on a 3-kg object produces an acceleration of 4 m/s², the same force applied to a 6-kg object produces an acceleration of 2 m/s². According to a huge number of similar observations, we conclude that the magnitude of the acceleration of an object is inversely proportional to its mass when acted on by a given force. If one object has a known mass, the mass of the other object can be obtained from acceleration measurements.

Mass is an inherent property of an object and is independent of the object's surroundings and of the method used to measure it. Also, mass is a scalar quantity and thus obeys the rules of ordinary arithmetic. For example, if you combine a 3-kg mass with a 5-kg mass, the total mass is 8 kg. This result can be verified experimentally by comparing the acceleration that a known force gives to several objects separately with the acceleration that the same force gives to the same objects combined as one unit.

Mass should not be confused with weight. Mass and weight are two different quantities. The weight of an object is equal to the magnitude of the gravitational force exerted on the object and varies with location (see Section 5.5). For example, a person weighing 180 lb on the Earth weighs only about 30 lb on the Moon. On the other hand, the mass of an object is the same everywhere: an object having a mass of 2 kg on the Earth also has a mass of 2 kg on the Moon.

◀ **Mass and weight are different quantities**

5.4 Newton's Second Law

Newton's first law explains what happens to an object when no forces act on it: it maintains its original motion; it either remains at rest or moves in a straight line with constant speed. Newton's second law answers the question of what happens to an object when one or more forces act on it.

Imagine performing an experiment in which you push a block of mass m across a frictionless, horizontal surface. When you exert some horizontal force $\vec{\mathbf{F}}$ on the block, it moves with some acceleration $\vec{\mathbf{a}}$. If you apply a force twice as great on the same block, experimental results show that the acceleration of the block doubles; if you increase the applied force to $3\vec{\mathbf{F}}$, the acceleration triples; and so on. From such observations, we conclude that the acceleration of an object is directly proportional to the force acting on it: $\vec{\mathbf{F}} \propto \vec{\mathbf{a}}$. This idea was first introduced in Section 2.4 when we discussed the direction of the acceleration of an object. We also know from the preceding section that the magnitude of the acceleration of an object is inversely proportional to its mass: $|\vec{\mathbf{a}}| \propto 1/m$.

These experimental observations are summarized in **Newton's second law:**

> When viewed from an inertial reference frame, the acceleration of an object is directly proportional to the net force acting on it and inversely proportional to its mass:
>
> $$\vec{\mathbf{a}} \propto \frac{\sum \vec{\mathbf{F}}}{m}$$

Pitfall Prevention 5.3

$m\vec{\mathbf{a}}$ **Is Not a Force** Equation 5.2 does *not* say that the product $m\vec{\mathbf{a}}$ is a force. All forces on an object are added vectorially to generate the net force on the left side of the equation. This net force is then equated to the product of the mass of the object and the acceleration that results from the net force. Do *not* include an "$m\vec{\mathbf{a}}$ force" in your analysis of the forces on an object.

If we choose a proportionality constant of 1, we can relate mass, acceleration, and force through the following mathematical statement of Newton's second law:[1]

$$\sum \vec{\mathbf{F}} = m\vec{\mathbf{a}}$$ (5.2)

◀ **Newton's second law**

---

[1]Equation 5.2 is valid only when the speed of the object is much less than the speed of light. We treat the relativistic situation in Chapter 39.

In both the textual and mathematical statements of Newton's second law, we have indicated that the acceleration is due to the *net force* $\sum \vec{F}$ acting on an object. The **net force** on an object is the vector sum of all forces acting on the object. (We sometimes refer to the net force as the *total force,* the *resultant force,* or the *unbalanced force.*) In solving a problem using Newton's second law, it is imperative to determine the correct net force on an object. Many forces may be acting on an object, but there is only one acceleration.

Equation 5.2 is a vector expression and hence is equivalent to three component equations:

$\blacktriangleright$ **Newton's second law: component form**

$$\sum F_x = ma_x \qquad \sum F_y = ma_y \qquad \sum F_z = ma_z \qquad (5.3)$$

> **Quick Quiz 5.2** An object experiences no acceleration. Which of the following *cannot* be true for the object? **(a)** A single force acts on the object. **(b)** No forces act on the object. **(c)** Forces act on the object, but the forces cancel.

> **Quick Quiz 5.3** You push an object, initially at rest, across a frictionless floor with a constant force for a time interval Δt, resulting in a final speed of v for the object. You then repeat the experiment, but with a force that is twice as large. What time interval is now required to reach the same final speed v?
> **(a)** $4\,\Delta t$ **(b)** $2\,\Delta t$ **(c)** Δt **(d)** $\Delta t/2$ **(e)** $\Delta t/4$

The SI unit of force is the **newton** (N). A force of 1 N is the force that, when acting on an object of mass 1 kg, produces an acceleration of 1 m/s². From this definition and Newton's second law, we see that the newton can be expressed in terms of the following fundamental units of mass, length, and time:

$\blacktriangleright$ **Definition of the newton**

$$1 \text{ N} \equiv 1 \text{ kg} \cdot \text{m/s}^2 \qquad (5.4)$$

In the U.S. customary system, the unit of force is the **pound** (lb). A force of 1 lb is the force that, when acting on a 1-slug mass,[2] produces an acceleration of 1 ft/s²:

$$1 \text{ lb} \equiv 1 \text{ slug} \cdot \text{ft/s}^2$$

A convenient approximation is $1 \text{ N} \approx \frac{1}{4}$ lb.

Example 5.1   **An Accelerating Hockey Puck**   AM

A hockey puck having a mass of 0.30 kg slides on the frictionless, horizontal surface of an ice rink. Two hockey sticks strike the puck simultaneously, exerting the forces on the puck shown in Figure 5.4. The force $\vec{F}_1$ has a magnitude of 5.0 N, and is directed at $\theta = 20°$ below the x axis. The force $\vec{F}_2$ has a magnitude of 8.0 N and its direction is $\phi = 60°$ above the x axis. Determine both the magnitude and the direction of the puck's acceleration.

Figure 5.4
(Example 5.1) A hockey puck moving on a frictionless surface is subject to two forces $\vec{F}_1$ and $\vec{F}_2$.

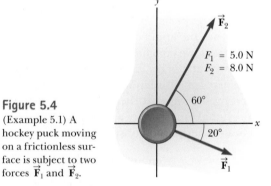

$F_1 = 5.0$ N
$F_2 = 8.0$ N

SOLUTION

Conceptualize Study Figure 5.4. Using your expertise in vector addition from Chapter 3, predict the approximate direction of the net force vector on the puck. The acceleration of the puck will be in the same direction.

Categorize Because we can determine a net force and we want an acceleration, this problem is categorized as one that may be solved using Newton's second law. In Section 5.7, we will formally introduce the *particle under a net force* analysis model to describe a situation such as this one.

Analyze Find the component of the net force acting on the puck in the x direction:

$$\sum F_x = F_{1x} + F_{2x} = F_1 \cos \theta + F_2 \cos \phi$$

---

[2]The *slug* is the unit of mass in the U.S. customary system and is that system's counterpart of the SI unit the *kilogram.* Because most of the calculations in our study of classical mechanics are in SI units, the slug is seldom used in this text.

▶ **5.1** continued

Find the component of the net force acting on the puck in the y direction:

$$\sum F_y = F_{1y} + F_{2y} = F_1 \sin \theta + F_2 \sin \phi$$

Use Newton's second law in component form (Eq. 5.3) to find the x and y components of the puck's acceleration:

$$a_x = \frac{\sum F_x}{m} = \frac{F_1 \cos \theta + F_2 \cos \phi}{m}$$

$$a_y = \frac{\sum F_y}{m} = \frac{F_1 \sin \theta + F_2 \sin \phi}{m}$$

Substitute numerical values:

$$a_x = \frac{(5.0 \text{ N}) \cos(-20°) + (8.0 \text{ N}) \cos(60°)}{0.30 \text{ kg}} = 29 \text{ m/s}^2$$

$$a_y = \frac{(5.0 \text{ N}) \sin(-20°) + (8.0 \text{ N}) \sin(60°)}{0.30 \text{ kg}} = 17 \text{ m/s}^2$$

Find the magnitude of the acceleration:

$$a = \sqrt{(29 \text{ m/s}^2)^2 + (17 \text{ m/s}^2)^2} = \boxed{34 \text{ m/s}^2}$$

Find the direction of the acceleration relative to the positive x axis:

$$\theta = \tan^{-1}\left(\frac{a_y}{a_x}\right) = \tan^{-1}\left(\frac{17}{29}\right) = \boxed{31°}$$

Finalize The vectors in Figure 5.4 can be added graphically to check the reasonableness of our answer. Because the acceleration vector is along the direction of the resultant force, a drawing showing the resultant force vector helps us check the validity of the answer. (Try it!)

WHAT IF? Suppose three hockey sticks strike the puck simultaneously, with two of them exerting the forces shown in Figure 5.4. The result of the three forces is that the hockey puck shows *no* acceleration. What must be the components of the third force?

Answer If there is zero acceleration, the net force acting on the puck must be zero. Therefore, the three forces must cancel. The components of the third force must be of equal magnitude and opposite sign compared to the components of the net force applied by the first two forces so that all the components add to zero. Therefore, $F_{3x} = -\sum F_x = -(0.30 \text{ kg})(29 \text{ m/s}^2) = -8.7 \text{ N}$ and $F_{3y} = -\sum F_y = -(0.30 \text{ kg})(17 \text{ m/s}^2) = -5.2 \text{ N}$.

5.5 The Gravitational Force and Weight

All objects are attracted to the Earth. The attractive force exerted by the Earth on an object is called the **gravitational force** $\vec{\mathbf{F}}_g$. This force is directed toward the center of the Earth,[3] and its magnitude is called the **weight** of the object.

We saw in Section 2.6 that a freely falling object experiences an acceleration $\vec{\mathbf{g}}$ acting toward the center of the Earth. Applying Newton's second law $\sum \vec{\mathbf{F}} = m\vec{\mathbf{a}}$ to a freely falling object of mass m, with $\vec{\mathbf{a}} = \vec{\mathbf{g}}$ and $\sum \vec{\mathbf{F}} = \vec{\mathbf{F}}_g$, gives

$$\vec{\mathbf{F}}_g = m\vec{\mathbf{g}} \tag{5.5}$$

Therefore, the weight of an object, being defined as the magnitude of $\vec{\mathbf{F}}_g$, is given by

$$F_g = mg \tag{5.6}$$

Because it depends on g, weight varies with geographic location. Because g decreases with increasing distance from the center of the Earth, objects weigh less at higher altitudes than at sea level. For example, a 1 000-kg pallet of bricks used in the construction of the Empire State Building in New York City weighed 9 800 N at street level, but weighed about 1 N less by the time it was lifted from sidewalk level to the top of the building. As another example, suppose a student has a mass

Pitfall Prevention 5.4

"Weight of an Object" We are familiar with the everyday phrase, the "weight of an object." Weight, however, is not an inherent property of an object; rather, it is a measure of the gravitational force between the object and the Earth (or other planet). Therefore, weight is a property of a *system* of items: the object and the Earth.

Pitfall Prevention 5.5

Kilogram Is Not a Unit of Weight You may have seen the "conversion" 1 kg = 2.2 lb. Despite popular statements of weights expressed in kilograms, the kilogram is not a unit of *weight*, it is a unit of *mass*. The conversion statement is not an equality; it is an *equivalence* that is valid only on the Earth's surface.

---

[3]This statement ignores that the mass distribution of the Earth is not perfectly spherical.

The life-support unit strapped to the back of astronaut Harrison Schmitt weighed 300 lb on the Earth and had a mass of 136 kg. During his training, a 50-lb mock-up with a mass of 23 kg was used. Although this strategy effectively simulated the reduced weight the unit would have on the Moon, it did not correctly mimic the unchanging mass. It was more difficult to accelerate the 136-kg unit (perhaps by jumping or twisting suddenly) on the Moon than it was to accelerate the 23-kg unit on the Earth.

of 70.0 kg. The student's weight in a location where $g = 9.80$ m/s^2 is 686 N (about 150 lb). At the top of a mountain, however, where $g = 9.77$ m/s^2, the student's weight is only 684 N. Therefore, if you want to lose weight without going on a diet, climb a mountain or weigh yourself at 30 000 ft during an airplane flight!

Equation 5.6 quantifies the gravitational force on the object, but notice that this equation does not require the object to be moving. Even for a stationary object or for an object on which several forces act, Equation 5.6 can be used to calculate the magnitude of the gravitational force. The result is a subtle shift in the interpretation of m in the equation. The mass m in Equation 5.6 determines the strength of the gravitational attraction between the object and the Earth. This role is completely different from that previously described for mass, that of measuring the resistance to changes in motion in response to an external force. In that role, mass is also called **inertial mass.** We call m in Equation 5.6 the **gravitational mass.** Even though this quantity is different in behavior from inertial mass, it is one of the experimental conclusions in Newtonian dynamics that gravitational mass and inertial mass have the same value.

Although this discussion has focused on the gravitational force on an object due to the Earth, the concept is generally valid on any planet. The value of g will vary from one planet to the next, but the magnitude of the gravitational force will always be given by the value of mg.

Quick Quiz 5.4 Suppose you are talking by interplanetary telephone to a friend who lives on the Moon. He tells you that he has just won a newton of gold in a contest. Excitedly, you tell him that you entered the Earth version of the same contest and also won a newton of gold! Who is richer? **(a)** You are. **(b)** Your friend is. **(c)** You are equally rich.

Conceptual Example 5.2   **How Much Do You Weigh in an Elevator?**

You have most likely been in an elevator that accelerates upward as it moves toward a higher floor. In this case, you feel heavier. In fact, if you are standing on a bathroom scale at the time, the scale measures a force having a magnitude that is greater than your weight. Therefore, you have tactile and measured evidence that leads you to believe you are heavier in this situation. *Are* you heavier?

SOLUTION

No; your weight is unchanged. Your experiences are due to your being in a noninertial reference frame. To provide the acceleration upward, the floor or scale must exert on your feet an upward force that is greater in magnitude than your weight. It is this greater force you feel, which you interpret as feeling heavier. The scale reads this upward force, not your weight, and so its reading increases.

5.6 Newton's Third Law

If you press against a corner of this textbook with your fingertip, the book pushes back and makes a small dent in your skin. If you push harder, the book does the same and the dent in your skin is a little larger. This simple activity illustrates that forces are *interactions* between two objects: when your finger pushes on the book, the book pushes back on your finger. This important principle is known as **Newton's third law:**

Newton's third law ▶

If two objects interact, the force $\vec{\mathbf{F}}_{12}$ exerted by object 1 on object 2 is equal in magnitude and opposite in direction to the force $\vec{\mathbf{F}}_{21}$ exerted by object 2 on object 1:

$$\vec{\mathbf{F}}_{12} = -\vec{\mathbf{F}}_{21} \tag{5.7}$$

When it is important to designate forces as interactions between two objects, we will use this subscript notation, where $\vec{F}_{ab}$ means "the force exerted *by* a *on* b." The third law is illustrated in Figure 5.5. The force that object 1 exerts on object 2 is popularly called the *action force,* and the force of object 2 on object 1 is called the *reaction force.* These italicized terms are not scientific terms; furthermore, either force can be labeled the action or reaction force. We will use these terms for convenience. In all cases, the action and reaction forces act on *different* objects and must be of the same type (gravitational, electrical, etc.). For example, the force acting on a freely falling projectile is the gravitational force exerted by the Earth on the projectile $\vec{F}_g = \vec{F}_{Ep}$ (E = Earth, p = projectile), and the magnitude of this force is *mg.* The reaction to this force is the gravitational force exerted by the projectile on the Earth $\vec{F}_{pE} = -\vec{F}_{Ep}$. The reaction force $\vec{F}_{pE}$ must accelerate the Earth toward the projectile just as the action force $\vec{F}_{Ep}$ accelerates the projectile toward the Earth. Because the Earth has such a large mass, however, its acceleration due to this reaction force is negligibly small.

Consider a computer monitor at rest on a table as in Figure 5.6a. The gravitational force on the monitor is $\vec{F}_g = \vec{F}_{Em}$. The reaction to this force is the force $\vec{F}_{mE} = -\vec{F}_{Em}$ exerted by the monitor on the Earth. The monitor does not accelerate because it is held up by the table. The table exerts on the monitor an upward force $\vec{n} = \vec{F}_{tm}$, called the **normal force.** (*Normal* in this context means *perpendicular.*) In general, whenever an object is in contact with a surface, the surface exerts a normal force on the object. The normal force on the monitor can have any value needed, up to the point of breaking the table. Because the monitor has zero acceleration, Newton's second law applied to the monitor gives us $\sum \vec{F} = \vec{n} + m\vec{g} = 0$, so $n\hat{j} - mg\hat{j} = 0$, or $n = mg$. The normal force balances the gravitational force on the monitor, so the net force on the monitor is zero. The reaction force to $\vec{n}$ is the force exerted by the monitor downward on the table, $\vec{F}_{mt} = -\vec{F}_{tm} = -\vec{n}$.

Notice that the forces acting on the monitor are $\vec{F}_g$ and $\vec{n}$ as shown in Figure 5.6b. The two forces $\vec{F}_{mE}$ and $\vec{F}_{mt}$ are exerted on objects other than the monitor.

Figure 5.6 illustrates an extremely important step in solving problems involving forces. Figure 5.6a shows many of the forces in the situation: those acting on the monitor, one acting on the table, and one acting on the Earth. Figure 5.6b, by contrast, shows only the forces acting on *one object,* the monitor, and is called a **force diagram** or a *diagram showing the forces on the object.* The important pictorial representation in Figure 5.6c is called a **free-body diagram.** In a free-body diagram, the particle model is used by representing the object as a dot and showing the forces that act on the object as being applied to the dot. When analyzing an object subject to forces, we are interested in the net force acting on one object, which we will model as a particle. Therefore, a free-body diagram helps us isolate only those forces on the object and eliminate the other forces from our analysis.

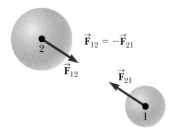

Figure 5.5 Newton's third law. The force $\vec{F}_{12}$ exerted by object 1 on object 2 is equal in magnitude and opposite in direction to the force $\vec{F}_{21}$ exerted by object 2 on object 1.

Pitfall Prevention 5.6

n Does Not Always Equal mg In the situation shown in Figure 5.6 and in many others, we find that $n = mg$ (the normal force has the same magnitude as the gravitational force). This result, however, is *not* generally true. If an object is on an incline, if there are applied forces with vertical components, or if there is a vertical acceleration of the system, then $n \neq mg$. *Always* apply Newton's second law to find the relationship between n and mg.

Pitfall Prevention 5.7

Newton's Third Law Remember that Newton's third-law action and reaction forces act on *different* objects. For example, in Figure 5.6, $\vec{n} = \vec{F}_{tm} = -m\vec{g} = -\vec{F}_{Em}$. The forces $\vec{n}$ and $m\vec{g}$ are equal in magnitude and opposite in direction, but they do not represent an action–reaction pair because both forces act on the *same* object, the monitor.

Pitfall Prevention 5.8

Free-Body Diagrams The *most important* step in solving a problem using Newton's laws is to draw a proper sketch, the free-body diagram. Be sure to draw *only* those forces that act on the object you are isolating. Be sure to draw *all* forces acting on the object, including any field forces, such as the gravitational force.

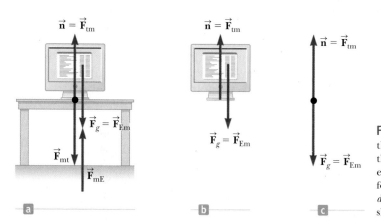

Figure 5.6 (a) When a computer monitor is at rest on a table, the forces acting on the monitor are the normal force $\vec{n}$ and the gravitational force $\vec{F}_g$. The reaction to $\vec{n}$ is the force $\vec{F}_{mt}$ exerted by the monitor on the table. The reaction to $\vec{F}_g$ is the force $\vec{F}_{mE}$ exerted by the monitor on the Earth. (b) A *force diagram* shows the forces on the monitor. (c) A *free-body diagram* shows the monitor as a black dot with the forces acting on it.

Quick Quiz 5.5 **(i)** If a fly collides with the windshield of a fast-moving bus, which experiences an impact force with a larger magnitude? (a) The fly. (b) The bus. (c) The same force is experienced by both. **(ii)** Which experiences the greater acceleration? (a) The fly. (b) The bus. (c) The same acceleration is experienced by both.

Conceptual Example 5.3   **You Push Me and I'll Push You**

A large man and a small boy stand facing each other on frictionless ice. They put their hands together and push against each other so that they move apart.

(A) Who moves away with the higher speed?

SOLUTION

This situation is similar to what we saw in Quick Quiz 5.5. According to Newton's third law, the force exerted by the man on the boy and the force exerted by the boy on the man are a third-law pair of forces, so they must be equal in magnitude. (A bathroom scale placed between their hands would read the same, regardless of which way it faced.) Therefore, the boy, having the smaller mass, experiences the greater acceleration. Both individuals accelerate for the same amount of time, but the greater acceleration of the boy over this time interval results in his moving away from the interaction with the higher speed.

(B) Who moves farther while their hands are in contact?

SOLUTION

Because the boy has the greater acceleration and therefore the greater average velocity, he moves farther than the man during the time interval during which their hands are in contact.

5.7 Analysis Models Using Newton's Second Law

In this section, we discuss two analysis models for solving problems in which objects are either in equilibrium ($\vec{a} = 0$) or accelerating under the action of constant external forces. Remember that when Newton's laws are applied to an object, we are interested only in external forces that act on the object. If the objects are modeled as particles, we need not worry about rotational motion. For now, we also neglect the effects of friction in those problems involving motion, which is equivalent to stating that the surfaces are *frictionless*. (The friction force is discussed in Section 5.8.)

We usually neglect the mass of any ropes, strings, or cables involved. In this approximation, the magnitude of the force exerted by any element of the rope on the adjacent element is the same for all elements along the rope. In problem statements, the synonymous terms *light* and *of negligible mass* are used to indicate that a mass is to be ignored when you work the problems. When a rope attached to an object is pulling on the object, the rope exerts a force on the object in a direction away from the object, parallel to the rope. The magnitude *T* of that force is called the **tension** in the rope. Because it is the magnitude of a vector quantity, tension is a scalar quantity.

Analysis Model: The Particle in Equilibrium

If the acceleration of an object modeled as a particle is zero, the object is treated with the **particle in equilibrium** model. In this model, the net force on the object is zero:

$$\sum \vec{F} = 0 \tag{5.8}$$

Consider a lamp suspended from a light chain fastened to the ceiling as in Figure 5.7a. The force diagram for the lamp (Fig. 5.7b) shows that the forces acting on the lamp are the downward gravitational force $\vec{\mathbf{F}}_g$ and the upward force $\vec{\mathbf{T}}$ exerted by the chain. Because there are no forces in the x direction, $\Sigma F_x = 0$ provides no helpful information. The condition $\Sigma F_y = 0$ gives

$$\sum F_y = T - F_g = 0 \text{ or } T = F_g$$

Again, notice that $\vec{\mathbf{T}}$ and $\vec{\mathbf{F}}_g$ are *not* an action–reaction pair because they act on the same object, the lamp. The reaction force to $\vec{\mathbf{T}}$ is a downward force exerted by the lamp on the chain.

Example 5.4 (page 100) shows an application of the particle in equilibrium model.

Analysis Model: The Particle Under a Net Force

If an object experiences an acceleration, its motion can be analyzed with the **particle under a net force** model. The appropriate equation for this model is Newton's second law, Equation 5.2:

$$\sum \vec{\mathbf{F}} = m\vec{\mathbf{a}} \tag{5.2}$$

Consider a crate being pulled to the right on a frictionless, horizontal floor as in Figure 5.8a. Of course, the floor directly under the boy must have friction; otherwise, his feet would simply slip when he tries to pull on the crate! Suppose you wish to find the acceleration of the crate and the force the floor exerts on it. The forces acting on the crate are illustrated in the free-body diagram in Figure 5.8b. Notice that the horizontal force $\vec{\mathbf{T}}$ being applied to the crate acts through the rope. The magnitude of $\vec{\mathbf{T}}$ is equal to the tension in the rope. In addition to the force $\vec{\mathbf{T}}$, the free-body diagram for the crate includes the gravitational force $\vec{\mathbf{F}}_g$ and the normal force $\vec{\mathbf{n}}$ exerted by the floor on the crate.

We can now apply Newton's second law in component form to the crate. The only force acting in the x direction is $\vec{\mathbf{T}}$. Applying $\Sigma F_x = ma_x$ to the horizontal motion gives

$$\sum F_x = T = ma_x \quad \text{or} \quad a_x = \frac{T}{m}$$

No acceleration occurs in the y direction because the crate moves only horizontally. Therefore, we use the particle in equilibrium model in the y direction. Applying the y component of Equation 5.8 yields

$$\sum F_y = n - F_g = 0 \quad \text{or} \quad n = F_g$$

That is, the normal force has the same magnitude as the gravitational force but acts in the opposite direction.

If $\vec{\mathbf{T}}$ is a constant force, the acceleration $a_x = T/m$ also is constant. Hence, the crate is also modeled as a particle under constant acceleration in the x direction, and the equations of kinematics from Chapter 2 can be used to obtain the crate's position x and velocity v_x as functions of time.

Notice from this discussion two concepts that will be important in future problem solving: (1) *In a given problem, it is possible to have different analysis models applied in different directions.* The crate in Figure 5.8 is a particle in equilibrium in the vertical direction and a particle under a net force in the horizontal direction. (2) *It is possible to describe an object by multiple analysis models.* The crate is a particle under a net force in the horizontal direction and is also a particle under constant acceleration in the same direction.

In the situation just described, the magnitude of the normal force $\vec{\mathbf{n}}$ is equal to the magnitude of $\vec{\mathbf{F}}_g$, but that is not always the case, as noted in Pitfall Prevention 5.6. For example, suppose a book is lying on a table and you push down on the book with a force $\vec{\mathbf{F}}$ as in Figure 5.9. Because the book is at rest and therefore not accelerating, $\Sigma F_y = 0$, which gives $n - F_g - F = 0$, or $n = F_g + F = mg + F$. In this situation, the normal force is *greater* than the gravitational force. Other examples in which $n \neq F_g$ are presented later.

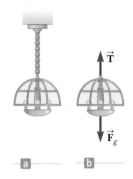

Figure 5.7 (a) A lamp suspended from a ceiling by a chain of negligible mass. (b) The forces acting on the lamp are the gravitational force $\vec{\mathbf{F}}_g$ and the force $\vec{\mathbf{T}}$ exerted by the chain.

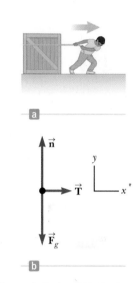

Figure 5.8 (a) A crate being pulled to the right on a frictionless floor. (b) The free-body diagram representing the external forces acting on the crate.

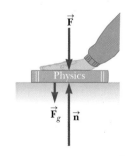

Figure 5.9 When a force $\vec{\mathbf{F}}$ pushes vertically downward on another object, the normal force $\vec{\mathbf{n}}$ on the object is greater than the gravitational force: $n = F_g + F$.

Several examples below demonstrate the use of the particle under a net force model.

Analysis Model	Particle in Equilibrium

Imagine an object that can be modeled as a particle. If it has several forces acting on it so that the forces all cancel, giving a net force of zero, the object will have an acceleration of zero. This condition is mathematically described as

$$\sum \vec{F} = 0 \qquad (5.8)$$

$$\vec{a} = 0$$
$$m$$
$$\xleftrightarrow{\hspace{2cm}}$$
$$\Sigma \vec{F} = 0$$

Examples

- a chandelier hanging over a dining room table
- an object moving at terminal speed through a viscous medium (Chapter 6)
- a steel beam in the frame of a building (Chapter 12)
- a boat floating on a body of water (Chapter 14)

Analysis Model	Particle Under a Net Force

Imagine an object that can be modeled as a particle. If it has one or more forces acting on it so that there is a net force on the object, it will accelerate in the direction of the net force. The relationship between the net force and the acceleration is

$$\sum \vec{F} = m\vec{a} \qquad (5.2)$$

$$m \quad \vec{a}$$
$$\circ \xrightarrow{\hspace{1.5cm}}$$
$$\Sigma \vec{F}$$

Examples

- a crate pushed across a factory floor
- a falling object acted upon by a gravitational force
- a piston in an automobile engine pushed by hot gases (Chapter 22)
- a charged particle in an electric field (Chapter 23)

Example 5.4	A Traffic Light at Rest	AM

A traffic light weighing 122 N hangs from a cable tied to two other cables fastened to a support as in Figure 5.10a. The upper cables make angles of $\theta_1 = 37.0°$ and $\theta_2 = 53.0°$ with the horizontal. These upper cables are not as strong as the vertical cable and will break if the tension in them exceeds 100 N. Does the traffic light remain hanging in this situation, or will one of the cables break?

SOLUTION

Conceptualize Inspect the drawing in Figure 5.10a. Let us assume the cables do not break and nothing is moving.

Categorize If nothing is moving, no part of the system is accelerating. We can now model the light as a *particle in equilibrium* on which the net force is zero. Similarly, the net force on the knot (Fig. 5.10c) is zero, so it is also modeled as a *particle in equilibrium*.

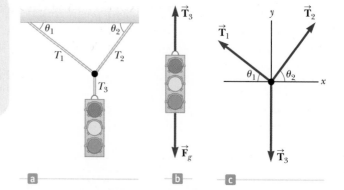

Figure 5.10 (Example 5.4) (a) A traffic light suspended by cables. (b) The forces acting on the traffic light. (c) The free-body diagram for the knot where the three cables are joined.

Analyze We construct a diagram of the forces acting on the traffic light, shown in Figure 5.10b, and a free-body diagram for the knot that holds the three cables together, shown in Figure 5.10c. This knot is a convenient object to choose because all the forces of interest act along lines passing through the knot.

From the particle in equilibrium model, apply Equation 5.8 for the traffic light in the *y* direction:

$$\sum F_y = 0 \;\rightarrow\; T_3 - F_g = 0$$
$$T_3 = F_g$$

▶ **5.4** continued

Choose the coordinate axes as shown in Figure 5.10c and resolve the forces acting on the knot into their components:

Force	x Component	y Component
$\vec{T}_1$	$-T_1 \cos \theta_1$	$T_1 \sin \theta_1$
$\vec{T}_2$	$T_2 \cos \theta_2$	$T_2 \sin \theta_2$
$\vec{T}_3$	0	$-F_g$

Apply the particle in equilibrium model to the knot:

(1) $\sum F_x = -T_1 \cos \theta_1 + T_2 \cos \theta_2 = 0$

(2) $\sum F_y = T_1 \sin \theta_1 + T_2 \sin \theta_2 + (-F_g) = 0$

Equation (1) shows that the horizontal components of $\vec{T}_1$ and $\vec{T}_2$ must be equal in magnitude, and Equation (2) shows that the sum of the vertical components of $\vec{T}_1$ and $\vec{T}_2$ must balance the downward force $\vec{T}_3$, which is equal in magnitude to the weight of the light.

Solve Equation (1) for T_2 in terms of T_1:

(3) $T_2 = T_1 \left(\dfrac{\cos \theta_1}{\cos \theta_2} \right)$

Substitute this value for T_2 into Equation (2):

$T_1 \sin \theta_1 + T_1 \left(\dfrac{\cos \theta_1}{\cos \theta_2} \right) (\sin \theta_2) - F_g = 0$

Solve for T_1:

$T_1 = \dfrac{F_g}{\sin \theta_1 + \cos \theta_1 \tan \theta_2}$

Substitute numerical values:

$T_1 = \dfrac{122 \text{ N}}{\sin 37.0° + \cos 37.0° \tan 53.0°} = 73.4 \text{ N}$

Using Equation (3), solve for T_2:

$T_2 = (73.4 \text{ N}) \left(\dfrac{\cos 37.0°}{\cos 53.0°} \right) = 97.4 \text{ N}$

Both values are less than 100 N (just barely for T_2), so the cables will not break .

....

Finalize Let us finalize this problem by imagining a change in the system, as in the following What If?

WHAT IF? Suppose the two angles in Figure 5.10a are equal. What would be the relationship between T_1 and T_2?

Answer We can argue from the symmetry of the problem that the two tensions T_1 and T_2 would be equal to each other. Mathematically, if the equal angles are called θ, Equation (3) becomes

$$T_2 = T_1 \left(\frac{\cos \theta}{\cos \theta} \right) = T_1$$

which also tells us that the tensions are equal. Without knowing the specific value of θ, we cannot find the values of T_1 and T_2. The tensions will be equal to each other, however, regardless of the value of θ.

Conceptual Example 5.5   **Forces Between Cars in a Train**

Train cars are connected by *couplers,* which are under tension as the locomotive pulls the train. Imagine you are on a train speeding up with a constant acceleration. As you move through the train from the locomotive to the last car, measuring the tension in each set of couplers, does the tension increase, decrease, or stay the same? When the engineer applies the brakes, the couplers are under compression. How does this compression force vary from the locomotive to the last car? (Assume only the brakes on the wheels of the engine are applied.)

SOLUTION

While the train is speeding up, tension decreases from the front of the train to the back. The coupler between the locomotive and the first car must apply enough force to accelerate the rest of the cars. As you move back along the

continued

▶ **5.5** continued

train, each coupler is accelerating less mass behind it. The last coupler has to accelerate only the last car, and so it is under the least tension.

When the brakes are applied, the force again decreases from front to back. The coupler connecting the locomotive to the first car must apply a large force to slow down the rest of the cars, but the final coupler must apply a force large enough to slow down only the last car.

Example 5.6 The Runaway Car AM

A car of mass m is on an icy driveway inclined at an angle θ as in Figure 5.11a.

(A) Find the acceleration of the car, assuming the driveway is frictionless.

SOLUTION

Conceptualize Use Figure 5.11a to conceptualize the situation. From everyday experience, we know that a car on an icy incline will accelerate down the incline. (The same thing happens to a car on a hill with its brakes not set.)

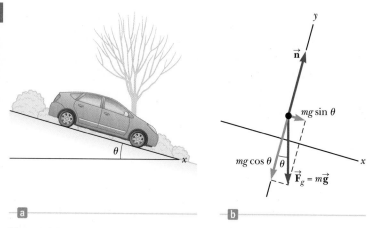

Figure 5.11 (Example 5.6) (a) A car on a frictionless incline. (b) The free-body diagram for the car. The black dot represents the position of the center of mass of the car. We will learn about center of mass in Chapter 9.

Categorize We categorize the car as a *particle under a net force* because it accelerates. Furthermore, this example belongs to a very common category of problems in which an object moves under the influence of gravity on an inclined plane.

....

Analyze Figure 5.11b shows the free-body diagram for the car. The only forces acting on the car are the normal force $\vec{\mathbf{n}}$ exerted by the inclined plane, which acts perpendicular to the plane, and the gravitational force $\vec{\mathbf{F}}_g = m\vec{\mathbf{g}}$, which acts vertically downward. For problems involving inclined planes, it is convenient to choose the coordinate axes with x along the incline and y perpendicular to it as in Figure 5.11b. With these axes, we represent the gravitational force by a component of magnitude $mg\sin\theta$ along the positive x axis and one of magnitude $mg\cos\theta$ along the negative y axis. Our choice of axes results in the car being modeled as a particle under a net force in the x direction and a particle in equilibrium in the y direction.

Apply these models to the car:

$$(1) \quad \sum F_x = mg\sin\theta = ma_x$$

$$(2) \quad \sum F_y = n - mg\cos\theta = 0$$

Solve Equation (1) for a_x:

$$(3) \quad a_x = \boxed{g\sin\theta}$$

....

Finalize Note that the acceleration component a_x is independent of the mass of the car! It depends only on the angle of inclination and on g.

From Equation (2), we conclude that the component of $\vec{\mathbf{F}}_g$ perpendicular to the incline is balanced by the normal force; that is, $n = mg\cos\theta$. This situation is a case in which the normal force is *not* equal in magnitude to the weight of the object (as discussed in Pitfall Prevention 5.6 on page 97).

It is possible, although inconvenient, to solve the problem with "standard" horizontal and vertical axes. You may want to try it, just for practice.

(B) Suppose the car is released from rest at the top of the incline and the distance from the car's front bumper to the bottom of the incline is d. How long does it take the front bumper to reach the bottom of the hill, and what is the car's speed as it arrives there?

▶ **5.6** continued

SOLUTION

Conceptualize Imagine the car is sliding down the hill and you use a stopwatch to measure the entire time interval until it reaches the bottom.

Categorize This part of the problem belongs to kinematics rather than to dynamics, and Equation (3) shows that the acceleration a_x is constant. Therefore, you should categorize the car in this part of the problem as a particle under constant acceleration.

Analyze Defining the initial position of the front bumper as $x_i = 0$ and its final position as $x_f = d$, and recognizing that $v_{xi} = 0$, choose Equation 2.16 from the particle under constant acceleration model, $x_f = x_i + v_{xi}t + \frac{1}{2}a_x t^2$:

$$d = \tfrac{1}{2}a_x t^2$$

Solve for t:

$$(4) \quad t = \sqrt{\frac{2d}{a_x}} = \sqrt{\frac{2d}{g\sin\theta}}$$

Use Equation 2.17, with $v_{xi} = 0$, to find the final velocity of the car:

$$v_{xf}^2 = 2a_x d$$

$$(5) \quad v_{xf} = \sqrt{2a_x d} = \sqrt{2gd\sin\theta}$$

Finalize We see from Equations (4) and (5) that the time t at which the car reaches the bottom and its final speed v_{xf} are independent of the car's mass, as was its acceleration. Notice that we have combined techniques from Chapter 2 with new techniques from this chapter in this example. As we learn more techniques in later chapters, this process of combining analysis models and information from several parts of the book will occur more often. In these cases, use the General Problem-Solving Strategy to help you identify what analysis models you will need.

WHAT IF? What previously solved problem does this situation become if $\theta = 90°$?

Answer Imagine θ going to 90° in Figure 5.11. The inclined plane becomes vertical, and the car is an object in free fall! Equation (3) becomes

$$a_x = g\sin\theta = g\sin 90° = g$$

which is indeed the free-fall acceleration. (We find $a_x = g$ rather than $a_x = -g$ because we have chosen positive x to be downward in Fig. 5.11.) Notice also that the condition $n = mg\cos\theta$ gives us $n = mg\cos 90° = 0$. That is consistent with the car falling downward *next to* the vertical plane, in which case there is no contact force between the car and the plane.

Example 5.7 **One Block Pushes Another** AM

Two blocks of masses m_1 and m_2, with $m_1 > m_2$, are placed in contact with each other on a frictionless, horizontal surface as in Figure 5.12a. A constant horizontal force $\vec{F}$ is applied to m_1 as shown.

(A) Find the magnitude of the acceleration of the system.

SOLUTION

Conceptualize Conceptualize the situation by using Figure 5.12a and realize that both blocks must experience the *same* acceleration because they are in contact with each other and remain in contact throughout the motion.

Categorize We categorize this problem as one involving a *particle under a net force* because a force is applied to a system of blocks and we are looking for the acceleration of the system.

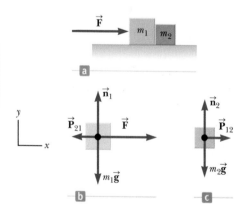

Figure 5.12 (Example 5.7) (a) A force is applied to a block of mass m_1, which pushes on a second block of mass m_2. (b) The forces acting on m_1. (c) The forces acting on m_2.

continued

▶ **5.7** continued

Analyze First model the combination of two blocks as a single particle under a net force. Apply Newton's second law to the combination in the x direction to find the acceleration:

$$\sum F_x = F = (m_1 + m_2)a_x$$

$$(1) \quad a_x = \frac{F}{m_1 + m_2}$$

Finalize The acceleration given by Equation (1) is the same as that of a single object of mass $m_1 + m_2$ and subject to the same force.

(B) Determine the magnitude of the contact force between the two blocks.

SOLUTION

Conceptualize The contact force is internal to the system of two blocks. Therefore, we cannot find this force by modeling the whole system (the two blocks) as a single particle.

Categorize Now consider each of the two blocks individually by categorizing each as a *particle under a net force*.

Analyze We construct a diagram of forces acting on the object for each block as shown in Figures 5.12b and 5.12c, where the contact force is denoted by $\vec{\mathbf{P}}$. From Figure 5.12c, we see that the only horizontal force acting on m_2 is the contact force $\vec{\mathbf{P}}_{12}$ (the force exerted by m_1 on m_2), which is directed to the right.

Apply Newton's second law to m_2:

$$(2) \quad \sum F_x = P_{12} = m_2 a_x$$

Substitute the value of the acceleration a_x given by Equation (1) into Equation (2):

$$(3) \quad P_{12} = m_2 a_x = \left(\frac{m_2}{m_1 + m_2}\right)F$$

Finalize This result shows that the contact force P_{12} is *less* than the applied force F. The force required to accelerate block 2 alone must be less than the force required to produce the same acceleration for the two-block system.

To finalize further, let us check this expression for P_{12} by considering the forces acting on m_1, shown in Figure 5.12b. The horizontal forces acting on m_1 are the applied force $\vec{\mathbf{F}}$ to the right and the contact force $\vec{\mathbf{P}}_{21}$ to the left (the force exerted by m_2 on m_1). From Newton's third law, $\vec{\mathbf{P}}_{21}$ is the reaction force to $\vec{\mathbf{P}}_{12}$, so $P_{21} = P_{12}$.

Apply Newton's second law to m_1:

$$(4) \quad \sum F_x = F - P_{21} = F - P_{12} = m_1 a_x$$

Solve for P_{12} and substitute the value of a_x from Equation (1):

$$P_{12} = F - m_1 a_x = F - m_1\left(\frac{F}{m_1 + m_2}\right) = \left(\frac{m_2}{m_1 + m_2}\right)F$$

This result agrees with Equation (3), as it must.

WHAT IF? Imagine that the force $\vec{\mathbf{F}}$ in Figure 5.12 is applied toward the left on the right-hand block of mass m_2. Is the magnitude of the force $\vec{\mathbf{P}}_{12}$ the same as it was when the force was applied toward the right on m_1?

Answer When the force is applied toward the left on m_2, the contact force must accelerate m_1. In the original situation, the contact force accelerates m_2. Because $m_1 > m_2$, more force is required, so the magnitude of $\vec{\mathbf{P}}_{12}$ is greater than in the original situation. To see this mathematically, modify Equation (4) appropriately and solve for $\vec{\mathbf{P}}_{12}$.

Example 5.8   **Weighing a Fish in an Elevator**

A person weighs a fish of mass m on a spring scale attached to the ceiling of an elevator as illustrated in Figure 5.13.

(A) Show that if the elevator accelerates either upward or downward, the spring scale gives a reading that is different from the weight of the fish.

▶ **5.8** continued

SOLUTION

Conceptualize The reading on the scale is related to the extension of the spring in the scale, which is related to the force on the end of the spring as in Figure 5.2. Imagine that the fish is hanging on a string attached to the end of the spring. In this case, the magnitude of the force exerted on the spring is equal to the tension T in the string. Therefore, we are looking for T. The force $\vec{T}$ pulls down on the string and pulls up on the fish.

Categorize We can categorize this problem by identifying the fish as a *particle in equilibrium* if the elevator is not accelerating or as a *particle under a net force* if the elevator is accelerating.

· ·

Analyze Inspect the diagrams of the forces acting on the fish in Figure 5.13 and notice that the external forces acting on the fish are the downward gravitational force $\vec{F}_g = m\vec{g}$ and the force $\vec{T}$ exerted by the string. If the elevator is either at rest or moving at constant velocity, the fish is a particle in equilibrium, so $\Sigma F_y = T - F_g = 0$ or $T = F_g = mg$. (Remember that the scalar mg is the weight of the fish.)

When the elevator accelerates upward, the spring scale reads a value greater than the weight of the fish.

When the elevator accelerates downward, the spring scale reads a value less than the weight of the fish.

Figure 5.13 (Example 5.8) A fish is weighed on a spring scale in an accelerating elevator car.

Now suppose the elevator is moving with an acceleration $\vec{a}$ relative to an observer standing outside the elevator in an inertial frame. The fish is now a particle under a net force.

Apply Newton's second law to the fish:

$$\Sigma F_y = T - mg = ma_y$$

Solve for T:

$$(1) \quad T = ma_y + mg = mg\left(\frac{a_y}{g} + 1\right) = F_g\left(\frac{a_y}{g} + 1\right)$$

where we have chosen upward as the positive y direction. We conclude from Equation (1) that the scale reading T is greater than the fish's weight mg if $\vec{a}$ is upward, so a_y is positive (Fig. 5.13a), and that the reading is less than mg if $\vec{a}$ is downward, so a_y is negative (Fig. 5.13b).

(B) Evaluate the scale readings for a 40.0-N fish if the elevator moves with an acceleration $a_y = \pm 2.00 \text{ m/s}^2$.

SOLUTION

Evaluate the scale reading from Equation (1) if $\vec{a}$ is upward:

$$T = (40.0 \text{ N})\left(\frac{2.00 \text{ m/s}^2}{9.80 \text{ m/s}^2} + 1\right) = \boxed{48.2 \text{ N}}$$

Evaluate the scale reading from Equation (1) if $\vec{a}$ is downward:

$$T = (40.0 \text{ N})\left(\frac{-2.00 \text{ m/s}^2}{9.80 \text{ m/s}^2} + 1\right) = \boxed{31.8 \text{ N}}$$

Finalize Take this advice: if you buy a fish in an elevator, make sure the fish is weighed while the elevator is either at rest or accelerating downward! Furthermore, notice that from the information given here, one cannot determine the direction of the velocity of the elevator.

WHAT IF? Suppose the elevator cable breaks and the elevator and its contents are in free fall. What happens to the reading on the scale?

Answer If the elevator falls freely, the fish's acceleration is $a_y = -g$. We see from Equation (1) that the scale reading T is zero in this case; that is, the fish *appears* to be weightless.

Example 5.9   The Atwood Machine   AM

When two objects of unequal mass are hung vertically over a frictionless pulley of negligible mass as in Figure 5.14a, the arrangement is called an *Atwood machine.* The device is sometimes used in the laboratory to determine the value of *g*. Determine the magnitude of the acceleration of the two objects and the tension in the lightweight string.

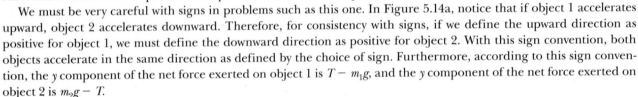

SOLUTION

Conceptualize Imagine the situation pictured in Figure 5.14a in action: as one object moves upward, the other object moves downward. Because the objects are connected by an inextensible string, their accelerations must be of equal magnitude.

Categorize The objects in the Atwood machine are subject to the gravitational force as well as to the forces exerted by the strings connected to them. Therefore, we can categorize this problem as one involving two *particles under a net force.*

Figure 5.14 (Example 5.9) The Atwood machine. (a) Two objects connected by a massless inextensible string over a frictionless pulley. (b) The free-body diagrams for the two objects.

Analyze The free-body diagrams for the two objects are shown in Figure 5.14b. Two forces act on each object: the upward force $\vec{T}$ exerted by the string and the downward gravitational force. In problems such as this one in which the pulley is modeled as massless and frictionless, the tension in the string on both sides of the pulley is the same. If the pulley has mass or is subject to friction, the tensions on either side are not the same and the situation requires techniques we will learn in Chapter 10.

We must be very careful with signs in problems such as this one. In Figure 5.14a, notice that if object 1 accelerates upward, object 2 accelerates downward. Therefore, for consistency with signs, if we define the upward direction as positive for object 1, we must define the downward direction as positive for object 2. With this sign convention, both objects accelerate in the same direction as defined by the choice of sign. Furthermore, according to this sign convention, the *y* component of the net force exerted on object 1 is $T - m_1 g$, and the *y* component of the net force exerted on object 2 is $m_2 g - T$.

From the particle under a net force model, apply Newton's second law to object 1:

$$(1) \quad \sum F_y = T - m_1 g = m_1 a_y$$

Apply Newton's second law to object 2:

$$(2) \quad \sum F_y = m_2 g - T = m_2 a_y$$

Add Equation (2) to Equation (1), noticing that *T* cancels:

$$-m_1 g + m_2 g = m_1 a_y + m_2 a_y$$

Solve for the acceleration:

$$(3) \quad a_y = \left(\frac{m_2 - m_1}{m_1 + m_2} \right) g$$

Substitute Equation (3) into Equation (1) to find *T*:

$$(4) \quad T = m_1 (g + a_y) = \left(\frac{2 m_1 m_2}{m_1 + m_2} \right) g$$

Finalize The acceleration given by Equation (3) can be interpreted as the ratio of the magnitude of the unbalanced force on the system $(m_2 - m_1)g$ to the total mass of the system $(m_1 + m_2)$, as expected from Newton's second law. Notice that the sign of the acceleration depends on the relative masses of the two objects.

WHAT IF?   Describe the motion of the system if the objects have equal masses, that is, $m_1 = m_2$.

Answer If we have the same mass on both sides, the system is balanced and should not accelerate. Mathematically, we see that if $m_1 = m_2$, Equation (3) gives us $a_y = 0$.

WHAT IF?   What if one of the masses is much larger than the other: $m_1 \gg m_2$?

Answer In the case in which one mass is infinitely larger than the other, we can ignore the effect of the smaller mass. Therefore, the larger mass should simply fall as if the smaller mass were not there. We see that if $m_1 \gg m_2$, Equation (3) gives us $a_y = -g$.

| Example 5.10 | **Acceleration of Two Objects Connected by a Cord** | **AM** |

A ball of mass m_1 and a block of mass m_2 are attached by a lightweight cord that passes over a frictionless pulley of negligible mass as in Figure 5.15a. The block lies on a frictionless incline of angle θ. Find the magnitude of the acceleration of the two objects and the tension in the cord.

SOLUTION

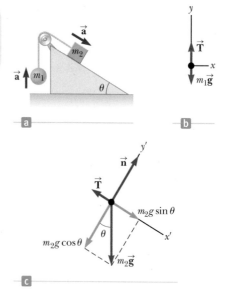

Conceptualize Imagine the objects in Figure 5.15 in motion. If m_2 moves down the incline, then m_1 moves upward. Because the objects are connected by a cord (which we assume does not stretch), their accelerations have the same magnitude. Notice the normal coordinate axes in Figure 5.15b for the ball and the "tilted" axes for the block in Figure 5.15c.

Categorize We can identify forces on each of the two objects and we are looking for an acceleration, so we categorize the objects as *particles under a net force*. For the block, this model is only valid for the x' direction. In the y' direction, we apply the *particle in equilibrium* model because the block does not accelerate in that direction.

Analyze Consider the free-body diagrams shown in Figures 5.15b and 5.15c.

Figure 5.15 (Example 5.10) (a) Two objects connected by a lightweight cord strung over a frictionless pulley. (b) The free-body diagram for the ball. (c) The free-body diagram for the block. (The incline is frictionless.)

Apply Newton's second law in the y direction to the ball, choosing the upward direction as positive:

$$(1) \quad \sum F_y = T - m_1 g = m_1 a_y = m_1 a$$

For the ball to accelerate upward, it is necessary that $T > m_1 g$. In Equation (1), we replaced a_y with a because the acceleration has only a y component.

For the block, we have chosen the x' axis along the incline as in Figure 5.15c. For consistency with our choice for the ball, we choose the positive x' direction to be down the incline.

Apply the particle under a net force model to the block in the x' direction and the particle in equilibrium model in the y' direction:

$$(2) \quad \sum F_{x'} = m_2 g \sin\theta - T = m_2 a_{x'} = m_2 a$$

$$(3) \quad \sum F_{y'} = n - m_2 g \cos\theta = 0$$

In Equation (2), we replaced $a_{x'}$ with a because the two objects have accelerations of equal magnitude a.

Solve Equation (1) for T:

$$(4) \quad T = m_1(g + a)$$

Substitute this expression for T into Equation (2):

$$m_2 g \sin\theta - m_1(g + a) = m_2 a$$

Solve for a:

$$(5) \quad a = \left(\frac{m_2 \sin\theta - m_1}{m_1 + m_2} \right) g$$

Substitute this expression for a into Equation (4) to find T:

$$(6) \quad T = \left(\frac{m_1 m_2 (\sin\theta + 1)}{m_1 + m_2} \right) g$$

Finalize The block accelerates down the incline only if $m_2 \sin\theta > m_1$. If $m_1 > m_2 \sin\theta$, the acceleration is up the incline for the block and downward for the ball. Also notice that the result for the acceleration, Equation (5), can be interpreted as the magnitude of the net external force acting on the ball–block system divided by the total mass of the system; this result is consistent with Newton's second law.

WHAT IF? What happens in this situation if $\theta = 90°$?

continued

▶ **5.10** continued

Answer If $\theta = 90°$, the inclined plane becomes vertical and there is no interaction between its surface and m_2. Therefore, this problem becomes the Atwood machine of Example 5.9. Letting $\theta \to 90°$ in Equations (5) and (6) causes them to reduce to Equations (3) and (4) of Example 5.9!

WHAT IF?   What if $m_1 = 0$?

Answer If $m_1 = 0$, then m_2 is simply sliding down an inclined plane without interacting with m_1 through the string. Therefore, this problem becomes the sliding car problem in Example 5.6. Letting $m_1 \to 0$ in Equation (5) causes it to reduce to Equation (3) of Example 5.6!

5.8 Forces of Friction

When an object is in motion either on a surface or in a viscous medium such as air or water, there is resistance to the motion because the object interacts with its surroundings. We call such resistance a **force of friction.** Forces of friction are very important in our everyday lives. They allow us to walk or run and are necessary for the motion of wheeled vehicles.

Imagine that you are working in your garden and have filled a trash can with yard clippings. You then try to drag the trash can across the surface of your concrete patio as in Figure 5.16a. This surface is *real*, not an idealized, frictionless surface. If we apply an external horizontal force $\vec{\mathbf{F}}$ to the trash can, acting to the right, the trash can remains stationary when $\vec{\mathbf{F}}$ is small. The force on the trash can that counteracts $\vec{\mathbf{F}}$ and keeps it from moving acts toward the left and is called the

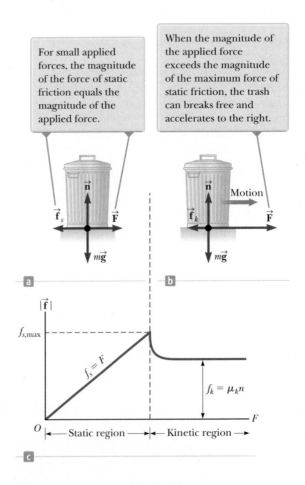

Figure 5.16 (a) and (b) When pulling on a trash can, the direction of the force of friction $\vec{\mathbf{f}}$ between the can and a rough surface is opposite the direction of the applied force $\vec{\mathbf{F}}$. (c) A graph of friction force versus applied force. Notice that $f_{s,\text{max}} > f_k$.

force of static friction $\vec{f}_s$. As long as the trash can is not moving, $f_s = F$. Therefore, if $\vec{F}$ is increased, $\vec{f}_s$ also increases. Likewise, if $\vec{F}$ decreases, $\vec{f}_s$ also decreases.

◀ **Force of static friction**

Experiments show that the friction force arises from the nature of the two surfaces: because of their roughness, contact is made only at a few locations where peaks of the material touch. At these locations, the friction force arises in part because one peak physically blocks the motion of a peak from the opposing surface and in part from chemical bonding ("spot welds") of opposing peaks as they come into contact. Although the details of friction are quite complex at the atomic level, this force ultimately involves an electrical interaction between atoms or molecules.

If we increase the magnitude of $\vec{F}$ as in Figure 5.16b, the trash can eventually slips. When the trash can is on the verge of slipping, f_s has its maximum value $f_{s,\text{max}}$ as shown in Figure 5.16c. When F exceeds $f_{s,\text{max}}$, the trash can moves and accelerates to the right. We call the friction force for an object in motion the **force of kinetic friction** $\vec{f}_k$. When the trash can is in motion, the force of kinetic friction on the can is less than $f_{s,\text{max}}$ (Fig. 5.16c). The net force $F - f_k$ in the x direction produces an acceleration to the right, according to Newton's second law. If $F = f_k$, the acceleration is zero and the trash can moves to the right with constant speed. If the applied force $\vec{F}$ is removed from the moving can, the friction force $\vec{f}_k$ acting to the left provides an acceleration of the trash can in the $-x$ direction and eventually brings it to rest, again consistent with Newton's second law.

◀ **Force of kinetic friction**

Experimentally, we find that, to a good approximation, both $f_{s,\text{max}}$ and f_k are proportional to the magnitude of the normal force exerted on an object by the surface. The following descriptions of the force of friction are based on experimental observations and serve as the simplification model we shall use for forces of friction in problem solving:

- The magnitude of the force of static friction between any two surfaces in contact can have the values

$$f_s \leq \mu_s n \tag{5.9}$$

where the dimensionless constant μ_s is called the **coefficient of static friction** and n is the magnitude of the normal force exerted by one surface on the other. The equality in Equation 5.9 holds when the surfaces are on the verge of slipping, that is, when $f_s = f_{s,\text{max}} = \mu_s n$. This situation is called *impending motion*. The inequality holds when the surfaces are not on the verge of slipping.

- The magnitude of the force of kinetic friction acting between two surfaces is

$$f_k = \mu_k n \tag{5.10}$$

where μ_k is the **coefficient of kinetic friction.** Although the coefficient of kinetic friction can vary with speed, we shall usually neglect any such variations in this text.

- The values of μ_k and μ_s depend on the nature of the surfaces, but μ_k is generally less than μ_s. Typical values range from around 0.03 to 1.0. Table 5.1 (page 110) lists some reported values.

- The direction of the friction force on an object is parallel to the surface with which the object is in contact and opposite to the actual motion (kinetic friction) or the impending motion (static friction) of the object relative to the surface.

- The coefficients of friction are nearly independent of the area of contact between the surfaces. We might expect that placing an object on the side having the most area might increase the friction force. Although this method provides more points in contact, the weight of the object is spread out over a larger area and the individual points are not pressed together as tightly. Because these effects approximately compensate for each other, the friction force is independent of the area.

Pitfall Prevention 5.9

The Equal Sign Is Used in Limited Situations In Equation 5.9, the equal sign is used *only* in the case in which the surfaces are just about to break free and begin sliding. Do not fall into the common trap of using $f_s = \mu_s n$ in *any* static situation.

Pitfall Prevention 5.10

Friction Equations Equations 5.9 and 5.10 are *not* vector equations. They are relationships between the *magnitudes* of the vectors representing the friction and normal forces. Because the friction and normal forces are perpendicular to each other, the vectors cannot be related by a multiplicative constant.

Pitfall Prevention 5.11

The Direction of the Friction Force Sometimes, an incorrect statement about the friction force between an object and a surface is made—"the friction force on an object is opposite to its motion or impending motion"—rather than the correct phrasing, "the friction force on an object is opposite to its motion or impending motion *relative to the surface.*"

Table 5.1	**Coefficients of Friction**	
	μ_s	μ_k
Rubber on concrete	1.0	0.8
Steel on steel	0.74	0.57
Aluminum on steel	0.61	0.47
Glass on glass	0.94	0.4
Copper on steel	0.53	0.36
Wood on wood	0.25–0.5	0.2
Waxed wood on wet snow	0.14	0.1
Waxed wood on dry snow	—	0.04
Metal on metal (lubricated)	0.15	0.06
Teflon on Teflon	0.04	0.04
Ice on ice	0.1	0.03
Synovial joints in humans	0.01	0.003

Note: All values are approximate. In some cases, the coefficient of friction can exceed 1.0.

a

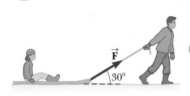

b

Figure 5.17 (Quick Quiz 5.7) A father slides his daughter on a sled either by (a) pushing down on her shoulders or (b) pulling up on a rope.

Quick Quiz 5.6 You press your physics textbook flat against a vertical wall with your hand. What is the direction of the friction force exerted by the wall on the book? **(a)** downward **(b)** upward **(c)** out from the wall **(d)** into the wall

Quick Quiz 5.7 You are playing with your daughter in the snow. She sits on a sled and asks you to slide her across a flat, horizontal field. You have a choice of **(a)** pushing her from behind by applying a force downward on her shoulders at 30° below the horizontal (Fig. 5.17a) or **(b)** attaching a rope to the front of the sled and pulling with a force at 30° above the horizontal (Fig. 5.17b). Which would be easier for you and why?

Example 5.11   **Experimental Determination of μ_s and μ_k**   AM

The following is a simple method of measuring coefficients of friction. Suppose a block is placed on a rough surface inclined relative to the horizontal as shown in Figure 5.18. The incline angle is increased until the block starts to move. Show that you can obtain μ_s by measuring the critical angle θ_c at which this slipping just occurs.

SOLUTION

Conceptualize Consider Figure 5.18 and imagine that the block tends to slide down the incline due to the gravitational force. To simulate the situation, place a coin on this book's cover and tilt the book until the coin begins to slide. Notice how this example differs from Example 5.6. When there is no friction on an incline, *any* angle of the incline will cause a stationary object to begin moving. When there is friction, however, there is no movement of the object for angles less than the critical angle.

Categorize The block is subject to various forces. Because we are raising the plane to the angle at which the block is just ready to begin to move but is not moving, we categorize the block as a *particle in equilibrium*.

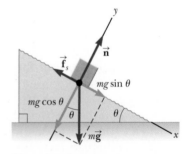

Figure 5.18 (Example 5.11) The external forces exerted on a block lying on a rough incline are the gravitational force $m\vec{\mathbf{g}}$, the normal force $\vec{\mathbf{n}}$, and the force of friction $\vec{\mathbf{f}}_s$. For convenience, the gravitational force is resolved into a component $mg\sin\theta$ along the incline and a component $mg\cos\theta$ perpendicular to the incline.

· ·

Analyze The diagram in Figure 5.18 shows the forces on the block: the gravitational force $m\vec{\mathbf{g}}$, the normal force $\vec{\mathbf{n}}$, and the force of static friction $\vec{\mathbf{f}}_s$. We choose x to be parallel to the plane and y perpendicular to it.

From the particle in equilibrium model, apply Equation 5.8 to the block in both the x and y directions:

(1) $\sum F_x = mg\sin\theta - f_s = 0$

(2) $\sum F_y = n - mg\cos\theta = 0$

▶ **5.11** continued

Substitute $mg = n/\cos\theta$ from Equation (2) into Equation (1):

$$(3) \quad f_s = mg\sin\theta = \left(\frac{n}{\cos\theta}\right)\sin\theta = n\tan\theta$$

When the incline angle is increased until the block is on the verge of slipping, the force of static friction has reached its maximum value $\mu_s n$. The angle θ in this situation is the critical angle θ_c. Make these substitutions in Equation (3):

$$\mu_s n = n\tan\theta_c$$

$$\mu_s = \tan\theta_c$$

We have shown, as requested, that the coefficient of static friction is related only to the critical angle. For example, if the block just slips at $\theta_c = 20.0°$, we find that $\mu_s = \tan 20.0° = 0.364$.

Finalize Once the block starts to move at $\theta \geq \theta_c$, it accelerates down the incline and the force of friction is $f_k = \mu_k n$. If θ is reduced to a value less than θ_c, however, it may be possible to find an angle θ_c' such that the block moves down the incline with constant speed as a particle in equilibrium again ($a_x = 0$). In this case, use Equations (1) and (2) with f_s replaced by f_k to find μ_k: $\mu_k = \tan\theta_c'$, where $\theta_c' < \theta_c$.

Example 5.12   **The Sliding Hockey Puck**   AM

A hockey puck on a frozen pond is given an initial speed of 20.0 m/s. If the puck always remains on the ice and slides 115 m before coming to rest, determine the coefficient of kinetic friction between the puck and ice.

SOLUTION

Conceptualize Imagine that the puck in Figure 5.19 slides to the right. The kinetic friction force acts to the left and slows the puck, which eventually comes to rest due to that force.

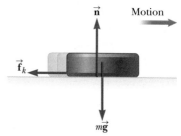

Figure 5.19 (Example 5.12) After the puck is given an initial velocity to the right, the only external forces acting on it are the gravitational force $m\vec{g}$, the normal force $\vec{n}$, and the force of kinetic friction $\vec{f}_k$.

Categorize The forces acting on the puck are identified in Figure 5.19, but the text of the problem provides kinematic variables. Therefore, we categorize the problem in several ways. First, it involves modeling the puck as a *particle under a net force* in the horizontal direction: kinetic friction causes the puck to accelerate. There is no acceleration of the puck in the vertical direction, so we use the *particle in equilibrium* model for that direction. Furthermore, because we model the force of kinetic friction as independent of speed, the acceleration of the puck is constant. So, we can also categorize this problem by modeling the puck as a *particle under constant acceleration*.

Analyze First, let's find the acceleration algebraically in terms of the coefficient of kinetic friction, using Newton's second law. Once we know the acceleration of the puck and the distance it travels, the equations of kinematics can be used to find the numerical value of the coefficient of kinetic friction. The diagram in Figure 5.19 shows the forces on the puck.

Apply the particle under a net force model in the x direction to the puck:

$$(1) \quad \sum F_x = -f_k = ma_x$$

Apply the particle in equilibrium model in the y direction to the puck:

$$(2) \quad \sum F_y = n - mg = 0$$

Substitute $n = mg$ from Equation (2) and $f_k = \mu_k n$ into Equation (1):

$$-\mu_k n = -\mu_k mg = ma_x$$

$$a_x = -\mu_k g$$

The negative sign means the acceleration is to the left in Figure 5.19. Because the velocity of the puck is to the right, the puck is slowing down. The acceleration is independent of the mass of the puck and is constant because we assume μ_k remains constant.

continued

▶ **5.12** continued

Apply the particle under constant acceleration model to the puck, choosing Equation 2.17 from the model, $v_{xf}^2 = v_{xi}^2 + 2a_x(x_f - x_i)$, with $x_i = 0$ and $v_{xf} = 0$:

$$0 = v_{xi}^2 + 2a_x x_f = v_{xi}^2 - 2\mu_k g x_f$$

Solve for the coefficient of kinetic friction:

$$\mu_k = \frac{v_{xi}^2}{2gx_f}$$

Substitute the numerical values:

$$\mu_k = \frac{(20.0 \text{ m/s})^2}{2(9.80 \text{ m/s}^2)(115 \text{ m})} = \boxed{0.177}$$

Finalize Notice that μ_k is dimensionless, as it should be, and that it has a low value, consistent with an object sliding on ice.

Example 5.13   Acceleration of Two Connected Objects When Friction Is Present   **AM**

A block of mass m_2 on a rough, horizontal surface is connected to a ball of mass m_1 by a lightweight cord over a lightweight, frictionless pulley as shown in Figure 5.20a. A force of magnitude F at an angle θ with the horizontal is applied to the block as shown, and the block slides to the right. The coefficient of kinetic friction between the block and surface is μ_k. Determine the magnitude of the acceleration of the two objects.

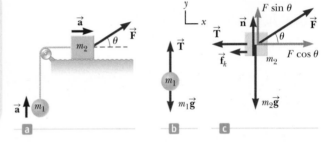

Figure 5.20   (Example 5.13) (a) The external force $\vec{\mathbf{F}}$ applied as shown can cause the block to accelerate to the right. (b, c) Diagrams showing the forces on the two objects, assuming the block accelerates to the right and the ball accelerates upward.

SOLUTION

Conceptualize Imagine what happens as $\vec{\mathbf{F}}$ is applied to the block. Assuming $\vec{\mathbf{F}}$ is large enough to break the block free from static friction but not large enough to lift the block, the block slides to the right and the ball rises.

Categorize We can identify forces and we want an acceleration, so we categorize this problem as one involving two *particles under a net force*, the ball and the block. Because we assume that the block does not rise into the air due to the applied force, we model the block as a *particle in equilibrium* in the vertical direction.

Analyze First draw force diagrams for the two objects as shown in Figures 5.20b and 5.20c. Notice that the string exerts a force of magnitude T on both objects. The applied force $\vec{\mathbf{F}}$ has x and y components $F\cos\theta$ and $F\sin\theta$, respectively. Because the two objects are connected, we can equate the magnitudes of the x component of the acceleration of the block and the y component of the acceleration of the ball and call them both a. Let us assume the motion of the block is to the right.

Apply the particle under a net force model to the block in the horizontal direction:

(1) $\sum F_x = F\cos\theta - f_k - T = m_2 a_x = m_2 a$

Because the block moves only horizontally, apply the particle in equilibrium model to the block in the vertical direction:

(2) $\sum F_y = n + F\sin\theta - m_2 g = 0$

Apply the particle under a net force model to the ball in the vertical direction:

(3) $\sum F_y = T - m_1 g = m_1 a_y = m_1 a$

Solve Equation (2) for n:

$n = m_2 g - F\sin\theta$

Substitute n into $f_k = \mu_k n$ from Equation 5.10:

(4) $f_k = \mu_k(m_2 g - F\sin\theta)$

▶ **5.13** continued

Substitute Equation (4) and the value of T from Equation (3) into Equation (1):

$$F\cos\theta - \mu_k(m_2 g - F\sin\theta) - m_1(a + g) = m_2 a$$

Solve for a:

$$(5)\quad a = \frac{F(\cos\theta + \mu_k\sin\theta) - (m_1 + \mu_k m_2)g}{m_1 + m_2}$$

Finalize The acceleration of the block can be either to the right or to the left depending on the sign of the numerator in Equation (5). If the velocity is to the left, we must reverse the sign of f_k in Equation (1) because the force of kinetic friction must oppose the motion of the block relative to the surface. In this case, the value of a is the same as in Equation (5), with the two plus signs in the numerator changed to minus signs.

What does Equation (5) reduce to if the force $\vec{\mathbf{F}}$ is removed and the surface becomes frictionless? Call this expression Equation (6). Does this algebraic expression match your intuition about the physical situation in this case? Now go back to Example 5.10 and let angle θ go to zero in Equation (5) of that example. How does the resulting equation compare with your Equation (6) here in Example 5.13? Should the algebraic expressions compare in this way based on the physical situations?

Summary

Definitions

An **inertial frame of reference** is a frame in which an object that does not interact with other objects experiences zero acceleration. Any frame moving with constant velocity relative to an inertial frame is also an inertial frame.

We define **force** as **that which causes a change in motion of an object.**

Concepts and Principles

Newton's first law states that it is possible to find an inertial frame in which an object that does not interact with other objects experiences zero acceleration, or, equivalently, in the absence of an external force, when viewed from an inertial frame, an object at rest remains at rest and an object in uniform motion in a straight line maintains that motion.

Newton's second law states that the acceleration of an object is directly proportional to the net force acting on it and inversely proportional to its mass.

Newton's third law states that if two objects interact, the force exerted by object 1 on object 2 is equal in magnitude and opposite in direction to the force exerted by object 2 on object 1.

The **gravitational force** exerted on an object is equal to the product of its mass (a scalar quantity) and the free-fall acceleration:

$$\vec{\mathbf{F}}_g = m\vec{\mathbf{g}} \qquad \textbf{(5.5)}$$

The **weight** of an object is the magnitude of the gravitational force acting on the object:

$$F_g = mg \qquad \textbf{(5.6)}$$

The maximum **force of static friction** $\vec{\mathbf{f}}_{s,\max}$ between an object and a surface is proportional to the normal force acting on the object. In general, $f_s \leq \mu_s n$, where μ_s is the **coefficient of static friction** and n is the magnitude of the normal force.

When an object slides over a surface, the magnitude of the **force of kinetic friction** $\vec{\mathbf{f}}_k$ is given by $f_k = \mu_k n$, where μ_k is the **coefficient of kinetic friction.**

continued

Analysis Models for Problem Solving

Particle Under a Net Force If a particle of mass m experiences a nonzero net force, its acceleration is related to the net force by Newton's second law:

$$\sum \vec{F} = m\vec{a} \qquad (5.2)$$

Particle in Equilibrium If a particle maintains a constant velocity (so that $\vec{a} = 0$), which could include a velocity of zero, the forces on the particle balance and Newton's second law reduces to

$$\sum \vec{F} = 0 \qquad (5.8)$$

Objective Questions

1. denotes answer available in *Student Solutions Manual/Study Guide*

1. The driver of a speeding empty truck slams on the brakes and skids to a stop through a distance d. On a second trial, the truck carries a load that doubles its mass. What will now be the truck's "skidding distance"? (a) $4d$ (b) $2d$ (c) $\sqrt{2}d$ (d) d (e) $d/2$

2. In Figure OQ5.2, a locomotive has broken through the wall of a train station. During the collision, what can be said about the force exerted by the locomotive on the wall? (a) The force exerted by the locomotive on the wall was larger than the force the wall could exert on the locomotive. (b) The force exerted by the locomotive on the wall was the same in magnitude as the force exerted by the wall on the locomotive. (c) The force exerted by the locomotive on the wall was less than the force exerted by the wall on the locomotive. (d) The wall cannot be said to "exert" a force; after all, it broke.

Figure OQ5.2

3. The third graders are on one side of a schoolyard, and the fourth graders are on the other. They are throwing snowballs at each other. Between them, snowballs of various masses are moving with different velocities as shown in Figure OQ5.3. Rank the snowballs (a) through (e) according to the magnitude of the total force exerted on each one. Ignore air resistance. If two snowballs rank together, make that fact clear.

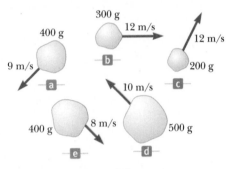

Figure OQ5.3

4. The driver of a speeding truck slams on the brakes and skids to a stop through a distance d. On another trial, the initial speed of the truck is half as large. What now will be the truck's skidding distance? (a) $2d$ (b) $\sqrt{2}d$ (c) d (d) $d/2$ (e) $d/4$

5. An experiment is performed on a puck on a level air hockey table, where friction is negligible. A constant horizontal force is applied to the puck, and the puck's acceleration is measured. Now the same puck is transported far into outer space, where both friction and gravity are negligible. The same constant force is applied to the puck (through a spring scale that stretches the same amount), and the puck's acceleration (relative to the distant stars) is measured. What is the puck's acceleration in outer space? (a) It is somewhat greater than its acceleration on the Earth. (b) It is the same as its acceleration on the Earth. (c) It is less than its acceleration on the Earth. (d) It is infinite because neither friction nor gravity constrains it. (e) It is very large because acceleration is inversely proportional to weight and the puck's weight is very small but not zero.

6. The manager of a department store is pushing horizontally with a force of magnitude 200 N on a box of shirts. The box is sliding across the horizontal floor with a forward acceleration. Nothing else touches the box. What must be true about the magnitude of the force of kinetic friction acting on the box (choose one)? (a) It is greater than 200 N. (b) It is less than 200 N. (c) It is equal to 200 N. (d) None of those statements is necessarily true.

7. Two objects are connected by a string that passes over a frictionless pulley as in Figure 5.14a, where $m_1 < m_2$ and a_1 and a_2 are the magnitudes of the respective accelerations. Which mathematical statement is true regarding the magnitude of the acceleration a_2 of the mass m_2? (a) $a_2 < g$ (b) $a_2 > g$ (c) $a_2 = g$ (d) $a_2 < a_1$ (e) $a_2 > a_1$

8. An object of mass m is sliding with speed v_i at some instant across a level tabletop, with which its coefficient of kinetic friction is μ. It then moves through a distance d and comes to rest. Which of the following equations for the speed v_i is reasonable? (a) $v_i = \sqrt{-2\mu mgd}$ (b) $v_i = \sqrt{2\mu mgd}$ (c) $v_i = \sqrt{-2\mu gd}$ (d) $v_i = \sqrt{2\mu gd}$ (e) $v_i = \sqrt{2\mu d}$

9. A truck loaded with sand accelerates along a highway. The driving force on the truck remains constant. What happens to the acceleration of the truck if its trailer leaks sand at a constant rate through a hole in its bottom? (a) It decreases at a steady rate. (b) It increases at a steady rate. (c) It increases and then decreases. (d) It decreases and then increases. (e) It remains constant.

10. A large crate of mass m is place on the flatbed of a truck but not tied down. As the truck accelerates forward with acceleration a, the crate remains at rest relative to the truck. What force causes the crate to accelerate? (a) the normal force (b) the gravitational force (c) the friction force (d) the ma force exerted by the crate (e) No force is required.

11. If an object is in equilibrium, which of the following statements is *not* true? (a) The speed of the object remains constant. (b) The acceleration of the object is zero. (c) The net force acting on the object is zero. (d) The object must be at rest. (e) There are at least two forces acting on the object.

12. A crate remains stationary after it has been placed on a ramp inclined at an angle with the horizontal. Which of the following statements is or are correct about the magnitude of the friction force that acts on the crate? Choose all that are true. (a) It is larger than the weight of the crate. (b) It is equal to $\mu_s n$. (c) It is greater than the component of the gravitational force acting down the ramp. (d) It is equal to the component of the gravitational force acting down the ramp. (e) It is less than the component of the gravitational force acting down the ramp.

13. An object of mass m moves with acceleration $\vec{a}$ down a rough incline. Which of the following forces should appear in a free-body diagram of the object? Choose all correct answers. (a) the gravitational force exerted by the planet (b) $m\vec{a}$ in the direction of motion (c) the normal force exerted by the incline (d) the friction force exerted by the incline (e) the force exerted by the object on the incline

---

Conceptual Questions   **1.** denotes answer available in *Student Solutions Manual/Study Guide*

1. If you hold a horizontal metal bar several centimeters above the ground and move it through grass, each leaf of grass bends out of the way. If you increase the speed of the bar, each leaf of grass will bend more quickly. How then does a rotary power lawn mower manage to cut grass? How can it exert enough force on a leaf of grass to shear it off?

2. Your hands are wet, and the restroom towel dispenser is empty. What do you do to get drops of water off your hands? How does the motion of the drops exemplify one of Newton's laws? Which one?

3. In the motion picture *It Happened One Night* (Columbia Pictures, 1934), Clark Gable is standing inside a stationary bus in front of Claudette Colbert, who is seated. The bus suddenly starts moving forward and Clark falls into Claudette's lap. Why did this happen?

4. If a car is traveling due westward with a constant speed of 20 m/s, what is the resultant force acting on it?

5. A passenger sitting in the rear of a bus claims that she was injured when the driver slammed on the brakes, causing a suitcase to come flying toward her from the front of the bus. If you were the judge in this case, what disposition would you make? Why?

6. A child tosses a ball straight up. She says that the ball is moving away from her hand because the ball feels an upward "force of the throw" as well as the gravitational force. (a) Can the "force of the throw" exceed the gravitational force? How would the ball move if it did? (b) Can the "force of the throw" be equal in magnitude to the gravitational force? Explain. (c) What strength can accurately be attributed to the "force of the throw"? Explain. (d) Why does the ball move away from the child's hand?

7. A person holds a ball in her hand. (a) Identify all the external forces acting on the ball and the Newton's third-law reaction force to each one. (b) If the ball is dropped, what force is exerted on it while it is falling? Identify the reaction force in this case. (Ignore air resistance.)

8. A spherical rubber balloon inflated with air is held stationary, with its opening, on the west side, pinched shut. (a) Describe the forces exerted by the air inside and outside the balloon on sections of the rubber. (b) After the balloon is released, it takes off toward the east, gaining speed rapidly. Explain this motion in terms of the forces now acting on the rubber. (c) Account for the motion of a skyrocket taking off from its launch pad.

9. A rubber ball is dropped onto the floor. What force causes the ball to bounce?

10. Twenty people participate in a tug-of-war. The two teams of ten people are so evenly matched that neither team wins. After the game they notice that a car is stuck in the mud. They attach the tug-of-war rope to the bumper of the car, and all the people pull on the rope. The heavy car has just moved a couple of decimeters when the rope breaks. Why did the rope break in this situation when it did not break when the same twenty people pulled on it in a tug-of-war?

11. Can an object exert a force on itself? Argue for your answer.

12. When you push on a box with a 200-N force instead of a 50-N force, you can feel that you are making a greater effort. When a table exerts a 200-N normal force instead of one of smaller magnitude, is the table really doing anything differently?

13. A weightlifter stands on a bathroom scale. He pumps a barbell up and down. What happens to the reading on the scale as he does so? **What If?** What if he is strong enough to actually *throw* the barbell upward? How does the reading on the scale vary now?

14. An athlete grips a light rope that passes over a low-friction pulley attached to the ceiling of a gym. A sack of sand precisely equal in weight to the athlete is tied to the other end of the rope. Both the sand and the athlete are initially at rest. The athlete climbs the rope, sometimes speeding up and slowing down as he does so. What happens to the sack of sand? Explain.

15. Suppose you are driving a classic car. Why should you avoid slamming on your brakes when you want to stop in the shortest possible distance? (Many modern cars have antilock brakes that avoid this problem.)

16. In Figure CQ5.16, the light, taut, unstretchable cord B joins block 1 and the larger-mass block 2. Cord A exerts a force on block 1 to make it accelerate forward. (a) How

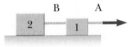

Figure CQ5.16

does the magnitude of the force exerted by cord A on block 1 compare with the magnitude of the force exerted by cord B on block 2? Is it larger, smaller, or equal? (b) How does the acceleration of block 1 compare with the acceleration (if any) of block 2? (c) Does cord B exert a force on block 1? If so, is it forward or backward? Is it larger, smaller, or equal in magnitude to the force exerted by cord B on block 2?

17. Describe two examples in which the force of friction exerted on an object is in the direction of motion of the object.

18. The mayor of a city reprimands some city employees because they will not remove the obvious sags from the cables that support the city traffic lights. What explanation can the employees give? How do you think the case will be settled in mediation?

19. Give reasons for the answers to each of the following questions: (a) Can a normal force be horizontal? (b) Can a normal force be directed vertically downward? (c) Consider a tennis ball in contact with a stationary floor and with nothing else. Can the normal force be different in magnitude from the gravitational force exerted on the ball? (d) Can the force exerted by the

floor on the ball be different in magnitude from the force the ball exerts on the floor?

20. Balancing carefully, three boys inch out onto a horizontal tree branch above a pond, each planning to dive in separately. The third boy in line notices that the branch is barely strong enough to support them. He decides to jump straight up and land back on the branch to break it, spilling all three into the pond. When he starts to carry out his plan, at what precise moment does the branch break? Explain. *Suggestion:* Pretend to be the third boy and imitate what he does in slow motion. If you are still unsure, stand on a bathroom scale and repeat the suggestion.

21. Identify action–reaction pairs in the following situations: (a) a man takes a step (b) a snowball hits a girl in the back (c) a baseball player catches a ball (d) a gust of wind strikes a window

22. As shown in Figure CQ5.22, student A, a 55-kg girl, sits on one chair with metal runners, at rest on a classroom floor. Student B, an 80-kg boy, sits on an identical chair. Both students keep their feet off the floor. A rope runs from student A's hands around a light pulley and then over her shoulder to the hands of a teacher standing on the floor behind her. The low-friction axle of the pulley is attached to a second rope held by student B. All ropes run parallel to the chair runners. (a) If student A pulls on her end of the rope, will her chair or will B's chair slide on the floor? Explain why. (b) If instead the teacher pulls on his rope end, which chair slides? Why this one? (c) If student B pulls on his rope, which chair slides? Why? (d) Now the teacher ties his end of the rope to student A's chair. Student A pulls on the end of the rope in her hands. Which chair slides and why?

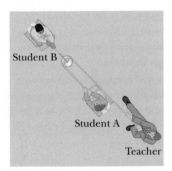

Figure CQ5.22

23. A car is moving forward slowly and is speeding up. A student claims that "the car exerts a force on itself" or that "the car's engine exerts a force on the car." (a) Argue that this idea cannot be accurate and that friction exerted by the road is the propulsive force on the car. Make your evidence and reasoning as persuasive as possible. (b) Is it static or kinetic friction? *Suggestions:* Consider a road covered with light gravel. Consider a sharp print of the tire tread on an asphalt road, obtained by coating the tread with dust.

Problems available in Access end-of-chapter problems online at **www.webassign.net**

Section 5.1 The Concept of Force
Section 5.2 Newton's First Law and Inertial Frames
Section 5.3 Mass
Section 5.4 Newton's Second Law
Section 5.5 The Gravitational Force and Weight
Section 5.6 Newton's Third Law
Problems 1–24

Section 5.7 Analysis Models Using Newton's Second Law
Problems 25–51

Section 5.8 Forces of Friction
Problems 52–71

Additional Problems
Problems 72–95

Challenge Problems
Problem 96–104

Solutions to the following Problems are available in the *Student Solutions Manual/Study Guide:*
5.3, 5.5, 5.8, 5.11, 5.19, 5.23, 5.31, 5.34, 5.37, 5.45, 5.47, 5.51, 5.61, 5.65, 5.74, 5.81, 5.85, 5.93, 5.95, and 5.101

List of Enhanced Problems

Problem Number	Targeted Feedback in Enhanced WebAssign	Analysis Model Tutorial in Enhanced WebAssign	Master It in Enhanced WebAssign	Watch It in Enhanced WebAssign
5.3	✓			✓
5.5	✓		✓	
5.11			✓	
5.16	✓		✓	
5.18	✓			✓
5.19	✓		✓	
5.22	✓			✓
5.28	✓			✓
5.29	✓		✓	
5.30	✓			✓
5.32	✓			✓
5.33	✓	✓		✓
5.40	✓	✓		✓
5.45			✓	
5.51	✓	✓	✓	
5.55	✓			✓
5.60	✓			✓
5.61	✓		✓	
5.63	✓			✓
5.65	✓	✓	✓	
5.77	✓		✓	
5.87	✓		✓	
5.95			✓	

Circular Motion and Other Applications of Newton's Laws

Kyle Busch, driver of the #18 Snickers Toyota, leads Jeff Gordon, driver of the #24 Dupont Chevrolet, during the NASCAR Sprint Cup Series Kobalt Tools 500 at the Atlanta Motor Speedway on March 9, 2008, in Hampton, Georgia. The cars travel on a banked roadway to help them undergo circular motion on the turns. *(Chris Graythen/Getty Images for NASCAR)*

In the preceding chapter, we introduced Newton's laws of motion and incorporated them into two analysis models involving linear motion. Now we discuss motion that is slightly more complicated. For example, we shall apply Newton's laws to objects traveling in circular paths. We shall also discuss motion observed from an accelerating frame of reference and motion of an object through a viscous medium. For the most part, this chapter consists of a series of examples selected to illustrate the application of Newton's laws to a variety of new circumstances.

6.1 Extending the Particle in Uniform Circular Motion Model

In Section 4.4, we discussed the analysis model of a particle in uniform circular motion, in which a particle moves with constant speed v in a circular path having a radius r. The particle experiences an acceleration that has a magnitude

$$a_c = \frac{v^2}{r}$$

118

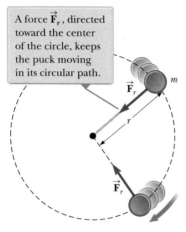

A force $\vec{\mathbf{F}}_r$, directed toward the center of the circle, keeps the puck moving in its circular path.

Figure 6.1 An overhead view of a puck moving in a circular path in a horizontal plane.

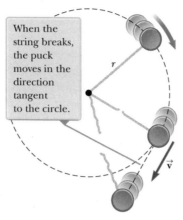

When the string breaks, the puck moves in the direction tangent to the circle.

Figure 6.2 The string holding the puck in its circular path breaks.

The acceleration is called *centripetal acceleration* because $\vec{\mathbf{a}}_c$ is directed toward the center of the circle. Furthermore, $\vec{\mathbf{a}}_c$ is *always* perpendicular to $\vec{\mathbf{v}}$. (If there were a component of acceleration parallel to $\vec{\mathbf{v}}$, the particle's speed would be changing.)

Let us now extend the particle in uniform circular motion model from Section 4.4 by incorporating the concept of force. Consider a puck of mass m that is tied to a string of length r and moves at constant speed in a horizontal, circular path as illustrated in Figure 6.1. Its weight is supported by a frictionless table, and the string is anchored to a peg at the center of the circular path of the puck. Why does the puck move in a circle? According to Newton's first law, the puck would move in a straight line if there were no force on it; the string, however, prevents motion along a straight line by exerting on the puck a radial force $\vec{\mathbf{F}}_r$ that makes it follow the circular path. This force is directed along the string toward the center of the circle as shown in Figure 6.1.

If Newton's second law is applied along the radial direction, the net force causing the centripetal acceleration can be related to the acceleration as follows:

$$\sum F = ma_c = m\frac{v^2}{r} \tag{6.1}$$

◀ **Force causing centripetal acceleration**

A force causing a centripetal acceleration acts toward the center of the circular path and causes a change in the direction of the velocity vector. If that force should vanish, the object would no longer move in its circular path; instead, it would move along a straight-line path tangent to the circle. This idea is illustrated in Figure 6.2 for the puck moving in a circular path at the end of a string in a horizontal plane. If the string breaks at some instant, the puck moves along the straight-line path that is tangent to the circle at the position of the puck at this instant.

Quick Quiz 6.1 You are riding on a Ferris wheel that is rotating with constant speed. The car in which you are riding always maintains its correct upward orientation; it does not invert. **(i)** What is the direction of the normal force on you from the seat when you are at the top of the wheel? (a) upward (b) downward (c) impossible to determine **(ii)** From the same choices, what is the direction of the net force on you when you are at the top of the wheel?

Pitfall Prevention 6.1
Direction of Travel When the String Is Cut Study Figure 6.2 very carefully. Many students (wrongly) think that the puck will move *radially* away from the center of the circle when the string is cut. The velocity of the puck is *tangent* to the circle. By Newton's first law, the puck continues to move in the same direction in which it is moving just as the force from the string disappears.

Analysis Model   Particle in Uniform Circular Motion (Extension)

Imagine a moving object that can be modeled as a particle. If it moves in a circular path of radius r at a constant speed v, it experiences a centripetal acceleration. Because the particle is accelerating, there must be a net force acting on the particle. That force is directed toward the center of the circular path and is given by

$$\sum F = ma_c = m\frac{v^2}{r} \quad (6.1)$$

Examples

- the tension in a string of constant length acting on a rock twirled in a circle
- the gravitational force acting on a planet traveling around the Sun in a perfectly circular orbit (Chapter 13)
- the magnetic force acting on a charged particle moving in a uniform magnetic field (Chapter 29)
- the electric force acting on an electron in orbit around a nucleus in the Bohr model of the hydrogen atom (Chapter 42)

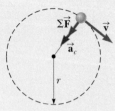

Example 6.1   The Conical Pendulum   AM

A small ball of mass m is suspended from a string of length L. The ball revolves with constant speed v in a horizontal circle of radius r as shown in Figure 6.3. (Because the string sweeps out the surface of a cone, the system is known as a *conical pendulum*.) Find an expression for v in terms of the geometry in Figure 6.3.

SOLUTION

Conceptualize Imagine the motion of the ball in Figure 6.3a and convince yourself that the string sweeps out a cone and that the ball moves in a horizontal circle.

Categorize The ball in Figure 6.3 does not accelerate vertically. Therefore, we model it as a *particle in equilibrium* in the vertical direction. It experiences a centripetal acceleration in the horizontal direction, so it is modeled as a *particle in uniform circular motion* in this direction.

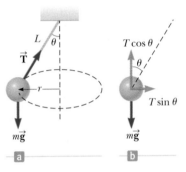

Figure 6.3 (Example 6.1) (a) A conical pendulum. The path of the ball is a horizontal circle. (b) The forces acting on the ball.

Analyze Let θ represent the angle between the string and the vertical. In the diagram of forces acting on the ball in Figure 6.3b, the force $\vec{T}$ exerted by the string on the ball is resolved into a vertical component $T\cos\theta$ and a horizontal component $T\sin\theta$ acting toward the center of the circular path.

Apply the particle in equilibrium model in the vertical direction:	$\sum F_y = T\cos\theta - mg = 0$ (1) $T\cos\theta = mg$
Use Equation 6.1 from the particle in uniform circular motion model in the horizontal direction:	(2) $\sum F_x = T\sin\theta = ma_c = \dfrac{mv^2}{r}$
Divide Equation (2) by Equation (1) and use $\sin\theta/\cos\theta = \tan\theta$:	$\tan\theta = \dfrac{v^2}{rg}$
Solve for v:	$v = \sqrt{rg\tan\theta}$
Incorporate $r = L\sin\theta$ from the geometry in Figure 6.3a:	$v = \boxed{\sqrt{Lg\sin\theta\tan\theta}}$

Finalize Notice that the speed is independent of the mass of the ball. Consider what happens when θ goes to 90° so that the string is horizontal. Because the tangent of 90° is infinite, the speed v is infinite, which tells us the string cannot possibly be horizontal. If it were, there would be no vertical component of the force $\vec{T}$ to balance the gravitational force on the ball. That is why we mentioned in regard to Figure 6.1 that the puck's weight in the figure is supported by a frictionless table.

Example 6.2 How Fast Can It Spin?

A puck of mass 0.500 kg is attached to the end of a cord 1.50 m long. The puck moves in a horizontal circle as shown in Figure 6.1. If the cord can withstand a maximum tension of 50.0 N, what is the maximum speed at which the puck can move before the cord breaks? Assume the string remains horizontal during the motion.

SOLUTION

Conceptualize It makes sense that the stronger the cord, the faster the puck can move before the cord breaks. Also, we expect a more massive puck to break the cord at a lower speed. (Imagine whirling a bowling ball on the cord!)

Categorize Because the puck moves in a circular path, we model it as a *particle in uniform circular motion*.

Analyze Incorporate the tension and the centripetal acceleration into Newton's second law as described by Equation 6.1:

$$T = m\frac{v^2}{r}$$

Solve for v:

$$(1) \quad v = \sqrt{\frac{Tr}{m}}$$

Find the maximum speed the puck can have, which corresponds to the maximum tension the string can withstand:

$$v_{max} = \sqrt{\frac{T_{max}r}{m}} = \sqrt{\frac{(50.0 \text{ N})(1.50 \text{ m})}{0.500 \text{ kg}}} = \boxed{12.2 \text{ m/s}}$$

Finalize Equation (1) shows that v increases with T and decreases with larger m, as we expected from our conceptualization of the problem.

WHAT IF? Suppose the puck moves in a circle of larger radius at the same speed v. Is the cord more likely or less likely to break?

Answer The larger radius means that the change in the direction of the velocity vector will be smaller in a given time interval. Therefore, the acceleration is smaller and the required tension in the string is smaller. As a result, the string is less likely to break when the puck travels in a circle of larger radius.

Example 6.3 What Is the Maximum Speed of the Car?

A 1 500-kg car moving on a flat, horizontal road negotiates a curve as shown in Figure 6.4a. If the radius of the curve is 35.0 m and the coefficient of static friction between the tires and dry pavement is 0.523, find the maximum speed the car can have and still make the turn successfully.

SOLUTION

Conceptualize Imagine that the curved roadway is part of a large circle so that the car is moving in a circular path.

Categorize Based on the Conceptualize step of the problem, we model the car as a *particle in uniform circular motion* in the horizontal direction. The car is not accelerating vertically, so it is modeled as a *particle in equilibrium* in the vertical direction.

Analyze Figure 6.4b shows the forces on the car. The force that enables the car to remain in its circular path is the force of static friction. (It is *static* because no slipping occurs at the point of contact between road and tires. If this force of static friction were zero—for example, if the car were on an icy road—the car would continue in a straight line and slide off the curved road.) The maximum speed v_{max} the car can have around the curve is the speed at which it is on the verge of skidding outward. At this point, the friction force has its maximum value $f_{s,max} = \mu_s n$.

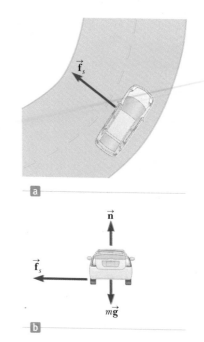

Figure 6.4 (Example 6.3) (a) The force of static friction directed toward the center of the curve keeps the car moving in a circular path. (b) The forces acting on the car.

continued

▶ **6.3** continued

Apply Equation 6.1 from the particle in uniform circular motion model in the radial direction for the maximum speed condition:

$$(1) \quad f_{s,\text{max}} = \mu_s n = m\frac{v_{\text{max}}^2}{r}$$

Apply the particle in equilibrium model to the car in the vertical direction:

$$\sum F_y = 0 \quad \rightarrow \quad n - mg = 0 \quad \rightarrow \quad n = mg$$

Solve Equation (1) for the maximum speed and substitute for n:

$$(2) \quad v_{\text{max}} = \sqrt{\frac{\mu_s n r}{m}} = \sqrt{\frac{\mu_s mg r}{m}} = \sqrt{\mu_s g r}$$

Substitute numerical values:

$$v_{\text{max}} = \sqrt{(0.523)(9.80 \text{ m/s}^2)(35.0 \text{ m})} = \boxed{13.4 \text{ m/s}}$$

Finalize This speed is equivalent to 30.0 mi/h. Therefore, if the speed limit on this roadway is higher than 30 mi/h, this roadway could benefit greatly from some banking, as in the next example! Notice that the maximum speed does not depend on the mass of the car, which is why curved highways do not need multiple speed limits to cover the various masses of vehicles using the road.

WHAT IF? Suppose a car travels this curve on a wet day and begins to skid on the curve when its speed reaches only 8.00 m/s. What can we say about the coefficient of static friction in this case?

Answer The coefficient of static friction between the tires and a wet road should be smaller than that between the tires and a dry road. This expectation is consistent with experience with driving because a skid is more likely on a wet road than a dry road.

To check our suspicion, we can solve Equation (2) for the coefficient of static friction:

$$\mu_s = \frac{v_{\text{max}}^2}{gr}$$

Substituting the numerical values gives

$$\mu_s = \frac{v_{\text{max}}^2}{gr} = \frac{(8.00 \text{ m/s})^2}{(9.80 \text{ m/s}^2)(35.0 \text{ m})} = 0.187$$

which is indeed smaller than the coefficient of 0.523 for the dry road.

Example 6.4   **The Banked Roadway**   AM

A civil engineer wishes to redesign the curved roadway in Example 6.3 in such a way that a car will not have to rely on friction to round the curve without skidding. In other words, a car moving at the designated speed can negotiate the curve even when the road is covered with ice. Such a road is usually *banked*, which means that the roadway is tilted toward the inside of the curve as seen in the opening photograph for this chapter. Suppose the designated speed for the road is to be 13.4 m/s (30.0 mi/h) and the radius of the curve is 35.0 m. At what angle should the curve be banked?

SOLUTION

Conceptualize The difference between this example and Example 6.3 is that the car is no longer moving on a flat roadway. Figure 6.5 shows the banked roadway, with the center of the circular path of the car far to the left of the figure. Notice that the horizontal component of the normal force participates in causing the car's centripetal acceleration.

Categorize As in Example 6.3, the car is modeled as a *particle in equilibrium* in the vertical direction and a *particle in uniform circular motion* in the horizontal direction.

Analyze On a level (unbanked) road, the force that causes the centripetal acceleration is the force of static friction between tires and the road as we saw in the preceding example. If the road is banked at an angle θ as in Figure 6.5, however, the

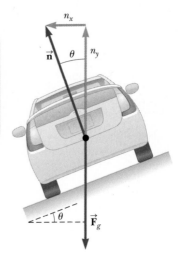

Figure 6.5 (Example 6.4) A car moves into the page and is rounding a curve on a road banked at an angle θ to the horizontal. When friction is neglected, the force that causes the centripetal acceleration and keeps the car moving in its circular path is the horizontal component of the normal force.

▶ **6.4** continued

normal force $\vec{\mathbf{n}}$ has a horizontal component toward the center of the curve. Because the road is to be designed so that the force of static friction is zero, the component $n_x = n \sin \theta$ is the only force that causes the centripetal acceleration.

Write Newton's second law for the car in the radial direction, which is the x direction:

$$(1) \quad \sum F_r = n \sin \theta = \frac{mv^2}{r}$$

Apply the particle in equilibrium model to the car in the vertical direction:

$$\sum F_y = n \cos \theta - mg = 0$$
$$(2) \quad n \cos \theta = mg$$

Divide Equation (1) by Equation (2):

$$(3) \quad \tan \theta = \frac{v^2}{rg}$$

Solve for the angle θ:

$$\theta = \tan^{-1}\left[\frac{(13.4 \text{ m/s})^2}{(35.0 \text{ m})(9.80 \text{ m/s}^2)}\right] = \boxed{27.6°}$$

Finalize Equation (3) shows that the banking angle is independent of the mass of the vehicle negotiating the curve. If a car rounds the curve at a speed less than 13.4 m/s, the centripetal acceleration decreases. Therefore, the normal force, which is unchanged, is sufficient to cause *two* accelerations: the lower centripetal acceleration and an acceleration of the car down the inclined roadway. Consequently, an additional friction force parallel to the roadway and upward is needed to keep the car from sliding down the bank (to the left in Fig. 6.5). Similarly, a driver attempting to negotiate the curve at a speed greater than 13.4 m/s has to depend on friction to keep from sliding up the bank (to the right in Fig. 6.5).

WHAT IF? Imagine that this same roadway were built on Mars in the future to connect different colony centers. Could it be traveled at the same speed?

Answer The reduced gravitational force on Mars would mean that the car is not pressed as tightly to the roadway. The reduced normal force results in a smaller component of the normal force toward the center of the circle. This smaller component would not be sufficient to provide the centripetal acceleration associated with the original speed. The centripetal acceleration must be reduced, which can be done by reducing the speed v.

Mathematically, notice that Equation (3) shows that the speed v is proportional to the square root of g for a roadway of fixed radius r banked at a fixed angle θ. Therefore, if g is smaller, as it is on Mars, the speed v with which the roadway can be safely traveled is also smaller.

Example 6.5 **Riding the Ferris Wheel** **AM**

A child of mass m rides on a Ferris wheel as shown in Figure 6.6a. The child moves in a vertical circle of radius 10.0 m at a constant speed of 3.00 m/s.

(A) Determine the force exerted by the seat on the child at the bottom of the ride. Express your answer in terms of the weight of the child, mg.

SOLUTION

Conceptualize Look carefully at Figure 6.6a. Based on experiences you may have had on a Ferris wheel or driving over small hills on a roadway, you would expect to feel lighter at the top of the path. Similarly, you would expect to feel heavier at the bottom of the path. At both the bottom of the path and the top, the normal and gravitational forces on the child act in *opposite* directions. The vector sum of these two forces gives a force of constant magnitude that keeps the child moving in a circular path at a constant speed. To yield net force vectors with the same magnitude, the normal force at the bottom must be greater than that at the top.

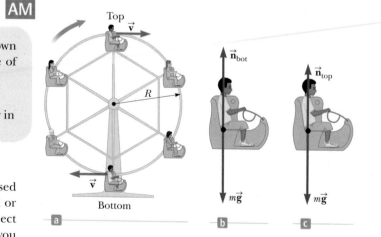

Figure 6.6 (Example 6.5) (a) A child rides on a Ferris wheel. (b) The forces acting on the child at the bottom of the path. (c) The forces acting on the child at the top of the path.

continued

▶ **6.5** continued

Categorize Because the speed of the child is constant, we can categorize this problem as one involving a *particle* (the child) *in uniform circular motion*, complicated by the gravitational force acting at all times on the child.

Analyze We draw a diagram of forces acting on the child at the bottom of the ride as shown in Figure 6.6b. The only forces acting on him are the downward gravitational force $\vec{F}_g = m\vec{g}$ and the upward force $\vec{n}_{bot}$ exerted by the seat. The net upward force on the child that provides his centripetal acceleration has a magnitude $n_{bot} - mg$.

Using the particle in uniform circular motion model, apply Newton's second law to the child in the radial direction when he is at the bottom of the ride:

$$\sum F = n_{bot} - mg = m\frac{v^2}{r}$$

Solve for the force exerted by the seat on the child:

$$n_{bot} = mg + m\frac{v^2}{r} = mg\left(1 + \frac{v^2}{rg}\right)$$

Substitute numerical values given for the speed and radius:

$$n_{bot} = mg\left[1 + \frac{(3.00 \text{ m/s})^2}{(10.0 \text{ m})(9.80 \text{ m/s}^2)}\right]$$

$$= 1.09\, mg$$

Hence, the magnitude of the force $\vec{n}_{bot}$ exerted by the seat on the child is *greater* than the weight of the child by a factor of 1.09. So, the child experiences an apparent weight that is greater than his true weight by a factor of 1.09.

(B) Determine the force exerted by the seat on the child at the top of the ride.

SOLUTION

Analyze The diagram of forces acting on the child at the top of the ride is shown in Figure 6.6c. The net downward force that provides the centripetal acceleration has a magnitude $mg - n_{top}$.

Apply Newton's second law to the child at this position:

$$\sum F = mg - n_{top} = m\frac{v^2}{r}$$

Solve for the force exerted by the seat on the child:

$$n_{top} = mg - m\frac{v^2}{r} = mg\left(1 - \frac{v^2}{rg}\right)$$

Substitute numerical values:

$$n_{top} = mg\left[1 - \frac{(3.00 \text{ m/s})^2}{(10.0 \text{ m})(9.80 \text{ m/s}^2)}\right]$$

$$= 0.908\, mg$$

In this case, the magnitude of the force exerted by the seat on the child is *less* than his true weight by a factor of 0.908, and the child feels lighter.

Finalize The variations in the normal force are consistent with our prediction in the Conceptualize step of the problem.

WHAT IF? Suppose a defect in the Ferris wheel mechanism causes the speed of the child to increase to 10.0 m/s. What does the child experience at the top of the ride in this case?

Answer If the calculation above is performed with $v = 10.0$ m/s, the magnitude of the normal force at the top of the ride is negative, which is impossible. We interpret it to mean that the required centripetal acceleration of the child is larger than that due to gravity. As a result, the child will lose contact with the seat and will only stay in his circular path if there is a safety bar or a seat belt that provides a downward force on him to keep him in his seat. At the bottom of the ride, the normal force is 2.02 *mg*, which would be uncomfortable.

6.2 Nonuniform Circular Motion

In Chapter 4, we found that if a particle moves with varying speed in a circular path, there is, in addition to the radial component of acceleration, a tangential component having magnitude $|dv/dt|$. Therefore, the force acting on the particle

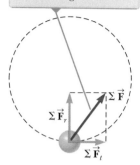

The net force exerted on the particle is the vector sum of the radial force and the tangential force.

Figure 6.7 When the net force acting on a particle moving in a circular path has a tangential component ΣF_t, the particle's speed changes.

must also have a tangential and a radial component. Because the total acceleration is $\vec{a} = \vec{a}_r + \vec{a}_t$, the total force exerted on the particle is $\Sigma \vec{F} = \Sigma \vec{F}_r + \Sigma \vec{F}_t$ as shown in Figure 6.7. (We express the radial and tangential forces as net forces with the summation notation because each force could consist of multiple forces that combine.) The vector $\Sigma \vec{F}_r$ is directed toward the center of the circle and is responsible for the centripetal acceleration. The vector $\Sigma \vec{F}_t$ tangent to the circle is responsible for the tangential acceleration, which represents a change in the particle's speed with time.

Q uick Quiz 6.2 A bead slides at constant speed along a curved wire lying on a horizontal surface as shown in Figure 6.8. **(a)** Draw the vectors representing the force exerted by the wire on the bead at points Ⓐ, Ⓑ, and Ⓒ. **(b)** Suppose the bead in Figure 6.8 speeds up with constant tangential acceleration as it moves toward the right. Draw the vectors representing the force on the bead at points Ⓐ, Ⓑ, and Ⓒ.

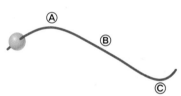

Figure 6.8 (Quick Quiz 6.2) A bead slides along a curved wire.

Example 6.6 **Keep Your Eye on the Ball** **AM**

A small sphere of mass m is attached to the end of a cord of length R and set into motion in a *vertical* circle about a fixed point O as illustrated in Figure 6.9. Determine the tangential acceleration of the sphere and the tension in the cord at any instant when the speed of the sphere is v and the cord makes an angle θ with the vertical.

SOLUTION

Conceptualize Compare the motion of the sphere in Figure 6.9 with that of the child in Figure 6.6a associated with Example 6.5. Both objects travel in a circular path. Unlike the child in Example 6.5, however, the speed of the sphere is *not* uniform in this example because, at most points along the path, a tangential component of acceleration arises from the gravitational force exerted on the sphere.

Categorize We model the sphere as a *particle under a net force* and moving in a circular path, but it is not a particle in *uniform* circular motion. We need to use the techniques discussed in this section on nonuniform circular motion.

Analyze From the force diagram in Figure 6.9, we see that the only forces acting on the sphere are the gravitational force

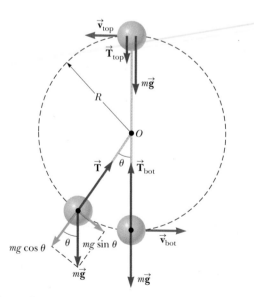

Figure 6.9 (Example 6.6) The forces acting on a sphere of mass m connected to a cord of length R and rotating in a vertical circle centered at O. Forces acting on the sphere are shown when the sphere is at the top and bottom of the circle and at an arbitrary location.

▶ **6.6** continued

$\vec{F}_g = m\vec{g}$ exerted by the Earth and the force $\vec{T}$ exerted by the cord. We resolve $\vec{F}_g$ into a tangential component $mg \sin\theta$ and a radial component $mg \cos\theta$.

From the particle under a net force model, apply Newton's second law to the sphere in the tangential direction:

$$\sum F_t = mg \sin\theta = ma_t$$
$$a_t = \boxed{g \sin\theta}$$

Apply Newton's second law to the forces acting on the sphere in the radial direction, noting that both $\vec{T}$ and $\vec{a}_r$ are directed toward O. As noted in Section 4.5, we can use Equation 4.14 for the centripetal acceleration of a particle even when it moves in a circular path in nonuniform motion:

$$\sum F_r = T - mg \cos\theta = \frac{mv^2}{R}$$
$$T = \boxed{mg\left(\frac{v^2}{Rg} + \cos\theta\right)}$$

Finalize Let us evaluate this result at the top and bottom of the circular path (Fig. 6.9):

$$T_{top} = mg\left(\frac{v_{top}^2}{Rg} - 1\right) \qquad T_{bot} = mg\left(\frac{v_{bot}^2}{Rg} + 1\right)$$

These results have similar mathematical forms as those for the normal forces n_{top} and n_{bot} on the child in Example 6.5, which is consistent with the normal force on the child playing a similar physical role in Example 6.5 as the tension in the string plays in this example. Keep in mind, however, that the normal force $\vec{n}$ on the child in Example 6.5 is always upward, whereas the force $\vec{T}$ in this example changes direction because it must always point inward along the string. Also note that v in the expressions above varies for different positions of the sphere, as indicated by the subscripts, whereas v in Example 6.5 is constant.

WHAT IF?   What if the ball is set in motion with a slower speed?

(A) What speed would the ball have as it passes over the top of the circle if the tension in the cord goes to zero instantaneously at this point?

Answer Let us set the tension equal to zero in the expression for T_{top}:

$$0 = mg\left(\frac{v_{top}^2}{Rg} - 1\right) \quad \to \quad v_{top} = \sqrt{gR}$$

(B) What if the ball is set in motion such that the speed at the top is less than this value? What happens?

Answer In this case, the ball never reaches the top of the circle. At some point on the way up, the tension in the string goes to zero and the ball becomes a projectile. It follows a segment of a parabolic path over the top of its motion, rejoining the circular path on the other side when the tension becomes nonzero again.

6.3 Motion in Accelerated Frames

Newton's laws of motion, which we introduced in Chapter 5, describe observations that are made in an inertial frame of reference. In this section, we analyze how Newton's laws are applied by an observer in a noninertial frame of reference, that is, one that is accelerating. For example, recall the discussion of the air hockey table on a train in Section 5.2. The train moving at constant velocity represents an inertial frame. An observer on the train sees the puck at rest remain at rest, and Newton's first law appears to be obeyed. The accelerating train is not an inertial frame. According to you as the observer on this train, there appears to be no force on the puck, yet it accelerates from rest toward the back of the train, appearing to violate Newton's first law. This property is a general property of observations made in noninertial frames: there appear to be unexplained accelerations of objects that are not "fastened" to the frame. Newton's first law is not violated, of course. It only appears to be violated because of observations made from a noninertial frame.

On the accelerating train, as you watch the puck accelerating toward the back of the train, you might conclude based on your belief in Newton's second law that a force has acted on the puck to cause it to accelerate. We call an apparent force such as this one a **fictitious force** because it is not a real force and is due only to observations made in an accelerated reference frame. A fictitious force appears to act on an object in the same way as a real force. Real forces are always interactions between two objects, however, and you cannot identify a second object for a fictitious force. (What second object is interacting with the puck to cause it to accelerate?) In general, simple fictitious forces appear to act in the direction *opposite* that of the acceleration of the noninertial frame. For example, the train accelerates forward and there appears to be a fictitious force causing the puck to slide toward the back of the train.

The train example describes a fictitious force due to a change in the train's speed. Another fictitious force is due to the change in the *direction* of the velocity vector. To understand the motion of a system that is noninertial because of a change in direction, consider a car traveling along a highway at a high speed and approaching a curved exit ramp on the left as shown in Figure 6.10a. As the car takes the sharp left turn on the ramp, a person sitting in the passenger seat leans or slides to the right and hits the door. At that point the force exerted by the door on the passenger keeps her from being ejected from the car. What causes her to move toward the door? A popular but incorrect explanation is that a force acting toward the right in Figure 6.10b pushes the passenger outward from the center of the circular path. Although often called the "centrifugal force," it is a fictitious force. The car represents a noninertial reference frame that has a centripetal acceleration toward the center of its circular path. As a result, the passenger feels an apparent force which is outward from the center of the circular path, or to the right in Figure 6.10b, in the direction opposite that of the acceleration.

Let us address this phenomenon in terms of Newton's laws. Before the car enters the ramp, the passenger is moving in a straight-line path. As the car enters the ramp and travels a curved path, the passenger tends to move along the original straight-line path, which is in accordance with Newton's first law: the natural tendency of an object is to continue moving in a straight line. If a sufficiently large force (toward the center of curvature) acts on the passenger as in Figure 6.10c, however, she moves in a curved path along with the car. This force is the force of friction between her and the car seat. If this friction force is not large enough, the seat follows a curved path while the passenger tends to continue in the straight-line path of the car before the car began the turn. Therefore, from the point of view of an observer in the car, the passenger leans or slides to the right relative to the seat. Eventually, she encounters the door, which provides a force large enough to enable her to follow the same curved path as the car.

Another interesting fictitious force is the "Coriolis force." It is an apparent force caused by changing the radial position of an object in a rotating coordinate system.

For example, suppose you and a friend are on opposite sides of a rotating circular platform and you decide to throw a baseball to your friend. Figure 6.11a on page 128 represents what an observer would see if the ball is viewed while the observer is hovering at rest above the rotating platform. According to this observer, who is in an inertial frame, the ball follows a straight line as it must according to Newton's first law. At $t = 0$ you throw the ball toward your friend, but by the time t_f when the ball has crossed the platform, your friend has moved to a new position and can't catch the ball. Now, however, consider the situation from your friend's viewpoint. Your friend is in a noninertial reference frame because he is undergoing a centripetal acceleration relative to the inertial frame of the Earth's surface. He starts off seeing the baseball coming toward him, but as it crosses the platform, it veers to one side as shown in Figure 6.11b. Therefore, your friend on the rotating platform states that the ball does not obey Newton's first law and claims that a sideways force is causing the ball to follow a curved path. This fictitious force is called the Coriolis force.

Fictitious forces may not be real forces, but they can have real effects. An object on your dashboard *really* slides off if you press the accelerator of your car. As you ride on a merry-go-round, you feel pushed toward the outside as if due to the fictitious "centrifugal force." You are likely to fall over and injure yourself due to the

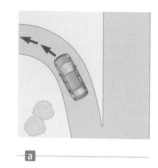

From the passenger's frame of reference, a force appears to push her toward the right door, but it is a fictitious force.

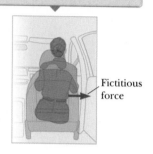

Relative to the reference frame of the Earth, the car seat applies a real force (friction) toward the left on the passenger, causing her to change direction along with the rest of the car.

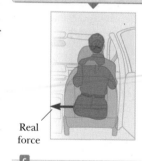

Figure 6.10 (a) A car approaching a curved exit ramp. What causes a passenger in the front seat to move toward the right-hand door? (b) Passenger's frame of reference. (c) Reference frame of the Earth.

By the time t_f that the ball arrives at the other side of the platform, your friend is no longer there to catch it. According to this observer, the ball follows a straight-line path, consistent with Newton's laws.

From your friend's point of view, the ball veers to one side during its flight. Your friend introduces a fictitious force to explain this deviation from the expected path.

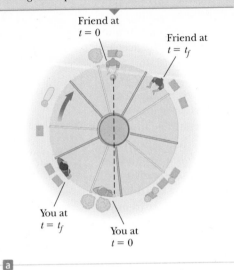

Friend at
$t = 0$

Friend at
$t = t_f$

You at
$t = t_f$

You at
$t = 0$

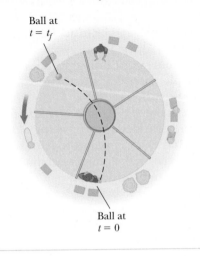

Ball at
$t = t_f$

Ball at
$t = 0$

a

b

Figure 6.11   You and your friend stand at the edge of a rotating circular platform. You throw the ball at $t = 0$ in the direction of your friend. (a) Overhead view observed by someone in an inertial reference frame attached to the Earth. The ground appears stationary, and the platform rotates clockwise. (b) Overhead view observed by someone in an inertial reference frame attached to the platform. The platform appears stationary, and the ground rotates counterclockwise.

Pitfall Prevention 6.2
Centrifugal Force   The commonly heard phrase "centrifugal force" is described as a force pulling *outward* on an object moving in a circular path. If you are feeling a "centrifugal force" on a rotating carnival ride, what is the other object with which you are interacting? You cannot identify another object because it is a fictitious force that occurs when you are in a noninertial reference frame.

Coriolis force if you walk along a radial line while a merry-go-round rotates. (One of the authors did so and suffered a separation of the ligaments from his ribs when he fell over.) The Coriolis force due to the rotation of the Earth is responsible for rotations of hurricanes and for large-scale ocean currents.

Quick **Quiz** 6.3   Consider the passenger in the car making a left turn in Figure 6.10. Which of the following is correct about forces in the horizontal direction if she is making contact with the right-hand door? **(a)** The passenger is in equilibrium between real forces acting to the right and real forces acting to the left. **(b)** The passenger is subject only to real forces acting to the right. **(c)** The passenger is subject only to real forces acting to the left. **(d)** None of those statements is true.

Example 6.7   **Fictitious Forces in Linear Motion**

A small sphere of mass m hangs by a cord from the ceiling of a boxcar that is accelerating to the right as shown in Figure 6.12. Both the inertial observer on the ground in Figure 6.12a and the noninertial observer on the train in Figure 6.12b agree that the cord makes an angle θ with respect to the vertical. The noninertial observer claims that a force, which we know to be fictitious, causes the observed deviation of the cord from the vertical. How is the magnitude of this force related to the boxcar's acceleration measured by the inertial observer in Figure 6.12a?

SOLUTION

Conceptualize   Place yourself in the role of each of the two observers in Figure 6.12. As the inertial observer on the ground, you see the boxcar accelerating and know that the deviation of the cord is due to this acceleration. As the noninertial observer on the boxcar, imagine that you ignore any effects of the car's motion so that you are not aware of its acceleration. Because you are unaware of this acceleration, you claim that a force is pushing sideways on the sphere to cause the deviation of the cord from the vertical. To make the conceptualization more real, try running from rest while holding a hanging object on a string and notice that the string is at an angle to the vertical while you are accelerating, as if a force is pushing the object backward.

▶ **6.7** c o n t i n u e d

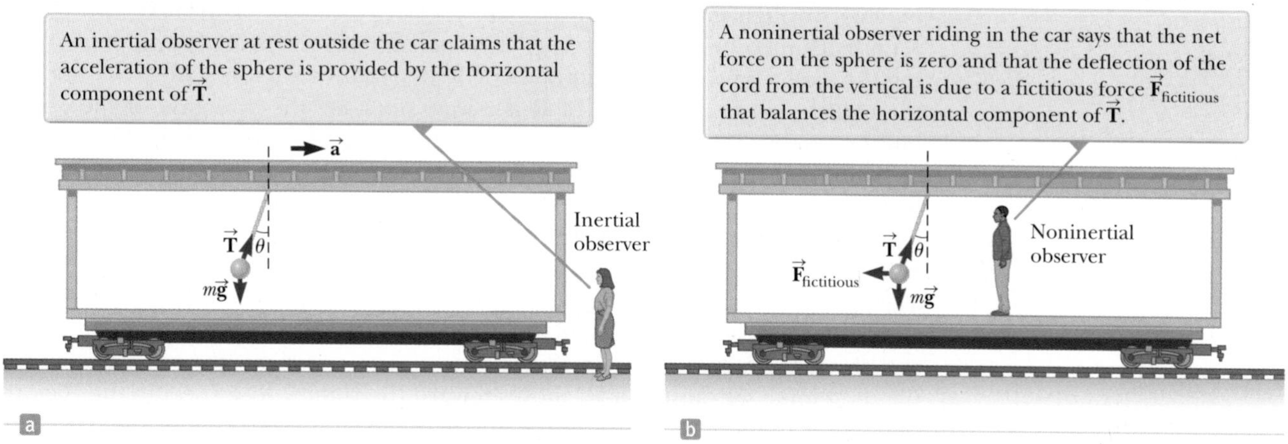

An inertial observer at rest outside the car claims that the acceleration of the sphere is provided by the horizontal component of $\vec{\mathbf{T}}$.

A noninertial observer riding in the car says that the net force on the sphere is zero and that the deflection of the cord from the vertical is due to a fictitious force $\vec{\mathbf{F}}_{\text{fictitious}}$ that balances the horizontal component of $\vec{\mathbf{T}}$.

Figure 6.12 (Example 6.7) A small sphere suspended from the ceiling of a boxcar accelerating to the right is deflected as shown.

Categorize For the inertial observer, we model the sphere as a *particle under a net force* in the horizontal direction and a *particle in equilibrium* in the vertical direction. For the noninertial observer, the sphere is modeled as a *particle in equilibrium* in both directions.

Analyze According to the inertial observer at rest (Fig. 6.12a), the forces on the sphere are the force $\vec{\mathbf{T}}$ exerted by the cord and the gravitational force. The inertial observer concludes that the sphere's acceleration is the same as that of the boxcar and that this acceleration is provided by the horizontal component of $\vec{\mathbf{T}}$.

For this observer, apply the particle under a net force and particle in equilibrium models:

$$\text{Inertial observer} \quad \begin{cases} (1) \quad \sum F_x = T \sin \theta = ma \\ (2) \quad \sum F_y = T \cos \theta - mg = 0 \end{cases}$$

According to the noninertial observer riding in the car (Fig. 6.12b), the cord also makes an angle θ with the vertical; to that observer, however, the sphere is at rest and so its acceleration is zero. Therefore, the noninertial observer introduces a force (which we know to be fictitious) in the horizontal direction to balance the horizontal component of $\vec{\mathbf{T}}$ and claims that the net force on the sphere is zero.

Apply the particle in equilibrium model for this observer in both directions:

$$\text{Noninertial observer} \quad \begin{cases} \sum F_x' = T \sin \theta - F_{\text{fictitious}} = 0 \\ \sum F_y' = T \cos \theta - mg = 0 \end{cases}$$

These expressions are equivalent to Equations (1) and (2) if $F_{\text{fictitious}} = ma$, where a is the acceleration according to the inertial observer.

Finalize If we make this substitution in the equation for $\sum F_x'$ above, we obtain the same mathematical results as the inertial observer. The physical interpretation of the cord's deflection, however, differs in the two frames of reference.

WHAT IF? Suppose the inertial observer wants to measure the acceleration of the train by means of the pendulum (the sphere hanging from the cord). How could she do so?

Answer Our intuition tells us that the angle θ the cord makes with the vertical should increase as the acceleration increases. By solving Equations (1) and (2) simultaneously for a, we find that $a = g \tan \theta$. Therefore, the inertial observer can determine the magnitude of the car's acceleration by measuring the angle θ and using that relationship. Because the deflection of the cord from the vertical serves as a measure of acceleration, *a simple pendulum can be used as an accelerometer.*

6.4 Motion in the Presence of Resistive Forces

In Chapter 5, we described the force of kinetic friction exerted on an object moving on some surface. We completely ignored any interaction between the object and the medium through which it moves. Now consider the effect of that medium, which can be either a liquid or a gas. The medium exerts a **resistive force** $\vec{R}$ on the object moving through it. Some examples are the air resistance associated with moving vehicles (sometimes called *air drag*) and the viscous forces that act on objects moving through a liquid. The magnitude of $\vec{R}$ depends on factors such as the speed of the object, and the direction of $\vec{R}$ is always opposite the direction of the object's motion relative to the medium. This direction may or may not be in the direction opposite the object's velocity according to the observer. For example, if a marble is dropped into a bottle of shampoo, the marble moves downward and the resistive force is upward, resisting the falling of the marble. In contrast, imagine the moment at which there is no wind and you are looking at a flag hanging limply on a flagpole. When a breeze begins to blow toward the right, the flag moves toward the right. In this case, the drag force on the flag from the moving air is to the right and the motion of the flag in response is also to the right, the *same* direction as the drag force. Because the air moves toward the right with respect to the flag, the flag moves to the left relative to the air. Therefore, the direction of the drag force is indeed opposite to the direction of the motion of the flag with respect to the air!

The magnitude of the resistive force can depend on speed in a complex way, and here we consider only two simplified models. In the first model, we assume the resistive force is proportional to the velocity of the moving object; this model is valid for objects falling slowly through a liquid and for very small objects, such as dust particles, moving through air. In the second model, we assume a resistive force that is proportional to the square of the speed of the moving object; large objects, such as skydivers moving through air in free fall, experience such a force.

Model 1: Resistive Force Proportional to Object Velocity

If we model the resistive force acting on an object moving through a liquid or gas as proportional to the object's velocity, the resistive force can be expressed as

$$\vec{R} = -b\vec{v} \qquad (6.2)$$

where b is a constant whose value depends on the properties of the medium and on the shape and dimensions of the object and $\vec{v}$ is the velocity of the object relative to the medium. The negative sign indicates that $\vec{R}$ is in the opposite direction to $\vec{v}$.

Consider a small sphere of mass m released from rest in a liquid as in Figure 6.13a. Assuming the only forces acting on the sphere are the resistive force $\vec{R} = -b\vec{v}$ and the gravitational force $\vec{F}_g$, let us describe its motion.[1] We model the sphere as a par-

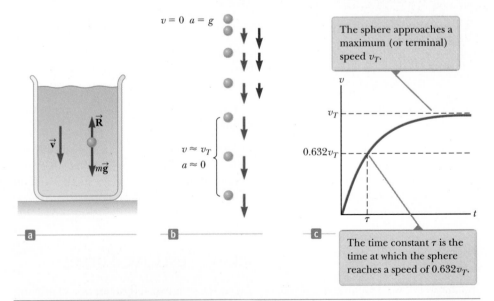

Figure 6.13 (a) A small sphere falling through a liquid. (b) A motion diagram of the sphere as it falls. Velocity vectors (red) and acceleration vectors (violet) are shown for each image after the first one. (c) A speed–time graph for the sphere.

[1] A *buoyant force* is also acting on the submerged object. This force is constant, and its magnitude is equal to the weight of the displaced liquid. This force can be modeled by changing the apparent weight of the sphere by a constant factor, so we will ignore the force here. We will discuss buoyant forces in Chapter 14.

ticle under a net force. Applying Newton's second law to the vertical motion of the sphere and choosing the downward direction to be positive, we obtain

$$\sum F_y = ma \quad \rightarrow \quad mg - bv = ma \tag{6.3}$$

where the acceleration of the sphere is downward. Noting that the acceleration a is equal to dv/dt gives

$$\frac{dv}{dt} = g - \frac{b}{m}v \tag{6.4}$$

This equation is called a *differential equation,* and the methods of solving it may not be familiar to you as yet. Notice, however, that initially when $v = 0$, the magnitude of the resistive force is also zero and the acceleration of the sphere is simply g. As t increases, the magnitude of the resistive force increases and the acceleration decreases. The acceleration approaches zero when the magnitude of the resistive force approaches the sphere's weight so that the net force on the sphere is zero. In this situation, the speed of the sphere approaches its **terminal speed** v_T.

◀ **Terminal speed**

The terminal speed is obtained from Equation 6.4 by setting $dv/dt = 0$, which gives

$$mg - bv_T = 0 \quad \text{or} \quad v_T = \frac{mg}{b} \tag{6.5}$$

Because you may not be familiar with differential equations yet, we won't show the details of the process that gives the expression for v for all times t. If $v = 0$ at $t = 0$, this expression is

$$v = \frac{mg}{b}(1 - e^{-bt/m}) = v_T(1 - e^{-t/\tau}) \tag{6.6}$$

This function is plotted in Figure 6.13c. The symbol e represents the base of the natural logarithm and is also called *Euler's number:* $e = 2.718\,28$. The **time constant** $\tau = m/b$ (Greek letter tau) is the time at which the sphere released from rest at $t = 0$ reaches 63.2% of its terminal speed; when $t = \tau$, Equation 6.6 yields $v = 0.632v_T$. (The number 0.632 is $1 - e^{-1}$.)

We can check that Equation 6.6 is a solution to Equation 6.4 by direct differentiation:

$$\frac{dv}{dt} = \frac{d}{dt}\left[\frac{mg}{b}(1 - e^{-bt/m})\right] = \frac{mg}{b}\left(0 + \frac{b}{m}e^{-bt/m}\right) = ge^{-bt/m}$$

(See Appendix Table B.4 for the derivative of e raised to some power.) Substituting into Equation 6.4 both this expression for dv/dt and the expression for v given by Equation 6.6 shows that our solution satisfies the differential equation.

Example 6.8 Sphere Falling in Oil AM

A small sphere of mass 2.00 g is released from rest in a large vessel filled with oil, where it experiences a resistive force proportional to its speed. The sphere reaches a terminal speed of 5.00 cm/s. Determine the time constant τ and the time at which the sphere reaches 90.0% of its terminal speed.

SOLUTION

Conceptualize With the help of Figure 6.13, imagine dropping the sphere into the oil and watching it sink to the bottom of the vessel. If you have some thick shampoo in a clear container, drop a marble in it and observe the motion of the marble.

Categorize We model the sphere as a *particle under a net force,* with one of the forces being a resistive force that depends on the speed of the sphere. This model leads to the result in Equation 6.5.

Analyze From Equation 6.5, evaluate the coefficient b: $\qquad b = \dfrac{mg}{v_T}$

continued

▶ **6.8** continued

Evaluate the time constant τ:

$$\tau = \frac{m}{b} = m\left(\frac{v_T}{mg}\right) = \frac{v_T}{g}$$

Substitute numerical values:

$$\tau = \frac{5.00 \text{ cm/s}}{980 \text{ cm/s}^2} = 5.10 \times 10^{-3} \text{ s}$$

Find the time t at which the sphere reaches a speed of $0.900v_T$ by setting $v = 0.900v_T$ in Equation 6.6 and solving for t:

$$0.900v_T = v_T(1 - e^{-t/\tau})$$

$$1 - e^{-t/\tau} = 0.900$$

$$e^{-t/\tau} = 0.100$$

$$-\frac{t}{\tau} = \ln(0.100) = -2.30$$

$$t = 2.30\tau = 2.30(5.10 \times 10^{-3} \text{ s}) = 11.7 \times 10^{-3} \text{ s}$$

$$= \boxed{11.7 \text{ ms}}$$

Finalize The sphere reaches 90.0% of its terminal speed in a very short time interval. You should have also seen this behavior if you performed the activity with the marble and the shampoo. Because of the short time interval required to reach terminal velocity, you may not have noticed the time interval at all. The marble may have appeared to immediately begin moving through the shampoo at a constant velocity.

Model 2: Resistive Force Proportional to Object Speed Squared

For objects moving at high speeds through air, such as airplanes, skydivers, cars, and baseballs, the resistive force is reasonably well modeled as proportional to the square of the speed. In these situations, the magnitude of the resistive force can be expressed as

$$R = \tfrac{1}{2}D\rho Av^2 \tag{6.7}$$

where D is a dimensionless empirical quantity called the *drag coefficient*, ρ is the density of air, and A is the cross-sectional area of the moving object measured in a plane perpendicular to its velocity. The drag coefficient has a value of about 0.5 for spherical objects but can have a value as great as 2 for irregularly shaped objects.

Let us analyze the motion of a falling object subject to an upward air resistive force of magnitude $R = \tfrac{1}{2}D\rho Av^2$. Suppose an object of mass m is released from rest. As Figure 6.14 shows, the object experiences two external forces:[2] the downward gravitational force $\vec{F}_g = m\vec{g}$ and the upward resistive force $\vec{R}$. Hence, the magnitude of the net force is

$$\sum F = mg - \tfrac{1}{2}D\rho Av^2 \tag{6.8}$$

where we have taken downward to be the positive vertical direction. Modeling the object as a particle under a net force, with the net force given by Equation 6.8, we find that the object has a downward acceleration of magnitude

$$a = g - \left(\frac{D\rho A}{2m}\right)v^2 \tag{6.9}$$

We can calculate the terminal speed v_T by noticing that when the gravitational force is balanced by the resistive force, the net force on the object is zero and therefore its acceleration is zero. Setting $a = 0$ in Equation 6.9 gives

$$g - \left(\frac{D\rho A}{2m}\right)v_T^2 = 0$$

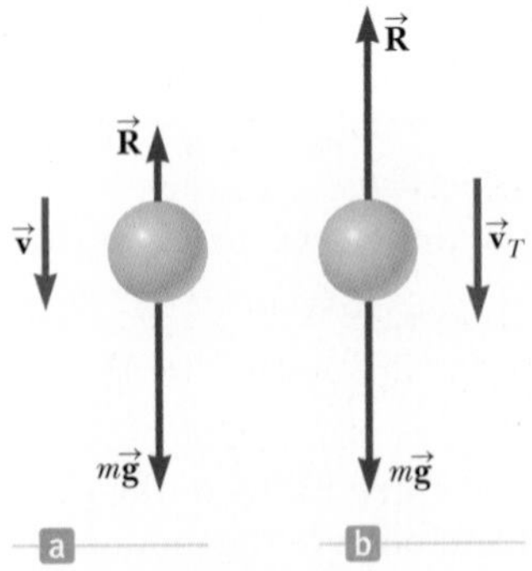

Figure 6.14 (a) An object falling through air experiences a resistive force $\vec{R}$ and a gravitational force $\vec{F}_g = m\vec{g}$. (b) The object reaches terminal speed when the net force acting on it is zero, that is, when $\vec{R} = -\vec{F}_g$ or $R = mg$.

[2]As with Model 1, there is also an upward buoyant force that we neglect.

Table 6.1	**Terminal Speed for Various Objects Falling Through Air**		
Object	Mass (kg)	Cross-Sectional Area (m²)	v_T (m/s)
Skydiver	75	0.70	60
Baseball (radius 3.7 cm)	0.145	4.2×10^{-3}	43
Golf ball (radius 2.1 cm)	0.046	1.4×10^{-3}	44
Hailstone (radius 0.50 cm)	4.8×10^{-4}	7.9×10^{-5}	14
Raindrop (radius 0.20 cm)	3.4×10^{-5}	1.3×10^{-5}	9.0

so

$$v_T = \sqrt{\frac{2mg}{D\rho A}} \tag{6.10}$$

Table 6.1 lists the terminal speeds for several objects falling through air.

Quick Quiz 6.4 A baseball and a basketball, having the same mass, are dropped through air from rest such that their bottoms are initially at the same height above the ground, on the order of 1 m or more. Which one strikes the ground first? **(a)** The baseball strikes the ground first. **(b)** The basketball strikes the ground first. **(c)** Both strike the ground at the same time.

Conceptual Example 6.9	The Skysurfer

Consider a skysurfer (Fig. 6.15) who jumps from a plane with his feet attached firmly to his surfboard, does some tricks, and then opens his parachute. Describe the forces acting on him during these maneuvers.

SOLUTION

When the surfer first steps out of the plane, he has no vertical velocity. The downward gravitational force causes him to accelerate toward the ground. As his downward speed increases, so does the upward resistive force exerted by the air on his body and the board. This upward force reduces their acceleration, and so their speed increases more slowly. Eventually, they are going so fast that the upward resistive force matches the downward gravitational force. Now the net force is zero and they no longer accelerate, but instead reach their terminal speed. At some point after reaching terminal speed, he opens his parachute, resulting in a drastic increase in the upward resistive force. The net force (and therefore the acceleration) is now upward, in the direction opposite the direction of the velocity. The downward velocity therefore decreases rapidly, and the resistive force on the parachute also decreases. Eventually, the upward resistive force and the downward gravitational force balance each other again and a much smaller terminal speed is reached, permitting a safe landing.

Figure 6.15 (Conceptual Example 6.9) A skysurfer.

Oliver Furrer/Jupiter Images

(Contrary to popular belief, the velocity vector of a skydiver never points upward. You may have seen a video in which a skydiver appears to "rocket" upward once the parachute opens. In fact, what happens is that the skydiver slows down but the person holding the camera continues falling at high speed.)

Example 6.10	Falling Coffee Filters

The dependence of resistive force on the square of the speed is a simplification model. Let's test the model for a specific situation. Imagine an experiment in which we drop a series of bowl-shaped, pleated coffee filters and measure their terminal speeds. Table 6.2 on page 134 presents typical terminal speed data from a real experiment using these coffee filters as

continued

▶ **6.10** continued

they fall through the air. The time constant τ is small, so a dropped filter quickly reaches terminal speed. Each filter has a mass of 1.64 g. When the filters are nested together, they combine in such a way that the front-facing surface area does not increase. Determine the relationship between the resistive force exerted by the air and the speed of the falling filters.

SOLUTION

Conceptualize Imagine dropping the coffee filters through the air. (If you have some coffee filters, try dropping them.) Because of the relatively small mass of the coffee filter, you probably won't notice the time interval during which there is an acceleration. The filters will appear to fall at constant velocity immediately upon leaving your hand.

Categorize Because a filter moves at constant velocity, we model it as a *particle in equilibrium.*

Analyze At terminal speed, the upward resistive force on the filter balances the downward gravitational force so that $R = mg$.

Evaluate the magnitude of the resistive force:
$$R = mg = (1.64 \text{ g})\left(\frac{1 \text{ kg}}{1\,000 \text{ g}}\right)(9.80 \text{ m/s}^2) = 0.016\,1 \text{ N}$$

Likewise, two filters nested together experience 0.032 2 N of resistive force, and so forth. These values of resistive force are shown in the far right column of Table 6.2. A graph of the resistive force on the filters as a function of terminal speed is shown in Figure 6.16a. A straight line is not a good fit, indicating that the resistive force is *not* proportional to the speed. The behavior is more clearly seen in Figure 6.16b, in which the resistive force is plotted as a function of the square of the terminal speed. This graph indicates that the resistive force is proportional to the *square* of the speed as suggested by Equation 6.7.

Finalize Here is a good opportunity for you to take some actual data at home on real coffee filters and see if you can reproduce the results shown in Figure 6.16. If you have shampoo and a marble as mentioned in Example 6.8, take data on that system too and see if the resistive force is appropriately modeled as being proportional to the speed.

Table 6.2 **Terminal Speed and Resistive Force for Nested Coffee Filters**

Number of Filters	v_T (m/s)ᵃ	R (N)
1	1.01	0.016 1
2	1.40	0.032 2
3	1.63	0.048 3
4	2.00	0.064 4
5	2.25	0.080 5
6	2.40	0.096 6
7	2.57	0.112 7
8	2.80	0.128 8
9	3.05	0.144 9
10	3.22	0.161 0

ᵃAll values of v_T are approximate.

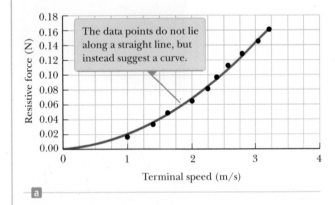

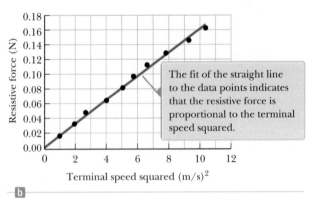

Figure 6.16 (Example 6.10) (a) Relationship between the resistive force acting on falling coffee filters and their terminal speed. (b) Graph relating the resistive force to the square of the terminal speed.

Example 6.11 **Resistive Force Exerted on a Baseball**

A pitcher hurls a 0.145-kg baseball past a batter at 40.2 m/s (= 90 mi/h). Find the resistive force acting on the ball at this speed.

▶ **6.11** continued

SOLUTION

Conceptualize This example is different from the previous ones in that the object is now moving horizontally through the air instead of moving vertically under the influence of gravity and the resistive force. The resistive force causes the ball to slow down, and gravity causes its trajectory to curve downward. We simplify the situation by assuming the velocity vector is exactly horizontal at the instant it is traveling at 40.2 m/s.

Categorize In general, the ball is a *particle under a net force*. Because we are considering only one instant of time, however, we are not concerned about acceleration, so the problem involves only finding the value of one of the forces.

Analyze To determine the drag coefficient D, imagine that we drop the baseball and allow it to reach terminal speed. Solve Equation 6.10 for D:

$$D = \frac{2mg}{v_T^2 \rho A}$$

Use this expression for D in Equation 6.7 to find an expression for the magnitude of the resistive force:

$$R = \tfrac{1}{2} D\rho A v^2 = \frac{1}{2}\left(\frac{2mg}{v_T^2 \rho A}\right)\rho A v^2 = mg\left(\frac{v}{v_T}\right)^2$$

Substitute numerical values, using the terminal speed from Table 6.1:

$$R = (0.145 \text{ kg})(9.80 \text{ m/s}^2)\left(\frac{40.2 \text{ m/s}}{43 \text{ m/s}}\right)^2 = 1.2 \text{ N}$$

Finalize The magnitude of the resistive force is similar in magnitude to the weight of the baseball, which is about 1.4 N. Therefore, air resistance plays a major role in the motion of the ball, as evidenced by the variety of curve balls, floaters, sinkers, and the like thrown by baseball pitchers.

Summary

Concepts and Principles

A particle moving in uniform circular motion has a centripetal acceleration; this acceleration must be provided by a net force directed toward the center of the circular path.

An observer in a noninertial (accelerating) frame of reference introduces **fictitious forces** when applying Newton's second law in that frame.

An object moving through a liquid or gas experiences a speed-dependent **resistive force.** This resistive force is in a direction opposite that of the velocity of the object relative to the medium and generally increases with speed. The magnitude of the resistive force depends on the object's size and shape and on the properties of the medium through which the object is moving. In the limiting case for a falling object, when the magnitude of the resistive force equals the object's weight, the object reaches its **terminal speed.**

Analysis Model for Problem-Solving

Particle in Uniform Circular Motion (Extension) With our new knowledge of forces, we can extend the model of a particle in uniform circular motion, first introduced in Chapter 4. Newton's second law applied to a particle moving in uniform circular motion states that the net force causing the particle to undergo a centripetal acceleration (Eq. 4.14) is related to the acceleration according to

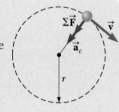

$$\sum F = ma_c = m\frac{v^2}{r} \qquad \textbf{(6.1)}$$

1. A child is practicing for a BMX race. His speed remains constant as he goes counterclockwise around a level track with two straight sections and two nearly semicircular sections as shown in the aerial view of Figure OQ6.1. (a) Rank

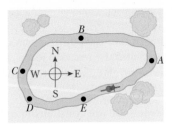

Figure OQ6.1

the magnitudes of his acceleration at the points *A*, *B*, *C*, *D*, and *E* from largest to smallest. If his acceleration is the same size at two points, display that fact in your ranking. If his acceleration is zero, display that fact. (b) What are the directions of his velocity at points *A*, *B*, and *C*? For each point, choose one: north, south, east, west, or nonexistent. (c) What are the directions of his acceleration at points *A*, *B*, and *C*?

2. Consider a skydiver who has stepped from a helicopter and is falling through air. Before she reaches terminal speed and long before she opens her parachute, does her speed (a) increase, (b) decrease, or (c) stay constant?

3. A door in a hospital has a pneumatic closer that pulls the door shut such that the doorknob moves with constant speed over most of its path. In this part of its motion, (a) does the doorknob experience a centripetal acceleration? (b) Does it experience a tangential acceleration?

4. A pendulum consists of a small object called a bob hanging from a light cord of fixed length, with the top end of the cord fixed, as represented in Figure OQ6.4. The bob moves without friction, swinging equally high on both sides. It moves from its turning point *A* through point *B* and reaches its maximum speed at point *C*. (a) Of these points, is there a point where the bob has nonzero radial acceleration and zero tangential acceleration? If so, which point? What is the

direction of its total acceleration at this point? (b) Of these points, is there a point where the bob has nonzero tangential acceleration and zero radial acceleration? If so, which point? What is the direction of its total acceleration at this point? (c) Is there a point where the bob has no acceleration? If so, which point? (d) Is there a point where the bob has both nonzero tangential and radial acceleration? If so, which point? What is the direction of its total acceleration at this point?

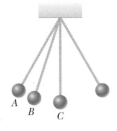

Figure OQ6.4

5. As a raindrop falls through the atmosphere, its speed initially changes as it falls toward the Earth. Before the raindrop reaches its terminal speed, does the magnitude of its acceleration (a) increase, (b) decrease, (c) stay constant at zero, (d) stay constant at 9.80 m/s^2, or (e) stay constant at some other value?

6. An office door is given a sharp push and swings open against a pneumatic device that slows the door down and then reverses its motion. At the moment the door is open the widest, (a) does the doorknob have a centripetal acceleration? (b) Does it have a tangential acceleration?

7. Before takeoff on an airplane, an inquisitive student on the plane dangles an iPod by its earphone wire. It hangs straight down as the plane is at rest waiting to take off. The plane then gains speed rapidly as it moves down the runway. **(i)** Relative to the student's hand, does the iPod (a) shift toward the front of the plane, (b) continue to hang straight down, or (c) shift toward the back of the plane? **(ii)** The speed of the plane increases at a constant rate over a time interval of several seconds. During this interval, does the angle the earphone wire makes with the vertical (a) increase, (b) stay constant, or (c) decrease?

1. What forces cause (a) an automobile, (b) a propeller-driven airplane, and (c) a rowboat to move?

2. A falling skydiver reaches terminal speed with her parachute closed. After the parachute is opened, what parameters change to decrease this terminal speed?

3. An object executes circular motion with constant speed whenever a net force of constant magnitude acts perpendicular to the velocity. What happens to the speed if the force is not perpendicular to the velocity?

4. Describe the path of a moving body in the event that (a) its acceleration is constant in magnitude at all times and perpendicular to the velocity, and (b) its accelera-

tion is constant in magnitude at all times and parallel to the velocity.

5. The observer in the accelerating elevator of Example 5.8 would claim that the "weight" of the fish is *T*, the scale reading, but this answer is obviously wrong. Why does this observation differ from that of a person outside the elevator, at rest with respect to the Earth?

6. If someone told you that astronauts are weightless in orbit because they are beyond the pull of gravity, would you accept the statement? Explain.

7. It has been suggested that rotating cylinders about 20 km in length and 8 km in diameter be placed in space and used as colonies. The purpose of the rotation is to simulate gravity for the inhabitants. Explain this concept for producing an effective imitation of gravity.

8. Consider a small raindrop and a large raindrop falling through the atmosphere. (a) Compare their terminal speeds. (b) What are their accelerations when they reach terminal speed?

9. Why does a pilot tend to black out when pulling out of a steep dive?

10. A pail of water can be whirled in a vertical path such that no water is spilled. Why does the water stay in the pail, even when the pail is above your head?

11. "If the current position and velocity of every particle in the Universe were known, together with the laws describing the forces that particles exert on one another, the whole future of the Universe could be calculated. The future is determinate and preordained. Free will is an illusion." Do you agree with this thesis? Argue for or against it.

Problems available in Access end-of-chapter problems online at www.webassign.net

Section 6.1 Extending the Particle in Uniform Circular Motion Model
Problems 1–11

Section 6.2 Nonuniform Circular Motion
Problems 12–19

Section 6.3 Motion in Accelerated Frames
Problems 20–25

Section 6.4 Motion in the Presence of Resistive Forces
Problems 26–36

Additional Problems
Problems 37–64

Challenge Problems
Problem 65–70

Solutions to the following Problems are available in the *Student Solutions Manual/Study Guide:*
6.1, 6.9, 6.11, 6.14, 6.19, 6.21, 6.23, 6.31, 6.35, 6.49, 6.55, 6.57, 6.59, and 6.63

List of Enhanced Problems

Problem Number	Targeted Feedback in Enhanced WebAssign	Analysis Model Tutorial in Enhanced WebAssign	Master It in Enhanced WebAssign	Watch It in Enhanced WebAssign
6.1		✓	✓	
6.6	✓			✓
6.8	✓			✓
6.9	✓		✓	
6.11	✓			✓
6.12	✓			✓
6.14	✓		✓	
6.16	✓	✓		✓
6.21	✓		✓	
6.23	✓		✓	
6.30	✓			✓
6.31			✓	
6.34		✓		
6.44	✓			✓
6.55			✓	
6.57	✓	✓		✓
6.62	✓			✓
6.63			✓	

Energy of a System

The definitions of quantities such as position, velocity, acceleration, and force and associated principles such as Newton's second law have allowed us to solve a variety of problems. Some problems that could theoretically be solved with Newton's laws, however, are very difficult in practice, but they can be made much simpler with a different approach. Here and in the following chapters, we will investigate this new approach, which will include definitions of quantities that may not be familiar to you. Other quantities may sound familiar, but they may have more specific meanings in physics than in everyday life. We begin this discussion by exploring the notion of *energy*.

The concept of energy is one of the most important topics in science and engineering. In everyday life, we think of energy in terms of fuel for transportation and heating, electricity for lights and appliances, and foods for consumption. These ideas, however, do not truly define energy. They merely tell us that fuels are needed to do a job and that those fuels provide us with something we call energy.

Energy is present in the Universe in various forms. *Every* physical process that occurs in the Universe involves energy and energy transfers or transformations. Unfortunately, despite its extreme importance, energy cannot be easily defined. The variables in previous chapters were relatively concrete; we have everyday experience with velocities and forces, for example. Although we have *experiences* with energy, such as running out of gasoline or losing our electrical service following a violent storm, the *notion* of energy is more abstract.

On a wind farm at the mouth of the River Mersey in Liverpool, England, the moving air does work on the blades of the windmills, causing the blades and the rotor of an electrical generator to rotate. Energy is transferred out of the system of the windmill by means of electricity. *(Christopher Furlong/Getty Images)*

The concept of energy can be applied to mechanical systems without resorting to Newton's laws. Furthermore, the energy approach allows us to understand thermal and electrical phenomena in later chapters of the book in terms of the same models that we will develop here in our study of mechanics.

Our analysis models presented in earlier chapters were based on the motion of a *particle* or an object that could be modeled as a particle. We begin our new approach by focusing our attention on a new simplification model, a *system*, and analysis models based on the model of a system. These analysis models will be formally introduced in Chapter 8. In this chapter, we introduce systems and three ways to store energy in a system.

7.1 Systems and Environments

In the system model, we focus our attention on a small portion of the Universe—the **system**—and ignore details of the rest of the Universe outside of the system. A critical skill in applying the system model to problems is *identifying the system*.

A valid system

Pitfall Prevention 7.1
Identify the System The most important *first* step to take in solving a problem using the energy approach is to identify the appropriate system of interest.

- may be a single object or particle
- may be a collection of objects or particles
- may be a region of space (such as the interior of an automobile engine combustion cylinder)
- may vary with time in size and shape (such as a rubber ball, which deforms upon striking a wall)

Identifying the need for a system approach to solving a problem (as opposed to a particle approach) is part of the Categorize step in the General Problem-Solving Strategy outlined in Chapter 2. Identifying the particular system is a second part of this step.

No matter what the particular system is in a given problem, we identify a **system boundary,** an imaginary surface (not necessarily coinciding with a physical surface) that divides the Universe into the system and the **environment** surrounding the system.

As an example, imagine a force applied to an object in empty space. We can define the object as the system and its surface as the system boundary. The force applied to it is an influence on the system from the environment that acts across the system boundary. We will see how to analyze this situation from a system approach in a subsequent section of this chapter.

Another example was seen in Example 5.10, where the system can be defined as the combination of the ball, the block, and the cord. The influence from the environment includes the gravitational forces on the ball and the block, the normal and friction forces on the block, and the force exerted by the pulley on the cord. The forces exerted by the cord on the ball and the block are internal to the system and therefore are not included as an influence from the environment.

There are a number of mechanisms by which a system can be influenced by its environment. The first one we shall investigate is *work*.

7.2 Work Done by a Constant Force

Almost all the terms we have used thus far—velocity, acceleration, force, and so on—convey a similar meaning in physics as they do in everyday life. Now, however, we encounter a term whose meaning in physics is distinctly different from its everyday meaning: work.

To understand what work as an influence on a system means to the physicist, consider the situation illustrated in Figure 7.1. A force $\vec{\mathbf{F}}$ is applied to a chalkboard

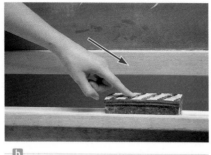

a **b** **c**

Figure 7.1 An eraser being pushed along a chalkboard tray by a force acting at different angles with respect to the horizontal direction.

eraser, which we identify as the system, and the eraser slides along the tray. If we want to know how effective the force is in moving the eraser, we must consider not only the magnitude of the force but also its direction. Notice that the finger in Figure 7.1 applies forces in three different directions on the eraser. Assuming the magnitude of the applied force is the same in all three photographs, the push applied in Figure 7.1b does more to move the eraser than the push in Figure 7.1a. On the other hand, Figure 7.1c shows a situation in which the applied force does not move the eraser at all, regardless of how hard it is pushed (unless, of course, we apply a force so great that we break the chalkboard tray!). These results suggest that when analyzing forces to determine the influence they have on the system, we must consider the vector nature of forces. We must also consider the magnitude of the force. Moving a force with a magnitude of $|\vec{\mathbf{F}}| = 2$ N through a displacement represents a greater influence on the system than moving a force of magnitude 1 N through the same displacement. The magnitude of the displacement is also important. Moving the eraser 3 m along the tray represents a greater influence than moving it 2 cm if the same force is used in both cases.

Let us examine the situation in Figure 7.2, where the object (the system) undergoes a displacement along a straight line while acted on by a constant force of magnitude F that makes an angle θ with the direction of the displacement.

> The **work** W done on a system by an agent exerting a constant force on the system is the product of the magnitude F of the force, the magnitude Δr of the displacement of the point of application of the force, and $\cos\theta$, where θ is the angle between the force and displacement vectors:
>
> $$W \equiv F\,\Delta r\cos\theta \qquad (7.1)$$

Notice in Equation 7.1 that work is a scalar, even though it is defined in terms of two vectors, a force $\vec{\mathbf{F}}$ and a displacement $\Delta\vec{\mathbf{r}}$. In Section 7.3, we explore how to combine two vectors to generate a scalar quantity.

Notice also that the displacement in Equation 7.1 is that of *the point of application of the force*. If the force is applied to a particle or a rigid object that can be modeled as a particle, this displacement is the same as that of the particle. For a deformable system, however, these displacements are not the same. For example, imagine pressing in on the sides of a balloon with both hands. The center of the balloon moves through zero displacement. The points of application of the forces from your hands on the sides of the balloon, however, do indeed move through a displacement as the balloon is compressed, and that is the displacement to be used in Equation 7.1. We will see other examples of deformable systems, such as springs and samples of gas contained in a vessel.

As an example of the distinction between the definition of work and our everyday understanding of the word, consider holding a heavy chair at arm's length for 3 min. At the end of this time interval, your tired arms may lead you to think you

Pitfall Prevention 7.2
Work Is Done by . . . on . . . Not only must you identify the system, you must also identify what agent in the environment is doing work on the system. When discussing work, always use the phrase, "the work done by . . . on . . . ." After "by," insert the part of the environment that is interacting directly with the system. After "on," insert the system. For example, "the work done by the hammer on the nail" identifies the nail as the system, and the force from the hammer represents the influence from the environment.

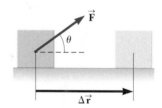

Figure 7.2 An object undergoes a displacement $\Delta\vec{\mathbf{r}}$ under the action of a constant force $\vec{\mathbf{F}}$.

◀ **Work done by a constant force**

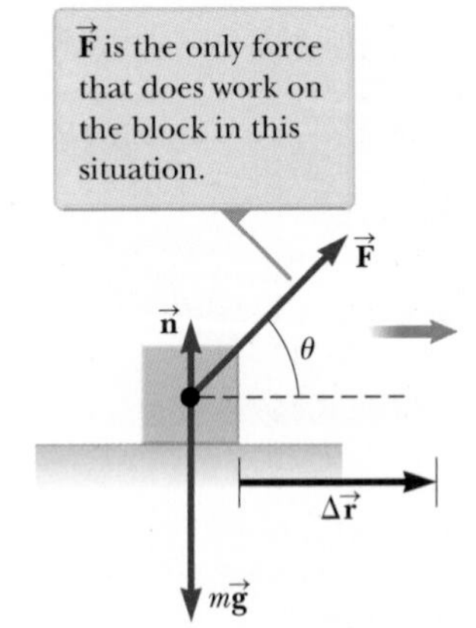

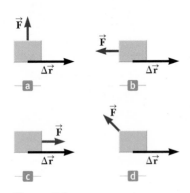

F is the only force that does work on the block in this situation.

Figure 7.3 An object is displaced on a frictionless, horizontal surface. The normal force $\vec{n}$ and the gravitational force $m\vec{g}$ do no work on the object.

Pitfall Prevention 7.3

Cause of the Displacement We can calculate the work done by a force on an object, but that force is *not* necessarily the cause of the object's displacement. For example, if you lift an object, (negative) work is done on the object by the gravitational force, although gravity is not the cause of the object moving upward!

Figure 7.4 (Quick Quiz 7.2) A block is pulled by a force in four different directions. In each case, the displacement of the block is to the right and of the same magnitude.

---

have done a considerable amount of work on the chair. According to our definition, however, you have done no work on it whatsoever. You exert a force to support the chair, but you do not move it. A force does no work on an object if the force does not move through a displacement. If $\Delta r = 0$, Equation 7.1 gives $W = 0$, which is the situation depicted in Figure 7.1c.

Also notice from Equation 7.1 that the work done by a force on a moving object is zero when the force applied is perpendicular to the displacement of its point of application. That is, if $\theta = 90°$, then $W = 0$ because $\cos 90° = 0$. For example, in Figure 7.3, the work done by the normal force on the object and the work done by the gravitational force on the object are both zero because both forces are perpendicular to the displacement and have zero components along an axis in the direction of $\Delta \vec{r}$.

The sign of the work also depends on the direction of $\vec{F}$ relative to $\Delta \vec{r}$. The work done by the applied force on a system is positive when the projection of $\vec{F}$ onto $\Delta \vec{r}$ is in the same direction as the displacement. For example, when an object is lifted, the work done by the applied force on the object is positive because the direction of that force is upward, in the same direction as the displacement of its point of application. When the projection of $\vec{F}$ onto $\Delta \vec{r}$ is in the direction opposite the displacement, W is negative. For example, as an object is lifted, the work done by the gravitational force on the object is negative. The factor $\cos \theta$ in the definition of W (Eq. 7.1) automatically takes care of the sign.

If an applied force $\vec{F}$ is in the same direction as the displacement $\Delta \vec{r}$, then $\theta = 0$ and $\cos 0 = 1$. In this case, Equation 7.1 gives

$$W = F \Delta r$$

The units of work are those of force multiplied by those of length. Therefore, the SI unit of work is the **newton · meter** (N · m = kg · m²/s²). This combination of units is used so frequently that it has been given a name of its own, the **joule** (J).

An important consideration for a system approach to problems is that **work is an energy transfer.** If W is the work done on a system and W is positive, energy is transferred *to* the system; if W is negative, energy is transferred *from* the system. Therefore, if a system interacts with its environment, this interaction can be described as a transfer of energy across the system boundary. The result is a change in the energy stored in the system. We will learn about the first type of energy storage in Section 7.5, after we investigate more aspects of work.

Quick Quiz 7.1 The gravitational force exerted by the Sun on the Earth holds the Earth in an orbit around the Sun. Let us assume that the orbit is perfectly circular. The work done by this gravitational force during a short time interval in which the Earth moves through a displacement in its orbital path is **(a)** zero **(b)** positive **(c)** negative **(d)** impossible to determine

Quick Quiz 7.2 Figure 7.4 shows four situations in which a force is applied to an object. In all four cases, the force has the same magnitude, and the displacement of the object is to the right and of the same magnitude. Rank the situations in order of the work done by the force on the object, from most positive to most negative.

Example 7.1   Mr. Clean

A man cleaning a floor pulls a vacuum cleaner with a force of magnitude $F = 50.0$ N at an angle of $30.0°$ with the horizontal (Fig. 7.5). Calculate the work done by the force on the vacuum cleaner as the vacuum cleaner is displaced 3.00 m to the right.

▶ **7.1** continued

SOLUTION

Figure 7.5 (Example 7.1) A vacuum cleaner being pulled at an angle of 30.0° from the horizontal.

Conceptualize Figure 7.5 helps conceptualize the situation. Think about an experience in your life in which you pulled an object across the floor with a rope or cord.

Categorize We are asked for the work done on an object by a force and are given the force on the object, the displacement of the object, and the angle between the two vectors, so we categorize this example as a substitution problem. We identify the vacuum cleaner as the system.

Use the definition of work (Eq. 7.1):

$$W = F\,\Delta r\cos\theta = (50.0\text{ N})(3.00\text{ m})(\cos 30.0°)$$
$$= \boxed{130\text{ J}}$$

Notice in this situation that the normal force $\vec{n}$ and the gravitational $\vec{F}_g = m\vec{g}$ do no work on the vacuum cleaner because these forces are perpendicular to the displacements of their points of application. Furthermore, there was no mention of whether there was friction between the vacuum cleaner and the floor. The presence or absence of friction is not important when calculating the work done by the applied force. In addition, this work does not depend on whether the vacuum moved at constant velocity or if it accelerated.

7.3 The Scalar Product of Two Vectors

Because of the way the force and displacement vectors are combined in Equation 7.1, it is helpful to use a convenient mathematical tool called the **scalar product** of two vectors. We write this scalar product of vectors $\vec{A}$ and $\vec{B}$ as $\vec{A} \cdot \vec{B}$. (Because of the dot symbol, the scalar product is often called the **dot product**.)

The scalar product of any two vectors $\vec{A}$ and $\vec{B}$ is defined as a scalar quantity equal to the product of the magnitudes of the two vectors and the cosine of the angle θ between them:

$$\boxed{\vec{A} \cdot \vec{B} \equiv AB\cos\theta} \tag{7.2}$$

As is the case with any multiplication, $\vec{A}$ and $\vec{B}$ need not have the same units.

By comparing this definition with Equation 7.1, we can express Equation 7.1 as a scalar product:

$$\boxed{W = F\,\Delta r\cos\theta = \vec{F} \cdot \Delta\vec{r}} \tag{7.3}$$

In other words, $\vec{F} \cdot \Delta\vec{r}$ is a shorthand notation for $F\,\Delta r\cos\theta$.

Before continuing with our discussion of work, let us investigate some properties of the dot product. Figure 7.6 shows two vectors $\vec{A}$ and $\vec{B}$ and the angle θ between them used in the definition of the dot product. In Figure 7.6, $B\cos\theta$ is the projection of $\vec{B}$ onto $\vec{A}$. Therefore, Equation 7.2 means that $\vec{A} \cdot \vec{B}$ is the product of the magnitude of $\vec{A}$ and the projection of $\vec{B}$ onto $\vec{A}$.[1]

From the right-hand side of Equation 7.2, we also see that the scalar product is **commutative.**[2] That is,

$$\vec{A} \cdot \vec{B} = \vec{B} \cdot \vec{A}$$

Pitfall Prevention 7.4
Work Is a Scalar Although Equation 7.3 defines the work in terms of two vectors, *work is a scalar;* there is no direction associated with it. *All* types of energy and energy transfer are scalars. This fact is a major advantage of the energy approach because we don't need vector calculations!

◀ Scalar product of any two vectors $\vec{A}$ and $\vec{B}$

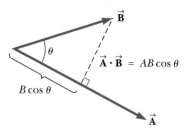

Figure 7.6 The scalar product $\vec{A} \cdot \vec{B}$ equals the magnitude of $\vec{A}$ multiplied by $B\cos\theta$, which is the projection of $\vec{B}$ onto $\vec{A}$.

---

[1]This statement is equivalent to stating that $\vec{A} \cdot \vec{B}$ equals the product of the magnitude of $\vec{B}$ and the projection of $\vec{A}$ onto $\vec{B}$.

[2]In Chapter 11, you will see another way of combining vectors that proves useful in physics and is not commutative.

Finally, the scalar product obeys the **distributive law of multiplication,** so

$$\vec{A} \cdot (\vec{B} + \vec{C}) = \vec{A} \cdot \vec{B} + \vec{A} \cdot \vec{C}$$

The scalar product is simple to evaluate from Equation 7.2 when $\vec{A}$ is either perpendicular or parallel to $\vec{B}$. If $\vec{A}$ is perpendicular to $\vec{B}$ ($\theta = 90°$), then $\vec{A} \cdot \vec{B} = 0$. (The equality $\vec{A} \cdot \vec{B} = 0$ also holds in the more trivial case in which either $\vec{A}$ or $\vec{B}$ is zero.) If vector $\vec{A}$ is parallel to vector $\vec{B}$ and the two point in the same direction ($\theta = 0$), then $\vec{A} \cdot \vec{B} = AB$. If vector $\vec{A}$ is parallel to vector $\vec{B}$ but the two point in opposite directions ($\theta = 180°$), then $\vec{A} \cdot \vec{B} = -AB$. The scalar product is negative when $90° < \theta \le 180°$.

The unit vectors $\hat{\mathbf{i}}$, $\hat{\mathbf{j}}$, and $\hat{\mathbf{k}}$, which were defined in Chapter 3, lie in the positive x, y, and z directions, respectively, of a right-handed coordinate system. Therefore, it follows from the definition of $\vec{A} \cdot \vec{B}$ that the scalar products of these unit vectors are

◀ **Scalar products of unit vectors**

$$\hat{\mathbf{i}} \cdot \hat{\mathbf{i}} = \hat{\mathbf{j}} \cdot \hat{\mathbf{j}} = \hat{\mathbf{k}} \cdot \hat{\mathbf{k}} = 1 \tag{7.4}$$

$$\hat{\mathbf{i}} \cdot \hat{\mathbf{j}} = \hat{\mathbf{i}} \cdot \hat{\mathbf{k}} = \hat{\mathbf{j}} \cdot \hat{\mathbf{k}} = 0 \tag{7.5}$$

Equations 3.18 and 3.19 state that two vectors $\vec{A}$ and $\vec{B}$ can be expressed in unit-vector form as

$$\vec{A} = A_x \hat{\mathbf{i}} + A_y \hat{\mathbf{j}} + A_z \hat{\mathbf{k}}$$

$$\vec{B} = B_x \hat{\mathbf{i}} + B_y \hat{\mathbf{j}} + B_z \hat{\mathbf{k}}$$

Using these expressions for the vectors and the information given in Equations 7.4 and 7.5 shows that the scalar product of $\vec{A}$ and $\vec{B}$ reduces to

$$\vec{A} \cdot \vec{B} = A_x B_x + A_y B_y + A_z B_z \tag{7.6}$$

(Details of the derivation are left for you in Problem 7 in Enhanced WebAssign.) In the special case in which $\vec{A} = \vec{B}$, we see that

$$\vec{A} \cdot \vec{A} = A_x^2 + A_y^2 + A_z^2 = A^2$$

Quick Quiz 7.3 Which of the following statements is true about the relationship between the dot product of two vectors and the product of the magnitudes of the vectors? **(a)** $\vec{A} \cdot \vec{B}$ is larger than AB. **(b)** $\vec{A} \cdot \vec{B}$ is smaller than AB. **(c)** $\vec{A} \cdot \vec{B}$ could be larger or smaller than AB, depending on the angle between the vectors. **(d)** $\vec{A} \cdot \vec{B}$ could be equal to AB.

Example 7.2   **The Scalar Product**

The vectors $\vec{A}$ and $\vec{B}$ are given by $\vec{A} = 2\hat{\mathbf{i}} + 3\hat{\mathbf{j}}$ and $\vec{B} = -\hat{\mathbf{i}} + 2\hat{\mathbf{j}}$.

(A) Determine the scalar product $\vec{A} \cdot \vec{B}$.

SOLUTION

Conceptualize There is no physical system to imagine here. Rather, it is purely a mathematical exercise involving two vectors.

Categorize Because we have a definition for the scalar product, we categorize this example as a substitution problem.

Substitute the specific vector expressions for $\vec{A}$ and $\vec{B}$:

$$\vec{A} \cdot \vec{B} = (2\hat{\mathbf{i}} + 3\hat{\mathbf{j}}) \cdot (-\hat{\mathbf{i}} + 2\hat{\mathbf{j}})$$

$$= -2\hat{\mathbf{i}} \cdot \hat{\mathbf{i}} + 2\hat{\mathbf{i}} \cdot 2\hat{\mathbf{j}} - 3\hat{\mathbf{j}} \cdot \hat{\mathbf{i}} + 3\hat{\mathbf{j}} \cdot 2\hat{\mathbf{j}}$$

$$= -2(1) + 4(0) - 3(0) + 6(1) = -2 + 6 = \boxed{4}$$

The same result is obtained when we use Equation 7.6 directly, where $A_x = 2$, $A_y = 3$, $B_x = -1$, and $B_y = 2$.

▶ **7.2** continued

(B) Find the angle θ between $\vec{\mathbf{A}}$ and $\vec{\mathbf{B}}$.

SOLUTION

Evaluate the magnitudes of $\vec{\mathbf{A}}$ and $\vec{\mathbf{B}}$ using the Pythagorean theorem:

$$A = \sqrt{A_x^2 + A_y^2} = \sqrt{(2)^2 + (3)^2} = \sqrt{13}$$

$$B = \sqrt{B_x^2 + B_y^2} = \sqrt{(-1)^2 + (2)^2} = \sqrt{5}$$

Use Equation 7.2 and the result from part (A) to find the angle:

$$\cos \theta = \frac{\vec{\mathbf{A}} \cdot \vec{\mathbf{B}}}{AB} = \frac{4}{\sqrt{13}\sqrt{5}} = \frac{4}{\sqrt{65}}$$

$$\theta = \cos^{-1} \frac{4}{\sqrt{65}} = \boxed{60.3°}$$

Example 7.3   **Work Done by a Constant Force**

A particle moving in the xy plane undergoes a displacement given by $\Delta \vec{\mathbf{r}} = (2.0\hat{\mathbf{i}} + 3.0\hat{\mathbf{j}})$ m as a constant force $\vec{\mathbf{F}} = (5.0\hat{\mathbf{i}} + 2.0\hat{\mathbf{j}})$ N acts on the particle. Calculate the work done by $\vec{\mathbf{F}}$ on the particle.

SOLUTION

Conceptualize Although this example is a little more physical than the previous one in that it identifies a force and a displacement, it is similar in terms of its mathematical structure.

Categorize Because we are given force and displacement vectors and asked to find the work done by this force on the particle, we categorize this example as a substitution problem.

Substitute the expressions for $\vec{\mathbf{F}}$ and $\Delta \vec{\mathbf{r}}$ into Equation 7.3 and use Equations 7.4 and 7.5:

$$W = \vec{\mathbf{F}} \cdot \Delta \vec{\mathbf{r}} = [(5.0\hat{\mathbf{i}} + 2.0\hat{\mathbf{j}}) \text{ N}] \cdot [(2.0\hat{\mathbf{i}} + 3.0\hat{\mathbf{j}}) \text{ m}]$$

$$= (5.0\hat{\mathbf{i}} \cdot 2.0\hat{\mathbf{i}} + 5.0\hat{\mathbf{i}} \cdot 3.0\hat{\mathbf{j}} + 2.0\hat{\mathbf{j}} \cdot 2.0\hat{\mathbf{i}} + 2.0\hat{\mathbf{j}} \cdot 3.0\hat{\mathbf{j}}) \text{ N} \cdot \text{m}$$

$$= [10 + 0 + 0 + 6] \text{ N} \cdot \text{m} = \boxed{16 \text{ J}}$$

7.4 Work Done by a Varying Force

Consider a particle being displaced along the x axis under the action of a force that varies with position. In such a situation, we cannot use Equation 7.1 to calculate the work done by the force because this relationship applies only when $\vec{\mathbf{F}}$ is constant in magnitude and direction. Figure 7.7a (page 146) shows a varying force applied on a particle that moves from initial position x_i to final position x_f. Imagine a particle undergoing a very small displacement Δx, shown in the figure. The x component F_x of the force is approximately constant over this small interval; for this small displacement, we can approximate the work done on the particle by the force using Equation 7.1 as

$$W \approx F_x \Delta x$$

which is the area of the shaded rectangle in Figure 7.7a. If the F_x versus x curve is divided into a large number of such intervals, the total work done for the displacement from x_i to x_f is approximately equal to the sum of a large number of such terms:

$$W \approx \sum_{x_i}^{x_f} F_x \Delta x$$

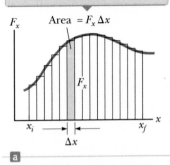

The total work done for the displacement from x_i to x_f is approximately equal to the sum of the areas of all the rectangles.

a

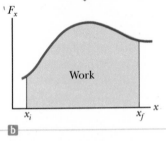

The work done by the component F_x of the varying force as the particle moves from x_i to x_f is *exactly* equal to the area under the curve.

b

Figure 7.7 (a) The work done on a particle by the force component F_x for the small displacement Δx is $F_x \Delta x$, which equals the area of the shaded rectangle. (b) The width Δx of each rectangle is shrunk to zero.

If the size of the small displacements is allowed to approach zero, the number of terms in the sum increases without limit but the value of the sum approaches a definite value equal to the area bounded by the F_x curve and the x axis:

$$\lim_{\Delta x \to 0} \sum_{x_i}^{x_f} F_x \, \Delta x = \int_{x_i}^{x_f} F_x \, dx$$

Therefore, we can express the work done by F_x on the system of the particle as it moves from x_i to x_f as

$$W = \int_{x_i}^{x_f} F_x \, dx \tag{7.7}$$

This equation reduces to Equation 7.1 when the component $F_x = F \cos \theta$ remains constant.

If more than one force acts on a system *and the system can be modeled as a particle,* the total work done on the system is just the work done by the net force. If we express the net force in the x direction as $\sum F_x$, the total work, or *net work*, done as the particle moves from x_i to x_f is

$$\sum W = W_{\text{ext}} = \int_{x_i}^{x_f} \left(\sum F_x\right) dx \quad \text{(particle)}$$

For the general case of a net force $\sum \vec{F}$ whose magnitude and direction may both vary, we use the scalar product,

$$\sum W = W_{\text{ext}} = \int \left(\sum \vec{F}\right) \cdot d\vec{r} \quad \text{(particle)} \tag{7.8}$$

where the integral is calculated over the path that the particle takes through space. The subscript "ext" on work reminds us that the net work is done by an *external* agent on the system. We will use this notation in this chapter as a reminder and to differentiate this work from an *internal* work to be described shortly.

If the system cannot be modeled as a particle (for example, if the system is deformable), we cannot use Equation 7.8 because different forces on the system may move through different displacements. In this case, we must evaluate the work done by each force separately and then add the works algebraically to find the net work done on the system:

$$\sum W = W_{\text{ext}} = \sum_{\text{forces}} \left(\int \vec{F} \cdot d\vec{r}\right) \quad \text{(deformable system)}$$

Example 7.4   **Calculating Total Work Done from a Graph**

A force acting on a particle varies with x as shown in Figure 7.8. Calculate the work done by the force on the particle as it moves from $x = 0$ to $x = 6.0$ m.

SOLUTION

Conceptualize Imagine a particle subject to the force in Figure 7.8. The force remains constant as the particle moves through the first 4.0 m and then decreases linearly to zero at 6.0 m. In terms of earlier discussions of motion, the particle could be modeled as a particle under constant acceleration for the first 4.0 m because the force is constant. Between 4.0 m and 6.0 m, however, the motion does not fit into one of our earlier analysis models because the acceleration of the particle is changing. If the particle starts from rest, its speed increases throughout the motion, and the particle is always moving in the positive x direction. These details about its speed and direction are not necessary for the calculation of the work done, however.

Categorize Because the force varies during the motion of the particle, we must use the techniques for work done by varying forces. In this case, the graphical representation in Figure 7.8 can be used to evaluate the work done.

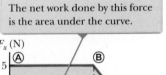

The net work done by this force is the area under the curve.

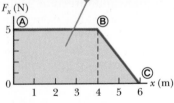

Figure 7.8 (Example 7.4) The force acting on a particle is constant for the first 4.0 m of motion and then decreases linearly with x from $x_{\text{B}} = 4.0$ m to $x_{\text{C}} = 6.0$ m.

▶ **7.4** continued

Analyze The work done by the force is equal to the area under the curve from $x_{Ⓐ} = 0$ to $x_{Ⓒ} = 6.0$ m. This area is equal to the area of the rectangular section from Ⓐ to Ⓑ plus the area of the triangular section from Ⓑ to Ⓒ.

Evaluate the area of the rectangle:

$$W_{Ⓐ \text{ to } Ⓑ} = (5.0 \text{ N})(4.0 \text{ m}) = 20 \text{ J}$$

Evaluate the area of the triangle:

$$W_{Ⓑ \text{ to } Ⓒ} = \tfrac{1}{2}(5.0 \text{ N})(2.0 \text{ m}) = 5.0 \text{ J}$$

Find the total work done by the force on the particle:

$$W_{Ⓐ \text{ to } Ⓒ} = W_{Ⓐ \text{ to } Ⓑ} + W_{Ⓑ \text{ to } Ⓒ} = 20 \text{ J} + 5.0 \text{ J} = \boxed{25 \text{ J}}$$

Finalize Because the graph of the force consists of straight lines, we can use rules for finding the areas of simple geometric models to evaluate the total work done in this example. If a force does not vary linearly as in Figure 7.7, such rules cannot be used and the force function must be integrated as in Equation 7.7 or 7.8.

Work Done by a Spring

A model of a common physical system on which the force varies with position is shown in Figure 7.9. The system is a block on a frictionless, horizontal surface and connected to a spring. For many springs, if the spring is either stretched or compressed a small distance from its unstretched (equilibrium) configuration, it exerts on the block a force that can be mathematically modeled as

$$F_s = -kx \qquad\qquad (7.9)$$ ◀ **Spring force**

where x is the position of the block relative to its equilibrium ($x = 0$) position and k is a positive constant called the **force constant** or the **spring constant** of the spring. In other words, the force required to stretch or compress a spring is proportional to the amount of stretch or compression x. This force law for springs is known as **Hooke's law.** The value of k is a measure of the *stiffness* of the spring. Stiff springs have large k values, and soft springs have small k values. As can be seen from Equation 7.9, the units of k are N/m.

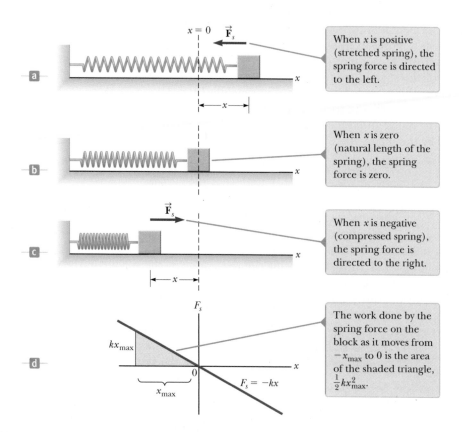

When x is positive (stretched spring), the spring force is directed to the left.

When x is zero (natural length of the spring), the spring force is zero.

When x is negative (compressed spring), the spring force is directed to the right.

The work done by the spring force on the block as it moves from $-x_{max}$ to 0 is the area of the shaded triangle, $\tfrac{1}{2}kx_{max}^2$.

Figure 7.9 The force exerted by a spring on a block varies with the block's position x relative to the equilibrium position $x = 0$. (a) x is positive. (b) x is zero. (c) x is negative. (d) Graph of F_s versus x for the block–spring system.

The vector form of Equation 7.9 is

$$\vec{\mathbf{F}}_s = F_s\hat{\mathbf{i}} = -kx\hat{\mathbf{i}} \qquad\qquad (7.10)$$

where we have chosen the x axis to lie along the direction the spring extends or compresses.

The negative sign in Equations 7.9 and 7.10 signifies that the force exerted by the spring is always directed *opposite* the displacement from equilibrium. When $x > 0$ as in Figure 7.9a so that the block is to the right of the equilibrium position, the spring force is directed to the left, in the negative x direction. When $x < 0$ as in Figure 7.9c, the block is to the left of equilibrium and the spring force is directed to the right, in the positive x direction. When $x = 0$ as in Figure 7.9b, the spring is unstretched and $F_s = 0$. Because the spring force always acts toward the equilibrium position ($x = 0$), it is sometimes called a *restoring force.*

If the spring is compressed until the block is at the point $-x_{max}$ and is then released, the block moves from $-x_{max}$ through zero to $+x_{max}$. It then reverses direction, returns to $-x_{max}$, and continues oscillating back and forth. We will study these oscillations in more detail in Chapter 15. For now, let's investigate the work done by the spring on the block over small portions of one oscillation.

Suppose the block has been pushed to the left to a position $-x_{max}$ and is then released. We identify the block as our system and calculate the work W_s done by the spring force on the block as the block moves from $x_i = -x_{max}$ to $x_f = 0$. Applying Equation 7.8 and assuming the block may be modeled as a particle, we obtain

$$W_s = \int \vec{\mathbf{F}}_s \cdot d\vec{\mathbf{r}} = \int_{x_i}^{x_f}(-kx\hat{\mathbf{i}}) \cdot (dx\hat{\mathbf{i}}) = \int_{-x_{max}}^{0}(-kx)\,dx = \tfrac{1}{2}kx_{max}^2 \qquad (7.11)$$

where we have used the integral $\int x^n\,dx = x^{n+1}/(n+1)$ with $n = 1$. The work done by the spring force is positive because the force is in the same direction as its displacement (both are to the right). Because the block arrives at $x = 0$ with some speed, it will continue moving until it reaches a position $+x_{max}$. The work done by the spring force on the block as it moves from $x_i = 0$ to $x_f = x_{max}$ is $W_s = -\tfrac{1}{2}kx_{max}^2$. The work is negative because for this part of the motion the spring force is to the left and its displacement is to the right. Therefore, the *net* work done by the spring force on the block as it moves from $x_i = -x_{max}$ to $x_f = x_{max}$ is *zero.*

Figure 7.9d is a plot of F_s versus x. The work calculated in Equation 7.11 is the area of the shaded triangle, corresponding to the displacement from $-x_{max}$ to 0. Because the triangle has base x_{max} and height kx_{max}, its area is $\tfrac{1}{2}kx_{max}^2$, agreeing with the work done by the spring as given by Equation 7.11.

If the block undergoes an arbitrary displacement from $x = x_i$ to $x = x_f$, the work done by the spring force on the block is

Work done by a spring ▶

$$W_s = \int_{x_i}^{x_f}(-kx)\,dx = \tfrac{1}{2}kx_i^2 - \tfrac{1}{2}kx_f^2 \qquad (7.12)$$

From Equation 7.12, we see that the work done by the spring force is zero for any motion that ends where it began ($x_i = x_f$). We shall make use of this important result in Chapter 8 when we describe the motion of this system in greater detail.

Equations 7.11 and 7.12 describe the work done by the spring on the block. Now let us consider the work done on the block by an *external agent* as the agent applies a force on the block and the block moves *very slowly* from $x_i = -x_{max}$ to $x_f = 0$ as in Figure 7.10. We can calculate this work by noting that at any value of the position, the *applied force* $\vec{\mathbf{F}}_{app}$ is equal in magnitude and opposite in direction to the spring force $\vec{\mathbf{F}}_s$, so $\vec{\mathbf{F}}_{app} = F_{app}\hat{\mathbf{i}} = -\vec{\mathbf{F}}_s = -(-kx\hat{\mathbf{i}}) = kx\hat{\mathbf{i}}$. Therefore, the work done by this applied force (the external agent) on the system of the block is

$$W_{ext} = \int \vec{\mathbf{F}}_{app} \cdot d\vec{\mathbf{r}} = \int_{x_i}^{x_f}(kx\hat{\mathbf{i}}) \cdot (dx\hat{\mathbf{i}}) = \int_{-x_{max}}^{0} kx\,dx = -\tfrac{1}{2}kx_{max}^2$$

This work is equal to the negative of the work done by the spring force for this displacement (Eq. 7.11). The work is negative because the external agent must push inward on the spring to prevent it from expanding, and this direction is opposite the direction of the displacement of the point of application of the force as the block moves from $-x_{max}$ to 0.

For an arbitrary displacement of the block, the work done on the system by the external agent is

$$W_{ext} = \int_{x_i}^{x_f} kx\,dx = \tfrac{1}{2}kx_f^2 - \tfrac{1}{2}kx_i^2 \qquad (7.13)$$

Notice that this equation is the negative of Equation 7.12.

Q uick Quiz 7.4 A dart is inserted into a spring-loaded dart gun by pushing the spring in by a distance x. For the next loading, the spring is compressed a distance $2x$. How much work is required to load the second dart compared with that required to load the first? **(a)** four times as much **(b)** two times as much **(c)** the same **(d)** half as much **(e)** one-fourth as much

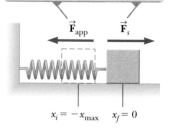

If the process of moving the block is carried out very slowly, then $\vec{F}_{app}$ is equal in magnitude and opposite in direction to $\vec{F}_s$ at all times.

$x_i = -x_{max}$ $x_f = 0$

Figure 7.10 A block moves from $x_i = -x_{max}$ to $x_f = 0$ on a frictionless surface as a force $\vec{F}_{app}$ is applied to the block.

Example 7.5 **Measuring k for a Spring** AM

A common technique used to measure the force constant of a spring is demonstrated by the setup in Figure 7.11. The spring is hung vertically (Fig. 7.11a), and an object of mass m is attached to its lower end. Under the action of the "load" mg, the spring stretches a distance d from its equilibrium position (Fig. 7.11b).

(A) If a spring is stretched 2.0 cm by a suspended object having a mass of 0.55 kg, what is the force constant of the spring?

SOLUTION

Conceptualize Figure 7.11b shows what happens to the spring when the object is attached to it. Simulate this situation by hanging an object on a rubber band.

Categorize The object in Figure 7.11b is at rest and not accelerating, so it is modeled as a *particle in equilibrium*.

Analyze Because the object is in equilibrium, the net force on it is zero and the upward spring force balances the downward gravitational force $m\vec{g}$ (Fig. 7.11c).

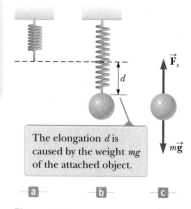

The elongation d is caused by the weight mg of the attached object.

a   b   c

Figure 7.11 (Example 7.5) Determining the force constant k of a spring.

Apply the particle in equilibrium model to the object: $\vec{F}_s + m\vec{g} = 0 \quad \to \quad F_s - mg = 0 \quad \to \quad F_s = mg$

Apply Hooke's law to give $F_s = kd$ and solve for k: $k = \dfrac{mg}{d} = \dfrac{(0.55\text{ kg})(9.80\text{ m/s}^2)}{2.0 \times 10^{-2}\text{ m}} = \boxed{2.7 \times 10^2\text{ N/m}}$

(B) How much work is done by the spring on the object as it stretches through this distance?

SOLUTION

Use Equation 7.12 to find the work done by the spring on the object: $W_s = 0 - \tfrac{1}{2}kd^2 = -\tfrac{1}{2}(2.7 \times 10^2\text{ N/m})(2.0 \times 10^{-2}\text{ m})^2$

$= -5.4 \times 10^{-2}\text{ J}$

Finalize This work is negative because the spring force acts upward on the object, but its point of application (where the spring attaches to the object) moves downward. As the object moves through the 2.0-cm distance, the gravitational force also does work on it. This work is positive because the gravitational force is downward and so is the displacement

continued

▶ **7.5** continued

of the point of application of this force. Would we expect the work done by the gravitational force, as the applied force in a direction opposite to the spring force, to be the negative of the answer above? Let's find out.

Evaluate the work done by the gravitational force on the object:

$$W = \vec{\mathbf{F}} \cdot \Delta \vec{\mathbf{r}} = (mg)(d) \cos 0 = mgd$$

$$= (0.55 \text{ kg})(9.80 \text{ m/s}^2)(2.0 \times 10^{-2} \text{ m}) = 1.1 \times 10^{-1} \text{ J}$$

If you expected the work done by gravity simply to be that done by the spring with a positive sign, you may be surprised by this result! To understand why that is not the case, we need to explore further, as we do in the next section.

7.5 Kinetic Energy and the Work–Kinetic Energy Theorem

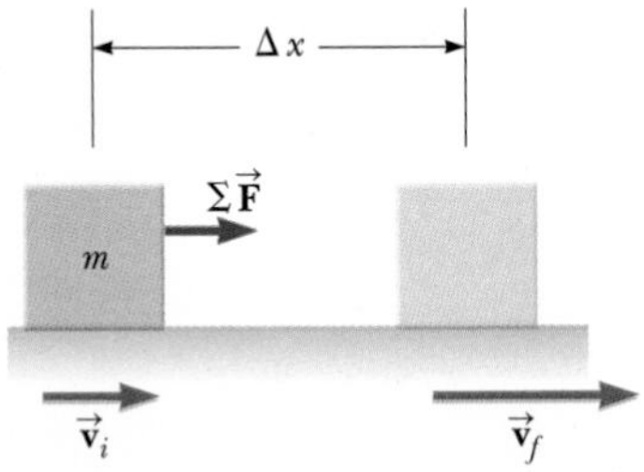

We have investigated work and identified it as a mechanism for transferring energy into a system. We have stated that work is an influence on a system from the environment, but we have not yet discussed the *result* of this influence on the system. One possible result of doing work on a system is that the system changes its speed. In this section, we investigate this situation and introduce our first type of energy that a system can possess, called *kinetic energy*.

Consider a system consisting of a single object. Figure 7.12 shows a block of mass m moving through a displacement directed to the right under the action of a net force $\Sigma \vec{\mathbf{F}}$, also directed to the right. We know from Newton's second law that the block moves with an acceleration $\vec{\mathbf{a}}$. If the block (and therefore the force) moves through a displacement $\Delta \vec{\mathbf{r}} = \Delta x \hat{\mathbf{i}} = (x_f - x_i)\hat{\mathbf{i}}$, the net work done on the block by the external net force $\Sigma \vec{\mathbf{F}}$ is

$$W_{\text{ext}} = \int_{x_i}^{x_f} \Sigma F \, dx \tag{7.14}$$

Figure 7.12 An object undergoing a displacement $\Delta \vec{\mathbf{r}} = \Delta x \hat{\mathbf{i}}$ and a change in velocity under the action of a constant net force $\Sigma \vec{\mathbf{F}}$.

Using Newton's second law, we substitute for the magnitude of the net force $\Sigma F = ma$ and then perform the following chain-rule manipulations on the integrand:

$$W_{\text{ext}} = \int_{x_i}^{x_f} ma \, dx = \int_{x_i}^{x_f} m \frac{dv}{dt} \, dx = \int_{x_i}^{x_f} m \frac{dv}{dx} \frac{dx}{dt} \, dx = \int_{v_i}^{v_f} mv \, dv$$

$$W_{\text{ext}} = \tfrac{1}{2}mv_f^2 - \tfrac{1}{2}mv_i^2 \tag{7.15}$$

where v_i is the speed of the block at $x = x_i$ and v_f is its speed at x_f.

Equation 7.15 was generated for the specific situation of one-dimensional motion, but it is a general result. It tells us that the work done by the net force on a particle of mass m is equal to the difference between the initial and final values of a quantity $\tfrac{1}{2}mv^2$. This quantity is so important that it has been given a special name, **kinetic energy:**

Kinetic energy ▶

$$K \equiv \tfrac{1}{2}mv^2 \tag{7.16}$$

Kinetic energy represents the energy associated with the motion of the particle. Note that kinetic energy is a scalar quantity and has the same units as work. For example, a 2.0-kg object moving with a speed of 4.0 m/s has a kinetic energy of 16 J. Table 7.1 lists the kinetic energies for various objects.

Equation 7.15 states that the work done on a particle by a net force $\Sigma \vec{\mathbf{F}}$ acting on it equals the change in kinetic energy of the particle. It is often convenient to write Equation 7.15 in the form

$$W_{\text{ext}} = K_f - K_i = \Delta K \tag{7.17}$$

Another way to write it is $K_f = K_i + W_{\text{ext}}$, which tells us that the final kinetic energy of an object is equal to its initial kinetic energy plus the change in energy due to the net work done on it.

Table 7.1	Kinetic Energies for Various Objects		
Object	**Mass (kg)**	**Speed (m/s)**	**Kinetic Energy (J)**
Earth orbiting the Sun	5.97×10^{24}	2.98×10^{4}	2.65×10^{33}
Moon orbiting the Earth	7.35×10^{22}	1.02×10^{3}	3.82×10^{28}
Rocket moving at escape speed[a]	500	1.12×10^{4}	3.14×10^{10}
Automobile at 65 mi/h	2 000	29	8.4×10^{5}
Running athlete	70	10	3 500
Stone dropped from 10 m	1.0	14	98
Golf ball at terminal speed	0.046	44	45
Raindrop at terminal speed	3.5×10^{-5}	9.0	1.4×10^{-3}
Oxygen molecule in air	5.3×10^{-26}	500	6.6×10^{-21}

[a]Escape speed is the minimum speed an object must reach near the Earth's surface to move infinitely far away from the Earth.

We have generated Equation 7.17 by imagining doing work on a particle. We could also do work on a deformable system, in which parts of the system move with respect to one another. In this case, we also find that Equation 7.17 is valid as long as the net work is found by adding up the works done by each force and adding, as discussed earlier with regard to Equation 7.8.

Equation 7.17 is an important result known as the **work–kinetic energy theorem:**

> When work is done on a system and the only change in the system is in its speed, the net work done on the system equals the change in kinetic energy of the system, as expressed by Equation 7.17: $W = \Delta K$.

◀ Work–kinetic energy theorem

The work–kinetic energy theorem indicates that the speed of a system *increases* if the net work done on it is *positive* because the final kinetic energy is greater than the initial kinetic energy. The speed *decreases* if the net work is *negative* because the final kinetic energy is less than the initial kinetic energy.

Because we have so far only investigated translational motion through space, we arrived at the work–kinetic energy theorem by analyzing situations involving translational motion. Another type of motion is *rotational motion*, in which an object spins about an axis. We will study this type of motion in Chapter 10. The work–kinetic energy theorem is also valid for systems that undergo a change in the rotational speed due to work done on the system. The windmill in the photograph at the beginning of this chapter is an example of work causing rotational motion.

The work–kinetic energy theorem will clarify a result seen earlier in this chapter that may have seemed odd. In Section 7.4, we arrived at a result of zero net work done when we let a spring push a block from $x_i = -x_{max}$ to $x_f = x_{max}$. Notice that because the speed of the block is continually changing, it may seem complicated to analyze this process. The quantity ΔK in the work–kinetic energy theorem, however, only refers to the initial and final points for the speeds; it does not depend on details of the path followed between these points. Therefore, because the speed is zero at both the initial and final points of the motion, the net work done on the block is zero. We will often see this concept of path independence in similar approaches to problems.

Let us also return to the mystery in the Finalize step at the end of Example 7.5. Why was the work done by gravity not just the value of the work done by the spring with a positive sign? Notice that the work done by gravity is larger than the magnitude of the work done by the spring. Therefore, the total work done by all forces on the object is positive. Imagine now how to create the situation in which the *only* forces on the object are the spring force and the gravitational force. You must support the object at the highest point and then remove your hand and let the object fall. If you do so, you know that when the object reaches a position 2.0 cm below your hand, it will be *moving*, which is consistent with Equation 7.17. Positive net

Pitfall Prevention 7.5

Conditions for the Work–Kinetic Energy Theorem The work–kinetic energy theorem is important but limited in its application; it is not a general principle. In many situations, other changes in the system occur besides its speed, and there are other interactions with the environment besides work. A more general principle involving energy is *conservation of energy* in Section 8.1.

Pitfall Prevention 7.6

The Work–Kinetic Energy Theorem: Speed, Not Velocity The work–kinetic energy theorem relates work to a change in the *speed* of a system, not a change in its velocity. For example, if an object is in uniform circular motion, its speed is constant. Even though its velocity is changing, no work is done on the object by the force causing the circular motion.

work is done on the object, and the result is that it has a kinetic energy as it passes through the 2.0-cm point.

The only way to prevent the object from having a kinetic energy after moving through 2.0 cm is to slowly lower it with your hand. Then, however, there is a third force doing work on the object, the normal force from your hand. If this work is calculated and added to that done by the spring force and the gravitational force, the net work done on the object is zero, which is consistent because it is not moving at the 2.0-cm point.

Earlier, we indicated that work can be considered as a mechanism for transferring energy into a system. Equation 7.17 is a mathematical statement of this concept. When work W_{ext} is done on a system, the result is a transfer of energy across the boundary of the system. The result on the system, in the case of Equation 7.17, is a change ΔK in kinetic energy. In the next section, we investigate another type of energy that can be stored in a system as a result of doing work on the system.

Quick Quiz 7.5 A dart is inserted into a spring-loaded dart gun by pushing the spring in by a distance x. For the next loading, the spring is compressed a distance $2x$. How much faster does the second dart leave the gun compared with the first? **(a)** four times as fast **(b)** two times as fast **(c)** the same **(d)** half as fast **(e)** one-fourth as fast

---

Example 7.6   **A Block Pulled on a Frictionless Surface**   **AM**

A 6.0-kg block initially at rest is pulled to the right along a frictionless, horizontal surface by a constant horizontal force of magnitude 12 N. Find the block's speed after it has moved through a horizontal distance of 3.0 m.

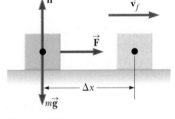

Figure 7.13 (Example 7.6) A block pulled to the right on a frictionless surface by a constant horizontal force.

SOLUTION

Conceptualize Figure 7.13 illustrates this situation. Imagine pulling a toy car across a table with a horizontal rubber band attached to the front of the car. The force is maintained constant by ensuring that the stretched rubber band always has the same length.

Categorize We could apply the equations of kinematics to determine the answer, but let us practice the energy approach. The block is the system, and three external forces act on the system. The normal force balances the gravitational force on the block, and neither of these vertically acting forces does work on the block because their points of application are horizontally displaced.

Analyze The net external force acting on the block is the horizontal 12-N force.

Use the work–kinetic energy theorem for the block, noting that its initial kinetic energy is zero:

$$W_{ext} = \Delta K = K_f - K_i = \tfrac{1}{2}mv_f^2 - 0 = \tfrac{1}{2}mv_f^2$$

Solve for v_f and use Equation 7.1 for the work done on the block by $\vec{F}$:

$$v_f = \sqrt{\frac{2W_{ext}}{m}} = \sqrt{\frac{2F\,\Delta x}{m}}$$

Substitute numerical values:

$$v_f = \sqrt{\frac{2(12\ \text{N})(3.0\ \text{m})}{6.0\ \text{kg}}} = 3.5\ \text{m/s}$$

Finalize You should solve this problem again by modeling the block as a *particle under a net force* to find its acceleration and then as a *particle under constant acceleration* to find its final velocity. In Chapter 8, we will see that the energy procedure followed above is an example of the analysis model of the *nonisolated system*.

WHAT IF?   Suppose the magnitude of the force in this example is doubled to $F' = 2F$. The 6.0-kg block accelerates to 3.5 m/s due to this applied force while moving through a displacement $\Delta x'$. How does the displacement $\Delta x'$ compare with the original displacement Δx?

▶ **7.6** continued

Answer If we pull harder, the block should accelerate to a given speed in a shorter distance, so we expect that $\Delta x' < \Delta x$. In both cases, the block experiences the same change in kinetic energy ΔK. Mathematically, from the work–kinetic energy theorem, we find that

$$W_{\text{ext}} = F'\Delta x' = \Delta K = F\,\Delta x$$

$$\Delta x' = \frac{F}{F'}\,\Delta x = \frac{F}{2F}\,\Delta x = \tfrac{1}{2}\,\Delta x$$

and the distance is shorter as suggested by our conceptual argument.

| **Conceptual Example 7.7** | **Does the Ramp Lessen the Work Required?** |

A man wishes to load a refrigerator onto a truck using a ramp at angle θ as shown in Figure 7.14. He claims that less work would be required to load the truck if the length L of the ramp were increased. Is his claim valid?

SOLUTION

No. Suppose the refrigerator is wheeled on a hand truck up the ramp at constant speed. In this case, for the system of the refrigerator and the hand truck, $\Delta K = 0$. The normal force exerted by the ramp on the system is directed at 90° to the displacement of its point of application and so does no work on the system. Because $\Delta K = 0$, the work–kinetic energy theorem gives

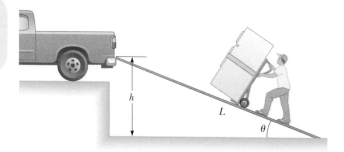

Figure 7.14 (Conceptual Example 7.7) A refrigerator attached to a frictionless, wheeled hand truck is moved up a ramp at constant speed.

$$W_{\text{ext}} = W_{\text{by man}} + W_{\text{by gravity}} = 0$$

The work done by the gravitational force equals the product of the weight mg of the system, the distance L through which the refrigerator is displaced, and $\cos(\theta + 90°)$. Therefore,

$$W_{\text{by man}} = -W_{\text{by gravity}} = -(mg)(L)[\cos(\theta + 90°)]$$

$$= mgL\sin\theta = mgh$$

where $h = L\sin\theta$ is the height of the ramp. Therefore, the man must do the same amount of work mgh on the system *regardless* of the length of the ramp. The work depends only on the height of the ramp. Although less force is required with a longer ramp, the point of application of that force moves through a greater displacement.

7.6 Potential Energy of a System

So far in this chapter, we have defined a system in general, but have focused our attention primarily on single particles or objects under the influence of external forces. Let us now consider systems of two or more particles or objects interacting via a force that is *internal* to the system. The kinetic energy of such a system is the algebraic sum of the kinetic energies of all members of the system. There may be systems, however, in which one object is so massive that it can be modeled as stationary and its kinetic energy can be neglected. For example, if we consider a ball–Earth system as the ball falls to the Earth, the kinetic energy of the system can be considered as just the kinetic energy of the ball. The Earth moves so slowly in this process that we can ignore its kinetic energy. On the other hand, the kinetic energy of a system of two electrons must include the kinetic energies of both particles.

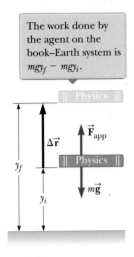

The work done by the agent on the book–Earth system is $mgy_f - mgy_i$.

Figure 7.15 An external agent lifts a book slowly from a height y_i to a height y_f.

Gravitational ▶ potential energy

Let us imagine a system consisting of a book and the Earth, interacting via the gravitational force. We do some work on the system by lifting the book slowly from rest through a vertical displacement $\Delta \vec{r} = (y_f - y_i)\hat{j}$ as in Figure 7.15. According to our discussion of work as an energy transfer, this work done on the system must appear as an increase in energy of the system. The book is at rest before we perform the work and is at rest after we perform the work. Therefore, there is no change in the kinetic energy of the system.

Because the energy change of the system is not in the form of kinetic energy, the work–kinetic energy theorem does not apply here and the energy change must appear as some form of energy storage other than kinetic energy. After lifting the book, we could release it and let it fall back to the position y_i. Notice that the book (and therefore, the system) now has kinetic energy and that its source is in the work that was done in lifting the book. While the book was at the highest point, the system had the *potential* to possess kinetic energy, but it did not do so until the book was allowed to fall. Therefore, we call the energy storage mechanism before the book is released **potential energy.** We will find that the potential energy of a system can only be associated with specific types of forces acting between members of a system. The amount of potential energy in the system is determined by the *configuration* of the system. Moving members of the system to different positions or rotating them may change the configuration of the system and therefore its potential energy.

Let us now derive an expression for the potential energy associated with an object at a given location above the surface of the Earth. Consider an external agent lifting an object of mass m from an initial height y_i above the ground to a final height y_f as in Figure 7.15. We assume the lifting is done slowly, with no acceleration, so the applied force from the agent is equal in magnitude to the gravitational force on the object: the object is modeled as a particle in equilibrium moving at constant velocity. The work done by the external agent on the system (object and the Earth) as the object undergoes this upward displacement is given by the product of the upward applied force $\vec{F}_{app}$ and the upward displacement of this force, $\Delta \vec{r} = \Delta y \hat{j}$:

$$W_{ext} = (\vec{F}_{app}) \cdot \Delta \vec{r} = (mg\hat{j}) \cdot [(y_f - y_i)\hat{j}] = mgy_f - mgy_i \qquad (7.18)$$

where this result is the net work done on the system because the applied force is the only force on the system from the environment. (Remember that the gravitational force is *internal* to the system.) Notice the similarity between Equation 7.18 and Equation 7.15. In each equation, the work done on a system equals a difference between the final and initial values of a quantity. In Equation 7.15, the work represents a transfer of energy into the system and the increase in energy of the system is kinetic in form. In Equation 7.18, the work represents a transfer of energy into the system and the system energy appears in a different form, which we have called potential energy.

Therefore, we can identify the quantity mgy as the **gravitational potential energy** U_g of the system of an object of mass m and the Earth:

$$U_g \equiv mgy \qquad (7.19)$$

The units of gravitational potential energy are joules, the same as the units of work and kinetic energy. Potential energy, like work and kinetic energy, is a scalar quantity. Notice that Equation 7.19 is valid only for objects near the surface of the Earth, where g is approximately constant.[3]

Using our definition of gravitational potential energy, Equation 7.18 can now be rewritten as

$$W_{ext} = \Delta U_g \qquad (7.20)$$

which mathematically describes that the net external work done on the system in this situation appears as a change in the gravitational potential energy of the system.

Equation 7.20 is similar in form to the work–kinetic energy theorem, Equation 7.17. In Equation 7.17, work is done on a system and energy appears in the system as

---

[3]The assumption that g is constant is valid as long as the vertical displacement of the object is small compared with the Earth's radius.

kinetic energy, representing *motion* of the members of the system. In Equation 7.20, work is done on the system and energy appears in the system as potential energy, representing a change in the *configuration* of the members of the system.

Gravitational potential energy depends only on the vertical height of the object above the surface of the Earth. The same amount of work must be done on an object–Earth system whether the object is lifted vertically from the Earth or is pushed starting from the same point up a frictionless incline, ending up at the same height. We verified this statement for a specific situation of rolling a refrigerator up a ramp in Conceptual Example 7.7. This statement can be shown to be true in general by calculating the work done on an object by an agent moving the object through a displacement having both vertical and horizontal components:

$$W_{\text{ext}} = (\vec{\mathbf{F}}_{\text{app}}) \cdot \Delta \vec{\mathbf{r}} = (mg\hat{\mathbf{j}}) \cdot [(x_f - x_i)\hat{\mathbf{i}} + (y_f - y_i)\hat{\mathbf{j}}] = mgy_f - mgy_i$$

where there is no term involving x in the final result because $\hat{\mathbf{j}} \cdot \hat{\mathbf{i}} = 0$.

In solving problems, you must choose a reference configuration for which the gravitational potential energy of the system is set equal to some reference value, which is normally zero. The choice of reference configuration is completely arbitrary because the important quantity is the *difference* in potential energy, and this difference is independent of the choice of reference configuration.

It is often convenient to choose as the reference configuration for zero gravitational potential energy the configuration in which an object is at the surface of the Earth, but this choice is not essential. Often, the statement of the problem suggests a convenient configuration to use.

Quick Quiz 7.6 Choose the correct answer. The gravitational potential energy of a system **(a)** is always positive **(b)** is always negative **(c)** can be negative or positive

---

Example 7.8 | The Proud Athlete and the Sore Toe

A trophy being shown off by a careless athlete slips from the athlete's hands and drops on his foot. Choosing floor level as the $y = 0$ point of your coordinate system, estimate the change in gravitational potential energy of the trophy–Earth system as the trophy falls. Repeat the calculation, using the top of the athlete's head as the origin of coordinates.

SOLUTION

Conceptualize The trophy changes its vertical position with respect to the surface of the Earth. Associated with this change in position is a change in the gravitational potential energy of the trophy–Earth system.

Categorize We evaluate a change in gravitational potential energy defined in this section, so we categorize this example as a substitution problem. Because there are no numbers provided in the problem statement, it is also an estimation problem.

The problem statement tells us that the reference configuration of the trophy–Earth system corresponding to zero potential energy is when the bottom of the trophy is at the floor. To find the change in potential energy for the system, we need to estimate a few values. Let's say the trophy has a mass of approximately 2 kg, and the top of a person's foot is about 0.05 m above the floor. Also, let's assume the trophy falls from a height of 1.4 m.

Calculate the gravitational potential energy of the trophy–Earth system just before the trophy is released:

$$U_i = mgy_i = (2 \text{ kg})(9.80 \text{ m/s}^2)(1.4 \text{ m}) = 27.4 \text{ J}$$

Calculate the gravitational potential energy of the trophy–Earth system when the trophy reaches the athlete's foot:

$$U_f = mgy_f = (2 \text{ kg})(9.80 \text{ m/s}^2)(0.05 \text{ m}) = 0.98 \text{ J}$$

Evaluate the change in gravitational potential energy of the trophy–Earth system:

$$\Delta U_g = 0.98 \text{ J} - 27.4 \text{ J} = -26.4 \text{ J}$$

continued

▶ **7.8** continued

We should probably keep only two digits because of the roughness of our estimates; therefore, we estimate that the change in gravitational potential energy is $-26\,\text{J}$. The system had about 27 J of gravitational potential energy before the trophy began its fall and approximately 1 J of potential energy as the trophy reaches the top of the foot.

The second case presented indicates that the reference configuration of the system for zero potential energy is chosen to be when the trophy is on the athlete's head (even though the trophy is never at this position in its motion). We estimate this position to be 2.0 m above the floor).

Calculate the gravitational potential energy of the trophy–Earth system just before the trophy is released from its position 0.6 m below the athlete's head:

$$U_i = mgy_i = (2\,\text{kg})(9.80\,\text{m/s}^2)(-0.6\,\text{m}) = -11.8\,\text{J}$$

Calculate the gravitational potential energy of the trophy–Earth system when the trophy reaches the athlete's foot located 1.95 m below its initial position:

$$U_f = mgy_f = (2\,\text{kg})(9.80\,\text{m/s}^2)(-1.95\,\text{m}) = -38.2\,\text{J}$$

Evaluate the change in gravitational potential energy of the trophy–Earth system:

$$\Delta U_g = -38.2\,\text{J} - (-11.8\,\text{J}) = -26.4\,\text{J} \approx -26\,\text{J}$$

This value is the same as before, as it must be. The change in potential energy is independent of the choice of configuration of the system representing the zero of potential energy. If we wanted to keep only one digit in our estimates, we could write the final result as $3 \times 10^1\,\text{J}$.

Elastic Potential Energy

Because members of a system can interact with one another by means of different types of forces, it is possible that there are different types of potential energy in a system. We have just become familiar with gravitational potential energy of a system in which members interact via the gravitational force. Let us explore a second type of potential energy that a system can possess.

Consider a system consisting of a block and a spring as shown in Figure 7.16. In Section 7.4, we identified *only* the block as the system. Now we include both the block and the spring in the system and recognize that the spring force is the interaction between the two members of the system. The force that the spring exerts on the block is given by $F_s = -kx$ (Eq. 7.9). The external work done by an applied force F_{app} on the block–spring system is given by Equation 7.13:

$$W_{ext} = \tfrac{1}{2}kx_f^{\,2} - \tfrac{1}{2}kx_i^{\,2} \tag{7.21}$$

In this situation, the initial and final x coordinates of the block are measured from its equilibrium position, $x = 0$. Again (as in the gravitational case, Eq. 7.18) the work done on the system is equal to the difference between the initial and final values of an expression related to the system's configuration. The **elastic potential energy** function associated with the block–spring system is defined by

Elastic potential energy ▶

$$U_s \equiv \tfrac{1}{2}kx^2 \tag{7.22}$$

Equation 7.21 can be expressed as

$$W_{ext} = \Delta U_s \tag{7.23}$$

Compare this equation to Equations 7.17 and 7.20. In all three situations, external work is done on a system and a form of energy storage in the system changes as a result.

The elastic potential energy of the system can be thought of as the energy stored in the deformed spring (one that is either compressed or stretched from its equilibrium position). The elastic potential energy stored in a spring is zero whenever the spring is undeformed ($x = 0$). Energy is stored in the spring only when the spring is

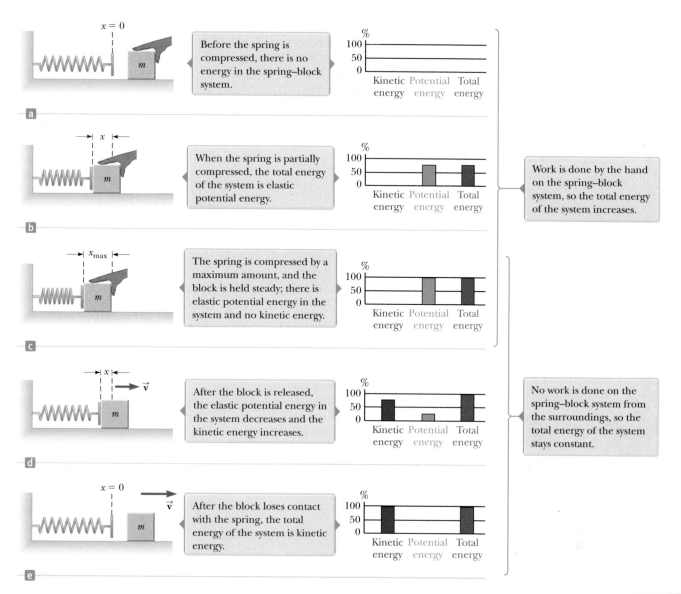

Figure 7.16 A spring on a frictionless, horizontal surface is compressed a distance x_{max} when a block of mass m is pushed against it. The block is then released and the spring pushes it to the right, where the block eventually loses contact with the spring. Parts (a) through (e) show various instants in the process. Energy bar charts on the right of each part of the figure help keep track of the energy in the system.

either stretched or compressed. Because the elastic potential energy is proportional to x^2, we see that U_s is always positive in a deformed spring. Everyday examples of the storage of elastic potential energy can be found in old-style clocks or watches that operate from a wound-up spring and small wind-up toys for children.

Consider Figure 7.16 once again, which shows a spring on a frictionless, horizontal surface. When a block is pushed against the spring by an external agent, the elastic potential energy and the total energy of the system increase as indicated in Figure 7.16b. When the spring is compressed a distance x_{max} (Fig. 7.16c), the elastic potential energy stored in the spring is $\frac{1}{2}kx_{max}^2$. When the block is released from rest, the spring exerts a force on the block and pushes the block to the right. The elastic potential energy of the system decreases, whereas the kinetic energy increases and the total energy remains fixed (Fig. 7.16d). When the spring returns to its original length, the stored elastic potential energy is completely transformed into kinetic energy of the block (Fig. 7.16e).

Figure 7.17 (Quick Quiz 7.7) A ball connected to a massless spring suspended vertically. What forms of potential energy are associated with the system when the ball is displaced downward?

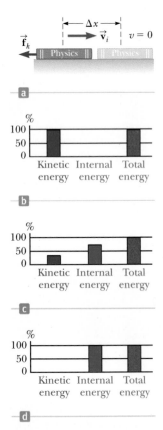

Figure 7.18 (a) A book sliding to the right on a horizontal surface slows down in the presence of a force of kinetic friction acting to the left. (b) An energy bar chart showing the energy in the system of the book and the surface at the initial instant of time. The energy of the system is all kinetic energy. (c) While the book is sliding, the kinetic energy of the system decreases as it is transformed to internal energy. (d) After the book has stopped, the energy of the system is all internal energy.

Quick Quiz 7.7 A ball is connected to a light spring suspended vertically as shown in Figure 7.17. When pulled downward from its equilibrium position and released, the ball oscillates up and down. **(i)** In the system of *the ball, the spring, and the Earth,* what forms of energy are there during the motion? (a) kinetic and elastic potential (b) kinetic and gravitational potential (c) kinetic, elastic potential, and gravitational potential (d) elastic potential and gravitational potential **(ii)** In the system of *the ball and the spring,* what forms of energy are there during the motion? Choose from the same possibilities (a) through (d).

Energy Bar Charts

Figure 7.16 shows an important graphical representation of information related to energy of systems called an **energy bar chart.** The vertical axis represents the amount of energy of a given type in the system. The horizontal axis shows the types of energy in the system. The bar chart in Figure 7.16a shows that the system contains zero energy because the spring is relaxed and the block is not moving. Between Figure 7.16a and Figure 7.16c, the hand does work on the system, compressing the spring and storing elastic potential energy in the system. In Figure 7.16d, the block has been released and is moving to the right while still in contact with the spring. The height of the bar for the elastic potential energy of the system decreases, the kinetic energy bar increases, and the total energy bar remains fixed. In Figure 7.16e, the spring has returned to its relaxed length and the system now contains only kinetic energy associated with the moving block.

Energy bar charts can be a very useful representation for keeping track of the various types of energy in a system. For practice, try making energy bar charts for the book–Earth system in Figure 7.15 when the book is dropped from the higher position. Figure 7.17 associated with Quick Quiz 7.7 shows another system for which drawing an energy bar chart would be a good exercise. We will show energy bar charts in some figures in this chapter. Some figures will not show a bar chart in the text but will include one in animated versions that appear in Enhanced WebAssign.

7.7 Conservative and Nonconservative Forces

We now introduce a third type of energy that a system can possess. Imagine that the book in Figure 7.18a has been accelerated by your hand and is now sliding to the right on the surface of a heavy table and slowing down due to the friction force. Suppose the *surface* is the system. Then the friction force from the sliding book does work on the surface. The force on the surface is to the right and the displacement of the point of application of the force is to the right because the book has moved to the right. The work done on the surface is therefore positive, but the surface is not moving after the book has stopped. Positive work has been done on the surface, yet there is no increase in the surface's kinetic energy or the potential energy of any system. So where is the energy?

From your everyday experience with sliding over surfaces with friction, you can probably guess that the surface will be *warmer* after the book slides over it. The work that was done on the surface has gone into warming the surface rather than increasing its speed or changing the configuration of a system. We call the energy associated with the temperature of a system its **internal energy,** symbolized E_{int}. (We will define internal energy more generally in Chapter 20.) In this case, the work done on the surface does indeed represent energy transferred into the system, but it appears in the system as internal energy rather than kinetic or potential energy.

Now consider the book and the surface in Figure 7.18a together as a system. Initially, the system has kinetic energy because the book is moving. While the book is sliding, the internal energy of the system increases: the book and the surface are warmer than before. When the book stops, the kinetic energy has been completely transformed to internal energy. We can consider the nonconservative force within the system—that is, between the book and the surface—as a *transformation mechanism* for energy. This nonconservative force transforms the kinetic energy of the system into internal energy. Rub your hands together briskly to experience this effect!

Figures 7.18b through 7.18d show energy bar charts for the situation in Figure 7.18a. In Figure 7.18b, the bar chart shows that the system contains kinetic energy at the instant the book is released by your hand. We define the reference amount of internal energy in the system as zero at this instant. Figure 7.18c shows the kinetic energy transforming to internal energy as the book slows down due to the friction force. In Figure 7.18d, after the book has stopped sliding, the kinetic energy is zero, and the system now contains only internal energy E_{int}. Notice that the total energy bar in red has not changed during the process. The amount of internal energy in the system after the book has stopped is equal to the amount of kinetic energy in the system at the initial instant. This equality is described by an important principle called *conservation of energy*. We will explore this principle in Chapter 8.

Now consider in more detail an object moving downward near the surface of the Earth. The work done by the gravitational force on the object does not depend on whether it falls vertically or slides down a sloping incline with friction. All that matters is the change in the object's elevation. The energy transformation to internal energy due to friction on that incline, however, depends very much on the distance the object slides. The longer the incline, the more potential energy is transformed to internal energy. In other words, the path makes no difference when we consider the work done by the gravitational force, but it does make a difference when we consider the energy transformation due to friction forces. We can use this varying dependence on path to classify forces as either *conservative* or *nonconservative*. Of the two forces just mentioned, the gravitational force is conservative and the friction force is nonconservative.

Conservative Forces

Conservative forces have these two equivalent properties:

1. The work done by a conservative force on a particle moving between any two points is independent of the path taken by the particle.
2. The work done by a conservative force on a particle moving through any closed path is zero. (A closed path is one for which the beginning point and the endpoint are identical.)

◀ **Properties of conservative forces**

The gravitational force is one example of a conservative force; the force that an ideal spring exerts on any object attached to the spring is another. The work done by the gravitational force on an object moving between any two points near the Earth's surface is $W_g = -mg\hat{\mathbf{j}} \cdot [(y_f - y_i)\hat{\mathbf{j}}] = mgy_i - mgy_f$. From this equation, notice that W_g depends only on the initial and final y coordinates of the object and hence is independent of the path. Furthermore, W_g is zero when the object moves over any closed path (where $y_i = y_f$).

For the case of the object–spring system, the work W_s done by the spring force is given by $W_s = \frac{1}{2}kx_i^2 - \frac{1}{2}kx_f^2$ (Eq. 7.12). We see that the spring force is conservative because W_s depends only on the initial and final x coordinates of the object and is zero for any closed path.

We can associate a potential energy for a system with a force acting between members of the system, but we can do so only if the force is conservative. In general, the work W_{int} done by a conservative force on an object that is a member of a system as the system changes from one configuration to another is equal to the initial value of the potential energy of the system minus the final value:

$$W_{int} = U_i - U_f = -\Delta U \tag{7.24}$$

The subscript "int" in Equation 7.24 reminds us that the work we are discussing is done by one member of the system on another member and is therefore *internal* to

Pitfall Prevention 7.9
Similar Equation Warning Compare Equation 7.24 with Equation 7.20. These equations are similar except for the negative sign, which is a common source of confusion. Equation 7.20 tells us that positive work done *by an outside agent* on a system causes an increase in the potential energy of the system (with no change in the kinetic or internal energy). Equation 7.24 states that positive work done on a component of a system by a conservative force *internal to the system* causes a decrease in the potential energy of the system.

the system. It is different from the work W_{ext} done *on* the system as a whole by an external agent. As an example, compare Equation 7.24 with the equation for the work done by an external agent on a block–spring system (Eq. 7.23) as the extension of the spring changes.

Nonconservative Forces

A force is **nonconservative** if it does not satisfy properties 1 and 2 above. The work done by a nonconservative force is path-dependent. We define the sum of the kinetic and potential energies of a system as the **mechanical energy** of the system:

$$E_{mech} \equiv K + U \tag{7.25}$$

where K includes the kinetic energy of all moving members of the system and U includes all types of potential energy in the system. For a book falling under the action of the gravitational force, the mechanical energy of the book–Earth system remains fixed; gravitational potential energy transforms to kinetic energy, and the total energy of the system remains constant. Nonconservative forces acting within a system, however, cause a *change* in the mechanical energy of the system. For example, for a book sent sliding on a horizontal surface that is not frictionless (Fig. 7.18a), the mechanical energy of the book–surface system is transformed to internal energy as we discussed earlier. Only part of the book's kinetic energy is transformed to internal energy in the book. The rest appears as internal energy in the surface. (When you trip and slide across a gymnasium floor, not only does the skin on your knees warm up, so does the floor!) Because the force of kinetic friction transforms the mechanical energy of a system into internal energy, it is a nonconservative force.

As an example of the path dependence of the work for a nonconservative force, consider Figure 7.19. Suppose you displace a book between two points on a table. If the book is displaced in a straight line along the blue path between points Ⓐ and Ⓑ in Figure 7.19, you do a certain amount of work against the kinetic friction force to keep the book moving at a constant speed. Now, imagine that you push the book along the brown semicircular path in Figure 7.19. You perform more work against friction along this curved path than along the straight path because the curved path is longer. The work done on the book depends on the path, so the friction force *cannot* be conservative.

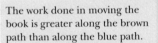

The work done in moving the book is greater along the brown path than along the blue path.

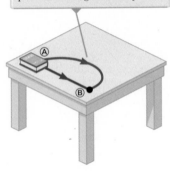

Figure 7.19 The work done against the force of kinetic friction depends on the path taken as the book is moved from Ⓐ to Ⓑ.

7.8 Relationship Between Conservative Forces and Potential Energy

In the preceding section, we found that the work done on a member of a system by a conservative force between the members of the system does not depend on the path taken by the moving member. The work depends only on the initial and final coordinates. For such a system, we can define a **potential energy function** U such that the work done within the system by the conservative force equals the negative of the change in the potential energy of the system according to Equation 7.24. Let us imagine a system of particles in which a conservative force $\vec{F}$ acts between the particles. Imagine also that the configuration of the system changes due to the motion of one particle along the x axis. Then we can evaluate the internal work done by this force as the particle moves along the x axis[4] using Equations 7.7 and 7.24:

$$W_{int} = \int_{x_i}^{x_f} F_x \, dx = -\Delta U \tag{7.26}$$

---

[4]For a general displacement, the work done in two or three dimensions also equals $-\Delta U$, where $U = U(x, y, z)$. We write this equation formally as $W_{int} = \int_i^f \vec{F} \cdot d\vec{r} = U_i - U_f$.

where F_x is the component of $\vec{F}$ in the direction of the displacement. We can also express Equation 7.26 as

$$\Delta U = U_f - U_i = -\int_{x_i}^{x_f} F_x \, dx \qquad (7.27)$$

Therefore, ΔU is negative when F_x and dx are in the same direction, as when an object is lowered in a gravitational field or when a spring pushes an object toward equilibrium.

It is often convenient to establish some particular location x_i of one member of a system as representing a reference configuration and measure all potential energy differences with respect to it. We can then define the potential energy function as

$$U_f(x) = -\int_{x_i}^{x_f} F_x \, dx + U_i \qquad (7.28)$$

The value of U_i is often taken to be zero for the reference configuration. It does not matter what value we assign to U_i because any nonzero value merely shifts $U_f(x)$ by a constant amount and only the *change* in potential energy is physically meaningful.

If the point of application of the force undergoes an infinitesimal displacement dx, we can express the infinitesimal change in the potential energy of the system dU as

$$dU = -F_x \, dx$$

Therefore, the conservative force is related to the potential energy function through the relationship[5]

$$\boxed{F_x = -\frac{dU}{dx}} \qquad (7.29)$$

◀ Relation of force between members of a system to the potential energy of the system

That is, the x component of a conservative force acting on a member within a system equals the negative derivative of the potential energy of the system with respect to x.

We can easily check Equation 7.29 for the two examples already discussed. In the case of the deformed spring, $U_s = \frac{1}{2}kx^2$; therefore,

$$F_s = -\frac{dU_s}{dx} = -\frac{d}{dx}\left(\tfrac{1}{2}kx^2\right) = -kx$$

which corresponds to the restoring force in the spring (Hooke's law). Because the gravitational potential energy function is $U_g = mgy$, it follows from Equation 7.29 that $F_g = -mg$ when we differentiate U_g with respect to y instead of x.

We now see that U is an important function because a conservative force can be derived from it. Furthermore, Equation 7.29 should clarify that adding a constant to the potential energy is unimportant because the derivative of a constant is zero.

Quick Quiz 7.8 What does the slope of a graph of $U(x)$ versus x represent? **(a)** the magnitude of the force on the object **(b)** the negative of the magnitude of the force on the object **(c)** the x component of the force on the object **(d)** the negative of the x component of the force on the object

7.9 Energy Diagrams and Equilibrium of a System

The motion of a system can often be understood qualitatively through a graph of its potential energy versus the position of a member of the system. Consider the potential

---

[5]In three dimensions, the expression is

$$\vec{F} = -\frac{\partial U}{\partial x}\hat{i} - \frac{\partial U}{\partial y}\hat{j} - \frac{\partial U}{\partial z}\hat{k}$$

where $(\partial U/\partial x)$ and so forth are partial derivatives. In the language of vector calculus, $\vec{F}$ equals the negative of the *gradient* of the scalar quantity $U(x, y, z)$.

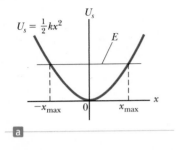

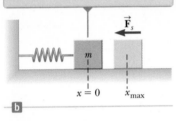

The restoring force exerted by the spring always acts toward $x = 0$, the position of stable equilibrium.

Figure 7.20 (a) Potential energy as a function of x for the friction-less block–spring system shown in (b). For a given energy E of the system, the block oscillates between the turning points, which have the coordinates $x = \pm x_{max}$.

Pitfall Prevention 7.10

Energy Diagrams A common mistake is to think that potential energy on the graph in an energy diagram represents the height of some object. For example, that is not the case in Figure 7.20, where the block is only moving horizontally.

energy function for a block–spring system, given by $U_s = \frac{1}{2}kx^2$. This function is plotted versus x in Figure 7.20a, where x is the position of the block. The force F_s exerted by the spring on the block is related to U_s through Equation 7.29:

$$F_s = -\frac{dU_s}{dx} = -kx$$

As we saw in Quick Quiz 7.8, the x component of the force is equal to the nega-tive of the slope of the U-versus-x curve. When the block is placed at rest at the equilibrium position of the spring ($x = 0$), where $F_s = 0$, it will remain there unless some external force F_{ext} acts on it. If this external force stretches the spring from equilibrium, x is positive and the slope dU/dx is positive; therefore, the force F_s exerted by the spring is negative and the block accelerates back toward $x = 0$ when released. If the external force compresses the spring, x is negative and the slope is negative; therefore, F_s is positive and again the mass accelerates toward $x = 0$ upon release.

From this analysis, we conclude that the $x = 0$ position for a block–spring sys-tem is one of **stable equilibrium.** That is, any movement away from this position results in a force directed back toward $x = 0$. In general, configurations of a sys-tem in stable equilibrium correspond to those for which $U(x)$ for the system is a minimum.

If the block in Figure 7.20 is moved to an initial position x_{max} and then released from rest, its total energy initially is the potential energy $\frac{1}{2}kx_{max}^2$ stored in the spring. As the block starts to move, the system acquires kinetic energy and loses potential energy. The block oscillates (moves back and forth) between the two points $x = -x_{max}$ and $x = +x_{max}$, called the *turning points.* In fact, because no energy is trans-formed to internal energy due to friction, the block oscillates between $-x_{max}$ and $+x_{max}$ forever. (We will discuss these oscillations further in Chapter 15.)

Another simple mechanical system with a configuration of stable equilibrium is a ball rolling about in the bottom of a bowl. Anytime the ball is displaced from its lowest position, it tends to return to that position when released.

Now consider a particle moving along the x axis under the influence of a conser-vative force F_x, where the U-versus-x curve is as shown in Figure 7.21. Once again, $F_x = 0$ at $x = 0$, and so the particle is in equilibrium at this point. This position, however, is one of **unstable equilibrium** for the following reason. Suppose the particle is displaced to the right ($x > 0$). Because the slope is negative for $x > 0$, $F_x = -dU/dx$ is positive and the particle accelerates away from $x = 0$. If instead the particle is at $x = 0$ and is displaced to the left ($x < 0$), the force is negative because the slope is positive for $x < 0$ and the particle again accelerates away from the equi-librium position. The position $x = 0$ in this situation is one of unstable equilibrium because for any displacement from this point, the force pushes the particle farther away from equilibrium and toward a position of lower potential energy. A pencil balanced on its point is in a position of unstable equilibrium. If the pencil is dis-placed slightly from its absolutely vertical position and is then released, it will surely fall over. In general, configurations of a system in unstable equilibrium correspond to those for which $U(x)$ for the system is a maximum.

Finally, a configuration called **neutral equilibrium** arises when U is constant over some region. Small displacements of an object from a position in this region produce neither restoring nor disrupting forces. A ball lying on a flat, horizontal surface is an example of an object in neutral equilibrium.

Figure 7.21 A plot of U versus x for a particle that has a position of unstable equilibrium located at $x = 0$. For any finite displace-ment of the particle, the force on the particle is directed away from $x = 0$.

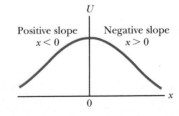

| Example 7.9 | Force and Energy on an Atomic Scale |

The potential energy associated with the force between two neutral atoms in a molecule can be modeled by the Lennard–Jones potential energy function:

$$U(x) = 4\epsilon\left[\left(\frac{\sigma}{x}\right)^{12} - \left(\frac{\sigma}{x}\right)^{6}\right]$$

where x is the separation of the atoms. The function $U(x)$ contains two parameters σ and ϵ that are determined from experiments. Sample values for the interaction between two atoms in a molecule are $\sigma = 0.263$ nm and $\epsilon = 1.51 \times 10^{-22}$ J. Using a spreadsheet or similar tool, graph this function and find the most likely distance between the two atoms.

SOLUTION

Conceptualize We identify the two atoms in the molecule as a system. Based on our understanding that stable molecules exist, we expect to find stable equilibrium when the two atoms are separated by some equilibrium distance.

Categorize Because a potential energy function exists, we categorize the force between the atoms as conservative. For a conservative force, Equation 7.29 describes the relationship between the force and the potential energy function.

Analyze Stable equilibrium exists for a separation distance at which the potential energy of the system of two atoms (the molecule) is a minimum.

Take the derivative of the function $U(x)$:

$$\frac{dU(x)}{dx} = 4\epsilon\frac{d}{dx}\left[\left(\frac{\sigma}{x}\right)^{12} - \left(\frac{\sigma}{x}\right)^{6}\right] = 4\epsilon\left[\frac{-12\sigma^{12}}{x^{13}} + \frac{6\sigma^{6}}{x^{7}}\right]$$

Minimize the function $U(x)$ by setting its derivative equal to zero:

$$4\epsilon\left[\frac{-12\sigma^{12}}{x_{eq}^{13}} + \frac{6\sigma^{6}}{x_{eq}^{7}}\right] = 0 \quad \rightarrow \quad x_{eq} = (2)^{1/6}\sigma$$

Evaluate x_{eq}, the equilibrium separation of the two atoms in the molecule:

$$x_{eq} = (2)^{1/6}(0.263 \text{ nm}) = \boxed{2.95 \times 10^{-10} \text{ m}}$$

We graph the Lennard–Jones function on both sides of this critical value to create our energy diagram as shown in Figure 7.22.

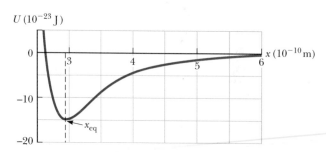

Finalize Notice that $U(x)$ is extremely large when the atoms are very close together, is a minimum when the atoms are at their critical separation, and then increases again as the atoms move apart. When $U(x)$ is a minimum, the atoms are in stable equilibrium, indicating that the most likely separation between them occurs at this point.

Figure 7.22 (Example 7.9) Potential energy curve associated with a molecule. The distance x is the separation between the two atoms making up the molecule.

Summary

Definitions

▪ A **system** is most often a single particle, a collection of particles, or a region of space, and may vary in size and shape. A **system boundary** separates the system from the **environment.**

▪ The **work** W done on a system by an agent exerting a constant force $\vec{\mathbf{F}}$ on the system is the product of the magnitude Δr of the displacement of the point of application of the force and the component $F\cos\theta$ of the force along the direction of the displacement $\Delta\vec{\mathbf{r}}$:

$$W \equiv F\Delta r\cos\theta \qquad (7.1)$$

continued

■ If a varying force does work on a particle as the particle moves along the x axis from x_i to x_f, the work done by the force on the particle is given by

$$W = \int_{x_i}^{x_f} F_x \, dx \qquad (7.7)$$

where F_x is the component of force in the x direction.

■ The **scalar product** (dot product) of two vectors $\vec{A}$ and $\vec{B}$ is defined by the relationship

$$\vec{A} \cdot \vec{B} \equiv AB \cos \theta \qquad (7.2)$$

where the result is a scalar quantity and θ is the angle between the two vectors. The scalar product obeys the commutative and distributive laws.

■ The **kinetic energy** of a particle of mass m moving with a speed v is

$$K \equiv \tfrac{1}{2}mv^2 \qquad (7.16)$$

■ If a particle of mass m is at a distance y above the Earth's surface, the **gravitational potential energy** of the particle–Earth system is

$$U_g \equiv mgy \qquad (7.19)$$

The **elastic potential energy** stored in a spring of force constant k is

$$U_s \equiv \tfrac{1}{2}kx^2 \qquad (7.22)$$

■ A force is **conservative** if the work it does on a particle that is a member of the system as the particle moves between two points is independent of the path the particle takes between the two points. Furthermore, a force is conservative if the work it does on a particle is zero when the particle moves through an arbitrary closed path and returns to its initial position. A force that does not meet these criteria is said to be **nonconservative.**

■ The **total mechanical energy of a system** is defined as the sum of the kinetic energy and the potential energy:

$$E_{\text{mech}} \equiv K + U \qquad (7.25)$$

Concepts and Principles

■ The **work–kinetic energy theorem** states that if work is done on a system by external forces and the only change in the system is in its speed,

$$W_{\text{ext}} = K_f - K_i = \Delta K = \tfrac{1}{2}mv_f^2 - \tfrac{1}{2}mv_i^2 \quad (7.15, 7.17)$$

■ A **potential energy function** U can be associated only with a conservative force. If a conservative force $\vec{F}$ acts between members of a system while one member moves along the x axis from x_i to x_f, the change in the potential energy of the system equals the negative of the work done by that force:

$$U_f - U_i = -\int_{x_i}^{x_f} F_x \, dx \qquad (7.27)$$

■ Systems can be in three types of equilibrium configurations when the net force on a member of the system is zero. Configurations of **stable equilibrium** correspond to those for which $U(x)$ is a minimum.

■ Configurations of **unstable equilibrium** correspond to those for which $U(x)$ is a maximum.

■ **Neutral equilibrium** arises when U is constant as a member of the system moves over some region.

Objective Questions   <small>**1.** denotes answer available in *Student Solutions Manual/Study Guide*</small>

1. Alex and John are loading identical cabinets onto a truck. Alex lifts his cabinet straight up from the ground to the bed of the truck, whereas John slides his cabinet up a rough ramp to the truck. Which statement is correct about the work done on the cabinet–Earth system? (a) Alex and John do the same amount of work. (b) Alex does more work than John. (c) John does more work than Alex. (d) None of those statements is necessarily true because the force of friction is unknown. (e) None of those statements is necessarily true because the angle of the incline is unknown.

2. If the net work done by external forces on a particle is zero, which of the following statements about the particle must be true? (a) Its velocity is zero. (b) Its velocity is decreased. (c) Its velocity is unchanged. (d) Its speed is unchanged. (e) More information is needed.

3. A worker pushes a wheelbarrow with a horizontal force of 50 N on level ground over a distance of 5.0 m. If a friction force of 43 N acts on the wheelbarrow in a direction opposite that of the worker, what work is done on the wheelbarrow by the worker? (a) 250 J (b) 215 J (c) 35 J (d) 10 J (e) None of those answers is correct.

4. A cart is set rolling across a level table, at the same speed on every trial. If it runs into a patch of sand, the cart exerts on the sand an average horizontal force of 6 N and travels a distance of 6 cm through the sand as it comes to a stop. If instead the cart runs into a patch of gravel on which the cart exerts an average horizontal force of 9 N, how far into the gravel will the cart roll before stopping? (a) 9 cm (b) 6 cm (c) 4 cm (d) 3 cm (e) none of those answers

5. Let $\hat{\mathbf{N}}$ represent the direction horizontally north, $\widehat{\mathbf{NE}}$ represent northeast (halfway between north and east), and so on. Each direction specification can be thought of as a unit vector. Rank from the largest to the smallest the following dot products. Note that zero is larger than a negative number. If two quantities are equal, display that fact in your ranking. (a) $\hat{\mathbf{N}} \cdot \hat{\mathbf{N}}$ (b) $\hat{\mathbf{N}} \cdot \widehat{\mathbf{NE}}$ (c) $\hat{\mathbf{N}} \cdot \hat{\mathbf{S}}$ (d) $\hat{\mathbf{N}} \cdot \hat{\mathbf{E}}$ (e) $\widehat{\mathbf{SE}} \cdot \hat{\mathbf{S}}$

6. Is the work required to be done by an external force on an object on a frictionless, horizontal surface to accelerate it from a speed v to a speed $2v$ (a) equal to the work required to accelerate the object from $v = 0$ to v, (b) twice the work required to accelerate the object from $v = 0$ to v, (c) three times the work required to accelerate the object from $v = 0$ to v, (d) four times the work required to accelerate the object from 0 to v, or (e) not known without knowledge of the acceleration?

7. A block of mass m is dropped from the fourth floor of an office building and hits the sidewalk below at speed v. From what floor should the block be dropped to double that impact speed? (a) the sixth floor (b) the eighth floor (c) the tenth floor (d) the twelfth floor (e) the sixteenth floor

8. As a simple pendulum swings back and forth, the forces acting on the suspended object are (a) the gravitational force, (b) the tension in the supporting cord, and (c) air resistance. **(i)** Which of these forces, if any, does no work on the pendulum at any time? **(ii)** Which of these forces does negative work on the pendulum at all times during its motion?

9. Bullet 2 has twice the mass of bullet 1. Both are fired so that they have the same speed. If the kinetic energy of bullet 1 is K, is the kinetic energy of bullet 2 (a) $0.25K$, (b) $0.5K$, (c) $0.71K$, (d) K, or (e) $2K$?

10. Figure OQ7.10 shows a light extended spring exerting a force F_s to the left on a block. **(i)** Does the block exert a force on the spring? Choose every correct answer. (a) No, it doesn't. (b) Yes, it does, to the left. (c) Yes, it does, to the right. (d) Yes, it does, and its magnitude is larger than F_s. (e) Yes, it does, and its magnitude is equal to F_s. **(ii)** Does the spring exert a force

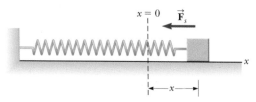

Figure OQ7.10

on the wall? Choose your answers from the same list (a) through (e).

11. If the speed of a particle is doubled, what happens to its kinetic energy? (a) It becomes four times larger. (b) It becomes two times larger. (c) It becomes $\sqrt{2}$ times larger. (d) It is unchanged. (e) It becomes half as large.

12. Mark and David are loading identical cement blocks onto David's pickup truck. Mark lifts his block straight up from the ground to the truck, whereas David slides his block up a ramp containing frictionless rollers. Which statement is true about the work done on the block–Earth system? (a) Mark does more work than David. (b) Mark and David do the same amount of work. (c) David does more work than Mark. (d) None of those statements is necessarily true because the angle of the incline is unknown. (e) None of those statements is necessarily true because the mass of one block is not given.

13. **(i)** Rank the gravitational accelerations you would measure for the following falling objects: (a) a 2-kg object 5 cm above the floor, (b) a 2-kg object 120 cm above the floor, (c) a 3-kg object 120 cm above the floor, and (d) a 3-kg object 80 cm above the floor. List the one with the largest magnitude of acceleration first. If any are equal, show their equality in your list. **(ii)** Rank the gravitational forces on the same four objects, listing the one with the largest magnitude first. **(iii)** Rank the gravitational potential energies (of the object–Earth system) for the same four objects, largest first, taking $y = 0$ at the floor.

14. A certain spring that obeys Hooke's law is stretched by an external agent. The work done in stretching the spring by 10 cm is 4 J. How much additional work is required to stretch the spring an additional 10 cm? (a) 2 J (b) 4 J (c) 8 J (d) 12 J (e) 16 J

15. A cart is set rolling across a level table, at the same speed on every trial. If it runs into a patch of sand, the cart exerts on the sand an average horizontal force of 6 N and travels a distance of 6 cm through the sand as it comes to a stop. If instead the cart runs into a patch of flour, it rolls an average of 18 cm before stopping. What is the average magnitude of the horizontal force the cart exerts on the flour? (a) 2 N (b) 3 N (c) 6 N (d) 18 N (e) none of those answers

16. An ice cube has been given a push and slides without friction on a level table. Which is correct? (a) It is in stable equilibrium. (b) It is in unstable equilibrium. (c) It is in neutral equilibrium. (d) It is not in equilibrium.

Conceptual Questions

1. denotes answer available in *Student Solutions Manual/Study Guide*

1. Can a normal force do work? If not, why not? If so, give an example.

2. Object 1 pushes on object 2 as the objects move together, like a bulldozer pushing a stone. Assume object 1 does 15.0 J of work on object 2. Does object 2 do work on object 1? Explain your answer. If possible, determine how much work and explain your reasoning.

3. A student has the idea that the total work done on an object is equal to its final kinetic energy. Is this idea true always, sometimes, or never? If it is sometimes true, under what circumstances? If it is always or never true, explain why.

4. (a) For what values of the angle θ between two vectors is their scalar product positive? (b) For what values of θ is their scalar product negative?

5. Can kinetic energy be negative? Explain.

6. Discuss the work done by a pitcher throwing a baseball. What is the approximate distance through which the force acts as the ball is thrown?

7. Discuss whether any work is being done by each of the following agents and, if so, whether the work is positive or negative. (a) a chicken scratching the ground (b) a person studying (c) a crane lifting a bucket of concrete (d) the gravitational force on the bucket in part (c) (e) the leg muscles of a person in the act of sitting down

8. If only one external force acts on a particle, does it necessarily change the particle's (a) kinetic energy? (b) Its velocity?

9. Preparing to clean them, you pop all the removable keys off a computer keyboard. Each key has the shape of a tiny box with one side open. By accident, you spill the keys onto the floor. Explain why many more keys land letter-side down than land open-side down.

10. You are reshelving books in a library. You lift a book from the floor to the top shelf. The kinetic energy of the book on the floor was zero and the kinetic energy of the book on the top shelf is zero, so no change occurs in the kinetic energy, yet you did some work in lifting the book. Is the work–kinetic energy theorem violated? Explain.

11. A certain uniform spring has spring constant k. Now the spring is cut in half. What is the relationship between k and the spring constant k' of each resulting smaller spring? Explain your reasoning.

12. What shape would the graph of U versus x have if a particle were in a region of neutral equilibrium?

13. Does the kinetic energy of an object depend on the frame of reference in which its motion is measured? Provide an example to prove this point.

14. Cite two examples in which a force is exerted on an object without doing any work on the object.

Problems available in

 Access end-of-chapter problems online at www.webassign.net

Section 7.2 Work Done by a Constant Force
Problems 1–6

Section 7.3 The Scalar Product of Two Vectors
Problems 7–13

Section 7.4 Work Done by a Varying Force
Problems 14–30

Section 7.5 Kinetic Energy and the Work–Kinetic Energy Theorem
Problems 31–39

Section 7.6 Potential Energy of a System
Problems 40–42

Section 7.7 Conservative and Nonconservative Forces
Problems 43–46

Section 7.8 Relationship Between Conservative Forces and Potential Energy
Problems 47–51

Section 7.9 Energy Diagrams and Equilibrium of a System
Problems 52–53

Additional Problems
Problems 54–65

Challenge Problems
Problem 66–67

Solutions to the following Problems are available in the *Student Solutions Manual/ Study Guide:*
7.5, 7.6, 7.11, 7.15, 7.17, 7.21, 7.35, 7.40, 7.43, 7.45, 7.47, 7.51, 7.63, and 7.67

List of Enhanced Problems

Problem Number	Targeted Feedback in Enhanced WebAssign	Analysis Model Tutorial in Enhanced WebAssign	Master It in Enhanced WebAssign	Watch It in Enhanced WebAssign
7.2	✓			✓
7.5	✓			✓
7.6			✓	
7.9	✓			✓
7.11			✓	
7.14	✓		✓	✓
7.15	✓			✓
7.17	✓	✓	✓	
7.29	✓			✓
7.31	✓			✓
7.32		✓		
7.33	✓			✓
7.34	✓			✓
7.35			✓	
7.36		✓		
7.42	✓			✓
7.43	✓		✓	
7.45	✓		✓	
7.51			✓	

Conservation of Energy

In Chapter 7, we introduced three methods for storing energy in a system: kinetic energy, associated with movement of members of the system; potential energy, determined by the configuration of the system; and internal energy, which is related to the temperature of the system.

We now consider analyzing physical situations using the energy approach for two types of systems: *nonisolated* and *isolated* systems. For nonisolated systems, we shall investigate ways that energy can cross the boundary of the system, resulting in a change in the system's total energy. This analysis leads to a critically important principle called *conservation of energy.* The conservation of energy principle extends well beyond physics and can be applied to biological organisms, technological systems, and engineering situations.

In isolated systems, energy does not cross the boundary of the system. For these systems, the total energy of the system is constant. If no nonconservative forces act within the system, we can use *conservation of mechanical energy* to solve a variety of problems.

Three youngsters enjoy the transformation of potential energy to kinetic energy on a waterslide. We can analyze processes such as these with the techniques developed in this chapter.
(Jade Lee/Asia Images/Getty Images)

169

Situations involving the transformation of mechanical energy to internal energy due to nonconservative forces require special handling. We investigate the procedures for these types of problems.

Finally, we recognize that energy can cross the boundary of a system at different rates. We describe the rate of energy transfer with the quantity *power*.

8.1 Analysis Model: Nonisolated System (Energy)

As we have seen, an object, modeled as a particle, can be acted on by various forces, resulting in a change in its kinetic energy according to the work–kinetic energy theorem from Chapter 7. If we choose the object as the system, this very simple situation is the first example of a *nonisolated system,* for which energy crosses the boundary of the system during some time interval due to an interaction with the environment. This scenario is common in physics problems. If a system does not interact with its environment, it is an *isolated system,* which we will study in Section 8.2.

The work–kinetic energy theorem is our first example of an energy equation appropriate for a nonisolated system. In the case of that theorem, the interaction of the system with its environment is the work done by the external force, and the quantity in the system that changes is the kinetic energy.

So far, we have seen only one way to transfer energy into a system: work. We mention below a few other ways to transfer energy into or out of a system. The details of these processes will be studied in other sections of the book. We illustrate mechanisms to transfer energy in Figure 8.1 and summarize them as follows.

Work, as we have learned in Chapter 7, is a method of transferring energy to a system by applying a force to the system such that the point of application of the force undergoes a displacement (Fig. 8.1a).

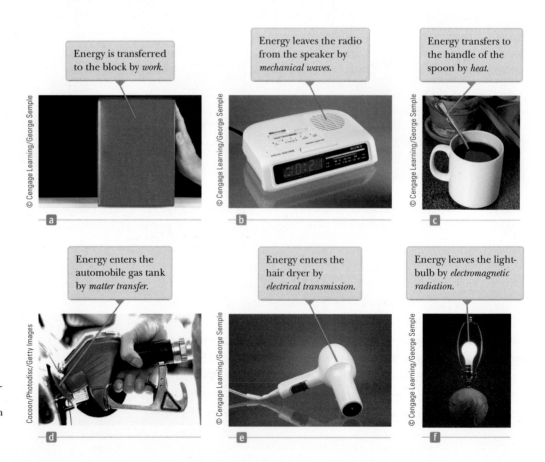

Energy is transferred to the block by *work.*

Energy leaves the radio from the speaker by *mechanical waves.*

Energy transfers to the handle of the spoon by *heat.*

Energy enters the automobile gas tank by *matter transfer.*

Energy enters the hair dryer by *electrical transmission.*

Energy leaves the light-bulb by *electromagnetic radiation.*

© Cengage Learning/George Semple

Cocoon/Photodisc/Getty Images

Figure 8.1 Energy transfer mechanisms. In each case, the system into which or from which energy is transferred is indicated.

Mechanical waves (Chapters 16–18) are a means of transferring energy by allowing a disturbance to propagate through air or another medium. It is the method by which energy (which you detect as sound) leaves the system of your clock radio through the loudspeaker and enters your ears to stimulate the hearing process (Fig. 8.1b). Other examples of mechanical waves are seismic waves and ocean waves.

Heat (Chapter 20) is a mechanism of energy transfer that is driven by a temperature difference between a system and its environment. For example, imagine dividing a metal spoon into two parts: the handle, which we identify as the system, and the portion submerged in a cup of coffee, which is part of the environment (Fig. 8.1c). The handle of the spoon becomes hot because fast-moving electrons and atoms in the submerged portion bump into slower ones in the nearby part of the handle. These particles move faster because of the collisions and bump into the next group of slow particles. Therefore, the internal energy of the spoon handle rises from energy transfer due to this collision process.

Matter transfer (Chapter 20) involves situations in which matter physically crosses the boundary of a system, carrying energy with it. Examples include filling your automobile tank with gasoline (Fig. 8.1d) and carrying energy to the rooms of your home by circulating warm air from the furnace, a process called *convection*.

Electrical transmission (Chapters 27 and 28) involves energy transfer into or out of a system by means of electric currents. It is how energy transfers into your hair dryer (Fig. 8.1e), home theater system, or any other electrical device.

Electromagnetic radiation (Chapter 34) refers to electromagnetic waves such as light (Fig. 8.1f), microwaves, and radio waves crossing the boundary of a system. Examples of this method of transfer include cooking a baked potato in your microwave oven and energy traveling from the Sun to the Earth by light through space.[1]

A central feature of the energy approach is the notion that we can neither create nor destroy energy, that energy is always *conserved*. This feature has been tested in countless experiments, and no experiment has ever shown this statement to be incorrect. Therefore, **if the total amount of energy in a system changes, it can *only* be because energy has crossed the boundary of the system by a transfer mechanism such as one of the methods listed above.**

Energy is one of several quantities in physics that are conserved. We will see other conserved quantities in subsequent chapters. There are many physical quantities that do not obey a conservation principle. For example, there is no conservation of force principle or conservation of velocity principle. Similarly, in areas other than physical quantities, such as in everyday life, some quantities are conserved and some are not. For example, the money in the system of your bank account is a conserved quantity. The only way the account balance changes is if money crosses the boundary of the system by deposits or withdrawals. On the other hand, the number of people in the system of a country is not conserved. Although people indeed cross the boundary of the system, which changes the total population, the population can also change by people dying and by giving birth to new babies. Even if no people cross the system boundary, the births and deaths will change the number of people in the system. There is no equivalent in the concept of energy to dying or giving birth. The general statement of the principle of **conservation of energy** can be described mathematically with the **conservation of energy equation** as follows:

$$\Delta E_{\text{system}} = \sum T \qquad \text{(8.1)}$$

◀ **Conservation of energy**

where E_{system} is the total energy of the system, including all methods of energy storage (kinetic, potential, and internal), and T (for *transfer*) is the amount of energy transferred across the system boundary by some mechanism. Two of our transfer mechanisms have well-established symbolic notations. For work, $T_{\text{work}} = W$ as discussed in Chapter 7, and for heat, $T_{\text{heat}} = Q$ as defined in Chapter 20. (Now that we

Pitfall Prevention 8.1
Heat Is Not a Form of Energy
The word *heat* is one of the most misused words in our popular language. Heat is a method of *transferring* energy, *not* a form of storing energy. Therefore, phrases such as "heat content," "the heat of the summer," and "the heat escaped" all represent uses of this word that are inconsistent with our physics definition. See Chapter 20.

---

[1]Electromagnetic radiation and work done by field forces are the only energy transfer mechanisms that do not require molecules of the environment to be available at the system boundary. Therefore, systems surrounded by a vacuum (such as planets) can only exchange energy with the environment by means of these two possibilities.

are familiar with work, we can simplify the appearance of equations by letting the simple symbol W represent the external work W_{ext} on a system. For internal work, we will *always* use W_{int} to differentiate it from W.) The other four members of our list do not have established symbols, so we will call them T_{MW} (mechanical waves), T_{MT} (matter transfer), T_{ET} (electrical transmission), and T_{ER} (electromagnetic radiation).

The full expansion of Equation 8.1 is

$$\Delta K + \Delta U + \Delta E_{int} = W + Q + T_{MW} + T_{MT} + T_{ET} + T_{ER} \qquad (8.2)$$

which is the primary mathematical representation of the energy version of the analysis model of the **nonisolated system.** (We will see other versions of the nonisolated system model, involving linear momentum and angular momentum, in later chapters.) In most cases, Equation 8.2 reduces to a much simpler one because some of the terms are zero for the specific situation. If, for a given system, all terms on the right side of the conservation of energy equation are zero, the system is an *isolated system*, which we study in the next section.

The conservation of energy equation is no more complicated in theory than the process of balancing your checking account statement. If your account is the system, the change in the account balance for a given month is the sum of all the transfers: deposits, withdrawals, fees, interest, and checks written. You may find it useful to think of energy as the *currency of nature!*

Suppose a force is applied to a nonisolated system and the point of application of the force moves through a displacement. Then suppose the only effect on the system is to change its speed. In this case, the only transfer mechanism is work (so that the right side of Eq. 8.2 reduces to just W) and the only kind of energy in the system that changes is the kinetic energy (so that the left side of Eq. 8.2 reduces to just ΔK). Equation 8.2 then becomes

$$\Delta K = W$$

which is the work–kinetic energy theorem. This theorem is a special case of the more general principle of conservation of energy. We shall see several more special cases in future chapters.

Quick Quiz 8.1 By what transfer mechanisms does energy enter and leave **(a)** your television set? **(b)** Your gasoline-powered lawn mower? **(c)** Your hand-cranked pencil sharpener?

Quick Quiz 8.2 Consider a block sliding over a horizontal surface with friction. Ignore any sound the sliding might make. **(i)** If the system is the *block*, this system is (a) isolated (b) nonisolated (c) impossible to determine **(ii)** If the system is the *surface*, describe the system from the same set of choices. **(iii)** If the system is the *block and the surface*, describe the system from the same set of choices.

Analysis Model   Nonisolated System (Energy)

Imagine you have identified a system to be analyzed and have defined a system boundary. Energy can exist in the system in three forms: kinetic, potential, and internal. The total of that energy can be changed when energy crosses the system boundary by any of six transfer methods shown in the diagram here. The total change in the energy in the system is equal to the total amount of energy that has crossed the system boundary. The mathematical statement of that concept is expressed in the **conservation of energy equation**:

$$\Delta E_{system} = \Sigma T \qquad (8.1)$$

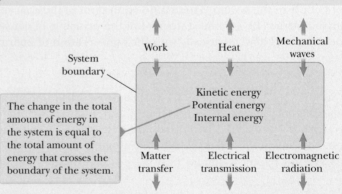

Analysis Model	**Nonisolated System (Energy) (continued)**

The full expansion of Equation 8.1 shows the specific types of energy storage and transfer:

$$\Delta K + \Delta U + \Delta E_{int} = W + Q + T_{MW} + T_{MT} + T_{ET} + T_{ER} \tag{8.2}$$

For a specific problem, this equation is generally reduced to a smaller number of terms by eliminating the terms that are equal to zero because they are not appropriate to the situation.

Examples:

- a force does work on a system of a single object, changing its speed: the work–kinetic energy theorem, $W = \Delta K$
- a gas contained in a vessel has work done on it and experiences a transfer of energy by heat, resulting in a change in its temperature: the first law of thermodynamics, $\Delta E_{int} = W + Q$ (Chapter 20)
- an incandescent light bulb is turned on, with energy entering the filament by electricity, causing its temperature to increase, and leaving by light: $\Delta E_{int} = T_{ET} + T_{ER}$ (Chapter 27)
- a photon enters a metal, causing an electron to be ejected from the metal: the photoelectric effect, $\Delta K + \Delta U = T_{ER}$ (Chapter 40)

8.2 Analysis Model: Isolated System (Energy)

In this section, we study another very common scenario in physics problems: a system is chosen such that no energy crosses the system boundary by any method. We begin by considering a gravitational situation. Think about the book–Earth system in Figure 7.15 in the preceding chapter. After we have lifted the book, there is gravitational potential energy stored in the system, which can be calculated from the work done by the external agent on the system, using $W = \Delta U_g$. (Check to see that this equation, which we've seen before, is contained within Eq. 8.2 above.)

Let us now shift our focus to the work done *on the book alone* by the gravitational force (Fig. 8.2) as the book falls back to its original height. As the book falls from y_i to y_f, the work done by the gravitational force on the book is

$$W_{on\ book} = (m\vec{\mathbf{g}}) \cdot \Delta \vec{\mathbf{r}} = (-mg\hat{\mathbf{j}}) \cdot [(y_f - y_i)\hat{\mathbf{j}}] = mgy_i - mgy_f \tag{8.3}$$

From the work–kinetic energy theorem of Chapter 7, the work done on the book is equal to the change in the kinetic energy of the book:

$$W_{on\ book} = \Delta K_{book}$$

We can equate these two expressions for the work done on the book:

$$\Delta K_{book} = mgy_i - mgy_f \tag{8.4}$$

Let us now relate each side of this equation to the *system* of the book and the Earth. For the right-hand side,

$$mgy_i - mgy_f = -(mgy_f - mgy_i) = -\Delta U_g$$

where $U_g = mgy$ is the gravitational potential energy of the system. For the left-hand side of Equation 8.4, because the book is the only part of the system that is moving, we see that $\Delta K_{book} = \Delta K$, where K is the kinetic energy of the system. Therefore, with each side of Equation 8.4 replaced with its system equivalent, the equation becomes

$$\Delta K = -\Delta U_g \tag{8.5}$$

This equation can be manipulated to provide a very important general result for solving problems. First, we move the change in potential energy to the left side of the equation:

$$\Delta K + \Delta U_g = 0$$

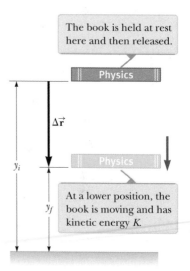

The book is held at rest here and then released.

Physics

$\Delta \vec{\mathbf{r}}$

y_i

Physics

y_f

At a lower position, the book is moving and has kinetic energy K.

Figure 8.2 A book is released from rest and falls due to work done by the gravitational force on the book.

The left side represents a sum of changes of the energy stored in the system. The right-hand side is zero because there are no transfers of energy across the boundary of the system; the book–Earth system is *isolated* from the environment. We developed this equation for a gravitational system, but it can be shown to be valid for a system with any type of potential energy. Therefore, for an isolated system,

$$\Delta K + \Delta U = 0 \qquad (8.6)$$

(Check to see that this equation is contained within Eq. 8.2.)

We defined in Chapter 7 the sum of the kinetic and potential energies of a system as its mechanical energy:

$$E_{\text{mech}} \equiv K + U \qquad (8.7)$$

◄ **Mechanical energy of a system**

where U represents the total of *all* types of potential energy. Because the system under consideration is isolated, Equations 8.6 and 8.7 tell us that the mechanical energy of the system is conserved:

$$\Delta E_{\text{mech}} = 0 \qquad (8.8)$$

◄ **The mechanical energy of an isolated system with no nonconservative forces acting is conserved.**

Equation 8.8 is a statement of **conservation of mechanical energy** for an isolated system with no nonconservative forces acting. The mechanical energy in such a system is conserved: the sum of the kinetic and potential energies remains constant:

Let us now write the changes in energy in Equation 8.6 explicitly:

$$(K_f - K_i) + (U_f - U_i) = 0$$
$$K_f + U_f = K_i + U_i \qquad (8.9)$$

For the gravitational situation of the falling book, Equation 8.9 can be written as

$$\tfrac{1}{2}mv_f^2 + mgy_f = \tfrac{1}{2}mv_i^2 + mgy_i$$

As the book falls to the Earth, the book–Earth system loses potential energy and gains kinetic energy such that the total of the two types of energy always remains constant: $E_{\text{total},i} = E_{\text{total},f}$.

If there are nonconservative forces acting within the system, mechanical energy is transformed to internal energy as discussed in Section 7.7. If nonconservative forces act in an isolated system, the total energy of the system is conserved although the mechanical energy is not. In that case, we can express the conservation of energy of the system as

◄ **The total energy of an isolated system is conserved.**

$$\Delta E_{\text{system}} = 0 \qquad (8.10)$$

where E_{system} includes all kinetic, potential, and internal energies. This equation is the most general statement of the energy version of the **isolated system** model. It is equivalent to Equation 8.2 with all terms on the right-hand side equal to zero.

Quick Quiz 8.3 A rock of mass m is dropped to the ground from a height h. A second rock, with mass $2m$, is dropped from the same height. When the second rock strikes the ground, what is its kinetic energy? **(a)** twice that of the first rock **(b)** four times that of the first rock **(c)** the same as that of the first rock **(d)** half as much as that of the first rock **(e)** impossible to determine

Quick Quiz 8.4 Three identical balls are thrown from the top of a building, all with the same initial speed. As shown in Figure 8.3, the first is thrown horizontally, the second at some angle above the horizontal, and the third at some angle below the horizontal. Neglecting air resistance, rank the speeds of the balls at the instant each hits the ground.

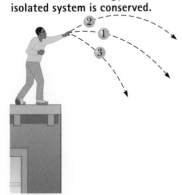

Figure 8.3 (Quick Quiz 8.4) Three identical balls are thrown with the same initial speed from the top of a building.

Analysis Model   Isolated System (Energy)

Imagine you have identified a system to be analyzed and have defined a system boundary. Energy can exist in the system in three forms: kinetic, potential, and internal. Imagine also a situation in which no energy crosses the boundary of the system by any method. Then, the system is isolated; energy transforms from one form to another and Equation 8.2 becomes

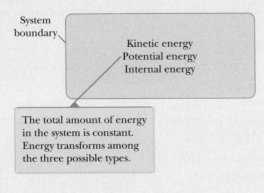

System boundary

Kinetic energy
Potential energy
Internal energy

The total amount of energy in the system is constant. Energy transforms among the three possible types.

$$\Delta E_{system} = 0 \qquad (8.10)$$

If no nonconservative forces act within the isolated system, the mechanical energy of the system is conserved, so

$$\Delta E_{mech} = 0 \qquad (8.8)$$

Examples:

- an object is in free-fall; gravitational potential energy transforms to kinetic energy: $\Delta K + \Delta U = 0$
- a basketball rolling across a gym floor comes to rest; kinetic energy transforms to internal energy: $\Delta K + \Delta E_{int} = 0$
- a pendulum is raised and released with an initial speed; its motion eventually stops due to air resistance; gravitational potential energy and kinetic energy transform to internal energy, $\Delta K + \Delta U + \Delta E_{int} = 0$ (Chapter 15)
- a battery is connected to a resistor; chemical potential energy in the battery transforms to internal energy in the resistor: $\Delta U + \Delta E_{int} = 0$ (Chapter 27)

Problem-Solving Strategy   Isolated and Nonisolated Systems with No Nonconservative Forces: Conservation of Energy

Many problems in physics can be solved using the principle of conservation of energy. The following procedure should be used when you apply this principle:

1. Conceptualize. Study the physical situation carefully and form a mental representation of what is happening. As you become more proficient working energy problems, you will begin to be comfortable imagining the types of energy that are changing in the system and the types of energy transfers occurring across the system boundary.

2. Categorize. Define your system, which may consist of more than one object and may or may not include springs or other possibilities for storing potential energy. Identify the time interval over which you will analyze the energy changes in the problem. Determine if any energy transfers occur across the boundary of your system during this time interval. If so, use the nonisolated system model, $\Delta E_{system} = \Sigma T$, from Section 8.1. If not, use the isolated system model, $\Delta E_{system} = 0$.

Determine whether any nonconservative forces are present within the system. If so, use the techniques of Sections 8.3 and 8.4. If not, use the principle of conservation of energy as outlined below.

3. Analyze. Choose configurations to represent the initial and final conditions of the system based on your choice of time interval. For each object that changes elevation, select a reference position for the object that defines the zero configuration of gravitational potential energy for the system. For an object on a spring, the zero configuration for elastic potential energy is when the object is at its equilibrium position. If there is more than one conservative force, write an expression for the potential energy associated with each force.

Begin with Equation 8.2 and retain only those terms in the equation that are appropriate for the situation in the problem. Express each change of energy stored in the system as the final value minus the initial value. Substitute appropriate expressions for each initial and final value of energy storage on the left side of the equation and for the energy transfers on the right side of the equation. Solve for the unknown quantity.

continued

▶ **Problem-Solving Strategy** continued

4. Finalize. Make sure your results are consistent with your mental representation. Also make sure the values of your results are reasonable and consistent with connections to everyday experience.

Example 8.1   Ball in Free Fall   AM

A ball of mass m is dropped from a height h above the ground as shown in Figure 8.4.

(A) Neglecting air resistance, determine the speed of the ball when it is at a height y above the ground. Choose the system as the ball and the Earth.

SOLUTION

Conceptualize Figure 8.4 and our everyday experience with falling objects allow us to conceptualize the situation. Although we can readily solve this problem with the techniques of Chapter 2, let us practice an energy approach.

Categorize As suggested in the problem, we identify the system as the ball and the Earth. Because there is neither air resistance nor any other interaction between the system and the environment, the system is isolated and we use the *isolated system* model. The only force between members of the system is the gravitational force, which is conservative.

$y_i = h$
$U_{gi} = mgh$
$K_i = 0$

$y_f = y$
$U_{gf} = mgy$
$K_f = \frac{1}{2}mv_f^2$

$\vec{v}_f$

$y = 0$
$U_g = 0$

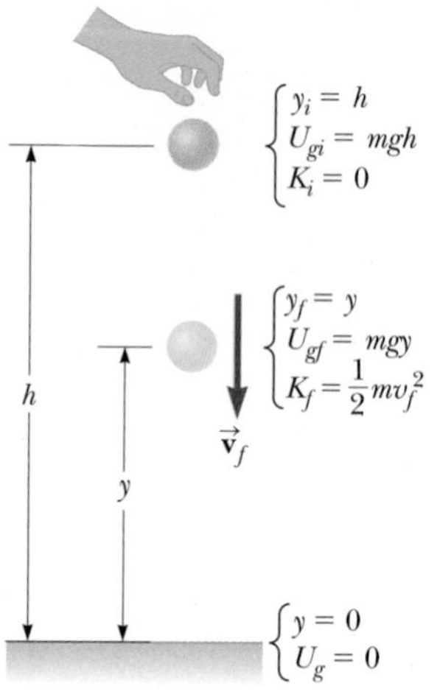

Figure 8.4 (Example 8.1) A ball is dropped from a height h above the ground. Initially, the total energy of the ball–Earth system is gravitational potential energy, equal to mgh relative to the ground. At the position y, the total energy is the sum of the kinetic and potential energies.

Analyze Because the system is isolated and there are no nonconservative forces acting within the system, we apply the principle of conservation of mechanical energy to the ball–Earth system. At the instant the ball is released, its kinetic energy is $K_i = 0$ and the gravitational potential energy of the system is $U_{gi} = mgh$. When the ball is at a position y above the ground, its kinetic energy is $K_f = \frac{1}{2}mv_f^2$ and the potential energy relative to the ground is $U_{gf} = mgy$.

Write the appropriate reduction of Equation 8.2, noting that the only types of energy in the system that change are kinetic energy and gravitational potential energy:

$$\Delta K + \Delta U_g = 0$$

Substitute for the energies:

$$\left(\tfrac{1}{2}mv_f^2 - 0\right) + \left(mgy - mgh\right) = 0$$

Solve for v_f:

$$v_f^2 = 2g(h - y) \quad \rightarrow \quad v_f = \sqrt{2g(h - y)}$$

The speed is always positive. If you had been asked to find the ball's velocity, you would use the negative value of the square root as the y component to indicate the downward motion.

(B) Find the speed of the ball again at height y by choosing the ball as the system.

SOLUTION

Categorize In this case, the only type of energy in the system that changes is kinetic energy. A single object that can be modeled as a particle cannot possess potential energy. The effect of gravity is to do work on the ball across the boundary of the system. We use the *nonisolated system* model.

Analyze Write the appropriate reduction of Equation 8.2:

$$\Delta K = W$$

Substitute for the initial and final kinetic energies and the work:

$$\left(\tfrac{1}{2}mv_f^2 - 0\right) = \vec{F}_g \cdot \Delta \vec{r} = -mg\hat{j} \cdot \Delta y\hat{j}$$
$$= -mg\Delta y = -mg(y - h) = mg(h - y)$$

Solve for v_f:

$$v_f^2 = 2g(h - y) \quad \rightarrow \quad v_f = \sqrt{2g(h - y)}$$

▶ **8.1** continued

Finalize The final result is the same, regardless of the choice of system. In your future problem solving, keep in mind that the choice of system is yours to make. Sometimes the problem is much easier to solve if a judicious choice is made as to the system to analyze.

WHAT IF? What if the ball were thrown downward from its highest position with a speed v_i? What would its speed be at height y?

Answer If the ball is thrown downward initially, we would expect its speed at height y to be larger than if simply dropped. Make your choice of system, either the ball alone or the ball and the Earth. You should find that either choice gives you the following result:

$$v_f = \sqrt{v_i^2 + 2g(h - y)}$$

Example 8.2 **A Grand Entrance** **AM**

You are designing an apparatus to support an actor of mass 65.0 kg who is to "fly" down to the stage during the performance of a play. You attach the actor's harness to a 130-kg sandbag by means of a lightweight steel cable running smoothly over two frictionless pulleys as in Figure 8.5a. You need 3.00 m of cable between the harness and the nearest pulley so that the pulley can be hidden behind a curtain. For the apparatus to work successfully, the sandbag must never lift above the floor as the actor swings from above the stage to the floor. Let us call the initial angle that the actor's cable makes with the vertical θ. What is the maximum value θ can have before the sandbag lifts off the floor?

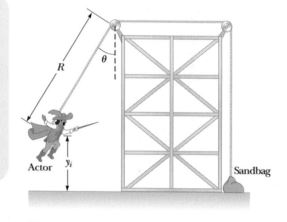

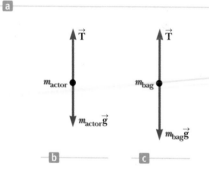

a

b   **c**

Figure 8.5 (Example 8.2) (a) An actor uses some clever staging to make his entrance. (b) The free-body diagram for the actor at the bottom of the circular path. (c) The free-body diagram for the sandbag if the normal force from the floor goes to zero.

SOLUTION

Conceptualize We must use several concepts to solve this problem. Imagine what happens as the actor approaches the bottom of the swing. At the bottom, the cable is vertical and must support his weight as well as provide centripetal acceleration of his body in the upward direction. At this point in his swing, the tension in the cable is the highest and the sandbag is most likely to lift off the floor.

Categorize Looking first at the swinging of the actor from the initial point to the lowest point, we model the actor and the Earth as an *isolated system*. We ignore air resistance, so there are no nonconservative forces acting. You might initially be tempted to model the system as nonisolated because of the interaction of the system with the cable, which is in the environment. The force applied to the actor by the cable, however, is always perpendicular to each element of the displacement of the actor and hence does no work. Therefore, in terms of energy transfers across the boundary, the system is isolated.

Analyze We first find the actor's speed as he arrives at the floor as a function of the initial angle θ and the radius R of the circular path through which he swings.

From the isolated system model, make the appropriate reduction of Equation 8.2 for the actor–Earth system:

$$\Delta K + \Delta U_g = 0$$

continued

▶ **8.2** continued

Let y_i be the initial height of the actor above the floor and v_f be his speed at the instant before he lands. (Notice that $K_i = 0$ because the actor starts from rest and that $U_f = 0$ because we define the configuration of the actor at the floor as having a gravitational potential energy of zero.)

(1) $(\frac{1}{2}m_{actor}v_f^2 - 0) + (0 - m_{actor}gy_i) = 0$

From the geometry in Figure 8.5a, notice that $y_f = 0$, so $y_i = R - R\cos\theta = R(1 - \cos\theta)$. Use this relationship in Equation (1) and solve for v_f^2:

(2) $v_f^2 = 2gR(1 - \cos\theta)$

....

Categorize Next, focus on the instant the actor is at the lowest point. Because the tension in the cable is transferred as a force applied to the sandbag, we model the actor at this instant as a *particle under a net force*. Because the actor moves along a circular arc, he experiences at the bottom of the swing a centripetal acceleration of v_f^2/r directed upward.

....

Analyze Apply Newton's second law from the particle under a net force model to the actor at the bottom of his path, using the free-body diagram in Figure 8.5b as a guide, and recognizing the acceleration as centripetal:

$\sum F_y = T - m_{actor}g = m_{actor}\dfrac{v_f^2}{R}$

(3) $T = m_{actor}g + m_{actor}\dfrac{v_f^2}{R}$

....

Categorize Finally, notice that the sandbag lifts off the floor when the upward force exerted on it by the cable exceeds the gravitational force acting on it; the normal force from the floor is zero when that happens. We do *not*, however, want the sandbag to lift off the floor. The sandbag must remain at rest, so we model it as a *particle in equilibrium*.

....

Analyze A force T of the magnitude given by Equation (3) is transmitted by the cable to the sandbag. If the sandbag remains at rest but is just ready to be lifted off the floor if any more force were applied by the cable, the normal force on it becomes zero and the particle in equilibrium model tells us that $T = m_{bag}g$ as in Figure 8.5c.

Substitute this condition and Equation (2) into Equation (3):

$m_{bag}g = m_{actor}g + m_{actor}\dfrac{2gR(1 - \cos\theta)}{R}$

Solve for $\cos\theta$ and substitute the given parameters:

$\cos\theta = \dfrac{3m_{actor} - m_{bag}}{2m_{actor}} = \dfrac{3(65.0 \text{ kg}) - 130 \text{ kg}}{2(65.0 \text{ kg})} = 0.500$

$\theta = \boxed{60.0°}$

....

Finalize Here we had to combine several analysis models from different areas of our study. Notice that the length R of the cable from the actor's harness to the leftmost pulley did not appear in the final algebraic equation for $\cos\theta$. Therefore, the final answer is independent of R.

Example 8.3   **The Spring-Loaded Popgun**   AM

The launching mechanism of a popgun consists of a trigger-released spring (Fig. 8.6a). The spring is compressed to a position $y_Ⓐ$, and the trigger is fired. The projectile of mass m rises to a position $y_Ⓒ$ above the position at which it leaves the spring, indicated in Figure 8.6b as position $y_Ⓑ = 0$. Consider a firing of the gun for which $m = 35.0$ g, $y_Ⓐ = -0.120$ m, and $y_Ⓒ = 20.0$ m.

(A) Neglecting all resistive forces, determine the spring constant.

SOLUTION

Conceptualize Imagine the process illustrated in parts (a) and (b) of Figure 8.6. The projectile starts from rest at Ⓐ, speeds up as the spring pushes upward on it, leaves the spring at Ⓑ, and then slows down as the gravitational force pulls downward on it, eventually coming to rest at point Ⓒ.

▶ **8.3** continued

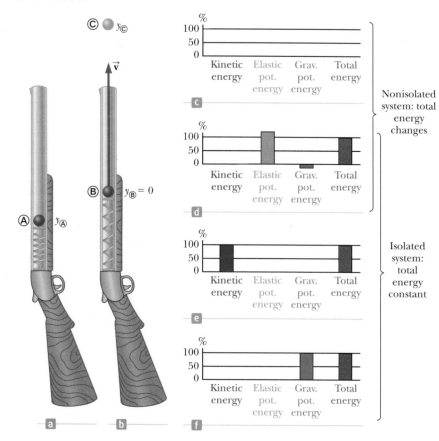

Figure 8.6 (Example 8.3) A spring-loaded popgun (a) before firing and (b) when the spring extends to its relaxed length. (c) An energy bar chart for the popgun–projectile–Earth system before the popgun is loaded. The energy in the system is zero. (d) The popgun is loaded by means of an external agent doing work on the system to push the spring downward. Therefore the system is nonisolated during this process. After the popgun is loaded, elastic potential energy is stored in the spring and the gravitational potential energy of the system is lower because the projectile is below point Ⓑ. (e) as the projectile passes through point Ⓑ, all of the energy of the isolated system is kinetic. (f) When the projectile reaches point Ⓒ, all of the energy of the isolated system is gravitational potential.

Categorize We identify the system as the projectile, the spring, and the Earth. We ignore both air resistance on the projectile and friction in the gun, so we model the system as isolated with no nonconservative forces acting.

Analyze Because the projectile starts from rest, its initial kinetic energy is zero. We choose the zero configuration for the gravitational potential energy of the system to be when the projectile leaves the spring at Ⓑ. For this configuration, the elastic potential energy is also zero.

After the gun is fired, the projectile rises to a maximum height $y_Ⓒ$. The final kinetic energy of the projectile is zero.

From the isolated system model, write a conservation of mechanical energy equation for the system between configurations when the projectile is at points Ⓐ and Ⓒ:

(1) $\quad \Delta K + \Delta U_g + \Delta U_s = 0$

Substitute for the initial and final energies:

$$(0 - 0) + (mgy_Ⓒ - mgy_Ⓐ) + (0 - \tfrac{1}{2}kx^2) = 0$$

Solve for k:

$$k = \frac{2mg(y_Ⓒ - y_Ⓐ)}{x^2}$$

Substitute numerical values:

$$k = \frac{2(0.035\,0\text{ kg})(9.80\text{ m/s}^2)[20.0\text{ m} - (-0.120\text{ m})]}{(0.120\text{ m})^2} = \boxed{958\text{ N/m}}$$

(B) Find the speed of the projectile as it moves through the equilibrium position Ⓑ of the spring as shown in Figure 8.6b.

SOLUTION

Analyze The energy of the system as the projectile moves through the equilibrium position of the spring includes only the kinetic energy of the projectile $\tfrac{1}{2}mv_Ⓑ^2$. Both types of potential energy are equal to zero for this configuration of the system.

continued

▶ **8.3** continued

Write Equation (1) again for the system between points Ⓐ and Ⓑ:	$\Delta K + \Delta U_g + \Delta U_s = 0$
Substitute for the initial and final energies:	$(\frac{1}{2}mv_{Ⓑ}^2 - 0) + (0 - mgy_{Ⓐ}) + (0 - \frac{1}{2}kx^2) = 0$
Solve for $v_{Ⓑ}$:	$v_{Ⓑ} = \sqrt{\dfrac{kx^2}{m} + 2gy_{Ⓐ}}$
Substitute numerical values:	$v_{Ⓑ} = \sqrt{\dfrac{(958\ \text{N/m})(0.120\ \text{m})^2}{(0.035\ 0\ \text{kg})} + 2(9.80\ \text{m/s}^2)(-0.120\ \text{m})} = \boxed{19.8\ \text{m/s}}$

Finalize This example is the first one we have seen in which we must include two different types of potential energy. Notice in part (A) that we never needed to consider anything about the speed of the ball between points Ⓐ and Ⓒ, which is part of the power of the energy approach: changes in kinetic and potential energy depend only on the initial and final values, not on what happens between the configurations corresponding to these values.

8.3 Situations Involving Kinetic Friction

Consider again the book in Figure 7.18a sliding to the right on the surface of a heavy table and slowing down due to the friction force. Work is done by the friction force on the book because there is a force and a displacement. Keep in mind, however, that our equations for work involve the displacement *of the point of application of the force*. A simple model of the friction force between the book and the surface is shown in Figure 8.7a. We have represented the entire friction force between the book and surface as being due to two identical teeth that have been spot-welded together.[2] One tooth projects upward from the surface, the other downward from the book, and they are welded at the points where they touch. The friction force acts at the junction of the two teeth. Imagine that the book slides a small distance d to the right as in Figure 8.7b. Because the teeth are modeled as identical, the junction of the teeth moves to the right by a distance $d/2$. Therefore, the displacement of the point of application of the friction force is $d/2$, but the displacement of the book is d!

In reality, the friction force is spread out over the entire contact area of an object sliding on a surface, so the force is not localized at a point. In addition, because the magnitudes of the friction forces at various points are constantly changing as individual spot welds occur, the surface and the book deform locally, and so on, the displacement of the point of application of the friction force is not at all the same as the displacement of the book. In fact, the displacement of the point of application of the friction force is not calculable and so neither is the work done by the friction force.

The work–kinetic energy theorem is valid for a particle or an object that can be modeled as a particle. When a friction force acts, however, we cannot calculate the work done by friction. For such situations, Newton's second law is still valid for the system even though the work–kinetic energy theorem is not. The case of a nondeformable object like our book sliding on the surface[3] can be handled in a relatively straightforward way.

Starting from a situation in which forces, including friction, are applied to the book, we can follow a similar procedure to that done in developing Equation 7.17. Let us start by writing Equation 7.8 for all forces on an object other than friction:

$$\sum W_{\text{other forces}} = \int \left(\sum \vec{\mathbf{F}}_{\text{other forces}}\right) \cdot d\vec{\mathbf{r}} \qquad \textbf{(8.11)}$$

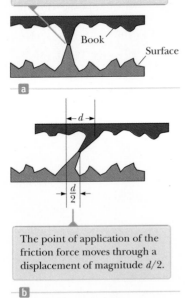

The entire friction force is modeled to be applied at the interface between two identical teeth projecting from the book and the surface.

Book

Surface

a

←d→

$\frac{d}{2}$

The point of application of the friction force moves through a displacement of magnitude $d/2$.

b

Figure 8.7 (a) A simplified model of friction between a book and a surface. (b) The book is moved to the right by a distance d.

[2]Figure 8.7 and its discussion are inspired by a classic article on friction: B. A. Sherwood and W. H. Bernard, "Work and heat transfer in the presence of sliding friction," *American Journal of Physics,* **52**:1001, 1984.

[3]The overall shape of the book remains the same, which is why we say it is nondeformable. On a microscopic level, however, there is deformation of the book's face as it slides over the surface.

The $d\vec{\mathbf{r}}$ in this equation is the displacement of the object because for forces other than friction, under the assumption that these forces do not deform the object, this displacement is the same as the displacement of the point of application of the forces. To each side of Equation 8.11 let us add the integral of the scalar product of the force of kinetic friction and $d\vec{\mathbf{r}}$. In doing so, we are not defining this quantity as work! We are simply saying that it is a quantity that can be calculated mathematically and will turn out to be useful to us in what follows.

$$\sum W_{\text{other forces}} + \int \vec{\mathbf{f}}_k \cdot d\vec{\mathbf{r}} = \int \left(\sum \vec{\mathbf{F}}_{\text{other forces}} \right) \cdot d\vec{\mathbf{r}} + \int \vec{\mathbf{f}}_k \cdot d\vec{\mathbf{r}}$$

$$= \int \left(\sum \vec{\mathbf{F}}_{\text{other forces}} + \vec{\mathbf{f}}_k \right) \cdot d\vec{\mathbf{r}}$$

The integrand on the right side of this equation is the net force $\sum \vec{\mathbf{F}}$ on the object, so

$$\sum W_{\text{other forces}} + \int \vec{\mathbf{f}}_k \cdot d\vec{\mathbf{r}} = \int \sum \vec{\mathbf{F}} \cdot d\vec{\mathbf{r}}$$

Incorporating Newton's second law $\sum \vec{\mathbf{F}} = m\vec{\mathbf{a}}$ gives

$$\sum W_{\text{other forces}} + \int \vec{\mathbf{f}}_k \cdot d\vec{\mathbf{r}} = \int m\vec{\mathbf{a}} \cdot d\vec{\mathbf{r}} = \int m\frac{d\vec{\mathbf{v}}}{dt} \cdot d\vec{\mathbf{r}} = \int_{t_i}^{t_f} m\frac{d\vec{\mathbf{v}}}{dt} \cdot \vec{\mathbf{v}}\, dt \quad \textbf{(8.12)}$$

where we have used Equation 4.3 to rewrite $d\vec{\mathbf{r}}$ as $\vec{\mathbf{v}}\,dt$. The scalar product obeys the product rule for differentiation (See Eq. B.30 in Appendix B.6), so the derivative of the scalar product of $\vec{\mathbf{v}}$ with itself can be written

$$\frac{d}{dt}(\vec{\mathbf{v}} \cdot \vec{\mathbf{v}}) = \frac{d\vec{\mathbf{v}}}{dt} \cdot \vec{\mathbf{v}} + \vec{\mathbf{v}} \cdot \frac{d\vec{\mathbf{v}}}{dt} = 2\frac{d\vec{\mathbf{v}}}{dt} \cdot \vec{\mathbf{v}}$$

We used the commutative property of the scalar product to justify the final expression in this equation. Consequently,

$$\frac{d\vec{\mathbf{v}}}{dt} \cdot \vec{\mathbf{v}} = \tfrac{1}{2}\frac{d}{dt}(\vec{\mathbf{v}} \cdot \vec{\mathbf{v}}) = \tfrac{1}{2}\frac{dv^2}{dt}$$

Substituting this result into Equation 8.12 gives

$$\sum W_{\text{other forces}} + \int \vec{\mathbf{f}}_k \cdot d\vec{\mathbf{r}} = \int_{t_i}^{t_f} m\left(\tfrac{1}{2}\frac{dv^2}{dt}\right) dt = \tfrac{1}{2}m \int_{v_i}^{v_f} d(v^2) = \tfrac{1}{2}mv_f^2 - \tfrac{1}{2}mv_i^2 = \Delta K$$

Looking at the left side of this equation, notice that in the inertial frame of the surface, $\vec{\mathbf{f}}_k$ and $d\vec{\mathbf{r}}$ will be in opposite directions for every increment $d\vec{\mathbf{r}}$ of the path followed by the object. Therefore, $\vec{\mathbf{f}}_k \cdot d\vec{\mathbf{r}} = -f_k\,dr$. The previous expression now becomes

$$\sum W_{\text{other forces}} - \int f_k\,dr = \Delta K$$

In our model for friction, the magnitude of the kinetic friction force is constant, so f_k can be brought out of the integral. The remaining integral $\int dr$ is simply the sum of increments of length along the path, which is the total path length d. Therefore,

$$\sum W_{\text{other forces}} - f_k d = \Delta K \quad \textbf{(8.13)}$$

Equation 8.13 can be used when a friction force acts on an object. The change in kinetic energy is equal to the work done by all forces other than friction minus a term $f_k d$ associated with the friction force.

Considering the sliding book situation again, let's identify the larger system of the book *and* the surface as the book slows down under the influence of a friction force alone. There is no work done across the boundary of this system by other forces because the system does not interact with the environment. There are no other types of energy transfer occurring across the boundary of the system, assuming we ignore the inevitable sound the sliding book makes! In this case, Equation 8.2 becomes

$$\Delta E_{\text{system}} = \Delta K + \Delta E_{\text{int}} = 0$$

The change in kinetic energy of this book–surface system is the same as the change in kinetic energy of the book alone because the book is the only part of the system that is moving. Therefore, incorporating Equation 8.13 with no work done by other forces gives

$$-f_k d + \Delta E_{int} = 0$$

**Change in internal energy ▶
due to a constant friction
force within the system**

$$\Delta E_{int} = f_k d \qquad (8.14)$$

Equation 8.14 tells us that the increase in internal energy of the system is equal to the product of the friction force and the path length through which the block moves. In summary, a friction force transforms kinetic energy in a system to internal energy. If work is done on the system by forces other than friction, Equation 8.13, with the help of Equation 8.14, can be written as

$$\sum W_{other\ forces} = W = \Delta K + \Delta E_{int} \qquad (8.15)$$

which is a reduced form of Equation 8.2 and represents the nonisolated system model for a system within which a nonconservative force acts.

Quick **Quiz** 8.5 You are traveling along a freeway at 65 mi/h. Your car has kinetic energy. You suddenly skid to a stop because of congestion in traffic. Where is the kinetic energy your car once had? **(a)** It is all in internal energy in the road. **(b)** It is all in internal energy in the tires. **(c)** Some of it has transformed to internal energy and some of it transferred away by mechanical waves. **(d)** It is all transferred away from your car by various mechanisms.

Example 8.4   **A Block Pulled on a Rough Surface**   **AM**

A 6.0-kg block initially at rest is pulled to the right along a horizontal surface by a constant horizontal force of 12 N.

(A) Find the speed of the block after it has moved 3.0 m if the surfaces in contact have a coefficient of kinetic friction of 0.15.

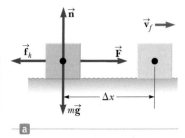

SOLUTION

Conceptualize This example is similar to Example 7.6 (page 152), but modified so that the surface is no longer frictionless. The rough surface applies a friction force on the block opposite to the applied force. As a result, we expect the speed to be lower than that found in Example 7.6.

Figure 8.8 (Example 8.4) (a) A block pulled to the right on a rough surface by a constant horizontal force. (b) The applied force is at an angle θ to the horizontal.

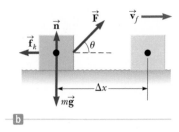

Categorize The block is pulled by a force and the surface is rough, so the block and the surface are modeled as a *nonisolated system* with a nonconservative force acting.

Analyze Figure 8.8a illustrates this situation. Neither the normal force nor the gravitational force does work on the system because their points of application are displaced horizontally.

Find the work done on the system by the applied force just as in Example 7.6:

$$\sum W_{other\ forces} = W_F = F\Delta x$$

Apply the *particle in equilibrium* model to the block in the vertical direction:

$$\sum F_y = 0 \rightarrow n - mg = 0 \rightarrow n = mg$$

Find the magnitude of the friction force:

$$f_k = \mu_k n = \mu_k mg = (0.15)(6.0\ \text{kg})(9.80\ \text{m/s}^2) = 8.82\ \text{N}$$

▶ **8.4** continued

Substitute the energies into Equation 8.15 and solve for the final speed of the block:

$$F\Delta x = \Delta K + \Delta E_{\text{int}} = (\tfrac{1}{2}mv_f^2 - 0) + f_k d$$

$$v_f = \sqrt{\frac{2}{m}(-f_k d + F\Delta x)}$$

Substitute numerical values:

$$v_f = \sqrt{\frac{2}{6.0 \text{ kg}}[-(8.82 \text{ N})(3.0 \text{ m}) + (12 \text{ N})(3.0 \text{ m})]} = \boxed{1.8 \text{ m/s}}$$

Finalize As expected, this value is less than the 3.5 m/s found in the case of the block sliding on a frictionless surface (see Example 7.6). The difference in kinetic energies between the block in Example 7.6 and the block in this example is equal to the increase in internal energy of the block–surface system in this example.

(B) Suppose the force $\vec{\mathbf{F}}$ is applied at an angle θ as shown in Figure 8.8b. At what angle should the force be applied to achieve the largest possible speed after the block has moved 3.0 m to the right?

SOLUTION

Conceptualize You might guess that $\theta = 0$ would give the largest speed because the force would have the largest component possible in the direction parallel to the surface. Think about $\vec{\mathbf{F}}$ applied at an arbitrary nonzero angle, however. Although the horizontal component of the force would be reduced, the vertical component of the force would reduce the normal force, in turn reducing the force of friction, which suggests that the speed could be maximized by pulling at an angle other than $\theta = 0$.

Categorize As in part (A), we model the block and the surface as a *nonisolated system* with a nonconservative force acting.

Analyze Find the work done by the applied force, noting that $\Delta x = d$ because the path followed by the block is a straight line:

$$(1) \quad \sum W_{\text{other forces}} = W_F = F\Delta x \cos\theta = Fd\cos\theta$$

Apply the particle in equilibrium model to the block in the vertical direction:

$$\sum F_y = n + F\sin\theta - mg = 0$$

Solve for n:

$$(2) \quad n = mg - F\sin\theta$$

Use Equation 8.15 to find the final kinetic energy for this situation:

$$W_F = \Delta K + \Delta E_{\text{int}} = (K_f - 0) + f_k d \quad \rightarrow \quad K_f = W_F - f_k d$$

Substitute the results in Equations (1) and (2):

$$K_f = Fd\cos\theta - \mu_k nd = Fd\cos\theta - \mu_k(mg - F\sin\theta)d$$

Maximizing the speed is equivalent to maximizing the final kinetic energy. Consequently, differentiate K_f with respect to θ and set the result equal to zero:

$$\frac{dK_f}{d\theta} = -Fd\sin\theta - \mu_k(0 - F\cos\theta)d = 0$$

$$-\sin\theta + \mu_k\cos\theta = 0$$

$$\tan\theta = \mu_k$$

Evaluate θ for $\mu_k = 0.15$:

$$\theta = \tan^{-1}(\mu_k) = \tan^{-1}(0.15) = \boxed{8.5°}$$

Finalize Notice that the angle at which the speed of the block is a maximum is indeed not $\theta = 0$. When the angle exceeds 8.5°, the horizontal component of the applied force is too small to be compensated by the reduced friction force and the speed of the block begins to decrease from its maximum value.

Conceptual Example 8.5   **Useful Physics for Safer Driving**

A car traveling at an initial speed v slides a distance d to a halt after its brakes lock. If the car's initial speed is instead $2v$ at the moment the brakes lock, estimate the distance it slides.

continued

▶ **8.5** c o n t i n u e d

SOLUTION

Let us assume the force of kinetic friction between the car and the road surface is constant and the same for both speeds. According to Equation 8.13, the friction force multiplied by the distance d is equal to the initial kinetic energy of the car (because $K_f = 0$ and there is no work done by other forces). If the speed is doubled, as it is in this example, the kinetic energy is quadrupled. For a given friction force, the distance traveled is four times as great when the initial speed is doubled, and so the estimated distance the car slides is $4d$.

| **Example 8.6** | **A Block–Spring System** | AM |

A block of mass 1.6 kg is attached to a horizontal spring that has a force constant of 1 000 N/m as shown in Figure 8.9a. The spring is compressed 2.0 cm and is then released from rest as in Figure 8.9b.

(A) Calculate the speed of the block as it passes through the equilibrium position $x = 0$ if the surface is frictionless.

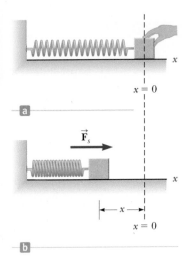

SOLUTION

Conceptualize This situation has been discussed before, and it is easy to visualize the block being pushed to the right by the spring and moving with some speed at $x = 0$.

Categorize We identify the system as the block and model the block as a *nonisolated system*.

Figure 8.9 (Example 8.6)
(a) A block attached to a spring is pushed inward from an initial position $x = 0$ by an external agent.
(b) At position x, the block is released from rest and the spring pushes it to the right.

Analyze In this situation, the block starts with $v_i = 0$ at $x_i = -2.0$ cm, and we want to find v_f at $x_f = 0$.

Use Equation 7.11 to find the work done by the spring on the system with $x_{max} = x_i$:

$$W_s = \tfrac{1}{2}kx_{max}^2$$

Work is done on the block, and its speed changes. The conservation of energy equation, Equation 8.2, reduces to the work–kinetic energy theorem. Use that theorem to find the speed at $x = 0$:

$$W_s = \tfrac{1}{2}mv_f^2 - \tfrac{1}{2}mv_i^2$$

$$v_f = \sqrt{v_i^2 + \frac{2}{m}W_s} = \sqrt{v_i^2 + \frac{2}{m}\left(\tfrac{1}{2}kx_{max}^2\right)}$$

Substitute numerical values:

$$v_f = \sqrt{0 + \frac{2}{1.6 \text{ kg}}\left[\tfrac{1}{2}(1\,000 \text{ N/m})(0.020 \text{ m})^2\right]} = \boxed{0.50 \text{ m/s}}$$

Finalize Although this problem could have been solved in Chapter 7, it is presented here to provide contrast with the following part (B), which requires the techniques of this chapter.

(B) Calculate the speed of the block as it passes through the equilibrium position if a constant friction force of 4.0 N retards its motion from the moment it is released.

SOLUTION

Conceptualize The correct answer must be less than that found in part (A) because the friction force retards the motion.

Categorize We identify the system as the block and the surface, a *nonisolated system* because of the work done by the spring. There is a nonconservative force acting within the system: the friction between the block and the surface.

▶ **8.6** continued

Analyze Write Equation 8.15:

$$W_s = \Delta K + \Delta E_{int} = (\tfrac{1}{2}mv_f^2 - 0) + f_k d$$

Solve for v_f:

$$v_f = \sqrt{\frac{2}{m}(W_s - f_k d)}$$

Substitute for the work done by the spring:

$$v_f = \sqrt{\frac{2}{m}(\tfrac{1}{2}kx_{max}^2 - f_k d)}$$

Substitute numerical values:

$$v_f = \sqrt{\frac{2}{1.6\ kg}[\tfrac{1}{2}(1\,000\ N/m)(0.020\ m)^2 - (4.0\ N)(0.020\ m)]} = 0.39\ m/s$$

Finalize As expected, this value is less than the 0.50 m/s found in part (A).

WHAT IF? What if the friction force were increased to 10.0 N? What is the block's speed at $x = 0$?

Answer In this case, the value of $f_k d$ as the block moves to $x = 0$ is

$$f_k d = (10.0\ N)(0.020\ m) = 0.20\ J$$

which is equal in magnitude to the kinetic energy at $x = 0$ for the frictionless case. (Verify it!). Therefore, all the kinetic energy has been transformed to internal energy by friction when the block arrives at $x = 0$, and its speed at this point is $v = 0$.

In this situation as well as that in part (B), the speed of the block reaches a maximum at some position other than $x = 0$. Problem 53 in Enhanced WebAssign asks you to locate these positions.

8.4 Changes in Mechanical Energy for Nonconservative Forces

Consider the book sliding across the surface in the preceding section. As the book moves through a distance d, the only force in the horizontal direction is the force of kinetic friction. This force causes a change $-f_k d$ in the kinetic energy of the book as described by Equation 8.13.

Now, however, suppose the book is part of a system that also exhibits a change in potential energy. In this case, $-f_k d$ is the amount by which the *mechanical* energy of the system changes because of the force of kinetic friction. For example, if the book moves on an incline that is not frictionless, there is a change in both the kinetic energy and the gravitational potential energy of the book–Earth system. Consequently,

$$\Delta E_{mech} = \Delta K + \Delta U_g = -f_k d = -\Delta E_{int}$$

In general, if a nonconservative force acts within an isolated system,

$$\Delta K + \Delta U + \Delta E_{int} = 0 \tag{8.16}$$

where ΔU is the change in all forms of potential energy. We recognize Equation 8.16 as Equation 8.2 with no transfers of energy across the boundary of the system.

If the system in which nonconservative forces act is nonisolated and the external influence on the system is by means of work, the generalization of Equation 8.13 is

$$\sum W_{other\ forces} - f_k d = \Delta E_{mech}$$

This equation, with the help of Equations 8.7 and 8.14, can be written as

$$\sum W_{other\ forces} = W = \Delta K + \Delta U + \Delta E_{int} \tag{8.17}$$

This reduced form of Equation 8.2 represents the nonisolated system model for a system that possesses potential energy and within which a nonconservative force acts.

Example 8.7   Crate Sliding Down a Ramp   AM

A 3.00-kg crate slides down a ramp. The ramp is 1.00 m in length and inclined at an angle of 30.0° as shown in Figure 8.10. The crate starts from rest at the top, experiences a constant friction force of magnitude 5.00 N, and continues to move a short distance on the horizontal floor after it leaves the ramp.

(A) Use energy methods to determine the speed of the crate at the bottom of the ramp.

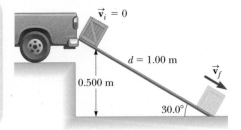

SOLUTION

Figure 8.10 (Example 8.7) A crate slides down a ramp under the influence of gravity. The potential energy of the system decreases, whereas the kinetic energy increases.

Conceptualize Imagine the crate sliding down the ramp in Figure 8.10. The larger the friction force, the more slowly the crate will slide.

Categorize We identify the crate, the surface, and the Earth as an *isolated system* with a nonconservative force acting.

Analyze Because $v_i = 0$, the initial kinetic energy of the system when the crate is at the top of the ramp is zero. If the y coordinate is measured from the bottom of the ramp (the final position of the crate, for which we choose the gravitational potential energy of the system to be zero) with the upward direction being positive, then $y_i = 0.500$ m.

Write the conservation of energy equation (Eq. 8.2) for this system:

$$\Delta K + \Delta U + \Delta E_{\text{int}} = 0$$

Substitute for the energies:

$$(\tfrac{1}{2}mv_f^2 - 0) + (0 - mgy_i) + f_k d = 0$$

Solve for v_f:

$$(1) \quad v_f = \sqrt{\frac{2}{m}(mgy_i - f_k d)}$$

Substitute numerical values:

$$v_f = \sqrt{\frac{2}{3.00 \text{ kg}} [(3.00 \text{ kg})(9.80 \text{ m/s}^2)(0.500 \text{ m}) - (5.00 \text{ N})(1.00 \text{ m})]} = \boxed{2.54 \text{ m/s}}$$

(B) How far does the crate slide on the horizontal floor if it continues to experience a friction force of magnitude 5.00 N?

SOLUTION

Analyze This part of the problem is handled in exactly the same way as part (A), but in this case we can consider the mechanical energy of the system to consist only of kinetic energy because the potential energy of the system remains fixed.

Write the conservation of energy equation for this situation:

$$\Delta K + \Delta E_{\text{int}} = 0$$

Substitute for the energies:

$$(0 - \tfrac{1}{2}mv_i^2) + f_k d = 0$$

Solve for the distance d and substitute numerical values:

$$d = \frac{mv_i^2}{2f_k} = \frac{(3.00 \text{ kg})(2.54 \text{ m/s})^2}{2(5.00 \text{ N})} = \boxed{1.94 \text{ m}}$$

Finalize For comparison, you may want to calculate the speed of the crate at the bottom of the ramp in the case in which the ramp is frictionless. Also notice that the increase in internal energy of the system as the crate slides down the ramp is $f_k d = (5.00 \text{ N})(1.00 \text{ m}) = 5.00$ J. This energy is shared between the crate and the surface, each of which is a bit warmer than before.

Also notice that the distance d the object slides on the horizontal surface is infinite if the surface is frictionless. Is that consistent with your conceptualization of the situation?

WHAT IF?   A cautious worker decides that the speed of the crate when it arrives at the bottom of the ramp may be so large that its contents may be damaged. Therefore, he replaces the ramp with a longer one such that the new

▶ **8.7** continued

ramp makes an angle of 25.0° with the ground. Does this new ramp reduce the speed of the crate as it reaches the ground?

Answer Because the ramp is longer, the friction force acts over a longer distance and transforms more of the mechanical energy into internal energy. The result is a reduction in the kinetic energy of the crate, and we expect a lower speed as it reaches the ground.

Find the length d of the new ramp:

$$\sin 25.0° = \frac{0.500 \text{ m}}{d} \quad \rightarrow \quad d = \frac{0.500 \text{ m}}{\sin 25.0°} = 1.18 \text{ m}$$

Find v_f from Equation (1) in part (A):

$$v_f = \sqrt{\frac{2}{3.00 \text{ kg}}[(3.00 \text{ kg})(9.80 \text{ m/s}^2)(0.500 \text{ m}) - (5.00 \text{ N})(1.18 \text{ m})]} = 2.42 \text{ m/s}$$

The final speed is indeed lower than in the higher-angle case.

Example 8.8   Block–Spring Collision   AM

A block having a mass of 0.80 kg is given an initial velocity $v_{Ⓐ} = 1.2$ m/s to the right and collides with a spring whose mass is negligible and whose force constant is $k = 50$ N/m as shown in Figure 8.11.

(A) Assuming the surface to be frictionless, calculate the maximum compression of the spring after the collision.

SOLUTION

Conceptualize The various parts of Figure 8.11 help us imagine what the block will do in this situation. All motion takes place in a horizontal plane, so we do not need to consider changes in gravitational potential energy.

Categorize We identify the system to be the block and the spring and model it as an *isolated system* with no nonconservative forces acting.

Analyze Before the collision, when the block is at Ⓐ, it has kinetic energy and the spring is uncompressed, so the elastic potential

Figure 8.11 (Example 8.8) A block sliding on a frictionless, horizontal surface collides with a light spring. (a) Initially, the mechanical energy is all kinetic energy. (b) The mechanical energy is the sum of the kinetic energy of the block and the elastic potential energy in the spring. (c) The energy is entirely potential energy. (d) The energy is transformed back to the kinetic energy of the block. The total energy of the system remains constant throughout the motion.

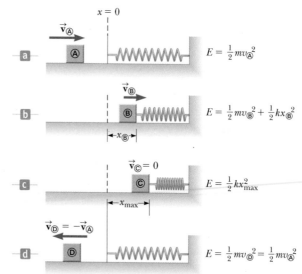

energy stored in the system is zero. Therefore, the total mechanical energy of the system before the collision is just $\frac{1}{2}mv_{Ⓐ}^2$. After the collision, when the block is at Ⓒ, the spring is fully compressed; now the block is at rest and so has zero kinetic energy. The elastic potential energy stored in the system, however, has its maximum value $\frac{1}{2}kx^2 = \frac{1}{2}kx_{max}^2$, where the origin of coordinates $x = 0$ is chosen to be the equilibrium position of the spring and x_{max} is the maximum compression of the spring, which in this case happens to be $x_{Ⓒ}$. The total mechanical energy of the system is conserved because no nonconservative forces act on objects within the isolated system.

Write the conservation of energy equation for this situation:

$$\Delta K + \Delta U = 0$$

Substitute for the energies:

$$\left(0 - \tfrac{1}{2}mv_{Ⓐ}^2\right) + \left(\tfrac{1}{2}kx_{max}^2 - 0\right) = 0$$

Solve for x_{max} and evaluate:

$$x_{max} = \sqrt{\frac{m}{k}}\,v_{Ⓐ} = \sqrt{\frac{0.80 \text{ kg}}{50 \text{ N/m}}}(1.2 \text{ m/s}) = \boxed{0.15 \text{ m}}$$

continued

▶ **8.8** continued

(B) Suppose a constant force of kinetic friction acts between the block and the surface, with $\mu_k = 0.50$. If the speed of the block at the moment it collides with the spring is $v_{\circledA} = 1.2$ m/s, what is the maximum compression $x_{\copyright}$ in the spring?

SOLUTION

Conceptualize Because of the friction force, we expect the compression of the spring to be smaller than in part (A) because some of the block's kinetic energy is transformed to internal energy in the block and the surface.

Categorize We identify the system as the block, the surface, and the spring. This is an *isolated system* but now involves a nonconservative force.

. . . .

Analyze In this case, the mechanical energy $E_{\text{mech}} = K + U_s$ of the system is *not* conserved because a friction force acts on the block. From the *particle in equilibrium* model in the vertical direction, we see that $n = mg$.

Evaluate the magnitude of the friction force:

$$f_k = \mu_k n = \mu_k mg$$

Write the conservation of energy equation for this situation:

$$\Delta K + \Delta U + \Delta E_{\text{int}} = 0$$

Substitute the initial and final energies:

$$\left(0 - \tfrac{1}{2}mv_{\circledA}^2\right) + \left(\tfrac{1}{2}kx_{\copyright}^2 - 0\right) + \mu_k mgx_{\copyright} = 0$$

Rearrange the terms into a qaudratic equation:

$$kx_{\copyright}^2 + 2\mu_k mgx_{\copyright} - mv_{\circledA}^2 = 0$$

Substitute numerical values:

$$50x_{\copyright}^2 + 2(0.50)(0.80)(9.80)x_{\copyright} - (0.80)(1.2)^2 = 0$$

$$50x_{\copyright}^2 + 7.84x_{\copyright} - 1.15 = 0$$

Solving the quadratic equation for $x_{\copyright}$ gives $x_{\copyright} = 0.092$ m and $x_{\copyright} = -0.25$ m. The physically meaningful root is $x_{\copyright} = \boxed{0.092 \text{ m}}$.

. . . .

Finalize The negative root does not apply to this situation because the block must be to the right of the origin (positive value of x) when it comes to rest. Notice that the value of 0.092 m is less than the distance obtained in the frictionless case of part (A) as we expected.

Example 8.9   Connected Blocks in Motion   AM

Two blocks are connected by a light string that passes over a frictionless pulley as shown in Figure 8.12. The block of mass m_1 lies on a horizontal surface and is connected to a spring of force constant k. The system is released from rest when the spring is unstretched. If the hanging block of mass m_2 falls a distance h before coming to rest, calculate the coefficient of kinetic friction between the block of mass m_1 and the surface.

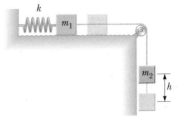

SOLUTION

Conceptualize The key word *rest* appears twice in the problem statement. This word suggests that the configurations of the system associated with rest are good candidates for the initial and final configurations because the kinetic energy of the system is zero for these configurations.

Categorize In this situation, the system consists of the two blocks, the spring, the surface, and the Earth. This is an *isolated system* with a nonconservative force acting. We also model the sliding block as a *particle in equilibrium* in the vertical direction, leading to $n = m_1 g$.

Figure 8.12 (Example 8.9) As the hanging block moves from its highest elevation to its lowest, the system loses gravitational potential energy but gains elastic potential energy in the spring. Some mechanical energy is transformed to internal energy because of friction between the sliding block and the surface.

. . . .

Analyze We need to consider two forms of potential energy for the system, gravitational and elastic: $\Delta U_g = U_{gf} - U_{gi}$ is the change in the system's gravitational potential energy, and $\Delta U_s = U_{sf} - U_{si}$ is the change in the system's elastic potential energy. The change in the gravitational potential energy of the system is associated with only the falling block

▶ **8.9** continued

because the vertical coordinate of the horizontally sliding block does not change. The initial and final kinetic energies of the system are zero, so $\Delta K = 0$.

Write the appropriate reduction of Equation 8.2:

$$(1) \quad \Delta U_g + \Delta U_s + \Delta E_{int} = 0$$

Substitute for the energies, noting that as the hanging block falls a distance h, the horizontally moving block moves the same distance h to the right, and the spring stretches by a distance h:

$$(0 - m_2gh) + (\tfrac{1}{2}kh^2 - 0) + f_kh = 0$$

Substitute for the friction force:

$$-m_2gh + \tfrac{1}{2}kh^2 + \mu_k m_1gh = 0$$

Solve for μ_k:

$$\mu_k = \frac{m_2g - \tfrac{1}{2}kh}{m_1g}$$

Finalize This setup represents a method of measuring the coefficient of kinetic friction between an object and some surface. Notice how we have solved the examples in this chapter using the energy approach. We begin with Equation 8.2 and then tailor it to the physical situation. This process may include deleting terms, such as the kinetic energy term and all terms on the right-hand side of Equation 8.2 in this example. It can also include expanding terms, such as rewriting ΔU due to two types of potential energy in this example.

Conceptual Example 8.10   **Interpreting the Energy Bars**

The energy bar charts in Figure 8.13 show three instants in the motion of the system in Figure 8.12 and described in Example 8.9. For each bar chart, identify the configuration of the system that corresponds to the chart.

SOLUTION

In Figure 8.13a, there is no kinetic energy in the system. Therefore, nothing in the system is moving. The bar chart shows that the system contains only gravitational potential energy and no internal energy yet, which corresponds to the configuration with the darker blocks in Figure 8.12 and represents the instant just after the system is released.

In Figure 8.13b, the system contains four types of energy. The height of the gravitational potential energy bar is at 50%, which tells us that the hanging block has moved halfway between its position corresponding to Figure 8.13a and the position defined as $y = 0$. Therefore, in this configuration, the hanging block is between the dark and light images of the hanging block in Figure 8.12. The system has gained kinetic energy because the blocks are moving, elastic potential energy because the spring is stretching, and internal energy because of friction between the block of mass m_1 and the surface.

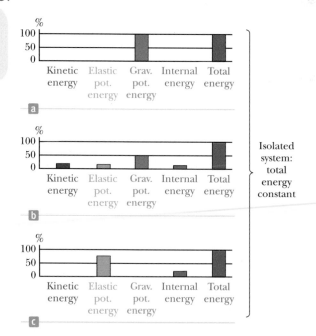

Figure 8.13 (Conceptual Example 8.10) Three energy bar charts are shown for the system in Figure 8.12.

In Figure 8.13c, the height of the gravitational potential energy bar is zero, telling us that the hanging block is at $y = 0$. In addition, the height of the kinetic energy bar is zero, indicating that the blocks have stopped moving momentarily. Therefore, the configuration of the system is that shown by the light images of the blocks in Figure 8.12. The height of the elastic potential energy bar is high because the spring is stretched its maximum amount. The height of the internal energy bar is higher than in Figure 8.13b because the block of mass m_1 has continued to slide over the surface after the configuration shown in Figure 8.13b.

8.5 Power

Consider Conceptual Example 7.7 again, which involved rolling a refrigerator up a ramp into a truck. Suppose the man is not convinced the work is the same regardless of the ramp's length and sets up a long ramp with a gentle rise. Although he does the same amount of work as someone using a shorter ramp, he takes longer to do the work because he has to move the refrigerator over a greater distance. Although the work done on both ramps is the same, there is *something* different about the tasks: the *time interval* during which the work is done.

The time rate of energy transfer is called the **instantaneous power** P and is defined as

◀ Definition of power

$$P \equiv \frac{dE}{dt} \qquad (8.18)$$

We will focus on work as the energy transfer method in this discussion, but keep in mind that the notion of power is valid for *any* means of energy transfer discussed in Section 8.1. If an external force is applied to an object (which we model as a particle) and if the work done by this force on the object in the time interval Δt is W, the **average power** during this interval is

$$P_{\text{avg}} = \frac{W}{\Delta t}$$

Therefore, in Conceptual Example 7.7, although the same work is done in rolling the refrigerator up both ramps, less power is required for the longer ramp.

In a manner similar to the way we approached the definition of velocity and acceleration, the instantaneous power is the limiting value of the average power as Δt approaches zero:

$$P = \lim_{\Delta t \to 0} \frac{W}{\Delta t} = \frac{dW}{dt}$$

where we have represented the infinitesimal value of the work done by dW. We find from Equation 7.3 that $dW = \vec{\mathbf{F}} \cdot d\vec{\mathbf{r}}$. Therefore, the instantaneous power can be written

$$P = \frac{dW}{dt} = \vec{\mathbf{F}} \cdot \frac{d\vec{\mathbf{r}}}{dt} = \vec{\mathbf{F}} \cdot \vec{\mathbf{v}} \qquad (8.19)$$

where $\vec{\mathbf{v}} = d\vec{\mathbf{r}}/dt$.

The SI unit of power is joules per second (J/s), also called the **watt** (W) after James Watt:

◀ The watt

$$1\ \text{W} = 1\ \text{J/s} = 1\ \text{kg} \cdot \text{m}^2/\text{s}^3$$

A unit of power in the U.S. customary system is the **horsepower** (hp):

$$1\ \text{hp} = 746\ \text{W}$$

A unit of energy (or work) can now be defined in terms of the unit of power. One **kilowatt-hour** (kWh) is the energy transferred in 1 h at the constant rate of 1 kW = 1 000 J/s. The amount of energy represented by 1 kWh is

$$1\ \text{kWh} = (10^3\ \text{W})(3\ 600\ \text{s}) = 3.60 \times 10^6\ \text{J}$$

A kilowatt-hour is a unit of energy, not power. When you pay your electric bill, you are buying energy, and the amount of energy transferred by electrical transmission into a home during the period represented by the electric bill is usually expressed in kilowatt-hours. For example, your bill may state that you used 900 kWh of energy during a month and that you are being charged at the rate of 10¢ per kilowatt-hour. Your obligation is then $90 for this amount of energy. As another example, suppose an electric bulb is rated at 100 W. In 1.00 h of operation, it would have energy transferred to it by electrical transmission in the amount of (0.100 kW)(1.00 h) = 0.100 kWh = 3.60 × 10⁵ J.

Example 8.11 | Power Delivered by an Elevator Motor AM

An elevator car (Fig. 8.14a) has a mass of 1 600 kg and is carrying passengers having a combined mass of 200 kg. A constant friction force of 4 000 N retards its motion.

(A) How much power must a motor deliver to lift the elevator car and its passengers at a constant speed of 3.00 m/s?

SOLUTION

Conceptualize The motor must supply the force of magnitude T that pulls the elevator car upward.

Categorize The friction force increases the power necessary to lift the elevator. The problem states that the speed of the elevator is constant, which tells us that $a = 0$. We model the elevator as a *particle in equilibrium*.

Analyze The free-body diagram in Figure 8.14b specifies the upward direction as positive. The *total* mass M of the system (car plus passengers) is equal to 1 800 kg.

Figure 8.14 (Example 8.11) (a) The motor exerts an upward force $\vec{T}$ on the elevator car. The magnitude of this force is the total tension T in the cables connecting the car and motor. The downward forces acting on the car are a friction force $\vec{f}$ and the gravitational force $\vec{F}_g = M\vec{g}$. (b) The free-body diagram for the elevator car.

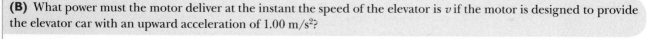

Using the particle in equilibrium model, apply Newton's second law to the car:	$\sum F_y = T - f - Mg = 0$
Solve for T:	$T = Mg + f$
Use Equation 8.19 and that $\vec{T}$ is in the same direction as $\vec{v}$ to find the power:	$P = \vec{T} \cdot \vec{v} = Tv = (Mg + f)v$
Substitute numerical values:	$P = [(1\ 800\ \text{kg})(9.80\ \text{m/s}^2) + (4\ 000\ \text{N})](3.00\ \text{m/s}) = \boxed{6.49 \times 10^4\ \text{W}}$

(B) What power must the motor deliver at the instant the speed of the elevator is v if the motor is designed to provide the elevator car with an upward acceleration of 1.00 m/s^2?

SOLUTION

Conceptualize In this case, the motor must supply the force of magnitude T that pulls the elevator car upward with an increasing speed. We expect that more power will be required to do that than in part (A) because the motor must now perform the additional task of accelerating the car.

Categorize In this case, we model the elevator car as a *particle under a net force* because it is accelerating.

Analyze Using the particle under a net force model, apply Newton's second law to the car:	$\sum F_y = T - f - Mg = Ma$
Solve for T:	$T = M(a + g) + f$
Use Equation 8.19 to obtain the required power:	$P = Tv = [M(a + g) + f]v$
Substitute numerical values:	$P = [(1\ 800\ \text{kg})(1.00\ \text{m/s}^2 + 9.80\ \text{m/s}^2) + 4\ 000\ \text{N}]v$
	$= (2.34 \times 10^4)v$

where v is the instantaneous speed of the car in meters per second and P is in watts.

Finalize To compare with part (A), let $v = 3.00$ m/s, giving a power of

$$P = (2.34 \times 10^4\ \text{N})(3.00\ \text{m/s}) = 7.02 \times 10^4\ \text{W}$$

which is larger than the power found in part (A), as expected.

Summary

Definitions

A **nonisolated system** is one for which energy crosses the boundary of the system. An **isolated system** is one for which no energy crosses the boundary of the system.

The **instantaneous power** P is defined as the time rate of energy transfer:

$$P \equiv \frac{dE}{dt} \tag{8.18}$$

Concepts and Principles

For a nonisolated system, we can equate the change in the total energy stored in the system to the sum of all the transfers of energy across the system boundary, which is a statement of **conservation of energy**. For an isolated system, the total energy is constant.

If a friction force of magnitude f_k acts over a distance d within a system, the change in internal energy of the system is

$$\Delta E_{int} = f_k d \tag{8.14}$$

Analysis Models for Problem Solving

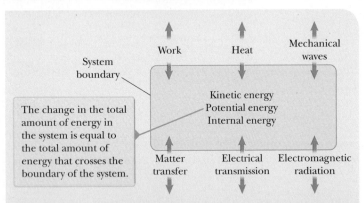

Nonisolated System (Energy). The most general statement describing the behavior of a nonisolated system is the **conservation of energy equation**:

$$\Delta E_{system} = \Sigma\, T \tag{8.1}$$

Including the types of energy storage and energy transfer that we have discussed gives

$$\Delta K + \Delta U + \Delta E_{int} = W + Q + T_{MW} + T_{MT} + T_{ET} + T_{ER} \tag{8.2}$$

For a specific problem, this equation is generally reduced to a smaller number of terms by eliminating the terms that are not appropriate to the situation.

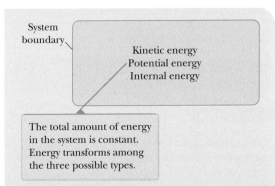

Isolated System (Energy). The total energy of an isolated system is conserved, so

$$\Delta E_{system} = 0 \tag{8.10}$$

which can be written as

$$\Delta K + \Delta U + \Delta E_{int} = 0 \tag{8.16}$$

If no nonconservative forces act within the isolated system, the mechanical energy of the system is conserved, so

$$\Delta E_{mech} = 0 \tag{8.8}$$

which can be written as

$$\Delta K + \Delta U = 0 \tag{8.6}$$

1. You hold a slingshot at arm's length, pull the light elastic band back to your chin, and release it to launch a pebble horizontally with speed 200 cm/s. With the same procedure, you fire a bean with speed 600 cm/s. What is the ratio of the mass of the bean to the mass of the pebble? (a) $\frac{1}{9}$ (b) $\frac{1}{3}$ (c) 1 (d) 3 (e) 9

2. Two children stand on a platform at the top of a curving slide next to a backyard swimming pool. At the same moment the smaller child hops off to jump straight down into the pool, the bigger child releases herself at the top of the frictionless slide. **(i)** Upon reaching the water, the kinetic energy of the smaller child compared with that of the larger child is (a) greater (b) less (c) equal. **(ii)** Upon reaching the water, the speed of the smaller child compared with that of the larger child is (a) greater (b) less (c) equal. **(iii)** During their motions from the platform to the water, the average acceleration of the smaller child compared with that of the larger child is (a) greater (b) less (c) equal.

3. At the bottom of an air track tilted at angle θ, a glider of mass m is given a push to make it coast a distance d up the slope as it slows down and stops. Then the glider comes back down the track to its starting point. Now the experiment is repeated with the same original speed but with a second identical glider set on top of the first. The airflow from the track is strong enough to support the stacked pair of gliders so that the combination moves over the track with negligible friction. Static friction holds the second glider stationary relative to the first glider throughout the motion. The coefficient of static friction between the two gliders is μ_s. What is the change in mechanical energy of the two-glider–Earth system in the up- and down-slope motion after the pair of gliders is released? Choose one. (a) $-2\mu_s mg$ (b) $-2mgd \cos \theta$ (c) $-2\mu_s mgd \cos \theta$ (d) 0 (e) $+2\mu_s mgd \cos \theta$

4. An athlete jumping vertically on a trampoline leaves the surface with a velocity of 8.5 m/s upward. What maximum height does she reach? (a) 13 m (b) 2.3 m (c) 3.7 m (d) 0.27 m (e) The answer can't be determined because the mass of the athlete isn't given.

5. Answer yes or no to each of the following questions. (a) Can an object–Earth system have kinetic energy and not gravitational potential energy? (b) Can it have gravitational potential energy and not kinetic energy? (c) Can it have both types of energy at the same moment? (d) Can it have neither?

6. In a laboratory model of cars skidding to a stop, data are measured for four trials using two blocks. The blocks have identical masses but different coefficients of kinetic friction with a table: $\mu_k = 0.2$ and 0.8. Each block is launched with speed $v_i = 1$ m/s and slides across the level table as the block comes to rest. This process represents the first two trials. For the next two trials, the procedure is repeated but the blocks are launched with speed $v_i = 2$ m/s. Rank the four trials (a) through (d) according to the stopping distance from largest to smallest. If the stopping distance is the same in two cases, give them equal rank. (a) $v_i = 1$ m/s, $\mu_k = 0.2$ (b) $v_i = 1$ m/s, $\mu_k = 0.8$ (c) $v_i = 2$ m/s, $\mu_k = 0.2$ (d) $v_i = 2$ m/s, $\mu_k = 0.8$

7. What average power is generated by a 70.0-kg mountain climber who climbs a summit of height 325 m in 95.0 min? (a) 39.1 W (b) 54.6 W (c) 25.5 W (d) 67.0 W (e) 88.4 W

8. A ball of clay falls freely to the hard floor. It does not bounce noticeably, and it very quickly comes to rest. What, then, has happened to the energy the ball had while it was falling? (a) It has been used up in producing the downward motion. (b) It has been transformed back into potential energy. (c) It has been transferred into the ball by heat. (d) It is in the ball and floor (and walls) as energy of invisible molecular motion. (e) Most of it went into sound.

9. A pile driver drives posts into the ground by repeatedly dropping a heavy object on them. Assume the object is dropped from the same height each time. By what factor does the energy of the pile driver–Earth system change when the mass of the object being dropped is doubled? (a) $\frac{1}{2}$ (b) 1; the energy is the same (c) 2 (d) 4

1. One person drops a ball from the top of a building while another person at the bottom observes its motion. Will these two people agree (a) on the value of the gravitational potential energy of the ball–Earth system? (b) On the change in potential energy? (c) On the kinetic energy of the ball at some point in its motion?

2. A car salesperson claims that a 300-hp engine is a necessary option in a compact car, in place of the conventional 130-hp engine. Suppose you intend to drive the car within speed limits (≤ 65 mi/h) on flat terrain. How would you counter this sales pitch?

3. Does everything have energy? Give the reasoning for your answer.

4. You ride a bicycle. In what sense is your bicycle solar-powered?

5. A bowling ball is suspended from the ceiling of a lecture hall by a strong cord. The ball is drawn away from its equilibrium position and released from rest at the

tip of the demonstrator's nose as shown in Figure CQ8.5. The demonstrator remains stationary. (a) Explain why the ball does not strike her on its return swing. (b) Would this demonstrator be safe if the ball were given a push from its starting position at her nose?

Figure CQ8.5

6. Can a force of static friction do work? If not, why not? If so, give an example.

7. In the general conservation of energy equation, state which terms predominate in describing each of the following devices and processes. For a process going on continuously, you may consider what happens in a 10-s time interval. State which terms in the equation represent original and final forms of energy, which would be inputs, and which outputs. (a) a slingshot firing a pebble (b) a fire burning (c) a portable radio operating (d) a car braking to a stop (e) the surface of the Sun shining visibly (f) a person jumping up onto a chair

8. Consider the energy transfers and transformations listed below in parts (a) through (e). For each part, (i) describe human-made devices designed to produce each of the energy transfers or transformations and, (ii) whenever possible, describe a natural process in which the energy transfer or transformation occurs. Give details to defend your choices, such as identifying the system and identifying other output energy if the device or natural process has limited efficiency. (a) Chemical potential energy transforms into internal energy. (b) Energy transferred by electrical transmission becomes gravitational potential energy. (c) Elastic potential energy transfers out of a system by heat. (d) Energy transferred by mechanical waves does work on a system. (e) Energy carried by electromagnetic waves becomes kinetic energy in a system.

9. A block is connected to a spring that is suspended from the ceiling. Assuming air resistance is ignored, describe the energy transformations that occur within the system consisting of the block, the Earth, and the spring when the block is set into vertical motion.

10. In Chapter 7, the work–kinetic energy theorem, $W = \Delta K$, was introduced. This equation states that work done on a system appears as a change in kinetic energy. It is a special-case equation, valid if there are no changes in any other type of energy such as potential or internal. Give two or three examples in which work is done on a system but the change in energy of the system is not a change in kinetic energy.

Problems available in **WebAssign** Access end-of-chapter problems online at www.webassign.net

Section 8.1 Analysis Model: Nonisolated System (Energy)
Problems 1–2

Section 8.2 Analysis Model: Isolated System (Energy)
Problems 3–11

Section 8.3 Situations Involving Kinetic Friction
Problems 12–17

Section 8.4 Changes in Mechanical Energy for Nonconservative Forces
Problems 18–27

Section 8.5 Power
Problems 28–41

Additional Problems
Problems 42–78

Challenge Problems
Problem 79–85

Solutions to the following Problems are available in the *Student Solutions Manual/Study Guide:*
8.5, 8.7, 8.12, 8.14, 8.16, 8.22, 8.23, 8.35, 8.29, 8.39, 8.44, 8.47, 8.51, 8.63, 8.64, 8.65, and 8.73

List of Enhanced Problems

Problem Number	Targeted Feedback in Enhanced WebAssign	Analysis Model Tutorial in Enhanced WebAssign	Master It in Enhanced WebAssign	Watch It in Enhanced WebAssign
8.3	✓			✓
8.4	✓			✓
8.5	✓	✓	✓	
8.6	✓			✓
8.7	✓		✓	
8.14	✓		✓	
8.15	✓			✓
8.21	✓			✓
8.22	✓	✓		✓
8.23	✓		✓	
8.25			✓	
8.29	✓			✓
8.41		✓	✓	
8.47			✓	
8.51	✓			
8.62	✓			✓
8.63	✓		✓	
8.64		✓	✓	

Linear Momentum and Collisions

Consider what happens when two cars collide as in the opening photograph for this chapter. Both cars change their motion from having a very large velocity to being at rest because of the collision. Because each car experiences a large change in velocity over a very short time interval, the average force on it is very large. By Newton's third law, each of the cars experiences a force of the same magnitude. By Newton's second law, the results of those forces on the motion of the car depends on the mass of the car.

One of the main objectives of this chapter is to enable you to understand and analyze such events in a simple way. First, we introduce the concept of *momentum*, which is useful for describing objects in motion. The momentum of an object is related to both its mass and its velocity. The concept of momentum leads us to a second conservation law, that of conservation of momentum. In turn, we identify new momentum versions of analysis models for isolated and nonisolated system. These models are especially useful for treating problems that involve collisions between objects and for analyzing rocket propulsion. This chapter also introduces the concept of the center of mass of a system of particles. We find that the motion of a system of particles can be described by the motion of one particle located at the center of mass that represents the entire system.

The concept of momentum allows the analysis of car collisions even without detailed knowledge of the forces involved. Such analysis can determine the relative velocity of the cars before the collision, and in addition aid engineers in designing safer vehicles. (The English translation of the German text on the side of the trailer in the background is: "Pit stop for your vehicle.") *(AP Photos/Keystone/ Regina Kuehne)*

9.1 Linear Momentum

In Chapter 8, we studied situations that are difficult to analyze with Newton's laws. We were able to solve problems involving these situations by identifying a system and applying a conservation principle, conservation of energy. Let us consider another situation and see if we can solve it with the models we have developed so far:

> A 60-kg archer stands at rest on frictionless ice and fires a 0.030-kg arrow horizontally at 85 m/s. With what velocity does the archer move across the ice after firing the arrow?

From Newton's third law, we know that the force that the bow exerts on the arrow is paired with a force in the opposite direction on the bow (and the archer). This force causes the archer to slide backward on the ice with the speed requested in the problem. We cannot determine this speed using motion models such as the particle under constant acceleration because we don't have any information about the acceleration of the archer. We cannot use force models such as the particle under a net force because we don't know anything about forces in this situation. Energy models are of no help because we know nothing about the work done in pulling the bowstring back or the elastic potential energy in the system related to the taut bowstring.

Despite our inability to solve the archer problem using models learned so far, this problem is very simple to solve if we introduce a new quantity that describes motion, *linear momentum*. To generate this new quantity, consider an isolated system of two particles (Fig. 9.1) with masses m_1 and m_2 moving with velocities $\vec{\mathbf{v}}_1$ and $\vec{\mathbf{v}}_2$ at an instant of time. Because the system is isolated, the only force on one particle is that from the other particle. If a force from particle 1 (for example, a gravitational force) acts on particle 2, there must be a second force—equal in magnitude but opposite in direction—that particle 2 exerts on particle 1. That is, the forces on the particles form a Newton's third law action–reaction pair, and $\vec{\mathbf{F}}_{12} = -\vec{\mathbf{F}}_{21}$. We can express this condition as

$$\vec{\mathbf{F}}_{21} + \vec{\mathbf{F}}_{12} = 0$$

From a system point of view, this equation says that if we add up the forces on the particles in an isolated system, the sum is zero.

Let us further analyze this situation by incorporating Newton's second law. At the instant shown in Figure 9.1, the interacting particles in the system have accelerations corresponding to the forces on them. Therefore, replacing the force on each particle with $m\vec{\mathbf{a}}$ for the particle gives

$$m_1\vec{\mathbf{a}}_1 + m_2\vec{\mathbf{a}}_2 = 0$$

Now we replace each acceleration with its definition from Equation 4.5:

$$m_1\frac{d\vec{\mathbf{v}}_1}{dt} + m_2\frac{d\vec{\mathbf{v}}_2}{dt} = 0$$

If the masses m_1 and m_2 are constant, we can bring them inside the derivative operation, which gives

$$\frac{d(m_1\vec{\mathbf{v}}_1)}{dt} + \frac{d(m_2\vec{\mathbf{v}}_2)}{dt} = 0$$

$$\frac{d}{dt}(m_1\vec{\mathbf{v}}_1 + m_2\vec{\mathbf{v}}_2) = 0 \tag{9.1}$$

Notice that the derivative of the sum $m_1\vec{\mathbf{v}}_1 + m_2\vec{\mathbf{v}}_2$ with respect to time is zero. Consequently, this sum must be constant. We learn from this discussion that the quantity $m\vec{\mathbf{v}}$ for a particle is important in that the sum of these quantities for an isolated system of particles is conserved. We call this quantity *linear momentum*:

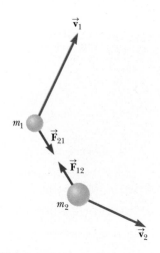

Figure 9.1 Two particles interact with each other. According to Newton's third law, we must have $\vec{\mathbf{F}}_{12} = -\vec{\mathbf{F}}_{21}$.

Definition of linear ▶
momentum of a particle

The **linear momentum** of a particle or an object that can be modeled as a particle of mass m moving with a velocity $\vec{\mathbf{v}}$ is defined to be the product of the mass and velocity of the particle:

$$\vec{\mathbf{p}} \equiv m\vec{\mathbf{v}} \tag{9.2}$$

Linear momentum is a vector quantity because it equals the product of a scalar quantity m and a vector quantity $\vec{\mathbf{v}}$. Its direction is along $\vec{\mathbf{v}}$, it has dimensions ML/T, and its SI unit is kg · m/s.

If a particle is moving in an arbitrary direction, $\vec{\mathbf{p}}$ has three components, and Equation 9.2 is equivalent to the component equations

$$p_x = mv_x \qquad p_y = mv_y \qquad p_z = mv_z$$

As you can see from its definition, the concept of momentum[1] provides a quantitative distinction between heavy and light particles moving at the same velocity. For example, the momentum of a bowling ball is much greater than that of a tennis ball moving at the same speed. Newton called the product $m\vec{\mathbf{v}}$ *quantity of motion;* this term is perhaps a more graphic description than our present-day word *momentum,* which comes from the Latin word for movement.

We have seen another quantity, kinetic energy, that is a combination of mass and speed. It would be a legitimate question to ask why we need another quantity, momentum, based on mass and velocity. There are clear differences between kinetic energy and momentum. First, kinetic energy is a scalar, whereas momentum is a vector. Consider a system of two equal-mass particles heading toward each other along a line with equal speeds. There is kinetic energy associated with this system because members of the system are moving. Because of the vector nature of momentum, however, the momentum of this system is zero. A second major difference is that kinetic energy can transform to other types of energy, such as potential energy or internal energy. There is only one type of linear momentum, so we see no such transformations when using a momentum approach to a problem. These differences are sufficient to make models based on momentum separate from those based on energy, providing an independent tool to use in solving problems.

Using Newton's second law of motion, we can relate the linear momentum of a particle to the resultant force acting on the particle. We start with Newton's second law and substitute the definition of acceleration:

$$\sum \vec{\mathbf{F}} = m\vec{\mathbf{a}} = m\,\frac{d\vec{\mathbf{v}}}{dt}$$

In Newton's second law, the mass m is assumed to be constant. Therefore, we can bring m inside the derivative operation to give us

$$\sum \vec{\mathbf{F}} = \frac{d(m\vec{\mathbf{v}})}{dt} = \frac{d\vec{\mathbf{p}}}{dt} \qquad (9.3)$$

◀ Newton's second law for a particle

This equation shows that **the time rate of change of the linear momentum of a particle is equal to the net force acting on the particle**. In Chapter 5, we identified force as that which causes a change in the motion of an object (Section 5.2). In Newton's second law (Eq. 5.2), we used acceleration $\vec{\mathbf{a}}$ to represent the change in motion. We see now in Equation 9.3 that we can use the derivative of momentum $\vec{\mathbf{p}}$ with respect to time to represent the change in motion.

This alternative form of Newton's second law is the form in which Newton presented the law, and it is actually more general than the form introduced in Chapter 5. In addition to situations in which the velocity vector varies with time, we can use Equation 9.3 to study phenomena in which the mass changes. For example, the mass of a rocket changes as fuel is burned and ejected from the rocket. We cannot use $\sum \vec{\mathbf{F}} = m\vec{\mathbf{a}}$ to analyze rocket propulsion; we must use a momentum approach, as we will show in Section 9.9.

---

[1] In this chapter, the terms *momentum* and *linear momentum* have the same meaning. Later, in Chapter 11, we shall use the term *angular momentum* for a different quantity when dealing with rotational motion.

Quick Quiz 9.1 Two objects have equal kinetic energies. How do the magnitudes of their momenta compare? **(a)** $p_1 < p_2$ **(b)** $p_1 = p_2$ **(c)** $p_1 > p_2$ **(d)** not enough information to tell

Quick Quiz 9.2 Your physical education teacher throws a baseball to you at a certain speed and you catch it. The teacher is next going to throw you a medicine ball whose mass is ten times the mass of the baseball. You are given the following choices: You can have the medicine ball thrown with **(a)** the same speed as the baseball, **(b)** the same momentum, or **(c)** the same kinetic energy. Rank these choices from easiest to hardest to catch.

9.2 Analysis Model: Isolated System (Momentum)

Using the definition of momentum, Equation 9.1 can be written

$$\frac{d}{dt}(\vec{\mathbf{p}}_1 + \vec{\mathbf{p}}_2) = 0$$

Because the time derivative of the total momentum $\vec{\mathbf{p}}_{tot} = \vec{\mathbf{p}}_1 + \vec{\mathbf{p}}_2$ is *zero*, we conclude that the *total* momentum of the isolated system of the two particles in Figure 9.1 must remain constant:

$$\vec{\mathbf{p}}_{tot} = \text{constant} \tag{9.4}$$

or, equivalently, over some time interval,

$$\Delta\vec{\mathbf{p}}_{tot} = 0 \tag{9.5}$$

Equation 9.5 can be written as

$$\vec{\mathbf{p}}_{1i} + \vec{\mathbf{p}}_{2i} = \vec{\mathbf{p}}_{1f} + \vec{\mathbf{p}}_{2f}$$

where $\vec{\mathbf{p}}_{1i}$ and $\vec{\mathbf{p}}_{2i}$ are the initial values and $\vec{\mathbf{p}}_{1f}$ and $\vec{\mathbf{p}}_{2f}$ are the final values of the momenta for the two particles for the time interval during which the particles interact. This equation in component form demonstrates that the total momenta in the x, y, and z directions are all independently conserved:

$$p_{1ix} + p_{2ix} = p_{1fx} + p_{2fx} \qquad p_{1iy} + p_{2iy} = p_{1fy} + p_{2fy} \qquad p_{1iz} + p_{2iz} = p_{1fz} + p_{2fz} \tag{9.6}$$

Equation 9.5 is the mathematical statement of a new analysis model, the **isolated system (momentum).** It can be extended to any number of particles in an isolated system as we show in Section 9.7. We studied the energy version of the isolated system model in Chapter 8 ($\Delta E_{system} = 0$) and now we have a momentum version. In general, Equation 9.5 can be stated in words as follows:

The momentum version of the isolated system model ▶

Whenever two or more particles in an isolated system interact, the total momentum of the system does not change.

This statement tells us that the total momentum of an isolated system at all times equals its initial momentum.

Notice that we have made no statement concerning the type of forces acting on the particles of the system. Furthermore, we have not specified whether the forces are conservative or nonconservative. We have also not indicated whether or not the forces are constant. The only requirement is that the forces must be *internal* to the system. This single requirement should give you a hint about the power of this new model.

Analysis Model **Isolated System (Momentum)**

Imagine you have identified a system to be analyzed and have defined a system boundary. If there are no external forces on the system, the system is *isolated*. In that case, the total momentum of the system, which is the vector sum of the momenta of all members of the system, is conserved:

$$\Delta \vec{\mathbf{p}}_{\text{tot}} = 0 \qquad \qquad \text{(9.5)}$$

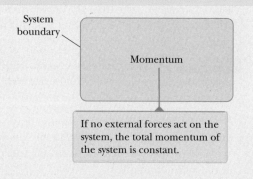

System boundary

Momentum

If no external forces act on the system, the total momentum of the system is constant.

Examples:

- a cue ball strikes another ball on a pool table
- a spacecraft fires its rockets and moves faster through space
- molecules in a gas at a specific temperature move about and strike each other (Chapter 21)
- an incoming particle strikes a nucleus, creating a new nucleus and a different outgoing particle (Chapter 44)
- an electron and a positron annihilate to form two outgoing photons (Chapter 46)

Example 9.1 **The Archer** AM

Let us consider the situation proposed at the beginning of Section 9.1. A 60-kg archer stands at rest on frictionless ice and fires a 0.030-kg arrow horizontally at 85 m/s (Fig. 9.2). With what velocity does the archer move across the ice after firing the arrow?

SOLUTION

Conceptualize You may have conceptualized this problem already when it was introduced at the beginning of Section 9.1. Imagine the arrow being fired one way and the archer recoiling in the opposite direction.

Categorize As discussed in Section 9.1, we cannot solve this problem with models based on motion, force, or energy. Nonetheless, we *can* solve this problem very easily with an approach involving momentum.

Let us take the system to consist of the archer (including the bow) and the arrow. The system is not isolated because the gravitational force and the normal force from the ice act on the system. These forces, however, are vertical and perpendicular to the motion of the system. There are no external forces in the horizontal direction, and we can apply the *isolated system (momentum)* model in terms of momentum components in this direction.

Figure 9.2 (Example 9.1) An archer fires an arrow horizontally to the right. Because he is standing on frictionless ice, he will begin to slide to the left across the ice.

Analyze The total horizontal momentum of the system before the arrow is fired is zero because nothing in the system is moving. Therefore, the total horizontal momentum of the system after the arrow is fired must also be zero. We choose the direction of firing of the arrow as the positive *x* direction. Identifying the archer as particle 1 and the arrow as particle 2, we have $m_1 = 60$ kg, $m_2 = 0.030$ kg, and $\vec{\mathbf{v}}_{2f} = 85\hat{\mathbf{i}}$ m/s.

Using the isolated system (momentum) model, begin with Equation 9.5:

$$\Delta \vec{\mathbf{p}} = 0 \;\; \rightarrow \;\; \vec{\mathbf{p}}_f - \vec{\mathbf{p}}_i = 0 \;\; \rightarrow \;\; \vec{\mathbf{p}}_f = \vec{\mathbf{p}}_i \;\; \rightarrow \;\; m_1\vec{\mathbf{v}}_{1f} + m_2\vec{\mathbf{v}}_{2f} = 0$$

Solve this equation for $\vec{\mathbf{v}}_{1f}$ and substitute numerical values:

$$\vec{\mathbf{v}}_{1f} = -\frac{m_2}{m_1}\vec{\mathbf{v}}_{2f} = -\left(\frac{0.030 \text{ kg}}{60 \text{ kg}}\right)(85\hat{\mathbf{i}} \text{ m/s}) = \boxed{-0.042\hat{\mathbf{i}} \text{ m/s}}$$

Finalize The negative sign for $\vec{\mathbf{v}}_{1f}$ indicates that the archer is moving to the left in Figure 9.2 after the arrow is fired, in the direction opposite the direction of motion of the arrow, in accordance with Newton's third law. Because the archer

continued

▶ **9.1** continued

is much more massive than the arrow, his acceleration and consequent velocity are much smaller than the acceleration and velocity of the arrow. Notice that this problem sounds very simple, but we could not solve it with models based on motion, force, or energy. Our new momentum model, however, shows us that it not only *sounds* simple, it *is* simple!

WHAT IF?   What if the arrow were fired in a direction that makes an angle θ with the horizontal? How will that change the recoil velocity of the archer?

Answer The recoil velocity should decrease in magnitude because only a component of the velocity of the arrow is in the x direction. Conservation of momentum in the x direction gives

$$m_1 v_{1f} + m_2 v_{2f} \cos \theta = 0$$

leading to

$$v_{1f} = -\frac{m_2}{m_1} v_{2f} \cos \theta$$

For $\theta = 0$, $\cos \theta = 1$ and the final velocity of the archer reduces to the value when the arrow is fired horizontally. For nonzero values of θ, the cosine function is less than 1 and the recoil velocity is less than the value calculated for $\theta = 0$. If $\theta = 90°$, then $\cos \theta = 0$ and $v_{1f} = 0$, so there is no recoil velocity. In this case, the archer is simply pushed downward harder against the ice as the arrow is fired.

Example 9.2   **Can We Really Ignore the Kinetic Energy of the Earth?**

In Section 7.6, we claimed that we can ignore the kinetic energy of the Earth when considering the energy of a system consisting of the Earth and a dropped ball. Verify this claim.

SOLUTION

Conceptualize Imagine dropping a ball at the surface of the Earth. From your point of view, the ball falls while the Earth remains stationary. By Newton's third law, however, the Earth experiences an upward force and therefore an upward acceleration while the ball falls. In the calculation below, we will show that this motion is extremely small and can be ignored.

Categorize We identify the system as the ball and the Earth. We assume there are no forces on the system from outer space, so the system is isolated. Let's use the *momentum* version of the *isolated system* model.

....

Analyze We begin by setting up a ratio of the kinetic energy of the Earth to that of the ball. We identify v_E and v_b as the speeds of the Earth and the ball, respectively, after the ball has fallen through some distance.

Use the definition of kinetic energy to set up this ratio:

$$(1) \quad \frac{K_E}{K_b} = \frac{\frac{1}{2} m_E v_E^2}{\frac{1}{2} m_b v_b^2} = \left(\frac{m_E}{m_b} \right) \left(\frac{v_E}{v_b} \right)^2$$

Apply the isolated system (momentum) model, recognizing that the initial momentum of the system is zero:

$$\Delta \vec{\mathbf{p}} = 0 \quad \rightarrow \quad p_i = p_f \quad \rightarrow \quad 0 = m_b v_b + m_E v_E$$

Solve the equation for the ratio of speeds:

$$\frac{v_E}{v_b} = -\frac{m_b}{m_E}$$

Substitute this expression for v_E/v_b in Equation (1):

$$\frac{K_E}{K_b} = \left(\frac{m_E}{m_b} \right) \left(-\frac{m_b}{m_E} \right)^2 = \frac{m_b}{m_E}$$

Substitute order-of-magnitude numbers for the masses:

$$\frac{K_E}{K_b} = \frac{m_b}{m_E} \sim \frac{1 \text{ kg}}{10^{25} \text{ kg}} \sim 10^{-25}$$

....

Finalize The kinetic energy of the Earth is a very small fraction of the kinetic energy of the ball, so we are justified in ignoring it in the kinetic energy of the system.

9.3 Analysis Model: Nonisolated System (Momentum)

According to Equation 9.3, the momentum of a particle changes if a net force acts on the particle. The same can be said about a net force applied to a system as we will show explicitly in Section 9.7: the momentum of a system changes if a net force from the environment acts on the system. This may sound similar to our discussion of energy in Chapter 8: the energy of a system changes if energy crosses the boundary of the system to or from the environment. In this section, we consider a *nonisolated system*. For energy considerations, a system is nonisolated if energy transfers across the boundary of the system by any of the means listed in Section 8.1. For momentum considerations, a system is nonisolated if a net force acts on the system for a time interval. In this case, we can imagine momentum being transferred to the system from the environment by means of the net force. Knowing the change in momentum caused by a force is useful in solving some types of problems. To build a better understanding of this important concept, let us assume a net force $\sum \vec{F}$ acts on a particle and this force may vary with time. According to Newton's second law, in the form expressed in Equation 9.3, $\sum \vec{F} = d\vec{p}/dt$, we can write

$$d\vec{p} = \sum \vec{F} \, dt \tag{9.7}$$

We can integrate[2] this expression to find the change in the momentum of a particle when the force acts over some time interval. If the momentum of the particle changes from $\vec{p}_i$ at time t_i to $\vec{p}_f$ at time t_f, integrating Equation 9.7 gives

$$\Delta\vec{p} = \vec{p}_f - \vec{p}_i = \int_{t_i}^{t_f} \sum \vec{F} \, dt \tag{9.8}$$

To evaluate the integral, we need to know how the net force varies with time. The quantity on the right side of this equation is a vector called the **impulse** of the net force $\sum \vec{F}$ acting on a particle over the time interval $\Delta t = t_f - t_i$:

$$\vec{I} \equiv \int_{t_i}^{t_f} \sum \vec{F} \, dt \tag{9.9}$$

◀ Impulse of a force

From its definition, we see that impulse $\vec{I}$ is a vector quantity having a magnitude equal to the area under the force–time curve as described in Figure 9.3a. It is assumed the force varies in time in the general manner shown in the figure and is nonzero in the time interval $\Delta t = t_f - t_i$. The direction of the impulse vector is the same as the direction of the change in momentum. Impulse has the dimensions of momentum, that is, ML/T. Impulse is *not* a property of a particle; rather, it is a measure of the degree to which an external force changes the particle's momentum.

Because the net force imparting an impulse to a particle can generally vary in time, it is convenient to define a time-averaged net force:

$$\left(\sum \vec{F}\right)_{\text{avg}} \equiv \frac{1}{\Delta t} \int_{t_i}^{t_f} \sum \vec{F} \, dt \tag{9.10}$$

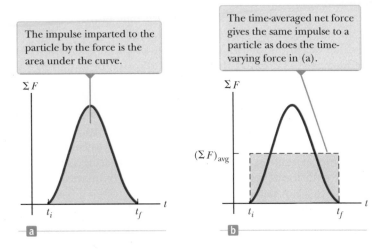

The impulse imparted to the particle by the force is the area under the curve.

The time-averaged net force gives the same impulse to a particle as does the time-varying force in (a).

Figure 9.3 (a) A net force acting on a particle may vary in time. (b) The value of the constant force $(\sum F)_{\text{avg}}$ (horizontal dashed line) is chosen so that the area $(\sum F)_{\text{avg}} \Delta t$ of the rectangle is the same as the area under the curve in (a).

---

[2]Here we are integrating force with respect to time. Compare this strategy with our efforts in Chapter 7, where we integrated force with respect to position to find the work done by the force.

where $\Delta t = t_f - t_i$. (This equation is an application of the mean value theorem of calculus.) Therefore, we can express Equation 9.9 as

$$\vec{\mathbf{I}} = \left(\sum \vec{\mathbf{F}} \right)_{\text{avg}} \Delta t \qquad (9.11)$$

This time-averaged force, shown in Figure 9.3b, can be interpreted as the constant force that would give to the particle in the time interval Δt the same impulse that the time-varying force gives over this same interval.

In principle, if $\sum \vec{\mathbf{F}}$ is known as a function of time, the impulse can be calculated from Equation 9.9. The calculation becomes especially simple if the force acting on the particle is constant. In this case, $\left(\sum \vec{\mathbf{F}} \right)_{\text{avg}} = \sum \vec{\mathbf{F}}$, where $\sum \vec{\mathbf{F}}$ is the constant net force, and Equation 9.11 becomes

$$\vec{\mathbf{I}} = \sum \vec{\mathbf{F}} \, \Delta t \qquad (9.12)$$

Combining Equations 9.8 and 9.9 gives us an important statement known as the **impulse–momentum theorem:**

Impulse–momentum theorem ▶ for a particle

> The change in the momentum of a particle is equal to the impulse of the net force acting on the particle:
>
> $$\Delta \vec{\mathbf{p}} = \vec{\mathbf{I}} \qquad (9.13)$$

This statement is equivalent to Newton's second law. When we say that an impulse is given to a particle, we mean that momentum is transferred from an external agent to that particle. Equation 9.13 is identical in form to the conservation of energy equation, Equation 8.1, and its full expansion, Equation 8.2. Equation 9.13 is the most general statement of the principle of **conservation of momentum** and is called the **conservation of momentum equation.** In the case of a momentum approach, isolated systems tend to appear in problems more often than nonisolated systems, so, in practice, the conservation of momentum equation is often identified as the special case of Equation 9.5.

The left side of Equation 9.13 represents the change in the momentum of the system, which in this case is a single particle. The right side is a measure of how much momentum crosses the boundary of the system due to the net force being applied to the system. Equation 9.13 is the mathematical statement of a new analysis model, the **nonisolated system (momentum)** model. Although this equation is similar in form to Equation 8.1, there are several differences in its application to problems. First, Equation 9.13 is a vector equation, whereas Equation 8.1 is a scalar equation. Therefore, directions are important for Equation 9.13. Second, there is only one type of momentum and therefore only one way to store momentum in a system. In contrast, as we see from Equation 8.2, there are three ways to store energy in a system: kinetic, potential, and internal. Third, there is only one way to transfer momentum into a system: by the application of a force on the system over a time interval. Equation 8.2 shows six ways we have identified as transferring energy into a system. Therefore, there is no expansion of Equation 9.13 analogous to Equation 8.2.

In many physical situations, we shall use what is called the **impulse approximation,** in which we assume one of the forces exerted on a particle acts for a short time but is much greater than any other force present. In this case, the net force $\sum \vec{\mathbf{F}}$ in Equation 9.9 is replaced with a single force $\vec{\mathbf{F}}$ to find the impulse on the particle. This approximation is especially useful in treating collisions in which the duration of the collision is very short. When this approximation is made, the single force is referred to as an *impulsive force*. For example, when a baseball is struck with a bat, the time of the collision is about 0.01 s and the average force that the bat exerts on the ball in this time is typically several thousand newtons. Because this contact force is much greater than the magnitude of the gravitational force, the impulse approximation justifies our ignoring the gravitational forces exerted on

Air bags in automobiles have saved countless lives in accidents. The air bag increases the time interval during which the passenger is brought to rest, thereby decreasing the force on (and resultant injury to) the passenger.

the ball and bat during the collision. When we use this approximation, it is important to remember that $\vec{p}_i$ and $\vec{p}_f$ represent the momenta *immediately* before and after the collision, respectively. Therefore, in any situation in which it is proper to use the impulse approximation, the particle moves very little during the collision.

Quick Quiz 9.3 Two objects are at rest on a frictionless surface. Object 1 has a greater mass than object 2. **(i)** When a constant force is applied to object 1, it accelerates through a distance *d* in a straight line. The force is removed from object 1 and is applied to object 2. At the moment when object 2 has accelerated through the same distance *d*, which statements are true? (a) $p_1 < p_2$ (b) $p_1 = p_2$ (c) $p_1 > p_2$ (d) $K_1 < K_2$ (e) $K_1 = K_2$ (f) $K_1 > K_2$ **(ii)** When a force is applied to object 1, it accelerates for a time interval Δt. The force is removed from object 1 and is applied to object 2. From the same list of choices, which statements are true after object 2 has accelerated for the same time interval Δt?

Quick Quiz 9.4 Rank an automobile dashboard, seat belt, and air bag, each used alone in separate collisions from the same speed, in terms of (a) the impulse and (b) the average force each delivers to a front-seat passenger, from greatest to least.

| **Analysis Model** | **Nonisolated System (Momentum)** |

Imagine you have identified a system to be analyzed and have defined a system boundary. If external forces are applied on the system, the system is *nonisolated*. In that case, the change in the total momentum of the system is equal to the impulse on the system, a statement known as the **impulse–momentum theorem**:

$$\Delta \vec{p} = \vec{I} \qquad (9.13)$$

Examples:

- a baseball is struck by a bat
- a spool sitting on a table is pulled by a string (Example 10.14 in Chapter 10)
- a gas molecule strikes the wall of the container holding the gas (Chapter 21)
- photons strike an absorbing surface and exert pressure on the surface (Chapter 34)

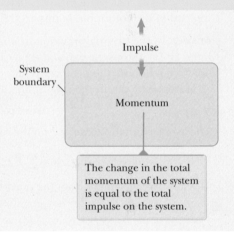

Impulse

System boundary

Momentum

The change in the total momentum of the system is equal to the total impulse on the system.

| **Example 9.3** | **How Good Are the Bumpers?** | **AM** |

In a particular crash test, a car of mass 1 500 kg collides with a wall as shown in Figure 9.4. The initial and final velocities of the car are $\vec{v}_i = -15.0\hat{i}$ m/s and $\vec{v}_f = 2.60\hat{i}$ m/s, respectively. If the collision lasts 0.150 s, find the impulse caused by the collision and the average net force exerted on the car.

SOLUTION

Conceptualize The collision time is short, so we can imagine the car being brought to rest very rapidly and then moving back in the opposite direction with a reduced speed.

Categorize Let us assume the net force exerted on the car by the wall and friction from the ground is large compared with other forces on the car (such as

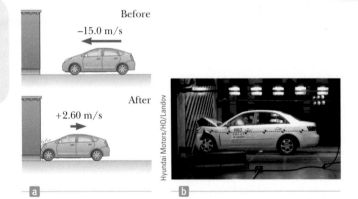

Before

−15.0 m/s

After

+2.60 m/s

Hyundai Motors/HO/Landov

a   b

Figure 9.4 (Example 9.3) (a) This car's momentum changes as a result of its collision with the wall. (b) In a crash test, much of the car's initial kinetic energy is transformed into energy associated with the damage to the car.

continued

▶ **9.3** continued

air resistance). Furthermore, the gravitational force and the normal force exerted by the road on the car are perpendicular to the motion and therefore do not affect the horizontal momentum. Therefore, we categorize the problem as one in which we can apply the impulse approximation in the horizontal direction. We also see that the car's momentum changes due to an impulse from the environment. Therefore, we can apply the *nonisolated system (momentum)* model.

Analyze

Use Equation 9.13 to find the impulse on the car:

$$\vec{I} = \Delta\vec{p} = \vec{p}_f - \vec{p}_i = m\vec{v}_f - m\vec{v}_i = m(\vec{v}_f - \vec{v}_i)$$

$$= (1\ 500\ \text{kg})[2.60\hat{i}\ \text{m/s} - (-15.0\hat{i}\ \text{m/s})] = \boxed{2.64 \times 10^4 \hat{i}\ \text{kg}\cdot\text{m/s}}$$

Use Equation 9.11 to evaluate the average net force exerted on the car:

$$\left(\sum \vec{F}\right)_{\text{avg}} = \frac{\vec{I}}{\Delta t} = \frac{2.64 \times 10^4 \hat{i}\ \text{kg}\cdot\text{m/s}}{0.150\ \text{s}} = \boxed{1.76 \times 10^5 \hat{i}\ \text{N}}$$

Finalize The net force found above is a combination of the normal force on the car from the wall and any friction force between the tires and the ground as the front of the car crumples. If the brakes are not operating while the crash occurs and the crumpling metal does not interfere with the free rotation of the tires, this friction force could be relatively small due to the freely rotating wheels. Notice that the signs of the velocities in this example indicate the reversal of directions. What would the mathematics be describing if both the initial and final velocities had the same sign?

WHAT IF? What if the car did not rebound from the wall? Suppose the final velocity of the car is zero and the time interval of the collision remains at 0.150 s. Would that represent a larger or a smaller net force on the car?

Answer In the original situation in which the car rebounds, the net force on the car does two things during the time interval: (1) it stops the car, and (2) it causes the car to move away from the wall at 2.60 m/s after the collision. If the car does not rebound, the net force is only doing the first of these steps—stopping the car—which requires a *smaller* force.

Mathematically, in the case of the car that does not rebound, the impulse is

$$\vec{I} = \Delta\vec{p} = \vec{p}_f - \vec{p}_i = 0 - (1\ 500\ \text{kg})(-15.0\hat{i}\ \text{m/s}) = 2.25 \times 10^4 \hat{i}\ \text{kg}\cdot\text{m/s}$$

The average net force exerted on the car is

$$\left(\sum \vec{F}\right)_{\text{avg}} = \frac{\vec{I}}{\Delta t} = \frac{2.25 \times 10^4 \hat{i}\ \text{kg}\cdot\text{m/s}}{0.150\ \text{s}} = 1.50 \times 10^5 \hat{i}\ \text{N}$$

which is indeed smaller than the previously calculated value, as was argued conceptually.

9.4 Collisions in One Dimension

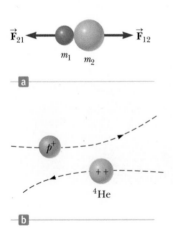

Figure 9.5 (a) The collision between two objects as the result of direct contact. (b) The "collision" between two charged particles.

In this section, we use the isolated system (momentum) model to describe what happens when two particles collide. The term **collision** represents an event during which two particles come close to each other and interact by means of forces. The interaction forces are assumed to be much greater than any external forces present, so we can use the impulse approximation.

A collision may involve physical contact between two macroscopic objects as described in Figure 9.5a, but the notion of what is meant by a collision must be generalized because "physical contact" on a submicroscopic scale is ill-defined and hence meaningless. To understand this concept, consider a collision on an atomic scale (Fig. 9.5b) such as the collision of a proton with an alpha particle (the nucleus of a helium atom). Because the particles are both positively charged, they repel each other due to the strong electrostatic force between them at close separations and never come into "physical contact."

When two particles of masses m_1 and m_2 collide as shown in Figure 9.5, the impulsive forces may vary in time in complicated ways, such as that shown in Figure 9.3. Regardless of the complexity of the time behavior of the impulsive force, however, this force is internal to the system of two particles. Therefore, the two particles form an isolated system and the momentum of the system must be conserved in *any* collision.

In contrast, the total kinetic energy of the system of particles may or may not be conserved, depending on the type of collision. In fact, collisions are categorized as being either *elastic* or *inelastic* depending on whether or not kinetic energy is conserved.

An **elastic collision** between two objects is one in which the total kinetic energy (as well as total momentum) of the system is the same before and after the collision. Collisions between certain objects in the macroscopic world, such as billiard balls, are only *approximately* elastic because some deformation and loss of kinetic energy take place. For example, you can hear a billiard ball collision, so you know that some of the energy is being transferred away from the system by sound. An elastic collision must be perfectly silent! *Truly* elastic collisions occur between atomic and subatomic particles. These collisions are described by the isolated system model for both energy and momentum. Furthermore, there must be no transformation of kinetic energy into other types of energy within the system.

An **inelastic collision** is one in which the total kinetic energy of the system is not the same before and after the collision (even though the momentum of the system is conserved). Inelastic collisions are of two types. When the objects stick together after they collide, as happens when a meteorite collides with the Earth, the collision is called **perfectly inelastic.** When the colliding objects do not stick together but some kinetic energy is transformed or transferred away, as in the case of a rubber ball colliding with a hard surface, the collision is called **inelastic** (with no modifying adverb). When the rubber ball collides with the hard surface, some of the ball's kinetic energy is transformed when the ball is deformed while it is in contact with the surface. Inelastic collisions are described by the momentum version of the isolated system model. The system could be isolated for energy, with kinetic energy transformed to potential or internal energy. If the system is nonisolated, there could be energy leaving the system by some means. In this latter case, there could also be some transformation of energy within the system. In either of these cases, the kinetic energy of the system changes.

In the remainder of this section, we investigate the mathematical details for collisions in one dimension and consider the two extreme cases, perfectly inelastic and elastic collisions.

Perfectly Inelastic Collisions

Consider two particles of masses m_1 and m_2 moving with initial velocities $\vec{\mathbf{v}}_{1i}$ and $\vec{\mathbf{v}}_{2i}$ along the same straight line as shown in Figure 9.6. The two particles collide head-on, stick together, and then move with some common velocity $\vec{\mathbf{v}}_f$ after the collision. Because the momentum of an isolated system is conserved in *any* collision, we can say that the total momentum before the collision equals the total momentum of the composite system after the collision:

$$\Delta\vec{\mathbf{p}} = 0 \quad \rightarrow \quad \vec{\mathbf{p}}_i = \vec{\mathbf{p}}_f \quad \rightarrow \quad m_1\vec{\mathbf{v}}_{1i} + m_2\vec{\mathbf{v}}_{2i} = (m_1 + m_2)\vec{\mathbf{v}}_f \tag{9.14}$$

Solving for the final velocity gives

$$\vec{\mathbf{v}}_f = \frac{m_1\vec{\mathbf{v}}_{1i} + m_2\vec{\mathbf{v}}_{2i}}{m_1 + m_2} \tag{9.15}$$

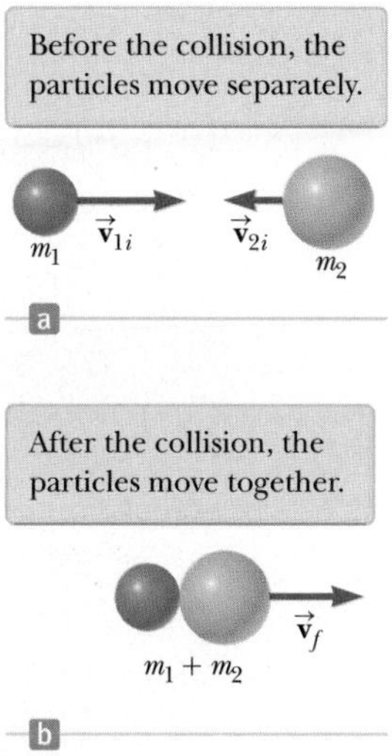

Elastic Collisions

Consider two particles of masses m_1 and m_2 moving with initial velocities $\vec{\mathbf{v}}_{1i}$ and $\vec{\mathbf{v}}_{2i}$ along the same straight line as shown in Figure 9.7 on page 206. The two particles collide head-on and then leave the collision site with different velocities, $\vec{\mathbf{v}}_{1f}$ and $\vec{\mathbf{v}}_{2f}$. In an elastic collision, both the momentum and kinetic energy of the system are conserved. Therefore, considering velocities along the horizontal direction in Figure 9.7, we have

$$p_i = p_f \quad \rightarrow \quad m_1 v_{1i} + m_2 v_{2i} = m_1 v_{1f} + m_2 v_{2f} \tag{9.16}$$

$$K_i = K_f \quad \rightarrow \quad \tfrac{1}{2}m_1 v_{1i}{}^2 + \tfrac{1}{2}m_2 v_{2i}{}^2 = \tfrac{1}{2}m_1 v_{1f}{}^2 + \tfrac{1}{2}m_2 v_{2f}{}^2 \tag{9.17}$$

Pitfall Prevention 9.2
Inelastic Collisions Generally, inelastic collisions are hard to analyze without additional information. Lack of this information appears in the mathematical representation as having more unknowns than equations.

Before the collision, the particles move separately.

After the collision, the particles move together.

Figure 9.6 Schematic representation of a perfectly inelastic head-on collision between two particles.

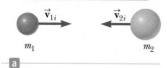

Before the collision, the particles move separately.

a

After the collision, the particles continue to move separately with new velocities.

$\vec{v}_{1f}$ $\vec{v}_{2f}$

b

Figure 9.7 Schematic representation of an elastic head-on collision between two particles.

Pitfall Prevention 9.3

Not a General Equation Equation 9.20 can only be used in a very *specific* situation, a one-dimensional, elastic collision between two objects. The *general* concept is conservation of momentum (and conservation of kinetic energy if the collision is elastic) for an isolated system.

Because all velocities in Figure 9.7 are either to the left or the right, they can be represented by the corresponding speeds along with algebraic signs indicating direction. We shall indicate v as positive if a particle moves to the right and negative if it moves to the left.

In a typical problem involving elastic collisions, there are two unknown quantities, and Equations 9.16 and 9.17 can be solved simultaneously to find them. An alternative approach, however—one that involves a little mathematical manipulation of Equation 9.17—often simplifies this process. To see how, let us cancel the factor $\frac{1}{2}$ in Equation 9.17 and rewrite it by gathering terms with subscript 1 on the left and 2 on the right:

$$m_1(v_{1i}^2 - v_{1f}^2) = m_2(v_{2f}^2 - v_{2i}^2)$$

Factoring both sides of this equation gives

$$m_1(v_{1i} - v_{1f})(v_{1i} + v_{1f}) = m_2(v_{2f} - v_{2i})(v_{2f} + v_{2i}) \tag{9.18}$$

Next, let us separate the terms containing m_1 and m_2 in Equation 9.16 in a similar way to obtain

$$m_1(v_{1i} - v_{1f}) = m_2(v_{2f} - v_{2i}) \tag{9.19}$$

To obtain our final result, we divide Equation 9.18 by Equation 9.19 and obtain

$$v_{1i} + v_{1f} = v_{2f} + v_{2i}$$

Now rearrange terms once again so as to have initial quantities on the left and final quantities on the right:

$$v_{1i} - v_{2i} = -(v_{1f} - v_{2f}) \tag{9.20}$$

This equation, in combination with Equation 9.16, can be used to solve problems dealing with elastic collisions. This pair of equations (Eqs. 9.16 and 9.20) is easier to handle than the pair of Equations 9.16 and 9.17 because there are no quadratic terms like there are in Equation 9.17. According to Equation 9.20, the *relative* velocity of the two particles before the collision, $v_{1i} - v_{2i}$, equals the negative of their relative velocity after the collision, $-(v_{1f} - v_{2f})$.

Suppose the masses and initial velocities of both particles are known. Equations 9.16 and 9.20 can be solved for the final velocities in terms of the initial velocities because there are two equations and two unknowns:

$$v_{1f} = \left(\frac{m_1 - m_2}{m_1 + m_2}\right)v_{1i} + \left(\frac{2m_2}{m_1 + m_2}\right)v_{2i} \tag{9.21}$$

$$v_{2f} = \left(\frac{2m_1}{m_1 + m_2}\right)v_{1i} + \left(\frac{m_2 - m_1}{m_1 + m_2}\right)v_{2i} \tag{9.22}$$

It is important to use the appropriate signs for v_{1i} and v_{2i} in Equations 9.21 and 9.22.

Let us consider some special cases. If $m_1 = m_2$, Equations 9.21 and 9.22 show that $v_{1f} = v_{2i}$ and $v_{2f} = v_{1i}$, which means that the particles exchange velocities if they have equal masses. That is approximately what one observes in head-on billiard ball collisions: the cue ball stops and the struck ball moves away from the collision with the same velocity the cue ball had.

If particle 2 is initially at rest, then $v_{2i} = 0$, and Equations 9.21 and 9.22 become

◄ **Elastic collision: particle 2 initially at rest**

$$v_{1f} = \left(\frac{m_1 - m_2}{m_1 + m_2}\right)v_{1i} \tag{9.23}$$

$$v_{2f} = \left(\frac{2m_1}{m_1 + m_2}\right)v_{1i} \tag{9.24}$$

If m_1 is much greater than m_2 and $v_{2i} = 0$, we see from Equations 9.23 and 9.24 that $v_{1f} \approx v_{1i}$ and $v_{2f} \approx 2v_{1i}$. That is, when a very heavy particle collides head-on with a very light one that is initially at rest, the heavy particle continues its motion unaltered after the collision and the light particle rebounds with a speed equal to about twice the initial speed of the heavy particle. An example of such a collision is that of a moving heavy atom, such as uranium, striking a light atom, such as hydrogen.

If m_2 is much greater than m_1 and particle 2 is initially at rest, then $v_{1f} \approx -v_{1i}$ and $v_{2f} \approx 0$. That is, when a very light particle collides head-on with a very heavy particle that is initially at rest, the light particle has its velocity reversed and the heavy one remains approximately at rest. For example, imagine what happens when you throw a table tennis ball at a bowling ball as in Quick Quiz 9.6 below.

Quick Quiz 9.5 In a perfectly inelastic one-dimensional collision between two moving objects, what condition alone is necessary so that the final kinetic energy of the system is zero after the collision? **(a)** The objects must have initial momenta with the same magnitude but opposite directions. **(b)** The objects must have the same mass. **(c)** The objects must have the same initial velocity. **(d)** The objects must have the same initial speed, with velocity vectors in opposite directions.

Quick Quiz 9.6 A table-tennis ball is thrown at a stationary bowling ball. The table-tennis ball makes a one-dimensional elastic collision and bounces back along the same line. Compared with the bowling ball after the collision, does the table-tennis ball have **(a)** a larger magnitude of momentum and more kinetic energy, **(b)** a smaller magnitude of momentum and more kinetic energy, **(c)** a larger magnitude of momentum and less kinetic energy, **(d)** a smaller magnitude of momentum and less kinetic energy, or **(e)** the same magnitude of momentum and the same kinetic energy?

Problem-Solving Strategy One-Dimensional Collisions

You should use the following approach when solving collision problems in one dimension:

1. Conceptualize. Imagine the collision occurring in your mind. Draw simple diagrams of the particles before and after the collision and include appropriate velocity vectors. At first, you may have to guess at the directions of the final velocity vectors.

2. Categorize. Is the system of particles isolated? If so, use the isolated system (momentum) model. Further categorize the collision as elastic, inelastic, or perfectly inelastic.

3. Analyze. Set up the appropriate mathematical representation for the problem. If the collision is perfectly inelastic, use Equation 9.15. If the collision is elastic, use Equations 9.16 and 9.20. If the collision is inelastic, use Equation 9.16. To find the final velocities in this case, you will need some additional information.

4. Finalize. Once you have determined your result, check to see if your answers are consistent with the mental and pictorial representations and that your results are realistic.

Example 9.4 The Executive Stress Reliever

An ingenious device that illustrates conservation of momentum and kinetic energy is shown in Figure 9.8 on page 208. It consists of five identical hard balls supported by strings of equal lengths. When ball 1 is pulled out and released, after the almost-elastic collision between it and ball 2, ball 1 stops and ball 5 moves out as shown in Figure 9.8b. If balls 1 and 2 are pulled out and released, they stop after the collision and balls 4 and 5 swing out, and so forth. Is it ever possible that when ball 1 is released, it stops after the collision and balls 4 and 5 will swing out on the opposite side and travel with half the speed of ball 1 as in Figure 9.8c?

continued

▶ **9.4** continued

SOLUTION

Conceptualize With the help of Figure 9.8c, imagine one ball coming in from the left and two balls exiting the collision on the right. That is the phenomenon we want to test to see if it could ever happen.

Categorize Because of the very short time interval between the arrival of the ball from the left and the departure of the ball(s) from the right, we can use the impulse approximation to ignore the gravitational forces on the balls and model the five balls as an *isolated system* in terms of both *momentum* and *energy*. Because the balls are hard, we can categorize the collisions between them as elastic for purposes of calculation.

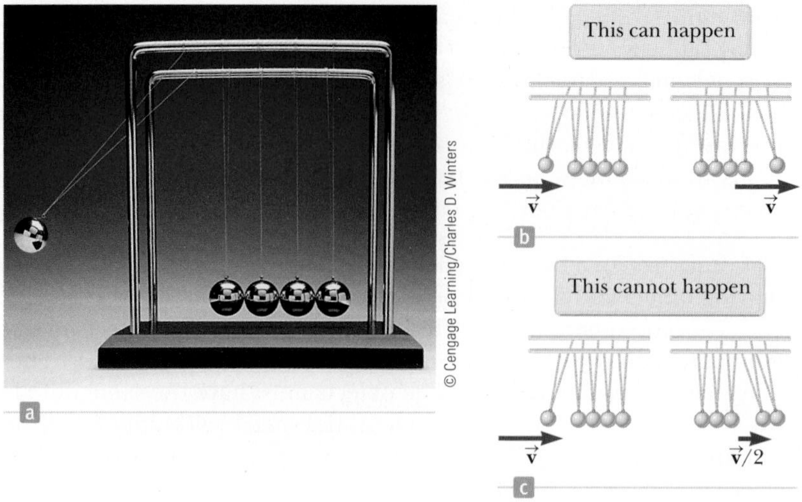

Figure 9.8 (Example 9.4) (a) An executive stress reliever. (b) If one ball swings down, we see one ball swing out at the other end. (c) Is it possible for one ball to swing down and two balls to leave the other end with half the speed of the first ball? In (b) and (c), the velocity vectors shown represent those of the balls immediately before and immediately after the collision.

Analyze Let's consider the situation shown in Figure 9.8c. The momentum of the system before the collision is mv, where m is the mass of ball 1 and v is its speed immediately before the collision. After the collision, we imagine that ball 1 stops and balls 4 and 5 swing out, each moving with speed $v/2$. The total momentum of the system after the collision would be $m(v/2) + m(v/2) = mv$. Therefore, the momentum of the system is conserved in the situation shown in Figure 9.8c!

The kinetic energy of the system immediately before the collision is $K_i = \frac{1}{2}mv^2$ and that after the collision is $K_f = \frac{1}{2}m(v/2)^2 + \frac{1}{2}m(v/2)^2 = \frac{1}{4}mv^2$. That shows that the kinetic energy of the system is *not* conserved, which is inconsistent with our assumption that the collisions are elastic.

Finalize Our analysis shows that it is *not* possible for balls 4 and 5 to swing out when only ball 1 is released. The only way to conserve both momentum and kinetic energy of the system is for one ball to move out when one ball is released, two balls to move out when two are released, and so on.

WHAT IF? Consider what would happen if balls 4 and 5 are glued together. Now what happens when ball 1 is pulled out and released?

Answer In this situation, balls 4 and 5 *must* move together as a single object after the collision. We have argued that both momentum and energy of the system cannot be conserved in this case. We assumed, however, ball 1 stopped after striking ball 2. What if we do not make this assumption? Consider the conservation equations with the assumption that ball 1 moves after the collision. For conservation of momentum,

$$p_i = p_f$$
$$mv_{1i} = mv_{1f} + 2mv_{4,5}$$

where $v_{4,5}$ refers to the final speed of the ball 4–ball 5 combination. Conservation of kinetic energy gives us

$$K_i = K_f$$
$$\tfrac{1}{2}mv_{1i}^2 = \tfrac{1}{2}mv_{1f}^2 + \tfrac{1}{2}(2m)v_{4,5}^2$$

Combining these equations gives

$$v_{4,5} = \tfrac{2}{3}v_{1i} \qquad v_{1f} = -\tfrac{1}{3}v_{1i}$$

Therefore, balls 4 and 5 move together as one object after the collision while ball 1 bounces back from the collision with one third of its original speed.

Example 9.5   **Carry Collision Insurance!**

An 1 800-kg car stopped at a traffic light is struck from the rear by a 900-kg car. The two cars become entangled, moving along the same path as that of the originally moving car. If the smaller car were moving at 20.0 m/s before the collision, what is the velocity of the entangled cars after the collision?

SOLUTION

Conceptualize This kind of collision is easily visualized, and one can predict that after the collision both cars will be moving in the same direction as that of the initially moving car. Because the initially moving car has only half the mass of the stationary car, we expect the final velocity of the cars to be relatively small.

Categorize We identify the two cars as an *isolated system* in terms of *momentum* in the horizontal direction and apply the impulse approximation during the short time interval of the collision. The phrase "become entangled" tells us to categorize the collision as perfectly inelastic.

......

Analyze The magnitude of the total momentum of the system before the collision is equal to that of the smaller car because the larger car is initially at rest.

Use the isolated system model for momentum:

$$\Delta \vec{p} = 0 \quad \rightarrow \quad p_i = p_f \quad \rightarrow \quad m_1 v_i = (m_1 + m_2) v_f$$

Solve for v_f and substitute numerical values:

$$v_f = \frac{m_1 v_i}{m_1 + m_2} = \frac{(900 \text{ kg})(20.0 \text{ m/s})}{900 \text{ kg} + 1\ 800 \text{ kg}} = \boxed{6.67 \text{ m/s}}$$

......

Finalize Because the final velocity is positive, the direction of the final velocity of the combination is the same as the velocity of the initially moving car as predicted. The speed of the combination is also much lower than the initial speed of the moving car.

WHAT IF? Suppose we reverse the masses of the cars. What if a stationary 900-kg car is struck by a moving 1 800-kg car? Is the final speed the same as before?

Answer Intuitively, we can guess that the final speed of the combination is higher than 6.67 m/s if the initially moving car is the more massive car. Mathematically, that should be the case because the system has a larger momentum if the initially moving car is the more massive one. Solving for the new final velocity, we find

$$v_f = \frac{m_1 v_i}{m_1 + m_2} = \frac{(1\ 800 \text{ kg})(20.0 \text{ m/s})}{1\ 800 \text{ kg} + 900 \text{ kg}} = 13.3 \text{ m/s}$$

which is two times greater than the previous final velocity.

Example 9.6   **The Ballistic Pendulum**

The ballistic pendulum (Fig. 9.9, page 210) is an apparatus used to measure the speed of a fast-moving projectile such as a bullet. A projectile of mass m_1 is fired into a large block of wood of mass m_2 suspended from some light wires. The projectile embeds in the block, and the entire system swings through a height h. How can we determine the speed of the projectile from a measurement of h?

SOLUTION

Conceptualize Figure 9.9a helps conceptualize the situation. Run the animation in your mind: the projectile enters the pendulum, which swings up to some height at which it momentarily comes to rest.

Categorize The projectile and the block form an *isolated system* in terms of *momentum* if we identify configuration A as immediately before the collision and configuration B as immediately after the collision. Because the projectile imbeds in the block, we can categorize the collision between them as perfectly inelastic.

......

Analyze To analyze the collision, we use Equation 9.15, which gives the speed of the system immediately after the collision when we assume the impulse approximation.

continued

▶ **9.6** continued

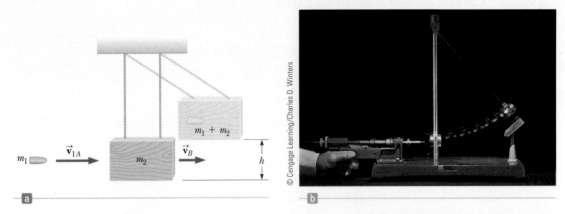

Figure 9.9 (Example 9.6) (a) Diagram of a ballistic pendulum. Notice that $\vec{v}_{1A}$ is the velocity of the projectile immediately before the collision and $\vec{v}_B$ is the velocity of the projectile–block system immediately after the perfectly inelastic collision. (b) Multiflash photograph of a ballistic pendulum used in the laboratory.

Noting that $v_{2A} = 0$, solve Equation 9.15 for v_B:

$$(1) \quad v_B = \frac{m_1 v_{1A}}{m_1 + m_2}$$

Categorize For the process during which the projectile–block combination swings upward to height h (ending at a configuration we'll call C), we focus on a *different* system, that of the projectile, the block, and the Earth. We categorize this part of the problem as one involving an *isolated system* for *energy* with no nonconservative forces acting.

Analyze Write an expression for the total kinetic energy of the system immediately after the collision:

$$(2) \quad K_B = \tfrac{1}{2}(m_1 + m_2)v_B^2$$

Substitute the value of v_B from Equation (1) into Equation (2):

$$K_B = \frac{m_1^2 v_{1A}^2}{2(m_1 + m_2)}$$

This kinetic energy of the system immediately after the collision is *less* than the initial kinetic energy of the projectile as is expected in an inelastic collision.

We define the gravitational potential energy of the system for configuration B to be zero. Therefore, $U_B = 0$, whereas $U_C = (m_1 + m_2)gh$.

Apply the isolated system model to the system:

$$\Delta K + \Delta U = 0 \quad \rightarrow \quad (K_C - K_B) + (U_C - U_B) = 0$$

Substitute the energies:

$$\left(0 - \frac{m_1^2 v_{1A}^2}{2(m_1 + m_2)}\right) + [(m_1 + m_2)gh - 0] = 0$$

Solve for v_{1A}:

$$v_{1A} = \left(\frac{m_1 + m_2}{m_1}\right)\sqrt{2gh}$$

Finalize We had to solve this problem in two steps. Each step involved a different system and a different analysis model: isolated system (momentum) for the first step and isolated system (energy) for the second. Because the collision was assumed to be perfectly inelastic, some mechanical energy was transformed to internal energy during the collision. Therefore, it would have been *incorrect* to apply the isolated system (energy) model to the entire process by equating the initial kinetic energy of the incoming projectile with the final gravitational potential energy of the projectile–block–Earth combination.

Example 9.7 **A Two-Body Collision with a Spring** AM

A block of mass $m_1 = 1.60$ kg initially moving to the right with a speed of 4.00 m/s on a frictionless, horizontal track collides with a light spring attached to a second block of mass $m_2 = 2.10$ kg initially moving to the left with a speed of 2.50 m/s as shown in Figure 9.10a. The spring constant is 600 N/m.

▶ **9.7** continued

(A) Find the velocities of the two blocks after the collision.

SOLUTION

Conceptualize With the help of Figure 9.10a, run an animation of the collision in your mind. Figure 9.10b shows an instant during the collision when the spring is compressed. Eventually, block 1 and the spring will again separate, so the system will look like Figure 9.10a again but with different velocity vectors for the two blocks.

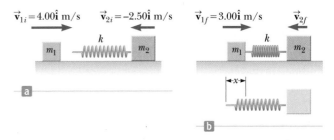

Categorize Because the spring force is conservative, kinetic energy in the system of two blocks and the spring is not transformed to internal energy during the

Figure 9.10 (Example 9.7) A moving block approaches a second moving block that is attached to a spring.

compression of the spring. Ignoring any sound made when the block hits the spring, we can categorize the collision as being elastic and the two blocks and the spring as an *isolated system* for both *energy* and *momentum*.

Analyze Because momentum of the system is conserved, apply Equation 9.16:

$$(1) \quad m_1 v_{1i} + m_2 v_{2i} = m_1 v_{1f} + m_2 v_{2f}$$

Because the collision is elastic, apply Equation 9.20:

$$(2) \quad v_{1i} - v_{2i} = -(v_{1f} - v_{2f})$$

Multiply Equation (2) by m_1:

$$(3) \quad m_1 v_{1i} - m_1 v_{2i} = -m_1 v_{1f} + m_1 v_{2f}$$

Add Equations (1) and (3):

$$2m_1 v_{1i} + (m_2 - m_1)v_{2i} = (m_1 + m_2)v_{2f}$$

Solve for v_{2f}:

$$v_{2f} = \frac{2m_1 v_{1i} + (m_2 - m_1)v_{2i}}{m_1 + m_2}$$

Substitute numerical values:

$$v_{2f} = \frac{2(1.60 \text{ kg})(4.00 \text{ m/s}) + (2.10 \text{ kg} - 1.60 \text{ kg})(-2.50 \text{ m/s})}{1.60 \text{ kg} + 2.10 \text{ kg}} = \boxed{3.12 \text{ m/s}}$$

Solve Equation (2) for v_{1f} and substitute numerical values:

$$v_{1f} = v_{2f} - v_{1i} + v_{2i} = 3.12 \text{ m/s} - 4.00 \text{ m/s} + (-2.50 \text{ m/s}) = \boxed{-3.38 \text{ m/s}}$$

(B) Determine the velocity of block 2 during the collision, at the instant block 1 is moving to the right with a velocity of +3.00 m/s as in Figure 9.10b.

SOLUTION

Conceptualize Focus your attention now on Figure 9.10b, which represents the final configuration of the system for the time interval of interest.

Categorize Because the momentum and mechanical energy of the *isolated system* of two blocks and the spring are conserved *throughout* the collision, the collision can be categorized as elastic for *any* final instant of time. Let us now choose the final instant to be when block 1 is moving with a velocity of +3.00 m/s.

Analyze Apply Equation 9.16:

$$m_1 v_{1i} + m_2 v_{2i} = m_1 v_{1f} + m_2 v_{2f}$$

Solve for v_{2f}:

$$v_{2f} = \frac{m_1 v_{1i} + m_2 v_{2i} - m_1 v_{1f}}{m_2}$$

Substitute numerical values:

$$v_{2f} = \frac{(1.60 \text{ kg})(4.00 \text{ m/s}) + (2.10 \text{ kg})(-2.50 \text{ m/s}) - (1.60 \text{ kg})(3.00 \text{ m/s})}{2.10 \text{ kg}}$$

$$= -1.74 \text{ m/s}$$

continued

▶ **9.7** continued

Finalize The negative value for v_{2f} means that block 2 is still moving to the left at the instant we are considering.

(C) Determine the distance the spring is compressed at that instant.

SOLUTION

Conceptualize Once again, focus on the configuration of the system shown in Figure 9.10b.

Categorize For the system of the spring and two blocks, no friction or other nonconservative forces act within the system. Therefore, we categorize the system as an *isolated system* in terms of *energy* with no nonconservative forces acting. The system also remains an *isolated system* in terms of *momentum*.

Analyze We choose the initial configuration of the system to be that existing immediately before block 1 strikes the spring and the final configuration to be that when block 1 is moving to the right at 3.00 m/s.

Write the appropriate reduction of
Equation 8.2:

$$\Delta K + \Delta U = 0$$

Evaluate the energies, recognizing that two objects in the system have kinetic energy and that the potential energy is elastic:

$$\left[\left(\tfrac{1}{2}m_1v_{1f}^2 + \tfrac{1}{2}m_2v_{2f}^2\right) - \left(\tfrac{1}{2}m_1v_{1i}^2 + \tfrac{1}{2}m_2v_{2i}^2\right)\right] + \left(\tfrac{1}{2}kx^2 - 0\right) = 0$$

Solve for x^2:

$$x^2 = \tfrac{1}{k}\left[m_1\left(v_{1i}^2 - v_{1f}^2\right) + m_2\left(v_{2i}^2 - v_{2f}^2\right)\right]$$

Substitute numerical values:

$$x^2 = \left(\frac{1}{600 \text{ N/m}}\right)\{(1.60 \text{ kg})[(4.00 \text{ m/s})^2 - (3.00 \text{ m/s})^2] + (2.10 \text{ kg})[(2.50 \text{ m/s})^2 - (1.74 \text{ m/s})^2]\}$$

$$\rightarrow x = \boxed{0.173 \text{ m}}$$

Finalize This answer is not the maximum compression of the spring because the two blocks are still moving toward each other at the instant shown in Figure 9.10b. Can you determine the maximum compression of the spring?

9.5 Collisions in Two Dimensions

In Section 9.2, we showed that the momentum of a system of two particles is conserved when the system is isolated. For any collision of two particles, this result implies that the momentum in each of the directions x, y, and z is conserved. An important subset of collisions takes place in a plane. The game of billiards is a familiar example involving multiple collisions of objects moving on a two-dimensional surface. For such two-dimensional collisions, we obtain two component equations for conservation of momentum:

$$m_1v_{1ix} + m_2v_{2ix} = m_1v_{1fx} + m_2v_{2fx}$$

$$m_1v_{1iy} + m_2v_{2iy} = m_1v_{1fy} + m_2v_{2fy}$$

where the three subscripts on the velocity components in these equations represent, respectively, the identification of the object (1, 2), initial and final values (i, f), and the velocity component (x, y).

Let us consider a specific two-dimensional problem in which particle 1 of mass m_1 collides with particle 2 of mass m_2 initially at rest as in Figure 9.11. After the collision (Fig. 9.11b), particle 1 moves at an angle θ with respect to the horizontal and particle 2 moves at an angle ϕ with respect to the horizontal. This event is called a *glancing* collision. Applying the law of conservation of momentum in component form and noting that the initial y component of the momentum of the two-particle system is zero gives

$$\Delta p_x = 0 \quad \rightarrow \quad p_{ix} = p_{fx} \quad \rightarrow \quad m_1v_{1i} = m_1v_{1f}\cos\theta + m_2v_{2f}\cos\phi \qquad \textbf{(9.25)}$$

$$\Delta p_y = 0 \quad \rightarrow \quad p_{iy} = p_{fy} \quad \rightarrow \quad 0 = m_1v_{1f}\sin\theta - m_2v_{2f}\sin\phi \qquad \textbf{(9.26)}$$

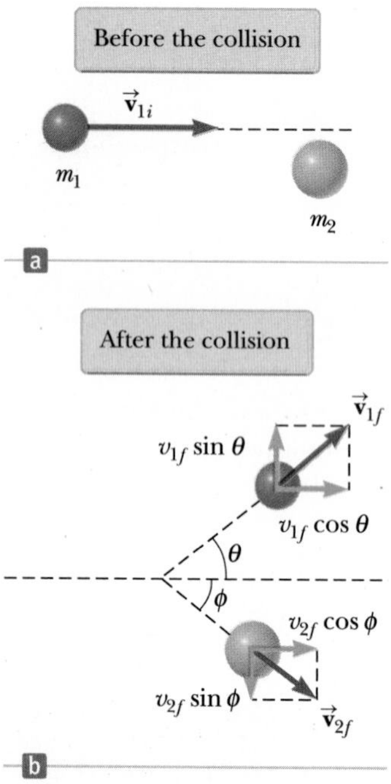

Before the collision

$\vec{v}_{1i}$

m_1

m_2

a

After the collision

$\vec{v}_{1f}$

$v_{1f}\sin\theta$

$v_{1f}\cos\theta$

θ

ϕ

$v_{2f}\cos\phi$

$v_{2f}\sin\phi$

$\vec{v}_{2f}$

b

Figure 9.11 An elastic, glancing collision between two particles.

where the minus sign in Equation 9.26 is included because after the collision particle 2 has a y component of velocity that is downward. (The symbols v in these particular equations are speeds, not velocity components. The direction of the component vector is indicated explicitly with plus or minus signs.) We now have two independent equations. As long as no more than two of the seven quantities in Equations 9.25 and 9.26 are unknown, we can solve the problem.

If the collision is elastic, we can also use Equation 9.17 (conservation of kinetic energy) with $v_{2i} = 0$:

$$K_i = K_f \quad \rightarrow \quad \tfrac{1}{2}m_1v_{1i}^2 = \tfrac{1}{2}m_1v_{1f}^2 + \tfrac{1}{2}m_2v_{2f}^2 \tag{9.27}$$

Knowing the initial speed of particle 1 and both masses, we are left with four unknowns (v_{1f}, v_{2f}, θ, and ϕ). Because we have only three equations, one of the four remaining quantities must be given to determine the motion after the elastic collision from conservation principles alone.

If the collision is inelastic, kinetic energy is *not* conserved and Equation 9.27 does *not* apply.

> **Pitfall Prevention 9.4**
> **Don't Use Equation 9.20** Equation 9.20, relating the initial and final relative velocities of two colliding objects, is only valid for one-dimensional elastic collisions. Do not use this equation when analyzing two-dimensional collisions.

Problem-Solving Strategy   Two-Dimensional Collisions

The following procedure is recommended when dealing with problems involving collisions between two particles in two dimensions.

1. Conceptualize. Imagine the collisions occurring and predict the approximate directions in which the particles will move after the collision. Set up a coordinate system and define your velocities in terms of that system. It is convenient to have the x axis coincide with one of the initial velocities. Sketch the coordinate system, draw and label all velocity vectors, and include all the given information.

2. Categorize. Is the system of particles truly isolated? If so, categorize the collision as elastic, inelastic, or perfectly inelastic.

3. Analyze. Write expressions for the x and y components of the momentum of each object before and after the collision. Remember to include the appropriate signs for the components of the velocity vectors and pay careful attention to signs throughout the calculation.

Apply the isolated system model for momentum $\Delta\vec{p} = 0$. When applied in each direction, this equation will generally reduce to $p_{ix} = p_{fx}$ and $p_{iy} = p_{fy}$, where each of these terms refer to the sum of the momenta of all objects in the system. Write expressions for the *total* momentum in the x direction *before* and *after* the collision and equate the two. Repeat this procedure for the total momentum in the y direction.

Proceed to solve the momentum equations for the unknown quantities. If the collision is inelastic, kinetic energy is *not* conserved and additional information is probably required. If the collision is perfectly inelastic, the final velocities of the two objects are equal.

If the collision is elastic, kinetic energy is conserved and you can equate the total kinetic energy of the system before the collision to the total kinetic energy after the collision, providing an additional relationship between the velocity magnitudes.

4. Finalize. Once you have determined your result, check to see if your answers are consistent with the mental and pictorial representations and that your results are realistic.

Example 9.8   Collision at an Intersection

A 1 500-kg car traveling east with a speed of 25.0 m/s collides at an intersection with a 2 500-kg truck traveling north at a speed of 20.0 m/s as shown in Figure 9.12 on page 214. Find the direction and magnitude of the velocity of the wreckage after the collision, assuming the vehicles stick together after the collision.

continued

▶ **9.8** continued

SOLUTION

Conceptualize Figure 9.12 should help you conceptualize the situation before and after the collision. Let us choose east to be along the positive x direction and north to be along the positive y direction.

Categorize Because we consider moments immediately before and immediately after the collision as defining our time interval, we ignore the small effect that friction would have on the wheels of the vehicles and model the two vehicles as an *isolated system* in terms of *momentum*. We also ignore the vehicles' sizes and model them as particles. The collision is perfectly inelastic because the car and the truck stick together after the collision.

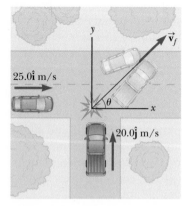

Figure 9.12 (Example 9.8) An eastbound car colliding with a north-bound truck.

Analyze Before the collision, the only object having momentum in the x direction is the car. Therefore, the magnitude of the total initial momentum of the system (car plus truck) in the x direction is that of only the car. Similarly, the total initial momentum of the system in the y direction is that of the truck. After the collision, let us assume the wreckage moves at an angle θ with respect to the x axis with speed v_f.

Apply the isolated system model for momentum in the x direction:

$$\Delta p_x = 0 \quad \rightarrow \quad \sum p_{xi} = \sum p_{xf} \quad \rightarrow \quad (1) \quad m_1 v_{1i} = (m_1 + m_2)v_f \cos \theta$$

Apply the isolated system model for momentum in the y direction:

$$\Delta p_y = 0 \quad \rightarrow \quad \sum p_{yi} = \sum p_{yf} \quad \rightarrow \quad (2) \quad m_2 v_{2i} = (m_1 + m_2)v_f \sin \theta$$

Divide Equation (2) by Equation (1):

$$\frac{m_2 v_{2i}}{m_1 v_{1i}} = \frac{\sin \theta}{\cos \theta} = \tan \theta$$

Solve for θ and substitute numerical values:

$$\theta = \tan^{-1}\left(\frac{m_2 v_{2i}}{m_1 v_{1i}}\right) = \tan^{-1}\left[\frac{(2\,500 \text{ kg})(20.0 \text{ m/s})}{(1\,500 \text{ kg})(25.0 \text{ m/s})}\right] = \boxed{53.1°}$$

Use Equation (2) to find the value of v_f and substitute numerical values:

$$v_f = \frac{m_2 v_{2i}}{(m_1 + m_2)\sin \theta} = \frac{(2\,500 \text{ kg})(20.0 \text{ m/s})}{(1\,500 \text{ kg} + 2\,500 \text{ kg})\sin 53.1°} = \boxed{15.6 \text{ m/s}}$$

Finalize Notice that the angle θ is qualitatively in agreement with Figure 9.12. Also notice that the final speed of the combination is less than the initial speeds of the two cars. This result is consistent with the kinetic energy of the system being reduced in an inelastic collision. It might help if you draw the momentum vectors of each vehicle before the collision and the two vehicles together after the collision.

Example 9.9 **Proton–Proton Collision**

A proton collides elastically with another proton that is initially at rest. The incoming proton has an initial speed of 3.50×10^5 m/s and makes a glancing collision with the second proton as in Figure 9.11. (At close separations, the protons exert a repulsive electrostatic force on each other.) After the collision, one proton moves off at an angle of 37.0° to the original direction of motion and the second deflects at an angle of ϕ to the same axis. Find the final speeds of the two protons and the angle ϕ.

SOLUTION

Conceptualize This collision is like that shown in Figure 9.11, which will help you conceptualize the behavior of the system. We define the x axis to be along the direction of the velocity vector of the initially moving proton.

Categorize The pair of protons form an *isolated system*. Both momentum and kinetic energy of the system are conserved in this glancing elastic collision.

▶ **9.9** continued

Analyze Using the isolated system model for both momentum and energy for a two-dimensional elastic collision, set up the mathematical representation with Equations 9.25 through 9.27:

(1) $v_{1i} = v_{1f}\cos\theta + v_{2f}\cos\phi$

(2) $0 = v_{1f}\sin\theta - v_{2f}\sin\phi$

(3) $v_{1i}{}^2 = v_{1f}{}^2 + v_{2f}{}^2$

Rearrange Equations (1) and (2):

$v_{2f}\cos\phi = v_{1i} - v_{1f}\cos\theta$

$v_{2f}\sin\phi = v_{1f}\sin\theta$

Square these two equations and add them:

$v_{2f}{}^2\cos^2\phi + v_{2f}{}^2\sin^2\phi =$
$v_{1i}{}^2 - 2v_{1i}v_{1f}\cos\theta + v_{1f}{}^2\cos^2\theta + v_{1f}{}^2\sin^2\theta$

Incorporate that the sum of the squares of sine and cosine for *any* angle is equal to 1:

(4) $v_{2f}{}^2 = v_{1i}{}^2 - 2v_{1i}v_{1f}\cos\theta + v_{1f}{}^2$

Substitute Equation (4) into Equation (3):

$v_{1f}{}^2 + (v_{1i}{}^2 - 2v_{1i}v_{1f}\cos\theta + v_{1f}{}^2) = v_{1i}{}^2$

(5) $v_{1f}{}^2 - v_{1i}v_{1f}\cos\theta = 0$

One possible solution of Equation (5) is $v_{1f} = 0$, which corresponds to a head-on, one-dimensional collision in which the first proton stops and the second continues with the same speed in the same direction. That is not the solution we want.

Divide both sides of Equation (5) by v_{1f} and solve for the remaining factor of v_{1f}:

$v_{1f} = v_{1i}\cos\theta = (3.50 \times 10^5\,\text{m/s})\cos 37.0° = \boxed{2.80 \times 10^5\,\text{m/s}}$

Use Equation (3) to find v_{2f}:

$v_{2f} = \sqrt{v_{1i}{}^2 - v_{1f}{}^2} = \sqrt{(3.50 \times 10^5\,\text{m/s})^2 - (2.80 \times 10^5\,\text{m/s})^2}$
$= \boxed{2.11 \times 10^5\,\text{m/s}}$

Use Equation (2) to find ϕ:

(2) $\phi = \sin^{-1}\left(\dfrac{v_{1f}\sin\theta}{v_{2f}}\right) = \sin^{-1}\left[\dfrac{(2.80 \times 10^5\,\text{m/s})\sin 37.0°}{(2.11 \times 10^5\,\text{m/s})}\right]$
$= \boxed{53.0°}$

Finalize It is interesting that $\theta + \phi = 90°$. This result is *not* accidental. Whenever two objects of equal mass collide elastically in a glancing collision and one of them is initially at rest, their final velocities are perpendicular to each other.

9.6 The Center of Mass

In this section, we describe the overall motion of a system in terms of a special point called the **center of mass** of the system. The system can be either a small number of particles or an extended, continuous object, such as a gymnast leaping through the air. We shall see that the translational motion of the center of mass of the system is the same as if all the mass of the system were concentrated at that point. That is, the system moves as if the net external force were applied to a single particle located at the center of mass. This model, the *particle model*, was introduced in Chapter 2. This behavior is independent of other motion, such as rotation or vibration of the system or deformation of the system (for instance, when a gymnast folds her body).

Consider a system consisting of a pair of particles that have different masses and are connected by a light, rigid rod (Fig. 9.13 on page 216). The position of the center of mass of a system can be described as being the *average position* of the system's mass. The center of mass of the system is located somewhere on the line joining the two particles and is closer to the particle having the larger mass. If a single force is applied at a point on the rod above the center of mass, the system rotates clockwise (see Fig. 9.13a). If the force is applied at a point on the rod below the center of mass, the system rotates counterclockwise (see Fig. 9.13b). If the force

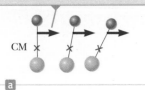

The system rotates clockwise when a force is applied above the center of mass.

a

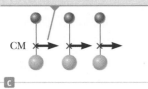

The system rotates counter-clockwise when a force is applied below the center of mass.

b

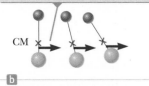

The system moves in the direction of the force without rotating when a force is applied at the center of mass.

c

Figure 9.13 A force is applied to a system of two particles of unequal mass connected by a light, rigid rod.

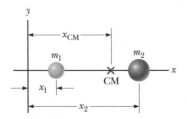

Figure 9.14 The center of mass of two particles of unequal mass on the x axis is located at x_{CM}, a point between the particles, closer to the one having the larger mass.

is applied at the center of mass, the system moves in the direction of the force without rotating (see Fig. 9.13c). The center of mass of an object can be located with this procedure.

The center of mass of the pair of particles described in Figure 9.14 is located on the x axis and lies somewhere between the particles. Its x coordinate is given by

$$x_{CM} \equiv \frac{m_1 x_1 + m_2 x_2}{m_1 + m_2} \tag{9.28}$$

For example, if $x_1 = 0$, $x_2 = d$, and $m_2 = 2m_1$, we find that $x_{CM} = \frac{2}{3}d$. That is, the center of mass lies closer to the more massive particle. If the two masses are equal, the center of mass lies midway between the particles.

We can extend this concept to a system of many particles with masses m_i in three dimensions. The x coordinate of the center of mass of n particles is defined to be

$$x_{CM} \equiv \frac{m_1 x_1 + m_2 x_2 + m_3 x_3 + \cdots + m_n x_n}{m_1 + m_2 + m_3 + \cdots + m_n} = \frac{\sum_i m_i x_i}{\sum_i m_i} = \frac{\sum_i m_i x_i}{M} = \frac{1}{M}\sum_i m_i x_i \tag{9.29}$$

where x_i is the x coordinate of the ith particle and the total mass is $M \equiv \sum_i m_i$ where the sum runs over all n particles. The y and z coordinates of the center of mass are similarly defined by the equations

$$y_{CM} \equiv \frac{1}{M}\sum_i m_i y_i \quad \text{and} \quad z_{CM} \equiv \frac{1}{M}\sum_i m_i z_i \tag{9.30}$$

The center of mass can be located in three dimensions by its position vector $\vec{r}_{CM}$. The components of this vector are x_{CM}, y_{CM}, and z_{CM}, defined in Equations 9.29 and 9.30. Therefore,

$$\vec{r}_{CM} = x_{CM}\hat{i} + y_{CM}\hat{j} + z_{CM}\hat{k} = \frac{1}{M}\sum_i m_i x_i\hat{i} + \frac{1}{M}\sum_i m_i y_i\hat{j} + \frac{1}{M}\sum_i m_i z_i\hat{k}$$

$$\vec{r}_{CM} \equiv \frac{1}{M}\sum_i m_i \vec{r}_i \tag{9.31}$$

where $\vec{r}_i$ is the position vector of the ith particle, defined by

$$\vec{r}_i \equiv x_i\hat{i} + y_i\hat{j} + z_i\hat{k}$$

Although locating the center of mass for an extended, continuous object is somewhat more cumbersome than locating the center of mass of a small number of particles, the basic ideas we have discussed still apply. Think of an extended object as a system containing a large number of small mass elements such as the cube in Figure 9.15. Because the separation between elements is very small, the object can be considered to have a continuous mass distribution. By dividing the object into elements of mass Δm_i with coordinates x_i, y_i, z_i, we see that the x coordinate of the center of mass is approximately

$$x_{CM} \approx \frac{1}{M}\sum_i x_i \Delta m_i$$

with similar expressions for y_{CM} and z_{CM}. If we let the number of elements n approach infinity, the size of each element approaches zero and x_{CM} is given precisely. In this limit, we replace the sum by an integral and Δm_i by the differential element dm:

$$x_{CM} = \lim_{\Delta m_i \to 0} \frac{1}{M}\sum_i x_i \Delta m_i = \frac{1}{M}\int x\, dm \tag{9.32}$$

Likewise, for y_{CM} and z_{CM} we obtain

$$y_{CM} = \frac{1}{M}\int y\, dm \quad \text{and} \quad z_{CM} = \frac{1}{M}\int z\, dm \tag{9.33}$$

We can express the vector position of the center of mass of an extended object in the form

$$\vec{\mathbf{r}}_{CM} = \frac{1}{M} \int \vec{\mathbf{r}}\, dm \qquad (9.34)$$

which is equivalent to the three expressions given by Equations 9.32 and 9.33.

The center of mass of any symmetric object of uniform density lies on an axis of symmetry and on any plane of symmetry. For example, the center of mass of a uniform rod lies in the rod, midway between its ends. The center of mass of a sphere or a cube lies at its geometric center.

Because an extended object is a continuous distribution of mass, each small mass element is acted upon by the gravitational force. The net effect of all these forces is equivalent to the effect of a single force $M\vec{\mathbf{g}}$ acting through a special point, called the **center of gravity**. If $\vec{\mathbf{g}}$ is constant over the mass distribution, the center of gravity coincides with the center of mass. If an extended object is pivoted at its center of gravity, it balances in any orientation.

The center of gravity of an irregularly shaped object such as a wrench can be determined by suspending the object first from one point and then from another. In Figure 9.16, a wrench is hung from point A and a vertical line AB (which can be established with a plumb bob) is drawn when the wrench has stopped swinging. The wrench is then hung from point C, and a second vertical line CD is drawn. The center of gravity is halfway through the thickness of the wrench, under the intersection of these two lines. In general, if the wrench is hung freely from any point, the vertical line through this point must pass through the center of gravity.

Quick Quiz 9.7 A baseball bat of uniform density is cut at the location of its center of mass as shown in Figure 9.17. Which piece has the smaller mass? **(a)** the piece on the right **(b)** the piece on the left **(c)** both pieces have the same mass **(d)** impossible to determine

Figure 9.17 (Quick Quiz 9.7) A baseball bat cut at the location of its center of mass.

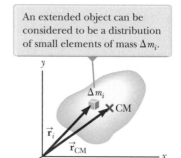

An extended object can be considered to be a distribution of small elements of mass Δm_i.

Figure 9.15 The center of mass is located at the vector position $\vec{\mathbf{r}}_{CM}$, which has coordinates x_{CM}, y_{CM}, and z_{CM}.

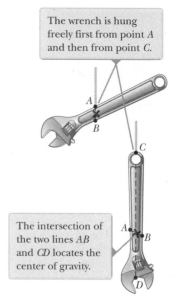

The wrench is hung freely first from point A and then from point C.

The intersection of the two lines AB and CD locates the center of gravity.

Figure 9.16 An experimental technique for determining the center of gravity of a wrench.

Example 9.10   **The Center of Mass of Three Particles**

A system consists of three particles located as shown in Figure 9.18. Find the center of mass of the system. The masses of the particles are $m_1 = m_2 = 1.0$ kg and $m_3 = 2.0$ kg.

SOLUTION

Conceptualize Figure 9.18 shows the three masses. Your intuition should tell you that the center of mass is located somewhere in the region between the blue particle and the pair of tan particles as shown in the figure.

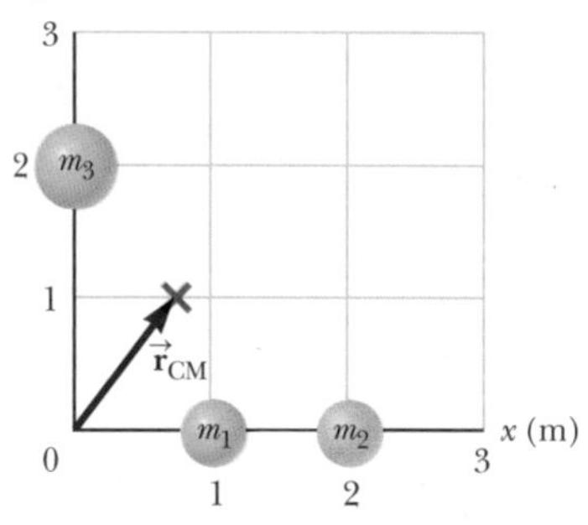

Figure 9.18 (Example 9.10) Two particles are located on the x axis, and a single particle is located on the y axis as shown. The vector indicates the location of the system's center of mass.

Categorize We categorize this example as a substitution problem because we will be using the equations for the center of mass developed in this section.

continued

▶ **9.10** continued

Use the defining equations for the coordinates of the center of mass and notice that $z_{CM} = 0$:

$$x_{CM} = \frac{1}{M} \sum_i m_i x_i = \frac{m_1 x_1 + m_2 x_2 + m_3 x_3}{m_1 + m_2 + m_3}$$

$$= \frac{(1.0 \text{ kg})(1.0 \text{ m}) + (1.0 \text{ kg})(2.0 \text{ m}) + (2.0 \text{ kg})(0)}{1.0 \text{ kg} + 1.0 \text{ kg} + 2.0 \text{ kg}} = \frac{3.0 \text{ kg} \cdot \text{m}}{4.0 \text{ kg}} = 0.75 \text{ m}$$

$$y_{CM} = \frac{1}{M} \sum_i m_i y_i = \frac{m_1 y_1 + m_2 y_2 + m_3 y_3}{m_1 + m_2 + m_3}$$

$$= \frac{(1.0 \text{ kg})(0) + (1.0 \text{ kg})(0) + (2.0 \text{ kg})(2.0 \text{ m})}{4.0 \text{ kg}} = \frac{4.0 \text{ kg} \cdot \text{m}}{4.0 \text{ kg}} = 1.0 \text{ m}$$

Write the position vector of the center of mass:

$$\vec{\mathbf{r}}_{CM} \equiv x_{CM}\hat{\mathbf{i}} + y_{CM}\hat{\mathbf{j}} = \boxed{(0.75\hat{\mathbf{i}} + 1.0\hat{\mathbf{j}}) \text{ m}}$$

Example 9.11 | The Center of Mass of a Rod

(A) Show that the center of mass of a rod of mass M and length L lies midway between its ends, assuming the rod has a uniform mass per unit length.

SOLUTION

Conceptualize The rod is shown aligned along the x axis in Figure 9.19, so $y_{CM} = z_{CM} = 0$. What is your prediction of the value of x_{CM}?

Categorize We categorize this example as an analysis problem because we need to divide the rod into small mass elements to perform the integration in Equation 9.32.

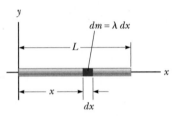

Figure 9.19 (Example 9.11) The geometry used to find the center of mass of a uniform rod.

Analyze The mass per unit length (this quantity is called the *linear mass density*) can be written as $\lambda = M/L$ for the uniform rod. If the rod is divided into elements of length dx, the mass of each element is $dm = \lambda \, dx$.

Use Equation 9.32 to find an expression for x_{CM}:

$$x_{CM} = \frac{1}{M} \int x \, dm = \frac{1}{M} \int_0^L x\lambda \, dx = \frac{\lambda}{M} \left.\frac{x^2}{2}\right|_0^L = \frac{\lambda L^2}{2M}$$

Substitute $\lambda = M/L$:

$$x_{CM} = \frac{L^2}{2M}\left(\frac{M}{L}\right) = \boxed{\tfrac{1}{2}L}$$

One can also use symmetry arguments to obtain the same result.

(B) Suppose a rod is *nonuniform* such that its mass per unit length varies linearly with x according to the expression $\lambda = \alpha x$, where α is a constant. Find the x coordinate of the center of mass as a fraction of L.

SOLUTION

Conceptualize Because the mass per unit length is not constant in this case but is proportional to x, elements of the rod to the right are more massive than elements near the left end of the rod.

Categorize This problem is categorized similarly to part (A), with the added twist that the linear mass density is not constant.

Analyze In this case, we replace dm in Equation 9.32 by $\lambda \, dx$, where $\lambda = \alpha x$.

Use Equation 9.32 to find an expression for x_{CM}:

$$x_{CM} = \frac{1}{M} \int x \, dm = \frac{1}{M} \int_0^L x\lambda \, dx = \frac{1}{M} \int_0^L x\alpha x \, dx$$

$$= \frac{\alpha}{M} \int_0^L x^2 \, dx = \frac{\alpha L^3}{3M}$$

▶ **9.11** continued

Find the total mass of the rod:

$$M = \int dm = \int_0^L \lambda \, dx = \int_0^L \alpha x \, dx = \frac{\alpha L^2}{2}$$

Substitute M into the expression for x_{CM}:

$$x_{CM} = \frac{\alpha L^3}{3\alpha L^2/2} = \boxed{\tfrac{2}{3}L}$$

· ·

Finalize Notice that the center of mass in part (B) is farther to the right than that in part (A). That result is reasonable because the elements of the rod become more massive as one moves to the right along the rod in part (B).

Example 9.12	**The Center of Mass of a Right Triangle**

You have been asked to hang a metal sign from a single vertical string. The sign has the triangular shape shown in Figure 9.20a. The bottom of the sign is to be parallel to the ground. At what distance from the left end of the sign should you attach the support string?

SOLUTION

Conceptualize Figure 9.20a shows the sign hanging from the string. The string must be attached at a point directly above the center of gravity of the sign, which is the same as its center of mass because it is in a uniform gravitational field.

Categorize As in the case of Example 9.11, we categorize this example as an analysis problem because it is necessary to identify infinitesimal mass elements of the sign to perform the integration in Equation 9.32.

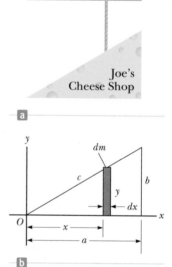

· ·

Analyze We assume the triangular sign has a uniform density and total mass M. Because the sign is a continuous distribution of mass, we must use the integral expression in Equation 9.32 to find the x coordinate of the center of mass.

We divide the triangle into narrow strips of width dx and height y as shown in Figure 9.20b, where y is the height of the hypotenuse of the triangle above the x axis for a given value of x. The mass of each strip is the product of the volume of the strip and the density ρ of the material from which the sign is made: $dm = \rho y t \, dx$, where t is the thickness of the metal sign. The density of the material is the total mass of the sign divided by its total volume (area of the triangle times thickness).

Figure 9.20 (Example 9.12) (a) A triangular sign to be hung from a single string. (b) Geometric construction for locating the center of mass.

Evaluate dm:

$$dm = \rho y t \, dx = \left(\frac{M}{\frac{1}{2}abt}\right) y t \, dx = \frac{2My}{ab} \, dx$$

Use Equation 9.32 to find the x coordinate of the center of mass:

$$(1) \quad x_{CM} = \frac{1}{M} \int x \, dm = \frac{1}{M} \int_0^a x \frac{2My}{ab} \, dx = \frac{2}{ab} \int_0^a xy \, dx$$

To proceed further and evaluate the integral, we must express y in terms of x. The line representing the hypotenuse of the triangle in Figure 9.20b has a slope of b/a and passes through the origin, so the equation of this line is $y = (b/a)x$.

Substitute for y in Equation (1):

$$x_{CM} = \frac{2}{ab} \int_0^a x\left(\frac{b}{a}x\right) dx = \frac{2}{a^2} \int_0^a x^2 \, dx = \frac{2}{a^2}\left[\frac{x^3}{3}\right]_0^a$$

$$= \tfrac{2}{3}a$$

Therefore, the string must be attached to the sign at a distance two-thirds of the length of the bottom edge from the left end.

· ·

continued

▶ **9.12** continued

Finalize This answer is identical to that in part (B) of Example 9.11. For the triangular sign, the linear increase in height y with position x means that elements in the sign increase in mass linearly along the x axis, just like the linear increase in mass density in Example 9.11. We could also find the y coordinate of the center of mass of the sign, but that is not needed to determine where the string should be attached. You might try cutting a right triangle out of cardboard and hanging it from a string so that the long base is horizontal. Does the string need to be attached at $\frac{2}{3}a$?

9.7 Systems of Many Particles

Consider a system of two or more particles for which we have identified the center of mass. We can begin to understand the physical significance and utility of the center of mass concept by taking the time derivative of the position vector for the center of mass given by Equation 9.31. From Section 4.1, we know that the time derivative of a position vector is by definition the velocity vector. Assuming M remains constant for a system of particles—that is, no particles enter or leave the system—we obtain the following expression for the **velocity of the center of mass** of the system:

◀ **Velocity of the center of mass of a system of particles**

$$\vec{\mathbf{v}}_{CM} = \frac{d\vec{\mathbf{r}}_{CM}}{dt} = \frac{1}{M}\sum_i m_i \frac{d\vec{\mathbf{r}}_i}{dt} = \frac{1}{M}\sum_i m_i\vec{\mathbf{v}}_i \qquad (9.35)$$

where $\vec{\mathbf{v}}_i$ is the velocity of the ith particle. Rearranging Equation 9.35 gives

◀ **Total momentum of a system of particles**

$$M\vec{\mathbf{v}}_{CM} = \sum_i m_i\vec{\mathbf{v}}_i = \sum_i \vec{\mathbf{p}}_i = \vec{\mathbf{p}}_{tot} \qquad (9.36)$$

Therefore, the total linear momentum of the system equals the total mass multiplied by the velocity of the center of mass. In other words, the total linear momentum of the system is equal to that of a single particle of mass M moving with a velocity $\vec{\mathbf{v}}_{CM}$.

Differentiating Equation 9.35 with respect to time, we obtain the **acceleration of the center of mass** of the system:

◀ **Acceleration of the center of mass of a system of particles**

$$\vec{\mathbf{a}}_{CM} = \frac{d\vec{\mathbf{v}}_{CM}}{dt} = \frac{1}{M}\sum_i m_i \frac{d\vec{\mathbf{v}}_i}{dt} = \frac{1}{M}\sum_i m_i\vec{\mathbf{a}}_i \qquad (9.37)$$

Rearranging this expression and using Newton's second law gives

$$M\vec{\mathbf{a}}_{CM} = \sum_i m_i\vec{\mathbf{a}}_i = \sum_i \vec{\mathbf{F}}_i \qquad (9.38)$$

where $\vec{\mathbf{F}}_i$ is the net force on particle i.

The forces on any particle in the system may include both external forces (from outside the system) and internal forces (from within the system). By Newton's third law, however, the internal force exerted by particle 1 on particle 2, for example, is equal in magnitude and opposite in direction to the internal force exerted by particle 2 on particle 1. Therefore, when we sum over all internal force vectors in Equation 9.38, they cancel in pairs and we find that the net force on the system is caused *only* by external forces. We can then write Equation 9.38 in the form

◀ **Newton's second law for a system of particles**

$$\sum \vec{\mathbf{F}}_{ext} = M\vec{\mathbf{a}}_{CM} \qquad (9.39)$$

That is, the net external force on a system of particles equals the total mass of the system multiplied by the acceleration of the center of mass. Comparing Equation 9.39 with Newton's second law for a single particle, we see that the particle model we have used in several chapters can be described in terms of the center of mass:

> The center of mass of a system of particles having combined mass M moves like an equivalent particle of mass M would move under the influence of the net external force on the system.

Let us integrate Equation 9.39 over a finite time interval:

$$\int \sum \vec{\mathbf{F}}_{\text{ext}}\, dt = \int M\vec{\mathbf{a}}_{\text{CM}}\, dt = \int M\frac{d\vec{\mathbf{v}}_{\text{CM}}}{dt}\, dt = M \int d\vec{\mathbf{v}}_{\text{CM}} = M\Delta\vec{\mathbf{v}}_{\text{CM}}$$

Notice that this equation can be written as

$$\Delta\vec{\mathbf{p}}_{\text{tot}} = \vec{\mathbf{I}} \tag{9.40}$$

◀ **Impulse–momentum theorem for a system of particles**

where $\vec{\mathbf{I}}$ is the impulse imparted to the system by external forces and $\vec{\mathbf{p}}_{\text{tot}}$ is the momentum of the system. Equation 9.40 is the generalization of the impulse–momentum theorem for a particle (Eq. 9.13) to a system of many particles. It is also the mathematical representation of the nonisolated system (momentum) model for a system of many particles.

Finally, if the net external force on a system is zero so that the system is isolated, it follows from Equation 9.39 that

$$M\vec{\mathbf{a}}_{\text{CM}} = M\frac{d\vec{\mathbf{v}}_{\text{CM}}}{dt} = 0$$

Therefore, the isolated system model for momentum for a system of many particles is described by

$$\Delta\vec{\mathbf{p}}_{\text{tot}} = 0 \tag{9.41}$$

which can be rewritten as

$$M\vec{\mathbf{v}}_{\text{CM}} = \vec{\mathbf{p}}_{\text{tot}} = \text{constant} \quad \left(\text{when } \sum \vec{\mathbf{F}}_{\text{ext}} = 0\right) \tag{9.42}$$

That is, the total linear momentum of a system of particles is conserved if no net external force is acting on the system. It follows that for an isolated system of particles, both the total momentum and the velocity of the center of mass are constant in time. This statement is a generalization of the isolated system (momentum) model for a many-particle system.

Suppose the center of mass of an isolated system consisting of two or more members is at rest. The center of mass of the system remains at rest if there is no net force on the system. For example, consider a system of a swimmer standing on a raft, with the system initially at rest. When the swimmer dives horizontally off the raft, the raft moves in the direction opposite that of the swimmer and the center of mass of the system remains at rest (if we neglect friction between raft and water). Furthermore, the linear momentum of the diver is equal in magnitude to that of the raft, but opposite in direction.

Quick Quiz 9.8 A cruise ship is moving at constant speed through the water. The vacationers on the ship are eager to arrive at their next destination. They decide to try to speed up the cruise ship by gathering at the bow (the front) and running together toward the stern (the back) of the ship. **(i)** While they are running toward the stern, is the speed of the ship (a) higher than it was before, (b) unchanged, (c) lower than it was before, or (d) impossible to determine? **(ii)** The vacationers stop running when they reach the stern of the ship. After they have all stopped running, is the speed of the ship (a) higher than it was before they started running, (b) unchanged from what it was before they started running, (c) lower than it was before they started running, or (d) impossible to determine?

Conceptual Example 9.13   **Exploding Projectile**

A projectile fired into the air suddenly explodes into several fragments (Fig. 9.21 on page 222).

(A) What can be said about the motion of the center of mass of the system made up of all the fragments after the explosion?

continued

▶ **9.13** continued

SOLUTION

Neglecting air resistance, the only external force on the projectile is the gravitational force. Therefore, if the projectile did not explode, it would continue to move along the parabolic path indicated by the dashed line in Figure 9.21. Because the forces caused by the explosion are internal, they do not affect the motion of the center of mass of the system (the fragments). Therefore, after the explosion, the center of mass of the fragments follows the same parabolic path the projectile would have followed if no explosion had occurred.

(B) If the projectile did not explode, it would land at a distance R from its launch point. Suppose the projectile explodes and splits into two pieces of equal mass. One piece lands at a distance $2R$ to the right of the launch point. Where does the other piece land?

SOLUTION

Figure 9.21 (Conceptual Example 9.13) When a projectile explodes into several fragments, the center of mass of the system made up of all the fragments follows the same parabolic path the projectile would have taken had there been no explosion.

As discussed in part (A), the center of mass of the two-piece system lands at a distance R from the launch point. One of the pieces lands at a farther distance R from the landing point (or a distance $2R$ from the launch point), to the right in Figure 9.21. Because the two pieces have the same mass, the other piece must land a distance R to the left of the landing point in Figure 9.21, which places this piece right back at the launch point!

Example 9.14   The Exploding Rocket   AM

A rocket is fired vertically upward. At the instant it reaches an altitude of 1 000 m and a speed of $v_i = 300$ m/s, it explodes into three fragments having equal mass. One fragment moves upward with a speed of $v_1 = 450$ m/s following the explosion. The second fragment has a speed of $v_2 = 240$ m/s and is moving east right after the explosion. What is the velocity of the third fragment immediately after the explosion?

SOLUTION

Conceptualize Picture the explosion in your mind, with one piece going upward and a second piece moving horizontally toward the east. Do you have an intuitive feeling about the direction in which the third piece moves?

Categorize This example is a two-dimensional problem because we have two fragments moving in perpendicular directions after the explosion as well as a third fragment moving in an unknown direction in the plane defined by the velocity vectors of the other two fragments. We assume the time interval of the explosion is very short, so we use the impulse approximation in which we ignore the gravitational force and air resistance. Because the forces of the explosion are internal to the system (the rocket), the rocket is an *isolated system* in terms of *momentum*. Therefore, the total momentum $\vec{\mathbf{p}}_i$ of the rocket immediately before the explosion must equal the total momentum $\vec{\mathbf{p}}_f$ of the fragments immediately after the explosion.

....

Analyze Because the three fragments have equal mass, the mass of each fragment is $M/3$, where M is the total mass of the rocket. We will let $\vec{\mathbf{v}}_3$ represent the unknown velocity of the third fragment.

Use the isolated system (momentum) model to equate the initial and final momenta of the system and express the momenta in terms of masses and velocities:

$$\Delta\vec{\mathbf{p}} = 0 \quad \rightarrow \quad \vec{\mathbf{p}}_i = \vec{\mathbf{p}}_f \quad \rightarrow \quad M\vec{\mathbf{v}}_i = \frac{M}{3}\vec{\mathbf{v}}_1 + \frac{M}{3}\vec{\mathbf{v}}_2 + \frac{M}{3}\vec{\mathbf{v}}_3$$

Solve for $\vec{\mathbf{v}}_3$:

$$\vec{\mathbf{v}}_3 = 3\vec{\mathbf{v}}_i - \vec{\mathbf{v}}_1 - \vec{\mathbf{v}}_2$$

Substitute the numerical values:

$$\vec{\mathbf{v}}_3 = 3(300\hat{\mathbf{j}}\ \text{m/s}) - (450\hat{\mathbf{j}}\ \text{m/s}) - (240\hat{\mathbf{i}}\ \text{m/s}) = \boxed{(-240\hat{\mathbf{i}} + 450\hat{\mathbf{j}})\ \text{m/s}}$$

....

Finalize Notice that this event is the reverse of a perfectly inelastic collision. There is one object before the collision and three objects afterward. Imagine running a movie of the event backward: the three objects would come together and become a single object. In a perfectly inelastic collision, the kinetic energy of the system decreases. If you were

▶ **9.14** continued

to calculate the kinetic energy before and after the event in this example, you would find that the kinetic energy of the system increases. (Try it!) This increase in kinetic energy comes from the potential energy stored in whatever fuel exploded to cause the breakup of the rocket.

9.8 Deformable Systems

So far in our discussion of mechanics, we have analyzed the motion of particles or nondeformable systems that can be modeled as particles. The discussion in Section 9.7 can be applied to an analysis of the motion of deformable systems. For example, suppose you stand on a skateboard and push off a wall, setting yourself in motion away from the wall. Your body has deformed during this event: your arms were bent before the event, and they straightened out while you pushed off the wall. How would we describe this event?

The force from the wall on your hands moves through no displacement; the force is always located at the interface between the wall and your hands. Therefore, the force does no work on the system, which is you and your skateboard. Pushing off the wall, however, does indeed result in a change in the kinetic energy of the system. If you try to use the work–kinetic energy theorem, $W = \Delta K$, to describe this event, you will notice that the left side of the equation is zero but the right side is not zero. The work–kinetic energy theorem is not valid for this event and is often not valid for systems that are deformable.

To analyze the motion of deformable systems, we appeal to Equation 8.2, the conservation of energy equation, and Equation 9.40, the impulse–momentum theorem. For the example of you pushing off the wall on your skateboard, identifying the system as you and the skateboard, Equation 8.2 gives

$$\Delta E_{\text{system}} = \sum T \quad \rightarrow \quad \Delta K + \Delta U = 0$$

where ΔK is the change in kinetic energy, which is related to the increased speed of the system, and ΔU is the decrease in potential energy stored in the body from previous meals. This equation tells us that the system transformed potential energy into kinetic energy by virtue of the muscular exertion necessary to push off the wall. Notice that the system is isolated in terms of energy but nonisolated in terms of momentum.

Applying Equation 9.40 to the system in this situation gives us

$$\Delta \vec{\mathbf{p}}_{\text{tot}} = \vec{\mathbf{I}} \quad \rightarrow \quad m \, \Delta \vec{\mathbf{v}} = \int \vec{\mathbf{F}}_{\text{wall}} \, dt$$

where $\vec{\mathbf{F}}_{\text{wall}}$ is the force exerted by the wall on your hands, m is the mass of you and the skateboard, and $\Delta \vec{\mathbf{v}}$ is the change in the velocity of the system during the event. To evaluate the right side of this equation, we would need to know how the force from the wall varies in time. In general, this process might be complicated. In the case of constant forces, or well-behaved forces, however, the integral on the right side of the equation can be evaluated.

Example 9.15 **Pushing on a Spring**[3] **AM**

As shown in Figure 9.22a (page 224), two blocks are at rest on a frictionless, level table. Both blocks have the same mass m, and they are connected by a spring of negligible mass. The separation distance of the blocks when the spring is relaxed is L. During a time interval Δt, a constant force of magnitude F is applied horizontally to the left block,

---

[3]Example 9.15 was inspired in part by C. E. Mungan, "A primer on work–energy relationships for introductory physics," *The Physics Teacher* **43**:10, 2005.

continued

▶ **9.15** continued

moving it through a distance x_1 as shown in Figure 9.22b. During this time interval, the right block moves through a distance x_2. At the end of this time interval, the force F is removed.

(A) Find the resulting speed $\vec{v}_{CM}$ of the center of mass of the system.

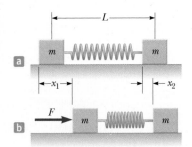

Figure 9.22 (Example 9.15) (a) Two blocks of equal mass are connected by a spring. (b) The left block is pushed with a constant force of magnitude F and moves a distance x_1 during some time interval. During this same time interval, the right block moves through a distance x_2.

SOLUTION

Conceptualize Imagine what happens as you push on the left block. It begins to move to the right in Figure 9.22, and the spring begins to compress. As a result, the spring pushes to the right on the right block, which begins to move to the right. At any given time, the blocks are generally moving with different velocities. As the center of mass of the system moves to the right with a constant speed after the force is removed, the two blocks oscillate back and forth with respect to the center of mass.

Categorize We apply three analysis models in this problem: the deformable system of two blocks and a spring is modeled as a *nonisolated system* in terms of *energy* because work is being done on it by the applied force. It is also modeled as a *nonisolated system* in terms of *momentum* because of the force acting on the system during a time interval. Because the applied force on the system is constant, the acceleration of its center of mass is constant and the center of mass is modeled as a *particle under constant acceleration.*

Analyze Using the nonisolated system (momentum) model, we apply the impulse–momentum theorem to the system of two blocks, recognizing that the force F is constant during the time interval Δt while the force is applied.

Write Equation 9.40 for the system:

$$\Delta p_x = I_x \quad \rightarrow \quad (2m)(v_{CM} - 0) = F\Delta t$$

$$(1) \quad 2mv_{CM} = F\Delta t$$

During the time interval Δt, the center of mass of the system moves a distance $\frac{1}{2}(x_1 + x_2)$. Use this fact to express the time interval in terms of $v_{CM,avg}$:

$$\Delta t = \frac{\frac{1}{2}(x_1 + x_2)}{v_{CM,avg}}$$

Because the center of mass is modeled as a particle under constant acceleration, the average velocity of the center of mass is the average of the initial velocity, which is zero, and the final velocity v_{CM}:

$$\Delta t = \frac{\frac{1}{2}(x_1 + x_2)}{\frac{1}{2}(0 + v_{CM})} = \frac{(x_1 + x_2)}{v_{CM}}$$

Substitute this expression into Equation (1):

$$2mv_{CM} = F\frac{(x_1 + x_2)}{v_{CM}}$$

Solve for v_{CM}:

$$v_{CM} = \sqrt{F\frac{(x_1 + x_2)}{2m}}$$

(B) Find the total energy of the system associated with vibration relative to its center of mass after the force F is removed.

SOLUTION

Analyze The vibrational energy is all the energy of the system other than the kinetic energy associated with translational motion of the center of mass. To find the vibrational energy, we apply the conservation of energy equation. The kinetic energy of the system can be expressed as $K = K_{CM} + K_{vib}$, where K_{vib} is the kinetic energy of the blocks relative to the center of mass due to their vibration. The potential energy of the system is U_{vib}, which is the potential energy stored in the spring when the separation of the blocks is some value other than L.

From the nonisolated system (energy) model, express Equation 8.2 for this system:

$$(2) \quad \Delta K_{CM} + \Delta K_{vib} + \Delta U_{vib} = W$$

▶ **9.15** c o n t i n u e d

Express Equation (2) in an alternate form, noting that $K_{vib} + U_{vib} = E_{vib}$:

$$\Delta K_{CM} + \Delta E_{vib} = W$$

The initial values of the kinetic energy of the center of mass and the vibrational energy of the system are zero. Use this fact and substitute for the work done on the system by the force F:

$$K_{CM} + E_{vib} = W = Fx_1$$

Solve for the vibrational energy and use the result from part (A):

$$E_{vib} = Fx_1 - K_{CM} = Fx_1 - \tfrac{1}{2}(2m)v_{CM}^2 = \boxed{F\,\frac{(x_1 - x_2)}{2}}$$

Finalize Neither of the two answers in this example depends on the spring length, the spring constant, or the time interval. Notice also that the magnitude x_1 of the displacement of the point of application of the applied force is different from the magnitude $\tfrac{1}{2}(x_1 + x_2)$ of the displacement of the center of mass of the system. This difference reminds us that the displacement in the definition of work (Eq. 7.1) is that of the point of application of the force.

9.9 Rocket Propulsion

When ordinary vehicles such as cars are propelled, the driving force for the motion is friction. In the case of the car, the driving force is the force exerted by the road on the car. We can model the car as a nonisolated system in terms of momentum. An impulse is applied to the car from the roadway, and the result is a change in the momentum of the car as described by Equation 9.40.

A rocket moving in space, however, has no road to push against. The rocket is an isolated system in terms of momentum. Therefore, the source of the propulsion of a rocket must be something other than an external force. The operation of a rocket depends on the law of conservation of linear momentum as applied to an isolated system, where the system is the rocket plus its ejected fuel.

Rocket propulsion can be understood by first considering our archer standing on frictionless ice in Example 9.1. Imagine the archer fires several arrows horizontally. For each arrow fired, the archer receives a compensating momentum in the opposite direction. As more arrows are fired, the archer moves faster and faster across the ice. In addition to this analysis in terms of momentum, we can also understand this phenomenon in terms of Newton's second and third laws. Every time the bow pushes an arrow forward, the arrow pushes the bow (and the archer) backward, and these forces result in an acceleration of the archer.

In a similar manner, as a rocket moves in free space, its linear momentum changes when some of its mass is ejected in the form of exhaust gases. Because the gases are given momentum when they are ejected out of the engine, the rocket receives a compensating momentum in the opposite direction. Therefore, the rocket is accelerated as a result of the "push," or thrust, from the exhaust gases. In free space, the center of mass of the system (rocket plus expelled gases) moves uniformly, independent of the propulsion process.[4]

Suppose at some time t the magnitude of the momentum of a rocket plus its fuel is $(M + \Delta m)v$, where v is the speed of the rocket relative to the Earth (Fig. 9.23a). Over a short time interval Δt, the rocket ejects fuel of mass Δm. At the end of this interval, the rocket's mass is M and its speed is $v + \Delta v$, where Δv is the change in speed of the rocket (Fig. 9.23b). If the fuel is ejected with a speed v_e relative to

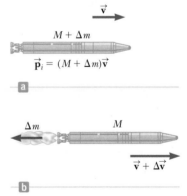

Courtesy of NASA

The force from a nitrogen-propelled hand-controlled device allows an astronaut to move about freely in space without restrictive tethers, using the thrust force from the expelled nitrogen.

$\vec{\mathbf{v}}$

$M + \Delta m$

$\vec{\mathbf{p}}_i = (M + \Delta m)\vec{\mathbf{v}}$

a

Δm M

$\vec{\mathbf{v}} + \Delta\vec{\mathbf{v}}$

b

Figure 9.23 Rocket propulsion. (a) The initial mass of the rocket plus all its fuel is $M + \Delta m$ at a time t, and its speed is v. (b) At a time $t + \Delta t$, the rocket's mass has been reduced to M and an amount of fuel Δm has been ejected. The rocket's speed increases by an amount Δv.

---

[4]The rocket and the archer represent cases of the reverse of a perfectly inelastic collision: momentum is conserved, but the kinetic energy of the rocket–exhaust gas system increases (at the expense of chemical potential energy in the fuel), as does the kinetic energy of the archer–arrow system (at the expense of potential energy from the archer's previous meals).

the rocket (the subscript e stands for *exhaust*, and v_e is usually called the *exhaust speed*), the velocity of the fuel relative to the Earth is $v - v_e$. Because the system of the rocket and the ejected fuel is isolated, we apply the isolated system model for momentum and obtain

$$\Delta p = 0 \quad \rightarrow \quad p_i = p_f \quad \rightarrow \quad (M + \Delta m)v = M(v + \Delta v) + \Delta m(v - v_e)$$

Simplifying this expression gives

$$M\,\Delta v = v_e\,\Delta m$$

If we now take the limit as Δt goes to zero, we let $\Delta v \rightarrow dv$ and $\Delta m \rightarrow dm$. Furthermore, the increase in the exhaust mass dm corresponds to an equal decrease in the rocket mass, so $dm = -dM$. Notice that dM is negative because it represents a decrease in mass, so $-dM$ is a positive number. Using this fact gives

$$M\,dv = v_e\,dm = -v_e\,dM \tag{9.43}$$

Now divide the equation by M and integrate, taking the initial mass of the rocket plus fuel to be M_i and the final mass of the rocket plus its remaining fuel to be M_f. The result is

$$\int_{v_i}^{v_f} dv = -v_e \int_{M_i}^{M_f} \frac{dM}{M}$$

◀ Expression for rocket propulsion

$$v_f - v_i = v_e \ln\left(\frac{M_i}{M_f}\right) \tag{9.44}$$

which is the basic expression for rocket propulsion. First, Equation 9.44 tells us that the increase in rocket speed is proportional to the exhaust speed v_e of the ejected gases. Therefore, the exhaust speed should be very high. Second, the increase in rocket speed is proportional to the natural logarithm of the ratio M_i/M_f. Therefore, this ratio should be as large as possible; that is, the mass of the rocket without its fuel should be as small as possible and the rocket should carry as much fuel as possible.

The **thrust** on the rocket is the force exerted on it by the ejected exhaust gases. We obtain the following expression for the thrust from Newton's second law and Equation 9.43:

$$\text{Thrust} = M\frac{dv}{dt} = \left| v_e \frac{dM}{dt} \right| \tag{9.45}$$

This expression shows that the thrust increases as the exhaust speed increases and as the rate of change of mass (called the *burn rate*) increases.

Example 9.16   **Fighting a Fire**

Two firefighters must apply a total force of 600 N to steady a hose that is discharging water at the rate of 3 600 L/min. Estimate the speed of the water as it exits the nozzle.

SOLUTION

Conceptualize As the water leaves the hose, it acts in a way similar to the gases being ejected from a rocket engine. As a result, a force (thrust) acts on the firefighters in a direction opposite the direction of motion of the water. In this case, we want the end of the hose to be modeled as a particle in equilibrium rather than to accelerate as in the case of the rocket. Consequently, the firefighters must apply a force of magnitude equal to the thrust in the opposite direction to keep the end of the hose stationary.

Categorize This example is a substitution problem in which we use given values in an equation derived in this section. The water exits at 3 600 L/min, which is 60 L/s. Knowing that 1 L of water has a mass of 1 kg, we estimate that about 60 kg of water leaves the nozzle each second.

▶ **9.16** continued

Use Equation 9.45 for the thrust:

$$\text{Thrust} = \left| v_e \frac{dM}{dt} \right|$$

Solve for the exhaust speed:

$$v_e = \frac{\text{Thrust}}{|dM/dt|}$$

Substitute numerical values:

$$v_e = \frac{600 \text{ N}}{60 \text{ kg/s}} = \boxed{10 \text{ m/s}}$$

Example 9.17   **A Rocket in Space**

A rocket moving in space, far from all other objects, has a speed of 3.0×10^3 m/s relative to the Earth. Its engines are turned on, and fuel is ejected in a direction opposite the rocket's motion at a speed of 5.0×10^3 m/s relative to the rocket.

(A) What is the speed of the rocket relative to the Earth once the rocket's mass is reduced to half its mass before ignition?

SOLUTION

Conceptualize Figure 9.23 shows the situation in this problem. From the discussion in this section and scenes from science fiction movies, we can easily imagine the rocket accelerating to a higher speed as the engine operates.

Categorize This problem is a substitution problem in which we use given values in the equations derived in this section.

Solve Equation 9.44 for the final velocity and substitute the known values:

$$v_f = v_i + v_e \ln\left(\frac{M_i}{M_f}\right)$$

$$= 3.0 \times 10^3 \text{ m/s} + (5.0 \times 10^3 \text{ m/s}) \ln\left(\frac{M_i}{0.50 M_i}\right)$$

$$= \boxed{6.5 \times 10^3 \text{ m/s}}$$

(B) What is the thrust on the rocket if it burns fuel at the rate of 50 kg/s?

SOLUTION

Use Equation 9.45, noting that $dM/dt = 50$ kg/s:

$$\text{Thrust} = \left| v_e \frac{dM}{dt} \right| = (5.0 \times 10^3 \text{ m/s})(50 \text{ kg/s}) = \boxed{2.5 \times 10^5 \text{ N}}$$

Summary

Definitions

▮ The **linear momentum** $\vec{\mathbf{p}}$ of a particle of mass m moving with a velocity $\vec{\mathbf{v}}$ is

$$\vec{\mathbf{p}} \equiv m\vec{\mathbf{v}} \qquad (9.2)$$

▮ The **impulse** imparted to a particle by a net force $\sum \vec{\mathbf{F}}$ is equal to the time integral of the force:

$$\vec{\mathbf{I}} \equiv \int_{t_i}^{t_f} \sum \vec{\mathbf{F}} \, dt \qquad (9.9)$$

continued

An **inelastic collision** is one for which the total kinetic energy of the system of colliding particles is not conserved. A **perfectly inelastic collision** is one in which the colliding particles stick together after the collision. An **elastic collision** is one in which the kinetic energy of the system is conserved.

The position vector of the **center of mass** of a system of particles is defined as

$$\vec{\mathbf{r}}_{CM} \equiv \frac{1}{M} \sum_i m_i \vec{\mathbf{r}}_i \qquad (9.31)$$

where $M = \sum_i m_i$ is the total mass of the system and $\vec{\mathbf{r}}_i$ is the position vector of the ith particle.

Concepts and Principles

The position vector of the center of mass of an extended object can be obtained from the integral expression

$$\vec{\mathbf{r}}_{CM} = \frac{1}{M} \int \vec{\mathbf{r}} \, dm \qquad (9.34)$$

The velocity of the center of mass for a system of particles is

$$\vec{\mathbf{v}}_{CM} = \frac{1}{M} \sum_i m_i \vec{\mathbf{v}}_i \qquad (9.35)$$

The total momentum of a system of particles equals the total mass multiplied by the velocity of the center of mass.

Newton's second law applied to a system of particles is

$$\sum \vec{\mathbf{F}}_{ext} = M \vec{\mathbf{a}}_{CM} \qquad (9.39)$$

where $\vec{\mathbf{a}}_{CM}$ is the acceleration of the center of mass and the sum is over all external forces. The center of mass moves like an imaginary particle of mass M under the influence of the resultant external force on the system.

Analysis Models for Problem Solving

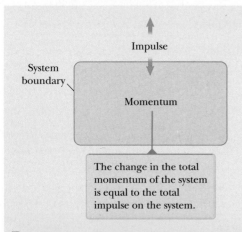

Nonisolated System (Momentum). If a system interacts with its environment in the sense that there is an external force on the system, the behavior of the system is described by the **impulse–momentum theorem:**

$$\Delta \vec{\mathbf{p}}_{tot} = \vec{\mathbf{I}} \qquad (9.40)$$

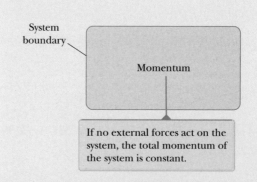

Isolated System (Momentum). The total momentum of an isolated system (no external forces) is conserved regardless of the nature of the forces between the members of the system:

$$\Delta \vec{\mathbf{p}}_{tot} = 0 \qquad (9.41)$$

The system may be isolated in terms of momentum but nonisolated in terms of energy, as in the case of inelastic collisions.

1. You are standing on a saucer-shaped sled at rest in the middle of a frictionless ice rink. Your lab partner throws you a heavy Frisbee. You take different actions in successive experimental trials. Rank the following situations according to your final speed from largest to smallest. If your final speed is the same in two cases, give them equal rank. (a) You catch the Frisbee and hold onto it. (b) You catch the Frisbee and throw it back to your partner. (c) You bobble the catch, just touching the Frisbee so that it continues in its original direction more slowly. (d) You catch the Frisbee and throw it so that it moves vertically upward above your head. (e) You catch the Frisbee and set it down so that it remains at rest on the ice.

2. A boxcar at a rail yard is set into motion at the top of a hump. The car rolls down quietly and without friction onto a straight, level track where it couples with a flatcar of smaller mass, originally at rest, so that the two cars then roll together without friction. Consider the two cars as a system from the moment of release of the boxcar until both are rolling together. Answer the following questions yes or no. (a) Is mechanical energy of the system conserved? (b) Is momentum of the system conserved? Next, consider only the process of the boxcar gaining speed as it rolls down the hump. For the boxcar and the Earth as a system, (c) is mechanical energy conserved? (d) Is momentum conserved? Finally, consider the two cars as a system as the boxcar is slowing down in the coupling process. (e) Is mechanical energy of this system conserved? (f) Is momentum of this system conserved?

3. A massive tractor is rolling down a country road. In a perfectly inelastic collision, a small sports car runs into the machine from behind. **(i)** Which vehicle experiences a change in momentum of larger magnitude? (a) The car does. (b) The tractor does. (c) Their momentum changes are the same size. (d) It could be either vehicle. **(ii)** Which vehicle experiences a larger change in kinetic energy? (a) The car does. (b) The tractor does. (c) Their kinetic energy changes are the same size. (d) It could be either vehicle.

4. A 2-kg object moving to the right with a speed of 4 m/s makes a head-on, elastic collision with a 1-kg object that is initially at rest. The velocity of the 1-kg object after the collision is (a) greater than 4 m/s, (b) less than 4 m/s, (c) equal to 4 m/s, (d) zero, or (e) impossible to say based on the information provided.

5. A 5-kg cart moving to the right with a speed of 6 m/s collides with a concrete wall and rebounds with a speed of 2 m/s. What is the change in momentum of the cart? (a) 0 (b) 40 kg · m/s (c) −40 kg · m/s (d) −30 kg · m/s (e) −10 kg · m/s

6. A 57.0-g tennis ball is traveling straight at a player at 21.0 m/s. The player volleys the ball straight back at 25.0 m/s. If the ball remains in contact with the racket for 0.060 0 s, what average force acts on the ball? (a) 22.6 N (b) 32.5 N (c) 43.7 N (d) 72.1 N (e) 102 N

7. The momentum of an object is increased by a factor of 4 in magnitude. By what factor is its kinetic energy changed? (a) 16 (b) 8 (c) 4 (d) 2 (e) 1

8. The kinetic energy of an object is increased by a factor of 4. By what factor is the magnitude of its momentum changed? (a) 16 (b) 8 (c) 4 (d) 2 (e) 1

9. If two particles have equal momenta, are their kinetic energies equal? (a) yes, always (b) no, never (c) no, except when their speeds are the same (d) yes, as long as they move along parallel lines

10. If two particles have equal kinetic energies, are their momenta equal? (a) yes, always (b) no, never (c) yes, as long as their masses are equal (d) yes, if both their masses and directions of motion are the same (e) yes, as long as they move along parallel lines

11. A 10.0-g bullet is fired into a 200-g block of wood at rest on a horizontal surface. After impact, the block slides 8.00 m before coming to rest. If the coefficient of friction between the block and the surface is 0.400, what is the speed of the bullet before impact? (a) 106 m/s (b) 166 m/s (c) 226 m/s (d) 286 m/s (e) none of those answers is correct

12. Two particles of different mass start from rest. The same net force acts on both of them as they move over equal distances. How do their final kinetic energies compare? (a) The particle of larger mass has more kinetic energy. (b) The particle of smaller mass has more kinetic energy. (c) The particles have equal kinetic energies. (d) Either particle might have more kinetic energy.

13. Two particles of different mass start from rest. The same net force acts on both of them as they move over equal distances. How do the magnitudes of their final momenta compare? (a) The particle of larger mass has more momentum. (b) The particle of smaller mass has more momentum. (c) The particles have equal momenta. (d) Either particle might have more momentum.

14. A basketball is tossed up into the air, falls freely, and bounces from the wooden floor. From the moment after the player releases it until the ball reaches the top of its bounce, what is the smallest system for which momentum is conserved? (a) the ball (b) the ball plus player (c) the ball plus floor (d) the ball plus the Earth (e) momentum is not conserved for any system

15. A 3-kg object moving to the right on a frictionless, horizontal surface with a speed of 2 m/s collides head-on and sticks to a 2-kg object that is initially moving to the left with a speed of 4 m/s. After the collision, which statement is true? (a) The kinetic energy of the system is 20 J. (b) The momentum of the system is 14 kg · m/s. (c) The kinetic energy of the system is greater than 5 J but less than 20 J. (d) The momentum of the system is −2 kg · m/s. (e) The momentum of the system is less than the momentum of the system before the collision.

16. A ball is suspended by a string that is tied to a fixed point above a wooden block standing on end. The ball is pulled back as shown in Figure OQ9.16 and released. In trial A, the ball rebounds elastically from the block. In trial B, two-sided tape causes the ball to stick to the block. In which case is the ball more likely to knock the block over? (a) It is more likely in trial A. (b) It is more likely in trial B. (c) It makes no difference. (d) It could be either case, depending on other factors.

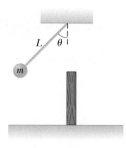

Figure OQ9.16

17. A car of mass m traveling at speed v crashes into the rear of a truck of mass $2m$ that is at rest and in neutral at an intersection. If the collision is perfectly inelastic, what is the speed of the combined car and truck after the collision? (a) v (b) $v/2$ (c) $v/3$ (d) $2v$ (e) None of those answers is correct.

18. A head-on, elastic collision occurs between two billiard balls of equal mass. If a red ball is traveling to the right with speed v and a blue ball is traveling to the left with speed $3v$ before the collision, what statement is true concerning their velocities subsequent to the collision? Neglect any effects of spin. (a) The red ball travels to the left with speed v, while the blue ball travels to the right with speed $3v$. (b) The red ball travels to the left with speed v, while the blue ball continues to move to the left with a speed $2v$. (c) The red ball travels to the left with speed $3v$, while the blue ball travels to the right with speed v. (d) Their final velocities cannot be determined because momentum is not conserved in the collision. (e) The velocities cannot be determined without knowing the mass of each ball.

Conceptual Questions   **1.** denotes answer available in *Student Solutions Manual/Study Guide*

1. An airbag in an automobile inflates when a collision occurs, which protects the passenger from serious injury (see the photo on page 202). Why does the airbag soften the blow? Discuss the physics involved in this dramatic photograph.

2. In golf, novice players are often advised to be sure to "follow through" with their swing. Why does this advice make the ball travel a longer distance? If a shot is taken near the green, very little follow-through is required. Why?

3. An open box slides across a frictionless, icy surface of a frozen lake. What happens to the speed of the box as water from a rain shower falls vertically downward into the box? Explain.

4. While in motion, a pitched baseball carries kinetic energy and momentum. (a) Can we say that it carries a force that it can exert on any object it strikes? (b) Can the baseball deliver more kinetic energy to the bat and batter than the ball carries initially? (c) Can the baseball deliver to the bat and batter more momentum than the ball carries initially? Explain each of your answers.

5. You are standing perfectly still and then take a step forward. Before the step, your momentum was zero, but afterward you have some momentum. Is the principle of conservation of momentum violated in this case? Explain your answer.

6. A sharpshooter fires a rifle while standing with the butt of the gun against her shoulder. If the forward momentum of a bullet is the same as the backward momentum of the gun, why isn't it as dangerous to be hit by the gun as by the bullet?

7. Two students hold a large bed sheet vertically between them. A third student, who happens to be the star pitcher on the school baseball team, throws a raw egg at the center of the sheet. Explain why the egg does not break when it hits the sheet, regardless of its initial speed.

8. A juggler juggles three balls in a continuous cycle. Any one ball is in contact with one of his hands for one fifth of the time. (a) Describe the motion of the center of mass of the three balls. (b) What average force does the juggler exert on one ball while he is touching it?

9. (a) Does the center of mass of a rocket in free space accelerate? Explain. (b) Can the speed of a rocket exceed the exhaust speed of the fuel? Explain.

10. On the subject of the following positions, state your own view and argue to support it. (a) The best theory of motion is that force causes acceleration. (b) The true measure of a force's effectiveness is the work it does, and the best theory of motion is that work done on an object changes its energy. (c) The true measure of a force's effect is impulse, and the best theory of motion is that impulse imparted to an object changes its momentum.

11. Does a larger net force exerted on an object always produce a larger change in the momentum of the object compared with a smaller net force? Explain.

12. Does a larger net force always produce a larger change in kinetic energy than a smaller net force? Explain.

13. A bomb, initially at rest, explodes into several pieces. (a) Is linear momentum of the system (the bomb before the explosion, the pieces after the explosion) conserved? Explain. (b) Is kinetic energy of the system conserved? Explain.

Problems available in ENHANCED WebAssign   Access end-of-chapter problems online at www.webassign.net

Section 9.1 Linear Momentum
Problems 1–5

Section 9.2 Analysis Model: Isolated System (Momentum)
Problems 6–11

Section 9.3 Analysis Model: Nonisolated System (Momentum)
Problems 12–21

Section 9.4 Collisions in One Dimension
Problems 22–34

Section 9.5 Collisions in Two Dimensions
Problems 35–44

Section 9.6 The Center of Mass
Problems 45–50

Section 9.7 Systems of Many Particles
Problems 51–55

Section 9.8 Deformable Systems
Problems 56–59

Section 9.9 Rocket Propulsion
Problems 60–64

Additional Problems
Problems 65–91

Challenge Problems
Problem 92–96

Solutions to the following Problems are available in the *Student Solutions Manual/Study Guide:*
9.6, 9.13, 9.18, 9.23, 9.27, 9.31, 9.37, 9.41, 9.43, 9.48, 9.51, 9.53, 9.56, 9.62, 9.67, 9.71, 9.79, 9.86, and 9.89

List of Enhanced Problems

Problem Number	Targeted Feedback in Enhanced WebAssign	Analysis Model Tutorial in Enhanced WebAssign	Master It in Enhanced WebAssign	Watch It in Enhanced WebAssign
9.6			✓	
9.11	✓			✓
9.13	✓			✓
9.17	✓		✓	
9.18	✓	✓		
9.23	✓			✓
9.27			✓	
9.29	✓		✓	
9.31		✓	✓	
9.33	✓	✓		✓
9.37	✓			✓
9.38	✓			✓
9.41	✓		✓	
9.42	✓		✓	
9.43	✓		✓	
9.45	✓			✓
9.48	✓			✓
9.51	✓			✓
9.53			✓	
9.67	✓		✓	
9.80	✓			✓
9.81	✓		✓	
9.89		✓	✓	
9.91	✓		✓	

Rotation of a Rigid Object About a Fixed Axis

When an extended object such as a wheel rotates about its axis, the motion cannot be analyzed by modeling the object as a particle because at any given time different parts of the object have different linear velocities and linear accelerations. We can, however, analyze the motion of an extended object by modeling it as a system of many particles, each of which has its own linear velocity and linear acceleration as discussed in Section 9.7.

In dealing with a rotating object, analysis is greatly simplified by assuming the object is rigid. A **rigid object** is one that is nondeformable; that is, the relative locations of all particles of which the object is composed remain constant. All real objects are deformable to some extent; our rigid-object model, however, is useful in many situations in which deformation is negligible. We have developed analysis models based on particles and systems. In this chapter, we introduce another class of analysis models based on the rigid-object model.

The Malaysian pastime of *gasing* involves the spinning of tops that can have masses up to 5 kg. Professional spinners can spin their tops so that they might rotate for more than an hour before stopping. We will study the rotational motion of objects such as these tops in this chapter. *(Courtesy Tourism Malaysia)*

10.1 Angular Position, Velocity, and Acceleration

We will develop our understanding of rotational motion in a manner parallel to that used for translational motion in previous chapters. We began in Chapter 2 by

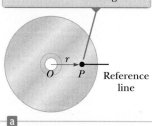

To define angular position for the disc, a fixed reference line is chosen. A particle at P is located at a distance r from the rotation axis through O.

a

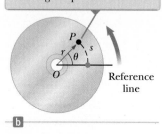

As the disc rotates, a particle at P moves through an arc length s on a circular path of radius r. The angular position of P is θ.

b

Figure 10.1 A compact disc rotating about a fixed axis through O perpendicular to the plane of the figure.

Pitfall Prevention 10.1
Remember the Radian In rotational equations, you *must* use angles expressed in radians. Don't fall into the trap of using angles measured in degrees in rotational equations.

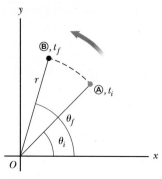

Figure 10.2 A particle on a rotating rigid object moves from Ⓐ to Ⓑ along the arc of a circle. In the time interval $\Delta t = t_f - t_i$, the radial line of length r moves through an angular displacement $\Delta\theta = \theta_f - \theta_i$.

Average angular speed ▶

defining kinematic variables: position, velocity, and acceleration. We do the same here for rotational motion.

Figure 10.1 illustrates an overhead view of a rotating compact disc, or CD. The disc rotates about a fixed axis perpendicular to the plane of the figure and passing through the center of the disc at O. A small element of the disc modeled as a particle at P is at a fixed distance r from the origin and rotates about it in a circle of radius r. (In fact, *every* element of the disc undergoes circular motion about O.) It is convenient to represent the position of P with its polar coordinates (r, θ), where r is the distance from the origin to P and θ is measured *counterclockwise* from some reference line fixed in space as shown in Figure 10.1a. In this representation, the angle θ changes in time while r remains constant. As the particle moves along the circle from the reference line, which is at angle θ = 0, it moves through an arc of length s as in Figure 10.1b. The arc length s is related to the angle θ through the relationship

$$s = r\theta \tag{10.1a}$$

$$\theta = \frac{s}{r} \tag{10.1b}$$

Because θ is the ratio of an arc length and the radius of the circle, it is a pure number. Usually, however, we give θ the artificial unit **radian** (rad), where one radian is the angle subtended by an arc length equal to the radius of the arc. Because the circumference of a circle is $2\pi r$, it follows from Equation 10.1b that 360° corresponds to an angle of $(2\pi r/r)$ rad $= 2\pi$ rad. Hence, 1 rad $= 360°/2\pi \approx 57.3°$. To convert an angle in degrees to an angle in radians, we use that π rad $= 180°$, so

$$\theta(\text{rad}) = \frac{\pi}{180°}\theta(\text{deg})$$

For example, 60° equals $\pi/3$ rad and 45° equals $\pi/4$ rad.

Because the disc in Figure 10.1 is a rigid object, as the particle moves through an angle θ from the reference line, every other particle on the object rotates through the same angle θ. Therefore, we can associate the angle θ with the entire rigid object as well as with an individual particle, which allows us to define the *angular position* of a rigid object in its rotational motion. We choose a reference line on the object, such as a line connecting O and a chosen particle on the object. The **angular position** of the rigid object is the angle θ between this reference line on the object and the fixed reference line in space, which is often chosen as the x axis. Such identification is similar to the way we define the position of an object in translational motion as the distance x between the object and the reference position, which is the origin, x = 0. Therefore, the angle θ plays the same role in rotational motion that the position x does in translational motion.

As the particle in question on our rigid object travels from position Ⓐ to position Ⓑ in a time interval Δt as in Figure 10.2, the reference line fixed to the object sweeps out an angle $\Delta\theta = \theta_f - \theta_i$. This quantity $\Delta\theta$ is defined as the **angular displacement** of the rigid object:

$$\Delta\theta \equiv \theta_f - \theta_i$$

The rate at which this angular displacement occurs can vary. If the rigid object spins rapidly, this displacement can occur in a short time interval. If it rotates slowly, this displacement occurs in a longer time interval. These different rotation rates can be quantified by defining the **average angular speed** ω_{avg} (Greek letter omega) as the ratio of the angular displacement of a rigid object to the time interval Δt during which the displacement occurs:

$$\omega_{\text{avg}} \equiv \frac{\theta_f - \theta_i}{t_f - t_i} = \frac{\Delta\theta}{\Delta t} \tag{10.2}$$

In analogy to translational speed, the **instantaneous angular speed** ω is defined as the limit of the average angular speed as Δt approaches zero:

$$\omega \equiv \lim_{\Delta t \to 0} \frac{\Delta \theta}{\Delta t} = \frac{d\theta}{dt} \qquad (10.3)$$

◀ **Instantaneous angular speed**

Angular speed has units of radians per second (rad/s), which can be written as s^{-1} because radians are not dimensional. We take ω to be positive when θ is increasing (counterclockwise motion in Fig. 10.2) and negative when θ is decreasing (clockwise motion in Fig. 10.2).

Quick Quiz 10.1 A rigid object rotates in a counterclockwise sense around a fixed axis. Each of the following pairs of quantities represents an initial angular position and a final angular position of the rigid object. **(i)** Which of the sets can *only* occur if the rigid object rotates through more than 180°? (a) 3 rad, 6 rad (b) −1 rad, 1 rad (c) 1 rad, 5 rad **(ii)** Suppose the change in angular position for each of these pairs of values occurs in 1 s. Which choice represents the lowest average angular speed?

If the instantaneous angular speed of an object changes from ω_i to ω_f in the time interval Δt, the object has an angular acceleration. The **average angular acceleration** α_{avg} (Greek letter alpha) of a rotating rigid object is defined as the ratio of the change in the angular speed to the time interval Δt during which the change in the angular speed occurs:

$$\alpha_{\text{avg}} \equiv \frac{\omega_f - \omega_i}{t_f - t_i} = \frac{\Delta \omega}{\Delta t} \qquad (10.4)$$

◀ **Average angular acceleration**

In analogy to translational acceleration, the **instantaneous angular acceleration** is defined as the limit of the average angular acceleration as Δt approaches zero:

$$\alpha \equiv \lim_{\Delta t \to 0} \frac{\Delta \omega}{\Delta t} = \frac{d\omega}{dt} \qquad (10.5)$$

◀ **Instantaneous angular acceleration**

Angular acceleration has units of radians per second squared (rad/s^2), or simply s^{-2}. Notice that α is positive when a rigid object rotating counterclockwise is speeding up or when a rigid object rotating clockwise is slowing down during some time interval.

When a rigid object is rotating about a *fixed* axis, every particle on the object rotates through the same angle in a given time interval and has the same angular speed and the same angular acceleration. Therefore, like the angular position θ, the quantities ω and α characterize the rotational motion of the entire rigid object as well as individual particles in the object.

Angular position (θ), angular speed (ω), and angular acceleration (α) are analogous to translational position (x), translational speed (v), and translational acceleration (a). The variables θ, ω, and α differ dimensionally from the variables x, v, and a only by a factor having the unit of length. (See Section 10.3.)

We have not specified any direction for angular speed and angular acceleration. Strictly speaking, ω and α are the magnitudes of the angular velocity and the angular acceleration vectors[1] $\vec{\boldsymbol{\omega}}$ and $\vec{\boldsymbol{\alpha}}$, respectively, and they should always be positive. Because we are considering rotation about a fixed axis, however, we can use non-vector notation and indicate the vectors' directions by assigning a positive or negative sign to ω and α as discussed earlier with regard to Equations 10.3 and 10.5. For rotation about a fixed axis, the only direction that uniquely specifies the rotational motion is the direction along the axis of rotation. Therefore, the directions of $\vec{\boldsymbol{\omega}}$ and $\vec{\boldsymbol{\alpha}}$ are along this axis. If a particle rotates in the xy plane as in Figure 10.2, the

Pitfall Prevention 10.2
Specify Your Axis In solving rotation problems, you must specify an axis of rotation. This new feature does not exist in our study of translational motion. The choice is arbitrary, but once you make it, you must maintain that choice consistently throughout the problem. In some problems, the physical situation suggests a natural axis, such as one along the axle of an automobile wheel. In other problems, there may not be an obvious choice, and you must exercise judgment.

---

[1]Although we do not verify it here, the instantaneous angular velocity and instantaneous angular acceleration are vector quantities, but the corresponding average values are not because angular displacements do not add as vector quantities for finite rotations.

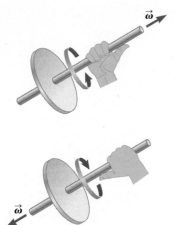

Figure 10.3 The right-hand rule for determining the direction of the angular velocity vector.

direction of $\vec{\omega}$ for the particle is out of the plane of the diagram when the rotation is counterclockwise and into the plane of the diagram when the rotation is clockwise. To illustrate this convention, it is convenient to use the *right-hand rule* demonstrated in Figure 10.3. When the four fingers of the right hand are wrapped in the direction of rotation, the extended right thumb points in the direction of $\vec{\omega}$. The direction of $\vec{\alpha}$ follows from its definition $\vec{\alpha} \equiv d\vec{\omega}/dt$. It is in the same direction as $\vec{\omega}$ if the angular speed is increasing in time, and it is antiparallel to $\vec{\omega}$ if the angular speed is decreasing in time.

10.2 Analysis Model: Rigid Object Under Constant Angular Acceleration

In our study of translational motion, after introducing the kinematic variables, we considered the special case of a particle under constant acceleration. We follow the same procedure here for a rigid object under constant angular acceleration.

Imagine a rigid object such as the CD in Figure 10.1 rotates about a fixed axis and has a constant angular acceleration. In parallel with our analysis model of the particle under constant acceleration, we generate a new analysis model for rotational motion called the **rigid object under constant angular acceleration.** We develop kinematic relationships for this model in this section. Writing Equation 10.5 in the form $d\omega = \alpha \, dt$ and integrating from $t_i = 0$ to $t_f = t$ gives

◀ Rotational kinematic equations

$$\omega_f = \omega_i + \alpha t \quad \text{(for constant } \alpha\text{)} \tag{10.6}$$

where ω_i is the angular speed of the rigid object at time $t = 0$. Equation 10.6 allows us to find the angular speed ω_f of the object at any later time t. Substituting Equation 10.6 into Equation 10.3 and integrating once more, we obtain

$$\theta_f = \theta_i + \omega_i t + \tfrac{1}{2}\alpha t^2 \quad \text{(for constant } \alpha\text{)} \tag{10.7}$$

where θ_i is the angular position of the rigid object at time $t = 0$. Equation 10.7 allows us to find the angular position θ_f of the object at any later time t. Eliminating t from Equations 10.6 and 10.7 gives

$$\omega_f^2 = \omega_i^2 + 2\alpha(\theta_f - \theta_i) \quad \text{(for constant } \alpha\text{)} \tag{10.8}$$

This equation allows us to find the angular speed ω_f of the rigid object for any value of its angular position θ_f. If we eliminate α between Equations 10.6 and 10.7, we obtain

$$\theta_f = \theta_i + \tfrac{1}{2}(\omega_i + \omega_f)t \quad \text{(for constant } \alpha\text{)} \tag{10.9}$$

Notice that these kinematic expressions for the rigid object under constant angular acceleration are of the same mathematical form as those for a particle under constant acceleration (Chapter 2). They can be generated from the equations for translational motion by making the substitutions $x \to \theta$, $v \to \omega$, and $a \to \alpha$. Table 10.1 compares the kinematic equations for the rigid object under constant angular acceleration and particle under constant acceleration models.

Pitfall Prevention 10.3

Just Like Translation? Equations 10.6 to 10.9 and Table 10.1 might suggest that rotational kinematics is just like translational kinematics. That is almost true, with two key differences. (1) In rotational kinematics, you must specify a rotation axis (per Pitfall Prevention 10.2). (2) In rotational motion, the object keeps returning to its original orientation; therefore, you may be asked for the number of revolutions made by a rigid object. This concept has no analog in translational motion.

Quick **Q**uiz 10.2 Consider again the pairs of angular positions for the rigid object in Quick Quiz 10.1. If the object starts from rest at the initial angular position, moves counterclockwise with constant angular acceleration, and arrives at the final angular position with the same angular speed in all three cases, for which choice is the angular acceleration the highest?

Table 10.1   Kinematic Equations for Rotational and Translational Motion

Rigid Object Under Constant Angular Acceleration		Particle Under Constant Acceleration	
$\omega_f = \omega_i + \alpha t$	(10.6)	$v_f = v_i + at$	(2.13)
$\theta_f = \theta_i + \omega_i t + \tfrac{1}{2}\alpha t^2$	(10.7)	$x_f = x_i + v_i t + \tfrac{1}{2}at^2$	(2.16)
$\omega_f^2 = \omega_i^2 + 2\alpha(\theta_f - \theta_i)$	(10.8)	$v_f^2 = v_i^2 + 2a(x_f - x_i)$	(2.17)
$\theta_f = \theta_i + \tfrac{1}{2}(\omega_i + \omega_f)t$	(10.9)	$x_f = x_i + \tfrac{1}{2}(v_i + v_f)t$	(2.15)

Analysis Model   Rigid Object Under Constant Angular Acceleration

Imagine an object that undergoes a spinning motion such that its angular acceleration is constant. The equations describing its angular position and angular speed are analogous to those for the particle under constant acceleration model:

$\alpha = $ constant

$$\omega_f = \omega_i + \alpha t \qquad \textbf{(10.6)}$$

$$\theta_f = \theta_i + \omega_i t + \tfrac{1}{2}\alpha t^2 \qquad \textbf{(10.7)}$$

$$\omega_f^2 = \omega_i^2 + 2\alpha(\theta_f - \theta_i) \qquad \textbf{(10.8)}$$

$$\theta_f = \theta_i + \tfrac{1}{2}(\omega_i + \omega_f)t \qquad \textbf{(10.9)}$$

Examples:

- during its spin cycle, the tub of a clothes washer begins from rest and accelerates up to its final spin speed
- a workshop grinding wheel is turned off and comes to rest under the action of a constant friction force in the bearings of the wheel
- a gyroscope is powered up and approaches its operating speed (Chapter 11)
- the crankshaft of a diesel engine changes to a higher angular speed (Chapter 22)

Example 10.1   Rotating Wheel   AM

A wheel rotates with a constant angular acceleration of 3.50 rad/s².

(A) If the angular speed of the wheel is 2.00 rad/s at $t_i = 0$, through what angular displacement does the wheel rotate in 2.00 s?

SOLUTION

Conceptualize Look again at Figure 10.1. Imagine that the compact disc rotates with its angular speed increasing at a constant rate. You start your stopwatch when the disc is rotating at 2.00 rad/s. This mental image is a model for the motion of the wheel in this example.

Categorize The phrase "with a constant angular acceleration" tells us to apply the *rigid object under constant angular acceleration* model to the wheel.

Analyze From the rigid object under constant angular acceleration model, choose Equation 10.7 and rearrange it so that it expresses the angular displacement of the wheel:

$$\Delta\theta = \theta_f - \theta_i = \omega_i t + \tfrac{1}{2}\alpha t^2$$

Substitute the known values to find the angular displacement at $t = 2.00$ s:

$$\Delta\theta = (2.00 \text{ rad/s})(2.00 \text{ s}) + \tfrac{1}{2}(3.50 \text{ rad/s}^2)(2.00 \text{ s})^2$$
$$= 11.0 \text{ rad} = (11.0 \text{ rad})(180°/\pi \text{ rad}) = 630°$$

(B) Through how many revolutions has the wheel turned during this time interval?

SOLUTION

Multiply the angular displacement found in part (A) by a conversion factor to find the number of revolutions:

$$\Delta\theta = 630°\left(\frac{1 \text{ rev}}{360°}\right) = 1.75 \text{ rev}$$

(C) What is the angular speed of the wheel at $t = 2.00$ s?

SOLUTION

Use Equation 10.6 from the rigid object under constant angular acceleration model to find the angular speed at $t = 2.00$ s:

$$\omega_f = \omega_i + \alpha t = 2.00 \text{ rad/s} + (3.50 \text{ rad/s}^2)(2.00 \text{ s})$$
$$= 9.00 \text{ rad/s}$$

Finalize We could also obtain this result using Equation 10.8 and the results of part (A). (Try it!)

WHAT IF? Suppose a particle moves along a straight line with a constant acceleration of 3.50 m/s². If the velocity of the particle is 2.00 m/s at $t_i = 0$, through what displacement does the particle move in 2.00 s? What is the velocity of the particle at $t = 2.00$ s?

continued

▶ **10.1** continued

Answer Notice that these questions are translational analogs to parts (A) and (C) of the original problem. The mathematical solution follows exactly the same form. For the displacement, from the particle under constant acceleration model,

$$\Delta x = x_f - x_i = v_i t + \tfrac{1}{2} a t^2$$

$$= (2.00 \text{ m/s})(2.00 \text{ s}) + \tfrac{1}{2}(3.50 \text{ m/s}^2)(2.00 \text{ s})^2 = 11.0 \text{ m}$$

and for the velocity,

$$v_f = v_i + at = 2.00 \text{ m/s} + (3.50 \text{ m/s}^2)(2.00 \text{ s}) = 9.00 \text{ m/s}$$

There is no translational analog to part (B) because translational motion under constant acceleration is not repetitive.

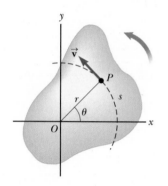

Figure 10.4 As a rigid object rotates about the fixed axis (the *z* axis) through *O*, the point *P* has a tangential velocity $\vec{v}$ that is always tangent to the circular path of radius *r*.

10.3 Angular and Translational Quantities

In this section, we derive some useful relationships between the angular speed and acceleration of a rotating rigid object and the translational speed and acceleration of a point in the object. To do so, we must keep in mind that when a rigid object rotates about a fixed axis as in Figure 10.4, every particle of the object moves in a circle whose center is on the axis of rotation.

Because point *P* in Figure 10.4 moves in a circle, the translational velocity vector $\vec{v}$ is always tangent to the circular path and hence is called *tangential velocity*. The magnitude of the tangential velocity of the point *P* is by definition the tangential speed $v = ds/dt$, where *s* is the distance traveled by this point measured along the circular path. Recalling that $s = r\theta$ (Eq. 10.1a) and noting that *r* is constant, we obtain

$$v = \frac{ds}{dt} = r\frac{d\theta}{dt}$$

Because $d\theta/dt = \omega$ (see Eq. 10.3), it follows that

▶ **Relation between tangential velocity and angular velocity**

$$v = r\omega \tag{10.10}$$

As we saw in Equation 4.17, the tangential speed of a point on a rotating rigid object equals the perpendicular distance of that point from the axis of rotation multiplied by the angular speed. Therefore, although every point on the rigid object has the same *angular* speed, not every point has the same *tangential* speed because *r* is not the same for all points on the object. Equation 10.10 shows that the tangential speed of a point on the rotating object increases as one moves outward from the center of rotation, as we would intuitively expect. For example, the outer end of a swinging golf club moves much faster than a point near the handle.

We can relate the angular acceleration of the rotating rigid object to the tangential acceleration of the point *P* by taking the time derivative of *v*:

$$a_t = \frac{dv}{dt} = r\frac{d\omega}{dt}$$

▶ **Relation between tangential acceleration and angular acceleration**

$$a_t = r\alpha \tag{10.11}$$

That is, the tangential component of the translational acceleration of a point on a rotating rigid object equals the point's perpendicular distance from the axis of rotation multiplied by the angular acceleration.

In Section 4.4, we found that a point moving in a circular path undergoes a radial acceleration a_r directed toward the center of rotation and whose magnitude is that of the centripetal acceleration v^2/r (Fig. 10.5). Because $v = r\omega$ for a point

P on a rotating object, we can express the centripetal acceleration at that point in terms of angular speed as we did in Equation 4.18:

$$a_c = \frac{v^2}{r} = r\omega^2 \qquad (10.12)$$

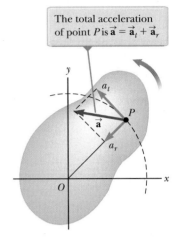

The total acceleration of point *P* is $\vec{a} = \vec{a}_t + \vec{a}_r$

The total acceleration vector at the point is $\vec{a} = \vec{a}_t + \vec{a}_r$, where the magnitude of $\vec{a}_r$ is the centripetal acceleration a_c. Because $\vec{a}$ is a vector having a radial and a tangential component, the magnitude of $\vec{a}$ at the point *P* on the rotating rigid object is

$$a = \sqrt{a_t^2 + a_r^2} = \sqrt{r^2\alpha^2 + r^2\omega^4} = r\sqrt{\alpha^2 + \omega^4} \qquad (10.13)$$

Quick Quiz 10.3 Ethan and Joseph are riding on a merry-go-round. Ethan rides on a horse at the outer rim of the circular platform, twice as far from the center of the circular platform as Joseph, who rides on an inner horse. **(i)** When the merry-go-round is rotating at a constant angular speed, what is Ethan's angular speed? (a) twice Joseph's (b) the same as Joseph's (c) half of Joseph's (d) impossible to determine **(ii)** When the merry-go-round is rotating at a constant angular speed, describe Ethan's tangential speed from the same list of choices.

Figure 10.5 As a rigid object rotates about a fixed axis (the *z* axis) through *O*, the point *P* experiences a tangential component of translational acceleration a_t and a radial component of translational acceleration a_r.

Example 10.2 **CD Player** AM

On a compact disc (Fig. 10.6), audio information is stored digitally in a series of pits and flat areas on the surface of the disc. The alternations between pits and flat areas on the surface represent binary ones and zeros to be read by the CD player and converted back to sound waves. The pits and flat areas are detected by a system consisting of a laser and lenses. The length of a string of ones and zeros representing one piece of information is the same everywhere on the disc, whether the information is near the center of the disc or near its outer edge. So that this length of ones and zeros always passes by the laser–lens system in the same time interval, the tangential speed of the disc surface at the location of the lens must be constant. According to Equation 10.10, the angular speed must therefore vary as the laser–lens system moves radially along the disc. In a typical CD player, the constant speed of the surface at the point of the laser–lens system is 1.3 m/s.

(A) Find the angular speed of the disc in revolutions per minute when information is being read from the innermost first track (*r* = 23 mm) and the outermost final track (*r* = 58 mm).

SOLUTION

Conceptualize Figure 10.6 shows a photograph of a compact disc. Trace your finger around the circle marked "23 mm" and mentally estimate the time interval to go around the circle once. Now trace your finger around the circle marked "58 mm," moving your finger across the surface of the page at the same speed as you did when tracing the smaller circle. Notice how much longer in time it takes your finger to go around the larger circle. If your finger represents the laser reading the disc, you can see that the disc rotates once in a longer time interval when the laser reads the information in the outer circle. Therefore, the disc must rotate more slowly when the laser is reading information from this part of the disc.

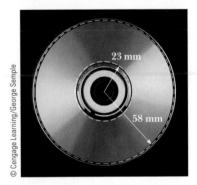

© Cengage Learning/George Semple

Figure 10.6 (Example 10.2) A compact disc.

Categorize This part of the example is categorized as a simple substitution problem. In later parts, we will need to identify analysis models.

Use Equation 10.10 to find the angular speed that gives the required tangential speed at the position of the inner track:

$$\omega_i = \frac{v}{r_i} = \frac{1.3 \text{ m/s}}{2.3 \times 10^{-2} \text{ m}} = 57 \text{ rad/s}$$

$$= (57 \text{ rad/s})\left(\frac{1 \text{ rev}}{2\pi \text{ rad}}\right)\left(\frac{60 \text{ s}}{1 \text{ min}}\right) = \boxed{5.4 \times 10^2 \text{ rev/min}}$$

continued

▶ **10.2** continued

Do the same for the outer track:

$$\omega_f = \frac{v}{r_f} = \frac{1.3 \text{ m/s}}{5.8 \times 10^{-2} \text{ m}} = 22 \text{ rad/s} = \boxed{2.1 \times 10^2 \text{ rev/min}}$$

The CD player adjusts the angular speed ω of the disc within this range so that information moves past the objective lens at a constant rate.

(B) The maximum playing time of a standard music disc is 74 min and 33 s. How many revolutions does the disc make during that time?

SOLUTION

Categorize From part (A), the angular speed decreases as the disc plays. Let us assume it decreases steadily, with α constant. We can then apply the *rigid object under constant angular acceleration* model to the disc.

Analyze If $t = 0$ is the instant the disc begins rotating, with angular speed of 57 rad/s, the final value of the time t is $(74 \text{ min})(60 \text{ s/min}) + 33 \text{ s} = 4\,473 \text{ s}$. We are looking for the angular displacement $\Delta\theta$ during this time interval.

Use Equation 10.9 to find the angular displacement of the disc at $t = 4\,473$ s:

$$\Delta\theta = \theta_f - \theta_i = \tfrac{1}{2}(\omega_i + \omega_f)t$$
$$= \tfrac{1}{2}(57 \text{ rad/s} + 22 \text{ rad/s})(4\,473 \text{ s}) = 1.8 \times 10^5 \text{ rad}$$

Convert this angular displacement to revolutions:

$$\Delta\theta = (1.8 \times 10^5 \text{ rad})\left(\frac{1 \text{ rev}}{2\pi \text{ rad}}\right) = \boxed{2.8 \times 10^4 \text{ rev}}$$

(C) What is the angular acceleration of the compact disc over the 4 473-s time interval?

SOLUTION

Categorize We again model the disc as a *rigid object under constant angular acceleration*. In this case, Equation 10.6 gives the value of the constant angular acceleration. Another approach is to use Equation 10.4 to find the average angular acceleration. In this case, we are not assuming the angular acceleration is constant. The answer is the same from both equations; only the interpretation of the result is different.

Analyze Use Equation 10.6 to find the angular acceleration:

$$\alpha = \frac{\omega_f - \omega_i}{t} = \frac{22 \text{ rad/s} - 57 \text{ rad/s}}{4\,473 \text{ s}} = \boxed{-7.6 \times 10^{-3} \text{ rad/s}^2}$$

Finalize The disc experiences a very gradual decrease in its rotation rate, as expected from the long time interval required for the angular speed to change from the initial value to the final value. In reality, the angular acceleration of the disc is not constant. Problem 90 in Enhanced WebAssign allows you to explore the actual time behavior of the angular acceleration.

The component $F \sin \phi$ tends to rotate the wrench about an axis through O.

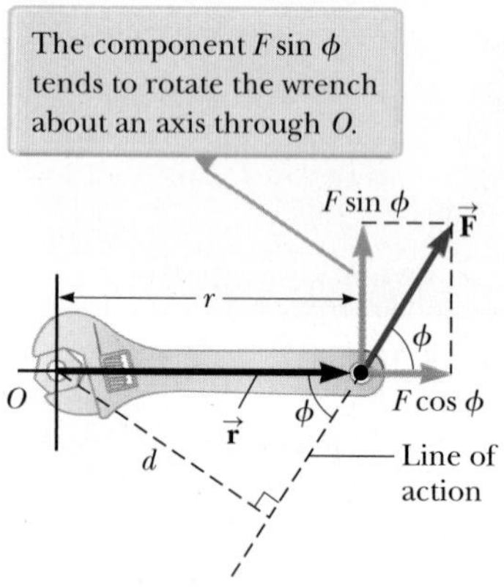

Figure 10.7 The force $\vec{F}$ has a greater rotating tendency about an axis through O as F increases and as the moment arm d increases.

10.4 Torque

In our study of translational motion, after investigating the description of motion, we studied the cause of changes in motion: force. We follow the same plan here: What is the cause of changes in rotational motion?

Imagine trying to rotate a door by applying a force of magnitude F perpendicular to the door surface near the hinges and then at various distances from the hinges. You will achieve a more rapid rate of rotation for the door by applying the force near the doorknob than by applying it near the hinges.

When a force is exerted on a rigid object pivoted about an axis, the object tends to rotate about that axis. The tendency of a force to rotate an object about some axis is measured by a quantity called **torque** $\vec{\tau}$ (Greek letter tau). Torque is a vector, but we will consider only its magnitude here; we will explore its vector nature in Chapter 11.

Consider the wrench in Figure 10.7 that we wish to rotate around an axis that is perpendicular to the page and passes through the center of the bolt. The applied force $\vec{F}$ acts at an angle ϕ to the horizontal. We define the magnitude of the torque associated with the force $\vec{F}$ around the axis passing through O by the expression

$$\tau \equiv rF \sin \phi = Fd \qquad (10.14)$$

where r is the distance between the rotation axis and the point of application of $\vec{F}$, and d is the perpendicular distance from the rotation axis to the line of action of $\vec{F}$. (The *line of action* of a force is an imaginary line extending out both ends of the vector representing the force. The dashed line extending from the tail of $\vec{F}$ in Fig. 10.7 is part of the line of action of $\vec{F}$.) From the right triangle in Figure 10.7 that has the wrench as its hypotenuse, we see that $d = r \sin \phi$. The quantity d is called the **moment arm** (or *lever arm*) of $\vec{F}$.

In Figure 10.7, the only component of $\vec{F}$ that tends to cause rotation of the wrench around an axis through O is $F \sin \phi$, the component perpendicular to a line drawn from the rotation axis to the point of application of the force. The horizontal component $F \cos \phi$, because its line of action passes through O, has no tendency to produce rotation about an axis passing through O. From the definition of torque, the rotating tendency increases as F increases and as d increases, which explains why it is easier to rotate a door if we push at the doorknob rather than at a point close to the hinges. We also want to apply our push as closely perpendicular to the door as we can so that ϕ is close to 90°. Pushing sideways on the doorknob ($\phi = 0$) will not cause the door to rotate.

If two or more forces act on a rigid object as in Figure 10.8, each tends to produce rotation about the axis through O. In this example, $\vec{F}_2$ tends to rotate the object clockwise and $\vec{F}_1$ tends to rotate it counterclockwise. We use the convention that the sign of the torque resulting from a force is positive if the turning tendency of the force is counterclockwise and negative if the turning tendency is clockwise. For Example, in Figure 10.8, the torque resulting from $\vec{F}_1$, which has a moment arm d_1, is positive and equal to $+F_1 d_1$; the torque from $\vec{F}_2$ is negative and equal to $-F_2 d_2$. Hence, the *net* torque about an axis through O is

$$\sum \tau = \tau_1 + \tau_2 = F_1 d_1 - F_2 d_2$$

Torque should not be confused with force. Forces can cause a change in translational motion as described by Newton's second law. Forces can also cause a change in rotational motion, but the effectiveness of the forces in causing this change depends on both the magnitudes of the forces and the moment arms of the forces, in the combination we call *torque*. Torque has units of force times length—newton meters (N · m) in SI units—and should be reported in these units. Do not confuse torque and work, which have the same units but are very different concepts.

Quick Quiz 10.4 (i) If you are trying to loosen a stubborn screw from a piece of wood with a screwdriver and fail, should you find a screwdriver for which the handle is (a) longer or (b) fatter? (ii) If you are trying to loosen a stubborn bolt from a piece of metal with a wrench and fail, should you find a wrench for which the handle is (a) longer or (b) fatter?

Example 10.3	The Net Torque on a Cylinder

A one-piece cylinder is shaped as shown in Figure 10.9, with a core section protruding from the larger drum. The cylinder is free to rotate about the central z axis shown in the drawing. A rope wrapped around the drum, which has radius R_1, exerts a force $\vec{T}_1$ to the right on the cylinder. A rope wrapped around the core, which has radius R_2, exerts a force $\vec{T}_2$ downward on the cylinder.

(A) What is the net torque acting on the cylinder about the rotation axis (which is the z axis in Fig. 10.9)?

continued

Unless otherwise noted, all content on this page is © Cengage Learning.

◀ **Moment arm**

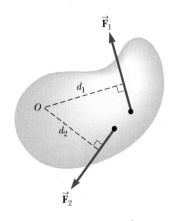

Figure 10.8 The force $\vec{F}_1$ tends to rotate the object counterclockwise about an axis through O, and $\vec{F}_2$ tends to rotate it clockwise.

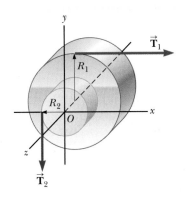

Figure 10.9 (Example 10.3) A solid cylinder pivoted about the z axis through O. The moment arm of $\vec{T}_1$ is R_1, and the moment arm of $\vec{T}_2$ is R_2.

▶ **10.3** continued

SOLUTION

Conceptualize Imagine that the cylinder in Figure 10.9 is a shaft in a machine. The force $\vec{T}_1$ could be applied by a drive belt wrapped around the drum. The force $\vec{T}_2$ could be applied by a friction brake at the surface of the core.

Categorize This example is a substitution problem in which we evaluate the net torque using Equation 10.14.

The torque due to $\vec{T}_1$ about the rotation axis is $-R_1 T_1$. (The sign is negative because the torque tends to produce clockwise rotation.) The torque due to $\vec{T}_2$ is $+R_2 T_2$. (The sign is positive because the torque tends to produce counterclockwise rotation of the cylinder.)

Evaluate the net torque about the rotation axis:

$$\sum \tau = \tau_1 + \tau_2 = \boxed{R_2 T_2 - R_1 T_1}$$

As a quick check, notice that if the two forces are of equal magnitude, the net torque is negative because $R_1 > R_2$. Starting from rest with both forces of equal magnitude acting on it, the cylinder would rotate clockwise because $\vec{T}_1$ would be more effective at turning it than would $\vec{T}_2$.

(B) Suppose $T_1 = 5.0$ N, $R_1 = 1.0$ m, $T_2 = 15$ N, and $R_2 = 0.50$ m. What is the net torque about the rotation axis, and which way does the cylinder rotate starting from rest?

SOLUTION

Substitute the given values:

$$\sum \tau = (0.50 \text{ m})(15 \text{ N}) - (1.0 \text{ m})(5.0 \text{ N}) = \boxed{2.5 \text{ N} \cdot \text{m}}$$

Because this net torque is positive, the cylinder begins to rotate in the counterclockwise direction.

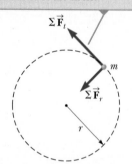

The tangential force on the particle results in a torque on the particle about an axis through the center of the circle.

$\sum \vec{F}_t$

m

$\sum \vec{F}_r$

r

Figure 10.10 A particle rotating in a circle under the influence of a tangential net force $\sum \vec{F}_t$. A radial net force $\sum \vec{F}_r$ also must be present to maintain the circular motion.

10.5 Analysis Model: Rigid Object Under a Net Torque

In Chapter 5, we learned that a net force on an object causes an acceleration of the object and that the acceleration is proportional to the net force. These facts are the basis of the particle under a net force model whose mathematical representation is Newton's second law. In this section, we show the rotational analog of Newton's second law: the angular acceleration of a rigid object rotating about a fixed axis is proportional to the net torque acting about that axis. Before discussing the more complex case of rigid-object rotation, however, it is instructive first to discuss the case of a particle moving in a circular path about some fixed point under the influence of an external force.

Consider a particle of mass m rotating in a circle of radius r under the influence of a tangential net force $\sum \vec{F}_t$ and a radial net force $\sum \vec{F}_r$ as shown in Figure 10.10. The radial net force causes the particle to move in the circular path with a centripetal acceleration. The tangential force provides a tangential acceleration $\vec{a}_t$, and

$$\sum F_t = ma_t$$

The magnitude of the net torque due to $\sum \vec{F}_t$ on the particle about an axis perpendicular to the page through the center of the circle is

$$\sum \tau = \sum F_t r = (ma_t)r$$

Because the tangential acceleration is related to the angular acceleration through the relationship $a_t = r\alpha$ (Eq. 10.11), the net torque can be expressed as

$$\sum \tau = (mr\alpha)r = (mr^2)\alpha \tag{10.15}$$

Let us denote the quantity mr^2 with the symbol I for now. We will say more about this quantity below. Using this notation, Equation 10.15 can be written as

$$\sum \tau = I\alpha \tag{10.16}$$

That is, the net torque acting on the particle is proportional to its angular acceleration. Notice that $\sum \tau = I\alpha$ has the same mathematical form as Newton's second law of motion, $\sum F = ma$.

Now let us extend this discussion to a rigid object of arbitrary shape rotating about a fixed axis passing through a point O as in Figure 10.11. The object can be regarded as a collection of particles of mass m_i. If we impose a Cartesian coordinate system on the object, each particle rotates in a circle about the origin and each has a tangential acceleration a_i produced by an external tangential force of magnitude F_i. For any given particle, we know from Newton's second law that

$$F_i = m_i a_i$$

The external torque $\vec{\tau}_i$ associated with the force $\vec{F}_i$ acts about the origin and its magnitude is given by

$$\tau_i = r_i F_i = r_i m_i a_i$$

Because $a_i = r_i \alpha$, the expression for τ_i becomes

$$\tau_i = m_i r_i^2 \alpha$$

Although each particle in the rigid object may have a different translational acceleration a_i, they all have the *same* angular acceleration α. With that in mind, we can add the torques on all of the particles making up the rigid object to obtain the net torque on the object about an axis through O due to all external forces:

$$\sum \tau_{\text{ext}} = \sum_i \tau_i = \sum_i m_i r_i^2 \alpha = \left(\sum_i m_i r_i^2 \right) \alpha \qquad \textbf{(10.17)}$$

where α can be taken outside the summation because it is common to all particles. Calling the quantity in parentheses I, the expression for $\sum \tau_{\text{ext}}$ becomes

$$\sum \tau_{\text{ext}} = I\alpha \qquad \textbf{(10.18)}$$

This equation for a rigid object is the same as that found for a particle moving in a circular path (Eq. 10.16). The net torque about the rotation axis is proportional to the angular acceleration of the object, with the proportionality factor being I, a quantity that we have yet to describe fully. Equation 10.18 is the mathematical representation of the analysis model of a **rigid object under a net torque,** the rotational analog to the particle under a net force.

Let us now address the quantity I, defined as follows:

$$I = \sum_i m_i r_i^2 \qquad \textbf{(10.19)}$$

This quantity is called the **moment of inertia** of the object, and depends on the masses of the particles making up the object and their distances from the rotation axis. Notice that Equation 10.19 reduces to $I = mr^2$ for a single particle, consistent with our use of the notation I that we used in going from Equation 10.15 to Equation 10.16. Note that moment of inertia has units of $\text{kg} \cdot \text{m}^2$ in SI units.

Equation 10.18 has the same form as Newton's second law for a system of particles as expressed in Equation 9.39:

$$\sum \vec{F}_{\text{ext}} = M\vec{a}_{\text{CM}}$$

Consequently, the moment of inertia I must play the same role in rotational motion as the role that mass plays in translational motion: the moment of inertia is the resistance to changes in rotational motion. This resistance depends not only on the mass of the object, but also on how the mass is distributed around the rotation axis. Table 10.2 on page 244 gives the moments of inertia[2] for a number of objects about specific axes. The moments of inertia of rigid objects with simple geometry (high symmetry) are relatively easy to calculate provided the rotation axis coincides with an axis of symmetry, as we show in the next section.

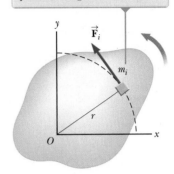

The particle of mass m_i of the rigid object experiences a torque in the same way that the particle in Figure 10.10 does.

Figure 10.11 A rigid object rotating about an axis through O. Each particle of mass m_i rotates about the axis with the same angular acceleration α.

◀ Torque on a rigid object is proportional to angular acceleration

Pitfall Prevention 10.5
No Single Moment of Inertia
There is one major difference between mass and moment of inertia. Mass is an inherent property of an object. The moment of inertia of an object depends on your choice of rotation axis. Therefore, there is no single value of the moment of inertia for an object. There is a *minimum* value of the moment of inertia, which is that calculated about an axis passing through the center of mass of the object.

---

[2]Civil engineers use moment of inertia to characterize the elastic properties (rigidity) of such structures as loaded beams. Hence, it is often useful even in a nonrotational context.

| Table 10.2 | **Moments of Inertia of Homogeneous Rigid Objects with Different Geometries** |

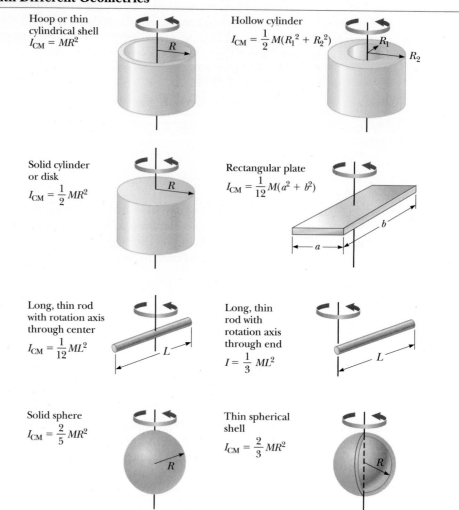

Hoop or thin cylindrical shell
$I_{CM} = MR^2$

Hollow cylinder
$I_{CM} = \frac{1}{2}M(R_1^2 + R_2^2)$

Solid cylinder or disk
$I_{CM} = \frac{1}{2}MR^2$

Rectangular plate
$I_{CM} = \frac{1}{12}M(a^2 + b^2)$

Long, thin rod with rotation axis through center
$I_{CM} = \frac{1}{12}ML^2$

Long, thin rod with rotation axis through end
$I = \frac{1}{3}ML^2$

Solid sphere
$I_{CM} = \frac{2}{5}MR^2$

Thin spherical shell
$I_{CM} = \frac{2}{3}MR^2$

> ⓠuick Quiz 10.5 You turn off your electric drill and find that the time interval
> for the rotating bit to come to rest due to frictional torque in the drill is Δt. You
> replace the bit with a larger one that results in a doubling of the moment of
> inertia of the drill's entire rotating mechanism. When this larger bit is rotated
> at the same angular speed as the first and the drill is turned off, the frictional
> torque remains the same as that for the previous situation. What is the time
> interval for this second bit to come to rest? (a) $4\Delta t$ (b) $2\Delta t$ (c) Δt (d) $0.5\Delta t$
> (e) $0.25\Delta t$ (f) impossible to determine

Analysis Model **Rigid Object Under a Net Torque**

Imagine you are analyzing the motion of an object that is free to rotate about a fixed axis. The cause
of changes in rotational motion of this object is torque applied to the object and, in parallel to Newton's second law for translation motion, the torque is equal to the product of the moment of inertia of
the object and the angular acceleration:

$$\sum \tau_{ext} = I\alpha \qquad \text{(10.18)}$$

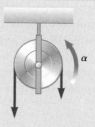

The torque, the moment of inertia, and the angular acceleration must all be evaluated around the
same rotation axis.

| Analysis Model | Rigid Object Under a Net Torque *(continued)* |

Examples:

- a bicycle chain around the sprocket of a bicycle causes the rear wheel of the bicycle to rotate
- an electric dipole moment in an electric field rotates due to the electric force from the field (Chapter 23)
- a magnetic dipole moment in a magnetic field rotates due to the magnetic force from the field (Chapter 30)
- the armature of a motor rotates due to the torque exerted by a surrounding magnetic field (Chapter 31)

| Example 10.4 | Rotating Rod | AM |

A uniform rod of length L and mass M is attached at one end to a frictionless pivot and is free to rotate about the pivot in the vertical plane as in Figure 10.12. The rod is released from rest in the horizontal position. What are the initial angular acceleration of the rod and the initial translational acceleration of its right end?

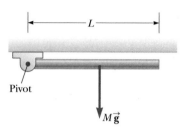

Figure 10.12 (Example 10.4) A rod is free to rotate around a pivot at the left end. The gravitational force on the rod acts at its center of mass.

SOLUTION

Conceptualize Imagine what happens to the rod in Figure 10.12 when it is released. It rotates clockwise around the pivot at the left end.

Categorize The rod is categorized as a *rigid object under a net torque*. The torque is due only to the gravitational force on the rod if the rotation axis is chosen to pass through the pivot in Figure 10.12. We *cannot* categorize the rod as a rigid object under constant angular acceleration because the torque exerted on the rod and therefore the angular acceleration of the rod vary with its angular position.

Analyze The only force contributing to the torque about an axis through the pivot is the gravitational force $M\vec{g}$ exerted on the rod. (The force exerted by the pivot on the rod has zero torque about the pivot because its moment arm is zero.) To compute the torque on the rod, we assume the gravitational force acts at the center of mass of the rod as shown in Figure 10.12.

Write an expression for the magnitude of the net external torque due to the gravitational force about an axis through the pivot:

$$\sum \tau_{ext} = Mg\left(\frac{L}{2}\right)$$

Use Equation 10.18 to obtain the angular acceleration of the rod, using the moment of inertia for the rod from Table 10.2:

$$(1) \quad \alpha = \frac{\sum \tau_{ext}}{I} = \frac{Mg(L/2)}{\frac{1}{3}ML^2} = \boxed{\frac{3g}{2L}}$$

Use Equation 10.11 with $r = L$ to find the initial translational acceleration of the right end of the rod:

$$a_t = L\alpha = \boxed{\tfrac{3}{2}g}$$

Finalize These values are the *initial* values of the angular and translational accelerations. Once the rod begins to rotate, the gravitational force is no longer perpendicular to the rod and the values of the two accelerations decrease, going to zero at the moment the rod passes through the vertical orientation.

WHAT IF? What if we were to place a penny on the end of the rod and then release the rod? Would the penny stay in contact with the rod?

Answer The result for the initial acceleration of a point on the end of the rod shows that $a_t > g$. An unsupported penny falls at acceleration g. So, if we place a penny on the end of the rod and then release the rod, the end of the rod falls faster than the penny does! The penny does not stay in contact with the rod. (Try this with a penny and a meterstick!)

The question now is to find the location on the rod at which we can place a penny that *will* stay in contact as both begin to fall. To find the translational acceleration of an arbitrary point on the rod at a distance $r < L$ from the pivot point, we combine Equation (1) with Equation 10.11:

$$a_t = r\alpha = \frac{3g}{2L}r$$

continued

▶ **10.4** continued

For the penny to stay in contact with the rod, the limiting case is that the translational acceleration must be equal to that due to gravity:

$$a_t = g = \frac{3g}{2L}\, r$$

$$r = \tfrac{2}{3}L$$

Therefore, a penny placed closer to the pivot than two-thirds of the length of the rod stays in contact with the falling rod, but a penny farther out than this point loses contact.

Conceptual Example 10.5   **Falling Smokestacks and Tumbling Blocks**

When a tall smokestack falls over, it often breaks somewhere along its length before it hits the ground as shown in Figure 10.13. Why?

SOLUTION

As the smokestack rotates around its base, each higher portion of the smokestack falls with a larger tangential acceleration than the portion below it according to Equation 10.11. The angular acceleration increases as the smokestack tips farther. Eventually, higher portions of the smokestack experience an acceleration greater than the acceleration that could result from gravity alone; this situation is similar to that described in Example 10.4. That can happen only if these portions are being pulled downward by a force in addition to the gravitational force. The force that causes that to occur is the shear force from lower portions of the smokestack. Eventually, the shear force that provides this acceleration is greater than the smokestack can withstand, and the smokestack breaks. The same thing happens with a tall tower of children's toy blocks. Borrow some blocks from a child and build such a tower. Push it over and watch it come apart at some point before it strikes the floor.

Figure 10.13 (Conceptual Example 10.5) A falling smokestack breaks at some point along its length.

Example 10.6   **Angular Acceleration of a Wheel**   **AM**

A wheel of radius R, mass M, and moment of inertia I is mounted on a frictionless, horizontal axle as in Figure 10.14. A light cord wrapped around the wheel supports an object of mass m. When the wheel is released, the object accelerates downward, the cord unwraps off the wheel, and the wheel rotates with an angular acceleration. Find expressions for the angular acceleration of the wheel, the translational acceleration of the object, and the tension in the cord.

SOLUTION

Conceptualize Imagine that the object is a bucket in an old-fashioned water well. It is tied to a cord that passes around a cylinder equipped with a crank for raising the bucket. After the bucket has been raised, the system is released and the bucket accelerates downward while the cord unwinds off the cylinder.

Categorize We apply two analysis models here. The object is modeled as a *particle under a net force*. The wheel is modeled as a *rigid object under a net torque*.

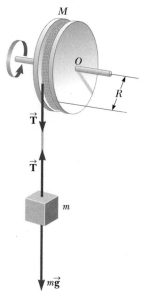

Figure 10.14 (Example 10.6) An object hangs from a cord wrapped around a wheel.

Analyze The magnitude of the torque acting on the wheel about its axis of rotation is $\tau = TR$, where T is the force exerted by the cord on the rim of the wheel. (The gravitational force exerted by the Earth on the wheel and the

▶ **10.6** continued

normal force exerted by the axle on the wheel both pass through the axis of rotation and therefore produce no torque.)

From the rigid object under a net torque model, write Equation 10.18:

$$\sum \tau_{\text{ext}} = I\alpha$$

Solve for α and substitute the net torque:

$$(1)\quad \alpha = \frac{\sum \tau_{\text{ext}}}{I} = \frac{TR}{I}$$

From the particle under a net force model, apply Newton's second law to the motion of the object, taking the downward direction to be positive:

$$\sum F_y = mg - T = ma$$

Solve for the acceleration a:

$$(2)\quad a = \frac{mg - T}{m}$$

Equations (1) and (2) have three unknowns: α, a, and T. Because the object and wheel are connected by a cord that does not slip, the translational acceleration of the suspended object is equal to the tangential acceleration of a point on the wheel's rim. Therefore, the angular acceleration α of the wheel and the translational acceleration of the object are related by $a = R\alpha$.

Use this fact together with Equations (1) and (2):

$$(3)\quad a = R\alpha = \frac{TR^2}{I} = \frac{mg - T}{m}$$

Solve for the tension T:

$$(4)\quad T = \frac{mg}{1 + (mR^2/I)}$$

Substitute Equation (4) into Equation (2) and solve for a:

$$(5)\quad a = \frac{g}{1 + (I/mR^2)}$$

Use $a = R\alpha$ and Equation (5) to solve for α:

$$\alpha = \frac{a}{R} = \frac{g}{R + (I/mR)}$$

Finalize We finalize this problem by imagining the behavior of the system in some extreme limits.

WHAT IF? What if the wheel were to become very massive so that I becomes very large? What happens to the acceleration a of the object and the tension T?

Answer If the wheel becomes infinitely massive, we can imagine that the object of mass m will simply hang from the cord without causing the wheel to rotate.

We can show that mathematically by taking the limit $I \to \infty$. Equation (5) then becomes

$$a = \frac{g}{1 + (I/mR^2)} \to 0$$

which agrees with our conceptual conclusion that the object will hang at rest. Also, Equation (4) becomes

$$T = \frac{mg}{1 + (mR^2/I)} \to mg$$

which is consistent because the object simply hangs at rest in equilibrium between the gravitational force and the tension in the string.

10.6 Calculation of Moments of Inertia

The moment of inertia of a system of discrete particles can be calculated in a straightforward way with Equation 10.19. We can evaluate the moment of inertia of a continuous rigid object by imagining the object to be divided into many small elements, each of which has mass Δm_i. We use the definition $I = \sum_i r_i^2 \Delta m_i$

and take the limit of this sum as $\Delta m_i \to 0$. In this limit, the sum becomes an integral over the volume of the object:

Moment of inertia ▶
of a rigid object

$$I = \lim_{\Delta m_i \to 0} \sum_i r_i^2 \, \Delta m_i = \int r^2 \, dm \tag{10.20}$$

It is usually easier to calculate moments of inertia in terms of the volume of the elements rather than their mass, and we can easily make that change by using Equation 1.1, $\rho \equiv m/V$, where ρ is the density of the object and V is its volume. From this equation, the mass of a small element is $dm = \rho \, dV$. Substituting this result into Equation 10.20 gives

$$I = \int \rho r^2 \, dV \tag{10.21}$$

If the object is homogeneous, ρ is constant and the integral can be evaluated for a known geometry. If ρ is not constant, its variation with position must be known to complete the integration.

The density given by $\rho = m/V$ sometimes is referred to as *volumetric mass density* because it represents mass per unit volume. Often we use other ways of expressing density. For instance, when dealing with a sheet of uniform thickness t, we can define a *surface mass density* $\sigma = \rho t$, which represents *mass per unit area*. Finally, when mass is distributed along a rod of uniform cross-sectional area A, we sometimes use *linear mass density* $\lambda = M/L = \rho A$, which is the *mass per unit length*.

Example 10.7   Uniform Rigid Rod

Calculate the moment of inertia of a uniform thin rod of length L and mass M (Fig. 10.15) about an axis perpendicular to the rod (the y' axis) and passing through its center of mass.

SOLUTION

Conceptualize Imagine twirling the rod in Figure 10.15 with your fingers around its midpoint. If you have a meterstick handy, use it to simulate the spinning of a thin rod and feel the resistance it offers to being spun.

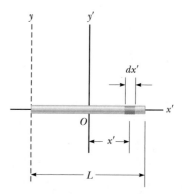

Figure 10.15 (Example 10.7) A uniform rigid rod of length L. The moment of inertia about the y' axis is less than that about the y axis. The latter axis is examined in Example 10.9.

Categorize This example is a substitution problem, using the definition of moment of inertia in Equation 10.20. As with any integration problem, the solution involves reducing the integrand to a single variable.

The shaded length element dx' in Figure 10.15 has a mass dm equal to the mass per unit length λ multiplied by dx'.

Express dm in terms of dx':

$$dm = \lambda \, dx' = \frac{M}{L} \, dx'$$

Substitute this expression into Equation 10.20, with $r^2 = (x')^2$:

$$I_{y'} = \int r^2 \, dm = \int_{-L/2}^{L/2} (x')^2 \frac{M}{L} \, dx' = \frac{M}{L} \int_{-L/2}^{L/2} (x')^2 \, dx'$$

$$= \frac{M}{L} \left[\frac{(x')^3}{3} \right]_{-L/2}^{L/2} = \tfrac{1}{12} M L^2$$

Check this result in Table 10.2.

Example 10.8   Uniform Solid Cylinder

A uniform solid cylinder has a radius R, mass M, and length L. Calculate its moment of inertia about its central axis (the z axis in Fig. 10.16).

▶ **10.8** continued

SOLUTION

Conceptualize To simulate this situation, imagine twirling a can of frozen juice around its central axis. Don't twirl a nonfrozen can of vegetable soup; it is not a rigid object! The liquid is able to move relative to the metal can.

Categorize This example is a substitution problem, using the definition of moment of inertia. As with Example 10.7, we must reduce the integrand to a single variable.

It is convenient to divide the cylinder into many cylindrical shells, each having radius r, thickness dr, and length L as shown in Figure 10.16. The density of the cylinder is ρ. The volume dV of each shell is its cross-sectional area multiplied by its length: $dV = L\,dA = L(2\pi r)\,dr$.

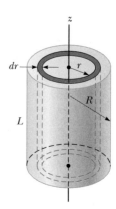

Figure 10.16 (Example 10.8) Calculating I about the z axis for a uniform solid cylinder.

Express dm in terms of dr:

$$dm = \rho\,dV = \rho L(2\pi r)\,dr$$

Substitute this expression into Equation 10.20:

$$I_z = \int r^2\,dm = \int r^2[\rho L(2\pi r)\,dr] = 2\pi\rho L\int_0^R r^3\,dr = \tfrac{1}{2}\pi\rho LR^4$$

Use the total volume $\pi R^2 L$ of the cylinder to express its density:

$$\rho = \frac{M}{V} = \frac{M}{\pi R^2 L}$$

Substitute this value into the expression for I_z:

$$I_z = \tfrac{1}{2}\pi\left(\frac{M}{\pi R^2 L}\right)LR^4 = \boxed{\tfrac{1}{2}MR^2}$$

Check this result in Table 10.2.

WHAT IF? What if the length of the cylinder in Figure 10.16 is increased to $2L$, while the mass M and radius R are held fixed? How does that change the moment of inertia of the cylinder?

Answer Notice that the result for the moment of inertia of a cylinder does not depend on L, the length of the cylinder. It applies equally well to a long cylinder and a flat disk having the same mass M and radius R. Therefore, the moment of inertia of the cylinder is not affected by how the mass is distributed along its length.

The calculation of moments of inertia of an object about an arbitrary axis can be cumbersome, even for a highly symmetric object. Fortunately, use of an important theorem, called the **parallel-axis theorem,** often simplifies the calculation.

To generate the parallel-axis theorem, suppose the object in Figure 10.17a on page 250 rotates about the z axis. The moment of inertia does not depend on how the mass is distributed along the z axis; as we found in Example 10.8, the moment of inertia of a cylinder is independent of its length. Imagine collapsing the three-dimensional object into a planar object as in Figure 10.17b. In this imaginary process, all mass moves parallel to the z axis until it lies in the xy plane. The coordinates of the object's center of mass are now x_{CM}, y_{CM}, and $z_{CM} = 0$. Let the mass element dm have coordinates $(x, y, 0)$ as shown in the view down the z axis in Figure 10.17c. Because this element is a distance $r = \sqrt{x^2 + y^2}$ from the z axis, the moment of inertia of the entire object about the z axis is

$$I = \int r^2\,dm = \int (x^2 + y^2)\,dm$$

We can relate the coordinates x, y of the mass element dm to the coordinates of this same element located in a coordinate system having the object's center of mass as its origin. If the coordinates of the center of mass are x_{CM}, y_{CM}, and $z_{CM} = 0$ in the original coordinate system centered on O, we see from Figure 10.17c that

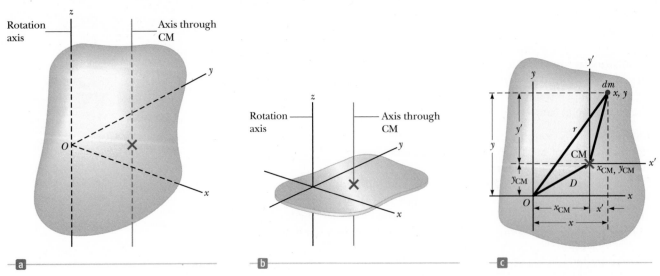

Figure 10.17 (a) An arbitrarily shaped rigid object. The origin of the coordinate system is not at the center of mass of the object. Imagine the object rotating about the z axis. (b) All mass elements of the object are collapsed parallel to the z axis to form a planar object. (c) An arbitrary mass element dm is indicated in blue in this view down the z axis. The parallel axis theorem can be used with the geometry shown to determine the moment of inertia of the original object around the z axis.

the relationships between the unprimed and primed coordinates are $x = x' + x_{CM}$, $y = y' + y_{CM}$, and $z = z' = 0$. Therefore,

$$I = \int [(x' + x_{CM})^2 + (y' + y_{CM})^2] \, dm$$

$$= \int [(x')^2 + (y')^2] \, dm + 2x_{CM} \int x' \, dm + 2y_{CM} \int y' \, dm + (x_{CM}^2 + y_{CM}^2) \int dm$$

The first integral is, by definition, the moment of inertia I_{CM} about an axis that is parallel to the z axis and passes through the center of mass. The second two integrals are zero because, by definition of the center of mass, $\int x' \, dm = \int y' \, dm = 0$. The last integral is simply MD^2 because $\int dm = M$ and $D^2 = x_{CM}^2 + y_{CM}^2$. Therefore, we conclude that

Parallel–axis theorem ▶

$$I = I_{CM} + MD^2 \tag{10.22}$$

Example 10.9   **Applying the Parallel-Axis Theorem**

Consider once again the uniform rigid rod of mass M and length L shown in Figure 10.15. Find the moment of inertia of the rod about an axis perpendicular to the rod through one end (the y axis in Fig. 10.15).

SOLUTION

Conceptualize Imagine twirling the rod around an endpoint rather than the midpoint. If you have a meterstick handy, try it and notice the degree of difficulty in rotating it around the end compared with rotating it around the center.

Categorize This example is a substitution problem, involving the parallel-axis theorem.

Intuitively, we expect the moment of inertia to be greater than the result $I_{CM} = \frac{1}{12}ML^2$ from Example 10.7 because there is mass up to a distance of L away from the rotation axis, whereas the farthest distance in Example 10.7 was only $L/2$. The distance between the center-of-mass axis and the y axis is $D = L/2$.

▶ **10.9** continued

Use the parallel-axis theorem:

$$I = I_{CM} + MD^2 = \tfrac{1}{12}ML^2 + M\left(\frac{L}{2}\right)^2 = \tfrac{1}{3}ML^2$$

Check this result in Table 10.2.

10.7 Rotational Kinetic Energy

After investigating the role of forces in our study of translational motion, we turned our attention to approaches involving energy. We do the same thing in our current study of rotational motion.

In Chapter 7, we defined the kinetic energy of an object as the energy associated with its motion through space. An object rotating about a fixed axis remains stationary in space, so there is no kinetic energy associated with translational motion. The individual particles making up the rotating object, however, are moving through space; they follow circular paths. Consequently, there is kinetic energy associated with rotational motion.

Let us consider an object as a system of particles and assume it rotates about a fixed z axis with an angular speed ω. Figure 10.18 shows the rotating object and identifies one particle on the object located at a distance r_i from the rotation axis. If the mass of the ith particle is m_i and its tangential speed is v_i, its kinetic energy is

$$K_i = \tfrac{1}{2}m_i v_i^2$$

To proceed further, recall that although every particle in the rigid object has the same angular speed ω, the individual tangential speeds depend on the distance r_i from the axis of rotation according to Equation 10.10. The *total* kinetic energy of the rotating rigid object is the sum of the kinetic energies of the individual particles:

$$K_R = \sum_i K_i = \sum_i \tfrac{1}{2}m_i v_i^2 = \tfrac{1}{2}\sum_i m_i r_i^2 \omega^2$$

We can write this expression in the form

$$K_R = \tfrac{1}{2}\left(\sum_i m_i r_i^2\right)\omega^2 \tag{10.23}$$

where we have factored ω^2 from the sum because it is common to every particle. We recognize the quantity in parentheses as the moment of inertia of the object, introduced in Section 10.5.

Therefore, Equation 10.23 can be written

$$K_R = \tfrac{1}{2}I\omega^2 \tag{10.24}$$

◀ **Rotational kinetic energy**

Although we commonly refer to the quantity $\tfrac{1}{2}I\omega^2$ as **rotational kinetic energy,** it is not a new form of energy. It is ordinary kinetic energy because it is derived from a sum over individual kinetic energies of the particles contained in the rigid object. The mathematical form of the kinetic energy given by Equation 10.24 is convenient when we are dealing with rotational motion, provided we know how to calculate I.

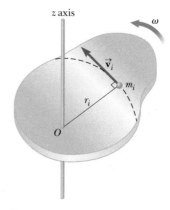

Figure 10.18 A rigid object rotating about the z axis with angular speed ω. The kinetic energy of the particle of mass m_i is $\tfrac{1}{2}m_i v_i^2$. The total kinetic energy of the object is called its rotational kinetic energy.

Q uick **Quiz** 10.6 A section of hollow pipe and a solid cylinder have the same radius, mass, and length. They both rotate about their long central axes with the same angular speed. Which object has the higher rotational kinetic energy? **(a)** The hollow pipe does. **(b)** The solid cylinder does. **(c)** They have the same rotational kinetic energy. **(d)** It is impossible to determine.

Example 10.10   An Unusual Baton

Four tiny spheres are fastened to the ends of two rods of negligible mass lying in the xy plane to form an unusual baton (Fig. 10.19). We shall assume the radii of the spheres are small compared with the dimensions of the rods.

(A) If the system rotates about the y axis (Fig. 10.19a) with an angular speed ω, find the moment of inertia and the rotational kinetic energy of the system about this axis.

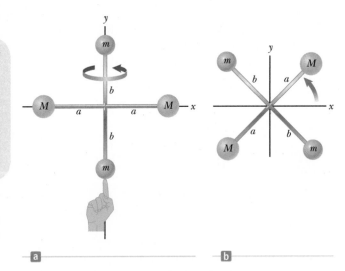

SOLUTION

Conceptualize Figure 10.19 is a pictorial representation that helps conceptualize the system of spheres and how it spins. Model the spheres as particles.

Categorize This example is a substitution problem because it is a straightforward application of the definitions discussed in this section.

Figure 10.19 (Example 10.10) Four spheres form an unusual baton. (a) The baton is rotated about the y axis. (b) The baton is rotated about the z axis.

Apply Equation 10.19 to the system:

$$I_y = \sum_i m_i r_i^2 = Ma^2 + Ma^2 = \boxed{2Ma^2}$$

Evaluate the rotational kinetic energy using Equation 10.24:

$$K_R = \tfrac{1}{2} I_y \omega^2 = \tfrac{1}{2}(2Ma^2)\omega^2 = \boxed{Ma^2\omega^2}$$

That the two spheres of mass m do not enter into this result makes sense because they have no motion about the axis of rotation; hence, they have no rotational kinetic energy. By similar logic, we expect the moment of inertia about the x axis to be $I_x = 2mb^2$ with a rotational kinetic energy about that axis of $K_R = mb^2\omega^2$.

(B) Suppose the system rotates in the xy plane about an axis (the z axis) through the center of the baton (Fig. 10.19b). Calculate the moment of inertia and rotational kinetic energy about this axis.

SOLUTION

Apply Equation 10.19 for this new rotation axis:

$$I_z = \sum_i m_i r_i^2 = Ma^2 + Ma^2 + mb^2 + mb^2 = \boxed{2Ma^2 + 2mb^2}$$

Evaluate the rotational kinetic energy using Equation 10.24:

$$K_R = \tfrac{1}{2} I_z \omega^2 = \tfrac{1}{2}(2Ma^2 + 2mb^2)\omega^2 = \boxed{(Ma^2 + mb^2)\omega^2}$$

Comparing the results for parts (A) and (B), we conclude that the moment of inertia and therefore the rotational kinetic energy associated with a given angular speed depend on the axis of rotation. In part (B), we expect the result to include all four spheres and distances because all four spheres are rotating in the xy plane. Based on the work–kinetic energy theorem, the smaller rotational kinetic energy in part (A) than in part (B) indicates it would require less work to set the system into rotation about the y axis than about the z axis.

WHAT IF?   What if the mass M is much larger than m? How do the answers to parts (A) and (B) compare?

Answer If $M \gg m$, then m can be neglected and the moment of inertia and the rotational kinetic energy in part (B) become

$$I_z = 2Ma^2 \quad \text{and} \quad K_R = Ma^2\omega^2$$

which are the same as the answers in part (A). If the masses m of the two tan spheres in Figure 10.19 are negligible, these spheres can be removed from the figure and rotations about the y and z axes are equivalent.

10.8 Energy Considerations in Rotational Motion

Having introduced rotational kinetic energy in Section 10.7, let us now see how an energy approach can be useful in solving rotational problems. We begin by considering the relationship between the torque acting on a rigid object and its resulting rotational motion so as to generate expressions for power and a rotational analog to the work–kinetic energy theorem. Consider the rigid object pivoted at O in Figure 10.20. Suppose a single external force $\vec{F}$ is applied at P, where $\vec{F}$ lies in the plane of the page. The work done on the object by $\vec{F}$ as its point of application rotates through an infinitesimal distance $ds = r\,d\theta$ is

$$dW = \vec{F} \cdot d\vec{s} = (F\sin\phi)r\,d\theta$$

where $F\sin\phi$ is the tangential component of $\vec{F}$, or, in other words, the component of the force along the displacement. Notice that the radial component vector of $\vec{F}$ does no work on the object because it is perpendicular to the displacement of the point of application of $\vec{F}$.

Because the magnitude of the torque due to $\vec{F}$ about an axis through O is defined as $rF\sin\phi$ by Equation 10.14, we can write the work done for the infinitesimal rotation as

$$dW = \tau\,d\theta \tag{10.25}$$

The rate at which work is being done by $\vec{F}$ as the object rotates about the fixed axis through the angle $d\theta$ in a time interval dt is

$$\frac{dW}{dt} = \tau\frac{d\theta}{dt}$$

Because dW/dt is the instantaneous power P (see Section 8.5) delivered by the force and $d\theta/dt = \omega$, this expression reduces to

$$P = \frac{dW}{dt} = \tau\omega \tag{10.26}$$

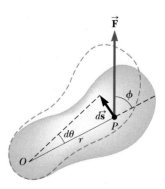

Figure 10.20 A rigid object rotates about an axis through O under the action of an external force $\vec{F}$ applied at P.

◀ **Power delivered to a rotating rigid object**

This equation is analogous to $P = Fv$ in the case of translational motion, and Equation 10.25 is analogous to $dW = F_x\,dx$.

In studying translational motion, we have seen that models based on an energy approach can be extremely useful in describing a system's behavior. From what we learned of translational motion, we expect that when a symmetric object rotates about a fixed axis, the work done by external forces equals the change in the rotational energy of the object.

To prove that fact, let us begin with the rigid object under a net torque model, whose mathematical representation is $\sum \tau_{\text{ext}} = I\alpha$. Using the chain rule from calculus, we can express the net torque as

$$\sum \tau_{\text{ext}} = I\alpha = I\frac{d\omega}{dt} = I\frac{d\omega}{d\theta}\frac{d\theta}{dt} = I\frac{d\omega}{d\theta}\omega$$

Rearranging this expression and noting that $\sum \tau_{\text{ext}}\,d\theta = dW$ gives

$$\sum \tau_{\text{ext}}\,d\theta = dW = I\omega\,d\omega$$

Integrating this expression, we obtain for the work W done by the net external force acting on a rotating system

$$W = \int_{\omega_i}^{\omega_f} I\omega\,d\omega = \tfrac{1}{2}I\omega_f^2 - \tfrac{1}{2}I\omega_i^2 \tag{10.27}$$

◀ **Work–kinetic energy theorem for rotational motion**

where the angular speed changes from ω_i to ω_f. Equation 10.27 is the **work–kinetic energy theorem for rotational motion.** Similar to the work–kinetic energy theorem in translational motion (Section 7.5), this theorem states that the net work done by external forces in rotating a symmetric rigid object about a fixed axis equals the change in the object's rotational energy.

This theorem is a form of the nonisolated system (energy) model discussed in Chapter 8. Work is done on the system of the rigid object, which represents a transfer of energy across the boundary of the system that appears as an increase in the object's rotational kinetic energy.

Table 10.3	Useful Equations in Rotational and Translational Motion	
Rotational Motion About a Fixed Axis	**Translational Motion**	
Angular speed $\omega = d\theta/dt$	Translational speed $v = dx/dt$	
Angular acceleration $\alpha = d\omega/dt$	Translational acceleration $a = dv/dt$	
Net torque $\Sigma\tau_{\text{ext}} = I\alpha$	Net force $\Sigma F = ma$	

If $\alpha = $ constant $\begin{cases} \omega_f = \omega_i + \alpha t \\ \theta_f = \theta_i + \omega_i t + \frac{1}{2}\alpha t^2 \\ \omega_f^2 = \omega_i^2 + 2\alpha(\theta_f - \theta_i) \end{cases}$	If $a = $ constant $\begin{cases} v_f = v_i + at \\ x_f = x_i + v_i t + \frac{1}{2}at^2 \\ v_f^2 = v_i^2 + 2a(x_f - x_i) \end{cases}$

Work $W = \int_{\theta_i}^{\theta_f} \tau \, d\theta$	Work $W = \int_{x_i}^{x_f} F_x \, dx$	
Rotational kinetic energy $K_R = \frac{1}{2}I\omega^2$	Kinetic energy $K = \frac{1}{2}mv^2$	
Power $P = \tau\omega$	Power $P = Fv$	
Angular momentum $L = I\omega$	Linear momentum $p = mv$	
Net torque $\Sigma\tau = dL/dt$	Net force $\Sigma F = dp/dt$	

In general, we can combine this theorem with the translational form of the work–kinetic energy theorem from Chapter 7. Therefore, the net work done by external forces on an object is the change in its *total* kinetic energy, which is the sum of the translational and rotational kinetic energies. For example, when a pitcher throws a baseball, the work done by the pitcher's hands appears as kinetic energy associated with the ball moving through space as well as rotational kinetic energy associated with the spinning of the ball.

In addition to the work–kinetic energy theorem, other energy principles can also be applied to rotational situations. For example, if a system involving rotating objects is isolated and no nonconservative forces act within the system, the isolated system model and the principle of conservation of mechanical energy can be used to analyze the system as in Example 10.11 below. In general, Equation 8.2, the conservation of energy equation, applies to rotational situations, with the recognition that the change in kinetic energy ΔK will include changes in both translational and rotational kinetic energies.

Finally, in some situations an energy approach does not provide enough information to solve the problem and it must be combined with a momentum approach. Such a case is illustrated in Example 10.14 in Section 10.9.

Table 10.3 lists the various equations we have discussed pertaining to rotational motion together with the analogous expressions for translational motion. Notice the similar mathematical forms of the equations. The last two equations in the left-hand column of Table 10.3, involving angular momentum L, are discussed in Chapter 11 and are included here only for the sake of completeness.

Example 10.11   Rotating Rod Revisited   AM

A uniform rod of length L and mass M is free to rotate on a frictionless pin passing through one end (Fig 10.21). The rod is released from rest in the horizontal position.

(A) What is its angular speed when the rod reaches its lowest position?

SOLUTION

Conceptualize Consider Figure 10.21 and imagine the rod rotating downward through a quarter turn about the pivot at the left end. Also look back at Example 10.8. This physical situation is the same.

Categorize As mentioned in Example 10.4, the angular acceleration of the rod is not constant. Therefore, the kinematic equations for rotation (Section 10.2) can-

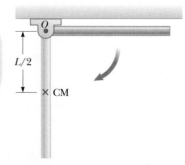

Figure 10.21 (Example 10.11)
A uniform rigid rod pivoted at O rotates in a vertical plane under the action of the gravitational force.

▶ **10.11** continued

not be used to solve this example. We categorize the system of the rod and the Earth as an *isolated system* in terms of *energy* with no nonconservative forces acting and use the principle of conservation of mechanical energy.

Analyze We choose the configuration in which the rod is hanging straight down as the reference configuration for gravitational potential energy and assign a value of zero for this configuration. When the rod is in the horizontal position, it has no rotational kinetic energy. The potential energy of the system in this configuration relative to the reference configuration is $MgL/2$ because the center of mass of the rod is at a height $L/2$ higher than its position in the reference configuration. When the rod reaches its lowest position, the energy of the system is entirely rotational energy $\frac{1}{2}I\omega^2$, where I is the moment of inertia of the rod about an axis passing through the pivot.

Using the isolated system (energy) model, write an appropriate reduction of Equation 8.2:

$$\Delta K + \Delta U = 0$$

Substitute for each of the final and initial energies:

$$\left(\tfrac{1}{2}I\omega^2 - 0\right) + \left(0 - \tfrac{1}{2}MgL\right) = 0$$

Solve for ω and use $I = \frac{1}{3}ML^2$ (see Table 10.2) for the rod:

$$\omega = \sqrt{\frac{MgL}{I}} = \sqrt{\frac{MgL}{\frac{1}{3}ML^2}} = \boxed{\sqrt{\frac{3g}{L}}}$$

(B) Determine the tangential speed of the center of mass and the tangential speed of the lowest point on the rod when it is in the vertical position.

SOLUTION

Use Equation 10.10 and the result from part (A):

$$v_{CM} = r\omega = \frac{L}{2}\omega = \boxed{\tfrac{1}{2}\sqrt{3gL}}$$

Because r for the lowest point on the rod is twice what it is for the center of mass, the lowest point has a tangential speed twice that of the center of mass:

$$v = 2v_{CM} = \boxed{\sqrt{3gL}}$$

Finalize The initial configuration in this example is the same as that in Example 10.4. In Example 10.4, however, we could only find the initial angular acceleration of the rod. Applying an energy approach in the current example allows us to find additional information, the angular speed of the rod at the lowest point. Convince yourself that you could find the angular speed of the rod at any angular position by knowing the location of the center of mass at this position.

WHAT IF? What if we want to find the angular speed of the rod when the angle it makes with the horizontal is 45.0°? Because this angle is half of 90.0°, for which we solved the problem above, is the angular speed at this configuration half the answer in the calculation above, that is, $\frac{1}{2}\sqrt{3g/L}$?

Answer Imagine the rod in Figure 10.21 at the 45.0° position. Use a pencil or a ruler to represent the rod at this position. Notice that the center of mass has dropped through more than half of the distance $L/2$ in this configuration. Therefore, more than half of the initial gravitational potential energy has been transformed to rotational kinetic energy. So, we should not expect the value of the angular speed to be as simple as proposed above.

Note that the center of mass of the rod drops through a distance of $0.500L$ as the rod reaches the vertical configuration. When the rod is at 45.0° to the horizontal, we can show that the center of mass of the rod drops through a distance of $0.354L$. Continuing the calculation, we find that the angular speed of the rod at this configuration is $0.841\sqrt{3g/L}$, (not $\frac{1}{2}\sqrt{3g/L}$).

Example 10.12 **Energy and the Atwood Machine** AM

Two blocks having different masses m_1 and m_2 are connected by a string passing over a pulley as shown in Figure 10.22 on page 256. The pulley has a radius R and moment of inertia I about its axis of rotation. The string does not slip on the pulley, and the system is released from rest. Find the translational speeds of the blocks after block 2 descends through a distance h and find the angular speed of the pulley at this time.

continued

▶ **10.12** continued

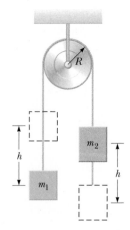

Conceptualize We have already seen examples involving the Atwood machine, so the motion of the objects in Figure 10.22 should be easy to visualize.

Categorize Because the string does not slip, the pulley rotates about the axle. We can neglect friction in the axle because the axle's radius is small relative to that of the pulley. Hence, the frictional torque is much smaller than the net torque applied by the two blocks provided that their masses are significantly different. Consequently, the system consisting of the two blocks, the pulley, and the Earth is an *isolated system* in terms of *energy* with no nonconservative forces acting; therefore, the mechanical energy of the system is conserved.

Figure 10.22 (Example 10.12) An Atwood machine with a massive pulley.

Analyze We define the zero configuration for gravitational potential energy as that which exists when the system is released. From Figure 10.22, we see that the descent of block 2 is associated with a decrease in system potential energy and that the rise of block 1 represents an increase in potential energy.

Using the isolated system (energy) model, write an appropriate reduction of the conservation of energy equation:

$$\Delta K + \Delta U = 0$$

Substitute for each of the energies:

$$\left[\left(\tfrac{1}{2}m_1 v_f^2 + \tfrac{1}{2}m_2 v_f^2 + \tfrac{1}{2}I\omega_f^2\right) - 0\right] + \left[(m_1 gh - m_2 gh) - 0\right] = 0$$

Use $v_f = R\omega_f$ to substitute for ω_f:

$$\tfrac{1}{2}m_1 v_f^2 + \tfrac{1}{2}m_2 v_f^2 + \tfrac{1}{2}I\frac{v_f^2}{R^2} = m_2 gh - m_1 gh$$

$$\tfrac{1}{2}\left(m_1 + m_2 + \frac{I}{R^2}\right)v_f^2 = (m_2 - m_1)gh$$

Solve for v_f:

$$(1)\quad v_f = \left[\frac{2(m_2 - m_1)gh}{m_1 + m_2 + I/R^2}\right]^{1/2}$$

Use $v_f = R\omega_f$ to solve for ω_f:

$$\omega_f = \frac{v_f}{R} = \frac{1}{R}\left[\frac{2(m_2 - m_1)gh}{m_1 + m_2 + I/R^2}\right]^{1/2}$$

Finalize Each block can be modeled as a *particle under constant acceleration* because it experiences a constant net force. Think about what you would need to do to use Equation (1) to find the acceleration of one of the blocks. Then imagine the pulley becoming massless and determine the acceleration of a block. How does this result compare with the result of Example 5.9?

10.9 Rolling Motion of a Rigid Object

In this section, we treat the motion of a rigid object rolling along a flat surface. In general, such motion is complex. For example, suppose a cylinder is rolling on a straight path such that the axis of rotation remains parallel to its initial orientation in space. As Figure 10.23 shows, a point on the rim of the cylinder moves in a complex path called a *cycloid*. We can simplify matters, however, by focusing on the center of mass rather than on a point on the rim of the rolling object. As shown in Figure 10.23, the center of mass moves in a straight line. If an object such as a cylinder rolls without slipping on the surface (called *pure rolling motion*), a simple relationship exists between its rotational and translational motions.

Consider a uniform cylinder of radius R rolling without slipping on a horizontal surface (Fig. 10.24). As the cylinder rotates through an angle θ, its center of mass

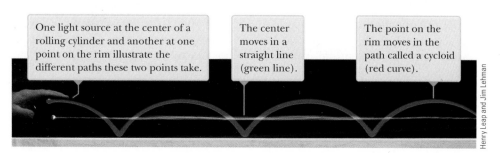

Figure 10.23 Two points on a rolling object take different paths through space.

One light source at the center of a rolling cylinder and another at one point on the rim illustrate the different paths these two points take.

The center moves in a straight line (green line).

The point on the rim moves in the path called a cycloid (red curve).

moves a linear distance $s = R\theta$ (see Eq. 10.1a). Therefore, the translational speed of the center of mass for pure rolling motion is given by

$$v_{CM} = \frac{ds}{dt} = R\frac{d\theta}{dt} = R\omega \qquad (10.28)$$

where ω is the angular speed of the cylinder. Equation 10.28 holds whenever a cylinder or sphere rolls without slipping and is the **condition for pure rolling motion.** The magnitude of the linear acceleration of the center of mass for pure rolling motion is

$$a_{CM} = \frac{dv_{CM}}{dt} = R\frac{d\omega}{dt} = R\alpha \qquad (10.29)$$

where α is the angular acceleration of the cylinder.

Imagine that you are moving along with a rolling object at speed v_{CM}, staying in a frame of reference at rest with respect to the center of mass of the object. As you observe the object, you will see the object in pure rotation around its center of mass. Figure 10.25a shows the velocities of points at the top, center, and bottom of the object as observed by you. In addition to these velocities, every point on the object moves in the same direction with speed v_{CM} relative to the surface on which it rolls. Figure 10.25b shows these velocities for a nonrotating object. In the reference frame at rest with respect to the surface, the velocity of a given point on the object is the sum of the velocities shown in Figures 10.25a and 10.25b. Figure 10.25c shows the results of adding these velocities.

Notice that the contact point between the surface and object in Figure 10.25c has a translational speed of zero. At this instant, the rolling object is moving in exactly the same way as if the surface were removed and the object were pivoted at point P and spun about an axis passing through P. We can express the total kinetic energy of this imagined spinning object as

$$K = \tfrac{1}{2}I_P\omega^2 \qquad (10.30)$$

where I_P is the moment of inertia about a rotation axis through P.

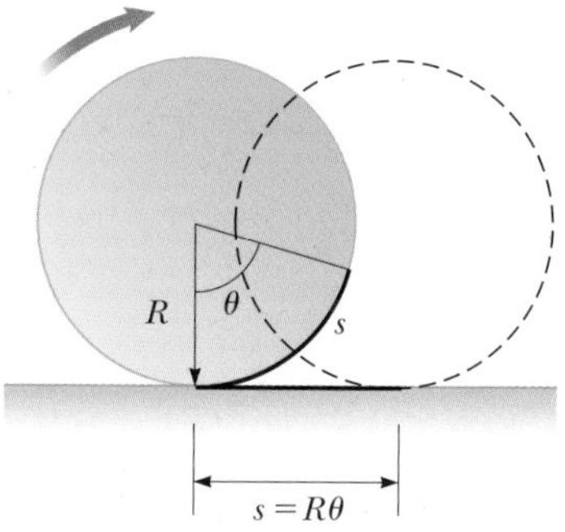

Figure 10.24 For pure rolling motion, as the cylinder rotates through an angle θ its center moves a linear distance $s = R\theta$.

Pitfall Prevention 10.6
Equation 10.28 Looks Familiar Equation 10.28 looks very similar to Equation 10.10, so be sure to be clear on the difference. Equation 10.10 gives the *tangential* speed of a point on a *rotating* object located a distance r from a fixed rotation axis if the object is rotating with angular speed ω. Equation 10.28 gives the *translational* speed of the center of mass of a *rolling* object of radius R rotating with angular speed ω.

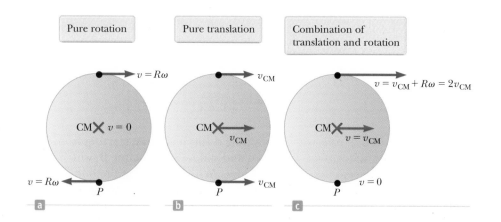

Pure rotation

Pure translation

Combination of translation and rotation

$v = R\omega$

v_{CM}

$v = v_{CM} + R\omega = 2v_{CM}$

$CM \times v = 0$

$CM \times v_{CM}$

$CM \times v = v_{CM}$

$v = R\omega$ P

v_{CM} P

$v = 0$ P

a   b   c

Figure 10.25 The motion of a rolling object can be modeled as a combination of pure translation and pure rotation.

Because the motion of the imagined spinning object is the same at this instant as our actual rolling object, Equation 10.30 also gives the kinetic energy of the rolling object. Applying the parallel-axis theorem, we can substitute $I_P = I_{CM} + MR^2$ into Equation 10.30 to obtain

$$K = \tfrac{1}{2}I_{CM}\omega^2 + \tfrac{1}{2}MR^2\omega^2$$

Using $v_{CM} = R\omega$, this equation can be expressed as

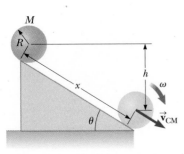

Total kinetic energy of a rolling object ▶

Figure 10.26 A sphere rolling down an incline. Mechanical energy of the sphere–Earth system is conserved if no slipping occurs.

$$K = \tfrac{1}{2}I_{CM}\omega^2 + \tfrac{1}{2}Mv_{CM}{}^2 \qquad \text{(10.31)}$$

The term $\tfrac{1}{2}I_{CM}\omega^2$ represents the rotational kinetic energy of the object about its center of mass, and the term $\tfrac{1}{2}Mv_{CM}{}^2$ represents the kinetic energy the object would have if it were just translating through space without rotating. Therefore, the total kinetic energy of a rolling object is the sum of the rotational kinetic energy *about* the center of mass and the translational kinetic energy *of* the center of mass. This statement is consistent with the situation illustrated in Figure 10.25, which shows that the velocity of a point on the object is the sum of the velocity of the center of mass and the tangential velocity around the center of mass.

Energy methods can be used to treat a class of problems concerning the rolling motion of an object on a rough incline. For example, consider Figure 10.26, which shows a sphere rolling without slipping after being released from rest at the top of the incline. Accelerated rolling motion is possible only if a friction force is present between the sphere and the incline to produce a net torque about the center of mass. Despite the presence of friction, no loss of mechanical energy occurs because the contact point is at rest relative to the surface at any instant. (On the other hand, if the sphere were to slip, mechanical energy of the sphere–incline–Earth system would decrease due to the nonconservative force of kinetic friction.)

In reality, *rolling friction* causes mechanical energy to transform to internal energy. Rolling friction is due to deformations of the surface and the rolling object. For example, automobile tires flex as they roll on a roadway, representing a transformation of mechanical energy to internal energy. The roadway also deforms a small amount, representing additional rolling friction. In our problem-solving models, we ignore rolling friction unless stated otherwise.

Using $v_{CM} = R\omega$ for pure rolling motion, we can express Equation 10.31 as

$$K = \tfrac{1}{2}I_{CM}\left(\frac{v_{CM}}{R}\right)^2 + \tfrac{1}{2}Mv_{CM}{}^2$$

$$K = \tfrac{1}{2}\left(\frac{I_{CM}}{R^2} + M\right)v_{CM}{}^2 \qquad \text{(10.32)}$$

For the sphere–Earth system in Figure 10.26, we define the zero configuration of gravitational potential energy to be when the sphere is at the bottom of the incline. Therefore, Equation 8.2 gives

$$\Delta K + \Delta U = 0$$

$$\left[\tfrac{1}{2}\left(\frac{I_{CM}}{R^2} + M\right)v_{CM}{}^2 - 0\right] + (0 - Mgh) = 0$$

$$v_{CM} = \left[\frac{2gh}{1 + (I_{CM}/MR^2)}\right]^{1/2} \qquad \text{(10.33)}$$

Quick **Quiz** 10.7 A ball rolls without slipping down incline A, starting from rest. At the same time, a box starts from rest and slides down incline B, which is identical to incline A except that it is frictionless. Which arrives at the bottom first? **(a)** The ball arrives first. **(b)** The box arrives first. **(c)** Both arrive at the same time. **(d)** It is impossible to determine.

Example 10.13 | Sphere Rolling Down an Incline

For the solid sphere shown in Figure 10.26, calculate the translational speed of the center of mass at the bottom of the incline and the magnitude of the translational acceleration of the center of mass.

SOLUTION

Conceptualize Imagine rolling the sphere down the incline. Compare it in your mind to a book sliding down a frictionless incline. You probably have experience with objects rolling down inclines and may be tempted to think that the sphere would move down the incline faster than the book. You do *not*, however, have experience with objects sliding down *frictionless* inclines! So, which object will reach the bottom first? (See Quick Quiz 10.7.)

Categorize We model the sphere and the Earth as an *isolated system* in terms of *energy* with no nonconservative forces acting. This model is the one that led to Equation 10.33, so we can use that result.

Analyze Evaluate the speed of the center of mass of the sphere from Equation 10.33:

$$(1) \quad v_{CM} = \left[\frac{2gh}{1 + (\frac{2}{5}MR^2/MR^2)} \right]^{1/2} = (\tfrac{10}{7}gh)^{1/2}$$

This result is less than $\sqrt{2gh}$, which is the speed an object would have if it simply slid down the incline without rotating. (Eliminate the rotation by setting $I_{CM} = 0$ in Eq. 10.33.)

To calculate the translational acceleration of the center of mass, notice that the vertical displacement of the sphere is related to the distance x it moves along the incline through the relationship $h = x \sin \theta$.

Use this relationship to rewrite Equation (1):

$$v_{CM}{}^2 = \tfrac{10}{7}gx \sin \theta$$

Write Equation 2.17 for an object starting from rest and moving through a distance x under constant acceleration:

$$v_{CM}{}^2 = 2a_{CM}x$$

Equate the preceding two expressions to find a_{CM}:

$$a_{CM} = \tfrac{5}{7}g \sin \theta$$

Finalize Both the speed and the acceleration of the center of mass are *independent* of the mass and the radius of the sphere. That is, all homogeneous solid spheres experience the same speed and acceleration on a given incline. Try to verify this statement experimentally with balls of different sizes, such as a marble and a croquet ball.

If we were to repeat the acceleration calculation for a hollow sphere, a solid cylinder, or a hoop, we would obtain similar results in which only the factor in front of $g \sin \theta$ would differ. The constant factors that appear in the expressions for v_{CM} and a_{CM} depend only on the moment of inertia about the center of mass for the specific object. In all cases, the acceleration of the center of mass is *less* than $g \sin \theta$, the value the acceleration would have if the incline were frictionless and no rolling occurred.

Example 10.14 | Pulling on a Spool[3]

A cylindrically symmetric spool of mass m and radius R sits at rest on a horizontal table with friction (Fig. 10.27). With your hand on a light string wrapped around the axle of radius r, you pull on the spool with a constant horizontal force of magnitude T to the right. As a result, the spool rolls without slipping a distance L along the table with no rolling friction.

(A) Find the final translational speed of the center of mass of the spool.

SOLUTION

Conceptualize Use Figure 10.27 to visualize the motion of the spool when you pull the string. For the spool to roll through a distance L, notice that your hand on the string must pull through a distance *different* from L.

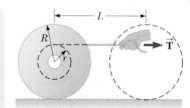

Figure 10.27 (Example 10.14) A spool rests on a horizontal table. A string is wrapped around the axle and is pulled to the right by a hand.

continued

---

[3]Example 10.14 was inspired in part by C. E. Mungan, "A primer on work–energy relationships for introductory physics," *The Physics Teacher,* **43**:10, 2005.

▶ **10.14** continued

Categorize The spool is a *rigid object under a net torque,* but the net torque includes that due to the friction force at the bottom of the spool, about which we know nothing. Therefore, an approach based on the rigid object under a net torque model will not be successful. Work is done by your hand on the spool and string, which form a nonisolated system in terms of energy. Let's see if an approach based on the *nonisolated system (energy)* model is fruitful.

Analyze The only type of energy that changes in the system is the kinetic energy of the spool. There is no rolling friction, so there is no change in internal energy. The only way that energy crosses the system's boundary is by the work done by your hand on the string. No work is done by the static force of friction on the bottom of the spool (to the left in Fig. 10.27) because the point of application of the force moves through no displacement.

Write the appropriate reduction of the conservation of energy equation, Equation 8.2:

$$(1) \quad W = \Delta K = \Delta K_{\text{trans}} + \Delta K_{\text{rot}}$$

where W is the work done on the string by your hand. To find this work, we need to find the displacement of your hand during the process.

We first find the length of string that has unwound off the spool. If the spool rolls through a distance L, the total angle through which it rotates is $\theta = L/R$. The axle also rotates through this angle.

Use Equation 10.1a to find the total arc length through which the axle turns:

$$\ell = r\theta = \frac{r}{R}L$$

This result also gives the length of string pulled off the axle. Your hand will move through this distance *plus* the distance L through which the spool moves. Therefore, the magnitude of the displacement of the point of application of the force applied by your hand is $\ell + L = L(1 + r/R)$.

Evaluate the work done by your hand on the string:

$$(2) \quad W = TL\left(1 + \frac{r}{R}\right)$$

Substitute Equation (2) into Equation (1):

$$TL\left(1 + \frac{r}{R}\right) = \tfrac{1}{2}mv_{\text{CM}}^2 + \tfrac{1}{2}I\omega^2$$

where I is the moment of inertia of the spool about its center of mass and v_{CM} and ω are the final values after the wheel rolls through the distance L.

Apply the nonslip rolling condition $\omega = v_{\text{CM}}/R$:

$$TL\left(1 + \frac{r}{R}\right) = \tfrac{1}{2}mv_{\text{CM}}^2 + \tfrac{1}{2}I\frac{v_{\text{CM}}^2}{R^2}$$

Solve for v_{CM}:

$$(3) \quad v_{\text{CM}} = \sqrt{\frac{2TL(1 + r/R)}{m(1 + I/mR^2)}}$$

(B) Find the value of the friction force f.

SOLUTION

Categorize Because the friction force does no work, we cannot evaluate it from an energy approach. We model the spool as a *nonisolated system,* but this time in terms of *momentum.* The string applies a force across the boundary of the system, resulting in an impulse on the system. Because the forces on the spool are constant, we can model the spool's center of mass as a *particle under constant acceleration.*

Analyze Write the impulse–momentum theorem (Eq. 9.40) for the spool:

$$m(v_{\text{CM}} - 0) = (T - f)\Delta t$$

$$(4) \quad mv_{\text{CM}} = (T - f)\Delta t$$

For a particle under constant acceleration starting from rest, Equation 2.14 tells us that the average velocity of the center of mass is half the final velocity.

Use Equation 2.2 to find the time interval for the center of mass of the spool to move a distance L from rest to a final speed v_{CM}:

$$(5) \quad \Delta t = \frac{L}{v_{\text{CM,avg}}} = \frac{2L}{v_{\text{CM}}}$$

▶ **10.14** continued

Substitute Equation (5) into Equation (4):

$$mv_{\text{CM}} = (T - f)\frac{2L}{v_{\text{CM}}}$$

Solve for the friction force f:

$$f = T - \frac{mv_{\text{CM}}^2}{2L}$$

Substitute v_{CM} from Equation (3):

$$f = T - \frac{m}{2L}\left[\frac{2TL(1 + r/R)}{m(1 + I/mR^2)}\right]$$

$$= T - T\frac{(1 + r/R)}{(1 + I/mR^2)} = \boxed{T\left[\frac{I - mrR}{I + mR^2}\right]}$$

Finalize Notice that we could use the impulse–momentum theorem for the translational motion of the spool while ignoring that the spool is rotating! This fact demonstrates the power of our growing list of approaches to solving problems.

Summary

Definitions

The **angular position** of a rigid object is defined as the angle θ between a reference line attached to the object and a reference line fixed in space. The **angular displacement** of a particle moving in a circular path or a rigid object rotating about a fixed axis is $\Delta\theta \equiv \theta_f - \theta_i$.

The **instantaneous angular speed** of a particle moving in a circular path or of a rigid object rotating about a fixed axis is

$$\omega \equiv \frac{d\theta}{dt} \qquad \textbf{(10.3)}$$

The **instantaneous angular acceleration** of a particle moving in a circular path or of a rigid object rotating about a fixed axis is

$$\alpha \equiv \frac{d\omega}{dt} \qquad \textbf{(10.5)}$$

When a rigid object rotates about a fixed axis, every part of the object has the same angular speed and the same angular acceleration.

The magnitude of the **torque** associated with a force $\vec{F}$ acting on an object at a distance r from the rotation axis is

$$\tau = rF\sin\phi = Fd \qquad \textbf{(10.14)}$$

where ϕ is the angle between the position vector of the point of application of the force and the force vector, and d is the moment arm of the force, which is the perpendicular distance from the rotation axis to the line of action of the force.

The **moment of inertia of a system of particles** is defined as

$$I \equiv \sum_i m_i r_i^2 \qquad \textbf{(10.19)}$$

where m_i is the mass of the ith particle and r_i is its distance from the rotation axis.

Concepts and Principles

When a rigid object rotates about a fixed axis, the angular position, angular speed, and angular acceleration are related to the translational position, translational speed, and translational acceleration through the relationships

$$s = r\theta \qquad \textbf{(10.1a)}$$

$$v = r\omega \qquad \textbf{(10.10)}$$

$$a_t = r\alpha \qquad \textbf{(10.11)}$$

If a rigid object rotates about a fixed axis with angular speed ω, its **rotational kinetic energy** can be written

$$K_R = \tfrac{1}{2}I\omega^2 \qquad \textbf{(10.24)}$$

where I is the moment of inertia of the object about the axis of rotation.

The **moment of inertia of a rigid object** is

$$I = \int r^2\, dm \qquad \textbf{(10.20)}$$

where r is the distance from the mass element dm to the axis of rotation.

continued

The rate at which work is done by an external force in rotating a rigid object about a fixed axis, or the **power** delivered, is

$$P = \tau\omega \qquad \textbf{(10.26)}$$

If work is done on a rigid object and the only result of the work is rotation about a fixed axis, the net work done by external forces in rotating the object equals the change in the rotational kinetic energy of the object:

$$W = \tfrac{1}{2}I\omega_f^2 - \tfrac{1}{2}I\omega_i^2 \qquad \textbf{(10.27)}$$

The **total kinetic energy** of a rigid object rolling on a rough surface without slipping equals the rotational kinetic energy about its center of mass plus the translational kinetic energy of the center of mass:

$$K = \tfrac{1}{2}I_{\text{CM}}\omega^2 + \tfrac{1}{2}Mv_{\text{CM}}^2 \qquad \textbf{(10.31)}$$

Analysis Models for Problem Solving

Rigid Object Under Constant Angular Acceleration. If a rigid object rotates about a fixed axis under constant angular acceleration, one can apply equations of kinematics that are analogous to those for translational motion of a particle under constant acceleration:

$$\omega_f = \omega_i + \alpha t \qquad \textbf{(10.6)}$$

$$\theta_f = \theta_i + \omega_i t + \tfrac{1}{2}\alpha t^2 \qquad \textbf{(10.7)}$$

$$\omega_f^2 = \omega_i^2 + 2\alpha(\theta_f - \theta_i) \qquad \textbf{(10.8)}$$

$$\theta_f = \theta_i + \tfrac{1}{2}(\omega_i + \omega_f)t \qquad \textbf{(10.9)}$$

Rigid Object Under a Net Torque. If a rigid object free to rotate about a fixed axis has a net external torque acting on it, the object undergoes an angular acceleration α, where

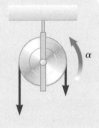

$$\sum \tau_{\text{ext}} = I\alpha \qquad \textbf{(10.18)}$$

This equation is the rotational analog to Newton's second law in the particle under a net force model.

Objective Questions

1. denotes answer available in *Student Solutions Manual/Study Guide*

1. A cyclist rides a bicycle with a wheel radius of 0.500 m across campus. A piece of plastic on the front rim makes a clicking sound every time it passes through the fork. If the cyclist counts 320 clicks between her apartment and the cafeteria, how far has she traveled? (a) 0.50 km (b) 0.80 km (c) 1.0 km (d) 1.5 km (e) 1.8 km

2. Consider an object on a rotating disk a distance r from its center, held in place on the disk by static friction. Which of the following statements is *not* true concerning this object? (a) If the angular speed is constant, the object must have constant tangential speed. (b) If the angular speed is constant, the object is not accelerated. (c) The object has a tangential acceleration only if the disk has an angular acceleration. (d) If the disk has an angular acceleration, the object has both a centripetal acceleration and a tangential acceleration. (e) The object always has a centripetal acceleration except when the angular speed is zero.

3. A wheel is rotating about a fixed axis with constant angular acceleration 3 rad/s². At different moments, its angular speed is −2 rad/s, 0, and +2 rad/s. For a point on the rim of the wheel, consider at these moments the magnitude of the tangential component of acceleration and the magnitude of the radial component of acceleration. Rank the following five items from largest to smallest: (a) $|a_t|$ when $\omega = -2$ rad/s, (b) $|a_r|$ when

$\omega = -2$ rad/s, (c) $|a_r|$ when $\omega = 0$, (d) $|a_t|$ when $\omega = 2$ rad/s, and (e) $|a_r|$ when $\omega = 2$ rad/s. If two items are equal, show them as equal in your ranking. If a quantity is equal to zero, show that fact in your ranking.

4. A grindstone increases in angular speed from 4.00 rad/s to 12.00 rad/s in 4.00 s. Through what angle does it turn during that time interval if the angular acceleration is constant? (a) 8.00 rad (b) 12.0 rad (c) 16.0 rad (d) 32.0 rad (e) 64.0 rad

5. Suppose a car's standard tires are replaced with tires 1.30 times larger in diameter. **(i)** Will the car's speedometer reading be (a) 1.69 times too high, (b) 1.30 times too high, (c) accurate, (d) 1.30 times too low, (e) 1.69 times too low, or (f) inaccurate by an unpredictable factor? **(ii)** Will the car's fuel economy in miles per gallon or km/L appear to be (a) 1.69 times better, (b) 1.30 times better, (c) essentially the same, (d) 1.30 times worse, or (e) 1.69 times worse?

6. Figure OQ10.6 shows a system of four particles joined by light, rigid rods. Assume $a = b$ and M is larger than m. About which of the coordinate axes does the system have **(i)** the smallest and **(ii)** the largest moment of inertia? (a) the x axis (b) the y axis (c) the z axis. (d) The moment of inertia has the same small value for two axes. (e) The moment of inertia is the same for all three axes.

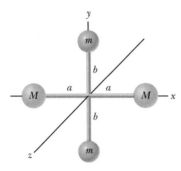

Figure OQ10.6

7. As shown in Figure OQ10.7, a cord is wrapped onto a cylindrical reel mounted on a fixed, frictionless, horizontal axle. When does the reel have a greater magnitude of angular acceleration? (a) When the cord is pulled down with a constant force of 50 N. (b) When an object of weight 50 N is hung from the cord and released. (c) The angular accelerations in parts (a) and (b) are equal. (d) It is impossible to determine.

Figure OQ10.7 Objective Question 7 and Conceptual Question 4.

8. A constant net torque is exerted on an object. Which of the following quantities for the object cannot be constant? Choose all that apply. (a) angular position (b) angular velocity (c) angular acceleration (d) moment of inertia (e) kinetic energy

9. A basketball rolls across a classroom floor without slipping, with its center of mass moving at a certain speed. A block of ice of the same mass is set sliding across the floor with the same speed along a parallel line. Which object has more **(i)** kinetic energy and **(ii)** momentum? (a) The basketball does. (b) The ice does. (c) The two quantities are equal. **(iii)** The two objects encounter a ramp sloping upward. Which object will travel farther up the ramp? (a) The basketball will. (b) The ice will. (c) They will travel equally far up the ramp.

10. A toy airplane hangs from the ceiling at the bottom end of a string. You turn the airplane many times to wind up the string clockwise and release it. The airplane starts to spin counterclockwise, slowly at first and then faster and faster. Take counterclockwise as the positive sense and assume friction is negligible. When the string is entirely unwound, the airplane has its maximum rate of rotation. **(i)** At this moment, is its angular acceleration (a) positive, (b) negative, or (c) zero? **(ii)** The airplane continues to spin, winding the string counterclockwise as it slows down. At the moment it momentarily stops, is its angular acceleration (a) positive, (b) negative, or (c) zero?

11. A solid aluminum sphere of radius R has moment of inertia I about an axis through its center. Will the moment of inertia about a central axis of a solid aluminum sphere of radius $2R$ be (a) $2I$, (b) $4I$, (c) $8I$, (d) $16I$, or (e) $32I$?

Conceptual Questions   1. denotes answer available in *Student Solutions Manual/Study Guide*

1. Is it possible to change the translational kinetic energy of an object without changing its rotational energy?

2. Must an object be rotating to have a nonzero moment of inertia?

3. Suppose just two external forces act on a stationary, rigid object and the two forces are equal in magnitude and opposite in direction. Under what condition does the object start to rotate?

4. Explain how you might use the apparatus described in Figure OQ10.7 to determine the moment of inertia of the wheel. *Note:* If the wheel does not have a uniform mass density, the moment of inertia is not necessarily equal to $\frac{1}{2}MR^2$.

5. Using the results from Example 10.6, how would you calculate the angular speed of the wheel and the linear speed of the hanging object at $t = 2$ s, assuming the system is released from rest at $t = 0$?

6. Explain why changing the axis of rotation of an object changes its moment of inertia.

7. Suppose you have two eggs, one hard-boiled and the other uncooked. You wish to determine which is the hard-boiled egg without breaking the eggs, which

can be done by spinning the two eggs on the floor and comparing the rotational motions. (a) Which egg spins faster? (b) Which egg rotates more uniformly? (c) Which egg begins spinning again after being stopped and then immediately released? Explain your answers to parts (a), (b), and (c).

8. Suppose you set your textbook sliding across a gymnasium floor with a certain initial speed. It quickly stops moving because of a friction force exerted on it by the floor. Next, you start a basketball rolling with the same initial speed. It keeps rolling from one end of the gym to the other. (a) Why does the basketball roll so far? (b) Does friction significantly affect the basketball's motion?

9. (a) What is the angular speed of the second hand of an analog clock? (b) What is the direction of $\vec{\boldsymbol{\omega}}$ as you view a clock hanging on a vertical wall? (c) What is the magnitude of the angular acceleration vector $\vec{\boldsymbol{\alpha}}$ of the second hand?

10. One blade of a pair of scissors rotates counterclockwise in the *xy* plane. (a) What is the direction of $\vec{\boldsymbol{\omega}}$ for the blade? (b) What is the direction of $\vec{\boldsymbol{\alpha}}$ if the magnitude of the angular velocity is decreasing in time?

11. If you see an object rotating, is there necessarily a net torque acting on it?

12. If a small sphere of mass M were placed at the end of the rod in Figure 10.21, would the result for ω be greater than, less than, or equal to the value obtained in Example 10.11?

13. Three objects of uniform density—a solid sphere, a solid cylinder, and a hollow cylinder—are placed at the top of an incline (Fig. CQ10.13). They are all released from rest at the same elevation and roll without slipping. (a) Which object reaches the bottom first? (b) Which reaches it last? *Note:* The result is independent of the masses and the radii of the objects. (Try this activity at home!)

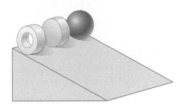

Figure CQ10.13

14. Which of the entries in Table 10.2 applies to finding the moment of inertia (a) of a long, straight sewer pipe rotating about its axis of symmetry? (b) Of an embroidery hoop rotating about an axis through its center and perpendicular to its plane? (c) Of a uniform door turning on its hinges? (d) Of a coin turning about an axis through its center and perpendicular to its faces?

15. Figure CQ10.15 shows a side view of a child's tricycle with rubber tires on a horizontal concrete sidewalk. If a string were attached to the upper pedal on the far side and pulled forward horizontally, the tricycle would start to roll forward. (a) Instead, assume a string is attached to the lower pedal on the near side and pulled forward horizontally as shown by A. Will the tricycle start to roll? If so, which way? Answer the same questions if (b) the string is pulled forward and upward as shown by B, (c) if the string is pulled straight down as shown by C, and (d) if the string is pulled forward and downward as shown by D. (e) **What If?** Suppose the string is instead attached to the rim of the front wheel and pulled upward and backward as shown by E. Which way does the tricycle roll? (f) Explain a pattern of reasoning, based on the figure, that makes it easy to answer questions such as these. What physical quantity must you evaluate?

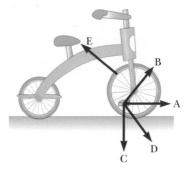

Figure CQ10.15

16. A person balances a meterstick in a horizontal position on the extended index fingers of her right and left hands. She slowly brings the two fingers together. The stick remains balanced, and the two fingers always meet at the 50-cm mark regardless of their original positions. (Try it!) Explain why that occurs.

Problems available in ENHANCED **WebAssign**   Access end-of-chapter problems online at **www.webassign.net**

Section 10.1 Angular Position, Velocity, and Acceleration
Problems 1–4

Section 10.2 Analysis Model: Rigid Object Under Constant Angular Acceleration
Problems 5–14

Section 10.3 Angular and Translational Quantities
Problems 15–26

Section 10.4 Torque
Problems 27–28

Section 10.5 Analysis Model: Rigid Object Under a Net Torque
Problems 29–37

Section 10.6 Calculation of Moments of Inertia
Problems 38–43

Section 10.7 Rotational Kinetic Energy
Problems 44–47

Section 10.8 Energy Considerations in Rotational Motion
Problems 48–58

Section 10.9 Rolling Motion of a Rigid Object
Problems 59–65

Additional Problems
Problems 66–86

Challenge Problems
Problem 87–94

Solutions to the following Problems are available in the *Student Solutions Manual/Study Guide:*
10.5, 10.7, 10.19, 10.21, 10.27, 10.33, 10.45, 10.55, 10.56, 10.57, 10.59, 10.61, 10.64, 10.67, 10.73, 10.77, 10.78, 10.82, and 10.89

List of Enhanced Problems

Problem Number	Targeted Feedback in Enhanced WebAssign	Analysis Model Tutorial in Enhanced WebAssign	Master It in Enhanced WebAssign	Watch It in Enhanced WebAssign
10.3	✓			✓
10.5	✓			✓
10.7	✓		✓	
10.11	✓	✓		✓
10.17	✓			✓
10.18	✓			✓
10.19			✓	
10.20	✓			✓
10.21	✓		✓	
10.23	✓			✓
10.27	✓		✓	
10.28	✓			✓
10.30	✓	✓		✓
10.31	✓		✓	
10.32	✓			✓
10.37	✓			✓
10.44	✓			✓
10.45	✓			✓
10.54		✓		
10.59	✓		✓	
10.67		✓		

Angular Momentum

The central topic of this chapter is angular momentum, a quantity that plays a key role in rotational dynamics. In analogy to the principle of conservation of linear momentum, there is also a principle of conservation of angular momentum. The angular momentum of an isolated system is constant. For angular momentum, an isolated system is one for which no external torques act on the system. If a net external torque acts on a system, it is nonisolated. Like the law of conservation of linear momentum, the law of conservation of angular momentum is a fundamental law of physics, equally valid for relativistic and quantum systems.

Two motorcycle racers lean precariously into a turn around a racetrack. The analysis of such a leaning turn is based on principles associated with angular momentum. *(Stuart Westmorland/The Image Bank/ Getty Images)*

11.1 The Vector Product and Torque

An important consideration in defining angular momentum is the process of multiplying two vectors by means of the operation called the *vector product*. We will introduce the vector product by considering the vector nature of torque.

Consider a force $\vec{F}$ acting on a particle located at point P and described by the vector position $\vec{r}$ (Fig. 11.1 on page 268). As we saw in Section 10.6, the *magnitude* of the torque due to this force about an axis through the origin is $rF \sin \phi$, where ϕ is the angle between $\vec{r}$ and $\vec{F}$. The axis about which $\vec{F}$ tends to produce rotation is perpendicular to the plane formed by $\vec{r}$ and $\vec{F}$.

The torque vector $\vec{\tau}$ is related to the two vectors $\vec{r}$ and $\vec{F}$. We can establish a mathematical relationship between $\vec{\tau}$, $\vec{r}$, and $\vec{F}$ using a mathematical operation called the **vector product:**

$$\vec{\tau} \equiv \vec{r} \times \vec{F} \qquad (11.1)$$

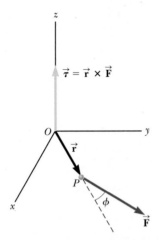

Figure 11.1 The torque vector $\vec{\tau}$ lies in a direction perpendicular to the plane formed by the position vector $\vec{r}$ and the applied force vector $\vec{F}$. In the situation shown, $\vec{r}$ and $\vec{F}$ lie in the xy plane, so the torque is along the z axis.

Properties of the ▶
vector product

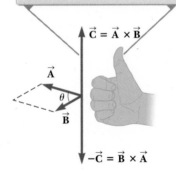

The direction of $\vec{C}$ is perpendicular to the plane formed by $\vec{A}$ and $\vec{B}$, and its direction is determined by the right-hand rule.

Figure 11.2 The vector product $\vec{A} \times \vec{B}$ is a third vector $\vec{C}$ having a magnitude $AB\sin\theta$ equal to the area of the parallelogram shown.

Cross products of ▶
unit vectors

We now give a formal definition of the vector product. Given any two vectors $\vec{A}$ and $\vec{B}$, the vector product $\vec{A} \times \vec{B}$ is defined as a third vector $\vec{C}$, which has a magnitude of $AB\sin\theta$, where θ is the angle between $\vec{A}$ and $\vec{B}$. That is, if $\vec{C}$ is given by

$$\vec{C} = \vec{A} \times \vec{B} \tag{11.2}$$

its magnitude is

$$C = AB\sin\theta \tag{11.3}$$

The quantity $AB\sin\theta$ is equal to the area of the parallelogram formed by $\vec{A}$ and $\vec{B}$ as shown in Figure 11.2. The *direction* of $\vec{C}$ is perpendicular to the plane formed by $\vec{A}$ and $\vec{B}$, and the best way to determine this direction is to use the right-hand rule illustrated in Figure 11.2. The four fingers of the right hand are pointed along $\vec{A}$ and then "wrapped" in the direction that would rotate $\vec{A}$ into $\vec{B}$ through the angle θ. The direction of the upright thumb is the direction of $\vec{A} \times \vec{B} = \vec{C}$. Because of the notation, $\vec{A} \times \vec{B}$ is often read "$\vec{A}$ cross $\vec{B}$," so the vector product is also called the **cross product.**

Some properties of the vector product that follow from its definition are as follows:

1. Unlike the scalar product, the vector product is *not* commutative. Instead, the order in which the two vectors are multiplied in a vector product is important:

$$\vec{A} \times \vec{B} = -\vec{B} \times \vec{A} \tag{11.4}$$

   Therefore, if you change the order of the vectors in a vector product, you must change the sign. You can easily verify this relationship with the right-hand rule.

2. If $\vec{A}$ is parallel to $\vec{B}$ ($\theta = 0$ or $180°$), then $\vec{A} \times \vec{B} = 0$; therefore, it follows that $\vec{A} \times \vec{A} = 0$.

3. If $\vec{A}$ is perpendicular to $\vec{B}$, then $|\vec{A} \times \vec{B}| = AB$.

4. The vector product obeys the distributive law:

$$\vec{A} \times (\vec{B} + \vec{C}) = \vec{A} \times \vec{B} + \vec{A} \times \vec{C} \tag{11.5}$$

5. The derivative of the vector product with respect to some variable such as t is

$$\frac{d}{dt}(\vec{A} \times \vec{B}) = \frac{d\vec{A}}{dt} \times \vec{B} + \vec{A} \times \frac{d\vec{B}}{dt} \tag{11.6}$$

   where it is important to preserve the multiplicative order of the terms on the right side in view of Equation 11.4.

It is left as an exercise (Problem 4 in Enhanced WebAssign) to show from Equations 11.3 and 11.4 and from the definition of unit vectors that the cross products of the unit vectors $\hat{i}$, $\hat{j}$, and $\hat{k}$ obey the following rules:

$$\hat{i} \times \hat{i} = \hat{j} \times \hat{j} = \hat{k} \times \hat{k} = 0 \tag{11.7a}$$

$$\hat{i} \times \hat{j} = -\hat{j} \times \hat{i} = \hat{k} \tag{11.7b}$$

$$\hat{j} \times \hat{k} = -\hat{k} \times \hat{j} = \hat{i} \tag{11.7c}$$

$$\hat{k} \times \hat{i} = -\hat{i} \times \hat{k} = \hat{j} \tag{11.7d}$$

Signs are interchangeable in cross products. For example, $\vec{A} \times (-\vec{B}) = -\vec{A} \times \vec{B}$ and $\hat{i} \times (-\hat{j}) = -\hat{i} \times \hat{j}$.

The cross product of any two vectors $\vec{A}$ and $\vec{B}$ can be expressed in the following determinant form:

$$\vec{A} \times \vec{B} = \begin{vmatrix} \hat{i} & \hat{j} & \hat{k} \\ A_x & A_y & A_z \\ B_x & B_y & B_z \end{vmatrix} = \begin{vmatrix} A_y & A_z \\ B_y & B_z \end{vmatrix}\hat{i} + \begin{vmatrix} A_z & A_x \\ B_z & B_x \end{vmatrix}\hat{j} + \begin{vmatrix} A_x & A_y \\ B_x & B_y \end{vmatrix}\hat{k}$$

Expanding these determinants gives the result

$$\vec{A} \times \vec{B} = (A_y B_z - A_z B_y)\,\hat{i} + (A_z B_x - A_x B_z)\,\hat{j} + (A_x B_y - A_y B_x)\,\hat{k} \quad \textbf{(11.8)}$$

Given the definition of the cross product, we can now assign a direction to the torque vector. If the force lies in the xy plane as in Figure 11.1, the torque $\vec{\tau}$ is represented by a vector parallel to the z axis. The force in Figure 11.1 creates a torque that tends to rotate the particle counterclockwise about the z axis; the direction of $\vec{\tau}$ is toward increasing z, and $\vec{\tau}$ is therefore in the positive z direction. If we reversed the direction of $\vec{F}$ in Figure 11.1, $\vec{\tau}$ would be in the negative z direction.

Quick Quiz 11.1 Which of the following statements about the relationship between the magnitude of the cross product of two vectors and the product of the magnitudes of the vectors is true? (a) $|\vec{A} \times \vec{B}|$ is larger than AB. (b) $|\vec{A} \times \vec{B}|$ is smaller than AB. (c) $|\vec{A} \times \vec{B}|$ could be larger or smaller than AB, depending on the angle between the vectors. (d) $|\vec{A} \times \vec{B}|$ could be equal to AB.

Example 11.1   The Vector Product

Two vectors lying in the xy plane are given by the equations $\vec{A} = 2\,\hat{i} + 3\,\hat{j}$ and $\vec{B} = -\hat{i} + 2\,\hat{j}$. Find $\vec{A} \times \vec{B}$ and verify that $\vec{A} \times \vec{B} = -\vec{B} \times \vec{A}$.

SOLUTION

Conceptualize Given the unit-vector notations of the vectors, think about the directions the vectors point in space. Draw them on graph paper and imagine the parallelogram shown in Figure 11.2 for these vectors.

Categorize Because we use the definition of the cross product discussed in this section, we categorize this example as a substitution problem.

Write the cross product of the two vectors:	$\vec{A} \times \vec{B} = (2\,\hat{i} + 3\,\hat{j}) \times (-\hat{i} + 2\,\hat{j})$
Perform the multiplication:	$\vec{A} \times \vec{B} = 2\,\hat{i} \times (-\hat{i}) + 2\,\hat{i} \times 2\,\hat{j} + 3\,\hat{j} \times (-\hat{i}) + 3\,\hat{j} \times 2\,\hat{j}$
Use Equations 11.7a through 11.7d to evaluate the various terms:	$\vec{A} \times \vec{B} = 0 + 4\hat{k} + 3\hat{k} + 0 = \boxed{7\hat{k}}$
To verify that $\vec{A} \times \vec{B} = -\vec{B} \times \vec{A}$, evaluate $\vec{B} \times \vec{A}$:	$\vec{B} \times \vec{A} = (-\hat{i} + 2\,\hat{j}) \times (2\,\hat{i} + 3\,\hat{j})$
Perform the multiplication:	$\vec{B} \times \vec{A} = (-\hat{i}) \times 2\,\hat{i} + (-\hat{i}) \times 3\,\hat{j} + 2\,\hat{j} \times 2\,\hat{i} + 2\,\hat{j} \times 3\,\hat{j}$
Use Equations 11.7a through 11.7d to evaluate the various terms:	$\vec{B} \times \vec{A} = 0 - 3\hat{k} - 4\hat{k} + 0 = \boxed{-7\hat{k}}$

Therefore, $\vec{A} \times \vec{B} = -\vec{B} \times \vec{A}$. As an alternative method for finding $\vec{A} \times \vec{B}$, you could use Equation 11.8. Try it!

Example 11.2   The Torque Vector

A force of $\vec{F} = (2.00\,\hat{i} + 3.00\,\hat{j})$ N is applied to an object that is pivoted about a fixed axis aligned along the z coordinate axis. The force is applied at a point located at $\vec{r} = (4.00\,\hat{i} + 5.00\,\hat{j})$ m. Find the torque $\vec{\tau}$ applied to the object.

SOLUTION

Conceptualize Given the unit-vector notations, think about the directions of the force and position vectors. If this force were applied at this position, in what direction would an object pivoted at the origin turn?

continued

▶ **11.2** continued

Categorize Because we use the definition of the cross product discussed in this section, we categorize this example as a substitution problem.

Set up the torque vector using Equation 11.1:

$$\vec{\tau} = \vec{r} \times \vec{F} = [(4.00\,\hat{i} + 5.00\,\hat{j})\,m] \times [(2.00\,\hat{i} + 3.00\,\hat{j})\,N]$$

Perform the multiplication:

$$\vec{\tau} = [(4.00)(2.00)\,\hat{i} \times \hat{i} + (4.00)(3.00)\,\hat{i} \times \hat{j}$$
$$+ (5.00)(2.00)\,\hat{j} \times \hat{i} + (5.00)(3.00)\,\hat{j} \times \hat{j}]\,N \cdot m$$

Use Equations 11.7a through 11.7d to evaluate the various terms:

$$\vec{\tau} = [0 + 12.0\,\hat{k} - 10.0\,\hat{k} + 0]\,N \cdot m = \boxed{2.0\,\hat{k}\,N \cdot m}$$

Notice that both $\vec{r}$ and $\vec{F}$ are in the xy plane. As expected, the torque vector is perpendicular to this plane, having only a z component. We have followed the rules for significant figures discussed in Section 1.6, which lead to an answer with two significant figures. We have lost some precision because we ended up subtracting two numbers that are close.

11.2 Analysis Model: Nonisolated System (Angular Momentum)

Figure 11.3 As the skater passes the pole, she grabs hold of it, which causes her to swing around the pole rapidly in a circular path.

Imagine a rigid pole sticking up through the ice on a frozen pond (Fig. 11.3). A skater glides rapidly toward the pole, aiming a little to the side so that she does not hit it. As she passes the pole, she reaches out to her side and grabs it, an action that causes her to move in a circular path around the pole. Just as the idea of linear momentum helps us analyze translational motion, a rotational analog—*angular momentum*—helps us analyze the motion of this skater and other objects undergoing rotational motion.

In Chapter 9, we developed the mathematical form of linear momentum and then proceeded to show how this new quantity was valuable in problem solving. We will follow a similar procedure for angular momentum.

Consider a particle of mass m located at the vector position $\vec{r}$ and moving with linear momentum $\vec{p}$ as in Figure 11.4. In describing translational motion, we found that the net force on the particle equals the time rate of change of its linear momentum, $\sum \vec{F} = d\vec{p}/dt$ (see Eq. 9.3). Let us take the cross product of each side of Equation 9.3 with $\vec{r}$, which gives the net torque on the particle on the left side of the equation:

$$\vec{r} \times \sum \vec{F} = \sum \vec{\tau} = \vec{r} \times \frac{d\vec{p}}{dt}$$

Now let's add to the right side the term $(d\vec{r}/dt) \times \vec{p}$, which is zero because $d\vec{r}/dt = \vec{v}$ and $\vec{v}$ and $\vec{p}$ are parallel. Therefore,

$$\sum \vec{\tau} = \vec{r} \times \frac{d\vec{p}}{dt} + \frac{d\vec{r}}{dt} \times \vec{p}$$

We recognize the right side of this equation as the derivative of $\vec{r} \times \vec{p}$ (see Eq. 11.6). Therefore,

$$\sum \vec{\tau} = \frac{d(\vec{r} \times \vec{p})}{dt} \tag{11.9}$$

which looks very similar in form to Equation 9.3, $\sum \vec{F} = d\vec{p}/dt$. Because torque plays the same role in rotational motion that force plays in translational motion, this result suggests that the combination $\vec{r} \times \vec{p}$ should play the same role in rotational motion that $\vec{\mathbf{p}}$ plays in translational motion. We call this combination the *angular momentum* of the particle:

> The instantaneous **angular momentum** $\vec{\mathbf{L}}$ of a particle relative to an axis through the origin O is defined by the cross product of the particle's instantaneous position vector $\vec{\mathbf{r}}$ and its instantaneous linear momentum $\vec{\mathbf{p}}$:
>
> $$\vec{\mathbf{L}} \equiv \vec{\mathbf{r}} \times \vec{\mathbf{p}} \qquad (11.10)$$

◀ **Angular momentum of a particle**

We can now write Equation 11.9 as

$$\sum \vec{\boldsymbol{\tau}} = \frac{d\vec{\mathbf{L}}}{dt} \qquad (11.11)$$

which is the rotational analog of Newton's second law, $\sum \vec{\mathbf{F}} = d\vec{\mathbf{p}}/dt$. Torque causes the angular momentum $\vec{\mathbf{L}}$ to change just as force causes linear momentum $\vec{\mathbf{p}}$ to change.

Notice that Equation 11.11 is valid only if $\sum \vec{\boldsymbol{\tau}}$ and $\vec{\mathbf{L}}$ are measured about the same axis. Furthermore, the expression is valid for any axis fixed in an inertial frame.

The SI unit of angular momentum is kg · m²/s. Notice also that both the magnitude and the direction of $\vec{\mathbf{L}}$ depend on the choice of axis. Following the right-hand rule, we see that the direction of $\vec{\mathbf{L}}$ is perpendicular to the plane formed by $\vec{\mathbf{r}}$ and $\vec{\mathbf{p}}$. In Figure 11.4, $\vec{\mathbf{r}}$ and $\vec{\mathbf{p}}$ are in the xy plane, so $\vec{\mathbf{L}}$ points in the z direction. Because $\vec{\mathbf{p}} = m\vec{\mathbf{v}}$, the magnitude of $\vec{\mathbf{L}}$ is

$$L = mvr \sin \phi \qquad (11.12)$$

where ϕ is the angle between $\vec{\mathbf{r}}$ and $\vec{\mathbf{p}}$. It follows that L is zero when $\vec{\mathbf{r}}$ is parallel to $\vec{\mathbf{p}}$ ($\phi = 0$ or $180°$). In other words, when the translational velocity of the particle is along a line that passes through the axis, the particle has zero angular momentum with respect to the axis. On the other hand, if $\vec{\mathbf{r}}$ is perpendicular to $\vec{\mathbf{p}}$ ($\phi = 90°$), then $L = mvr$. At that instant, the particle moves exactly as if it were on the rim of a wheel rotating about the axis in a plane defined by $\vec{\mathbf{r}}$ and $\vec{\mathbf{p}}$.

Quick Quiz 11.2 Recall the skater described at the beginning of this section. Let her mass be m. **(i)** What would be her angular momentum relative to the pole at the instant she is a distance d from the pole if she were skating directly toward it at speed v? (a) zero (b) mvd (c) impossible to determine **(ii)** What would be her angular momentum relative to the pole at the instant she is a distance d from the pole if she were skating at speed v along a straight path that is a perpendicular distance a from the pole? (a) zero (b) mvd (c) mva (d) impossible to determine

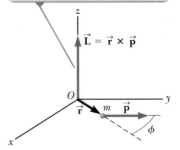

The angular momentum $\vec{\mathbf{L}}$ of a particle about an axis is a vector perpendicular to both the particle's position $\vec{\mathbf{r}}$ relative to the axis and its momentum $\vec{\mathbf{p}}$.

Figure 11.4 The angular momentum $\vec{\mathbf{L}}$ of a particle is a vector given by $\vec{\mathbf{L}} = \vec{\mathbf{r}} \times \vec{\mathbf{p}}$.

Pitfall Prevention 11.2
Is Rotation Necessary for Angular Momentum? We can define angular momentum even if the particle is not moving in a circular path. A particle moving in a straight line has angular momentum about any axis displaced from the path of the particle.

Example 11.3	**Angular Momentum of a Particle in Circular Motion**

A particle moves in the xy plane in a circular path of radius r as shown in Figure 11.5. Find the magnitude and direction of its angular momentum relative to an axis through O when its velocity is $\vec{\mathbf{v}}$.

SOLUTION

Conceptualize The linear momentum of the particle is always changing in direction (but not in magnitude). You might therefore be tempted to conclude that the angular momentum of the particle is always changing. In this situation, however, that is not the case. Let's see why.

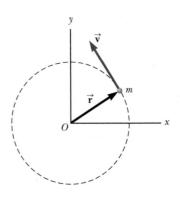

Figure 11.5 (Example 11.3) A particle moving in a circle of radius r has an angular momentum about an axis through O that has magnitude mvr. The vector $\vec{\mathbf{L}} = \vec{\mathbf{r}} \times \vec{\mathbf{p}}$ points *out* of the page.

continued

▶ **11.3** continued

Categorize We use the definition of the angular momentum of a particle discussed in this section, so we categorize this example as a substitution problem.

Use Equation 11.12 to evaluate the magnitude of $\vec{L}$:

$$L = mvr\sin 90° = \boxed{mvr}$$

This value of L is constant because all three factors on the right are constant. The direction of $\vec{L}$ also is constant, even though the direction of $\vec{p} = m\vec{v}$ keeps changing. To verify this statement, apply the right-hand rule to find the direction of $\vec{L} = \vec{r} \times \vec{p} = m\vec{r} \times \vec{v}$ in Figure 11.5. Your thumb points out of the page, so that is the direction of $\vec{L}$. Hence, we can write the vector expression $\vec{L} = (mvr)\hat{k}$. If the particle were to move clockwise, $\vec{L}$ would point downward and into the page and $\vec{L} = -(mvr)\hat{k}$. A particle in uniform circular motion has a constant angular momentum about an axis through the center of its path.

Angular Momentum of a System of Particles

Using the techniques of Section 9.7, we can show that Newton's second law for a system of particles is

$$\sum \vec{F}_{ext} = \frac{d\vec{p}_{tot}}{dt}$$

This equation states that the net external force on a system of particles is equal to the time rate of change of the total linear momentum of the system. Let's see if a similar statement can be made for rotational motion. The total angular momentum of a system of particles about some axis is defined as the vector sum of the angular momenta of the individual particles:

$$\vec{L}_{tot} = \vec{L}_1 + \vec{L}_2 + \cdots + \vec{L}_n = \sum_i \vec{L}_i$$

where the vector sum is over all n particles in the system.

Differentiating this equation with respect to time gives

$$\frac{d\vec{L}_{tot}}{dt} = \sum_i \frac{d\vec{L}_i}{dt} = \sum_i \vec{\tau}_i$$

where we have used Equation 11.11 to replace the time rate of change of the angular momentum of each particle with the net torque on the particle.

The torques acting on the particles of the system are those associated with internal forces between particles and those associated with external forces. The net torque associated with all internal forces, however, is zero. Recall that Newton's third law tells us that internal forces between particles of the system are equal in magnitude and opposite in direction. If we assume these forces lie along the line of separation of each pair of particles, the total torque around some axis passing through an origin O due to each action–reaction force pair is zero (that is, the moment arm d from O to the line of action of the forces is equal for both particles, and the forces are in opposite directions). In the summation, therefore, the net internal torque is zero. We conclude that the total angular momentum of a system can vary with time only if a net external torque is acting on the system:

▶ The net external torque on a system equals the time rate of change of angular momentum of the system

$$\sum \vec{\tau}_{ext} = \frac{d\vec{L}_{tot}}{dt} \tag{11.13}$$

This equation is indeed the rotational analog of $\sum \vec{F}_{ext} = d\vec{p}_{tot}/dt$ for a system of particles. Equation 11.13 is the mathematical representation of the **angular momentum version of the nonisolated system model.** If a system is nonisolated in the sense that there is a net torque on it, the torque is equal to the time rate of change of angular momentum.

Although we do not prove it here, this statement is true regardless of the motion of the center of mass. It applies even if the center of mass is accelerating, provided the torque and angular momentum are evaluated relative to an axis through the center of mass.

Equation 11.13 can be rearranged and integrated to give

$$\Delta \vec{\mathbf{L}}_{tot} = \int \left(\sum \vec{\boldsymbol{\tau}}_{ext} \right) dt$$

This equation represents the *angular impulse–angular momentum theorem*. Compare this equation to the translational version, Equation 9.40.

Analysis Model | Nonisolated System (Angular Momentum)

Imagine a system that rotates about an axis. If there is a net external torque acting on the system, the time rate of change of the angular momentum of the system is equal to the net external torque:

$$\sum \vec{\boldsymbol{\tau}}_{ext} = \frac{d\vec{\mathbf{L}}_{tot}}{dt} \qquad \text{(11.13)}$$

Examples:

- a flywheel in an automobile engine increases its angular momentum when the engine applies torque to it
- the tub of a washing machine decreases in angular momentum due to frictional torque after the machine is turned off
- the axis of the Earth undergoes a precessional motion due to the torque exerted on the Earth by the gravitational force from the Sun
- the armature of a motor increases its angular momentum due to the torque exerted by a surrounding magnetic field (Chapter 31)

The rate of change in the angular momentum of the nonisolated system is equal to the net external torque on the system.

Example 11.4 | A System of Objects | AM

A sphere of mass m_1 and a block of mass m_2 are connected by a light cord that passes over a pulley as shown in Figure 11.6. The radius of the pulley is R, and the mass of the thin rim is M. The spokes of the pulley have negligible mass. The block slides on a frictionless, horizontal surface. Find an expression for the linear acceleration of the two objects, using the concepts of angular momentum and torque.

SOLUTION

Conceptualize When the system is released, the block slides to the left, the sphere drops downward, and the pulley rotates counterclockwise. This situation is similar to problems we have solved earlier except that now we want to use an angular momentum approach.

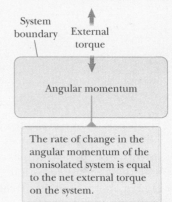

Figure 11.6 (Example 11.4) When the system is released, the sphere moves downward and the block moves to the left.

Categorize We identify the block, pulley, and sphere as a *nonisolated system* for *angular momentum*, subject to the external torque due to the gravitational force on the sphere. We shall calculate the angular momentum about an axis that coincides with the axle of the pulley. The angular momentum of the system includes that of two objects moving translationally (the sphere and the block) and one object undergoing pure rotation (the pulley).

Analyze At any instant of time, the sphere and the block have a common speed v, so the angular momentum of the sphere about the pulley axle is m_1vR and that of the block is m_2vR. At the same instant, all points on the rim of the pulley also move with speed v, so the angular momentum of the pulley is MvR.

Now let's address the total external torque acting on the system about the pulley axle. Because it has a moment arm of zero, the force exerted by the axle on the pulley does not contribute to the torque. Furthermore, the normal force

continued

▶ **11.4** continued

acting on the block is balanced by the gravitational force $m_2\vec{\mathbf{g}}$, so these forces do not contribute to the torque. The gravitational force $m_1\vec{\mathbf{g}}$ acting on the sphere produces a torque about the axle equal in magnitude to m_1gR, where R is the moment arm of the force about the axle. This result is the total external torque about the pulley axle; that is, $\sum \tau_{\text{ext}} = m_1gR$.

Write an expression for the total angular momentum of the system:

(1) $L = m_1vR + m_2vR + MvR = (m_1 + m_2 + M)vR$

Substitute this expression and the total external torque into Equation 11.13, the mathematical representation of the nonisolated system model for angular momentum:

$$\sum \tau_{\text{ext}} = \frac{dL}{dt}$$

$$m_1gR = \frac{d}{dt}[(m_1 + m_2 + M)vR]$$

(2) $m_1gR = (m_1 + m_2 + M)R\dfrac{dv}{dt}$

Recognizing that $dv/dt = a$, solve Equation (2) for a:

(3) $a = \dfrac{m_1g}{m_1 + m_2 + M}$

Finalize When we evaluated the net torque about the axle, we did not include the forces that the cord exerts on the objects because these forces are internal to the system under consideration. Instead, we analyzed the system as a whole. Only *external* torques contribute to the change in the system's angular momentum. Let $M \to 0$ in Equation (3) and call the result Equation A. Now go back to Equation (5) in Example 5.10, let $\theta \to 0$, and call the result Equation B. Do Equations A and B match? Looking at Figures 5.15 and 11.6 in these limits, *should* the two equations match?

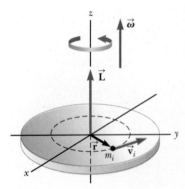

Figure 11.7 When a rigid object rotates about an axis, the angular momentum $\vec{\mathbf{L}}$ is in the same direction as the angular velocity $\vec{\boldsymbol{\omega}}$ according to the expression $\vec{\mathbf{L}} = I\vec{\boldsymbol{\omega}}$.

11.3 Angular Momentum of a Rotating Rigid Object

In Example 11.4, we considered the angular momentum of a deformable system of particles. Let us now restrict our attention to a nondeformable system, a rigid object. Consider a rigid object rotating about a fixed axis that coincides with the z axis of a coordinate system as shown in Figure 11.7. Let's determine the angular momentum of this object. Each *particle* of the object rotates in the xy plane about the z axis with an angular speed ω. The magnitude of the angular momentum of a particle of mass m_i about the z axis is $m_iv_ir_i$. Because $v_i = r_i\omega$ (Eq. 10.10), we can express the magnitude of the angular momentum of this particle as

$$L_i = m_ir_i^2\omega$$

The vector $\vec{\mathbf{L}}_i$ for this particle is directed along the z axis, as is the vector $\vec{\boldsymbol{\omega}}$.

We can now find the angular momentum (which in this situation has only a z component) of the whole object by taking the sum of L_i over all particles:

$$L_z = \sum_i L_i = \sum_i m_ir_i^2\omega = \left(\sum_i m_ir_i^2\right)\omega$$

$$L_z = I\omega \tag{11.14}$$

where we have recognized $\sum_i m_ir_i^2$ as the moment of inertia I of the object about the z axis (Eq. 10.19). Notice that Equation 11.14 is mathematically similar in form to Equation 9.2 for linear momentum: $\vec{\mathbf{p}} = m\vec{\mathbf{v}}$.

Now let's differentiate Equation 11.14 with respect to time, noting that I is constant for a rigid object:

$$\frac{dL_z}{dt} = I\frac{d\omega}{dt} = I\alpha \tag{11.15}$$

where α is the angular acceleration relative to the axis of rotation. Because dL_z/dt is equal to the net external torque (see Eq. 11.13), we can express Equation 11.15 as

$$\sum \tau_{ext} = I\alpha \qquad (11.16)$$

◀ Rotational form of Newton's second law

That is, the net external torque acting on a rigid object rotating about a fixed axis equals the moment of inertia about the rotation axis multiplied by the object's angular acceleration relative to that axis. This result is the same as Equation 10.18, which was derived using a force approach, but we derived Equation 11.16 using the concept of angular momentum. As we saw in Section 10.7, Equation 11.16 is the mathematical representation of the rigid object under a net torque analysis model. This equation is also valid for a rigid object rotating about a moving axis, provided the moving axis (1) passes through the center of mass and (2) is a symmetry axis.

If a symmetrical object rotates about a fixed axis passing through its center of mass, you can write Equation 11.14 in vector form as $\vec{L} = I\vec{\omega}$, where $\vec{L}$ is the total angular momentum of the object measured with respect to the axis of rotation. Furthermore, the expression is valid for any object, regardless of its symmetry, if $\vec{L}$ stands for the component of angular momentum along the axis of rotation.[1]

Quick Quiz 11.3 A solid sphere and a hollow sphere have the same mass and radius. They are rotating with the same angular speed. Which one has the higher angular momentum? **(a)** the solid sphere **(b)** the hollow sphere **(c)** both have the same angular momentum **(d)** impossible to determine

---

Example 11.5   Bowling Ball

Estimate the magnitude of the angular momentum of a bowling ball spinning at 10 rev/s as shown in Figure 11.8.

SOLUTION

Conceptualize Imagine spinning a bowling ball on the smooth floor of a bowling alley. Because a bowling ball is relatively heavy, the angular momentum should be relatively large.

Categorize We evaluate the angular momentum using Equation 11.14, so we categorize this example as a substitution problem.

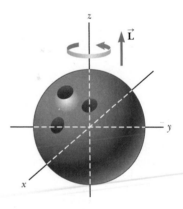

Figure 11.8 (Example 11.5) A bowling ball that rotates about the z axis in the direction shown has an angular momentum $\vec{L}$ in the positive z direction. If the direction of rotation is reversed, then $\vec{L}$ points in the negative z direction.

We start by making some estimates of the relevant physical parameters and model the ball as a uniform solid sphere. A typical bowling ball might have a mass of 7.0 kg and a radius of 12 cm.

Evaluate the moment of inertia of the ball about an axis through its center from Table 10.2:

$$I = \tfrac{2}{5}MR^2 = \tfrac{2}{5}(7.0 \text{ kg})(0.12 \text{ m})^2 = 0.040 \text{ kg} \cdot \text{m}^2$$

Evaluate the magnitude of the angular momentum from Equation 11.14:

$$L_z = I\omega = (0.040 \text{ kg} \cdot \text{m}^2)(10 \text{ rev/s})(2\pi \text{ rad/rev}) = 2.53 \text{ kg} \cdot \text{m}^2/\text{s}$$

Because of the roughness of our estimates, we should keep only one significant figure, so $L_z = \boxed{3 \text{ kg} \cdot \text{m}^2/\text{s}.}$

---

[1]In general, the expression $\vec{L} = I\vec{\omega}$ is not always valid. If a rigid object rotates about an *arbitrary* axis, then $\vec{L}$ and $\vec{\omega}$ may point in different directions. In this case, the moment of inertia cannot be treated as a scalar. Strictly speaking, $\vec{L} = I\vec{\omega}$ applies only to rigid objects of any shape that rotate about one of three mutually perpendicular axes (called *principal axes*) through the center of mass. This concept is discussed in more advanced texts on mechanics.

Example 11.6   The Seesaw   AM

A father of mass m_f and his daughter of mass m_d sit on opposite ends of a seesaw at equal distances from the pivot at the center (Fig. 11.9). The seesaw is modeled as a rigid rod of mass M and length ℓ and is pivoted without friction. At a given moment, the combination rotates in a vertical plane with an angular speed ω.

(A) Find an expression for the magnitude of the system's angular momentum.

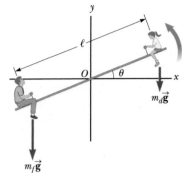

Figure 11.9 (Example 11.6) A father and daughter demonstrate angular momentum on a seesaw.

SOLUTION

Conceptualize Identify the z axis through O as the axis of rotation in Figure 11.9. The rotating system has angular momentum about that axis.

Categorize Ignore any movement of arms or legs of the father and daughter and model them both as particles. The system is therefore modeled as a rigid object. This first part of the example is categorized as a substitution problem.

The moment of inertia of the system equals the sum of the moments of inertia of the three components: the seesaw and the two individuals. We can refer to Table 10.2 to obtain the expression for the moment of inertia of the rod and use the particle expression $I = mr^2$ for each person.

Find the total moment of inertia of the system about the z axis through O:

$$I = \tfrac{1}{12}M\ell^2 + m_f\left(\frac{\ell}{2}\right)^2 + m_d\left(\frac{\ell}{2}\right)^2 = \frac{\ell^2}{4}\left(\frac{M}{3} + m_f + m_d\right)$$

Find the magnitude of the angular momentum of the system:

$$L = I\omega = \frac{\ell^2}{4}\left(\frac{M}{3} + m_f + m_d\right)\omega$$

(B) Find an expression for the magnitude of the angular acceleration of the system when the seesaw makes an angle θ with the horizontal.

SOLUTION

Conceptualize Generally, fathers are more massive than daughters, so the system is not in equilibrium and has an angular acceleration. We expect the angular acceleration to be positive in Figure 11.9.

Categorize The combination of the board, father, and daughter is a *rigid object under a net torque* because of the external torque associated with the gravitational forces on the father and daughter. We again identify the axis of rotation as the z axis in Figure 11.9.

Analyze To find the angular acceleration of the system at any angle θ, we first calculate the net torque on the system and then use $\Sigma \tau_{\text{ext}} = I\alpha$ from the rigid object under a net torque model to obtain an expression for α.

Evaluate the torque due to the gravitational force on the father:

$$\tau_f = m_f g \frac{\ell}{2} \cos\theta \quad (\vec{\tau}_f \text{ out of page})$$

Evaluate the torque due to the gravitational force on the daughter:

$$\tau_d = -m_d g \frac{\ell}{2} \cos\theta \quad (\vec{\tau}_d \text{ into page})$$

Evaluate the net external torque exerted on the system:

$$\sum \tau_{\text{ext}} = \tau_f + \tau_d = \tfrac{1}{2}(m_f - m_d)g\ell \cos\theta$$

Use Equation 11.16 and I from part (A) to find α:

$$\alpha = \frac{\sum \tau_{\text{ext}}}{I} = \frac{2(m_f - m_d)g\cos\theta}{\ell\,[(M/3) + m_f + m_d]}$$

Finalize For a father more massive than his daughter, the angular acceleration is positive as expected. If the seesaw begins in a horizontal orientation ($\theta = 0$) and is released, the rotation is counterclockwise in Figure 11.9 and the father's end of the seesaw drops, which is consistent with everyday experience.

WHAT IF? Imagine the father moves inward on the seesaw to a distance d from the pivot to try to balance the two sides. What is the angular acceleration of the system in this case when it is released from an arbitrary angle θ?

▶ **11.6** continued

Answer The angular acceleration of the system should decrease if the system is more balanced.

Find the total moment of inertia about the z axis through O for the modified system:

$$I = \tfrac{1}{12}M\ell^2 + m_f d^2 + m_d\left(\frac{\ell}{2}\right)^2 = \frac{\ell^2}{4}\left(\frac{M}{3} + m_d\right) + m_f d^2$$

Find the net torque exerted on the system about an axis through O:

$$\sum \tau_{ext} = \tau_f + \tau_d = m_f g d \cos\theta - \tfrac{1}{2}m_d g \ell \cos\theta$$

Find the new angular acceleration of the system:

$$\alpha = \frac{\sum \tau_{ext}}{I} = \frac{(m_f d - \tfrac{1}{2}m_d \ell)g\cos\theta}{(\ell^2/4)\,[(M/3) + m_d] + m_f d^2}$$

The seesaw is balanced when the angular acceleration is zero. In this situation, both father and daughter can push off the ground and rise to the highest possible point.

Find the required position of the father by setting $\alpha = 0$:

$$\alpha = \frac{(m_f d - \tfrac{1}{2}m_d \ell)g\cos\theta}{(\ell^2/4)[(M/3) + m_d] + m_f d^2} = 0$$

$$m_f d - \tfrac{1}{2}m_d \ell = 0 \quad \rightarrow \quad d = \left(\frac{m_d}{m_f}\right)\frac{\ell}{2}$$

In the rare case that the father and daughter have the same mass, the father is located at the end of the seesaw, $d = \ell/2$.

11.4 Analysis Model: Isolated System (Angular Momentum)

In Chapter 9, we found that the total linear momentum of a system of particles remains constant if the system is isolated, that is, if the net external force acting on the system is zero. We have an analogous conservation law in rotational motion:

> The total angular momentum of a system is constant in both magnitude and direction if the net external torque acting on the system is zero, that is, if the system is isolated.

◀ **Conservation of angular momentum**

This statement is often called[2] the principle of **conservation of angular momentum** and is the basis of the **angular momentum version of the isolated system model**. This principle follows directly from Equation 11.13, which indicates that if

$$\sum \vec{\tau}_{ext} = \frac{d\vec{L}_{tot}}{dt} = 0 \tag{11.17}$$

then

$$\Delta \vec{L}_{tot} = 0 \tag{11.18}$$

Equation 11.18 can be written as

$$\vec{L}_{tot} = \text{constant} \quad \text{or} \quad \vec{L}_i = \vec{L}_f$$

For an isolated system consisting of a small number of particles, we write this conservation law as $\vec{L}_{tot} = \sum \vec{L}_n = \text{constant}$, where the index n denotes the nth particle in the system.

If an isolated rotating system is deformable so that its mass undergoes redistribution in some way, the system's moment of inertia changes. Because the magnitude of the angular momentum of the system is $L = I\omega$ (Eq. 11.14), conservation

---

[2]The most general conservation of angular momentum equation is Equation 11.13, which describes how the system interacts with its environment.

When his arms and legs are close to his body, the skater's moment of inertia is small and his angular speed is large.

Clive Rose/Getty Images

To slow down for the finish of his spin, the skater moves his arms and legs outward, increasing his moment of inertia.

Al Bello/Getty Images

Figure 11.10 Angular momentum is conserved as Russian gold medalist Evgeni Plushenko performs during the Turin 2006 Winter Olympic Games.

of angular momentum requires that the product of I and ω must remain constant. Therefore, a change in I for an isolated system requires a change in ω. In this case, we can express the principle of conservation of angular momentum as

$$I_i\omega_i = I_f\omega_f = \text{constant} \tag{11.19}$$

This expression is valid both for rotation about a fixed axis and for rotation about an axis through the center of mass of a moving system as long as that axis remains fixed in direction. We require only that the net external torque be zero.

Many examples demonstrate conservation of angular momentum for a deformable system. You may have observed a figure skater spinning in the finale of a program (Fig. 11.10). The angular speed of the skater is large when his hands and feet are close to the trunk of his body. (Notice the skater's hair!) Ignoring friction between skater and ice, there are no external torques on the skater. The moment of inertia of his body increases as his hands and feet are moved away from his body at the finish of the spin. According to the isolated system model for angular momentum, his angular speed must decrease. In a similar way, when divers or acrobats wish to make several somersaults, they pull their hands and feet close to their bodies to rotate at a higher rate. In these cases, the external force due to gravity acts through the center of mass and hence exerts no torque about an axis through this point. Therefore, the angular momentum about the center of mass must be conserved; that is, $I_i\omega_i = I_f\omega_f$. For example, when divers wish to double their angular speed, they must reduce their moment of inertia to half its initial value.

In Equation 11.18, we have a third version of the isolated system model. We can now state that the energy, linear momentum, and angular momentum of an isolated system are all constant:

$$\Delta E_{\text{system}} = 0 \quad \text{(if there are no energy transfers across the system boundary)}$$

$$\Delta \vec{\mathbf{p}}_{\text{tot}} = 0 \quad \text{(if the net external force on the system is zero)}$$

$$\Delta \vec{\mathbf{L}}_{\text{tot}} = 0 \quad \text{(if the net external torque on the system is zero)}$$

A system may be isolated in terms of one of these quantities but not in terms of another. If a system is nonisolated in terms of momentum or angular momentum, it will often be nonisolated also in terms of energy because the system has a net force or torque on it and the net force or torque will do work on the system. We can, however, identify systems that are nonisolated in terms of energy but isolated in terms of momentum. For example, imagine pushing inward on a balloon (the system) between your hands. Work is done in compressing the balloon, so the system is nonisolated in terms of energy, but there is zero net force on the system, so the system is isolated in terms of momentum. A similar statement could be made about twisting the ends of a long, springy piece of metal with both hands. Work is done on the metal (the system), so energy is stored in the nonisolated system as elastic potential energy, but the net torque on the system is zero. Therefore, the system is isolated in terms of angular momentum. Other examples are collisions of macroscopic objects, which represent isolated systems in terms of momentum but nonisolated systems in terms of energy because of the output of energy from the system by mechanical waves (sound).

Quick Quiz 11.4 A competitive diver leaves the diving board and falls toward the water with her body straight and rotating slowly. She pulls her arms and legs into a tight tuck position. What happens to her rotational kinetic energy? **(a)** It increases. **(b)** It decreases. **(c)** It stays the same. **(d)** It is impossible to determine.

Analysis Model | Isolated System (Angular Momentum)

Imagine a system rotates about an axis. If there is no net external torque on the system, there is no change in the angular momentum of the system:

$$\Delta \vec{L}_{tot} = 0 \qquad \text{(11.18)}$$

Applying this law of conservation of angular momentum to a system whose moment of inertia changes gives

$$I_i\omega_i = I_f\omega_f = \text{constant} \qquad \text{(11.19)}$$

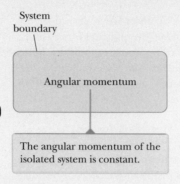

System boundary

Angular momentum

The angular momentum of the isolated system is constant.

Examples:

- after a supernova explosion, the core of a star collapses to a small radius and spins at a much higher rate
- the square of the orbital period of a planet is proportional to the cube of its semimajor axis; Kepler's third law (Chapter 13)
- in atomic transitions, selection rules on the quantum numbers must be obeyed in order to conserve angular momentum (Chapter 42)
- in beta decay of a radioactive nucleus, a neutrino must be emitted in order to conserve angular momentum (Chapter 44)

Example 11.7 | Formation of a Neutron Star AM

A star rotates with a period of 30 days about an axis through its center. The period is the time interval required for a point on the star's equator to make one complete revolution around the axis of rotation. After the star undergoes a supernova explosion, the stellar core, which had a radius of 1.0×10^4 km, collapses into a neutron star of radius 3.0 km. Determine the period of rotation of the neutron star.

SOLUTION

Conceptualize The change in the neutron star's motion is similar to that of the skater described earlier, but in the reverse direction. As the mass of the star moves closer to the rotation axis, we expect the star to spin faster.

Categorize Let us assume that during the collapse of the stellar core, (1) no external torque acts on it, (2) it remains spherical with the same relative mass distribution, and (3) its mass remains constant. We categorize the star as an *isolated system* in terms of *angular momentum*. We do not know the mass distribution of the star, but we have assumed the distribution is symmetric, so the moment of inertia can be expressed as kMR^2, where k is some numerical constant. (From Table 10.2, for example, we see that $k = \frac{2}{5}$ for a solid sphere and $k = \frac{2}{3}$ for a spherical shell.)

. . . .

Analyze Let's use the symbol T for the period, with T_i being the initial period of the star and T_f being the period of the neutron star. The star's angular speed is given by $\omega = 2\pi/T$.

From the isolated system model for angular momentum, write Equation 11.19 for the star:

$$I_i\omega_i = I_f\omega_f$$

Use $\omega = 2\pi/T$ to rewrite this equation in terms of the initial and final periods:

$$I_i\left(\frac{2\pi}{T_i}\right) = I_f\left(\frac{2\pi}{T_f}\right)$$

Substitute the moments of inertia in the preceding equation:

$$kMR_i^2\left(\frac{2\pi}{T_i}\right) = kMR_f^2\left(\frac{2\pi}{T_f}\right)$$

Solve for the final period of the star:

$$T_f = \left(\frac{R_f}{R_i}\right)^2 T_i$$

Substitute numerical values:

$$T_f = \left(\frac{3.0 \text{ km}}{1.0 \times 10^4 \text{ km}}\right)^2 (30 \text{ days}) = 2.7 \times 10^{-6} \text{ days} = \boxed{0.23 \text{ s}}$$

. . . .

Finalize The neutron star does indeed rotate faster after it collapses, as predicted. It moves very fast, in fact, rotating about four times each second!

Example 11.8   The Merry-Go-Round

A horizontal platform in the shape of a circular disk rotates freely in a horizontal plane about a frictionless, vertical axle (Fig. 11.11). The platform has a mass $M = 100$ kg and a radius $R = 2.0$ m. A student whose mass is $m = 60$ kg walks slowly from the rim of the disk toward its center. If the angular speed of the system is 2.0 rad/s when the student is at the rim, what is the angular speed when she reaches a point $r = 0.50$ m from the center?

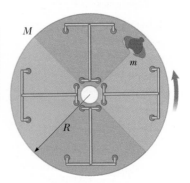

Figure 11.11 (Example 11.8) As the student walks toward the center of the rotating platform, the angular speed of the system increases because the angular momentum of the system remains constant.

SOLUTION

Conceptualize The speed change here is similar to those of the spinning skater and the neutron star in preceding discussions. This problem is different because part of the moment of inertia of the system changes (that of the student) while part remains fixed (that of the platform).

Categorize Because the platform rotates on a frictionless axle, we identify the system of the student and the platform as an *isolated system* in terms of *angular momentum.*

Analyze Let us denote the moment of inertia of the platform as I_p and that of the student as I_s. We model the student as a particle.

Find the initial moment of inertia I_i of the system (student plus platform) about the axis of rotation:

$$I_i = I_{pi} + I_{si} = \tfrac{1}{2}MR^2 + mR^2$$

Find the moment of inertia of the system when the student walks to the position $r < R$:

$$I_f = I_{pf} + I_{sf} = \tfrac{1}{2}MR^2 + mr^2$$

Write Equation 11.19 for the system:

$$I_i \omega_i = I_f \omega_f$$

Substitute the moments of inertia:

$$(\tfrac{1}{2}MR^2 + mR^2)\omega_i = (\tfrac{1}{2}MR^2 + mr^2)\omega_f$$

Solve for the final angular speed:

$$\omega_f = \left(\frac{\tfrac{1}{2}MR^2 + mR^2}{\tfrac{1}{2}MR^2 + mr^2} \right)\omega_i$$

Substitute numerical values:

$$\omega_f = \left[\frac{\tfrac{1}{2}(100 \text{ kg})(2.0 \text{ m})^2 + (60 \text{ kg})(2.0 \text{ m})^2}{\tfrac{1}{2}(100 \text{ kg})(2.0 \text{ m})^2 + (60 \text{ kg})(0.50 \text{ m})^2} \right](2.0 \text{ rad/s}) = \boxed{4.1 \text{ rad/s}}$$

Finalize As expected, the angular speed increases. The fastest that this system could spin would be when the student moves to the center of the platform. Do this calculation to show that this maximum angular speed is 4.4 rad/s. Notice that the activity described in this problem is dangerous as discussed with regard to the Coriolis force in Section 6.3.

WHAT IF?   What if you measured the kinetic energy of the system before and after the student walks inward? Are the initial kinetic energy and the final kinetic energy the same?

Answer You may be tempted to say yes because the system is isolated. Remember, however, that energy can be transformed among several forms, so we have to handle an energy question carefully.

Find the initial kinetic energy:

$$K_i = \tfrac{1}{2}I_i \omega_i^2 = \tfrac{1}{2}(440 \text{ kg} \cdot \text{m}^2)(2.0 \text{ rad/s})^2 = 880 \text{ J}$$

Find the final kinetic energy:

$$K_f = \tfrac{1}{2}I_f \omega_f^2 = \tfrac{1}{2}(215 \text{ kg} \cdot \text{m}^2)(4.1 \text{ rad/s})^2 = 1.80 \times 10^3 \text{ J}$$

Therefore, the kinetic energy of the system *increases.* The student must perform muscular activity to move herself closer to the center of rotation, so this extra kinetic energy comes from potential energy stored in the student's body from previous meals. The system is isolated in terms of energy, but a transformation process within the system changes potential energy to kinetic energy.

Example 11.9   Disk and Stick Collision   AM

A 2.0-kg disk traveling at 3.0 m/s strikes a 1.0-kg stick of length 4.0 m that is lying flat on nearly frictionless ice as shown in the overhead view of Figure 11.12a. The disk strikes at the endpoint of the stick, at a distance $r = 2.0$ m from the stick's center. Assume the collision is elastic and the disk does not deviate from its original line of motion. Find the translational speed of the disk, the translational speed of the stick, and the angular speed of the stick after the collision. The moment of inertia of the stick about its center of mass is 1.33 kg · m².

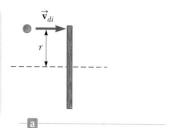

Before

SOLUTION

Conceptualize Examine Figure 11.12a and imagine what happens after the disk hits the stick. Figure 11.12b shows what you might expect: the disk continues to move at a slower speed, and the stick is in both translational and rotational motion. We assume the disk does not deviate from its original line of motion because the force exerted by the stick on the disk is parallel to the original path of the disk.

Figure 11.12 (Example 11.9) Overhead view of a disk striking a stick in an elastic collision. (a) Before the collision, the disk moves toward the stick. (b) The collision causes the stick to rotate and move to the right.

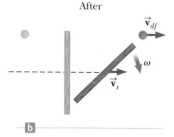

After

Categorize Because the ice is frictionless, the disk and stick form an *isolated system* in terms of *momentum* and *angular momentum*. Ignoring the sound made in the collision, we also model the system as an *isolated system* in terms of *energy*. In addition, because the collision is assumed to be elastic, the kinetic energy of the system is constant.

Analyze First notice that we have three unknowns, so we need three equations to solve simultaneously.

Apply the isolated system model for momentum to the system and then rearrange the result:

$$\Delta \vec{p}_{tot} = 0 \quad \rightarrow \quad (m_d v_{df} + m_s v_s) - m_d v_{di} = 0$$
$$(1) \quad m_d(v_{di} - v_{df}) = m_s v_s$$

Apply the isolated system model for angular momentum to the system and rearrange the result. Use an axis passing through the center of the stick as the rotation axis so that the path of the disk is a distance $r = 2.0$ m from the rotation axis:

$$\Delta \vec{L}_{tot} = 0 \quad \rightarrow \quad (-rm_d v_{df} + I\omega) - (-rm_d v_{di}) = 0$$
$$(2) \quad -rm_d(v_{di} - v_{df}) = I\omega$$

Apply the isolated system model for energy to the system, rearrange the equation, and factor the combination of terms related to the disk:

$$\Delta K = 0 \quad \rightarrow \quad (\tfrac{1}{2}m_d v_{df}^2 + \tfrac{1}{2}m_s v_s^2 + \tfrac{1}{2}I\omega^2) - \tfrac{1}{2}m_d v_{di}^2 = 0$$
$$(3) \quad m_d(v_{di} - v_{df})(v_{di} + v_{df}) = m_s v_s^2 + I\omega^2$$

Multiply Equation (1) by r and add to Equation (2):

$$rm_d(v_{di} - v_{df}) = rm_s v_s$$
$$-rm_d(v_{di} - v_{df}) = I\omega$$
$$0 = rm_s v_s + I\omega$$

Solve for ω:

$$(4) \quad \omega = -\frac{rm_s v_s}{I}$$

Divide Equation (3) by Equation (1):

$$\frac{m_d(v_{di} - v_{df})(v_{di} + v_{df})}{m_d(v_{di} - v_{df})} = \frac{m_s v_s^2 + I\omega^2}{m_s v_s}$$
$$(5) \quad v_{di} + v_{df} = v_s + \frac{I\omega^2}{m_s v_s}$$

Substitute Equation (4) into Equation (5):

$$(6) \quad v_{di} + v_{df} = v_s\left(1 + \frac{r^2 m_s}{I}\right)$$

Substitute v_{df} from Equation (1) into Equation (6):

$$v_{di} + \left(v_{di} - \frac{m_s}{m_d}v_s\right) = v_s\left(1 + \frac{r^2 m_s}{I}\right)$$

continued

▶ **11.9** continued

Solve for v_s and substitute numerical values:

$$v_s = \frac{2v_{di}}{1 + (m_s/m_d) + (r^2 m_s/I)}$$

$$= \frac{2(3.0 \text{ m/s})}{1 + (1.0 \text{ kg}/2.0 \text{ kg}) + [(2.0 \text{ m})^2(1.0 \text{ kg})/1.33 \text{ kg} \cdot \text{m}^2]} = \boxed{1.3 \text{ m/s}}$$

Substitute numerical values into Equation (4):

$$\omega = -\frac{(2.0 \text{ m})(1.0 \text{ kg})(1.3 \text{ m/s})}{1.33 \text{ kg} \cdot \text{m}^2} = \boxed{-2.0 \text{ rad/s}}$$

Solve Equation (1) for v_{df} and substitute numerical values:

$$v_{df} = v_{di} - \frac{m_s}{m_d}v_s = 3.0 \text{ m/s} - \frac{1.0 \text{ kg}}{2.0 \text{ kg}}(1.3 \text{ m/s}) = \boxed{2.3 \text{ m/s}}$$

Finalize These values seem reasonable. The disk is moving more slowly after the collision than it was before the collision, and the stick has a small translational speed. Table 11.1 summarizes the initial and final values of variables for the disk and the stick, and it verifies the conservation of linear momentum, angular momentum, and kinetic energy for the isolated system.

Table 11.1   **Comparison of Values in Example 11.9 Before and After the Collision**

	v (m/s)	ω (rad/s)	p (kg · m/s)	L (kg · m²/s)	K_{trans} (J)	K_{rot} (J)
Before						
Disk	3.0	—	6.0	−12	9.0	—
Stick	0	0	0	0	0	0
Total for system	—	—	6.0	−12	9.0	0
After						
Disk	2.3	—	4.7	−9.3	5.4	—
Stick	1.3	−2.0	1.3	−2.7	0.9	2.7
Total for system	—	—	6.0	−12	6.3	2.7

Note: Linear momentum, angular momentum, and total kinetic energy of the system are all conserved.

11.5 The Motion of Gyroscopes and Tops

An unusual and fascinating type of motion you have probably observed is that of a top spinning about its axis of symmetry as shown in Figure 11.13a. If the top spins rapidly, the symmetry axis rotates about the z axis, sweeping out a cone (see Fig. 11.13b). The motion of the symmetry axis about the vertical—known as **precessional motion**—is usually slow relative to the spinning motion of the top.

It is quite natural to wonder why the top does not fall over. Because the center of mass is not directly above the pivot point O, a net torque is acting on the top about an axis passing through O, a torque resulting from the gravitational force $M\vec{\mathbf{g}}$. The top would certainly fall over if it were not spinning. Because it is spinning, however, it has an angular momentum $\vec{\mathbf{L}}$ directed along its symmetry axis. We shall show that this symmetry axis moves about the z axis (precessional motion occurs) because the torque produces a change in the *direction* of the symmetry axis. This illustration is an excellent example of the importance of the vector nature of angular momentum.

The essential features of precessional motion can be illustrated by considering the simple gyroscope shown in Figure 11.14a. The two forces acting on the gyroscope are shown in Figure 11.14b: the downward gravitational force $M\vec{\mathbf{g}}$ and the normal force $\vec{\mathbf{n}}$ acting upward at the pivot point O. The normal force produces no torque about an axis passing through the pivot because its moment arm through that point is zero. The gravitational force, however, produces a torque $\vec{\boldsymbol{\tau}} = \vec{\mathbf{r}} \times M\vec{\mathbf{g}}$ about an axis passing through O, where the direction of $\vec{\boldsymbol{\tau}}$ is perpendicular to the plane formed by $\vec{\mathbf{r}}$ and $M\vec{\mathbf{g}}$. By necessity, the vector $\vec{\boldsymbol{\tau}}$ lies in a horizontal xy plane perpendicular to the angular momentum vector. The net torque and angular momentum of the gyroscope are related through Equation 11.13:

$$\sum \vec{\tau}_{ext} = \frac{d\vec{L}}{dt}$$

This expression shows that in the infinitesimal time interval dt, the nonzero torque produces a change in angular momentum $d\vec{L}$, a change that is in the same direction as $\vec{\tau}$. Therefore, like the torque vector, $d\vec{L}$ must also be perpendicular to $\vec{L}$. Figure 11.14c illustrates the resulting precessional motion of the symmetry axis of the gyroscope. In a time interval dt, the change in angular momentum is $d\vec{L} = \vec{L}_f - \vec{L}_i = \vec{\tau}\, dt$. Because $d\vec{L}$ is perpendicular to $\vec{L}$, the magnitude of $\vec{L}$ does not change ($|\vec{L}_i| = |\vec{L}_f|$). Rather, what is changing is the *direction* of $\vec{L}$. Because the change in angular momentum $d\vec{L}$ is in the direction of $\vec{\tau}$, which lies in the xy plane, the gyroscope undergoes precessional motion.

To simplify the description of the system, we assume the total angular momentum of the precessing wheel is the sum of the angular momentum $I\vec{\omega}$ due to the spinning and the angular momentum due to the motion of the center of mass about the pivot. In our treatment, we shall neglect the contribution from the center-of-mass motion and take the total angular momentum to be simply $I\vec{\omega}$. In practice, this approximation is good if $\vec{\omega}$ is made very large.

The vector diagram in Figure 11.14c shows that in the time interval dt, the angular momentum vector rotates through an angle $d\phi$, which is also the angle through which the gyroscope axle rotates. From the vector triangle formed by the vectors $\vec{L}_i$, $\vec{L}_f$, and $d\vec{L}$, we see that

$$d\phi = \frac{dL}{L} = \frac{\sum \tau_{ext}\, dt}{L} = \frac{(Mgr_{CM})\, dt}{L}$$

Dividing through by dt and using the relationship $L = I\omega$, we find that the rate at which the axle rotates about the vertical axis is

$$\omega_p = \frac{d\phi}{dt} = \frac{Mgr_{CM}}{I\omega} \qquad \textbf{(11.20)}$$

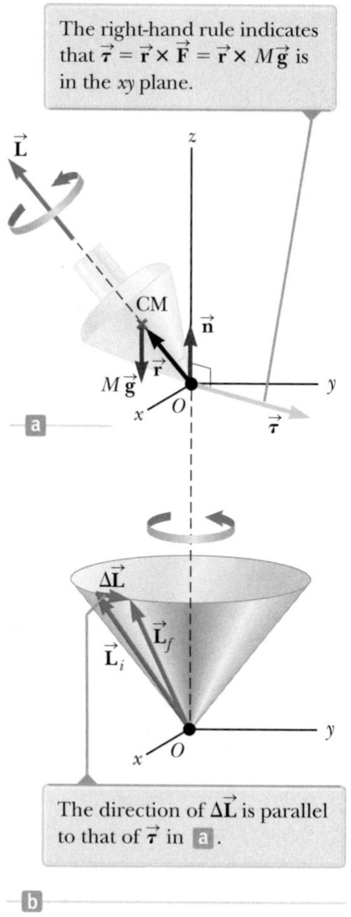

The right-hand rule indicates that $\vec{\tau} = \vec{r} \times \vec{F} = \vec{r} \times M\vec{g}$ is in the xy plane.

The direction of $\Delta\vec{L}$ is parallel to that of $\vec{\tau}$ in a.

Figure 11.13 Precessional motion of a top spinning about its symmetry axis. (a) The only external forces acting on the top are the normal force $\vec{n}$ and the gravitational force $M\vec{g}$. The direction of the angular momentum $\vec{L}$ is along the axis of symmetry. (b) Because $\vec{L}_f = \Delta\vec{L} + \vec{L}_i$, the top precesses about the z axis.

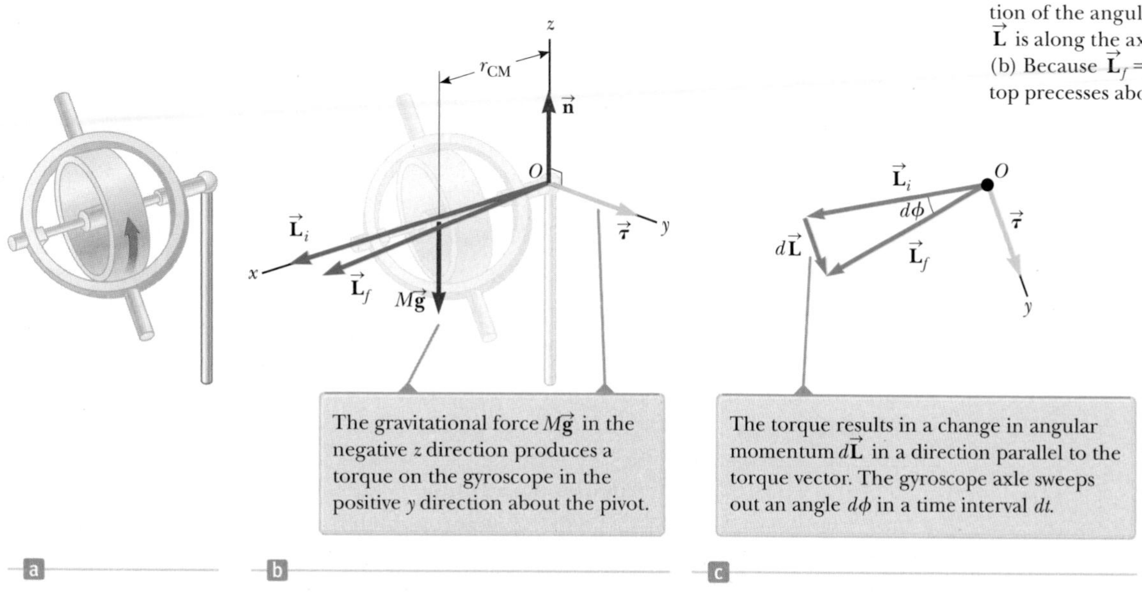

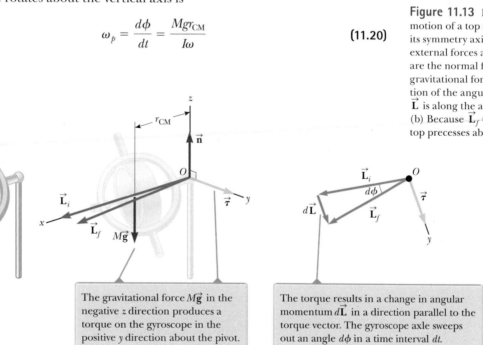

The gravitational force $M\vec{g}$ in the negative z direction produces a torque on the gyroscope in the positive y direction about the pivot.

The torque results in a change in angular momentum $d\vec{L}$ in a direction parallel to the torque vector. The gyroscope axle sweeps out an angle $d\phi$ in a time interval dt.

Figure 11.14 (a) A spinning gyroscope is placed on a pivot at the right end. (b) Diagram for the spinning gyroscope showing forces, torque, and angular momentum. (c) Overhead view (looking down the z axis) of the gyroscope's initial and final angular momentum vectors for an infinitesimal time interval dt.

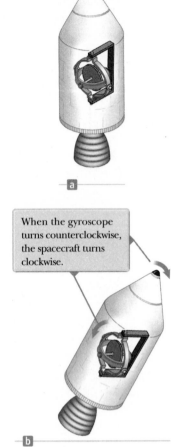

Figure 11.15 (a) A spacecraft carries a gyroscope that is not spinning. (b) The gyroscope is set into rotation.

The angular speed ω_p is called the **precessional frequency**. This result is valid only when $\omega_p \ll \omega$. Otherwise, a much more complicated motion is involved. As you can see from Equation 11.20, the condition $\omega_p \ll \omega$ is met when ω is large, that is, when the wheel spins rapidly. Furthermore, notice that the precessional frequency decreases as ω increases, that is, as the wheel spins faster about its axis of symmetry.

As an example of the usefulness of gyroscopes, suppose you are in a spacecraft in deep space and you need to alter your trajectory. To fire the engines in the correct direction, you need to turn the spacecraft. How, though, do you turn a spacecraft in empty space? One way is to have small rocket engines that fire perpendicularly out the side of the spacecraft, providing a torque around its center of mass. Such a setup is desirable, and many spacecraft have such rockets.

Let us consider another method, however, that does not require the consumption of rocket fuel. Suppose the spacecraft carries a gyroscope that is not rotating as in Figure 11.15a. In this case, the angular momentum of the spacecraft about its center of mass is zero. Suppose the gyroscope is set into rotation, giving the gyroscope a nonzero angular momentum. There is no external torque on the isolated system (spacecraft and gyroscope), so the angular momentum of this system must remain zero according to the isolated system (angular momentum) model. The zero value can be satisfied if the spacecraft rotates in the direction opposite that of the gyroscope so that the angular momentum vectors of the gyroscope and the spacecraft cancel, resulting in no angular momentum of the system. The result of rotating the gyroscope, as in Figure 11.15b, is that the spacecraft turns around! By including three gyroscopes with mutually perpendicular axles, any desired rotation in space can be achieved.

This effect created an undesirable situation with the *Voyager 2* spacecraft during its flight. The spacecraft carried a tape recorder whose reels rotated at high speeds. Each time the tape recorder was turned on, the reels acted as gyroscopes and the spacecraft started an undesirable rotation in the opposite direction. This rotation had to be counteracted by Mission Control by using the sideward-firing jets to *stop* the rotation!

Summary

Definitions

Given two vectors $\vec{A}$ and $\vec{B}$, the **vector product** $\vec{A} \times \vec{B}$ is a vector $\vec{C}$ having a magnitude

$$C = AB \sin \theta \qquad \text{(11.3)}$$

where θ is the angle between $\vec{A}$ and $\vec{B}$. The direction of the vector $\vec{C} = \vec{A} \times \vec{B}$ is perpendicular to the plane formed by $\vec{A}$ and $\vec{B}$, and this direction is determined by the right-hand rule.

The **torque** $\vec{\tau}$ on a particle due to a force $\vec{F}$ about an axis through the origin in an inertial frame is defined to be

$$\vec{\tau} \equiv \vec{r} \times \vec{F} \qquad \text{(11.1)}$$

The **angular momentum** $\vec{L}$ about an axis through the origin of a particle having linear momentum $\vec{p} = m\vec{v}$ is

$$\vec{L} \equiv \vec{r} \times \vec{p} \qquad \text{(11.10)}$$

where $\vec{r}$ is the vector position of the particle relative to the origin.

Concepts and Principles

The z component of angular momentum of a rigid object rotating about a fixed z axis is

$$L_z = I\omega \tag{11.14}$$

where I is the moment of inertia of the object about the axis of rotation and ω is its angular speed.

Analysis Models for Problem Solving

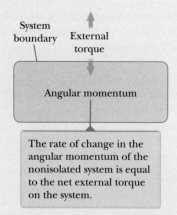

Nonisolated System (Angular Momentum). If a system interacts with its environment in the sense that there is an external torque on the system, the net external torque acting on a system is equal to the time rate of change of its angular momentum:

$$\sum \vec{\tau}_{\text{ext}} = \frac{d\vec{L}_{\text{tot}}}{dt} \tag{11.13}$$

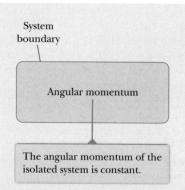

Isolated System (Angular Momentum). If a system experiences no external torque from the environment, the total angular momentum of the system is conserved:

$$\Delta \vec{L}_{\text{tot}} = 0 \tag{11.18}$$

Applying this law of conservation of angular momentum to a system whose moment of inertia changes gives

$$I_i\omega_i = I_f\omega_f = \text{constant} \tag{11.19}$$

Objective Questions

[1.] denotes answer available in *Student Solutions Manual/Study Guide*

1. An ice skater starts a spin with her arms stretched out to the sides. She balances on the tip of one skate to turn without friction. She then pulls her arms in so that her moment of inertia decreases by a factor of 2. In the process of her doing so, what happens to her kinetic energy? (a) It increases by a factor of 4. (b) It increases by a factor of 2. (c) It remains constant. (d) It decreases by a factor of 2. (e) It decreases by a factor of 4.

2. A pet mouse sleeps near the eastern edge of a stationary, horizontal turntable that is supported by a frictionless, vertical axle through its center. The mouse wakes up and starts to walk north on the turntable. (i) As it takes its first steps, what is the direction of the mouse's displacement relative to the stationary ground below? (a) north (b) south (c) no displacement. (ii) In this process, the spot on the turntable where the mouse had been snoozing undergoes a displacement in what direction relative to the ground below? (a) north (b) south (c) no displacement. Answer yes or no for the following questions. (iii) In this process, is the mechanical energy of the mouse–turntable system constant? (iv) Is the momentum of the system constant? (v) Is the angular momentum of the system constant?

[3.] Let us name three perpendicular directions as right, up, and toward you as you might name them when you are facing a television screen that lies in a vertical plane. Unit vectors for these directions are $\hat{\mathbf{r}}$, $\hat{\mathbf{u}}$, and $\hat{\mathbf{t}}$, respectively. Consider the quantity $(-3\hat{\mathbf{u}} \times 2\hat{\mathbf{t}})$. (i) Is the magnitude of this vector (a) 6, (b) 3, (c) 2, or (d) 0? (ii) Is the direction of this vector (a) down, (b) toward you, (c) up, (d) away from you, or (e) left?

4. Let the four compass directions north, east, south, and west be represented by unit vectors $\hat{\mathbf{n}}$, $\hat{\mathbf{e}}$, $\hat{\mathbf{s}}$, and $\hat{\mathbf{w}}$, respectively. Vertically up and down are represented as $\hat{\mathbf{u}}$ and $\hat{\mathbf{d}}$. Let us also identify unit vectors that are halfway between these directions such as $\hat{\mathbf{ne}}$ for northeast. Rank the magnitudes of the following cross products from largest to smallest. If any are equal in magnitude

or are equal to zero, show that in your ranking. (a) $\hat{\mathbf{n}} \times \hat{\mathbf{n}}$ (b) $\hat{\mathbf{w}} \times \hat{\mathbf{ne}}$ (c) $\hat{\mathbf{u}} \times \hat{\mathbf{ne}}$ (d) $\hat{\mathbf{n}} \times \hat{\mathbf{nw}}$ (e) $\hat{\mathbf{n}} \times \hat{\mathbf{e}}$

5. Answer yes or no to the following questions. (a) Is it possible to calculate the torque acting on a rigid object without specifying an axis of rotation? (b) Is the torque independent of the location of the axis of rotation?

6. Vector $\vec{\mathbf{A}}$ is in the negative y direction, and vector $\vec{\mathbf{B}}$ is in the negative x direction. (i) What is the direction of $\vec{\mathbf{A}} \times \vec{\mathbf{B}}$? (a) no direction because it is a scalar (b) x (c) $-y$ (d) z (e) $-z$ (ii) What is the direction of $\vec{\mathbf{B}} \times \vec{\mathbf{A}}$? Choose from the same possibilities (a) through (e).

7. Two ponies of equal mass are initially at diametrically opposite points on the rim of a large horizontal turntable that is turning freely on a frictionless, vertical axle through its center. The ponies simultaneously start walking toward each other across the turntable. (i) As they walk, what happens to the angular speed of the turntable? (a) It increases. (b) It decreases. (c) It stays constant. Consider the ponies–turntable system in this process and answer yes or no for the following questions. (ii) Is the mechanical energy of the system conserved? (iii) Is the momentum of the system conserved? (iv) Is the angular momentum of the system conserved?

8. Consider an isolated system moving through empty space. The system consists of objects that interact with each other and can change location with respect to one another. Which of the following quantities can change in time? (a) The angular momentum of the system. (b) The linear momentum of the system. (c) Both the angular momentum and linear momentum of the system. (d) Neither the angular momentum nor linear momentum of the system.

Conceptual Questions   1. denotes answer available in *Student Solutions Manual/Study Guide*

1. Stars originate as large bodies of slowly rotating gas. Because of gravity, these clumps of gas slowly decrease in size. What happens to the angular speed of a star as it shrinks? Explain.

2. A scientist arriving at a hotel asks a bellhop to carry a heavy suitcase. When the bellhop rounds a corner, the suitcase suddenly swings away from him for some unknown reason. The alarmed bellhop drops the suitcase and runs away. What might be in the suitcase?

3. Why does a long pole help a tightrope walker stay balanced?

4. Two children are playing with a roll of paper towels. One child holds the roll between the index fingers of her hands so that it is free to rotate, and the second child pulls at constant speed on the free end of the paper towels. As the child pulls the paper towels, the radius of the roll of remaining towels decreases. (a) How does the torque on the roll change with time? (b) How does the angular speed of the roll change in time? (c) If the child suddenly jerks the end paper towel with a large force, is the towel more likely to break from the others when it is being pulled from a nearly full roll or from a nearly empty roll?

5. Both torque and work are products of force and displacement. How are they different? Do they have the same units?

6. In some motorcycle races, the riders drive over small hills and the motorcycle becomes airborne for a short time interval. If the motorcycle racer keeps the throttle open while leaving the hill and going into the air, the motorcycle tends to nose upward. Why?

7. If the torque acting on a particle about an axis through a certain origin is zero, what can you say about its angular momentum about that axis?

8. A ball is thrown in such a way that it does not spin about its own axis. Does this statement imply that the angular momentum is zero about an arbitrary axis? Explain.

9. If global warming continues over the next one hundred years, it is likely that some polar ice will melt and the water will be distributed closer to the equator. (a) How would that change the moment of inertia of the Earth? (b) Would the duration of the day (one revolution) increase or decrease?

10. A cat usually lands on its feet regardless of the position from which it is dropped. A slow-motion film of a cat falling shows that the upper half of its body twists in one direction while the lower half twists in the opposite direction. (See Fig. CQ11.10.) Why does this type of rotation occur?

Agence Nature/Photo Researchers, Inc.

Figure CQ11.10

11. In Chapters 7 and 8, we made use of energy bar charts to analyze physical situations. Why have we not used bar charts for angular momentum in this chapter?

Problems available in ☑ **WebAssign** Access end-of-chapter problems online at www.webassign.net

Section 11.1 The Vector Product and Torque
Problems 1–10

Section 11.2 Analysis Model: Nonisolated System (Angular Momentum)
Problems 11–21

Section 11.3 Angular Momentum of a Rotating Rigid Object
Problems 22–29

Section 11.4 Analysis Model: Isolated System (Angular Momentum)
Problems 30–41

Section 11.5 The Motion of Gyroscopes and Tops
Problems 42–43

Additional Problems
Problems 44–60

Challenge Problems
Problem 61–64

Solutions to the following Problems are available in the *Student Solutions Manual/Study Guide:*
11.3, 11.7, 11.11, 11.15, 11.19, 11.27, 11.33, 11.37, 11.39, 11.41, 11.52, 11.55, and 11.61

List of Enhanced Problems

Problem Number	Targeted Feedback in Enhanced WebAssign	Analysis Model Tutorial in Enhanced WebAssign	Master It in Enhanced WebAssign	Watch It in Enhanced WebAssign
11.1	✓			✓
11.3	✓		✓	
11.11	✓		✓	
11.12	✓			✓
11.18	✓	✓		✓
11.19			✓	
11.25	✓			✓
11.27	✓		✓	
11.29		✓		
11.30	✓			✓
11.31	✓	✓		✓
11.33			✓	
11.34	✓			✓
11.37	✓			
11.52		✓	✓	
11.55	✓		✓	

Static Equilibrium and Elasticity

In Chapters 10 and 11, we studied the dynamics of rigid objects. Part of this chapter addresses the conditions under which a rigid object is in equilibrium. The term *equilibrium* implies that the object moves with both constant velocity and constant angular velocity relative to an observer in an inertial reference frame. We deal here only with the special case in which both of these velocities are equal to zero. In this case, the object is in what is called *static equilibrium*. Static equilibrium represents a common situation in engineering practice, and the principles it involves are of special interest to civil engineers, architects, and mechanical engineers. If you are an engineering student, you will undoubtedly take an advanced course in statics in the near future.

The last section of this chapter deals with how objects deform under load conditions. An *elastic* object returns to its original shape when the deforming forces are removed. Several elastic constants are defined, each corresponding to a different type of deformation.

Balanced Rock in Arches National Park, Utah, is a 3 000 000-kg boulder that has been in stable equilibrium for several millennia. It had a smaller companion nearby, called "Chip Off the Old Block," that fell during the winter of 1975. Balanced Rock appeared in an early scene of the movie *Indiana Jones and the Last Crusade.* We will study the conditions under which an object is in equilibrium in this chapter. *(John W. Jewett, Jr.)*

12.1 Analysis Model: Rigid Object in Equilibrium

In Chapter 5, we discussed the particle in equilibrium model, in which a particle moves with constant velocity because the net force acting on it is zero. The situation with real (extended) objects is more complex because these objects often cannot be modeled as particles. For an extended object to be in equilibrium, a second condition must be satisfied. This second condition involves the rotational motion of the extended object.

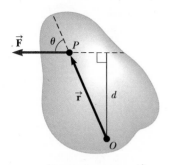

Figure 12.1 A single force $\vec{\mathbf{F}}$ acts on a rigid object at the point P.

Consider a single force $\vec{\mathbf{F}}$ acting on a rigid object as shown in Figure 12.1. Recall that the torque associated with the force $\vec{\mathbf{F}}$ about an axis through O is given by Equation 11.1:

$$\vec{\boldsymbol{\tau}} = \vec{\mathbf{r}} \times \vec{\mathbf{F}}$$

The magnitude of $\vec{\boldsymbol{\tau}}$ is Fd (see Equation 10.14), where d is the moment arm shown in Figure 12.1. According to Equation 10.18, the net torque on a rigid object causes it to undergo an angular acceleration.

In this discussion, we investigate those rotational situations in which the angular acceleration of a rigid object is zero. Such an object is in **rotational equilibrium.** Because $\Sigma\,\tau_{\text{ext}} = I\alpha$ for rotation about a fixed axis, the necessary condition for rotational equilibrium is that the net torque about any axis must be zero. We now have two necessary conditions for equilibrium of a rigid object:

Pitfall Prevention 12.1

Zero Torque Zero net torque does not mean an absence of rotational motion. An object that is rotating at a constant angular speed can be under the influence of a net torque of zero. This possibility is analogous to the translational situation: zero net force does not mean an absence of translational motion.

1. The net external force on the object must equal zero:

$$\sum \vec{\mathbf{F}}_{\text{ext}} = 0 \qquad (12.1)$$

2. The net external torque on the object about *any* axis must be zero:

$$\sum \vec{\boldsymbol{\tau}}_{\text{ext}} = 0 \qquad (12.2)$$

These conditions describe the **rigid object in equilibrium** analysis model. The first condition is a statement of translational equilibrium; it states that the translational acceleration of the object's center of mass must be zero when viewed from an inertial reference frame. The second condition is a statement of rotational equilibrium; it states that the angular acceleration about any axis must be zero. In the special case of **static equilibrium,** which is the main subject of this chapter, the object in equilibrium is at rest relative to the observer and so has no translational or angular speed (that is, $v_{\text{CM}} = 0$ and $\omega = 0$).

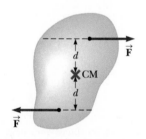

Figure 12.2 (Quick Quiz 12.1) Two forces of equal magnitude are applied at equal distances from the center of mass of a rigid object.

Quick Quiz 12.1 Consider the object subject to the two forces of equal magnitude in Figure 12.2. Choose the correct statement with regard to this situation. **(a)** The object is in force equilibrium but not torque equilibrium. **(b)** The object is in torque equilibrium but not force equilibrium. **(c)** The object is in both force equilibrium and torque equilibrium. **(d)** The object is in neither force equilibrium nor torque equilibrium.

Quick Quiz 12.2 Consider the object subject to the three forces in Figure 12.3. Choose the correct statement with regard to this situation. **(a)** The object is in force equilibrium but not torque equilibrium. **(b)** The object is in torque equilibrium but not force equilibrium. **(c)** The object is in both force equilibrium and torque equilibrium. **(d)** The object is in neither force equilibrium nor torque equilibrium.

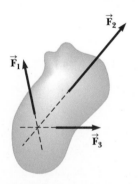

Figure 12.3 (Quick Quiz 12.2) Three forces act on an object. Notice that the lines of action of all three forces pass through a common point.

The two vector expressions given by Equations 12.1 and 12.2 are equivalent, in general, to six scalar equations: three from the first condition for equilibrium and three from the second (corresponding to x, y, and z components). Hence, in a complex system involving several forces acting in various directions, you could be faced with solving a set of equations with many unknowns. Here, we restrict our discussion to situations in which all the forces lie in the xy plane. (Forces whose vector representations are in the same plane are said to be *coplanar.*) With this restriction, we must deal with only three scalar equations. Two come from balancing the forces in the x and y directions. The third comes from the torque equation, namely that the net torque about a perpendicular axis through *any* point in the xy plane must be zero. This perpendicular axis will necessarily be parallel to the z axis, so the two conditions of the rigid object in equilibrium model provide the equations

$$\sum F_x = 0 \quad \sum F_y = 0 \quad \sum \tau_z = 0 \qquad \textbf{(12.3)}$$

where the location of the axis of the torque equation is arbitrary.

Analysis Model **Rigid Object in Equilibrium**

Imagine an object that can rotate, but is exhibiting no translational acceleration a and no rotational acceleration α. Such an object is in both translational *and* rotational equilibrium, so the net force *and* the net torque about any axis are both equal to zero:

$$\sum \vec{\mathbf{F}}_{\text{ext}} = 0 \qquad \textbf{(12.1)}$$

$$\sum \vec{\boldsymbol{\tau}}_{\text{ext}} = 0 \qquad \textbf{(12.2)}$$

Examples:

- a balcony juts out from a building and must support the weight of several humans without collapsing
- a gymnast performs the difficult *iron cross* maneuver in an Olympic event
- a ship moves at constant speed through calm water and maintains a perfectly level orientation (Chapter 14)
- polarized molecules in a dielectric material in a constant electric field take on an average equilibrium orientation that remains fixed in time (Chapter 26)

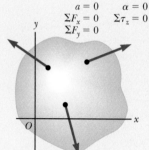

12.2 More on the Center of Gravity

Whenever we deal with a rigid object, one of the forces we must consider is the gravitational force acting on it, and we must know the point of application of this force. As we learned in Section 9.5, associated with every object is a special point called its center of gravity. The combination of the various gravitational forces acting on all the various mass elements of the object is equivalent to a single gravitational force acting through this point. Therefore, to compute the torque due to the gravitational force on an object of mass M, we need only consider the force $M\vec{\mathbf{g}}$ acting at the object's center of gravity.

How do we find this special point? As mentioned in Section 9.5, if we assume $\vec{\mathbf{g}}$ is uniform over the object, the center of gravity of the object coincides with its center of mass. To see why, consider an object of arbitrary shape lying in the xy plane as illustrated in Figure 12.4. Suppose the object is divided into a large number of particles of masses $m_1, m_2, m_3, \ldots$ having coordinates $(x_1, y_1), (x_2, y_2), (x_3, y_3), \ldots$. In Equation 9.29, we defined the x coordinate of the center of mass of such an object to be

$$x_{\text{CM}} = \frac{m_1 x_1 + m_2 x_2 + m_3 x_3 + \cdots}{m_1 + m_2 + m_3 + \cdots} = \frac{\displaystyle\sum_i m_i x_i}{\displaystyle\sum_i m_i}$$

We use a similar equation to define the y coordinate of the center of mass, replacing each x with its y counterpart.

Let us now examine the situation from another point of view by considering the gravitational force exerted on each particle as shown in Figure 12.5. Each particle contributes a torque about an axis through the origin equal in magnitude to the particle's weight mg multiplied by its moment arm. For example, the magnitude of the torque due to the force $m_1\vec{\mathbf{g}}_1$ is $m_1 g_1 x_1$, where g_1 is the value of the gravitational acceleration at the position of the particle of mass m_1. We wish to locate the center of gravity, the point at which application of the single gravitational force $M\vec{\mathbf{g}}_{\text{CG}}$ (where $M = m_1 + m_2 + m_3 + \cdots$ is the total mass of the object and $\vec{\mathbf{g}}_{\text{CG}}$ is the acceleration due to gravity at the location of the center of gravity) has the same effect on

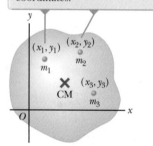

Each particle of the object has a specific mass and specific coordinates.

Figure 12.4 An object can be divided into many small particles. These particles can be used to locate the center of mass.

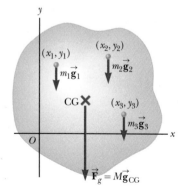

Figure 12.5 By dividing an object into many particles, we can find its center of gravity.

rotation as does the combined effect of all the individual gravitational forces $m_i\vec{\mathbf{g}}_i$. Equating the torque resulting from $M\vec{\mathbf{g}}_{CG}$ acting at the center of gravity to the sum of the torques acting on the individual particles gives

$$(m_1 + m_2 + m_3 + \cdots)g_{CG}\, x_{CG} = m_1 g_1 x_1 + m_2 g_2 x_2 + m_3 g_3 x_3 + \cdots$$

This expression accounts for the possibility that the value of g can in general vary over the object. If we assume uniform g over the object (as is usually the case), the g factors cancel and we obtain

$$x_{CG} = \frac{m_1 x_1 + m_2 x_2 + m_3 x_3 + \cdots}{m_1 + m_2 + m_3 + \cdots} \tag{12.4}$$

Comparing this result with Equation 9.29 shows that the center of gravity is located at the center of mass as long as $\vec{\mathbf{g}}$ is uniform over the entire object. Several examples in the next section deal with homogeneous, symmetric objects. The center of gravity for any such object coincides with its geometric center.

> **Q**uick Quiz 12.3 A meterstick of uniform density is hung from a string tied at
> the 25-cm mark. A 0.50-kg object is hung from the zero end of the meterstick,
> and the meterstick is balanced horizontally. What is the mass of the meterstick?
> (a) 0.25 kg (b) 0.50 kg (c) 0.75 kg (d) 1.0 kg (e) 2.0 kg (f) impossible to
> determine

12.3 Examples of Rigid Objects in Static Equilibrium

The photograph of the one-bottle wine holder in Figure 12.6 shows one example of a balanced mechanical system that seems to defy gravity. For the system (wine holder plus bottle) to be in equilibrium, the net external force must be zero (see Eq. 12.1) and the net external torque must be zero (see Eq. 12.2). The second condition can be satisfied only when the center of gravity of the system is directly over the support point.

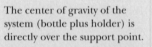

The center of gravity of the system (bottle plus holder) is directly over the support point.

Figure 12.6 This one-bottle wine holder is a surprising display of static equilibrium.

© Cengage Learning/Charles D. Winters

| **Problem-Solving Strategy** | **Rigid Object in Equilibrium** |

When analyzing a rigid object in equilibrium under the action of several external forces, use the following procedure.

1. Conceptualize. Think about the object that is in equilibrium and identify all the forces on it. Imagine what effect each force would have on the rotation of the object if it were the only force acting.

2. Categorize. Confirm that the object under consideration is indeed a rigid object in equilibrium. The object must have zero translational acceleration and zero angular acceleration.

3. Analyze. Draw a diagram and label all external forces acting on the object. Try to guess the correct direction for any forces that are not specified. When using the particle under a net force model, the object on which forces act can be represented in a free-body diagram with a dot because it does not matter where on the object the forces are applied. When using the rigid object in equilibrium model, however, we cannot use a dot to represent the object because the location where forces act is important in the calculation. Therefore, in a diagram showing the forces on an object, we must show the actual object or a simplified version of it.

Resolve all forces into rectangular components, choosing a convenient coordinate system. Then apply the first condition for equilibrium, Equation 12.1. Remember to keep track of the signs of the various force components.

▶ **Problem-Solving Strategy** continued

Choose a convenient axis for calculating the net torque on the rigid object. Remember that the choice of the axis for the torque equation is arbitrary; therefore, choose an axis that simplifies your calculation as much as possible. Usually, the most convenient axis for calculating torques is one through a point through which the lines of action of several forces pass, so their torques around this axis are zero. If you don't know a force or don't need to know a force, it is often beneficial to choose an axis through the point at which this force acts. Apply the second condition for equilibrium, Equation 12.2.

Solve the simultaneous equations for the unknowns in terms of the known quantities.

4. Finalize. Make sure your results are consistent with your diagram. If you selected a direction that leads to a negative sign in your solution for a force, do not be alarmed; it merely means that the direction of the force is the opposite of what you guessed. Add up the vertical and horizontal forces on the object and confirm that each set of components adds to zero. Add up the torques on the object and confirm that the sum equals zero.

Example 12.1   The Seesaw Revisited   AM

A seesaw consisting of a uniform board of mass M and length ℓ supports at rest a father and daughter with masses m_f and m_d, respectively, as shown in Figure 12.7. The support (called the *fulcrum*) is under the center of gravity of the board, the father is a distance d from the center, and the daughter is a distance $\ell/2$ from the center.

(A) Determine the magnitude of the upward force $\vec{n}$ exerted by the support on the board.

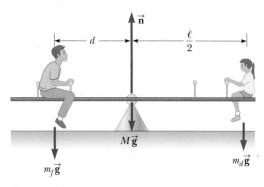

Figure 12.7 (Example 12.1) A balanced system.

SOLUTION

Conceptualize Let us focus our attention on the board and consider the gravitational forces on the father and daughter as forces applied directly to the board. The daughter would cause a clockwise rotation of the board around the support, whereas the father would cause a counterclockwise rotation.

Categorize Because the text of the problem states that the system is at rest, we model the board as a *rigid object in equilibrium*. Because we will only need the first condition of equilibrium to solve this part of the problem, however, we could also simply model the board as a *particle in equilibrium*.

⋯⋯

Analyze Define upward as the positive y direction and substitute the forces on the board into Equation 12.1:

$$n - m_f g - m_d g - Mg = 0$$

Solve for the magnitude of the force $\vec{n}$:

$$(1) \quad n = m_f g + m_d g + Mg = \boxed{(m_f + m_d + M)g}$$

(B) Determine where the father should sit to balance the system at rest.

SOLUTION

Categorize This part of the problem requires the introduction of torque to find the position of the father, so we model the board as a *rigid object in equilibrium*.

⋯⋯

Analyze The board's center of gravity is at its geometric center because we are told that the board is uniform. If we choose a rotation axis perpendicular to the page through the center of gravity of the board, the torques produced by $\vec{n}$ and the gravitational force on the board about this axis are zero.

continued

▶ **12.1** continued

Substitute expressions for the torques on the board due to the father and daughter into Equation 12.2:

$$(m_f g)(d) - (m_d g)\frac{\ell}{2} = 0$$

Solve for d:

$$d = \left(\frac{m_d}{m_f}\right)\frac{\ell}{2}$$

....

Finalize This result is the same one we obtained in Example 11.6 by evaluating the angular acceleration of the system and setting the angular acceleration equal to zero.

WHAT IF? Suppose we had chosen another point through which the rotation axis were to pass. For example, suppose the axis is perpendicular to the page and passes through the location of the father. Does that change the results to parts (A) and (B)?

Answer Part (A) is unaffected because the calculation of the net force does not involve a rotation axis. In part (B), we would conceptually expect there to be no change if a different rotation axis is chosen because the second condition of equilibrium claims that the torque is zero about *any* rotation axis.

Let's verify this answer mathematically. Recall that the sign of the torque associated with a force is positive if that force tends to rotate the system counterclockwise, whereas the sign of the torque is negative if the force tends to rotate the system clockwise. Let's choose a rotation axis perpendicular to the page and passing through the location of the father.

Substitute expressions for the torques on the board around this axis into Equation 12.2:

$$n(d) - (Mg)(d) - (m_d g)\left(d + \frac{\ell}{2}\right) = 0$$

Substitute from Equation (1) in part (A) and solve for d:

$$(m_f + m_d + M)g(d) - (Mg)(d) - (m_d g)\left(d + \frac{\ell}{2}\right) = 0$$

$$(m_f g)(d) - (m_d g)\left(\frac{\ell}{2}\right) = 0 \quad \rightarrow \quad d = \left(\frac{m_d}{m_f}\right)\frac{\ell}{2}$$

This result is in agreement with the one obtained in part (B).

Example 12.2   Standing on a Horizontal Beam   AM

A uniform horizontal beam with a length of $\ell = 8.00$ m and a weight of $W_b = 200$ N is attached to a wall by a pin connection. Its far end is supported by a cable that makes an angle of $\phi = 53.0°$ with the beam (Fig. 12.8a). A person of weight $W_p = 600$ N stands a distance $d = 2.00$ m from the wall. Find the tension in the cable as well as the magnitude and direction of the force exerted by the wall on the beam.

SOLUTION

Conceptualize Imagine the person in Figure 12.8a moving outward on the beam. It seems reasonable that the farther he moves outward, the larger the torque he applies about the pivot and the larger the tension in the cable must be to balance this torque.

Categorize Because the system is at rest, we categorize the beam as a *rigid object in equilibrium*.

....

Analyze We identify all the external forces acting on the beam: the 200-N gravitational force, the

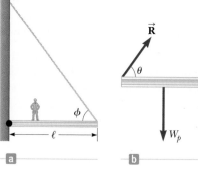

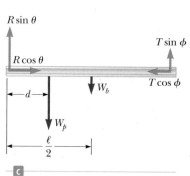

Figure 12.8 (Example 12.2) (a) A uniform beam supported by a cable. A person walks outward on the beam. (b) The force diagram for the beam. (c) The force diagram for the beam showing the components of $\vec{R}$ and $\vec{T}$.

▶ **12.2** continued

force $\vec{\mathbf{T}}$ exerted by the cable, the force $\vec{\mathbf{R}}$ exerted by the wall at the pivot, and the 600-N force that the person exerts on the beam. These forces are all indicated in the force diagram for the beam shown in Figure 12.8b. When we assign directions for forces, it is sometimes helpful to imagine what would happen if a force were suddenly removed. For example, if the wall were to vanish suddenly, the left end of the beam would move to the left as it begins to fall. This scenario tells us that the wall is not only holding the beam up but is also pressing outward against it. Therefore, we draw the vector $\vec{\mathbf{R}}$ in the direction shown in Figure 12.8b. Figure 12.8c shows the horizontal and vertical components of $\vec{\mathbf{T}}$ and $\vec{\mathbf{R}}$.

Applying the first condition of equilibrium, substitute expressions for the forces on the beam into component equations from Equation 12.1:

$$(1) \quad \sum F_x = R \cos\theta - T\cos\phi = 0$$

$$(2) \quad \sum F_y = R \sin\theta + T\sin\phi - W_p - W_b = 0$$

where we have chosen rightward and upward as our positive directions. Because R, T, and θ are all unknown, we cannot obtain a solution from these expressions alone. (To solve for the unknowns, the number of simultaneous equations must generally equal the number of unknowns.)

Now let's invoke the condition for rotational equilibrium. A convenient axis to choose for our torque equation is the one that passes through the pin connection. The feature that makes this axis so convenient is that the force $\vec{\mathbf{R}}$ and the horizontal component of $\vec{\mathbf{T}}$ both have a moment arm of zero; hence, these forces produce no torque about this axis.

Substitute expressions for the torques on the beam into Equation 12.2:

$$\sum \tau_z = (T\sin\phi)(\ell) - W_p d - W_b\left(\frac{\ell}{2}\right) = 0$$

This equation contains only T as an unknown because of our choice of rotation axis. Solve for T and substitute numerical values:

$$T = \frac{W_p d + W_b(\ell/2)}{\ell \sin\phi} = \frac{(600\text{ N})(2.00\text{ m}) + (200\text{ N})(4.00\text{ m})}{(8.00\text{ m})\sin 53.0°} = \boxed{313\text{ N}}$$

Rearrange Equations (1) and (2) and then divide:

$$\frac{R\sin\theta}{R\cos\theta} = \tan\theta = \frac{W_p + W_b - T\sin\phi}{T\cos\phi}$$

Solve for θ and substitute numerical values:

$$\theta = \tan^{-1}\left(\frac{W_p + W_b - T\sin\phi}{T\cos\phi}\right)$$

$$= \tan^{-1}\left[\frac{600\text{ N} + 200\text{ N} - (313\text{ N})\sin 53.0°}{(313\text{ N})\cos 53.0°}\right] = \boxed{71.1°}$$

Solve Equation (1) for R and substitute numerical values:

$$R = \frac{T\cos\phi}{\cos\theta} = \frac{(313\text{ N})\cos 53.0°}{\cos 71.1°} = \boxed{581\text{ N}}$$

Finalize The positive value for the angle θ indicates that our estimate of the direction of $\vec{\mathbf{R}}$ was accurate.

Had we selected some other axis for the torque equation, the solution might differ in the details but the answers would be the same. For example, had we chosen an axis through the center of gravity of the beam, the torque equation would involve both T and R. This equation, coupled with Equations (1) and (2), however, could still be solved for the unknowns. Try it!

WHAT IF? What if the person walks farther out on the beam? Does T change? Does R change? Does θ change?

Answer T must increase because the gravitational force on the person exerts a larger torque about the pin connection, which must be countered by a larger torque in the opposite direction due to an increased value of T. If T increases, the vertical component of $\vec{\mathbf{R}}$ decreases to maintain force equilibrium in the vertical direction. Force equilibrium in the horizontal direction, however, requires an increased horizontal component of $\vec{\mathbf{R}}$ to balance the horizontal component of the increased $\vec{\mathbf{T}}$. This fact suggests that θ becomes smaller, but it is hard to predict what happens to R. Problem 66 in Enhanced WebAssign asks you to explore the behavior of R.

A uniform ladder of length ℓ rests against a smooth, vertical wall (Fig. 12.9a). The mass of the ladder is m, and the coefficient of static friction between the ladder and the ground is $\mu_s = 0.40$. Find the minimum angle θ_{min} at which the ladder does not slip.

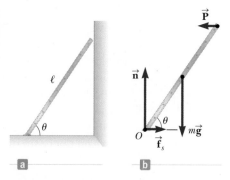

S O L U T I O N

Conceptualize Think about any ladders you have climbed. Do you want a large friction force between the bottom of the ladder and the surface or a small one? If the friction force is zero, will the ladder stay up? Simulate a ladder with a ruler leaning against a vertical surface. Does the ruler slip at some angles and stay up at others?

Figure 12.9 (Example 12.3) (a) A uniform ladder at rest, leaning against a smooth wall. The ground is rough. (b) The forces on the ladder.

Categorize We do not wish the ladder to slip, so we model it as a *rigid object in equilibrium*.

Analyze A diagram showing all the external forces acting on the ladder is illustrated in Figure 12.9b. The force exerted by the ground on the ladder is the vector sum of a normal force $\vec{n}$ and the force of static friction $\vec{f}_s$. The wall exerts a normal force $\vec{P}$ on the top of the ladder, but there is no friction force here because the wall is smooth. So the net force on the top of the ladder is perpendicular to the wall and of magnitude P.

Apply the first condition for equilibrium to the ladder in both the x and the y directions:

(1) $\sum F_x = f_s - P = 0$

(2) $\sum F_y = n - mg = 0$

Solve Equation (1) for P:

(3) $P = f_s$

Solve Equation (2) for n:

(4) $n = mg$

When the ladder is on the verge of slipping, the force of static friction must have its maximum value, which is given by $f_{s,max} = \mu_s n$. Combine this equation with Equations (3) and (4):

(5) $P_{max} = f_{s,max} = \mu_s n = \mu_s mg$

Apply the second condition for equilibrium to the ladder, evaluating torques about an axis perpendicular to the page through O:

$$\sum \tau_O = P\ell \sin\theta - mg\frac{\ell}{2}\cos\theta = 0$$

Solve for $\tan\theta$:

$$\frac{\sin\theta}{\cos\theta} = \tan\theta = \frac{mg}{2P} \rightarrow \theta = \tan^{-1}\left(\frac{mg}{2P}\right)$$

Under the conditions that the ladder is just ready to slip, θ becomes θ_{min} and P_{max} is given by Equation (5). Substitute:

$$\theta_{min} = \tan^{-1}\left(\frac{mg}{2P_{max}}\right) = \tan^{-1}\left(\frac{1}{2\mu_s}\right) = \tan^{-1}\left[\frac{1}{2(0.40)}\right] = \boxed{51°}$$

Finalize Notice that the angle depends only on the coefficient of friction, not on the mass or length of the ladder.

(A) Estimate the magnitude of the force $\vec{F}$ a person must apply to a wheelchair's main wheel to roll up over a sidewalk curb (Fig. 12.10a). This main wheel that comes in contact with the curb has a radius r, and the height of the curb is h.

▶ **12.4** continued

SOLUTION

Conceptualize Think about wheelchair access to buildings. Generally, there are ramps built for individuals in wheelchairs. Steplike structures such as curbs are serious barriers to a wheelchair.

Categorize Imagine the person exerts enough force so that the bottom of the main wheel just loses contact with the lower surface and hovers at rest. We model the wheel in this situation as a *rigid object in equilibrium*.

Analyze Usually, the person's hands supply the required force to a slightly smaller wheel that is concentric with the main wheel. For simplicity, let's assume the radius of this second wheel is the same as the radius of the main wheel. Let's estimate a combined gravitational force of magnitude $mg = 1\ 400$ N for the person and the wheelchair, acting along a line of action passing through the axle of the main wheel, and choose a wheel radius of $r = 30$ cm. We also pick a curb height of $h = 10$ cm. Let's also assume the wheelchair and occupant are symmetric and each wheel supports a weight of 700 N. We then proceed to analyze only one of the main wheels. Figure 12.10b shows the geometry for a single wheel.

When the wheel is just about to be raised from the street, the normal force exerted by the ground on the wheel at point B goes to zero. Hence, at this time only three forces act on the wheel as shown in the force diagram in Figure 12.10c. The force $\vec{\mathbf{R}}$, which is the force exerted by the curb on the wheel, acts at point A, so if we choose to have our axis of rotation be perpendicular to the page and pass through point A, we do not need to include $\vec{\mathbf{R}}$ in our torque equation. The moment arm of $\vec{\mathbf{F}}$ relative to an axis through A is given by $2r - h$ (see Fig. 12.10c).

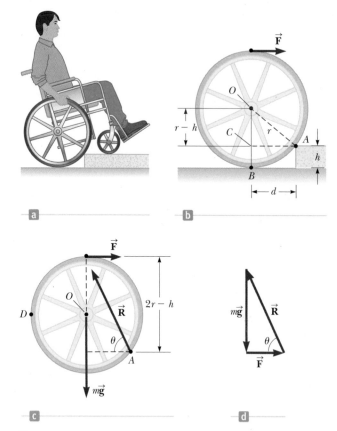

Figure 12.10 (Example 12.4) (a) A person in a wheelchair attempts to roll up over a curb. (b) Details of the wheel and curb. The person applies a force $\vec{\mathbf{F}}$ to the top of the wheel. (c) A force diagram for the wheel when it is just about to be raised. Three forces act on the wheel at this instant: $\vec{\mathbf{F}}$, which is exerted by the hand; $\vec{\mathbf{R}}$, which is exerted by the curb; and the gravitational force $m\vec{\mathbf{g}}$. (d) The vector sum of the three external forces acting on the wheel is zero.

Use the triangle *OAC* in Figure 12.10b to find the moment arm d of the gravitational force $m\vec{\mathbf{g}}$ acting on the wheel relative to an axis through point A:

$$(1) \quad d = \sqrt{r^2 - (r-h)^2} = \sqrt{2rh - h^2}$$

Apply the second condition for equilibrium to the wheel, taking torques about an axis through A:

$$(2) \quad \sum \tau_A = mgd - F(2r - h) = 0$$

Substitute for d from Equation (1):

$$mg\sqrt{2rh - h^2} - F(2r - h) = 0$$

Solve for F:

$$(3) \quad F = \frac{mg\sqrt{2rh - h^2}}{2r - h}$$

Simplify:

$$F = mg\frac{\sqrt{h}\sqrt{2r - h}}{2r - h} = mg\sqrt{\frac{h}{2r - h}}$$

Substitute the known values:

$$F = (700\ \text{N})\sqrt{\frac{0.1\ \text{m}}{2(0.3\ \text{m}) - 0.1\ \text{m}}}$$

$$= 3 \times 10^2\ \text{N}$$

continued

▶ **12.4** continued

(B) Determine the magnitude and direction of $\vec{R}$.

SOLUTION

Apply the first condition for equilibrium to the x and y components of the forces on the wheel:

$$\text{(4)} \quad \sum F_x = F - R\cos\theta = 0$$

$$\text{(5)} \quad \sum F_y = R\sin\theta - mg = 0$$

Divide Equation (5) by Equation (4):

$$\frac{R\sin\theta}{R\cos\theta} = \tan\theta = \frac{mg}{F}$$

Solve for the angle θ:

$$\theta = \tan^{-1}\left(\frac{mg}{F}\right) = \tan^{-1}\left(\frac{700\text{ N}}{300\text{ N}}\right) = \boxed{70°}$$

Solve Equation (5) for R and substitute numerical values:

$$R = \frac{mg}{\sin\theta} = \frac{700\text{ N}}{\sin 70°} = \boxed{8 \times 10^2\text{ N}}$$

Finalize Notice that we have kept only one digit as significant. (We have written the angle as 70° because $7 \times 10^{1\circ}$ is awkward!) The results indicate that the force that must be applied to each wheel is substantial. You may want to estimate the force required to roll a wheelchair up a typical sidewalk accessibility ramp for comparison.

WHAT IF? Would it be easier to negotiate the curb if the person grabbed the wheel at point D in Figure 12.10c and pulled *upward*?

Answer If the force $\vec{F}$ in Figure 12.10c is rotated counterclockwise by 90° and applied at D, its moment arm about an axis through A is $d + r$. Let's call the magnitude of this new force F'.

Modify Equation (2) for this situation:

$$\sum \tau_A = mgd - F'(d + r) = 0$$

Solve this equation for F' and substitute for d:

$$F' = \frac{mgd}{d + r} = \frac{mg\sqrt{2rh - h^2}}{\sqrt{2rh - h^2} + r}$$

Take the ratio of this force to the original force from Equation (3) and express the result in terms of h/r, the ratio of the curb height to the wheel radius:

$$\frac{F'}{F} = \frac{\dfrac{mg\sqrt{2rh - h^2}}{\sqrt{2rh - h^2} + r}}{\dfrac{mg\sqrt{2rh - h^2}}{2r - h}} = \frac{2r - h}{\sqrt{2rh - h^2} + r} = \frac{2 - \left(\dfrac{h}{r}\right)}{\sqrt{2\left(\dfrac{h}{r}\right) - \left(\dfrac{h}{r}\right)^2} + 1}$$

Substitute the ratio $h/r = 0.33$ from the given values:

$$\frac{F'}{F} = \frac{2 - 0.33}{\sqrt{2(0.33) - (0.33)^2} + 1} = 0.96$$

This result tells us that, *for these values*, it is slightly easier to pull upward at D than horizontally at the top of the wheel. For very high curbs, so that h/r is close to 1, the ratio F'/F drops to about 0.5 because point A is located near the right edge of the wheel in Figure 12.10b. The force at D is applied at a distance of about $2r$ from A, whereas the force at the top of the wheel has a moment arm of only about r. For high curbs, then, it is best to pull upward at D, although a large value of the force is required. For small curbs, it is best to apply the force at the top of the wheel. The ratio F'/F becomes larger than 1 at about $h/r = 0.3$ because point A is now close to the bottom of the wheel and the force applied at the top of the wheel has a larger moment arm than when applied at D.

Finally, let's comment on the validity of these mathematical results. Consider Figure 12.10d and imagine that the vector $\vec{F}$ is upward instead of to the right. There is no way the three vectors can add to equal zero as required by the first equilibrium condition. Therefore, our results above may be qualitatively valid, but not exact quantitatively. To cancel the horizontal component of $\vec{R}$, the force at D must be applied at an angle to the vertical rather than straight upward. This feature makes the calculation more complicated and requires both conditions of equilibrium.

12.4 Elastic Properties of Solids

Except for our discussion about springs in earlier chapters, we have assumed objects remain rigid when external forces act on them. In Section 9.8, we explored deformable systems. In reality, all objects are deformable to some extent. That is, it is possible to change the shape or the size (or both) of an object by applying external forces. As these changes take place, however, internal forces in the object resist the deformation.

We shall discuss the deformation of solids in terms of the concepts of *stress* and *strain*. **Stress** is a quantity that is proportional to the force causing a deformation; more specifically, stress is the external force acting on an object per unit cross-sectional area. The result of a stress is **strain,** which is a measure of the degree of deformation. It is found that, for sufficiently small stresses, stress is proportional to strain; the constant of proportionality depends on the material being deformed and on the nature of the deformation. We call this proportionality constant the **elastic modulus.** The elastic modulus is therefore defined as the ratio of the stress to the resulting strain:

$$\text{Elastic modulus} \equiv \frac{\text{stress}}{\text{strain}} \tag{12.5}$$

The elastic modulus in general relates what is done to a solid object (a force is applied) to how that object responds (it deforms to some extent). It is similar to the spring constant k in Hooke's law (Eq. 7.9) that relates a force applied to a spring and the resultant deformation of the spring, measured by its extension or compression.

We consider three types of deformation and define an elastic modulus for each:

1. **Young's modulus** measures the resistance of a solid to a change in its length.
2. **Shear modulus** measures the resistance to motion of the planes within a solid parallel to each other.
3. **Bulk modulus** measures the resistance of solids or liquids to changes in their volume.

Young's Modulus: Elasticity in Length

Consider a long bar of cross-sectional area A and initial length L_i that is clamped at one end as in Figure 12.11. When an external force is applied perpendicular to the cross section, internal molecular forces in the bar resist distortion ("stretching"), but the bar reaches an equilibrium situation in which its final length L_f is greater than L_i and in which the external force is exactly balanced by the internal forces. In such a situation, the bar is said to be stressed. We define the **tensile stress** as the ratio of the magnitude of the external force F to the cross-sectional area A, where the cross section is perpendicular to the force vector. The **tensile strain** in this case is defined as the ratio of the change in length ΔL to the original length L_i. We define **Young's modulus** by a combination of these two ratios:

$$Y \equiv \frac{\text{tensile stress}}{\text{tensile strain}} = \frac{F/A}{\Delta L / L_i} \tag{12.6}$$

◀ Young's modulus

Young's modulus is typically used to characterize a rod or wire stressed under either tension or compression. Because strain is a dimensionless quantity, Y has units of force per unit area. Typical values are given in Table 12.1 on page 300.

For relatively small stresses, the bar returns to its initial length when the force is removed. The **elastic limit** of a substance is defined as the maximum stress that can be applied to the substance before it becomes permanently deformed and does not return to its initial length. It is possible to exceed the elastic limit of a substance by

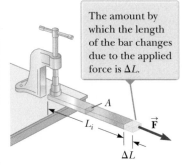

The amount by which the length of the bar changes due to the applied force is ΔL.

Figure 12.11 A force $\vec{F}$ is applied to the free end of a bar clamped at the other end.

Table 12.1	Typical Values for Elastic Moduli		
Substance	Young's Modulus (N/m²)	Shear Modulus (N/m²)	Bulk Modulus (N/m²)
Tungsten	35×10^{10}	14×10^{10}	20×10^{10}
Steel	20×10^{10}	8.4×10^{10}	6×10^{10}
Copper	11×10^{10}	4.2×10^{10}	14×10^{10}
Brass	9.1×10^{10}	3.5×10^{10}	6.1×10^{10}
Aluminum	7.0×10^{10}	2.5×10^{10}	7.0×10^{10}
Glass	$6.5–7.8 \times 10^{10}$	$2.6–3.2 \times 10^{10}$	$5.0–5.5 \times 10^{10}$
Quartz	5.6×10^{10}	2.6×10^{10}	2.7×10^{10}
Water	—	—	0.21×10^{10}
Mercury	—	—	2.8×10^{10}

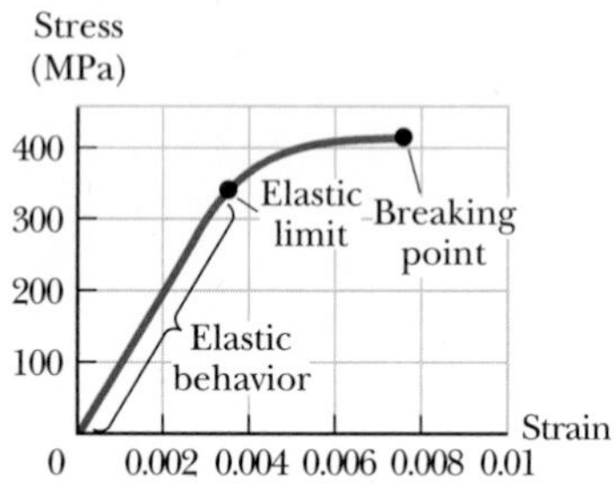

Figure 12.12 Stress-versus-strain curve for an elastic solid.

applying a sufficiently large stress as seen in Figure 12.12. Initially, a stress-versus-strain curve is a straight line. As the stress increases, however, the curve is no longer a straight line. When the stress exceeds the elastic limit, the object is permanently distorted and does not return to its original shape after the stress is removed. As the stress is increased even further, the material ultimately breaks.

Shear Modulus: Elasticity of Shape

Another type of deformation occurs when an object is subjected to a force parallel to one of its faces while the opposite face is held fixed by another force (Fig. 12.13a). The stress in this case is called a *shear stress*. If the object is originally a rectangular block, a shear stress results in a shape whose cross section is a parallelogram. A book pushed sideways as shown in Figure 12.13b is an example of an object subjected to a shear stress. To a first approximation (for small distortions), no change in volume occurs with this deformation.

We define the **shear stress** as F/A, the ratio of the tangential force to the area A of the face being sheared. The **shear strain** is defined as the ratio $\Delta x/h$, where Δx is the horizontal distance that the sheared face moves and h is the height of the object. In terms of these quantities, the **shear modulus** is

◀ Shear modulus

$$S \equiv \frac{\text{shear stress}}{\text{shear strain}} = \frac{F/A}{\Delta x/h} \qquad (12.7)$$

Values of the shear modulus for some representative materials are given in Table 12.1. Like Young's modulus, the unit of shear modulus is the ratio of that for force to that for area.

Bulk Modulus: Volume Elasticity

Bulk modulus characterizes the response of an object to changes in a force of uniform magnitude applied perpendicularly over the entire surface of the object as shown in Figure 12.14. (We assume here the object is made of a single substance.)

Figure 12.13 (a) A shear deformation in which a rectangular block is distorted by two forces of equal magnitude but opposite directions applied to two parallel faces. (b) A book is under shear stress when a hand placed on the cover applies a horizontal force away from the spine.

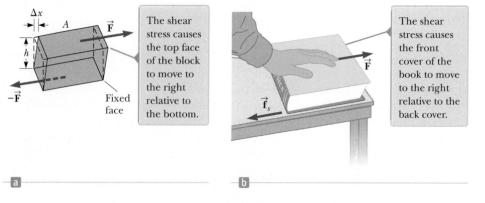

The shear stress causes the top face of the block to move to the right relative to the bottom.

The shear stress causes the front cover of the book to move to the right relative to the back cover.

As we shall see in Chapter 14, such a uniform distribution of forces occurs when an object is immersed in a fluid. An object subject to this type of deformation undergoes a change in volume but no change in shape. The **volume stress** is defined as the ratio of the magnitude of the total force F exerted on a surface to the area A of the surface. The quantity $P = F/A$ is called **pressure,** which we shall study in more detail in Chapter 14. If the pressure on an object changes by an amount $\Delta P = \Delta F/A$, the object experiences a volume change ΔV. The **volume strain** is equal to the change in volume ΔV divided by the initial volume V_i. Therefore, from Equation 12.5, we can characterize a volume ("bulk") compression in terms of the **bulk modulus,** which is defined as

$$B \equiv \frac{\text{volume stress}}{\text{volume strain}} = -\frac{\Delta F/A}{\Delta V/V_i} = -\frac{\Delta P}{\Delta V/V_i} \qquad (12.8)$$

◀ **Bulk modulus**

A negative sign is inserted in this defining equation so that B is a positive number. This maneuver is necessary because an increase in pressure (positive ΔP) causes a decrease in volume (negative ΔV) and vice versa.

Table 12.1 lists bulk moduli for some materials. If you look up such values in a different source, you may find the reciprocal of the bulk modulus listed. The reciprocal of the bulk modulus is called the **compressibility** of the material.

Notice from Table 12.1 that both solids and liquids have a bulk modulus. No shear modulus and no Young's modulus are given for liquids, however, because a liquid does not sustain a shearing stress or a tensile stress. If a shearing force or a tensile force is applied to a liquid, the liquid simply flows in response.

> ⓠuick Quiz 12.4 For the three parts of this Quick Quiz, choose from the fol-
> lowing choices the correct answer for the elastic modulus that describes the
> relationship between stress and strain for the system of interest, which is in ital-
> ics: (a) Young's modulus (b) shear modulus (c) bulk modulus (d) none of those
> choices (i) A *block of iron* is sliding across a horizontal floor. The friction force
> between the sliding block and the floor causes the block to deform. (ii) A tra-
> peze artist swings through a circular arc. At the bottom of the swing, the *wires*
> supporting the trapeze are longer than when the trapeze artist simply hangs
> from the trapeze due to the increased tension in them. (iii) A spacecraft carries
> a *steel sphere* to a planet on which atmospheric pressure is much higher than on
> the Earth. The higher pressure causes the radius of the sphere to decrease.

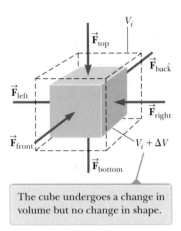

The cube undergoes a change in volume but no change in shape.

Figure 12.14 A cube is under uniform pressure and is therefore compressed on all sides by forces normal to its six faces. The arrowheads of force vectors on the sides of the cube that are not visible are hidden by the cube.

Prestressed Concrete

If the stress on a solid object exceeds a certain value, the object fractures. The maximum stress that can be applied before fracture occurs—called the *tensile strength, compressive strength,* or *shear strength*—depends on the nature of the material and on the type of applied stress. For example, concrete has a tensile strength of about 2×10^6 N/m², a compressive strength of 20×10^6 N/m², and a shear strength of 2×10^6 N/m². If the applied stress exceeds these values, the concrete fractures. It is common practice to use large safety factors to prevent failure in concrete structures.

Concrete is normally very brittle when it is cast in thin sections. Therefore, concrete slabs tend to sag and crack at unsupported areas as shown in Figure 12.15a. The slab can be strengthened by the use of steel rods to reinforce the concrete as illustrated in Figure 12.15b. Because concrete is much stronger under compression (squeezing) than under tension (stretching) or shear, vertical columns of concrete can support

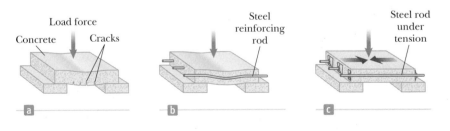

Figure 12.15 (a) A concrete slab with no reinforcement tends to crack under a heavy load. (b) The strength of the concrete is increased by using steel reinforcement rods. (c) The concrete is further strengthened by prestressing it with steel rods under tension.

very heavy loads, whereas horizontal beams of concrete tend to sag and crack. A significant increase in shear strength is achieved, however, if the reinforced concrete is prestressed as shown in Figure 12.15c. As the concrete is being poured, the steel rods are held under tension by external forces. The external forces are released after the concrete cures; the result is a permanent tension in the steel and hence a compressive stress on the concrete. The concrete slab can now support a much heavier load.

Example 12.5   Stage Design

In Example 8.2, we analyzed a cable used to support an actor as he swings onto the stage. Now suppose the tension in the cable is 940 N as the actor reaches the lowest point. What diameter should a 10-m-long steel cable have if we do not want it to stretch more than 0.50 cm under these conditions?

SOLUTION

Conceptualize Look back at Example 8.2 to recall what is happening in this situation. We ignored any stretching of the cable there, but we wish to address this phenomenon in this example.

Categorize We perform a simple calculation involving Equation 12.6, so we categorize this example as a substitution problem.

Solve Equation 12.6 for the cross-sectional area of the cable:

$$A = \frac{FL_i}{Y\,\Delta L}$$

Assuming the cross section is circular, find the diameter of the cable from $d = 2r$ and $A = \pi r^2$:

$$d = 2r = 2\sqrt{\frac{A}{\pi}} = 2\sqrt{\frac{FL_i}{\pi Y \Delta L}}$$

Substitute numerical values:

$$d = 2\sqrt{\frac{(940\ \text{N})(10\ \text{m})}{\pi(20 \times 10^{10}\ \text{N/m}^2)(0.005\,0\ \text{m})}} = 3.5 \times 10^{-3}\ \text{m} = 3.5\ \text{mm}$$

To provide a large margin of safety, you would probably use a flexible cable made up of many smaller wires having a total cross-sectional area substantially greater than our calculated value.

Example 12.6   Squeezing a Brass Sphere

A solid brass sphere is initially surrounded by air, and the air pressure exerted on it is $1.0 \times 10^5\ \text{N/m}^2$ (normal atmospheric pressure). The sphere is lowered into the ocean to a depth where the pressure is $2.0 \times 10^7\ \text{N/m}^2$. The volume of the sphere in air is $0.50\ \text{m}^3$. By how much does this volume change once the sphere is submerged?

SOLUTION

Conceptualize Think about movies or television shows you have seen in which divers go to great depths in the water in submersible vessels. These vessels must be very strong to withstand the large pressure under water. This pressure squeezes the vessel and reduces its volume.

Categorize We perform a simple calculation involving Equation 12.8, so we categorize this example as a substitution problem.

Solve Equation 12.8 for the volume change of the sphere:

$$\Delta V = -\frac{V_i \Delta P}{B}$$

Substitute numerical values:

$$\Delta V = -\frac{(0.50\ \text{m}^3)(2.0 \times 10^7\ \text{N/m}^2 - 1.0 \times 10^5\ \text{N/m}^2)}{6.1 \times 10^{10}\ \text{N/m}^2}$$

$$= -1.6 \times 10^{-4}\ \text{m}^3$$

The negative sign indicates that the volume of the sphere decreases.

Summary

Definitions

The gravitational force exerted on an object can be considered as acting at a single point called the **center of gravity.** An object's center of gravity coincides with its center of mass if the object is in a uniform gravitational field.

We can describe the elastic properties of a substance using the concepts of stress and strain. **Stress** is a quantity proportional to the force producing a deformation; **strain** is a measure of the degree of deformation. Stress is proportional to strain, and the constant of proportionality is the **elastic modulus:**

$$\text{Elastic modulus} \equiv \frac{\text{stress}}{\text{strain}} \qquad (12.5)$$

Concepts and Principles

Three common types of deformation are represented by (1) the resistance of a solid to elongation under a load, characterized by **Young's modulus** Y; (2) the resistance of a solid to the motion of internal planes sliding past each other, characterized by the **shear modulus** S; and (3) the resistance of a solid or fluid to a volume change, characterized by the **bulk modulus** B.

Analysis Model for Problem Solving

Rigid Object in Equilibrium A rigid object in equilibrium exhibits no translational or angular acceleration. The net external force acting on it is zero, and the net external torque on it is zero about any axis:

$$\sum \vec{\mathbf{F}}_{\text{ext}} = 0 \qquad (12.1)$$

$$\sum \vec{\boldsymbol{\tau}}_{\text{ext}} = 0 \qquad (12.2)$$

The first condition is the condition for translational equilibrium, and the second is the condition for rotational equilibrium.

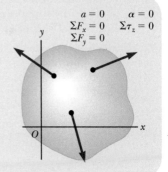

Objective Questions

1. The acceleration due to gravity becomes weaker by about three parts in ten million for each meter of increased elevation above the Earth's surface. Suppose a skyscraper is 100 stories tall, with the same floor plan for each story and with uniform average density. Compare the location of the building's center of mass and the location of its center of gravity. Choose one: (a) Its center of mass is higher by a distance of several meters. (b) Its center of mass is higher by a distance of several millimeters. (c) Its center of mass and its center of gravity are in the same location. (d) Its center of gravity is higher by a distance of several millimeters. (e) Its center of gravity is higher by a distance of several meters.

2. A rod 7.0 m long is pivoted at a point 2.0 m from the left end. A downward force of 50 N acts at the left end, and a downward force of 200 N acts at the right end. At what distance to the right of the pivot can a third force of 300 N acting upward be placed to produce rotational equilibrium? *Note:* Neglect the weight of the rod. (a) 1.0 m (b) 2.0 m (c) 3.0 m (d) 4.0 m (e) 3.5 m

3. Consider the object in Figure OQ12.3. A single force is exerted on the object. The line of action of the force does not pass through the object's center of mass. The acceleration of the object's center of mass due to this force (a) is the same as if the force were applied at the

center of mass, (b) is larger than the acceleration would be if the force were applied at the center of mass, (c) is smaller than the acceleration would be if the force were applied at the center of mass, or (d) is zero because the force causes only angular acceleration about the center of mass.

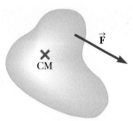

Figure OQ12.3

4. Two forces are acting on an object. Which of the following statements is correct? (a) The object is in equilibrium if the forces are equal in magnitude and opposite in direction. (b) The object is in equilibrium if the net torque on the object is zero. (c) The object is in equilibrium if the forces act at the same point on the object. (d) The object is in equilibrium if the net force and the net torque on the object are both zero. (e) The object cannot be in equilibrium because more than one force acts on it.

5. In the cabin of a ship, a soda can rests in a saucer-shaped indentation in a built-in counter. The can tilts as the ship slowly rolls. In which case is the can most stable against tipping over? (a) It is most stable when it is full. (b) It is most stable when it is half full. (c) It is most stable when it is empty. (d) It is most stable in two of these cases. (e) It is equally stable in all cases.

6. A 20.0-kg horizontal plank 4.00 m long rests on two supports, one at the left end and a second 1.00 m from the right end. What is the magnitude of the force exerted on the plank by the support near the right end? (a) 32.0 N (b) 45.2 N (c) 112 N (d) 131 N (e) 98.2 N

7. Assume a single 300-N force is exerted on a bicycle frame as shown in Figure OQ12.7. Consider the torque produced by this force about axes perpendicular to the plane of the paper and through each of the points

A through *E*, where *E* is the center of mass of the frame. Rank the torques τ_A, τ_B, τ_C, τ_D, and τ_E from largest to smallest, noting that zero is greater than a negative quantity. If two torques are equal, note their equality in your ranking.

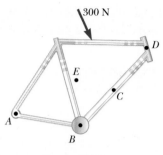

Figure OQ12.7

8. In analyzing the equilibrium of a flat, rigid object, you are about to choose an axis about which you will calculate torques. Which of the following describes the choice you should make? (a) The axis should pass through the object's center of mass. (b) The axis should pass through one end of the object. (c) The axis should be either the *x* axis or the *y* axis. (d) The axis should pass through any point within the object. (e) Any axis within or outside the object can be chosen.

9. A certain wire, 3 m long, stretches by 1.2 mm when under tension 200 N. **(i)** Does an equally thick wire 6 m long, made of the same material and under the same tension, stretch by (a) 4.8 mm, (b) 2.4 mm, (c) 1.2 mm, (d) 0.6 mm, or (e) 0.3 mm? **(ii)** A wire with twice the diameter, 3 m long, made of the same material and under the same tension, stretches by what amount? Choose from the same possibilities (a) through (e).

10. The center of gravity of an ax is on the centerline of the handle, close to the head. Assume you saw across the handle through the center of gravity and weigh the two parts. What will you discover? (a) The handle side is heavier than the head side. (b) The head side is heavier than the handle side. (c) The two parts are equally heavy. (d) Their comparative weights cannot be predicted.

Conceptual Questions | 1. denotes answer available in *Student Solutions Manual/Study Guide*

1. A ladder stands on the ground, leaning against a wall. Would you feel safer climbing up the ladder if you were told that the ground is frictionless but the wall is rough or if you were told that the wall is frictionless but the ground is rough? Explain your answer.

2. The center of gravity of an object may be located outside the object. Give two examples for which that is the case.

3. (a) Give an example in which the net force acting on an object is zero and yet the net torque is nonzero. (b) Give an example in which the net torque acting on an object is zero and yet the net force is nonzero.

4. Stand with your back against a wall. Why can't you put your heels firmly against the wall and then bend forward without falling?

5. An arbitrarily shaped piece of plywood can be suspended from a string attached to the ceiling. Explain how you could use a plumb bob to find its center of gravity.

6. A girl has a large, docile dog she wishes to weigh on a small bathroom scale. She reasons that she can determine her dog's weight with the following method. First she puts the dog's two front feet on the scale and records the scale reading. Then she places only the dog's two back feet on the scale and records the reading. She thinks that the sum of the readings will be the dog's weight. Is she correct? Explain your answer.

7. Can an object be in equilibrium if it is in motion? Explain.

8. What kind of deformation does a cube of Jell-O exhibit when it jiggles?

Problems available in Access end-of-chapter problems online at **www.webassign.net**

Section 12.1 Analysis Model: Rigid Object in Equilibrium
Problems 1–2

Section 12.2 More on the Center of Gravity
Problems 3–7

Section 12.3 Examples of Rigid Objects in Static Equilibrium
Problems 8–25

Section 12.4 Elastic Properties of Solids
Problems 26–36

Additional Problems
Problems 37–64

Challenge Problems
Problem 65–68

Solutions to the following Problems are available in the *Student Solutions Manual/Study Guide:*
12.4, 12.5, 12.8, 12.13, 12.31, 12.32, 12.33, 12.37, 12.45, 12.49, 12.51, 12.56, 12.60, and 12.62

List of Enhanced Problems

Problem Number	Targeted Feedback in Enhanced WebAssign	Analysis Model Tutorial in Enhanced WebAssign	Master It in Enhanced WebAssign	Watch It in Enhanced WebAssign
12.3	✓			✓
12.4			✓	
12.8	✓	✓	✓	
12.9	✓			✓
12.10	✓			✓
12.13	✓	✓	✓	
12.17	✓			✓
12.18	✓			✓
12.19	✓			
12.23	✓			✓
12.31			✓	
12.33	✓		✓	
12.36		✓		
12.37	✓		✓	
12.39	✓			✓
12.43		✓		
12.49	✓		✓	
12.56	✓		✓	
12.63	✓		✓	

Universal Gravitation

Hubble Space Telescope image of
the Whirlpool Galaxy, M51, taken
in 2005. The arms of this spiral
galaxy compress hydrogen gas
and create new clusters of stars.
Some astronomers believe that the
arms are prominent due to a close
encounter with the small, yellow
galaxy, NGC 5195, at the tip of one
of its arms. *(NASA, Hubble Heritage Team,
(STScI/AURA), ESA, S. Beckwith (STScI).
Additional Processing: Robert Gendler)*

Before 1687, a large amount of data had been collected on the motions of the Moon and
the planets, but a clear understanding of the forces related to these motions was not available.
In that year, Isaac Newton provided the key that unlocked the secrets of the heavens. He knew,
from his first law, that a net force had to be acting on the Moon because without such a force
the Moon would move in a straight-line path rather than in its almost circular orbit. Newton
reasoned that this force was the gravitational attraction exerted by the Earth on the Moon. He
realized that the forces involved in the Earth–Moon attraction and in the Sun–planet attraction were not something special to those systems, but rather were particular cases of a general
and universal attraction between objects. In other words, Newton saw that the same force of
attraction that causes the Moon to follow its path around the Earth also causes an apple to
fall from a tree. It was the first time that "earthly" and "heavenly" motions were unified.

In this chapter, we study the law of universal gravitation. We emphasize a description of
planetary motion because astronomical data provide an important test of this law's validity.
We then show that the laws of planetary motion developed by Johannes Kepler follow from

the law of universal gravitation and the principle of conservation of angular momentum for an isolated system. We conclude by deriving a general expression for the gravitational potential energy of a system and examining the energetics of planetary and satellite motion.

13.1 Newton's Law of Universal Gravitation

You may have heard the legend that, while napping under a tree, Newton was struck on the head by a falling apple. This alleged accident supposedly prompted him to imagine that perhaps all objects in the Universe were attracted to each other in the same way the apple was attracted to the Earth. Newton analyzed astronomical data on the motion of the Moon around the Earth. From that analysis, he made the bold assertion that the force law governing the motion of planets was the *same* as the force law that attracted a falling apple to the Earth.

In 1687, Newton published his work on the law of gravity in his treatise *Mathematical Principles of Natural Philosophy*. **Newton's law of universal gravitation** states that

> every particle in the Universe attracts every other particle with a force that is directly proportional to the product of their masses and inversely proportional to the square of the distance between them.

◀ The law of universal gravitation

If the particles have masses m_1 and m_2 and are separated by a distance r, the magnitude of this gravitational force is

$$F_g = G \frac{m_1 m_2}{r^2} \tag{13.1}$$

where G is a constant, called the *universal gravitational constant*. Its value in SI units is

$$G = 6.674 \times 10^{-11} \, \text{N} \cdot \text{m}^2/\text{kg}^2 \tag{13.2}$$

The universal gravitational constant G was first evaluated in the late nineteenth century, based on results of an important experiment by Sir Henry Cavendish (1731–1810) in 1798. The law of universal gravitation was not expressed by Newton in the form of Equation 13.1, and Newton did not mention a constant such as G. In fact, even by the time of Cavendish, a unit of force had not yet been included in the existing system of units. Cavendish's goal was to measure the density of the Earth. His results were then used by other scientists 100 years later to generate a value for G.

Cavendish's apparatus consists of two small spheres, each of mass m, fixed to the ends of a light, horizontal rod suspended by a fine fiber or thin metal wire as illustrated in Figure 13.1. When two large spheres, each of mass M, are placed near the smaller ones, the attractive force between smaller and larger spheres causes the rod to rotate and twist the wire suspension to a new equilibrium orientation. The angle of rotation is measured by the deflection of a light beam reflected from a mirror attached to the vertical suspension.

The form of the force law given by Equation 13.1 is often referred to as an **inverse-square law** because the magnitude of the force varies as the inverse square of the separation of the particles.[1] We shall see other examples of this type of force law in subsequent chapters. We can express this force in vector form by defining a unit vector $\hat{\mathbf{r}}_{12}$ (Fig. 13.2). Because this unit vector is directed from particle 1 toward particle 2, the force exerted by particle 1 on particle 2 is

$$\vec{\mathbf{F}}_{12} = -G \frac{m_1 m_2}{r^2} \hat{\mathbf{r}}_{12} \tag{13.3}$$

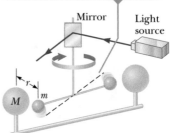

The dashed line represents the original position of the rod.

Mirror   Light source

Figure 13.1 Cavendish apparatus for measuring gravitational forces.

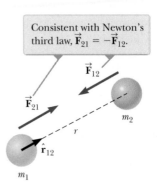

Consistent with Newton's third law, $\vec{\mathbf{F}}_{21} = -\vec{\mathbf{F}}_{12}$.

$\vec{\mathbf{F}}_{12}$

$\vec{\mathbf{F}}_{21}$

m_2

r

$\hat{\mathbf{r}}_{12}$

m_1

Figure 13.2 The gravitational force between two particles is attractive. The unit vector $\hat{\mathbf{r}}_{12}$ is directed from particle 1 toward particle 2.

---

[1]An *inverse* proportionality between two quantities x and y is one in which $y = k/x$, where k is a constant. A *direct* proportion between x and y exists when $y = kx$.

where the negative sign indicates that particle 2 is attracted to particle 1; hence, the force on particle 2 must be directed toward particle 1. By Newton's third law, the force exerted by particle 2 on particle 1, designated $\vec{F}_{21}$, is equal in magnitude to $\vec{F}_{12}$ and in the opposite direction. That is, these forces form an action–reaction pair, and $\vec{F}_{21} = -\vec{F}_{12}$.

Two features of Equation 13.3 deserve mention. First, the gravitational force is a field force that always exists between two particles, regardless of the medium that separates them. Second, because the force varies as the inverse square of the distance between the particles, it decreases rapidly with increasing separation.

Equation 13.3 can also be used to show that the gravitational force exerted by a finite-size, spherically symmetric mass distribution on a particle outside the distribution is the same as if the entire mass of the distribution were concentrated at the center. For example, the magnitude of the force exerted by the Earth on a particle of mass m near the Earth's surface is

$$F_g = G\frac{M_E m}{R_E{}^2} \tag{13.4}$$

where M_E is the Earth's mass and R_E its radius. This force is directed toward the center of the Earth.

Quick Quiz 13.1 A planet has two moons of equal mass. Moon 1 is in a circular orbit of radius r. Moon 2 is in a circular orbit of radius $2r$. What is the magnitude of the gravitational force exerted by the planet on Moon 2? **(a)** four times as large as that on Moon 1 **(b)** twice as large as that on Moon 1 **(c)** equal to that on Moon 1 **(d)** half as large as that on Moon 1 **(e)** one-fourth as large as that on Moon 1

Example 13.1 **Billiards, Anyone?**

Three 0.300-kg billiard balls are placed on a table at the corners of a right triangle as shown in Figure 13.3. The sides of the triangle are of lengths $a = 0.400$ m, $b = 0.300$ m, and $c = 0.500$ m. Calculate the gravitational force vector on the cue ball (designated m_1) resulting from the other two balls as well as the magnitude and direction of this force.

SOLUTION

Conceptualize Notice in Figure 13.3 that the cue ball is attracted to both other balls by the gravitational force. We can see graphically that the net force should point upward and toward the right. We locate our coordinate axes as shown in Figure 13.3, placing our origin at the position of the cue ball.

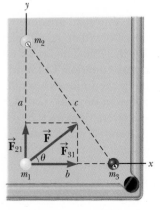

Figure 13.3 (Example 13.1) The resultant gravitational force acting on the cue ball is the vector sum $\vec{F}_{21} + \vec{F}_{31}$.

Categorize This problem involves evaluating the gravitational forces on the cue ball using Equation 13.3. Once these forces are evaluated, it becomes a vector addition problem to find the net force.

Analyze Find the force exerted by m_2 on the cue ball:

$$\vec{F}_{21} = G\frac{m_2 m_1}{a^2}\hat{\mathbf{j}}$$

$$= (6.674 \times 10^{-11}\,\text{N} \cdot \text{m}^2/\text{kg}^2)\frac{(0.300\,\text{kg})(0.300\,\text{kg})}{(0.400\,\text{m})^2}\hat{\mathbf{j}}$$

$$= 3.75 \times 10^{-11}\hat{\mathbf{j}}\,\text{N}$$

Find the force exerted by m_3 on the cue ball:

$$\vec{F}_{31} = G\frac{m_3 m_1}{b^2}\hat{\mathbf{i}}$$

$$= (6.674 \times 10^{-11}\,\text{N} \cdot \text{m}^2/\text{kg}^2)\frac{(0.300\,\text{kg})(0.300\,\text{kg})}{(0.300\,\text{m})^2}\hat{\mathbf{i}}$$

$$= 6.67 \times 10^{-11}\hat{\mathbf{i}}\,\text{N}$$

▶ **13.1** continued

Find the net gravitational force on the cue ball by adding these force vectors:

$$\vec{\mathbf{F}} = \vec{\mathbf{F}}_{31} + \vec{\mathbf{F}}_{21} = \boxed{(6.67\,\hat{\mathbf{i}} + 3.75\,\hat{\mathbf{j}}) \times 10^{-11}\,\text{N}}$$

Find the magnitude of this force:

$$F = \sqrt{F_{31}^2 + F_{21}^2} = \sqrt{(6.67)^2 + (3.75)^2} \times 10^{-11}\,\text{N}$$

$$= \boxed{7.66 \times 10^{-11}\,\text{N}}$$

Find the tangent of the angle θ for the net force vector:

$$\tan\theta = \frac{F_y}{F_x} = \frac{F_{21}}{F_{31}} = \frac{3.75 \times 10^{-11}\,\text{N}}{6.67 \times 10^{-11}\,\text{N}} = 0.562$$

Evaluate the angle θ:

$$\theta = \tan^{-1}(0.562) = \boxed{29.4°}$$

Finalize The result for F shows that the gravitational forces between everyday objects have extremely small magnitudes.

13.2 Free-Fall Acceleration and the Gravitational Force

We have called the magnitude of the gravitational force on an object near the Earth's surface the *weight* of the object, where the weight is given by Equation 5.6. Equation 13.4 is another expression for this force. Therefore, we can set Equations 5.6 and 13.4 equal to each other to obtain

$$mg = G\frac{M_E m}{R_E^2}$$

$$g = G\frac{M_E}{R_E^2} \tag{13.5}$$

Equation 13.5 relates the free-fall acceleration g to physical parameters of the Earth—its mass and radius—and explains the origin of the value of 9.80 m/s² that we have used in earlier chapters. Now consider an object of mass m located a distance h above the Earth's surface or a distance r from the Earth's center, where $r = R_E + h$. The magnitude of the gravitational force acting on this object is

$$F_g = G\frac{M_E m}{r^2} = G\frac{M_E m}{(R_E + h)^2}$$

The magnitude of the gravitational force acting on the object at this position is also $F_g = mg$, where g is the value of the free-fall acceleration at the altitude h. Substituting this expression for F_g into the last equation shows that g is given by

$$g = \frac{GM_E}{r^2} = \frac{GM_E}{(R_E + h)^2} \tag{13.6}$$

◀ Variation of g with altitude

Therefore, it follows that g *decreases* with *increasing altitude*. Values of g for the Earth at various altitudes are listed in Table 13.1. Because an object's weight is mg, we see that as $r \to \infty$, the weight of the object approaches zero.

Table 13.1 Free-Fall Acceleration g at Various Altitudes Above the Earth's Surface

Altitude h (km)	g (m/s²)
1 000	7.33
2 000	5.68
3 000	4.53
4 000	3.70
5 000	3.08
6 000	2.60
7 000	2.23
8 000	1.93
9 000	1.69
10 000	1.49
50 000	0.13
∞	0

Ⓠ**uick Quiz 13.2** Superman stands on top of a very tall mountain and throws a baseball horizontally with a speed such that the baseball goes into a circular orbit around the Earth. While the baseball is in orbit, what is the magnitude of the acceleration of the ball? **(a)** It depends on how fast the baseball is thrown. **(b)** It is zero because the ball does not fall to the ground. **(c)** It is slightly less than 9.80 m/s². **(d)** It is equal to 9.80 m/s².

Example 13.2 **The Density of the Earth**

Using the known radius of the Earth and that $g = 9.80$ m/s^2 at the Earth's surface, find the average density of the Earth.

SOLUTION

Conceptualize Assume the Earth is a perfect sphere. The density of material in the Earth varies, but let's adopt a simplified model in which we assume the density to be uniform throughout the Earth. The resulting density is the average density of the Earth.

Categorize This example is a relatively simple substitution problem.

Using Equation 13.5, solve for the mass of the Earth:

$$M_E = \frac{gR_E^2}{G}$$

Substitute this mass and the volume of a sphere into the definition of density (Eq. 1.1):

$$\rho_E = \frac{M_E}{V_E} = \frac{gR_E^2/G}{\frac{4}{3}\pi R_E^3} = \frac{3}{4}\frac{g}{\pi G R_E}$$

$$= \frac{3}{4}\frac{9.80 \text{ m/s}^2}{\pi(6.674 \times 10^{-11} \text{ N} \cdot \text{m}^2/\text{kg}^2)(6.37 \times 10^6 \text{ m})} = \boxed{5.50 \times 10^3 \text{ kg/m}^3}$$

WHAT IF? What if you were told that a typical density of granite at the Earth's surface is 2.75×10^3 kg/m^3? What would you conclude about the density of the material in the Earth's interior?

Answer Because this value is about half the density we calculated as an average for the entire Earth, we would conclude that the inner core of the Earth has a density much higher than the average value. It is most amazing that the Cavendish experiment—which can be used to determine G and can be done today on a tabletop—combined with simple free-fall measurements of g provides information about the core of the Earth!

13.3 Analysis Model: Particle in a Field (Gravitational)

When Newton published his theory of universal gravitation, it was considered a success because it satisfactorily explained the motion of the planets. It represented strong evidence that the same laws that describe phenomena on the Earth can be used on large objects like planets and throughout the Universe. Since 1687, Newton's theory has been used to account for the motions of comets, the deflection of a Cavendish balance, the orbits of binary stars, and the rotation of galaxies. Nevertheless, both Newton's contemporaries and his successors found it difficult to accept the concept of a force that acts at a distance. They asked how it was possible for two objects such as the Sun and the Earth to interact when they were not in contact with each other. Newton himself could not answer that question.

An approach to describing interactions between objects that are not in contact came well after Newton's death. This approach enables us to look at the gravitational interaction in a different way, using the concept of a **gravitational field** that exists at every point in space. When a particle is placed at a point where the gravitational field exists, the particle experiences a gravitational force. In other words, we imagine that the field exerts a force on the particle rather than consider a direct interaction between two particles. The gravitational field $\vec{g}$ is defined as

Gravitational field ▶

$$\vec{g} \equiv \frac{\vec{F}_g}{m_0} \tag{13.7}$$

That is, the gravitational field at a point in space equals the gravitational force $\vec{F}_g$ experienced by a *test particle* placed at that point divided by the mass m_0 of the test particle. We call the object creating the field the *source particle*. (Although the Earth is not a particle, it is possible to show that we can model the Earth as a particle for the purpose of finding the gravitational field that it creates.) Notice that the presence of the test particle is not necessary for the field to exist: the source particle creates the gravitational field. We can detect the presence of the field and measure its strength by placing a test particle in the field and noting the force exerted on it. In essence, we are describing the "effect" that any object (in this case, the Earth) has on the empty space around itself in terms of the force that *would* be present *if* a second object were somewhere in that space.[2]

The concept of a field is at the heart of the **particle in a field** analysis model. In the general version of this model, a particle resides in an area of space in which a field exists. Because of the existence of the field and a property of the particle, the particle experiences a force. In the gravitational version of the particle in a field model discussed here, the type of field is gravitational, and the property of the particle that results in the force is the particle's mass m. The mathematical representation of the gravitational version of the particle in a field model is Equation 5.5:

$$\vec{\mathbf{F}}_g = m\vec{\mathbf{g}} \tag{5.5}$$

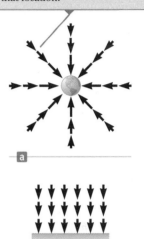

The field vectors point in the direction of the acceleration a particle would experience if it were placed in the field. The magnitude of the field vector at any location is the magnitude of the free-fall acceleration at that location.

In future chapters, we will see two other versions of the particle in a field model. In the electric version, the property of a particle that results in a force is *electric charge:* when a charged particle is placed in an *electric field*, it experiences a force. The magnitude of the force is the product of the electric charge and the field, in analogy with the gravitational force in Equation 5.5. In the magnetic version of the particle in a field model, a charged particle is placed in a *magnetic field*. One other property of this particle is required for the particle to experience a force: the particle must have a *velocity* at some nonzero angle to the magnetic field. The electric and magnetic versions of the particle in a field model are critical to the understanding of the principles of *electromagnetism*, which we will study in Chapters 23–34.

Because the gravitational force acting on the object has a magnitude $GM_E m/r^2$ (see Eq. 13.4), the gravitational field $\vec{\mathbf{g}}$ at a distance r from the center of the Earth is

$$\vec{\mathbf{g}} = \frac{\vec{\mathbf{F}}_g}{m} = -\frac{GM_E}{r^2}\hat{\mathbf{r}} \tag{13.8}$$

Figure 13.4 (a) The gravitational field vectors in the vicinity of a uniform spherical mass such as the Earth vary in both direction and magnitude. (b) The gravitational field vectors in a small region near the Earth's surface are uniform in both direction and magnitude.

where $\hat{\mathbf{r}}$ is a unit vector pointing radially outward from the Earth and the negative sign indicates that the field points toward the center of the Earth as illustrated in Figure 13.4a. The field vectors at different points surrounding the Earth vary in both direction and magnitude. In a small region near the Earth's surface, the downward field $\vec{\mathbf{g}}$ is approximately constant and uniform as indicated in Figure 13.4b. Equation 13.8 is valid at all points *outside* the Earth's surface, assuming the Earth is spherical. At the Earth's surface, where $r = R_E$, $\vec{\mathbf{g}}$ has a magnitude of 9.80 N/kg. (The unit N/kg is the same as m/s².)

| Analysis Model | **Particle in a Field (Gravitational)** |

Imagine an object with mass that we call a *source particle*. The source particle establishes a **gravitational field** $\vec{\mathbf{g}}$ throughout space. The gravitational field is evaluated by measuring the force on a test particle of mass m_0 and then using Equation 13.7. Now imagine a particle of mass m is placed in that field. The particle interacts with the gravitational field so that it experiences a gravitational force given by

$$\vec{\mathbf{F}}_g = m\vec{\mathbf{g}} \tag{5.5}$$

continued

---

[2]We shall return to this idea of mass affecting the space around it when we discuss Einstein's theory of gravitation in Chapter 39.

Analysis Model   **Particle in a Field (Gravitational)** *(continued)*

Examples:

- an object of mass m near the surface of the Earth has a *weight*, which is the result of the gravitational field established in space by the Earth
- a planet in the solar system is in orbit around the Sun, due to the gravitational force on the planet exerted by the gravitational field established by the Sun
- an object near a black hole is drawn into the black hole, never to escape, due to the tremendous gravitational field established by the black hole (Section 13.6)
- in the general theory of relativity, the gravitational field of a massive object is imagined to be described by a *curvature of space–time* (Chapter 39)
- the gravitational field of a massive object is imagined to be mediated by particles called *gravitons*, which have never been detected (Chapter 46)

Example 13.3   **The Weight of the Space Station**   AM

The International Space Station operates at an altitude of 350 km. Plans for the final construction show that material of weight 4.22×10^6 N, measured at the Earth's surface, will have been lifted off the surface by various spacecraft during the construction process. What is the weight of the space station when in orbit?

SOLUTION

Conceptualize The mass of the space station is fixed; it is independent of its location. Based on the discussions in this section and Section 13.2, we realize that the value of g will be reduced at the height of the space station's orbit. Therefore, the weight of the Space Station will be smaller than that at the surface of the Earth.

Categorize We model the Space Station as a *particle in a gravitational field*.

Analyze From the particle in a field model, find the mass of the space station from its weight at the surface of the Earth:

$$m = \frac{F_g}{g} = \frac{4.22 \times 10^6 \text{ N}}{9.80 \text{ m/s}^2} = 4.31 \times 10^5 \text{ kg}$$

Use Equation 13.6 with $h = 350$ km to find the magnitude of the gravitational field at the orbital location:

$$g = \frac{GM_E}{(R_E + h)^2}$$

$$= \frac{(6.674 \times 10^{-11} \text{ N} \cdot \text{m}^2/\text{kg}^2)(5.97 \times 10^{24} \text{ kg})}{(6.37 \times 10^6 \text{ m} + 0.350 \times 10^6 \text{ m})^2} = 8.82 \text{ m/s}^2$$

Use the particle in a field model again to find the space station's weight in orbit:

$$F_g = mg = (4.31 \times 10^5 \text{ kg})(8.82 \text{ m/s}^2) = \boxed{3.80 \times 10^6 \text{ N}}$$

Finalize Notice that the weight of the Space Station is less when it is in orbit, as we expected. It has about 10% less weight than it has when on the Earth's surface, representing a 10% decrease in the magnitude of the gravitational field.

13.4 Kepler's Laws and the Motion of Planets

Humans have observed the movements of the planets, stars, and other celestial objects for thousands of years. In early history, these observations led scientists to regard the Earth as the center of the Universe. This *geocentric model* was elaborated and formalized by the Greek astronomer Claudius Ptolemy (c. 100–c. 170) in the second century and was accepted for the next 1400 years. In 1543, Polish astronomer Nicolaus Copernicus (1473–1543) suggested that the Earth and the other planets revolved in circular orbits around the Sun (the *heliocentric model*).

Danish astronomer Tycho Brahe (1546–1601) wanted to determine how the heavens were constructed and pursued a project to determine the positions of both stars and planets. Those observations of the planets and 777 stars visible to the naked eye were carried out with only a large sextant and a compass. (The telescope had not yet been invented.)

German astronomer Johannes Kepler was Brahe's assistant for a short while before Brahe's death, whereupon he acquired his mentor's astronomical data and spent 16 years trying to deduce a mathematical model for the motion of the planets. Such data are difficult to sort out because the moving planets are observed from a moving Earth. After many laborious calculations, Kepler found that Brahe's data on the revolution of Mars around the Sun led to a successful model.

Kepler's complete analysis of planetary motion is summarized in three statements known as **Kepler's laws:**

◀ **Kepler's laws**

1. All planets move in elliptical orbits with the Sun at one focus.
2. The radius vector drawn from the Sun to a planet sweeps out equal areas in equal time intervals.
3. The square of the orbital period of any planet is proportional to the cube of the semimajor axis of the elliptical orbit.

Kepler's First Law

The geocentric and original heliocentric models of the solar system both suggested circular orbits for heavenly bodies. Kepler's first law indicates that the circular orbit is a very special case and elliptical orbits are the general situation. This notion was difficult for scientists of the time to accept because they believed that perfect circular orbits of the planets reflected the perfection of heaven.

Figure 13.5 shows the geometry of an ellipse, which serves as our model for the elliptical orbit of a planet. An ellipse is mathematically defined by choosing two points F_1 and F_2, each of which is a called a **focus,** and then drawing a curve through points for which the sum of the distances r_1 and r_2 from F_1 and F_2, respectively, is a constant. The longest distance through the center between points on the ellipse (and passing through each focus) is called the **major axis,** and this distance is $2a$. In Figure 13.5, the major axis is drawn along the x direction. The distance a is called the **semimajor axis.** Similarly, the shortest distance through the center between points on the ellipse is called the **minor axis** of length $2b$, where the distance b is the **semiminor axis.** Either focus of the ellipse is located at a distance c from the center of the ellipse, where $a^2 = b^2 + c^2$. In the elliptical orbit of a planet around the Sun, the Sun is at one focus of the ellipse. There is nothing at the other focus.

The **eccentricity** of an ellipse is defined as $e = c/a$, and it describes the general shape of the ellipse. For a circle, $c = 0$, and the eccentricity is therefore zero. The smaller b is compared with a, the shorter the ellipse is along the y direction compared with its extent in the x direction in Figure 13.5. As b decreases, c increases and the eccentricity e increases. Therefore, higher values of eccentricity correspond to longer and thinner ellipses. The range of values of the eccentricity for an ellipse is $0 < e < 1$.

Eccentricities for planetary orbits vary widely in the solar system. The eccentricity of the Earth's orbit is 0.017, which makes it nearly circular. On the other hand, the eccentricity of Mercury's orbit is 0.21, the highest of the eight planets. Figure 13.6a on page 314 shows an ellipse with an eccentricity equal to that of Mercury's orbit. Notice that even this highest-eccentricity orbit is difficult to distinguish from a circle, which is one reason Kepler's first law is an admirable accomplishment. The eccentricity of the orbit of Comet Halley is 0.97, describing an orbit whose major axis is much longer than its minor axis, as shown in Figure 13.6b. As a result, Comet Halley spends much of its 76-year period far from the Sun and invisible from the Earth. It is only visible to the naked eye during a small part of its orbit when it is near the Sun.

Now imagine a planet in an elliptical orbit such as that shown in Figure 13.5, with the Sun at focus F_2. When the planet is at the far left in the diagram, the distance

Johannes Kepler
German astronomer (1571–1630)
Kepler is best known for developing the laws of planetary motion based on the careful observations of Tycho Brahe.

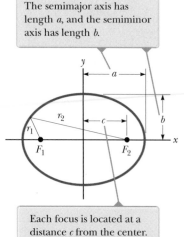

The semimajor axis has length a, and the semiminor axis has length b.

Each focus is located at a distance c from the center.

Figure 13.5 Plot of an ellipse.

Pitfall Prevention 13.2
Where Is the Sun? The Sun is located at one focus of the elliptical orbit of a planet. It is *not* located at the center of the ellipse.

Figure 13.6 (a) The shape of the orbit of Mercury, which has the highest eccentricity ($e = 0.21$) among the eight planets in the solar system. (b) The shape of the orbit of Comet Halley. The shape of the orbit is correct; the comet and the Sun are shown larger than in reality for clarity.

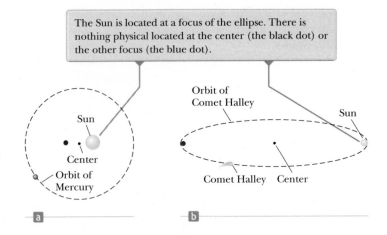

The Sun is located at a focus of the ellipse. There is nothing physical located at the center (the black dot) or the other focus (the blue dot).

Sun

Center

Orbit of Mercury

Orbit of Comet Halley

Sun

Comet Halley   Center

a

b

between the planet and the Sun is $a + c$. At this point, called the *aphelion*, the planet is at its maximum distance from the Sun. (For an object in orbit around the Earth, this point is called the *apogee*.) Conversely, when the planet is at the right end of the ellipse, the distance between the planet and the Sun is $a - c$. At this point, called the *perihelion* (for an Earth orbit, the *perigee*), the planet is at its minimum distance from the Sun.

Kepler's first law is a direct result of the inverse-square nature of the gravitational force. Circular and elliptical orbits correspond to objects that are *bound* to the gravitational force center. These objects include planets, asteroids, and comets that move repeatedly around the Sun as well as moons orbiting a planet. There are also *unbound* objects, such as a meteoroid from deep space that might pass by the Sun once and then never return. The gravitational force between the Sun and these objects also varies as the inverse square of the separation distance, and the allowed paths for these objects include parabolas ($e = 1$) and hyperbolas ($e > 1$).

Kepler's Second Law

Kepler's second law can be shown to be a result of the isolated system model for angular momentum. Consider a planet of mass M_p moving about the Sun in an elliptical orbit (Fig. 13.7a). Let's consider the planet as a system. We model the Sun to be so much more massive than the planet that the Sun does not move. The gravitational force exerted by the Sun on the planet is a central force, always along the radius vector, directed toward the Sun (Fig. 13.7a). The torque on the planet due to this central force about an axis through the Sun is zero because $\vec{\mathbf{F}}_g$ is parallel to $\vec{\mathbf{r}}$.

Therefore, because the external torque on the planet is zero, it is modeled as an isolated system for angular momentum, and the angular momentum $\vec{\mathbf{L}}$ of the planet is a constant of the motion:

$$\Delta \vec{\mathbf{L}} = 0 \quad \rightarrow \quad \vec{\mathbf{L}} = \text{constant}$$

Evaluating $\vec{\mathbf{L}}$ for the planet,

$$\vec{\mathbf{L}} = \vec{\mathbf{r}} \times \vec{\mathbf{p}} = M_p \vec{\mathbf{r}} \times \vec{\mathbf{v}} \quad \rightarrow \quad L = M_p |\vec{\mathbf{r}} \times \vec{\mathbf{v}}| \tag{13.9}$$

We can relate this result to the following geometric consideration. In a time interval dt, the radius vector $\vec{\mathbf{r}}$ in Figure 13.7b sweeps out the area dA, which equals half the area $|\vec{\mathbf{r}} \times d\vec{\mathbf{r}}|$ of the parallelogram formed by the vectors $\vec{\mathbf{r}}$ and $d\vec{\mathbf{r}}$. Because the displacement of the planet in the time interval dt is given by $d\vec{\mathbf{r}} = \vec{\mathbf{v}} \, dt$,

$$dA = \tfrac{1}{2} |\vec{\mathbf{r}} \times d\vec{\mathbf{r}}| = \tfrac{1}{2} |\vec{\mathbf{r}} \times \vec{\mathbf{v}} \, dt| = \tfrac{1}{2} |\vec{\mathbf{r}} \times \vec{\mathbf{v}}| \, dt$$

Substitute for the absolute value of the cross product from Equation 13.9:

$$dA = \tfrac{1}{2} \left(\frac{L}{M_p} \right) dt$$

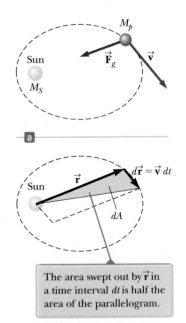

M_p

Sun

M_S

$\vec{\mathbf{F}}_g$

$\vec{\mathbf{v}}$

a

Sun

$\vec{\mathbf{r}}$

$d\vec{\mathbf{r}} = \vec{\mathbf{v}} \, dt$

dA

The area swept out by $\vec{\mathbf{r}}$ in a time interval dt is half the area of the parallelogram.

b

Figure 13.7 (a) The gravitational force acting on a planet is directed toward the Sun. (b) During a time interval dt, a parallelogram is formed by the vectors $\vec{\mathbf{r}}$ and $d\vec{\mathbf{r}} = \vec{\mathbf{v}} \, dt$.

Divide both sides by dt to obtain

$$\frac{dA}{dt} = \frac{L}{2M_p}$$

(13.10)

where L and M_p are both constants. This result shows that that the derivative dA/dt is constant—the radius vector from the Sun to any planet sweeps out equal areas in equal time intervals as stated in Kepler's second law.

This conclusion is a result of the gravitational force being a central force, which in turn implies that angular momentum of the planet is constant. Therefore, the law applies to *any* situation that involves a central force, whether inverse square or not.

Kepler's Third Law

Kepler's third law can be predicted from the inverse-square law for circular orbits and our analysis models. Consider a planet of mass M_p that is assumed to be moving about the Sun (mass M_S) in a circular orbit as in Figure 13.8. Because the gravitational force provides the centripetal acceleration of the planet as it moves in a circle, we model the planet as a particle under a net force and as a particle in uniform circular motion and incorporate Newton's law of universal gravitation,

$$F_g = M_p a \quad \rightarrow \quad \frac{GM_S M_p}{r^2} = M_p \left(\frac{v^2}{r}\right)$$

The orbital speed of the planet is $2\pi r/T$, where T is the period; therefore, the preceding expression becomes

$$\frac{GM_S}{r^2} = \frac{(2\pi r/T)^2}{r}$$

$$T^2 = \left(\frac{4\pi^2}{GM_S}\right)r^3 = K_S r^3$$

where K_S is a constant given by

$$K_S = \frac{4\pi^2}{GM_S} = 2.97 \times 10^{-19} \text{ s}^2/\text{m}^3$$

This equation is also valid for elliptical orbits if we replace r with the length a of the semimajor axis (Fig. 13.5):

$$T^2 = \left(\frac{4\pi^2}{GM_S}\right)a^3 = K_S a^3$$

(13.11)

◀ **Kepler's third law**

Equation 13.11 is Kepler's third law: the square of the period is proportional to the cube of the semimajor axis. Because the semimajor axis of a circular orbit is its radius, this equation is valid for both circular and elliptical orbits. Notice that the constant of proportionality K_S is independent of the mass of the planet.[3] Equation 13.11 is therefore valid for *any* planet. If we were to consider the orbit of a satellite such as the Moon about the Earth, the constant would have a different value, with the Sun's mass replaced by the Earth's mass; that is, $K_E = 4\pi^2/GM_E$.

Table 13.2 on page 316 is a collection of useful data for planets and other objects in the solar system. The far-right column verifies that the ratio T^2/r^3 is constant for all objects orbiting the Sun. The small variations in the values in this column are the result of uncertainties in the data measured for the periods and semimajor axes of the objects.

Recent astronomical work has revealed the existence of a large number of solar system objects beyond the orbit of Neptune. In general, these objects lie in the *Kuiper belt*, a region that extends from about 30 AU (the orbital radius of Neptune) to 50 AU. (An AU is an *astronomical unit*, equal to the radius of the Earth's orbit.) Current

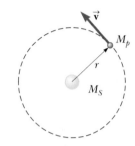

Figure 13.8 A planet of mass M_p moving in a circular orbit around the Sun. The orbits of all planets except Mercury are nearly circular.

---

[3]Equation 13.11 is indeed a proportion because the ratio of the two quantities T^2 and a^3 is a constant. The variables in a proportion are not required to be limited to the first power only.

Table 13.2   **Useful Planetary Data**

Body	Mass (kg)	Mean Radius (m)	Period of Revolution (s)	Mean Distance from the Sun (m)	$\frac{T^2}{r^3}$ (s²/m³)
Mercury	3.30×10^{23}	2.44×10^6	7.60×10^6	5.79×10^{10}	2.98×10^{-19}
Venus	4.87×10^{24}	6.05×10^6	1.94×10^7	1.08×10^{11}	2.99×10^{-19}
Earth	5.97×10^{24}	6.37×10^6	3.156×10^7	1.496×10^{11}	2.97×10^{-19}
Mars	6.42×10^{23}	3.39×10^6	5.94×10^7	2.28×10^{11}	2.98×10^{-19}
Jupiter	1.90×10^{27}	6.99×10^7	3.74×10^8	7.78×10^{11}	2.97×10^{-19}
Saturn	5.68×10^{26}	5.82×10^7	9.29×10^8	1.43×10^{12}	2.95×10^{-19}
Uranus	8.68×10^{25}	2.54×10^7	2.65×10^9	2.87×10^{12}	2.97×10^{-19}
Neptune	1.02×10^{26}	2.46×10^7	5.18×10^9	4.50×10^{12}	2.94×10^{-19}
Pluto[a]	1.25×10^{22}	1.20×10^6	7.82×10^9	5.91×10^{12}	2.96×10^{-19}
Moon	7.35×10^{22}	1.74×10^6	—	—	—
Sun	1.989×10^{30}	6.96×10^8	—	—	—

[a]In August 2006, the International Astronomical Union adopted a definition of a planet that separates Pluto from the other eight planets. Pluto is now defined as a "dwarf planet" like the asteroid Ceres.

estimates identify at least 70 000 objects in this region with diameters larger than 100 km. The first Kuiper belt object (KBO) is Pluto, discovered in 1930 and formerly classified as a planet. Starting in 1992, many more have been detected. Several have diameters in the 1 000-km range, such as Varuna (discovered in 2000), Ixion (2001), Quaoar (2002), Sedna (2003), Haumea (2004), Orcus (2004), and Makemake (2005). One KBO, Eris, discovered in 2005, is believed to be significantly larger than Pluto. Other KBOs do not yet have names, but are currently indicated by their year of discovery and a code, such as 2009 YE7 and 2010 EK139.

A subset of about 1 400 KBOs are called "Plutinos" because, like Pluto, they exhibit a resonance phenomenon, orbiting the Sun two times in the same time interval as Neptune revolves three times. The contemporary application of Kepler's laws and such exotic proposals as planetary angular momentum exchange and migrating planets suggest the excitement of this active area of current research.

Quick Quiz 13.3 An asteroid is in a highly eccentric elliptical orbit around the Sun. The period of the asteroid's orbit is 90 days. Which of the following statements is true about the possibility of a collision between this asteroid and the Earth? **(a)** There is no possible danger of a collision. **(b)** There is a possibility of a collision. **(c)** There is not enough information to determine whether there is danger of a collision.

Example 13.4   **The Mass of the Sun**

Calculate the mass of the Sun, noting that the period of the Earth's orbit around the Sun is 3.156×10^7 s and its distance from the Sun is 1.496×10^{11} m.

SOLUTION

Conceptualize Based on the mathematical representation of Kepler's third law expressed in Equation 13.11, we realize that the mass of the central object in a gravitational system is related to the orbital size and period of objects in orbit around the central object.

Categorize This example is a relatively simple substitution problem.

Solve Equation 13.11 for the mass of the Sun:
$$M_S = \frac{4\pi^2 r^3}{GT^2}$$

Substitute the known values:
$$M_S = \frac{4\pi^2(1.496 \times 10^{11}\ \text{m})^3}{(6.674 \times 10^{-11}\ \text{N} \cdot \text{m}^2/\text{kg}^2)(3.156 \times 10^7\ \text{s})^2} = 1.99 \times 10^{30}\ \text{kg}$$

▶ **13.4** continued

In Example 13.2, an understanding of gravitational forces enabled us to find out something about the density of the Earth's core, and now we have used this understanding to determine the mass of the Sun!

| **Example 13.5** | **A Geosynchronous Satellite** | AM |

Consider a satellite of mass m moving in a circular orbit around the Earth at a constant speed v and at an altitude h above the Earth's surface as illustrated in Figure 13.9.

(A) Determine the speed of satellite in terms of G, h, R_E (the radius of the Earth), and M_E (the mass of the Earth).

SOLUTION

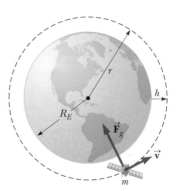

Conceptualize Imagine the satellite moving around the Earth in a circular orbit under the influence of the gravitational force. This motion is similar to that of the International Space Station, the Hubble Space Telescope, and other objects in orbit around the Earth.

Categorize The satellite moves in a circular orbit at a constant speed. Therefore, we categorize the satellite as a *particle in uniform circular motion* as well as a *particle under a net force.*

Figure 13.9 (Example 13.5) A satellite of mass m moving around the Earth in a circular orbit of radius r with constant speed v. The only force acting on the satellite is the gravitational force $\vec{\mathbf{F}}_g$. (Not drawn to scale.)

Analyze The only external force acting on the satellite is the gravitational force from the Earth, which acts toward the center of the Earth and keeps the satellite in its circular orbit.

Apply the particle under a net force and particle in uniform circular motion models to the satellite:

$$F_g = ma \quad \rightarrow \quad G\frac{M_E m}{r^2} = m\left(\frac{v^2}{r}\right)$$

Solve for v, noting that the distance r from the center of the Earth to the satellite is $r = R_E + h$:

$$(1) \quad v = \sqrt{\frac{GM_E}{r}} = \sqrt{\frac{GM_E}{R_E + h}}$$

(B) If the satellite is to be *geosynchronous* (that is, appearing to remain over a fixed position on the Earth), how fast is it moving through space?

SOLUTION

To appear to remain over a fixed position on the Earth, the period of the satellite must be 24 h = 86 400 s and the satellite must be in orbit directly over the equator.

Solve Kepler's third law (Equation 13.11, with $a = r$ and $M_S \rightarrow M_E$) for r:

$$r = \left(\frac{GM_E T^2}{4\pi^2}\right)^{1/3}$$

Substitute numerical values:

$$r = \left[\frac{(6.674 \times 10^{-11}\ \text{N} \cdot \text{m}^2/\text{kg}^2)(5.97 \times 10^{24}\ \text{kg})(86\ 400\ \text{s})^2}{4\pi^2}\right]^{1/3}$$

$$= 4.22 \times 10^7\ \text{m}$$

Use Equation (1) to find the speed of the satellite:

$$v = \sqrt{\frac{(6.674 \times 10^{-11}\ \text{N} \cdot \text{m}^2/\text{kg}^2)(5.97 \times 10^{24}\ \text{kg})}{4.22 \times 10^7\ \text{m}}}$$

$$= 3.07 \times 10^3\ \text{m/s}$$

Finalize The value of r calculated here translates to a height of the satellite above the surface of the Earth of almost 36 000 km. Therefore, geosynchronous satellites have the advantage of allowing an earthbound antenna to be aimed

continued

▶ **13.5** continued

in a fixed direction, but there is a disadvantage in that the signals between the Earth and the satellite must travel a long distance. It is difficult to use geosynchronous satellites for optical observation of the Earth's surface because of their high altitude.

WHAT IF? What if the satellite motion in part (A) were taking place at height h above the surface of another planet more massive than the Earth but of the same radius? Would the satellite be moving at a higher speed or a lower speed than it does around the Earth?

Answer If the planet exerts a larger gravitational force on the satellite due to its larger mass, the satellite must move with a higher speed to avoid moving toward the surface. This conclusion is consistent with the predictions of Equation (1), which shows that because the speed v is proportional to the square root of the mass of the planet, the speed increases as the mass of the planet increases.

13.5 Gravitational Potential Energy

In Chapter 8, we introduced the concept of gravitational potential energy, which is the energy associated with the configuration of a system of objects interacting via the gravitational force. We emphasized that the gravitational potential energy function $U = mgy$ for a particle–Earth system is valid only when the particle of mass m is near the Earth's surface, where the gravitational force is independent of y. This expression for the gravitational potential energy is also restricted to situations where a very massive object (such as the Earth) establishes a gravitational field of magnitude g and a particle of much smaller mass m resides in that field. Because the gravitational force between two particles varies as $1/r^2$, we expect that a more general potential energy function—one that is valid without the restrictions mentioned above—will be different from $U = mgy$.

Recall from Equation 7.27 that the change in the potential energy of a system associated with a given displacement of a member of the system is defined as the negative of the internal work done by the force on that member during the displacement:

$$\Delta U = U_f - U_i = -\int_{r_i}^{r_f} F(r)\, dr \qquad \text{(13.12)}$$

We can use this result to evaluate the general gravitational potential energy function. Consider a particle of mass m moving between two points Ⓐ and Ⓑ above the Earth's surface (Fig. 13.10). The particle is subject to the gravitational force given by Equation 13.1. We can express this force as

$$F(r) = -\frac{GM_E m}{r^2}$$

where the negative sign indicates that the force is attractive. Substituting this expression for $F(r)$ into Equation 13.12, we can compute the change in the gravitational potential energy function for the particle–Earth system as the separation distance r changes:

$$U_f - U_i = GM_E m \int_{r_i}^{r_f} \frac{dr}{r^2} = GM_E m \left[-\frac{1}{r} \right]_{r_i}^{r_f}$$

$$U_f - U_i = -GM_E m \left(\frac{1}{r_f} - \frac{1}{r_i} \right) \qquad \text{(13.13)}$$

As always, the choice of a reference configuration for the potential energy is completely arbitrary. It is customary to choose the reference configuration for zero

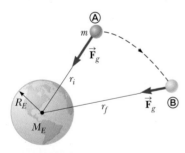

Figure 13.10 As a particle of mass m moves from Ⓐ to Ⓑ above the Earth's surface, the gravitational potential energy of the particle–Earth system changes according to Equation 13.12.

potential energy to be the same as that for which the force is zero. Taking $U_i = 0$ at $r_i = \infty$, we obtain the important result

$$U(r) = -\frac{GM_E m}{r} \qquad (13.14)$$

◀ **Gravitational potential energy of the Earth–particle system**

This expression applies when the particle is separated from the center of the Earth by a distance r, provided that $r \geq R_E$. The result is not valid for particles inside the Earth, where $r < R_E$. Because of our choice of U_i, the function U is always negative (Fig. 13.11).

Although Equation 13.14 was derived for the particle–Earth system, a similar form of the equation can be applied to any two particles. That is, the gravitational potential energy associated with any pair of particles of masses m_1 and m_2 separated by a distance r is

$$U = -\frac{Gm_1 m_2}{r} \qquad (13.15)$$

This expression shows that the gravitational potential energy for any pair of particles varies as $1/r$, whereas the force between them varies as $1/r^2$. Furthermore, the potential energy is negative because the force is attractive and we have chosen the potential energy as zero when the particle separation is infinite. Because the force between the particles is attractive, an external agent must do positive work to increase the separation between the particles. The work done by the external agent produces an increase in the potential energy as the two particles are separated. That is, U becomes less negative as r increases.

When two particles are at rest and separated by a distance r, an external agent has to supply an energy at least equal to $+Gm_1 m_2/r$ to separate the particles to an infinite distance. It is therefore convenient to think of the absolute value of the potential energy as the *binding energy* of the system. If the external agent supplies an energy greater than the binding energy, the excess energy of the system is in the form of kinetic energy of the particles when the particles are at an infinite separation.

We can extend this concept to three or more particles. In this case, the total potential energy of the system is the sum over all pairs of particles. Each pair contributes a term of the form given by Equation 13.15. For example, if the system contains three particles as in Figure 13.12,

$$U_{\text{total}} = U_{12} + U_{13} + U_{23} = -G\left(\frac{m_1 m_2}{r_{12}} + \frac{m_1 m_3}{r_{13}} + \frac{m_2 m_3}{r_{23}}\right)$$

The absolute value of U_{total} represents the work needed to separate the particles by an infinite distance.

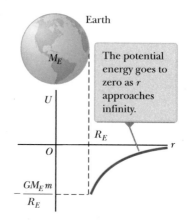

Figure 13.11 Graph of the gravitational potential energy U versus r for the system of an object above the Earth's surface.

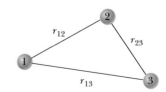

Figure 13.12 Three interacting particles.

Example 13.6 **The Change in Potential Energy**

A particle of mass m is displaced through a small vertical distance Δy near the Earth's surface. Show that in this situation the general expression for the change in gravitational potential energy given by Equation 13.13 reduces to the familiar relationship $\Delta U = mg\,\Delta y$.

SOLUTION

Conceptualize Compare the two different situations for which we have developed expressions for gravitational potential energy: (1) a planet and an object that are far apart for which the energy expression is Equation 13.14 and (2) a small object at the surface of a planet for which the energy expression is Equation 7.19. We wish to show that these two expressions are equivalent.

continued

▶ **13.6** continued

Categorize This example is a substitution problem.

Combine the fractions in Equation 13.13:

$$(1) \quad \Delta U = -GM_E m\left(\frac{1}{r_f} - \frac{1}{r_i}\right) = GM_E m\left(\frac{r_f - r_i}{r_i r_f}\right)$$

Evaluate $r_f - r_i$ and $r_i r_f$ if both the initial and final positions of the particle are close to the Earth's surface:

$$r_f - r_i = \Delta y \quad r_i r_f \approx R_E^2$$

Substitute these expressions into Equation (1):

$$\Delta U \approx \frac{GM_E m}{R_E^2} \Delta y = mg\,\Delta y$$

where $g = GM_E/R_E^2$ (Eq. 13.5).

WHAT IF? Suppose you are performing upper-atmosphere studies and are asked by your supervisor to find the height in the Earth's atmosphere at which the "surface equation" $\Delta U = mg\,\Delta y$ gives a 1.0% error in the change in the potential energy. What is this height?

Answer Because the surface equation assumes a constant value for g, it will give a ΔU value that is larger than the value given by the general equation, Equation 13.13.

Set up a ratio reflecting a 1.0% error:

$$\frac{\Delta U_{\text{surface}}}{\Delta U_{\text{general}}} = 1.010$$

Substitute the expressions for each of these changes ΔU:

$$\frac{mg\,\Delta y}{GM_E m(\Delta y/r_i r_f)} = \frac{g r_i r_f}{GM_E} = 1.010$$

Substitute for r_i, r_f, and g from Equation 13.5:

$$\frac{(GM_E/R_E^2)R_E(R_E + \Delta y)}{GM_E} = \frac{R_E + \Delta y}{R_E} = 1 + \frac{\Delta y}{R_E} = 1.010$$

Solve for Δy:

$$\Delta y = 0.010 R_E = 0.010(6.37 \times 10^6 \text{ m}) = 6.37 \times 10^4 \text{ m} = 63.7 \text{ km}$$

13.6 Energy Considerations in Planetary and Satellite Motion

Given the general expression for gravitational potential energy developed in Section 13.5, we can now apply our energy analysis models to gravitational systems. Consider an object of mass m moving with a speed v in the vicinity of a massive object of mass M, where $M \gg m$. The system might be a planet moving around the Sun, a satellite in orbit around the Earth, or a comet making a one-time flyby of the Sun. If we assume the object of mass M is at rest in an inertial reference frame, the total mechanical energy E of the two-object system when the objects are separated by a distance r is the sum of the kinetic energy of the object of mass m and the potential energy of the system, given by Equation 13.15:

$$E = K + U$$

$$E = \tfrac{1}{2}mv^2 - \frac{GMm}{r} \tag{13.16}$$

If the system of objects of mass m and M is isolated, and there are no nonconservative forces acting within the system, the mechanical energy of the system given by Equation 13.16 is the total energy of the system and this energy is conserved:

$$\Delta E_{\text{system}} = 0 \quad \rightarrow \quad \Delta K + \Delta U_g = 0 \quad \rightarrow \quad E_i = E_f$$

Therefore, as the object of mass m moves from Ⓐ to Ⓑ in Figure 13.10, the total energy remains constant and Equation 13.16 gives

$$\tfrac{1}{2}mv_i^2 - \frac{GMm}{r_i} = \tfrac{1}{2}mv_f^2 - \frac{GMm}{r_f} \tag{13.17}$$

Combining this statement of energy conservation with our earlier discussion of conservation of angular momentum, we see that both the total energy and the total angular momentum of a gravitationally bound, two-object system are constants of the motion.

Equation 13.16 shows that E may be positive, negative, or zero, depending on the value of v. For a bound system such as the Earth–Sun system, however, E is necessarily *less than zero* because we have chosen the convention that $U \to 0$ as $r \to \infty$.

We can easily establish that $E < 0$ for the system consisting of an object of mass m moving in a circular orbit about an object of mass $M \gg m$ (Fig. 13.13). Modeling the object of mass m as a particle under a net force and a particle in uniform circular motion gives

$$F_g = ma \quad \to \quad \frac{GMm}{r^2} = \frac{mv^2}{r}$$

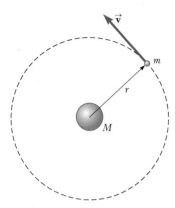

Figure 13.13 An object of mass m moving in a circular orbit about a much larger object of mass M.

Multiplying both sides by r and dividing by 2 gives

$$\tfrac{1}{2}mv^2 = \frac{GMm}{2r} \tag{13.18}$$

Substituting this equation into Equation 13.16, we obtain

$$E = \frac{GMm}{2r} - \frac{GMm}{r}$$

$$E = -\frac{GMm}{2r} \quad \text{(circular orbits)} \tag{13.19}$$

◀ Total energy for circular orbits of an object of mass m around an object of mass $M \gg m$

This result shows that the total mechanical energy is negative in the case of circular orbits. Notice that the kinetic energy is positive and equal to half the absolute value of the potential energy. The absolute value of E is also equal to the binding energy of the system because this amount of energy must be provided to the system to move the two objects infinitely far apart.

The total mechanical energy is also negative in the case of elliptical orbits. The expression for E for elliptical orbits is the same as Equation 13.19 with r replaced by the semimajor axis length a:

$$E = -\frac{GMm}{2a} \quad \text{(elliptical orbits)} \tag{13.20}$$

◀ Total energy for elliptical orbits of an object of mass m around an object of mass $M \gg m$

Quick Quiz 13.4 A comet moves in an elliptical orbit around the Sun. Which point in its orbit (perihelion or aphelion) represents the highest value of (a) the speed of the comet, (b) the potential energy of the comet–Sun system, (c) the kinetic energy of the comet, and (d) the total energy of the comet–Sun system?

Example 13.7   **Changing the Orbit of a Satellite**

A space transportation vehicle releases a 470-kg communications satellite while in an orbit 280 km above the surface of the Earth. A rocket engine on the satellite boosts it into a geosynchronous orbit. How much energy does the engine have to provide?

SOLUTION

Conceptualize Notice that the height of 280 km is much lower than that for a geosynchronous satellite, 36 000 km, as mentioned in Example 13.5. Therefore, energy must be expended to raise the satellite to this much higher position.

Categorize This example is a substitution problem.

Find the initial radius of the satellite's orbit when it is still in the vehicle's cargo bay: $\quad r_i = R_E + 280 \text{ km} = 6.65 \times 10^6 \text{ m}$

continued

▶ **13.7** continued

Use Equation 13.19 to find the difference in energies for the satellite–Earth system with the satellite at the initial and final radii:

$$\Delta E = E_f - E_i = -\frac{GM_E m}{2r_f} - \left(-\frac{GM_E m}{2r_i}\right) = -\frac{GM_E m}{2}\left(\frac{1}{r_f} - \frac{1}{r_i}\right)$$

Substitute numerical values, using $r_f = 4.22 \times 10^7$ m from Example 13.5:

$$\Delta E = -\frac{(6.674 \times 10^{-11}\ \text{N} \cdot \text{m}^2/\text{kg}^2)(5.97 \times 10^{24}\ \text{kg})(470\ \text{kg})}{2} \times$$
$$\left(\frac{1}{4.22 \times 10^7\ \text{m}} - \frac{1}{6.65 \times 10^6\ \text{m}}\right) = \boxed{1.19 \times 10^{10}\ \text{J}}$$

which is the energy equivalent of 89 gal of gasoline. NASA engineers must account for the changing mass of the spacecraft as it ejects burned fuel, something we have not done here. Would you expect the calculation that includes the effect of this changing mass to yield a greater or a lesser amount of energy required from the engine?

Escape Speed

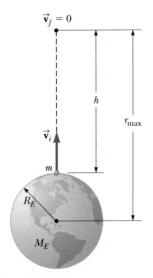

Figure 13.14 An object of mass m projected upward from the Earth's surface with an initial speed v_i reaches a maximum altitude h.

Suppose an object of mass m is projected vertically upward from the Earth's surface with an initial speed v_i as illustrated in Figure 13.14. We can use energy considerations to find the value of the initial speed needed to allow the object to reach a certain distance away from the center of the Earth. Equation 13.16 gives the total energy of the system for any configuration. As the object is projected upward from the surface of the Earth, $v = v_i$ and $r = r_i = R_E$. When the object reaches its maximum altitude, $v = v_f = 0$ and $r = r_f = r_{max}$. Because the object–Earth system is isolated, we substitute these values into the isolated-system model expression given by Equation 13.17:

$$\tfrac{1}{2}mv_i^2 - \frac{GM_E m}{R_E} = -\frac{GM_E m}{r_{max}}$$

Solving for v_i^2 gives

$$v_i^2 = 2GM_E\left(\frac{1}{R_E} - \frac{1}{r_{max}}\right) \tag{13.21}$$

For a given maximum altitude $h = r_{max} - R_E$, we can use this equation to find the required initial speed.

We are now in a position to calculate the **escape speed,** which is the minimum speed the object must have at the Earth's surface to approach an infinite separation distance from the Earth. Traveling at this minimum speed, the object continues to move farther and farther away from the Earth as its speed asymptotically approaches zero. Letting $r_{max} \rightarrow \infty$ in Equation 13.21 and identifying v_i as v_{esc} gives

Escape speed from ▶ the Earth

$$v_{esc} = \sqrt{\frac{2GM_E}{R_E}} \tag{13.22}$$

Pitfall Prevention 13.3
You Can't Really Escape Although Equation 13.22 provides the "escape speed" from the Earth, *complete* escape from the Earth's gravitational influence is impossible because the gravitational force is of infinite range.

This expression for v_{esc} is independent of the mass of the object. In other words, a spacecraft has the same escape speed as a molecule. Furthermore, the result is independent of the direction of the velocity and ignores air resistance.

If the object is given an initial speed equal to v_{esc}, the total energy of the system is equal to zero. Notice that when $r \rightarrow \infty$, the object's kinetic energy and the potential energy of the system are both zero. If v_i is greater than v_{esc}, however, the total energy of the system is greater than zero and the object has some residual kinetic energy as $r \rightarrow \infty$.

Example 13.8 **Escape Speed of a Rocket**

Calculate the escape speed from the Earth for a 5 000-kg spacecraft and determine the kinetic energy it must have at the Earth's surface to move infinitely far away from the Earth.

▶ **13.8** continued

SOLUTION

Conceptualize Imagine projecting the spacecraft from the Earth's surface so that it moves farther and farther away, traveling more and more slowly, with its speed approaching zero. Its speed will never reach zero, however, so the object will never turn around and come back.

Categorize This example is a substitution problem.

Use Equation 13.22 to find the escape speed:

$$v_{esc} = \sqrt{\frac{2GM_E}{R_E}} = \sqrt{\frac{2(6.674 \times 10^{-11}\ N \cdot m^2/kg^2)(5.97 \times 10^{24}\ kg)}{6.37 \times 10^6\ m}}$$

$$= \boxed{1.12 \times 10^4\ m/s}$$

Evaluate the kinetic energy of the spacecraft from Equation 7.16:

$$K = \tfrac{1}{2}mv_{esc}^2 = \tfrac{1}{2}(5.00 \times 10^3\ kg)(1.12 \times 10^4\ m/s)^2$$

$$= \boxed{3.13 \times 10^{11}\ J}$$

The calculated escape speed corresponds to about 25 000 mi/h. The kinetic energy of the spacecraft is equivalent to the energy released by the combustion of about 2 300 gal of gasoline.

WHAT IF? What if you want to launch a 1 000-kg spacecraft at the escape speed? How much energy would that require?

Answer In Equation 13.22, the mass of the object moving with the escape speed does not appear. Therefore, the escape speed for the 1 000-kg spacecraft is the same as that for the 5 000-kg spacecraft. The only change in the kinetic energy is due to the mass, so the 1 000-kg spacecraft requires one-fifth of the energy of the 5 000-kg spacecraft:

$$K = \tfrac{1}{5}(3.13 \times 10^{11}\ J) = 6.25 \times 10^{10}\ J$$

Equations 13.21 and 13.22 can be applied to objects projected from any planet. That is, in general, the escape speed from the surface of any planet of mass M and radius R is

$$v_{esc} = \sqrt{\frac{2GM}{R}} \qquad \text{(13.23)}$$

◀ Escape speed from the surface of a planet of mass M and radius R

Escape speeds for the planets, the Moon, and the Sun are provided in Table 13.3. The values vary from 2.3 km/s for the Moon to about 618 km/s for the Sun. These results, together with some ideas from the kinetic theory of gases (see Chapter 21), explain why some planets have atmospheres and others do not. As we shall see later, at a given temperature the average kinetic energy of a gas molecule depends only on the mass of the molecule. Lighter molecules, such as hydrogen and helium, have a higher average speed than heavier molecules at the same temperature. When the average speed of the lighter molecules is not much less than the escape speed of a planet, a significant fraction of them have a chance to escape.

This mechanism also explains why the Earth does not retain hydrogen molecules and helium atoms in its atmosphere but does retain heavier molecules, such as oxygen and nitrogen. On the other hand, the very large escape speed for Jupiter enables that planet to retain hydrogen, the primary constituent of its atmosphere.

Black Holes

In Example 11.7, we briefly described a rare event called a supernova, the catastrophic explosion of a very massive star. The material that remains in the central core of such an object continues to collapse, and the core's ultimate fate depends on its mass. If the core has a mass less than 1.4 times the mass of our Sun, it gradually cools down and ends its life as a white dwarf star. If the core's mass is greater than this value, however, it may collapse further due to gravitational forces. What

Table 13.3 **Escape Speeds from the Surfaces of the Planets, Moon, and Sun**

Planet	v_{esc} (km/s)
Mercury	4.3
Venus	10.3
Earth	11.2
Mars	5.0
Jupiter	60
Saturn	36
Uranus	22
Neptune	24
Moon	2.3
Sun	618

remains is a neutron star, discussed in Example 11.7, in which the mass of a star is compressed to a radius of about 10 km. (On the Earth, a teaspoon of this material would weigh about 5 billion tons!)

An even more unusual star death may occur when the core has a mass greater than about three solar masses. The collapse may continue until the star becomes a very small object in space, commonly referred to as a **black hole.** In effect, black holes are remains of stars that have collapsed under their own gravitational force. If an object such as a spacecraft comes close to a black hole, the object experiences an extremely strong gravitational force and is trapped forever.

The escape speed for a black hole is very high because of the concentration of the star's mass into a sphere of very small radius (see Eq. 13.23). If the escape speed exceeds the speed of light c, radiation from the object (such as visible light) cannot escape and the object appears to be black (hence the origin of the terminology "black hole"). The critical radius R_S at which the escape speed is c is called the **Schwarzschild radius** (Fig. 13.15). The imaginary surface of a sphere of this radius surrounding the black hole is called the **event horizon,** which is the limit of how close you can approach the black hole and hope to escape.

There is evidence that supermassive black holes exist at the centers of galaxies, with masses very much larger than the Sun. (There is strong evidence of a supermassive black hole of mass 2–3 million solar masses at the center of our galaxy.)

Dark Matter

Equation (1) in Example 13.5 shows that the speed of an object in orbit around the Earth decreases as the object is moved farther away from the Earth:

$$v = \sqrt{\frac{GM_E}{r}} \tag{13.24}$$

Using data in Table 13.2 to find the speeds of planets in their orbits around the Sun, we find the same behavior for the planets. Figure 13.16 shows this behavior for the eight planets of our solar system. The theoretical prediction of the planet speed as a function of distance from the Sun is shown by the red-brown curve, using Equation 13.24 with the mass of the Earth replaced by the mass of the Sun. Data for the individual planets lie right on this curve. This behavior results from the vast majority of the mass of the solar system being concentrated in a small space, i.e., the Sun.

Extending this concept further, we might expect the same behavior in a galaxy. Much of the visible galactic mass, including that of a supermassive black hole, is near the central core of a galaxy. The opening photograph for this chapter shows the central core of the Whirlpool galaxy as a very bright area surrounded by the "arms" of the galaxy, which contain material in orbit around the central core. Based on this distribution of matter in the galaxy, the speed of an object in the outer part of the galaxy would be smaller than that for objects closer to the center, just like for the planets of the solar system.

That is *not* what is observed, however. Figure 13.17 shows the results of measurements of the speeds of objects in the Andromeda galaxy as a function of distance from the galaxy's center.[4] The red-brown curve shows the expected speeds for these objects if they were traveling in circular orbits around the mass concentrated in the central core. The data for the individual objects in the galaxy shown by the black dots are all well above the theoretical curve. These data, as well as an extensive amount of data taken over the past half century, show that for objects outside the central core of the galaxy, the curve of speed versus distance from the center of the galaxy is approximately flat rather than decreasing at larger distances. Therefore, these objects (including our own Solar System in the Milky Way) are rotating faster than can be accounted for by gravity due to the visible galaxy! This surprising

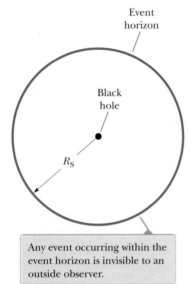

Event horizon

Black hole

R_S

Any event occurring within the event horizon is invisible to an outside observer.

Figure 13.15 A black hole. The distance R_S equals the Schwarzschild radius.

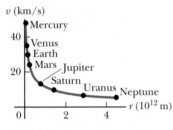

v (km/s)

Mercury
Venus
Earth
Mars
Jupiter
Saturn
Uranus Neptune
$r\ (10^{12}\,\text{m})$

Figure 13.16 The orbital speed v as a function of distance r from the Sun for the eight planets of the solar system. The theoretical curve is in red-brown, and the data points for the planets are in black.

[4]V. C. Rubin and W. K. Ford, "Rotation of the Andromeda Nebula from a Spectroscopic Survey of Emission Regions," *Astrophysical Journal* **159:** 379–403 (1970).

result means that there must be additional mass in a more extended distribution, causing these objects to orbit so fast, and has led scientists to propose the existence of **dark matter.** This matter is proposed to exist in a large halo around each galaxy (with a radius up to 10 times as large as the visible galaxy's radius). Because it is not luminous (i.e., does not emit electromagnetic radiation) it must be either very cold or electrically neutral. Therefore, we cannot "see" dark matter, except through its gravitational effects.

The proposed existence of dark matter is also implied by earlier observations made on larger gravitationally bound structures known as galaxy clusters.[5] These observations show that the orbital speeds of galaxies in a cluster are, on average, too large to be explained by the luminous matter in the cluster alone. The speeds of the individual galaxies are so high, they suggest that there is 50 times as much dark matter in galaxy clusters as in the galaxies themselves!

Why doesn't dark matter affect the orbital speeds of planets like it does those of a galaxy? It seems that a solar system is too small a structure to contain enough dark matter to affect the behavior of orbital speeds. A galaxy or galaxy cluster, on the other hand, contains huge amounts of dark matter, resulting in the surprising behavior.

What, though, *is* dark matter? At this time, no one knows. One theory claims that dark matter is based on a particle called a weakly interacting massive particle, or WIMP. If this theory is correct, calculations show that about 200 WIMPs pass through a human body at any given time. The new Large Hadron Collider in Europe (see Chapter 46) is the first particle accelerator with enough energy to possibly generate and detect the existence of WIMPs, which has generated much current interest in dark matter. Keeping an eye on this research in the future should be exciting.

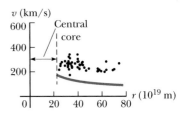

Figure 13.17 The orbital speed v of a galaxy object as a function of distance r from the center of the central core of the Andromeda galaxy. The theoretical curve is in red-brown, and the data points for the galaxy objects are in black. No data are provided on the left because the behavior inside the central core of the galaxy is more complicated.

Summary

Definitions

The **gravitational field** at a point in space is defined as the gravitational force $\vec{F}_g$ experienced by any test particle located at that point divided by the mass m_0 of the test particle:

$$\vec{g} \equiv \frac{\vec{F}_g}{m_0} \qquad (13.7)$$

Concepts and Principles

Newton's law of universal gravitation states that the gravitational force of attraction between any two particles of masses m_1 and m_2 separated by a distance r has the magnitude

$$F_g = G \frac{m_1 m_2}{r^2} \qquad (13.1)$$

where $G = 6.674 \times 10^{-11}$ N · m²/kg² is the **universal gravitational constant.** This equation enables us to calculate the force of attraction between masses under many circumstances.

An object at a distance h above the Earth's surface experiences a gravitational force of magnitude mg, where g is the free-fall acceleration at that elevation:

$$g = \frac{GM_E}{r^2} = \frac{GM_E}{(R_E + h)^2} \qquad (13.6)$$

In this expression, M_E is the mass of the Earth and R_E is its radius. Therefore, the weight of an object decreases as the object moves away from the Earth's surface.

---

[5]F. Zwicky, "On the Masses of Nebulae and of Clusters of Nebulae," *Astrophysical Journal* **86**: 217–246 (1937).

Kepler's laws of planetary motion state:

1. All planets move in elliptical orbits with the Sun at one focus.
2. The radius vector drawn from the Sun to a planet sweeps out equal areas in equal time intervals.
3. The square of the orbital period of any planet is proportional to the cube of the semimajor axis of the elliptical orbit.

Kepler's third law can be expressed as

$$T^2 = \left(\frac{4\pi^2}{GM_S}\right) a^3 \qquad (13.11)$$

where M_S is the mass of the Sun and a is the semimajor axis. For a circular orbit, a can be replaced in Equation 13.11 by the radius r. Most planets have nearly circular orbits around the Sun.

The **gravitational potential energy** associated with a system of two particles of mass m_1 and m_2 separated by a distance r is

$$U = -\frac{Gm_1 m_2}{r} \qquad (13.15)$$

where U is taken to be zero as $r \to \infty$.

If an isolated system consists of an object of mass m moving with a speed v in the vicinity of a massive object of mass M, the total energy E of the system is the sum of the kinetic and potential energies:

$$E = \tfrac{1}{2}mv^2 - \frac{GMm}{r} \qquad (13.16)$$

The total energy of the system is a constant of the motion. If the object moves in an elliptical orbit of semimajor axis a around the massive object and $M \gg m$, the total energy of the system is

$$E = -\frac{GMm}{2a} \qquad (13.20)$$

For a circular orbit, this same equation applies with $a = r$.

The **escape speed** for an object projected from the surface of a planet of mass M and radius R is

$$v_{esc} = \sqrt{\frac{2GM}{R}} \qquad (13.23)$$

Analysis Model for Problem Solving

Particle in a Field (Gravitational) A source particle with some mass establishes a **gravitational field $\vec{g}$** throughout space. When a particle of mass m is placed in that field, it experiences a gravitational force given by

$$\vec{F}_g = m\vec{g} \qquad (5.5)$$

Objective Questions

1. denotes answer available in *Student Solutions Manual/Study Guide*

1. A system consists of five particles. How many terms appear in the expression for the total gravitational potential energy of the system? (a) 4 (b) 5 (c) 10 (d) 20 (e) 25

2. Rank the following quantities of energy from largest to smallest. State if any are equal. (a) the absolute value of the average potential energy of the Sun–Earth system (b) the average kinetic energy of the Earth in its orbital motion relative to the Sun (c) the absolute value of the total energy of the Sun–Earth system

3. A satellite moves in a circular orbit at a constant speed around the Earth. Which of the following statements is true? (a) No force acts on the satellite. (b) The satellite moves at constant speed and hence doesn't accelerate. (c) The satellite has an acceleration directed away from the Earth. (d) The satellite has an acceleration directed toward the Earth. (e) Work is done on the satellite by the gravitational force.

4. Suppose the gravitational acceleration at the surface of a certain moon A of Jupiter is 2 m/s². Moon B has twice the mass and twice the radius of moon A. What is the gravitational acceleration at its surface? Neglect the gravitational acceleration due to Jupiter. (a) 8 m/s² (b) 4 m/s² (c) 2 m/s² (d) 1 m/s² (e) 0.5 m/s²

5. Imagine that nitrogen and other atmospheric gases were more soluble in water so that the atmosphere of the Earth is entirely absorbed by the oceans. Atmospheric pressure would then be zero, and outer space would start at the planet's surface. Would the Earth then have a gravitational field? (a) Yes, and at the surface it would be larger in magnitude than 9.8 N/kg. (b) Yes, and it would be essentially the same as the current value. (c) Yes, and it would be somewhat less than 9.8 N/kg. (d) Yes, and it would be much less than 9.8 N/kg. (e) No, it would not.

6. An object of mass m is located on the surface of a spherical planet of mass M and radius R. The escape speed from the planet does not depend on which of the following? (a) M (b) m (c) the density of the planet (d) R (e) the acceleration due to gravity on that planet

7. A satellite originally moves in a circular orbit of radius R around the Earth. Suppose it is moved into a circular orbit of radius $4R$. **(i)** What does the force exerted on the satellite then become? (a) eight times larger (b) four times larger (c) one-half as large (d) one-eighth as large (e) one-sixteenth as large **(ii)** What happens to the satellite's speed? Choose from the same possibilities (a) through (e). **(iii)** What happens to its period? Choose from the same possibilities (a) through (e).

8. The vernal equinox and the autumnal equinox are associated with two points 180° apart in the Earth's orbit. That is, the Earth is on precisely opposite sides of the Sun when it passes through these two points. From the vernal equinox, 185.4 days elapse before the autumnal equinox. Only 179.8 days elapse from the autumnal equinox until the next vernal equinox. Why is the interval from the March (vernal) to the September (autumnal) equinox (which contains the summer solstice) longer than the interval from the September to the March equinox rather than being equal to that interval? Choose one of the following reasons. (a) They are really the same, but the Earth spins faster during the "summer" interval, so the days are shorter. (b) Over the "summer" interval, the Earth moves slower because it is farther from the Sun. (c) Over the March-to-September interval, the Earth moves slower because it is closer to the Sun. (d) The Earth has less kinetic energy when it is warmer. (e) The Earth has less orbital angular momentum when it is warmer.

9. Rank the magnitudes of the following gravitational forces from largest to smallest. If two forces are equal, show their equality in your list. (a) the force exerted by a 2-kg object on a 3-kg object 1 m away (b) the force exerted by a 2-kg object on a 9-kg object 1 m away (c) the force exerted by a 2-kg object on a 9-kg object 2 m away (d) the force exerted by a 9-kg object on a 2-kg object 2 m away (e) the force exerted by a 4-kg object on another 4-kg object 2 m away

10. The gravitational force exerted on an astronaut on the Earth's surface is 650 N directed downward. When she is in the space station in orbit around the Earth, is the gravitational force on her (a) larger, (b) exactly the same, (c) smaller, (d) nearly but not exactly zero, or (e) exactly zero?

11. Halley's comet has a period of approximately 76 years, and it moves in an elliptical orbit in which its distance from the Sun at closest approach is a small fraction of its maximum distance. Estimate the comet's maximum distance from the Sun in astronomical units (AUs) (the distance from the Earth to the Sun). (a) 6 AU (b) 12 AU (c) 20 AU (d) 28 AU (e) 35 AU

Conceptual Questions   **1.** denotes answer available in *Student Solutions Manual/Study Guide*

1. Each *Voyager* spacecraft was accelerated toward escape speed from the Sun by the gravitational force exerted by Jupiter on the spacecraft. (a) Is the gravitational force a conservative or a nonconservative force? (b) Does the interaction of the spacecraft with Jupiter meet the definition of an elastic collision? (c) How could the spacecraft be moving faster after the collision?

2. In his 1798 experiment, Cavendish was said to have "weighed the Earth." Explain this statement.

3. Why don't we put a geosynchronous weather satellite in orbit around the 45th parallel? Wouldn't such a satellite be more useful in the United States than one in orbit around the equator?

4. (a) Explain why the force exerted on a particle by a uniform sphere must be directed toward the center of the sphere. (b) Would this statement be true if the mass distribution of the sphere were not spherically symmetric? Explain.

5. (a) At what position in its elliptical orbit is the speed of a planet a maximum? (b) At what position is the speed a minimum?

6. You are given the mass and radius of planet X. How would you calculate the free-fall acceleration on this planet's surface?

7. (a) If a hole could be dug to the center of the Earth, would the force on an object of mass m still obey Equation 13.1 there? (b) What do you think the force on m would be at the center of the Earth?

8. Explain why it takes more fuel for a spacecraft to travel from the Earth to the Moon than for the return trip. Estimate the difference.

9. A satellite in low-Earth orbit is not truly traveling through a vacuum. Rather, it moves through very thin air. Does the resulting air friction cause the satellite to slow down?

Problems available in Access end-of-chapter problems online at www.webassign.net

Section 13.1 Newton's Law of Universal Gravitation
Problems 1–10

Section 13.2 Free-Fall Acceleration and the Gravitational Force
Problems 11–13

Section 13.3 Analysis Model: Particle in a Field (Gravitational)
Problems 14–16

Section 13.4 Kepler's Laws and the Motion of Planets
Problems 17–29

Section 13.5 Gravitational Potential Energy
Problems 30–35

Section 13.6 Energy Considerations in Planetary and Satellite Motion
Problems 36–49

Additional Problems
Problems 50–77

Challenge Problems
Problems 78–80

Solutions to the following Problems are available in the *Student Solutions Manual/Study Guide:*
13.1, 13.11, 13.12, 13.14, 13.18, 13.21, 13.27, 13.33, 13.36, 13.38, 13.61, 13.62, 13.64, 13.69, 13.74, and 13.80

List of Enhanced Problems

Problem Number	Targeted Feedback in Enhanced WebAssign	Analysis Model Tutorial in Enhanced WebAssign	Master It in Enhanced WebAssign	Watch It in Enhanced WebAssign
13.1			✓	
13.3	✓			✓
13.6	✓			✓
13.9	✓			✓
13.11	✓		✓	
13.12	✓			✓
13.16	✓	✓		✓
13.21	✓		✓	
13.23	✓			✓
13.26	✓			✓
13.30	✓			✓
13.36		✓	✓	
13.39	✓			✓
13.45	✓			✓
13.51				✓
13.53	✓		✓	
13.61		✓	✓	
13.65		✓		

Fluid Mechanics

Matter is normally classified as being in one of three states: solid, liquid, or gas. From everyday experience we know that a solid has a definite volume and shape, a liquid has a definite volume but no definite shape, and an unconfined gas has neither a definite volume nor a definite shape. These descriptions help us picture the states of matter, but they are somewhat artificial. For example, asphalt and plastics are normally considered solids, but over long time intervals they tend to flow like liquids. Likewise, most substances can be a solid, a liquid, or a gas (or a combination of any of these three), depending on the temperature and pressure. In general, the time interval required for a particular substance to change its shape in response to an external force determines whether we treat the substance as a solid, a liquid, or a gas.

A **fluid** is a collection of molecules that are randomly arranged and held together by weak cohesive forces and by forces exerted by the walls of a container. Both liquids and gases are fluids.

In our treatment of the mechanics of fluids, we'll be applying principles and analysis models that we have already discussed. First, we consider the mechanics of a fluid at rest, that is, *fluid statics*, and then study fluids in motion, that is, *fluid dynamics*.

14.1 Pressure

Fluids do not sustain shearing stresses or tensile stresses such as those discussed in Chapter 12; therefore, the only stress that can be exerted on an object submerged in a static fluid is one that tends to compress the object from all sides. In other words, the force exerted by a static fluid on an object is always perpendicular to the surfaces of the object as shown in Figure 14.1. We discussed this situation in Section 12.4.

Fish congregate around a reef in Hawaii searching for food. How do fish such as the lined butterflyfish (*Chaetodon lineolatus*) at the upper left control their movements up and down in the water? We'll find out in this chapter. *(Vlad61/Shutterstock.com)*

At any point on the surface of the object, the force exerted by the fluid is perpendicular to the surface of the object.

Figure 14.1 The forces exerted by a fluid on the surfaces of a submerged object.

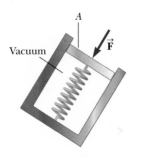

Figure 14.2 A simple device for measuring the pressure exerted by a fluid.

The pressure in a fluid can be measured with the device pictured in Figure 14.2. The device consists of an evacuated cylinder that encloses a light piston connected to a spring. As the device is submerged in a fluid, the fluid presses on the top of the piston and compresses the spring until the inward force exerted by the fluid is balanced by the outward force exerted by the spring. The fluid pressure can be measured directly if the spring is calibrated in advance. If F is the magnitude of the force exerted on the piston and A is the surface area of the piston, the **pressure** P of the fluid at the level to which the device has been submerged is defined as the ratio of the force to the area:

$$P \equiv \frac{F}{A} \tag{14.1}$$

Pressure is a scalar quantity because it is proportional to the magnitude of the force on the piston.

If the pressure varies over an area, the infinitesimal force dF on an infinitesimal surface element of area dA is

$$dF = P\, dA \tag{14.2}$$

where P is the pressure at the location of the area dA. To calculate the total force exerted on a surface of a container, we must integrate Equation 14.2 over the surface.

The units of pressure are newtons per square meter (N/m^2) in the SI system. Another name for the SI unit of pressure is the **pascal** (Pa):

$$1\ \text{Pa} \equiv 1\ N/m^2 \tag{14.3}$$

Pitfall Prevention 14.1

Force and Pressure Equations 14.1 and 14.2 make a clear distinction between force and pressure. Another important distinction is that *force is a vector* and *pressure is a scalar*. There is no direction associated with pressure, but the direction of the force associated with the pressure is perpendicular to the surface on which the pressure acts.

For a tactile demonstration of the definition of pressure, hold a tack between your thumb and forefinger, with the point of the tack on your thumb and the head of the tack on your forefinger. Now *gently* press your thumb and forefinger together. Your thumb will begin to feel pain immediately while your forefinger will not. The tack is exerting the same force on both your thumb and forefinger, but the pressure on your thumb is much larger because of the small area over which the force is applied.

Quick Quiz 14.1 Suppose you are standing directly behind someone who steps back and accidentally stomps on your foot with the heel of one shoe. Would you be better off if that person were **(a)** a large, male professional basketball player wearing sneakers or **(b)** a petite woman wearing spike-heeled shoes?

Example 14.1 | **The Water Bed**

The mattress of a water bed is 2.00 m long by 2.00 m wide and 30.0 cm deep.

(A) Find the weight of the water in the mattress.

SOLUTION

Conceptualize Think about carrying a jug of water and how heavy it is. Now imagine a sample of water the size of a water bed. We expect the weight to be relatively large.

Categorize This example is a substitution problem.

Find the volume of the water filling the mattress: $V = (2.00\ \text{m})(2.00\ \text{m})(0.300\ \text{m}) = 1.20\ \text{m}^3$

Use Equation 1.1 and the density of fresh water (see Table 14.1) to find the mass of the water bed: $M = \rho V = (1\,000\ \text{kg/m}^3)(1.20\ \text{m}^3) = 1.20 \times 10^3\ \text{kg}$

Find the weight of the bed: $Mg = (1.20 \times 10^3\ \text{kg})(9.80\ \text{m/s}^2) = \boxed{1.18 \times 10^4\ \text{N}}$

which is approximately 2 650 lb. (A regular bed, including mattress, box spring, and metal frame, weighs approximately 300 lb.) Because this load is so great, it is best to place a water bed in the basement or on a sturdy, well-supported floor.

▶ **14.1** continued

(B) Find the pressure exerted by the water bed on the floor when the bed rests in its normal position. Assume the entire lower surface of the bed makes contact with the floor.

SOLUTION

When the water bed is in its normal position, the area in contact with the floor is 4.00 m². Use Equation 14.1 to find the pressure:

$$P = \frac{1.18 \times 10^4 \text{ N}}{4.00 \text{ m}^2} = \boxed{2.94 \times 10^3 \text{ Pa}}$$

WHAT IF? What if the water bed is replaced by a 300-lb regular bed that is supported by four legs? Each leg has a circular cross section of radius 2.00 cm. What pressure does this bed exert on the floor?

Answer The weight of the regular bed is distributed over four circular cross sections at the bottom of the legs. Therefore, the pressure is

$$P = \frac{F}{A} = \frac{mg}{4(\pi r^2)} = \frac{300 \text{ lb}}{4\pi(0.020\ 0 \text{ m})^2}\left(\frac{1 \text{ N}}{0.225 \text{ lb}}\right)$$

$$= 2.65 \times 10^5 \text{ Pa}$$

This result is almost 100 times larger than the pressure due to the water bed! The weight of the regular bed, even though it is much less than the weight of the water bed, is applied over the very small area of the four legs. The high pressure on the floor at the feet of a regular bed could cause dents in wood floors or permanently crush carpet pile.

14.2 Variation of Pressure with Depth

As divers well know, water pressure increases with depth. Likewise, atmospheric pressure decreases with increasing altitude; for this reason, aircraft flying at high altitudes must have pressurized cabins for the comfort of the passengers.

We now show how the pressure in a liquid increases with depth. As Equation 1.1 describes, the *density* of a substance is defined as its mass per unit volume; Table 14.1 lists the densities of various substances. These values vary slightly with temperature because the volume of a substance is dependent on temperature (as shown in Chapter 19). Under standard conditions (at 0°C and at atmospheric pressure), the densities of gases are about $\frac{1}{1\ 000}$ the densities of solids and liquids. This difference in densities implies that the average molecular spacing in a gas under these conditions is about ten times greater than that in a solid or liquid.

Table 14.1 **Densities of Some Common Substances at Standard Temperature (0°C) and Pressure (Atmospheric)**

Substance	ρ (kg/m³)	Substance	ρ (kg/m³)
Air	1.29	Iron	7.86×10^3
Air (at 20°C and		Lead	11.3×10^3
atmospheric pressure)	1.20	Mercury	13.6×10^3
Aluminum	2.70×10^3	Nitrogen gas	1.25
Benzene	0.879×10^3	Oak	0.710×10^3
Brass	8.4×10^3	Osmium	22.6×10^3
Copper	8.92×10^3	Oxygen gas	1.43
Ethyl alcohol	0.806×10^3	Pine	0.373×10^3
Fresh water	1.00×10^3	Platinum	21.4×10^3
Glycerin	1.26×10^3	Seawater	1.03×10^3
Gold	19.3×10^3	Silver	10.5×10^3
Helium gas	1.79×10^{-1}	Tin	7.30×10^3
Hydrogen gas	8.99×10^{-2}	Uranium	19.1×10^3
Ice	0.917×10^3		

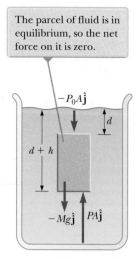

The parcel of fluid is in equilibrium, so the net force on it is zero.

$-P_0 A\hat{\mathbf{j}}$

d

$d + h$

$-Mg\hat{\mathbf{j}}$ $PA\hat{\mathbf{j}}$

Figure 14.3 A parcel of fluid in a larger volume of fluid is singled out.

Variation of pressure ▶ with depth

Pascal's law ▶

Now consider a liquid of density ρ at rest as shown in Figure 14.3. We assume ρ is uniform throughout the liquid, which means the liquid is incompressible. Let us select a parcel of the liquid contained within an imaginary block of cross-sectional area A extending from depth d to depth $d + h$. The liquid external to our parcel exerts forces at all points on the surface of the parcel, perpendicular to the surface. The pressure exerted by the liquid on the bottom face of the parcel is P, and the pressure on the top face is P_0. Therefore, the upward force exerted by the outside fluid on the bottom of the parcel has a magnitude PA, and the downward force exerted on the top has a magnitude P_0A. The mass of liquid in the parcel is $M = \rho V = \rho Ah$; therefore, the weight of the liquid in the parcel is $Mg = \rho Ahg$. Because the parcel is at rest and remains at rest, it can be modeled as a particle in equilibrium, so that the net force acting on it must be zero. Choosing upward to be the positive y direction, we see that

$$\sum \vec{\mathbf{F}} = PA\hat{\mathbf{j}} - P_0A\hat{\mathbf{j}} - Mg\hat{\mathbf{j}} = 0$$

or

$$PA - P_0A - \rho Ahg = 0$$

$$\boxed{P = P_0 + \rho gh} \tag{14.4}$$

That is, the pressure P at a depth h below a point in the liquid at which the pressure is P_0 is greater by an amount ρgh. If the liquid is open to the atmosphere and P_0 is the pressure at the surface of the liquid, then P_0 is **atmospheric pressure.** In our calculations and working of end-of-chapter problems, we usually take atmospheric pressure to be

$$P_0 = 1.00 \text{ atm} = 1.013 \times 10^5 \text{ Pa}$$

Equation 14.4 implies that the pressure is the same at all points having the same depth, independent of the shape of the container.

Because the pressure in a fluid depends on depth and on the value of P_0, any increase in pressure at the surface must be transmitted to every other point in the fluid. This concept was first recognized by French scientist Blaise Pascal (1623–1662) and is called **Pascal's law: a change in the pressure applied to a fluid is transmitted undiminished to every point of the fluid and to the walls of the container.**

An important application of Pascal's law is the hydraulic press illustrated in Figure 14.4a. A force of magnitude F_1 is applied to a small piston of surface area A_1. The pressure is transmitted through an incompressible liquid to a larger piston of surface area A_2. Because the pressure must be the same on both sides, $P = F_1/A_1 = F_2/A_2$. Therefore, the force F_2 is greater than the force F_1 by a factor of A_2/A_1. By designing a hydraulic press with appropriate areas A_1 and A_2, a large out-

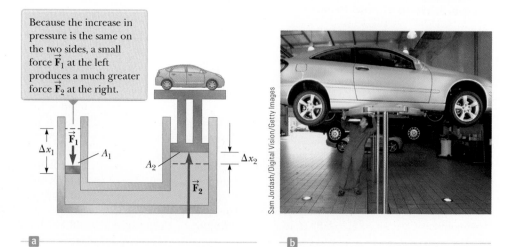

Because the increase in pressure is the same on the two sides, a small force $\vec{\mathbf{F}}_1$ at the left produces a much greater force $\vec{\mathbf{F}}_2$ at the right.

Δx_1 $\vec{\mathbf{F}}_1$ A_1 A_2 Δx_2 $\vec{\mathbf{F}}_2$

Sam Jordash/Digital Vision/Getty Images

Figure 14.4 (a) Diagram of a hydraulic press. (b) A vehicle undergoing repair is supported by a hydraulic lift in a garage.

a   b

put force can be applied by means of a small input force. Hydraulic brakes, car lifts, hydraulic jacks, and forklifts all make use of this principle (Fig. 14.4b).

Because liquid is neither added to nor removed from the system, the volume of liquid pushed down on the left in Figure 14.4a as the piston moves downward through a displacement Δx_1 equals the volume of liquid pushed up on the right as the right piston moves upward through a displacement Δx_2. That is, $A_1 \Delta x_1 = A_2 \Delta x_2$; therefore, $A_2/A_1 = \Delta x_1/\Delta x_2$. We have already shown that $A_2/A_1 = F_2/F_1$. Therefore, $F_2/F_1 = \Delta x_1/\Delta x_2$, so $F_1 \Delta x_1 = F_2 \Delta x_2$. Each side of this equation is the work done by the force on its respective piston. Therefore, the work done by $\vec{F}_1$ on the input piston equals the work done by $\vec{F}_2$ on the output piston, as it must to conserve energy. (The process can be modeled as a special case of the nonisolated system model: the *nonisolated system in steady state*. There is energy transfer into and out of the system, but these energy transfers balance, so that there is no net change in the energy of the system.)

Quick Quiz 14.2 The pressure at the bottom of a filled glass of water ($\rho = 1\,000$ kg/m^3) is P. The water is poured out, and the glass is filled with ethyl alcohol ($\rho = 806$ kg/m^3). What is the pressure at the bottom of the glass? **(a)** smaller than P **(b)** equal to P **(c)** larger than P **(d)** indeterminate

---

Example 14.2   The Car Lift

In a car lift used in a service station, compressed air exerts a force on a small piston that has a circular cross section of radius 5.00 cm. This pressure is transmitted by a liquid to a piston that has a radius of 15.0 cm.

(A) What force must the compressed air exert to lift a car weighing 13 300 N?

SOLUTION

Conceptualize Review the material just discussed about Pascal's law to understand the operation of a car lift.

Categorize This example is a substitution problem.

Solve $F_1/A_1 = F_2/A_2$ for F_1:

$$F_1 = \left(\frac{A_1}{A_2}\right)F_2 = \frac{\pi(5.00 \times 10^{-2}\ \mathrm{m})^2}{\pi(15.0 \times 10^{-2}\ \mathrm{m})^2}\,(1.33 \times 10^4\ \mathrm{N})$$

$$= 1.48 \times 10^3\ \mathrm{N}$$

(B) What air pressure produces this force?

SOLUTION

Use Equation 14.1 to find the air pressure that produces this force:

$$P = \frac{F_1}{A_1} = \frac{1.48 \times 10^3\ \mathrm{N}}{\pi(5.00 \times 10^{-2}\ \mathrm{m})^2}$$

$$= 1.88 \times 10^5\ \mathrm{Pa}$$

This pressure is approximately twice atmospheric pressure.

Example 14.3   A Pain in Your Ear

Estimate the force exerted on your eardrum due to the water when you are swimming at the bottom of a pool that is 5.0 m deep.

SOLUTION

Conceptualize As you descend in the water, the pressure increases. You may have noticed this increased pressure in your ears while diving in a swimming pool, a lake, or the ocean. We can find the pressure difference exerted on the

continued

▶ **14.3** continued

eardrum from the depth given in the problem; then, after estimating the ear drum's surface area, we can determine the net force the water exerts on it.

Categorize This example is a substitution problem.
The air inside the middle ear is normally at atmospheric pressure P_0. Therefore, to find the net force on the eardrum, we must consider the difference between the total pressure at the bottom of the pool and atmospheric pressure. Let's estimate the surface area of the eardrum to be approximately $1 \text{ cm}^2 = 1 \times 10^{-4} \text{ m}^2$.

Use Equation 14.4 to find this pressure difference:

$$P_{\text{bot}} - P_0 = \rho g h$$
$$= (1.00 \times 10^3 \text{ kg/m}^3)(9.80 \text{ m/s}^2)(5.0 \text{ m}) = 4.9 \times 10^4 \text{ Pa}$$

Use Equation 14.1 to find the magnitude of the net force on the ear:

$$F = (P_{\text{bot}} - P_0)A = (4.9 \times 10^4 \text{ Pa})(1 \times 10^{-4} \text{ m}^2) \approx \boxed{5 \text{ N}}$$

Because a force of this magnitude on the eardrum is extremely uncomfortable, swimmers often "pop their ears" while under water, an action that pushes air from the lungs into the middle ear. Using this technique equalizes the pressure on the two sides of the eardrum and relieves the discomfort.

Example 14.4   The Force on a Dam

Water is filled to a height H behind a dam of width w (Fig. 14.5). Determine the resultant force exerted by the water on the dam.

SOLUTION

Conceptualize Because pressure varies with depth, we cannot calculate the force simply by multiplying the area by the pressure. As the pressure in the water increases with depth, the force on the adjacent portion of the dam also increases.

Categorize Because of the variation of pressure with depth, we must use integration to solve this example, so we categorize it as an analysis problem.

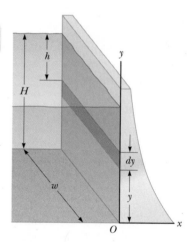

Figure 14.5 (Example 14.4) Water exerts a force on a dam.

Analyze Let's imagine a vertical y axis, with $y = 0$ at the bottom of the dam. We divide the face of the dam into narrow horizontal strips at a distance y above the bottom, such as the red strip in Figure 14.5. The pressure on each such strip is due only to the water; atmospheric pressure acts on both sides of the dam.

Use Equation 14.4 to calculate the pressure due to the water at the depth h:

$$P = \rho g h = \rho g (H - y)$$

Use Equation 14.2 to find the force exerted on the shaded strip of area $dA = w \, dy$:

$$dF = P \, dA = \rho g (H - y) w \, dy$$

Integrate to find the total force on the dam:

$$F = \int P \, dA = \int_0^H \rho g (H - y) w \, dy = \boxed{\tfrac{1}{2}\rho g w H^2}$$

Finalize Notice that the thickness of the dam shown in Figure 14.5 increases with depth. This design accounts for the greater force the water exerts on the dam at greater depths.

WHAT IF? What if you were asked to find this force without using calculus? How could you determine its value?

Answer We know from Equation 14.4 that pressure varies linearly with depth. Therefore, the average pressure due to the water over the face of the dam is the average of the pressure at the top and the pressure at the bottom:

$$P_{\text{avg}} = \frac{P_{\text{top}} + P_{\text{bottom}}}{2} = \frac{0 + \rho g H}{2} = \tfrac{1}{2}\rho g H$$

▶ **14.4** continued

The total force on the dam is equal to the product of the average pressure and the area of the face of the dam:

$$F = P_{avg} A = (\tfrac{1}{2}\rho g H)(Hw) = \tfrac{1}{2}\rho g w H^2$$

which is the same result we obtained using calculus.

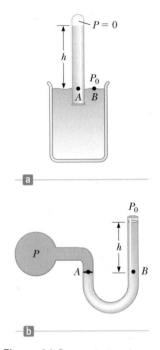

Figure 14.6 Two devices for measuring pressure: (a) a mercury barometer and (b) an open-tube manometer.

14.3 Pressure Measurements

During the weather report on a television news program, the *barometric pressure* is often provided. This reading is the current local pressure of the atmosphere, which varies over a small range from the standard value provided earlier. How is this pressure measured?

One instrument used to measure atmospheric pressure is the common barometer, invented by Evangelista Torricelli (1608–1647). A long tube closed at one end is filled with mercury and then inverted into a dish of mercury (Fig. 14.6a). The closed end of the tube is nearly a vacuum, so the pressure at the top of the mercury column can be taken as zero. In Figure 14.6a, the pressure at point A, due to the column of mercury, must equal the pressure at point B, due to the atmosphere. If that were not the case, there would be a net force that would move mercury from one point to the other until equilibrium is established. Therefore, $P_0 = \rho_{Hg} g h$, where ρ_{Hg} is the density of the mercury and h is the height of the mercury column. As atmospheric pressure varies, the height of the mercury column varies, so the height can be calibrated to measure atmospheric pressure. Let us determine the height of a mercury column for one atmosphere of pressure, $P_0 = 1$ atm $= 1.013 \times 10^5$ Pa:

$$P_0 = \rho_{Hg} g h \quad \rightarrow \quad h = \frac{P_0}{\rho_{Hg} g} = \frac{1.013 \times 10^5 \text{ Pa}}{(13.6 \times 10^3 \text{ kg/m}^3)(9.80 \text{ m/s}^2)} = 0.760 \text{ m}$$

Based on such a calculation, one atmosphere of pressure is defined to be the pressure equivalent of a column of mercury that is exactly 0.760 0 m in height at 0°C.

A device for measuring the pressure of a gas contained in a vessel is the open-tube manometer illustrated in Figure 14.6b. One end of a U-shaped tube containing a liquid is open to the atmosphere, and the other end is connected to a container of gas at pressure P. In an equilibrium situation, the pressures at points A and B must be the same (otherwise, the curved portion of the liquid would experience a net force and would accelerate), and the pressure at A is the unknown pressure of the gas. Therefore, equating the unknown pressure P to the pressure at point B, we see that $P = P_0 + \rho g h$. Again, we can calibrate the height h to the pressure P.

The difference in the pressures in each part of Figure 14.6 (that is, $P - P_0$) is equal to $\rho g h$. The pressure P is called the **absolute pressure,** and the difference $P - P_0$ is called the **gauge pressure.** For example, the pressure you measure in your bicycle tire is gauge pressure.

Ouick Quiz 14.3 Several common barometers are built, with a variety of fluids. For which of the following fluids will the column of fluid in the barometer be the highest? **(a)** mercury **(b)** water **(c)** ethyl alcohol **(d)** benzene

14.4 Buoyant Forces and Archimedes's Principle

Have you ever tried to push a beach ball down under water (Fig. 14.7a, p. 424)? It is extremely difficult to do because of the large upward force exerted by the water on the ball. The upward force exerted by a fluid on any immersed object is called

Archimedes
Greek Mathematician, Physicist, and Engineer (c. 287–212 BC)
Archimedes was perhaps the greatest scientist of antiquity. He was the first to compute accurately the ratio of a circle's circumference to its diameter, and he also showed how to calculate the volume and surface area of spheres, cylinders, and other geometric shapes. He is well known for discovering the nature of the buoyant force and was also a gifted inventor. One of his practical inventions, still in use today, is Archimedes's screw, an inclined, rotating, coiled tube used originally to lift water from the holds of ships. He also invented the catapult and devised systems of levers, pulleys, and weights for raising heavy loads. Such inventions were successfully used to defend his native city, Syracuse, during a two-year siege by Romans.

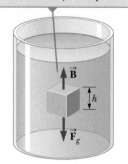

The buoyant force on the cube is the resultant of the forces exerted on its top and bottom faces by the liquid.

Figure 14.8 The external forces acting on an immersed cube are the gravitational force $\vec{F}_g$ and the buoyant force $\vec{B}$.

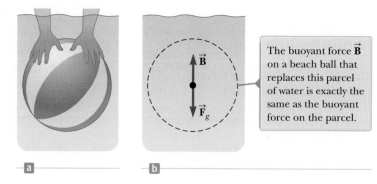

The buoyant force $\vec{B}$ on a beach ball that replaces this parcel of water is exactly the same as the buoyant force on the parcel.

Figure 14.7 (a) A swimmer pushes a beach ball under water. (b) The forces on a beach ball–sized parcel of water.

a **buoyant force.** We can determine the magnitude of a buoyant force by applying some logic. Imagine a beach ball–sized parcel of water beneath the water surface as in Figure 14.7b. Because this parcel is in equilibrium, there must be an upward force that balances the downward gravitational force on the parcel. This upward force is the buoyant force, and its magnitude is equal to the weight of the water in the parcel. The buoyant force is the resultant force on the parcel due to all forces applied by the fluid surrounding the parcel.

Now imagine replacing the beach ball–sized parcel of water with a beach ball of the same size. The net force applied by the fluid surrounding the beach ball is the same, regardless of whether it is applied to a beach ball or to a parcel of water. Consequently, **the magnitude of the buoyant force on an object always equals the weight of the fluid displaced by the object.** This statement is known as **Archimedes's principle.**

With the beach ball under water, the buoyant force, equal to the weight of a beach ball–sized parcel of water, is much larger than the weight of the beach ball. Therefore, there is a large net upward force, which explains why it is so hard to hold the beach ball under the water. Note that Archimedes's principle does not refer to the makeup of the object experiencing the buoyant force. The object's composition is not a factor in the buoyant force because the buoyant force is exerted by the surrounding fluid.

To better understand the origin of the buoyant force, consider a cube of solid material immersed in a liquid as in Figure 14.8. According to Equation 14.4, the pressure P_{bot} at the bottom of the cube is greater than the pressure P_{top} at the top by an amount $\rho_{fluid}gh$, where h is the height of the cube and ρ_{fluid} is the density of the fluid. The pressure at the bottom of the cube causes an *upward* force equal to $P_{bot}A$, where A is the area of the bottom face. The pressure at the top of the cube causes a *downward* force equal to $P_{top}A$. The resultant of these two forces is the buoyant force $\vec{B}$ with magnitude

$$B = (P_{bot} - P_{top})A = (\rho_{fluid}gh)A$$

$$B = \rho_{fluid}gV_{disp} \tag{14.5}$$

where $V_{disp} = Ah$ is the volume of the fluid displaced by the cube. Because the product $\rho_{fluid}V_{disp}$ is equal to the mass of fluid displaced by the object,

$$B = Mg$$

where Mg is the weight of the fluid displaced by the cube. This result is consistent with our initial statement about Archimedes's principle above, based on the discussion of the beach ball.

Under normal conditions, the weight of a fish in the opening photograph for this chapter is slightly greater than the buoyant force on the fish. Hence, the fish would sink if it did not have some mechanism for adjusting the buoyant force. The fish accomplishes that by internally regulating the size of its air-filled swim bladder to increase its volume and the magnitude of the buoyant force acting on it, according to Equation 14.5. In this manner, fish are able to swim to various depths.

Before we proceed with a few examples, it is instructive to discuss two common situations: a totally submerged object and a floating (partly submerged) object.

Case 1: Totally Submerged Object When an object is totally submerged in a fluid of density ρ_{fluid}, the volume V_{disp} of the displaced fluid is equal to the volume V_{obj} of the object; so, from Equation 14.5, the magnitude of the upward buoyant force is $B = \rho_{\text{fluid}}gV_{\text{obj}}$. If the object has a mass M and density ρ_{obj}, its weight is equal to $F_g = Mg = \rho_{\text{obj}}gV_{\text{obj}}$, and the net force on the object is $B - F_g = (\rho_{\text{fluid}} - \rho_{\text{obj}})gV_{\text{obj}}$. Hence, if the density of the object is less than the density of the fluid, the downward gravitational force is less than the buoyant force and the unsupported object accelerates upward (Fig. 14.9a). If the density of the object is greater than the density of the fluid, the upward buoyant force is less than the downward gravitational force and the unsupported object sinks (Fig. 14.9b). If the density of the submerged object equals the density of the fluid, the net force on the object is zero and the object remains in equilibrium. Therefore, the direction of motion of an object submerged in a fluid is determined *only* by the densities of the object and the fluid.

Case 2: Floating Object Now consider an object of volume V_{obj} and density $\rho_{\text{obj}} < \rho_{\text{fluid}}$ in static equilibrium floating on the surface of a fluid, that is, an object that is only *partially* submerged (Fig. 14.10). In this case, the upward buoyant force is balanced by the downward gravitational force acting on the object. If V_{disp} is the volume of the fluid displaced by the object (this volume is the same as the volume of that part of the object beneath the surface of the fluid), the buoyant force has a magnitude $B = \rho_{\text{fluid}}gV_{\text{disp}}$. Because the weight of the object is $F_g = Mg = \rho_{\text{obj}}gV_{\text{obj}}$ and because $F_g = B$, we see that $\rho_{\text{fluid}}gV_{\text{disp}} = \rho_{\text{obj}}gV_{\text{obj}}$, or

$$\frac{V_{\text{disp}}}{V_{\text{obj}}} = \frac{\rho_{\text{obj}}}{\rho_{\text{fluid}}} \tag{14.6}$$

This equation shows that the fraction of the volume of a floating object that is below the fluid surface is equal to the ratio of the density of the object to that of the fluid.

> **Q**uick Quiz 14.4 You are shipwrecked and floating in the middle of the ocean on
> a raft. Your cargo on the raft includes a treasure chest full of gold that you found
> before your ship sank, and the raft is just barely afloat. To keep you floating as
> high as possible in the water, should you **(a)** leave the treasure chest on top of
> the raft, **(b)** secure the treasure chest to the underside of the raft, or **(c)** hang
> the treasure chest in the water with a rope attached to the raft? (Assume throw-
> ing the treasure chest overboard is not an option you wish to consider.)

> **Pitfall Prevention 14.2**
> **Buoyant Force Is Exerted by the Fluid** Remember that **the buoyant force is exerted by the fluid.** It is not determined by properties of the object except for the amount of fluid displaced by the object. Therefore, if several objects of different densities but the same volume are immersed in a fluid, they will all experience the same buoyant force. Whether they sink or float is determined by the relationship between the buoyant force and the gravitational force.

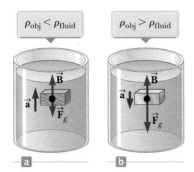

Figure 14.9 (a) A totally submerged object that is less dense than the fluid in which it is submerged experiences a net upward force and rises to the surface after it is released. (b) A totally submerged object that is denser than the fluid experiences a net downward force and sinks.

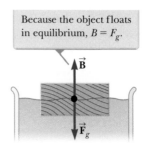

Figure 14.10 An object floating on the surface of a fluid experiences two forces, the gravitational force $\vec{F}_g$ and the buoyant force $\vec{B}$.

Example 14.5   Eureka! **AM**

Archimedes supposedly was asked to determine whether a crown made for the king consisted of pure gold. According to legend, he solved this problem by weighing the crown first in air and then in water as shown in Figure 14.11. Suppose the scale read 7.84 N when the crown was in air and 6.84 N when it was in water. What should Archimedes have told the king?

SOLUTION

Conceptualize Figure 14.11 helps us imagine what is happening in this example. Because of the buoyant force, the scale reading is smaller in Figure 14.11b than in Figure 14.11a.

Categorize This problem is an example of Case 1 discussed earlier because the crown is completely submerged. The scale reading is a measure of one of the forces on the crown, and the crown is stationary. Therefore, we can categorize the crown as a *particle in equilibrium*.

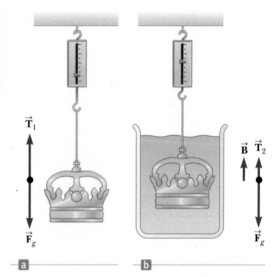

Figure 14.11 (Example 14.5) (a) When the crown is suspended in air, the scale reads its true weight because $T_1 = F_g$ (the buoyancy of air is negligible). (b) When the crown is immersed in water, the buoyant force $\vec{B}$ changes the scale reading to a lower value $T_2 = F_g - B$.

Analyze When the crown is suspended in air, the scale reads the true weight $T_1 = F_g$ (neglecting the small buoyant force due to the surrounding air). When the crown is immersed in water, the buoyant force $\vec{B}$ reduces the scale reading to an *apparent* weight of $T_2 = F_g - B$.

Apply the particle in equilibrium model to the crown in water:

$$\sum F = B + T_2 - F_g = 0$$

Solve for B:

$$B = F_g - T_2$$

Because this buoyant force is equal in magnitude to the weight of the displaced water, $B = \rho_w g V_{\text{disp}}$, where V_{disp} is the volume of the displaced water and ρ_w is its density. Also, the volume of the crown V_c is equal to the volume of the displaced water because the crown is completely submerged, so $B = \rho_w g V_c$.

Find the density of the crown from Equation 1.1:

$$\rho_c = \frac{m_c}{V_c} = \frac{m_c g}{V_c g} = \frac{m_c g}{(B/\rho_w)} = \frac{m_c g \rho_w}{B} = \frac{m_c g \rho_w}{F_g - T_2}$$

Substitute numerical values:

$$\rho_c = \frac{(7.84\ \text{N})(1\,000\ \text{kg/m}^3)}{7.84\ \text{N} - 6.84\ \text{N}} = 7.84 \times 10^3\ \text{kg/m}^3$$

Finalize From Table 14.1, we see that the density of gold is $19.3 \times 10^3\ \text{kg/m}^3$. Therefore, Archimedes should have reported that the king had been cheated. Either the crown was hollow, or it was not made of pure gold.

WHAT IF?   Suppose the crown has the same weight but is indeed pure gold and not hollow. What would the scale reading be when the crown is immersed in water?

Answer Find the buoyant force on the crown:

$$B = \rho_w g V_w = \rho_w g V_c = \rho_w g \left(\frac{m_c}{\rho_c}\right) = \rho_w \left(\frac{m_c g}{\rho_c}\right)$$

Substitute numerical values:

$$B = (1.00 \times 10^3\ \text{kg/m}^3)\frac{7.84\ \text{N}}{19.3 \times 10^3\ \text{kg/m}^3} = 0.406\ \text{N}$$

Find the tension in the string hanging from the scale:

$$T_2 = F_g - B = 7.84\ \text{N} - 0.406\ \text{N} = 7.43\ \text{N}$$

Example 14.6 A Titanic Surprise

An iceberg floating in seawater as shown in Figure 14.12a is extremely dangerous because most of the ice is below the surface. This hidden ice can damage a ship that is still a considerable distance from the visible ice. What fraction of the iceberg lies below the water level?

SOLUTION

Conceptualize You are likely familiar with the phrase, "That's only the tip of the iceberg." The origin of this popular saying is that most of the volume of a floating iceberg is beneath the surface of the water (Fig. 14.12b).

Categorize This example corresponds to Case 2 because only part of the iceberg is underneath the water. It is also a simple substitution problem involving Equation 14.6.

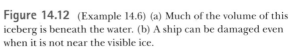

Figure 14.12 (Example 14.6) (a) Much of the volume of this iceberg is beneath the water. (b) A ship can be damaged even when it is not near the visible ice.

Evaluate Equation 14.6 using the densities of ice and seawater (Table 14.1):

$$f = \frac{V_{\text{disp}}}{V_{\text{ice}}} = \frac{\rho_{\text{ice}}}{\rho_{\text{seawater}}} = \frac{917 \text{ kg/m}^3}{1\,030 \text{ kg/m}^3} = \boxed{0.890 \text{ or } 89.0\%}$$

Therefore, the visible fraction of ice above the water's surface is about 11%. It is the unseen 89% below the water that represents the danger to a passing ship.

14.5 Fluid Dynamics

Thus far, our study of fluids has been restricted to fluids at rest. We now turn our attention to fluids in motion. When fluid is in motion, its flow can be characterized as being one of two main types. The flow is said to be **steady,** or **laminar,** if each particle of the fluid follows a smooth path such that the paths of different particles never cross each other as shown in Figure 14.13. In steady flow, every fluid particle arriving at a given point in space has the same velocity.

Above a certain critical speed, fluid flow becomes **turbulent.** Turbulent flow is irregular flow characterized by small whirlpool-like regions as shown in Figure 14.14.

The term **viscosity** is commonly used in the description of fluid flow to characterize the degree of internal friction in the fluid. This internal friction, or *viscous force*, is associated with the resistance that two adjacent layers of fluid have to moving relative to each other. Viscosity causes part of the fluid's kinetic energy to be transformed to internal energy. This mechanism is similar to the one by which the kinetic energy of an object sliding over a rough, horizontal surface decreases as discussed in Sections 8.3 and 8.4.

Because the motion of real fluids is very complex and not fully understood, we make some simplifying assumptions in our approach. In our simplification model of **ideal fluid flow,** we make the following four assumptions:

1. **The fluid is nonviscous.** In a nonviscous fluid, internal friction is neglected. An object moving through the fluid experiences no viscous force.
2. **The flow is steady.** In steady (laminar) flow, all particles passing through a point have the same velocity.
3. **The fluid is incompressible.** The density of an incompressible fluid is constant.
4. **The flow is irrotational.** In irrotational flow, the fluid has no angular momentum about any point. If a small paddle wheel placed anywhere in the fluid does not rotate about the wheel's center of mass, the flow is irrotational.

Figure 14.13 Laminar flow around an automobile in a test wind tunnel.

Figure 14.14 Hot gases from a cigarette made visible by smoke particles. The smoke first moves in laminar flow at the bottom and then in turbulent flow above.

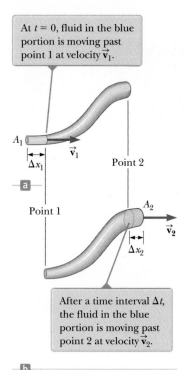

At $t = 0$, fluid in the blue portion is moving past point 1 at velocity $\vec{v}_1$.

A_1

Δx_1 $\vec{v}_1$

Point 2

a

Point 1 A_2

$\vec{v}_2$

Δx_2

After a time interval Δt, the fluid in the blue portion is moving past point 2 at velocity $\vec{v}_2$.

b

Figure 14.16 A fluid moving with steady flow through a pipe of varying cross-sectional area. (a) At $t = 0$, the small blue-colored portion of the fluid at the left is moving through area A_1. (b) After a time interval Δt, the blue-colored portion shown here is that fluid that has moved through area A_2.

Equation of Continuity ▶
for Fluids

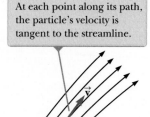

At each point along its path, the particle's velocity is tangent to the streamline.

$\vec{v}$

Figure 14.15 A particle in laminar flow follows a streamline.

The path taken by a fluid particle under steady flow is called a **streamline.** The velocity of the particle is always tangent to the streamline as shown in Figure 14.15. A set of streamlines like the ones shown in Figure 14.15 form a *tube of flow.* Fluid particles cannot flow into or out of the sides of this tube; if they could, the streamlines would cross one another.

Consider ideal fluid flow through a pipe of nonuniform size as illustrated in Figure 14.16. Let's focus our attention on a segment of fluid in the pipe. Figure 14.16a shows the segment at time $t = 0$ consisting of the gray portion between point 1 and point 2 and the short blue portion to the left of point 1. At this time, the fluid in the short blue portion is flowing through a cross section of area A_1 at speed v_1. During the time interval Δt, the small length Δx_1 of fluid in the blue portion moves past point 1. During the same time interval, fluid at the right end of the segment moves past point 2 in the pipe. Figure 14.16b shows the situation at the end of the time interval Δt. The blue portion at the right end represents the fluid that has moved past point 2 through an area A_2 at a speed v_2.

The mass of fluid contained in the blue portion in Figure 14.16a is given by $m_1 = \rho A_1 \, \Delta x_1 = \rho A_1 v_1 \, \Delta t$, where ρ is the (unchanging) density of the ideal fluid. Similarly, the fluid in the blue portion in Figure 14.16b has a mass $m_2 = \rho A_2 \, \Delta x_2 = \rho A_2 v_2 \, \Delta t$. Because the fluid is incompressible and the flow is steady, however, the mass of fluid that passes point 1 in a time interval Δt must equal the mass that passes point 2 in the same time interval. That is, $m_1 = m_2$ or $\rho A_1 v_1 \, \Delta t = \rho A_2 v_2 \, \Delta t$, which means that

$$A_1 v_1 = A_2 v_2 = \text{constant} \tag{14.7}$$

This expression is called the **equation of continuity for fluids.** It states that the product of the area and the fluid speed at all points along a pipe is constant for an incompressible fluid. Equation 14.7 shows that the speed is high where the tube is constricted (small A) and low where the tube is wide (large A). The product Av, which has the dimensions of volume per unit time, is called either the *volume flux* or the *flow rate.* The condition $Av = \text{constant}$ is equivalent to the statement that the volume of fluid that enters one end of a tube in a given time interval equals the volume leaving the other end of the tube in the same time interval if no leaks are present.

You demonstrate the equation of continuity each time you water your garden with your thumb over the end of a garden hose as in Figure 14.17. By partially block-

Figure 14.17 The speed of water spraying from the end of a garden hose increases as the size of the opening is decreased with the thumb.

ing the opening with your thumb, you reduce the cross-sectional area through which the water passes. As a result, the speed of the water increases as it exits the hose, and the water can be sprayed over a long distance.

Example 14.7 Watering a Garden

A gardener uses a water hose to fill a 30.0-L bucket. The gardener notes that it takes 1.00 min to fill the bucket. A nozzle with an opening of cross-sectional area 0.500 cm² is then attached to the hose. The nozzle is held so that water is projected horizontally from a point 1.00 m above the ground. Over what horizontal distance can the water be projected?

SOLUTION

Conceptualize Imagine any past experience you have with projecting water from a horizontal hose or a pipe using either your thumb or a nozzle, which can be attached to the end of the hose. The faster the water is traveling as it leaves the hose, the farther it will land on the ground from the end of the hose.

Categorize Once the water leaves the hose, it is in free fall. Therefore, we categorize a given element of the water as a *projectile*. The element is modeled as a *particle under constant acceleration* (due to gravity) in the vertical direction and a *particle under constant velocity* in the horizontal direction. The horizontal distance over which the element is projected depends on the speed with which it is projected. This example involves a change in area for the pipe, so we also categorize it as one in which we use the continuity equation for fluids.

Analyze

Express the volume flow rate R in terms of area and speed of the water in the hose:

$$R = A_1 v_1$$

Solve for the speed of the water in the hose:

$$v_1 = \frac{R}{A_1}$$

We have labeled this speed v_1 because we identify point 1 within the hose. We identify point 2 in the air just outside the nozzle. We must find the speed $v_2 = v_{xi}$ with which the water exits the nozzle. The subscript i anticipates that it will be the *initial* velocity component of the water projected from the hose, and the subscript x indicates that the initial velocity vector of the projected water is horizontal.

Solve the continuity equation for fluids for v_2:

$$(1) \quad v_2 = v_{xi} = \frac{A_1}{A_2} v_1 = \frac{A_1}{A_2}\left(\frac{R}{A_1}\right) = \frac{R}{A_2}$$

We now shift our thinking away from fluids and to projectile motion. In the vertical direction, an element of the water starts from rest and falls through a vertical distance of 1.00 m.

Write Equation 2.16 for the vertical position of an element of water, modeled as a particle under constant acceleration:

$$y_f = y_i + v_{yi}t - \tfrac{1}{2}gt^2$$

Call the initial position of the water $y_i = 0$ and recognize that the water begins with a vertical velocity component of zero. Solve for the time at which the water reaches the ground:

$$(2) \quad y_f = 0 + 0 - \tfrac{1}{2}gt^2 \;\rightarrow\; t = \sqrt{\frac{-2y_f}{g}}$$

Use Equation 2.7 to find the horizontal position of the element at this time, modeled as a particle under constant velocity:

$$x_f = x_i + v_{xi}t = 0 + v_2 t = v_2 t$$

Substitute from Equations (1) and (2):

$$x_f = \frac{R}{A_2}\sqrt{\frac{-2y_f}{g}}$$

Substitute numerical values:

$$x_f = \frac{30.0\ \text{L/min}}{0.500\ \text{cm}^2}\sqrt{\frac{-2(-1.00\ \text{m})}{9.80\ \text{m/s}^2}}\left(\frac{10^3\ \text{cm}^3}{1\ \text{L}}\right)\left(\frac{1\ \text{min}}{60\ \text{s}}\right) = 452\ \text{cm} = 4.52\ \text{m}$$

continued

▶ **14.7** continued

Finalize The time interval for the element of water to fall to the ground is unchanged if the projection speed is changed because the projection is horizontal. Increasing the projection speed results in the water hitting the ground farther from the end of the hose, but requires the same time interval to strike the ground.

Daniel Bernoulli
Swiss physicist (1700–1782)
Bernoulli made important discoveries in fluid dynamics. Bernoulli's most famous work, *Hydrodynamica*, was published in 1738; it is both a theoretical and a practical study of equilibrium, pressure, and speed in fluids. He showed that as the speed of a fluid increases, its pressure decreases. Referred to as "Bernoulli's principle," Bernoulli's work is used to produce a partial vacuum in chemical laboratories by connecting a vessel to a tube through which water is running rapidly.

© iStockphoto.com/ZU_09

14.6 Bernoulli's Equation

You have probably experienced driving on a highway and having a large truck pass you at high speed. In this situation, you may have had the frightening feeling that your car was being pulled in toward the truck as it passed. We will investigate the origin of this effect in this section.

As a fluid moves through a region where its speed or elevation above the Earth's surface changes, the pressure in the fluid varies with these changes. The relationship between fluid speed, pressure, and elevation was first derived in 1738 by Swiss physicist Daniel Bernoulli. Consider the flow of a segment of an ideal fluid through a nonuniform pipe in a time interval Δt as illustrated in Figure 14.18. This figure is very similar to Figure 14.16, which we used to develop the continuity equation. We have added two features: the forces on the outer ends of the blue portions of fluid and the heights of these portions above the reference position $y = 0$.

The force exerted on the segment by the fluid to the left of the blue portion in Figure 14.18a has a magnitude $P_1 A_1$. The work done by this force on the segment in a time interval Δt is $W_1 = F_1 \Delta x_1 = P_1 A_1 \Delta x_1 = P_1 V$, where V is the volume of the blue portion of fluid passing point 1 in Figure 14.18a. In a similar manner, the work done on the segment by the fluid to the right of the segment in the same time interval Δt is $W_2 = -P_2 A_2 \Delta x_2 = -P_2 V$, where V is the volume of the blue portion of fluid passing point 2 in Figure 14.18b. (The volumes of the blue portions of fluid in Figures 14.18a and 14.18b are equal because the fluid is incompressible.) This work is negative because the force on the segment of fluid is to the left and the displacement of the point of application of the force is to the right. Therefore, the net work done on the segment by these forces in the time interval Δt is

$$W = (P_1 - P_2)V$$

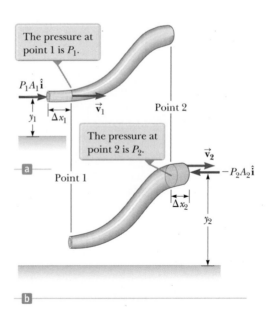

Figure 14.18 A fluid in laminar flow through a pipe. (a) A segment of the fluid at time $t = 0$. A small portion of the blue-colored fluid is at height y_1 above a reference position. (b) After a time interval Δt, the entire segment has moved to the right. The blue-colored portion of the fluid is that which has passed point 2 and is at height y_2.

Part of this work goes into changing the kinetic energy of the segment of fluid, and part goes into changing the gravitational potential energy of the segment–Earth system. Because we are assuming streamline flow, the kinetic energy K_{gray} of the gray portion of the segment is the same in both parts of Figure 14.18. Therefore, the change in the kinetic energy of the segment of fluid is

$$\Delta K = \left(\tfrac{1}{2}mv_2^2 + K_{\text{gray}}\right) - \left(\tfrac{1}{2}mv_1^2 + K_{\text{gray}}\right) = \tfrac{1}{2}mv_2^2 - \tfrac{1}{2}mv_1^2$$

where m is the mass of the blue portions of fluid in both parts of Figure 14.18. (Because the volumes of both portions are the same, they also have the same mass.)

Considering the gravitational potential energy of the segment–Earth system, once again there is no change during the time interval for the gravitational potential energy U_{gray} associated with the gray portion of the fluid. Consequently, the change in gravitational potential energy of the system is

$$\Delta U = \left(mgy_2 + U_{\text{gray}}\right) - \left(mgy_1 + U_{\text{gray}}\right) = mgy_2 - mgy_1$$

From Equation 8.2, the total work done on the system by the fluid outside the segment is equal to the change in mechanical energy of the system: $W = \Delta K + \Delta U$. Substituting for each of these terms gives

$$(P_1 - P_2)V = \tfrac{1}{2}mv_2^2 - \tfrac{1}{2}mv_1^2 + mgy_2 - mgy_1$$

If we divide each term by the portion volume V and recall that $\rho = m/V$, this expression reduces to

$$P_1 - P_2 = \tfrac{1}{2}\rho v_2^2 - \tfrac{1}{2}\rho v_1^2 + \rho g y_2 - \rho g y_1$$

Rearranging terms gives

$$P_1 + \tfrac{1}{2}\rho v_1^2 + \rho g y_1 = P_2 + \tfrac{1}{2}\rho v_2^2 + \rho g y_2 \tag{14.8}$$

which is **Bernoulli's equation** as applied to an ideal fluid. This equation is often expressed as

$$P + \tfrac{1}{2}\rho v^2 + \rho g y = \text{constant} \tag{14.9}$$

◀ **Bernoulli's equation**

Bernoulli's equation shows that the pressure of a fluid decreases as the speed of the fluid increases. In addition, the pressure decreases as the elevation increases. This latter point explains why water pressure from faucets on the upper floors of a tall building is weak unless measures are taken to provide higher pressure for these upper floors.

When the fluid is at rest, $v_1 = v_2 = 0$ and Equation 14.8 becomes

$$P_1 - P_2 = \rho g(y_2 - y_1) = \rho g h$$

This result is in agreement with Equation 14.4.

Although Equation 14.9 was derived for an incompressible fluid, the general behavior of pressure with speed is true even for gases: as the speed increases, the pressure decreases. This *Bernoulli effect* explains the experience with the truck on the highway at the opening of this section. As air passes between you and the truck, it must pass through a relatively narrow channel. According to the continuity equation, the speed of the air is higher. According to the Bernoulli effect, this higher-speed air exerts less pressure on your car than the slower-moving air on the other side of your car. Therefore, there is a net force pushing you toward the truck!

Quick Quiz 14.5 You observe two helium balloons floating next to each other at the ends of strings secured to a table. The facing surfaces of the balloons are separated by 1–2 cm. You blow through the small space between the balloons. What happens to the balloons? **(a)** They move toward each other. **(b)** They move away from each other. **(c)** They are unaffected.

Example 14.8 The Venturi Tube

The horizontal constricted pipe illustrated in Figure 14.19, known as a *Venturi tube*, can be used to measure the flow speed of an incompressible fluid. Determine the flow speed at point 2 of Figure 14.19a if the pressure difference $P_1 - P_2$ is known.

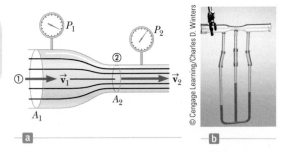

SOLUTION

Conceptualize Bernoulli's equation shows how the pressure of an ideal fluid decreases as its speed increases. Therefore, we should be able to calibrate a device to give us the fluid speed if we can measure pressure.

Categorize Because the problem states that the fluid is incompressible, we can categorize it as one in which we can use the equation of continuity for fluids and Bernoulli's equation.

Figure 14.19 (Example 14.8) (a) Pressure P_1 is greater than pressure P_2 because $v_1 < v_2$. This device can be used to measure the speed of fluid flow. (b) A Venturi tube, located at the top of the photograph. The higher level of fluid in the middle column shows that the pressure at the top of the column, which is in the constricted region of the Venturi tube, is lower.

Analyze Apply Equation 14.8 to points 1 and 2, noting that $y_1 = y_2$ because the pipe is horizontal:

$$(1) \quad P_1 + \tfrac{1}{2}\rho v_1^2 = P_2 + \tfrac{1}{2}\rho v_2^2$$

Solve the equation of continuity for v_1:

$$v_1 = \frac{A_2}{A_1} v_2$$

Substitute this expression into Equation (1):

$$P_1 + \tfrac{1}{2}\rho \left(\frac{A_2}{A_1}\right)^2 v_2^2 = P_2 + \tfrac{1}{2}\rho v_2^2$$

Solve for v_2:

$$v_2 = A_1 \sqrt{\frac{2(P_1 - P_2)}{\rho(A_1^2 - A_2^2)}}$$

Finalize From the design of the tube (areas A_1 and A_2) and measurements of the pressure difference $P_1 - P_2$, we can calculate the speed of the fluid with this equation. To see the relationship between fluid speed and pressure difference, place two empty soda cans on their sides about 2 cm apart on a table. Gently blow a stream of air horizontally between the cans and watch them roll together slowly due to a modest pressure difference between the stagnant air on their outside edges and the moving air between them. Now blow more strongly and watch the increased pressure difference move the cans together more rapidly.

Example 14.9 Torricelli's Law **AM**

An enclosed tank containing a liquid of density ρ has a hole in its side at a distance y_1 from the tank's bottom (Fig. 14.20). The hole is open to the atmosphere, and its diameter is much smaller than the diameter of the tank. The air above the liquid is maintained at a pressure P. Determine the speed of the liquid as it leaves the hole when the liquid's level is a distance h above the hole.

SOLUTION

Conceptualize Imagine that the tank is a fire extinguisher. When the hole is opened, liquid leaves the hole with a certain speed. If the pressure P at the top of the liquid is increased, the liquid leaves with a higher speed. If the pressure P falls too low, the liquid leaves with a low speed and the extinguisher must be replaced.

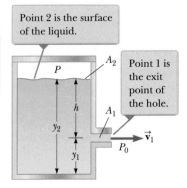

Point 2 is the surface of the liquid.

Point 1 is the exit point of the hole.

Figure 14.20 (Example 14.9) A liquid leaves a hole in a tank at speed v_1.

▶ **14.9** continued

Categorize Looking at Figure 14.20, we know the pressure at two points and the velocity at one of those points. We wish to find the velocity at the second point. Therefore, we can categorize this example as one in which we can apply Bernoulli's equation.

. . . .

Analyze Because $A_2 \gg A_1$, the liquid is approximately at rest at the top of the tank, where the pressure is P. At the hole, P_1 is equal to atmospheric pressure P_0.

Apply Bernoulli's equation between points 1 and 2:

$$P_0 + \tfrac{1}{2}\rho v_1^{\,2} + \rho g y_1 = P + \rho g y_2$$

Solve for v_1, noting that $y_2 - y_1 = h$:

$$v_1 = \sqrt{\frac{2(P - P_0)}{\rho} + 2gh}$$

. . . .

Finalize When P is much greater than P_0 (so that the term $2gh$ can be neglected), the exit speed of the water is mainly a function of P. If the tank is open to the atmosphere, then $P = P_0$ and $v_1 = \sqrt{2gh}$. In other words, for an open tank, the speed of the liquid leaving a hole a distance h below the surface is equal to that acquired by an object falling freely through a vertical distance h. This phenomenon is known as *Torricelli's law*.

WHAT IF? What if the position of the hole in Figure 14.20 could be adjusted vertically? If the tank is open to the atmosphere and sitting on a table, what position of the hole would cause the water to land on the table at the farthest distance from the tank?

Answer Model a parcel of water exiting the hole as a projectile. From the *particle under constant acceleration* model, find the time at which the parcel strikes the table from a hole at an arbitrary position y_1:

$$y_f = y_i + v_{yi}t - \tfrac{1}{2}gt^2$$
$$0 = y_1 + 0 - \tfrac{1}{2}gt^2$$
$$t = \sqrt{\frac{2y_1}{g}}$$

From the *particle under constant velocity* model, find the horizontal position of the parcel at the time it strikes the table:

$$x_f = x_i + v_{xi}t = 0 + \sqrt{2g(y_2 - y_1)}\sqrt{\frac{2y_1}{g}}$$
$$= 2\sqrt{\left(y_2 y_1 - y_1^{\,2}\right)}$$

Maximize the horizontal position by taking the derivative of x_f with respect to y_1 (because y_1, the height of the hole, is the variable that can be adjusted) and setting it equal to zero:

$$\frac{dx_f}{dy_1} = \tfrac{1}{2}(2)\left(y_2 y_1 - y_1^{\,2}\right)^{-1/2}(y_2 - 2y_1) = 0$$

Solve for y_1:

$$y_1 = \tfrac{1}{2}y_2$$

Therefore, to maximize the horizontal distance, the hole should be halfway between the bottom of the tank and the upper surface of the water. Below this location, the water is projected at a higher speed but falls for a short time interval, reducing the horizontal range. Above this point, the water is in the air for a longer time interval but is projected with a smaller horizontal speed.

14.7 Other Applications of Fluid Dynamics

Consider the streamlines that flow around an airplane wing as shown in Figure 14.21 on page 346. Let's assume the airstream approaches the wing horizontally from the right with a velocity $\vec{\mathbf{v}}_1$. The tilt of the wing causes the airstream to be deflected downward with a velocity $\vec{\mathbf{v}}_2$. Because the airstream is deflected by the wing, the wing must exert a force on the airstream. According to Newton's third law, the airstream exerts a force $\vec{\mathbf{F}}$ on the wing that is equal in magnitude and

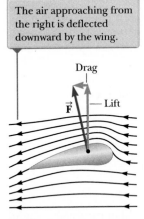

> The air approaching from the right is deflected downward by the wing.

Figure 14.21 Streamline flow around a moving airplane wing. By Newton's third law, the air deflected by the wing results in an upward force on the wing from the air: *lift*. Because of air resistance, there is also a force opposite the velocity of the wing: *drag*.

opposite in direction. This force has a vertical component called **lift** (or aerodynamic lift) and a horizontal component called **drag.** The lift depends on several factors, such as the speed of the airplane, the area of the wing, the wing's curvature, and the angle between the wing and the horizontal. The curvature of the wing surfaces causes the pressure above the wing to be lower than that below the wing due to the Bernoulli effect. This pressure difference assists with the lift on the wing. As the angle between the wing and the horizontal increases, turbulent flow can set in above the wing to reduce the lift.

In general, an object moving through a fluid experiences lift as the result of any effect that causes the fluid to change its direction as it flows past the object. Some factors that influence lift are the shape of the object, its orientation with respect to the fluid flow, any spinning motion it might have, and the texture of its surface. For example, a golf ball struck with a club is given a rapid backspin due to the slant of the club. The dimples on the ball increase the friction force between the ball and the air so that air adheres to the ball's surface. Figure 14.22 shows air adhering to the ball and being deflected downward as a result. Because the ball pushes the air down, the air must push up on the ball. Without the dimples, the friction force is lower and the golf ball does not travel as far. It may seem counterintuitive to increase the range by increasing the friction force, but the lift gained by spinning the ball more than compensates for the loss of range due to the effect of friction on the translational motion of the ball. For the same reason, a baseball's cover helps the spinning ball "grab" the air rushing by and helps deflect it when a "curve ball" is thrown.

A number of devices operate by means of the pressure differentials that result from differences in a fluid's speed. For example, a stream of air passing over one end of an open tube, the other end of which is immersed in a liquid, reduces the pressure above the tube as illustrated in Figure 14.23. This reduction in pressure causes the liquid to rise into the airstream. The liquid is then dispersed into a fine spray of droplets. You might recognize that this *atomizer* is used in perfume bottles and paint sprayers.

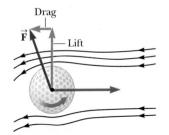

Figure 14.22 Because of the deflection of air, a spinning golf ball experiences a lifting force that allows it to travel much farther than it would if it were not spinning.

Figure 14.23 A stream of air passing over a tube dipped into a liquid causes the liquid to rise in the tube.

Summary

Definitions

The **pressure** *P* in a fluid is the force per unit area exerted by the fluid on a surface:

$$P \equiv \frac{F}{A} \qquad (14.1)$$

In the SI system, pressure has units of newtons per square meter (N/m²), and 1 N/m² = 1 **pascal** (Pa).

Concepts and Principles

The pressure in a fluid at rest varies with depth h in the fluid according to the expression

$$P = P_0 + \rho g h \qquad (14.4)$$

where P_0 is the pressure at $h = 0$ and ρ is the density of the fluid, assumed uniform.

Pascal's law states that when pressure is applied to an enclosed fluid, the pressure is transmitted undiminished to every point in the fluid and to every point on the walls of the container.

When an object is partially or fully submerged in a fluid, the fluid exerts on the object an upward force called the **buoyant force**. According to **Archimedes's principle,** the magnitude of the buoyant force is equal to the weight of the fluid displaced by the object:

$$B = \rho_{\text{fluid}} g V_{\text{disp}} \qquad (14.5)$$

The flow rate (volume flux) through a pipe that varies in cross-sectional area is constant; that is equivalent to stating that the product of the cross-sectional area A and the speed v at any point is a constant. This result is expressed in the **equation of continuity for fluids:**

$$A_1 v_1 = A_2 v_2 = \text{constant} \qquad (14.7)$$

The sum of the pressure, kinetic energy per unit volume, and gravitational potential energy per unit volume has the same value at all points along a streamline for an ideal fluid. This result is summarized in **Bernoulli's equation:**

$$P + \tfrac{1}{2}\rho v^2 + \rho g y = \text{constant} \qquad (14.9)$$

Objective Questions   1. denotes answer available in *Student Solutions Manual/Study Guide*

1. Figure OQ14.1 shows aerial views from directly above two dams. Both dams are equally wide (the vertical dimension in the diagram) and equally high (into the page in the diagram). The dam on the left holds back a very large lake, and the dam on the right holds back a narrow river. Which dam has to be built more strongly? (a) the dam on the left (b) the dam on the right (c) both the same (d) cannot be predicted

of the following statements are valid? (Choose all correct statements.) (a) The buoyant force on the steel object is equal to its weight. (b) The buoyant force on the block is equal to its weight. (c) The tension in the string is equal to the weight of the steel object. (d) The tension in the string is less than the weight of the steel object. (e) The buoyant force on the block is equal to the volume of water it displaces.

Dam   Dam

Figure OQ14.1

Figure OQ14.3

2. A beach ball filled with air is pushed about 1 m below the surface of a swimming pool and released from rest. Which of the following statements are valid, assuming the size of the ball remains the same? (Choose all correct statements.) (a) As the ball rises in the pool, the buoyant force on it increases. (b) When the ball is released, the buoyant force exceeds the gravitational force, and the ball accelerates upward. (c) The buoyant force on the ball decreases as the ball approaches the surface of the pool. (d) The buoyant force on the ball equals its weight and remains constant as the ball rises. (e) The buoyant force on the ball while it is submerged is approximately equal to the weight of a volume of water that could fill the ball.

3. A wooden block floats in water, and a steel object is attached to the bottom of the block by a string as in Figure OQ14.3. If the block remains floating, which

4. An apple is held completely submerged just below the surface of water in a container. The apple is then moved to a deeper point in the water. Compared with the force needed to hold the apple just below the surface, what is the force needed to hold it at the deeper point? (a) larger (b) the same (c) smaller (d) impossible to determine

5. A beach ball is made of thin plastic. It has been inflated with air, but the plastic is not stretched. By swimming with fins on, you manage to take the ball from the surface of a pool to the bottom. Once the ball is completely submerged, what happens to the buoyant force exerted on the beach ball as you take it deeper? (a) It increases. (b) It remains constant. (c) It decreases. (d) It is impossible to determine.

6. A solid iron sphere and a solid lead sphere of the same size are each suspended by strings and are submerged in a tank of water. (Note that the density of lead is greater than that of iron.) Which of the following statements are valid? (Choose all correct statements.) (a) The buoyant force on each is the same. (b) The buoyant force on the lead sphere is greater than the buoyant force on the iron sphere because lead has the greater density. (c) The tension in the string supporting the lead sphere is greater than the tension in the string supporting the iron sphere. (d) The buoyant force on the iron sphere is greater than the buoyant force on the lead sphere because lead displaces more water. (e) None of those statements is true.

7. Three vessels of different shapes are filled to the same level with water as in Figure OQ14.7. The area of the base is the same for all three vessels. Which of the following statements are valid? (Choose all correct statements.) (a) The pressure at the top surface of vessel A is greatest because it has the largest surface area. (b) The pressure at the bottom of vessel A is greatest because it contains the most water. (c) The pressure at the bottom of each vessel is the same. (d) The force on the bottom of each vessel is not the same. (e) At a given depth below the surface of each vessel, the pressure on the side of vessel A is greatest because of its slope.

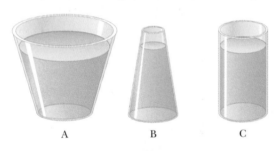

Figure OQ14.7

8. One of the predicted problems due to global warming is that ice in the polar ice caps will melt and raise sea levels everywhere in the world. Is that more of a worry for ice (a) at the north pole, where most of the ice floats on water; (b) at the south pole, where most of the ice sits on land; (c) both at the north and south pole equally; or (d) at neither pole?

9. A boat develops a leak and, after its passengers are rescued, eventually sinks to the bottom of a lake. When the boat is at the bottom, what is the force of the lake bottom on the boat? (a) greater than the weight of the boat (b) equal to the weight of the boat (c) less than the weight of the boat (d) equal to the weight of the displaced water (e) equal to the buoyant force on the boat

10. A small piece of steel is tied to a block of wood. When the wood is placed in a tub of water with the steel on top, half of the block is submerged. Now the block is inverted so that the steel is under water. **(i)** Does the amount of the block submerged (a) increase, (b) decrease, or (c) remain the same? **(ii)** What happens to the water level in the tub when the block is inverted? (a) It rises. (b) It falls. (c) It remains the same.

11. A piece of unpainted porous wood barely floats in an open container partly filled with water. The container is then sealed and pressurized above atmospheric pressure. What happens to the wood? (a) It rises in the water. (b) It sinks lower in the water. (c) It remains at the same level.

12. A person in a boat floating in a small pond throws an anchor overboard. What happens to the level of the pond? (a) It rises. (b) It falls. (c) It remains the same.

13. Rank the buoyant forces exerted on the following five objects of equal volume from the largest to the smallest. Assume the objects have been dropped into a swimming pool and allowed to come to mechanical equilibrium. If any buoyant forces are equal, state that in your ranking. (a) a block of solid oak (b) an aluminum block (c) a beach ball made of thin plastic and inflated with air (d) an iron block (e) a thin-walled, sealed bottle of water

14. A water supply maintains a constant rate of flow for water in a hose. You want to change the opening of the nozzle so that water leaving the nozzle will reach a height that is four times the current maximum height the water reaches with the nozzle vertical. To do so, should you (a) decrease the area of the opening by a factor of 16, (b) decrease the area by a factor of 8, (c) decrease the area by a factor of 4, (d) decrease the area by a factor of 2, or (e) give up because it cannot be done?

15. A glass of water contains floating ice cubes. When the ice melts, does the water level in the glass (a) go up, (b) go down, or (c) remain the same?

16. An ideal fluid flows through a horizontal pipe whose diameter varies along its length. Measurements would indicate that the sum of the kinetic energy per unit volume and pressure at different sections of the pipe would (a) decrease as the pipe diameter increases, (b) increase as the pipe diameter increases, (c) increase as the pipe diameter decreases, (d) decrease as the pipe diameter decreases, or (e) remain the same as the pipe diameter changes.

Conceptual Questions

1. denotes answer available in *Student Solutions Manual/Study Guide*

1. When an object is immersed in a liquid at rest, why is the net force on the object in the horizontal direction equal to zero?

2. Two thin-walled drinking glasses having equal base areas but different shapes, with very different cross-sectional areas above the base, are filled to the same level with water. According to the expression $P = P_0 + \rho g h$, the pressure is the same at the bottom of both glasses. In view of this equality, why does one weigh more than the other?

3. Because atmospheric pressure is about 10^5 N/m² and the area of a person's chest is about 0.13 m², the force of the atmosphere on one's chest is around 13 000 N. In view of this enormous force, why don't our bodies collapse?

4. A fish rests on the bottom of a bucket of water while the bucket is being weighed on a scale. When the fish begins to swim around, does the scale reading change? Explain your answer.

5. You are a passenger on a spacecraft. For your survival and comfort, the interior contains air just like that at the surface of the Earth. The craft is coasting through a very empty region of space. That is, a nearly perfect vacuum exists just outside the wall. Suddenly, a meteoroid pokes a hole, about the size of a large coin, right through the wall next to your seat. (a) What happens? (b) Is there anything you can or should do about it?

6. If the airstream from a hair dryer is directed over a table-tennis ball, the ball can be levitated. Explain.

7. A water tower is a common sight in many communities. Figure CQ14.7 shows a collection of colorful water towers in Kuwait City, Kuwait. Notice that the large weight of the water results in the center of mass of the system being high above the ground. Why is it desirable for a water tower to have this highly unstable shape rather than being shaped as a tall cylinder?

Figure CQ14.7

8. If you release a ball while inside a freely falling elevator, the ball remains in front of you rather than falling to the floor because the ball, the elevator, and you all experience the same downward gravitational acceleration. What happens if you repeat this experiment with a helium-filled balloon?

9. (a) Is the buoyant force a conservative force? (b) Is a potential energy associated with the buoyant force? (c) Explain your answers to parts (a) and (b).

10. An empty metal soap dish barely floats in water. A bar of Ivory soap floats in water. When the soap is stuck in the soap dish, the combination sinks. Explain why.

11. How would you determine the density of an irregularly shaped rock?

12. Place two cans of soft drinks, one regular and one diet, in a container of water. You will find that the diet drink floats while the regular one sinks. Use Archimedes's principle to devise an explanation.

13. The water supply for a city is often provided from reservoirs built on high ground. Water flows from the reservoir, through pipes, and into your home when you turn the tap on your faucet. Why does water flow more rapidly out of a faucet on the first floor of a building than in an apartment on a higher floor?

14. Does a ship float higher in the water of an inland lake or in the ocean? Why?

15. When ski jumpers are airborne (Fig. CQ14.15), they bend their bodies forward and keep their hands at their sides. Why?

Figure CQ14.15

16. Why do airplane pilots prefer to take off with the airplane facing into the wind?

17. Prairie dogs ventilate their burrows by building a mound around one entrance, which is open to a stream of air when wind blows from any direction. A second entrance at ground level is open to almost stagnant air. How does this construction create an airflow through the burrow?

18. In Figure CQ14.18, an airstream moves from right to left through a tube that is constricted at the middle. Three table-tennis balls are levitated in equilibrium above the vertical columns through which the air escapes. (a) Why is the ball at the right higher than the one in the middle? (b) Why is the ball at the left lower than the ball at the right even though the horizontal tube has the same dimensions at these two points?

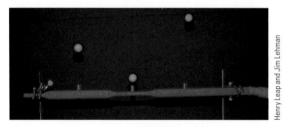

Figure CQ14.18

19. A typical silo on a farm has many metal bands wrapped around its perimeter for support as shown in Figure CQ14.19. Why is the spacing between successive bands smaller for the lower portions of the silo on the left, and why are double bands used at lower portions of the silo on the right?

Figure CQ14.19

Problems available in ENHANCED WebAssign Access end-of-chapter problems online at www.webassign.net

Section 14.1 Pressure
Problems 1–5

Section 14.2 Variation of Pressure with Depth
Problems 6–18

Section 14.3 Pressure Measurements
Problems 19–24

Section 14.4 Buoyant Forces and Archimedes's Principle
Problems 25–39

Section 14.5 Fluid Dynamics
Section 14.6 Bernoulli's Equation
Problems 40–50

Section 14.7 Other Applications of Fluid Dynamics
Problems 51–55

Additional Problems
Problems 56–84

Challenge Problems
Problems 85–87

Solutions to the following Problems are available in the *Student Solutions Manual/Study Guide:*
14.3, 14.5, 14.7, 14.9, 14.15, 14.21, 14.25, 14.29, 14.31, 14.39, 14.41, 15.55, 14.62, 14.70, 14.73, 14.81, and 14.85

List of Enhanced Problems

Problem Number	Targeted Feedback in Enhanced WebAssign	Analysis Model Tutorial in Enhanced WebAssign	Master It in Enhanced WebAssign	Watch It in Enhanced WebAssign
14.3	✓			✓
14.5			✓	
14.7			✓	
14.8	✓			✓
14.9		✓	✓	
14.10	✓			
14.14				✓
14.22	✓			✓
14.28	✓			✓
14.29	✓	✓	✓	
14.31			✓	
14.39			✓	
14.41	✓		✓	
14.55			✓	
14.57	✓			✓
14.61		✓		
14.65		✓		
14.84	✓		✓	

Oscillations and Mechanical Waves

Falling drops of water cause a water surface to oscillate. These oscillations are associated with circular waves moving away from the point at which the drops fall. In Part 2 of the text, we will explore the principles related to oscillations and waves. *(Marga Buschbell Steeger/Photographer's Choice/ Getty Images)*

We begin this new part of the text by studying a special type of motion called *periodic* motion, the repeating motion of an object in which it continues to return to a given position after a fixed time interval. The repetitive movements of such an object are called *oscillations.* We will focus our attention on a special case of periodic motion called *simple harmonic motion.* All periodic motions can be modeled as combinations of simple harmonic motions.

Simple harmonic motion also forms the basis for our understanding of *mechanical waves.* Sound waves, seismic waves, waves on stretched strings, and water waves are all produced by some source of oscillation. As a sound wave travels through the air, elements of the air oscillate back and forth; as a water wave travels across a pond, elements of the water oscillate up and down and backward and forward. The motion of the elements of the medium bears a strong resemblance to the periodic motion of an oscillating pendulum or an object attached to a spring.

To explain many other phenomena in nature, we must understand the concepts of oscillations and waves. For instance, although skyscrapers and bridges appear to be rigid, they actually oscillate, something the architects and engineers who design and build them must take into account. To understand how radio and television work, we must understand the origin and nature of electromagnetic waves and how they propagate through space. Finally, much of what scientists have learned about atomic structure has come from information carried by waves. Therefore, we must first study oscillations and waves if we are to understand the concepts and theories of atomic physics. ■

Oscillatory Motion

The London Millennium Bridge over the River Thames in London. On opening day of the bridge, pedestrians noticed a swinging motion of the bridge, leading to its being named the "Wobbly Bridge." The bridge was closed after two days and remained closed for two years. Over 50 *tuned mass dampers* were added to the bridge: the pairs of spring-loaded structures on top of the cross members (arrow). We will study both oscillations and damping of oscillations in this chapter. *(Monkey Business Images/ Shutterstock.com)*

Periodic motion is motion of an object that regularly returns to a given position after a fixed time interval. With a little thought, we can identify several types of periodic motion in everyday life. Your car returns to the driveway each afternoon. You return to the dinner table each night to eat. A bumped chandelier swings back and forth, returning to the same position at a regular rate. The Earth returns to the same position in its orbit around the Sun each year, resulting in the variation among the four seasons.

A special kind of periodic motion occurs in mechanical systems when the force acting on an object is proportional to the position of the object relative to some equilibrium position. If this force is always directed toward the equilibrium position, the motion is called *simple harmonic motion*, which is the primary focus of this chapter.

15.1 Motion of an Object Attached to a Spring

As a model for simple harmonic motion, consider a block of mass m attached to the end of a spring, with the block free to move on a frictionless, horizontal surface

(Fig. 15.1). When the spring is neither stretched nor compressed, the block is at rest at the position called the **equilibrium position** of the system, which we identify as $x = 0$ (Fig. 15.1b). We know from experience that such a system oscillates back and forth if disturbed from its equilibrium position.

We can understand the oscillating motion of the block in Figure 15.1 qualitatively by first recalling that when the block is displaced to a position x, the spring exerts on the block a force that is proportional to the position and given by **Hooke's law** (see Section 7.4):

$$F_s = -kx \qquad (15.1)$$

◀ Hooke's law

We call F_s a **restoring force** because it is always directed toward the equilibrium position and therefore *opposite* the displacement of the block from equilibrium. That is, when the block is displaced to the right of $x = 0$ in Figure 15.1a, the position is positive and the restoring force is directed to the left. When the block is displaced to the left of $x = 0$ as in Figure 15.1c, the position is negative and the restoring force is directed to the right.

When the block is displaced from the equilibrium point and released, it is a particle under a net force and consequently undergoes an acceleration. Applying the particle under a net force model to the motion of the block, with Equation 15.1 providing the net force in the x direction, we obtain

$$\sum F_x = ma_x \;\; \rightarrow \;\; -kx = ma_x$$

$$a_x = -\frac{k}{m}x \qquad (15.2)$$

That is, the acceleration of the block is proportional to its position, and the direction of the acceleration is opposite the direction of the displacement of the block from equilibrium. Systems that behave in this way are said to exhibit **simple harmonic motion.** An object moves with simple harmonic motion whenever its acceleration is proportional to its position and is oppositely directed to the displacement from equilibrium.

If the block in Figure 15.1 is displaced to a position $x = A$ and released from rest, its *initial* acceleration is $-kA/m$. When the block passes through the equilibrium position $x = 0$, its acceleration is zero. At this instant, its speed is a maximum because the acceleration changes sign. The block then continues to travel to the left of equilibrium with a positive acceleration and finally reaches $x = -A$, at which time its acceleration is $+kA/m$ and its speed is again zero as discussed in Sections 7.4 and 7.9. The block completes a full cycle of its motion by returning to the original position, again passing through $x = 0$ with maximum speed. Therefore, the block oscillates between the turning points $x = \pm A$. In the absence of

Pitfall Prevention 15.1
The Orientation of the Spring Figure 15.1 shows a *horizontal* spring, with an attached block sliding on a frictionless surface. Another possibility is a block hanging from a *vertical* spring. All the results we discuss for the horizontal spring are the same for the vertical spring with one exception: when the block is placed on the vertical spring, its weight causes the spring to extend. If the resting position of the block is defined as $x = 0$, the results of this chapter also apply to this vertical system.

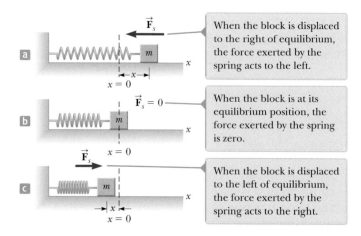

When the block is displaced to the right of equilibrium, the force exerted by the spring acts to the left.

When the block is at its equilibrium position, the force exerted by the spring is zero.

When the block is displaced to the left of equilibrium, the force exerted by the spring acts to the right.

Figure 15.1 A block attached to a spring moving on a frictionless surface.

friction, this idealized motion will continue forever because the force exerted by the spring is conservative. Real systems are generally subject to friction, so they do not oscillate forever. We shall explore the details of the situation with friction in Section 15.6.

Quick Quiz 15.1 A block on the end of a spring is pulled to position $x = A$ and released from rest. In one full cycle of its motion, through what total distance does it travel? **(a)** $A/2$ **(b)** A **(c)** $2A$ **(d)** $4A$

15.2 Analysis Model: Particle in Simple Harmonic Motion

The motion described in the preceding section occurs so often that we identify the **particle in simple harmonic motion** model to represent such situations. To develop a mathematical representation for this model, we will generally choose x as the axis along which the oscillation occurs; hence, we will drop the subscript-x notation in this discussion. Recall that, by definition, $a = dv/dt = d^2x/dt^2$, so we can express Equation 15.2 as

$$\frac{d^2x}{dt^2} = -\frac{k}{m}x \tag{15.3}$$

If we denote the ratio k/m with the symbol ω^2 (we choose ω^2 rather than ω so as to make the solution we develop below simpler in form), then

$$\omega^2 = \frac{k}{m} \tag{15.4}$$

and Equation 15.3 can be written in the form

$$\frac{d^2x}{dt^2} = -\omega^2 x \tag{15.5}$$

Let's now find a mathematical solution to Equation 15.5, that is, a function $x(t)$ that satisfies this second-order differential equation and is a mathematical representation of the position of the particle as a function of time. We seek a function whose second derivative is the same as the original function with a negative sign and multiplied by ω^2. The trigonometric functions sine and cosine exhibit this behavior, so we can build a solution around one or both of them. The following cosine function is a solution to the differential equation:

Position versus time for ▶
a particle in simple
harmonic motion

$$x(t) = A\cos(\omega t + \phi) \tag{15.6}$$

where A, ω, and ϕ are constants. To show explicitly that this solution satisfies Equation 15.5, notice that

$$\frac{dx}{dt} = A\frac{d}{dt}\cos(\omega t + \phi) = -\omega A\sin(\omega t + \phi) \tag{15.7}$$

$$\frac{d^2x}{dt^2} = -\omega A\frac{d}{dt}\sin(\omega t + \phi) = -\omega^2 A\cos(\omega t + \phi) \tag{15.8}$$

Comparing Equations 15.6 and 15.8, we see that $d^2x/dt^2 = -\omega^2 x$ and Equation 15.5 is satisfied.

The parameters A, ω, and ϕ are constants of the motion. To give physical significance to these constants, it is convenient to form a graphical representation of the motion by plotting x as a function of t as in Figure 15.2a. First, A, called the **amplitude** of the motion, is simply the maximum value of the position of the particle in

Pitfall Prevention 15.2
A Nonconstant Acceleration The acceleration of a particle in simple harmonic motion is not constant. Equation 15.3 shows that its acceleration varies with position x. Therefore, we *cannot* apply the kinematic equations of Chapter 2 in this situation.

Pitfall Prevention 15.3
Where's the Triangle? Equation 15.6 includes a trigonometric function, a *mathematical function* that can be used whether it refers to a triangle or not. In this case, the cosine function happens to have the correct behavior for representing the position of a particle in simple harmonic motion.

either the positive or negative x direction. The constant ω is called the **angular frequency,** and it has units[1] of radians per second. It is a measure of how rapidly the oscillations are occurring; the more oscillations per unit time, the higher the value of ω. From Equation 15.4, the angular frequency is

$$\omega = \sqrt{\frac{k}{m}} \qquad \text{(15.9)}$$

The constant angle ϕ is called the **phase constant** (or initial phase angle) and, along with the amplitude A, is determined uniquely by the position and velocity of the particle at $t = 0$. If the particle is at its maximum position $x = A$ at $t = 0$, the phase constant is $\phi = 0$ and the graphical representation of the motion is as shown in Figure 15.2b. The quantity $(\omega t + \phi)$ is called the **phase** of the motion. Notice that the function $x(t)$ is periodic and its value is the same each time ωt increases by 2π radians.

Equations 15.1, 15.5, and 15.6 form the basis of the mathematical representation of the particle in simple harmonic motion model. If you are analyzing a situation and find that the force on an object modeled as a particle is of the mathematical form of Equation 15.1, you know the motion is that of a simple harmonic oscillator and the position of the particle is described by Equation 15.6. If you analyze a system and find that it is described by a differential equation of the form of Equation 15.5, the motion is that of a simple harmonic oscillator. If you analyze a situation and find that the position of a particle is described by Equation 15.6, you know the particle undergoes simple harmonic motion.

Quick Quiz 15.2 Consider a graphical representation (Fig. 15.3) of simple harmonic motion as described mathematically in Equation 15.6. When the particle is at point Ⓐ on the graph, what can you say about its position and velocity? **(a)** The position and velocity are both positive. **(b)** The position and velocity are both negative. **(c)** The position is positive, and the velocity is zero. **(d)** The position is negative, and the velocity is zero. **(e)** The position is positive, and the velocity is negative. **(f)** The position is negative, and the velocity is positive.

Quick Quiz 15.3 Figure 15.4 shows two curves representing particles undergoing simple harmonic motion. The correct description of these two motions is that the simple harmonic motion of particle B is **(a)** of larger angular frequency and larger amplitude than that of particle A, **(b)** of larger angular frequency and smaller amplitude than that of particle A, **(c)** of smaller angular frequency and larger amplitude than that of particle A, or **(d)** of smaller angular frequency and smaller amplitude than that of particle A.

Let us investigate further the mathematical description of simple harmonic motion. The **period** T of the motion is the time interval required for the particle to go through one full cycle of its motion (Fig. 15.2a). That is, the values of x and v for the particle at time t equal the values of x and v at time $t + T$. Because the phase increases by 2π radians in a time interval of T,

$$[\omega(t + T) + \phi] - (\omega t + \phi) = 2\pi$$

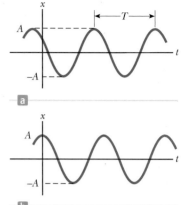

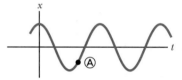

Figure 15.2 (a) An x–t graph for a particle undergoing simple harmonic motion. The amplitude of the motion is A, and the period (defined in Eq. 15.10) is T. (b) The x–t graph for the special case in which $x = A$ at $t = 0$ and hence $\phi = 0$.

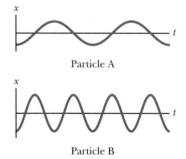

Figure 15.3 (Quick Quiz 15.2) An x–t graph for a particle undergoing simple harmonic motion. At a particular time, the particle's position is indicated by Ⓐ in the graph.

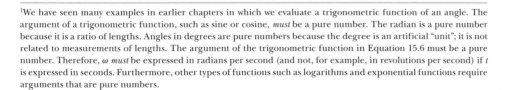

Particle A

Particle B

Figure 15.4 (Quick Quiz 15.3) Two x–t graphs for particles undergoing simple harmonic motion. The amplitudes and frequencies are different for the two particles.

---

[1] We have seen many examples in earlier chapters in which we evaluate a trigonometric function of an angle. The argument of a trigonometric function, such as sine or cosine, *must* be a pure number. The radian is a pure number because it is a ratio of lengths. Angles in degrees are pure numbers because the degree is an artificial "unit"; it is not related to measurements of lengths. The argument of the trigonometric function in Equation 15.6 must be a pure number. Therefore, ω *must* be expressed in radians per second (and not, for example, in revolutions per second) if t is expressed in seconds. Furthermore, other types of functions such as logarithms and exponential functions require arguments that are pure numbers.

Simplifying this expression gives $\omega T = 2\pi$, or

$$T = \frac{2\pi}{\omega} \tag{15.10}$$

The inverse of the period is called the **frequency** f of the motion. Whereas the period is the time interval per oscillation, the frequency represents the number of oscillations the particle undergoes per unit time interval:

$$f = \frac{1}{T} = \frac{\omega}{2\pi} \tag{15.11}$$

The units of f are cycles per second, or **hertz** (Hz). Rearranging Equation 15.11 gives

$$\omega = 2\pi f = \frac{2\pi}{T} \tag{15.12}$$

Equations 15.9 through 15.11 can be used to express the period and frequency of the motion for the particle in simple harmonic motion in terms of the characteristics m and k of the system as

Period ▶
$$T = \frac{2\pi}{\omega} = 2\pi\sqrt{\frac{m}{k}} \tag{15.13}$$

Frequency ▶
$$f = \frac{1}{T} = \frac{1}{2\pi}\sqrt{\frac{k}{m}} \tag{15.14}$$

That is, the period and frequency depend *only* on the mass of the particle and the force constant of the spring and *not* on the parameters of the motion, such as A or ϕ. As we might expect, the frequency is larger for a stiffer spring (larger value of k) and decreases with increasing mass of the particle.

We can obtain the velocity and acceleration[2] of a particle undergoing simple harmonic motion from Equations 15.7 and 15.8:

Velocity of a particle in ▶
simple harmonic motion
$$v = \frac{dx}{dt} = -\omega A \sin(\omega t + \phi) \tag{15.15}$$

Acceleration of a particle in ▶
simple harmonic motion
$$a = \frac{d^2 x}{dt^2} = -\omega^2 A \cos(\omega t + \phi) \tag{15.16}$$

From Equation 15.15, we see that because the sine and cosine functions oscillate between ± 1, the extreme values of the velocity v are $\pm \omega A$. Likewise, Equation 15.16 shows that the extreme values of the acceleration a are $\pm \omega^2 A$. Therefore, the *maximum* values of the magnitudes of the velocity and acceleration are

Maximum magnitudes of ▶
velocity and acceleration in
simple harmonic motion
$$v_{\max} = \omega A = \sqrt{\frac{k}{m}}\, A \tag{15.17}$$

$$a_{\max} = \omega^2 A = \frac{k}{m}\, A \tag{15.18}$$

Figure 15.5a plots position versus time for an arbitrary value of the phase constant. The associated velocity–time and acceleration–time curves are illustrated in Figures 15.5b and 15.5c, respectively. They show that the phase of the velocity differs from the phase of the position by $\pi/2$ rad, or 90°. That is, when x is a maximum or a minimum, the velocity is zero. Likewise, when x is zero, the speed is a maximum. Furthermore, notice that the phase of the acceleration differs from the phase of the position by π radians, or 180°. For example, when x is a maximum, a has a maximum magnitude in the opposite direction.

---

[2]Because the motion of a simple harmonic oscillator takes place in one dimension, we denote velocity as v and acceleration as a, with the direction indicated by a positive or negative sign as in Chapter 2.

Quick Quiz 15.4 An object of mass m is hung from a spring and set into oscillation. The period of the oscillation is measured and recorded as T. The object
of mass m is removed and replaced with an object of mass $2m$. When this object
is set into oscillation, what is the period of the motion? **(a)** $2T$ **(b)** $\sqrt{2}\,T$ **(c)** T
(d) $T/\sqrt{2}$ **(e)** $T/2$

Equation 15.6 describes simple harmonic motion of a particle in general. Let's
now see how to evaluate the constants of the motion. The angular frequency ω is
evaluated using Equation 15.9. The constants A and ϕ are evaluated from the initial conditions, that is, the state of the oscillator at $t = 0$.

Suppose a block is set into motion by pulling it from equilibrium by a distance A
and releasing it from rest at $t = 0$ as in Figure 15.6. We must then require our solutions for $x(t)$ and $v(t)$ (Eqs. 15.6 and 15.15) to obey the initial conditions that $x(0) =$
A and $v(0) = 0$:

$$x(0) = A \cos \phi = A$$

$$v(0) = -\omega A \sin \phi = 0$$

These conditions are met if $\phi = 0$, giving $x = A \cos \omega t$ as our solution. To check this
solution, notice that it satisfies the condition that $x(0) = A$ because $\cos 0 = 1$.

The position, velocity, and acceleration of the block versus time are plotted in
Figure 15.7a for this special case. The acceleration reaches extreme values of $\mp\omega^2 A$
when the position has extreme values of $\pm A$. Furthermore, the velocity has extreme
values of $\pm\omega A$, which both occur at $x = 0$. Hence, the quantitative solution agrees
with our qualitative description of this system.

Let's consider another possibility. Suppose the system is oscillating and we define
$t = 0$ as the instant the block passes through the unstretched position of the spring
while moving to the right (Fig. 15.8). In this case, our solutions for $x(t)$ and $v(t)$
must obey the initial conditions that $x(0) = 0$ and $v(0) = v_i$:

$$x(0) = A \cos \phi = 0$$

$$v(0) = -\omega A \sin \phi = v_i$$

The first of these conditions tells us that $\phi = \pm\pi/2$. With these choices for ϕ, the
second condition tells us that $A = \mp v_i/\omega$. Because the initial velocity is positive and
the amplitude must be positive, we must have $\phi = -\pi/2$. Hence, the solution is

$$x = \frac{v_i}{\omega} \cos\left(\omega t - \frac{\pi}{2}\right)$$

The graphs of position, velocity, and acceleration versus time for this choice of $t = 0$
are shown in Figure 15.7b. Notice that these curves are the same as those in Figure

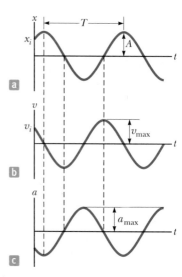

Figure 15.5 Graphical representation of simple harmonic
motion. (a) Position versus time.
(b) Velocity versus time. (c) Acceleration versus time. Notice that at
any specified time the velocity is
90° out of phase with the position
and the acceleration is 180° out of
phase with the position.

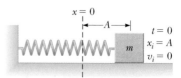

Figure 15.6 A block–spring
system that begins its motion from
rest with the block at $x = A$ at $t = 0$.

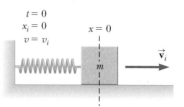

Figure 15.8 The block–spring
system is undergoing oscillation,
and $t = 0$ is defined at an instant
when the block passes through the
equilibrium position $x = 0$ and is
moving to the right with speed v_i.

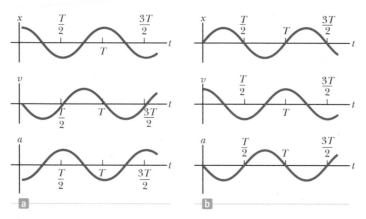

Figure 15.7 (a) Position, velocity, and acceleration versus time for the block in Figure 15.6 under
the initial conditions that at $t = 0$, $x(0) = A$, and $v(0) = 0$. (b) Position, velocity, and acceleration versus time for the block in Figure 15.8 under the initial conditions that at $t = 0$, $x(0) = 0$, and $v(0) = v_i$.

15.7a, but shifted to the right by one-fourth of a cycle. This shift is described mathematically by the phase constant $\phi = -\pi/2$, which is one-fourth of a full cycle of 2π.

Analysis Model   Particle in Simple Harmonic Motion

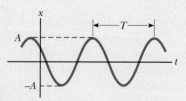

Imagine an object that is subject to a force that is proportional to the negative of the object's position, $F = -kx$. Such a force equation is known as Hooke's law, and it describes the force applied to an object attached to an ideal spring. The parameter k in Hooke's law is called the *spring constant* or the *force constant*. The position of an object acted on by a force described by Hooke's law is given by

$$x(t) = A \cos (\omega t + \phi) \qquad \text{(15.6)}$$

where A is the **amplitude** of the motion, ω is the **angular frequency**, and ϕ is the **phase constant**. The values of A and ϕ depend on the initial position and initial velocity of the particle.

The **period** of the oscillation of the particle is

$$T = \frac{2\pi}{\omega} = 2\pi\sqrt{\frac{m}{k}} \qquad \text{(15.13)}$$

and the inverse of the period is the **frequency**.

Examples:

- a bungee jumper hangs from a bungee cord and oscillates up and down
- a guitar string vibrates back and forth in a standing wave, with each element of the string moving in simple harmonic motion (Chapter 18)
- a piston in a gasoline engine oscillates up and down within the cylinder of the engine (Chapter 22)
- an atom in a diatomic molecule vibrates back and forth as if it is connected by a spring to the other atom in the molecule (Chapter 43)

Example 15.1   A Block–Spring System   AM

A 200-g block connected to a light spring for which the force constant is 5.00 N/m is free to oscillate on a frictionless, horizontal surface. The block is displaced 5.00 cm from equilibrium and released from rest as in Figure 15.6.

(A) Find the period of its motion.

SOLUTION

Conceptualize Study Figure 15.6 and imagine the block moving back and forth in simple harmonic motion once it is released. Set up an experimental model in the vertical direction by hanging a heavy object such as a stapler from a strong rubber band.

Categorize The block is modeled as a *particle in simple harmonic motion*.

Analyze

Use Equation 15.9 to find the angular frequency of the block–spring system:

$$\omega = \sqrt{\frac{k}{m}} = \sqrt{\frac{5.00 \text{ N/m}}{200 \times 10^{-3} \text{ kg}}} = 5.00 \text{ rad/s}$$

Use Equation 15.13 to find the period of the system:

$$T = \frac{2\pi}{\omega} = \frac{2\pi}{5.00 \text{ rad/s}} = \boxed{1.26 \text{ s}}$$

(B) Determine the maximum speed of the block.

SOLUTION

Use Equation 15.17 to find v_{max}:

$$v_{max} = \omega A = (5.00 \text{ rad/s})(5.00 \times 10^{-2} \text{ m}) = \boxed{0.250 \text{ m/s}}$$

(C) What is the maximum acceleration of the block?

▶ **15.1** continued

SOLUTION

Use Equation 15.18 to find a_{max}:

$a_{max} = \omega^2 A = (5.00 \text{ rad/s})^2 (5.00 \times 10^{-2} \text{ m}) = \boxed{1.25 \text{ m/s}^2}$

(D) Express the position, velocity, and acceleration as functions of time in SI units.

SOLUTION

Find the phase constant from the initial condition that $x = A$ at $t = 0$:

$x(0) = A \cos \phi = A \rightarrow \phi = 0$

Use Equation 15.6 to write an expression for $x(t)$:

$x = A \cos (\omega t + \phi) = \boxed{0.050\ 0 \cos 5.00t}$

Use Equation 15.15 to write an expression for $v(t)$:

$v = -\omega A \sin (\omega t + \phi) = \boxed{-0.250 \sin 5.00t}$

Use Equation 15.16 to write an expression for $a(t)$:

$a = -\omega^2 A \cos (\omega t + \phi) = \boxed{-1.25 \cos 5.00t}$

Finalize Consider part (a) of Figure 15.7, which shows the graphical representations of the motion of the block in this problem. Make sure that the mathematical representations found above in part (D) are consistent with these graphical representations.

WHAT IF? What if the block were released from the same initial position, $x_i = 5.00$ cm, but with an initial velocity of $v_i = -0.100$ m/s? Which parts of the solution change, and what are the new answers for those that do change?

Answers Part (A) does not change because the period is independent of how the oscillator is set into motion. Parts (B), (C), and (D) will change.

Write position and velocity expressions for the initial conditions:

(1) $x(0) = A \cos \phi = x_i$

(2) $v(0) = -\omega A \sin \phi = v_i$

Divide Equation (2) by Equation (1) to find the phase constant:

$\dfrac{-\omega A \sin \phi}{A \cos \phi} = \dfrac{v_i}{x_i}$

$\tan \phi = -\dfrac{v_i}{\omega x_i} = -\dfrac{-0.100 \text{ m/s}}{(5.00 \text{ rad/s})(0.050\ 0 \text{ m})} = 0.400$

$\phi = \tan^{-1} (0.400) = 0.121\pi$

Use Equation (1) to find A:

$A = \dfrac{x_i}{\cos \phi} = \dfrac{0.050\ 0 \text{ m}}{\cos (0.121\pi)} = 0.053\ 9 \text{ m}$

Find the new maximum speed:

$v_{max} = \omega A = (5.00 \text{ rad/s})(5.39 \times 10^{-2} \text{ m}) = 0.269 \text{ m/s}$

Find the new magnitude of the maximum acceleration:

$a_{max} = \omega^2 A = (5.00 \text{ rad/s})^2 (5.39 \times 10^{-2} \text{ m}) = 1.35 \text{ m/s}^2$

Find new expressions for position, velocity, and acceleration in SI units:

$x = 0.053\ 9 \cos (5.00t + 0.121\pi)$

$v = -0.269 \sin (5.00t + 0.121\pi)$

$a = -1.35 \cos (5.00t + 0.121\pi)$

As we saw in Chapters 7 and 8, many problems are easier to solve using an energy approach rather than one based on variables of motion. This particular What If? is easier to solve from an energy approach. Therefore, we shall investigate the energy of the simple harmonic oscillator in the next section.

Example 15.2 **Watch Out for Potholes!** AM

A car with a mass of 1 300 kg is constructed so that its frame is supported by four springs. Each spring has a force constant of 20 000 N/m. Two people riding in the car have a combined mass of 160 kg. Find the frequency of vibration of the car after it is driven over a pothole in the road.

continued

▶ **15.2** continued

SOLUTION

Conceptualize Think about your experiences with automobiles. When you sit in a car, it moves downward a small distance because your weight is compressing the springs further. If you push down on the front bumper and release it, the front of the car oscillates a few times.

Categorize We imagine the car as being supported by a single spring and model the car as a *particle in simple harmonic motion*.

....

Analyze First, let's determine the effective spring constant of the four springs combined. For a given extension x of the springs, the combined force on the car is the sum of the forces from the individual springs.

Find an expression for the total force on the car:

$$F_{\text{total}} = \sum (-kx) = -\left(\sum k\right) x$$

In this expression, x has been factored from the sum because it is the same for all four springs. The effective spring constant for the combined springs is the sum of the individual spring constants.

Evaluate the effective spring constant:

$$k_{\text{eff}} = \sum k = 4 \times 20\ 000\ \text{N/m} = 80\ 000\ \text{N/m}$$

Use Equation 15.14 to find the frequency of vibration:

$$f = \frac{1}{2\pi}\sqrt{\frac{k_{\text{eff}}}{m}} = \frac{1}{2\pi}\sqrt{\frac{80\ 000\ \text{N/m}}{1\ 460\ \text{kg}}} = \boxed{1.18\ \text{Hz}}$$

....

Finalize The mass we used here is that of the car plus the people because that is the total mass that is oscillating. Also notice that we have explored only up-and-down motion of the car. If an oscillation is established in which the car rocks back and forth such that the front end goes up when the back end goes down, the frequency will be different.

WHAT IF? Suppose the car stops on the side of the road and the two people exit the car. One of them pushes downward on the car and releases it so that it oscillates vertically. Is the frequency of the oscillation the same as the value we just calculated?

Answer The suspension system of the car is the same, but the mass that is oscillating is smaller: it no longer includes the mass of the two people. Therefore, the frequency should be higher. Let's calculate the new frequency, taking the mass to be 1 300 kg:

$$f = \frac{1}{2\pi}\sqrt{\frac{k_{\text{eff}}}{m}} = \frac{1}{2\pi}\sqrt{\frac{80\ 000\ \text{N/m}}{1\ 300\ \text{kg}}} = 1.25\ \text{Hz}$$

As predicted, the new frequency is a bit higher.

15.3 Energy of the Simple Harmonic Oscillator

As we have done before, after studying the the motion of an object modeled as a particle in a new situation and investigating the forces involved in influencing that motion, we turn our attention to *energy*. Let us examine the mechanical energy of a system in which a particle undergoes simple harmonic motion, such as the block–spring system illustrated in Figure 15.1. Because the surface is frictionless, the system is isolated and we expect the total mechanical energy of the system to be constant. We assume a massless spring, so the kinetic energy of the system corresponds only to that of the block. We can use Equation 15.15 to express the kinetic energy of the block as

Kinetic energy of a simple ▶
harmonic oscillator

$$K = \tfrac{1}{2}mv^2 = \tfrac{1}{2}m\omega^2 A^2 \sin^2(\omega t + \phi) \qquad (15.19)$$

The elastic potential energy stored in the spring for any elongation x is given by $\tfrac{1}{2}kx^2$ (see Eq. 7.22). Using Equation 15.6 gives

Potential energy of a simple ▶
harmonic oscillator

$$U = \tfrac{1}{2}kx^2 = \tfrac{1}{2}kA^2 \cos^2(\omega t + \phi) \qquad (15.20)$$

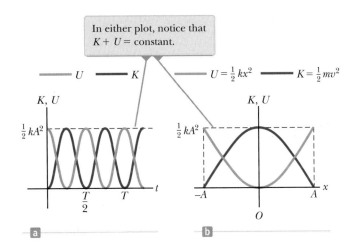

In either plot, notice that $K + U =$ constant.

Figure 15.9 (a) Kinetic energy and potential energy versus time for a simple harmonic oscillator with $\phi = 0$. (b) Kinetic energy and potential energy versus position for a simple harmonic oscillator.

We see that K and U are *always* positive quantities or zero. Because $\omega^2 = k/m$, we can express the total mechanical energy of the simple harmonic oscillator as

$$E = K + U = \tfrac{1}{2}kA^2[\sin^2(\omega t + \phi) + \cos^2(\omega t + \phi)]$$

From the identity $\sin^2\theta + \cos^2\theta = 1$, we see that the quantity in square brackets is unity. Therefore, this equation reduces to

$$E = \tfrac{1}{2}kA^2 \qquad\qquad \text{(15.21)}$$

◀ Total energy of a simple harmonic oscillator

That is, the total mechanical energy of a simple harmonic oscillator is a constant of the motion and is proportional to the square of the amplitude. The total mechanical energy is equal to the maximum potential energy stored in the spring when $x = \pm A$ because $v = 0$ at these points and there is no kinetic energy. At the equilibrium position, where $U = 0$ because $x = 0$, the total energy, all in the form of kinetic energy, is again $\tfrac{1}{2}kA^2$.

Plots of the kinetic and potential energies versus time appear in Figure 15.9a, where we have taken $\phi = 0$. At all times, the sum of the kinetic and potential energies is a constant equal to $\tfrac{1}{2}kA^2$, the total energy of the system.

The variations of K and U with the position x of the block are plotted in Figure 15.9b. Energy is continuously being transformed between potential energy stored in the spring and kinetic energy of the block.

Figure 15.10 on page 362 illustrates the position, velocity, acceleration, kinetic energy, and potential energy of the block–spring system for one full period of the motion. Most of the ideas discussed so far are incorporated in this important figure. Study it carefully.

Finally, we can obtain the velocity of the block at an arbitrary position by expressing the total energy of the system at some arbitrary position x as

$$E = K + U = \tfrac{1}{2}mv^2 + \tfrac{1}{2}kx^2 = \tfrac{1}{2}kA^2$$

$$v = \pm\sqrt{\frac{k}{m}(A^2 - x^2)} = \pm\omega\sqrt{A^2 - x^2} \qquad\qquad \text{(15.22)}$$

◀ Velocity as a function of position for a simple harmonic oscillator

When you check Equation 15.22 to see whether it agrees with known cases, you find that it verifies that the speed is a maximum at $x = 0$ and is zero at the turning points $x = \pm A$.

You may wonder why we are spending so much time studying simple harmonic oscillators. We do so because they are good models of a wide variety of physical phenomena. For example, recall the Lennard–Jones potential discussed in Example 7.9. This complicated function describes the forces holding atoms together. Figure 15.11a on page 362 shows that for small displacements from the equilibrium

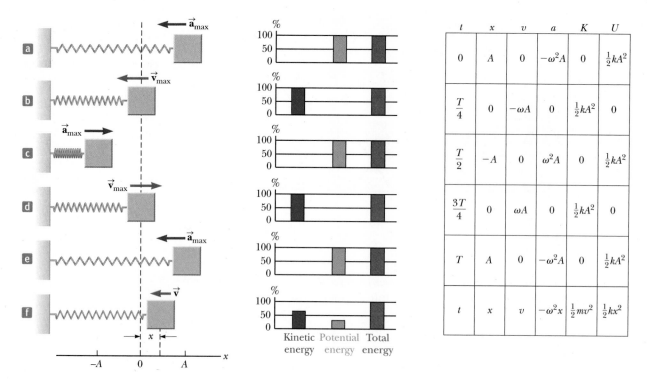

Figure 15.10 (a) through (e) Several instants in the simple harmonic motion for a block–spring system. Energy bar graphs show the distribution of the energy of the system at each instant. The parameters in the table at the right refer to the block–spring system, assuming at $t = 0$, $x = A$; hence, $x = A \cos \omega t$. For these five special instants, one of the types of energy is zero. (f) An arbitrary point in the motion of the oscillator. The system possesses both kinetic energy and potential energy at this instant as shown in the bar graph.

position, the potential energy curve for this function approximates a parabola, which represents the potential energy function for a simple harmonic oscillator. Therefore, we can model the complex atomic binding forces as being due to tiny springs as depicted in Figure 15.11b.

The ideas presented in this chapter apply not only to block–spring systems and atoms, but also to a wide range of situations that include bungee jumping, playing a musical instrument, and viewing the light emitted by a laser. You will see more examples of simple harmonic oscillators as you work through this book.

Figure 15.11 (a) If the atoms in a molecule do not move too far from their equilibrium positions, a graph of potential energy versus separation distance between atoms is similar to the graph of potential energy versus position for a simple harmonic oscillator (dashed black curve). (b) The forces between atoms in a solid can be modeled by imagining springs between neighboring atoms.

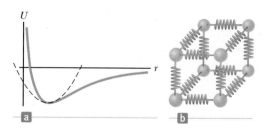

Example 15.3   Oscillations on a Horizontal Surface   AM

A 0.500-kg cart connected to a light spring for which the force constant is 20.0 N/m oscillates on a frictionless, horizontal air track.

(A) Calculate the maximum speed of the cart if the amplitude of the motion is 3.00 cm.

SOLUTION

Conceptualize The system oscillates in exactly the same way as the block in Figure 15.10, so use that figure in your mental image of the motion.

▶ **15.3** continued

Categorize The cart is modeled as a *particle in simple harmonic motion.*

. . . .

Analyze Use Equation 15.21 to express the total energy of the oscillator system and equate it to the kinetic energy of the system when the cart is at $x = 0$:

$$E = \tfrac{1}{2}kA^2 = \tfrac{1}{2}mv_{max}^2$$

Solve for the maximum speed and substitute numerical values:

$$v_{max} = \sqrt{\frac{k}{m}}\,A = \sqrt{\frac{20.0 \text{ N/m}}{0.500 \text{ kg}}}(0.030\ 0 \text{ m}) = \boxed{0.190 \text{ m/s}}$$

(B) What is the velocity of the cart when the position is 2.00 cm?

SOLUTION

Use Equation 15.22 to evaluate the velocity:

$$v = \pm\sqrt{\frac{k}{m}(A^2 - x^2)}$$

$$= \pm\sqrt{\frac{20.0 \text{ N/m}}{0.500 \text{ kg}}[(0.030\ 0 \text{ m})^2 - (0.020\ 0 \text{ m})^2]}$$

$$= \boxed{\pm 0.141 \text{ m/s}}$$

The positive and negative signs indicate that the cart could be moving to either the right or the left at this instant.

(C) Compute the kinetic and potential energies of the system when the position of the cart is 2.00 cm.

SOLUTION

Use the result of part (B) to evaluate the kinetic energy at $x = 0.020\ 0$ m:

$$K = \tfrac{1}{2}mv^2 = \tfrac{1}{2}(0.500 \text{ kg})(0.141 \text{ m/s})^2 = \boxed{5.00 \times 10^{-3} \text{ J}}$$

Evaluate the elastic potential energy at $x = 0.020\ 0$ m:

$$U = \tfrac{1}{2}kx^2 = \tfrac{1}{2}(20.0 \text{ N/m})(0.0200 \text{ m})^2 = \boxed{4.00 \times 10^{-3} \text{ J}}$$

Finalize The sum of the kinetic and potential energies in part (C) is equal to the total energy, which can be found from Equation 15.21. That must be true for *any* position of the cart.

WHAT IF? The cart in this example could have been set into motion by releasing the cart from rest at $x = 3.00$ cm. What if the cart were released from the same position, but with an initial velocity of $v = -0.100$ m/s? What are the new amplitude and maximum speed of the cart?

Answer This question is of the same type we asked at the end of Example 15.1, but here we apply an energy approach.

First calculate the total energy of the system at $t = 0$:

$$E = \tfrac{1}{2}mv^2 + \tfrac{1}{2}kx^2$$
$$= \tfrac{1}{2}(0.500 \text{ kg})(-0.100 \text{ m/s})^2 + \tfrac{1}{2}(20.0 \text{ N/m})(0.030\ 0 \text{ m})^2$$
$$= 1.15 \times 10^{-2} \text{ J}$$

Equate this total energy to the potential energy of the system when the cart is at the endpoint of the motion:

$$E = \tfrac{1}{2}kA^2$$

Solve for the amplitude A:

$$A = \sqrt{\frac{2E}{k}} = \sqrt{\frac{2(1.15 \times 10^{-2} \text{ J})}{20.0 \text{ N/m}}} = 0.033\ 9 \text{ m}$$

Equate the total energy to the kinetic energy of the system when the cart is at the equilibrium position:

$$E = \tfrac{1}{2}mv_{max}^2$$

Solve for the maximum speed:

$$v_{max} = \sqrt{\frac{2E}{m}} = \sqrt{\frac{2(1.15 \times 10^{-2} \text{ J})}{0.500 \text{ kg}}} = 0.214 \text{ m/s}$$

The amplitude and maximum velocity are larger than the previous values because the cart was given an initial velocity at $t = 0$.

The back edge of the treadle goes up and down as one's feet rock the treadle.

The oscillation of the treadle causes circular motion of the drive wheel, eventually resulting in additional up and down motion—of the sewing needle.

John W. Jewett, Jr.

Figure 15.12 The bottom of a treadle-style sewing machine from the early twentieth century. The treadle is the wide, flat foot pedal with the metal grillwork.

15.4 Comparing Simple Harmonic Motion with Uniform Circular Motion

Some common devices in everyday life exhibit a relationship between oscillatory motion and circular motion. For example, consider the drive mechanism for a non-electric sewing machine in Figure 15.12. The operator of the machine places her feet on the treadle and rocks them back and forth. This oscillatory motion causes the large wheel at the right to undergo circular motion. The red drive belt seen in the photograph transfers this circular motion to the sewing machine mechanism (above the photo) and eventually results in the oscillatory motion of the sewing needle. In this section, we explore this interesting relationship between these two types of motion.

Figure 15.13 is a view of an experimental arrangement that shows this relationship. A ball is attached to the rim of a turntable of radius A, which is illuminated from above by a lamp. The ball casts a shadow on a screen. As the turntable rotates with constant angular speed, the shadow of the ball moves back and forth in simple harmonic motion.

Consider a particle located at point P on the circumference of a circle of radius A as in Figure 15.14a, with the line OP making an angle ϕ with the x axis at $t = 0$. We call this circle a *reference circle* for comparing simple harmonic motion with uniform circular motion, and we choose the position of P at $t = 0$ as our reference position. If the particle moves along the circle with constant angular speed ω until OP makes an angle θ with the x axis as in Figure 15.14b, at some time $t > 0$ the angle between OP and the x axis is $\theta = \omega t + \phi$. As the particle moves along the circle, the projection of P on the x axis, labeled point Q, moves back and forth along the x axis between the limits $x = \pm A$.

Notice that points P and Q always have the same x coordinate. From the right triangle OPQ, we see that this x coordinate is

$$x(t) = A \cos (\omega t + \phi) \tag{15.23}$$

This expression is the same as Equation 15.6 and shows that the point Q moves with simple harmonic motion along the x axis. Therefore, the motion of an object described by the analysis model of a particle in simple harmonic motion along a straight line can be represented by the projection of an object that can be modeled as a particle in uniform circular motion along a diameter of a reference circle.

This geometric interpretation shows that the time interval for one complete revolution of the point P on the reference circle is equal to the period of motion T for simple harmonic motion between $x = \pm A$. Therefore, the angular speed ω of P is the same as the angular frequency ω of simple harmonic motion along the x axis

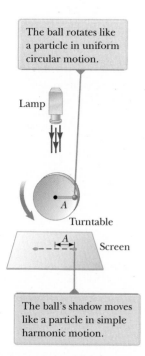

The ball rotates like a particle in uniform circular motion.

Lamp

A

Turntable

A Screen

The ball's shadow moves like a particle in simple harmonic motion.

Figure 15.13 An experimental setup for demonstrating the connection between a particle in simple harmonic motion and a corresponding particle in uniform circular motion.

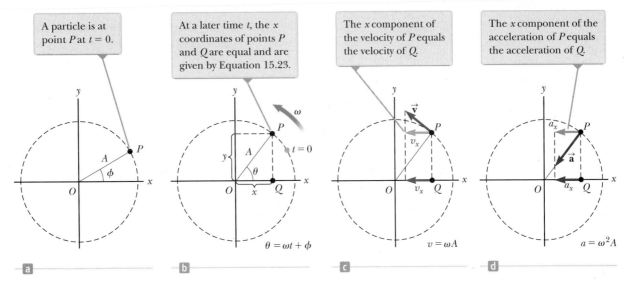

Figure 15.14 Relationship between the uniform circular motion of a point P and the simple harmonic motion of a point Q. A particle at P moves in a circle of radius A with constant angular speed ω.

(which is why we use the same symbol). The phase constant ϕ for simple harmonic motion corresponds to the initial angle OP makes with the x axis. The radius A of the reference circle equals the amplitude of the simple harmonic motion.

Because the relationship between linear and angular speed for circular motion is $v = r\omega$ (see Eq. 10.10), the particle moving on the reference circle of radius A has a velocity of magnitude ωA. From the geometry in Figure 15.14c, we see that the x component of this velocity is $-\omega A \sin(\omega t + \phi)$. By definition, point Q has a velocity given by dx/dt. Differentiating Equation 15.23 with respect to time, we find that the velocity of Q is the same as the x component of the velocity of P.

The acceleration of P on the reference circle is directed radially inward toward O and has a magnitude $v^2/A = \omega^2 A$. From the geometry in Figure 15.14d, we see that the x component of this acceleration is $-\omega^2 A \cos(\omega t + \phi)$. This value is also the acceleration of the projected point Q along the x axis, as you can verify by taking the second derivative of Equation 15.23.

Quick **Quiz** 15.5 Figure 15.15 shows the position of an object in uniform circular motion at $t = 0$. A light shines from above and projects a shadow of the object on a screen below the circular motion. What are the correct values for the *amplitude* and *phase constant* (relative to an x axis to the right) of the simple harmonic motion of the shadow? (a) 0.50 m and 0 (b) 1.00 m and 0 (c) 0.50 m and π (d) 1.00 m and π

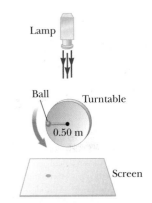

Figure 15.15 (Quick Quiz 15.5) An object moves in circular motion, casting a shadow on the screen below. Its position at an instant of time is shown.

Example 15.4   **Circular Motion with Constant Angular Speed**   AM

The ball in Figure 15.13 rotates counterclockwise in a circle of radius 3.00 m with a constant angular speed of 8.00 rad/s. At $t = 0$, its shadow has an x coordinate of 2.00 m and is moving to the right.

(A) Determine the x coordinate of the shadow as a function of time in SI units.

SOLUTION

Conceptualize Be sure you understand the relationship between circular motion of the ball and simple harmonic motion of its shadow as described in Figure 15.13. Notice that the shadow is *not* at is maximum position at $t = 0$.

Categorize The ball on the turntable is a *particle in uniform circular motion*. The shadow is modeled as a *particle in simple harmonic motion*.

continued

▶ **15.4** c o n t i n u e d

Analyze Use Equation 15.23 to write an expression for the x coordinate of the rotating ball:

$$x = A \cos(\omega t + \phi)$$

Solve for the phase constant:

$$\phi = \cos^{-1}\left(\frac{x}{A}\right) - \omega t$$

Substitute numerical values for the initial conditions:

$$\phi = \cos^{-1}\left(\frac{2.00 \text{ m}}{3.00 \text{ m}}\right) - 0 = \pm 48.2° = \pm 0.841 \text{ rad}$$

If we were to take $\phi = +0.841$ rad as our answer, the shadow would be moving to the left at $t = 0$. Because the shadow is moving to the right at $t = 0$, we must choose $\phi = -0.841$ rad.

Write the x coordinate as a function of time:

$$x = \boxed{3.00 \cos(8.00t - 0.841)}$$

(B) Find the x components of the shadow's velocity and acceleration at any time t.

SOLUTION

Differentiate the x coordinate with respect to time to find the velocity at any time in m/s:

$$v_x = \frac{dx}{dt} = (-3.00 \text{ m})(8.00 \text{ rad/s}) \sin(8.00t - 0.841)$$

$$= \boxed{-24.0 \sin(8.00t - 0.841)}$$

Differentiate the velocity with respect to time to find the acceleration at any time in m/s²:

$$a_x = \frac{dv_x}{dt} = (-24.0 \text{ m/s})(8.00 \text{ rad/s}) \cos(8.00t - 0.841)$$

$$= \boxed{-192 \cos(8.00t - 0.841)}$$

Finalize These results are equally valid for the ball moving in uniform circular motion and the shadow moving in simple harmonic motion. Notice that the value of the phase constant puts the ball in the fourth quadrant of the xy coordinate system of Figure 15.14, which is consistent with the shadow having a positive value for x and moving toward the right.

15.5 The Pendulum

The **simple pendulum** is another mechanical system that exhibits periodic motion. It consists of a particle-like bob of mass m suspended by a light string of length L that is fixed at the upper end as shown in Figure 15.16. The motion occurs in the vertical plane and is driven by the gravitational force. We shall show that, provided the angle θ is small (less than about 10°), the motion is very close to that of a simple harmonic oscillator.

The forces acting on the bob are the force $\vec{T}$ exerted by the string and the gravitational force $m\vec{g}$. The tangential component $mg \sin\theta$ of the gravitational force always acts toward $\theta = 0$, opposite the displacement of the bob from the lowest position. Therefore, the tangential component is a restoring force, and we can apply Newton's second law for motion in the tangential direction:

$$F_t = ma_t \rightarrow -mg \sin\theta = m\frac{d^2s}{dt^2}$$

where the negative sign indicates that the tangential force acts toward the equilibrium (vertical) position and s is the bob's position measured along the arc. We have expressed the tangential acceleration as the second derivative of the position s. Because $s = L\theta$ (Eq. 10.1a with $r = L$) and L is constant, this equation reduces to

$$\frac{d^2\theta}{dt^2} = -\frac{g}{L}\sin\theta$$

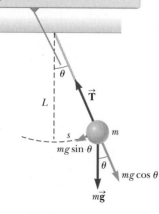

When θ is small, a simple pendulum's motion can be modeled as simple harmonic motion about the equilibrium position $\theta = 0$.

Figure 15.16 A simple pendulum.

Considering θ as the position, let us compare this equation with Equation 15.3. Does it have the same mathematical form? No! The right side is proportional to $\sin \theta$ rather than to θ; hence, we would not expect simple harmonic motion because this expression is not of the same mathematical form as Equation 15.3. If we assume θ is *small* (less than about 10° or 0.2 rad), however, we can use the **small angle approximation,** in which $\sin \theta \approx \theta$, where θ is measured in radians. Table 15.1 shows angles in degrees and radians and the sines of these angles. As long as θ is less than approximately 10°, the angle in radians and its sine are the same to within an accuracy of less than 1.0%.

Therefore, for small angles, the equation of motion becomes

$$\frac{d^2\theta}{dt^2} = -\frac{g}{L}\theta \quad \text{(for small values of } \theta\text{)} \tag{15.24}$$

Equation 15.24 has the same mathematical form as Equation 15.3, so we conclude that the motion for small amplitudes of oscillation can be modeled as simple harmonic motion. Therefore, the solution of Equation 15.24 is modeled after Equation 15.6 and is given by $\theta = \theta_{max} \cos(\omega t + \phi)$, where θ_{max} is the *maximum angular position* and the angular frequency ω is

$$\omega = \sqrt{\frac{g}{L}} \tag{15.25}$$

◀ **Angular frequency for a simple pendulum**

The period of the motion is

$$T = \frac{2\pi}{\omega} = 2\pi\sqrt{\frac{L}{g}} \tag{15.26}$$

◀ **Period of a simple pendulum**

In other words, the period and frequency of a simple pendulum depend only on the length of the string and the acceleration due to gravity. Because the period is independent of the mass, we conclude that all simple pendula that are of equal length and are at the same location (so that g is constant) oscillate with the same period.

The simple pendulum can be used as a timekeeper because its period depends only on its length and the local value of g. It is also a convenient device for making precise measurements of the free-fall acceleration. Such measurements are important because variations in local values of g can provide information on the location of oil and other valuable underground resources.

> **Q**uick Quiz 15.6 A grandfather clock depends on the period of a pendulum to
> keep correct time. **(i)** Suppose a grandfather clock is calibrated correctly and
> then a mischievous child slides the bob of the pendulum downward on the oscil-
> lating rod. Does the grandfather clock run (a) slow, (b) fast, or (c) correctly?
> **(ii)** Suppose a grandfather clock is calibrated correctly at sea level and is then
> taken to the top of a very tall mountain. Does the grandfather clock now run
> (a) slow, (b) fast, or (c) correctly?

> **Pitfall Prevention 15.5**
> **Not True Simple Harmonic Motion**
> The pendulum *does not* exhibit true simple harmonic motion for *any* angle. If the angle is less than about 10°, the motion is close to and can be *modeled* as simple harmonic.

Table 15.1	Angles and Sines of Angles		
Angle in Degrees	**Angle in Radians**	**Sine of Angle**	**Percent Difference**
0°	0.000 0	0.000 0	0.0%
1°	0.017 5	0.017 5	0.0%
2°	0.034 9	0.034 9	0.0%
3°	0.052 4	0.052 3	0.0%
5°	0.087 3	0.087 2	0.1%
10°	0.174 5	0.173 6	0.5%
15°	0.261 8	0.258 8	1.2%
20°	0.349 1	0.342 0	2.1%
30°	0.523 6	0.500 0	4.7%

A Connection Between Length and Time

Christian Huygens (1629–1695), the greatest clockmaker in history, suggested that an international unit of length could be defined as the length of a simple pendulum having a period of exactly 1 s. How much shorter would our length unit be if his suggestion had been followed?

SOLUTION

Conceptualize Imagine a pendulum that swings back and forth in exactly 1 second. Based on your experience in observing swinging objects, can you make an estimate of the required length? Hang a small object from a string and simulate the 1-s pendulum.

Categorize This example involves a simple pendulum, so we categorize it as a substitution problem that applies the concepts introduced in this section.

Solve Equation 15.26 for the length and substitute the known values:

$$L = \frac{T^2 g}{4\pi^2} = \frac{(1.00\ \text{s})^2 (9.80\ \text{m/s}^2)}{4\pi^2} = \boxed{0.248\ \text{m}}$$

The meter's length would be slightly less than one-fourth of its current length. Also, the number of significant digits depends only on how precisely we know g because the time has been defined to be exactly 1 s.

WHAT IF? What if Huygens had been born on another planet? What would the value for g have to be on that planet such that the meter based on Huygens's pendulum would have the same value as our meter?

Answer Solve Equation 15.26 for g:

$$g = \frac{4\pi^2 L}{T^2} = \frac{4\pi^2 (1.00\ \text{m})}{(1.00\ \text{s})^2} = 4\pi^2\ \text{m/s}^2 = 39.5\ \text{m/s}^2$$

No planet in our solar system has an acceleration due to gravity that large.

Physical Pendulum

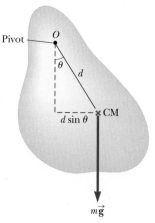

Figure 15.17 A physical pendulum pivoted at O.

Suppose you balance a wire coat hanger so that the hook is supported by your extended index finger. When you give the hanger a small angular displacement with your other hand and then release it, it oscillates. If a hanging object oscillates about a fixed axis that does not pass through its center of mass and the object cannot be approximated as a point mass, we cannot treat the system as a simple pendulum. In this case, the system is called a **physical pendulum.**

Consider a rigid object pivoted at a point O that is a distance d from the center of mass (Fig. 15.17). The gravitational force provides a torque about an axis through O, and the magnitude of that torque is $mgd \sin\theta$, where θ is as shown in Figure 15.17. We apply the rigid object under a net torque analysis model to the object and use the rotational form of Newton's second law, $\Sigma \tau_{\text{ext}} = I\alpha$, where I is the moment of inertia of the object about the axis through O. The result is

$$-mgd \sin\theta = I \frac{d^2\theta}{dt^2}$$

The negative sign indicates that the torque about O tends to decrease θ. That is, the gravitational force produces a restoring torque. If we again assume θ is small, the approximation $\sin\theta \approx \theta$ is valid and the equation of motion reduces to

$$\frac{d^2\theta}{dt^2} = -\left(\frac{mgd}{I}\right)\theta = -\omega^2\theta \qquad (15.27)$$

Because this equation is of the same mathematical form as Equation 15.3, its solution is modeled after that of the simple harmonic oscillator. That is, the solution of Equation 15.27 is given by $\theta = \theta_{max} \cos(\omega t + \phi)$, where θ_{max} is the maximum angular position and

$$\omega = \sqrt{\frac{mgd}{I}}$$

The period is

$$T = \frac{2\pi}{\omega} = 2\pi\sqrt{\frac{I}{mgd}} \qquad (15.28)$$

◀ **Period of a physical pendulum**

This result can be used to measure the moment of inertia of a flat, rigid object. If the location of the center of mass—and hence the value of d—is known, the moment of inertia can be obtained by measuring the period. Finally, notice that Equation 15.28 reduces to the period of a simple pendulum (Eq. 15.26) when $I = md^2$, that is, when all the mass is concentrated at the center of mass.

Example 15.6 **A Swinging Rod**

A uniform rod of mass M and length L is pivoted about one end and oscillates in a vertical plane (Fig. 15.18). Find the period of oscillation if the amplitude of the motion is small.

SOLUTION

Conceptualize Imagine a rod swinging back and forth when pivoted at one end. Try it with a meterstick or a scrap piece of wood.

Categorize Because the rod is not a point particle, we categorize it as a physical pendulum.

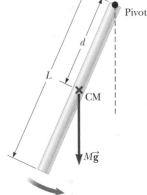

Figure 15.18 (Example 15.6) A rigid rod oscillating about a pivot through one end is a physical pendulum with $d = L/2$.

Analyze In Chapter 10, we found that the moment of inertia of a uniform rod about an axis through one end is $\frac{1}{3}ML^2$. The distance d from the pivot to the center of mass of the rod is $L/2$.

Substitute these quantities into Equation 15.28:

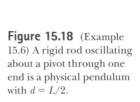

$$T = 2\pi\sqrt{\frac{\frac{1}{3}ML^2}{Mg(L/2)}} = 2\pi\sqrt{\frac{2L}{3g}}$$

Finalize In one of the Moon landings, an astronaut walking on the Moon's surface had a belt hanging from his space suit, and the belt oscillated as a physical pendulum. A scientist on the Earth observed this motion on television and used it to estimate the free-fall acceleration on the Moon. How did the scientist make this calculation?

Torsional Pendulum

Figure 15.19 on page 370 shows a rigid object such as a disk suspended by a wire attached at the top to a fixed support. When the object is twisted through some angle θ, the twisted wire exerts on the object a restoring torque that is proportional to the angular position. That is,

$$\tau = -\kappa\theta$$

where κ (Greek letter kappa) is called the *torsion constant* of the support wire and is a rotational analog to the force constant k for a spring. The value of κ can be obtained by applying a known torque to twist the wire through a measurable angle θ. Applying Newton's second law for rotational motion, we find that

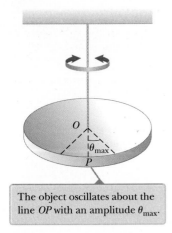

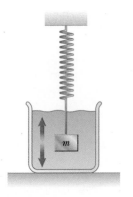

The object oscillates about the line OP with an amplitude θ_{max}.

Figure 15.19 A torsional pendulum.

$$\sum \tau = I\alpha \quad \rightarrow \quad -\kappa\theta = I\frac{d^2\theta}{dt^2}$$

$$\frac{d^2\theta}{dt^2} = -\frac{\kappa}{I}\theta \tag{15.29}$$

Again, this result is the equation of motion for a simple harmonic oscillator, with $\omega = \sqrt{\kappa/I}$ and a period

$$T = 2\pi\sqrt{\frac{I}{\kappa}} \tag{15.30}$$

This system is called a *torsional pendulum*. There is no small-angle restriction in this situation as long as the elastic limit of the wire is not exceeded.

15.6 Damped Oscillations

The oscillatory motions we have considered so far have been for ideal systems, that is, systems that oscillate indefinitely under the action of only one force, a linear restoring force. In many real systems, nonconservative forces such as friction or air resistance also act and retard the motion of the system. Consequently, the mechanical energy of the system diminishes in time, and the motion is said to be *damped*. The mechanical energy of the system is transformed into internal energy in the object and the retarding medium. Figure 15.20 depicts one such system: an object attached to a spring and submersed in a viscous liquid. Another example is a simple pendulum oscillating in air. After being set into motion, the pendulum eventually stops oscillating due to air resistance. The opening photograph for this chapter depicts damped oscillations in practice. The spring-loaded devices mounted below the bridge are dampers that transform mechanical energy of the oscillating bridge into internal energy.

One common type of retarding force is that discussed in Section 6.4, where the force is proportional to the speed of the moving object and acts in the direction opposite the velocity of the object with respect to the medium. This retarding force is often observed when an object moves through air, for instance. Because the retarding force can be expressed as $\vec{\mathbf{R}} = -b\vec{\mathbf{v}}$ (where b is a constant called the *damping coefficient*) and the restoring force of the system is $-kx$, we can write Newton's second law as

$$\sum F_x = -kx - bv_x = ma_x$$

$$-kx - b\frac{dx}{dt} = m\frac{d^2x}{dt^2} \tag{15.31}$$

The solution to this equation requires mathematics that may be unfamiliar to you; we simply state it here without proof. When the retarding force is small compared with the maximum restoring force—that is, when the damping coefficient b is small—the solution to Equation 15.31 is

$$x = Ae^{-(b/2m)t}\cos(\omega t + \phi) \tag{15.32}$$

where the angular frequency of oscillation is

$$\omega = \sqrt{\frac{k}{m} - \left(\frac{b}{2m}\right)^2} \tag{15.33}$$

This result can be verified by substituting Equation 15.32 into Equation 15.31. It is convenient to express the angular frequency of a damped oscillator in the form

$$\omega = \sqrt{\omega_0^2 - \left(\frac{b}{2m}\right)^2}$$

where $\omega_0 = \sqrt{k/m}$ represents the angular frequency in the absence of a retarding force (the undamped oscillator) and is called the **natural frequency** of the system.

Figure 15.20 One example of a damped oscillator is an object attached to a spring and submersed in a viscous liquid.

Figure 15.21 shows the position as a function of time for an object oscillating in the presence of a retarding force. When the retarding force is small, the oscillatory character of the motion is preserved but the amplitude decreases exponentially in time, with the result that the motion ultimately becomes undetectable. Any system that behaves in this way is known as a **damped oscillator.** The dashed black lines in Figure 15.21, which define the *envelope* of the oscillatory curve, represent the exponential factor in Equation 15.32. This envelope shows that the amplitude decays exponentially with time. For motion with a given spring constant and object mass, the oscillations dampen more rapidly for larger values of the retarding force.

When the magnitude of the retarding force is small such that $b/2m < \omega_0$, the system is said to be **underdamped.** The resulting motion is represented by Figure 15.21 and the the blue curve in Figure 15.22. As the value of b increases, the amplitude of the oscillations decreases more and more rapidly. When b reaches a critical value b_c such that $b_c/2m = \omega_0$, the system does not oscillate and is said to be **critically damped.** In this case, the system, once released from rest at some nonequilibrium position, approaches but does not pass through the equilibrium position. The graph of position versus time for this case is the red curve in Figure 15.22.

If the medium is so viscous that the retarding force is large compared with the restoring force—that is, if $b/2m > \omega_0$—the system is **overdamped.** Again, the displaced system, when free to move, does not oscillate but rather simply returns to its equilibrium position. As the damping increases, the time interval required for the system to approach equilibrium also increases as indicated by the black curve in Figure 15.22. For critically damped and overdamped systems, there is no angular frequency ω and the solution in Equation 15.32 is not valid.

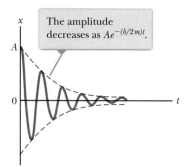

Figure 15.21 Graph of position versus time for a damped oscillator.

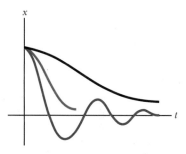

Figure 15.22 Graphs of position versus time for an underdamped oscillator (blue curve), a critically damped oscillator (red curve), and an overdamped oscillator (black curve).

15.7 Forced Oscillations

We have seen that the mechanical energy of a damped oscillator decreases in time as a result of the retarding force. It is possible to compensate for this energy decrease by applying a periodic external force that does positive work on the system. At any instant, energy can be transferred into the system by an applied force that acts in the direction of motion of the oscillator. For example, a child on a swing can be kept in motion by appropriately timed "pushes." The amplitude of motion remains constant if the energy input per cycle of motion exactly equals the decrease in mechanical energy in each cycle that results from retarding forces.

A common example of a forced oscillator is a damped oscillator driven by an external force that varies periodically, such as $F(t) = F_0 \sin \omega t$, where F_0 is a constant and ω is the angular frequency of the driving force. In general, the frequency ω of the driving force is variable, whereas the natural frequency ω_0 of the oscillator is fixed by the values of k and m. Modeling an oscillator with both retarding and driving forces as a particle under a net force, Newton's second law in this situation gives

$$\sum F_x = ma_x \quad \rightarrow \quad F_0 \sin \omega t - b\frac{dx}{dt} - kx = m\frac{d^2x}{dt^2} \qquad \textbf{(15.34)}$$

Again, the solution of this equation is rather lengthy and will not be presented. After the driving force on an initially stationary object begins to act, the amplitude of the oscillation will increase. The system of the oscillator and the surrounding medium is a nonisolated system: work is done by the driving force, such that the vibrational energy of the system (kinetic energy of the object, elastic potential energy in the spring) and internal energy of the object and the medium increase. After a sufficiently long period of time, when the energy input per cycle from the driving force equals the amount of mechanical energy transformed to internal energy for each cycle, a steady-state condition is reached in which the oscillations proceed with constant amplitude. In this situation, the solution of Equation 15.34 is

$$x = A \cos (\omega t + \phi) \qquad \textbf{(15.35)}$$

where

$$A = \frac{F_0/m}{\sqrt{(\omega^2 - \omega_0^2)^2 + \left(\frac{b\omega}{m}\right)^2}} \qquad (15.36)$$

and where $\omega_0 = \sqrt{k/m}$ is the natural frequency of the undamped oscillator ($b = 0$).

Equations 15.35 and 15.36 show that the forced oscillator vibrates at the frequency of the driving force and that the amplitude of the oscillator is constant for a given driving force because it is being driven in steady-state by an external force. For small damping, the amplitude is large when the frequency of the driving force is near the natural frequency of oscillation, or when $\omega \approx \omega_0$. The dramatic increase in amplitude near the natural frequency is called **resonance,** and the natural frequency ω_0 is also called the **resonance frequency** of the system.

The reason for large-amplitude oscillations at the resonance frequency is that energy is being transferred to the system under the most favorable conditions. We can better understand this concept by taking the first time derivative of x in Equation 15.35, which gives an expression for the velocity of the oscillator. We find that v is proportional to $\sin(\omega t + \phi)$, which is the same trigonometric function as that describing the driving force. Therefore, the applied force $\vec{F}$ is in phase with the velocity. The rate at which work is done on the oscillator by $\vec{F}$ equals the dot product $\vec{F} \cdot \vec{v}$; this rate is the power delivered to the oscillator. Because the product $\vec{F} \cdot \vec{v}$ is a maximum when $\vec{F}$ and $\vec{v}$ are in phase, we conclude that at resonance, the applied force is in phase with the velocity and the power transferred to the oscillator is a maximum.

Figure 15.23 is a graph of amplitude as a function of driving frequency for a forced oscillator with and without damping. Notice that the amplitude increases with decreasing damping ($b \to 0$) and that the resonance curve broadens as the damping increases. In the absence of a damping force ($b = 0$), we see from Equation 15.36 that the steady-state amplitude approaches infinity as ω approaches ω_0. In other words, if there are no losses in the system and we continue to drive an initially motionless oscillator with a periodic force that is in phase with the velocity, the amplitude of motion builds without limit (see the red-brown curve in Fig. 15.23). This limitless building does not occur in practice because some damping is always present in reality.

Later in this book we shall see that resonance appears in other areas of physics. For example, certain electric circuits have natural frequencies and can be set into strong resonance by a varying voltage applied at a given frequency. A bridge has natural frequencies that can be set into resonance by an appropriate driving force. A dramatic example of such resonance occurred in 1940 when the Tacoma Narrows Bridge in the state of Washington was destroyed by resonant vibrations. Although the winds were not particularly strong on that occasion, the "flapping" of the wind across the roadway (think of the "flapping" of a flag in a strong wind) provided a periodic driving force whose frequency matched that of the bridge. The resulting oscillations of the bridge caused it to ultimately collapse (Fig. 15.24) because the bridge design had inadequate built-in safety features.

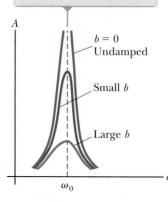

When the frequency ω of the driving force equals the natural frequency ω_0 of the oscillator, resonance occurs.

$b = 0$
Undamped

Small b

Large b

ω_0

ω

Figure 15.23 Graph of amplitude versus frequency for a damped oscillator when a periodic driving force is present. Notice that the shape of the resonance curve depends on the size of the damping coefficient b.

Figure 15.24 (a) In 1940, turbulent winds set up torsional vibrations in the Tacoma Narrows Bridge, causing it to oscillate at a frequency near one of the natural frequencies of the bridge structure. (b) Once established, this resonance condition led to the bridge's collapse. (Mathematicians and physicists are currently challenging some aspects of this interpretation.)

a

b

AP Photos

© Topham/The Image Works

Many other examples of resonant vibrations can be cited. A resonant vibration you may have experienced is the "singing" of telephone wires in the wind. Machines often break if one vibrating part is in resonance with some other moving part. Soldiers marching in cadence across a bridge have been known to set up resonant vibrations in the structure and thereby cause it to collapse. Whenever any real physical system is driven near its resonance frequency, you can expect oscillations of very large amplitudes.

Summary

Concepts and Principles

The kinetic energy and potential energy for an object of mass m oscillating at the end of a spring of force constant k vary with time and are given by

$$K = \tfrac{1}{2}mv^2 = \tfrac{1}{2}m\omega^2 A^2 \sin^2(\omega t + \phi) \quad \textbf{(15.19)}$$

$$U = \tfrac{1}{2}kx^2 = \tfrac{1}{2}kA^2 \cos^2(\omega t + \phi) \quad \textbf{(15.20)}$$

The total energy of a simple harmonic oscillator is a constant of the motion and is given by

$$E = \tfrac{1}{2}kA^2 \quad \textbf{(15.21)}$$

A **simple pendulum** of length L can be modeled to move in simple harmonic motion for small angular displacements from the vertical. Its period is

$$T = 2\pi\sqrt{\frac{L}{g}} \quad \textbf{(15.26)}$$

A **physical pendulum** is an extended object that, for small angular displacements, can be modeled to move in simple harmonic motion about a pivot that does not go through the center of mass. The period of this motion is

$$T = 2\pi\sqrt{\frac{I}{mgd}} \quad \textbf{(15.28)}$$

where I is the moment of inertia of the object about an axis through the pivot and d is the distance from the pivot to the center of mass of the object.

If an oscillator experiences a damping force $\vec{\mathbf{R}} = -b\vec{\mathbf{v}}$, its position for small damping is described by

$$x = Ae^{-(b/2m)t}\cos(\omega t + \phi) \quad \textbf{(15.32)}$$

where

$$\omega = \sqrt{\frac{k}{m} - \left(\frac{b}{2m}\right)^2} \quad \textbf{(15.33)}$$

If an oscillator is subject to a sinusoidal driving force that is described by $F(t) = F_0 \sin \omega t$, it exhibits **resonance**, in which the amplitude is largest when the driving frequency ω matches the natural frequency $\omega_0 = \sqrt{k/m}$ of the oscillator.

Analysis Model for Problem Solving

Particle in Simple Harmonic Motion If a particle is subject to a force of the form of Hooke's law $F = -kx$, the particle exhibits **simple harmonic motion.** Its position is described by

$$x(t) = A\cos(\omega t + \phi) \quad \textbf{(15.6)}$$

where A is the **amplitude** of the motion, ω is the **angular frequency**, and ϕ is the **phase constant.** The value of ϕ depends on the initial position and initial velocity of the particle.

The **period** of the oscillation of the particle is

$$T = \frac{2\pi}{\omega} = 2\pi\sqrt{\frac{m}{k}} \quad \textbf{(15.13)}$$

and the inverse of the period is the **frequency.**

1. If a simple pendulum oscillates with small amplitude and its length is doubled, what happens to the frequency of its motion? (a) It doubles. (b) It becomes $\sqrt{2}$ times as large. (c) It becomes half as large. (d) It becomes $1/\sqrt{2}$ times as large. (e) It remains the same.

2. You attach a block to the bottom end of a spring hanging vertically. You slowly let the block move down and find that it hangs at rest with the spring stretched by 15.0 cm. Next, you lift the block back up to the initial position and release it from rest with the spring unstretched. What maximum distance does it move down? (a) 7.5 cm (b) 15.0 cm (c) 30.0 cm (d) 60.0 cm (e) The distance cannot be determined without knowing the mass and spring constant.

3. A block–spring system vibrating on a frictionless, horizontal surface with an amplitude of 6.0 cm has an energy of 12 J. If the block is replaced by one whose mass is twice the mass of the original block and the amplitude of the motion is again 6.0 cm, what is the energy of the system? (a) 12 J (b) 24 J (c) 6 J (d) 48 J (e) none of those answers

4. An object–spring system moving with simple harmonic motion has an amplitude A. When the kinetic energy of the object equals twice the potential energy stored in the spring, what is the position x of the object? (a) A (b) $\frac{1}{3}A$ (c) $A/\sqrt{3}$ (d) 0 (e) none of those answers

5. An object of mass 0.40 kg, hanging from a spring with a spring constant of 8.0 N/m, is set into an up-and-down simple harmonic motion. What is the magnitude of the acceleration of the object when it is at its maximum displacement of 0.10 m? (a) zero (b) 0.45 m/s² (c) 1.0 m/s² (d) 2.0 m/s² (e) 2.4 m/s²

6. A runaway railroad car, with mass 3.0×10^5 kg, coasts across a level track at 2.0 m/s when it collides elastically with a spring-loaded bumper at the end of the track. If the spring constant of the bumper is 2.0×10^6 N/m, what is the maximum compression of the spring during the collision? (a) 0.77 m (b) 0.58 m (c) 0.34 m (d) 1.07 m (e) 1.24 m

7. The position of an object moving with simple harmonic motion is given by $x = 4 \cos(6\pi t)$, where x is in meters and t is in seconds. What is the period of the oscillating system? (a) 4 s (b) $\frac{1}{6}$ s (c) $\frac{1}{3}$ s (d) 6π s (e) impossible to determine from the information given

8. If an object of mass m attached to a light spring is replaced by one of mass $9m$, the frequency of the vibrating system changes by what factor? (a) $\frac{1}{9}$ (b) $\frac{1}{3}$ (c) 3.0 (d) 9.0 (e) 6.0

9. You stand on the end of a diving board and bounce to set it into oscillation. You find a maximum response in terms of the amplitude of oscillation of the end of the board when you bounce at frequency f. You now move to the middle of the board and repeat the experiment. Is the resonance frequency for forced oscillations at this point (a) higher, (b) lower, or (c) the same as f?

10. A mass–spring system moves with simple harmonic motion along the x axis between turning points at $x_1 = 20$ cm and $x_2 = 60$ cm. For parts (i) through (iii), choose from the same five possibilities. (i) At which position does the particle have the greatest magnitude of momentum? (a) 20 cm (b) 30 cm (c) 40 cm (d) some other position (e) The greatest value occurs at multiple points. (ii) At which position does the particle have greatest kinetic energy? (iii) At which position does the particle-spring system have the greatest total energy?

11. A block with mass $m = 0.1$ kg oscillates with amplitude $A = 0.1$ m at the end of a spring with force constant $k = 10$ N/m on a frictionless, horizontal surface. Rank the periods of the following situations from greatest to smallest. If any periods are equal, show their equality in your ranking. (a) The system is as described above. (b) The system is as described in situation (a) except the amplitude is 0.2 m. (c) The situation is as described in situation (a) except the mass is 0.2 kg. (d) The situation is as described in situation (a) except the spring has force constant 20 N/m. (e) A small resistive force makes the motion underdamped.

12. For a simple harmonic oscillator, answer yes or no to the following questions. (a) Can the quantities position and velocity have the same sign? (b) Can velocity and acceleration have the same sign? (c) Can position and acceleration have the same sign?

13. The top end of a spring is held fixed. A block is hung on the bottom end as in Figure OQ15.13a, and the frequency f of the oscillation of the system is measured. The block, a second identical block, and the spring are carried up in a space shuttle to Earth orbit. The two blocks are attached to the ends of the spring. The spring is compressed without making adjacent coils touch (Fig. OQ15.13b), and the system is released to oscillate while floating within the shuttle cabin (Fig. OQ15.13c). What is the frequency of oscillation for this system in terms of f? (a) $f/2$ (b) $f/\sqrt{2}$ (c) f (d) $\sqrt{2}f$ (e) $2f$

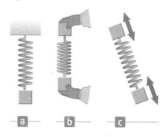

Figure OQ15.13

14. Which of the following statements is *not* true regarding a mass–spring system that moves with simple harmonic motion in the absence of friction? (a) The total energy of the system remains constant. (b) The energy of the system is continually transformed between kinetic and potential energy. (c) The total energy of the system is proportional to the square of the amplitude. (d) The potential energy stored in the system is greatest when the mass passes through the equilibrium position. (e) The velocity of the oscillating mass has its maximum value when the mass passes through the equilibrium position.

15. A simple pendulum has a period of 2.5 s. **(i)** What is its period if its length is made four times larger? (a) 1.25 s (b) 1.77 s (c) 2.5 s (d) 3.54 s (e) 5 s **(ii)** What is its period if the length is held constant at its initial value and the mass of the suspended bob is made four times larger? Choose from the same possibilities.

16. A simple pendulum is suspended from the ceiling of a stationary elevator, and the period is determined. **(i)** When the elevator accelerates upward, is the period (a) greater, (b) smaller, or (c) unchanged? **(ii)** When the elevator has a downward acceleration, is the period (a) greater, (b) smaller, or (c) unchanged? **(iii)** When the elevator moves with constant upward velocity, is the period of the pendulum (a) greater, (b) smaller, or (c) unchanged?

17. A particle on a spring moves in simple harmonic motion along the x axis between turning points at $x_1 = 100$ cm and $x_2 = 140$ cm. **(i)** At which of the following positions does the particle have maximum speed? (a) 100 cm (b) 110 cm (c) 120 cm (d) at none of those positions **(ii)** At which position does it have maximum acceleration? Choose from the same possibilities as in part (i). **(iii)** At which position is the greatest net force exerted on the particle? Choose from the same possibilities as in part (i).

Conceptual Questions

1. denotes answer available in *Student Solutions Manual/Study Guide*

1. You are looking at a small, leafy tree. You do not notice any breeze, and most of the leaves on the tree are motionless. One leaf, however, is fluttering back and forth wildly. After a while, that leaf stops moving and you notice a different leaf moving much more than all the others. Explain what could cause the large motion of one particular leaf.

2. The equations listed together on page 34 give position as a function of time, velocity as a function of time, and velocity as a function of position for an object moving in a straight line with constant acceleration. The quantity v_{xi} appears in every equation. (a) Do any of these equations apply to an object moving in a straight line with simple harmonic motion? (b) Using a similar format, make a table of equations describing simple harmonic motion. Include equations giving acceleration as a function of time and acceleration as a function of position. State the equations in such a form that they apply equally to a block–spring system, to a pendulum, and to other vibrating systems. (c) What quantity appears in every equation?

3. (a) If the coordinate of a particle varies as $x = -A \cos \omega t$, what is the phase constant in Equation 15.6? (b) At what position is the particle at $t = 0$?

4. A pendulum bob is made from a sphere filled with water. What would happen to the frequency of vibration of this pendulum if there were a hole in the sphere that allowed the water to leak out slowly?

5. Figure CQ15.5 shows graphs of the potential energy of four different systems versus the position of a particle in each system. Each particle is set into motion with a push at an arbitrarily chosen location. Describe its subsequent motion in each case (a), (b), (c), and (d).

6. A student thinks that any real vibration must be damped. Is the student correct? If so, give convincing reasoning. If not, give an example of a real vibration that keeps constant amplitude forever if the system is isolated.

7. The mechanical energy of an undamped block–spring system is constant as kinetic energy transforms to elastic potential energy and vice versa. For comparison, explain what happens to the energy of a damped oscillator in terms of the mechanical, potential, and kinetic energies.

8. Is it possible to have damped oscillations when a system is at resonance? Explain.

9. Will damped oscillations occur for any values of b and k? Explain.

10. If a pendulum clock keeps perfect time at the base of a mountain, will it also keep perfect time when it is moved to the top of the mountain? Explain.

11. Is a bouncing ball an example of simple harmonic motion? Is the daily movement of a student from home to school and back simple harmonic motion? Why or why not?

12. A simple pendulum can be modeled as exhibiting simple harmonic motion when θ is small. Is the motion periodic when θ is large?

13. Consider the simplified single-piston engine in Figure CQ15.13. Assuming the wheel rotates with constant angular speed, explain why the piston rod oscillates in simple harmonic motion.

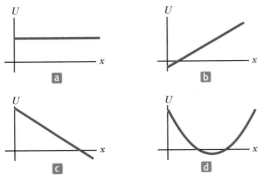

Figure CQ15.5

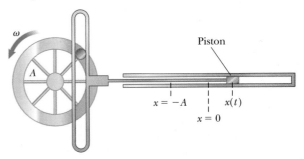

Figure CQ15.13

Section 15.1 Motion of an Object Attached to a Spring
Problems 1–2

Section 15.2 Analysis Model: Particle in Simple Harmonic Motion
Problems 3–20

Section 15.3 Energy of the Simple Harmonic Oscillator
Problems 21–32

Section 15.4 Comparing Simple Harmonic Motion with Uniform Circular Motion
Problem 33

Section 15.5 The Pendulum
Problems 34–45

Section 15.6 Damped Oscillations
Problems 46–49

Section 15.7 Forced Oscillations
Problems 50–55

Additional Problems
Problems 56–83

Challenge Problems
Problems 84–89

Solutions to the following Problems are available in the *Student Solutions Manual/Study Guide:*
15.5, 15.9, 15.12, 15.15, 15.19, 15.21, 15.29, 15.36, 15.37, 15.41, 15.55, 15.59, 15.65, 15.67, 15.72, 15.82, and 15.84

List of Enhanced Problems

Problem Number	Targeted Feedback in Enhanced WebAssign	Analysis Model Tutorial in Enhanced WebAssign	Master It in Enhanced WebAssign	Watch It in Enhanced WebAssign
15.3			✓	
15.4	✓			✓
15.5	✓		✓	
15.8	✓			✓
15.9	✓	✓		✓
15.10	✓		✓	
15.18	✓			✓
15.19	✓		✓	
15.21		✓	✓	
15.27	✓			✓
15.32		✓		
15.37	✓		✓	
15.43	✓			✓
15.46	✓			✓
15.55			✓	
15.59			✓	
15.63	✓		✓	
15.65			✓	
15.75	✓	✓		✓

Wave Motion

Many of us experienced waves as children when we dropped a pebble into a pond. At the point the pebble hits the water's surface, circular waves are created. These waves move outward from the creation point in expanding circles until they reach the shore. If you were to examine carefully the motion of a small object floating on the disturbed water, you would see that the object moves vertically and horizontally about its original position but does not undergo any net displacement away from or toward the point at which the pebble hit the water. The small elements of water in contact with the object, as well as all the other water elements on the pond's surface, behave in the same way. That is, the water *wave* moves from the point of origin to the shore, but the water is not carried with it.

The world is full of waves, the two main types being *mechanical* waves and *electromagnetic* waves. In the case of mechanical waves, some physical medium is being disturbed; in our pebble example, elements of water are disturbed. Electromagnetic waves do not require a medium to propagate; some examples of electromagnetic waves are visible light, radio waves, television signals, and x-rays. Here, in this part of the book, we study only mechanical waves.

Consider again the small object floating on the water. We have caused the object to move at one point in the water by dropping a pebble at another location. The object has gained kinetic energy from our action, so energy must have transferred from the point at

Lifeguards in New South Wales, Australia, practice taking their boat over large water waves breaking near the shore. A wave moving over the surface of water is one example of a mechanical wave. *(Travel Ink/Gallo Images/Getty Images)*

377

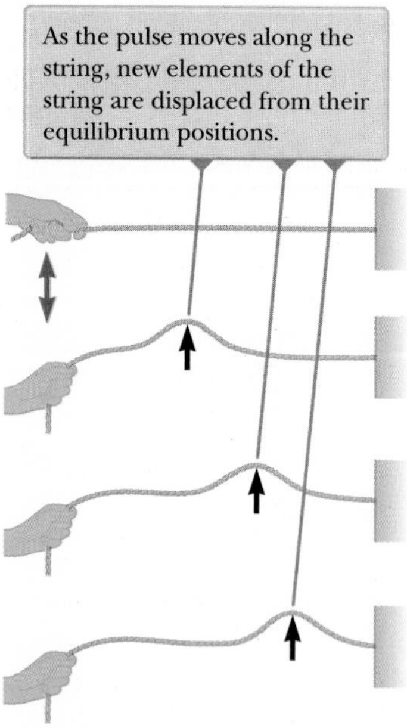

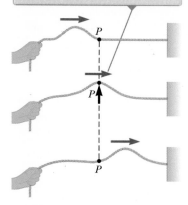

Figure 16.1 A hand moves the end of a stretched string up and down once (red arrow), causing a pulse to travel along the string.

Figure 16.2 The displacement of a particular string element for a transverse pulse traveling on a stretched string.

Figure 16.3 A longitudinal pulse along a stretched spring.

which the pebble is dropped to the position of the object. This feature is central to wave motion: *energy* is transferred over a distance, but *matter* is not.

16.1 Propagation of a Disturbance

The introduction to this chapter alluded to the essence of wave motion: the transfer of energy through space without the accompanying transfer of matter. In the list of energy transfer mechanisms in Chapter 8, two mechanisms—mechanical waves and electromagnetic radiation—depend on waves. By contrast, in another mechanism, matter transfer, the energy transfer is accompanied by a movement of matter through space with no wave character in the process.

All mechanical waves require (1) some source of disturbance, (2) a medium containing elements that can be disturbed, and (3) some physical mechanism through which elements of the medium can influence each other. One way to demonstrate wave motion is to flick one end of a long string that is under tension and has its opposite end fixed as shown in Figure 16.1. In this manner, a single bump (called a *pulse*) is formed and travels along the string with a definite speed. Figure 16.1 represents four consecutive "snapshots" of the creation and propagation of the traveling pulse. The hand is the source of the disturbance. The string is the medium through which the pulse travels—individual elements of the string are disturbed from their equilibrium position. Furthermore, the elements of the string are connected together so they influence each other. The pulse has a definite height and a definite speed of propagation along the medium. The shape of the pulse changes very little as it travels along the string.[1]

We shall first focus on a pulse traveling through a medium. Once we have explored the behavior of a pulse, we will then turn our attention to a *wave*, which is a *periodic* disturbance traveling through a medium. We create a pulse on our string by flicking the end of the string once as in Figure 16.1. If we were to move the end of the string up and down repeatedly, we would create a traveling wave, which has characteristics a pulse does not have. We shall explore these characteristics in Section 16.2.

As the pulse in Figure 16.1 travels, each disturbed element of the string moves in a direction *perpendicular* to the direction of propagation. Figure 16.2 illustrates this point for one particular element, labeled P. Notice that no part of the string ever moves in the direction of the propagation. A traveling wave or pulse that causes the elements of the disturbed medium to move perpendicular to the direction of propagation is called a **transverse wave.**

Compare this wave with another type of pulse, one moving down a long, stretched spring as shown in Figure 16.3. The left end of the spring is pushed briefly to the right and then pulled briefly to the left. This movement creates a sudden compression of a region of the coils. The compressed region travels along the spring (to the right in Fig. 16.3). Notice that the direction of the displacement of the coils is *parallel* to the direction of propagation of the compressed region. A traveling wave or pulse that causes the elements of the medium to move parallel to the direction of propagation is called a **longitudinal wave.**

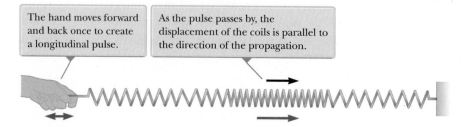

---

[1]In reality, the pulse changes shape and gradually spreads out during the motion. This effect, called *dispersion,* is common to many mechanical waves as well as to electromagnetic waves. We do not consider dispersion in this chapter.

Sound waves, which we shall discuss in Chapter 17, are another example of longitudinal waves. The disturbance in a sound wave is a series of high-pressure and low-pressure regions that travel through air.

Some waves in nature exhibit a combination of transverse and longitudinal displacements. Surface-water waves are a good example. When a water wave travels on the surface of deep water, elements of water at the surface move in nearly circular paths as shown in Figure 16.4. The disturbance has both transverse and longitudinal components. The transverse displacements seen in Figure 16.4 represent the variations in vertical position of the water elements. The longitudinal displacements represent elements of water moving back and forth in a horizontal direction.

The three-dimensional waves that travel out from a point under the Earth's surface at which an earthquake occurs are of both types, transverse and longitudinal. The longitudinal waves are the faster of the two, traveling at speeds in the range of 7 to 8 km/s near the surface. They are called **P waves,** with "P" standing for *primary*, because they travel faster than the transverse waves and arrive first at a seismograph (a device used to detect waves due to earthquakes). The slower transverse waves, called **S waves,** with "S" standing for *secondary*, travel through the Earth at 4 to 5 km/s near the surface. By recording the time interval between the arrivals of these two types of waves at a seismograph, the distance from the seismograph to the point of origin of the waves can be determined. This distance is the radius of an imaginary sphere centered on the seismograph. The origin of the waves is located somewhere on that sphere. The imaginary spheres from three or more monitoring stations located far apart from one another intersect at one region of the Earth, and this region is where the earthquake occurred.

Consider a pulse traveling to the right on a long string as shown in Figure 16.5. Figure 16.5a represents the shape and position of the pulse at time $t = 0$. At this time, the shape of the pulse, whatever it may be, can be represented by some mathematical function that we will write as $y(x, 0) = f(x)$. This function describes the transverse position y of the element of the string located at each value of x at time $t = 0$. Because the speed of the pulse is v, the pulse has traveled to the right a distance vt at the time t (Fig. 16.5b). We assume the shape of the pulse does not change with time. Therefore, at time t, the shape of the pulse is the same as it was at time $t = 0$ as in Figure 16.5a. Consequently, an element of the string at x at this time has the same y position as an element located at $x - vt$ had at time $t = 0$:

$$y(x, t) = y(x - vt, 0)$$

In general, then, we can represent the transverse position y for all positions and times, measured in a stationary frame with the origin at O, as

$$y(x, t) = f(x - vt) \tag{16.1}$$

Similarly, if the pulse travels to the left, the transverse positions of elements of the string are described by

$$y(x, t) = f(x + vt) \tag{16.2}$$

The function y, sometimes called the **wave function,** depends on the two variables x and t. For this reason, it is often written $y(x, t)$, which is read "y as a function of x and t."

It is important to understand the meaning of y. Consider an element of the string at point P in Figure 16.5, identified by a particular value of its x coordinate. As the pulse passes through P, the y coordinate of this element increases, reaches a maximum, and then decreases to zero. The wave function $y(x, t)$ represents the y coordinate—the transverse position—of any element located at position x at any time t. Furthermore, if t is fixed (as, for example, in the case of taking a snapshot of the pulse), the wave function $y(x)$, sometimes called the **waveform,** defines a curve representing the geometric shape of the pulse at that time.

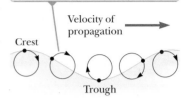

The elements at the surface move in nearly circular paths. Each element is displaced both horizontally and vertically from its equilibrium position.

Figure 16.4 The motion of water elements on the surface of deep water in which a wave is propagating is a combination of transverse and longitudinal displacements.

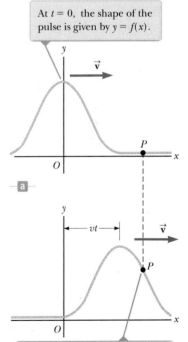

At $t = 0$, the shape of the pulse is given by $y = f(x)$.

At some later time t, the shape of the pulse remains unchanged and the vertical position of an element of the medium at any point P is given by $y = f(x - vt)$.

Figure 16.5 A one-dimensional pulse traveling to the right on a string with a speed v.

Quick Quiz 16.1 **(i)** In a long line of people waiting to buy tickets, the first person leaves and a pulse of motion occurs as people step forward to fill the gap. As each person steps forward, the gap moves through the line. Is the propagation of this gap (a) transverse or (b) longitudinal? **(ii)** Consider "the wave" at a baseball game: people stand up and raise their arms as the wave arrives at their location, and the resultant pulse moves around the stadium. Is this wave (a) transverse or (b) longitudinal?

Example 16.1 | A Pulse Moving to the Right

A pulse moving to the right along the x axis is represented by the wave function

$$y(x, t) = \frac{2}{(x - 3.0t)^2 + 1}$$

where x and y are measured in centimeters and t is measured in seconds. Find expressions for the wave function at $t = 0$, $t = 1.0$ s, and $t = 2.0$ s.

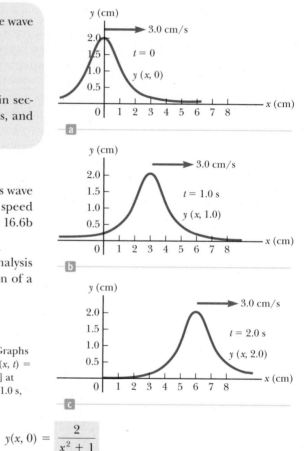

SOLUTION

Conceptualize Figure 16.6a shows the pulse represented by this wave function at $t = 0$. Imagine this pulse moving to the right at a speed of 3.0 cm/s and maintaining its shape as suggested by Figures 16.6b and 16.6c.

Categorize We categorize this example as a relatively simple analysis problem in which we interpret the mathematical representation of a pulse.

Analyze The wave function is of the form $y = f(x - vt)$. Inspection of the expression for $y(x, t)$ and comparison to Equation 16.1 reveal that the wave speed is $v = 3.0$ cm/s. Furthermore, by letting $x - 3.0t = 0$, we find that the maximum value of y is given by $A = 2.0$ cm.

Figure 16.6
(Example 16.1) Graphs of the function $y(x, t) = 2/[(x - 3.0t)^2 + 1]$ at (a) $t = 0$, (b) $t = 1.0$ s, and (c) $t = 2.0$ s.

Write the wave function expression at $t = 0$:
$$y(x, 0) = \frac{2}{x^2 + 1}$$

Write the wave function expression at $t = 1.0$ s:
$$y(x, 1.0) = \frac{2}{(x - 3.0)^2 + 1}$$

Write the wave function expression at $t = 2.0$ s:
$$y(x, 2.0) = \frac{2}{(x - 6.0)^2 + 1}$$

For each of these expressions, we can substitute various values of x and plot the wave function. This procedure yields the wave functions shown in the three parts of Figure 16.6.

Finalize These snapshots show that the pulse moves to the right without changing its shape and that it has a constant speed of 3.0 cm/s.

WHAT IF? What if the wave function were

$$y(x, t) = \frac{4}{(x + 3.0t)^2 + 1}$$

How would that change the situation?

Answer One new feature in this expression is the plus sign in the denominator rather than the minus sign. The new expression represents a pulse with a similar shape as that in Figure 16.6, but moving to the left as time progresses.

▶ **16.1** continued

Another new feature here is the numerator of 4 rather than 2. Therefore, the new expression represents a pulse with twice the height of that in Figure 16.6.

16.2 Analysis Model: Traveling Wave

In this section, we introduce an important wave function whose shape is shown in Figure 16.7. The wave represented by this curve is called a **sinusoidal wave** because the curve is the same as that of the function sin θ plotted against θ. A sinusoidal wave could be established on the rope in Figure 16.1 by shaking the end of the rope up and down in simple harmonic motion.

The sinusoidal wave is the simplest example of a periodic continuous wave and can be used to build more complex waves (see Section 18.8). The brown curve in Figure 16.7 represents a snapshot of a traveling sinusoidal wave at $t = 0$, and the blue curve represents a snapshot of the wave at some later time t. Imagine two types of motion that can occur. First, the entire waveform in Figure 16.7 moves to the right so that the brown curve moves toward the right and eventually reaches the position of the blue curve. This movement is the motion of the *wave*. If we focus on one element of the medium, such as the element at $x = 0$, we see that each element moves up and down along the y axis in simple harmonic motion. This movement is the motion of the *elements of the medium*. It is important to differentiate between the motion of the wave and the motion of the elements of the medium.

In the early chapters of this book, we developed several analysis models based on three simplification models: the particle, the system, and the rigid object. With our introduction to waves, we can develop a new simplification model, the **wave**, that will allow us to explore more analysis models for solving problems. An ideal particle has zero size. We can build physical objects with nonzero size as combinations of particles. Therefore, the particle can be considered a basic building block. An ideal wave has a single frequency and is infinitely long; that is, the wave exists throughout the Universe. (A wave of finite length must necessarily have a mixture of frequencies.) When this concept is explored in Section 18.8, we will find that ideal waves can be combined to build complex waves, just as we combined particles.

In what follows, we will develop the principal features and mathematical representations of the analysis model of a **traveling wave.** This model is used in situations in which a wave moves through space without interacting with other waves or particles.

Figure 16.8a shows a snapshot of a traveling wave moving through a medium. Figure 16.8b shows a graph of the position of one element of the medium as a function of time. A point in Figure 16.8a at which the displacement of the element from its normal position is highest is called the **crest** of the wave. The lowest point is called the **trough.** The distance from one crest to the next is called the **wavelength** λ (Greek letter lambda). More generally, the wavelength is the minimum distance between any two identical points on adjacent waves as shown in Figure 16.8a.

If you count the number of seconds between the arrivals of two adjacent crests at a given point in space, you measure the **period** T of the waves. In general, the period is the time interval required for two identical points of adjacent waves to pass by a point as shown in Figure 16.8b. The period of the wave is the same as the period of the simple harmonic oscillation of one element of the medium.

The same information is more often given by the inverse of the period, which is called the **frequency** f. In general, the frequency of a periodic wave is the number of crests (or troughs, or any other point on the wave) that pass a given point in a unit time interval. The frequency of a sinusoidal wave is related to the period by the expression

$$f = \frac{1}{T} \tag{16.3}$$

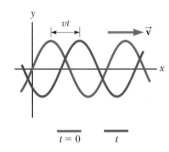

Figure 16.7 A one-dimensional sinusoidal wave traveling to the right with a speed v. The brown curve represents a snapshot of the wave at $t = 0$, and the blue curve represents a snapshot at some later time t.

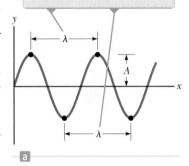

The wavelength λ of a wave is the distance between adjacent crests or adjacent troughs.

The period T of a wave is the time interval required for the element to complete one cycle of its oscillation and for the wave to travel one wavelength.

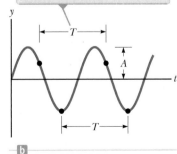

Figure 16.8 (a) A snapshot of a sinusoidal wave. (b) The position of one element of the medium as a function of time.

The frequency of the wave is the same as the frequency of the simple harmonic oscillation of one element of the medium. The most common unit for frequency, as we learned in Chapter 15, is s^{-1}, or **hertz** (Hz). The corresponding unit for T is seconds.

The maximum position of an element of the medium relative to its equilibrium position is called the **amplitude** A of the wave as indicated in Figure 16.8.

Waves travel with a specific speed, and this speed depends on the properties of the medium being disturbed. For instance, sound waves travel through room-temperature air with a speed of about 343 m/s (781 mi/h), whereas they travel through most solids with a speed greater than 343 m/s.

Consider the sinusoidal wave in Figure 16.8a, which shows the position of the wave at $t = 0$. Because the wave is sinusoidal, we expect the wave function at this instant to be expressed as $y(x, 0) = A \sin ax$, where A is the amplitude and a is a constant to be determined. At $x = 0$, we see that $y(0, 0) = A \sin a(0) = 0$, consistent with Figure 16.8a. The next value of x for which y is zero is $x = \lambda/2$. Therefore,

$$y\left(\frac{\lambda}{2}, 0\right) = A \sin\left(a\frac{\lambda}{2}\right) = 0$$

For this equation to be true, we must have $a\lambda/2 = \pi$, or $a = 2\pi/\lambda$. Therefore, the function describing the positions of the elements of the medium through which the sinusoidal wave is traveling can be written

$$y(x, 0) = A \sin\left(\frac{2\pi}{\lambda} x\right) \tag{16.4}$$

where the constant A represents the wave amplitude and the constant λ is the wavelength. Notice that the vertical position of an element of the medium is the same whenever x is increased by an integral multiple of λ. Based on our discussion of Equation 16.1, if the wave moves to the right with a speed v, the wave function at some later time t is

$$y(x, t) = A \sin\left[\frac{2\pi}{\lambda}(x - vt)\right] \tag{16.5}$$

If the wave were traveling to the left, the quantity $x - vt$ would be replaced by $x + vt$ as we learned when we developed Equations 16.1 and 16.2.

By definition, the wave travels through a displacement Δx equal to one wavelength λ in a time interval Δt of one period T. Therefore, the wave speed, wavelength, and period are related by the expression

$$v = \frac{\Delta x}{\Delta t} = \frac{\lambda}{T} \tag{16.6}$$

Substituting this expression for v into Equation 16.5 gives

$$y = A \sin\left[2\pi\left(\frac{x}{\lambda} - \frac{t}{T}\right)\right] \tag{16.7}$$

This form of the wave function shows the *periodic* nature of y. Note that we will often use y rather than $y(x, t)$ as a shorthand notation. At any given time t, y has the *same* value at the positions x, $x + \lambda$, $x + 2\lambda$, and so on. Furthermore, at any given position x, the value of y is the same at times t, $t + T$, $t + 2T$, and so on.

We can express the wave function in a convenient form by defining two other quantities, the **angular wave number** k (usually called simply the **wave number**) and the **angular frequency** ω:

Angular wave number ▶

$$k \equiv \frac{2\pi}{\lambda} \tag{16.8}$$

Angular frequency ▶

$$\omega \equiv \frac{2\pi}{T} = 2\pi f \tag{16.9}$$

Using these definitions, Equation 16.7 can be written in the more compact form

$$y = A \sin (kx - \omega t) \qquad (16.10)$$

◀ **Wave function for a sinusoidal wave**

Using Equations 16.3, 16.8, and 16.9, the wave speed v originally given in Equation 16.6 can be expressed in the following alternative forms:

$$v = \frac{\omega}{k} \qquad (16.11)$$

$$v = \lambda f \qquad (16.12)$$

◀ **Speed of a sinusoidal wave**

The wave function given by Equation 16.10 assumes the vertical position y of an element of the medium is zero at $x = 0$ and $t = 0$. That need not be the case. If it is not, we generally express the wave function in the form

$$y = A \sin (kx - \omega t + \phi) \qquad (16.13)$$

◀ **General expression for a sinusoidal wave**

where ϕ is the **phase constant,** just as we learned in our study of periodic motion in Chapter 15. This constant can be determined from the initial conditions. The primary equations in the mathematical representation of the traveling wave analysis model are Equations 16.3, 16.10, and 16.12.

Quick Quiz 16.2 A sinusoidal wave of frequency f is traveling along a stretched
string. The string is brought to rest, and a second traveling wave of frequency
$2f$ is established on the string. **(i)** What is the wave speed of the second wave?
(a) twice that of the first wave **(b)** half that of the first wave **(c)** the same as
that of the first wave **(d)** impossible to determine **(ii)** From the same choices,
describe the wavelength of the second wave. **(iii)** From the same choices,
describe the amplitude of the second wave.

Example 16.2   **A Traveling Sinusoidal Wave**   AM

A sinusoidal wave traveling in the positive x direction has an amplitude of 15.0 cm, a wavelength of 40.0 cm, and a frequency of 8.00 Hz. The vertical position of an element of the medium at $t = 0$ and $x = 0$ is also 15.0 cm as shown in Figure 16.9.

(A) Find the wave number k, period T, angular frequency ω, and speed v of the wave.

SOLUTION

Conceptualize Figure 16.9 shows the wave at $t = 0$.
Imagine this wave moving to the right and maintaining its shape.

Categorize From the description in the problem statement, we see that we are analyzing a mechanical wave moving through a medium, so we categorize the problem with the *traveling wave* model.

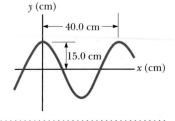

Figure 16.9 (Example 16.2) A sinusoidal wave of wavelength $\lambda = 40.0$ cm and amplitude $A = 15.0$ cm.

Analyze

Evaluate the wave number from Equation 16.8:

$$k = \frac{2\pi}{\lambda} = \frac{2\pi \text{ rad}}{40.0 \text{ cm}} = \boxed{15.7 \text{ rad/m}}$$

Evaluate the period of the wave from Equation 16.3:

$$T = \frac{1}{f} = \frac{1}{8.00 \text{ s}^{-1}} = \boxed{0.125 \text{ s}}$$

Evaluate the angular frequency of the wave from Equation 16.9:

$$\omega = 2\pi f = 2\pi(8.00 \text{ s}^{-1}) = \boxed{50.3 \text{ rad/s}}$$

Evaluate the wave speed from Equation 16.12:

$$v = \lambda f = (40.0 \text{ cm})(8.00 \text{ s}^{-1}) = \boxed{3.20 \text{ m/s}}$$

continued

▶ **16.2** continued

(B) Determine the phase constant ϕ and write a general expression for the wave function.

SOLUTION

Substitute $A = 15.0$ cm, $y = 15.0$ cm, $x = 0$, and $t = 0$ into Equation 16.13:

$$15.0 = (15.0) \sin \phi \rightarrow \sin \phi = 1 \rightarrow \phi = \frac{\pi}{2} \text{ rad}$$

Write the wave function:

$$y = A \sin\left(kx - \omega t + \frac{\pi}{2}\right) = A \cos(kx - \omega t)$$

Substitute the values for A, k, and ω in SI units into this expression:

$$y = \boxed{0.150 \cos(15.7x - 50.3t)}$$

Finalize Review the results carefully and make sure you understand them. How would the graph in Figure 16.9 change if the phase angle were zero? How would the graph change if the amplitude were 30.0 cm? How would the graph change if the wavelength were 10.0 cm?

Sinusoidal Waves on Strings

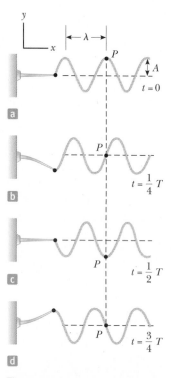

Figure 16.10 One method for producing a sinusoidal wave on a string. The left end of the string is connected to a blade that is set into oscillation. Every element of the string, such as that at point P, oscillates with simple harmonic motion in the vertical direction.

In Figure 16.1, we demonstrated how to create a pulse by jerking a taut string up and down once. To create a series of such pulses—a wave—let's replace the hand with an oscillating blade vibrating in simple harmonic motion. Figure 16.10 represents snapshots of the wave created in this way at intervals of $T/4$. Because the end of the blade oscillates in simple harmonic motion, each element of the string, such as that at P, also oscillates vertically with simple harmonic motion. Therefore, every element of the string can be treated as a simple harmonic oscillator vibrating with a frequency equal to the frequency of oscillation of the blade.[2] Notice that while each element oscillates in the y direction, the wave travels to the right in the $+x$ direction with a speed v. Of course, that is the definition of a transverse wave.

If we define $t = 0$ as the time for which the configuration of the string is as shown in Figure 16.10a, the wave function can be written as

$$y = A \sin(kx - \omega t)$$

We can use this expression to describe the motion of any element of the string. An element at point P (or any other element of the string) moves only vertically, and so its x coordinate remains constant. Therefore, the **transverse speed** v_y (not to be confused with the wave speed v) and the **transverse acceleration** a_y of elements of the string are

$$v_y = \frac{dy}{dt}\bigg]_{x=\text{constant}} = \frac{\partial y}{\partial t} = -\omega A \cos(kx - \omega t) \tag{16.14}$$

$$a_y = \frac{dv_y}{dt}\bigg]_{x=\text{constant}} = \frac{\partial v_y}{\partial t} = -\omega^2 A \sin(kx - \omega t) \tag{16.15}$$

These expressions incorporate partial derivatives because y depends on both x and t. In the operation $\partial y/\partial t$, for example, we take a derivative with respect to t while holding x constant. The maximum magnitudes of the transverse speed and transverse acceleration are simply the absolute values of the coefficients of the cosine and sine functions:

$$v_{y,\text{max}} = \omega A \tag{16.16}$$

$$a_{y,\text{max}} = \omega^2 A \tag{16.17}$$

The transverse speed and transverse acceleration of elements of the string do not reach their maximum values simultaneously. The transverse speed reaches its maximum value (ωA) when $y = 0$, whereas the magnitude of the transverse acceleration

---

[2]In this arrangement, we are assuming that a string element always oscillates in a vertical line. The tension in the string would vary if an element were allowed to move sideways. Such motion would make the analysis very complex.

reaches its maximum value ($\omega^2 A$) when $y = \pm A$. Finally, Equations 16.16 and 16.17 are identical in mathematical form to the corresponding equations for simple harmonic motion, Equations 15.17 and 15.18.

> **Q**uick Quiz 16.3 The amplitude of a wave is doubled, with no other changes made to the wave. As a result of this doubling, which of the following statements is correct? **(a)** The speed of the wave changes. **(b)** The frequency of the wave changes. **(c)** The maximum transverse speed of an element of the medium changes. **(d)** Statements **(a)** through **(c)** are all true. **(e)** None of statements **(a)** through **(c)** is true.

> **Pitfall Prevention 16.2**
>
> **Two Kinds of Speed/Velocity**
> Do not confuse v, the speed of the wave as it propagates along the string, with v_y, the transverse velocity of a point on the string. The speed v is constant for a uniform medium, whereas v_y varies sinusoidally.

Analysis Model · Traveling Wave

Imagine a source vibrating such that it influences the medium that is in contact with the source. Such a source creates a disturbance that propagates through the medium. If the source vibrates in simple harmonic motion with period T, sinusoidal waves propagate through the medium at a speed given by

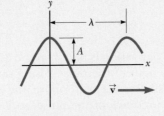

$$v = \frac{\lambda}{T} = \lambda f \qquad (16.6, 16.12)$$

where λ is the **wavelength** of the wave and f is its **frequency**. A sinusoidal wave can be expressed as

$$y = A \sin (kx - \omega t) \qquad (16.10)$$

where A is the **amplitude** of the wave, k is its **wave number**, and ω is its **angular frequency.**

Examples:

- a vibrating blade sends a sinusoidal wave down a string attached to the blade
- a loudspeaker vibrates back and forth, emitting sound waves into the air (Chapter 17)
- a guitar body vibrates, emitting sound waves into the air (Chapter 18)
- a vibrating electric charge creates an electromagnetic wave that propagates into space at the speed of light (Chapter 34)

16.3 The Speed of Waves on Strings

One aspect of the behavior of *linear* mechanical waves is that the wave speed depends only on the properties of the medium through which the wave travels. Waves for which the amplitude A is small relative to the wavelength λ can be represented as linear waves. (See Section 16.6.) In this section, we determine the speed of a transverse wave traveling on a stretched string.

Let us use a mechanical analysis to derive the expression for the speed of a pulse traveling on a stretched string under tension T. Consider a pulse moving to the right with a uniform speed v, measured relative to a stationary (with respect to the Earth) inertial reference frame as shown in Figure 16.11a. Newton's laws are valid in any inertial reference frame. Therefore, let us view this pulse from a different inertial reference frame, one that moves along with the pulse at the same speed so that the pulse appears to be at rest in the frame as in Figure 16.11b. In this reference frame, the pulse remains fixed and each element of the string moves to the left through the pulse shape.

A short element of the string, of length Δs, forms an approximate arc of a circle of radius R as shown in the magnified view in Figure 16.11b. In our moving frame of reference, the element of the string moves to the left with speed v. As it travels through the arc, we can model the element as a particle in uniform circular motion. This element has a centripetal acceleration of v^2/R, which is supplied by components of the force $\vec{T}$ whose magnitude is the tension in the string. The force $\vec{T}$ acts on each side of the element, tangent to the arc, as in Figure 16.11b. The horizontal components of $\vec{T}$ cancel, and each vertical component $T \sin \theta$ acts downward. Hence, the magnitude of the total radial force on the element is $2T \sin \theta$.

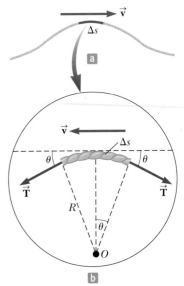

Figure 16.11 (a) In the reference frame of the Earth, a pulse moves to the right on a string with speed v. (b) In a frame of reference moving to the right with the pulse, the small element of length Δs moves to the left with speed v.

Because the element is small, θ is small and we can use the small-angle approximation $\sin \theta \approx \theta$. Therefore, the magnitude of the total radial force is

$$F_r = 2T \sin \theta \approx 2T\theta$$

The element has mass $m = \mu \Delta s$, where μ is the mass per unit length of the string. Because the element forms part of a circle and subtends an angle of 2θ at the center, $\Delta s = R(2\theta)$, and

$$m = \mu \Delta s = 2\mu R\theta$$

The element of the string is modeled as a particle under a net force. Therefore, applying Newton's second law to this element in the radial direction gives

$$F_r = \frac{mv^2}{R} \quad \rightarrow \quad 2T\theta = \frac{2\mu R\theta v^2}{R} \quad \rightarrow \quad T = \mu v^2$$

Solving for v gives

▶ **Speed of a wave on a stretched string**

$$v = \sqrt{\frac{T}{\mu}} \tag{16.18}$$

Notice that this derivation is based on the assumption that the pulse height is small relative to the length of the pulse. Using this assumption, we were able to use the approximation $\sin \theta \approx \theta$. Furthermore, the model assumes that the tension T is not affected by the presence of the pulse, so T is the same at all points on the pulse. Finally, this proof does *not* assume any particular shape for the pulse. We therefore conclude that a pulse of *any shape* will travel on the string with speed $v = \sqrt{T/\mu}$, without any change in pulse shape.

Pitfall Prevention 16.3

Multiple T's Do not confuse the T in Equation 16.18 for the tension with the symbol T used in this chapter for the period of a wave. The context of the equation should help you identify which quantity is meant. There simply aren't enough letters in the alphabet to assign a unique letter to each variable!

Quick Quiz 16.4 Suppose you create a pulse by moving the free end of a taut string up and down once with your hand beginning at $t = 0$. The string is attached at its other end to a distant wall. The pulse reaches the wall at time t. Which of the following actions, taken by itself, decreases the time interval required for the pulse to reach the wall? More than one choice may be correct. **(a)** moving your hand more quickly, but still only up and down once by the same amount **(b)** moving your hand more slowly, but still only up and down once by the same amount **(c)** moving your hand a greater distance up and down in the same amount of time **(d)** moving your hand a lesser distance up and down in the same amount of time **(e)** using a heavier string of the same length and under the same tension **(f)** using a lighter string of the same length and under the same tension **(g)** using a string of the same linear mass density but under decreased tension **(h)** using a string of the same linear mass density but under increased tension

Example 16.3 The Speed of a Pulse on a Cord   AM

A uniform string has a mass of 0.300 kg and a length of 6.00 m (Fig. 16.12). The string passes over a pulley and supports a 2.00-kg object. Find the speed of a pulse traveling along this string.

SOLUTION

Conceptualize In Figure 16.12, the hanging block establishes a tension in the horizontal string. This tension determines the speed with which waves move on the string.

Categorize To find the tension in the string, we model the hanging block as a *particle in equilibrium*. Then we use the tension to evaluate the wave speed on the string using Equation 16.18.

Figure 16.12 (Example 16.3) The tension T in the cord is maintained by the suspended object. The speed of any wave traveling along the cord is given by $v = \sqrt{T/\mu}$.

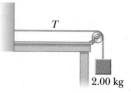

Analyze Apply the particle in equilibrium model to the block:

$$\sum F_y = T - m_{block}g = 0$$

Solve for the tension in the string:

$$T = m_{block}g$$

▶ **16.3** continued

Use Equation 16.18 to find the wave speed, using $\mu = m_{string}/\ell$ for the linear mass density of the string:

$$v = \sqrt{\frac{T}{\mu}} = \sqrt{\frac{m_{block}\,g\ell}{m_{string}}}$$

Evaluate the wave speed:

$$v = \sqrt{\frac{(2.00 \text{ kg})(9.80 \text{ m/s}^2)(6.00 \text{ m})}{0.300 \text{ kg}}} = \boxed{19.8 \text{ m/s}}$$

Finalize The calculation of the tension neglects the small mass of the string. Strictly speaking, the string can never be exactly straight; therefore, the tension is not uniform.

WHAT IF? What if the block were swinging back and forth with respect to the vertical like a pendulum? How would that affect the wave speed on the string?

Answer The swinging block is categorized as a *particle under a net force*. The magnitude of one of the forces on the block is the tension in the string, which determines the wave speed. As the block swings, the tension changes, so the wave speed changes.

When the block is at the bottom of the swing, the string is vertical and the tension is larger than the weight of the block because the net force must be upward to provide the centripetal acceleration of the block. Therefore, the wave speed must be greater than 19.8 m/s.

When the block is at its highest point at the end of a swing, it is momentarily at rest, so there is no centripetal acceleration at that instant. The block is a particle in equilibrium in the radial direction. The tension is balanced by a component of the gravitational force on the block. Therefore, the tension is smaller than the weight and the wave speed is less than 19.8 m/s. With what frequency does the speed of the wave vary? Is it the same frequency as the pendulum?

Example 16.4 **Rescuing the Hiker**

An 80.0-kg hiker is trapped on a mountain ledge following a storm. A helicopter rescues the hiker by hovering above him and lowering a cable to him. The mass of the cable is 8.00 kg, and its length is 15.0 m. A sling of mass 70.0 kg is attached to the end of the cable. The hiker attaches himself to the sling, and the helicopter then accelerates upward. Terrified by hanging from the cable in midair, the hiker tries to signal the pilot by sending transverse pulses up the cable. A pulse takes 0.250 s to travel the length of the cable. What is the acceleration of the helicopter? Assume the tension in the cable is uniform.

SOLUTION

Conceptualize Imagine the effect of the acceleration of the helicopter on the cable. The greater the upward acceleration, the larger the tension in the cable. In turn, the larger the tension, the higher the speed of pulses on the cable.

Categorize This problem is a combination of one involving the speed of pulses on a string and one in which the hiker and sling are modeled as a *particle under a net force*.

Analyze Use the time interval for the pulse to travel from the hiker to the helicopter to find the speed of the pulses on the cable:

$$v = \frac{\Delta x}{\Delta t} = \frac{15.0 \text{ m}}{0.250 \text{ s}} = 60.0 \text{ m/s}$$

Solve Equation 16.18 for the tension in the cable:

$$(1) \quad v = \sqrt{\frac{T}{\mu}} \;\rightarrow\; T = \mu v^2$$

Model the hiker and sling as a particle under a net force, noting that the acceleration of this particle of mass m is the same as the acceleration of the helicopter:

$$\sum F = ma \;\rightarrow\; T - mg = ma$$

Solve for the acceleration and substitute the tension from Equation (1):

$$a = \frac{T}{m} - g = \frac{\mu v^2}{m} - g = \frac{m_{cable}\,v^2}{\ell_{cable}\,m} - g$$

continued

▶ **16.4** continued

Substitute numerical values:

$$a = \frac{(8.00 \text{ kg})(60.0 \text{ m/s})^2}{(15.0 \text{ m})(150.0 \text{ kg})} - 9.80 \text{ m/s}^2 = \boxed{3.00 \text{ m/s}^2}$$

Finalize A real cable has stiffness in addition to tension. Stiffness tends to return a wire to its original straight-line shape even when it is not under tension. For example, a piano wire straightens if released from a curved shape; package-wrapping string does not.

Stiffness represents a restoring force in addition to tension and increases the wave speed. Consequently, for a real cable, the speed of 60.0 m/s that we determined is most likely associated with a smaller acceleration of the helicopter.

16.4 Reflection and Transmission

The traveling wave model describes waves traveling through a uniform medium without interacting with anything along the way. We now consider how a traveling wave is affected when it encounters a change in the medium. For example, consider a pulse traveling on a string that is rigidly attached to a support at one end as in Figure 16.13. When the pulse reaches the support, a severe change in the medium occurs: the string ends. As a result, the pulse undergoes **reflection;** that is, the pulse moves back along the string in the opposite direction.

Notice that the reflected pulse is *inverted*. This inversion can be explained as follows. When the pulse reaches the fixed end of the string, the string produces an upward force on the support. By Newton's third law, the support must exert an equal-magnitude and oppositely directed (downward) reaction force on the string. This downward force causes the pulse to invert upon reflection.

Now consider another case. This time, the pulse arrives at the end of a string that is free to move vertically as in Figure 16.14. The tension at the free end is maintained because the string is tied to a ring of negligible mass that is free to slide vertically on a smooth post without friction. Again, the pulse is reflected, but this time it is not inverted. When it reaches the post, the pulse exerts a force on the free end of the string, causing the ring to accelerate upward. The ring rises as high as the incoming pulse, and then the downward component of the tension force pulls the ring back down. This movement of the ring produces a reflected pulse that is not inverted and that has the same amplitude as the incoming pulse.

Finally, consider a situation in which the boundary is intermediate between these two extremes. In this case, part of the energy in the incident pulse is reflected and part undergoes **transmission;** that is, some of the energy passes through the boundary. For instance, suppose a light string is attached to a heavier string as in Figure 16.15. When a pulse traveling on the light string reaches the boundary between the two strings, part of the pulse is reflected and inverted and part is transmitted to the heavier string. The reflected pulse is inverted for the same reasons described earlier in the case of the string rigidly attached to a support.

The reflected pulse has a smaller amplitude than the incident pulse. In Section 16.5, we show that the energy carried by a wave is related to its amplitude. According to the principle of conservation of energy, when the pulse breaks up into a reflected pulse and a transmitted pulse at the boundary, the sum of the energies of these two pulses must equal the energy of the incident pulse. Because the reflected pulse contains only part of the energy of the incident pulse, its amplitude must be smaller.

When a pulse traveling on a heavy string strikes the boundary between the heavy string and a lighter one as in Figure 16.16, again part is reflected and part is transmitted. In this case, the reflected pulse is not inverted.

In either case, the relative heights of the reflected and transmitted pulses depend on the relative densities of the two strings. If the strings are identical, there is no discontinuity at the boundary and no reflection takes place.

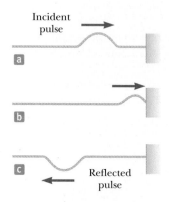

Figure 16.13 The reflection of a traveling pulse at the fixed end of a stretched string. The reflected pulse is inverted, but its shape is otherwise unchanged.

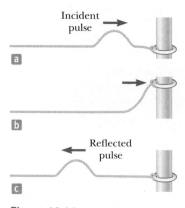

Figure 16.14 The reflection of a traveling pulse at the free end of a stretched string. The reflected pulse is not inverted.

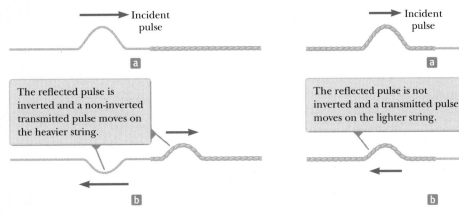

Figure 16.15 (a) A pulse traveling to the right on a light string approaches the junction with a heavier string. (b) The situation after the pulse reaches the junction.

Figure 16.16 (a) A pulse traveling to the right on a heavy string approaches the junction with a lighter string. (b) The situation after the pulse reaches the junction.

According to Equation 16.18, the speed of a wave on a string increases as the mass per unit length of the string decreases. In other words, a wave travels more rapidly on a light string than on a heavy string if both are under the same tension. The following general rules apply to reflected waves: When a wave or pulse travels from medium A to medium B and $v_A > v_B$ (that is, when B is denser than A), it is inverted upon reflection. When a wave or pulse travels from medium A to medium B and $v_A < v_B$ (that is, when A is denser than B), it is not inverted upon reflection.

16.5 Rate of Energy Transfer by Sinusoidal Waves on Strings

Waves transport energy through a medium as they propagate. For example, suppose an object is hanging on a stretched string and a pulse is sent down the string as in Figure 16.17a. When the pulse meets the suspended object, the object is momentarily displaced upward as in Figure 16.17b. In the process, energy is transferred to the object and appears as an increase in the gravitational potential energy of the object–Earth system. This section examines the rate at which energy is transported along a string. We shall assume a one-dimensional sinusoidal wave in the calculation of the energy transferred.

Consider a sinusoidal wave traveling on a string (Fig. 16.18). The source of the energy is some external agent at the left end of the string. We can consider the string to be a nonisolated system. As the external agent performs work on the end of the string, moving it up and down, energy enters the system of the string and propagates along its length. Let's focus our attention on an infinitesimal element of the string of length dx and mass dm. Each such element oscillates vertically with its position described by Equation 15.6. Therefore, we can model each element of the string as a particle in simple harmonic motion, with the oscillation in the y direction. All elements have the same angular frequency ω and the same amplitude A. The kinetic energy K associated with a moving particle is $K = \frac{1}{2}mv^2$. If we apply this equation to the infinitesimal element, the kinetic energy dK associated with the up and down motion of this element is

$$dK = \tfrac{1}{2}(dm)v_y^2$$

where v_y is the transverse speed of the element. If μ is the mass per unit length of the string, the mass dm of the element of length dx is equal to $\mu\, dx$. Hence, we can express the kinetic energy of an element of the string as

$$dK = \tfrac{1}{2}(\mu\, dx)v_y^2 \qquad\qquad \textbf{(16.19)}$$

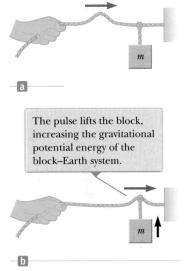

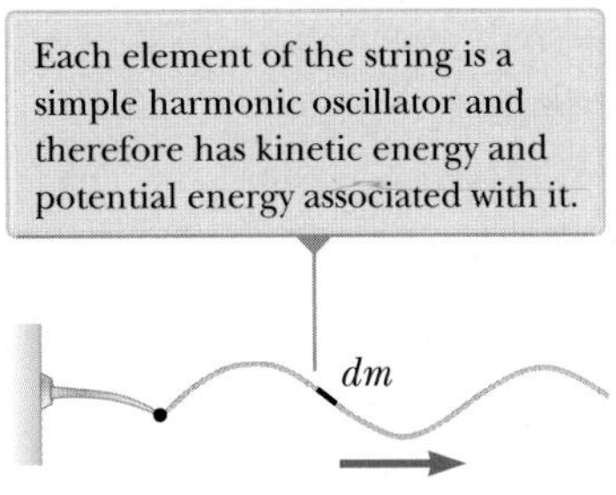

The pulse lifts the block, increasing the gravitational potential energy of the block–Earth system.

Figure 16.17 (a) A pulse travels to the right on a stretched string, carrying energy with it. (b) The energy of the pulse arrives at the hanging block.

Each element of the string is a simple harmonic oscillator and therefore has kinetic energy and potential energy associated with it.

Figure 16.18 A sinusoidal wave traveling along the x axis on a stretched string.

Substituting for the general transverse speed of an element of the medium using Equation 16.14 gives

$$dK = \tfrac{1}{2}\mu[-\omega A \cos(kx - \omega t)]^2 \, dx = \tfrac{1}{2}\mu\omega^2 A^2 \cos^2(kx - \omega t) \, dx$$

If we take a snapshot of the wave at time $t = 0$, the kinetic energy of a given element is

$$dK = \tfrac{1}{2}\mu\omega^2 A^2 \cos^2 kx \, dx$$

Integrating this expression over all the string elements in a wavelength of the wave gives the total kinetic energy K_λ in one wavelength:

$$K_\lambda = \int dK = \int_0^\lambda \tfrac{1}{2}\mu\omega^2 A^2 \cos^2 kx \, dx = \tfrac{1}{2}\mu\omega^2 A^2 \int_0^\lambda \cos^2 kx \, dx$$

$$= \tfrac{1}{2}\mu\omega^2 A^2 \left[\tfrac{1}{2}x + \frac{1}{4k}\sin 2kx \right]_0^\lambda = \tfrac{1}{2}\mu\omega^2 A^2 \left[\tfrac{1}{2}\lambda \right] = \tfrac{1}{4}\mu\omega^2 A^2 \lambda$$

In addition to kinetic energy, there is potential energy associated with each element of the string due to its displacement from the equilibrium position and the restoring forces from neighboring elements. A similar analysis to that above for the total potential energy U_λ in one wavelength gives exactly the same result:

$$U_\lambda = \tfrac{1}{4}\mu\omega^2 A^2 \lambda$$

The total energy in one wavelength of the wave is the sum of the potential and kinetic energies:

$$E_\lambda = U_\lambda + K_\lambda = \tfrac{1}{2}\mu\omega^2 A^2 \lambda \tag{16.20}$$

As the wave moves along the string, this amount of energy passes by a given point on the string during a time interval of one period of the oscillation. Therefore, the power P, or rate of energy transfer T_{MW} associated with the mechanical wave, is

$$P = \frac{T_{MW}}{\Delta t} = \frac{E_\lambda}{T} = \frac{\tfrac{1}{2}\mu\omega^2 A^2 \lambda}{T} = \tfrac{1}{2}\mu\omega^2 A^2 \left(\frac{\lambda}{T}\right)$$

Power of a wave ▶

$$P = \tfrac{1}{2}\mu\omega^2 A^2 v \tag{16.21}$$

Equation 16.21 shows that the rate of energy transfer by a sinusoidal wave on a string is proportional to (a) the square of the frequency, (b) the square of the amplitude, and (c) the wave speed. In fact, the rate of energy transfer in *any* sinusoidal wave is proportional to the square of the angular frequency and to the square of the amplitude.

Quick Quiz 16.5 Which of the following, taken by itself, would be most effective in increasing the rate at which energy is transferred by a wave traveling along a string? **(a)** reducing the linear mass density of the string by one half **(b)** doubling the wavelength of the wave **(c)** doubling the tension in the string **(d)** doubling the amplitude of the wave

---

Example 16.5 Power Supplied to a Vibrating String

A taut string for which $\mu = 5.00 \times 10^{-2}$ kg/m is under a tension of 80.0 N. How much power must be supplied to the string to generate sinusoidal waves at a frequency of 60.0 Hz and an amplitude of 6.00 cm?

SOLUTION

Conceptualize Consider Figure 16.10 again and notice that the vibrating blade supplies energy to the string at a certain rate. This energy then propagates to the right along the string.

▶ **16.5** continued

Categorize We evaluate quantities from equations developed in the chapter, so we categorize this example as a substitution problem.

Use Equation 16.21 to evaluate the power:

$$P = \tfrac{1}{2}\mu\omega^2 A^2 v$$

Use Equations 16.9 and 16.18 to substitute for ω and v:

$$P = \tfrac{1}{2}\mu(2\pi f)^2 A^2 \left(\sqrt{\frac{T}{\mu}}\right) = 2\pi^2 f^2 A^2 \sqrt{\mu T}$$

Substitute numerical values:

$$P = 2\pi^2 (60.0 \text{ Hz})^2 (0.060\ 0 \text{ m})^2 \sqrt{(0.050\ 0 \text{ kg/m})(80.0 \text{ N})} = \boxed{512 \text{ W}}$$

WHAT IF? What if the string is to transfer energy at a rate of 1 000 W? What must be the required amplitude if all other parameters remain the same?

Answer Let us set up a ratio of the new and old power, reflecting only a change in the amplitude:

$$\frac{P_{new}}{P_{old}} = \frac{\tfrac{1}{2}\mu\omega^2 A^2_{new} v}{\tfrac{1}{2}\mu\omega^2 A^2_{old} v} = \frac{A^2_{new}}{A^2_{old}}$$

Solving for the new amplitude gives

$$A_{new} = A_{old}\sqrt{\frac{P_{new}}{P_{old}}} = (6.00 \text{ cm})\sqrt{\frac{1\ 000 \text{ W}}{512 \text{ W}}} = 8.39 \text{ cm}$$

16.6 The Linear Wave Equation

In Section 16.1, we introduced the concept of the wave function to represent waves traveling on a string. All wave functions $y(x, t)$ represent solutions of an equation called the *linear wave equation*. This equation gives a complete description of the wave motion, and from it one can derive an expression for the wave speed. Furthermore, the linear wave equation is basic to many forms of wave motion. In this section, we derive this equation as applied to waves on strings.

Suppose a traveling wave is propagating along a string that is under a tension T. Let's consider one small string element of length Δx (Fig. 16.19). The ends of the element make small angles θ_A and θ_B with the x axis. Forces act on the string at its ends where it connects to neighboring elements. Therefore, the element is modeled as a particle under a net force. The net force acting on the element in the vertical direction is

$$\sum F_y = T\sin\theta_B - T\sin\theta_A = T(\sin\theta_B - \sin\theta_A)$$

Because the angles are small, we can use the approximation $\sin\theta \approx \tan\theta$ to express the net force as

$$\sum F_y \approx T(\tan\theta_B - \tan\theta_A) \tag{16.22}$$

Imagine undergoing an infinitesimal displacement outward from the right end of the rope element in Figure 16.19 along the blue line representing the force $\vec{T}$. This displacement has infinitesimal x and y components and can be represented by the vector $dx\,\hat{\mathbf{i}} + dy\,\hat{\mathbf{j}}$. The tangent of the angle with respect to the x axis for this displacement is dy/dx. Because we evaluate this tangent at a particular instant of time, we must express it in partial form as $\partial y/\partial x$. Substituting for the tangents in Equation 16.22 gives

$$\sum F_y \approx T\left[\left(\frac{\partial y}{\partial x}\right)_B - \left(\frac{\partial y}{\partial x}\right)_A\right] \tag{16.23}$$

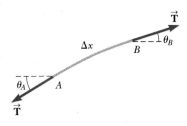

Figure 16.19 An element of a string under tension T.

Now, from the particle under a net force model, let's apply Newton's second law to the element, with the mass of the element given by $m = \mu\,\Delta x$:

$$\sum F_y = ma_y = \mu\,\Delta x\left(\frac{\partial^2 y}{\partial t^2}\right) \tag{16.24}$$

Combining Equation 16.23 with Equation 16.24 gives

$$\mu\,\Delta x\left(\frac{\partial^2 y}{\partial t^2}\right) = T\left[\left(\frac{\partial y}{\partial x}\right)_B - \left(\frac{\partial y}{\partial x}\right)_A\right]$$

$$\frac{\mu}{T}\frac{\partial^2 y}{\partial t^2} = \frac{(\partial y/\partial x)_B - (\partial y/dx)_A}{\Delta x} \tag{16.25}$$

The right side of Equation 16.25 can be expressed in a different form if we note that the partial derivative of any function is defined as

$$\frac{\partial f}{\partial x} \equiv \lim_{\Delta x \to 0}\frac{f(x + \Delta x) - f(x)}{\Delta x}$$

Associating $f(x + \Delta x)$ with $(\partial y/\partial x)_B$ and $f(x)$ with $(\partial y/\partial x)_A$, we see that, in the limit $\Delta x \to 0$, Equation 16.25 becomes

$$\frac{\mu}{T}\frac{\partial^2 y}{\partial t^2} = \frac{\partial^2 y}{\partial x^2} \tag{16.26}$$

◀ **Linear wave equation for a string**

This expression is the linear wave equation as it applies to waves on a string.

The linear wave equation (Eq. 16.26) is often written in the form

$$\frac{\partial^2 y}{\partial x^2} = \frac{1}{v^2}\frac{\partial^2 y}{\partial t^2} \tag{16.27}$$

◀ **Linear wave equation in general**

Equation 16.27 applies in general to various types of traveling waves. For waves on strings, y represents the vertical position of elements of the string. For sound waves propagating through a gas, y corresponds to longitudinal position of elements of the gas from equilibrium or variations in either the pressure or the density of the gas. In the case of electromagnetic waves, y corresponds to electric or magnetic field components.

We have shown that the sinusoidal wave function (Eq. 16.10) is one solution of the linear wave equation (Eq. 16.27). Although we do not prove it here, the linear wave equation is satisfied by *any* wave function having the form $y = f(x \pm vt)$. Furthermore, we have seen that the linear wave equation is a direct consequence of the particle under a net force model applied to any element of a string carrying a traveling wave.

Summary

Definitions

▪ A one-dimensional **sinusoidal wave** is one for which the positions of the elements of the medium vary sinusoidally. A sinusoidal wave traveling to the right can be expressed with a **wave function**

$$y(x, t) = A\sin\left[\frac{2\pi}{\lambda}(x - vt)\right] \tag{16.5}$$

where A is the **amplitude**, λ is the **wavelength**, and v is the **wave speed**.

▪ The **angular wave number** k and **angular frequency** ω of a wave are defined as follows:

$$k \equiv \frac{2\pi}{\lambda} \tag{16.8}$$

$$\omega \equiv \frac{2\pi}{T} = 2\pi f \tag{16.9}$$

where T is the **period** of the wave and f is its **frequency**.

A **transverse wave** is one in which the elements of the medium move in a direction *perpendicular* to the direction of propagation.

A **longitudinal wave** is one in which the elements of the medium move in a direction *parallel* to the direction of propagation.

Concepts and Principles

Any one-dimensional wave traveling with a speed v in the x direction can be represented by a wave function of the form

$$y(x, t) = f(x \pm vt) \qquad \textbf{(16.1, 16.2)}$$

where the positive sign applies to a wave traveling in the negative x direction and the negative sign applies to a wave traveling in the positive x direction. The shape of the wave at any instant in time (a snapshot of the wave) is obtained by holding t constant.

The speed of a wave traveling on a taut string of mass per unit length μ and tension T is

$$v = \sqrt{\frac{T}{\mu}} \qquad \textbf{(16.18)}$$

A wave is totally or partially reflected when it reaches the end of the medium in which it propagates or when it reaches a boundary where its speed changes discontinuously. If a wave traveling on a string meets a fixed end, the wave is reflected and inverted. If the wave reaches a free end, it is reflected but not inverted.

The **power** transmitted by a sinusoidal wave on a stretched string is

$$P = \tfrac{1}{2}\mu\omega^2 A^2 v \qquad \textbf{(16.21)}$$

Wave functions are solutions to a differential equation called the **linear wave equation:**

$$\frac{\partial^2 y}{\partial x^2} = \frac{1}{v^2}\frac{\partial^2 y}{\partial t^2} \qquad \textbf{(16.27)}$$

Analysis Model for Problem Solving

Traveling Wave. The wave speed of a sinusoidal wave is

$$v = \frac{\lambda}{T} = \lambda f \qquad \textbf{(16.6, 16.12)}$$

A sinusoidal wave can be expressed as

$$y = A \sin(kx - \omega t) \qquad \textbf{(16.10)}$$

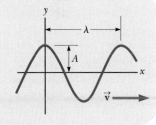

Objective Questions

> **1.** denotes answer available in *Student Solutions Manual/Study Guide*

1. If one end of a heavy rope is attached to one end of a lightweight rope, a wave can move from the heavy rope into the lighter one. **(i)** What happens to the speed of the wave? (a) It increases. (b) It decreases. (c) It is constant. (d) It changes unpredictably. **(ii)** What happens to the frequency? Choose from the same possibilities. **(iii)** What happens to the wavelength? Choose from the same possibilities.

2. If you stretch a rubber hose and pluck it, you can observe a pulse traveling up and down the hose. **(i)** What happens to the speed of the pulse if you stretch the hose more tightly? (a) It increases. (b) It decreases. (c) It is constant. (d) It changes unpredictably. **(ii)** What hap-

pens to the speed if you fill the hose with water? Choose from the same possibilities.

3. Rank the waves represented by the following functions from the largest to the smallest according to **(i)** their amplitudes, **(ii)** their wavelengths, **(iii)** their frequencies, **(iv)** their periods, and **(v)** their speeds. If the values of a quantity are equal for two waves, show them as having equal rank. For all functions, x and y are in meters and t is in seconds. (a) $y = 4 \sin(3x - 15t)$ (b) $y = 6 \cos(3x + 15t - 2)$ (c) $y = 8 \sin(2x + 15t)$ (d) $y = 8 \cos(4x + 20t)$ (e) $y = 7 \sin(6x - 24t)$

4. By what factor would you have to multiply the tension in a stretched string so as to double the wave speed?

Assume the string does not stretch. (a) a factor of 8 (b) a factor of 4 (c) a factor of 2 (d) a factor of 0.5 (e) You could not change the speed by a predictable factor by changing the tension.

5. When all the strings on a guitar (Fig. OQ16.5) are stretched to the same tension, will the speed of a wave along the most massive bass string be (a) faster, (b) slower, or (c) the same as the speed of a wave on the lighter strings? Alternatively, (d) is the speed on the bass string not necessarily any of these answers?

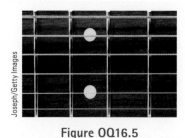

Figure OQ16.5

6. Which of the following statements is not necessarily true regarding mechanical waves? (a) They are formed by some source of disturbance. (b) They are sinusoidal in nature. (c) They carry energy. (d) They require a medium through which to propagate. (e) The wave speed depends on the properties of the medium in which they travel.

7. (a) Can a wave on a string move with a wave speed that is greater than the maximum transverse speed $v_{y,max}$ of an element of the string? (b) Can the wave speed be much greater than the maximum element speed? (c) Can the wave speed be equal to the maximum element speed? (d) Can the wave speed be less than $v_{y,max}$?

8. A source vibrating at constant frequency generates a sinusoidal wave on a string under constant tension. If the power delivered to the string is doubled, by what factor does the amplitude change? (a) a factor of 4 (b) a factor of 2 (c) a factor of $\sqrt{2}$ (d) a factor of 0.707 (e) cannot be predicted

9. The distance between two successive peaks of a sinusoidal wave traveling along a string is 2 m. If the frequency of this wave is 4 Hz, what is the speed of the wave? (a) 4 m/s (b) 1 m/s (c) 8 m/s (d) 2 m/s (e) impossible to answer from the information given

Conceptual Questions

1. denotes answer available in *Student Solutions Manual/Study Guide*

1. Why is a solid substance able to transport both longitudinal waves and transverse waves, but a homogeneous fluid is able to transport only longitudinal waves?

2. (a) How would you create a longitudinal wave in a stretched spring? (b) Would it be possible to create a transverse wave in a spring?

3. When a pulse travels on a taut string, does it always invert upon reflection? Explain.

4. In mechanics, massless strings are often assumed. Why is that not a good assumption when discussing waves on strings?

5. If you steadily shake one end of a taut rope three times each second, what would be the period of the sinusoidal wave set up in the rope?

6. (a) If a long rope is hung from a ceiling and waves are sent up the rope from its lower end, why does the speed of the waves change as they ascend? (b) Does the speed of the ascending waves increase or decrease? Explain.

7. Why is a pulse on a string considered to be transverse?

8. Does the vertical speed of an element of a horizontal, taut string, through which a wave is traveling, depend on the wave speed? Explain.

9. In an earthquake, both S (transverse) and P (longitudinal) waves propagate from the focus of the earthquake. The focus is in the ground radially below the epicenter on the surface (Fig. CQ16.9). Assume the waves move in straight lines through uniform material. The S waves travel through the Earth more slowly than the P waves (at about 5 km/s versus 8 km/s). By detecting the time of arrival of the waves at a seismograph, (a) how can one determine the distance to the focus of the earthquake? (b) How many detection stations are necessary to locate the focus unambiguously?

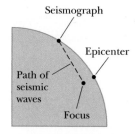

Figure CQ16.9

Section 16.1 Propagation of a Disturbance
Problems 1–4

Section 16.2 Analysis Model: Traveling Wave
Problems 5–20

Section 16.3 The Speed of Waves on Strings
Problems 21–31

Section 16.5 Rate of Energy Transfer by Sinusoidal Waves on Strings
Problems 32–40

Section 16.6 The Linear Wave Equation
Problems 41–44

Additional Problems
Problems 45–63

Challenge Problems
Problems 64–67

Solutions to the following Problems are available in the *Student Solutions Manual/Study Guide:*
16.3, 16.5, 16.7, 16.9, 16.19, 16.23, 16.27, 16.34, 16.35, 16.41, 16.53, 16.60, 16.63, and 16.65

List of Enhanced Problems

Problem Number	Targeted Feedback in Enhanced WebAssign	Analysis Model Tutorial in Enhanced WebAssign	Master It in Enhanced WebAssign	Watch It in Enhanced WebAssign
16.1	✓			✓
16.5	✓		✓	
16.7	✓		✓	
16.9			✓	
16.10	✓			✓
16.11	✓			✓
16.15	✓			✓
16.17	✓			✓
16.22	✓			✓
16.23			✓	
16.25	✓			✓
16.27		✓	✓	
16.29		✓		
16.31	✓			✓
16.34			✓	
16.35			✓	
16.36	✓			✓
16.37	✓	✓		
16.55		✓		
16.63			✓	

Sound Waves

17.1 Pressure Variations in Sound Waves

17.2 Speed of Sound Waves

17.3 Intensity of Periodic Sound Waves

17.4 The Doppler Effect

Most of the waves we studied in Chapter 16 are constrained to move along a one- dimensional medium. For example, the wave in Figure 16.7 is a purely mathematical construct moving along the *x* axis. The wave in Figure 16.10 is constrained to move along the length of the string. We have also seen waves moving through a two-dimensional medium, such as the ripples on the water surface in the introduction to Part 2 on page 351 and the waves moving over the surface of the ocean in Figure 16.4. In this chapter, we investigate mechanical waves that move through three-dimensional bulk media. For example, seismic waves leaving the focus of an earthquake travel through the three-dimensional interior of the Earth.

We will focus our attention on **sound waves,** which travel through any material, but are most commonly experienced as the mechanical waves traveling through air that result in the human perception of hearing. As sound waves travel through air, elements of air are disturbed from their equilibrium positions. Accompanying these movements are changes in density and pressure of the air along the direction of wave motion. If the source of the sound waves vibrates sinusoidally, the density and pressure variations are also sinusoidal. The mathematical description of sinusoidal sound waves is very similar to that of sinusoidal waves on strings, as discussed in Chapter 16.

Sound waves are divided into three categories that cover different frequency ranges. (1) *Audible waves* lie within the range of sensitivity of the human ear. They can be generated in a variety of ways, such as by musical instruments, human voices, or loudspeakers. (2) *Infrasonic waves* have frequencies below the audible range. Elephants can use infrasonic waves to communicate with one another, even when separated by many kilometers. (3) *Ultrasonic waves* have frequencies above the audible range. You may have used a "silent" whistle to retrieve your dog. Dogs easily hear the ultrasonic sound this whistle emits, although humans cannot detect it at all. Ultrasonic waves are also used in medical imaging.

Three musicians play the alpenhorn in Valais, Switzerland. In this chapter, we explore the behavior of sound waves such as those coming from these large musical instruments. *(Stefano Cellai/AGE fotostock)*

This chapter begins with a discussion of the pressure variations in a sound wave, the speed of sound waves, and wave intensity, which is a function of wave amplitude. We then provide an alternative description of the intensity of sound waves that compresses the wide range of intensities to which the ear is sensitive into a smaller range for convenience. The effects of the motion of sources and listeners on the frequency of a sound are also investigated.

17.1 Pressure Variations in Sound Waves

In Chapter 16, we began our investigation of waves by imagining the creation of a single pulse that traveled down a string (Figure 16.1) or a spring (Figure 16.3). Let's do something similar for sound. We describe pictorially the motion of a one-dimensional longitudinal sound pulse moving through a long tube containing a compressible gas as shown in Figure 17.1. A piston at the left end can be quickly moved to the right to compress the gas and create the pulse. Before the piston is moved, the gas is undisturbed and of uniform density as represented by the uniformly shaded region in Figure 17.1a. When the piston is pushed to the right (Fig. 17.1b), the gas just in front of it is compressed (as represented by the more heavily shaded region); the pressure and density in this region are now higher than they were before the piston moved. When the piston comes to rest (Fig. 17.1c), the compressed region of the gas continues to move to the right, corresponding to a longitudinal pulse traveling through the tube with speed v.

One can produce a one-dimensional *periodic* sound wave in the tube of gas in Figure 17.1 by causing the piston to move in simple harmonic motion. The results are shown in Figure 17.2. The darker parts of the colored areas in this figure represent regions in which the gas is compressed and the density and pressure are above their equilibrium values. A compressed region is formed whenever the pis-

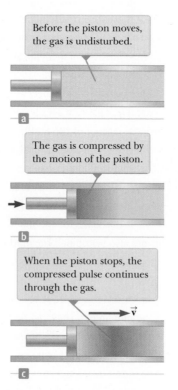

Before the piston moves, the gas is undisturbed.

a

The gas is compressed by the motion of the piston.

b

When the piston stops, the compressed pulse continues through the gas.

$\vec{v}$

c

Figure 17.1 Motion of a longitudinal pulse through a compressible gas. The compression (darker region) is produced by the moving piston.

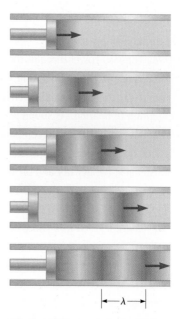

$\longmapsto \lambda \longmapsto$

Figure 17.2 A longitudinal wave propagating through a gas-filled tube. The source of the wave is an oscillating piston at the left.

ton is pushed into the tube. This compressed region, called a **compression,** moves through the tube, continuously compressing the region just in front of itself. When the piston is pulled back, the gas in front of it expands and the pressure and density in this region fall below their equilibrium values (represented by the lighter parts of the colored areas in Fig. 17.2). These low-pressure regions, called **rarefactions,** also propagate along the tube, following the compressions. Both regions move at the speed of sound in the medium.

As the piston oscillates sinusoidally, regions of compression and rarefaction are continuously set up. The distance between two successive compressions (or two successive rarefactions) equals the wavelength λ of the sound wave. Because the sound wave is longitudinal, as the compressions and rarefactions travel through the tube, any small element of the gas moves with simple harmonic motion parallel to the direction of the wave. If $s(x, t)$ is the position of a small element relative to its equilibrium position,[1] we can express this harmonic position function as

$$s(x, t) = s_{max} \cos (kx - \omega t) \tag{17.1}$$

where s_{max} is the maximum position of the element relative to equilibrium. This parameter is often called the **displacement amplitude** of the wave. The parameter k is the wave number, and ω is the angular frequency of the wave. Notice that the displacement of the element is along x, in the direction of propagation of the sound wave.

The variation in the gas pressure ΔP measured from the equilibrium value is also periodic with the same wave number and angular frequency as for the displacement in Equation 17.1. Therefore, we can write

$$\Delta P = \Delta P_{max} \sin (kx - \omega t) \tag{17.2}$$

where **the pressure amplitude ΔP_{max}** is the maximum change in pressure from the equilibrium value.

Notice that we have expressed the displacement by means of a cosine function and the pressure by means of a sine function. We will justify this choice in the procedure that follows and relate the pressure amplitude P_{max} to the displacement amplitude s_{max}. Consider the piston–tube arrangement of Figure 17.1 once again. In Figure 17.3a, we focus our attention on a small cylindrical element of undisturbed gas of length Δx and area A. The volume of this element is $V_i = A\,\Delta x$.

Figure 17.3b shows this element of gas after a sound wave has moved it to a new position. The cylinder's two flat faces move through different distances s_1 and s_2. The change in volume ΔV of the element in the new position is equal to $A\,\Delta s$, where $\Delta s = s_1 - s_2$.

From the definition of bulk modulus (see Eq. 12.8), we express the pressure variation in the element of gas as a function of its change in volume:

$$\Delta P = -B \frac{\Delta V}{V_i}$$

Let's substitute for the initial volume and the change in volume of the element:

$$\Delta P = -B \frac{A\,\Delta s}{A\,\Delta x}$$

Let the length Δx of the cylinder approach zero so that the ratio $\Delta s/\Delta x$ becomes a partial derivative:

$$\Delta P = -B \frac{\partial s}{\partial x} \tag{17.3}$$

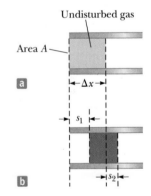

Figure 17.3 (a) An undisturbed element of gas of length Δx in a tube of cross-sectional area A. (b) When a sound wave propagates through the gas, the element is moved to a new position and has a different length. The parameters s_1 and s_2 describe the displacements of the ends of the element from their equilibrium positions.

---

[1] We use $s(x, t)$ here instead of $y(x, t)$ because the displacement of elements of the medium is not perpendicular to the x direction.

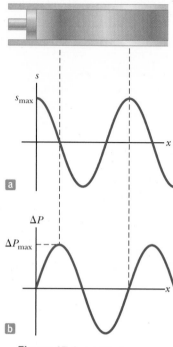

Figure 17.4 (a) Displacement amplitude and (b) pressure amplitude versus position for a sinusoidal longitudinal wave.

Substitute the position function given by Equation 17.1:

$$\Delta P = -B\frac{\partial}{\partial x}\left[s_{max}\cos\left(kx - \omega t\right)\right] = Bs_{max}k\,\sin\left(kx - \omega t\right)$$

From this result, we see that a displacement described by a cosine function leads to a pressure described by a sine function. We also see that the displacement and pressure amplitudes are related by

$$\Delta P_{max} = Bs_{max}k \tag{17.4}$$

This relationship depends on the bulk modulus of the gas, which is not as readily available as is the density of the gas. Once we determine the speed of sound in a gas in Section 17.2, we will be able to provide an expression that relates ΔP_{max} and s_{max} in terms of the density of the gas.

This discussion shows that a sound wave may be described equally well in terms of either pressure or displacement. A comparison of Equations 17.1 and 17.2 shows that the pressure wave is 90° out of phase with the displacement wave. Graphs of these functions are shown in Figure 17.4. The pressure variation is a maximum when the displacement from equilibrium is zero, and the displacement from equilibrium is a maximum when the pressure variation is zero.

Quick **Quiz 17.1** If you blow across the top of an empty soft-drink bottle, a pulse of sound travels down through the air in the bottle. At the moment the pulse reaches the bottom of the bottle, what is the correct description of the displacement of elements of air from their equilibrium positions and the pressure of the air at this point? **(a)** The displacement and pressure are both at a maximum. **(b)** The displacement and pressure are both at a minimum. **(c)** The displacement is zero, and the pressure is a maximum. **(d)** The displacement is zero, and the pressure is a minimum.

17.2 Speed of Sound Waves

We now extend the discussion begun in Section 17.1 to evaluate the speed of sound in a gas. In Figure 17.5a, consider the cylindrical element of gas between the piston and the dashed line. This element of gas is in equilibrium under the influence of forces of equal magnitude, from the piston on the left and from the rest of the gas on the right. The magnitude of these forces is PA, where P is the pressure in the gas and A is the cross-sectional area of the tube.

Figure 17.5b shows the situation after a time interval Δt during which the piston moves to the right at a constant speed v_x due to a force from the left on the piston that has increased in magnitude to $(P + \Delta P)A$. By the end of the time interval Δt,

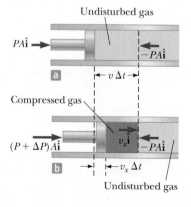

Figure 17.5 (a) An undisturbed element of gas of length $v\ \Delta t$ in a tube of cross-sectional area A. The element is in equilibrium between forces on either end. (b) When the piston moves inward at constant velocity v_x due to an increased force on the left, the element also moves with the same velocity.

every bit of gas in the element is moving with speed v_x. That will not be true in general for a macroscopic element of gas, but it will become true if we shrink the length of the element to an infinitesimal value.

The length of the undisturbed element of gas is chosen to be $v \, \Delta t$, where v is the speed of sound in the gas and Δt is the time interval between the configurations in Figures 17.5a and 17.5b. Therefore, at the end of the time interval Δt, the sound wave will just reach the right end of the cylindrical element of gas. The gas to the right of the element is undisturbed because the sound wave has not reached it yet.

The element of gas is modeled as a nonisolated system in terms of momentum. The force from the piston has provided an impulse to the element, which in turn exhibits a change in momentum. Therefore, we evaluate both sides of the impulse–momentum theorem:

$$\Delta \vec{\mathbf{p}} = \vec{\mathbf{I}} \tag{17.5}$$

On the right, the impulse is provided by the constant force due to the increased pressure on the piston:

$$\vec{\mathbf{I}} = \sum \vec{\mathbf{F}} \, \Delta t = (A \, \Delta P \, \Delta t)\hat{\mathbf{i}}$$

The pressure change ΔP can be related to the volume change and then to the speeds v and v_x through the bulk modulus:

$$\Delta P = -B \frac{\Delta V}{V_i} = -B \frac{(-v_x A \, \Delta t)}{v A \, \Delta t} = B \frac{v_x}{v}$$

Therefore, the impulse becomes

$$\vec{\mathbf{I}} = \left(AB \frac{v_x}{v} \Delta t \right)\hat{\mathbf{i}} \tag{17.6}$$

On the left-hand side of the impulse–momentum theorem, Equation 17.5, the change in momentum of the element of gas of mass m is as follows:

$$\Delta \vec{\mathbf{p}} = m \, \Delta \vec{\mathbf{v}} = (\rho V_i)(v_x \hat{\mathbf{i}} - 0) = (\rho v v_x A \, \Delta t)\hat{\mathbf{i}} \tag{17.7}$$

Substituting Equations 17.6 and 17.7 into Equation 17.5, we find

$$\rho v v_x A \, \Delta t = AB \frac{v_x}{v} \Delta t$$

which reduces to an expression for the speed of sound in a gas:

$$v = \sqrt{\frac{B}{\rho}} \tag{17.8}$$

It is interesting to compare this expression with Equation 16.18 for the speed of transverse waves on a string, $v = \sqrt{T/\mu}$. In both cases, the wave speed depends on an elastic property of the medium (bulk modulus B or string tension T) and on an inertial property of the medium (volume density ρ or linear density μ). In fact, the speed of all mechanical waves follows an expression of the general form

$$v = \sqrt{\frac{\text{elastic property}}{\text{inertial property}}}$$

For longitudinal sound waves in a solid rod of material, for example, the speed of sound depends on Young's modulus Y and the density ρ. Table 17.1 (page 402) provides the speed of sound in several different materials.

The speed of sound also depends on the temperature of the medium. For sound traveling through air, the relationship between wave speed and air temperature is

$$v = 331 \sqrt{1 + \frac{T_C}{273}} \tag{17.9}$$

Table 17.1		Speed of Sound in Various Media				
Medium	**v (m/s)**	**Medium**	**v (m/s)**	**Medium**	**v (m/s)**
Gases		**Liquids at 25°C**		**Solids**[a]	
Hydrogen (0°C)	1 286	Glycerol	1 904	Pyrex glass	5 640
Helium (0°C)	972	Seawater	1 533	Iron	5 950
Air (20°C)	343	Water	1 493	Aluminum	6 420
Air (0°C)	331	Mercury	1 450	Brass	4 700
Oxygen (0°C)	317	Kerosene	1 324	Copper	5 010
		Methyl alcohol	1 143	Gold	3 240
		Carbon tetrachloride	926	Lucite	2 680
				Lead	1 960
				Rubber	1 600

[a]Values given are for propagation of longitudinal waves in bulk media. Speeds for longitudinal waves in thin rods are smaller, and speeds of transverse waves in bulk are smaller yet.

where v is in meters/second, 331 m/s is the speed of sound in air at 0°C, and T_C is the air temperature in degrees Celsius. Using this equation, one finds that at 20°C, the speed of sound in air is approximately 343 m/s.

This information provides a convenient way to estimate the distance to a thunderstorm. First count the number of seconds between seeing the flash of lightning and hearing the thunder. Dividing this time interval by 3 gives the approximate distance to the lightning in kilometers because 343 m/s is approximately $\frac{1}{3}$ km/s. Dividing the time interval in seconds by 5 gives the approximate distance to the lightning in miles because the speed of sound is approximately $\frac{1}{5}$ mi/s.

Having an expression (Eq. 17.8) for the speed of sound, we can now express the relationship between pressure amplitude and displacement amplitude for a sound wave (Eq. 17.4) as

$$\Delta P_{max} = B s_{max} k = (\rho v^2) s_{max}\left(\frac{\omega}{v}\right) = \rho v \omega s_{max} \tag{17.10}$$

This expression is a bit more useful than Equation 17.4 because the density of a gas is more readily available than is the bulk modulus.

17.3 Intensity of Periodic Sound Waves

In Chapter 16, we showed that a wave traveling on a taut string transports energy, consistent with the notion of energy transfer by mechanical waves in Equation 8.2. Naturally, we would expect sound waves to also represent a transfer of energy. Consider the element of gas acted on by the piston in Figure 17.5. Imagine that the piston is moving back and forth in simple harmonic motion at angular frequency ω. Imagine also that the length of the element becomes very small so that the entire element moves with the same velocity as the piston. Then we can model the element as a particle on which the piston is doing work. The rate at which the piston is doing work on the element at any instant of time is given by Equation 8.19:

$$Power = \vec{\mathbf{F}} \cdot \vec{\mathbf{v}}_x$$

where we have used *Power* rather than P so that we don't confuse power P with pressure P! The force $\vec{\mathbf{F}}$ on the element of gas is related to the pressure and the velocity $\vec{\mathbf{v}}_x$ of the element is the derivative of the displacement function, so we find

$$Power = [\Delta P(x, t)A]\hat{\mathbf{i}} \cdot \frac{\partial}{\partial t}[s(x, t)\hat{\mathbf{i}}]$$

$$= [\rho v \omega A s_{max} \sin(kx - \omega t)]\left\{\frac{\partial}{\partial t}[s_{max} \cos(kx - \omega t)]\right\}$$

$$= \rho v \omega A s_{max} \sin (kx - \omega t)][\omega s_{max} \sin (kx - \omega t)]$$

$$= \rho v \omega^2 A s_{max}^2 \sin^2 (kx - \omega t)$$

We now find the time average power over one period of the oscillation. For any given value of x, which we can choose to be $x = 0$, the average value of $\sin^2 (kx - \omega t)$ over one period T is

$$\frac{1}{T}\int_0^T \sin^2 (0 - \omega t)\, dt = \frac{1}{T}\int_0^T \sin^2 \omega t\, dt = \frac{1}{T}\left(\frac{t}{2} + \frac{\sin 2\omega t}{2\omega}\right)\Big|_0^T = \tfrac{1}{2}$$

Therefore,

$$(Power)_{avg} = \tfrac{1}{2}\rho v \omega^2 A s_{max}^2$$

We define the **intensity** I of a wave, or the power per unit area, as the rate at which the energy transported by the wave transfers through a unit area A perpendicular to the direction of travel of the wave:

$$I \equiv \frac{(Power)_{avg}}{A} \qquad\qquad \textbf{(17.11)}$$

◀ **Intensity of a sound wave**

In this case, the intensity is therefore

$$I = \tfrac{1}{2}\rho v (\omega s_{max})^2$$

Hence, the intensity of a periodic sound wave is proportional to the square of the displacement amplitude and to the square of the angular frequency. This expression can also be written in terms of the pressure amplitude ΔP_{max}; in this case, we use Equation 17.10 to obtain

$$I = \frac{(\Delta P_{max})^2}{2\rho v} \qquad\qquad \textbf{(17.12)}$$

The string waves we studied in Chapter 16 are constrained to move along the one-dimensional string, as discussed in the introduction to this chapter. The sound waves we have studied with regard to Figures 17.1 through 17.3 and 17.5 are constrained to move in one dimension along the length of the tube. As we mentioned in the introduction, however, sound waves can move through three-dimensional bulk media, so let's place a sound source in the open air and study the results.

Consider the special case of a point source emitting sound waves equally in all directions. If the air around the source is perfectly uniform, the sound power radiated in all directions is the same, and the speed of sound in all directions is the same. The result in this situation is called a **spherical wave.** Figure 17.6 shows these spherical waves as a series of circular arcs concentric with the source. Each arc represents a surface over which the phase of the wave is constant. We call such a surface of constant phase a **wave front.** The radial distance between adjacent wave fronts that have the same phase is the wavelength λ of the wave. The radial lines pointing outward from the source, representing the direction of propagation of the waves, are called **rays.**

The average power emitted by the source must be distributed uniformly over each spherical wave front of area $4\pi r^2$. Hence, the wave intensity at a distance r from the source is

$$I = \frac{(Power)_{avg}}{A} = \frac{(Power)_{avg}}{4\pi r^2} \qquad\qquad \textbf{(17.13)}$$

The intensity decreases as the square of the distance from the source. This inverse-square law is reminiscent of the behavior of gravity in Chapter 13.

The rays are radial lines pointing outward from the source, perpendicular to the wave fronts.

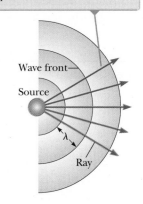

Wave front

Source

λ

Ray

Figure 17.6 Spherical waves emitted by a point source. The circular arcs represent the spherical wave fronts that are concentric with the source.

Quick Quiz 17.2 A vibrating guitar string makes very little sound if it is not mounted on the guitar body. Why does the sound have greater intensity if the string is attached to the guitar body? (a) The string vibrates with more energy. (b) The energy leaves the guitar at a greater rate. (c) The sound power is spread over a larger area at the listener's position. (d) The sound power is concentrated over a smaller area at the listener's position. (e) The speed of sound is higher in the material of the guitar body. (f) None of these answers is correct.

Example 17.1   **Hearing Limits**

The faintest sounds the human ear can detect at a frequency of 1 000 Hz correspond to an intensity of about 1.00×10^{-12} W/m², which is called *threshold of hearing*. The loudest sounds the ear can tolerate at this frequency correspond to an intensity of about 1.00 W/m², the *threshold of pain*. Determine the pressure amplitude and displacement amplitude associated with these two limits.

SOLUTION

Conceptualize Think about the quietest environment you have ever experienced. It is likely that the intensity of sound in even this quietest environment is higher than the threshold of hearing.

Categorize Because we are given intensities and asked to calculate pressure and displacement amplitudes, this problem is an analysis problem requiring the concepts discussed in this section.

Analyze To find the amplitude of the pressure variation at the threshold of hearing, use Equation 17.12, taking the speed of sound waves in air to be $v = 343$ m/s and the density of air to be $\rho = 1.20$ kg/m³:

$$\Delta P_{max} = \sqrt{2\rho v I}$$
$$= \sqrt{2(1.20 \text{ kg/m}^3)(343 \text{ m/s})(1.00 \times 10^{-12} \text{ W/m}^2)}$$
$$= 2.87 \times 10^{-5} \text{ N/m}^2$$

Calculate the corresponding displacement amplitude using Equation 17.10, recalling that $\omega = 2\pi f$ (Eq. 16.9):

$$s_{max} = \frac{\Delta P_{max}}{\rho v \omega} = \frac{2.87 \times 10^{-5} \text{ N/m}^2}{(1.20 \text{ kg/m}^3)(343 \text{ m/s})(2\pi \times 1\,000 \text{ Hz})}$$
$$= 1.11 \times 10^{-11} \text{ m}$$

In a similar manner, one finds that the loudest sounds the human ear can tolerate (the threshold of pain) correspond to a pressure amplitude of 28.7 N/m² and a displacement amplitude equal to 1.11×10^{-5} m .

Finalize Because atmospheric pressure is about 10^5 N/m², the result for the pressure amplitude tells us that the ear is sensitive to pressure fluctuations as small as 3 parts in 10^{10}! The displacement amplitude is also a remarkably small number! If we compare this result for s_{max} to the size of an atom (about 10^{-10} m), we see that the ear is an extremely sensitive detector of sound waves.

Example 17.2   **Intensity Variations of a Point Source**

A point source emits sound waves with an average power output of 80.0 W.

(A) Find the intensity 3.00 m from the source.

SOLUTION

Conceptualize Imagine a small loudspeaker sending sound out at an average rate of 80.0 W uniformly in all directions. You are standing 3.00 m away from the speakers. As the sound propagates, the energy of the sound waves is spread out over an ever-expanding sphere, so the intensity of the sound falls off with distance.

Categorize We evaluate the intensity from an equation generated in this section, so we categorize this example as a substitution problem.

▶ **17.2** continued

Because a point source emits energy in the form of spherical waves, use Equation 17.13 to find the intensity:

$$I = \frac{(Power)_{avg}}{4\pi r^2} = \frac{80.0 \text{ W}}{4\pi(3.00 \text{ m})^2} = \boxed{0.707 \text{ W/m}^2}$$

This intensity is close to the threshold of pain.

(B) Find the distance at which the intensity of the sound is 1.00×10^{-8} W/m².

SOLUTION

Solve for r in Equation 17.13 and use the given value for I:

$$r = \sqrt{\frac{(Power)_{avg}}{4\pi I}} = \sqrt{\frac{80.0 \text{ W}}{4\pi(1.00 \times 10^{-8} \text{ W/m}^2)}}$$

$$= \boxed{2.52 \times 10^4 \text{ m}}$$

Sound Level in Decibels

Example 17.1 illustrates the wide range of intensities the human ear can detect. Because this range is so wide, it is convenient to use a logarithmic scale, where the **sound level** β (Greek letter beta) is defined by the equation

$$\beta \equiv 10 \log \left(\frac{I}{I_0}\right) \tag{17.14}$$

The constant I_0 is the *reference intensity*, taken to be at the threshold of hearing ($I_0 = 1.00 \times 10^{-12}$ W/m²), and I is the intensity in watts per square meter to which the sound level β corresponds, where β is measured[2] in **decibels** (dB). On this scale, the threshold of pain ($I = 1.00$ W/m²) corresponds to a sound level of $\beta = 10 \log [(1 \text{ W/m}^2)/(10^{-12} \text{ W/m}^2)] = 10 \log (10^{12}) = 120$ dB, and the threshold of hearing corresponds to $\beta = 10 \log [(10^{-12} \text{ W/m}^2)/(10^{-12} \text{ W/m}^2)] = 0$ dB.

Prolonged exposure to high sound levels may seriously damage the human ear. Ear plugs are recommended whenever sound levels exceed 90 dB. Recent evidence suggests that "noise pollution" may be a contributing factor to high blood pressure, anxiety, and nervousness. Table 17.2 gives some typical sound levels.

Table 17.2

Sound Levels

Source of Sound	β (dB)
Nearby jet airplane	150
Jackhammer; machine gun	130
Siren; rock concert	120
Subway; power lawn mower	100
Busy traffic	80
Vacuum cleaner	70
Normal conversation	60
Mosquito buzzing	40
Whisper	30
Rustling leaves	10
Threshold of hearing	0

Quick Quiz 17.3 Increasing the intensity of a sound by a factor of 100 causes the sound level to increase by what amount? **(a)** 100 dB **(b)** 20 dB **(c)** 10 dB **(d)** 2 dB

Example 17.3 **Sound Levels**

Two identical machines are positioned the same distance from a worker. The intensity of sound delivered by each operating machine at the worker's location is 2.0×10^{-7} W/m².

(A) Find the sound level heard by the worker when one machine is operating.

SOLUTION

Conceptualize Imagine a situation in which one source of sound is active and is then joined by a second identical source, such as one person speaking and then a second person speaking at the same time or one musical instrument playing and then being joined by a second instrument.

Categorize This example is a relatively simple analysis problem requiring Equation 17.14.

continued

---

[2]The unit *bel* is named after the inventor of the telephone, Alexander Graham Bell (1847–1922). The prefix *deci-* is the SI prefix that stands for 10^{-1}.

▶ **17.3** continued

. . . .

Analyze Use Equation 17.14 to calculate the sound level at the worker's location with one machine operating:

$$\beta_1 = 10 \log \left(\frac{2.0 \times 10^{-7} \, \text{W/m}^2}{1.00 \times 10^{-12} \, \text{W/m}^2} \right) = 10 \log \left(2.0 \times 10^5 \right) = \boxed{53 \, \text{dB}}$$

(B) Find the sound level heard by the worker when two machines are operating.

SOLUTION

Use Equation 17.14 to calculate the sound level at the worker's location with double the intensity:

$$\beta_2 = 10 \log \left(\frac{4.0 \times 10^{-7} \, \text{W/m}^2}{1.00 \times 10^{-12} \, \text{W/m}^2} \right) = 10 \log \left(4.0 \times 10^5 \right) = \boxed{56 \, \text{dB}}$$

Finalize These results show that when the intensity is doubled, the sound level increases by only 3 dB. This 3-dB increase is independent of the original sound level. (Prove this to yourself!)

WHAT IF? *Loudness* is a psychological response to a sound. It depends on both the intensity and the frequency of the sound. As a rule of thumb, a doubling in loudness is approximately associated with an increase in sound level of 10 dB. (This rule of thumb is relatively inaccurate at very low or very high frequencies.) If the loudness of the machines in this example is to be doubled, how many machines at the same distance from the worker must be running?

Answer Using the rule of thumb, a doubling of loudness corresponds to a sound level increase of 10 dB. Therefore,

$$\beta_2 - \beta_1 = 10 \, \text{dB} = 10 \log \left(\frac{I_2}{I_0} \right) - 10 \log \left(\frac{I_1}{I_0} \right) = 10 \log \left(\frac{I_2}{I_1} \right)$$

$$\log \left(\frac{I_2}{I_1} \right) = 1 \quad \rightarrow \quad I_2 = 10 I_1$$

Therefore, ten machines must be operating to double the loudness.

Loudness and Frequency

The discussion of sound level in decibels relates to a *physical* measurement of the strength of a sound. Let us now extend our discussion from the What If? section of Example 17.3 concerning the *psychological* "measurement" of the strength of a sound.

Of course, we don't have instruments in our bodies that can display numerical values of our reactions to stimuli. We have to "calibrate" our reactions somehow by comparing different sounds to a reference sound, but that is not easy to accomplish. For example, earlier we mentioned that the threshold intensity is 10^{-12} W/m², corresponding to an intensity level of 0 dB. In reality, this value is the threshold only for a sound of frequency 1 000 Hz, which is a standard reference frequency in acoustics. If we perform an experiment to measure the threshold intensity at other frequencies, we find a distinct variation of this threshold as a function of frequency. For example, at 100 Hz, a barely audible sound must have an intensity level of about 30 dB! Unfortunately, there is no simple relationship between physical measurements and psychological "measurements." The 100-Hz, 30-dB sound is psychologically "equal" in loudness to the 1 000-Hz, 0-dB sound (both are just barely audible), but they are not physically equal in sound level (30 dB ≠ 0 dB).

By using test subjects, the human response to sound has been studied, and the results are shown in the white area of Figure 17.7 along with the approximate frequency and sound-level ranges of other sound sources. The lower curve of the white area corresponds to the threshold of hearing. Its variation with frequency is clear from this diagram. Notice that humans are sensitive to frequencies ranging from about 20 Hz to about 20 000 Hz. The upper bound of the white area is the thresh-

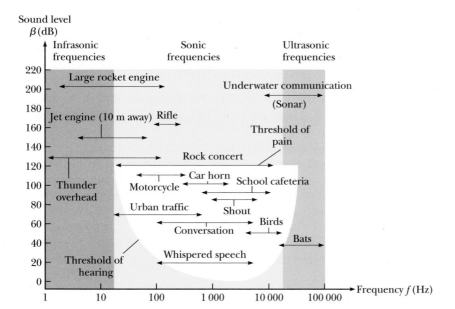

Figure 17.7 Approximate ranges of frequency and sound level of various sources and that of normal human hearing, shown by the white area. (From R. L. Reese, *University Physics*, Pacific Grove, Brooks/Cole, 2000.)

old of pain. Here the boundary of the white area appears straight because the psychological response is relatively independent of frequency at this high sound level.

The most dramatic change with frequency is in the lower left region of the white area, for low frequencies and low intensity levels. Our ears are particularly insensitive in this region. If you are listening to your home entertainment system and the bass (low frequencies) and treble (high frequencies) sound balanced at a high volume, try turning the volume down and listening again. You will probably notice that the bass seems weak, which is due to the insensitivity of the ear to low frequencies at low sound levels as shown in Figure 17.7.

17.4 The Doppler Effect

Perhaps you have noticed how the sound of a vehicle's horn changes as the vehicle moves past you. The frequency of the sound you hear as the vehicle approaches you is higher than the frequency you hear as it moves away from you. This experience is one example of the **Doppler effect.**[3]

To see what causes this apparent frequency change, imagine you are in a boat that is lying at anchor on a gentle sea where the waves have a period of $T = 3.0$ s. Hence, every 3.0 s a crest hits your boat. Figure 17.8a shows this situation, with the water waves moving toward the left. If you set your watch to $t = 0$ just as one crest hits, the watch reads 3.0 s when the next crest hits, 6.0 s when the third crest

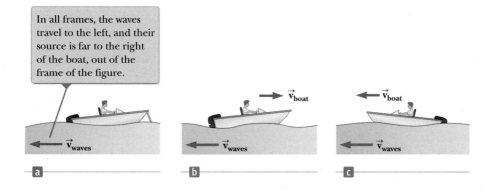

In all frames, the waves travel to the left, and their source is far to the right of the boat, out of the frame of the figure.

Figure 17.8 (a) Waves moving toward a stationary boat. (b) The boat moving toward the wave source. (c) The boat moving away from the wave source.

[3]Named after Austrian physicist Christian Johann Doppler (1803–1853), who in 1842 predicted the effect for both sound waves and light waves.

hits, and so on. From these observations, you conclude that the wave frequency is $f = 1/T = 1/(3.0 \text{ s}) = 0.33$ Hz. Now suppose you start your motor and head directly into the oncoming waves as in Figure 17.8b. Again you set your watch to $t = 0$ as a crest hits the front (the bow) of your boat. Now, however, because you are moving toward the next wave crest as it moves toward you, it hits you less than 3.0 s after the first hit. In other words, the period you observe is shorter than the 3.0-s period you observed when you were stationary. Because $f = 1/T$, you observe a higher wave frequency than when you were at rest.

If you turn around and move in the same direction as the waves (Fig. 17.8c), you observe the opposite effect. You set your watch to $t = 0$ as a crest hits the back (the stern) of the boat. Because you are now moving away from the next crest, more than 3.0 s has elapsed on your watch by the time that crest catches you. Therefore, you observe a lower frequency than when you were at rest.

These effects occur because the *relative* speed between your boat and the waves depends on the direction of travel and on the speed of your boat. (See Section 4.6.) When you are moving toward the right in Figure 17.8b, this relative speed is higher than that of the wave speed, which leads to the observation of an increased frequency. When you turn around and move to the left, the relative speed is lower, as is the observed frequency of the water waves.

Let's now examine an analogous situation with sound waves in which the water waves become sound waves, the water becomes the air, and the person on the boat becomes an observer listening to the sound. In this case, an observer O is moving and a sound source S is stationary. For simplicity, we assume the air is also stationary and the observer moves directly toward the source (Fig. 17.9). The observer moves with a speed v_O toward a stationary point source ($v_S = 0$), where *stationary* means at rest with respect to the medium, air.

If a point source emits sound waves and the medium is uniform, the waves move at the same speed in all directions radially away from the source; the result is a spherical wave as mentioned in Section 17.3. The distance between adjacent wave fronts equals the wavelength λ. In Figure 17.9, the circles are the intersections of these three-dimensional wave fronts with the two-dimensional paper.

We take the frequency of the source in Figure 17.9 to be f, the wavelength to be λ, and the speed of sound to be v. If the observer were also stationary, he would detect wave fronts at a frequency f. (That is, when $v_O = 0$ and $v_S = 0$, the observed frequency equals the source frequency.) When the observer moves toward the source, the speed of the waves relative to the observer is $v' = v + v_O$, as in the case of the boat in Figure 17.8, but the wavelength λ is unchanged. Hence, using Equation 16.12, $v = \lambda f$, we can say that the frequency f' heard by the observer is *increased* and is given by

$$f' = \frac{v'}{\lambda} = \frac{v + v_O}{\lambda}$$

Because $\lambda = v/f$, we can express f' as

$$f' = \left(\frac{v + v_O}{v}\right)f \quad \text{(observer moving toward source)} \tag{17.15}$$

If the observer is moving away from the source, the speed of the wave relative to the observer is $v' = v - v_O$. The frequency heard by the observer in this case is *decreased* and is given by

$$f' = \left(\frac{v - v_O}{v}\right)f \quad \text{(observer moving away from source)} \tag{17.16}$$

These last two equations can be reduced to a single equation by adopting a sign convention. Whenever an observer moves with a speed v_O relative to a stationary source, the frequency heard by the observer is given by Equation 17.15, with v_O interpreted as follows: a positive value is substituted for v_O when the observer moves

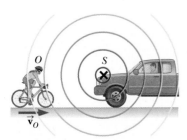

Figure 17.9 An observer O (the cyclist) moves with a speed v_O toward a stationary point source S, the horn of a parked truck. The observer hears a frequency f' that is greater than the source frequency.

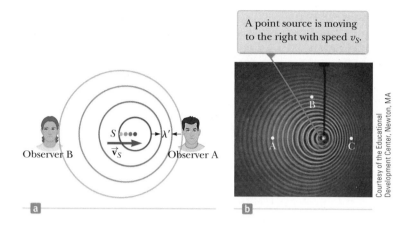

A point source is moving to the right with speed v_S.

a

b

Courtesy of the Educational Development Center, Newton, MA

Observer B

S ●●●● λ'

$\vec{v}_S$

Observer A

Figure 17.10 (a) A source S moving with a speed v_S toward a stationary observer A and away from a stationary observer B. Observer A hears an increased frequency, and observer B hears a decreased frequency. (b) The Doppler effect in water, observed in a ripple tank. Letters shown in the photo refer to Quick Quiz 17.4.

toward the source, and a negative value is substituted when the observer moves away from the source.

Now suppose the source is in motion and the observer is at rest. If the source moves directly toward observer A in Figure 17.10a, each new wave is emitted from a position to the right of the origin of the previous wave. As a result, the wave fronts heard by the observer are closer together than they would be if the source were not moving. (Fig. 17.10b shows this effect for waves moving on the surface of water.) As a result, the wavelength λ' measured by observer A is shorter than the wavelength λ of the source. During each vibration, which lasts for a time interval T (the period), the source moves a distance $v_S T = v_S/f$ and the wavelength is *shortened* by this amount. Therefore, the observed wavelength λ' is

$$\lambda' = \lambda - \Delta\lambda = \lambda - \frac{v_S}{f}$$

Because $\lambda = v/f$, the frequency f' heard by observer A is

$$f' = \frac{v}{\lambda'} = \frac{v}{\lambda - (v_S/f)} = \frac{v}{(v/f) - (v_S/f)}$$

$$f' = \left(\frac{v}{v - v_S}\right)f \quad \text{(source moving toward observer)} \qquad \textbf{(17.17)}$$

That is, the observed frequency is *increased* whenever the source is moving toward the observer.

When the source moves away from a stationary observer, as is the case for observer B in Figure 17.10a, the observer measures a wavelength λ' that is *greater* than λ and hears a *decreased* frequency:

$$f' = \left(\frac{v}{v + v_S}\right)f \quad \text{(source moving away from observer)} \qquad \textbf{(17.18)}$$

We can express the general relationship for the observed frequency when a source is moving and an observer is at rest as Equation 17.17, with the same sign convention applied to v_S as was applied to v_O: a positive value is substituted for v_S when the source moves toward the observer, and a negative value is substituted when the source moves away from the observer.

Finally, combining Equations 17.15 and 17.17 gives the following general relationship for the observed frequency that includes all four conditions described by Equations 17.15 through 17.18:

$$f' = \left(\frac{v + v_O}{v - v_S}\right)f \qquad \textbf{(17.19)}$$

◀ General Doppler–shift expression

Pitfall Prevention 17.1
Doppler Effect Does Not Depend on Distance Some people think that the Doppler effect depends on the distance between the source and the observer. Although the *intensity* of a sound varies as the distance changes, the apparent *frequency* depends only on the relative speed of source and observer. As you listen to an approaching source, you will detect increasing intensity but constant frequency. As the source passes, you will hear the frequency suddenly drop to a new constant value and the intensity begin to decrease.

In this expression, the signs for the values substituted for v_O and v_S depend on the direction of the velocity. A positive value is used for motion of the observer or the source *toward* the other (associated with an *increase* in observed frequency), and a negative value is used for motion of one *away from* the other (associated with a *decrease* in observed frequency).

Although the Doppler effect is most typically experienced with sound waves, it is a phenomenon common to all waves. For example, the relative motion of source and observer produces a frequency shift in light waves. The Doppler effect is used in police radar systems to measure the speeds of motor vehicles. Likewise, astronomers use the effect to determine the speeds of stars, galaxies, and other celestial objects relative to the Earth.

Quick Quiz 17.4 Consider detectors of water waves at three locations A, B, and C in Figure 17.10b. Which of the following statements is true? **(a)** The wave speed is highest at location A. **(b)** The wave speed is highest at location C. **(c)** The detected wavelength is largest at location B. **(d)** The detected wavelength is largest at location C. **(e)** The detected frequency is highest at location C. **(f)** The detected frequency is highest at location A.

Quick Quiz 17.5 You stand on a platform at a train station and listen to a train approaching the station at a constant velocity. While the train approaches, but before it arrives, what do you hear? **(a)** the intensity and the frequency of the sound both increasing **(b)** the intensity and the frequency of the sound both decreasing **(c)** the intensity increasing and the frequency decreasing **(d)** the intensity decreasing and the frequency increasing **(e)** the intensity increasing and the frequency remaining the same **(f)** the intensity decreasing and the frequency remaining the same

---

Example 17.4   **The Broken Clock Radio**

Your clock radio awakens you with a steady and irritating sound of frequency 600 Hz. One morning, it malfunctions and cannot be turned off. In frustration, you drop the clock radio out of your fourth-story dorm window, 15.0 m from the ground. Assume the speed of sound is 343 m/s. As you listen to the falling clock radio, what frequency do you hear just before you hear it striking the ground?

SOLUTION

Conceptualize The speed of the clock radio increases as it falls. Therefore, it is a source of sound moving away from you with an increasing speed so the frequency you hear should be less than 600 Hz.

Categorize We categorize this problem as one in which we combine the *particle under constant acceleration* model for the falling radio with our understanding of the frequency shift of sound due to the Doppler effect.

Analyze Because the clock radio is modeled as a particle under constant acceleration due to gravity, use Equation 2.13 to express the speed of the source of sound:

$$(1)\quad v_S = v_{yi} + a_y t = 0 - gt = -gt$$

From Equation 2.16, find the time at which the clock radio strikes the ground:

$$y_f = y_i + v_{yi}t - \tfrac{1}{2}gt^2 = 0 + 0 - \tfrac{1}{2}gt^2 \quad \rightarrow \quad t = \sqrt{-\frac{2y_f}{g}}$$

Substitute into Equation (1):

$$v_S = (-g)\sqrt{-\frac{2y_f}{g}} = -\sqrt{-2gy_f}$$

Use Equation 17.19 to determine the Doppler-shifted frequency heard from the falling clock radio:

$$f' = \left[\frac{v + 0}{v - (-\sqrt{-2gy_f})}\right]f = \left(\frac{v}{v + \sqrt{-2gy_f}}\right)f$$

▶ **17.4** continued

Substitute numerical values:

$$f' = \left[\frac{343 \text{ m/s}}{343 \text{ m/s} + \sqrt{-2(9.80 \text{ m/s}^2)(-15.0 \text{ m})}} \right](600 \text{ Hz})$$

$$= \boxed{571 \text{ Hz}}$$

Finalize The frequency is lower than the actual frequency of 600 Hz because the clock radio is moving away from you. If it were to fall from a higher floor so that it passes below $y = -15.0$ m, the clock radio would continue to accelerate and the frequency would continue to drop.

Example 17.5   Doppler Submarines

A submarine (sub A) travels through water at a speed of 8.00 m/s, emitting a sonar wave at a frequency of 1 400 Hz. The speed of sound in the water is 1 533 m/s. A second submarine (sub B) is located such that both submarines are traveling directly toward each other. The second submarine is moving at 9.00 m/s.

(A) What frequency is detected by an observer riding on sub B as the subs approach each other?

SOLUTION

Conceptualize Even though the problem involves subs moving in water, there is a Doppler effect just like there is when you are in a moving car and listening to a sound moving through the air from another car.

Categorize Because both subs are moving, we categorize this problem as one involving the Doppler effect for both a moving source and a moving observer.

Analyze Use Equation 17.19 to find the Doppler-shifted frequency heard by the observer in sub B, being careful with the signs assigned to the source and observer speeds:

$$f' = \left(\frac{v + v_O}{v - v_S} \right) f$$

$$f' = \left[\frac{1\ 533 \text{ m/s} + (+9.00 \text{ m/s})}{1\ 533 \text{ m/s} - (+8.00 \text{ m/s})} \right](1\ 400 \text{ Hz}) = \boxed{1\ 416 \text{ Hz}}$$

(B) The subs barely miss each other and pass. What frequency is detected by an observer riding on sub B as the subs recede from each other?

SOLUTION

Use Equation 17.19 to find the Doppler-shifted frequency heard by the observer in sub B, again being careful with the signs assigned to the source and observer speeds:

$$f' = \left(\frac{v + v_O}{v - v_S} \right) f$$

$$f' = \left[\frac{1\ 533 \text{ m/s} + (-9.00 \text{ m/s})}{1\ 533 \text{ m/s} - (-8.00 \text{ m/s})} \right](1\ 400 \text{ Hz}) = \boxed{1\ 385 \text{ Hz}}$$

Notice that the frequency drops from 1 416 Hz to 1 385 Hz as the subs pass. This effect is similar to the drop in frequency you hear when a car passes by you while blowing its horn.

(C) While the subs are approaching each other, some of the sound from sub A reflects from sub B and returns to sub A. If this sound were to be detected by an observer on sub A, what is its frequency?

SOLUTION

The sound of apparent frequency 1 416 Hz found in part (A) is reflected from a moving source (sub B) and then detected by a moving observer (sub A). Find the frequency detected by sub A:

$$f'' = \left(\frac{v + v_O}{v - v_S} \right) f'$$

$$= \left[\frac{1\ 533 \text{ m/s} + (+8.00 \text{ m/s})}{1\ 533 \text{ m/s} - (+9.00 \text{ m/s})} \right](1\ 416 \text{ Hz}) = \boxed{1\ 432 \text{ Hz}}$$

continued

▶ **17.5** c o n t i n u e d

Finalize This technique is used by police officers to measure the speed of a moving car. Microwaves are emitted from the police car and reflected by the moving car. By detecting the Doppler-shifted frequency of the reflected microwaves, the police officer can determine the speed of the moving car.

Figure 17.11 (a) A representation of a shock wave produced when a source moves from S_0 to the right with a speed v_S that is greater than the wave speed v in the medium. (b) A stroboscopic photograph of a bullet moving at supersonic speed through the hot air above a candle.

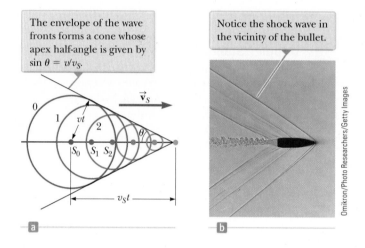

The envelope of the wave fronts forms a cone whose apex half-angle is given by $\sin \theta = v/v_S$.

Notice the shock wave in the vicinity of the bullet.

Shock Waves

Now consider what happens when the speed v_S of a source *exceeds* the wave speed v. This situation is depicted graphically in Figure 17.11a. The circles represent spherical wave fronts emitted by the source at various times during its motion. At $t = 0$, the source is at S_0 and moving toward the right. At later times, the source is at S_1, and then S_2, and so on. At the time t, the wave front centered at S_0 reaches a radius of vt. In this same time interval, the source travels a distance $v_S t$. Notice in Figure 17.11a that a straight line can be drawn tangent to all the wave fronts generated at various times. Therefore, the envelope of these wave fronts is a cone whose apex half-angle θ (the "Mach angle") is given by

$$\sin \theta = \frac{vt}{v_S t} = \frac{v}{v_S}$$

The ratio v_S/v is referred to as the *Mach number*, and the conical wave front produced when $v_S > v$ (supersonic speeds) is known as a *shock wave*. An interesting analogy to shock waves is the V-shaped wave fronts produced by a boat (the bow wave) when the boat's speed exceeds the speed of the surface-water waves (Fig. 17.12).

Jet airplanes traveling at supersonic speeds produce shock waves, which are responsible for the loud "sonic boom" one hears. The shock wave carries a great deal of energy concentrated on the surface of the cone, with correspondingly great pressure variations. Such shock waves are unpleasant to hear and can cause damage to buildings when aircraft fly supersonically at low altitudes. In fact, an airplane flying at supersonic speeds produces a double boom because two shock waves are formed, one from the nose of the plane and one from the tail. People near the path of a space shuttle as it glides toward its landing point have reported hearing what sounds like two very closely spaced cracks of thunder.

Figure 17.12 The V-shaped bow wave of a boat is formed because the boat speed is greater than the speed of the water waves it generates. A bow wave is analogous to a shock wave formed by an airplane traveling faster than sound.

ⓠuick Quiz 17.6 An airplane flying with a constant velocity moves from a cold air mass into a warm air mass. Does the Mach number **(a)** increase, **(b)** decrease, or **(c)** stay the same?

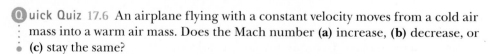

Summary

Definitions

The **intensity** of a periodic sound wave, which is the power per unit area, is

$$I \equiv \frac{(Power)_{avg}}{A} = \frac{(\Delta P_{max})^2}{2\rho v} \quad \text{(17.11, 17.12)}$$

The **sound level** of a sound wave in decibels is

$$\beta \equiv 10 \log \left(\frac{I}{I_0} \right) \quad \text{(17.14)}$$

The constant I_0 is a reference intensity, usually taken to be at the threshold of hearing (1.00×10^{-12} W/m²), and I is the intensity of the sound wave in watts per square meter.

Concepts and Principles

Sound waves are longitudinal and travel through a compressible medium with a speed that depends on the elastic and inertial properties of that medium. The speed of sound in a gas having a bulk modulus B and density ρ is

$$v = \sqrt{\frac{B}{\rho}} \quad \text{(17.8)}$$

For sinusoidal sound waves, the variation in the position of an element of the medium is

$$s(x, t) = s_{max} \cos (kx - \omega t) \quad \text{(17.1)}$$

and the variation in pressure from the equilibrium value is

$$\Delta P = \Delta P_{max} \sin (kx - \omega t) \quad \text{(17.2)}$$

where ΔP_{max} is the **pressure amplitude.** The pressure wave is 90° out of phase with the displacement wave. The relationship between s_{max} and ΔP_{max} is

$$\Delta P_{max} = \rho v \omega s_{max} \quad \text{(17.10)}$$

The change in frequency heard by an observer whenever there is relative motion between a source of sound waves and the observer is called the **Doppler effect.** The observed frequency is

$$f' = \left(\frac{v + v_O}{v - v_S} \right) f \quad \text{(17.19)}$$

In this expression, the signs for the values substituted for v_O and v_S depend on the direction of the velocity. A positive value for the speed of the observer or source is substituted if the velocity of one is toward the other, whereas a negative value represents a velocity of one away from the other.

Objective Questions

1. Table 17.1 shows the speed of sound is typically an order of magnitude larger in solids than in gases. To what can this higher value be most directly attributed? (a) the difference in density between solids and gases (b) the difference in compressibility between solids and gases (c) the limited size of a solid object compared to a free gas (d) the impossibility of holding a gas under significant tension

2. Two sirens A and B are sounding so that the frequency from A is twice the frequency from B. Compared with the speed of sound from A, is the speed of sound from B (a) twice as fast, (b) half as fast, (c) four times as fast, (d) one-fourth as fast, or (e) the same?

3. As you travel down the highway in your car, an ambulance approaches you from the rear at a high speed (Fig. OQ17.3) sounding its siren at a frequency of 500 Hz. Which statement is correct? (a) You hear a frequency less than 500 Hz. (b) You hear a frequency equal to 500 Hz. (c) You hear a frequency greater

Anthony Redpath/Corbis

Figure OQ17.3

than 500 Hz. (d) You hear a frequency greater than 500 Hz, whereas the ambulance driver hears a frequency lower than 500 Hz. (e) You hear a frequency less than 500 Hz, whereas the ambulance driver hears a frequency of 500 Hz.

4. What happens to a sound wave as it travels from air into water? (a) Its intensity increases. (b) Its wavelength decreases. (c) Its frequency increases. (d) Its frequency remains the same. (e) Its velocity decreases.

5. A church bell in a steeple rings once. At 300 m in front of the church, the maximum sound intensity is 2 μW/m². At 950 m behind the church, the maximum intensity is 0.2 μW/m². What is the main reason for the difference in the intensity? (a) Most of the sound is absorbed by the air before it gets far away from the source. (b) Most of the sound is absorbed by the ground as it travels away from the source. (c) The bell broadcasts the sound mostly toward the front. (d) At a larger distance, the power is spread over a larger area.

6. If a 1.00-kHz sound source moves at a speed of 50.0 m/s toward a listener who moves at a speed of 30.0 m/s in a direction away from the source, what is the apparent frequency heard by the listener? (a) 796 Hz (b) 949 Hz (c) 1 000 Hz (d) 1 068 Hz (e) 1 273 Hz

7. A sound wave can be characterized as (a) a transverse wave, (b) a longitudinal wave, (c) a transverse wave or a longitudinal wave, depending on the nature of its source, (d) one that carries no energy, or (e) a wave that does not require a medium to be transmitted from one place to the other.

8. Assume a change at the source of sound reduces the wavelength of a sound wave in air by a factor of 2. **(i)** What happens to its frequency? (a) It increases by a factor of 4. (b) It increases by a factor of 2. (c) It is unchanged. (d) It decreases by a factor of 2. (e) It changes by an unpredictable factor. **(ii)** What happens to its speed? Choose from the same possibilities as in part (i).

9. A point source broadcasts sound into a uniform medium. If the distance from the source is tripled, how does the intensity change? (a) It becomes one-ninth as large. (b) It becomes one-third as large. (c) It is unchanged. (d) It becomes three times larger. (e) It becomes nine times larger.

10. Suppose an observer and a source of sound are both at rest relative to the ground and a strong wind is blowing away from the source toward the observer. **(i)** What effect does the wind have on the observed frequency? (a) It causes an increase. (b) It causes a decrease. (c) It causes no change. **(ii)** What effect does the wind have on the observed wavelength? Choose from the same possibilities as in part (i). **(iii)** What effect does the wind have on the observed speed of the wave? Choose from the same possibilities as in part (i).

11. A source of sound vibrates with constant frequency. Rank the frequency of sound observed in the following cases from highest to the lowest. If two frequencies are equal, show their equality in your ranking. All the motions mentioned have the same speed, 25 m/s. (a) The source and observer are stationary. (b) The source is moving toward a stationary observer. (c) The source is moving away from a stationary observer. (d) The observer is moving toward a stationary source. (e) The observer is moving away from a stationary source.

12. With a sensitive sound-level meter, you measure the sound of a running spider as −10 dB. What does the negative sign imply? (a) The spider is moving away from you. (b) The frequency of the sound is too low to be audible to humans. (c) The intensity of the sound is too faint to be audible to humans. (d) You have made a mistake; negative signs do not fit with logarithms.

13. Doubling the power output from a sound source emitting a single frequency will result in what increase in decibel level? (a) 0.50 dB (b) 2.0 dB (c) 3.0 dB (d) 4.0 dB (e) above 20 dB

14. Of the following sounds, which one is most likely to have a sound level of 60 dB? (a) a rock concert (b) the turning of a page in this textbook (c) dinner-table conversation (d) a cheering crowd at a football game

Conceptual Questions

1. denotes answer available in *Student Solutions Manual/Study Guide*

1. How can an object move with respect to an observer so that the sound from it is not shifted in frequency?

2. Older auto-focus cameras sent out a pulse of sound and measured the time interval required for the pulse to reach an object, reflect off of it, and return to be detected. Can air temperature affect the camera's focus? New cameras use a more reliable infrared system.

3. A friend sitting in her car far down the road waves to you and beeps her horn at the same moment. How far away must she be for you to calculate the speed of sound to two significant figures by measuring the time interval required for the sound to reach you?

4. How can you determine that the speed of sound is the same for all frequencies by listening to a band or orchestra?

5. Explain how the distance to a lightning bolt (Fig. CQ17.5) can be determined by counting the seconds between the flash and the sound of thunder.

6. You are driving toward a cliff and honk your horn. Is there a Doppler shift of the sound when you hear the echo? If so, is it like a moving source or a moving observer? What if the reflection occurs not from a cliff, but from the forward edge of a huge alien spacecraft moving toward you as you drive?

Figure CQ17.5

© iStockphoto.com/Colin Orthner

7. The radar systems used by police to detect speeders are sensitive to the Doppler shift of a pulse of microwaves. Discuss how this sensitivity can be used to measure the speed of a car.

8. *The Tunguska event.* On June 30, 1908, a meteor burned up and exploded in the atmosphere above the Tunguska River valley in Siberia. It knocked down trees over thousands of square kilometers and started a forest fire, but produced no crater and apparently caused no human casualties. A witness sitting on his doorstep outside the zone of falling trees recalled events in the following sequence. He saw a moving light in the sky, brighter than the Sun and descending at a low angle to the horizon. He felt his face become warm. He felt the ground shake. An invisible agent picked him up and immediately dropped him about a meter from where he had been seated. He heard a very loud protracted rumbling. Suggest an explanation for these observations and for the order in which they happened.

9. A sonic ranger is a device that determines the distance to an object by sending out an ultrasonic sound pulse and measuring the time interval required for the wave to return by reflection from the object. Typically, these devices cannot reliably detect an object that is less than half a meter from the sensor. Why is that?

Problems available in Access end-of-chapter problems online at www.webassign.net

Section 17.1 Pressure Variations in Sound Waves
Problems 1–3

Section 17.2 Speed of Sound Waves
Problems 4–18

Section 17.3 Intensity of Periodic Sound Waves
Problems 19–36

Section 17.4 The Doppler Effect
Problems 37–47

Additional Problems
Problems 48–70

Challenge Problems
Problems 71–73

Solutions to the following Problems are available in the *Student Solutions Manual/Study Guide:*

List of Enhanced Problems

Problem Number	Targeted Feedback in Enhanced WebAssign	Analysis Model Tutorial in Enhanced WebAssign	Master It in Enhanced WebAssign	Watch It in Enhanced WebAssign
17.1	✓			✓
17.2	✓			✓
17.4	✓		✓	
17.10	✓			✓
17.11	✓			✓
17.12	✓			✓
17.13	✓	✓		✓
17.19	✓			
17.25	✓			✓
17.27	✓		✓	
17.31			✓	
17.32	✓			✓
17.33			✓	
17.41		✓		
17.45	✓		✓	
17.47		✓	✓	
17.57		✓		
17.61			✓	
17.69			✓	

Superposition and Standing Waves

The wave model was introduced in the previous two chapters. We have seen that waves are very different from particles. A particle is of zero size, whereas a wave has a characteristic size, its wavelength. Another important difference between waves and particles is that we can explore the possibility of two or more waves combining at one point in the same medium. Particles can be combined to form extended objects, but the particles must be at *different* locations. In contrast, two waves can both be present at the same location. The ramifications of this possibility are explored in this chapter.

When waves are combined in systems with boundary conditions, only certain allowed frequencies can exist and we say the frequencies are *quantized.* Quantization is a notion that is at the heart of quantum mechanics, a subject introduced formally in Chapter 40. There we show that analysis of waves under boundary conditions explains many of the quantum phenomena. In this chapter, we use quantization to understand the behavior of the wide array of musical instruments that are based on strings and air columns.

Blues master B. B. King takes advantage of standing waves on strings. He changes to higher notes on the guitar by pushing the strings against the frets on the fingerboard, shortening the lengths of the portions of the strings that vibrate. *(AP Photo/Danny Moloshok)*

We also consider the combination of waves having different frequencies. When two sound waves having nearly the same frequency interfere, we hear variations in the loudness called *beats*. Finally, we discuss how any nonsinusoidal periodic wave can be described as a sum of sine and cosine functions.

18.1 Analysis Model: Waves in Interference

Many interesting wave phenomena in nature cannot be described by a single traveling wave. Instead, one must analyze these phenomena in terms of a combination of traveling waves. As noted in the introduction, waves have a remarkable difference from particles in that waves can be combined at the *same* location in space. To analyze such wave combinations, we make use of the **superposition principle:**

Superposition principle ▶

> If two or more traveling waves are moving through a medium, the resultant value of the wave function at any point is the algebraic sum of the values of the wave functions of the individual waves.

Waves that obey this principle are called *linear waves*. (See Section 16.6.) In the case of mechanical waves, linear waves are generally characterized by having amplitudes much smaller than their wavelengths. Waves that violate the superposition principle are called *nonlinear waves* and are often characterized by large amplitudes. In this book, we deal only with linear waves.

One consequence of the superposition principle is that two traveling waves can pass through each other without being destroyed or even altered. For instance, when two pebbles are thrown into a pond and hit the surface at different locations, the expanding circular surface waves from the two locations simply pass through each other with no permanent effect. The resulting complex pattern can be viewed as two independent sets of expanding circles.

Figure 18.1 is a pictorial representation of the superposition of two pulses. The wave function for the pulse moving to the right is y_1, and the wave function for the pulse moving to the left is y_2. The pulses have the same speed but different shapes, and the displacement of the elements of the medium is in the positive y direction for both pulses. When the waves overlap (Fig. 18.1b), the wave function for the resulting complex wave is given by $y_1 + y_2$. When the crests of the pulses coincide (Fig. 18.1c), the resulting wave given by $y_1 + y_2$ has a larger amplitude than that of the individual pulses. The two pulses finally separate and continue moving in their original directions (Fig. 18.1d). Notice that the pulse shapes remain unchanged after the interaction, as if the two pulses had never met!

The combination of separate waves in the same region of space to produce a resultant wave is called **interference.** For the two pulses shown in Figure 18.1, the displacement of the elements of the medium is in the positive y direction for both pulses, and the resultant pulse (created when the individual pulses overlap) exhibits an amplitude greater than that of either individual pulse. Because the displacements caused by the two pulses are in the same direction, we refer to their superposition as **constructive interference.**

Constructive interference ▶

Now consider two pulses traveling in opposite directions on a taut string where one pulse is inverted relative to the other as illustrated in Figure 18.2. When these pulses begin to overlap, the resultant pulse is given by $y_1 + y_2$, but the values of the function y_2 are negative. Again, the two pulses pass through each other; because the displacements caused by the two pulses are in opposite directions, however, we refer to their superposition as **destructive interference.**

Destructive interference ▶

The superposition principle is the centerpiece of the analysis model called **waves in interference.** In many situations, both in acoustics and optics, waves combine according to this principle and exhibit interesting phenomena with practical applications.

Pitfall Prevention 18.1
Do Waves Actually *Interfere*? In popular usage, the term *interfere* implies that an agent affects a situation in some way so as to preclude something from happening. For example, in American football, *pass interference* means that a defending player has affected the receiver so that the receiver is unable to catch the ball. This usage is very different from its use in physics, where waves pass through each other and interfere, but do not affect each other in any way. In physics, interference is similar to the notion of *combination* as described in this chapter.

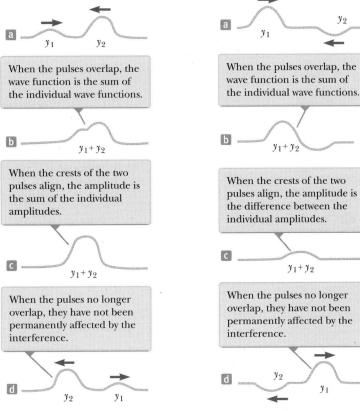

Figure 18.1 Constructive interference. Two positive pulses travel on a stretched string in opposite directions and overlap.

Figure 18.2 Destructive interference. Two pulses, one positive and one negative, travel on a stretched string in opposite directions and overlap.

Q uick **Quiz** 18.1 Two pulses move in opposite directions on a string and are identical in shape except that one has positive displacements of the elements of the string and the other has negative displacements. At the moment the two pulses completely overlap on the string, what happens? **(a)** The energy associated with the pulses has disappeared. **(b)** The string is not moving. **(c)** The string forms a straight line. **(d)** The pulses have vanished and will not reappear.

Superposition of Sinusoidal Waves

Let us now apply the principle of superposition to two sinusoidal waves traveling in the same direction in a linear medium. If the two waves are traveling to the right and have the same frequency, wavelength, and amplitude but differ in phase, we can express their individual wave functions as

$$y_1 = A \sin (kx - \omega t) \quad y_2 = A \sin (kx - \omega t + \phi)$$

where, as usual, $k = 2\pi/\lambda$, $\omega = 2\pi f$, and ϕ is the phase constant as discussed in Section 16.2. Hence, the resultant wave function y is

$$y = y_1 + y_2 = A [\sin (kx - \omega t) + \sin (kx - \omega t + \phi)]$$

To simplify this expression, we use the trigonometric identity

$$\sin a + \sin b = 2 \cos \left(\frac{a - b}{2}\right) \sin \left(\frac{a + b}{2}\right)$$

Figure 18.3 The superposition of two identical waves y_1 and y_2 (blue and green, respectively) to yield a resultant wave (red-brown).

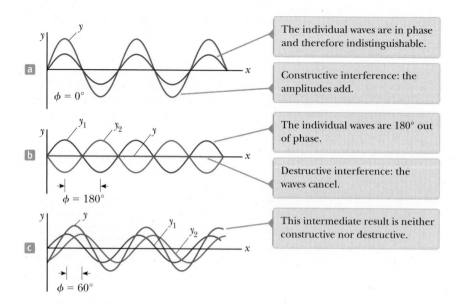

The individual waves are in phase and therefore indistinguishable.

Constructive interference: the amplitudes add.

$\phi = 0°$

The individual waves are 180° out of phase.

Destructive interference: the waves cancel.

$\phi = 180°$

This intermediate result is neither constructive nor destructive.

$\phi = 60°$

Letting $a = kx - \omega t$ and $b = kx - \omega t + \phi$, we find that the resultant wave function y reduces to

◀ Resultant of two traveling sinusoidal waves

$$y = 2A \cos\left(\frac{\phi}{2}\right) \sin\left(kx - \omega t + \frac{\phi}{2}\right)$$

This result has several important features. The resultant wave function y also is sinusoidal and has the same frequency and wavelength as the individual waves because the sine function incorporates the same values of k and ω that appear in the original wave functions. The amplitude of the resultant wave is $2A \cos(\phi/2)$, and its phase constant is $\phi/2$. If the phase constant ϕ of the original wave equals 0, then $\cos(\phi/2) = \cos 0 = 1$ and the amplitude of the resultant wave is $2A$, twice the amplitude of either individual wave. In this case, the crests of the two waves are at the same locations in space and the waves are said to be everywhere *in phase* and therefore interfere constructively. The individual waves y_1 and y_2 combine to form the red-brown curve y of amplitude $2A$ shown in Figure 18.3a. Because the individual waves are in phase, they are indistinguishable in Figure 18.3a, where they appear as a single blue curve. In general, constructive interference occurs when $\cos(\phi/2) = \pm 1$. That is true, for example, when $\phi = 0, 2\pi, 4\pi, \ldots$ rad, that is, when ϕ is an *even* multiple of π.

When ϕ is equal to π rad or to any *odd* multiple of π, then $\cos(\phi/2) = \cos(\pi/2) = 0$ and the crests of one wave occur at the same positions as the troughs of the second wave (Fig. 18.3b). Therefore, as a consequence of destructive interference, the resultant wave has *zero* amplitude everywhere as shown by the straight red-brown line in Figure 18.3b. Finally, when the phase constant has an arbitrary value other than 0 or an integer multiple of π rad (Fig. 18.3c), the resultant wave has an amplitude whose value is somewhere between 0 and $2A$.

In the more general case in which the waves have the same wavelength but different amplitudes, the results are similar with the following exceptions. In the in-phase case, the amplitude of the resultant wave is not twice that of a single wave, but rather is the sum of the amplitudes of the two waves. When the waves are π rad out of phase, they do not completely cancel as in Figure 18.3b. The result is a wave whose amplitude is the difference in the amplitudes of the individual waves.

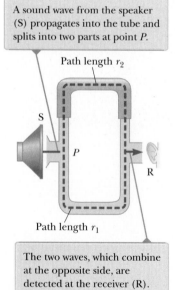

A sound wave from the speaker (S) propagates into the tube and splits into two parts at point P.

Path length r_2

S

P

R

Path length r_1

The two waves, which combine at the opposite side, are detected at the receiver (R).

Figure 18.4 An acoustical system for demonstrating interference of sound waves. The upper path length r_2 can be varied by sliding the upper section.

Interference of Sound Waves

One simple device for demonstrating interference of sound waves is illustrated in Figure 18.4. Sound from a loudspeaker S is sent into a tube at point P, where there is a T-shaped junction. Half the sound energy travels in one direction, and half travels in the opposite direction. Therefore, the sound waves that reach the receiver R can travel along either of the two paths. The distance along any path from speaker to receiver is called the **path length** r. The lower path length r_1 is fixed, but the upper path length r_2 can be varied by sliding the U-shaped tube, which is similar to that on a slide trombone. When the difference in the path lengths $\Delta r = |r_2 - r_1|$ is either zero or some integer multiple of the wavelength λ (that is, $\Delta r = n\lambda$, where $n = 0, 1, 2, 3, \ldots$), the two waves reaching the receiver at any instant are in phase and interfere constructively as shown in Figure 18.3a. For this case, a maximum in the sound intensity is detected at the receiver. If the path length r_2 is adjusted such that the path difference $\Delta r = \lambda/2, 3\lambda/2, \ldots, n\lambda/2$ (for n odd), the two waves are exactly π rad, or 180°, out of phase at the receiver and hence cancel each other. In this case of destructive interference, no sound is detected at the receiver. This simple experiment demonstrates that a phase difference may arise between two waves generated by the same source when they travel along paths of unequal lengths. This important phenomenon will be indispensable in our investigation of the interference of light waves in Chapter 37.

Analysis Model   Waves in Interference

Imagine two waves traveling in the same location through a medium. The displacement of elements of the medium is affected by both waves. According to the **principle of superposition**, the displacement is the sum of the individual displacements that would be caused by each wave. When the waves are in phase, **constructive interference** occurs and the resultant displacement is larger than the individual displacements. **Destructive interference** occurs when the waves are out of phase.

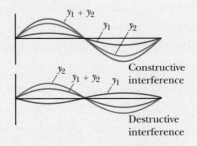

Examples:

- a piano tuner listens to a piano string and a tuning fork vibrating together and notices beats (Section 18.7)
- light waves from two coherent sources combine to form an interference pattern on a screen (Chapter 37)
- a thin film of oil on top of water shows swirls of color (Chapter 37)
- x-rays passing through a crystalline solid combine to form a Laue pattern (Chapter 38)

Example 18.1   Two Speakers Driven by the Same Source   AM

Two identical loudspeakers placed 3.00 m apart are driven by the same oscillator (Fig. 18.5). A listener is originally at point O, located 8.00 m from the center of the line connecting the two speakers. The listener then moves to point P, which is a perpendicular distance 0.350 m from O, and she experiences the *first minimum* in sound intensity. What is the frequency of the oscillator?

SOLUTION

Conceptualize In Figure 18.4, a sound wave enters a tube and is then *acoustically* split into two different paths before recombining at the other end. In this example, a signal representing the sound is *electrically* split and sent to two different loudspeakers. After leaving the speakers, the sound waves recombine at the position of the listener. Despite the difference in how the splitting occurs, the path difference discussion related to Figure 18.4 can be applied here.

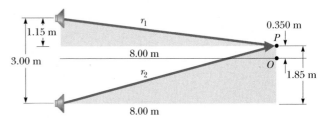

Figure 18.5 (Example 18.1) Two identical loudspeakers emit sound waves to a listener at P.

Categorize Because the sound waves from two separate sources combine, we apply the *waves in interference* analysis model.

continued

▶ **18.1** continued

Analyze Figure 18.5 shows the physical arrangement of the speakers, along with two shaded right triangles that can be drawn on the basis of the lengths described in the problem. The first minimum occurs when the two waves reaching the listener at point P are 180° out of phase, in other words, when their path difference Δr equals $\lambda/2$.

From the shaded triangles, find the path lengths from the speakers to the listener:

$$r_1 = \sqrt{(8.00 \text{ m})^2 + (1.15 \text{ m})^2} = 8.08 \text{ m}$$

$$r_2 = \sqrt{(8.00 \text{ m})^2 + (1.85 \text{ m})^2} = 8.21 \text{ m}$$

Hence, the path difference is $r_2 - r_1 = 0.13$ m. Because this path difference must equal $\lambda/2$ for the first minimum, $\lambda = 0.26$ m.

To obtain the oscillator frequency, use Equation 16.12, $v = \lambda f$, where v is the speed of sound in air, 343 m/s:

$$f = \frac{v}{\lambda} = \frac{343 \text{ m/s}}{0.26 \text{ m}} = \boxed{1.3 \text{ kHz}}$$

Finalize This example enables us to understand why the speaker wires in a stereo system should be connected properly. When connected the wrong way—that is, when the positive (or red) wire is connected to the negative (or black) terminal on one of the speakers and the other is correctly wired—the speakers are said to be "out of phase," with one speaker moving outward while the other moves inward. As a consequence, the sound wave coming from one speaker destructively interferes with the wave coming from the other at point O in Figure 18.5. A rarefaction region due to one speaker is superposed on a compression region from the other speaker. Although the two sounds probably do not completely cancel each other (because the left and right stereo signals are usually not identical), a substantial loss of sound quality occurs at point O.

WHAT IF? What if the speakers were connected out of phase? What happens at point P in Figure 18.5?

Answer In this situation, the path difference of $\lambda/2$ combines with a phase difference of $\lambda/2$ due to the incorrect wiring to give a full phase difference of λ. As a result, the waves are in phase and there is a *maximum* intensity at point P.

Figure 18.6 Two identical loudspeakers emit sound waves toward each other. When they overlap, identical waves traveling in opposite directions will combine to form standing waves.

18.2 Standing Waves

The sound waves from the pair of loudspeakers in Example 18.1 leave the speakers in the forward direction, and we considered interference at a point in front of the speakers. Suppose we turn the speakers so that they face each other and then have them emit sound of the same frequency and amplitude. In this situation, two identical waves travel in opposite directions in the same medium as in Figure 18.6. These waves combine in accordance with the waves in interference model.

We can analyze such a situation by considering wave functions for two transverse sinusoidal waves having the same amplitude, frequency, and wavelength but traveling in opposite directions in the same medium:

$$y_1 = A \sin(kx - \omega t) \qquad y_2 = A \sin(kx + \omega t)$$

where y_1 represents a wave traveling in the positive x direction and y_2 represents one traveling in the negative x direction. Adding these two functions gives the resultant wave function y:

$$y = y_1 + y_2 = A \sin(kx - \omega t) + A \sin(kx + \omega t)$$

When we use the trigonometric identity $\sin(a \pm b) = \sin a \cos b \pm \cos a \sin b$, this expression reduces to

$$y = (2A \sin kx) \cos \omega t \qquad \text{(18.1)}$$

Equation 18.1 represents the wave function of a **standing wave.** A standing wave such as the one on a string shown in Figure 18.7 is an oscillation pattern *with a stationary outline* that results from the superposition of two identical waves traveling in opposite directions.

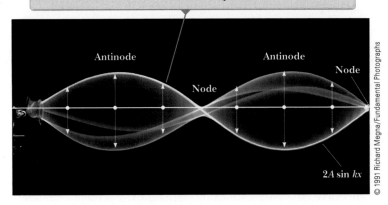

The amplitude of the vertical oscillation of any element of the string depends on the horizontal position of the element. Each element vibrates within the confines of the envelope function 2A sin kx.

Antinode   Antinode

Node   Node

2A sin kx

© 1991 Richard Megna/Fundamental Photographs

Figure 18.7 Multiflash photograph of a standing wave on a string. The time behavior of the vertical displacement from equilibrium of an individual element of the string is given by cos ωt. That is, each element vibrates at an angular frequency ω.

Notice that Equation 18.1 does not contain a function of $kx - \omega t$. Therefore, it is not an expression for a single traveling wave. When you observe a standing wave, there is no sense of motion in the direction of propagation of either original wave. Comparing Equation 18.1 with Equation 15.6, we see that it describes a special kind of simple harmonic motion. Every element of the medium oscillates in simple harmonic motion with the same angular frequency ω (according to the cos ωt factor in the equation). The amplitude of the simple harmonic motion of a given element (given by the factor 2A sin kx, the coefficient of the cosine function) depends on the location x of the element in the medium, however.

If you can find a noncordless telephone with a coiled cord connecting the handset to the base unit, you can see the difference between a standing wave and a traveling wave. Stretch the coiled cord out and flick it with a finger. You will see a pulse traveling along the cord. Now shake the handset up and down and adjust your shaking frequency until every coil on the cord is moving up at the same time and then down. That is a standing wave, formed from the combination of waves moving away from your hand and reflected from the base unit toward your hand. Notice that there is no sense of traveling along the cord like there was for the pulse. You only see up-and-down motion of the elements of the cord.

Equation 18.1 shows that the amplitude of the simple harmonic motion of an element of the medium has a minimum value of zero when x satisfies the condition sin kx = 0, that is, when

$$kx = 0, \pi, 2\pi, 3\pi, \ldots$$

Because $k = 2\pi/\lambda$, these values for kx give

$$x = 0, \frac{\lambda}{2}, \lambda, \frac{3\lambda}{2}, \ldots = \frac{n\lambda}{2} \quad n = 0, 1, 2, 3, \ldots \quad (18.2)$$ ◀ Positions of nodes

These points of zero amplitude are called **nodes.**

The element of the medium with the *greatest* possible displacement from equilibrium has an amplitude of 2A, which we define as the amplitude of the standing wave. The positions in the medium at which this maximum displacement occurs are called **antinodes.** The antinodes are located at positions for which the coordinate x satisfies the condition sin kx = ±1, that is, when

$$kx = \frac{\pi}{2}, \frac{3\pi}{2}, \frac{5\pi}{2}, \ldots$$

Therefore, the positions of the antinodes are given by

$$x = \frac{\lambda}{4}, \frac{3\lambda}{4}, \frac{5\lambda}{4}, \ldots = \frac{n\lambda}{4} \quad n = 1, 3, 5, \ldots \quad (18.3)$$ ◀ Positions of antinodes

Pitfall Prevention 18.2
Three Types of Amplitude We need to distinguish carefully here between the **amplitude of the individual waves,** which is A, and the **amplitude of the simple harmonic motion of the elements of the medium,** which is 2A sin kx. A given element in a standing wave vibrates within the constraints of the *envelope* function 2A sin kx, where x is that element's position in the medium. Such vibration is in contrast to traveling sinusoidal waves, in which all elements oscillate with the same amplitude and the same frequency and the amplitude A of the wave is the same as the amplitude A of the simple harmonic motion of the elements. Furthermore, we can identify the **amplitude of the standing wave** as 2A.

Figure 18.8 Standing-wave patterns produced at various times by two waves of equal amplitude traveling in opposite directions. For the resultant wave y, the nodes (N) are points of zero displacement and the antinodes (A) are points of maximum displacement.

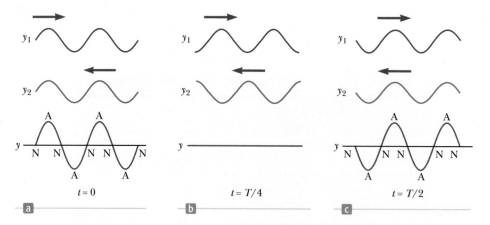

$t = 0$ a

$t = T/4$ b

$t = T/2$ c

Two nodes and two antinodes are labeled in the standing wave in Figure 18.7. The light blue curve labeled $2A \sin kx$ in Figure 18.7 represents one wavelength of the traveling waves that combine to form the standing wave. Figure 18.7 and Equations 18.2 and 18.3 provide the following important features of the locations of nodes and antinodes:

> The distance between adjacent antinodes is equal to $\lambda/2$.
> The distance between adjacent nodes is equal to $\lambda/2$.
> The distance between a node and an adjacent antinode is $\lambda/4$.

Wave patterns of the elements of the medium produced at various times by two transverse traveling waves moving in opposite directions are shown in Figure 18.8. The blue and green curves are the wave patterns for the individual traveling waves, and the red-brown curves are the wave patterns for the resultant standing wave. At $t = 0$ (Fig. 18.8a), the two traveling waves are in phase, giving a wave pattern in which each element of the medium is at rest and experiencing its maximum displacement from equilibrium. One-quarter of a period later, at $t = T/4$ (Fig. 18.8b), the traveling waves have moved one-fourth of a wavelength (one to the right and the other to the left). At this time, the traveling waves are out of phase, and each element of the medium is passing through the equilibrium position in its simple harmonic motion. The result is zero displacement for elements at all values of x; that is, the wave pattern is a straight line. At $t = T/2$ (Fig. 18.8c), the traveling waves are again in phase, producing a wave pattern that is inverted relative to the $t = 0$ pattern. In the standing wave, the elements of the medium alternate in time between the extremes shown in Figures 18.8a and 18.8c.

Quick Quiz 18.2 Consider the waves in Figure 18.8 to be waves on a stretched string. Define the velocity of elements of the string as positive if they are moving upward in the figure. **(i)** At the moment the string has the shape shown by the red-brown curve in Figure 18.8a, what is the instantaneous velocity of elements along the string? (a) zero for all elements (b) positive for all elements (c) negative for all elements (d) varies with the position of the element **(ii)** From the same choices, at the moment the string has the shape shown by the red-brown curve in Figure 18.8b, what is the instantaneous velocity of elements along the string?

Example 18.2   **Formation of a Standing Wave**

Two waves traveling in opposite directions produce a standing wave. The individual wave functions are

$$y_1 = 4.0 \sin (3.0x - 2.0t)$$
$$y_2 = 4.0 \sin (3.0x + 2.0t)$$

where x and y are measured in centimeters and t is in seconds.

(A) Find the amplitude of the simple harmonic motion of the element of the medium located at $x = 2.3$ cm.

▶ **18.2** c o n t i n u e d

SOLUTION

Conceptualize The waves described by the given equations are identical except for their directions of travel, so they indeed combine to form a standing wave as discussed in this section. We can represent the waves graphically by the blue and green curves in Figure 18.8.

Categorize We will substitute values into equations developed in this section, so we categorize this example as a substitution problem.

From the equations for the waves, we see that $A = 4.0$ cm, $k = 3.0$ rad/cm, and $\omega = 2.0$ rad/s. Use Equation 18.1 to write an expression for the standing wave:

$$y = (2A \sin kx) \cos \omega t = 8.0 \sin 3.0x \cos 2.0t$$

Find the amplitude of the simple harmonic motion of the element at the position $x = 2.3$ cm by evaluating the sine function at this position:

$$y_{max} = (8.0 \text{ cm}) \sin 3.0x \,|_{x = 2.3}$$
$$= (8.0 \text{ cm}) \sin (6.9 \text{ rad}) = \boxed{4.6 \text{ cm}}$$

(B) Find the positions of the nodes and antinodes if one end of the string is at $x = 0$.

SOLUTION

Find the wavelength of the traveling waves:

$$k = \frac{2\pi}{\lambda} = 3.0 \text{ rad/cm} \quad \rightarrow \quad \lambda = \frac{2\pi}{3.0} \text{ cm}$$

Use Equation 18.2 to find the locations of the nodes:

$$x = n\frac{\lambda}{2} = n\left(\frac{\pi}{3.0}\right) \text{ cm} \quad n = 0, 1, 2, 3, \ldots$$

Use Equation 18.3 to find the locations of the antinodes:

$$x = n\frac{\lambda}{4} = n\left(\frac{\pi}{6.0}\right) \text{ cm} \quad n = 1, 3, 5, 7, \ldots$$

18.3 Analysis Model: Waves Under Boundary Conditions

Consider a string of length L fixed at both ends as shown in Figure 18.9. We will use this system as a model for a guitar string or piano string. Waves can travel in both directions on the string. Therefore, standing waves can be set up in the string by a continuous superposition of waves incident on and reflected from the ends. Notice that there is a *boundary condition* for the waves on the string: because the ends of the string are fixed, they must necessarily have zero displacement and are therefore nodes by definition. The condition that both ends of the string must be nodes fixes the wavelength of the standing wave on the string according to Equation 18.2, which, in turn, determines the frequency of the wave. The boundary condition results in the string having a number of discrete natural patterns of oscillation, called **normal modes,** each of which has a characteristic frequency that is easily calculated. This situation in which only certain frequencies of oscillation are allowed is called **quantization.** Quantization is a common occurrence when waves are subject to boundary conditions and is a central feature in our discussions of quantum physics in the extended version of this text. Notice in Figure 18.8 that there are no boundary conditions, so standing waves of *any* frequency can be established; there is no quantization without boundary conditions. Because boundary conditions occur so often for waves, we identify an analysis model called **waves under boundary conditions** for the discussion that follows.

The normal modes of oscillation for the string in Figure 18.9 can be described by imposing the boundary conditions that the ends be nodes and that the nodes be separated by one-half of a wavelength with antinodes halfway between the nodes. The first normal mode that is consistent with these requirements, shown in Figure 18.10a (page 426), has nodes at its ends and one antinode in the middle. This normal

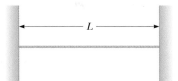

Figure 18.9 A string of length L fixed at both ends.

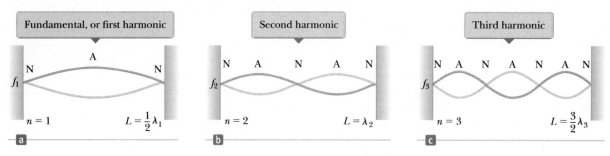

Figure 18.10 The normal modes of vibration of the string in Figure 18.9 form a harmonic series. The string vibrates between the extremes shown.

mode is the longest-wavelength mode that is consistent with our boundary conditions. The first normal mode occurs when the wavelength λ_1 is equal to twice the length of the string, or $\lambda_1 = 2L$. The section of a standing wave from one node to the next node is called a *loop*. In the first normal mode, the string is vibrating in one loop. In the second normal mode (see Fig. 18.10b), the string vibrates in two loops. When the left half of the string is moving upward, the right half is moving downward. In this case, the wavelength λ_2 is equal to the length of the string, as expressed by $\lambda_2 = L$. The third normal mode (see Fig. 18.10c) corresponds to the case in which $\lambda_3 = 2L/3$, and the string vibrates in three loops. In general, the wavelengths of the various normal modes for a string of length L fixed at both ends are

Wavelengths of normal modes ▶

$$\lambda_n = \frac{2L}{n} \quad n = 1, 2, 3, \ldots \quad (18.4)$$

where the index n refers to the nth normal mode of oscillation. These modes are *possible*. The *actual* modes that are excited on a string are discussed shortly.

The natural frequencies associated with the modes of oscillation are obtained from the relationship $f = v/\lambda$, where the wave speed v is the same for all frequencies. Using Equation 18.4, we find that the natural frequencies f_n of the normal modes are

Natural frequencies of normal modes as functions of wave speed and length of string ▶

$$f_n = \frac{v}{\lambda_n} = n\frac{v}{2L} \quad n = 1, 2, 3, \ldots \quad (18.5)$$

These natural frequencies are also called the *quantized frequencies* associated with the vibrating string fixed at both ends.

Because $v = \sqrt{T/\mu}$ (see Eq. 16.18) for waves on a string, where T is the tension in the string and μ is its linear mass density, we can also express the natural frequencies of a taut string as

Natural frequencies of normal modes as functions of string tension and linear mass density ▶

$$f_n = \frac{n}{2L}\sqrt{\frac{T}{\mu}} \quad n = 1, 2, 3, \ldots \quad (18.6)$$

The lowest frequency f_1, which corresponds to $n = 1$, is called either the **fundamental** or the **fundamental frequency** and is given by

Fundamental frequency of a taut string ▶

$$f_1 = \frac{1}{2L}\sqrt{\frac{T}{\mu}} \quad (18.7)$$

The frequencies of the remaining normal modes are integer multiples of the fundamental frequency (Eq. 18.5). Frequencies of normal modes that exhibit such an integer-multiple relationship form a **harmonic series,** and the normal modes are called **harmonics.** The fundamental frequency f_1 is the frequency of the first harmonic; the frequency $f_2 = 2f_1$ is that of the second harmonic; and the frequency $f_n = nf_1$ is that of the nth harmonic. Other oscillating systems, such as a drumhead, exhibit normal modes, but the frequencies are not related as integer multiples of a fundamental (see Section 18.6). Therefore, we do not use the term *harmonic* in association with those types of systems.

Let us examine further how the various harmonics are created in a string. To excite only a single harmonic, the string would have to be distorted into a shape that corresponds to that of the desired harmonic. After being released, the string would vibrate at the frequency of that harmonic. This maneuver is difficult to perform, however, and is not how a string of a musical instrument is excited. If the string is distorted into a general, nonsinusoidal shape, the resulting vibration includes a combination of various harmonics. Such a distortion occurs in musical instruments when the string is plucked (as in a guitar), bowed (as in a cello), or struck (as in a piano). When the string is distorted into a nonsinusoidal shape, only waves that satisfy the boundary conditions can persist on the string. These waves are the harmonics.

The frequency of a string that defines the musical note that it plays is that of the fundamental, even though other harmonics are present. The string's frequency can be varied by changing the string's tension or its length. For example, the tension in guitar and violin strings is varied by a screw adjustment mechanism or by tuning pegs located on the neck of the instrument. As the tension is increased, the frequency of the normal modes increases in accordance with Equation 18.6. Once the instrument is "tuned," players vary the frequency by moving their fingers along the neck, thereby changing the length of the oscillating portion of the string. As the length is shortened, the frequency increases because, as Equation 18.6 specifies, the normal-mode frequencies are inversely proportional to string length.

> **Quick Quiz 18.3** When a standing wave is set up on a string fixed at both ends, which of the following statements is true? **(a)** The number of nodes is equal to the number of antinodes. **(b)** The wavelength is equal to the length of the string divided by an integer. **(c)** The frequency is equal to the number of nodes times the fundamental frequency. **(d)** The shape of the string at any instant shows a symmetry about the midpoint of the string.

Analysis Model **Waves Under Boundary Conditions**

Imagine a wave that is not free to travel throughout all space as in the traveling wave model. If the wave is subject to boundary conditions, such that certain requirements must be met at specific locations in space, the wave is limited to a set of **normal modes** with quantized wavelengths and quantized natural frequencies.

For waves on a string fixed at both ends, the natural frequencies are

$$f_n = \frac{n}{2L}\sqrt{\frac{T}{\mu}} \quad n = 1, 2, 3, \ldots \quad (18.6)$$

where T is the tension in the string and μ is its linear mass density.

Examples:

- waves traveling back and forth on a guitar string combine to form a standing wave
- sound waves traveling back and forth in a clarinet combine to form standing waves (Section 18.5)
- a microscopic particle confined to a small region of space is modeled as a wave and exhibits quantized energies (Chapter 41)
- the Fermi energy of a metal is determined by modeling electrons as wave-like particles in a box (Chapter 43)

Example 18.3 **Give Me a C Note!**

The middle C string on a piano has a fundamental frequency of 262 Hz, and the string for the first A above middle C has a fundamental frequency of 440 Hz.

(A) Calculate the frequencies of the next two harmonics of the C string.

continued

▶ **18.3** c o n t i n u e d

SOLUTION

Conceptualize Remember that the harmonics of a vibrating string have frequencies that are related by integer multiples of the fundamental.

Categorize This first part of the example is a simple substitution problem.

Knowing that the fundamental frequency is $f_1 = 262$ Hz, find the frequencies of the next harmonics by multiplying by integers:

$$f_2 = 2f_1 = \boxed{524 \text{ Hz}}$$

$$f_3 = 3f_1 = \boxed{786 \text{ Hz}}$$

(B) If the A and C strings have the same linear mass density μ and length L, determine the ratio of tensions in the two strings.

SOLUTION

Categorize This part of the example is more of an analysis problem than is part (A) and uses the *waves under boundary conditions* model.

Analyze Use Equation 18.7 to write expressions for the fundamental frequencies of the two strings:

$$f_{1A} = \frac{1}{2L}\sqrt{\frac{T_A}{\mu}} \quad \text{and} \quad f_{1C} = \frac{1}{2L}\sqrt{\frac{T_C}{\mu}}$$

Divide the first equation by the second and solve for the ratio of tensions:

$$\frac{f_{1A}}{f_{1C}} = \sqrt{\frac{T_A}{T_C}} \rightarrow \frac{T_A}{T_C} = \left(\frac{f_{1A}}{f_{1C}}\right)^2 = \left(\frac{440}{262}\right)^2 = \boxed{2.82}$$

Finalize If the frequencies of piano strings were determined solely by tension, this result suggests that the ratio of tensions from the lowest string to the highest string on the piano would be enormous. Such large tensions would make it difficult to design a frame to support the strings. In reality, the frequencies of piano strings vary due to additional parameters, including the mass per unit length and the length of the string. The What If? below explores a variation in length.

WHAT IF? If you look inside a real piano, you'll see that the assumption made in part (B) is only partially true. The strings are not likely to have the same length. The string densities for the given notes might be equal, but suppose the length of the A string is only 64% of the length of the C string. What is the ratio of their tensions?

Answer Using Equation 18.7 again, we set up the ratio of frequencies:

$$\frac{f_{1A}}{f_{1C}} = \frac{L_C}{L_A}\sqrt{\frac{T_A}{T_C}} \rightarrow \frac{T_A}{T_C} = \left(\frac{L_A}{L_C}\right)^2\left(\frac{f_{1A}}{f_{1C}}\right)^2$$

$$\frac{T_A}{T_C} = (0.64)^2\left(\frac{440}{262}\right)^2 = 1.16$$

Notice that this result represents only a 16% increase in tension, compared with the 182% increase in part (B).

Example 18.4 **Changing String Vibration with Water** **AM**

One end of a horizontal string is attached to a vibrating blade, and the other end passes over a pulley as in Figure 18.11a. A sphere of mass 2.00 kg hangs on the end of the string. The string is vibrating in its second harmonic. A container of water is raised under the sphere so that the sphere is completely submerged. In this configuration, the string vibrates in its fifth harmonic as shown in Figure 18.11b. What is the radius of the sphere?

SOLUTION

Conceptualize Imagine what happens when the sphere is immersed in the water. The buoyant force acts upward on the sphere, reducing the tension in the string. The change in tension causes a change in the speed of waves on the

▶ **18.4** c o n t i n u e d

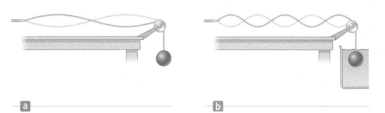

Figure 18.11 (Example 18.4) (a) When the sphere hangs in air, the string vibrates in its second harmonic. (b) When the sphere is immersed in water, the string vibrates in its fifth harmonic.

string, which in turn causes a change in the wavelength. This altered wavelength results in the string vibrating in its fifth normal mode rather than the second.

Categorize The hanging sphere is modeled as a *particle in equilibrium*. One of the forces acting on it is the buoyant force from the water. We also apply the *waves under boundary conditions* model to the string.

Analyze Apply the particle in equilibrium model to the sphere in Figure 18.11a, identifying T_1 as the tension in the string as the sphere hangs in air:

$$\sum F = T_1 - mg = 0$$
$$T_1 = mg$$

Apply the particle in equilibrium model to the sphere in Figure 18.11b, where T_2 is the tension in the string as the sphere is immersed in water:

$$T_2 + B - mg = 0$$
$$(1) \quad B = mg - T_2$$

The desired quantity, the radius of the sphere, will appear in the expression for the buoyant force B. Before proceeding in this direction, however, we must evaluate T_2 from the information about the standing wave.

Write the equation for the frequency of a standing wave on a string (Eq. 18.6) twice, once before the sphere is immersed and once after. Notice that the frequency f is the same in both cases because it is determined by the vibrating blade. In addition, the linear mass density μ and the length L of the vibrating portion of the string are the same in both cases. Divide the equations:

$$f = \frac{n_1}{2L}\sqrt{\frac{T_1}{\mu}} \quad \rightarrow \quad 1 = \frac{n_1}{n_2}\sqrt{\frac{T_1}{T_2}}$$
$$f = \frac{n_2}{2L}\sqrt{\frac{T_2}{\mu}}$$

Solve for T_2:

$$T_2 = \left(\frac{n_1}{n_2}\right)^2 T_1 = \left(\frac{n_1}{n_2}\right)^2 mg$$

Substitute this result into Equation (1):

$$(2) \quad B = mg - \left(\frac{n_1}{n_2}\right)^2 mg = mg\left[1 - \left(\frac{n_1}{n_2}\right)^2\right]$$

Using Equation 14.5, express the buoyant force in terms of the radius of the sphere:

$$B = \rho_{water} g V_{sphere} = \rho_{water} g\left(\tfrac{4}{3}\pi r^3\right)$$

Solve for the radius of the sphere and substitute from Equation (2):

$$r = \left(\frac{3B}{4\pi\rho_{water}g}\right)^{1/3} = \left\{\frac{3m}{4\pi\rho_{water}}\left[1 - \left(\frac{n_1}{n_2}\right)^2\right]\right\}^{1/3}$$

Substitute numerical values:

$$r = \left\{\frac{3(2.00 \text{ kg})}{4\pi(1\,000 \text{ kg/m}^3)}\left[1 - \left(\frac{2}{5}\right)^2\right]\right\}^{1/3}$$
$$= 0.073\,7 \text{ m} = \boxed{7.37 \text{ cm}}$$

Finalize Notice that only certain radii of the sphere will result in the string vibrating in a normal mode; the speed of waves on the string must be changed to a value such that the length of the string is an integer multiple of half wavelengths. This limitation is a feature of the *quantization* that was introduced earlier in this chapter: the sphere radii that cause the string to vibrate in a normal mode are *quantized*.

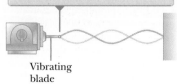

Vibrating blade

Figure 18.12 Standing waves are set up in a string when one end is connected to a vibrating blade.

18.4 Resonance

We have seen that a system such as a taut string is capable of oscillating in one or more normal modes of oscillation. Suppose we drive such a string with a vibrating blade as in Figure 18.12. We find that if a periodic force is applied to such a system, the amplitude of the resulting motion of the string is greatest when the frequency of the applied force is equal to one of the natural frequencies of the system. This phenomenon, known as *resonance,* was discussed in Section 15.7 with regard to a simple harmonic oscillator. Although a block–spring system or a simple pendulum has only one natural frequency, standing-wave systems have a whole set of natural frequencies, such as that given by Equation 18.6 for a string. Because an oscillating system exhibits a large amplitude when driven at any of its natural frequencies, these frequencies are often referred to as **resonance frequencies.**

Consider the string in Figure 18.12 again. The fixed end is a node, and the end connected to the blade is very nearly a node because the amplitude of the blade's motion is small compared with that of the elements of the string. As the blade oscillates, transverse waves sent down the string are reflected from the fixed end. As we learned in Section 18.3, the string has natural frequencies that are determined by its length, tension, and linear mass density (see Eq. 18.6). When the frequency of the blade equals one of the natural frequencies of the string, standing waves are produced and the string oscillates with a large amplitude. In this resonance case, the wave generated by the oscillating blade is in phase with the reflected wave and the string absorbs energy from the blade. If the string is driven at a frequency that is not one of its natural frequencies, the oscillations are of low amplitude and exhibit no stable pattern.

Resonance is very important in the excitation of musical instruments based on air columns. We shall discuss this application of resonance in Section 18.5.

18.5 Standing Waves in Air Columns

The waves under boundary conditions model can also be applied to sound waves in a column of air such as that inside an organ pipe or a clarinet. Standing waves in this case are the result of interference between longitudinal sound waves traveling in opposite directions.

In a pipe closed at one end, the closed end is a **displacement node** because the rigid barrier at this end does not allow longitudinal motion of the air. Because the pressure wave is 90° out of phase with the displacement wave (see Section 17.1), the closed end of an air column corresponds to a **pressure antinode** (that is, a point of maximum pressure variation).

The open end of an air column is approximately a **displacement antinode**[1] and a pressure node. We can understand why no pressure variation occurs at an open end by noting that the end of the air column is open to the atmosphere; therefore, the pressure at this end must remain constant at atmospheric pressure.

You may wonder how a sound wave can reflect from an open end because there may not appear to be a change in the medium at this point: the medium through which the sound wave moves is air both inside and outside the pipe. Sound can be represented as a pressure wave, however, and a compression region of the sound wave is constrained by the sides of the pipe as long as the region is inside the pipe. As the compression region exits at the open end of the pipe, the constraint of the pipe is removed and the compressed air is free to expand into the atmosphere. Therefore, there is a change in the *character* of the medium between the inside

---

[1]Strictly speaking, the open end of an air column is not exactly a displacement antinode. A compression reaching an open end does not reflect until it passes beyond the end. For a tube of circular cross section, an end correction equal to approximately $0.6R$, where R is the tube's radius, must be added to the length of the air column. Hence, the effective length of the air column is longer than the true length L. We ignore this end correction in this discussion.

Figure 18.13 Graphical representations of the motion of elements of air in standing longitudinal waves in (a) a column open at both ends and (b) a column closed at one end.

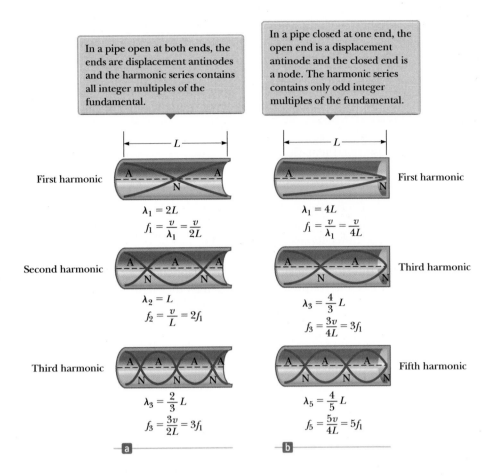

In a pipe open at both ends, the ends are displacement antinodes and the harmonic series contains all integer multiples of the fundamental.

In a pipe closed at one end, the open end is a displacement antinode and the closed end is a node. The harmonic series contains only odd integer multiples of the fundamental.

First harmonic

$\lambda_1 = 2L$

$f_1 = \dfrac{v}{\lambda_1} = \dfrac{v}{2L}$

Second harmonic

$\lambda_2 = L$

$f_2 = \dfrac{v}{L} = 2f_1$

Third harmonic

$\lambda_3 = \dfrac{2}{3}L$

$f_3 = \dfrac{3v}{2L} = 3f_1$

First harmonic

$\lambda_1 = 4L$

$f_1 = \dfrac{v}{\lambda_1} = \dfrac{v}{4L}$

Third harmonic

$\lambda_3 = \dfrac{4}{3}L$

$f_3 = \dfrac{3v}{4L} = 3f_1$

Fifth harmonic

$\lambda_5 = \dfrac{4}{5}L$

$f_5 = \dfrac{5v}{4L} = 5f_1$

of the pipe and the outside even though there is no change in the *material* of the medium. This change in character is sufficient to allow some reflection.

With the boundary conditions of nodes or antinodes at the ends of the air column, we have a set of normal modes of oscillation as is the case for the string fixed at both ends. Therefore, the air column has quantized frequencies.

The first three normal modes of oscillation of a pipe open at both ends are shown in Figure 18.13a. Notice that both ends are displacement antinodes (approximately). In the first normal mode, the standing wave extends between two adjacent antinodes, which is a distance of half a wavelength. Therefore, the wavelength is twice the length of the pipe, and the fundamental frequency is $f_1 = v/2L$. As Figure 18.13a shows, the frequencies of the higher harmonics are $2f_1, 3f_1, \ldots$.

> In a pipe open at both ends, the natural frequencies of oscillation form a harmonic series that includes all integral multiples of the fundamental frequency.

Because all harmonics are present and because the fundamental frequency is given by the same expression as that for a string (see Eq. 18.5), we can express the natural frequencies of oscillation as

$$f_n = n\frac{v}{2L} \quad n = 1, 2, 3, \ldots \qquad \textbf{(18.8)}$$

Despite the similarity between Equations 18.5 and 18.8, you must remember that v in Equation 18.5 is the speed of waves on the string, whereas v in Equation 18.8 is the speed of sound in air.

If a pipe is closed at one end and open at the other, the closed end is a displacement node (see Fig. 18.13b). In this case, the standing wave for the fundamental mode extends from an antinode to the adjacent node, which is one-fourth of a wavelength. Hence, the wavelength for the first normal mode is $4L$, and the fundamental

Pitfall Prevention 18.3

Sound Waves in Air Are Longitudinal, Not Transverse The standing longitudinal waves are drawn as transverse waves in Figure 18.13. Because they are in the same direction as the propagation, it is difficult to draw longitudinal displacements. Therefore, it is best to interpret the red-brown curves in Figure 18.13 as a graphical representation of the waves (our diagrams of string waves are pictorial representations), with the vertical axis representing the horizontal displacement $s(x, t)$ of the elements of the medium.

Natural frequencies of a pipe ◀ **open at both ends**

frequency is $f_1 = v/4L$. As Figure 18.13b shows, the higher-frequency waves that satisfy our conditions are those that have a node at the closed end and an antinode at the open end; hence, the higher harmonics have frequencies $3f_1, 5f_1, \ldots$.

> In a pipe closed at one end, the natural frequencies of oscillation form a harmonic series that includes only odd integral multiples of the fundamental frequency.

We express this result mathematically as

Natural frequencies of ▶
a pipe closed at one end
and open at the other

$$f_n = n\frac{v}{4L} \quad n = 1, 3, 5, \ldots \qquad (18.9)$$

It is interesting to investigate what happens to the frequencies of instruments based on air columns and strings during a concert as the temperature rises. The sound emitted by a flute, for example, becomes sharp (increases in frequency) as the flute warms up because the speed of sound increases in the increasingly warmer air inside the flute (consider Eq. 18.8). The sound produced by a violin becomes flat (decreases in frequency) as the strings thermally expand because the expansion causes their tension to decrease (see Eq. 18.6).

Musical instruments based on air columns are generally excited by resonance. The air column is presented with a sound wave that is rich in many frequencies. The air column then responds with a large-amplitude oscillation to the frequencies that match the quantized frequencies in its set of harmonics. In many woodwind instruments, the initial rich sound is provided by a vibrating reed. In brass instruments, this excitation is provided by the sound coming from the vibration of the player's lips. In a flute, the initial excitation comes from blowing over an edge at the mouthpiece of the instrument in a manner similar to blowing across the opening of a bottle with a narrow neck. The sound of the air rushing across the bottle opening has many frequencies, including one that sets the air cavity in the bottle into resonance.

Quick Quiz 18.4 A pipe open at both ends resonates at a fundamental frequency f_{open}. When one end is covered and the pipe is again made to resonate, the fundamental frequency is f_{closed}. Which of the following expressions describes how these two resonant frequencies compare? (a) $f_{closed} = f_{open}$ (b) $f_{closed} = \frac{1}{2}f_{open}$ (c) $f_{closed} = 2f_{open}$ (d) $f_{closed} = \frac{3}{2}f_{open}$

Quick Quiz 18.5 Balboa Park in San Diego has an outdoor organ. When the air temperature increases, the fundamental frequency of one of the organ pipes (a) stays the same, (b) goes down, (c) goes up, or (d) is impossible to determine.

Example 18.5 **Wind in a Culvert**

A section of drainage culvert 1.23 m in length makes a howling noise when the wind blows across its open ends.

(A) Determine the frequencies of the first three harmonics of the culvert if it is cylindrical in shape and open at both ends. Take $v = 343$ m/s as the speed of sound in air.

SOLUTION

Conceptualize The sound of the wind blowing across the end of the pipe contains many frequencies, and the culvert responds to the sound by vibrating at the natural frequencies of the air column.

Categorize This example is a relatively simple substitution problem.

Find the frequency of the first harmonic of the culvert, modeling it as an air column open at both ends:

$$f_1 = \frac{v}{2L} = \frac{343 \text{ m/s}}{2(1.23 \text{ m})} = \boxed{139 \text{ Hz}}$$

Find the next harmonics by multiplying by integers:

$$f_2 = 2f_1 = \boxed{279 \text{ Hz}}$$

$$f_3 = 3f_1 = \boxed{418 \text{ Hz}}$$

▶ **18.5** continued

(B) What are the three lowest natural frequencies of the culvert if it is blocked at one end?

SOLUTION

Find the frequency of the first harmonic of the culvert, modeling it as an air column closed at one end:

$$f_1 = \frac{v}{4L} = \frac{343 \text{ m/s}}{4(1.23 \text{ m})} = \boxed{69.7 \text{ Hz}}$$

Find the next two harmonics by multiplying by odd integers:

$$f_3 = 3f_1 = \boxed{209 \text{ Hz}}$$
$$f_5 = 5f_1 = \boxed{349 \text{ Hz}}$$

Example 18.6 Measuring the Frequency of a Tuning Fork AM

A simple apparatus for demonstrating resonance in an air column is depicted in Figure 18.14. A vertical pipe open at both ends is partially submerged in water, and a tuning fork vibrating at an unknown frequency is placed near the top of the pipe. The length L of the air column can be adjusted by moving the pipe vertically. The sound waves generated by the fork are reinforced when L corresponds to one of the resonance frequencies of the pipe. For a certain pipe, the smallest value of L for which a peak occurs in the sound intensity is 9.00 cm.

(A) What is the frequency of the tuning fork?

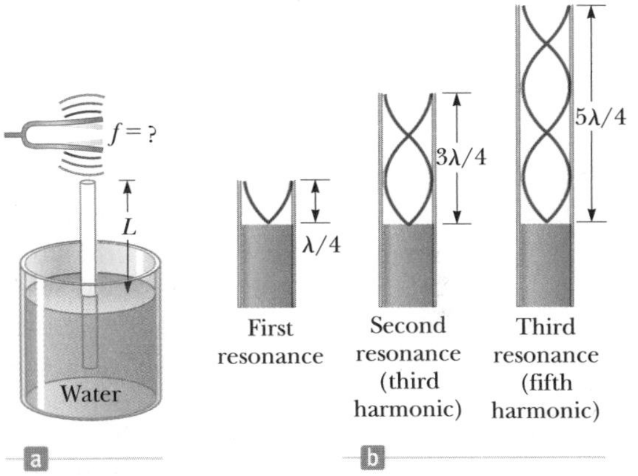

Figure 18.14 (Example 18.6) (a) Apparatus for demonstrating the resonance of sound waves in a pipe closed at one end. The length L of the air column is varied by moving the pipe vertically while it is partially submerged in water. (b) The first three normal modes of the system shown in (a).

SOLUTION

Conceptualize Sound waves from the tuning fork enter the pipe at its upper end. Although the pipe is open at its lower end to allow the water to enter, the water's surface acts like a barrier. The waves reflect from the water surface and combine with those moving downward to form a standing wave.

Categorize Because of the reflection of the sound waves from the water surface, we can model the pipe as open at the upper end and closed at the lower end. Therefore, we can apply the *waves under boundary conditions* model to this situation.

Analyze

Use Equation 18.9 to find the fundamental frequency for $L = 0.090\ 0$ m:

$$f_1 = \frac{v}{4L} = \frac{343 \text{ m/s}}{4(0.090\ 0 \text{ m})} = \boxed{953 \text{ Hz}}$$

Because the tuning fork causes the air column to resonate at this frequency, this frequency must also be that of the tuning fork.

(B) What are the values of L for the next two resonance conditions?

SOLUTION

Use Equation 16.12 to find the wavelength of the sound wave from the tuning fork:

$$\lambda = \frac{v}{f} = \frac{343 \text{ m/s}}{953 \text{ Hz}} = 0.360 \text{ m}$$

Notice from Figure 18.14b that the length of the air column for the second resonance is $3\lambda/4$:

$$L = 3\lambda/4 = \boxed{0.270 \text{ m}}$$

Notice from Figure 18.14b that the length of the air column for the third resonance is $5\lambda/4$:

$$L = 5\lambda/4 = \boxed{0.450 \text{ m}}$$

Finalize Consider how this problem differs from the preceding example. In the culvert, the length was fixed and the air column was presented with a mixture of many frequencies. The pipe in this example is presented with one single frequency from the tuning fork, and the length of the pipe is varied until resonance is achieved.

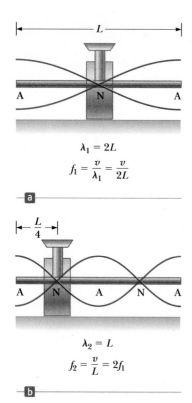

$$\lambda_1 = 2L$$

$$f_1 = \frac{v}{\lambda_1} = \frac{v}{2L}$$

a

$$\lambda_2 = L$$

$$f_2 = \frac{v}{L} = 2f_1$$

b

Figure 18.15 Normal-mode longitudinal vibrations of a rod of length L (a) clamped at the middle to produce the first normal mode and (b) clamped at a distance $L/4$ from one end to produce the second normal mode. Notice that the red-brown curves are graphical representations of oscillations parallel to the rod (longitudinal waves).

18.6 Standing Waves in Rods and Membranes

Standing waves can also be set up in rods and membranes. A rod clamped in the middle and stroked parallel to the rod at one end oscillates as depicted in Figure 18.15a. The oscillations of the elements of the rod are longitudinal, and so the red-brown curves in Figure 18.15 represent *longitudinal* displacements of various parts of the rod. For clarity, the displacements have been drawn in the transverse direction as they were for air columns. The midpoint is a displacement node because it is fixed by the clamp, whereas the ends are displacement antinodes because they are free to oscillate. The oscillations in this setup are analogous to those in a pipe open at both ends. The red-brown lines in Figure 18.15a represent the first normal mode, for which the wavelength is $2L$ and the frequency is $f = v/2L$, where v is the speed of longitudinal waves in the rod. Other normal modes may be excited by clamping the rod at different points. For example, the second normal mode (Fig. 18.15b) is excited by clamping the rod a distance $L/4$ away from one end.

It is also possible to set up transverse standing waves in rods. Musical instruments that depend on transverse standing waves in rods or bars include triangles, marimbas, xylophones, glockenspiels, chimes, and vibraphones. Other devices that make sounds from vibrating bars include music boxes and wind chimes.

Two-dimensional oscillations can be set up in a flexible membrane stretched over a circular hoop such as that in a drumhead. As the membrane is struck at some point, waves that arrive at the fixed boundary are reflected many times. The resulting sound is not harmonic because the standing waves have frequencies that are *not* related by integer multiples. Without this relationship, the sound may be more correctly described as *noise* rather than as music. The production of noise is in contrast to the situation in wind and stringed instruments, which produce sounds that we describe as musical.

Some possible normal modes of oscillation for a two-dimensional circular membrane are shown in Figure 18.16. Whereas nodes are *points* in one-dimensional standing waves on strings and in air columns, a two-dimensional oscillator has *curves* along which there is no displacement of the elements of the medium. The lowest normal mode, which has a frequency f_1, contains only one nodal curve; this curve runs around the outer edge of the membrane. The other possible normal modes show additional nodal curves that are circles and straight lines across the diameter of the membrane.

18.7 Beats: Interference in Time

The interference phenomena we have studied so far involve the superposition of two or more waves having the same frequency. Because the amplitude of the oscil-

Figure 18.16 Representation of some of the normal modes possible in a circular membrane fixed at its perimeter. The pair of numbers above each pattern corresponds to the number of radial nodes and the number of circular nodes, respectively. In each diagram, elements of the membrane on either side of a nodal line move in opposite directions, as indicated by the colors. (*Adapted from T. D. Rossing,* The Science of Sound, *3rd ed., Reading, Massachusetts, Addison-Wesley Publishing Co., 2001*)

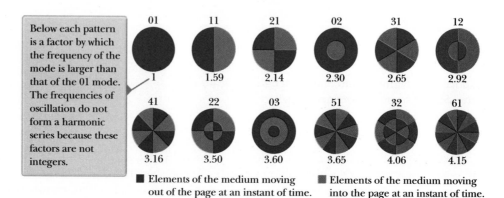

Below each pattern is a factor by which the frequency of the mode is larger than that of the 01 mode. The frequencies of oscillation do not form a harmonic series because these factors are not integers.

■ Elements of the medium moving out of the page at an instant of time.
■ Elements of the medium moving into the page at an instant of time.

lation of elements of the medium varies with the position in space of the element in such a wave, we refer to the phenomenon as *spatial interference*. Standing waves in strings and pipes are common examples of spatial interference.

Now let's consider another type of interference, one that results from the superposition of two waves having slightly *different* frequencies. In this case, when the two waves are observed at a point in space, they are periodically in and out of phase. That is, there is a *temporal* (time) alternation between constructive and destructive interference. As a consequence, we refer to this phenomenon as *interference in time* or *temporal interference*. For example, if two tuning forks of slightly different frequencies are struck, one hears a sound of periodically varying amplitude. This phenomenon is called **beating.**

> Beating is the periodic variation in amplitude at a given point due to the superposition of two waves having slightly different frequencies.

◀ **Definition of beating**

The number of amplitude maxima one hears per second, or the *beat frequency*, equals the difference in frequency between the two sources as we shall show below. The maximum beat frequency that the human ear can detect is about 20 beats/s. When the beat frequency exceeds this value, the beats blend indistinguishably with the sounds producing them.

Consider two sound waves of equal amplitude and slightly different frequencies f_1 and f_2 traveling through a medium. We use equations similar to Equation 16.13 to represent the wave functions for these two waves at a point that we identify as $x = 0$. We also choose the phase angle in Equation 16.13 as $\phi = \pi/2$:

$$y_1 = A \sin\left(\frac{\pi}{2} - \omega_1 t\right) = A \cos\left(2\pi f_1 t\right)$$

$$y_2 = A \sin\left(\frac{\pi}{2} - \omega_2 t\right) = A \cos\left(2\pi f_2 t\right)$$

Using the superposition principle, we find that the resultant wave function at this point is

$$y = y_1 + y_2 = A\left(\cos 2\pi f_1 t + \cos 2\pi f_2 t\right)$$

The trigonometric identity

$$\cos a + \cos b = 2 \cos\left(\frac{a - b}{2}\right) \cos\left(\frac{a + b}{2}\right)$$

allows us to write the expression for y as

$$y = \left[2A \cos 2\pi\left(\frac{f_1 - f_2}{2}\right)t\right] \cos 2\pi\left(\frac{f_1 + f_2}{2}\right)t \qquad \text{(18.10)}$$

◀ **Resultant of two waves of different frequencies but equal amplitude**

Graphs of the individual waves and the resultant wave are shown in Figure 18.17. From the factors in Equation 18.10, we see that the resultant wave has an effective

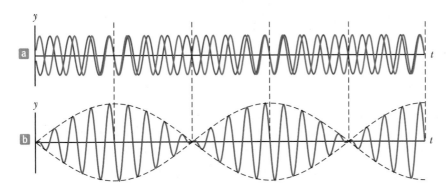

Figure 18.17 Beats are formed by the combination of two waves of slightly different frequencies. (a) The individual waves. (b) The combined wave. The envelope wave (dashed line) represents the beating of the combined sounds.

frequency equal to the average frequency $(f_1 + f_2)/2$. This wave is multiplied by an envelope wave given by the expression in the square brackets:

$$y_{\text{envelope}} = 2A \cos 2\pi\left(\frac{f_1 - f_2}{2}\right)t \tag{18.11}$$

That is, the amplitude and therefore the intensity of the resultant sound vary in time. The dashed black line in Figure 18.17b is a graphical representation of the envelope wave in Equation 18.11 and is a sine wave varying with frequency $(f_1 - f_2)/2$.

A maximum in the amplitude of the resultant sound wave is detected whenever

$$\cos 2\pi\left(\frac{f_1 - f_2}{2}\right)t = \pm 1$$

Hence, there are *two* maxima in each period of the envelope wave. Because the amplitude varies with frequency as $(f_1 - f_2)/2$, the number of beats per second, or the **beat frequency** f_{beat}, is twice this value. That is,

Beat frequency ▶

$$f_{\text{beat}} = |f_1 - f_2| \tag{18.12}$$

For instance, if one tuning fork vibrates at 438 Hz and a second one vibrates at 442 Hz, the resultant sound wave of the combination has a frequency of 440 Hz (the musical note A) and a beat frequency of 4 Hz. A listener would hear a 440-Hz sound wave go through an intensity maximum four times every second.

Example 18.7   **The Mistuned Piano Strings**

Two identical piano strings of length 0.750 m are each tuned exactly to 440 Hz. The tension in one of the strings is then increased by 1.0%. If they are now struck, what is the beat frequency between the fundamentals of the two strings?

SOLUTION

Conceptualize As the tension in one of the strings is changed, its fundamental frequency changes. Therefore, when both strings are played, they will have different frequencies and beats will be heard.

Categorize We must combine our understanding of the *waves under boundary conditions* model for strings with our new knowledge of beats.

· ·

Analyze Set up a ratio of the fundamental frequencies of the two strings using Equation 18.5:

$$\frac{f_2}{f_1} = \frac{(v_2/2L)}{(v_1/2L)} = \frac{v_2}{v_1}$$

Use Equation 16.18 to substitute for the wave speeds on the strings:

$$\frac{f_2}{f_1} = \frac{\sqrt{T_2/\mu}}{\sqrt{T_1/\mu}} = \sqrt{\frac{T_2}{T_1}}$$

Incorporate that the tension in one string is 1.0% larger than the other; that is, $T_2 = 1.010T_1$:

$$\frac{f_2}{f_1} = \sqrt{\frac{1.010T_1}{T_1}} = 1.005$$

Solve for the frequency of the tightened string:

$$f_2 = 1.005f_1 = 1.005(440 \text{ Hz}) = 442 \text{ Hz}$$

Find the beat frequency using Equation 18.12:

$$f_{\text{beat}} = 442 \text{ Hz} - 440 \text{ Hz} = \boxed{2 \text{ Hz}}$$

· ·

Finalize Notice that a 1.0% mistuning in tension leads to an easily audible beat frequency of 2 Hz. A piano tuner can use beats to tune a stringed instrument by "beating" a note against a reference tone of known frequency. The tuner can then adjust the string tension until the frequency of the sound it emits equals the frequency of the reference tone. The tuner does so by tightening or loosening the string until the beats produced by it and the reference source become too infrequent to notice.

18.8 Nonsinusoidal Wave Patterns

It is relatively easy to distinguish the sounds coming from a violin and a saxophone even when they are both playing the same note. On the other hand, a person untrained in music may have difficulty distinguishing a note played on a clarinet from the same note played on an oboe. We can use the pattern of the sound waves from various sources to explain these effects.

When frequencies that are integer multiples of a fundamental frequency are combined to make a sound, the result is a *musical* sound. A listener can assign a pitch to the sound based on the fundamental frequency. Pitch is a psychological reaction to a sound that allows the listener to place the sound on a scale from low to high (bass to treble). Combinations of frequencies that are not integer multiples of a fundamental result in a *noise* rather than a musical sound. It is much harder for a listener to assign a pitch to a noise than to a musical sound.

The wave patterns produced by a musical instrument are the result of the superposition of frequencies that are integer multiples of a fundamental. This superposition results in the corresponding richness of musical tones. The human perceptive response associated with various mixtures of harmonics is the *quality* or *timbre* of the sound. For instance, the sound of the trumpet is perceived to have a "brassy" quality (that is, we have learned to associate the adjective *brassy* with that sound); this quality enables us to distinguish the sound of the trumpet from that of the saxophone, whose quality is perceived as "reedy." The clarinet and oboe, however, both contain air columns excited by reeds; because of this similarity, they have similar mixtures of frequencies and it is more difficult for the human ear to distinguish them on the basis of their sound quality.

The sound wave patterns produced by the majority of musical instruments are nonsinusoidal. Characteristic patterns produced by a tuning fork, a flute, and a clarinet, each playing the same note, are shown in Figure 18.18. Each instrument has its own characteristic pattern. Notice, however, that despite the differences in the patterns, each pattern is periodic. This point is important for our analysis of these waves.

The problem of analyzing nonsinusoidal wave patterns appears at first sight to be a formidable task. If the wave pattern is periodic, however, it can be represented as closely as desired by the combination of a sufficiently large number of sinusoidal waves that form a harmonic series. In fact, we can represent any periodic function as a series of sine and cosine terms by using a mathematical technique based on **Fourier's theorem.**[2] The corresponding sum of terms that represents the periodic wave pattern is called a **Fourier series.** Let $y(t)$ be any function that is periodic in time with period T such that $y(t + T) = y(t)$. Fourier's theorem states that this function can be written as

$$y(t) = \sum (A_n \sin 2\pi f_n t + B_n \cos 2\pi f_n t) \qquad \text{(18.13)}$$

where the lowest frequency is $f_1 = 1/T$. The higher frequencies are integer multiples of the fundamental, $f_n = nf_1$, and the coefficients A_n and B_n represent the amplitudes of the various waves. Figure 18.19 on page 438 represents a harmonic analysis of the wave patterns shown in Figure 18.18. Each bar in the graph represents one of the terms in the series in Equation 18.13 up to $n = 9$. Notice that a struck tuning fork produces only one harmonic (the first), whereas the flute and clarinet produce the first harmonic and many higher ones.

Notice the variation in relative intensity of the various harmonics for the flute and the clarinet. In general, any musical sound consists of a fundamental frequency f plus other frequencies that are integer multiples of f, all having different intensities.

Pitfall Prevention 18.4
Pitch Versus Frequency Do not confuse the term *pitch* with *frequency*. Frequency is the physical measurement of the number of oscillations per second. Pitch is a psychological reaction to sound that enables a person to place the sound on a scale from high to low or from treble to bass. Therefore, frequency is the stimulus and pitch is the response. Although pitch is related mostly (but not completely) to frequency, they are not the same. A phrase such as "the pitch of the sound" is incorrect because pitch is not a physical property of the sound.

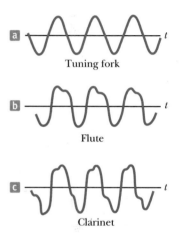

Figure 18.18 Sound wave patterns produced by (a) a tuning fork, (b) a flute, and (c) a clarinet, each at approximately the same frequency.

◀ **Fourier's theorem**

---

[2] Developed by Jean Baptiste Joseph Fourier (1786–1830).

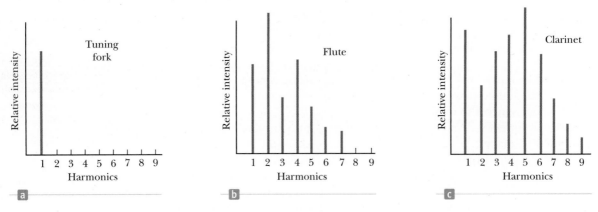

Figure 18.19 Harmonics of the wave patterns shown in Figure 18.18. Notice the variations in intensity of the various harmonics. Parts (a), (b), and (c) correspond to those in Figure 18.18.

We have discussed the *analysis* of a wave pattern using Fourier's theorem. The analysis involves determining the coefficients of the harmonics in Equation 18.13 from a knowledge of the wave pattern. The reverse process, called *Fourier synthesis*, can also be performed. In this process, the various harmonics are added together to form a resultant wave pattern. As an example of Fourier synthesis, consider the building of a square wave as shown in Figure 18.20. The symmetry of the square wave results in only odd multiples of the fundamental frequency combining in its synthesis. In Figure 18.20a, the blue curve shows the combination of f and $3f$. In Figure 18.20b, we have added $5f$ to the combination and obtained the green curve. Notice how the general shape of the square wave is approximated, even though the upper and lower portions are not flat as they should be.

Figure 18.20c shows the result of adding odd frequencies up to $9f$. This approximation (red-brown curve) to the square wave is better than the approximations in Figures 18.20a and 18.20b. To approximate the square wave as closely as possible, we must add all odd multiples of the fundamental frequency, up to infinite frequency.

Using modern technology, musical sounds can be generated electronically by mixing different amplitudes of any number of harmonics. These widely used electronic music synthesizers are capable of producing an infinite variety of musical tones.

Figure 18.20 Fourier synthesis of a square wave, represented by the sum of odd multiples of the first harmonic, which has frequency f.

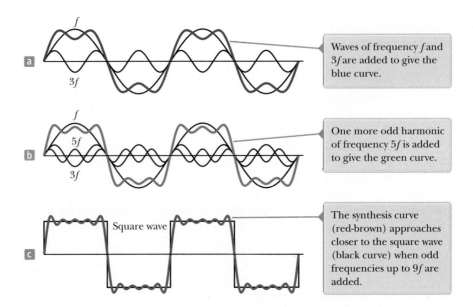

Waves of frequency f and $3f$ are added to give the blue curve.

One more odd harmonic of frequency $5f$ is added to give the green curve.

The synthesis curve (red-brown) approaches closer to the square wave (black curve) when odd frequencies up to $9f$ are added.

Summary

Concepts and Principles

The **superposition principle** specifies that when two or more waves move through a medium, the value of the resultant wave function equals the algebraic sum of the values of the individual wave functions.

The phenomenon of **beating** is the periodic variation in intensity at a given point due to the superposition of two waves having slightly different frequencies. The **beat frequency** is

$$f_{\text{beat}} = |f_1 - f_2| \tag{18.12}$$

where f_1 and f_2 are the frequencies of the individual waves.

Standing waves are formed from the combination of two sinusoidal waves having the same frequency, amplitude, and wavelength but traveling in opposite directions. The resultant standing wave is described by the wave function

$$y = (2A \sin kx) \cos \omega t \tag{18.1}$$

Hence, the amplitude of the standing wave is $2A$, and the amplitude of the simple harmonic motion of any element of the medium varies according to its position as $2A \sin kx$. The points of zero amplitude (called **nodes**) occur at $x = n\lambda/2$ ($n = 0, 1, 2, 3, \ldots$). The maximum amplitude points (called **antinodes**) occur at $x = n\lambda/4$ ($n = 1, 3, 5, \ldots$). Adjacent antinodes are separated by a distance $\lambda/2$. Adjacent nodes also are separated by a distance $\lambda/2$.

Analysis Models for Problem Solving

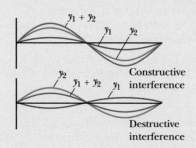

Waves in Interference. When two traveling waves having equal frequencies superimpose, the resultant wave is described by the **principle of superposition** and has an amplitude that depends on the phase angle ϕ between the two waves. **Constructive interference** occurs when the two waves are in phase, corresponding to $\phi = 0, 2\pi, 4\pi, \ldots$ rad. **Destructive interference** occurs when the two waves are 180° out of phase, corresponding to $\phi = \pi, 3\pi, 5\pi, \ldots$ rad.

Waves Under Boundary Conditions. When a wave is subject to boundary conditions, only certain natural frequencies are allowed; we say that the frequencies are quantized.

For waves on a string fixed at both ends, the natural frequencies are

$$f_n = \frac{n}{2L}\sqrt{\frac{T}{\mu}} \quad n = 1, 2, 3, \ldots \tag{18.6}$$

where T is the tension in the string and μ is its linear mass density.

For sound waves with speed v in an air column of length L open at both ends, the natural frequencies are

$$f_n = n\frac{v}{2L} \quad n = 1, 2, 3, \ldots \tag{18.8}$$

If an air column is open at one end and closed at the other, only odd harmonics are present and the natural frequencies are

$$f_n = n\frac{v}{4L} \quad n = 1, 3, 5, \ldots \tag{18.9}$$

Objective Questions   **1.** denotes answer available in *Student Solutions Manual/Study Guide*

1. In Figure OQ18.1 (page 440), a sound wave of wavelength 0.8 m divides into two equal parts that recombine to interfere constructively, with the original difference between their path lengths being $|r_2 - r_1| = 0.8$ m. Rank the following situations according to the intensity of sound at the receiver from the highest to the lowest. Assume the tube walls absorb no sound energy. Give equal ranks to situations in which the intensity is equal.

(a) From its original position, the sliding section is moved out by 0.1 m. (b) Next it slides out an additional 0.1 m. (c) It slides out still another 0.1 m. (d) It slides out 0.1 m more.

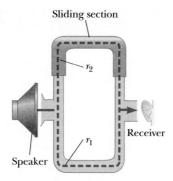

Figure OQ18.1

2. A string of length L, mass per unit length μ, and tension T is vibrating at its fundamental frequency. **(i)** If the length of the string is doubled, with all other factors held constant, what is the effect on the fundamental frequency? (a) It becomes two times larger. (b) It becomes $\sqrt{2}$ times larger. (c) It is unchanged. (d) It becomes $1/\sqrt{2}$ times as large. (e) It becomes one-half as large. **(ii)** If the mass per unit length is doubled, with all other factors held constant, what is the effect on the fundamental frequency? Choose from the same possibilities as in part (i). **(iii)** If the tension is doubled, with all other factors held constant, what is the effect on the fundamental frequency? Choose from the same possibilities as in part (i).

3. In Example 18.1, we investigated an oscillator at 1.3 kHz driving two identical side-by-side speakers. We found that a listener at point O hears sound with maximum intensity, whereas a listener at point P hears a minimum. What is the intensity at P? (a) less than but close to the intensity at O (b) half the intensity at O (c) very low but not zero (d) zero (e) indeterminate

4. A series of pulses, each of amplitude 0.1 m, is sent down a string that is attached to a post at one end. The pulses are reflected at the post and travel back along the string without loss of amplitude. **(i)** What is the net displacement at a point on the string where two pulses are crossing? Assume the string is rigidly attached to the post. (a) 0.4 m (b) 0.3 m (c) 0.2 m (d) 0.1 m (e) 0 **(ii)** Next assume the end at which reflection occurs is free to slide up and down. Now what is the net displacement at a point on the string where two pulses are crossing? Choose your answer from the same possibilities as in part (i).

5. A flute has a length of 58.0 cm. If the speed of sound in air is 343 m/s, what is the fundamental frequency of the flute, assuming it is a tube closed at one end and open at the other? (a) 148 Hz (b) 296 Hz (c) 444 Hz (d) 591 Hz (e) none of those answers

6. When two tuning forks are sounded at the same time, a beat frequency of 5 Hz occurs. If one of the tuning forks has a frequency of 245 Hz, what is the frequency of the other tuning fork? (a) 240 Hz (b) 242.5 Hz (c) 247.5 Hz (d) 250 Hz (e) More than one answer could be correct.

7. A tuning fork is known to vibrate with frequency 262 Hz. When it is sounded along with a mandolin string, four beats are heard every second. Next, a bit of tape is put onto each tine of the tuning fork, and the tuning fork now produces five beats per second with the same mandolin string. What is the frequency of the string? (a) 257 Hz (b) 258 Hz (c) 262 Hz (d) 266 Hz (e) 267 Hz

8. An archer shoots an arrow horizontally from the center of the string of a bow held vertically. After the arrow leaves it, the string of the bow will vibrate as a superposition of what standing-wave harmonics? (a) It vibrates only in harmonic number 1, the fundamental. (b) It vibrates only in the second harmonic. (c) It vibrates only in the odd-numbered harmonics 1, 3, 5, 7, ... (d) It vibrates only in the even-numbered harmonics 2, 4, 6, 8, ... (e) It vibrates in all harmonics.

9. As oppositely moving pulses of the same shape (one upward, one downward) on a string pass through each other, at one particular instant the string shows no displacement from the equilibrium position at any point. What has happened to the energy carried by the pulses at this instant of time? (a) It was used up in producing the previous motion. (b) It is all potential energy. (c) It is all internal energy. (d) It is all kinetic energy. (e) The positive energy of one pulse adds to zero with the negative energy of the other pulse.

10. A standing wave having three nodes is set up in a string fixed at both ends. If the frequency of the wave is doubled, how many antinodes will there be? (a) 2 (b) 3 (c) 4 (d) 5 (e) 6

11. Suppose all six equal-length strings of an acoustic guitar are played without fingering, that is, without being pressed down at any frets. What quantities are the same for all six strings? Choose all correct answers. (a) the fundamental frequency (b) the fundamental wavelength of the string wave (c) the fundamental wavelength of the sound emitted (d) the speed of the string wave (e) the speed of the sound emitted

12. Assume two identical sinusoidal waves are moving through the same medium in the same direction. Under what condition will the amplitude of the resultant wave be greater than either of the two original waves? (a) in all cases (b) only if the waves have no difference in phase (c) only if the phase difference is less than 90° (d) only if the phase difference is less than 120° (e) only if the phase difference is less than 180°

Conceptual Questions

1. denotes answer available in *Student Solutions Manual/Study Guide*

1. A crude model of the human throat is that of a pipe open at both ends with a vibrating source to introduce the sound into the pipe at one end. Assuming the vibrating source produces a range of frequencies, discuss the effect of changing the pipe's length.

2. When two waves interfere constructively or destructively, is there any gain or loss in energy in the system of the waves? Explain.

3. Explain how a musical instrument such as a piano may be tuned by using the phenomenon of beats.

4. What limits the amplitude of motion of a real vibrating system that is driven at one of its resonant frequencies?

5. A tuning fork by itself produces a faint sound. Explain how each of the following methods can be used to obtain a louder sound from it. Explain also any effect on the time interval for which the fork vibrates audibly. (a) holding the edge of a sheet of paper against one vibrating tine (b) pressing the handle of the tuning fork against a chalkboard or a tabletop (c) holding the tuning fork above a column of air of properly chosen length as in Example 18.6 (d) holding the tuning fork close to an open slot cut in a sheet of foam plastic or cardboard (with the slot similar in size and shape to one tine of the fork and the motion of the tines perpendicular to the sheet)

6. An airplane mechanic notices that the sound from a twin-engine aircraft rapidly varies in loudness when both engines are running. What could be causing this variation from loud to soft?

7. Despite a reasonably steady hand, a person often spills his coffee when carrying it to his seat. Discuss resonance as a possible cause of this difficulty and devise a means for preventing the spills.

8. A soft-drink bottle resonates as air is blown across its top. What happens to the resonance frequency as the level of fluid in the bottle decreases?

9. Does the phenomenon of wave interference apply only to sinusoidal waves?

Problems available in Access end-of-chapter problems online at www.webassign.net

---

Section 18.1 Analysis Model: Waves in Interference
Problems 1–13

Section 18.2 Standing Waves
Problems 14–19

Section 18.3 Analysis Model: Waves Under Boundary Conditions
Problems 20–33

Section 18.4 Resonance
Problems 34–36

Section 18.5 Standing Waves in Air Columns
Problems 37–53

Section 18.6 Standing Waves in Rods and Membranes
Problems 54–55

Section 18.7 Beats: Interference in Time
Problems 56–59

Section 18.8 Nonsinusoidal Wave Patterns
Problems 60–61

Additional Problems
Problems 62–86

Challenge Problems
Problems 87–88

Solutions to the following Problems are available in the *Student Solutions Manual/Study Guide:*
18.7, 18.9, 18.11, 18.17, 18.19, 18.23, 18.26, 18.39, 18.43, 18.46, 18.51, 18.55, 18.57, 18.59, 18.61, 18.77, 18.80, and 18.82

List of Enhanced Problems

Problem Number	Targeted Feedback in Enhanced WebAssign	Analysis Model Tutorial in Enhanced WebAssign	Master It in Enhanced WebAssign	Watch It in Enhanced WebAssign
18.1	✓			✓
18.3	✓			✓
18.7	✓			
18.8	✓	✓		✓
18.9			✓	
18.11			✓	
18.15	✓			✓
18.17	✓		✓	
18.19	✓		✓	
18.23	✓			✓
18.27	✓	✓		✓
18.28	✓		✓	
18.40	✓			✓
18.51		✓	✓	
18.56	✓			✓
18.57			✓	
18.59			✓	
18.62	✓		✓	
18.77			✓	
18.80			✓	
18.85		✓		

Thermodynamics

A bubble in one of the many mud pots in Yellowstone National Park is caught just at the moment of popping. A mud pot is a pool of bubbling hot mud that demonstrates the existence of thermodynamic processes below the Earth's surface. (© Adambooth/ Dreamstime.com)

We now direct our attention to the study of thermodynamics, which involves situations in which the temperature or state (solid, liquid, gas) of a system changes due to energy transfers. As we shall see, thermodynamics is very successful in explaining the bulk properties of matter and the correlation between these properties and the mechanics of atoms and molecules.

Historically, the development of thermodynamics paralleled the development of the atomic theory of matter. By the 1820s, chemical experiments had provided solid evidence for the existence of atoms. At that time, scientists recognized that a connection between thermodynamics and the structure of matter must exist. In 1827, botanist Robert Brown reported that grains of pollen suspended in a liquid move erratically from one place to another as if under constant agitation. In 1905, Albert Einstein used kinetic theory to explain the cause of this erratic motion, known today as *Brownian motion*. Einstein explained this phenomenon by assuming the grains are under constant bombardment by "invisible" molecules in the liquid, which themselves move erratically. This explanation gave scientists insight into the concept of molecular motion and gave credence to the idea that matter is made up of atoms. A connection was thus forged between the everyday world and the tiny, invisible building blocks that make up this world.

Thermodynamics also addresses more practical questions. Have you ever wondered how a refrigerator is able to cool its contents, or what types of transformations occur in a power plant or in the engine of your automobile, or what happens to the kinetic energy of a moving object when the object comes to rest? The laws of thermodynamics can be used to provide explanations for these and other phenomena. ■

CHAPTER

19

Temperature

Why would someone designing a pipeline include these strange loops? Pipelines carrying liquids often contain such loops to allow for expansion and contraction as the temperature changes. We will study thermal expansion in this chapter.
(© Lowell Georgia/CORBIS)

In our study of mechanics, we carefully defined such concepts as *mass, force,* and *kinetic energy* to facilitate our quantitative approach. Likewise, a quantitative description of thermal phenomena requires careful definitions of such important terms as *temperature, heat,* and *internal energy.* This chapter begins with a discussion of temperature.

Next, when studying thermal phenomena, we consider the importance of the particular substance we are investigating. For example, gases expand appreciably when heated, whereas liquids and solids expand only slightly.

This chapter concludes with a study of ideal gases on the macroscopic scale. Here, we are concerned with the relationships among such quantities as pressure, volume, and temperature of a gas. In Chapter 21, we shall examine gases on a microscopic scale, using a model that represents the components of a gas as small particles.

19.1 Temperature and the Zeroth Law of Thermodynamics

We often associate the concept of temperature with how hot or cold an object feels when we touch it. In this way, our senses provide us with a qualitative indication of temperature. Our senses, however, are unreliable and often mislead us. For exam-

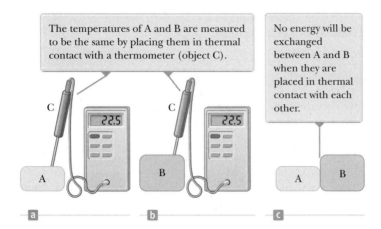

The temperatures of A and B are measured to be the same by placing them in thermal contact with a thermometer (object C).

No energy will be exchanged between A and B when they are placed in thermal contact with each other.

Figure 19.1 The zeroth law of thermodynamics.

ple, if you stand in bare feet with one foot on carpet and the other on an adjacent tile floor, the tile feels colder than the carpet *even though both are at the same temperature.* The two objects feel different because tile transfers energy by heat at a higher rate than carpet does. Your skin "measures" the rate of energy transfer by heat rather than the actual temperature. What we need is a reliable and reproducible method for measuring the relative hotness or coldness of objects rather than the rate of energy transfer. Scientists have developed a variety of thermometers for making such quantitative measurements.

Two objects at different initial temperatures eventually reach some intermediate temperature when placed in contact with each other. For example, when hot water and cold water are mixed in a bathtub, energy is transferred from the hot water to the cold water and the final temperature of the mixture is somewhere between the initial hot and cold temperatures.

Imagine that two objects are placed in an insulated container such that they interact with each other but not with the environment. If the objects are at different temperatures, energy is transferred between them, even if they are initially not in physical contact with each other. The energy-transfer mechanisms from Chapter 8 that we will focus on are heat and electromagnetic radiation. For purposes of this discussion, let's assume two objects are in **thermal contact** with each other if energy can be exchanged between them by these processes due to a temperature difference. **Thermal equilibrium** is a situation in which two objects would not exchange energy by heat or electromagnetic radiation if they were placed in thermal contact.

Let's consider two objects A and B, which are not in thermal contact, and a third object C, which is our thermometer. We wish to determine whether A and B are in thermal equilibrium with each other. The thermometer (object C) is first placed in thermal contact with object A until thermal equilibrium is reached[1] as shown in Figure 19.1a. From that moment on, the thermometer's reading remains constant and we record this reading. The thermometer is then removed from object A and placed in thermal contact with object B as shown in Figure 19.1b. The reading is again recorded after thermal equilibrium is reached. If the two readings are the same, we can conclude that object A and object B are in thermal equilibrium with each other. If they are placed in contact with each other as in Figure 19.1c, there is no exchange of energy between them.

---

[1]We assume a negligible amount of energy transfers between the thermometer and object A in the time interval during which they are in thermal contact. Without this assumption, which is also made for the thermometer and object B, the measurement of the temperature of an object disturbs the system so that the measured temperature is different from the initial temperature of the object. In practice, whenever you measure a temperature with a thermometer, you measure the disturbed system, not the original system.

We can summarize these results in a statement known as the **zeroth law of thermodynamics** (the law of equilibrium):

Zeroth law ▶
of thermodynamics

> If objects A and B are separately in thermal equilibrium with a third object C, then A and B are in thermal equilibrium with each other.

This statement can easily be proved experimentally and is very important because it enables us to define temperature. We can think of **temperature** as the property that determines whether an object is in thermal equilibrium with other objects. Two objects in thermal equilibrium with each other are at the same temperature. Conversely, if two objects have different temperatures, they are not in thermal equilibrium with each other. We now know that temperature is something that determines whether or not energy will transfer between two objects in thermal contact. In Chapter 21, we will relate temperature to the mechanical behavior of molecules.

Quick Quiz 19.1 Two objects, with different sizes, masses, and temperatures, are placed in thermal contact. In which direction does the energy travel? **(a)** Energy travels from the larger object to the smaller object. **(b)** Energy travels from the object with more mass to the one with less mass. **(c)** Energy travels from the object at higher temperature to the object at lower temperature.

19.2 Thermometers and the Celsius Temperature Scale

Thermometers are devices used to measure the temperature of a system. All thermometers are based on the principle that some physical property of a system changes as the system's temperature changes. Some physical properties that change with temperature are (1) the volume of a liquid, (2) the dimensions of a solid, (3) the pressure of a gas at constant volume, (4) the volume of a gas at constant pressure, (5) the electric resistance of a conductor, and (6) the color of an object.

A common thermometer in everyday use consists of a mass of liquid—usually mercury or alcohol—that expands into a glass capillary tube when heated (Fig. 19.2). In this case, the physical property that changes is the volume of a liquid. Any temperature change in the range of the thermometer can be defined as being proportional to the change in length of the liquid column. The thermometer can be calibrated by placing it in thermal contact with a natural system that remains

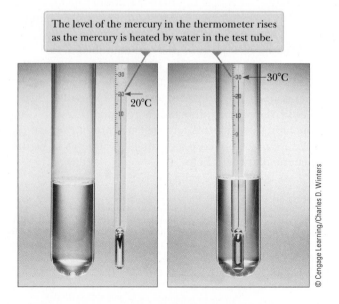

The level of the mercury in the thermometer rises as the mercury is heated by water in the test tube.

Figure 19.2 A mercury thermometer before and after increasing its temperature.

© Cengage Learning/Charles D. Winters

at constant temperature. One such system is a mixture of water and ice in thermal equilibrium at atmospheric pressure. On the **Celsius temperature scale,** this mixture is defined to have a temperature of zero degrees Celsius, which is written as 0°C; this temperature is called the *ice point* of water. Another commonly used system is a mixture of water and steam in thermal equilibrium at atmospheric pressure; its temperature is defined as 100°C, which is the *steam point* of water. Once the liquid levels in the thermometer have been established at these two points, the length of the liquid column between the two points is divided into 100 equal segments to create the Celsius scale. Therefore, each segment denotes a change in temperature of one Celsius degree.

Thermometers calibrated in this way present problems when extremely accurate readings are needed. For instance, the readings given by an alcohol thermometer calibrated at the ice and steam points of water might agree with those given by a mercury thermometer only at the calibration points. Because mercury and alcohol have different thermal expansion properties, when one thermometer reads a temperature of, for example, 50°C, the other may indicate a slightly different value. The discrepancies between thermometers are especially large when the temperatures to be measured are far from the calibration points.[2]

An additional practical problem of any thermometer is the limited range of temperatures over which it can be used. A mercury thermometer, for example, cannot be used below the freezing point of mercury, which is −39°C, and an alcohol thermometer is not useful for measuring temperatures above 85°C, the boiling point of alcohol. To surmount this problem, we need a universal thermometer whose readings are independent of the substance used in it. The gas thermometer, discussed in the next section, approaches this requirement.

19.3 The Constant-Volume Gas Thermometer and the Absolute Temperature Scale

One version of a gas thermometer is the constant-volume apparatus shown in Figure 19.3. The physical change exploited in this device is the variation of pressure of a fixed volume of gas with temperature. The flask is immersed in an ice-water bath, and mercury reservoir B is raised or lowered until the top of the mercury in column A is at the zero point on the scale. The height h, the difference between the mercury levels in reservoir B and column A, indicates the pressure in the flask at 0°C by means of Equation 14.4, $P = P_0 + \rho gh$.

The flask is then immersed in water at the steam point. Reservoir B is readjusted until the top of the mercury in column A is again at zero on the scale, which ensures that the gas's volume is the same as it was when the flask was in the ice bath (hence the designation "constant-volume"). This adjustment of reservoir B gives a value for the gas pressure at 100°C. These two pressure and temperature values are then plotted as shown in Figure 19.4. The line connecting the two points serves as a calibration curve for unknown temperatures. (Other experiments show that a linear relationship between pressure and temperature is a very good assumption.) To measure the temperature of a substance, the gas flask of Figure 19.3 is placed in thermal contact with the substance and the height of reservoir B is adjusted until the top of the mercury column in A is at zero on the scale. The height of the mercury column in B indicates the pressure of the gas; knowing the pressure, the temperature of the substance is found using the graph in Figure 19.4.

Now suppose temperatures of different gases at different initial pressures are measured with gas thermometers. Experiments show that the thermometer readings are nearly independent of the type of gas used as long as the gas pressure is low and the temperature is well above the point at which the gas liquefies

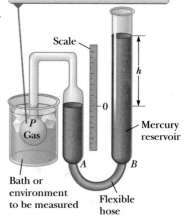

The volume of gas in the flask is kept constant by raising or lowering reservoir B to keep the mercury level in column A constant.

Figure 19.3 A constant-volume gas thermometer measures the pressure of the gas contained in the flask immersed in the bath.

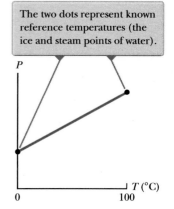

The two dots represent known reference temperatures (the ice and steam points of water).

Figure 19.4 A typical graph of pressure versus temperature taken with a constant-volume gas thermometer.

---

[2]Two thermometers that use the same liquid may also give different readings, due in part to difficulties in constructing uniform-bore glass capillary tubes.

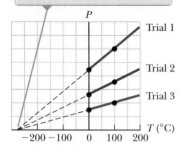

Figure 19.5 Pressure versus temperature for experimental trials in which gases have different pressures in a constant-volume gas thermometer.

Pitfall Prevention 19.1

A Matter of Degree Notations for temperatures in the Kelvin scale do not use the degree sign. The unit for a Kelvin temperature is simply "kelvins" and not "degrees Kelvin."

Note that the scale is logarithmic.

Temperature (K)

10^9

10^8 ←— Hydrogen bomb

10^7 ←— Interior of the Sun

10^6 ←— Solar corona

10^5

10^4 ←— Surface of the Sun

10^3 ←— Copper melts

10^2 ←— Water freezes
   ←— Liquid nitrogen
10 ←— Liquid hydrogen

1 ←— Liquid helium

Lowest temperature achieved ~ 10^{-9} K

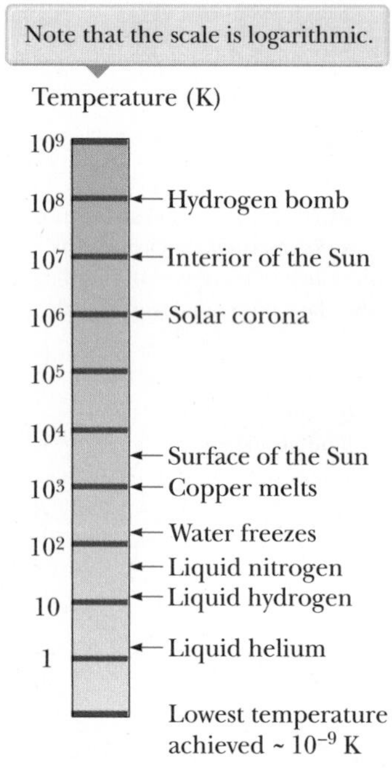

Figure 19.6 Absolute temperatures at which various physical processes occur.

(Fig. 19.5). The agreement among thermometers using various gases improves as the pressure is reduced.

If we extend the straight lines in Figure 19.5 toward negative temperatures, we find a remarkable result: **in every case, the pressure is zero when the temperature is −273.15°C!** This finding suggests some special role that this particular temperature must play. It is used as the basis for the **absolute temperature scale,** which sets −273.15°C as its zero point. This temperature is often referred to as **absolute zero.** It is indicated as a zero because at a lower temperature, the pressure of the gas would become negative, which is meaningless. The size of one degree on the absolute temperature scale is chosen to be identical to the size of one degree on the Celsius scale. Therefore, the conversion between these temperatures is

$$T_C = T - 273.15 \tag{19.1}$$

where T_C is the Celsius temperature and T is the absolute temperature.

Because the ice and steam points are experimentally difficult to duplicate and depend on atmospheric pressure, an absolute temperature scale based on two new fixed points was adopted in 1954 by the International Committee on Weights and Measures. The first point is absolute zero. The second reference temperature for this new scale was chosen as the **triple point of water,** which is the single combination of temperature and pressure at which liquid water, gaseous water, and ice (solid water) coexist in equilibrium. This triple point occurs at a temperature of 0.01°C and a pressure of 4.58 mm of mercury. On the new scale, which uses the unit *kelvin*, the temperature of water at the triple point was set at 273.16 kelvins, abbreviated 273.16 K. This choice was made so that the old absolute temperature scale based on the ice and steam points would agree closely with the new scale based on the triple point. This new absolute temperature scale (also called the **Kelvin scale**) employs the SI unit of absolute temperature, the **kelvin,** which is defined to be 1/273.16 of the difference between absolute zero and the temperature of the triple point of water.

Figure 19.6 gives the absolute temperature for various physical processes and structures. The temperature of absolute zero (0 K) cannot be achieved, although laboratory experiments have come very close, reaching temperatures of less than one nanokelvin.

The Celsius, Fahrenheit, and Kelvin Temperature Scales[3]

Equation 19.1 shows that the Celsius temperature T_C is shifted from the absolute (Kelvin) temperature T by 273.15°. Because the size of one degree is the same on the two scales, a temperature difference of 5°C is equal to a temperature difference of 5 K. The two scales differ only in the choice of the zero point. Therefore, the ice-point temperature on the Kelvin scale, 273.15 K, corresponds to 0.00°C, and the Kelvin-scale steam point, 373.15 K, is equivalent to 100.00°C.

A common temperature scale in everyday use in the United States is the **Fahrenheit scale.** This scale sets the temperature of the ice point at 32°F and the temperature of the steam point at 212°F. The relationship between the Celsius and Fahrenheit temperature scales is

$$T_F = \tfrac{9}{5}T_C + 32°F \tag{19.2}$$

We can use Equations 19.1 and 19.2 to find a relationship between changes in temperature on the Celsius, Kelvin, and Fahrenheit scales:

$$\Delta T_C = \Delta T = \tfrac{5}{9}\Delta T_F \tag{19.3}$$

Of these three temperature scales, only the Kelvin scale is based on a true zero value of temperature. The Celsius and Fahrenheit scales are based on an arbitrary zero associated with one particular substance, water, on one particular planet, the

---

[3]Named after Anders Celsius (1701–1744), Daniel Gabriel Fahrenheit (1686–1736), and William Thomson, Lord Kelvin (1824–1907), respectively.

Earth. Therefore, if you encounter an equation that calls for a temperature T or that involves a ratio of temperatures, you *must* convert all temperatures to kelvins. If the equation contains a change in temperature ΔT, using Celsius temperatures will give you the correct answer, in light of Equation 19.3, but it is always *safest* to convert temperatures to the Kelvin scale.

Quick Quiz 19.2 Consider the following pairs of materials. Which pair represents two materials, one of which is twice as hot as the other? **(a)** boiling water at 100°C, a glass of water at 50°C **(b)** boiling water at 100°C, frozen methane at −50°C **(c)** an ice cube at −20°C, flames from a circus fire-eater at 233°C **(d)** none of those pairs

---

Example 19.1   **Converting Temperatures**

On a day when the temperature reaches 50°F, what is the temperature in degrees Celsius and in kelvins?

SOLUTION

Conceptualize In the United States, a temperature of 50°F is well understood. In many other parts of the world, however, this temperature might be meaningless because people are familiar with the Celsius temperature scale.

Categorize This example is a simple substitution problem.

Solve Equation 19.2 for the Celsius temperature and substitute numerical values:

$$T_C = \tfrac{5}{9}(T_F - 32) = \tfrac{5}{9}(50 - 32) = \boxed{10°C}$$

Use Equation 19.1 to find the Kelvin temperature:

$$T = T_C + 273.15 = 10°C + 273.15 = \boxed{283 \text{ K}}$$

A convenient set of weather-related temperature equivalents to keep in mind is that 0°C is (literally) freezing at 32°F, 10°C is cool at 50°F, 20°C is room temperature, 30°C is warm at 86°F, and 40°C is a hot day at 104°F.

19.4 Thermal Expansion of Solids and Liquids

Our discussion of the liquid thermometer makes use of one of the best-known changes in a substance: as its temperature increases, its volume increases. This phenomenon, known as **thermal expansion,** plays an important role in numerous engineering applications. For example, thermal-expansion joints such as those shown in Figure 19.7 must be included in buildings, concrete highways, railroad tracks,

Without these joints to separate sections of roadway on bridges, the surface would buckle due to thermal expansion on very hot days or crack due to contraction on very cold days.

The long, vertical joint is filled with a soft material that allows the wall to expand and contract as the temperature of the bricks changes.

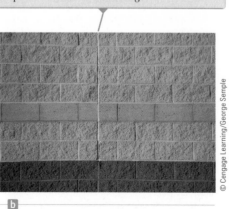

© Cengage Learning/George Semple

Figure 19.7 Thermal-expansion joints in (a) bridges and (b) walls.

brick walls, and bridges to compensate for dimensional changes that occur as the temperature changes.

Thermal expansion is a consequence of the change in the *average* separation between the atoms in an object. To understand this concept, let's model the atoms as being connected by stiff springs as discussed in Section 15.3 and shown in Figure 15.11b. At ordinary temperatures, the atoms in a solid oscillate about their equilibrium positions with an amplitude of approximately 10^{-11} m and a frequency of approximately 10^{13} Hz. The average spacing between the atoms is about 10^{-10} m. As the temperature of the solid increases, the atoms oscillate with greater amplitudes; as a result, the average separation between them increases.[4] Consequently, the object expands.

If thermal expansion is sufficiently small relative to an object's initial dimensions, the change in any dimension is, to a good approximation, proportional to the first power of the temperature change. Suppose an object has an initial length L_i along some direction at some temperature and the length changes by an amount ΔL for a change in temperature ΔT. Because it is convenient to consider the fractional change in length per degree of temperature change, we define the **average coefficient of linear expansion** as

$$\alpha \equiv \frac{\Delta L / L_i}{\Delta T}$$

Experiments show that α is constant for small changes in temperature. For purposes of calculation, this equation is usually rewritten as

◀ **Thermal expansion in one dimension**

$$\Delta L = \alpha L_i \, \Delta T \qquad (19.4)$$

or as

$$L_f - L_i = \alpha L_i (T_f - T_i) \qquad (19.5)$$

where L_f is the final length, T_i and T_f are the initial and final temperatures, respectively, and the proportionality constant α is the average coefficient of linear expansion for a given material and has units of $(\degree C)^{-1}$. Equation 19.4 can be used for both thermal expansion, when the temperature of the material increases, and thermal contraction, when its temperature decreases.

It may be helpful to think of thermal expansion as an effective magnification or as a photographic enlargement of an object. For example, as a metal washer is heated (Fig. 19.8), all dimensions, including the radius of the hole, increase according to Equation 19.4. A cavity in a piece of material expands in the same way as if the cavity were filled with the material.

Table 19.1 lists the average coefficients of linear expansion for various materials. For these materials, α is positive, indicating an increase in length with increasing temperature. That is not always the case, however. Some substances—calcite ($CaCO_3$) is one example—expand along one dimension (positive α) and contract along another (negative α) as their temperatures are increased.

Because the linear dimensions of an object change with temperature, it follows that surface area and volume change as well. The change in volume is proportional to the initial volume V_i and to the change in temperature according to the relationship

◀ **Thermal expansion in three dimensions**

$$\Delta V = \beta V_i \, \Delta T \qquad (19.6)$$

where β is the **average coefficient of volume expansion.** To find the relationship between β and α, assume the average coefficient of linear expansion of the solid is the same in all directions; that is, assume the material is *isotropic.* Consider a solid box of dimensions ℓ, w, and h. Its volume at some temperature T_i is $V_i = \ell w h$. If the

---

[4]More precisely, thermal expansion arises from the *asymmetrical* nature of the potential energy curve for the atoms in a solid as shown in Figure 15.11a. If the oscillators were truly harmonic, the average atomic separations would not change regardless of the amplitude of vibration.

Table 19.1		Average Expansion Coefficients for Some Materials Near Room Temperature		
Material (Solids)	**Average Linear Expansion Coefficient $(\alpha)(°C)^{-1}$**	**Material (Liquids and Gases)**	**Average Volume Expansion Coefficient $(\beta)(°C)^{-1}$**	
Aluminum	24×10^{-6}	Acetone	1.5×10^{-4}	
Brass and bronze	19×10^{-6}	Alcohol, ethyl	1.12×10^{-4}	
Concrete	12×10^{-6}	Benzene	1.24×10^{-4}	
Copper	17×10^{-6}	Gasoline	9.6×10^{-4}	
Glass (ordinary)	9×10^{-6}	Glycerin	4.85×10^{-4}	
Glass (Pyrex)	3.2×10^{-6}	Mercury	1.82×10^{-4}	
Invar (Ni–Fe alloy)	0.9×10^{-6}	Turpentine	9.0×10^{-4}	
Lead	29×10^{-6}	Air[a] at 0°C	3.67×10^{-3}	
Steel	11×10^{-6}	Helium[a]	3.665×10^{-3}	

[a]Gases do not have a specific value for the volume expansion coefficient because the amount of expansion depends on the type of process through which the gas is taken. The values given here assume the gas undergoes an expansion at constant pressure.

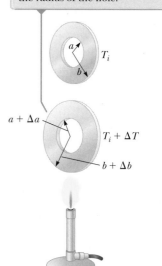

As the washer is heated, all dimensions increase, including the radius of the hole.

Figure 19.8 Thermal expansion of a homogeneous metal washer. (The expansion is exaggerated in this figure.)

temperature changes to $T_i + \Delta T$, its volume changes to $V_i + \Delta V$, where each dimension changes according to Equation 19.4. Therefore,

$$V_i + \Delta V = (\ell + \Delta\ell)(w + \Delta w)(h + \Delta h)$$

$$= (\ell + \alpha\ell \, \Delta T)(w + \alpha w \, \Delta T)(h + \alpha h \, \Delta T)$$

$$= \ell w h (1 + \alpha \, \Delta T)^3$$

$$= V_i[1 + 3\alpha \, \Delta T + 3(\alpha \, \Delta T)^2 + (\alpha \, \Delta T)^3]$$

Dividing both sides by V_i and isolating the term $\Delta V/V_i$, we obtain the fractional change in volume:

$$\frac{\Delta V}{V_i} = 3\alpha \, \Delta T + 3(\alpha \, \Delta T)^2 + (\alpha \, \Delta T)^3$$

Because $\alpha \, \Delta T \ll 1$ for typical values of ΔT ($< \sim 100°C$), we can neglect the terms $3(\alpha \, \Delta T)^2$ and $(\alpha \, \Delta T)^3$. Upon making this approximation, we see that

$$\frac{\Delta V}{V_i} = 3\alpha \, \Delta T \quad \rightarrow \quad \Delta V = (3\alpha) V_i \, \Delta T$$

Comparing this expression to Equation 19.6 shows that

$$\beta = 3\alpha$$

In a similar way, you can show that the change in area of a rectangular plate is given by $\Delta A = 2\alpha A_i \, \Delta T$ (see Problem 61 in Enhanced WebAssign).

A simple mechanism called a *bimetallic strip*, found in practical devices such as mechanical thermostats, uses the difference in coefficients of expansion for different materials. It consists of two thin strips of dissimilar metals bonded together. As the temperature of the strip increases, the two metals expand by different amounts and the strip bends as shown in Figure 19.9.

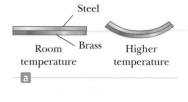

Quick **Quiz 19.3** If you are asked to make a very sensitive glass thermometer, which of the following working liquids would you choose? **(a)** mercury **(b)** alcohol **(c)** gasoline **(d)** glycerin

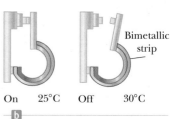

Figure 19.9 (a) A bimetallic strip bends as the temperature changes because the two metals have different expansion coefficients. (b) A bimetallic strip used in a thermostat to break or make electrical contact.

Quick **Quiz 19.4** Two spheres are made of the same metal and have the same radius, but one is hollow and the other is solid. The spheres are taken through the same temperature increase. Which sphere expands more? **(a)** The solid sphere expands more. **(b)** The hollow sphere expands more. **(c)** They expand by the same amount. **(d)** There is not enough information to say.

| Example 19.2 | Expansion of a Railroad Track |

A segment of steel railroad track has a length of 30.000 m when the temperature is 0.0°C.

(A) What is its length when the temperature is 40.0°C?

SOLUTION

Conceptualize Because the rail is relatively long, we expect to obtain a measurable increase in length for a 40°C temperature increase.

Categorize We will evaluate a length increase using the discussion of this section, so this part of the example is a substitution problem.

Use Equation 19.4 and the value of the coefficient of linear expansion from Table 19.1:

$$\Delta L = \alpha L_i \Delta T = [11 \times 10^{-6} \ (°C)^{-1}](30.000 \text{ m})(40.0°C) = 0.013 \text{ m}$$

Find the new length of the track:

$$L_f = 30.000 \text{ m} + 0.013 \text{ m} = \boxed{30.013 \text{ m}}$$

(B) Suppose the ends of the rail are rigidly clamped at 0.0°C so that expansion is prevented. What is the thermal stress set up in the rail if its temperature is raised to 40.0°C?

SOLUTION

Categorize This part of the example is an analysis problem because we need to use concepts from another chapter.

Analyze The thermal stress is the same as the tensile stress in the situation in which the rail expands freely and is then compressed with a mechanical force F back to its original length.

Find the tensile stress from Equation 12.6 using Young's modulus for steel from Table 12.1:

$$\text{Tensile stress} = \frac{F}{A} = Y\frac{\Delta L}{L_i}$$

$$\frac{F}{A} = (20 \times 10^{10} \text{ N/m}^2)\left(\frac{0.013 \text{ m}}{30.000 \text{ m}}\right) = \boxed{8.7 \times 10^7 \text{ N/m}^2}$$

Finalize The expansion in part (A) is 1.3 cm. This expansion is indeed measurable as predicted in the Conceptualize step. The thermal stress in part (B) can be avoided by leaving small expansion gaps between the rails.

WHAT IF? What if the temperature drops to −40.0°C? What is the length of the unclamped segment?

Answer The expression for the change in length in Equation 19.4 is the same whether the temperature increases or decreases. Therefore, if there is an increase in length of 0.013 m when the temperature increases by 40°C, there is a decrease in length of 0.013 m when the temperature decreases by 40°C. (We assume α is constant over the entire range of temperatures.) The new length at the colder temperature is 30.000 m − 0.013 m = 29.987 m.

| Example 19.3 | The Thermal Electrical Short |

A poorly designed electronic device has two bolts attached to different parts of the device that almost touch each other in its interior as in Figure 19.10. The steel and brass bolts are at different electric potentials, and if they touch, a short circuit will develop, damaging the device. (We will study electric potential in Chapter 25.) The initial gap between the ends of the bolts is $d = 5.0 \ \mu m$ at 27°C. At what temperature will the bolts touch? Assume the distance between the walls of the device is not affected by the temperature change.

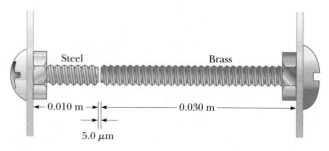

Figure 19.10 (Example 19.3) Two bolts attached to different parts of an electrical device are almost touching when the temperature is 27°C. As the temperature increases, the ends of the bolts move toward each other.

SOLUTION

Conceptualize Imagine the ends of both bolts expanding into the gap between them as the temperature rises.

▶ **19.3** continued

Categorize We categorize this example as a thermal expansion problem in which the *sum* of the changes in length of the two bolts must equal the length of the initial gap between the ends.

....

Analyze Set the sum of the length changes equal to the width of the gap:

$$\Delta L_{br} + \Delta L_{st} = \alpha_{br} L_{i,br}\, \Delta T + \alpha_{st} L_{i,st}\, \Delta T = d$$

Solve for ΔT:

$$\Delta T = \frac{d}{\alpha_{br} L_{i,br} + \alpha_{st} L_{i,st}}$$

Substitute numerical values:

$$\Delta T = \frac{5.0 \times 10^{-6}\ \text{m}}{[19 \times 10^{-6}\ (°C)^{-1}](0.030\ \text{m}) + [11 \times 10^{-6}\ (°C)^{-1}](0.010\ \text{m})} = 7.4°C$$

Find the temperature at which the bolts touch:

$$T = 27°C + 7.4°C = \boxed{34°C}$$

....

Finalize This temperature is possible if the air conditioning in the building housing the device fails for a long period on a very hot summer day.

The Unusual Behavior of Water

Liquids generally increase in volume with increasing temperature and have average coefficients of volume expansion about ten times greater than those of solids. Cold water is an exception to this rule as you can see from its density-versus-temperature curve shown in Figure 19.11. As the temperature increases from 0°C to 4°C, water contracts and its density therefore increases. Above 4°C, water expands with increasing temperature and so its density decreases. Therefore, the density of water reaches a maximum value of 1.000 g/cm³ at 4°C.

We can use this unusual thermal-expansion behavior of water to explain why a pond begins freezing at the surface rather than at the bottom. When the air temperature drops from, for example, 7°C to 6°C, the surface water also cools and consequently decreases in volume. The surface water is denser than the water below it, which has not cooled and decreased in volume. As a result, the surface water sinks, and warmer water from below moves to the surface. When the air temperature is between 4°C and 0°C, however, the surface water expands as it cools, becoming less dense than the water below it. The mixing process stops, and eventually the surface water freezes. As the water freezes, the ice remains on the surface because ice is less dense than water. The ice continues to build up at the surface, while water near the

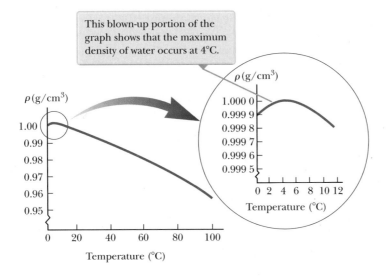

This blown-up portion of the graph shows that the maximum density of water occurs at 4°C.

Figure 19.11 The variation in the density of water at atmospheric pressure with temperature.

bottom remains at 4°C. If that were not the case, fish and other forms of marine life would not survive.

19.5 Macroscopic Description of an Ideal Gas

The volume expansion equation $\Delta V = \beta V_i \Delta T$ is based on the assumption that the material has an initial volume V_i before the temperature change occurs. Such is the case for solids and liquids because they have a fixed volume at a given temperature.

The case for gases is completely different. The interatomic forces within gases are very weak, and, in many cases, we can imagine these forces to be nonexistent and still make very good approximations. Therefore, *there is no equilibrium separation* for the atoms and no "standard" volume at a given temperature; the volume depends on the size of the container. As a result, we cannot express changes in volume ΔV in a process on a gas with Equation 19.6 because we have no defined volume V_i at the beginning of the process. Equations involving gases contain the volume V, rather than a *change* in the volume from an initial value, as a variable.

For a gas, it is useful to know how the quantities volume V, pressure P, and temperature T are related for a sample of gas of mass m. In general, the equation that interrelates these quantities, called the *equation of state*, is very complicated. If the gas is maintained at a very low pressure (or low density), however, the equation of state is quite simple and can be determined from experimental results. Such a low-density gas is commonly referred to as an **ideal gas.**[5] We can use the **ideal gas model** to make predictions that are adequate to describe the behavior of real gases at low pressures.

It is convenient to express the amount of gas in a given volume in terms of the number of moles n. One **mole** of any substance is that amount of the substance that contains **Avogadro's number** $N_A = 6.022 \times 10^{23}$ of constituent particles (atoms or molecules). The number of moles n of a substance is related to its mass m through the expression

$$n = \frac{m}{M} \tag{19.7}$$

where M is the molar mass of the substance. The molar mass of each chemical element is the atomic mass (from the periodic table; see Appendix C) expressed in grams per mole. For example, the mass of one He atom is 4.00 u (atomic mass units), so the molar mass of He is 4.00 g/mol.

Now suppose an ideal gas is confined to a cylindrical container whose volume can be varied by means of a movable piston as in Figure 19.12. If we assume the cylinder does not leak, the mass (or the number of moles) of the gas remains constant. For such a system, experiments provide the following information:

Figure 19.12 An ideal gas confined to a cylinder whose volume can be varied by means of a movable piston.

- When the gas is kept at a constant temperature, its pressure is inversely proportional to the volume. (This behavior is described historically as Boyle's law.)
- When the pressure of the gas is kept constant, the volume is directly proportional to the temperature. (This behavior is described historically as Charles's law.)
- When the volume of the gas is kept constant, the pressure is directly proportional to the temperature. (This behavior is described historically as Gay–Lussac's law.)

These observations are summarized by the **equation of state for an ideal gas:**

Equation of state for ▶ an ideal gas

$$PV = nRT \tag{19.8}$$

[5] To be more specific, the assumptions here are that the temperature of the gas must not be too low (the gas must not condense into a liquid) or too high and that the pressure must be low. The concept of an ideal gas implies that the gas molecules do not interact except upon collision and that the molecular volume is negligible compared with the volume of the container. In reality, an ideal gas does not exist. The concept of an ideal gas is nonetheless very useful because real gases at low pressures are well-modeled as ideal gases.

In this expression, also known as the **ideal gas law,** n is the number of moles of gas in the sample and R is a constant. Experiments on numerous gases show that as the pressure approaches zero, the quantity PV/nT approaches the same value R for all gases. For this reason, R is called the **universal gas constant.** In SI units, in which pressure is expressed in pascals (1 Pa = 1 N/m^2) and volume in cubic meters, the product PV has units of newton $\cdot$ meters, or joules, and R has the value

$$R = 8.314 \, \text{J/mol} \cdot \text{K} \qquad \textbf{(19.9)}$$

If the pressure is expressed in atmospheres and the volume in liters (1 L = 10^3 cm^3 = 10^{-3} m^3), then R has the value

$$R = 0.082\,06 \, \text{L} \cdot \text{atm/mol} \cdot \text{K}$$

Using this value of R and Equation 19.8 shows that the volume occupied by 1 mol of *any* gas at atmospheric pressure and at 0°C (273 K) is 22.4 L.

The ideal gas law states that if the volume and temperature of a fixed amount of gas do not change, the pressure also remains constant. Consider a bottle of champagne that is shaken and then spews liquid when opened as shown in Figure 19.13. A common misconception is that the pressure inside the bottle is increased when the bottle is shaken. On the contrary, because the temperature of the bottle and its contents remains constant as long as the bottle is sealed, so does the pressure, as can be shown by replacing the cork with a pressure gauge. The correct explanation is as follows. Carbon dioxide gas resides in the volume between the liquid surface and the cork. The pressure of the gas in this volume is set higher than atmospheric pressure in the bottling process. Shaking the bottle displaces some of the carbon dioxide gas into the liquid, where it forms bubbles, and these bubbles become attached to the inside of the bottle. (No new gas is generated by shaking.) When the bottle is opened, the pressure is reduced to atmospheric pressure, which causes the volume of the bubbles to increase suddenly. If the bubbles are attached to the bottle (beneath the liquid surface), their rapid expansion expels liquid from the bottle. If the sides and bottom of the bottle are first tapped until no bubbles remain beneath the surface, however, the drop in pressure does not force liquid from the bottle when the champagne is opened.

The ideal gas law is often expressed in terms of the total number of molecules N. Because the number of moles n equals the ratio of the total number of molecules and Avogadro's number N_A, we can write Equation 19.8 as

$$PV = nRT = \frac{N}{N_A} RT$$

$$PV = Nk_B T \qquad \textbf{(19.10)}$$

where k_B is **Boltzmann's constant,** which has the value

$$k_B = \frac{R}{N_A} = 1.38 \times 10^{-23} \, \text{J/K} \qquad \textbf{(19.11)}$$

◀ Boltzmann's constant

It is common to call quantities such as P, V, and T the **thermodynamic variables** of an ideal gas. If the equation of state is known, one of the variables can always be expressed as some function of the other two.

Quick Quiz 19.5 A common material for cushioning objects in packages is made by trapping bubbles of air between sheets of plastic. Is this material more effective at keeping the contents of the package from moving around inside the package on **(a)** a hot day, **(b)** a cold day, or **(c)** either hot or cold days?

Quick Quiz 19.6 On a winter day, you turn on your furnace and the temperature of the air inside your home increases. Assume your home has the normal amount of leakage between inside air and outside air. Is the number of moles of air in your room at the higher temperature **(a)** larger than before, **(b)** smaller than before, or **(c)** the same as before?

Figure 19.13 A bottle of champagne is shaken and opened. Liquid spews out of the opening. A common misconception is that the pressure inside the bottle is increased by the shaking.

Pitfall Prevention 19.3
So Many ks There are a variety of physical quantities for which the letter k is used. Two we have seen previously are the force constant for a spring (Chapter 15) and the wave number for a mechanical wave (Chapter 16). Boltzmann's constant is another k, and we will see k used for thermal conductivity in Chapter 20 and for an electrical constant in Chapter 23. To make some sense of this confusing state of affairs, we use a subscript B for Boltzmann's constant to help us recognize it. In this book, you will see Boltzmann's constant as k_B, but you may see Boltzmann's constant in other resources as simply k.

Example 19.4 Heating a Spray Can

A spray can containing a propellant gas at twice atmospheric pressure (202 kPa) and having a volume of 125.00 cm³ is at 22°C. It is then tossed into an open fire. (*Warning:* Do not do this experiment; it is very dangerous.) When the temperature of the gas in the can reaches 195°C, what is the pressure inside the can? Assume any change in the volume of the can is negligible.

SOLUTION

Conceptualize Intuitively, you should expect that the pressure of the gas in the container increases because of the increasing temperature.

Categorize We model the gas in the can as ideal and use the ideal gas law to calculate the new pressure.

Analyze Rearrange Equation 19.8:

$$(1) \quad \frac{PV}{T} = nR$$

No air escapes during the compression, so n, and therefore nR, remains constant. Hence, set the initial value of the left side of Equation (1) equal to the final value:

$$(2) \quad \frac{P_i V_i}{T_i} = \frac{P_f V_f}{T_f}$$

Because the initial and final volumes of the gas are assumed to be equal, cancel the volumes:

$$(3) \quad \frac{P_i}{T_i} = \frac{P_f}{T_f}$$

Solve for P_f:

$$P_f = \left(\frac{T_f}{T_i}\right)P_i = \left(\frac{468 \text{ K}}{295 \text{ K}}\right)(202 \text{ kPa}) = \boxed{320 \text{ kPa}}$$

Finalize The higher the temperature, the higher the pressure exerted by the trapped gas as expected. If the pressure increases sufficiently, the can may explode. Because of this possibility, you should never dispose of spray cans in a fire.

WHAT IF? Suppose we include a volume change due to thermal expansion of the steel can as the temperature increases. Does that alter our answer for the final pressure significantly?

Answer Because the thermal expansion coefficient of steel is very small, we do not expect much of an effect on our final answer.

Find the change in the volume of the can using Equation 19.6 and the value for α for steel from Table 19.1:

$$\Delta V = \beta V_i \Delta T = 3\alpha V_i \Delta T$$
$$= 3[11 \times 10^{-6} \text{ (°C)}^{-1}](125.00 \text{ cm}^3)(173°C) = 0.71 \text{ cm}^3$$

Start from Equation (2) again and find an equation for the final pressure:

$$P_f = \left(\frac{T_f}{T_i}\right)\left(\frac{V_i}{V_f}\right)P_i$$

This result differs from Equation (3) only in the factor V_i/V_f. Evaluate this factor:

$$\frac{V_i}{V_f} = \frac{125.00 \text{ cm}^3}{(125.00 \text{ cm}^3 + 0.71 \text{ cm}^3)} = 0.994 = 99.4\%$$

Therefore, the final pressure will differ by only 0.6% from the value calculated without considering the thermal expansion of the can. Taking 99.4% of the previous final pressure, the final pressure including thermal expansion is 318 kPa.

Summary

Definitions

Two objects are in **thermal equilibrium** with each other if they do not exchange energy when in thermal contact.

Temperature is the property that determines whether an object is in thermal equilibrium with other objects. Two objects in thermal equilibrium with each other are at the same temperature. The SI unit of absolute temperature is the **kelvin,** which is defined to be 1/273.16 of the difference between absolute zero and the temperature of the triple point of water.

Concepts and Principles

The **zeroth law of thermodynamics** states that if objects A and B are separately in thermal equilibrium with a third object C, then objects A and B are in thermal equilibrium with each other.

When the temperature of an object is changed by an amount ΔT, its length changes by an amount ΔL that is proportional to ΔT and to its initial length L_i:

$$\Delta L = \alpha L_i \Delta T \qquad (19.4)$$

where the constant α is the **average coefficient of linear expansion.** The **average coefficient of volume expansion** β for a solid is approximately equal to 3α.

An **ideal gas** is one for which PV/nT is constant. An ideal gas is described by the **equation of state,**

$$PV = nRT \qquad (19.8)$$

where n equals the number of moles of the gas, P is its pressure, V is its volume, R is the universal gas constant $(8.314\ \mathrm{J/mol \cdot K})$, and T is the absolute temperature of the gas. A real gas behaves approximately as an ideal gas if it has a low density.

Objective Questions   1. denotes answer available in *Student Solutions Manual/Study Guide*

1. Markings to indicate length are placed on a steel tape in a room that is at a temperature of 22°C. Measurements are then made with the same tape on a day when the temperature is 27°C. Assume the objects you are measuring have a smaller coefficient of linear expansion than steel. Are the measurements (a) too long, (b) too short, or (c) accurate?

2. When a certain gas under a pressure of 5.00×10^6 Pa at 25.0°C is allowed to expand to 3.00 times its original volume, its final pressure is 1.07×10^6 Pa. What is its final temperature? (a) 450 K (b) 233 K (c) 212 K (d) 191 K (e) 115 K

3. If the volume of an ideal gas is doubled while its temperature is quadrupled, does the pressure (a) remain the same, (b) decrease by a factor of 2, (c) decrease by a factor of 4, (d) increase by a factor of 2, or (e) increase by a factor of 4

4. The pendulum of a certain pendulum clock is made of brass. When the temperature increases, what happens to the period of the clock? (a) It increases. (b) It decreases. (c) It remains the same.

5. A temperature of 162°F is equivalent to what temperature in kelvins? (a) 373 K (b) 288 K (c) 345 K (d) 201 K (e) 308 K

6. A cylinder with a piston holds 0.50 m^3 of oxygen at an absolute pressure of 4.0 atm. The piston is pulled outward, increasing the volume of the gas until the pressure drops to 1.0 atm. If the temperature stays con-

stant, what new volume does the gas occupy? (a) 1.0 m^3 (b) 1.5 m^3 (c) 2.0 m^3 (d) 0.12 m^3 (e) 2.5 m^3

7. What would happen if the glass of a thermometer expanded more on warming than did the liquid in the tube? (a) The thermometer would break. (b) It could be used only for temperatures below room temperature. (c) You would have to hold it with the bulb on top. (d) The scale on the thermometer is reversed so that higher temperature values would be found closer to the bulb. (e) The numbers would not be evenly spaced.

8. A cylinder with a piston contains a sample of a thin gas. The kind of gas and the sample size can be changed. The cylinder can be placed in different constant-temperature baths, and the piston can be held in different positions. Rank the following cases according to the pressure of the gas from the highest to the lowest, displaying any cases of equality. (a) A 0.002-mol sample of oxygen is held at 300 K in a 100-cm^3 container. (b) A 0.002-mol sample of oxygen is held at 600 K in a 200-cm^3 container. (c) A 0.002-mol sample of oxygen is held at 600 K in a 300-cm^3 container. (d) A 0.004-mol sample of helium is held at 300 K in a 200-cm^3 container. (e) A 0.004-mol sample of helium is held at 250 K in a 200-cm^3 container.

9. Two cylinders A and B at the same temperature contain the same quantity of the same kind of gas. Cylinder A has three times the volume of cylinder B. What can you conclude about the pressures the gases exert? (a) We can conclude nothing about the pressures.

(b) The pressure in A is three times the pressure in B. (c) The pressures must be equal. (d) The pressure in A must be one-third the pressure in B.

10. A rubber balloon is filled with 1 L of air at 1 atm and 300 K and is then put into a cryogenic refrigerator at 100 K. The rubber remains flexible as it cools. **(i)** What happens to the volume of the balloon? (a) It decreases to $\frac{1}{3}$ L. (b) It decreases to $1/\sqrt{3}$ L. (c) It is constant. (d) It increases to $\sqrt{3}$ L. (e) It increases to 3 L. **(ii)** What happens to the pressure of the air in the balloon? (a) It decreases to $\frac{1}{3}$ atm. (b) It decreases to $1/\sqrt{3}$ atm. (c) It is constant. (d) It increases to $\sqrt{3}$ atm. (e) It increases to 3 atm.

11. The average coefficient of linear expansion of copper is 17×10^{-6} (°C)$^{-1}$. The Statue of Liberty is 93 m tall on a summer morning when the temperature is 25°C. Assume the copper plates covering the statue are mounted edge to edge without expansion joints and do not buckle or bind on the framework supporting them as the day grows hot. What is the order of magnitude of the statue's increase in height? (a) 0.1 mm (b) 1 mm (c) 1 cm (d) 10 cm (e) 1 m

12. Suppose you empty a tray of ice cubes into a bowl partly full of water and cover the bowl. After one-half hour, the contents of the bowl come to thermal equilibrium, with more liquid water and less ice than you started with. Which of the following is true? (a) The temperature of the liquid water is higher than the temperature of the remaining ice. (b) The temperature of the liquid water is the same as that of the ice. (c) The temperature of the liquid water is less than that of the ice. (d) The comparative temperatures of the liquid water and ice depend on the amounts present.

13. A hole is drilled in a metal plate. When the metal is raised to a higher temperature, what happens to the diameter of the hole? (a) It decreases. (b) It increases. (c) It remains the same. (d) The answer depends on the initial temperature of the metal. (e) None of those answers is correct.

14. On a very cold day in upstate New York, the temperature is −25°C, which is equivalent to what Fahrenheit temperature? (a) −46°F (b) −77°F (c) 18°F (d) −25°F (e) −13°F

Conceptual Questions

1. denotes answer available in *Student Solutions Manual/Study Guide*

1. Common thermometers are made of a mercury column in a glass tube. Based on the operation of these thermometers, which has the larger coefficient of linear expansion, glass or mercury? (Don't answer the question by looking in a table.)

2. A piece of copper is dropped into a beaker of water. (a) If the water's temperature rises, what happens to the temperature of the copper? (b) Under what conditions are the water and copper in thermal equilibrium?

3. (a) What does the ideal gas law predict about the volume of a sample of gas at absolute zero? (b) Why is this prediction incorrect?

4. Some picnickers stop at a convenience store to buy some food, including bags of potato chips. They then drive up into the mountains to their picnic site. When they unload the food, they notice that the bags of chips are puffed up like balloons. Why did that happen?

5. In describing his upcoming trip to the Moon, and as portrayed in the movie *Apollo 13* (Universal, 1995), astronaut Jim Lovell said, "I'll be walking in a place where there's a 400-degree difference between sunlight and shadow." Suppose an astronaut standing on the Moon holds a thermometer in his gloved hand. (a) Is the thermometer reading the temperature of the vacuum at the Moon's surface? (b) Does it read any temperature? If so, what object or substance has that temperature?

6. Metal lids on glass jars can often be loosened by running hot water over them. Why does that work?

7. An automobile radiator is filled to the brim with water when the engine is cool. (a) What happens to the water when the engine is running and the water has been raised to a high temperature? (b) What do modern automobiles have in their cooling systems to prevent the loss of coolants?

8. When the metal ring and metal sphere in Figure CQ19.8 are both at room temperature, the sphere can barely be passed through the ring. (a) After the sphere is warmed in a flame, it cannot be passed through the ring. Explain. (b) **What If?** What if the ring is warmed and the sphere is left at room temperature? Does the sphere pass through the ring?

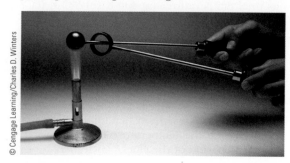
© Cengage Learning/Charles D. Winters

Figure CQ19.8

9. Is it possible for two objects to be in thermal equilibrium if they are not in contact with each other? Explain.

10. Use a periodic table of the elements (see Appendix C) to determine the number of grams in one mole of (a) hydrogen, which has diatomic molecules; (b) helium; and (c) carbon monoxide.

Section 19.2 Thermometers and the Celsius Temperature Scale
Section 19.3 The Constant-Volume Gas Thermometer and the Absolute Temperature Scale
Problems 1–7

Section 19.4 Thermal Expansion of Solids and Liquids
Problems 8–25

Section 19.5 Macroscopic Description of an Ideal Gas
Problems 26–44

Additional Problems
Problems 45–71

Challenge Problems
Problems 72–79

Solutions to the following Problems are available in the *Student Solutions Manual/Study Guide:*
19.5, 19.7, 19.9, 19.11, 19.19, 19.31, 19.32, 19.37, 19.39, 19.51, 19.57, 19.61, 19.64, 19.71, 19.73, and 19.76

List of Enhanced Problems

Problem Number	Targeted Feedback in Enhanced WebAssign	Analysis Model Tutorial in Enhanced WebAssign	Master It in Enhanced WebAssign	Watch It in Enhanced WebAssign
19.7			✓	
19.8	✓			✓
19.9	✓		✓	
19.11	✓		✓	
19.17	✓			✓
19.18	✓			✓
19.25	✓		✓	
19.29	✓			✓
19.31			✓	
19.32			✓	
19.36	✓			✓
19.37			✓	
19.39	✓	✓	✓	
19.41	✓			
19.43	✓			✓
19.51		✓		
19.53		✓		
19.64		✓		
19.70		✓		
19.74	✓			✓

The First Law of Thermodynamics

In this photograph of the Mt. Baker area near Bellingham, Washington, we see evidence of water in all three phases. In the lake is liquid water, and solid water in the form of snow appears on the ground. The clouds in the sky consist of liquid water droplets that have condensed from the gaseous water vapor in the air. Changes of a substance from one phase to another are a result of energy transfer. (©iStockphoto.com/ KingWu)

Until about 1850, the fields of thermodynamics and mechanics were considered to be two distinct branches of science. The principle of conservation of energy seemed to describe only certain kinds of mechanical systems. Mid-19th-century experiments performed by Englishman James Joule and others, however, showed a strong connection between the transfer of energy by heat in thermal processes and the transfer of energy by work in mechanical processes. Today we know that mechanical energy can be transformed to internal energy, which is formally defined in this chapter. Once the concept of energy was generalized from mechanics to include internal energy, the principle of conservation of energy as discussed in Chapter 8 emerged as a universal law of nature.

This chapter focuses on the concept of internal energy, the first law of thermodynamics, and some important applications of the first law. The first law of thermodynamics describes systems in which the only energy change is that of internal energy and the transfers of energy are by heat and work. A major difference in our discussion of work in this chapter from that in most of the chapters on mechanics is that we will consider work done on *deformable* systems.

20.1 Heat and Internal Energy

At the outset, it is important to make a major distinction between internal energy and heat, terms that are often incorrectly used interchangeably in popular language.

Internal energy is all the energy of a system that is associated with its microscopic components—atoms and molecules—when viewed from a reference frame at rest with respect to the center of mass of the system.

The last part of this sentence ensures that any bulk kinetic energy of the system due to its motion through space is not included in internal energy. Internal energy includes kinetic energy of random translational, rotational, and vibrational motion of molecules; vibrational potential energy associated with forces between atoms in molecules; and electric potential energy associated with forces between molecules. It is useful to relate internal energy to the temperature of an object, but this relationship is limited. We show in Section 20.3 that internal energy changes can also occur in the absence of temperature changes. In that discussion, we will investigate the internal energy of the system when there is a *physical change*, most often related to a phase change, such as melting or boiling. We assign energy associated with *chemical changes*, related to chemical reactions, to the potential energy term in Equation 8.2, not to internal energy. Therefore, we discuss the *chemical potential energy* in, for example, a human body (due to previous meals), the gas tank of a car (due to an earlier transfer of fuel), and a battery of an electric circuit (placed in the battery during its construction in the manufacturing process).

Heat is defined as a process of transferring energy across the boundary of a system because of a temperature difference between the system and its surroundings. It is also the amount of energy Q transferred by this process.

When you *heat* a substance, you are transferring energy into it by placing it in contact with surroundings that have a higher temperature. Such is the case, for example, when you place a pan of cold water on a stove burner. The burner is at a higher temperature than the water, and so the water gains energy by heat.

Read this definition of heat (Q in Eq. 8.2) very carefully. In particular, notice what heat is *not* in the following common quotes. (1) Heat is *not* energy in a hot substance. For example, "The boiling water has a lot of heat" is incorrect; the boiling water has *internal energy* E_{int}. (2) Heat is *not* radiation. For example, "It was so hot because the sidewalk was radiating heat" is incorrect; energy is leaving the sidewalk by *electromagnetic radiation*, T_{ER} in Equation 8.2. (3) Heat is *not* warmth of an environment. For example, "The heat in the air was so oppressive" is incorrect; on a hot day, the air has a high *temperature T*.

As an analogy to the distinction between heat and internal energy, consider the distinction between work and mechanical energy discussed in Chapter 7. The work done on a system is a measure of the amount of energy transferred to the system from its surroundings, whereas the mechanical energy (kinetic energy plus potential energy) of a system is a consequence of the motion and configuration of the system. Therefore, when a person does work on a system, energy is transferred from the person to the system. It makes no sense to talk about the work *of* a system; one can refer only to the work done *on* or *by* a system when some process has occurred in which energy has been transferred to or from the system. Likewise, it makes no sense to talk about the heat *of* a system; one can refer to heat only when energy has been transferred as a result of a temperature difference. Both heat and work are ways of transferring energy between a system and its surroundings.

Units of Heat

Early studies of heat focused on the resultant increase in temperature of a substance, which was often water. Initial notions of heat were based on a fluid called *caloric* that flowed from one substance to another and caused changes in temperature. From the name of this mythical fluid came an energy unit related to thermal processes, the **calorie (cal),** which is defined as the amount of energy transfer

Pitfall Prevention 20.1

Internal Energy, Thermal Energy, and Bond Energy When reading other physics books, you may see terms such as *thermal energy* and *bond energy.* Thermal energy can be interpreted as that part of the internal energy associated with random motion of molecules and therefore related to temperature. Bond energy is the intermolecular potential energy. Therefore,

Internal energy =
   thermal energy + bond energy

Although this breakdown is presented here for clarification with regard to other books, we will not use these terms because there is no need for them.

Pitfall Prevention 20.2

Heat, Temperature, and Internal Energy Are Different As you read the newspaper or explore the Internet, be alert for incorrectly used phrases including the word *heat* and think about the proper word to be used in place of *heat.* Incorrect examples include "As the truck braked to a stop, a large amount of heat was generated by friction" and "The heat of a hot summer day . . . ."

James Prescott Joule
British physicist (1818–1889)
Joule received some formal education in mathematics, philosophy, and chemistry from John Dalton but was in large part self-educated. Joule's research led to the establishment of the principle of conservation of energy. His study of the quantitative relationship among electrical, mechanical, and chemical effects of heat culminated in his announcement in 1843 of the amount of work required to produce a unit of energy, called the mechanical equivalent of heat.

necessary to raise the temperature of 1 g of water from 14.5°C to 15.5°C.[1] (The "Calorie," written with a capital "C" and used in describing the energy content of foods, is actually a kilocalorie.) The unit of energy in the U.S. customary system is the **British thermal unit (Btu),** which is defined as the amount of energy transfer required to raise the temperature of 1 lb of water from 63°F to 64°F.

Once the relationship between energy in thermal and mechanical processes became clear, there was no need for a separate unit related to thermal processes. The *joule* has already been defined as an energy unit based on mechanical processes. Scientists are increasingly turning away from the calorie and the Btu and are using the joule when describing thermal processes. In this textbook, heat, work, and internal energy are usually measured in joules.

The Mechanical Equivalent of Heat

In Chapters 7 and 8, we found that whenever friction is present in a mechanical system, the mechanical energy in the system decreases; in other words, mechanical energy is not conserved in the presence of nonconservative forces. Various experiments show that this mechanical energy does not simply disappear but is transformed into internal energy. You can perform such an experiment at home by hammering a nail into a scrap piece of wood. What happens to all the kinetic energy of the hammer once you have finished? Some of it is now in the nail as internal energy, as demonstrated by the nail being measurably warmer. Notice that there is *no* transfer of energy by heat in this process. For the nail and board as a nonisolated system, Equation 8.2 becomes $\Delta E_{\text{int}} = W + T_{\text{MW}}$, where W is the work done by the hammer on the nail and T_{MW} is the energy leaving the system by sound waves when the nail is struck. Although this connection between mechanical and internal energy was first suggested by Benjamin Thompson, it was James Prescott Joule who established the equivalence of the decrease in mechanical energy and the increase in internal energy.

A schematic diagram of Joule's most famous experiment is shown in Figure 20.1. The system of interest is the Earth, the two blocks, and the water in a thermally insulated container. Work is done within the system on the water by a rotating paddle wheel, which is driven by heavy blocks falling at a constant speed. If the energy transformed in the bearings and the energy passing through the walls by heat are neglected, the decrease in potential energy of the system as the blocks fall equals the work done by the paddle wheel on the water and, in turn, the increase in internal energy of the water. If the two blocks fall through a distance h, the decrease in potential energy of the system is $2mgh$, where m is the mass of one block; this energy causes the temperature of the water to increase. By varying the conditions of the experiment, Joule found that the decrease in mechanical energy is proportional to the product of the mass of the water and the increase in water temperature. The proportionality constant was found to be approximately 4.18 J/g · °C. Hence, 4.18 J of mechanical energy raises the temperature of 1 g of water by 1°C. More precise measurements taken later demonstrated the proportionality to be 4.186 J/g · °C when the temperature of the water was raised from 14.5°C to 15.5°C. We adopt this "15-degree calorie" value:

$$1 \text{ cal} = 4.186 \text{ J} \tag{20.1}$$

This equality is known, for purely historical reasons, as the **mechanical equivalent of heat.** A more proper name would be *equivalence between mechanical energy and internal energy,* but the historical name is well entrenched in our language, despite the incorrect use of the word *heat.*

The falling blocks rotate the paddles, causing the temperature of the water to increase.

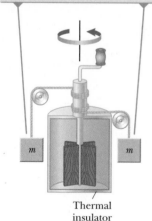

Thermal
insulator

Figure 20.1 Joule's experiment for determining the mechanical equivalent of heat.

---

[1]Originally, the calorie was defined as the energy transfer necessary to raise the temperature of 1 g of water by 1°C. Careful measurements, however, showed that the amount of energy required to produce a 1°C change depends somewhat on the initial temperature; hence, a more precise definition evolved.

Example 20.1	**Losing Weight the Hard Way**

A student eats a dinner rated at 2 000 Calories. He wishes to do an equivalent amount of work in the gymnasium by lifting a 50.0-kg barbell. How many times must he raise the barbell to expend this much energy? Assume he raises the barbell 2.00 m each time he lifts it and he regains no energy when he lowers the barbell.

SOLUTION

Conceptualize Imagine the student raising the barbell. He is doing work on the system of the barbell and the Earth, so energy is leaving his body. The total amount of work that the student must do is 2 000 Calories.

Categorize We model the system of the barbell and the Earth as a *nonisolated system* for *energy.*

....

Analyze Reduce the conservation of energy equation, Equation 8.2, to the appropriate expression for the system of the barbell and the Earth:

$$(1) \quad \Delta U_{total} = W_{total}$$

Express the change in gravitational potential energy of the system after the barbell is raised once:

$$\Delta U = mgh$$

Express the total amount of energy that must be transferred into the system by work for lifting the barbell n times, assuming energy is not regained when the barbell is lowered:

$$(2) \quad \Delta U_{total} = nmgh$$

Substitute Equation (2) into Equation (1):

$$nmgh = W_{total}$$

Solve for n:

$$n = \frac{W_{total}}{mgh}$$

Substitute numerical values:

$$n = \frac{(2\ 000\ \text{Cal})}{(50.0\ \text{kg})(9.80\ \text{m/s}^2)(2.00\ \text{m})}\left(\frac{1.00 \times 10^3\ \text{cal}}{\text{Calorie}}\right)\left(\frac{4.186\ \text{J}}{1\ \text{cal}}\right)$$

$$= 8.54 \times 10^3\ \text{times}$$

....

Finalize If the student is in good shape and lifts the barbell once every 5 s, it will take him about 12 h to perform this feat. Clearly, it is much easier for this student to lose weight by dieting.

In reality, the human body is not 100% efficient. Therefore, not all the energy transformed within the body from the dinner transfers out of the body by work done on the barbell. Some of this energy is used to pump blood and perform other functions within the body. Therefore, the 2 000 Calories can be worked off in less time than 12 h when these other energy processes are included.

20.2 Specific Heat and Calorimetry

When energy is added to a system and there is no change in the kinetic or potential energy of the system, the temperature of the system usually rises. (An exception to this statement is the case in which a system undergoes a change of state—also called a *phase transition*—as discussed in the next section.) If the system consists of a sample of a substance, we find that the quantity of energy required to raise the temperature of a given mass of the substance by some amount varies from one substance to another. For example, the quantity of energy required to raise the temperature of 1 kg of water by 1°C is 4 186 J, but the quantity of energy required to raise the temperature of 1 kg of copper by 1°C is only 387 J. In the discussion that follows, we shall use heat as our example of energy transfer, but keep in mind that the temperature of the system could be changed by means of any method of energy transfer.

The **heat capacity** C of a particular sample is defined as the amount of energy needed to raise the temperature of that sample by 1°C. From this definition, we see that if energy Q produces a change ΔT in the temperature of a sample, then

$$Q = C\Delta T \qquad \qquad \textbf{(20.2)}$$

Table 20.1	Specific Heats of Some Substances at 25°C and Atmospheric Pressure			
Substance	Specific Heat (J/kg · °C)		Substance	Specific Heat (J/kg · °C)
Elemental solids			*Other solids*	
Aluminum	900		Brass	380
Beryllium	1 830		Glass	837
Cadmium	230		Ice (−5°C)	2 090
Copper	387		Marble	860
Germanium	322		Wood	1 700
Gold	129			
Iron	448		*Liquids*	
Lead	128		Alcohol (ethyl)	2 400
Silicon	703		Mercury	140
Silver	234		Water (15°C)	4 186
			Gas	
			Steam (100°C)	2 010

Note: To convert values to units of cal/g · °C, divide by 4 186.

The **specific heat** c of a substance is the heat capacity per unit mass. Therefore, if energy Q transfers to a sample of a substance with mass m and the temperature of the sample changes by ΔT, the specific heat of the substance is

▶ Specific heat

$$c \equiv \frac{Q}{m \, \Delta T} \tag{20.3}$$

Specific heat is essentially a measure of how thermally insensitive a substance is to the addition of energy. The greater a material's specific heat, the more energy must be added to a given mass of the material to cause a particular temperature change. Table 20.1 lists representative specific heats.

From this definition, we can relate the energy Q transferred between a sample of mass m of a material and its surroundings to a temperature change ΔT as

$$Q = mc \, \Delta T \tag{20.4}$$

For example, the energy required to raise the temperature of 0.500 kg of water by 3.00°C is $Q = (0.500 \text{ kg})(4\,186 \text{ J/kg} \cdot {}^\circ\text{C})(3.00{}^\circ\text{C}) = 6.28 \times 10^3$ J. Notice that when the temperature increases, Q and ΔT are taken to be positive and energy transfers into the system. When the temperature decreases, Q and ΔT are negative and energy transfers out of the system.

We can identify $mc \, \Delta T$ as the change in internal energy of the system if we ignore any thermal expansion or contraction of the system. (Thermal expansion or contraction would result in a very small amount of work being done on the system by the surrounding air.) Then, Equation 20.4 is a reduced form of Equation 8.2: $\Delta E_{int} = Q$. The internal energy of the system can be changed by transferring energy into the system by any mechanism. For example, if the system is a baked potato in a microwave oven, Equation 8.2 reduces to the following analog to Equation 20.4: $\Delta E_{int} = T_{ER} = mc \, \Delta T$, where T_{ER} is the energy transferred to the potato from the microwave oven by electromagnetic radiation. If the system is the air in a bicycle pump, which becomes hot when the pump is operated, Equation 8.2 reduces to the following analog to Equation 20.4: $\Delta E_{int} = W = mc \, \Delta T$, where W is the work done on the pump by the operator. By identifying $mc \, \Delta T$ as ΔE_{int}, we have taken a step toward a better understanding of temperature: temperature is related to the energy of the molecules of a system. We will learn more details of this relationship in Chapter 21.

Specific heat varies with temperature. If, however, temperature intervals are not too great, the temperature variation can be ignored and c can be treated as a constant.[2]

Pitfall Prevention 20.3

An Unfortunate Choice of Terminology The name *specific heat* is an unfortunate holdover from the days when thermodynamics and mechanics developed separately. A better name would be *specific energy transfer*, but the existing term is too entrenched to be replaced.

Pitfall Prevention 20.4

Energy Can Be Transferred by Any Method The symbol Q represents the amount of energy transferred, but keep in mind that the energy transfer in Equation 20.4 could be by *any* of the methods introduced in Chapter 8; it does not have to be heat. For example, repeatedly bending a wire coat hanger raises the temperature at the bending point by *work*.

---

[2]The definition given by Equation 20.4 assumes the specific heat does not vary with temperature over the interval $\Delta T = T_f - T_i$. In general, if c varies with temperature over the interval, the correct expression for Q is $Q = m \int_{T_i}^{T_f} c \, dT$.

For example, the specific heat of water varies by only about 1% from 0°C to 100°C at atmospheric pressure. Unless stated otherwise, we shall neglect such variations.

Quick Quiz 20.1 Imagine you have 1 kg each of iron, glass, and water, and all three samples are at 10°C. **(a)** Rank the samples from highest to lowest temperature after 100 J of energy is added to each sample. **(b)** Rank the samples from greatest to least amount of energy transferred by heat if each sample increases in temperature by 20°C.

Notice from Table 20.1 that water has the highest specific heat of common materials. This high specific heat is in part responsible for the moderate climates found near large bodies of water. As the temperature of a body of water decreases during the winter, energy is transferred from the cooling water to the air by heat, increasing the internal energy of the air. Because of the high specific heat of water, a relatively large amount of energy is transferred to the air for even modest temperature changes of the water. The prevailing winds on the West Coast of the United States are toward the land (eastward). Hence, the energy liberated by the Pacific Ocean as it cools keeps coastal areas much warmer than they would otherwise be. As a result, West Coast states generally have more favorable winter weather than East Coast states, where the prevailing winds do not tend to carry the energy toward land.

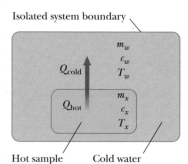

Figure 20.2 In a calorimetry experiment, a hot sample whose specific heat is unknown is placed in cold water in a container that isolates the system from the environment.

Calorimetry

One technique for measuring specific heat involves heating a sample to some known temperature T_x, placing it in a vessel containing water of known mass and temperature $T_w < T_x$, and measuring the temperature of the water after equilibrium has been reached. This technique is called **calorimetry,** and devices in which this energy transfer occurs are called **calorimeters.** Figure 20.2 shows the hot sample in the cold water and the resulting energy transfer by heat from the high-temperature part of the system to the low-temperature part. If the system of the sample and the water is isolated, the principle of conservation of energy requires that the amount of energy Q_{hot} that leaves the sample (of unknown specific heat) equal the amount of energy Q_{cold} that enters the water.[3] Conservation of energy allows us to write the mathematical representation of this energy statement as

$$Q_{cold} = -Q_{hot} \tag{20.5}$$

Suppose m_x is the mass of a sample of some substance whose specific heat we wish to determine. Let's call its specific heat c_x and its initial temperature T_x as shown in Figure 20.2. Likewise, let m_w, c_w, and T_w represent corresponding values for the water. If T_f is the final temperature after the system comes to equilibrium, Equation 20.4 shows that the energy transfer for the water is $m_w c_w(T_f - T_w)$, which is positive because $T_f > T_w$, and that the energy transfer for the sample of unknown specific heat is $m_x c_x(T_f - T_x)$, which is negative. Substituting these expressions into Equation 20.5 gives

$$m_w c_w(T_f - T_w) = -m_x c_x(T_f - T_x)$$

This equation can be solved for the unknown specific heat c_x.

Pitfall Prevention 20.5
Remember the Negative Sign It is *critical* to include the negative sign in Equation 20.5. The negative sign in the equation is necessary for consistency with our sign convention for energy transfer. The energy transfer Q_{hot} has a negative value because energy is leaving the hot substance. The negative sign in the equation ensures that the right side is a positive number, consistent with the left side, which is positive because energy is entering the cold water.

Example 20.2　　**Cooling a Hot Ingot**

A 0.050 0-kg ingot of metal is heated to 200.0°C and then dropped into a calorimeter containing 0.400 kg of water initially at 20.0°C. The final equilibrium temperature of the mixed system is 22.4°C. Find the specific heat of the metal.

continued

---

[3]For precise measurements, the water container should be included in our calculations because it also exchanges energy with the sample. Doing so would require that we know the container's mass and composition, however. If the mass of the water is much greater than that of the container, we can neglect the effects of the container.

▶ **20.2** continued

SOLUTION

Conceptualize Imagine the process occurring in the isolated system of Figure 20.2. Energy leaves the hot ingot and goes into the cold water, so the ingot cools off and the water warms up. Once both are at the same temperature, the energy transfer stops.

Categorize We use an equation developed in this section, so we categorize this example as a substitution problem.

Use Equation 20.4 to evaluate each side of Equation 20.5:

$$m_w c_w (T_f - T_w) = -m_x c_x (T_f - T_x)$$

Solve for c_x:

$$c_x = \frac{m_w c_w (T_f - T_w)}{m_x (T_x - T_f)}$$

Substitute numerical values:

$$c_x = \frac{(0.400 \text{ kg})(4\,186 \text{ J/kg} \cdot {}^\circ\text{C})(22.4{}^\circ\text{C} - 20.0{}^\circ\text{C})}{(0.050\,0 \text{ kg})(200.0{}^\circ\text{C} - 22.4{}^\circ\text{C})}$$

$$= \boxed{453 \text{ J/kg} \cdot {}^\circ\text{C}}$$

The ingot is most likely iron as you can see by comparing this result with the data given in Table 20.1. The temperature of the ingot is initially above the steam point. Therefore, some of the water may vaporize when the ingot is dropped into the water. We assume the system is sealed and this steam cannot escape. Because the final equilibrium temperature is lower than the steam point, any steam that does result recondenses back into water.

WHAT IF? Suppose you are performing an experiment in the laboratory that uses this technique to determine the specific heat of a sample and you wish to decrease the overall uncertainty in your final result for c_x. Of the data given in this example, changing which value would be most effective in decreasing the uncertainty?

Answer The largest experimental uncertainty is associated with the small difference in temperature of 2.4°C for the water. For example, using the rules for propagation of uncertainty in Appendix Section B.8, an uncertainty of 0.1°C in each of T_f and T_w leads to an 8% uncertainty in their difference. For this temperature difference to be larger experimentally, the most effective change is to *decrease the amount of water*.

Example 20.3 **Fun Time for a Cowboy**

A cowboy fires a silver bullet with a muzzle speed of 200 m/s into the pine wall of a saloon. Assume all the internal energy generated by the impact remains with the bullet. What is the temperature change of the bullet?

SOLUTION

Conceptualize Imagine similar experiences you may have had in which mechanical energy is transformed to internal energy when a moving object is stopped. For example, as mentioned in Section 20.1, a nail becomes warm after it is hit a few times with a hammer.

Categorize The bullet is modeled as an *isolated system*. No work is done on the system because the force from the wall moves through no displacement. This example is similar to the skateboarder pushing off a wall in Section 9.7. There, no work is done on the skateboarder by the wall, and potential energy stored in the body from previous meals is transformed to kinetic energy. Here, no work is done by the wall on the bullet, and kinetic energy is transformed to internal energy.

Analyze Reduce the conservation of energy equation, Equation 8.2, to the appropriate expression for the system of the bullet:

(1) $\quad \Delta K + \Delta E_{int} = 0$

The change in the bullet's internal energy is related to its change in temperature:

(2) $\quad \Delta E_{int} = mc\,\Delta T$

Substitute Equation (2) into Equation (1):

$(0 - \frac{1}{2}mv^2) + mc\,\Delta T = 0$

▶ **20.3** continued

Solve for ΔT, using 234 J/kg · °C as the specific heat of silver (see Table 20.1):

$$(3) \quad \Delta T = \frac{\frac{1}{2}mv^2}{mc} = \frac{v^2}{2c} = \frac{(200 \text{ m/s})^2}{2(234 \text{ J/kg} \cdot °C)} = \boxed{85.5°C}$$

. . . .

Finalize Notice that the result does not depend on the mass of the bullet.

WHAT IF? Suppose the cowboy runs out of silver bullets and fires a lead bullet at the same speed into the wall. Will the temperature change of the bullet be larger or smaller?

Answer Table 20.1 shows that the specific heat of lead is 128 J/kg · °C, which is smaller than that for silver. Therefore, a given amount of energy input or transformation raises lead to a higher temperature than silver and the final temperature of the lead bullet will be larger. In Equation (3), let's substitute the new value for the specific heat:

$$\Delta T = \frac{v^2}{2c} = \frac{(200 \text{ m/s})^2}{2(128 \text{ J/kg} \cdot °C)} = 156°C$$

There is no requirement that the silver and lead bullets have the same mass to determine this change in temperature. The only requirement is that they have the same speed.

20.3 Latent Heat

As we have seen in the preceding section, a substance can undergo a change in temperature when energy is transferred between it and its surroundings. In some situations, however, the transfer of energy does not result in a change in temperature. That is the case whenever the physical characteristics of the substance change from one form to another; such a change is commonly referred to as a **phase change.** Two common phase changes are from solid to liquid (melting) and from liquid to gas (boiling); another is a change in the crystalline structure of a solid. All such phase changes involve a change in the system's internal energy but no change in its temperature. The increase in internal energy in boiling, for example, is represented by the breaking of bonds between molecules in the liquid state; this bond breaking allows the molecules to move farther apart in the gaseous state, with a corresponding increase in intermolecular potential energy.

As you might expect, different substances respond differently to the addition or removal of energy as they change phase because their internal molecular arrangements vary. Also, the amount of energy transferred during a phase change depends on the amount of substance involved. (It takes less energy to melt an ice cube than it does to thaw a frozen lake.) When discussing two phases of a material, we will use the term *higher-phase material* to mean the material existing at the higher temperature. So, for example, if we discuss water and ice, water is the higher-phase material, whereas steam is the higher-phase material in a discussion of steam and water. Consider a system containing a substance in two phases in equilibrium such as water and ice. The initial amount of the higher-phase material, water, in the system is m_i. Now imagine that energy Q enters the system. As a result, the final amount of water is m_f due to the melting of some of the ice. Therefore, the amount of ice that melted, equal to the amount of *new* water, is $\Delta m = m_f - m_i$. We define the **latent heat** for this phase change as

$$L \equiv \frac{Q}{\Delta m} \tag{20.6}$$

This parameter is called latent heat (literally, the "hidden" heat) because this added or removed energy does not result in a temperature change. The value of L for a substance depends on the nature of the phase change as well as on the properties of the substance. If the entire amount of the lower-phase material undergoes a phase change, the change in mass Δm of the higher-phase material is equal to the initial mass of the lower-phase material. For example, if an ice cube of mass m on a

Table 20.2	**Latent Heats of Fusion and Vaporization**			
Substance	Melting Point (°C)	Latent Heat of Fusion (J/kg)	Boiling Point (°C)	Latent Heat of Vaporization (J/kg)
Helium[a]	−272.2	5.23×10^3	−268.93	2.09×10^4
Oxygen	−218.79	1.38×10^4	−182.97	2.13×10^5
Nitrogen	−209.97	2.55×10^4	−195.81	2.01×10^5
Ethyl alcohol	−114	1.04×10^5	78	8.54×10^5
Water	0.00	3.33×10^5	100.00	2.26×10^6
Sulfur	119	3.81×10^4	444.60	3.26×10^5
Lead	327.3	2.45×10^4	1 750	8.70×10^5
Aluminum	660	3.97×10^5	2 450	1.14×10^7
Silver	960.80	8.82×10^4	2 193	2.33×10^6
Gold	1 063.00	6.44×10^4	2 660	1.58×10^6
Copper	1 083	1.34×10^5	1 187	5.06×10^6

[a]Helium does not solidify at atmospheric pressure. The melting point given here corresponds to a pressure of 2.5 MPa.

plate melts completely, the change in mass of the water is $m_f - 0 = m$, which is the mass of new water and is also equal to the initial mass of the ice cube.

From the definition of latent heat, and again choosing heat as our energy transfer mechanism, the energy required to change the phase of a pure substance is

Energy transferred to ▶
a substance during
a phase change

$$Q = L \, \Delta m \tag{20.7}$$

where Δm is the change in mass of the higher-phase material.

Latent heat of fusion L_f is the term used when the phase change is from solid to liquid (*to fuse* means "to combine by melting"), and **latent heat of vaporization** L_v is the term used when the phase change is from liquid to gas (the liquid "vaporizes").[4] The latent heats of various substances vary considerably as data in Table 20.2 show. When energy enters a system, causing melting or vaporization, the amount of the higher-phase material increases, so Δm is positive and Q is positive, consistent with our sign convention. When energy is extracted from a system, causing freezing or condensation, the amount of the higher-phase material decreases, so Δm is negative and Q is negative, again consistent with our sign convention. Keep in mind that Δm in Equation 20.7 always refers to the higher-phase material.

To understand the role of latent heat in phase changes, consider the energy required to convert a system consisting of a 1.00-g cube of ice at −30.0°C to steam at 120.0°C. Figure 20.3 indicates the experimental results obtained when energy is gradually added to the ice. The results are presented as a graph of temperature of the system versus energy added to the system. Let's examine each portion of the red-brown curve, which is divided into parts A through E.

Part A. On this portion of the curve, the temperature of the system changes from −30.0°C to 0.0°C. Equation 20.4 indicates that the temperature varies linearly with the energy added, so the experimental result is a straight line on the graph. Because the specific heat of ice is 2 090 J/kg · °C, we can calculate the amount of energy added by using Equation 20.4:

$$Q = m_i c_i \, \Delta T = (1.00 \times 10^{-3} \text{ kg})(2\,090 \text{ J/kg} \cdot °\text{C})(30.0°\text{C}) = 62.7 \text{ J}$$

Part B. When the temperature of the system reaches 0.0°C, the ice–water mixture remains at this temperature—even though energy is being added—until all the ice melts. The energy required to melt 1.00 g of ice at 0.0°C is, from Equation 20.7,

$$Q = L_f \, \Delta m_w = L_f m_i = (3.33 \times 10^5 \text{ J/kg})(1.00 \times 10^{-3} \text{ kg}) = 333 \text{ J}$$

Pitfall Prevention 20.6

Signs Are Critical Sign errors occur very often when students apply calorimetry equations. For phase changes, remember that Δm in Equation 20.7 is always the change in mass of the higher-phase material. In Equation 20.4, be sure your ΔT is *always* the final temperature minus the initial temperature. In addition, you must *always* include the negative sign on the right side of Equation 20.5.

---

[4]When a gas cools, it eventually *condenses;* that is, it returns to the liquid phase. The energy given up per unit mass is called the *latent heat of condensation* and is numerically equal to the latent heat of vaporization. Likewise, when a liquid cools, it eventually solidifies, and the *latent heat of solidification* is numerically equal to the latent heat of fusion.

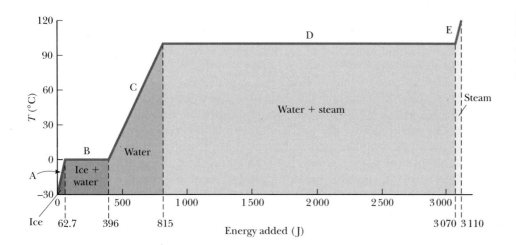

Figure 20.3 A plot of temperature versus energy added when a system initially consisting of 1.00 g of ice at −30.0°C is converted to steam at 120.0°C.

At this point, we have moved to the 396 J (= 62.7 J + 333 J) mark on the energy axis in Figure 20.3.

Part C. Between 0.0°C and 100.0°C, nothing surprising happens. No phase change occurs, and so all energy added to the system, which is now water, is used to increase its temperature. The amount of energy necessary to increase the temperature from 0.0°C to 100.0°C is

$$Q = m_w c_w \Delta T = (1.00 \times 10^{-3} \text{ kg})(4.19 \times 10^3 \text{ J/kg} \cdot \text{°C})(100.0\text{°C}) = 419 \text{ J}$$

where m_w is the mass of the water in the system, which is the same as the mass m_i of the original ice.

Part D. At 100.0°C, another phase change occurs as the system changes from water at 100.0°C to steam at 100.0°C. Similar to the ice–water mixture in part B, the water–steam mixture remains at 100.0°C—even though energy is being added—until all the liquid has been converted to steam. The energy required to convert 1.00 g of water to steam at 100.0°C is

$$Q = L_v \Delta m_s = L_v m_w = (2.26 \times 10^6 \text{ J/kg})(1.00 \times 10^{-3} \text{ kg}) = 2.26 \times 10^3 \text{ J}$$

Part E. On this portion of the curve, as in parts A and C, no phase change occurs; therefore, all energy added is used to increase the temperature of the system, which is now steam. The energy that must be added to raise the temperature of the steam from 100.0°C to 120.0°C is

$$Q = m_s c_s \Delta T = (1.00 \times 10^{-3} \text{ kg})(2.01 \times 10^3 \text{ J/kg} \cdot \text{°C})(20.0\text{°C}) = 40.2 \text{ J}$$

The total amount of energy that must be added to the system to change 1 g of ice at −30.0°C to steam at 120.0°C is the sum of the results from all five parts of the curve, which is 3.11×10^3 J. Conversely, to cool 1 g of steam at 120.0°C to ice at −30.0°C, we must remove 3.11×10^3 J of energy.

Notice in Figure 20.3 the relatively large amount of energy that is transferred into the water to vaporize it to steam. Imagine reversing this process, with a large amount of energy transferred out of steam to condense it into water. That is why a burn to your skin from steam at 100°C is much more damaging than exposure of your skin to water at 100°C. A very large amount of energy enters your skin from the steam, and the steam remains at 100°C for a long time while it condenses. Conversely, when your skin makes contact with water at 100°C, the water immediately begins to drop in temperature as energy transfers from the water to your skin.

If liquid water is held perfectly still in a very clean container, it is possible for the water to drop below 0°C without freezing into ice. This phenomenon, called **supercooling,** arises because the water requires a disturbance of some sort for the molecules to move apart and start forming the large, open ice structure that makes the density of ice lower than that of water as discussed in Section 19.4. If supercooled water is disturbed, it suddenly freezes. The system drops into the lower-energy configuration of bound molecules of the ice structure, and the energy released raises the temperature back to 0°C.

Commercial hand warmers consist of liquid sodium acetate in a sealed plastic pouch. The solution in the pouch is in a stable supercooled state. When a disk in the pouch is clicked by your fingers, the liquid solidifies and the temperature increases, just like the supercooled water just mentioned. In this case, however, the freezing point of the liquid is higher than body temperature, so the pouch feels warm to the touch. To reuse the hand warmer, the pouch must be boiled until the solid liquefies. Then, as it cools, it passes below its freezing point into the supercooled state.

It is also possible to create **superheating.** For example, clean water in a very clean cup placed in a microwave oven can sometimes rise in temperature beyond 100°C without boiling because the formation of a bubble of steam in the water requires scratches in the cup or some type of impurity in the water to serve as a nucleation site. When the cup is removed from the microwave oven, the superheated water can become explosive as bubbles form immediately and the hot water is forced upward out of the cup.

Quick Quiz 20.2 Suppose the same process of adding energy to the ice cube is performed as discussed above, but instead we graph the internal energy of the system as a function of energy input. What would this graph look like?

Example 20.4 **Cooling the Steam** AM

What mass of steam initially at 130°C is needed to warm 200 g of water in a 100-g glass container from 20.0°C to 50.0°C?

SOLUTION

Conceptualize Imagine placing water and steam together in a closed insulated container. The system eventually reaches a uniform state of water with a final temperature of 50.0°C.

Categorize Based on our conceptualization of this situation, we categorize this example as one involving calorimetry in which a phase change occurs. The calorimeter is an *isolated system* for *energy:* energy transfers between the components of the system but does not cross the boundary between the system and the environment.

Analyze Write Equation 20.5 to describe the calorimetry process:

$$(1) \quad Q_{\text{cold}} = -Q_{\text{hot}}$$

The steam undergoes three processes: first a decrease in temperature to 100°C, then condensation into liquid water, and finally a decrease in temperature of the water to 50.0°C. Find the energy transfer in the first process using the unknown mass m_s of the steam:

$$Q_1 = m_s c_s \Delta T_s$$

Find the energy transfer in the second process:

$$Q_2 = L_v \Delta m_s = L_v(0 - m_s) = -m_s L_v$$

Find the energy transfer in the third process:

$$Q_3 = m_s c_w \Delta T_{\text{hot water}}$$

Add the energy transfers in these three stages:

$$(2) \quad Q_{\text{hot}} = Q_1 + Q_2 + Q_3 = m_s(c_s \Delta T_s - L_v + c_w \Delta T_{\text{hot water}})$$

The 20.0°C water and the glass undergo only one process, an increase in temperature to 50.0°C. Find the energy transfer in this process:

$$(3) \quad Q_{\text{cold}} = m_w c_w \Delta T_{\text{cold water}} + m_g c_g \Delta T_{\text{glass}}$$

Substitute Equations (2) and (3) into Equation (1):

$$m_w c_w \Delta T_{\text{cold water}} + m_g c_g \Delta T_{\text{glass}} = -m_s(c_s \Delta T_s - L_v + c_w \Delta T_{\text{hot water}})$$

Solve for m_s:

$$m_s = -\frac{m_w c_w \Delta T_{\text{cold water}} + m_g c_g \Delta T_{\text{glass}}}{c_s \Delta T_s - L_v + c_w \Delta T_{\text{hot water}}}$$

▶ **20.4** continued

Substitute
numerical
values:

$$m_s = -\frac{(0.200 \text{ kg})(4\,186 \text{ J/kg} \cdot {}^{\circ}\text{C})(50.0{}^{\circ}\text{C} - 20.0{}^{\circ}\text{C}) + (0.100 \text{ kg})(837 \text{ J/kg} \cdot {}^{\circ}\text{C})(50.0{}^{\circ}\text{C} - 20.0{}^{\circ}\text{C})}{(2\,010 \text{ J/kg} \cdot {}^{\circ}\text{C})(100{}^{\circ}\text{C} - 130{}^{\circ}\text{C}) - (2.26 \times 10^6 \text{ J/kg}) + (4\,186 \text{ J/kg} \cdot {}^{\circ}\text{C})(50.0{}^{\circ}\text{C} - 100{}^{\circ}\text{C})}$$

$$= 1.09 \times 10^{-2} \text{ kg} = \boxed{10.9 \text{ g}}$$

WHAT IF? What if the final state of the system is water at 100°C? Would we need more steam or less steam? How would the analysis above change?

Answer More steam would be needed to raise the temperature of the water and glass to 100°C instead of 50.0°C. There would be two major changes in the analysis. First, we would not have a term Q_3 for the steam because the water that condenses from the steam does not cool below 100°C. Second, in Q_{cold}, the temperature change would be 80.0°C instead of 30.0°C. For practice, show that the result is a required mass of steam of 31.8 g.

20.4 Work and Heat in Thermodynamic Processes

In thermodynamics, we describe the *state* of a system using such variables as pressure, volume, temperature, and internal energy. As a result, these quantities belong to a category called **state variables.** For any given configuration of the system, we can identify values of the state variables. (For mechanical systems, the state variables include kinetic energy K and potential energy U.) A state of a system can be specified only if the system is in thermal equilibrium internally. In the case of a gas in a container, internal thermal equilibrium requires that every part of the gas be at the same pressure and temperature.

A second category of variables in situations involving energy is **transfer variables.** These variables are those that appear on the right side of the conservation of energy equation, Equation 8.2. Such a variable has a nonzero value if a process occurs in which energy is transferred across the system's boundary. The transfer variable is positive or negative, depending on whether energy is entering or leaving the system. Because a transfer of energy across the boundary represents a change in the system, transfer variables are not associated with a given state of the system, but rather with a *change* in the state of the system.

In the previous sections, we discussed heat as a transfer variable. In this section, we study another important transfer variable for thermodynamic systems, work. Work performed on particles was studied extensively in Chapter 7, and here we investigate the work done on a deformable system, a gas. Consider a gas contained in a cylinder fitted with a movable piston (Fig. 20.4). At equilibrium, the gas occupies a volume V and exerts a uniform pressure P on the cylinder's walls and on the piston. If the piston has a cross-sectional area A, the magnitude of the force exerted by the gas on the piston is $F = PA$. By Newton's third law, the magnitude of the force exerted by the piston on the gas is also PA. Now let's assume we push the piston inward and compress the gas **quasi-statically,** that is, slowly enough to allow the system to remain essentially in internal thermal equilibrium at all times. The point of application of the force on the gas is the bottom face of the piston. As the piston is pushed downward by an external force $\vec{\mathbf{F}} = -F\hat{\mathbf{j}}$ through a displacement of $d\vec{\mathbf{r}} = dy\hat{\mathbf{j}}$ (Fig. 20.4b), the work done on the gas is, according to our definition of work in Chapter 7,

$$dW = \vec{\mathbf{F}} \cdot d\vec{\mathbf{r}} = -F\hat{\mathbf{j}} \cdot dy\hat{\mathbf{j}} = -F\,dy = -PA\,dy$$

The mass of the piston is assumed to be negligible in this discussion. Because $A\,dy$ is the change in volume of the gas dV, we can express the work done on the gas as

$$dW = -P\,dV \tag{20.8}$$

If the gas is compressed, dV is negative and the work done on the gas is positive. If the gas expands, dV is positive and the work done on the gas is negative. If the

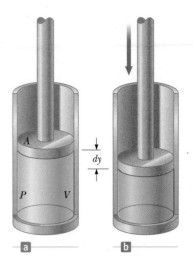

Figure 20.4 Work is done on a gas contained in a cylinder at a pressure P as the piston is pushed downward so that the gas is compressed.

volume remains constant, the work done on the gas is zero. The total work done on the gas as its volume changes from V_i to V_f is given by the integral of Equation 20.8:

Work done on a gas ▶

$$W = -\int_{V_i}^{V_f} P \, dV \qquad (20.9)$$

To evaluate this integral, you must know how the pressure varies with volume during the process.

In general, the pressure is not constant during a process followed by a gas, but depends on the volume and temperature. If the pressure and volume are known at each step of the process, the state of the gas at each step can be plotted on an important graphical representation called a **PV diagram** as in Figure 20.5. This type of diagram allows us to visualize a process through which a gas is progressing. The curve on a PV diagram is called the *path* taken between the initial and final states.

Notice that the integral in Equation 20.9 is equal to the area under a curve on a PV diagram. Therefore, we can identify an important use for PV diagrams:

> The work done on a gas in a quasi-static process that takes the gas from an initial state to a final state is the negative of the area under the curve on a PV diagram, evaluated between the initial and final states.

For the process of compressing a gas in a cylinder, the work done depends on the particular path taken between the initial and final states as Figure 20.5 suggests. To illustrate this important point, consider several different paths connecting i and f (Fig. 20.6). In the process depicted in Figure 20.6a, the volume of the gas is first reduced from V_i to V_f at constant pressure P_i and the pressure of the gas then increases from P_i to P_f by heating at constant volume V_f. The work done on the gas along this path is $-P_i(V_f - V_i)$. In Figure 20.6b, the pressure of the gas is increased from P_i to P_f at constant volume V_i and then the volume of the gas is reduced from V_i to V_f at constant pressure P_f. The work done on the gas is $-P_f(V_f - V_i)$. This value is greater than that for the process described in Figure 20.6a because the piston is moved through the same displacement by a larger force. Finally, for the process described in Figure 20.6c, where both P and V change continuously, the work done on the gas has some value between the values obtained in the first two processes. To evaluate the work in this case, the function $P(V)$ must be known so that we can evaluate the integral in Equation 20.9.

The energy transfer Q into or out of a system by heat also depends on the process. Consider the situations depicted in Figure 20.7. In each case, the gas has the same initial volume, temperature, and pressure, and is assumed to be ideal. In Figure 20.7a, the gas is thermally insulated from its surroundings except at the bottom of the gas-filled region, where it is in thermal contact with an energy reservoir. An *energy reservoir* is a source of energy that is considered to be so great that a finite transfer of energy to or from the reservoir does not change its temperature. The piston is held

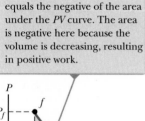

The work done on a gas equals the negative of the area under the PV curve. The area is negative here because the volume is decreasing, resulting in positive work.

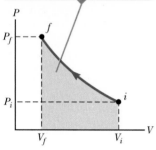

Figure 20.5 A gas is compressed quasi-statically (slowly) from state i to state f. An outside agent must do positive work on the gas to compress it.

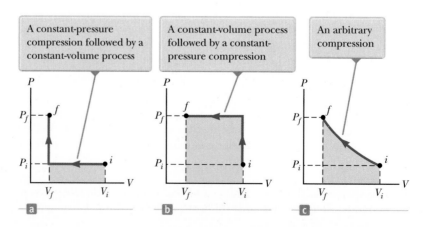

A constant-pressure compression followed by a constant-volume process

A constant-volume process followed by a constant-pressure compression

An arbitrary compression

Figure 20.6 The work done on a gas as it is taken from an initial state to a final state depends on the path between these states.

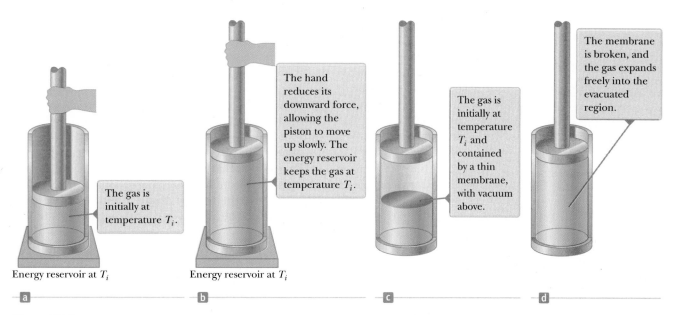

Figure 20.7 Gas in a cylinder. (a) The gas is in contact with an energy reservoir. The walls of the cylinder are perfectly insulating, but the base in contact with the reservoir is conducting. (b) The gas expands slowly to a larger volume. (c) The gas is contained by a membrane in half of a volume, with vacuum in the other half. The entire cylinder is perfectly insulating. (d) The gas expands freely into the larger volume.

at its initial position by an external agent such as a hand. When the force holding the piston is reduced slightly, the piston rises very slowly to its final position shown in Figure 20.7b. Because the piston is moving upward, the gas is doing work on the piston. During this expansion to the final volume V_f, just enough energy is transferred by heat from the reservoir to the gas to maintain a constant temperature T_i.

Now consider the completely thermally insulated system shown in Figure 20.7c. When the membrane is broken, the gas expands rapidly into the vacuum until it occupies a volume V_f and is at a pressure P_f. The final state of the gas is shown in Figure 20.7d. In this case, the gas does no work because it does not apply a force; no force is required to expand into a vacuum. Furthermore, no energy is transferred by heat through the insulating wall.

As we discuss in Section 20.5, experiments show that the temperature of the ideal gas does not change in the process indicated in Figures 20.7c and 20.7d. Therefore, the initial and final states of the ideal gas in Figures 20.7a and 20.7b are identical to the initial and final states in Figures 20.7c and 20.7d, but the paths are different. In the first case, the gas does work on the piston and energy is transferred slowly to the gas by heat. In the second case, no energy is transferred by heat and the value of the work done is zero. Therefore, energy transfer by heat, like work done, depends on the particular process occurring in the system. In other words, because heat and work both depend on the path followed on a *PV* diagram between the initial and final states, neither quantity is determined solely by the endpoints of a thermodynamic process.

20.5 The First Law of Thermodynamics

When we introduced the law of conservation of energy in Chapter 8, we stated that the change in the energy of a system is equal to the sum of all transfers of energy across the system's boundary (Eq. 8.2). The **first law of thermodynamics** is a special case of the law of conservation of energy that describes processes in which only the internal energy[5] changes and the only energy transfers are by heat and work:

$$\Delta E_{int} = Q + W \qquad \text{(20.10)}$$

◀ **First law of thermodynamics**

---

[5]It is an unfortunate accident of history that the traditional symbol for internal energy is U, which is also the traditional symbol for potential energy as introduced in Chapter 7. To avoid confusion between potential energy and internal energy, we use the symbol E_{int} for internal energy in this book. If you take an advanced course in thermodynamics, however, be prepared to see U used as the symbol for internal energy in the first law.

Figure 20.8 The first law of thermodynamics equates the change in
internal energy E_{int} in a system to
the net energy transfer to the system by heat Q and work W. In the
situation shown here, the internal
energy of the gas increases.

Look back at Equation 8.2 to see that the first law of thermodynamics is contained
within that more general equation.

Let us investigate some special cases in which the first law can be applied. First,
consider an *isolated system,* that is, one that does not interact with its surroundings,
as we have seen before. In this case, no energy transfer by heat takes place and the
work done on the system is zero; hence, the internal energy remains constant. That
is, because $Q = W = 0$, it follows that $\Delta E_{int} = 0$; therefore, $E_{int,i} = E_{int,f}$. We conclude that the internal energy E_{int} of an isolated system remains constant.

Next, consider the case of a system that can exchange energy with its surroundings and is taken through a **cyclic process,** that is, a process that starts and ends at
the same state. In this case, the change in the internal energy must again be zero
because E_{int} is a state variable; therefore, the energy Q added to the system must
equal the negative of the work W done on the system during the cycle. That is, in a
cyclic process,

$$\Delta E_{int} = 0 \quad \text{and} \quad Q = -W \quad \text{(cyclic process)}$$

On a PV diagram for a gas, a cyclic process appears as a closed curve. (The processes described in Figure 20.6 are represented by open curves because the initial
and final states differ.) It can be shown that in a cyclic process for a gas, the net
work done on the system per cycle equals the area enclosed by the path representing the process on a PV diagram.

20.6 Some Applications of the First Law of Thermodynamics

In this section, we consider additional applications of the first law to processes
through which a gas is taken. As a model, let's consider the sample of gas contained
in the piston–cylinder apparatus in Figure 20.8. This figure shows work being done
on the gas and energy transferring in by heat, so the internal energy of the gas is
rising. In the following discussion of various processes, refer back to this figure
and mentally alter the directions of the transfer of energy to reflect what is happening in the process.

Before we apply the first law of thermodynamics to specific systems, it is useful
to first define some idealized thermodynamic processes. An **adiabatic process** is
one during which no energy enters or leaves the system by heat; that is, $Q = 0$. An
adiabatic process can be achieved either by thermally insulating the walls of the
system or by performing the process rapidly so that there is negligible time for
energy to transfer by heat. Applying the first law of thermodynamics to an adiabatic process gives

$$\Delta E_{int} = W \quad \text{(adiabatic process)} \tag{20.11}$$

This result shows that if a gas is compressed adiabatically such that W is positive,
then ΔE_{int} is positive and the temperature of the gas increases. Conversely, the temperature of a gas decreases when the gas expands adiabatically.

Adiabatic processes are very important in engineering practice. Some common
examples are the expansion of hot gases in an internal combustion engine, the
liquefaction of gases in a cooling system, and the compression stroke in a diesel
engine.

The process described in Figures 20.7c and 20.7d, called an **adiabatic free
expansion,** is unique. The process is adiabatic because it takes place in an insulated
container. Because the gas expands into a vacuum, it does not apply a force on a
piston as does the gas in Figures 20.7a and 20.7b, so no work is done on or by the
gas. Therefore, in this adiabatic process, both $Q = 0$ and $W = 0$. As a result, $\Delta E_{int} =
0$ for this process as can be seen from the first law. That is, the initial and final
internal energies of a gas are equal in an adiabatic free expansion. As we shall see

in Chapter 21, the internal energy of an ideal gas depends only on its temperature. Therefore, we expect no change in temperature during an adiabatic free expansion. This prediction is in accord with the results of experiments performed at low pressures. (Experiments performed at high pressures for real gases show a slight change in temperature after the expansion due to intermolecular interactions, which represent a deviation from the model of an ideal gas.)

A process that occurs at constant pressure is called an **isobaric process.** In Figure 20.8, an isobaric process could be established by allowing the piston to move freely so that it is always in equilibrium between the net force from the gas pushing upward and the weight of the piston plus the force due to atmospheric pressure pushing downward. The first process in Figure 20.6a and the second process in Figure 20.6b are both isobaric.

In such a process, the values of the heat and the work are both usually nonzero. The work done on the gas in an isobaric process is simply

$$W = -P(V_f - V_i) \quad \text{(isobaric process)} \tag{20.12}$$

◀ Isobaric process

where P is the constant pressure of the gas during the process.

A process that takes place at constant volume is called an **isovolumetric process.** In Figure 20.8, clamping the piston at a fixed position would ensure an isovolumetric process. The second process in Figure 20.6a and the first process in Figure 20.6b are both isovolumetric.

Because the volume of the gas does not change in such a process, the work given by Equation 20.9 is zero. Hence, from the first law we see that in an isovolumetric process, because $W = 0$,

$$\Delta E_{int} = Q \quad \text{(isovolumetric process)} \tag{20.13}$$

◀ Isovolumetric process

This expression specifies that if energy is added by heat to a system kept at constant volume, all the transferred energy remains in the system as an increase in its internal energy. For example, when a can of spray paint is thrown into a fire, energy enters the system (the gas in the can) by heat through the metal walls of the can. Consequently, the temperature, and therefore the pressure, in the can increases until the can possibly explodes.

A process that occurs at constant temperature is called an **isothermal process.** This process can be established by immersing the cylinder in Figure 20.8 in an ice–water bath or by putting the cylinder in contact with some other constant-temperature reservoir. A plot of P versus V at constant temperature for an ideal gas yields a hyperbolic curve called an *isotherm*. The internal energy of an ideal gas is a function of temperature only. Hence, because the temperature does not change in an isothermal process involving an ideal gas, we must have $\Delta E_{int} = 0$. For an isothermal process, we conclude from the first law that the energy transfer Q must be equal to the negative of the work done on the gas; that is, $Q = -W$. Any energy that enters the system by heat is transferred out of the system by work; as a result, no change in the internal energy of the system occurs in an isothermal process.

◀ Isothermal process

Pitfall Prevention 20.9

$Q \neq 0$ in an Isothermal Process
Do not fall into the common trap of thinking there must be no transfer of energy by heat if the temperature does not change as is the case in an isothermal process. Because the cause of temperature change can be either heat *or* work, the temperature can remain constant even if energy enters the gas by heat, which can only happen if the energy entering the gas by heat leaves by work.

Ⓠuick Quiz 20.3 In the last three columns of the following table, fill in the boxes with the correct signs (−, +, or 0) for Q, W, and ΔE_{int}. For each situation, the system to be considered is identified.

Situation	System	Q	W	ΔE_{int}
(a) Rapidly pumping up a bicycle tire	Air in the pump			
(b) Pan of room-temperature water sitting on a hot stove	Water in the pan			
(c) Air quickly leaking out of a balloon	Air originally in the balloon			

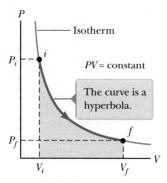

Figure 20.9 The *PV* diagram for an isothermal expansion of an ideal gas from an initial state to a final state.

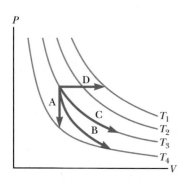

Figure 20.10 (Quick Quiz 20.4) Identify the nature of paths A, B, C, and D.

Isothermal Expansion of an Ideal Gas

Suppose an ideal gas is allowed to expand quasi-statically at constant temperature. This process is described by the *PV* diagram shown in Figure 20.9. The curve is a hyperbola (see Appendix B, Eq. B.23), and the ideal gas law (Eq. 19.8) with *T* constant indicates that the equation of this curve is $PV = nRT$ = constant.

Let's calculate the work done on the gas in the expansion from state *i* to state *f*. The work done on the gas is given by Equation 20.9. Because the gas is ideal and the process is quasi-static, the ideal gas law is valid for each point on the path. Therefore,

$$W = -\int_{V_i}^{V_f} P \, dV = -\int_{V_i}^{V_f} \frac{nRT}{V} \, dV$$

Because *T* is constant in this case, it can be removed from the integral along with *n* and *R*:

$$W = -nRT \int_{V_i}^{V_f} \frac{dV}{V} = -nRT \ln V \Big|_{V_i}^{V_f}$$

To evaluate the integral, we used $\int (dx/x) = \ln x$. (See Appendix B.) Evaluating the result at the initial and final volumes gives

$$W = nRT \ln \left(\frac{V_i}{V_f} \right) \tag{20.14}$$

Numerically, this work *W* equals the negative of the shaded area under the *PV* curve shown in Figure 20.9. Because the gas expands, $V_f > V_i$ and the value for the work done on the gas is negative as we expect. If the gas is compressed, then $V_f < V_i$ and the work done on the gas is positive.

Quick Quiz 20.4 Characterize the paths in Figure 20.10 as isobaric, isovolumetric, isothermal, or adiabatic. For path B, $Q = 0$. The blue curves are isotherms.

Example 20.5 **An Isothermal Expansion**

A 1.0-mol sample of an ideal gas is kept at 0.0°C during an expansion from 3.0 L to 10.0 L.

(A) How much work is done on the gas during the expansion?

SOLUTION

Conceptualize Run the process in your mind: the cylinder in Figure 20.8 is immersed in an ice-water bath, and the piston moves outward so that the volume of the gas increases. You can also use the graphical representation in Figure 20.9 to conceptualize the process.

Categorize We will evaluate parameters using equations developed in the preceding sections, so we categorize this example as a substitution problem. Because the temperature of the gas is fixed, the process is isothermal.

Substitute the given values into Equation 20.14:

$$W = nRT \ln \left(\frac{V_i}{V_f} \right)$$

$$= (1.0 \text{ mol})(8.31 \text{ J/mol} \cdot \text{K})(273 \text{ K}) \ln \left(\frac{3.0 \text{ L}}{10.0 \text{ L}} \right)$$

$$= \boxed{-2.7 \times 10^3 \text{ J}}$$

(B) How much energy transfer by heat occurs between the gas and its surroundings in this process?

SOLUTION

Find the heat from the first law:

$$\Delta E_{\text{int}} = Q + W$$
$$0 = Q + W$$
$$Q = -W = \boxed{2.7 \times 10^3 \text{ J}}$$

▶ **20.5** continued

(C) If the gas is returned to the original volume by means of an isobaric process, how much work is done on the gas?

SOLUTION

Use Equation 20.12. The pressure is not given, so incorporate the ideal gas law:

$$W = -P(V_f - V_i) = -\frac{nRT_i}{V_i}(V_f - V_i)$$

$$= -\frac{(1.0 \text{ mol})(8.31 \text{ J/mol} \cdot \text{K})(273 \text{ K})}{10.0 \times 10^{-3} \text{ m}^3}(3.0 \times 10^{-3} \text{ m}^3 - 10.0 \times 10^{-3} \text{ m}^3)$$

$$= \boxed{1.6 \times 10^3 \text{ J}}$$

We used the initial temperature and volume to calculate the work done because the final temperature was unknown. The work done on the gas is positive because the gas is being compressed.

Example 20.6 Boiling Water

Suppose 1.00 g of water vaporizes isobarically at atmospheric pressure (1.013×10^5 Pa). Its volume in the liquid state is $V_i = V_{liquid} = 1.00 \text{ cm}^3$, and its volume in the vapor state is $V_f = V_{vapor} = 1\,671 \text{ cm}^3$. Find the work done in the expansion and the change in internal energy of the system. Ignore any mixing of the steam and the surrounding air; imagine that the steam simply pushes the surrounding air out of the way.

SOLUTION

Conceptualize Notice that the temperature of the system does not change. There is a phase change occurring as the water evaporates to steam.

Categorize Because the expansion takes place at constant pressure, we categorize the process as isobaric. We will use equations developed in the preceding sections, so we categorize this example as a substitution problem.

Use Equation 20.12 to find the work done on the system as the air is pushed out of the way:

$$W = -P(V_f - V_i)$$
$$= -(1.013 \times 10^5 \text{ Pa})(1\,671 \times 10^{-6} \text{ m}^3 - 1.00 \times 10^{-6} \text{ m}^3)$$
$$= \boxed{-169 \text{ J}}$$

Use Equation 20.7 and the latent heat of vaporization for water to find the energy transferred into the system by heat:

$$Q = L_v \, \Delta m_s = m_s L_v = (1.00 \times 10^{-3} \text{ kg})(2.26 \times 10^6 \text{ J/kg})$$
$$= 2\,260 \text{ J}$$

Use the first law to find the change in internal energy of the system:

$$\Delta E_{int} = Q + W = 2\,260 \text{ J} + (-169 \text{ J}) = \boxed{2.09 \text{ kJ}}$$

The positive value for ΔE_{int} indicates that the internal energy of the system increases. The largest fraction of the energy ($2\,090 \text{ J} / 2260 \text{ J} = 93\%$) transferred to the liquid goes into increasing the internal energy of the system. The remaining 7% of the energy transferred leaves the system by work done by the steam on the surrounding atmosphere.

Example 20.7 Heating a Solid

A 1.0-kg bar of copper is heated at atmospheric pressure so that its temperature increases from 20°C to 50°C.

(A) What is the work done on the copper bar by the surrounding atmosphere?

SOLUTION

Conceptualize This example involves a solid, whereas the preceding two examples involved liquids and gases. For a solid, the change in volume due to thermal expansion is very small.

continued

▶ **20.7** continued

Categorize Because the expansion takes place at constant atmospheric pressure, we categorize the process as isobaric.

Analyze Find the work done on the copper bar using Equation 20.12:

$$W = -P \Delta V$$

Express the change in volume using Equation 19.6 and that $\beta = 3\alpha$:

$$W = -P(\beta V_i \, \Delta T) = -P(3\alpha V_i \, \Delta T) = -3\alpha P V_i \, \Delta T$$

Substitute for the volume in terms of the mass and density of the copper:

$$W = -3\alpha P \left(\frac{m}{\rho}\right) \Delta T$$

Substitute numerical values:

$$W = -3[1.7 \times 10^{-5} \, (\degree\text{C})^{-1}](1.013 \times 10^5 \, \text{N/m}^2)\left(\frac{1.0 \text{ kg}}{8.92 \times 10^3 \text{ kg/m}^3}\right)(50\degree\text{C} - 20\degree\text{C})$$

$$= \boxed{-1.7 \times 10^{-2} \text{ J}}$$

Because this work is negative, work is done *by* the copper bar on the atmosphere.

(B) How much energy is transferred to the copper bar by heat?

SOLUTION

Use Equation 20.4 and the specific heat of copper from Table 20.1:

$$Q = mc \, \Delta T = (1.0 \text{ kg})(387 \text{ J/kg} \cdot \degree\text{C})(50\degree\text{C} - 20\degree\text{C})$$

$$= \boxed{1.2 \times 10^4 \text{ J}}$$

(C) What is the increase in internal energy of the copper bar?

SOLUTION

Use the first law of thermodynamics:

$$\Delta E_{\text{int}} = Q + W = 1.2 \times 10^4 \text{ J} + (-1.7 \times 10^{-2} \text{ J})$$

$$= \boxed{1.2 \times 10^4 \text{ J}}$$

Finalize Most of the energy transferred into the system by heat goes into increasing the internal energy of the copper bar. The fraction of energy used to do work on the surrounding atmosphere is only about 10^{-6}. Hence, when the thermal expansion of a solid or a liquid is analyzed, the small amount of work done on or by the system is usually ignored.

20.7 Energy Transfer Mechanisms in Thermal Processes

In Chapter 8, we introduced a global approach to the energy analysis of physical processes through Equation 8.1, $\Delta E_{\text{system}} = \Sigma \, T$, where T represents energy transfer, which can occur by several mechanisms. Earlier in this chapter, we discussed two of the terms on the right side of this equation, work W and heat Q. In this section, we explore more details about heat as a means of energy transfer and two other energy transfer methods often related to temperature changes: convection (a form of matter transfer T_{MT}) and electromagnetic radiation T_{ER}.

Thermal Conduction

The process of energy transfer by heat (Q in Eq. 8.2) can also be called **conduction** or **thermal conduction**. In this process, the transfer can be represented on an atomic scale as an exchange of kinetic energy between microscopic particles—molecules, atoms, and free electrons—in which less-energetic particles gain energy in collisions with more-energetic particles. For example, if you hold one end of a long metal bar and insert the other end into a flame, you will find that the temperature of the metal in your hand soon increases. The energy reaches your hand by means of conduction. Initially, before the rod is inserted into the flame, the microscopic particles in the metal are vibrating about their equilibrium positions. As the flame raises the temperature of the rod, the particles near the flame begin to vibrate with greater and greater amplitudes. These particles, in turn, collide with their neighbors and transfer some of their energy in the collisions. Slowly, the amplitudes of vibration of metal atoms and electrons farther and farther from the flame increase until eventually those in the metal near your hand are affected. This increased vibration is detected by an increase in the temperature of the metal and of your potentially burned hand.

The rate of thermal conduction depends on the properties of the substance being heated. For example, it is possible to hold a piece of asbestos in a flame indefinitely, which implies that very little energy is conducted through the asbestos. In general, metals are good thermal conductors and materials such as asbestos, cork, paper, and fiberglass are poor conductors. Gases also are poor conductors because the separation distance between the particles is so great. Metals are good thermal conductors because they contain large numbers of electrons that are relatively free to move through the metal and so can transport energy over large distances. Therefore, in a good conductor such as copper, conduction takes place by means of both the vibration of atoms and the motion of free electrons.

Conduction occurs only if there is a difference in temperature between two parts of the conducting medium. Consider a slab of material of thickness Δx and cross-sectional area A. One face of the slab is at a temperature T_c, and the other face is at a temperature $T_h > T_c$ (Fig. 20.11). Experimentally, it is found that energy Q transfers in a time interval Δt from the hotter face to the colder one. The rate $P = Q/\Delta t$ at which this energy transfer occurs is found to be proportional to the cross-sectional area and the temperature difference $\Delta T = T_h - T_c$ and inversely proportional to the thickness:

$$P = \frac{Q}{\Delta t} \propto A \frac{\Delta T}{\Delta x}$$

Notice that P has units of watts when Q is in joules and Δt is in seconds. That is not surprising because P is power, the rate of energy transfer by heat. For a slab of infinitesimal thickness dx and temperature difference dT, we can write the **law of thermal conduction** as

$$P = kA \left| \frac{dT}{dx} \right| \tag{20.15}$$

where the proportionality constant k is the **thermal conductivity** of the material and $|dT/dx|$ is the **temperature gradient** (the rate at which temperature varies with position).

Substances that are good thermal conductors have large thermal conductivity values, whereas good thermal insulators have low thermal conductivity values. Table 20.3 lists thermal conductivities for various substances. Notice that metals are generally better thermal conductors than nonmetals.

Suppose a long, uniform rod of length L is thermally insulated so that energy cannot escape by heat from its surface except at the ends as shown in Figure 20.12 (page 480). One end is in thermal contact with an energy reservoir at temperature T_c, and the other end is in thermal contact with a reservoir at temperature $T_h > T_c$. When a steady state has been reached, the temperature at each point along the rod is constant in time. In this case, if we assume k is not a function of temperature, the temperature gradient is the same everywhere along the rod and is

$$\left| \frac{dT}{dx} \right| = \frac{T_h - T_c}{L}$$

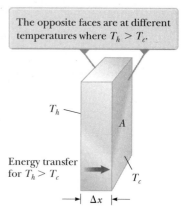

The opposite faces are at different temperatures where $T_h > T_c$.

Energy transfer for $T_h > T_c$

Figure 20.11 Energy transfer through a conducting slab with a cross-sectional area A and a thickness Δx.

Table 20.3

Thermal Conductivities

Substance	Thermal Conductivity (W/m · °C)
Metals (at 25°C)	
Aluminum	238
Copper	397
Gold	314
Iron	79.5
Lead	34.7
Silver	427
Nonmetals (approximate values)	
Asbestos	0.08
Concrete	0.8
Diamond	2 300
Glass	0.8
Ice	2
Rubber	0.2
Water	0.6
Wood	0.08
Gases (at 20°C)	
Air	0.023 4
Helium	0.138
Hydrogen	0.172
Nitrogen	0.023 4
Oxygen	0.023 8

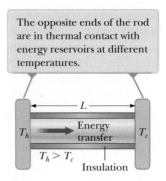

Figure 20.12 Conduction of energy through a uniform, insulated rod of length L.

Therefore, the rate of energy transfer by conduction through the rod is

$$P = kA\left(\frac{T_h - T_c}{L}\right) \tag{20.16}$$

For a compound slab containing several materials of thicknesses L_1, L_2, ... and thermal conductivities k_1, k_2, ..., the rate of energy transfer through the slab at steady state is

$$P = \frac{A(T_h - T_c)}{\sum\limits_i (L_i / k_i)} \tag{20.17}$$

where T_h and T_c are the temperatures of the outer surfaces (which are held constant) and the summation is over all slabs. Example 20.8 shows how Equation 20.17 results from a consideration of two thicknesses of materials.

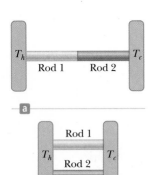

Figure 20.13 (Quick Quiz 20.5) In which case is the rate of energy transfer larger?

Quick Quiz 20.5 You have two rods of the same length and diameter, but they are formed from different materials. The rods are used to connect two regions at different temperatures so that energy transfers through the rods by heat. They can be connected in series as in Figure 20.13a or in parallel as in Figure 20.13b. In which case is the rate of energy transfer by heat larger? **(a)** The rate is larger when the rods are in series. **(b)** The rate is larger when the rods are in parallel. **(c)** The rate is the same in both cases.

Example 20.8   Energy Transfer Through Two Slabs

Two slabs of thickness L_1 and L_2 and thermal conductivities k_1 and k_2 are in thermal contact with each other as shown in Figure 20.14. The temperatures of their outer surfaces are T_c and T_h, respectively, and $T_h > T_c$. Determine the temperature at the interface and the rate of energy transfer by conduction through an area A of the slabs in the steady-state condition.

SOLUTION

Conceptualize Notice the phrase "in the steady-state condition." We interpret this phrase to mean that energy transfers through the compound slab at the same rate at all points. Otherwise, energy would be building up or disappearing at some point. Furthermore, the temperature varies with position in the two slabs, most likely at different rates in each part of the compound slab. When the system is in steady state, the interface is at some fixed temperature T.

Categorize We categorize this example as a thermal conduction problem and impose the condition that the power is the same in both slabs of material.

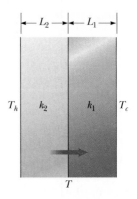

Figure 20.14 (Example 20.8) Energy transfer by conduction through two slabs in thermal contact with each other. At steady state, the rate of energy transfer through slab 1 equals the rate of energy transfer through slab 2.

▶ **20.8** continued

Analyze Use Equation 20.16 to express the rate at which energy is transferred through an area A of slab 1:

$$(1) \quad P_1 = k_1 A \left(\frac{T - T_c}{L_1} \right)$$

Express the rate at which energy is transferred through the same area of slab 2:

$$(2) \quad P_2 = k_2 A \left(\frac{T_h - T}{L_2} \right)$$

Set these two rates equal to represent the steady-state situation:

$$k_1 A \left(\frac{T - T_c}{L_1} \right) = k_2 A \left(\frac{T_h - T}{L_2} \right)$$

Solve for T:

$$(3) \quad T = \frac{k_1 L_2 T_c + k_2 L_1 T_h}{k_1 L_2 + k_2 L_1}$$

Substitute Equation (3) into either Equation (1) or Equation (2):

$$(4) \quad P = \frac{A(T_h - T_c)}{(L_1/k_1) + (L_2/k_2)}$$

Finalize Extension of this procedure to several slabs of materials leads to Equation 20.17.

WHAT IF? Suppose you are building an insulated container with two layers of insulation and the rate of energy transfer determined by Equation (4) turns out to be too high. You have enough room to increase the thickness of one of the two layers by 20%. How would you decide which layer to choose?

Answer To decrease the power as much as possible, you must increase the denominator in Equation (4) as much as possible. Whichever thickness you choose to increase, L_1 or L_2, you increase the corresponding term L/k in the denominator by 20%. For this percentage change to represent the largest absolute change, you want to take 20% of the larger term. Therefore, you should increase the thickness of the layer that has the larger value of L/k.

Home Insulation

In engineering practice, the term L/k for a particular substance is referred to as the **R-value** of the material. Therefore, Equation 20.17 reduces to

$$P = \frac{A(T_h - T_c)}{\sum_i R_i} \tag{20.18}$$

where $R_i = L_i / k_i$. The R-values for a few common building materials are given in Table 20.4. In the United States, the insulating properties of materials used in buildings are usually expressed in U.S. customary units, not SI units. Therefore, in

Table 20.4 **R-Values for Some Common Building Materials**

Material	R-value (ft² · °F · h/Btu)
Hardwood siding (1 in. thick)	0.91
Wood shingles (lapped)	0.87
Brick (4 in. thick)	4.00
Concrete block (filled cores)	1.93
Fiberglass insulation (3.5 in. thick)	10.90
Fiberglass insulation (6 in. thick)	18.80
Fiberglass board (1 in. thick)	4.35
Cellulose fiber (1 in. thick)	3.70
Flat glass (0.125 in. thick)	0.89
Insulating glass (0.25-in. space)	1.54
Air space (3.5 in. thick)	1.01
Stagnant air layer	0.17
Drywall (0.5 in. thick)	0.45
Sheathing (0.5 in. thick)	1.32

Table 20.4, *R*-values are given as a combination of British thermal units, feet, hours, and degrees Fahrenheit.

At any vertical surface open to the air, a very thin stagnant layer of air adheres to the surface. One must consider this layer when determining the *R*-value for a wall. The thickness of this stagnant layer on an outside wall depends on the speed of the wind. Energy transfer through the walls of a house on a windy day is greater than that on a day when the air is calm. A representative *R*-value for this stagnant layer of air is given in Table 20.4.

Example 20.9   The *R*-Value of a Typical Wall

Calculate the total *R*-value for a wall constructed as shown in Figure 20.15a. Starting outside the house (toward the front in the figure) and moving inward, the wall consists of 4 in. of brick, 0.5 in. of sheathing, an air space 3.5 in. thick, and 0.5 in. of drywall.

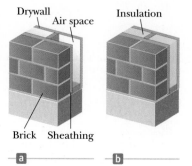

Figure 20.15 (Example 20.9) An exterior house wall containing (a) an air space and (b) insulation.

SOLUTION

Conceptualize Use Figure 20.15 to help conceptualize the structure of the wall. Do not forget the stagnant air layers inside and outside the house.

Categorize We will use specific equations developed in this section on home insulation, so we categorize this example as a substitution problem.

Use Table 20.4 to find the *R*-value of each layer:

$$R_1 \text{ (outside stagnant air layer)} = 0.17 \text{ ft}^2 \cdot {}^\circ\text{F} \cdot \text{h/Btu}$$

$$R_2 \text{ (brick)} = 4.00 \text{ ft}^2 \cdot {}^\circ\text{F} \cdot \text{h/Btu}$$

$$R_3 \text{ (sheathing)} = 1.32 \text{ ft}^2 \cdot {}^\circ\text{F} \cdot \text{h/Btu}$$

$$R_4 \text{ (air space)} = 1.01 \text{ ft}^2 \cdot {}^\circ\text{F} \cdot \text{h/Btu}$$

$$R_5 \text{ (drywall)} = 0.45 \text{ ft}^2 \cdot {}^\circ\text{F} \cdot \text{h/Btu}$$

$$R_6 \text{ (inside stagnant air layer)} = 0.17 \text{ ft}^2 \cdot {}^\circ\text{F} \cdot \text{h/Btu}$$

Add the *R*-values to obtain the total *R*-value for the wall:

$$R_{\text{total}} = R_1 + R_2 + R_3 + R_4 + R_5 + R_6 = \boxed{7.12 \text{ ft}^2 \cdot {}^\circ\text{F} \cdot \text{h/Btu}}$$

WHAT IF?   Suppose you are not happy with this total *R*-value for the wall. You cannot change the overall structure, but you can fill the air space as in Figure 20.15b. To *maximize* the total *R*-value, what material should you choose to fill the air space?

Answer Looking at Table 20.4, we see that 3.5 in. of fiberglass insulation is more than ten times as effective as 3.5 in. of air. Therefore, we should fill the air space with fiberglass insulation. The result is that we add 10.90 ft² · °F · h/Btu of *R*-value, and we lose 1.01 ft² · °F · h/Btu due to the air space we have replaced. The new total *R*-value is equal to 7.12 ft² · °F · h/Btu + 9.89 ft² · °F · h/Btu = 17.01 ft² · °F · h/Btu.

Convection

At one time or another, you probably have warmed your hands by holding them over an open flame. In this situation, the air directly above the flame is heated and expands. As a result, the density of this air decreases and the air rises. This hot air warms your hands as it flows by. Energy transferred by the movement of a warm substance is said to have been transferred by **convection,** which is a form of matter transfer, T_{MT} in Equation 8.2. When resulting from differences in density, as with air around a fire, the process is referred to as *natural convection*. Airflow at a beach is an example of natural convection, as is the mixing that occurs as surface water in a lake cools and sinks (see Section 19.4). When the heated substance is forced to move by a fan or pump, as in some hot-air and hot-water heating systems, the process is called *forced convection.*

If it were not for convection currents, it would be very difficult to boil water. As water is heated in a teakettle, the lower layers are warmed first. This water expands and rises to the top because its density is lowered. At the same time, the denser, cool water at the surface sinks to the bottom of the kettle and is heated.

The same process occurs when a room is heated by a radiator. The hot radiator warms the air in the lower regions of the room. The warm air expands and rises to the ceiling because of its lower density. The denser, cooler air from above sinks, and the continuous air current pattern shown in Figure 20.16 is established.

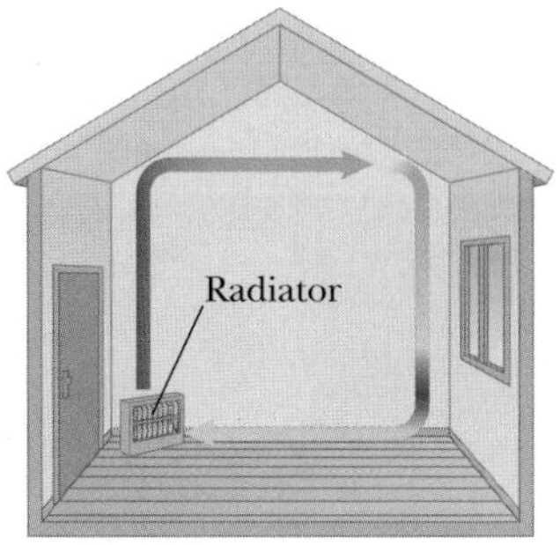

Figure 20.16 Convection currents are set up in a room warmed by a radiator.

Radiation

The third means of energy transfer we shall discuss is **thermal radiation,** T_{ER} in Equation 8.2. All objects radiate energy continuously in the form of electromagnetic waves (see Chapter 34) produced by thermal vibrations of the molecules. You are likely familiar with electromagnetic radiation in the form of the orange glow from an electric stove burner, an electric space heater, or the coils of a toaster.

The rate at which the surface of an object radiates energy is proportional to the fourth power of the absolute temperature of the surface. Known as **Stefan's law,** this behavior is expressed in equation form as

$$P = \sigma A e T^4 \qquad\qquad (20.19)$$

◀ Stefan's law

where P is the power in watts of electromagnetic waves radiated from the surface of the object, σ is a constant equal to $5.669\ 6 \times 10^{-8}\ \text{W/m}^2 \cdot \text{K}^4$, A is the surface area of the object in square meters, e is the **emissivity,** and T is the surface temperature in kelvins. The value of e can vary between zero and unity depending on the properties of the surface of the object. The emissivity is equal to the **absorptivity,** which is the fraction of the incoming radiation that the surface absorbs. A mirror has very low absorptivity because it reflects almost all incident light. Therefore, a mirror surface also has a very low emissivity. At the other extreme, a black surface has high absorptivity and high emissivity. An **ideal absorber** is defined as an object that absorbs all the energy incident on it, and for such an object, $e = 1$. An object for which $e = 1$ is often referred to as a **black body.** We shall investigate experimental and theoretical approaches to radiation from a black body in Chapter 40.

Every second, approximately 1 370 J of electromagnetic radiation from the Sun passes perpendicularly through each 1 m^2 at the top of the Earth's atmosphere. This radiation is primarily visible and infrared light accompanied by a significant amount of ultraviolet radiation. We shall study these types of radiation in detail in Chapter 34. Enough energy arrives at the surface of the Earth each day to supply all our energy needs on this planet hundreds of times over, if only it could be captured and used efficiently. The growth in the number of solar energy–powered houses and proposals for solar energy "farms" in the United States reflects the increasing efforts being made to use this abundant energy.

What happens to the atmospheric temperature at night is another example of the effects of energy transfer by radiation. If there is a cloud cover above the Earth, the water vapor in the clouds absorbs part of the infrared radiation emitted by the Earth and re-emits it back to the surface. Consequently, temperature levels at the surface remain moderate. In the absence of this cloud cover, there is less in the way to prevent this radiation from escaping into space; therefore, the temperature decreases more on a clear night than on a cloudy one.

As an object radiates energy at a rate given by Equation 20.19, it also absorbs electromagnetic radiation from the surroundings, which consist of other objects that radiate energy. If the latter process did not occur, an object would eventually radiate all its energy and its temperature would reach absolute zero. If an object is at a temperature T and its surroundings are at an average temperature T_0, the net rate of energy gained or lost by the object as a result of radiation is

$$P_{net} = \sigma Ae(T^4 - T_0^{\,4}) \qquad (20.20)$$

When an object is in equilibrium with its surroundings, it radiates and absorbs energy at the same rate and its temperature remains constant. When an object is hotter than its surroundings, it radiates more energy than it absorbs and its temperature decreases.

The Dewar Flask

The *Dewar flask*[6] is a container designed to minimize energy transfers by conduction, convection, and radiation. Such a container is used to store cold or hot liquids for long periods of time. (An insulated bottle, such as a Thermos, is a common household equivalent of a Dewar flask.) The standard construction (Fig. 20.17) consists of a double-walled Pyrex glass vessel with silvered walls. The space between the walls is evacuated to minimize energy transfer by conduction and convection. The silvered surfaces minimize energy transfer by radiation because silver is a very good reflector and has very low emissivity. A further reduction in energy loss is obtained by reducing the size of the neck. Dewar flasks are commonly used to store liquid nitrogen (boiling point 77 K) and liquid oxygen (boiling point 90 K).

To confine liquid helium (boiling point 4.2 K), which has a very low heat of vaporization, it is often necessary to use a double Dewar system in which the Dewar flask containing the liquid is surrounded by a second Dewar flask. The space between the two flasks is filled with liquid nitrogen.

Newer designs of storage containers use "superinsulation" that consists of many layers of reflecting material separated by fiberglass. All this material is in a vacuum, and no liquid nitrogen is needed with this design.

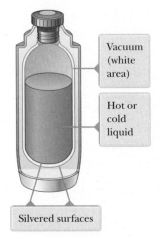

Vacuum (white area)

Hot or cold liquid

Silvered surfaces

Figure 20.17 A cross-sectional view of a Dewar flask, which is used to store hot or cold substances.

[6]Invented by Sir James Dewar (1842–1923).

Summary

Definitions

Internal energy is a system's energy associated with its temperature and its physical state (solid, liquid, gas). Internal energy includes kinetic energy of random translation, rotation, and vibration of molecules; vibrational potential energy within molecules; and potential energy between molecules.

Heat is the process of energy transfer across the boundary of a system resulting from a temperature difference between the system and its surroundings. The symbol Q represents the amount of energy transferred by this process.

A **calorie** is the amount of energy necessary to raise the temperature of 1 g of water from 14.5°C to 15.5°C.

The **heat capacity** C of any sample is the amount of energy needed to raise the temperature of the sample by 1°C.

The **specific heat** c of a substance is the heat capacity per unit mass:

$$c \equiv \frac{Q}{m\,\Delta T} \qquad (20.3)$$

The **latent heat** of a substance is defined as the ratio of the energy input to a substance to the change in mass of the higher-phase material:

$$L \equiv \frac{Q}{\Delta m} \qquad (20.6)$$

Concepts and Principles

The energy Q required to change the temperature of a mass m of a substance by an amount ΔT is

$$Q = mc\,\Delta T \qquad \textbf{(20.4)}$$

where c is the specific heat of the substance.

The energy required to change the phase of a pure substance is

$$Q = L\,\Delta m \qquad \textbf{(20.7)}$$

where L is the latent heat of the substance, which depends on the nature of the phase change and the substance, and Δm is the change in mass of the higher-phase material.

The **work** done on a gas as its volume changes from some initial value V_i to some final value V_f is

$$W = -\int_{V_i}^{V_f} P\,dV \qquad \textbf{(20.9)}$$

where P is the pressure of the gas, which may vary during the process. To evaluate W, the process must be fully specified; that is, P and V must be known during each step. The work done depends on the path taken between the initial and final states.

The **first law of thermodynamics** is a specific reduction of the conservation of energy equation (Eq. 8.2) and states that when a system undergoes a change from one state to another, the change in its internal energy is

$$\Delta E_{\text{int}} = Q + W \qquad \textbf{(20.10)}$$

where Q is the energy transferred into the system by heat and W is the work done on the system. Although Q and W both depend on the path taken from the initial state to the final state, the quantity ΔE_{int} does not depend on the path.

In a **cyclic process** (one that originates and terminates at the same state), $\Delta E_{\text{int}} = 0$ and therefore $Q = -W$. That is, the energy transferred into the system by heat equals the negative of the work done on the system during the process.

In an **adiabatic process,** no energy is transferred by heat between the system and its surroundings ($Q = 0$). In this case, the first law gives $\Delta E_{\text{int}} = W$. In the **adiabatic free expansion** of a gas, $Q = 0$ and $W = 0$, so $\Delta E_{\text{int}} = 0$. That is, the internal energy of the gas does not change in such a process.

An **isobaric process** is one that occurs at constant pressure. The work done on a gas in such a process is $W = -P(V_f - V_i)$.

An **isovolumetric process** is one that occurs at constant volume. No work is done in such a process, so $\Delta E_{\text{int}} = Q$.

An **isothermal process** is one that occurs at constant temperature. The work done on an ideal gas during an isothermal process is

$$W = nRT \ln\!\left(\frac{V_i}{V_f}\right) \qquad \textbf{(20.14)}$$

Conduction can be viewed as an exchange of kinetic energy between colliding molecules or electrons. The rate of energy transfer by conduction through a slab of area A is

$$P = kA\left|\frac{dT}{dx}\right| \qquad \textbf{(20.15)}$$

where k is the **thermal conductivity** of the material from which the slab is made and $|dT/dx|$ is the **temperature gradient.**

In **convection,** a warm substance transfers energy from one location to another.

All objects emit **thermal radiation** in the form of electromagnetic waves at the rate

$$P = \sigma AeT^4 \qquad \textbf{(20.19)}$$

Objective Questions

1. denotes answer available in *Student Solutions Manual/Study Guide*

1. An ideal gas is compressed to half its initial volume by means of several possible processes. Which of the following processes results in the most work done on the gas? (a) isothermal (b) adiabatic (c) isobaric (d) The work done is independent of the process.

2. A poker is a stiff, nonflammable rod used to push burning logs around in a fireplace. For safety and comfort of use, should the poker be made from a material with (a) high specific heat and high thermal conductivity, (b) low specific heat and low thermal conductivity,

(c) low specific heat and high thermal conductivity, or (d) high specific heat and low thermal conductivity?

3. Assume you are measuring the specific heat of a sample of originally hot metal by using a calorimeter containing water. Because your calorimeter is not perfectly insulating, energy can transfer by heat between the contents of the calorimeter and the room. To obtain the most accurate result for the specific heat of the metal, you should use water with which initial temperature? (a) slightly lower than room temperature (b) the same as room temperature (c) slightly higher than room temperature (d) whatever you like because the initial temperature makes no difference

4. An amount of energy is added to ice, raising its temperature from $-10°C$ to $-5°C$. A larger amount of energy is added to the same mass of water, raising its temperature from $15°C$ to $20°C$. From these results, what would you conclude? (a) Overcoming the latent heat of fusion of ice requires an input of energy. (b) The latent heat of fusion of ice delivers some energy to the system. (c) The specific heat of ice is less than that of water. (d) The specific heat of ice is greater than that of water. (e) More information is needed to draw any conclusion.

5. How much energy is required to raise the temperature of 5.00 kg of lead from $20.0°C$ to its melting point of $327°C$? The specific heat of lead is $128 \text{ J/kg} \cdot °C$. (a) 4.04×10^5 J (b) 1.07×10^5 J (c) 8.15×10^4 J (d) 2.13×10^4 J (e) 1.96×10^5 J

6. Ethyl alcohol has about one-half the specific heat of water. Assume equal amounts of energy are transferred by heat into equal-mass liquid samples of alcohol and water in separate insulated containers. The water rises in temperature by $25°C$. How much will the alcohol rise in temperature? (a) It will rise by $12°C$. (b) It will rise by $25°C$. (c) It will rise by $50°C$. (d) It depends on the rate of energy transfer. (e) It will not rise in temperature.

7. The specific heat of substance A is greater than that of substance B. Both A and B are at the same initial temperature when equal amounts of energy are added to them. Assuming no melting or vaporization occurs, which of the following can be concluded about the final temperature T_A of substance A and the final temperature T_B of substance B? (a) $T_A > T_B$ (b) $T_A < T_B$ (c) $T_A = T_B$ (d) More information is needed.

8. Beryllium has roughly one-half the specific heat of water (H_2O). Rank the quantities of energy input required to produce the following changes from the largest to the smallest. In your ranking, note any cases of equality. (a) raising the temperature of 1 kg of H_2O from $20°C$ to $26°C$ (b) raising the temperature of 2 kg of H_2O from $20°C$ to $23°C$ (c) raising the temperature of 2 kg of H_2O from $1°C$ to $4°C$ (d) raising the temperature of 2 kg of beryllium from $-1°C$ to $2°C$ (e) raising the temperature of 2 kg of H_2O from $-1°C$ to $2°C$

9. A person shakes a sealed insulated bottle containing hot coffee for a few minutes. (i) What is the change in the temperature of the coffee? (a) a large decrease (b) a slight decrease (c) no change (d) a slight increase (e) a large increase (ii) What is the change in the internal energy of the coffee? Choose from the same possibilities.

10. A 100-g piece of copper, initially at $95.0°C$, is dropped into 200 g of water contained in a 280-g aluminum can; the water and can are initially at $15.0°C$. What is the final temperature of the system? (Specific heats of copper and aluminum are 0.092 and 0.215 cal/g · °C, respectively.) (a) $16°C$ (b) $18°C$ (c) $24°C$ (d) $26°C$ (e) none of those answers

11. Star A has twice the radius and twice the absolute surface temperature of star B. The emissivity of both stars can be assumed to be 1. What is the ratio of the power output of star A to that of star B? (a) 4 (b) 8 (c) 16 (d) 32 (e) 64

12. If a gas is compressed isothermally, which of the following statements is true? (a) Energy is transferred into the gas by heat. (b) No work is done on the gas. (c) The temperature of the gas increases. (d) The internal energy of the gas remains constant. (e) None of those statements is true.

13. When a gas undergoes an adiabatic expansion, which of the following statements is true? (a) The temperature of the gas does not change. (b) No work is done by the gas. (c) No energy is transferred to the gas by heat. (d) The internal energy of the gas does not change. (e) The pressure increases.

14. If a gas undergoes an isobaric process, which of the following statements is true? (a) The temperature of the gas doesn't change. (b) Work is done on or by the gas. (c) No energy is transferred by heat to or from the gas. (d) The volume of the gas remains the same. (e) The pressure of the gas decreases uniformly.

15. How long would it take a 1 000 W heater to melt 1.00 kg of ice at $-20.0°C$, assuming all the energy from the heater is absorbed by the ice? (a) 4.18 s (b) 41.8 s (c) 5.55 min (d) 6.25 min (e) 38.4 min

Conceptual Questions 1. denotes answer available in *Student Solutions Manual/Study Guide*

1. Rub the palm of your hand on a metal surface for about 30 seconds. Place the palm of your other hand on an unrubbed portion of the surface and then on the rubbed portion. The rubbed portion will feel warmer. Now repeat this process on a wood surface. Why does the temperature difference between the rubbed and unrubbed portions of the wood surface seem larger than for the metal surface?

2. You need to pick up a very hot cooking pot in your kitchen. You have a pair of cotton oven mitts. To pick up the pot most comfortably, should you soak them in cold water or keep them dry?

3. What is wrong with the following statement: "Given any two bodies, the one with the higher temperature contains more heat."

4. Why is a person able to remove a piece of dry aluminum foil from a hot oven with bare fingers, whereas a burn results if there is moisture on the foil?

5. Using the first law of thermodynamics, explain why the *total* energy of an isolated system is always constant.

6. In 1801, Humphry Davy rubbed together pieces of ice inside an icehouse. He made sure that nothing in the environment was at a higher temperature than the rubbed pieces. He observed the production of drops of liquid water. Make a table listing this and other experiments or processes to illustrate each of the following situations. (a) A system can absorb energy by heat, increase in internal energy, and increase in temperature. (b) A system can absorb energy by heat and increase in internal energy without an increase in temperature. (c) A system can absorb energy by heat without increasing in temperature or in internal energy. (d) A system can increase in internal energy and in temperature without absorbing energy by heat. (e) A system can increase in internal energy without absorbing energy by heat or increasing in temperature.

7. It is the morning of a day that will become hot. You just purchased drinks for a picnic and are loading them, with ice, into a chest in the back of your car. (a) You wrap a wool blanket around the chest. Does doing so help to keep the beverages cool, or should you expect the wool blanket to warm them up? Explain your answer. (b) Your younger sister suggests you wrap her up in another wool blanket to keep her cool on the hot day like the ice chest. Explain your response to her.

8. In usually warm climates that experience a hard freeze, fruit growers will spray the fruit trees with water, hoping that a layer of ice will form on the fruit. Why would such a layer be advantageous?

9. Suppose you pour hot coffee for your guests, and one of them wants it with cream. He wants the coffee to be as warm as possible several minutes later when he drinks it. To have the warmest coffee, should the person add the cream just after the coffee is poured or just before drinking? Explain.

10. When camping in a canyon on a still night, a camper notices that as soon as the sun strikes the surrounding peaks, a breeze begins to stir. What causes the breeze?

11. Pioneers stored fruits and vegetables in underground cellars. In winter, why did the pioneers place an open barrel of water alongside their produce?

12. Is it possible to convert internal energy to mechanical energy? Explain with examples.

Problems available in Access end-of-chapter problems online at **www.webassign.net**

Section 20.1 Heat and Internal Energy
Problem 1

Section 20.2 Specific Heat and Calorimetry
Problems 2–15

Section 20.3 Latent Heat
Problems 16–24

Section 20.4 Work and Heat in Thermodynamic Processes
Problems 25–29

Section 20.5 The First Law of Thermodynamics
Problems 30–34

Section 20.6 Some Applications of the First Law of Thermodynamics
Problems 35–42

Section 20.7 Energy Transfer Mechanisms in Thermal Processes
Problems 43–57

Additional Problems
Problems 58–80

Challenge Problems
Problems 81–84

Solutions to the following Problems are available in the *Student Solutions Manual/Study Guide:*
20.1, 20.4, 20.6, 20.11, 20.20, 20.23, 20.25, 20.29, 20.33, 20.35, 20.36, 20.37, 20.55, 20.61, 20.72, 20.77, and 20.84

List of Enhanced Problems

Problem Number	Targeted Feedback in Enhanced WebAssign	Analysis Model Tutorial in Enhanced WebAssign	Master It in Enhanced WebAssign	Watch It in Enhanced WebAssign
20.2	✓	✓		✓
20.6	✓		✓	
20.11	✓		✓	
20.13	✓			✓
20.17	✓		✓	
20.18	✓			✓
20.20	✓	✓	✓	
20.22	✓			✓
20.28	✓			✓
20.29	✓		✓	
20.30	✓			✓
20.34	✓			✓
20.35			✓	
20.37	✓		✓	
20.38	✓			✓
20.41	✓			✓
20.51	✓		✓	
20.58	✓		✓	
20.59	✓		✓	
20.61			✓	
20.62		✓		
20.65		✓		
20.71	✓		✓	
20.77			✓	

The Kinetic Theory of Gases

A boy inflates his bicycle tire with a hand-operated pump. Kinetic theory helps to describe the details of the air in the pump. (© Cengage Learning/ George Semple)

In Chapter 19, we discussed the properties of an ideal gas by using such macroscopic variables as pressure, volume, and temperature. Such large-scale properties can be related to a description on a microscopic scale, where matter is treated as a collection of molecules. Applying Newton's laws of motion in a statistical manner to a collection of particles provides a reasonable description of thermodynamic processes. To keep the mathematics relatively simple, we shall consider primarily the behavior of gases because in gases the interactions between molecules are much weaker than they are in liquids or solids.

We shall begin by relating pressure and temperature directly to the details of molecular motion in a sample of gas. Based on these results, we will make predictions of molar specific heats of gases. Some of these predictions will be correct and some will not. We will extend our model to explain those values that are not predicted correctly by the simpler model. Finally, we discuss the distribution of molecular speeds in a gas.

21.1 Molecular Model of an Ideal Gas

In this chapter, we will investigate a *structural model* for an ideal gas. A **structural model** is a theoretical construct designed to represent a system that cannot be observed directly because it is too large or too small. For example, we can only observe the solar system from the inside; we cannot travel outside the solar system and look back to see how it works. This restricted vantage point has led to different historical structural models of the solar system: the *geocentric model*, with the Earth at the center, and the *heliocentric model*, with the Sun at the center. Of course, the latter has been shown to be correct. An example of a system too small to observe directly is the hydrogen atom. Various structural models of this system have been developed, including the *Bohr model* (Section 42.3) and the *quantum model* (Section 42.4). Once a structural model is developed, various predictions are made for experimental observations. For example, the geocentric model of the solar system makes predictions of how the movement of Mars should appear from the Earth. It turns out that those predictions do not match the actual observations. When that occurs with a structural model, the model must be modified or replaced with another model.

The structural model that we will develop for an ideal gas is called **kinetic theory**. This model treats an ideal gas as a collection of molecules with the following properties:

1. *Physical components:*

   The gas consists of a number of identical molecules within a cubic container of side length d. The number of molecules in the gas is large, and the average separation between them is large compared with their dimensions. Therefore, the molecules occupy a negligible volume in the container. This assumption is consistent with the ideal gas model, in which we imagine the molecules to be point-like.

2. *Behavior of the components:*

   (a) The molecules obey Newton's laws of motion, but as a whole their motion is isotropic: any molecule can move in any direction with any speed.

   (b) The molecules interact only by short-range forces during elastic collisions. This assumption is consistent with the ideal gas model, in which the molecules exert no long-range forces on one another.

   (c) The molecules make elastic collisions with the walls.

Although we often picture an ideal gas as consisting of single atoms, the behavior of molecular gases approximates that of ideal gases rather well at low pressures. Usually, molecular rotations or vibrations have no effect on the motions considered here.

For our first application of kinetic theory, let us relate the macroscope variable of pressure P to microscopic quantities. Consider a collection of N molecules of an ideal gas in a container of volume V. As indicated above, the container is a cube with edges of length d (Fig. 21.1). We shall first focus our attention on one of these molecules of mass m_0 and assume it is moving so that its component of velocity in the x direction is v_{xi} as in Figure 21.2. (The subscript i here refers to the ith molecule in the collection, not to an initial value. We will combine the effects of all the molecules shortly.) As the molecule collides elastically with any wall (property 2(c) above), its velocity component perpendicular to the wall is reversed because the mass of the wall is far greater than the mass of the molecule. The molecule is modeled as a nonisolated system for which the impulse from the wall causes a change in the molecule's momentum. Because the momentum component p_{xi} of the molecule is $m_0 v_{xi}$ before the collision and $-m_0 v_{xi}$ after the collision, the change in the x component of the momentum of the molecule is

$$\Delta p_{xi} = -m_0 v_{xi} - (m_0 v_{xi}) = -2 m_0 v_{xi} \tag{21.1}$$

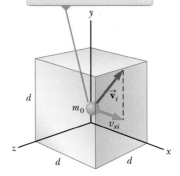

One molecule of the gas moves with velocity $\vec{v}$ on its way toward a collision with the wall.

Figure 21.1 A cubical box with sides of length d containing an ideal gas.

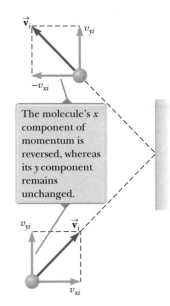

The molecule's x component of momentum is reversed, whereas its y component remains unchanged.

Figure 21.2 A molecule makes an elastic collision with the wall of the container. In this construction, we assume the molecule moves in the xy plane.

From the nonisolated system model for momentum, we can apply the impulse-momentum theorem (Eqs. 9.11 and 9.13) to the molecule to give

$$\overline{F}_{i,\text{on molecule}}\, \Delta t_{\text{collision}} = \Delta p_{xi} = -2m_0 v_{xi} \qquad (21.2)$$

where $\overline{F}_{i,\text{on molecule}}$ is the x component of the average force[1] the wall exerts on the molecule during the collision and $\Delta t_{\text{collision}}$ is the duration of the collision. For the molecule to make another collision with the same wall after this first collision, it must travel a distance of $2d$ in the x direction (across the cube and back). Therefore, the time interval between two collisions with the same wall is

$$\Delta t = \frac{2d}{v_{xi}} \qquad (21.3)$$

The force that causes the change in momentum of the molecule in the collision with the wall occurs only during the collision. We can, however, find the long-term average force for many back-and-forth trips across the cube by averaging the force in Equation 21.2 over the time interval for the molecule to move across the cube and back once, Equation 21.3. The average change in momentum per trip for the time interval for many trips is the same as that for the short duration of the collision. Therefore, we can rewrite Equation 21.2 as

$$\overline{F}_i\, \Delta t = -2m_0 v_{xi} \qquad (21.4)$$

where $\overline{F}_i$ is the average force component over the time interval for the molecule to move across the cube and back. Because exactly one collision occurs for each such time interval, this result is also the long-term average force on the molecule over long time intervals containing any number of multiples of Δt.

Equation 21.3 and 21.4 enable us to express the x component of the long-term average force exerted by the wall on the molecule as

$$\overline{F}_i = -\frac{2m_0 v_{xi}}{\Delta t} = -\frac{2m_0 v_{xi}^2}{2d} = -\frac{m_0 v_{xi}^2}{d} \qquad (21.5)$$

Now, by Newton's third law, the x component of the long-term average force exerted by the *molecule* on the *wall* is equal in magnitude and opposite in direction:

$$\overline{F}_{i,\text{on wall}} = -\overline{F}_i = -\left(-\frac{m_0 v_{xi}^2}{d}\right) = \frac{m_0 v_{xi}^2}{d} \qquad (21.6)$$

The total average force $\overline{F}$ exerted by the gas on the wall is found by adding the average forces exerted by the individual molecules. Adding terms such as those in Equation 21.6 for all molecules gives

$$\overline{F} = \sum_{i=1}^{N} \frac{m_0 v_{xi}^2}{d} = \frac{m_0}{d} \sum_{i=1}^{N} v_{xi}^2 \qquad (21.7)$$

where we have factored out the length of the box and the mass m_0 because property 1 tells us that all the molecules are the same. We now impose an additional feature from property 1, that the number of molecules is large. For a small number of molecules, the actual force on the wall would vary with time. It would be nonzero during the short interval of a collision of a molecule with the wall and zero when no molecule happens to be hitting the wall. For a very large number of molecules such as Avogadro's number, however, these variations in force are smoothed out so that the average force given above is the same over *any* time interval. Therefore, the *constant* force F on the wall due to the molecular collisions is

$$F = \frac{m_0}{d} \sum_{i=1}^{N} v_{xi}^2 \qquad (21.8)$$

---

[1]For this discussion, we use a bar over a variable to represent the average value of the variable, such as $\overline{F}$ for the average force, rather than the subscript "avg" that we have used before. This notation is to save confusion because we already have a number of subscripts on variables.

To proceed further, let's consider how to express the average value of the square of the x component of the velocity for N molecules. The traditional average of a set of values is the sum of the values over the number of values:

$$\overline{v_x^2} = \frac{\sum_{i=1}^{N} v_{xi}^2}{N} \rightarrow \sum_{i=1}^{N} v_{xi}^2 = N\overline{v_x^2} \tag{21.9}$$

Using Equation 21.9 to substitute for the sum in Equation 21.8 gives

$$F = \frac{m_0}{d} N\overline{v_x^2} \tag{21.10}$$

Now let's focus again on one molecule with velocity components v_{xi}, v_{yi}, and v_{zi}. The Pythagorean theorem relates the square of the speed of the molecule to the squares of the velocity components:

$$v_i^2 = v_{xi}^2 + v_{yi}^2 + v_{zi}^2 \tag{21.11}$$

Hence, the average value of v^2 for all the molecules in the container is related to the average values of v_x^2, v_y^2, and v_z^2 according to the expression

$$\overline{v^2} = \overline{v_x^2} + \overline{v_y^2} + \overline{v_z^2} \tag{21.12}$$

Because the motion is isotropic (property 2(a) above), the average values $\overline{v_x^2}$, $\overline{v_y^2}$, and $\overline{v_z^2}$ are equal to one another. Using this fact and Equation 21.12, we find that

$$\overline{v^2} = 3\overline{v_x^2} \tag{21.13}$$

Therefore, from Equation 21.10, the total force exerted on the wall is

$$F = \tfrac{1}{3}N \frac{m_0\overline{v^2}}{d} \tag{21.14}$$

Using this expression, we can find the total pressure exerted on the wall:

$$P = \frac{F}{A} = \frac{F}{d^2} = \tfrac{1}{3}N \frac{m_0\overline{v^2}}{d^3} = \tfrac{1}{3}\left(\frac{N}{V}\right)m_0\overline{v^2}$$

$$P = \tfrac{2}{3}\left(\frac{N}{V}\right)(\tfrac{1}{2}m_0\overline{v^2}) \tag{21.15}$$

◀ Relationship between pressure and molecular kinetic energy

where we have recognized the volume V of the cube as d^3.

Equation 21.15 indicates that the pressure of a gas is proportional to (1) the number of molecules per unit volume and (2) the average translational kinetic energy of the molecules, $\tfrac{1}{2}m_0\overline{v^2}$. In analyzing this structural model of an ideal gas, we obtain an important result that relates the macroscopic quantity of pressure to a microscopic quantity, the average value of the square of the molecular speed. Therefore, a key link between the molecular world and the large-scale world has been established.

Notice that Equation 21.15 verifies some features of pressure with which you are probably familiar. One way to increase the pressure inside a container is to increase the number of molecules per unit volume N/V in the container. That is what you do when you add air to a tire. The pressure in the tire can also be raised by increasing the average translational kinetic energy of the air molecules in the tire. That can be accomplished by increasing the temperature of that air, which is why the pressure inside a tire increases as the tire warms up during long road trips. The continuous flexing of the tire as it moves along the road surface results in work done on the rubber as parts of the tire distort, causing an increase in internal energy of the rubber. The increased temperature of the rubber results in the transfer of energy by heat into the air inside the tire. This transfer increases the air's temperature, and this increase in temperature in turn produces an increase in pressure.

Molecular Interpretation of Temperature

Let's now consider another macroscopic variable, the temperature T of the gas. We can gain some insight into the meaning of temperature by first writing Equation 21.15 in the form

$$PV = \tfrac{2}{3}N(\tfrac{1}{2}m_0\overline{v^2})   \tag{21.16}$$

Let's now compare this expression with the equation of state for an ideal gas (Eq. 19.10):

$$PV = Nk_{\mathrm{B}}T   \tag{21.17}$$

Equating the right sides of Equations 21.16 and 21.17 and solving for T gives

▶ **Relationship between temperature and molecular kinetic energy**

$$T = \frac{2}{3k_{\mathrm{B}}}\left(\tfrac{1}{2}m_0\overline{v^2}\right)   \tag{21.18}$$

This result tells us that temperature is a direct measure of average molecular kinetic energy. By rearranging Equation 21.18, we can relate the translational molecular kinetic energy to the temperature:

▶ **Average kinetic energy per molecule**

$$\tfrac{1}{2}m_0\overline{v^2} = \tfrac{3}{2}k_{\mathrm{B}}T   \tag{21.19}$$

That is, the average translational kinetic energy per molecule is $\tfrac{3}{2}k_{\mathrm{B}}T$. Because $\overline{v_x^2} = \tfrac{1}{3}\overline{v^2}$ (Eq. 21.13), it follows that

$$\tfrac{1}{2}m_0\overline{v_x^2} = \tfrac{1}{2}k_{\mathrm{B}}T   \tag{21.20}$$

In a similar manner, for the y and z directions,

$$\tfrac{1}{2}m_0\overline{v_y^2} = \tfrac{1}{2}k_{\mathrm{B}}T \ \text{ and } \ \tfrac{1}{2}m_0\overline{v_z^2} = \tfrac{1}{2}k_{\mathrm{B}}T$$

Therefore, each translational degree of freedom contributes an equal amount of energy, $\tfrac{1}{2}k_{\mathrm{B}}T$, to the gas. (In general, a "degree of freedom" refers to an independent means by which a molecule can possess energy.) A generalization of this result, known as the **theorem of equipartition of energy,** is as follows:

▶ **Theorem of equipartition of energy**

Each degree of freedom contributes $\tfrac{1}{2}k_{\mathrm{B}}T$ to the energy of a system, where possible degrees of freedom are those associated with translation, rotation, and vibration of molecules.

The total translational kinetic energy of N molecules of gas is simply N times the average energy per molecule, which is given by Equation 21.19:

▶ **Total translational kinetic energy of N molecules**

$$K_{\mathrm{tot\,trans}} = N(\tfrac{1}{2}m_0\overline{v^2}) = \tfrac{3}{2}Nk_{\mathrm{B}}T = \tfrac{3}{2}nRT   \tag{21.21}$$

where we have used $k_{\mathrm{B}} = R/N_{\mathrm{A}}$ for Boltzmann's constant and $n = N/N_{\mathrm{A}}$ for the number of moles of gas. If the gas molecules possess only translational kinetic energy, Equation 21.21 represents the internal energy of the gas. This result implies that the internal energy of an ideal gas depends *only* on the temperature. We will follow up on this point in Section 21.2.

The square root of $\overline{v^2}$ is called the **root-mean-square (rms) speed** of the molecules. From Equation 21.19, we find that the rms speed is

▶ **Root-mean-square speed**

$$v_{\mathrm{rms}} = \sqrt{\overline{v^2}} = \sqrt{\frac{3k_{\mathrm{B}}T}{m_0}} = \sqrt{\frac{3RT}{M}}   \tag{21.22}$$

where M is the molar mass in kilograms per mole and is equal to $m_0 N_{\mathrm{A}}$. This expression shows that, at a given temperature, lighter molecules move faster, on the average, than do heavier molecules. For example, at a given temperature, hydrogen molecules, whose molar mass is 2.02×10^{-3} kg/mol, have an average speed approximately four times that of oxygen molecules, whose molar mass is 32.0×10^{-3} kg/mol. Table 21.1 lists the rms speeds for various molecules at 20°C.

Table 21.1	**Some Root-Mean-Square (rms) Speeds**				
Gas	Molar Mass (g/mol)	v_{rms} at 20°C (m/s)	Gas	Molar Mass (g/mol)	v_{rms} at 20°C (m/s)
H_2	2.02	1902	NO	30.0	494
He	4.00	1352	O_2	32.0	478
H_2O	18.0	637	CO_2	44.0	408
Ne	20.2	602	SO_2	64.1	338
N_2 or CO	28.0	511			

Quick Quiz 21.1 Two containers hold an ideal gas at the same temperature and pressure. Both containers hold the same type of gas, but container B has twice the volume of container A. **(i)** What is the average translational kinetic energy per molecule in container B? (a) twice that of container A (b) the same as that of container A (c) half that of container A (d) impossible to determine **(ii)** From the same choices, describe the internal energy of the gas in container B.

> **Pitfall Prevention 21.1**
> **The Square Root of the Square?**
> Taking the square root of $\overline{v^2}$ does not "undo" the square because we have taken an average *between* squaring and taking the square root. Although the square root of $(\overline{v})^2$ is $\overline{v} = v_{avg}$ because the squaring is done after the averaging, the square root of $\overline{v^2}$ is *not* v_{avg}, but rather v_{rms}.

Example 21.1	**A Tank of Helium**

A tank used for filling helium balloons has a volume of 0.300 m³ and contains 2.00 mol of helium gas at 20.0°C. Assume the helium behaves like an ideal gas.

(A) What is the total translational kinetic energy of the gas molecules?

SOLUTION

Conceptualize Imagine a microscopic model of a gas in which you can watch the molecules move about the container more rapidly as the temperature increases. Because the gas is monatomic, the total translational kinetic energy of the molecules is the internal energy of the gas.

Categorize We evaluate parameters with equations developed in the preceding discussion, so this example is a substitution problem.

Use Equation 21.21 with $n = 2.00$ mol and $T = 293$ K:

$$E_{int} = \tfrac{3}{2}nRT = \tfrac{3}{2}(2.00 \text{ mol})(8.31 \text{ J/mol} \cdot \text{K})(293 \text{ K})$$

$$= \boxed{7.30 \times 10^3 \text{ J}}$$

(B) What is the average kinetic energy per molecule?

SOLUTION

Use Equation 21.19:

$$\tfrac{1}{2}m_0\overline{v^2} = \tfrac{3}{2}k_BT = \tfrac{3}{2}(1.38 \times 10^{-23} \text{ J/K})(293 \text{ K})$$

$$= \boxed{6.07 \times 10^{-21} \text{ J}}$$

WHAT IF? What if the temperature is raised from 20.0°C to 40.0°C? Because 40.0 is twice as large as 20.0, is the total translational energy of the molecules of the gas twice as large at the higher temperature?

Answer The expression for the total translational energy depends on the temperature, and the value for the temperature must be expressed in kelvins, not in degrees Celsius. Therefore, the ratio of 40.0 to 20.0 is *not* the appropriate ratio. Converting the Celsius temperatures to kelvins, 20.0°C is 293 K and 40.0°C is 313 K. Therefore, the total translational energy increases by a factor of only 313 K/293 K = 1.07.

21.2 Molar Specific Heat of an Ideal Gas

Consider an ideal gas undergoing several processes such that the change in temperature is $\Delta T = T_f - T_i$ for all processes. The temperature change can be achieved

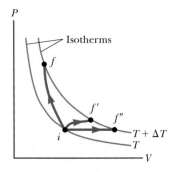

Figure 21.3 An ideal gas is taken from one isotherm at temperature T to another at temperature $T + \Delta T$ along three different paths.

by taking a variety of paths from one isotherm to another as shown in Figure 21.3. Because ΔT is the same for all paths, the change in internal energy ΔE_{int} is the same for all paths. The work W done on the gas (the negative of the area under the curves), however, is different for each path. Therefore, from the first law of thermodynamics, we can argue that the heat $Q = \Delta E_{\text{int}} - W$ associated with a given change in temperature does *not* have a unique value as discussed in Section 20.4.

We can address this difficulty by defining specific heats for two special processes that we have studied: isovolumetric and isobaric. Because the number of moles n is a convenient measure of the amount of gas, we define the **molar specific heats** associated with these processes as follows:

$$Q = nC_V\Delta T \quad \text{(constant volume)} \tag{21.23}$$

$$Q = nC_P\Delta T \quad \text{(constant pressure)} \tag{21.24}$$

where C_V is the **molar specific heat at constant volume** and C_P is the **molar specific heat at constant pressure.** When energy is added to a gas by heat at constant pressure, not only does the internal energy of the gas increase, but (negative) work is done on the gas because of the change in volume required to keep the pressure constant. Therefore, the heat Q in Equation 21.24 must account for both the increase in internal energy and the transfer of energy out of the system by work. For this reason, Q is greater in Equation 21.24 than in Equation 21.23 for given values of n and ΔT. Therefore, C_P is greater than C_V.

In the previous section, we found that the temperature of a gas is a measure of the average translational kinetic energy of the gas molecules. This kinetic energy is associated with the motion of the center of mass of each molecule. It does not include the energy associated with the internal motion of the molecule, namely, vibrations and rotations about the center of mass. That should not be surprising because the simple kinetic theory model assumes a structureless molecule.

So, let's first consider the simplest case of an ideal monatomic gas, that is, a gas containing one atom per molecule such as helium, neon, or argon. When energy is added to a monatomic gas in a container of fixed volume, all the added energy goes into increasing the translational kinetic energy of the atoms. There is no other way to store the energy in a monatomic gas. Therefore, from Equation 21.21, we see that the internal energy E_{int} of N molecules (or n mol) of an ideal monatomic gas is

$$\blacktriangleright \text{Internal energy of an ideal monatomic gas}$$

$$E_{\text{int}} = K_{\text{tot trans}} = \tfrac{3}{2}Nk_BT = \tfrac{3}{2}nRT \tag{21.25}$$

For a monatomic ideal gas, E_{int} is a function of T only and the functional relationship is given by Equation 21.25. In general, the internal energy of any ideal gas is a function of T only and the exact relationship depends on the type of gas.

If energy is transferred by heat to a system at constant volume, no work is done on the system. That is, $W = -\int P\,dV = 0$ for a constant-volume process. Hence, from the first law of thermodynamics,

$$Q = \Delta E_{\text{int}} \tag{21.26}$$

In other words, all the energy transferred by heat goes into increasing the internal energy of the system. A constant-volume process from i to f for an ideal gas is described in Figure 21.4, where ΔT is the temperature difference between the two isotherms. Substituting the expression for Q given by Equation 21.23 into Equation 21.26, we obtain

$$\Delta E_{\text{int}} = nC_V\,\Delta T \tag{21.27}$$

This equation applies to all ideal gases, those gases having more than one atom per molecule as well as monatomic ideal gases.

In the limit of infinitesimal changes, we can use Equation 21.27 to express the molar specific heat at constant volume as

$$C_V = \frac{1}{n}\frac{dE_{\text{int}}}{dT} \tag{21.28}$$

Let's now apply the results of this discussion to a monatomic gas. Substituting the internal energy from Equation 21.25 into Equation 21.28 gives

$$C_V = \tfrac{3}{2}R = 12.5 \text{ J/mol} \cdot \text{K} \tag{21.29}$$

This expression predicts a value of $C_V = \tfrac{3}{2}R$ for *all* monatomic gases. This prediction is in excellent agreement with measured values of molar specific heats for such gases as helium, neon, argon, and xenon over a wide range of temperatures (Table 21.2). Small variations in Table 21.2 from the predicted values are because real gases are not ideal gases. In real gases, weak intermolecular interactions occur, which are not addressed in our ideal gas model.

Now suppose the gas is taken along the constant-pressure path $i \to f'$ shown in Figure 21.4. Along this path, the temperature again increases by ΔT. The energy that must be transferred by heat to the gas in this process is $Q = nC_P \Delta T$. Because the volume changes in this process, the work done on the gas is $W = -P \Delta V$, where P is the constant pressure at which the process occurs. Applying the first law of thermodynamics to this process, we have

$$\Delta E_{int} = Q + W = nC_P \Delta T + (-P \Delta V) \tag{21.30}$$

In this case, the energy added to the gas by heat is channeled as follows. Part of it leaves the system by work (that is, the gas moves a piston through a displacement), and the remainder appears as an increase in the internal energy of the gas. The change in internal energy for the process $i \to f'$, however, is equal to that for the process $i \to f$ because E_{int} depends only on temperature for an ideal gas and ΔT is the same for both processes. In addition, because $PV = nRT$, note that for a constant-pressure process, $P \Delta V = nR \Delta T$. Substituting this value for $P \Delta V$ into Equation 21.30 with $\Delta E_{int} = nC_V \Delta T$ (Eq. 21.27) gives

$$nC_V \Delta T = nC_P \Delta T - nR \Delta T$$

$$C_P - C_V = R \tag{21.31}$$

This expression applies to *any* ideal gas. It predicts that the molar specific heat of an ideal gas at constant pressure is greater than the molar specific heat at constant volume by an amount R, the universal gas constant (which has the value 8.31 J/mol · K). This expression is applicable to real gases as the data in Table 21.2 show.

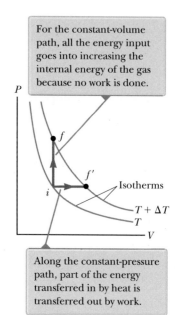

For the constant-volume path, all the energy input goes into increasing the internal energy of the gas because no work is done.

Along the constant-pressure path, part of the energy transferred in by heat is transferred out by work.

Figure 21.4 Energy is transferred by heat to an ideal gas in two ways.

Molar Specific Heats of Various Gases

Gas	C_P	C_V	$C_P - C_V$	$\gamma = C_P/C_V$
Monatomic gases				
He	20.8	12.5	8.33	1.67
Ar	20.8	12.5	8.33	1.67
Ne	20.8	12.7	8.12	1.64
Kr	20.8	12.3	8.49	1.69
Diatomic gases				
H_2	28.8	20.4	8.33	1.41
N_2	29.1	20.8	8.33	1.40
O_2	29.4	21.1	8.33	1.40
CO	29.3	21.0	8.33	1.40
Cl_2	34.7	25.7	8.96	1.35
Polyatomic gases				
CO_2	37.0	28.5	8.50	1.30
SO_2	40.4	31.4	9.00	1.29
H_2O	35.4	27.0	8.37	1.30
CH_4	35.5	27.1	8.41	1.31

The header of the table reads: **Table 21.2** Molar Specific Heats of Various Gases, with the spanning header "Molar Specific Heat (J/mol · K)[a]".

[a] All values except that for water were obtained at 300 K.

Because $C_V = \frac{3}{2}R$ for a monatomic ideal gas, Equation 21.31 predicts a value $C_P = \frac{5}{2}R = 20.8\,\text{J/mol} \cdot \text{K}$ for the molar specific heat of a monatomic gas at constant pressure. The ratio of these molar specific heats is a dimensionless quantity γ (Greek letter gamma):

Ratio of molar specific heats ▶
for a monatomic ideal gas

$$\gamma = \frac{C_P}{C_V} = \frac{5R/2}{3R/2} = \frac{5}{3} = 1.67 \tag{21.32}$$

Theoretical values of C_V, C_P, and γ are in excellent agreement with experimental values obtained for monatomic gases, but they are in serious disagreement with the values for the more complex gases (see Table 21.2). That is not surprising; the value $C_V = \frac{3}{2}R$ was derived for a monatomic ideal gas, and we expect some additional contribution to the molar specific heat from the internal structure of the more complex molecules. In Section 21.3, we describe the effect of molecular structure on the molar specific heat of a gas. The internal energy—and hence the molar specific heat—of a complex gas must include contributions from the rotational and the vibrational motions of the molecule.

In the case of solids and liquids heated at constant pressure, very little work is done during such a process because the thermal expansion is small. Consequently, C_P and C_V are approximately equal for solids and liquids.

Quick Quiz 21.2 (i) How does the internal energy of an ideal gas change as it follows path $i \rightarrow f$ in Figure 21.4? (a) E_{int} increases. (b) E_{int} decreases. (c) E_{int} stays the same. (d) There is not enough information to determine how E_{int} changes. (ii) From the same choices, how does the internal energy of an ideal gas change as it follows path $f \rightarrow f'$ along the isotherm labeled $T + \Delta T$ in Figure 21.4?

Example 21.2   **Heating a Cylinder of Helium**

A cylinder contains 3.00 mol of helium gas at a temperature of 300 K.

(A) If the gas is heated at constant volume, how much energy must be transferred by heat to the gas for its temperature to increase to 500 K?

SOLUTION

Conceptualize Run the process in your mind with the help of the piston–cylinder arrangement in Figure 19.12. Imagine that the piston is clamped in position to maintain the constant volume of the gas.

Categorize We evaluate parameters with equations developed in the preceding discussion, so this example is a substitution problem.

Use Equation 21.23 to find the energy transfer:
$$Q_1 = nC_V\,\Delta T$$

Substitute the given values:
$$Q_1 = (3.00\ \text{mol})(12.5\ \text{J/mol} \cdot \text{K})(500\ \text{K} - 300\ \text{K})$$
$$= \boxed{7.50 \times 10^3\,\text{J}}$$

(B) How much energy must be transferred by heat to the gas at constant pressure to raise the temperature to 500 K?

SOLUTION

Use Equation 21.24 to find the energy transfer:
$$Q_2 = nC_P\,\Delta T$$

Substitute the given values:
$$Q_2 = (3.00\ \text{mol})(20.8\ \text{J/mol} \cdot \text{K})(500\ \text{K} - 300\ \text{K})$$
$$= \boxed{12.5 \times 10^3\,\text{J}}$$

This value is larger than Q_1 because of the transfer of energy out of the gas by work to raise the piston in the constant pressure process.

21.3 The Equipartition of Energy

Predictions based on our model for molar specific heat agree quite well with the behavior of monatomic gases, but not with the behavior of complex gases (see Table 21.2). The value predicted by the model for the quantity $C_P - C_V = R$, however, is the same for all gases. This similarity is not surprising because this difference is the result of the work done on the gas, which is independent of its molecular structure.

To clarify the variations in C_V and C_P in gases more complex than monatomic gases, let's explore further the origin of molar specific heat. So far, we have assumed the sole contribution to the internal energy of a gas is the translational kinetic energy of the molecules. The internal energy of a gas, however, includes contributions from the translational, vibrational, and rotational motion of the molecules. The rotational and vibrational motions of molecules can be activated by collisions and therefore are "coupled" to the translational motion of the molecules. The branch of physics known as *statistical mechanics* has shown that, for a large number of particles obeying the laws of Newtonian mechanics, the available energy is, on average, shared equally by each independent degree of freedom. Recall from Section 21.1 that the equipartition theorem states that, at equilibrium, each degree of freedom contributes $\frac{1}{2}k_B T$ of energy per molecule.

Let's consider a diatomic gas whose molecules have the shape of a dumbbell (Fig. 21.5). In this model, the center of mass of the molecule can translate in the x, y, and z directions (Fig. 21.5a). In addition, the molecule can rotate about three mutually perpendicular axes (Fig. 21.5b). The rotation about the y axis can be neglected because the molecule's moment of inertia I_y and its rotational energy $\frac{1}{2}I_y\omega^2$ about this axis are negligible compared with those associated with the x and z axes. (If the two atoms are modeled as particles, then I_y is identically zero.) Therefore, there are five degrees of freedom for translation and rotation: three associated with the translational motion and two associated with the rotational motion. Because each degree of freedom contributes, on average, $\frac{1}{2}k_B T$ of energy per molecule, the internal energy for a system of N molecules, ignoring vibration for now, is

$$E_{\text{int}} = 3N(\tfrac{1}{2}k_B T) + 2N(\tfrac{1}{2}k_B T) = \tfrac{5}{2}Nk_B T = \tfrac{5}{2}nRT$$

We can use this result and Equation 21.28 to find the molar specific heat at constant volume:

$$C_V = \frac{1}{n}\frac{dE_{\text{int}}}{dT} = \frac{1}{n}\frac{d}{dT}(\tfrac{5}{2}nRT) = \tfrac{5}{2}R = 20.8 \text{ J/mol} \cdot \text{K} \qquad \textbf{(21.33)}$$

From Equations 21.31 and 21.32, we find that

$$C_P = C_V + R = \tfrac{7}{2}R = 29.1 \text{ J/mol} \cdot \text{K}$$

$$\gamma = \frac{C_P}{C_V} = \frac{\tfrac{7}{2}R}{\tfrac{5}{2}R} = \frac{7}{5} = 1.40$$

These results agree quite well with most of the data for diatomic molecules given in Table 21.2. That is rather surprising because we have not yet accounted for the possible vibrations of the molecule.

In the model for vibration, the two atoms are joined by an imaginary spring (see Fig. 21.5c). The vibrational motion adds two more degrees of freedom, which correspond to the kinetic energy and the potential energy associated with vibrations along the length of the molecule. Hence, a model that includes all three types of motion predicts a total internal energy of

$$E_{\text{int}} = 3N(\tfrac{1}{2}k_B T) + 2N(\tfrac{1}{2}k_B T) + 2N(\tfrac{1}{2}k_B T) = \tfrac{7}{2}Nk_B T = \tfrac{7}{2}nRT$$

and a molar specific heat at constant volume of

$$C_V = \frac{1}{n}\frac{dE_{\text{int}}}{dT} = \frac{1}{n}\frac{d}{dT}(\tfrac{7}{2}nRT) = \tfrac{7}{2}R = 29.1 \text{ J/mol} \cdot \text{K} \qquad \textbf{(21.34)}$$

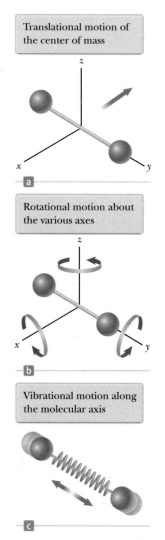

Translational motion of the center of mass

a

Rotational motion about the various axes

b

Vibrational motion along the molecular axis

c

Figure 21.5 Possible motions of a diatomic molecule.

Figure 21.6 The molar specific heat of hydrogen as a function of temperature.

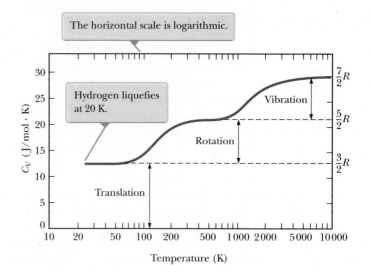

This value is inconsistent with experimental data for molecules such as H_2 and N_2 (see Table 21.2) and suggests a breakdown of our model based on classical physics.

It might seem that our model is a failure for predicting molar specific heats for diatomic gases. We can claim some success for our model, however, if measurements of molar specific heat are made over a wide temperature range rather than at the single temperature that gives us the values in Table 21.2. Figure 21.6 shows the molar specific heat of hydrogen as a function of temperature. The remarkable feature about the three plateaus in the graph's curve is that they are at the values of the molar specific heat predicted by Equations 21.29, 21.33, and 21.34! For low temperatures, the diatomic hydrogen gas behaves like a monatomic gas. As the temperature rises to room temperature, its molar specific heat rises to a value for a diatomic gas, consistent with the inclusion of rotation but not vibration. For high temperatures, the molar specific heat is consistent with a model including all types of motion.

Before addressing the reason for this mysterious behavior, let's make some brief remarks about polyatomic gases. For molecules with more than two atoms, three axes of rotation are available. The vibrations are more complex than for diatomic molecules. Therefore, the number of degrees of freedom is even larger. The result is an even higher predicted molar specific heat, which is in qualitative agreement with experiment. The molar specific heats for the polyatomic gases in Table 21.2 are higher than those for diatomic gases. The more degrees of freedom available to a molecule, the more "ways" there are to store energy, resulting in a higher molar specific heat.

A Hint of Energy Quantization

Our model for molar specific heats has been based so far on purely classical notions. It predicts a value of the specific heat for a diatomic gas that, according to Figure 21.6, only agrees with experimental measurements made at high temperatures. To explain why this value is only true at high temperatures and why the plateaus in Figure 21.6 exist, we must go beyond classical physics and introduce some quantum physics into the model. In Chapter 18, we discussed quantization of frequency for vibrating strings and air columns; only certain frequencies of standing waves can exist. That is a natural result whenever waves are subject to boundary conditions.

Quantum physics (Chapters 40 through 43) shows that atoms and molecules can be described by the waves under boundary conditions analysis model. Consequently, these waves have quantized frequencies. Furthermore, in quantum physics, the energy of a system is proportional to the frequency of the wave representing the system. Hence, **the energies of atoms and molecules are quantized.**

For a molecule, quantum physics tells us that the rotational and vibrational energies are quantized. Figure 21.7 shows an **energy-level diagram** for the rotational and vibrational quantum states of a diatomic molecule. The lowest allowed state is called the **ground state.** The black lines show the energies allowed for the molecule. Notice that allowed vibrational states are separated by larger energy gaps than are rotational states.

At low temperatures, the energy a molecule gains in collisions with its neighbors is generally not large enough to raise it to the first excited state of either rotation or vibration. Therefore, even though rotation and vibration are allowed according to classical physics, they do not occur in reality at low temperatures. All molecules are in the ground state for rotation and vibration. The only contribution to the molecules' average energy is from translation, and the specific heat is that predicted by Equation 21.29.

As the temperature is raised, the average energy of the molecules increases. In some collisions, a molecule may have enough energy transferred to it from another molecule to excite the first rotational state. As the temperature is raised further, more molecules can be excited to this state. The result is that rotation begins to contribute to the internal energy, and the molar specific heat rises. At about room temperature in Figure 21.6, the second plateau has been reached and rotation contributes fully to the molar specific heat. The molar specific heat is now equal to the value predicted by Equation 21.33.

There is no contribution at room temperature from vibration because the molecules are still in the ground vibrational state. The temperature must be raised even further to excite the first vibrational state, which happens in Figure 21.6 between 1 000 K and 10 000 K. At 10 000 K on the right side of the figure, vibration is contributing fully to the internal energy and the molar specific heat has the value predicted by Equation 21.34.

The predictions of this model are supportive of the theorem of equipartition of energy. In addition, the inclusion in the model of energy quantization from quantum physics allows a full understanding of Figure 21.6.

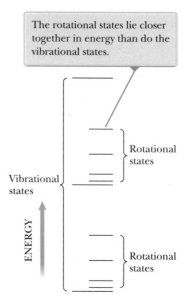

The rotational states lie closer together in energy than do the vibrational states.

Figure 21.7 An energy-level diagram for vibrational and rotational states of a diatomic molecule.

Quick Quiz 21.3 The molar specific heat of a diatomic gas is measured at constant volume and found to be 29.1 J/mol · K. What are the types of energy that are contributing to the molar specific heat? **(a)** translation only **(b)** translation and rotation only **(c)** translation and vibration only **(d)** translation, rotation, and vibration

Quick Quiz 21.4 The molar specific heat of a gas is measured at constant volume and found to be $11R/2$. Is the gas most likely to be **(a)** monatomic, **(b)** diatomic, or **(c)** polyatomic?

21.4 Adiabatic Processes for an Ideal Gas

As noted in Section 20.6, an **adiabatic process** is one in which no energy is transferred by heat between a system and its surroundings. For example, if a gas is compressed (or expanded) rapidly, very little energy is transferred out of (or into) the system by heat, so the process is nearly adiabatic. Such processes occur in the cycle of a gasoline engine, which is discussed in detail in Chapter 22. Another example of an adiabatic process is the slow expansion of a gas that is thermally insulated from its surroundings. All three variables in the ideal gas law—P, V, and T—change during an adiabatic process.

Let's imagine an adiabatic gas process involving an infinitesimal change in volume dV and an accompanying infinitesimal change in temperature dT. The work done on the gas is $-P\,dV$. Because the internal energy of an ideal gas depends only on temperature, the change in the internal energy in an adiabatic process is the same as that for an isovolumetric process between the same temperatures, $dE_{int} = nC_V\,dT$ (Eq. 21.27). Hence, the first law of thermodynamics, $\Delta E_{int} = Q + W$, with $Q = 0$, becomes the infinitesimal form

$$dE_{int} = nC_V\,dT = -P\,dV \qquad \textbf{(21.35)}$$

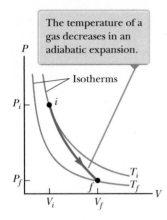

Figure 21.8 The *PV* diagram for an adiabatic expansion of an ideal gas.

Taking the total differential of the equation of state of an ideal gas, $PV = nRT$, gives

$$P\,dV + V\,dP = nR\,dT \qquad (21.36)$$

Eliminating dT from Equations 21.35 and 21.36, we find that

$$P\,dV + V\,dP = -\frac{R}{C_V}P\,dV$$

Substituting $R = C_P - C_V$ and dividing by PV gives

$$\frac{dV}{V} + \frac{dP}{P} = -\left(\frac{C_P - C_V}{C_V}\right)\frac{dV}{V} = (1-\gamma)\frac{dV}{V}$$

$$\frac{dP}{P} + \gamma\frac{dV}{V} = 0$$

Integrating this expression, we have

$$\ln P + \gamma \ln V = \text{constant}$$

which is equivalent to

▶ **Relationship between *P* and *V* for an adiabatic process involving an ideal gas**

$$PV^{\gamma} = \text{constant} \qquad (21.37)$$

The *PV* diagram for an adiabatic expansion is shown in Figure 21.8. Because $\gamma > 1$, the *PV* curve is steeper than it would be for an isothermal expansion, for which $PV = \text{constant}$. By the definition of an adiabatic process, no energy is transferred by heat into or out of the system. Hence, from the first law, we see that ΔE_{int} is negative (work is done *by* the gas, so its internal energy decreases) and so ΔT also is negative. Therefore, the temperature of the gas decreases ($T_f < T_i$) during an adiabatic expansion.[2] Conversely, the temperature increases if the gas is compressed adiabatically. Applying Equation 21.37 to the initial and final states, we see that

$$P_i V_i^{\gamma} = P_f V_f^{\gamma} \qquad (21.38)$$

Using the ideal gas law, we can express Equation 21.37 as

▶ **Relationship between *T* and *V* for an adiabatic process involving an ideal gas**

$$TV^{\gamma-1} = \text{constant} \qquad (21.39)$$

Example 21.3 A Diesel Engine Cylinder

Air at 20.0°C in the cylinder of a diesel engine is compressed from an initial pressure of 1.00 atm and volume of 800.0 cm³ to a volume of 60.0 cm³. Assume air behaves as an ideal gas with $\gamma = 1.40$ and the compression is adiabatic. Find the final pressure and temperature of the air.

SOLUTION

Conceptualize Imagine what happens if a gas is compressed into a smaller volume. Our discussion above and Figure 21.8 tell us that the pressure and temperature both increase.

Categorize We categorize this example as a problem involving an adiabatic process.

··

Analyze Use Equation 21.38 to find the final pressure:

$$P_f = P_i\left(\frac{V_i}{V_f}\right)^{\gamma} = (1.00\text{ atm})\left(\frac{800.0\text{ cm}^3}{60.0\text{ cm}^3}\right)^{1.40}$$

$$= \boxed{37.6\text{ atm}}$$

---

[2]In the adiabatic free expansion discussed in Section 20.6, the temperature remains constant. In this unique process, no work is done because the gas expands into a vacuum. In general, the temperature decreases in an adiabatic expansion in which work is done.

▶ **21.3** continued

Use the ideal gas law to find the final temperature:

$$\frac{P_i V_i}{T_i} = \frac{P_f V_f}{T_f}$$

$$T_f = \frac{P_f V_f}{P_i V_i} T_i = \frac{(37.6 \text{ atm})(60.0 \text{ cm}^3)}{(1.00 \text{ atm})(800.0 \text{ cm}^3)}(293 \text{ K})$$

$$= 826 \text{ K} = \boxed{553°\text{C}}$$

Finalize The temperature of the gas increases by a factor of 826 K/293 K = 2.82. The high compression in a diesel engine raises the temperature of the gas enough to cause the combustion of fuel without the use of spark plugs.

21.5 Distribution of Molecular Speeds

Thus far, we have considered only average values of the energies of all the molecules in a gas and have not addressed the distribution of energies among individual molecules. The motion of the molecules is extremely chaotic. Any individual molecule collides with others at an enormous rate, typically a billion times per second. Each collision results in a change in the speed and direction of motion of each of the participant molecules. Equation 21.22 shows that rms molecular speeds increase with increasing temperature. At a given time, what is the relative number of molecules that possess some characteristic such as energy within a certain range?

We shall address this question by considering the **number density** $n_V(E)$. This quantity, called a *distribution function,* is defined so that $n_V(E)\, dE$ is the number of molecules per unit volume with energy between E and $E + dE$. (The ratio of the number of molecules that have the desired characteristic to the total number of molecules is the probability that a particular molecule has that characteristic.) In general, the number density is found from statistical mechanics to be

$$n_V(E) = n_0 e^{-E/k_B T} \qquad (21.40)$$

◀ Boltzmann distribution law

where n_0 is defined such that $n_0\, dE$ is the number of molecules per unit volume having energy between $E = 0$ and $E = dE$. This equation, known as the **Boltzmann distribution law,** is important in describing the statistical mechanics of a large number of molecules. It states that the probability of finding the molecules in a particular energy state varies exponentially as the negative of the energy divided by $k_B T$. All the molecules would fall into the lowest energy level if the thermal agitation at a temperature T did not excite the molecules to higher energy levels.

Pitfall Prevention 21.2
The Distribution Function The distribution function $n_V(E)$ is defined in terms of the number of molecules with energy in the range E to $E + dE$ rather than in terms of the number of molecules with energy E. Because the number of molecules is finite and the number of possible values of the energy is infinite, the number of molecules with an *exact* energy E may be zero.

Example 21.4 **Thermal Excitation of Atomic Energy Levels**

As discussed in Section 21.4, atoms can occupy only certain discrete energy levels. Consider a gas at a temperature of 2 500 K whose atoms can occupy only two energy levels separated by 1.50 eV, where 1 eV (electron volt) is an energy unit equal to 1.60×10^{-19} J (Fig. 21.9). Determine the ratio of the number of atoms in the higher energy level to the number in the lower energy level.

SOLUTION

Conceptualize In your mental representation of this example, remember that only two possible states are allowed for the system of the atom. Figure 21.9 helps you visualize the two states on an energy-level diagram. In this case, the atom has two possible energies, E_1 and E_2, where $E_1 < E_2$.

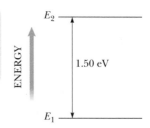

Figure 21.9 (Example 21.4) Energy-level diagram for a gas whose atoms can occupy two energy states.

continued

▶ **21.4** continued

Categorize We categorize this example as one in which we focus on particles in a two-state quantized system. We will apply the Boltzmann distribution law to this system.

Analyze Set up the ratio of the number of atoms in the higher energy level to the number in the lower energy level and use Equation 21.40 to express each number:

$$(1) \quad \frac{n_V(E_2)}{n_V(E_1)} = \frac{n_0 e^{-E_2/k_B T}}{n_0 e^{-E_1/k_B T}} = e^{-(E_2-E_1)/k_B T}$$

Evaluate $k_B T$ in the exponent:

$$k_B T = (1.38 \times 10^{-23} \text{ J/K})(2\,500 \text{ K})\left(\frac{1 \text{ eV}}{1.60 \times 10^{-19} \text{ J}}\right) = 0.216 \text{ eV}$$

Substitute this value into Equation (1):

$$\frac{n_V(E_2)}{n_V(E_1)} = e^{-1.50 \text{ eV}/0.216 \text{ eV}} = e^{-6.96} = \boxed{9.52 \times 10^{-4}}$$

Finalize This result indicates that at $T = 2\,500$ K, only a small fraction of the atoms are in the higher energy level. In fact, for every atom in the higher energy level, there are about 1 000 atoms in the lower level. The number of atoms in the higher level increases at even higher temperatures, but the distribution law specifies that at equilibrium there are always more atoms in the lower level than in the higher level.

WHAT IF? What if the energy levels in Figure 21.9 were closer together in energy? Would that increase or decrease the fraction of the atoms in the upper energy level?

Answer If the excited level is lower in energy than that in Figure 21.9, it would be easier for thermal agitation to excite atoms to this level and the fraction of atoms in this energy level would be larger, which we can see mathematically by expressing Equation (1) as

$$r_2 = e^{-(E_2-E_1)/k_B T}$$

where r_2 is the ratio of atoms having energy E_2 to those with energy E_1. Differentiating with respect to E_2, we find

$$\frac{dr_2}{dE_2} = \frac{d}{dE_2}\left[e^{-(E_2-E_1)/k_B T}\right] = -\frac{1}{k_B T} e^{-(E_2-E_1)/k_B T} < 0$$

Because the derivative has a negative value, as E_2 decreases, r_2 increases.

Ludwig Boltzmann
Austrian physicist (1844–1906)
Boltzmann made many important contributions to the development of the kinetic theory of gases, electromagnetism, and thermodynamics. His pioneering work in the field of kinetic theory led to the branch of physics known as statistical mechanics.

© INTERFOTO/Alamy

Now that we have discussed the distribution of energies among molecules in a gas, let's think about the distribution of molecular speeds. In 1860, James Clerk Maxwell (1831–1879) derived an expression that describes the distribution of molecular speeds in a very definite manner. His work and subsequent developments by other scientists were highly controversial because direct detection of molecules could not be achieved experimentally at that time. About 60 years later, however, experiments were devised that confirmed Maxwell's predictions.

Let's consider a container of gas whose molecules have some distribution of speeds. Suppose we want to determine how many gas molecules have a speed in the range from, for example, 400 to 401 m/s. Intuitively, we expect the speed distribution to depend on temperature. Furthermore, we expect the distribution to peak in the vicinity of v_{rms}. That is, few molecules are expected to have speeds much less than or much greater than v_{rms} because these extreme speeds result only from an unlikely chain of collisions.

The observed speed distribution of gas molecules in thermal equilibrium is shown in Figure 21.10. The quantity N_v, called the **Maxwell–Boltzmann speed distribution function,** is defined as follows. If N is the total number of molecules, the number of molecules with speeds between v and $v + dv$ is $dN = N_v \, dv$. This number is also equal to the area of the shaded rectangle in Figure 21.10. Furthermore, the fraction of molecules with speeds between v and $v + dv$ is $(N_v \, dv)/N$. This fraction is also equal to the probability that a molecule has a speed in the range v to $v + dv$.

The fundamental expression that describes the distribution of speeds of N gas molecules is

$$N_v = 4\pi N \left(\frac{m_0}{2\pi k_B T}\right)^{3/2} v^2 e^{-m_0 v^2/2 k_B T} \tag{21.41}$$

where m_0 is the mass of a gas molecule, k_B is Boltzmann's constant, and T is the absolute temperature.[3] Observe the appearance of the Boltzmann factor $e^{-E/k_B T}$ with $E = \frac{1}{2}m_0 v^2$.

As indicated in Figure 21.10, the average speed is somewhat lower than the rms speed. The *most probable speed* v_{mp} is the speed at which the distribution curve reaches a peak. Using Equation 21.41, we find that

$$v_{rms} = \sqrt{\overline{v^2}} = \sqrt{\frac{3k_B T}{m_0}} = 1.73\sqrt{\frac{k_B T}{m_0}} \tag{21.42}$$

$$v_{avg} = \sqrt{\frac{8k_B T}{\pi m_0}} = 1.60\sqrt{\frac{k_B T}{m_0}} \tag{21.43}$$

$$v_{mp} = \sqrt{\frac{2k_B T}{m_0}} = 1.41\sqrt{\frac{k_B T}{m_0}} \tag{21.44}$$

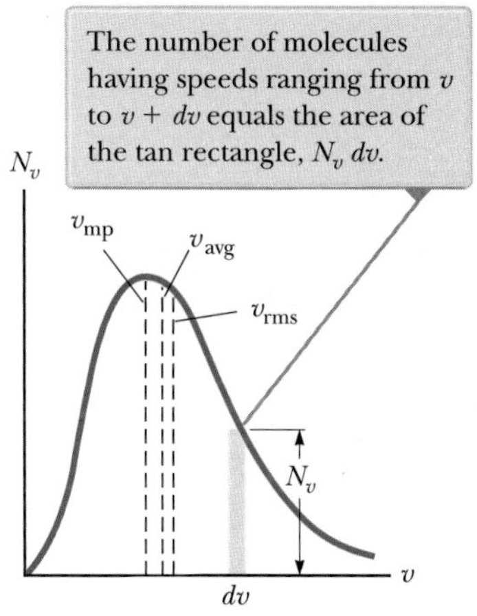

Equation 21.42 has previously appeared as Equation 21.22. The details of the derivations of these equations from Equation 21.41 are left for the problems in Enhanced WebAssign (see Problems 42 and 69). From these equations, we see that

$$v_{rms} > v_{avg} > v_{mp}$$

Figure 21.11 represents speed distribution curves for nitrogen, N_2. The curves were obtained by using Equation 21.41 to evaluate the distribution function at various speeds and at two temperatures. Notice that the peak in each curve shifts to the right as T increases, indicating that the average speed increases with increasing temperature, as expected. Because the lowest speed possible is zero and the upper classical limit of the speed is infinity, the curves are asymmetrical. (In Chapter 39, we show that the actual upper limit is the speed of light.)

Equation 21.41 shows that the distribution of molecular speeds in a gas depends both on mass and on temperature. At a given temperature, the fraction of molecules with speeds exceeding a fixed value increases as the mass decreases. Hence,

The number of molecules having speeds ranging from v to $v + dv$ equals the area of the tan rectangle, $N_v\, dv$.

Figure 21.10 The speed distribution of gas molecules at some temperature. The function N_v approaches zero as v approaches infinity.

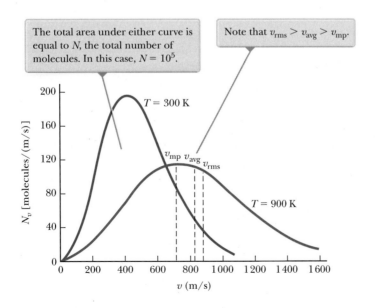

The total area under either curve is equal to N, the total number of molecules. In this case, $N = 10^5$.

Note that $v_{rms} > v_{avg} > v_{mp}$.

Figure 21.11 The speed distribution function for 10^5 nitrogen molecules at 300 K and 900 K.

---

[3] For the derivation of this expression, see an advanced textbook on thermodynamics.

lighter molecules such as H_2 and He escape into space more readily from the Earth's atmosphere than do heavier molecules such as N_2 and O_2. (See the discussion of escape speed in Chapter 13. Gas molecules escape even more readily from the Moon's surface than from the Earth's because the escape speed on the Moon is lower than that on the Earth.)

The speed distribution curves for molecules in a liquid are similar to those shown in Figure 21.11. We can understand the phenomenon of evaporation of a liquid from this distribution in speeds, given that some molecules in the liquid are more energetic than others. Some of the faster-moving molecules in the liquid penetrate the surface and even leave the liquid at temperatures well below the boiling point. The molecules that escape the liquid by evaporation are those that have sufficient energy to overcome the attractive forces of the molecules in the liquid phase. Consequently, the molecules left behind in the liquid phase have a lower average kinetic energy; as a result, the temperature of the liquid decreases. Hence, evaporation is a cooling process. For example, an alcohol-soaked cloth can be placed on a feverish head to cool and comfort a patient.

Example 21.5   A System of Nine Particles

Nine particles have speeds of 5.00, 8.00, 12.0, 12.0, 12.0, 14.0, 14.0, 17.0, and 20.0 m/s.

(A) Find the particles' average speed.

SOLUTION

Conceptualize Imagine a small number of particles moving in random directions with the few speeds listed. This situation is not representative of the large number of molecules in a gas, so we should not expect the results to be consistent with those from statistical mechanics.

Categorize Because we are dealing with a small number of particles, we can calculate the average speed directly.

Analyze Find the average speed of the particles by dividing the sum of the speeds by the total number of particles:

$$v_{avg} = \frac{(5.00 + 8.00 + 12.0 + 12.0 + 12.0 + 14.0 + 14.0 + 17.0 + 20.0) \text{ m/s}}{9}$$

$$= 12.7 \text{ m/s}$$

(B) What is the rms speed of the particles?

SOLUTION

Find the average speed squared of the particles by dividing the sum of the speeds squared by the total number of particles:

$$\overline{v^2} = \frac{(5.00^2 + 8.00^2 + 12.0^2 + 12.0^2 + 12.0^2 + 14.0^2 + 14.0^2 + 17.0^2 + 20.0^2) \text{ m}^2/\text{s}^2}{9}$$

$$= 178 \text{ m}^2/\text{s}^2$$

Find the rms speed of the particles by taking the square root:

$$v_{rms} = \sqrt{\overline{v^2}} = \sqrt{178 \text{ m}^2/\text{s}^2} = 13.3 \text{ m/s}$$

(C) What is the most probable speed of the particles?

SOLUTION

Three of the particles have a speed of 12.0 m/s, two have a speed of 14.0 m/s, and the remaining four have different speeds. Hence, the most probable speed v_{mp} is 12.0 m/s.

Finalize Compare this example, in which the number of particles is small and we know the individual particle speeds, with the next example.

| Example 21.6 | Molecular Speeds in a Hydrogen Gas |

A 0.500-mol sample of hydrogen gas is at 300 K.

(A) Find the average speed, the rms speed, and the most probable speed of the hydrogen molecules.

SOLUTION

Conceptualize Imagine a huge number of particles in a real gas, all moving in random directions with different speeds.

Categorize We cannot calculate the averages as was done in Example 21.5 because the individual speeds of the particles are not known. We are dealing with a very large number of particles, however, so we can use the Maxwell-Boltzmann speed distribution function.

....

Analyze Use Equation 21.43 to find the average speed:

$$v_{avg} = 1.60\sqrt{\frac{k_B T}{m_0}} = 1.60\sqrt{\frac{(1.38 \times 10^{-23}\,\text{J/K})(300\,\text{K})}{2(1.67 \times 10^{-27}\,\text{kg})}}$$

$$= \boxed{1.78 \times 10^3\,\text{m/s}}$$

Use Equation 21.42 to find the rms speed:

$$v_{rms} = 1.73\sqrt{\frac{k_B T}{m_0}} = 1.73\sqrt{\frac{(1.38 \times 10^{-23}\,\text{J/K})(300\,\text{K})}{2(1.67 \times 10^{-27}\,\text{kg})}}$$

$$= \boxed{1.93 \times 10^3\,\text{m/s}}$$

Use Equation 21.44 to find the most probable speed:

$$v_{mp} = 1.41\sqrt{\frac{k_B T}{m_0}} = 1.41\sqrt{\frac{(1.38 \times 10^{-23}\,\text{J/K})(300\,\text{K})}{2(1.67 \times 10^{-27}\,\text{kg})}}$$

$$= \boxed{1.57 \times 10^3\,\text{m/s}}$$

(B) Find the number of molecules with speeds between 400 m/s and 401 m/s.

SOLUTION

Use Equation 21.41 to evaluate the number of molecules in a narrow speed range between v and $v + dv$:

$$(1)\quad N_v\,dv = 4\pi N\left(\frac{m_0}{2\pi k_B T}\right)^{3/2} v^2 e^{-m_0 v^2/2k_B T}\,dv$$

Evaluate the constant in front of v^2:

$$4\pi N\left(\frac{m_0}{2\pi k_B T}\right)^{3/2} = 4\pi n N_A\left(\frac{m_0}{2\pi k_B T}\right)^{3/2}$$

$$= 4\pi(0.500\,\text{mol})(6.02 \times 10^{23}\,\text{mol}^{-1})\left[\frac{2(1.67 \times 10^{-27}\,\text{kg})}{2\pi(1.38 \times 10^{-23}\,\text{J/K})(300\,\text{K})}\right]^{3/2}$$

$$= 1.74 \times 10^{14}\,\text{s}^3/\text{m}^3$$

Evaluate the exponent of e that appears in Equation (1):

$$-\frac{m_0 v^2}{2k_B T} = -\frac{2(1.67 \times 10^{-27}\,\text{kg})(400\,\text{m/s})^2}{2(1.38 \times 10^{-23}\,\text{J/K})(300\,\text{K})} = -0.064\,5$$

Evaluate $N_v\,dv$ using these values in Equation (1):

$$N_v\,dv = (1.74 \times 10^{14}\,\text{s}^3/\text{m}^3)(400\,\text{m/s})^2 e^{-0.064\,5}(1\,\text{m/s})$$

$$= \boxed{2.61 \times 10^{19}\,\text{molecules}}$$

Finalize In this evaluation, we could calculate the result without integration because $dv = 1$ m/s is much smaller than $v = 400$ m/s. Had we sought the number of particles between, say, 400 m/s and 500 m/s, we would need to integrate Equation (1) between these speed limits.

Summary

Concepts and Principles

▌ The pressure of N molecules of an ideal gas contained in a volume V is

$$P = \frac{2}{3}\left(\frac{N}{V}\right)\left(\frac{1}{2}m_0\overline{v^2}\right) \qquad \text{(21.15)}$$

The average translational kinetic energy per molecule of a gas, $\frac{1}{2}m_0\overline{v^2}$, is related to the temperature T of the gas through the expression

$$\frac{1}{2}m_0\overline{v^2} = \frac{3}{2}k_B T \qquad \text{(21.19)}$$

where k_B is Boltzmann's constant. Each translational degree of freedom (x, y, or z) has $\frac{1}{2}k_B T$ of energy associated with it.

▌ The internal energy of N molecules (or n mol) of an ideal monatomic gas is

$$E_{int} = \frac{3}{2}Nk_B T = \frac{3}{2}nRT \qquad \text{(21.25)}$$

The change in internal energy for n mol of any ideal gas that undergoes a change in temperature ΔT is

$$\Delta E_{int} = nC_V\,\Delta T \qquad \text{(21.27)}$$

where C_V is the **molar specific heat at constant volume.**

▌ The molar specific heat of an ideal monatomic gas at constant volume is $C_V = \frac{3}{2}R$; the molar specific heat at constant pressure is $C_P = \frac{5}{2}R$. The ratio of specific heats is given by $\gamma = C_P/C_V = \frac{5}{3}$.

▌ If an ideal gas undergoes an adiabatic expansion or compression, the first law of thermodynamics, together with the equation of state, shows that

$$PV^\gamma = \text{constant} \qquad \text{(21.37)}$$

▌ The **Boltzmann distribution law** describes the distribution of particles among available energy states. The relative number of particles having energy between E and $E + dE$ is $n_V(E)\, dE$, where

$$n_V(E) = n_0 e^{-E/k_B T} \qquad \text{(21.40)}$$

The **Maxwell–Boltzmann speed distribution function** describes the distribution of speeds of molecules in a gas:

$$N_v = 4\pi N\left(\frac{m_0}{2\pi k_B T}\right)^{3/2} v^2 e^{-m_0 v^2/2k_B T} \qquad \text{(21.41)}$$

▌ Equation 21.41 enables us to calculate the **root-mean-square speed**, the **average speed,** and the **most probable speed** of molecules in a gas:

$$v_{rms} = \sqrt{\overline{v^2}} = \sqrt{\frac{3k_B T}{m_0}} = 1.73\sqrt{\frac{k_B T}{m_0}} \qquad \text{(21.42)}$$

$$v_{avg} = \sqrt{\frac{8k_B T}{\pi m_0}} = 1.60\sqrt{\frac{k_B T}{m_0}} \qquad \text{(21.43)}$$

$$v_{mp} = \sqrt{\frac{2k_B T}{m_0}} = 1.41\sqrt{\frac{k_B T}{m_0}} \qquad \text{(21.44)}$$

Objective Questions   1. denotes answer available in *Student Solutions Manual/Study Guide*

1. Cylinder A contains oxygen (O_2) gas, and cylinder B contains nitrogen (N_2) gas. If the molecules in the two cylinders have the same rms speeds, which of the following statements is *false*? (a) The two gases have different temperatures. (b) The temperature of cylinder B is less than the temperature of cylinder A. (c) The temperature of cylinder B is greater than the temperature of cylinder A. (d) The average kinetic energy of the nitrogen molecules is less than the average kinetic energy of the oxygen molecules.

2. An ideal gas is maintained at constant pressure. If the temperature of the gas is increased from 200 K to 600 K, what happens to the rms speed of the molecules? (a) It increases by a factor of 3. (b) It remains the same. (c) It is one-third the original speed. (d) It is $\sqrt{3}$ times the original speed. (e) It increases by a factor of 6.

3. Two samples of the same ideal gas have the same pressure and density. Sample B has twice the volume of sample A. What is the rms speed of the molecules in sample B? (a) twice that in sample A (b) equal to that in sample A (c) half that in sample A (d) impossible to determine

4. A helium-filled latex balloon initially at room temperature is placed in a freezer. The latex remains flexible.

(i) Does the balloon's volume (a) increase, (b) decrease, or (c) remain the same? **(ii)** Does the pressure of the helium gas (a) increase significantly, (b) decrease significantly, or (c) remain approximately the same?

5. A gas is at 200 K. If we wish to double the rms speed of the molecules of the gas, to what value must we raise its temperature? (a) 283 K (b) 400 K (c) 566 K (d) 800 K (e) 1130 K

6. Rank the following from largest to smallest, noting any cases of equality. (a) the average speed of molecules in a particular sample of ideal gas (b) the most probable speed (c) the root-mean-square speed (d) the average vector velocity of the molecules

7. A sample of gas with a thermometer immersed in the gas is held over a hot plate. A student is asked to give a step-by-step account of what makes our observation of the temperature of the gas increase. His response includes the following steps. (a) The molecules speed up. (b) Then the molecules collide with one another more often. (c) Internal friction makes the collisions inelastic. (d) Heat is produced in the collisions. (e) The molecules of the gas transfer more energy to the thermometer when they strike it, so we observe that the temperature has gone up. (f) The same process can take place without the use of a hot plate if

you quickly push in the piston in an insulated cylinder containing the gas. **(i)** Which of the parts (a) through (f) of this account are correct statements necessary for a clear and complete explanation? **(ii)** Which are correct statements that are not necessary to account for the higher thermometer reading? **(iii)** Which are incorrect statements?

8. An ideal gas is contained in a vessel at 300 K. The temperature of the gas is then increased to 900 K. **(i)** By what factor does the average kinetic energy of the molecules change, (a) a factor of 9, (b) a factor of 3, (c) a factor of $\sqrt{3}$, (d) a factor of 1, or (e) a factor of $\frac{1}{3}$? Using the same choices as in part (i), by what factor does each of the following change: **(ii)** the rms molecular speed of the molecules, **(iii)** the average momentum change that one molecule undergoes in a collision with one particular wall, **(iv)** the rate of collisions of molecules with walls, and **(v)** the pressure of the gas.

9. Which of the assumptions below is *not* made in the kinetic theory of gases? (a) The number of molecules is very large. (b) The molecules obey Newton's laws of motion. (c) The forces between molecules are long range. (d) The gas is a pure substance. (e) The average separation between molecules is large compared to their dimensions.

Conceptual Questions

1. denotes answer available in *Student Solutions Manual/Study Guide*

1. Hot air rises, so why does it generally become cooler as you climb a mountain? *Note:* Air has low thermal conductivity.

2. Why does a diatomic gas have a greater energy content per mole than a monatomic gas at the same temperature?

3. When alcohol is rubbed on your body, it lowers your skin temperature. Explain this effect.

4. What happens to a helium-filled latex balloon released into the air? Does it expand or contract? Does it stop rising at some height?

5. Which is denser, dry air or air saturated with water vapor? Explain.

6. One container is filled with helium gas and another with argon gas. Both containers are at the same temperature. Which molecules have the higher rms speed? Explain.

7. Dalton's law of partial pressures states that the total pressure of a mixture of gases is equal to the sum of the pressures that each gas in the mixture would exert if it were alone in the container. Give a convincing argument for this law based on the kinetic theory of gases.

Problems available in Access end-of-chapter problems online at www.webassign.net

Section 21.1 Molecular Model of an Ideal Gas
Problems 1–13

Section 21.2 Molar Specific Heat of an Ideal Gas
Problems 14–21

Section 21.3 The Equipartition of Energy
Problems 22–25

Section 21.4 Adiabatic Processes for an Ideal Gas
Problems 26–34

Section 21.5 Distribution of Molecular Speeds
Problems 35–43

Additional Problems
Problems 44–73

Challenge Problems
Problems 74–75

Solutions to the following Problems are available in the *Student Solutions Manual/ Study Guide:*
21.1, 21.2, 21.4, 21.5, 21.9, 21.17, 21.21, 21.25, 21.26, 21.29, 21.36, 21.42, 21.52, 21.58, and 21.67

List of Enhanced Problems

Problem Number	Targeted Feedback in Enhanced WebAssign	Analysis Model Tutorial in Enhanced WebAssign	Master It in Enhanced WebAssign	Watch It in Enhanced WebAssign
21.1	✓		✓	
21.2			✓	
21.3	✓			✓
21.5	✓		✓	
21.7	✓			✓
21.13	✓		✓	✓
21.14	✓			✓
21.17	✓		✓	
21.25			✓	
21.26	✓		✓	
21.27	✓			✓
21.28	✓			✓
21.29			✓	
21.36			✓	
21.37	✓			✓
21.48		✓		
21.49		✓		
21.63		✓		
21.75		✓		

Heat Engines, Entropy, and the Second Law of Thermodynamics

The first law of thermodynamics, which we studied in Chapter 20, is a statement of conservation of energy and is a special-case reduction of Equation 8.2. This law states that a change in internal energy in a system can occur as a result of energy transfer by heat, by work, or by both. Although the first law of thermodynamics is very important, it makes no distinction between processes that occur spontaneously and those that do not. Only certain types of energy transformation and energy transfer processes actually take place in nature, however. The *second law of thermodynamics,* the major topic in this chapter, establishes which processes do and do not occur. The following are examples

A Stirling engine from the early nineteenth century. Air is heated in the lower cylinder using an external source. As this happens, the air expands and pushes against a piston, causing it to move. The air is then cooled, allowing the cycle to begin again. This is one example of a heat engine, which we study in this chapter. *(© SSPL/The Image Works)*

509

Lord Kelvin
British physicist and mathematician
(1824–1907)
Born William Thomson in Belfast, Kelvin was the first to propose the use of an absolute scale of temperature. The Kelvin temperature scale is named in his honor. Kelvin's work in thermodynamics led to the idea that energy cannot pass spontaneously from a colder object to a hotter object.

of processes that do not violate the first law of thermodynamics if they proceed in either direction, but are observed in reality to proceed in only one direction:

- When two objects at different temperatures are placed in thermal contact with each other, the net transfer of energy by heat is always from the warmer object to the cooler object, never from the cooler to the warmer.
- A rubber ball dropped to the ground bounces several times and eventually comes to rest, but a ball lying on the ground never gathers internal energy from the ground and begins bouncing on its own.
- An oscillating pendulum eventually comes to rest because of collisions with air molecules and friction at the point of suspension. The mechanical energy of the system is converted to internal energy in the air, the pendulum, and the suspension; the reverse conversion of energy never occurs.

All these processes are *irreversible;* that is, they are processes that occur naturally in one direction only. No irreversible process has ever been observed to run backward. If it were to do so, it would violate the second law of thermodynamics.[1]

22.1 Heat Engines and the Second Law of Thermodynamics

A **heat engine** is a device that takes in energy by heat[2] and, operating in a cyclic process, expels a fraction of that energy by means of work. For instance, in a typical process by which a power plant produces electricity, a fuel such as coal is burned and the high-temperature gases produced are used to convert liquid water to steam. This steam is directed at the blades of a turbine, setting it into rotation. The mechanical energy associated with this rotation is used to drive an electric generator. Another device that can be modeled as a heat engine is the internal combustion engine in an automobile. This device uses energy from a burning fuel to perform work on pistons that results in the motion of the automobile.

Let us consider the operation of a heat engine in more detail. A heat engine carries some working substance through a cyclic process during which (1) the working substance absorbs energy by heat from a high-temperature energy reservoir, (2) work is done by the engine, and (3) energy is expelled by heat to a lower-temperature reservoir. As an example, consider the operation of a steam engine (Fig. 22.1), which uses water as the working substance. The water in a boiler absorbs energy from burning fuel and evaporates to steam, which then does work by expanding against a piston. After the steam cools and condenses, the liquid water produced returns to the boiler and the cycle repeats.

It is useful to represent a heat engine schematically as in Figure 22.2. The engine absorbs a quantity of energy $|Q_h|$ from the hot reservoir. For the mathematical discussion of heat engines, we use absolute values to make all energy transfers by heat positive, and the direction of transfer is indicated with an explicit positive or negative sign. The engine does work W_{eng} (so that *negative* work $W = -W_{eng}$ is done *on* the engine) and then gives up a quantity of energy $|Q_c|$ to the cold reservoir.

Figure 22.1 A steam-driven locomotive obtains its energy by burning wood or coal. The generated energy vaporizes water into steam, which powers the locomotive. Modern locomotives use diesel fuel instead of wood or coal. Whether old-fashioned or modern, such locomotives can be modeled as heat engines, which extract energy from a burning fuel and convert a fraction of it to mechanical energy.

---

[1]Although a process occurring in the time-reversed sense has never been *observed*, it is *possible* for it to occur. As we shall see later in this chapter, however, the probability of such a process occurring is infinitesimally small. From this viewpoint, processes occur with a vastly greater probability in one direction than in the opposite direction.

[2]We use heat as our model for energy transfer into a heat engine. Other methods of energy transfer are possible in the model of a heat engine, however. For example, the Earth's atmosphere can be modeled as a heat engine in which the input energy transfer is by means of electromagnetic radiation from the Sun. The output of the atmospheric heat engine causes the wind structure in the atmosphere.

Because the working substance goes through a cycle, its initial and final internal energies are equal: $\Delta E_{int} = 0$. Hence, from the first law of thermodynamics, $\Delta E_{int} = Q + W = Q - W_{eng} = 0$, and the net work W_{eng} done by a heat engine is equal to the net energy Q_{net} transferred to it. As you can see from Figure 22.2, $Q_{net} = |Q_h| - |Q_c|$; therefore,

$$W_{eng} = |Q_h| - |Q_c| \qquad (22.1)$$

The **thermal efficiency** e of a heat engine is defined as the ratio of the net work done by the engine during one cycle to the energy input at the higher temperature during the cycle:

$$e \equiv \frac{W_{eng}}{|Q_h|} = \frac{|Q_h| - |Q_c|}{|Q_h|} = 1 - \frac{|Q_c|}{|Q_h|} \qquad (22.2)$$

◀ Thermal efficiency of a heat engine

You can think of the efficiency as the ratio of what you gain (work) to what you give (energy transfer at the higher temperature). In practice, all heat engines expel only a fraction of the input energy Q_h by mechanical work; consequently, their efficiency is always less than 100%. For example, a good automobile engine has an efficiency of about 20%, and diesel engines have efficiencies ranging from 35% to 40%.

Equation 22.2 shows that a heat engine has 100% efficiency ($e = 1$) only if $|Q_c| = 0$, that is, if no energy is expelled to the cold reservoir. In other words, a heat engine with perfect efficiency would have to expel all the input energy by work. Because efficiencies of real engines are well below 100%, the **Kelvin–Planck form of the second law of thermodynamics** states the following:

> It is impossible to construct a heat engine that, operating in a cycle, produces no effect other than the input of energy by heat from a reservoir and the performance of an equal amount of work.

This statement of the second law means that during the operation of a heat engine, W_{eng} can never be equal to $|Q_h|$ or, alternatively, that some energy $|Q_c|$ *must* be rejected to the environment. Figure 22.3 is a schematic diagram of the impossible "perfect" heat engine.

Quick Quiz 22.1 The energy input to an engine is 4.00 times greater than the work it performs. **(i)** What is its thermal efficiency? (a) 4.00 (b) 1.00 (c) 0.250 (d) impossible to determine **(ii)** What fraction of the energy input is expelled to the cold reservoir? (a) 0.250 (b) 0.750 (c) 1.00 (d) impossible to determine

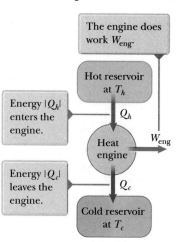

Figure 22.2 Schematic representation of a heat engine.

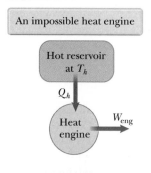

Figure 22.3 Schematic diagram of a heat engine that takes in energy from a hot reservoir and does an equivalent amount of work. It is impossible to construct such a perfect engine.

Pitfall Prevention 22.1

The First and Second Laws Notice the distinction between the first and second laws of thermodynamics. If a gas undergoes a *one-time isothermal process*, then $\Delta E_{int} = Q + W = 0$ and $W = -Q$. Therefore, the first law allows *all* energy input by heat to be expelled by work. In a heat engine, however, in which a substance undergoes a *cyclic* process, only a *portion* of the energy input by heat can be expelled by work according to the second law.

| **Example 22.1** | **The Efficiency of an Engine** |

An engine transfers 2.00×10^3 J of energy from a hot reservoir during a cycle and transfers 1.50×10^3 J as exhaust to a cold reservoir.

(A) Find the efficiency of the engine.

SOLUTION

Conceptualize Review Figure 22.2; think about energy going into the engine from the hot reservoir and splitting, with part coming out by work and part by heat into the cold reservoir.

Categorize This example involves evaluation of quantities from the equations introduced in this section, so we categorize it as a substitution problem.

Find the efficiency of the engine from Equation 22.2:

$$e = 1 - \frac{|Q_c|}{|Q_h|} = 1 - \frac{1.50 \times 10^3 \text{ J}}{2.00 \times 10^3 \text{ J}} = \boxed{0.250, \text{ or } 25.0\%}$$

(B) How much work does this engine do in one cycle?

SOLUTION

Find the work done by the engine by taking the difference between the input and output energies:

$$W_{eng} = |Q_h| - |Q_c| = 2.00 \times 10^3 \text{ J} - 1.50 \times 10^3 \text{ J}$$
$$= \boxed{5.0 \times 10^2 \text{ J}}$$

WHAT IF? Suppose you were asked for the power output of this engine. Do you have sufficient information to answer this question?

Answer No, you do not have enough information. The power of an engine is the *rate* at which work is done by the engine. You know how much work is done per cycle, but you have no information about the time interval associated with one cycle. If you were told that the engine operates at 2 000 rpm (revolutions per minute), however, you could relate this rate to the period of rotation T of the mechanism of the engine. Assuming there is one thermodynamic cycle per revolution, the power is

$$P = \frac{W_{eng}}{T} = \frac{5.0 \times 10^2 \text{ J}}{\left(\frac{1}{2\,000} \text{ min}\right)} \left(\frac{1 \text{ min}}{60 \text{ s}}\right) = 1.7 \times 10^4 \text{ W}$$

22.2 Heat Pumps and Refrigerators

In a heat engine, the direction of energy transfer is from the hot reservoir to the cold reservoir, which is the natural direction. The role of the heat engine is to process the energy from the hot reservoir so as to do useful work. What if we wanted to transfer energy from the cold reservoir to the hot reservoir? Because that is not the natural direction of energy transfer, we must put some energy into a device to be successful. Devices that perform this task are called **heat pumps** and **refrigerators.** For example, homes in summer are cooled using heat pumps called *air conditioners.* The air conditioner transfers energy from the cool room in the home to the warm air outside.

In a refrigerator or a heat pump, the engine takes in energy $|Q_c|$ from a cold reservoir and expels energy $|Q_h|$ to a hot reservoir (Fig. 22.4), which can be accomplished only if work is done *on* the engine. From the first law, we know that the energy given up to the hot reservoir must equal the sum of the work done and the energy taken in from the cold reservoir. Therefore, the refrigerator or heat pump transfers energy from a colder body (for example, the contents of a kitchen refrigerator or the winter air outside a building) to a hotter body (the air in the kitchen or a room in the building). In practice, it is desirable to carry out this process with

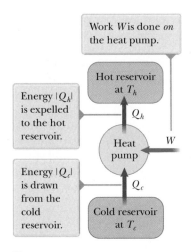

Work W is done *on* the heat pump.

Energy $|Q_h|$ is expelled to the hot reservoir.

Hot reservoir at T_h

Q_h

Heat pump

W

Energy $|Q_c|$ is drawn from the cold reservoir.

Q_c

Cold reservoir at T_c

Figure 22.4 Schematic representation of a heat pump.

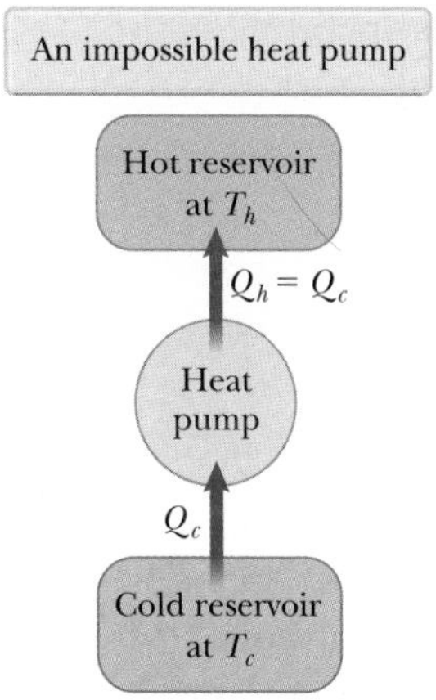

An impossible heat pump

Hot reservoir at T_h

$Q_h = Q_c$

Heat pump

Q_c

Cold reservoir at T_c

Figure 22.5 Schematic diagram of an impossible heat pump or refrigerator, that is, one that takes in energy from a cold reservoir and expels an equivalent amount of energy to a hot reservoir without the input of energy by work.

a minimum of work. If the process could be accomplished without doing any work, the refrigerator or heat pump would be "perfect" (Fig. 22.5). Again, the existence of such a device would be in violation of the second law of thermodynamics, which in the form of the **Clausius statement**[3] states:

> It is impossible to construct a cyclical machine whose sole effect is to transfer energy continuously by heat from one object to another object at a higher temperature without the input of energy by work.

In simpler terms, energy does not transfer spontaneously by heat from a cold object to a hot object. Work input is required to run a refrigerator.

The Clausius and Kelvin–Planck statements of the second law of thermodynamics appear at first sight to be unrelated, but in fact they are equivalent in all respects. Although we do not prove so here, if either statement is false, so is the other.[4]

In practice, a heat pump includes a circulating fluid that passes through two sets of metal coils that can exchange energy with the surroundings. The fluid is cold and at low pressure when it is in the coils located in a cool environment, where it absorbs energy by heat. The resulting warm fluid is then compressed and enters the other coils as a hot, high-pressure fluid. There it releases its stored energy to the warm surroundings. In an air conditioner, energy is absorbed into the fluid in coils located in a building's interior; after the fluid is compressed, energy leaves the fluid through coils located outdoors. In a refrigerator, the external coils are behind the unit (Fig. 22.6) or underneath the unit. The internal coils are in the walls of the refrigerator and absorb energy from the food.

The effectiveness of a heat pump is described in terms of a number called the **coefficient of performance** (COP). The COP is similar to the thermal efficiency for a heat engine in that it is a ratio of what you gain (energy transferred to or from a reservoir) to what you give (work input). For a heat pump operating in the cooling mode, "what you gain" is energy removed from the cold reservoir. The most effective refrigerator or air conditioner is one that removes the greatest amount of energy

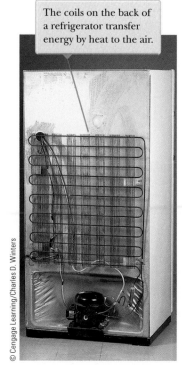

The coils on the back of a refrigerator transfer energy by heat to the air.

Figure 22.6 The back of a household refrigerator. The air surrounding the coils is the hot reservoir.

© Cengage Learning/Charles D. Winters

---

[3]First expressed by Rudolf Clausius (1822–1888).

[4]See an advanced textbook on thermodynamics for this proof.

from the cold reservoir in exchange for the least amount of work. Therefore, for these devices operating in the cooling mode, we define the COP in terms of $|Q_c|$:

$$\text{COP (cooling mode)} = \frac{\text{energy transferred at low temperature}}{\text{work done on heat pump}} = \frac{|Q_c|}{W} \quad \textbf{(22.3)}$$

A good refrigerator should have a high COP, typically 5 or 6.

In addition to cooling applications, heat pumps are becoming increasingly popular for heating purposes. The energy-absorbing coils for a heat pump are located outside a building, in contact with the air or buried in the ground. The other set of coils are in the building's interior. The circulating fluid flowing through the coils absorbs energy from the outside and releases it to the interior of the building from the interior coils.

In the heating mode, the COP of a heat pump is defined as the ratio of the energy transferred to the hot reservoir to the work required to transfer that energy:

$$\text{COP (heating mode)} = \frac{\text{energy transferred at high temperature}}{\text{work done on heat pump}} = \frac{|Q_h|}{W} \quad \textbf{(22.4)}$$

If the outside temperature is 25°F (−4°C) or higher, a typical value of the COP for a heat pump is about 4. That is, the amount of energy transferred to the building is about four times greater than the work done by the motor in the heat pump. As the outside temperature decreases, however, it becomes more difficult for the heat pump to extract sufficient energy from the air and so the COP decreases. Therefore, the use of heat pumps that extract energy from the air, although satisfactory in moderate climates, is not appropriate in areas where winter temperatures are very low. It is possible to use heat pumps in colder areas by burying the external coils deep in the ground. In that case, the energy is extracted from the ground, which tends to be warmer than the air in the winter.

Quick Quiz 22.2 The energy entering an electric heater by electrical transmission can be converted to internal energy with an efficiency of 100%. By what factor does the cost of heating your home change when you replace your electric heating system with an electric heat pump that has a COP of 4.00? Assume the motor running the heat pump is 100% efficient. **(a)** 4.00 **(b)** 2.00 **(c)** 0.500 **(d)** 0.250

Example 22.2   Freezing Water

A certain refrigerator has a COP of 5.00. When the refrigerator is running, its power input is 500 W. A sample of water of mass 500 g and temperature 20.0°C is placed in the freezer compartment. How long does it take to freeze the water to ice at 0°C? Assume all other parts of the refrigerator stay at the same temperature and there is no leakage of energy from the exterior, so the operation of the refrigerator results only in energy being extracted from the water.

SOLUTION

Conceptualize Energy leaves the water, reducing its temperature and then freezing it into ice. The time interval required for this entire process is related to the rate at which energy is withdrawn from the water, which, in turn, is related to the power input of the refrigerator.

Categorize We categorize this example as one that combines our understanding of temperature changes and phase changes from Chapter 20 and our understanding of heat pumps from this chapter.

Analyze Use the power rating of the refrigerator to find the time interval Δt required for the freezing process to occur:

$$P = \frac{W}{\Delta t} \quad \rightarrow \quad \Delta t = \frac{W}{P}$$

▶ **22.2** continued

Use Equation 22.3 to relate the work W done on the heat pump to the energy $|Q_c|$ extracted from the water:

$$\Delta t = \frac{|Q_c|}{P(\text{COP})}$$

Use Equations 20.4 and 20.7 to substitute the amount of energy $|Q_c|$ that must be extracted from the water of mass m:

$$\Delta t = \frac{|mc\,\Delta T + L_f\,\Delta m|}{P(\text{COP})}$$

Recognize that the amount of water that freezes is $\Delta m = -m$ because all the water freezes:

$$\Delta t = \frac{|m(c\,\Delta T - L_f)|}{P(\text{COP})}$$

Substitute numerical values:

$$\Delta t = \frac{|(0.500\ \text{kg})[(4\ 186\ \text{J/kg}\cdot{}^\circ\text{C})(-20.0^\circ\text{C}) - 3.33\times10^5\ \text{J/kg}]|}{(500\ \text{W})(5.00)}$$

$$= \boxed{83.3\ \text{s}}$$

Finalize In reality, the time interval for the water to freeze in a refrigerator is much longer than 83.3 s, which suggests that the assumptions of our model are not valid. Only a small part of the energy extracted from the refrigerator interior in a given time interval comes from the water. Energy must also be extracted from the container in which the water is placed, and energy that continuously leaks into the interior from the exterior must be extracted.

22.3 Reversible and Irreversible Processes

In the next section, we will discuss a theoretical heat engine that is the most efficient possible. To understand its nature, we must first examine the meaning of reversible and irreversible processes. In a **reversible** process, the system undergoing the process can be returned to its initial conditions along the same path on a PV diagram, and every point along this path is an equilibrium state. A process that does not satisfy these requirements is **irreversible.**

All natural processes are known to be irreversible. Let's examine the adiabatic free expansion of a gas, which was already discussed in Section 20.6, and show that it cannot be reversible. Consider a gas in a thermally insulated container as shown in Figure 22.7. A membrane separates the gas from a vacuum. When the membrane is punctured, the gas expands freely into the vacuum. As a result of the puncture, the system has changed because it occupies a greater volume after the expansion. Because the gas does not exert a force through a displacement, it does no work on the surroundings as it expands. In addition, no energy is transferred to or from the gas by heat because the container is insulated from its surroundings. Therefore, in this adiabatic process, the system has changed but the surroundings have not.

For this process to be reversible, we must return the gas to its original volume and temperature without changing the surroundings. Imagine trying to reverse the process by compressing the gas to its original volume. To do so, we fit the container with a piston and use an engine to force the piston inward. During this process, the surroundings change because work is being done by an outside agent on the system. In addition, the system changes because the compression increases the temperature of the gas. The temperature of the gas can be lowered by allowing it to come into contact with an external energy reservoir. Although this step returns the gas to its original conditions, the surroundings are again affected because energy is being added to the surroundings from the gas. If this

Pitfall Prevention 22.2
All Real Processes Are Irreversible
The reversible process is an idealization; all real processes on the Earth are irreversible.

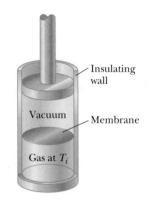

Figure 22.7 Adiabatic free expansion of a gas.

The gas is compressed slowly as individual grains of sand drop onto the piston.

Energy reservoir

Figure 22.8 A method for compressing a gas in an almost reversible isothermal process.

energy could be used to drive the engine that compressed the gas, the net energy transfer to the surroundings would be zero. In this way, the system and its surroundings could be returned to their initial conditions and we could identify the process as reversible. The Kelvin–Planck statement of the second law, however, specifies that the energy removed from the gas to return the temperature to its original value cannot be completely converted to mechanical energy by the process of work done by the engine in compressing the gas. Therefore, we must conclude that the process is irreversible.

We could also argue that the adiabatic free expansion is irreversible by relying on the portion of the definition of a reversible process that refers to equilibrium states. For example, during the sudden expansion, significant variations in pressure occur throughout the gas. Therefore, there is no well-defined value of the pressure for the entire system at any time between the initial and final states. In fact, the process cannot even be represented as a path on a *PV* diagram. The *PV* diagram for an adiabatic free expansion would show the initial and final conditions as points, but these points would not be connected by a path. Therefore, because the intermediate conditions between the initial and final states are not equilibrium states, the process is irreversible.

Although all real processes are irreversible, some are almost reversible. If a real process occurs very slowly such that the system is always very nearly in an equilibrium state, the process can be approximated as being reversible. Suppose a gas is compressed isothermally in a piston–cylinder arrangement in which the gas is in thermal contact with an energy reservoir and we continuously transfer just enough energy from the gas to the reservoir to keep the temperature constant. For example, imagine that the gas is compressed very slowly by dropping grains of sand onto a frictionless piston as shown in Figure 22.8. As each grain lands on the piston and compresses the gas a small amount, the system deviates from an equilibrium state, but it is so close to one that it achieves a new equilibrium state in a relatively short time interval. Each grain added represents a change to a new equilibrium state, but the differences between states are so small that the entire process can be approximated as occurring through continuous equilibrium states. The process can be reversed by slowly removing grains from the piston.

A general characteristic of a reversible process is that no nonconservative effects (such as turbulence or friction) that transform mechanical energy to internal energy can be present. Such effects can be impossible to eliminate completely. Hence, it is not surprising that real processes in nature are irreversible.

Pitfall Prevention 22.3
Don't Shop for a Carnot Engine
The Carnot engine is an idealization; do not expect a Carnot engine to be developed for commercial use. We explore the Carnot engine only for theoretical considerations.

22.4 The Carnot Engine

In 1824, a French engineer named Sadi Carnot described a theoretical engine, now called a **Carnot engine,** that is of great importance from both practical and theoretical viewpoints. He showed that a heat engine operating in an ideal, reversible cycle—called a **Carnot cycle**—between two energy reservoirs is the most efficient engine possible. Such an ideal engine establishes an upper limit on the efficiencies of all other engines. That is, the net work done by a working substance taken through the Carnot cycle is the greatest amount of work possible for a given amount of energy supplied to the substance at the higher temperature. **Carnot's theorem** can be stated as follows:

> No real heat engine operating between two energy reservoirs can be more efficient than a Carnot engine operating between the same two reservoirs.

In this section, we will show that the efficiency of a Carnot engine depends only on the temperatures of the reservoirs. In turn, that efficiency represents the

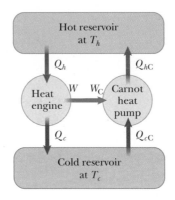

Figure 22.9 A Carnot engine operated as a heat pump and another engine with a proposed higher efficiency operate between two energy reservoirs. The work output and input are matched.

Sadi Carnot
French engineer (1796–1832)
Carnot was the first to show the quantitative relationship between work and heat. In 1824, he published his only work, *Reflections on the Motive Power of Heat*, which reviewed the industrial, political, and economic importance of the steam engine. In it, he defined work as "weight lifted through a height."

maximum possible efficiency for real engines. Let us confirm that the Carnot engine is the most efficient. We imagine a hypothetical engine with an efficiency greater than that of the Carnot engine. Consider Figure 22.9, which shows the hypothetical engine with $e > e_C$ on the left connected between hot and cold reservoirs. In addition, let us attach a Carnot engine between the same reservoirs. Because the Carnot cycle is reversible, the Carnot engine can be run in reverse as a Carnot heat pump as shown on the right in Figure 22.9. We match the output work of the engine to the input work of the heat pump, $W = W_C$, so there is no exchange of energy by work between the surroundings and the engine–heat pump combination.

Because of the proposed relation between the efficiencies, we must have

$$e > e_C \quad \rightarrow \quad \frac{|W|}{|Q_h|} > \frac{|W_C|}{|Q_{hC}|}$$

The numerators of these two fractions cancel because the works have been matched. This expression requires that

$$|Q_{hC}| > |Q_h| \tag{22.5}$$

From Equation 22.1, the equality of the works gives us

$$|W| = |W_C| \quad \rightarrow \quad |Q_h| - |Q_c| = |Q_{hC}| - |Q_{cC}|$$

which can be rewritten to put the energies exchanged with the cold reservoir on the left and those with the hot reservoir on the right:

$$|Q_{hC}| - |Q_h| = |Q_{cC}| - |Q_c| \tag{22.6}$$

Note that the left side of Equation 22.6 is positive, so the right side must be positive also. We see that the net energy exchange with the hot reservoir is equal to the net energy exchange with the cold reservoir. As a result, for the combination of the heat engine and the heat pump, energy is transferring from the cold reservoir to the hot reservoir by heat with no input of energy by work.

This result is in violation of the Clausius statement of the second law. Therefore, our original assumption that $e > e_C$ must be incorrect, and we must conclude that the Carnot engine represents the highest possible efficiency for an engine. The key feature of the Carnot engine that makes it the most efficient is its *reversibility;* it can be run in reverse as a heat pump. All real engines are less efficient than the Carnot engine because they do not operate through a reversible cycle. The efficiency of a real engine is further reduced by such practical difficulties as friction and energy losses by conduction.

To describe the Carnot cycle taking place between temperatures T_c and T_h, let's assume the working substance is an ideal gas contained in a cylinder fitted with a movable piston at one end. The cylinder's walls and the piston are thermally nonconducting. Four stages of the Carnot cycle are shown in Figure 22.10(page 518),

Figure 22.10 The Carnot cycle. The letters A, B, C, and D refer to the states of the gas shown in Figure 22.11. The arrows on the piston indicate the direction of its motion during each process.

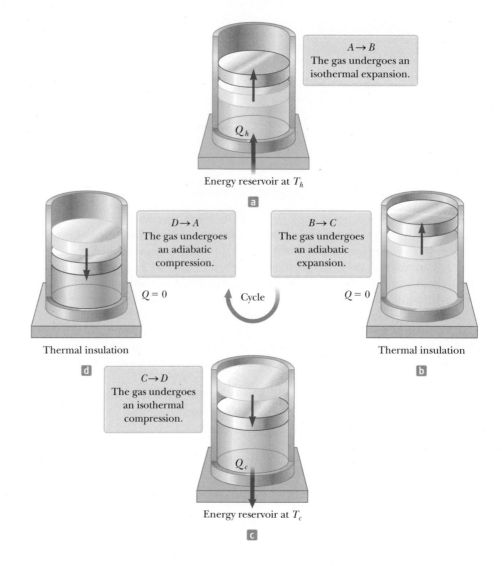

$A \rightarrow B$
The gas undergoes an isothermal expansion.

Q_h

Energy reservoir at T_h

a

$D \rightarrow A$
The gas undergoes an adiabatic compression.

$B \rightarrow C$
The gas undergoes an adiabatic expansion.

$Q = 0$ Cycle $Q = 0$

Thermal insulation   Thermal insulation

d   b

$C \rightarrow D$
The gas undergoes an isothermal compression.

Q_c

Energy reservoir at T_c

c

and the PV diagram for the cycle is shown in Figure 22.11. The Carnot cycle consists of two adiabatic processes and two isothermal processes, all reversible:

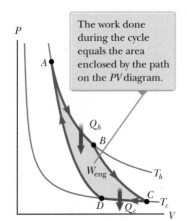

The work done during the cycle equals the area enclosed by the path on the PV diagram.

Figure 22.11 PV diagram for the Carnot cycle. The net work done W_{eng} equals the net energy transferred into the Carnot engine in one cycle, $|Q_h| - |Q_c|$.

1. Process $A \rightarrow B$ (Fig. 22.10a) is an isothermal expansion at temperature T_h. The gas is placed in thermal contact with an energy reservoir at temperature T_h. During the expansion, the gas absorbs energy $|Q_h|$ from the reservoir through the base of the cylinder and does work W_{AB} in raising the piston.

2. In process $B \rightarrow C$ (Fig. 22.10b), the base of the cylinder is replaced by a thermally nonconducting wall and the gas expands adiabatically; that is, no energy enters or leaves the system by heat. During the expansion, the temperature of the gas decreases from T_h to T_c and the gas does work W_{BC} in raising the piston.

3. In process $C \rightarrow D$ (Fig. 22.10c), the gas is placed in thermal contact with an energy reservoir at temperature T_c and is compressed isothermally at temperature T_c. During this time, the gas expels energy $|Q_c|$ to the reservoir and the work done by the piston on the gas is W_{CD}.

4. In the final process $D \rightarrow A$ (Fig. 22.10d), the base of the cylinder is replaced by a nonconducting wall and the gas is compressed adiabatically. The temperature of the gas increases to T_h, and the work done by the piston on the gas is W_{DA}.

The thermal efficiency of the engine is given by Equation 22.2:

$$e = 1 - \frac{|Q_c|}{|Q_h|}$$

In Example 22.3, we show that for a Carnot cycle,

$$\frac{|Q_c|}{|Q_h|} = \frac{T_c}{T_h} \qquad (22.7)$$

Hence, the thermal efficiency of a Carnot engine is

$$e_C = 1 - \frac{T_c}{T_h} \qquad (22.8)$$

◀ **Efficiency of a Carnot engine**

This result indicates that all Carnot engines operating between the same two temperatures have the same efficiency.[5]

Equation 22.8 can be applied to any working substance operating in a Carnot cycle between two energy reservoirs. According to this equation, the efficiency is zero if $T_c = T_h$, as one would expect. The efficiency increases as T_c is lowered and T_h is raised. The efficiency can be unity (100%), however, only if $T_c = 0$ K. Such reservoirs are not available; therefore, the maximum efficiency is always less than 100%. In most practical cases, T_c is near room temperature, which is about 300 K. Therefore, one usually strives to increase the efficiency by raising T_h.

Theoretically, a Carnot-cycle heat engine run in reverse constitutes the most effective heat pump possible, and it determines the maximum COP for a given combination of hot and cold reservoir temperatures. Using Equations 22.1 and 22.4, we see that the maximum COP for a heat pump in its heating mode is

$$\text{COP}_C \, (\text{heating mode}) = \frac{|Q_h|}{W}$$

$$= \frac{|Q_h|}{|Q_h| - |Q_c|} = \frac{1}{1 - \dfrac{|Q_c|}{|Q_h|}} = \frac{1}{1 - \dfrac{T_c}{T_h}} = \frac{T_h}{T_h - T_c}$$

The Carnot COP for a heat pump in the cooling mode is

$$\text{COP}_C \, (\text{cooling mode}) = \frac{T_c}{T_h - T_c}$$

As the difference between the temperatures of the two reservoirs approaches zero in this expression, the theoretical COP approaches infinity. In practice, the low temperature of the cooling coils and the high temperature at the compressor limit the COP to values below 10.

Quick **Quiz** 22.3 Three engines operate between reservoirs separated in temperature by 300 K. The reservoir temperatures are as follows: Engine A: $T_h = 1\,000$ K, $T_c = 700$ K; Engine B: $T_h = 800$ K, $T_c = 500$ K; Engine C: $T_h = 600$ K, $T_c = 300$ K. Rank the engines in order of theoretically possible efficiency from highest to lowest.

---

[5]For the processes in the Carnot cycle to be reversible, they must be carried out infinitesimally slowly. Therefore, although the Carnot engine is the most efficient engine possible, it has zero power output because it takes an infinite time interval to complete one cycle! For a real engine, the short time interval for each cycle results in the working substance reaching a high temperature lower than that of the hot reservoir and a low temperature higher than that of the cold reservoir. An engine undergoing a Carnot cycle between this narrower temperature range was analyzed by F. L. Curzon and B. Ahlborn ("Efficiency of a Carnot engine at maximum power output," *Am. J. Phys.* **43**(1), 22, 1975), who found that the efficiency at maximum power output depends only on the reservoir temperatures T_c and T_h and is given by $e_{C\text{-}A} = 1 - (T_c/T_h)^{1/2}$. The Curzon–Ahlborn efficiency $e_{C\text{-}A}$ provides a closer approximation to the efficiencies of real engines than does the Carnot efficiency.

Example 22.3   Efficiency of the Carnot Engine

Show that the ratio of energy transfers by heat in a Carnot engine is equal to the ratio of reservoir temperatures, as given by Equation 22.7.

SOLUTION

Conceptualize Make use of Figures 22.10 and 22.11 to help you visualize the processes in the Carnot cycle.

Categorize Because of our understanding of the Carnot cycle, we can categorize the processes in the cycle as isothermal and adiabatic.

Analyze For the isothermal expansion (process $A \to B$ in Fig. 22.10), find the energy transfer by heat from the hot reservoir using Equation 20.14 and the first law of thermodynamics:

$$|Q_h| = |\Delta E_{int} - W_{AB}| = |0 - W_{AB}| = nRT_h \ln \frac{V_B}{V_A}$$

In a similar manner, find the energy transfer to the cold reservoir during the isothermal compression $C \to D$:

$$|Q_c| = |\Delta E_{int} - W_{CD}| = |0 - W_{CD}| = nRT_c \ln \frac{V_C}{V_D}$$

Divide the second expression by the first:

$$(1) \quad \frac{|Q_c|}{|Q_h|} = \frac{T_c \ln (V_C/V_D)}{T_h \ln (V_B/V_A)}$$

Apply Equation 21.39 to the adiabatic processes $B \to C$ and $D \to A$:

$$T_h V_B^{\gamma-1} = T_c V_C^{\gamma-1}$$
$$T_h V_A^{\gamma-1} = T_c V_D^{\gamma-1}$$

Divide the first equation by the second:

$$\left(\frac{V_B}{V_A}\right)^{\gamma-1} = \left(\frac{V_C}{V_D}\right)^{\gamma-1}$$

$$(2) \quad \frac{V_B}{V_A} = \frac{V_C}{V_D}$$

Substitute Equation (2) into Equation (1):

$$\frac{|Q_c|}{|Q_h|} = \frac{T_c \ln (V_C/V_D)}{T_h \ln (V_B/V_A)} = \frac{T_c \ln (V_C/V_D)}{T_h \ln (V_C/V_D)} = \frac{T_c}{T_h}$$

Finalize This last equation is Equation 22.7, the one we set out to prove.

Example 22.4   The Steam Engine

A steam engine has a boiler that operates at 500 K. The energy from the burning fuel changes water to steam, and this steam then drives a piston. The cold reservoir's temperature is that of the outside air, approximately 300 K. What is the maximum thermal efficiency of this steam engine?

SOLUTION

Conceptualize In a steam engine, the gas pushing on the piston in Figure 22.10 is steam. A real steam engine does not operate in a Carnot cycle, but, to find the maximum possible efficiency, imagine a Carnot steam engine.

Categorize We calculate an efficiency using Equation 22.8, so we categorize this example as a substitution problem.

Substitute the reservoir temperatures into Equation 22.8: $e_C = 1 - \dfrac{T_c}{T_h} = 1 - \dfrac{300 \text{ K}}{500 \text{ K}} = \boxed{0.400} \quad \text{or} \quad \boxed{40.0\%}$

This result is the highest *theoretical* efficiency of the engine. In practice, the efficiency is considerably lower.

▶ **22.4** continued

WHAT IF? Suppose we wished to increase the theoretical efficiency of this engine. This increase can be achieved by raising T_h by ΔT or by decreasing T_c by the same ΔT. Which would be more effective?

Answer A given ΔT would have a larger fractional effect on a smaller temperature, so you would expect a larger change in efficiency if you alter T_c by ΔT. Let's test that numerically. Raising T_h by 50 K, corresponding to $T_h = 550$ K, would give a maximum efficiency of

$$e_C = 1 - \frac{T_c}{T_h} = 1 - \frac{300 \text{ K}}{550 \text{ K}} = 0.455$$

Decreasing T_c by 50 K, corresponding to $T_c = 250$ K, would give a maximum efficiency of

$$e_C = 1 - \frac{T_c}{T_h} = 1 - \frac{250 \text{ K}}{500 \text{ K}} = 0.500$$

Although changing T_c is *mathematically* more effective, often changing T_h is *practically* more feasible.

22.5 Gasoline and Diesel Engines

In a gasoline engine, six processes occur in each cycle; they are illustrated in Figure 22.12. In this discussion, let's consider the interior of the cylinder above the piston to be the system that is taken through repeated cycles in the engine's operation. For a given cycle, the piston moves up and down twice, which represents a four-stroke cycle consisting of two upstrokes and two downstrokes. The processes in the cycle can be approximated by the **Otto cycle** shown in the *PV* diagram in Figure 22.13 (page 522). In the following discussion, refer to Figure 22.12 for the pictorial representation of the strokes and Figure 22.13 for the significance on the *PV* diagram of the letter designations below:

1. During the *intake stroke* (Fig. 22.12a and $O \rightarrow A$ in Figure 22.13), the piston moves downward and a gaseous mixture of air and fuel is drawn into the

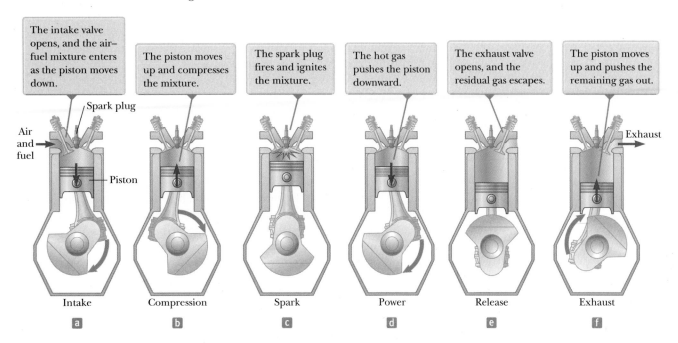

Figure 22.12 The four-stroke cycle of a conventional gasoline engine. The arrows on the piston indicate the direction of its motion during each process.

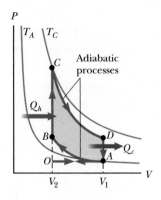

Figure 22.13 *PV* diagram for the Otto cycle, which approximately represents the processes occurring in an internal combustion engine.

cylinder at atmospheric pressure. That is the energy input part of the cycle: energy enters the system (the interior of the cylinder) by matter transfer as potential energy stored in the fuel. In this process, the volume increases from V_2 to V_1. This apparent backward numbering is based on the compression stroke (process 2 below), in which the air–fuel mixture is compressed from V_1 to V_2.

2. During the *compression stroke* (Fig. 22.12b and $A \rightarrow B$ in Fig. 22.13), the piston moves upward, the air–fuel mixture is compressed adiabatically from volume V_1 to volume V_2, and the temperature increases from T_A to T_B. The work done on the gas is positive, and its value is equal to the negative of the area under the curve AB in Figure 22.13.

3. Combustion occurs when the spark plug fires (Fig. 22.12c and $B \rightarrow C$ in Fig. 22.13). That is not one of the strokes of the cycle because it occurs in a very short time interval while the piston is at its highest position. The combustion represents a rapid energy transformation from potential energy stored in chemical bonds in the fuel to internal energy associated with molecular motion, which is related to temperature. During this time interval, the mixture's pressure and temperature increase rapidly, with the temperature rising from T_B to T_C. The volume, however, remains approximately constant because of the short time interval. As a result, approximately no work is done on or by the gas. We can model this process in the *PV* diagram (Fig. 22.13) as that process in which the energy $|Q_h|$ enters the system. (In reality, however, this process is a *transformation* of energy already in the cylinder from process $O \rightarrow A$.)

4. In the *power stroke* (Fig. 22.12d and $C \rightarrow D$ in Fig. 22.13), the gas expands adiabatically from V_2 to V_1. This expansion causes the temperature to drop from T_C to T_D. Work is done by the gas in pushing the piston downward, and the value of this work is equal to the area under the curve CD.

5. Release of the residual gases occurs when an exhaust valve is opened (Fig. 22.12e and $D \rightarrow A$ in Fig. 22.13). The pressure suddenly drops for a short time interval. During this time interval, the piston is almost stationary and the volume is approximately constant. Energy is expelled from the interior of the cylinder and continues to be expelled during the next process.

6. In the final process, the *exhaust stroke* (Fig. 22.12e and $A \rightarrow O$ in Fig. 22.13), the piston moves upward while the exhaust valve remains open. Residual gases are exhausted at atmospheric pressure, and the volume decreases from V_1 to V_2. The cycle then repeats.

If the air–fuel mixture is assumed to be an ideal gas, the efficiency of the Otto cycle is

$$e = 1 - \frac{1}{(V_1/V_2)^{\gamma-1}} \quad \text{(Otto cycle)} \qquad \text{(22.9)}$$

where V_1/V_2 is the **compression ratio** and γ is the ratio of the molar specific heats C_P/C_V for the air–fuel mixture. Equation 22.9, which is derived in Example 22.5, shows that the efficiency increases as the compression ratio increases. For a typical compression ratio of 8 and with $\gamma = 1.4$, Equation 22.9 predicts a theoretical efficiency of 56% for an engine operating in the idealized Otto cycle. This value is much greater than that achieved in real engines (15% to 20%) because of such effects as friction, energy transfer by conduction through the cylinder walls, and incomplete combustion of the air–fuel mixture.

Diesel engines operate on a cycle similar to the Otto cycle, but they do not employ a spark plug. The compression ratio for a diesel engine is much greater than that for a gasoline engine. Air in the cylinder is compressed to a very small volume, and, as a consequence, the cylinder temperature at the end of the compression stroke is very high. At this point, fuel is injected into the cylinder. The temperature is high enough for the air–fuel mixture to ignite without the assistance of a spark plug. Diesel engines are more efficient than gasoline engines because of their greater compression ratios and resulting higher combustion temperatures.

Example 22.5 Efficiency of the Otto Cycle

Show that the thermal efficiency of an engine operating in an idealized Otto cycle (see Figs. 22.12 and 22.13) is given by Equation 22.9. Treat the working substance as an ideal gas.

SOLUTION

Conceptualize Study Figures 22.12 and 22.13 to make sure you understand the working of the Otto cycle.

Categorize As seen in Figure 22.13, we categorize the processes in the Otto cycle as isovolumetric and adiabatic.

Analyze Model the energy input and output as occurring by heat in processes $B \to C$ and $D \to A$. (In reality, most of the energy enters and leaves by matter transfer as the air–fuel mixture enters and leaves the cylinder.) Use Equation 21.23 to find the energy transfers by heat for these processes, which take place at constant volume:

$$B \to C \quad |Q_h| = nC_V (T_C - T_B)$$
$$D \to A \quad |Q_c| = nC_V (T_D - T_A)$$

Substitute these expressions into Equation 22.2:

$$(1) \quad e = 1 - \frac{|Q_c|}{|Q_h|} = 1 - \frac{T_D - T_A}{T_C - T_B}$$

Apply Equation 21.39 to the adiabatic processes $A \to B$ and $C \to D$:

$$A \to B \quad T_A V_A^{\gamma-1} = T_B V_B^{\gamma-1}$$
$$C \to D \quad T_C V_C^{\gamma-1} = T_D V_D^{\gamma-1}$$

Solve these equations for the temperatures T_A and T_D, noting that $V_A = V_D = V_1$ and $V_B = V_C = V_2$:

$$(2) \quad T_A = T_B \left(\frac{V_B}{V_A}\right)^{\gamma-1} = T_B \left(\frac{V_2}{V_1}\right)^{\gamma-1}$$

$$(3) \quad T_D = T_C \left(\frac{V_C}{V_D}\right)^{\gamma-1} = T_C \left(\frac{V_2}{V_1}\right)^{\gamma-1}$$

Subtract Equation (2) from Equation (3) and rearrange:

$$(4) \quad \frac{T_D - T_A}{T_C - T_B} = \left(\frac{V_2}{V_1}\right)^{\gamma-1}$$

Substitute Equation (4) into Equation (1):

$$e = 1 - \frac{1}{(V_1/V_2)^{\gamma-1}}$$

Finalize This final expression is Equation 22.9.

22.6 Entropy

The zeroth law of thermodynamics involves the concept of temperature, and the first law involves the concept of internal energy. Temperature and internal energy are both state variables; that is, the value of each depends only on the thermodynamic state of a system, not on the process that brought it to that state. Another state variable—this one related to the second law of thermodynamics—is *entropy*.

Entropy was originally formulated as a useful concept in thermodynamics. Its importance grew, however, as the field of statistical mechanics developed because the analytical techniques of statistical mechanics provide an alternative means of interpreting entropy and a more global significance to the concept. In statistical

Pitfall Prevention 22.4

Entropy Is Abstract Entropy is one of the most abstract notions in physics, so follow the discussion in this and the subsequent sections very carefully. Do not confuse energy with entropy. Even though the names sound similar, they are very different concepts. On the other hand, energy and entropy are intimately related, as we shall see in this discussion.

mechanics, the behavior of a substance is described in terms of the statistical behavior of its atoms and molecules.

We will develop our understanding of entropy by first considering some nonthermodynamic systems, such as a pair of dice and poker hands. We will then expand on these ideas and use them to understand the concept of entropy as applied to thermodynamic systems.

We begin this process by distinguishing between *microstates* and *macrostates* of a system. A **microstate** is a particular configuration of the individual constituents of the system. A **macrostate** is a description of the system's conditions from a macroscopic point of view.

For any given macrostate of the system, a number of microstates are possible. For example, the macrostate of a 4 on a pair of dice can be formed from the possible microstates 1–3, 2–2, and 3–1. The macrostate of 2 has only one microstate, 1–1. It is assumed all microstates are equally probable. We can compare these two macrostates in three ways: (1) *Uncertainty:* If we know that a macrostate of 4 exists, there is some uncertainty as to the microstate that exists, because there are multiple microstates that will result in a 4. In comparison, there is lower uncertainty (in fact, *zero* uncertainty) for a macrostate of 2 because there is only one microstate. (2) *Choice:* There are more choices of microstates for a 4 than for a 2. (3) *Probability:* The macrostate of 4 has a higher probability than a macrostate of 2 because there are more ways (microstates) of achieving a 4. The notions of uncertainty, choice, and probability are central to the concept of entropy, as we discuss below.

Let's look at another example related to a poker hand. There is only one microstate associated with the macrostate of a royal flush of five spades, laid out in order from ten to ace (Fig. 22.14a). Figure 22.14b shows another poker hand. The macrostate here is "worthless hand." The *particular* hand (the microstate) in Figure 22.14b and the hand in Figure 22.14a are equally probable. There are, however, *many* other hands similar to that in Figure 22.14b; that is, there are many microstates that also qualify as worthless hands. If you, as a poker player, are told your opponent holds a macrostate of a royal flush in spades, there is *zero uncertainty* as to what five cards are in the hand, only *one choice* of what those cards are, and *low probability* that the hand actually occurred. In contrast, if you are told that your opponent has the macrostate of "worthless hand," there is *high uncertainty* as to what the five cards are, *many choices* of what they could be, and a *high probability* that a worthless hand occurred. Another variable in poker, of course, is the value of the hand, related to the probability: the higher the probability, the lower the value. The important point to take away from this discussion is that uncertainty, choice, and probability are related in these situations: if one is high, the others are high, and vice versa.

Another way of describing macrostates is by means of "missing information." For high-probability macrostates with many microstates, there is a large amount

Figure 22.14 (a) A royal flush has low probability of occurring. (b) A worthless poker hand, one of many.

of missing information, meaning we have very little information about what microstate actually exists. For a macrostate of a 2 on a pair of dice, we have no missing information; we *know* the microstate is 1–1. For a macrostate of a worthless poker hand, however, we have lots of missing information, related to the large number of choices we could make as to the actual hand that is held.

Quick Quiz 22.4 **(a)** Suppose you select four cards at random from a standard deck of playing cards and end up with a macrostate of four deuces. How many microstates are associated with this macrostate? **(b)** Suppose you pick up two cards and end up with a macrostate of two aces. How many microstates are associated with this macrostate?

For thermodynamic systems, the variable **entropy** S is used to represent the level of uncertainty, choice, probability, or missing information in the system. Consider a configuration (a macrostate) in which all the oxygen molecules in your room are located in the west half of the room and the nitrogen molecules in the east half. Compare that macrostate to the more common configuration of the air molecules distributed uniformly throughout the room. The latter configuration has the higher uncertainty and more missing information as to where the molecules are located because they could be anywhere, not just in one half of the room according to the type of molecule. The configuration with a uniform distribution also represents more choices as to where to locate molecules. It also has a much higher probability of occurring; have you ever noticed your half of the room suddenly being empty of oxygen? Therefore, the latter configuration represents a higher entropy.

For systems of dice and poker hands, the comparisons between probabilities for various macrostates involve relatively small numbers. For example, a macrostate of a 4 on a pair of dice is only three times as probable as a macrostate of 2. The ratio of probabilities of a worthless hand and a royal flush is significantly larger. When we are talking about a macroscopic thermodynamic system containing on the order of Avogadro's number of molecules, however, the ratios of probabilities can be astronomical.

Let's explore this concept by considering 100 molecules in a container. Half of the molecules are oxygen and the other half are nitrogen. At any given moment, the probability of one molecule being in the left part of the container shown in Figure 22.15a as a result of random motion is $\frac{1}{2}$. If there are two molecules as shown in Figure 22.15b, the probability of both being in the left part is $(\frac{1}{2})^2$, or 1 in 4. If there are three molecules (Fig. 22.15c), the probability of them all being in the left portion at the same moment is $(\frac{1}{2})^3$, or 1 in 8. For 100 independently moving molecules, the probability that the 50 oxygen molecules will be found in the left part at any moment is $(\frac{1}{2})^{50}$. Likewise, the probability that the remaining 50 nitrogen molecules will be found in the right part at any moment is $(\frac{1}{2})^{50}$. Therefore, the probability of

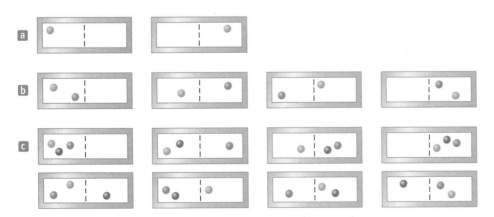

Figure 22.15 Possible distributions of identical molecules in a container. The colors used here exist only to allow us to distinguish among the molecules. (a) One molecule in a container has a 1-in-2 chance of being on the left side. (b) Two molecules have a 1-in-4 chance of being on the left side at the same time. (c) Three molecules have a 1-in-8 chance of being on the left side at the same time.

finding this oxygen–nitrogen separation as a result of random motion is the product $(\frac{1}{2})^{50}(\frac{1}{2})^{50} = (\frac{1}{2})^{100}$, which corresponds to about 1 in 10^{30}. When this calculation is extrapolated from 100 molecules to the number in 1 mol of gas (6.02×10^{23}), the separated arrangement is found to be *extremely* improbable!

Conceptual Example 22.6 **Let's Play Marbles!**

Suppose you have a bag of 100 marbles of which 50 are red and 50 are green. You are allowed to draw four marbles from the bag according to the following rules. Draw one marble, record its color, and return it to the bag. Shake the bag and then draw another marble. Continue this process until you have drawn and returned four marbles. What are the possible macrostates for this set of events? What is the most likely macrostate? What is the least likely macrostate?

SOLUTION

Because each marble is returned to the bag before the next one is drawn and the bag is then shaken, the probability of drawing a red marble is always the same as the probability of drawing a green one. All the possible microstates and macrostates are shown in Table 22.1. As this table indicates, there is only one way to draw a macrostate of four red marbles, so there is only one microstate for that macrostate. There are, however, four

Table 22.1 **Possible Results of Drawing Four Marbles from a Bag**

Macrostate	Possible Microstates	Total Number of Microstates
All R	RRRR	1
1G, 3R	RRRG, RRGR, RGRR, GRRR	4
2G, 2R	RRGG, RGRG, GRRG, RGGR, GRGR, GGRR	6
3G, 1R	GGGR, GGRG, GRGG, RGGG	4
All G	GGGG	1

possible microstates that correspond to the macrostate of one green marble and three red marbles, six microstates that correspond to two green marbles and two red marbles, four microstates that correspond to three green marbles and one red marble, and one microstate that corresponds to four green marbles. The most likely macrostate—two red marbles and two green marbles—corresponds to the largest number of choices of microstates, and, therefore, the most uncertainty as to what the exact microstate is. The least likely macrostates—four red marbles or four green marbles—correspond to only one choice of microstate and, therefore, zero uncertainty. There is no missing information for the least likely states: we know the colors of all four marbles.

We have investigated the notions of uncertainty, number of choices, probability, and missing information for some non-thermodynamic systems and have argued that the concept of entropy can be related to these notions for thermodynamic systems. We have not yet indicated how to evaluate entropy numerically for a thermodynamic system. This evaluation was done through statistical means by Boltzmann in the 1870s and appears in its currently accepted form as

$$S = k_B \ln W \tag{22.10}$$

where k_B is Boltzmann's constant. Boltzmann intended W, standing for *Wahrscheinlichkeit*, the German word for probability, to be proportional to the probability that a given macrostate exists. It is equivalent to let W be the number of microstates associated with the macrostate, so we can interpret W as representing the number of "ways" of achieving the macrostate. Therefore, macrostates with larger numbers of microstates have higher probability and, equivalently, higher entropy.

In the kinetic theory of gases, gas molecules are represented as particles moving randomly. Suppose the gas is confined to a volume V. For a uniform distribution of gas in the volume, there are a large number of equivalent microstates, and the entropy of the gas can be related to the number of microstates corresponding to a given macrostate. Let us count the number of microstates by considering the variety of molecular locations available to the molecules. Let us assume each molecule occupies some microscopic volume V_m. The total number of possible locations of a single molecule in a macroscopic volume V is the ratio $w = V/V_m$, which is a huge number. We use lowercase w here to represent the number of ways a single molecule can be placed in the volume or the number of microstates for a single molecule, which is equivalent to the number of available locations. We assume the probabilities of a molecule occupying any of these locations are equal. As more molecules are added to the system, the number of possible ways the molecules can be positioned in the volume multiplies, as we saw in Figure 22.15. For example, if you consider two molecules, for every possible placement of the first, all possible placements of the second are available. Therefore, there are w ways of locating the first molecule, and for each way, there are w ways of locating the second molecule. The total number of ways of locating the two molecules is $W = w \times w = w^2 = (V/V_m)^2$. (Uppercase W represents the number of ways of putting multiple molecules into the volume and is not to be confused with work.)

Now consider placing N molecules of gas in the volume V. Neglecting the very small probability of having two molecules occupy the same location, each molecule may go into any of the V/V_m locations, and so the number of ways of locating N molecules in the volume becomes $W = w^N = (V/V_m)^N$. Therefore, the spatial part of the entropy of the gas, from Equation 22.10, is

$$S = k_B \ln W = k_B \ln \left(\frac{V}{V_m}\right)^N = Nk_B \ln \left(\frac{V}{V_m}\right) = nR \ln \left(\frac{V}{V_m}\right) \qquad \textbf{(22.11)}$$

We will use this expression in the next section as we investigate changes in entropy for processes occurring in thermodynamic systems.

Notice that we have indicated Equation 22.11 as representing only the *spatial* portion of the entropy of the gas. There is also a temperature-dependent portion of the entropy that the discussion above does not address. For example, imagine an isovolumetric process in which the temperature of the gas increases. Equation 22.11 above shows no change in the spatial portion of the entropy for this situation. There *is* a change in entropy, however, associated with the increase in temperature. We can understand this by appealing again to a bit of quantum physics. Recall from Section 21.3 that the energies of the gas molecules are quantized. When the temperature of a gas changes, the distribution of energies of the gas molecules changes according to the Boltzmann distribution law, discussed in Section 21.5. Therefore, as the temperature of the gas increases, there is more uncertainty about the particular microstate that exists as gas molecules distribute themselves into higher available quantum states. We will see the entropy change associated with an isovolumetric process in Example 22.8.

22.7 Changes in Entropy for Thermodynamic Systems

Thermodynamic systems are constantly in flux, changing continuously from one microstate to another. If the system is in equilibrium, a given macrostate exists, and the system fluctuates from one microstate associated with that macrostate to another. This change is unobservable because we are only able to detect the macrostate. Equilibrium states have tremendously higher probability than nonequilibrium states, so it is highly unlikely that an equilibrium state will spontaneously change to a nonequilibrium state. For example, we do not observe a spontaneous split into the oxygen–nitrogen separation discussed in Section 22.6.

What if the system begins in a low-probability macrostate, however? What if the room *begins* with an oxygen–nitrogen separation? In this case, the system will progress from this low-probability macrostate to the much-higher probability state: the gases will disperse and mix throughout the room. Because entropy is related to probability, a spontaneous increase in entropy, such as in the latter situation, is natural. If the oxygen and nitrogen molecules were initially spread evenly throughout the room, a decrease in entropy would occur if the spontaneous splitting of molecules occurred.

One way of conceptualizing a change in entropy is to relate it to *energy spreading.* A natural tendency is for energy to undergo spatial spreading in time, representing an increase in entropy. If a basketball is dropped onto a floor, it bounces several times and eventually comes to rest. The initial gravitational potential energy in the basketball–Earth system has been transformed to internal energy in the ball and the floor. That energy is spreading outward by heat into the air and into regions of the floor farther from the drop point. In addition, some of the energy has spread throughout the room by sound. It would be unnatural for energy in the room and floor to reverse this motion and concentrate into the stationary ball so that it spontaneously begins to bounce again.

In the adiabatic free expansion of Section 22.3, the spreading of energy accompanies the spreading of the molecules as the gas rushes into the evacuated half of the container. If a warm object is placed in thermal contact with a cool object, energy transfers from the warm object to the cool one by heat, representing a spread of energy until it is distributed more evenly between the two objects.

Now consider a mathematical representation of this spreading of energy or, equivalently, the change in entropy. The original formulation of entropy in thermodynamics involves the transfer of energy by heat during a reversible process. Consider any infinitesimal process in which a system changes from one equilibrium state to another. If dQ_r is the amount of energy transferred by heat when the system follows a reversible path between the states, the change in entropy dS is equal to this amount of energy divided by the absolute temperature of the system:

Change in entropy for ▶
an infinitesimal process

$$dS = \frac{dQ_r}{T} \tag{22.12}$$

We have assumed the temperature is constant because the process is infinitesimal. Because entropy is a state variable, the change in entropy during a process depends only on the endpoints and therefore is independent of the actual path followed. Consequently, the entropy change for an irreversible process can be determined by calculating the entropy change for a *reversible* process that connects the same initial and final states.

The subscript r on the quantity dQ_r is a reminder that the transferred energy is to be measured along a reversible path even though the system may actually have followed some irreversible path. When energy is absorbed by the system, dQ_r is positive and the entropy of the system increases. When energy is expelled by the system, dQ_r is negative and the entropy of the system decreases. Notice that Equation 22.12 does not define entropy but rather the *change* in entropy. Hence, the meaningful quantity in describing a process is the *change* in entropy.

To calculate the change in entropy for a *finite* process, first recognize that T is generally not constant during the process. Therefore, we must integrate Equation 22.12:

Change in entropy for ▶
a finite process

$$\Delta S = \int_i^f dS = \int_i^f \frac{dQ_r}{T} \tag{22.13}$$

As with an infinitesimal process, the change in entropy ΔS of a system going from one state to another has the same value for *all* paths connecting the two states. That is, the finite change in entropy ΔS of a system depends only on the properties of the initial and final equilibrium states. Therefore, we are free to choose any convenient reversible path over which to evaluate the entropy in place of the actual path as long as the initial and final states are the same for both paths. This point is explored further on in this section.

From Equation 22.10, we see that a change in entropy is represented in the Boltzmann formulation as

$$\Delta S = k_B \ln\left(\frac{W_f}{W_i}\right) \tag{22.14}$$

where W_i and W_f represent the inital and final numbers of microstates, respectively, for the initial and final configurations of the system. If $W_f > W_i$, the final state is more probable than the the initial state (there are more choices of microstates), and the entropy increases.

Quick Quiz 22.5 An ideal gas is taken from an initial temperature T_i to a higher final temperature T_f along two different reversible paths. Path A is at constant pressure, and path B is at constant volume. What is the relation between the entropy changes of the gas for these paths? (a) $\Delta S_A > \Delta S_B$ (b) $\Delta S_A = \Delta S_B$ (c) $\Delta S_A < \Delta S_B$

Quick Quiz 22.6 True or False: The entropy change in an adiabatic process must be zero because $Q = 0$.

Example 22.7 **Change in Entropy: Melting**

A solid that has a latent heat of fusion L_f melts at a temperature T_m. Calculate the change in entropy of this substance when a mass m of the substance melts.

SOLUTION

Conceptualize We can choose any convenient reversible path to follow that connects the initial and final states. It is not necessary to identify the process or the path because, whatever it is, the effect is the same: energy enters the substance by heat and the substance melts. The mass m of the substance that melts is equal to Δm, the change in mass of the higher-phase (liquid) substance.

Categorize Because the melting takes place at a fixed temperature, we categorize the process as isothermal.

Analyze Use Equation 20.7 in Equation 22.13, noting that the temperature remains fixed:

$$\Delta S = \int \frac{dQ_r}{T} = \frac{1}{T_m} \int dQ_r = \frac{Q_r}{T_m} = \frac{L_f \Delta m}{T_m} = \boxed{\frac{L_f m}{T_m}}$$

Finalize Notice that Δm is positive so that ΔS is positive, representing that energy is added to the substance.

Entropy Change in a Carnot Cycle

Let's consider the changes in entropy that occur in a Carnot heat engine that operates between the temperatures T_c and T_h. In one cycle, the engine takes in energy $|Q_h|$ from the hot reservoir and expels energy $|Q_c|$ to the cold reservoir. These energy transfers occur only during the isothermal portions of the Carnot cycle; therefore, the constant temperature can be brought out in front of the integral sign in Equation 22.13. The integral then simply has the value of the total amount of energy transferred by heat. Therefore, the total change in entropy for one cycle is

$$\Delta S = \frac{|Q_h|}{T_h} - \frac{|Q_c|}{T_c} \tag{22.15}$$

where the minus sign represents that energy is leaving the engine at temperature T_c. In Example 22.3, we showed that for a Carnot engine,

$$\frac{|Q_c|}{|Q_h|} = \frac{T_c}{T_h}$$

Using this result in Equation 22.15, we find that the total change in entropy for a Carnot engine operating in a cycle is *zero:*

$$\Delta S = 0$$

Now consider a system taken through an arbitrary (non-Carnot) reversible cycle. Because entropy is a state variable—and hence depends only on the properties of a given equilibrium state—we conclude that $\Delta S = 0$ for *any* reversible cycle. In general, we can write this condition as

$$\oint \frac{dQ_r}{T} = 0 \quad \text{(reversible cycle)} \tag{22.16}$$

where the symbol $\oint$ indicates that the integration is over a closed path.

Entropy Change in a Free Expansion

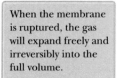

When the membrane is ruptured, the gas will expand freely and irreversibly into the full volume.

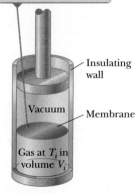

Insulating wall

Vacuum

Membrane

Gas at T_i in volume V_i

Figure 22.16 Adiabatic free expansion of a gas. The container is thermally insulated from its surroundings; therefore, $Q = 0$.

Let's again consider the adiabatic free expansion of a gas occupying an initial volume V_i (Fig. 22.16). In this situation, a membrane separating the gas from an evacuated region is broken and the gas expands to a volume V_f. This process is irreversible; the gas would not spontaneously crowd into half the volume after filling the entire volume. What is the change in entropy of the gas during this process? The process is neither reversible nor quasi-static. As shown in Section 20.6, the initial and final temperatures of the gas are the same.

To apply Equation 22.13, we cannot take $Q = 0$, the value for the irreversible process, but must instead find Q_r; that is, we must find an equivalent reversible path that shares the same initial and final states. A simple choice is an isothermal, reversible expansion in which the gas pushes slowly against a piston while energy enters the gas by heat from a reservoir to hold the temperature constant. Because T is constant in this process, Equation 22.13 gives

$$\Delta S = \int_i^f \frac{dQ_r}{T} = \frac{1}{T} \int_i^f dQ_r$$

For an isothermal process, the first law of thermodynamics specifies that $\int_i^f dQ_r$ is equal to the negative of the work done on the gas during the expansion from V_i to V_f, which is given by Equation 20.14. Using this result, we find that the entropy change for the gas is

$$\Delta S = nR \ln \left(\frac{V_f}{V_i} \right) \tag{22.17}$$

Because $V_f > V_i$, we conclude that ΔS is positive. This positive result indicates that the entropy of the gas *increases* as a result of the irreversible, adiabatic expansion.

It is easy to see that the energy has spread after the expansion. Instead of being concentrated in a relatively small space, the molecules and the energy associated with them are scattered over a larger region.

Entropy Change in Thermal Conduction

Let us now consider a system consisting of a hot reservoir and a cold reservoir that are in thermal contact with each other and isolated from the rest of the Universe. A process occurs during which energy Q is transferred by heat from the hot reservoir at temperature T_h to the cold reservoir at temperature T_c. The process as described is irreversible (energy would not spontaneously flow from cold to hot), so we must find an equivalent reversible process. The overall process is a combination of two processes: energy leaving the hot reservoir and energy entering the cold reservoir. We will calculate the entropy change for the reservoir in each process and add to obtain the overall entropy change.

Consider first the process of energy entering the cold reservoir. Although the reservoir has absorbed some energy, the temperature of the reservoir has not changed. The energy that has entered the reservoir is the same as that which would enter by means of a reversible, isothermal process. The same is true for energy leaving the hot reservoir.

Because the cold reservoir absorbs energy Q, its entropy increases by Q/T_c. At the same time, the hot reservoir loses energy Q, so its entropy change is $-Q/T_h$. Therefore, the change in entropy of the system is

$$\Delta S = \frac{Q}{T_c} + \frac{-Q}{T_h} = Q\left(\frac{1}{T_c} - \frac{1}{T_h}\right) > 0 \qquad \textbf{(22.18)}$$

This increase is consistent with our interpretation of entropy changes as representing the spreading of energy. In the initial configuration, the hot reservoir has excess internal energy relative to the cold reservoir. The process that occurs spreads the energy into a more equitable distribution between the two reservoirs.

Example 22.8 Adiabatic Free Expansion: Revisited

Let's verify that the macroscopic and microscopic approaches to the calculation of entropy lead to the same conclusion for the adiabatic free expansion of an ideal gas. Suppose the ideal gas in Figure 22.16 expands to four times its initial volume. As we have seen for this process, the initial and final temperatures are the same.

(A) Using a macroscopic approach, calculate the entropy change for the gas.

SOLUTION

Conceptualize Look back at Figure 22.16, which is a diagram of the system before the adiabatic free expansion. Imagine breaking the membrane so that the gas moves into the evacuated area. The expansion is irreversible.

Categorize We can replace the irreversible process with a reversible isothermal process between the same initial and final states. This approach is macroscopic, so we use a thermodynamic variable, in particular, the volume V.

....

Analyze Use Equation 22.17 to evaluate the entropy change:

$$\Delta S = nR \ln\left(\frac{V_f}{V_i}\right) = nR \ln\left(\frac{4V_i}{V_i}\right) = \boxed{nR \ln 4}$$

(B) Using statistical considerations, calculate the change in entropy for the gas and show that it agrees with the answer you obtained in part (A).

SOLUTION

Categorize This approach is microscopic, so we use variables related to the individual molecules.

....

Analyze As in the discussion leading to Equation 22.11, the number of microstates available to a single molecule in the initial volume V_i is $w_i = V_i/V_m$, where V_i is the initial volume of the gas and V_m is the microscopic volume occupied by the molecule. Use this number to find the number of available microstates for N molecules:

$$W_i = w_i^N = \left(\frac{V_i}{V_m}\right)^N$$

Find the number of available microstates for N molecules in the final volume $V_f = 4V_i$:

$$W_f = \left(\frac{V_f}{V_m}\right)^N = \left(\frac{4V_i}{V_m}\right)^N$$

continued

▶ **22.8** continued

Use Equation 22.14 to find the entropy change:

$$\Delta S = k_B \ln\left(\frac{W_f}{W_i}\right)$$

$$= k_B \ln\left(\frac{4V_i}{V_i}\right)^N = k_B \ln(4^N) = Nk_B \ln 4 = \boxed{nR \ln 4}$$

Finalize The answer is the same as that for part (A), which dealt with macroscopic parameters.

WHAT IF? In part (A), we used Equation 22.17, which was based on a reversible isothermal process connecting the initial and final states. Would you arrive at the same result if you chose a different reversible process?

Answer You *must* arrive at the same result because entropy is a state variable. For example, consider the two-step process in Figure 22.17: a reversible adiabatic expansion from V_i to $4V_i$ ($A \rightarrow B$) during which the temperature drops from T_1 to T_2 and a reversible isovolumetric process ($B \rightarrow C$) that takes the gas back to the initial temperature T_1. During the reversible adiabatic process, $\Delta S = 0$ because $Q_r = 0$.

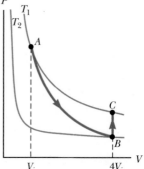

Figure 22.17 (Example 22.8) A gas expands to four times its initial volume and back to the initial temperature by means of a two-step process.

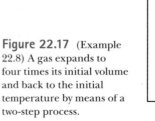

For the reversible isovolumetric process ($B \rightarrow C$), use Equation 22.13:

$$\Delta S = \int_i^f \frac{dQ_r}{T} = \int_{T_2}^{T_1} \frac{nC_V dT}{T} = nC_V \ln\left(\frac{T_1}{T_2}\right)$$

Find the ratio of temperature T_1 to T_2 from Equation 21.39 for the adiabatic process:

$$\frac{T_1}{T_2} = \left(\frac{4V_i}{V_i}\right)^{\gamma-1} = (4)^{\gamma-1}$$

Substitute to find ΔS:

$$\Delta S = nC_V \ln(4)^{\gamma-1} = nC_V(\gamma-1)\ln 4$$

$$= nC_V\left(\frac{C_P}{C_V} - 1\right)\ln 4 = n(C_P - C_V)\ln 4 = nR \ln 4$$

We do indeed obtain the exact same result for the entropy change.

22.8 Entropy and the Second Law

If we consider a system and its surroundings to include the entire Universe, the Universe is always moving toward a higher-probability macrostate, corresponding to the continuous spreading of energy. An alternative way of stating this behavior is as follows:

Entropy statement of ▶
the second law of
thermodynamics

The entropy of the Universe increases in all real processes.

This statement is yet another wording of the second law of thermodynamics that can be shown to be equivalent to the Kelvin-Planck and Clausius statements.

Let us show this equivalence first for the Clausius statement. Looking at Figure 22.5, we see that, if the heat pump operates in this manner, energy is spontaneously flowing from the cold reservoir to the hot reservoir without an input of energy by work. As a result, the energy in the system is not spreading evenly between the two reservoirs, but is *concentrating* in the hot reservoir. Consequently, if the Clausius statement of the second law is not true, then the entropy statement is also not true, demonstrating their equivalence.

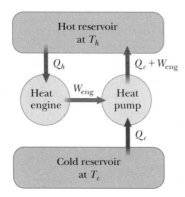

Figure 22.18 The impossible engine of Figure 22.3 transfers energy by work to a heat pump operating between two energy reservoirs. This situation is forbidden by the Clausius statement of the second law of thermodynamics.

For the equivalence of the Kelvin–Planck statement, consider Figure 22.18, which shows the impossible engine of Figure 22.3 connected to a heat pump operating between the same reservoirs. The output work of the engine is used to drive the heat pump. The net effect of this combination is that energy leaves the cold reservoir and is delivered to the hot reservoir without the input of work. (The work done by the engine on the heat pump is *internal* to the system of both devices.) This is forbidden by the Clausius statement of the second law, which we have shown to be equivalent to the entropy statement. Therefore, the Kelvin–Planck statement of the second law is also equivalent to the entropy statement.

When dealing with a system that is not isolated from its surroundings, remember that the increase in entropy described in the second law is that of the system *and* its surroundings. When a system and its surroundings interact in an irreversible process, the increase in entropy of one is greater than the decrease in entropy of the other. Hence, the change in entropy of the Universe must be greater than zero for an irreversible process and equal to zero for a reversible process.

We can check this statement of the second law for the calculations of entropy change that we made in Section 22.7. Consider first the entropy change in a free expansion, described by Equation 22.17. Because the free expansion takes place in an insulated container, no energy is transferred by heat from the surroundings. Therefore, Equation 22.17 represents the entropy change of the entire Universe. Because $V_f > V_i$, the entropy change of the Universe is positive, consistent with the second law.

Now consider the entropy change in thermal conduction, described by Equation 22.18. Let each reservoir be half the Universe. (The larger the reservoir, the better is the assumption that its temperature remains constant!) Then the entropy change of the Universe is represented by Equation 22.18. Because $T_h > T_c$, this entropy change is positive, again consistent with the second law. The positive entropy change is also consistent with the notion of energy spreading. The warm portion of the Universe has excess internal energy relative to the cool portion. Thermal conduction represents a spreading of the energy more equitably throughout the Universe.

Finally, let us look at the entropy change in a Carnot cycle, given by Equation 22.15. The entropy change of the engine itself is zero. The entropy change of the reservoirs is

$$\Delta S = \frac{|Q_c|}{T_c} - \frac{|Q_h|}{T_h}$$

In light of Equation 22.7, this entropy change is also zero. Therefore, the entropy change of the Universe is only that associated with the work done by the engine. A portion of that work will be used to change the mechanical energy of a system external to the engine: speed up the shaft of a machine, raise a weight, and so on. There is no change in internal energy of the external system due to this portion of the work, or, equivalently, no energy spreading, so the entropy change is again zero. The other portion of the work will be used to overcome various friction forces or other nonconservative forces in the external system. This process will cause an increase in internal energy of that system. That same increase in internal energy could have happened via a reversible thermodynamic process in which energy Q_r is transferred by heat, so the entropy change associated with that part of the work is positive. As a result, the overall entropy change of the Universe for the operation of the Carnot engine is positive, again consistent with the second law.

Ultimately, because real processes are irreversible, the entropy of the Universe should increase steadily and eventually reach a maximum value. At this value, assuming that the second law of thermodynamics, as formulated here on Earth, applies to the entire expanding Universe, the Universe will be in a state of uniform temperature and density. The total energy of the Universe will have spread more evenly throughout the Universe. All physical, chemical, and biological processes will have ceased at this time. This gloomy state of affairs is sometimes referred to as the *heat death* of the Universe.

Summary

Definitions

The **thermal efficiency** e of a heat engine is

$$e \equiv \frac{W_{\text{eng}}}{|Q_h|} = \frac{|Q_h| - |Q_c|}{|Q_h|} = 1 - \frac{|Q_c|}{|Q_h|} \quad \text{(22.2)}$$

The **microstate** of a system is the description of its individual components. The **macrostate** is a description of the system from a macroscopic point of view. A given macrostate can have many microstates.

From a microscopic viewpoint, the **entropy** of a given macrostate is defined as

$$S \equiv k_B \ln W \quad \text{(22.10)}$$

where k_B is Boltzmann's constant and W is the number of microstates of the system corresponding to the macrostate.

In a **reversible** process, the system can be returned to its initial conditions along the same path on a PV diagram, and every point along this path is an equilibrium state. A process that does not satisfy these requirements is **irreversible.**

Concepts and Principles

A **heat engine** is a device that takes in energy by heat and, operating in a cyclic process, expels a fraction of that energy by means of work. The net work done by a heat engine in carrying a working substance through a cyclic process ($\Delta E_{\text{int}} = 0$) is

$$W_{\text{eng}} = |Q_h| - |Q_c| \quad \text{(22.1)}$$

where $|Q_h|$ is the energy taken in from a hot reservoir and $|Q_c|$ is the energy expelled to a cold reservoir.

Two ways the **second law of thermodynamics** can be stated are as follows:

- It is impossible to construct a heat engine that, operating in a cycle, produces no effect other than the input of energy by heat from a reservoir and the performance of an equal amount of work (the Kelvin–Planck statement).
- It is impossible to construct a cyclical machine whose sole effect is to transfer energy continuously by heat from one object to another object at a higher temperature without the input of energy by work (the Clausius statement).

■ **Carnot's theorem** states that no real heat engine operating (irreversibly) between the temperatures T_c and T_h can be more efficient than an engine operating reversibly in a Carnot cycle between the same two temperatures.

■ The thermal efficiency of a heat engine operating in the Carnot cycle is

$$e_C = 1 - \frac{T_c}{T_h} \tag{22.8}$$

■ The macroscopic state of a system that has a large number of microstates has four qualities that are all related: (1) *uncertainty:* because of the large number of microstates, there is a large uncertainty as to which one actually exists; (2) *choice:* again because of the large number of microstates, there is a large number of choices from which to select as to which one exists; (3) *probability:* a macrostate with a large number of microstates is more likely to exist than one with a small number of microstates; (4) *missing information:* because of the large number of microstates, there is a high amount of missing information as to which one exists. For a thermodynamic system, all four of these can be related to the state variable of **entropy.**

■ The second law of thermodynamics states that when real (irreversible) processes occur, there is a spatial spreading of energy. This spreading of energy is related to a thermodynamic state variable called **entropy** S. Therefore, yet another way the second law can be stated is as follows:

• The entropy of the Universe increases in all real processes.

■ The **change in entropy** dS of a system during a process between two infinitesimally separated equilibrium states is

$$dS = \frac{dQ_r}{T} \tag{22.12}$$

where dQ_r is the energy transfer by heat for the system for a reversible process that connects the initial and final states.

■ The change in entropy of a system during an arbitrary finite process between an initial state and a final state is

$$\Delta S = \int_i^f \frac{dQ_r}{T} \tag{22.13}$$

The value of ΔS for the system is the same for all paths connecting the initial and final states. The change in entropy for a system undergoing any reversible, cyclic process is zero.

Objective Questions   **1.** denotes answer available in *Student Solutions Manual/Study Guide*

1. The second law of thermodynamics implies that the coefficient of performance of a refrigerator must be what? (a) less than 1 (b) less than or equal to 1 (c) greater than or equal to 1 (d) finite (e) greater than 0

2. Assume a sample of an ideal gas is at room temperature. What action will *necessarily* make the entropy of the sample increase? (a) Transfer energy into it by heat. (b) Transfer energy into it irreversibly by heat. (c) Do work on it. (d) Increase either its temperature or its volume, without letting the other variable decrease. (e) None of those choices is correct.

3. A refrigerator has 18.0 kJ of work done on it while 115 kJ of energy is transferred from inside its interior. What is its coefficient of performance? (a) 3.40 (b) 2.80 (c) 8.90 (d) 6.40 (e) 5.20

4. Of the following, which is *not* a statement of the second law of thermodynamics? (a) No heat engine operating in a cycle can absorb energy from a reservoir and use it entirely to do work. (b) No real engine operating between two energy reservoirs can be more efficient than a Carnot engine operating between the same two reservoirs. (c) When a system undergoes a change in state, the change in the internal energy of the system is the sum of the energy transferred to the system by heat and the work done on the system. (d) The entropy of the Universe increases in all natural processes. (e) Energy will not spontaneously transfer by heat from a cold object to a hot object.

5. Consider cyclic processes completely characterized by each of the following net energy inputs and outputs. In each case, the energy transfers listed are the *only* ones occurring. Classify each process as (a) possible, (b) impossible according to the first law of thermodynamics, (c) impossible according to the second law of thermodynamics, or (d) impossible according to both the first and second laws. (i) Input is 5 J of work, and output is 4 J of work. (ii) Input is 5 J of work, and output is 5 J of energy transferred by heat. (iii) Input is 5 J of energy transferred by electrical transmission, and output is 6 J of work. (iv) Input is 5 J of energy transferred by heat, and output is 5 J of energy transferred

by heat. **(v)** Input is 5 J of energy transferred by heat, and output is 5 J of work. **(vi)** Input is 5 J of energy transferred by heat, and output is 3 J of work plus 2 J of energy transferred by heat.

6. A compact air-conditioning unit is placed on a table inside a well-insulated apartment and is plugged in and turned on. What happens to the average temperature of the apartment? (a) It increases. (b) It decreases. (c) It remains constant. (d) It increases until the unit warms up and then decreases. (e) The answer depends on the initial temperature of the apartment.

7. A steam turbine operates at a boiler temperature of 450 K and an exhaust temperature of 300 K. What is the maximum theoretical efficiency of this system? (a) 0.240 (b) 0.500 (c) 0.333 (d) 0.667 (e) 0.150

8. A thermodynamic process occurs in which the entropy of a system changes by -8 J/K. According to the second law of thermodynamics, what can you conclude about the entropy change of the environment? (a) It must be $+8$ J/K or less. (b) It must be between $+8$ J/K and 0. (c) It must be equal to $+8$ J/K. (d) It must be $+8$ J/K or more. (e) It must be zero.

9. A sample of a monatomic ideal gas is contained in a cylinder with a piston. Its state is represented by the dot in the *PV* diagram shown in Figure OQ22.9. Arrows *A* through *E* represent isobaric, isothermal, adiabatic, and isovolumetric processes that the sample can undergo. In each process except *D*, the volume changes by a factor of 2. All five processes are reversible. Rank the processes according to the change in entropy of the gas from the largest positive value to the

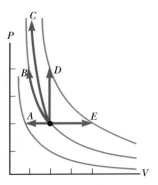

Figure OQ22.9

largest-magnitude negative value. In your rankings, display any cases of equality.

10. An engine does 15.0 kJ of work while exhausting 37.0 kJ to a cold reservoir. What is the efficiency of the engine? (a) 0.150 (b) 0.288 (c) 0.333 (d) 0.450 (e) 1.20

11. The arrow *OA* in the *PV* diagram shown in Figure OQ22.11 represents a reversible adiabatic expansion of an ideal gas. The same sample of gas, starting from the same state *O*, now undergoes an adiabatic free expansion to the same final volume. What point on the diagram could represent the final state of the gas? (a) the same point *A* as for the reversible expansion (b) point *B* (c) point *C* (d) any of those choices (e) none of those choices

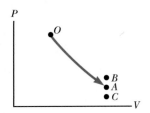

Figure OQ22.11

Conceptual Questions

1. denotes answer available in *Student Solutions Manual/Study Guide*

1. The energy exhaust from a certain coal-fired electric generating station is carried by "cooling water" into Lake Ontario. The water is warm from the viewpoint of living things in the lake. Some of them congregate around the outlet port and can impede the water flow. (a) Use the theory of heat engines to explain why this action can reduce the electric power output of the station. (b) An engineer says that the electric output is reduced because of "higher back pressure on the turbine blades." Comment on the accuracy of this statement.

2. Discuss three different common examples of natural processes that involve an increase in entropy. Be sure to account for all parts of each system under consideration.

3. Does the second law of thermodynamics contradict or correct the first law? Argue for your answer.

4. "The first law of thermodynamics says you can't really win, and the second law says you can't even break even." Explain how this statement applies to a particular device or process; alternatively, argue against the statement.

5. "Energy is the mistress of the Universe, and entropy is her shadow." Writing for an audience of general readers, argue for this statement with at least two examples. Alternatively, argue for the view that entropy is like an executive who instantly determines what will happen, whereas energy is like a bookkeeper telling us how little we can afford. (Arnold Sommerfeld suggested the idea for this question.)

6. (a) Give an example of an irreversible process that occurs in nature. (b) Give an example of a process in nature that is nearly reversible.

7. The device shown in Figure CQ22.7, called a thermoelectric converter, uses a series of semiconductor cells to transform internal energy to electric potential energy, which we will study in Chapter 25. In the photograph on the left, both legs of the device are at the same temperature and no electric potential energy is produced. When one leg is at a higher temperature than the other as shown in the photograph on the right, however, electric potential energy is produced as

the device extracts energy from the hot reservoir and drives a small electric motor. (a) Why is the difference in temperature necessary to produce electric potential energy in this demonstration? (b) In what sense does this intriguing experiment demonstrate the second law of thermodynamics?

Courtesy of PASCO Scientific Company

Figure CQ22.7

8. A steam-driven turbine is one major component of an electric power plant. Why is it advantageous to have the temperature of the steam as high as possible?

9. Discuss the change in entropy of a gas that expands (a) at constant temperature and (b) adiabatically.

10. Suppose your roommate cleans and tidies up your messy room after a big party. Because she is creating more order, does this process represent a violation of the second law of thermodynamics?

11. Is it possible to construct a heat engine that creates no thermal pollution? Explain.

12. (a) If you shake a jar full of jelly beans of different sizes, the larger beans tend to appear near the top and the smaller ones tend to fall to the bottom. Why? (b) Does this process violate the second law of thermodynamics?

13. What are some factors that affect the efficiency of automobile engines?

Problems available in ENHANCED **WebAssign**   Access end-of-chapter problems online at www.webassign.net

Section 22.1 Heat Engines and the Second Law of Thermodynamics
Problems 1–7

Section 22.2 Heat Pumps and Refrigerators
Problems 8–13

Section 22.3 Reversible and Irreversible Processes
Section 22.4 The Carnot Engine
Problems 14–35

Section 22.5 Gasoline and Diesel Engines
Problems 36–38

Section 22.6 Entropy
Problems 39–41

Section 22.7 Changes in Entropy for Thermodynamic Systems
Section 22.8 Entropy and the Second Law
Problems 42–55

Additional Problems
Problems 56–80

Challenge Problems
Problems 81–82

Solutions to the following Problems are available in the *Student Solutions Manual/Study Guide:*
22.1, 22.15, 22.20, 22.22, 22.27, 22.37, 22.38, 22.39, 22.43, 22.48, 22.51, 22.56, 22.63, 22.67, 22.74, 22.76, and 22.81

List of Enhanced Problems

Problem Number	Targeted Feedback in Enhanced WebAssign	Analysis Model Tutorial in Enhanced WebAssign	Master It in Enhanced WebAssign	Watch It in Enhanced WebAssign
22.1	✓		✓	
22.3	✓			✓
22.7	✓			✓
22.8	✓			✓
22.15	✓		✓	
22.17	✓			✓
22.22			✓	
22.27	✓		✓	
22.28	✓			✓
22.29	✓		✓	
22.31	✓			✓
22.37			✓	
22.42	✓			✓
22.45		✓		
22.46		✓		
22.47		✓		
22.55	✓		✓	✓
22.60		✓		
22.63			✓	
22.64	✓		✓	

Electricity and Magnetism

A Transrapid maglev train pulls into a station in Shanghai, China. The word *maglev* is an abbreviated form of *magnetic levitation*. This train makes no physical contact with its rails; its weight is totally supported by electromagnetic forces. In this part of the book, we will study these forces. *(OTHK/Asia Images/Jupiterimages)*

We now study the branch of physics concerned with electric and magnetic phenomena. The laws of electricity and magnetism play a central role in the operation of such devices as smartphones, televisions, electric motors, computers, high-energy accelerators, and other electronic devices. More fundamentally, the interatomic and intermolecular forces responsible for the formation of solids and liquids are electric in origin.

Evidence in Chinese documents suggests magnetism was observed as early as 2000 BC. The ancient Greeks observed electric and magnetic phenomena possibly as early as 700 BC. The Greeks knew about magnetic forces from observations that the naturally occurring stone *magnetite* (Fe_3O_4) is attracted to iron. (The word *electric* comes from *elecktron*, the Greek word for "amber." The word *magnetic* comes from *Magnesia*, the name of the district of Greece where magnetite was first found.)

Not until the early part of the nineteenth century did scientists establish that electricity and magnetism are related phenomena. In 1819, Hans Oersted discovered that a compass needle is deflected when placed near a circuit carrying an electric current. In 1831, Michael Faraday and, almost simultaneously, Joseph Henry showed that when a wire is moved near a magnet (or, equivalently, when a magnet is moved near a wire), an electric current is established in the wire. In 1873, James Clerk Maxwell used these observations and other experimental facts as a basis for formulating the laws of electromagnetism as we know them today. (*Electromagnetism* is a name given to the combined study of electricity and magnetism.)

Maxwell's contributions to the field of electromagnetism were especially significant because the laws he formulated are basic to *all* forms of electromagnetic phenomena. His work is as important as Newton's work on the laws of motion and the theory of gravitation. ■

Electric Fields

This young woman is enjoying the effects of electrically charging her body. Each individual hair on her head becomes charged and exerts a repulsive force on the other hairs, resulting in the "stand-up" hairdo seen here. *(Ted Kinsman / Photo Researchers, Inc.)*

In this chapter, we begin the study of electromagnetism. The first link that we will make to our previous study is through the concept of *force*. The electromagnetic force between charged particles is one of the fundamental forces of nature. We begin by describing some basic properties of one manifestation of the electromagnetic force, the electric force. We then discuss Coulomb's law, which is the fundamental law governing the electric force between any two charged particles. Next, we introduce the concept of an electric field associated with a charge distribution and describe its effect on other charged particles. We then show how to use Coulomb's law to calculate the electric field for a given charge distribution. The chapter concludes with a discussion of the motion of a charged particle in a uniform electric field.

The second link between electromagnetism and our previous study is through the concept of *energy*. We will discuss that connection in Chapter 25.

23.1 Properties of Electric Charges

A number of simple experiments demonstrate the existence of electric forces. For example, after rubbing a balloon on your hair on a dry day, you will find that the balloon attracts bits of paper. The attractive force is often strong enough to suspend the paper from the balloon.

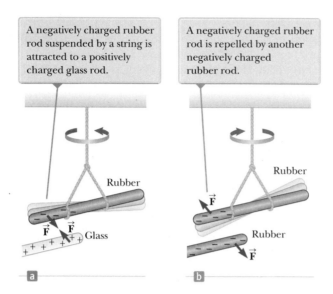

A negatively charged rubber rod suspended by a string is attracted to a positively charged glass rod.

A negatively charged rubber rod is repelled by another negatively charged rubber rod.

Rubber

$\vec{F}$

Glass

$+ + + + + + +$

Rubber

$\vec{F}$

Rubber

$\vec{F}$

a

b

Figure 23.1 The electric force between (a) oppositely charged objects and (b) like-charged objects.

When materials behave in this way, they are said to be *electrified* or to have become **electrically charged.** You can easily electrify your body by vigorously rubbing your shoes on a wool rug. Evidence of the electric charge on your body can be detected by lightly touching (and startling) a friend. Under the right conditions, you will see a spark when you touch and both of you will feel a slight tingle. (Experiments such as these work best on a dry day because an excessive amount of moisture in the air can cause any charge you build up to "leak" from your body to the Earth.)

In a series of simple experiments, it was found that there are two kinds of electric charges, which were given the names **positive** and **negative** by Benjamin Franklin (1706–1790). Electrons are identified as having negative charge, and protons are positively charged. To verify that there are two types of charge, suppose a hard rubber rod that has been rubbed on fur is suspended by a string as shown in Figure 23.1. When a glass rod that has been rubbed on silk is brought near the rubber rod, the two attract each other (Fig. 23.1a). On the other hand, if two charged rubber rods (or two charged glass rods) are brought near each other as shown in Figure 23.1b, the two repel each other. This observation shows that the rubber and glass have two different types of charge on them. On the basis of these observations, we conclude that **charges of the same sign repel one another and charges with opposite signs attract one another.**

Using the convention suggested by Franklin, the electric charge on the glass rod is called positive and that on the rubber rod is called negative. Therefore, any charged object attracted to a charged rubber rod (or repelled by a charged glass rod) must have a positive charge, and any charged object repelled by a charged rubber rod (or attracted to a charged glass rod) must have a negative charge.

Another important aspect of electricity that arises from experimental observations is that **electric charge is always conserved** in an isolated system. That is, when one object is rubbed against another, charge is not created in the process. The electrified state is due to a *transfer* of charge from one object to the other. One object gains some amount of negative charge while the other gains an equal amount of positive charge. For example, when a glass rod is rubbed on silk as in Figure 23.2, the silk obtains a negative charge equal in magnitude to the positive charge on the glass rod. We now know from our understanding of atomic structure that electrons are transferred in the rubbing process from the glass to the silk. Similarly, when rubber is rubbed on fur, electrons are transferred from the fur to the rubber, giving the rubber a net negative charge and the fur a net positive charge. This process works because neutral, uncharged matter contains as many positive charges (protons within atomic nuclei)

Because of conservation of charge, each electron adds negative charge to the silk and an equal positive charge is left on the glass rod.

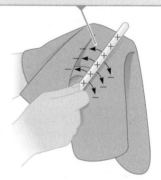

Figure 23.2 When a glass rod is rubbed with silk, electrons are transferred from the glass to the silk.

◀ Electric charge is conserved

The neutral sphere has equal numbers of positive and negative charges.

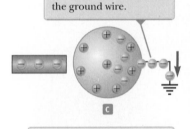

a

Electrons redistribute when a charged rod is brought close.

b

Some electrons leave the grounded sphere through the ground wire.

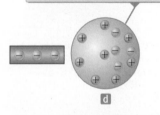

c

The excess positive charge is nonuniformly distributed.

d

The remaining electrons redistribute uniformly, and there is a net uniform distribution of positive charge on the sphere.

e

Figure 23.3 Charging a metallic object by *induction*. (a) A neutral metallic sphere. (b) A charged rubber rod is placed near the sphere. (c) The sphere is grounded. (d) The ground connection is removed. (e) The rod is removed.

as negative charges (electrons). Conservation of electric charge for an isolated system is like conservation of energy, momentum, and angular momentum, but we don't identify an analysis model for this conservation principle because it is not used often enough in the mathematical solution to problems.

In 1909, Robert Millikan (1868–1953) discovered that electric charge always occurs as integral multiples of a fundamental amount of charge e (see Section 25.7). In modern terms, the electric charge q is said to be **quantized,** where q is the standard symbol used for charge as a variable. That is, electric charge exists as discrete "packets," and we can write $q = \pm Ne$, where N is some integer. Other experiments in the same period showed that the electron has a charge $-e$ and the proton has a charge of equal magnitude but opposite sign $+e$. Some particles, such as the neutron, have no charge.

> **Quick Quiz 23.1** Three objects are brought close to each other, two at a time. When objects A and B are brought together, they repel. When objects B and C are brought together, they also repel. Which of the following are true? **(a)** Objects A and C possess charges of the same sign. **(b)** Objects A and C possess charges of opposite sign. **(c)** All three objects possess charges of the same sign. **(d)** One object is neutral. **(e)** Additional experiments must be performed to determine the signs of the charges.

23.2 Charging Objects by Induction

It is convenient to classify materials in terms of the ability of electrons to move through the material:

> Electrical **conductors** are materials in which some of the electrons are free electrons[1] that are not bound to atoms and can move relatively freely through the material; electrical **insulators** are materials in which all electrons are bound to atoms and cannot move freely through the material.

Materials such as glass, rubber, and dry wood fall into the category of electrical insulators. When such materials are charged by rubbing, only the area rubbed becomes charged and the charged particles are unable to move to other regions of the material.

In contrast, materials such as copper, aluminum, and silver are good electrical conductors. When such materials are charged in some small region, the charge readily distributes itself over the entire surface of the material.

Semiconductors are a third class of materials, and their electrical properties are somewhere between those of insulators and those of conductors. Silicon and germanium are well-known examples of semiconductors commonly used in the fabrication of a variety of electronic chips used in computers, cellular telephones, and home theater systems. The electrical properties of semiconductors can be changed over many orders of magnitude by the addition of controlled amounts of certain atoms to the materials.

To understand how to charge a conductor by a process known as **induction,** consider a neutral (uncharged) conducting sphere insulated from the ground as shown in Figure 23.3a. There are an equal number of electrons and protons in the sphere if the charge on the sphere is exactly zero. When a negatively charged rubber rod is brought near the sphere, electrons in the region nearest the rod experience a repulsive force and migrate to the opposite side of the sphere. This migration leaves

---

[1]A metal atom contains one or more outer electrons, which are weakly bound to the nucleus. When many atoms combine to form a metal, the *free electrons* are these outer electrons, which are not bound to any one atom. These electrons move about the metal in a manner similar to that of gas molecules moving in a container.

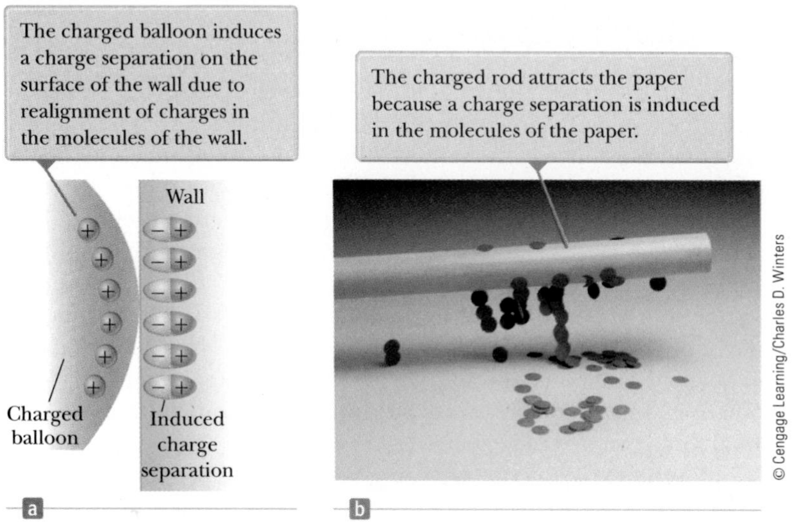

The charged balloon induces a charge separation on the surface of the wall due to realignment of charges in the molecules of the wall.

The charged rod attracts the paper because a charge separation is induced in the molecules of the paper.

Figure 23.4 (a) A charged balloon is brought near an insulating wall. (b) A charged rod is brought close to bits of paper.

Wall

Charged balloon

Induced charge separation

© Cengage Learning/Charles D. Winters

a

b

the side of the sphere near the rod with an effective positive charge because of the diminished number of electrons as in Figure 23.3b. (The left side of the sphere in Figure 23.3b is positively charged *as if* positive charges moved into this region, but remember that only electrons are free to move.) This process occurs even if the rod never actually touches the sphere. If the same experiment is performed with a conducting wire connected from the sphere to the Earth (Fig. 23.3c), some of the electrons in the conductor are so strongly repelled by the presence of the negative charge in the rod that they move out of the sphere through the wire and into the Earth. The symbol ⏚ at the end of the wire in Figure 23.3c indicates that the wire is connected to **ground**, which means a reservoir, such as the Earth, that can accept or provide electrons freely with negligible effect on its electrical characteristics. If the wire to ground is then removed (Fig. 23.3d), the conducting sphere contains an excess of *induced* positive charge because it has fewer electrons than it needs to cancel out the positive charge of the protons. When the rubber rod is removed from the vicinity of the sphere (Fig. 23.3e), this induced positive charge remains on the ungrounded sphere. Notice that the rubber rod loses none of its negative charge during this process.

Charging an object by induction requires no contact with the object inducing the charge. That is in contrast to charging an object by rubbing (that is, by *conduction*), which does require contact between the two objects.

A process similar to induction in conductors takes place in insulators. In most neutral molecules, the center of positive charge coincides with the center of negative charge. In the presence of a charged object, however, these centers inside each molecule in an insulator may shift slightly, resulting in more positive charge on one side of the molecule than on the other. This realignment of charge within individual molecules produces a layer of charge on the surface of the insulator as shown in Figure 23.4a. The proximity of the positive charges on the surface of the object and the negative charges on the surface of the insulator results in an attractive force between the object and the insulator. Your knowledge of induction in insulators should help you explain why a charged rod attracts bits of electrically neutral paper as shown in Figure 23.4b.

Quick Quiz 23.2 Three objects are brought close to one another, two at a time. When objects A and B are brought together, they attract. When objects B and C are brought together, they repel. Which of the following are necessarily true? **(a)** Objects A and C possess charges of the same sign. **(b)** Objects A and C possess charges of opposite sign. **(c)** All three objects possess charges of the same sign. **(d)** One object is neutral. **(e)** Additional experiments must be performed to determine information about the charges on the objects.

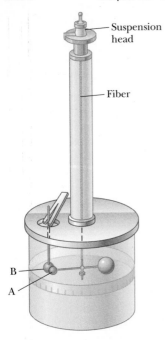

Figure 23.5 Coulomb's balance, used to establish the inverse-square law for the electric force.

23.3 Coulomb's Law

Charles Coulomb measured the magnitudes of the electric forces between charged objects using the torsion balance, which he invented (Fig. 23.5). The operating principle of the torsion balance is the same as that of the apparatus used by Cavendish to measure the density of the Earth (see Section 13.1), with the electrically neutral spheres replaced by charged ones. The electric force between charged spheres A and B in Figure 23.5 causes the spheres to either attract or repel each other, and the resulting motion causes the suspended fiber to twist. Because the restoring torque of the twisted fiber is proportional to the angle through which the fiber rotates, a measurement of this angle provides a quantitative measure of the electric force of attraction or repulsion. Once the spheres are charged by rubbing, the electric force between them is very large compared with the gravitational attraction, and so the gravitational force can be neglected.

From Coulomb's experiments, we can generalize the properties of the **electric force** (sometimes called the *electrostatic force*) between two stationary charged particles. We use the term **point charge** to refer to a charged particle of zero size. The electrical behavior of electrons and protons is very well described by modeling them as point charges. From experimental observations, we find that the magnitude of the electric force (sometimes called the *Coulomb force*) between two point charges is given by **Coulomb's law.**

◀ Coulomb's law

$$F_e = k_e \frac{|q_1||q_2|}{r^2} \tag{23.1}$$

where k_e is a constant called the **Coulomb constant.** In his experiments, Coulomb was able to show that the value of the exponent of r was 2 to within an uncertainty of a few percent. Modern experiments have shown that the exponent is 2 to within an uncertainty of a few parts in 10^{16}. Experiments also show that the electric force, like the gravitational force, is conservative.

The value of the Coulomb constant depends on the choice of units. The SI unit of charge is the **coulomb** (C). The Coulomb constant k_e in SI units has the value

◀ Coulomb constant

$$k_e = 8.987\ 6 \times 10^9\ \text{N} \cdot \text{m}^2/\text{C}^2 \tag{23.2}$$

This constant is also written in the form

$$k_e = \frac{1}{4\pi\epsilon_0} \tag{23.3}$$

where the constant ϵ_0 (Greek letter epsilon) is known as the **permittivity of free space** and has the value

$$\epsilon_0 = 8.854\ 2 \times 10^{-12}\ \text{C}^2/\text{N} \cdot \text{m}^2 \tag{23.4}$$

The smallest unit of free charge e known in nature,[2] the charge on an electron $(-e)$ or a proton $(+e)$, has a magnitude

$$e = 1.602\ 18 \times 10^{-19}\ \text{C} \tag{23.5}$$

Therefore, 1 C of charge is approximately equal to the charge of 6.24×10^{18} electrons or protons. This number is very small when compared with the number of free electrons in 1 cm³ of copper, which is on the order of 10^{23}. Nevertheless, 1 C is a substantial amount of charge. In typical experiments in which a rubber or glass rod is charged by friction, a net charge on the order of 10^{-6} C is obtained. In other

Charles Coulomb
French physicist (1736–1806)
Coulomb's major contributions to science were in the areas of electrostatics and magnetism. During his lifetime, he also investigated the strengths of materials, thereby contributing to the field of structural mechanics. In ergonomics, his research provided an understanding of the ways in which people and animals can best do work.

---

[2]No unit of charge smaller than e has been detected on a free particle; current theories, however, propose the existence of particles called *quarks* having charges $-e/3$ and $2e/3$. Although there is considerable experimental evidence for such particles inside nuclear matter, *free* quarks have never been detected. We discuss other properties of quarks in Chapter 46.

Table 23.1	**Charge and Mass of the Electron, Proton, and Neutron**	
Particle	Charge (C)	Mass (kg)
Electron (e)	$-1.602\ 176\ 5 \times 10^{-19}$	$9.109\ 4 \times 10^{-31}$
Proton (p)	$+1.602\ 176\ 5 \times 10^{-19}$	$1.672\ 62 \times 10^{-27}$
Neutron (n)	0	$1.674\ 93 \times 10^{-27}$

words, only a very small fraction of the total available charge is transferred between the rod and the rubbing material.

The charges and masses of the electron, proton, and neutron are given in Table 23.1. Notice that the electron and proton are identical in the magnitude of their charge but vastly different in mass. On the other hand, the proton and neutron are similar in mass but vastly different in charge. Chapter 46 will help us understand these interesting properties.

Example 23.1   The Hydrogen Atom

The electron and proton of a hydrogen atom are separated (on the average) by a distance of approximately 5.3×10^{-11} m. Find the magnitudes of the electric force and the gravitational force between the two particles.

SOLUTION

Conceptualize Think about the two particles separated by the very small distance given in the problem statement. In Chapter 13, we mentioned that the gravitational force between an electron and a proton is very small compared to the electric force between them, so we expect this to be the case with the results of this example.

Categorize The electric and gravitational forces will be evaluated from universal force laws, so we categorize this example as a substitution problem.

Use Coulomb's law to find the magnitude of the electric force:

$$F_e = k_e \frac{|e||-e|}{r^2} = (8.988 \times 10^9 \ \text{N} \cdot \text{m}^2/\text{C}^2) \frac{(1.60 \times 10^{-19} \ \text{C})^2}{(5.3 \times 10^{-11} \ \text{m})^2}$$

$$= \boxed{8.2 \times 10^{-8} \ \text{N}}$$

Use Newton's law of universal gravitation and Table 23.1 (for the particle masses) to find the magnitude of the gravitational force:

$$F_g = G \frac{m_e m_p}{r^2}$$

$$= (6.674 \times 10^{-11} \ \text{N} \cdot \text{m}^2/\text{kg}^2) \frac{(9.11 \times 10^{-31} \ \text{kg})(1.67 \times 10^{-27} \ \text{kg})}{(5.3 \times 10^{-11} \ \text{m})^2}$$

$$= \boxed{3.6 \times 10^{-47} \ \text{N}}$$

The ratio $F_e/F_g \approx 2 \times 10^{39}$. Therefore, the gravitational force between charged atomic particles is negligible when compared with the electric force. Notice the similar forms of Newton's law of universal gravitation and Coulomb's law of electric forces. Other than the magnitude of the forces between elementary particles, what is a fundamental difference between the two forces?

When dealing with Coulomb's law, remember that force is a vector quantity and must be treated accordingly. Coulomb's law expressed in vector form for the electric force exerted by a charge q_1 on a second charge q_2, written $\vec{\mathbf{F}}_{12}$, is

$$\vec{\mathbf{F}}_{12} = k_e \frac{q_1 q_2}{r^2} \hat{\mathbf{r}}_{12} \qquad\qquad (23.6)$$

◀ Vector form of Coulomb's law

where $\hat{\mathbf{r}}_{12}$ is a unit vector directed from q_1 toward q_2 as shown in Figure 23.6a (page 546). Because the electric force obeys Newton's third law, the electric force exerted by q_2 on q_1 is equal in magnitude to the force exerted by q_1 on q_2 and in the opposite direction; that is, $\vec{\mathbf{F}}_{21} = -\vec{\mathbf{F}}_{12}$. Finally, Equation 23.6 shows that if q_1 and q_2 have the

Figure 23.6 Two point charges separated by a distance r exert a force on each other that is given by Coulomb's law. The force $\vec{\mathbf{F}}_{21}$ exerted by q_2 on q_1 is equal in magnitude and opposite in direction to the force $\vec{\mathbf{F}}_{12}$ exerted by q_1 on q_2.

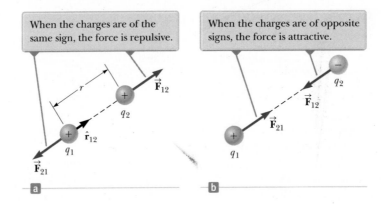

When the charges are of the same sign, the force is repulsive.

When the charges are of opposite signs, the force is attractive.

same sign as in Figure 23.6a, the product $q_1 q_2$ is positive and the electric force on one particle is directed away from the other particle. If q_1 and q_2 are of opposite sign as shown in Figure 23.6b, the product $q_1 q_2$ is negative and the electric force on one particle is directed toward the other particle. These signs describe the *relative* direction of the force but not the *absolute* direction. A negative product indicates an attractive force, and a positive product indicates a repulsive force. The *absolute* direction of the force on a charge depends on the location of the other charge. For example, if an x axis lies along the two charges in Figure 23.6a, the product $q_1 q_2$ is positive, but $\vec{\mathbf{F}}_{12}$ points in the positive x direction and $\vec{\mathbf{F}}_{21}$ points in the negative x direction.

When more than two charges are present, the force between any pair of them is given by Equation 23.6. Therefore, the resultant force on any one of them equals the vector sum of the forces exerted by the other individual charges. For example, if four charges are present, the resultant force exerted by particles 2, 3, and 4 on particle 1 is

$$\vec{\mathbf{F}}_1 = \vec{\mathbf{F}}_{21} + \vec{\mathbf{F}}_{31} + \vec{\mathbf{F}}_{41}$$

Quick Quiz 23.3 Object A has a charge of $+2\ \mu\text{C}$, and object B has a charge of $+6\ \mu\text{C}$. Which statement is true about the electric forces on the objects?
(a) $\vec{\mathbf{F}}_{AB} = -3\vec{\mathbf{F}}_{BA}$ **(b)** $\vec{\mathbf{F}}_{AB} = -\vec{\mathbf{F}}_{BA}$ **(c)** $3\vec{\mathbf{F}}_{AB} = -\vec{\mathbf{F}}_{BA}$ **(d)** $\vec{\mathbf{F}}_{AB} = 3\vec{\mathbf{F}}_{BA}$
(e) $\vec{\mathbf{F}}_{AB} = \vec{\mathbf{F}}_{BA}$ **(f)** $3\vec{\mathbf{F}}_{AB} = \vec{\mathbf{F}}_{BA}$

Example 23.2   **Find the Resultant Force**

Consider three point charges located at the corners of a right triangle as shown in Figure 23.7, where $q_1 = q_3 = 5.00\ \mu\text{C}$, $q_2 = -2.00\ \mu\text{C}$, and $a = 0.100$ m. Find the resultant force exerted on q_3.

SOLUTION

Conceptualize Think about the net force on q_3. Because charge q_3 is near two other charges, it will experience two electric forces. These forces are exerted in different directions as shown in Figure 23.7. Based on the forces shown in the figure, estimate the direction of the net force vector.

Categorize Because two forces are exerted on charge q_3, we categorize this example as a vector addition problem.

Analyze The directions of the individual forces exerted by q_1 and q_2 on q_3 are shown in Figure 23.7. The force $\vec{\mathbf{F}}_{23}$ exerted by q_2 on q_3 is attractive because q_2 and q_3 have opposite signs. In the coordinate system shown in Figure 23.7, the attractive force $\vec{\mathbf{F}}_{23}$ is to the left (in the negative x direction).

The force $\vec{\mathbf{F}}_{13}$ exerted by q_1 on q_3 is repulsive because both charges are positive. The repulsive force $\vec{\mathbf{F}}_{13}$ makes an angle of $45.0°$ with the x axis.

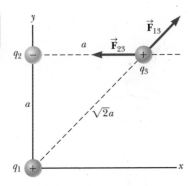

Figure 23.7 (Example 23.2) The force exerted by q_1 on q_3 is $\vec{\mathbf{F}}_{13}$. The force exerted by q_2 on q_3 is $\vec{\mathbf{F}}_{23}$. The resultant force $\vec{\mathbf{F}}_3$ exerted on q_3 is the vector sum $\vec{\mathbf{F}}_{13} + \vec{\mathbf{F}}_{23}$.

▶ **23.2** continued

Use Equation 23.1 to find the magnitude of $\vec{\mathbf{F}}_{23}$:

$$F_{23} = k_e \frac{|q_2||q_3|}{a^2}$$

$$= (8.988 \times 10^9 \text{ N} \cdot \text{m}^2/\text{C}^2) \frac{(2.00 \times 10^{-6} \text{ C})(5.00 \times 10^{-6} \text{ C})}{(0.100 \text{ m})^2} = 8.99 \text{ N}$$

Find the magnitude of the force $\vec{\mathbf{F}}_{13}$:

$$F_{13} = k_e \frac{|q_1||q_3|}{(\sqrt{2}\, a)^2}$$

$$= (8.988 \times 10^9 \text{ N} \cdot \text{m}^2/\text{C}^2) \frac{(5.00 \times 10^{-6} \text{ C})(5.00 \times 10^{-6} \text{ C})}{2(0.100 \text{ m})^2} = 11.2 \text{ N}$$

Find the x and y components of the force $\vec{\mathbf{F}}_{13}$:

$$F_{13x} = (11.2 \text{ N}) \cos 45.0° = 7.94 \text{ N}$$
$$F_{13y} = (11.2 \text{ N}) \sin 45.0° = 7.94 \text{ N}$$

Find the components of the resultant force acting on q_3:

$$F_{3x} = F_{13x} + F_{23x} = 7.94 \text{ N} + (-8.99 \text{ N}) = -1.04 \text{ N}$$
$$F_{3y} = F_{13y} + F_{23y} = 7.94 \text{ N} + 0 = 7.94 \text{ N}$$

Express the resultant force acting on q_3 in unit-vector form:

$$\vec{\mathbf{F}}_3 = (-1.04\hat{\mathbf{i}} + 7.94\hat{\mathbf{j}}) \text{ N}$$

....

Finalize The net force on q_3 is upward and toward the left in Figure 23.7. If q_3 moves in response to the net force, the distances between q_3 and the other charges change, so the net force changes. Therefore, if q_3 is free to move, it can be modeled as a particle under a net force as long as it is recognized that the force exerted on q_3 is *not* constant. As a reminder, we display most numerical values to three significant figures, which leads to operations such as 7.94 N + (−8.99 N) = −1.04 N above. If you carry all intermediate results to more significant figures, you will see that this operation is correct.

WHAT IF? What if the signs of all three charges were changed to the opposite signs? How would that affect the result for $\vec{\mathbf{F}}_3$?

Answer The charge q_3 would still be attracted toward q_2 and repelled from q_1 with forces of the same magnitude. Therefore, the final result for $\vec{\mathbf{F}}_3$ would be the same.

Example 23.3   **Where Is the Net Force Zero?**   **AM**

Three point charges lie along the x axis as shown in Figure 23.8. The positive charge $q_1 = 15.0\ \mu\text{C}$ is at $x = 2.00$ m, the positive charge $q_2 = 6.00\ \mu\text{C}$ is at the origin, and the net force acting on q_3 is zero. What is the x coordinate of q_3?

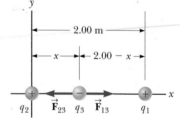

Figure 23.8 (Example 23.3) Three point charges are placed along the x axis. If the resultant force acting on q_3 is zero, the force $\vec{\mathbf{F}}_{13}$ exerted by q_1 on q_3 must be equal in magnitude and opposite in direction to the force $\vec{\mathbf{F}}_{23}$ exerted by q_2 on q_3.

SOLUTION

Conceptualize Because q_3 is near two other charges, it experiences two electric forces. Unlike the preceding example, however, the forces lie along the same line in this problem as indicated in Figure 23.8. Because q_3 is negative and q_1 and q_2 are positive, the forces $\vec{\mathbf{F}}_{13}$ and $\vec{\mathbf{F}}_{23}$ are both attractive. Because q_2 is the smaller charge, the position of q_3 at which the force is zero should be closer to q_2 than to q_1.

Categorize Because the net force on q_3 is zero, we model the point charge as a *particle in equilibrium*.

....

Analyze Write an expression for the net force on charge q_3 when it is in equilibrium:

$$\vec{\mathbf{F}}_3 = \vec{\mathbf{F}}_{23} + \vec{\mathbf{F}}_{13} = -k_e \frac{|q_2||q_3|}{x^2}\hat{\mathbf{i}} + k_e \frac{|q_1|}{(9}$$

Move the second term to the right side of the equation and set the coefficients of the unit vector $\hat{\mathbf{i}}$ equal:

$$k_e \frac{|q_2||q_3|}{x^2} = k_e \frac{|q_1||q_3|}{(2.00 - x)^2}$$

▶ **23.3** continued

Eliminate k_e and $|q_3|$ and rearrange the equation:	$(2.00 - x)^2 |q_2| = x^2 |q_1|$
Take the square root of both sides of the equation:	$(2.00 - x)\sqrt{|q_2|} = \pm x\sqrt{|q_1|}$
Solve for x:	$x = \dfrac{2.00\sqrt{|q_2|}}{\sqrt{|q_2|} \pm \sqrt{|q_1|}}$
Substitute numerical values, choosing the plus sign:	$x = \dfrac{2.00\sqrt{6.00 \times 10^{-6}\,\text{C}}}{\sqrt{6.00 \times 10^{-6}\,\text{C}} + \sqrt{15.0 \times 10^{-6}\,\text{C}}} = 0.775\,\text{m}$

Finalize Notice that the movable charge is indeed closer to q_2 as we predicted in the Conceptualize step. The second solution to the equation (if we choose the negative sign) is $x = -3.44$ m. That is another location where the *magnitudes* of the forces on q_3 are equal, but both forces are in the same direction, so they do not cancel.

WHAT IF? Suppose q_3 is constrained to move only along the x axis. From its initial position at $x = 0.775$ m, it is pulled a small distance along the x axis. When released, does it return to equilibrium, or is it pulled farther from equilibrium? That is, is the equilibrium stable or unstable?

Answer If q_3 is moved to the right, $\vec{F}_{13}$ becomes larger and $\vec{F}_{23}$ becomes smaller. The result is a net force to the right, in the same direction as the displacement. Therefore, the charge q_3 would continue to move to the right and the equilibrium is *unstable*. (See Section 7.9 for a review of stable and unstable equilibria.)

If q_3 is constrained to stay at a *fixed x* coordinate but allowed to move up and down in Figure 23.8, the equilibrium is stable. In this case, if the charge is pulled upward (or downward) and released, it moves back toward the equilibrium position and oscillates about this point.

Example 23.4   Find the Charge on the Spheres   AM

Two identical small charged spheres, each having a mass of 3.00×10^{-2} kg, hang in equilibrium as shown in Figure 23.9a. The length L of each string is 0.150 m, and the angle θ is 5.00°. Find the magnitude of the charge on each sphere.

SOLUTION

Conceptualize Figure 23.9a helps us conceptualize this example. The two spheres exert repulsive forces on each other. If they are held close to each other and released, they move outward from the center and settle into the configuration in Figure 23.9a after the oscillations have vanished due to air resistance.

Categorize The key phrase "in equilibrium" helps us model each sphere as a *particle in equilibrium*. This example is similar to the particle in equilibrium problems in Chapter 5 with the added feature that one of the forces on a sphere is an electric force.

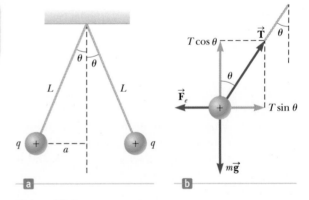

Figure 23.9 (Example 23.4) (a) Two identical spheres, each carrying the same charge q, suspended in equilibrium. (b) Diagram of the forces acting on the sphere on the left part of (a).

Analyze The force diagram for the left-hand sphere is shown in Figure 23.9b. The sphere is in equilibrium under the application of the force $\vec{T}$ from the string, the electric force $\vec{F}_e$ from the other sphere, and the gravitational force $m\vec{g}$.

From the particle in equilibrium model, set the net force on the left-hand sphere equal to zero for each component:

$$(1) \quad \sum F_x = T\sin\theta - F_e = 0 \;\to\; T\sin\theta = F_e$$

$$(2) \quad \sum F_y = T\cos\theta - mg = 0 \;\to\; T\cos\theta = mg$$

[Divid]e Equation (1) by Equation (2) to find F_e:

$$(3) \quad \tan\theta = \frac{F_e}{mg} \;\to\; F_e = mg\tan\theta$$

▶ **23.4** continued

Use the geometry of the right triangle in Figure 23.9a to find a relationship between a, L, and θ:

$$(4) \quad \sin \theta = \frac{a}{L} \rightarrow a = L \sin \theta$$

Solve Coulomb's law (Eq. 23.1) for the charge $|q|$ on each sphere and substitute from Equations (3) and (4):

$$|q| = \sqrt{\frac{F_e r^2}{k_e}} = \sqrt{\frac{F_e (2a)^2}{k_e}} = \sqrt{\frac{mg \tan \theta (2L \sin \theta)^2}{k_e}}$$

Substitute numerical values:

$$|q| = \sqrt{\frac{(3.00 \times 10^{-2}\ \text{kg})(9.80\ \text{m/s}^2) \tan (5.00°)[2(0.150\ \text{m}) \sin (5.00°)]^2}{8.988 \times 10^9\ \text{N} \cdot \text{m}^2/\text{C}^2}}$$

$$= 4.42 \times 10^{-8}\ \text{C}$$

Finalize If the sign of the charges were not given in Figure 23.9, we could not determine them. In fact, the sign of the charge is not important. The situation is the same whether both spheres are positively charged or negatively charged.

WHAT IF? Suppose your roommate proposes solving this problem without the assumption that the charges are of equal magnitude. She claims the symmetry of the problem is destroyed if the charges are not equal, so the strings would make two different angles with the vertical and the problem would be much more complicated. How would you respond?

Answer The symmetry is not destroyed and the angles are not different. Newton's third law requires the magnitudes of the electric forces on the two spheres to be the same, regardless of the equality or nonequality of the charges. The solution to the example remains the same with one change: the value of $|q|$ in the solution is replaced by $\sqrt{|q_1 q_2|}$ in the new situation, where q_1 and q_2 are the values of the charges on the two spheres. The symmetry of the problem would be destroyed if the *masses* of the spheres were not the same. In this case, the strings would make different angles with the vertical and the problem would be more complicated.

23.4 Analysis Model: Particle in a Field (Electric)

In Section 5.1, we discussed the differences between contact forces and field forces. Two field forces—the gravitational force in Chapter 13 and the electric force here—have been introduced into our discussions so far. As pointed out earlier, field forces can act through space, producing an effect even when no physical contact occurs between interacting objects. Such an interaction can be modeled as a two-step process: a source particle establishes a field, and then a charged particle interacts with the field and experiences a force. The gravitational field $\vec{g}$ at a point in space due to a source particle was defined in Section 13.4 to be equal to the gravitational force $\vec{F}_g$ acting on a test particle of mass m divided by that mass: $\vec{g} \equiv \vec{F}_g / m$. Then the force exerted by the field is $\vec{F} = m\vec{g}$ (Eq. 5.5).

The concept of a field was developed by Michael Faraday (1791–1867) in the context of electric forces and is of such practical value that we shall devote much attention to it in the next several chapters. In this approach, an **electric field** is said to exist in the region of space around a charged object, the **source charge.** The presence of the electric field can be detected by placing a **test charge** in the field and noting the electric force on it. As an example, consider Figure 23.10, which shows a small positive test charge q_0 placed near a second object carrying a much greater positive charge Q. We define the electric field due to the source charge at the location of the test charge to be the electric force on the test charge *per unit charge*, or, to be more specific, the **electric field vector** $\vec{E}$ at a point in space is defined as the electric force $\vec{F}_e$ acting on a positive test charge q_0 placed at that point divided by the test charge:[3]

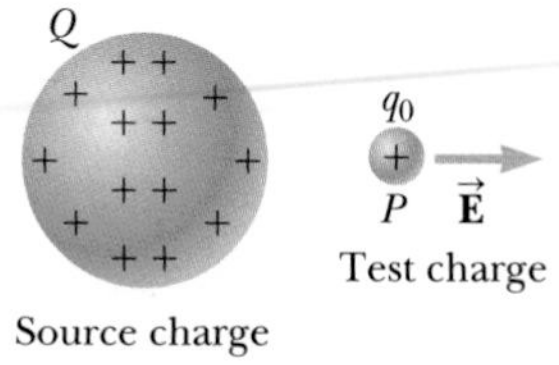

Figure 23.10 A small positive test charge q_0 placed at point P near an object carrying a much larger positive charge Q experiences an electric field $\vec{E}$ at point P established by the source charge Q. We will *always* assume that the test charge is so small that the field of the source charge is unaffected by its presence.

$$\vec{E} \equiv \frac{\vec{F}_e}{q_0}$$

(23.7) ◀ **Definition of electric field**

[3]When using Equation 23.7, we must assume the test charge q_0 is small enough that it does not disturb the charge distribution responsible for the electric field. If the test charge is great enough, the charge on the metallic sphere is redistributed and the electric field it sets up is different from the field it sets up in the presence of the much smaller test charge.

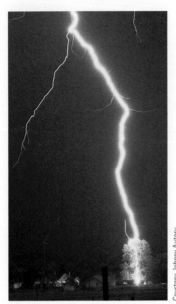

Courtesy Johnny Autery

This dramatic photograph captures a lightning bolt striking a tree near some rural homes. Lightning is associated with very strong electric fields in the atmosphere.

Pitfall Prevention 23.1

Particles Only Equation 23.8 is valid only for a *particle* of charge q, that is, an object of zero size. For a charged *object* of finite size in an electric field, the field may vary in magnitude and direction over the size of the object, so the corresponding force equation may be more complicated.

The vector $\vec{\mathbf{E}}$ has the SI units of newtons per coulomb (N/C). The direction of $\vec{\mathbf{E}}$ as shown in Figure 23.10 is the direction of the force a positive test charge experiences when placed in the field. Note that $\vec{\mathbf{E}}$ is the field produced by some charge or charge distribution *separate from* the test charge; it is not the field produced by the test charge itself. Also note that the existence of an electric field is a property of its source; the presence of the test charge is not necessary for the field to exist. The test charge serves as a *detector* of the electric field: an electric field exists at a point if a test charge at that point experiences an electric force.

If an arbitrary charge q is placed in an electric field $\vec{\mathbf{E}}$, it experiences an electric force given by

$$\vec{\mathbf{F}}_e = q\vec{\mathbf{E}} \tag{23.8}$$

This equation is the mathematical representation of the electric version of the **particle in a field** analysis model. If q is positive, the force is in the same direction as the field. If q is negative, the force and the field are in opposite directions. Notice the similarity between Equation 23.8 and the corresponding equation from the gravitational version of the particle in a field model, $\vec{\mathbf{F}}_g = m\vec{\mathbf{g}}$ (Section 5.5). Once the magnitude and direction of the electric field are known at some point, the electric force exerted on *any* charged particle placed at that point can be calculated from Equation 23.8.

To determine the direction of an electric field, consider a point charge q as a source charge. This charge creates an electric field at all points in space surrounding it. A test charge q_0 is placed at point P, a distance r from the source charge, as in Figure 23.11a. We imagine using the test charge to determine the direction of the electric force and therefore that of the electric field. According to Coulomb's law, the force exerted by q on the test charge is

$$\vec{\mathbf{F}}_e = k_e \frac{q q_0}{r^2} \hat{\mathbf{r}}$$

where $\hat{\mathbf{r}}$ is a unit vector directed from q toward q_0. This force in Figure 23.11a is directed away from the source charge q. Because the electric field at P, the position of the test charge, is defined by $\vec{\mathbf{E}} = \vec{\mathbf{F}}_e / q_0$, the electric field at P created by q is

$$\vec{\mathbf{E}} = k_e \frac{q}{r^2} \hat{\mathbf{r}} \tag{23.9}$$

If the source charge q is positive, Figure 23.11b shows the situation with the test charge removed: the source charge sets up an electric field at P, directed away from q. If q is

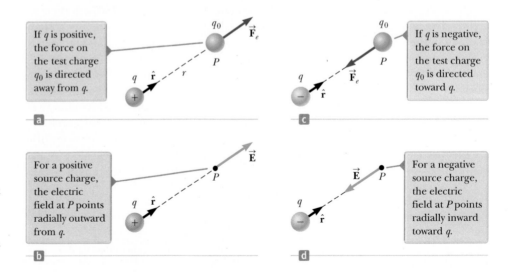

Figure 23.11 (a), (c) When a test charge q_0 is placed near a source charge q, the test charge experiences a force. (b), (d) At a point P near a source charge q, there exists an electric field.

If q is positive, the force on the test charge q_0 is directed away from q.

If q is negative, the force on the test charge q_0 is directed toward q.

For a positive source charge, the electric field at P points radially outward from q.

For a negative source charge, the electric field at P points radially inward toward q.

negative as in Figure 23.11c, the force on the test charge is toward the source charge, so the electric field at P is directed toward the source charge as in Figure 23.11d.

To calculate the electric field at a point P due to a small number of point charges, we first calculate the electric field vectors at P individually using Equation 23.9 and then add them vectorially. In other words, at any point P, the total electric field due to a group of source charges equals the vector sum of the electric fields of all the charges. This superposition principle applied to fields follows directly from the vector addition of electric forces. Therefore, the electric field at point P due to a group of source charges can be expressed as the vector sum

$$\vec{E} = k_e \sum_i \frac{q_i}{r_i^2} \hat{r}_i \qquad (23.10)$$

◀ Electric field due to a finite number of point charges

where r_i is the distance from the ith source charge q_i to the point P and $\hat{r}_i$ is a unit vector directed from q_i toward P.

In Example 23.6, we explore the electric field due to two charges using the superposition principle. Part (B) of the example focuses on an **electric dipole,** which is defined as a positive charge q and a negative charge $-q$ separated by a distance $2a$. The electric dipole is a good model of many molecules, such as hydrochloric acid (HCl). Neutral atoms and molecules behave as dipoles when placed in an external electric field. Furthermore, many molecules, such as HCl, are permanent dipoles. The effect of such dipoles on the behavior of materials subjected to electric fields is discussed in Chapter 26.

Ⓠuick Quiz 23.4 A test charge of $+3\ \mu\text{C}$ is at a point P where an external electric field is directed to the right and has a magnitude of 4×10^6 N/C. If the test charge is replaced with another test charge of $-3\ \mu\text{C}$, what happens to the external electric field at P? **(a)** It is unaffected. **(b)** It reverses direction. **(c)** It changes in a way that cannot be determined.

Analysis Model **Particle in a Field (Electric)**

Imagine an object with charge that we call a *source charge.* The source charge establishes an **electric field $\vec{E}$** throughout space. Now imagine a particle with charge q is placed in that field. The particle interacts with the electric field so that the particle experiences an electric force given by

$$\vec{F}_e = q\vec{E} \qquad (23.8)$$

Examples:

- an electron moves between the deflection plates of a cathode ray oscilloscope and is deflected from its original path
- charged ions experience an electric force from the electric field in a velocity selector before entering a mass spectrometer (Chapter 29)
- an electron moves around the nucleus in the electric field established by the proton in a hydrogen atom as modeled by the Bohr theory (Chapter 42)
- a hole in a semiconducting material moves in response to the electric field established by applying a voltage to the material (Chapter 43)

Example 23.5 **A Suspended Water Droplet**

A water droplet of mass 3.00×10^{-12} kg is located in the air near the ground during a stormy day. An atmospheric electric field of magnitude 6.00×10^3 N/C points vertically downward in the vicinity of the water droplet. The droplet remains suspended at rest in the air. What is the electric charge on the droplet?

SOLUTION

Conceptualize Imagine the water droplet hovering at rest in the air. This situation is not what is normally observed, so something must be holding the water droplet up.

continued

▶ **23.5** continued

Categorize The droplet can be modeled as a particle and is described by two analysis models associated with fields: the *particle in a field (gravitational)* and the *particle in a field (electric)*. Furthermore, because the droplet is subject to forces but remains at rest, it is also described by the *particle in equilibrium* model.

Analyze

Write Newton's second law from the particle in equilibrium model in the vertical direction:

(1) $\sum F_y = 0 \rightarrow F_e - F_g = 0$

Using the two particle in a field models mentioned in the Categorize step, substitute for the forces in Equation (1), recognizing that the vertical component of the electric field is negative:

$q(-E) - mg = 0$

Solve for the charge on the water droplet:

$q = -\dfrac{mg}{E}$

Substitute numerical values:

$q = -\dfrac{(3.00 \times 10^{-12}\,\text{kg})(9.80\,\text{m/s}^2)}{6.00 \times 10^3\,\text{N/C}} = -4.90 \times 10^{-15}\,\text{C}$

Finalize Noting the smallest unit of free charge in Equation 23.5, the charge on the water droplet is a large number of these units. Notice that the electric *force* is upward to balance the downward gravitational force. The problem statement claims that the electric *field* is in the downward direction. Therefore, the charge found above is negative so that the electric force is in the direction opposite to the electric field.

Example 23.6   Electric Field Due to Two Charges

Charges q_1 and q_2 are located on the x axis, at distances a and b, respectively, from the origin as shown in Figure 23.12.

(A) Find the components of the net electric field at the point P, which is at position $(0, y)$.

SOLUTION

Conceptualize Compare this example with Example 23.2. There, we add vector forces to find the net force on a charged particle. Here, we add electric field vectors to find the net electric field at a point in space. If a charged particle were placed at P, we could use the particle in a field model to find the electric force on the particle.

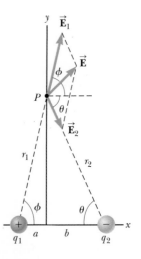

Figure 23.12 (Example 23.6) The total electric field $\vec{\mathbf{E}}$ at P equals the vector sum $\vec{\mathbf{E}}_1 + \vec{\mathbf{E}}_2$, where $\vec{\mathbf{E}}_1$ is the field due to the positive charge q_1 and $\vec{\mathbf{E}}_2$ is the field due to the negative charge q_2.

Categorize We have two source charges and wish to find the resultant electric field, so we categorize this example as one in which we can use the superposition principle represented by Equation 23.10.

Analyze Find the magnitude of the electric field at P due to charge q_1:

$E_1 = k_e \dfrac{|q_1|}{r_1^2} = k_e \dfrac{|q_1|}{a^2 + y^2}$

Find the magnitude of the electric field at P due to charge q_2:

$E_2 = k_e \dfrac{|q_2|}{r_2^2} = k_e \dfrac{|q_2|}{b^2 + y^2}$

Write the electric field vectors for each charge in unit-vector form:

$\vec{\mathbf{E}}_1 = k_e \dfrac{|q_1|}{a^2 + y^2} \cos\phi\,\hat{\mathbf{i}} + k_e \dfrac{|q_1|}{a^2 + y^2} \sin\phi\,\hat{\mathbf{j}}$

$\vec{\mathbf{E}}_2 = k_e \dfrac{|q_2|}{b^2 + y^2} \cos\theta\,\hat{\mathbf{i}} - k_e \dfrac{|q_2|}{b^2 + y^2} \sin\theta\,\hat{\mathbf{j}}$

▶ **23.6** continued

Write the components of the net electric field vector:

$$(1) \quad E_x = E_{1x} + E_{2x} = k_e \frac{|q_1|}{a^2 + y^2} \cos \phi + k_e \frac{|q_2|}{b^2 + y^2} \cos \theta$$

$$(2) \quad E_y = E_{1y} + E_{2y} = k_e \frac{|q_1|}{a^2 + y^2} \sin \phi - k_e \frac{|q_2|}{b^2 + y^2} \sin \theta$$

(B) Evaluate the electric field at point P in the special case that $|q_1| = |q_2|$ and $a = b$.

SOLUTION

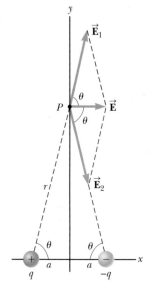

Conceptualize Figure 23.13 shows the situation in this special case. Notice the symmetry in the situation and that the charge distribution is now an electric dipole.

Categorize Because Figure 23.13 is a special case of the general case shown in Figure 23.12, we can categorize this example as one in which we can take the result of part (A) and substitute the appropriate values of the variables.

Figure 23.13 (Example 23.6) When the charges in Figure 23.12 are of equal magnitude and equidistant from the origin, the situation becomes symmetric as shown here.

Analyze Based on the symmetry in Figure 23.13, evaluate Equations (1) and (2) from part (A) with $a = b$, $|q_1| = |q_2| = q$, and $\phi = \theta$:

$$(3) \quad E_x = k_e \frac{q}{a^2 + y^2} \cos \theta + k_e \frac{q}{a^2 + y^2} \cos \theta = 2k_e \frac{q}{a^2 + y^2} \cos \theta$$

$$E_y = k_e \frac{q}{a^2 + y^2} \sin \theta - k_e \frac{q}{a^2 + y^2} \sin \theta = 0$$

From the geometry in Figure 23.13, evaluate $\cos \theta$:

$$(4) \quad \cos \theta = \frac{a}{r} = \frac{a}{(a^2 + y^2)^{1/2}}$$

Substitute Equation (4) into Equation (3):

$$E_x = 2k_e \frac{q}{a^2 + y^2} \left[\frac{a}{(a^2 + y^2)^{1/2}} \right] = k_e \frac{2aq}{(a^2 + y^2)^{3/2}}$$

(C) Find the electric field due to the electric dipole when point P is a distance $y \gg a$ from the origin.

SOLUTION

In the solution to part (B), because $y \gg a$, neglect a^2 compared with y^2 and write the expression for E in this case:

$$(5) \quad E \approx k_e \frac{2aq}{y^3}$$

Finalize From Equation (5), we see that at points far from a dipole but along the perpendicular bisector of the line joining the two charges, the magnitude of the electric field created by the dipole varies as $1/r^3$, whereas the more slowly varying field of a point charge varies as $1/r^2$ (see Eq. 23.9). That is because at distant points, the fields of the two charges of equal magnitude and opposite sign almost cancel each other. The $1/r^3$ variation in E for the dipole also is obtained for a distant point along the x axis and for any general distant point.

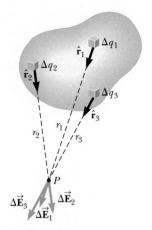

Figure 23.14 The electric field at P due to a continuous charge distribution is the vector sum of the fields $\Delta\vec{\mathbf{E}}_i$ due to all the elements Δq_i of the charge distribution. Three sample elements are shown.

23.5 Electric Field of a Continuous Charge Distribution

Equation 23.10 is useful for calculating the electric field due to a small number of charges. In many cases, we have a continuous distribution of charge rather than a collection of discrete charges. The charge in these situations can be described as continuously distributed along some line, over some surface, or throughout some volume.

To set up the process for evaluating the electric field created by a continuous charge distribution, let's use the following procedure. First, divide the charge distribution into small elements, each of which contains a small charge Δq as shown in Figure 23.14. Next, use Equation 23.9 to calculate the electric field due to one of these elements at a point P. Finally, evaluate the total electric field at P due to the charge distribution by summing the contributions of all the charge elements (that is, by applying the superposition principle).

The electric field at P due to one charge element carrying charge Δq is

$$\Delta\vec{\mathbf{E}} = k_e \frac{\Delta q}{r^2}\hat{\mathbf{r}}$$

where r is the distance from the charge element to point P and $\hat{\mathbf{r}}$ is a unit vector directed from the element toward P. The total electric field at P due to all elements in the charge distribution is approximately

$$\vec{\mathbf{E}} \approx k_e \sum_i \frac{\Delta q_i}{r_i^2}\hat{\mathbf{r}}_i$$

where the index i refers to the ith element in the distribution. Because the number of elements is very large and the charge distribution is modeled as continuous, the total field at P in the limit $\Delta q_i \rightarrow 0$ is

Electric field due to a continuous charge distribution ▶

$$\vec{\mathbf{E}} = k_e \lim_{\Delta q_i \to 0} \sum_i \frac{\Delta q_i}{r_i^2}\hat{\mathbf{r}}_i = k_e \int \frac{dq}{r^2}\hat{\mathbf{r}} \qquad (23.11)$$

where the integration is over the entire charge distribution. The integration in Equation 23.11 is a vector operation and must be treated appropriately.

Let's illustrate this type of calculation with several examples in which the charge is distributed on a line, on a surface, or throughout a volume. When performing such calculations, it is convenient to use the concept of a *charge density* along with the following notations:

- If a charge Q is uniformly distributed throughout a volume V, the **volume charge density** ρ is defined by

Volume charge density ▶

$$\rho \equiv \frac{Q}{V}$$

where ρ has units of coulombs per cubic meter (C/m^3).

- If a charge Q is uniformly distributed on a surface of area A, the **surface charge density** σ (Greek letter sigma) is defined by

Surface charge density ▶

$$\sigma \equiv \frac{Q}{A}$$

where σ has units of coulombs per square meter (C/m^2).

- If a charge Q is uniformly distributed along a line of length ℓ, the **linear charge density** λ is defined by

Linear charge density ▶

$$\lambda \equiv \frac{Q}{\ell}$$

where λ has units of coulombs per meter (C/m).

- If the charge is nonuniformly distributed over a volume, surface, or line, the amounts of charge dq in a small volume, surface, or length element are

$$dq = \rho \, dV \qquad dq = \sigma \, dA \qquad dq = \lambda \, d\ell$$

Problem-Solving Strategy   Calculating the Electric Field

The following procedure is recommended for solving problems that involve the determination of an electric field due to individual charges or a charge distribution.

1. Conceptualize. Establish a mental representation of the problem: think carefully about the individual charges or the charge distribution and imagine what type of electric field it would create. Appeal to any symmetry in the arrangement of charges to help you visualize the electric field.

2. Categorize. Are you analyzing a group of individual charges or a continuous charge distribution? The answer to this question tells you how to proceed in the Analyze step.

3. Analyze.

(a) If you are analyzing a group of individual charges, use the superposition principle: when several point charges are present, the resultant field at a point in space is the *vector sum* of the individual fields due to the individual charges (Eq. 23.10). Be very careful in the manipulation of vector quantities. It may be useful to review the material on vector addition in Chapter 3. Example 23.6 demonstrated this procedure.

(b) If you are analyzing a continuous charge distribution, the superposition principle is applied by replacing the vector sums for evaluating the total electric field from individual charges by vector integrals. The charge distribution is divided into infinitesimal pieces, and the vector sum is carried out by integrating over the entire charge distribution (Eq. 23.11). Examples 23.7 through 23.9 demonstrate such procedures.

Consider symmetry when dealing with either a distribution of point charges or a continuous charge distribution. Take advantage of any symmetry in the system you observed in the Conceptualize step to simplify your calculations. The cancellation of field components perpendicular to the axis in Example 23.8 is an example of the application of symmetry.

4. Finalize. Check to see if your electric field expression is consistent with the mental representation and if it reflects any symmetry that you noted previously. Imagine varying parameters such as the distance of the observation point from the charges or the radius of any circular objects to see if the mathematical result changes in a reasonable way.

Example 23.7   The Electric Field Due to a Charged Rod

A rod of length ℓ has a uniform positive charge per unit length λ and a total charge Q. Calculate the electric field at a point P that is located along the long axis of the rod and a distance a from one end (Fig. 23.15).

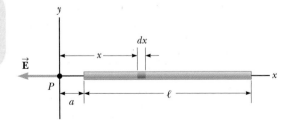

Figure 23.15 (Example 23.7) The electric field at P due to a uniformly charged rod lying along the x axis.

SOLUTION

Conceptualize The field $d\vec{\mathbf{E}}$ at P due to each segment of charge on the rod is in the negative x direction because every segment carries a positive charge. Figure 23.15 shows the appropriate geometry. In our result, we expect the electric field to become smaller as the distance a becomes larger because point P is farther from the charge distribution.

continued

▶ **23.7** continued

Categorize Because the rod is continuous, we are evaluating the field due to a continuous charge distribution rather than a group of individual charges. Because every segment of the rod produces an electric field in the negative x direction, the sum of their contributions can be handled without the need to add vectors.

....

Analyze Let's assume the rod is lying along the x axis, dx is the length of one small segment, and dq is the charge on that segment. Because the rod has a charge per unit length λ, the charge dq on the small segment is $dq = \lambda\, dx$.

Find the magnitude of the electric field at P due to one segment of the rod having a charge dq:

$$dE = k_e \frac{dq}{x^2} = k_e \frac{\lambda\, dx}{x^2}$$

Find the total field at P using[4] Equation 23.11:

$$E = \int_a^{\ell+a} k_e \lambda \frac{dx}{x^2}$$

Noting that k_e and $\lambda = Q/\ell$ are constants and can be removed from the integral, evaluate the integral:

$$E = k_e \lambda \int_a^{\ell+a} \frac{dx}{x^2} = k_e \lambda \left[-\frac{1}{x} \right]_a^{\ell+a}$$

$$(1) \quad E = k_e \frac{Q}{\ell} \left(\frac{1}{a} - \frac{1}{\ell + a} \right) = \boxed{\frac{k_e Q}{a(\ell + a)}}$$

....

Finalize We see that our prediction is correct; if a becomes larger, the denominator of the fraction grows larger, and E becomes smaller. On the other hand, if $a \to 0$, which corresponds to sliding the bar to the left until its left end is at the origin, then $E \to \infty$. That represents the condition in which the observation point P is at zero distance from the charge at the end of the rod, so the field becomes infinite. We explore large values of a below.

WHAT IF? Suppose point P is very far away from the rod. What is the nature of the electric field at such a point?

Answer If P is far from the rod ($a \gg \ell$), then ℓ in the denominator of Equation (1) can be neglected and $E \approx k_e Q/a^2$. That is exactly the form you would expect for a point charge. Therefore, at large values of a/ℓ, the charge distribution appears to be a point charge of magnitude Q; the point P is so far away from the rod we cannot distinguish that it has a size. The use of the limiting technique ($a/\ell \to \infty$) is often a good method for checking a mathematical expression.

Example 23.8 The Electric Field of a Uniform Ring of Charge

A ring of radius a carries a uniformly distributed positive total charge Q. Calculate the electric field due to the ring at a point P lying a distance x from its center along the central axis perpendicular to the plane of the ring (Fig. 23.16a).

SOLUTION

Conceptualize Figure 23.16a shows the electric field contribution $d\vec{\mathbf{E}}$ at P due to a single segment of charge at the top of the ring. This field vector can be resolved into components dE_x parallel to

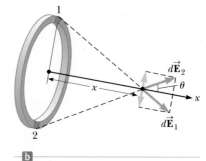

Figure 23.16 (Example 23.8) A uniformly charged ring of radius a. (a) The field at P on the x axis due to an element of charge dq. (b) The total electric field at P is along the x axis. The perpendicular component of the field at P due to segment 1 is canceled by the perpendicular component due to segment 2.

---

[4]To carry out integrations such as this one, first express the charge element dq in terms of the other variables in the integral. (In this example, there is one variable, x, so we made the change $dq = \lambda\, dx$.) The integral must be over scalar quantities; therefore, express the electric field in terms of components, if necessary. (In this example, the field has only an x component, so this detail is of no concern.) Then, reduce your expression to an integral over a single variable (or to multiple integrals, each over a single variable). In examples that have spherical or cylindrical symmetry, the single variable is a radial coordinate.

▶ **23.8** continued

the axis of the ring and $dE_\perp$ perpendicular to the axis. Figure 23.16b shows the electric field contributions from two segments on opposite sides of the ring. Because of the symmetry of the situation, the perpendicular components of the field cancel. That is true for all pairs of segments around the ring, so we can ignore the perpendicular component of the field and focus solely on the parallel components, which simply add.

Categorize Because the ring is continuous, we are evaluating the field due to a continuous charge distribution rather than a group of individual charges.

Analyze Evaluate the parallel component of an electric field contribution from a segment of charge dq on the ring:

$$(1) \quad dE_x = k_e \frac{dq}{r^2} \cos\theta = k_e \frac{dq}{a^2 + x^2} \cos\theta$$

From the geometry in Figure 23.16a, evaluate $\cos\theta$:

$$(2) \quad \cos\theta = \frac{x}{r} = \frac{x}{(a^2 + x^2)^{1/2}}$$

Substitute Equation (2) into Equation (1):

$$dE_x = k_e \frac{dq}{a^2 + x^2}\left[\frac{x}{(a^2 + x^2)^{1/2}}\right] = \frac{k_e x}{(a^2 + x^2)^{3/2}} \, dq$$

All segments of the ring make the same contribution to the field at P because they are all equidistant from this point. Integrate over the circumference of the ring to obtain the total field at P:

$$E_x = \int \frac{k_e x}{(a^2 + x^2)^{3/2}} \, dq = \frac{k_e x}{(a^2 + x^2)^{3/2}} \int dq$$

$$(3) \quad E = \frac{k_e x}{(a^2 + x^2)^{3/2}} \, Q$$

Finalize This result shows that the field is zero at $x = 0$. Is that consistent with the symmetry in the problem? Furthermore, notice that Equation (3) reduces to $k_e Q/x^2$ if $x \gg a$, so the ring acts like a point charge for locations far away from the ring. From a faraway point, we cannot distinguish the ring shape of the charge.

WHAT IF? Suppose a negative charge is placed at the center of the ring in Figure 23.16 and displaced slightly by a distance $x \ll a$ along the x axis. When the charge is released, what type of motion does it exhibit?

Answer In the expression for the field due to a ring of charge, let $x \ll a$, which results in

$$E_x = \frac{k_e Q}{a^3} \, x$$

Therefore, from Equation 23.8, the force on a charge $-q$ placed near the center of the ring is

$$F_x = -\frac{k_e q Q}{a^3} \, x$$

Because this force has the form of Hooke's law (Eq. 15.1), the motion of the negative charge is described with the *particle in simple harmonic motion model!*

Example 23.9   **The Electric Field of a Uniformly Charged Disk**

A disk of radius R has a uniform surface charge density σ. Calculate the electric field at a point P that lies along the central perpendicular axis of the disk and a distance x from the center of the disk (Fig. 23.17).

SOLUTION

Conceptualize If the disk is considered to be a set of concentric rings, we can use our result from Example 23.8—which gives the field created by a single ring of radius a—and sum the contributions of all rings making up the disk. By symmetry, the field at an axial point must be along the central axis.

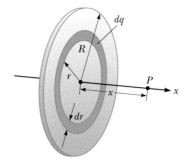

Figure 23.17 (Example 23.9) A uniformly charged disk of radius R. The electric field at an axial point P is directed along the central axis, perpendicular to the plane of the disk.

continued

▶ **23.9** continued

Categorize Because the disk is continuous, we are evaluating the field due to a continuous charge distribution rather than a group of individual charges.

. . . .

Analyze Find the amount of charge dq on the surface area of a ring of radius r and width dr as shown in Figure 23.17:

$$dq = \sigma \, dA = \sigma(2\pi r \, dr) = 2\pi\sigma r \, dr$$

Use this result in the equation given for E_x in Example 23.8 (with a replaced by r and Q replaced by dq) to find the field due to the ring:

$$dE_x = \frac{k_e x}{(r^2 + x^2)^{3/2}} \, (2\pi\sigma r \, dr)$$

To obtain the total field at P, integrate this expression over the limits $r = 0$ to $r = R$, noting that x is a constant in this situation:

$$E_x = k_e x \pi \sigma \int_0^R \frac{2r \, dr}{(r^2 + x^2)^{3/2}}$$

$$= k_e x \pi \sigma \int_0^R (r^2 + x^2)^{-3/2} d(r^2)$$

$$= k_e x \pi \sigma \left[\frac{(r^2 + x^2)^{-1/2}}{-1/2} \right]_0^R = 2\pi k_e \sigma \left[1 - \frac{x}{(R^2 + x^2)^{1/2}} \right]$$

. . . .

Finalize This result is valid for all values of $x > 0$. For large values of x, the result above can be evaluated by a series expansion and shown to be equivalent to the electric field of a point charge Q. We can calculate the field close to the disk along the axis by assuming $x \ll R$; in this case, the expression in brackets reduces to unity to give us the near-field approximation

$$E = 2\pi k_e \sigma = \frac{\sigma}{2\epsilon_0}$$

where ϵ_0 is the permittivity of free space. In Chapter 24, we obtain the same result for the field created by an infinite plane of charge with uniform surface charge density.

WHAT IF? What if we let the radius of the disk grow so that the disk becomes an infinite plane of charge?

Answer The result of letting $R \to \infty$ in the final result of the example is that the magnitude of the electric field becomes

$$E = 2\pi k_e \sigma = \frac{\sigma}{2\epsilon_0}$$

This is the same expression that we obtained for $x \ll R$. If $R \to \infty$, *everywhere* is near-field—the result is independent of the position at which you measure the electric field. Therefore, the electric field due to an infinite plane of charge is uniform throughout space.

An infinite plane of charge is impossible in practice. If two planes of charge are placed close to each other, however, with one plane positively charged, and the other negatively, the electric field between the plates is very close to uniform at points far from the edges. Such a configuration will be investigated in Chapter 26.

23.6 Electric Field Lines

We have defined the electric field in the mathematical representation with Equation 23.7. Let's now explore a means of visualizing the electric field in a pictorial representation. A convenient way of visualizing electric field patterns is to draw lines, called **electric field lines** and first introduced by Faraday, that are related to the electric field in a region of space in the following manner:

- The electric field vector $\vec{\mathbf{E}}$ is tangent to the electric field line at each point. The line has a direction, indicated by an arrowhead, that is the same as that of the electric field vector. The direction of the line is that of the force on a positive charge placed in the field according to the particle in a field model.

- The number of lines per unit area through a surface perpendicular to the lines is proportional to the magnitude of the electric field in that region. Therefore, the field lines are close together where the electric field is strong and far apart where the field is weak.

These properties are illustrated in Figure 23.18. The density of field lines through surface A is greater than the density of lines through surface B. Therefore, the magnitude of the electric field is larger on surface A than on surface B. Furthermore, because the lines at different locations point in different directions, the field is nonuniform.

Is this relationship between strength of the electric field and the density of field lines consistent with Equation 23.9, the expression we obtained for E using Coulomb's law? To answer this question, consider an imaginary spherical surface of radius r concentric with a point charge. From symmetry, we see that the magnitude of the electric field is the same everywhere on the surface of the sphere. The number of lines N that emerge from the charge is equal to the number that penetrate the spherical surface. Hence, the number of lines per unit area on the sphere is $N/4\pi r^2$ (where the surface area of the sphere is $4\pi r^2$). Because E is proportional to the number of lines per unit area, we see that E varies as $1/r^2$; this finding is consistent with Equation 23.9.

Representative electric field lines for the field due to a single positive point charge are shown in Figure 23.19a. This two-dimensional drawing shows only the field lines that lie in the plane containing the point charge. The lines are actually directed radially outward from the charge in all directions; therefore, instead of the flat "wheel" of lines shown, you should picture an entire spherical distribution of lines. Because a positive charge placed in this field would be repelled by the positive source charge, the lines are directed radially away from the source charge. The electric field lines representing the field due to a single negative point charge are directed toward the charge (Fig. 23.19b). In either case, the lines are along the radial direction and extend all the way to infinity. Notice that the lines become closer together as they approach the charge, indicating that the strength of the field increases as we move toward the source charge.

The rules for drawing electric field lines are as follows:

- The lines must begin on a positive charge and terminate on a negative charge. In the case of an excess of one type of charge, some lines will begin or end infinitely far away.

The magnitude of the field is greater on surface A than on surface B.

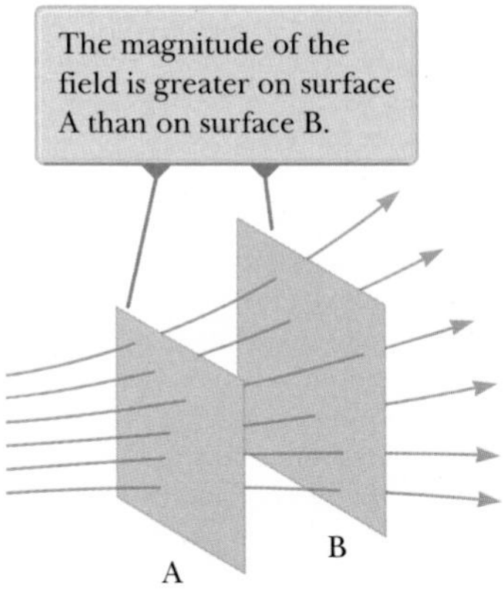

Figure 23.18 Electric field lines penetrating two surfaces.

Pitfall Prevention 23.2
Electric Field Lines Are Not Paths of Particles! Electric field lines represent the field at various locations. Except in very special cases, they *do not* represent the path of a charged particle moving in an electric field.

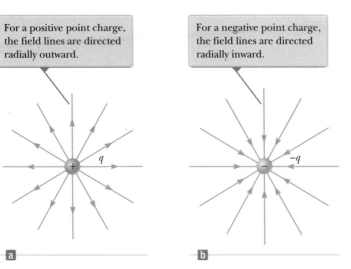

For a positive point charge, the field lines are directed radially outward.

q

For a negative point charge, the field lines are directed radially inward.

$-q$

a

b

Figure 23.19 The electric field lines for a point charge. Notice that the figures show only those field lines that lie in the plane of the page.

Pitfall Prevention 23.3
Electric Field Lines Are Not Real
Electric field lines are not material objects. They are used only as a pictorial representation to provide a qualitative description of the electric field. Only a finite number of lines from each charge can be drawn, which makes it appear as if the field were quantized and exists only in certain parts of space. The field, in fact, is continuous, existing at every point. You should avoid obtaining the wrong impression from a two-dimensional drawing of field lines used to describe a three-dimensional situation.

- The number of lines drawn leaving a positive charge or approaching a negative charge is proportional to the magnitude of the charge.
- No two field lines can cross.

We choose the number of field lines starting from any object with a positive charge q_+ to be Cq_+ and the number of lines ending on any object with a negative charge q_- to be $C|q_-|$, where C is an arbitrary proportionality constant. Once C is chosen, the number of lines is fixed. For example, in a two-charge system, if object 1 has charge Q_1 and object 2 has charge Q_2, the ratio of number of lines in contact with the charges is $N_2/N_1 = |Q_2/Q_1|$. The electric field lines for two point charges of equal magnitude but opposite signs (an electric dipole) are shown in Figure 23.20. Because the charges are of equal magnitude, the number of lines that begin at the positive charge must equal the number that terminate at the negative charge. At points very near the charges, the lines are nearly radial, as for a single isolated charge. The high density of lines between the charges indicates a region of strong electric field.

Figure 23.21 shows the electric field lines in the vicinity of two equal positive point charges. Again, the lines are nearly radial at points close to either charge, and the same number of lines emerges from each charge because the charges are equal in magnitude. Because there are no negative charges available, the electric field lines end infinitely far away. At great distances from the charges, the field is approximately equal to that of a single point charge of magnitude $2q$.

Finally, in Figure 23.22, we sketch the electric field lines associated with a positive charge $+2q$ and a negative charge $-q$. In this case, the number of lines leaving $+2q$ is twice the number terminating at $-q$. Hence, only half the lines that leave the positive charge reach the negative charge. The remaining half terminate on a negative charge we assume to be at infinity. At distances much greater than the charge separation, the electric field lines are equivalent to those of a single charge $+q$.

Quick Quiz 23.5 Rank the magnitudes of the electric field at points A, B, and C shown in Figure 23.21 (greatest magnitude first).

The number of field lines leaving the positive charge equals the number terminating at the negative charge.

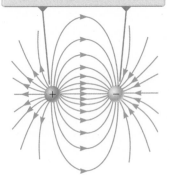

Figure 23.20 The electric field lines for two point charges of equal magnitude and opposite sign (an electric dipole).

Two field lines leave $+2q$ for every one that terminates on $-q$.

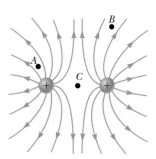

Figure 23.21 The electric field lines for two positive point charges. (The locations A, B, and C are discussed in Quick Quiz 23.5.)

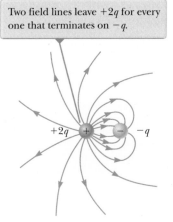

Figure 23.22 The electric field lines for a point charge $+2q$ and a second point charge $-q$.

23.7 Motion of a Charged Particle in a Uniform Electric Field

When a particle of charge q and mass m is placed in an electric field $\vec{E}$, the electric force exerted on the charge is $q\vec{E}$ according to Equation 23.8 in the particle in a field model. If that is the only force exerted on the particle, it must be the net force, and it causes the particle to accelerate according to the particle under a net force model. Therefore,

$$\vec{F}_e = q\vec{E} = m\vec{a}$$

and the acceleration of the particle is

$$\vec{a} = \frac{q\vec{E}}{m} \qquad (23.12)$$

If $\vec{E}$ is uniform (that is, constant in magnitude and direction), and the particle is free to move, the electric force on the particle is constant and we can apply the particle under constant acceleration model to the motion of the particle. Therefore, the particle in this situation is described by *three* analysis models: particle in a field, particle under a net force, and particle under constant acceleration! If the particle has a positive charge, its acceleration is in the direction of the electric field. If the particle has a negative charge, its acceleration is in the direction opposite the electric field.

> **Pitfall Prevention 23.4**
> **Just Another Force** Electric forces and fields may seem abstract to you. Once $\vec{F}_e$ is evaluated, however, it causes a particle to move according to our well-established models of forces and motion from Chapters 2 through 6. Keeping this link with the past in mind should help you solve problems in this chapter.

Example 23.10 | An Accelerating Positive Charge: Two Models AM

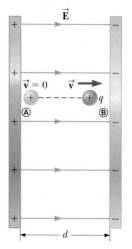

A uniform electric field $\vec{E}$ is directed along the x axis between parallel plates of charge separated by a distance d as shown in Figure 23.23. A positive point charge q of mass m is released from rest at a point Ⓐ next to the positive plate and accelerates to a point Ⓑ next to the negative plate.

(A) Find the speed of the particle at Ⓑ by modeling it as a particle under constant acceleration.

SOLUTION

Conceptualize When the positive charge is placed at Ⓐ, it experiences an electric force toward the right in Figure 23.23 due to the electric field directed toward the right. As a result, it will accelerate to the right and arrive at Ⓑ with some speed.

Figure 23.23 (Example 23.10) A positive point charge q in a uniform electric field $\vec{E}$ undergoes constant acceleration in the direction of the field.

Categorize Because the electric field is uniform, a constant electric force acts on the charge. Therefore, as suggested in the discussion preceding the example and in the problem statement, the point charge can be modeled as a charged *particle under constant acceleration.*

. . . .

Analyze Use Equation 2.17 to express the velocity of the particle as a function of position:

$$v_f^2 = v_i^2 + 2a(x_f - x_i) = 0 + 2a(d - 0) = 2ad$$

Solve for v_f and substitute for the magnitude of the acceleration from Equation 23.12:

$$v_f = \sqrt{2ad} = \sqrt{2\left(\frac{qE}{m}\right)d} = \boxed{\sqrt{\frac{2qEd}{m}}}$$

(B) Find the speed of the particle at Ⓑ by modeling it as a nonisolated system in terms of energy.

SOLUTION

Categorize The problem statement tells us that the charge is a *nonisolated system* for *energy.* The electric force, like any force, can do work on a system. Energy is transferred to the system of the charge by work done by the electric force exerted on the charge. The initial configuration of the system is when the particle is at rest at Ⓐ, and the final configuration is when it is moving with some speed at Ⓑ.

continued

▶ **23.10** continued

Analyze Write the appropriate reduction of the conservation of energy equation, Equation 8.2, for the system of the charged particle:

$$W = \Delta K$$

Replace the work and kinetic energies with values appropriate for this situation:

$$F_e \, \Delta x = K_\circledB - K_\circledA = \tfrac{1}{2} m v_f^2 - 0 \quad \rightarrow \quad v_f = \sqrt{\frac{2 F_e \, \Delta x}{m}}$$

Substitute for the magnitude of the electric force F_e from the particle in a field model and the displacement Δx:

$$v_f = \sqrt{\frac{2(qE)(d)}{m}} = \sqrt{\frac{2qEd}{m}}$$

Finalize The answer to part (B) is the same as that for part (A), as we expect. This problem can be solved with different approaches. We saw the same possibilities with mechanical problems.

Example 23.11   An Accelerated Electron   AM

An electron enters the region of a uniform electric field as shown in Figure 23.24, with $v_i = 3.00 \times 10^6$ m/s and $E = 200$ N/C. The horizontal length of the plates is $\ell = 0.100$ m.

(A) Find the acceleration of the electron while it is in the electric field.

SOLUTION

Conceptualize This example differs from the preceding one because the velocity of the charged particle is initially perpendicular to the electric field lines. (In Example 23.10, the velocity of the charged particle is always parallel to the electric field lines.) As a result, the electron in this example follows a curved path as shown in Figure 23.24. The motion of the electron is the same as that of a massive particle projected horizontally in a gravitational field near the surface of the Earth.

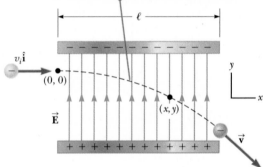

The electron undergoes a downward acceleration (opposite $\vec{\mathbf{E}}$), and its motion is parabolic while it is between the plates.

Figure 23.24 (Example 23.11) An electron is projected horizontally into a uniform electric field produced by two charged plates.

Categorize The electron is a *particle in a field (electric)*. Because the electric field is uniform, a constant electric force is exerted on the electron. To find the acceleration of the electron, we can model it as a *particle under a net force*.

Analyze From the particle in a field model, we know that the direction of the electric force on the electron is downward in Figure 23.24, opposite the direction of the electric field lines. From the particle under a net force model, therefore, the acceleration of the electron is downward.

The particle under a net force model was used to develop Equation 23.12 in the case in which the electric force on a particle is the only force. Use this equation to evaluate the y component of the acceleration of the electron:

$$a_y = -\frac{eE}{m_e}$$

Substitute numerical values:

$$a_y = -\frac{(1.60 \times 10^{-19} \text{ C})(200 \text{ N/C})}{9.11 \times 10^{-31} \text{ kg}} = -3.51 \times 10^{13} \text{ m/s}^2$$

(B) Assuming the electron enters the field at time $t = 0$, find the time at which it leaves the field.

SOLUTION

Categorize Because the electric force acts only in the vertical direction in Figure 23.24, the motion of the particle in the horizontal direction can be analyzed by modeling it as a *particle under constant velocity*.

▶ **23.11** continued

Analyze Solve Equation 2.7 for the time at which the electron arrives at the right edges of the plates:

$$x_f = x_i + v_x t \rightarrow t = \frac{x_f - x_i}{v_x}$$

Substitute numerical values:

$$t = \frac{\ell - 0}{v_x} = \frac{0.100 \text{ m}}{3.00 \times 10^6 \text{ m/s}} = \boxed{3.33 \times 10^{-8} \text{ s}}$$

(C) Assuming the vertical position of the electron as it enters the field is $y_i = 0$, what is its vertical position when it leaves the field?

SOLUTION

Categorize Because the electric force is constant in Figure 23.24, the motion of the particle in the vertical direction can be analyzed by modeling it as a *particle under constant acceleration.*

Analyze Use Equation 2.16 to describe the position of the particle at any time t:

$$y_f = y_i + v_{yi}t + \tfrac{1}{2}a_y t^2$$

Substitute numerical values:

$$y_f = 0 + 0 + \tfrac{1}{2}(-3.51 \times 10^{13} \text{ m/s}^2)(3.33 \times 10^{-8} \text{ s})^2$$

$$= -0.019\,5 \text{ m} = \boxed{-1.95 \text{ cm}}$$

Finalize If the electron enters just below the negative plate in Figure 23.24 and the separation between the plates is less than the value just calculated, the electron will strike the positive plate.

Notice that we have used *four* analysis models to describe the electron in the various parts of this problem. We have neglected the gravitational force acting on the electron, which represents a good approximation when dealing with atomic particles. For an electric field of 200 N/C, the ratio of the magnitude of the electric force eE to the magnitude of the gravitational force mg is on the order of 10^{12} for an electron and on the order of 10^9 for a proton.

Summary

Definitions

The **electric field** $\vec{\mathbf{E}}$ at some point in space is defined as the electric force $\vec{\mathbf{F}}_e$ that acts on a small positive test charge placed at that point divided by the magnitude q_0 of the test charge:

$$\vec{\mathbf{E}} \equiv \frac{\vec{\mathbf{F}}_e}{q_0} \tag{23.7}$$

Concepts and Principles

Electric charges have the following important properties:

- Charges of opposite sign attract one another, and charges of the same sign repel one another.
- The total charge in an isolated system is conserved.
- Charge is quantized.

Conductors are materials in which electrons move freely. **Insulators** are materials in which electrons do not move freely.

continued

Coulomb's law states that the electric force exerted by a point charge q_1 on a second point charge q_2 is

$$\vec{F}_{12} = k_e \frac{q_1 q_2}{r^2} \hat{r}_{12} \qquad (23.6)$$

where r is the distance between the two charges and $\hat{r}_{12}$ is a unit vector directed from q_1 toward q_2. The constant k_e, which is called the **Coulomb constant,** has the value $k_e = 8.988 \times 10^9 \text{ N} \cdot \text{m}^2/\text{C}^2$.

At a distance r from a point charge q, the electric field due to the charge is

$$\vec{E} = k_e \frac{q}{r^2} \hat{r} \qquad (23.9)$$

where $\hat{r}$ is a unit vector directed from the charge toward the point in question. The electric field is directed radially outward from a positive charge and radially inward toward a negative charge.

The electric field due to a group of point charges can be obtained by using the superposition principle. That is, the total electric field at some point equals the vector sum of the electric fields of all the charges:

$$\vec{E} = k_e \sum_i \frac{q_i}{r_i^2} \hat{r}_i \qquad (23.10)$$

The electric field at some point due to a continuous charge distribution is

$$\vec{E} = k_e \int \frac{dq}{r^2} \hat{r} \qquad (23.11)$$

where dq is the charge on one element of the charge distribution and r is the distance from the element to the point in question.

Analysis Models for Problem Solving

Particle in a Field (Electric) A source particle with some electric charge establishes an **electric field** $\vec{E}$ throughout space. When a particle with charge q is placed in that field, it experiences an electric force given by

$$\vec{F}_e = q\vec{E} \qquad (23.8)$$

Objective Questions

1. A free electron and a free proton are released in identical electric fields. **(i)** How do the magnitudes of the electric force exerted on the two particles compare? (a) It is millions of times greater for the electron. (b) It is thousands of times greater for the electron. (c) They are equal. (d) It is thousands of times smaller for the electron. (e) It is millions of times smaller for the electron. **(ii)** Compare the magnitudes of their accelerations. Choose from the same possibilities as in part (i).

2. What prevents gravity from pulling you through the ground to the center of the Earth? Choose the best answer. (a) The density of matter is too great. (b) The positive nuclei of your body's atoms repel the positive nuclei of the atoms of the ground. (c) The density of the ground is greater than the density of your body. (d) Atoms are bound together by chemical bonds. (e) Electrons on the ground's surface and the surface of your feet repel one another.

3. A very small ball has a mass of 5.00×10^{-3} kg and a charge of 4.00 μC. What magnitude electric field directed upward will balance the weight of the ball so that the ball is suspended motionless above the ground? (a) 8.21×10^2 N/C (b) 1.22×10^4 N/C (c) 2.00×10^{-2} N/C (d) 5.11×10^6 N/C (e) 3.72×10^3 N/C

4. An electron with a speed of 3.00×10^6 m/s moves into a uniform electric field of magnitude 1.00×10^3 N/C. The field lines are parallel to the electron's velocity and pointing in the same direction as the velocity. How far does the electron travel before it is brought to rest? (a) 2.56 cm (b) 5.12 cm (c) 11.2 cm (d) 3.34 m (e) 4.24 m

5. A point charge of -4.00 nC is located at $(0, 1.00)$ m. What is the x component of the electric field due to the point charge at $(4.00, -2.00)$ m? (a) 1.15 N/C (b) -0.864 N/C (c) 1.44 N/C (d) -1.15 N/C (e) 0.864 N/C

6. A circular ring of charge with radius b has total charge q uniformly distributed around it. What is the magnitude of the electric field at the center of the ring? (a) 0 (b) $k_e q/b^2$ (c) $k_e q^2/b^2$ (d) $k_e q^2/b$ (e) none of those answers

7. What happens when a charged insulator is placed near an uncharged metallic object? (a) They repel each other. (b) They attract each other. (c) They may attract or repel each other, depending on whether the charge on the insulator is positive or negative. (d) They exert no electrostatic force on each other. (e) The charged insulator always spontaneously discharges.

8. Estimate the magnitude of the electric field due to the proton in a hydrogen atom at a distance of 5.29×10^{-11} m, the expected position of the electron in the atom. (a) 10^{-11} N/C (b) 10^8 N/C (c) 10^{14} N/C (d) 10^6 N/C (e) 10^{12} N/C

9. (i) A metallic coin is given a positive electric charge. Does its mass (a) increase measurably, (b) increase by an amount too small to measure directly, (c) remain unchanged, (d) decrease by an amount too small to measure directly, or (e) decrease measurably? **(ii)** Now the coin is given a negative electric charge. What happens to its mass? Choose from the same possibilities as in part (i).

10. Assume the charged objects in Figure OQ23.10 are fixed. Notice that there is no sight line from the location of q_2 to the location of q_1. If you were at q_1, you would be unable to see q_2 because it is behind q_3. How would you calculate the electric force exerted on the object with charge q_1? (a) Find only the force exerted by q_2 on charge q_1. (b) Find only the force exerted by q_3 on charge q_1. (c) Add the force that q_2 would exert by itself on charge q_1 to the force that q_3 would exert by itself on charge q_1. (d) Add the force that q_3 would exert by itself to a certain fraction of the force that q_2 would exert by itself. (e) There is no definite way to find the force on charge q_1.

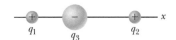

Figure OQ23.10

11. Three charged particles are arranged on corners of a square as shown in Figure OQ23.11, with charge $-Q$ on both the particle at the upper left corner and the particle at the lower right corner and with charge $+2Q$ on the particle at the lower left corner. **(i)** What is the direction of the electric field at the upper right corner, which is a point in empty space? (a) It is upward and to the right. (b) It is straight to the right. (c) It is straight downward. (d) It is downward and to the left. (e) It is perpendicular to the plane of the picture and outward. **(ii)** Suppose the $+2Q$ charge at the lower left corner is removed. Then does the magnitude of the field at the upper right corner (a) become larger, (b) become smaller, (c) stay the same, or (d) change unpredictably?

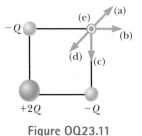

Figure OQ23.11

12. Two point charges attract each other with an electric force of magnitude F. If the charge on one of the particles is reduced to one-third its original value and the distance between the particles is doubled, what is the resulting magnitude of the electric force between them? (a) $\frac{1}{12}F$ (b) $\frac{1}{3}F$ (c) $\frac{1}{6}F$ (d) $\frac{3}{4}F$ (e) $\frac{3}{2}F$

13. Assume a uniformly charged ring of radius R and charge Q produces an electric field E_{ring} at a point P on its axis, at distance x away from the center of the ring as in Figure OQ23.13a. Now the same charge Q is spread uniformly over the circular area the ring encloses, forming a flat disk of charge with the same radius as in Figure OQ23.13b. How does the field E_{disk} produced by the disk at P compare with the field produced by the ring at the same point? (a) $E_{disk} < E_{ring}$ (b) $E_{disk} = E_{ring}$ (c) $E_{disk} > E_{ring}$ (d) impossible to determine

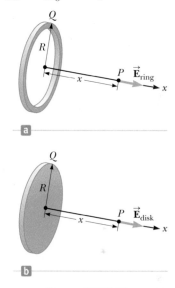

Figure OQ23.13

14. An object with negative charge is placed in a region of space where the electric field is directed vertically upward. What is the direction of the electric force exerted on this charge? (a) It is up. (b) It is down. (c) There is no force. (d) The force can be in any direction.

15. The magnitude of the electric force between two protons is 2.30×10^{-26} N. How far apart are they? (a) 0.100 m (b) 0.022 0 m (c) 3.10 m (d) 0.005 70 m (e) 0.480 m

Conceptual Questions | **1.** denotes answer available in *Student Solutions Manual/Study Guide*

1. (a) Would life be different if the electron were positively charged and the proton were negatively charged? (b) Does the choice of signs have any bearing on physical and chemical interactions? Explain your answers.

2. A charged comb often attracts small bits of dry paper that then fly away when they touch the comb. Explain why that occurs.

3. A person is placed in a large, hollow, metallic sphere that is insulated from ground. If a large charge is placed on the sphere, will the person be harmed upon touching the inside of the sphere?

4. A student who grew up in a tropical country and is studying in the United States may have no experience with static electricity sparks and shocks until his or her first American winter. Explain.

5. If a suspended object A is attracted to a charged object B, can we conclude that A is charged? Explain.

6. Consider point *A* in Figure CQ23.6 located an arbitrary distance from two positive point charges in otherwise empty space. (a) Is it possible for an electric field to exist at point *A* in empty space? Explain. (b) Does charge exist at this point? Explain. (c) Does a force exist at this point? Explain.

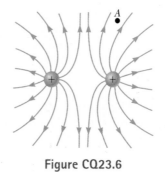

Figure CQ23.6

7. In fair weather, there is an electric field at the surface of the Earth, pointing down into the ground. What is the sign of the electric charge on the ground in this situation?

8. Why must hospital personnel wear special conducting shoes while working around oxygen in an operating room? What might happen if the personnel wore shoes with rubber soles?

9. A balloon clings to a wall after it is negatively charged by rubbing. (a) Does that occur because the wall is positively charged? (b) Why does the balloon eventually fall?

10. Consider two electric dipoles in empty space. Each dipole has zero net charge. (a) Does an electric force exist between the dipoles; that is, can two objects with zero net charge exert electric forces on each other? (b) If so, is the force one of attraction or of repulsion?

11. A glass object receives a positive charge by rubbing it with a silk cloth. In the rubbing process, have protons been added to the object or have electrons been removed from it?

Problems available in ENHANCED **WebAssign**   Access end-of-chapter problems online at **www.webassign.net**

Section 23.1 Properties of Electric Charges
Problems 1–2

Section 23.2 Charging Objects by Induction
Section 23.3 Coulomb's Law
Problems 3–22

Section 23.4 Analysis Model: Particle in a Field (Electric)
Problems 23–36

Section 23.5 Electric Field of a Continuous Charge Distribution
Problems 37–46

Section 23.6 Electric Field Lines
Problems 47–50

Section 23.7 Motion of a Charged Particle in a Uniform Electric Field
Problems 51–57

Additional Problems
Problems 58–83

Challenge Problems
Problems 84–91

Solutions to the following Problems are available in the *Student Solutions Manual/Study Guide:*
23.8, 23.11, 23.15, 23.25, 23.29, 23.39, 23.44, 23.45, 23.47, 23.50, 23.51, 23.55, 23.57, 23.61, 23.67, 23.73, 23.81, and 23.82

List of Enhanced Problems

Problem Number	Targeted Feedback in Enhanced WebAssign	Analysis Model Tutorial in Enhanced WebAssign	Master It in Enhanced WebAssign	Watch It in Enhanced WebAssign
23.2	✓			✓
23.10	✓			✓
23.11			✓	
23.13	✓			✓
23.15	✓		✓	
23.21	✓			✓
23.29	✓		✓	
23.30	✓			✓
23.33		✓		
23.37	✓			✓
23.39	✓		✓	
23.43				✓
23.45			✓	
23.49	✓			✓
23.51		✓	✓	
23.52	✓			✓
23.57			✓	
23.61		✓		
23.65		✓		
23.67			✓	

Gauss's Law

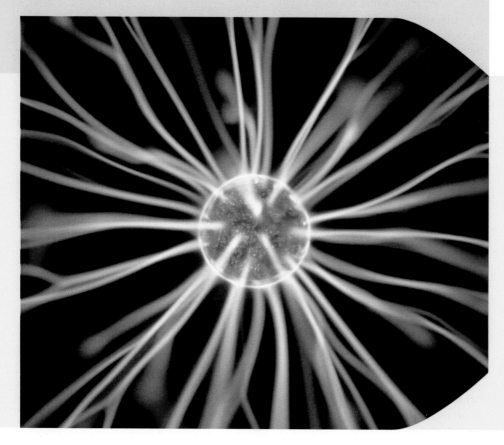

In Chapter 23, we showed how to calculate the electric field due to a given charge distribution by integrating over the distribution. In this chapter, we describe *Gauss's law* and an alternative procedure for calculating electric fields. Gauss's law is based on the inverse-square behavior of the electric force between point charges. Although Gauss's law is a direct consequence of Coulomb's law, it is more convenient for calculating the electric fields of highly symmetric charge distributions and makes it possible to deal with complicated problems using qualitative reasoning. As we show in this chapter, Gauss's law is important in understanding and verifying the properties of conductors in electrostatic equilibrium.

In a tabletop plasma ball, the colorful lines emanating from the sphere give evidence of strong electric fields. Using Gauss's law, we show in this chapter that the electric field surrounding a uniformly charged sphere is identical to that of a point charge. *(Steve Cole/Getty Images)*

24.1 Electric Flux

The concept of electric field lines was described qualitatively in Chapter 23. We now treat electric field lines in a more quantitative way.

Consider an electric field that is uniform in both magnitude and direction as shown in Figure 24.1. The field lines penetrate a rectangular surface of area A, whose plane is oriented perpendicular to the field. Recall from Section 23.6 that the number of lines per unit area (in other words, the *line density*) is proportional to the magnitude of the electric field. Therefore, the total number of lines penetrating the surface is proportional to the product EA. This product of the magnitude of the electric field E and surface area A perpendicular to the field is called the **electric flux** Φ_E (uppercase Greek letter phi):

$$\Phi_E = EA \tag{24.1}$$

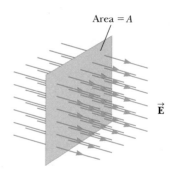

Area = A

$\vec{E}$

Figure 24.1 Field lines representing a uniform electric field penetrating a plane of area A perpendicular to the field.

567

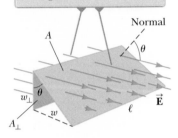

The number of field lines that go through the area $A_\perp$ is the same as the number that go through area A.

Figure 24.2 Field lines representing a uniform electric field penetrating an area A whose normal is at an angle θ to the field.

From the SI units of E and A, we see that Φ_E has units of newton meters squared per coulomb $(\text{N} \cdot \text{m}^2/\text{C})$. Electric flux is proportional to the number of electric field lines penetrating some surface.

If the surface under consideration is not perpendicular to the field, the flux through it must be less than that given by Equation 24.1. Consider Figure 24.2, where the normal to the surface of area A is at an angle θ to the uniform electric field. Notice that the number of lines that cross this area A is equal to the number of lines that cross the area $A_\perp$, which is a projection of area A onto a plane oriented perpendicular to the field. The area A is the product of the length and the width of the surface: $A = \ell w$. At the left edge of the figure, we see that the widths of the surfaces are related by $w_\perp = w \cos \theta$. The area $A_\perp$ is given by $A_\perp = \ell w_\perp = \ell w \cos \theta$ and we see that the two areas are related by $A_\perp = A \cos \theta$. Because the flux through A equals the flux through $A_\perp$, the flux through A is

$$\Phi_E = EA_\perp = EA \cos \theta \qquad (24.2)$$

From this result, we see that the flux through a surface of fixed area A has a maximum value EA when the surface is perpendicular to the field (when the normal to the surface is parallel to the field, that is, when $\theta = 0°$ in Fig. 24.2); the flux is zero when the surface is parallel to the field (when the normal to the surface is perpendicular to the field, that is, when $\theta = 90°$).

In this discussion, the angle θ is used to describe the orientation of the surface of area A. We can also interpret the angle as that between the electric field vector and the normal to the surface. In this case, the product $E \cos \theta$ in Equation 24.2 is the component of the electric field perpendicular to the surface. The flux through the surface can then be written $\Phi_E = (E \cos \theta)A = E_n A$, where we use E_n as the component of the electric field normal to the surface.

We assumed a uniform electric field in the preceding discussion. In more general situations, the electric field may vary over a large surface. Therefore, the definition of flux given by Equation 24.2 has meaning only for a small element of area over which the field is approximately constant. Consider a general surface divided into a large number of small elements, each of area ΔA_i. It is convenient to define a vector $\Delta \vec{\mathbf{A}}_i$ whose magnitude represents the area of the ith element of the large surface and whose direction is defined to be *perpendicular* to the surface element as shown in Figure 24.3. The electric field $\vec{\mathbf{E}}_i$ at the location of this element makes an angle θ_i with the vector $\Delta \vec{\mathbf{A}}_i$. The electric flux $\Phi_{E,i}$ through this element is

$$\Phi_{E,i} = E_i \, \Delta A_i \, \cos \theta_i = \vec{\mathbf{E}}_i \cdot \Delta \vec{\mathbf{A}}_i$$

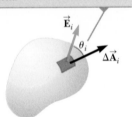

The electric field makes an angle θ_i with the vector $\Delta \vec{\mathbf{A}}_i$, defined as being normal to the surface element.

Figure 24.3 A small element of surface area ΔA_i in an electric field.

where we have used the definition of the scalar product of two vectors $(\vec{\mathbf{A}} \cdot \vec{\mathbf{B}} \equiv AB \cos \theta$; see Chapter 7). Summing the contributions of all elements gives an approximation to the total flux through the surface:

$$\Phi_E \approx \sum \vec{\mathbf{E}}_i \cdot \Delta \vec{\mathbf{A}}_i$$

If the area of each element approaches zero, the number of elements approaches infinity and the sum is replaced by an integral. Therefore, the general definition of electric flux is

◀ **Definition of electric flux**

$$\Phi_E \equiv \int_{\text{surface}} \vec{\mathbf{E}} \cdot d\vec{\mathbf{A}} \qquad (24.3)$$

Equation 24.3 is a *surface integral*, which means it must be evaluated over the surface in question. In general, the value of Φ_E depends both on the field pattern and on the surface.

We are often interested in evaluating the flux through a *closed surface*, defined as a surface that divides space into an inside and an outside region so that one cannot move from one region to the other without crossing the surface. The surface of a sphere, for example, is a closed surface. By convention, if the area element in Equa-

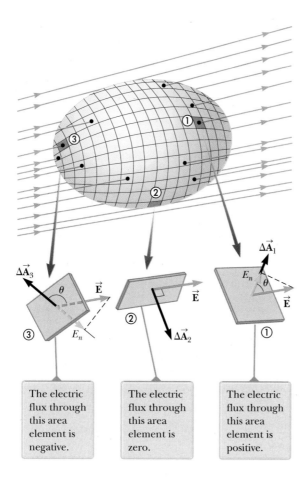

Figure 24.4 A closed surface in an electric field. The area vectors are, by convention, normal to the surface and point outward.

The electric flux through this area element is negative.

The electric flux through this area element is zero.

The electric flux through this area element is positive.

tion 24.3 is part of a closed surface, the direction of the area vector is chosen so that the vector points outward from the surface. If the area element is not part of a closed surface, the direction of the area vector is chosen so that the angle between the area vector and the electric field vector is less than or equal to 90°.

Consider the closed surface in Figure 24.4. The vectors $\Delta \vec{\mathbf{A}}_i$ point in different directions for the various surface elements, but for each element they are normal to the surface and point outward. At the element labeled ①, the field lines are crossing the surface from the inside to the outside and $\theta < 90°$; hence, the flux $\Phi_{E,1} = \vec{\mathbf{E}} \cdot \Delta \vec{\mathbf{A}}_1$ through this element is positive. For element ②, the field lines graze the surface (perpendicular to $\Delta \vec{\mathbf{A}}_2$); therefore, $\theta = 90°$ and the flux is zero. For elements such as ③, where the field lines are crossing the surface from outside to inside, $180° > \theta > 90°$ and the flux is negative because $\cos \theta$ is negative. The *net* flux through the surface is proportional to the net number of lines leaving the surface, where the net number means *the number of lines leaving the surface minus the number of lines entering the surface.* If more lines are leaving than entering, the net flux is positive. If more lines are entering than leaving, the net flux is negative. Using the symbol $\oint$ to represent an integral over a closed surface, we can write the net flux Φ_E through a closed surface as

$$\Phi_E = \oint \vec{\mathbf{E}} \cdot d\vec{\mathbf{A}} = \oint E_n \, dA \qquad \textbf{(24.4)}$$

where E_n represents the component of the electric field normal to the surface.

Quick Quiz 24.1 Suppose a point charge is located at the center of a spherical surface. The electric field at the surface of the sphere and the total flux through the sphere are determined. Now the radius of the sphere is halved.

What happens to the flux through the sphere and the magnitude of the electric field at the surface of the sphere? **(a)** The flux and field both increase. **(b)** The flux and field both decrease. **(c)** The flux increases, and the field decreases. **(d)** The flux decreases, and the field increases. **(e)** The flux remains the same, and the field increases. **(f)** The flux decreases, and the field remains the same.

Example 24.1 Flux Through a Cube

Consider a uniform electric field $\vec{\mathbf{E}}$ oriented in the x direction in empty space. A cube of edge length ℓ is placed in the field, oriented as shown in Figure 24.5. Find the net electric flux through the surface of the cube.

SOLUTION

Conceptualize Examine Figure 24.5 carefully. Notice that the electric field lines pass through two faces perpendicularly and are parallel to four other faces of the cube.

Categorize We evaluate the flux from its definition, so we categorize this example as a substitution problem.

The flux through four of the faces (③, ④, and the unnumbered faces) is zero because $\vec{\mathbf{E}}$ is parallel to the four faces and therefore perpendicular to $d\vec{\mathbf{A}}$ on these faces.

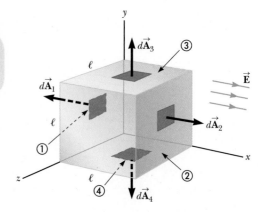

Figure 24.5 (Example 24.1) A closed surface in the shape of a cube in a uniform electric field oriented parallel to the x axis. Side ④ is the bottom of the cube, and side ① is opposite side ②.

Write the integrals for the net flux through faces ① and ②:

$$\Phi_E = \int_1 \vec{\mathbf{E}} \cdot d\vec{\mathbf{A}} + \int_2 \vec{\mathbf{E}} \cdot d\vec{\mathbf{A}}$$

For face ①, $\vec{\mathbf{E}}$ is constant and directed inward but $d\vec{\mathbf{A}}_1$ is directed outward ($\theta = 180°$). Find the flux through this face:

$$\int_1 \vec{\mathbf{E}} \cdot d\vec{\mathbf{A}} = \int_1 E(\cos 180°)\, dA = -E \int_1 dA = -EA = -E\ell^2$$

For face ②, $\vec{\mathbf{E}}$ is constant and outward and in the same direction as $d\vec{\mathbf{A}}_2$ ($\theta = 0°$). Find the flux through this face:

$$\int_2 \vec{\mathbf{E}} \cdot d\vec{\mathbf{A}} = \int_2 E(\cos 0°)\, dA = E \int_2 dA = +EA = E\ell^2$$

Find the net flux by adding the flux over all six faces:

$$\Phi_E = -E\ell^2 + E\ell^2 + 0 + 0 + 0 + 0 = \boxed{0}$$

24.2 Gauss's Law

When the charge is at the center of the sphere, the electric field is everywhere normal to the surface and constant in magnitude.

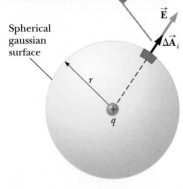

Figure 24.6 A spherical gaussian surface of radius r surrounding a positive point charge q.

In this section, we describe a general relationship between the net electric flux through a closed surface (often called a *gaussian surface*) and the charge enclosed by the surface. This relationship, known as *Gauss's law*, is of fundamental importance in the study of electric fields.

Consider a positive point charge q located at the center of a sphere of radius r as shown in Figure 24.6. From Equation 23.9, we know that the magnitude of the electric field everywhere on the surface of the sphere is $E = k_e q/r^2$. The field lines are directed radially outward and hence are perpendicular to the surface at every point on the surface. That is, at each surface point, $\vec{\mathbf{E}}$ is parallel to the vector $\Delta \vec{\mathbf{A}}_i$ representing a local element of area ΔA_i surrounding the surface point. Therefore,

$$\vec{\mathbf{E}} \cdot \Delta \vec{\mathbf{A}}_i = E\, \Delta A_i$$

and, from Equation 24.4, we find that the net flux through the gaussian surface is

$$\Phi_E = \oint \vec{\mathbf{E}} \cdot d\vec{\mathbf{A}} = \oint E\, dA = E \oint dA$$

where we have moved E outside of the integral because, by symmetry, E is constant over the surface. The value of E is given by $E = k_e q/r^2$. Furthermore, because the surface is spherical, $\oint dA = A = 4\pi r^2$. Hence, the net flux through the gaussian surface is

$$\Phi_E = k_e \frac{q}{r^2} (4\pi r^2) = 4\pi k_e q$$

Recalling from Equation 23.3 that $k_e = 1/4\pi\epsilon_0$, we can write this equation in the form

$$\Phi_E = \frac{q}{\epsilon_0} \tag{24.5}$$

Equation 24.5 shows that the net flux through the spherical surface is proportional to the charge inside the surface. The flux is independent of the radius r because the area of the spherical surface is proportional to r^2, whereas the electric field is proportional to $1/r^2$. Therefore, in the product of area and electric field, the dependence on r cancels.

Now consider several closed surfaces surrounding a charge q as shown in Figure 24.7. Surface S_1 is spherical, but surfaces S_2 and S_3 are not. From Equation 24.5, the flux that passes through S_1 has the value q/ϵ_0. As discussed in the preceding section, flux is proportional to the number of electric field lines passing through a surface. The construction shown in Figure 24.7 shows that the number of lines through S_1 is equal to the number of lines through the nonspherical surfaces S_2 and S_3. Therefore,

> the net flux through *any* closed surface surrounding a point charge q is given by q/ϵ_0 and is independent of the shape of that surface.

Now consider a point charge located *outside* a closed surface of arbitrary shape as shown in Figure 24.8. As can be seen from this construction, any electric field line entering the surface leaves the surface at another point. The number of electric field lines entering the surface equals the number leaving the surface. Therefore, the net electric flux through a closed surface that surrounds no charge is zero. Applying this result to Example 24.1, we see that the net flux through the cube is zero because there is no charge inside the cube.

Let's extend these arguments to two generalized cases: (1) that of many point charges and (2) that of a continuous distribution of charge. We once again use the superposition principle, which states that the electric field due to many charges is

Karl Friedrich Gauss
German mathematician and astronomer (1777–1855)
Gauss received a doctoral degree in mathematics from the University of Helmstedt in 1799. In addition to his work in electromagnetism, he made contributions to mathematics and science in number theory, statistics, non-Euclidean geometry, and cometary orbital mechanics. He was a founder of the German Magnetic Union, which studies the Earth's magnetic field on a continual basis.

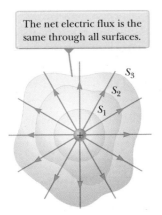

Figure 24.7 Closed surfaces of various shapes surrounding a positive charge.

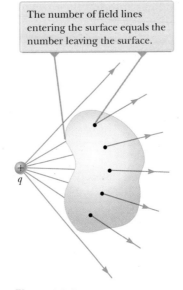

Figure 24.8 A point charge located *outside* a closed surface.

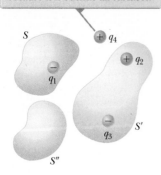

Charge q_4 does not contribute to the flux through any surface because it is outside all surfaces.

Figure 24.9 The net electric flux through any closed surface depends only on the charge *inside* that surface. The net flux through surface S is q_1/ϵ_0, the net flux through surface S' is $(q_2 + q_3)/\epsilon_0$, and the net flux through surface S'' is zero.

Pitfall Prevention 24.1

Zero Flux Is Not Zero Field
In two situations, there is zero flux through a closed surface: either (1) there are no charged particles enclosed by the surface or (2) there are charged particles enclosed, but the net charge inside the surface is zero. For either situation, it is *incorrect* to conclude that the electric field on the surface is zero. Gauss's law states that the electric *flux* is proportional to the enclosed charge, not the electric *field*.

---

the vector sum of the electric fields produced by the individual charges. Therefore, the flux through any closed surface can be expressed as

$$\oint \vec{E} \cdot d\vec{A} = \oint (\vec{E}_1 + \vec{E}_2 + \cdots) \cdot d\vec{A}$$

where $\vec{E}$ is the total electric field at any point on the surface produced by the vector addition of the electric fields at that point due to the individual charges. Consider the system of charges shown in Figure 24.9. The surface S surrounds only one charge, q_1; hence, the net flux through S is q_1/ϵ_0. The flux through S due to charges q_2, q_3, and q_4 outside it is zero because each electric field line from these charges that enters S at one point leaves it at another. The surface S' surrounds charges q_2 and q_3; hence, the net flux through it is $(q_2 + q_3)/\epsilon_0$. Finally, the net flux through surface S'' is zero because there is no charge inside this surface. That is, *all* the electric field lines that enter S'' at one point leave at another. Charge q_4 does not contribute to the net flux through any of the surfaces.

The mathematical form of **Gauss's law** is a generalization of what we have just described and states that the net flux through *any* closed surface is

$$\Phi_E = \oint \vec{E} \cdot d\vec{A} = \frac{q_{in}}{\epsilon_0} \qquad \text{(24.6)}$$

where $\vec{E}$ represents the electric field at any point on the surface and q_{in} represents the net charge inside the surface.

When using Equation 24.6, you should note that although the charge q_{in} is the net charge inside the gaussian surface, $\vec{E}$ represents the *total electric field*, which includes contributions from charges both inside and outside the surface.

In principle, Gauss's law can be solved for $\vec{E}$ to determine the electric field due to a system of charges or a continuous distribution of charge. In practice, however, this type of solution is applicable only in a limited number of highly symmetric situations. In the next section, we use Gauss's law to evaluate the electric field for charge distributions that have spherical, cylindrical, or planar symmetry. If one chooses the gaussian surface surrounding the charge distribution carefully, the integral in Equation 24.6 can be simplified and the electric field determined.

Quick Quiz 24.2 If the net flux through a gaussian surface is *zero*, the following four statements *could be true*. Which of the statements *must be true*? **(a)** There are no charges inside the surface. **(b)** The net charge inside the surface is zero. **(c)** The electric field is zero everywhere on the surface. **(d)** The number of electric field lines entering the surface equals the number leaving the surface.

Conceptual Example 24.2 **Flux Due to a Point Charge**

A spherical gaussian surface surrounds a point charge q. Describe what happens to the total flux through the surface if **(A)** the charge is tripled, **(B)** the radius of the sphere is doubled, **(C)** the surface is changed to a cube, and **(D)** the charge is moved to another location inside the surface.

SOLUTION

(A) The flux through the surface is tripled because flux is proportional to the amount of charge inside the surface.

(B) The flux does not change because all electric field lines from the charge pass through the sphere, regardless of its radius.

(C) The flux does not change when the shape of the gaussian surface changes because all electric field lines from the charge pass through the surface, regardless of its shape.

(D) The flux does not change when the charge is moved to another location inside that surface because Gauss's law refers to the total charge enclosed, regardless of where the charge is located inside the surface.

24.3 Application of Gauss's Law to Various Charge Distributions

As mentioned earlier, Gauss's law is useful for determining electric fields when the charge distribution is highly symmetric. The following examples demonstrate ways of choosing the gaussian surface over which the surface integral given by Equation 24.6 can be simplified and the electric field determined. In choosing the surface, always take advantage of the symmetry of the charge distribution so that E can be removed from the integral. The goal in this type of calculation is to determine a surface for which each portion of the surface satisfies one or more of the following conditions:

1. The value of the electric field can be argued by symmetry to be constant over the portion of the surface.
2. The dot product in Equation 24.6 can be expressed as a simple algebraic product $E\, dA$ because $\vec{\mathbf{E}}$ and $d\vec{\mathbf{A}}$ are parallel.
3. The dot product in Equation 24.6 is zero because $\vec{\mathbf{E}}$ and $d\vec{\mathbf{A}}$ are perpendicular.
4. The electric field is zero over the portion of the surface.

Different portions of the gaussian surface can satisfy different conditions as long as every portion satisfies at least one condition. All four conditions are used in examples throughout the remainder of this chapter and will be identified by number. If the charge distribution does not have sufficient symmetry such that a gaussian surface that satisfies these conditions can be found, Gauss's law is still true, but is not useful for determining the electric field for that charge distribution.

> **Pitfall Prevention 24.2**
> **Gaussian Surfaces Are Not Real**
> A gaussian surface is an imaginary surface you construct to satisfy the conditions listed here. It does not have to coincide with a physical surface in the situation.

Example 24.3   A Spherically Symmetric Charge Distribution

An insulating solid sphere of radius a has a uniform volume charge density ρ and carries a total positive charge Q (Fig. 24.10).

(A) Calculate the magnitude of the electric field at a point outside the sphere.

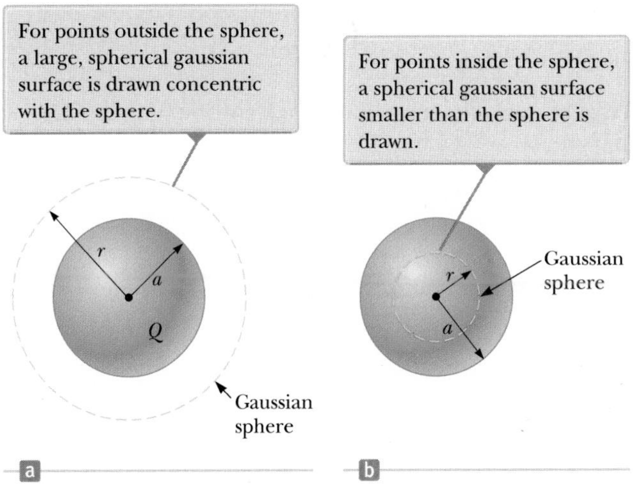

SOLUTION

Conceptualize Notice how this problem differs from our previous discussion of Gauss's law. The electric field due to point charges was discussed in Section 24.2. Now we are considering the electric field due to a distribution of charge. We found the field for various distributions of charge in Chapter 23 by integrating over the distribution. This example demonstrates a difference from our discussions in Chapter 23. In this chapter, we find the electric field using Gauss's law.

Categorize Because the charge is distributed uniformly throughout the sphere, the charge distribution has spherical symmetry and we can apply Gauss's law to find the electric field.

For points outside the sphere, a large, spherical gaussian surface is drawn concentric with the sphere.

For points inside the sphere, a spherical gaussian surface smaller than the sphere is drawn.

Figure 24.10 (Example 24.3) A uniformly charged insulating sphere of radius a and total charge Q. In diagrams such as this one, the dotted line represents the intersection of the gaussian surface with the plane of the page.

Analyze To reflect the spherical symmetry, let's choose a spherical gaussian surface of radius r, concentric with the sphere, as shown in Figure 24.10a. For this choice, condition (2) is satisfied everywhere on the surface and $\vec{\mathbf{E}} \cdot d\vec{\mathbf{A}} = E\, dA$.

continued

▶ **24.3** continued

Replace $\vec{\mathbf{E}} \cdot d\vec{\mathbf{A}}$ in Gauss's law with $E \, dA$:

$$\Phi_E = \oint \vec{\mathbf{E}} \cdot d\vec{\mathbf{A}} = \oint E \, dA = \frac{Q}{\epsilon_0}$$

By symmetry, E has the same value everywhere on the surface, which satisfies condition (1), so we can remove E from the integral:

$$\oint E \, dA = E \oint dA = E(4\pi r^2) = \frac{Q}{\epsilon_0}$$

Solve for E:

$$(1) \quad E = \frac{Q}{4\pi\epsilon_0 r^2} = k_e \frac{Q}{r^2} \quad (\text{for } r > a)$$

Finalize This field is identical to that for a point charge. Therefore, **the electric field due to a uniformly charged sphere in the region external to the sphere is** *equivalent* **to that of a point charge located at the center of the sphere.**

(B) Find the magnitude of the electric field at a point inside the sphere.

SOLUTION

Analyze In this case, let's choose a spherical gaussian surface having radius $r < a$, concentric with the insulating sphere (Fig. 24.10b). Let V' be the volume of this smaller sphere. To apply Gauss's law in this situation, recognize that the charge q_{in} within the gaussian surface of volume V' is less than Q.

Calculate q_{in} by using $q_{in} = \rho V'$:

$$q_{in} = \rho V' = \rho(\tfrac{4}{3}\pi r^3)$$

Notice that conditions (1) and (2) are satisfied everywhere on the gaussian surface in Figure 24.10b. Apply Gauss's law in the region $r < a$:

$$\oint E \, dA = E \oint dA = E(4\pi r^2) = \frac{q_{in}}{\epsilon_0}$$

Solve for E and substitute for q_{in}:

$$E = \frac{q_{in}}{4\pi\epsilon_0 r^2} = \frac{\rho(\tfrac{4}{3}\pi r^3)}{4\pi\epsilon_0 r^2} = \frac{\rho}{3\epsilon_0} r$$

Substitute $\rho = Q/\tfrac{4}{3}\pi a^3$ and $\epsilon_0 = 1/4\pi k_e$:

$$(2) \quad E = \frac{Q/\tfrac{4}{3}\pi a^3}{3(1/4\pi k_e)} r = k_e \frac{Q}{a^3} r \quad (\text{for } r < a)$$

Finalize This result for E differs from the one obtained in part (A). It shows that $E \to 0$ as $r \to 0$. Therefore, the result eliminates the problem that would exist at $r = 0$ if E varied as $1/r^2$ inside the sphere as it does outside the sphere. That is, if $E \propto 1/r^2$ for $r < a$, the field would be infinite at $r = 0$, which is physically impossible.

WHAT IF? Suppose the radial position $r = a$ is approached from inside the sphere and from outside. Do we obtain the same value of the electric field from both directions?

Answer Equation (1) shows that the electric field approaches a value from the outside given by

$$E = \lim_{r \to a} \left(k_e \frac{Q}{r^2} \right) = k_e \frac{Q}{a^2}$$

From the inside, Equation (2) gives

$$E = \lim_{r \to a} \left(k_e \frac{Q}{a^3} r \right) = k_e \frac{Q}{a^3} a = k_e \frac{Q}{a^2}$$

Therefore, the value of the field is the same as the surface is approached from both directions. A plot of E versus r is shown in Figure 24.11. Notice that the magnitude of the field is continuous.

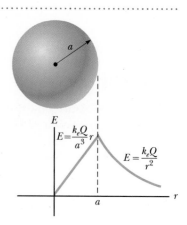

Figure 24.11 (Example 24.3) A plot of E versus r for a uniformly charged insulating sphere. The electric field inside the sphere ($r < a$) varies linearly with r. The field outside the sphere ($r > a$) is the same as that of a point charge Q located at $r = 0$.

| Example 24.4 | **A Cylindrically Symmetric Charge Distribution** |

Find the electric field a distance r from a line of positive charge of infinite length and constant charge per unit length λ (Fig. 24.12a).

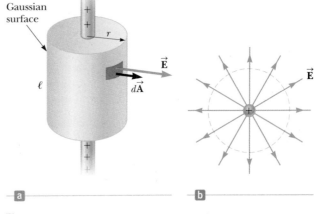

SOLUTION

Conceptualize The line of charge is *infinitely* long. Therefore, the field is the same at all points equidistant from the line, regardless of the vertical position of the point in Figure 24.12a. We expect the field to become weaker as we move farther away from the line of charge.

Categorize Because the charge is distributed uniformly along the line, the charge distribution has cylindrical symmetry and we can apply Gauss's law to find the electric field.

Figure 24.12 (Example 24.4) (a) An infinite line of charge surrounded by a cylindrical gaussian surface concentric with the line. (b) An end view shows that the electric field at the cylindrical surface is constant in magnitude and perpendicular to the surface.

Analyze The symmetry of the charge distribution requires that $\vec{E}$ be perpendicular to the line charge and directed outward as shown in Figure 24.12b. To reflect the symmetry of the charge distribution, let's choose a cylindrical gaussian surface of radius r and length ℓ that is coaxial with the line charge. For the curved part of this surface, $\vec{E}$ is constant in magnitude and perpendicular to the surface at each point, satisfying conditions (1) and (2). Furthermore, the flux through the ends of the gaussian cylinder is zero because $\vec{E}$ is parallel to these surfaces. That is the first application we have seen of condition (3).

We must take the surface integral in Gauss's law over the entire gaussian surface. Because $\vec{E} \cdot d\vec{A}$ is zero for the flat ends of the cylinder, however, we restrict our attention to only the curved surface of the cylinder.

Apply Gauss's law and conditions (1) and (2) for the curved surface, noting that the total charge inside our gaussian surface is $\lambda \ell$:

$$\Phi_E = \oint \vec{E} \cdot d\vec{A} = E \oint dA = EA = \frac{q_{in}}{\epsilon_0} = \frac{\lambda \ell}{\epsilon_0}$$

Substitute the area $A = 2\pi r \ell$ of the curved surface:

$$E(2\pi r \ell) = \frac{\lambda \ell}{\epsilon_0}$$

Solve for the magnitude of the electric field:

$$E = \frac{\lambda}{2\pi \epsilon_0 r} = \boxed{2k_e \frac{\lambda}{r}} \qquad \textbf{(24.7)}$$

Finalize This result shows that the electric field due to a cylindrically symmetric charge distribution varies as $1/r$, whereas the field external to a spherically symmetric charge distribution varies as $1/r^2$. Equation 24.7 can also be derived by direct integration over the charge distribution. (See Problem 44 in Chapter 23 in Enhanced WebAssign.)

| WHAT IF? | What if the line segment in this example were not infinitely long? |

Answer If the line charge in this example were of finite length, the electric field would not be given by Equation 24.7. A finite line charge does not possess sufficient symmetry to make use of Gauss's law because the magnitude of the electric field is no longer constant over the surface of the gaussian cylinder: the field near the ends of the line would be different from that far from the ends. Therefore, condition (1) would not be satisfied in this situation. Furthermore, $\vec{E}$ is not perpendicular to the cylindrical surface at all points: the field vectors near the ends would have a component parallel to the line. Therefore, condition (2) would not be satisfied. For points close to a finite line charge and far from the ends, Equation 24.7 gives a good approximation of the value of the field.

It is left for you to show (see Problem 33 in Enhanced WebAssign) that the electric field inside a uniformly charged rod of finite radius and infinite length is proportional to r.

Example 24.5 A Plane of Charge

Find the electric field due to an infinite plane of positive charge with uniform surface charge density σ.

SOLUTION

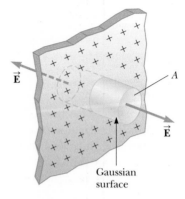

Conceptualize Notice that the plane of charge is *infinitely* large. Therefore, the electric field should be the same at all points equidistant from the plane. How would you expect the electric field to depend on the distance from the plane?

Categorize Because the charge is distributed uniformly on the plane, the charge distribution is symmetric; hence, we can use Gauss's law to find the electric field.

Figure 24.13 (Example 24.5) A cylindrical gaussian surface penetrating an infinite plane of charge. The flux is EA through each end of the gaussian surface and zero through its curved surface.

Analyze By symmetry, $\vec{E}$ must be perpendicular to the plane at all points. The direction of $\vec{E}$ is away from positive charges, indicating that the direction of $\vec{E}$ on one side of the plane must be opposite its direction on the other side as shown in Figure 24.13. A gaussian surface that reflects the symmetry is a small cylinder whose axis is perpendicular to the plane and whose ends each have an area A and are equidistant from the plane. Because $\vec{E}$ is parallel to the curved surface of the cylinder—and therefore perpendicular to $d\vec{A}$ at all points on this surface— condition (3) is satisfied and there is no contribution to the surface integral from this surface. For the flat ends of the cylinder, conditions (1) and (2) are satisfied. The flux through each end of the cylinder is EA; hence, the total flux through the entire gaussian surface is just that through the ends, $\Phi_E = 2EA$.

Write Gauss's law for this surface, noting that the enclosed charge is $q_{in} = \sigma A$:

$$\Phi_E = 2EA = \frac{q_{in}}{\epsilon_0} = \frac{\sigma A}{\epsilon_0}$$

Solve for E:

$$E = \frac{\sigma}{2\epsilon_0} \qquad (24.8)$$

Finalize Because the distance from each flat end of the cylinder to the plane does not appear in Equation 24.8, we conclude that $E = \sigma/2\epsilon_0$ at *any* distance from the plane. That is, the field is uniform everywhere. Figure 24.14 shows this uniform field due to an infinite plane of charge, seen edge-on.

WHAT IF? Suppose two infinite planes of charge are parallel to each other, one positively charged and the other negatively charged. The surface charge densities of both planes are of the same magnitude. What does the electric field look like in this situation?

Answer We first addressed this configuration in the **What If?** section of Example 23.9. The electric fields due to the two planes add in the region between the planes, resulting in a uniform field of magnitude σ/ϵ_0, and cancel elsewhere to give a field of zero. Figure 24.15 shows the field lines for such a configuration. This method is a practical way to achieve uniform electric fields with finite-sized planes placed close to each other.

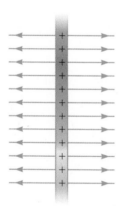

Figure 24.14 (Example 24.5) The electric field lines due to an infinite plane of positive charge.

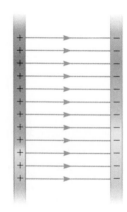

Figure 24.15 (Example 24.5) The electric field lines between two infinite planes of charge, one positive and one negative. In practice, the field lines near the edges of finite-sized sheets of charge will curve outward.

Conceptual Example 24.6 Don't Use Gauss's Law Here!

Explain why Gauss's law cannot be used to calculate the electric field near an electric dipole, a charged disk, or a triangle with a point charge at each corner.

▶ **24.6** continued

The charge distributions of all these configurations do not have sufficient symmetry to make the use of Gauss's law practical. We cannot find a closed surface surrounding any of these distributions for which all portions of the surface satisfy one or more of conditions (1) through (4) listed at the beginning of this section.

24.4 Conductors in Electrostatic Equilibrium

As we learned in Section 23.2, a good electrical conductor contains charges (electrons) that are not bound to any atom and therefore are free to move about within the material. When there is no net motion of charge within a conductor, the conductor is in **electrostatic equilibrium.** A conductor in electrostatic equilibrium has the following properties:

1. The electric field is zero everywhere inside the conductor, whether the conductor is solid or hollow.
2. If the conductor is isolated and carries a charge, the charge resides on its surface.
3. The electric field at a point just outside a charged conductor is perpendicular to the surface of the conductor and has a magnitude σ/ϵ_0, where σ is the surface charge density at that point.
4. On an irregularly shaped conductor, the surface charge density is greatest at locations where the radius of curvature of the surface is smallest.

◀ **Properties of a conductor in electrostatic equilibrium**

We verify the first three properties in the discussion that follows. The fourth property is presented here (but not verified until we have studied the appropriate material in Chapter 25) to provide a complete list of properties for conductors in electrostatic equilibrium.

We can understand the first property by considering a conducting slab placed in an external field $\vec{\mathbf{E}}$ (Fig. 24.16). The electric field inside the conductor *must* be zero, assuming electrostatic equilibrium exists. If the field were not zero, free electrons in the conductor would experience an electric force ($\vec{\mathbf{F}} = q\vec{\mathbf{E}}$) and would accelerate due to this force. This motion of electrons, however, would mean that the conductor is not in electrostatic equilibrium. Therefore, the existence of electrostatic equilibrium is consistent only with a zero field in the conductor.

Let's investigate how this zero field is accomplished. Before the external field is applied, free electrons are uniformly distributed throughout the conductor. When the external field is applied, the free electrons accelerate to the left in Figure 24.16, causing a plane of negative charge to accumulate on the left surface. The movement of electrons to the left results in a plane of positive charge on the right surface. These planes of charge create an additional electric field inside the conductor that opposes the external field. As the electrons move, the surface charge densities on the left and right surfaces increase until the magnitude of the internal field equals that of the external field, resulting in a net field of zero inside the conductor. The time it takes a good conductor to reach equilibrium is on the order of 10^{-16} s, which for most purposes can be considered instantaneous.

If the conductor is hollow, the electric field inside the conductor is also zero, whether we consider points in the conductor or in the cavity within the conductor. The zero value of the electric field in the cavity is easiest to argue with the concept of electric potential, so we will address this issue in Section 25.6.

Gauss's law can be used to verify the second property of a conductor in electrostatic equilibrium. Figure 24.17 shows an arbitrarily shaped conductor. A gaussian

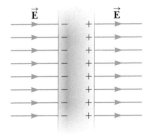

Figure 24.16 A conducting slab in an external electric field $\vec{\mathbf{E}}$. The charges induced on the two surfaces of the slab produce an electric field that opposes the external field, giving a resultant field of zero inside the slab.

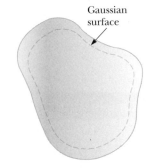

Figure 24.17 A conductor of arbitrary shape. The broken line represents a gaussian surface that can be just inside the conductor's surface.

surface is drawn inside the conductor and can be very close to the conductor's surface. As we have just shown, the electric field everywhere inside the conductor is zero when it is in electrostatic equilibrium. Therefore, the electric field must be zero at every point on the gaussian surface, in accordance with condition (4) in Section 24.3, and the net flux through this gaussian surface is zero. From this result and Gauss's law, we conclude that the net charge inside the gaussian surface is zero. Because there can be no net charge inside the gaussian surface (which is arbitrarily close to the conductor's surface), any net charge on the conductor must reside on its surface. Gauss's law does not indicate how this excess charge is distributed on the conductor's surface, only that it resides exclusively on the surface.

To verify the third property, let's begin with the perpendicularity of the field to the surface. If the field vector $\vec{E}$ had a component parallel to the conductor's surface, free electrons would experience an electric force and move along the surface; in such a case, the conductor would not be in equilibrium. Therefore, the field vector must be perpendicular to the surface.

To determine the magnitude of the electric field, we use Gauss's law and draw a gaussian surface in the shape of a small cylinder whose end faces are parallel to the conductor's surface (Fig. 24.18). Part of the cylinder is just outside the conductor, and part is inside. The field is perpendicular to the conductor's surface from the condition of electrostatic equilibrium. Therefore, condition (3) in Section 24.3 is satisfied for the curved part of the cylindrical gaussian surface: there is no flux through this part of the gaussian surface because $\vec{E}$ is parallel to the surface. There is no flux through the flat face of the cylinder inside the conductor because here $\vec{E} = 0$, which satisfies condition (4). Hence, the net flux through the gaussian surface is equal to that through only the flat face outside the conductor, where the field is perpendicular to the gaussian surface. Using conditions (1) and (2) for this face, the flux is EA, where E is the electric field just outside the conductor and A is the area of the cylinder's face. Applying Gauss's law to this surface gives

$$\Phi_E = \oint E\, dA = EA = \frac{q_{in}}{\epsilon_0} = \frac{\sigma A}{\epsilon_0}$$

where we have used $q_{in} = \sigma A$. Solving for E gives for the electric field immediately outside a charged conductor:

$$E = \frac{\sigma}{\epsilon_0} \tag{24.9}$$

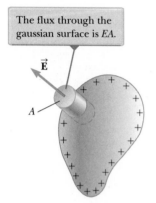

The flux through the gaussian surface is EA.

$\vec{E}$

A

Figure 24.18 A gaussian surface in the shape of a small cylinder is used to calculate the electric field immediately outside a charged conductor.

Quick Quiz 24.3 Your younger brother likes to rub his feet on the carpet and then touch you to give you a shock. While you are trying to escape the shock treatment, you discover a hollow metal cylinder in your basement, large enough to climb inside. In which of the following cases will you *not* be shocked? **(a)** You climb inside the cylinder, making contact with the inner surface, and your charged brother touches the outer metal surface. **(b)** Your charged brother is inside touching the inner metal surface and you are outside, touching the outer metal surface. **(c)** Both of you are outside the cylinder, touching its outer metal surface but not touching each other directly.

Example 24.7 A Sphere Inside a Spherical Shell

A solid insulating sphere of radius a carries a net positive charge Q uniformly distributed throughout its volume. A conducting spherical shell of inner radius b and outer radius c is concentric with the solid sphere and carries a net charge $-2Q$. Using Gauss's law, find the electric field in the regions labeled ①, ②, ③, and ④ in Figure 24.19 and the charge distribution on the shell when the entire system is in electrostatic equilibrium.

▶ **24.7** continued

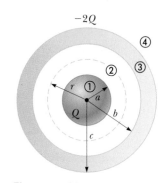

Figure 24.19 (Example 24.7) An insulating sphere of radius a and carrying a charge Q surrounded by a conducting spherical shell carrying a charge $-2Q$.

SOLUTION

Conceptualize Notice how this problem differs from Example 24.3. The charged sphere in Figure 24.10 appears in Figure 24.19, but it is now surrounded by a shell carrying a charge $-2Q$. Think about how the presence of the shell will affect the electric field of the sphere.

Categorize The charge is distributed uniformly throughout the sphere, and we know that the charge on the conducting shell distributes itself uniformly on the surfaces. Therefore, the system has spherical symmetry and we can apply Gauss's law to find the electric field in the various regions.

. . . .

Analyze In region ②—between the surface of the solid sphere and the inner surface of the shell—we construct a spherical gaussian surface of radius r, where $a < r < b$, noting that the charge inside this surface is $+Q$ (the charge on the solid sphere). Because of the spherical symmetry, the electric field lines must be directed radially outward and be constant in magnitude on the gaussian surface.

The charge on the conducting shell creates zero electric field in the region $r < b$, so the shell has no effect on the field in region ② due to the sphere. Therefore, write an expression for the field in region ② as that due to the sphere from part (A) of Example 24.3:

$$E_2 = k_e \frac{Q}{r^2} \quad (\text{for } a < r < b)$$

Because the conducting shell creates zero field inside itself, it also has no effect on the field inside the sphere. Therefore, write an expression for the field in region ① as that due to the sphere from part (B) of Example 24.3:

$$E_1 = k_e \frac{Q}{a^3} r \quad (\text{for } r < a)$$

In region ④, where $r > c$, construct a spherical gaussian surface; this surface surrounds a total charge $q_{in} = Q + (-2Q) = -Q$. Therefore, model the charge distribution as a sphere with charge $-Q$ and write an expression for the field in region ④ from part (A) of Example 24.3:

$$E_4 = -k_e \frac{Q}{r^2} \quad (\text{for } r > c)$$

In region ③, the electric field must be zero because the spherical shell is a conductor in equilibrium:

$$E_3 = 0 \quad (\text{for } b < r < c)$$

Construct a gaussian surface of radius r in region ③, where $b < r < c$, and note that q_{in} must be zero because $E_3 = 0$. Find the amount of charge q_{inner} on the inner surface of the shell:

$$q_{in} = q_{sphere} + q_{inner}$$
$$q_{inner} = q_{in} - q_{sphere} = 0 - Q = -Q$$

. . . .

Finalize The charge on the inner surface of the spherical shell must be $-Q$ to cancel the charge $+Q$ on the solid sphere and give zero electric field in the material of the shell. Because the net charge on the shell is $-2Q$, its outer surface must carry a charge $-Q$.

WHAT IF? How would the results of this problem differ if the sphere were conducting instead of insulating?

Answer The only change would be in region ①, where $r < a$. Because there can be no charge inside a conductor in electrostatic equilibrium, $q_{in} = 0$ for a gaussian surface of radius $r < a$; therefore, on the basis of Gauss's law and symmetry, $E_1 = 0$. In regions ②, ③, and ④, there would be no way to determine from observations of the electric field whether the sphere is conducting or insulating.

Summary

Definition

Electric flux is proportional to the number of electric field lines that penetrate a surface. If the electric field is uniform and makes an angle θ with the normal to a surface of area A, the electric flux through the surface is

$$\Phi_E = EA \cos \theta \qquad \text{(24.2)}$$

In general, the electric flux through a surface is

$$\Phi_E \equiv \int_{\text{surface}} \vec{E} \cdot d\vec{A} \qquad \text{(24.3)}$$

Concepts and Principles

Gauss's law says that the net electric flux Φ_E through any closed gaussian surface is equal to the *net* charge q_{in} inside the surface divided by ϵ_0:

$$\Phi_E = \oint \vec{E} \cdot d\vec{A} = \frac{q_{\text{in}}}{\epsilon_0} \qquad \text{(24.6)}$$

Using Gauss's law, you can calculate the electric field due to various symmetric charge distributions.

A conductor in **electrostatic equilibrium** has the following properties:

1. The electric field is zero everywhere inside the conductor, whether the conductor is solid or hollow.
2. If the conductor is isolated and carries a charge, the charge resides on its surface.
3. The electric field at a point just outside a charged conductor is perpendicular to the surface of the conductor and has a magnitude σ/ϵ_0, where σ is the surface charge density at that point.
4. On an irregularly shaped conductor, the surface charge density is greatest at locations where the radius of curvature of the surface is smallest.

Objective Questions   1. denotes answer available in *Student Solutions Manual/Study Guide*

1. A cubical gaussian surface surrounds a long, straight, charged filament that passes perpendicularly through two opposite faces. No other charges are nearby. **(i)** Over how many of the cube's faces is the electric field zero? (a) 0 (b) 2 (c) 4 (d) 6 **(ii)** Through how many of the cube's faces is the electric flux zero? Choose from the same possibilities as in part (i).

2. A coaxial cable consists of a long, straight filament surrounded by a long, coaxial, cylindrical conducting shell. Assume charge Q is on the filament, zero net charge is on the shell, and the electric field is $E_1\hat{\mathbf{i}}$ at a particular point P midway between the filament and the inner surface of the shell. Next, you place the cable into a uniform external field $-E\hat{\mathbf{i}}$. What is the x component of the electric field at P then? (a) 0 (b) between 0 and E_1 (c) E_1 (d) between 0 and $-E_1$ (e) $-E_1$

3. In which of the following contexts can Gauss's law *not* be readily applied to find the electric field? (a) near a long, uniformly charged wire (b) above a large, uniformly charged plane (c) inside a uniformly charged ball (d) outside a uniformly charged sphere (e) Gauss's law can be readily applied to find the electric field in all these contexts.

4. A particle with charge q is located inside a cubical gaussian surface. No other charges are nearby. **(i)** If the particle is at the center of the cube, what is the flux through each one of the faces of the cube? (a) 0 (b) $q/2\epsilon_0$ (c) $q/6\epsilon_0$ (d) $q/8\epsilon_0$ (e) depends on the size of the cube **(ii)** If the particle can be moved to any point within the cube, what maximum value can the flux through one face approach? Choose from the same possibilities as in part (i).

5. Charges of 3.00 nC, −2.00 nC, −7.00 nC, and 1.00 nC are contained inside a rectangular box with length 1.00 m, width 2.00 m, and height 2.50 m. Outside the box are charges of 1.00 nC and 4.00 nC. What is the electric flux through the surface of the box? (a) 0 (b) -5.64×10^2 N · m²/C (c) -1.47×10^3 N · m²/C (d) 1.47×10^3 N · m²/C (e) 5.64×10^2 N · m²/C

6. A large, metallic, spherical shell has no net charge. It is supported on an insulating stand and has a small hole at the top. A small tack with charge Q is lowered on a silk thread through the hole into the interior of the shell. **(i)** What is the charge on the inner surface of the shell, (a) Q (b) $Q/2$ (c) 0 (d) $-Q/2$ or (e) $-Q$? Choose your answers to the following questions from

the same possibilities. **(ii)** What is the charge on the outer surface of the shell? **(iii)** The tack is now allowed to touch the interior surface of the shell. After this contact, what is the charge on the tack? **(iv)** What is the charge on the inner surface of the shell now? **(v)** What is the charge on the outer surface of the shell now?

7. Two solid spheres, both of radius 5 cm, carry identical total charges of 2 μC. Sphere A is a good conductor. Sphere B is an insulator, and its charge is distributed uniformly throughout its volume. **(i)** How do the magnitudes of the electric fields they separately create at a radial distance of 6 cm compare? (a) $E_A > E_B = 0$ (b) $E_A > E_B > 0$ (c) $E_A = E_B > 0$ (d) $0 < E_A < E_B$ (e) $0 = E_A < E_B$ **(ii)** How do the magnitudes of the electric fields they separately create at radius 4 cm compare? Choose from the same possibilities as in part (i).

8. A uniform electric field of 1.00 N/C is set up by a uniform distribution of charge in the xy plane. What is the electric field inside a metal ball placed 0.500 m above the xy plane? (a) 1.00 N/C (b) −1.00 N/C (c) 0 (d) 0.250 N/C (e) varies depending on the position inside the ball

9. A solid insulating sphere of radius 5 cm carries electric charge uniformly distributed throughout its volume. Concentric with the sphere is a conducting spherical shell with no net charge as shown in Figure OQ24.9. The inner radius of the shell is 10 cm, and the outer radius is 15 cm. No other charges are nearby. (a) Rank

the magnitude of the electric field at points A (at radius 4 cm), B (radius 8 cm), C (radius 12 cm), and D (radius 16 cm) from largest to smallest. Display any cases of equality in your ranking. (b) Similarly rank the electric flux through concentric spherical surfaces through points A, B, C, and D.

Figure OQ24.9

10. A cubical gaussian surface is bisected by a large sheet of charge, parallel to its top and bottom faces. No other charges are nearby. **(i)** Over how many of the cube's faces is the electric field zero? (a) 0 (b) 2 (c) 4 (d) 6 **(ii)** Through how many of the cube's faces is the electric flux zero? Choose from the same possibilities as in part (i).

11. Rank the electric fluxes through each gaussian surface shown in Figure OQ24.11 from largest to smallest. Display any cases of equality in your ranking.

Figure OQ24.11

Conceptual Questions

1. denotes answer available in *Student Solutions Manual/Study Guide*

1. Consider an electric field that is uniform in direction throughout a certain volume. Can it be uniform in magnitude? Must it be uniform in magnitude? Answer these questions (a) assuming the volume is filled with an insulating material carrying charge described by a volume charge density and (b) assuming the volume is empty space. State reasoning to prove your answers.

2. A cubical surface surrounds a point charge q. Describe what happens to the total flux through the surface if (a) the charge is doubled, (b) the volume of the cube is doubled, (c) the surface is changed to a sphere, (d) the charge is moved to another location inside the surface, and (e) the charge is moved outside the surface.

3. A uniform electric field exists in a region of space containing no charges. What can you conclude about the net electric flux through a gaussian surface placed in this region of space?

4. If the total charge inside a closed surface is known but the distribution of the charge is unspecified, can you use Gauss's law to find the electric field? Explain.

5. Explain why the electric flux through a closed surface with a given enclosed charge is independent of the size or shape of the surface.

6. If more electric field lines leave a gaussian surface than enter it, what can you conclude about the net charge enclosed by that surface?

7. A person is placed in a large, hollow, metallic sphere that is insulated from ground. (a) If a large charge is placed on the sphere, will the person be harmed upon touching the inside of the sphere? (b) Explain what will happen if the person also has an initial charge whose sign is opposite that of the charge on the sphere.

8. Consider two identical conducting spheres whose surfaces are separated by a small distance. One sphere is given a large net positive charge, and the other is given a small net positive charge. It is found that the force between the spheres is attractive even though they both have net charges of the same sign. Explain how this attraction is possible.

9. A common demonstration involves charging a rubber balloon, which is an insulator, by rubbing it on your hair and then touching the balloon to a ceiling or wall, which is also an insulator. Because of the electrical attraction between the charged balloon and the neutral wall, the balloon sticks to the wall. Imagine now that we have two infinitely large, flat sheets of insulating

material. One is charged, and the other is neutral. If these sheets are brought into contact, does an attractive force exist between them as there was for the balloon and the wall?

10. On the basis of the repulsive nature of the force between like charges and the freedom of motion of charge within a conductor, explain why excess charge on an isolated conductor must reside on its surface.

11. The Sun is lower in the sky during the winter than it is during the summer. (a) How does this change affect the flux of sunlight hitting a given area on the surface of the Earth? (b) How does this change affect the weather?

Problems available in Access end-of-chapter problems online at www.webassign.net

Section 24.1 Electric Flux
Problems 1–6

Section 24.2 Gauss's Law
Problems 7–22

Section 24.3 Application of Gauss's Law to Various Charge Distributions
Problems 23–36

Section 24.4 Conductors in Electrostatic Equilibrium
Problems 37–47

Additional Problems
Problems 48–60

Challenge Problems
Problems 61–69

Solutions to the following Problems are available in the *Student Solutions Manual/Study Guide:*
24.3, 24.9, 24.21, 24.27, 24.29, 24.31, 24.33, 24.45, 24.46, 24.52, 24.54, 24.56, 24.66, and 24.69

List of Enhanced Problems

Problem Number	Targeted Feedback in Enhanced WebAssign	Analysis Model Tutorial in Enhanced WebAssign	Master It in Enhanced WebAssign	Watch It in Enhanced WebAssign
24.2	✓			✓
24.3	✓		✓	
24.4	✓			✓
24.5			✓	
24.9	✓		✓	
24.10	✓			✓
24.11				✓
24.14	✓			✓
24.24	✓			✓
24.25		✓		
24.27			✓	
24.29	✓		✓	
24.30	✓			✓
24.31			✓	
24.34	✓			✓
24.35	✓			✓
24.36		✓		
24.37			✓	
24.39	✓			✓
24.43		✓		
24.46			✓	
24.47	✓		✓	
24.57	✓			✓
24.62		✓		

Electric Potential

Processes occurring during thunderstorms cause large differences in electric potential between a thundercloud and the ground. The result of this potential difference is an electrical discharge that we call lightning, such as this display. Notice at the left that a downward channel of lightning (a *stepped leader*) is about to make contact with a channel coming up from the ground (a *return stroke*). (Costazzurra/Shutterstock.com)

In Chapter 23, we linked our new study of electromagnetism to our earlier studies of *force*. Now we make a new link to our earlier investigations into *energy*. The concept of potential energy was introduced in Chapter 7 in connection with such conservative forces as the gravitational force and the elastic force exerted by a spring. By using the law of conservation of energy, we could solve various problems in mechanics that were not solvable with an approach using forces. The concept of potential energy is also of great value in the study of electricity. Because the electrostatic force is conservative, electrostatic phenomena can be conveniently described in terms of an electric potential energy. This idea enables us to define a quantity known as *electric potential*. Because the electric potential at any point in an electric field is a scalar quantity, we can use it to describe electrostatic phenomena more simply than if we were to rely only on the electric field and electric forces. The concept of electric potential is of great practical value in the operation of electric circuits and devices that we will study in later chapters.

25.1 Electric Potential and Potential Difference

When a charge q is placed in an electric field $\vec{\mathbf{E}}$ created by some source charge distribution, the particle in a field model tells us that there is an electric force $q\vec{\mathbf{E}}$

acting on the charge. This force is conservative because the force between charges described by Coulomb's law is conservative. Let us identify the charge and the field as a system. If the charge is free to move, it will do so in response to the electric force. Therefore, the electric field will be doing work on the charge. This work is *internal* to the system. This situation is similar to that in a gravitational system: When an object is released near the surface of the Earth, the gravitational force does work on the object. This work is internal to the object–Earth system as discussed in Sections 7.7 and 7.8.

When analyzing electric and magnetic fields, it is common practice to use the notation $d\vec{\mathbf{s}}$ to represent an infinitesimal displacement vector that is oriented tangent to a path through space. This path may be straight or curved, and an integral performed along this path is called either a *path integral* or a *line integral* (the two terms are synonymous).

For an infinitesimal displacement $d\vec{\mathbf{s}}$ of a point charge q immersed in an electric field, the work done within the charge–field system by the electric field on the charge is $W_{int} = \vec{\mathbf{F}}_e \cdot d\vec{\mathbf{s}} = q\vec{\mathbf{E}} \cdot d\vec{\mathbf{s}}$. Recall from Equation 7.26 that internal work done in a system is equal to the negative of the change in the potential energy of the system: $W_{int} = -\Delta U$. Therefore, as the charge q is displaced, the electric potential energy of the charge–field system is changed by an amount $dU = -W_{int} = -q\vec{\mathbf{E}} \cdot d\vec{\mathbf{s}}$. For a finite displacement of the charge from some point Ⓐ in space to some other point Ⓑ, the change in electric potential energy of the system is

$$\Delta U = -q\int_{Ⓐ}^{Ⓑ} \vec{\mathbf{E}} \cdot d\vec{\mathbf{s}} \tag{25.1}$$

◀ **Change in electric potential energy of a system**

The integration is performed along the path that q follows as it moves from Ⓐ to Ⓑ. Because the force $q\vec{\mathbf{E}}$ is conservative, this line integral does not depend on the path taken from Ⓐ to Ⓑ.

For a given position of the charge in the field, the charge–field system has a potential energy U relative to the configuration of the system that is defined as $U = 0$. Dividing the potential energy by the charge gives a physical quantity that depends only on the source charge distribution and has a value at every point in an electric field. This quantity is called the **electric potential** (or simply the **potential**) V:

$$V = \frac{U}{q} \tag{25.2}$$

Because potential energy is a scalar quantity, electric potential also is a scalar quantity.

The **potential difference** $\Delta V = V_{Ⓑ} - V_{Ⓐ}$ between two points Ⓐ and Ⓑ in an electric field is defined as the change in electric potential energy of the system when a charge q is moved between the points (Eq. 25.1) divided by the charge:

$$\Delta V \equiv \frac{\Delta U}{q} = -\int_{Ⓐ}^{Ⓑ} \vec{\mathbf{E}} \cdot d\vec{\mathbf{s}} \tag{25.3}$$

◀ **Potential difference between two points**

In this definition, the infinitesimal displacement $d\vec{\mathbf{s}}$ is interpreted as the displacement between two points in space rather than the displacement of a point charge as in Equation 25.1.

Just as with potential energy, only *differences* in electric potential are meaningful. We often take the value of the electric potential to be zero at some convenient point in an electric field.

Potential difference should not be confused with difference in potential energy. The potential *difference* between Ⓐ and Ⓑ exists solely because of a source charge and depends on the source charge distribution (consider points Ⓐ and Ⓑ in the discussion above *without* the presence of the charge q). For a potential *energy* to exist, we must have a system of two or more charges. The potential

Pitfall Prevention 25.1
Potential and Potential Energy
The *potential is characteristic of the field only*, independent of a charged particle that may be placed in the field. *Potential energy is characteristic of the charge-field system* due to an interaction between the field and a charged particle placed in the field.

energy belongs to the system and changes only if a charge is moved relative to the rest of the system. This situation is similar to that for the electric field. An electric *field* exists solely because of a source charge. An electric *force* requires two charges: the source charge to set up the field and another charge placed within that field.

Let's now consider the situation in which an external agent moves the charge in the field. If the agent moves the charge from Ⓐ to Ⓑ without changing the kinetic energy of the charge, the agent performs work that changes the potential energy of the system: $W = \Delta U$. From Equation 25.3, the work done by an external agent in moving a charge q through an electric field at constant velocity is

$$W = q\,\Delta V \qquad (25.4)$$

Because electric potential is a measure of potential energy per unit charge, the SI unit of both electric potential and potential difference is joules per coulomb, which is defined as a **volt** (V):

$$1\ \text{V} \equiv 1\ \text{J/C}$$

That is, as we can see from Equation 25.4, 1 J of work must be done to move a 1-C charge through a potential difference of 1 V.

Equation 25.3 shows that potential difference also has units of electric field times distance. It follows that the SI unit of electric field (N/C) can also be expressed in volts per meter:

$$1\ \text{N/C} = 1\ \text{V/m}$$

Therefore, we can state a new interpretation of the electric field:

> The electric field is a measure of the rate of change of the electric potential with respect to position.

A unit of energy commonly used in atomic and nuclear physics is the **electron volt** (eV), which is defined as the energy a charge–field system gains or loses when a charge of magnitude e (that is, an electron or a proton) is moved through a potential difference of 1 V. Because 1 V = 1 J/C and the fundamental charge is equal to 1.60×10^{-19} C, the electron volt is related to the joule as follows:

$$1\ \text{eV} = 1.60 \times 10^{-19}\ \text{C} \cdot \text{V} = 1.60 \times 10^{-19}\ \text{J} \qquad (25.5)$$

For instance, an electron in the beam of a typical dental x-ray machine may have a speed of 1.4×10^8 m/s. This speed corresponds to a kinetic energy 1.1×10^{-14} J (using relativistic calculations as discussed in Chapter 39), which is equivalent to 6.7×10^4 eV. Such an electron has to be accelerated from rest through a potential difference of 67 kV to reach this speed.

Quick Quiz 25.1 In Figure 25.1, two points Ⓐ and Ⓑ are located within a region in which there is an electric field. **(i)** How would you describe the potential difference $\Delta V = V_{Ⓑ} - V_{Ⓐ}$? (a) It is positive. (b) It is negative. (c) It is zero. **(ii)** A negative charge is placed at Ⓐ and then moved to Ⓑ. How would you describe the change in potential energy of the charge–field system for this process? Choose from the same possibilities.

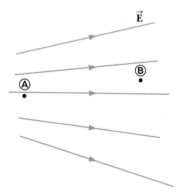

Figure 25.1 (Quick Quiz 25.1) Two points in an electric field.

25.2 Potential Difference in a Uniform Electric Field

Equations 25.1 and 25.3 hold in all electric fields, whether uniform or varying, but they can be simplified for the special case of a uniform field. First, consider a uniform electric field directed along the negative *y* axis as shown in Figure 25.2a. Let's calculate the potential difference between two points Ⓐ and Ⓑ separated by a dis-

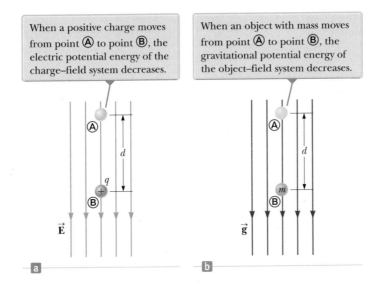

Figure 25.2 (a) When the electric field $\vec{\mathbf{E}}$ is directed downward, point Ⓑ is at a lower electric potential than point Ⓐ. (b) A gravitational analog to the situation in (a).

tance d, where the displacement $\vec{\mathbf{s}}$ points from Ⓐ toward Ⓑ and is parallel to the field lines. Equation 25.3 gives

$$V_Ⓑ - V_Ⓐ = \Delta V = -\int_Ⓐ^Ⓑ \vec{\mathbf{E}} \cdot d\vec{\mathbf{s}} = -\int_Ⓐ^Ⓑ E\, ds\, (\cos 0°) = -\int_Ⓐ^Ⓑ E\, ds$$

Because E is constant, it can be removed from the integral sign, which gives

$$\Delta V = -E \int_Ⓐ^Ⓑ ds$$

$$\Delta V = -Ed \tag{25.6}$$

◀ **Potential difference between two points in a uniform electric field**

The negative sign indicates that the electric potential at point Ⓑ is lower than at point Ⓐ; that is, $V_Ⓑ < V_Ⓐ$. Electric field lines *always* point in the direction of decreasing electric potential as shown in Figure 25.2a.

Now suppose a charge q moves from Ⓐ to Ⓑ. We can calculate the change in the potential energy of the charge–field system from Equations 25.3 and 25.6:

$$\Delta U = q\, \Delta V = -qEd \tag{25.7}$$

This result shows that if q is positive, then ΔU is negative. Therefore, in a system consisting of a positive charge and an electric field, the electric potential energy of the system decreases when the charge moves in the direction of the field. If a positive charge is released from rest in this electric field, it experiences an electric force $q\vec{\mathbf{E}}$ in the direction of $\vec{\mathbf{E}}$ (downward in Fig. 25.2a). Therefore, it accelerates downward, gaining kinetic energy. As the charged particle gains kinetic energy, the electric potential energy of the charge–field system decreases by an equal amount. This equivalence should not be surprising; it is simply conservation of mechanical energy in an isolated system as introduced in Chapter 8.

Figure 25.2b shows an analogous situation with a gravitational field. When a particle with mass m is released in a gravitational field, it accelerates downward, gaining kinetic energy. At the same time, the gravitational potential energy of the object–field system decreases.

The comparison between a system of a positive charge residing in an electrical field and an object with mass residing in a gravitational field in Figure 25.2 is useful for conceptualizing electrical behavior. The electrical situation, however, has one feature that the gravitational situation does not: the charge can be negative. If q is negative, then ΔU in Equation 25.7 is positive and the situation is reversed.

Pitfall Prevention 25.4
The Sign of ΔV The negative sign in Equation 25.6 is due to the fact that we started at point Ⓐ and moved to a new point in the *same* direction as the electric field lines. If we started from Ⓑ and moved to Ⓐ, the potential difference would be $+Ed$. In a uniform electric field, the magnitude of the potential difference is Ed and the sign can be determined by the direction of travel.

Figure 25.3 A uniform electric field directed along the positive x axis. Three points in the electric field are labeled.

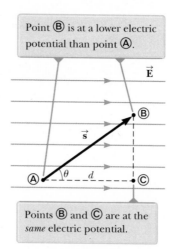

Point Ⓑ is at a lower electric potential than point Ⓐ.

Points Ⓑ and Ⓒ are at the *same* electric potential.

A system consisting of a negative charge and an electric field *gains* electric potential energy when the charge moves in the direction of the field. If a negative charge is released from rest in an electric field, it accelerates in a direction *opposite* the direction of the field. For the negative charge to move in the direction of the field, an external agent must apply a force and do positive work on the charge.

Now consider the more general case of a charged particle that moves between Ⓐ and Ⓑ in a uniform electric field such that the vector $\vec{s}$ is *not* parallel to the field lines as shown in Figure 25.3. In this case, Equation 25.3 gives

◀ **Change in potential between two points in a uniform electric field**

$$\Delta V = -\int_Ⓐ^Ⓑ \vec{E} \cdot d\vec{s} = -\vec{E} \cdot \int_Ⓐ^Ⓑ d\vec{s} = -\vec{E} \cdot \vec{s} \tag{25.8}$$

where again $\vec{E}$ was removed from the integral because it is constant. The change in potential energy of the charge–field system is

$$\Delta U = q\Delta V = -q\vec{E} \cdot \vec{s} \tag{25.9}$$

Finally, we conclude from Equation 25.8 that all points in a plane perpendicular to a uniform electric field are at the same electric potential. We can see that in Figure 25.3, where the potential difference $V_Ⓑ - V_Ⓐ$ is equal to the potential difference $V_Ⓒ - V_Ⓐ$. (Prove this fact to yourself by working out two dot products for $\vec{E} \cdot \vec{s}$: one for $\vec{s}_{Ⓐ\rightarrowⒷ}$, where the angle θ between $\vec{E}$ and $\vec{s}$ is arbitrary as shown in Figure 25.3, and one for $\vec{s}_{Ⓐ\rightarrowⒸ}$, where $\theta = 0$.) Therefore, $V_Ⓑ = V_Ⓒ$. The name **equipotential surface** is given to any surface consisting of a continuous distribution of points having the same electric potential.

The equipotential surfaces associated with a uniform electric field consist of a family of parallel planes that are all perpendicular to the field. Equipotential surfaces associated with fields having other symmetries are described in later sections.

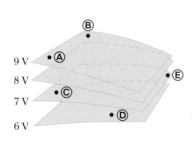

Figure 25.4 (Quick Quiz 25.2) Four equipotential surfaces.

Ⓠ**uick Quiz 25.2** The labeled points in Figure 25.4 are on a series of equipotential surfaces associated with an electric field. Rank (from greatest to least) the work done by the electric field on a positively charged particle that moves from Ⓐ to Ⓑ, from Ⓑ to Ⓒ, from Ⓒ to Ⓓ, and from Ⓓ to Ⓔ.

Example 25.1 | **The Electric Field Between Two Parallel Plates of Opposite Charge**

A battery has a specified potential difference ΔV between its terminals and establishes that potential difference between conductors attached to the terminals. A 12-V battery is connected between two parallel plates as shown in Figure 25.5. The separation between the plates is $d = 0.30$ cm, and we assume the electric field between the plates to be uniform. (This assumption is reasonable if the plate separation is small relative to the plate dimensions and we do not consider locations near the plate edges.) Find the magnitude of the electric field between the plates.

▶ **25.1** continued

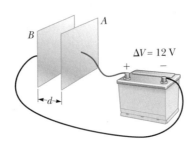

Figure 25.5 (Example 25.1) A 12-V battery connected to two parallel plates. The electric field between the plates has a magnitude given by the potential difference ΔV divided by the plate separation d.

$\Delta V = 12$ V

Conceptualize In Example 24.5, we illustrated the uniform electric field between parallel plates. The new feature to this problem is that the electric field is related to the new concept of electric potential.

Categorize The electric field is evaluated from a relationship between field and potential given in this section, so we categorize this example as a substitution problem.

Use Equation 25.6 to evaluate the magnitude of the electric field between the plates:

$$E = \frac{|V_B - V_A|}{d} = \frac{12 \text{ V}}{0.30 \times 10^{-2} \text{ m}} = 4.0 \times 10^3 \text{ V/m}$$

The configuration of plates in Figure 25.5 is called a *parallel-plate capacitor* and is examined in greater detail in Chapter 26.

Example 25.2 | **Motion of a Proton in a Uniform Electric Field** AM

A proton is released from rest at point Ⓐ in a uniform electric field that has a magnitude of 8.0×10^4 V/m (Fig. 25.6). The proton undergoes a displacement of magnitude $d = 0.50$ m to point Ⓑ in the direction of $\vec{E}$. Find the speed of the proton after completing the displacement.

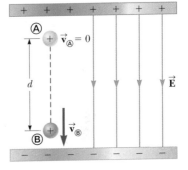

Conceptualize Visualize the proton in Figure 25.6 moving downward through the potential difference. The situation is analogous to an object falling through a gravitational field. Also compare this example to Example 23.10 where a positive charge was moving in a uniform electric field. In that example, we applied the particle under constant acceleration and nonisolated system models. Now that we have investigated electric potential energy, what model can we use here?

Figure 25.6 (Example 25.2) A proton accelerates from Ⓐ to Ⓑ in the direction of the electric field.

Categorize The system of the proton and the two plates in Figure 25.6 does not interact with the environment, so we model it as an *isolated system* for *energy*.

Analyze

Write the appropriate reduction of Equation 8.2, the conservation of energy equation, for the isolated system of the charge and the electric field:

$\Delta K + \Delta U = 0$

Substitute the changes in energy for both terms:

$(\frac{1}{2}mv^2 - 0) + e\Delta V = 0$

Solve for the final speed of the proton and substitute for ΔV from Equation 25.6:

$v = \sqrt{\frac{-2e\,\Delta V}{m}} = \sqrt{\frac{-2e(-Ed)}{m}} = \sqrt{\frac{2eEd}{m}}$

Substitute numerical values:

$v = \sqrt{\dfrac{2(1.6 \times 10^{-19} \text{ C})(8.0 \times 10^4 \text{ V})(0.50 \text{ m})}{1.67 \times 10^{-27} \text{ kg}}}$

$= 2.8 \times 10^6$ m/s

continued

▶ **25.2** continued

Finalize Because ΔV is negative for the field, ΔU is also negative for the proton–field system. The negative value of ΔU means the potential energy of the system decreases as the proton moves in the direction of the electric field. As the proton accelerates in the direction of the field, it gains kinetic energy while the electric potential energy of the system decreases at the same time.

Figure 25.6 is oriented so that the proton moves downward. The proton's motion is analogous to that of an object falling in a gravitational field. Although the gravitational field is always downward at the surface of the Earth, an electric field can be in any direction, depending on the orientation of the plates creating the field. Therefore, Figure 25.6 could be rotated 90° or 180° and the proton could move horizontally or upward in the electric field!

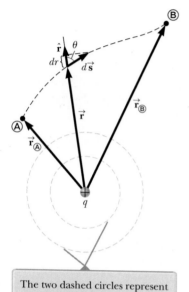

The two dashed circles represent intersections of spherical equipotential surfaces with the page.

Figure 25.7 The potential difference between points Ⓐ and Ⓑ due to a point charge q depends *only* on the initial and final radial coordinates $r_Ⓐ$ and $r_Ⓑ$.

Pitfall Prevention 25.5

Similar Equation Warning Do not confuse Equation 25.11 for the electric potential of a point charge with Equation 23.9 for the electric field of a point charge. Potential is proportional to $1/r$, whereas the magnitude of the field is proportional to $1/r^2$. The effect of a charge on the space surrounding it can be described in two ways. The charge sets up a vector electric field $\vec{E}$, which is related to the force experienced by a charge placed in the field. It also sets up a scalar potential V, which is related to the potential energy of the two-charge system when a charge is placed in the field.

25.3 Electric Potential and Potential Energy Due to Point Charges

As discussed in Section 23.4, an isolated positive point charge q produces an electric field directed radially outward from the charge. To find the electric potential at a point located a distance r from the charge, let's begin with the general expression for potential difference, Equation 25.3,

$$V_Ⓑ - V_Ⓐ = -\int_Ⓐ^Ⓑ \vec{E} \cdot d\vec{s}$$

where Ⓐ and Ⓑ are the two arbitrary points shown in Figure 25.7. At any point in space, the electric field due to the point charge is $\vec{E} = (k_e q/r^2)\hat{r}$ (Eq. 23.9), where $\hat{r}$ is a unit vector directed radially outward from the charge. Therefore, the quantity $\vec{E} \cdot d\vec{s}$ can be expressed as

$$\vec{E} \cdot d\vec{s} = k_e \frac{q}{r^2} \hat{r} \cdot d\vec{s}$$

Because the magnitude of $\hat{r}$ is 1, the dot product $\hat{r} \cdot d\vec{s} = ds \cos\theta$, where θ is the angle between $\hat{r}$ and $d\vec{s}$. Furthermore, $ds \cos\theta$ is the projection of $d\vec{s}$ onto $\hat{r}$; therefore, $ds \cos\theta = dr$. That is, any displacement $d\vec{s}$ along the path from point Ⓐ to point Ⓑ produces a change dr in the magnitude of $\vec{r}$, the position vector of the point relative to the charge creating the field. Making these substitutions, we find that $\vec{E} \cdot d\vec{s} = (k_e q/r^2)dr$; hence, the expression for the potential difference becomes

$$V_Ⓑ - V_Ⓐ = -k_e q \int_{r_Ⓐ}^{r_Ⓑ} \frac{dr}{r^2} = k_e \frac{q}{r}\Big|_{r_Ⓐ}^{r_Ⓑ}$$

$$V_Ⓑ - V_Ⓐ = k_e q \left[\frac{1}{r_Ⓑ} - \frac{1}{r_Ⓐ}\right] \tag{25.10}$$

Equation 25.10 shows us that the integral of $\vec{E} \cdot d\vec{s}$ is *independent* of the path between points Ⓐ and Ⓑ. Multiplying by a charge q_0 that moves between points Ⓐ and Ⓑ, we see that the integral of $q_0\vec{E} \cdot d\vec{s}$ is also independent of path. This latter integral, which is the work done by the electric force on the charge q_0, shows that the electric force is conservative (see Section 7.7). We define a field that is related to a conservative force as a **conservative field**. Therefore, Equation 25.10 tells us that the electric field of a fixed point charge q is conservative. Furthermore, Equation 25.10 expresses the important result that the potential difference between any two points Ⓐ and Ⓑ in a field created by a point charge depends only on the radial coordinates $r_Ⓐ$ and $r_Ⓑ$. It is customary to choose the reference of electric potential for a point charge to be $V = 0$ at $r_Ⓐ = \infty$. With this reference choice, the electric potential due to a point charge at any distance r from the charge is

$$V = k_e \frac{q}{r} \tag{25.11}$$

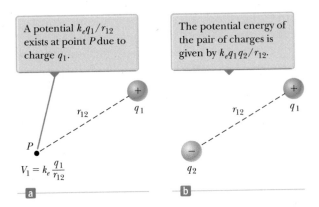

Figure 25.8 (a) Charge q_1 establishes an electric potential V_1 at point P. (b) Charge q_2 is brought from infinity to point P.

We obtain the electric potential resulting from two or more point charges by applying the superposition principle. That is, the total electric potential at some point P due to several point charges is the sum of the potentials due to the individual charges. For a group of point charges, we can write the total electric potential at P as

$$V = k_e \sum_i \frac{q_i}{r_i} \qquad (25.12)$$

◀ **Electric potential due to several point charges**

Figure 25.8a shows a charge q_1, which sets up an electric field throughout space. The charge also establishes an electric potential at all points, including point P, where the electric potential is V_1. Now imagine that an external agent brings a charge q_2 from infinity to point P. The work that must be done to do this is given by Equation 25.4, $W = q_2 \Delta V$. This work represents a transfer of energy across the boundary of the two-charge system, and the energy appears in the system as potential energy U when the particles are separated by a distance r_{12} as in Figure 25.8b. From Equation 8.2, we have $W = \Delta U$. Therefore, the **electric potential energy** of a pair of point charges[1] can be found as follows:

$$\Delta U = W = q_2 \Delta V \;\; \rightarrow \;\; U - 0 = q_2 \left(k_e \frac{q_1}{r_{12}} - 0 \right)$$

$$U = k_e \frac{q_1 q_2}{r_{12}} \qquad (25.13)$$

If the charges are of the same sign, then U is positive. Positive work must be done by an external agent on the system to bring the two charges near each other (because charges of the same sign repel). If the charges are of opposite sign, as in Figure 25.8b, then U is negative. Negative work is done by an external agent against the attractive force between the charges of opposite sign as they are brought near each other; a force must be applied opposite the displacement to prevent q_2 from accelerating toward q_1.

If the system consists of more than two charged particles, we can obtain the total potential energy of the system by calculating U for every *pair* of charges and summing the terms algebraically. For example, the total potential energy of the system of three charges shown in Figure 25.9 is

$$U = k_e \left(\frac{q_1 q_2}{r_{12}} + \frac{q_1 q_3}{r_{13}} + \frac{q_2 q_3}{r_{23}} \right) \qquad (25.14)$$

The potential energy of this system of charges is given by Equation 25.14.

Physically, this result can be interpreted as follows. Imagine q_1 is fixed at the position shown in Figure 25.9 but q_2 and q_3 are at infinity. The work an external agent must do to bring q_2 from infinity to its position near q_1 is $k_e q_1 q_2 / r_{12}$, which is the first term in Equation 25.14. The last two terms represent the work required to bring q_3 from infinity to its position near q_1 and q_2. (The result is independent of the order in which the charges are transported.)

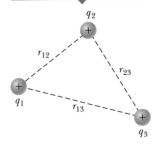

Figure 25.9 Three point charges are fixed at the positions shown.

---

[1]The expression for the electric potential energy of a system made up of two point charges, Equation 25.13, is of the *same* form as the equation for the gravitational potential energy of a system made up of two point masses, $-G m_1 m_2 / r$ (see Chapter 13). The similarity is not surprising considering that both expressions are derived from an inverse-square force law.

Quiz 25.3 In Figure 25.8b, take q_2 to be a negative source charge and q_1 to be a second charge whose sign can be changed. (i) If q_1 is initially positive and is changed to a charge of the same magnitude but negative, what happens to the potential at the position of q_1 due to q_2? (a) It increases. (b) It decreases. (c) It remains the same. (ii) When q_1 is changed from positive to negative, what happens to the potential energy of the two-charge system? Choose from the same possibilities.

Example 25.3   The Electric Potential Due to Two Point Charges

As shown in Figure 25.10a, a charge $q_1 = 2.00$ μC is located at the origin and a charge $q_2 = -6.00$ μC is located at (0, 3.00) m.

(A) Find the total electric potential due to these charges at the point P, whose coordinates are (4.00, 0) m.

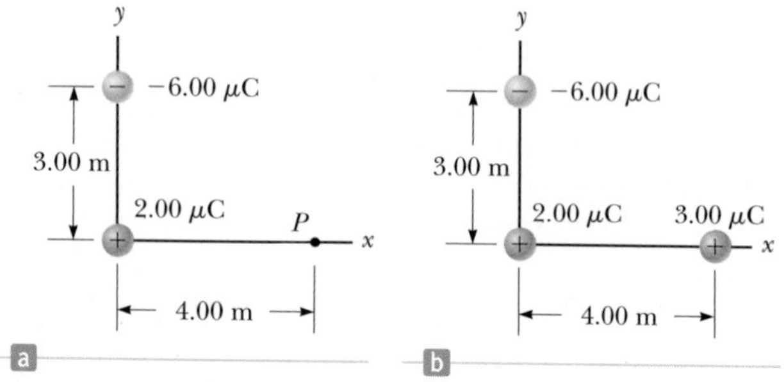

SOLUTION

Conceptualize Recognize first that the 2.00-μC and −6.00-μC charges are source charges and set up an electric field as well as a potential at all points in space, including point P.

Categorize The potential is evaluated using an equation developed in this chapter, so we categorize this example as a substitution problem.

Figure 25.10 (Example 25.3) (a) The electric potential at P due to the two charges q_1 and q_2 is the algebraic sum of the potentials due to the individual charges. (b) A third charge $q_3 = 3.00$ μC is brought from infinity to point P.

Use Equation 25.12 for the system of two source charges:

$$V_P = k_e \left(\frac{q_1}{r_1} + \frac{q_2}{r_2} \right)$$

Substitute numerical values:

$$V_P = (8.988 \times 10^9 \text{ N} \cdot \text{m}^2/\text{C}^2) \left(\frac{2.00 \times 10^{-6} \text{ C}}{4.00 \text{ m}} + \frac{-6.00 \times 10^{-6} \text{ C}}{5.00 \text{ m}} \right)$$

$$= \boxed{-6.29 \times 10^3 \text{ V}}$$

(B) Find the change in potential energy of the system of two charges plus a third charge $q_3 = 3.00$ μC as the latter charge moves from infinity to point P (Fig. 25.10b).

SOLUTION

Assign $U_i = 0$ for the system to the initial configuration in which the charge q_3 is at infinity. Use Equation 25.2 to evaluate the potential energy for the configuration in which the charge is at P:

$$U_f = q_3 V_P$$

Substitute numerical values to evaluate ΔU:

$$\Delta U = U_f - U_i = q_3 V_P - 0 = (3.00 \times 10^{-6} \text{ C})(-6.29 \times 10^3 \text{ V})$$

$$= \boxed{-1.89 \times 10^{-2} \text{ J}}$$

Therefore, because the potential energy of the system has decreased, an external agent has to do positive work to remove the charge q_3 from point P back to infinity.

WHAT IF? You are working through this example with a classmate and she says, "Wait a minute! In part (B), we ignored the potential energy associated with the pair of charges q_1 and q_2!" How would you respond?

Answer Given the statement of the problem, it is not necessary to include this potential energy because part (B) asks for the *change* in potential energy of the system as q_3 is brought in from infinity. Because the configuration of charges q_1 and q_2 does not change in the process, there is no ΔU associated with these charges. Had part (B) asked to find the change in potential energy when *all three* charges start out infinitely far apart and are then brought to the positions in Figure 25.10b, however, you would have to calculate the change using Equation 25.14.

25.4 Obtaining the Value of the Electric Field from the Electric Potential

The electric field $\vec{E}$ and the electric potential V are related as shown in Equation 25.3, which tells us how to find ΔV if the electric field $\vec{E}$ is known. What if the situation is reversed? How do we calculate the value of the electric field if the electric potential is known in a certain region?

From Equation 25.3, the potential difference dV between two points a distance ds apart can be expressed as

$$dV = -\vec{E} \cdot d\vec{s} \tag{25.15}$$

If the electric field has only one component E_x, then $\vec{E} \cdot d\vec{s} = E_x\, dx$. Therefore, Equation 25.15 becomes $dV = -E_x\, dx$, or

$$E_x = -\frac{dV}{dx} \tag{25.16}$$

That is, the x component of the electric field is equal to the negative of the derivative of the electric potential with respect to x. Similar statements can be made about the y and z components. Equation 25.16 is the mathematical statement of the electric field being a measure of the rate of change with position of the electric potential as mentioned in Section 25.1.

Experimentally, electric potential and position can be measured easily with a voltmeter (a device for measuring potential difference) and a meterstick. Consequently, an electric field can be determined by measuring the electric potential at several positions in the field and making a graph of the results. According to Equation 25.16, the slope of a graph of V versus x at a given point provides the magnitude of the electric field at that point.

Imagine starting at a point and then moving through a displacement $d\vec{s}$ along an equipotential surface. For this motion, $dV = 0$ because the potential is constant along an equipotential surface. From Equation 25.15, we see that $dV = -\vec{E} \cdot d\vec{s} = 0$; therefore, because the dot product is zero, $\vec{E}$ must be perpendicular to the displacement along the equipotential surface. This result shows that the equipotential surfaces must always be perpendicular to the electric field lines passing through them.

As mentioned at the end of Section 25.2, the equipotential surfaces associated with a uniform electric field consist of a family of planes perpendicular to the field lines. Figure 25.11a shows some representative equipotential surfaces for this situation.

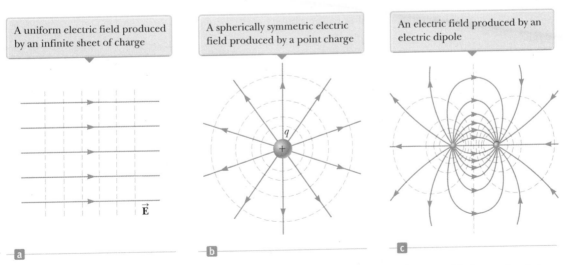

Figure 25.11 Equipotential surfaces (the dashed blue lines are intersections of these surfaces with the page) and electric field lines. In all cases, the equipotential surfaces are *perpendicular* to the electric field lines at every point.

If the charge distribution creating an electric field has spherical symmetry such that the volume charge density depends only on the radial distance r, the electric field is radial. In this case, $\vec{E} \cdot d\vec{s} = E_r \, dr$, and we can express dV as $dV = -E_r \, dr$. Therefore,

$$E_r = -\frac{dV}{dr} \tag{25.17}$$

For example, the electric potential of a point charge is $V = k_e q/r$. Because V is a function of r only, the potential function has spherical symmetry. Applying Equation 25.17, we find that the magnitude of the electric field due to the point charge is $E_r = k_e q/r^2$, a familiar result. Notice that the potential changes only in the radial direction, not in any direction perpendicular to r. Therefore, V (like E_r) is a function only of r, which is again consistent with the idea that equipotential surfaces are perpendicular to field lines. In this case, the equipotential surfaces are a family of spheres concentric with the spherically symmetric charge distribution (Fig. 25.11b). The equipotential surfaces for an electric dipole are sketched in Figure 25.11c.

In general, the electric potential is a function of all three spatial coordinates. If $V(r)$ is given in terms of the Cartesian coordinates, the electric field components E_x, E_y, and E_z can readily be found from $V(x, y, z)$ as the partial derivatives[2]

Finding the electric field ▶ from the potential

$$E_x = -\frac{\partial V}{\partial x} \qquad E_y = -\frac{\partial V}{\partial y} \qquad E_z = -\frac{\partial V}{\partial z} \tag{25.18}$$

Quick **Quiz** 25.4 In a certain region of space, the electric potential is zero everywhere along the x axis. **(i)** From this information, you can conclude that the x component of the electric field in this region is (a) zero, (b) in the positive x direction, or (c) in the negative x direction. **(ii)** Suppose the electric potential is $+2$ V everywhere along the x axis. From the same choices, what can you conclude about the x component of the electric field now?

25.5 Electric Potential Due to Continuous Charge Distributions

In Section 25.3, we found how to determine the electric potential due to a small number of charges. What if we wish to find the potential due to a continuous distribution of charge? The electric potential in this situation can be calculated using two different methods. The first method is as follows. If the charge distribution is known, we consider the potential due to a small charge element dq, treating this element as a point charge (Fig. 25.12). From Equation 25.11, the electric potential dV at some point P due to the charge element dq is

$$dV = k_e \frac{dq}{r} \tag{25.19}$$

where r is the distance from the charge element to point P. To obtain the total potential at point P, we integrate Equation 25.19 to include contributions from all elements of the charge distribution. Because each element is, in general, a different distance from point P and k_e is constant, we can express V as

Electric potential due to ▶ a continuous charge distribution

$$V = k_e \int \frac{dq}{r} \tag{25.20}$$

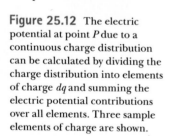

Figure 25.12 The electric potential at point P due to a continuous charge distribution can be calculated by dividing the charge distribution into elements of charge dq and summing the electric potential contributions over all elements. Three sample elements of charge are shown.

---

[2]In vector notation, $\vec{E}$ is often written in Cartesian coordinate systems as

$$\vec{E} = -\nabla V = -\left(\hat{\mathbf{i}} \frac{\partial}{\partial x} + \hat{\mathbf{j}} \frac{\partial}{\partial y} + \hat{\mathbf{k}} \frac{\partial}{\partial z}\right) V$$

where ∇ is called the *gradient operator*.

In effect, we have replaced the sum in Equation 25.12 with an integral. In this expression for V, the electric potential is taken to be zero when point P is infinitely far from the charge distribution.

The second method for calculating the electric potential is used if the electric field is already known from other considerations such as Gauss's law. If the charge distribution has sufficient symmetry, we first evaluate $\vec{\mathbf{E}}$ using Gauss's law and then substitute the value obtained into Equation 25.3 to determine the potential difference ΔV between any two points. We then choose the electric potential V to be zero at some convenient point.

Problem-Solving Strategy   Calculating Electric Potential

The following procedure is recommended for solving problems that involve the determination of an electric potential due to a charge distribution.

1. Conceptualize. Think carefully about the individual charges or the charge distribution you have in the problem and imagine what type of potential would be created. Appeal to any symmetry in the arrangement of charges to help you visualize the potential.

2. Categorize. Are you analyzing a group of individual charges or a continuous charge distribution? The answer to this question will tell you how to proceed in the *Analyze* step.

3. Analyze. When working problems involving electric potential, remember that it is a *scalar quantity*, so there are no components to consider. Therefore, when using the superposition principle to evaluate the electric potential at a point, simply take the algebraic sum of the potentials due to each charge. You must keep track of signs, however.

As with potential energy in mechanics, only *changes* in electric potential are significant; hence, the point where the potential is set at zero is arbitrary. When dealing with point charges or a finite-sized charge distribution, we usually define $V = 0$ to be at a point infinitely far from the charges. If the charge distribution itself extends to infinity, however, some other nearby point must be selected as the reference point.

(a) *If you are analyzing a group of individual charges:* Use the superposition principle, which states that when several point charges are present, the resultant potential at a point P in space is the *algebraic sum* of the individual potentials at P due to the individual charges (Eq. 25.12). Example 25.4 below demonstrates this procedure.

(b) *If you are analyzing a continuous charge distribution:* Replace the sums for evaluating the total potential at some point P from individual charges by integrals (Eq. 25.20). The total potential at P is obtained by integrating over the entire charge distribution. For many problems, it is possible in performing the integration to express dq and r in terms of a single variable. To simplify the integration, give careful consideration to the geometry involved in the problem. Examples 25.5 through 25.7 demonstrate such a procedure.

To obtain the potential from the electric field: Another method used to obtain the potential is to start with the definition of the potential difference given by Equation 25.3. If $\vec{\mathbf{E}}$ is known or can be obtained easily (such as from Gauss's law), the line integral of $\vec{\mathbf{E}} \cdot d\vec{\mathbf{s}}$ can be evaluated.

4. Finalize. Check to see if your expression for the potential is consistent with your mental representation and reflects any symmetry you noted previously. Imagine varying parameters such as the distance of the observation point from the charges or the radius of any circular objects to see if the mathematical result changes in a reasonable way.

Example 25.4 The Electric Potential Due to a Dipole

An electric dipole consists of two charges of equal magnitude and opposite sign separated by a distance $2a$ as shown in Figure 25.13. The dipole is along the x axis and is centered at the origin.

(A) Calculate the electric potential at point P on the y axis.

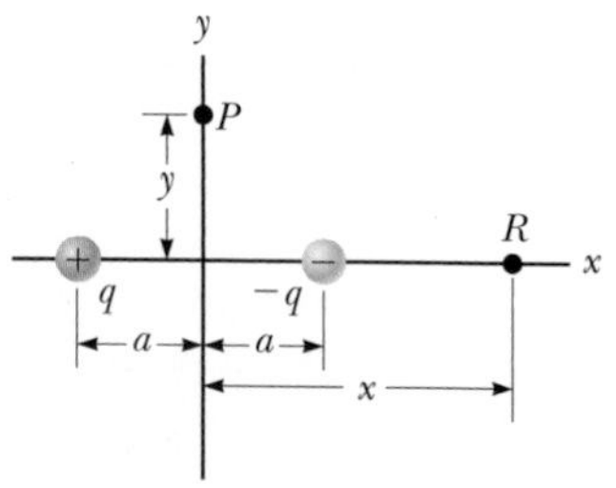

Figure 25.13 (Example 25.4) An electric dipole located on the x axis.

SOLUTION

Conceptualize Compare this situation to that in part (B) of Example 23.6. It is the same situation, but here we are seeking the electric potential rather than the electric field.

Categorize We categorize the problem as one in which we have a small number of particles rather than a continuous distribution of charge. The electric potential can be evaluated by summing the potentials due to the individual charges.

Analyze Use Equation 25.12 to find the electric potential at P due to the two charges:

$$V_P = k_e \sum_i \frac{q_i}{r_i} = k_e \left(\frac{q}{\sqrt{a^2 + y^2}} + \frac{-q}{\sqrt{a^2 + y^2}} \right) = \boxed{0}$$

(B) Calculate the electric potential at point R on the positive x axis.

SOLUTION

Use Equation 25.12 to find the electric potential at R due to the two charges:

$$V_R = k_e \sum_i \frac{q_i}{r_i} = k_e \left(\frac{-q}{x - a} + \frac{q}{x + a} \right) = \boxed{-\frac{2k_e qa}{x^2 - a^2}}$$

(C) Calculate V and E_x at a point on the x axis far from the dipole.

SOLUTION

For point R far from the dipole such that $x \gg a$, neglect a^2 in the denominator of the answer to part (B) and write V in this limit:

$$V_R = \lim_{x \gg a} \left(-\frac{2k_e qa}{x^2 - a^2} \right) \approx \boxed{-\frac{2k_e qa}{x^2}} \quad (x \gg a)$$

Use Equation 25.16 and this result to calculate the x component of the electric field at a point on the x axis far from the dipole:

$$E_x = -\frac{dV}{dx} = -\frac{d}{dx} \left(-\frac{2k_e qa}{x^2} \right)$$

$$= 2k_e qa \frac{d}{dx} \left(\frac{1}{x^2} \right) = \boxed{-\frac{4k_e qa}{x^3}} \quad (x \gg a)$$

Finalize The potentials in parts (B) and (C) are negative because points on the positive x axis are closer to the negative charge than to the positive charge. For the same reason, the x component of the electric field is negative. Notice that we have a $1/r^3$ falloff of the electric field with distance far from the dipole, similar to the behavior of the electric field on the y axis in Example 23.6.

WHAT IF? Suppose you want to find the electric field at a point P on the y axis. In part (A), the electric potential was found to be zero for all values of y. Is the electric field zero at all points on the y axis?

Answer No. That there is no change in the potential along the y axis tells us only that the y component of the electric field is zero. Look back at Figure 23.13 in Example 23.6. We showed there that the electric field of a dipole on the y axis has only an x component. We could not find the x component in the current example because we do not have an expression for the potential near the y axis as a function of x.

| Example 25.5 | **Electric Potential Due to a Uniformly Charged Ring** |

(A) Find an expression for the electric potential at a point P located on the perpendicular central axis of a uniformly charged ring of radius a and total charge Q.

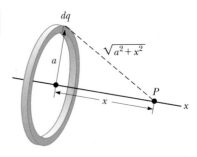

S O L U T I O N

Conceptualize Study Figure 25.14, in which the ring is oriented so that its plane is perpendicular to the x axis and its center is at the origin. Notice that the symmetry of the situation means that all the charges on the ring are the same distance from point P. Compare this example to Example 23.8. Notice that no vector considerations are necessary here because electric potential is a scalar.

Figure 25.14 (Example 25.5) A uniformly charged ring of radius a lies in a plane perpendicular to the x axis. All elements dq of the ring are the same distance from a point P lying on the x axis.

Categorize Because the ring consists of a continuous distribution of charge rather than a set of discrete charges, we must use the integration technique represented by Equation 25.20 in this example.

. . . .

Analyze We take point P to be at a distance x from the center of the ring as shown in Figure 25.14.

Use Equation 25.20 to express V in terms of the geometry:

$$V = k_e \int \frac{dq}{r} = k_e \int \frac{dq}{\sqrt{a^2 + x^2}}$$

Noting that a and x do not vary for an integration over the ring, bring $\sqrt{a^2 + x^2}$ in front of the integral sign and integrate over the ring:

$$V = \frac{k_e}{\sqrt{a^2 + x^2}} \int dq = \boxed{\frac{k_e Q}{\sqrt{a^2 + x^2}}} \qquad (25.21)$$

(B) Find an expression for the magnitude of the electric field at point P.

S O L U T I O N

From symmetry, notice that along the x axis $\vec{E}$ can have only an x component. Therefore, apply Equation 25.16 to Equation 25.21:

$$E_x = -\frac{dV}{dx} = -k_e Q \frac{d}{dx}(a^2 + x^2)^{-1/2}$$

$$= -k_e Q(-\tfrac{1}{2})(a^2 + x^2)^{-3/2}(2x)$$

$$E_x = \boxed{\frac{k_e x}{(a^2 + x^2)^{3/2}} Q} \qquad (25.22)$$

Finalize The only variable in the expressions for V and E_x is x. That is not surprising because our calculation is valid only for points along the x axis, where y and z are both zero. This result for the electric field agrees with that obtained by direct integration (see Example 23.8). For practice, use the result of part (B) in Equation 25.3 to verify that the potential is given by the expression in part (A).

| Example 25.6 | **Electric Potential Due to a Uniformly Charged Disk** |

A uniformly charged disk has radius R and surface charge density σ.

(A) Find the electric potential at a point P along the perpendicular central axis of the disk.

S O L U T I O N

Conceptualize If we consider the disk to be a set of concentric rings, we can use our result from Example 25.5—which gives the potential due to a ring of radius a—and sum the contributions of all rings making up the disk. Figure

continued

▶ **25.6** continued

25.15 shows one such ring. Because point P is on the central axis of the disk, symmetry again tells us that all points in a given ring are the same distance from P.

Categorize Because the disk is continuous, we evaluate the potential due to a continuous charge distribution rather than a group of individual charges.

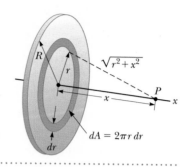

Figure 25.15 (Example 25.6) A uniformly charged disk of radius R lies in a plane perpendicular to the x axis. The calculation of the electric potential at any point P on the x axis is simplified by dividing the disk into many rings of radius r and width dr, with area $2\pi r\, dr$.

Analyze Find the amount of charge dq on a ring of radius r and width dr as shown in Figure 25.15:

$$dq = \sigma\, dA = \sigma(2\pi r\, dr) = 2\pi\sigma r\, dr$$

Use this result in Equation 25.21 in Example 25.5 (with a replaced by the variable r and Q replaced by the differential dq) to find the potential due to the ring:

$$dV = \frac{k_e\, dq}{\sqrt{r^2 + x^2}} = \frac{k_e 2\pi\sigma r\, dr}{\sqrt{r^2 + x^2}}$$

To obtain the total potential at P, integrate this expression over the limits $r = 0$ to $r = R$, noting that x is a constant:

$$V = \pi k_e \sigma \int_0^R \frac{2r\, dr}{\sqrt{r^2 + x^2}} = \pi k_e \sigma \int_0^R (r^2 + x^2)^{-1/2}\, 2r\, dr$$

This integral is of the common form $\int u^n\, du$, where $n = -\frac{1}{2}$ and $u = r^2 + x^2$, and has the value $u^{n+1}/(n + 1)$. Use this result to evaluate the integral:

$$V = 2\pi k_e \sigma [(R^2 + x^2)^{1/2} - x] \qquad (25.23)$$

(B) Find the x component of the electric field at a point P along the perpendicular central axis of the disk.

SOLUTION

As in Example 25.5, use Equation 25.16 to find the electric field at any axial point:

$$E_x = -\frac{dV}{dx} = 2\pi k_e \sigma \left[1 - \frac{x}{(R^2 + x^2)^{1/2}} \right] \qquad (25.24)$$

Finalize Compare Equation 25.24 with the result of Example 23.9. They are the same. The calculation of V and $\vec{E}$ for an arbitrary point off the x axis is more difficult to perform because of the absence of symmetry and we do not treat that situation in this book.

Example 25.7   **Electric Potential Due to a Finite Line of Charge**

A rod of length ℓ located along the x axis has a total charge Q and a uniform linear charge density λ. Find the electric potential at a point P located on the y axis a distance a from the origin (Fig. 25.16).

SOLUTION

Conceptualize The potential at P due to every segment of charge on the rod is positive because every segment carries a positive charge. Notice that we have no symmetry to appeal to here, but the simple geometry should make the problem solvable.

Categorize Because the rod is continuous, we evaluate the potential due to a continuous charge distribution rather than a group of individual charges.

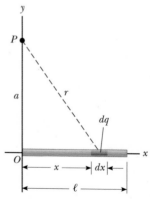

Figure 25.16 (Example 25.7) A uniform line charge of length ℓ located along the x axis. To calculate the electric potential at P, the line charge is divided into segments each of length dx and each carrying a charge $dq = \lambda\, dx$.

Analyze In Figure 25.16, the rod lies along the x axis, dx is the length of one small segment, and dq is the charge on that segment. Because the rod has a charge per unit length λ, the charge dq on the small segment is $dq = \lambda\, dx$.

▶ **25.7** continued

Find the potential at P due to one segment of the rod at an arbitrary position x:

$$dV = k_e \frac{dq}{r} = k_e \frac{\lambda \, dx}{\sqrt{a^2 + x^2}}$$

Find the total potential at P by integrating this expression over the limits $x = 0$ to $x = \ell$:

$$V = \int_0^\ell k_e \frac{\lambda \, dx}{\sqrt{a^2 + x^2}}$$

Noting that k_e and $\lambda = Q/\ell$ are constants and can be removed from the integral, evaluate the integral with the help of Appendix B:

$$V = k_e \lambda \int_0^\ell \frac{dx}{\sqrt{a^2 + x^2}} = k_e \frac{Q}{\ell} \ln \left(x + \sqrt{a^2 + x^2} \right) \Big|_0^\ell$$

Evaluate the result between the limits:

$$V = k_e \frac{Q}{\ell} [\ln (\ell + \sqrt{a^2 + \ell^2}) - \ln a] = k_e \frac{Q}{\ell} \ln \left(\frac{\ell + \sqrt{a^2 + \ell^2}}{a} \right) \quad \textbf{(25.25)}$$

Finalize If $\ell \ll a$, the potential at P should approach that of a point charge because the rod is very short compared to the distance from the rod to P. By using a series expansion for the natural logarithm from Appendix B.5, it is easy to show that Equation 25.25 becomes $V = k_e Q/a$.

WHAT IF? What if you were asked to find the electric field at point P? Would that be a simple calculation?

Answer Calculating the electric field by means of Equation 23.11 would be a little messy. There is no symmetry to appeal to, and the integration over the line of charge would represent a vector addition of electric fields at point P. Using Equation 25.18, you could find E_y by replacing a with y in Equation 25.25 and performing the differentiation with respect to y. Because the charged rod in Figure 25.16 lies entirely to the right of $x = 0$, the electric field at point P would have an x component to the left if the rod is charged positively. You cannot use Equation 25.18 to find the x component of the field, however, because the potential due to the rod was evaluated at a specific value of x ($x = 0$) rather than a general value of x. You would have to find the potential as a function of both x and y to be able to find the x and y components of the electric field using Equation 25.18.

25.6 Electric Potential Due to a Charged Conductor

In Section 24.4, we found that when a solid conductor in equilibrium carries a net charge, the charge resides on the conductor's outer surface. Furthermore, the electric field just outside the conductor is perpendicular to the surface and the field inside is zero.

We now generate another property of a charged conductor, related to electric potential. Consider two points Ⓐ and Ⓑ on the surface of a charged conductor as shown in Figure 25.17. Along a surface path connecting these points, $\vec{E}$ is always

Pitfall Prevention 25.6

Potential May Not Be Zero The electric potential inside the conductor is not necessarily zero in Figure 25.17, even though the electric field is zero. Equation 25.15 shows that a zero value of the field results in no *change* in the potential from one point to another inside the conductor. Therefore, the potential everywhere inside the conductor, including the surface, has the same value, which may or may not be zero, depending on where the zero of potential is defined.

Notice from the spacing of the positive signs that the surface charge density is nonuniform.

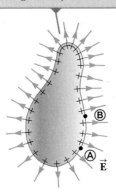

Figure 25.17 An arbitrarily shaped conductor carrying a positive charge. When the conductor is in electrostatic equilibrium, all the charge resides at the surface, $\vec{E} = 0$ inside the conductor, and the direction of $\vec{E}$ immediately outside the conductor is perpendicular to the surface. The electric potential is constant inside the conductor and is equal to the potential at the surface.

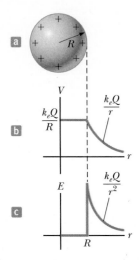

Figure 25.18 (a) The excess charge on a conducting sphere of radius R is uniformly distributed on its surface. (b) Electric potential versus distance r from the center of the charged conducting sphere. (c) Electric field magnitude versus distance r from the center of the charged conducting sphere.

perpendicular to the displacement $d\vec{\mathbf{s}}$; therefore, $\vec{\mathbf{E}} \cdot d\vec{\mathbf{s}} = 0$. Using this result and Equation 25.3, we conclude that the potential difference between Ⓐ and Ⓑ is necessarily zero:

$$V_{Ⓑ} - V_{Ⓐ} = -\int_{Ⓐ}^{Ⓑ} \vec{\mathbf{E}} \cdot d\vec{\mathbf{s}} = 0$$

This result applies to any two points on the surface. Therefore, V is constant everywhere on the surface of a charged conductor in equilibrium. That is,

> the surface of any charged conductor in electrostatic equilibrium is an equipotential surface: every point on the surface of a charged conductor in equilibrium is at the same electric potential. Furthermore, because the electric field is zero inside the conductor, the electric potential is constant everywhere inside the conductor and equal to its value at the surface.

Because of the constant value of the potential, no work is required to move a charge from the interior of a charged conductor to its surface.

Consider a solid metal conducting sphere of radius R and total positive charge Q as shown in Figure 25.18a. As determined in part (A) of Example 24.3, the electric field outside the sphere is $k_e Q/r^2$ and points radially outward. Because the field outside a spherically symmetric charge distribution is identical to that of a point charge, we expect the potential to also be that of a point charge, $k_e Q/r$. At the surface of the conducting sphere in Figure 25.18a, the potential must be $k_e Q/R$. Because the entire sphere must be at the same potential, the potential at any point within the sphere must also be $k_e Q/R$. Figure 25.18b is a plot of the electric potential as a function of r, and Figure 25.18c shows how the electric field varies with r.

When a net charge is placed on a spherical conductor, the surface charge density is uniform as indicated in Figure 25.18a. If the conductor is nonspherical as in Figure 25.17, however, the surface charge density is high where the radius of curvature is small (as noted in Section 24.4) and low where the radius of curvature is large. Because the electric field immediately outside the conductor is proportional to the surface charge density, the electric field is large near convex points having small radii of curvature and reaches very high values at sharp points. In Example 25.8, the relationship between electric field and radius of curvature is explored mathematically.

Example 25.8   **Two Connected Charged Spheres**

Two spherical conductors of radii r_1 and r_2 are separated by a distance much greater than the radius of either sphere. The spheres are connected by a conducting wire as shown in Figure 25.19. The charges on the spheres in equilibrium are q_1 and q_2, respectively, and they are uniformly charged. Find the ratio of the magnitudes of the electric fields at the surfaces of the spheres.

Figure 25.19 (Example 25.8) Two charged spherical conductors connected by a conducting wire. The spheres are at the *same* electric potential V.

SOLUTION

Conceptualize Imagine the spheres are much farther apart than shown in Figure 25.19. Because they are so far apart, the field of one does not affect the charge distribution on the other. The conducting wire between them ensures that both spheres have the same electric potential.

Categorize Because the spheres are so far apart, we model the charge distribution on them as spherically symmetric, and we can model the field and potential outside the spheres to be that due to point charges.

....

Analyze Set the electric potentials at the surfaces of the spheres equal to each other:

$$V = k_e \frac{q_1}{r_1} = k_e \frac{q_2}{r_2}$$

▶ **25.8** continued

Solve for the ratio of charges on the spheres:

$$(1) \quad \frac{q_1}{q_2} = \frac{r_1}{r_2}$$

Write expressions for the magnitudes of the electric fields at the surfaces of the spheres:

$$E_1 = k_e \frac{q_1}{r_1^2} \quad \text{and} \quad E_2 = k_e \frac{q_2}{r_2^2}$$

Evaluate the ratio of these two fields:

$$\frac{E_1}{E_2} = \frac{q_1}{q_2} \frac{r_2^2}{r_1^2}$$

Substitute for the ratio of charges from Equation (1):

$$(2) \quad \frac{E_1}{E_2} = \frac{r_1}{r_2} \frac{r_2^2}{r_1^2} = \frac{r_2}{r_1}$$

Finalize The field is stronger in the vicinity of the smaller sphere even though the electric potentials at the surfaces of both spheres are the same. If $r_2 \to 0$, then $E_2 \to \infty$, verifying the statement above that the electric field is very large at sharp points.

A Cavity Within a Conductor

Suppose a conductor of arbitrary shape contains a cavity as shown in Figure 25.20. Let's assume no charges are inside the cavity. In this case, the electric field inside the cavity must be *zero* regardless of the charge distribution on the outside surface of the conductor as we mentioned in Section 24.4. Furthermore, the field in the cavity is zero even if an electric field exists outside the conductor.

To prove this point, remember that every point on the conductor is at the same electric potential; therefore, any two points Ⓐ and Ⓑ on the cavity's surface must be at the same potential. Now imagine a field $\vec{\mathbf{E}}$ exists in the cavity and evaluate the potential difference $V_{Ⓑ} - V_{Ⓐ}$ defined by Equation 25.3:

$$V_{Ⓑ} - V_{Ⓐ} = -\int_{Ⓐ}^{Ⓑ} \vec{\mathbf{E}} \cdot d\vec{\mathbf{s}}$$

Because $V_{Ⓑ} - V_{Ⓐ} = 0$, the integral of $\vec{\mathbf{E}} \cdot d\vec{\mathbf{s}}$ must be zero for all paths between any two points Ⓐ and Ⓑ on the conductor. The only way that can be true for *all* paths is if $\vec{\mathbf{E}}$ is zero *everywhere* in the cavity. Therefore, a cavity surrounded by conducting walls is a field-free region as long as no charges are inside the cavity.

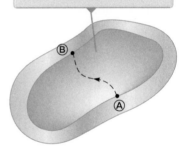

The electric field in the cavity is zero regardless of the charge on the conductor.

Figure 25.20 A conductor in electrostatic equilibrium containing a cavity.

Corona Discharge

A phenomenon known as **corona discharge** is often observed near a conductor such as a high-voltage power line. When the electric field in the vicinity of the conductor is sufficiently strong, electrons resulting from random ionizations of air molecules near the conductor accelerate away from their parent molecules. These rapidly moving electrons can ionize additional molecules near the conductor, creating more free electrons. The observed glow (or corona discharge) results from the recombination of these free electrons with the ionized air molecules. If a conductor has an irregular shape, the electric field can be very high near sharp points or edges of the conductor; consequently, the ionization process and corona discharge are most likely to occur around such points.

Corona discharge is used in the electrical transmission industry to locate broken or faulty components. For example, a broken insulator on a transmission tower has sharp edges where corona discharge is likely to occur. Similarly, corona discharge will occur at the sharp end of a broken conductor strand. Observation of these discharges is difficult because the visible radiation emitted is weak and most of the radiation is in the ultraviolet. (We will discuss ultraviolet radiation and other portions of the electromagnetic spectrum in Section 34.7.) Even use of traditional ultraviolet cameras is of little help because the radiation from the corona

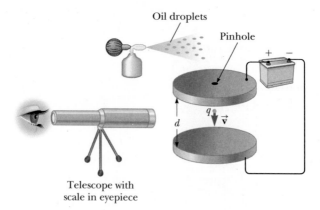

Oil droplets

Pinhole

q $\vec{v}$

d

Telescope with scale in eyepiece

discharge is overwhelmed by ultraviolet radiation from the Sun. Newly developed dual-spectrum devices combine a narrow-band ultraviolet camera with a visible-light camera to show a daylight view of the corona discharge in the actual location on the transmission tower or cable. The ultraviolet part of the camera is designed to operate in a wavelength range in which radiation from the Sun is very weak.

25.7 The Millikan Oil-Drop Experiment

Robert Millikan performed a brilliant set of experiments from 1909 to 1913 in which he measured e, the magnitude of the elementary charge on an electron, and demonstrated the quantized nature of this charge. His apparatus, diagrammed in Figure 25.21, contains two parallel metallic plates. Oil droplets from an atomizer are allowed to pass through a small hole in the upper plate. Millikan used x-rays to ionize the air in the chamber so that freed electrons would adhere to the oil drops, giving them a negative charge. A horizontally directed light beam is used to illuminate the oil droplets, which are viewed through a telescope whose long axis is perpendicular to the light beam. When viewed in this manner, the droplets appear as shining stars against a dark background and the rate at which individual drops fall can be determined.

Let's assume a single drop having a mass m and carrying a charge q is being viewed and its charge is negative. If no electric field is present between the plates, the two forces acting on the charge are the gravitational force $m\vec{g}$ acting downward[3] and a viscous drag force $\vec{F}_D$ acting upward as indicated in Figure 25.22a. The drag force is proportional to the drop's speed as discussed in Section 6.4. When the drop reaches its terminal speed v_T the two forces balance each other ($mg = F_D$).

Now suppose a battery connected to the plates sets up an electric field between the plates such that the upper plate is at the higher electric potential. In this case, a third force $q\vec{E}$ acts on the charged drop. The particle in a field model applies twice to the particle: it is in a gravitational field and an electric field. Because q is negative and $\vec{E}$ is directed downward, this electric force is directed upward as shown in Figure 25.22b. If this upward force is strong enough, the drop moves upward and the drag force $\vec{F}'_D$ acts downward. When the upward electric force $q\vec{E}$ balances the sum of the gravitational force and the downward drag force $\vec{F}'_D$, the drop reaches a new terminal speed v'_T in the upward direction.

With the field turned on, a drop moves slowly upward, typically at rates of hundredths of a centimeter per second. The rate of fall in the absence of a field is comparable. Hence, one can follow a single droplet for hours, alternately rising and falling, by simply turning the electric field on and off.

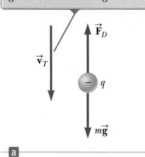

With the electric field off, the droplet falls at terminal velocity $\vec{v}_T$ under the influence of the gravitational and drag forces.

$\vec{F}_D$

$\vec{v}_T$

q

$m\vec{g}$

a

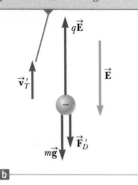

When the electric field is turned on, the droplet moves upward at terminal velocity $\vec{v}'_T$ under the influence of the electric, gravitational, and drag forces.

$q\vec{E}$

$\vec{E}$

$\vec{v}'_T$

$m\vec{g}$ $\vec{F}'_D$

b

Figure 25.22 The forces acting on a negatively charged oil droplet in the Millikan experiment.

[3]There is also a buoyant force on the oil drop due to the surrounding air. This force can be incorporated as a correction in the gravitational force $m\vec{g}$ on the drop, so we will not consider it in our analysis.

After recording measurements on thousands of droplets, Millikan and his coworkers found that all droplets, to within about 1% precision, had a charge equal to some integer multiple of the elementary charge e:

$$q = ne \quad n = 0, -1, -2, -3, \ldots$$

where $e = 1.60 \times 10^{-19}$ C. Millikan's experiment yields conclusive evidence that charge is quantized. For this work, he was awarded the Nobel Prize in Physics in 1923.

25.8 Applications of Electrostatics

The practical application of electrostatics is represented by such devices as lightning rods and electrostatic precipitators and by such processes as xerography and the painting of automobiles. Scientific devices based on the principles of electrostatics include electrostatic generators, the field-ion microscope, and ion-drive rocket engines. Details of two devices are given below.

The Van de Graaff Generator

Experimental results show that when a charged conductor is placed in contact with the inside of a hollow conductor, all the charge on the charged conductor is transferred to the hollow conductor. In principle, the charge on the hollow conductor and its electric potential can be increased without limit by repetition of the process.

In 1929, Robert J. Van de Graaff (1901–1967) used this principle to design and build an electrostatic generator, and a schematic representation of it is given in Figure 25.23. This type of generator was once used extensively in nuclear physics research. Charge is delivered continuously to a high-potential electrode by means of a moving belt of insulating material. The high-voltage electrode is a hollow metal dome mounted on an insulating column. The belt is charged at point Ⓐ by means of a corona discharge between comb-like metallic needles and a grounded grid. The needles are maintained at a positive electric potential of typically 10^4 V. The positive charge on the moving belt is transferred to the dome by a second comb of needles at point Ⓑ. Because the electric field inside the dome is negligible, the positive charge on the belt is easily transferred to the conductor regardless of its potential. In practice, it is possible to increase the electric potential of the dome until electrical discharge occurs through the air. Because the "breakdown" electric field in air is about 3×10^6 V/m, a sphere 1.00 m in radius can be raised to a maximum potential of 3×10^6 V. The potential can be increased further by increasing the dome's radius and placing the entire system in a container filled with high-pressure gas.

Van de Graaff generators can produce potential differences as large as 20 million volts. Protons accelerated through such large potential differences receive enough energy to initiate nuclear reactions between themselves and various target nuclei. Smaller generators are often seen in science classrooms and museums. If a person insulated from the ground touches the sphere of a Van de Graaff generator, his or her body can be brought to a high electric potential. The person's hair acquires a net positive charge, and each strand is repelled by all the others as in the opening photograph of Chapter 23.

The Electrostatic Precipitator

One important application of electrical discharge in gases is the *electrostatic precipitator*. This device removes particulate matter from combustion gases, thereby reducing air pollution. Precipitators are especially useful in coal-burning power plants and industrial operations that generate large quantities of smoke. Current systems are able to eliminate more than 99% of the ash from smoke.

Figure 25.24a (page 604) shows a schematic diagram of an electrostatic precipitator. A high potential difference (typically 40 to 100 kV) is maintained between

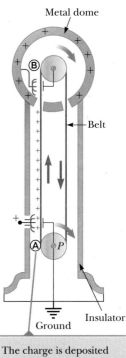

The charge is deposited on the belt at point Ⓐ and transferred to the hollow conductor at point Ⓑ.

Figure 25.23 Schematic diagram of a Van de Graaff generator. Charge is transferred to the metal dome at the top by means of a moving belt.

The high negative electric potential maintained on the central wire creates a corona discharge in the vicinity of the wire.

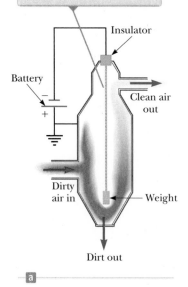

By Courtesy of Tenova TAKRAF

Figure 25.24 (a) Schematic diagram of an electrostatic precipitator. Compare the air pollution when the electrostatic precipitator is (b) operating and (c) turned off.

a wire running down the center of a duct and the walls of the duct, which are grounded. The wire is maintained at a negative electric potential with respect to the walls, so the electric field is directed toward the wire. The values of the field near the wire become high enough to cause a corona discharge around the wire; the air near the wire contains positive ions, electrons, and such negative ions as O_2^-. The air to be cleaned enters the duct and moves near the wire. As the electrons and negative ions created by the discharge are accelerated toward the outer wall by the electric field, the dirt particles in the air become charged by collisions and ion capture. Because most of the charged dirt particles are negative, they too are drawn to the duct walls by the electric field. When the duct is periodically shaken, the particles break loose and are collected at the bottom.

In addition to reducing the level of particulate matter in the atmosphere (compare Figs. 25.24b and c), the electrostatic precipitator recovers valuable materials in the form of metal oxides.

Summary

Definitions

The **potential difference** ΔV between points Ⓐ and Ⓑ in an electric field $\vec{\mathbf{E}}$ is defined as

$$\Delta V \equiv \frac{\Delta U}{q} = -\int_{Ⓐ}^{Ⓑ} \vec{\mathbf{E}} \cdot d\vec{\mathbf{s}} \qquad \textbf{(25.3)}$$

where ΔU is given by Equation 25.1 on page 585. The **electric potential** $V = U/q$ is a scalar quantity and has the units of joules per coulomb, where $1 \text{ J/C} \equiv 1 \text{ V}$.

An **equipotential surface** is one on which all points are at the same electric potential. Equipotential surfaces are perpendicular to electric field lines.

Concepts and Principles

When a positive charge q is moved between points Ⓐ and Ⓑ in an electric field $\vec{\mathbf{E}}$, the change in the potential energy of the charge–field system is

$$\Delta U = -q \int_{Ⓐ}^{Ⓑ} \vec{\mathbf{E}} \cdot d\vec{\mathbf{s}} \qquad \textbf{(25.1)}$$

The potential difference between two points separated by a distance d in a uniform electric field $\vec{\mathbf{E}}$ is

$$\Delta V = -Ed \qquad \textbf{(25.6)}$$

if the direction of travel between the points is in the same direction as the electric field.

If we define $V = 0$ at $r = \infty$, the electric potential due to a point charge at any distance r from the charge is

$$V = k_e \frac{q}{r} \qquad \textbf{(25.11)}$$

The electric potential associated with a group of point charges is obtained by summing the potentials due to the individual charges.

The **electric potential energy** associated with a pair of point charges separated by a distance r_{12} is

$$U = k_e \frac{q_1 q_2}{r_{12}} \qquad \textbf{(25.13)}$$

We obtain the potential energy of a distribution of point charges by summing terms like Equation 25.13 over all pairs of particles.

If the electric potential is known as a function of coordinates x, y, and z, we can obtain the components of the electric field by taking the negative derivative of the electric potential with respect to the coordinates. For example, the x component of the electric field is

$$E_x = -\frac{dV}{dx} \qquad \textbf{(25.16)}$$

The electric potential due to a continuous charge distribution is

$$V = k_e \int \frac{dq}{r} \qquad \textbf{(25.20)}$$

Every point on the surface of a charged conductor in electrostatic equilibrium is at the same electric potential. The potential is constant everywhere inside the conductor and equal to its value at the surface.

Objective Questions

[1.] denotes answer available in *Student Solutions Manual/Study Guide*

1. In a certain region of space, the electric field is zero. From this fact, what can you conclude about the electric potential in this region? (a) It is zero. (b) It does not vary with position. (c) It is positive. (d) It is negative. (e) None of those answers is necessarily true.

2. Consider the equipotential surfaces shown in Figure 25.4. In this region of space, what is the approximate direction of the electric field? (a) It is out of the page. (b) It is into the page. (c) It is toward the top of the page. (d) It is toward the bottom of the page. (e) The field is zero.

3. **(i)** A metallic sphere A of radius 1.00 cm is several centimeters away from a metallic spherical shell B of radius 2.00 cm. Charge 450 nC is placed on A, with no charge on B or anywhere nearby. Next, the two objects are joined by a long, thin, metallic wire (as shown in Fig. 25.19), and finally the wire is removed. How is the charge shared between A and B? (a) 0 on A, 450 nC on B (b) 90.0 nC on A and 360 nC on B, with equal surface charge densities (c) 150 nC on A and 300 nC on B (d) 225 nC on A and 225 nC on B (e) 450 nC on A and 0 on B **(ii)** A metallic sphere A of radius 1 cm with charge 450 nC hangs on an insulating thread inside an uncharged thin metallic spherical shell B of radius 2 cm. Next, A is made temporarily to touch the inner surface of B. How is the charge then shared between

them? Choose from the same possibilities. Arnold Arons, the only physics teacher yet to have his picture on the cover of *Time* magazine, suggested the idea for this question.

4. The electric potential at $x = 3.00$ m is 120 V, and the electric potential at $x = 5.00$ m is 190 V. What is the x component of the electric field in this region, assuming the field is uniform? (a) 140 N/C (b) −140 N/C (c) 35.0 N/C (d) −35.0 N/C (e) 75.0 N/C

5. Rank the potential energies of the four systems of particles shown in Figure OQ25.5 from largest to smallest. Include equalities if appropriate.

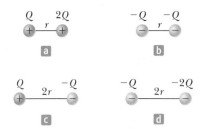

Figure OQ25.5

6. In a certain region of space, a uniform electric field is in the x direction. A particle with negative charge is carried from $x = 20.0$ cm to $x = 60.0$ cm. **(i)** Does

the electric potential energy of the charge–field system (a) increase, (b) remain constant, (c) decrease, or (d) change unpredictably? **(ii)** Has the particle moved to a position where the electric potential is (a) higher than before, (b) unchanged, (c) lower than before, or (d) unpredictable?

7. Rank the electric potentials at the four points shown in Figure OQ25.7 from largest to smallest.

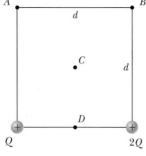

Figure OQ25.7

8. An electron in an x-ray machine is accelerated through a potential difference of 1.00×10^4 V before it hits the target. What is the kinetic energy of the electron in electron volts? (a) 1.00×10^4 eV (b) 1.60×10^{-15} eV (c) 1.60×10^{-22} eV (d) 6.25×10^{22} eV (e) 1.60×10^{-19} eV

9. Rank the electric potential energies of the systems of charges shown in Figure OQ25.9 from largest to smallest. Indicate equalities if appropriate.

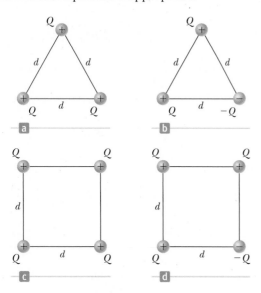

Figure OQ25.9

10. Four particles are positioned on the rim of a circle. The charges on the particles are $+0.500\ \mu C$, $+1.50\ \mu C$, $-1.00\ \mu C$, and $-0.500\ \mu C$. If the electric potential at the center of the circle due to the $+0.500\ \mu C$ charge alone is 4.50×10^4 V, what is the total electric potential at the center due to the four charges? (a) 18.0×10^4 V (b) 4.50×10^4 V (c) 0 (d) -4.50×10^4 V (e) 9.00×10^4 V

11. A proton is released from rest at the origin in a uniform electric field in the positive x direction with magnitude 850 N/C. What is the change in the electric potential energy of the proton–field system when the proton travels to $x = 2.50$ m? (a) 3.40×10^{-16} J (b) -3.40×10^{-16} J (c) 2.50×10^{-16} J (d) -2.50×10^{-16} J (e) -1.60×10^{-19} J

12. A particle with charge -40.0 nC is on the x axis at the point with coordinate $x = 0$. A second particle, with charge -20.0 nC, is on the x axis at $x = 0.500$ m. **(i)** Is the point at a finite distance where the electric field is zero (a) to the left of $x = 0$, (b) between $x = 0$ and $x = 0.500$ m, or (c) to the right of $x = 0.500$ m? **(ii)** Is the electric potential zero at this point? (a) No; it is positive. (b) Yes. (c) No; it is negative. **(iii)** Is there a point at a finite distance where the electric potential is zero? (a) Yes; it is to the left of $x = 0$. (b) Yes; it is between $x = 0$ and $x = 0.500$ m. (c) Yes; it is to the right of $x = 0.500$ m. (d) No.

13. A filament running along the x axis from the origin to $x = 80.0$ cm carries electric charge with uniform density. At the point P with coordinates ($x = 80.0$ cm, $y = 80.0$ cm), this filament creates electric potential 100 V. Now we add another filament along the y axis, running from the origin to $y = 80.0$ cm, carrying the same amount of charge with the same uniform density. At the same point P, is the electric potential created by the pair of filaments (a) greater than 200 V, (b) 200 V, (c) 100 V, (d) between 0 and 200 V, or (e) 0?

14. In different experimental trials, an electron, a proton, or a doubly charged oxygen atom (O^{--}), is fired within a vacuum tube. The particle's trajectory carries it through a point where the electric potential is 40.0 V and then through a point at a different potential. Rank each of the following cases according to the change in kinetic energy of the particle over this part of its flight from the largest increase to the largest decrease in kinetic energy. In your ranking, display any cases of equality. (a) An electron moves from 40.0 V to 60.0 V. (b) An electron moves from 40.0 V to 20.0 V. (c) A proton moves from 40.0 V to 20.0 V. (d) A proton moves from 40.0 V to 10.0 V. (e) An O^{--} ion moves from 40.0 V to 60.0 V.

15. A helium nucleus (charge = $2e$, mass = 6.63×10^{-27} kg) traveling at 6.20×10^5 m/s enters an electric field, traveling from point Ⓐ, at a potential of 1.50×10^3 V, to point Ⓑ, at 4.00×10^3 V. What is its speed at point Ⓑ? (a) 7.91×10^5 m/s (b) 3.78×10^5 m/s (c) 2.13×10^5 m/s (d) 2.52×10^6 m/s (e) 3.01×10^8 m/s

Conceptual Questions

1. denotes answer available in *Student Solutions Manual/Study Guide*

1. What determines the maximum electric potential to which the dome of a Van de Graaff generator can be raised?

2. Describe the motion of a proton (a) after it is released from rest in a uniform electric field. Describe the changes (if any) in (b) its kinetic energy and (c) the electric potential energy of the proton–field system.

3. When charged particles are separated by an infinite distance, the electric potential energy of the pair is zero. When the particles are brought close, the elec-

tric potential energy of a pair with the same sign is positive, whereas the electric potential energy of a pair with opposite signs is negative. Give a physical explanation of this statement.

4. Study Figure 23.3 and the accompanying text discussion of charging by induction. When the grounding wire is touched to the rightmost point on the sphere in Figure 23.3c, electrons are drained away from the sphere to leave the sphere positively charged. Suppose the grounding wire is touched to the leftmost point on the sphere instead. (a) Will electrons still drain away, moving closer to the negatively charged rod as they do so? (b) What kind of charge, if any, remains on the sphere?

5. Distinguish between electric potential and electric potential energy.

6. Describe the equipotential surfaces for (a) an infinite line of charge and (b) a uniformly charged sphere.

Problems available in Access end-of-chapter problems online at www.webassign.net

Section 25.1 Electric Potential and Potential Difference
Section 25.2 Potential Difference in a Uniform Electric Field
Problems 1–11

Section 25.3 Electric Potential and Potential Energy Due to Point Charges
Problems 12–35

Section 25.4 Obtaining the Value of the Electric Field from the Electric Potential
Problems 36–42

Section 25.5 Electric Potential Due to Continuous Charge Distributions
Problems 43–47

Section 25.6 Electric Potential Due to a Charged Conductor
Problems 48–52

Additional Problems
Problems 53–70

Challenge Problems
Problems 71–77

Solutions to the following Problems are available in the *Student Solutions Manual/Study Guide:*
25.3, 25.7, 25.20, 25.22, 25.24, 25.39, 25.42, 25.45, 25.50, 25.52, 25.57, 25.62, 25.65, 25.71, and 25.73

List of Enhanced Problems

Problem Number	Targeted Feedback in Enhanced WebAssign	Analysis Model Tutorial in Enhanced WebAssign	Master It in Enhanced WebAssign	Watch It in Enhanced WebAssign
25.1	✓		✓	
25.3	✓		✓	
25.4	✓			✓
25.5	✓			✓
25.7		✓	✓	
25.9		✓		
25.16	✓		✓	
25.19	✓			✓
25.20			✓	
25.22	✓		✓	
25.27	✓			✓
25.31		✓		
25.35		✓		
25.37	✓			✓
25.39	✓			✓
25.44	✓			✓
25.47				✓
25.50	✓		✓	
25.52			✓	
25.57			✓	

Capacitance and Dielectrics

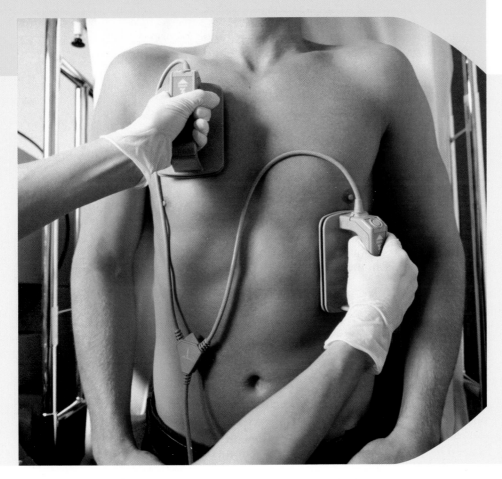

When a patient receives a shock from a defibrillator, the energy delivered to the patient is initially stored in a *capacitor*. We will study capacitors and capacitance in this chapter. *(Andrew Olney/Getty Images)*

In this chapter, we introduce the first of three simple circuit elements that can be connected with wires to form an electric circuit. Electric circuits are the basis for the vast majority of the devices used in our society. Here we shall discuss *capacitors*, devices that store electric charge. This discussion is followed by the study of *resistors* in Chapter 27 and *inductors* in Chapter 32. In later chapters, we will study more sophisticated circuit elements such as *diodes* and *transistors*.

Capacitors are commonly used in a variety of electric circuits. For instance, they are used to tune the frequency of radio receivers, as filters in power supplies, to eliminate sparking in automobile ignition systems, and as energy-storing devices in electronic flash units.

26.1 Definition of Capacitance

Consider two conductors as shown in Figure 26.1 (page 610). Such a combination of two conductors is called a **capacitor.** The conductors are called *plates.* If the conductors carry charges of equal magnitude and opposite sign, a potential difference ΔV exists between them.

609

Figure 26.1 A capacitor consists of two conductors.

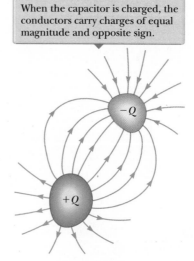

When the capacitor is charged, the conductors carry charges of equal magnitude and opposite sign.

What determines how much charge is on the plates of a capacitor for a given voltage? Experiments show that the quantity of charge Q on a capacitor[1] is linearly proportional to the potential difference between the conductors; that is, $Q \propto \Delta V$. The proportionality constant depends on the shape and separation of the conductors.[2] This relationship can be written as $Q = C\Delta V$ if we define capacitance as follows:

> The **capacitance** C of a capacitor is defined as the ratio of the magnitude of the charge on either conductor to the magnitude of the potential difference between the conductors:
>
> $$C \equiv \frac{Q}{\Delta V} \tag{26.1}$$

◀ **Definition of capacitance**

By definition *capacitance is always a positive quantity*. Furthermore, the charge Q and the potential difference ΔV are always expressed in Equation 26.1 as positive quantities.

From Equation 26.1, we see that capacitance has SI units of coulombs per volt. Named in honor of Michael Faraday, the SI unit of capacitance is the **farad** (F):

$$1\text{ F} = 1\text{ C/V}$$

The farad is a very large unit of capacitance. In practice, typical devices have capacitances ranging from microfarads (10^{-6} F) to picofarads (10^{-12} F). We shall use the symbol μF to represent microfarads. In practice, to avoid the use of Greek letters, physical capacitors are often labeled "mF" for microfarads and "mmF" for micromicrofarads or, equivalently, "pF" for picofarads.

Let's consider a capacitor formed from a pair of parallel plates as shown in Figure 26.2. Each plate is connected to one terminal of a battery, which acts as a source of potential difference. If the capacitor is initially uncharged, the battery establishes an electric field in the connecting wires when the connections are made. Let's focus on the plate connected to the negative terminal of the battery. The electric field in the wire applies a force on electrons in the wire immediately outside this plate; this force causes the electrons to move onto the plate. The movement continues until the plate, the wire, and the terminal are all at the same electric potential. Once this equilibrium situation is attained, a potential difference no longer exists between the terminal and the plate; as a result, no electric field is present in the wire and

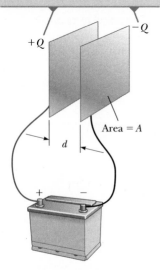

When the capacitor is connected to the terminals of a battery, electrons transfer between the plates and the wires so that the plates become charged.

Area = A

$+Q$ $-Q$ d

Figure 26.2 A parallel-plate capacitor consists of two parallel conducting plates, each of area A, separated by a distance d.

---

[1]Although the total charge on the capacitor is zero (because there is as much excess positive charge on one conductor as there is excess negative charge on the other), it is common practice to refer to the magnitude of the charge on either conductor as "the charge on the capacitor."

[2]The proportionality between Q and ΔV can be proven from Coulomb's law or by experiment.

the electrons stop moving. The plate now carries a negative charge. A similar process occurs at the other capacitor plate, where electrons move from the plate to the wire, leaving the plate positively charged. In this final configuration, the potential difference across the capacitor plates is the same as that between the terminals of the battery.

Quick Quiz 26.1 A capacitor stores charge Q at a potential difference ΔV. What happens if the voltage applied to the capacitor by a battery is doubled to $2\,\Delta V$? **(a)** The capacitance falls to half its initial value, and the charge remains the same. **(b)** The capacitance and the charge both fall to half their initial values. **(c)** The capacitance and the charge both double. **(d)** The capacitance remains the same, and the charge doubles.

26.2 Calculating Capacitance

We can derive an expression for the capacitance of a pair of oppositely charged conductors having a charge of magnitude Q in the following manner. First we calculate the potential difference using the techniques described in Chapter 25. We then use the expression $C = Q/\Delta V$ to evaluate the capacitance. The calculation is relatively easy if the geometry of the capacitor is simple.

Although the most common situation is that of two conductors, a single conductor also has a capacitance. For example, imagine a single spherical, charged conductor. The electric field lines around this conductor are exactly the same as if there were a conducting, spherical shell of infinite radius, concentric with the sphere and carrying a charge of the same magnitude but opposite sign. Therefore, we can identify the imaginary shell as the second conductor of a two-conductor capacitor. The electric potential of the sphere of radius a is simply $k_e Q/a$ (see Section 25.6), and setting $V = 0$ for the infinitely large shell gives

$$C = \frac{Q}{\Delta V} = \frac{Q}{k_e Q/a} = \frac{a}{k_e} = 4\pi\epsilon_0 a \qquad (26.2)$$

> **Pitfall Prevention 26.3**
> **Too Many Cs** Do not confuse an italic C for capacitance with a non-italic C for the unit coulomb.

◀ Capacitance of an isolated charged sphere

This expression shows that the capacitance of an isolated, charged sphere is proportional to its radius and is independent of both the charge on the sphere and its potential, as is the case with all capacitors. Equation 26.1 is the general definition of capacitance in terms of electrical parameters, but the capacitance of a given capacitor will depend only on the geometry of the plates.

The capacitance of a pair of conductors is illustrated below with three familiar geometries, namely, parallel plates, concentric cylinders, and concentric spheres. In these calculations, we assume the charged conductors are separated by a vacuum.

Parallel–Plate Capacitors

Two parallel, metallic plates of equal area A are separated by a distance d as shown in Figure 26.2. One plate carries a charge $+Q$, and the other carries a charge $-Q$. The surface charge density on each plate is $\sigma = Q/A$. If the plates are very close together (in comparison with their length and width), we can assume the electric field is uniform between the plates and zero elsewhere. According to the What If? feature of Example 24.5, the value of the electric field between the plates is

$$E = \frac{\sigma}{\epsilon_0} = \frac{Q}{\epsilon_0 A}$$

Because the field between the plates is uniform, the magnitude of the potential difference between the plates equals Ed (see Eq. 25.6); therefore,

$$\Delta V = Ed = \frac{Qd}{\epsilon_0 A}$$

Substituting this result into Equation 26.1, we find that the capacitance is

$$C = \frac{Q}{\Delta V} = \frac{Q}{Qd/\epsilon_0 A}$$

$$\boxed{C = \frac{\epsilon_0 A}{d}} \tag{26.3}$$

◀ Capacitance of parallel plates

That is, the capacitance of a parallel-plate capacitor is proportional to the area of its plates and inversely proportional to the plate separation.

Let's consider how the geometry of these conductors influences the capacity of the pair of plates to store charge. As a capacitor is being charged by a battery, electrons flow into the negative plate and out of the positive plate. If the capacitor plates are large, the accumulated charges are able to distribute themselves over a substantial area and the amount of charge that can be stored on a plate for a given potential difference increases as the plate area is increased. Therefore, it is reasonable that the capacitance is proportional to the plate area A as in Equation 26.3.

Now consider the region that separates the plates. Imagine moving the plates closer together. Consider the situation before any charges have had a chance to move in response to this change. Because no charges have moved, the electric field between the plates has the same value but extends over a shorter distance. Therefore, the magnitude of the potential difference between the plates $\Delta V = Ed$ (Eq. 25.6) is smaller. The difference between this new capacitor voltage and the terminal voltage of the battery appears as a potential difference across the wires connecting the battery to the capacitor, resulting in an electric field in the wires that drives more charge onto the plates and increases the potential difference between the plates. When the potential difference between the plates again matches that of the battery, the flow of charge stops. Therefore, moving the plates closer together causes the charge on the capacitor to increase. If d is increased, the charge decreases. As a result, the inverse relationship between C and d in Equation 26.3 is reasonable.

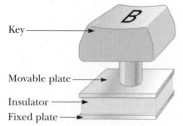

Key

Movable plate

Insulator

Fixed plate

Figure 26.3 (Quick Quiz 26.2) One type of computer keyboard button.

Ⓠuick Quiz 26.2 Many computer keyboard buttons are constructed of capacitors as shown in Figure 26.3. When a key is pushed down, the soft insulator between the movable plate and the fixed plate is compressed. When the key is pressed, what happens to the capacitance? **(a)** It increases. **(b)** It decreases. **(c)** It changes in a way you cannot determine because the electric circuit connected to the keyboard button may cause a change in ΔV.

Example 26.1 **The Cylindrical Capacitor**

A solid cylindrical conductor of radius a and charge Q is coaxial with a cylindrical shell of negligible thickness, radius $b > a$, and charge $-Q$ (Fig. 26.4a). Find the capacitance of this cylindrical capacitor if its length is ℓ.

SOLUTION

Conceptualize Recall that any pair of conductors qualifies as a capacitor, so the system described in this example therefore qualifies. Figure 26.4b helps visualize the electric field between the conductors. We expect the capacitance to depend only on geometric factors, which, in this case, are a, b, and ℓ.

Categorize Because of the cylindrical symmetry of the system, we can use results from previous studies of cylindrical systems to find the capacitance.

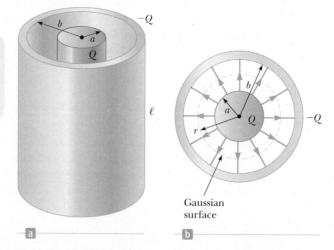

Figure 26.4 (Example 26.1) (a) A cylindrical capacitor consists of a solid cylindrical conductor of radius a and length ℓ surrounded by a coaxial cylindrical shell of radius b. (b) End view. The electric field lines are radial. The dashed line represents the end of a cylindrical gaussian surface of radius r and length ℓ.

▶ **26.1** continued

Analyze Assuming ℓ is much greater than a and b, we can neglect end effects. In this case, the electric field is perpendicular to the long axis of the cylinders and is confined to the region between them (Fig. 26.4b).

Write an expression for the potential difference between the two cylinders from Equation 25.3:

$$V_b - V_a = -\int_a^b \vec{\mathbf{E}} \cdot d\vec{\mathbf{s}}$$

Apply Equation 24.7 for the electric field outside a cylindrically symmetric charge distribution and notice from Figure 26.4b that $\vec{\mathbf{E}}$ is parallel to $d\vec{\mathbf{s}}$ along a radial line:

$$V_b - V_a = -\int_a^b E_r\, dr = -2k_e\lambda \int_a^b \frac{dr}{r} = -2k_e\lambda \ln\left(\frac{b}{a}\right)$$

Substitute the absolute value of ΔV into Equation 26.1 and use $\lambda = Q/\ell$:

$$C = \frac{Q}{\Delta V} = \frac{Q}{(2k_e Q/\ell)\ln(b/a)} = \frac{\ell}{2k_e \ln(b/a)} \quad \textbf{(26.4)}$$

Finalize The capacitance depends on the radii a and b and is proportional to the length of the cylinders. Equation 26.4 shows that the capacitance per unit length of a combination of concentric cylindrical conductors is

$$\frac{C}{\ell} = \frac{1}{2k_e \ln(b/a)} \quad \textbf{(26.5)}$$

An example of this type of geometric arrangement is a *coaxial cable,* which consists of two concentric cylindrical conductors separated by an insulator. You probably have a coaxial cable attached to your television set if you are a subscriber to cable television. The coaxial cable is especially useful for shielding electrical signals from any possible external influences.

WHAT IF? Suppose $b = 2.00a$ for the cylindrical capacitor. You would like to increase the capacitance, and you can do so by choosing to increase either ℓ by 10% or a by 10%. Which choice is more effective at increasing the capacitance?

Answer According to Equation 26.4, C is proportional to ℓ, so increasing ℓ by 10% results in a 10% increase in C. For the result of the change in a, let's use Equation 26.4 to set up a ratio of the capacitance C' for the enlarged cylinder radius a' to the original capacitance:

$$\frac{C'}{C} = \frac{\ell/2k_e \ln(b/a')}{\ell/2k_e \ln(b/a)} = \frac{\ln(b/a)}{\ln(b/a')}$$

We now substitute $b = 2.00a$ and $a' = 1.10a$, representing a 10% increase in a:

$$\frac{C'}{C} = \frac{\ln(2.00a/a)}{\ln(2.00a/1.10a)} = \frac{\ln 2.00}{\ln 1.82} = 1.16$$

which corresponds to a 16% increase in capacitance. Therefore, it is more effective to increase a than to increase ℓ.

Note two more extensions of this problem. First, it is advantageous to increase a only for a range of relationships between a and b. If $b > 2.85a$, increasing ℓ by 10% is more effective than increasing a (see Problem 70 in Enhanced WebAssign). Second, if b decreases, the capacitance increases. Increasing a or decreasing b has the effect of bringing the plates closer together, which increases the capacitance.

Example 26.2 The Spherical Capacitor

A spherical capacitor consists of a spherical conducting shell of radius b and charge $-Q$ concentric with a smaller conducting sphere of radius a and charge Q (Fig. 26.5, page 614). Find the capacitance of this device.

SOLUTION

Conceptualize As with Example 26.1, this system involves a pair of conductors and qualifies as a capacitor. We expect the capacitance to depend on the spherical radii a and b.

continued

▶ **26.2** continued

Categorize Because of the spherical symmetry of the system, we can use results from previous studies of spherical systems to find the capacitance.

Analyze As shown in Chapter 24, the direction of the electric field outside a spherically symmetric charge distribution is radial and its magnitude is given by the expression $E = k_e Q / r^2$. In this case, this result applies to the field *between* the spheres ($a < r < b$).

Write an expression for the potential difference between the two conductors from Equation 25.3:

Apply the result of Example 24.3 for the electric field outside a spherically symmetric charge distribution and note that $\vec{\mathbf{E}}$ is parallel to $d\vec{\mathbf{s}}$ along a radial line:

Substitute the absolute value of ΔV into Equation 26.1:

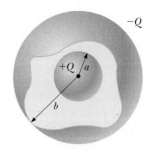

Figure 26.5 (Example 26.2) A spherical capacitor consists of an inner sphere of radius a surrounded by a concentric spherical shell of radius b. The electric field between the spheres is directed radially outward when the inner sphere is positively charged.

$$V_b - V_a = -\int_a^b \vec{\mathbf{E}} \cdot d\vec{\mathbf{s}}$$

$$V_b - V_a = -\int_a^b E_r\, dr = -k_e Q \int_a^b \frac{dr}{r^2} = k_e Q \left[\frac{1}{r}\right]_a^b$$

$$(1) \quad V_b - V_a = k_e Q \left(\frac{1}{b} - \frac{1}{a}\right) = k_e Q \frac{a - b}{ab}$$

$$C = \frac{Q}{\Delta V} = \frac{Q}{|V_b - V_a|} = \boxed{\frac{ab}{k_e(b - a)}} \qquad (26.6)$$

Finalize The capacitance depends on a and b as expected. The potential difference between the spheres in Equation (1) is negative because Q is positive and $b > a$. Therefore, in Equation 26.6, when we take the absolute value, we change $a - b$ to $b - a$. The result is a positive number.

WHAT IF? If the radius b of the outer sphere approaches infinity, what does the capacitance become?

Answer In Equation 26.6, we let $b \to \infty$:

$$C = \lim_{b \to \infty} \frac{ab}{k_e(b - a)} = \frac{ab}{k_e(b)} = \frac{a}{k_e} = 4\pi\epsilon_0 a$$

Notice that this expression is the same as Equation 26.2, the capacitance of an isolated spherical conductor.

26.3 Combinations of Capacitors

Two or more capacitors often are combined in electric circuits. We can calculate the equivalent capacitance of certain combinations using methods described in this section. Throughout this section, we assume the capacitors to be combined are initially uncharged.

In studying electric circuits, we use a simplified pictorial representation called a **circuit diagram.** Such a diagram uses **circuit symbols** to represent various circuit elements. The circuit symbols are connected by straight lines that represent the wires between the circuit elements. The circuit symbols for capacitors, batteries, and switches as well as the color codes used for them in this text are given in Figure 26.6. The symbol for the capacitor reflects the geometry of the most common model for a capacitor, a pair of parallel plates. The positive terminal of the battery is at the higher potential and is represented in the circuit symbol by the longer line.

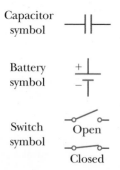

Figure 26.6 Circuit symbols for capacitors, batteries, and switches. Notice that capacitors are in blue, batteries are in green, and switches are in red. The closed switch can carry current, whereas the open one cannot.

Parallel Combination

Two capacitors connected as shown in Figure 26.7a are known as a **parallel combination** of capacitors. Figure 26.7b shows a circuit diagram for this combination of capacitors. The left plates of the capacitors are connected to the positive terminal of the battery by a conducting wire and are therefore both at the same electric potential

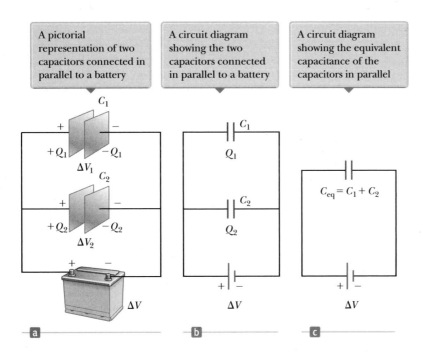

Figure 26.7 Two capacitors connected in parallel. All three diagrams are equivalent.

A pictorial representation of two capacitors connected in parallel to a battery

A circuit diagram showing the two capacitors connected in parallel to a battery

A circuit diagram showing the equivalent capacitance of the capacitors in parallel

as the positive terminal. Likewise, the right plates are connected to the negative terminal and so are both at the same potential as the negative terminal. Therefore, the individual potential differences across capacitors connected in parallel are the same and are equal to the potential difference applied across the combination. That is,

$$\Delta V_1 = \Delta V_2 = \Delta V$$

where ΔV is the battery terminal voltage.

After the battery is attached to the circuit, the capacitors quickly reach their maximum charge. Let's call the maximum charges on the two capacitors Q_1 and Q_2, where $Q_1 = C_1 \Delta V_1$ and $Q_2 = C_2 \Delta V_2$. The *total charge* Q_{tot} stored by the two capacitors is the sum of the charges on the individual capacitors:

$$Q_{tot} = Q_1 + Q_2 = C_1 \Delta V_1 + C_2 \Delta V_2 \qquad \textbf{(26.7)}$$

Suppose you wish to replace these two capacitors by one *equivalent capacitor* having a capacitance C_{eq} as in Figure 26.7c. The effect this equivalent capacitor has on the circuit must be exactly the same as the effect of the combination of the two individual capacitors. That is, the equivalent capacitor must store charge Q_{tot} when connected to the battery. Figure 26.7c shows that the voltage across the equivalent capacitor is ΔV because the equivalent capacitor is connected directly across the battery terminals. Therefore, for the equivalent capacitor,

$$Q_{tot} = C_{eq} \Delta V$$

Substituting this result into Equation 26.7 gives

$$C_{eq} \Delta V = C_1 \Delta V_1 + C_2 \Delta V_2$$

$$C_{eq} = C_1 + C_2 \quad \text{(parallel combination)}$$

where we have canceled the voltages because they are all the same. If this treatment is extended to three or more capacitors connected in parallel, the **equivalent capacitance** is found to be

$$C_{eq} = C_1 + C_2 + C_3 + \cdots \quad \text{(parallel combination)} \qquad \textbf{(26.8)}$$

◀ Equivalent capacitance for capacitors in parallel

Therefore, the equivalent capacitance of a parallel combination of capacitors is (1) the algebraic sum of the individual capacitances and (2) greater than any of

Figure 26.8 Two capacitors connected in series. All three diagrams are equivalent.

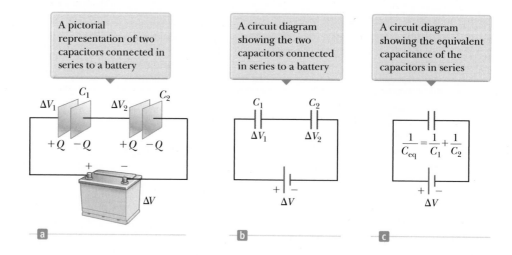

A pictorial representation of two capacitors connected in series to a battery

A circuit diagram showing the two capacitors connected in series to a battery

A circuit diagram showing the equivalent capacitance of the capacitors in series

the individual capacitances. Statement (2) makes sense because we are essentially combining the areas of all the capacitor plates when they are connected with conducting wire, and capacitance of parallel plates is proportional to area (Eq. 26.3).

Series Combination

Two capacitors connected as shown in Figure 26.8a and the equivalent circuit diagram in Figure 26.8b are known as a **series combination** of capacitors. The left plate of capacitor 1 and the right plate of capacitor 2 are connected to the terminals of a battery. The other two plates are connected to each other and to nothing else; hence, they form an isolated system that is initially uncharged and must continue to have zero net charge. To analyze this combination, let's first consider the uncharged capacitors and then follow what happens immediately after a battery is connected to the circuit. When the battery is connected, electrons are transferred out of the left plate of C_1 and into the right plate of C_2. As this negative charge accumulates on the right plate of C_2, an equivalent amount of negative charge is forced off the left plate of C_2, and this left plate therefore has an excess positive charge. The negative charge leaving the left plate of C_2 causes negative charges to accumulate on the right plate of C_1. As a result, both right plates end up with a charge $-Q$ and both left plates end up with a charge $+Q$. Therefore, the charges on capacitors connected in series are the same:

$$Q_1 = Q_2 = Q$$

where Q is the charge that moved between a wire and the connected outside plate of one of the capacitors.

Figure 26.8a shows the individual voltages ΔV_1 and ΔV_2 across the capacitors. These voltages add to give the total voltage ΔV_{tot} across the combination:

$$\Delta V_{\text{tot}} = \Delta V_1 + \Delta V_2 = \frac{Q_1}{C_1} + \frac{Q_2}{C_2} \qquad \textbf{(26.9)}$$

In general, the total potential difference across any number of capacitors connected in series is the sum of the potential differences across the individual capacitors.

Suppose the equivalent single capacitor in Figure 26.8c has the same effect on the circuit as the series combination when it is connected to the battery. After it is fully charged, the equivalent capacitor must have a charge of $-Q$ on its right plate and a charge of $+Q$ on its left plate. Applying the definition of capacitance to the circuit in Figure 26.8c gives

$$\Delta V_{\text{tot}} = \frac{Q}{C_{\text{eq}}}$$

Substituting this result into Equation 26.9, we have

$$\frac{Q}{C_{eq}} = \frac{Q_1}{C_1} + \frac{Q_2}{C_2}$$

Canceling the charges because they are all the same gives

$$\frac{1}{C_{eq}} = \frac{1}{C_1} + \frac{1}{C_2} \quad \text{(series combination)}$$

When this analysis is applied to three or more capacitors connected in series, the relationship for the **equivalent capacitance** is

$$\frac{1}{C_{eq}} = \frac{1}{C_1} + \frac{1}{C_2} + \frac{1}{C_3} + \cdots \quad \text{(series combination)} \qquad \text{(26.10)}$$ ◀ Equivalent capacitance for capacitors in series

This expression shows that (1) the inverse of the equivalent capacitance is the algebraic sum of the inverses of the individual capacitances and (2) the equivalent capacitance of a series combination is always less than any individual capacitance in the combination.

Quick Quiz 26.3 Two capacitors are identical. They can be connected in series or in parallel. If you want the *smallest* equivalent capacitance for the combination, how should you connect them? **(a)** in series **(b)** in parallel **(c)** either way because both combinations have the same capacitance

---

Example 26.3 | Equivalent Capacitance

Find the equivalent capacitance between a and b for the combination of capacitors shown in Figure 26.9a. All capacitances are in microfarads.

SOLUTION

Conceptualize Study Figure 26.9a carefully and make sure you understand how the capacitors are connected. Verify that there are only series and parallel connections between capacitors.

Categorize Figure 26.9a shows that the circuit contains both series and parallel connections, so we use the rules for series and parallel combinations discussed in this section.

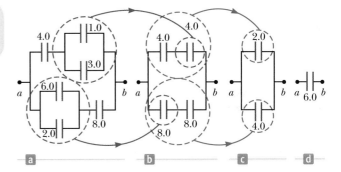

Figure 26.9 (Example 26.3) To find the equivalent capacitance of the capacitors in (a), we reduce the various combinations in steps as indicated in (b), (c), and (d), using the series and parallel rules described in the text. All capacitances are in microfarads.

Analyze Using Equations 26.8 and 26.10, we reduce the combination step by step as indicated in the figure. As you follow along below, notice that in each step we replace the combination of two capacitors in the circuit diagram with a single capacitor having the equivalent capacitance.

The 1.0-μF and 3.0-μF capacitors (upper red-brown circle in Fig. 26.9a) are in parallel. Find the equivalent capacitance from Equation 26.8:

$$C_{eq} = C_1 + C_2 = 4.0 \ \mu F$$

The 2.0-μF and 6.0-μF capacitors (lower red-brown circle in Fig. 26.9a) are also in parallel:

$$C_{eq} = C_1 + C_2 = 8.0 \ \mu F$$

The circuit now looks like Figure 26.9b. The two 4.0-μF capacitors (upper green circle in Fig. 26.9b) are in series. Find the equivalent capacitance from Equation 26.10:

$$\frac{1}{C_{eq}} = \frac{1}{C_1} + \frac{1}{C_2} = \frac{1}{4.0 \ \mu F} + \frac{1}{4.0 \ \mu F} = \frac{1}{2.0 \ \mu F}$$

$$C_{eq} = 2.0 \ \mu F$$

continued

▶ **26.3** continued

The two 8.0-μF capacitors (lower green circle in Fig. 26.9b) are also in series. Find the equivalent capacitance from Equation 26.10:

$$\frac{1}{C_{eq}} = \frac{1}{C_1} + \frac{1}{C_2} = \frac{1}{8.0 \ \mu F} + \frac{1}{8.0 \ \mu F} = \frac{1}{4.0 \ \mu F}$$

$$C_{eq} = 4.0 \ \mu F$$

The circuit now looks like Figure 26.9c. The 2.0-μF and 4.0-μF capacitors are in parallel:

$$C_{eq} = C_1 + C_2 = \boxed{6.0 \ \mu F}$$

Finalize This final value is that of the single equivalent capacitor shown in Figure 26.9d. For further practice in treating circuits with combinations of capacitors, imagine a battery is connected between points a and b in Figure 26.9a so that a potential difference ΔV is established across the combination. Can you find the voltage across and the charge on each capacitor?

26.4 Energy Stored in a Charged Capacitor

Because positive and negative charges are separated in the system of two conductors in a capacitor, electric potential energy is stored in the system. Many of those who work with electronic equipment have at some time verified that a capacitor can store energy. If the plates of a charged capacitor are connected by a conductor such as a wire, charge moves between each plate and its connecting wire until the capacitor is uncharged. The discharge can often be observed as a visible spark. If you accidentally touch the opposite plates of a charged capacitor, your fingers act as a pathway for discharge and the result is an electric shock. The degree of shock you receive depends on the capacitance and the voltage applied to the capacitor. Such a shock could be dangerous if high voltages are present as in the power supply of a home theater system. Because the charges can be stored in a capacitor even when the system is turned off, unplugging the system does not make it safe to open the case and touch the components inside.

Figure 26.10a shows a battery connected to a single parallel-plate capacitor with a switch in the circuit. Let us identify the circuit as a system. When the switch is closed (Fig. 26.10b), the battery establishes an electric field in the wires and charges

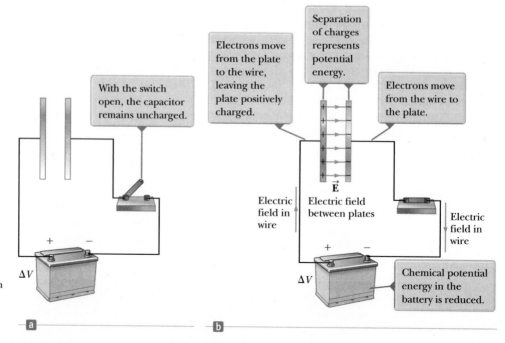

With the switch open, the capacitor remains uncharged.

Electrons move from the plate to the wire, leaving the plate positively charged.

Separation of charges represents potential energy.

Electrons move from the wire to the plate.

Electric field in wire

Electric field between plates

Electric field in wire

$\vec{E}$

ΔV

ΔV

Chemical potential energy in the battery is reduced.

Figure 26.10 (a) A circuit consisting of a capacitor, a battery, and a switch. (b) When the switch is closed, the battery establishes an electric field in the wire and the capacitor becomes charged.

flow between the wires and the capacitor. As that occurs, there is a transformation of energy within the system. Before the switch is closed, energy is stored as chemical potential energy in the battery. This energy is transformed during the chemical reaction that occurs within the battery when it is operating in an electric circuit. When the switch is closed, some of the chemical potential energy in the battery is transformed to electric potential energy associated with the separation of positive and negative charges on the plates.

To calculate the energy stored in the capacitor, we shall assume a charging process that is different from the actual process described in Section 26.1 but that gives the same final result. This assumption is justified because the energy in the final configuration does not depend on the actual charge-transfer process.[3] Imagine the plates are disconnected from the battery and you transfer the charge mechanically through the space between the plates as follows. You grab a small amount of positive charge on one plate and apply a force that causes this positive charge to move over to the other plate. Therefore, you do work on the charge as it is transferred from one plate to the other. At first, no work is required to transfer a small amount of charge dq from one plate to the other,[4] but once this charge has been transferred, a small potential difference exists between the plates. Therefore, work must be done to move additional charge through this potential difference. As more and more charge is transferred from one plate to the other, the potential difference increases in proportion and more work is required. The overall process is described by the nonisolated system model for energy. Equation 8.2 reduces to $W = \Delta U_E$; the work done on the system by the external agent appears as an increase in electric potential energy in the system.

Suppose q is the charge on the capacitor at some instant during the charging process. At the same instant, the potential difference across the capacitor is $\Delta V = q/C$. This relationship is graphed in Figure 26.11. From Section 25.1, we know that the work necessary to transfer an increment of charge dq from the plate carrying charge $-q$ to the plate carrying charge q (which is at the higher electric potential) is

$$dW = \Delta V \, dq = \frac{q}{C} \, dq$$

The work required to transfer the charge dq is the area of the tan rectangle in Figure 26.11. Because 1 V = 1 J/C, the unit for the area is the joule. The total work required to charge the capacitor from $q = 0$ to some final charge $q = Q$ is

$$W = \int_0^Q \frac{q}{C} \, dq = \frac{1}{C} \int_0^Q q \, dq = \frac{Q^2}{2C}$$

The work done in charging the capacitor appears as electric potential energy U_E stored in the capacitor. Using Equation 26.1, we can express the potential energy stored in a charged capacitor as

$$U_E = \frac{Q^2}{2C} = \tfrac{1}{2}Q\,\Delta V = \tfrac{1}{2}C(\Delta V)^2 \qquad \text{(26.11)}$$

◀ **Energy stored in a charged capacitor**

Because the curve in Figure 26.11 is a straight line, the total area under the curve is that of a triangle of base Q and height ΔV.

Equation 26.11 applies to any capacitor, regardless of its geometry. For a given capacitance, the stored energy increases as the charge and the potential difference increase. In practice, there is a limit to the maximum energy (or charge) that can be stored because, at a sufficiently large value of ΔV, discharge ultimately occurs

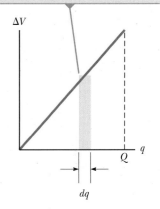

The work required to move charge dq through the potential difference ΔV across the capacitor plates is given approximately by the area of the shaded rectangle.

Figure 26.11 A plot of potential difference versus charge for a capacitor is a straight line having slope $1/C$.

---

[3]This discussion is similar to that of state variables in thermodynamics. The change in a state variable such as temperature is independent of the path followed between the initial and final states. The potential energy of a capacitor (or any system) is also a state variable, so its change does not depend on the process followed to charge the capacitor.

[4]We shall use lowercase q for the time-varying charge on the capacitor while it is charging to distinguish it from uppercase Q, which is the total charge on the capacitor after it is completely charged.

between the plates. For this reason, capacitors are usually labeled with a maximum operating voltage.

We can consider the energy in a capacitor to be stored in the electric field created between the plates as the capacitor is charged. This description is reasonable because the electric field is proportional to the charge on the capacitor. For a parallel-plate capacitor, the potential difference is related to the electric field through the relationship $\Delta V = Ed$. Furthermore, its capacitance is $C = \epsilon_0 A/d$ (Eq. 26.3). Substituting these expressions into Equation 26.11 gives

$$U_E = \tfrac{1}{2}\left(\frac{\epsilon_0 A}{d}\right)(Ed)^2 = \tfrac{1}{2}(\epsilon_0 Ad)E^2 \tag{26.12}$$

Because the volume occupied by the electric field is Ad, the *energy per unit volume* $u_E = U_E/Ad$, known as the *energy density*, is

▶ Energy density in an electric field

$$u_E = \tfrac{1}{2}\epsilon_0 E^2 \tag{26.13}$$

Although Equation 26.13 was derived for a parallel-plate capacitor, the expression is generally valid regardless of the source of the electric field. That is, the energy density in any electric field is proportional to the square of the magnitude of the electric field at a given point.

Quick Quiz 26.4 You have three capacitors and a battery. In which of the following combinations of the three capacitors is the maximum possible energy stored when the combination is attached to the battery? **(a)** series **(b)** parallel **(c)** no difference because both combinations store the same amount of energy

---

Example 26.4   **Rewiring Two Charged Capacitors**

Two capacitors C_1 and C_2 (where $C_1 > C_2$) are charged to the same initial potential difference ΔV_i. The charged capacitors are removed from the battery, and their plates are connected with opposite polarity as in Figure 26.12a. The switches S_1 and S_2 are then closed as in Figure 26.12b.

(A) Find the final potential difference ΔV_f between a and b after the switches are closed.

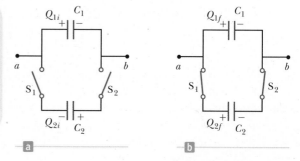

Figure 26.12 (Example 26.4) (a) Two capacitors are charged to the same initial potential difference and connected together with plates of opposite sign to be in contact when the switches are closed. (b) When the switches are closed, the charges redistribute.

SOLUTION

Conceptualize Figure 26.12 helps us understand the initial and final configurations of the system. When the switches are closed, the charge on the system will redistribute between the capacitors until both capacitors have the same potential difference. Because $C_1 > C_2$, more charge exists on C_1 than on C_2, so the final configuration will have positive charge on the left plates as shown in Figure 26.12b.

Categorize In Figure 26.12b, it might appear as if the capacitors are connected in parallel, but there is no battery in this circuit to apply a voltage across the combination. Therefore, we *cannot* categorize this problem as one in which capacitors are connected in parallel. We *can* categorize it as a problem involving an isolated system for electric charge. The left-hand plates of the capacitors form an isolated system because they are not connected to the right-hand plates by conductors.

Analyze Write an expression for the total charge on the left-hand plates of the system before the switches are closed, noting that a negative sign for Q_{2i} is necessary because the charge on the left plate of capacitor C_2 is negative:

(1) $Q_i = Q_{1i} + Q_{2i} = C_1\,\Delta V_i - C_2\,\Delta V_i = (C_1 - C_2)\Delta V_i$

▶ **26.4** continued

After the switches are closed, the charges on the individual capacitors change to new values Q_{1f} and Q_{2f} such that the potential difference is again the same across both capacitors, with a value of ΔV_f. Write an expression for the total charge on the left-hand plates of the system after the switches are closed:

$$(2) \quad Q_f = Q_{1f} + Q_{2f} = C_1 \, \Delta V_f + C_2 \, \Delta V_f = (C_1 + C_2)\Delta V_f$$

Because the system is isolated, the initial and final charges on the system must be the same. Use this condition and Equations (1) and (2) to solve for ΔV_f:

$$Q_f = Q_i \;\rightarrow\; (C_1 + C_2)\,\Delta V_f = (C_1 - C_2)\,\Delta V_i$$

$$(3) \quad \Delta V_f = \left(\frac{C_1 - C_2}{C_1 + C_2}\right)\Delta V_i$$

(B) Find the total energy stored in the capacitors before and after the switches are closed and determine the ratio of the final energy to the initial energy.

SOLUTION

Use Equation 26.11 to find an expression for the total energy stored in the capacitors before the switches are closed:

$$(4) \quad U_i = \tfrac{1}{2}C_1(\Delta V_i)^2 + \tfrac{1}{2}C_2(\Delta V_i)^2 = \tfrac{1}{2}(C_1 + C_2)(\Delta V_i)^2$$

Write an expression for the total energy stored in the capacitors after the switches are closed:

$$U_f = \tfrac{1}{2}C_1(\Delta V_f)^2 + \tfrac{1}{2}C_2(\Delta V_f)^2 = \tfrac{1}{2}(C_1 + C_2)(\Delta V_f)^2$$

Use the results of part (A) to rewrite this expression in terms of ΔV_i:

$$(5) \quad U_f = \tfrac{1}{2}(C_1 + C_2)\left[\left(\frac{C_1 - C_2}{C_1 + C_2}\right)\Delta V_i\right]^2 = \tfrac{1}{2}\frac{(C_1 - C_2)^2(\Delta V_i)^2}{C_1 + C_2}$$

Divide Equation (5) by Equation (4) to obtain the ratio of the energies stored in the system:

$$\frac{U_f}{U_i} = \frac{\tfrac{1}{2}(C_1 - C_2)^2(\Delta V_i)^2/(C_1 + C_2)}{\tfrac{1}{2}(C_1 + C_2)(\Delta V_i)^2}$$

$$(6) \quad \frac{U_f}{U_i} = \left(\frac{C_1 - C_2}{C_1 + C_2}\right)^2$$

Finalize The ratio of energies is *less* than unity, indicating that the final energy is *less* than the initial energy. At first, you might think the law of energy conservation has been violated, but that is not the case. The "missing" energy is transferred out of the system by the mechanism of electromagnetic waves (T_{ER} in Eq. 8.2), as we shall see in Chapter 34. Therefore, this system is isolated for electric charge, but nonisolated for energy.

WHAT IF? What if the two capacitors have the same capacitance? What would you expect to happen when the switches are closed?

Answer Because both capacitors have the same initial potential difference applied to them, the charges on the identical capacitors have the same magnitude. When the capacitors with opposite polarities are connected together, the equal-magnitude charges should cancel each other, leaving the capacitors uncharged.

Let's test our results to see if that is the case mathematically. In Equation (1), because the capacitances are equal, the initial charge Q_i on the system of left-hand plates is zero. Equation (3) shows that $\Delta V_f = 0$, which is consistent with uncharged capacitors. Finally, Equation (5) shows that $U_f = 0$, which is also consistent with uncharged capacitors.

One device in which capacitors have an important role is the portable *defibrillator* (see the chapter-opening photo on page 609). When cardiac fibrillation (random contractions) occurs, the heart produces a rapid, irregular pattern of beats. A fast discharge of energy through the heart can return the organ to its normal beat pattern. Emergency medical teams use portable defibrillators that contain batteries capable of charging a capacitor to a high voltage. (The circuitry actually permits the capacitor to be charged to a much higher voltage than that of the battery.) Up to 360 J is stored in the electric field of a large capacitor in a defibrillator when it is fully charged. The stored energy is released through the heart by conducting electrodes, called paddles, which are placed on both sides of the victim's chest. The defibrillator can deliver the energy to a patient in about 2 ms (roughly equivalent to 3 000 times the power delivered to a 60-W lightbulb!). The paramedics must wait between applications of the energy because of the time interval necessary for the capacitors to become fully charged. In this application and others (e.g., camera flash units and lasers used for fusion experiments), capacitors serve as energy reservoirs that can be slowly charged and then quickly discharged to provide large amounts of energy in a short pulse.

26.5 Capacitors with Dielectrics

A **dielectric** is a nonconducting material such as rubber, glass, or waxed paper. We can perform the following experiment to illustrate the effect of a dielectric in a capacitor. Consider a parallel-plate capacitor that without a dielectric has a charge Q_0 and a capacitance C_0. The potential difference across the capacitor is $\Delta V_0 = Q_0/C_0$. Figure 26.13a illustrates this situation. The potential difference is measured by a device called a *voltmeter*. Notice that no battery is shown in the figure; also, we must assume no charge can flow through an ideal voltmeter. Hence, there is no path by which charge can flow and alter the charge on the capacitor. If a dielectric is now inserted between the plates as in Figure 26.13b, the voltmeter indicates that the voltage between the plates decreases to a value ΔV. The voltages with and without the dielectric are related by a factor κ as follows:

$$\Delta V = \frac{\Delta V_0}{\kappa}$$

Because $\Delta V < \Delta V_0$, we see that $\kappa > 1$. The dimensionless factor κ is called the **dielectric constant** of the material. The dielectric constant varies from one material to another. In this section, we analyze this change in capacitance in terms of electrical parameters such as electric charge, electric field, and potential difference; Section 26.7 describes the microscopic origin of these changes.

Because the charge Q_0 on the capacitor does not change, the capacitance must change to the value

$$C = \frac{Q_0}{\Delta V} = \frac{Q_0}{\Delta V_0/\kappa} = \kappa \frac{Q_0}{\Delta V_0}$$

◀ Capacitance of a capacitor filled with a material of dielectric constant κ

$$C = \kappa C_0 \tag{26.14}$$

The potential difference across the charged capacitor is initially ΔV_0.

After the dielectric is inserted between the plates, the charge remains the same, but the potential difference decreases and the capacitance increases.

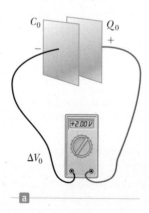

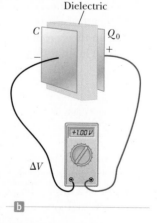

Figure 26.13 A charged capacitor (a) before and (b) after insertion of a dielectric between the plates.

That is, the capacitance *increases* by the factor κ when the dielectric completely fills the region between the plates.[5] Because $C_0 = \epsilon_0 A/d$ (Eq. 26.3) for a parallel-plate capacitor, we can express the capacitance of a parallel-plate capacitor filled with a dielectric as

$$C = \kappa \frac{\epsilon_0 A}{d} \qquad \textbf{(26.15)}$$

From Equation 26.15, it would appear that the capacitance could be made very large by inserting a dielectric between the plates and decreasing d. In practice, the lowest value of d is limited by the electric discharge that could occur through the dielectric medium separating the plates. For any given separation d, the maximum voltage that can be applied to a capacitor without causing a discharge depends on the **dielectric strength** (maximum electric field) of the dielectric. If the magnitude of the electric field in the dielectric exceeds the dielectric strength, the insulating properties break down and the dielectric begins to conduct.

Physical capacitors have a specification called by a variety of names, including *working voltage, breakdown voltage,* and *rated voltage.* This parameter represents the largest voltage that can be applied to the capacitor without exceeding the dielectric strength of the dielectric material in the capacitor. Consequently, when selecting a capacitor for a given application, you must consider its capacitance as well as the expected voltage across the capacitor in the circuit, making sure the expected voltage is smaller than the rated voltage of the capacitor.

Insulating materials have values of κ greater than unity and dielectric strengths greater than that of air as Table 26.1 indicates. Therefore, a dielectric provides the following advantages:

- An increase in capacitance
- An increase in maximum operating voltage
- Possible mechanical support between the plates, which allows the plates to be close together without touching, thereby decreasing d and increasing C

Table 26.1 **Approximate Dielectric Constants and Dielectric Strengths of Various Materials at Room Temperature**

Material	Dielectric Constant κ	Dielectric Strength[a] (10^6 V/m)
Air (dry)	1.000 59	3
Bakelite	4.9	24
Fused quartz	3.78	8
Mylar	3.2	7
Neoprene rubber	6.7	12
Nylon	3.4	14
Paper	3.7	16
Paraffin-impregnated paper	3.5	11
Polystyrene	2.56	24
Polyvinyl chloride	3.4	40
Porcelain	6	12
Pyrex glass	5.6	14
Silicone oil	2.5	15
Strontium titanate	233	8
Teflon	2.1	60
Vacuum	1.000 00	—
Water	80	—

[a]The dielectric strength equals the maximum electric field that can exist in a dielectric without electrical breakdown. These values depend strongly on the presence of impurities and flaws in the materials.

---

[5] If the dielectric is introduced while the potential difference is held constant by a battery, the charge increases to a value $Q = \kappa Q_0$. The additional charge comes from the wires attached to the capacitor, and the capacitance again increases by the factor κ.

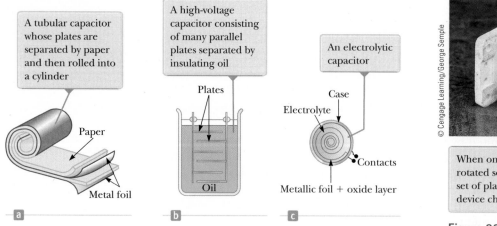

A tubular capacitor whose plates are separated by paper and then rolled into a cylinder

Paper

Metal foil

A high-voltage capacitor consisting of many parallel plates separated by insulating oil

Plates

Oil

An electrolytic capacitor

Case

Electrolyte

Contacts

Metallic foil + oxide layer

a   **b**   **c**

Figure 26.14 Three commercial capacitor designs.

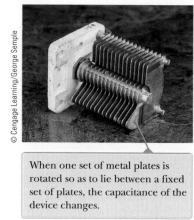

When one set of metal plates is rotated so as to lie between a fixed set of plates, the capacitance of the device changes.

Figure 26.15 A variable capacitor.

Types of Capacitors

Many capacitors are built into integrated circuit chips, but some electrical devices still use stand-alone capacitors. Commercial capacitors are often made from metallic foil interlaced with thin sheets of either paraffin-impregnated paper or Mylar as the dielectric material. These alternate layers of metallic foil and dielectric are rolled into a cylinder to form a small package (Fig. 26.14a). High-voltage capacitors commonly consist of a number of interwoven metallic plates immersed in silicone oil (Fig. 26.14b). Small capacitors are often constructed from ceramic materials.

Often, an *electrolytic capacitor* is used to store large amounts of charge at relatively low voltages. This device, shown in Figure 26.14c, consists of a metallic foil in contact with an *electrolyte,* a solution that conducts electricity by virtue of the motion of ions contained in the solution. When a voltage is applied between the foil and the electrolyte, a thin layer of metal oxide (an insulator) is formed on the foil, and this layer serves as the dielectric. Very large values of capacitance can be obtained in an electrolytic capacitor because the dielectric layer is very thin and therefore the plate separation is very small.

Electrolytic capacitors are not reversible as are many other capacitors. They have a polarity, which is indicated by positive and negative signs marked on the device. When electrolytic capacitors are used in circuits, the polarity must be correct. If the polarity of the applied voltage is the opposite of what is intended, the oxide layer is removed and the capacitor conducts electricity instead of storing charge.

Variable capacitors (typically 10 to 500 pF) usually consist of two interwoven sets of metallic plates, one fixed and the other movable, and contain air as the dielectric (Fig. 26.15). These types of capacitors are often used in radio tuning circuits.

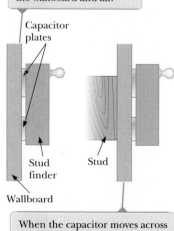

The materials between the plates of the capacitor are the wallboard and air.

Capacitor plates

Stud finder

Stud

Wallboard

When the capacitor moves across a stud in the wall, the materials between the plates are the wallboard and the wood stud. The change in the dielectric constant causes a signal light to illuminate.

a   **b**

Figure 26.16 (Quick Quiz 26.5) A stud finder.

 uick Quiz 26.5 If you have ever tried to hang a picture or a mirror, you know it can be difficult to locate a wooden stud in which to anchor your nail or screw. A carpenter's stud finder is a capacitor with its plates arranged side by side instead of facing each other as shown in Figure 26.16. When the device is moved over a stud, does the capacitance **(a)** increase or **(b)** decrease?

Example 26.5   **Energy Stored Before and After**   **AM**

A parallel-plate capacitor is charged with a battery to a charge Q_0. The battery is then removed, and a slab of material that has a dielectric constant κ is inserted between the plates. Identify the system as the capacitor and the dielectric. Find the energy stored in the system before and after the dielectric is inserted.

▶ **26.5** continued

SOLUTION

Conceptualize Think about what happens when the dielectric is inserted between the plates. Because the battery has been removed, the charge on the capacitor must remain the same. We know from our earlier discussion, however, that the capacitance must change. Therefore, we expect a change in the energy of the system.

Categorize Because we expect the energy of the system to change, we model it as a *nonisolated system* for *energy* involving a capacitor and a dielectric.

Analyze From Equation 26.11, find the energy stored in the absence of the dielectric:

$$U_0 = \frac{Q_0^2}{2C_0}$$

Find the energy stored in the capacitor after the dielectric is inserted between the plates:

$$U = \frac{Q_0^2}{2C}$$

Use Equation 26.14 to replace the capacitance C:

$$U = \frac{Q_0^2}{2\kappa C_0} = \frac{U_0}{\kappa}$$

Finalize Because $\kappa > 1$, the final energy is less than the initial energy. We can account for the decrease in energy of the system by performing an experiment and noting that the dielectric, when inserted, is pulled into the device. To keep the dielectric from accelerating, an external agent must do negative work on the dielectric. Equation 8.2 becomes $\Delta U = W$, where both sides of the equation are negative.

26.6 Electric Dipole in an Electric Field

We have discussed the effect on the capacitance of placing a dielectric between the plates of a capacitor. In Section 26.7, we shall describe the microscopic origin of this effect. Before we can do so, however, let's expand the discussion of the electric dipole introduced in Section 23.4 (see Example 23.6). The electric dipole consists of two charges of equal magnitude and opposite sign separated by a distance $2a$ as shown in Figure 26.17. The **electric dipole moment** of this configuration is defined as the vector $\vec{\mathbf{p}}$ directed from $-q$ toward $+q$ along the line joining the charges and having magnitude

$$p \equiv 2aq \tag{26.16}$$

Now suppose an electric dipole is placed in a uniform electric field $\vec{\mathbf{E}}$ and makes an angle θ with the field as shown in Figure 26.18. We identify $\vec{\mathbf{E}}$ as the field *external* to the dipole, established by some other charge distribution, to distinguish it from the field *due to* the dipole, which we discussed in Section 23.4.

Each of the charges is modeled as a particle in an electric field. The electric forces acting on the two charges are equal in magnitude ($F = qE$) and opposite in direction as shown in Figure 26.18. Therefore, the net force on the dipole is zero. The two forces produce a net torque on the dipole, however; the dipole is therefore described by the rigid object under a net torque model. As a result, the dipole rotates in the direction that brings the dipole moment vector into greater alignment with the field. The torque due to the force on the positive charge about an axis through O in Figure 26.18 has magnitude $Fa \sin \theta$, where $a \sin \theta$ is the moment arm of F about O. This force tends to produce a clockwise rotation. The torque about O on the negative charge is also of magnitude $Fa \sin \theta$; here again, the force tends to produce a clockwise rotation. Therefore, the magnitude of the net torque about O is

$$\tau = 2Fa \sin \theta$$

Because $F = qE$ and $p = 2aq$, we can express τ as

$$\tau = 2aqE \sin \theta = pE \sin \theta \tag{26.17}$$

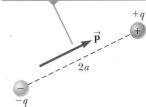

The electric dipole moment $\vec{\mathbf{p}}$ is directed from $-q$ toward $+q$.

Figure 26.17 An electric dipole consists of two charges of equal magnitude and opposite sign separated by a distance of $2a$.

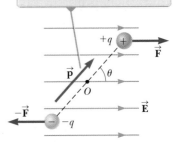

The dipole moment $\vec{\mathbf{p}}$ is at an angle θ to the field, causing the dipole to experience a torque.

Figure 26.18 An electric dipole in a uniform external electric field.

Based on this expression, it is convenient to express the torque in vector form as the cross product of the vectors $\vec{p}$ and $\vec{E}$:

Torque on an electric dipole ▶
in an external electric field

$$\vec{\tau} = \vec{p} \times \vec{E} \qquad (26.18)$$

We can also model the system of the dipole and the external electric field as an isolated system for energy. Let's determine the potential energy of the system as a function of the dipole's orientation with respect to the field. To do so, recognize that work must be done by an external agent to rotate the dipole through an angle so as to cause the dipole moment vector to become less aligned with the field. The work done is then stored as electric potential energy in the system. Notice that this potential energy is associated with a *rotational* configuration of the system. Previously, we have seen potential energies associated with *translational* configurations: an object with mass was moved in a gravitational field, a charge was moved in an electric field, or a spring was extended. The work dW required to rotate the dipole through an angle $d\theta$ is $dW = \tau \, d\theta$ (see Eq. 10.25). Because $\tau = pE \sin \theta$ and the work results in an increase in the electric potential energy U, we find that for a rotation from θ_i to θ_f, the change in potential energy of the system is

$$U_f - U_i = \int_{\theta_i}^{\theta_f} \tau \, d\theta = \int_{\theta_i}^{\theta_f} pE \sin \theta \, d\theta = pE \int_{\theta_i}^{\theta_f} \sin \theta \, d\theta$$

$$= pE\left[-\cos \theta\right]_{\theta_i}^{\theta_f} = pE(\cos \theta_i - \cos \theta_f)$$

The term that contains $\cos \theta_i$ is a constant that depends on the initial orientation of the dipole. It is convenient to choose a reference angle of $\theta_i = 90°$ so that $\cos \theta_i = \cos 90° = 0$. Furthermore, let's choose $U_i = 0$ at $\theta_i = 90°$ as our reference value of potential energy. Hence, we can express a general value of $U_E = U_f$ as

$$U_E = -pE \cos \theta \qquad (26.19)$$

We can write this expression for the potential energy of a dipole in an electric field as the dot product of the vectors $\vec{p}$ and $\vec{E}$:

Potential energy of the ▶
system of an electric dipole
in an external electric field

$$\boxed{U_E = -\vec{p} \cdot \vec{E}} \qquad (26.20)$$

To develop a conceptual understanding of Equation 26.19, compare it with the expression for the potential energy of the system of an object in the Earth's gravitational field, $U_g = mgy$ (Eq. 7.19). First, both expressions contain a parameter of the entity placed in the field: mass for the object, dipole moment for the dipole. Second, both expressions contain the field, g for the object, E for the dipole. Finally, both expressions contain a configuration description: translational position y for the object, rotational position θ for the dipole. In both cases, once the configuration is changed, the system tends to return to the original configuration when the object is released: the object of mass m falls toward the ground, and the dipole begins to rotate back toward the configuration in which it is aligned with the field.

Molecules are said to be *polarized* when a separation exists between the average position of the negative charges and the average position of the positive charges in the molecule. In some molecules such as water, this condition is always present; such molecules are called **polar molecules.** Molecules that do not possess a permanent polarization are called **nonpolar molecules.**

We can understand the permanent polarization of water by inspecting the geometry of the water molecule. The oxygen atom in the water molecule is bonded to the hydrogen atoms such that an angle of 105° is formed between the two bonds (Fig. 26.19). The center of the negative charge distribution is near the oxygen atom, and the center of the positive charge distribution lies at a point midway along the line joining the hydrogen atoms (the point labeled ✕ in Fig. 26.19). We can model the water molecule and other polar molecules as dipoles because the average positions of the positive and negative charges act as point charges. As a result, we can apply our discussion of dipoles to the behavior of polar molecules.

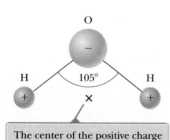

The center of the positive charge distribution is at the point ✕.

Figure 26.19 The water molecule, H_2O, has a permanent polarization resulting from its nonlinear geometry.

Washing with soap and water is a household scenario in which the dipole structure of water is exploited. Grease and oil are made up of nonpolar molecules, which are generally not attracted to water. Plain water is not very useful for removing this type of grime. Soap contains long molecules called *surfactants*. In a long molecule, the polarity characteristics of one end of the molecule can be different from those at the other end. In a surfactant molecule, one end acts like a nonpolar molecule and the other acts like a polar molecule. The nonpolar end can attach to a grease or oil molecule, and the polar end can attach to a water molecule. Therefore, the soap serves as a chain, linking the dirt and water molecules together. When the water is rinsed away, the grease and oil go with it.

A symmetric molecule (Fig. 26.20a) has no permanent polarization, but polarization can be induced by placing the molecule in an electric field. A field directed to the left as in Figure 26.20b causes the center of the negative charge distribution to shift to the right relative to the positive charges. This *induced polarization* is the effect that predominates in most materials used as dielectrics in capacitors.

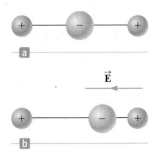

Figure 26.20 (a) A linear symmetric molecule has no permanent polarization. (b) An external electric field induces a polarization in the molecule.

Example 26.6 **The H$_2$O Molecule** AM

The water (H$_2$O) molecule has an electric dipole moment of 6.3×10^{-30} C · m. A sample contains 10^{21} water molecules, with the dipole moments all oriented in the direction of an electric field of magnitude 2.5×10^{5} N/C. How much work is required to rotate the dipoles from this orientation ($\theta = 0°$) to one in which all the moments are perpendicular to the field ($\theta = 90°$)?

SOLUTION

Conceptualize When all the dipoles are aligned with the electric field, the dipoles–electric field system has the minimum potential energy. This energy has a negative value given by the product of the right side of Equation 26.19, evaluated at 0°, and the number N of dipoles.

Categorize The combination of the dipoles and the electric field is identified as a system. We use the *nonisolated system* model because an external agent performs work on the system to change its potential energy.

....

Analyze Write the appropriate reduction of the conservation of energy equation, Equation 8.2, for this situation:

(1) $\Delta U_E = W$

Use Equation 26.19 to evaluate the initial and final potential energies of the system and Equation (1) to calculate the work required to rotate the dipoles:

$W = U_{90°} - U_{0°} = (-NpE \cos 90°) - (-NpE \cos 0°)$

$= NpE = (10^{21})(6.3 \times 10^{-30} \text{ C} \cdot \text{m})(2.5 \times 10^{5} \text{ N/C})$

$= \boxed{1.6 \times 10^{-3} \text{ J}}$

....

Finalize Notice that the work done on the system is positive because the potential energy of the system has been raised from a negative value to a value of zero.

26.7 An Atomic Description of Dielectrics

In Section 26.5, we found that the potential difference ΔV_0 between the plates of a capacitor is reduced to $\Delta V_0 / \kappa$ when a dielectric is introduced. The potential difference is reduced because the magnitude of the electric field decreases between the plates. In particular, if $\vec{\mathbf{E}}_0$ is the electric field without the dielectric, the field in the presence of a dielectric is

$$\vec{\mathbf{E}} = \frac{\vec{\mathbf{E}}_0}{\kappa} \tag{26.21}$$

First consider a dielectric made up of polar molecules placed in the electric field between the plates of a capacitor. The dipoles (that is, the polar molecules making

Figure 26.21 (a) Polar molecules in a dielectric. (b) An electric field is applied to the dielectric. (c) Details of the electric field inside the dielectric.

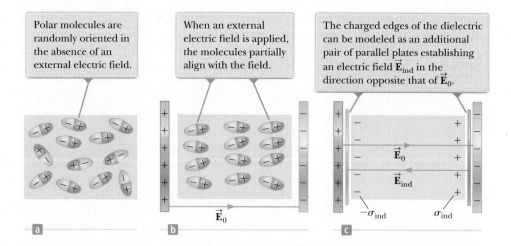

Polar molecules are randomly oriented in the absence of an external electric field.

When an external electric field is applied, the molecules partially align with the field.

The charged edges of the dielectric can be modeled as an additional pair of parallel plates establishing an electric field $\vec{E}_{ind}$ in the direction opposite that of $\vec{E}_0$.

up the dielectric) are randomly oriented in the absence of an electric field as shown in Figure 26.21a. When an external field $\vec{E}_0$ due to charges on the capacitor plates is applied, a torque is exerted on the dipoles, causing them to partially align with the field as shown in Figure 26.21b. The dielectric is now polarized. The degree of alignment of the molecules with the electric field depends on temperature and the magnitude of the field. In general, the alignment increases with decreasing temperature and with increasing electric field.

If the molecules of the dielectric are nonpolar, the electric field due to the plates produces an induced polarization in the molecule. These induced dipole moments tend to align with the external field, and the dielectric is polarized. Therefore, a dielectric can be polarized by an external field regardless of whether the molecules in the dielectric are polar or nonpolar.

With these ideas in mind, consider a slab of dielectric material placed between the plates of a capacitor so that it is in a uniform electric field $\vec{E}_0$ as shown in Figure 26.21b. The electric field due to the plates is directed to the right and polarizes the dielectric. The net effect on the dielectric is the formation of an *induced* positive surface charge density σ_{ind} on the right face and an equal-magnitude negative surface charge density $-\sigma_{ind}$ on the left face as shown in Figure 26.21c. Because we can model these surface charge distributions as being due to charged parallel plates, the induced surface charges on the dielectric give rise to an induced electric field $\vec{E}_{ind}$ in the direction opposite the external field $\vec{E}_0$. Therefore, the net electric field $\vec{E}$ in the dielectric has a magnitude

$$E = E_0 - E_{ind} \qquad (26.22)$$

In the parallel-plate capacitor shown in Figure 26.22, the external field E_0 is related to the charge density σ on the plates through the relationship $E_0 = \sigma/\epsilon_0$. The induced electric field in the dielectric is related to the induced charge density σ_{ind} through the relationship $E_{ind} = \sigma_{ind}/\epsilon_0$. Because $E = E_0/\kappa = \sigma/\kappa\epsilon_0$, substitution into Equation 26.22 gives

$$\frac{\sigma}{\kappa\epsilon_0} = \frac{\sigma}{\epsilon_0} - \frac{\sigma_{ind}}{\epsilon_0}$$

$$\sigma_{ind} = \left(\frac{\kappa - 1}{\kappa}\right)\sigma \qquad (26.23)$$

The induced charge density σ_{ind} on the dielectric is *less* than the charge density σ on the plates.

Figure 26.22 Induced charge on a dielectric placed between the plates of a charged capacitor.

Because $\kappa > 1$, this expression shows that the charge density σ_{ind} induced on the dielectric is less than the charge density σ on the plates. For instance, if $\kappa = 3$, the induced charge density is two-thirds the charge density on the plates. If no dielectric is present, then $\kappa = 1$ and $\sigma_{ind} = 0$ as expected. If the dielectric is replaced by an electrical conductor for which $E = 0$, however, Equation 26.22 indicates that $E_0 = E_{ind}$, which corresponds to $\sigma_{ind} = \sigma$. That is, the surface charge induced on the conductor is equal in magnitude but opposite in sign to that on the plates, resulting in a net electric field of zero in the conductor (see Fig. 24.16).

Example 26.7 Effect of a Metallic Slab

A parallel-plate capacitor has a plate separation d and plate area A. An uncharged metallic slab of thickness a is inserted midway between the plates.

(A) Find the capacitance of the device.

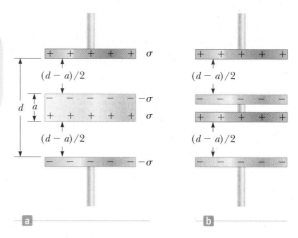

Figure 26.23 (Example 26.7) (a) A parallel-plate capacitor of plate separation d partially filled with a metallic slab of thickness a. (b) The equivalent circuit of the device in (a) consists of two capacitors in series, each having a plate separation $(d - a)/2$.

SOLUTION

Conceptualize Figure 26.23a shows the metallic slab between the plates of the capacitor. Any charge that appears on one plate of the capacitor must induce a charge of equal magnitude and opposite sign on the near side of the slab as shown in Figure 26.23a. Consequently, the net charge on the slab remains zero and the electric field inside the slab is zero.

Categorize The planes of charge on the metallic slab's upper and lower edges are identical to the distribution of charges on the plates of a capacitor. The metal between the slab's edges serves only to make an electrical connection between the edges. Therefore, we can model the edges of the slab as conducting planes and the bulk of the slab as a wire. As a result, the capacitor in Figure 26.23a is equivalent to two capacitors in series, each having a plate separation $(d - a)/2$ as shown in Figure 26.23b.

Analyze Use Equation 26.3 and the rule for adding two capacitors in series (Eq. 26.10) to find the equivalent capacitance in Figure 26.23b:

$$\frac{1}{C} = \frac{1}{C_1} + \frac{1}{C_2} = \frac{1}{\dfrac{\epsilon_0 A}{(d - a)/2}} + \frac{1}{\dfrac{\epsilon_0 A}{(d - a)/2}}$$

$$C = \frac{\epsilon_0 A}{d - a}$$

(B) Show that the capacitance of the original capacitor is unaffected by the insertion of the metallic slab if the slab is infinitesimally thin.

SOLUTION

In the result for part (A), let $a \to 0$:

$$C = \lim_{a \to 0} \left(\frac{\epsilon_0 A}{d - a} \right) = \frac{\epsilon_0 A}{d}$$

Finalize The result of part (B) is the original capacitance before the slab is inserted, which tells us that we can insert an infinitesimally thin metallic sheet between the plates of a capacitor without affecting the capacitance. We use this fact in the next example.

WHAT IF? What if the metallic slab in part (A) is not midway between the plates? How would that affect the capacitance?

Answer Let's imagine moving the slab in Figure 26.23a upward so that the distance between the upper edge of the slab and the upper plate is b. Then, the distance between the lower edge of the slab and the lower plate is $d - b - a$. As in part (A), we find the total capacitance of the series combination:

$$\frac{1}{C} = \frac{1}{C_1} + \frac{1}{C_2} = \frac{1}{\epsilon_0 A / b} + \frac{1}{\epsilon_0 A / (d - b - a)}$$

$$= \frac{b}{\epsilon_0 A} + \frac{d - b - a}{\epsilon_0 A} = \frac{d - a}{\epsilon_0 A} \quad \to \quad C = \frac{\epsilon_0 A}{d - a}$$

which is the same result as found in part (A). The capacitance is independent of the value of b, so it does not matter where the slab is located. In Figure 26.23b, when the central structure is moved up or down, the decrease in plate separation of one capacitor is compensated by the increase in plate separation for the other.

Example 26.8   A Partially Filled Capacitor

A parallel-plate capacitor with a plate separation d has a capacitance C_0 in the absence of a dielectric. What is the capacitance when a slab of dielectric material of dielectric constant κ and thickness fd is inserted between the plates (Fig. 26.24a), where f is a fraction between 0 and 1?

SOLUTION

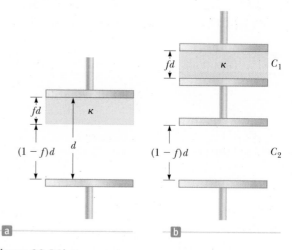

Conceptualize In our previous discussions of dielectrics between the plates of a capacitor, the dielectric filled the volume between the plates. In this example, only part of the volume between the plates contains the dielectric material.

Categorize In Example 26.7, we found that an infinitesimally thin metallic sheet inserted between the plates of a capacitor does not affect the capacitance. Imagine sliding an infinitesimally thin metallic slab along the bottom face of the dielectric shown in Figure 26.24a. We can model this system as a series combination of two capacitors as shown in Figure 26.24b. One capacitor has a plate separation fd and is filled with a dielectric; the other has a plate separation $(1 - f)d$ and has air between its plates.

Figure 26.24 (Example 26.8) (a) A parallel-plate capacitor of plate separation d partially filled with a dielectric of thickness fd. (b) The equivalent circuit of the capacitor consists of two capacitors connected in series.

Analyze Evaluate the two capacitances in Figure 26.24b from Equation 26.15:

$$C_1 = \frac{\kappa \epsilon_0 A}{fd} \quad \text{and} \quad C_2 = \frac{\epsilon_0 A}{(1 - f)d}$$

Find the equivalent capacitance C from Equation 26.10 for two capacitors combined in series:

$$\frac{1}{C} = \frac{1}{C_1} + \frac{1}{C_2} = \frac{fd}{\kappa \epsilon_0 A} + \frac{(1 - f)d}{\epsilon_0 A}$$

$$\frac{1}{C} = \frac{fd}{\kappa \epsilon_0 A} + \frac{\kappa(1 - f)d}{\kappa \epsilon_0 A} = \frac{f + \kappa(1 - f)}{\kappa} \frac{d}{\epsilon_0 A}$$

Invert and substitute for the capacitance without the dielectric, $C_0 = \epsilon_0 A/d$:

$$C = \frac{\kappa}{f + \kappa(1 - f)} \frac{\epsilon_0 A}{d} = \frac{\kappa}{f + \kappa(1 - f)} C_0$$

Finalize Let's test this result for some known limits. If $f \rightarrow 0$, the dielectric should disappear. In this limit, $C \rightarrow C_0$, which is consistent with a capacitor with air between the plates. If $f \rightarrow 1$, the dielectric fills the volume between the plates. In this limit, $C \rightarrow \kappa C_0$, which is consistent with Equation 26.14.

Summary

Definitions

A **capacitor** consists of two conductors carrying charges of equal magnitude and opposite sign. The **capacitance** C of any capacitor is the ratio of the charge Q on either conductor to the potential difference ΔV between them:

$$C \equiv \frac{Q}{\Delta V} \quad \text{(26.1)}$$

The capacitance depends only on the geometry of the conductors and not on an external source of charge or potential difference. The SI unit of capacitance is coulombs per volt, or the **farad** (F): 1 F = 1 C/V.

The **electric dipole moment** $\vec{p}$ of an electric dipole has a magnitude

$$p \equiv 2aq \quad \text{(26.16)}$$

where $2a$ is the distance between the charges q and $-q$. The direction of the electric dipole moment vector is from the negative charge toward the positive charge.

Concepts and Principles

If two or more capacitors are connected in parallel, the potential difference is the same across all capacitors. The equivalent capacitance of a **parallel combination** of capacitors is

$$C_{eq} = C_1 + C_2 + C_3 + \cdots \qquad \textbf{(26.8)}$$

If two or more capacitors are connected in series, the charge is the same on all capacitors, and the equivalent capacitance of the **series combination** is given by

$$\frac{1}{C_{eq}} = \frac{1}{C_1} + \frac{1}{C_2} + \frac{1}{C_3} + \cdots \qquad \textbf{(26.10)}$$

These two equations enable you to simplify many electric circuits by replacing multiple capacitors with a single equivalent capacitance.

Energy is stored in a charged capacitor because the charging process is equivalent to the transfer of charges from one conductor at a lower electric potential to another conductor at a higher potential. The energy stored in a capacitor of capacitance C with charge Q and potential difference ΔV is

$$U_E = \frac{Q^2}{2C} = \tfrac{1}{2} Q \Delta V = \tfrac{1}{2} C (\Delta V)^2 \qquad \textbf{(26.11)}$$

When a dielectric material is inserted between the plates of a capacitor, the capacitance increases by a dimensionless factor κ, called the **dielectric constant:**

$$C = \kappa C_0 \qquad \textbf{(26.14)}$$

where C_0 is the capacitance in the absence of the dielectric.

The torque acting on an electric dipole in a uniform electric field $\vec{\mathbf{E}}$ is

$$\vec{\boldsymbol{\tau}} = \vec{\mathbf{p}} \times \vec{\mathbf{E}} \qquad \textbf{(26.18)}$$

The potential energy of the system of an electric dipole in a uniform external electric field $\vec{\mathbf{E}}$ is

$$U_E = -\vec{\mathbf{p}} \cdot \vec{\mathbf{E}} \qquad \textbf{(26.20)}$$

Objective Questions

1. denotes answer available in *Student Solutions Manual/Study Guide*

1. A fully charged parallel-plate capacitor remains connected to a battery while you slide a dielectric between the plates. Do the following quantities (a) increase, (b) decrease, or (c) stay the same? (i) C (ii) Q (iii) ΔV (iv) the energy stored in the capacitor

2. By what factor is the capacitance of a metal sphere multiplied if its volume is tripled? (a) 3 (b) $3^{1/3}$ (c) 1 (d) $3^{-1/3}$ (e) $\tfrac{1}{3}$

3. An electronics technician wishes to construct a parallel-plate capacitor using rutile ($\kappa = 100$) as the dielectric. The area of the plates is 1.00 cm². What is the capacitance if the rutile thickness is 1.00 mm? (a) 88.5 pF (b) 177 pF (c) 8.85 μF (d) 100 μF (e) 35.4 μF

4. A parallel-plate capacitor is connected to a battery. What happens to the stored energy if the plate separation is doubled while the capacitor remains connected to the battery? (a) It remains the same. (b) It is doubled. (c) It decreases by a factor of 2. (d) It decreases by a factor of 4. (e) It increases by a factor of 4.

5. If three unequal capacitors, initially uncharged, are connected in series across a battery, which of the following statements is true? (a) The equivalent capacitance is greater than any of the individual capacitances. (b) The largest voltage appears across the smallest capacitance. (c) The largest voltage appears across the largest capacitance. (d) The capacitor with the largest capacitance has the greatest charge. (e) The capacitor with the smallest capacitance has the smallest charge.

6. Assume a device is designed to obtain a large potential difference by first charging a bank of capacitors connected in parallel and then activating a switch arrangement that in effect disconnects the capacitors from the charging source and from each other and reconnects them all in a series arrangement. The group of charged capacitors is then discharged in series. What is the maximum potential difference that can be obtained in this manner by using ten 500-μF capacitors and an 800-V charging source? (a) 500 V (b) 8.00 kV (c) 400 kV (d) 800 V (e) 0

7. (i) What happens to the magnitude of the charge on each plate of a capacitor if the potential difference between the conductors is doubled? (a) It becomes four times larger. (b) It becomes two times larger. (c) It is unchanged. (d) It becomes one-half as large. (e) It becomes one-fourth as large. (ii) If the potential difference across a capacitor is doubled, what happens to the energy stored? Choose from the same possibilities as in part (i).

8. A capacitor with very large capacitance is in series with another capacitor with very small capacitance. What is the equivalent capacitance of the combination? (a) slightly greater than the capacitance of the large capacitor (b) slightly less than the capacitance of the large capacitor (c) slightly greater than the capacitance of the small capacitor (d) slightly less than the capacitance of the small capacitor

9. A parallel-plate capacitor filled with air carries a charge Q. The battery is disconnected, and a slab of material with dielectric constant $\kappa = 2$ is inserted between the plates. Which of the following statements is true? (a) The voltage across the capacitor decreases by a factor of 2. (b) The voltage across the capacitor is doubled. (c) The charge on the plates is doubled. (d) The charge on the plates decreases by a factor of 2. (e) The electric field is doubled.

10. **(i)** A battery is attached to several different capacitors connected in parallel. Which of the following statements is true? (a) All capacitors have the same charge, and the equivalent capacitance is greater than the capacitance of any of the capacitors in the group. (b) The capacitor with the largest capacitance carries the smallest charge. (c) The potential difference across each capacitor is the same, and the equivalent capacitance is greater than any of the capacitors in the group. (d) The capacitor with the smallest capacitance carries the largest charge. (e) The potential differences across the capacitors are the same only if the capacitances are the same. **(ii)** The capacitors are reconnected in series, and the combination is again connected to the battery. From the same choices, choose the one that is true.

11. A parallel-plate capacitor is charged and then is disconnected from the battery. By what factor does the stored energy change when the plate separation is then doubled? (a) It becomes four times larger. (b) It becomes two times larger. (c) It stays the same. (d) It becomes one-half as large. (e) It becomes one-fourth as large.

12. **(i)** Rank the following five capacitors from greatest to smallest capacitance, noting any cases of equality. (a) a 20-μF capacitor with a 4-V potential difference between its plates (b) a 30-μF capacitor with charges of magnitude 90 μC on each plate (c) a capacitor with charges of magnitude 80 μC on its plates, differing by 2 V in potential, (d) a 10-μF capacitor storing energy 125 μJ (e) a capacitor storing energy 250 μJ with a 10-V potential difference **(ii)** Rank the same capacitors in part (i) from largest to smallest according to the potential difference between the plates. **(iii)** Rank the capacitors in part (i) in the order of the magnitudes of the charges on their plates. **(iv)** Rank the capacitors in part (i) in the order of the energy they store.

13. True or False? (a) From the definition of capacitance $C = Q/\Delta V$, it follows that an uncharged capacitor has a capacitance of zero. (b) As described by the definition of capacitance, the potential difference across an uncharged capacitor is zero.

14. You charge a parallel-plate capacitor, remove it from the battery, and prevent the wires connected to the plates from touching each other. When you increase the plate separation, do the following quantities (a) increase, (b) decrease, or (c) stay the same? **(i)** C **(ii)** Q **(iii)** E between the plates **(iv)** ΔV

Conceptual Questions

1. denotes answer available in *Student Solutions Manual/Study Guide*

1. (a) Why is it dangerous to touch the terminals of a high-voltage capacitor even after the voltage source that charged the capacitor is disconnected from the capacitor? (b) What can be done to make the capacitor safe to handle after the voltage source has been removed?

2. Assume you want to increase the maximum operating voltage of a parallel-plate capacitor. Describe how you can do that with a fixed plate separation.

3. If you were asked to design a capacitor in which small size and large capacitance were required, what would be the two most important factors in your design?

4. Explain why a dielectric increases the maximum operating voltage of a capacitor even though the physical size of the capacitor doesn't change.

5. Explain why the work needed to move a particle with charge Q through a potential difference ΔV is $W = Q\Delta V$, whereas the energy stored in a charged capacitor is $U_E = \frac{1}{2}Q\Delta V$. Where does the factor $\frac{1}{2}$ come from?

6. An air-filled capacitor is charged, then disconnected from the power supply, and finally connected to a voltmeter. Explain how and why the potential difference changes when a dielectric is inserted between the plates of the capacitor.

7. The sum of the charges on both plates of a capacitor is zero. What does a capacitor store?

8. Because the charges on the plates of a parallel-plate capacitor are opposite in sign, they attract each other. Hence, it would take positive work to increase the plate separation. What type of energy in the system changes due to the external work done in this process?

Section 26.1 Definition of Capacitance
Problems 1–3

Section 26.2 Calculating Capacitance
Problems 4–12

Section 26.3 Combinations of Capacitors
Problems 13–29

Section 26.4 Energy Stored in a Charged Capacitor
Problems 30–41

Section 26.5 Capacitors with Dielectrics
Problems 42–49

Section 26.6 Electric Dipole in an Electric Field
Problems 50–52

Section 26.7 An Atomic Description of Dielectrics
Problem 53

Additional Problems
Problems 54–71

Challenge Problems
Problems 72–78

Solutions to the following Problems are available in the *Student Solutions Manual/Study Guide:*
26.4, 26.5, 26.7, 26.9, 26.11, 26.21, 26.23, 26.24, 26.38, 26.46, 26.50, 26.59, 26.67, 26.68, 26.72, and 26.78

List of Enhanced Problems

Problem Number	Targeted Feedback in Enhanced WebAssign	Analysis Model Tutorial in Enhanced WebAssign	Master It in Enhanced WebAssign	Watch It in Enhanced WebAssign
26.2	✓			
26.3	✓			
26.4			✓	
26.5			✓	
26.6	✓			✓
26.9	✓		✓	
26.13	✓			✓
26.14	✓			✓
26.21			✓	
26.22	✓			✓
26.23	✓		✓	
26.24	✓		✓	
26.32	✓			✓
26.39		✓		
26.41		✓		
26.43	✓			✓
26.45	✓			✓
26.49		✓		
26.50			✓	
26.51	✓			
26.59		✓		

Current and Resistance

These two lightbulbs provide similar power output by visible light (electromagnetic radiation). The compact fluorescent bulb on the left, however, produces this light output with far less input by electrical transmission than the incandescent bulb on the right. The fluorescent bulb, therefore, is less costly to operate and saves valuable resources needed to generate electricity. *(Christina Richards/ Shutterstock.com)*

We now consider situations involving electric charges that are in motion through some region of space. We use the term *electric current*, or simply *current*, to describe the rate of flow of charge. Most practical applications of electricity deal with electric currents, including a variety of home appliances. For example, the voltage from a wall plug produces a current in the coils of a toaster when it is turned on. In these common situations, current exists in a conductor such as a copper wire. Currents can also exist outside a conductor. For instance, a beam of electrons in a particle accelerator constitutes a current.

This chapter begins with the definition of current. A microscopic description of current is given, and some factors that contribute to the opposition to the flow of charge in conductors are discussed. A classical model is used to describe electrical conduction in metals, and some limitations of this model are cited. We also define electrical resistance and introduce a new circuit element, the resistor. We conclude by discussing the rate at which energy is transferred to a device in an electric circuit. The energy transfer mechanism in Equation 8.2 that corresponds to this process is electrical transmission T_{ET}.

27.1 Electric Current

In this section, we study the flow of electric charges through a piece of material. The amount of flow depends on both the material through which the charges are passing and the potential difference across the material. Whenever there is a net flow of charge through some region, an electric *current* is said to exist.

It is instructive to draw an analogy between water flow and current. The flow of water in a plumbing pipe can be quantified by specifying the amount of water that emerges from a faucet during a given time interval, often measured in liters per minute. A river current can be characterized by describing the rate at which the water flows past a particular location. For example, the flow over the brink at Niagara Falls is maintained at rates between 1 400 m³/s and 2 800 m³/s.

There is also an analogy between thermal conduction and current. In Section 20.7, we discussed the flow of energy by heat through a sample of material. The rate of energy flow is determined by the material as well as the temperature difference across the material as described by Equation 20.15.

To define current quantitatively, suppose charges are moving perpendicular to a surface of area A as shown in Figure 27.1. (This area could be the cross-sectional area of a wire, for example.) The **current** is defined as the rate at which charge flows through this surface. If ΔQ is the amount of charge that passes through this surface in a time interval Δt, the **average current** I_{avg} is equal to the charge that passes through A per unit time:

$$I_{avg} = \frac{\Delta Q}{\Delta t} \qquad (27.1)$$

If the rate at which charge flows varies in time, the current varies in time; we define the **instantaneous current** I as the limit of the average current as $\Delta t \to 0$:

$$I \equiv \frac{dQ}{dt} \qquad (27.2)$$ ◀ **Electric current**

The SI unit of current is the **ampere** (A):

$$1 \text{ A} = 1 \text{ C/s} \qquad (27.3)$$

That is, 1 A of current is equivalent to 1 C of charge passing through a surface in 1 s.

The charged particles passing through the surface in Figure 27.1 can be positive, negative, or both. It is conventional to assign to the current the same direction as the flow of positive charge. In electrical conductors such as copper or aluminum, the current results from the motion of negatively charged electrons. Therefore, in an ordinary conductor, the direction of the current is opposite the direction of flow of electrons. For a beam of positively charged protons in an accelerator, however, the current is in the direction of motion of the protons. In some cases—such as those involving gases and electrolytes, for instance—the current is the result of the flow of both positive and negative charges. It is common to refer to a moving charge (positive or negative) as a mobile **charge carrier.**

If the ends of a conducting wire are connected to form a loop, all points on the loop are at the same electric potential; hence, the electric field is zero within and at the surface of the conductor. Because the electric field is zero, there is no net transport of charge through the wire; therefore, there is no current. If the ends of the conducting wire are connected to a battery, however, all points on the loop are not at the same potential. The battery sets up a potential difference between the ends of the loop, creating an electric field within the wire. The electric field exerts forces on the electrons in the wire, causing them to move in the wire and therefore creating a current.

Microscopic Model of Current

We can relate current to the motion of the charge carriers by describing a microscopic model of conduction in a metal. Consider the current in a cylindrical

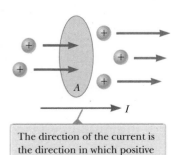

The direction of the current is the direction in which positive charges flow when free to do so.

Figure 27.1 Charges in motion through an area A. The time rate at which charge flows through the area is defined as the current I.

Pitfall Prevention 27.1

"Current Flow" Is Redundant
The phrase *current flow* is commonly used, although it is technically incorrect because current *is* a flow (of charge). This wording is similar to the phrase *heat transfer*, which is also redundant because heat *is* a transfer (of energy). We will avoid this phrase and speak of *flow of charge* or *charge flow.*

Pitfall Prevention 27.2

Batteries Do Not Supply Electrons
A battery does not supply electrons to the circuit. It establishes the electric field that exerts a force on electrons already in the wires and elements of the circuit.

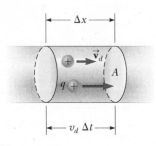

Figure 27.2 A segment of a uniform conductor of cross-sectional area A.

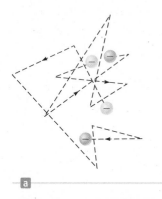

a

The random motion of the charge carriers is modified by the field, and they have a drift velocity opposite the direction of the electric field.

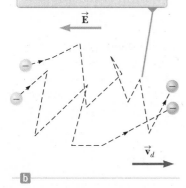

b

Figure 27.3 (a) A schematic diagram of the random motion of two charge carriers in a conductor in the absence of an electric field. The drift velocity is zero. (b) The motion of the charge carriers in a conductor in the presence of an electric field. Because of the acceleration of the charge carriers due to the electric force, the paths are actually parabolic. The drift speed, however, is much smaller than the average speed, so the parabolic shape is not visible on this scale.

conductor of cross-sectional area A (Fig. 27.2). The volume of a segment of the conductor of length Δx (between the two circular cross sections shown in Fig. 27.2) is $A\,\Delta x$. If n represents the number of mobile charge carriers per unit volume (in other words, the charge carrier density), the number of carriers in the segment is $nA\,\Delta x$. Therefore, the total charge ΔQ in this segment is

$$\Delta Q = (nA\,\Delta x)q$$

where q is the charge on each carrier. If the carriers move with a velocity $\vec{v}_d$ parallel to the axis of the cylinder, the magnitude of the displacement they experience in the x direction in a time interval Δt is $\Delta x = v_d\,\Delta t$. Let Δt be the time interval required for the charge carriers in the segment to move through a displacement whose magnitude is equal to the length of the segment. This time interval is also the same as that required for all the charge carriers in the segment to pass through the circular area at one end. With this choice, we can write ΔQ as

$$\Delta Q = (nAv_d\,\Delta t)q$$

Dividing both sides of this equation by Δt, we find that the average current in the conductor is

$$I_{\text{avg}} = \frac{\Delta Q}{\Delta t} = nqv_dA \qquad \textbf{(27.4)}$$

In reality, the speed of the charge carriers v_d is an average speed called the **drift speed.** To understand the meaning of drift speed, consider a conductor in which the charge carriers are free electrons. If the conductor is isolated—that is, the potential difference across it is zero—these electrons undergo random motion that is analogous to the motion of gas molecules. The electrons collide repeatedly with the metal atoms, and their resultant motion is complicated and zigzagged as in Figure 27.3a. As discussed earlier, when a potential difference is applied across the conductor (for example, by means of a battery), an electric field is set up in the conductor; this field exerts an electric force on the electrons, producing a current. In addition to the zigzag motion due to the collisions with the metal atoms, the electrons move slowly along the conductor (in a direction opposite that of $\vec{E}$) at the **drift velocity** $\vec{v}_d$ as shown in Figure 27.3b.

You can think of the atom–electron collisions in a conductor as an effective internal friction (or drag force) similar to that experienced by a liquid's molecules flowing through a pipe stuffed with steel wool. The energy transferred from the electrons to the metal atoms during collisions causes an increase in the atom's vibrational energy and a corresponding increase in the conductor's temperature.

Quick Quiz 27.1 Consider positive and negative charges moving horizontally through the four regions shown in Figure 27.4. Rank the current in these four regions from highest to lowest.

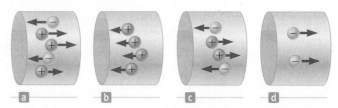

Figure 27.4 (Quick Quiz 27.1) Charges move through four regions.

| Example 27.1 | Drift Speed in a Copper Wire |

The 12-gauge copper wire in a typical residential building has a cross-sectional area of 3.31×10^{-6} m². It carries a constant current of 10.0 A. What is the drift speed of the electrons in the wire? Assume each copper atom contributes one free electron to the current. The density of copper is 8.92 g/cm³.

SOLUTION

Conceptualize Imagine electrons following a zigzag motion such as that in Figure 27.3a, with a drift velocity parallel to the wire superimposed on the motion as in Figure 27.3b. As mentioned earlier, the drift speed is small, and this example helps us quantify the speed.

Categorize We evaluate the drift speed using Equation 27.4. Because the current is constant, the average current during any time interval is the same as the constant current: $I_{avg} = I$.

Analyze The periodic table of the elements in Appendix C shows that the molar mass of copper is $M = 63.5$ g/mol. Recall that 1 mol of any substance contains Avogadro's number of atoms ($N_A = 6.02 \times 10^{23}$ mol⁻¹).

Use the molar mass and the density of copper to find the volume of 1 mole of copper:

$$V = \frac{M}{\rho}$$

From the assumption that each copper atom contributes one free electron to the current, find the electron density in copper:

$$n = \frac{N_A}{V} = \frac{N_A \rho}{M}$$

Solve Equation 27.4 for the drift speed and substitute for the electron density:

$$v_d = \frac{I_{avg}}{nqA} = \frac{I}{nqA} = \frac{IM}{qAN_A\rho}$$

Substitute numerical values:

$$v_d = \frac{(10.0 \text{ A})(0.063\,5 \text{ kg/mol})}{(1.60 \times 10^{-19} \text{ C})(3.31 \times 10^{-6} \text{ m}^2)(6.02 \times 10^{23} \text{ mol}^{-1})(8\,920 \text{ kg/m}^3)}$$

$$= 2.23 \times 10^{-4} \text{ m/s}$$

Finalize This result shows that typical drift speeds are very small. For instance, electrons traveling with a speed of 2.23×10^{-4} m/s would take about 75 min to travel 1 m! You might therefore wonder why a light turns on almost instantaneously when its switch is thrown. In a conductor, changes in the electric field that drives the free electrons according to the particle in a field model travel through the conductor with a speed close to that of light. So, when you flip on a light switch, electrons already in the filament of the lightbulb experience electric forces and begin moving after a time interval on the order of nanoseconds.

27.2 Resistance

In Section 24.4, we argued that the electric field inside a conductor is zero. This statement is true, however, *only* if the conductor is in static equilibrium as stated in that discussion. The purpose of this section is to describe what happens when there is a nonzero electric field in the conductor. As we saw in Section 27.1, a current exists in the wire in this case.

Consider a conductor of cross-sectional area A carrying a current I. The **current density** J in the conductor is defined as the current per unit area. Because the current $I = nqv_dA$, the current density is

$$J \equiv \frac{I}{A} = nqv_d \qquad \qquad \textbf{(27.5)} \quad \blacktriangleleft \text{ Current density}$$

Georg Simon Ohm
German physicist (1789–1854)
Ohm, a high school teacher and later a professor at the University of Munich, formulated the concept of resistance and discovered the proportionalities expressed in Equations 27.6 and 27.7.

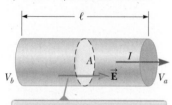

A potential difference $\Delta V = V_b - V_a$ maintained across the conductor sets up an electric field $\vec{E}$, and this field produces a current I that is proportional to the potential difference.

Figure 27.5 A uniform conductor of length ℓ and cross-sectional area A.

Pitfall Prevention 27.3

Equation 27.7 Is Not Ohm's Law
Many individuals call Equation 27.7 Ohm's law, but that is incorrect. This equation is simply the definition of resistance, and it provides an important relationship between voltage, current, and resistance. Ohm's law is related to a proportionality of J to E (Eq. 27.6) or, equivalently, of I to ΔV, which, from Equation 27.7, indicates that the resistance is constant, independent of the applied voltage. We will see some devices for which Equation 27.7 correctly describes their resistance, but that do *not* obey Ohm's law.

where J has SI units of amperes per meter squared. This expression is valid only if the current density is uniform and only if the surface of cross-sectional area A is perpendicular to the direction of the current.

A current density and an electric field are established in a conductor whenever a potential difference is maintained across the conductor. In some materials, the current density is proportional to the electric field:

$$J = \sigma E \qquad (27.6)$$

where the constant of proportionality σ is called the **conductivity** of the conductor.[1] Materials that obey Equation 27.6 are said to follow **Ohm's law,** named after Georg Simon Ohm. More specifically, Ohm's law states the following:

> For many materials (including most metals), the ratio of the current density to the electric field is a constant σ that is independent of the electric field producing the current.

Materials and devices that obey Ohm's law and hence demonstrate this simple relationship between E and J are said to be *ohmic*. Experimentally, however, it is found that not all materials and devices have this property. Those that do not obey Ohm's law are said to be *nonohmic*. Ohm's law is not a fundamental law of nature; rather, it is an empirical relationship valid only for certain situations.

We can obtain an equation useful in practical applications by considering a segment of straight wire of uniform cross-sectional area A and length ℓ as shown in Figure 27.5. A potential difference $\Delta V = V_b - V_a$ is maintained across the wire, creating in the wire an electric field and a current. If the field is assumed to be uniform, the magnitude of the potential difference across the wire is related to the field within the wire through Equation 25.6,

$$\Delta V = E\ell$$

Therefore, we can express the current density (Eq. 27.6) in the wire as

$$J = \sigma \frac{\Delta V}{\ell}$$

Because $J = I/A$, the potential difference across the wire is

$$\Delta V = \frac{\ell}{\sigma} J = \left(\frac{\ell}{\sigma A}\right) I = R I$$

The quantity $R = \ell/\sigma A$ is called the **resistance** of the conductor. We define the resistance as the ratio of the potential difference across a conductor to the current in the conductor:

$$R \equiv \frac{\Delta V}{I} \qquad (27.7)$$

We will use this equation again and again when studying electric circuits. This result shows that resistance has SI units of volts per ampere. One volt per ampere is defined to be one **ohm** (Ω):

$$1\,\Omega \equiv 1\,\text{V/A} \qquad (27.8)$$

Equation 27.7 shows that if a potential difference of 1 V across a conductor causes a current of 1 A, the resistance of the conductor is 1 Ω. For example, if an electrical appliance connected to a 120-V source of potential difference carries a current of 6 A, its resistance is 20 Ω.

Most electric circuits use circuit elements called **resistors** to control the current in the various parts of the circuit. As with capacitors in Chapter 26, many resistors are built into integrated circuit chips, but stand-alone resistors are still available and

---

[1]Do not confuse conductivity σ with surface charge density, for which the same symbol is used.

Color	Number	Multiplier	Tolerance
Black	0	1	
Brown	1	10^1	
Red	2	10^2	
Orange	3	10^3	
Yellow	4	10^4	
Green	5	10^5	
Blue	6	10^6	
Violet	7	10^7	
Gray	8	10^8	
White	9	10^9	
Gold		10^{-1}	5%
Silver		10^{-2}	10%
Colorless			20%

Table 27.1 **Color Coding for Resistors**

The colored bands on this resistor are yellow, violet, black, and gold.

Figure 27.6 A close-up view of a circuit board shows the color coding on a resistor. The gold band on the left tells us that the resistor is oriented "backward" in this view and we need to read the colors from right to left.

dexns/Shutterstock.com

widely used. Two common types are the *composition resistor*, which contains carbon, and the *wire-wound resistor*, which consists of a coil of wire. Values of resistors in ohms are normally indicated by color coding as shown in Figure 27.6 and Table 27.1. The first two colors on a resistor give the first two digits in the resistance value, with the decimal place to the right of the second digit. The third color represents the power of 10 for the multiplier of the resistance value. The last color is the tolerance of the resistance value. As an example, the four colors on the resistor at the bottom of Figure 27.6 are yellow ($= 4$), violet ($= 7$), black ($= 10^0$), and gold ($= 5\%$), and so the resistance value is $47 \times 10^0 = 47\ \Omega$ with a tolerance value of $5\% = 2\ \Omega$.

The inverse of conductivity is **resistivity**[2] ρ:

$$\rho = \frac{1}{\sigma} \tag{27.9}$$

◀ **Resistivity is the inverse of conductivity**

where ρ has the units ohm $\cdot$ meters ($\Omega \cdot \mathrm{m}$). Because $R = \ell/\sigma A$, we can express the resistance of a uniform block of material along the length ℓ as

$$R = \rho \frac{\ell}{A} \tag{27.10}$$

◀ **Resistance of a uniform material along the length ℓ**

Every ohmic material has a characteristic resistivity that depends on the properties of the material and on temperature. In addition, as you can see from Equation 27.10, the resistance of a sample of the material depends on the geometry of the sample as well as on the resistivity of the material. Table 27.2 (page 640) gives the resistivities of a variety of materials at 20°C. Notice the enormous range, from very low values for good conductors such as copper and silver to very high values for good insulators such as glass and rubber. An ideal conductor would have zero resistivity, and an ideal insulator would have infinite resistivity.

Equation 27.10 shows that the resistance of a given cylindrical conductor such as a wire is proportional to its length and inversely proportional to its cross-sectional area. If the length of a wire is doubled, its resistance doubles. If its cross-sectional area is doubled, its resistance decreases by one half. The situation is analogous to the flow of a liquid through a pipe. As the pipe's length is increased, the resistance to flow increases. As the pipe's cross-sectional area is increased, more liquid crosses a given cross section of the pipe per unit time interval. Therefore, more liquid flows for the same pressure differential applied to the pipe, and the resistance to flow decreases.

Ohmic materials and devices have a linear current–potential difference relationship over a broad range of applied potential differences (Fig. 27.7a, page 640). The slope of the I-versus-ΔV curve in the linear region yields a value for $1/R$. Nonohmic

Pitfall Prevention 27.4

Resistance and Resistivity Resistivity is a property of a *substance*, whereas resistance is a property of an *object*. We have seen similar pairs of variables before. For example, density is a property of a substance, whereas mass is a property of an object. Equation 27.10 relates resistance to resistivity, and Equation 1.1 relates mass to density.

---

[2]Do not confuse resistivity ρ with mass density or charge density, for which the same symbol is used.

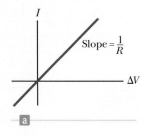

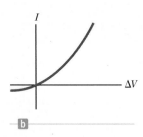

Figure 27.7 (a) The current–potential difference curve for an ohmic material. The curve is linear, and the slope is equal to the inverse of the resistance of the conductor. (b) A nonlinear current–potential difference curve for a junction diode. This device does not obey Ohm's law.

Table 27.2 **Resistivities and Temperature Coefficients of Resistivity for Various Materials**

Material	Resistivity[a] $(\Omega \cdot m)$	Temperature Coefficient[b] α $[(°C)^{-1}]$
Silver	1.59×10^{-8}	3.8×10^{-3}
Copper	1.7×10^{-8}	3.9×10^{-3}
Gold	2.44×10^{-8}	3.4×10^{-3}
Aluminum	2.82×10^{-8}	3.9×10^{-3}
Tungsten	5.6×10^{-8}	4.5×10^{-3}
Iron	10×10^{-8}	5.0×10^{-3}
Platinum	11×10^{-8}	3.92×10^{-3}
Lead	22×10^{-8}	3.9×10^{-3}
Nichrome[c]	1.00×10^{-6}	0.4×10^{-3}
Carbon	3.5×10^{-5}	-0.5×10^{-3}
Germanium	0.46	-48×10^{-3}
Silicon[d]	2.3×10^{3}	-75×10^{-3}
Glass	10^{10} to 10^{14}	
Hard rubber	$\sim 10^{13}$	
Sulfur	10^{15}	
Quartz (fused)	75×10^{16}	

[a] All values at 20°C. All elements in this table are assumed to be free of impurities.
[b] See Section 27.4.
[c] A nickel–chromium alloy commonly used in heating elements. The resistivity of Nichrome varies with composition and ranges between 1.00×10^{-6} and $1.50 \times 10^{-6} \, \Omega \cdot m$.
[d] The resistivity of silicon is very sensitive to purity. The value can be changed by several orders of magnitude when it is doped with other atoms.

materials have a nonlinear current–potential difference relationship. One common semiconducting device with nonlinear *I*-versus-ΔV characteristics is the *junction diode* (Fig. 27.7b). The resistance of this device is low for currents in one direction (positive ΔV) and high for currents in the reverse direction (negative ΔV). In fact, most modern electronic devices, such as transistors, have nonlinear current–potential difference relationships; their proper operation depends on the particular way they violate Ohm's law.

Quick Quiz 27.2 A cylindrical wire has a radius r and length ℓ. If both r and ℓ are doubled, does the resistance of the wire **(a)** increase, **(b)** decrease, or **(c)** remain the same?

Quick Quiz 27.3 In Figure 27.7b, as the applied voltage increases, does the resistance of the diode **(a)** increase, **(b)** decrease, or **(c)** remain the same?

Example 27.2 **The Resistance of Nichrome Wire**

The radius of 22-gauge Nichrome wire is 0.32 mm.

(A) Calculate the resistance per unit length of this wire.

SOLUTION

Conceptualize Table 27.2 shows that Nichrome has a resistivity two orders of magnitude larger than the best conductors in the table. Therefore, we expect it to have some special practical applications that the best conductors may not have.

Categorize We model the wire as a cylinder so that a simple geometric analysis can be applied to find the resistance.

Analyze Use Equation 27.10 and the resistivity of Nichrome from Table 27.2 to find the resistance per unit length:

$$\frac{R}{\ell} = \frac{\rho}{A} = \frac{\rho}{\pi r^2} = \frac{1.0 \times 10^{-6} \, \Omega \cdot m}{\pi (0.32 \times 10^{-3} \, m)^2} = \boxed{3.1 \, \Omega/m}$$

▶ **27.2** continued

(B) If a potential difference of 10 V is maintained across a 1.0-m length of the Nichrome wire, what is the current in the wire?

SOLUTION

Analyze Use Equation 27.7 to find the current:

$$I = \frac{\Delta V}{R} = \frac{\Delta V}{(R/\ell)\ell} = \frac{10\text{ V}}{(3.1\ \Omega/\text{m})(1.0\text{ m})} = \boxed{3.2\text{ A}}$$

Finalize Because of its high resistivity and resistance to oxidation, Nichrome is often used for heating elements in toasters, irons, and electric heaters.

WHAT IF? What if the wire were composed of copper instead of Nichrome? How would the values of the resistance per unit length and the current change?

Answer Table 27.2 shows us that copper has a resistivity two orders of magnitude smaller than that for Nichrome. Therefore, we expect the answer to part (A) to be smaller and the answer to part (B) to be larger. Calculations show that a copper wire of the same radius would have a resistance per unit length of only 0.053 Ω/m. A 1.0-m length of copper wire of the same radius would carry a current of 190 A with an applied potential difference of 10 V.

Example 27.3   **The Radial Resistance of a Coaxial Cable**

Coaxial cables are used extensively for cable television and other electronic applications. A coaxial cable consists of two concentric cylindrical conductors. The region between the conductors is completely filled with polyethylene plastic as shown in Figure 27.8a. Current leakage through the plastic, in the *radial* direction, is unwanted. (The cable is designed to conduct current along its length, but that is *not* the current being considered here.) The radius of the inner conductor is $a = 0.500$ cm, the radius of the outer conductor is $b = 1.75$ cm, and the length is $L = 15.0$ cm. The resistivity of the plastic is $1.0 \times 10^{13}\ \Omega \cdot$ m. Calculate the resistance of the plastic between the two conductors.

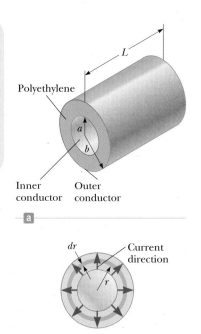

Figure 27.8 (Example 27.3) A coaxial cable. (a) Polyethylene plastic fills the gap between the two conductors. (b) End view, showing current leakage.

SOLUTION

Conceptualize Imagine two currents as suggested in the text of the problem. The desired current is along the cable, carried within the conductors. The undesired current corresponds to leakage through the plastic, and its direction is radial.

Categorize Because the resistivity and the geometry of the plastic are known, we categorize this problem as one in which we find the resistance of the plastic from these parameters. Equation 27.10, however, represents the resistance of a block of material. We have a more complicated geometry in this situation. Because the area through which the charges pass depends on the radial position, we must use integral calculus to determine the answer.

Analyze We divide the plastic into concentric cylindrical shells of infinitesimal thickness dr (Fig. 27.8b). Any charge passing from the inner to the outer conductor must move radially through this shell. Use a differential form of Equation 27.10, replacing ℓ with dr for the length variable: $dR = \rho\, dr/A$, where dR is the resistance of a shell of plastic of thickness dr and surface area A.

Write an expression for the resistance of our hollow cylindrical shell of plastic representing the area as the surface area of the shell:

$$dR = \frac{\rho\, dr}{A} = \frac{\rho}{2\pi rL}\, dr$$

continued

▶ **27.3** c o n t i n u e d

Integrate this expression from $r = a$ to $r = b$:

$$(1) \quad R = \int dR = \frac{\rho}{2\pi L} \int_a^b \frac{dr}{r} = \frac{\rho}{2\pi L} \ln\left(\frac{b}{a}\right)$$

Substitute the values given:

$$R = \frac{1.0 \times 10^{13} \,\Omega \cdot \text{m}}{2\pi(0.150 \,\text{m})} \ln\left(\frac{1.75 \,\text{cm}}{0.500 \,\text{cm}}\right) = \boxed{1.33 \times 10^{13} \,\Omega}$$

Finalize Let's compare this resistance to that of the inner copper conductor of the cable along the 15.0-cm length.

Use Equation 27.10 to find the resistance of the copper cylinder:

$$R_{Cu} = \rho \frac{\ell}{A} = (1.7 \times 10^{-8} \,\Omega \cdot \text{m}) \left[\frac{0.150 \,\text{m}}{\pi(5.00 \times 10^{-3} \,\text{m})^2}\right]$$

$$= 3.2 \times 10^{-5} \,\Omega$$

This resistance is 18 orders of magnitude smaller than the radial resistance. Therefore, almost all the current corresponds to charge moving along the length of the cable, with a very small fraction leaking in the radial direction.

WHAT IF? Suppose the coaxial cable is enlarged to twice the overall diameter with two possible choices: (1) the ratio b/a is held fixed, or (2) the difference $b - a$ is held fixed. For which choice does the leakage current between the inner and outer conductors increase when the voltage is applied between them?

Answer For the current to increase, the resistance must decrease. For choice (1), in which b/a is held fixed, Equa-

tion (1) shows that the resistance is unaffected. For choice (2), we do not have an equation involving the difference $b - a$ to inspect. Looking at Figure 27.8b, however, we see that increasing b and a while holding the difference constant results in charge flowing through the same thickness of plastic but through a larger area perpendicular to the flow. This larger area results in lower resistance and a higher current.

27.3 A Model for Electrical Conduction

In this section, we describe a structural model of electrical conduction in metals that was first proposed by Paul Drude (1863–1906) in 1900. (See Section 21.1 for a review of structural models.) This model leads to Ohm's law and shows that resistivity can be related to the motion of electrons in metals. Although the Drude model described here has limitations, it introduces concepts that are applied in more elaborate treatments.

Following the outline of structural models from Section 21.1, the Drude model for electrical conduction has the following properties:

1. *Physical components:*
   Consider a conductor as a regular array of atoms plus a collection of free electrons, which are sometimes called *conduction* electrons. We identify the system as the combination of the atoms and the conduction electrons. The conduction electrons, although bound to their respective atoms when the atoms are not part of a solid, become free when the atoms condense into a solid.

2. *Behavior of the components:*
   (a) In the absence of an electric field, the conduction electrons move in random directions through the conductor (Fig. 27.3a). The situation is similar to the motion of gas molecules confined in a vessel. In fact, some scientists refer to conduction electrons in a metal as an *electron gas.*
   (b) When an electric field is applied to the system, the free electrons drift slowly in a direction opposite that of the electric field (Fig. 27.3b), with an average drift speed v_d that is much smaller (typically 10^{-4} m/s) than their average speed v_{avg} between collisions (typically 10^6 m/s).
   (c) The electron's motion after a collision is independent of its motion before the collision. The excess energy acquired by the electrons due to

the work done on them by the electric field is transferred to the atoms of the conductor when the electrons and atoms collide.

With regard to property 2(c) above, the energy transferred to the atoms causes the internal energy of the system and, therefore, the temperature of the conductor to increase.

We are now in a position to derive an expression for the drift velocity, using several of our analysis models. When a free electron of mass m_e and charge q $(= -e)$ is subjected to an electric field $\vec{\mathbf{E}}$, it is described by the particle in a field model and experiences a force $\vec{\mathbf{F}} = q\vec{\mathbf{E}}$. The electron is a particle under a net force, and its acceleration can be found from Newton's second law, $\sum \vec{\mathbf{F}} = m\vec{\mathbf{a}}$:

$$\vec{\mathbf{a}} = \frac{\sum \vec{\mathbf{F}}}{m} = \frac{q\vec{\mathbf{E}}}{m_e} \tag{27.11}$$

Because the electric field is uniform, the electron's acceleration is constant, so the electron can be modeled as a particle under constant acceleration. If $\vec{\mathbf{v}}_i$ is the electron's initial velocity the instant after a collision (which occurs at a time defined as $t = 0$), the velocity of the electron at a very short time t later (immediately before the next collision occurs) is, from Equation 4.8,

$$\vec{\mathbf{v}}_f = \vec{\mathbf{v}}_i + \vec{\mathbf{a}}t = \vec{\mathbf{v}}_i + \frac{q\vec{\mathbf{E}}}{m_e}t \tag{27.12}$$

Let's now take the average value of $\vec{\mathbf{v}}_f$ for all the electrons in the wire over all possible collision times t and all possible values of $\vec{\mathbf{v}}_i$. Assuming the initial velocities are randomly distributed over all possible directions (property 2(a) above), the average value of $\vec{\mathbf{v}}_i$ is zero. The average value of the second term of Equation 27.12 is $\left(q\vec{\mathbf{E}}/m_e\right)\tau$, where τ is the *average time interval between successive collisions*. Because the average value of $\vec{\mathbf{v}}_f$ is equal to the drift velocity,

$$\vec{\mathbf{v}}_{f,\mathrm{avg}} = \vec{\mathbf{v}}_d = \frac{q\vec{\mathbf{E}}}{m_e}\tau \tag{27.13}$$

◀ **Drift velocity in terms of microscopic quantities**

The value of τ depends on the size of the metal atoms and the number of electrons per unit volume. We can relate this expression for drift velocity in Equation 27.13 to the current in the conductor. Substituting the magnitude of the velocity from Equation 27.13 into Equation 27.4, the average current in the conductor is given by

$$I_{\mathrm{avg}} = nq\left(\frac{qE}{m_e}\tau\right)A = \frac{nq^2E}{m_e}\tau A \tag{27.14}$$

Because the current density J is the current divided by the area A,

$$J = \frac{nq^2E}{m_e}\tau$$

◀ **Current density in terms of microscopic quantities**

where n is the number of electrons per unit volume. Comparing this expression with Ohm's law, $J = \sigma E$, we obtain the following relationships for conductivity and resistivity of a conductor:

$$\sigma = \frac{nq^2\tau}{m_e} \tag{27.15}$$

◀ **Conductivity in terms of microscopic quantities**

$$\rho = \frac{1}{\sigma} = \frac{m_e}{nq^2\tau} \tag{27.16}$$

◀ **Resistivity in terms of microscopic quantities**

According to this classical model, conductivity and resistivity do not depend on the strength of the electric field. This feature is characteristic of a conductor obeying Ohm's law.

The model shows that the resistivity can be calculated from a knowledge of the density of the electrons, their charge and mass, and the average time interval τ between collisions. This time interval is related to the average distance between collisions ℓ_{avg} (the *mean free path*) and the average speed v_{avg} through the expression[3]

$$\tau = \frac{\ell_{avg}}{v_{avg}} \tag{27.17}$$

Although this structural model of conduction is consistent with Ohm's law, it does not correctly predict the values of resistivity or the behavior of the resistivity with temperature. For example, the results of classical calculations for v_{avg} using the ideal gas model for the electrons are about a factor of ten smaller than the actual values, which results in incorrect predictions of values of resistivity from Equation 27.16. Furthermore, according to Equations 27.16 and 27.17, the resistivity is predicted to vary with temperature as does v_{avg}, which, according to an ideal-gas model (Chapter 21, Eq. 21.43), is proportional to $\sqrt{T}$. This behavior is in disagreement with the experimentally observed linear dependence of resistivity with temperature for pure metals. (See Section 27.4.) Because of these incorrect predictions, we must modify our structural model. We shall call the model that we have developed so far the *classical* model for electrical conduction. To account for the incorrect predictions of the classical model, we develop it further into a *quantum mechanical* model, which we shall describe briefly.

We discussed two important simplification models in earlier chapters, the particle model and the wave model. Although we discussed these two simplification models separately, quantum physics tells us that this separation is not so clear-cut. As we shall discuss in detail in Chapter 40, particles have wave-like properties. The predictions of some models can only be matched to experimental results if the model includes the wave-like behavior of particles. The structural model for electrical conduction in metals is one of these cases.

Let us imagine that the electrons moving through the metal have wave-like properties. If the array of atoms in a conductor is regularly spaced (that is, periodic), the wave-like character of the electrons makes it possible for them to move freely through the conductor and a collision with an atom is unlikely. For an idealized conductor, no collisions would occur, the mean free path would be infinite, and the resistivity would be zero. Electrons are scattered only if the atomic arrangement is irregular (not periodic), as a result of structural defects or impurities, for example. At low temperatures, the resistivity of metals is dominated by scattering caused by collisions between the electrons and impurities. At high temperatures, the resistivity is dominated by scattering caused by collisions between the electrons and the atoms of the conductor, which are continuously displaced as a result of thermal agitation, destroying the perfect periodicity. The thermal motion of the atoms makes the structure irregular (compared with an atomic array at rest), thereby reducing the electron's mean free path.

Although it is beyond the scope of this text to show this modification in detail, the classical model modified with the wave-like character of the electrons results in predictions of resistivity values that are in agreement with measured values and predicts a linear temperature dependence. Quantum notions had to be introduced in Chapter 21 to understand the temperature behavior of molar specific heats of gases. Here we have another case in which quantum physics is necessary for the model to agree with experiment. Although classical physics can explain a tremendous range of phenomena, we continue to see hints that quantum physics must be incorporated into our models. We shall study quantum physics in detail in Chapters 40 through 46.

---

[3]Recall that the average speed of a group of particles depends on the temperature of the group (Chapter 21) and is not the same as the drift speed v_d.

27.4 Resistance and Temperature

Over a limited temperature range, the resistivity of a conductor varies approximately linearly with temperature according to the expression

$$\rho = \rho_0[1 + \alpha(T - T_0)] \tag{27.18}$$

◀ Variation of ρ with temperature

where ρ is the resistivity at some temperature T (in degrees Celsius), ρ_0 is the resistivity at some reference temperature T_0 (usually taken to be 20°C), and α is the **temperature coefficient of resistivity.** From Equation 27.18, the temperature coefficient of resistivity can be expressed as

$$\alpha = \frac{1}{\rho_0}\frac{\Delta\rho}{\Delta T} \tag{27.19}$$

◀ Temperature coefficient of resistivity

where $\Delta\rho = \rho - \rho_0$ is the change in resistivity in the temperature interval $\Delta T = T - T_0$.

The temperature coefficients of resistivity for various materials are given in Table 27.2. Notice that the unit for α is degrees Celsius^{-1} [(°C)$^{-1}$]. Because resistance is proportional to resistivity (Eq. 27.10), the variation of resistance of a sample is

$$R = R_0[1 + \alpha(T - T_0)] \tag{27.20}$$

where R_0 is the resistance at temperature T_0. Use of this property enables precise temperature measurements through careful monitoring of the resistance of a probe made from a particular material.

For some metals such as copper, resistivity is nearly proportional to temperature as shown in Figure 27.9. A nonlinear region always exists at very low temperatures, however, and the resistivity usually reaches some finite value as the temperature approaches absolute zero. This residual resistivity near absolute zero is caused primarily by the collision of electrons with impurities and imperfections in the metal. In contrast, high-temperature resistivity (the linear region) is predominantly characterized by collisions between electrons and metal atoms.

Notice that three of the α values in Table 27.2 are negative, indicating that the resistivity of these materials decreases with increasing temperature. This behavior is indicative of a class of materials called *semiconductors,* first introduced in Section 23.2, and is due to an increase in the density of charge carriers at higher temperatures.

Because the charge carriers in a semiconductor are often associated with impurity atoms (as we discuss in more detail in Chapter 43), the resistivity of these materials is very sensitive to the type and concentration of such impurities.

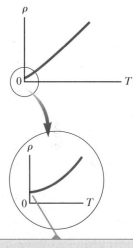

As T approaches absolute zero, the resistivity approaches a nonzero value.

Figure 27.9 Resistivity versus temperature for a metal such as copper. The curve is linear over a wide range of temperatures, and ρ increases with increasing temperature.

Quick Quiz 27.4 When does an incandescent lightbulb carry more current, **(a)** immediately after it is turned on and the glow of the metal filament is increasing or **(b)** after it has been on for a few milliseconds and the glow is steady?

27.5 Superconductors

There is a class of metals and compounds whose resistance decreases to zero when they are below a certain temperature T_c, known as the **critical temperature.** These materials are known as **superconductors.** The resistance–temperature graph for a superconductor follows that of a normal metal at temperatures above T_c (Fig. 27.10). When the temperature is at or below T_c, the resistivity drops suddenly to zero. This phenomenon was discovered in 1911 by Dutch physicist Heike Kamerlingh-Onnes (1853–1926) as he worked with mercury, which is a superconductor below 4.2 K. Measurements have shown that the resistivities of superconductors below their T_c values are less than 4×10^{-25} $\Omega \cdot$ m, or approximately 10^{17} times smaller than the resistivity of copper. In practice, these resistivities are considered to be zero.

The resistance drops discontinuously to zero at T_c, which is 4.15 K for mercury.

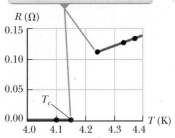

Figure 27.10 Resistance versus temperature for a sample of mercury (Hg). The graph follows that of a normal metal above the critical temperature T_c.

A small permanent magnet levitated above a disk of the superconductor $YBa_2Cu_3O_7$, which is in liquid nitrogen at 77 K.

Table 27.3	Critical Temperatures for Various Superconductors
Material	T_c (K)
$HgBa_2Ca_2Cu_3O_8$	134
Tl—Ba—Ca—Cu—O	125
Bi—Sr—Ca—Cu—O	105
$YBa_2Cu_3O_7$	92
Nb_3Ge	23.2
Nb_3Sn	18.05
Nb	9.46
Pb	7.18
Hg	4.15
Sn	3.72
Al	1.19
Zn	0.88

Today, thousands of superconductors are known, and as Table 27.3 illustrates, the critical temperatures of recently discovered superconductors are substantially higher than initially thought possible. Two kinds of superconductors are recognized. The more recently identified ones are essentially ceramics with high critical temperatures, whereas superconducting materials such as those observed by Kamerlingh-Onnes are metals. If a room-temperature superconductor is ever identified, its effect on technology could be tremendous.

The value of T_c is sensitive to chemical composition, pressure, and molecular structure. Copper, silver, and gold, which are excellent conductors, do not exhibit superconductivity.

One truly remarkable feature of superconductors is that once a current is set up in them, it persists *without any applied potential difference* (because $R = 0$). Steady currents have been observed to persist in superconducting loops for several years with no apparent decay!

An important and useful application of superconductivity is in the development of superconducting magnets, in which the magnitudes of the magnetic field are approximately ten times greater than those produced by the best normal electromagnets. Such superconducting magnets are being considered as a means of storing energy. Superconducting magnets are currently used in medical magnetic resonance imaging, or MRI, units, which produce high-quality images of internal organs without the need for excessive exposure of patients to x-rays or other harmful radiation.

27.6 Electrical Power

The direction of the effective flow of positive charge is clockwise.

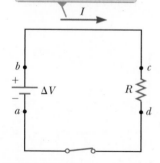

Figure 27.11 A circuit consisting of a resistor of resistance R and a battery having a potential difference ΔV across its terminals.

In typical electric circuits, energy T_{ET} is transferred by electrical transmission from a source such as a battery to some device such as a lightbulb or a radio receiver. Let's determine an expression that will allow us to calculate the rate of this energy transfer. First, consider the simple circuit in Figure 27.11, where energy is delivered to a resistor. (Resistors are designated by the circuit symbol —⋀⋀⋀—.) Because the connecting wires also have resistance, some energy is delivered to the wires and some to the resistor. Unless noted otherwise, we shall assume the resistance of the wires is small compared with the resistance of the circuit element so that the energy delivered to the wires is negligible.

Imagine following a positive quantity of charge Q moving clockwise around the circuit in Figure 27.11 from point a through the battery and resistor back to point a. We identify the entire circuit as our system. As the charge moves from a to b through the battery, the electric potential energy of the system *increases* by an amount $Q \Delta V$

while the chemical potential energy in the battery *decreases* by the same amount. (Recall from Eq. 25.3 that $\Delta U = q \Delta V$.) As the charge moves from c to d through the resistor, however, the electric potential energy of the system decreases due to collisions of electrons with atoms in the resistor. In this process, the electric potential energy is transformed to internal energy corresponding to increased vibrational motion of the atoms in the resistor. Because the resistance of the interconnecting wires is neglected, no energy transformation occurs for paths bc and da. When the charge returns to point a, the net result is that some of the chemical potential energy in the battery has been delivered to the resistor and resides in the resistor as internal energy E_{int} associated with molecular vibration.

The resistor is normally in contact with air, so its increased temperature results in a transfer of energy by heat Q into the air. In addition, the resistor emits thermal radiation T_{ER}, representing another means of escape for the energy. After some time interval has passed, the resistor reaches a constant temperature. At this time, the input of energy from the battery is balanced by the output of energy from the resistor by heat and radiation, and the resistor is a nonisolated system in steady state. Some electrical devices include *heat sinks*[4] connected to parts of the circuit to prevent these parts from reaching dangerously high temperatures. Heat sinks are pieces of metal with many fins. Because the metal's high thermal conductivity provides a rapid transfer of energy by heat away from the hot component and the large number of fins provides a large surface area in contact with the air, energy can transfer by radiation and into the air by heat at a high rate.

Let's now investigate the rate at which the electric potential energy of the system decreases as the charge Q passes through the resistor:

$$\frac{dU}{dt} = \frac{d}{dt}(Q\,\Delta V) = \frac{dQ}{dt}\Delta V = I\,\Delta V$$

where I is the current in the circuit. The system regains this potential energy when the charge passes through the battery, at the expense of chemical energy in the battery. The rate at which the potential energy of the system decreases as the charge passes through the resistor is equal to the rate at which the system gains internal energy in the resistor. Therefore, the power P, representing the rate at which energy is delivered to the resistor, is

$$P = I\,\Delta V \qquad (27.21)$$

We derived this result by considering a battery delivering energy to a resistor. Equation 27.21, however, can be used to calculate the power delivered by a voltage source to *any* device carrying a current I and having a potential difference ΔV between its terminals.

Using Equation 27.21 and $\Delta V = IR$ for a resistor, we can express the power delivered to the resistor in the alternative forms

$$P = I^2 R = \frac{(\Delta V)^2}{R} \qquad (27.22)$$

When I is expressed in amperes, ΔV in volts, and R in ohms, the SI unit of power is the watt, as it was in Chapter 8 in our discussion of mechanical power. The process by which energy is transformed to internal energy in a conductor of resistance R is often called *joule heating*;[5] this transformation is also often referred to as an I^2R loss.

---

[4]This usage is another misuse of the word *heat* that is ingrained in our common language.

[5]It is commonly called *joule heating* even though the process of heat does not occur when energy delivered to a resistor appears as internal energy. It is another example of incorrect usage of the word *heat* that has become entrenched in our language.

Pitfall Prevention 27.5

Charges Do Not Move All the Way Around a Circuit in a Short Time In terms of understanding the energy transfer in a circuit, it is useful to *imagine* a charge moving all the way around the circuit even though it would take hours to do so.

Pitfall Prevention 27.6

Misconceptions About Current Several common misconceptions are associated with current in a circuit like that in Figure 27.11. One is that current comes out of one terminal of the battery and is then "used up" as it passes through the resistor, leaving current in only one part of the circuit. The current is actually the same *everywhere* in the circuit. A related misconception has the current coming out of the resistor being smaller than that going in because some of the current is "used up." Yet another misconception has current coming out of both terminals of the battery, in opposite directions, and then "clashing" in the resistor, delivering the energy in this manner. That is not the case; charges flow in the same rotational sense at *all* points in the circuit.

Pitfall Prevention 27.7

Energy Is Not "Dissipated" In some books, you may see Equation 27.22 described as the power "dissipated in" a resistor, suggesting that energy disappears. Instead, we say energy is "delivered to" a resistor.

Lester Lefkowitz/Taxi/Getty Images

Figure 27.12 These power lines transfer energy from the electric company to homes and businesses. The energy is transferred at a very high voltage, possibly hundreds of thousands of volts in some cases. Even though it makes power lines very dangerous, the high voltage results in less loss of energy due to resistance in the wires.

When transporting energy by electricity through power lines (Fig. 27.12), you should not assume the lines have zero resistance. Real power lines do indeed have resistance, and power is delivered to the resistance of these wires. Utility companies seek to minimize the energy transformed to internal energy in the lines and maximize the energy delivered to the consumer. Because $P = I \Delta V$, the same amount of energy can be transported either at high currents and low potential differences or at low currents and high potential differences. Utility companies choose to transport energy at low currents and high potential differences primarily for economic reasons. Copper wire is very expensive, so it is cheaper to use high-resistance wire (that is, wire having a small cross-sectional area; see Eq. 27.10). Therefore, in the expression for the power delivered to a resistor, $P = I^2R$, the resistance of the wire is fixed at a relatively high value for economic considerations. The I^2R loss can be reduced by keeping the current I as low as possible, which means transferring the energy at a high voltage. In some instances, power is transported at potential differences as great as 765 kV. At the destination of the energy, the potential difference is usually reduced to 4 kV by a device called a *transformer*. Another transformer drops the potential difference to 240 V for use in your home. Of course, each time the potential difference decreases, the current increases by the same factor and the power remains the same. We shall discuss transformers in greater detail in Chapter 33.

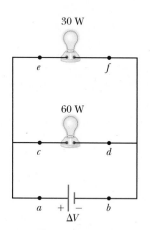

Figure 27.13 (Quick Quiz 27.5) Two lightbulbs connected across the same potential difference.

Quick **Quiz** 27.5 For the two lightbulbs shown in Figure 27.13, rank the current values at points *a* through *f* from greatest to least.

Example 27.4   **Power in an Electric Heater**

An electric heater is constructed by applying a potential difference of 120 V across a Nichrome wire that has a total resistance of 8.00 Ω. Find the current carried by the wire and the power rating of the heater.

SOLUTION

Conceptualize As discussed in Example 27.2, Nichrome wire has high resistivity and is often used for heating elements in toasters, irons, and electric heaters. Therefore, we expect the power delivered to the wire to be relatively high.

Categorize We evaluate the power from Equation 27.22, so we categorize this example as a substitution problem.

Use Equation 27.7 to find the current in the wire:

$$I = \frac{\Delta V}{R} = \frac{120 \text{ V}}{8.00 \ \Omega} = \boxed{15.0 \text{ A}}$$

Find the power rating using the expression $P = I^2R$ from Equation 27.22:

$$P = I^2R = (15.0 \text{ A})^2(8.00 \ \Omega) = 1.80 \times 10^3 \text{ W} = \boxed{1.80 \text{ kW}}$$

WHAT IF? What if the heater were accidentally connected to a 240-V supply? (That is difficult to do because the shape and orientation of the metal contacts in 240-V plugs are different from those in 120-V plugs.) How would that affect the current carried by the heater and the power rating of the heater, assuming the resistance remains constant?

Answer If the applied potential difference were doubled, Equation 27.7 shows that the current would double. According to Equation 27.22, $P = (\Delta V)^2/R$, the power would be four times larger.

| Example 27.5 | **Linking Electricity and Thermodynamics** | AM |

An immersion heater must increase the temperature of 1.50 kg of water from 10.0°C to 50.0°C in 10.0 min while operating at 110 V.

(A) What is the required resistance of the heater?

SOLUTION

Conceptualize An immersion heater is a resistor that is inserted into a container of water. As energy is delivered to the immersion heater, raising its temperature, energy leaves the surface of the resistor by heat, going into the water. When the immersion heater reaches a constant temperature, the rate of energy delivered to the resistance by electrical transmission (T_{ET}) is equal to the rate of energy delivered by heat (Q) to the water.

Categorize This example allows us to link our new understanding of power in electricity with our experience with specific heat in thermodynamics (Chapter 20). The water is a *nonisolated system*. Its internal energy is rising because of energy transferred into the water by heat from the resistor, so Equation 8.2 reduces to $\Delta E_{int} = Q$. In our model, we assume the energy that enters the water from the heater remains in the water.

....

Analyze To simplify the analysis, let's ignore the initial period during which the temperature of the resistor increases and also ignore any variation of resistance with temperature. Therefore, we imagine a constant rate of energy transfer for the entire 10.0 min.

Set the rate of energy delivered to the resistor equal to the rate of energy Q entering the water by heat:

$$P = \frac{(\Delta V)^2}{R} = \frac{Q}{\Delta t}$$

Use Equation 20.4, $Q = mc\,\Delta T$, to relate the energy input by heat to the resulting temperature change of the water and solve for the resistance:

$$\frac{(\Delta V)^2}{R} = \frac{mc\,\Delta T}{\Delta t} \rightarrow R = \frac{(\Delta V)^2\,\Delta t}{mc\,\Delta T}$$

Substitute the values given in the statement of the problem:

$$R = \frac{(110\text{ V})^2(600\text{ s})}{(1.50\text{ kg})(4\,186\text{ J/kg}\cdot\text{°C})(50.0\text{°C} - 10.0\text{°C})} = \boxed{28.9\ \Omega}$$

(B) Estimate the cost of heating the water.

SOLUTION

Multiply the power by the time interval to find the amount of energy transferred to the resistor:

$$T_{ET} = P\,\Delta t = \frac{(\Delta V)^2}{R}\,\Delta t = \frac{(110\text{ V})^2}{28.9\ \Omega}(10.0\text{ min})\left(\frac{1\text{ h}}{60.0\text{ min}}\right)$$

$$= 69.8\text{ Wh} = 0.069\,8\text{ kWh}$$

Find the cost knowing that energy is purchased at an estimated price of 11¢ per kilowatt-hour:

$$\text{Cost} = (0.069\,8\text{ kWh})(\$0.11/\text{kWh}) = \$0.008 = \boxed{0.8\text{¢}}$$

Finalize The cost to heat the water is very low, less than one cent. In reality, the cost is higher because some energy is transferred from the water into the surroundings by heat and electromagnetic radiation while its temperature is increasing. If you have electrical devices in your home with power ratings on them, use this power rating and an approximate time interval of use to estimate the cost for one use of the device.

Summary

Definitions

The electric **current** I in a conductor is defined as

$$I \equiv \frac{dQ}{dt} \qquad \textbf{(27.2)}$$

where dQ is the charge that passes through a cross section of the conductor in a time interval dt. The SI unit of current is the **ampere** (A), where 1 A = 1 C/s.

continued

The **current density** J in a conductor is the current per unit area:

$$J \equiv \frac{I}{A} \quad \textbf{(27.5)}$$

The **resistance** R of a conductor is defined as

$$R \equiv \frac{\Delta V}{I} \quad \textbf{(27.7)}$$

where ΔV is the potential difference across the conductor and I is the current it carries. The SI unit of resistance is volts per ampere, which is defined to be 1 **ohm** (Ω); that is, $1\ \Omega = 1\ \text{V/A}$.

Concepts and Principles

The average current in a conductor is related to the motion of the charge carriers through the relationship

$$I_{\text{avg}} = nqv_d A \quad \textbf{(27.4)}$$

where n is the density of charge carriers, q is the charge on each carrier, v_d is the drift speed, and A is the cross-sectional area of the conductor.

The current density in an ohmic conductor is proportional to the electric field according to the expression

$$J = \sigma E \quad \textbf{(27.6)}$$

The proportionality constant σ is called the **conductivity** of the material of which the conductor is made. The inverse of σ is known as **resistivity** ρ (that is, $\rho = 1/\sigma$). Equation 27.6 is known as **Ohm's law,** and a material is said to obey this law if the ratio of its current density to its applied electric field is a constant that is independent of the applied field.

For a uniform block of material of cross-sectional area A and length ℓ, the resistance over the length ℓ is

$$R = \rho \frac{\ell}{A} \quad \textbf{(27.10)}$$

where ρ is the resistivity of the material.

In a classical model of electrical conduction in metals, the electrons are treated as molecules of a gas. In the absence of an electric field, the average velocity of the electrons is zero. When an electric field is applied, the electrons move (on average) with a **drift velocity** $\vec{\mathbf{v}}_d$ that is opposite the electric field. The drift velocity is given by

$$\vec{\mathbf{v}}_d = \frac{q\vec{\mathbf{E}}}{m_e}\tau \quad \textbf{(27.13)}$$

where q is the electron's charge, m_e is the mass of the electron, and τ is the average time interval between electron–atom collisions. According to this model, the resistivity of the metal is

$$\rho = \frac{m_e}{nq^2\tau} \quad \textbf{(27.16)}$$

where n is the number of free electrons per unit volume.

The resistivity of a conductor varies approximately linearly with temperature according to the expression

$$\rho = \rho_0[1 + \alpha(T - T_0)] \quad \textbf{(27.18)}$$

where ρ_0 is the resistivity at some reference temperature T_0 and α is the **temperature coefficient of resistivity.**

If a potential difference ΔV is maintained across a circuit element, the **power,** or rate at which energy is supplied to the element, is

$$P = I\,\Delta V \quad \textbf{(27.21)}$$

Because the potential difference across a resistor is given by $\Delta V = IR$, we can express the power delivered to a resistor as

$$P = I^2 R = \frac{(\Delta V)^2}{R} \quad \textbf{(27.22)}$$

The energy delivered to a resistor by electrical transmission T_{ET} appears in the form of internal energy E_{int} in the resistor.

Objective Questions   1. denotes answer available in *Student Solutions Manual/Study Guide*

1. Car batteries are often rated in ampere-hours. Does this information designate the amount of (a) current, (b) power, (c) energy, (d) charge, or (e) potential the battery can supply?

2. Two wires A and B with circular cross sections are made of the same metal and have equal lengths, but the resistance of wire A is three times greater than that of wire B. (i) What is the ratio of the cross-sectional

area of A to that of B? (a) 3 (b) $\sqrt{3}$ (c) 1 (d) $1/\sqrt{3}$ (e) $\frac{1}{3}$ **(ii)** What is the ratio of the radius of A to that of B? Choose from the same possibilities as in part (i).

3. A cylindrical metal wire at room temperature is carrying electric current between its ends. One end is at potential $V_A = 50$ V, and the other end is at potential $V_B = 0$ V. Rank the following actions in terms of the change that each one separately would produce in the current from the greatest increase to the greatest decrease. In your ranking, note any cases of equality. (a) Make $V_A = 150$ V with $V_B = 0$ V. (b) Adjust V_A to triple the power with which the wire converts electrically transmitted energy into internal energy. (c) Double the radius of the wire. (d) Double the length of the wire. (e) Double the Celsius temperature of the wire.

4. A current-carrying ohmic metal wire has a cross-sectional area that gradually becomes smaller from one end of the wire to the other. The current has the same value for each section of the wire, so charge does not accumulate at any one point. **(i)** How does the drift speed vary along the wire as the area becomes smaller? (a) It increases. (b) It decreases. (c) It remains constant. **(ii)** How does the resistance per unit length vary along the wire as the area becomes smaller? Choose from the same possibilities as in part (i).

5. A potential difference of 1.00 V is maintained across a 10.0-Ω resistor for a period of 20.0 s. What total charge passes by a point in one of the wires connected to the resistor in this time interval? (a) 200 C (b) 20.0 C (c) 2.00 C (d) 0.005 00 C (e) 0.050 0 C

6. Three wires are made of copper having circular cross sections. Wire 1 has a length L and radius r. Wire 2 has a length L and radius $2r$. Wire 3 has a length $2L$ and radius $3r$. Which wire has the smallest resistance? (a) wire 1 (b) wire 2 (c) wire 3 (d) All have the same resistance. (e) Not enough information is given to answer the question.

7. A metal wire of resistance R is cut into three equal pieces that are then placed together side by side to form a new cable with a length equal to one-third the original length. What is the resistance of this new cable? (a) $\frac{1}{9}R$ (b) $\frac{1}{3}R$ (c) R (d) $3R$ (e) $9R$

8. A metal wire has a resistance of 10.0 Ω at a temperature of 20.0°C. If the same wire has a resistance of 10.6 Ω at 90.0°C, what is the resistance of this wire when its temperature is −20.0°C? (a) 0.700 Ω (b) 9.66 Ω (c) 10.3 Ω (d) 13.8 Ω (e) 6.59 Ω

9. The current-versus-voltage behavior of a certain electrical device is shown in Figure OQ27.9. When the potential difference across the device is 2 V, what is its resistance? (a) 1 Ω (b) $\frac{3}{4}$ Ω (c) $\frac{4}{3}$ Ω (d) undefined (e) none of those answers

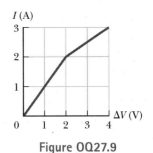

Figure OQ27.9

10. Two conductors made of the same material are connected across the same potential difference. Conductor A has twice the diameter and twice the length of conductor B. What is the ratio of the power delivered to A to the power delivered to B? (a) 8 (b) 4 (c) 2 (d) 1 (e) $\frac{1}{2}$

11. Two conducting wires A and B of the same length and radius are connected across the same potential difference. Conductor A has twice the resistivity of conductor B. What is the ratio of the power delivered to A to the power delivered to B? (a) 2 (b) $\sqrt{2}$ (c) 1 (d) $1/\sqrt{2}$ (e) $\frac{1}{2}$

12. Two lightbulbs both operate on 120 V. One has a power of 25 W and the other 100 W. **(i)** Which lightbulb has higher resistance? (a) The dim 25-W lightbulb does. (b) The bright 100-W lightbulb does. (c) Both are the same. **(ii)** Which lightbulb carries more current? Choose from the same possibilities as in part (i).

13. Wire B has twice the length and twice the radius of wire A. Both wires are made from the same material. If wire A has a resistance R, what is the resistance of wire B? (a) $4R$ (b) $2R$ (c) R (d) $\frac{1}{2}R$ (e) $\frac{1}{4}R$

Conceptual Questions   1. denotes answer available in *Student Solutions Manual/Study Guide*

1. If you were to design an electric heater using Nichrome wire as the heating element, what parameters of the wire could you vary to meet a specific power output such as 1 000 W?

2. What factors affect the resistance of a conductor?

3. When the potential difference across a certain conductor is doubled, the current is observed to increase by a factor of 3. What can you conclude about the conductor?

4. Over the time interval after a difference in potential is applied between the ends of a wire, what would happen to the drift velocity of the electrons in a wire and to the current in the wire if the electrons could move freely without resistance through the wire?

5. How does the resistance for copper and for silicon change with temperature? Why are the behaviors of these two materials different?

6. Use the atomic theory of matter to explain why the resistance of a material should increase as its temperature increases.

7. If charges flow very slowly through a metal, why does it not require several hours for a light to come on when you throw a switch?

8. Newspaper articles often contain statements such as "10 000 volts of electricity surged through the victim's body." What is wrong with this statement?

Problems available in Access end-of-chapter problems online at www.webassign.net

Section 27.1 Electric Current
Problems 1–13

Section 27.2 Resistance
Problems 14–21

Section 27.3 A Model for Electrical Conduction
Problems 22–25

Section 27.4 Resistance and Temperature
Problems 26–35

Section 27.6 Electrical Power
Problems 36–56

Additional Problems
Problems 57–81

Challenge Problems
Problem 82–85

Solutions to the following Problems are available in the *Student Solutions Manual/Study Guide:*
27.7, 27.11, 27.16, 27.19, 27.25, 27.27, 27.33, 27.42, 27.43, 27.47, 27.53, 27.62, 27.63, 27.66, 27.67, 27.82, and 27.84

List of Enhanced Problems

Problem Number	Targeted Feedback in Enhanced WebAssign	Analysis Model Tutorial in Enhanced WebAssign	Master It in Enhanced WebAssign	Watch It in Enhanced WebAssign
27.1	✓	✓	✓	
27.3	✓			✓
27.4		✓		
27.8	✓			✓
27.9	✓			✓
27.11			✓	
27.12	✓			✓
27.13	✓			✓
27.14	✓			✓
27.15	✓		✓	
27.19	✓		✓	
27.25			✓	
27.31	✓		✓	
27.33			✓	
27.42		✓	✓	
27.45	✓			✓
27.46	✓			✓
27.47			✓	
27.53			✓	
27.57	✓		✓	
27.61	✓			✓
27.66		✓	✓	
27.69	✓			✓

Direct-Current Circuits

In this chapter, we analyze simple electric circuits that contain batteries, resistors, and capacitors in various combinations. Some circuits contain resistors that can be combined using simple rules. The analysis of more complicated circuits is simplified using *Kirchhoff's rules,* which follow from the laws of conservation of energy and conservation of electric charge for isolated systems. Most of the circuits analyzed are assumed to be in *steady state,* which means that currents in the circuit are constant in magnitude and direction. A current that is constant in direction is called a *direct current* (DC). We will study *alternating current* (AC), in which the current changes direction periodically, in Chapter 33. Finally, we discuss electrical circuits in the home.

A technician repairs a connection on a circuit board from a computer. In our lives today, we use various items containing electric circuits, including many with circuit boards much smaller than the board shown in the photograph. These include handheld game players, cell phones, and digital cameras. In this chapter, we study simple types of circuits and learn how to analyze them. *(Trombax/Shutterstock.com)*

28.1 Electromotive Force

In Section 27.6, we discussed a circuit in which a battery produces a current. We will generally use a battery as a source of energy for circuits in our discussion. Because the potential difference at the battery terminals is constant in a particular circuit, the current in the circuit is constant in magnitude and direction and is called **direct current.** A battery is called either a *source of electromotive force* or, more commonly, a *source of emf.* (The phrase *electromotive force* is an unfortunate historical term, describing not a force, but rather a potential difference in volts.) The **emf $\mathcal{E}$** of a battery is **the maximum possible voltage the battery can provide between its terminals.** You can think of a source of emf as a "charge pump." When an electric potential difference exists between two points, the source moves charges "uphill" from the lower potential to the higher.

We shall generally assume the connecting wires in a circuit have no resistance. The positive terminal of a battery is at a higher potential than the negative terminal.

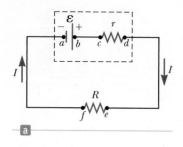

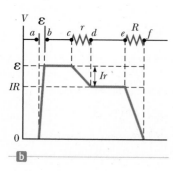

b

Figure 28.1 (a) Circuit diagram of a source of emf $\mathcal{E}$ (in this case, a battery), of internal resistance r, connected to an external resistor of resistance R. (b) Graphical representation showing how the electric potential changes as the circuit in (a) is traversed clockwise.

Pitfall Prevention 28.1

What Is Constant in a Battery?
It is a common misconception that a battery is a source of constant current. Equation 28.3 shows that is not true. The current in the circuit depends on the resistance R connected to the battery. It is also not true that a battery is a source of constant terminal voltage as shown by Equation 28.1. **A battery is a source of constant emf.**

Because a real battery is made of matter, there is resistance to the flow of charge within the battery. This resistance is called **internal resistance** r. For an idealized battery with zero internal resistance, the potential difference across the battery (called its *terminal voltage*) equals its emf. For a real battery, however, the terminal voltage is *not* equal to the emf for a battery in a circuit in which there is a current. To understand why, consider the circuit diagram in Figure 28.1a. We model the battery as shown in the diagram; it is represented by the dashed rectangle containing an ideal, resistance-free emf $\mathcal{E}$ in series with an internal resistance r. A resistor of resistance R is connected across the terminals of the battery. Now imagine moving through the battery from a to d and measuring the electric potential at various locations. Passing from the negative terminal to the positive terminal, the potential increases by an amount $\mathcal{E}$. As we move through the resistance r, however, the potential *decreases* by an amount Ir, where I is the current in the circuit. Therefore, the terminal voltage of the battery $\Delta V = V_d - V_a$ is

$$\Delta V = \mathcal{E} - Ir \qquad (28.1)$$

From this expression, notice that $\mathcal{E}$ is equivalent to the **open-circuit voltage,** that is, the terminal voltage when the current is zero. The emf is the voltage labeled on a battery; for example, the emf of a D cell is 1.5 V. The actual potential difference between a battery's terminals depends on the current in the battery as described by Equation 28.1. Figure 28.1b is a graphical representation of the changes in electric potential as the circuit is traversed in the clockwise direction.

Figure 28.1a shows that the terminal voltage ΔV must equal the potential difference across the external resistance R, often called the **load resistance.** The load resistor might be a simple resistive circuit element as in Figure 28.1a, or it could be the resistance of some electrical device (such as a toaster, electric heater, or lightbulb) connected to the battery (or, in the case of household devices, to the wall outlet). The resistor represents a *load* on the battery because the battery must supply energy to operate the device containing the resistance. The potential difference across the load resistance is $\Delta V = IR$. Combining this expression with Equation 28.1, we see that

$$\mathcal{E} = IR + Ir \qquad (28.2)$$

Figure 28.1a shows a graphical representation of this equation. Solving for the current gives

$$I = \frac{\mathcal{E}}{R + r} \qquad (28.3)$$

Equation 28.3 shows that the current in this simple circuit depends on both the load resistance R external to the battery and the internal resistance r. If R is much greater than r, as it is in many real-world circuits, we can neglect r.

Multiplying Equation 28.2 by the current I in the circuit gives

$$I\mathcal{E} = I^2R + I^2r \qquad (28.4)$$

Equation 28.4 indicates that because power $P = I\Delta V$ (see Eq. 27.21), the total power output $I\mathcal{E}$ associated with the emf of the battery is delivered to the external load resistance in the amount I^2R and to the internal resistance in the amount I^2r.

Quick Quiz 28.1 To maximize the percentage of the power from the emf of a battery that is delivered to a device external to the battery, what should the internal resistance of the battery be? **(a)** It should be as low as possible. **(b)** It should be as high as possible. **(c)** The percentage does not depend on the internal resistance.

Example 28.1   **Terminal Voltage of a Battery**

A battery has an emf of 12.0 V and an internal resistance of 0.050 0 Ω. Its terminals are connected to a load resistance of 3.00 Ω.

▶ **28.1** continued

(A) Find the current in the circuit and the terminal voltage of the battery.

SOLUTION

Conceptualize Study Figure 28.1a, which shows a circuit consistent with the problem statement. The battery delivers energy to the load resistor.

Categorize This example involves simple calculations from this section, so we categorize it as a substitution problem.

Use Equation 28.3 to find the current in the circuit:
$$I = \frac{\mathcal{E}}{R + r} = \frac{12.0 \text{ V}}{3.00 \ \Omega + 0.050 \ 0 \ \Omega} = \boxed{3.93 \text{ A}}$$

Use Equation 28.1 to find the terminal voltage:
$$\Delta V = \mathcal{E} - Ir = 12.0 \text{ V} - (3.93 \text{ A})(0.050 \ 0 \ \Omega) = \boxed{11.8 \text{ V}}$$

To check this result, calculate the voltage across the load resistance R:
$$\Delta V = IR = (3.93 \text{ A})(3.00 \ \Omega) = 11.8 \text{ V}$$

(B) Calculate the power delivered to the load resistor, the power delivered to the internal resistance of the battery, and the power delivered by the battery.

SOLUTION

Use Equation 27.22 to find the power delivered to the load resistor:
$$P_R = I^2 R = (3.93 \text{ A})^2(3.00 \ \Omega) = \boxed{46.3 \text{ W}}$$

Find the power delivered to the internal resistance:
$$P_r = I^2 r = (3.93 \text{ A})^2(0.050 \ 0 \ \Omega) = \boxed{0.772 \text{ W}}$$

Find the power delivered by the battery by adding these quantities:
$$P = P_R + P_r = 46.3 \text{ W} + 0.772 \text{ W} = \boxed{47.1 \text{ W}}$$

WHAT IF? As a battery ages, its internal resistance increases. Suppose the internal resistance of this battery rises to 2.00 Ω toward the end of its useful life. How does that alter the battery's ability to deliver energy?

Answer Let's connect the same 3.00-Ω load resistor to the battery.

Find the new current in the battery:
$$I = \frac{\mathcal{E}}{R + r} = \frac{12.0 \text{ V}}{3.00 \ \Omega + 2.00 \ \Omega} = 2.40 \text{ A}$$

Find the new terminal voltage:
$$\Delta V = \mathcal{E} - Ir = 12.0 \text{ V} - (2.40 \text{ A})(2.00 \ \Omega) = 7.2 \text{ V}$$

Find the new powers delivered to the load resistor and internal resistance:
$$P_R = I^2 R = (2.40 \text{ A})^2(3.00 \ \Omega) = 17.3 \text{ W}$$
$$P_r = I^2 r = (2.40 \text{ A})^2(2.00 \ \Omega) = 11.5 \text{ W}$$

In this situation, the terminal voltage is only 60% of the emf. Notice that 40% of the power from the battery is delivered to the internal resistance when r is 2.00 Ω. When r is 0.050 0 Ω as in part (B), this percentage is only 1.6%. Consequently, even though the emf remains fixed, the increasing internal resistance of the battery significantly reduces the battery's ability to deliver energy to an external load.

Example 28.2 **Matching the Load**

Find the load resistance R for which the maximum power is delivered to the load resistance in Figure 28.1a.

SOLUTION

Conceptualize Think about varying the load resistance in Figure 28.1a and the effect on the power delivered to the load resistance. When R is large, there is very little current, so the power I^2R delivered to the load resistor is small.

continued

▶ **28.2** continued

When R is small, let's say $R \ll r$, the current is large and the power delivered to the internal resistance is $I^2 r \gg I^2 R$. Therefore, the power delivered to the load resistor is small compared to that delivered to the internal resistance. For some intermediate value of the resistance R, the power must maximize.

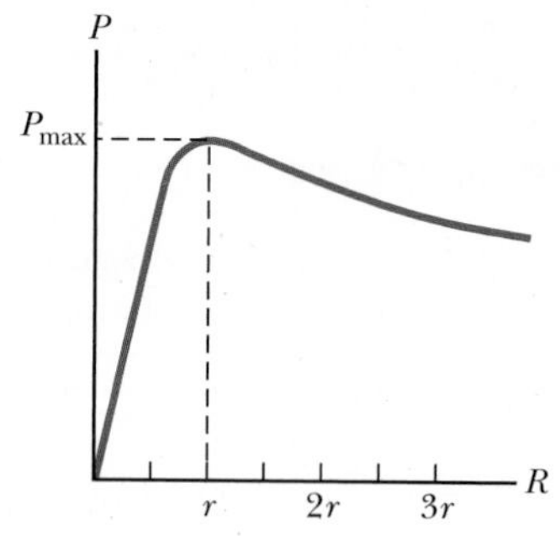

Categorize We categorize this example as an analysis problem because we must undertake a procedure to maximize the power. The circuit is the same as that in Example 28.1. The load resistance R in this case, however, is a variable.

Figure 28.2 (Example 28.2) Graph of the power P delivered by a battery to a load resistor of resistance R as a function of R.

Analyze Find the power delivered to the load resistance using Equation 27.22, with I given by Equation 28.3:

$$(1) \quad P = I^2 R = \frac{\mathcal{E}^2 R}{(R + r)^2}$$

Differentiate the power with respect to the load resistance R and set the derivative equal to zero to maximize the power:

$$\frac{dP}{dR} = \frac{d}{dR}\left[\frac{\mathcal{E}^2 R}{(R + r)^2}\right] = \frac{d}{dR}\left[\mathcal{E}^2 R(R + r)^{-2}\right] = 0$$

$$[\mathcal{E}^2(R + r)^{-2}] + [\mathcal{E}^2 R(-2)(R + r)^{-3}] = 0$$

$$\frac{\mathcal{E}^2(R + r)}{(R + r)^3} - \frac{2\mathcal{E}^2 R}{(R + r)^3} = \frac{\mathcal{E}^2(r - R)}{(R + r)^3} = 0$$

Solve for R:

$$R = \boxed{r}$$

Finalize To check this result, let's plot P versus R as in Figure 28.2. The graph shows that P reaches a maximum value at $R = r$. Equation (1) shows that this maximum value is $P_{max} = \mathcal{E}^2/4r$.

28.2 Resistors in Series and Parallel

When two or more resistors are connected together as are the incandescent light-bulbs in Figure 28.3a, they are said to be in a **series combination.** Figure 28.3b is the circuit diagram for the lightbulbs, shown as resistors, and the battery. What if you wanted to replace the series combination with a single resistor that would draw the same current from the battery? What would be its value? In a series connection, if an amount of charge Q exits resistor R_1, charge Q must also enter the second resistor R_2. Otherwise, charge would accumulate on the wire between the resistors. Therefore, the same amount of charge passes through both resistors in a given time interval and the currents are the same in both resistors:

$$I = I_1 = I_2$$

where I is the current leaving the battery, I_1 is the current in resistor R_1, and I_2 is the current in resistor R_2.

The potential difference applied across the series combination of resistors divides between the resistors. In Figure 28.3b, because the voltage drop[1] from a to b equals $I_1 R_1$ and the voltage drop from b to c equals $I_2 R_2$, the voltage drop from a to c is

$$\Delta V = \Delta V_1 + \Delta V_2 = I_1 R_1 + I_2 R_2$$

The potential difference across the battery is also applied to the **equivalent resistance** R_{eq} in Figure 28.3c:

$$\Delta V = I R_{eq}$$

---

[1]The term *voltage drop* is synonymous with a decrease in electric potential across a resistor. It is often used by individuals working with electric circuits.

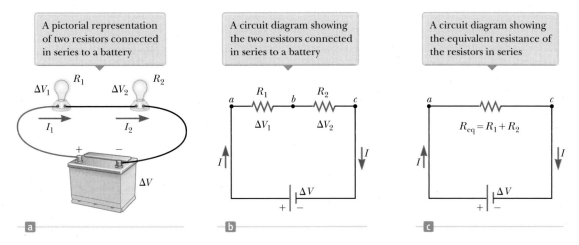

A pictorial representation of two resistors connected in series to a battery

A circuit diagram showing the two resistors connected in series to a battery

A circuit diagram showing the equivalent resistance of the resistors in series

Figure 28.3 Two lightbulbs with resistances R_1 and R_2 connected in series. All three diagrams are equivalent.

where the equivalent resistance has the same effect on the circuit as the series combination because it results in the same current I in the battery. Combining these equations for ΔV gives

$$IR_{eq} = I_1 R_1 + I_2 R_2 \quad \rightarrow \quad R_{eq} = R_1 + R_2 \qquad \textbf{(28.5)}$$

where we have canceled the currents I, I_1, and I_2 because they are all the same. We see that we can replace the two resistors in series with a single equivalent resistance whose value is the *sum* of the individual resistances.

The equivalent resistance of three or more resistors connected in series is

$$R_{eq} = R_1 + R_2 + R_3 + \cdots \qquad \textbf{(28.6)}$$

This relationship indicates that the equivalent resistance of a series combination of resistors is the numerical sum of the individual resistances and is always greater than any individual resistance.

Looking back at Equation 28.3, we see that the denominator of the right-hand side is the simple algebraic sum of the external and internal resistances. That is consistent with the internal and external resistances being in series in Figure 28.1a.

If the filament of one lightbulb in Figure 28.3 were to fail, the circuit would no longer be complete (resulting in an open-circuit condition) and the second lightbulb would also go out. This fact is a general feature of a series circuit: if one device in the series creates an open circuit, all devices are inoperative.

◄ The equivalent resistance of a series combination of resistors

Pitfall Prevention 28.2
Lightbulbs Don't Burn We will describe the end of the life of an incandescent lightbulb by saying *the filament fails* rather than by saying the lightbulb "burns out." The word *burn* suggests a combustion process, which is not what occurs in a lightbulb. The failure of a lightbulb results from the slow sublimation of tungsten from the very hot filament over the life of the lightbulb. The filament eventually becomes very thin because of this process. The mechanical stress from a sudden temperature increase when the lightbulb is turned on causes the thin filament to break.

Quick Quiz 28.2 With the switch in the circuit of Figure 28.4a closed, there is no current in R_2 because the current has an alternate zero-resistance path through the switch. There is current in R_1, and this current is measured with the ammeter (a device for measuring current) at the bottom of the circuit. If the switch is opened (Fig. 28.4b), there is current in R_2. What happens to the reading on the ammeter when the switch is opened? **(a)** The reading goes up. **(b)** The reading goes down. **(c)** The reading does not change.

Pitfall Prevention 28.3
Local and Global Changes A local change in one part of a circuit may result in a global change throughout the circuit. For example, if a single resistor is changed in a circuit containing several resistors and batteries, the currents in all resistors and batteries, the terminal voltages of all batteries, and the voltages across all resistors may change as a result.

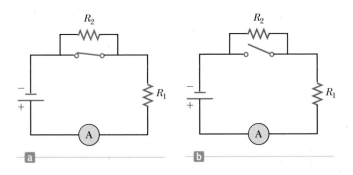

Figure 28.4 (Quick Quiz 28.2) What happens when the switch is opened?

Figure 28.5 Two lightbulbs with resistances R_1 and R_2 connected in parallel. All three diagrams are equivalent.

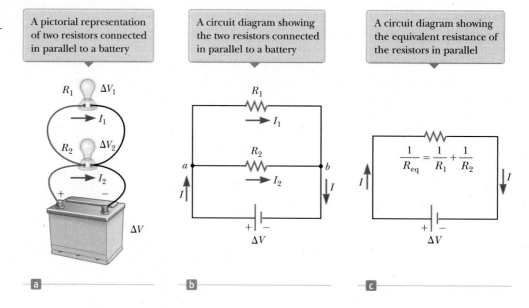

A pictorial representation of two resistors connected in parallel to a battery

A circuit diagram showing the two resistors connected in parallel to a battery

A circuit diagram showing the equivalent resistance of the resistors in parallel

Pitfall Prevention 28.4

Current Does Not Take the Path of Least Resistance You may have heard the phrase "current takes the path of least resistance" (or similar wording) in reference to a parallel combination of current paths such that there are two or more paths for the current to take. Such wording is incorrect. The current takes *all* paths. Those paths with lower resistance have larger currents, but even very high resistance paths carry *some* of the current. In theory, if current has a choice between a zero-resistance path and a finite resistance path, all the current takes the path of zero resistance; a path with zero resistance, however, is an idealization.

Now consider two resistors in a **parallel combination** as shown in Figure 28.5. As with the series combination, what is the value of the single resistor that could replace the combination and draw the same current from the battery? Notice that both resistors are connected directly across the terminals of the battery. Therefore, the potential differences across the resistors are the same:

$$\Delta V = \Delta V_1 = \Delta V_2$$

where ΔV is the terminal voltage of the battery.

When charges reach point a in Figure 28.5b, they split into two parts, with some going toward R_1 and the rest going toward R_2. A **junction** is any such point in a circuit where a current can split. This split results in less current in each individual resistor than the current leaving the battery. Because electric charge is conserved, the current I that enters point a must equal the total current leaving that point:

$$I = I_1 + I_2 = \frac{\Delta V_1}{R_1} + \frac{\Delta V_2}{R_2}$$

where I_1 is the current in R_1 and I_2 is the current in R_2.

The current in the **equivalent resistance** R_{eq} in Figure 28.5c is

$$I = \frac{\Delta V}{R_{eq}}$$

where the equivalent resistance has the same effect on the circuit as the two resistors in parallel; that is, the equivalent resistance draws the same current I from the battery. Combining these equations for I, we see that the equivalent resistance of two resistors in parallel is given by

$$\frac{\Delta V}{R_{eq}} = \frac{\Delta V_1}{R_1} + \frac{\Delta V_2}{R_2} \quad \rightarrow \quad \frac{1}{R_{eq}} = \frac{1}{R_1} + \frac{1}{R_2} \tag{28.7}$$

where we have canceled ΔV, ΔV_1, and ΔV_2 because they are all the same.

An extension of this analysis to three or more resistors in parallel gives

$$\frac{1}{R_{eq}} = \frac{1}{R_1} + \frac{1}{R_2} + \frac{1}{R_3} + \cdots \tag{28.8}$$

◄ **The equivalent resistance of a parallel combination of resistors**

This expression shows that the inverse of the equivalent resistance of two or more resistors in a parallel combination is equal to the sum of the inverses of the individual resistances. Furthermore, the equivalent resistance is always less than the smallest resistance in the group.

Household circuits are always wired such that the appliances are connected in parallel. Each device operates independently of the others so that if one is switched off, the others remain on. In addition, in this type of connection, all the devices operate on the same voltage.

Let's consider two examples of practical applications of series and parallel circuits. Figure 28.6 illustrates how a three-way incandescent lightbulb is constructed to provide three levels of light intensity.[2] The socket of the lamp is equipped with a three-way switch for selecting different light intensities. The lightbulb contains two filaments. When the lamp is connected to a 120-V source, one filament receives 100 W of power and the other receives 75 W. The three light intensities are made possible by applying the 120 V to one filament alone, to the other filament alone, or to the two filaments in parallel. When switch S_1 is closed and switch S_2 is opened, current exists only in the 75-W filament. When switch S_1 is open and switch S_2 is closed, current exists only in the 100-W filament. When both switches are closed, current exists in both filaments and the total power is 175 W.

If the filaments were connected in series and one of them were to break, no charges could pass through the lightbulb and it would not glow, regardless of the switch position. If, however, the filaments were connected in parallel and one of them (for example, the 75-W filament) were to break, the lightbulb would continue to glow in two of the switch positions because current exists in the other (100-W) filament.

As a second example, consider strings of incandescent lights that are used for many ornamental purposes such as decorating Christmas trees. Over the years, both parallel and series connections have been used for strings of lights. Because series-wired lightbulbs operate with less energy per bulb and at a lower temperature, they are safer than parallel-wired lightbulbs for indoor Christmas-tree use. If, however, the filament of a single lightbulb in a series-wired string were to fail (or if the lightbulb were removed from its socket), all the lights on the string would go out. The popularity of series-wired light strings diminished because troubleshooting a failed lightbulb is a tedious, time-consuming chore that involves trial-and-error substitution of a good lightbulb in each socket along the string until the defective one is found.

In a parallel-wired string, each lightbulb operates at 120 V. By design, the lightbulbs are brighter and hotter than those on a series-wired string. As a result, they are inherently more dangerous (more likely to start a fire, for instance), but if one lightbulb in a parallel-wired string fails or is removed, the rest of the lightbulbs continue to glow.

To prevent the failure of one lightbulb from causing the entire string to go out, a new design was developed for so-called miniature lights wired in series. When the filament breaks in one of these miniature lightbulbs, the break in the filament represents the largest resistance in the series, much larger than that of the intact filaments. As a result, most of the applied 120 V appears across the lightbulb with the broken filament. Inside the lightbulb, a small jumper loop covered by an insulating material is wrapped around the filament leads. When the filament fails and 120 V appears across the lightbulb, an arc burns the insulation on the jumper and connects the filament leads. This connection now completes the circuit through the lightbulb even though its filament is no longer active (Fig. 28.7, page 660).

When a lightbulb fails, the resistance across its terminals is reduced to almost zero because of the alternate jumper connection mentioned in the preceding paragraph. All the other lightbulbs not only stay on, but they glow more brightly because

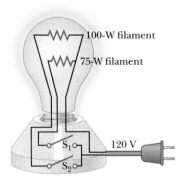

Figure 28.6 A three-way incandescent lightbulb.

---

[2]The three-way lightbulb and other household devices actually operate on alternating current (AC), to be introduced in Chapter 33.

Figure 28.7 (a) Schematic diagram of a modern "miniature" incandescent holiday lightbulb, with a jumper connection to provide a current path if the filament breaks. (b) A holiday lightbulb with a broken filament. (c) A Christmas-tree lightbulb.

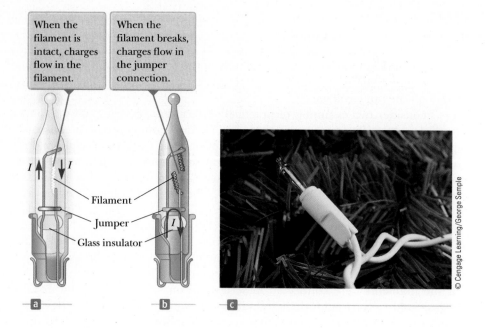

When the filament is intact, charges flow in the filament.

When the filament breaks, charges flow in the jumper connection.

Filament

Jumper

Glass insulator

a · b · c

© Cengage Learning/George Semple

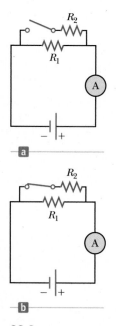

Figure 28.8 (Quick Quiz 28.3) What happens when the switch is closed?

the total resistance of the string is reduced and consequently the current in each remaining lightbulb increases. Each lightbulb operates at a slightly higher temperature than before. As more lightbulbs fail, the current keeps rising, the filament of each remaining lightbulb operates at a higher temperature, and the lifetime of the lightbulb is reduced. For this reason, you should check for failed (nonglowing) lightbulbs in such a series-wired string and replace them as soon as possible, thereby maximizing the lifetimes of all the lightbulbs.

Quick Quiz 28.3 With the switch in the circuit of Figure 28.8a open, there is no current in R_2. There is current in R_1, however, and it is measured with the ammeter at the right side of the circuit. If the switch is closed (Fig. 28.8b), there is current in R_2. What happens to the reading on the ammeter when the switch is closed? **(a)** The reading increases. **(b)** The reading decreases. **(c)** The reading does not change.

Quick Quiz 28.4 Consider the following choices: (a) increases, (b) decreases, (c) remains the same. From these choices, choose the best answer for the following situations. **(i)** In Figure 28.3, a third resistor is added in series with the first two. What happens to the current in the battery? **(ii)** What happens to the terminal voltage of the battery? **(iii)** In Figure 28.5, a third resistor is added in parallel with the first two. What happens to the current in the battery? **(iv)** What happens to the terminal voltage of the battery?

Conceptual Example 28.3   **Landscape Lights**

A homeowner wishes to install low-voltage landscape lighting in his back yard. To save money, he purchases inexpensive 18-gauge cable, which has a relatively high resistance per unit length. This cable consists of two side-by-side wires separated by insulation, like the cord on an appliance. He runs a 200-foot length of this cable from the power supply to the farthest point at which he plans to position a light fixture. He attaches light fixtures across the two wires on the cable at 10-foot intervals so that the light fixtures are in parallel. Because of the cable's resistance, the brightness of the lightbulbs in the fixtures is not as desired. Which of the following problems does the homeowner have? (a) All the lightbulbs glow equally less brightly than they would if lower-resistance cable had been used. (b) The brightness of the lightbulbs decreases as you move farther from the power supply.

▶ **28.3** continued

SOLUTION

A circuit diagram for the system appears in Figure 28.9. The horizontal resistors with letter subscripts (such as R_A) represent the resistance of the wires in the cable between the light fixtures, and the vertical resistors with number subscripts (such as R_1) represent the resistance of the light fixtures themselves. Part of the terminal voltage of the power supply is dropped across resistors R_A and R_B. Therefore, the voltage across light fixture R_1 is less than the terminal voltage. There is a further voltage drop across resistors R_C and R_D. Consequently, the voltage across light fixture R_2 is smaller than that across R_1. This pattern continues down the line of light fixtures, so the correct choice is (b). Each successive light fixture has a smaller voltage across it and glows less brightly than the one before.

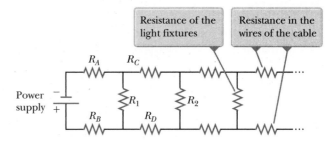

Figure 28.9 (Conceptual Example 28.3) The circuit diagram for a set of landscape light fixtures connected in parallel across the two wires of a two-wire cable.

Example 28.4 Find the Equivalent Resistance

Four resistors are connected as shown in Figure 28.10a.

(A) Find the equivalent resistance between points a and c.

SOLUTION

Conceptualize Imagine charges flowing into and through this combination from the left. All charges must pass from a to b through the first two resistors, but the charges split at b into two different paths when encountering the combination of the 6.0-Ω and the 3.0-Ω resistors.

Categorize Because of the simple nature of the combination of resistors in Figure 28.10, we categorize this example as one for which we can use the rules for series and parallel combinations of resistors.

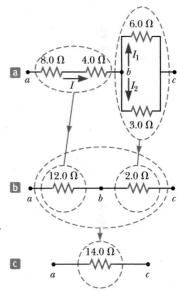

Figure 28.10 (Example 28.4) The original network of resistors is reduced to a single equivalent resistance.

Analyze The combination of resistors can be reduced in steps as shown in Figure 28.10.

Find the equivalent resistance between a and b of the 8.0-Ω and 4.0-Ω resistors, which are in series (left-hand red-brown circles):

$$R_{eq} = 8.0\ \Omega + 4.0\ \Omega = 12.0\ \Omega$$

Find the equivalent resistance between b and c of the 6.0-Ω and 3.0-Ω resistors, which are in parallel (right-hand red-brown circles):

$$\frac{1}{R_{eq}} = \frac{1}{6.0\ \Omega} + \frac{1}{3.0\ \Omega} = \frac{3}{6.0\ \Omega}$$

$$R_{eq} = \frac{6.0\ \Omega}{3} = 2.0\ \Omega$$

The circuit of equivalent resistances now looks like Figure 28.10b. The 12.0-Ω and 2.0-Ω resistors are in series (green circles). Find the equivalent resistance from a to c:

$$R_{eq} = 12.0\ \Omega + 2.0\ \Omega = \boxed{14.0\ \Omega}$$

This resistance is that of the single equivalent resistor in Figure 28.10c.

(B) What is the current in each resistor if a potential difference of 42 V is maintained between a and c?

continued

▶ **28.4** continued

SOLUTION

The currents in the 8.0-Ω and 4.0-Ω resistors are the same because they are in series. In addition, they carry the same current that would exist in the 14.0-Ω equivalent resistor subject to the 42-V potential difference.

Use Equation 27.7 ($R = \Delta V/I$) and the result from part (A) to find the current in the 8.0-Ω and 4.0-Ω resistors:

$$I = \frac{\Delta V_{ac}}{R_{eq}} = \frac{42\text{ V}}{14.0\ \Omega} = \boxed{3.0\text{ A}}$$

Set the voltages across the resistors in parallel in Figure 28.10a equal to find a relationship between the currents:

$$\Delta V_1 = \Delta V_2 \quad \rightarrow \quad (6.0\ \Omega)I_1 = (3.0\ \Omega)I_2 \quad \rightarrow \quad I_2 = 2I_1$$

Use $I_1 + I_2 = 3.0$ A to find I_1:

$$I_1 + I_2 = 3.0\text{ A} \quad \rightarrow \quad I_1 + 2I_1 = 3.0\text{ A} \quad \rightarrow \quad I_1 = \boxed{1.0\text{ A}}$$

Find I_2:

$$I_2 = 2I_1 = 2(1.0\text{ A}) = \boxed{2.0\text{ A}}$$

Finalize As a final check of our results, note that $\Delta V_{bc} = (6.0\ \Omega)I_1 = (3.0\ \Omega)I_2 = 6.0$ V and $\Delta V_{ab} = (12.0\ \Omega)I = 36$ V; therefore, $\Delta V_{ac} = \Delta V_{ab} + \Delta V_{bc} = 42$ V, as it must.

Example 28.5　**Three Resistors in Parallel**

Three resistors are connected in parallel as shown in Figure 28.11a. A potential difference of 18.0 V is maintained between points a and b.

(A) Calculate the equivalent resistance of the circuit.

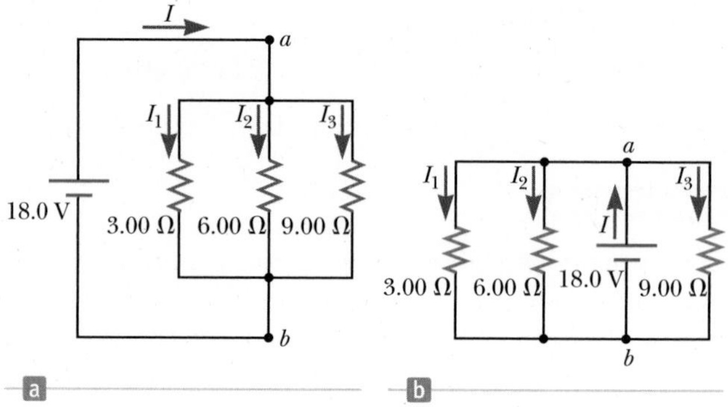

SOLUTION

Conceptualize Figure 28.11a shows that we are dealing with a simple parallel combination of three resistors. Notice that the current I splits into three currents I_1, I_2, and I_3 in the three resistors.

Figure 28.11 (Example 28.5) (a) Three resistors connected in parallel. The voltage across each resistor is 18.0 V. (b) Another circuit with three resistors and a battery. Is it equivalent to the circuit in (a)?

Categorize This problem can be solved with rules developed in this section, so we categorize it as a substitution problem. Because the three resistors are connected in parallel, we can use the rule for resistors in parallel, Equation 28.8, to evaluate the equivalent resistance.

Use Equation 28.8 to find R_{eq}:

$$\frac{1}{R_{eq}} = \frac{1}{3.00\ \Omega} + \frac{1}{6.00\ \Omega} + \frac{1}{9.00\ \Omega} = \frac{11}{18.0\ \Omega}$$

$$R_{eq} = \frac{18.0\ \Omega}{11} = \boxed{1.64\ \Omega}$$

(B) Find the current in each resistor.

SOLUTION

The potential difference across each resistor is 18.0 V. Apply the relationship $\Delta V = IR$ to find the currents:

$$I_1 = \frac{\Delta V}{R_1} = \frac{18.0\text{ V}}{3.00\ \Omega} = \boxed{6.00\text{ A}}$$

$$I_2 = \frac{\Delta V}{R_2} = \frac{18.0\text{ V}}{6.00\ \Omega} = \boxed{3.00\text{ A}}$$

$$I_3 = \frac{\Delta V}{R_3} = \frac{18.0\text{ V}}{9.00\ \Omega} = \boxed{2.00\text{ A}}$$

(C) Calculate the power delivered to each resistor and the total power delivered to the combination of resistors.

▶ **28.5** continued

SOLUTION

Apply the relationship $P = I^2R$ to each resistor using the currents calculated in part (B):

3.00-Ω: $P_1 = I_1^2R_1 = (6.00 \text{ A})^2(3.00 \text{ Ω}) = \boxed{108 \text{ W}}$

6.00-Ω: $P_2 = I_2^2R_2 = (3.00 \text{ A})^2(6.00 \text{ Ω}) = \boxed{54 \text{ W}}$

9.00-Ω: $P_3 = I_3^2R_3 = (2.00 \text{ A})^2(9.00 \text{ Ω}) = \boxed{36 \text{ W}}$

These results show that the smallest resistor receives the most power. Summing the three quantities gives a total power of $\boxed{198 \text{ W}}$. We could have calculated this final result from part (A) by considering the equivalent resistance as follows: $P = (\Delta V)^2/R_{eq} = (18.0 \text{ V})^2/1.64 \text{ Ω} = 198 \text{ W}$.

WHAT IF? What if the circuit were as shown in Figure 28.11b instead of as in Figure 28.11a? How would that affect the calculation?

Answer There would be no effect on the calculation. The physical placement of the battery is not important. Only the electrical arrangement is important. In Figure 28.11b, the battery still maintains a potential difference of 18.0 V between points a and b, so the two circuits in the figure are electrically identical.

28.3 Kirchhoff's Rules

As we saw in the preceding section, combinations of resistors can be simplified and analyzed using the expression $\Delta V = IR$ and the rules for series and parallel combinations of resistors. Very often, however, it is not possible to reduce a circuit to a single loop using these rules. The procedure for analyzing more complex circuits is made possible by using the following two principles, called **Kirchhoff's rules.**

> **1. Junction rule.** At any junction, the sum of the currents must equal zero:
>
> $$\sum_{junction} I = 0 \qquad \textbf{(28.9)}$$
>
> **2. Loop rule.** The sum of the potential differences across all elements around any closed circuit loop must be zero:
>
> $$\sum_{closed\ loop} \Delta V = 0 \qquad \textbf{(28.10)}$$

Kirchhoff's first rule is a statement of conservation of electric charge. All charges that enter a given point in a circuit must leave that point because charge cannot build up or disappear at a point. Currents directed into the junction are entered into the sum in the junction rule as $+I$, whereas currents directed out of a junction are entered as $-I$. Applying this rule to the junction in Figure 28.12a gives

$$I_1 - I_2 - I_3 = 0$$

Figure 28.12b represents a mechanical analog of this situation, in which water flows through a branched pipe having no leaks. Because water does not build up anywhere in the pipe, the flow rate into the pipe on the left equals the total flow rate out of the two branches on the right.

Kirchhoff's second rule follows from the law of conservation of energy for an isolated system. Let's imagine moving a charge around a closed loop of a circuit. When the charge returns to the starting point, the charge–circuit system must have the same total energy as it had before the charge was moved. The sum of the increases in energy as the charge passes through some circuit elements must equal the sum of the decreases in energy as it passes through other elements. The potential energy of the system decreases whenever the charge moves through a potential drop $-IR$ across a resistor or whenever it moves in the reverse direction through a

The amount of charge flowing out of the branches on the right must equal the amount flowing into the single branch on the left.

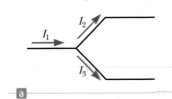

The amount of water flowing out of the branches on the right must equal the amount flowing into the single branch on the left.

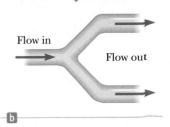

Figure 28.12 (a) Kirchhoff's junction rule. (b) A mechanical analog of the junction rule.

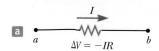

In each diagram, $\Delta V = V_b - V_a$ and the circuit element is traversed from a to b, left to right.

Figure 28.13 Rules for determining the signs of the potential differences across a resistor and a battery. (The battery is assumed to have no internal resistance.)

source of emf. The potential energy increases whenever the charge passes through a battery from the negative terminal to the positive terminal.

When applying Kirchhoff's second rule, imagine *traveling* around the loop and consider changes in *electric potential* rather than the changes in *potential energy* described in the preceding paragraph. Imagine traveling through the circuit elements in Figure 28.13 toward the right. The following sign conventions apply when using the second rule:

- Charges move from the high-potential end of a resistor toward the low-potential end, so if a resistor is traversed in the direction of the current, the potential difference ΔV across the resistor is $-IR$ (Fig. 28.13a).
- If a resistor is traversed in the direction *opposite* the current, the potential difference ΔV across the resistor is $+IR$ (Fig. 28.13b).
- If a source of emf (assumed to have zero internal resistance) is traversed in the direction of the emf (from negative to positive), the potential difference ΔV is $+\mathcal{E}$ (Fig. 28.13c).
- If a source of emf (assumed to have zero internal resistance) is traversed in the direction opposite the emf (from positive to negative), the potential difference ΔV is $-\mathcal{E}$ (Fig. 28.13d).

There are limits on the number of times you can usefully apply Kirchhoff's rules in analyzing a circuit. You can use the junction rule as often as you need as long as you include in it a current that has not been used in a preceding junction-rule equation. In general, the number of times you can use the junction rule is one fewer than the number of junction points in the circuit. You can apply the loop rule as often as needed as long as a new circuit element (resistor or battery) or a new current appears in each new equation. In general, to solve a particular circuit problem, the number of independent equations you need to obtain from the two rules equals the number of unknown currents.

Complex networks containing many loops and junctions generate a great number of independent linear equations and a correspondingly great number of unknowns. Such situations can be handled formally through the use of matrix algebra. Computer software can also be used to solve for the unknowns.

The following examples illustrate how to use Kirchhoff's rules. In all cases, it is assumed the circuits have reached steady-state conditions; in other words, the currents in the various branches are constant. Any capacitor acts as an open branch in a circuit; that is, the current in the branch containing the capacitor is zero under steady-state conditions.

Gustav Kirchhoff
German Physicist (1824–1887)
Kirchhoff, a professor at Heidelberg, and Robert Bunsen invented the spectroscope and founded the science of spectroscopy, which we shall study in Chapter 42. They discovered the elements cesium and rubidium and invented astronomical spectroscopy.

Problem-Solving Strategy Kirchhoff's Rules

The following procedure is recommended for solving problems that involve circuits that cannot be reduced by the rules for combining resistors in series or parallel.

1. Conceptualize. Study the circuit diagram and make sure you recognize all elements in the circuit. Identify the polarity of each battery and try to imagine the directions in which the current would exist in the batteries.

2. Categorize. Determine whether the circuit can be reduced by means of combining series and parallel resistors. If so, use the techniques of Section 28.2. If not, apply Kirchhoff's rules according to the *Analyze* step below.

3. Analyze. Assign labels to all known quantities and symbols to all unknown quantities. You must assign *directions* to the currents in each part of the circuit. Although the assignment of current directions is arbitrary, you must adhere *rigorously* to the directions you assign when you apply Kirchhoff's rules.

Apply the junction rule (Kirchhoff's first rule) to all junctions in the circuit except one. Now apply the loop rule (Kirchhoff's second rule) to as many loops in

▶ **Problem-Solving Strategy** continued

the circuit as are needed to obtain, in combination with the equations from the junction rule, as many equations as there are unknowns. To apply this rule, you must choose a direction in which to travel around the loop (either clockwise or counterclockwise) and correctly identify the change in potential as you cross each element. Be careful with signs!

Solve the equations simultaneously for the unknown quantities.

4. Finalize. Check your numerical answers for consistency. Do not be alarmed if any of the resulting currents have a negative value. That only means you have guessed the direction of that current incorrectly, but *its magnitude will be correct.*

| **Example 28.6** | **A Single-Loop Circuit** |

A single-loop circuit contains two resistors and two batteries as shown in Figure 28.14. (Neglect the internal resistances of the batteries.) Find the current in the circuit.

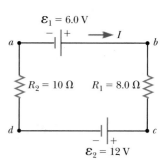

SOLUTION

Conceptualize Figure 28.14 shows the polarities of the batteries and a guess at the direction of the current. The 12-V battery is the stronger of the two, so the current should be counterclockwise. Therefore, we expect our guess for the direction of the current to be wrong, but we will continue and see how this incorrect guess is represented by our final answer.

Categorize We do not need Kirchhoff's rules to analyze this simple circuit, but let's use them anyway simply to see how they are applied. There are no junctions in this single-loop circuit; therefore, the current is the same in all elements.

Figure 28.14 (Example 28.6) A series circuit containing two batteries and two resistors, where the polarities of the batteries are in opposition.

Analyze Let's assume the current is clockwise as shown in Figure 28.14. Traversing the circuit in the clockwise direction, starting at a, we see that $a \to b$ represents a potential difference of $+\mathcal{E}_1$, $b \to c$ represents a potential difference of $-IR_1$, $c \to d$ represents a potential difference of $-\mathcal{E}_2$, and $d \to a$ represents a potential difference of $-IR_2$.

Apply Kirchhoff's loop rule to the single loop in the circuit:

$$\sum \Delta V = 0 \quad \to \quad \mathcal{E}_1 - IR_1 - \mathcal{E}_2 - IR_2 = 0$$

Solve for I and use the values given in Figure 28.14:

$$(1) \quad I = \frac{\mathcal{E}_1 - \mathcal{E}_2}{R_1 + R_2} = \frac{6.0 \text{ V} - 12 \text{ V}}{8.0 \ \Omega + 10 \ \Omega} = \boxed{-0.33 \text{ A}}$$

Finalize The negative sign for I indicates that the direction of the current is opposite the assumed direction. The emfs in the numerator subtract because the batteries in Figure 28.14 have opposite polarities. The resistances in the denominator add because the two resistors are in series.

WHAT IF? What if the polarity of the 12.0-V battery were reversed? How would that affect the circuit?

Answer Although we could repeat the Kirchhoff's rules calculation, let's instead examine Equation (1) and modify it accordingly. Because the polarities of the two batteries are now in the same direction, the signs of $\mathcal{E}_1$ and $\mathcal{E}_2$ are the same and Equation (1) becomes

$$I = \frac{\mathcal{E}_1 + \mathcal{E}_2}{R_1 + R_2} = \frac{6.0 \text{ V} + 12 \text{ V}}{8.0 \ \Omega + 10 \ \Omega} = 1.0 \text{ A}$$

| **Example 28.7** | **A Multiloop Circuit** |

Find the currents I_1, I_2, and I_3 in the circuit shown in Figure 28.15 on page 666.

continued

▶ **28.7** continued

SOLUTION

Conceptualize Imagine physically rearranging the circuit while keeping it electrically the same. Can you rearrange it so that it consists of simple series or parallel combinations of resistors? You should find that you cannot. (If the 10.0-V battery were removed and replaced by a wire from b to the 6.0-Ω resistor, the circuit would consist of only series and parallel combinations.)

Categorize We cannot simplify the circuit by the rules associated with combining resistances in series and in parallel. Therefore, this problem is one in which we must use Kirchhoff's rules.

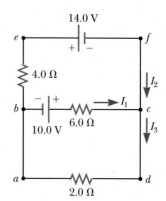

Figure 28.15 (Example 28.7) A circuit containing different branches.

Analyze We arbitrarily choose the directions of the currents as labeled in Figure 28.15.

Apply Kirchhoff's junction rule to junction c:

(1) $I_1 + I_2 - I_3 = 0$

We now have one equation with three unknowns: I_1, I_2, and I_3. There are three loops in the circuit: $abcda$, $befcb$, and $aefda$. We need only two loop equations to determine the unknown currents. (The third equation would give no new information.) Let's choose to traverse these loops in the clockwise direction. Apply Kirchhoff's loop rule to loops $abcda$ and $befcb$:

$abcda$: (2) $10.0\ \text{V} - (6.0\ \Omega)I_1 - (2.0\ \Omega)I_3 = 0$

$befcb$: $-(4.0\ \Omega)I_2 - 14.0\ \text{V} + (6.0\ \Omega)I_1 - 10.0\ \text{V} = 0$

(3) $-24.0\ \text{V} + (6.0\ \Omega)I_1 - (4.0\ \Omega)I_2 = 0$

Solve Equation (1) for I_3 and substitute into Equation (2):

$10.0\ \text{V} - (6.0\ \Omega)I_1 - (2.0\ \Omega)(I_1 + I_2) = 0$

(4) $10.0\ \text{V} - (8.0\ \Omega)I_1 - (2.0\ \Omega)I_2 = 0$

Multiply each term in Equation (3) by 4 and each term in Equation (4) by 3:

(5) $-96.0\ \text{V} + (24.0\ \Omega)I_1 - (16.0\ \Omega)I_2 = 0$

(6) $30.0\ \text{V} - (24.0\ \Omega)I_1 - (6.0\ \Omega)I_2 = 0$

Add Equation (6) to Equation (5) to eliminate I_1 and find I_2:

$-66.0\ \text{V} - (22.0\ \Omega)I_2 = 0$

$I_2 = \boxed{-3.0\ \text{A}}$

Use this value of I_2 in Equation (3) to find I_1:

$-24.0\ \text{V} + (6.0\ \Omega)I_1 - (4.0\ \Omega)(-3.0\ \text{A}) = 0$

$-24.0\ \text{V} + (6.0\ \Omega)I_1 + 12.0\ \text{V} = 0$

$I_1 = \boxed{2.0\ \text{A}}$

Use Equation (1) to find I_3:

$I_3 = I_1 + I_2 = 2.0\ \text{A} - 3.0\ \text{A} = \boxed{-1.0\ \text{A}}$

Finalize Because our values for I_2 and I_3 are negative, the directions of these currents are opposite those indicated in Figure 28.15. The numerical values for the currents are correct. Despite the incorrect direction, we *must* continue to use these negative values in subsequent calculations because our equations were established with our original choice of direction. What would have happened had we left the current directions as labeled in Figure 28.15 but traversed the loops in the opposite direction?

28.4 *RC* Circuits

So far, we have analyzed direct-current circuits in which the current is constant. In DC circuits containing capacitors, the current is always in the same direction but may vary in magnitude at different times. A circuit containing a series combination of a resistor and a capacitor is called an ***RC* circuit.**

Charging a Capacitor

Figure 28.16 shows a simple series *RC* circuit. Let's assume the capacitor in this circuit is initially uncharged. There is no current while the switch is open (Fig. 28.16a). If the switch is thrown to position *a* at *t* = 0 (Fig. 28.16b), however, charge begins to flow, setting up a current in the circuit, and the capacitor begins to charge.[3] Notice that during charging, charges do not jump across the capacitor plates because the gap between the plates represents an open circuit. Instead, charge is transferred between each plate and its connecting wires due to the electric field established in the wires by the battery until the capacitor is fully charged. As the plates are being charged, the potential difference across the capacitor increases. The value of the maximum charge on the plates depends on the voltage of the battery. Once the maximum charge is reached, the current in the circuit is zero because the potential difference across the capacitor matches that supplied by the battery.

To analyze this circuit quantitatively, let's apply Kirchhoff's loop rule to the circuit after the switch is thrown to position *a*. Traversing the loop in Figure 28.16b clockwise gives

$$\mathcal{E} - \frac{q}{C} - iR = 0 \qquad \textbf{(28.11)}$$

where q/C is the potential difference across the capacitor and iR is the potential difference across the resistor. We have used the sign conventions discussed earlier for the signs on $\mathcal{E}$ and iR. The capacitor is traversed in the direction from the positive plate to the negative plate, which represents a decrease in potential. Therefore, we use a negative sign for this potential difference in Equation 28.11. Note that lowercase q and i are *instantaneous* values that depend on time (as opposed to steady-state values) as the capacitor is being charged.

We can use Equation 28.11 to find the initial current I_i in the circuit and the maximum charge Q_{max} on the capacitor. At the instant the switch is thrown to position *a* (t = 0), the charge on the capacitor is zero. Equation 28.11 shows that the initial current I_i in the circuit is a maximum and is given by

$$I_i = \frac{\mathcal{E}}{R} \quad \text{(current at } t = 0\text{)} \qquad \textbf{(28.12)}$$

At this time, the potential difference from the battery terminals appears entirely across the resistor. Later, when the capacitor is charged to its maximum value Q_{max}, charges cease to flow, the current in the circuit is zero, and the potential difference from the battery terminals appears entirely across the capacitor. Substituting $i = 0$ into Equation 28.11 gives the maximum charge on the capacitor:

$$Q_{max} = C\mathcal{E} \quad \text{(maximum charge)} \qquad \textbf{(28.13)}$$

To determine analytical expressions for the time dependence of the charge and current, we must solve Equation 28.11, a single equation containing two variables q and i. The current in all parts of the series circuit must be the same. Therefore, the current in the resistance R must be the same as the current between each capacitor plate and the wire connected to it. This current is equal to the time rate of change of the charge on the capacitor plates. Therefore, we substitute $i = dq/dt$ into Equation 28.11 and rearrange the equation:

$$\frac{dq}{dt} = \frac{\mathcal{E}}{R} - \frac{q}{RC}$$

To find an expression for q, we solve this separable differential equation as follows. First combine the terms on the right-hand side:

$$\frac{dq}{dt} = \frac{C\mathcal{E}}{RC} - \frac{q}{RC} = -\frac{q - C\mathcal{E}}{RC}$$

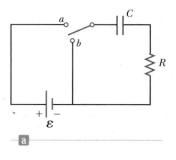

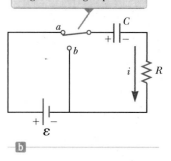

When the switch is thrown to position *a*, the capacitor begins to charge up.

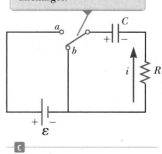

When the switch is thrown to position *b*, the capacitor discharges.

Figure 28.16 A capacitor in series with a resistor, switch, and battery.

---

[3]In previous discussions of capacitors, we assumed a steady-state situation, in which no current was present in any branch of the circuit containing a capacitor. Now we are considering the case *before* the steady-state condition is realized; in this situation, charges are moving and a current exists in the wires connected to the capacitor.

Multiply this equation by dt and divide by $q - C\mathcal{E}$:

$$\frac{dq}{q - C\mathcal{E}} = -\frac{1}{RC}\,dt$$

Integrate this expression, using $q = 0$ at $t = 0$:

$$\int_0^q \frac{dq}{q - C\mathcal{E}} = -\frac{1}{RC}\int_0^t dt$$

$$\ln\left(\frac{q - C\mathcal{E}}{-C\mathcal{E}}\right) = -\frac{t}{RC}$$

From the definition of the natural logarithm, we can write this expression as

Charge as a function of time for a capacitor being charged ▶

$$q(t) = C\mathcal{E}(1 - e^{-t/RC}) = Q_{\max}(1 - e^{-t/RC}) \qquad (28.14)$$

where e is the base of the natural logarithm and we have made the substitution from Equation 28.13.

We can find an expression for the charging current by differentiating Equation 28.14 with respect to time. Using $i = dq/dt$, we find that

Current as a function of time for a capacitor being charged ▶

$$i(t) = \frac{\mathcal{E}}{R}\,e^{-t/RC} \qquad (28.15)$$

Plots of capacitor charge and circuit current versus time are shown in Figure 28.17. Notice that the charge is zero at $t = 0$ and approaches the maximum value $C\mathcal{E}$ as $t \to \infty$. The current has its maximum value $I_i = \mathcal{E}/R$ at $t = 0$ and decays exponentially to zero as $t \to \infty$. The quantity RC, which appears in the exponents of Equations 28.14 and 28.15, is called the **time constant** τ of the circuit:

$$\tau = RC \qquad (28.16)$$

The time constant represents the time interval during which the current decreases to $1/e$ of its initial value; that is, after a time interval τ, the current decreases to $i = e^{-1}I_i = 0.368I_i$. After a time interval 2τ, the current decreases to $i = e^{-2}I_i = 0.135I_i$, and so forth. Likewise, in a time interval τ, the charge increases from zero to $C\mathcal{E}[1 - e^{-1}] = 0.632C\mathcal{E}$.

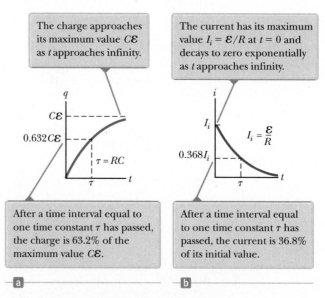

Figure 28.17 (a) Plot of capacitor charge versus time for the circuit shown in Figure 28.16b. (b) Plot of current versus time for the circuit shown in Figure 28.16b.

The following dimensional analysis shows that τ has units of time:

$$[\tau] = [RC] = \left[\left(\frac{\Delta V}{I}\right)\left(\frac{Q}{\Delta V}\right)\right] = \left[\frac{Q}{Q/\Delta t}\right] = [\Delta t] = \text{T}$$

Because $\tau = RC$ has units of time, the combination t/RC is dimensionless, as it must be to be an exponent of e in Equations 28.14 and 28.15.

The energy supplied by the battery during the time interval required to fully charge the capacitor is $Q_{max}\mathcal{E} = C\mathcal{E}^2$. After the capacitor is fully charged, the energy stored in the capacitor is $\frac{1}{2}Q_{max}\mathcal{E} = \frac{1}{2}C\mathcal{E}^2$, which is only half the energy output of the battery. It is left as a problem (Problem 68 in Enhanced WebAssign) to show that the remaining half of the energy supplied by the battery appears as internal energy in the resistor.

Discharging a Capacitor

Imagine that the capacitor in Figure 28.16b is completely charged. An initial potential difference Q_i/C exists across the capacitor, and there is zero potential difference across the resistor because $i = 0$. If the switch is now thrown to position b at $t = 0$ (Fig. 28.16c), the capacitor begins to discharge through the resistor. At some time t during the discharge, the current in the circuit is i and the charge on the capacitor is q. The circuit in Figure 28.16c is the same as the circuit in Figure 28.16b except for the absence of the battery. Therefore, we eliminate the emf $\mathcal{E}$ from Equation 28.11 to obtain the appropriate loop equation for the circuit in Figure 28.16c:

$$-\frac{q}{C} - iR = 0 \tag{28.17}$$

When we substitute $i = dq/dt$ into this expression, it becomes

$$-R\frac{dq}{dt} = \frac{q}{C}$$

$$\frac{dq}{q} = -\frac{1}{RC}\,dt$$

Integrating this expression using $q = Q_i$ at $t = 0$ gives

$$\int_{Q_i}^{q}\frac{dq}{q} = -\frac{1}{RC}\int_0^t dt$$

$$\ln\left(\frac{q}{Q_i}\right) = -\frac{t}{RC}$$

$$q(t) = Q_i e^{-t/RC} \tag{28.18}$$

◀ **Charge as a function of time for a discharging capacitor**

Differentiating Equation 28.18 with respect to time gives the instantaneous current as a function of time:

$$i(t) = -\frac{Q_i}{RC}e^{-t/RC} \tag{28.19}$$

◀ **Current as a function of time for a discharging capacitor**

where $Q_i/RC = I_i$ is the initial current. The negative sign indicates that as the capacitor discharges, the current direction is opposite its direction when the capacitor was being charged. (Compare the current directions in Figs. 28.16b and 28.16c.) Both the charge on the capacitor and the current decay exponentially at a rate characterized by the time constant $\tau = RC$.

Quick Quiz 28.5 Consider the circuit in Figure 28.18 and assume the battery has no internal resistance. **(i)** Just after the switch is closed, what is the current in the battery? (a) 0 (b) $\mathcal{E}/2R$ (c) $2\mathcal{E}/R$ (d) $\mathcal{E}/R$ (e) impossible to determine **(ii)** After a very long time, what is the current in the battery? Choose from the same choices.

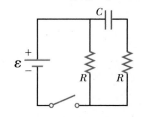

Figure 28.18 (Quick Quiz 28.5) How does the current vary after the switch is closed?

Conceptual Example 28.8 **Intermittent Windshield Wipers**

Many automobiles are equipped with windshield wipers that can operate intermittently during a light rainfall. How does the operation of such wipers depend on the charging and discharging of a capacitor?

SOLUTION

The wipers are part of an *RC* circuit whose time constant can be varied by selecting different values of *R* through a multiposition switch. As the voltage across the capacitor increases, the capacitor reaches a point at which it discharges and triggers the wipers. The circuit then begins another charging cycle. The time interval between the individual sweeps of the wipers is determined by the value of the time constant.

Example 28.9 **Charging a Capacitor in an *RC* Circuit**

An uncharged capacitor and a resistor are connected in series to a battery as shown in Figure 28.16, where $\mathcal{E} = 12.0$ V, $C = 5.00$ μF, and $R = 8.00 \times 10^5$ Ω. The switch is thrown to position *a*. Find the time constant of the circuit, the maximum charge on the capacitor, the maximum current in the circuit, and the charge and current as functions of time.

SOLUTION

Conceptualize Study Figure 28.16 and imagine throwing the switch to position *a* as shown in Figure 28.16b. Upon doing so, the capacitor begins to charge.

Categorize We evaluate our results using equations developed in this section, so we categorize this example as a substitution problem.

Evaluate the time constant of the circuit from Equation 28.16:

$$\tau = RC = (8.00 \times 10^5 \ \Omega)(5.00 \times 10^{-6} \ \text{F}) = \boxed{4.00 \ \text{s}}$$

Evaluate the maximum charge on the capacitor from Equation 28.13:

$$Q_{max} = C\mathcal{E} = (5.00 \ \mu\text{F})(12.0 \ \text{V}) = \boxed{60.0 \ \mu\text{C}}$$

Evaluate the maximum current in the circuit from Equation 28.12:

$$I_i = \frac{\mathcal{E}}{R} = \frac{12.0 \ \text{V}}{8.00 \times 10^5 \ \Omega} = \boxed{15.0 \ \mu\text{A}}$$

Use these values in Equations 28.14 and 28.15 to find the charge and current as functions of time:

(1) $q(t) = \boxed{60.0(1 - e^{-t/4.00})}$

(2) $i(t) = \boxed{15.0e^{-t/4.00}}$

In Equations (1) and (2), *q* is in microcoulombs, *i* is in microamperes, and *t* is in seconds.

Example 28.10 **Discharging a Capacitor in an *RC* Circuit**

Consider a capacitor of capacitance *C* that is being discharged through a resistor of resistance *R* as shown in Figure 28.16c.

(A) After how many time constants is the charge on the capacitor one-fourth its initial value?

SOLUTION

Conceptualize Study Figure 28.16 and imagine throwing the switch to position *b* as shown in Figure 28.16c. Upon doing so, the capacitor begins to discharge.

Categorize We categorize the example as one involving a discharging capacitor and use the appropriate equations.

▶ **28.10** continued

Analyze Substitute $q(t) = Q_i/4$ into Equation 28.18:

$$\frac{Q_i}{4} = Q_i e^{-t/RC}$$

$$\tfrac{1}{4} = e^{-t/RC}$$

Take the logarithm of both sides of the equation and solve for t:

$$-\ln 4 = -\frac{t}{RC}$$

$$t = RC \ln 4 = 1.39 RC = \boxed{1.39\tau}$$

(B) The energy stored in the capacitor decreases with time as the capacitor discharges. After how many time constants is this stored energy one-fourth its initial value?

SOLUTION

Use Equations 26.11 and 28.18 to express the energy stored in the capacitor at any time t:

$$(1) \quad U(t) = \frac{q^2}{2C} = \frac{Q_i^{\,2}}{2C} e^{-2t/RC}$$

Substitute $U(t) = \tfrac{1}{4}(Q_i^{\,2}/2C)$ into Equation (1):

$$\frac{1}{4}\frac{Q_i^{\,2}}{2C} = \frac{Q_i^{\,2}}{2C} e^{-2t/RC}$$

$$\tfrac{1}{4} = e^{-2t/RC}$$

Take the logarithm of both sides of the equation and solve for t:

$$-\ln 4 = -\frac{2t}{RC}$$

$$t = \tfrac{1}{2} RC \ln 4 = 0.693 RC = \boxed{0.693\tau}$$

Finalize Notice that because the energy depends on the square of the charge, the energy in the capacitor drops more rapidly than the charge on the capacitor.

WHAT IF?   What if you want to describe the circuit in terms of the time interval required for the charge to fall to one-half its original value rather than by the time constant τ? That would give a parameter for the circuit called its *half-life* $t_{1/2}$. How is the half-life related to the time constant?

Answer In one half-life, the charge falls from Q_i to $Q_i/2$. Therefore, from Equation 28.18,

$$\frac{Q_i}{2} = Q_i e^{-t_{1/2}/RC} \quad \rightarrow \quad \tfrac{1}{2} = e^{-t_{1/2}/RC}$$

which leads to

$$t_{1/2} = 0.693\tau$$

The concept of half-life will be important to us when we study nuclear decay in Chapter 44. The radioactive decay of an unstable sample behaves in a mathematically similar manner to a discharging capacitor in an *RC* circuit.

Example 28.11   **Energy Delivered to a Resistor**   **AM**

A 5.00-μF capacitor is charged to a potential difference of 800 V and then discharged through a resistor. How much energy is delivered to the resistor in the time interval required to fully discharge the capacitor?

SOLUTION

Conceptualize In Example 28.10, we considered the energy decrease in a discharging capacitor to a value of one-fourth the initial energy. In this example, the capacitor fully discharges.

Categorize We solve this example using two approaches. The first approach is to model the circuit as an *isolated system* for *energy*. Because energy in an isolated system is conserved, the initial electric potential energy U_E stored in the

continued

▶ **28.11** continued

capacitor is transformed into internal energy $E_{int} = E_R$ in the resistor. The second approach is to model the resistor as a *nonisolated system* for *energy*. Energy enters the resistor by electrical transmission from the capacitor, causing an increase in the resistor's internal energy.

Analyze We begin with the isolated system approach.

Write the appropriate reduction of the conservation of energy equation, Equation 8.2:

$$\Delta U + \Delta E_{int} = 0$$

Substitute the initial and final values of the energies:

$$(0 - U_E) + (E_{int} - 0) = 0 \quad \rightarrow \quad E_R = U_E$$

Use Equation 26.11 for the electric potential energy in the capacitor:

$$E_R = \tfrac{1}{2}C\mathcal{E}^2$$

Substitute numerical values:

$$E_R = \tfrac{1}{2}(5.00 \times 10^{-6}\,\text{F})(800\,\text{V})^2 = \boxed{1.60\,\text{J}}$$

The second approach, which is more difficult but perhaps more instructive, is to note that as the capacitor discharges through the resistor, the rate at which energy is delivered to the resistor by electrical transmission is i^2R, where i is the instantaneous current given by Equation 28.19.

Evaluate the energy delivered to the resistor by integrating the power over all time because it takes an infinite time interval for the capacitor to completely discharge:

$$P = \frac{dE}{dt} \quad \rightarrow \quad E_R = \int_0^\infty P\,dt$$

Substitute for the power delivered to the resistor:

$$E_R = \int_0^\infty i^2 R\,dt$$

Substitute for the current from Equation 28.19:

$$E_R = \int_0^\infty \left(-\frac{Q_i}{RC}\,e^{-t/RC}\right)^2 R\,dt = \frac{Q_i^2}{RC^2}\int_0^\infty e^{-2t/RC}\,dt = \frac{\mathcal{E}^2}{R}\int_0^\infty e^{-2t/RC}\,dt$$

Substitute the value of the integral, which is $RC/2$ (see Problem 44 in Enhanced WebAssign):

$$E_R = \frac{\mathcal{E}^2}{R}\left(\frac{RC}{2}\right) = \tfrac{1}{2}C\mathcal{E}^2$$

Finalize This result agrees with that obtained using the isolated system approach, as it must. We can use this second approach to find the total energy delivered to the resistor at *any* time after the switch is closed by simply replacing the upper limit in the integral with that specific value of t.

28.5 Household Wiring and Electrical Safety

Many considerations are important in the design of an electrical system of a home that will provide adequate electrical service for the occupants while maximizing their safety. We discuss some aspects of a home electrical system in this section.

Household Wiring

Household circuits represent a practical application of some of the ideas presented in this chapter. In our world of electrical appliances, it is useful to understand the power requirements and limitations of conventional electrical systems and the safety measures that prevent accidents.

In a conventional installation, the utility company distributes electric power to individual homes by means of a pair of wires, with each home connected in parallel to these wires. One wire is called the *live wire*[4] as illustrated in Figure 28.19, and the other is called the *neutral wire*. The neutral wire is grounded; that is, its electric potential is taken to be zero. The potential difference between the live and neutral wires is approximately 120 V. This voltage alternates in time, and the potential of the live wire oscillates relative to ground. Much of what we have learned so far for the constant-emf situation (direct current) can also be applied to the alternating current that power companies supply to businesses and households. (Alternating voltage and current are discussed in Chapter 33.)

To record a household's energy consumption, a meter is connected in series with the live wire entering the house. After the meter, the wire splits so that there are several separate circuits in parallel distributed throughout the house. Each circuit contains a circuit breaker (or, in older installations, a fuse). A circuit breaker is a special switch that opens if the current exceeds the rated value for the circuit breaker. The wire and circuit breaker for each circuit are carefully selected to meet the current requirements for that circuit. If a circuit is to carry currents as large as 30 A, a heavy wire and an appropriate circuit breaker must be selected to handle this current. A circuit used to power only lamps and small appliances often requires only 20 A. Each circuit has its own circuit breaker to provide protection for that part of the entire electrical system of the house.

As an example, consider a circuit in which a toaster oven, a microwave oven, and a coffee maker are connected (corresponding to R_1, R_2, and R_3 in Fig. 28.19). We can calculate the current in each appliance by using the expression $P = I \Delta V$. The toaster oven, rated at 1 000 W, draws a current of 1 000 W/120 V = 8.33 A. The microwave oven, rated at 1 300 W, draws 10.8 A, and the coffee maker, rated at 800 W, draws 6.67 A. When the three appliances are operated simultaneously, they draw a total current of 25.8 A. Therefore, the circuit must be wired to handle at least this much current. If the rating of the circuit breaker protecting the circuit is too small—say, 20 A—the breaker will be tripped when the third appliance is turned on, preventing all three appliances from operating. To avoid this situation, the toaster oven and coffee maker can be operated on one 20-A circuit and the microwave oven on a separate 20-A circuit.

Many heavy-duty appliances such as electric ranges and clothes dryers require 240 V for their operation. The power company supplies this voltage by providing a third wire that is 120 V below ground potential (Fig. 28.20). The potential difference between this live wire and the other live wire (which is 120 V above ground potential) is 240 V. An appliance that operates from a 240-V line requires half as much current compared with operating it at 120 V; therefore, smaller wires can be used in the higher-voltage circuit without overheating.

Electrical Safety

When the live wire of an electrical outlet is connected directly to ground, the circuit is completed and a *short-circuit condition* exists. A short circuit occurs when almost zero resistance exists between two points at different potentials, and the result is a very large current. When that happens accidentally, a properly operating circuit breaker opens the circuit and no damage is done. A person in contact with ground, however, can be electrocuted by touching the live wire of a frayed cord or other exposed conductor. An exceptionally effective (and dangerous!) ground contact is made when the person either touches a water pipe (normally at ground potential) or stands on the ground with wet feet. The latter situation represents effective ground contact because normal, nondistilled water is a conductor due to the large number of ions associated with impurities. This situation should be avoided at all cost.

The electrical meter measures the power in watts.

Figure 28.19 Wiring diagram for a household circuit. The resistances represent appliances or other electrical devices that operate with an applied voltage of 120 V.

© Cengage Learning/George Semple

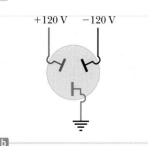

Figure 28.20 (a) An outlet for connection to a 240-V supply. (b) The connections for each of the openings in a 240-V outlet.

---

[4]*Live wire* is a common expression for a conductor whose electric potential is above or below ground potential.

Figure 28.21 (a) A diagram of the circuit for an electric drill with only two connecting wires. The normal current path is from the live wire through the motor connections and back to ground through the neutral wire. (b) This shock can be avoided by connecting the drill case to ground through a third ground wire. The wire colors represent electrical standards in the United States: the "hot" wire is black, the ground wire is green, and the neutral wire is white (shown as gray in the figure).

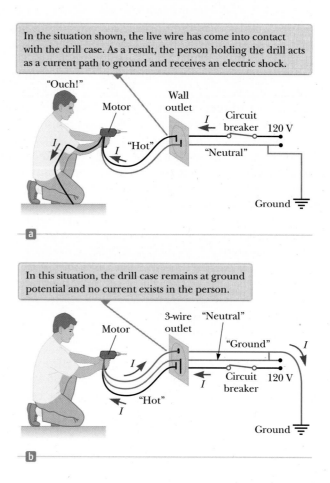

In the situation shown, the live wire has come into contact with the drill case. As a result, the person holding the drill acts as a current path to ground and receives an electric shock.

In this situation, the drill case remains at ground potential and no current exists in the person.

Electric shock can result in fatal burns or can cause the muscles of vital organs such as the heart to malfunction. The degree of damage to the body depends on the magnitude of the current, the length of time it acts, the part of the body touched by the live wire, and the part of the body in which the current exists. Currents of 5 mA or less cause a sensation of shock, but ordinarily do little or no damage. If the current is larger than about 10 mA, the muscles contract and the person may be unable to release the live wire. If the body carries a current of about 100 mA for only a few seconds, the result can be fatal. Such a large current paralyzes the respiratory muscles and prevents breathing. In some cases, currents of approximately 1 A can produce serious (and sometimes fatal) burns. In practice, no contact with live wires is regarded as safe whenever the voltage is greater than 24 V.

Many 120-V outlets are designed to accept a three-pronged power cord. (This feature is required in all new electrical installations.) One of these prongs is the live wire at a nominal potential of 120 V. The second is the neutral wire, nominally at 0 V, which carries current to ground. Figure 28.21a shows a connection to an electric drill with only these two wires. If the live wire accidentally makes contact with the casing of the electric drill (which can occur if the wire insulation wears off), current can be carried to ground by way of the person, resulting in an electric shock. The third wire in a three-pronged power cord, the round prong, is a safety ground wire that normally carries no current. It is both grounded and connected directly to the casing of the appliance. If the live wire is accidentally shorted to the casing in this situation, most of the current takes the low-resistance path through the appliance to ground as shown in Figure 28.21b.

Special power outlets called *ground-fault circuit interrupters,* or GFCIs, are used in kitchens, bathrooms, basements, exterior outlets, and other hazardous areas of homes. These devices are designed to protect persons from electric shock by sensing small currents (< 5 mA) leaking to ground. (The principle of their operation is described in Chapter 31.) When an excessive leakage current is detected, the current is shut off in less than 1 ms.

Summary

Definition

▎ The **emf** of a battery is equal to the voltage across its terminals when the current is zero. That is, the emf is equivalent to the **open-circuit voltage** of the battery.

Concepts and Principles

▎ The **equivalent resistance** of a set of resistors connected in a **series combination** is

$$R_{eq} = R_1 + R_2 + R_3 + \cdots \quad \text{(28.6)}$$

The **equivalent resistance** of a set of resistors connected in a **parallel combination** is found from the relationship

$$\frac{1}{R_{eq}} = \frac{1}{R_1} + \frac{1}{R_2} + \frac{1}{R_3} + \cdots \quad \text{(28.8)}$$

▎ Circuits involving more than one loop are conveniently analyzed with the use of **Kirchhoff's rules:**

1. **Junction rule.** At any junction, the sum of the currents must equal zero:

$$\sum_{\text{junction}} I = 0 \quad \text{(28.9)}$$

2. **Loop rule.** The sum of the potential differences across all elements around any circuit loop must be zero:

$$\sum_{\text{closed loop}} \Delta V = 0 \quad \text{(28.10)}$$

When a resistor is traversed in the direction of the current, the potential difference ΔV across the resistor is $-IR$. When a resistor is traversed in the direction opposite the current, $\Delta V = +IR$. When a source of emf is traversed in the direction of the emf (negative terminal to positive terminal), the potential difference is $+\varepsilon$. When a source of emf is traversed opposite the emf (positive to negative), the potential difference is $-\varepsilon$.

▎ If a capacitor is charged with a battery through a resistor of resistance R, the charge on the capacitor and the current in the circuit vary in time according to the expressions

$$q(t) = Q_{max}(1 - e^{-t/RC}) \quad \text{(28.14)}$$

$$i(t) = \frac{\varepsilon}{R} e^{-t/RC} \quad \text{(28.15)}$$

where $Q_{max} = C\varepsilon$ is the maximum charge on the capacitor. The product RC is called the **time constant** τ of the circuit.

▎ If a charged capacitor of capacitance C is discharged through a resistor of resistance R, the charge and current decrease exponentially in time according to the expressions

$$q(t) = Q_i e^{-t/RC} \quad \text{(28.18)}$$

$$i(t) = -\frac{Q_i}{RC} e^{-t/RC} \quad \text{(28.19)}$$

where Q_i is the initial charge on the capacitor and Q_i/RC is the initial current in the circuit.

Objective Questions

▎1.▎ denotes answer available in *Student Solutions Manual/Study Guide*

1. Is a circuit breaker wired (a) in series with the device it is protecting, (b) in parallel, or (c) neither in series or in parallel, or (d) is it impossible to tell?

2. A battery has some internal resistance. (i) Can the potential difference across the terminals of the battery be equal to its emf? (a) no (b) yes, if the battery is absorbing energy by electrical transmission (c) yes, if more than one wire is connected to each terminal (d) yes, if the current in the battery is zero (e) yes, with no special condition required. (ii) Can the terminal voltage exceed the emf? Choose your answer from the same possibilities as in part (i).

3. The terminals of a battery are connected across two resistors in series. The resistances of the resistors are not the same. Which of the following statements are correct? Choose all that are correct. (a) The resistor with the smaller resistance carries more current than the other resistor. (b) The resistor with the larger resistance carries less current than the other resistor. (c) The current in each resistor is the same. (d) The potential difference across each resistor is the same. (e) The potential difference is greatest across the resistor closest to the positive terminal.

4. When operating on a 120-V circuit, an electric heater receives 1.30×10^3 W of power, a toaster receives 1.00×10^3 W, and an electric oven receives 1.54×10^3 W. If all three appliances are connected in parallel on a 120-V circuit and turned on, what is the total current drawn from an external source? (a) 24.0 A (b) 32.0 A (c) 40.0 A (d) 48.0 A (e) none of those answers

5. If the terminals of a battery with zero internal resistance are connected across two identical resistors in series, the total power delivered by the battery is 8.00 W. If the same battery is connected across the same resistors in parallel, what is the total power delivered by the battery? (a) 16.0 W (b) 32.0 W (c) 2.00 W (d) 4.00 W (e) none of those answers

6. Several resistors are connected in series. Which of the following statements is correct? Choose all that are correct. (a) The equivalent resistance is greater than any of the resistances in the group. (b) The equivalent resistance is less than any of the resistances in the group. (c) The equivalent resistance depends on the voltage applied across the group. (d) The equivalent resistance is equal to the sum of the resistances in the group. (e) None of those statements is correct.

7. What is the time constant of the circuit shown in Figure OQ28.7? Each of the five resistors has resistance R, and each of the five capacitors has capacitance C. The internal resistance of the battery is negligible. (a) RC (b) $5RC$ (c) $10RC$ (d) $25RC$ (e) none of those answers

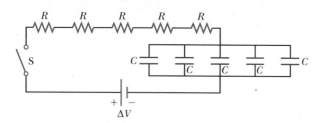

Figure OQ28.7

8. When resistors with different resistances are connected in series, which of the following must be the same for each resistor? Choose all correct answers. (a) potential difference (b) current (c) power delivered (d) charge entering each resistor in a given time interval (e) none of those answers

9. When resistors with different resistances are connected in parallel, which of the following must be the same for each resistor? Choose all correct answers. (a) potential difference (b) current (c) power delivered (d) charge entering each resistor in a given time interval (e) none of those answers

10. The terminals of a battery are connected across two resistors in parallel. The resistances of the resistors are not the same. Which of the following statements is correct? Choose all that are correct. (a) The resistor with the larger resistance carries more current than the other resistor. (b) The resistor with the larger resistance carries less current than the other resistor. (c) The potential difference across each resistor is the same. (d) The potential difference across the larger resistor is greater than the potential difference across the smaller resistor. (e) The potential difference is greater across the resistor closer to the battery.

11. Are the two headlights of a car wired (a) in series with each other, (b) in parallel, or (c) neither in series nor in parallel, or (d) is it impossible to tell?

12. In the circuit shown in Figure OQ28.12, each battery is delivering energy to the circuit by electrical transmission. All the resistors have equal resistance. **(i)** Rank the electric potentials at points a, b, c, d, and e from highest to lowest, noting any cases of equality in the ranking. **(ii)** Rank the magnitudes of the currents at the same points from greatest to least, noting any cases of equality.

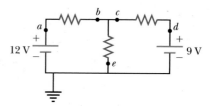

Figure OQ28.12

13. Several resistors are connected in parallel. Which of the following statements are correct? Choose all that are correct. (a) The equivalent resistance is greater than any of the resistances in the group. (b) The equivalent resistance is less than any of the resistances in the group. (c) The equivalent resistance depends on the voltage applied across the group. (d) The equivalent resistance is equal to the sum of the resistances in the group. (e) None of those statements is correct.

14. A circuit consists of three identical lamps connected to a battery as in Figure OQ28.14. The battery has some internal resistance. The switch S, originally open, is closed. **(i)** What then happens to the brightness of lamp B? (a) It increases. (b) It decreases somewhat. (c) It does not change. (d) It drops to zero. For parts (ii) to (vi), choose from the same possibilities (a) through (d). **(ii)** What happens to the brightness of lamp C? **(iii)** What happens to the current in the battery? **(iv)** What happens to the potential difference across lamp A? **(v)** What happens to the potential difference

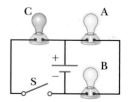

Figure OQ28.14

across lamp C? **(vi)** What happens to the total power delivered to the lamps by the battery?

15. A series circuit consists of three identical lamps connected to a battery as shown in Figure OQ28.15. The switch S, originally open, is closed. **(i)** What then happens to the brightness of lamp B? (a) It increases. (b) It decreases somewhat. (c) It does not change. (d) It drops to zero. For parts (ii) to (vi), choose from the same possibilities (a) through (d). **(ii)** What happens to the brightness of lamp C? **(iii)** What happens to the current in the battery? **(iv)** What happens to the potential difference across

lamp A? **(v)** What happens to the potential difference across lamp C? **(vi)** What happens to the total power delivered to the lamps by the battery?

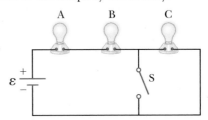

Figure OQ28.15

Conceptual Questions

1. denotes answer available in *Student Solutions Manual/Study Guide*

1. Suppose a parachutist lands on a high-voltage wire and grabs the wire as she prepares to be rescued. (a) Will she be electrocuted? (b) If the wire then breaks, should she continue to hold onto the wire as she falls to the ground? Explain.

2. A student claims that the second of two lightbulbs in series is less bright than the first because the first lightbulb uses up some of the current. How would you respond to this statement?

3. Why is it possible for a bird to sit on a high-voltage wire without being electrocuted?

4. Given three lightbulbs and a battery, sketch as many different electric circuits as you can.

5. A ski resort consists of a few chairlifts and several interconnected downhill runs on the side of a mountain, with a lodge at the bottom. The chairlifts are analogous to batteries, and the runs are analogous to resistors. Describe how two runs can be in series. Describe how three runs can be in parallel. Sketch a junction between one chairlift and two runs. State Kirchhoff's junction rule for ski resorts. One of the skiers happens to be carrying a skydiver's altimeter. She never takes the same set of chairlifts and runs twice, but keeps passing you at the fixed location where you are working. State Kirchhoff's loop rule for ski resorts.

6. Referring to Figure CQ28.6, describe what happens to the lightbulb after the switch is closed. Assume the capacitor has a large capacitance and is initially uncharged. Also assume the light illuminates when connected directly across the battery terminals.

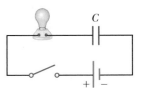

Figure CQ28.6

7. So that your grandmother can listen to *A Prairie Home Companion*, you take her bedside radio to the hospital where she is staying. You are required to have a maintenance worker test the radio for electrical safety. Finding that it develops 120 V on one of its knobs, he does not let you take it to your grandmother's room. Your grandmother complains that she has had the radio for many years and nobody has ever gotten a shock from it. You end up having to buy a new plastic radio. (a) Why is your grandmother's old radio dangerous in a hospital room? (b) Will the old radio be safe back in her bedroom?

8. (a) What advantage does 120-V operation offer over 240 V? (b) What disadvantages does it have?

9. Is the direction of current in a battery always from the negative terminal to the positive terminal? Explain.

10. Compare series and parallel resistors to the series and parallel rods in Figure 20.13 on page 480. How are the situations similar?

Problems available in ENHANCED WebAssign   Access end-of-chapter problems online at **www.webassign.net**

Section 28.1 Electromotive Force
Problems 1–4

Section 28.2 Resistors in Series and Parallel
Problems 5–21

Section 28.3 Kirchhoff's Rules
Problems 22–36

Section 28.4 *RC* Circuits
Problems 37–45

Section 28.5 Household Wiring and Electrical Safety
Problems 43–46

28.6 Relative Velocity and Relative Acceleration
Problems 44–48

Additional Problems
Problems 49–81

Challenge Problems
Problem 82–83

Solutions to the following Problems are available in the *Student Solutions Manual/Study Guide*:
28.1, 28.9, 28.19, 28.22, 28.23, 28.24, 28.27, 28.38, 28.41, 28.43, 28.46, 28.59, 28.66, 28.67, 28.71, 28.72, and 28.75

List of Enhanced Problems

Problem Number	Targeted Feedback in Enhanced WebAssign	Analysis Model Tutorial in Enhanced WebAssign	Master It in Enhanced WebAssign	Watch It in Enhanced WebAssign
28.1			✓	
28.2		✓		
28.3	✓			✓
28.5	✓			✓
28.9	✓		✓	
28.13	✓		✓	
28.19	✓			✓
28.23	✓		✓	
28.25			✓	
28.29	✓			✓
28.30	✓			✓
28.31	✓		✓	
28.35	✓		✓	
28.38	✓			✓
28.39	✓			✓
28.41	✓			✓
28.43			✓	
28.46			✓	
28.47	✓		✓	
28.59			✓	
28.67		✓	✓	
28.72			✓	

Magnetic Fields

An engineer performs a test on the electronics associated with one of the superconducting magnets in the Large Hadron Collider at the European Laboratory for Particle Physics, operated by the European Organization for Nuclear Research (CERN). The magnets are used to control the motion of charged particles in the accelerator. We will study the effects of magnetic fields on moving charged particles in this chapter. *(CERN)*

Many historians of science believe that the compass, which uses a magnetic needle, was used in China as early as the 13th century BC, its invention being of Arabic or Indian origin. The early Greeks knew about magnetism as early as 800 BC. They discovered that the stone magnetite (Fe_3O_4) attracts pieces of iron. Legend ascribes the name *magnetite* to the shepherd Magnes, the nails of whose shoes and the tip of whose staff stuck fast to chunks of magnetite while he pastured his flocks.

In 1269, Pierre de Maricourt of France found that the directions of a needle near a spherical natural magnet formed lines that encircled the sphere and passed through two points diametrically opposite each other, which he called the *poles* of the magnet. Subsequent experiments showed that every magnet, regardless of its shape, has two poles, called *north* (N) and *south* (S) poles, that exert forces on other magnetic poles similar to the way electric charges exert forces on one another. That is, like poles (N–N or S–S) repel each other, and opposite poles (N–S) attract each other.

The poles received their names because of the way a magnet, such as that in a compass, behaves in the presence of the Earth's magnetic field. If a bar magnet is suspended from its midpoint and can swing freely in a horizontal plane, it will rotate until its north pole points to the Earth's geographic North Pole and its south pole points to the Earth's geographic South Pole.[1]

In 1600, William Gilbert (1540–1603) extended de Maricourt's experiments to a variety of materials. He knew that a compass needle orients in preferred directions, so he suggested that the Earth itself is a large, permanent magnet. In 1750, experimenters used a torsion balance to show that magnetic poles exert attractive or repulsive forces on each other and that these forces vary as the inverse square of the distance between interacting poles. Although the force between two magnetic poles is otherwise similar to the force between two electric charges, electric charges can be isolated (witness the electron and proton), whereas a single magnetic pole has never been isolated. That is, magnetic poles are always found in pairs. All attempts thus far to detect an isolated magnetic pole have been unsuccessful. No matter how many times a permanent magnet is cut in two, each piece always has a north and a south pole.[2]

The relationship between magnetism and electricity was discovered in 1819 when, during a lecture demonstration, Hans Christian Oersted found that an electric current in a wire deflected a nearby compass needle.[3] In the 1820s, further connections between electricity and magnetism were demonstrated independently by Faraday and Joseph Henry (1797–1878). They showed that an electric current can be produced in a circuit either by moving a magnet near the circuit or by changing the current in a nearby circuit. These observations demonstrate that a changing magnetic field creates an electric field. Years later, theoretical work by Maxwell showed that the reverse is also true: a changing electric field creates a magnetic field.

This chapter examines the forces that act on moving charges and on current-carrying wires in the presence of a magnetic field. The source of the magnetic field is described in Chapter 30.

Hans Christian Oersted
Danish Physicist and Chemist (1777–1851)
Oersted is best known for observing that a compass needle deflects when placed near a wire carrying a current. This important discovery was the first evidence of the connection between electric and magnetic phenomena. Oersted was also the first to prepare pure aluminum.

29.1 Analysis Model: Particle in a Field (Magnetic)

In our study of electricity, we described the interactions between charged objects in terms of electric fields. Recall that an electric field surrounds any electric charge. In addition to containing an electric field, the region of space surrounding any *moving* electric charge also contains a **magnetic field.** A magnetic field also surrounds a magnetic substance making up a permanent magnet.

Historically, the symbol $\vec{\mathbf{B}}$ has been used to represent a magnetic field, and we use this notation in this book. The direction of the magnetic field $\vec{\mathbf{B}}$ at any location is the direction in which a compass needle points at that location. As with the electric field, we can represent the magnetic field by means of drawings with *magnetic field lines*.

Figure 29.1 shows how the magnetic field lines of a bar magnet can be traced with the aid of a compass. Notice that the magnetic field lines outside the magnet

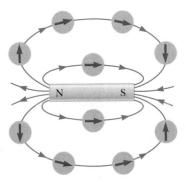

Figure 29.1 Compass needles can be used to trace the magnetic field lines in the region outside a bar magnet.

---

[1]The Earth's geographic North Pole is magnetically a south pole, whereas the Earth's geographic South Pole is magnetically a north pole. Because *opposite* magnetic poles attract each other, the pole on a magnet that is attracted to the Earth's geographic North Pole is the magnet's *north* pole and the pole attracted to the Earth's geographic South Pole is the magnet's *south* pole.

[2]There is some theoretical basis for speculating that magnetic *monopoles*—isolated north or south poles—may exist in nature, and attempts to detect them are an active experimental field of investigation.

[3]The same discovery was reported in 1802 by an Italian jurist, Gian Domenico Romagnosi, but was overlooked, probably because it was published in an obscure journal.

Figure 29.2 Magnetic field patterns can be displayed with iron filings sprinkled on paper near magnets.

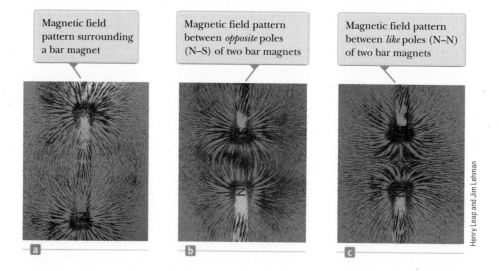

Magnetic field pattern surrounding a bar magnet

Magnetic field pattern between *opposite* poles (N–S) of two bar magnets

Magnetic field pattern between *like* poles (N–N) of two bar magnets

Henry Leap and Jim Lehman

point away from the north pole and toward the south pole. One can display magnetic field patterns of a bar magnet using small iron filings as shown in Figure 29.2.

When we speak of a compass magnet having a north pole and a south pole, it is more proper to say that it has a "north-seeking" pole and a "south-seeking" pole. This wording means that the north-seeking pole points to the north geographic pole of the Earth, whereas the south-seeking pole points to the south geographic pole. Because the north pole of a magnet is attracted toward the north geographic pole of the Earth, the Earth's south magnetic pole is located near the north geographic pole and the Earth's north magnetic pole is located near the south geographic pole. In fact, the configuration of the Earth's magnetic field, pictured in Figure 29.3, is very much like the one that would be achieved by burying a gigantic bar magnet deep in the Earth's interior. If a compass needle is supported by bearings that allow it to rotate in the vertical plane as well as in the horizontal plane, the needle is horizontal with respect to the Earth's surface only near the equator. As the compass is moved northward, the needle rotates so that it points more and more toward the Earth's surface. Finally, at a point near Hudson Bay in Canada, the north pole of the needle points directly downward. This site, first found in 1832, is considered to be the location of the south magnetic pole of the Earth. It is approximately 1 300 mi from the Earth's geographic

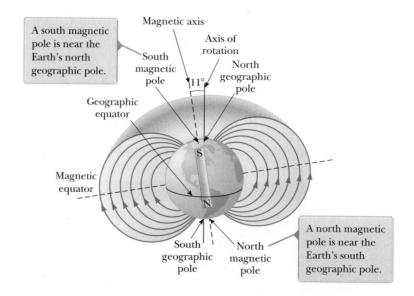

A south magnetic pole is near the Earth's north geographic pole.

Magnetic axis

Axis of rotation

South magnetic pole

North geographic pole

11°

Geographic equator

Magnetic equator

S

N

A north magnetic pole is near the Earth's south geographic pole.

South geographic pole

North magnetic pole

Figure 29.3 The Earth's magnetic field lines.

North Pole, and its exact position varies slowly with time. Similarly, the north magnetic pole of the Earth is about 1 200 mi away from the Earth's geographic South Pole.

Although the Earth's magnetic field pattern is similar to the one that would be set up by a bar magnet deep within the Earth, it is easy to understand why the source of this magnetic field cannot be large masses of permanently magnetized material. The Earth does have large deposits of iron ore deep beneath its surface, but the high temperatures in the Earth's core prevent the iron from retaining any permanent magnetization. Scientists consider it more likely that the source of the Earth's magnetic field is convection currents in the Earth's core. Charged ions or electrons circulating in the liquid interior could produce a magnetic field just like a current loop does, as we shall see in Chapter 30. There is also strong evidence that the magnitude of a planet's magnetic field is related to the planet's rate of rotation. For example, Jupiter rotates faster than the Earth, and space probes indicate that Jupiter's magnetic field is stronger than the Earth's. Venus, on the other hand, rotates more slowly than the Earth, and its magnetic field is found to be weaker. Investigation into the cause of the Earth's magnetism is ongoing.

The direction of the Earth's magnetic field has reversed several times during the last million years. Evidence for this reversal is provided by basalt, a type of rock that contains iron. Basalt forms from material spewed forth by volcanic activity on the ocean floor. As the lava cools, it solidifies and retains a picture of the Earth's magnetic field direction. The rocks are dated by other means to provide a time line for these periodic reversals of the magnetic field.

We can quantify the magnetic field $\vec{B}$ by using our model of a particle in a field, like the model discussed for gravity in Chapter 13 and for electricity in Chapter 23. The existence of a magnetic field at some point in space can be determined by measuring the **magnetic force** $\vec{F}_B$ exerted on an appropriate test particle placed at that point. This process is the same one we followed in defining the electric field in Chapter 23. If we perform such an experiment by placing a particle with charge q in the magnetic field, we find the following results that are similar to those for experiments on electric forces:

- The magnetic force is proportional to the charge q of the particle.
- The magnetic force on a negative charge is directed opposite to the force on a positive charge moving in the same direction.
- The magnetic force is proportional to the magnitude of the magnetic field vector $\vec{B}$.

We also find the following results, which are *totally different* from those for experiments on electric forces:

- The magnetic force is proportional to the speed v of the particle.
- If the velocity vector makes an angle θ with the magnetic field, the magnitude of the magnetic force is proportional to $\sin \theta$.
- When a charged particle moves *parallel* to the magnetic field vector, the magnetic force on the charge is zero.
- When a charged particle moves in a direction *not* parallel to the magnetic field vector, the magnetic force acts in a direction perpendicular to both $\vec{v}$ and $\vec{B}$; that is, the magnetic force is perpendicular to the plane formed by $\vec{v}$ and $\vec{B}$.

These results show that the magnetic force on a particle is more complicated than the electric force. The magnetic force is distinctive because it depends on the velocity of the particle and because its direction is perpendicular to both $\vec{v}$ and $\vec{B}$. Figure 29.4 (page 684) shows the details of the direction of the magnetic force on a charged

Figure 29.4 (a) The direction of the magnetic force $\vec{\mathbf{F}}_B$ acting on a charged particle moving with a velocity $\vec{\mathbf{v}}$ in the presence of a magnetic field $\vec{\mathbf{B}}$. (b) Magnetic forces on positive and negative charges. The dashed lines show the paths of the particles, which are investigated in Section 29.2.

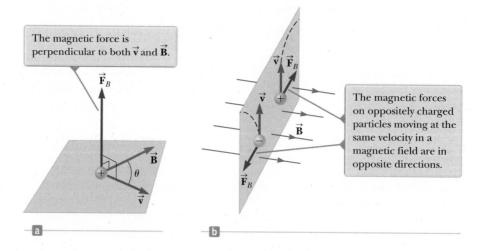

particle. Despite this complicated behavior, these observations can be summarized in a compact way by writing the magnetic force in the form

◄ Vector expression for the magnetic force on a charged particle moving in a magnetic field

$$\vec{\mathbf{F}}_B = q\vec{\mathbf{v}} \times \vec{\mathbf{B}} \qquad (29.1)$$

which by definition of the cross product (see Section 11.1) is perpendicular to both $\vec{\mathbf{v}}$ and $\vec{\mathbf{B}}$. We can regard this equation as an operational definition of the magnetic field at some point in space. That is, the magnetic field is defined in terms of the force acting on a moving charged particle. Equation 29.1 is the mathematical representation of the magnetic version of the **particle in a field** analysis model.

Figure 29.5 reviews two right-hand rules for determining the direction of the cross product $\vec{\mathbf{v}} \times \vec{\mathbf{B}}$ and determining the direction of $\vec{\mathbf{F}}_B$. The rule in Figure 29.5a depends on our right-hand rule for the cross product in Figure 11.2. Point the four fingers of your right hand along the direction of $\vec{\mathbf{v}}$ with the palm facing $\vec{\mathbf{B}}$ and curl them toward $\vec{\mathbf{B}}$. Your extended thumb, which is at a right angle to your fingers, points in the direction of $\vec{\mathbf{v}} \times \vec{\mathbf{B}}$. Because $\vec{\mathbf{F}}_B = q\vec{\mathbf{v}} \times \vec{\mathbf{B}}$, $\vec{\mathbf{F}}_B$ is in the direction of your thumb if q is positive and is opposite the direction of your thumb if q is negative. (If you need more help understanding the cross product, you should review Section 11.1, including Fig. 11.2.)

An alternative rule is shown in Figure 29.5b. Here the thumb points in the direction of $\vec{\mathbf{v}}$ and the extended fingers in the direction of $\vec{\mathbf{B}}$. Now, the force $\vec{\mathbf{F}}_B$ on a positive charge extends outward from the palm. The advantage of this rule is that the force on the charge is in the direction you would push on something with your

Figure 29.5 Two right-hand rules for determining the direction of the magnetic force $\vec{\mathbf{F}}_B = q\vec{\mathbf{v}} \times \vec{\mathbf{B}}$ acting on a particle with charge q moving with a velocity $\vec{\mathbf{v}}$ in a magnetic field $\vec{\mathbf{B}}$. (a) In this rule, the magnetic force is in the direction in which your thumb points. (b) In this rule, the magnetic force is in the direction of your palm, as if you are pushing the particle with your hand.

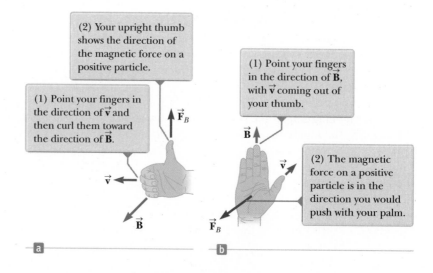

hand: outward from your palm. The force on a negative charge is in the opposite direction. You can use either of these two right-hand rules.

The magnitude of the magnetic force on a charged particle is

$$F_B = |q|vB \sin \theta \tag{29.2}$$

where θ is the smaller angle between $\vec{v}$ and $\vec{B}$. From this expression, we see that F_B is zero when $\vec{v}$ is parallel or antiparallel to $\vec{B}$ ($\theta = 0$ or $180°$) and maximum when $\vec{v}$ is perpendicular to $\vec{B}$ ($\theta = 90°$).

◀ **Magnitude of the magnetic force on a charged particle moving in a magnetic field**

Let's compare the important differences between the electric and magnetic versions of the particle in a field model:

- The electric force vector is along the direction of the electric field, whereas the magnetic force vector is perpendicular to the magnetic field.
- The electric force acts on a charged particle regardless of whether the particle is moving, whereas the magnetic force acts on a charged particle only when the particle is in motion.
- The electric force does work in displacing a charged particle, whereas the magnetic force associated with a steady magnetic field does no work when a particle is displaced because the force is perpendicular to the displacement of its point of application.

From the last statement and on the basis of the work–kinetic energy theorem, we conclude that the kinetic energy of a charged particle moving through a magnetic field cannot be altered by the magnetic field alone. The field can alter the direction of the velocity vector, but it cannot change the speed or kinetic energy of the particle.

From Equation 29.2, we see that the SI unit of magnetic field is the newton per coulomb-meter per second, which is called the **tesla** (T):

$$1 \text{ T} = 1 \frac{\text{N}}{\text{C} \cdot \text{m/s}}$$

◀ **The tesla**

Because a coulomb per second is defined to be an ampere,

$$1 \text{ T} = 1 \frac{\text{N}}{\text{A} \cdot \text{m}}$$

A non-SI magnetic-field unit in common use, called the *gauss* (G), is related to the tesla through the conversion $1 \text{ T} = 10^4 \text{ G}$. Table 29.1 shows some typical values of magnetic fields.

Quick **Quiz** 29.1 An electron moves in the plane of this paper toward the top of the page. A magnetic field is also in the plane of the page and directed toward the right. What is the direction of the magnetic force on the electron? **(a)** toward the top of the page **(b)** toward the bottom of the page **(c)** toward the left edge of the page **(d)** toward the right edge of the page **(e)** upward out of the page **(f)** downward into the page

Table 29.1 Some Approximate Magnetic Field Magnitudes	
Source of Field	**Field Magnitude (T)**
Strong superconducting laboratory magnet	30
Strong conventional laboratory magnet	2
Medical MRI unit	1.5
Bar magnet	10^{-2}
Surface of the Sun	10^{-2}
Surface of the Earth	0.5×10^{-4}
Inside human brain (due to nerve impulses)	10^{-13}

Analysis Model Particle in a Field (Magnetic)

Imagine some source (which we will investigate later) establishes a **magnetic field $\vec{B}$** throughout space. Now imagine a particle with charge q is placed in that field. The particle interacts with the magnetic field so that the particle experiences a magnetic force given by

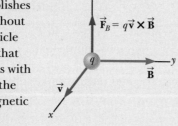

$$\vec{F}_B = q\vec{v} \times \vec{B} \qquad (29.1)$$

Examples:

- an ion moves in a circular path in the magnetic field of a mass spectrometer (Section 29.3)
- a coil in a motor rotates in response to the magnetic field in the motor (Chapter 31)
- a magnetic field is used to separate particles emitted by radioactive sources (Chapter 44)
- in a bubble chamber, particles created in collisions follow curved paths in a magnetic field, allowing the particles to be identified (Chapter 46)

Example 29.1 An Electron Moving in a Magnetic Field AM

An electron in an old-style television picture tube moves toward the front of the tube with a speed of 8.0×10^6 m/s along the x axis (Fig. 29.6). Surrounding the neck of the tube are coils of wire that create a magnetic field of magnitude 0.025 T, directed at an angle of 60° to the x axis and lying in the xy plane. Calculate the magnetic force on the electron.

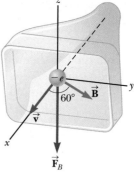

Figure 29.6 (Example 29.1)
The magnetic force $\vec{F}_B$ acting on the electron is in the negative z direction when $\vec{v}$ and $\vec{B}$ lie in the xy plane.

SOLUTION

Conceptualize Recall that the magnetic force on a charged particle is perpendicular to the plane formed by the velocity and magnetic field vectors. Use one of the right-hand rules in Figure 29.5 to convince yourself that the direction of the force on the electron is downward in Figure 29.6.

Categorize We evaluate the magnetic force using the *magnetic* version of the *particle in a field* model.

Analyze Use Equation 29.2 to find the magnitude of the magnetic force:

$$F_B = |q|vB\sin\theta$$
$$= (1.6 \times 10^{-19}\,\text{C})(8.0 \times 10^6\,\text{m/s})(0.025\,\text{T})(\sin 60°)$$
$$= \boxed{2.8 \times 10^{-14}\,\text{N}}$$

Finalize For practice using the vector product, evaluate this force in vector notation using Equation 29.1. The magnitude of the magnetic force may seem small to you, but remember that it is acting on a very small particle, the electron. To convince yourself that this is a substantial force for an electron, calculate the initial acceleration of the electron due to this force.

29.2 Motion of a Charged Particle in a Uniform Magnetic Field

Before we continue our discussion, some explanation of the notation used in this book is in order. To indicate the direction of $\vec{B}$ in illustrations, we sometimes present perspective views such as those in Figure 29.6. If $\vec{B}$ lies in the plane of the page or is present in a perspective drawing, we use green vectors or green field lines with arrowheads. In nonperspective illustrations, we depict a magnetic field perpendicular to and directed out of the page with a series of green dots, which represent the tips of arrows coming toward you (see Fig. 29.7a). In this case, the field is labeled

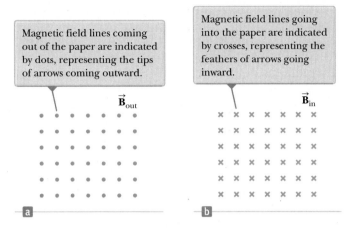

Magnetic field lines coming out of the paper are indicated by dots, representing the tips of arrows coming outward.

Magnetic field lines going into the paper are indicated by crosses, representing the feathers of arrows going inward.

$\vec{\mathbf{B}}_{out}$

$\vec{\mathbf{B}}_{in}$

Figure 29.7 Representations of magnetic field lines perpendicular to the page.

$\vec{\mathbf{B}}_{out}$. If $\vec{\mathbf{B}}$ is directed perpendicularly into the page, we use green crosses, which represent the feathered tails of arrows fired away from you, as in Figure 29.7b. In this case, the field is labeled $\vec{\mathbf{B}}_{in}$, where the subscript "in" indicates "into the page." The same notation with crosses and dots is also used for other quantities that might be perpendicular to the page such as forces and current directions.

In Section 29.1, we found that the magnetic force acting on a charged particle moving in a magnetic field is perpendicular to the particle's velocity and consequently the work done by the magnetic force on the particle is zero. Now consider the special case of a positively charged particle moving in a uniform magnetic field with the initial velocity vector of the particle perpendicular to the field. Let's assume the direction of the magnetic field is into the page as in Figure 29.8. The particle in a field model tells us that the magnetic force on the particle is perpendicular to both the magnetic field lines and the velocity of the particle. The fact that there is a force on the particle tells us to apply the particle under a net force model to the particle. As the particle changes the direction of its velocity in response to the magnetic force, the magnetic force remains perpendicular to the velocity. As we found in Section 6.1, if the force is always perpendicular to the velocity, the path of the particle is a circle! Figure 29.8 shows the particle moving in a circle in a plane perpendicular to the magnetic field. Although magnetism and magnetic forces may be new and unfamiliar to you now, we see a magnetic effect that results in something with which we are familiar: the particle in uniform circular motion model!

The particle moves in a circle because the magnetic force $\vec{\mathbf{F}}_B$ is perpendicular to $\vec{\mathbf{v}}$ and $\vec{\mathbf{B}}$ and has a constant magnitude qvB. As Figure 29.8 illustrates, the

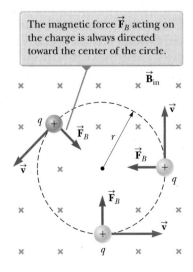

The magnetic force $\vec{\mathbf{F}}_B$ acting on the charge is always directed toward the center of the circle.

$\vec{\mathbf{B}}_{in}$

Figure 29.8 When the velocity of a charged particle is perpendicular to a uniform magnetic field, the particle moves in a circular path in a plane perpendicular to $\vec{\mathbf{B}}$.

rotation is counterclockwise for a positive charge in a magnetic field directed into the page. If q were negative, the rotation would be clockwise. We use the particle under a net force model to write Newton's second law for the particle:

$$\sum F = F_B = ma$$

Because the particle moves in a circle, we also model it as a particle in uniform circular motion and we replace the acceleration with centripetal acceleration:

$$F_B = qvB = \frac{mv^2}{r}$$

This expression leads to the following equation for the radius of the circular path:

$$r = \frac{mv}{qB} \tag{29.3}$$

That is, the radius of the path is proportional to the linear momentum mv of the particle and inversely proportional to the magnitude of the charge on the particle and to the magnitude of the magnetic field. The angular speed of the particle (from Eq. 10.10) is

$$\omega = \frac{v}{r} = \frac{qB}{m} \tag{29.4}$$

The period of the motion (the time interval the particle requires to complete one revolution) is equal to the circumference of the circle divided by the speed of the particle:

$$T = \frac{2\pi r}{v} = \frac{2\pi}{\omega} = \frac{2\pi m}{qB} \tag{29.5}$$

These results show that the angular speed of the particle and the period of the circular motion do not depend on the speed of the particle or on the radius of the orbit. The angular speed ω is often referred to as the **cyclotron frequency** because charged particles circulate at this angular frequency in the type of accelerator called a *cyclotron*, which is discussed in Section 29.3.

If a charged particle moves in a uniform magnetic field with its velocity at some arbitrary angle with respect to $\vec{\mathbf{B}}$, its path is a helix. For example, if the field is directed in the x direction as shown in Figure 29.9, there is no component of force in the x direction. As a result, $a_x = 0$, and the x component of velocity remains constant. The charged particle is a particle in equilibrium in this direction. The magnetic force $q\vec{\mathbf{v}} \times \vec{\mathbf{B}}$ causes the components v_y and v_z to change in time, however, and the resulting motion is a helix whose axis is parallel to the magnetic field. The projection of the path onto the yz plane (viewed along the x axis) is a circle. (The projections of the path onto the xy and xz planes are sinusoids!) Equations 29.3 to 29.5 still apply provided v is replaced by $v_\perp = \sqrt{v_y^2 + v_z^2}$.

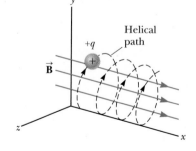

Figure 29.9 A charged particle having a velocity vector that has a component parallel to a uniform magnetic field moves in a helical path.

Quick Quiz 29.2 A charged particle is moving perpendicular to a magnetic field in a circle with a radius r. **(i)** An identical particle enters the field, with $\vec{v}$ perpendicular to $\vec{B}$, but with a higher speed than the first particle. Compared with the radius of the circle for the first particle, is the radius of the circular path for the second particle (a) smaller, (b) larger, or (c) equal in size? **(ii)** The magnitude of the magnetic field is increased. From the same choices, compare the radius of the new circular path of the first particle with the radius of its initial path.

Example 29.2   **A Proton Moving Perpendicular to a Uniform Magnetic Field**   `AM`

A proton is moving in a circular orbit of radius 14 cm in a uniform 0.35-T magnetic field perpendicular to the velocity of the proton. Find the speed of the proton.

SOLUTION

Conceptualize From our discussion in this section, we know the proton follows a circular path when moving perpendicular to a uniform magnetic field. In Chapter 39, we will learn that the highest possible speed for a particle is the speed of light, 3.00×10^8 m/s, so the speed of the particle in this problem must come out to be smaller than that value.

Categorize The proton is described by both the *particle in a field* model and the *particle in uniform circular motion* model. These models led to Equation 29.3.

Analyze

Solve Equation 29.3 for the speed of the particle:

$$v = \frac{qBr}{m_p}$$

Substitute numerical values:

$$v = \frac{(1.60 \times 10^{-19}\text{ C})(0.35\text{ T})(0.14\text{ m})}{1.67 \times 10^{-27}\text{ kg}}$$

$$= 4.7 \times 10^6 \text{ m/s}$$

Finalize The speed is indeed smaller than the speed of light, as required.

WHAT IF? What if an electron, rather than a proton, moves in a direction perpendicular to the same magnetic field with this same speed? Will the radius of its orbit be different?

Answer An electron has a much smaller mass than a proton, so the magnetic force should be able to change its velocity much more easily than that for the proton. Therefore, we expect the radius to be smaller. Equation 29.3 shows that r is proportional to m with q, B, and v the same for the electron as for the proton. Consequently, the radius will be smaller by the same factor as the ratio of masses m_e/m_p.

Example 29.3   **Bending an Electron Beam**   `AM`

In an experiment designed to measure the magnitude of a uniform magnetic field, electrons are accelerated from rest through a potential difference of 350 V and then enter a uniform magnetic field that is perpendicular to the velocity vector of the electrons. The electrons travel along a curved path because of the magnetic force exerted on them, and the radius of the path is measured to be 7.5 cm. (Such a curved beam of electrons is shown in Fig. 29.10.)

(A) What is the magnitude of the magnetic field?

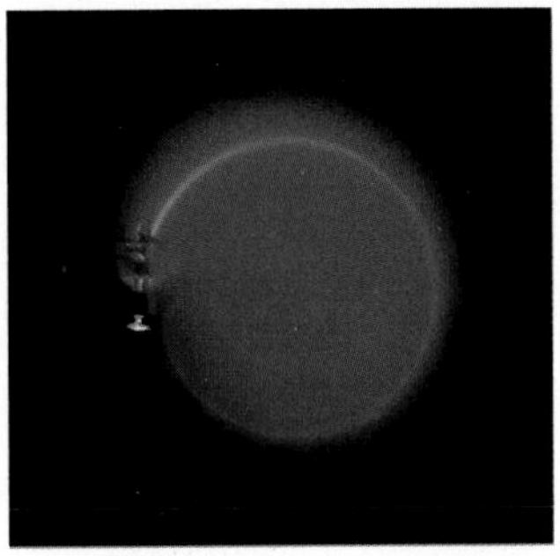

Henry Leap and Jim Lehman

Figure 29.10 (Example 29.3) The bending of an electron beam in a magnetic field.

continued

▶ **29.3** continued

SOLUTION

Conceptualize This example involves electrons accelerating from rest due to an electric force and then moving in a circular path due to a magnetic force. With the help of Figures 29.8 and 29.10, visualize the circular motion of the electrons.

Categorize Equation 29.3 shows that we need the speed v of the electron to find the magnetic field magnitude, and v is not given. Consequently, we must find the speed of the electron based on the potential difference through which it is accelerated. To do so, we categorize the first part of the problem by modeling an electron and the electric field as an *isolated system* in terms of *energy*. Once the electron enters the magnetic field, we categorize the second part of the problem as one involving a *particle in a field* and a *particle in uniform circular motion*, as we have done in this section.

Analyze Write the appropriate reduction of the conservation of energy equation, Equation 8.2, for the electron–electric field system:

$$\Delta K + \Delta U = 0$$

Substitute the appropriate initial and final energies:

$$\left(\tfrac{1}{2}m_e v^2 - 0\right) + (q\,\Delta V) = 0$$

Solve for the speed of the electron:

$$v = \sqrt{\frac{-2q\,\Delta V}{m_e}}$$

Substitute numerical values:

$$v = \sqrt{\frac{-2(-1.60 \times 10^{-19}\,\text{C})(350\,\text{V})}{9.11 \times 10^{-31}\,\text{kg}}} = 1.11 \times 10^7\,\text{m/s}$$

Now imagine the electron entering the magnetic field with this speed. Solve Equation 29.3 for the magnitude of the magnetic field:

$$B = \frac{m_e v}{er}$$

Substitute numerical values:

$$B = \frac{(9.11 \times 10^{-31}\,\text{kg})(1.11 \times 10^7\,\text{m/s})}{(1.60 \times 10^{-19}\,\text{C})(0.075\,\text{m})} = \boxed{8.4 \times 10^{-4}\,\text{T}}$$

(B) What is the angular speed of the electrons?

SOLUTION

Use Equation 10.10:

$$\omega = \frac{v}{r} = \frac{1.11 \times 10^7\,\text{m/s}}{0.075\,\text{m}} = \boxed{1.5 \times 10^8\,\text{rad/s}}$$

Finalize The angular speed can be represented as $\omega = (1.5 \times 10^8\,\text{rad/s})(1\,\text{rev}/2\pi\,\text{rad}) = 2.4 \times 10^7\,\text{rev/s}$. The electrons travel around the circle 24 million times per second! This answer is consistent with the very high speed found in part (A).

WHAT IF? What if a sudden voltage surge causes the accelerating voltage to increase to 400 V? How does that affect the angular speed of the electrons, assuming the magnetic field remains constant?

Answer The increase in accelerating voltage ΔV causes the electrons to enter the magnetic field with a higher speed v. This higher speed causes them to travel in a circle with a larger radius r. The angular speed is the ratio of v to r. Both v and r increase by the same factor, so the effects can-

cel and the angular speed remains the same. Equation 29.4 is an expression for the cyclotron frequency, which is the same as the angular speed of the electrons. The cyclotron frequency depends only on the charge q, the magnetic field B, and the mass m_e, none of which have changed. Therefore, the voltage surge has no effect on the angular speed. (In reality, however, the voltage surge may also increase the magnetic field if the magnetic field is powered by the same source as the accelerating voltage. In that case, the angular speed increases according to Eq. 29.4.)

When charged particles move in a nonuniform magnetic field, the motion is complex. For example, in a magnetic field that is strong at the ends and weak in the middle such as that shown in Figure 29.11, the particles can oscillate between two positions. A charged particle starting at one end spirals along the field lines until it reaches the other end, where it reverses its path and spirals back. This configura-

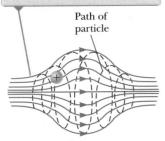

The magnetic force exerted on the particle near either end of the bottle has a component that causes the particle to spiral back toward the center.

Path of particle

Figure 29.11 A charged particle moving in a nonuniform magnetic field (a magnetic bottle) spirals about the field and oscillates between the endpoints.

Figure 29.12 The Van Allen belts are made up of charged particles trapped by the Earth's nonuniform magnetic field. The magnetic field lines are in green, and the particle paths are dashed black lines.

tion is known as a *magnetic bottle* because charged particles can be trapped within it. The magnetic bottle has been used to confine a *plasma*, a gas consisting of ions and electrons. Such a plasma-confinement scheme could fulfill a crucial role in the control of nuclear fusion, a process that could supply us in the future with an almost endless source of energy. Unfortunately, the magnetic bottle has its problems. If a large number of particles are trapped, collisions between them cause the particles to eventually leak from the system.

The Van Allen radiation belts consist of charged particles (mostly electrons and protons) surrounding the Earth in doughnut-shaped regions (Fig. 29.12). The particles, trapped by the Earth's nonuniform magnetic field, spiral around the field lines from pole to pole, covering the distance in only a few seconds. These particles originate mainly from the Sun, but some come from stars and other heavenly objects. For this reason, the particles are called *cosmic rays*. Most cosmic rays are deflected by the Earth's magnetic field and never reach the atmosphere. Some of the particles become trapped, however, and it is these particles that make up the Van Allen belts. When the particles are located over the poles, they sometimes collide with atoms in the atmosphere, causing the atoms to emit visible light. Such collisions are the origin of the beautiful aurora borealis, or northern lights, in the northern hemisphere and the aurora australis in the southern hemisphere. Auroras are usually confined to the polar regions because the Van Allen belts are nearest the Earth's surface there. Occasionally, though, solar activity causes larger numbers of charged particles to enter the belts and significantly distort the normal magnetic field lines associated with the Earth. In these situations, an aurora can sometimes be seen at lower latitudes.

29.3 Applications Involving Charged Particles Moving in a Magnetic Field

A charge moving with a velocity $\vec{v}$ in the presence of both an electric field $\vec{E}$ and a magnetic field $\vec{B}$ is described by two particle in a field models. It experiences both an electric force $q\vec{E}$ and a magnetic force $q\vec{v} \times \vec{B}$. The total force (called the Lorentz force) acting on the charge is

$$\vec{F} = q\vec{E} + q\vec{v} \times \vec{B} \qquad \text{(29.6)}$$

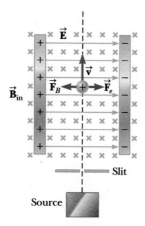

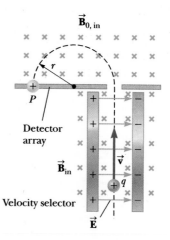

Figure 29.13 A velocity selector. When a positively charged particle is moving with velocity $\vec{v}$ in the presence of a magnetic field directed into the page and an electric field directed to the right, it experiences an electric force $q\vec{E}$ to the right and a magnetic force $q\vec{v} \times \vec{B}$ to the left.

Figure 29.14 A mass spectrometer. Positively charged particles are sent first through a velocity selector and then into a region where the magnetic field $\vec{B}_0$ causes the particles to move in a semicircular path and strike a detector array at P.

Velocity Selector

In many experiments involving moving charged particles, it is important that all particles move with essentially the same velocity, which can be achieved by applying a combination of an electric field and a magnetic field oriented as shown in Figure 29.13. A uniform electric field is directed to the right (in the plane of the page in Fig. 29.13), and a uniform magnetic field is applied in the direction perpendicular to the electric field (into the page in Fig. 29.13). If q is positive and the velocity $\vec{v}$ is upward, the magnetic force $q\vec{v} \times \vec{B}$ is to the left and the electric force $q\vec{E}$ is to the right. When the magnitudes of the two fields are chosen so that $qE = qvB$, the forces cancel. The charged particle is modeled as a particle in equilibrium and moves in a straight vertical line through the region of the fields. From the expression $qE = qvB$, we find that

$$v = \frac{E}{B} \tag{29.7}$$

Only those particles having this speed pass undeflected through the mutually perpendicular electric and magnetic fields. The magnetic force exerted on particles moving at speeds greater than that is stronger than the electric force, and the particles are deflected to the left. Those moving at slower speeds are deflected to the right.

The Mass Spectrometer

A **mass spectrometer** separates ions according to their mass-to-charge ratio. In one version of this device, known as the *Bainbridge mass spectrometer*, a beam of ions first passes through a velocity selector and then enters a second uniform magnetic field $\vec{B}_0$ that has the same direction as the magnetic field in the selector (Fig. 29.14). Upon entering the second magnetic field, the ions are described by the particle in uniform circular motion model. They move in a semicircle of radius r before striking a detector array at P. If the ions are positively charged, the beam deflects to the left as Figure 29.14 shows. If the ions are negatively charged, the beam deflects to the right. From Equation 29.3, we can express the ratio m/q as

$$\frac{m}{q} = \frac{rB_0}{v}$$

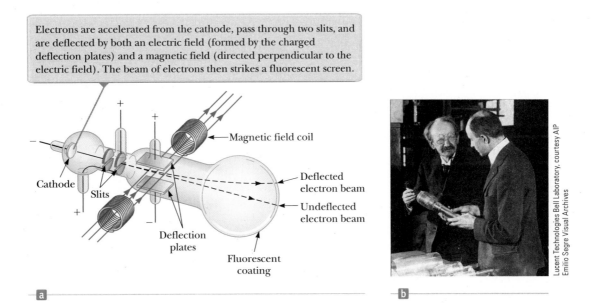

Electrons are accelerated from the cathode, pass through two slits, and are deflected by both an electric field (formed by the charged deflection plates) and a magnetic field (directed perpendicular to the electric field). The beam of electrons then strikes a fluorescent screen.

Figure 29.15 (a) Thomson's apparatus for measuring e/m_e. (b) J. J. Thomson (*left*) in the Cavendish Laboratory, University of Cambridge. The man on the right, Frank Baldwin Jewett, is a distant relative of John W. Jewett, Jr., coauthor of this text.

Using Equation 29.7 gives

$$\frac{m}{q} = \frac{rB_0 B}{E} \tag{29.8}$$

Therefore, we can determine m/q by measuring the radius of curvature and knowing the field magnitudes B, B_0, and E. In practice, one usually measures the masses of various isotopes of a given ion, with the ions all carrying the same charge q. In this way, the mass ratios can be determined even if q is unknown.

A variation of this technique was used by J. J. Thomson (1856–1940) in 1897 to measure the ratio e/m_e for electrons. Figure 29.15a shows the basic apparatus he used. Electrons are accelerated from the cathode and pass through two slits. They then drift into a region of perpendicular electric and magnetic fields. The magnitudes of the two fields are first adjusted to produce an undeflected beam. When the magnetic field is turned off, the electric field produces a measurable beam deflection that is recorded on the fluorescent screen. From the size of the deflection and the measured values of E and B, the charge-to-mass ratio can be determined. The results of this crucial experiment represent the discovery of the electron as a fundamental particle of nature.

The Cyclotron

A **cyclotron** is a device that can accelerate charged particles to very high speeds. The energetic particles produced are used to bombard atomic nuclei and thereby produce nuclear reactions of interest to researchers. A number of hospitals use cyclotron facilities to produce radioactive substances for diagnosis and treatment.

Both electric and magnetic forces play key roles in the operation of a cyclotron, a schematic drawing of which is shown in Figure 29.16a (page 694). The charges move inside two semicircular containers D_1 and D_2, referred to as *dees* because of their shape like the letter D. A high-frequency alternating potential difference is applied to the dees, and a uniform magnetic field is directed perpendicular to them. A positive ion released at P near the center of the magnet in one dee moves in a semicircular path (indicated by the dashed black line in the drawing) and arrives back at the gap in a time interval $T/2$, where T is the time interval needed to make one complete trip around the two dees, given by Equation 29.5. The frequency

Pitfall Prevention 29.1

The Cyclotron Is Not the Only Type of Particle Accelerator The cyclotron is important historically because it was the first particle accelerator to produce particles with very high speeds. Cyclotrons still play important roles in medical applications and some research activities. Many other research activities make use of a different type of accelerator called a *synchrotron*.

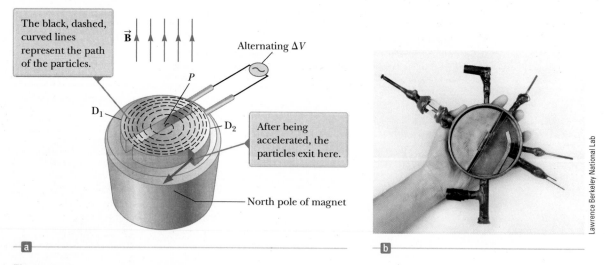

The black, dashed, curved lines represent the path of the particles.

Alternating ΔV

After being accelerated, the particles exit here.

North pole of magnet

a

b

Lawrence Berkeley National Lab

Figure 29.16 (a) A cyclotron consists of an ion source at P, two dees D_1 and D_2 across which an alternating potential difference is applied, and a uniform magnetic field. (The south pole of the magnet is not shown.) (b) The first cyclotron, invented by E. O. Lawrence and M. S. Livingston in 1934.

of the applied potential difference is adjusted so that the polarity of the dees is reversed in the same time interval during which the ion travels around one dee. If the applied potential difference is adjusted such that D_1 is at a lower electric potential than D_2 by an amount ΔV, the ion accelerates across the gap to D_1 and its kinetic energy increases by an amount $q\,\Delta V$. It then moves around D_1 in a semicircular path of greater radius (because its speed has increased). After a time interval $T/2$, it again arrives at the gap between the dees. By this time, the polarity across the dees has again been reversed and the ion is given another "kick" across the gap. The motion continues so that for each half-circle trip around one dee, the ion gains additional kinetic energy equal to $q\,\Delta V$. When the radius of its path is nearly that of the dees, the energetic ion leaves the system through the exit slit. The cyclotron's operation depends on T being independent of the speed of the ion and of the radius of the circular path (Eq. 29.5).

We can obtain an expression for the kinetic energy of the ion when it exits the cyclotron in terms of the radius R of the dees. From Equation 29.3, we know that $v = qBR/m$. Hence, the kinetic energy is

$$K = \tfrac{1}{2}mv^2 = \frac{q^2 B^2 R^2}{2m} \qquad (29.9)$$

When the energy of the ions in a cyclotron exceeds about 20 MeV, relativistic effects come into play. (Such effects are discussed in Chapter 39.) Observations show that T increases and the moving ions do not remain in phase with the applied potential difference. Some accelerators overcome this problem by modifying the period of the applied potential difference so that it remains in phase with the moving ions.

29.4 Magnetic Force Acting on a Current-Carrying Conductor

If a magnetic force is exerted on a single charged particle when the particle moves through a magnetic field, it should not surprise you that a current-carrying wire also experiences a force when placed in a magnetic field. The current is a collection of many charged particles in motion; hence, the resultant force exerted by the field on the wire is the vector sum of the individual forces exerted on all the charged particles making up the current. The force exerted on the particles is transmitted to the wire when the particles collide with the atoms making up the wire.

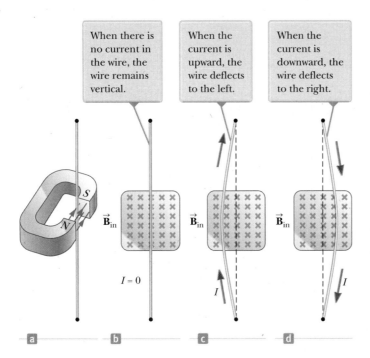

When there is no current in the wire, the wire remains vertical.

When the current is upward, the wire deflects to the left.

When the current is downward, the wire deflects to the right.

$\vec{\textbf{B}}_{in}$ $\vec{\textbf{B}}_{in}$ $\vec{\textbf{B}}_{in}$

$I = 0$ I I

a   b   c   d

Figure 29.17 (a) A wire suspended vertically between the poles of a magnet. (b)–(d) The setup shown in (a) as seen looking at the south pole of the magnet so that the magnetic field (green crosses) is directed into the page.

One can demonstrate the magnetic force acting on a current-carrying conductor by hanging a wire between the poles of a magnet as shown in Figure 29.17a. For ease in visualization, part of the horseshoe magnet in part (a) is removed to show the end face of the south pole in parts (b) through (d) of Figure 29.17. The magnetic field is directed into the page and covers the region within the shaded squares. When the current in the wire is zero, the wire remains vertical as in Figure 29.17b. When the wire carries a current directed upward as in Figure 29.17c, however, the wire deflects to the left. If the current is reversed as in Figure 29.17d, the wire deflects to the right.

Let's quantify this discussion by considering a straight segment of wire of length L and cross-sectional area A carrying a current I in a uniform magnetic field $\vec{\textbf{B}}$ as in Figure 29.18. According to the magnetic version of the particle in a field model, the magnetic force exerted on a charge q moving with a drift velocity $\vec{\textbf{v}}_d$ is $q\vec{\textbf{v}}_d \times \vec{\textbf{B}}$. To find the total force acting on the wire, we multiply the force $q\vec{\textbf{v}}_d \times \vec{\textbf{B}}$ exerted on one charge by the number of charges in the segment. Because the volume of the segment is AL, the number of charges in the segment is nAL, where n is the number of mobile charge carriers per unit volume. Hence, the total magnetic force on the segment of wire of length L is

$$\vec{\textbf{F}}_B = (q\vec{\textbf{v}}_d \times \vec{\textbf{B}})nAL$$

We can write this expression in a more convenient form by noting that, from Equation 27.4, the current in the wire is $I = nqv_dA$. Therefore,

$$\vec{\textbf{F}}_B = I\vec{\textbf{L}} \times \vec{\textbf{B}} \qquad (29.10)$$

where $\vec{\textbf{L}}$ is a vector that points in the direction of the current I and has a magnitude equal to the length L of the segment. This expression applies only to a straight segment of wire in a uniform magnetic field.

Now consider an arbitrarily shaped wire segment of uniform cross section in a magnetic field as shown in Figure 29.19 (page 696). It follows from Equation 29.10 that the magnetic force exerted on a small segment of vector length $d\vec{\textbf{s}}$ in the presence of a field $\vec{\textbf{B}}$ is

$$d\vec{\textbf{F}}_B = I\,d\vec{\textbf{s}} \times \vec{\textbf{B}} \qquad (29.11)$$

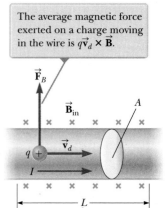

The average magnetic force exerted on a charge moving in the wire is $q\vec{\textbf{v}}_d \times \vec{\textbf{B}}$.

$\vec{\textbf{F}}_B$

$\vec{\textbf{B}}_{in}$ A

q + $\vec{\textbf{v}}_d$

I

L

The magnetic force on the wire segment of length L is $I\vec{\textbf{L}} \times \vec{\textbf{B}}$.

Figure 29.18 A segment of a current-carrying wire in a magnetic field $\vec{\textbf{B}}$.

◀ Force on a segment of current-carrying wire in a uniform magnetic field

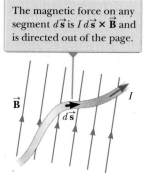

Figure 29.19 A wire segment of arbitrary shape carrying a current I in a magnetic field $\vec{\mathbf{B}}$ experiences a magnetic force.

The magnetic force on any segment $d\vec{\mathbf{s}}$ is $I\,d\vec{\mathbf{s}} \times \vec{\mathbf{B}}$ and is directed out of the page.

where $d\vec{\mathbf{F}}_B$ is directed out of the page for the directions of $\vec{\mathbf{B}}$ and $d\vec{\mathbf{s}}$ in Figure 29.19. Equation 29.11 can be considered as an alternative definition of $\vec{\mathbf{B}}$. That is, we can define the magnetic field $\vec{\mathbf{B}}$ in terms of a measurable force exerted on a current element, where the force is a maximum when $\vec{\mathbf{B}}$ is perpendicular to the element and zero when $\vec{\mathbf{B}}$ is parallel to the element.

To calculate the total force $\vec{\mathbf{F}}_B$ acting on the wire shown in Figure 29.19, we integrate Equation 29.11 over the length of the wire:

$$\vec{\mathbf{F}}_B = I \int_a^b d\vec{\mathbf{s}} \times \vec{\mathbf{B}} \tag{29.12}$$

where a and b represent the endpoints of the wire. When this integration is carried out, the magnitude of the magnetic field and the direction the field makes with the vector $d\vec{\mathbf{s}}$ may differ at different points.

Quick Quiz 29.3 A wire carries current in the plane of this paper toward the top of the page. The wire experiences a magnetic force toward the right edge of the page. Is the direction of the magnetic field causing this force **(a)** in the plane of the page and toward the left edge, **(b)** in the plane of the page and toward the bottom edge, **(c)** upward out of the page, or **(d)** downward into the page?

Example 29.4   **Force on a Semicircular Conductor**

A wire bent into a semicircle of radius R forms a closed circuit and carries a current I. The wire lies in the xy plane, and a uniform magnetic field is directed along the positive y axis as in Figure 29.20. Find the magnitude and direction of the magnetic force acting on the straight portion of the wire and on the curved portion.

SOLUTION

Conceptualize Using the right-hand rule for cross products, we see that the force $\vec{\mathbf{F}}_1$ on the straight portion of the wire is out of the page and the force $\vec{\mathbf{F}}_2$ on the curved portion is into the page. Is $\vec{\mathbf{F}}_2$ larger in magnitude than $\vec{\mathbf{F}}_1$ because the length of the curved portion is longer than that of the straight portion?

Categorize Because we are dealing with a current-carrying wire in a magnetic field rather than a single charged particle, we must use Equation 29.12 to find the total force on each portion of the wire.

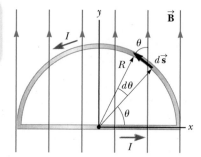

Figure 29.20 (Example 29.4) The magnetic force on the straight portion of the loop is directed out of the page, and the magnetic force on the curved portion is directed into the page.

Analyze Notice that $d\vec{\mathbf{s}}$ is perpendicular to $\vec{\mathbf{B}}$ everywhere on the straight portion of the wire. Use Equation 29.12 to find the force on this portion:

$$\vec{\mathbf{F}}_1 = I \int_a^b d\vec{\mathbf{s}} \times \vec{\mathbf{B}} = I \int_{-R}^{R} B\,dx\,\hat{\mathbf{k}} = \boxed{2IRB\,\hat{\mathbf{k}}}$$

▶ **29.4** continued

To find the magnetic force on the curved part, first write an expression for the magnetic force $d\vec{\mathbf{F}}_2$ on the element $d\vec{\mathbf{s}}$ in Figure 29.20:

(1) $d\vec{\mathbf{F}}_2 = I\,d\vec{\mathbf{s}} \times \vec{\mathbf{B}} = -IB\sin\theta\,ds\,\hat{\mathbf{k}}$

From the geometry in Figure 29.20, write an expression for ds:

(2) $ds = R\,d\theta$

Substitute Equation (2) into Equation (1) and integrate over the angle θ from 0 to π:

$$\vec{\mathbf{F}}_2 = -\int_0^\pi IRB\sin\theta\,d\theta\,\hat{\mathbf{k}} = -IRB\int_0^\pi \sin\theta\,d\theta\,\hat{\mathbf{k}} = -IRB[-\cos\theta]_0^\pi\,\hat{\mathbf{k}}$$
$$= IRB(\cos\pi - \cos 0)\hat{\mathbf{k}} = IRB(-1-1)\hat{\mathbf{k}} = \boxed{-2IRB\,\hat{\mathbf{k}}}$$

Finalize Two very important general statements follow from this example. First, the force on the curved portion is the same in magnitude as the force on a straight wire between the same two points. In general, the magnetic force on a curved current-carrying wire in a uniform magnetic field is equal to that on a straight wire connecting the endpoints and carrying the same current. Furthermore, $\vec{\mathbf{F}}_1 + \vec{\mathbf{F}}_2 = 0$ is also a general result: the net magnetic force acting on any closed current loop in a uniform magnetic field is zero.

29.5 Torque on a Current Loop in a Uniform Magnetic Field

In Section 29.4, we showed how a magnetic force is exerted on a current-carrying conductor placed in a magnetic field. With that as a starting point, we now show that a torque is exerted on a current loop placed in a magnetic field.

Consider a rectangular loop carrying a current I in the presence of a uniform magnetic field directed parallel to the plane of the loop as shown in Figure 29.21a. No magnetic forces act on sides ① and ③ because these wires are parallel to the field; hence, $\vec{\mathbf{L}} \times \vec{\mathbf{B}} = 0$ for these sides. Magnetic forces do, however, act on sides ② and ④ because these sides are oriented perpendicular to the field. The magnitude of these forces is, from Equation 29.10,

$$F_2 = F_4 = IaB$$

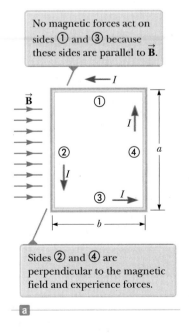

No magnetic forces act on sides ① and ③ because these sides are parallel to $\vec{\mathbf{B}}$.

Sides ② and ④ are perpendicular to the magnetic field and experience forces.

The magnetic forces $\vec{\mathbf{F}}_2$ and $\vec{\mathbf{F}}_4$ exerted on sides ② and ④ create a torque that tends to rotate the loop clockwise.

Figure 29.21 (a) Overhead view of a rectangular current loop in a uniform magnetic field. (b) Edge view of the loop sighting down sides ② and ④. The purple dot in the left circle represents current in wire ② coming toward you; the purple cross in the right circle represents current in wire ④ moving away from you.

The direction of $\vec{\mathbf{F}}_2$, the magnetic force exerted on wire ②, is out of the page in the view shown in Figure 29.20a and that of $\vec{\mathbf{F}}_4$, the magnetic force exerted on wire ④, is into the page in the same view. If we view the loop from side ③ and sight along sides ② and ④, we see the view shown in Figure 29.21b, and the two magnetic forces $\vec{\mathbf{F}}_2$ and $\vec{\mathbf{F}}_4$ are directed as shown. Notice that the two forces point in opposite directions but are *not* directed along the same line of action. If the loop is pivoted so that it can rotate about point O, these two forces produce about O a torque that rotates the loop clockwise. The magnitude of this torque τ_{max} is

$$\tau_{max} = F_2 \frac{b}{2} + F_4 \frac{b}{2} = (IaB)\frac{b}{2} + (IaB)\frac{b}{2} = IabB$$

where the moment arm about O is $b/2$ for each force. Because the area enclosed by the loop is $A = ab$, we can express the maximum torque as

$$\tau_{max} = IAB \tag{29.13}$$

This maximum-torque result is valid only when the magnetic field is parallel to the plane of the loop. The sense of the rotation is clockwise when viewed from side ③ as indicated in Figure 29.21b. If the current direction were reversed, the force directions would also reverse and the rotational tendency would be counterclockwise.

Now suppose the uniform magnetic field makes an angle $\theta < 90°$ with a line perpendicular to the plane of the loop as in Figure 29.22. For convenience, let's assume $\vec{\mathbf{B}}$ is perpendicular to sides ② and ④. In this case, the magnetic forces $\vec{\mathbf{F}}_1$ and $\vec{\mathbf{F}}_3$ exerted on sides ① and ③ cancel each other and produce no torque because they act along the same line. The magnetic forces $\vec{\mathbf{F}}_2$ and $\vec{\mathbf{F}}_4$ acting on sides ② and ④, however, produce a torque about *any point*. Referring to the edge view shown in Figure 29.22, we see that the moment arm of $\vec{\mathbf{F}}_2$ about the point O is equal to $(b/2)\sin\theta$. Likewise, the moment arm of $\vec{\mathbf{F}}_4$ about O is also equal to $(b/2)\sin\theta$. Because $F_2 = F_4 = IaB$, the magnitude of the net torque about O is

$$\tau = F_2 \frac{b}{2} \sin\theta + F_4 \frac{b}{2} \sin\theta$$

$$= IaB\left(\frac{b}{2}\sin\theta\right) + IaB\left(\frac{b}{2}\sin\theta\right) = IabB\sin\theta$$

$$= IAB\sin\theta$$

where $A = ab$ is the area of the loop. This result shows that the torque has its maximum value IAB when the field is perpendicular to the normal to the plane of the loop ($\theta = 90°$) as discussed with regard to Figure 29.21 and is zero when the field is parallel to the normal to the plane of the loop ($\theta = 0$).

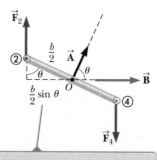

Figure 29.22 An edge view of the loop in Figure 29.21 with the normal to the loop at an angle θ with respect to the magnetic field.

When the normal to the loop makes an angle θ with the magnetic field, the moment arm for the torque is $(b/2)\sin\theta$.

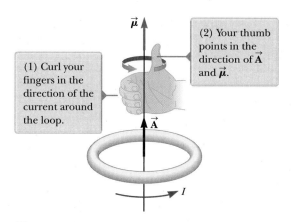

Figure 29.23 Right-hand rule for determining the direction of the vector $\vec{A}$ for a current loop. The direction of the magnetic moment $\vec{\mu}$ is the same as the direction of $\vec{A}$.

A convenient vector expression for the torque exerted on a loop placed in a uniform magnetic field $\vec{B}$ is

$$\vec{\tau} = I\vec{A} \times \vec{B} \qquad (29.14)$$

◀ **Torque on a current loop in a magnetic field**

where $\vec{A}$, the vector shown in Figure 29.22, is perpendicular to the plane of the loop and has a magnitude equal to the area of the loop. To determine the direction of $\vec{A}$, use the right-hand rule described in Figure 29.23. When you curl the fingers of your right hand in the direction of the current in the loop, your thumb points in the direction of $\vec{A}$. Figure 29.22 shows that the loop tends to rotate in the direction of decreasing values of θ (that is, such that the area vector $\vec{A}$ rotates toward the direction of the magnetic field).

The product $I\vec{A}$ is defined to be the **magnetic dipole moment** $\vec{\mu}$ (often simply called the "magnetic moment") of the loop:

$$\vec{\mu} \equiv I\vec{A} \qquad (29.15)$$

◀ **Magnetic dipole moment of a current loop**

The SI unit of magnetic dipole moment is the ampere-meter^2 (A · m^2). If a coil of wire contains N loops of the same area, the magnetic moment of the coil is

$$\vec{\mu}_{\text{coil}} = NI\vec{A} \qquad (29.16)$$

Using Equation 29.15, we can express the torque exerted on a current-carrying loop in a magnetic field $\vec{B}$ as

$$\vec{\tau} = \vec{\mu} \times \vec{B} \qquad (29.17)$$

◀ **Torque on a magnetic moment in a magnetic field**

This result is analogous to Equation 26.18, $\vec{\tau} = \vec{p} \times \vec{E}$, for the torque exerted on an electric dipole in the presence of an electric field $\vec{E}$, where $\vec{p}$ is the electric dipole moment.

Although we obtained the torque for a particular orientation of $\vec{B}$ with respect to the loop, the equation $\vec{\tau} = \vec{\mu} \times \vec{B}$ is valid for any orientation. Furthermore, although we derived the torque expression for a rectangular loop, the result is valid for a loop of any shape. The torque on an N-turn coil is given by Equation 29.17 by using Equation 29.16 for the magnetic moment.

In Section 26.6, we found that the potential energy of a system of an electric dipole in an electric field is given by $U_E = -\vec{p} \cdot \vec{E}$. This energy depends on the orientation of the dipole in the electric field. Likewise, the potential energy of a system of a magnetic dipole in a magnetic field depends on the orientation of the dipole in the magnetic field and is given by

Potential energy of a system of a magnetic moment in a magnetic field

$$U_B = -\vec{\mu} \cdot \vec{B} \qquad (29.18)$$

◀

This expression shows that the system has its lowest energy $U_{min} = -\mu B$ when $\vec{\mu}$ points in the same direction as $\vec{B}$. The system has its highest energy $U_{max} = +\mu B$ when $\vec{\mu}$ points in the direction opposite $\vec{B}$.

Imagine the loop in Figure 29.22 is pivoted at point O on sides ① and ③, so that it is free to rotate. If the loop carries current and the magnetic field is turned on, the loop is modeled as a rigid object under a net torque, with the torque given by Equation 29.17. The torque on the current loop causes the loop to rotate; this effect is exploited practically in a **motor.** Energy enters the motor by electrical transmission, and the rotating coil can do work on some device external to the motor. For example, the motor in a car's electrical window system does work on the windows, applying a force on them and moving them up or down through some displacement. We will discuss motors in more detail in Section 31.5.

Quick Quiz 29.4 (i) Rank the magnitudes of the torques acting on the rectangular loops (a), (b), and (c) shown edge-on in Figure 29.24 from highest to lowest. All loops are identical and carry the same current. (ii) Rank the magnitudes of the net forces acting on the rectangular loops shown in Figure 29.24 from highest to lowest.

Figure 29.24 (Quick Quiz 29.4) Which current loop (seen edge-on) experiences the greatest torque, (a), (b), or (c)? Which experiences the greatest net force?

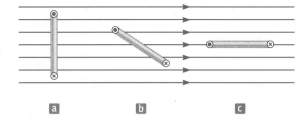

| a | b | c |

Example 29.5 The Magnetic Dipole Moment of a Coil

A rectangular coil of dimensions 5.40 cm × 8.50 cm consists of 25 turns of wire and carries a current of 15.0 mA. A 0.350-T magnetic field is applied parallel to the plane of the coil.

(A) Calculate the magnitude of the magnetic dipole moment of the coil.

SOLUTION

Conceptualize The magnetic moment of the coil is independent of any magnetic field in which the loop resides, so it depends only on the geometry of the loop and the current it carries.

Categorize We evaluate quantities based on equations developed in this section, so we categorize this example as a substitution problem.

Use Equation 29.16 to calculate the magnetic moment associated with a coil consisting of N turns:

$$\mu_{coil} = NIA = (25)(15.0 \times 10^{-3}\,\text{A})(0.054\,0\,\text{m})(0.085\,0\,\text{m})$$
$$= 1.72 \times 10^{-3}\,\text{A} \cdot \text{m}^2$$

(B) What is the magnitude of the torque acting on the loop?

SOLUTION

Use Equation 29.17, noting that $\vec{B}$ is perpendicular to $\vec{\mu}_{coil}$:

$$\tau = \mu_{coil}B = (1.72 \times 10^{-3}\,\text{A} \cdot \text{m}^2)(0.350\,\text{T})$$
$$= 6.02 \times 10^{-4}\,\text{N} \cdot \text{m}$$

Example 29.6 Rotating a Coil AM

Consider the loop of wire in Figure 29.25a. Imagine it is pivoted along side ④, which is parallel to the z axis and fastened so that side ④ remains fixed and the rest of the loop hangs vertically in the gravitational field of the Earth but can rotate around side ④ (Fig. 29.25b). The mass of the loop is 50.0 g, and the sides are of lengths a = 0.200 m and b = 0.100 m. The loop carries a current of 3.50 A and is immersed in a vertical uniform magnetic field of magnitude 0.010 0 T in the positive y direction (Fig. 29.25c). What angle does the plane of the loop make with the vertical?

SOLUTION

Conceptualize In the edge view of Figure 29.25b, notice that the magnetic moment of the loop is to the left. Therefore, when the loop is in the magnetic field, the magnetic torque on the loop causes it to rotate in a clockwise direction around side ④, which we choose as the rotation axis. Imagine the loop making this clockwise rotation so that the plane of the loop is at some angle θ to the vertical as in Figure 29.25c. The gravitational force on the loop exerts a torque that would cause a rotation in the counterclockwise direction if the magnetic field were turned off.

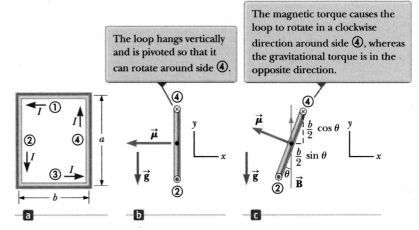

The loop hangs vertically and is pivoted so that it can rotate around side ④.

The magnetic torque causes the loop to rotate in a clockwise direction around side ④, whereas the gravitational torque is in the opposite direction.

Figure 29.25 (Example 29.6) (a) The dimensions of a rectangular current loop. (b) Edge view of the loop sighting down sides ② and ④. (c) An edge view of the loop in (b) rotated through an angle with respect to the horizontal when it is placed in a magnetic field.

Categorize At some angle of the loop, the two torques described in the Conceptualize step are equal in magnitude and the loop is at rest. We therefore model the loop as a *rigid object in equilibrium*.

- -

Analyze Evaluate the magnetic torque on the loop about side ④ from Equation 29.17:

$$\vec{\boldsymbol{\tau}}_B = \vec{\boldsymbol{\mu}} \times \vec{\mathbf{B}} = -\mu B \sin(90° - \theta)\hat{\mathbf{k}} = -IAB\cos\theta\,\hat{\mathbf{k}} = -IabB\cos\theta\,\hat{\mathbf{k}}$$

Evaluate the gravitational torque on the loop, noting that the gravitational force can be modeled to act at the center of the loop:

$$\vec{\boldsymbol{\tau}}_g = \vec{\mathbf{r}} \times m\vec{\mathbf{g}} = mg\frac{b}{2}\sin\theta\,\hat{\mathbf{k}}$$

From the rigid body in equilibrium model, add the torques and set the net torque equal to zero:

$$\sum\vec{\boldsymbol{\tau}} = -IabB\cos\theta\,\hat{\mathbf{k}} + mg\frac{b}{2}\sin\theta\,\hat{\mathbf{k}} = 0$$

Solve for θ:

$$IabB\cos\theta = mg\frac{b}{2}\sin\theta \quad \rightarrow \quad \tan\theta = \frac{2IaB}{mg}$$

$$\theta = \tan^{-1}\left(\frac{2IaB}{mg}\right)$$

Substitute numerical values:

$$\theta = \tan^{-1}\left[\frac{2(3.50\text{ A})(0.200\text{ m})(0.010\ 0\text{ T})}{(0.050\ 0\text{ kg})(9.80\text{ m/s}^2)}\right] = \boxed{1.64°}$$

- -

Finalize The angle is relatively small, so the loop still hangs almost vertically. If the current I or the magnetic field B is increased, however, the angle increases as the magnetic torque becomes stronger.

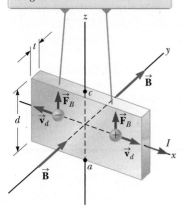

Figure 29.26 To observe the Hall effect, a magnetic field is applied to a current-carrying conductor. The Hall voltage is measured between points a and c.

29.6 The Hall Effect

When a current-carrying conductor is placed in a magnetic field, a potential difference is generated in a direction perpendicular to both the current and the magnetic field. This phenomenon, first observed by Edwin Hall (1855–1938) in 1879, is known as the *Hall effect*. The arrangement for observing the Hall effect consists of a flat conductor carrying a current I in the x direction as shown in Figure 29.26. A uniform magnetic field $\vec{\mathbf{B}}$ is applied in the y direction. If the charge carriers are electrons moving in the negative x direction with a drift velocity $\vec{\mathbf{v}}_d$, they experience an upward magnetic force $\vec{\mathbf{F}}_B = q\vec{\mathbf{v}}_d \times \vec{\mathbf{B}}$, are deflected upward, and accumulate at the upper edge of the flat conductor, leaving an excess of positive charge at the lower edge (Fig. 29.27a). This accumulation of charge at the edges establishes an electric field in the conductor and increases until the electric force on carriers remaining in the bulk of the conductor balances the magnetic force acting on the carriers. The electrons can now be described by the particle in equilibrium model, and they are no longer deflected upward. A sensitive voltmeter connected across the sample as shown in Figure 29.27 can measure the potential difference, known as the **Hall voltage** ΔV_H, generated across the conductor.

If the charge carriers are positive and hence move in the positive x direction (for rightward current) as shown in Figures 29.26 and 29.27b, they also experience an upward magnetic force $q\vec{\mathbf{v}}_d \times \vec{\mathbf{B}}$, which produces a buildup of positive charge on the upper edge and leaves an excess of negative charge on the lower edge. Hence, the sign of the Hall voltage generated in the sample is opposite the sign of the Hall voltage resulting from the deflection of electrons. The sign of the charge carriers can therefore be determined from measuring the polarity of the Hall voltage.

In deriving an expression for the Hall voltage, first note that the magnetic force exerted on the carriers has magnitude qv_dB. In equilibrium, this force is balanced by the electric force qE_H, where E_H is the magnitude of the electric field due to the charge separation (sometimes referred to as the *Hall field*). Therefore,

$$qv_dB = qE_H$$

$$E_H = v_dB$$

If d is the width of the conductor, the Hall voltage is

$$\Delta V_H = E_Hd = v_dBd \tag{29.19}$$

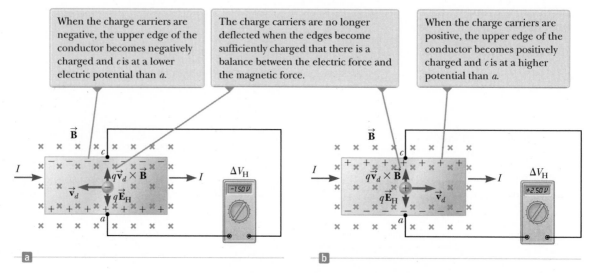

Figure 29.27 The sign of the Hall voltage depends on the sign of the charge carriers.

Therefore, the measured Hall voltage gives a value for the drift speed of the charge carriers if d and B are known.

We can obtain the charge-carrier density n by measuring the current in the sample. From Equation 27.4, we can express the drift speed as

$$v_d = \frac{I}{nqA} \qquad\qquad \text{(29.20)}$$

where A is the cross-sectional area of the conductor. Substituting Equation 29.20 into Equation 29.19 gives

$$\Delta V_H = \frac{IBd}{nqA} \qquad\qquad \text{(29.21)}$$

Because $A = td$, where t is the thickness of the conductor, we can also express Equation 29.21 as

$$\Delta V_H = \frac{IB}{nqt} = \frac{R_H IB}{t} \qquad\qquad \text{(29.22)} \qquad \blacktriangleleft \text{ The Hall voltage}$$

where $R_H = 1/nq$ is called the **Hall coefficient.** This relationship shows that a properly calibrated conductor can be used to measure the magnitude of an unknown magnetic field.

Because all quantities in Equation 29.22 other than nq can be measured, a value for the Hall coefficient is readily obtainable. The sign and magnitude of R_H give the sign of the charge carriers and their number density. In most metals, the charge carriers are electrons and the charge-carrier density determined from Hall-effect measurements is in good agreement with calculated values for such metals as lithium (Li), sodium (Na), copper (Cu), and silver (Ag), whose atoms each give up one electron to act as a current carrier. In this case, n is approximately equal to the number of conducting electrons per unit volume. This classical model, however, is not valid for metals such as iron (Fe), bismuth (Bi), and cadmium (Cd) or for semiconductors. These discrepancies can be explained only by using a model based on the quantum nature of solids.

Example 29.7 The Hall Effect for Copper

A rectangular copper strip 1.5 cm wide and 0.10 cm thick carries a current of 5.0 A. Find the Hall voltage for a 1.2-T magnetic field applied in a direction perpendicular to the strip.

SOLUTION

Conceptualize Study Figures 29.26 and 29.27 carefully and make sure you understand that a Hall voltage is developed between the top and bottom edges of the strip.

Categorize We evaluate the Hall voltage using an equation developed in this section, so we categorize this example as a substitution problem.

Assuming one electron per atom is available for conduction, find the charge-carrier density in terms of the molar mass M and density ρ of copper:

$$n = \frac{N_A}{V} = \frac{N_A \rho}{M}$$

Substitute this result into Equation 29.22:

$$\Delta V_H = \frac{IB}{nqt} = \frac{MIB}{N_A \rho qt}$$

Substitute numerical values:

$$\Delta V_H = \frac{(0.063\,5 \text{ kg/mol})(5.0 \text{ A})(1.2 \text{ T})}{(6.02 \times 10^{23} \text{ mol}^{-1})(8\,920 \text{ kg/m}^3)(1.60 \times 10^{-19} \text{ C})(0.001\,0 \text{ m})}$$

$$= 0.44 \ \mu\text{V}$$

continued

▶ **29.7** continued

Such an extremely small Hall voltage is expected in good conductors. (Notice that the width of the conductor is not needed in this calculation.)

WHAT IF? What if the strip has the same dimensions but is made of a semiconductor? Will the Hall voltage be smaller or larger?

Answer In semiconductors, n is much smaller than it is in metals that contribute one electron per atom to the current; hence, the Hall voltage is usually larger because it varies as the inverse of n. Currents on the order of 0.1 mA are generally used for such materials. Consider a piece of silicon that has the same dimensions as the copper strip in this example and whose value for n is 1.0×10^{20} electrons/m³. Taking $B = 1.2$ T and $I = 0.10$ mA, we find that $\Delta V_H = 7.5$ mV. A potential difference of this magnitude is readily measured.

Summary

Definition

The **magnetic dipole moment** $\vec{\mu}$ of a loop carrying a current I is

$$\vec{\mu} \equiv I\vec{A} \tag{29.15}$$

where the area vector $\vec{A}$ is perpendicular to the plane of the loop and $|\vec{A}|$ is equal to the area of the loop. The SI unit of $\vec{\mu}$ is A · m².

Concepts and Principles

If a charged particle moves in a uniform magnetic field so that its initial velocity is perpendicular to the field, the particle moves in a circle, the plane of which is perpendicular to the magnetic field. The radius of the circular path is

$$r = \frac{mv}{qB} \tag{29.3}$$

where m is the mass of the particle and q is its charge. The angular speed of the charged particle is

$$\omega = \frac{qB}{m} \tag{29.4}$$

If a straight conductor of length L carries a current I, the force exerted on that conductor when it is placed in a uniform magnetic field $\vec{B}$ is

$$\vec{F}_B = I\vec{L} \times \vec{B} \tag{29.10}$$

where the direction of $\vec{L}$ is in the direction of the current and $|\vec{L}| = L$.

If an arbitrarily shaped wire carrying a current I is placed in a magnetic field, the magnetic force exerted on a very small segment $d\vec{s}$ is

$$d\vec{F}_B = I\,d\vec{s} \times \vec{B} \tag{29.11}$$

To determine the total magnetic force on the wire, one must integrate Equation 29.11 over the wire, keeping in mind that both $\vec{B}$ and $d\vec{s}$ may vary at each point.

The torque $\vec{\tau}$ on a current loop placed in a uniform magnetic field $\vec{B}$ is

$$\vec{\tau} = \vec{\mu} \times \vec{B} \tag{29.17}$$

The potential energy of the system of a magnetic dipole in a magnetic field is

$$U_B = -\vec{\mu} \cdot \vec{B} \tag{29.18}$$

Analysis Models for Problem Solving

Particle in a Field (Magnetic) A source (to be discussed in Chapter 30) establishes a **magnetic field** $\vec{\mathbf{B}}$ throughout space. When a particle with charge q and moving with velocity $\vec{\mathbf{v}}$ is placed in that field, it experiences a magnetic force given by

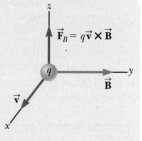

$$\vec{\mathbf{F}}_B = q\vec{\mathbf{v}} \times \vec{\mathbf{B}} \qquad (29.1)$$

The direction of this magnetic force is perpendicular both to the velocity of the particle and to the magnetic field. The magnitude of this force is

$$F_B = |q|vB \sin \theta \qquad (29.2)$$

where θ is the smaller angle between $\vec{\mathbf{v}}$ and $\vec{\mathbf{B}}$. The SI unit of $\vec{\mathbf{B}}$ is the **tesla** (T), where $1 \text{ T} = 1 \text{ N/A} \cdot \text{m}$.

Objective Questions   **1.** denotes answer available in *Student Solutions Manual/Study Guide*

Objective Questions 3, 4, and 6 in Chapter 11 can be assigned with this chapter as review for the vector product.

1. A spatially uniform magnetic field cannot exert a magnetic force on a particle in which of the following circumstances? There may be more than one correct statement. (a) The particle is charged. (b) The particle moves perpendicular to the magnetic field. (c) The particle moves parallel to the magnetic field. (d) The magnitude of the magnetic field changes with time. (e) The particle is at rest.

2. Rank the magnitudes of the forces exerted on the following particles from largest to smallest. In your ranking, display any cases of equality. (a) an electron moving at 1 Mm/s perpendicular to a 1-mT magnetic field (b) an electron moving at 1 Mm/s parallel to a 1-mT magnetic field (c) an electron moving at 2 Mm/s perpendicular to a 1-mT magnetic field (d) a proton moving at 1 Mm/s perpendicular to a 1-mT magnetic field (e) a proton moving at 1 Mm/s at a 45° angle to a 1-mT magnetic field

3. A particle with electric charge is fired into a region of space where the electric field is zero. It moves in a straight line. Can you conclude that the magnetic field in that region is zero? (a) Yes, you can. (b) No; the field might be perpendicular to the particle's velocity. (c) No; the field might be parallel to the particle's velocity. (d) No; the particle might need to have charge of the opposite sign to have a force exerted on it. (e) No; an observation of an object with *electric* charge gives no information about a *magnetic* field.

4. A proton moving horizontally enters a region where a uniform magnetic field is directed perpendicular to the proton's velocity as shown in Figure OQ29.4. After the proton enters the field, does it (a) deflect downward, with its speed remaining constant; (b) deflect upward, moving in a semicircular path with constant speed, and exit the field moving to the left; (c) continue to move in the horizontal direction with constant velocity; (d) move in a circular orbit and become trapped by the field; or (e) deflect out of the plane of the paper?

Figure OQ29.4

5. At a certain instant, a proton is moving in the positive x direction through a magnetic field in the negative z direction. What is the direction of the magnetic force exerted on the proton? (a) positive z direction (b) negative z direction (c) positive y direction (d) negative y direction (e) The force is zero.

6. A thin copper rod 1.00 m long has a mass of 50.0 g. What is the minimum current in the rod that would allow it to levitate above the ground in a magnetic field of magnitude 0.100 T? (a) 1.20 A (b) 2.40 A (c) 4.90 A (d) 9.80 A (e) none of those answers

7. Electron A is fired horizontally with speed 1.00 Mm/s into a region where a vertical magnetic field exists. Electron B is fired along the same path with speed 2.00 Mm/s. (i) Which electron has a larger magnetic force exerted on it? (a) A does. (b) B does. (c) The forces have the same nonzero magnitude. (d) The forces are both zero. (ii) Which electron has a path that curves more sharply? (a) A does. (b) B does. (c) The particles follow the same curved path. (d) The particles continue to go straight.

8. Classify each of the following statements as a characteristic (a) of electric forces only, (b) of magnetic forces only, (c) of both electric and magnetic forces, or (d) of neither electric nor magnetic forces. (i) The force is proportional to the magnitude of the field exerting it. (ii) The force is proportional to the magnitude of the charge of the object on which the force is exerted. (iii) The force exerted on a negatively charged object is opposite in direction to the force on a positive charge. (iv) The force exerted on a stationary charged object is nonzero. (v) The force exerted on a moving charged

object is zero. **(vi)** The force exerted on a charged object is proportional to its speed. **(vii)** The force exerted on a charged object cannot alter the object's speed. **(viii)** The magnitude of the force depends on the charged object's direction of motion.

9. An electron moves horizontally across the Earth's equator at a speed of 2.50×10^6 m/s and in a direction $35.0°$ N of E. At this point, the Earth's magnetic field has a direction due north, is parallel to the surface, and has a value of 3.00×10^{-5} T. What is the force acting on the electron due to its interaction with the Earth's magnetic field? (a) 6.88×10^{-18} N due west (b) 6.88×10^{-18} N toward the Earth's surface (c) 9.83×10^{-18} N toward the Earth's surface (d) 9.83×10^{-18} N away from the Earth's surface (e) 4.00×10^{-18} N away from the Earth's surface

10. A charged particle is traveling through a uniform magnetic field. Which of the following statements are true of the magnetic field? There may be more than one correct statement. (a) It exerts a force on the particle parallel to the field. (b) It exerts a force on the particle along the direction of its motion. (c) It increases the kinetic energy of the particle. (d) It exerts a force that is perpendicular to the direction of motion. (e) It does not change the magnitude of the momentum of the particle.

11. In the velocity selector shown in Figure 29.13, electrons with speed $v = E/B$ follow a straight path. Electrons moving significantly faster than this speed through the same selector will move along what kind of path? (a) a circle (b) a parabola (c) a straight line (d) a more complicated trajectory

12. Answer each question yes or no. Assume the motions and currents mentioned are along the x axis and fields are in the y direction. (a) Does an electric field exert a force on a stationary charged object? (b) Does a magnetic field do so? (c) Does an electric field exert a force on a moving charged object? (d) Does a magnetic field do so? (e) Does an electric field exert a force on a straight current-carrying wire? (f) Does a magnetic field do so? (g) Does an electric field exert a force on a beam of moving electrons? (h) Does a magnetic field do so?

13. A magnetic field exerts a torque on each of the current-carrying single loops of wire shown in Figure OQ29.13. The loops lie in the xy plane, each carrying the same magnitude current, and the uniform magnetic field points in the positive x direction. Rank the loops by the magnitude of the torque exerted on them by the field from largest to smallest.

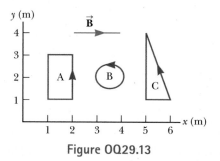

Figure OQ29.13

Conceptual Questions | **1.** denotes answer available in *Student Solutions Manual/Study Guide*

1. Can a constant magnetic field set into motion an electron initially at rest? Explain your answer.

2. Explain why it is not possible to determine the charge and the mass of a charged particle separately by measuring accelerations produced by electric and magnetic forces on the particle.

3. Is it possible to orient a current loop in a uniform magnetic field such that the loop does not tend to rotate? Explain.

4. How can the motion of a moving charged particle be used to distinguish between a magnetic field and an electric field? Give a specific example to justify your argument.

5. How can a current loop be used to determine the presence of a magnetic field in a given region of space?

6. Charged particles from outer space, called cosmic rays, strike the Earth more frequently near the poles than near the equator. Why?

7. Two charged particles are projected in the same direction into a magnetic field perpendicular to their velocities. If the particles are deflected in opposite directions, what can you say about them?

Problems available in ENHANCED WebAssign Access end-of-chapter problems online at www.webassign.net

Section 29.1 Analysis Model: Particle in a Field (Magnetic)
Problems 1–12

Section 29.2 Motion of a Charged Particle in a Uniform Magnetic Field
Problems 13–23

Section 29.3 Applications Involving Charged Particles Moving in a Magnetic Field
Problems 24–31

Section 29.4 Magnetic Force Acting on a Current-Carrying Conductor
Problems 32–44

Section 29.5 Torque on a Current Loop in a Uniform Magnetic Field
Problems 45–53

Section 29.6 The Hall Effect
Problems 54–55

Additional Problems
Problems 56–76

Challenge Problems
Problem 77–80

Solutions to the following Problems are available in the *Student Solutions Manual/Study Guide:*
29.2, 29.6, 29.8, 29.11, 29.15, 29.21, 29.24, 29.31, 29.39, 29.42, 29.51, 29.55, 29.58, 29.59, 29.72, and 29.77

List of Enhanced Problems

Problem Number	Targeted Feedback in Enhanced WebAssign	Analysis Model Tutorial in Enhanced WebAssign	Master It in Enhanced WebAssign	Watch It in Enhanced WebAssign
29.2	✓			✓
29.6		✓		
29.7	✓			✓
29.8	✓			✓
29.9		✓		
29.11			✓	
29.17	✓			
29.21			✓	
29.24			✓	
29.25	✓			✓
29.29	✓			✓
29.34	✓			✓
29.35	✓			✓
29.37	✓	✓		✓
29.39			✓	
29.48	✓			✓
29.49	✓		✓	
29.51	✓		✓	
29.53	✓			✓
29.55			✓	
29.59			✓	
29.61		✓		
29.65		✓		
29.69		✓		

CHAPTER 30

Sources of the Magnetic Field

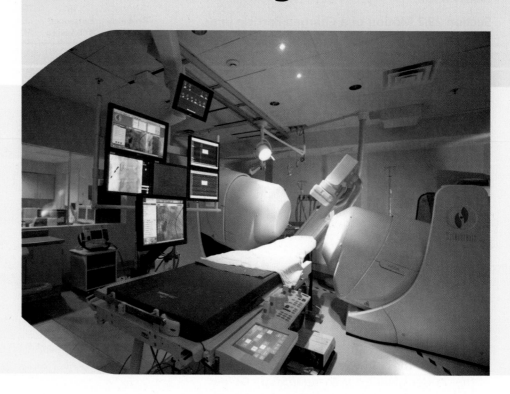

A cardiac catheterization laboratory stands ready to receive a patient suffering from atrial fibrillation. The large white objects on either side of the operating table are strong magnets that place the patient in a magnetic field. The electrophysiologist performing a catheter ablation procedure sits at a computer in the room to the left. With guidance from the magnetic field, he or she uses a joystick and other controls to thread the magnetically sensitive tip of a cardiac catheter through blood vessels and into the chambers of the heart. *(© Courtesy of Stereotaxis, Inc.)*

In Chapter 29, we discussed the magnetic force exerted on a charged particle moving in a magnetic field. To complete the description of the magnetic interaction, this chapter explores the origin of the magnetic field, moving charges. We begin by showing how to use the law of Biot and Savart to calculate the magnetic field produced at some point in space by a small current element. This formalism is then used to calculate the total magnetic field due to various current distributions. Next, we show how to determine the force between two current-carrying conductors, leading to the definition of the ampere. We also introduce Ampère's law, which is useful in calculating the magnetic field of a highly symmetric configuration carrying a steady current.

This chapter is also concerned with the complex processes that occur in magnetic materials. All magnetic effects in matter can be explained on the basis of atomic magnetic moments, which arise both from the orbital motion of electrons and from an intrinsic property of electrons known as spin.

30.1 The Biot–Savart Law

Shortly after Oersted's discovery in 1819 that a compass needle is deflected by a current-carrying conductor, Jean-Baptiste Biot (1774–1862) and Félix Savart (1791–1841) performed quantitative experiments on the force exerted by an electric current on a nearby magnet. From their experimental results, Biot and Savart arrived at a mathematical expression that gives the magnetic field at some point in space in terms of the current that produces the field. That expression is based on the following experimental observations for the magnetic field $d\vec{\mathbf{B}}$ at a point P associated with a length element $d\vec{\mathbf{s}}$ of a wire carrying a steady current I (Fig. 30.1):

- The vector $d\vec{\mathbf{B}}$ is perpendicular both to $d\vec{\mathbf{s}}$ (which points in the direction of the current) and to the unit vector $\hat{\mathbf{r}}$ directed from $d\vec{\mathbf{s}}$ toward P.
- The magnitude of $d\vec{\mathbf{B}}$ is inversely proportional to r^2, where r is the distance from $d\vec{\mathbf{s}}$ to P.
- The magnitude of $d\vec{\mathbf{B}}$ is proportional to the current I and to the magnitude ds of the length element $d\vec{\mathbf{s}}$.
- The magnitude of $d\vec{\mathbf{B}}$ is proportional to $\sin\theta$, where θ is the angle between the vectors $d\vec{\mathbf{s}}$ and $\hat{\mathbf{r}}$.

These observations are summarized in the mathematical expression known today as the **Biot–Savart law**:

$$d\vec{\mathbf{B}} = \frac{\mu_0}{4\pi}\frac{I\,d\vec{\mathbf{s}} \times \hat{\mathbf{r}}}{r^2} \qquad (30.1)$$

◀ Biot–Savart law

where μ_0 is a constant called the **permeability of free space**:

$$\mu_0 = 4\pi \times 10^{-7}\,\mathrm{T \cdot m/A} \qquad (30.2)$$

◀ Permeability of free space

Notice that the field $d\vec{\mathbf{B}}$ in Equation 30.1 is the field created at a point by the current in only a small length element $d\vec{\mathbf{s}}$ of the conductor. To find the *total* magnetic field $\vec{\mathbf{B}}$ created at some point by a current of finite size, we must sum up contributions from all current elements $I\,d\vec{\mathbf{s}}$ that make up the current. That is, we must evaluate $\vec{\mathbf{B}}$ by integrating Equation 30.1:

$$\vec{\mathbf{B}} = \frac{\mu_0 I}{4\pi}\int\frac{d\vec{\mathbf{s}} \times \hat{\mathbf{r}}}{r^2} \qquad (30.3)$$

where the integral is taken over the entire current distribution. This expression must be handled with special care because the integrand is a cross product and therefore a vector quantity. We shall see one case of such an integration in Example 30.1.

Although the Biot–Savart law was discussed for a current-carrying wire, it is also valid for a current consisting of charges flowing through space such as the particle beam in an accelerator. In that case, $d\vec{\mathbf{s}}$ represents the length of a small segment of space in which the charges flow.

Interesting similarities and differences exist between Equation 30.1 for the magnetic field due to a current element and Equation 23.9 for the electric field due to a point charge. The magnitude of the magnetic field varies as the inverse square of the distance from the source, as does the electric field due to a point charge. The directions of the two fields are quite different, however. The electric field created by a point charge is radial, but the magnetic field created by a current element is perpendicular to both the length element $d\vec{\mathbf{s}}$ and the unit vector $\hat{\mathbf{r}}$ as described by the cross product in Equation 30.1. Hence, if the conductor lies in the plane of the page as shown in Figure 30.1, $d\vec{\mathbf{B}}$ points out of the page at P and into the page at P'.

Another difference between electric and magnetic fields is related to the source of the field. An electric field is established by an isolated electric charge. The Biot–Savart law gives the magnetic field of an isolated current element at some point, but such an isolated current element cannot exist the way an isolated electric charge can. A current element *must* be part of an extended current distribution because a complete circuit is needed for charges to flow. Therefore,

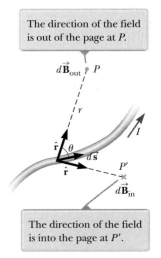

The direction of the field is out of the page at P.

The direction of the field is into the page at P'.

Figure 30.1 The magnetic field $d\vec{\mathbf{B}}$ at a point due to the current I through a length element $d\vec{\mathbf{s}}$ is given by the Biot–Savart law.

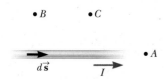

Figure 30.2 (Quick Quiz 30.1) Where is the magnetic field due to the current element the greatest?

the Biot–Savart law (Eq. 30.1) is only the first step in a calculation of a magnetic field; it must be followed by an integration over the current distribution as in Equation 30.3.

Quick Quiz 30.1 Consider the magnetic field due to the current in the wire shown in Figure 30.2. Rank the points A, B, and C in terms of magnitude of the magnetic field that is due to the current in just the length element $d\vec{s}$ shown from greatest to least.

Example 30.1   **Magnetic Field Surrounding a Thin, Straight Conductor**

Consider a thin, straight wire of finite length carrying a constant current I and placed along the x axis as shown in Figure 30.3. Determine the magnitude and direction of the magnetic field at point P due to this current.

SOLUTION

Conceptualize From the Biot–Savart law, we expect that the magnitude of the field is proportional to the current in the wire and decreases as the distance a from the wire to point P increases. We also expect the field to depend on the angles θ_1 and θ_2 in Figure 30.3b. We place the origin at O and let point P be along the positive y axis, with $\hat{\mathbf{k}}$ being a unit vector pointing out of the page.

Categorize We are asked to find the magnetic field due to a simple current distribution, so this example is a typical problem for which the Biot–Savart law is appropriate. We must find the field contribution from a small element of current and then integrate over the current distribution.

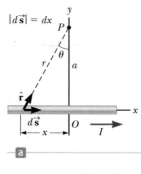

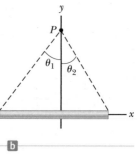

Figure 30.3 (Example 30.1) (a) A thin, straight wire carrying a current I. (b) The angles θ_1 and θ_2 used for determining the net field.

Analyze Let's start by considering a length element $d\vec{s}$ located a distance r from P. The direction of the magnetic field at point P due to the current in this element is out of the page because $d\vec{s} \times \hat{\mathbf{r}}$ is out of the page. In fact, because *all* the current elements $I\,d\vec{s}$ lie in the plane of the page, they all produce a magnetic field directed out of the page at point P. Therefore, the direction of the magnetic field at point P is out of the page and we need only find the magnitude of the field.

Evaluate the cross product in the Biot–Savart law:

$$d\vec{s} \times \hat{\mathbf{r}} = |d\vec{s} \times \hat{\mathbf{r}}|\hat{\mathbf{k}} = \left[dx \sin\left(\frac{\pi}{2} - \theta \right) \right]\hat{\mathbf{k}} = (dx \cos\theta)\hat{\mathbf{k}}$$

Substitute into Equation 30.1:

$$(1) \quad d\vec{B} = (dB)\hat{\mathbf{k}} = \frac{\mu_0 I}{4\pi} \frac{dx \cos\theta}{r^2}\hat{\mathbf{k}}$$

From the geometry in Figure 30.3a, express r in terms of θ:

$$(2) \quad r = \frac{a}{\cos\theta}$$

Notice that $\tan\theta = -x/a$ from the right triangle in Figure 30.3a (the negative sign is necessary because $d\vec{s}$ is located at a negative value of x) and solve for x:

$$x = -a \tan\theta$$

Find the differential dx:

$$(3) \quad dx = -a \sec^2\theta\, d\theta = -\frac{a\, d\theta}{\cos^2\theta}$$

Substitute Equations (2) and (3) into the expression for the z component of the field from Equation (1):

$$(4) \quad dB = -\frac{\mu_0 I}{4\pi}\left(\frac{a\, d\theta}{\cos^2\theta} \right)\left(\frac{\cos^2\theta}{a^2} \right)\cos\theta = -\frac{\mu_0 I}{4\pi a}\cos\theta\, d\theta$$

▶ **30.1** continued

Integrate Equation (4) over all length elements on the wire, where the subtending angles range from θ_1 to θ_2 as defined in Figure 30.3b:

$$B = -\frac{\mu_0 I}{4\pi a} \int_{\theta_1}^{\theta_2} \cos\theta \; d\theta = \frac{\mu_0 I}{4\pi a}(\sin\theta_1 - \sin\theta_2) \tag{30.4}$$

....

Finalize We can use this result to find the magnitude of the magnetic field of *any* straight current-carrying wire if we know the geometry and hence the angles θ_1 and θ_2. Consider the special case of an infinitely long, straight wire. If the wire in Figure 30.3b becomes infinitely long, we see that $\theta_1 = \pi/2$ and $\theta_2 = -\pi/2$ for length elements ranging between positions $x = -\infty$ and $x = +\infty$. Because $(\sin\theta_1 - \sin\theta_2) = [\sin \pi/2 - \sin(-\pi/2)] = 2$, Equation 30.4 becomes

$$B = \frac{\mu_0 I}{2\pi a} \tag{30.5}$$

Equations 30.4 and 30.5 both show that the magnitude of the magnetic field is proportional to the current and decreases with increasing distance from the wire, as expected. Equation 30.5 has the same mathematical form as the expression for the magnitude of the electric field due to a long charged wire (see Eq. 24.7).

| **Example 30.2** | **Magnetic Field Due to a Curved Wire Segment** |

Calculate the magnetic field at point O for the current-carrying wire segment shown in Figure 30.4. The wire consists of two straight portions and a circular arc of radius a, which subtends an angle θ.

SOLUTION

Conceptualize The magnetic field at O due to the current in the straight segments AA' and CC' is zero because $d\vec{s}$ is parallel to $\hat{\mathbf{r}}$ along these paths, which means that $d\vec{s} \times \hat{\mathbf{r}} = 0$ for these paths. Therefore, we expect the magnetic field at O to be due only to the current in the curved portion of the wire.

Categorize Because we can ignore segments AA' and CC', this example is categorized as an application of the Biot–Savart law to the curved wire segment AC.

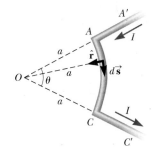

Figure 30.4 (Example 30.2) The length of the curved segment AC is s.

....

Analyze Each length element $d\vec{s}$ along path AC is at the same distance a from O, and the current in each contributes a field element $d\vec{B}$ directed into the page at O. Furthermore, at every point on AC, $d\vec{s}$ is perpendicular to $\hat{\mathbf{r}}$; hence, $|d\vec{s} \times \hat{\mathbf{r}}| = ds$.

From Equation 30.1, find the magnitude of the field at O due to the current in an element of length ds:

$$dB = \frac{\mu_0}{4\pi}\frac{I\,ds}{a^2}$$

Integrate this expression over the curved path AC, noting that I and a are constants:

$$B = \frac{\mu_0 I}{4\pi a^2}\int ds = \frac{\mu_0 I}{4\pi a^2}s$$

From the geometry, note that $s = a\theta$ and substitute:

$$B = \frac{\mu_0 I}{4\pi a^2}(a\theta) = \frac{\mu_0 I}{4\pi a}\theta \tag{30.6}$$

....

Finalize Equation 30.6 gives the magnitude of the magnetic field at O. The direction of $\vec{B}$ is into the page at O because $d\vec{s} \times \hat{\mathbf{r}}$ is into the page for every length element.

WHAT IF? What if you were asked to find the magnetic field at the center of a circular wire loop of radius R that carries a current I? Can this question be answered at this point in our understanding of the source of magnetic fields?

continued

▶ **30.2** continued

Answer Yes, it can. The straight wires in Figure 30.4 do not contribute to the magnetic field. The only contribution is from the curved segment. As the angle θ increases, the curved segment becomes a full circle when $\theta = 2\pi$. Therefore, you can find the magnetic field at the center of a wire loop by letting $\theta = 2\pi$ in Equation 30.6:

$$B = \frac{\mu_0 I}{4\pi a} 2\pi = \frac{\mu_0 I}{2a}$$

This result is a limiting case of a more general result discussed in Example 30.3.

Example 30.3   **Magnetic Field on the Axis of a Circular Current Loop**

Consider a circular wire loop of radius a located in the yz plane and carrying a steady current I as in Figure 30.5. Calculate the magnetic field at an axial point P a distance x from the center of the loop.

SOLUTION

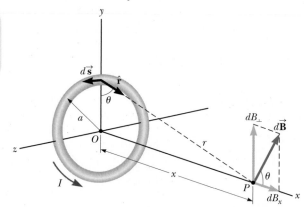

Figure 30.5 (Example 30.3) Geometry for calculating the magnetic field at a point P lying on the axis of a current loop. By symmetry, the total field $\vec{\mathbf{B}}$ is along this axis.

Conceptualize Compare this problem to Example 23.8 for the electric field due to a ring of charge. Figure 30.5 shows the magnetic field contribution $d\vec{\mathbf{B}}$ at P due to a single current element at the top of the ring. This field vector can be resolved into components dB_x parallel to the axis of the ring and $dB_\perp$ perpendicular to the axis. Think about the magnetic field contributions from a current element at the bottom of the loop. Because of the symmetry of the situation, the perpendicular components of the field due to elements at the top and bottom of the ring cancel. This cancellation occurs for all pairs of segments around the ring, so we can ignore the perpendicular component of the field and focus solely on the parallel components, which simply add.

Categorize We are asked to find the magnetic field due to a simple current distribution, so this example is a typical problem for which the Biot–Savart law is appropriate.

. . . .

Analyze In this situation, every length element $d\vec{\mathbf{s}}$ is perpendicular to the vector $\hat{\mathbf{r}}$ at the location of the element. Therefore, for any element, $|d\vec{\mathbf{s}} \times \hat{\mathbf{r}}| = (ds)(1) \sin 90° = ds$. Furthermore, all length elements around the loop are at the same distance r from P, where $r^2 = a^2 + x^2$.

Use Equation 30.1 to find the magnitude of $d\vec{\mathbf{B}}$ due to the current in any length element $d\vec{\mathbf{s}}$:

$$dB = \frac{\mu_0 I}{4\pi} \frac{|d\vec{\mathbf{s}} \times \hat{\mathbf{r}}|}{r^2} = \frac{\mu_0 I}{4\pi} \frac{ds}{(a^2 + x^2)}$$

Find the x component of the field element:

$$dB_x = \frac{\mu_0 I}{4\pi} \frac{ds}{(a^2 + x^2)} \cos\theta$$

Integrate over the entire loop:

$$B_x = \oint dB_x = \frac{\mu_0 I}{4\pi} \oint \frac{ds \cos\theta}{a^2 + x^2}$$

From the geometry, evaluate $\cos\theta$:

$$\cos\theta = \frac{a}{(a^2 + x^2)^{1/2}}$$

Substitute this expression for $\cos\theta$ into the integral and note that x and a are both constant:

$$B_x = \frac{\mu_0 I}{4\pi} \oint \frac{ds}{a^2 + x^2} \left[\frac{a}{(a^2 + x^2)^{1/2}} \right] = \frac{\mu_0 I}{4\pi} \frac{a}{(a^2 + x^2)^{3/2}} \oint ds$$

Integrate around the loop:

$$B_x = \frac{\mu_0 I}{4\pi} \frac{a}{(a^2 + x^2)^{3/2}} (2\pi a) = \frac{\mu_0 I a^2}{2(a^2 + x^2)^{3/2}} \qquad \textbf{(30.7)}$$

. . . .

▶ **30.3** continued

Finalize To find the magnetic field at the center of the loop, set $x = 0$ in Equation 30.7. At this special point,

$$B = \frac{\mu_0 I}{2a} \quad \text{(at } x = 0)$$ **(30.8)**

which is consistent with the result of the **What If?** feature of Example 30.2.

The pattern of magnetic field lines for a circular current loop is shown in Figure 30.6a. For clarity, the lines are drawn for only the plane that contains the axis of the loop. The field-line pattern is axially symmetric and looks like the pattern around a bar magnet, which is shown in Figure 30.6b.

WHAT IF?   What if we consider points on the x axis very far from the loop? How does the magnetic field behave at these distant points?

Answer In this case, in which $x \gg a$, we can neglect the term a^2 in the denominator of Equation 30.7 and obtain

$$B \approx \frac{\mu_0 I a^2}{2x^3} \quad \text{(for } x \gg a)$$ **(30.9)**

The magnitude of the magnetic moment μ of the loop is defined as the product of current and loop area (see Eq. 29.15): $\mu = I(\pi a^2)$ for our circular loop. We can express Equation 30.9 as

$$B \approx \frac{\mu_0}{2\pi} \frac{\mu}{x^3}$$ **(30.10)**

This result is similar in form to the expression for the electric field due to an electric dipole, $E = k_e(p/y^3)$ (see Example 23.6), where $p = 2aq$ is the electric dipole moment as defined in Equation 26.16.

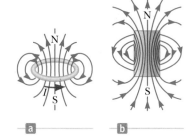

Figure 30.6 (Example 30.3) (a) Magnetic field lines surrounding a current loop. (b) Magnetic field lines surrounding a bar magnet. Notice the similarity between this line pattern and that of a current loop.

30.2 The Magnetic Force Between Two Parallel Conductors

In Chapter 29, we described the magnetic force that acts on a current-carrying conductor placed in an external magnetic field. Because a current in a conductor sets up its own magnetic field, it is easy to understand that two current-carrying conductors exert magnetic forces on each other. One wire establishes the magnetic field and the other wire is modeled as a collection of particles in a magnetic field. Such forces between wires can be used as the basis for defining the ampere and the coulomb.

Consider two long, straight, parallel wires separated by a distance a and carrying currents I_1 and I_2 in the same direction as in Figure 30.7. Let's determine the force exerted on one wire due to the magnetic field set up by the other wire. Wire 2, which carries a current I_2 and is identified arbitrarily as the source wire, creates a magnetic field $\vec{\mathbf{B}}_2$ at the location of wire 1, the test wire. The magnitude of this magnetic field is the same at all points on wire 1. The direction of $\vec{\mathbf{B}}_2$ is perpendicular to wire 1 as shown in Figure 30.7. According to Equation 29.10, the magnetic force on a length ℓ of wire 1 is $\vec{\mathbf{F}}_1 = I_1\vec{\ell} \times \vec{\mathbf{B}}_2$. Because $\vec{\ell}$ is perpendicular to $\vec{\mathbf{B}}_2$ in this situation, the magnitude of $\vec{\mathbf{F}}_1$ is $F_1 = I_1\ell B_2$. Because the magnitude of $\vec{\mathbf{B}}_2$ is given by Equation 30.5,

$$F_1 = I_1 \ell B_2 = I_1 \ell \left(\frac{\mu_0 I_2}{2\pi a}\right) = \frac{\mu_0 I_1 I_2}{2\pi a}\ell$$ **(30.11)**

The direction of $\vec{\mathbf{F}}_1$ is toward wire 2 because $\vec{\ell} \times \vec{\mathbf{B}}_2$ is in that direction. When the field set up at wire 2 by wire 1 is calculated, the force $\vec{\mathbf{F}}_2$ acting on wire 2 is found to be equal in magnitude and opposite in direction to $\vec{\mathbf{F}}_1$, which is what we expect because Newton's third law must be obeyed. When the currents are in opposite directions (that is, when one of the currents is reversed in Fig. 30.7), the forces

The field $\vec{\mathbf{B}}_2$ due to the current in wire 2 exerts a magnetic force of magnitude $F_1 = I_1\ell B_2$ on wire 1.

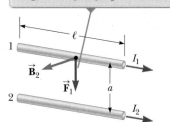

Figure 30.7 Two parallel wires that each carry a steady current exert a magnetic force on each other. The force is attractive if the currents are parallel (as shown) and repulsive if the currents are antiparallel.

are reversed and the wires repel each other. Hence, parallel conductors carrying currents in the *same* direction *attract* each other, and parallel conductors carrying currents in *opposite* directions *repel* each other.

Because the magnitudes of the forces are the same on both wires, we denote the magnitude of the magnetic force between the wires as simply F_B. We can rewrite this magnitude in terms of the force per unit length:

$$\frac{F_B}{\ell} = \frac{\mu_0 I_1 I_2}{2\pi a} \qquad \text{(30.12)}$$

The force between two parallel wires is used to define the **ampere** as follows:

Definition of the ampere ▶ When the magnitude of the force per unit length between two long, parallel wires that carry identical currents and are separated by 1 m is 2×10^{-7} N/m, the current in each wire is defined to be 1 A.

The value 2×10^{-7} N/m is obtained from Equation 30.12 with $I_1 = I_2 = 1$ A and $a = 1$ m. Because this definition is based on a force, a mechanical measurement can be used to standardize the ampere. For instance, the National Institute of Standards and Technology uses an instrument called a *current balance* for primary current measurements. The results are then used to standardize other, more conventional instruments such as ammeters.

The SI unit of charge, the **coulomb,** is defined in terms of the ampere: When a conductor carries a steady current of 1 A, the quantity of charge that flows through a cross section of the conductor in 1 s is 1 C.

In deriving Equations 30.11 and 30.12, we assumed both wires are long compared with their separation distance. In fact, only one wire needs to be long. The equations accurately describe the forces exerted on each other by a long wire and a straight, parallel wire of limited length ℓ.

Quick Quiz 30.2 A loose spiral spring carrying no current is hung from a ceiling. When a switch is thrown so that a current exists in the spring, do the coils (a) move closer together, (b) move farther apart, or (c) not move at all?

Example 30.4   Suspending a Wire   AM

Two infinitely long, parallel wires are lying on the ground a distance $a = 1.00$ cm apart as shown in Figure 30.8a. A third wire, of length $L = 10.0$ m and mass 400 g, carries a current of $I_1 = 100$ A and is levitated above the first two wires, at a horizontal position midway between them. The infinitely long wires carry equal currents I_2 in the same direction, but in the direction opposite that in the levitated wire. What current must the infinitely long wires carry so that the three wires form an equilateral triangle?

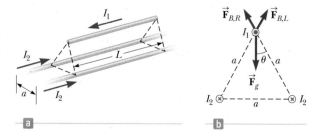

Figure 30.8 (Example 30.4) (a) Two current-carrying wires lie on the ground and suspend a third wire in the air by magnetic forces. (b) End view. In the situation described in the example, the three wires form an equilateral triangle. The two magnetic forces on the levitated wire are $\vec{\mathbf{F}}_{B,L}$, the force due to the left-hand wire on the ground, and $\vec{\mathbf{F}}_{B,R}$, the force due to the right-hand wire. The gravitational force $\vec{\mathbf{F}}_g$ on the levitated wire is also shown.

SOLUTION

Conceptualize Because the current in the short wire is opposite those in the long wires, the short wire is repelled from both of the others. Imagine the currents in the long wires in Figure 30.8a are increased. The repulsive force becomes stronger, and the levitated wire rises to the point at which the wire is once again levitated in equilibrium at a higher position. Figure 30.8b shows the desired situation with the three wires forming an equilateral triangle.

Categorize Because the levitated wire is subject to forces but does not accelerate, it is modeled as a *particle in equilibrium.*

▶ **30.4** continued

Analyze The horizontal components of the magnetic forces on the levitated wire cancel. The vertical components are both positive and add together. Choose the z axis to be upward through the top wire in Figure 30.8b and in the plane of the page.

Find the total magnetic force in the upward direction on the levitated wire:

$$\vec{\mathbf{F}}_B = 2\left(\frac{\mu_0 I_1 I_2}{2\pi a}\ell\right)\cos\theta\,\hat{\mathbf{k}} = \frac{\mu_0 I_1 I_2}{\pi a}\ell\cos\theta\,\hat{\mathbf{k}}$$

Find the gravitational force on the levitated wire:

$$\vec{\mathbf{F}}_g = -mg\hat{\mathbf{k}}$$

Apply the particle in equilibrium model by adding the forces and setting the net force equal to zero:

$$\sum \vec{\mathbf{F}} = \vec{\mathbf{F}}_B + \vec{\mathbf{F}}_g = \frac{\mu_0 I_1 I_2}{\pi a}\ell\cos\theta\,\hat{\mathbf{k}} - mg\hat{\mathbf{k}} = 0$$

Solve for the current in the wires on the ground:

$$I_2 = \frac{mg\pi a}{\mu_0 I_1 \ell \cos\theta}$$

Substitute numerical values:

$$I_2 = \frac{(0.400\ \text{kg})(9.80\ \text{m/s}^2)\pi(0.010\ 0\ \text{m})}{(4\pi \times 10^{-7}\ \text{T}\cdot\text{m/A})(100\ \text{A})(10.0\ \text{m})\cos 30.0°}$$

$$= \boxed{113\ \text{A}}$$

Finalize The currents in all wires are on the order of 10^2 A. Such large currents would require specialized equipment. Therefore, this situation would be difficult to establish in practice. Is the equilibrium of wire 1 stable or unstable?

30.3 Ampère's Law

Looking back, we can see that the result of Example 30.1 is important because a current in the form of a long, straight wire occurs often. Figure 30.9 is a perspective view of the magnetic field surrounding a long, straight, current-carrying wire. Because of the wire's symmetry, the magnetic field lines are circles concentric with the wire and lie in planes perpendicular to the wire. The magnitude of $\vec{\mathbf{B}}$ is constant on any circle of radius a and is given by Equation 30.5. A convenient rule for determining the direction of $\vec{\mathbf{B}}$ is to grasp the wire with the right hand, positioning the thumb along the direction of the current. The four fingers wrap in the direction of the magnetic field.

Figure 30.9 also shows that the magnetic field line has no beginning and no end. Rather, it forms a closed loop. That is a major difference between magnetic field lines and electric field lines, which begin on positive charges and end on negative charges. We will explore this feature of magnetic field lines further in Section 30.5.

Oersted's 1819 discovery about deflected compass needles demonstrates that a current-carrying conductor produces a magnetic field. Figure 30.10a (page 716) shows how this effect can be demonstrated in the classroom. Several compass needles are placed in a horizontal plane near a long, vertical wire. When no current is present in the wire, all the needles point in the same direction (that of the horizontal component of the Earth's magnetic field) as expected. When the wire carries a strong, steady current, the needles all deflect in a direction tangent to the circle as in Figure 30.10b. These observations demonstrate that the direction of the magnetic field produced by the current in the wire is consistent with the right-hand rule described in Figure 30.9. When the current is reversed, the needles in Figure 30.10b also reverse.

Now let's evaluate the product $\vec{\mathbf{B}} \cdot d\vec{\mathbf{s}}$ for a small length element $d\vec{\mathbf{s}}$ on the circular path defined by the compass needles and sum the products for all elements

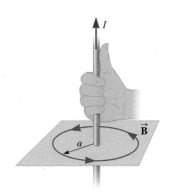

Figure 30.9 The right-hand rule for determining the direction of the magnetic field surrounding a long, straight wire carrying a current. Notice that the magnetic field lines form circles around the wire.

Andre-Marie Ampère
French Physicist (1775–1836)
Ampère is credited with the discovery of electromagnetism, which is the relationship between electric currents and magnetic fields. Ampère's genius, particularly in mathematics, became evident by the time he was 12 years old; his personal life, however, was filled with tragedy. His father, a wealthy city official, was guillotined during the French Revolution, and his wife died young, in 1803. Ampère died at the age of 61 of pneumonia.

Pitfall Prevention 30.2

Avoiding Problems with Signs When using Ampère's law, apply the following right-hand rule. Point your thumb in the direction of the current through the amperian loop. Your curled fingers then point in the direction that you should integrate when traversing the loop to avoid having to define the current as negative.

Ampère's law ▶

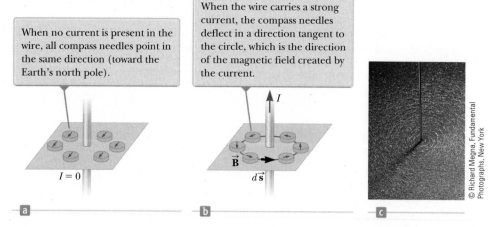

When no current is present in the wire, all compass needles point in the same direction (toward the Earth's north pole).

When the wire carries a strong current, the compass needles deflect in a direction tangent to the circle, which is the direction of the magnetic field created by the current.

$I = 0$

I

$\vec{\mathbf{B}}$ $d\vec{\mathbf{s}}$

a   **b**   **c**

Figure 30.10 (a) and (b) Compasses show the effects of the current in a nearby wire. (c) Circular magnetic field lines surrounding a current-carrying conductor, displayed with iron filings.

over the closed circular path.[1] Along this path, the vectors $d\vec{\mathbf{s}}$ and $\vec{\mathbf{B}}$ are parallel at each point (see Fig. 30.10b), so $\vec{\mathbf{B}} \cdot d\vec{\mathbf{s}} = B \, ds$. Furthermore, the magnitude of $\vec{\mathbf{B}}$ is constant on this circle and is given by Equation 30.5. Therefore, the sum of the products $B \, ds$ over the closed path, which is equivalent to the line integral of $\vec{\mathbf{B}} \cdot d\vec{\mathbf{s}}$, is

$$\oint \vec{\mathbf{B}} \cdot d\vec{\mathbf{s}} = B \oint ds = \frac{\mu_0 I}{2\pi r} (2\pi r) = \mu_0 I$$

where $\oint ds = 2\pi r$ is the circumference of the circular path of radius r. Although this result was calculated for the special case of a circular path surrounding a wire of infinite length, it holds for a closed path of *any* shape (an *amperian loop*) surrounding a current that exists in an unbroken circuit. The general case, known as **Ampère's law,** can be stated as follows:

> The line integral of $\vec{\mathbf{B}} \cdot d\vec{\mathbf{s}}$ around any closed path equals $\mu_0 I$, where I is the total steady current passing through any surface bounded by the closed path:
>
> $$\oint \vec{\mathbf{B}} \cdot d\vec{\mathbf{s}} = \mu_0 I \qquad (30.13)$$

Ampère's law describes the creation of magnetic fields by all continuous current configurations, but at our mathematical level it is useful only for calculating the magnetic field of current configurations having a high degree of symmetry. Its use is similar to that of Gauss's law in calculating electric fields for highly symmetric charge distributions.

Quick Quiz 30.3 Rank the magnitudes of $\oint \vec{\mathbf{B}} \cdot d\vec{\mathbf{s}}$ for the closed paths *a* through *d* in Figure 30.11 from greatest to least.

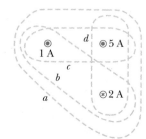

d

$\odot 5\,\text{A}$

$1\,\text{A}$

c

b

a

$\otimes 2\,\text{A}$

Figure 30.11 (Quick Quiz 30.3) Four closed paths around three current-carrying wires.

---

[1] You may wonder why we would choose to evaluate this scalar product. The origin of Ampère's law is in 19th-century science, in which a "magnetic charge" (the supposed analog to an isolated electric charge) was imagined to be moved around a circular field line. The work done on the charge was related to $\vec{\mathbf{B}} \cdot d\vec{\mathbf{s}}$, just as the work done moving an electric charge in an electric field is related to $\vec{\mathbf{E}} \cdot d\vec{\mathbf{s}}$. Therefore, Ampère's law, a valid and useful principle, arose from an erroneous and abandoned work calculation!

Quick Quiz 30.4 Rank the magnitudes of $\oint \vec{\mathbf{B}} \cdot d\vec{\mathbf{s}}$ for the closed paths a through d in Figure 30.12 from greatest to least.

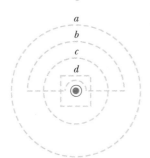

Figure 30.12 (Quick Quiz 30.4) Several closed paths near a single current-carrying wire.

Example 30.5   **The Magnetic Field Created by a Long Current-Carrying Wire**

A long, straight wire of radius R carries a steady current I that is uniformly distributed through the cross section of the wire (Fig. 30.13). Calculate the magnetic field a distance r from the center of the wire in the regions $r \geq R$ and $r < R$.

SOLUTION

Conceptualize Study Figure 30.13 to understand the structure of the wire and the current in the wire. The current creates magnetic fields everywhere, both inside and outside the wire. Based on our discussions about long, straight wires, we expect the magnetic field lines to be circles centered on the central axis of the wire.

Categorize Because the wire has a high degree of symmetry, we categorize this example as an Ampère's law problem. For the $r \geq R$ case, we should arrive at the same result as was obtained in Example 30.1, where we applied the Biot–Savart law to the same situation.

Figure 30.13 (Example 30.5) A long, straight wire of radius R carrying a steady current I uniformly distributed across the cross section of the wire. The magnetic field at any point can be calculated from Ampère's law using a circular path of radius r, concentric with the wire.

Analyze For the magnetic field exterior to the wire, let us choose for our path of integration circle 1 in Figure 30.13. From symmetry, $\vec{\mathbf{B}}$ must be constant in magnitude and parallel to $d\vec{\mathbf{s}}$ at every point on this circle.

Note that the total current passing through the plane of the circle is I and apply Ampère's law:

$$\oint \vec{\mathbf{B}} \cdot d\vec{\mathbf{s}} = B \oint ds = B(2\pi r) = \mu_0 I$$

Solve for B:

$$B = \frac{\mu_0 I}{2\pi r} \quad \text{(for } r \geq R) \tag{30.14}$$

Now consider the interior of the wire, where $r < R$. Here the current I' passing through the plane of circle 2 is less than the total current I.

Set the ratio of the current I' enclosed by circle 2 to the entire current I equal to the ratio of the area πr^2 enclosed by circle 2 to the cross-sectional area πR^2 of the wire:

$$\frac{I'}{I} = \frac{\pi r^2}{\pi R^2}$$

Solve for I':

$$I' = \frac{r^2}{R^2} I$$

Apply Ampère's law to circle 2:

$$\oint \vec{\mathbf{B}} \cdot d\vec{\mathbf{s}} = B(2\pi r) = \mu_0 I' = \mu_0 \left(\frac{r^2}{R^2} I\right)$$

Solve for B:

$$B = \left(\frac{\mu_0 I}{2\pi R^2}\right) r \quad \text{(for } r < R) \tag{30.15}$$

continued

▶ **30.5** c o n t i n u e d

Finalize The magnetic field exterior to the wire is identical in form to Equation 30.5. As is often the case in highly symmetric situations, it is much easier to use Ampère's law than the Biot–Savart law (Example 30.1). The magnetic field interior to the wire is similar in form to the expression for the electric field inside a uniformly charged sphere (see Example 24.3). The magnitude of the magnetic field versus *r* for this configuration is plotted in Figure 30.14. Inside the

Figure 30.14 (Example 30.5) Magnitude of the magnetic field versus *r* for the wire shown in Figure 30.13. The field is proportional to *r* inside the wire and varies as $1/r$ outside the wire.

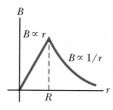

wire, $B \to 0$ as $r \to 0$. Furthermore, Equations 30.14 and 30.15 give the same value of the magnetic field at $r = R$, demonstrating that the magnetic field is continuous at the surface of the wire.

Example 30.6 The Magnetic Field Created by a Toroid

A device called a *toroid* (Fig. 30.15) is often used to create an almost uniform magnetic field in some enclosed area. The device consists of a conducting wire wrapped around a ring (a *torus*) made of a nonconducting material. For a toroid having *N* closely spaced turns of wire, calculate the magnetic field in the region occupied by the torus, a distance *r* from the center.

SOLUTION

Conceptualize Study Figure 30.15 carefully to understand how the wire is wrapped around the torus. The torus could be a solid material or it could be air, with a stiff wire wrapped into the shape shown in Figure 30.15 to form an empty toroid. Imagine each turn of the wire to be a circular loop as in Example 30.3. The magnetic field at the center of the loop is perpendicular to the plane of the loop. Therefore, the magnetic field lines of the collection of loops will form circles within the toroid such as suggested by loop 1 in Figure 30.15.

Categorize Because the toroid has a high degree of symmetry, we categorize this example as an Ampère's law problem.

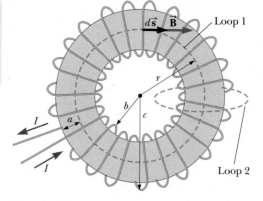

Figure 30.15 (Example 30.6) A toroid consisting of many turns of wire. If the turns are closely spaced, the magnetic field in the interior of the toroid is tangent to the dashed circle (loop 1) and varies as $1/r$. The dimension *a* is the cross-sectional radius of the torus. The field outside the toroid is very small and can be described by using the amperian loop (loop 2) at the right side, perpendicular to the page.

Analyze Consider the circular amperian loop (loop 1) of radius *r* in the plane of Figure 30.15. By symmetry, the magnitude of the field is constant on this circle and tangent to it, so $\vec{B} \cdot d\vec{s} = B\,ds$. Furthermore, the wire passes through the loop *N* times, so the total current through the loop is *NI*.

Apply Ampère's law to loop 1:

$$\oint \vec{B} \cdot d\vec{s} = B \oint ds = B(2\pi r) = \mu_0 NI$$

Solve for *B*:

$$B = \frac{\mu_0 NI}{2\pi r} \tag{30.16}$$

Finalize This result shows that *B* varies as $1/r$ and hence is *nonuniform* in the region occupied by the torus. If, however, *r* is very large compared with the cross-sectional radius *a* of the torus, the field is approximately uniform inside the torus.

For an ideal toroid, in which the turns are closely spaced, the external magnetic field is close to zero, but it is not exactly zero. In Figure 30.15, imagine the radius *r*

of amperian loop 1 to be either smaller than *b* or larger than *c*. In either case, the loop encloses zero net current, so $\oint \vec{B} \cdot d\vec{s} = 0$. You might think this result proves that $\vec{B} = 0$, but it does not. Consider the amperian loop (loop 2) on the right side of the toroid in Figure 30.15. The plane of this loop is perpendicular to the page, and the toroid passes through the loop. As charges enter the toroid as indicated by the current directions in Figure 30.15,

▶ **30.6** continued

they work their way counterclockwise around the toroid. Therefore, there is a counterclockwise current around the toroid, so that a current passes through amperian loop 2! This current is small, but not zero. As a result, the toroid acts as a current loop and produces a weak external field of the form shown in Figure 30.6. The reason $\oint \vec{\mathbf{B}} \cdot d\vec{\mathbf{s}} = 0$ for amperian loop 1 of radius $r < b$ or $r > c$ is that the field lines are perpendicular to $d\vec{\mathbf{s}}$, *not* because $\vec{\mathbf{B}} = 0$.

30.4 The Magnetic Field of a Solenoid

A **solenoid** is a long wire wound in the form of a helix. With this configuration, a reasonably uniform magnetic field can be produced in the space surrounded by the turns of wire—which we shall call the *interior* of the solenoid—when the solenoid carries a current. When the turns are closely spaced, each can be approximated as a circular loop; the net magnetic field is the vector sum of the fields resulting from all the turns.

Figure 30.16 shows the magnetic field lines surrounding a loosely wound solenoid. The field lines in the interior are nearly parallel to one another, are uniformly distributed, and are close together, indicating that the field in this space is strong and almost uniform.

If the turns are closely spaced and the solenoid is of finite length, the external magnetic field lines are as shown in Figure 30.17a. This field line distribution is similar to that surrounding a bar magnet (Fig. 30.17b). Hence, one end of the solenoid behaves like the north pole of a magnet and the opposite end behaves like the south pole. As the length of the solenoid increases, the interior field becomes more uniform and the exterior field becomes weaker. An *ideal solenoid* is approached when the turns are closely spaced and the length is much greater than the radius of the turns. Figure 30.18 (page 720) shows a longitudinal cross section of part of such a solenoid carrying a current I. In this case, the external field is close to zero and the interior field is uniform over a great volume.

Consider the amperian loop (loop 1) perpendicular to the page in Figure 30.18 (page 720), surrounding the ideal solenoid. This loop encloses a small

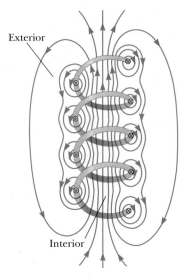

Figure 30.16 The magnetic field lines for a loosely wound solenoid.

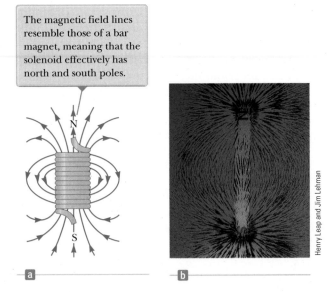

The magnetic field lines resemble those of a bar magnet, meaning that the solenoid effectively has north and south poles.

Figure 30.17 (a) Magnetic field lines for a tightly wound solenoid of finite length, carrying a steady current. The field in the interior space is strong and nearly uniform. (b) The magnetic field pattern of a bar magnet, displayed with small iron filings on a sheet of paper.

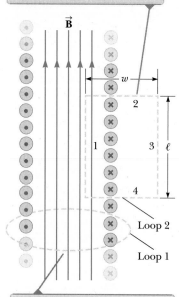

Figure 30.18 Cross-sectional view of an ideal solenoid, where the interior magnetic field is uniform and the exterior field is close to zero.

▶ Magnetic field inside a solenoid

current as the charges in the wire move coil by coil along the length of the solenoid. Therefore, there is a nonzero magnetic field outside the solenoid. It is a weak field, with circular field lines, like those due to a line of current as in Figure 30.9. For an ideal solenoid, this weak field is the only field external to the solenoid.

We can use Ampère's law to obtain a quantitative expression for the interior magnetic field in an ideal solenoid. Because the solenoid is ideal, $\vec{B}$ in the interior space is uniform and parallel to the axis and the magnetic field lines in the exterior space form circles around the solenoid. The planes of these circles are perpendicular to the page. Consider the rectangular path (loop 2) of length ℓ and width w shown in Figure 30.18. Let's apply Ampère's law to this path by evaluating the integral of $\vec{B} \cdot d\vec{s}$ over each side of the rectangle. The contribution along side 3 is zero because the external magnetic field lines are perpendicular to the path in this region. The contributions from sides 2 and 4 are both zero, again because $\vec{B}$ is perpendicular to $d\vec{s}$ along these paths, both inside and outside the solenoid. Side 1 gives a contribution to the integral because along this path $\vec{B}$ is uniform and parallel to $d\vec{s}$. The integral over the closed rectangular path is therefore

$$\oint \vec{B} \cdot d\vec{s} = \int_{\text{path 1}} \vec{B} \cdot d\vec{s} = B \int_{\text{path 1}} ds = B\ell$$

The right side of Ampère's law involves the total current I through the area bounded by the path of integration. In this case, the total current through the rectangular path equals the current through each turn multiplied by the number of turns. If N is the number of turns in the length ℓ, the total current through the rectangle is NI. Therefore, Ampère's law applied to this path gives

$$\oint \vec{B} \cdot d\vec{s} = B\ell = \mu_0 NI$$

$$B = \mu_0 \frac{N}{\ell} I = \mu_0 nI \tag{30.17}$$

where $n = N/\ell$ is the number of turns per unit length.

We also could obtain this result by reconsidering the magnetic field of a toroid (see Example 30.6). If the radius r of the torus in Figure 30.15 containing N turns is much greater than the toroid's cross-sectional radius a, a short section of the toroid approximates a solenoid for which $n = N/2\pi r$. In this limit, Equation 30.16 agrees with Equation 30.17.

Equation 30.17 is valid only for points near the center (that is, far from the ends) of a very long solenoid. As you might expect, the field near each end is smaller than the value given by Equation 30.17. As the length of a solenoid increases, the magnitude of the field at the end approaches half the magnitude at the center (see Problem 69 in Enhanced WebAssign).

Quick Quiz 30.5 Consider a solenoid that is very long compared with its radius. Of the following choices, what is the most effective way to increase the magnetic field in the interior of the solenoid? **(a)** double its length, keeping the number of turns per unit length constant **(b)** reduce its radius by half, keeping the number of turns per unit length constant **(c)** overwrap the entire solenoid with an additional layer of current-carrying wire

30.5 Gauss's Law in Magnetism

The flux associated with a magnetic field is defined in a manner similar to that used to define electric flux (see Eq. 24.3). Consider an element of area dA on an arbitrarily shaped surface as shown in Figure 30.19. If the magnetic field at this element is $\vec{B}$, the magnetic flux through the element is $\vec{B} \cdot d\vec{A}$, where $d\vec{A}$ is a vector that is perpendicular to the surface and has a magnitude equal to the area dA. Therefore, the total magnetic flux Φ_B through the surface is

$$\Phi_B \equiv \int \vec{B} \cdot d\vec{A} \qquad (30.18)$$

◀ **Definition of magnetic flux**

Consider the special case of a plane of area A in a uniform field $\vec{B}$ that makes an angle θ with $d\vec{A}$. The magnetic flux through the plane in this case is

$$\Phi_B = BA \cos \theta \qquad (30.19)$$

If the magnetic field is parallel to the plane as in Figure 30.20a, then $\theta = 90°$ and the flux through the plane is zero. If the field is perpendicular to the plane as in Figure 30.20b, then $\theta = 0$ and the flux through the plane is BA (the maximum value).

The unit of magnetic flux is $T \cdot m^2$, which is defined as a *weber* (Wb); 1 Wb = $1 \, T \cdot m^2$.

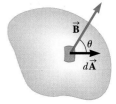

Figure 30.19 The magnetic flux through an area element dA is $\vec{B} \cdot d\vec{A} = B \, dA \cos \theta$, where $d\vec{A}$ is a vector perpendicular to the surface.

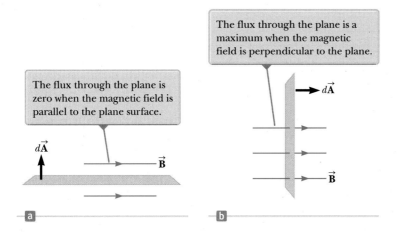

The flux through the plane is zero when the magnetic field is parallel to the plane surface.

The flux through the plane is a maximum when the magnetic field is perpendicular to the plane.

Figure 30.20 Magnetic flux through a plane lying in a magnetic field.

Example 30.7   Magnetic Flux Through a Rectangular Loop

A rectangular loop of width a and length b is located near a long wire carrying a current I (Fig. 30.21). The distance between the wire and the closest side of the loop is c. The wire is parallel to the long side of the loop. Find the total magnetic flux through the loop due to the current in the wire.

SOLUTION

Conceptualize As we saw in Section 30.3, the magnetic field lines due to the wire will be circles, many of which will pass through the rectangular loop. We know that the magnetic field is a function of distance r from a long wire. Therefore, the magnetic field varies over the area of the rectangular loop.

Categorize Because the magnetic field varies over the area of the loop, we must integrate over this area to find the total flux. That identifies this as an analysis problem.

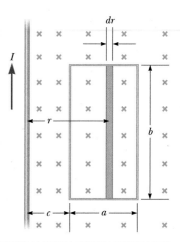

Figure 30.21 (Example 30.7) The magnetic field due to the wire carrying a current I is not uniform over the rectangular loop.

Analyze Noting that $\vec{B}$ is parallel to $d\vec{A}$ at any point within the loop, find the magnetic flux through the rectangular area using Equation 30.18 and incorporate Equation 30.14 for the magnetic field:

$$\Phi_B = \int \vec{B} \cdot d\vec{A} = \int B \, dA = \int \frac{\mu_0 I}{2\pi r} \, dA$$

continued

▶ **30.7** c o n t i n u e d

Express the area element (the tan strip in Fig. 30.21) as $dA = b\,dr$ and substitute:

$$\Phi_B = \int \frac{\mu_0 I}{2\pi r}\, b\, dr = \frac{\mu_0 Ib}{2\pi}\int \frac{dr}{r}$$

Integrate from $r = c$ to $r = a + c$:

$$\Phi_B = \frac{\mu_0 Ib}{2\pi}\int_c^{a+c}\frac{dr}{r} = \frac{\mu_0 Ib}{2\pi}\ln r\,\Big|_c^{a+c}$$

$$= \frac{\mu_0 Ib}{2\pi}\ln\left(\frac{a+c}{c}\right) = \frac{\mu_0 Ib}{2\pi}\ln\left(1+\frac{a}{c}\right)$$

Finalize Notice how the flux depends on the size of the loop. Increasing either a or b increases the flux as expected. If c becomes large such that the loop is very far from the wire, the flux approaches zero, also as expected. If c goes to zero, the flux becomes infinite. In principle, this infinite value occurs because the field becomes infinite at $r = 0$ (assuming an infinitesimally thin wire). That will not happen in reality because the thickness of the wire prevents the left edge of the loop from reaching $r = 0$.

In Chapter 24, we found that the electric flux through a closed surface surrounding a net charge is proportional to that charge (Gauss's law). In other words, the number of electric field lines leaving the surface depends only on the net charge within it. This behavior exists because electric field lines originate and terminate on electric charges.

The situation is quite different for magnetic fields, which are continuous and form closed loops. In other words, as illustrated by the magnetic field lines of a current in Figure 30.9 and of a bar magnet in Figure 30.22, magnetic field lines do not begin or end at any point. For any closed surface such as the one outlined by the dashed line in Figure 30.22, the number of lines entering the surface equals the number leaving the surface; therefore, the net magnetic flux is zero. In contrast, for a closed surface surrounding one charge of an electric dipole (Fig. 30.23), the net electric flux is not zero.

Gauss's law in magnetism states that

the net magnetic flux through any closed surface is always zero:

Gauss's law in magnetism ▶

$$\oint \vec{\mathbf{B}} \cdot d\vec{\mathbf{A}} = 0 \tag{30.20}$$

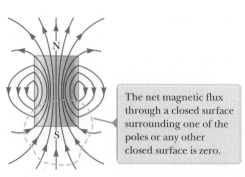

The net magnetic flux through a closed surface surrounding one of the poles or any other closed surface is zero.

Figure 30.22 The magnetic field lines of a bar magnet form closed loops. (The dashed line represents the intersection of a closed surface with the page.)

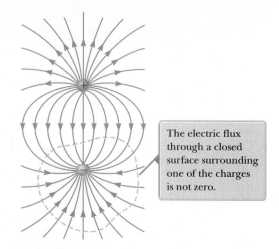

The electric flux through a closed surface surrounding one of the charges is not zero.

Figure 30.23 The electric field lines surrounding an electric dipole begin on the positive charge and terminate on the negative charge.

This statement represents that isolated magnetic poles (monopoles) have never been detected and perhaps do not exist. Nonetheless, scientists continue the search because certain theories that are otherwise successful in explaining fundamental physical behavior suggest the possible existence of magnetic monopoles.

30.6 Magnetism in Matter

The magnetic field produced by a current in a coil of wire gives us a hint as to what causes certain materials to exhibit strong magnetic properties. Earlier we found that a solenoid like the one shown in Figure 30.17a has a north pole and a south pole. In general, *any* current loop has a magnetic field and therefore has a magnetic dipole moment, including the atomic-level current loops described in some models of the atom.

The Magnetic Moments of Atoms

Let's begin our discussion with a classical model of the atom in which electrons move in circular orbits around the much more massive nucleus. In this model, an orbiting electron constitutes a tiny current loop (because it is a moving charge), and the magnetic moment of the electron is associated with this orbital motion. Although this model has many deficiencies, some of its predictions are in good agreement with the correct theory, which is expressed in terms of quantum physics.

In our classical model, we assume an electron is a particle in uniform circular motion: it moves with constant speed v in a circular orbit of radius r about the nucleus as in Figure 30.24. The current I associated with this orbiting electron is its charge e divided by its period T. Using Equation 4.15 from the particle in uniform circular motion model, $T = 2\pi r/v$, gives

$$I = \frac{e}{T} = \frac{ev}{2\pi r}$$

The magnitude of the magnetic moment associated with this current loop is given by $\mu = IA$, where $A = \pi r^2$ is the area enclosed by the orbit. Therefore,

$$\mu = IA = \left(\frac{ev}{2\pi r}\right)\pi r^2 = \tfrac{1}{2}evr \tag{30.21}$$

Because the magnitude of the orbital angular momentum of the electron is given by $L = m_e vr$ (Eq. 11.12 with $\phi = 90°$), the magnetic moment can be written as

$$\mu = \left(\frac{e}{2m_e}\right)L \tag{30.22}$$

◀ **Orbital magnetic moment**

This result demonstrates that the magnetic moment of the electron is proportional to its orbital angular momentum. Because the electron is negatively charged, the vectors $\vec{\mu}$ and $\vec{L}$ point in *opposite* directions. Both vectors are perpendicular to the plane of the orbit as indicated in Figure 30.24.

A fundamental outcome of quantum physics is that orbital angular momentum is quantized and is equal to multiples of $\hbar = h/2\pi = 1.05 \times 10^{-34}$ J · s, where h is Planck's constant (see Chapter 40). The smallest nonzero value of the electron's magnetic moment resulting from its orbital motion is

$$\mu = \sqrt{2}\,\frac{e}{2m_e}\,\hbar \tag{30.23}$$

We shall see in Chapter 42 how expressions such as Equation 30.23 arise.

Because all substances contain electrons, you may wonder why most substances are not magnetic. The main reason is that, in most substances, the magnetic

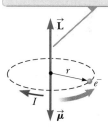

The electron has an angular momentum $\vec{L}$ in one direction and a magnetic moment $\vec{\mu}$ in the opposite direction.

Figure 30.24 An electron moving in the direction of the gray arrow in a circular orbit of radius r. Because the electron carries a negative charge, the direction of the current due to its motion about the nucleus is opposite the direction of that motion.

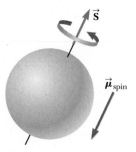

Figure 30.25 Classical model of a spinning electron. We can adopt this model to remind ourselves that electrons have an intrinsic angular momentum. The model should not be pushed too far, however; it gives an incorrect magnitude for the magnetic moment, incorrect quantum numbers, and too many degrees of freedom.

Table 30.1 Magnetic Moments of Some Atoms and Ions	
Atom or Ion	**Magnetic Moment (10^{-24} J/T)**
H	9.27
He	0
Ne	0
Ce^{3+}	19.8
Yb^{3+}	37.1

moment of one electron in an atom is canceled by that of another electron orbiting in the opposite direction. The net result is that, for most materials, the magnetic effect produced by the orbital motion of the electrons is either zero or very small.

In addition to its orbital magnetic moment, an electron (as well as protons, neutrons, and other particles) has an intrinsic property called **spin** that also contributes to its magnetic moment. Classically, the electron might be viewed as spinning about its axis as shown in Figure 30.25, but you should be very careful with the classical interpretation. The magnitude of the angular momentum $\vec{S}$ associated with spin is on the same order of magnitude as the magnitude of the angular momentum $\vec{L}$ due to the orbital motion. The magnitude of the spin angular momentum of an electron predicted by quantum theory is

$$S = \frac{\sqrt{3}}{2}\hbar$$

The magnetic moment characteristically associated with the spin of an electron has the value

$$\mu_{spin} = \frac{e\hbar}{2m_e} \qquad (30.24)$$

This combination of constants is called the **Bohr magneton** μ_B:

$$\mu_B = \frac{e\hbar}{2m_e} = 9.27 \times 10^{-24}\,\text{J/T} \qquad (30.25)$$

Therefore, atomic magnetic moments can be expressed as multiples of the Bohr magneton. (Note that $1\,\text{J/T} = 1\,\text{A}\cdot\text{m}^2$.)

In atoms containing many electrons, the electrons usually pair up with their spins opposite each other; therefore, the spin magnetic moments cancel. Atoms containing an odd number of electrons, however, must have at least one unpaired electron and therefore some spin magnetic moment. The total magnetic moment of an atom is the vector sum of the orbital and spin magnetic moments, and a few examples are given in Table 30.1. Notice that helium and neon have zero moments because their individual spin and orbital moments cancel.

The nucleus of an atom also has a magnetic moment associated with its constituent protons and neutrons. The magnetic moment of a proton or neutron, however, is much smaller than that of an electron and can usually be neglected. We can understand this smaller value by inspecting Equation 30.25 and replacing the mass of the electron with the mass of a proton or a neutron. Because the masses of the proton and neutron are much greater than that of the electron, their magnetic moments are on the order of 10^3 times smaller than that of the electron.

Ferromagnetism

A small number of crystalline substances exhibit strong magnetic effects called **ferromagnetism.** Some examples of ferromagnetic substances are iron, cobalt, nickel, gadolinium, and dysprosium. These substances contain permanent atomic magnetic moments that tend to align parallel to each other even in a weak external magnetic field. Once the moments are aligned, the substance remains magnetized after the external field is removed. This permanent alignment is due to a strong coupling between neighboring moments, a coupling that can be understood only in quantum-mechanical terms.

All ferromagnetic materials are made up of microscopic regions called **domains,** regions within which all magnetic moments are aligned. These domains have volumes of about 10^{-12} to 10^{-8} m^3 and contain 10^{17} to 10^{21} atoms. The boundaries between the various domains having different orientations are called **domain walls.** In an unmagnetized sample, the magnetic moments in the domains are randomly oriented so that the net magnetic moment is zero as in Figure 30.26a. When the sample is placed in an external magnetic field $\vec{B}$, the size of those domains with magnetic moments aligned with the field grows, which results in a magnetized sample as in Figure 30.26b. As the external field becomes very strong as in Figure 30.26c, the domains in which the magnetic moments are not aligned with the field become very small. When the external field is removed, the sample may retain a net magnetization in the direction of the original field. At ordinary temperatures, thermal agitation is not sufficient to disrupt this preferred orientation of magnetic moments.

When the temperature of a ferromagnetic substance reaches or exceeds a critical temperature called the **Curie temperature,** the substance loses its residual magnetization. Below the Curie temperature, the magnetic moments are aligned and the substance is ferromagnetic. Above the Curie temperature, the thermal agitation is great enough to cause a random orientation of the moments and the substance becomes paramagnetic. Curie temperatures for several ferromagnetic substances are given in Table 30.2.

Paramagnetism

Paramagnetic substances have a weak magnetism resulting from the presence of atoms (or ions) that have permanent magnetic moments. These moments interact only weakly with one another and are randomly oriented in the absence of an external magnetic field. When a paramagnetic substance is placed in an external magnetic field, its atomic moments tend to line up with the field. This alignment process, however, must compete with thermal motion, which tends to randomize the magnetic moment orientations.

Diamagnetism

When an external magnetic field is applied to a diamagnetic substance, a weak magnetic moment is induced in the direction opposite the applied field, causing diamagnetic substances to be weakly repelled by a magnet. Although diamagnetism is present in all matter, its effects are much smaller than those of paramagnetism or ferromagnetism and are evident only when those other effects do not exist.

We can attain some understanding of diamagnetism by considering a classical model of two atomic electrons orbiting the nucleus in opposite directions but with the same speed. The electrons remain in their circular orbits because of the attractive electrostatic force exerted by the positively charged nucleus. Because the magnetic moments of the two electrons are equal in magnitude and opposite in direction, they cancel each other and the magnetic moment of the atom is zero. When an external magnetic field is applied, the electrons experience an additional magnetic force $q\vec{v} \times \vec{B}$. This added magnetic force combines with the electrostatic force to increase the orbital speed of the electron whose magnetic moment is antiparallel to the field and to decrease the speed of the electron whose magnetic moment is parallel to the field. As a result, the two magnetic moments of the electrons no longer cancel and the substance acquires a net magnetic moment that is opposite the applied field.

In an unmagnetized substance, the atomic magnetic dipoles are randomly oriented.

a

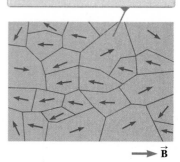

When an external field $\vec{B}$ is applied, the domains with components of magnetic moment in the same direction as $\vec{B}$ grow larger, giving the sample a net magnetization.

$\longrightarrow \vec{B}$

b

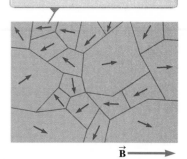

As the field is made even stronger, the domains with magnetic moment vectors not aligned with the external field become very small.

$\vec{B} \longrightarrow$

c

Figure 30.26 Orientation of magnetic dipoles before and after a magnetic field is applied to a ferromagnetic substance.

Table 30.2	Curie Temperatures for Several Ferromagnetic Substances
Substance	T_{Curie} **(K)**
Iron	1 043
Cobalt	1 394
Nickel	631
Gadolinium	317
Fe_2O_3	893

In the Meissner effect, the small magnet at the top induces currents in the superconducting disk below, which is cooled to −321°F (77 K). The currents create a repulsive magnetic force on the magnet causing it to levitate above the superconducting disk.

Liquid oxygen, a paramagnetic material, is attracted to the poles of a magnet.

The levitation force is exerted on the diamagnetic water molecules in the frog's body.

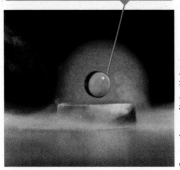

(*Left*) Paramagnetism. (*Right*) Diamagnetism: a frog is levitated in a 16-T magnetic field at the Nijmegen High Field Magnet Laboratory in the Netherlands.

Figure 30.27 An illustration of the Meissner effect, shown by this magnet suspended above a cooled ceramic superconductor disk, has become our most visual image of high-temperature superconductivity. Superconductivity is the loss of all resistance to electrical current and is a key to more-efficient energy use.

As you recall from Chapter 27, a superconductor is a substance in which the electrical resistance is zero below some critical temperature. Certain types of superconductors also exhibit perfect diamagnetism in the superconducting state. As a result, an applied magnetic field is expelled by the superconductor so that the field is zero in its interior. This phenomenon is known as the **Meissner effect.** If a permanent magnet is brought near a superconductor, the two objects repel each other. This repulsion is illustrated in Figure 30.27, which shows a small permanent magnet levitated above a superconductor maintained at 77 K.

Summary

Definition

The **magnetic flux** Φ_B through a surface is defined by the surface integral

$$\Phi_B \equiv \int \vec{\mathbf{B}} \cdot d\vec{\mathbf{A}} \tag{30.18}$$

Concepts and Principles

The **Biot–Savart law** says that the magnetic field $d\vec{\mathbf{B}}$ at a point P due to a length element $d\vec{\mathbf{s}}$ that carries a steady current I is

$$d\vec{\mathbf{B}} = \frac{\mu_0}{4\pi} \frac{I \, d\vec{\mathbf{s}} \times \hat{\mathbf{r}}}{r^2} \tag{30.1}$$

where μ_0 is the **permeability of free space,** r is the distance from the element to the point P, and $\hat{\mathbf{r}}$ is a unit vector pointing from $d\vec{\mathbf{s}}$ toward point P. We find the total field at P by integrating this expression over the entire current distribution.

The magnetic force per unit length between two parallel wires separated by a distance a and carrying currents I_1 and I_2 has a magnitude

$$\frac{F_B}{\ell} = \frac{\mu_0 I_1 I_2}{2\pi a} \tag{30.12}$$

The force is attractive if the currents are in the same direction and repulsive if they are in opposite directions.

Ampère's law says that the line integral of $\vec{\mathbf{B}} \cdot d\vec{\mathbf{s}}$ around any closed path equals $\mu_0 I$, where I is the total steady current through any surface bounded by the closed path:

$$\oint \vec{\mathbf{B}} \cdot d\vec{\mathbf{s}} = \mu_0 I \qquad (30.13)$$

The magnitude of the magnetic field at a distance r from a long, straight wire carrying an electric current I is

$$B = \frac{\mu_0 I}{2\pi r} \qquad (30.14)$$

The field lines are circles concentric with the wire.

The magnitudes of the fields inside a toroid and solenoid are

$$B = \frac{\mu_0 N I}{2\pi r} \quad (\text{toroid}) \qquad (30.16)$$

$$B = \mu_0 \frac{N}{\ell} I = \mu_0 n I \quad (\text{solenoid}) \qquad (30.17)$$

where N is the total number of turns.

Gauss's law of magnetism states that the net magnetic flux through any closed surface is zero:

$$\oint \vec{\mathbf{B}} \cdot d\vec{\mathbf{A}} = 0 \qquad (30.20)$$

Substances can be classified into one of three categories that describe their magnetic behavior. **Diamagnetic** substances are those in which the magnetic moment is weak and opposite the applied magnetic field. **Paramagnetic** substances are those in which the magnetic moment is weak and in the same direction as the applied magnetic field. In **ferromagnetic** substances, interactions between atoms cause magnetic moments to align and create a strong magnetization that remains after the external field is removed.

Objective Questions   **1.** denotes answer available in *Student Solutions Manual/Study Guide*

1. **(i)** What happens to the magnitude of the magnetic field inside a long solenoid if the current is doubled? (a) It becomes four times larger. (b) It becomes twice as large. (c) It is unchanged. (d) It becomes one-half as large. (e) It becomes one-fourth as large. **(ii)** What happens to the field if instead the length of the solenoid is doubled, with the number of turns remaining the same? Choose from the same possibilities as in part (i). **(iii)** What happens to the field if the number of turns is doubled, with the length remaining the same? Choose from the same possibilities as in part (i). **(iv)** What happens to the field if the radius is doubled? Choose from the same possibilities as in part (i).

2. In Figure 30.7, assume $I_1 = 2.00$ A and $I_2 = 6.00$ A. What is the relationship between the magnitude F_1 of the force exerted on wire 1 and the magnitude F_2 of the force exerted on wire 2? (a) $F_1 = 6F_2$ (b) $F_1 = 3F_2$ (c) $F_1 = F_2$ (d) $F_1 = \frac{1}{3}F_2$ (e) $F_1 = \frac{1}{6}F_2$

3. Answer each question yes or no. (a) Is it possible for each of three stationary charged particles to exert a force of attraction on the other two? (b) Is it possible for each of three stationary charged particles to repel both of the other particles? (c) Is it possible for each of three current-carrying metal wires to attract the other two wires? (d) Is it possible for each of three current-carrying metal wires to repel the other two wires? André-Marie Ampère's experiments on electromagnetism are models of logical precision and included observation of the phenomena referred to in this question.

4. Two long, parallel wires each carry the same current I in the same direction (Fig. OQ30.4). Is the total magnetic field at the point P midway between the wires (a) zero, (b) directed into the page, (c) directed out of the page, (d) directed to the left, or (e) directed to the right?

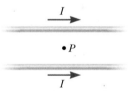

Figure OQ30.4

5. Two long, straight wires cross each other at a right angle, and each carries the same current I (Fig. OQ30.5). Which of the following statements is true regarding the total magnetic field due to the two wires at the various points in the figure? More than one statement may be correct. (a) The field is strongest at points B and D. (b) The field is strongest at points A and C. (c) The field is out of the page at point B and

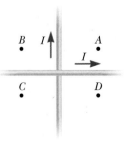

Figure OQ30.5

into the page at point *D*. (d) The field is out of the page at point *C* and out of the page at point *D*. (e) The field has the same magnitude at all four points.

6. A long, vertical, metallic wire carries downward electric current. **(i)** What is the direction of the magnetic field it creates at a point 2 cm horizontally east of the center of the wire? (a) north (b) south (c) east (d) west (e) up **(ii)** What would be the direction of the field if the current consisted of positive charges moving downward instead of electrons moving upward? Choose from the same possibilities as in part (i).

7. Suppose you are facing a tall makeup mirror on a vertical wall. Fluorescent tubes framing the mirror carry a clockwise electric current. **(i)** What is the direction of the magnetic field created by that current at the center of the mirror? (a) left (b) right (c) horizontally toward you (d) horizontally away from you (e) no direction because the field has zero magnitude **(ii)** What is the direction of the field the current creates at a point on the wall outside the frame to the right? Choose from the same possibilities as in part (i).

8. A long, straight wire carries a current *I* (Fig. OQ30.8). Which of the following statements is true regarding the magnetic field due to the wire? More than one statement may be correct. (a) The magnitude is proportional to I/r, and the direction is out of the page at *P*. (b) The magnitude is proportional to I/r^2, and the direction is out of the page at *P*. (c) The magnitude is proportional to I/r, and the direction is into the page at *P*. (d) The magnitude is proportional to I/r^2, and the direction is into the page at *P*. (e) The magnitude is proportional to *I*, but does not depend on *r*.

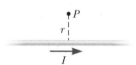

Figure OQ30.8

9. Two long, parallel wires carry currents of 20.0 A and 10.0 A in opposite directions (Fig. OQ30.9). Which of the following statements is true? More than one state-

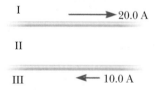

Figure OQ30.9 Objective Questions 9 and 10.

ment may be correct. (a) In region I, the magnetic field is into the page and is never zero. (b) In region II, the field is into the page and can be zero. (c) In region III, it is possible for the field to be zero. (d) In region I, the magnetic field is out of the page and is never zero. (e) There are no points where the field is zero.

10. Consider the two parallel wires carrying currents in opposite directions in Figure OQ30.9. Due to the magnetic interaction between the wires, does the lower wire experience a magnetic force that is (a) upward, (b) downward, (c) to the left, (d) to the right, or (e) into the paper?

11. What creates a magnetic field? More than one answer may be correct. (a) a stationary object with electric charge (b) a moving object with electric charge (c) a stationary conductor carrying electric current (d) a difference in electric potential (e) a charged capacitor disconnected from a battery and at rest *Note:* In Chapter 34, we will see that a changing electric field also creates a magnetic field.

12. A long solenoid with closely spaced turns carries electric current. Does each turn of wire exert (a) an attractive force on the next adjacent turn, (b) a repulsive force on the next adjacent turn, (c) zero force on the next adjacent turn, or (d) either an attractive or a repulsive force on the next turn, depending on the direction of current in the solenoid?

13. A uniform magnetic field is directed along the *x* axis. For what orientation of a flat, rectangular coil is the flux through the rectangle a maximum? (a) It is a maximum in the *xy* plane. (b) It is a maximum in the *xz* plane. (c) It is a maximum in the *yz* plane. (d) The flux has the same nonzero value for all these orientations. (e) The flux is zero in all cases.

14. Rank the magnitudes of the following magnetic fields from largest to smallest, noting any cases of equality. (a) the field 2 cm away from a long, straight wire carrying a current of 3 A (b) the field at the center of a flat, compact, circular coil, 2 cm in radius, with 10 turns, carrying a current of 0.3 A (c) the field at the center of a solenoid 2 cm in radius and 200 cm long, with 1 000 turns, carrying a current of 0.3 A (d) the field at the center of a long, straight, metal bar, 2 cm in radius, carrying a current of 300 A (e) a field of 1 mT

15. Solenoid A has length *L* and *N* turns, solenoid B has length 2*L* and *N* turns, and solenoid C has length *L*/2 and 2*N* turns. If each solenoid carries the same current, rank the magnitudes of the magnetic fields in the centers of the solenoids from largest to smallest.

Conceptual Questions   1. denotes answer available in *Student Solutions Manual/Study Guide*

1. Is the magnetic field created by a current loop uniform? Explain.

2. One pole of a magnet attracts a nail. Will the other pole of the magnet attract the nail? Explain. Also explain how a magnet sticks to a refrigerator door.

3. Compare Ampère's law with the Biot–Savart law. Which is more generally useful for calculating $\vec{\mathbf{B}}$ for a current-carrying conductor?

4. A hollow copper tube carries a current along its length. Why is $B = 0$ inside the tube? Is *B* nonzero outside the tube?

5. Imagine you have a compass whose needle can rotate vertically as well as horizontally. Which way would the compass needle point if you were at the Earth's north magnetic pole?

6. Is Ampère's law valid for all closed paths surrounding a conductor? Why is it not useful for calculating $\vec{B}$ for all such paths?

7. A magnet attracts a piece of iron. The iron can then attract another piece of iron. On the basis of domain alignment, explain what happens in each piece of iron.

8. Why does hitting a magnet with a hammer cause the magnetism to be reduced?

9. The quantity $\int \vec{B} \cdot d\vec{s}$ in Ampère's law is called *magnetic circulation*. Figures 30.10 and 30.13 show paths around which the magnetic circulation is evaluated. Each of these paths encloses an area. What is the magnetic flux through each area? Explain your answer.

10. Figure CQ30.10 shows four permanent magnets, each having a hole through its center. Notice that the blue and yellow magnets are levitated above the red ones. (a) How does this levitation occur? (b) What purpose do the rods serve? (c) What can you say about the poles of the magnets from this observation? (d) If the blue magnet were inverted, what do you suppose would happen?

Figure CQ30.10

© Cengage Learning/Charles D. Winters

11. Explain why two parallel wires carrying currents in opposite directions repel each other.

12. Consider a magnetic field that is uniform in direction throughout a certain volume. (a) Can the field be uniform in magnitude? (b) Must it be uniform in magnitude? Give evidence for your answers.

Problems available in ENHANCED **WebAssign**   Access end-of-chapter problems online at **www.webassign.net**

Section 30.1 The Biot–Savart Law
Problems 1–20

Section 30.2 The Magnetic Force Between Two Parallel Conductors
Problems 21–29

Section 30.3 Ampère's Law
Problems 30–39

Section 30.4 The Magnetic Field of a Solenoid
Problems 40–45

Section 30.5 Gauss's Law in Magnetism
Problems 46–48

Section 30.6 Magnetism in Matter
Problems 49–50

Additional Problems
Problems 51–68

Challenge Problems
Problem 69–77

Solutions to the following Problems are available in the *Student Solutions Manual/Study Guide:*
30.5, 30.10, 30.14, 30.25, 30.31, 30.36, 30.38, 30.39, 30.41, 30.47, 30.49, 30.55, 30.56, 30.57, 30.76, and 30.78

List of Enhanced Problems

Problem Number	Targeted Feedback in Enhanced WebAssign	Analysis Model Tutorial in Enhanced WebAssign	Master It in Enhanced WebAssign	Watch It in Enhanced WebAssign
30.3	✓			✓
30.5	✓		✓	
30.6				✓
30.14		✓	✓	
30.21	✓			✓
30.25	✓		✓	
30.29		✓		
30.31	✓			✓
30.32	✓			✓
30.35	✓			✓
30.39	✓		✓	
30.41	✓		✓	
30.43	✓			✓
30.47	✓		✓	
30.48	✓			✓
30.49			✓	
30.52	✓		✓	
30.55			✓	
30.62		✓		
30.66		✓		

Faraday's Law

So far, our studies in electricity and magnetism have focused on the electric fields produced by stationary charges and the magnetic fields produced by moving charges. This chapter explores the effects produced by magnetic fields that vary in time.

Experiments conducted by Michael Faraday in England in 1831 and independently by Joseph Henry in the United States that same year showed that an emf can be induced in a circuit by a changing magnetic field. The results of these experiments led to a very basic and important law of electromagnetism known as *Faraday's law of induction*. An emf (and therefore a current as well) can be induced in various processes that involve a change in a magnetic flux.

31.1 Faraday's Law of Induction

To see how an emf can be induced by a changing magnetic field, consider the experimental results obtained when a loop of wire is connected to a sensitive ammeter as illustrated in Figure 31.1 (page 732). When a magnet is moved toward the loop, the reading on the ammeter changes from zero to a nonzero value, arbitrarily shown as negative in Figure 31.1a. When the magnet is brought to rest and held stationary relative to the loop (Fig. 31.1b), a reading of zero is observed. When the magnet is moved away from the loop, the reading on the ammeter changes to a positive value as shown in Figure 31.1c. Finally, when the magnet is held stationary and the loop

An artist's impression of the Skerries SeaGen Array, a tidal energy generator under development near the island of Anglesey, North Wales. When it is brought online, it will offer 10.5 MW of power from generators turned by tidal streams. The image shows the underwater blades that are driven by the tidal currents. The second blade system has been raised from the water for servicing. We will study generators in this chapter. *(Marine Current Turbines TM Ltd.)*

Michael Faraday
British Physicist and Chemist
(1791–1867)
Faraday is often regarded as the greatest experimental scientist of the 1800s. His many contributions to the study of electricity include the invention of the electric motor, electric generator, and transformer as well as the discovery of electromagnetic induction and the laws of electrolysis. Greatly influenced by religion, he refused to work on the development of poison gas for the British military.

is moved either toward or away from it, the reading changes from zero. From these observations, we conclude that the loop detects that the magnet is moving relative to it and we relate this detection to a change in magnetic field. Therefore, it seems that a relationship exists between a current and a changing magnetic field.

These results are quite remarkable because a current is set up even though no batteries are present in the circuit! We call such a current an *induced current* and say that it is produced by an *induced emf.*

Now let's describe an experiment conducted by Faraday and illustrated in Figure 31.2. A primary coil is wrapped around an iron ring and connected to a switch and a battery. A current in the coil produces a magnetic field when the switch is closed. A secondary coil also is wrapped around the ring and is connected to a sensitive ammeter. No battery is present in the secondary circuit, and the secondary coil is not electrically connected to the primary coil. Any current detected in the secondary circuit must be induced by some external agent.

Initially, you might guess that no current is ever detected in the secondary circuit. Something quite amazing happens when the switch in the primary circuit is either opened or thrown closed, however. At the instant the switch is closed, the ammeter reading changes from zero momentarily and then returns to zero. At the instant the switch is opened, the ammeter changes to a reading with the opposite sign and again returns to zero. Finally, the ammeter reads zero when there is either a steady current or no current in the primary circuit. To understand what happens in this experiment, note that when the switch is closed, the current in the primary circuit produces a magnetic field that penetrates the secondary circuit. Furthermore, when the switch is thrown closed, the magnetic field produced by the current in the primary circuit changes from zero to some value over some finite time, and this changing field induces a current in the secondary circuit. Notice that no current is induced in the secondary coil even when a steady current exists in the primary coil. It is a *change* in the current in the primary coil that induces a current in the secondary coil, not just the *existence* of a current.

As a result of these observations, Faraday concluded that an electric current can be induced in a loop by a changing magnetic field. The induced current exists only while the magnetic field through the loop is changing. Once the magnetic field reaches a steady value, the current in the loop disappears. In effect, the loop behaves as though a source of emf were connected to it for a short time. It is customary to say that an induced emf is produced in the loop by the changing magnetic field.

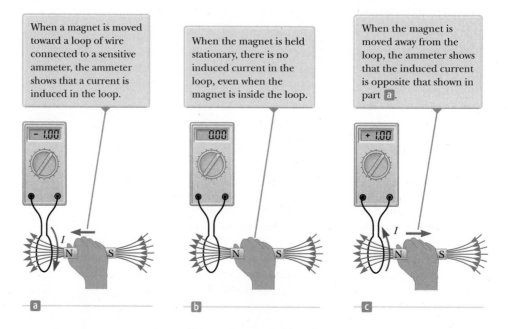

When a magnet is moved toward a loop of wire connected to a sensitive ammeter, the ammeter shows that a current is induced in the loop.

When the magnet is held stationary, there is no induced current in the loop, even when the magnet is inside the loop.

When the magnet is moved away from the loop, the ammeter shows that the induced current is opposite that shown in part **a**.

Figure 31.1 A simple experiment showing that a current is induced in a loop when a magnet is moved toward or away from the loop.

a　　　　**b**　　　　**c**

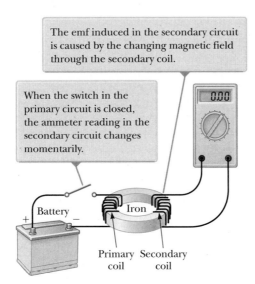

The emf induced in the secondary circuit is caused by the changing magnetic field through the secondary coil.

When the switch in the primary circuit is closed, the ammeter reading in the secondary circuit changes momentarily.

Battery   Iron

Primary Secondary
coil   coil

Figure 31.2 Faraday's experiment.

The experiments shown in Figures 31.1 and 31.2 have one thing in common: in each case, an emf is induced in a loop when the magnetic flux through the loop changes with time. In general, this emf is directly proportional to the time rate of change of the magnetic flux through the loop. This statement can be written mathematically as **Faraday's law of induction:**

$$\mathcal{E} = -\frac{d\Phi_B}{dt}$$ (31.1)

◀ **Faraday's law of induction**

where $\Phi_B = \int \vec{\mathbf{B}} \cdot d\vec{\mathbf{A}}$ is the magnetic flux through the loop. (See Section 30.5.)

If a coil consists of N loops with the same area and Φ_B is the magnetic flux through one loop, an emf is induced in every loop. The loops are in series, so their emfs add; therefore, the total induced emf in the coil is given by

$$\mathcal{E} = -N\frac{d\Phi_B}{dt}$$ (31.2)

The negative sign in Equations 31.1 and 31.2 is of important physical significance and will be discussed in Section 31.3.

Suppose a loop enclosing an area A lies in a uniform magnetic field $\vec{\mathbf{B}}$ as in Figure 31.3. The magnetic flux through the loop is equal to $BA \cos\theta$, where θ is the angle between the magnetic field and the normal to the loop; hence, the induced emf can be expressed as

$$\mathcal{E} = -\frac{d}{dt}(BA \cos\theta)$$ (31.3)

From this expression, we see that an emf can be induced in the circuit in several ways:

- The magnitude of $\vec{\mathbf{B}}$ can change with time.
- The area enclosed by the loop can change with time.
- The angle θ between $\vec{\mathbf{B}}$ and the normal to the loop can change with time.
- Any combination of the above can occur.

Quick Quiz 31.1 A circular loop of wire is held in a uniform magnetic field, with the plane of the loop perpendicular to the field lines. Which of the following will *not* cause a current to be induced in the loop? **(a)** crushing the loop **(b)** rotating the loop about an axis perpendicular to the field lines **(c)** keeping the orientation of the loop fixed and moving it along the field lines **(d)** pulling the loop out of the field

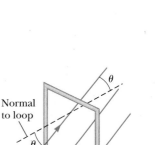

Normal
to loop

θ

θ

Loop of
area A

$\vec{\mathbf{B}}$

Figure 31.3 A conducting loop that encloses an area A in the presence of a uniform magnetic field $\vec{\mathbf{B}}$. The angle between $\vec{\mathbf{B}}$ and the normal to the loop is θ.

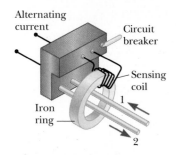

Figure 31.4 Essential components of a ground fault circuit interrupter.

Some Applications of Faraday's Law

The ground fault circuit interrupter (GFCI) is an interesting safety device that protects users of electrical appliances against electric shock. Its operation makes use of Faraday's law. In the GFCI shown in Figure 31.4, wire 1 leads from the wall outlet to the appliance to be protected and wire 2 leads from the appliance back to the wall outlet. An iron ring surrounds the two wires, and a sensing coil is wrapped around part of the ring. Because the currents in the wires are in opposite directions and of equal magnitude, there is zero net current flowing through the ring and the net magnetic flux through the sensing coil is zero. Now suppose the return current in wire 2 changes so that the two currents are not equal in magnitude. (That can happen if, for example, the appliance becomes wet, enabling current to leak to ground.) Then the net current through the ring is not zero and the magnetic flux through the sensing coil is no longer zero. Because household current is alternating (meaning that its direction keeps reversing), the magnetic flux through the sensing coil changes with time, inducing an emf in the coil. This induced emf is used to trigger a circuit breaker, which stops the current before it is able to reach a harmful level.

Another interesting application of Faraday's law is the production of sound in an electric guitar. The coil in this case, called the *pickup coil*, is placed near the vibrating guitar string, which is made of a metal that can be magnetized. A permanent magnet inside the coil magnetizes the portion of the string nearest the coil (Fig. 31.5a). When the string vibrates at some frequency, its magnetized segment produces a changing magnetic flux through the coil. The changing flux induces an emf in the coil that is fed to an amplifier. The output of the amplifier is sent to the loudspeakers, which produce the sound waves we hear.

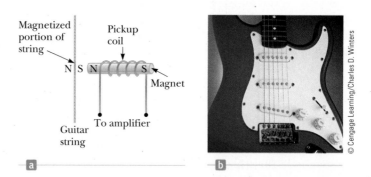

Figure 31.5 (a) In an electric guitar, a vibrating magnetized string induces an emf in a pickup coil. (b) The pickups (the circles beneath the metallic strings) of this electric guitar detect the vibrations of the strings and send this information through an amplifier and into speakers. (A switch on the guitar allows the musician to select which set of six pickups is used.)

Example 31.1   Inducing an emf in a Coil

A coil consists of 200 turns of wire. Each turn is a square of side $d = 18$ cm, and a uniform magnetic field directed perpendicular to the plane of the coil is turned on. If the field changes linearly from 0 to 0.50 T in 0.80 s, what is the magnitude of the induced emf in the coil while the field is changing?

SOLUTION

Conceptualize From the description in the problem, imagine magnetic field lines passing through the coil. Because the magnetic field is changing in magnitude, an emf is induced in the coil.

Categorize We will evaluate the emf using Faraday's law from this section, so we categorize this example as a substitution problem.

▶ **31.1** continued

Evaluate Equation 31.2 for the situation described here, noting that the magnetic field changes linearly with time:

$$|\mathcal{E}| = N\frac{\Delta \Phi_B}{\Delta t} = N\frac{\Delta(BA)}{\Delta t} = NA\frac{\Delta B}{\Delta t} = Nd^2\frac{B_f - B_i}{\Delta t}$$

Substitute numerical values:

$$|\mathcal{E}| = (200)(0.18 \text{ m})^2 \frac{(0.50 \text{ T} - 0)}{0.80 \text{ s}} = \boxed{4.0 \text{ V}}$$

WHAT IF? What if you were asked to find the magnitude of the induced current in the coil while the field is changing? Can you answer that question?

Answer If the ends of the coil are not connected to a circuit, the answer to this question is easy: the current is zero! (Charges move within the wire of the coil, but they cannot move into or out of the ends of the coil.) For a steady current to exist, the ends of the coil must be connected to an external circuit. Let's assume the coil is connected to a circuit and the total resistance of the coil and the circuit is 2.0 Ω. Then, the magnitude of the induced current in the coil is

$$I = \frac{|\mathcal{E}|}{R} = \frac{4.0 \text{ V}}{2.0 \text{ Ω}} = 2.0 \text{ A}$$

Example 31.2 **An Exponentially Decaying Magnetic Field**

A loop of wire enclosing an area A is placed in a region where the magnetic field is perpendicular to the plane of the loop. The magnitude of $\vec{B}$ varies in time according to the expression $B = B_{max}e^{-at}$, where a is some constant. That is, at $t = 0$, the field is B_{max}, and for $t > 0$, the field decreases exponentially (Fig. 31.6). Find the induced emf in the loop as a function of time.

Figure 31.6 (Example 31.2) Exponential decrease in the magnitude of the magnetic field through a loop with time. The induced emf and induced current in a conducting path attached to the loop vary with time in the same way.

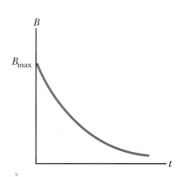

SOLUTION

Conceptualize The physical situation is similar to that in Example 31.1 except for two things: there is only one loop, and the field varies exponentially with time rather than linearly.

Categorize We will evaluate the emf using Faraday's law from this section, so we categorize this example as a substitution problem.

Evaluate Equation 31.1 for the situation described here:

$$\mathcal{E} = -\frac{d\Phi_B}{dt} = -\frac{d}{dt}(AB_{max}e^{-at}) = -AB_{max}\frac{d}{dt}e^{-at} = \boxed{aAB_{max}e^{-at}}$$

This expression indicates that the induced emf decays exponentially in time. The maximum emf occurs at $t = 0$, where $\mathcal{E}_{max} = aAB_{max}$. The plot of $\mathcal{E}$ versus t is similar to the B-versus-t curve shown in Figure 31.6.

31.2 Motional emf

In Examples 31.1 and 31.2, we considered cases in which an emf is induced in a stationary circuit placed in a magnetic field when the field changes with time. In this section, we describe **motional emf,** the emf induced in a conductor moving through a constant magnetic field.

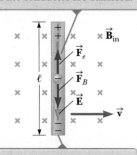

In steady state, the electric and magnetic forces on an electron in the conductor are balanced.

Due to the magnetic force on electrons, the ends of the conductor become oppositely charged, which establishes an electric field in the conductor.

Figure 31.7 A straight electrical conductor of length ℓ moving with a velocity $\vec{v}$ through a uniform magnetic field $\vec{B}$ directed perpendicular to $\vec{v}$.

The straight conductor of length ℓ shown in Figure 31.7 is moving through a uniform magnetic field directed into the page. For simplicity, let's assume the conductor is moving in a direction perpendicular to the field with constant velocity under the influence of some external agent. From the magnetic version of the particle in a field model, the electrons in the conductor experience a force $\vec{F}_B = q\vec{v} \times \vec{B}$ (Eq. 29.1) that is directed along the length ℓ, perpendicular to both $\vec{v}$ and $\vec{B}$. Under the influence of this force, the electrons move to the lower end of the conductor and accumulate there, leaving a net positive charge at the upper end. As a result of this charge separation, an electric field $\vec{E}$ is produced inside the conductor. Therefore, the electrons are also described by the electric version of the particle in a field model. The charges accumulate at both ends until the downward magnetic force qvB on charges remaining in the conductor is balanced by the upward electric force qE. The electrons are then described by the particle in equilibrium model. The condition for equilibrium requires that the forces on the electrons balance:

$$qE = qvB \quad \text{or} \quad E = vB$$

The magnitude of the electric field produced in the conductor is related to the potential difference across the ends of the conductor according to the relationship $\Delta V = E\ell$ (Eq. 25.6). Therefore, for the equilibrium condition,

$$\Delta V = E\ell = B\ell v \tag{31.4}$$

where the upper end of the conductor in Figure 31.7 is at a higher electric potential than the lower end. Therefore, a potential difference is maintained between the ends of the conductor as long as the conductor continues to move through the uniform magnetic field. If the direction of the motion is reversed, the polarity of the potential difference is also reversed.

A more interesting situation occurs when the moving conductor is part of a closed conducting path. This situation is particularly useful for illustrating how a changing magnetic flux causes an induced current in a closed circuit. Consider a circuit consisting of a conducting bar of length ℓ sliding along two fixed, parallel conducting rails as shown in Figure 31.8a. For simplicity, let's assume the bar has zero resistance and the stationary part of the circuit has a resistance R. A uniform and constant magnetic field $\vec{B}$ is applied perpendicular to the plane of the circuit. As the bar is pulled to the right with a velocity $\vec{v}$ under the influence of an applied force $\vec{F}_{app}$, free charges in the bar are moving particles in a magnetic field that experience a magnetic force directed along the length of the bar. This force sets up an induced current because the charges are free to move in the closed conducting path. In this case, the rate of change of magnetic flux through the circuit and the corresponding

A counterclockwise current I is induced in the loop. The magnetic force $\vec{F}_B$ on the bar carrying this current opposes the motion.

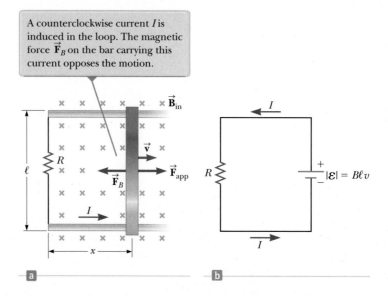

Figure 31.8 (a) A conducting bar sliding with a velocity $\vec{v}$ along two conducting rails under the action of an applied force $\vec{F}_{app}$. (b) The equivalent circuit diagram for the setup shown in (a).

induced motional emf across the moving bar are proportional to the change in area of the circuit.

Because the area enclosed by the circuit at any instant is ℓx, where x is the position of the bar, the magnetic flux through that area is

$$\Phi_B = B\ell x$$

Using Faraday's law and noting that x changes with time at a rate $dx/dt = v$, we find that the induced motional emf is

$$\varepsilon = -\frac{d\Phi_B}{dt} = -\frac{d}{dt}(B\ell x) = -B\ell\frac{dx}{dt}$$

$$\varepsilon = -B\ell v \qquad\qquad \text{(31.5)} \qquad \blacktriangleleft \text{ Motional emf}$$

Because the resistance of the circuit is R, the magnitude of the induced current is

$$I = \frac{|\varepsilon|}{R} = \frac{B\ell v}{R} \qquad\qquad \text{(31.6)}$$

The equivalent circuit diagram for this example is shown in Figure 31.8b.

Let's examine the system using energy considerations. Because no battery is in the circuit, you might wonder about the origin of the induced current and the energy delivered to the resistor. We can understand the source of this current and energy by noting that the applied force does work on the conducting bar. Therefore, we model the circuit as a nonisolated system. The movement of the bar through the field causes charges to move along the bar with some average drift velocity; hence, a current is established. The change in energy in the system during some time interval must be equal to the transfer of energy into the system by work, consistent with the general principle of conservation of energy described by Equation 8.2. The appropriate reduction of Equation 8.2 is $W = \Delta E_{int}$, because the input energy appears as internal energy in the resistor.

Let's verify this equality mathematically. As the bar moves through the uniform magnetic field $\vec{\mathbf{B}}$, it experiences a magnetic force $\vec{\mathbf{F}}_B$ of magnitude $I\ell B$ (see Section 29.4). Because the bar moves with constant velocity, it is modeled as a particle in equilibrium and the magnetic force must be equal in magnitude and opposite in direction to the applied force, or to the left in Figure 31.8a. (If $\vec{\mathbf{F}}_B$ acted in the direction of motion, it would cause the bar to accelerate, violating the principle of conservation of energy.) Using Equation 31.6 and $F_{app} = F_B = I\ell B$, the power delivered by the applied force is

$$P = F_{app}v = (I\ell B)v = \frac{B^2\ell^2 v^2}{R} = \frac{\varepsilon^2}{R} \qquad\qquad \text{(31.7)}$$

From Equation 27.22, we see that this power input is equal to the rate at which energy is delivered to the resistor, consistent with the principle of conservation of energy.

Ⓠuick Quiz 31.2 In Figure 31.8a, a given applied force of magnitude F_{app} results in a constant speed v and a power input P. Imagine that the force is increased so that the constant speed of the bar is doubled to $2v$. Under these conditions, what are the new force and the new power input? **(a)** $2F$ and $2P$ **(b)** $4F$ and $2P$ **(c)** $2F$ and $4P$ **(d)** $4F$ and $4P$

Example 31.3   Magnetic Force Acting on a Sliding Bar   **AM**

The conducting bar illustrated in Figure 31.9 (page 738) moves on two frictionless, parallel rails in the presence of a uniform magnetic field directed into the page. The bar has mass m, and its length is ℓ. The bar is given an initial velocity $\vec{\mathbf{v}}_i$ to the right and is released at $t = 0$.

continued

▶ **31.3** c o n t i n u e d

(A) Using Newton's laws, find the velocity of the bar as a function of time.

SOLUTION

Conceptualize As the bar slides to the right in Figure 31.9, a counterclockwise current is established in the circuit consisting of the bar, the rails, and the resistor. The upward current in the bar results in a magnetic force to the left on the bar as shown in the figure. Therefore, the bar must slow down, so our mathematical solution should demonstrate that.

Categorize The text already categorizes this problem as one that uses Newton's laws. We model the bar as a *particle under a net force*.

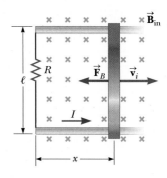

Figure 31.9 (Example 31.3) A conducting bar of length ℓ on two fixed conducting rails is given an initial velocity $\vec{v}_i$ to the right.

Analyze From Equation 29.10, the magnetic force is $F_B = -I\ell B$, where the negative sign indicates that the force is to the left. The magnetic force is the *only* horizontal force acting on the bar.

Using the particle under a net force model, apply Newton's second law to the bar in the horizontal direction:

$$F_x = ma \quad \rightarrow \quad -I\ell B = m\frac{dv}{dt}$$

Substitute $I = B\ell v/R$ from Equation 31.6:

$$m\frac{dv}{dt} = -\frac{B^2\ell^2}{R}v$$

Rearrange the equation so that all occurrences of the variable v are on the left and those of t are on the right:

$$\frac{dv}{v} = -\left(\frac{B^2\ell^2}{mR}\right)dt$$

Integrate this equation using the initial condition that $v = v_i$ at $t = 0$ and noting that $(B^2\ell^2/mR)$ is a constant:

$$\int_{v_i}^{v}\frac{dv}{v} = -\frac{B^2\ell^2}{mR}\int_{0}^{t}dt$$

$$\ln\left(\frac{v}{v_i}\right) = -\left(\frac{B^2\ell^2}{mR}\right)t$$

Define the constant $\tau = mR/B^2\ell^2$ and solve for the velocity:

$$(1) \quad v = \boxed{v_i e^{-t/\tau}}$$

Finalize This expression for v indicates that the velocity of the bar decreases with time under the action of the magnetic force as expected from our conceptualization of the problem.

(B) Show that the same result is found by using an energy approach.

SOLUTION

Categorize The text of this part of the problem tells us to use an energy approach for the same situation. We model the entire circuit in Figure 31.9 as an *isolated system*.

Analyze Consider the sliding bar as one system component possessing kinetic energy, which decreases because energy is transferring *out* of the bar by electrical transmission through the rails. The resistor is another system component possessing internal energy, which rises because energy is transferring *into* the resistor. Because energy is not leaving the system, the rate of energy transfer out of the bar equals the rate of energy transfer into the resistor.

Equate the power entering the resistor to that leaving the bar:

$$P_{\text{resistor}} = -P_{\text{bar}}$$

Substitute for the electrical power delivered to the resistor and the time rate of change of kinetic energy for the bar:

$$I^2 R = -\frac{d}{dt}\left(\tfrac{1}{2}mv^2\right)$$

Use Equation 31.6 for the current and carry out the derivative:

$$\frac{B^2\ell^2 v^2}{R} = -mv\frac{dv}{dt}$$

▶ **31.3** continued

Rearrange terms:
$$\frac{dv}{v} = -\left(\frac{B^2\ell^2}{mR}\right)dt$$

Finalize This result is the same expression to be integrated that we found in part (A).

WHAT IF? Suppose you wished to increase the distance through which the bar moves between the time it is initially projected and the time it essentially comes to rest. You can do so by changing one of three variables—v_i, R, or B—by a factor of 2 or $\frac{1}{2}$. Which variable should you change to maximize the distance, and would you double it or halve it?

Answer Increasing v_i would make the bar move farther. Increasing R would decrease the current and therefore the magnetic force, making the bar move farther. Decreasing B would decrease the magnetic force and make the bar move farther. Which method is most effective, though?

Use Equation (1) to find the distance the bar moves by integration:

$$v = \frac{dx}{dt} = v_i e^{-t/\tau}$$

$$x = \int_0^\infty v_i e^{-t/\tau}\, dt = -v_i \tau e^{-t/\tau}\Big|_0^\infty$$

$$= -v_i \tau(0-1) = v_i \tau = v_i\left(\frac{mR}{B^2\ell^2}\right)$$

This expression shows that doubling v_i or R will double the distance. Changing B by a factor of $\frac{1}{2}$, however, causes the distance to be four times as great!

Example 31.4   Motional emf Induced in a Rotating Bar

A conducting bar of length ℓ rotates with a constant angular speed ω about a pivot at one end. A uniform magnetic field $\vec{B}$ is directed perpendicular to the plane of rotation as shown in Figure 31.10. Find the motional emf induced between the ends of the bar.

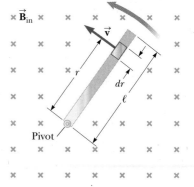

Figure 31.10 (Example 31.4) A conducting bar rotating around a pivot at one end in a uniform magnetic field that is perpendicular to the plane of rotation. A motional emf is induced between the ends of the bar.

SOLUTION

Conceptualize The rotating bar is different in nature from the sliding bar in Figure 31.8. Consider a small segment of the bar, however. It is a short length of conductor moving in a magnetic field and has an emf generated in it like the sliding bar. By thinking of each small segment as a source of emf, we see that all segments are in series and the emfs add.

Categorize Based on the conceptualization of the problem, we approach this example as we did in the discussion leading to Equation 31.5, with the added feature that the short segments of the bar are traveling in circular paths.

Analyze Evaluate the magnitude of the emf induced in a segment of the bar of length dr having a velocity $\vec{v}$ from Equation 31.5:

$$d\mathcal{E} = Bv\, dr$$

Find the total emf between the ends of the bar by adding the emfs induced across all segments:

$$\mathcal{E} = \int Bv\, dr$$

The tangential speed v of an element is related to the angular speed ω through the relationship $v = r\omega$ (Eq. 10.10); use that fact and integrate:

$$\mathcal{E} = B\int v\, dr = B\omega\int_0^\ell r\, dr = \tfrac{1}{2}B\omega\ell^2$$

continued

▶ **31.4** continued

Finalize In Equation 31.5 for a sliding bar, we can increase ε by increasing B, ℓ, or v. Increasing any one of these variables by a given factor increases ε by the same factor. Therefore, you would choose whichever of these three variables is most convenient to increase. For the rotating rod, however, there is an advantage to increasing the length of the rod to raise the emf because ℓ is squared. Doubling the length gives four times the emf, whereas doubling the angular speed only doubles the emf.

WHAT IF?   Suppose, after reading through this example, you come up with a brilliant idea. A Ferris wheel has radial metallic spokes between the hub and the circular rim. These spokes move in the magnetic field of the Earth, so each spoke acts like the bar in Figure 31.10. You plan to use the emf generated by the rotation of the Ferris wheel to power the lightbulbs on the wheel. Will this idea work?

Answer Let's estimate the emf that is generated in this situation. We know the magnitude of the magnetic field of the Earth from Table 29.1: $B = 0.5 \times 10^{-4}$ T. A typical spoke on a Ferris wheel might have a length on the order of 10 m. Suppose the period of rotation is on the order of 10 s.

Determine the angular speed of the spoke:

$$\omega = \frac{2\pi}{T} = \frac{2\pi}{10 \text{ s}} = 0.63 \text{ s}^{-1} \sim 1 \text{ s}^{-1}$$

Assume the magnetic field lines of the Earth are horizontal at the location of the Ferris wheel and perpendicular to the spokes. Find the emf generated:

$$\varepsilon = \tfrac{1}{2}B\omega\ell^2 = \tfrac{1}{2}(0.5 \times 10^{-4} \text{ T})(1 \text{ s}^{-1})(10 \text{ m})^2$$

$$= 2.5 \times 10^{-3} \text{ V} \sim 1 \text{ mV}$$

This value is a tiny emf, far smaller than that required to operate lightbulbs.

An additional difficulty is related to energy. Even assuming you could find lightbulbs that operate using a potential difference on the order of millivolts, a spoke must be part of a circuit to provide a voltage to the lightbulbs. Consequently, the spoke must carry a current. Because this current-carrying spoke is in a magnetic field, a magnetic force is exerted on the spoke in the direction opposite its direction of motion. As a result, the motor of the Ferris wheel must supply more energy to perform work against this magnetic drag force. The motor must ultimately provide the energy that is operating the lightbulbs, and you have not gained anything for free!

31.3 Lenz's Law

Faraday's law (Eq. 31.1) indicates that the induced emf and the change in flux have opposite algebraic signs. This feature has a very real physical interpretation that has come to be known as **Lenz's law:**[1]

Lenz's law ▶

> The induced current in a loop is in the direction that creates a magnetic field that opposes the change in magnetic flux through the area enclosed by the loop.

That is, the induced current tends to keep the original magnetic flux through the loop from changing. We shall show that this law is a consequence of the law of conservation of energy.

To understand Lenz's law, let's return to the example of a bar moving to the right on two parallel rails in the presence of a uniform magnetic field (the *external* magnetic field, shown by the green crosses in Fig. 31.11a). As the bar moves to the right, the magnetic flux through the area enclosed by the circuit increases with time because the area increases. Lenz's law states that the induced current must be directed so that the magnetic field it produces opposes the change in the external magnetic flux. Because the magnetic flux due to an external field directed into the page is increasing, the induced current—if it is to oppose this change—must

---

[1]Developed by German physicist Heinrich Lenz (1804–1865).

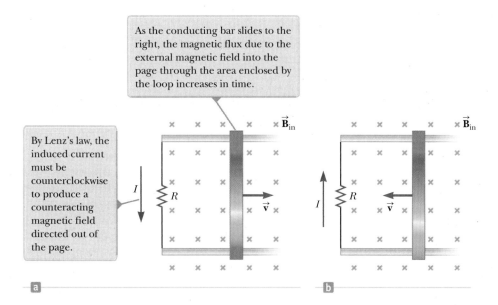

As the conducting bar slides to the right, the magnetic flux due to the external magnetic field into the page through the area enclosed by the loop increases in time.

By Lenz's law, the induced current must be counterclockwise to produce a counteracting magnetic field directed out of the page.

Figure 31.11 (a) Lenz's law can be used to determine the direction of the induced current. (b) When the bar moves to the left, the induced current must be clockwise. Why?

produce a field directed out of the page. Hence, the induced current must be directed counterclockwise when the bar moves to the right. (Use the right-hand rule to verify this direction.) If the bar is moving to the left as in Figure 31.11b, the external magnetic flux through the area enclosed by the loop decreases with time. Because the field is directed into the page, the direction of the induced current must be clockwise if it is to produce a field that also is directed into the page. In either case, the induced current attempts to maintain the original flux through the area enclosed by the current loop.

Let's examine this situation using energy considerations. Suppose the bar is given a slight push to the right. In the preceding analysis, we found that this motion sets up a counterclockwise current in the loop. What happens if we assume the current is clockwise such that the direction of the magnetic force exerted on the bar is to the right? This force would accelerate the rod and increase its velocity, which in turn would cause the area enclosed by the loop to increase more rapidly. The result would be an increase in the induced current, which would cause an increase in the force, which would produce an increase in the current, and so on. In effect, the system would acquire energy with no input of energy. This behavior is clearly inconsistent with all experience and violates the law of conservation of energy. Therefore, the current must be counterclockwise.

Quick Quiz 31.3 Figure 31.12 shows a circular loop of wire falling toward a wire carrying a current to the left. What is the direction of the induced current in the loop of wire? **(a)** clockwise **(b)** counterclockwise **(c)** zero **(d)** impossible to determine

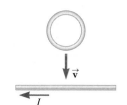

Figure 31.12 (Quick Quiz 31.3)

Conceptual Example 31.5 **Application of Lenz's Law**

A magnet is placed near a metal loop as shown in Figure 31.13a (page 742).

(A) Find the direction of the induced current in the loop when the magnet is pushed toward the loop.

SOLUTION

As the magnet moves to the right toward the loop, the external magnetic flux through the loop increases with time. To counteract this increase in flux due to a field toward the right, the induced current produces its own magnetic field to the left as illustrated in Figure 31.13b; hence, the induced current is in the direction shown. Knowing that like
continued

▶ **31.5** continued

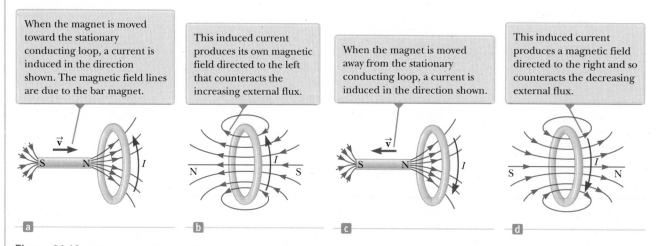

When the magnet is moved toward the stationary conducting loop, a current is induced in the direction shown. The magnetic field lines are due to the bar magnet.

This induced current produces its own magnetic field directed to the left that counteracts the increasing external flux.

When the magnet is moved away from the stationary conducting loop, a current is induced in the direction shown.

This induced current produces a magnetic field directed to the right and so counteracts the decreasing external flux.

Figure 31.13 (Conceptual Example 31.5) A moving bar magnet induces a current in a conducting loop.

magnetic poles repel each other, we conclude that the left face of the current loop acts like a north pole and the right face acts like a south pole.

(B) Find the direction of the induced current in the loop when the magnet is pulled away from the loop.

SOLUTION

If the magnet moves to the left as in Figure 31.13c, its flux through the area enclosed by the loop decreases in time. Now the induced current in the loop is in the direction shown in Figure 31.13d because this current direction produces a magnetic field in the same direction as the external field. In this case, the left face of the loop is a south pole and the right face is a north pole.

Conceptual Example 31.6   **A Loop Moving Through a Magnetic Field**

A rectangular metallic loop of dimensions ℓ and w and resistance R moves with constant speed v to the right as in Figure 31.14a. The loop passes through a uniform magnetic field $\vec{\mathbf{B}}$ directed into the page and extending a distance $3w$ along the x axis. Define x as the position of the right side of the loop along the x axis.

(A) Plot the magnetic flux through the area enclosed by the loop as a function of x.

SOLUTION

Figure 31.14b shows the flux through the area enclosed by the loop as a function of x. Before the loop enters the field, the flux through the loop is zero. As the loop enters the field, the flux increases linearly with position until the left edge of the loop is just inside the field. Finally, the flux through the loop decreases linearly to zero as the loop leaves the field.

(B) Plot the induced motional emf in the loop as a function of x.

SOLUTION

Before the loop enters the field, no motional emf is induced in it because no field is present (Fig. 31.14c). As the right side of the loop enters the field, the magnetic flux directed into the page increases. Hence, according to Lenz's law, the induced current is counterclockwise because it must produce its own magnetic field directed out of the page. The motional emf $-B\ell v$ (from Eq. 31.5) arises from the magnetic force experienced by charges in the right side of the loop. When the loop is entirely in the field, the change in magnetic flux through the loop is zero; hence, the motional emf vanishes. That happens because once the left side of the loop enters the field, the motional emf induced in it

▶ **31.6** continued

cancels the motional emf present in the right side of the loop. As the right side of the loop leaves the field, the flux through the loop begins to decrease, a clockwise current is induced, and the induced emf is $B\ell v$. As soon as the left side leaves the field, the emf decreases to zero.

(C) Plot the external applied force necessary to counter the magnetic force and keep v constant as a function of x.

SOLUTION

The external force that must be applied to the loop to maintain this motion is plotted in Figure 31.14d. Before the loop enters the field, no magnetic force acts on it; hence, the applied force must be zero if v is constant. When the right side of the loop enters the field, the applied force necessary to maintain constant speed must be equal in magnitude and opposite in direction to the magnetic force exerted on that side, so that the loop is a particle in equilibrium. When the loop is entirely in the field, the

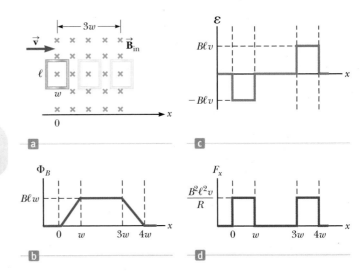

Figure 31.14 (Conceptual Example 31.6) (a) A conducting rectangular loop of width w and length ℓ moving with a velocity $\vec{v}$ through a uniform magnetic field extending a distance $3w$. (b) Magnetic flux through the area enclosed by the loop as a function of loop position. (c) Induced emf as a function of loop position. (d) Applied force required for constant velocity as a function of loop position.

flux through the loop is not changing with time. Hence, the net emf induced in the loop is zero and the current also is zero. Therefore, no external force is needed to maintain the motion. Finally, as the right side leaves the field, the applied force must be equal in magnitude and opposite in direction to the magnetic force acting on the left side of the loop.

From this analysis, we conclude that power is supplied only when the loop is either entering or leaving the field. Furthermore, this example shows that the motional emf induced in the loop can be zero even when there is motion through the field! A motional emf is induced *only* when the magnetic flux through the loop *changes in time*.

31.4 Induced emf and Electric Fields

We have seen that a changing magnetic flux induces an emf and a current in a conducting loop. In our study of electricity, we related a current to an electric field that applies electric forces on charged particles. In the same way, we can relate an induced current in a conducting loop to an electric field by claiming that an electric field is created in the conductor as a result of the changing magnetic flux.

We also noted in our study of electricity that the existence of an electric field is independent of the presence of any test charges. This independence suggests that even in the absence of a conducting loop, a changing magnetic field generates an electric field in empty space.

This induced electric field is *nonconservative*, unlike the electrostatic field produced by stationary charges. To illustrate this point, consider a conducting loop of radius r situated in a uniform magnetic field that is perpendicular to the plane of the loop as in Figure 31.15. If the magnetic field changes with time, an emf $\varepsilon = -d\Phi_B/dt$ is, according to Faraday's law (Eq. 31.1), induced in the loop. The induction of a current in the loop implies the presence of an induced electric field $\vec{E}$, which must be tangent to the loop because that is the direction in which the charges in the wire move in response to the electric force. The work done by the electric field in moving a charge q once around the loop is equal to $q\varepsilon$. Because the electric force acting on the charge is $q\vec{E}$, the work done by the electric field in

If $\vec{B}$ changes in time, an electric field is induced in a direction tangent to the circumference of the loop.

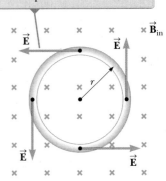

Figure 31.15 A conducting loop of radius r in a uniform magnetic field perpendicular to the plane of the loop.

Pitfall Prevention 31.1
Induced Electric Fields The changing magnetic field does *not* need to exist at the location of the induced electric field. In Figure 31.15, even a loop outside the region of magnetic field experiences an induced electric field.

moving the charge once around the loop is $qE(2\pi r)$, where $2\pi r$ is the circumference of the loop. These two expressions for the work done must be equal; therefore,

$$q\mathcal{E} = qE(2\pi r)$$

$$E = \frac{\mathcal{E}}{2\pi r}$$

Using this result along with Equation 31.1 and that $\Phi_B = BA = B\pi r^2$ for a circular loop, the induced electric field can be expressed as

$$E = -\frac{1}{2\pi r}\frac{d\Phi_B}{dt} = -\frac{r}{2}\frac{dB}{dt} \tag{31.8}$$

If the time variation of the magnetic field is specified, the induced electric field can be calculated from Equation 31.8.

The emf for any closed path can be expressed as the line integral of $\vec{E} \cdot d\vec{s}$ over that path: $\mathcal{E} = \oint \vec{E} \cdot d\vec{s}$. In more general cases, E may not be constant and the path may not be a circle. Hence, Faraday's law of induction, $\mathcal{E} = -d\Phi_B/dt$, can be written in the general form

Faraday's law in general form ▶

$$\oint \vec{E} \cdot d\vec{s} = -\frac{d\Phi_B}{dt} \tag{31.9}$$

The induced electric field $\vec{E}$ in Equation 31.9 is a nonconservative field that is generated by a changing magnetic field. The field $\vec{E}$ that satisfies Equation 31.9 cannot possibly be an electrostatic field because were the field electrostatic and hence conservative, the line integral of $\vec{E} \cdot d\vec{s}$ over a closed loop would be zero (Section 25.1), which would be in contradiction to Equation 31.9.

Example 31.7 **Electric Field Induced by a Changing Magnetic Field in a Solenoid**

A long solenoid of radius R has n turns of wire per unit length and carries a time-varying current that varies sinusoidally as $I = I_{max} \cos \omega t$, where I_{max} is the maximum current and ω is the angular frequency of the alternating current source (Fig. 31.16).

(A) Determine the magnitude of the induced electric field outside the solenoid at a distance $r > R$ from its long central axis.

SOLUTION

Conceptualize Figure 31.16 shows the physical situation. As the current in the coil changes, imagine a changing magnetic field at all points in space as well as an induced electric field.

Categorize In this analysis problem, because the current varies in time, the magnetic field is changing, leading to an induced electric field as opposed to the electrostatic electric fields due to stationary electric charges.

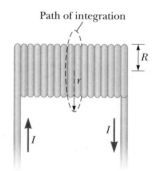

Path of integration

Figure 31.16 (Example 31.7) A long solenoid carrying a time-varying current given by $I = I_{max} \cos \omega t$. An electric field is induced both inside and outside the solenoid.

Analyze First consider an external point and take the path for the line integral to be a circle of radius r centered on the solenoid as illustrated in Figure 31.16.

Evaluate the right side of Equation 31.9, noting that the magnetic field $\vec{B}$ inside the solenoid is perpendicular to the circle bounded by the path of integration:

$$(1) \quad -\frac{d\Phi_B}{dt} = -\frac{d}{dt}(B\pi R^2) = -\pi R^2 \frac{dB}{dt}$$

Evaluate the magnetic field inside the solenoid from Equation 30.17:

$$(2) \quad B = \mu_0 nI = \mu_0 nI_{max} \cos \omega t$$

▶ **31.7** continued

Substitute Equation (2) into Equation (1):

$$(3) \quad -\frac{d\Phi_B}{dt} = -\pi R^2 \mu_0 n I_{max} \frac{d}{dt}(\cos \omega t) = \pi R^2 \mu_0 n I_{max} \omega \sin \omega t$$

Evaluate the left side of Equation 31.9, noting that the magnitude of $\vec{E}$ is constant on the path of integration and $\vec{E}$ is tangent to it:

$$(4) \quad \oint \vec{E} \cdot d\vec{s} = E(2\pi r)$$

Substitute Equations (3) and (4) into Equation 31.9:

$$E(2\pi r) = \pi R^2 \mu_0 n I_{max} \omega \sin \omega t$$

Solve for the magnitude of the electric field:

$$E = \frac{\mu_0 n I_{max} \omega R^2}{2r} \sin \omega t \quad \text{(for } r > R)$$

Finalize This result shows that the amplitude of the electric field outside the solenoid falls off as $1/r$ and varies sinusoidally with time. It is proportional to the current I as well as to the frequency ω, consistent with the fact that a larger value of ω means more change in magnetic flux per unit time. As we will learn in Chapter 34, the time-varying electric field creates an additional contribution to the magnetic field. The magnetic field can be somewhat stronger than we first stated, both inside and outside the solenoid. The correction to the magnetic field is small if the angular frequency ω is small. At high frequencies, however, a new phenomenon can dominate: The electric and magnetic fields, each re-creating the other, constitute an electromagnetic wave radiated by the solenoid as we will study in Chapter 34.

(B) What is the magnitude of the induced electric field inside the solenoid, a distance r from its axis?

SOLUTION

Analyze For an interior point $(r < R)$, the magnetic flux through an integration loop is given by $\Phi_B = B\pi r^2$.

Evaluate the right side of Equation 31.9:

$$(5) \quad -\frac{d\Phi_B}{dt} = -\frac{d}{dt}(B\pi r^2) = -\pi r^2 \frac{dB}{dt}$$

Substitute Equation (2) into Equation (5):

$$(6) \quad -\frac{d\Phi_B}{dt} = -\pi r^2 \mu_0 n I_{max} \frac{d}{dt}(\cos \omega t) = \pi r^2 \mu_0 n I_{max} \omega \sin \omega t$$

Substitute Equations (4) and (6) into Equation 31.9:

$$E(2\pi r) = \pi r^2 \mu_0 n I_{max} \omega \sin \omega t$$

Solve for the magnitude of the electric field:

$$E = \frac{\mu_0 n I_{max} \omega}{2} r \sin \omega t \quad \text{(for } r < R)$$

Finalize This result shows that the amplitude of the electric field induced inside the solenoid by the changing magnetic flux through the solenoid increases linearly with r and varies sinusoidally with time. As with the field outside the solenoid, the field inside is proportional to the current I and the frequency ω.

31.5 Generators and Motors

Electric generators are devices that take in energy by work and transfer it out by electrical transmission. To understand how they operate, let us consider the **alternating-current (AC) generator.** In its simplest form, it consists of a loop of wire rotated by some external means in a magnetic field (Fig. 31.17a, page 746).

In commercial power plants, the energy required to rotate the loop can be derived from a variety of sources. For example, in a hydroelectric plant, falling water directed against the blades of a turbine produces the rotary motion; in a coal-fired plant, the energy released by burning coal is used to convert water to steam, and this steam is directed against the turbine blades.

Figure 31.17 (a) Schematic diagram of an AC generator. (b) The alternating emf induced in the loop plotted as a function of time.

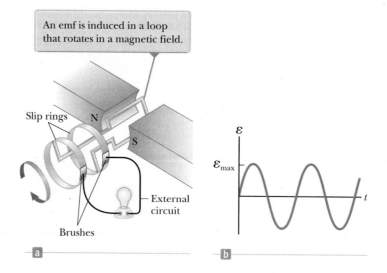

An emf is induced in a loop that rotates in a magnetic field.

Slip rings   N   S

External circuit

Brushes

$\mathcal{E}$

$\mathcal{E}_{max}$

t

a

b

As a loop rotates in a magnetic field, the magnetic flux through the area enclosed by the loop changes with time, and this change induces an emf and a current in the loop according to Faraday's law. The ends of the loop are connected to slip rings that rotate with the loop. Connections from these slip rings, which act as output terminals of the generator, to the external circuit are made by stationary metallic brushes in contact with the slip rings.

Instead of a single turn, suppose a coil with N turns (a more practical situation), with the same area A, rotates in a magnetic field with a constant angular speed ω. If θ is the angle between the magnetic field and the normal to the plane of the coil as in Figure 31.18, the magnetic flux through the coil at any time t is

$$\Phi_B = BA \cos \theta = BA \cos \omega t$$

where we have used the relationship $\theta = \omega t$ between angular position and angular speed (see Eq. 10.3). (We have set the clock so that $t = 0$ when $\theta = 0$.) Hence, the induced emf in the coil is

$$\mathcal{E} = -N\frac{d\Phi_B}{dt} = -NBA\frac{d}{dt}(\cos \omega t) = NBA\omega \sin \omega t \qquad \textbf{(31.10)}$$

This result shows that the emf varies sinusoidally with time as plotted in Figure 31.17b. Equation 31.10 shows that the maximum emf has the value

$$\mathcal{E}_{max} = NBA\omega \qquad \textbf{(31.11)}$$

which occurs when $\omega t = 90°$ or $270°$. In other words, $\mathcal{E} = \mathcal{E}_{max}$ when the magnetic field is in the plane of the coil and the time rate of change of flux is a maximum. Furthermore, the emf is zero when $\omega t = 0$ or $180°$, that is, when $\vec{\textbf{B}}$ is perpendicular to the plane of the coil and the time rate of change of flux is zero.

The frequency for commercial generators in the United States and Canada is 60 Hz, whereas in some European countries it is 50 Hz. (Recall that $\omega = 2\pi f$, where f is the frequency in hertz.)

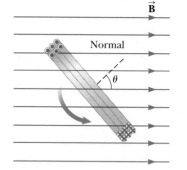

$\vec{\textbf{B}}$

Normal

θ

Figure 31.18 A cutaway view of a loop enclosing an area A and containing N turns, rotating with constant angular speed ω in a magnetic field. The emf induced in the loop varies sinusoidally in time.

Quick Quiz 31.4 In an AC generator, a coil with N turns of wire spins in a magnetic field. Of the following choices, which does *not* cause an increase in the emf generated in the coil? **(a)** replacing the coil wire with one of lower resistance **(b)** spinning the coil faster **(c)** increasing the magnetic field **(d)** increasing the number of turns of wire on the coil

| Example 31.8 | emf Induced in a Generator |

The coil in an AC generator consists of 8 turns of wire, each of area $A = 0.090\ 0$ m^2, and the total resistance of the wire is 12.0 Ω. The coil rotates in a 0.500-T magnetic field at a constant frequency of 60.0 Hz.

(A) Find the maximum induced emf in the coil.

SOLUTION

Conceptualize Study Figure 31.17 to make sure you understand the operation of an AC generator.

Categorize We evaluate parameters using equations developed in this section, so we categorize this example as a substitution problem.

Use Equation 31.11 to find the maximum induced emf:

$$\mathcal{E}_{max} = NBA\omega = NBA(2\pi f)$$

Substitute numerical values:

$$\mathcal{E}_{max} = 8(0.500\text{ T})(0.090\ 0\text{ m}^2)(2\pi)(60.0\text{ Hz}) = \boxed{136\text{ V}}$$

(B) What is the maximum induced current in the coil when the output terminals are connected to a low-resistance conductor?

SOLUTION

Use Equation 27.7 and the result to part (A):

$$I_{max} = \frac{\mathcal{E}_{max}}{R} = \frac{136\text{ V}}{12.0\ \Omega} = \boxed{11.3\text{ A}}$$

The **direct-current (DC) generator** is illustrated in Figure 31.19a. Such generators are used, for instance, in older cars to charge the storage batteries. The components are essentially the same as those of the AC generator except that the contacts to the rotating coil are made using a split ring called a *commutator*.

In this configuration, the output voltage always has the same polarity and pulsates with time as shown in Figure 31.19b. We can understand why by noting that the contacts to the split ring reverse their roles every half cycle. At the same time, the polarity of the induced emf reverses; hence, the polarity of the split ring (which is the same as the polarity of the output voltage) remains the same.

A pulsating DC current is not suitable for most applications. To obtain a steadier DC current, commercial DC generators use many coils and commutators distributed so that the sinusoidal pulses from the various coils are out of phase. When these pulses are superimposed, the DC output is almost free of fluctuations.

A **motor** is a device into which energy is transferred by electrical transmission while energy is transferred out by work. A motor is essentially a generator operating

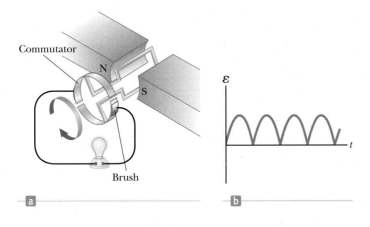

Commutator

N

S

Brush

$\mathcal{E}$

t

a

b

Figure 31.19 (a) Schematic diagram of a DC generator. (b) The magnitude of the emf varies in time, but the polarity never changes.

Figure 31.20 The engine compartment of a Toyota Prius, a hybrid vehicle.

in reverse. Instead of generating a current by rotating a coil, a current is supplied to the coil by a battery, and the torque acting on the current-carrying coil (Section 29.5) causes it to rotate.

Useful mechanical work can be done by attaching the rotating coil of a motor to some external device. As the coil rotates in a magnetic field, however, the changing magnetic flux induces an emf in the coil; consistent with Lenz's law, this induced emf always acts to reduce the current in the coil. The back emf increases in magnitude as the rotational speed of the coil increases. (The phrase *back emf* is used to indicate an emf that tends to reduce the supplied current.) Because the voltage available to supply current equals the difference between the supply voltage and the back emf, the current in the rotating coil is limited by the back emf.

When a motor is turned on, there is initially no back emf, and the current is very large because it is limited only by the resistance of the coil. As the coil begins to rotate, the induced back emf opposes the applied voltage and the current in the coil decreases. If the mechanical load increases, the motor slows down, which causes the back emf to decrease. This reduction in the back emf increases the current in the coil and therefore also increases the power needed from the external voltage source. For this reason, the power requirements for running a motor are greater for heavy loads than for light ones. If the motor is allowed to run under no mechanical load, the back emf reduces the current to a value just large enough to overcome energy losses due to internal energy and friction. If a very heavy load jams the motor so that it cannot rotate, the lack of a back emf can lead to dangerously high current in the motor's wire. This dangerous situation is explored in the What If? section of Example 31.9.

A modern application of motors in automobiles is seen in the development of *hybrid drive systems.* In these automobiles, a gasoline engine and an electric motor are combined to increase the fuel economy of the vehicle and reduce its emissions. Figure 31.20 shows the engine compartment of a Toyota Prius, one of the hybrids available in the United States. In this automobile, power to the wheels can come from either the gasoline engine or the electric motor. In normal driving, the electric motor accelerates the vehicle from rest until it is moving at a speed of about 15 mi/h (24 km/h). During this acceleration period, the engine is not running, so gasoline is not used and there is no emission. At higher speeds, the motor and engine work together so that the engine always operates at or near its most efficient speed. The result is a significantly higher gasoline mileage than that obtained by a traditional gasoline-powered automobile. When a hybrid vehicle brakes, the motor acts as a generator and returns some of the vehicle's kinetic energy back to the battery as stored energy. In a normal vehicle, this kinetic energy is not recovered because it is transformed to internal energy in the brakes and roadway.

Example 31.9   The Induced Current in a Motor

A motor contains a coil with a total resistance of 10 Ω and is supplied by a voltage of 120 V. When the motor is running at its maximum speed, the back emf is 70 V.

(A) Find the current in the coil at the instant the motor is turned on.

SOLUTION

Conceptualize Think about the motor just after it is turned on. It has not yet moved, so there is no back emf generated. As a result, the current in the motor is high. After the motor begins to turn, a back emf is generated and the current decreases.

Categorize We need to combine our new understanding about motors with the relationship between current, voltage, and resistance in this substitution problem.

▶ **31.9** continued

Evaluate the current in the coil from Equation 27.7 with no back emf generated:

$$I = \frac{\mathcal{E}}{R} = \frac{120 \text{ V}}{10 \text{ }\Omega} = \boxed{12 \text{ A}}$$

(B) Find the current in the coil when the motor has reached maximum speed.

SOLUTION

Evaluate the current in the coil with the maximum back emf generated:

$$I = \frac{\mathcal{E} - \mathcal{E}_{\text{back}}}{R} = \frac{120 \text{ V} - 70 \text{ V}}{10 \text{ }\Omega} = \frac{50 \text{ V}}{10 \text{ }\Omega} = \boxed{5.0 \text{ A}}$$

The current drawn by the motor when operating at its maximum speed is significantly less than that drawn before it begins to turn.

WHAT IF? Suppose this motor is in a circular saw. When you are operating the saw, the blade becomes jammed in a piece of wood and the motor cannot turn. By what percentage does the power input to the motor increase when it is jammed?

Answer You may have everyday experiences with motors becoming warm when they are prevented from turning. That is due to the increased power input to the motor. The higher rate of energy transfer results in an increase in the internal energy of the coil, an undesirable effect.

Set up the ratio of power input to the motor when jammed, using the current calculated in part (A), to that when it is not jammed, part (B):

$$\frac{P_{\text{jammed}}}{P_{\text{not jammed}}} = \frac{I_A^2 R}{I_B^2 R} = \frac{I_A^2}{I_B^2}$$

Substitute numerical values:

$$\frac{P_{\text{jammed}}}{P_{\text{not jammed}}} = \frac{(12 \text{ A})^2}{(5.0 \text{ A})^2} = 5.76$$

That represents a 476% increase in the input power! Such a high power input can cause the coil to become so hot that it is damaged.

31.6 Eddy Currents

As we have seen, an emf and a current are induced in a circuit by a changing magnetic flux. In the same manner, circulating currents called **eddy currents** are induced in bulk pieces of metal moving through a magnetic field. This phenomenon can be demonstrated by allowing a flat copper or aluminum plate attached at the end of a rigid bar to swing back and forth through a magnetic field (Fig. 31.21). As the plate enters the field, the changing magnetic flux induces an emf in the plate, which in turn causes the free electrons in the plate to move, producing the swirling eddy currents. According to Lenz's law, the direction of the eddy currents is such that they create magnetic fields that oppose the change that causes the currents. For this reason, the eddy currents must produce effective magnetic poles on the plate, which are repelled by the poles of the magnet; this situation gives rise to a repulsive force that opposes the motion of the plate. (If the opposite were true, the plate would accelerate and its energy would increase after each swing, in violation of the law of conservation of energy.)

As indicated in Figure 31.22a (page 750), with $\vec{\mathbf{B}}$ directed into the page, the induced eddy current is counterclockwise as the swinging plate enters the field at position 1 because the flux due to the external magnetic field into the page through the plate is increasing. Hence, by Lenz's law, the induced current must provide its own magnetic field out of the page. The opposite is true as the plate

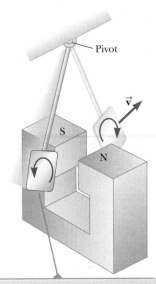

As the plate enters or leaves the field, the changing magnetic flux induces an emf, which causes eddy currents in the plate.

Figure 31.21 Formation of eddy currents in a conducting plate moving through a magnetic field.

Figure 31.22 When a conducting plate swings through a magnetic field, eddy currents are induced and the magnetic force $\vec{\mathbf{F}}_B$ on the plate opposes its velocity, causing it to eventually come to rest.

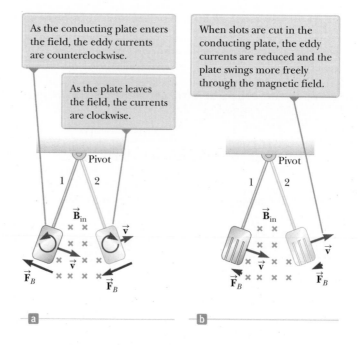

As the conducting plate enters the field, the eddy currents are counterclockwise.

As the plate leaves the field, the currents are clockwise.

When slots are cut in the conducting plate, the eddy currents are reduced and the plate swings more freely through the magnetic field.

leaves the field at position 2, where the current is clockwise. Because the induced eddy current always produces a magnetic retarding force $\vec{\mathbf{F}}_B$ when the plate enters or leaves the field, the swinging plate eventually comes to rest.

If slots are cut in the plate as shown in Figure 31.22b, the eddy currents and the corresponding retarding force are greatly reduced. We can understand this reduction in force by realizing that the cuts in the plate prevent the formation of any large current loops.

The braking systems on many subway and rapid-transit cars make use of electromagnetic induction and eddy currents. An electromagnet attached to the train is positioned near the steel rails. (An electromagnet is essentially a solenoid with an iron core.) The braking action occurs when a large current is passed through the electromagnet. The relative motion of the magnet and rails induces eddy currents in the rails, and the direction of these currents produces a drag force on the moving train. Because the eddy currents decrease steadily in magnitude as the train slows down, the braking effect is quite smooth. As a safety measure, some power tools use eddy currents to stop rapidly spinning blades once the device is turned off.

Eddy currents are often undesirable because they represent a transformation of mechanical energy to internal energy. To reduce this energy loss, conducting parts are often laminated; that is, they are built up in thin layers separated by a nonconducting material such as lacquer or a metal oxide. This layered structure prevents large current loops and effectively confines the currents to small loops in individual layers. Such a laminated structure is used in transformer cores (see Section 33.8) and motors to minimize eddy currents and thereby increase the efficiency of these devices.

Quick Quiz 31.5 In an equal-arm balance from the early 20th century (Fig. 31.23), an aluminum sheet hangs from one of the arms and passes between the poles of a magnet, causing the oscillations of the balance to decay rapidly. In the absence of such magnetic braking, the oscillation might continue for a long time, and the experimenter would have to wait to take a reading. Why do the oscillations decay? **(a)** because the aluminum sheet is attracted to the magnet

(b) because currents in the aluminum sheet set up a magnetic field that opposes the oscillations **(c)** because aluminum is paramagnetic

John W. Jewett, Jr.

Figure 31.23 (Quick Quiz 31.5) In an old-fashioned equal-arm balance, an aluminum sheet hangs between the poles of a magnet.

Summary

Concepts and Principles

▪ **Faraday's law of induction** states that the emf induced in a loop is directly proportional to the time rate of change of magnetic flux through the loop, or

$$\mathcal{E} = -\frac{d\Phi_B}{dt} \qquad \text{(31.1)}$$

where $\Phi_B = \int \vec{\mathbf{B}} \cdot d\vec{\mathbf{A}}$ is the magnetic flux through the loop.

▪ When a conducting bar of length ℓ moves at a velocity $\vec{\mathbf{v}}$ through a magnetic field $\vec{\mathbf{B}}$, where $\vec{\mathbf{B}}$ is perpendicular to the bar and to $\vec{\mathbf{v}}$, the **motional emf** induced in the bar is

$$\mathcal{E} = -B\ell v \qquad \text{(31.5)}$$

▪ **Lenz's law** states that the induced current and induced emf in a conductor are in such a direction as to set up a magnetic field that opposes the change that produced them.

▪ A general form of **Faraday's law of induction** is

$$\oint \vec{\mathbf{E}} \cdot d\vec{\mathbf{s}} = -\frac{d\Phi_B}{dt} \qquad \text{(31.9)}$$

where $\vec{\mathbf{E}}$ is the nonconservative electric field that is produced by the changing magnetic flux.

Objective Questions   1. denotes answer available in *Student Solutions Manual/Study Guide*

1. Figure OQ31.1 is a graph of the magnetic flux through a certain coil of wire as a function of time during an interval while the radius of the coil is increased, the coil is rotated through 1.5 revolutions, and the external source of the magnetic field is turned off, in that order. Rank the emf induced in the coil at the instants marked A through E from the largest positive value to the largest-magnitude negative value. In your ranking,

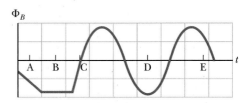

Figure OQ31.1

note any cases of equality and also any instants when the emf is zero.

2. A flat coil of wire is placed in a uniform magnetic field that is in the y direction. **(i)** The magnetic flux through the coil is a maximum if the plane of the coil is where? More than one answer may be correct. (a) in the xy plane (b) in the yz plane (c) in the xz plane (d) in any orientation, because it is a constant **(ii)** For what orientation is the flux zero? Choose from the same possibilities as in part (i).

3. A rectangular conducting loop is placed near a long wire carrying a current I as shown in Figure OQ31.3. If I decreases in time, what can be said of the current induced in the loop? (a) The direction of the current depends on the size of the loop. (b) The current is clockwise. (c) The current is counterclockwise. (d) The current is zero. (e) Nothing can be said about the current in the loop without more information.

Figure OQ31.3

4. A circular loop of wire with a radius of 4.0 cm is in a uniform magnetic field of magnitude 0.060 T. The plane of the loop is perpendicular to the direction of the magnetic field. In a time interval of 0.50 s, the magnetic field changes to the opposite direction with a magnitude of 0.040 T. What is the magnitude of the average emf induced in the loop?   (a) 0.20 V   (b) 0.025 V   (c) 5.0 mV   (d) 1.0 mV   (e) 0.20 mV

5. A square, flat loop of wire is pulled at constant velocity through a region of uniform magnetic field directed perpendicular to the plane of the loop as shown in Figure OQ31.5. Which of the following statements are correct? More than one statement may be correct. (a) Current is induced in the loop in the clockwise direction. (b) Current is induced in the loop in the counterclockwise direction. (c) No current is induced in the loop. (d) Charge separation occurs in the loop, with the top edge positive. (e) Charge separation occurs in the loop, with the top edge negative.

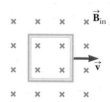

Figure OQ31.5

6. The bar in Figure OQ31.6 moves on rails to the right with a velocity $\vec{v}$, and a uniform, constant magnetic field is directed out of the page. Which of the following statements are correct? More than one statement may be correct. (a) The induced current in the loop is zero. (b) The induced current in the loop is clockwise. (c) The induced current in the loop is counterclockwise. (d) An external force is required to keep the bar moving at constant speed. (e) No force is required to keep the bar moving at constant speed.

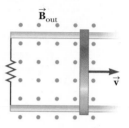

Figure OQ31.6

7. A bar magnet is held in a vertical orientation above a loop of wire that lies in the horizontal plane as shown in Figure OQ31.7. The south end of the magnet is toward the loop. After the magnet is dropped, what is true of the induced current in the loop as viewed from above? (a) It is clockwise as the magnet falls toward the loop. (b) It is counterclockwise as the magnet falls toward the loop. (c) It is clockwise after the magnet has moved through the loop and moves away from it. (d) It is always clockwise. (e) It is first counterclockwise as the magnet approaches the loop and then clockwise after it has passed through the loop.

Figure OQ31.7

8. What happens to the amplitude of the induced emf when the rate of rotation of a generator coil is doubled? (a) It becomes four times larger. (b) It becomes two times larger. (c) It is unchanged. (d) It becomes one-half as large. (e) It becomes one-fourth as large.

9. Two coils are placed near each other as shown in Figure OQ31.9. The coil on the left is connected to a battery and a switch, and the coil on the right is connected to a resistor. What is the direction of the cur-

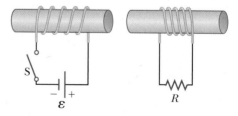

Figure OQ31.9

rent in the resistor **(i)** at an instant immediately after the switch is thrown closed, **(ii)** after the switch has been closed for several seconds, and **(iii)** at an instant after the switch has then been thrown open? Choose each answer from the possibilities (a) left, (b) right, or (c) the current is zero.

10. A circuit consists of a conducting movable bar and a lightbulb connected to two conducting rails as shown in Figure OQ31.10. An external magnetic field is directed perpendicular to the plane of the circuit. Which of the following actions will make the bulb light up? More than one statement may be correct. (a) The bar is

moved to the left. (b) The bar is moved to the right. (c) The magnitude of the magnetic field is increased. (d) The magnitude of the magnetic field is decreased. (e) The bar is lifted off the rails.

11. Two rectangular loops of wire lie in the same plane as shown in Figure OQ31.11. If the current I in the outer loop is counterclockwise and increases with time, what is true of the current induced in the inner loop? More than one statement may be correct. (a) It is zero. (b) It is clockwise. (c) It is counterclockwise. (d) Its magnitude depends on the dimensions of the loops. (e) Its direction depends on the dimensions of the loops.

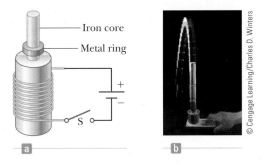

Figure OQ31.10

Figure OQ31.11

Conceptual Questions

1. denotes answer available in *Student Solutions Manual/Study Guide*

1. In Section 7.7, we defined conservative and nonconservative forces. In Chapter 23, we stated that an electric charge creates an electric field that produces a conservative force. Argue now that induction creates an electric field that produces a nonconservative force.

2. A spacecraft orbiting the Earth has a coil of wire in it. An astronaut measures a small current in the coil, although there is no battery connected to it and there are no magnets in the spacecraft. What is causing the current?

3. In a hydroelectric dam, how is energy produced that is then transferred out by electrical transmission? That is, how is the energy of motion of the water converted to energy that is transmitted by AC electricity?

4. A bar magnet is dropped toward a conducting ring lying on the floor. As the magnet falls toward the ring, does it move as a freely falling object? Explain.

5. A circular loop of wire is located in a uniform and constant magnetic field. Describe how an emf can be induced in the loop in this situation.

6. A piece of aluminum is dropped vertically downward between the poles of an electromagnet. Does the magnetic field affect the velocity of the aluminum?

7. What is the difference between magnetic flux and magnetic field?

8. When the switch in Figure CQ31.8a is closed, a current is set up in the coil and the metal ring springs upward (Fig. CQ31.8b). Explain this behavior.

Figure CQ31.8 Conceptual Questions 8 and 9.

9. Assume the battery in Figure CQ31.8a is replaced by an AC source and the switch is held closed. If held down, the metal ring on top of the solenoid becomes hot. Why?

10. A loop of wire is moving near a long, straight wire carrying a constant current I as shown in Figure CQ31.10. (a) Determine the direction of the induced current in the loop as it moves away from the wire. (b) What would be the direction of the induced current in the loop if it were moving toward the wire?

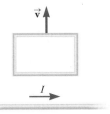

Figure CQ31.10

Problems available in ~~ENHANCED~~ **WebAssign**   Access end-of-chapter problems online at **www.webassign.net**

Section 31.1 Faraday's Law of Induction
Problems 1–20

Section 31.2 Motional emf
Section 31.3 Lenz's Law
Problems 21–38

Section 31.4 Induced emf and Electric Fields
Problems 39–41

Section 31.5 Generators and Motors
Problems 42–49

Section 31.6 Eddy Currents
Problem 50

Additional Problems
Problems 51–78

Challenge Problems
Problem 79–83

Solutions to the following Problems are available in the *Student Solutions Manual/Study Guide:*
31.8, 31.11, 31.16, 31.17, 31.27, 31.33, 31.36, 31.40, 31.42, 31.44, 31.60, 31.65, 31.67, 31.71, 31.75, 31.77, and 31.81

List of Enhanced Problems

Problem Number	Targeted Feedback in Enhanced WebAssign	Analysis Model Tutorial in Enhanced WebAssign	Master It in Enhanced WebAssign	Watch It in Enhanced WebAssign
31.4	✓			✓
31.8	✓			✓
31.9	✓			✓
31.11			✓	
31.13	✓			✓
31.14				✓
31.16			✓	
31.22				✓
31.27	✓		✓	
31.31		✓		
31.33			✓	
31.35		✓		
31.39	✓			
31.40			✓	
31.45	✓			✓
31.53	✓		✓	
31.60		✓		
31.61	✓			✓
31.65			✓	
31.69		✓		
31.75			✓	
31.77			✓	

Inductance

A treasure hunter uses a metal detector to search for buried objects at a beach. At the end of the metal detector is a coil of wire that is part of a circuit. When the coil comes near a metal object, the inductance of the coil is affected and the current in the circuit changes. This change triggers a signal in the earphones worn by the treasure hunter. We investigate inductance in this chapter. *(Andy Ryan/Stone/Getty Images)*

In Chapter 31, we saw that an emf and a current are induced in a loop of wire when the magnetic flux through the area enclosed by the loop changes with time. This phenomenon of electromagnetic induction has some practical consequences. In this chapter, we first describe an effect known as *self-induction,* in which a time-varying current in a circuit produces an induced emf opposing the emf that initially set up the time-varying current. Self-induction is the basis of the *inductor,* an electrical circuit element. We discuss the energy stored in the magnetic field of an inductor and the energy density associated with the magnetic field.

Next, we study how an emf is induced in a coil as a result of a changing magnetic flux produced by a second coil, which is the basic principle of *mutual induction.* Finally, we examine the characteristics of circuits that contain inductors, resistors, and capacitors in various combinations.

32.1 Self-Induction and Inductance

In this chapter, we need to distinguish carefully between emfs and currents that are caused by physical sources such as batteries and those that are induced by changing magnetic fields. When we use a term (such as *emf* or *current*) without an adjective, we are describing the parameters associated with a physical source. We use the adjective *induced* to describe those emfs and currents caused by a changing magnetic field.

Consider a circuit consisting of a switch, a resistor, and a source of emf as shown in Figure 32.1. The circuit diagram is represented in perspective to show the orientations of some of the magnetic field lines due to the current in the circuit. When the switch is thrown to its closed position, the current does not immediately jump from zero to its maximum value $\mathcal{E}/R$. Faraday's law of electromagnetic induction (Eq. 31.1) can be used to describe this effect as follows. As the current increases with time, the magnetic field lines surrounding the wires pass through the loop represented by the circuit itself. This magnetic field passing through the loop causes a magnetic flux through the loop. This increasing flux creates an induced emf in the circuit. The direction of the induced emf is such that it would cause an induced current in the loop (if the loop did not already carry a current), which would establish a magnetic field opposing the change in the original magnetic field. Therefore, the direction of the induced emf is opposite the direction of the emf of the battery, which results in a gradual rather than instantaneous increase in the current to its final equilibrium value. Because of the direction of the induced emf, it is also called a *back emf*, similar to that in a motor as discussed in Chapter 31. This effect is called **self-induction** because the changing flux through the circuit and the resultant induced emf arise from the circuit itself. The emf $\mathcal{E}_L$ set up in this case is called a **self-induced emf.**

To obtain a quantitative description of self-induction, recall from Faraday's law that the induced emf is equal to the negative of the time rate of change of the magnetic flux. The magnetic flux is proportional to the magnetic field, which in turn is proportional to the current in the circuit. Therefore, a self-induced emf is always proportional to the time rate of change of the current. For any loop of wire, we can write this proportionality as

$$\mathcal{E}_L = -L \frac{di}{dt} \tag{32.1}$$

where L is a proportionality constant—called the **inductance** of the loop—that depends on the geometry of the loop and other physical characteristics. If we consider a closely spaced coil of N turns (a toroid or an ideal solenoid) carrying a current i and containing N turns, Faraday's law tells us that $\mathcal{E}_L = -N\, d\Phi_B/dt$. Combining this expression with Equation 32.1 gives

$$L = \frac{N\Phi_B}{i} \tag{32.2}$$

◀ **Inductance of an N-turn coil**

where it is assumed the same magnetic flux passes through each turn and L is the inductance of the entire coil.

From Equation 32.1, we can also write the inductance as the ratio

$$L = -\frac{\mathcal{E}_L}{di/dt} \tag{32.3}$$

Recall that resistance is a measure of the opposition to current as given by Equation 27.7, $R = \Delta V/I$; in comparison, Equation 32.3, being of the same mathematical form as Equation 27.7, shows us that inductance is a measure of the opposition to a *change* in current.

The SI unit of inductance is the **henry** (H), which as we can see from Equation 32.3 is 1 volt-second per ampere: $1\ \text{H} = 1\ \text{V} \cdot \text{s/A}$.

As shown in Example 32.1, the inductance of a coil depends on its geometry. This dependence is analogous to the capacitance of a capacitor depending on the geometry of its plates as we found in Equation 26.3 and the resistance of a resistor depending on the length and area of the conducting material in Equation 27.10. Inductance calculations can be quite difficult to perform for complicated geometries, but the examples below involve simple situations for which inductances are easily evaluated.

Joseph Henry
American Physicist (1797–1878)
Henry became the first director of the Smithsonian Institution and first president of the Academy of Natural Science. He improved the design of the electromagnet and constructed one of the first motors. He also discovered the phenomenon of self-induction, but he failed to publish his findings. The unit of inductance, the henry, is named in his honor.

After the switch is closed, the current produces a magnetic flux through the area enclosed by the loop. As the current increases toward its equilibrium value, this magnetic flux changes in time and induces an emf in the loop.

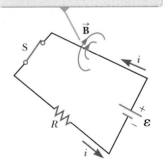

Figure 32.1 Self-induction in a simple circuit.

Quick Quiz 32.1 A coil with zero resistance has its ends labeled a and b. The potential at a is higher than at b. Which of the following could be consistent with this situation? **(a)** The current is constant and is directed from a to b. **(b)** The current is constant and is directed from b to a. **(c)** The current is increasing and is directed from a to b. **(d)** The current is decreasing and is directed from a to b. **(e)** The current is increasing and is directed from b to a. **(f)** The current is decreasing and is directed from b to a.

Example 32.1   Inductance of a Solenoid

Consider a uniformly wound solenoid having N turns and length ℓ. Assume ℓ is much longer than the radius of the windings and the core of the solenoid is air.

(A) Find the inductance of the solenoid.

SOLUTION

Conceptualize The magnetic field lines from each turn of the solenoid pass through all the turns, so an induced emf in each coil opposes changes in the current.

Categorize We categorize this example as a substitution problem. Because the solenoid is long, we can use the results for an ideal solenoid obtained in Chapter 30.

Find the magnetic flux through each turn of area A in the solenoid, using the expression for the magnetic field from Equation 30.17:

$$\Phi_B = BA = \mu_0 niA = \mu_0 \frac{N}{\ell} iA$$

Substitute this expression into Equation 32.2:

$$L = \frac{N\Phi_B}{i} = \mu_0 \frac{N^2}{\ell} A \qquad \textbf{(32.4)}$$

(B) Calculate the inductance of the solenoid if it contains 300 turns, its length is 25.0 cm, and its cross-sectional area is 4.00 cm^2.

SOLUTION

Substitute numerical values into Equation 32.4:

$$L = (4\pi \times 10^{-7}\,\text{T}\cdot\text{m/A}) \frac{300^2}{25.0 \times 10^{-2}\,\text{m}} (4.00 \times 10^{-4}\,\text{m}^2)$$

$$= 1.81 \times 10^{-4}\,\text{T}\cdot\text{m}^2/\text{A} = \boxed{0.181\ \text{mH}}$$

(C) Calculate the self-induced emf in the solenoid if the current it carries decreases at the rate of 50.0 A/s.

SOLUTION

Substitute $di/dt = -50.0$ A/s and the answer to part (B) into Equation 32.1:

$$\varepsilon_L = -L\frac{di}{dt} = -(1.81 \times 10^{-4}\,\text{H})(-50.0\,\text{A/s})$$

$$= \boxed{9.05\ \text{mV}}$$

The result for part (A) shows that L depends on geometry and is proportional to the square of the number of turns. Because $N = n\ell$, we can also express the result in the form

$$L = \mu_0 \frac{(n\ell)^2}{\ell} A = \mu_0 n^2 A\ell = \mu_0 n^2 V \qquad \textbf{(32.5)}$$

where $V = A\ell$ is the interior volume of the solenoid.

32.2 RL Circuits

If a circuit contains a coil such as a solenoid, the inductance of the coil prevents the current in the circuit from increasing or decreasing instantaneously. A circuit element that has a large inductance is called an **inductor** and has the circuit symbol —⦚⦚⦚—. We always assume the inductance of the remainder of a circuit is negligible compared with that of the inductor. Keep in mind, however, that even a circuit without a coil has some inductance that can affect the circuit's behavior.

Because the inductance of an inductor results in a back emf, an inductor in a circuit opposes changes in the current in that circuit. The inductor attempts to keep the current the same as it was before the change occurred. If the battery voltage in the circuit is increased so that the current rises, the inductor opposes this change and the rise is not instantaneous. If the battery voltage is decreased, the inductor causes a slow drop in the current rather than an immediate drop. Therefore, the inductor causes the circuit to be "sluggish" as it reacts to changes in the voltage.

Consider the circuit shown in Figure 32.2, which contains a battery of negligible internal resistance. This circuit is an ***RL* circuit** because the elements connected to the battery are a resistor and an inductor. The curved lines on switch S_2 suggest this switch can never be open; it is always set to either *a* or *b*. (If the switch is connected to neither *a* nor *b*, any current in the circuit suddenly stops.) Suppose S_2 is set to *a* and switch S_1 is open for $t < 0$ and then thrown closed at $t = 0$. The current in the circuit begins to increase, and a back emf (Eq. 32.1) that opposes the increasing current is induced in the inductor.

With this point in mind, let's apply Kirchhoff's loop rule to this circuit, traversing the circuit in the clockwise direction:

$$\mathcal{E} - iR - L\frac{di}{dt} = 0 \qquad \text{(32.6)}$$

where iR is the voltage drop across the resistor. (Kirchhoff's rules were developed for circuits with steady currents, but they can also be applied to a circuit in which the current is changing if we imagine them to represent the circuit at one *instant* of time.) Now let's find a solution to this differential equation, which is similar to that for the *RC* circuit (see Section 28.4).

A mathematical solution of Equation 32.6 represents the current in the circuit as a function of time. To find this solution, we change variables for convenience, letting $x = (\mathcal{E}/R) - i$, so $dx = -di$. With these substitutions, Equation 32.6 becomes

$$x + \frac{L}{R}\frac{dx}{dt} = 0$$

Rearranging and integrating this last expression gives

$$\int_{x_0}^{x}\frac{dx}{x} = -\frac{R}{L}\int_{0}^{t} dt$$

$$\ln\frac{x}{x_0} = -\frac{R}{L}t$$

where x_0 is the value of x at time $t = 0$. Taking the antilogarithm of this result gives

$$x = x_0 e^{-Rt/L}$$

Because $i = 0$ at $t = 0$, note from the definition of x that $x_0 = \mathcal{E}/R$. Hence, this last expression is equivalent to

$$\frac{\mathcal{E}}{R} - i = \frac{\mathcal{E}}{R}e^{-Rt/L}$$

$$i = \frac{\mathcal{E}}{R}\left(1 - e^{-Rt/L}\right)$$

This expression shows how the inductor affects the current. The current does not increase instantly to its final equilibrium value when the switch is closed, but instead increases according to an exponential function. If the inductance is removed from the circuit, which corresponds to letting L approach zero, the exponential term

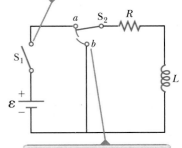

When switch S_1 is thrown closed, the current increases and an emf that opposes the increasing current is induced in the inductor.

When the switch S_2 is thrown to position *b*, the battery is no longer part of the circuit and the current decreases.

Figure 32.2 An *RL* circuit. When switch S_2 is in position *a*, the battery is in the circuit.

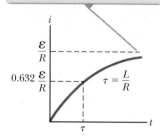

After switch S_1 is thrown closed at $t = 0$, the current increases toward its maximum value $\mathcal{E}/R$.

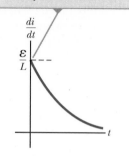

The time rate of change of current is a maximum at $t = 0$, which is the instant at which switch S_1 is thrown closed.

Figure 32.3 Plot of the current versus time for the *RL* circuit shown in Figure 32.2. The time constant τ is the time interval required for i to reach 63.2% of its maximum value.

Figure 32.4 Plot of di/dt versus time for the *RL* circuit shown in Figure 32.2. The rate decreases exponentially with time as i increases toward its maximum value.

becomes zero and there is no time dependence of the current in this case; the current increases instantaneously to its final equilibrium value in the absence of the inductance.

We can also write this expression as

$$i = \frac{\mathcal{E}}{R}\left(1 - e^{-t/\tau}\right) \tag{32.7}$$

where the constant τ is the **time constant** of the *RL* circuit:

$$\tau = \frac{L}{R} \tag{32.8}$$

Physically, τ is the time interval required for the current in the circuit to reach $(1 - e^{-1}) = 0.632 = 63.2\%$ of its final value $\mathcal{E}/R$. The time constant is a useful parameter for comparing the time responses of various circuits.

Figure 32.3 shows a graph of the current versus time in the *RL* circuit. Notice that the equilibrium value of the current, which occurs as t approaches infinity, is $\mathcal{E}/R$. That can be seen by setting di/dt equal to zero in Equation 32.6 and solving for the current i. (At equilibrium, the change in the current is zero.) Therefore, the current initially increases very rapidly and then gradually approaches the equilibrium value $\mathcal{E}/R$ as t approaches infinity.

Let's also investigate the time rate of change of the current. Taking the first time derivative of Equation 32.7 gives

$$\frac{di}{dt} = \frac{\mathcal{E}}{L}e^{-t/\tau} \tag{32.9}$$

This result shows that the time rate of change of the current is a maximum (equal to $\mathcal{E}/L$) at $t = 0$ and falls off exponentially to zero as t approaches infinity (Fig. 32.4).

Now consider the *RL* circuit in Figure 32.2 again. Suppose switch S_2 has been set at position a long enough (and switch S_1 remains closed) to allow the current to reach its equilibrium value $\mathcal{E}/R$. In this situation, the circuit is described by the outer loop in Figure 32.2. If S_2 is thrown from a to b, the circuit is now described by only the right-hand loop in Figure 32.2. Therefore, the battery has been eliminated from the circuit. Setting $\mathcal{E} = 0$ in Equation 32.6 gives

$$iR + L\frac{di}{dt} = 0$$

It is left as a problem (Problem 22 in Enhanced WebAssign) to show that the solution of this differential equation is

$$i = \frac{\mathcal{E}}{R} e^{-t/\tau} = I_i e^{-t/\tau}$$ (32.10)

where $\mathcal{E}$ is the emf of the battery and $I_i = \mathcal{E}/R$ is the initial current at the instant the switch is thrown to b.

If the circuit did not contain an inductor, the current would immediately decrease to zero when the battery is removed. When the inductor is present, it opposes the decrease in the current and causes the current to decrease exponentially. A graph of the current in the circuit versus time (Fig. 32.5) shows that the current is continuously decreasing with time.

Quick Quiz 32.2 Consider the circuit in Figure 32.2 with S_1 open and S_2 at position a. Switch S_1 is now thrown closed. **(i)** At the instant it is closed, across which circuit element is the voltage equal to the emf of the battery? (a) the resistor (b) the inductor (c) both the inductor and resistor **(ii)** After a very long time, across which circuit element is the voltage equal to the emf of the battery? Choose from among the same answers.

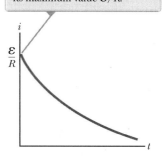

At $t = 0$, the switch is thrown to position b and the current has its maximum value $\mathcal{E}/R$.

Figure 32.5 Current versus time for the right-hand loop of the circuit shown in Figure 32.2. For $t < 0$, switch S_2 is at position a.

| **Example 32.2** | **Time Constant of an *RL* Circuit** |

Consider the circuit in Figure 32.2 again. Suppose the circuit elements have the following values: $\mathcal{E} = 12.0$ V, $R = 6.00\ \Omega$, and $L = 30.0$ mH.

(A) Find the time constant of the circuit.

SOLUTION

Conceptualize You should understand the operation and behavior of the circuit in Figure 32.2 from the discussion in this section.

Categorize We evaluate the results using equations developed in this section, so this example is a substitution problem.

Evaluate the time constant from Equation 32.8:

$$\tau = \frac{L}{R} = \frac{30.0 \times 10^{-3}\ \text{H}}{6.00\ \Omega} = 5.00\ \text{ms}$$

(B) Switch S_2 is at position a, and switch S_1 is thrown closed at $t = 0$. Calculate the current in the circuit at $t = 2.00$ ms.

SOLUTION

Evaluate the current at $t = 2.00$ ms from Equation 32.7:

$$i = \frac{\mathcal{E}}{R}(1 - e^{-t/\tau}) = \frac{12.0\ \text{V}}{6.00\ \Omega}(1 - e^{-2.00\ \text{ms}/5.00\ \text{ms}}) = 2.00\ \text{A}\,(1 - e^{-0.400})$$

$$= 0.659\ \text{A}$$

(C) Compare the potential difference across the resistor with that across the inductor.

SOLUTION

At the instant the switch is closed, there is no current and therefore no potential difference across the resistor. At this instant, the battery voltage appears entirely across the inductor in the form of a back emf of 12.0 V as the inductor tries to maintain the zero-current condition. (The top end of the inductor in Fig. 32.2 is at a higher electric potential than the bottom end.) As time passes, the emf across the inductor decreases and the current in the resistor (and hence the voltage across it) increases as shown in Figure 32.6 (page 762). The sum of the two voltages at all times is 12.0 V.

WHAT IF? In Figure 32.6, the voltages across the resistor and inductor are equal at 3.4 ms. What if you wanted to delay the condition in which the voltages are equal to some later instant, such as $t = 10.0$ ms? Which parameter, L or R, would require the least adjustment, in terms of a percentage change, to achieve that? *continued*

▶ **32.2** continued

Answer Figure 32.6 shows that the voltages are equal when the voltage across the inductor has fallen to half its original value. Therefore, the time interval required for the voltages to become equal is the *half-life* $t_{1/2}$ of the decay. We introduced the half-life in the What If? section of Example 28.10 to describe the exponential decay in *RC* circuits, where $t_{1/2} = 0.693\tau$.

Figure 32.6 (Example 32.2) The time behavior of the voltages across the resistor and inductor in Figure 32.2 given the values provided in this example.

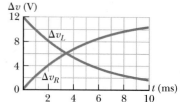

From the desired half-life of 10.0 ms, use the result from Example 28.10 to find the time constant of the circuit:

$$\tau = \frac{t_{1/2}}{0.693} = \frac{10.0\ \text{ms}}{0.693} = 14.4\ \text{ms}$$

Hold *L* fixed and find the value of *R* that gives this time constant:

$$\tau = \frac{L}{R} \;\rightarrow\; R = \frac{L}{\tau} = \frac{30.0 \times 10^{-3}\ \text{H}}{14.4\ \text{ms}} = 2.08\ \Omega$$

Now hold *R* fixed and find the appropriate value of *L*:

$$\tau = \frac{L}{R} \;\rightarrow\; L = \tau R = (14.4\ \text{ms})(6.00\ \Omega) = 86.4 \times 10^{-3}\ \text{H}$$

The change in *R* corresponds to a 65% decrease compared with the initial resistance. The change in *L* represents a 188% increase in inductance! Therefore, a much smaller percentage adjustment in *R* can achieve the desired effect than would an adjustment in *L*.

Pitfall Prevention 32.1

Capacitors, Resistors, and Inductors Store Energy Differently Different energy-storage mechanisms are at work in capacitors, inductors, and resistors. A charged capacitor stores energy as electrical potential energy. An inductor stores energy as what we could call magnetic potential energy when it carries current. Energy delivered to a resistor is transformed to internal energy.

32.3 Energy in a Magnetic Field

A battery in a circuit containing an inductor must provide more energy than one in a circuit without the inductor. Consider Figure 32.2 with switch S_2 in position *a*. When switch S_1 is thrown closed, part of the energy supplied by the battery appears as internal energy in the resistance in the circuit, and the remaining energy is stored in the magnetic field of the inductor. Multiplying each term in Equation 32.6 by *i* and rearranging the expression gives

$$i\mathcal{E} = i^2 R + Li\frac{di}{dt} \tag{32.11}$$

Recognizing $i\mathcal{E}$ as the rate at which energy is supplied by the battery and $i^2 R$ as the rate at which energy is delivered to the resistor, we see that $Li(di/dt)$ must represent the rate at which energy is being stored in the inductor. If U_B is the energy stored in the inductor at any time, we can write the rate dU_B/dt at which energy is stored as

$$\frac{dU_B}{dt} = Li\frac{di}{dt}$$

To find the total energy stored in the inductor at any instant, let's rewrite this expression as $dU_B = Li\,di$ and integrate:

$$U_B = \int dU_B = \int_0^i Li\,di = L\int_0^i i\,di$$

◀ **Energy stored in an inductor**

$$U_B = \tfrac{1}{2}Li^2 \tag{32.12}$$

where *L* is constant and has been removed from the integral. Equation 32.12 represents the energy stored in the magnetic field of the inductor when the current is *i*. It is similar in form to Equation 26.11 for the energy stored in the electric field of a capacitor, $U_E = \tfrac{1}{2}C(\Delta V)^2$. In either case, energy is required to establish a field.

We can also determine the energy density of a magnetic field. For simplicity, consider a solenoid whose inductance is given by Equation 32.5:

$$L = \mu_0 n^2 V$$

The magnetic field of a solenoid is given by Equation 30.17:

$$B = \mu_0 n i$$

Substituting the expression for L and $i = B/\mu_0 n$ into Equation 32.12 gives

$$U_B = \tfrac{1}{2}Li^2 = \tfrac{1}{2}\mu_0 n^2 V \left(\frac{B}{\mu_0 n}\right)^2 = \frac{B^2}{2\mu_0} V \qquad \textbf{(32.13)}$$

The magnetic energy density, or the energy stored per unit volume in the magnetic field of the inductor, is $u_B = U_B/V$, or

$$u_B = \frac{B^2}{2\mu_0} \qquad \textbf{(32.14)} \qquad \blacktriangleleft \ \textbf{Magnetic energy density}$$

Although this expression was derived for the special case of a solenoid, it is valid for any region of space in which a magnetic field exists. Equation 32.14 is similar in form to Equation 26.13 for the energy per unit volume stored in an electric field, $u_E = \tfrac{1}{2}\epsilon_0 E^2$. In both cases, the energy density is proportional to the square of the field magnitude.

Quick Quiz 32.3 You are performing an experiment that requires the highest-possible magnetic energy density in the interior of a very long current-carrying solenoid. Which of the following adjustments increases the energy density? (More than one choice may be correct.) **(a)** increasing the number of turns per unit length on the solenoid **(b)** increasing the cross-sectional area of the solenoid **(c)** increasing only the length of the solenoid while keeping the number of turns per unit length fixed **(d)** increasing the current in the solenoid

---

Example 32.3   **What Happens to the Energy in the Inductor?**   AM

Consider once again the RL circuit shown in Figure 32.2, with switch S_2 at position a and the current having reached its steady-state value. When S_2 is thrown to position b, the current in the right-hand loop decays exponentially with time according to the expression $i = I_i e^{-t/\tau}$, where $I_i = \mathcal{E}/R$ is the initial current in the circuit and $\tau = L/R$ is the time constant. Show that all the energy initially stored in the magnetic field of the inductor appears as internal energy in the resistor as the current decays to zero.

SOLUTION

Conceptualize Before S_2 is thrown to b, energy is being delivered at a constant rate to the resistor from the battery and energy is stored in the magnetic field of the inductor. After $t = 0$, when S_2 is thrown to b, the battery can no longer provide energy and energy is delivered to the resistor only from the inductor.

Categorize We model the right-hand loop of the circuit as an *isolated system* so that energy is transferred between components of the system but does not leave the system.

Analyze We begin by evaluating the energy delivered to the resistor, which appears as internal energy in the resistor.

Begin with Equation 27.22 and recognize that the rate of change of internal energy in the resistor is the power delivered to the resistor:

$$\frac{dE_{int}}{dt} = P = i^2 R$$

Substitute the current given by Equation 32.10 into this equation:

$$\frac{dE_{int}}{dt} = i^2 R = (I_i e^{-Rt/L})^2 R = I_i^2 R e^{-2Rt/L}$$

Solve for dE_{int} and integrate this expression over the limits $t = 0$ to $t \to \infty$:

$$E_{int} = \int_0^\infty I_i^2 R e^{-2Rt/L}\, dt = I_i^2 R \int_0^\infty e^{-2Rt/L}\, dt$$

The value of the definite integral can be shown to be $L/2R$ (see Problem 36 in Enhanced WebAssign). Use this result to evaluate E_{int}:

$$E_{int} = I_i^2 R \left(\frac{L}{2R}\right) = \tfrac{1}{2} L I_i^2$$

continued

▶ **32.3** continued

Finalize This result is equal to the initial energy stored in the magnetic field of the inductor, given by Equation 32.12, as we set out to prove.

Example 32.4 **The Coaxial Cable**

Coaxial cables are often used to connect electrical devices, such as your video system, and in receiving signals in television cable systems. Model a long coaxial cable as a thin, cylindrical conducting shell of radius b concentric with a solid cylinder of radius a as in Figure 32.7. The conductors carry the same current I in opposite directions. Calculate the inductance L of a length ℓ of this cable.

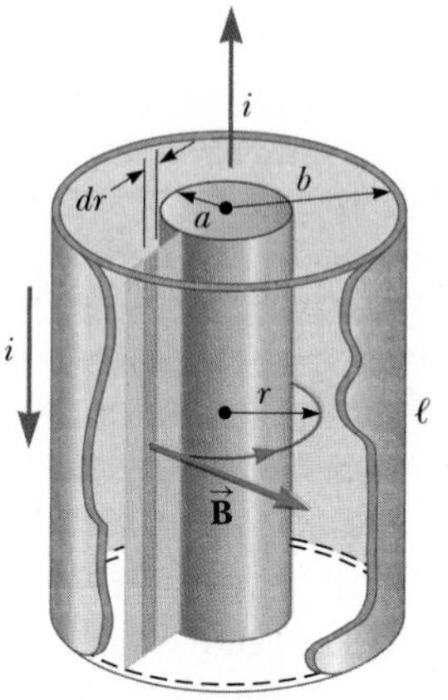

SOLUTION

Conceptualize Consider Figure 32.7. Although we do not have a visible coil in this geometry, imagine a thin, radial slice of the coaxial cable such as the light gold rectangle in Figure 32.7. If the inner and outer conductors are connected at the ends of the cable (above and below the figure), this slice represents one large conducting loop. The current in the loop sets up a magnetic field between the inner and outer conductors that passes through this loop. If the current changes, the magnetic field changes and the induced emf opposes the original change in the current in the conductors.

Categorize We categorize this situation as one in which we must return to the fundamental definition of inductance, Equation 32.2.

Figure 32.7 (Example 32.4) Section of a long coaxial cable. The inner and outer conductors carry equal currents in opposite directions.

Analyze We must find the magnetic flux through the light gold rectangle in Figure 32.7. Ampère's law (see Section 30.3) tells us that the magnetic field in the region between the conductors is due to the inner conductor alone and that its magnitude is $B = \mu_0 i/2\pi r$, where r is measured from the common center of the cylinders. A sample circular field line is shown in Figure 32.7, along with a field vector tangent to the field line. The magnetic field is zero outside the outer shell because the net current passing through the area enclosed by a circular path surrounding the cable is zero; hence, from Ampère's law, $\oint \vec{B} \cdot d\vec{s} = 0$.

The magnetic field is perpendicular to the light gold rectangle of length ℓ and width $b - a$, the cross section of interest. Because the magnetic field varies with radial position across this rectangle, we must use calculus to find the total magnetic flux.

Divide the light gold rectangle into strips of width dr such as the darker strip in Figure 32.7. Evaluate the magnetic flux through such a strip:

$$d\Phi_B = B\, dA = B\ell\, dr$$

Substitute for the magnetic field and integrate over the entire light gold rectangle:

$$\Phi_B = \int_a^b \frac{\mu_0 i}{2\pi r}\, \ell\, dr = \frac{\mu_0 i \ell}{2\pi} \int_a^b \frac{dr}{r} = \frac{\mu_0 i \ell}{2\pi} \ln\left(\frac{b}{a}\right)$$

Use Equation 32.2 to find the inductance of the cable:

$$L = \frac{\Phi_B}{i} = \frac{\mu_0 \ell}{2\pi} \ln\left(\frac{b}{a}\right)$$

Finalize The inductance depends only on geometric factors related to the cable. It increases if ℓ increases, if b increases, or if a decreases. This result is consistent with our conceptualization: any of these changes increases the size of the loop represented by our radial slice and through which the magnetic field passes, increasing the inductance.

32.4 Mutual Inductance

Very often, the magnetic flux through the area enclosed by a circuit varies with time because of time-varying currents in nearby circuits. This condition induces an emf through a process known as *mutual induction,* so named because it depends on the interaction of two circuits.

Consider the two closely wound coils of wire shown in cross-sectional view in Figure 32.8. The current i_1 in coil 1, which has N_1 turns, creates a magnetic field. Some of the magnetic field lines pass through coil 2, which has N_2 turns. The magnetic flux caused by the current in coil 1 and passing through coil 2 is represented by Φ_{12}. In analogy to Equation 32.2, we can identify the **mutual inductance** M_{12} of coil 2 with respect to coil 1:

$$M_{12} = \frac{N_2 \Phi_{12}}{i_1} \tag{32.15}$$

Mutual inductance depends on the geometry of both circuits and on their orientation with respect to each other. As the circuit separation distance increases, the mutual inductance decreases because the flux linking the circuits decreases.

If the current i_1 varies with time, we see from Faraday's law and Equation 32.15 that the emf induced by coil 1 in coil 2 is

$$\mathcal{E}_2 = -N_2 \frac{d\Phi_{12}}{dt} = -N_2 \frac{d}{dt}\left(\frac{M_{12} i_1}{N_2}\right) = -M_{12}\frac{di_1}{dt} \tag{32.16}$$

In the preceding discussion, it was assumed the current is in coil 1. Let's also imagine a current i_2 in coil 2. The preceding discussion can be repeated to show that there is a mutual inductance M_{21}. If the current i_2 varies with time, the emf induced by coil 2 in coil 1 is

$$\mathcal{E}_1 = -M_{21}\frac{di_2}{dt} \tag{32.17}$$

In mutual induction, the emf induced in one coil is always proportional to the rate at which the current in the other coil is changing. Although the proportionality constants M_{12} and M_{21} have been treated separately, it can be shown that they are equal. Therefore, with $M_{12} = M_{21} = M$, Equations 32.16 and 32.17 become

$$\mathcal{E}_2 = -M\frac{di_1}{dt} \quad \text{and} \quad \mathcal{E}_1 = -M\frac{di_2}{dt}$$

These two equations are similar in form to Equation 32.1 for the self-induced emf $\mathcal{E} = -L\,(di/dt)$. The unit of mutual inductance is the henry.

Quick Quiz 32.4 In Figure 32.8, coil 1 is moved closer to coil 2, with the orientation of both coils remaining fixed. Because of this movement, the mutual induction of the two coils **(a)** increases, **(b)** decreases, or **(c)** is unaffected.

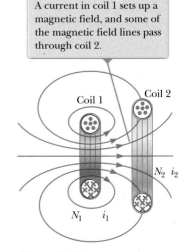

A current in coil 1 sets up a magnetic field, and some of the magnetic field lines pass through coil 2.

Figure 32.8 A cross-sectional view of two adjacent coils.

| Example 32.5 | "Wireless" Battery Charger |

An electric toothbrush has a base designed to hold the toothbrush handle when not in use. As shown in Figure 32.9a, the handle has a cylindrical hole that fits loosely over a matching cylinder on the base. When the handle is placed on the base, a changing current in a solenoid inside the base cylinder induces a current in a coil inside the handle. This induced current charges the battery in the handle.

We can model the base as a solenoid of length ℓ with N_B turns (Fig. 32.9b), carrying a current i, and having a cross-sectional area A. The handle coil contains N_H turns and completely surrounds the base coil. Find the mutual inductance of the system.

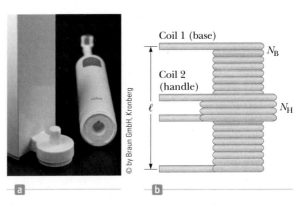

Figure 32.9 (Example 32.5) (a) This electric toothbrush uses the mutual induction of solenoids as part of its battery-charging system. (b) A coil of N_H turns wrapped around the center of a solenoid of N_B turns.

continued

▶ **32.5** continued

Conceptualize Be sure you can identify the two coils in the situation and understand that a changing current in one coil induces a current in the second coil.

Categorize We will determine the result using concepts discussed in this section, so we categorize this example as a substitution problem.

Use Equation 30.17 to express the magnetic field in the interior of the base solenoid:

$$B = \mu_0 \frac{N_B}{\ell} i$$

Find the mutual inductance, noting that the magnetic flux Φ_{BH} through the handle's coil caused by the magnetic field of the base coil is BA:

$$M = \frac{N_H \Phi_{BH}}{i} = \frac{N_H\, BA}{i} = \boxed{\mu_0 \frac{N_B N_H}{\ell} A}$$

Wireless charging is used in a number of other "cordless" devices. One significant example is the inductive charging used by some manufacturers of electric cars that avoids direct metal-to-metal contact between the car and the charging apparatus.

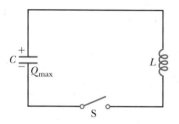

Figure 32.10 A simple LC circuit. The capacitor has an initial charge Q_{max}, and the switch is open for $t < 0$ and then closed at $t = 0$.

32.5 Oscillations in an *LC* Circuit

When a capacitor is connected to an inductor as illustrated in Figure 32.10, the combination is an **LC circuit.** If the capacitor is initially charged and the switch is then closed, both the current in the circuit and the charge on the capacitor oscillate between maximum positive and negative values. If the resistance of the circuit is zero, no energy is transformed to internal energy. In the following analysis, the resistance in the circuit is neglected. We also assume an idealized situation in which energy is not radiated away from the circuit. This radiation mechanism is discussed in Chapter 34.

Assume the capacitor has an initial charge Q_{max} (the maximum charge) and the switch is open for $t < 0$ and then closed at $t = 0$. Let's investigate what happens from an energy viewpoint.

When the capacitor is fully charged, the energy U in the circuit is stored in the capacitor's electric field and is equal to $Q_{max}^2/2C$ (Eq. 26.11). At this time, the current in the circuit is zero; therefore, no energy is stored in the inductor. After the switch is closed, the rate at which charges leave or enter the capacitor plates (which is also the rate at which the charge on the capacitor changes) is equal to the current in the circuit. After the switch is closed and the capacitor begins to discharge, the energy stored in its electric field decreases. The capacitor's discharge represents a current in the circuit, and some energy is now stored in the magnetic field of the inductor. Therefore, energy is transferred from the electric field of the capacitor to the magnetic field of the inductor. When the capacitor is fully discharged, it stores no energy. At this time, the current reaches its maximum value and all the energy in the circuit is stored in the inductor. The current continues in the same direction, decreasing in magnitude, with the capacitor eventually becoming fully charged again but with the polarity of its plates now opposite the initial polarity. This process is followed by another discharge until the circuit returns to its original state of maximum charge Q_{max} and the plate polarity shown in Figure 32.10. The energy continues to oscillate between inductor and capacitor.

The oscillations of the LC circuit are an electromagnetic analog to the mechanical oscillations of the particle in simple harmonic motion studied in Chapter 15. Much of what was discussed there is applicable to LC oscillations. For example, we investigated the effect of driving a mechanical oscillator with an external force,

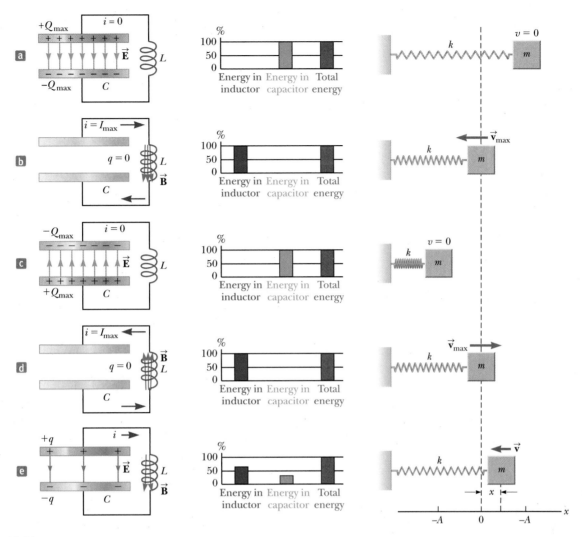

Figure 32.11 Energy transfer in a resistanceless, nonradiating *LC* circuit. The capacitor has a charge Q_{max} at $t = 0$, the instant at which the switch in Figure 32.10 is closed. The mechanical analog of this circuit is the particle in simple harmonic motion, represented by the block–spring system at the right of the figure. (a)–(d) At these special instants, all of the energy in the circuit resides in one of the circuit elements. (e) At an arbitrary instant, the energy is split between the capacitor and the inductor.

which leads to the phenomenon of *resonance*. The same phenomenon is observed in the *LC* circuit. (See Section 33.7.)

A representation of the energy transfer in an *LC* circuit is shown in Figure 32.11. As mentioned, the behavior of the circuit is analogous to that of the particle in simple harmonic motion studied in Chapter 15. For example, consider the block–spring system shown in Figure 15.10. The oscillations of this system are shown at the right of Figure 32.11. The potential energy $\frac{1}{2}kx^2$ stored in the stretched spring is analogous to the potential energy $Q_{max}^2/2C$ stored in the capacitor in Figure 32.11. The kinetic energy $\frac{1}{2}mv^2$ of the moving block is analogous to the magnetic energy $\frac{1}{2}Li^2$ stored in the inductor, which requires the presence of moving charges. In Figure 32.11a, all the energy is stored as electric potential energy in the capacitor at $t = 0$ (because $i = 0$), just as all the energy in a block–spring system is initially stored as potential energy in the spring if it is stretched and released at $t = 0$. In Figure 32.11b, all the energy is stored as magnetic energy $\frac{1}{2}LI_{max}^2$ in the inductor, where I_{max} is the maximum current. Figures 32.11c and 32.11d show subsequent quarter-cycle situations in which the energy is all electric or all magnetic. At intermediate points, part of the energy is electric and part is magnetic.

Consider some arbitrary time t after the switch is closed so that the capacitor has a charge $q < Q_{max}$ and the current is $i < I_{max}$. At this time, both circuit elements store energy, as shown in Figure 32.11e, but the sum of the two energies must equal the total initial energy U stored in the fully charged capacitor at $t = 0$:

Total energy stored in ▶
an *LC* circuit

$$U = U_E + U_B = \frac{q^2}{2C} + \tfrac{1}{2}Li^2 = \frac{Q_{max}^2}{2C} \tag{32.18}$$

Because we have assumed the circuit resistance to be zero and we ignore electromagnetic radiation, no energy is transformed to internal energy and none is transferred out of the system of the circuit. Therefore, with these assumptions, the system of the circuit is isolated: *the total energy of the system must remain constant in time.* We describe the constant energy of the system mathematically by setting $dU/dt = 0$. Therefore, by differentiating Equation 32.18 with respect to time while noting that q and i vary with time gives

$$\frac{dU}{dt} = \frac{d}{dt}\left(\frac{q^2}{2C} + \tfrac{1}{2}Li^2\right) = \frac{q}{C}\frac{dq}{dt} + Li\frac{di}{dt} = 0 \tag{32.19}$$

We can reduce this result to a differential equation in one variable by remembering that the current in the circuit is equal to the rate at which the charge on the capacitor changes: $i = dq/dt$. It then follows that $di/dt = d^2q/dt^2$. Substitution of these relationships into Equation 32.19 gives

$$\frac{q}{C} + L\frac{d^2q}{dt^2} = 0$$

$$\frac{d^2q}{dt^2} = -\frac{1}{LC}q \tag{32.20}$$

Let's solve for q by noting that this expression is of the same form as the analogous Equations 15.3 and 15.5 for a particle in simple harmonic motion:

$$\frac{d^2x}{dt^2} = -\frac{k}{m}x = -\omega^2 x$$

where k is the spring constant, m is the mass of the block, and $\omega = \sqrt{k/m}$. The solution of this mechanical equation has the general form (Eq. 15.6):

$$x = A\cos(\omega t + \phi)$$

where A is the amplitude of the simple harmonic motion (the maximum value of x), ω is the angular frequency of this motion, and ϕ is the phase constant; the values of A and ϕ depend on the initial conditions. Because Equation 32.20 is of the same mathematical form as the differential equation of the simple harmonic oscillator, it has the solution

Charge as a function of time ▶
for an ideal *LC* circuit

$$q = Q_{max}\cos(\omega t + \phi) \tag{32.21}$$

where Q_{max} is the maximum charge of the capacitor and the angular frequency ω is

Angular frequency of ▶
oscillation in an *LC* circuit

$$\omega = \frac{1}{\sqrt{LC}} \tag{32.22}$$

Note that the angular frequency of the oscillations depends solely on the inductance and capacitance of the circuit. Equation 32.22 gives the *natural frequency* of oscillation of the *LC* circuit.

Because q varies sinusoidally with time, the current in the circuit also varies sinusoidally with time. We can show that by differentiating Equation 32.21 with respect to time:

Current as a function of ▶
time for an ideal *LC* current

$$i = \frac{dq}{dt} = -\omega Q_{max}\sin(\omega t + \phi) \tag{32.23}$$

To determine the value of the phase angle ϕ, let's examine the initial conditions, which in our situation require that at $t = 0$, $i = 0$, and $q = Q_{max}$. Setting $i = 0$ at $t = 0$ in Equation 32.23 gives

$$0 = -\omega Q_{max} \sin \phi$$

which shows that $\phi = 0$. This value for ϕ also is consistent with Equation 32.21 and the condition that $q = Q_{max}$ at $t = 0$. Therefore, in our case, the expressions for q and i are

$$q = Q_{max} \cos \omega t \tag{32.24}$$

$$i = -\omega Q_{max} \sin \omega t = -I_{max} \sin \omega t \tag{32.25}$$

Graphs of q versus t and i versus t are shown in Figure 32.12. The charge on the capacitor oscillates between the extreme values Q_{max} and $-Q_{max}$, and the current oscillates between I_{max} and $-I_{max}$. Furthermore, the current is 90° out of phase with the charge. That is, when the charge is a maximum, the current is zero, and when the charge is zero, the current has its maximum value.

Let's return to the energy discussion of the *LC* circuit. Substituting Equations 32.24 and 32.25 in Equation 32.18, we find that the total energy is

$$U = U_E + U_B = \frac{Q_{max}^2}{2C} \cos^2 \omega t + \tfrac{1}{2}LI_{max}^2 \sin^2 \omega t \tag{32.26}$$

This expression contains all the features described qualitatively at the beginning of this section. It shows that the energy of the *LC* circuit continuously oscillates between energy stored in the capacitor's electric field and energy stored in the inductor's magnetic field. When the energy stored in the capacitor has its maximum value $Q_{max}^2/2C$, the energy stored in the inductor is zero. When the energy stored in the inductor has its maximum value $\tfrac{1}{2}LI_{max}^2$, the energy stored in the capacitor is zero.

Plots of the time variations of U_E and U_B are shown in Figure 32.13. The sum $U_E + U_B$ is a constant and is equal to the total energy $Q_{max}^2/2C$, or $\tfrac{1}{2}LI_{max}^2$. Analytical verification is straightforward. The amplitudes of the two graphs in Figure 32.13 must be equal because the maximum energy stored in the capacitor (when $I = 0$) must equal the maximum energy stored in the inductor (when $q = 0$). This equality is expressed mathematically as

$$\frac{Q_{max}^2}{2C} = \frac{LI_{max}^2}{2}$$

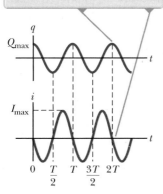

The charge q and the current i are 90° out of phase with each other.

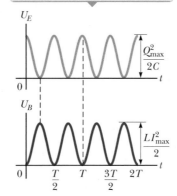

The sum of the two curves is a constant and is equal to the total energy stored in the circuit.

Figure 32.12 Graphs of charge versus time and current versus time for a resistanceless, nonradiating *LC* circuit.

Figure 32.13 Plots of U_E versus t and U_B versus t for a resistanceless, nonradiating *LC* circuit.

Using this expression in Equation 32.26 for the total energy gives

$$U = \frac{Q_{max}^2}{2C}(\cos^2 \omega t + \sin^2 \omega t) = \frac{Q_{max}^2}{2C} \qquad \textbf{(32.27)}$$

because $\cos^2 \omega t + \sin^2 \omega t = 1$.

In our idealized situation, the oscillations in the circuit persist indefinitely; the total energy U of the circuit, however, remains constant only if energy transfers and transformations are neglected. In actual circuits, there is always some resistance and some energy is therefore transformed to internal energy. We mentioned at the beginning of this section that we are also ignoring radiation from the circuit. In reality, radiation is inevitable in this type of circuit, and the total energy in the circuit continuously decreases as a result of this process.

Quick Quiz 32.5 **(i)** At an instant of time during the oscillations of an LC circuit, the current is at its maximum value. At this instant, what happens to the voltage across the capacitor? (a) It is different from that across the inductor. (b) It is zero. (c) It has its maximum value. (d) It is impossible to determine. **(ii)** Now consider an instant when the current is momentarily zero. From the same choices, describe the magnitude of the voltage across the capacitor at this instant.

Example 32.6   **Oscillations in an *LC* Circuit**

In Figure 32.14, the battery has an emf of 12.0 V, the inductance is 2.81 mH, and the capacitance is 9.00 pF. The switch has been set to position *a* for a long time so that the capacitor is charged. The switch is then thrown to position *b*, removing the battery from the circuit and connecting the capacitor directly across the inductor.

(A) Find the frequency of oscillation of the circuit.

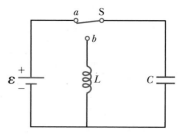

Figure 32.14 (Example 32.6) First the capacitor is fully charged with the switch set to position *a*. Then the switch is thrown to position *b*, and the battery is no longer in the circuit.

SOLUTION

Conceptualize When the switch is thrown to position *b*, the active part of the circuit is the right-hand loop, which is an LC circuit.

Categorize We use equations developed in this section, so we categorize this example as a substitution problem.

Use Equation 32.22 to find the frequency:

$$f = \frac{\omega}{2\pi} = \frac{1}{2\pi\sqrt{LC}}$$

Substitute numerical values:

$$f = \frac{1}{2\pi[(2.81 \times 10^{-3}\,\text{H})(9.00 \times 10^{-12}\,\text{F})]^{1/2}} = \boxed{1.00 \times 10^6\,\text{Hz}}$$

(B) What are the maximum values of charge on the capacitor and current in the circuit?

SOLUTION

Find the initial charge on the capacitor, which equals the maximum charge:

$$Q_{max} = C\,\Delta V = (9.00 \times 10^{-12}\,\text{F})(12.0\,\text{V}) = \boxed{1.08 \times 10^{-10}\,\text{C}}$$

Use Equation 32.25 to find the maximum current from the maximum charge:

$$I_{max} = \omega Q_{max} = 2\pi f Q_{max} = (2\pi \times 10^6\,\text{s}^{-1})(1.08 \times 10^{-10}\,\text{C})$$

$$= \boxed{6.79 \times 10^{-4}\,\text{A}}$$

32.6 The *RLC* Circuit

Let's now turn our attention to a more realistic circuit consisting of a resistor, an inductor, and a capacitor connected in series as shown in Figure 32.15. We assume the resistance of the resistor represents all the resistance in the circuit. Suppose the switch is at position *a* so that the capacitor has an initial charge Q_{max}. The switch is now thrown to position *b*. At this instant, the total energy stored in the capacitor and inductor is $Q_{max}^2/2C$. This total energy, however, is no longer constant as it was in the *LC* circuit because the resistor causes transformation to internal energy. (We continue to ignore electromagnetic radiation from the circuit in this discussion.) Because the rate of energy transformation to internal energy within a resistor is i^2R,

$$\frac{dU}{dt} = -i^2R$$

where the negative sign signifies that the energy *U* of the circuit is decreasing in time. Substituting $U = U_E + U_B$ gives

$$\frac{q}{C}\frac{dq}{dt} + Li\frac{di}{dt} = -i^2R \tag{32.28}$$

To convert this equation into a form that allows us to compare the electrical oscillations with their mechanical analog, we first use $i = dq/dt$ and move all terms to the left-hand side to obtain

$$Li\frac{d^2q}{dt^2} + i^2R + \frac{q}{C}i = 0$$

Now divide through by *i*:

$$L\frac{d^2q}{dt^2} + iR + \frac{q}{C} = 0$$

$$L\frac{d^2q}{dt^2} + R\frac{dq}{dt} + \frac{q}{C} = 0 \tag{32.29}$$

The *RLC* circuit is analogous to the damped harmonic oscillator discussed in Section 15.6 and illustrated in Figure 15.20. The equation of motion for a damped block–spring system is, from Equation 15.31,

$$m\frac{d^2x}{dt^2} + b\frac{dx}{dt} + kx = 0 \tag{32.30}$$

Comparing Equations 32.29 and 32.30, we see that *q* corresponds to the position *x* of the block at any instant, *L* to the mass *m* of the block, *R* to the damping coefficient *b*, and *C* to $1/k$, where *k* is the force constant of the spring. These and other relationships are listed in Table 32.1 on page 772.

Because the analytical solution of Equation 32.29 is cumbersome, we give only a qualitative description of the circuit behavior. In the simplest case, when $R = 0$, Equation 32.29 reduces to that of a simple *LC* circuit as expected, and the charge and the current oscillate sinusoidally in time. This situation is equivalent to removing all damping in the mechanical oscillator.

When *R* is small, a situation that is analogous to light damping in the mechanical oscillator, the solution of Equation 32.29 is

$$q = Q_{max}e^{-Rt/2L}\cos\omega_d t \tag{32.31}$$

where ω_d, the angular frequency at which the circuit oscillates, is given by

$$\omega_d = \left[\frac{1}{LC} - \left(\frac{R}{2L}\right)^2\right]^{1/2} \tag{32.32}$$

That is, the value of the charge on the capacitor undergoes a damped harmonic oscillation in analogy with a block–spring system moving in a viscous medium. Equation 32.32 shows that when $R \ll \sqrt{4L/C}$ (so that the second term in the

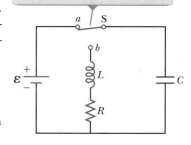

Figure 32.15 A series *RLC* circuit.

Table 32.1	Analogies Between the *RLC* Circuit and the Particle in Simple Harmonic Motion	
RLC Circuit		**One-Dimensional Particle in Simple Harmonic Motion**
Charge	$q \leftrightarrow x$	Position
Current	$i \leftrightarrow v_x$	Velocity
Potential difference	$\Delta V \leftrightarrow F_x$	Force
Resistance	$R \leftrightarrow b$	Viscous damping coefficient
Capacitance	$C \leftrightarrow 1/k$	(k = spring constant)
Inductance	$L \leftrightarrow m$	Mass
Current = time derivative of charge	$i = \dfrac{dq}{dt} \leftrightarrow v_x = \dfrac{dx}{dt}$	Velocity = time derivative of position
Rate of change of current = second time derivative of charge	$\dfrac{di}{dt} = \dfrac{d^2q}{dt^2} \leftrightarrow a_x = \dfrac{dv_x}{dt} = \dfrac{d^2x}{dt^2}$	Acceleration = second time derivative of position
Energy in inductor	$U_B = \frac{1}{2}Li^2 \leftrightarrow K = \frac{1}{2}mv^2$	Kinetic energy of moving object
Energy in capacitor	$U_E = \frac{1}{2}\dfrac{q^2}{C} \leftrightarrow U = \frac{1}{2}kx^2$	Potential energy stored in a spring
Rate of energy loss due to resistance	$i^2R \leftrightarrow bv^2$	Rate of energy loss due to friction
RLC circuit	$L\dfrac{d^2q}{dt^2} + R\dfrac{dq}{dt} + \dfrac{q}{C} = 0 \leftrightarrow m\dfrac{d^2x}{dt^2} + b\dfrac{dx}{dt} + kx = 0$	Damped object on a spring

brackets is much smaller than the first), the frequency ω_d of the damped oscillator is close to that of the undamped oscillator, $1/\sqrt{LC}$. Because $i = dq/dt$, it follows that the current also undergoes damped harmonic oscillation. A plot of the charge versus time for the damped oscillator is shown in Figure 32.16a, and an oscilloscope trace for a real *RLC* circuit is shown in Figure 32.16b. The maximum value of q decreases after each oscillation, just as the amplitude of a damped block–spring system decreases in time.

For larger values of R, the oscillations damp out more rapidly; in fact, there exists a critical resistance value $R_c = \sqrt{4L/C}$ above which no oscillations occur. A system with $R = R_c$ is said to be *critically damped*. When R exceeds R_c, the system is said to be *overdamped*.

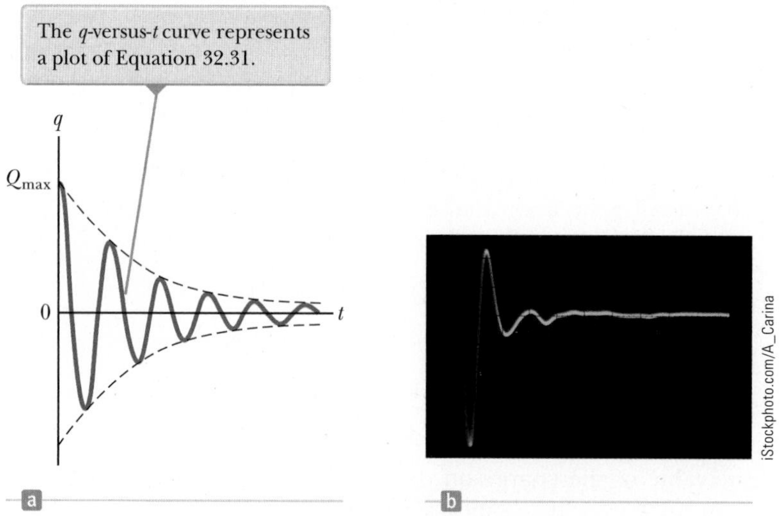

The q-versus-t curve represents a plot of Equation 32.31.

Figure 32.16 (a) Charge versus time for a damped *RLC* circuit. The charge decays in this way when $R < \sqrt{4L/C}$. (b) Oscilloscope pattern showing the decay in the oscillations of an *RLC* circuit.

iStockphoto.com/A_Carina

Summary

Concepts and Principles

When the current in a loop of wire changes with time, an emf is induced in the loop according to Faraday's law. The **self-induced emf** is

$$\mathcal{E}_L = -L \frac{di}{dt} \qquad (32.1)$$

where L is the **inductance** of the loop. Inductance is a measure of how much opposition a loop offers to a change in the current in the loop. Inductance has the SI unit of **henry** (H), where $1\ \text{H} = 1\ \text{V} \cdot \text{s/A}$.

The inductance of any coil is

$$L = \frac{N\Phi_B}{i} \qquad (32.2)$$

where N is the total number of turns and Φ_B is the magnetic flux through the coil. The inductance of a device depends on its geometry. For example, the inductance of an air-core solenoid is

$$L = \mu_0 \frac{N^2}{\ell} A \qquad (32.4)$$

where ℓ is the length of the solenoid and A is the cross-sectional area.

If a resistor and inductor are connected in series to a battery of emf $\mathcal{E}$ at time $t = 0$, the current in the circuit varies in time according to the expression

$$i = \frac{\mathcal{E}}{R}(1 - e^{-t/\tau}) \qquad (32.7)$$

where $\tau = L/R$ is the **time constant** of the RL circuit. If we replace the battery in the circuit by a resistanceless wire, the current decays exponentially with time according to the expression

$$i = \frac{\mathcal{E}}{R} e^{-t/\tau} \qquad (32.10)$$

where $\mathcal{E}/R$ is the initial current in the circuit.

The energy stored in the magnetic field of an inductor carrying a current i is

$$U_B = \tfrac{1}{2} L i^2 \qquad (32.12)$$

This energy is the magnetic counterpart to the energy stored in the electric field of a charged capacitor.

The energy density at a point where the magnetic field is B is

$$u_B = \frac{B^2}{2\mu_0} \qquad (32.14)$$

The **mutual inductance** of a system of two coils is

$$M_{12} = \frac{N_2 \Phi_{12}}{i_1} = M_{21} = \frac{N_1 \Phi_{21}}{i_2} = M \qquad (32.15)$$

This mutual inductance allows us to relate the induced emf in a coil to the changing source current in a nearby coil using the relationships

$$\mathcal{E}_2 = -M_{12} \frac{di_1}{dt} \quad \text{and} \quad \mathcal{E}_1 = -M_{21} \frac{di_2}{dt} \qquad (32.16, 32.17)$$

In an LC circuit that has zero resistance and does not radiate electromagnetically (an idealization), the values of the charge on the capacitor and the current in the circuit vary sinusoidally in time at an angular frequency given by

$$\omega = \frac{1}{\sqrt{LC}} \qquad (32.22)$$

The energy in an LC circuit continuously transfers between energy stored in the capacitor and energy stored in the inductor.

In an RLC circuit with small resistance, the charge on the capacitor varies with time according to

$$q = Q_{\text{max}} e^{-Rt/2L} \cos \omega_d t \qquad (32.31)$$

where

$$\omega_d = \left[\frac{1}{LC} - \left(\frac{R}{2L} \right)^2 \right]^{1/2} \qquad (32.32)$$

1. The centers of two circular loops are separated by a fixed distance. **(i)** For what relative orientation of the loops is their mutual inductance a maximum? (a) coaxial and lying in parallel planes (b) lying in the same plane (c) lying in perpendicular planes, with the center of one on the axis of the other (d) The orientation makes no difference. **(ii)** For what relative orientation is their mutual inductance a minimum? Choose from the same possibilities as in part (i).

2. A long, fine wire is wound into a coil with inductance 5 mH. The coil is connected across the terminals of a battery, and the current is measured a few seconds after the connection is made. The wire is unwound and wound again into a different coil with $L = 10$ mH. This second coil is connected across the same battery, and the current is measured in the same way. Compared with the current in the first coil, is the current in the second coil (a) four times as large, (b) twice as large, (c) unchanged, (d) half as large, or (e) one-fourth as large?

3. A solenoidal inductor for a printed circuit board is being redesigned. To save weight, the number of turns is reduced by one-half, with the geometric dimensions kept the same. By how much must the current change if the energy stored in the inductor is to remain the same? (a) It must be four times larger. (b) It must be two times larger. (c) It should be left the same. (d) It should be one-half as large. (e) No change in the current can compensate for the reduction in the number of turns.

4. In Figure OQ32.4, the switch is left in position *a* for a long time interval and is then quickly thrown to position *b*. Rank the magnitudes of the voltages across the four circuit elements a short time thereafter from the largest to the smallest.

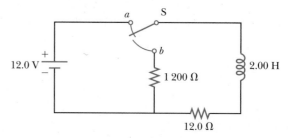

Figure OQ32.4

5. Two solenoids, A and B, are wound using equal lengths of the same kind of wire. The length of the axis of each solenoid is large compared with its diameter. The axial length of A is twice as large as that of B, and A has twice as many turns as B. What is the ratio of the inductance of solenoid A to that of solenoid B? (a) 4 (b) 2 (c) 1 (d) $\frac{1}{2}$ (e) $\frac{1}{4}$

6. If the current in an inductor is doubled, by what factor is the stored energy multiplied? (a) 4 (b) 2 (c) 1 (d) $\frac{1}{2}$ (e) $\frac{1}{4}$

7. Initially, an inductor with no resistance carries a constant current. Then the current is brought to a new constant value twice as large. *After* this change, when the current is constant at its higher value, what has happened to the emf in the inductor? (a) It is larger than before the change by a factor of 4. (b) It is larger by a factor of 2. (c) It has the same nonzero value. (d) It continues to be zero. (e) It has decreased.

1. Consider this thesis: "Joseph Henry, America's first professional physicist, caused a basic change in the human view of the Universe when he discovered self-induction during a school vacation at the Albany Academy about 1830. Before that time, one could think of the Universe as composed of only one thing: matter. The energy that temporarily maintains the current after a battery is removed from a coil, on the other hand, is not energy that belongs to any chunk of matter. It is energy in the massless magnetic field surrounding the coil. With Henry's discovery, Nature forced us to admit that the Universe consists of fields as well as matter." (a) Argue for or against the statement. (b) In your view, what makes up the Universe?

2. (a) What parameters affect the inductance of a coil? (b) Does the inductance of a coil depend on the current in the coil?

3. A switch controls the current in a circuit that has a large inductance. The electric arc at the switch (Fig.

CQ32.3) can melt and oxidize the contact surfaces, resulting in high resistivity of the contacts and eventual destruction of the switch. Is a spark more likely to be produced at the switch when the switch is being closed, when it is being opened, or does it not matter?

Alexandra Héder

Figure CQ32.3

4. Consider the four circuits shown in Figure CQ32.4, each consisting of a battery, a switch, a lightbulb, a

resistor, and either a capacitor or an inductor. Assume the capacitor has a large capacitance and the inductor has a large inductance but no resistance. The lightbulb has high efficiency, glowing whenever it carries electric current. **(i)** Describe what the lightbulb does in each of circuits (a) through (d) after the switch is thrown closed. **(ii)** Describe what the lightbulb does in each of circuits (a) through (d) when, having been closed for a long time interval, the switch is opened.

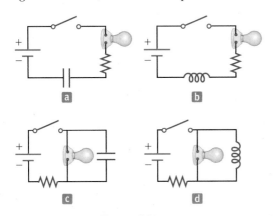

Figure CQ32.4

5. The current in a circuit containing a coil, a resistor, and a battery has reached a constant value. (a) Does the coil have an inductance? (b) Does the coil affect the value of the current?

6. (a) Can an object exert a force on itself? (b) When a coil induces an emf in itself, does it exert a force on itself?

7. The open switch in Figure CQ32.7 is thrown closed at $t = 0$. Before the switch is closed, the capacitor is uncharged and all currents are zero. Determine the currents in L, C, and R, the emf across L, and the potential differences across C and R (a) at the instant after the switch is closed and (b) long after it is closed.

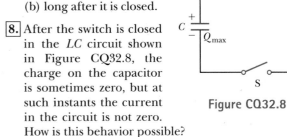

Figure CQ32.7

8. After the switch is closed in the LC circuit shown in Figure CQ32.8, the charge on the capacitor is sometimes zero, but at such instants the current in the circuit is not zero. How is this behavior possible?

Figure CQ32.8

9. How can you tell whether an RLC circuit is overdamped or underdamped?

10. Discuss the similarities between the energy stored in the electric field of a charged capacitor and the energy stored in the magnetic field of a current-carrying coil.

Problems available in ENHANCED **WebAssign**   Access end-of-chapter problems online at **www.webassign.net**

Section 32.1 Self-Induction and Inductance
Problems 1–14

Section 32.2 *RL* Circuits
Problems 15–31

Section 32.3 Energy in a Magnetic Field
Problems 32–39

Section 32.4 Mutual Inductance
Problems 40–46

Section 32.5 Oscillations in an *LC* Circuit
Problems 47–55

Section 32.6 The *RLC* Circuit
Problems 56–59

Additional Problems
Problems 60–77

Challenge Problems
Problem 78–83

Solutions to the following Problems are available in the *Student Solutions Manual/Study Guide:*
32.3, 32.10, 32.13, 32.16, 32.27, 32.31, 32.33, 32.35, 32.38, 32.43, 32.48, 32.54, 32.58, 32.70, 32.71, 32.75, 32.80, and 32.81

List of Enhanced Problems

Problem Number	Targeted Feedback in Enhanced WebAssign	Analysis Model Tutorial in Enhanced WebAssign	Master It in Enhanced WebAssign	Watch It in Enhanced WebAssign
32.4		✓		
32.9	✓			✓
32.10	✓	✓		
32.13		✓		
32.16		✓		
32.23	✓			✓
32.24	✓			
32.31		✓		
32.33	✓	✓		
32.34	✓			✓
32.37	✓	✓		
32.38	✓			✓
32.43		✓		
32.49	✓			
32.53	✓	✓		✓
32.55	✓	✓		✓
32.58		✓		
32.64		✓		

Alternating-Current Circuits

These large transformers are used to increase the voltage at a power plant for distribution of energy by electrical transmission to the power grid. Voltages can be changed relatively easily because power is distributed by alternating current rather than direct current. *(©Lester Lefkowitz/Getty Images)*

In this chapter, we describe alternating-current (AC) circuits. Every time you turn on a television set, a computer, or any of a multitude of other electrical appliances in a home, you are calling on alternating currents to provide the power to operate them. We begin our study by investigating the characteristics of simple series circuits that contain resistors, inductors, and capacitors and that are driven by a sinusoidal voltage. The primary aim of this chapter can be summarized as follows: if an AC source applies an alternating voltage to a series circuit containing resistors, inductors, and capacitors, we want to know the amplitude and time characteristics of the alternating current. We conclude this chapter with two sections concerning transformers, power transmission, and electrical filters.

33.1 AC Sources

An AC circuit consists of circuit elements and a power source that provides an alternating voltage Δv. This time-varying voltage from the source is described by

$$\Delta v = \Delta V_{\max} \sin \omega t$$

where $\Delta V_{\max}$ is the maximum output voltage of the source, or the **voltage amplitude.** There are various possibilities for AC sources, including generators as discussed in Section 31.5 and electrical oscillators. In a home, each electrical outlet serves as an AC source. Because the output voltage of an AC source varies sinusoidally with time, the voltage is positive during one half of the cycle and negative during the other half as in Figure 33.1. Likewise, the current in any circuit driven by an AC source is an alternating current that also varies sinusoidally with time.

From Equation 15.12, the angular frequency of the AC voltage is

$$\omega = 2\pi f = \frac{2\pi}{T}$$

where f is the frequency of the source and T is the period. The source determines the frequency of the current in any circuit connected to it. Commercial electric-power plants in the United States use a frequency of 60.0 Hz, which corresponds to an angular frequency of 377 rad/s.

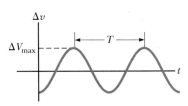

Figure 33.1 The voltage supplied by an AC source is sinusoidal with a period T.

33.2 Resistors in an AC Circuit

Consider a simple AC circuit consisting of a resistor and an AC source as shown in Figure 33.2. At any instant, the algebraic sum of the voltages around a closed loop in a circuit must be zero (Kirchhoff's loop rule). Therefore, $\Delta v + \Delta v_R = 0$ or, using Equation 27.7 for the voltage across the resistor,

$$\Delta v - i_R R = 0$$

If we rearrange this expression and substitute $\Delta V_{max} \sin \omega t$ for Δv, the instantaneous current in the resistor is

$$i_R = \frac{\Delta v}{R} = \frac{\Delta V_{max}}{R} \sin \omega t = I_{max} \sin \omega t \qquad (33.1)$$

where I_{max} is the maximum current:

$$I_{max} = \frac{\Delta V_{max}}{R} \qquad (33.2)$$

Equation 33.1 shows that the instantaneous voltage across the resistor is

$$\Delta v_R = i_R R = I_{max} R \sin \omega t \qquad (33.3)$$

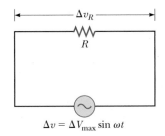

Figure 33.2 A circuit consisting of a resistor of resistance R connected to an AC source, designated by the symbol

◀ **Maximum current in a resistor**

◀ **Voltage across a resistor**

A plot of voltage and current versus time for this circuit is shown in Figure 33.3a on page 778. At point a, the current has a maximum value in one direction, arbitrarily called the positive direction. Between points a and b, the current is decreasing in magnitude but is still in the positive direction. At point b, the current is momentarily zero; it then begins to increase in the negative direction between points b and c. At point c, the current has reached its maximum value in the negative direction.

The current and voltage are in step with each other because they vary identically with time. Because i_R and Δv_R both vary as $\sin \omega t$ and reach their maximum values at the same time as shown in Figure 33.3a, they are said to be **in phase**, similar to the way two waves can be in phase as discussed in our study of wave motion in Chapter 18. Therefore, for a sinusoidal applied voltage, the current in a resistor is always in phase with the voltage across the resistor. For resistors in AC circuits, there are no new concepts to learn. Resistors behave essentially the same way in both DC and AC circuits. That, however, is not the case for capacitors and inductors.

To simplify our analysis of circuits containing two or more elements, we use a graphical representation called a *phasor diagram*. A **phasor** is a vector whose length is proportional to the maximum value of the variable it represents (ΔV_{max} for voltage and I_{max} for current in this discussion). The phasor rotates counterclockwise at an angular speed equal to the angular frequency associated with the variable. The

Pitfall Prevention 33.1
Time-Varying Values We continue to use lowercase symbols Δv and i to indicate the instantaneous values of time-varying voltages and currents. We will add a subscript to indicate the appropriate circuit element. Capital letters represent fixed values of voltage and current such as ΔV_{max} and I_{max}.

Figure 33.3 (a) Plots of the instantaneous current i_R and instantaneous voltage Δv_R across a resistor as functions of time. At time $t = T$, one cycle of the time-varying voltage and current has been completed. (b) Phasor diagram for the resistive circuit showing that the current is in phase with the voltage.

Pitfall Prevention 33.2

A Phasor Is Like a Graph An alternating voltage can be presented in different representations. One graphical representation is shown in Figure 33.1 in which the voltage is drawn in rectangular coordinates, with voltage on the vertical axis and time on the horizontal axis. Figure 33.3b shows another graphical representation. The phase space in which the phasor is drawn is similar to polar coordinate graph paper. The radial coordinate represents the amplitude of the voltage. The angular coordinate is the phase angle. The vertical-axis coordinate of the tip of the phasor represents the instantaneous value of the voltage. The horizontal coordinate represents nothing at all. As shown in Figure 33.3b, alternating currents can also be represented by phasors.

To help with this discussion of phasors, review Section 15.4, where we represented the simple harmonic motion of a real object by the projection of an imaginary object's uniform circular motion onto a coordinate axis. A phasor is a direct analog to this representation.

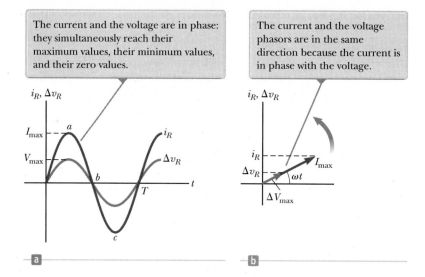

The current and the voltage are in phase: they simultaneously reach their maximum values, their minimum values, and their zero values.

The current and the voltage phasors are in the same direction because the current is in phase with the voltage.

projection of the phasor onto the vertical axis represents the instantaneous value of the quantity it represents.

Figure 33.3b shows voltage and current phasors for the circuit of Figure 33.2 at some instant of time. The projections of the phasor arrows onto the vertical axis are determined by a sine function of the angle of the phasor with respect to the horizontal axis. For example, the projection of the current phasor in Figure 33.3b is $I_{max} \sin \omega t$. Notice that this expression is the same as Equation 33.1. Therefore, the projections of phasors represent current values that vary sinusoidally in time. We can do the same with time-varying voltages. The advantage of this approach is that the phase relationships among currents and voltages can be represented as vector additions of phasors using the vector addition techniques discussed in Chapter 3.

In the case of the single-loop resistive circuit of Figure 33.2, the current and voltage phasors are in the same direction in Figure 33.3b because i_R and Δv_R are in phase. The current and voltage in circuits containing capacitors and inductors have different phase relationships.

Quick Quiz 33.1 Consider the voltage phasor in Figure 33.4, shown at three instants of time. **(i)** Choose the part of the figure, (a), (b), or (c), that represents the instant of time at which the instantaneous value of the voltage has the largest magnitude. **(ii)** Choose the part of the figure that represents the instant of time at which the instantaneous value of the voltage has the smallest magnitude.

Figure 33.4 (Quick Quiz 33.1) A voltage phasor is shown at three instants of time, (a), (b), and (c).

For the simple resistive circuit in Figure 33.2, notice that the average value of the current over one cycle is zero. That is, the current is maintained in the positive direction for the same amount of time and at the same magnitude as it is maintained in the negative direction. The direction of the current, however, has no effect on the behavior of the resistor. We can understand this concept by realizing that collisions between electrons and the fixed atoms of the resistor result in an

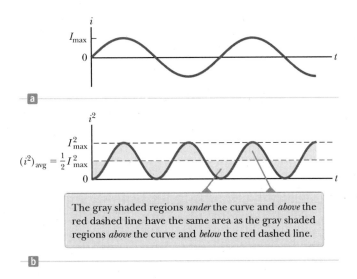

Figure 33.5 (a) Graph of the current in a resistor as a function of time. (b) Graph of the current squared in a resistor as a function of time, showing that the red dashed line is the average of $I_{max}^2 \sin^2 \omega t$. In general, the average value of $\sin^2 \omega t$ or $\cos^2 \omega t$ over one cycle is $\frac{1}{2}$.

The gray shaded regions *under* the curve and *above* the red dashed line have the same area as the gray shaded regions *above* the curve and *below* the red dashed line.

increase in the resistor's temperature. Although this temperature increase depends on the magnitude of the current, it is independent of the current's direction.

We can make this discussion quantitative by recalling that the rate at which energy is delivered to a resistor is the power $P = i^2 R$, where i is the instantaneous current in the resistor. Because this rate is proportional to the square of the current, it makes no difference whether the current is direct or alternating, that is, whether the sign associated with the current is positive or negative. The temperature increase produced by an alternating current having a maximum value I_{max}, however, is not the same as that produced by a direct current equal to I_{max} because the alternating current has this maximum value for only an instant during each cycle (Fig. 33.5a). What is of importance in an AC circuit is an average value of current, referred to as the **rms current.** As we learned in Section 21.1, the notation *rms* stands for *root-mean-square*, which in this case means the square root of the mean (average) value of the square of the current: $I_{rms} = \sqrt{(i^2)_{avg}}$. Because i^2 varies as $\sin^2 \omega t$ and because the average value of i^2 is $\frac{1}{2}I_{max}^2$ (see Fig. 33.5b), the rms current is

$$I_{rms} = \frac{I_{max}}{\sqrt{2}} = 0.707 I_{max} \tag{33.4}$$ ◀ rms current

This equation states that an alternating current whose maximum value is 2.00 A delivers to a resistor the same power as a direct current that has a value of $(0.707)(2.00\ A) = 1.41\ A$. The average power delivered to a resistor that carries an alternating current is

$$P_{avg} = I_{rms}^2 R$$ ◀ **Average power delivered to a resistor**

Alternating voltage is also best discussed in terms of rms voltage, and the relationship is identical to that for current:

$$\Delta V_{rms} = \frac{\Delta V_{max}}{\sqrt{2}} = 0.707\ \Delta V_{max} \tag{33.5}$$ ◀ rms voltage

When we speak of measuring a 120-V alternating voltage from an electrical outlet, we are referring to an rms voltage of 120 V. A calculation using Equation 33.5 shows that such an alternating voltage has a maximum value of about 170 V. One reason rms values are often used when discussing alternating currents and voltages is that AC ammeters and voltmeters are designed to read rms values. Furthermore, with rms values, many of the equations we use have the same form as their direct-current counterparts.

Example 33.1   What Is the rms Current?

The voltage output of an AC source is given by the expression $\Delta v = 200 \sin \omega t$, where Δv is in volts. Find the rms current in the circuit when this source is connected to a 100-Ω resistor.

SOLUTION

Conceptualize Figure 33.2 shows the physical situation for this problem.

Categorize We evaluate the current with an equation developed in this section, so we categorize this example as a substitution problem.

Combine Equations 33.2 and 33.4 to find the rms current:

$$I_{\text{rms}} = \frac{I_{\text{max}}}{\sqrt{2}} = \frac{\Delta V_{\text{max}}}{\sqrt{2}R}$$

Comparing the expression for voltage output with the general form $\Delta v = \Delta V_{\text{max}} \sin \omega t$ shows that $\Delta V_{\text{max}} = 200$ V. Substitute numerical values:

$$I_{\text{rms}} = \frac{200 \text{ V}}{\sqrt{2}\,(100 \text{ }\Omega)} = \boxed{1.41 \text{ A}}$$

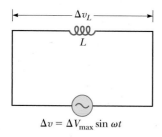

$$\Delta v = \Delta V_{\text{max}} \sin \omega t$$

Figure 33.6 A circuit consisting of an inductor of inductance L connected to an AC source.

33.3 Inductors in an AC Circuit

Now consider an AC circuit consisting only of an inductor connected to the terminals of an AC source as shown in Figure 33.6. Because $\Delta v_L = -L(di_L/dt)$ is the self-induced instantaneous voltage across the inductor (see Eq. 32.1), Kirchhoff's loop rule applied to this circuit gives $\Delta v + \Delta v_L = 0$, or

$$\Delta v - L \frac{di_L}{dt} = 0$$

Substituting $\Delta V_{\text{max}} \sin \omega t$ for Δv and rearranging gives

$$\Delta v = L \frac{di_L}{dt} = \Delta V_{\text{max}} \sin \omega t \tag{33.6}$$

Solving this equation for di_L gives

$$di_L = \frac{\Delta V_{\text{max}}}{L} \sin \omega t \, dt$$

Integrating this expression[1] gives the instantaneous current i_L in the inductor as a function of time:

$$i_L = \frac{\Delta V_{\text{max}}}{L} \int \sin \omega t \, dt = -\frac{\Delta V_{\text{max}}}{\omega L} \cos \omega t \tag{33.7}$$

Using the trigonometric identity $\cos \omega t = -\sin(\omega t - \pi/2)$, we can express Equation 33.7 as

Current in an inductor ▶

$$i_L = \frac{\Delta V_{\text{max}}}{\omega L} \sin\left(\omega t - \frac{\pi}{2}\right) \tag{33.8}$$

Comparing this result with Equation 33.6 shows that the instantaneous current i_L in the inductor and the instantaneous voltage Δv_L across the inductor are *out of* phase by $\pi/2$ rad = 90°.

A plot of voltage and current versus time is shown in Figure 33.7a. When the current i_L in the inductor is a maximum (point b in Fig. 33.7a), it is momentarily

---

[1]We neglect the constant of integration here because it depends on the initial conditions, which are not important for this situation.

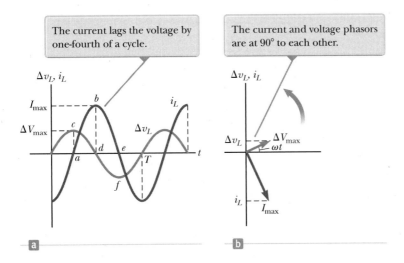

The current lags the voltage by one-fourth of a cycle.

The current and voltage phasors are at 90° to each other.

Figure 33.7 (a) Plots of the instantaneous current i_L and instantaneous voltage Δv_L across an inductor as functions of time. (b) Phasor diagram for the inductive circuit.

not changing, so the voltage across the inductor is zero (point d). At points such as a and e, the current is zero and the rate of change of current is at a maximum. Therefore, the voltage across the inductor is also at a maximum (points c and f). Notice that the voltage reaches its maximum value one-quarter of a period before the current reaches its maximum value. Therefore, for a sinusoidal applied voltage, the current in an inductor always *lags* behind the voltage across the inductor by 90° (one-quarter cycle in time).

As with the relationship between current and voltage for a resistor, we can represent this relationship for an inductor with a phasor diagram as in Figure 33.7b. The phasors are at 90° to each other, representing the 90° phase difference between current and voltage.

Equation 33.7 shows that the current in an inductive circuit reaches its maximum value when $\cos \omega t = \pm 1$:

$$I_{max} = \frac{\Delta V_{max}}{\omega L}$$ (33.9)

◀ **Maximum current in an inductor**

This expression is similar to the relationship between current, voltage, and resistance in a DC circuit, $I = \Delta V/R$ (Eq. 27.7). Because I_{max} has units of amperes and ΔV_{max} has units of volts, ωL must have units of ohms. Therefore, ωL has the same units as resistance and is related to current and voltage in the same way as resistance. It must behave in a manner similar to resistance in the sense that it represents opposition to the flow of charge. Because ωL depends on the applied frequency ω, the inductor *reacts* differently, in terms of offering opposition to current, for different frequencies. For this reason, we define ωL as the **inductive reactance** X_L:

$$X_L \equiv \omega L$$ (33.10)

◀ **Inductive reactance**

Therefore, we can write Equation 33.9 as

$$I_{max} = \frac{\Delta V_{max}}{X_L}$$ (33.11)

The expression for the rms current in an inductor is similar to Equation 33.11, with I_{max} replaced by I_{rms} and ΔV_{max} replaced by ΔV_{rms}.

Equation 33.10 indicates that, for a given applied voltage, the inductive reactance increases as the frequency increases. This conclusion is consistent with Faraday's law: the greater the rate of change of current in the inductor, the larger the back emf. The larger back emf translates to an increase in the reactance and a decrease in the current.

Using Equations 33.6 and 33.11, we find that the instantaneous voltage across the inductor is

Voltage across an inductor ▶

$$\Delta v_L = -L\frac{di_L}{dt} = -\Delta V_{max}\sin \omega t = -I_{max}X_L\sin \omega t \qquad \text{(33.12)}$$

Quick Quiz 33.2 Consider the AC circuit in Figure 33.8. The frequency of the AC source is adjusted while its voltage amplitude is held constant. When does the lightbulb glow the brightest? **(a)** It glows brightest at high frequencies. **(b)** It glows brightest at low frequencies. **(c)** The brightness is the same at all frequencies.

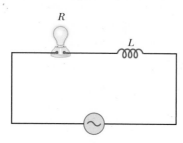

Figure 33.8 (Quick Quiz 33.2) At what frequencies does the lightbulb glow the brightest?

Example 33.2   **A Purely Inductive AC Circuit**

In a purely inductive AC circuit, $L = 25.0$ mH and the rms voltage is 150 V. Calculate the inductive reactance and rms current in the circuit if the frequency is 60.0 Hz.

SOLUTION

Conceptualize Figure 33.6 shows the physical situation for this problem. Keep in mind that inductive reactance increases with increasing frequency of the applied voltage.

Categorize We determine the reactance and the current from equations developed in this section, so we categorize this example as a substitution problem.

Use Equation 33.10 to find the inductive reactance:

$$X_L = \omega L = 2\pi f L = 2\pi(60.0 \text{ Hz})(25.0 \times 10^{-3} \text{ H})$$
$$= \boxed{9.42 \ \Omega}$$

Use an rms version of Equation 33.11 to find the rms current:

$$I_{rms} = \frac{\Delta V_{rms}}{X_L} = \frac{150 \text{ V}}{9.42 \ \Omega} = \boxed{15.9 \text{ A}}$$

WHAT IF?   If the frequency increases to 6.00 kHz, what happens to the rms current in the circuit?

Answer If the frequency increases, the inductive reactance also increases because the current is changing at a higher rate. The increase in inductive reactance results in a lower current.

Let's calculate the new inductive reactance and the new rms current:

$$X_L = 2\pi(6.00 \times 10^3 \text{ Hz})(25.0 \times 10^{-3} \text{ H}) = 942 \ \Omega$$

$$I_{rms} = \frac{150 \text{ V}}{942 \ \Omega} = 0.159 \text{ A}$$

33.4 Capacitors in an AC Circuit

Figure 33.9 shows an AC circuit consisting of a capacitor connected across the terminals of an AC source. Kirchhoff's loop rule applied to this circuit gives $\Delta v + \Delta v_C = 0$, or

$$\Delta v - \frac{q}{C} = 0 \qquad \text{(33.13)}$$

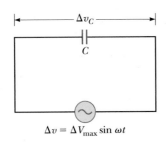

Figure 33.9 A circuit consisting of a capacitor of capacitance C connected to an AC source.

Substituting $\Delta V_{max} \sin \omega t$ for Δv and rearranging gives

$$q = C \Delta V_{max} \sin \omega t \tag{33.14}$$

where q is the instantaneous charge on the capacitor. Differentiating Equation 33.14 with respect to time gives the instantaneous current in the circuit:

$$i_C = \frac{dq}{dt} = \omega C \Delta V_{max} \cos \omega t \tag{33.15}$$

Using the trigonometric identity

$$\cos \omega t = \sin \left(\omega t + \frac{\pi}{2} \right)$$

we can express Equation 33.15 in the alternative form

$$i_C = \omega C \Delta V_{max} \sin \left(\omega t + \frac{\pi}{2} \right) \tag{33.16}$$ ◀ **Current in a capacitor**

Comparing this expression with $\Delta v = \Delta V_{max} \sin \omega t$ shows that the current is $\pi/2$ rad $= 90°$ out of phase with the voltage across the capacitor. A plot of current and voltage versus time (Fig. 33.10a) shows that the current reaches its maximum value one-quarter of a cycle sooner than the voltage reaches its maximum value.

Consider a point such as b in Figure 33.10a where the current is zero at this instant. That occurs when the capacitor reaches its maximum charge so that the voltage across the capacitor is a maximum (point d). At points such as a and e, the current is a maximum, which occurs at those instants when the charge on the capacitor reaches zero and the capacitor begins to recharge with the opposite polarity. When the charge is zero, the voltage across the capacitor is zero (points c and f).

As with inductors, we can represent the current and voltage for a capacitor on a phasor diagram. The phasor diagram in Figure 33.10b shows that for a sinusoidally applied voltage, the current always *leads* the voltage across a capacitor by 90°.

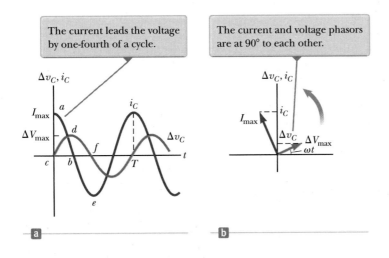

Figure 33.10 (a) Plots of the instantaneous current i_C and instantaneous voltage Δv_C across a capacitor as functions of time. (b) Phasor diagram for the capacitive circuit.

Equation 33.15 shows that the current in the circuit reaches its maximum value when $\cos \omega t = \pm 1$:

$$I_{max} = \omega C \, \Delta V_{max} = \frac{\Delta V_{max}}{(1/\omega C)} \tag{33.17}$$

As in the case with inductors, this looks like Equation 27.7, so the denominator plays the role of resistance, with units of ohms. We give the combination $1/\omega C$ the symbol X_C, and because this function varies with frequency, we define it as the **capacitive reactance:**

Capacitive reactance ▶
$$X_C \equiv \frac{1}{\omega C} \tag{33.18}$$

We can now write Equation 33.17 as

Maximum current ▶
in a capacitor
$$I_{max} = \frac{\Delta V_{max}}{X_C} \tag{33.19}$$

The rms current is given by an expression similar to Equation 33.19, with I_{max} replaced by I_{rms} and ΔV_{max} replaced by ΔV_{rms}.

Using Equation 33.19, we can express the instantaneous voltage across the capacitor as

Voltage across a capacitor ▶
$$\Delta v_C = \Delta V_{max} \sin \omega t = I_{max} X_C \sin \omega t \tag{33.20}$$

Equations 33.18 and 33.19 indicate that as the frequency of the voltage source increases, the capacitive reactance decreases and the maximum current therefore increases. The frequency of the current is determined by the frequency of the voltage source driving the circuit. As the frequency approaches zero, the capacitive reactance approaches infinity and the current therefore approaches zero. This conclusion makes sense because the circuit approaches direct current conditions as ω approaches zero and the capacitor represents an open circuit.

Quick Quiz 33.3 Consider the AC circuit in Figure 33.11. The frequency of the AC source is adjusted while its voltage amplitude is held constant. When does the lightbulb glow the brightest? **(a)** It glows brightest at high frequencies. **(b)** It glows brightest at low frequencies. **(c)** The brightness is the same at all frequencies.

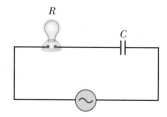

Figure 33.11 (Quick Quiz 33.3)

Quick Quiz 33.4 Consider the AC circuit in Figure 33.12. The frequency of the AC source is adjusted while its voltage amplitude is held constant. When does the lightbulb glow the brightest? **(a)** It glows brightest at high frequencies. **(b)** It glows brightest at low frequencies. **(c)** The brightness is the same at all frequencies.

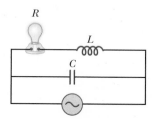

Figure 33.12 (Quick Quiz 33.4)

| Example 33.3 | A Purely Capacitive AC Circuit |

An 8.00-μF capacitor is connected to the terminals of a 60.0-Hz AC source whose rms voltage is 150 V. Find the capacitive reactance and the rms current in the circuit.

SOLUTION

Conceptualize Figure 33.9 shows the physical situation for this problem. Keep in mind that capacitive reactance decreases with increasing frequency of the applied voltage.

Categorize We determine the reactance and the current from equations developed in this section, so we categorize this example as a substitution problem.

Use Equation 33.18 to find the capacitive reactance:

$$X_C = \frac{1}{\omega C} = \frac{1}{2\pi f C} = \frac{1}{2\pi (60.0 \text{ Hz})(8.00 \times 10^{-6} \text{ F})} = 332 \ \Omega$$

Use an rms version of Equation 33.19 to find the rms current:

$$I_{\text{rms}} = \frac{\Delta V_{\text{rms}}}{X_C} = \frac{150 \text{ V}}{332 \ \Omega} = 0.452 \text{ A}$$

WHAT IF? What if the frequency is doubled? What happens to the rms current in the circuit?

Answer If the frequency increases, the capacitive reactance decreases, which is just the opposite from the case of an inductor. The decrease in capacitive reactance results in an increase in the current.

Let's calculate the new capacitive reactance and the new rms current:

$$X_C = \frac{1}{\omega C} = \frac{1}{2\pi (120 \text{ Hz})(8.00 \times 10^{-6} \text{ F})} = 166 \ \Omega$$

$$I_{\text{rms}} = \frac{150 \text{ V}}{166 \ \Omega} = 0.904 \text{ A}$$

33.5 The *RLC* Series Circuit

In the previous sections, we considered individual circuit elements connected to an AC source. Figure 33.13a shows a circuit that contains a combination of circuit elements: a resistor, an inductor, and a capacitor connected in series across an alternating-voltage source. If the applied voltage varies sinusoidally with time, the instantaneous applied voltage is

$$\Delta v = \Delta V_{\text{max}} \sin \omega t$$

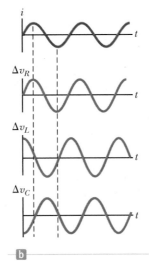

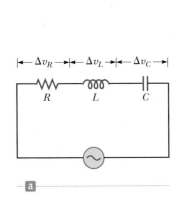

Figure 33.13 (a) A series circuit consisting of a resistor, an inductor, and a capacitor connected to an AC source. (b) Phase relationships between the current and the voltages in the individual circuit elements if they were connected alone to the AC source.

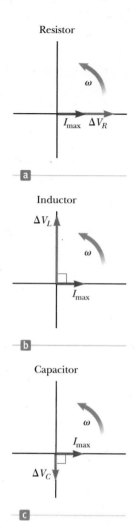

Figure 33.14 Phase relationships between the voltage and current phasors for (a) a resistor, (b) an inductor, and (c) a capacitor connected in series.

Figure 33.13b shows the voltage versus time across each element in the circuit and its phase relationships to the current if it were connected individually to the AC source, as discussed in Sections 33.2–33.4.

When the circuit elements are all connected together to the AC source, as in Figure 33.13a, the current in the circuit is given by

$$i = I_{max} \sin (\omega t - \phi)$$

where ϕ is some **phase angle** between the current and the applied voltage. Based on our discussions of phase in Sections 33.3 and 33.4, we expect that the current will generally not be in phase with the voltage in an *RLC* circuit.

Because the circuit elements in Figure 33.13a are in series, the current everywhere in the circuit must be the same at any instant. That is, the current at all points in a series AC circuit has the same amplitude and phase. Based on the preceding sections, we know that the voltage across each element has a different amplitude and phase. In particular, the voltage across the resistor is in phase with the current, the voltage across the inductor leads the current by 90°, and the voltage across the capacitor lags behind the current by 90°. Using these phase relationships, we can express the instantaneous voltages across the three circuit elements as

$$\Delta v_R = I_{max} R \sin \omega t = \Delta V_R \sin \omega t \qquad \text{(33.21)}$$

$$\Delta v_L = I_{max} X_L \sin \left(\omega t + \frac{\pi}{2} \right) = \Delta V_L \cos \omega t \qquad \text{(33.22)}$$

$$\Delta v_C = I_{max} X_C \sin \left(\omega t - \frac{\pi}{2} \right) = -\Delta V_C \cos \omega t \qquad \text{(33.23)}$$

The sum of these three voltages must equal the instantaneous voltage from the AC source, but it is important to recognize that because the three voltages have different phase relationships with the current, they cannot be added directly. Figure 33.14 represents the phasors at an instant at which the current in all three elements is momentarily zero. The zero current is represented by the current phasor along the horizontal axis in each part of the figure. Next the voltage phasor is drawn at the appropriate phase angle to the current for each element.

Because phasors are rotating vectors, the voltage phasors in Figure 33.14 can be combined using vector addition as in Figure 33.15. In Figure 33.15a, the voltage phasors in Figure 33.14 are combined on the same coordinate axes. Figure 33.15b shows the vector addition of the voltage phasors. The voltage phasors ΔV_L and ΔV_C are in *opposite* directions along the same line, so we can construct the difference phasor $\Delta V_L - \Delta V_C$, which is perpendicular to the phasor ΔV_R. This diagram shows that the vector sum of the voltage amplitudes ΔV_R, ΔV_L, and ΔV_C equals a phasor whose length is the maximum applied voltage ΔV_{max} and which makes an angle ϕ with the current phasor I_{max}. From the right triangle in Figure 33.15b, we see that

Figure 33.15 (a) Phasor diagram for the series *RLC* circuit shown in Figure 33.13a. (b) The inductance and capacitance phasors are added together and then added vectorially to the resistance phasor.

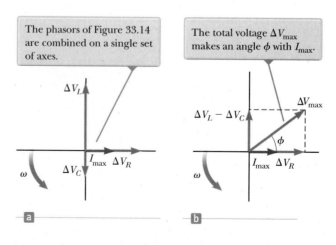

$$\Delta V_{\max} = \sqrt{\Delta V_R^2 + (\Delta V_L - \Delta V_C)^2} = \sqrt{(I_{\max}R)^2 + (I_{\max}X_L - I_{\max}X_C)^2}$$

$$\Delta V_{\max} = I_{\max}\sqrt{R^2 + (X_L - X_C)^2}$$

Therefore, we can express the maximum current as

$$I_{\max} = \frac{\Delta V_{\max}}{\sqrt{R^2 + (X_L - X_C)^2}} \qquad (33.24)$$ ◀ **Maximum current in an *RLC* circuit**

Once again, this expression has the same mathematical form as Equation 27.7. The denominator of the fraction plays the role of resistance and is called the **impedance *Z*** of the circuit:

$$Z \equiv \sqrt{R^2 + (X_L - X_C)^2} \qquad (33.25)$$ ◀ **Impedance**

where impedance also has units of ohms. Therefore, Equation 33.24 can be written in the form

$$I_{\max} = \frac{\Delta V_{\max}}{Z} \qquad (33.26)$$

Equation 33.26 is the AC equivalent of Equation 27.7. Note that the impedance and therefore the current in an AC circuit depend on the resistance, the inductance, the capacitance, and the frequency (because the reactances are frequency dependent).

From the right triangle in the phasor diagram in Figure 33.15b, the phase angle ϕ between the current and the voltage is found as follows:

$$\phi = \tan^{-1}\left(\frac{\Delta V_L - \Delta V_C}{\Delta V_R}\right) = \tan^{-1}\left(\frac{I_{\max}X_L - I_{\max}X_C}{I_{\max}R}\right)$$

$$\phi = \tan^{-1}\left(\frac{X_L - X_C}{R}\right) \qquad (33.27)$$ ◀ **Phase angle**

When $X_L > X_C$ (which occurs at high frequencies), the phase angle is positive, signifying that the current lags the applied voltage as in Figure 33.15b. We describe this situation by saying that the circuit is *more inductive than capacitive*. When $X_L < X_C$, the phase angle is negative, signifying that the current leads the applied voltage, and the circuit is *more capacitive than inductive*. When $X_L = X_C$, the phase angle is zero and the circuit is *purely resistive*.

Quick Quiz 33.5 Label each part of Figure 33.16, (a), (b), and (c), as representing $X_L > X_C$, $X_L = X_C$, or $X_L < X_C$.

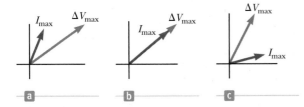

Figure 33.16 (Quick Quiz 33.5) Match the phasor diagrams to the relationships between the reactances.

Example 33.4   **Analyzing a Series *RLC* Circuit**

A series *RLC* circuit has $R = 425\ \Omega$, $L = 1.25\ H$, and $C = 3.50\ \mu F$. It is connected to an AC source with $f = 60.0\ Hz$ and $\Delta V_{\max} = 150\ V$.

(A) Determine the inductive reactance, the capacitive reactance, and the impedance of the circuit.   *continued*

▶ **33.4** continued

SOLUTION

Conceptualize The circuit of interest in this example is shown in Figure 33.13a. The current in the combination of the resistor, inductor, and capacitor oscillates at a particular phase angle with respect to the applied voltage.

Categorize The circuit is a simple series RLC circuit, so we can use the approach discussed in this section.

Analyze Find the angular frequency:

$$\omega = 2\pi f = 2\pi(60.0 \text{ Hz}) = 377 \text{ s}^{-1}$$

Use Equation 33.10 to find the inductive reactance:

$$X_L = \omega L = (377 \text{ s}^{-1})(1.25 \text{ H}) = \boxed{471 \ \Omega}$$

Use Equation 33.18 to find the capacitive reactance:

$$X_C = \frac{1}{\omega C} = \frac{1}{(377 \text{ s}^{-1})(3.50 \times 10^{-6} \text{ F})} = \boxed{758 \ \Omega}$$

Use Equation 33.25 to find the impedance:

$$Z = \sqrt{R^2 + (X_L - X_C)^2}$$

$$= \sqrt{(425 \ \Omega)^2 + (471 \ \Omega - 758 \ \Omega)^2} = \boxed{513 \ \Omega}$$

(B) Find the maximum current in the circuit.

SOLUTION

Use Equation 33.26 to find the maximum current:

$$I_{max} = \frac{\Delta V_{max}}{Z} = \frac{150 \text{ V}}{513 \ \Omega} = \boxed{0.293 \text{ A}}$$

(C) Find the phase angle between the current and voltage.

SOLUTION

Use Equation 33.27 to calculate the phase angle:

$$\phi = \tan^{-1}\left(\frac{X_L - X_C}{R}\right) = \tan^{-1}\left(\frac{471 \ \Omega - 758 \ \Omega}{425 \ \Omega}\right) = \boxed{-34.0°}$$

(D) Find the maximum voltage across each element.

SOLUTION

Use Equations 33.2, 33.11, and 33.19 to calculate the maximum voltages:

$$\Delta V_R = I_{max} R = (0.293 \text{ A})(425 \ \Omega) = \boxed{124 \text{ V}}$$

$$\Delta V_L = I_{max} X_L = (0.293 \text{ A})(471 \ \Omega) = \boxed{138 \text{ V}}$$

$$\Delta V_C = I_{max} X_C = (0.293 \text{ A})(758 \ \Omega) = \boxed{222 \text{ V}}$$

(E) What replacement value of L should an engineer analyzing the circuit choose such that the current leads the applied voltage by 30.0° rather than 34.0°? All other values in the circuit stay the same.

SOLUTION

Solve Equation 33.27 for the inductive reactance:

$$X_L = X_C + R \tan \phi$$

Substitute Equations 33.10 and 33.18 into this expression:

$$\omega L = \frac{1}{\omega C} + R \tan \phi$$

Solve for L:

$$L = \frac{1}{\omega}\left(\frac{1}{\omega C} + R \tan \phi\right)$$

Substitute the given values:

$$L = \frac{1}{(377 \text{ s}^{-1})}\left[\frac{1}{(377 \text{ s}^{-1})(3.50 \times 10^{-6} \text{ F})} + (425 \ \Omega) \tan(-30.0°)\right]$$

$$L = \boxed{1.36 \text{ H}}$$

Finalize Because the capacitive reactance is larger than the inductive reactance, the circuit is more capacitive than inductive. In this case, the phase angle ϕ is negative, so the current leads the applied voltage.

▶ **33.4** continued

Using Equations 33.21, 33.22, and 33.23, the instantaneous voltages across the three elements are

$$\Delta v_R = (124\text{ V}) \sin 377t$$

$$\Delta v_L = (138\text{ V}) \cos 377t$$

$$\Delta v_C = (-222\text{ V}) \cos 377t$$

WHAT IF? What if you added up the maximum voltages across the three circuit elements? Is that a physically meaningful quantity?

Answer The sum of the maximum voltages across the elements is $\Delta V_R + \Delta V_L + \Delta V_C = 484$ V. This sum is much greater than the maximum voltage of the source, 150 V. The sum of the maximum voltages is a meaningless quantity because when sinusoidally varying quantities are added, *both their amplitudes and their phases* must be taken into account. The maximum voltages across the various elements occur at different times. Therefore, the voltages must be added in a way that takes account of the different phases as shown in Figure 33.15.

33.6 Power in an AC Circuit

Now let's take an energy approach to analyzing AC circuits and consider the transfer of energy from the AC source to the circuit. The power delivered by a battery to an external DC circuit is equal to the product of the current and the terminal voltage of the battery. Likewise, the instantaneous power delivered by an AC source to a circuit is the product of the current and the applied voltage. For the *RLC* circuit shown in Figure 33.13a, we can express the instantaneous power *P* as

$$P = i\,\Delta v = I_{max} \sin (\omega t - \phi)\, \Delta V_{max} \sin \omega t$$

$$P = I_{max}\, \Delta V_{max} \sin \omega t \sin (\omega t - \phi) \tag{33.28}$$

This result is a complicated function of time and is therefore not very useful from a practical viewpoint. What is generally of interest is the average power over one or more cycles. Such an average can be computed by first using the trigonometric identity $\sin (\omega t - \phi) = \sin \omega t \cos \phi - \cos \omega t \sin \phi$. Substituting this identity into Equation 33.28 gives

$$P = I_{max}\, \Delta V_{max} \sin^2 \omega t \cos \phi - I_{max}\, \Delta V_{max} \sin \omega t \cos \omega t \sin \phi \tag{33.29}$$

Let's now take the time average of *P* over one or more cycles, noting that I_{max}, ΔV_{max}, ϕ, and ω are all constants. The time average of the first term on the right of the equal sign in Equation 33.29 involves the average value of $\sin^2 \omega t$, which is $\frac{1}{2}$. The time average of the second term on the right of the equal sign is identically zero because $\sin \omega t \cos \omega t = \frac{1}{2} \sin 2\omega t$, and the average value of $\sin 2\omega t$ is zero. Therefore, we can express the **average power** P_{avg} as

$$P_{avg} = \tfrac{1}{2} I_{max}\, \Delta V_{max} \cos \phi \tag{33.30}$$

It is convenient to express the average power in terms of the rms current and rms voltage defined by Equations 33.4 and 33.5:

$$P_{avg} = I_{rms}\, \Delta V_{rms} \cos \phi \tag{33.31}$$

◀ **Average power delivered to an *RLC* circuit**

where the quantity $\cos \phi$ is called the **power factor.** Figure 33.15b shows that the maximum voltage across the resistor is given by $\Delta V_R = \Delta V_{max} \cos \phi = I_{max} R$. Therefore, $\cos \phi = I_{max}R/\Delta V_{max} = R/Z$, and we can express P_{avg} as

$$P_{avg} = I_{rms}\, \Delta V_{rms} \cos \phi = I_{rms}\, \Delta V_{rms}\left(\frac{R}{Z}\right) = I_{rms}\left(\frac{\Delta V_{rms}}{Z}\right)R$$

Recognizing that $\Delta V_{rms}/Z = I_{rms}$ gives

$$P_{avg} = I_{rms}^2 R \qquad (33.32)$$

The average power delivered by the source is converted to internal energy in the resistor, just as in the case of a DC circuit. When the load is purely resistive, $\phi = 0$, $\cos \phi = 1$, and, from Equation 33.31, we see that

$$P_{avg} = I_{rms} \, \Delta V_{rms}$$

Note that no power losses are associated with pure capacitors and pure inductors in an AC circuit. To see why that is true, let's first analyze the power in an AC circuit containing only a source and a capacitor. When the current begins to increase in one direction in an AC circuit, charge begins to accumulate on the capacitor and a voltage appears across it. When this voltage reaches its maximum value, the energy stored in the capacitor as electric potential energy is $\frac{1}{2}C(\Delta V_{max})^2$. This energy storage, however, is only momentary. The capacitor is charged and discharged twice during each cycle: charge is delivered to the capacitor during two quarters of the cycle and is returned to the voltage source during the remaining two quarters. Therefore, the average power supplied by the source is zero. In other words, no power losses occur in a capacitor in an AC circuit.

Now consider the case of an inductor. When the current in an inductor reaches its maximum value, the energy stored in the inductor is a maximum and is given by $\frac{1}{2}LI_{max}^2$. When the current begins to decrease in the circuit, this stored energy in the inductor returns to the source as the inductor attempts to maintain the current in the circuit.

Equation 33.31 shows that the power delivered by an AC source to any circuit depends on the phase, a result that has many interesting applications. For example, a factory that uses large motors in machines, generators, or transformers has a large inductive load (because of all the windings). To deliver greater power to such devices in the factory without using excessively high voltages, technicians introduce capacitance in the circuits to shift the phase.

Ⓠuick Quiz 33.6 An AC source drives an *RLC* circuit with a fixed voltage amplitude. If the driving frequency is ω_1, the circuit is more capacitive than inductive and the phase angle is $-10°$. If the driving frequency is ω_2, the circuit is more inductive than capacitive and the phase angle is $+10°$. At what frequency is the largest amount of power delivered to the circuit? **(a)** It is largest at ω_1. **(b)** It is largest at ω_2. **(c)** The same amount of power is delivered at both frequencies.

Example 33.5   Average Power in an *RLC* Series Circuit

Calculate the average power delivered to the series *RLC* circuit described in Example 33.4.

SOLUTION

Conceptualize Consider the circuit in Figure 33.13a and imagine energy being delivered to the circuit by the AC source. Review Example 33.4 for other details about this circuit.

Categorize We find the result by using equations developed in this section, so we categorize this example as a substitution problem.

Use Equation 33.5 and the maximum voltage from Example 33.4 to find the rms voltage from the source:

$$\Delta V_{rms} = \frac{\Delta V_{max}}{\sqrt{2}} = \frac{150 \text{ V}}{\sqrt{2}} = 106 \text{ V}$$

Similarly, find the rms current in the circuit:

$$I_{rms} = \frac{I_{max}}{\sqrt{2}} = \frac{0.293 \text{ A}}{\sqrt{2}} = 0.207 \text{ A}$$

▶ **33.5** continued

Use Equation 33.31 to find the power delivered by the source:

$$P_{avg} = I_{rms} V_{rms} \cos \phi = (0.207 \text{ A})(106 \text{ V}) \cos(-34.0°)$$

$$= \boxed{18.2 \text{ W}}$$

33.7 Resonance in a Series *RLC* Circuit

We investigated resonance in mechanical oscillating systems in Chapter 15. As shown in Chapter 32, a series *RLC* circuit is an electrical oscillating system. Such a circuit is said to be **in resonance** when the driving frequency is such that the rms current has its maximum value. In general, the rms current can be written

$$I_{rms} = \frac{\Delta V_{rms}}{Z} \tag{33.33}$$

where *Z* is the impedance. Substituting the expression for *Z* from Equation 33.25 into Equation 33.33 gives

$$I_{rms} = \frac{\Delta V_{rms}}{\sqrt{R^2 + (X_L - X_C)^2}} \tag{33.34}$$

Because the impedance depends on the frequency of the source, the current in the *RLC* circuit also depends on the frequency. The angular frequency ω_0 at which $X_L - X_C = 0$ is called the **resonance frequency** of the circuit. To find ω_0, we set $X_L = X_C$, which gives $\omega_0 L = 1/\omega_0 C$, or

$$\omega_0 = \frac{1}{\sqrt{LC}} \tag{33.35}$$

◀ **Resonance frequency**

This frequency also corresponds to the natural frequency of oscillation of an *LC* circuit (see Section 32.5). Therefore, the rms current in a series *RLC* circuit has its maximum value when the frequency of the applied voltage matches the natural oscillator frequency, which depends only on *L* and *C*. Furthermore, at the resonance frequency, the current is in phase with the applied voltage.

Quick Quiz 33.7 What is the impedance of a series *RLC* circuit at resonance? **(a)** larger than *R* **(b)** less than *R* **(c)** equal to *R* **(d)** impossible to determine

A plot of rms current versus angular frequency for a series *RLC* circuit is shown in Figure 33.17a on page 792. The data assume a constant $\Delta V_{rms} = 5.0$ mV, $L = 5.0$ μH, and $C = 2.0$ nF. The three curves correspond to three values of *R*. In each case, the rms current has its maximum value at the resonance frequency ω_0. Furthermore, the curves become narrower and taller as the resistance decreases.

Equation 33.34 shows that when $R = 0$, the current becomes infinite at resonance. Real circuits, however, always have some resistance, which limits the value of the current to some finite value.

We can also calculate the average power as a function of frequency for a series *RLC* circuit. Using Equations 33.32, 33.33, and 33.25 gives

$$P_{avg} = I_{rms}^2 R = \frac{(\Delta V_{rms})^2}{Z^2} R = \frac{(\Delta V_{rms})^2 R}{R^2 + (X_L - X_C)^2} \tag{33.36}$$

Because $X_L = \omega L$, $X_C = 1/\omega C$, and $\omega_0^2 = 1/LC$, the term $(X_L - X_C)^2$ can be expressed as

$$(X_L - X_C)^2 = \left(\omega L - \frac{1}{\omega C}\right)^2 = \frac{L^2}{\omega^2}(\omega^2 - \omega_0^2)^2$$

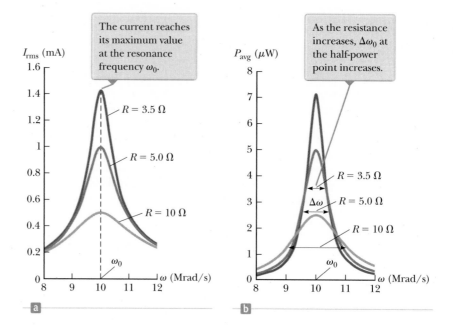

Using this result in Equation 33.36 gives

$$P_{avg} = \frac{(\Delta V_{rms})^2 \, R\omega^2}{R^2\,\omega^2 + L^2(\omega^2 - \omega_0^2)^2} \tag{33.37}$$

◀ **Average power as a function of frequency in an *RLC* circuit**

Equation 33.37 shows that at resonance, when $\omega = \omega_0$, the average power is a maximum and has the value $(\Delta V_{rms})^2/R$. Figure 33.17b is a plot of average power versus frequency for three values of R in a series *RLC* circuit. As the resistance is made smaller, the curve becomes sharper in the vicinity of the resonance frequency. This curve sharpness is usually described by a dimensionless parameter known as the **quality factor,**[2] denoted by Q:

◀ **Quality factor**

$$Q = \frac{\omega_0}{\Delta\omega}$$

where $\Delta\omega$ is the width of the curve measured between the two values of ω for which P_{avg} has one-half its maximum value, called the *half-power points* (see Fig. 33.17b.) It is left as a problem (Problem 76 in Enhanced WebAssign) to show that the width at the half-power points has the value $\Delta\omega = R/L$ so that

$$Q = \frac{\omega_0 L}{R} \tag{33.38}$$

A radio's receiving circuit is an important application of a resonant circuit. The radio is tuned to a particular station (which transmits an electromagnetic wave or signal of a specific frequency) by varying a capacitor, which changes the receiving circuit's resonance frequency. When the circuit is driven by the electromagnetic oscillations a radio signal produces in an antenna, the tuner circuit responds with a large amplitude of electrical oscillation only for the station frequency that matches the resonance frequency. Therefore, only the signal from one radio station is passed on to the amplifier and loudspeakers even though signals from all stations are driving the circuit at the same time. Because many signals are often present over a range of frequencies, it is important to design a high-Q circuit to eliminate unwanted signals. In this manner, stations whose frequencies are near but not equal to the resonance frequency have a response at the receiver that is negligibly small relative to the signal that matches the resonance frequency.

---

[2]The quality factor is also defined as the ratio $2\pi E/\Delta E$, where E is the energy stored in the oscillating system and ΔE is the energy decrease per cycle of oscillation due to the resistance.

| Example 33.6 | A Resonating Series *RLC* Circuit |

Consider a series *RLC* circuit for which $R = 150\ \Omega$, $L = 20.0$ mH, $\Delta V_{rms} = 20.0$ V, and $\omega = 5\ 000$ s^{-1}. Determine the value of the capacitance for which the current is a maximum.

SOLUTION

Conceptualize Consider the circuit in Figure 33.13a and imagine varying the frequency of the AC source. The current in the circuit has its maximum value at the resonance frequency ω_0.

Categorize We find the result by using equations developed in this section, so we categorize this example as a substitution problem.

Use Equation 33.35 to solve for the required capacitance in terms of the resonance frequency:

$$\omega_0 = \frac{1}{\sqrt{LC}} \rightarrow C = \frac{1}{\omega_0^2 L}$$

Substitute numerical values:

$$C = \frac{1}{(5.00 \times 10^3\ \text{s}^{-1})^2 (20.0 \times 10^{-3}\ \text{H})} = \boxed{2.00\ \mu\text{F}}$$

33.8 The Transformer and Power Transmission

As discussed in Section 27.6, it is economical to use a high voltage and a low current to minimize the I^2R loss in transmission lines when electric power is transmitted over great distances. Consequently, 350-kV lines are common, and in many areas, even higher-voltage (765-kV) lines are used. At the receiving end of such lines, the consumer requires power at a low voltage (for safety and for efficiency in design). In practice, the voltage is decreased to approximately 20 000 V at a distribution substation, then to 4 000 V for delivery to residential areas, and finally to 120 V and 240 V at the customer's site. Therefore, a device is needed that can change the alternating voltage and current without causing appreciable changes in the power delivered. The AC transformer is that device.

In its simplest form, the **AC transformer** consists of two coils of wire wound around a core of iron as illustrated in Figure 33.18. (Compare this arrangement to Faraday's experiment in Figure 31.2.) The coil on the left, which is connected to the input alternating-voltage source and has N_1 turns, is called the *primary winding* (or the *primary*). The coil on the right, consisting of N_2 turns and connected to a load resistor R_L, is called the *secondary winding* (or the *secondary*). The purposes of the iron core are to increase the magnetic flux through the coil and to provide a medium in which nearly all the magnetic field lines through one coil pass through the other coil. Eddy-current losses are reduced by using a laminated core. Transformation of energy to internal energy in the finite resistance of the coil wires is usually quite small. Typical transformers have power efficiencies from 90% to

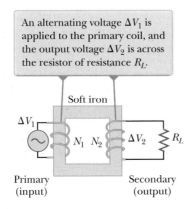

An alternating voltage ΔV_1 is applied to the primary coil, and the output voltage ΔV_2 is across the resistor of resistance R_L.

ΔV_1 Soft iron N_1 N_2 ΔV_2 R_L

Primary (input)   Secondary (output)

Figure 33.18 An ideal transformer consists of two coils wound on the same iron core.

99%. In the discussion that follows, let's assume we are working with an *ideal transformer,* one in which the energy losses in the windings and core are zero.

Faraday's law states that the voltage Δv_1 across the primary is

$$\Delta v_1 = -N_1 \frac{d\Phi_B}{dt} \tag{33.39}$$

where Φ_B is the magnetic flux through each turn. If we assume all magnetic field lines remain within the iron core, the flux through each turn of the primary equals the flux through each turn of the secondary. Hence, the voltage across the secondary is

$$\Delta v_2 = -N_2 \frac{d\Phi_B}{dt} \tag{33.40}$$

Solving Equation 33.39 for $d\Phi_B/dt$ and substituting the result into Equation 33.40 gives

$$\Delta v_2 = \frac{N_2}{N_1} \Delta v_1 \tag{33.41}$$

When $N_2 > N_1$, the output voltage Δv_2 exceeds the input voltage Δv_1. This configuration is referred to as a *step-up transformer.* When $N_2 < N_1$, the output voltage is less than the input voltage, and we have a *step-down transformer.* A circuit diagram for a transformer connected to a load resistance is shown in Figure 33.19.

When a current I_1 exists in the primary circuit, a current I_2 is induced in the secondary. (In this discussion, uppercase I and ΔV refer to rms values.) If the load in the secondary circuit is a pure resistance, the induced current is in phase with the induced voltage. The power supplied to the secondary circuit must be provided by the AC source connected to the primary circuit. In an ideal transformer where there are no losses, the power $I_1 \Delta V_1$ supplied by the source is equal to the power $I_2 \Delta V_2$ in the secondary circuit. That is,

$$I_1 \Delta V_1 = I_2 \Delta V_2 \tag{33.42}$$

The value of the load resistance R_L determines the value of the secondary current because $I_2 = \Delta V_2/R_L$. Furthermore, the current in the primary is $I_1 = \Delta V_1/R_{eq}$, where

$$R_{eq} = \left(\frac{N_1}{N_2}\right)^2 R_L \tag{33.43}$$

is the equivalent resistance of the load resistance when viewed from the primary side. We see from this analysis that a transformer may be used to match resistances between the primary circuit and the load. In this manner, maximum power transfer can be achieved between a given power source and the load resistance. For example, a transformer connected between the 1-kΩ output of an audio amplifier and an 8-Ω speaker ensures that as much of the audio signal as possible is transferred into the speaker. In stereo terminology, this process is called *impedance matching.*

To operate properly, many common household electronic devices require low voltages. A small transformer that plugs directly into the wall like the one illustrated in Figure 33.20 can provide the proper voltage. The photograph shows the two windings wrapped around a common iron core that is found inside all these

Nikola Tesla
American Physicist (1856–1943)
Tesla was born in Croatia, but he spent most of his professional life as an inventor in the United States. He was a key figure in the development of alternating-current electricity, high-voltage transformers, and the transport of electrical power using AC transmission lines. Tesla's viewpoint was at odds with the ideas of Thomas Edison, who committed himself to the use of direct current in power transmission. Tesla's AC approach won out.

Figure 33.19　Circuit diagram for a transformer.

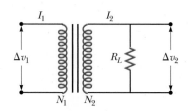

This transformer is smaller than the one in the opening photograph of this chapter. In addition, it is a step-down transformer. It drops the voltage from 4 000 V to 240 V for delivery to a group of residences.

The primary winding in this transformer is attached to the prongs of the plug, whereas the secondary winding is connected to the power cord on the right.

Figure 33.20 Electronic devices are often powered by AC adaptors containing transformers such as this one. These adaptors alter the AC voltage. In many applications, the adaptors also convert alternating current to direct current.

little "black boxes." This particular transformer converts the 120-V AC in the wall socket to 12.5-V AC. (Can you determine the ratio of the numbers of turns in the two coils?) Some black boxes also make use of diodes to convert the alternating current to direct current. (See Section 33.9.)

Example 33.7 The Economics of AC Power

An electricity-generating station needs to deliver energy at a rate of 20 MW to a city 1.0 km away. A common voltage for commercial power generators is 22 kV, but a step-up transformer is used to boost the voltage to 230 kV before transmission.

(A) If the resistance of the wires is 2.0 Ω and the energy costs are about 11¢/kWh, estimate the cost of the energy converted to internal energy in the wires during one day.

SOLUTION

Conceptualize The resistance of the wires is in series with the resistance representing the load (homes and businesses). Therefore, there is a voltage drop in the wires, which means that some of the transmitted energy is converted to internal energy in the wires and never reaches the load.

Categorize This problem involves finding the power delivered to a resistive load in an AC circuit. Let's ignore any capacitive or inductive characteristics of the load and set the power factor equal to 1.

Analyze Calculate I_{rms} in the wires from Equation 33.31:

$$I_{rms} = \frac{P_{avg}}{\Delta V_{rms}} = \frac{20 \times 10^6 \text{ W}}{230 \times 10^3 \text{ V}} = 87 \text{ A}$$

Determine the rate at which energy is delivered to the resistance in the wires from Equation 33.32:

$$P_{wires} = I_{rms}^2 R = (87 \text{ A})^2 (2.0 \text{ }\Omega) = 15 \text{ kW}$$

Calculate the energy T_{ET} delivered to the wires over the course of a day:

$$T_{ET} = P_{wires} \Delta t = (15 \text{ kW})(24 \text{ h}) = \boxed{363 \text{ kWh}}$$

Find the cost of this energy at a rate of 11¢/kWh:

$$\text{Cost} = (363 \text{ kWh})(\$0.11/\text{kWh}) = \boxed{\$40}$$

(B) Repeat the calculation for the situation in which the power plant delivers the energy at its original voltage of 22 kV.

continued

▶ **33.7** continued

SOLUTION

Calculate I_{rms} in the wires from Equation 33.31:	$I_{rms} = \dfrac{P_{avg}}{\Delta V_{rms}} = \dfrac{20 \times 10^6 \text{ W}}{22 \times 10^3 \text{ V}} = 909 \text{ A}$
From Equation 33.32, determine the rate at which energy is delivered to the resistance in the wires:	$P_{wires} = I_{rms}^2 R = (909 \text{ A})^2 (2.0 \text{ }\Omega) = 1.7 \times 10^3 \text{ kW}$
Calculate the energy delivered to the wires over the course of a day:	$T_{ET} = P_{wires} \Delta t = (1.7 \times 10^3 \text{ kW})(24 \text{ h}) = 4.0 \times 10^4 \text{ kWh}$
Find the cost of this energy at a rate of 11¢/kWh:	Cost $= (4.0 \times 10^4 \text{ kWh})(\$0.11/\text{kWh}) = $ **\$4.4 $\times 10^3$**

Finalize Notice the tremendous savings that are possible through the use of transformers and high-voltage transmission lines. Such savings in combination with the efficiency of using alternating current to operate motors led to the universal adoption of alternating current instead of direct current for commercial power grids.

33.9 Rectifiers and Filters

Portable electronic devices such as radios and laptop computers are often powered by direct current supplied by batteries. Many devices come with AC–DC converters such as that shown in Figure 33.20. Such a converter contains a transformer that steps the voltage down from 120 V to, typically, 6 V or 9 V and a circuit that converts alternating current to direct current. The AC–DC converting process is called **rectification,** and the converting device is called a **rectifier.**

The most important element in a rectifier circuit is a **diode,** a circuit element that conducts current in one direction but not the other. Most diodes used in modern electronics are semiconductor devices. The circuit symbol for a diode is ──▶│──, where the arrow indicates the direction of the current in the diode. A diode has low resistance to current in one direction (the direction of the arrow) and high resistance to current in the opposite direction. To understand how a diode rectifies a current, consider Figure 33.21a, which shows a diode and a resistor connected to the secondary of a transformer. The transformer reduces the voltage from 120-V AC to the lower voltage that is needed for the device having a resistance R (the load

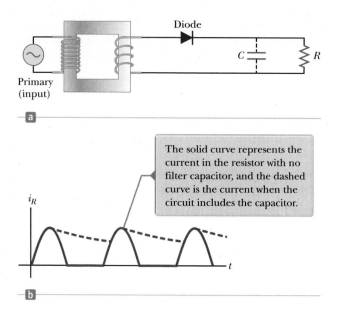

Figure 33.21 (a) A half-wave rectifier with an optional filter capacitor. (b) Current versus time in the resistor.

The solid curve represents the current in the resistor with no filter capacitor, and the dashed curve is the current when the circuit includes the capacitor.

resistance). Because the diode conducts current in only one direction, the alternating current in the load resistor is reduced to the form shown by the solid curve in Figure 33.21b. The diode conducts current only when the side of the symbol containing the arrowhead has a positive potential relative to the other side. In this situation, the diode acts as a *half-wave rectifier* because current is present in the circuit only during half of each cycle.

When a capacitor is added to the circuit as shown by the dashed lines and the capacitor symbol in Figure 33.21a, the circuit is a simple DC power supply. The time variation of the current in the load resistor (the dashed curve in Fig. 33.21b) is close to being zero, as determined by the *RC* time constant of the circuit. As the current in the circuit begins to rise at $t = 0$ in Figure 33.21b, the capacitor charges up. When the current begins to fall, however, the capacitor discharges through the resistor, so the current in the resistor does not fall as quickly as the current from the transformer.

The *RC* circuit in Figure 33.21a is one example of a **filter circuit,** which is used to smooth out or eliminate a time-varying signal. For example, radios are usually powered by a 60-Hz alternating voltage. After rectification, the voltage still contains a small AC component at 60 Hz (sometimes called *ripple*), which must be filtered. By "filtered," we mean that the 60-Hz ripple must be reduced to a value much less than that of the audio signal to be amplified because without filtering, the resulting audio signal includes an annoying hum at 60 Hz.

We can also design filters that respond differently to different frequencies. Consider the simple series *RC* circuit shown in Figure 33.22a. The input voltage is across the series combination of the two elements. The output is the voltage across the resistor. A plot of the ratio of the output voltage to the input voltage as a function of the logarithm of angular frequency (see Fig. 33.22b) shows that at low frequencies, ΔV_{out} is much smaller than ΔV_{in}, whereas at high frequencies, the two voltages are equal. Because the circuit preferentially passes signals of higher frequency while blocking low-frequency signals, the circuit is called an ***RC* high-pass filter.** (See Problem 54 in Enhanced WebAssign for an analysis of this filter.)

Physically, a high-pass filter works because a capacitor "blocks out" direct current and AC current at low frequencies. At low frequencies, the capacitive reactance is large and much of the applied voltage appears across the capacitor rather than across the output resistor. As the frequency increases, the capacitive reactance drops and more of the applied voltage appears across the resistor.

Now consider the circuit shown in Figure 33.23a on page 798, where we have interchanged the resistor and capacitor and where the output voltage is taken across the capacitor. At low frequencies, the reactance of the capacitor and the voltage across the capacitor is high. As the frequency increases, the voltage across the capacitor drops. Therefore, this filter is an ***RC* low-pass filter.** The ratio of output voltage to input voltage (see Problem 56 in Enhanced WebAssign), plotted as a function of the logarithm of ω in Figure 33.23b, shows this behavior.

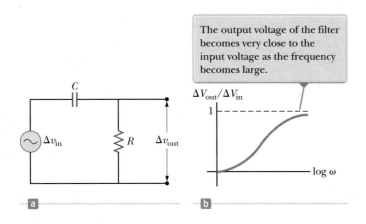

The output voltage of the filter becomes very close to the input voltage as the frequency becomes large.

Figure 33.22 (a) A simple *RC* high-pass filter. (b) Ratio of output voltage to input voltage for an *RC* high-pass filter as a function of the angular frequency of the AC source.

Figure 33.23 (a) A simple *RC* low-pass filter. (b) Ratio of output voltage to input voltage for an *RC* low-pass filter as a function of the angular frequency of the AC source.

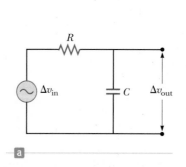

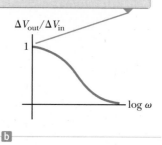

The output voltage of the filter becomes very close to the input voltage as the frequency becomes small.

You may be familiar with crossover networks, which are an important part of the speaker systems for high-quality audio systems. These networks use low-pass filters to direct low frequencies to a special type of speaker, the "woofer," which is designed to reproduce the low notes accurately. The high frequencies are sent by a high-pass filter to the "tweeter" speaker.

Summary

Definitions

In AC circuits that contain inductors and capacitors, it is useful to define the **inductive reactance** X_L and the **capacitive reactance** X_C as

$$X_L \equiv \omega L \qquad (33.10)$$

$$X_C \equiv \frac{1}{\omega C} \qquad (33.18)$$

where ω is the angular frequency of the AC source. The SI unit of reactance is the ohm.

The **impedance** Z of an *RLC* series AC circuit is

$$Z \equiv \sqrt{R^2 + (X_L - X_C)^2} \qquad (33.25)$$

This expression illustrates that we cannot simply add the resistance and reactances in a circuit. We must account for the applied voltage and current being out of phase, with the **phase angle** ϕ between the current and voltage being

$$\phi = \tan^{-1}\left(\frac{X_L - X_C}{R}\right) \qquad (33.27)$$

The sign of ϕ can be positive or negative, depending on whether X_L is greater or less than X_C. The phase angle is zero when $X_L = X_C$.

Concepts and Principles

The **rms current** and **rms voltage** in an AC circuit in which the voltages and current vary sinusoidally are given by

$$I_{rms} = \frac{I_{max}}{\sqrt{2}} = 0.707 I_{max} \qquad (33.4)$$

$$\Delta V_{rms} = \frac{\Delta V_{max}}{\sqrt{2}} = 0.707 \Delta V_{max} \qquad (33.5)$$

where I_{max} and ΔV_{max} are the maximum values.

If an AC circuit consists of a source and a resistor, the current is in phase with the voltage. That is, the current and voltage reach their maximum values at the same time.

If an AC circuit consists of a source and an inductor, the current lags the voltage by 90°. That is, the voltage reaches its maximum value one-quarter of a period before the current reaches its maximum value.

If an AC circuit consists of a source and a capacitor, the current leads the voltage by 90°. That is, the current reaches its maximum value one-quarter of a period before the voltage reaches its maximum value.

■ The **average power** delivered by the source in an *RLC* circuit is

$$P_{avg} = I_{rms} \Delta V_{rms} \cos \phi \tag{33.31}$$

An equivalent expression for the average power is

$$P_{avg} = I_{rms}^2 R \tag{33.32}$$

The average power delivered by the source results in increasing internal energy in the resistor. No power loss occurs in an ideal inductor or capacitor.

■ A series *RLC* circuit is in resonance when the inductive reactance equals the capacitive reactance. When this condition is met, the rms current given by Equation 33.34 has its maximum value. The **resonance frequency** ω_0 of the circuit is

$$\omega_0 = \frac{1}{\sqrt{LC}} \tag{33.35}$$

The rms current in a series *RLC* circuit has its maximum value when the frequency of the source equals ω_0, that is, when the "driving" frequency matches the resonance frequency.

■ The rms current in a series *RLC* circuit is

$$I_{rms} = \frac{\Delta V_{rms}}{\sqrt{R^2 + (X_L - X_C)^2}} \tag{33.34}$$

■ **AC transformers** allow for easy changes in alternating voltage according to

$$\Delta v_2 = \frac{N_2}{N_1} \Delta v_1 \tag{33.41}$$

where N_1 and N_2 are the numbers of windings on the primary and secondary coils, respectively, and Δv_1 and Δv_2 are the voltages on these coils.

Objective Questions   **1.** denotes answer available in *Student Solutions Manual/Study Guide*

1. An inductor and a resistor are connected in series across an AC source as in Figure OQ33.1. Immediately after the switch is closed, which of the following statements is true? (a) The current in the circuit is $\Delta V/R$. (b) The voltage across the inductor is zero. (c) The current in the circuit is zero. (d) The voltage across the resistor is ΔV. (e) The voltage across the inductor is half its maximum value.

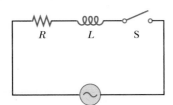

Figure OQ33.1

2. **(i)** When a particular inductor is connected to a source of sinusoidally varying emf with constant amplitude and a frequency of 60.0 Hz, the rms current is 3.00 A. What is the rms current if the source frequency is doubled? (a) 12.0 A (b) 6.00 A (c) 4.24 A (d) 3.00 A (e) 1.50 A **(ii)** Repeat part (i) assuming the load is a capacitor instead of an inductor. **(iii)** Repeat part (i) assuming the load is a resistor instead of an inductor.

3. A capacitor and a resistor are connected in series across an AC source as shown in Figure OQ33.3. After the switch is closed, which of the following statements is true? (a) The voltage across the capacitor lags the current by 90°. (b) The voltage across the resistor is out of phase with the current. (c) The voltage across the capacitor leads the current by 90°. (d) The current decreases as the frequency of the source is increased,

but its peak voltage remains the same. (e) None of those statements is correct.

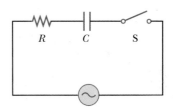

Figure OQ33.3

4. **(i)** What is the time average of the "square-wave" potential shown in Figure OQ33.4? (a) $\sqrt{2}\,\Delta V_{max}$ (b) ΔV_{max} (c) $\Delta V_{max}/\sqrt{2}$ (d) $\Delta V_{max}/2$ (e) $\Delta V_{max}/4$ **(ii)** What is the rms voltage? Choose from the same possibilities as in part (i).

Figure OQ33.4

5. If the voltage across a circuit element has its maximum value when the current in the circuit is zero, which of the following statements *must* be true? (a) The circuit element is a resistor. (b) The circuit element is a capacitor. (c) The circuit element is an inductor. (d) The current and voltage are 90° out of phase. (e) The current and voltage are 180° out of phase.

6. A sinusoidally varying potential difference has amplitude 170 V. **(i)** What is its minimum instantaneous

value? (a) 170 V (b) 120 V (c) 0 (d) −120 V (e) −170 V **(ii)** What is its average value? **(iii)** What is its rms value? Choose from the same possibilities as in part (i) in each case.

7. A series *RLC* circuit contains a 20.0-Ω resistor, a 0.750-μF capacitor, and a 120-mH inductor. **(i)** If a sinusoidally varying rms voltage of 120 V at f = 500 Hz is applied across this combination of elements, what is the rms current in the circuit? (a) 2.33 A (b) 6.00 A (c) 10.0 A (d) 17.0 A (e) none of those answers **(ii) What If?** What is the rms current in the circuit when operating at its resonance frequency? Choose from the same possibilities as in part (i).

8. A resistor, a capacitor, and an inductor are connected in series across an AC source. Which of the following statements is *false*? (a) The instantaneous voltage across the capacitor lags the current by 90°. (b) The instantaneous voltage across the inductor leads the current by 90°. (c) The instantaneous voltage across the resistor is in phase with the current. (d) The voltages across the resistor, capacitor, and inductor are not in phase. (e) The rms voltage across the combination of the three elements equals the algebraic sum of the rms voltages across each element separately.

9. Under what conditions is the impedance of a series *RLC* circuit equal to the resistance in the circuit? (a) The driving frequency is lower than the resonance frequency. (b) The driving frequency is equal to the resonance frequency. (c) The driving frequency is higher than the resonance frequency. (d) always (e) never

10. What is the phase angle in a series *RLC* circuit at resonance? (a) 180° (b) 90° (c) 0 (d) −90° (e) None of those answers is necessarily correct.

11. A circuit containing an AC source, a capacitor, an inductor, and a resistor has a high-Q resonance at 1 000 Hz. From greatest to least, rank the following contributions to the impedance of the circuit at that frequency and at lower and higher frequencies. Note any cases of equality in your ranking. (a) X_C at 500 Hz (b) X_C at 1 500 Hz (c) X_L at 500 Hz (d) X_L at 1 500 Hz (e) R at 1 000 Hz

12. A 6.00-V battery is connected across the primary coil of a transformer having 50 turns. If the secondary coil of the transformer has 100 turns, what voltage appears across the secondary? (a) 24.0 V (b) 12.0 V (c) 6.00 V (d) 3.00 V (e) none of those answers

13. Do AC ammeters and voltmeters read (a) peak-to-valley, (b) maximum, (c) rms, or (d) average values?

Conceptual Questions   **1.** denotes answer available in *Student Solutions Manual/Study Guide*

1. (a) Explain how the quality factor is related to the response characteristics of a radio receiver. (b) Which variable most strongly influences the quality factor?

2. (a) Explain how the mnemonic "ELI the ICE man" can be used to recall whether current leads voltage or voltage leads current in *RLC* circuits. Note that E represents emf $\mathcal{E}$. (b) Explain how "CIVIL" works as another mnemonic device, where V represents voltage.

3. Why is the sum of the maximum voltages across each element in a series *RLC* circuit usually greater than the maximum applied voltage? Doesn't that inequality violate Kirchhoff's loop rule?

4. (a) Does the phase angle in an *RLC* series circuit depend on frequency? (b) What is the phase angle for the circuit when the inductive reactance equals the capacitive reactance?

5. Do some research to answer these questions: Who invented the metal detector? Why? What are its limitations?

6. As shown in Figure CQ33.6, a person pulls a vacuum cleaner at speed v across a horizontal floor, exerting

on it a force of magnitude F directed upward at an angle θ with the horizontal. (a) At what rate is the person doing work on the cleaner? (b) State as completely as you can the analogy between power in this situation and in an electric circuit.

7. A certain power supply can be modeled as a source of emf in series with both a resistance of 10 Ω and an inductive reactance of 5 Ω. To obtain maximum power delivered to the load, it is found that the load should have a resistance of R_L = 10 Ω, an inductive reactance of zero, and a capacitive reactance of 5 Ω. (a) With this load, is the circuit in resonance? (b) With this load, what fraction of the average power put out by the source of emf is delivered to the load? (c) To increase the fraction of the power delivered to the load, how could the load be changed? You may wish to review Example 28.2 and Problem 4 in Chapter 28 in Enhanced WebAssign on maximum power transfer in DC circuits.

8. Will a transformer operate if a battery is used for the input voltage across the primary? Explain.

9. (a) Why does a capacitor act as a short circuit at high frequencies? (b) Why does a capacitor act as an open circuit at low frequencies?

10. An ice storm breaks a transmission line and interrupts electric power to a town. A homeowner starts a gasoline-powered 120-V generator and clips its output terminals to "hot" and "ground" terminals of the electrical panel for his house. On a power pole down the block is a transformer designed to step down the voltage for household use. It has a ratio of turns N_1/N_2 of 100 to 1. A repairman climbs the pole. What voltage

Figure CQ33.6

will he encounter on the input side of the transformer? As this question implies, safety precautions must be taken in the use of home generators and during power failures in general.

Problems available in  **WebAssign** Access end-of-chapter problems online at www.webassign.net

Section 33.1 AC Sources
Section 33.2 Resistors in an AC Circuit
Problems 1–8

Section 33.3 Inductors in an AC Circuit
Problems 9–16

Section 33.4 Capacitors in an AC Circuit
Problems 17–23

Section 33.5 The *RLC* Series Circuit
Problems 24–33

Section 33.6 Power in an AC Circuit
Problems 34–41

Section 33.7 Resonance in a Series *RLC* Circuit
Problems 42–47

Section 33.8 The Transformer and Power Transmission
Problems 48–52

Section 33.9 Rectifiers and Filters
Problems 53–56

Additional Problems
Problems 57–76

Challenge Problems
Problem 77–81

Solutions to the following Problems are available in the *Student Solutions Manual/Study Guide:*
33.5, 33.10, 33.11, 33.21, 33.29, 33.31, 33.36, 33.39, 33.43, 33.49, 33.53, 33.58, 33.60, 33.65, 33.75, and 33.76

List of Enhanced Problems

Problem Number	Targeted Feedback in Enhanced WebAssign	Analysis Model Tutorial in Enhanced WebAssign	Master It in Enhanced WebAssign	Watch It in Enhanced WebAssign
33.10			✓	
33.11	✓			✓
33.14	✓			✓
33.21			✓	
33.23	✓			✓
33.24	✓			✓
33.26	✓			✓
33.27	✓			✓
33.31			✓	
33.35	✓			✓
33.36	✓			✓
33.39	✓			✓
33.40	✓			✓
33.42	✓			✓
33.43	✓	✓	✓	
33.48	✓			✓
33.49			✓	
33.50		✓		
33.53			✓	
33.57	✓			✓
33.60			✓	
33.67	✓			✓
33.69		✓		
33.75			✓	

Electromagnetic Waves

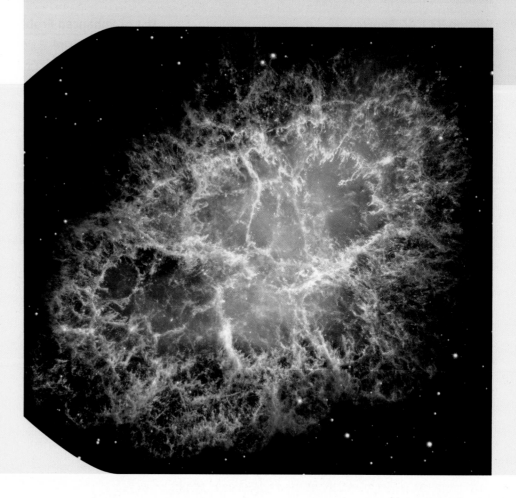

This image of the Crab Nebula taken with visible light shows a variety of colors, with each color representing a different wavelength of visible light. *(NASA, ESA, J. Hester, A. Loll (ASU))*

The waves described in Chapters 16, 17, and 18 are mechanical waves. By definition, the propagation of mechanical disturbances—such as sound waves, water waves, and waves on a string—requires the presence of a medium. This chapter is concerned with the properties of electromagnetic waves, which (unlike mechanical waves) can propagate through empty space.

We begin by considering Maxwell's contributions in modifying Ampère's law, which we studied in Chapter 30. We then discuss Maxwell's equations, which form the theoretical basis of all electromagnetic phenomena. These equations predict the existence of electromagnetic waves that propagate through space at the speed of light c according to the traveling wave analysis model. Heinrich Hertz confirmed Maxwell's prediction when he generated and detected electromagnetic waves in 1887. That discovery has led to many practical communication systems, including radio, television, cell phone systems, wireless Internet connectivity, and optoelectronics.

Next, we learn how electromagnetic waves are generated by oscillating electric charges. The waves radiated from the oscillating charges can be detected at great distances. Furthermore, because electromagnetic waves carry energy (T_{ER} in Eq. 8.2) and momentum, they can exert pressure on a surface. The chapter concludes with a description of the various frequency ranges in the electromagnetic spectrum.

34.1 Displacement Current and the General Form of Ampère's Law

In Chapter 30, we discussed using Ampère's law (Eq. 30.13) to analyze the magnetic fields created by currents:

$$\oint \vec{\mathbf{B}} \cdot d\vec{\mathbf{s}} = \mu_0 I$$

In this equation, the line integral is over any closed path through which conduction current passes, where conduction current is defined by the expression $I = dq/dt$. (In this section, we use the term *conduction current* to refer to the current carried by charge carriers in the wire to distinguish it from a new type of current we shall introduce shortly.) We now show that Ampère's law in this form is valid only if any electric fields present are constant in time. James Clerk Maxwell recognized this limitation and modified Ampère's law to include time-varying electric fields.

Consider a capacitor being charged as illustrated in Figure 34.1. When a conduction current is present, the charge on the positive plate changes, but no conduction current exists in the gap between the plates because there are no charge carriers in the gap. Now consider the two surfaces S_1 and S_2 in Figure 34.1, bounded by the same path P. Ampère's law states that $\oint \vec{\mathbf{B}} \cdot d\vec{\mathbf{s}}$ around this path must equal $\mu_0 I$, where I is the total current through *any* surface bounded by the path P.

When the path P is considered to be the boundary of S_1, $\oint \vec{\mathbf{B}} \cdot d\vec{\mathbf{s}} = \mu_0 I$ because the conduction current I passes through S_1. When the path is considered to be the boundary of S_2, however, $\oint \vec{\mathbf{B}} \cdot d\vec{\mathbf{s}} = 0$ because no conduction current passes through S_2. Therefore, we have a contradictory situation that arises from the discontinuity of the current! Maxwell solved this problem by postulating an additional term on the right side of Ampère's law, which includes a factor called the **displacement current** I_d defined as[1]

$$I_d \equiv \epsilon_0 \frac{d\Phi_E}{dt} \qquad\qquad (34.1)$$

◀ **Displacement current**

James Clerk Maxwell
Scottish Theoretical Physicist (1831–1879)
Maxwell developed the electromagnetic theory of light and the kinetic theory of gases, and explained the nature of Saturn's rings and color vision. Maxwell's successful interpretation of the electromagnetic field resulted in the field equations that bear his name. Formidable mathematical ability combined with great insight enabled him to lead the way in the study of electromagnetism and kinetic theory. He died of cancer before he was 50.

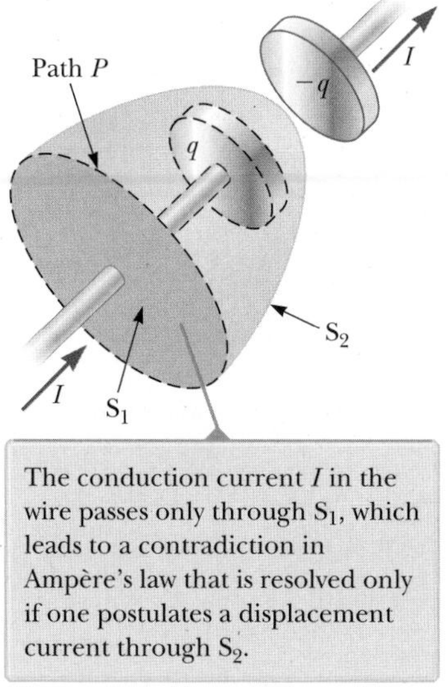

The conduction current I in the wire passes only through S_1, which leads to a contradiction in Ampère's law that is resolved only if one postulates a displacement current through S_2.

Figure 34.1 Two surfaces S_1 and S_2 near the plate of a capacitor are bounded by the same path P.

---

[1] *Displacement* in this context does not have the meaning it does in Chapter 2. Despite the inaccurate implications, the word is historically entrenched in the language of physics, so we continue to use it.

where ϵ_0 is the permittivity of free space (see Section 23.3) and $\Phi_E \equiv \int \vec{\mathbf{E}} \cdot d\vec{\mathbf{A}}$ is the electric flux (see Eq. 24.3) through the surface bounded by the path of integration.

As the capacitor is being charged (or discharged), the changing electric field between the plates may be considered equivalent to a current that acts as a continuation of the conduction current in the wire. When the expression for the displacement current given by Equation 34.1 is added to the conduction current on the right side of Ampère's law, the difficulty represented in Figure 34.1 is resolved. No matter which surface bounded by the path P is chosen, either a conduction current or a displacement current passes through it. With this new term I_d, we can express the general form of Ampère's law (sometimes called the **Ampère–Maxwell law**) as

◀ Ampère–Maxwell law

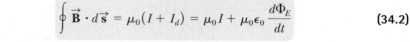

$$\oint \vec{\mathbf{B}} \cdot d\vec{\mathbf{s}} = \mu_0(I + I_d) = \mu_0 I + \mu_0\epsilon_0 \frac{d\Phi_E}{dt} \qquad (34.2)$$

We can understand the meaning of this expression by referring to Figure 34.2. The electric flux through surface S is $\Phi_E = \int \vec{\mathbf{E}} \cdot d\vec{\mathbf{A}} = EA$, where A is the area of the capacitor plates and E is the magnitude of the uniform electric field between the plates. If q is the charge on the plates at any instant, then $E = q/(\epsilon_0 A)$ (see Section 26.2). Therefore, the electric flux through S is

$$\Phi_E = EA = \frac{q}{\epsilon_0}$$

Hence, the displacement current through S is

$$I_d = \epsilon_0 \frac{d\Phi_E}{dt} = \frac{dq}{dt} \qquad (34.3)$$

That is, the displacement current I_d through S is precisely equal to the conduction current I in the wires connected to the capacitor!

By considering surface S, we can identify the displacement current as the source of the magnetic field on the surface boundary. The displacement current has its physical origin in the time-varying electric field. The central point of this formalism is that magnetic fields are produced *both* by conduction currents *and* by time-varying electric fields. This result was a remarkable example of theoretical work by Maxwell, and it contributed to major advances in the understanding of electromagnetism.

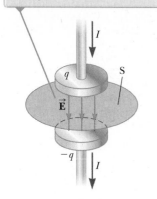

The electric field lines between the plates create an electric flux through surface S.

Figure 34.2 When a conduction current exists in the wires, a changing electric field $\vec{\mathbf{E}}$ exists between the plates of the capacitor.

Quick Quiz 34.1 In an *RC* circuit, the capacitor begins to discharge. **(i)** During the discharge, in the region of space between the plates of the capacitor, is there (a) conduction current but no displacement current, (b) displacement current but no conduction current, (c) both conduction and displacement current, or (d) no current of any type? **(ii)** In the same region of space, is there (a) an electric field but no magnetic field, (b) a magnetic field but no electric field, (c) both electric and magnetic fields, or (d) no fields of any type?

Example 34.1   **Displacement Current in a Capacitor**

A sinusoidally varying voltage is applied across a capacitor as shown in Figure 34.3. The capacitance is $C = 8.00 \ \mu F$, the frequency of the applied voltage is $f = 3.00$ kHz, and the voltage amplitude is $\Delta V_{max} = 30.0$ V. Find the displacement current in the capacitor.

SOLUTION

Conceptualize Figure 34.3 represents the circuit diagram for this situation. Figure 34.2 shows a close-up of the capacitor and the electric field between the plates.

Categorize We determine results using equations discussed in this section, so we categorize this example as a substitution problem.

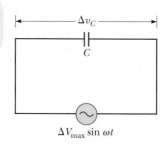

Figure 34.3 (Example 34.1)

▶ **34.1** continued

Evaluate the angular frequency of the source from Equation 15.12:

$$\omega = 2\pi f = 2\pi(3.00 \times 10^3 \text{ Hz}) = 1.88 \times 10^4 \text{ s}^{-1}$$

Use Equation 33.20 to express the potential difference in volts across the capacitor as a function of time in seconds:

$$\Delta v_C = \Delta V_{max} \sin \omega t = 30.0 \sin (1.88 \times 10^4 \, t)$$

Use Equation 34.3 to find the displacement current in amperes as a function of time. Note that the charge on the capacitor is $q = C \Delta v_C$:

$$i_d = \frac{dq}{dt} = \frac{d}{dt}(C \, \Delta v_C) = C \frac{d}{dt}(\Delta V_{max} \sin \omega t)$$

$$= \omega C \Delta V_{max} \cos \omega t$$

Substitute numerical values:

$$i_d = (1.88 \times 10^4 \text{ s}^{-1})(8.00 \times 10^{-6} \text{ C})(30.0 \text{ V}) \cos (1.88 \times 10^4 \, t)$$

$$= 4.51 \cos (1.88 \times 10^4 \, t)$$

34.2 Maxwell's Equations and Hertz's Discoveries

We now present four equations that are regarded as the basis of all electrical and magnetic phenomena. These equations, developed by Maxwell, are as fundamental to electromagnetic phenomena as Newton's laws are to mechanical phenomena. In fact, the theory that Maxwell developed was more far-reaching than even he imagined because it turned out to be in agreement with the special theory of relativity, as Einstein showed in 1905.

Maxwell's equations represent the laws of electricity and magnetism that we have already discussed, but they have additional important consequences. For simplicity, we present **Maxwell's equations** as applied to free space, that is, in the absence of any dielectric or magnetic material. The four equations are

$$\oint \vec{\mathbf{E}} \cdot d\vec{\mathbf{A}} = \frac{q}{\epsilon_0} \qquad (34.4) \qquad \blacktriangleleft \text{ Gauss's law}$$

$$\oint \vec{\mathbf{B}} \cdot d\vec{\mathbf{A}} = 0 \qquad (34.5) \qquad \blacktriangleleft \text{ Gauss's law in magnetism}$$

$$\oint \vec{\mathbf{E}} \cdot d\vec{\mathbf{s}} = -\frac{d\Phi_B}{dt} \qquad (34.6) \qquad \blacktriangleleft \text{ Faraday's law}$$

$$\oint \vec{\mathbf{B}} \cdot d\vec{\mathbf{s}} = \mu_0 I + \epsilon_0 \mu_0 \frac{d\Phi_E}{dt} \qquad (34.7) \qquad \blacktriangleleft \text{ Ampère–Maxwell law}$$

Equation 34.4 is Gauss's law: the total electric flux through any closed surface equals the net charge inside that surface divided by ϵ_0. This law relates an electric field to the charge distribution that creates it.

Equation 34.5 is Gauss's law in magnetism, and it states that the net magnetic flux through a closed surface is zero. That is, the number of magnetic field lines that enter a closed volume must equal the number that leave that volume, which implies that magnetic field lines cannot begin or end at any point. If they did, it would mean that isolated magnetic monopoles existed at those points. That isolated magnetic monopoles have not been observed in nature can be taken as a confirmation of Equation 34.5.

Equation 34.6 is Faraday's law of induction, which describes the creation of an electric field by a changing magnetic flux. This law states that the emf, which is the line integral of the electric field around any closed path, equals the rate of change of magnetic flux through any surface bounded by that path. One consequence of Faraday's law is the current induced in a conducting loop placed in a time-varying magnetic field.

Equation 34.7 is the Ampère–Maxwell law, discussed in Section 34.1, and it describes the creation of a magnetic field by a changing electric field and by electric current: the line integral of the magnetic field around any closed path is the sum of μ_0 multiplied by the net current through that path and $\epsilon_0\mu_0$ multiplied by the rate of change of electric flux through any surface bounded by that path.

Once the electric and magnetic fields are known at some point in space, the force acting on a particle of charge q can be calculated from the electric and magnetic versions of the particle in a field model:

Lorentz force law ▶

$$\vec{F} = q\vec{E} + q\vec{v} \times \vec{B} \tag{34.8}$$

This relationship is called the **Lorentz force law.** (We saw this relationship earlier as Eq. 29.6.) Maxwell's equations, together with this force law, completely describe all classical electromagnetic interactions in a vacuum.

Notice the symmetry of Maxwell's equations. Equations 34.4 and 34.5 are symmetric, apart from the absence of the term for magnetic monopoles in Equation 34.5. Furthermore, Equations 34.6 and 34.7 are symmetric in that the line integrals of $\vec{E}$ and $\vec{B}$ around a closed path are related to the rate of change of magnetic flux and electric flux, respectively. Maxwell's equations are of fundamental importance not only to electromagnetism, but to all science. Hertz once wrote, "One cannot escape the feeling that these mathematical formulas have an independent existence and an intelligence of their own, that they are wiser than we are, wiser even than their discoverers, that we get more out of them than we put into them."

In the next section, we show that Equations 34.6 and 34.7 can be combined to obtain a wave equation for both the electric field and the magnetic field. In empty space, where $q = 0$ and $I = 0$, the solution to these two equations shows that the speed at which electromagnetic waves travel equals the measured speed of light. This result led Maxwell to predict that light waves are a form of electromagnetic radiation.

Hertz performed experiments that verified Maxwell's prediction. The experimental apparatus Hertz used to generate and detect electromagnetic waves is shown schematically in Figure 34.4. An induction coil is connected to a transmitter made up of two spherical electrodes separated by a narrow gap. The coil provides short voltage surges to the electrodes, making one positive and the other negative. A spark is generated between the spheres when the electric field near either electrode surpasses the dielectric strength for air (3×10^6 V/m; see Table 26.1). Free electrons in a strong electric field are accelerated and gain enough energy to ionize any molecules they strike. This ionization provides more electrons, which can accelerate and cause further ionizations. As the air in the gap is ionized, it becomes a much better conductor and the discharge between the electrodes exhibits an oscillatory behavior at a very high frequency. From an electric-circuit viewpoint, this experimental apparatus is equivalent to an LC circuit in which the inductance is that of the coil and the capacitance is due to the spherical electrodes.

Because L and C are small in Hertz's apparatus, the frequency of oscillation is high, on the order of 100 MHz. (Recall from Eq. 32.22 that $\omega = 1/\sqrt{LC}$ for an LC circuit.) Electromagnetic waves are radiated at this frequency as a result of the oscillation of free charges in the transmitter circuit. Hertz was able to detect these waves using a single loop of wire with its own spark gap (the receiver). Such a receiver loop, placed several meters from the transmitter, has its own effective inductance, capacitance, and natural frequency of oscillation. In Hertz's experiment, sparks were induced across the gap of the receiving electrodes when the receiver's frequency was adjusted to match that of the transmitter. In this way,

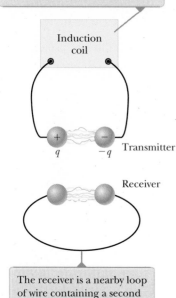

The transmitter consists of two spherical electrodes connected to an induction coil, which provides short voltage surges to the spheres, setting up oscillations in the discharge between the electrodes.

Induction coil

q $-q$ Transmitter

Receiver

The receiver is a nearby loop of wire containing a second spark gap.

Figure 34.4 Schematic diagram of Hertz's apparatus for generating and detecting electromagnetic waves.

Hertz demonstrated that the oscillating current induced in the receiver was produced by electromagnetic waves radiated by the transmitter. His experiment is analogous to the mechanical phenomenon in which a tuning fork responds to acoustic vibrations from an identical tuning fork that is oscillating nearby.

In addition, Hertz showed in a series of experiments that the radiation generated by his spark-gap device exhibited the wave properties of interference, diffraction, reflection, refraction, and polarization, which are all properties exhibited by light as we shall see in Part 5. Therefore, it became evident that the radio-frequency waves Hertz was generating had properties similar to those of light waves and that they differed only in frequency and wavelength. Perhaps his most convincing experiment was the measurement of the speed of this radiation. Waves of known frequency were reflected from a metal sheet and created a standing-wave interference pattern whose nodal points could be detected. The measured distance between the nodal points enabled determination of the wavelength λ. Using the relationship $v = \lambda f$ (Eq. 16.12) from the traveling wave model, Hertz found that v was close to 3×10^8 m/s, the known speed c of visible light.

Heinrich Rudolf Hertz
German Physicist (1857–1894)
Hertz made his most important discovery of electromagnetic waves in 1887. After finding that the speed of an electromagnetic wave was the same as that of light, Hertz showed that electromagnetic waves, like light waves, could be reflected, refracted, and diffracted. The hertz, equal to one complete vibration or cycle per second, is named after him.

34.3 Plane Electromagnetic Waves

The properties of electromagnetic waves can be deduced from Maxwell's equations. One approach to deriving these properties is to solve the second-order differential equation obtained from Maxwell's third and fourth equations. A rigorous mathematical treatment of that sort is beyond the scope of this text. To circumvent this problem, let's assume the vectors for the electric field and magnetic field in an electromagnetic wave have a specific space–time behavior that is simple but consistent with Maxwell equations.

To understand the prediction of electromagnetic waves more fully, let's focus our attention on an electromagnetic wave that travels in the x direction (the *direction of propagation*). For this wave, the electric field $\vec{\mathbf{E}}$ is in the y direction and the magnetic field $\vec{\mathbf{B}}$ is in the z direction as shown in Figure 34.5. Such waves, in which the electric and magnetic fields are restricted to being parallel to a pair of perpendicular axes, are said to be **linearly polarized waves.** Furthermore, let's assume the field magnitudes E and B depend on x and t only, not on the y or z coordinate.

Let's also imagine that the source of the electromagnetic waves is such that a wave radiated from *any* position in the yz plane (not only from the origin as might be suggested by Fig. 34.5) propagates in the x direction and all such waves are emitted in phase. If we define a **ray** as the line along which the wave travels, all rays for these waves are parallel. This entire collection of waves is often called a **plane wave.** A surface connecting points of equal phase on all waves is a geometric plane called a **wave front,** introduced in Chapter 17. In comparison, a point source of radiation sends waves out radially in all directions. A surface connecting points of equal phase for this situation is a sphere, so this wave is called a **spherical wave.**

To generate the prediction of plane electromagnetic waves, we start with Faraday's law, Equation 34.6:

$$\oint \vec{\mathbf{E}} \cdot d\vec{\mathbf{s}} = -\frac{d\Phi_B}{dt}$$

To apply this equation to the wave in Figure 34.5, consider a rectangle of width dx and height ℓ lying in the xy plane as shown in Figure 34.6 (page 808). Let's first evaluate the line integral of $\vec{\mathbf{E}} \cdot d\vec{\mathbf{s}}$ around this rectangle in the counterclockwise direction at an instant of time when the wave is passing through the rectangle. The contributions from the top and bottom of the rectangle are zero because $\vec{\mathbf{E}}$ is perpendicular to $d\vec{\mathbf{s}}$ for these paths. We can express the electric field on the right side of the rectangle as

$$E(x + dx) \approx E(x) + \frac{dE}{dx}\bigg|_{t\,\text{constant}} dx = E(x) + \frac{\partial E}{\partial x} dx$$

Pitfall Prevention 34.1
What Is "a" Wave? What do we mean by a *single* wave? The word *wave* represents both the emission from a *single point* ("wave radiated from *any* position in the *yz* plane" in the text) and the collection of waves from *all points* on the source ("**plane wave**" in the text). You should be able to use this term in both ways and understand its meaning from the context.

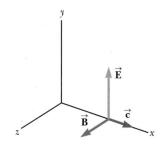

Figure 34.5 Electric and magnetic fields of an electromagnetic wave traveling at velocity $\vec{\mathbf{c}}$ in the positive x direction. The field vectors are shown at one instant of time and at one position in space. These fields depend on x and t.

According to Equation 34.11, this spatial variation in $\vec{\mathbf{E}}$ gives rise to a time-varying magnetic field along the z direction.

According to Equation 34.14, this spatial variation in $\vec{\mathbf{B}}$ gives rise to a time-varying electric field along the y direction.

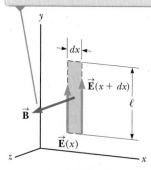

Figure 34.6 At an instant when a plane wave moving in the positive x direction passes through a rectangular path of width dx lying in the xy plane, the electric field in the y direction varies from $\vec{\mathbf{E}}(x)$ to $\vec{\mathbf{E}}(x + dx)$.

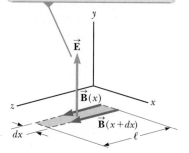

Figure 34.7 At an instant when a plane wave passes through a rectangular path of width dx lying in the xz plane, the magnetic field in the z direction varies from $\vec{\mathbf{B}}(x)$ to $\vec{\mathbf{B}}(x + dx)$.

where $E(x)$ is the field on the left side of the rectangle at this instant.[2] Therefore, the line integral over this rectangle is approximately

$$\oint \vec{\mathbf{E}} \cdot d\vec{\mathbf{s}} = [E(x + dx)]\ell - [E(x)]\ell \approx \ell \left(\frac{\partial E}{\partial x}\right) dx \tag{34.9}$$

Because the magnetic field is in the z direction, the magnetic flux through the rectangle of area $\ell\,dx$ is approximately $\Phi_B = B\ell\,dx$ (assuming dx is very small compared with the wavelength of the wave). Taking the time derivative of the magnetic flux gives

$$\frac{d\Phi_B}{dt} = \ell\,dx\,\frac{dB}{dt}\bigg|_{x\,\text{constant}} = \ell\,dx\,\frac{\partial B}{\partial t} \tag{34.10}$$

Substituting Equations 34.9 and 34.10 into Equation 34.6 gives

$$\ell \left(\frac{\partial E}{\partial x}\right) dx = -\ell\,dx\,\frac{\partial B}{\partial t}$$

$$\frac{\partial E}{\partial x} = -\frac{\partial B}{\partial t} \tag{34.11}$$

In a similar manner, we can derive a second equation by starting with Maxwell's fourth equation in empty space (Eq. 34.7). In this case, the line integral of $\vec{\mathbf{B}} \cdot d\vec{\mathbf{s}}$ is evaluated around a rectangle lying in the xz plane and having width dx and length ℓ as in Figure 34.7. Noting that the magnitude of the magnetic field changes from $B(x)$ to $B(x + dx)$ over the width dx and that the direction for taking the line integral is counterclockwise when viewed from above in Figure 34.7, the line integral over this rectangle is found to be approximately

$$\oint \vec{\mathbf{B}} \cdot d\vec{\mathbf{s}} = [B(x)]\ell - [B(x + dx)]\ell \approx -\ell \left(\frac{\partial B}{\partial x}\right) dx \tag{34.12}$$

---

[2]Because dE/dx in this equation is expressed as the change in E with x at a given instant t, dE/dx is equivalent to the partial derivative $\partial E/\partial x$. Likewise, dB/dt means the change in B with time at a particular position x; therefore, in Equation 34.10, we can replace dB/dt with $\partial B/\partial t$.

The electric flux through the rectangle is $\Phi_E = E\ell\,dx$, which, when differentiated with respect to time, gives

$$\frac{\partial \Phi_E}{\partial t} = \ell\,dx\,\frac{\partial E}{\partial t} \tag{34.13}$$

Substituting Equations 34.12 and 34.13 into Equation 34.7 gives

$$-\ell\left(\frac{\partial B}{\partial x}\right)dx = \mu_0\,\epsilon_0\,\ell\,dx\left(\frac{\partial E}{\partial t}\right)$$

$$\frac{\partial B}{\partial x} = -\mu_0\epsilon_0\,\frac{\partial E}{\partial t} \tag{34.14}$$

Taking the derivative of Equation 34.11 with respect to x and combining the result with Equation 34.14 gives

$$\frac{\partial^2 E}{\partial x^2} = -\frac{\partial}{\partial x}\left(\frac{\partial B}{\partial t}\right) = -\frac{\partial}{\partial t}\left(\frac{\partial B}{\partial x}\right) = -\frac{\partial}{\partial t}\left(-\mu_0\epsilon_0\,\frac{\partial E}{\partial t}\right)$$

$$\frac{\partial^2 E}{\partial x^2} = \mu_0\epsilon_0\,\frac{\partial^2 E}{\partial t^2} \tag{34.15}$$

In the same manner, taking the derivative of Equation 34.14 with respect to x and combining it with Equation 34.11 gives

$$\frac{\partial^2 B}{\partial x^2} = \mu_0\epsilon_0\,\frac{\partial^2 B}{\partial t^2} \tag{34.16}$$

Equations 34.15 and 34.16 both have the form of the linear wave equation[3] with the wave speed v replaced by c, where

$$c = \frac{1}{\sqrt{\mu_0\epsilon_0}} \tag{34.17}$$

◀ **Speed of electromagnetic waves**

Let's evaluate this speed numerically:

$$c = \frac{1}{\sqrt{(4\pi \times 10^{-7}\ \text{T}\cdot\text{m/A})(8.854\ 19 \times 10^{-12}\ \text{C}^2/\text{N}\cdot\text{m}^2)}}$$

$$= 2.997\ 92 \times 10^8\ \text{m/s}$$

Because this speed is precisely the same as the speed of light in empty space, we are led to believe (correctly) that light is an electromagnetic wave.

The simplest solution to Equations 34.15 and 34.16 is a sinusoidal wave for which the field magnitudes E and B vary with x and t according to the expressions

$$E = E_{\text{max}} \cos{(kx - \omega t)} \tag{34.18}$$

$$B = B_{\text{max}} \cos{(kx - \omega t)} \tag{34.19}$$

◀ **Sinusoidal electric and magnetic fields**

where E_{max} and B_{max} are the maximum values of the fields. The angular wave number is $k = 2\pi/\lambda$, where λ is the wavelength. The angular frequency is $\omega = 2\pi f$, where f is the wave frequency. According to the traveling wave model, the ratio ω/k equals the speed of an electromagnetic wave, c:

$$\frac{\omega}{k} = \frac{2\pi f}{2\pi/\lambda} = \lambda f = c$$

---

[3]The linear wave equation is of the form $(\partial^2 y/\partial x^2) = (1/v^2)(\partial^2 y/\partial t^2)$, where v is the speed of the wave and y is the wave function. The linear wave equation was introduced as Equation 16.27, and we suggest you review Section 16.6.

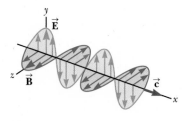

Figure 34.8 A sinusoidal electromagnetic wave moves in the positive x direction with a speed c.

where we have used Equation 16.12, $v = c = \lambda f$, which relates the speed, frequency, and wavelength of a sinusoidal wave. Therefore, for electromagnetic waves, the wavelength and frequency of these waves are related by

$$\lambda = \frac{c}{f} = \frac{3.00 \times 10^8 \text{ m/s}}{f} \qquad (34.20)$$

Figure 34.8 is a pictorial representation, at one instant, of a sinusoidal, linearly polarized electromagnetic wave moving in the positive x direction.

We can generate other mathematical representations of the traveling wave model for electromagnetic waves. Taking partial derivatives of Equations 34.18 (with respect to x) and 34.19 (with respect to t) gives

$$\frac{\partial E}{\partial x} = -kE_{max} \sin (kx - \omega t)$$

$$\frac{\partial B}{\partial t} = \omega B_{max} \sin (kx - \omega t)$$

Substituting these results into Equation 34.11 shows that, at any instant,

$$kE_{max} = \omega B_{max}$$

$$\frac{E_{max}}{B_{max}} = \frac{\omega}{k} = c$$

Using these results together with Equations 34.18 and 34.19 gives

$$\frac{E_{max}}{B_{max}} = \frac{E}{B} = c \qquad (34.21)$$

That is, at every instant, the ratio of the magnitude of the electric field to the magnitude of the magnetic field in an electromagnetic wave equals the speed of light.

Finally, note that electromagnetic waves obey the superposition principle as described in the waves in interference analysis model (which we discussed in Section 18.1 with respect to mechanical waves) because the differential equations involving E and B are linear equations. For example, we can add two waves with the same frequency and polarization simply by adding the magnitudes of the two electric fields algebraically.

Pitfall Prevention 34.2
$\vec{E}$ **Stronger Than** $\vec{B}$? Because the value of c is so large, some students incorrectly interpret Equation 34.21 as meaning that the electric field is much stronger than the magnetic field. Electric and magnetic fields are measured in different units, however, so they cannot be directly compared. In Section 34.4, we find that the electric and magnetic fields contribute equally to the wave's energy.

Quick Quiz 34.2 What is the phase difference between the sinusoidal oscillations of the electric and magnetic fields in Figure 34.8? **(a)** 180° **(b)** 90° **(c)** 0 **(d)** impossible to determine

Quick Quiz 34.3 An electromagnetic wave propagates in the negative y direction. The electric field at a point in space is momentarily oriented in the positive x direction. In which direction is the magnetic field at that point momentarily oriented? **(a)** the negative x direction **(b)** the positive y direction **(c)** the positive z direction **(d)** the negative z direction

Example 34.2 An Electromagnetic Wave

A sinusoidal electromagnetic wave of frequency 40.0 MHz travels in free space in the x direction as in Figure 34.9.

(A) Determine the wavelength and period of the wave.

▶ **34.2** continued

SOLUTION

Conceptualize Imagine the wave in Figure 34.9 moving to the right along the x axis, with the electric and magnetic fields oscillating in phase.

Categorize We use the mathematical representation of the *traveling wave* model for electromagnetic waves.

Figure 34.9 (Example 34.2) At some instant, a plane electromagnetic wave moving in the x direction has a maximum electric field of 750 N/C in the positive y direction.

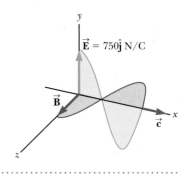

Analyze

Use Equation 34.20 to find the wavelength of the wave:

$$\lambda = \frac{c}{f} = \frac{3.00 \times 10^8 \text{ m/s}}{40.0 \times 10^6 \text{ Hz}} = 7.50 \text{ m}$$

Find the period T of the wave as the inverse of the frequency:

$$T = \frac{1}{f} = \frac{1}{40.0 \times 10^6 \text{ Hz}} = 2.50 \times 10^{-8} \text{ s}$$

(B) At some point and at some instant, the electric field has its maximum value of 750 N/C and is directed along the y axis. Calculate the magnitude and direction of the magnetic field at this position and time.

SOLUTION

Use Equation 34.21 to find the magnitude of the magnetic field:

$$B_{max} = \frac{E_{max}}{c} = \frac{750 \text{ N/C}}{3.00 \times 10^8 \text{ m/s}} = 2.50 \times 10^{-6} \text{ T}$$

Because $\vec{E}$ and $\vec{B}$ must be perpendicular to each other and perpendicular to the direction of wave propagation (x in this case), we conclude that $\vec{B}$ is in the z direction.

Finalize Notice that the wavelength is several meters. This is relatively long for an electromagnetic wave. As we will see in Section 34.7, this wave belongs to the radio range of frequencies.

34.4 Energy Carried by Electromagnetic Waves

In our discussion of the nonisolated system model for energy in Section 8.1, we identified electromagnetic radiation as one method of energy transfer across the boundary of a system. The amount of energy transferred by electromagnetic waves is symbolized as T_{ER} in Equation 8.2. The rate of transfer of energy by an electromagnetic wave is described by a vector $\vec{S}$, called the **Poynting vector,** which is defined by the expression

$$\vec{S} \equiv \frac{1}{\mu_0} \vec{E} \times \vec{B} \qquad (34.22)$$

◀ **Poynting vector**

The magnitude of the Poynting vector represents the rate at which energy passes through a unit surface area perpendicular to the direction of wave propagation. Therefore, the magnitude of $\vec{S}$ represents *power per unit area*. The direction of the vector is along the direction of wave propagation (Fig. 34.10, page 812). The SI units of $\vec{S}$ are J/s · m² = W/m².

As an example, let's evaluate the magnitude of $\vec{S}$ for a plane electromagnetic wave where $|\vec{E} \times \vec{B}| = EB$. In this case,

$$S = \frac{EB}{\mu_0} \qquad (34.23)$$

Pitfall Prevention 34.3

An Instantaneous Value The Poynting vector given by Equation 34.22 is time dependent. Its magnitude varies in time, reaching a maximum value at the same instant the magnitudes of $\vec{E}$ and $\vec{B}$ do. The *average* rate of energy transfer is given by Equation 34.24 on the next page.

Because $B = E/c$, we can also express this result as

$$S = \frac{E^2}{\mu_0 c} = \frac{cB^2}{\mu_0}$$

These equations for S apply at any instant of time and represent the *instantaneous* rate at which energy is passing through a unit area in terms of the instantaneous values of E and B.

What is of greater interest for a sinusoidal plane electromagnetic wave is the time average of S over one or more cycles, which is called the *wave intensity I*. (We discussed the intensity of sound waves in Chapter 17.) When this average is taken, we obtain an expression involving the time average of $\cos^2(kx - \omega t)$, which equals $\frac{1}{2}$. Hence, the average value of S (in other words, the intensity of the wave) is

Wave intensity ▶

$$I = S_{avg} = \frac{E_{max} B_{max}}{2\mu_0} = \frac{E_{max}^2}{2\mu_0 c} = \frac{cB_{max}^2}{2\mu_0} \qquad (34.24)$$

Recall that the energy per unit volume associated with an electric field, which is the instantaneous energy density u_E, is given by Equation 26.13:

$$u_E = \tfrac{1}{2}\epsilon_0 E^2$$

Also recall that the instantaneous energy density u_B associated with a magnetic field is given by Equation 32.14:

$$u_B = \frac{B^2}{2\mu_0}$$

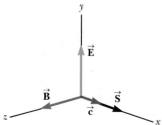

Figure 34.10 The Poynting vector $\vec{S}$ for a plane electromagnetic wave is along the direction of wave propagation.

Because E and B vary with time for an electromagnetic wave, the energy densities also vary with time. Using the relationships $B = E/c$ and $c = 1/\sqrt{\mu_0 \epsilon_0}$, the expression for u_B becomes

$$u_B = \frac{(E/c)^2}{2\mu_0} = \frac{\mu_0 \epsilon_0}{2\mu_0} E^2 = \tfrac{1}{2}\epsilon_0 E^2$$

Comparing this result with the expression for u_E, we see that

$$u_B = u_E = \tfrac{1}{2}\epsilon_0 E^2 = \frac{B^2}{2\mu_0}$$

That is, the instantaneous energy density associated with the magnetic field of an electromagnetic wave equals the instantaneous energy density associated with the electric field. Hence, in a given volume, the energy is equally shared by the two fields.

The **total instantaneous energy density** u is equal to the sum of the energy densities associated with the electric and magnetic fields:

Total instantaneous ▶
energy density of an
electromagnetic wave

$$u = u_E + u_B = \epsilon_0 E^2 = \frac{B^2}{\mu_0}$$

When this total instantaneous energy density is averaged over one or more cycles of an electromagnetic wave, we again obtain a factor of $\frac{1}{2}$. Hence, for any electromagnetic wave, the total average energy per unit volume is

Average energy density of ▶
an electromagnetic wave

$$u_{avg} = \epsilon_0 (E^2)_{avg} = \tfrac{1}{2}\epsilon_0 E_{max}^2 = \frac{B_{max}^2}{2\mu_0} \qquad (34.25)$$

Comparing this result with Equation 34.24 for the average value of S, we see that

$$I = S_{avg} = c u_{avg} \qquad (34.26)$$

In other words, the intensity of an electromagnetic wave equals the average energy density multiplied by the speed of light.

The Sun delivers about 10^3 W/m^2 of energy to the Earth's surface via electromagnetic radiation. Let's calculate the total power that is incident on the roof of a home. The roof's dimensions are 8.00 m $\times$ 20.0 m. We assume the average magnitude of the Poynting vector for solar radiation at the surface of the Earth is $S_{avg} = 1\,000$ W/m^2. This average value represents the power per unit area, or the light intensity. Assuming the radiation is incident normal to the roof, we obtain

$$P_{avg} = S_{avg}A = (1\,000 \text{ W/m}^2)(8.00 \text{ m} \times 20.0 \text{ m}) = 1.60 \times 10^5 \text{ W}$$

This power is large compared with the power requirements of a typical home. If this power could be absorbed and made available to electrical devices, it would provide more than enough energy for the average home. Solar energy is not easily harnessed, however, and the prospects for large-scale conversion are not as bright as may appear from this calculation. For example, the efficiency of conversion from solar energy is typically 12–18% for photovoltaic cells, reducing the available power by an order of magnitude. Other considerations reduce the power even further. Depending on location, the radiation is most likely not incident normal to the roof and, even if it is, this situation exists for only a short time near the middle of the day. No energy is available for about half of each day during the nighttime hours, and cloudy days further reduce the available energy. Finally, while energy is arriving at a large rate during the middle of the day, some of it must be stored for later use, requiring batteries or other storage devices. All in all, complete solar operation of homes is not currently cost effective for most homes.

Example 34.3   Fields on the Page

Estimate the maximum magnitudes of the electric and magnetic fields of the light that is incident on this page because of the visible light coming from your desk lamp. Treat the lightbulb as a point source of electromagnetic radiation that is 5% efficient at transforming energy coming in by electrical transmission to energy leaving by visible light.

SOLUTION

Conceptualize The filament in your lightbulb emits electromagnetic radiation. The brighter the light, the larger the magnitudes of the electric and magnetic fields.

Categorize Because the lightbulb is to be treated as a point source, it emits equally in all directions, so the outgoing electromagnetic radiation can be modeled as a spherical wave.

Analyze Recall from Equation 17.13 that the wave intensity I a distance r from a point source is $I = P_{avg}/4\pi r^2$, where P_{avg} is the average power output of the source and $4\pi r^2$ is the area of a sphere of radius r centered on the source.

Set this expression for I equal to the intensity of an electromagnetic wave given by Equation 34.24:

$$I = \frac{P_{avg}}{4\pi r^2} = \frac{E_{max}^2}{2\mu_0 c}$$

Solve for the electric field magnitude:

$$E_{max} = \sqrt{\frac{\mu_0 c P_{avg}}{2\pi r^2}}$$

Let's make some assumptions about numbers to enter in this equation. The visible light output of a 60-W lightbulb operating at 5% efficiency is approximately 3.0 W by visible light. (The remaining energy transfers out of the lightbulb by thermal conduction and invisible radiation.) A reasonable distance from the lightbulb to the page might be 0.30 m.

Substitute these values:

$$E_{max} = \sqrt{\frac{(4\pi \times 10^{-7} \text{ T} \cdot \text{m/A})(3.00 \times 10^8 \text{ m/s})(3.0 \text{ W})}{2\pi(0.30 \text{ m})^2}}$$

$$= 45 \text{ V/m}$$

continued

▶ **34.3** continued

Use Equation 34.21 to find the magnetic field magnitude:

$$B_{\text{max}} = \frac{E_{\text{max}}}{c} = \frac{45 \text{ V/m}}{3.00 \times 10^8 \text{ m/s}} = \boxed{1.5 \times 10^{-7} \text{ T}}$$

Finalize This value of the magnetic field magnitude is two orders of magnitude smaller than the Earth's magnetic field.

34.5 Momentum and Radiation Pressure

Electromagnetic waves transport linear momentum as well as energy. As this momentum is absorbed by some surface, pressure is exerted on the surface. Therefore, the surface is a nonisolated system for momentum. In this discussion, let's assume the electromagnetic wave strikes the surface at normal incidence and transports a total energy T_{ER} to the surface in a time interval Δt. Maxwell showed that if the surface absorbs all the incident energy T_{ER} in this time interval (as does a black body, introduced in Section 20.7), the total momentum $\vec{\mathbf{p}}$ transported to the surface has a magnitude

| Momentum transported to a perfectly absorbing surface ▶ | $$p = \frac{T_{\text{ER}}}{c} \qquad \text{(complete absorption)}$$ | (34.27) |

The pressure P exerted on the surface is defined as force per unit area F/A, which when combined with Newton's second law gives

$$P = \frac{F}{A} = \frac{1}{A}\frac{dp}{dt}$$

Substituting Equation 34.27 into this expression for pressure P gives

$$P = \frac{1}{A}\frac{dp}{dt} = \frac{1}{A}\frac{d}{dt}\left(\frac{T_{\text{ER}}}{c}\right) = \frac{1}{c}\frac{(dT_{\text{ER}}/dt)}{A}$$

We recognize $(dT_{\text{ER}}/dt)/A$ as the rate at which energy is arriving at the surface per unit area, which is the magnitude of the Poynting vector. Therefore, the radiation pressure P exerted on the perfectly absorbing surface is

| Radiation pressure exerted on ▶ a perfectly absorbing surface | $$P = \frac{S}{c} \qquad \text{(complete absorption)}$$ | (34.28) |

If the surface is a perfect reflector (such as a mirror) and incidence is normal, the momentum transported to the surface in a time interval Δt is twice that given by Equation 34.27. That is, the momentum transferred to the surface by the incoming light is $p = T_{\text{ER}}/c$ and that transferred by the reflected light is also $p = T_{\text{ER}}/c$. Therefore,

$$p = \frac{2T_{\text{ER}}}{c} \qquad \text{(complete reflection)} \tag{34.29}$$

The radiation pressure exerted on a perfectly reflecting surface for normal incidence of the wave is

| Radiation pressure exerted on ▶ a perfectly reflecting surface | $$P = \frac{2S}{c} \qquad \text{(complete reflection)}$$ | (34.30) |

The pressure on a surface having a reflectivity somewhere between these two extremes has a value between S/c and $2S/c$, depending on the properties of the surface.

Although radiation pressures are very small (about 5×10^{-6} N/m² for direct sunlight), *solar sailing* is a low-cost means of sending spacecraft to the planets. Large sheets experience radiation pressure from sunlight and are used in much the way canvas sheets are used on earthbound sailboats. In 2010, the Japan Aerospace Exploration Agency (JAXA) launched the first spacecraft to use solar sailing as its primary propulsion, *IKAROS* (Interplanetary Kite-craft Accelerated by Radiation of the Sun). Successful testing of this spacecraft would lead to a larger effort to send a spacecraft to Jupiter by radiation pressure later in the present decade.

Quick Quiz 34.4 To maximize the radiation pressure on the sails of a spacecraft using solar sailing, should the sheets be **(a)** very black to absorb as much sunlight as possible or **(b)** very shiny to reflect as much sunlight as possible?

Conceptual Example 34.4   **Sweeping the Solar System**

A great amount of dust exists in interplanetary space. Although in theory these dust particles can vary in size from molecular size to a much larger size, very little of the dust in our solar system is smaller than about $0.2~\mu m$. Why?

SOLUTION

The dust particles are subject to two significant forces: the gravitational force that draws them toward the Sun and the radiation-pressure force that pushes them away from the Sun. The gravitational force is proportional to the cube of the radius of a spherical dust particle because it is proportional to the mass and therefore to the volume $4\pi r^3/3$ of the particle. The radiation pressure is proportional to the square of the radius because it depends on the planar cross section of the particle. For large particles, the gravitational force is greater than the force from radiation pressure. For particles having radii less than about $0.2~\mu m$, the radiation-pressure force is greater than the gravitational force. As a result, these particles are swept out of our solar system by sunlight.

Example 34.5   **Pressure from a Laser Pointer**

When giving presentations, many people use a laser pointer to direct the attention of the audience to information on a screen. If a 3.0-mW pointer creates a spot on a screen that is 2.0 mm in diameter, determine the radiation pressure on a screen that reflects 70% of the light that strikes it. The power 3.0 mW is a time-averaged value.

SOLUTION

Conceptualize Imagine the waves striking the screen and exerting a radiation pressure on it. The pressure should not be very large.

Categorize This problem involves a calculation of radiation pressure using an approach like that leading to Equation 34.28 or Equation 34.30, but it is complicated by the 70% reflection.

Analyze We begin by determining the magnitude of the beam's Poynting vector.

Divide the time-averaged power delivered via the electromagnetic wave by the cross-sectional area of the beam:

$$S_{avg} = \frac{(Power)_{avg}}{A} = \frac{(Power)_{avg}}{\pi r^2} = \frac{3.0 \times 10^{-3}~\text{W}}{\pi \left(\dfrac{2.0 \times 10^{-3}~\text{m}}{2}\right)^2} = 955~\text{W/m}^2$$

Now let's determine the radiation pressure from the laser beam. Equation 34.30 indicates that a completely reflected beam would apply an average pressure of $P_{avg} = 2S_{avg}/c$. We can model the actual reflection as follows. Imagine that the surface absorbs the beam, resulting in pressure $P_{avg} = S_{avg}/c$. Then the surface emits the beam, resulting in additional pressure $P_{avg} = S_{avg}/c$. If the surface emits only a fraction f of the beam (so that f is the amount of the incident beam reflected), the pressure due to the emitted beam is $P_{avg} = fS_{avg}/c$.

Use this model to find the total pressure on the surface due to absorption and re-emission (reflection):

$$P_{avg} = \frac{S_{avg}}{c} + f\frac{S_{avg}}{c} = (1 + f)\frac{S_{avg}}{c}$$

continued

▶ **34.5** continued

Evaluate this pressure for a beam that is 70% reflected:

$$P_{avg} = (1 + 0.70)\frac{955 \text{ W/m}^2}{3.0 \times 10^8 \text{ m/s}} = 5.4 \times 10^{-6} \text{ N/m}^2$$

Finalize The pressure has an extremely small value, as expected. (Recall from Section 14.2 that atmospheric pressure is approximately 10^5 N/m².) Consider the magnitude of the Poynting vector, $S_{avg} = 955$ W/m². It is about the same as the intensity of sunlight at the Earth's surface. For this reason, it is not safe to shine the beam of a laser pointer into a person's eyes, which may be more dangerous than looking directly at the Sun.

WHAT IF? What if the laser pointer is moved twice as far away from the screen? Does that affect the radiation pressure on the screen?

Answer Because a laser beam is popularly represented as a beam of light with constant cross section, you might think that the intensity of radiation, and therefore the radiation pressure, is independent of distance from the screen. A laser beam, however, does not have a constant cross section at all distances from the source; rather, there is a small but measurable divergence of the beam. If the laser is moved farther away from the screen, the area of illumination on the screen increases, decreasing the intensity. In turn, the radiation pressure is reduced.

In addition, the doubled distance from the screen results in more loss of energy from the beam due to scattering from air molecules and dust particles as the light travels from the laser to the screen. This energy loss further reduces the radiation pressure on the screen.

34.6 Production of Electromagnetic Waves by an Antenna

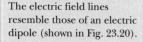

The electric field lines resemble those of an electric dipole (shown in Fig. 23.20).

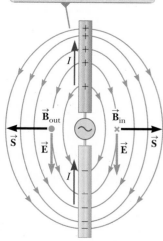

Figure 34.11 A half-wave antenna consists of two metal rods connected to an alternating voltage source. This diagram shows $\vec{E}$ and $\vec{B}$ at an arbitrary instant when the current is upward.

Stationary charges and steady currents cannot produce electromagnetic waves. If the current in a wire changes with time, however, the wire emits electromagnetic waves. The fundamental mechanism responsible for this radiation is the acceleration of a charged particle. **Whenever a charged particle accelerates, energy is transferred away from the particle by electromagnetic radiation.**

Let's consider the production of electromagnetic waves by a *half-wave antenna*. In this arrangement, two conducting rods are connected to a source of alternating voltage (such as an *LC* oscillator) as shown in Figure 34.11. The length of each rod is equal to one-quarter the wavelength of the radiation emitted when the oscillator operates at frequency *f*. The oscillator forces charges to accelerate back and forth between the two rods. Figure 34.11 shows the configuration of the electric and magnetic fields at some instant when the current is upward. The separation of charges in the upper and lower portions of the antenna make the electric field lines resemble those of an electric dipole. (As a result, this type of antenna is sometimes called a *dipole antenna*.) Because these charges are continuously oscillating between the two rods, the antenna can be approximated by an oscillating electric dipole. The current representing the movement of charges between the ends of the antenna produces magnetic field lines forming concentric circles around the antenna that are perpendicular to the electric field lines at all points. The magnetic field is zero at all points along the axis of the antenna. Furthermore, $\vec{E}$ and $\vec{B}$ are 90° out of phase in time; for example, the current is zero when the charges at the outer ends of the rods are at a maximum.

At the two points where the magnetic field is shown in Figure 34.11, the Poynting vector $\vec{S}$ is directed radially outward, indicating that energy is flowing away from the antenna at this instant. At later times, the fields and the Poynting vector reverse direction as the current alternates. Because $\vec{E}$ and $\vec{B}$ are 90° out of phase at points near the dipole, the net energy flow is zero. From this fact, you might conclude (incorrectly) that no energy is radiated by the dipole.

Energy is indeed radiated, however. Because the dipole fields fall off as $1/r^3$ (as shown in Example 23.6 for the electric field of a static dipole), they are negligible at great distances from the antenna. At these great distances, something else causes a type of radiation different from that close to the antenna. The source of this radiation is the continuous induction of an electric field by the time-varying magnetic field and the induction of a magnetic field by the time-varying electric field, predicted by Equations 34.6 and 34.7. The electric and magnetic fields produced in this manner are in phase with each other and vary as $1/r$. The result is an outward flow of energy at all times.

The angular dependence of the radiation intensity produced by a dipole antenna is shown in Figure 34.12. Notice that the intensity and the power radiated are a maximum in a plane that is perpendicular to the antenna and passing through its midpoint. Furthermore, the power radiated is zero along the antenna's axis. A mathematical solution to Maxwell's equations for the dipole antenna shows that the intensity of the radiation varies as $(\sin^2 \theta)/r^2$, where θ is measured from the axis of the antenna.

Electromagnetic waves can also induce currents in a receiving antenna. The response of a dipole receiving antenna at a given position is a maximum when the antenna axis is parallel to the electric field at that point and zero when the axis is perpendicular to the electric field.

Quick Quiz 34.5 If the antenna in Figure 34.11 represents the source of a distant radio station, what would be the best orientation for your portable radio antenna located to the right of the figure? **(a)** up-down along the page **(b)** left-right along the page **(c)** perpendicular to the page

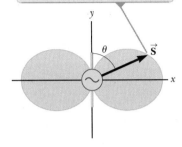

The distance from the origin to a point on the edge of the tan shape is proportional to the magnitude of the Poynting vector and the intensity of radiation in that direction.

Figure 34.12 Angular dependence of the intensity of radiation produced by an oscillating electric dipole.

34.7 The Spectrum of Electromagnetic Waves

The various types of electromagnetic waves are listed in Figure 34.13 (page 818), which shows the **electromagnetic spectrum.** Notice the wide ranges of frequencies and wavelengths. No sharp dividing point exists between one type of wave and the next. Remember that all forms of the various types of radiation are produced by the same phenomenon: acceleration of electric charges. The names given to the types of waves are simply a convenient way to describe the region of the spectrum in which they lie.

Radio waves, whose wavelengths range from more than 10^4 m to about 0.1 m, are the result of charges accelerating through conducting wires. They are generated by such electronic devices as *LC* oscillators and are used in radio and television communication systems.

Microwaves have wavelengths ranging from approximately 0.3 m to 10^{-4} m and are also generated by electronic devices. Because of their short wavelengths, they are well suited for radar systems and for studying the atomic and molecular properties of matter. Microwave ovens are an interesting domestic application of these waves. It has been suggested that solar energy could be harnessed by beaming microwaves to the Earth from a solar collector in space.

Infrared waves have wavelengths ranging from approximately 10^{-3} m to the longest wavelength of visible light, 7×10^{-7} m. These waves, produced by molecules and room-temperature objects, are readily absorbed by most materials. The infrared (IR) energy absorbed by a substance appears as internal energy because the energy agitates the object's atoms, increasing their vibrational or translational motion, which results in a temperature increase. Infrared radiation has practical and scientific applications in many areas, including physical therapy, IR photography, and vibrational spectroscopy.

Visible light, the most familiar form of electromagnetic waves, is the part of the electromagnetic spectrum the human eye can detect. Light is produced by the rearrangement of electrons in atoms and molecules. The various wavelengths of visible light, which correspond to different colors, range from red ($\lambda \approx 7 \times 10^{-7}$ m) to violet ($\lambda \approx 4 \times 10^{-7}$ m). The sensitivity of the human eye is a function of wavelength, being a maximum at a wavelength of about 5.5×10^{-7} m. With that in mind, why do you suppose tennis balls often have a yellow-green color? Table 34.1 provides

Pitfall Prevention 34.6
"Heat Rays" Infrared rays are often called "heat rays," but this terminology is a misnomer. Although infrared radiation is used to raise or maintain temperature as in the case of keeping food warm with "heat lamps" at a fast-food restaurant, all wavelengths of electromagnetic radiation carry energy that can cause the temperature of a system to increase. As an example, consider a potato baking in your microwave oven.

Table 34.1 **Approximate Correspondence Between Wavelengths of Visible Light and Color**

Wavelength Range (nm)	Color Description
400–430	Violet
430–485	Blue
485–560	Green
560–590	Yellow
590–625	Orange
625–700	Red

Note: The wavelength ranges here are approximate. Different people will describe colors differently.

Figure 34.13 The electromagnetic spectrum.

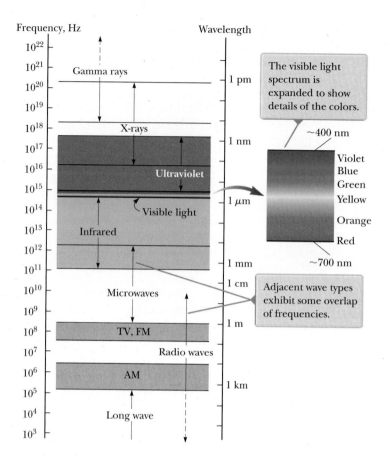

Wearing sunglasses that do not block ultraviolet (UV) light is worse for your eyes than wearing no sunglasses at all. The lenses of any sunglasses absorb some visible light, thereby causing the wearer's pupils to dilate. If the glasses do not also block UV light, more damage may be done to the lens of the eye because of the dilated pupils. If you wear no sunglasses at all, your pupils are contracted, you squint, and much less UV light enters your eyes. High-quality sunglasses block nearly all the eye-damaging UV light.

approximate correspondences between the wavelength of visible light and the color assigned to it by humans. Light is the basis of the science of optics and optical instruments, to be discussed in Chapters 35 through 38.

Ultraviolet waves cover wavelengths ranging from approximately 4×10^{-7} m to 6×10^{-10} m. The Sun is an important source of ultraviolet (UV) light, which is the main cause of sunburn. Sunscreen lotions are transparent to visible light but absorb most UV light. The higher a sunscreen's solar protection factor, or SPF, the greater the percentage of UV light absorbed. Ultraviolet rays have also been implicated in the formation of cataracts, a clouding of the lens inside the eye.

Most of the UV light from the Sun is absorbed by ozone (O_3) molecules in the Earth's upper atmosphere, in a layer called the stratosphere. This ozone shield converts lethal high-energy UV radiation to IR radiation, which in turn warms the stratosphere.

X-rays have wavelengths in the range from approximately 10^{-8} m to 10^{-12} m. The most common source of x-rays is the stopping of high-energy electrons upon bombarding a metal target. X-rays are used as a diagnostic tool in medicine and as a treatment for certain forms of cancer. Because x-rays can damage or destroy living tissues and organisms, care must be taken to avoid unnecessary exposure or overexposure. X-rays are also used in the study of crystal structure because x-ray wavelengths are comparable to the atomic separation distances in solids (about 0.1 nm).

Gamma rays are electromagnetic waves emitted by radioactive nuclei and during certain nuclear reactions. High-energy gamma rays are a component of cosmic rays that enter the Earth's atmosphere from space. They have wavelengths ranging from approximately 10^{-10} m to less than 10^{-14} m. Gamma rays are highly penetrating and produce serious damage when absorbed by living tissues. Consequently, those working near such dangerous radiation must be protected with heavily absorbing materials such as thick layers of lead.

Quick Quiz 34.6 In many kitchens, a microwave oven is used to cook food. The frequency of the microwaves is on the order of 10^{10} Hz. Are the wavelengths of these microwaves on the order of (a) kilometers, (b) meters, (c) centimeters, or (d) micrometers?

Quick Quiz 34.7 A radio wave of frequency on the order of 10^5 Hz is used to carry a sound wave with a frequency on the order of 10^3 Hz. Is the wavelength of this radio wave on the order of (a) kilometers, (b) meters, (c) centimeters, or (d) micrometers?

Summary

Definitions

In a region of space in which there is a changing electric field, there is a **displacement current** defined as

$$I_d \equiv \epsilon_0 \frac{d\Phi_E}{dt} \qquad (34.1)$$

where ϵ_0 is the permittivity of free space (see Section 23.3) and $\Phi_E = \int \vec{E} \cdot d\vec{A}$ is the electric flux.

The rate at which energy passes through a unit area by electromagnetic radiation is described by the **Poynting vector** $\vec{S}$, where

$$\vec{S} \equiv \frac{1}{\mu_0} \vec{E} \times \vec{B} \qquad (34.22)$$

Concepts and Principles

When used with the **Lorentz force law**, $\vec{F} = q\vec{E} + q\vec{v} \times \vec{B}$, **Maxwell's equations** describe all electromagnetic phenomena:

$$\oint \vec{E} \cdot d\vec{A} = \frac{q}{\epsilon_0} \qquad (34.4)$$

$$\oint \vec{B} \cdot d\vec{A} = 0 \qquad (34.5)$$

$$\oint \vec{E} \cdot d\vec{s} = -\frac{d\Phi_B}{dt} \qquad (34.6)$$

$$\oint \vec{B} \cdot d\vec{s} = \mu_0 I + \epsilon_0 \mu_0 \frac{d\Phi_E}{dt} \qquad (34.7)$$

Electromagnetic waves, which are predicted by Maxwell's equations, have the following properties and are described by the following mathematical representations of the traveling wave model for electromagnetic waves.

- The electric field and the magnetic field each satisfy a wave equation. These two wave equations, which can be obtained from Maxwell's third and fourth equations, are

$$\frac{\partial^2 E}{\partial x^2} = \mu_0 \epsilon_0 \frac{\partial^2 E}{\partial t^2} \qquad (34.15)$$

$$\frac{\partial^2 B}{\partial x^2} = \mu_0 \epsilon_0 \frac{\partial^2 B}{\partial t^2} \qquad (34.16)$$

- The waves travel through a vacuum with the speed of light c, where

$$c = \frac{1}{\sqrt{\mu_0 \epsilon_0}} \qquad (34.17)$$

- Numerically, the speed of electromagnetic waves in a vacuum is 3.00×10^8 m/s.
- The wavelength and frequency of electromagnetic waves are related by

$$\lambda = \frac{c}{f} = \frac{3.00 \times 10^8 \text{ m/s}}{f} \qquad (34.20)$$

- The electric and magnetic fields are perpendicular to each other and perpendicular to the direction of wave propagation.
- The instantaneous magnitudes of $\vec{E}$ and $\vec{B}$ in an electromagnetic wave are related by the expression

$$\frac{E}{B} = c \qquad (34.21)$$

- Electromagnetic waves carry energy.
- Electromagnetic waves carry momentum.

continued

Because electromagnetic waves carry momentum, they exert pressure on surfaces. If an electromagnetic wave whose Poynting vector is $\vec{S}$ is completely absorbed by a surface upon which it is normally incident, the radiation pressure on that surface is

$$P = \frac{S}{c} \quad (\text{complete absorption}) \qquad (34.28)$$

If the surface totally reflects a normally incident wave, the pressure is doubled.

The electric and magnetic fields of a sinusoidal plane electromagnetic wave propagating in the positive x direction can be written as

$$E = E_{max} \cos (kx - \omega t) \qquad (34.18)$$
$$B = B_{max} \cos (kx - \omega t) \qquad (34.19)$$

where k is the angular wave number and ω is the angular frequency of the wave. These equations represent special solutions to the wave equations for E and B.

The average value of the Poynting vector for a plane electromagnetic wave has a magnitude

$$S_{avg} = \frac{E_{max} B_{max}}{2\mu_0} = \frac{E_{max}^2}{2\mu_0 c} = \frac{cB_{max}^2}{2\mu_0} \qquad (34.24)$$

The intensity of a sinusoidal plane electromagnetic wave equals the average value of the Poynting vector taken over one or more cycles.

The electromagnetic spectrum includes waves covering a broad range of wavelengths, from long radio waves at more than 10^4 m to gamma rays at less than 10^{-14} m.

Objective Questions

1. denotes answer available in *Student Solutions Manual/Study Guide*

1. A spherical interplanetary grain of dust of radius 0.2 μm is at a distance r_1 from the Sun. The gravitational force exerted by the Sun on the grain just balances the force due to radiation pressure from the Sun's light. **(i)** Assume the grain is moved to a distance $2r_1$ from the Sun and released. At this location, what is the net force exerted on the grain? (a) toward the Sun (b) away from the Sun (c) zero (d) impossible to determine without knowing the mass of the grain **(ii)** Now assume the grain is moved back to its original location at r_1, compressed so that it crystallizes into a sphere with significantly higher density, and then released. In this situation, what is the net force exerted on the grain? Choose from the same possibilities as in part (i).

2. A small source radiates an electromagnetic wave with a single frequency into vacuum, equally in all directions. **(i)** As the wave moves, does its frequency (a) increase, (b) decrease, or (c) stay constant? Using the same choices, answer the same question about **(ii)** its wavelength, **(iii)** its speed, **(iv)** its intensity, and **(v)** the amplitude of its electric field.

3. A typical microwave oven operates at a frequency of 2.45 GHz. What is the wavelength associated with the electromagnetic waves in the oven? (a) 8.20 m (b) 12.2 cm (c) 1.20×10^8 m (d) 8.20×10^{-9} m (e) none of those answers

4. A student working with a transmitting apparatus like Heinrich Hertz's wishes to adjust the electrodes to generate electromagnetic waves with a frequency half as large as before. **(i)** How large should she make the effective capacitance of the pair of electrodes? (a) four times larger than before (b) two times larger than before (c) one-half as large as before (d) one-fourth as large as before (e) none of those answers **(ii)** After she makes the required adjustment, what will the wavelength of the transmitted wave be? Choose from the same possibilities as in part (i).

5. Assume you charge a comb by running it through your hair and then hold the comb next to a bar magnet. Do the electric and magnetic fields produced constitute an electromagnetic wave? (a) Yes they do, necessarily. (b) Yes they do because charged particles are moving inside the bar magnet. (c) They can, but only if the electric field of the comb and the magnetic field of the magnet are perpendicular. (d) They can, but only if both the comb and the magnet are moving. (e) They can, if either the comb or the magnet or both are accelerating.

6. Which of the following statements are true regarding electromagnetic waves traveling through a vacuum? More than one statement may be correct. (a) All waves have the same wavelength. (b) All waves have the same frequency. (c) All waves travel at 3.00×10^8 m/s. (d) The electric and magnetic fields associated with the waves are perpendicular to each other and to the direction of wave propagation. (e) The speed of the waves depends on their frequency.

7. A plane electromagnetic wave with a single frequency moves in vacuum in the positive x direction. Its amplitude is uniform over the yz plane. **(i)** As the wave moves, does its frequency (a) increase, (b) decrease, or (c) stay constant? Using the same choices, answer the same question about **(ii)** its wavelength, **(iii)** its speed, **(iv)** its intensity, and **(v)** the amplitude of its magnetic field.

8. Assume the amplitude of the electric field in a plane electromagnetic wave is E_1 and the amplitude of the magnetic field is B_1. The source of the wave is then adjusted so that the amplitude of the electric field doubles to become $2E_1$. **(i)** What happens to the amplitude of the magnetic field in this process? (a) It becomes four times larger. (b) It becomes two times larger. (c) It can stay constant. (d) It becomes one-half as large. (e) It becomes one-fourth as large. **(ii)** What happens to the intensity of the wave? Choose from the same possibilities as in part (i).

9. An electromagnetic wave with a peak magnetic field magnitude of 1.50×10^{-7} T has an associated peak electric field of what magnitude? (a) 0.500×10^{-15} N/C (b) 2.00×10^{-5} N/C (c) 2.20×10^4 N/C (d) 45.0 N/C (e) 22.0 N/C

10. **(i)** Rank the following kinds of waves according to their wavelength ranges from those with the largest typical or average wavelength to the smallest, noting any cases of equality: (a) gamma rays (b) microwaves (c) radio waves (d) visible light (e) x-rays **(ii)** Rank the kinds of waves according to their frequencies from highest to lowest. **(iii)** Rank the kinds of waves according to their speeds in vacuum from fastest to slowest.

11. Consider an electromagnetic wave traveling in the positive y direction. The magnetic field associated with the wave at some location at some instant points in the negative x direction as shown in Figure OQ34.11. What is the direction of the electric field at this position and at this instant? (a) the positive x direction (b) the positive y direction (c) the positive z direction (d) the negative z direction (e) the negative y direction

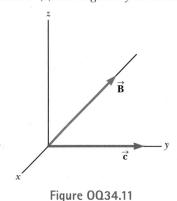

Figure OQ34.11

Conceptual Questions   **1.** denotes answer available in *Student Solutions Manual/Study Guide*

1. Suppose a creature from another planet has eyes that are sensitive to infrared radiation. Describe what the alien would see if it looked around your library. In particular, what would appear bright and what would appear dim?

2. For a given incident energy of an electromagnetic wave, why is the radiation pressure on a perfectly reflecting surface twice as great as that on a perfectly absorbing surface?

3. Radio stations often advertise "instant news." If that means you can hear the news the instant the radio announcer speaks it, is the claim true? What approximate time interval is required for a message to travel from Maine to California by radio waves? (Assume the waves can be detected at this range.)

4. List at least three differences between sound waves and light waves.

5. If a high-frequency current exists in a solenoid containing a metallic core, the core becomes warm due to induction. Explain why the material rises in temperature in this situation.

6. When light (or other electromagnetic radiation) travels across a given region, (a) what is it that oscillates? (b) What is it that is transported?

7. Why should an infrared photograph of a person look different from a photograph taken with visible light?

8. Do Maxwell's equations allow for the existence of magnetic monopoles? Explain.

9. Despite the advent of digital television, some viewers still use "rabbit ears" atop their sets (Fig. CQ34.9) instead of purchasing cable television service or satellite dishes. Certain orientations of the receiving antenna on a television set give better reception than others. Furthermore, the best orientation varies from station to station. Explain.

Figure CQ34.9

10. What does a radio wave do to the charges in the receiving antenna to provide a signal for your car radio?

11. Describe the physical significance of the Poynting vector.

12. An empty plastic or glass dish being removed from a microwave oven can be cool to the touch, even when food on an adjoining dish is hot. How is this phenomenon possible?

13. What new concept did Maxwell's generalized form of Ampère's law include?

Problems available in ~~ENHANCED~~ WebAssign   Access end-of-chapter problems online at www.webassign.net

Section 34.1 Displacement Current and the General Form of Ampère's Law
Problems 1–3

Section 34.2 Maxwell's Equations and Hertz's Discoveries
Problems 4–6

Section 34.3 Plane Electromagnetic Waves
Problems 7–19

Section 34.4 Energy Carried by Electromagnetic Waves
Problems 20–34

Section 34.5 Momentum and Radiation Pressure
Problems 35–43

Section 34.6 Production of Electromagnetic Waves by an Antenna
Problems 44–49

Section 34.7 The Spectrum of Electromagnetic Waves
Problems 50–53

Additional Problems
Problems 54–75

Challenge Problems
Problem 76–79

Solutions to the following Problems are available in the *Student Solutions Manual/Study Guide:*
34.3, 34.5, 34.15, 34.19, 34.23, 34.29, 34.36, 34.37, 34.49, 34.51, 34.56, 34.63, 34.65, 34.77, 34.78, and 34.79

List of Enhanced Problems

Problem Number	Targeted Feedback in Enhanced WebAssign	Analysis Model Tutorial in Enhanced WebAssign	Master It in Enhanced WebAssign	Watch It in Enhanced WebAssign
34.2	✓			✓
34.3			✓	
34.4	✓	✓		✓
34.5			✓	
34.12	✓			✓
34.15	✓		✓	
34.17		✓		
34.19	✓		✓	
34.21	✓			✓
34.23	✓		✓	
34.27	✓			✓
34.29			✓	
34.31	✓			✓
34.37	✓		✓	
34.39		✓		
34.43	✓	✓		✓
34.50	✓			✓
34.63		✓		
34.65	✓		✓	
34.79		✓		

Light and Optics

The Grand Tetons in western Wyoming are reflected in a smooth lake at sunset. The optical principles we study in this part of the book will explain the nature of the reflected image of the mountains and why the sky appears red. *(David Muench/Terra/Corbis)*

Light is basic to almost all life on the Earth. For example, plants convert the energy transferred by sunlight to chemical energy through photosynthesis. In addition, light is the principal means by which we are able to transmit and receive information to and from objects around us and throughout the Universe. Light is a form of electromagnetic radiation and represents energy transfer from the source to the observer.

Many phenomena in our everyday life depend on the properties of light. When you watch a television or view photos on a computer monitor, you are seeing millions of colors formed from combinations of only three colors that are physically on the screen: red, blue, and green. The blue color of the daytime sky is a result of the optical phenomenon of *scattering* of light by air molecules, as are the red and orange colors of sunrises and sunsets. You see your image in your bathroom mirror in the morning or the images of other cars in your rearview mirror when you are driving. These images result from *reflection* of light. If you wear glasses or contact lenses, you are depending on *refraction* of light for clear vision. The colors of a rainbow result from *dispersion* of light as it passes through raindrops hovering in the sky after a rainstorm. If you have ever seen the colored circles of the glory surrounding the shadow of your airplane on clouds as you fly above them, you are seeing an effect that results from *interference* of light. The phenomena mentioned here have been studied by scientists and are well understood.

In the introduction to Chapter 35, we discuss the dual nature of light. In some cases, it is best to model light as a stream of particles; in others, a wave model works better. Chapters 35 through 38 concentrate on those aspects of light that are best understood through the wave model of light. In Part 6, we will investigate the particle nature of light. ■

The Nature of Light and the Principles of Ray Optics

This photograph of a rainbow shows the range of colors from red on the top to violet on the bottom. The appearance of the rainbow depends on three optical phenomena discussed in this chapter: reflection, refraction, and dispersion. The faint pastel-colored bows beneath the main rainbow are called supernumerary bows. They are formed by interference between rays of light leaving raindrops below those causing the main rainbow. (John W. Jewett, Jr.)

This first chapter on optics begins by introducing two historical models for light and discussing early methods for measuring the speed of light. Next we study the fundamental phenomena of geometric optics: reflection of light from a surface and refraction as the light crosses the boundary between two media. We also study the dispersion of light as it refracts into materials, resulting in visual displays such as the rainbow. Finally, we investigate the phenomenon of total internal reflection, which is the basis for the operation of optical fibers and the technology of fiber optics.

35.1 The Nature of Light

Before the beginning of the 19th century, light was considered to be a stream of particles that either was emitted by the object being viewed or emanated from the eyes of the viewer. Newton, the chief architect of the particle model of light, held that particles were emitted from a light source and that these particles stimulated the sense of sight upon entering the eye. Using this idea, he was able to explain reflection and refraction.

Most scientists accepted Newton's particle model. During Newton's lifetime, however, another model was proposed, one that argued that light might be some sort of wave motion. In 1678, Dutch physicist and astronomer Christian Huygens showed that a wave model of light could also explain reflection and refraction.

In 1801, Thomas Young (1773–1829) provided the first clear experimental demonstration of the wave nature of light. Young showed that under appropriate conditions light rays interfere with one another according to the waves in interference model, just like mechanical waves (Chapter 18). Such behavior could not be explained at that time by a particle model because there was no conceivable way in which two or more particles could come together and cancel one another. Additional developments during the 19th century led to the general acceptance of the wave model of light, the most important resulting from the work of Maxwell, who in 1873 asserted that light was a form of high-frequency electromagnetic wave. As discussed in Chapter 34, Hertz provided experimental confirmation of Maxwell's theory in 1887 by producing and detecting electromagnetic waves.

Although the wave model and the classical theory of electricity and magnetism were able to explain most known properties of light, they could not explain some subsequent experiments. The most striking phenomenon is the photoelectric effect, also discovered by Hertz: when light strikes a metal surface, electrons are sometimes ejected from the surface. As one example of the difficulties that arose, experiments showed that the kinetic energy of an ejected electron is independent of the light intensity. This finding contradicted the wave model, which held that a more intense beam of light should add more energy to the electron. Einstein proposed an explanation of the photoelectric effect in 1905 using a model based on the concept of quantization developed by Max Planck (1858–1947) in 1900. The quantization model assumes the energy of a light wave is present in particles called *photons;* hence, the energy is said to be quantized. According to Einstein's theory, the energy of a photon is proportional to the frequency of the electromagnetic wave:

$$E = hf \tag{35.1}$$

◀ Energy of a photon

where the constant of proportionality $h = 6.63 \times 10^{-34}$ J · s is called *Planck's constant.* We study this theory in Chapter 40.

In view of these developments, light must be regarded as having a dual nature. Light exhibits the characteristics of a wave in some situations and the characteristics of a particle in other situations. Light is light, to be sure. The question "Is light a wave or a particle?" is inappropriate, however. Sometimes light acts like a wave, and other times it acts like a particle. In the next few chapters, we investigate the wave nature of light.

Christian Huygens
Dutch Physicist and Astronomer (1629–1695)
Huygens is best known for his contributions to the fields of optics and dynamics. To Huygens, light was a type of vibratory motion, spreading out and producing the sensation of light when impinging on the eye. On the basis of this theory, he deduced the laws of reflection and refraction and explained the phenomenon of double refraction.

35.2 Measurements of the Speed of Light

Light travels at such a high speed (to three digits, $c = 3.00 \times 10^8$ m/s) that early attempts to measure its speed were unsuccessful. Galileo attempted to measure the speed of light by positioning two observers in towers separated by approximately 10 km. Each observer carried a shuttered lantern. One observer would open his lantern first, and then the other would open his lantern at the moment he saw the light from the first lantern. Galileo reasoned that by knowing the transit time of the light beams from one lantern to the other and the distance between the two lanterns, he could obtain the speed. His results were inconclusive. Today, we realize (as Galileo concluded) that it is impossible to measure the speed of light in this manner because the transit time for the light is so much less than the reaction time of the observers.

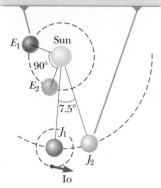

In the time interval during which the Earth travels 90° around the Sun (three months), Jupiter travels only about 7.5°.

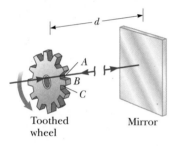

Figure 35.1 Roemer's method for measuring the speed of light (drawing not to scale).

Roemer's Method

In 1675, Danish astronomer Ole Roemer (1644–1710) made the first successful estimate of the speed of light. Roemer's technique involved astronomical observations of Io, one of the moons of Jupiter. Io has a period of revolution around Jupiter of approximately 42.5 h. The period of revolution of Jupiter around the Sun is about 12 yr; therefore, as the Earth moves through 90° around the Sun, Jupiter revolves through only $(\frac{1}{12})90° = 7.5°$ (Fig. 35.1).

An observer using the orbital motion of Io as a clock would expect the orbit to have a constant period. After collecting data for more than a year, however, Roemer observed a systematic variation in Io's period. He found that the periods were longer than average when the Earth was receding from Jupiter and shorter than average when the Earth was approaching Jupiter. Roemer attributed this variation in period to the distance between the Earth and Jupiter changing from one observation to the next.

Using Roemer's data, Huygens estimated the lower limit for the speed of light to be approximately 2.3×10^8 m/s. This experiment is important historically because it demonstrated that light does have a finite speed and gave an estimate of this speed.

Fizeau's Method

The first successful method for measuring the speed of light by means of purely terrestrial techniques was developed in 1849 by French physicist Armand H. L. Fizeau (1819–1896). Figure 35.2 represents a simplified diagram of Fizeau's apparatus. The basic procedure is to measure the total time interval during which light travels from some point to a distant mirror and back. If d is the distance between the light source (considered to be at the location of the wheel) and the mirror and if the time interval for one round trip is Δt, the speed of light is $c = 2d/\Delta t$.

To measure the transit time, Fizeau used a rotating toothed wheel, which converts a continuous beam of light into a series of light pulses. The rotation of such a wheel controls what an observer at the light source sees. For example, if the pulse traveling toward the mirror and passing the opening at point A in Figure 35.2 should return to the wheel at the instant tooth B had rotated into position to cover the return path, the pulse would not reach the observer. At a greater rate of rotation, the opening at point C could move into position to allow the reflected pulse to reach the observer. Knowing the distance d, the number of teeth in the wheel, and the angular speed of the wheel, Fizeau arrived at a value of 3.1×10^8 m/s. Similar measurements made by subsequent investigators yielded more precise values for c, which led to the currently accepted value of $2.997\ 924\ 58 \times 10^8$ m/s.

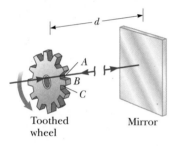

Toothed wheel   Mirror

Figure 35.2 Fizeau's method for measuring the speed of light using a rotating toothed wheel. The light source is considered to be at the location of the wheel; therefore, the distance d is known.

Example 35.1   Measuring the Speed of Light with Fizeau's Wheel   AM

Assume Fizeau's wheel has 360 teeth and rotates at 27.5 rev/s when a pulse of light passing through opening A in Figure 35.2 is blocked by tooth B on its return. If the distance to the mirror is 7 500 m, what is the speed of light?

SOLUTION

Conceptualize Imagine a pulse of light passing through opening A in Figure 35.2 and reflecting from the mirror. By the time the pulse arrives back at the wheel, tooth B has rotated into the position previously occupied by opening A.

Categorize The wheel is a rigid object rotating at constant angular speed. We model the pulse of light as a *particle under constant speed*.

Analyze The wheel has 360 teeth, so it must have 360 openings. Therefore, because the light passes through opening A but is blocked by the tooth immediately adjacent to A, the wheel must rotate through an angular displacement of $\frac{1}{720}$ rev in the time interval during which the light pulse makes its round trip.

Use Equation 10.2, with the angular speed constant, to find the time interval for the pulse's round trip:

$$\Delta t = \frac{\Delta \theta}{\omega} = \frac{\frac{1}{720}\ \text{rev}}{27.5\ \text{rev/s}} = 5.05 \times 10^{-5}\ \text{s}$$

▶ **35.1** continued

From the particle under constant speed model, find the speed of the pulse of light:

$$c = \frac{2d}{\Delta t} = \frac{2(7\,500 \text{ m})}{5.05 \times 10^{-5} \text{ s}} = \boxed{2.97 \times 10^8 \text{ m/s}}$$

Finalize This result is very close to the actual value of the speed of light.

35.3 The Ray Approximation in Ray Optics

The field of **ray optics** (sometimes called *geometric optics*) involves the study of the propagation of light. Ray optics assumes light travels in a fixed direction in a straight line as it passes through a uniform medium and changes its direction when it meets the surface of a different medium or if the optical properties of the medium are nonuniform in either space or time. In our study of ray optics here and in Chapter 36, we use what is called the **ray approximation.** To understand this approximation, first notice that the rays of a given wave are straight lines perpendicular to the wave fronts as illustrated in Figure 35.3 for a plane wave. In the ray approximation, a wave moving through a medium travels in a straight line in the direction of its rays.

If the wave meets a barrier in which there is a circular opening whose diameter is much larger than the wavelength as in Figure 35.4a, the wave emerging from the opening continues to move in a straight line (apart from some small edge effects); hence, the ray approximation is valid. If the diameter of the opening is on the order of the wavelength as in Figure 35.4b, the waves spread out from the opening in all directions. This effect, called *diffraction,* will be studied in Chapter 37. Finally, if the opening is much smaller than the wavelength, the opening can be approximated as a point source of waves as shown in Fig. 35.4c.

Similar effects are seen when waves encounter an opaque object of dimension d. In that case, when $\lambda \ll d$, the object casts a sharp shadow.

The ray approximation and the assumption that $\lambda \ll d$ are used in this chapter and in Chapter 36, both of which deal with ray optics. This approximation is very good for the study of mirrors, lenses, prisms, and associated optical instruments such as telescopes, cameras, and eyeglasses.

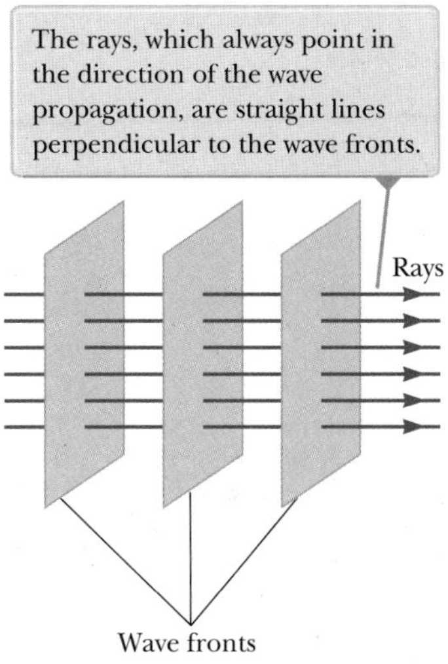

The rays, which always point in the direction of the wave propagation, are straight lines perpendicular to the wave fronts.

Rays

Wave fronts

Figure 35.3 A plane wave propagating to the right.

35.4 Analysis Model: Wave Under Reflection

We introduced the concept of reflection of waves in a discussion of waves on strings in Section 16.4. As with waves on strings, when a light ray traveling in one medium encounters a boundary with another medium, part of the incident light

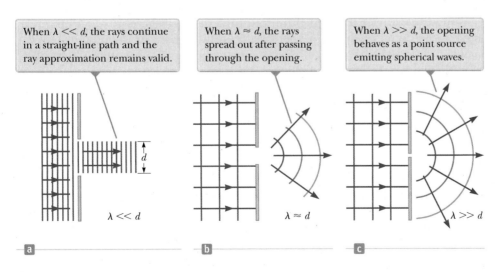

When $\lambda \ll d$, the rays continue in a straight-line path and the ray approximation remains valid.

When $\lambda \approx d$, the rays spread out after passing through the opening.

When $\lambda \gg d$, the opening behaves as a point source emitting spherical waves.

$\lambda \ll d$ $\lambda \approx d$ $\lambda \gg d$

a   b   c

Figure 35.4 A plane wave of wavelength λ is incident on a barrier in which there is an opening of diameter d.

Figure 35.5 Schematic representation of (a) specular reflection, where the reflected rays are all parallel to one another, and (b) diffuse reflection, where the reflected rays travel in random directions. (c) and (d) Photographs of specular and diffuse reflection using laser light.

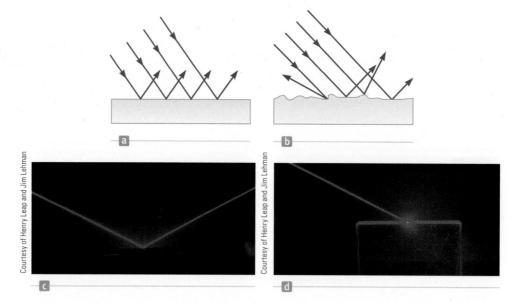

Courtesy of Henry Leap and Jim Lehman

Courtesy of Henry Leap and Jim Lehman

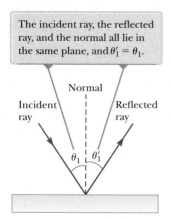

The incident ray, the reflected ray, and the normal all lie in the same plane, and $\theta_1' = \theta_1$.

Normal

Incident ray

Reflected ray

$\theta_1 \quad \theta_1'$

Figure 35.6 The wave under reflection model.

is reflected. For waves on a one-dimensional string, the reflected wave must necessarily be restricted to a direction along the string. For light waves traveling in three-dimensional space, no such restriction applies and the reflected light waves can be in directions different from the direction of the incident waves. Figure 35.5a shows several rays of a beam of light incident on a smooth, mirror-like, reflecting surface. The reflected rays are parallel to one another as indicated in the figure. The direction of a reflected ray is in the plane perpendicular to the reflecting surface that contains the incident ray. Reflection of light from such a smooth surface is called **specular reflection.** If the reflecting surface is rough as in Figure 35.5b, the surface reflects the rays not as a parallel set but in various directions. Reflection from any rough surface is known as **diffuse reflection.** A surface behaves as a smooth surface as long as the surface variations are much smaller than the wavelength of the incident light.

The difference between these two kinds of reflection explains why it is more difficult to see while driving on a rainy night than on a dry night. If the road is wet, the smooth surface of the water specularly reflects most of your headlight beams away from your car (and perhaps into the eyes of oncoming drivers). When the road is dry, its rough surface diffusely reflects part of your headlight beam back toward you, allowing you to see the road more clearly. Your bathroom mirror exhibits specular reflection, whereas light reflecting from this page experiences diffuse reflection. In this book, we restrict our study to specular reflection and use the term *reflection* to mean specular reflection.

Consider a light ray traveling in air and incident at an angle on a flat, smooth surface as shown in Figure 35.6. The incident and reflected rays make angles θ_1 and θ_1', respectively, where the angles are measured between the normal and the rays. (The normal is a line drawn perpendicular to the surface at the point where the incident ray strikes the surface.) Experiments and theory show that the angle of reflection equals the angle of incidence:

Law of reflection ▶

$$\theta_1' = \theta_1 \tag{35.2}$$

This relationship is called the **law of reflection.** Because reflection of waves from an interface between two media is a common phenomenon, we identify an analysis model for this situation: the **wave under reflection.** Equation 35.2 is the mathematical representation of this model.

Quick Quiz 35.1 In the movies, you sometimes see an actor looking in a mirror and you can see his face in the mirror. It can be said with certainty that during the filming of such a scene, the actor sees in the mirror: **(a)** his face **(b)** your face **(c)** the director's face **(d)** the movie camera **(e)** impossible to determine

Example 35.2 The Double-Reflected Light Ray AM

Two mirrors make an angle of 120° with each other as illustrated in Figure 35.7a. A ray is incident on mirror M_1 at an angle of 65° to the normal. Find the direction of the ray after it is reflected from mirror M_2.

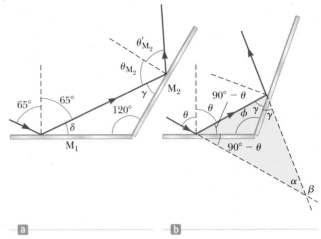

SOLUTION

Conceptualize Figure 35.7a helps conceptualize this situation. The incoming ray reflects from the first mirror, and the reflected ray is directed toward the second mirror. Therefore, there is a second reflection from the second mirror.

Categorize Because the interactions with both mirrors are simple reflections, we apply the *wave under reflection* model and some geometry.

Figure 35.7 (Example 35.2) (a) Mirrors M_1 and M_2 make an angle of 120° with each other. (b) The geometry for an arbitrary mirror angle.

Analyze From the law of reflection, the first reflected ray makes an angle of 65° with the normal.

Find the angle the first reflected ray makes with the horizontal:

$$\delta = 90° - 65° = 25°$$

From the triangle made by the first reflected ray and the two mirrors, find the angle the reflected ray makes with M_2:

$$\gamma = 180° - 25° - 120° = 35°$$

Find the angle the first reflected ray makes with the normal to M_2:

$$\theta_{M_2} = 90° - 35° = 55°$$

From the law of reflection, find the angle the second reflected ray makes with the normal to M_2:

$$\theta'_{M_2} = \theta_{M_2} = \boxed{55°}$$

Finalize Let's explore variations in the angle between the mirrors as follows.

WHAT IF? If the incoming and outgoing rays in Figure 35.7a are extended behind the mirror, they cross at an angle of 60° and the overall change in direction of the light ray is 120°. This angle is the same as that between the mirrors. What if the angle between the mirrors is changed? Is the overall change in the direction of the light ray always equal to the angle between the mirrors?

Answer Making a general statement based on one data point or one observation is always a dangerous practice! Let's investigate the change in direction for a general situation. Figure 35.7b shows the mirrors at an arbitrary angle ϕ and the incoming light ray striking the mirror at an arbitrary angle θ with respect to the normal to the mirror surface. In accordance with the law of reflection and the sum of the interior angles of a triangle, the angle γ is given by $\gamma = 180° - (90° - \theta) - \phi = 90° + \theta - \phi$.

Consider the triangle highlighted in yellow in Figure 35.7b and determine α:

$$\alpha + 2\gamma + 2(90° - \theta) = 180° \quad \rightarrow \quad \alpha = 2(\theta - \gamma)$$

Notice from Figure 35.7b that the change in direction of the light ray is angle β. Use the geometry in the figure to solve for β:

$$\beta = 180° - \alpha = 180° - 2(\theta - \gamma)$$
$$= 180° - 2[\theta - (90° + \theta - \phi)] = 360° - 2\phi$$

Notice that β is not equal to ϕ. For $\phi = 120°$, we obtain $\beta = 120°$, which happens to be the same as the mirror angle; that is true only for this special angle between the mirrors, however. For example, if $\phi = 90°$, we obtain $\beta = 180°$. In that case, the light is reflected straight back to its origin.

If the angle between two mirrors is 90°, the reflected beam returns to the source parallel to its original path as discussed in the What If? section of the preceding example. This phenomenon, called *retroreflection*, has many practical applications. If a third mirror is placed perpendicular to the first two so that the three form the

Figure 35.8 Applications of retroreflection.

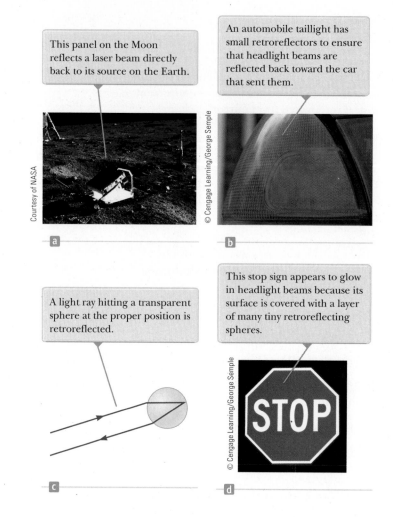

This panel on the Moon reflects a laser beam directly back to its source on the Earth.

Courtesy of NASA

a

An automobile taillight has small retroreflectors to ensure that headlight beams are reflected back toward the car that sent them.

© Cengage Learning/George Semple

b

A light ray hitting a transparent sphere at the proper position is retroreflected.

c

This stop sign appears to glow in headlight beams because its surface is covered with a layer of many tiny retroreflecting spheres.

© Cengage Learning/George Semple

STOP

d

This leg of an ant gives a scale for the size of the mirrors.

Courtesy of Texas Instruments, Inc.

a

The mirror on the left is "on," and the one on the right is "off."

Courtesy of Texas Instruments, Inc.

b

Figure 35.9 (a) An array of mirrors on the surface of a digital micromirror device. Each mirror has an area of approximately $16\ \mu\text{m}^2$. (b) A close-up view of two single micromirrors.

corner of a cube, retroreflection works in three dimensions. In 1969, a panel of many small reflectors was placed on the Moon by the *Apollo 11* astronauts (Fig. 35.8a). A laser beam from the Earth is reflected directly back on itself, and its transit time is measured. This information is used to determine the distance to the Moon with an uncertainty of 15 cm. (Imagine how difficult it would be to align a regular flat mirror on the Moon so that the reflected laser beam would hit a particular location on the Earth!) A more everyday application is found in automobile taillights. Part of the plastic making up the taillight is formed into many tiny cube corners (Fig. 35.8b) so that headlight beams from cars approaching from the rear are reflected back to the drivers. Instead of cube corners, small spherical bumps are sometimes used (Fig. 35.8c). Tiny clear spheres are used in a coating material found on many road signs. Due to retroreflection from these spheres, the stop sign in Figure 35.8d appears much brighter than it would if it were simply a flat, shiny surface. Retroreflectors are also used for reflective panels on running shoes and running clothing to allow joggers to be seen at night.

Another practical application of the law of reflection is the digital projection of movies, television shows, and computer presentations. A digital projector uses an optical semiconductor chip called a *digital micromirror device*. This device contains an array of tiny mirrors (Fig. 35.9a) that can be individually tilted by means of signals to an address electrode underneath the edge of the mirror. Each mirror corresponds to a pixel in the projected image. When the pixel corresponding to a given mirror is to be bright, the mirror is in the "on" position and is oriented so as to reflect light from a source illuminating the array to the screen (Fig. 35.9b). When the pixel for this mirror is to be dark, the mirror is "off" and is tilted so that the light is reflected away from the screen. The brightness of the pixel is determined by the total time interval during which the mirror is in the "on" position during the display of one image.

Digital movie projectors use three micromirror devices, one for each of the primary colors red, blue, and green, so that movies can be displayed with up to 35 trillion colors. Because information is stored as binary data, a digital movie does not degrade with time as does film. Furthermore, because the movie is entirely in the form of computer software, it can be delivered to theaters by means of satellites, optical discs, or optical fiber networks.

Analysis Model | **Wave Under Reflection**

Imagine a wave (electromagnetic or mechanical) traveling through space and striking a flat surface at an angle θ_1 with respect to the normal to the surface. The wave will reflect from the surface in a direction described by the **law of reflection**—the angle of reflection θ_1' equals the angle of incidence θ_1:

$$\theta_1' = \theta_1 \qquad \textbf{(35.2)}$$

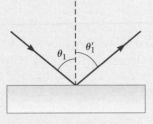

Examples:

- sound waves from an orchestra reflect from a bandshell out to the audience
- a mirror is used to deflect a laser beam in a laser light show
- your bathroom mirror reflects light from your face back to you to form an image of your face (Chapter 36)
- x-rays reflected from a crystalline material create an optical pattern that can be used to understand the structure of the solid (Chapter 38)

35.5 Analysis Model: Wave Under Refraction

In addition to the phenomenon of reflection discussed for waves on strings in Section 16.4, we also found that some of the energy of the incident wave transmits into the new medium. For example, consider Figures 16.15 and 16.16, in which a pulse on a string approaching a junction with another string both reflects from and transmits past the junction and into the second string. Similarly, when a ray of light traveling through a transparent medium encounters a boundary leading into another transparent medium as shown in Figure 35.10, part of the energy is reflected and part enters the second medium. As with reflection, the direction of the transmitted wave exhibits an interesting behavior because of the three-dimensional nature of the light waves. The ray that enters the second medium changes its direction of propagation at the boundary and is said to be **refracted.** The incident ray, the reflected ray, and the refracted ray all lie in the same plane. The **angle of refraction,** θ_2 in Figure 35.10a, depends on the properties of the two media and on the angle of incidence θ_1 through the relationship

$$\frac{\sin \theta_2}{\sin \theta_1} = \frac{v_2}{v_1} \qquad \textbf{(35.3)}$$

where v_1 is the speed of light in the first medium and v_2 is the speed of light in the second medium.

The path of a light ray through a refracting surface is reversible. For example, the ray shown in Figure 35.10a travels from point A to point B. If the ray originated at B, it would travel along line BA to reach point A and the reflected ray would point downward and to the left in the glass.

Quick Quiz 35.2 If beam ① is the incoming beam in Figure 35.10b, which of the other four red lines are reflected beams and which are refracted beams?

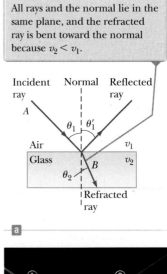

All rays and the normal lie in the same plane, and the refracted ray is bent toward the normal because $v_2 < v_1$.

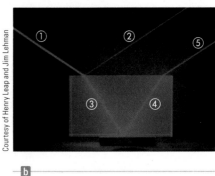

Figure 35.10 (a) The wave under refraction model. (b) Light incident on the Lucite block refracts both when it enters the block and when it leaves the block.

Figure 35.11 The refraction of light as it (a) moves from air into glass and (b) moves from glass into air.

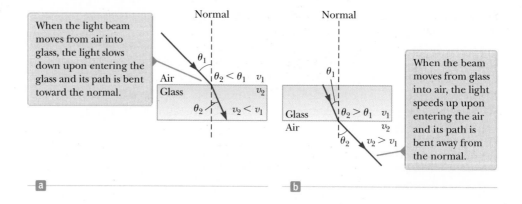

When the light beam moves from air into glass, the light slows down upon entering the glass and its path is bent toward the normal.

When the beam moves from glass into air, the light speeds up upon entering the air and its path is bent away from the normal.

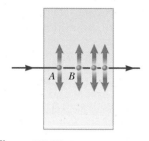

Figure 35.12 Light passing from one atom to another in a medium. The blue spheres are electrons, and the vertical arrows represent their oscillations.

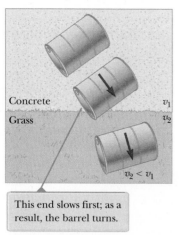

This end slows first; as a result, the barrel turns.

Figure 35.13 Overhead view of a barrel rolling from concrete onto grass.

From Equation 35.3, we can infer that when light moves from a material in which its speed is high to a material in which its speed is lower as shown in Figure 35.11a, the angle of refraction θ_2 is less than the angle of incidence θ_1 and the ray is bent *toward* the normal. If the ray moves from a material in which light moves slowly to a material in which it moves more rapidly as illustrated in Figure 35.11b, then θ_2 is greater than θ_1 and the ray is bent *away* from the normal.

The behavior of light as it passes from air into another substance and then re-emerges into air is often a source of confusion to students. When light travels in air, its speed is 3.00×10^8 m/s, but this speed is reduced to approximately 2×10^8 m/s when the light enters a block of glass. When the light re-emerges into air, its speed instantaneously increases to its original value of 3.00×10^8 m/s. This effect is far different from what happens, for example, when a bullet is fired through a block of wood. In that case, the speed of the bullet decreases as it moves through the wood because some of its original energy is used to tear apart the wood fibers. When the bullet enters the air once again, it emerges at a speed lower than it had when it entered the wood.

To see why light behaves as it does, consider Figure 35.12, which represents a beam of light entering a piece of glass from the left. Once inside the glass, the light may encounter an electron bound to an atom, indicated as point A. Let's assume light is absorbed by the atom, which causes the electron to oscillate (a detail represented by the double-headed vertical arrows). The oscillating electron then acts as an antenna and radiates the beam of light toward an atom at B, where the light is again absorbed. The details of these absorptions and radiations are best explained in terms of quantum mechanics (Chapter 42). For now, it is sufficient to think of light passing from one atom to another through the glass. Although light travels from one atom to another at 3.00×10^8 m/s, the absorption and radiation that take place cause the *average* light speed through the material to fall to approximately 2×10^8 m/s. Once the light emerges into the air, absorption and radiation cease and the light travels at a constant speed of 3.00×10^8 m/s.

A mechanical analog of refraction is shown in Figure 35.13. When the left end of the rolling barrel reaches the grass, it slows down, whereas the right end remains on the concrete and moves at its original speed. This difference in speeds causes the barrel to pivot, which changes the direction of travel.

Index of Refraction

In general, the speed of light in any material is *less* than its speed in vacuum. In fact, *light travels at its maximum speed c in vacuum.* It is convenient to define the **index of refraction** n of a medium to be the ratio

Index of refraction ▶

$$n \equiv \frac{\text{speed of light in vacuum}}{\text{speed of light in a medium}} \equiv \frac{c}{v} \qquad (35.4)$$

This definition shows that the index of refraction is a dimensionless number greater than unity because v is always less than c. Furthermore, n is equal to unity for vacuum. The indices of refraction for various substances are listed in Table 35.1.

As light travels from one medium to another, its frequency does not change but its wavelength does. To see why that is true, consider Figure 35.14. Waves pass an observer at point A in medium 1 with a certain frequency and are incident on the boundary between medium 1 and medium 2. The frequency with which the waves pass an observer at point B in medium 2 must equal the frequency at which they pass point A. If that were not the case, energy would be piling up or disappearing at the boundary. Because there is no mechanism for that to occur, the frequency must be a constant as a light ray passes from one medium into another. Therefore, because the relationship $v = \lambda f$ (Eq. 16.12) from the traveling wave model must be valid in both media and because $f_1 = f_2 = f$, we see that

$$v_1 = \lambda_1 f \quad \text{and} \quad v_2 = \lambda_2 f \tag{35.5}$$

Because $v_1 \neq v_2$, it follows that $\lambda_1 \neq \lambda_2$ as shown in Figure 35.14.

We can obtain a relationship between index of refraction and wavelength by dividing the first Equation 35.5 by the second and then using Equation 35.4:

$$\frac{\lambda_1}{\lambda_2} = \frac{v_1}{v_2} = \frac{c/n_1}{c/n_2} = \frac{n_2}{n_1} \tag{35.6}$$

This expression gives

$$\lambda_1 n_1 = \lambda_2 n_2$$

If medium 1 is vacuum or, for all practical purposes, air, then $n_1 = 1$. Hence, it follows from Equation 35.6 that the index of refraction of any medium can be expressed as the ratio

$$n = \frac{\lambda}{\lambda_n} \tag{35.7}$$

where λ is the wavelength of light in vacuum and λ_n is the wavelength of light in the medium whose index of refraction is n. From Equation 35.7, we see that because $n > 1$, $\lambda_n < \lambda$.

We are now in a position to express Equation 35.3 in an alternative form. Replacing the v_2/v_1 term in Equation 35.3 with n_1/n_2 from Equation 35.6 gives

$$n_1 \sin \theta_1 = n_2 \sin \theta_2 \tag{35.8}$$

The experimental discovery of this relationship is usually credited to Willebrord Snell (1591–1626) and it is therefore known as **Snell's law of refraction.** We shall

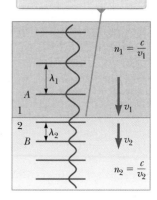

As a wave moves between the media, its wavelength changes but its frequency remains constant.

$n_1 = \dfrac{c}{v_1}$

$n_2 = \dfrac{c}{v_2}$

Figure 35.14 A wave travels from medium 1 to medium 2, in which it moves with lower speed.

Pitfall Prevention 35.2

An Inverse Relationship The index of refraction is *inversely* proportional to the wave speed. As the wave speed v decreases, the index of refraction n increases. Therefore, the higher the index of refraction of a material, the more it *slows down* light from its speed in vacuum. The more the light slows down, the more θ_2 differs from θ_1 in Equation 35.8.

◀ Snell's law of refraction

Pitfall Prevention 35.3

n **Is Not an Integer Here** The symbol n has been used several times as an integer, such as in Chapter 18 to indicate the standing wave mode on a string or in an air column. The index of refraction n is *not* an integer.

Table 35.1		Indices of Refraction	
Substance	**Index of Refraction**	**Substance**	**Index of Refraction**
Solids at 20°C		*Liquids at 20°C*	
Cubic zirconia	2.20	Benzene	1.501
Diamond (C)	2.419	Carbon disulfide	1.628
Fluorite (CaF₂)	1.434	Carbon tetrachloride	1.461
Fused quartz (SiO₂)	1.458	Ethyl alcohol	1.361
Gallium phosphide	3.50	Glycerin	1.473
Glass, crown	1.52	Water	1.333
Glass, flint	1.66		
Ice (H₂O)	1.309	*Gases at 0°C, 1 atm*	
Polystyrene	1.49	Air	1.000 293
Sodium chloride (NaCl)	1.544	Carbon dioxide	1.000 45

Note: All values are for light having a wavelength of 589 nm in vacuum.

examine this equation further in Section 35.6. Refraction of waves at an interface between two media is a common phenomenon, so we identify an analysis model for this situation: the **wave under refraction.** Equation 35.8 is the mathematical representation of this model for electromagnetic radiation. Other waves, such as seismic waves and sound waves, also exhibit refraction according to this model, and the mathematical representation of the model for these waves is Equation 35.3.

Quick Quiz 35.3 Light passes from a material with index of refraction 1.3 into one with index of refraction 1.2. Compared to the incident ray, what happens to the refracted ray? **(a)** It bends toward the normal. **(b)** It is undeflected. **(c)** It bends away from the normal.

Analysis Model **Wave Under Refraction**

Imagine a wave (electromagnetic or mechanical) traveling through space and striking a flat surface at an angle θ_1 with respect to the normal to the surface. Some of the energy of the wave refracts into the medium below the surface in a direction θ_2 described by the **law of refraction—**

$$\frac{\sin \theta_2}{\sin \theta_1} = \frac{v_2}{v_1} \quad (35.3)$$

where v_1 and v_2 are the speeds of the wave in medium 1 and medium 2, respectively.

For light waves, **Snell's law of refraction** states that

$$n_1 \sin \theta_1 = n_2 \sin \theta_2 \quad (35.8)$$

where n_1 and n_2 are the indices of refraction in the two media.

Examples:

- sound waves moving upward from the shore of a lake refract in warmer layers of air higher above the lake and travel downward to a listener in a boat, making sounds from the shore louder than expected
- light from the sky approaches a hot roadway at a grazing angle and refracts upward so as to miss the roadway and enter a driver's eye, giving the illusion of a pool of water on the distant roadway
- light is sent over long distances in an optical fiber because of a difference in index of refraction between the fiber and the surrounding material (Section 35.8)
- a magnifying glass forms an enlarged image of a postage stamp due to refraction of light through the lens (Chapter 36)

Example 35.3 **Angle of Refraction for Glass** **AM**

A light ray of wavelength 589 nm traveling through air is incident on a smooth, flat slab of crown glass at an angle of 30.0° to the normal.

(A) Find the angle of refraction.

SOLUTION

Conceptualize Study Figure 35.11a, which illustrates the refraction process occurring in this problem. We expect that $\theta_2 < \theta_1$ because the speed of light is lower in the glass.

Categorize This is a typical problem in which we apply the *wave under refraction* model.

....

Analyze Rearrange Snell's law of refraction to find $\sin \theta_2$:

$$\sin \theta_2 = \frac{n_1}{n_2} \sin \theta_1$$

Solve for θ_2:

$$\theta_2 = \sin^{-1}\left(\frac{n_1}{n_2} \sin \theta_1\right)$$

Substitute indices of refraction from Table 35.1 and the incident angle:

$$\theta_2 = \sin^{-1}\left(\frac{1.00}{1.52} \sin 30.0°\right) = \boxed{19.2°}$$

(B) Find the speed of this light once it enters the glass.

▶ **35.3** c o n t i n u e d

SOLUTION

Solve Equation 35.4 for the speed of light in the glass:

$$v = \frac{c}{n}$$

Substitute numerical values:

$$v = \frac{3.00 \times 10^8 \text{ m/s}}{1.52} = \boxed{1.97 \times 10^8 \text{ m/s}}$$

(C) What is the wavelength of this light in the glass?

SOLUTION

Use Equation 35.7 to find the wavelength in the glass:

$$\lambda_n = \frac{\lambda}{n} = \frac{589 \text{ nm}}{1.52} = \boxed{388 \text{ nm}}$$

Finalize In part (A), note that $\theta_2 < \theta_1$, consistent with the slower speed of the light found in part (B). In part (C), we see that the wavelength of the light is shorter in the glass than in the air.

Example 35.4	**Light Passing Through a Slab** AM

A light beam passes from medium 1 to medium 2, with the latter medium being a thick slab of material whose index of refraction is n_2 (Fig. 35.15). Show that the beam emerging into medium 1 from the other side is parallel to the incident beam.

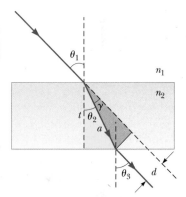

SOLUTION

Conceptualize Follow the path of the light beam as it enters and exits the slab of material in Figure 35.15, where we have assumed that $n_2 > n_1$. The ray bends toward the normal upon entering and away from the normal upon leaving.

Figure 35.15 (Example 35.4) The dashed line drawn parallel to the ray coming out the bottom of the slab represents the path the light would take were the slab not there.

Categorize Like Example 35.3, this is another typical problem in which we apply the *wave under refraction* model.

Analyze Apply Snell's law of refraction to the upper surface:

$$(1) \quad \sin \theta_2 = \frac{n_1}{n_2} \sin \theta_1$$

Apply Snell's law to the lower surface:

$$(2) \quad \sin \theta_3 = \frac{n_2}{n_1} \sin \theta_2$$

Substitute Equation (1) into Equation (2):

$$\sin \theta_3 = \frac{n_2}{n_1} \left(\frac{n_1}{n_2} \sin \theta_1 \right) = \sin \theta_1$$

Finalize Therefore, $\theta_3 = \theta_1$ and the slab does not alter the direction of the beam. It does, however, offset the beam parallel to itself by the distance d shown in Figure 35.15.

WHAT IF? What if the thickness t of the slab is doubled? Does the offset distance d also double?

Answer Consider the region of the light path within the slab in Figure 35.15. The distance a is the hypotenuse of two right triangles.

Find an expression for a from the yellow triangle:

$$a = \frac{t}{\cos \theta_2}$$

Find an expression for d from the red triangle:

$$d = a \sin \gamma = a \sin (\theta_1 - \theta_2)$$

Combine these equations:

$$d = \frac{t}{\cos \theta_2} \sin (\theta_1 - \theta_2)$$

For a given incident angle θ_1, the refracted angle θ_2 is determined solely by the index of refraction, so the offset distance d is proportional to t. If the thickness doubles, so does the offset distance.

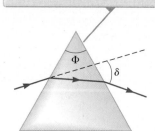

The apex angle Φ is the angle between the sides of the prism through which the light enters and leaves.

Figure 35.16 A prism refracts a single-wavelength light ray through an angle of deviation δ.

In Example 35.4, the light passes through a slab of material with parallel sides. What happens when light strikes a prism with nonparallel sides as shown in Figure 35.16? In this case, the outgoing ray does not propagate in the same direction as the incoming ray. A ray of single-wavelength light incident on the prism from the left emerges at angle δ from its original direction of travel. This angle δ is called the **angle of deviation.** The **apex angle** Φ of the prism, shown in the figure, is defined as the angle between the surface at which the light enters the prism and the second surface that the light encounters.

Example 35.5   **Measuring *n* Using a Prism**   **AM**

Although we do not prove it here, the minimum angle of deviation δ_{min} for a prism occurs when the angle of incidence θ_1 is such that the refracted ray inside the prism makes the same angle with the normal to the two prism faces[1] as shown in Figure 35.17. Obtain an expression for the index of refraction of the prism material in terms of the minimum angle of deviation and the apex angle Φ.

SOLUTION

Conceptualize Study Figure 35.17 carefully and be sure you understand why the light ray comes out of the prism traveling in a different direction.

Categorize In this example, light enters a material through one surface and leaves the material at another surface. Let's apply the *wave under refraction* model to the light passing through the prism.

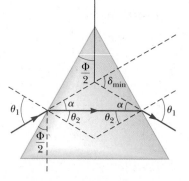

Figure 35.17 (Example 35.5) A light ray passing through a prism at the minimum angle of deviation δ_{min}.

Analyze Consider the geometry in Figure 35.17, where we have used symmetry to label several angles. The reproduction of the angle Φ/2 at the location of the incoming light ray shows that $\theta_2 = \Phi/2$. The theorem that an exterior angle of any triangle equals the sum of the two opposite interior angles shows that $\delta_{min} = 2\alpha$. The geometry also shows that $\theta_1 = \theta_2 + \alpha$.

Combine these three geometric results:

$$\theta_1 = \theta_2 + \alpha = \frac{\Phi}{2} + \frac{\delta_{min}}{2} = \frac{\Phi + \delta_{min}}{2}$$

Apply the wave under refraction model at the left surface and solve for *n*:

$$(1.00) \sin \theta_1 = n \sin \theta_2 \quad \rightarrow \quad n = \frac{\sin \theta_1}{\sin \theta_2}$$

Substitute for the incident and refracted angles:

$$n = \frac{\sin \left(\dfrac{\Phi + \delta_{min}}{2} \right)}{\sin (\Phi/2)} \qquad \textbf{(35.9)}$$

---

[1]The details of this proof are available in texts on optics.

▶ **35.5** continued

Finalize Knowing the apex angle Φ of the prism and measuring δ_{min}, you can calculate the index of refraction of the prism material. Furthermore, a hollow prism can be used to determine the values of n for various liquids filling the prism.

35.6 Huygens's Principle

The laws of reflection and refraction were stated earlier in this chapter without proof. In this section, we develop these laws by using a geometric method proposed by Huygens in 1678. **Huygens's principle** is a geometric construction for using knowledge of an earlier wave front to determine the position of a new wave front at some instant:

> All points on a given wave front are taken as point sources for the production of spherical secondary waves, called wavelets, that propagate outward through a medium with speeds characteristic of waves in that medium. After some time interval has passed, the new position of the wave front is the surface tangent to the wavelets.

First, consider a plane wave moving through free space as shown in Figure 35.18a. At $t = 0$, the wave front is indicated by the plane labeled AA'. In Huygens's construction, each point on this wave front is considered a point source. For clarity, only three point sources on AA' are shown. With these sources for the wavelets, we draw circular arcs, each of radius $c \Delta t$, where c is the speed of light in vacuum and Δt is some time interval during which the wave propagates. The surface drawn tangent to these wavelets is the plane BB', which is the wave front at a later time, and is parallel to AA'. In a similar manner, Figure 35.18b shows Huygens's construction for a spherical wave.

Huygens's Principle Applied to Reflection and Refraction

We now derive the laws of reflection and refraction, using Huygens's principle.

For the law of reflection, refer to Figure 35.19. The line AB represents a plane wave front of the incident light just as ray 1 strikes the surface. At this instant, the wave at A sends out a Huygens wavelet (appearing at a later time as the light brown circular arc passing through D); the reflected light makes an angle γ' with the surface. At the

Pitfall Prevention 35.4
Of What Use Is Huygens's Principle? At this point, the importance of Huygens's principle may not be evident. Predicting the position of a future wave front may not seem to be very critical. We will use Huygens's principle here to generate the laws of reflection and refraction and in later chapters to explain additional wave phenomena for light.

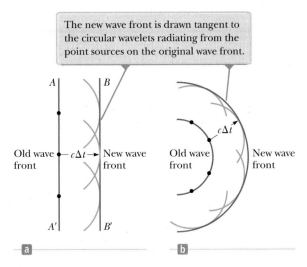

Figure 35.18 gives the text: "The new wave front is drawn tangent to the circular wavelets radiating from the point sources on the original wave front."

Figure 35.18 Huygens's construction for (a) a plane wave propagating to the right and (b) a spherical wave propagating to the right.

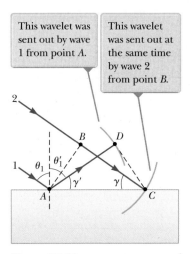

Figure 35.19 gives the text: "This wavelet was sent out by wave 1 from point A." and "This wavelet was sent out at the same time by wave 2 from point B."

Figure 35.19 Huygens's construction for proving the law of reflection.

same time, the wave at B emits a Huygens wavelet (the light brown circular arc passing through C) with the incident light making an angle γ with the surface. Figure 35.19 shows these wavelets after a time interval Δt, after which ray 2 strikes the surface. Because both rays 1 and 2 move with the same speed, we must have $AD = BC = c\,\Delta t$.

The remainder of our analysis depends on geometry. Notice that the two triangles ABC and ADC are congruent because they have the same hypotenuse AC and because $AD = BC$. Figure 35.19 shows that

$$\cos \gamma = \frac{BC}{AC} \quad \text{and} \quad \cos \gamma' = \frac{AD}{AC}$$

where $\gamma = 90° - \theta_1$ and $\gamma' = 90° - \theta_1'$. Because $AD = BC$,

$$\cos \gamma = \cos \gamma'$$

Therefore,

$$\gamma = \gamma'$$
$$90° - \theta_1 = 90° - \theta_1'$$

and

$$\theta_1 = \theta_1'$$

which is the law of reflection.

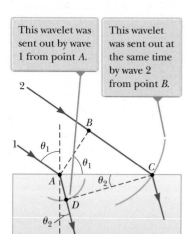

This wavelet was sent out by wave 1 from point A.

This wavelet was sent out at the same time by wave 2 from point B.

Figure 35.20 Huygens's construction for proving Snell's law of refraction.

Now let's use Huygens's principle to derive Snell's law of refraction. We focus our attention on the instant ray 1 strikes the surface and the subsequent time interval until ray 2 strikes the surface as in Figure 35.20. During this time interval, the wave at A sends out a Huygens wavelet (the light brown arc passing through D) and the light refracts into the material, making an angle θ_2 with the normal to the surface. In the same time interval, the wave at B sends out a Huygens wavelet (the light brown arc passing through C) and the light continues to propagate in the same direction. Because these two wavelets travel through different media, the radii of the wavelets are different. The radius of the wavelet from A is $AD = v_2\,\Delta t$, where v_2 is the wave speed in the second medium. The radius of the wavelet from B is $BC = v_1\,\Delta t$, where v_1 is the wave speed in the original medium.

From triangles ABC and ADC, we find that

$$\sin \theta_1 = \frac{BC}{AC} = \frac{v_1\,\Delta t}{AC} \quad \text{and} \quad \sin \theta_2 = \frac{AD}{AC} = \frac{v_2\,\Delta t}{AC}$$

Dividing the first equation by the second gives

$$\frac{\sin \theta_1}{\sin \theta_2} = \frac{v_1}{v_2}$$

From Equation 35.4, however, we know that $v_1 = c/n_1$ and $v_2 = c/n_2$. Therefore,

$$\frac{\sin \theta_1}{\sin \theta_2} = \frac{c/n_1}{c/n_2} = \frac{n_2}{n_1}$$

and

$$n_1 \sin \theta_1 = n_2 \sin \theta_2$$

which is Snell's law of refraction.

35.7 Dispersion

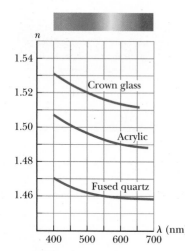

Figure 35.21 Variation of index of refraction with vacuum wavelength for three materials.

An important property of the index of refraction n is that, for a given material, the index varies with the wavelength of the light passing through the material as Figure 35.21 shows. This behavior is called **dispersion.** Because n is a function of wavelength, Snell's law of refraction indicates that light of different wavelengths is refracted at different angles when incident on a material.

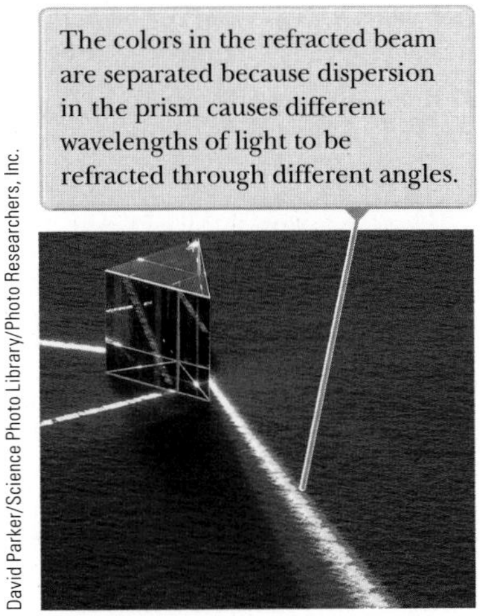

The colors in the refracted beam are separated because dispersion in the prism causes different wavelengths of light to be refracted through different angles.

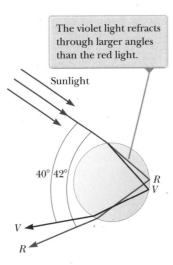

Figure 35.22 White light enters a glass prism at the upper left.

The violet light refracts through larger angles than the red light.

Figure 35.23 Path of sunlight through a spherical raindrop. Light following this path contributes to the visible rainbow.

Pitfall Prevention 35.5
A Rainbow of Many Light Rays
Pictorial representations such as Figure 35.23 are subject to misinterpretation. The figure shows one ray of light entering the raindrop and undergoing reflection and refraction, exiting the raindrop in a range of 40° to 42° from the entering ray. This illustration might be interpreted incorrectly as meaning that *all* light entering the raindrop exits in this small range of angles. In reality, light exits the raindrop over a much larger range of angles, from 0° to 42°. A careful analysis of the reflection and refraction from the spherical raindrop shows that the range of 40° to 42° is where the *highest-intensity light* exits the raindrop.

Figure 35.21 shows that the index of refraction generally decreases with increasing wavelength. For example, violet light refracts more than red light does when passing into a material.

Now suppose a beam of *white light* (a combination of all visible wavelengths) is incident on a prism as illustrated in Figure 35.22. Clearly, the angle of deviation δ depends on wavelength. The rays that emerge spread out in a series of colors known as the **visible spectrum.** These colors, in order of decreasing wavelength, are red, orange, yellow, green, blue, and violet. Newton showed that each color has a particular angle of deviation and that the colors can be recombined to form the original white light.

The dispersion of light into a spectrum is demonstrated most vividly in nature by the formation of a rainbow, which is often seen by an observer positioned between the Sun and a rain shower. To understand how a rainbow is formed, consider Figure 35.23. We will need to apply both the wave under reflection and wave under refraction models. A ray of sunlight (which is white light) passing overhead strikes a drop of water in the atmosphere and is refracted and reflected as follows. It is first refracted at the front surface of the drop, with the violet light deviating the most and the red light the least. At the back surface of the drop, the light is reflected and returns to the front surface, where it again undergoes refraction as it moves from water into air. The rays leave the drop such that the angle between the incident white light and the most intense returning violet ray is 40° and the angle between the incident white light and the most intense returning red ray is 42°. This small angular difference between the returning rays causes us to see a colored bow.

Now suppose an observer is viewing a rainbow as shown in Figure 35.24. If a raindrop high in the sky is being observed, the most intense red light returning from the drop reaches the observer because it is deviated the least; the most intense violet light, however, passes over the observer because it is deviated the most. Hence, the observer sees red light coming from this drop. Similarly, a drop lower in the sky directs the most intense violet light toward the observer and appears violet to the observer. (The most intense red light from this drop passes below the observer's eye and is not seen.) The most intense light from other colors of the spectrum reaches the observer from raindrops lying between these two extreme positions.

Figure 35.25 (page 840) shows a *double rainbow.* The secondary rainbow is fainter than the primary rainbow, and the colors are reversed. The secondary rainbow arises from light that makes two reflections from the interior surface before exiting

The highest-intensity light traveling from higher raindrops toward the eyes of the observer is red, whereas the most intense light from lower drops is violet.

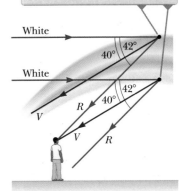

Figure 35.24 The formation of a rainbow seen by an observer standing with the Sun behind his back.

Figure 35.25 This photograph of a rainbow shows a distinct secondary rainbow with the colors reversed.

the raindrop. In the laboratory, rainbows have been observed in which the light makes more than 30 reflections before exiting the water drop. Because each reflection involves some loss of light due to refraction of part of the incident light out of the water drop, the intensity of these higher-order rainbows is small compared with that of the primary rainbow.

Quick Quiz 35.4 In photography, lenses in a camera use refraction to form an image on a light-sensitive surface. Ideally, you want all the colors in the light from the object being photographed to be refracted by the same amount. Of the materials shown in Figure 35.21, which would you choose for a single-element camera lens? **(a)** crown glass **(b)** acrylic **(c)** fused quartz **(d)** impossible to determine

35.8 Total Internal Reflection

An interesting effect called **total internal reflection** can occur when light is directed from a medium having a given index of refraction toward one having a lower index of refraction. Consider Figure 35.26a, in which a light ray travels in medium 1 and meets the boundary between medium 1 and medium 2, where n_1 is greater than n_2. In the figure, labels 1 through 5 indicate various possible directions of the ray consistent with the wave under refraction model. The refracted rays are bent away from the normal because n_1 is greater than n_2. At some particular angle of incidence θ_c, called the **critical angle**, the refracted light ray moves parallel to the boundary so that $\theta_2 = 90°$ (Fig. 35.26b). For angles of incidence greater than θ_c, the ray is entirely reflected at the boundary as shown by ray 5 in Figure 35.26a.

We can use Snell's law of refraction to find the critical angle. When $\theta_1 = \theta_c$, $\theta_2 = 90°$ and Equation 35.8 gives

$$n_1 \sin \theta_c = n_2 \sin 90° = n_2$$

◀ Critical angle for total internal reflection

$$\sin \theta_c = \frac{n_2}{n_1} \quad (\text{for } n_1 > n_2) \tag{35.10}$$

This equation can be used only when n_1 is greater than n_2. That is, total internal reflection occurs only when light is directed from a medium of a given index of refraction toward a medium of lower index of refraction. If n_1 were less than n_2,

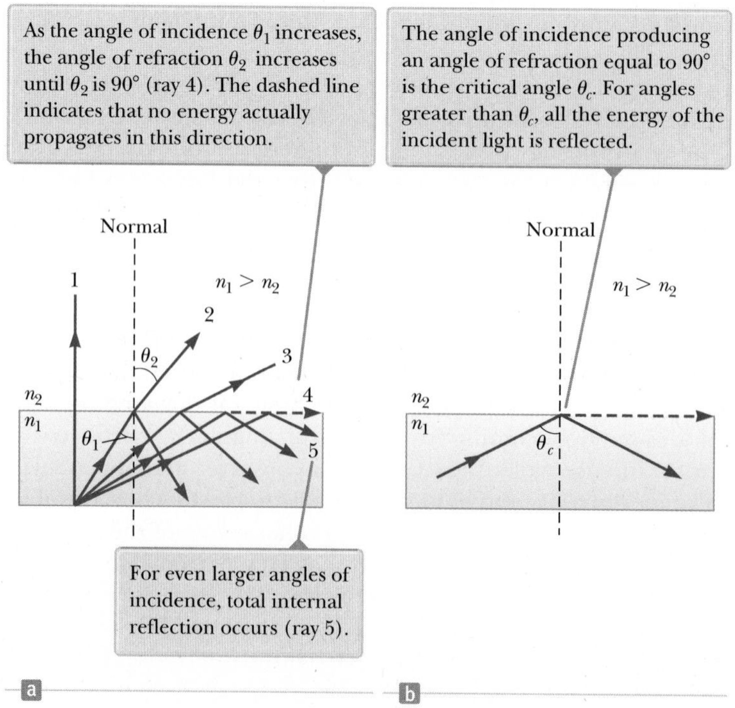

As the angle of incidence θ_1 increases, the angle of refraction θ_2 increases until θ_2 is 90° (ray 4). The dashed line indicates that no energy actually propagates in this direction.

The angle of incidence producing an angle of refraction equal to 90° is the critical angle θ_c. For angles greater than θ_c, all the energy of the incident light is reflected.

Normal

$n_1 > n_2$

1

2

θ_2

3

n_2

n_1

θ_1

4

5

Normal

$n_1 > n_2$

n_2

n_1

θ_c

For even larger angles of incidence, total internal reflection occurs (ray 5).

Figure 35.26 (a) Rays travel from a medium of index of refraction n_1 into a medium of index of refraction n_2, where $n_2 < n_1$. (b) Ray 4 is singled out.

a

b

Equation 35.10 would give $\sin \theta_c > 1$, which is a meaningless result because the sine of an angle can never be greater than unity.

The critical angle for total internal reflection is small when n_1 is considerably greater than n_2. For example, the critical angle for a diamond in air is 24°. Any ray inside the diamond that approaches the surface at an angle greater than 24° is completely reflected back into the crystal. This property, combined with proper faceting, causes diamonds to sparkle. The angles of the facets are cut so that light is "caught" inside the crystal through multiple internal reflections. These multiple reflections give the light a long path through the medium, and substantial dispersion of colors occurs. By the time the light exits through the top surface of the crystal, the rays associated with different colors have been fairly widely separated from one another.

Cubic zirconia also has a high index of refraction and can be made to sparkle very much like a diamond. If a suspect jewel is immersed in corn syrup, the difference in n for the cubic zirconia and that for the corn syrup is small and the critical angle is therefore great. Hence, more rays escape sooner; as a result, the sparkle completely disappears. A real diamond does not lose all its sparkle when placed in corn syrup.

Quick Quiz 35.5 In Figure 35.27, five light rays enter a glass prism from the left. **(i)** How many of these rays undergo total internal reflection at the slanted surface of the prism? (a) one (b) two (c) three (d) four (e) five **(ii)** Suppose the prism in Figure 35.27 can be rotated in the plane of the paper. For *all five* rays to experience total internal reflection from the slanted surface, should the prism be rotated (a) clockwise or (b) counterclockwise?

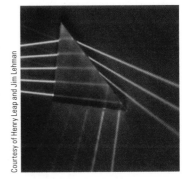

Courtesy of Henry Leap and Jim Lehman

Figure 35.27 (Quick Quiz 35.5) Five nonparallel light rays enter a glass prism from the left.

Example 35.6 **A View from the Fish's Eye**

Find the critical angle for an air–water boundary. (Assume the index of refraction of water is 1.33.)

SOLUTION

Conceptualize Study Figure 35.26 to understand the concept of total internal reflection and the significance of the critical angle.

Categorize We use concepts developed in this section, so we categorize this example as a substitution problem.

Apply Equation 35.10 to the air–water interface:

$$\sin \theta_c = \frac{n_2}{n_1} = \frac{1.00}{1.33} = 0.752$$

$$\theta_c = \boxed{48.8°}$$

WHAT IF? What if a fish in a still pond looks upward toward the water's surface at different angles relative to the surface as in Figure 35.28? What does it see?

Answer Because the path of a light ray is reversible, light traveling from medium 2 into medium 1 in Figure 35.26a follows the paths shown, but in the *opposite* direction. A fish looking upward toward the water surface as in Figure 35.28 can see out of the water if it looks toward the surface at an angle less than the critical angle. Therefore, when the fish's line of vision makes an angle of $\theta = 40°$ with the normal to the surface, for example, light from above the water reaches the fish's eye. At $\theta = 48.8°$, the critical angle for water, the light has to skim along the water's surface before being refracted to the fish's eye; at this angle, the fish can, in principle, see the entire shore of the pond. At angles greater than the critical angle, the light reaching the fish comes by means of total internal reflection at the surface. Therefore, at $\theta = 60°$, the fish sees a reflection of the bottom of the pond.

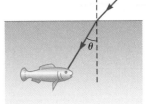

Figure 35.28 (Example 35.6) **What If?** A fish looks upward toward the water surface.

Optical Fibers

Another interesting application of total internal reflection is the use of glass or transparent plastic rods to "pipe" light from one place to another. As indicated in Figure 35.29 (page 842), light is confined to traveling within a rod, even around curves, as the result of successive total internal reflections. Such a light pipe is flexible

Figure 35.29 Light travels in a curved transparent rod by multiple internal reflections.

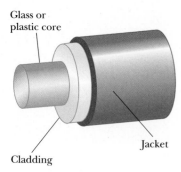

Glass or plastic core

Jacket

Cladding

Figure 35.30 The construction of an optical fiber. Light travels in the core, which is surrounded by a cladding and a protective jacket.

if thin fibers are used rather than thick rods. A flexible light pipe is called an **optical fiber.** If a bundle of parallel fibers is used to construct an optical transmission line, images can be transferred from one point to another. Part of the 2009 Nobel Prize in Physics was awarded to Charles K. Kao (b. 1933) for his discovery of how to transmit light signals over long distances through thin glass fibers. This discovery has led to the development of a sizable industry known as *fiber optics.*

A practical optical fiber consists of a transparent core surrounded by a *cladding,* a material that has a lower index of refraction than the core. The combination may be surrounded by a plastic *jacket* to prevent mechanical damage. Figure 35.30 shows a cutaway view of this construction. Because the index of refraction of the cladding is less than that of the core, light traveling in the core experiences total internal reflection if it arrives at the interface between the core and the cladding at an angle of incidence that exceeds the critical angle. In this case, light "bounces" along the core of the optical fiber, losing very little of its intensity as it travels.

Any loss in intensity in an optical fiber is essentially due to reflections from the two ends and absorption by the fiber material. Optical fiber devices are particularly useful for viewing an object at an inaccessible location. For example, physicians often use such devices to examine internal organs of the body or to perform surgery without making large incisions. Optical fiber cables are replacing copper wiring and coaxial cables for telecommunications because the fibers can carry a much greater volume of telephone calls or other forms of communication than electrical wires can.

Figure 35.31a shows a bundle of optical fibers gathered into an optical cable that can be used to carry communication signals. Figure 35.31b shows laser light following the curves of a coiled bundle by total internal reflection. Many computers and other electronic equipment now have optical ports as well as electrical ports for transferring information.

Figure 35.31 (a) Strands of glass optical fibers are used to carry voice, video, and data signals in telecommunication networks. (b) A bundle of optical fibers is illuminated by a laser.

Dennis O'Clair/Getty Images

a

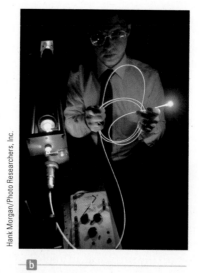

Hank Morgan/Photo Researchers, Inc.

b

Summary

Definition

The **index of refraction** n of a medium is defined by the ratio

$$n \equiv \frac{c}{v}$$

(35.4)

where c is the speed of light in vacuum and v is the speed of light in the medium.

Concepts and Principles

In geometric optics, we use the **ray approximation,** in which a wave travels through a uniform medium in straight lines in the direction of the rays.

Total internal reflection occurs when light travels from a medium of high index of refraction to one of lower index of refraction. The **critical angle** θ_c for which total internal reflection occurs at an interface is given by

$$\sin \theta_c = \frac{n_2}{n_1} \quad (\text{for } n_1 > n_2) \qquad \textbf{(35.10)}$$

Analysis Models for Problem Solving

Wave Under Reflection. The **law of reflection** states that for a light ray (or other type of wave) incident on a smooth surface, the angle of reflection θ_1' equals the angle of incidence θ_1:

$$\theta_1' = \theta_1 \qquad \textbf{(35.2)}$$

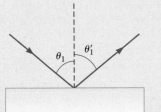

Wave Under Refraction. A wave crossing a boundary as it travels from medium 1 to medium 2 is **refracted.** The angle of refraction θ_2 is related to the incident angle θ_1 by the relationship

$$\frac{\sin \theta_2}{\sin \theta_1} = \frac{v_2}{v_1} \qquad \textbf{(35.3)}$$

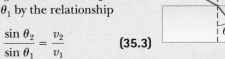

where v_1 and v_2 are the speeds of the wave in medium 1 and medium 2, respectively. The incident ray, the reflected ray, the refracted ray, and the normal to the surface all lie in the same plane.

For light waves, **Snell's law of refraction** states that

$$n_1 \sin \theta_1 = n_2 \sin \theta_2 \qquad \textbf{(35.8)}$$

where n_1 and n_2 are the indices of refraction in the two media.

Objective Questions   [1.] denotes answer available in *Student Solutions Manual/Study Guide*

1. In each of the following situations, a wave passes through an opening in an absorbing wall. Rank the situations in order from the one in which the wave is best described by the ray approximation to the one in which the wave coming through the opening spreads out most nearly equally in all directions in the hemisphere beyond the wall. (a) The sound of a low whistle at 1 kHz passes through a doorway 1 m wide. (b) Red light passes through the pupil of your eye. (c) Blue light passes through the pupil of your eye. (d) The wave broadcast by an AM radio station passes through a doorway 1 m wide. (e) An x-ray passes through the space between bones in your elbow joint.

2. A source emits monochromatic light of wavelength 495 nm in air. When the light passes through a liquid, its wavelength reduces to 434 nm. What is the liquid's index of refraction? (a) 1.26 (b) 1.49 (c) 1.14 (d) 1.33 (e) 2.03

3. Carbon disulfide ($n = 1.63$) is poured into a container made of crown glass ($n = 1.52$). What is the critical angle for total internal reflection of a light ray in the liquid when it is incident on the liquid-to-glass surface? (a) 89.2° (b) 68.8° (c) 21.2° (d) 1.07° (e) 43.0°

4. A light wave moves between medium 1 and medium 2. Which of the following are correct statements relating its speed, frequency, and wavelength in the two media, the indices of refraction of the media, and the angles of incidence and refraction? More than one statement may be correct. (a) $v_1/\sin \theta_1 = v_2/\sin \theta_2$ (b) $\csc \theta_1/n_1 = \csc \theta_2/n_2$ (c) $\lambda_1/\sin \theta_1 = \lambda_2/\sin \theta_2$ (d) $f_1/\sin \theta_1 = f_2/\sin \theta_2$ (e) $n_1/\cos \theta_1 = n_2/\cos \theta_2$

5. What happens to a light wave when it travels from air into glass? (a) Its speed remains the same. (b) Its speed increases. (c) Its wavelength increases. (d) Its wavelength remains the same. (e) Its frequency remains the same.

6. The index of refraction for water is about $\frac{4}{3}$. What happens as a beam of light travels from air into water? (a) Its speed increases to $\frac{4}{3}c$, and its frequency decreases. (b) Its speed decreases to $\frac{3}{4}c$, and its wavelength decreases by a factor of $\frac{3}{4}$. (c) Its speed decreases to $\frac{3}{4}c$, and its wavelength increases by a factor of $\frac{4}{3}$. (d) Its speed and frequency remain the same. (e) Its speed decreases to $\frac{3}{4}c$, and its frequency increases.

7. Light can travel from air into water. Some possible paths for the light ray in the water are shown in Figure

OQ35.7. Which path will the light most likely follow? (a) A (b) B (c) C (d) D (e) E

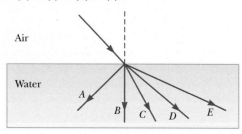

Figure OQ35.7

8. What is the order of magnitude of the time interval required for light to travel 10 km as in Galileo's attempt to measure the speed of light? (a) several seconds (b) several milliseconds (c) several microseconds (d) several nanoseconds

9. A light ray containing both blue and red wavelengths is incident at an angle on a slab of glass. Which of the sketches in Figure OQ35.9 represents the most likely outcome? (a) A (b) B (c) C (d) D (e) none of them

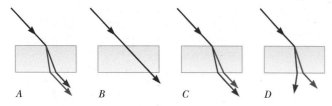

Figure OQ35.9

10. For the following questions, choose from the following possibilities: (a) yes; water (b) no; water (c) yes; air (d) no; air. **(i)** Can light undergo total internal reflection at a smooth interface between air and water? If so, in which medium must it be traveling originally? **(ii)** Can sound undergo total internal reflection at a smooth interface between air and water? If so, in which medium must it be traveling originally?

11. A light ray travels from vacuum into a slab of material with index of refraction n_1 at incident angle θ with respect to the surface. It subsequently passes into a second slab of material with index of refraction n_2 before passing back into vacuum again. The surfaces of the different materials are all parallel to one another. As the light exits the second slab, what can be said of the final angle ϕ that the outgoing light makes with the normal? (a) $\phi > \theta$ (b) $\phi < \theta$ (c) $\phi = \theta$ (d) The angle depends on the magnitudes of n_1 and n_2. (e) The angle depends on the wavelength of the light.

12. Suppose you find experimentally that two colors of light, A and B, originally traveling in the same direction in air, are sent through a glass prism, and A changes direction more than B. Which travels more slowly in the prism, A or B? Alternatively, is there insufficient information to determine which moves more slowly?

13. The core of an optical fiber transmits light with minimal loss if it is surrounded by what? (a) water (b) diamond (c) air (d) glass (e) fused quartz

14. Which color light refracts the most when entering crown glass from air at some incident angle θ with respect to the normal? (a) violet (b) blue (c) green (d) yellow (e) red

15. Light traveling in a medium of index of refraction n_1 is incident on another medium having an index of refraction n_2. Under which of the following conditions can total internal reflection occur at the interface of the two media? (a) The indices of refraction have the relation $n_2 > n_1$. (b) The indices of refraction have the relation $n_1 > n_2$. (c) Light travels slower in the second medium than in the first. (d) The angle of incidence is less than the critical angle. (e) The angle of incidence must equal the angle of refraction.

Conceptual Questions

1. denotes answer available in *Student Solutions Manual/Study Guide*

1. The level of water in a clear, colorless glass can easily be observed with the naked eye. The level of liquid helium in a clear glass vessel is extremely difficult to see with the naked eye. Explain.

2. A complete circle of a rainbow can sometimes be seen from an airplane. With a stepladder, a lawn sprinkler, and a sunny day, how can you show the complete circle to children?

3. You take a child for walks around the neighborhood. She loves to listen to echoes from houses when she shouts or when you clap loudly. A house with a large, flat front wall can produce an echo if you stand straight in front of it and reasonably far away. (a) Draw a bird's-eye view of the situation to explain the production of the echo. Shade the area where you can stand to hear the echo. For parts (b) through (e), explain your answers with diagrams. (b) **What If?** The

child helps you discover that a house with an L-shaped floor plan can produce echoes if you are standing in a wider range of locations. You can be standing at any reasonably distant location from which you can see the inside corner. Explain the echo in this case and compare with your diagram in part (a). (c) **What If?** What if the two wings of the house are not perpendicular? Will you and the child, standing close together, hear echoes? (d) **What If?** What if a rectangular house and its garage have perpendicular walls that would form an inside corner but have a breezeway between them so that the walls do not meet? Will the structure produce strong echoes for people in a wide range of locations?

4. The F-117A stealth fighter (Fig. CQ35.4) is specifically designed to be a *non*retroreflector of radar. What aspects of its design help accomplish this purpose?

Courtesy U.S. Air Force

Figure CQ35.4

5. Retroreflection by transparent spheres, mentioned in Section 35.4, can be observed with dewdrops. To do so, look at your head's shadow where it falls on dewy grass. The optical display around the shadow of your head is called *heiligenschein*, which is German for *holy light*. Renaissance artist Benvenuto Cellini described the phenomenon and his reaction in his *Autobiography*, at the end of Part One, and American philosopher Henry David Thoreau did the same in *Walden,* "Baker Farm," second paragraph. Do some Internet research to find out more about the heiligenschein.

6. Sound waves have much in common with light waves, including the properties of reflection and refraction. Give an example of each of these phenomena for sound waves.

7. Total internal reflection is applied in the periscope of a submerged submarine to let the user observe events above the water surface. In this device, two prisms are arranged as shown in Figure CQ35.7 so that an incident beam of light follows the path shown. Parallel tilted, silvered mirrors could be used, but glass prisms with no silvered surfaces give higher light throughput. Propose a reason for the higher efficiency.

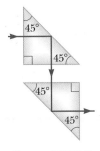

Figure CQ35.7

8. Explain why a diamond sparkles more than a glass crystal of the same shape and size.

9. A laser beam passing through a nonhomogeneous sugar solution follows a curved path. Explain.

10. The display windows of some department stores are slanted slightly inward at the bottom. This tilt is to decrease the glare from streetlights and the Sun, which would make it difficult for shoppers to see the display inside. Sketch a light ray reflecting from such a window to show how this design works.

11. At one restaurant, a worker uses colored chalk to write the daily specials on a blackboard illuminated with a spotlight. At another restaurant, a worker writes with colored grease pencils on a flat, smooth sheet of transparent acrylic plastic with an index of refraction 1.55. The panel hangs in front of a piece of black felt. Small, bright fluorescent tube lights are installed all along the edges of the sheet, inside an opaque channel. Figure CQ35.11 shows a cutaway view of the sign. (a) Explain why viewers at both restaurants see the letters shining against a black background. (b) Explain why the sign at the second restaurant may use less energy from the electric company than the illuminated blackboard at the first restaurant. (c) What would be a good choice for the index of refraction of the material in the grease pencils?

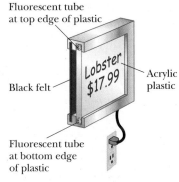

Fluorescent tube at top edge of plastic

Acrylic plastic

Black felt

Fluorescent tube at bottom edge of plastic

Figure CQ35.11

12. (a) Under what conditions is a mirage formed? While driving on a hot day, sometimes you see what appears to be water on the road far ahead. When you arrive at the location of the water, however, the road is perfectly dry. Explain this phenomenon. (b) The mirage called *fata morgana* often occurs over water or in cold regions covered with snow or ice. It can cause islands to sometimes become visible, even though they are not normally visible because they are below the horizon due to the curvature of the Earth. Explain this phenomenon.

13. Figure CQ35.13 shows a pencil partially immersed in a cup of water. Why does the pencil appear to be bent?

© Cengage Learning/Charles D. Winters

Figure CQ35.13

14. A scientific supply catalog advertises a material having an index of refraction of 0.85. Is that a good product to buy? Why or why not?

15. Why do astronomers looking at distant galaxies talk about looking backward in time?

16. Try this simple experiment on your own. Take two opaque cups, place a coin at the bottom of each cup near the edge, and fill one cup with water. Next, view the cups at some angle from the side so that the coin in water is just visible as shown on the left in Figure CQ35.16. Notice that the coin in air is not visible as shown on the right in Figure CQ35.16. Explain this observation.

© Cengage Learning/Ed Dodd

Figure CQ35.16

17. Figure CQ35.17a shows a desk ornament globe containing a photograph. The flat photograph is in air, inside a vertical slot located behind a water-filled compart-

ment having the shape of one half of a cylinder. Suppose you are looking at the center of the photograph and then rotate the globe about a vertical axis. You find that the center of the photograph disappears when you rotate the globe beyond a certain maximum angle (Fig. CQ35.17b). (a) Account for this phenomenon and (b) describe what you see when you turn the globe beyond this angle.

Courtesy of Edwin Lo

a   b

Figure CQ35.17

Problems available in **ENHANCED WebAssign** Access end-of-chapter problems online at **www.webassign.net**

Section 35.1 The Nature of Light
Section 35.2 Measurements of the Speed of Light
Problems 1–4

Section 35.3 The Ray Approximation in Ray Optics
Section 35.4 Analysis Model: Wave Under Reflection
Section 35.5 Analysis Model: Wave Under Refraction
Problems 5–35

Section 35.6 Huygens's Principle
Section 35.7 Dispersion
Problems 36–40

Section 35.8 Total Internal Reflection
Problems 41–50

Additional Problems
Problems 51–80

Challenge Problems
Problem 81–87

Solutions to the following Problems are available in the *Student Solutions Manual/Study Guide*:
35.3, 35.6, 35.12, 35.13, 35.35, 35.39, 35.43, 35.44, 35.45, 35.47, 35.53, 35.59, 35.65, 35.68, 35.80, 35.81, and 35.85

List of Enhanced Problems

Problem Number	Targeted Feedback in Enhanced WebAssign	Analysis Model Tutorial in Enhanced WebAssign	Master It in Enhanced WebAssign	Watch It in Enhanced WebAssign
35.1	✓		✓	
35.3		✓	✓	
35.5	✓			✓
35.6	✓			✓
35.8	✓		✓	✓
35.13			✓	
35.14	✓			✓
35.17	✓		✓	
35.21	✓			✓
35.22	✓			✓
35.27	✓	✓	✓	✓
35.39			✓	
35.42	✓			✓
35.43	✓		✓	
35.47			✓	
35.51	✓		✓	
35.53			✓	
35.56		✓		
35.59			✓	
35.65			✓	
35.67	✓			✓
35.69		✓		
35.77	✓			✓

Image Formation

The light rays coming from the leaves in the background of this scene did not form a focused image in the camera that took this photograph. Consequently, the background appears very blurry. Light rays passing though the raindrop, however, have been altered so as to form a focused image of the background leaves for the camera. In this chapter, we investigate the formation of images as light rays reflect from mirrors and refract through lenses. *(Don Hammond Photography Ltd. RF)*

This chapter is concerned with the images that result when light rays encounter flat or curved surfaces between two media. Images can be formed by either reflection or refraction due to these surfaces. We can design mirrors and lenses to form images with desired characteristics. In this chapter, we continue to use the ray approximation and assume light travels in straight lines. We first study the formation of images by mirrors and lenses and techniques for locating an image and determining its size. Then we investigate how to combine these elements into several useful optical instruments such as microscopes and telescopes.

36.1 Images Formed by Flat Mirrors

Image formation by mirrors can be understood through the behavior of light rays as described by the wave under reflection analysis model. We begin by considering the simplest possible mirror, the flat mirror. Consider a point source of light placed at O in Figure 36.1, a distance p in front of a flat mirror. The distance p is called the **object distance.** Diverging light rays leave the source and are reflected from the mirror. Upon reflection, the rays continue to diverge. The dashed lines in Figure 36.1 are extensions of the diverging rays back to a point of intersection at *I*. The diverging rays appear to the viewer to originate at the point *I* behind the mirror. Point *I*, which is a distance *q* behind the mirror, is called the **image** of the object at *O*. The distance *q* is called the **image distance.** Regardless of the system under study, images can always be located by extending diverging rays back to a point at which they intersect. Images are located either at a point from which rays of light *actually* diverge or at a point from which they *appear* to diverge.

Images are classified as *real* or *virtual*. A **real image** is formed when all light rays pass through and diverge from the image point; a **virtual image** is formed when most if not all of the light rays do *not* pass through the image point but only appear to diverge from that point. The image formed by the mirror in Figure 36.1 is virtual. No light rays from the object exist behind the mirror, at the location of the image, so the light rays in front of the mirror only seem to be diverging from *I*. The image of an object seen in a flat mirror is *always* virtual. Real images can be displayed on a screen (as at a movie theater), but virtual images cannot be displayed on a screen. We shall see an example of a real image in Section 36.2.

We can use the simple geometry in Figure 36.2 to examine the properties of the images of extended objects formed by flat mirrors. Even though there are an infinite number of choices of direction in which light rays could leave each point on the object (represented by a gray arrow), we need to choose only two rays to determine where an image is formed. One of those rays starts at *P*, follows a path perpendicular to the mirror to *Q*, and reflects back on itself. The second ray follows the oblique path *PR* and reflects as shown in Figure 36.2 according to the law of reflection. An observer in front of the mirror would extend the two reflected rays back to the point at which they appear to have originated, which is point *P'* behind the mirror. A continuation of this process for points other than *P* on the object would result in a virtual image (represented by a pink arrow) of the entire object behind the mirror. Because triangles *PQR* and *P'QR* are congruent, *PQ* = *P'Q*, so $|p| = |q|$. Therefore, the image formed of an object placed in front of a flat mirror is as far behind the mirror as the object is in front of the mirror.

The geometry in Figure 36.2 also reveals that the object height *h* equals the image height *h'*. Let us define **lateral magnification** *M* of an image as follows:

$$M \equiv \frac{\text{image height}}{\text{object height}} = \frac{h'}{h} \tag{36.1}$$

This general definition of the lateral magnification for an image from any type of mirror is also valid for images formed by lenses, which we study in Section 36.4. For a flat mirror, *M* = +1 for any image because *h'* = *h*. The positive value of the magnification signifies that the image is upright. (By upright we mean that if the object arrow points upward as in Figure 36.2, so does the image arrow.)

A flat mirror produces an image that has an *apparent* left–right reversal. You can see this reversal by standing in front of a mirror and raising your right hand as shown in Figure 36.3. The image you see raises its left hand. Likewise, your hair appears to be parted on the side opposite your real part, and a mole on your right cheek appears to be on your left cheek.

This reversal is not *actually* a left–right reversal. Imagine, for example, lying on your left side on the floor with your body parallel to the mirror surface. Now your head is on the left and your feet are on the right. If you shake your feet, the image does not shake its head! If you raise your right hand, however, the image again raises its left hand. Therefore, the mirror again appears to produce a left–right reversal but in the up–down direction!

The reversal is actually a *front–back reversal,* caused by the light rays going forward toward the mirror and then reflecting back from it. An interesting

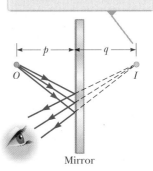

The image point *I* is located behind the mirror a distance *q* from the mirror. The image is virtual.

Figure 36.1 An image formed by reflection from a flat mirror.

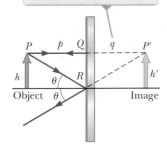

Because the triangles *PQR* and *P'QR* are congruent, $|p| = |q|$ and *h* = *h'*.

Figure 36.2 A geometric construction that is used to locate the image of an object placed in front of a flat mirror.

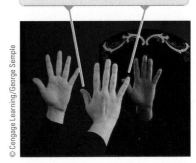

The thumb is on the left side of both real hands and on the left side of the image. That the thumb is not on the right side of the image indicates that there is no left-to-right reversal.

Figure 36.3 The image in the mirror of a person's right hand is reversed front to back, which makes the right hand appear to be a left hand.

exercise is to stand in front of a mirror while holding an overhead transparency in front of you so that you can read the writing on the transparency. You will also be able to read the writing on the image of the transparency. You may have had a similar experience if you have attached a transparent decal with words on it to the rear window of your car. If the decal can be read from outside the car, you can also read it when looking into your rearview mirror from inside the car.

Quick Quiz 36.1 You are standing approximately 2 m away from a mirror. The mirror has water spots on its surface. True or False: It is possible for you to see the water spots and your image both in focus at the same time.

Conceptual Example 36.1   Multiple Images Formed by Two Mirrors

Two flat mirrors are perpendicular to each other as in Figure 36.4, and an object is placed at point O. In this situation, multiple images are formed. Locate the positions of these images.

SOLUTION

The image of the object is at I_1 in mirror 1 (green rays) and at I_2 in mirror 2 (red rays). In addition, a third image is formed at I_3 (blue rays). This third image is the image of I_1 in mirror 2 or, equivalently, the image of I_2 in mirror 1. That is, the image at I_1 (or I_2) serves as the object for I_3. To form this image at I_3, the rays reflect twice after leaving the object at O.

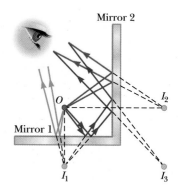

Figure 36.4 (Conceptual Example 36.1) When an object is placed in front of two mutually perpendicular mirrors as shown, three images are formed. Follow the different-colored light rays to understand the formation of each image.

Conceptual Example 36.2   The Tilting Rearview Mirror

Most rearview mirrors in cars have a day setting and a night setting. The night setting greatly diminishes the intensity of the image so that lights from trailing vehicles do not temporarily blind the driver. How does such a mirror work?

SOLUTION

Figure 36.5 shows a cross-sectional view of a rearview mirror for each setting. The unit consists of a reflective coating on the back of a wedge of glass. In the day setting (Fig. 36.5a), the light from an object behind the car strikes the glass wedge at point 1. Most of the light enters the wedge, refracting as it crosses the front surface, and reflects from the back surface to return to the front surface, where it is refracted again as it re-enters the air as ray B (for *bright*). In addition, a small portion of the light is reflected at the front surface of the glass as indicated by ray D (for *dim*).

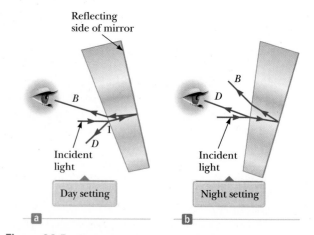

Figure 36.5 (Conceptual Example 36.2) Cross-sectional views of a rearview mirror.

This dim reflected light is responsible for the image observed when the mirror is in the night setting (Fig. 36.5b). In that case, the wedge is rotated so that the path followed by the bright light (ray B) does not lead to the eye. Instead, the dim light reflected from the front surface of the wedge travels to the eye, and the brightness of trailing headlights does not become a hazard.

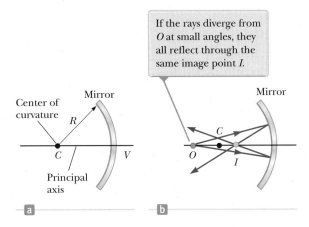

If the rays diverge from *O* at small angles, they all reflect through the same image point *I*.

Figure 36.6 (a) A concave mirror of radius *R*. The center of curvature *C* is located on the principal axis. (b) A point object placed at *O* in front of a concave spherical mirror of radius *R*, where *O* is any point on the principal axis farther than *R* from the mirror surface, forms a real image at *I*.

36.2 Images Formed by Spherical Mirrors

In the preceding section, we considered images formed by flat mirrors. Now we study images formed by curved mirrors. Although a variety of curvatures are possible, we will restrict our investigation to spherical mirrors. As its name implies, a **spherical mirror** has the shape of a section of a sphere.

Concave Mirrors

We first consider reflection of light from the inner, concave surface of a spherical mirror as shown in Figure 36.6. This type of reflecting surface is called a **concave mirror.** Figure 36.6a shows that the mirror has a radius of curvature *R*, and its center of curvature is point *C*. Point *V* is the center of the spherical section, and a line through *C* and *V* is called the **principal axis** of the mirror. Figure 36.6a shows a cross section of a spherical mirror, with its surface represented by the solid, curved dark blue line. (The lighter blue band represents the structural support for the mirrored surface, such as a curved piece of glass on which a silvered reflecting surface is deposited.) This type of mirror focuses incoming parallel rays to a point as demonstrated by the yellow light rays in Figure 36.7.

Now consider a point source of light placed at point *O* in Figure 36.6b, where *O* is any point on the principal axis to the left of *C*. Two diverging light rays that originate at *O* are shown. After reflecting from the mirror, these rays converge and cross at the image point *I*. They then continue to diverge from *I* as if an object were there. As a result, the image at point *I* is real.

In this section, we shall consider only rays that diverge from the object and make a small angle with the principal axis. Such rays are called **paraxial rays.** All paraxial rays reflect through the image point as shown in Figure 36.6b. Rays that are far from the principal axis such as those shown in Figure 36.8 converge to other points on the principal axis, producing a blurred image. This effect, called *spherical aberration,* is present to some extent for any spherical mirror and is discussed in Section 36.5.

If the object distance *p* and radius of curvature *R* are known, we can use Figure 36.9 (page 852) to calculate the image distance *q*. By convention, these distances are measured from point *V*. Figure 36.9 shows two rays leaving the tip of the object. The red ray passes through the center of curvature *C* of the mirror, hitting the mirror perpendicular to the mirror surface and reflecting back on itself. The blue ray strikes the mirror at its center (point *V*) and reflects as shown, obeying the law of reflection. The image of the tip of the arrow is located at the point where these two rays intersect. From the large, red right triangle in Figure 36.9, we see that $\tan \theta = h/p$, and from the yellow right triangle, we see that $\tan \theta = -h'/q$. The

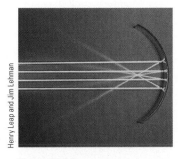

Figure 36.7 Reflection of parallel rays from a concave mirror.

The reflected rays intersect at different points on the principal axis.

Figure 36.8 A spherical concave mirror exhibits spherical aberration when light rays make large angles with the principal axis.

Henry Leap and Jim Lehman

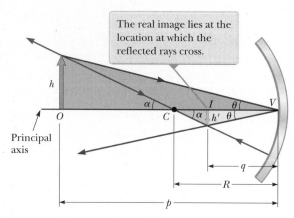

Figure 36.9 The image formed by a spherical concave mirror when the object *O* lies outside the center of curvature *C*. This geometric construction is used to derive Equation 36.4.

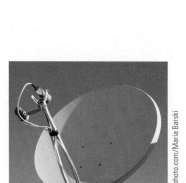

A satellite-dish antenna is a concave reflector for television signals from a satellite in orbit around the Earth. Because the satellite is so far away, the signals are carried by microwaves that are parallel when they arrive at the dish. These waves reflect from the dish and are focused on the receiver.

negative sign is introduced because the image is inverted, so h' is taken to be negative. Therefore, from Equation 36.1 and these results, we find that the magnification of the image is

$$M = \frac{h'}{h} = -\frac{q}{p}$$ (36.2)

Also notice from the green right triangle in Figure 36.9 and the smaller red right triangle that

$$\tan \alpha = \frac{-h'}{R - q} \quad \text{and} \quad \tan \alpha = \frac{h}{p - R}$$

from which it follows that

$$\frac{h'}{h} = -\frac{R - q}{p - R}$$ (36.3)

Comparing Equations 36.2 and 36.3 gives

$$\frac{R - q}{p - R} = \frac{q}{p}$$

Simple algebra reduces this expression to

$$\frac{1}{p} + \frac{1}{q} = \frac{2}{R}$$ (36.4)

which is called the *mirror equation*. We present a modified version of this equation shortly.

If the object is very far from the mirror—that is, if p is so much greater than R that p can be said to approach infinity—then $1/p \approx 0$, and Equation 36.4 shows that $q \approx R/2$. That is, when the object is very far from the mirror, the image point is halfway between the center of curvature and the center point on the mirror as shown in Figure 36.10. The incoming rays from the object are essentially parallel in this figure because the source is assumed to be very far from the mirror. The image point in this special case is called the **focal point** *F*, and the image distance is called the **focal length** *f*, where

$$f = \frac{R}{2}$$ (36.5)

The focal point is a distance f from the mirror, as noted in Figure 36.10. In Figure 36.7, the beams are traveling parallel to the principal axis and the mirror reflects all beams to the focal point.

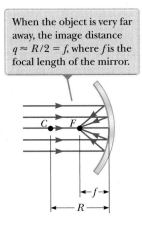

When the object is very far away, the image distance $q \approx R/2 = f$, where f is the focal length of the mirror.

Figure 36.10 Light rays from a distant object ($p \rightarrow \infty$) reflect from a concave mirror through the focal point F.

Pitfall Prevention 36.2

The *Focal* Point Is Not the *Focus* Point The focal point *is usually not* the point at which the light rays focus to form an image. The focal point is determined solely by the curvature of the mirror; it does not depend on the location of the object. In general, an image forms at a point different from the focal point of a mirror (or a lens), as in Figure 36.9. The *only* exception is when the object is located infinitely far away from the mirror.

Because the focal length is a parameter particular to a given mirror, it can be used to compare one mirror with another. Combining Equations 36.4 and 36.5, the **mirror equation** can be expressed in terms of the focal length:

$$\frac{1}{p} + \frac{1}{q} = \frac{1}{f}$$ (36.6)

◀ **Mirror equation in terms of focal length**

Notice that the focal length of a mirror depends only on the curvature of the mirror and not on the material from which the mirror is made because the formation of the image results from rays reflected from the surface of the material. The situation is different for lenses; in that case, the light actually passes through the material and the focal length depends on the type of material from which the lens is made. (See Section 36.4.)

Convex Mirrors

Figure 36.11 shows the formation of an image by a **convex mirror,** that is, one silvered so that light is reflected from the outer, convex surface. It is sometimes called a **diverging mirror** because the rays from any point on an object diverge after reflection as though they were coming from some point behind the mirror. The image in Figure 36.11 is virtual because the reflected rays only appear to originate at the image point as indicated by the dashed lines. Furthermore, the image is always upright and smaller than the object. This type of mirror is often used in stores to foil shoplifters. A single mirror can be used to survey a large field of view because it forms a smaller image of the interior of the store.

We do not derive any equations for convex spherical mirrors because Equations 36.2, 36.4, and 36.6 can be used for either concave or convex mirrors if we adhere to a strict sign convention. We will refer to the region in which light rays originate and move toward the mirror as the *front side* of the mirror and the other side as the

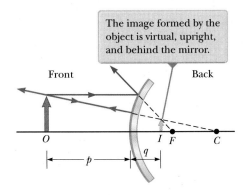

The image formed by the object is virtual, upright, and behind the mirror.

Figure 36.11 Formation of an image by a spherical convex mirror.

Table 36.1	**Sign Conventions for Mirrors**	
Quantity	Positive When . . .	Negative When . . .
Object location (p)	object is in front of mirror (real object).	object is in back of mirror (virtual object).
Image location (q)	image is in front of mirror (real image).	image is in back of mirror (virtual image).
Image height (h')	image is upright.	image is inverted.
Focal length (f) and radius (R)	mirror is concave.	mirror is convex.
Magnification (M)	image is upright.	image is inverted.

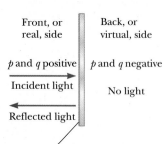

Front, or real, side | Back, or virtual, side
p and q positive | p and q negative
Incident light | No light
Reflected light |

Flat, convex, or concave mirrored surface

Figure 36.12 Signs of p and q for all types of mirrors.

back side. For example, in Figures 36.9 and 36.11, the side to the left of the mirrors is the front side and the side to the right of the mirrors is the back side. Figure 36.12 states the sign conventions for object and image distances for any type of mirror, and Table 36.1 summarizes the sign conventions for all quantities. One entry in the table, a *virtual object*, is formally introduced in Section 36.4.

Ray Diagrams for Mirrors

The positions and sizes of images formed by mirrors can be conveniently determined with *ray diagrams*. These pictorial representations reveal the nature of the image and can be used to check results calculated from the mathematical representation using the mirror and magnification equations. To draw a ray diagram, you must know the position of the object and the locations of the mirror's focal point and center of curvature. You then draw three rays to locate the image as shown by the examples in Figure 36.13. These rays all start from the same object point and are drawn as follows. You may choose any point on the object; here, let's choose the top of the object for simplicity. For concave mirrors (see Figs. 36.13a and 36.13b), draw the following three rays:

- Ray 1 is drawn from the top of the object parallel to the principal axis and is reflected through the focal point *F*.
- Ray 2 is drawn from the top of the object through the focal point (or as if coming from the focal point if $p < f$) and is reflected parallel to the principal axis.
- Ray 3 is drawn from the top of the object through the center of curvature *C* (or as if coming from the center *C* if $p < 2f$) and is reflected back on itself.

The intersection of any two of these rays locates the image. The third ray serves as a check of the construction. The image point obtained in this fashion must always agree with the value of *q* calculated from the mirror equation. With concave mirrors, notice what happens as the object is moved closer to the mirror. The real, inverted image in Figure 36.13a moves to the left and becomes larger as the object approaches the focal point. When the object is at the focal point, the image is infinitely far to the left. When the object lies between the focal point and the mirror surface as shown in Figure 36.13b, however, the image is to the right, behind the object, and virtual, upright, and enlarged. This latter situation applies when you use a shaving mirror or a makeup mirror, both of which are concave. Your face is closer to the mirror than the focal point, and you see an upright, enlarged image of your face.

For convex mirrors (see Fig. 36.13c), draw the following three rays:

- Ray 1 is drawn from the top of the object parallel to the principal axis and is reflected *away from* the focal point *F*.

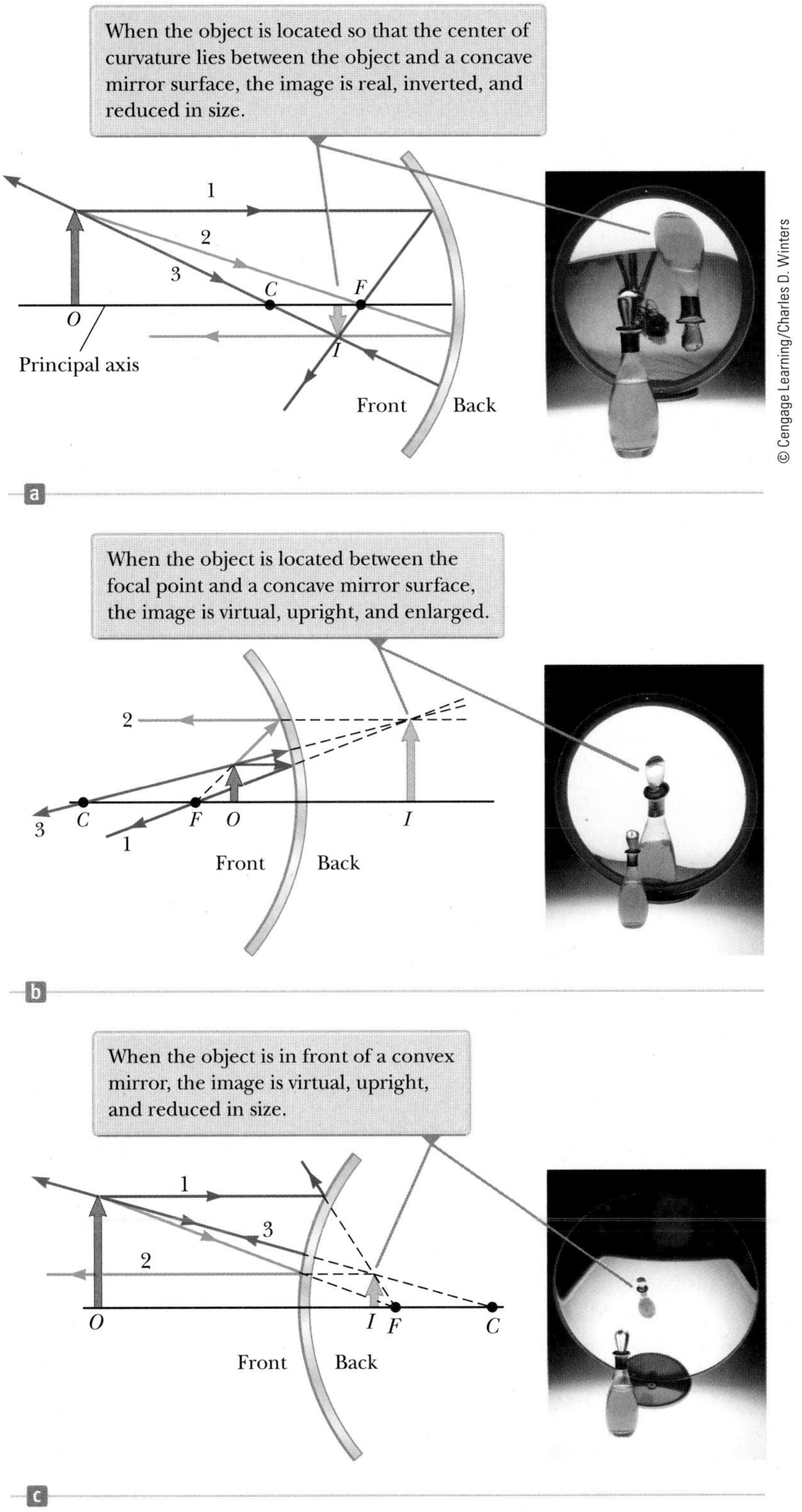

Figure 36.13 Ray diagrams for spherical mirrors along with corresponding photographs of the images of bottles.

When the object is located so that the center of curvature lies between the object and a concave mirror surface, the image is real, inverted, and reduced in size.

Principal axis

Front   Back

© Cengage Learning/Charles D. Winters

a

When the object is located between the focal point and a concave mirror surface, the image is virtual, upright, and enlarged.

Front   Back

b

When the object is in front of a convex mirror, the image is virtual, upright, and reduced in size.

Front   Back

c

- Ray 2 is drawn from the top of the object toward the focal point on the back side of the mirror and is reflected parallel to the principal axis.
- Ray 3 is drawn from the top of the object toward the center of curvature *C* on the back side of the mirror and is reflected back on itself.

In a convex mirror, the image of an object is always virtual, upright, and reduced in size as shown in Figure 36.13c. In this case, as the object distance decreases, the virtual image increases in size and moves away from the focal point toward the mirror as the object approaches the mirror. You should construct other diagrams to verify how image position varies with object position.

Quick Quiz 36.2 You wish to start a fire by reflecting sunlight from a mirror onto some paper under a pile of wood. Which would be the best choice for the type of mirror? **(a)** flat **(b)** concave **(c)** convex

Quick Quiz 36.3 Consider the image in the mirror in Figure 36.14. Based on the appearance of this image, would you conclude that **(a)** the mirror is concave and the image is real, **(b)** the mirror is concave and the image is virtual, **(c)** the mirror is convex and the image is real, or **(d)** the mirror is convex and the image is virtual?

NASA

Figure 36.14 (Quick Quiz 36.3) What type of mirror is shown here?

Example 36.3 **The Image Formed by a Concave Mirror**

A spherical mirror has a focal length of +10.0 cm.

(A) Locate and describe the image for an object distance of 25.0 cm.

SOLUTION

Conceptualize Because the focal length of the mirror is positive, it is a concave mirror (see Table 36.1). We expect the possibilities of both real and virtual images.

Categorize Because the object distance in this part of the problem is larger than the focal length, we expect the image to be real. This situation is analogous to that in Figure 36.13a.

Analyze Find the image distance by using Equation 36.6:

$$\frac{1}{q} = \frac{1}{f} - \frac{1}{p}$$

$$\frac{1}{q} = \frac{1}{10.0 \text{ cm}} - \frac{1}{25.0 \text{ cm}}$$

$$q = \boxed{16.7 \text{ cm}}$$

Find the magnification of the image from Equation 36.2:

$$M = -\frac{q}{p} = -\frac{16.7 \text{ cm}}{25.0 \text{ cm}} = \boxed{-0.667}$$

Finalize The absolute value of M is less than unity, so the image is smaller than the object, and the negative sign for M tells us that the image is inverted. Because q is positive, the image is located on the front side of the mirror and is real. Look into the bowl of a shiny spoon or stand far away from a shaving mirror to see this image.

(B) Locate and describe the image for an object distance of 10.0 cm.

SOLUTION

Categorize Because the object is at the focal point, we expect the image to be infinitely far away.

Analyze Find the image distance by using Equation 36.6:

$$\frac{1}{q} = \frac{1}{f} - \frac{1}{p}$$

$$\frac{1}{q} = \frac{1}{10.0 \text{ cm}} - \frac{1}{10.0 \text{ cm}}$$

$$q = \boxed{\infty}$$

▶ **36.3** continued

Finalize This result means that rays originating from an object positioned at the focal point of a mirror are reflected so that the image is formed at an infinite distance from the mirror; that is, the rays travel parallel to one another after reflection. Such is the situation in a flashlight or an automobile headlight, where the bulb filament is placed at the focal point of a reflector, producing a parallel beam of light.

(C) Locate and describe the image for an object distance of 5.00 cm.

SOLUTION

Categorize Because the object distance is smaller than the focal length, we expect the image to be virtual. This situation is analogous to that in Figure 36.13b.

Analyze Find the image distance by using Equation 36.6:

$$\frac{1}{q} = \frac{1}{f} - \frac{1}{p}$$

$$\frac{1}{q} = \frac{1}{10.0 \text{ cm}} - \frac{1}{5.00 \text{ cm}}$$

$$q = -10.0 \text{ cm}$$

Find the magnification of the image from Equation 36.2:

$$M = -\frac{q}{p} = -\left(\frac{-10.0 \text{ cm}}{5.00 \text{ cm}}\right) = +2.00$$

Finalize The image is twice as large as the object, and the positive sign for M indicates that the image is upright (see Fig. 36.13b). The negative value of the image distance tells us that the image is virtual, as expected. Put your face close to a shaving mirror to see this type of image.

WHAT IF? Suppose you set up the bottle and mirror apparatus illustrated in Figure 36.13a and described here in part (A). While adjusting the apparatus, you accidentally bump the bottle and it begins to slide toward the mirror at speed v_p. How fast does the image of the bottle move?

Answer Solve the mirror equation, Equation 36.6, for q:

$$q = \frac{fp}{p - f}$$

Differentiate this equation with respect to time to find the velocity of the image:

$$(1) \quad v_q = \frac{dq}{dt} = \frac{d}{dt}\left(\frac{fp}{p - f}\right) = -\frac{f^2}{(p - f)^2}\frac{dp}{dt} = -\frac{f^2 v_p}{(p - f)^2}$$

Substitute numerical values from part (A):

$$v_q = -\frac{(10.0 \text{ cm})^2 v_p}{(25.0 \text{ cm} - 10.0 \text{ cm})^2} = -0.444 v_p$$

Therefore, the speed of the image is less than that of the object in this case.

We can see two interesting behaviors of the function for v_q in Equation (1). First, the velocity is negative regardless of the value of p or f. Therefore, if the object moves toward the mirror, the image moves toward the left in Figure 36.13 without regard for the side of the focal point at which the object is located or whether the mirror is concave or convex. Second, in the limit of $p \to 0$, the velocity v_q approaches $-v_p$. As the object moves very close to the mirror, the mirror looks like a plane mirror, the image is as far behind the mirror as the object is in front, and both the object and the image move with the same speed.

Example 36.4   **The Image Formed by a Convex Mirror**

An automobile rearview mirror as shown in Figure 36.15 (page 858) shows an image of a truck located 10.0 m from the mirror. The focal length of the mirror is −0.60 m.

(A) Find the position of the image of the truck.

continued

▶ **36.4** continued

SOLUTION

Conceptualize This situation is depicted in Figure 36.13c.

Categorize Because the mirror is convex, we expect it to form an upright, reduced, virtual image for any object position.

Figure 36.15 (Example 36.4) An approaching truck is seen in a convex mirror on the right side of an automobile. Notice that the image of the truck is in focus, but the frame of the mirror is not, which demonstrates that the image is not at the same location as the mirror surface.

Analyze Find the image distance by using Equation 36.6:

$$\frac{1}{q} = \frac{1}{f} - \frac{1}{p}$$

$$\frac{1}{q} = \frac{1}{-0.60 \text{ m}} - \frac{1}{10.0 \text{ m}}$$

$$q = \boxed{-0.57 \text{ m}}$$

(B) Find the magnification of the image.

SOLUTION

Analyze Use Equation 36.2:

$$M = -\frac{q}{p} = -\left(\frac{-0.57 \text{ m}}{10.0 \text{ m}}\right) = \boxed{+0.057}$$

Finalize The negative value of q in part (A) indicates that the image is virtual, or behind the mirror, as shown in Figure 36.13c. The magnification in part (B) indicates that the image is much smaller than the truck and is upright because M is positive. The image is reduced in size, so the truck appears to be farther away than it actually is. Because of the image's small size, these mirrors carry the inscription, "Objects in this mirror are closer than they appear." Look into your rearview mirror or the back side of a shiny spoon to see an image of this type.

36.3 Images Formed by Refraction

In this section, we describe how images are formed when light rays follow the wave under refraction model at the boundary between two transparent materials. Consider two transparent media having indices of refraction n_1 and n_2, where the boundary between the two media is a spherical surface of radius R (Fig. 36.16). We assume the object at O is in the medium for which the index of refraction is n_1. Let's consider the paraxial rays leaving O. As we shall see, all such rays are refracted at the spherical surface and focus at a single point I, the image point.

Figure 36.17 shows a single ray leaving point O and refracting to point I. Snell's law of refraction applied to this ray gives

$$n_1 \sin \theta_1 = n_2 \sin \theta_2$$

Because θ_1 and θ_2 are assumed to be small, we can use the small-angle approximation $\sin \theta \approx \theta$ (with angles in radians) and write Snell's law as

$$n_1 \theta_1 = n_2 \theta_2$$

We know that an exterior angle of any triangle equals the sum of the two opposite interior angles, so applying this rule to triangles OPC and PIC in Figure 36.17 gives

$$\theta_1 = \alpha + \beta$$

$$\beta = \theta_2 + \gamma$$

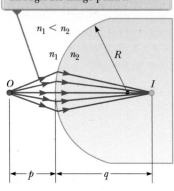

Rays making small angles with the principal axis diverge from a point object at O and are refracted through the image point I.

Figure 36.16 An image formed by refraction at a spherical surface.

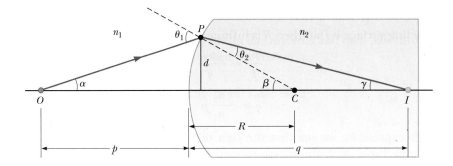

Figure 36.17 Geometry used to derive Equation 36.8, assuming $n_1 < n_2$.

Combining all three expressions and eliminating θ_1 and θ_2 gives

$$n_1\alpha + n_2\gamma = (n_2 - n_1)\beta \qquad (36.7)$$

Figure 36.17 shows three right triangles that have a common vertical leg of length d. For paraxial rays (unlike the relatively large-angle ray shown in Fig. 36.17), the horizontal legs of these triangles are approximately p for the triangle containing angle α, R for the triangle containing angle β, and q for the triangle containing angle γ. In the small-angle approximation, $\tan\theta \approx \theta$, so we can write the approximate relationships from these triangles as follows:

$$\tan\alpha \approx \alpha \approx \frac{d}{p} \qquad \tan\beta \approx \beta \approx \frac{d}{R} \qquad \tan\gamma \approx \gamma \approx \frac{d}{q}$$

Substituting these expressions into Equation 36.7 and dividing through by d gives

$$\frac{n_1}{p} + \frac{n_2}{q} = \frac{n_2 - n_1}{R} \qquad (36.8)$$

◀ **Relation between object and image distance for a refracting surface**

For a fixed object distance p, the image distance q is independent of the angle the ray makes with the axis. This result tells us that all paraxial rays focus at the same point I.

As with mirrors, we must use a sign convention to apply Equation 36.8 to a variety of cases. We define the side of the surface in which light rays originate as the front side. The other side is called the back side. In contrast with mirrors, where real images are formed in front of the reflecting surface, real images are formed by refraction of light rays to the back of the surface. Because of the difference in location of real images, the refraction sign conventions for q and R are opposite the reflection sign conventions. For example, q and R are both positive in Figure 36.17. The sign conventions for spherical refracting surfaces are summarized in Table 36.2.

We derived Equation 36.8 from an assumption that $n_1 < n_2$ in Figure 36.17. This assumption is not necessary, however. Equation 36.8 is valid regardless of which index of refraction is greater.

Table 36.2 **Sign Conventions for Refracting Surfaces**

Quantity	Positive When . . .	Negative When . . .
Object location (p)	object is in front of surface (real object).	object is in back of surface (virtual object).
Image location (q)	image is in back of surface (real image).	image is in front of surface (virtual image).
Image height (h')	image is upright.	image is inverted.
Radius (R)	center of curvature is in back of surface.	center of curvature is in front of surface.

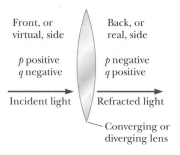

Front, or virtual, side

Back, or real, side

p positive
q negative

p negative
q positive

Incident light   Refracted light

Converging or diverging lens

Figure 36.24 A diagram for obtaining the signs of p and q for a thin lens. (This diagram also applies to a refracting surface.)

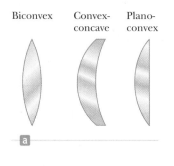

Biconvex   Convex-concave   Plano-convex

a

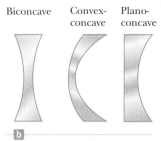

Biconcave   Convex-concave   Plano-concave

b

Figure 36.25 Various lens shapes. (a) Converging lenses have a positive focal length and are thickest at the middle. (b) Diverging lenses have a negative focal length and are thickest at the edges.

Table 36.3	**Sign Conventions for Thin Lenses**	
Quantity	**Positive When . . .**	**Negative When . . .**
Object location (p)	object is in front of lens (real object).	object is in back of lens (virtual object).
Image location (q)	image is in back of lens (real image).	image is in front of lens (virtual image).
Image height (h')	image is upright.	image is inverted.
R_1 and R_2	center of curvature is in back of lens.	center of curvature is in front of lens.
Focal length (f)	a converging lens.	a diverging lens.

Figure 36.24 is useful for obtaining the signs of p and q, and Table 36.3 gives the sign conventions for thin lenses. These sign conventions are the *same* as those for refracting surfaces (see Table 36.2).

Various lens shapes are shown in Figure 36.25. Notice that a converging lens is thicker at the center than at the edge, whereas a diverging lens is thinner at the center than at the edge.

Magnification of Images

Consider a thin lens through which light rays from an object pass. As with mirrors (Eq. 36.2), a geometric construction shows that the lateral magnification of the image is

$$M = \frac{h'}{h} = -\frac{q}{p} \tag{36.17}$$

From this expression, it follows that when M is positive, the image is upright and on the same side of the lens as the object. When M is negative, the image is inverted and on the side of the lens opposite the object.

Ray Diagrams for Thin Lenses

Ray diagrams are convenient for locating the images formed by thin lenses or systems of lenses. They also help clarify our sign conventions. Figure 36.26 shows such diagrams for three single-lens situations.

To locate the image of a *converging* lens (Figs. 36.26a and 36.26b), the following three rays are drawn from the top of the object:

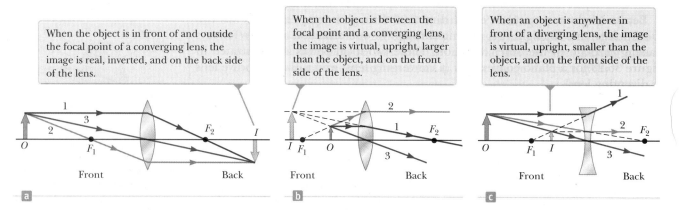

When the object is in front of and outside the focal point of a converging lens, the image is real, inverted, and on the back side of the lens.

When the object is between the focal point and a converging lens, the image is virtual, upright, larger than the object, and on the front side of the lens.

When an object is anywhere in front of a diverging lens, the image is virtual, upright, smaller than the object, and on the front side of the lens.

Figure 36.26 Ray diagrams for locating the image formed by a thin lens.

- Ray 1 is drawn parallel to the principal axis. After being refracted by the lens, this ray passes through the focal point on the back side of the lens.
- Ray 2 is drawn through the focal point on the front side of the lens (or as if coming from the focal point if $p < f$) and emerges from the lens parallel to the principal axis.
- Ray 3 is drawn through the center of the lens and continues in a straight line.

To locate the image of a *diverging* lens (Fig. 36.26c), the following three rays are drawn from the top of the object:

- Ray 1 is drawn parallel to the principal axis. After being refracted by the lens, this ray emerges directed away from the focal point on the front side of the lens.
- Ray 2 is drawn in the direction toward the focal point on the back side of the lens and emerges from the lens parallel to the principal axis.
- Ray 3 is drawn through the center of the lens and continues in a straight line.

For the converging lens in Figure 36.26a, where the object is to the left of the focal point ($p > f$), the image is real and inverted. When the object is between the focal point and the lens ($p < f$) as in Figure 36.26b, the image is virtual and upright. In that case, the lens acts as a magnifying glass, which we study in more detail in Section 36.8. For a diverging lens (Fig. 36.26c), the image is always virtual and upright, regardless of where the object is placed. These geometric constructions are reasonably accurate only if the distance between the rays and the principal axis is much less than the radii of the lens surfaces.

Refraction occurs only at the surfaces of the lens. A certain lens design takes advantage of this behavior to produce the *Fresnel lens*, a powerful lens without great thickness. Because only the surface curvature is important in the refracting qualities of the lens, material in the middle of a Fresnel lens is removed as shown in the cross sections of lenses in Figure 36.27. Because the edges of the curved segments cause some distortion, Fresnel lenses are generally used only in situations in which image quality is less important than reduction of weight. A classroom overhead projector often uses a Fresnel lens; the circular edges between segments of the lens can be seen by looking closely at the light projected onto a screen.

Quick Quiz 36.6 What is the focal length of a pane of window glass? **(a)** zero **(b)** infinity **(c)** the thickness of the glass **(d)** impossible to determine

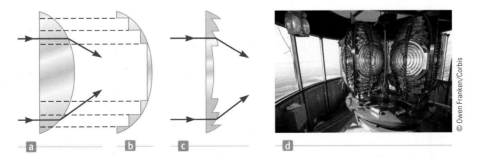

Figure 36.27 A side view of the construction of a Fresnel lens. (a) The thick lens refracts a light ray as shown. (b) Lens material in the bulk of the lens is cut away, leaving only the material close to the curved surface. (c) The small pieces of remaining material are moved to the left to form a flat surface on the left of the Fresnel lens with ridges on the right surface. From a front view, these ridges would be circular in shape. This new lens refracts light in the same way as the lens in (a). (d) A Fresnel lens used in a lighthouse shows several segments with the ridges discussed in (c).

© Owen Franken/Corbis

Example 36.8   Images Formed by a Converging Lens

A converging lens has a focal length of 10.0 cm.

(A) An object is placed 30.0 cm from the lens. Construct a ray diagram, find the image distance, and describe the image.

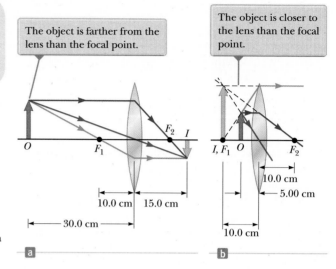

The object is farther from the lens than the focal point.

The object is closer to the lens than the focal point.

Figure 36.28
(Example 36.8) An image is formed by a converging lens.

SOLUTION

Conceptualize Because the lens is converging, the focal length is positive (see Table 36.3). We expect the possibilities of both real and virtual images.

Categorize Because the object distance is larger than the focal length, we expect the image to be real. The ray diagram for this situation is shown in Figure 36.28a.

Analyze Find the image distance by using Equation 36.16:

$$\frac{1}{q} = \frac{1}{f} - \frac{1}{p}$$

$$\frac{1}{q} = \frac{1}{10.0 \text{ cm}} - \frac{1}{30.0 \text{ cm}}$$

$$q = \boxed{+15.0 \text{ cm}}$$

Find the magnification of the image from Equation 36.17:

$$M = -\frac{q}{p} = -\frac{15.0 \text{ cm}}{30.0 \text{ cm}} = \boxed{-0.500}$$

Finalize The positive sign for the image distance tells us that the image is indeed real and on the back side of the lens. The magnification of the image tells us that the image is reduced in height by one half, and the negative sign for M tells us that the image is inverted.

(B) An object is placed 10.0 cm from the lens. Find the image distance and describe the image.

SOLUTION

Categorize Because the object is at the focal point, we expect the image to be infinitely far away.

Analyze Find the image distance by using Equation 36.16:

$$\frac{1}{q} = \frac{1}{f} - \frac{1}{p}$$

$$\frac{1}{q} = \frac{1}{10.0 \text{ cm}} - \frac{1}{10.0 \text{ cm}}$$

$$q = \boxed{\infty}$$

Finalize This result means that rays originating from an object positioned at the focal point of a lens are refracted so that the image is formed at an infinite distance from the lens; that is, the rays travel parallel to one another after refraction.

(C) An object is placed 5.00 cm from the lens. Construct a ray diagram, find the image distance, and describe the image.

SOLUTION

Categorize Because the object distance is smaller than the focal length, we expect the image to be virtual. The ray diagram for this situation is shown in Figure 36.28b.

▶ **36.8** continued

Analyze Find the image distance by using Equation 36.16:

$$\frac{1}{q} = \frac{1}{f} - \frac{1}{p}$$

$$\frac{1}{q} = \frac{1}{10.0 \text{ cm}} - \frac{1}{5.00 \text{ cm}}$$

$$q = \boxed{-10.0 \text{ cm}}$$

Find the magnification of the image from Equation 36.17:

$$M = -\frac{q}{p} = -\left(\frac{-10.0 \text{ cm}}{5.00 \text{ cm}}\right) = \boxed{+2.00}$$

Finalize The negative image distance tells us that the image is virtual and formed on the side of the lens from which the light is incident, the front side. The image is enlarged, and the positive sign for M tells us that the image is upright.

WHAT IF? What if the object moves right up to the lens surface so that $p \rightarrow 0$? Where is the image?

Answer In this case, because $p << R$, where R is either of the radii of the surfaces of the lens, the curvature of the lens can be ignored. The lens should appear to have the same effect as a flat piece of material, which suggests that the image is just on the front side of the lens, at $q = 0$. This conclusion can be verified mathematically by rearranging the thin lens equation:

$$\frac{1}{q} = \frac{1}{f} - \frac{1}{p}$$

If we let $p \rightarrow 0$, the second term on the right becomes very large compared with the first and we can neglect $1/f$. The equation becomes

$$\frac{1}{q} = -\frac{1}{p} \quad \rightarrow \quad q = -p = 0$$

Therefore, q is on the front side of the lens (because it has the opposite sign as p) and right at the lens surface.

Example 36.9 **Images Formed by a Diverging Lens**

A diverging lens has a focal length of 10.0 cm.

(A) An object is placed 30.0 cm from the lens. Construct a ray diagram, find the image distance, and describe the image.

SOLUTION

Conceptualize Because the lens is diverging, the focal length is negative (see Table 36.3). The ray diagram for this situation is shown in Figure 36.29a.

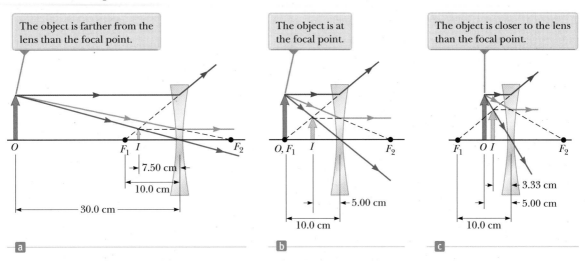

The object is farther from the lens than the focal point.

The object is at the focal point.

The object is closer to the lens than the focal point.

Figure 36.29 (Example 36.9) An image is formed by a diverging lens.

continued

▶ **36.9** continued

Categorize Because the lens is diverging, we expect it to form an upright, reduced, virtual image for any object position.

Analyze Find the image distance by using Equation 36.16:

$$\frac{1}{q} = \frac{1}{f} - \frac{1}{p}$$

$$\frac{1}{q} = \frac{1}{-10.0 \text{ cm}} - \frac{1}{30.0 \text{ cm}}$$

$$q = -7.50 \text{ cm}$$

Find the magnification of the image from Equation 36.17:

$$M = -\frac{q}{p} = -\left(\frac{-7.50 \text{ cm}}{30.0 \text{ cm}}\right) = +0.250$$

Finalize This result confirms that the image is virtual, smaller than the object, and upright. Look through the diverging lens in a door peephole to see this type of image.

(B) An object is placed 10.0 cm from the lens. Construct a ray diagram, find the image distance, and describe the image.

SOLUTION

The ray diagram for this situation is shown in Figure 36.29b.

Analyze Find the image distance by using Equation 36.16:

$$\frac{1}{q} = \frac{1}{f} - \frac{1}{p}$$

$$\frac{1}{q} = \frac{1}{-10.0 \text{ cm}} - \frac{1}{10.0 \text{ cm}}$$

$$q = -5.00 \text{ cm}$$

Find the magnification of the image from Equation 36.17:

$$M = -\frac{q}{p} = -\left(\frac{-5.00 \text{ cm}}{10.0 \text{ cm}}\right) = +0.500$$

Finalize Notice the difference between this situation and that for a converging lens. For a diverging lens, an object at the focal point does not produce an image infinitely far away.

(C) An object is placed 5.00 cm from the lens. Construct a ray diagram, find the image distance, and describe the image.

SOLUTION

The ray diagram for this situation is shown in Figure 36.29c.

Analyze Find the image distance by using Equation 36.16:

$$\frac{1}{q} = \frac{1}{f} - \frac{1}{p}$$

$$\frac{1}{q} = \frac{1}{-10.0 \text{ cm}} - \frac{1}{5.0 \text{ cm}}$$

$$q = -3.33 \text{ cm}$$

Find the magnification of the image from Equation 36.17:

$$M = -\left(\frac{-3.33 \text{ cm}}{5.00 \text{ cm}}\right) = +0.667$$

Finalize For all three object positions, the image position is negative and the magnification is a positive number smaller than 1, which confirms that the image is virtual, smaller than the object, and upright.

Combinations of Thin Lenses

If two thin lenses are used to form an image, the system can be treated in the following manner. First, the image formed by the first lens is located as if the second lens were not present. Then a ray diagram is drawn for the second lens, with the image formed by the first lens now serving as the object for the second lens. The second image formed is the final image of the system. If the image formed by the first lens lies on the back side of the second lens, that image is treated as a virtual object for the second lens (that is, in the thin lens equation, p is negative). The same procedure can be extended to a system of three or more lenses. Because the magnification due to the second lens is performed on the magnified image due to the first lens, the overall magnification of the image due to the combination of lenses is the product of the individual magnifications:

$$M = M_1 M_2 \tag{36.18}$$

This equation can be used for combinations of any optical elements such as a lens and a mirror. For more than two optical elements, the magnifications due to all elements are multiplied together.

Let's consider the special case of a system of two lenses of focal lengths f_1 and f_2 in contact with each other. If $p_1 = p$ is the object distance for the combination, application of the thin lens equation (Eq. 36.16) to the first lens gives

$$\frac{1}{p} + \frac{1}{q_1} = \frac{1}{f_1}$$

where q_1 is the image distance for the first lens. Treating this image as the object for the second lens, we see that the object distance for the second lens must be $p_2 = -q_1$. (The distances are the same because the lenses are in contact and assumed to be infinitesimally thin. The object distance is negative because the object is virtual if the image from the first lens is real.) Therefore, for the second lens,

$$\frac{1}{p_2} + \frac{1}{q_2} = \frac{1}{f_2} \quad \rightarrow \quad -\frac{1}{q_1} + \frac{1}{q} = \frac{1}{f_2}$$

where $q = q_2$ is the final image distance from the second lens, which is the image distance for the combination. Adding the equations for the two lenses eliminates q_1 and gives

$$\frac{1}{p} + \frac{1}{q} = \frac{1}{f_1} + \frac{1}{f_2}$$

If the combination is replaced with a single lens that forms an image at the same location, its focal length must be related to the individual focal lengths by the expression

$$\frac{1}{f} = \frac{1}{f_1} + \frac{1}{f_2} \tag{36.19}$$

◀ **Focal length for a combination of two thin lenses in contact**

Therefore, two thin lenses in contact with each other are equivalent to a single thin lens having a focal length given by Equation 36.19.

Example 36.10   Where Is the Final Image?

Two thin converging lenses of focal lengths $f_1 = 10.0$ cm and $f_2 = 20.0$ cm are separated by 20.0 cm as illustrated in Figure 36.30. An object is placed 30.0 cm to the left of lens 1. Find the position and the magnification of the final image.

SOLUTION

Conceptualize Imagine light rays passing through the first lens and forming a real image (because $p > f$) in the absence of a second lens. Figure 36.30 shows these light rays forming the inverted image I_1. Once the light rays converge to the image point, they do not stop. They continue through the image point and interact with the

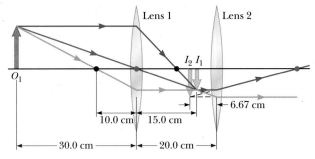

Figure 36.30 (Example 36.10) A combination of two converging lenses. The ray diagram shows the location of the final image (I_2) due to the combination of lenses. The black dots are the focal points of lens 1, and the red dots are the focal points of lens 2.

continued

▶ **36.10** c o n t i n u e d

second lens. The rays leaving the image point behave in the same way as the rays leaving an object. Therefore, the image of the first lens serves as the object of the second lens.

Categorize We categorize this problem as one in which the thin lens equation is applied in a stepwise fashion to the two lenses.

....

Analyze Find the location of the image formed by lens 1 from the thin lens equation:

$$\frac{1}{q_1} = \frac{1}{f} - \frac{1}{p_1}$$

$$\frac{1}{q_1} = \frac{1}{10.0 \text{ cm}} - \frac{1}{30.0 \text{ cm}}$$

$$q_1 = +15.0 \text{ cm}$$

Find the magnification of the image from Equation 36.17:

$$M_1 = -\frac{q_1}{p_1} = -\frac{15.0 \text{ cm}}{30.0 \text{ cm}} = -0.500$$

The image formed by this lens acts as the object for the second lens. Therefore, the object distance for the second lens is 20.0 cm − 15.0 cm = 5.00 cm.

Find the location of the image formed by lens 2 from the thin lens equation:

$$\frac{1}{q_2} = \frac{1}{20.0 \text{ cm}} - \frac{1}{5.00 \text{ cm}}$$

$$q_2 = \boxed{-6.67 \text{ cm}}$$

Find the magnification of the image from Equation 36.17:

$$M_2 = -\frac{q_2}{p_2} = -\frac{(-6.67 \text{ cm})}{5.00 \text{ cm}} = +1.33$$

Find the overall magnification of the system from Equation 36.18:

$$M = M_1 M_2 = (-0.500)(1.33) = \boxed{-0.667}$$

....

Finalize The negative sign on the overall magnification indicates that the final image is inverted with respect to the initial object. Because the absolute value of the magnification is less than 1, the final image is smaller than the object.

Because q_2 is negative, the final image is on the front, or left, side of lens 2. These conclusions are consistent with the ray diagram in Figure 36.30.

WHAT IF? Suppose you want to create an upright image with this system of two lenses. How must the second lens be moved?

Answer Because the object is farther from the first lens than the focal length of that lens, the first image is inverted. Consequently, the second lens must invert the image once again so that the final image is upright. An inverted image is only formed by a converging lens if the object is outside the focal point. Therefore, the image formed by the first lens must be to the left of the focal point of the second lens in Figure 36.30. To make that happen, you must move the second lens at least as far away from the first lens as the sum $q_1 + f_2 = 15.0 \text{ cm} + 20.0 \text{ cm} = 35.0 \text{ cm}$.

36.5 Lens Aberrations

Our analysis of mirrors and lenses assumes rays make small angles with the principal axis and the lenses are thin. In this simple model, all rays leaving a point source focus at a single point, producing a sharp image. Clearly, that is not always true. When the approximations used in this analysis do not hold, imperfect images are formed.

A precise analysis of image formation requires tracing each ray, using Snell's law at each refracting surface and the law of reflection at each reflecting surface. This procedure shows that the rays from a point object do not focus at a single point, with the result that the image is blurred. The departures of actual images from the ideal predicted by our simplified model are called **aberrations.**

Spherical Aberration

Spherical aberration occurs because the focal points of rays far from the principal axis of a spherical lens (or mirror) are different from the focal points of rays of the same wavelength passing near the axis. Figure 36.31 illustrates spherical aberration for parallel rays passing through a converging lens. Rays passing through points near the center of the lens are imaged farther from the lens than rays passing through points near the edges. Figure 36.8 earlier in the chapter shows spherical aberration for light rays leaving a point object and striking a spherical mirror.

Many cameras have an adjustable aperture to control light intensity and reduce spherical aberration. (An aperture is an opening that controls the amount of light passing through the lens.) Sharper images are produced as the aperture size is reduced; with a small aperture, only the central portion of the lens is exposed to the light and therefore a greater percentage of the rays are paraxial. At the same time, however, less light passes through the lens. To compensate for this lower light intensity, a longer exposure time is used.

In the case of mirrors, spherical aberration can be minimized through the use of a parabolic reflecting surface rather than a spherical surface. Parabolic surfaces are not used often, however, because those with high-quality optics are very expensive to make. Parallel light rays incident on a parabolic surface focus at a common point, regardless of their distance from the principal axis. Parabolic reflecting surfaces are used in many astronomical telescopes to enhance image quality.

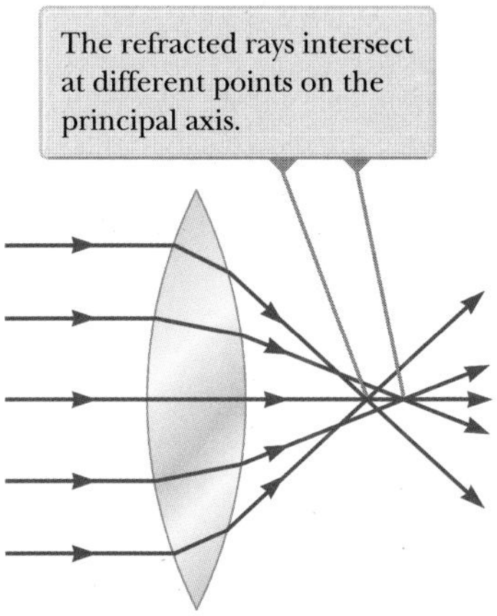

Chromatic Aberration

In Chapter 35, we described dispersion, whereby a material's index of refraction varies with wavelength. Because of this phenomenon, violet rays are refracted more than red rays when white light passes through a lens (Fig. 36.32). The figure shows that the focal length of a lens is greater for red light than for violet light. Other wavelengths (not shown in Fig. 36.32) have focal points intermediate between those of red and violet, which causes a blurred image and is called **chromatic aberration.**

Chromatic aberration for a diverging lens also results in a shorter focal length for violet light than for red light, but on the front side of the lens. Chromatic aberration can be greatly reduced by combining a converging lens made of one type of glass and a diverging lens made of another type of glass.

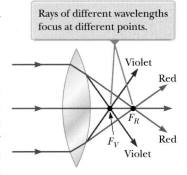

Figure 36.31 Spherical aberration caused by a converging lens. Does a diverging lens cause spherical aberration?

The refracted rays intersect at different points on the principal axis.

Rays of different wavelengths focus at different points.

Figure 36.32 Chromatic aberration caused by a converging lens.

36.6 The Camera

The photographic **camera** is a simple optical instrument whose essential features are shown in Figure 36.33. It consists of a light-tight chamber, a converging lens that produces a real image, and a light-sensitive component behind the lens on which the image is formed.

The image in a digital camera is formed on a *charge-coupled device* (CCD), which digitizes the image, turning it into binary code. (A CCD is described in Section 40.2.) The digital information is then stored on a memory chip for playback on the camera's display screen, or it can be downloaded to a computer. Film cameras are similar to digital cameras except that the light forms an image on light-sensitive film rather than on a CCD. The film must then be chemically processed to produce the image on paper. In the discussion that follows, we assume the camera is digital.

A camera is focused by varying the distance between the lens and the CCD. For proper focusing—which is necessary for the formation of sharp images—the lens-to-CCD distance depends on the object distance as well as the focal length of the lens.

The shutter, positioned behind the lens, is a mechanical device that is opened for selected time intervals, called *exposure times.* You can photograph moving objects by using short exposure times or photograph dark scenes (with low light levels) by using long exposure times. If this adjustment were not available, it would be impossible

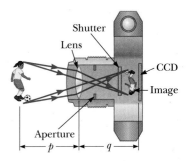

Figure 36.33 Cross-sectional view of a simple digital camera. The CCD is the light-sensitive component of the camera. In a nondigital camera, the light from the lens falls onto photographic film. In reality, $p \gg q$.

to take stop-action photographs. For example, a rapidly moving vehicle could move enough in the time interval during which the shutter is open to produce a blurred image. Another major cause of blurred images is the movement of the camera while the shutter is open. To prevent such movement, either short exposure times or a tripod should be used, even for stationary objects. Typical shutter speeds (that is, exposure times) are $\frac{1}{30}$ s, $\frac{1}{60}$ s, $\frac{1}{125}$ s, and $\frac{1}{250}$ s. In practice, stationary objects are normally shot with an intermediate shutter speed of $\frac{1}{60}$ s.

The intensity I of the light reaching the CCD is proportional to the area of the lens. Because this area is proportional to the square of the diameter D, it follows that I is also proportional to D^2. Light intensity is a measure of the rate at which energy is received by the CCD per unit area of the image. Because the area of the image is proportional to q^2 and $q \approx f$ (when $p \gg f$, so p can be approximated as infinite), we conclude that the intensity is also proportional to $1/f^2$ and therefore that $I \propto D^2/f^2$.

The ratio f/D is called the **f-number** of a lens:

$$f\text{-number} \equiv \frac{f}{D} \qquad (36.20)$$

Hence, the intensity of light incident on the CCD varies according to the following proportionality:

$$I \propto \frac{1}{(f/D)^2} \propto \frac{1}{(f\text{-number})^2} \qquad (36.21)$$

The f-number is often given as a description of the lens's "speed." The lower the f-number, the wider the aperture and the higher the rate at which energy from the light exposes the CCD; therefore, a lens with a low f-number is a "fast" lens. The conventional notation for an f-number is "$f/$" followed by the actual number. For example, "$f/4$" means an f-number of 4; it *does not* mean to divide f by 4! Extremely fast lenses, which have f-numbers as low as approximately $f/1.2$, are expensive because it is very difficult to keep aberrations acceptably small with light rays passing through a large area of the lens. Camera lens systems (that is, combinations of lenses with adjustable apertures) are often marked with multiple f-numbers, usually $f/2.8$, $f/4$, $f/5.6$, $f/8$, $f/11$, and $f/16$. Any one of these settings can be selected by adjusting the aperture, which changes the value of D. Increasing the setting from one f-number to the next higher value (for example, from $f/2.8$ to $f/4$) decreases the area of the aperture by a factor of 2. The lowest f-number setting on a camera lens corresponds to a wide-open aperture and the use of the maximum possible lens area.

Simple cameras usually have a fixed focal length and a fixed aperture size, with an f-number of about $f/11$. This high value for the f-number allows for a large **depth of field,** meaning that objects at a wide range of distances from the lens form reasonably sharp images on the CCD. In other words, the camera does not have to be focused.

> **Q**uick **Quiz** 36.7 A camera can be modeled as a simple converging lens that focuses an image on the CCD, acting as the screen. A camera is initially focused on a distant object. To focus the image of an object close to the camera, must the lens be **(a)** moved away from the CCD, **(b)** left where it is, or **(c)** moved toward the CCD?

Example 36.11   **Finding the Correct Exposure Time**

The lens of a digital camera has a focal length of 55 mm and a speed (an f-number) of $f/1.8$. The correct exposure time for this speed under certain conditions is known to be $\frac{1}{500}$ s.

(A) Determine the diameter of the lens.

SOLUTION

Conceptualize Remember that the f-number for a lens relates its focal length to its diameter.

▶ **36.11** continued

Categorize We determine results using equations developed in this section, so we categorize this example as a substitution problem.

Solve Equation 36.20 for D and substitute numerical values:

$$D = \frac{f}{f\text{-number}} = \frac{55 \text{ mm}}{1.8} = \boxed{31 \text{ mm}}$$

(B) Calculate the correct exposure time if the f-number is changed to $f/4$ under the same lighting conditions.

SOLUTION

The total light energy hitting the CCD is proportional to the product of the intensity and the exposure time. If I is the light intensity reaching the CCD, the energy per unit area received by the CCD in a time interval Δt is proportional to $I \Delta t$. Comparing the two situations, we require that $I_1 \Delta t_1 = I_2 \Delta t_2$, where Δt_1 is the correct exposure time for $f/1.8$ and Δt_2 is the correct exposure time for $f/4$.

Use this result and substitute for I from Equation 36.21:

$$I_1 \Delta t_1 = I_2 \Delta t_2 \quad \rightarrow \quad \frac{\Delta t_1}{(f_1\text{-number})^2} = \frac{\Delta t_2}{(f_2\text{-number})^2}$$

Solve for Δt_2 and substitute numerical values:

$$\Delta t_2 = \left(\frac{f_2\text{-number}}{f_1\text{-number}}\right)^2 \Delta t_1 = \left(\frac{4}{1.8}\right)^2 \left(\tfrac{1}{500} \text{ s}\right) \approx \boxed{\tfrac{1}{100} \text{ s}}$$

As the aperture size is reduced, the exposure time must increase.

36.7 The Eye

Like a camera, a normal eye focuses light and produces a sharp image. The mechanisms by which the eye controls the amount of light admitted and adjusts to produce correctly focused images, however, are far more complex, intricate, and effective than those in even the most sophisticated camera. In all respects, the eye is a physiological wonder.

Figure 36.34 shows the basic parts of the human eye. Light entering the eye passes through a transparent structure called the *cornea* (Fig. 36.35), behind which are a clear liquid (the *aqueous humor*), a variable aperture (the *pupil*, which is an opening in the *iris*), and the *crystalline lens*. Most of the refraction occurs at the outer surface of the eye, where the cornea is covered with a film of tears. Relatively little refraction occurs in the crystalline lens because the aqueous humor in contact with the lens has an average index of refraction close to that of the lens. The iris, which is the colored portion of the eye, is a muscular diaphragm that controls pupil size. The iris regulates the amount of light entering the eye by dilating, or

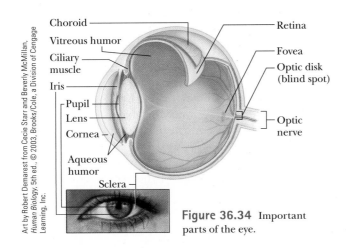

Choroid
Vitreous humor
Ciliary muscle
Iris
Pupil
Lens
Cornea
Aqueous humor
Sclera
Retina
Fovea
Optic disk (blind spot)
Optic nerve

Figure 36.34 Important parts of the eye.

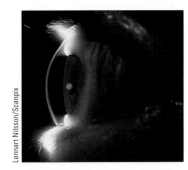

Figure 36.35 Close-up photograph of the cornea of the human eye.

opening, the pupil in low-light conditions and contracting, or closing, the pupil in high-light conditions. The f-number range of the human eye is approximately $f/2.8$ to $f/16$.

The cornea–lens system focuses light onto the back surface of the eye, the *retina*, which consists of millions of sensitive receptors called *rods* and *cones*. When stimulated by light, these receptors send impulses via the optic nerve to the brain, where an image is perceived. By this process, a distinct image of an object is observed when the image falls on the retina.

The eye focuses on an object by varying the shape of the pliable crystalline lens through a process called **accommodation.** The lens adjustments take place so swiftly that we are not even aware of the change. Accommodation is limited in that objects very close to the eye produce blurred images. The **near point** is the closest distance for which the lens can accommodate to focus light on the retina. This distance usually increases with age and has an average value of 25 cm. At age 10, the near point of the eye is typically approximately 18 cm. It increases to approximately 25 cm at age 20, to 50 cm at age 40, and to 500 cm or greater at age 60. The **far point** of the eye represents the greatest distance for which the lens of the relaxed eye can focus light on the retina. A person with normal vision can see very distant objects and therefore has a far point that can be approximated as infinity.

The retina is covered with two types of light-sensitive cells, called **rods** and **cones.** The rods are not sensitive to color but are more light sensitive than the cones. The rods are responsible for *scotopic vision,* or dark-adapted vision. Rods are spread throughout the retina and allow good peripheral vision for all light levels and motion detection in the dark. The cones are concentrated in the fovea. These cells are sensitive to different wavelengths of light. The three categories of these cells are called red, green, and blue cones because of the peaks of the color ranges to which they respond (Fig. 36.36). If the red and green cones are stimulated simultaneously (as would be the case if yellow light were shining on them), the brain interprets what is seen as yellow. If all three types of cones are stimulated by the separate colors red, blue, and green, white light is seen. If all three types of cones are stimulated by light that contains *all* colors, such as sunlight, again white light is seen.

Televisions and computer monitors take advantage of this visual illusion by having only red, green, and blue dots on the screen. With specific combinations of brightness in these three primary colors, our eyes can be made to see any color in the rainbow. Therefore, the yellow lemon you see in a television commercial is not actually yellow, it is red and green! The paper on which this page is printed is made of tiny, matted, translucent fibers that scatter light in all directions, and the resultant mixture of colors appears white to the eye. Snow, clouds, and white hair are not actually white. In fact, there is no such thing as a white pigment. The appearance of these things is a consequence of the scattering of light containing all colors, which we interpret as white.

Conditions of the Eye

When the eye suffers a mismatch between the focusing range of the lens–cornea system and the length of the eye, with the result that light rays from a near object reach the retina before they converge to form an image as shown in Figure 36.37a, the condition is known as **farsightedness** (or *hyperopia*). A farsighted person can usually see faraway objects clearly but not nearby objects. Although the near point of a normal eye is approximately 25 cm, the near point of a farsighted person is much farther away. The refracting power in the cornea and lens is insufficient to focus the light from all but distant objects satisfactorily. The condition can be corrected by placing a converging lens in front of the eye as shown in Figure 36.37b. The lens refracts the incoming rays more toward the principal axis before entering the eye, allowing them to converge and focus on the retina.

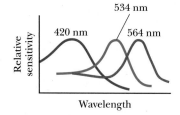

Figure 36.36 Approximate color sensitivity of the three types of cones in the retina.

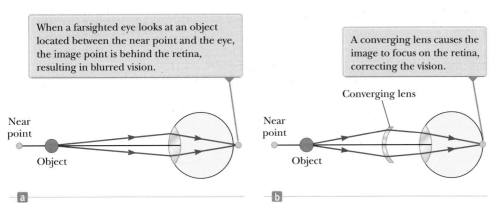

When a farsighted eye looks at an object located between the near point and the eye, the image point is behind the retina, resulting in blurred vision.

A converging lens causes the image to focus on the retina, correcting the vision.

Figure 36.37 (a) An uncorrected farsighted eye. (b) A farsighted eye corrected with a converging lens.

A person with **nearsightedness** (or *myopia*), another mismatch condition, can focus on nearby objects but not on faraway objects. The far point of the nearsighted eye is not infinity and may be less than 1 m. The maximum focal length of the nearsighted eye is insufficient to produce a sharp image on the retina, and rays from a distant object converge to a focus in front of the retina. They then continue past that point, diverging before they finally reach the retina and causing blurred vision (Fig. 36.38a). Nearsightedness can be corrected with a diverging lens as shown in Figure 36.38b. The lens refracts the rays away from the principal axis before they enter the eye, allowing them to focus on the retina.

Beginning in middle age, most people lose some of their accommodation ability as their visual muscles weaken and the lens hardens. Unlike farsightedness, which is a mismatch between focusing power and eye length, **presbyopia** (literally, "old-age vision") is due to a reduction in accommodation ability. The cornea and lens do not have sufficient focusing power to bring nearby objects into focus on the retina. The symptoms are the same as those of farsightedness, and the condition can be corrected with converging lenses.

In eyes having a defect known as **astigmatism,** light from a point source produces a line image on the retina. This condition arises when the cornea, the lens, or both are not perfectly symmetric. Astigmatism can be corrected with lenses that have different curvatures in two mutually perpendicular directions.

Optometrists and ophthalmologists usually prescribe lenses[1] measured in **diopters:** the **power** P of a lens in diopters equals the inverse of the focal length in meters: $P = 1/f$. For example, a converging lens of focal length $+20$ cm has a power of $+5.0$ diopters, and a diverging lens of focal length -40 cm has a power of -2.5 diopters.

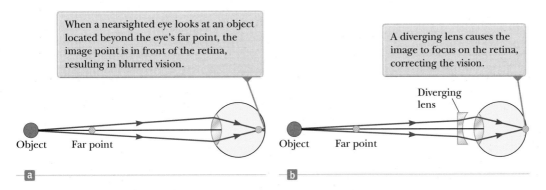

When a nearsighted eye looks at an object located beyond the eye's far point, the image point is in front of the retina, resulting in blurred vision.

A diverging lens causes the image to focus on the retina, correcting the vision.

Figure 36.38 (a) An uncorrected nearsighted eye. (b) A nearsighted eye corrected with a diverging lens.

---

[1]The word *lens* comes from *lentil,* the name of an Italian legume. (You may have eaten lentil soup.) Early eyeglasses were called "glass lentils" because the biconvex shape of their lenses resembled the shape of a lentil. The first lenses for farsightedness and presbyopia appeared around 1280; concave eyeglasses for correcting nearsightedness did not appear until more than 100 years later.

36.8 The Simple Magnifier

The simple magnifier, or magnifying glass, consists of a single converging lens. This device increases the apparent size of an object.

Suppose an object is viewed at some distance p from the eye as illustrated in Figure 36.39. The size of the image formed at the retina depends on the angle θ subtended by the object at the eye. As the object moves closer to the eye, θ increases and a larger image is observed. An average normal human eye, however, cannot focus on an object closer than about 25 cm, the near point (Fig. 36.40a). Therefore, θ is maximum at the near point.

To further increase the apparent angular size of an object, a converging lens can be placed in front of the eye as in Figure 36.40b, with the object located at point O, immediately inside the focal point of the lens. At this location, the lens forms a virtual, upright, enlarged image. We define **angular magnification** m as the ratio of the angle subtended by an object with a lens in use (angle θ in Fig. 36.40b) to the angle subtended by the object placed at the near point with no lens in use (angle θ_0 in Fig. 36.40a):

$$m \equiv \frac{\theta}{\theta_0} \tag{36.22}$$

The angular magnification is a maximum when the image is at the near point of the eye, that is, when $q = -25$ cm. The object distance corresponding to this image distance can be calculated from the thin lens equation:

$$\frac{1}{p} + \frac{1}{-25 \text{ cm}} = \frac{1}{f} \rightarrow p = \frac{25f}{25 + f}$$

where f is the focal length of the magnifier in centimeters. If we make the small-angle approximations

$$\tan \theta_0 \approx \theta_0 \approx \frac{h}{25} \quad \text{and} \quad \tan \theta \approx \theta \approx \frac{h}{p} \tag{36.23}$$

Equation 36.22 becomes

$$m_{\text{max}} = \frac{\theta}{\theta_0} = \frac{h/p}{h/25} = \frac{25}{p} = \frac{25}{25f/(25 + f)}$$

$$m_{\text{max}} = 1 + \frac{25 \text{ cm}}{f} \tag{36.24}$$

Although the eye can focus on an image formed anywhere between the near point and infinity, it is most relaxed when the image is at infinity. For the image formed by the magnifying lens to appear at infinity, the object has to be at the focal point of the lens. In this case, Equations 36.23 become

$$\theta_0 \approx \frac{h}{25} \quad \text{and} \quad \theta \approx \frac{h}{f}$$

and the magnification is

$$m_{\text{min}} = \frac{\theta}{\theta_0} = \frac{25 \text{ cm}}{f} \tag{36.25}$$

With a single lens, it is possible to obtain angular magnifications up to about 4 without serious aberrations. Magnifications up to about 20 can be achieved by using one or two additional lenses to correct for aberrations.

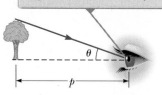

The size of the image formed on the retina depends on the angle θ subtended at the eye.

Figure 36.39 An observer looks at an object at distance p.

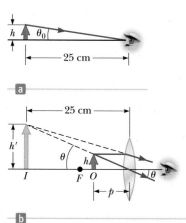

a

b

Figure 36.40 (a) An object placed at the near point of the eye ($p = 25$ cm) subtends an angle $\theta_0 \approx h/25$ cm at the eye. (b) An object placed near the focal point of a converging lens produces a magnified image that subtends an angle $\theta \approx h'/25$ cm at the eye.

© Cengage Learning/George Semple

A simple magnifier, also called a magnifying glass, is used to view an enlarged image of a portion of a map.

| Example 36.12 | **Magnification of a Lens** |

What is the maximum magnification that is possible with a lens having a focal length of 10 cm, and what is the magnification of this lens when the eye is relaxed?

SOLUTION

Conceptualize Study Figure 36.40b for the situation in which a magnifying glass forms an enlarged image of an object placed inside the focal point. The maximum magnification occurs when the image is located at the near point of the eye. When the eye is relaxed, the image is at infinity.

Categorize We determine results using equations developed in this section, so we categorize this example as a substitution problem.

Evaluate the maximum magnification from Equation 36.24:

$$m_{max} = 1 + \frac{25 \text{ cm}}{f} = 1 + \frac{25 \text{ cm}}{10 \text{ cm}} = \boxed{3.5}$$

Evaluate the minimum magnification, when the eye is relaxed, from Equation 36.25:

$$m_{min} = \frac{25 \text{ cm}}{f} = \frac{25 \text{ cm}}{10 \text{ cm}} = \boxed{2.5}$$

36.9 The Compound Microscope

A simple magnifier provides only limited assistance in inspecting minute details of an object. Greater magnification can be achieved by combining two lenses in a device called a **compound microscope** shown in Figure 36.41a. It consists of one lens, the *objective*, that has a very short focal length $f_o < 1$ cm and a second lens, the *eyepiece*, that has a focal length f_e of a few centimeters. The two lenses are separated by a distance L that is much greater than either f_o or f_e. The object, which is placed just outside the focal point of the objective, forms a real, inverted image at I_1, and this image is located at or close to the focal point of the eyepiece. The eyepiece, which serves as a simple magnifier, produces at I_2 a virtual, enlarged image of I_1. The lateral magnification M_1 of the first image is $-q_1/p_1$. Notice from Figure 36.41a that q_1 is approximately equal to L and that the object is very close

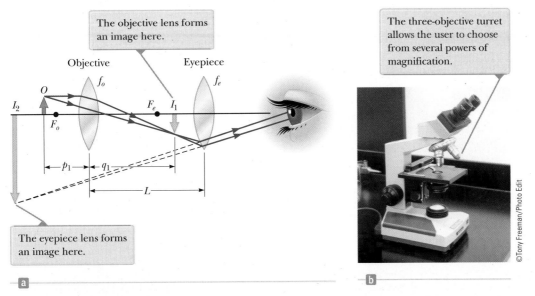

The objective lens forms an image here.

Objective   Eyepiece

The three-objective turret allows the user to choose from several powers of magnification.

The eyepiece lens forms an image here.

©Tony Freeman/Photo Edit

a   b

Figure 36.41 (a) Diagram of a compound microscope, which consists of an objective lens and an eyepiece lens. (b) A compound microscope.

to the focal point of the objective: $p_1 \approx f_o$. Therefore, the lateral magnification by the objective is

$$M_o \approx -\frac{L}{f_o}$$

The angular magnification by the eyepiece for an object (corresponding to the image at I_1) placed at the focal point of the eyepiece is, from Equation 36.25,

$$m_e = \frac{25 \text{ cm}}{f_e}$$

The overall magnification of the image formed by a compound microscope is defined as the product of the lateral and angular magnifications:

$$M = M_o m_e = -\frac{L}{f_o}\left(\frac{25 \text{ cm}}{f_e}\right) \tag{36.26}$$

The negative sign indicates that the image is inverted.

The microscope has extended human vision to the point where we can view previously unknown details of incredibly small objects. The capabilities of this instrument have steadily increased with improved techniques for precision grinding of lenses. A question often asked about microscopes is, "If one were extremely patient and careful, would it be possible to construct a microscope that would enable the human eye to see an atom?" The answer is no, as long as light is used to illuminate the object. For an object under an optical microscope (one that uses visible light) to be seen, the object must be at least as large as a wavelength of light. Because the diameter of any atom is many times smaller than the wavelengths of visible light, the mysteries of the atom must be probed using other types of "microscopes."

36.10 The Telescope

Two fundamentally different types of **telescopes** exist; both are designed to aid in viewing distant objects such as the planets in our solar system. The first type, the **refracting telescope,** uses a combination of lenses to form an image.

Like the compound microscope, the refracting telescope shown in Figure 36.42a has an objective and an eyepiece. The two lenses are arranged so that the objective

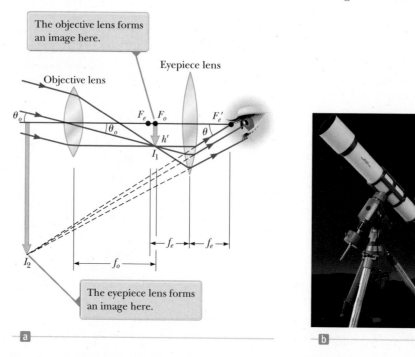

The objective lens forms an image here.

Eyepiece lens

Objective lens

θ_o

F_e F_o

θ_o

h'

θ

F_e'

I_1

f_e f_e

I_2

f_o

The eyepiece lens forms an image here.

Figure 36.42 (a) Lens arrangement in a refracting telescope, with the object at infinity. (b) A refracting telescope.

a   b

forms a real, inverted image of a distant object very near the focal point of the eyepiece. Because the object is essentially at infinity, this point at which I_1 forms is the focal point of the objective. The eyepiece then forms, at I_2, an enlarged, inverted image of the image at I_1. To provide the largest possible magnification, the image distance for the eyepiece is infinite. Therefore, the image due to the objective lens, which acts as the object for the eyepiece lens, must be located at the focal point of the eyepiece. Hence, the two lenses are separated by a distance $f_o + f_e$, which corresponds to the length of the telescope tube.

The angular magnification of the telescope is given by θ/θ_o, where θ_o is the angle subtended by the object at the objective and θ is the angle subtended by the final image at the viewer's eye. Consider Figure 36.42a, in which the object is a very great distance to the left of the figure. The angle θ_o (to the *left* of the objective) subtended by the object at the objective is the same as the angle (to the *right* of the objective) subtended by the first image at the objective. Therefore,

$$\tan \theta_o \approx \theta_o \approx -\frac{h'}{f_o}$$

where the negative sign indicates that the image is inverted.

The angle θ subtended by the final image at the eye is the same as the angle that a ray coming from the tip of I_1 and traveling parallel to the principal axis makes with the principal axis after it passes through the lens. Therefore,

$$\tan \theta \approx \theta \approx \frac{h'}{f_e}$$

We have not used a negative sign in this equation because the final image is not inverted; the object creating this final image I_2 is I_1, and both it and I_2 point in the same direction. Therefore, the angular magnification of the telescope can be expressed as

$$m = \frac{\theta}{\theta_o} = \frac{h'/f_e}{-h'/f_o} = -\frac{f_o}{f_e} \qquad \text{(36.27)}$$

This result shows that the angular magnification of a telescope equals the ratio of the objective focal length to the eyepiece focal length. The negative sign indicates that the image is inverted.

When you look through a telescope at such relatively nearby objects as the Moon and the planets, magnification is important. Individual stars in our galaxy, however, are so far away that they always appear as small points of light no matter how great the magnification. To gather as much light as possible, large research telescopes used to study very distant objects must have a large diameter. It is difficult and expensive to manufacture large lenses for refracting telescopes. Another difficulty with large lenses is that their weight leads to sagging, which is an additional source of aberration.

These problems associated with large lenses can be partially overcome by replacing the objective with a concave mirror, which results in the second type of telescope, the **reflecting telescope.** Because light is reflected from the mirror and does not pass through a lens, the mirror can have rigid supports on the back side. Such supports eliminate the problem of sagging.

Figure 36.43a shows the design for a typical reflecting telescope. The incoming light rays are reflected by a parabolic mirror at the base. These reflected rays converge toward point A in the figure, where an image would be formed. Before this image is formed, however, a small, flat mirror M reflects the light toward an opening in the tube's side and it passes into an eyepiece. This particular design is said to have a Newtonian focus because Newton developed it. Figure 36.43b shows such a telescope. Notice that the light never passes through glass (except through the small eyepiece) in the reflecting telescope. As a result, problems associated with chromatic aberration are virtually eliminated. The reflecting telescope can be made even shorter by orienting the flat mirror so that it reflects the light back

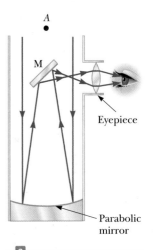

Figure 36.43 (a) A Newtonian-focus reflecting telescope. (b) A reflecting telescope. This type of telescope is shorter than that in Figure 36.42b.

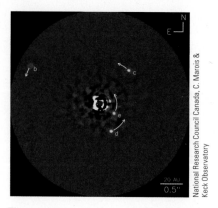

National Research Council Canada, C. Marois & Keck Observatory

Figure 36.44 A direct optical image of a solar system around the star HR8799, developed at the Keck Observatory in Hawaii.

toward the objective mirror and the light enters an eyepiece in a hole in the middle of the mirror.

The largest reflecting telescopes in the world are at the Keck Observatory on Mauna Kea, Hawaii. The site includes two telescopes with diameters of 10 m, each containing 36 hexagonally shaped, computer-controlled mirrors that work together to form a large reflecting surface. In addition, the two telescopes can work together to provide a telescope with an effective diameter of 85 m. In contrast, the largest refracting telescope in the world, at the Yerkes Observatory in Williams Bay, Wisconsin, has a diameter of only 1 m.

Figure 36.44 shows a remarkable optical image from the Keck Observatory of a solar system around the star HR8799, located 129 light-years from the Earth. The planets labeled b, c, and d were seen in 2008 and the innermost planet, labeled e, was observed in December 2010. This photograph represents the first direct image of another solar system and was made possible by the adaptive optics technology used in the Keck Observatory.

Summary

Definitions

The **lateral magnification** M of the image due to a mirror or lens is defined as the ratio of the image height h' to the object height h. It is equal to the negative of the ratio of the image distance q to the object distance p:

$$M \equiv \frac{\text{image height}}{\text{object height}} = \frac{h'}{h} = -\frac{q}{p} \quad \text{(36.1, 36.2, 36.17)}$$

The **angular magnification** m is the ratio of the angle subtended by an object with a lens in use (angle θ in Fig. 36.40b) to the angle subtended by the object placed at the near point with no lens in use (angle θ_0 in Fig. 36.40a):

$$m \equiv \frac{\theta}{\theta_0} \quad \text{(36.22)}$$

The ratio of the focal length of a camera lens to the diameter of the lens is called the **f-number** of the lens:

$$f\text{-number} \equiv \frac{f}{D} \quad \text{(36.20)}$$

Concepts and Principles

In the paraxial ray approximation, the object distance p and image distance q for a spherical mirror of radius R are related by the **mirror equation:**

$$\frac{1}{p} + \frac{1}{q} = \frac{2}{R} = \frac{1}{f} \quad \text{(36.4, 36.6)}$$

where $f = R/2$ is the **focal length** of the mirror.

An image can be formed by refraction from a spherical surface of radius R. The object and image distances for refraction from such a surface are related by

$$\frac{n_1}{p} + \frac{n_2}{q} = \frac{n_2 - n_1}{R} \quad \text{(36.8)}$$

where the light is incident in the medium for which the index of refraction is n_1 and is refracted in the medium for which the index of refraction is n_2.

The inverse of the **focal length** f of a thin lens surrounded by air is given by the **lens-makers' equation:**

$$\frac{1}{f} = (n - 1)\left(\frac{1}{R_1} - \frac{1}{R_2}\right) \quad \text{(36.15)}$$

Converging lenses have positive focal lengths, and **diverging lenses** have negative focal lengths.

For a thin lens, and in the paraxial ray approximation, the object and image distances are related by the **thin lens equation:**

$$\frac{1}{p} + \frac{1}{q} = \frac{1}{f} \quad \text{(36.16)}$$

The maximum magnification of a single lens of focal length f used as a simple magnifier is

$$m_{max} = 1 + \frac{25 \text{ cm}}{f} \qquad (36.24)$$

The overall magnification of the image formed by a compound microscope is

$$M = -\frac{L}{f_o}\left(\frac{25 \text{ cm}}{f_e}\right) \qquad (36.26)$$

where f_o and f_e are the focal lengths of the objective and eyepiece lenses, respectively, and L is the distance between the lenses.

The angular magnification of a refracting telescope can be expressed as

$$m = -\frac{f_o}{f_e} \qquad (36.27)$$

where f_o and f_e are the focal lengths of the objective and eyepiece lenses, respectively. The angular magnification of a reflecting telescope is given by the same expression where f_o is the focal length of the objective mirror.

Objective Questions

1. denotes answer available in *Student Solutions Manual/Study Guide*

1. The faceplate of a diving mask can be ground into a corrective lens for a diver who does not have perfect vision. The proper design allows the person to see clearly both under water and in the air. Normal eyeglasses have lenses with both the front and back surfaces curved. Should the lenses of a diving mask be curved (a) on the outer surface only, (b) on the inner surface only, or (c) on both surfaces?

2. Lulu looks at her image in a makeup mirror. It is enlarged when she is close to the mirror. As she backs away, the image becomes larger, then impossible to identify when she is 30.0 cm from the mirror, then upside down when she is beyond 30.0 cm, and finally small, clear, and upside down when she is much farther from the mirror. **(i)** Is the mirror (a) convex, (b) plane, or (c) concave? **(ii)** Is the magnitude of its focal length (a) 0, (b) 15.0 cm, (c) 30.0 cm, (d) 60.0 cm, or (e) ∞?

3. An object is located 50.0 cm from a converging lens having a focal length of 15.0 cm. Which of the following statements is true regarding the image formed by the lens? (a) It is virtual, upright, and larger than the object. (b) It is real, inverted, and smaller than the object. (c) It is virtual, inverted, and smaller than the object. (d) It is real, inverted, and larger than the object. (e) It is real, upright, and larger than the object.

4. **(i)** When an image of an object is formed by a converging lens, which of the following statements is *always* true? More than one statement may be correct. (a) The image is virtual. (b) The image is real. (c) The image is upright. (d) The image is inverted. (e) None of those statements is always true. **(ii)** When the image of an object is formed by a diverging lens, which of the statements is *always* true?

5. A converging lens in a vertical plane receives light from an object and forms an inverted image on a screen. An opaque card is then placed next to the lens, covering only the upper half of the lens. What happens to the image on the screen? (a) The upper half of the image disappears. (b) The lower half of the image disappears. (c) The entire image disappears. (d) The entire image is still visible, but is dimmer. (e) No change in the image occurs.

6. If Josh's face is 30.0 cm in front of a concave shaving mirror creating an upright image 1.50 times as large as the object, what is the mirror's focal length? (a) 12.0 cm (b) 20.0 cm (c) 70.0 cm (d) 90.0 cm (e) none of those answers

7. Two thin lenses of focal lengths $f_1 = 15.0$ and $f_2 = 10.0$ cm, respectively, are separated by 35.0 cm along a common axis. The f_1 lens is located to the left of the f_2 lens. An object is now placed 50.0 cm to the left of the f_1 lens, and a final image due to light passing though both lenses forms. By what factor is the final image different in size from the object? (a) 0.600 (b) 1.20 (c) 2.40 (d) 3.60 (e) none of those answers

8. If you increase the aperture diameter of a camera by a factor of 3, how is the intensity of the light striking the film affected? (a) It increases by factor of 3. (b) It decreases by a factor of 3. (c) It increases by a factor of 9. (d) It decreases by a factor of 9. (e) Increasing the aperture size doesn't affect the intensity.

9. A person spearfishing from a boat sees a stationary fish a few meters away in a direction about 30° below the horizontal. To spear the fish, and assuming the spear does not change direction when it enters the water, should the person (a) aim above where he sees the fish, (b) aim below the fish, or (c) aim precisely at the fish?

10. Model each of the following devices in use as consisting of a single converging lens. Rank the cases according to the ratio of the distance from the object to the lens to the focal length of the lens, from the largest ratio to the smallest. (a) a film-based movie projector showing a movie (b) a magnifying glass being used to examine a postage stamp (c) an astronomical refracting telescope being used to make a sharp image of stars on an electronic detector (d) a searchlight being used to produce a beam of parallel rays from a point source (e) a camera lens being used to photograph a soccer game

11. A converging lens made of crown glass has a focal length of 15.0 cm when used in air. If the lens is immersed in water, what is its focal length? (a) negative

(b) less than 15.0 cm (c) equal to 15.0 cm (d) greater than 15.0 cm (e) none of those answers

12. A converging lens of focal length 8 cm forms a sharp image of an object on a screen. What is the smallest possible distance between the object and the screen? (a) 0 (b) 4 cm (c) 8 cm (d) 16 cm (e) 32 cm

13. **(i)** When an image of an object is formed by a plane mirror, which of the following statements is *always* true? More than one statement may be correct. (a) The image is virtual. (b) The image is real. (c) The image is upright. (d) The image is inverted. (e) None of those statements is always true. **(ii)** When the image of an object is formed by a concave mirror, which of the preceding statements are *always* true? **(iii)** When the image of an object is formed by a convex mirror, which of the preceding statements are *always* true?

14. An object, represented by a gray arrow, is placed in front of a plane mirror. Which of the diagrams in Figure OQ36.14 correctly describes the image, represented by the pink arrow?

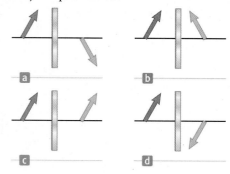

Figure OQ36.14

Conceptual Questions

1. denotes answer available in *Student Solutions Manual/Study Guide*

1. A converging lens of short focal length can take light diverging from a small source and refract it into a beam of parallel rays. A Fresnel lens as shown in Figure 36.27 is used in a lighthouse for this purpose. A concave mirror can take light diverging from a small source and reflect it into a beam of parallel rays. (a) Is it possible to make a Fresnel mirror? (b) Is this idea original, or has it already been done?

2. Explain this statement: "The focal point of a lens is the location of the image of a point object at infinity." (a) Discuss the notion of infinity in real terms as it applies to object distances. (b) Based on this statement, can you think of a simple method for determining the focal length of a converging lens?

3. Why do some emergency vehicles have the symbol ƎƆИAↃUꓭMA written on the front?

4. Explain why a mirror cannot give rise to chromatic aberration.

5. (a) Can a converging lens be made to diverge light if it is placed into a liquid? (b) **What If?** What about a converging mirror?

6. Explain why a fish in a spherical goldfish bowl appears larger than it really is.

7. In Figure 36.26a, assume the gray object arrow is replaced by one that is much taller than the lens. (a) How many rays from the top of the object will strike the lens? (b) How many principal rays can be drawn in a ray diagram?

8. Lenses used in eyeglasses, whether converging or diverging, are always designed so that the middle of the lens curves away from the eye like the center lenses of Figures 36.25a and 36.25b. Why?

9. Suppose you want to use a converging lens to project the image of two trees onto a screen. As shown in Figure CQ36.9, one tree is a distance *x* from the lens and the other is at 2*x*. You adjust the screen so that the near tree

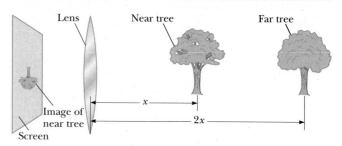

Figure CQ36.9

is in focus. If you now want the far tree to be in focus, do you move the screen toward or away from the lens?

10. Consider a spherical concave mirror with the object located to the left of the mirror beyond the focal point. Using ray diagrams, show that the image moves to the left as the object approaches the focal point.

11. In Figures CQ36.11a and CQ36.11b, which glasses correct nearsightedness and which correct farsightedness?

Figure CQ36.11 Conceptual Questions 11 and 12.

12. Bethany tries on either her hyperopic grandfather's or her myopic brother's glasses and complains, "Everything looks blurry." Why do the eyes of a person wearing glasses not look blurry? (See Fig. CQ36.11.)

13. In a Jules Verne novel, a piece of ice is shaped to form a magnifying lens, focusing sunlight to start a fire. Is that possible?

14. A solar furnace can be constructed by using a concave mirror to reflect and focus sunlight into a furnace enclosure. What factors in the design of the reflecting mirror would guarantee very high temperatures?

15. Figure CQ36.15 shows a lithograph by M. C. Escher titled *Hand with Reflection Sphere (Self-Portrait in Spherical Mirror)*. Escher said about the work: "The picture shows a spherical mirror, resting on a left hand. But as a print is the reverse of the original drawing on stone, it was my right hand that you see depicted. (Being left-handed, I needed my left hand to make the drawing.) Such a globe reflection collects almost

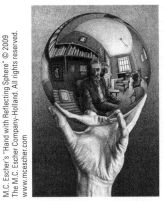

Figure CQ36.15

one's whole surroundings in one disk-shaped image. The whole room, four walls, the floor, and the ceiling, everything, albeit distorted, is compressed into that one small circle. Your own head, or more exactly the point between your eyes, is the absolute center. No matter how you turn or twist yourself, you can't get out of that central point. You are immovably the focus, the unshakable core, of your world." Comment on the accuracy of Escher's description.

16. If a cylinder of solid glass or clear plastic is placed above the words LEAD OXIDE and viewed from the side as shown in Figure CQ36.16, the word LEAD appears inverted, but the word OXIDE does not. Explain.

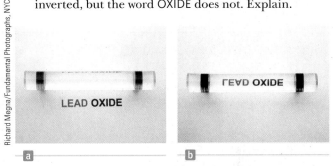

Figure CQ36.16

17. Do the equations $1/p + 1/q = 1/f$ and $M = -q/p$ apply to the image formed by a flat mirror? Explain your answer.

Problems available in **ENHANCED** **WebAssign** Access end-of-chapter problems online at www.webassign.net

Section 36.1 Images Formed by Flat Mirrors
Problems 1–7

Section 36.2 Images Formed by Spherical Mirrors
Problems 8–28

Section 36.3 Images Formed by Refraction
Problems 29–37

Section 36.4 Images Formed by Thin lenses
Problems 38–53

Section 36.5 Lens Aberrations
Problems 54–55

Section 36.6 The Camera
Problems 56–57

Section 36.7 The Eye
Problems 58–65

Section 36.8 The Simple Magnifier
Problem 66

Section 36.9 The Compound Microscope
Problem 67

Section 36.10 The Telescope
Problems 68–70

Additional Problems
Problems 71–92

Challenge Problems
Problem 93–97

Solutions to the following Problems are available in the *Student Solutions Manual/Study Guide:*
36.1, 36.9, 36.11, 36.25, 36.30, 36.35, 36.40, 36.47, 36.49, 36.58, 36.63, 36.65, 36.68, 36.81, 36.84, 36.87, 36.90, 36.91, and 36.96

List of Enhanced Problems

Problem Number	Targeted Feedback in Enhanced WebAssign	Analysis Model Tutorial in Enhanced WebAssign	Master It in Enhanced WebAssign	Watch It in Enhanced WebAssign
36.1	✓	✓		✓
36.2		✓		
36.9	✓		✓	
36.11	✓		✓	
36.12	✓		✓	✓
36.19	✓			✓
36.21	✓			✓
36.23	✓			✓
36.25			✓	
36.26		✓		
36.28	✓		✓	
36.35			✓	
36.37	✓			✓
36.40			✓	
36.41	✓			✓
36.43	✓			✓
36.51	✓			✓
36.53	✓		✓	
36.61	✓		✓	✓
36.63	✓			
36.68			✓	
36.75	✓		✓	
36.87			✓	
36.90			✓	

Wave Optics

The colors in many of a hummingbird's feathers are not due to pigment. The *iridescence* that makes the brilliant colors that often appear on the bird's throat and belly is due to an interference effect caused by structures in the feathers. The colors will vary with the viewing angle. *(Dec Hogan/ Shutterstock.com)*

In Chapter 36, we studied light rays passing through a lens or reflecting from a mirror to describe the formation of images. This discussion completed our study of *ray optics*. In this chapter and in Chapter 38, we are concerned with *wave optics*, sometimes called *physical optics*, the study of interference, diffraction, and polarization of light. These phenomena cannot be adequately explained with the ray optics used in Chapters 35 and 36. We now learn how treating light as waves rather than as rays leads to a satisfying description of such phenomena.

37.1 Young's Double-Slit Experiment

In Chapter 18, we studied the waves in interference model and found that the superposition of two mechanical waves can be constructive or destructive. In constructive interference, the amplitude of the resultant wave is greater than that of either individual wave, whereas in destructive interference, the resultant amplitude is less than that of the larger wave. Light waves also interfere with one another. Fundamentally, all interference associated with light waves arises when the electromagnetic fields that constitute the individual waves combine.

Interference in light waves from two sources was first demonstrated by Thomas Young in 1801. A schematic diagram of the apparatus Young used is shown in Figure 37.1a. Plane light waves arrive at a barrier that contains two slits S_1 and S_2. The light from S_1 and S_2 produces on a viewing screen a visible pattern of bright and dark parallel bands called **fringes** (Fig. 37.1b). When the light from S_1 and that from S_2 both arrive at a point on the screen such that constructive interference occurs at

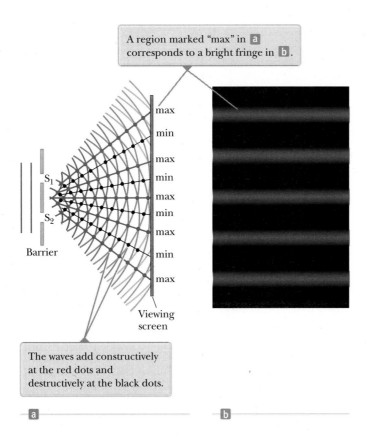

A region marked "max" in **a** corresponds to a bright fringe in **b**.

S₁

S₂

Barrier

max
min
max
min
max
min
max
min
max

Viewing screen

The waves add constructively at the red dots and destructively at the black dots.

a   **b**

Figure 37.1 (a) Schematic diagram of Young's double-slit experiment. Slits S₁ and S₂ behave as coherent sources of light waves that produce an interference pattern on the viewing screen (drawing not to scale). (b) A simulation of an enlargement of the center of a fringe pattern formed on the viewing screen.

that location, a bright fringe appears. When the light from the two slits combines destructively at any location on the screen, a dark fringe results.

Figure 37.2 is a photograph looking down on an interference pattern produced on the surface of a water tank by two vibrating sources. The linear regions of constructive interference, such as at *A*, and destructive interference, such as at *B*, radiating from the area between the sources are analogous to the red and black lines in Figure 37.1a.

Figure 37.3 on page 888 shows some of the ways in which two waves can combine at the screen. In Figure 37.3a, the two waves, which leave the two slits in phase, strike the screen at the central point *O*. Because both waves travel the same distance, they arrive at *O* in phase. As a result, constructive interference occurs at this location and a bright fringe is observed. In Figure 37.3b, the two waves also start in phase, but here the lower wave has to travel one wavelength farther than the upper wave to reach point *P*. Because the lower wave falls behind

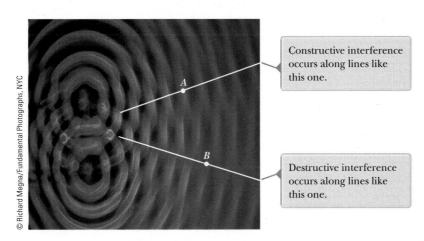

© Richard Megna/Fundamental Photographs, NYC

Constructive interference occurs along lines like this one.

Destructive interference occurs along lines like this one.

Figure 37.2 An interference pattern involving water waves is produced by two vibrating sources at the water's surface.

Figure 37.3 Waves leave the slits and combine at various points on the viewing screen. (All figures not to scale.)

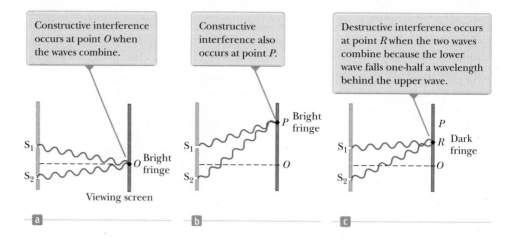

Constructive interference occurs at point *O* when the waves combine.

Constructive interference also occurs at point *P*.

Destructive interference occurs at point *R* when the two waves combine because the lower wave falls one-half a wavelength behind the upper wave.

the upper one by exactly one wavelength, they still arrive in phase at *P* and a second bright fringe appears at this location. At point *R* in Figure 37.3c, however, between points *O* and *P*, the lower wave has fallen half a wavelength behind the upper wave and a crest of the upper wave overlaps a trough of the lower wave, giving rise to destructive interference at point *R*. A dark fringe is therefore observed at this location.

If two lightbulbs are placed side by side so that light from both bulbs combines, no interference effects are observed because the light waves from one bulb are emitted independently of those from the other bulb. The emissions from the two lightbulbs do not maintain a constant phase relationship with each other over time. Light waves from an ordinary source such as a lightbulb undergo random phase changes in time intervals of less than a nanosecond. Therefore, the conditions for constructive interference, destructive interference, or some intermediate state are maintained only for such short time intervals. Because the eye cannot follow such rapid changes, no interference effects are observed. Such light sources are said to be **incoherent.**

To observe interference of waves from two sources, the following conditions must be met:

Conditions for interference ▶

- The sources must be **coherent;** that is, they must maintain a constant phase with respect to each other.
- The sources should be **monochromatic;** that is, they should be of a single wavelength.

As an example, single-frequency sound waves emitted by two side-by-side loudspeakers driven by a single amplifier can interfere with each other because the two speakers are coherent. In other words, they respond to the amplifier in the same way at the same time.

A common method for producing two coherent light sources is to use a monochromatic source to illuminate a barrier containing two small openings, usually in the shape of slits, as in the case of Young's experiment illustrated in Figure 37.1. The light emerging from the two slits is coherent because a single source produces the original light beam and the two slits serve only to separate the original beam into two parts (which, after all, is what is done to the sound signal from two side-by-side loudspeakers). Any random change in the light emitted by the source occurs in both beams at the same time. As a result, interference effects can be observed when the light from the two slits arrives at a viewing screen.

If the light traveled only in its original direction after passing through the slits as shown in Figure 37.4a, the waves would not overlap and no interference pattern would be seen. Instead, as we have discussed in our treatment of Huygens's principle (Section 35.6), the waves spread out from the slits as shown in Figure 37.4b. In other words, the light deviates from a straight-line path and enters the region that

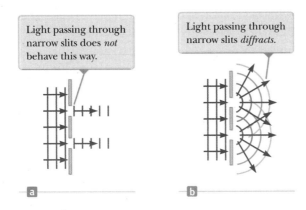

Figure 37.4 (a) If light waves did not spread out after passing through the slits, no interference would occur. (b) The light waves from the two slits overlap as they spread out, filling what we expect to be shadowed regions with light and producing interference fringes on a screen placed to the right of the slits.

would otherwise be shadowed. As noted in Section 35.3, this divergence of light from its initial line of travel is called **diffraction.**

37.2 Analysis Model: Waves in Interference

We discussed the superposition principle for waves on strings in Section 18.1, leading to a one-dimensional version of the waves in interference analysis model. In Example 18.1 on page 421, we briefly discussed a two-dimensional interference phenomenon for sound from two loudspeakers. In walking from point O to point P in Figure 18.5, the listener experienced a maximum in sound intensity at O and a minimum at P. This experience is exactly analogous to an observer looking at point O in Figure 37.3 and seeing a bright fringe and then sweeping his eyes upward to point R, where there is a minimum in light intensity.

Let's look in more detail at the two-dimensional nature of Young's experiment with the help of Figure 37.5. The viewing screen is located a perpendicular distance L from the barrier containing two slits, S_1 and S_2 (Fig. 37.5a). These slits are separated by a distance d, and the source is monochromatic. To reach any arbitrary point P in the upper half of the screen, a wave from the lower slit must travel farther than a wave from the upper slit. The extra distance traveled from the lower slit is the **path difference** δ (Greek letter delta). If we assume the rays labeled r_1 and r_2 are parallel (Fig. 37.5b), which is approximately true if L is much greater than d, then δ is given by

$$\delta = r_2 - r_1 = d \sin \theta \tag{37.1}$$

The value of δ determines whether the two waves are in phase when they arrive at point P. If δ is either zero or some integer multiple of the wavelength, the two waves

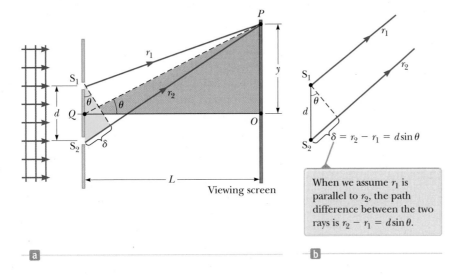

When we assume r_1 is parallel to r_2, the path difference between the two rays is $r_2 - r_1 = d \sin \theta$.

Figure 37.5 (a) Geometric construction for describing Young's double-slit experiment (not to scale). (b) The slits are represented as sources, and the outgoing light rays are assumed to be parallel as they travel to P. To achieve that in practice, it is essential that $L \gg d$.

are in phase at point P and constructive interference results. Therefore, the condition for bright fringes, or **constructive interference,** at point P is

Condition for constructive ▶
interference

$$d \sin \theta_{\text{bright}} = m\lambda \qquad m = 0, \pm 1, \pm 2, \ldots \qquad \text{(37.2)}$$

The number m is called the **order number.** For constructive interference, the order number is the same as the number of wavelengths that represents the path difference between the waves from the two slits. The central bright fringe at $\theta_{\text{bright}} = 0$ is called the *zeroth-order maximum.* The first maximum on either side, where $m = \pm 1$, is called the *first-order maximum,* and so forth.

When δ is an odd multiple of $\lambda/2$, the two waves arriving at point P are 180° out of phase and give rise to destructive interference. Therefore, the condition for dark fringes, or **destructive interference,** at point P is

Condition for destructive ▶
interference

$$d \sin \theta_{\text{dark}} = \left(m + \tfrac{1}{2}\right)\lambda \qquad m = 0, \pm 1, \pm 2, \ldots \qquad \text{(37.3)}$$

These equations provide the *angular* positions of the fringes. It is also useful to obtain expressions for the *linear* positions measured along the screen from O to P. From the triangle OPQ in Figure 37.5a, we see that

$$\tan \theta = \frac{y}{L} \qquad \text{(37.4)}$$

Using this result, the linear positions of bright and dark fringes are given by

$$y_{\text{bright}} = L \tan \theta_{\text{bright}} \qquad \text{(37.5)}$$

$$y_{\text{dark}} = L \tan \theta_{\text{dark}} \qquad \text{(37.6)}$$

where θ_{bright} and θ_{dark} are given by Equations 37.2 and 37.3.

When the angles to the fringes are small, the positions of the fringes are linear near the center of the pattern. That can be verified by noting that for small angles, $\tan \theta \approx \sin \theta$, so Equation 37.5 gives the positions of the bright fringes as $y_{\text{bright}} = L \sin \theta_{\text{bright}}$. Incorporating Equation 37.2 gives

$$y_{\text{bright}} = L \frac{m\lambda}{d} \quad \text{(small angles)} \qquad \text{(37.7)}$$

This result shows that y_{bright} is linear in the order number m, so the fringes are equally spaced for small angles. Similarly, for dark fringes,

$$y_{\text{dark}} = L \frac{\left(m + \tfrac{1}{2}\right)\lambda}{d} \quad \text{(small angles)} \qquad \text{(37.8)}$$

As demonstrated in Example 37.1, Young's double-slit experiment provides a method for measuring the wavelength of light. In fact, Young used this technique to do precisely that. In addition, his experiment gave the wave model of light a great deal of credibility. It was inconceivable that particles of light coming through the slits could cancel one another in a way that would explain the dark fringes.

The principles discussed in this section are the basis of the **waves in interference** analysis model. This model was applied to mechanical waves in one dimension in Chapter 18. Here we see the details of applying this model in three dimensions to light.

Ⓠ**uick Quiz** 37.1 Which of the following causes the fringes in a two-slit interference pattern to move farther apart? **(a)** decreasing the wavelength of the light **(b)** decreasing the screen distance L **(c)** decreasing the slit spacing d **(d)** immersing the entire apparatus in water

| Analysis Model | Waves in Interference |

Imagine a broad beam of light that illuminates a double slit in an otherwise opaque material. An interference pattern of bright and dark fringes is created on a distant screen. The condition for bright fringes (**constructive interference**) is

$$d \sin \theta_{bright} = m\lambda \quad m = 0, \pm1, \pm2, \ldots \qquad (37.2)$$

The condition for dark fringes (**destructive interference**) is

$$d \sin \theta_{dark} = (m + \tfrac{1}{2})\lambda \quad m = 0, \pm1, \pm2, \ldots \qquad (37.3)$$

The number m is called the **order number** of the fringe.

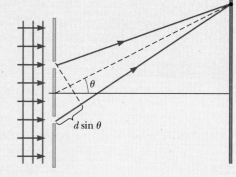

Examples:

- a thin film of oil on top of water shows swirls of color (Section 37.5)
- x-rays passing through a crystalline solid combine to form a Laue pattern (Chapter 38)
- a Michelson interferometer (Section 37.6) is used to search for the ether representing the medium through which light travels (Chapter 39)
- electrons exhibit interference just like light waves when they pass through a double slit (Chapter 40)

| Example 37.1 | Measuring the Wavelength of a Light Source | |

A viewing screen is separated from a double slit by 4.80 m. The distance between the two slits is 0.030 0 mm. Monochromatic light is directed toward the double slit and forms an interference pattern on the screen. The first dark fringe is 4.50 cm from the center line on the screen.

(A) Determine the wavelength of the light.

SOLUTION

Conceptualize Study Figure 37.5 to be sure you understand the phenomenon of interference of light waves. The distance of 4.50 cm is y in Figure 37.5. Because $L \gg y$, the angles for the fringes are small.

Categorize This problem is a simple application of the *waves in interference* model.

. . . .

Analyze

Solve Equation 37.8 for the wavelength and substitute numerical values, taking $m = 0$ for the first dark fringe:

$$\lambda = \frac{y_{dark}d}{(m + \frac{1}{2})L} = \frac{(4.50 \times 10^{-2}\ \text{m})(3.00 \times 10^{-5}\ \text{m})}{(0 + \frac{1}{2})(4.80\ \text{m})}$$

$$= 5.62 \times 10^{-7}\ \text{m} = \boxed{562\ \text{nm}}$$

(B) Calculate the distance between adjacent bright fringes.

SOLUTION

Find the distance between adjacent bright fringes from Equation 37.7 and the results of part (A):

$$y_{m+1} - y_m = L \frac{(m + 1)\lambda}{d} - L \frac{m\lambda}{d}$$

$$= L \frac{\lambda}{d} = 4.80\ \text{m} \left(\frac{5.62 \times 10^{-7}\ \text{m}}{3.00 \times 10^{-5}\ \text{m}} \right)$$

$$= 9.00 \times 10^{-2}\ \text{m} = \boxed{9.00\ \text{cm}}$$

Finalize For practice, find the wavelength of the sound in Example 18.1 using the procedure in part (A) of this example.

Example 37.2 Separating Double-Slit Fringes of Two Wavelengths **AM**

A light source emits visible light of two wavelengths: $\lambda = 430$ nm and $\lambda' = 510$ nm. The source is used in a double-slit interference experiment in which $L = 1.50$ m and $d = 0.025\ 0$ mm. Find the separation distance between the third-order bright fringes for the two wavelengths.

SOLUTION

Conceptualize In Figure 37.5a, imagine light of two wavelengths incident on the slits and forming two interference patterns on the screen. At some points, the fringes of the two colors might overlap, but at most points, they will not.

Categorize This problem is an application of the mathematical representation of the *waves in interference* analysis model.

Analyze

Use Equation 37.7 to find the fringe positions corresponding to these two wavelengths and subtract them:

$$\Delta y = y'_{bright} - y_{bright} = L\frac{m\lambda'}{d} - L\frac{m\lambda}{d} = \frac{Lm}{d}(\lambda' - \lambda)$$

Substitute numerical values:

$$\Delta y = \frac{(1.50\ \text{m})(3)}{0.025\ 0 \times 10^{-3}\ \text{m}}(510 \times 10^{-9}\ \text{m} - 430 \times 10^{-9}\ \text{m})$$

$$= 0.014\ 4\ \text{m} = \boxed{1.44\ \text{cm}}$$

Finalize Let's explore further details of the interference pattern in the following **What If?**

WHAT IF? What if we examine the entire interference pattern due to the two wavelengths and look for overlapping fringes? Are there any locations on the screen where the bright fringes from the two wavelengths overlap exactly?

Answer Find such a location by setting the location of any bright fringe due to λ equal to one due to λ', using Equation 37.7:

$$L\frac{m\lambda}{d} = L\frac{m'\lambda'}{d} \rightarrow \frac{m'}{m} = \frac{\lambda}{\lambda'}$$

Substitute the wavelengths:

$$\frac{m'}{m} = \frac{430\ \text{nm}}{510\ \text{nm}} = \frac{43}{51}$$

Therefore, the 51st fringe of the 430-nm light overlaps with the 43rd fringe of the 510-nm light.

Use Equation 37.7 to find the value of y for these fringes:

$$y = (1.50\ \text{m})\left[\frac{51(430 \times 10^{-9}\ \text{m})}{0.025\ 0 \times 10^{-3}\ \text{m}}\right] = 1.32\ \text{m}$$

This value of y is comparable to L, so the small-angle approximation used for Equation 37.7 is *not* valid. This conclusion suggests we should not expect Equation 37.7 to give us the correct result. If you use Equation 37.5, you can show that the bright fringes do indeed overlap when the same condition, $m'/m = \lambda/\lambda'$, is met (see Problem 48 in Enhanced WebAssign). Therefore, the 51st fringe of the 430-nm light does overlap with the 43rd fringe of the 510-nm light, but not at the location of 1.32 m. You are asked to find the correct location as part of Problem 48.

37.3 Intensity Distribution of the Double-Slit Interference Pattern

Notice that the edges of the bright fringes in Figure 37.1b are not sharp; rather, there is a gradual change from bright to dark. So far, we have discussed the locations of only the centers of the bright and dark fringes on a distant screen. Let's now direct our attention to the intensity of the light at other points between the positions of maximum constructive and destructive interference. In other words, we now calculate the distribution of light intensity associated with the double-slit interference pattern.

Again, suppose the two slits represent coherent sources of sinusoidal waves such that the two waves from the slits have the same angular frequency ω and are in phase. The total magnitude of the electric field at point P on the screen in Figure 37.5 is the superposition of the two waves. Assuming the two waves have the same amplitude E_0, we can write the magnitude of the electric field at point P due to each wave separately as

$$E_1 = E_0 \sin \omega t \quad \text{and} \quad E_2 = E_0 \sin (\omega t + \phi) \tag{37.9}$$

Although the waves are in phase at the slits, their phase difference ϕ at P depends on the path difference $\delta = r_2 - r_1 = d \sin \theta$. A path difference of λ (for constructive interference) corresponds to a phase difference of 2π rad. A path difference of δ is the same fraction of λ as the phase difference ϕ is of 2π. We can describe this fraction mathematically with the ratio

$$\frac{\delta}{\lambda} = \frac{\phi}{2\pi}$$

which gives

$$\phi = \frac{2\pi}{\lambda} \delta = \frac{2\pi}{\lambda} d \sin \theta \tag{37.10}$$ ◀ **Phase difference**

This equation shows how the phase difference ϕ depends on the angle θ in Figure 37.5.

Using the superposition principle and Equation 37.9, we obtain the following expression for the magnitude of the resultant electric field at point P:

$$E_P = E_1 + E_2 = E_0[\sin \omega t + \sin (\omega t + \phi)] \tag{37.11}$$

We can simplify this expression by using the trigonometric identity

$$\sin A + \sin B = 2 \sin \left(\frac{A + B}{2} \right) \cos \left(\frac{A - B}{2} \right)$$

Taking $A = \omega t + \phi$ and $B = \omega t$, Equation 37.11 becomes

$$E_P = 2E_0 \cos \left(\frac{\phi}{2} \right) \sin \left(\omega t + \frac{\phi}{2} \right) \tag{37.12}$$

This result indicates that the electric field at point P has the same frequency ω as the light at the slits but that the amplitude of the field is multiplied by the factor $2 \cos (\phi/2)$. To check the consistency of this result, note that if $\phi = 0, 2\pi, 4\pi, \ldots$, the magnitude of the electric field at point P is $2E_0$, corresponding to the condition for maximum constructive interference. These values of ϕ are consistent with Equation 37.2 for constructive interference. Likewise, if $\phi = \pi, 3\pi, 5\pi, \ldots$, the magnitude of the electric field at point P is zero, which is consistent with Equation 37.3 for total destructive interference.

Finally, to obtain an expression for the light intensity at point P, recall from Section 34.4 that the intensity of a wave is proportional to the square of the resultant electric field magnitude at that point (Eq. 34.24). Using Equation 37.12, we can therefore express the light intensity at point P as

$$I \propto E_P^2 = 4E_0^2 \cos^2 \left(\frac{\phi}{2} \right) \sin^2 \left(\omega t + \frac{\phi}{2} \right)$$

Most light-detecting instruments measure time-averaged light intensity, and the time-averaged value of $\sin^2 (\omega t + \phi/2)$ over one cycle is $\frac{1}{2}$. (See Fig. 33.5.) Therefore, we can write the average light intensity at point P as

$$I = I_{max} \cos^2 \left(\frac{\phi}{2} \right) \tag{37.13}$$

where I_{max} is the maximum intensity on the screen and the expression represents the time average. Substituting the value for ϕ given by Equation 37.10 into this expression gives

$$I = I_{max} \cos^2\left(\frac{\pi d \sin\theta}{\lambda}\right) \tag{37.14}$$

Alternatively, because $\sin\theta \approx y/L$ for small values of θ in Figure 37.5, we can write Equation 37.14 in the form

$$I = I_{max} \cos^2\left(\frac{\pi d}{\lambda L} y\right) \quad \text{(small angles)} \tag{37.15}$$

Constructive interference, which produces light intensity maxima, occurs when the quantity $\pi dy/\lambda L$ is an integral multiple of π, corresponding to $y = (\lambda L/d)m$, where m is the order number. This result is consistent with Equation 37.7.

A plot of light intensity versus $d\sin\theta$ is given in Figure 37.6. The interference pattern consists of equally spaced fringes of equal intensity.

Figure 37.7 shows similar plots of light intensity versus $d\sin\theta$ for light passing through multiple slits. For more than two slits, we would add together more electric field magnitudes than the two in Equation 37.9. In this case, the pattern contains primary and secondary maxima. For three slits, notice that the primary maxima are nine times more intense than the secondary maxima as measured by the height of the curve because the intensity varies as E^2. For N slits, the intensity of the primary maxima is N^2 times greater than that for the secondary maxima. As the number of slits increases, the primary maxima increase in intensity and become narrower, while the secondary maxima decrease in intensity relative to the primary maxima. Figure 37.7 also shows that as the number of slits increases, the number of secondary maxima also increases. In fact, the number of secondary maxima is always $N - 2$, where N is the number of slits. In Section 38.4, we shall investigate the pattern for a very large number of slits in a device called a *diffraction grating*.

Quick Quiz 37.2 Using Figure 37.7 as a model, sketch the interference pattern from six slits.

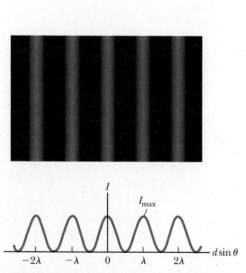

Figure 37.6 Light intensity versus $d\sin\theta$ for a double-slit interference pattern when the screen is far from the two slits ($L \gg d$).

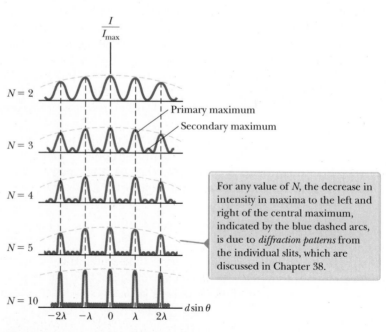

Figure 37.7 Multiple-slit interference patterns. As N, the number of slits, is increased, the primary maxima (the tallest peaks in each graph) become narrower but remain fixed in position and the number of secondary maxima increases.

37.4 Change of Phase Due to Reflection

Young's method for producing two coherent light sources involves illuminating a pair of slits with a single source. Another simple, yet ingenious, arrangement for producing an interference pattern with a single light source is known as *Lloyd's mirror*[1] (Fig. 37.8). A point light source S is placed close to a mirror, and a viewing screen is positioned some distance away and perpendicular to the mirror. Light waves can reach point P on the screen either directly from S to P or by the path involving reflection from the mirror. The reflected ray can be treated as a ray originating from a virtual source S'. As a result, we can think of this arrangement as a double-slit source where the distance d between sources S and S' in Figure 37.8 is analogous to length d in Figure 37.5. Hence, at observation points far from the source (L >> d), we expect waves from S and S' to form an interference pattern exactly like the one formed by two real coherent sources. An interference pattern is indeed observed. The positions of the dark and bright fringes, however, are reversed relative to the pattern created by two real coherent sources (Young's experiment). Such a reversal can only occur if the coherent sources S and S' differ in phase by 180°.

To illustrate further, consider point P', the point where the mirror intersects the screen. This point is equidistant from sources S and S'. If path difference alone were responsible for the phase difference, we would see a bright fringe at P' (because the path difference is zero for this point), corresponding to the central bright fringe of the two-slit interference pattern. Instead, a dark fringe is observed at P'. We therefore conclude that a 180° phase change must be produced by reflection from the mirror. In general, an electromagnetic wave undergoes a phase change of 180° upon reflection from a medium that has a higher index of refraction than the one in which the wave is traveling.

It is useful to draw an analogy between reflected light waves and the reflections of a transverse pulse on a stretched string (Section 16.4). The reflected pulse on a string undergoes a phase change of 180° when reflected from the boundary of a denser string or a rigid support, but no phase change occurs when the pulse is reflected from the boundary of a less dense string or a freely-supported end. Similarly, an electromagnetic wave undergoes a 180° phase change when reflected from a boundary leading to an optically denser medium (defined as a medium with a higher index of refraction), but no phase change occurs when the wave is reflected from a boundary leading to a less dense medium. These rules, summarized in Figure 37.9, can be deduced from Maxwell's equations, but the treatment is beyond the scope of this text.

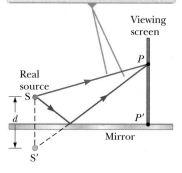

An interference pattern is produced on the screen as a result of the combination of the direct ray (red) and the reflected ray (blue).

Figure 37.8 Lloyd's mirror. The reflected ray undergoes a phase change of 180°.

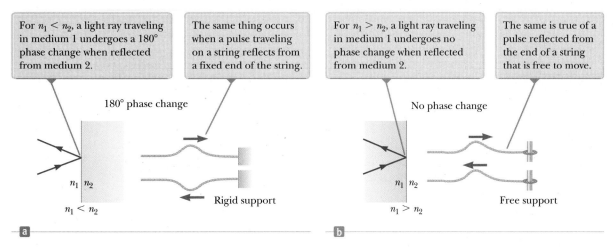

For $n_1 < n_2$, a light ray traveling in medium 1 undergoes a 180° phase change when reflected from medium 2.

The same thing occurs when a pulse traveling on a string reflects from a fixed end of the string.

For $n_1 > n_2$, a light ray traveling in medium 1 undergoes no phase change when reflected from medium 2.

The same is true of a pulse reflected from the end of a string that is free to move.

180° phase change

No phase change

n_1 | n_2

$n_1 < n_2$

Rigid support

n_1 | n_2

$n_1 > n_2$

Free support

a

b

Figure 37.9 Comparisons of reflections of light waves and waves on strings.

[1]Developed in 1834 by Humphrey Lloyd (1800–1881), Professor of Natural and Experimental Philosophy, Trinity College, Dublin.

37.5 Interference in Thin Films

Interference effects are commonly observed in thin films, such as thin layers of oil on water or the thin surface of a soap bubble. The varied colors observed when white light is incident on such films result from the interference of waves reflected from the two surfaces of the film.

Consider a film of uniform thickness t and index of refraction n. The wavelength of light λ_n in the film (see Section 35.5) is

$$\lambda_n = \frac{\lambda}{n}$$

where λ is the wavelength of the light in free space and n is the index of refraction of the film material. Let's assume light rays traveling in air are nearly normal to the two surfaces of the film as shown in Figure 37.10.

Reflected ray 1, which is reflected from the upper surface (A) in Figure 37.10, undergoes a phase change of 180° with respect to the incident wave. Reflected ray 2, which is reflected from the lower film surface (B), undergoes no phase change because it is reflected from a medium (air) that has a lower index of refraction. Therefore, ray 1 is 180° out of phase with ray 2, which is equivalent to a path difference of $\lambda_n/2$. We must also consider, however, that ray 2 travels an extra distance $2t$ before the waves recombine in the air above surface A. (Remember that we are considering light rays that are close to normal to the surface. If the rays are not close to normal, the path difference is larger than $2t$.) If $2t = \lambda_n/2$, rays 1 and 2 recombine in phase and the result is constructive interference. In general, the condition for *constructive* interference in thin films is[2]

$$2t = \left(m + \tfrac{1}{2}\right)\lambda_n \quad m = 0, 1, 2, \ldots \tag{37.16}$$

This condition takes into account two factors: (1) the difference in path length for the two rays (the term $m\lambda_n$) and (2) the 180° phase change upon reflection (the term $\tfrac{1}{2}\lambda_n$). Because $\lambda_n = \lambda/n$, we can write Equation 37.16 as

$$2nt = \left(m + \tfrac{1}{2}\right)\lambda \quad m = 0, 1, 2, \ldots \tag{37.17}$$

If the extra distance $2t$ traveled by ray 2 corresponds to a multiple of λ_n, the two waves combine out of phase and the result is destructive interference. The general equation for *destructive* interference in thin films is

$$2nt = m\lambda \quad m = 0, 1, 2, \ldots \tag{37.18}$$

The foregoing conditions for constructive and destructive interference are valid when the medium above the top surface of the film is the same as the medium below the bottom surface or, if there are different media above and below the film, the index of refraction of both is less than n. If the film is placed between two different media, one with $n < n_{\text{film}}$ and the other with $n > n_{\text{film}}$, the conditions for constructive and destructive interference are reversed. In that case, either there is a phase change of 180° for both ray 1 reflecting from surface A and ray 2 reflecting from surface B or there is no phase change for either ray; hence, the net change in relative phase due to the reflections is zero.

Rays 3 and 4 in Figure 37.10 lead to interference effects in the light transmitted through the thin film. The analysis of these effects is similar to that of the reflected light. You are asked to explore the transmitted light in Problems 35, 36, and 38 in Enhanced WebAssign.

Quick Quiz 37.3 One microscope slide is placed on top of another with their left edges in contact and a human hair under the right edge of the upper slide. As a result, a wedge of air exists between the slides. An interference pattern results

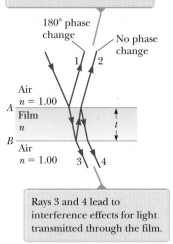

Interference in light reflected from a thin film is due to a combination of rays 1 and 2 reflected from the upper and lower surfaces of the film.

180° phase change

No phase change

1 2

Air
$n = 1.00$

A

Film
n

B

Air
$n = 1.00$

t

3 4

Rays 3 and 4 lead to interference effects for light transmitted through the film.

Figure 37.10 Light paths through a thin film.

Pitfall Prevention 37.1

Be Careful with Thin Films Be sure to include *both* effects—path length and phase change—when analyzing an interference pattern resulting from a thin film. The possible phase change is a new feature we did not need to consider for double-slit interference. Also think carefully about the material on either side of the film. If there are different materials on either side of the film, you may have a situation in which there is a 180° phase change at *both* surfaces or at *neither* surface.

---

[2]The full interference effect in a thin film requires an analysis of an infinite number of reflections back and forth between the top and bottom surfaces of the film. We focus here only on a single reflection from the bottom of the film, which provides the largest contribution to the interference effect.

when monochromatic light is incident on the wedge. What is at the left edges of the slides? **(a)** a dark fringe **(b)** a bright fringe **(c)** impossible to determine

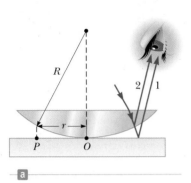

a

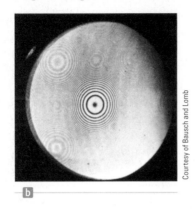

b

Courtesy of Bausch and Lomb

Figure 37.11 (a) The combination of rays reflected from the flat plate and the curved lens surface gives rise to an interference pattern known as Newton's rings. (b) Photograph of Newton's rings.

Newton's Rings

Another method for observing interference in light waves is to place a plano-convex lens on top of a flat glass surface as shown in Figure 37.11a. With this arrangement, the air film between the glass surfaces varies in thickness from zero at the point of contact to some nonzero value at point P. If the radius of curvature R of the lens is much greater than the distance r and the system is viewed from above, a pattern of light and dark rings is observed as shown in Figure 37.11b. These circular fringes, discovered by Newton, are called **Newton's rings.**

The interference effect is due to the combination of ray 1, reflected from the flat plate, with ray 2, reflected from the curved surface of the lens. Ray 1 undergoes a phase change of 180° upon reflection (because it is reflected from a medium of higher index of refraction), whereas ray 2 undergoes no phase change (because it is reflected from a medium of lower index of refraction). Hence, the conditions for constructive and destructive interference are given by Equations 37.17 and 37.18, respectively, with $n = 1$ because the film is air. Because there is no path difference and the total phase change is due only to the 180° phase change upon reflection, the contact point at O is dark as seen in Figure 37.11b.

Using the geometry shown in Figure 37.11a, we can obtain expressions for the radii of the bright and dark bands in terms of the radius of curvature R and wavelength λ. For example, the dark rings have radii given by the expression $r \approx \sqrt{m\lambda R/n}$. The details are left as a problem (see Problem 66 in Enhanced WebAssign). We can obtain the wavelength of the light causing the interference pattern by measuring the radii of the rings, provided R is known. Conversely, we can use a known wavelength to obtain R.

One important use of Newton's rings is in the testing of optical lenses. A circular pattern like that pictured in Figure 37.11b is obtained only when the lens is ground to a perfectly symmetric curvature. Variations from such symmetry produce a pattern with fringes that vary from a smooth, circular shape. These variations indicate how the lens must be reground and repolished to remove imperfections.

(a) A thin film of oil floating on water displays interference, shown by the pattern of colors when white light is incident on the film. Variations in film thickness produce the interesting color pattern. The razor blade gives you an idea of the size of the colored bands. (b) Interference in soap bubbles. The colors are due to interference between light rays reflected from the inner and outer surfaces of the thin film of soap making up the bubble. The color depends on the thickness of the film, ranging from black, where the film is thinnest, to magenta, where it is thickest.

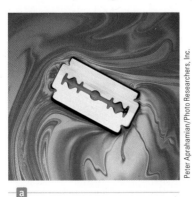

a

Peter Aprahamian/Photo Researchers, Inc.

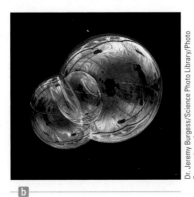

b

Dr. Jeremy Burgess/Science Photo Library/Photo Researchers, Inc.

Problem-Solving Strategy Thin-Film Interference

The following features should be kept in mind when working thin-film interference problems.

1. Conceptualize. Think about what is going on physically in the problem. Identify the light source and the location of the observer.

2. Categorize. Confirm that you should use the techniques for thin-film interference by identifying the thin film causing the interference.

3. Analyze. The type of interference that occurs is determined by the phase relationship between the portion of the wave reflected at the upper surface of the film and the portion reflected at the lower surface. Phase differences between the two portions of the wave have two causes: differences in the distances traveled by the two portions and phase changes occurring on reflection. *Both* causes must be considered when determining which type of interference occurs. If the media above and below the film both have index of refraction larger than that of the film or if both indices are smaller, use Equation 37.17 for constructive interference and Equation 37.18 for destructive interference. If the film is located between two different media, one with $n < n_{film}$ and the other with $n > n_{film}$, reverse these two equations for constructive and destructive interference.

4. Finalize. Inspect your final results to see if they make sense physically and are of an appropriate size.

Example 37.3   Interference in a Soap Film

Calculate the minimum thickness of a soap-bubble film that results in constructive interference in the reflected light if the film is illuminated with light whose wavelength in free space is $\lambda = 600$ nm. The index of refraction of the soap film is 1.33.

SOLUTION

Conceptualize Imagine that the film in Figure 37.10 is soap, with air on both sides.

Categorize We determine the result using an equation from this section, so we categorize this example as a substitution problem.

The minimum film thickness for constructive interference in the reflected light corresponds to $m = 0$ in Equation 37.17. Solve this equation for t and substitute numerical values:

$$t = \frac{(0 + \frac{1}{2})\lambda}{2n} = \frac{\lambda}{4n} = \frac{(600 \text{ nm})}{4(1.33)} = \boxed{113 \text{ nm}}$$

WHAT IF? What if the film is twice as thick? Does this situation produce constructive interference?

Answer Using Equation 37.17, we can solve for the thicknesses at which constructive interference occurs:

$$t = \left(m + \tfrac{1}{2}\right)\frac{\lambda}{2n} = (2m + 1)\frac{\lambda}{4n} \quad m = 0, 1, 2, \ldots$$

The allowed values of m show that constructive interference occurs for *odd* multiples of the thickness corresponding to $m = 0$, $t = 113$ nm. Therefore, constructive interference does *not* occur for a film that is twice as thick.

Example 37.4   Nonreflective Coatings for Solar Cells

Solar cells—devices that generate electricity when exposed to sunlight—are often coated with a transparent, thin film of silicon monoxide (SiO, $n = 1.45$) to minimize reflective losses from the surface. Suppose a silicon solar cell ($n = 3.5$) is coated with a thin film of silicon monoxide for this purpose (Fig. 37.12a). Determine the minimum film thickness that produces the least reflection at a wavelength of 550 nm, near the center of the visible spectrum.

▶ **37.4** continued

SOLUTION

Conceptualize Figure 37.12a helps us visualize the path of the rays in the SiO film that result in interference in the reflected light.

Categorize Based on the geometry of the SiO layer, we categorize this example as a thin-film interference problem.

· ·

Analyze The reflected light is a minimum when rays 1 and 2 in Figure 37.12a meet the condition of destructive interference. In this situation, *both* rays undergo a 180° phase change upon reflection: ray 1 from the upper SiO surface and ray 2 from the lower SiO surface. The net change in phase due to reflection is therefore zero, and the condition for a reflection minimum requires a path difference of $\lambda_n/2$, where λ_n is the wavelength of the light in SiO. Hence, $2nt = \lambda/2$, where λ is the wavelength in air and n is the index of refraction of SiO.

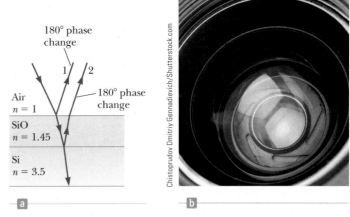

Figure 37.12 (Example 37.4) (a) Reflective losses from a silicon solar cell are minimized by coating the surface of the cell with a thin film of silicon monoxide. (b) The reflected light from a coated camera lens often has a reddish-violet appearance.

Solve the equation $2nt = \lambda/2$ for t and substitute numerical values:

$$t = \frac{\lambda}{4n} = \frac{550 \text{ nm}}{4(1.45)} = \boxed{94.8 \text{ nm}}$$

· ·

Finalize A typical uncoated solar cell has reflective losses as high as 30%, but a coating of SiO can reduce this value to about 10%. This significant decrease in reflective losses increases the cell's efficiency because less reflection means that more sunlight enters the silicon to create charge carriers in the cell. No coating can ever be made perfectly nonreflecting because the required thickness is wavelength-dependent and the incident light covers a wide range of wavelengths.

Glass lenses used in cameras and other optical instruments are usually coated with a transparent thin film to reduce or eliminate unwanted reflection and to enhance the transmission of light through the lenses. The camera lens in Figure 37.12b has several coatings (of different thicknesses) to minimize reflection of light waves having wavelengths near the center of the visible spectrum. As a result, the small amount of light that is reflected by the lens has a greater proportion of the far ends of the spectrum and often appears reddish violet.

37.6 The Michelson Interferometer

The **interferometer**, invented by American physicist A. A. Michelson (1852–1931), splits a light beam into two parts and then recombines the parts to form an interference pattern. The device can be used to measure wavelengths or other lengths with great precision because a large and precisely measurable displacement of one of the mirrors is related to an exactly countable number of wavelengths of light.

A schematic diagram of the interferometer is shown in Figure 37.13 (page 900). A ray of light from a monochromatic source is split into two rays by mirror M_0, which is inclined at 45° to the incident light beam. Mirror M_0, called a *beam splitter*, transmits half the light incident on it and reflects the rest. One ray is reflected from M_0 to the right toward mirror M_1, and the second ray is transmitted vertically through M_0 toward mirror M_2. Hence, the two rays travel separate paths L_1 and L_2. After reflecting from M_1 and M_2, the two rays eventually recombine at M_0 to produce an interference pattern, which can be viewed through a telescope.

The interference condition for the two rays is determined by the difference in their path length. When the two mirrors are exactly perpendicular to each other, the interference pattern is a target pattern of bright and dark circular fringes. As M_1 is moved, the fringe pattern collapses or expands, depending on the direction in which M_1 is moved. For example, if a dark circle appears at the center of the

Figure 37.13 Diagram of the Michelson interferometer.

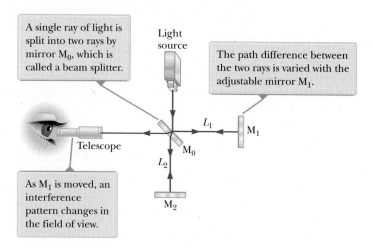

A single ray of light is split into two rays by mirror M_0, which is called a beam splitter.

Light source

The path difference between the two rays is varied with the adjustable mirror M_1.

Telescope

L_1 M_1

M_0

L_2

As M_1 is moved, an interference pattern changes in the field of view.

M_2

target pattern (corresponding to destructive interference) and M_1 is then moved a distance $\lambda/4$ toward M_0, the path difference changes by $\lambda/2$. What was a dark circle at the center now becomes a bright circle. As M_1 is moved an additional distance $\lambda/4$ toward M_0, the bright circle becomes a dark circle again. Therefore, the fringe pattern shifts by one-half fringe each time M_1 is moved a distance $\lambda/4$. The wavelength of light is then measured by counting the number of fringe shifts for a given displacement of M_1. If the wavelength is accurately known, mirror displacements can be measured to within a fraction of the wavelength.

We will see an important historical use of the Michelson interferometer in our discussion of relativity in Chapter 39. Modern uses include the following two applications, Fourier transform infrared spectroscopy and the laser interferometer gravitational-wave observatory.

Fourier Transform Infrared Spectroscopy

Spectroscopy is the study of the wavelength distribution of radiation from a sample that can be used to identify the characteristics of atoms or molecules in the sample. Infrared spectroscopy is particularly important to organic chemists when analyzing organic molecules. Traditional spectroscopy involves the use of an optical element, such as a prism (Section 35.5) or a diffraction grating (Section 38.4), which spreads out various wavelengths in a complex optical signal from the sample into different angles. In this way, the various wavelengths of radiation and their intensities in the signal can be determined. These types of devices are limited in their resolution and effectiveness because they must be scanned through the various angular deviations of the radiation.

The technique of *Fourier transform infrared* (FTIR) *spectroscopy* is used to create a higher-resolution spectrum in a time interval of 1 second that may have required 30 minutes with a standard spectrometer. In this technique, the radiation from a sample enters a Michelson interferometer. The movable mirror is swept through the zero-path-difference condition, and the intensity of radiation at the viewing position is recorded. The result is a complex set of data relating light intensity as a function of mirror position, called an *interferogram*. Because there is a relationship between mirror position and light intensity for a given wavelength, the interferogram contains information about all wavelengths in the signal.

In Section 18.8, we discussed Fourier analysis of a waveform. The waveform is a function that contains information about all the individual frequency components that make up the waveform.[3] Equation 18.13 shows how the waveform is generated from the individual frequency components. Similarly, the interferogram can be

---

[3]In acoustics, it is common to talk about the components of a complex signal in terms of frequency. In optics, it is more common to identify the components by wavelength.

Figure 37.14 The Laser Interferometer Gravitational-Wave
Observatory (LIGO) near Richland, Washington. Notice the two
perpendicular arms of the Michelson interferometer.

analyzed by computer, in a process called a *Fourier transform*, to provide all the wavelength components. This information is the same as that generated by traditional
spectroscopy, but the resolution of FTIR spectroscopy is much higher.

Laser Interferometer Gravitational-Wave Observatory

Einstein's general theory of relativity (Section 39.9) predicts the existence of *gravitational waves.* These waves propagate from the site of any gravitational disturbance,
which could be periodic and predictable, such as the rotation of a double star around
a center of mass, or unpredictable, such as the supernova explosion of a massive star.

In Einstein's theory, gravitation is equivalent to a distortion of space. Therefore,
a gravitational disturbance causes an additional distortion that propagates through
space in a manner similar to mechanical or electromagnetic waves. When gravitational waves from a disturbance pass by the Earth, they create a distortion of the local
space. The laser interferometer gravitational-wave observatory (LIGO) apparatus is
designed to detect this distortion. The apparatus employs a Michelson interferometer that uses laser beams with an effective path length of several kilometers. At the
end of an arm of the interferometer, a mirror is mounted on a massive pendulum.
When a gravitational wave passes by, the pendulum and the attached mirror move
and the interference pattern due to the laser beams from the two arms changes.

Two sites for interferometers have been developed in the United States—in
Richland, Washington, and in Livingston, Louisiana—to allow coincidence studies of gravitational waves. Figure 37.14 shows the Washington site. The two arms of
the Michelson interferometer are evident in the photograph. Six data runs have
been performed as of 2010. These runs have been coordinated with other gravitational wave detectors, such as GEO in Hannover, Germany, TAMA in Mitaka,
Japan, and VIRGO in Cascina, Italy. So far, gravitational waves have not yet been
detected, but the data runs have provided critical information for modifications
and design features for the next generation of detectors. The original detectors are
currently being dismantled, in preparation for the installation of Advanced LIGO,
an upgrade that should increase the sensitivity of the observatory by a factor of 10.
The target date for the beginning of scientific operation of Advanced LIGO is 2014.

Summary

Concepts and Principles

Interference in light waves occurs whenever two or more waves overlap at a given
point. An interference pattern is observed
if (1) the sources are coherent and (2) the
sources have identical wavelengths.

The **intensity** at a point in a double-slit interference pattern is

$$I = I_{\max} \cos^2 \left(\frac{\pi d \sin \theta}{\lambda} \right) \qquad \textbf{(37.14)}$$

where $I_{\max}$ is the maximum intensity on the screen and the
expression represents the time average.   *continued*

A wave traveling from a medium of index of refraction n_1 toward a medium of index of refraction n_2 undergoes a 180° phase change upon reflection when $n_2 > n_1$ and undergoes no phase change when $n_2 < n_1$.

The condition for constructive interference in a film of thickness t and index of refraction n surrounded by air is

$$2nt = (m + \tfrac{1}{2})\lambda \quad m = 0, 1, 2, \ldots \tag{37.17}$$

where λ is the wavelength of the light in free space.

Similarly, the condition for destructive interference in a thin film surrounded by air is

$$2nt = m\lambda \quad m = 0, 1, 2, \ldots \tag{37.18}$$

Analysis Models for Problem Solving

Waves in Interference. Young's double-slit experiment serves as a prototype for interference phenomena involving electromagnetic radiation. In this experiment, two slits separated by a distance d are illuminated by a single-wavelength light source. The condition for bright fringes (**constructive interference**) is

$$d \sin \theta_{\text{bright}} = m\lambda \quad m = 0, \pm 1, \pm 2, \ldots \tag{37.2}$$

The condition for dark fringes (**destructive interference**) is

$$d \sin \theta_{\text{dark}} = (m + \tfrac{1}{2})\lambda \quad m = 0, \pm 1, \pm 2, \ldots \tag{37.3}$$

The number m is called the **order number** of the fringe.

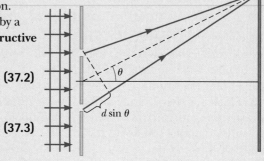

Objective Questions

1. denotes answer available in *Student Solutions Manual/Study Guide*

1. While using a Michelson interferometer (shown in Fig. 37.13), you see a dark circle at the center of the interference pattern. **(i)** As you gradually move the light source toward the central mirror M_0, through a distance $\lambda/2$, what do you see? (a) There is no change in the pattern. (b) The dark circle changes into a bright circle. (c) The dark circle changes into a bright circle and then back into a dark circle. (d) The dark circle changes into a bright circle, then into a dark circle, and then into a bright circle. **(ii)** As you gradually move the moving mirror toward the central mirror M_0, through a distance $\lambda/2$, what do you see? Choose from the same possibilities.

2. Four trials of Young's double-slit experiment are conducted. (a) In the first trial, blue light passes through two fine slits 400 μm apart and forms an interference pattern on a screen 4 m away. (b) In a second trial, red light passes through the same slits and falls on the same screen. (c) A third trial is performed with red light and the same screen, but with slits 800 μm apart. (d) A final trial is performed with red light, slits 800 μm apart, and a screen 8 m away. **(i)** Rank the trials (a) through (d) from the largest to the smallest value of the angle between the central maximum and the first-order side maximum. In your ranking, note any cases of equality. **(ii)** Rank the same trials according to the distance between the central maximum and the first-order side maximum on the screen.

3. Suppose Young's double-slit experiment is performed in air using red light and then the apparatus is immersed in water. What happens to the interference pattern on the screen? (a) It disappears. (b) The bright and dark fringes stay in the same locations, but the contrast is reduced. (c) The bright fringes are closer together. (d) The bright fringes are farther apart. (e) No change happens in the interference pattern.

4. Green light has a wavelength of 500 nm in air. **(i)** Assume green light is reflected from a mirror with angle of incidence 0°. The incident and reflected waves together constitute a standing wave with what distance from one node to the next node? (a) 1 000 nm (b) 500 nm (c) 250 nm (d) 125 nm (e) 62.5 nm **(ii).** The green light is sent into a Michelson interferometer that is adjusted to produce a central bright circle. How far must the interferometer's moving mirror be shifted to change the center of the pattern into a dark circle? Choose from the same possibilities as in part (i). **(iii).** The green light is reflected perpendicularly from a thin film of a plastic with an index of refraction 2.00. The film appears bright in the reflected light. How much additional thickness would make the film appear dark?

5. A thin layer of oil ($n = 1.25$) is floating on water ($n = 1.33$). What is the minimum nonzero thickness of the oil in the region that strongly reflects green light ($\lambda = 530$ nm)? (a) 500 nm (b) 313 nm (c) 404 nm (d) 212 nm (e) 285 nm

6. A monochromatic beam of light of wavelength 500 nm illuminates a double slit having a slit separation of 2.00×10^{-5} m. What is the angle of the second-order bright fringe? (a) 0.050 0 rad (b) 0.025 0 rad (c) 0.100 rad (d) 0.250 rad (e) 0.010 0 rad

7. According to Table 35.1, the index of refraction of flint glass is 1.66 and the index of refraction of crown glass is 1.52. **(i)** A film formed by one drop of sassafras oil, on a horizontal surface of a flint glass block, is viewed by reflected light. The film appears brightest at its outer margin, where it is thinnest. A film of the same oil on crown glass appears dark at its outer margin. What can you say about the index of refraction of the oil? (a) It must be less than 1.52. (b) It must be between 1.52 and 1.66. (c) It must be greater than 1.66. (d) None of those statements is necessarily true. **(ii)** Could a very thin film of some other liquid appear bright by reflected light on both of the glass blocks? **(iii)** Could it appear dark on both? **(iv)** Could it appear dark on crown glass and bright on flint glass? Experiments described by Thomas Young suggested this question.

8. Suppose you perform Young's double-slit experiment with the slit separation slightly smaller than the wavelength of the light. As a screen, you use a large half-cylinder with its axis along the midline between the slits. What interference pattern will you see on the interior surface of the cylinder? (a) bright and dark fringes so closely spaced as to be indistinguishable (b) one central bright fringe and two dark fringes only (c) a completely bright screen with no dark fringes (d) one central dark fringe and two bright fringes only (e) a completely dark screen with no bright fringes

9. A plane monochromatic light wave is incident on a double slit as illustrated in Figure 37.1. **(i)** As the viewing screen is moved away from the double slit, what happens to the separation between the interference fringes on the screen? (a) It increases. (b) It decreases. (c) It remains the same. (d) It may increase or decrease, depending on the wavelength of the light. (e) More information is required. **(ii)** As the slit separation increases, what happens to the separation between the interference fringes on the screen? Select from the same choices.

10. A film of oil on a puddle in a parking lot shows a variety of bright colors in swirled patches. What can you say about the thickness of the oil film? (a) It is much less than the wavelength of visible light. (b) It is on the same order of magnitude as the wavelength of visible light. (c) It is much greater than the wavelength of visible light. (d) It might have any relationship to the wavelength of visible light.

Conceptual Questions

1. denotes answer available in *Student Solutions Manual/Study Guide*

1. Why is the lens on a good-quality camera coated with a thin film?

2. A soap film is held vertically in air and is viewed in reflected light as in Figure CQ37.2. Explain why the film appears to be dark at the top.

3. Explain why two flashlights held close together do not produce an interference pattern on a distant screen.

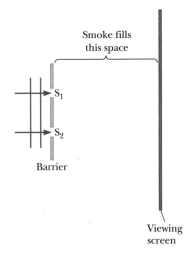

Figure CQ37.2

4. A lens with outer radius of curvature R and index of refraction n rests on a flat glass plate. The combination is illuminated with white light from above and observed from above. (a) Is there a dark spot or a light spot at the center of the lens? (b) What does it mean if the observed rings are noncircular?

5. Consider a dark fringe in a double-slit interference pattern at which almost no light energy is arriving. Light from both slits is arriving at the location of the dark fringe, but the waves cancel. Where does the energy at the positions of dark fringes go?

6. (a) In Young's double-slit experiment, why do we use monochromatic light? (b) If white light is used, how would the pattern change?

7. What is the necessary condition on the path length difference between two waves that interfere (a) constructively and (b) destructively?

8. In a laboratory accident, you spill two liquids onto different parts of a water surface. Neither of the liquids mixes with the water. Both liquids form thin films on the water surface. As the films spread and become very thin, you notice that one film becomes brighter and the other darker in reflected light. Why?

9. A theatrical smoke machine fills the space between the barrier and the viewing screen in the Young's double-slit experiment shown in Figure CQ37.9. Would the smoke show evidence of interference within this space? Explain your answer.

Smoke fills this space

S_1

S_2

Barrier

Viewing screen

Figure CQ37.9

Problems available in ENHANCED WebAssign   Access end-of-chapter problems online at www.webassign.net

Section 37.1 Young's Double-Slit Experiment
Section 37.2 Analysis Model: Waves in Interference
Problems 1–22

Section 37.3 Intensity Distribution of the Double-Slit Interference Pattern
Problems 23–29

Section 37.4 Change of Phase Due to Reflection
Section 37.5 Interference in Thin Films
Problems 30–41

Section 37.6 The Michelson Interferometer
Problems 42–44

Additional Problems
Problems 45–70

Challenge Problems
Problem 71–76

Solutions to the following Problems are available in the *Student Solutions Manual/Study Guide:*
37.5, 37.9, 37.13, 37.18, 37.24, 37.25, 37.29, 37.36, 37.37, 37.42, 37.57, 37.58, 37.60, 37.69, and 37.72

List of Enhanced Problems

Problem Number	Targeted Feedback in Enhanced WebAssign	Analysis Model Tutorial in Enhanced WebAssign	Master It in Enhanced WebAssign	Watch It in Enhanced WebAssign
37.4	✓			✓
37.5	✓			✓
37.9	✓	✓	✓	
37.11		✓	✓	
37.13	✓	✓	✓	
37.14	✓			✓
37.18			✓	
37.21	✓			✓
37.25			✓	
37.29	✓			✓
37.31	✓			✓
37.33			✓	
37.35	✓			✓
37.36			✓	
37.37	✓		✓	
37.39	✓			✓
37.42			✓	
37.43	✓		✓	
37.54		✓		
37.60	✓			✓
37.69			✓	

Diffraction Patterns and Polarization

The Hubble Space Telescope does its viewing above the atmosphere and does not suffer from the atmospheric blurring, caused by air turbulence, that plagues ground-based telescopes. Despite this advantage, it does have limitations due to diffraction effects. In this chapter, we show how the wave nature of light limits the ability of any optical system to distinguish between closely spaced objects. *(NASA Hubble Space Telescope Collection)*

When plane light waves pass through a small aperture in an opaque barrier, the aperture acts as if it were a point source of light, with waves entering the shadow region behind the barrier. This phenomenon, known as diffraction, was first mentioned in Section 35.3, and can be described only with a wave model for light. In this chapter, we investigate the features of the *diffraction pattern* that occurs when the light from the aperture is allowed to fall upon a screen.

In Chapter 34, we learned that electromagnetic waves are transverse. That is, the electric and magnetic field vectors associated with electromagnetic waves are perpendicular to the direction of wave propagation. In this chapter, we show that under certain conditions these transverse waves with electric field vectors in all possible transverse directions can be *polarized* in various ways. In other words, only certain directions of the electric field vectors are present in the polarized wave.

38.1 Introduction to Diffraction Patterns

In Sections 35.3 and 37.1, we discussed that light of wavelength comparable to or larger than the width of a slit spreads out in all forward directions upon passing through the slit. This phenomenon is called *diffraction*. When light passes through a narrow slit, it spreads beyond the narrow path defined by the slit into regions that would be in shadow if light traveled in straight lines. Other waves, such as sound waves and water waves, also have this property of spreading when passing through apertures or by sharp edges.

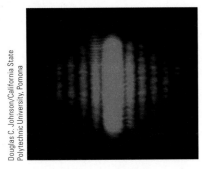

Figure 38.1 The diffraction pattern that appears on a screen when light passes through a narrow vertical slit. The pattern consists of a broad central fringe and a series of less intense and narrower side fringes.

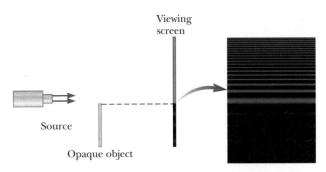

Figure 38.2 Light from a small source passes by the edge of an opaque object and continues on to a screen. A diffraction pattern consisting of bright and dark fringes appears on the screen in the region above the edge of the object.

You might expect that the light passing through a small opening would simply result in a broad region of light on a screen due to the spreading of the light as it passes through the opening. We find something more interesting, however. A **diffraction pattern** consisting of light and dark areas is observed, somewhat similar to the interference patterns discussed earlier. For example, when a narrow slit is placed between a distant light source (or a laser beam) and a screen, the light produces a diffraction pattern like that shown in Figure 38.1. The pattern consists of a broad, intense central band (called the **central maximum**) flanked by a series of narrower, less intense additional bands (called **side maxima** or **secondary maxima**) and a series of intervening dark bands (or **minima**). Figure 38.2 shows a diffraction pattern associated with light passing by the edge of an object. Again we see bright and dark fringes, which is reminiscent of an interference pattern.

Figure 38.3 shows a diffraction pattern associated with the shadow of a penny. A bright spot occurs at the center, and circular fringes extend outward from the shadow's edge. We can explain the central bright spot by using the wave theory of light, which predicts constructive interference at this point. From the viewpoint of ray optics (in which light is viewed as rays traveling in straight lines), we expect the center of the shadow to be dark because that part of the viewing screen is completely shielded by the penny.

Shortly before the central bright spot was first observed, one of the supporters of ray optics, Simeon Poisson, argued that if Augustin Fresnel's wave theory of light were valid, a central bright spot should be observed in the shadow of a circular object illuminated by a point source of light. To Poisson's astonishment, the spot was observed by Dominique Arago shortly thereafter. Therefore, Poisson's prediction reinforced the wave theory rather than disproving it.

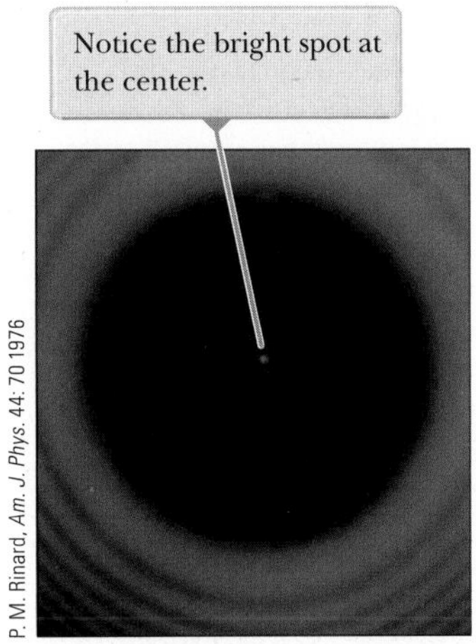

> Notice the bright spot at the center.

Figure 38.3 Diffraction pattern created by the illumination of a penny, with the penny positioned midway between the screen and light source.

38.2 Diffraction Patterns from Narrow Slits

Let's consider a common situation, that of light passing through a narrow opening modeled as a slit and projected onto a screen. To simplify our analysis, we assume the observing screen is far from the slit and the rays reaching the screen are approximately parallel. (This situation can also be achieved experimentally by using a converging lens to focus the parallel rays on a nearby screen.) In this model, the pattern on the screen is called a **Fraunhofer diffraction pattern.**[1]

Figure 38.4a (page 908) shows light entering a single slit from the left and diffracting as it propagates toward a screen. Figure 38.4b shows the fringe structure of

---

[1]If the screen is brought close to the slit (and no lens is used), the pattern is a *Fresnel* diffraction pattern. The Fresnel pattern is more difficult to analyze, so we shall restrict our discussion to Fraunhofer diffraction.

Figure 38.4 (a) Geometry for analyzing the Fraunhofer diffraction pattern of a single slit. (Drawing not to scale.) (b) Simulation of a single-slit Fraunhofer diffraction pattern.

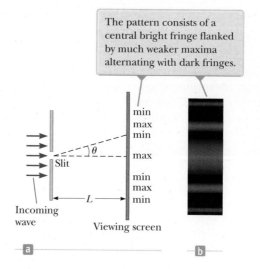

The pattern consists of a central bright fringe flanked by much weaker maxima alternating with dark fringes.

Slit

Incoming wave

Viewing screen

a

b

a Fraunhofer diffraction pattern. A bright fringe is observed along the axis at $\theta = 0$, with alternating dark and bright fringes on each side of the central bright fringe.

Until now, we have assumed slits are point sources of light. In this section, we abandon that assumption and see how the finite width of slits is the basis for understanding Fraunhofer diffraction. We can explain some important features of this phenomenon by examining waves coming from various portions of the slit as shown in Figure 38.5. According to Huygens's principle, each portion of the slit acts as a source of light waves. Hence, light from one portion of the slit can interfere with light from another portion, and the resultant light intensity on a viewing screen depends on the direction θ. Based on this analysis, we recognize that a diffraction pattern is actually an interference pattern in which the different sources of light are different portions of the single slit! Therefore, the diffraction patterns we discuss in this chapter are applications of the waves in interference analysis model.

To analyze the diffraction pattern, let's divide the slit into two halves as shown in Figure 38.5. Keeping in mind that all the waves are in phase as they leave the slit, consider rays 1 and 3. As these two rays travel toward a viewing screen far to the right of the figure, ray 1 travels farther than ray 3 by an amount equal to the path difference $(a/2) \sin \theta$, where a is the width of the slit. Similarly, the path difference between rays 2 and 4 is also $(a/2) \sin \theta$, as is that between rays 3 and 5. If this path difference is exactly half a wavelength (corresponding to a phase difference of 180°), the pairs of waves cancel each other and destructive interference results. This cancellation occurs for any two rays that originate at points separated by half the slit width because the phase difference between two such points is 180°. Therefore, waves from the upper half of the slit interfere destructively with waves from the lower half when

$$\frac{a}{2} \sin \theta = \frac{\lambda}{2}$$

or, if we consider waves at angle θ both above the dashed line in Figure 38.5 and below,

$$\sin \theta = \pm \frac{\lambda}{a}$$

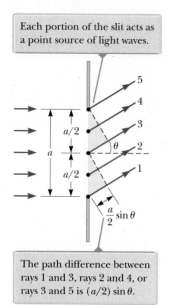

Each portion of the slit acts as a point source of light waves.

The path difference between rays 1 and 3, rays 2 and 4, or rays 3 and 5 is $(a/2) \sin \theta$.

Figure 38.5 Paths of light rays that encounter a narrow slit of width a and diffract toward a screen in the direction described by angle θ (not to scale).

Dividing the slit into four equal parts and using similar reasoning, we find that the viewing screen is also dark when

$$\sin \theta = \pm 2 \frac{\lambda}{a}$$

Likewise, dividing the slit into six equal parts shows that darkness occurs on the screen when

$$\sin \theta = \pm 3 \frac{\lambda}{a}$$

Therefore, the general condition for destructive interference is

$$\sin \theta_{dark} = m \frac{\lambda}{a} \qquad m = \pm 1, \pm 2, \pm 3, \ldots \qquad \textbf{(38.1)}$$

◀ **Condition for destructive interference for a single slit**

This equation gives the values of θ_{dark} for which the diffraction pattern has zero light intensity, that is, when a dark fringe is formed. It tells us nothing, however, about the variation in light intensity along the screen. The general features of the intensity distribution are shown in Figure 38.4. A broad, central bright fringe is observed; this fringe is flanked by much weaker bright fringes alternating with dark fringes. The various dark fringes occur at the values of θ_{dark} that satisfy Equation 38.1. Each bright-fringe peak lies approximately halfway between its bordering dark-fringe minima. Notice that the central bright maximum is twice as wide as the secondary maxima. There is no central dark fringe, represented by the absence of $m = 0$ in Equation 38.1.

Quick Quiz 38.1 Suppose the slit width in Figure 38.4 is made half as wide. Does the central bright fringe (a) become wider, (b) remain the same, or (c) become narrower?

> **Pitfall Prevention 38.2**
> **Similar Equation Warning!** Equation 38.1 has exactly the same form as Equation 37.2, with d, the slit separation, used in Equation 37.2 and a, the slit width, used in Equation 38.1. Equation 37.2, however, describes the *bright* regions in a two-slit interference pattern, whereas Equation 38.1 describes the *dark* regions in a single-slit diffraction pattern.

Example 38.1   **Where Are the Dark Fringes?**   AM

Light of wavelength 580 nm is incident on a slit having a width of 0.300 mm. The viewing screen is 2.00 m from the slit. Find the width of the central bright fringe.

SOLUTION

Conceptualize Based on the problem statement, we imagine a single-slit diffraction pattern similar to that in Figure 38.4.

Categorize We categorize this example as a straightforward application of our discussion of single-slit diffraction patterns, which comes from the *waves in interference* analysis model.

. . . .

Analyze Evaluate Equation 38.1 for the two dark fringes that flank the central bright fringe, which correspond to $m = \pm 1$:

$$\sin \theta_{dark} = \pm \frac{\lambda}{a}$$

Let y represent the vertical position along the viewing screen in Figure 38.4a, measured from the point on the screen directly behind the slit. Then, $\tan \theta_{dark} = y_1/L$, where the subscript 1 refers to the first dark fringe. Because θ_{dark} is very small, we can use the approximation $\sin \theta_{dark} \approx \tan \theta_{dark}$; therefore, $y_1 = L \sin \theta_{dark}$.

The width of the central bright fringe is twice the absolute value of y_1:

$$2|y_1| = 2|L \sin \theta_{dark}| = 2 \left| \pm L \frac{\lambda}{a} \right| = 2L \frac{\lambda}{a} = 2(2.00 \text{ m}) \frac{580 \times 10^{-9} \text{ m}}{0.300 \times 10^{-3} \text{ m}}$$

$$= 7.73 \times 10^{-3} \text{ m} = \boxed{7.73 \text{ mm}}$$

. . . .

Finalize Notice that this value is much greater than the width of the slit. Let's explore below what happens if we change the slit width.

WHAT IF? What if the slit width is increased by an order of magnitude to 3.00 mm? What happens to the diffraction pattern?

Answer Based on Equation 38.1, we expect that the angles at which the dark bands appear will decrease as a increases. Therefore, the diffraction pattern narrows.

Repeat the calculation with the larger slit width:

$$2|y_1| = 2L \frac{\lambda}{a} = 2(2.00 \text{ m}) \frac{580 \times 10^{-9} \text{ m}}{3.00 \times 10^{-3} \text{ m}} = 7.73 \times 10^{-4} \text{ m} = \boxed{0.773 \text{ mm}}$$

Notice that this result is *smaller* than the width of the slit. In general, for large values of a, the various maxima and minima are so closely spaced that only a large, central bright area resembling the geometric image of the slit is observed. This concept is very important in the performance of optical instruments such as telescopes.

Intensity of Single-Slit Diffraction Patterns

Analysis of the intensity variation in a diffraction pattern from a single slit of width a shows that the intensity is given by

Intensity of a single-slit ▶ Fraunhofer diffraction pattern

$$I = I_{max} \left[\frac{\sin{(\pi a \sin{\theta}/\lambda)}}{\pi a \sin{\theta}/\lambda} \right]^2 \tag{38.2}$$

where I_{max} is the intensity at $\theta = 0$ (the central maximum) and λ is the wavelength of light used to illuminate the slit. This expression shows that *minima* occur when

$$\frac{\pi a \sin{\theta_{dark}}}{\lambda} = m\pi$$

or

Condition for intensity ▶ minima for a single slit

$$\sin{\theta_{dark}} = m\frac{\lambda}{a} \quad m = \pm 1, \pm 2, \pm 3, \ldots$$

in agreement with Equation 38.1.

Figure 38.6a represents a plot of the intensity in the single-slit pattern as given by Equation 38.2, and Figure 38.6b is a simulation of a single-slit Fraunhofer diffraction pattern. Notice that most of the light intensity is concentrated in the central bright fringe.

Intensity of Two-Slit Diffraction Patterns

When more than one slit is present, we must consider not only diffraction patterns due to the individual slits but also the interference patterns due to the waves coming from different slits. Notice the curved dashed lines in Figure 37.7 in Chapter 37, which indicate a decrease in intensity of the interference maxima as θ increases. This decrease is due to a diffraction pattern. The interference patterns in that figure are located entirely within the central bright fringe of the diffraction pattern, so the only hint of the diffraction pattern we see is the falloff in intensity toward the outside of the pattern. To determine the effects of both two-slit interference and a single-slit diffraction pattern from each slit from a wider viewpoint than that in Figure 37.7, we combine Equations 37.14 and 38.2:

$$I = I_{max} \cos^2{\left(\frac{\pi d \sin{\theta}}{\lambda} \right)} \left[\frac{\sin{(\pi a \sin{\theta}/\lambda)}}{\pi a \sin{\theta}/\lambda} \right]^2 \tag{38.3}$$

Although this expression looks complicated, it merely represents the single-slit diffraction pattern (the factor in square brackets) acting as an "envelope" for a two-slit interference pattern (the cosine-squared factor) as shown in Figure 38.7. The broken

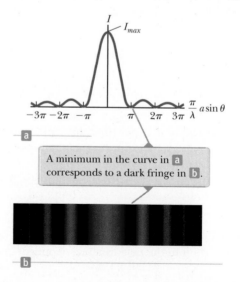

A minimum in the curve in **a** corresponds to a dark fringe in **b**.

Figure 38.6 (a) A plot of light intensity I versus $(\pi/\lambda)a \sin{\theta}$ for the single-slit Fraunhofer diffraction pattern. (b) Simulation of a single-slit Fraunhofer diffraction pattern.

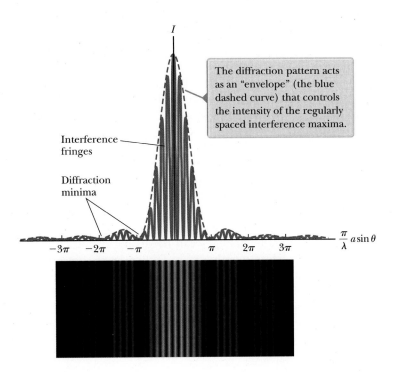

Figure 38.7 The combined effects of two-slit and single-slit interference. This pattern is produced when 650-nm light waves pass through two 3.0-μm slits that are 18 μm apart.

blue curve in Figure 38.7 represents the factor in square brackets in Equation 38.3. The cosine-squared factor by itself would give a series of peaks all with the same height as the highest peak of the red-brown curve in Figure 38.7. Because of the effect of the square-bracket factor, however, these peaks vary in height as shown.

Equation 37.2 indicates the conditions for interference maxima as $d \sin \theta = m\lambda$, where d is the distance between the two slits. Equation 38.1 specifies that the first diffraction minimum occurs when $a \sin \theta = \lambda$, where a is the slit width. Dividing Equation 37.2 by Equation 38.1 (with $m = 1$) allows us to determine which interference maximum coincides with the first diffraction minimum:

$$\frac{d \sin \theta}{a \sin \theta} = \frac{m\lambda}{\lambda}$$

$$\frac{d}{a} = m \qquad\qquad \textbf{(38.4)}$$

In Figure 38.7, $d/a = 18\ \mu\text{m}/3.0\ \mu\text{m} = 6$. Therefore, the sixth interference maximum (if we count the central maximum as $m = 0$) is aligned with the first diffraction minimum and is dark.

Quick Quiz 38.2 Consider the central peak in the diffraction envelope in Figure 38.7 and look closely at the horizontal scale. Suppose the wavelength of the light is changed to 450 nm. What happens to this central peak? **(a)** The width of the peak decreases, and the number of interference fringes it encloses decreases. **(b)** The width of the peak decreases, and the number of interference fringes it encloses increases. **(c)** The width of the peak decreases, and the number of interference fringes it encloses remains the same. **(d)** The width of the peak increases, and the number of interference fringes it encloses decreases. **(e)** The width of the peak increases, and the number of interference fringes it encloses increases. **(f)** The width of the peak increases, and the number of interference fringes it encloses remains the same. **(g)** The width of the peak remains the same, and the number of interference fringes it encloses decreases. **(h)** The width of the peak remains the same, and the number of interference fringes it encloses increases. **(i)** The width of the peak remains the same, and the number of interference fringes it encloses remains the same.

38.3 Resolution of Single-Slit and Circular Apertures

The ability of optical systems to distinguish between closely spaced objects is limited because of the wave nature of light. To understand this limitation, consider Figure 38.8, which shows two light sources far from a narrow slit of width a. The sources can be two noncoherent point sources S_1 and S_2; for example, they could be two distant stars. If no interference occurred between light passing through different parts of the slit, two distinct bright spots (or images) would be observed on the viewing screen. Because of such interference, however, each source is imaged as a bright central region flanked by weaker bright and dark fringes, a diffraction pattern. What is observed on the screen is the sum of two diffraction patterns: one from S_1 and the other from S_2.

If the two sources are far enough apart to keep their central maxima from overlapping as in Figure 38.8a, their images can be distinguished and are said to be *resolved*. If the sources are close together as in Figure 38.8b, however, the two central maxima overlap and the images are not resolved. To determine whether two images are resolved, the following condition is often used:

> When the central maximum of one image falls on the first minimum of another image, the images are said to be just resolved. This limiting condition of resolution is known as **Rayleigh's criterion.**

From Rayleigh's criterion, we can determine the minimum angular separation θ_{min} subtended by the sources at the slit in Figure 38.8 for which the images are just resolved. Equation 38.1 indicates that the first minimum in a single-slit diffraction pattern occurs at the angle for which

$$\sin \theta = \frac{\lambda}{a}$$

where a is the width of the slit. According to Rayleigh's criterion, this expression gives the smallest angular separation for which the two images are resolved. Because $\lambda \ll a$ in most situations, $\sin \theta$ is small and we can use the approximation $\sin \theta \approx \theta$. Therefore, the limiting angle of resolution for a slit of width a is

$$\theta_{min} = \frac{\lambda}{a} \tag{38.5}$$

where θ_{min} is expressed in radians. Hence, the angle subtended by the two sources at the slit must be greater than λ/a if the images are to be resolved.

Many optical systems use circular apertures rather than slits. The diffraction pattern of a circular aperture as shown in the photographs of Figure 38.9 consists of

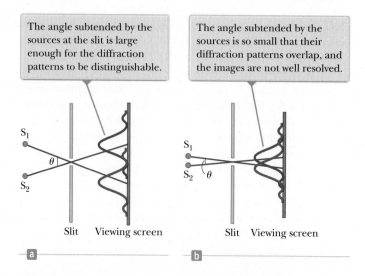

The angle subtended by the sources at the slit is large enough for the diffraction patterns to be distinguishable.

The angle subtended by the sources is so small that their diffraction patterns overlap, and the images are not well resolved.

S_1

θ

S_2

Slit   Viewing screen

a

S_1

S_2 θ

Slit   Viewing screen

b

Figure 38.8 Two point sources far from a narrow slit each produce a diffraction pattern. (a) The sources are separated by a large angle. (b) The sources are separated by a small angle. (Notice that the angles are greatly exaggerated. The drawing is not to scale.)

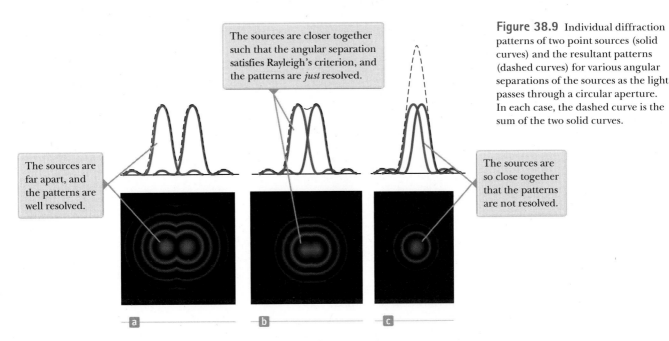

The sources are closer together such that the angular separation satisfies Rayleigh's criterion, and the patterns are *just* resolved.

The sources are far apart, and the patterns are well resolved.

The sources are so close together that the patterns are not resolved.

Figure 38.9 Individual diffraction patterns of two point sources (solid curves) and the resultant patterns (dashed curves) for various angular separations of the sources as the light passes through a circular aperture. In each case, the dashed curve is the sum of the two solid curves.

a central circular bright disk surrounded by progressively fainter bright and dark rings. Figure 38.9 shows diffraction patterns for three situations in which light from two point sources passes through a circular aperture. When the sources are far apart, their images are well resolved (Fig. 38.9a). When the angular separation of the sources satisfies Rayleigh's criterion, the images are just resolved (Fig. 38.9b). Finally, when the sources are close together, the images are said to be unresolved (Fig. 38.9c) and the pattern looks like that of a single source.

Analysis shows that the limiting angle of resolution of the circular aperture is

$$\theta_{min} = 1.22 \frac{\lambda}{D} \tag{38.6}$$

◀ **Limiting angle of resolution for a circular aperture**

where D is the diameter of the aperture. This expression is similar to Equation 38.5 except for the factor 1.22, which arises from a mathematical analysis of diffraction from the circular aperture.

Ⓠuick Quiz 38.3 Cat's eyes have pupils that can be modeled as vertical slits. At night, would cats be more successful in resolving **(a)** headlights on a distant car or **(b)** vertically separated lights on the mast of a distant boat?

Ⓠuick Quiz 38.4 Suppose you are observing a binary star with a telescope and are having difficulty resolving the two stars. You decide to use a colored filter to maximize the resolution. (A filter of a given color transmits only that color of light.) What color filter should you choose? **(a)** blue **(b)** green **(c)** yellow **(d)** red

---

Example 38.2   **Resolution of the Eye**

Light of wavelength 500 nm, near the center of the visible spectrum, enters a human eye. Although pupil diameter varies from person to person, let's estimate a daytime diameter of 2 mm.

(A) Estimate the limiting angle of resolution for this eye, assuming its resolution is limited only by diffraction.

SOLUTION

Conceptualize Identify the pupil of the eye as the aperture through which the light travels. Light passing through this small aperture causes diffraction patterns to occur on the retina.

Categorize We determine the result using equations developed in this section, so we categorize this example as a substitution problem.

continued

▶ **38.2** continued

Use Equation 38.6, taking $\lambda = 500$ nm and $D = 2$ mm:

$$\theta_{min} = 1.22 \frac{\lambda}{D} = 1.22 \left(\frac{5.00 \times 10^{-7} \text{ m}}{2 \times 10^{-3} \text{ m}} \right)$$

$$= \boxed{3 \times 10^{-4} \text{ rad}} \approx \boxed{1 \text{ min of arc}}$$

(B) Determine the minimum separation distance d between two point sources that the eye can distinguish if the point sources are a distance $L = 25$ cm from the observer (Fig. 38.10).

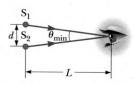

Figure 38.10 (Example 38.2) Two point sources separated by a distance d as observed by the eye.

SOLUTION

Noting that θ_{min} is small, find d:

$$\sin \theta_{min} \approx \theta_{min} \approx \frac{d}{L} \rightarrow d = L\theta_{min}$$

Substitute numerical values:

$$d = (25 \text{ cm})(3 \times 10^{-4} \text{ rad}) = \boxed{8 \times 10^{-3} \text{ cm}}$$

This result is approximately equal to the thickness of a human hair.

Example 38.3   Resolution of a Telescope

Each of the two telescopes at the Keck Observatory on the dormant Mauna Kea volcano in Hawaii has an effective diameter of 10 m. What is its limiting angle of resolution for 600-nm light?

SOLUTION

Conceptualize Identify the aperture through which the light travels as the opening of the telescope. Light passing through this aperture causes diffraction patterns to occur in the final image.

Categorize We determine the result using equations developed in this section, so we categorize this example as a substitution problem.

Use Equation 38.6, taking $\lambda = 6.00 \times 10^{-7}$ m and $D = 10$ m:

$$\theta_{min} = 1.22 \frac{\lambda}{D} = 1.22 \left(\frac{6.00 \times 10^{-7} \text{ m}}{10 \text{ m}} \right)$$

$$= \boxed{7.3 \times 10^{-8} \text{ rad}} \approx \boxed{0.015 \text{ s of arc}}$$

Any two stars that subtend an angle greater than or equal to this value are resolved (if atmospheric conditions are ideal).

WHAT IF? What if we consider radio telescopes? They are much larger in diameter than optical telescopes, but do they have better angular resolutions than optical telescopes? For example, the radio telescope at Arecibo, Puerto Rico, has a diameter of 305 m and is designed to detect radio waves of 0.75-m wavelength. How does its resolution compare with that of one of the Keck telescopes?

Answer The increase in diameter might suggest that radio telescopes would have better resolution than a Keck telescope, but Equation 38.6 shows that θ_{min} depends on *both* diameter and wavelength. Calculating the minimum angle of resolution for the radio telescope, we find

$$\theta_{min} = 1.22 \frac{\lambda}{D} = 1.22 \left(\frac{0.75 \text{ m}}{305 \text{ m}} \right)$$

$$= 3.0 \times 10^{-3} \text{ rad} \approx 10 \text{ min of arc}$$

This limiting angle of resolution is measured in *minutes* of arc rather than the *seconds* of arc for the optical telescope. Therefore, the change in wavelength more than compensates for the increase in diameter. The limiting angle of resolution for the Arecibo radio telescope is more than 40 000 times larger (that is, *worse*) than the Keck minimum.

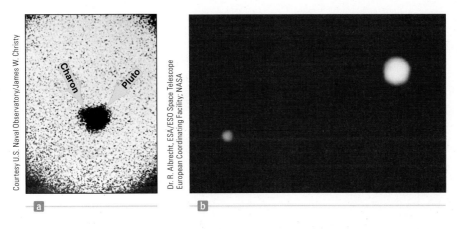

Figure 38.11 (a) The photograph on which Charon, the moon of Pluto, was discovered in 1978. From an Earth-based telescope, atmospheric blurring results in Charon appearing only as a subtle bump on the edge of Pluto. (b) A Hubble Space Telescope photo of Pluto and Charon, clearly resolving the two objects.

A telescope such as the one discussed in Example 38.3 can never reach its diffraction limit because the limiting angle of resolution is always set by atmospheric blurring at optical wavelengths. This seeing limit is usually about 1 s of arc and is never smaller than about 0.1 s of arc. The atmospheric blurring is caused by variations in index of refraction with temperature variations in the air. This blurring is one reason for the superiority of photographs from orbiting telescopes, which view celestial objects from a position above the atmosphere.

As an example of the effects of atmospheric blurring, consider telescopic images of Pluto and its moon, Charon. Figure 38.11a, an image taken in 1978, represents the discovery of Charon. In this photograph, taken from an Earth-based telescope, atmospheric turbulence causes the image of Charon to appear only as a bump on the edge of Pluto. In comparison, Figure 38.11b shows a photograph taken from the Hubble Space Telescope. Without the problems of atmospheric turbulence, Pluto and its moon are clearly resolved.

38.4 The Diffraction Grating

The **diffraction grating,** a useful device for analyzing light sources, consists of a large number of equally spaced parallel slits. A *transmission grating* can be made by cutting parallel grooves on a glass plate with a precision ruling machine. The spaces between the grooves are transparent to the light and hence act as separate slits. A *reflection grating* can be made by cutting parallel grooves on the surface of a reflective material. The reflection of light from the spaces between the grooves is specular, and the reflection from the grooves cut into the material is diffuse. Therefore, the spaces between the grooves act as parallel sources of reflected light like the slits in a transmission grating. Current technology can produce gratings that have very small slit spacings. For example, a typical grating ruled with 5 000 grooves/cm has a slit spacing $d = (1/5\,000)$ cm $= 2.00 \times 10^{-4}$ cm.

A section of a diffraction grating is illustrated in Figure 38.12 (page 916). A plane wave is incident from the left, normal to the plane of the grating. The pattern observed on the screen far to the right of the grating is the result of the combined effects of interference and diffraction. Each slit produces diffraction, and the diffracted beams interfere with one another to produce the final pattern.

The waves from all slits are in phase as they leave the slits. For an arbitrary direction θ measured from the horizontal, however, the waves must travel different path lengths before reaching the screen. Notice in Figure 38.12 that the path difference δ between rays from any two adjacent slits is equal to $d \sin \theta$. If this path difference equals one wavelength or some integral multiple of a wavelength, waves from all slits are in phase at the screen and a bright fringe is observed. Therefore, the condition for *maxima* in the interference pattern at the angle θ_{bright} is

$$d \sin \theta_{\text{bright}} = m\lambda \quad m = 0, \pm 1, \pm 2, \pm 3, \dots \qquad (38.7)$$

◀ Condition for interference maxima for a grating

Pitfall Prevention 38.3
A Diffraction Grating Is an Interference Grating As with *diffraction pattern*, *diffraction grating* is a misnomer, but is deeply entrenched in the language of physics. The diffraction grating depends on diffraction in the same way as the double slit, spreading the light so that light from different slits can interfere. It would be more correct to call it an *interference grating*, but *diffraction grating* is the name in use.

Figure 38.12 Side view of a diffraction grating. The slit separation is d, and the path difference between adjacent slits is $d \sin \theta$.

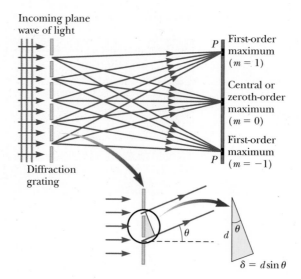

Incoming plane wave of light

First-order maximum ($m = 1$)

Central or zeroth-order maximum ($m = 0$)

First-order maximum ($m = -1$)

Diffraction grating

$\delta = d \sin \theta$

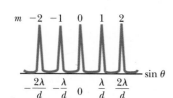

Figure 38.13 Intensity versus $\sin \theta$ for a diffraction grating. The zeroth-, first-, and second-order maxima are shown.

We can use this expression to calculate the wavelength if we know the grating spacing d and the angle θ_{bright}. If the incident radiation contains several wavelengths, the mth-order maximum for each wavelength occurs at a specific angle. All wavelengths are seen at $\theta = 0$, corresponding to $m = 0$, the zeroth-order maximum. The first-order maximum ($m = 1$) is observed at an angle that satisfies the relationship $\sin \theta_{bright} = \lambda/d$, the second-order maximum ($m = 2$) is observed at a larger angle θ_{bright}, and so on. For the small values of d typical in a diffraction grating, the angles θ_{bright} are large, as we see in Example 38.5.

The intensity distribution for a diffraction grating obtained with the use of a monochromatic source is shown in Figure 38.13. Notice the sharpness of the principal maxima and the broadness of the dark areas compared with the broad bright fringes characteristic of the two-slit interference pattern (see Fig. 37.6). You should also review Figure 37.7, which shows that the width of the intensity maxima decreases as the number of slits increases. Because the principal maxima are so sharp, they are much brighter than two-slit interference maxima.

Quick Quiz 38.5 Ultraviolet light of wavelength 350 nm is incident on a diffraction grating with slit spacing d and forms an interference pattern on a screen a distance L away. The angular positions θ_{bright} of the interference maxima are large. The locations of the bright fringes are marked on the screen. Now red light of wavelength 700 nm is used with a diffraction grating to form another diffraction pattern on the screen. Will the bright fringes of this pattern be located at the marks on the screen if **(a)** the screen is moved to a distance $2L$ from the grating, **(b)** the screen is moved to a distance $L/2$ from the grating, **(c)** the grating is replaced with one of slit spacing $2d$, **(d)** the grating is replaced with one of slit spacing $d/2$, or **(e)** nothing is changed?

Conceptual Example 38.4   **A Compact Disc Is a Diffraction Grating**

Light reflected from the surface of a compact disc is multicolored as shown in Figure 38.14. The colors and their intensities depend on the orientation of the CD relative to the eye and relative to the light source. Explain how that works.

Figure 38.14 (Conceptual Example 38.4) A compact disc observed under white light. The colors observed in the reflected light and their intensities depend on the orientation of the CD relative to the eye and relative to the light source.

Carlos E. Santa Maria/Shutterstock.com

SOLUTION

The surface of a CD has a spiral grooved track (with adjacent grooves having a separation on

▶ **38.4** continued

the order of 1 μm). Therefore, the surface acts as a reflection grating. The light reflecting from the regions between these closely spaced grooves interferes constructively only in certain directions that depend on the wavelength and the direction of the incident light. Any section of the CD serves as a diffraction grating for white light, sending different colors in different directions. The different colors you see upon viewing one section change when the light source, the CD, or you change position. This change in position causes the angle of incidence or the angle of the diffracted light to be altered.

Example 38.5   The Orders of a Diffraction Grating

Monochromatic light from a helium–neon laser (λ = 632.8 nm) is incident normally on a diffraction grating containing 6 000 grooves per centimeter. Find the angles at which the first- and second-order maxima are observed.

SOLUTION

Conceptualize Study Figure 38.12 and imagine that the light coming from the left originates from the helium–neon laser. Let's evaluate the possible values of the angle θ for constructive interference.

Categorize We determine results using equations developed in this section, so we categorize this example as a substitution problem.

Calculate the slit separation as the inverse of the number of grooves per centimeter:

$$d = \frac{1}{6\,000}\text{ cm} = 1.667 \times 10^{-4}\text{ cm} = 1\,667\text{ nm}$$

Solve Equation 38.7 for sin θ and substitute numerical values for the first-order maximum ($m = 1$) to find θ_1:

$$\sin\theta_1 = \frac{(1)\lambda}{d} = \frac{632.8\text{ nm}}{1\,667\text{ nm}} = 0.379\,7$$

$$\theta_1 = \boxed{22.31°}$$

Repeat for the second-order maximum ($m = 2$):

$$\sin\theta_2 = \frac{(2)\lambda}{d} = \frac{2(632.8\text{ nm})}{1\,667\text{ nm}} = 0.759\,4$$

$$\theta_2 = \boxed{49.41°}$$

WHAT IF? What if you looked for the third-order maximum? Would you find it?

Answer For $m = 3$, we find sin $\theta_3 = 1.139$. Because sin θ cannot exceed unity, this result does not represent a realistic solution. Hence, only zeroth-, first-, and second-order maxima can be observed for this situation.

Applications of Diffraction Gratings

A schematic drawing of a simple apparatus used to measure angles in a diffraction pattern is shown in Figure 38.15 (page 918). This apparatus is a *diffraction grating spectrometer*. The light to be analyzed passes through a slit, and a collimated beam of light is incident on the grating. The diffracted light leaves the grating at angles that satisfy Equation 38.7, and a telescope is used to view the image of the slit. The wavelength can be determined by measuring the precise angles at which the images of the slit appear for the various orders.

The spectrometer is a useful tool in *atomic spectroscopy,* in which the light from an atom is analyzed to find the wavelength components. These wavelength components can be used to identify the atom. We shall investigate atomic spectra in Chapter 42 of the extended version of this text.

Another application of diffraction gratings is the *grating light valve* (GLV), which competes in some video display applications with the digital micromirror devices (DMDs) discussed in Section 35.4. A GLV is a silicon microchip fitted with an array

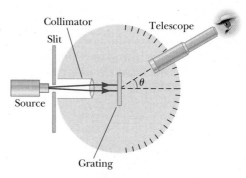

Figure 38.15 Diagram of a diffraction grating spectrometer. The collimated beam incident on the grating is spread into its various wavelength components with constructive interference for a particular wavelength occurring at the angles θ_{bright} that satisfy the equation $d \sin \theta_{bright} = m\lambda$, where $m = 0, \pm 1, \pm 2, \ldots$.

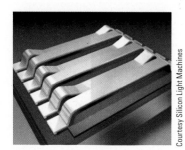

Figure 38.16 A small portion of a grating light valve. The alternating reflective ribbons at different levels act as a diffraction grating, offering very high-speed control of the direction of light toward a digital display device.

of parallel silicon nitride ribbons coated with a thin layer of aluminum (Fig. 38.16). Each ribbon is approximately 20 μm long and 5 μm wide and is separated from the silicon substrate by an air gap on the order of 100 nm. With no voltage applied, all ribbons are at the same level. In this situation, the array of ribbons acts as a flat surface, specularly reflecting incident light.

When a voltage is applied between a ribbon and the electrode on the silicon substrate, an electric force pulls the ribbon downward, closer to the substrate. Alternate ribbons can be pulled down, while those in between remain in an elevated configuration. As a result, the array of ribbons acts as a diffraction grating such that the constructive interference for a particular wavelength of light can be directed toward a screen or other optical display system. If one uses three such devices—one each for red, blue, and green light—full-color display is possible.

In addition to its use in video display, the GLV has found applications in laser optical navigation sensor technology, computer-to-plate commercial printing, and other types of imaging.

Another interesting application of diffraction gratings is **holography,** the production of three-dimensional images of objects. The physics of holography was developed by Dennis Gabor (1900–1979) in 1948 and resulted in the Nobel Prize in Physics for Gabor in 1971. The requirement of coherent light for holography delayed the realization of holographic images from Gabor's work until the development of lasers in the 1960s. Figure 38.17 shows a single hologram viewed from two different positions and the three-dimensional character of its image. Notice in particular the difference in the view through the magnifying glass in Figures 38.17a and 38.17b.

Figure 38.18 shows how a hologram is made. Light from the laser is split into two parts by a half-silvered mirror at *B*. One part of the beam reflects off the object to be photographed and strikes an ordinary photographic film. The other half of the beam is diverged by lens L_2, reflects from mirrors M_1 and M_2, and finally

Figure 38.17 In this hologram, a circuit board is shown from two different views. Notice the difference in the appearance of the measuring tape and the view through the magnifying lens in (a) and (b).

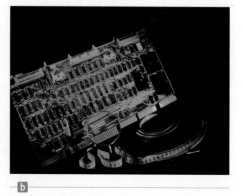

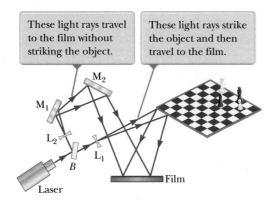

These light rays travel to the film without striking the object.

These light rays strike the object and then travel to the film.

M₂

M₁

L₂

L₁

B

Laser

Film

Figure 38.18 Experimental arrangement for producing a hologram.

strikes the film. The two beams overlap to form an extremely complicated interference pattern on the film. Such an interference pattern can be produced only if the phase relationship of the two waves is constant throughout the exposure of the film. This condition is met by illuminating the scene with light coming through a pinhole or with coherent laser radiation. The hologram records not only the intensity of the light scattered from the object (as in a conventional photograph), but also the phase difference between the reference beam and the beam scattered from the object. Because of this phase difference, an interference pattern is formed that produces an image in which all three-dimensional information available from the perspective of any point on the hologram is preserved.

In a normal photographic image, a lens is used to focus the image so that each point on the object corresponds to a single point on the photograph. Notice that there is no lens used in Figure 38.18 to focus the light onto the film. Therefore, light from each point on the object reaches *all* points on the film. As a result, each region of the photographic film on which the hologram is recorded contains information about all illuminated points on the object, which leads to a remarkable result: if a small section of the hologram is cut from the film, the complete image can be formed from the small piece! (The quality of the image is reduced, but the entire image is present.)

A hologram is best viewed by allowing coherent light to pass through the developed film as one looks back along the direction from which the beam comes. The interference pattern on the film acts as a diffraction grating. Figure 38.19 shows two rays of light striking and passing through the film. For each ray, the $m = 0$ and $m = \pm 1$ rays in the diffraction pattern are shown emerging from the right side of the film. The $m = +1$ rays converge to form a real image of the scene, which is not the image that is normally viewed. By extending the light rays corresponding to $m = -1$ behind the film, we see that there is a virtual image located there, with light coming from it in exactly the same way that light came from the actual object

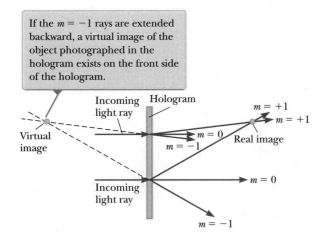

If the $m = -1$ rays are extended backward, a virtual image of the object photographed in the hologram exists on the front side of the hologram.

Incoming light ray

Hologram

$m = +1$

$m = +1$

Virtual image

$m = 0$

$m = -1$

Real image

$m = 0$

Incoming light ray

$m = -1$

Figure 38.19 Two light rays strike a hologram at normal incidence. For each ray, outgoing rays corresponding to $m = 0$ and $m = \pm 1$ are shown.

Figure 38.20 Schematic diagram of the technique used to observe the diffraction of x-rays by a crystal. The array of spots formed on the film is called a Laue pattern.

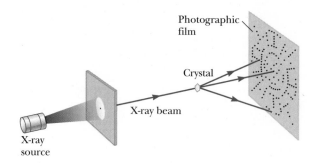

when the film was exposed. This image is what one sees when looking through the holographic film.

Holograms are finding a number of applications. You may have a hologram on your credit card. This special type of hologram is called a *rainbow hologram* and is designed to be viewed in reflected white light.

38.5 Diffraction of X-Rays by Crystals

In principle, the wavelength of any electromagnetic wave can be determined if a grating of the proper spacing (on the order of λ) is available. X-rays, discovered by Wilhelm Roentgen (1845–1923) in 1895, are electromagnetic waves of very short wavelength (on the order of 0.1 nm). It would be impossible to construct a grating having such a small spacing by the cutting process described at the beginning of Section 38.4. The atomic spacing in a solid is known to be about 0.1 nm, however. In 1913, Max von Laue (1879–1960) suggested that the regular array of atoms in a crystal could act as a three-dimensional diffraction grating for x-rays. Subsequent experiments confirmed this prediction. The diffraction patterns from crystals are complex because of the three-dimensional nature of the crystal structure. Nevertheless, x-ray diffraction has proved to be an invaluable technique for elucidating these structures and for understanding the structure of matter.

Figure 38.20 shows one experimental arrangement for observing x-ray diffraction from a crystal. A collimated beam of monochromatic x-rays is incident on a crystal. The diffracted beams are very intense in certain directions, corresponding to constructive interference from waves reflected from layers of atoms in the crystal. The diffracted beams, which can be detected by a photographic film, form an array of spots known as a *Laue pattern* as in Figure 38.21a. One can deduce the crystalline structure by analyzing the positions and intensities of the various spots in the pattern. Figure 38.21b shows a Laue pattern from a crystalline enzyme, using a wide range of wavelengths so that a swirling pattern results.

Figure 38.21 (a) A Laue pattern of a single crystal of the mineral beryl (beryllium aluminum silicate). Each dot represents a point of constructive interference. (b) A Laue pattern of the enzyme Rubisco, produced with a wideband x-ray spectrum. This enzyme is present in plants and takes part in the process of photosynthesis. The Laue pattern is used to determine the crystal structure of Rubisco.

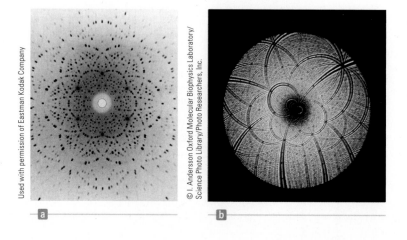

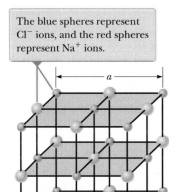

The blue spheres represent Cl⁻ ions, and the red spheres represent Na⁺ ions.

Figure 38.22 Crystalline structure of sodium chloride (NaCl). The length of the cube edge is $a = 0.562\ 737$ nm.

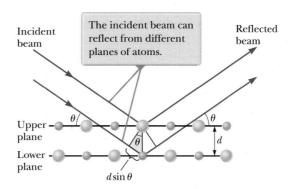

The incident beam can reflect from different planes of atoms.

Figure 38.23 A two-dimensional description of the reflection of an x-ray beam from two parallel crystalline planes separated by a distance d. The beam reflected from the lower plane travels farther than the beam reflected from the upper plane by a distance $2d \sin \theta$.

The arrangement of atoms in a crystal of sodium chloride (NaCl) is shown in Figure 38.22. Each unit cell (the geometric solid that repeats throughout the crystal) is a cube having an edge length a. A careful examination of the NaCl structure shows that the ions lie in discrete planes (the shaded areas in Fig. 38.22). Now suppose an incident x-ray beam makes an angle θ with one of the planes as in Figure 38.23. The beam can be reflected from both the upper plane and the lower one, but the beam reflected from the lower plane travels farther than the beam reflected from the upper plane. The effective path difference is $2d \sin \theta$. The two beams reinforce each other (constructive interference) when this path difference equals some integer multiple of λ. The same is true for reflection from the entire family of parallel planes. Hence, the condition for *constructive* interference (maxima in the reflected beam) is

$$2d \sin \theta = m\lambda \quad m = 1, 2, 3, \ldots \tag{38.8}$$

◀ **Bragg's law**

This condition is known as **Bragg's law,** after W. L. Bragg (1890–1971), who first derived the relationship. If the wavelength and diffraction angle are measured, Equation 38.8 can be used to calculate the spacing between atomic planes.

Pitfall Prevention 38.4
Different Angles Notice in Figure 38.23 that the angle θ is measured from the reflecting surface rather than from the normal as in the case of the law of reflection in Chapter 35. With slits and diffraction gratings, we also measured the angle θ from the normal to the array of slits. Because of historical tradition, the angle is measured differently in Bragg diffraction, so interpret Equation 38.8 with care.

38.6 Polarization of Light Waves

In Chapter 34, we described the transverse nature of light and all other electromagnetic waves. Polarization, discussed in this section, is firm evidence of this transverse nature.

An ordinary beam of light consists of a large number of waves emitted by the atoms of the light source. Each atom produces a wave having some particular orientation of the electric field vector $\vec{E}$, corresponding to the direction of atomic vibration. The *direction of polarization* of each individual wave is defined to be the direction in which the electric field is vibrating. In Figure 38.24, this direction happens to lie along the y axis. All individual electromagnetic waves traveling in the x direction have an $\vec{E}$ vector parallel to the yz plane, but this vector could be at any possible angle with respect to the y axis. Because all directions of vibration from a wave source are possible, the resultant electromagnetic wave is a superposition of waves vibrating in many different directions. The result is an **unpolarized** light beam, represented in Figure 38.25a (page 922). The direction of wave propagation in this figure is perpendicular to the page. The arrows show a few possible

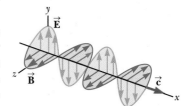

Figure 38.24 Schematic diagram of an electromagnetic wave propagating at velocity $\vec{c}$ in the x direction. The electric field vibrates in the xy plane, and the magnetic field vibrates in the xz plane.

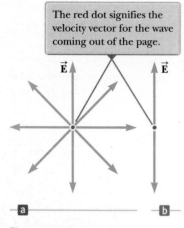

The red dot signifies the velocity vector for the wave coming out of the page.

Figure 38.25 (a) A representation of an unpolarized light beam viewed along the direction of propagation. The transverse electric field can vibrate in any direction in the plane of the page with equal probability. (b) A linearly polarized light beam with the electric field vibrating in the vertical direction.

directions of the electric field vectors for the individual waves making up the resultant beam. At any given point and at some instant of time, all these individual electric field vectors add to give one resultant electric field vector.

As noted in Section 34.3, a wave is said to be **linearly polarized** if the resultant electric field $\vec{E}$ vibrates in the same direction *at all times* at a particular point as shown in Figure 38.25b. (Sometimes, such a wave is described as *plane-polarized,* or simply *polarized.*) The plane formed by $\vec{E}$ and the direction of propagation is called the *plane of polarization* of the wave. If the wave in Figure 38.24 represents the resultant of all individual waves, the plane of polarization is the *xy* plane.

A linearly polarized beam can be obtained from an unpolarized beam by removing all waves from the beam except those whose electric field vectors oscillate in a single plane. We now discuss four processes for producing polarized light from unpolarized light.

Polarization by Selective Absorption

The most common technique for producing polarized light is to use a material that transmits waves whose electric fields vibrate in a plane parallel to a certain direction and that absorbs waves whose electric fields vibrate in all other directions.

In 1938, E. H. Land (1909–1991) discovered a material, which he called *Polaroid,* that polarizes light through selective absorption. This material is fabricated in thin sheets of long-chain hydrocarbons. The sheets are stretched during manufacture so that the long-chain molecules align. After a sheet is dipped into a solution containing iodine, the molecules become good electrical conductors. Conduction takes place primarily along the hydrocarbon chains because electrons can move easily only along the chains. If light whose electric field vector is parallel to the chains is incident on the material, the electric field accelerates electrons along the chains and energy is absorbed from the radiation. Therefore, the light does not pass through the material. Light whose electric field vector is perpendicular to the chains passes through the material because electrons cannot move from one molecule to the next. As a result, when unpolarized light is incident on the material, the exiting light is polarized perpendicular to the molecular chains.

It is common to refer to the direction perpendicular to the molecular chains as the *transmission axis*. In an ideal polarizer, all light with $\vec{E}$ parallel to the transmission axis is transmitted and all light with $\vec{E}$ perpendicular to the transmission axis is absorbed.

Figure 38.26 represents an unpolarized light beam incident on a first polarizing sheet, called the *polarizer*. Because the transmission axis is oriented vertically in the figure, the light transmitted through this sheet is polarized vertically. A second polarizing sheet, called the *analyzer,* intercepts the beam. In Figure 38.26, the analyzer transmission axis is set at an angle θ to the polarizer axis. We call the electric field vector of the first transmitted beam $\vec{E}_0$. The component of $\vec{E}_0$ perpendicular to the analyzer axis is completely absorbed. The component of $\vec{E}_0$ parallel to the

Figure 38.26 Two polarizing sheets whose transmission axes make an angle θ with each other. Only a fraction of the polarized light incident on the analyzer is transmitted through it.

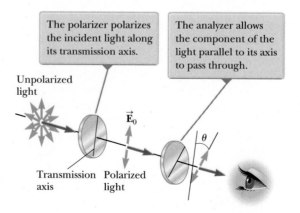

The polarizer polarizes the incident light along its transmission axis.

The analyzer allows the component of the light parallel to its axis to pass through.

Unpolarized light

$\vec{E}_0$

Transmission axis

Polarized light

θ

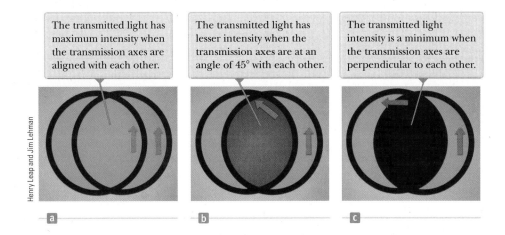

The transmitted light has maximum intensity when the transmission axes are aligned with each other.

The transmitted light has lesser intensity when the transmission axes are at an angle of 45° with each other.

The transmitted light intensity is a minimum when the transmission axes are perpendicular to each other.

Figure 38.27 The intensity of light transmitted through two polarizers depends on the relative orientation of their transmission axes. The red arrows indicate the transmission axes of the polarizers.

analyzer axis, which is transmitted through the analyzer, is $E_0 \cos \theta$. Because the intensity of the transmitted beam varies as the square of its magnitude, we conclude that the intensity I of the (polarized) beam transmitted through the analyzer varies as

$$I = I_{max} \cos^2 \theta \qquad \text{(38.9)}$$

◀ Malus's law

where I_{max} is the intensity of the polarized beam incident on the analyzer. This expression, known as **Malus's law,**[2] applies to any two polarizing materials whose transmission axes are at an angle θ to each other. This expression shows that the intensity of the transmitted beam is maximum when the transmission axes are parallel ($\theta = 0$ or $180°$) and is zero (complete absorption by the analyzer) when the transmission axes are perpendicular to each other. This variation in transmitted intensity through a pair of polarizing sheets is illustrated in Figure 38.27. Because the average value of $\cos^2 \theta$ is $\frac{1}{2}$, the intensity of initially unpolarized light is reduced by a factor of one-half as the light passes through a single ideal polarizer.

Polarization by Reflection

When an unpolarized light beam is reflected from a surface, the polarization of the reflected light depends on the angle of incidence. If the angle of incidence is 0°, the reflected beam is unpolarized. For other angles of incidence, the reflected light is polarized to some extent, and for one particular angle of incidence, the reflected light is completely polarized. Let's now investigate reflection at that special angle.

Suppose an unpolarized light beam is incident on a surface as in Figure 38.28a (page 924). Each individual electric field vector can be resolved into two components: one parallel to the surface (and perpendicular to the page in Fig. 38.28, represented by the dots) and the other (represented by the orange arrows) perpendicular both to the first component and to the direction of propagation. Therefore, the polarization of the entire beam can be described by two electric field components in these directions. It is found that the parallel component represented by the dots reflects more strongly than the other component represented by the arrows, resulting in a partially polarized reflected beam. Furthermore, the refracted beam is also partially polarized.

Now suppose the angle of incidence θ_1 is varied until the angle between the reflected and refracted beams is 90° as in Figure 38.28b. At this particular angle of incidence, the reflected beam is completely polarized (with its electric field vector parallel to the surface) and the refracted beam is still only partially polarized. The angle of incidence at which this polarization occurs is called the **polarizing angle** θ_p.

---

[2]Named after its discoverer, E. L. Malus (1775–1812). Malus discovered that reflected light was polarized by viewing it through a calcite ($CaCO_3$) crystal.

Figure 38.28 (a) When unpolarized light is incident on a reflecting surface, the reflected and refracted beams are partially polarized. (b) The reflected beam is completely polarized when the angle of incidence equals the polarizing angle θ_p, which satisfies the equation $n_2/n_1 = \tan\theta_p$. At this incident angle, the reflected and refracted rays are perpendicular to each other.

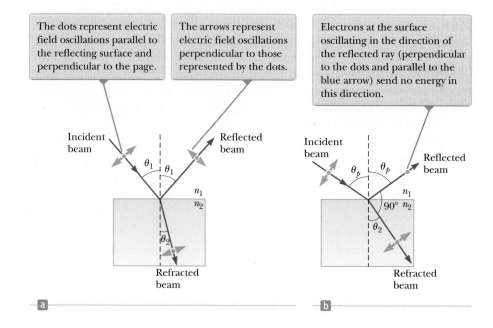

We can obtain an expression relating the polarizing angle to the index of refraction of the reflecting substance by using Figure 38.28b. From this figure, we see that $\theta_p + 90° + \theta_2 = 180°$; therefore, $\theta_2 = 90° - \theta_p$. Using Snell's law of refraction (Eq. 35.8) gives

$$\frac{n_2}{n_1} = \frac{\sin\theta_1}{\sin\theta_2} = \frac{\sin\theta_p}{\sin\theta_2}$$

Because $\sin\theta_2 = \sin(90° - \theta_p) = \cos\theta_p$, we can write this expression as $n_2/n_1 = \sin\theta_p/\cos\theta_p$, which means that

Brewster's law ▶

$$\tan\theta_p = \frac{n_2}{n_1} \tag{38.10}$$

This expression is called **Brewster's law,** and the polarizing angle θ_p is sometimes called **Brewster's angle,** after its discoverer, David Brewster (1781–1868). Because n varies with wavelength for a given substance, Brewster's angle is also a function of wavelength.

We can understand polarization by reflection by imagining that the electric field in the incident light sets electrons at the surface of the material in Figure 38.28b into oscillation. The component directions of oscillation are (1) parallel to the arrows shown on the refracted beam of light and therefore parallel to the reflected beam and (2) perpendicular to the page. The oscillating electrons act as dipole antennas radiating light with a polarization parallel to the direction of oscillation. Consult Figure 34.12, which shows the pattern of radiation from a dipole antenna. Notice that there is no radiation at an angle of $\theta = 0$, that is, along the oscillation direction of the antenna. Therefore, for the oscillations in direction 1, there is no radiation in the direction along the reflected ray. For oscillations in direction 2, the electrons radiate light with a polarization perpendicular to the page. Therefore, the light reflected from the surface at this angle is completely polarized parallel to the surface.

Polarization by reflection is a common phenomenon. Sunlight reflected from water, glass, and snow is partially polarized. If the surface is horizontal, the electric field vector of the reflected light has a strong horizontal component. Sunglasses made of polarizing material reduce the glare of reflected light. The transmission axes of such lenses are oriented vertically so that they absorb the strong horizontal component of the reflected light. If you rotate sunglasses through 90°, they are not as effective at blocking the glare from shiny horizontal surfaces.

Polarization by Double Refraction

Solids can be classified on the basis of internal structure. Those in which the atoms are arranged in a specific order are called *crystalline;* the NaCl structure of Figure 38.22 is one example of a crystalline solid. Those solids in which the atoms are distributed randomly are called *amorphous.* When light travels through an amorphous material such as glass, it travels with a speed that is the same in all directions. That is, glass has a single index of refraction. In certain crystalline materials such as calcite and quartz, however, the speed of light is not the same in all directions. In these materials, the speed of light depends on the direction of propagation *and* on the plane of polarization of the light. Such materials are characterized by two indices of refraction. Hence, they are often referred to as **double-refracting** or **birefringent** materials.

When unpolarized light enters a birefringent material, it may split into an **ordinary (O) ray** and an **extraordinary (E) ray.** These two rays have mutually perpendicular polarizations and travel at different speeds through the material. The two speeds correspond to two indices of refraction, n_O for the ordinary ray and n_E for the extraordinary ray.

There is one direction, called the **optic axis,** along which the ordinary and extraordinary rays have the same speed. If light enters a birefringent material at an angle to the optic axis, however, the different indices of refraction will cause the two polarized rays to split and travel in different directions as shown in Figure 38.29.

The index of refraction n_O for the ordinary ray is the same in all directions. If one could place a point source of light inside the crystal as in Figure 38.30, the ordinary waves would spread out from the source as spheres. The index of refraction n_E varies with the direction of propagation. A point source sends out an extraordinary wave having wave fronts that are elliptical in cross section. The difference in speed for the two rays is a maximum in the direction perpendicular to the optic axis. For example, in calcite, $n_O = 1.658$ at a wavelength of 589.3 nm and n_E varies from 1.658 along the optic axis to 1.486 perpendicular to the optic axis. Values for n_O and the extreme value of n_E for various double-refracting crystals are given in Table 38.1.

If you place a calcite crystal on a sheet of paper and then look through the crystal at any writing on the paper, you would see two images as shown in Figure 38.31. As can be seen from Figure 38.29, these two images correspond to one formed by the ordinary ray and one formed by the extraordinary ray. If the two images are viewed through a sheet of rotating polarizing glass, they alternately appear and disappear because the ordinary and extraordinary rays are plane-polarized along mutually perpendicular directions.

Some materials such as glass and plastic become birefringent when stressed. Suppose an unstressed piece of plastic is placed between a polarizer and an analyzer so that light passes from polarizer to plastic to analyzer. When the plastic is unstressed and the analyzer axis is perpendicular to the polarizer axis, none of the polarized light passes through the analyzer. In other words, the unstressed plastic has no effect on the light passing through it. If the plastic is stressed, however, regions of greatest stress become birefringent and the polarization of the light passing through the plastic changes. Hence, a series of bright and dark bands is observed in the transmitted light, with the bright bands corresponding to regions of greatest stress.

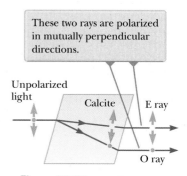

Figure 38.29 Unpolarized light incident at an angle to the optic axis in a calcite crystal splits into an ordinary (O) ray and an extraordinary (E) ray (not to scale).

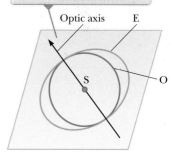

Figure 38.30 A point source S inside a double-refracting crystal produces a spherical wave front corresponding to the ordinary (O) ray and an elliptical wave front corresponding to the extraordinary (E) ray.

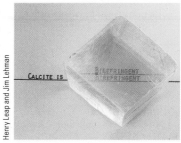

Henry Leap and Jim Lehman

Figure 38.31 A calcite crystal produces a double image because it is a birefringent (double-refracting) material.

Table 38.1	Indices of Refraction for Some Double-Refracting Crystals at a Wavelength of 589.3 nm		
Crystal	n_O	n_E	n_O/n_E
Calcite ($CaCO_3$)	1.658	1.486	1.116
Quartz (SiO_2)	1.544	1.553	0.994
Sodium nitrate ($NaNO_3$)	1.587	1.336	1.188
Sodium sulfite ($NaSO_3$)	1.565	1.515	1.033
Zinc chloride ($ZnCl_2$)	1.687	1.713	0.985
Zinc sulfide (ZnS)	2.356	2.378	0.991

Figure 38.32 A plastic model of an arch structure under load conditions. The pattern is produced when the plastic model is viewed between a polarizer and analyzer oriented perpendicular to each other. Such patterns are useful in the optimal design of architectural components.

Peter Aprahamian/Science Photo Library, Photo Researchers, Inc.

Engineers often use this technique, called *optical stress analysis,* in designing structures ranging from bridges to small tools. They build a plastic model and analyze it under different load conditions to determine regions of potential weakness and failure under stress. An example of a plastic model under stress is shown in Figure 38.32.

Polarization by Scattering

When light is incident on any material, the electrons in the material can absorb and reradiate part of the light. Such absorption and reradiation of light by electrons in the gas molecules that make up air is what causes sunlight reaching an observer on the Earth to be partially polarized. You can observe this effect—called **scattering**—by looking directly up at the sky through a pair of sunglasses whose lenses are made of polarizing material. Less light passes through at certain orientations of the lenses than at others.

Figure 38.33 illustrates how sunlight becomes polarized when it is scattered. The phenomenon is similar to that creating completely polarized light upon reflection from a surface at Brewster's angle. An unpolarized beam of sunlight traveling in the horizontal direction (parallel to the ground) strikes a molecule of one of the gases that make up air, setting the electrons of the molecule into vibration. These vibrating charges act like the vibrating charges in an antenna. The horizontal component of the electric field vector in the incident wave results in a horizontal component of the vibration of the charges, and the vertical component of the vector results in a vertical component of vibration. If the observer in Figure 38.33 is looking straight up (perpendicular to the original direction of propagation of the light), the vertical oscillations of the charges send no radiation toward the observer. Therefore, the observer sees light that is completely polarized in the horizontal direction as indicated by the orange arrows. If the observer looks in other directions, the light is partially polarized in the horizontal direction.

Variations in the color of scattered light in the atmosphere can be understood as follows. When light of various wavelengths λ is incident on gas molecules of diameter d, where $d \ll \lambda$, the relative intensity of the scattered light varies as $1/\lambda^4$. The condition $d \ll \lambda$ is satisfied for scattering from oxygen (O_2) and nitrogen (N_2) molecules in the atmosphere, whose diameters are about 0.2 nm. Hence, short wavelengths (violet light) are scattered more efficiently than long wavelengths (red light). Therefore, when sunlight is scattered by gas molecules in the air, the short-wavelength radiation (violet) is scattered more intensely than the long-wavelength radiation (red).

When you look up into the sky in a direction that is not toward the Sun, you see the scattered light, which is predominantly violet. Your eyes, however, are not very sensitive to violet light. Light of the next color in the spectrum, blue, is scattered with less intensity than violet, but your eyes are far more sensitive to blue light than to violet light. Hence, you see a blue sky. If you look toward the west at sunset (or toward the east at sunrise), you are looking in a direction toward the Sun and are seeing light that has passed through a large distance of air. Most of the blue light has been scattered by the air between you and the Sun. The light that survives this

The scattered light traveling perpendicular to the incident light is plane-polarized because the vertical vibrations of the charges in the air molecule send no light in this direction.

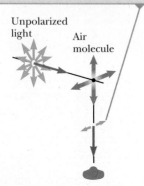

Unpolarized light
Air molecule

Figure 38.33 The scattering of unpolarized sunlight by air molecules.

trip through the air to you has had much of its blue component scattered and is therefore heavily weighted toward the red end of the spectrum; as a result, you see the red and orange colors of sunset (or sunrise).

Optical Activity

Many important applications of polarized light involve materials that display **optical activity.** A material is said to be optically active if it rotates the plane of polarization of any light transmitted through the material. The angle through which the light is rotated by a specific material depends on the length of the path through the material and on concentration if the material is in solution. One optically active material is a solution of the common sugar dextrose. A standard method for determining the concentration of sugar solutions is to measure the rotation produced by a fixed length of the solution.

Molecular asymmetry determines whether a material is optically active. For example, some proteins are optically active because of their spiral shape.

The liquid crystal displays found in most calculators have their optical activity changed by the application of electric potential across different parts of the display. Try using a pair of polarizing sunglasses to investigate the polarization used in the display of your calculator.

Quick Quiz 38.6 A polarizer for microwaves can be made as a grid of parallel metal wires approximately 1 cm apart. Is the electric field vector for microwaves transmitted through this polarizer **(a)** parallel or **(b)** perpendicular to the metal wires?

Quick Quiz 38.7 You are walking down a long hallway that has many light fixtures in the ceiling and a very shiny, newly waxed floor. When looking at the floor, you see reflections of every light fixture. Now you put on sunglasses that are polarized. Some of the reflections of the light fixtures can no longer be seen. (Try it!) Are the reflections that disappear those **(a)** nearest to you, **(b)** farthest from you, or **(c)** at an intermediate distance from you?

Summary

Concepts and Principles

Diffraction is the deviation of light from a straight-line path when the light passes through an aperture or around an obstacle. Diffraction is due to the wave nature of light.

The **Fraunhofer diffraction pattern** produced by a single slit of width a on a distant screen consists of a central bright fringe and alternating bright and dark fringes of much lower intensities. The angles θ_{dark} at which the diffraction pattern has zero intensity, corresponding to destructive interference, are given by

$$\sin \theta_{dark} = m \frac{\lambda}{a} \quad m = \pm 1, \pm 2, \pm 3, \ldots \quad \textbf{(38.1)}$$

Rayleigh's criterion, which is a limiting condition of resolution, states that two images formed by an aperture are just distinguishable if the central maximum of the diffraction pattern for one image falls on the first minimum of the diffraction pattern for the other image. The limiting angle of resolution for a slit of width a is $\theta_{min} = \lambda/a$, and the limiting angle of resolution for a circular aperture of diameter D is given by $\theta_{min} = 1.22\lambda/D$.

A **diffraction grating** consists of a large number of equally spaced, identical slits. The condition for intensity maxima in the interference pattern of a diffraction grating for normal incidence is

$$d \sin \theta_{bright} = m\lambda \quad m = 0, \pm 1, \pm 2, \pm 3, \ldots \quad \textbf{(38.7)}$$

where d is the spacing between adjacent slits and m is the order number of the intensity maximum.

continued

When polarized light of intensity I_{max} is emitted by a polarizer and then is incident on an analyzer, the light transmitted through the analyzer has an intensity equal to $I_{max} \cos^2 \theta$, where θ is the angle between the polarizer and analyzer transmission axes.

In general, reflected light is partially polarized. Reflected light, however, is completely polarized when the angle of incidence is such that the angle between the reflected and refracted beams is 90°. This angle of incidence, called the **polarizing angle** θ_p, satisfies **Brewster's law**:

$$\tan \theta_p = \frac{n_2}{n_1} \tag{38.10}$$

where n_1 is the index of refraction of the medium in which the light initially travels and n_2 is the index of refraction of the reflecting medium.

Objective Questions

1. denotes answer available in *Student Solutions Manual/Study Guide*

1. Certain sunglasses use a polarizing material to reduce the intensity of light reflected as glare from water or automobile windshields. What orientation should the polarizing filters have to be most effective? (a) The polarizers should absorb light with its electric field horizontal. (b) The polarizers should absorb light with its electric field vertical. (c) The polarizers should absorb both horizontal and vertical electric fields. (d) The polarizers should not absorb either horizontal or vertical electric fields.

2. What is most likely to happen to a beam of light when it reflects from a shiny metallic surface at an arbitrary angle? Choose the best answer. (a) It is totally absorbed by the surface. (b) It is totally polarized. (c) It is unpolarized. (d) It is partially polarized. (e) More information is required.

3. In Figure 38.4, assume the slit is in a barrier that is opaque to x-rays as well as to visible light. The photograph in Figure 38.4b shows the diffraction pattern produced with visible light. What will happen if the experiment is repeated with x-rays as the incoming wave and with no other changes? (a) The diffraction pattern is similar. (b) There is no noticeable diffraction pattern but rather a projected shadow of high intensity on the screen, having the same width as the slit. (c) The central maximum is much wider, and the minima occur at larger angles than with visible light. (d) No x-rays reach the screen.

4. A Fraunhofer diffraction pattern is produced on a screen located 1.00 m from a single slit. If a light source of wavelength 5.00×10^{-7} m is used and the distance from the center of the central bright fringe to the first dark fringe is 5.00×10^{-3} m, what is the slit width? (a) 0.010 0 mm (b) 0.100 mm (c) 0.200 mm (d) 1.00 mm (e) 0.005 00 mm

5. Consider a wave passing through a single slit. What happens to the width of the central maximum of its diffraction pattern as the slit is made half as wide? (a) It becomes one-fourth as wide. (b) It becomes one-half as wide. (c) Its width does not change. (d) It becomes twice as wide. (e) It becomes four times as wide.

6. Assume Figure 38.1 was photographed with red light of a single wavelength λ_0. The light passed through a single slit of width a and traveled distance L to the screen where the photograph was made. Consider the width of the central bright fringe, measured between the centers of the dark fringes on both sides of it. Rank from largest to smallest the widths of the central fringe in the following situations and note any cases of equality. (a) The experiment is performed as photographed. (b) The experiment is performed with light whose frequency is increased by 50%. (c) The experiment is performed with light whose wavelength is increased by 50%. (d) The experiment is performed with the original light and with a slit of width $2a$. (e) The experiment is performed with the original light and slit and with distance $2L$ to the screen.

7. If plane polarized light is sent through two polarizers, the first at 45° to the original plane of polarization and the second at 90° to the original plane of polarization, what fraction of the original polarized intensity passes through the last polarizer? (a) 0 (b) $\frac{1}{4}$ (c) $\frac{1}{2}$ (d) $\frac{1}{8}$ (e) $\frac{1}{10}$

8. Why is it advantageous to use a large-diameter objective lens in a telescope? (a) It diffracts the light more effectively than smaller-diameter objective lenses. (b) It increases its magnification. (c) It enables you to see more objects in the field of view. (d) It reflects unwanted wavelengths. (e) It increases its resolution.

9. What combination of optical phenomena causes the bright colored patterns sometimes seen on wet streets covered with a layer of oil? Choose the best answer. (a) diffraction and polarization (b) interference and diffraction (c) polarization and reflection (d) refraction and diffraction (e) reflection and interference

10. When you receive a chest x-ray at a hospital, the x-rays pass through a set of parallel ribs in your chest. Do your ribs act as a diffraction grating for x-rays? (a) Yes. They produce diffracted beams that can be observed separately. (b) Not to a measurable extent. The ribs are too far apart. (c) Essentially not. The ribs are too close together. (d) Essentially not. The ribs are too few in number. (e) Absolutely not. X-rays cannot diffract.

11. When unpolarized light passes through a diffraction grating, does it become polarized? (a) No, it does not. (b) Yes, it does, with the transmission axis parallel to the slits or grooves in the grating. (c) Yes, it does, with the transmission axis perpendicular to the slits or grooves in the grating. (d) It possibly does because an electric field above some threshold is blocked out by the grating if the field is perpendicular to the slits.

12. Off in the distance, you see the headlights of a car, but they are indistinguishable from the single headlight of a motorcycle. Assume the car's headlights are now switched from low beam to high beam so that the light intensity you receive becomes three times greater. What then happens to your ability to resolve the two light sources? (a) It increases by a factor of 9. (b) It increases by a factor of 3. (c) It remains the same. (d) It becomes one-third as good. (e) It becomes one-ninth as good.

Conceptual Questions   **1.** denotes answer available in *Student Solutions Manual/Study Guide*

1. The atoms in a crystal lie in planes separated by a few tenths of a nanometer. Can they produce a diffraction pattern for visible light as they do for x-rays? Explain your answer with reference to Bragg's law.

2. Holding your hand at arm's length, you can readily block sunlight from reaching your eyes. Why can you not block sound from reaching your ears this way?

3. How could the index of refraction of a flat piece of opaque obsidian glass be determined?

4. (a) Is light from the sky polarized? (b) Why is it that clouds seen through Polaroid glasses stand out in bold contrast to the sky?

5. A laser beam is incident at a shallow angle on a horizontal machinist's ruler that has a finely calibrated scale. The engraved rulings on the scale give rise to a diffraction pattern on a vertical screen. Discuss how you can use this technique to obtain a measure of the wavelength of the laser light.

6. If a coin is glued to a glass sheet and this arrangement is held in front of a laser beam, the projected shadow has diffraction rings around its edge and a bright spot in the center. How are these effects possible?

7. Fingerprints left on a piece of glass such as a windowpane often show colored spectra like that from a diffraction grating. Why?

8. A laser produces a beam a few millimeters wide, with uniform intensity across its width. A hair is stretched vertically across the front of the laser to cross the beam. (a) How is the diffraction pattern it produces on a distant screen related to that of a vertical slit equal in width to the hair? (b) How could you determine the width of the hair from measurements of its diffraction pattern?

9. A radio station serves listeners in a city to the northeast of its broadcast site. It broadcasts from three adjacent towers on a mountain ridge, along a line running east to west, in what's called a *phased array*. Show that by introducing time delays among the signals the individual towers radiate, the station can maximize net intensity in the direction toward the city (and in the opposite direction) and minimize the signal transmitted in other directions.

10. John William Strutt, Lord Rayleigh (1842–1919), invented an improved foghorn. To warn ships of a coastline, a foghorn should radiate sound in a wide horizontal sheet over the ocean's surface. It should not waste energy by broadcasting sound upward or downward. Rayleigh's foghorn trumpet is shown in two possible configurations, horizontal and vertical, in Figure CQ38.10. Which is the correct orientation? Decide whether the long dimension of the rectangular opening should be horizontal or vertical and argue for your decision.

Figure CQ38.10

11. Why can you hear around corners, but not see around corners?

12. Figure CQ38.12 shows a megaphone in use. Construct a theoretical description of how a megaphone works. You may assume the sound of your voice radiates just through the opening of your mouth. Most of the information in speech is carried not in a signal at the fundamental frequency, but in noises and in harmonics, with frequencies of a few thousand hertz. Does your theory allow any prediction that is simple to test?

Figure CQ38.12

Problems available in WebAssign   Access end-of-chapter problems online at www.webassign.net

Section 38.2 Diffraction Patterns from Narrow Slits
Problems 1–13

Section 38.3 Resolution of Single-Slit and Circular Apertures
Problems 14–24

Section 38.4 The Diffraction Grating
Problems 25–37

Section 38.5 Diffraction of X-Rays by Crystals
Problems 38–41

Section 38.6 Polarization of Light Waves
Problems 42–52

Additional Problems
Problems 53–74

Challenge Problems

Problem 75–79

Solutions to the following Problems are available in the *Student Solutions Manual/Study Guide:*

38.7, 38.11, 38.20, 38.23, 38.26, 38.30, 38.32, 38.35, 38.38, 38.43, 38.49, 38.53, 38.59, 38.65, 38.71, and 38.77

List of Enhanced Problems

Problem Number	Targeted Feedback in Enhanced WebAssign	Analysis Model Tutorial in Enhanced WebAssign	Master It in Enhanced WebAssign	Watch It in Enhanced WebAssign
38.1	✓		✓	
38.2	✓			✓
38.7	✓		✓	
38.11			✓	
38.20	✓		✓	
38.21	✓		✓	
38.23			✓	
38.24	✓			✓
38.26	✓		✓	
38.28	✓			✓
38.32	✓		✓	
38.33	✓			✓
38.38			✓	
38.39		✓	✓	
38.43			✓	
38.44	✓	✓	✓	✓
38.45	✓			✓
38.51	✓		✓	
38.58	✓			✓
38.65		✓		

Modern Physics

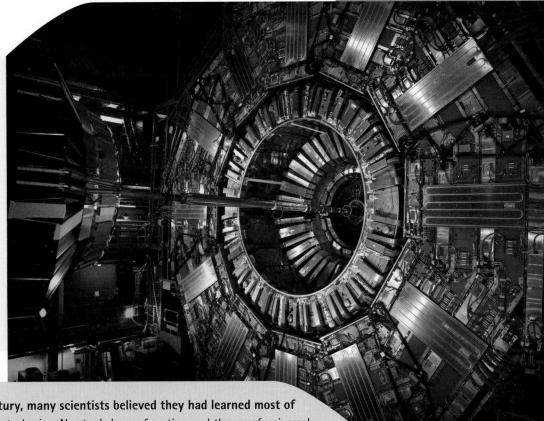

The Compact Muon Solenoid (CMS) Detector is part of the Large Hadron Collider at the European Laboratory for Particle Physics operated by CERN. It is one of several detectors that search for elementary particles. For a sense of scale, the green structure to the left of the detector and extending to the top is five stories high. *(CERN)*

At the end of the 19th century, many scientists believed they had learned most of what there was to know about physics. Newton's laws of motion and theory of universal gravitation, Maxwell's theoretical work in unifying electricity and magnetism, the laws of thermodynamics and kinetic theory, and the principles of optics were highly successful in explaining a variety of phenomena.

At the turn of the 20th century, however, a major revolution shook the world of physics. In 1900, Max Planck provided the basic ideas that led to the formulation of the quantum theory, and in 1905, Albert Einstein formulated his special theory of relativity. The excitement of the times is captured in Einstein's own words: "It was a marvelous time to be alive." Both theories were to have a profound effect on our understanding of nature. Within a few decades, they inspired new developments in the fields of atomic physics, nuclear physics, and condensed-matter physics.

In Chapter 39, we shall introduce the special theory of relativity. The theory provides us with a new and deeper view of physical laws. Although the predictions of this theory often violate our common sense, the theory correctly describes the results of experiments involving speeds near the speed of light. The extended version of this textbook, *Physics for Scientists and Engineers with Modern Physics,* covers the basic concepts of quantum mechanics and their application to atomic and molecular physics. In addition, we introduce condensed matter physics, nuclear physics, particle physics, and cosmology in the extended version.

Even though the physics that was developed during the 20th century has led to a multitude of important technological achievements, the story is still incomplete. Discoveries will continue to evolve during our lifetimes, and many of these discoveries will deepen or refine our understanding of nature and the Universe around us. It is still a "marvelous time to be alive." ■

Relativity

Standing on the shoulders of a giant. David Serway, son of one of the authors, watches over two of his children, Nathan and Kaitlyn, as they frolic in the arms of Albert Einstein's statue at the Einstein memorial in Washington, D.C. It is well known that Einstein, the principal architect of relativity, was very fond of children. *(Emily Serway)*

Our everyday experiences and observations involve objects that move at speeds much less than the speed of light. Newtonian mechanics was formulated by observing and describing the motion of such objects, and this formalism is very successful in describing a wide range of phenomena that occur at low speeds. Nonetheless, it fails to describe properly the motion of objects whose speeds approach that of light.

Experimentally, the predictions of Newtonian theory can be tested at high speeds by accelerating electrons or other charged particles through a large electric potential difference. For example, it is possible to accelerate an electron to a speed of $0.99c$ (where c is the speed of light) by using a potential difference of several million volts. According to Newtonian mechanics, if the potential difference is increased by a factor of 4, the electron's kinetic energy is four times greater and its speed should double to $1.98c$. Experiments show, however, that the speed of the electron—as well as the speed of any other object in the Universe—always remains less than the speed of light, regardless of the size of the accelerating voltage. Because it places no upper limit on speed, Newtonian mechanics is contrary to modern experimental results and is clearly a limited theory.

In 1905, at the age of only 26, Einstein published his special theory of relativity. Regarding the theory, Einstein wrote:

The relativity theory arose from necessity, from serious and deep contradictions in the old theory from which there seemed no escape. The strength of the new theory lies in the consistency and simplicity with which it solves all these difficulties.[1]

Although Einstein made many other important contributions to science, the special theory of relativity alone represents one of the greatest intellectual achievements of all time. With this theory, experimental observations can be correctly predicted over the range of speeds from $v = 0$ to speeds approaching the speed of light. At low speeds, Einstein's theory reduces to Newtonian mechanics as a limiting situation. It is important to recognize that Einstein was working on electromagnetism when he developed the special theory of relativity. He was convinced that Maxwell's equations were correct, and to reconcile them with one of his postulates, he was forced into the revolutionary notion of assuming that space and time are not absolute.

This chapter gives an introduction to the special theory of relativity, with emphasis on some of its predictions. In addition to its well-known and essential role in theoretical physics, the special theory of relativity has practical applications, including the design of nuclear power plants and modern global positioning system (GPS) units. These devices depend on relativistic principles for proper design and operation.

39.1 The Principle of Galilean Relativity

To describe a physical event, we must establish a frame of reference. You should recall from Chapter 5 that an inertial frame of reference is one in which an object is observed to have no acceleration when no forces act on it. Furthermore, any frame moving with constant velocity with respect to an inertial frame must also be an inertial frame.

There is no absolute inertial reference frame. Therefore, the results of an experiment performed in a vehicle moving with uniform velocity must be identical to the results of the same experiment performed in a stationary vehicle. The formal statement of this result is called the **principle of Galilean relativity:**

> The laws of mechanics must be the same in all inertial frames of reference.

◀ **Principle of Galilean relativity**

Let's consider an observation that illustrates the equivalence of the laws of mechanics in different inertial frames. The pickup truck in Figure 39.1a moves with a

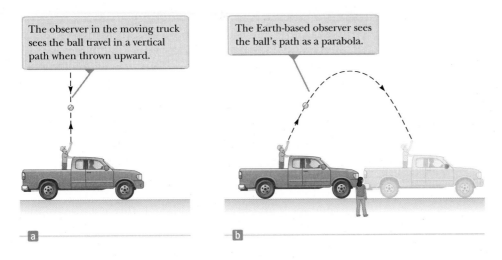

The observer in the moving truck sees the ball travel in a vertical path when thrown upward.

The Earth-based observer sees the ball's path as a parabola.

Figure 39.1 Two observers watch the path of a thrown ball and obtain different results.

[1] A. Einstein and L. Infield, *The Evolution of Physics* (New York: Simon and Schuster, 1961).

constant velocity with respect to the ground. If a passenger in the truck throws a ball straight up and if air resistance is neglected, the passenger observes that the ball moves in a vertical path. The motion of the ball appears to be precisely the same as if the ball were thrown by a person at rest on the Earth. The law of universal gravitation and the equations of motion under constant acceleration are obeyed whether the truck is at rest or in uniform motion.

Consider also an observer on the ground as in Figure 39.1b. Both observers agree on the laws of physics: the observer in the truck throws a ball straight up, and it rises and falls back into his hand according to the particle under constant acceleration model. Do the observers agree on the path of the ball thrown by the observer in the truck? The observer on the ground sees the path of the ball as a parabola as illustrated in Figure 39.1b, whereas, as mentioned earlier, the observer in the truck sees the ball move in a vertical path. Furthermore, according to the observer on the ground, the ball has a horizontal component of velocity equal to the velocity of the truck, and the horizontal motion of the ball is described by the particle under constant velocity model. Although the two observers disagree on certain aspects of the situation, they agree on the validity of Newton's laws and on the results of applying appropriate analysis models that we have learned. This agreement implies that no mechanical experiment can detect any difference between the two inertial frames. The only thing that can be detected is the relative motion of one frame with respect to the other.

Ⓠuick Quiz 39.1 Which observer in Figure 39.1 sees the ball's *correct* path? **(a)** the observer in the truck **(b)** the observer on the ground **(c)** both observers

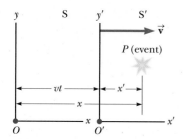

Figure 39.2 An event occurs at a point P. The event is seen by two observers in inertial frames S and S′, where S′ moves with a velocity $\vec{v}$ relative to S.

Suppose some physical phenomenon, which we call an *event*, occurs and is observed by an observer at rest in an inertial reference frame. The wording "in a frame" means that the observer is at rest with respect to the origin of that frame. The event's location and time of occurrence can be specified by the four coordinates (x, y, z, t). We would like to be able to transform these coordinates from those of an observer in one inertial frame to those of another observer in a frame moving with uniform relative velocity compared with the first frame.

Consider two inertial frames S and S′ (Fig. 39.2). The S′ frame moves with a constant velocity $\vec{v}$ along the common x and x' axes, where $\vec{v}$ is measured relative to S. We assume the origins of S and S′ coincide at $t = 0$ and an event occurs at point P in space at some instant of time. For simplicity, we show the observer O in the S frame and the observer O' in the S′ frame as blue dots at the origins of their coordinate frames in Figure 39.2, but that is not necessary: either observer could be at any fixed location in his or her frame. Observer O describes the event with space–time coordinates (x, y, z, t), whereas observer O' in S′ uses the coordinates (x', y', z', t') to describe the same event. Model the origin of S′ as a particle under constant velocity relative to the origin of S. As we see from the geometry in Figure 39.2, the relationships among these various coordinates can be written

Galilean transformation ▶ equations

$$x' = x - vt \qquad y' = y \qquad z' = z \qquad t' = t \qquad \text{(39.1)}$$

These equations are the **Galilean space–time transformation equations.** Note that time is assumed to be the same in both inertial frames. That is, within the framework of classical mechanics, all clocks run at the same rate, regardless of their velocity, so the time at which an event occurs for an observer in S is the same as the time for the same event in S′. Consequently, the time interval between two successive events should be the same for both observers. Although this assumption may seem obvious, it turns out to be incorrect in situations where v is comparable to the speed of light.

Now suppose a particle moves through a displacement of magnitude dx along the x axis in a time interval dt as measured by an observer in S. It follows from Equations 39.1 that the corresponding displacement dx' measured by an observer in S′ is

$dx' = dx - v\,dt$, where frame S′ is moving with speed v in the x direction relative to frame S. Because $dt = dt'$, we find that

$$\frac{dx'}{dt'} = \frac{dx}{dt} - v$$

or

$$u'_x = u_x - v \qquad\qquad \textbf{(39.2)}$$

where u_x and u'_x are the x components of the velocity of the particle measured by observers in S and S′, respectively. (We use the symbol $\vec{\mathbf{u}}$ rather than $\vec{\mathbf{v}}$ for particle velocity because $\vec{\mathbf{v}}$ is already used for the relative velocity of two reference frames.) Equation 39.2 is the **Galilean velocity transformation equation.** It is consistent with our intuitive notion of time and space as well as with our discussions in Section 4.6. As we shall soon see, however, it leads to serious contradictions when applied to electromagnetic waves.

> **Q**uick Quiz 39.2 A baseball pitcher with a 90-mi/h fastball throws a ball while standing on a railroad flatcar moving at 110 mi/h. The ball is thrown in the same direction as that of the velocity of the train. If you apply the Galilean velocity transformation equation to this situation, is the speed of the ball relative to the Earth **(a)** 90 mi/h, **(b)** 110 mi/h, **(c)** 20 mi/h, **(d)** 200 mi/h, or **(e)** impossible to determine?

The Speed of Light

It is quite natural to ask whether the principle of Galilean relativity also applies to electricity, magnetism, and optics. Experiments indicate that the answer is no. Recall from Chapter 34 that Maxwell showed that the speed of light in free space is $c = 3.00 \times 10^8$ m/s. Physicists of the late 1800s thought light waves move through a medium called the *ether* and the speed of light is c only in a special, absolute frame at rest with respect to the ether. The Galilean velocity transformation equation was expected to hold for observations of light made by an observer in any frame moving at speed v relative to the absolute ether frame. That is, if light travels along the x axis and an observer moves with velocity $\vec{\mathbf{v}}$ along the x axis, the observer measures the light to have speed $c \pm v$, depending on the directions of travel of the observer and the light.

Because the existence of a preferred, absolute ether frame would show that light is similar to other classical waves and that Newtonian ideas of an absolute frame are true, considerable importance was attached to establishing the existence of the ether frame. Prior to the late 1800s, experiments involving light traveling in media moving at the highest laboratory speeds attainable at that time were not capable of detecting differences as small as that between c and $c \pm v$. Starting in about 1880, scientists decided to use the Earth as the moving frame in an attempt to improve their chances of detecting these small changes in the speed of light.

Observers fixed on the Earth can take the view that they are stationary and that the absolute ether frame containing the medium for light propagation moves past them with speed v. Determining the speed of light under these circumstances is similar to determining the speed of an aircraft traveling in a moving air current, or wind; consequently, we speak of an "ether wind" blowing through our apparatus fixed to the Earth.

A direct method for detecting an ether wind would use an apparatus fixed to the Earth to measure the ether wind's influence on the speed of light. If v is the speed of the ether relative to the Earth, light should have its maximum speed $c + v$ when propagating downwind as in Figure 39.3a. Likewise, the speed of light should have its minimum value $c - v$ when the light is propagating upwind as in Figure 39.3b and an intermediate value $(c^2 - v^2)^{1/2}$ when the light is directed such that it travels perpendicular to the ether wind as in Figure 39.3c. In this latter case, the vector $\vec{\mathbf{c}}$

Pitfall Prevention 39.1

The Relationship Between the S and S′ Frames Many of the mathematical representations in this chapter are true *only* for the specified relationship between the S and S′ frames. The x and x' axes coincide, except their origins are different. The y and y' axes (and the z and z' axes) are parallel, but they only coincide at one instant due to the time-varying position of the origin of S′ with respect to that of S. We choose the time $t = 0$ to be the instant at which the origins of the two coordinate systems coincide. If the S′ frame is moving in the positive x direction relative to S, then v is positive; otherwise, it is negative.

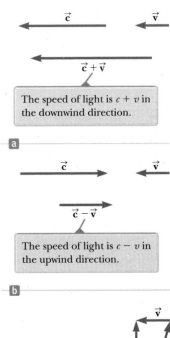

The speed of light is $c + v$ in the downwind direction.

a

The speed of light is $c - v$ in the upwind direction.

b

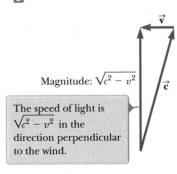

Magnitude: $\sqrt{c^2 - v^2}$

The speed of light is $\sqrt{c^2 - v^2}$ in the direction perpendicular to the wind.

c

Figure 39.3 If the velocity of the ether wind relative to the Earth is $\vec{\mathbf{v}}$ and the velocity of light relative to the ether is $\vec{\mathbf{c}}$, the speed of light relative to the Earth depends on the direction of the Earth's velocity.

must be aimed upstream so that the resultant velocity is perpendicular to the wind, like the boat in Figure 4.21b. If the Sun is assumed to be at rest in the ether, the velocity of the ether wind would be equal to the orbital velocity of the Earth around the Sun, which has a magnitude of approximately 30 km/s or 3×10^4 m/s. Because $c = 3 \times 10^8$ m/s, it is necessary to detect a change in speed of approximately 1 part in 10^4 for measurements in the upwind or downwind directions. Although such a change is experimentally measurable, all attempts to detect such changes and establish the existence of the ether wind (and hence the absolute frame) proved futile! We shall discuss the classic experimental search for the ether in Section 39.2.

The principle of Galilean relativity refers only to the laws of mechanics. If it is assumed the laws of electricity and magnetism are the same in all inertial frames, a paradox concerning the speed of light immediately arises. That can be understood by recognizing that Maxwell's equations imply that the speed of light always has the fixed value 3.00×10^8 m/s in all inertial frames, a result in direct contradiction to what is expected based on the Galilean velocity transformation equation. According to Galilean relativity, the speed of light should *not* be the same in all inertial frames.

To resolve this contradiction in theories, we must conclude that either (1) the laws of electricity and magnetism are not the same in all inertial frames or (2) the Galilean velocity transformation equation is incorrect. If we assume the first alternative, a preferred reference frame in which the speed of light has the value c must exist and the measured speed must be greater or less than this value in any other reference frame, in accordance with the Galilean velocity transformation equation. If we assume the second alternative, we must abandon the notions of absolute time and absolute length that form the basis of the Galilean space–time transformation equations.

39.2 The Michelson–Morley Experiment

The most famous experiment designed to detect small changes in the speed of light was first performed in 1881 by A. A. Michelson (see Section 37.6) and later repeated under various conditions by Michelson and Edward W. Morley (1838–1923). As we shall see, the outcome of the experiment contradicted the ether hypothesis.

The experiment was designed to determine the velocity of the Earth relative to that of the hypothetical ether. The experimental tool used was the Michelson interferometer, which was discussed in Section 37.6 and is shown again in Figure 39.4. Arm 2 is aligned along the direction of the Earth's motion through space. The Earth moving through the ether at speed v is equivalent to the ether flowing past the Earth in the opposite direction with speed v. This ether wind blowing in the direction opposite the direction of the Earth's motion should cause the speed of light measured in the Earth frame to be $c - v$ as the light approaches mirror M_2 and $c + v$ after reflection, where c is the speed of light in the ether frame.

The two light beams reflect from M_1 and M_2 and recombine, and an interference pattern is formed as discussed in Section 37.6. The interference pattern is then observed while the interferometer is rotated through an angle of 90°. This rotation interchanges the speed of the ether wind between the arms of the interferometer. The rotation should cause the fringe pattern to shift slightly but measurably. Measurements failed, however, to show any change in the interference pattern! The Michelson–Morley experiment was repeated at different times of the year when the ether wind was expected to change direction and magnitude, but the results were always the same: no fringe shift of the magnitude required was *ever* observed.[2]

The negative results of the Michelson–Morley experiment not only contradicted the ether hypothesis, but also showed that it is impossible to measure the absolute

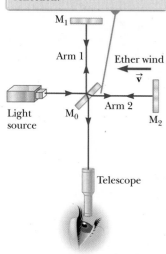

According to the ether wind theory, the speed of light should be $c - v$ as the beam approaches mirror M_2 and $c + v$ after reflection.

Figure 39.4 A Michelson interferometer is used in an attempt to detect the ether wind.

---

[2]From an Earth-based observer's point of view, changes in the Earth's speed and direction of motion in the course of a year are viewed as ether wind shifts. Even if the speed of the Earth with respect to the ether were zero at some time, six months later the speed of the Earth would be 60 km/s with respect to the ether and as a result a fringe shift should be noticed. No shift has ever been observed, however.

velocity of the Earth with respect to the ether frame. Einstein, however, offered a postulate for his special theory of relativity that places quite a different interpretation on these null results. In later years, when more was known about the nature of light, the idea of an ether that permeates all of space was abandoned. Light is now understood to be an electromagnetic wave, which requires no medium for its propagation. As a result, the idea of an ether in which these waves travel became unnecessary.

Details of the Michelson–Morley Experiment

To understand the outcome of the Michelson–Morley experiment, let's assume the two arms of the interferometer in Figure 39.4 are of equal length L. We shall analyze the situation as if there were an ether wind because that is what Michelson and Morley expected to find. As noted above, the speed of the light beam along arm 2 should be $c - v$ as the beam approaches M_2 and $c + v$ after the beam is reflected. We model a pulse of light as a particle under constant speed. Therefore, the time interval for travel to the right for the pulse is $\Delta t = L/(c - v)$ and the time interval for travel to the left is $\Delta t = L/(c + v)$. The total time interval for the round trip along arm 2 is

$$\Delta t_{\text{arm 2}} = \frac{L}{c + v} + \frac{L}{c - v} = \frac{2Lc}{c^2 - v^2} = \frac{2L}{c}\left(1 - \frac{v^2}{c^2}\right)^{-1}$$

Now consider the light beam traveling along arm 1, perpendicular to the ether wind. Because the speed of the beam relative to the Earth is $(c^2 - v^2)^{1/2}$ in this case (see Fig. 39.3c), the time interval for travel for each half of the trip is $\Delta t = L/(c^2 - v^2)^{1/2}$ and the total time interval for the round trip is

$$\Delta t_{\text{arm 1}} = \frac{2L}{(c^2 - v^2)^{1/2}} = \frac{2L}{c}\left(1 - \frac{v^2}{c^2}\right)^{-1/2}$$

The time difference Δt between the horizontal round trip (arm 2) and the vertical round trip (arm 1) is

$$\Delta t = \Delta t_{\text{arm 2}} - \Delta t_{\text{arm 1}} = \frac{2L}{c}\left[\left(1 - \frac{v^2}{c^2}\right)^{-1} - \left(1 - \frac{v^2}{c^2}\right)^{-1/2}\right]$$

Because $v^2/c^2 << 1$, we can simplify this expression by using the following binomial expansion after dropping all terms higher than second order:

$$(1 - x)^n \approx 1 - nx \quad \text{(for } x << 1)$$

In our case, $x = v^2/c^2$, and we find that

$$\Delta t = \Delta t_{\text{arm 2}} - \Delta t_{\text{arm 1}} \approx \frac{Lv^2}{c^3} \tag{39.3}$$

This time difference between the two instants at which the reflected beams arrive at the viewing telescope gives rise to a phase difference between the beams, producing an interference pattern when they combine at the position of the telescope. A shift in the interference pattern should be detected when the interferometer is rotated through 90° in a horizontal plane so that the two beams exchange roles. This rotation results in a time difference twice that given by Equation 39.3. Therefore, the path difference that corresponds to this time difference is

$$\Delta d = c(2\,\Delta t) = \frac{2Lv^2}{c^2}$$

Because a change in path length of one wavelength corresponds to a shift of one fringe, the corresponding fringe shift is equal to this path difference divided by the wavelength of the light:

$$\text{Shift} = \frac{2Lv^2}{\lambda c^2} \tag{39.4}$$

In the experiments by Michelson and Morley, each light beam was reflected by mirrors many times to give an effective path length L of approximately 11 m. Using this value, taking v to be equal to 3.0×10^4 m/s (the speed of the Earth around the Sun), and using 500 nm for the wavelength of the light, we expect a fringe shift of

$$\text{Shift} = \frac{2(11 \text{ m})(3.0 \times 10^4 \text{ m/s})^2}{(5.0 \times 10^{-7} \text{ m})(3.0 \times 10^8 \text{ m/s})^2} = 0.44$$

The instrument used by Michelson and Morley could detect shifts as small as 0.01 fringe, but it detected no shift whatsoever in the fringe pattern! The experiment has been repeated many times since by different scientists under a wide variety of conditions, and no fringe shift has ever been detected. Therefore, it was concluded that the motion of the Earth with respect to the postulated ether cannot be detected.

Many efforts were made to explain the null results of the Michelson–Morley experiment and to save the ether frame concept and the Galilean velocity transformation equation for light. All proposals resulting from these efforts have been shown to be wrong. No experiment in the history of physics received such valiant efforts to explain the absence of an expected result as did the Michelson–Morley experiment. The stage was set for Einstein, who solved the problem in 1905 with his special theory of relativity.

39.3 Einstein's Principle of Relativity

In the previous section, we noted the impossibility of measuring the speed of the ether with respect to the Earth and the failure of the Galilean velocity transformation equation in the case of light. Einstein proposed a theory that boldly removed these difficulties and at the same time completely altered our notion of space and time.[3] He based his special theory of relativity on two postulates:

1. **The principle of relativity:** The laws of physics must be the same in all inertial reference frames.
2. **The constancy of the speed of light:** The speed of light in vacuum has the same value, $c = 3.00 \times 10^8$ m/s, in all inertial frames, regardless of the velocity of the observer or the velocity of the source emitting the light.

The first postulate asserts that *all* the laws of physics—those dealing with mechanics, electricity and magnetism, optics, thermodynamics, and so on—are the same in all reference frames moving with constant velocity relative to one another. This postulate is a generalization of the principle of Galilean relativity, which refers only to the laws of mechanics. From an experimental point of view, Einstein's principle of relativity means that any kind of experiment (measuring the speed of light, for example) performed in a laboratory at rest must give the same result when performed in a laboratory moving at a constant velocity with respect to the first one. Hence, no preferred inertial reference frame exists, and it is impossible to detect absolute motion.

Note that postulate 2 is required by postulate 1: if the speed of light were not the same in all inertial frames, measurements of different speeds would make it possible to distinguish between inertial frames. As a result, a preferred, absolute frame could be identified, in contradiction to postulate 1.

Although the Michelson–Morley experiment was performed before Einstein published his work on relativity, it is not clear whether or not Einstein was aware of the details of the experiment. Nonetheless, the null result of the experiment can be readily understood within the framework of Einstein's theory. According to

Albert Einstein
German-American Physicist
(1879–1955)
Einstein, one of the greatest physicists of all time, was born in Ulm, Germany. In 1905, at age 26, he published four scientific papers that revolutionized physics. Two of these papers were concerned with what is now considered his most important contribution: the special theory of relativity.

In 1916, Einstein published his work on the general theory of relativity. The most dramatic prediction of this theory is the degree to which light is deflected by a gravitational field. Measurements made by astronomers on bright stars in the vicinity of the eclipsed Sun in 1919 confirmed Einstein's prediction, and Einstein became a world celebrity as a result. Einstein was deeply disturbed by the development of quantum mechanics in the 1920s despite his own role as a scientific revolutionary. In particular, he could never accept the probabilistic view of events in nature that is a central feature of quantum theory. The last few decades of his life were devoted to an unsuccessful search for a unified theory that would combine gravitation and electromagnetism.

© Mary Evans Picture Library/Alamy

---

[3]A. Einstein, "On the Electrodynamics of Moving Bodies," *Ann. Physik* **17**:891, 1905. For an English translation of this article and other publications by Einstein, see the book by H. Lorentz, A. Einstein, H. Minkowski, and H. Weyl, *The Principle of Relativity* (New York: Dover, 1958).

his principle of relativity, the premises of the Michelson–Morley experiment were incorrect. In the process of trying to explain the expected results, we stated that when light traveled against the ether wind, its speed was $c - v$, in accordance with the Galilean velocity transformation equation. If the state of motion of the observer or of the source has no influence on the value found for the speed of light, however, one always measures the value to be c. Likewise, the light makes the return trip after reflection from the mirror at speed c, not at speed $c + v$. Therefore, the motion of the Earth does not influence the interference pattern observed in the Michelson–Morley experiment, and a null result should be expected.

If we accept Einstein's theory of relativity, we must conclude that relative motion is unimportant when measuring the speed of light. At the same time, we must alter our commonsense notion of space and time and be prepared for some surprising consequences. As you read the pages ahead, keep in mind that our commonsense ideas are based on a lifetime of everyday experiences and not on observations of objects moving at hundreds of thousands of kilometers per second. Therefore, these results may seem strange, but that is only because we have no experience with them.

39.4 Consequences of the Special Theory of Relativity

As we examine some of the consequences of relativity in this section, we restrict our discussion to the concepts of simultaneity, time intervals, and lengths, all three of which are quite different in relativistic mechanics from what they are in Newtonian mechanics. In relativistic mechanics, for example, the distance between two points and the time interval between two events depend on the frame of reference in which they are measured.

Simultaneity and the Relativity of Time

A basic premise of Newtonian mechanics is that a universal time scale exists that is the same for all observers. Newton and his followers took simultaneity for granted. In his special theory of relativity, Einstein abandoned this assumption.

Einstein devised the following thought experiment to illustrate this point. A boxcar moves with uniform velocity, and two bolts of lightning strike its ends as illustrated in Figure 39.5a, leaving marks on the boxcar and on the ground. The marks on the boxcar are labeled A' and B', and those on the ground are labeled A and B. An observer O' moving with the boxcar is midway between A' and B', and a ground observer O is midway between A and B. The events recorded by the observers are the striking of the boxcar by the two lightning bolts.

The light signals emitted from A and B at the instant at which the two bolts strike later reach observer O at the same time as indicated in Figure 39.5b. This observer

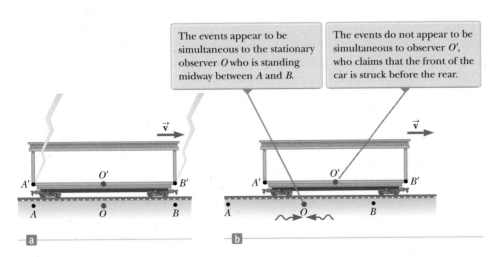

The events appear to be simultaneous to the stationary observer O who is standing midway between A and B.

The events do not appear to be simultaneous to observer O', who claims that the front of the car is struck before the rear.

Figure 39.5 (a) Two lightning bolts strike the ends of a moving boxcar. (b) The leftward-traveling light signal has already passed O', but the rightward-traveling signal has not yet reached O'.

realizes that the signals traveled at the same speed over equal distances and so concludes that the events at *A* and *B* occurred simultaneously. Now consider the same events as viewed by observer *O′*. By the time the signals have reached observer *O*, observer *O′* has moved as indicated in Figure 39.5b. Therefore, the signal from *B′* has already swept past *O′*, but the signal from *A′* has not yet reached *O′*. In other words, *O′* sees the signal from *B′* before seeing the signal from *A′*. According to Einstein, *the two observers must find that light travels at the same speed.* Therefore, observer *O′* concludes that one lightning bolt strikes the front of the boxcar *before* the other one strikes the back.

This thought experiment clearly demonstrates that the two events that appear to be simultaneous to observer *O* do *not* appear to be simultaneous to observer *O′*. Simultaneity is not an absolute concept but rather one that depends on the state of motion of the observer. Einstein's thought experiment demonstrates that two observers can disagree on the simultaneity of two events. This disagreement, however, depends on the transit time of light to the observers and therefore does *not* demonstrate the deeper meaning of relativity. In relativistic analyses of high-speed situations, simultaneity is relative even when the transit time is subtracted out. In fact, in all the relativistic effects we discuss, we ignore differences caused by the transit time of light to the observers.

Time Dilation

To illustrate that observers in different inertial frames can measure different time intervals between a pair of events, consider a vehicle moving to the right with a speed *v* such as the boxcar shown in Figure 39.6a. A mirror is fixed to the ceiling of the vehicle, and observer *O′* at rest in the frame attached to the vehicle holds a flashlight a distance *d* below the mirror. At some instant, the flashlight emits a pulse of light directed toward the mirror (event 1), and at some later time after reflecting from the mirror, the pulse arrives back at the flashlight (event 2). Observer *O′* carries a clock and uses it to measure the time interval Δt_p between these two events. (The subscript *p* stands for *proper*, as we shall see in a moment.) We model the pulse of light as a particle under constant speed. Because the light pulse has a speed *c*, the time interval required for the pulse to travel from *O′* to the mirror and back is

$$\Delta t_p = \frac{\text{distance traveled}}{\text{speed}} = \frac{2d}{c} \qquad (39.5)$$

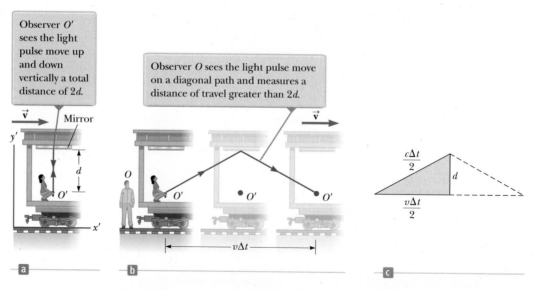

Figure 39.6 (a) A mirror is fixed to a moving vehicle, and a light pulse is sent out by observer *O′* at rest in the vehicle. (b) Relative to a stationary observer *O* standing alongside the vehicle, the mirror and *O′* move with a speed *v*. (c) The right triangle for calculating the relationship between Δt and Δt_p.

Now consider the same pair of events as viewed by observer O in a second frame at rest with respect to the ground as shown in Figure 39.6b. According to this observer, the mirror and the flashlight are moving to the right with a speed v, and as a result, the sequence of events appears entirely different. By the time the light from the flashlight reaches the mirror, the mirror has moved to the right a distance $v\,\Delta t/2$, where Δt is the time interval required for the light to travel from O' to the mirror and back to O' as measured by O. Observer O concludes that because of the motion of the vehicle, if the light is to hit the mirror, it must leave the flashlight at an angle with respect to the vertical direction. Comparing Figure 39.6a with Figure 39.6b, we see that the light must travel farther in part (b) than in part (a). (Notice that neither observer "knows" that he or she is moving. Each is at rest in his or her own inertial frame.)

According to the second postulate of the special theory of relativity, both observers must measure c for the speed of light. Because the light travels farther according to O, the time interval Δt measured by O is longer than the time interval Δt_p measured by O'. To obtain a relationship between these two time intervals, let's use the right triangle shown in Figure 39.6c. The Pythagorean theorem gives

$$\left(\frac{c\,\Delta t}{2}\right)^2 = \left(\frac{v\,\Delta t}{2}\right)^2 + d^2$$

Solving for Δt gives

$$\Delta t = \frac{2d}{\sqrt{c^2 - v^2}} = \frac{2d}{c\sqrt{1 - \dfrac{v^2}{c^2}}} \qquad \textbf{(39.6)}$$

Because $\Delta t_p = 2d/c$, we can express this result as

$$\Delta t = \frac{\Delta t_p}{\sqrt{1 - \dfrac{v^2}{c^2}}} = \gamma\,\Delta t_p \qquad \textbf{(39.7)} \qquad \blacktriangleleft \textbf{ Time dilation}$$

where

$$\gamma = \frac{1}{\sqrt{1 - \dfrac{v^2}{c^2}}} \qquad \textbf{(39.8)}$$

Because γ is always greater than unity, Equation 39.7 shows that the time interval Δt measured by an observer moving with respect to a clock is longer than the time interval Δt_p measured by an observer at rest with respect to the clock. This effect is known as **time dilation.**

Time dilation is not observed in our everyday lives, which can be understood by considering the factor γ. This factor deviates significantly from a value of 1 only for very high speeds as shown in Figure 39.7 and Table 39.1. For example, for a speed of $0.1c$, the value of γ is 1.005. Therefore, there is a time dilation of only 0.5% at

Table 39.1 Approximate Values for γ at Various Speeds	
v/c	γ
0	1
0.001 0	1.000 000 5
0.010	1.000 05
0.10	1.005
0.20	1.021
0.30	1.048
0.40	1.091
0.50	1.155
0.60	1.250
0.70	1.400
0.80	1.667
0.90	2.294
0.92	2.552
0.94	2.931
0.96	3.571
0.98	5.025
0.99	7.089
0.995	10.01
0.999	22.37

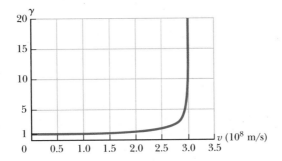

Figure 39.7 Graph of γ versus v. As the speed approaches that of light, γ increases rapidly.

one-tenth the speed of light. Speeds encountered on an everyday basis are far slower than $0.1c$, so we do not experience time dilation in normal situations.

The time interval Δt_p in Equations 39.5 and 39.7 is called the **proper time interval.** (Einstein used the German term *Eigenzeit*, which means "own-time.") In general, the proper time interval is the time interval between two events measured by an observer *who sees the events occur at the same point in space.*

If a clock is moving with respect to you, the time interval between ticks of the moving clock is observed to be longer than the time interval between ticks of an identical clock in your reference frame. Therefore, it is often said that a moving clock is measured to run more slowly than a clock in your reference frame by a factor γ. We can generalize this result by stating that all physical processes, including mechanical, chemical, and biological ones, are measured to slow down when those processes occur in a frame moving with respect to the observer. For example, the heartbeat of an astronaut moving through space keeps time with a clock inside the spacecraft. Both the astronaut's clock and heartbeat are measured to slow down relative to a clock back on the Earth (although the astronaut would have no sensation of life slowing down in the spacecraft).

Quick Quiz 39.3 Suppose the observer O' on the train in Figure 39.6 aims her flashlight at the far wall of the boxcar and turns it on and off, sending a pulse of light toward the far wall. Both O' and O measure the time interval between when the pulse leaves the flashlight and when it hits the far wall. Which observer measures the proper time interval between these two events? **(a)** O' **(b)** O **(c)** both observers **(d)** neither observer

Quick Quiz 39.4 A crew on a spacecraft watches a movie that is two hours long. The spacecraft is moving at high speed through space. Does an Earth-based observer watching the movie screen on the spacecraft through a powerful telescope measure the duration of the movie to be **(a)** longer than, **(b)** shorter than, or **(c)** equal to two hours?

> **Pitfall Prevention 39.3**
> **The Proper Time Interval** It is *very* important in relativistic calculations to correctly identify the observer who measures the proper time interval. The proper time interval between two events is always the time interval measured by an observer for whom the two events take place at the same position.

Time dilation is a very real phenomenon that has been verified by various experiments involving natural clocks. One experiment reported by J. C. Hafele and R. E. Keating provided direct evidence of time dilation.[4] Time intervals measured with four cesium atomic clocks in jet flight were compared with time intervals measured by Earth-based reference atomic clocks. To compare these results with theory, many factors had to be considered, including periods of speeding up and slowing down relative to the Earth, variations in direction of travel, and the weaker gravitational field experienced by the flying clocks than that experienced by the Earth-based clock. The results were in good agreement with the predictions of the special theory of relativity and were explained in terms of the relative motion between the Earth and the jet aircraft. In their paper, Hafele and Keating stated that "relative to the atomic time scale of the U.S. Naval Observatory, the flying clocks lost 59 ± 10 ns during the eastward trip and gained 273 ± 7 ns during the westward trip."

Another interesting example of time dilation involves the observation of *muons,* unstable elementary particles that have a charge equal to that of the electron and a mass 207 times that of the electron. Muons can be produced by the collision of cosmic radiation with atoms high in the atmosphere. Slow-moving muons in the laboratory have a lifetime that is measured to be the proper time interval $\Delta t_p = 2.2\ \mu s$. If we take $2.2\ \mu s$ as the average lifetime of a muon and assume that muons created by cosmic radiation have a speed close to the speed of light, we find that these particles can travel a distance of approximately $(3.0 \times 10^8\ \text{m/s})(2.2 \times 10^{-6}\ \text{s}) \approx 6.6 \times 10^2$ m before they decay (Fig. 39.8a). Hence, they are unlikely to reach the

---

[4]J. C. Hafele and R. E. Keating, "Around the World Atomic Clocks: Relativistic Time Gains Observed," *Science* **177**:168, 1972.

Figure 39.8 Travel of muons according to an Earth-based observer.

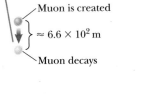

Without relativistic considerations, according to an observer on the Earth, muons created in the atmosphere and traveling downward with a speed close to c travel only about 6.6×10^2 m before decaying with an average lifetime of 2.2 μs. Therefore, very few muons would reach the surface of the Earth.

With relativistic considerations, the muon's lifetime is dilated according to an observer on the Earth. Hence, according to this observer, the muon can travel about 4.8×10^3 m before decaying. The result is many of them arriving at the surface.

Muon is created

$\approx 6.6 \times 10^2$ m

Muon decays

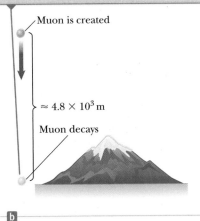

Muon is created

$\approx 4.8 \times 10^3$ m

Muon decays

a **b**

surface of the Earth from high in the atmosphere where they are produced. Experiments show, however, that a large number of muons *do* reach the surface. The phenomenon of time dilation explains this effect. As measured by an observer on the Earth, the muons have a dilated lifetime equal to $\gamma \, \Delta t_p$. For example, for $v = 0.99c$, $\gamma \approx 7.1$, and $\gamma \, \Delta t_p \approx 16$ μs. Hence, the average distance traveled by the muons in this time interval as measured by an observer on the Earth is approximately $(0.99)(3.0 \times 10^8 \text{ m/s})(16 \times 10^{-6} \text{ s}) \approx 4.8 \times 10^3$ m as indicated in Figure 39.8b.

In 1976, at the laboratory of the European Council for Nuclear Research (CERN) in Geneva, muons injected into a large storage ring reached speeds of approximately $0.999\,4c$. Electrons produced by the decaying muons were detected by counters around the ring, enabling scientists to measure the decay rate and hence the muon lifetime. The lifetime of the moving muons was measured to be approximately 30 times as long as that of the stationary muon, in agreement with the prediction of relativity to within two parts in a thousand.

Example 39.1 **What Is the Period of the Pendulum?**

The period of a pendulum is measured to be 3.00 s in the reference frame of the pendulum. What is the period when measured by an observer moving at a speed of $0.960c$ relative to the pendulum?

SOLUTION

Conceptualize Let's change frames of reference. Instead of the observer moving at $0.960c$, we can take the equivalent point of view that the observer is at rest and the pendulum is moving at $0.960c$ past the stationary observer. Hence, the pendulum is an example of a clock moving at high speed with respect to an observer.

Categorize Based on the Conceptualize step, we can categorize this example as a substitution problem involving relativistic time dilation.

The proper time interval, measured in the rest frame of the pendulum, is $\Delta t_p = 3.00$ s.

Use Equation 39.7 to find the dilated time interval:

$$\Delta t = \gamma \, \Delta t_p = \frac{1}{\sqrt{1 - \dfrac{(0.960c)^2}{c^2}}} \, \Delta t_p = \frac{1}{\sqrt{1 - 0.921\,6}} \, \Delta t_p$$

$$= 3.57(3.00 \text{ s}) = \boxed{10.7 \text{ s}}$$

continued

▶ **39.1** c o n t i n u e d

This result shows that a moving pendulum is indeed measured to take longer to complete a period than a pendulum at rest does. The period increases by a factor of $\gamma = 3.57$.

WHAT IF? What if the speed of the observer increases by 4.00%? Does the dilated time interval increase by 4.00%?

Answer Based on the highly nonlinear behavior of γ as a function of v in Figure 39.7, we would guess that the increase in Δt would be different from 4.00%.

Find the new speed if it increases by 4.00%:

$$v_{\text{new}} = (1.040\ 0)(0.960c) = 0.998\ 4c$$

Perform the time dilation calculation again:

$$\Delta t = \gamma\,\Delta t_p = \frac{1}{\sqrt{1 - \dfrac{(0.998\ 4c)^2}{c^2}}}\,\Delta t_p = \frac{1}{\sqrt{1 - 0.996\ 8}}\,\Delta t_p$$

$$= 17.68(3.00\text{ s}) = 53.1\text{ s}$$

Therefore, the 4.00% increase in speed results in almost a 400% increase in the dilated time!

Example 39.2 How Long Was Your Trip?

Suppose you are driving your car on a business trip and are traveling at 30 m/s. Your boss, who is waiting at your destination, expects the trip to take 5.0 h. When you arrive late, your excuse is that the clock in your car registered the passage of 5.0 h but that you were driving fast and so your clock ran more slowly than the clock in your boss's office. If your car clock actually did indicate a 5.0-h trip, how much time passed on your boss's clock, which was at rest on the Earth?

SOLUTION

Conceptualize The observer is your boss standing stationary on the Earth. The clock is in your car, moving at 30 m/s with respect to your boss.

Categorize The low speed of 30 m/s suggests we might categorize this problem as one in which we use classical concepts and equations. Based on the problem statement that the moving clock runs more slowly than a stationary clock, however, we categorize this problem as one involving time dilation.

Analyze The proper time interval, measured in the rest frame of the car, is $\Delta t_p = 5.0$ h.

Use Equation 39.8 to evaluate γ:

$$\gamma = \frac{1}{\sqrt{1 - \dfrac{v^2}{c^2}}} = \frac{1}{\sqrt{1 - \dfrac{(3.0 \times 10^1\text{ m/s})^2}{(3.0 \times 10^8\text{ m/s})^2}}} = \frac{1}{\sqrt{1 - 10^{-14}}}$$

If you try to determine this value on your calculator, you will probably obtain $\gamma = 1$. Instead, perform a binomial expansion:

$$\gamma = (1 - 10^{-14})^{-1/2} \approx 1 + \tfrac{1}{2}(10^{-14}) = 1 + 5.0 \times 10^{-15}$$

Use Equation 39.7 to find the dilated time interval measured by your boss:

$$\Delta t = \gamma\,\Delta t_p = (1 + 5.0 \times 10^{-15})(5.0\text{ h})$$

$$= 5.0\text{ h} + 2.5 \times 10^{-14}\text{ h} = \boxed{5.0\text{ h} + 0.090\text{ ns}}$$

Finalize Your boss's clock would be only 0.090 ns ahead of your car clock. You might want to think of another excuse!

The Twin Paradox

An intriguing consequence of time dilation is the *twin paradox* (Fig. 39.9). Consider an experiment involving a set of twins named Speedo and Goslo. When they are 20 years old, Speedo, the more adventuresome of the two, sets out on an epic journey from the Earth to Planet X, located 20 light-years away. One light-year (ly) is the distance light travels through free space in 1 year. Furthermore, Speedo's

As Speedo (on the left) leaves his brother on Earth, both twins are the same age.

When Speedo returns from his journey, Goslo (on the right) is much older than Speedo.

Figure 39.9 The twin paradox. Speedo takes a journey to a star 20 light-years away and returns to the Earth.

a

b

spacecraft is capable of reaching a speed of $0.95c$ relative to the inertial frame of his twin brother back home on the Earth. After reaching Planet X, Speedo becomes homesick and immediately returns to the Earth at the same speed $0.95c$. Upon his return, Speedo is shocked to discover that Goslo has aged 42 years and is now 62 years old. Speedo, on the other hand, has aged only 13 years.

The paradox is *not* that the twins have aged at different rates. Here is the apparent paradox. From Goslo's frame of reference, he was at rest while his brother traveled at a high speed away from him and then came back. According to Speedo, however, he himself remained stationary while Goslo and the Earth raced away from him and then headed back. Therefore, we might expect Speedo to claim that Goslo ages more slowly than himself. The situation appears to be symmetrical from either twin's point of view. Which twin *actually* ages more slowly?

The situation is actually not symmetrical. Consider a third observer moving at a constant speed relative to Goslo. According to the third observer, Goslo never changes inertial frames. Goslo's speed relative to the third observer is always the same. The third observer notes, however, that Speedo accelerates during his journey when he slows down and starts moving back toward the Earth, *changing reference frames in the process.* From the third observer's perspective, there is something very different about the motion of Goslo when compared to Speedo. Therefore, there is no paradox: only Goslo, who is always in a single inertial frame, can make correct predictions based on special relativity. Goslo finds that instead of aging 42 years, Speedo ages only $(1 - v^2/c^2)^{1/2}(42 \text{ years}) = 13 \text{ years}$. Of these 13 years, Speedo spends 6.5 years traveling to Planet X and 6.5 years returning.

Quick Quiz 39.5 Suppose astronauts are paid according to the amount of time they spend traveling in space. After a long voyage traveling at a speed approaching c, would a crew rather be paid according to **(a)** an Earth-based clock, **(b)** their spacecraft's clock, or **(c)** either clock?

Length Contraction

The measured distance between two points in space also depends on the frame of reference of the observer. The **proper length** L_p of an object is the length measured by an observer *at rest relative to the object.* The length of an object measured by someone in a reference frame that is moving with respect to the object is always less than the proper length. This effect is known as **length contraction.**

To understand length contraction, consider a spacecraft traveling with a speed v from one star to another. There are two observers: one on the Earth and the other in the spacecraft. The observer at rest on the Earth (and also assumed to be at rest with respect to the two stars) measures the distance between the stars to be the proper length L_p. According to this observer, the time interval required for the spacecraft to complete the voyage is given by the particle under constant velocity model as $\Delta t = L_p/v$. The passages of the two stars by the spacecraft occur at the same position for the space traveler. Therefore, the space traveler measures the proper time interval Δt_p. Because of time dilation, the proper time interval is related to the Earth-measured time interval by $\Delta t_p = \Delta t/\gamma$. Because the space traveler reaches the second star in the time Δt_p, he or she concludes that the distance L between the stars is

$$L = v\,\Delta t_p = v\,\frac{\Delta t}{\gamma}$$

Because the proper length is $L_p = v\,\Delta t$, we see that

Length contraction ▶

$$L = \frac{L_p}{\gamma} = L_p\sqrt{1 - \frac{v^2}{c^2}} \qquad (39.9)$$

where $\sqrt{1 - v^2/c^2}$ is a factor less than unity. If an object has a proper length L_p when it is measured by an observer at rest with respect to the object, its length L when it moves with speed v in a direction parallel to its length is measured to be shorter according to Equation 39.9.

For example, suppose a meterstick moves past a stationary Earth-based observer with speed v as in Figure 39.10. The length of the meterstick as measured by an observer in a frame attached to the stick is the proper length L_p shown in Figure 39.10a. The length of the stick L measured by the Earth observer is shorter than L_p by the factor $(1 - v^2/c^2)^{1/2}$ as suggested in Figure 39.10b. Notice that length contraction takes place only along the direction of motion.

The proper length and the proper time interval are defined differently. The proper length is measured by an observer for whom the endpoints of the length remain fixed in space. The proper time interval is measured by someone for whom the two events take place at the same position in space. As an example of this point, let's return to the decaying muons moving at speeds close to the speed of light. An observer in the muon's reference frame measures the proper lifetime, whereas an Earth-based observer measures the proper length (the distance between the creation point and the decay point in Fig. 39.8b). In the muon's reference frame, there is no time dilation, but the distance of travel to the surface is shorter when measured in this frame. Likewise, in the Earth observer's reference frame, there is time dilation, but the distance of travel is measured to be the proper length. Therefore, when calculations on the muon are performed in both frames, the outcome of the experiment in one frame is the same as the outcome in the other frame: more muons reach the surface than would be predicted without relativistic effects.

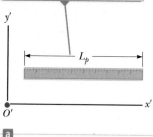

A meterstick measured by an observer in a frame attached to the stick has its proper length L_p.

y'

L_p

O' — x'

a

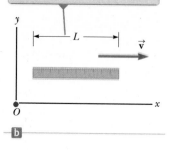

A meterstick measured by an observer in a frame in which the stick has a velocity relative to the frame is measured to be shorter than its proper length.

y

L

$\vec{v}$

O — x

b

Figure 39.10 The length of a meterstick is measured by two observers.

Quick Quiz 39.6 You are packing for a trip to another star. During the journey, you will be traveling at $0.99c$. You are trying to decide whether you should buy smaller sizes of your clothing because you will be thinner on your trip due to length contraction. You also plan to save money by reserving a smaller cabin to sleep in because you will be shorter when you lie down. Should you **(a)** buy smaller sizes of clothing, **(b)** reserve a smaller cabin, **(c)** do neither of these things, or **(d)** do both of these things?

Quick Quiz 39.7 You are observing a spacecraft moving away from you. You measure it to be shorter than when it was at rest on the ground next to you. You also see a clock through the spacecraft window, and you observe that the passage of time on the clock is measured to be slower than that of the watch on your wrist. Compared with when the spacecraft was on the ground, what do you measure if the spacecraft turns around and comes *toward* you at the same speed? **(a)** The spacecraft is measured to be longer, and the clock runs faster. **(b)** The spacecraft is measured to be longer, and the clock runs slower. **(c)** The spacecraft is measured to be shorter, and the clock runs faster. **(d)** The spacecraft is measured to be shorter, and the clock runs slower.

Space–Time Graphs

It is sometimes helpful to represent a physical situation with a **space–time graph,** in which ct is the ordinate and position x is the abscissa. The twin paradox is displayed in such a graph in Figure 39.11 from Goslo's point of view. A path through space–time is called a **world-line.** At the origin, the world-lines of Speedo (blue) and Goslo (green) coincide because the twins are in the same location at the same time. After Speedo leaves on his trip, his world-line diverges from that of his brother. Goslo's world-line is vertical because he remains fixed in location. At Goslo and Speedo's reunion, the two world-lines again come together. It would be impossible for Speedo to have a world-line that crossed the path of a light beam that left the Earth when he did. To do so would require him to have a speed greater than c (which, as shown in Sections 39.6 and 39.7, is not possible).

World-lines for light beams are diagonal lines on space–time graphs, typically drawn at 45° to the right or left of vertical (assuming the x and ct axes have the same scales), depending on whether the light beam is traveling in the direction of increasing or decreasing x. All possible future events for Goslo and Speedo lie above the x axis and between the red-brown lines in Figure 39.11 because neither twin can travel faster than light. The only past events that Goslo and Speedo could have experienced occur between two similar 45° world-lines that approach the origin from below the x axis.

If Figure 39.11 is rotated about the ct axis, the red-brown lines sweep out a cone, called the *light cone,* which generalizes Figure 39.11 to two space dimensions. The y axis can be imagined coming out of the page. All future events for an observer at the origin must lie within the light cone. We can imagine another rotation that would generalize the light cone to three space dimensions to include z, but because of the requirement for four dimensions (three space dimensions and time), we cannot represent this situation in a two-dimensional drawing on paper.

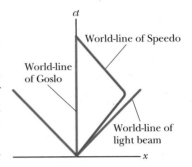

Figure 39.11 The twin paradox on a space–time graph. The twin who stays on the Earth has a world-line along the ct axis (green). The path of the traveling twin through space–time is represented by a world-line that changes direction (blue). The red-brown lines are world-lines for light beams traveling in the positive x direction (on the right) or the negative x direction (on the left).

Example 39.3 | **A Voyage to Sirius** AM

An astronaut takes a trip to Sirius, which is located a distance of 8 light-years from the Earth. The astronaut measures the time of the one-way journey to be 6 years. If the spaceship moves at a constant speed of $0.8c$, how can the 8-ly distance be reconciled with the 6-year trip time measured by the astronaut?

SOLUTION

Conceptualize An observer on the Earth measures light to require 8 years to travel between Sirius and the Earth. The astronaut measures a time interval for his travel of only 6 years. Is the astronaut traveling faster than light?

Categorize Because the astronaut is measuring a length of space between the Earth and Sirius that is in motion with respect to her, we categorize this example as a length contraction problem. We also model the astronaut as a *particle under constant velocity.*

....

Analyze The distance of 8 ly represents the proper length from the Earth to Sirius measured by an observer on the Earth seeing both objects nearly at rest.

Calculate the contracted length measured by the astronaut using Equation 39.9:

$$L = \frac{8 \text{ ly}}{\gamma} = (8 \text{ ly})\sqrt{1 - \frac{v^2}{c^2}} = (8 \text{ ly})\sqrt{1 - \frac{(0.8c)^2}{c^2}} = 5 \text{ ly}$$

Use the particle under constant velocity model to find the travel time measured on the astronaut's clock:

$$\Delta t = \frac{L}{v} = \frac{5 \text{ ly}}{0.8c} = \frac{5 \text{ ly}}{0.8(1 \text{ ly/yr})} = 6 \text{ yr}$$

continued

▶ **39.3** continued

Finalize Notice that we have used the value for the speed of light as $c = 1$ ly/yr. The trip takes a time interval shorter than 8 years for the astronaut because, to her, the distance between the Earth and Sirius is measured to be shorter.

WHAT IF? What if this trip is observed with a very powerful telescope by a technician in Mission Control on the Earth? At what time will this technician *see* that the astronaut has arrived at Sirius?

Answer The time interval the technician measures for the astronaut to arrive is

$$\Delta t = \frac{L_p}{v} = \frac{8 \text{ ly}}{0.8c} = 10 \text{ yr}$$

For the technician to *see* the arrival, the light from the scene of the arrival must travel back to the Earth and enter the telescope. This travel requires a time interval of

$$\Delta t = \frac{L_p}{v} = \frac{8 \text{ ly}}{c} = 8 \text{ yr}$$

Therefore, the technician sees the arrival after 10 yr + 8 yr = 18 yr. If the astronaut immediately turns around and comes back home, she arrives, according to the technician, 20 years after leaving, only 2 years *after the technician saw her arrive!* In addition, the astronaut would have aged by only 12 years.

Example 39.4   The Pole-in-the-Barn Paradox

The twin paradox, discussed earlier, is a classic "paradox" in relativity. Another classic "paradox" is as follows. Suppose a runner moving at 0.75c carries a horizontal pole 15 m long toward a barn that is 10 m long. The barn has front and rear doors that are initially open. An observer on the ground can instantly and simultaneously close and open the two doors by remote control. When the runner and the pole are inside the barn, the ground observer closes and then opens both doors so that the runner and pole are momentarily captured inside the barn and then proceed to exit the barn from the back doorway. Do both the runner and the ground observer agree that the runner makes it safely through the barn?

SOLUTION

Conceptualize From your everyday experience, you would be surprised to see a 15-m pole fit inside a 10-m barn, but we are becoming used to surprising results in relativistic situations.

Categorize The pole is in motion with respect to the ground observer so that the observer measures its length to be contracted, whereas the stationary barn has a proper length of 10 m. We categorize this example as a length contraction problem. The runner carrying the pole is modeled as a *particle under constant velocity*.

Analyze Use Equation 39.9 to find the contracted length of the pole according to the ground observer:

$$L_{\text{pole}} = L_p \sqrt{1 - \frac{v^2}{c^2}} = (15 \text{ m}) \sqrt{1 - (0.75)^2} = 9.9 \text{ m}$$

Therefore, the ground observer measures the pole to be slightly shorter than the barn and there is no problem with momentarily capturing the pole inside it. The "paradox" arises when we consider the runner's point of view.

Use Equation 39.9 to find the contracted length of the barn according to the running observer:

$$L_{\text{barn}} = L_p \sqrt{1 - \frac{v^2}{c^2}} = (10 \text{ m}) \sqrt{1 - (0.75)^2} = 6.6 \text{ m}$$

Because the pole is in the rest frame of the runner, the runner measures it to have its proper length of 15 m. Now the situation looks even worse: How can a 15-m pole fit inside a *6.6-m* barn? Although this question is the classic one that is often asked, it is not the question we have asked because it is not the important one. We asked, "*Does the runner make it safely through the barn?*"

The resolution of the "paradox" lies in the relativity of simultaneity. The closing of the two doors is measured to be simultaneous by the ground observer. Because the doors are at different positions, however, they do not close simulta-

▶ **39.4** continued

neously as measured by the runner. The rear door closes and then opens first, allowing the leading end of the pole to exit. The front door of the barn does not close until the trailing end of the pole passes by.

We can analyze this "paradox" using a space–time graph. Figure 39.12a is a space–time graph from the ground observer's point of view. We choose $x = 0$ as the position of the front doorway of the barn and $t = 0$ as the instant at which the leading end of the pole is located at the front doorway of the barn. The world-lines for the two doorways of the barn are separated by 10 m and are vertical because the barn is not moving relative to this observer. For the pole, we follow two tilted world-lines, one for each end of the moving pole. These world-lines are 9.9 m apart horizontally, which is the contracted length seen by the ground observer. As seen in Figure 39.12a, the pole is entirely within the barn at some time.

Figure 39.12b shows the space–time graph according to the runner. Here, the world-lines for the pole are separated by 15 m and are vertical because the pole is at rest in the runner's frame of reference. The barn is hurtling *toward* the runner, so the world-lines for the front and rear doorways of the barn are tilted to the left. The world-lines for the barn are separated by 6.6 m, the contracted length as seen by the runner. The leading end of the pole leaves the rear doorway of the barn long before the trailing end of the pole enters the barn. Therefore, the opening of the rear door occurs before the closing of the front door.

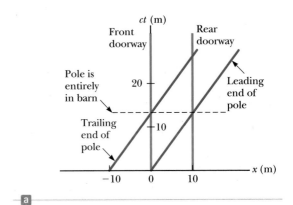

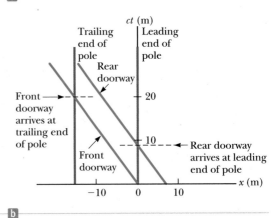

Figure 39.12 (Example 39.4) Space–time graphs for the pole-in-the-barn paradox (a) from the ground observer's point of view and (b) from the runner's point of view.

From the ground observer's point of view, use the particle under constant velocity model to find the time after $t = 0$ at which the trailing end of the pole enters the barn:

$$(1) \quad t = \frac{\Delta x}{v} = \frac{9.9 \text{ m}}{0.75c} = \frac{13.2 \text{ m}}{c}$$

From the runner's point of view, use the particle under constant velocity model to find the time at which the leading end of the pole leaves the barn:

$$(2) \quad t = \frac{\Delta x}{v} = \frac{6.6 \text{ m}}{0.75c} = \frac{8.8 \text{ m}}{c}$$

Find the time at which the trailing end of the pole enters the front door of the barn:

$$(3) \quad t = \frac{\Delta x}{v} = \frac{15 \text{ m}}{0.75c} = \frac{20 \text{ m}}{c}$$

Finalize From Equation (1), the pole should be completely inside the barn at a time corresponding to $ct = 13.2$ m. This situation is consistent with the point on the ct axis in Figure 39.12a where the pole is inside the barn. From Equation (2), the leading end of the pole leaves the barn at $ct = 8.8$ m. This situation is consistent with the point on the ct axis in Figure 39.12b where the rear doorway of the barn arrives at the leading end of the pole. Equation (3) gives $ct = 20$ m, which agrees with the instant shown in Figure 39.12b at which the front doorway of the barn arrives at the trailing end of the pole.

The Relativistic Doppler Effect

Another important consequence of time dilation is the shift in frequency observed for light emitted by atoms in motion as opposed to light emitted by atoms at rest. This phenomenon, known as the Doppler effect, was introduced in Chapter 17 as it pertains to sound waves. In the case of sound, the velocity v_S of the source with respect to the medium of propagation can be distinguished from the velocity v_O of the observer with respect to the medium (the air). Light waves must be analyzed differently, however, because *they require no medium of propagation,* and no method exists for distinguishing the velocity of a light source from the velocity of the observer. The only measurable velocity is the *relative velocity v* between the source and the observer.

If a light source and an observer approach each other with a relative speed v, the frequency f' measured by the observer is

$$f' = \frac{\sqrt{1 + v/c}}{\sqrt{1 - v/c}} f \tag{39.10}$$

where f is the frequency of the source measured in its rest frame. This relativistic Doppler shift equation, unlike the Doppler shift equation for sound, depends only on the relative speed v of the source and observer and holds for relative speeds as great as c. As you might expect, the equation predicts that $f' > f$ when the source and observer approach each other. We obtain the expression for the case in which the source and observer recede from each other by substituting negative values for v in Equation 39.10.

The most spectacular and dramatic use of the relativistic Doppler effect is the measurement of shifts in the frequency of light emitted by a moving astronomical object such as a galaxy. Light emitted by atoms and normally found in the extreme violet region of the spectrum is shifted toward the red end of the spectrum for atoms in other galaxies, indicating that these galaxies are *receding* from us. American astronomer Edwin Hubble (1889–1953) performed extensive measurements of this *red shift* to confirm that most galaxies are moving away from us, indicating that the Universe is expanding.

39.5 The Lorentz Transformation Equations

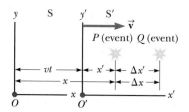

Figure 39.13 Events occur at points P and Q and are observed by an observer at rest in the S frame and another in the S' frame, which is moving to the right with a speed v.

Suppose two events occur at points P and Q and are reported by two observers, one at rest in a frame S and another in a frame S' that is moving to the right with speed v as in Figure 39.13. The observer in S reports the events with space–time coordinates (x, y, z, t), and the observer in S' reports the same events using the coordinates (x', y', z', t'). Equation 39.1 predicts that the distance between the two points in space at which the events occur does not depend on motion of the observer: $\Delta x = \Delta x'$. Because this prediction is contradictory to the notion of length contraction, the Galilean transformation is not valid when v approaches the speed of light. In this section, we present the correct transformation equations that apply for all speeds in the range $0 < v < c$.

The equations that are valid for all speeds and that enable us to transform coordinates from S to S' are the **Lorentz transformation equations:**

Lorentz transformation ▶
for S → S'

$$x' = \gamma(x - vt) \qquad y' = y \qquad z' = z \qquad t' = \gamma\left(t - \frac{v}{c^2}x\right) \tag{39.11}$$

These transformation equations were developed by Hendrik A. Lorentz (1853–1928) in 1890 in connection with electromagnetism. It was Einstein, however, who recognized their physical significance and took the bold step of interpreting them within the framework of the special theory of relativity.

Notice the difference between the Galilean and Lorentz time equations. In the Galilean case, $t = t'$. In the Lorentz case, however, the value for t' assigned to an event by an observer O' in the S' frame in Figure 39.13 depends both on the time t and on the coordinate x as measured by an observer O in the S frame, which is consistent with the notion that an event is characterized by four space–time coordinates (x, y, z, t). In other words, in relativity, space and time are *not* separate concepts but rather are closely interwoven with each other.

If you wish to transform coordinates in the S′ frame to coordinates in the S frame, simply replace v by $-v$ and interchange the primed and unprimed coordinates in Equations 39.11:

$$x = \gamma(x' + vt') \qquad y = y' \qquad z = z' \qquad t = \gamma\left(t' + \frac{v}{c^2}x'\right) \qquad \textbf{(39.12)}$$

◀ Inverse Lorentz transformation for S′ → S

When $v \ll c$, the Lorentz transformation equations should reduce to the Galilean equations. As v approaches zero, $v/c \ll 1$; therefore, $\gamma \to 1$ and Equations 39.11 indeed reduce to the Galilean space–time transformation equations in Equation 39.1.

In many situations, we would like to know the difference in coordinates between two events or the time interval between two events as seen by observers O and O'. From Equations 39.11 and 39.12, we can express the differences between the four variables x, x', t, and t' in the form

$$\left.\begin{aligned}\Delta x' &= \gamma(\Delta x - v\,\Delta t) \\ \Delta t' &= \gamma\left(\Delta t - \frac{v}{c^2}\Delta x\right)\end{aligned}\right\}\text{S} \to \text{S}' \qquad \textbf{(39.13)}$$

$$\left.\begin{aligned}\Delta x &= \gamma(\Delta x' + v\,\Delta t') \\ \Delta t &= \gamma\left(\Delta t' + \frac{v}{c^2}\Delta x'\right)\end{aligned}\right\}\text{S}' \to \text{S} \qquad \textbf{(39.14)}$$

where $\Delta x' = x_2' - x_1'$ and $\Delta t' = t_2' - t_1'$ are the differences measured by observer O' and $\Delta x = x_2 - x_1$ and $\Delta t = t_2 - t_1$ are the differences measured by observer O. (We have not included the expressions for relating the y and z coordinates because they are unaffected by motion along the x direction.[5])

Example 39.5　　Simultaneity and Time Dilation Revisited

(A) Use the Lorentz transformation equations in difference form to show that simultaneity is not an absolute concept.

SOLUTION

Conceptualize Imagine two events that are simultaneous and separated in space as measured in the S′ frame such that $\Delta t' = 0$ and $\Delta x' \neq 0$. These measurements are made by an observer O' who is moving with speed v relative to O.

Categorize The statement of the problem tells us to categorize this example as one involving the use of the Lorentz transformation.

Analyze From the expression for Δt given in Equation 39.14, find the time interval Δt measured by observer O:

$$\Delta t = \gamma\left(\Delta t' + \frac{v}{c^2}\Delta x'\right) = \gamma\left(0 + \frac{v}{c^2}\Delta x'\right) = \gamma\frac{v}{c^2}\Delta x'$$

Finalize The time interval for the same two events as measured by O is nonzero, so the events do not appear to be simultaneous to O.

(B) Use the Lorentz transformation equations in difference form to show that a moving clock is measured to run more slowly than a clock that is at rest with respect to an observer.

SOLUTION

Conceptualize Imagine that observer O' carries a clock that he uses to measure a time interval $\Delta t'$. He finds that two events occur at the same place in his reference frame ($\Delta x' = 0$) but at different times ($\Delta t' \neq 0$). Observer O' is moving with speed v relative to O.

continued

---

[5]Although relative motion of the two frames along the x axis does not change the y and z coordinates of an object, it does change the y and z velocity components of an object moving in either frame as noted in Section 39.6.

▶ **39.5** continued

Categorize The statement of the problem tells us to categorize this example as one involving the use of the Lorentz transformation.

....

Analyze From the expression for Δt given in Equation 39.14, find the time interval Δt measured by observer O:

$$\Delta t = \gamma\left(\Delta t' + \frac{v}{c^2}\Delta x'\right) = \gamma\left[\Delta t' + \frac{v}{c^2}(0)\right] = \gamma\,\Delta t'$$

....

Finalize This result is the equation for time dilation found earlier (Eq. 39.7), where $\Delta t' = \Delta t_p$ is the proper time interval measured by the clock carried by observer O'. Therefore, O measures the moving clock to run slow.

39.6 The Lorentz Velocity Transformation Equations

Suppose two observers in relative motion with respect to each other are both observing an object's motion. Previously, we defined an event as occurring at an instant of time. Now let's interpret the "event" as the object's motion. We know that the Galilean velocity transformation (Eq. 39.2) is valid for low speeds. How do the observers' measurements of the velocity of the object relate to each other if the speed of the object or the relative speed of the observers is close to that of light? Once again, S' is our frame moving at a speed v relative to S. Suppose an object has a velocity component u_x' measured in the S' frame, where

$$u_x' = \frac{dx'}{dt'} \tag{39.15}$$

Using Equation 39.11, we have

$$dx' = \gamma(dx - v\,dt)$$

$$dt' = \gamma\left(dt - \frac{v}{c^2}\,dx\right)$$

Substituting these values into Equation 39.15 gives

$$u_x' = \frac{dx - v\,dt}{dt - \frac{v}{c^2}\,dx} = \frac{\dfrac{dx}{dt} - v}{1 - \dfrac{v}{c^2}\dfrac{dx}{dt}}$$

The term dx/dt, however, is simply the velocity component u_x of the object measured by an observer in S, so this expression becomes

Lorentz velocity trans- ▶
formation for S → S'

$$u_x' = \frac{u_x - v}{1 - \dfrac{u_x v}{c^2}} \tag{39.16}$$

If the object has velocity components along the y and z axes, the components as measured by an observer in S' are

$$u_y' = \frac{u_y}{\gamma\left(1 - \dfrac{u_x v}{c^2}\right)} \quad \text{and} \quad u_z' = \frac{u_z}{\gamma\left(1 - \dfrac{u_x v}{c^2}\right)} \tag{39.17}$$

Notice that u_y' and u_z' do not contain the parameter v in the numerator because the relative velocity is along the x axis.

When v is much smaller than c (the nonrelativistic case), the denominator of Equation 39.16 approaches unity and so $u_x' \approx u_x - v$, which is the Galilean velocity transformation equation. In another extreme, when $u_x = c$, Equation 39.16 becomes

$$u'_x = \frac{c - v}{1 - \frac{cv}{c^2}} = \frac{c\left(1 - \frac{v}{c}\right)}{1 - \frac{v}{c}} = c$$

This result shows that a speed measured as c by an observer in S is also measured as c by an observer in S′, independent of the relative motion of S and S′. This conclusion is consistent with Einstein's second postulate: the speed of light must be c relative to all inertial reference frames. Furthermore, we find that the speed of an object can never be measured as larger than c. That is, the speed of light is the ultimate speed. We shall return to this point later.

To obtain u_x in terms of u'_x, we replace v by $-v$ in Equation 39.16 and interchange the roles of u_x and u'_x:

$$u_x = \frac{u'_x + v}{1 + \frac{u'_x v}{c^2}} \qquad (39.18)$$

Quick Quiz 39.8 You are driving on a freeway at a relativistic speed. **(i)** Straight ahead of you, a technician standing on the ground turns on a searchlight and a beam of light moves exactly vertically upward as seen by the technician. As you observe the beam of light, do you measure the magnitude of the vertical component of its velocity as (a) equal to c, (b) greater than c, or (c) less than c? **(ii)** If the technician aims the searchlight directly at you instead of upward, do you measure the magnitude of the horizontal component of its velocity as (a) equal to c, (b) greater than c, or (c) less than c?

> **Pitfall Prevention 39.5**
> **What Can the Observers Agree On?** We have seen several measurements that the two observers O and O' do *not* agree on: (1) the time interval between events that take place in the same position in one of their frames, (2) the distance between two points that remain fixed in one of their frames, (3) the velocity components of a moving particle, and (4) whether two events occurring at different locations in both frames are simultaneous or not. The two observers *can* agree on (1) their relative speed of motion v with respect to each other, (2) the speed c of any ray of light, and (3) the simultaneity of two events that take place at the same position *and* time in some frame.

Example 39.6 Relative Velocity of Two Spacecraft

Two spacecraft A and B are moving in opposite directions as shown in Figure 39.14. An observer on the Earth measures the speed of spacecraft A to be $0.750c$ and the speed of spacecraft B to be $0.850c$. Find the velocity of spacecraft B as observed by the crew on spacecraft A.

Figure 39.14 (Example 39.6) Two spacecraft A and B move in opposite directions. The speed of spacecraft B relative to spacecraft A is *less* than c and is obtained from the relativistic velocity transformation equation.

SOLUTION

Conceptualize There are two observers, one (O) on the Earth and one (O') on spacecraft A. The event is the motion of spacecraft B.

Categorize Because the problem asks to find an observed velocity, we categorize this example as one requiring the Lorentz velocity transformation.

Analyze The Earth-based observer at rest in the S frame makes two measurements, one of each spacecraft. We want to find the velocity of spacecraft B as measured by the crew on spacecraft A. Therefore, $u_x = -0.850c$. The velocity of spacecraft A is also the velocity of the observer at rest in spacecraft A (the S′ frame) relative to the observer at rest on the Earth. Therefore, $v = 0.750c$.

Obtain the velocity u'_x of spacecraft B relative to spacecraft A using Equation 39.16:

$$u'_x = \frac{u_x - v}{1 - \frac{u_x v}{c^2}} = \frac{-0.850c - 0.750c}{1 - \frac{(-0.850c)(0.750c)}{c^2}} = \boxed{-0.977c}$$

Finalize The negative sign indicates that spacecraft B is moving in the negative x direction as observed by the crew on spacecraft A. Is that consistent with your expectation from Figure 39.14? Notice that the speed is less than c. That is, an

continued

▶ **39.6** continued

object whose speed is less than c in one frame of reference must have a speed less than c in any other frame. (Had you used the Galilean velocity transformation equation in this example, you would have found that $u'_x = u_x - v = -0.850c - 0.750c = -1.60c$, which is impossible. The Galilean transformation equation does not work in relativistic situations.)

WHAT IF?   What if the two spacecraft pass each other? What is their relative speed now?

Answer The calculation using Equation 39.16 involves only the velocities of the two spacecraft and does not depend on their locations. After they pass each other, they have the same velocities, so the velocity of spacecraft B as observed by the crew on spacecraft A is the same, $-0.977c$. The only difference after they pass is that spacecraft B is receding from spacecraft A, whereas it was approaching spacecraft A before it passed.

Example 39.7   **Relativistic Leaders of the Pack**

Two motorcycle pack leaders named David and Emily are racing at relativistic speeds along perpendicular paths as shown in Figure 39.15. How fast does Emily recede as seen by David over his right shoulder?

SOLUTION

Conceptualize The two observers are David and the police officer in Figure 39.15. The event is the motion of Emily. Figure 39.15 represents the situation as seen by the police officer at rest in frame S. Frame S′ moves along with David.

Categorize Because the problem asks to find an observed velocity, we categorize this problem as one requiring the Lorentz velocity transformation. The motion takes place in two dimensions.

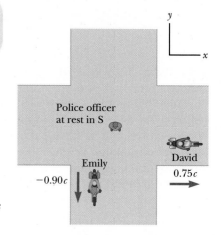

Figure 39.15 (Example 39.7) David moves east with a speed $0.75c$ relative to the police officer, and Emily travels south at a speed $0.90c$ relative to the officer.

Police officer at rest in S

Emily $-0.90c$

David $0.75c$

....

Analyze Identify the velocity components for David and Emily according to the police officer:

David: $v_x = v = 0.75c$ $v_y = 0$
Emily: $u_x = 0$ $u_y = -0.90c$

Using Equations 39.16 and 39.17, calculate u'_x and u'_y for Emily as measured by David:

$$u'_x = \frac{u_x - v}{1 - \dfrac{u_x v}{c^2}} = \frac{0 - 0.75c}{1 - \dfrac{(0)(0.75c)}{c^2}} = -0.75c$$

$$u'_y = \frac{u_y}{\gamma\left(1 - \dfrac{u_x v}{c^2}\right)} = \frac{\sqrt{1 - \dfrac{(0.75c)^2}{c^2}}\,(-0.90c)}{1 - \dfrac{(0)(0.75c)}{c^2}} = -0.60c$$

Using the Pythagorean theorem, find the speed of Emily as measured by David:

$$u' = \sqrt{(u'_x)^2 + (u'_y)^2} = \sqrt{(-0.75c)^2 + (-0.60c)^2} = \boxed{0.96c}$$

....

Finalize This speed is less than c, as required by the special theory of relativity.

39.7 Relativistic Linear Momentum

To describe the motion of particles within the framework of the special theory of relativity properly, you must replace the Galilean transformation equations by the Lorentz transformation equations. Because the laws of physics must remain unchanged under the Lorentz transformation, we must generalize Newton's laws and the definitions of linear momentum and energy to conform to the Lorentz transformation equations and the principle of relativity. These generalized definitions should reduce to the classical (nonrelativistic) definitions for $v \ll c$.

First, recall from the isolated system model that when two particles (or objects that can be modeled as particles) collide, the total momentum of the isolated system of the two particles remains constant. Suppose we observe this collision in a reference frame S and confirm that the momentum of the system is conserved. Now imagine that the momenta of the particles are measured by an observer in a second reference frame S′ moving with velocity $\vec{v}$ relative to the first frame. Using the Lorentz velocity transformation equation and the classical definition of linear momentum, $\vec{p} = m\vec{u}$ (where $\vec{u}$ is the velocity of a particle), we find that linear momentum of the system is *not* measured to be conserved by the observer in S′. Because the laws of physics are the same in all inertial frames, however, linear momentum of the system must be conserved in all frames. We have a contradiction. In view of this contradiction and assuming the Lorentz velocity transformation equation is correct, we must modify the definition of linear momentum so that the momentum of an isolated system is conserved for all observers. For any particle, the correct relativistic equation for linear momentum that satisfies this condition is

$$\vec{p} \equiv \frac{m\vec{u}}{\sqrt{1 - \dfrac{u^2}{c^2}}} = \gamma m \vec{u} \tag{39.19}$$

◀ **Definition of relativistic linear momentum**

Pitfall Prevention 39.6
Watch Out for "Relativistic Mass" Some older treatments of relativity maintained the conservation of momentum principle at high speeds by using a model in which a particle's mass increases with speed. You might still encounter this notion of "relativistic mass" in your outside reading, especially in older books. Be aware that this notion is no longer widely accepted; today, mass is considered as *invariant,* independent of speed. The mass of an object in all frames is considered to be the mass as measured by an observer at rest with respect to the object.

where m is the mass of the particle and $\vec{u}$ is the velocity of the particle. When u is much less than c, $\gamma = (1 - u^2/c^2)^{-1/2}$ approaches unity and $\vec{p}$ approaches $m\vec{u}$. Therefore, the relativistic equation for $\vec{p}$ reduces to the classical expression when u is much smaller than c, as it should.

The relativistic force $\vec{F}$ acting on a particle whose linear momentum is $\vec{p}$ is defined as

$$\vec{F} \equiv \frac{d\vec{p}}{dt} \tag{39.20}$$

where $\vec{p}$ is given by Equation 39.19. This expression, which is the relativistic form of Newton's second law, is reasonable because it preserves classical mechanics in the limit of low velocities and is consistent with conservation of linear momentum for an isolated system ($\vec{F}_{ext} = 0$) both relativistically and classically.

It is left as an end-of-chapter problem (Problem 88 in Enhanced WebAssign) to show that under relativistic conditions, the acceleration $\vec{a}$ of a particle decreases under the action of a constant force, in which case $a \propto (1 - u^2/c^2)^{3/2}$. This proportionality shows that as the particle's speed approaches c, the acceleration caused by any finite force approaches zero. Hence, it is impossible to accelerate a particle from rest to a speed $u \geq c$. This argument reinforces that the speed of light is the ultimate speed, the speed limit of the Universe. It is the maximum possible speed for energy transfer and for information transfer. Any object with mass must move at a lower speed.

Example 39.8   **Linear Momentum of an Electron**

An electron, which has a mass of 9.11×10^{-31} kg, moves with a speed of $0.750c$. Find the magnitude of its relativistic momentum and compare this value with the momentum calculated from the classical expression.

SOLUTION

Conceptualize Imagine an electron moving with high speed. The electron carries momentum, but the magnitude of its momentum is not given by $p = mu$ because the speed is relativistic.

Categorize We categorize this example as a substitution problem involving a relativistic equation.

continued

▶ **39.8** continued

Use Equation 39.19 with $u = 0.750c$ to find the magnitude of the momentum:

$$p = \frac{m_e u}{\sqrt{1 - \dfrac{u^2}{c^2}}}$$

$$p = \frac{(9.11 \times 10^{-31}\ \text{kg})(0.750)(3.00 \times 10^8\ \text{m/s})}{\sqrt{1 - \dfrac{(0.750c)^2}{c^2}}}$$

$$= 3.10 \times 10^{-22}\ \text{kg} \cdot \text{m/s}$$

The classical expression (used incorrectly here) gives $p_{\text{classical}} = m_e u = 2.05 \times 10^{-22}\ \text{kg} \cdot \text{m/s}$. Hence, the correct relativistic result is 50% greater than the classical result!

39.8 Relativistic Energy

We have seen that the definition of linear momentum requires generalization to make it compatible with Einstein's postulates. This conclusion implies that the definition of kinetic energy must most likely be modified also.

To derive the relativistic form of the work–kinetic energy theorem, imagine a particle moving in one dimension along the x axis. A force in the x direction causes the momentum of the particle to change according to Equation 39.20. In what follows, we assume the particle is accelerated from rest to some final speed u. The work done by the force F on the particle is

$$W = \int_{x_1}^{x_2} F\,dx = \int_{x_1}^{x_2} \frac{dp}{dt}\,dx \tag{39.21}$$

To perform this integration and find the work done on the particle and the relativistic kinetic energy as a function of u, we first evaluate dp/dt:

$$\frac{dp}{dt} = \frac{d}{dt}\frac{mu}{\sqrt{1 - \dfrac{u^2}{c^2}}} = \frac{m}{\left(1 - \dfrac{u^2}{c^2}\right)^{3/2}}\frac{du}{dt}$$

Substituting this expression for dp/dt and $dx = u\,dt$ into Equation 39.21 gives

$$W = \int_0^t \frac{m}{\left(1 - \dfrac{u^2}{c^2}\right)^{3/2}}\frac{du}{dt}(u\,dt) = m\int_0^u \frac{u}{\left(1 - \dfrac{u^2}{c^2}\right)^{3/2}}\,du$$

where we use the limits 0 and u in the integral because the integration variable has been changed from t to u. Evaluating the integral gives

$$W = \frac{mc^2}{\sqrt{1 - \dfrac{u^2}{c^2}}} - mc^2 \tag{39.22}$$

Recall from Chapter 7 that the work done by a force acting on a system consisting of a single particle equals the change in kinetic energy of the particle: $W = \Delta K$. Because we assumed the initial speed of the particle is zero, its initial kinetic energy is zero, so $W = K - K_i = K - 0 = K$. Therefore, the work W in Equation 39.22 is equivalent to the relativistic kinetic energy K:

Relativistic kinetic energy ▶

$$K = \frac{mc^2}{\sqrt{1 - \dfrac{u^2}{c^2}}} - mc^2 = \gamma mc^2 - mc^2 = (\gamma - 1)mc^2 \tag{39.23}$$

This equation is routinely confirmed by experiments using high-energy particle accelerators.

At low speeds, where $u/c \ll 1$, Equation 39.23 should reduce to the classical expression $K = \frac{1}{2}mu^2$. We can check that by using the binomial expansion $(1 - \beta^2)^{-1/2} \approx 1 + \frac{1}{2}\beta^2 + \cdots$ for $\beta \ll 1$, where the higher-order powers of β are neglected in the expansion. (In treatments of relativity, β is a common symbol used to represent u/c or v/c.) In our case, $\beta = u/c$, so

$$\gamma = \frac{1}{\sqrt{1 - \dfrac{u^2}{c^2}}} = \left(1 - \frac{u^2}{c^2}\right)^{-1/2} \approx 1 + \frac{1}{2}\frac{u^2}{c^2}$$

Substituting this result into Equation 39.23 gives

$$K \approx \left[\left(1 + \frac{1}{2}\frac{u^2}{c^2}\right) - 1\right]mc^2 = \frac{1}{2}mu^2 \quad (\text{for } u/c \ll 1)$$

which is the classical expression for kinetic energy. A graph comparing the relativistic and nonrelativistic expressions is given in Figure 39.16. In the relativistic case, the particle speed never exceeds c, regardless of the kinetic energy. The two curves are in good agreement when $u \ll c$.

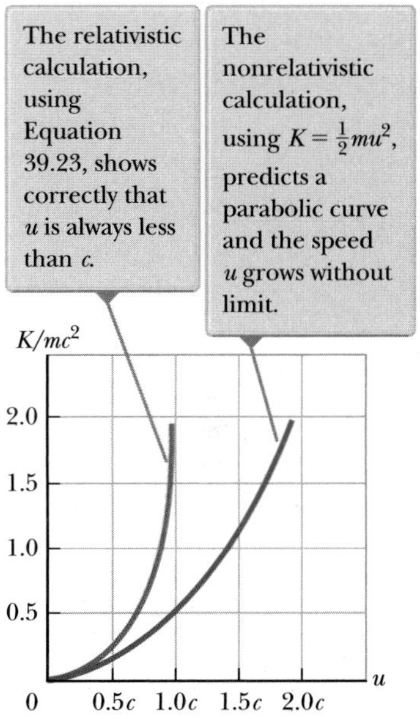

The constant term mc^2 in Equation 39.23, which is independent of the speed of the particle, is called the **rest energy** E_R of the particle:

$$E_R = mc^2 \tag{39.24}$$

Equation 39.24 shows that **mass is a form of energy,** where c^2 is simply a constant conversion factor. This expression also shows that a small mass corresponds to an enormous amount of energy, a concept fundamental to nuclear and elementary-particle physics.

The term γmc^2 in Equation 39.23, which depends on the particle speed, is the sum of the kinetic and rest energies. It is called the **total energy** E:

Total energy = kinetic energy + rest energy

$$E = K + mc^2 \tag{39.25}$$

or

$$E = \frac{mc^2}{\sqrt{1 - \dfrac{u^2}{c^2}}} = \gamma mc^2 \tag{39.26}$$

◀ **Total energy of a relativistic particle**

In many situations, the linear momentum or energy of a particle rather than its speed is measured. It is therefore useful to have an expression relating the total energy E to the relativistic linear momentum p, which is accomplished by using the expressions $E = \gamma mc^2$ and $p = \gamma mu$. By squaring these equations and subtracting, we can eliminate u (Problem 58 in Enhanced WebAssign). The result, after some algebra, is[6]

$$E^2 = p^2c^2 + (mc^2)^2 \tag{39.27}$$

◀ **Energy–momentum relationship for a relativistic particle**

When the particle is at rest, $p = 0$, so $E = E_R = mc^2$.

In Section 35.1, we introduced the concept of a particle of light, called a **photon.** For particles that have zero mass, such as photons, we set $m = 0$ in Equation 39.27 and find that

$$E = pc \tag{39.28}$$

The relativistic calculation, using Equation 39.23, shows correctly that u is always less than c.

The nonrelativistic calculation, using $K = \frac{1}{2}mu^2$, predicts a parabolic curve and the speed u grows without limit.

Figure 39.16 A graph comparing relativistic and nonrelativistic kinetic energy of a moving particle. The energies are plotted as a function of particle speed u.

---

[6]One way to remember this relationship is to draw a right triangle having a hypotenuse of length E and legs of lengths pc and mc^2.

This equation is an exact expression relating total energy and linear momentum for photons, which always travel at the speed of light (in vacuum).

Finally, because the mass m of a particle is independent of its motion, m must have the same value in all reference frames. For this reason, m is often called the **invariant mass.** On the other hand, because the total energy and linear momentum of a particle both depend on velocity, these quantities depend on the reference frame in which they are measured.

When dealing with subatomic particles, it is convenient to express their energy in electron volts (Section 25.1) because the particles are usually given this energy by acceleration through a potential difference. The conversion factor, as you recall from Equation 25.5, is

$$1 \text{ eV} = 1.602 \times 10^{-19} \text{ J}$$

For example, the mass of an electron is 9.109×10^{-31} kg. Hence, the rest energy of the electron is

$$m_e c^2 = (9.109 \times 10^{-31} \text{ kg})(2.998 \times 10^8 \text{ m/s})^2 = 8.187 \times 10^{-14} \text{ J}$$

$$= (8.187 \times 10^{-14} \text{ J})(1 \text{ eV}/1.602 \times 10^{-19} \text{ J}) = 0.511 \text{ MeV}$$

Quick Quiz 39.9 The following *pairs* of energies—particle 1: E, $2E$; particle 2: E, $3E$; particle 3: $2E$, $4E$—represent the rest energy and total energy of three different particles. Rank the particles from greatest to least according to their **(a)** mass, **(b)** kinetic energy, and **(c)** speed.

Example 39.9   The Energy of a Speedy Proton

(A) Find the rest energy of a proton in units of electron volts.

SOLUTION

Conceptualize Even if the proton is not moving, it has energy associated with its mass. If it moves, the proton possesses more energy, with the total energy being the sum of its rest energy and its kinetic energy.

Categorize The phrase "rest energy" suggests we must take a relativistic rather than a classical approach to this problem.

Analyze Use Equation 39.24 to find the rest energy:

$$E_R = m_p c^2 = (1.672\ 6 \times 10^{-27} \text{ kg})(2.998 \times 10^8 \text{ m/s})^2$$

$$= (1.504 \times 10^{-10} \text{ J})\left(\frac{1.00 \text{ eV}}{1.602 \times 10^{-19} \text{ J}}\right) = \boxed{938 \text{ MeV}}$$

(B) If the total energy of a proton is three times its rest energy, what is the speed of the proton?

SOLUTION

Use Equation 39.26 to relate the total energy of the proton to the rest energy:

$$E = 3m_p c^2 = \frac{m_p c^2}{\sqrt{1 - \dfrac{u^2}{c^2}}} \rightarrow 3 = \frac{1}{\sqrt{1 - \dfrac{u^2}{c^2}}}$$

Solve for u:

$$1 - \frac{u^2}{c^2} = \tfrac{1}{9} \rightarrow \frac{u^2}{c^2} = \tfrac{8}{9}$$

$$u = \frac{\sqrt{8}}{3}c = 0.943c = \boxed{2.83 \times 10^8 \text{ m/s}}$$

(C) Determine the kinetic energy of the proton in units of electron volts.

▶ **39.9** continued

SOLUTION

Use Equation 39.25 to find the kinetic energy of the proton:

$$K = E - m_p c^2 = 3m_p c^2 - m_p c^2 = 2m_p c^2$$
$$= 2(938 \text{ MeV}) = \boxed{1.88 \times 10^3 \text{ MeV}}$$

(D) What is the proton's momentum?

SOLUTION

Use Equation 39.27 to calculate the momentum:

$$E^2 = p^2 c^2 + (m_p c^2)^2 = (3m_p c^2)^2$$
$$p^2 c^2 = 9(m_p c^2)^2 - (m_p c^2)^2 = 8(m_p c^2)^2$$
$$p = \sqrt{8} \, \frac{m_p c^2}{c} = \sqrt{8} \, \frac{938 \text{ MeV}}{c} = \boxed{2.65 \times 10^3 \text{ MeV}/c}$$

Finalize The unit of momentum in part (D) is written MeV/c, which is a common unit in particle physics. For comparison, you might want to solve this example using classical equations.

WHAT IF?   In classical physics, if the momentum of a particle doubles, the kinetic energy increases by a factor of 4. What happens to the kinetic energy of the proton in this example if its momentum doubles?

Answer Based on what we have seen so far in relativity, it is likely you would predict that its kinetic energy does not increase by a factor of 4.

Find the new doubled momentum:

$$p_{\text{new}} = 2\left(\sqrt{8} \, \frac{m_p c^2}{c}\right) = 4\sqrt{2} \, \frac{m_p c^2}{c}$$

Use this result in Equation 39.27 to find the new total energy:

$$E_{\text{new}}^2 = p_{\text{new}}^2 c^2 + (m_p c^2)^2$$
$$E_{\text{new}}^2 = \left(4\sqrt{2} \, \frac{m_p c^2}{c}\right)^2 c^2 + (m_p c^2)^2 = 33(m_p c^2)^2$$
$$E_{\text{new}} = \sqrt{33} \, m_p c^2 = 5.7 m_p c^2$$

Use Equation 39.25 to find the new kinetic energy:

$$K_{\text{new}} = E_{\text{new}} - m_p c^2 = 5.7 m_p c^2 - m_p c^2 = 4.7 m_p c^2$$

This value is a little more than twice the kinetic energy found in part (C), not four times. In general, the factor by which the kinetic energy increases if the momentum doubles depends on the initial momentum, but it approaches 4 as the momentum approaches zero. In this latter situation, classical physics correctly describes the situation.

Equation 39.26, $E = \gamma mc^2$, represents the total energy of a particle. This important equation suggests that even when a particle is at rest ($\gamma = 1$), it still possesses enormous energy through its mass. The clearest experimental proof of the equivalence of mass and energy occurs in nuclear and elementary-particle interactions in which the conversion of mass into kinetic energy takes place. Consequently, we cannot use the principle of conservation of energy in relativistic situations as it was outlined in Chapter 8. We must modify the principle by including rest energy as another form of energy storage.

This concept is important in atomic and nuclear processes, in which the change in mass is a relatively large fraction of the initial mass. In a conventional nuclear reactor, for example, the uranium nucleus undergoes *fission*, a reaction that results in several lighter fragments having considerable kinetic energy. In the case of ^{235}U, which is used as fuel in nuclear power plants, the fragments are two lighter nuclei and a few neutrons. The total mass of the fragments is less than that of the ^{235}U by an amount Δm. The corresponding energy Δmc^2 associated with this mass difference is exactly equal to the sum of the kinetic energies of the fragments. The kinetic energy is absorbed as the fragments move through water, raising the internal energy of the water. This internal energy is used to produce steam for the generation of electricity.

Next, consider a basic *fusion* reaction in which two deuterium atoms combine to form one helium atom. The decrease in mass that results from the creation of one helium atom from two deuterium atoms is $\Delta m = 4.25 \times 10^{-29}$ kg. Hence, the corresponding energy that results from one fusion reaction is $\Delta mc^2 = 3.83 \times 10^{-12}$ J = 23.9 MeV. To appreciate the magnitude of this result, consider that if only 1 g of deuterium were converted to helium, the energy released would be on the order of 10^{12} J! In 2013's cost of electrical energy, this energy would be worth approximately $35 000. We shall present more details of these nuclear processes in Chapter 45 of the extended version of this textbook.

Example 39.10   Mass Change in a Radioactive Decay

The ^{216}Po nucleus is unstable and exhibits radioactivity (Chapter 44). It decays to ^{212}Pb by emitting an alpha particle, which is a helium nucleus, ^{4}He. The relevant masses, in atomic mass units (see Table A.1 in Appendix A), are $m_i = m(^{216}\text{Po}) = 216.001\ 915$ u and $m_f = m(^{212}\text{Pb}) + m(^{4}\text{He}) = 211.991\ 898$ u + 4.002 603 u.

(A) Find the mass change of the system in this decay.

SOLUTION

Conceptualize The initial system is the ^{216}Po nucleus. Imagine the mass of the system decreasing during the decay and transforming to kinetic energy of the alpha particle and the ^{212}Pb nucleus after the decay.

Categorize We use concepts discussed in this section, so we categorize this example as a substitution problem.

Calculate the change in mass using the mass values given in the problem statement.

$$\Delta m = 216.001\ 915\ \text{u} - (211.991\ 898\ \text{u} + 4.002\ 603\ \text{u})$$

$$= 0.007\ 414\ \text{u} = \boxed{1.23 \times 10^{-29}\ \text{kg}}$$

(B) Find the energy this mass change represents.

SOLUTION

Use Equation 39.24 to find the energy associated with this mass change:

$$E = \Delta mc^2 = (1.23 \times 10^{-29}\ \text{kg})(3.00 \times 10^8\ \text{m/s})^2$$

$$= 1.11 \times 10^{-12}\ \text{J} = \boxed{6.92\ \text{MeV}}$$

39.9 The General Theory of Relativity

Up to this point, we have sidestepped a curious puzzle. Mass has two seemingly different properties: a *gravitational attraction* for other masses and an *inertial* property that represents a resistance to acceleration. We first discussed these two attributes for mass in Section 5.5. To designate these two attributes, we use the subscripts g and i and write

$$\text{Gravitational property:} \qquad F_g = m_g g$$

$$\text{Inertial property:} \qquad \sum F = m_i a$$

The value for the gravitational constant G was chosen to make the magnitudes of m_g and m_i numerically equal. Regardless of how G is chosen, however, the strict proportionality of m_g and m_i has been established experimentally to an extremely high degree: a few parts in 10^{12}. Therefore, it appears that gravitational mass and inertial mass may indeed be exactly proportional.

Why, though? They seem to involve two entirely different concepts: a force of mutual gravitational attraction between two masses and the resistance of a single mass to being accelerated. This question, which puzzled Newton and many other physicists over the years, was answered by Einstein in 1916 when he published his theory of gravitation, known as the *general theory of relativity*. Because it is a mathematically complex theory, we offer merely a hint of its elegance and insight.

In Einstein's view, the dual behavior of mass was evidence for a very intimate and basic connection between the two behaviors. He pointed out that no mechanical experiment (such as dropping an object) could distinguish between the two situations illustrated in Figures 39.17a and 39.17b. In Figure 39.17a, a person standing in an elevator on the surface of a planet feels pressed into the floor due to the gravitational force. If he releases his briefcase, he observes it moving toward the floor with acceleration $\vec{g} = -g\hat{j}$. In Figure 39.17b, the person is in an elevator in empty space accelerating upward with $\vec{a}_{el} = +g\hat{j}$. The person feels pressed into the floor with the same force as in Figure 39.17a. If he releases his briefcase, he observes it moving toward the floor with acceleration g, exactly as in the previous situation. In each situation, an object released by the observer undergoes a downward acceleration of magnitude g relative to the floor. In Figure 39.17a, the person is at rest in an inertial frame in a gravitational field due to the planet. In Figure 39.17b, the person is in a noninertial frame accelerating in gravity-free space. Einstein's claim is that these two situations are completely equivalent.

Einstein carried this idea further and proposed that *no* experiment, mechanical or otherwise, could distinguish between the two situations. This extension to include all phenomena (not just mechanical ones) has interesting consequences. For example, suppose a light pulse is sent horizontally across the elevator as in Figure 39.17c, in which the elevator is accelerating upward in empty space. From the point of view of an observer in an inertial frame outside the elevator, the light travels in a straight line while the floor of the elevator accelerates upward. According to the observer on the elevator, however, the trajectory of the light pulse bends downward as the floor of the elevator (and the observer) accelerates upward. Therefore, based on the equality of parts (a) and (b) of the figure, Einstein proposed that a

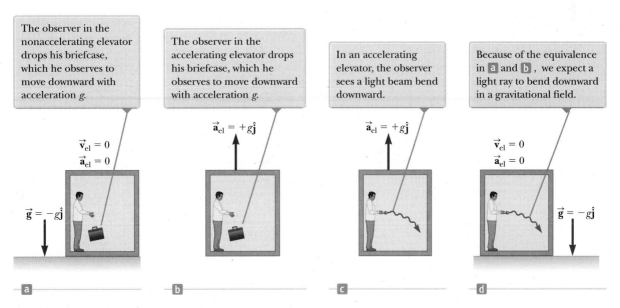

Figure 39.17 (a) The observer is at rest in an elevator in a uniform gravitational field $\vec{g} = -g\hat{j}$, directed downward. (b) The observer is in a region where gravity is negligible, but the elevator moves upward with an acceleration $\vec{a}_{el} = +g\hat{j}$. According to Einstein, the frames of reference in (a) and (b) are equivalent in every way. No local experiment can distinguish any difference between the two frames. (c) An observer watches a beam of light in an accelerating elevator. (d) Einstein's prediction of the behavior of a beam of light in a gravitational field.

beam of light should also be bent downward by a gravitational field as in Figure 39.17d. Experiments have verified the effect, although the bending is small. A laser aimed at the horizon falls less than 1 cm after traveling 6 000 km. (No such bending is predicted in Newton's theory of gravitation.)

Einstein's **general theory of relativity** has two postulates:

- All the laws of nature have the same form for observers in any frame of reference, whether accelerated or not.
- In the vicinity of any point, a gravitational field is equivalent to an accelerated frame of reference in gravity-free space (the **principle of equivalence**).

One interesting effect predicted by the general theory is that time is altered by gravity. A clock in the presence of gravity runs slower than one located where gravity is negligible. Consequently, the frequencies of radiation emitted by atoms in the presence of a strong gravitational field are *redshifted* to lower frequencies when compared with the same emissions in the presence of a weak field. This gravitational redshift has been detected in spectral lines emitted by atoms in massive stars. It has also been verified on the Earth by comparing the frequencies of gamma rays emitted from nuclei separated vertically by about 20 m.

The second postulate suggests a gravitational field may be "transformed away" at any point if we choose an appropriate accelerated frame of reference, a freely falling one. Einstein developed an ingenious method of describing the acceleration necessary to make the gravitational field "disappear." He specified a concept, the *curvature of space–time,* that describes the gravitational effect at every point. In fact, the curvature of space–time completely replaces Newton's gravitational theory. According to Einstein, there is no such thing as a gravitational force. Rather, the presence of a mass causes a curvature of space–time in the vicinity of the mass, and this curvature dictates the space–time path that all freely moving objects must follow.

As an example of the effects of curved space–time, imagine two travelers moving on parallel paths a few meters apart on the surface of the Earth and maintaining an exact northward heading along two longitude lines. As they observe each other near the equator, they will claim that their paths are exactly parallel. As they approach the North Pole, however, they notice that they are moving closer together and will meet at the North Pole. Therefore, they claim that they moved along parallel paths, but moved toward each other, *as if there were an attractive force between them.* The travelers make this conclusion based on their everyday experience of moving on flat surfaces. From our mental representation, however, we realize they are walking on a curved surface, and it is the geometry of the curved surface, rather than an attractive force, that causes them to converge. In a similar way, general relativity replaces the notion of forces with the movement of objects through curved space–time.

One prediction of the general theory of relativity is that a light ray passing near the Sun should be deflected in the curved space–time created by the Sun's mass. This prediction was confirmed when astronomers detected the bending of starlight near the Sun during a total solar eclipse that occurred shortly after World War I (Fig. 39.18). When this discovery was announced, Einstein became an international celebrity.

Courtesy of NASA

Einstein's cross. The four outer bright spots are images of the same galaxy that have been bent around a massive object located between the galaxy and the Earth. The massive object acts like a lens, causing the rays of light that were diverging from the distant galaxy to converge on the Earth. (If the intervening massive object had a uniform mass distribution, we would see a bright ring instead of four spots.)

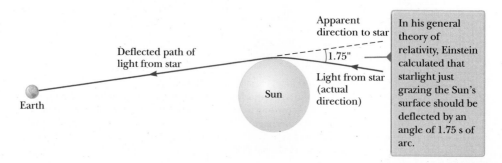

Figure 39.18 Deflection of starlight passing near the Sun. Because of this effect, the Sun or some other remote object can act as a *gravitational lens.*

Apparent direction to star

Deflected path of light from star

1.75"

Light from star (actual direction)

Earth

Sun

In his general theory of relativity, Einstein calculated that starlight just grazing the Sun's surface should be deflected by an angle of 1.75 s of arc.

If the concentration of mass becomes very great as is believed to occur when a large star exhausts its nuclear fuel and collapses to a very small volume, a **black hole** may form as discussed in Chapter 13. Here, the curvature of space–time is so extreme that within a certain distance from the center of the black hole all matter and light become trapped as discussed in Section 13.6.

Summary

Definitions

The relativistic expression for the **linear momentum** of a particle moving with a velocity $\vec{\mathbf{u}}$ is

$$\vec{\mathbf{p}} \equiv \frac{m\vec{\mathbf{u}}}{\sqrt{1 - \dfrac{u^2}{c^2}}} = \gamma m\vec{\mathbf{u}} \tag{39.19}$$

The relativistic force $\vec{\mathbf{F}}$ acting on a particle whose linear momentum is $\vec{\mathbf{p}}$ is defined as

$$\vec{\mathbf{F}} \equiv \frac{d\vec{\mathbf{p}}}{dt} \tag{39.20}$$

Concepts and Principles

The two basic postulates of the special theory of relativity are as follows:

- The laws of physics must be the same in all inertial reference frames.
- The speed of light in vacuum has the same value, $c = 3.00 \times 10^8$ m/s, in all inertial frames, regardless of the velocity of the observer or the velocity of the source emitting the light.

Three consequences of the special theory of relativity are as follows:

- Events that are measured to be simultaneous for one observer are not necessarily measured to be simultaneous for another observer who is in motion relative to the first.
- Clocks in motion relative to an observer are measured to run slower by a factor $\gamma = (1 - v^2/c^2)^{-1/2}$. This phenomenon is known as **time dilation.**
- The lengths of objects in motion are measured to be shorter in the direction of motion by a factor $1/\gamma = (1 - v^2/c^2)^{1/2}$. This phenomenon is known as **length contraction.**

To satisfy the postulates of special relativity, the Galilean transformation equations must be replaced by the **Lorentz transformation equations:**

$$x' = \gamma(x - vt) \quad y' = y \quad z' = z \quad t' = \gamma\left(t - \frac{v}{c^2}x\right) \tag{39.11}$$

where $\gamma = (1 - v^2/c^2)^{-1/2}$ and the S′ frame moves in the x direction at speed v relative to the S frame.

The relativistic form of the **Lorentz velocity transformation equation** is

$$u'_x = \frac{u_x - v}{1 - \dfrac{u_x v}{c^2}} \tag{39.16}$$

where u'_x is the x component of the velocity of an object as measured in the S′ frame and u_x is its component as measured in the S frame.

The relativistic expression for the **kinetic energy** of a particle is

$$K = \frac{mc^2}{\sqrt{1 - \dfrac{u^2}{c^2}}} - mc^2 = (\gamma - 1)mc^2 \tag{39.23}$$

The constant term mc^2 in Equation 39.23 is called the **rest energy** E_R of the particle:

$$E_R = mc^2 \tag{39.24}$$

The **total energy** E of a particle is given by

$$E = \frac{mc^2}{\sqrt{1 - \dfrac{u^2}{c^2}}} = \gamma mc^2 \tag{39.26}$$

The relativistic linear momentum of a particle is related to its total energy through the equation

$$E^2 = p^2c^2 + (mc^2)^2 \tag{39.27}$$

1. **(i)** Does the speed of an electron have an upper limit? (a) yes, the speed of light c (b) yes, with another value (c) no **(ii)** Does the magnitude of an electron's momentum have an upper limit? (a) yes, $m_e c$ (b) yes, with another value (c) no **(iii)** Does the electron's kinetic energy have an upper limit? (a) yes, $m_e c^2$ (b) yes, $\frac{1}{2} m_e c^2$ (c) yes, with another value (d) no

2. A spacecraft zooms past the Earth with a constant velocity. An observer on the Earth measures that an undamaged clock on the spacecraft is ticking at one-third the rate of an identical clock on the Earth. What does an observer on the spacecraft measure about the Earth-based clock's ticking rate? (a) It runs more than three times faster than his own clock. (b) It runs three times faster than his own. (c) It runs at the same rate as his own. (d) It runs at one-third the rate of his own. (e) It runs at less than one-third the rate of his own.

3. As a car heads down a highway traveling at a speed v away from a ground observer, which of the following statements are true about the measured speed of the light beam from the car's headlights? More than one statement may be correct. (a) The ground observer measures the light speed to be $c + v$. (b) The driver measures the light speed to be c. (c) The ground observer measures the light speed to be c. (d) The driver measures the light speed to be $c - v$. (e) The ground observer measures the light speed to be $c - v$.

4. A spacecraft built in the shape of a sphere moves past an observer on the Earth with a speed of $0.500c$. What shape does the observer measure for the spacecraft as it goes by? (a) a sphere (b) a cigar shape, elongated along the direction of motion (c) a round pillow shape, flattened along the direction of motion (d) a conical shape, pointing in the direction of motion

5. An astronaut is traveling in a spacecraft in outer space in a straight line at a constant speed of $0.500c$. Which of the following effects would she experience? (a) She would feel heavier. (b) She would find it harder to breathe. (c) Her heart rate would change. (d) Some

of the dimensions of her spacecraft would be shorter. (e) None of those answers is correct.

6. You measure the volume of a cube at rest to be V_0. You then measure the volume of the same cube as it passes you in a direction parallel to one side of the cube. The speed of the cube is $0.980c$, so $\gamma \approx 5$. Is the volume you measure close to (a) $V_0/25$, (b) $V_0/5$, (c) V_0, (d) $5V_0$, or (e) $25V_0$?

7. Two identical clocks are set side by side and synchronized. One remains on the Earth. The other is put into orbit around the Earth moving rapidly toward the east. **(i)** As measured by an observer on the Earth, does the orbiting clock (a) run faster than the Earth-based clock, (b) run at the same rate, or (c) run slower? **(ii)** The orbiting clock is returned to its original location and brought to rest relative to the Earth-based clock. Thereafter, what happens? (a) Its reading lags farther and farther behind the Earth-based clock. (b) It lags behind the Earth-based clock by a constant amount. (c) It is synchronous with the Earth-based clock. (d) It is ahead of the Earth-based clock by a constant amount. (e) It gets farther and farther ahead of the Earth-based clock.

8. The following three particles all have the same total energy E: (a) a photon, (b) a proton, and (c) an electron. Rank the magnitudes of the particles' momenta from greatest to smallest.

9. Which of the following statements are fundamental postulates of the special theory of relativity? More than one statement may be correct. (a) Light moves through a substance called the ether. (b) The speed of light depends on the inertial reference frame in which it is measured. (c) The laws of physics depend on the inertial reference frame in which they are used. (d) The laws of physics are the same in all inertial reference frames. (e) The speed of light is independent of the inertial reference frame in which it is measured.

10. A distant astronomical object (a quasar) is moving away from us at half the speed of light. What is the speed of the light we receive from this quasar? (a) greater than c (b) c (c) between $c/2$ and c (d) $c/2$ (e) between 0 and $c/2$

1. In several cases, a nearby star has been found to have a large planet orbiting about it, although light from the planet could not be seen separately from the starlight. Using the ideas of a system rotating about its center of mass and of the Doppler shift for light, explain how an astronomer could determine the presence of the invisible planet.

2. Explain why, when defining the length of a rod, it is necessary to specify that the positions of the ends of the rod are to be measured simultaneously.

3. A train is approaching you at very high speed as you stand next to the tracks. Just as an observer on the train passes you, you both begin to play the same recorded

version of a Beethoven symphony on identical iPods. (a) According to you, whose iPod finishes the symphony first? (b) **What If?** According to the observer on the train, whose iPod finishes the symphony first? (c) Whose iPod actually finishes the symphony first?

4. List three ways our day-to-day lives would change if the speed of light were only 50 m/s.

5. How is acceleration indicated on a space–time graph?

6. (a) "Newtonian mechanics correctly describes objects moving at ordinary speeds, and relativistic mechanics correctly describes objects moving very fast." (b) "Relativistic mechanics must make a smooth transition as it reduces to Newtonian mechanics in a case in which

the speed of an object becomes small compared with the speed of light." Argue for or against statements (a) and (b).

7. The speed of light in water is 230 Mm/s. Suppose an electron is moving through water at 250 Mm/s. Does that violate the principle of relativity? Explain.

8. A particle is moving at a speed less than $c/2$. If the speed of the particle is doubled, what happens to its momentum?

9. Give a physical argument that shows it is impossible to accelerate an object of mass m to the speed of light, even with a continuous force acting on it.

10. Explain how the Doppler effect with microwaves is used to determine the speed of an automobile.

11. It is said that Einstein, in his teenage years, asked the question, "What would I see in a mirror if I carried it in my hands and ran at a speed near that of light?" How would you answer this question?

12. (i) An object is placed at a position $p > f$ from a concave mirror as shown in Figure CQ39.12a, where f is the focal length of the mirror. In a finite time interval, the object is moved to the right to a position at the focal point F of the mirror. Show that the image of the object moves at a speed greater than the speed of light. (ii) A laser pointer is suspended in a horizontal plane and set into rapid rotation as shown in Figure CQ39.12b. Show that the spot of light it produces on a distant screen can move across the screen at a speed greater than the speed of light. (If you carry out this experiment, make sure the direct laser light cannot enter a person's eyes.) (iii) Argue that the experiments in parts (i) and (ii) do not invalidate the principle that no material, no energy, and no information can move faster than light moves in a vacuum.

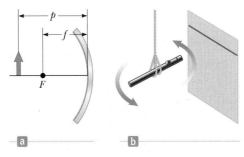

Figure CQ39.12

13. With regard to reference frames, how does general relativity differ from special relativity?

14. Two identical clocks are in the same house, one upstairs in a bedroom and the other downstairs in the kitchen. Which clock runs slower? Explain.

Problems available in **WebAssign** Access end-of-chapter problems online at www.webassign.net

Section 39.1 The Principle of Galilean Relativity
Problems 1–4

Section 39.2 The Michelson–Morley Experiment
Section 39.3 Einstein's Principle of Relativity
Section 39.4 Consequences of the Special Theory of Relativity
Problems 5–26

Section 39.5 The Lorentz Transformation Equations
Problems 27–31

Section 39.6 The Lorentz Velocity Transformation Equations
Problems 32–35

Section 39.7 Relativistic Linear Momentum
Problems 36–43

Section 39.8 Relativistic Energy
Problems 44–64

Section 39.9 The General Theory of Relativity
Problem 65

Additional Problems
Problems 66–87

Challenge Problems
Problem 88–91

Solutions to the following Problems are available in the Student Solutions Manual/Study Guide:
39.2, 39.15, 39.19, 39.25, 39.32, 39.33, 39.43, 39.47, 39.49, 39.51, 39.55, 39.57, 39.58, 39.61, 39.67, 39.77, 39.78, and 39.88

List of Enhanced Problems

Problem Number	Targeted Feedback in Enhanced WebAssign	Analysis Model Tutorial in Enhanced WebAssign	Master It in Enhanced WebAssign	Watch It in Enhanced WebAssign
39.4		✓		
39.7	✓			✓
39.9	✓			✓
39.10	✓			
39.12	✓			✓
39.16	✓		✓	
39.19	✓	✓	✓	
39.27	✓			✓
39.32	✓		✓	
39.35	✓		✓	
39.36	✓			✓
39.39	✓			
39.42			✓	
39.43	✓		✓	
39.49		✓	✓	
39.53	✓			✓
39.54	✓			✓
39.61			✓	
39.67			✓	
39.71	✓		✓	
39.73	✓		✓	
39.77			✓	
39.78			✓	
39.85		✓		

CHAPTER 40

Introduction to Quantum Physics

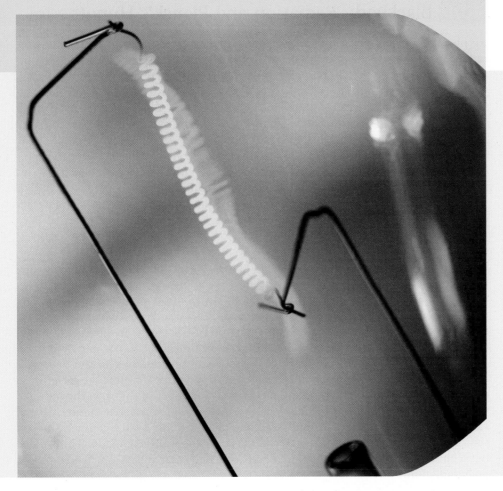

In Chapter 39, we discussed that Newtonian mechanics must be replaced by Einstein's special theory of relativity when dealing with particle speeds comparable to the speed of light. As the 20th century progressed, many experimental and theoretical problems were resolved by the special theory of relativity. For many other problems, however, neither relativity nor classical physics could provide a theoretical answer. Attempts to apply the laws of classical physics to explain the behavior of matter on the atomic scale were consistently unsuccessful. For example, the emission of discrete wavelengths of light from atoms in a high-temperature gas could not be explained within the framework of classical physics.

As physicists sought new ways to solve these puzzles, another revolution took place in physics between 1900 and 1930. A new theory called *quantum mechanics* was highly successful in explaining the behavior of particles of microscopic size. Like the special theory of relativity, the quantum theory requires a modification of our ideas concerning the physical world.

The first explanation of a phenomenon using quantum theory was introduced by Max Planck. Many subsequent mathematical developments and interpretations were made by a number of distinguished physicists, including Einstein, Bohr, de Broglie, Schrödinger, and

This lightbulb filament glows with an orange color. Why? Classical physics is unable to explain the experimentally observed wavelength distribution of electromagnetic radiation from a hot object. A theory proposed in 1900 and describing the radiation from such objects represents the dawn of quantum physics. *(Steve Cole/Getty Images)*

967

Heisenberg. Despite the great success of the quantum theory, Einstein frequently played the role of its critic, especially with regard to the manner in which the theory was interpreted.

Because an extensive study of quantum theory is beyond the scope of this book, this chapter is simply an introduction to its underlying principles.

40.1 Blackbody Radiation and Planck's Hypothesis

An object at any temperature emits electromagnetic waves in the form of **thermal radiation** from its surface as discussed in Section 20.7. The characteristics of this radiation depend on the temperature and properties of the object's surface. Careful study shows that the radiation consists of a continuous distribution of wavelengths from all portions of the electromagnetic spectrum. If the object is at room temperature, the wavelengths of thermal radiation are mainly in the infrared region and hence the radiation is not detected by the human eye. As the surface temperature of the object increases, the object eventually begins to glow visibly red, like the coils of a toaster. At sufficiently high temperatures, the glowing object appears white, as in the hot tungsten filament of an incandescent lightbulb.

From a classical viewpoint, thermal radiation originates from accelerated charged particles in the atoms near the surface of the object; those charged particles emit radiation much as small antennas do. The thermally agitated particles can have a distribution of energies, which accounts for the continuous spectrum of radiation emitted by the object. By the end of the 19th century, however, it became apparent that the classical theory of thermal radiation was inadequate. The basic problem was in understanding the observed distribution of wavelengths in the radiation emitted by a black body. As defined in Section 20.7, a **black body** is an ideal system that absorbs all radiation incident on it. The electromagnetic radiation emitted by the black body is called **blackbody radiation.**

A good approximation of a black body is a small hole leading to the inside of a hollow object as shown in Figure 40.1. Any radiation incident on the hole from outside the cavity enters the hole and is reflected a number of times on the interior walls of the cavity; hence, the hole acts as a perfect absorber. The nature of the radiation leaving the cavity through the hole depends only on the temperature of the cavity walls and not on the material of which the walls are made. The spaces between lumps of hot charcoal (Fig. 40.2) emit light that is very much like blackbody radiation.

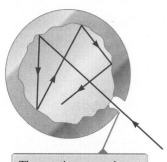

The opening to a cavity inside a hollow object is a good approximation of a black body: the hole acts as a perfect absorber.

Figure 40.1 A physical model of a black body.

The radiation emitted by oscillators in the cavity walls in Figure 40.1 experiences boundary conditions and can be analyzed using the waves under boundary conditions analysis model. As the radiation reflects from the cavity's walls, standing electromagnetic waves are established within the three-dimensional interior of the cavity. Many standing-wave modes are possible, and the distribution of the energy in the cavity among these modes determines the wavelength distribution of the radiation leaving the cavity through the hole.

The wavelength distribution of radiation from cavities was studied experimentally in the late 19th century. Figure 40.3 shows how the intensity of blackbody radiation varies with temperature and wavelength. The following two consistent experimental findings were seen as especially significant:

Figure 40.2 The glow emanating from the spaces between these hot charcoal briquettes is, to a close approximation, blackbody radiation. The color of the light depends only on the temperature of the briquettes.

1. **The total power of the emitted radiation increases with temperature.**
   We discussed this behavior briefly in Chapter 20, where we introduced Stefan's law:

Stefan's law ▶

$$P = \sigma A e T^4 \tag{40.1}$$

where P is the power in watts radiated at all wavelengths from the surface of an object, $\sigma = 5.670 \times 10^{-8}$ W/m$^2 \cdot$ K^4 is the Stefan–Boltzmann constant, A is the surface area of the object in square meters, e is the emissivity of the surface, and T is the surface temperature in kelvins. For a black body, the emissivity is $e = 1$ exactly.

2. **The peak of the wavelength distribution shifts to shorter wavelengths as the temperature increases.** This behavior is described by the following relationship, called **Wien's displacement law:**

$$\lambda_{\max} T = 2.898 \times 10^{-3} \text{ m} \cdot \text{K}$$ (40.2)

◀ **Wien's displacement law**

where $\lambda_{\max}$ is the wavelength at which the curve peaks and T is the absolute temperature of the surface of the object emitting the radiation. The wavelength at the curve's peak is inversely proportional to the absolute temperature; that is, as the temperature increases, the peak is "displaced" to shorter wavelengths (Fig. 40.3).

Wien's displacement law is consistent with the behavior of the object mentioned at the beginning of this section. At room temperature, the object does not appear to glow because the peak is in the infrared region of the electromagnetic spectrum. At higher temperatures, it glows red because the peak is in the near infrared with some radiation at the red end of the visible spectrum, and at still higher temperatures, it glows white because the peak is in the visible so that all colors are emitted.

Ⓠuick Quiz 40.1 Figure 40.4 shows two stars in the constellation Orion. Betelgeuse appears to glow red, whereas Rigel looks blue in color. Which star has a higher surface temperature? **(a)** Betelgeuse **(b)** Rigel **(c)** both the same **(d)** impossible to determine

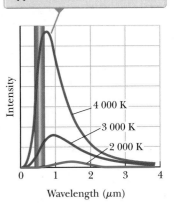

The 4 000-K curve has a peak near the visible range. This curve represents an object that would glow with a yellowish-white appearance.

Figure 40.3 Intensity of blackbody radiation versus wavelength at three temperatures. The visible range of wavelengths is between 0.4 μm and 0.7 μm. At approximately 6 000 K, the peak is in the center of the visible wavelengths and the object appears white.

Figure 40.4 (Quick Quiz 40.1) Which star is hotter, Betelgeuse or Rigel?

A successful theory for blackbody radiation must predict the shape of the curves in Figure 40.3, the temperature dependence expressed in Stefan's law, and the shift of the peak with temperature described by Wien's displacement law. Early attempts to use classical ideas to explain the shapes of the curves in Figure 40.3 failed.

Let's consider one of these early attempts. To describe the distribution of energy from a black body, we define $I(\lambda,T)\, d\lambda$ to be the intensity, or power per unit area, emitted in the wavelength interval $d\lambda$. The result of a calculation based on a classical theory of blackbody radiation known as the **Rayleigh–Jeans law** is

$$I(\lambda,T) = \frac{2\pi c k_B T}{\lambda^4}$$ (40.3)

◀ **Rayleigh–Jeans law**

where k_B is Boltzmann's constant. The black body is modeled as the hole leading into a cavity (Fig. 40.1), resulting in many modes of oscillation of the electromagnetic field caused by accelerated charges in the cavity walls and the emission of electromagnetic waves at all wavelengths. In the classical theory used to derive

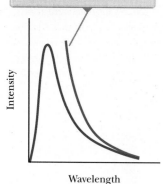

The classical theory (red-brown curve) shows intensity growing without bound for short wavelengths, unlike the experimental data (blue curve).

Figure 40.5 Comparison of experimental results and the curve predicted by the Rayleigh–Jeans law for the distribution of blackbody radiation.

Max Planck
German Physicist (1858–1947)
Planck introduced the concept of "quantum of action" (Planck's constant, h) in an attempt to explain the spectral distribution of blackbody radiation, which laid the foundations for quantum theory. In 1918, he was awarded the Nobel Prize in Physics for this discovery of the quantized nature of energy.

Equation 40.3, the average energy for each wavelength of the standing-wave modes is assumed to be proportional to $k_B T$, based on the theorem of equipartition of energy discussed in Section 21.1.

An experimental plot of the blackbody radiation spectrum, together with the theoretical prediction of the Rayleigh–Jeans law, is shown in Figure 40.5. At long wavelengths, the Rayleigh–Jeans law is in reasonable agreement with experimental data, but at short wavelengths, major disagreement is apparent.

As λ approaches zero, the function $I(\lambda,T)$ given by Equation 40.3 approaches infinity. Hence, according to classical theory, not only should short wavelengths predominate in a blackbody spectrum, but also the energy emitted by any black body should become infinite in the limit of zero wavelength. In contrast to this prediction, the experimental data plotted in Figure 40.5 show that as λ approaches zero, $I(\lambda,T)$ also approaches zero. This mismatch of theory and experiment was so disconcerting that scientists called it the *ultraviolet catastrophe*. (This "catastrophe"—infinite energy—occurs as the wavelength approaches zero; the word *ultraviolet* was applied because ultraviolet wavelengths are short.)

In 1900, Max Planck developed a theory of blackbody radiation that leads to an equation for $I(\lambda,T)$ that is in complete agreement with experimental results at all wavelengths. In discussing this theory, we use the outline of properties of structural models introduced in Chapter 21:

1. *Physical components:*
   Planck assumed the cavity radiation came from atomic oscillators in the cavity walls in Figure 40.1.
2. *Behavior of the components:*
   (a) The energy of an oscillator can have only certain *discrete* values E_n:

$$E_n = nhf \tag{40.4}$$

   where n is a positive integer called a **quantum number,**[1] f is the oscillator's frequency, and h is a parameter Planck introduced that is now called **Planck's constant.** Because the energy of each oscillator can have only discrete values given by Equation 40.4, we say the energy is **quantized.** Each discrete energy value corresponds to a different **quantum state,** represented by the quantum number n. When the oscillator is in the $n = 1$ quantum state, its energy is hf; when it is in the $n = 2$ quantum state, its energy is $2hf$; and so on.

   (b) The oscillators emit or absorb energy when making a transition from one quantum state to another. The entire energy difference between the initial and final states in the transition is emitted or absorbed as a single quantum of radiation. If the transition is from one state to a lower adjacent state—say, from the $n = 3$ state to the $n = 2$ state—Equation 40.4 shows that the amount of energy emitted by the oscillator and carried by the quantum of radiation is

$$E = hf \tag{40.5}$$

According to property 2(b), an oscillator emits or absorbs energy only when it changes quantum states. If it remains in one quantum state, no energy is absorbed or emitted. Figure 40.6 is an **energy-level diagram** showing the quantized energy levels and allowed transitions proposed by Planck. This important semigraphical representation is used often in quantum physics.[2] The vertical axis is linear in energy, and the allowed energy levels are represented as horizontal lines. The quantized system can have only the energies represented by the horizontal lines.

---

[1]A quantum number is generally an integer (although half-integer quantum numbers can occur) that describes an allowed state of a system, such as the values of n describing the normal modes of oscillation of a string fixed at both ends, as discussed in Section 18.3.

[2]We first saw an energy-level diagram in Section 21.3.

The key point in Planck's theory is the radical assumption of quantized energy states. This development—a clear deviation from classical physics—marked the birth of the quantum theory.

In the Rayleigh–Jeans model, the average energy associated with a particular wavelength of standing waves in the cavity is the same for all wavelengths and is equal to $k_B T$. Planck used the same classical ideas as in the Rayleigh–Jeans model to arrive at the energy density as a product of constants and the average energy for a given wavelength, but the average energy is not given by the equipartition theorem. A wave's average energy is the average energy difference between levels of the oscillator, *weighted according to the probability of the wave being emitted*. This weighting is based on the occupation of higher-energy states as described by the Boltzmann distribution law, which was discussed in Section 21.5. According to this law, the probability of a state being occupied is proportional to the factor $e^{-E/k_B T}$, where E is the energy of the state.

At low frequencies (long wavelengths), according to property 2(a), the energy levels are close together as on the right in Figure 40.7, and many of the energy states are excited because the Boltzmann factor $e^{-E/k_B T}$ is relatively large for these states. Therefore, there are many contributions to the outgoing radiation, although each contribution has very low energy. Now, consider high-frequency radiation, that is, radiation with short wavelength. To obtain this radiation, the allowed energies are very far apart as on the left in Figure 40.7. The probability of thermal agitation exciting these high energy levels is small because of the small value of the Boltzmann factor for large values of E. At high frequencies, the low probability of excitation results in very little contribution to the total energy, even though each quantum is of large energy. This low probability "turns the curve over" and brings it down to zero again at short wavelengths.

Using this approach, Planck generated a theoretical expression for the wavelength distribution that agreed remarkably well with the experimental curves in Figure 40.3:

$$I(\lambda, T) = \frac{2\pi h c^2}{\lambda^5 (e^{hc/\lambda k_B T} - 1)} \qquad (40.6)$$

Pitfall Prevention 40.2

n **Is Again an Integer** In the preceding chapters on optics, we used the symbol n for the index of refraction, which was not an integer. Here we are again using n as we did in Chapter 18 to indicate the standing-wave mode on a string or in an air column. In quantum physics, n is often used as an integer quantum number to identify a particular quantum state of a system.

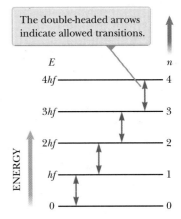

Figure 40.6 Allowed energy levels for an oscillator with frequency f.

◀ **Planck's wavelength distribution function**

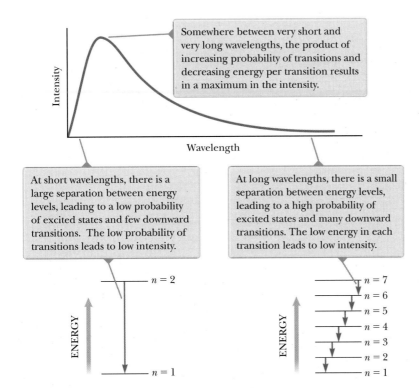

Figure 40.7 In Planck's model, the average energy associated with a given wavelength is the product of the energy of a transition and a factor related to the probability of the transition occurring.

This function includes the parameter h, which Planck adjusted so that his curve matched the experimental data at all wavelengths. The value of this parameter is found to be independent of the material of which the black body is made and independent of the temperature; it is a fundamental constant of nature. The value of h, Planck's constant, which was first introduced in Chapter 35, is

Planck's constant ▶

$$h = 6.626 \times 10^{-34} \, \text{J} \cdot \text{s} \tag{40.7}$$

At long wavelengths, Equation 40.6 reduces to the Rayleigh–Jeans expression, Equation 40.3 (see Problem 14 in Enhanced WebAssign), and at short wavelengths, it predicts an exponential decrease in $I(\lambda, T)$ with decreasing wavelength, in agreement with experimental results.

When Planck presented his theory, most scientists (including Planck!) did not consider the quantum concept to be realistic. They believed it was a mathematical trick that happened to predict the correct results. Hence, Planck and others continued to search for a more "rational" explanation of blackbody radiation. Subsequent developments, however, showed that a theory based on the quantum concept (rather than on classical concepts) had to be used to explain not only blackbody radiation but also a number of other phenomena at the atomic level.

In 1905, Einstein rederived Planck's results by assuming the oscillations of the electromagnetic field were themselves quantized. In other words, he proposed that quantization is a fundamental property of light and other electromagnetic radiation, which led to the concept of photons as shall be discussed in Section 40.2. Critical to the success of the quantum or photon theory was the relation between energy and frequency, which classical theory completely failed to predict.

You may have had your body temperature measured at the doctor's office by an *ear thermometer*, which can read your temperature very quickly (Fig. 40.8). In a fraction of a second, this type of thermometer measures the amount of infrared radiation emitted by the eardrum. It then converts the amount of radiation into a temperature reading. This thermometer is very sensitive because temperature is raised to the fourth power in Stefan's law. Suppose you have a fever 1°C above normal. Because absolute temperatures are found by adding 273 to Celsius temperatures, the ratio of your fever temperature to normal body temperature of 37°C is

$$\frac{T_{\text{fever}}}{T_{\text{normal}}} = \frac{38°\text{C} + 273°\text{C}}{37°\text{C} + 273°\text{C}} = 1.003\ 2$$

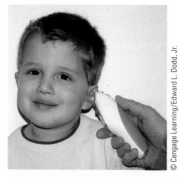

Figure 40.8 An ear thermometer measures a patient's temperature by detecting the intensity of infrared radiation leaving the eardrum.

© Cengage Learning/Edward L. Dodd, Jr.

which is only a 0.32% increase in temperature. The increase in radiated power, however, is proportional to the fourth power of temperature, so

$$\frac{P_{\text{fever}}}{P_{\text{normal}}} = \left(\frac{38°\text{C} + 273°\text{C}}{37°\text{C} + 273°\text{C}}\right)^4 = 1.013$$

The result is a 1.3% increase in radiated power, which is easily measured by modern infrared radiation sensors.

Example 40.1   Thermal Radiation from Different Objects

(A) Find the peak wavelength of the blackbody radiation emitted by the human body when the skin temperature is 35°C.

SOLUTION

Conceptualize Thermal radiation is emitted from the surface of any object. The peak wavelength is related to the surface temperature through Wien's displacement law (Eq. 40.2).

Categorize We evaluate results using an equation developed in this section, so we categorize this example as a substitution problem.

▶ **40.1** continued

Solve Equation 40.2 for λ_{max}:

$$(1) \quad \lambda_{max} = \frac{2.898 \times 10^{-3} \text{ m} \cdot \text{K}}{T}$$

Substitute the surface temperature:

$$\lambda_{max} = \frac{2.898 \times 10^{-3} \text{ m} \cdot \text{K}}{308 \text{ K}} = \boxed{9.41 \ \mu\text{m}}$$

This radiation is in the infrared region of the spectrum and is invisible to the human eye. Some animals (pit vipers, for instance) are able to detect radiation of this wavelength and therefore can locate warm-blooded prey even in the dark.

(B) Find the peak wavelength of the blackbody radiation emitted by the tungsten filament of a lightbulb, which operates at 2 000 K.

SOLUTION

Substitute the filament temperature into Equation (1):

$$\lambda_{max} = \frac{2.898 \times 10^{-3} \text{ m} \cdot \text{K}}{2 \ 000 \text{ K}} = \boxed{1.45 \ \mu\text{m}}$$

This radiation is also in the infrared, meaning that most of the energy emitted by a lightbulb is not visible to us.

(C) Find the peak wavelength of the blackbody radiation emitted by the Sun, which has a surface temperature of approximately 5 800 K.

SOLUTION

Substitute the surface temperature into Equation (1):

$$\lambda_{max} = \frac{2.898 \times 10^{-3} \text{ m} \cdot \text{K}}{5 \ 800 \text{ K}} = \boxed{0.500 \ \mu\text{m}}$$

This radiation is near the center of the visible spectrum, near the color of a yellow-green tennis ball. Because it is the most prevalent color in sunlight, our eyes have evolved to be most sensitive to light of approximately this wavelength.

Example 40.2   **The Quantized Oscillator**   AM

A 2.00-kg block is attached to a massless spring that has a force constant of $k = 25.0$ N/m. The spring is stretched 0.400 m from its equilibrium position and released from rest.

(A) Find the total energy of the system and the frequency of oscillation according to classical calculations.

SOLUTION

Conceptualize We understand the details of the block's motion from our study of simple harmonic motion in Chapter 15. Review that material if you need to.

Categorize The phrase "according to classical calculations" tells us to categorize this part of the problem as a classical analysis of the oscillator. We model the block as a *particle in simple harmonic motion*.

....

Analyze Based on the way the block is set into motion, its amplitude is 0.400 m.

Evaluate the total energy of the block–spring system using Equation 15.21:

$$E = \tfrac{1}{2}kA^2 = \tfrac{1}{2}(25.0 \text{ N/m})(0.400 \text{ m})^2 = \boxed{2.00 \text{ J}}$$

Evaluate the frequency of oscillation from Equation 15.14:

$$f = \frac{1}{2\pi}\sqrt{\frac{k}{m}} = \frac{1}{2\pi}\sqrt{\frac{25.0 \text{ N/m}}{2.00 \text{ kg}}} = \boxed{0.563 \text{ Hz}}$$

(B) Assuming the energy of the oscillator is quantized, find the quantum number n for the system oscillating with this amplitude.

continued

▶ **40.2** continued

SOLUTION

Categorize This part of the problem is categorized as a quantum analysis of the oscillator. We model the block–spring system as a Planck oscillator.

Analyze Solve Equation 40.4 for the quantum number n:

$$n = \frac{E_n}{hf}$$

Substitute numerical values:

$$n = \frac{2.00\text{ J}}{(6.626 \times 10^{-34}\text{ J} \cdot \text{s})(0.563\text{ Hz})} = 5.36 \times 10^{33}$$

Finalize Notice that 5.36×10^{33} is a very large quantum number, which is typical for macroscopic systems. Changes between quantum states for the oscillator are explored next.

WHAT IF? Suppose the oscillator makes a transition from the $n = 5.36 \times 10^{33}$ state to the state corresponding to $n = 5.36 \times 10^{33} - 1$. By how much does the energy of the oscillator change in this one-quantum change?

Answer From Equation 40.5 and the result to part (A), the energy carried away due to the transition between states differing in n by 1 is

$$E = hf = (6.626 \times 10^{-34}\text{ J} \cdot \text{s})(0.563\text{ Hz}) = 3.73 \times 10^{-34}\text{ J}$$

This energy change due to a one-quantum change is fractionally equal to 3.73×10^{-34} J/2.00 J, or on the order of one part in 10^{34}! It is such a small fraction of the total energy of the oscillator that it cannot be detected. Therefore, even though the energy of a macroscopic block–spring system is quantized and does indeed decrease by small quantum jumps, our senses perceive the decrease as continuous. Quantum effects become important and detectable only on the submicroscopic level of atoms and molecules.

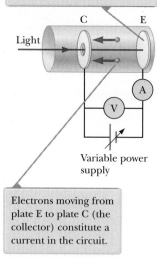

When light strikes plate E (the emitter), photoelectrons are ejected from the plate.

Electrons moving from plate E to plate C (the collector) constitute a current in the circuit.

Figure 40.9 A circuit diagram for studying the photoelectric effect.

40.2 The Photoelectric Effect

Blackbody radiation was the first phenomenon to be explained with a quantum model. In the latter part of the 19th century, at the same time that data were taken on thermal radiation, experiments showed that light incident on certain metallic surfaces causes electrons to be emitted from those surfaces. This phenomenon, which was first discussed in Section 35.1, is known as the **photoelectric effect,** and the emitted electrons are called **photoelectrons.**[3]

Figure 40.9 is a diagram of an apparatus for studying the photoelectric effect. An evacuated glass or quartz tube contains a metallic plate E (the emitter) connected to the negative terminal of a battery and another metallic plate C (the collector) that is connected to the positive terminal of the battery. When the tube is kept in the dark, the ammeter reads zero, indicating no current in the circuit. However, when plate E is illuminated by light having an appropriate wavelength, a current is detected by the ammeter, indicating a flow of charges across the gap between plates E and C. This current arises from photoelectrons emitted from plate E and collected at plate C.

Figure 40.10 is a plot of photoelectric current versus potential difference ΔV applied between plates E and C for two light intensities. At large values of ΔV, the current reaches a maximum value; all the electrons emitted from E are collected at C, and the current cannot increase further. In addition, the maximum current increases as the intensity of the incident light increases, as you might expect,

---

[3]Photoelectrons are not different from other electrons. They are given this name solely because of their ejection from a metal by light in the photoelectric effect.

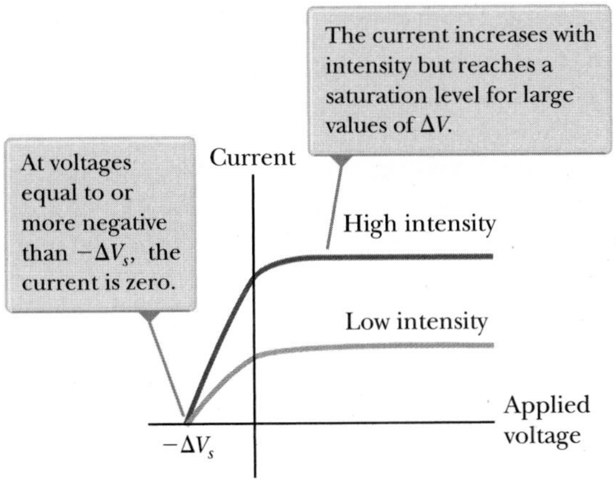

At voltages equal to or more negative than $-\Delta V_s$, the current is zero.

The current increases with intensity but reaches a saturation level for large values of ΔV.

Current

High intensity

Low intensity

Applied voltage

$-\Delta V_s$

Figure 40.10 Photoelectric current versus applied potential difference for two light intensities.

because more electrons are ejected by the higher-intensity light. Finally, when ΔV is negative—that is, when the battery in the circuit is reversed to make plate E positive and plate C negative—the current drops because many of the photoelectrons emitted from E are repelled by the now negative plate C. In this situation, only those photoelectrons having a kinetic energy greater than $e|\Delta V|$ reach plate C, where e is the magnitude of the charge on the electron. When ΔV is equal to or more negative than $-\Delta V_s$, where ΔV_s is the **stopping potential,** no photoelectrons reach C and the current is zero.

Let's model the combination of the electric field between the plates and an electron ejected from plate E as an isolated system. Suppose this electron stops just as it reaches plate C. Because the system is isolated, the appropriate reduction of Equation 8.2 is

$$\Delta K + \Delta U = 0$$

where the initial configuration is at the instant the electron leaves the metal with kinetic energy K_i and the final configuration is when the electron stops just before touching plate C. If we define the electric potential energy of the system in the initial configuration to be zero, we have

$$(0 - K_i) + [(q)(\Delta V) - 0] = 0 \quad \rightarrow \quad K_i = q\Delta V = -e\Delta V$$

Now suppose the potential difference ΔV is increased in the negative direction just until the current is zero at $\Delta V = -\Delta V_s$. In this case, the electron that stops immediately before reaching plate C has the maximum possible kinetic energy upon leaving the metal surface. The previous equation can then be written as

$$K_{max} = e\,\Delta V_s \tag{40.8}$$

This equation allows us to measure K_{max} experimentally by determining the magnitude of the voltage ΔV_s at which the current drops to zero.

Several features of the photoelectric effect are listed below. For each feature, we compare the predictions made by a classical approach, using the wave model for light, with the experimental results.

1. Dependence of photoelectron kinetic energy on light intensity

   Classical prediction: Electrons should absorb energy continuously from the electromagnetic waves. As the light intensity incident on a metal is increased, energy should be transferred into the metal at a higher rate and the electrons should be ejected with more kinetic energy.
   Experimental result: The maximum kinetic energy of photoelectrons is *independent* of light intensity as shown in Figure 40.10 with both curves falling to zero at the *same* negative voltage. (According to Equation 40.8, the maximum kinetic energy is proportional to the stopping potential.)

2. Time interval between incidence of light and ejection of photoelectrons

Classical prediction: At low light intensities, a measurable time interval should pass between the instant the light is turned on and the time an electron is ejected from the metal. This time interval is required for the electron to absorb the incident radiation before it acquires enough energy to escape from the metal.

Experimental result: Electrons are emitted from the surface of the metal almost *instantaneously* (less than 10^{-9} s after the surface is illuminated), even at very low light intensities.

3. Dependence of ejection of electrons on light frequency

Classical prediction: Electrons should be ejected from the metal at any incident light frequency, as long as the light intensity is high enough, because energy is transferred to the metal regardless of the incident light frequency.

Experimental result: No electrons are emitted if the incident light frequency falls below some **cutoff frequency** f_c, whose value is characteristic of the material being illuminated. No electrons are ejected below this cutoff frequency *regardless* of the light intensity.

4. Dependence of photoelectron kinetic energy on light frequency

Classical prediction: There should be *no* relationship between the frequency of the light and the electron kinetic energy. The kinetic energy should be related to the intensity of the light.

Experimental result: The maximum kinetic energy of the photoelectrons increases with increasing light frequency.

For these features, experimental results contradict *all four* classical predictions. A successful explanation of the photoelectric effect was given by Einstein in 1905, the same year he published his special theory of relativity. As part of a general paper on electromagnetic radiation, for which he received a Nobel Prize in Physics in 1921, Einstein extended Planck's concept of quantization to electromagnetic waves as mentioned in Section 40.1. Einstein assumed light (or any other electromagnetic wave) of frequency f from *any* source can be considered a stream of quanta. Today we call these quanta **photons.** Each photon has an energy E given by Equation 40.5, $E = hf$, and each moves in a vacuum at the speed of light c, where $c = 3.00 \times 10^8$ m/s.

Quick Quiz 40.2 While standing outdoors one evening, you are exposed to the following four types of electromagnetic radiation: yellow light from a sodium street lamp, radio waves from an AM radio station, radio waves from an FM radio station, and microwaves from an antenna of a communications system. Rank these types of waves in terms of photon energy from highest to lowest.

Let us organize Einstein's model for the photoelectric effect using the properties of structural models:

1. *Physical components:*

We imagine the system to consist of two physical components: (1) an electron that is to be ejected by an incoming photon and (2) the remainder of the metal.

2. *Behavior of the components:*

(a) In Einstein's model, a photon of the incident light gives *all* its energy hf to a *single* electron in the metal. Therefore, the absorption of energy by the electrons is not a continuous process as envisioned in the wave model, but rather a discontinuous process in which energy is delivered to the electrons in bundles. The energy transfer is accomplished via a one-photon/one-electron event.[4]

---

[4]In principle, two photons could combine to provide an electron with their combined energy. That is highly improbable, however, without the high intensity of radiation available from very strong lasers.

(b) We can describe the time evolution of the system by applying the non-isolated system model for energy over a time interval that includes the absorption of one photon and the ejection of the corresponding electron. Energy is transferred into the system by electromagnetic radiation, the photon. The system has two types of energy: the potential energy of the metal–electron system and the kinetic energy of the ejected electron. Therefore, we can write the conservation of energy equation (Eq. 8.2) as

$$\Delta K + \Delta U = T_{ER} \tag{40.9}$$

The energy transfer into the system is that of the photon, $T_{ER} = hf$. During the process, the kinetic energy of the electron increases from zero to its final value, which we assume to be the maximum possible value K_{max}. The potential energy of the system increases because the electron is pulled away from the metal to which it is attracted. We define the potential energy of the system when the electron is outside the metal as zero. The potential energy of the system when the electron is in the metal is $U = -\phi$, where ϕ is called the **work function** of the metal. The work function represents the minimum energy with which an electron is bound in the metal and is on the order of a few electron volts. Table 40.1 lists selected values. The increase in potential energy of the system when the electron is removed from the metal is the work function ϕ. Substituting these energies into Equation 40.9, we have

$$(K_{max} - 0) + [0 - (-\phi)] = hf$$

$$K_{max} + \phi = hf \tag{40.10}$$

If the electron makes collisions with other electrons or metal ions as it is being ejected, some of the incoming energy is transferred to the metal and the electron is ejected with less kinetic energy than K_{max}.

The prediction made by Einstein is an equation for the maximum kinetic energy of an ejected electron as a function of frequency of the illuminating radiation. This equation can be found by rearranging Equation 40.10:

$$K_{max} = hf - \phi \tag{40.11}$$

◀ Photoelectric effect equation

With Einstein's structural model, one can explain the observed features of the photoelectric effect that cannot be understood using classical concepts:

1. **Dependence of photoelectron kinetic energy on light intensity**

   Equation 40.11 shows that K_{max} is independent of the light intensity. The maximum kinetic energy of any one electron, which equals $hf - \phi$, depends only on the light frequency and the work function. If the light intensity is doubled, the number of photons arriving per unit time is doubled, which doubles the rate at which photoelectrons are emitted. The maximum kinetic energy of any one photoelectron, however, is unchanged.

2. **Time interval between incidence of light and ejection of photoelectrons**

   Near-instantaneous emission of electrons is consistent with the photon model of light. The incident energy appears in small packets, and there is a one-to-one interaction between photons and electrons. If the incident light has very low intensity, there are very few photons arriving per unit time interval; each photon, however, can have sufficient energy to eject an electron immediately.

3. **Dependence of ejection of electrons on light frequency**

   Because the photon must have energy greater than the work function ϕ to eject an electron, the photoelectric effect cannot be observed below a

Table 40.1 Work Functions of Selected Metals

Metal	ϕ (eV)
Na	2.46
Al	4.08
Fe	4.50
Cu	4.70
Zn	4.31
Ag	4.73
Pt	6.35
Pb	4.14

Note: Values are typical for metals listed. Actual values may vary depending on whether the metal is a single crystal or polycrystalline. Values may also depend on the face from which electrons are ejected from crystalline metals. Furthermore, different experimental procedures may produce differing values.

Figure 40.11 A plot of K_{max} for photoelectrons versus frequency of incident light in a typical photo-electric effect experiment.

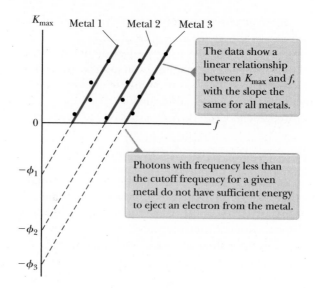

certain cutoff frequency. If the energy of an incoming photon does not satisfy this requirement, an electron cannot be ejected from the surface, even though many photons per unit time are incident on the metal in a very intense light beam.

4. Dependence of photoelectron kinetic energy on light frequency

A photon of higher frequency carries more energy and therefore ejects a photoelectron with more kinetic energy than does a photon of lower frequency.

Einstein's model predicts a linear relationship (Eq. 40.11) between the maximum electron kinetic energy K_{max} and the light frequency f. Experimental observation of a linear relationship between K_{max} and f would be a final confirmation of Einstein's theory. Indeed, such a linear relationship was observed experimentally within a few years of Einstein's theory and is sketched in Figure 40.11. The slope of the lines in such a plot is Planck's constant h. The intercept on the horizontal axis gives the cutoff frequency below which no photoelectrons are emitted. The cutoff frequency is related to the work function through the relationship $f_c = \phi/h$. The cutoff frequency corresponds to a **cutoff wavelength** λ_c, where

Cutoff wavelength ▶

$$\lambda_c = \frac{c}{f_c} = \frac{c}{\phi/h} = \frac{hc}{\phi} \qquad \text{(40.12)}$$

and c is the speed of light. Wavelengths greater than λ_c incident on a material having a work function ϕ do not result in the emission of photoelectrons.

The combination hc in Equation 40.12 often occurs when relating a photon's energy to its wavelength. A common shortcut when solving problems is to express this combination in useful units according to the following approximation:

$$hc = 1\ 240\ \text{eV} \cdot \text{nm}$$

One of the first practical uses of the photoelectric effect was as the detector in a camera's light meter. Light reflected from the object to be photographed strikes a photoelectric surface in the meter, causing it to emit photoelectrons that then pass through a sensitive ammeter. The magnitude of the current in the ammeter depends on the light intensity.

The phototube, another early application of the photoelectric effect, acts much like a switch in an electric circuit. It produces a current in the circuit when light of sufficiently high frequency falls on a metal plate in the phototube, but produces no current in the dark. Phototubes were used in burglar alarms and in the detection of the soundtrack on motion picture film. Modern semiconductor devices have now replaced older devices based on the photoelectric effect.

Today, the photoelectric effect is used in the operation of photomultiplier tubes. Figure 40.12 shows the structure of such a device. A photon striking the photocathode ejects an electron by means of the photoelectric effect. This electron accelerates across the potential difference between the photocathode and the first *dynode,* shown as being at +200 V relative to the photocathode in Figure 40.12. This high-energy electron strikes the dynode and ejects several more electrons. The same process is repeated through a series of dynodes at ever higher potentials until an electrical pulse is produced as millions of electrons strike the last dynode. The tube is therefore called a *multiplier:* one photon at the input has resulted in millions of electrons at the output.

The photomultiplier tube is used in nuclear detectors to detect photons produced by the interaction of energetic charged particles or gamma rays with certain materials. It is also used in astronomy in a technique called *photoelectric photometry.* In that technique, the light collected by a telescope from a single star is allowed to fall on a photomultiplier tube for a time interval. The tube measures the total energy transferred by light during the time interval, which can then be converted to a luminosity of the star.

The photomultiplier tube is being replaced in many astronomical observations with a *charge-coupled device* (CCD), which is the same device used in a digital camera (Section 36.6). Half of the 2009 Nobel Prize in Physics was awarded to Willard S. Boyle (b. 1924) and George E. Smith (b. 1930) for their 1969 invention of the charge-coupled device. In a CCD, an array of pixels is formed on the silicon surface of an integrated circuit (Section 43.7). When the surface is exposed to light from an astronomical scene through a telescope or a terrestrial scene through a digital camera, electrons generated by the photoelectric effect are caught in "traps" beneath the surface. The number of electrons is related to the intensity of the light striking the surface. A signal processor measures the number of electrons associated with each pixel and converts this information into a digital code that a computer can use to reconstruct and display the scene.

The *electron bombardment CCD camera* allows higher sensitivity than a conventional CCD. In this device, electrons ejected from a photocathode by the photoelectric effect are accelerated through a high voltage before striking a CCD array. The higher energy of the electrons results in a very sensitive detector of low-intensity radiation.

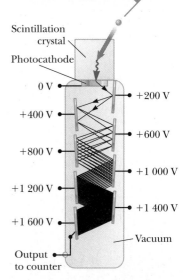

An incoming particle enters the scintillation crystal, where a collision results in a photon. The photon strikes the photocathode, which emits an electron by the photoelectric effect.

Figure 40.12 The multiplication of electrons in a photomultiplier tube.

Quick Quiz 40.3 Consider one of the curves in Figure 40.10. Suppose the intensity of the incident light is held fixed but its frequency is increased. Does the stopping potential in Figure 40.10 (a) remain fixed, (b) move to the right, or (c) move to the left?

Quick Quiz 40.4 Suppose classical physicists had the idea of plotting K_{max} versus f as in Figure 40.11. Draw a graph of what the expected plot would look like, based on the wave model for light.

| Example 40.3 | **The Photoelectric Effect for Sodium** |

A sodium surface is illuminated with light having a wavelength of 300 nm. As indicated in Table 40.1, the work function for sodium metal is 2.46 eV.

(A) Find the maximum kinetic energy of the ejected photoelectrons.

SOLUTION

Conceptualize Imagine a photon striking the metal surface and ejecting an electron. The electron with the maximum energy is one near the surface that experiences no interactions with other particles in the metal that would reduce its energy on its way out of the metal.

continued

▶ **40.3** c o n t i n u e d

Categorize We evaluate the results using equations developed in this section, so we categorize this example as a substitution problem.

Find the energy of each photon in the illuminating light beam from Equation 40.5:

$$E = hf = \frac{hc}{\lambda}$$

From Equation 40.11, find the maximum kinetic energy of an electron:

$$K_{max} = \frac{hc}{\lambda} - \phi = \frac{1\,240\ \text{eV} \cdot \text{nm}}{300\ \text{nm}} - 2.46\ \text{eV} = \boxed{1.67\ \text{eV}}$$

(B) Find the cutoff wavelength λ_c for sodium.

SOLUTION

Calculate λ_c using Equation 40.12:

$$\lambda_c = \frac{hc}{\phi} = \frac{1\,240\ \text{eV} \cdot \text{nm}}{2.46\ \text{eV}} = \boxed{504\ \text{nm}}$$

Arthur Holly Compton
American Physicist (1892–1962)
Compton was born in Wooster, Ohio, and attended Wooster College and Princeton University. He became the director of the laboratory at the University of Chicago, where experimental work concerned with sustained nuclear chain reactions was conducted. This work was of central importance to the construction of the first nuclear weapon. His discovery of the Compton effect led to his sharing of the 1927 Nobel Prize in Physics with Charles Wilson.

40.3 The Compton Effect

In 1919, Einstein concluded that a photon of energy E travels in a single direction and carries a momentum equal to $E/c = hf/c$. In 1923, Arthur Holly Compton (1892–1962) and Peter Debye (1884–1966) independently carried Einstein's idea of photon momentum further.

Prior to 1922, Compton and his coworkers had accumulated evidence showing that the classical wave theory of light failed to explain the scattering of x-rays from electrons. According to classical theory, electromagnetic waves of frequency f incident on electrons should have two effects: (1) radiation pressure (see Section 34.5) should cause the electrons to accelerate in the direction of propagation of the waves, and (2) the oscillating electric field of the incident radiation should set the electrons into oscillation at the apparent frequency f', where f' is the frequency in the frame of the moving electrons. This apparent frequency is different from the frequency f of the incident radiation because of the Doppler effect (see Section 17.4). Each electron first absorbs radiation as a moving particle and then reradiates as a moving particle, thereby exhibiting two Doppler shifts in the frequency of radiation.

Because different electrons move at different speeds after the interaction, depending on the amount of energy absorbed from the electromagnetic waves, the scattered wave frequency at a given angle to the incoming radiation should show a distribution of Doppler-shifted values. Contrary to this prediction, Compton's experiments showed that at a given angle only *one* frequency of radiation is observed. Compton and his coworkers explained these experiments by treating photons not as waves but rather as point-like particles having energy hf and momentum hf/c and by assuming the energy and momentum of the isolated system of the colliding photon–electron pair are conserved. Compton adopted a particle model for something that was well known as a wave, and today this scattering phenomenon is known as the **Compton effect.** Figure 40.13 shows the quantum picture of the collision between an individual x-ray photon of frequency f_0 and an electron. In the quantum model, the electron is scattered through an angle ϕ with respect to this direction as in a billiard-ball type of collision. (The symbol ϕ used here is an angle and is not to be confused with the work function, which was discussed in the preceding section.) Compare Figure 40.13 with the two-dimensional collision shown in Figure 9.11.

Figure 40.14 is a schematic diagram of the apparatus used by Compton. The x-rays, scattered from a carbon target, were diffracted by a rotating crystal spectrometer, and the intensity was measured with an ionization chamber that generated a current proportional to the intensity. The incident beam consisted of monochromatic x-rays of wavelength $\lambda_0 = 0.071$ nm. The experimental intensity-

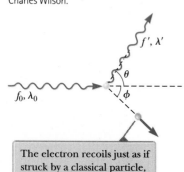

The electron recoils just as if struck by a classical particle, revealing the particle-like nature of the photon.

Figure 40.13 The quantum model for x-ray scattering from an electron.

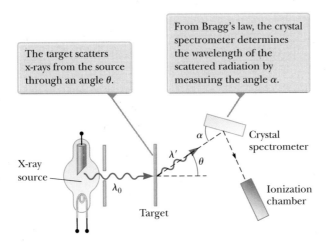

The target scatters x-rays from the source through an angle θ.

From Bragg's law, the crystal spectrometer determines the wavelength of the scattered radiation by measuring the angle α.

Figure 40.14 Schematic diagram of Compton's apparatus.

versus-wavelength plots observed by Compton for four scattering angles (corresponding to θ in Fig. 40.13) are shown in Figure 40.15. The graphs for the three nonzero angles show two peaks, one at λ_0 and one at $\lambda' > \lambda_0$. The shifted peak at λ' is caused by the scattering of x-rays from free electrons, which was predicted by Compton to depend on scattering angle as

$$\lambda' - \lambda_0 = \frac{h}{m_e c}(1 - \cos\theta) \tag{40.13}$$

◀ **Compton shift equation**

where m_e is the mass of the electron. This expression is known as the **Compton shift equation** and correctly describes the positions of the peaks in Figure 40.15. The factor $h/m_e c$, called the **Compton wavelength** of the electron, has a currently accepted value of

$$\lambda_C = \frac{h}{m_e c} = 0.002\ 43\ \text{nm}$$

◀ **Compton wavelength**

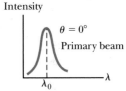

The unshifted peak at λ_0 in Figure 40.15 is caused by x-rays scattered from electrons tightly bound to the target atoms. This unshifted peak also is predicted by Equation 40.13 if the electron mass is replaced with the mass of a carbon atom, which is approximately 23 000 times the mass of the electron. Therefore, there is a wavelength shift for scattering from an electron bound to an atom, but it is so small that it was undetectable in Compton's experiment.

Compton's measurements were in excellent agreement with the predictions of Equation 40.13. These results were the first to convince many physicists of the fundamental validity of quantum theory.

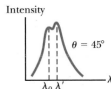

Quick Quiz 40.5 For any given scattering angle θ, Equation 40.13 gives the same value for the Compton shift for any wavelength. Keeping that in mind, for which of the following types of radiation is the fractional shift in wavelength at a given scattering angle the largest? **(a)** radio waves **(b)** microwaves **(c)** visible light **(d)** x-rays

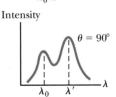

Derivation of the Compton Shift Equation

We can derive the Compton shift equation by assuming the photon behaves like a particle and collides elastically with a free electron initially at rest as shown in Figure 40.13. The photon is treated as a particle having energy $E = hf = hc/\lambda$ and zero rest energy. We apply the isolated system analysis models for energy and momentum to the photon and the electron. In the scattering process, the total energy and total linear momentum of the system are conserved. Applying the isolated system model for energy to this process gives

$$\Delta K_{photon} + \Delta K_e = 0 \rightarrow \frac{hc}{\lambda_0} = \frac{hc}{\lambda'} + K_e$$

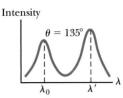

Figure 40.15 Scattered x-ray intensity versus wavelength for Compton scattering at θ = 0°, 45°, 90°, and 135°.

where hc/λ_0 is the energy of the incident photon, hc/λ' is the energy of the scattered photon, and K_e is the kinetic energy of the recoiling electron. Because the electron may recoil at a speed comparable to that of light, we must use the relativistic expression $K_e = (\gamma - 1)m_e c^2$ (Eq. 39.23). Therefore,

$$\frac{hc}{\lambda_0} = \frac{hc}{\lambda'} + (\gamma - 1)m_e c^2 \tag{40.14}$$

where $\gamma = 1/\sqrt{1 - (u^2/c^2)}$ and u is the speed of the electron.

Next, let's apply the isolated system model for momentum to this collision, noting that the x and y components of momentum are each conserved independently. Equation 39.28 shows that the momentum of a photon has a magnitude $p = E/c$, and we know from Equation 40.5 that $E = hf$. Therefore, $p = hf/c$. Substituting λf for c (Eq. 34.20) in this expression gives $p = h/\lambda$. Because the relativistic expression for the momentum of the recoiling electron is $p_e = \gamma m_e u$ (Eq. 39.19), we obtain the following expressions for the x and y components of linear momentum, where the angles are as described in Figure 40.13:

$$x \text{ component:} \quad \frac{h}{\lambda_0} = \frac{h}{\lambda'} \cos\theta + \gamma m_e u \cos\phi \tag{40.15}$$

$$y \text{ component:} \quad 0 = \frac{h}{\lambda'} \sin\theta - \gamma m_e u \sin\phi \tag{40.16}$$

Eliminating u and ϕ from Equations 40.14 through 40.16 gives a single expression that relates the remaining three variables (λ', λ_0, and θ). After some algebra (see Problem 64 in Enhanced WebAssign), we obtain Equation 40.13.

Example 40.4   **Compton Scattering at 45°**

X-rays of wavelength $\lambda_0 = 0.200\,000$ nm are scattered from a block of material. The scattered x-rays are observed at an angle of 45.0° to the incident beam. Calculate their wavelength.

SOLUTION

Conceptualize Imagine the process in Figure 40.13, with the photon scattered at 45° to its original direction.

Categorize We evaluate the result using an equation developed in this section, so we categorize this example as a substitution problem.

Solve Equation 40.13 for the wavelength of the scattered x-ray:

$$(1) \quad \lambda' = \lambda_0 + \frac{h(1 - \cos\theta)}{m_e c}$$

Substitute numerical values:

$$\lambda' = 0.200\,000 \times 10^{-9}\text{ m} + \frac{(6.626 \times 10^{-34}\text{ J}\cdot\text{s})(1 - \cos 45.0°)}{(9.11 \times 10^{-31}\text{ kg})(3.00 \times 10^8\text{ m/s})}$$

$$= 0.200\,000 \times 10^{-9}\text{ m} + 7.10 \times 10^{-13}\text{ m} = \boxed{0.200\,710\text{ nm}}$$

WHAT IF?   What if the detector is moved so that scattered x-rays are detected at an angle larger than 45°? Does the wavelength of the scattered x-rays increase or decrease as the angle θ increases?

Answer In Equation (1), if the angle θ increases, $\cos\theta$ decreases. Consequently, the factor $(1 - \cos\theta)$ increases. Therefore, the scattered wavelength increases.

We could also apply an energy argument to achieve this same result. As the scattering angle increases, more energy is transferred from the incident photon to the electron. As a result, the energy of the scattered photon decreases with increasing scattering angle. Because $E = hf$, the frequency of the scattered photon decreases, and because $\lambda = c/f$, the wavelength increases.

40.4 The Nature of Electromagnetic Waves

In Section 35.1, we introduced the notion of competing models of light: particles and waves. Let's expand on that earlier discussion. Phenomena such as the photoelectric effect and the Compton effect offer ironclad evidence that when light (or other forms of electromagnetic radiation) and matter interact, the light behaves as if it were composed of particles having energy hf and momentum h/λ. How can light be considered a photon (in other words, a particle) when we know it is a wave? On the one hand, we describe light in terms of photons having energy and momentum. On the other hand, light and other electromagnetic waves exhibit interference and diffraction effects, which are consistent only with a wave interpretation.

Which model is correct? Is light a wave or a particle? The answer depends on the phenomenon being observed. Some experiments can be explained either better or solely with the photon model, whereas others are explained either better or solely with the wave model. We must accept both models and admit that the true nature of light is not describable in terms of any single classical picture. The same light beam that can eject photoelectrons from a metal (meaning that the beam consists of photons) can also be diffracted by a grating (meaning that the beam is a wave). In other words, the particle model and the wave model of light complement each other.

The success of the particle model of light in explaining the photoelectric effect and the Compton effect raises many other questions. If light is a particle, what is the meaning of the "frequency" and "wavelength" of the particle, and which of these two properties determines its energy and momentum? Is light *simultaneously* a wave and a particle? Although photons have no rest energy (a nonobservable quantity because a photon cannot be at rest), is there a simple expression for the *effective mass* of a moving photon? If photons have effective mass, do they experience gravitational attraction? What is the spatial extent of a photon, and how does an electron absorb or scatter one photon? Some of these questions can be answered, but others demand a view of atomic processes that is too pictorial and literal. Many of them stem from classical analogies such as colliding billiard balls and ocean waves breaking on a seashore. Quantum mechanics gives light a more flexible nature by treating the particle model and the wave model of light as both necessary and complementary. Neither model can be used exclusively to describe all properties of light. A complete understanding of the observed behavior of light can be attained only if the two models are combined in a complementary manner.

40.5 The Wave Properties of Particles

Students introduced to the dual nature of light often find the concept difficult to accept. In the world around us, we are accustomed to regarding such things as baseballs solely as particles and other things such as sound waves solely as forms of wave motion. Every large-scale observation can be interpreted by considering either a wave explanation or a particle explanation, but in the world of photons and electrons, such distinctions are not as sharply drawn.

Even more disconcerting is that, under certain conditions, the things we unambiguously call "particles" exhibit wave characteristics. In his 1923 doctoral dissertation, Louis de Broglie postulated that because photons have both wave and particle characteristics, perhaps all forms of matter have both properties. This highly revolutionary idea had no experimental confirmation at the time. According to de Broglie, electrons, just like light, have a dual particle–wave nature.

In Section 40.3, we found that the momentum of a photon can be expressed as

$$p = \frac{h}{\lambda}$$

Louis de Broglie
French Physicist (1892–1987)
De Broglie was born in Dieppe, France. At the Sorbonne in Paris, he studied history in preparation for what he hoped would be a career in the diplomatic service. The world of science is lucky he changed his career path to become a theoretical physicist. De Broglie was awarded the Nobel Prize in Physics in 1929 for his prediction of the wave nature of electrons.

This equation shows that the photon wavelength can be specified by its momentum: $\lambda = h/p$. De Broglie suggested that material particles of momentum p have a characteristic wavelength that is given by the *same expression*. Because the magnitude of the momentum of a particle of mass m and speed u is $p = mu$, the **de Broglie wavelength** of that particle is[5]

$$\lambda = \frac{h}{p} = \frac{h}{mu} \tag{40.17}$$

Furthermore, in analogy with photons, de Broglie postulated that particles obey the Einstein relation $E = hf$, where E is the total energy of the particle. The frequency of a particle is then

$$f = \frac{E}{h} \tag{40.18}$$

The dual nature of matter is apparent in Equations 40.17 and 40.18 because each contains both particle quantities (p and E) and wave quantities (λ and f).

The problem of understanding the dual nature of matter and radiation is conceptually difficult because the two models seem to contradict each other. This problem as it applies to light was discussed earlier. The **principle of complementarity** states that

> the wave and particle models of either matter or radiation complement each other.

Neither model can be used exclusively to describe matter or radiation adequately. Because humans tend to generate mental images based on their experiences from the everyday world, we use both descriptions in a complementary manner to explain any given set of data from the quantum world.

The Davisson–Germer Experiment

De Broglie's 1923 proposal that matter exhibits both wave and particle properties was regarded as pure speculation. If particles such as electrons had wave properties, under the correct conditions they should exhibit diffraction effects. Only three years later, C. J. Davisson (1881–1958) and L. H. Germer (1896–1971) succeeded in observing electron diffraction and measuring the wavelength of electrons. Their important discovery provided the first experimental confirmation of the waves proposed by de Broglie.

Interestingly, the intent of the initial Davisson–Germer experiment was not to confirm the de Broglie hypothesis. In fact, their discovery was made by accident (as is often the case). The experiment involved the scattering of low-energy electrons (approximately 54 eV) from a nickel target in a vacuum. During one experiment, the nickel surface was badly oxidized because of an accidental break in the vacuum system. After the target was heated in a flowing stream of hydrogen to remove the oxide coating, electrons scattered by it exhibited intensity maxima and minima at specific angles. The experimenters finally realized that the nickel had formed large crystalline regions upon heating and that the regularly spaced planes of atoms in these regions served as a diffraction grating for electrons. (See the discussion of diffraction of x-rays by crystals in Section 38.5.)

Shortly thereafter, Davisson and Germer performed more extensive diffraction measurements on electrons scattered from single-crystal targets. Their results showed conclusively the wave nature of electrons and confirmed the de Broglie relationship $p = h/\lambda$. In the same year, G. P. Thomson (1892–1975) of Scotland also observed electron diffraction patterns by passing electrons through very thin gold

Pitfall Prevention 40.3

What's Waving? If particles have wave properties, what's waving? You are familiar with waves on strings, which are very concrete. Sound waves are more abstract, but you are likely comfortable with them. Electromagnetic waves are even more abstract, but at least they can be described in terms of physical variables and electric and magnetic fields. In contrast, waves associated with particles are completely abstract and cannot be associated with a physical variable. In Chapter 41, we describe the wave associated with a particle in terms of probability.

---

[5]The de Broglie wavelength for a particle moving at *any* speed u is $\lambda = h/\gamma mu$, where $\gamma = [1 - (u^2/c^2)]^{-1/2}$.

foils. Diffraction patterns were subsequently observed in the scattering of helium atoms, hydrogen atoms, and neutrons. Hence, the wave nature of particles has been established in various ways.

Quick Quiz 40.6 An electron and a proton both moving at nonrelativistic speeds have the same de Broglie wavelength. Which of the following quantities are also the same for the two particles? **(a)** speed **(b)** kinetic energy **(c)** momentum **(d)** frequency

---

Example 40.5   **Wavelengths for Microscopic and Macroscopic Objects**

(A) Calculate the de Broglie wavelength for an electron ($m_e = 9.11 \times 10^{-31}$ kg) moving at 1.00×10^7 m/s.

SOLUTION

Conceptualize Imagine the electron moving through space. From a classical viewpoint, it is a particle under constant velocity. From the quantum viewpoint, the electron has a wavelength associated with it.

Categorize We evaluate the result using an equation developed in this section, so we categorize this example as a substitution problem.

Evaluate the de Broglie wavelength using Equation 40.17:

$$\lambda = \frac{h}{m_e u} = \frac{6.626 \times 10^{-34}\, \text{J} \cdot \text{s}}{(9.11 \times 10^{-31}\, \text{kg})(1.00 \times 10^7\, \text{m/s})} = \boxed{7.27 \times 10^{-11}\, \text{m}}$$

The wave nature of this electron could be detected by diffraction techniques such as those in the Davisson–Germer experiment.

(B) A rock of mass 50 g is thrown with a speed of 40 m/s. What is its de Broglie wavelength?

SOLUTION

Evaluate the de Broglie wavelength using Equation 40.17:

$$\lambda = \frac{h}{mu} = \frac{6.626 \times 10^{-34}\, \text{J} \cdot \text{s}}{(50 \times 10^{-3}\, \text{kg})(40\, \text{m/s})} = \boxed{3.3 \times 10^{-34}\, \text{m}}$$

This wavelength is much smaller than any aperture through which the rock could possibly pass. Hence, we could not observe diffraction effects, and as a result, the wave properties of large-scale objects cannot be observed.

The Electron Microscope

A practical device that relies on the wave characteristics of electrons is the **electron microscope.** A *transmission* electron microscope, used for viewing flat, thin samples, is shown in Figure 40.16 on page 986. In many respects, it is similar to an optical microscope; the electron microscope, however, has a much greater resolving power because it can accelerate electrons to very high kinetic energies, giving them very short wavelengths. No microscope can resolve details that are significantly smaller than the wavelength of the waves used to illuminate the object. The shorter wavelengths of electrons gives an electron microscope a resolution that can be 1 000 times better than that from the visible light used in optical microscopes. As a result, an electron microscope with ideal lenses would be able to distinguish details approximately 1 000 times smaller than those distinguished by an optical microscope. (Electromagnetic radiation of the same wavelength as the electrons in an electron microscope is in the x-ray region of the spectrum.)

The electron beam in an electron microscope is controlled by electrostatic or magnetic deflection, which acts on the electrons to focus the beam and form an image. Rather than examining the image through an eyepiece as in an optical microscope, the viewer looks at an image formed on a monitor or other type of

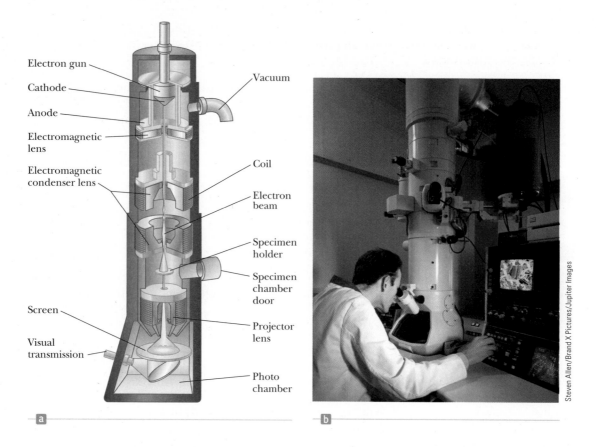

Electron gun

Cathode

Anode

Electromagnetic lens

Electromagnetic condenser lens

Screen

Visual transmission

Vacuum

Coil

Electron beam

Specimen holder

Specimen chamber door

Projector lens

Photo chamber

Steven Allen/Brand X Pictures/Jupiter Images

a

b

Figure 40.16 (a) Diagram of a transmission electron microscope for viewing a thinly sectioned sample. The "lenses" that control the electron beam are magnetic deflection coils. (b) An electron microscope in use.

© Robert Harding Picture Library Ltd./Alamy

Figure 40.17 A scanning electron microscope photograph shows significant detail of a cheese mite, *Tyrolichus casei*. The mite is so small, with a maximum length of 0.70 mm, that ordinary microscopes do not reveal minute anatomical details.

display screen. Figure 40.17 shows the amazing detail available with an electron microscope.

40.6 A New Model: The Quantum Particle

Because in the past we considered the particle and wave models to be distinct, the discussions presented in previous sections may be quite disturbing. The notion that both light and material particles have both particle and wave properties does not fit with this distinction. Experimental evidence shows, however, that this conclusion is exactly what we must accept. The recognition of this dual nature leads to a new model, the **quantum particle,** which is a combination of the particle model introduced in Chapter 2 and the wave model discussed in Chapter 16. In this new model, entities have both particle and wave characteristics, and we must choose one appropriate behavior—particle or wave—to understand a particular phenomenon.

In this section, we shall explore this model in a way that might make you more comfortable with this idea. We shall do so by demonstrating that an entity that exhibits properties of a particle can be constructed from waves.

Let's first recall some characteristics of ideal particles and ideal waves. An ideal particle has zero size. Therefore, an essential feature of a particle is that it is *localized* in space. An ideal wave has a single frequency and is infinitely long as suggested by Figure 40.18a. Therefore, an ideal wave is *unlocalized* in space. A localized entity can be built from infinitely long waves as follows. Imagine drawing one wave along the *x* axis, with one of its crests located at *x* = 0, as at the top of Figure 40.18b. Now draw a second wave, of the same amplitude but a different frequency, with one of its crests also at $x = 0$. As a result of the superposition of these two waves, *beats* exist as the waves are alternately in phase and out of phase. (Beats were discussed in Section 18.7.) The bottom curve in Figure 40.18b shows the results of superposing these two waves.

Notice that we have already introduced some localization by superposing the two waves. A single wave has the same amplitude everywhere in space; no point in space is any different from any other point. By adding a second wave, however, there is something different about the in-phase points compared with the out-of-phase points.

Now imagine that more and more waves are added to our original two, each new wave having a new frequency. Each new wave is added so that one of its crests is at $x = 0$ with the result that all the waves add constructively at $x = 0$. When we add a large number of waves, the probability of a positive value of a wave function at any point $x \neq 0$ is equal to the probability of a negative value, and there is destructive interference *everywhere* except near $x = 0$, where all the crests are superposed. The result is shown in Figure 40.19. The small region of constructive interference is called a **wave packet.** This localized region of space is different from all other regions. We can identify the wave packet as a particle because it has the localized nature of a particle! The location of the wave packet corresponds to the particle's position.

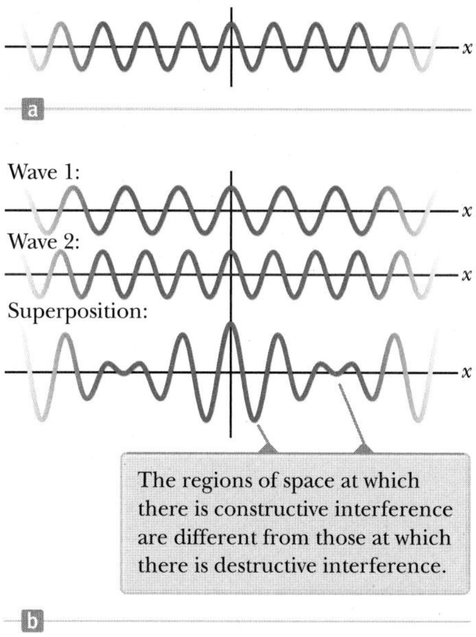

The localized nature of this entity is the *only* characteristic of a particle that was generated with this process. We have not addressed how the wave packet might achieve such particle characteristics as mass, electric charge, and spin. Therefore, you may not be completely convinced that we have built a particle. As further evidence that the wave packet can represent the particle, let's show that the wave packet has another characteristic of a particle.

To simplify the mathematical representation, we return to our combination of two waves. Consider two waves with equal amplitudes but different angular frequencies ω_1 and ω_2. We can represent the waves mathematically as

$$y_1 = A \cos (k_1 x - \omega_1 t) \quad \text{and} \quad y_2 = A \cos (k_2 x - \omega_2 t)$$

where, as in Chapter 16, $k = 2\pi/\lambda$ and $\omega = 2\pi f$. Using the superposition principle, let's add the waves:

$$y = y_1 + y_2 = A \cos (k_1 x - \omega_1 t) + A \cos (k_2 x - \omega_2 t)$$

It is convenient to write this expression in a form that uses the trigonometric identity

$$\cos a + \cos b = 2 \cos \left(\frac{a - b}{2} \right) \cos \left(\frac{a + b}{2} \right)$$

Letting $a = k_1 x - \omega_1 t$ and $b = k_2 x - \omega_2 t$ gives

$$y = 2A \cos \left[\frac{(k_1 x - \omega_1 t) - (k_2 x - \omega_2 t)}{2} \right] \cos \left[\frac{(k_1 x - \omega_1 t) + (k_2 x - \omega_2 t)}{2} \right]$$

$$y = \left[2A \cos \left(\frac{\Delta k}{2} x - \frac{\Delta \omega}{2} t \right) \right] \cos \left(\frac{k_1 + k_2}{2} x - \frac{\omega_1 + \omega_2}{2} t \right) \quad \textbf{(40.19)}$$

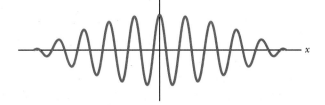

Wave 1:

Wave 2:

Superposition:

The regions of space at which there is constructive interference are different from those at which there is destructive interference.

Figure 40.18 (a) An idealized wave of an exact single frequency is the same throughout space and time. (b) If two ideal waves with slightly different frequencies are combined, beats result (Section 18.7).

Figure 40.19 If a large number of waves are combined, the result is a wave packet, which represents a particle.

Figure 40.20 The beat pattern of Figure 40.18b, with an envelope function (dashed curve) superimposed.

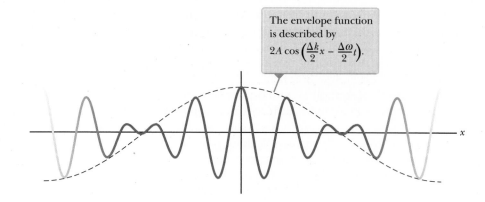

The envelope function is described by $2A \cos\left(\frac{\Delta k}{2}x - \frac{\Delta \omega}{2}t\right)$.

where $\Delta k = k_1 - k_2$ and $\Delta \omega = \omega_1 - \omega_2$. The second cosine factor represents a wave with a wave number and frequency that are equal to the averages of the values for the individual waves.

In Equation 40.19, the factor in square brackets represents the envelope of the wave as shown by the dashed curve in Figure 40.20. This factor also has the mathematical form of a wave. This envelope of the combination can travel through space with a different speed than the individual waves. As an extreme example of this possibility, imagine combining two identical waves moving in opposite directions. The two waves move with the same speed, but the envelope has a speed of *zero* because we have built a standing wave, which we studied in Section 18.2.

For an individual wave, the speed is given by Equation 16.11,

Phase speed of a wave ▶
in a wave packet

$$v_{\text{phase}} = \frac{\omega}{k} \qquad (40.20)$$

This speed is called the **phase speed** because it is the rate of advance of a crest on a single wave, which is a point of fixed phase. Equation 40.20 can be interpreted as follows: the phase speed of a wave is the ratio of the coefficient of the time variable t to the coefficient of the space variable x in the equation representing the wave, $y = A \cos (kx - \omega t)$.

The factor in brackets in Equation 40.19 is of the form of a wave, so it moves with a speed given by this same ratio:

$$v_g = \frac{\text{coefficient of time variable } t}{\text{coefficient of space variable } x} = \frac{(\Delta \omega / 2)}{(\Delta k / 2)} = \frac{\Delta \omega}{\Delta k}$$

The subscript g on the speed indicates that it is commonly called the **group speed,** or the speed of the wave packet (the *group* of waves) we have built. We have generated this expression for a simple addition of two waves. When a large number of waves are superposed to form a wave packet, this ratio becomes a derivative:

Group speed of a wave packet ▶

$$v_g = \frac{d\omega}{dk} \qquad (40.21)$$

Multiplying the numerator and the denominator by $\hbar$, where $\hbar = h/2\pi$, gives

$$v_g = \frac{\hbar \, d\omega}{\hbar \, dk} = \frac{d(\hbar \omega)}{d(\hbar k)} \qquad (40.22)$$

Let's look at the terms in the parentheses of Equation 40.22 separately. For the numerator,

$$\hbar \omega = \frac{h}{2\pi}(2\pi f) = hf = E$$

For the denominator,

$$\hbar k = \frac{h}{2\pi}\left(\frac{2\pi}{\lambda}\right) = \frac{h}{\lambda} = p$$

Therefore, Equation 40.22 can be written as

$$v_g = \frac{d(\hbar \omega)}{d(\hbar k)} = \frac{dE}{dp} \qquad \textbf{(40.23)}$$

Because we are exploring the possibility that the envelope of the combined waves represents the particle, consider a free particle moving with a speed u that is small compared with the speed of light. The energy of the particle is its kinetic energy:

$$E = \tfrac{1}{2}mu^2 = \frac{p^2}{2m}$$

Differentiating this equation with respect to p gives

$$v_g = \frac{dE}{dp} = \frac{d}{dp}\left(\frac{p^2}{2m}\right) = \frac{1}{2m}(2p) = u \qquad \textbf{(40.24)}$$

Therefore, the group speed of the wave packet is identical to the speed of the particle that it is modeled to represent, giving us further confidence that the wave packet is a reasonable way to build a particle.

Ⓠuick Quiz 40.7 As an analogy to wave packets, consider an "automobile packet" that occurs near the scene of an accident on a freeway. The phase speed is analogous to the speed of individual automobiles as they move through the backup caused by the accident. The group speed can be identified as the speed of the leading edge of the packet of cars. For the automobile packet, is the group speed (a) the same as the phase speed, (b) less than the phase speed, or (c) greater than the phase speed?

40.7 The Double-Slit Experiment Revisited

Wave–particle duality is now a firmly accepted concept reinforced by experimental results, including those of the Davisson–Germer experiment. As with the postulates of special relativity, however, this concept often leads to clashes with familiar thought patterns we hold from everyday experience.

One way to crystallize our ideas about the electron's wave–particle duality is through an experiment in which electrons are fired at a double slit. Consider a parallel beam of mono-energetic electrons incident on a double slit as in Figure 40.21. Let's assume the slit widths are small compared with the electron wavelength so that we need not worry about diffraction maxima and minima as discussed for light in Section 38.2. An electron detector screen is positioned far from the slits at a distance much greater than d, the separation distance of the slits. If the detector screen collects electrons for a long enough time, we find a typical wave interference pattern for the counts per minute, or probability of arrival of electrons. Such an interference

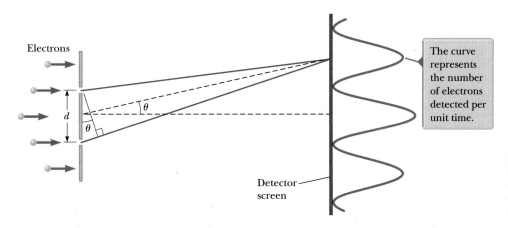

Figure 40.21 Electron interference. The slit separation d is much greater than the individual slit widths and much less than the distance between the slit and the detector screen.

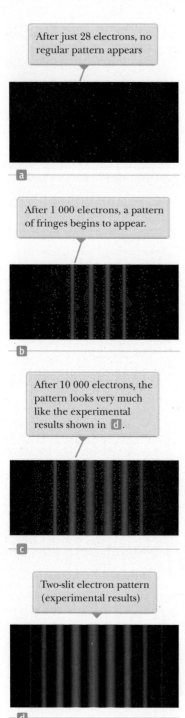

After just 28 electrons, no regular pattern appears

a

After 1 000 electrons, a pattern of fringes begins to appear.

b

After 10 000 electrons, the pattern looks very much like the experimental results shown in **d**.

c

Two-slit electron pattern (experimental results)

d

Figure 40.22 (a)–(c) Computer-simulated interference patterns for a small number of electrons incident on a double slit. (d) Computer simulation of a double-slit interference pattern produced by many electrons.

pattern would not be expected if the electrons behaved as classical particles, giving clear evidence that electrons are interfering, a distinct wave-like behavior.

If we measure the angles θ at which the maximum intensity of electrons arrives at the detector screen in Figure 40.21, we find they are described by exactly the same equation as that for light, $d \sin \theta = m\lambda$ (Eq. 37.2), where m is the order number and λ is the electron wavelength. Therefore, the dual nature of the electron is clearly shown in this experiment: the electrons are detected as particles at a localized spot on the detector screen at some instant of time, but the probability of arrival at that spot is determined by finding the intensity of two interfering waves.

Now imagine that we lower the beam intensity so that one electron at a time arrives at the double slit. It is tempting to assume the electron goes through either slit 1 or slit 2. You might argue that there are no interference effects because there is not a second electron going through the other slit to interfere with the first. This assumption places too much emphasis on the particle model of the electron, however. The interference pattern is still observed if the time interval for the measurement is sufficiently long for many electrons to pass one at a time through the slits and arrive at the detector screen! This situation is illustrated by the computer-simulated patterns in Figure 40.22 where the interference pattern becomes clearer as the number of electrons reaching the detector screen increases. Hence, our assumption that the electron is localized and goes through only one slit when both slits are open must be wrong (a painful conclusion!).

To interpret these results, we are forced to conclude that an electron interacts with both slits *simultaneously*. If you try to determine experimentally which slit the electron goes through, the act of measuring destroys the interference pattern. It is impossible to determine which slit the electron goes through. In effect, we can say only that the electron passes through *both* slits! The same arguments apply to photons.

If we restrict ourselves to a pure particle model, it is an uncomfortable notion that the electron can be present at both slits at once. From the quantum particle model, however, the particle can be considered to be built from waves that exist throughout space. Therefore, the wave components of the electron are present at both slits at the same time, and this model leads to a more comfortable interpretation of this experiment.

40.8 The Uncertainty Principle

Whenever one measures the position or velocity of a particle at any instant, experimental uncertainties are built into the measurements. According to classical mechanics, there is no fundamental barrier to an ultimate refinement of the apparatus or experimental procedures. In other words, it is possible, in principle, to make such measurements with arbitrarily small uncertainty. Quantum theory predicts, however, that it is fundamentally impossible to make simultaneous measurements of a particle's position and momentum with infinite accuracy.

In 1927, Werner Heisenberg (1901–1976) introduced this notion, which is now known as the **Heisenberg uncertainty principle:**

> If a measurement of the position of a particle is made with uncertainty Δx and a simultaneous measurement of its x component of momentum is made with uncertainty Δp_x, the product of the two uncertainties can never be smaller than $\hbar/2$:
>
> $$\Delta x \, \Delta p_x \geq \frac{\hbar}{2} \qquad \text{(40.25)}$$

That is, it is physically impossible to measure simultaneously the exact position and exact momentum of a particle. Heisenberg was careful to point out that the inescapable uncertainties Δx and Δp_x do not arise from imperfections in practical measuring instruments. Rather, the uncertainties arise from the quantum structure of matter.

To understand the uncertainty principle, imagine that a particle has a single wavelength that is known *exactly*. According to the de Broglie relation, $\lambda = h/p$, we would therefore know the momentum to be precisely $p = h/\lambda$. In reality, a single-wavelength wave would exist throughout space. Any region along this wave is the same as any other region (Fig. 40.18a). Suppose we ask, Where is the particle this wave represents? No special location in space along the wave could be identified with the particle; all points along the wave are the same. Therefore, we have *infinite* uncertainty in the position of the particle, and we know nothing about its location. Perfect knowledge of the particle's momentum has cost us all information about its location.

In comparison, now consider a particle whose momentum is uncertain so that it has a range of possible values of momentum. According to the de Broglie relation, the result is a range of wavelengths. Therefore, the particle is not represented by a single wavelength, but rather by a combination of wavelengths within this range. This combination forms a wave packet as we discussed in Section 40.6 and illustrated in Figure 40.19. If you were asked to determine the location of the particle, you could only say that it is somewhere in the region defined by the wave packet because there is a distinct difference between this region and the rest of space. Therefore, by losing some information about the momentum of the particle, we have gained information about its position.

If you were to lose *all* information about the momentum, you would be adding together waves of all possible wavelengths, resulting in a wave packet of zero length. Therefore, if you know nothing about the momentum, you know exactly where the particle is.

The mathematical form of the uncertainty principle states that the product of the uncertainties in position and momentum is always larger than some minimum value. This value can be calculated from the types of arguments discussed above, and the result is the value of $\hbar/2$ in Equation 40.25.

Another form of the uncertainty principle can be generated by reconsidering Figure 40.19. Imagine that the horizontal axis is time rather than spatial position x. We can then make the same arguments that were made about knowledge of wavelength and position in the time domain. The corresponding variables would be frequency and time. Because frequency is related to the energy of the particle by $E = hf$, the uncertainty principle in this form is

$$\Delta E\, \Delta t \geq \frac{\hbar}{2} \qquad (40.26)$$

The form of the uncertainty principle given in Equation 40.26 suggests that energy conservation can appear to be violated by an amount ΔE as long as it is only for a short time interval Δt consistent with that equation. We shall use this notion to estimate the rest energies of particles in Chapter 46.

Ⓠuick Quiz 40.8 A particle's location is measured and specified as being exactly at $x = 0$, with *zero* uncertainty in the x direction. How does that location affect the uncertainty of its velocity component in the y direction? **(a)** It does not affect it. **(b)** It makes it infinite. **(c)** It makes it zero.

Werner Heisenberg
German Theoretical Physicist
(1901–1976)
Heisenberg obtained his Ph.D. in 1923 at the University of Munich. While other physicists tried to develop physical models of quantum phenomena, Heisenberg developed an abstract mathematical model called *matrix mechanics*. The more widely accepted physical models were shown to be equivalent to matrix mechanics. Heisenberg made many other significant contributions to physics, including his famous uncertainty principle for which he received a Nobel Prize in Physics in 1932, the prediction of two forms of molecular hydrogen, and theoretical models of the nucleus.

Pitfall Prevention 40.4
The Uncertainty Principle Some students incorrectly interpret the uncertainty principle as meaning that a measurement interferes with the system. For example, if an electron is observed in a hypothetical experiment using an optical microscope, the photon used to see the electron collides with it and makes it move, giving it an uncertainty in momentum. This scenario does *not* represent the basis of the uncertainty principle. The uncertainty principle is independent of the measurement process and is based on the wave nature of matter.

Example 40.6   **Locating an Electron**

The speed of an electron is measured to be 5.00×10^3 m/s to an accuracy of 0.003 00%. Find the minimum uncertainty in determining the position of this electron.

SOLUTION

Conceptualize The fractional value given for the accuracy of the electron's speed can be interpreted as the fractional uncertainty in its momentum. This uncertainty corresponds to a minimum uncertainty in the electron's position through the uncertainty principle.

continued

▶ **40.6** continued

Categorize We evaluate the result using concepts developed in this section, so we categorize this example as a substitution problem.

Assume the electron is moving along the x axis and find the uncertainty in p_x, letting f represent the accuracy of the measurement of its speed:

$$\Delta p_x = m\,\Delta v_x = mfv_x$$

Solve Equation 40.25 for the uncertainty in the electron's position and substitute numerical values:

$$\Delta x \geq \frac{\hbar}{2\,\Delta p_x} = \frac{\hbar}{2mfv_x} = \frac{1.055 \times 10^{-34}\,\text{J}\cdot\text{s}}{2(9.11 \times 10^{-31}\,\text{kg})(0.000\,030\,0)(5.00 \times 10^3\,\text{m/s})}$$

$$= 3.86 \times 10^{-4}\,\text{m} = \boxed{0.386\,\text{mm}}$$

Example 40.7   **The Line Width of Atomic Emissions**

Atoms have quantized energy levels similar to those of Planck's oscillators, although the energy levels of an atom are usually not evenly spaced. When an atom makes a transition between states separated in energy by ΔE, energy is emitted in the form of a photon of frequency $f = \Delta E/h$. Although an excited atom can radiate at any time from $t = 0$ to $t = \infty$, the average time interval after excitation during which an atom radiates is called the **lifetime** τ. If $\tau = 1.0 \times 10^{-8}$ s, use the uncertainty principle to compute the line width Δf produced by this finite lifetime.

SOLUTION

Conceptualize The lifetime τ given for the excited state can be interpreted as the uncertainty Δt in the time at which the transition occurs. This uncertainty corresponds to a minimum uncertainty in the frequency of the radiated photon through the uncertainty principle.

Categorize We evaluate the result using concepts developed in this section, so we categorize this example as a substitution problem.

Use Equation 40.5 to relate the uncertainty in the photon's frequency to the uncertainty in its energy:

$$E = hf \;\rightarrow\; \Delta E = h\,\Delta f \;\rightarrow\; \Delta f = \frac{\Delta E}{h}$$

Use Equation 40.26 to substitute for the uncertainty in the photon's energy, giving the minimum value of Δf:

$$\Delta f \geq \frac{1}{h}\frac{\hbar}{2\,\Delta t} = \frac{1}{h}\frac{h/2\pi}{2\,\Delta t} = \frac{1}{4\pi\,\Delta t} = \frac{1}{4\pi\tau}$$

Substitute for the lifetime of the excited state:

$$\Delta f \geq \frac{1}{4\pi(1.0 \times 10^{-8}\,\text{s})} = \boxed{8.0 \times 10^6\,\text{Hz}}$$

WHAT IF?   What if this same lifetime were associated with a transition that emits a radio wave rather than a visible light wave from an atom? Is the fractional line width $\Delta f/f$ larger or smaller than for the visible light?

Answer Because we are assuming the same lifetime for both transitions, Δf is independent of the frequency of radiation. Radio waves have lower frequencies than light waves, so the ratio $\Delta f/f$ will be larger for the radio waves. Assuming a light-wave frequency f of 6.00×10^{14} Hz, the fractional line width is

$$\frac{\Delta f}{f} = \frac{8.0 \times 10^6\,\text{Hz}}{6.00 \times 10^{14}\,\text{Hz}} = 1.3 \times 10^{-8}$$

This narrow fractional line width can be measured with a sensitive interferometer. Usually, however, temperature and pressure effects overshadow the natural line width and broaden the line through mechanisms associated with the Doppler effect and collisions.

Assuming a radio-wave frequency f of 94.7×10^6 Hz, the fractional line width is

$$\frac{\Delta f}{f} = \frac{8.0 \times 10^6\,\text{Hz}}{94.7 \times 10^6\,\text{Hz}} = 8.4 \times 10^{-2}$$

Therefore, for the radio wave, this same absolute line width corresponds to a fractional line width of more than 8%.

Summary

Concepts and Principles

The characteristics of **blackbody radiation** cannot be explained using classical concepts. Planck introduced the quantum concept and Planck's constant h when he assumed atomic oscillators existing only in discrete energy states were responsible for this radiation. In Planck's model, radiation is emitted in single quantized packets whenever an oscillator makes a transition between discrete energy states. The energy of a packet is

$$E = hf \qquad (40.5)$$

where f is the frequency of the oscillator. Einstein successfully extended Planck's quantum hypothesis to the standing waves of electromagnetic radiation in a cavity used in the blackbody radiation model.

The **photoelectric effect** is a process whereby electrons are ejected from a metal surface when light is incident on that surface. In Einstein's model, light is viewed as a stream of particles, or **photons,** each having energy $E = hf$, where h is Planck's constant and f is the frequency. The maximum kinetic energy of the ejected photoelectron is

$$K_{max} = hf - \phi \qquad (40.11)$$

where ϕ is the **work function** of the metal.

X-rays are scattered at various angles by electrons in a target. In such a scattering event, a shift in wavelength is observed for the scattered x-rays, a phenomenon known as the **Compton effect.** Classical physics does not predict the correct behavior in this effect. If the x-ray is treated as a photon, conservation of energy and linear momentum applied to the photon–electron collisions yields, for the Compton shift,

$$\lambda' - \lambda_0 = \frac{h}{m_e c}(1 - \cos\theta) \qquad (40.13)$$

where m_e is the mass of the electron, c is the speed of light, and θ is the scattering angle.

Light has a dual nature in that it has both wave and particle characteristics. Some experiments can be explained either better or solely by the particle model, whereas others can be explained either better or solely by the wave model.

Every object of mass m and momentum $p = mu$ has wave properties, with a **de Broglie wavelength** given by

$$\lambda = \frac{h}{p} = \frac{h}{mu} \qquad (40.17)$$

By combining a large number of waves, a single region of constructive interference, called a **wave packet,** can be created. The wave packet carries the characteristic of localization like a particle does, but it has wave properties because it is built from waves. For an individual wave in the wave packet, the **phase speed** is

$$v_{phase} = \frac{\omega}{k} \qquad (40.20)$$

For the wave packet as a whole, the **group speed** is

$$v_g = \frac{d\omega}{dk} \qquad (40.21)$$

For a wave packet representing a particle, the group speed can be shown to be the same as the speed of the particle.

The **Heisenberg uncertainty principle** states that if a measurement of the position of a particle is made with uncertainty Δx and a simultaneous measurement of its linear momentum is made with uncertainty Δp_x, the product of the two uncertainties is restricted to

$$\Delta x\, \Delta p_x \geq \frac{\hbar}{2} \qquad (40.25)$$

Another form of the uncertainty principle relates measurements of energy and time:

$$\Delta E\, \Delta t \geq \frac{\hbar}{2} \qquad (40.26)$$

Objective Questions   **1.** denotes answer available in *Student Solutions Manual/Study Guide*

1. Rank the wavelengths of the following quantum particles from the largest to the smallest. If any have equal wavelengths, display the equality in your ranking. (a) a photon with energy 3 eV (b) an electron with kinetic energy 3 eV (c) a proton with kinetic energy 3 eV (d) a photon with energy 0.3 eV (e) an electron with momentum 3 eV/c

2. An x-ray photon is scattered by an originally stationary electron. Relative to the frequency of the incident photon, is the frequency of the scattered photon (a) lower, (b) higher, or (c) unchanged?

3. In a Compton scattering experiment, a photon of energy E is scattered from an electron at rest. After the scattering event occurs, which of the following statements is true? (a) The frequency of the photon is greater than E/h. (b) The energy of the photon is less than E. (c) The wavelength of the photon is less than hc/E. (d) The momentum of the photon increases. (e) None of those statements is true.

4. In a certain experiment, a filament in an evacuated lightbulb carries a current I_1 and you measure the spectrum of light emitted by the filament, which behaves as a black body at temperature T_1. The wavelength emitted with highest intensity (symbolized by λ_{max}) has the value λ_1. You then increase the potential difference across the filament by a factor of 8, and the current increases by a factor of 2. (i) After this change, what is the new value of the temperature of the filament? (a) $16T_1$ (b) $8T_1$ (c) $4T_1$ (d) $2T_1$ (e) still T_1 (ii) What is the new value of the wavelength emitted with highest intensity? (a) $4\lambda_1$ (b) $2\lambda_1$ (c) λ_1 (d) $\frac{1}{2}\lambda_1$ (e) $\frac{1}{4}\lambda_1$

5. Which of the following statements are true according to the uncertainty principle? More than one statement may be correct. (a) It is impossible to simultaneously determine both the position and the momentum of a particle along the same axis with arbitrary accuracy. (b) It is impossible to simultaneously determine both the energy and momentum of a particle with arbitrary accuracy. (c) It is impossible to determine a particle's energy with arbitrary accuracy in a finite amount of time. (d) It is impossible to measure the position of a particle with arbitrary accuracy in a finite amount of time. (e) It is impossible to simultaneously measure both the energy and position of a particle with arbitrary accuracy.

6. A monochromatic light beam is incident on a barium target that has a work function of 2.50 eV. If a potential difference of 1.00 V is required to turn back all the ejected electrons, what is the wavelength of the light

beam? (a) 355 nm (b) 497 nm (c) 744 nm (d) 1.42 pm (e) none of those answers

7. Which of the following is most likely to cause sunburn by delivering more energy to individual molecules in skin cells? (a) infrared light (b) visible light (c) ultraviolet light (d) microwaves (e) Choices (a) through (d) are equally likely.

8. Which of the following phenomena most clearly demonstrates the wave nature of electrons? (a) the photoelectric effect (b) blackbody radiation (c) the Compton effect (d) diffraction of electrons by crystals (e) none of those answers

9. What is the de Broglie wavelength of an electron accelerated from rest through a potential difference of 50.0 V? (a) 0.100 nm (b) 0.139 nm (c) 0.174 nm (d) 0.834 nm (e) none of those answers

10. A proton, an electron, and a helium nucleus all move at speed v. Rank their de Broglie wavelengths from largest to smallest.

11. Consider (a) an electron, (b) a photon, and (c) a proton, all moving in vacuum. Choose all correct answers for each question. (i) Which of the three possess rest energy? (ii) Which have charge? (iii) Which carry energy? (iv) Which carry momentum? (v) Which move at the speed of light? (vi) Which have a wavelength characterizing their motion?

12. An electron and a proton, moving in opposite directions, are accelerated from rest through the same potential difference. Which particle has the longer wavelength? (a) The electron does. (b) The proton does. (c) Both are the same. (d) Neither has a wavelength.

13. Which of the following phenomena most clearly demonstrates the particle nature of light? (a) diffraction (b) the photoelectric effect (c) polarization (d) interference (e) refraction

14. Both an electron and a proton are accelerated to the same speed, and the experimental uncertainty in the speed is the same for the two particles. The positions of the two particles are also measured. Is the minimum possible uncertainty in the electron's position (a) less than the minimum possible uncertainty in the proton's position, (b) the same as that for the proton, (c) more than that for the proton, or (d) impossible to tell from the given information?

Conceptual Questions   1. denotes answer available in *Student Solutions Manual/Study Guide*

1. The opening photograph for this chapter shows a filament of a lightbulb in operation. Look carefully at the last turns of wire at the upper and lower ends of the filament. Why are these turns dimmer than the others?

2. How does the Compton effect differ from the photoelectric effect?

3. If matter has a wave nature, why is this wave-like characteristic not observable in our daily experiences?

4. If the photoelectric effect is observed for one metal, can you conclude that the effect will also be observed

for another metal under the same conditions? Explain.

5. In the photoelectric effect, explain why the stopping potential depends on the frequency of light but not on the intensity.

6. Why does the existence of a cutoff frequency in the photoelectric effect favor a particle theory for light over a wave theory?

7. Which has more energy, a photon of ultraviolet radiation or a photon of yellow light? Explain.

8. All objects radiate energy. Why, then, are we not able to see all objects in a dark room?

9. Is an electron a wave or a particle? Support your answer by citing some experimental results.

10. Suppose a photograph were made of a person's face using only a few photons. Would the result be simply a very faint image of the face? Explain your answer.

11. Why is an electron microscope more suitable than an optical microscope for "seeing" objects less than 1 μm in size?

12. Is light a wave or a particle? Support your answer by citing specific experimental evidence.

13. (a) What does the slope of the lines in Figure 40.11 represent? (b) What does the y intercept represent? (c) How would such graphs for different metals compare with one another?

14. Why was the demonstration of electron diffraction by Davisson and Germer an important experiment?

15. *Iridescence* is the phenomenon that gives shining colors to the feathers of peacocks, hummingbirds (see page 886), resplendent quetzals, and even ducks and grackles. Without pigments, it colors Morpho butterflies (Fig. CQ40.15), Urania moths, some beetles and flies, rainbow trout, and mother-of-pearl in abalone shells. Iridescent colors change as you turn an object. They are produced by a wide variety of intricate structures in different species. Problem 64 in Chapter 38 describes the structures that produce iridescence in a peacock feather. These structures were all unknown until the

invention of the electron microscope. Explain why light microscopes cannot reveal them.

Figure CQ40.15

16. In describing the passage of electrons through a slit and arriving at a screen, physicist Richard Feynman said that "electrons arrive in lumps, like particles, but the probability of arrival of these lumps is determined as the intensity of the waves would be. It is in this sense that the electron behaves sometimes like a particle and sometimes like a wave." Elaborate on this point in your own words. For further discussion, see R. Feynman, *The Character of Physical Law* (Cambridge, MA: MIT Press, 1980), chap. 6.

17. The classical model of blackbody radiation given by the Rayleigh–Jeans law has two major flaws. (a) Identify the flaws and (b) explain how Planck's law deals with them.

Problems available in ENHANCED **WebAssign** Access end-of-chapter problems online at www.webassign.net

Section 40.1 Blackbody Radiation and Planck's Hypothesis
Problems 1–16

Section 40.2 The Photoelectric Effect
Problems 17–24

Section 40.3 The Compton Effect
Problems 25–36

Section 40.4 The Nature of Electromagnetic Waves
Problems 37–38

Section 40.5 The Wave Properties of Particles
Problems 39–48

Section 40.6 A New Model: The Quantum Particle
Problems 49–50

Section 40.7 The Double-Slit Experiment Revisited
Problems 51–53

Section 40.8 The Uncertainty Principle
Problems 54–59

Additional Problems
Problems 60–71

Challenge Problems
Problems 72–76

Solutions to the following Problems are available in the *Student Solutions Manual/Study Guide:*
40.6, 40.8, 40.9, 40.19, 40.29, 40.38, 40.41, 40.44, 40.49, 40.51, 40.55, 40.60, 40.67, 40.68, and 40.75

List of Enhanced Problems

Problem Number	Targeted Feedback in Enhanced WebAssign	Analysis Model Tutorial in Enhanced WebAssign	Master It in Enhanced WebAssign	Watch It in Enhanced WebAssign
40.3	✓			✓
40.5	✓			✓
40.6			✓	
40.8			✓	
40.9	✓			✓
40.10	✓			✓
40.11	✓			✓
40.20	✓			✓
40.23		✓		
40.29			✓	
40.31	✓			✓
40.38	✓			✓
40.51	✓		✓	
40.53	✓			✓
40.55			✓	
40.61		✓	✓	

Quantum Mechanics

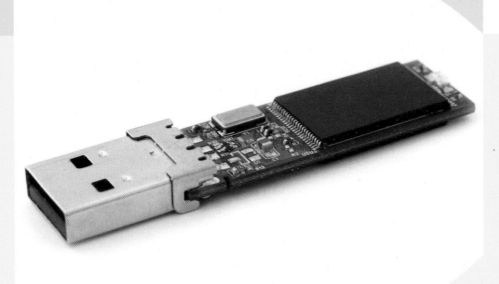

In this chapter, we introduce quantum mechanics, an extremely successful theory for explaining the behavior of microscopic particles. This theory, developed in the 1920s by Erwin Schrödinger, Werner Heisenberg, and others, enables us to understand a host of phenomena involving atoms, molecules, nuclei, and solids. The discussion in this chapter follows from the quantum particle model that was developed in Chapter 40 and incorporates some of the features of the waves under boundary conditions model that was explored in Chapter 18. We also discuss practical applications of quantum mechanics, including the scanning tunneling microscope and nanoscale devices that may be used in future quantum computers. Finally, we shall return to the simple harmonic oscillator that was introduced in Chapter 15 and examine it from a quantum mechanical point of view.

An opened flash drive of the type used as an external data storage device for a computer. Flash drives are employed extensively in computers, digital cameras, cell phones, and other devices. Writing data to and erasing data from flash drives incorporate the phenomenon of quantum tunneling, which we explore in this chapter. *(Image copyright Vasilius, 2009. Used under license from Shutterstock.com)*

41.1 The Wave Function

In Chapter 40, we introduced some new and strange ideas. In particular, we concluded on the basis of experimental evidence that both matter and electromagnetic radiation are sometimes best modeled as particles and sometimes as waves, depending on the phenomenon being observed. We can improve our understanding of quantum physics by making another connection between particles and waves using the notion of probability, a concept that was introduced in Chapter 40.

We begin by discussing electromagnetic radiation using the particle model. The probability per unit volume of finding a photon in a given region of space at an instant of time is proportional to the number of photons per unit volume at that time:

$$\frac{\text{Probability}}{V} \propto \frac{N}{V}$$

The number of photons per unit volume is proportional to the intensity of the radiation:

$$\frac{N}{V} \propto I$$

Now, let's form a connection between the particle model and the wave model by recalling that the intensity of electromagnetic radiation is proportional to the square of the electric field amplitude E for the electromagnetic wave (Eq. 34.24):

$$I \propto E^2$$

Equating the beginning and the end of this series of proportionalities gives

$$\frac{\text{Probability}}{V} \propto E^2 \tag{41.1}$$

Therefore, for electromagnetic radiation, the probability per unit volume of finding a particle associated with this radiation (the photon) is proportional to the square of the amplitude of the associated electromagnetic wave.

Recognizing the wave–particle duality of both electromagnetic radiation and matter, we should suspect a parallel proportionality for a material particle: the probability per unit volume of finding the particle is proportional to the square of the amplitude of a wave representing the particle. In Chapter 40, we learned that there is a de Broglie wave associated with every particle. The amplitude of the de Broglie wave associated with a particle is not a measurable quantity because the wave function representing a particle is generally a complex function as we discuss below. In contrast, the electric field for an electromagnetic wave is a real function. The matter analog to Equation 41.1 relates the square of the amplitude of the wave to the probability per unit volume of finding the particle. Hence, the amplitude of the wave associated with the particle is called the **probability amplitude,** or the **wave function,** and it has the symbol Ψ.

In general, the complete wave function Ψ for a system depends on the positions of all the particles in the system and on time; therefore, it can be written $\Psi(\vec{\mathbf{r}}_1, \vec{\mathbf{r}}_2, \vec{\mathbf{r}}_3, \ldots, \vec{\mathbf{r}}_j, \ldots, t)$, where $\vec{\mathbf{r}}_j$ is the position vector of the jth particle in the system. For many systems of interest, including all those we study in this text, the wave function Ψ is mathematically separable in space and time and can be written as a product of a space function ψ for one particle of the system and a complex time function:[1]

$$\Psi(\vec{\mathbf{r}}_1, \vec{\mathbf{r}}_2, \vec{\mathbf{r}}_3, \ldots, \vec{\mathbf{r}}_j, \ldots, t) = \psi(\vec{\mathbf{r}}_j)e^{-i\omega t} \tag{41.2}$$

▶ Space- and time-dependent wave function Ψ

where ω ($= 2\pi f$) is the angular frequency of the wave function and $i = \sqrt{-1}$.

For any system in which the potential energy is time-independent and depends only on the positions of particles within the system, the important information about the system is contained within the space part of the wave function. The time part is simply the factor $e^{-i\omega t}$. Therefore, an understanding of ψ is the critical aspect of a given problem.

The wave function ψ is often complex-valued. The absolute square $|\psi|^2 = \psi^*\psi$, where ψ^* is the complex conjugate[2] of ψ, is always real and positive and is proportional to the probability per unit volume of finding a particle at a given point at some instant. The wave function contains within it all the information that can be known about the particle.

---

[1]The standard form of a complex number is $a + ib$. The notation $e^{i\theta}$ is equivalent to the standard form as follows:

$$e^{i\theta} = \cos\theta + i\sin\theta$$

Therefore, the notation $e^{-i\omega t}$ in Equation 41.2 is equivalent to $\cos(-\omega t) + i\sin(-\omega t) = \cos\omega t - i\sin\omega t$.

[2]For a complex number $z = a + ib$, the complex conjugate is found by changing i to $-i$: $z^* = a - ib$. The product of a complex number and its complex conjugate is always real and positive. That is, $z^*z = (a - ib)(a + ib) = a^2 - (ib)^2 = a^2 - (i)^2b^2 = a^2 + b^2$.

Although ψ cannot be measured, we can measure the real quantity $|\psi|^2$, which can be interpreted as follows. If ψ represents a single particle, then $|\psi|^2$—called the **probability density**—is the relative probability per unit volume that the particle will be found at any given point in the volume. This interpretation can also be stated in the following manner. If dV is a small volume element surrounding some point, the probability of finding the particle in that volume element is

$$P(x, y, z)\ dV = |\psi|^2\ dV \qquad (41.3)$$

◀ **Probability density** $|\psi|^2$

This probabilistic interpretation of the wave function was first suggested by Max Born (1882–1970) in 1928. In 1926, Erwin Schrödinger proposed a wave equation that describes the manner in which the wave function changes in space and time. The *Schrödinger wave equation*, which we shall examine in Section 41.3, represents a key element in the theory of quantum mechanics.

The concepts of quantum mechanics, strange as they sometimes may seem, developed from classical ideas. In fact, when the techniques of quantum mechanics are applied to macroscopic systems, the results are essentially identical to those of classical physics. This blending of the two approaches occurs when the de Broglie wavelength is small compared with the dimensions of the system. The situation is similar to the agreement between relativistic mechanics and classical mechanics when $v \ll c$.

In Section 40.5, we found that the de Broglie equation relates the momentum of a particle to its wavelength through the relation $p = h/\lambda$. If an ideal free particle has a precisely known momentum p_x, its wave function is an infinitely long sinusoidal wave of wavelength $\lambda = h/p_x$ and the particle has equal probability of being at any point along the x axis (Fig. 40.18a). The wave function ψ for such a free particle moving along the x axis can be written as

$$\psi(x) = Ae^{ikx} \qquad (41.4)$$

◀ **Wave function for a free particle**

where A is a constant amplitude and $k = 2\pi/\lambda$ is the angular wave number (Eq. 16.8) of the wave representing the particle.[3]

One-Dimensional Wave Functions and Expectation Values

This section discusses only one-dimensional systems, where the particle must be located along the x axis, so the probability $|\psi|^2\ dV$ in Equation 41.3 is modified to become $|\psi|^2\ dx$. The probability that the particle will be found in the infinitesimal interval dx around the point x is

$$P(x)\ dx = |\psi|^2\ dx \qquad (41.5)$$

Although it is not possible to specify the position of a particle with complete certainty, it is possible through $|\psi|^2$ to specify the probability of observing it in a region surrounding a given point x. The probability of finding the particle in the arbitrary interval $a \le x \le b$ is

$$P_{ab} = \int_{a}^{b} |\psi|^2\ dx \qquad (41.6)$$

The probability P_{ab} is the area under the curve of $|\psi|^2$ versus x between the points $x = a$ and $x = b$ as in Figure 41.1.

Experimentally, there is a finite probability of finding a particle in an interval near some point at some instant. The value of that probability must lie between the

> **Pitfall Prevention 41.1**
> **The Wave Function Belongs to a System** The common language in quantum mechanics is to associate a wave function with a particle. The wave function, however, is determined by the particle *and* its interaction with its environment, so it more rightfully belongs to a system. In many cases, the particle is the only part of the system that experiences a change, which is why the common language has developed. You will see examples in the future in which it is more proper to think of the system wave function rather than the particle wave function.

> The probability of a particle being in the interval $a \le x \le b$ is the area under the probability density curve from a to b.

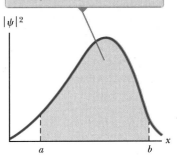

Figure 41.1 An arbitrary probability density curve for a particle.

---

[3]For the free particle, the full wave function, based on Equation 41.2, is

$$\Psi(x, t) = Ae^{ikx}e^{-i\omega t} = Ae^{i(kx - \omega t)} = A[\cos(kx - \omega t) + i\sin(kx - \omega t)]$$

The real part of this wave function has the same form as the waves we added together to form wave packets in Section 40.6.

limits 0 and 1. For example, if the probability is 0.30, there is a 30% chance of finding the particle in the interval.

Because the particle must be somewhere along the x axis, the sum of the probabilities over all values of x must be 1:

Normalization condition on ψ ▶

$$\int_{-\infty}^{\infty} |\psi|^2 \, dx = 1 \qquad (41.7)$$

Any wave function satisfying Equation 41.7 is said to be **normalized.** Normalization is simply a statement that the particle exists at some point in space.

Once the wave function for a particle is known, it is possible to calculate the average position at which you would expect to find the particle after many measurements. This average position is called the **expectation value** of x and is defined by the equation

Expectation value ▶
for position x

$$\langle x \rangle \equiv \int_{-\infty}^{\infty} \psi^* x \psi \, dx \qquad (41.8)$$

(Brackets, $\langle \ldots \rangle$, are used to denote expectation values.) Furthermore, one can find the expectation value of any function $f(x)$ associated with the particle by using the following equation:[4]

Expectation value for ▶
a function $f(x)$

$$\langle f(x) \rangle \equiv \int_{-\infty}^{\infty} \psi^* f(x) \psi \, dx \qquad (41.9)$$

Quick Quiz 41.1 Consider the wave function for the free particle, Equation 41.4. At what value of x is the particle most likely to be found at a given time? **(a)** at $x = 0$ **(b)** at small nonzero values of x **(c)** at large values of x **(d)** anywhere along the x axis

Example 41.1   A Wave Function for a Particle

Consider a particle whose wave function is graphed in Figure 41.2 and is given by

$$\psi(x) = Ae^{-ax^2}$$

(A) What is the value of A if this wave function is normalized?

SOLUTION

Conceptualize The particle is not a free particle because the wave function is not a sinusoidal function. Figure 41.2 indicates that the particle is constrained to remain close to $x = 0$ at all times. Think of a physical system in which the particle always stays close to a given point. Examples of such systems are a block on a spring, a marble at the bottom of a bowl, and the bob of a simple pendulum.

Categorize Because the statement of the problem describes the wave nature of a particle, this example requires a quantum approach rather than a classical approach.

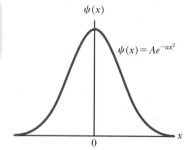

Figure 41.2 (Example 41.1) A symmetric wave function for a particle, given by $\psi(x) = Ae^{-ax^2}$.

Analyze Apply the normalization condition, Equation 41.7, to the wave function:

$$\int_{-\infty}^{\infty} |\psi|^2 \, dx = \int_{-\infty}^{\infty} (Ae^{-ax^2})^2 \, dx = A^2 \int_{-\infty}^{\infty} e^{-2ax^2} \, dx = 1$$

---

[4]Expectation values are analogous to "weighted averages," in which each possible value of a function is multiplied by the probability of the occurrence of that value before summing over all possible values. We write the expectation value as $\int_{-\infty}^{\infty} \psi^* f(x) \psi \, dx$ rather than $\int_{-\infty}^{\infty} f(x) \psi^2 \, dx$ because $f(x)$ may be represented by an operator (such as a derivative) rather than a simple multiplicative function in more advanced treatments of quantum mechanics. In these situations, the operator is applied only to ψ and not to ψ^*.

▶ **41.1** continued

Express the integral as the sum of two integrals:	$$(1)\quad A^2\int_{-\infty}^{\infty}e^{-2ax^2}\,dx=A^2\left(\int_{0}^{\infty}e^{-2ax^2}\,dx+\int_{-\infty}^{0}e^{-2ax^2}\,dx\right)=1$$
Change the integration variable from x to $-x$ in the second integral:	$$\int_{-\infty}^{0}e^{-2ax^2}\,dx=\int_{\infty}^{0}e^{-2a(-x)^2}\,(-dx)=-\int_{\infty}^{0}e^{-2ax^2}\,dx$$
Reverse the order of the limits, which introduces a negative sign:	$$-\int_{\infty}^{0}e^{-2ax^2}\,dx=\int_{0}^{\infty}e^{-2ax^2}\,dx$$
Substitute this expression for the second integral in Equation (1):	$$A^2\left(\int_{0}^{\infty}e^{-2ax^2}\,dx+\int_{0}^{\infty}e^{-2ax^2}\,dx\right)=1$$ $$(2)\quad 2A^2\int_{0}^{\infty}e^{-2ax^2}\,dx=1$$
Evaluate the integral with the help of Table B.6 in Appendix B:	$$\int_{0}^{\infty}e^{-2ax^2}\,dx=\tfrac{1}{2}\sqrt{\frac{\pi}{2a}}$$
Substitute this result into Equation (2) and solve for A:	$$2A^2\left(\tfrac{1}{2}\sqrt{\frac{\pi}{2a}}\right)=1\;\rightarrow\;A=\left(\frac{2a}{\pi}\right)^{1/4}$$

(B) What is the expectation value of x for this particle?

SOLUTION

Evaluate the expectation value using Equation 41.8:	$$\langle x\rangle\equiv\int_{-\infty}^{\infty}\psi^*x\psi\,dx=\int_{-\infty}^{\infty}(Ae^{-ax^2})x(Ae^{-ax^2})\,dx$$ $$=A^2\int_{-\infty}^{\infty}xe^{-2ax^2}\,dx$$
As in part (A), express the integral as a sum of two integrals:	$$(3)\quad \langle x\rangle=A^2\left(\int_{0}^{\infty}xe^{-2ax^2}\,dx+\int_{-\infty}^{0}xe^{-2ax^2}\,dx\right)$$
Change the integration variable from x to $-x$ in the second integral:	$$\int_{-\infty}^{0}xe^{-2ax^2}\,dx=\int_{\infty}^{0}-xe^{-2a(-x)^2}\,(-dx)=\int_{\infty}^{0}xe^{-2ax^2}\,dx$$
Reverse the order of the limits, which introduces a negative sign:	$$\int_{\infty}^{0}xe^{-2ax^2}\,dx=-\int_{0}^{\infty}xe^{-2ax^2}\,dx$$
Substitute this expression for the second integral in Equation (3):	$$\langle x\rangle=A^2\left(\int_{0}^{\infty}xe^{-2ax^2}\,dx-\int_{0}^{\infty}xe^{-2ax^2}\,dx\right)=\;0$$

Finalize Given the symmetry of the wave function around $x=0$ in Figure 41.2, it is not surprising that the average position of the particle is at $x=0$. In Section 41.7, we show that the wave function studied in this example represents the lowest-energy state of the quantum harmonic oscillator.

41.2 Analysis Model: Quantum Particle Under Boundary Conditions

The free particle discussed in Section 41.1 has no boundary conditions; it can be anywhere in space. The particle in Example 41.1 is not a free particle. Figure 41.2 shows that the particle is always restricted to positions near $x=0$. In this section, we shall investigate the effects of restrictions on the motion of a quantum particle.

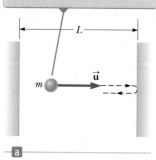

a

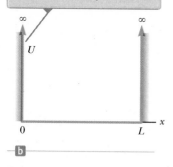

b

Figure 41.3 (a) The particle in a box. (b) The potential energy function for the system.

A Particle in a Box

We begin by applying some of the ideas we have developed to a simple physical problem, a particle confined to a one-dimensional region of space, called the *particle-in-a-box* problem (even though the "box" is one-dimensional!). From a classical viewpoint, if a particle is bouncing elastically back and forth along the x axis between two impenetrable walls separated by a distance L as in Figure 41.3a, it can be modeled as a particle under constant speed. If the speed of the particle is u, the magnitude of its momentum mu remains constant as does its kinetic energy. (Recall that in Chapter 39 we used u for particle speed to distinguish it from v, the speed of a reference frame.) Classical physics places no restrictions on the values of a particle's momentum and energy. The quantum-mechanical approach to this problem is quite different and requires that we find the appropriate wave function consistent with the conditions of the situation.

Because the walls are impenetrable, there is zero probability of finding the particle outside the box, so the wave function $\psi(x)$ must be zero for $x < 0$ and $x > L$. To be a mathematically well-behaved function, $\psi(x)$ must be continuous in space. There must be no discontinuous jumps in the value of the wave function at any point.[5] Therefore, if ψ is zero outside the walls, it must also be zero *at* the walls; that is, $\psi(0) = 0$ and $\psi(L) = 0$. Only those wave functions that satisfy these boundary conditions are allowed.

Figure 41.3b, a graphical representation of the particle-in-a-box problem, shows the potential energy of the particle–environment system as a function of the position of the particle. As long as the particle is inside the box, the potential energy of the system does not depend on the location of the particle and we can choose its constant value to be zero. Outside the box, we must ensure that the wave function is zero. We can do so by defining the system's potential energy as infinitely large if the particle were outside the box. Therefore, the only way a particle could be outside the box is if the system has an infinite amount of energy, which is impossible.

The wave function for a particle in the box can be expressed as a real sinusoidal function:[6]

$$\psi(x) = A \sin\left(\frac{2\pi x}{\lambda}\right) \tag{41.10}$$

where λ is the de Broglie wavelength associated with the particle. This wave function must satisfy the boundary conditions at the walls. The boundary condition $\psi(0) = 0$ is satisfied already because the sine function is zero when $x = 0$. The boundary condition $\psi(L) = 0$ gives

$$\psi(L) = 0 = A \sin\left(\frac{2\pi L}{\lambda}\right)$$

which can only be true if

$$\frac{2\pi L}{\lambda} = n\pi \quad \rightarrow \quad \lambda = \frac{2L}{n} \tag{41.11}$$

where $n = 1, 2, 3, \ldots$. Therefore, only certain wavelengths for the particle are allowed! Each of the allowed wavelengths corresponds to a quantum state for the system, and n is the quantum number. Incorporating Equation 41.11 in Equation 41.10 gives

◀ **Wave functions for a particle in a box**

$$\psi_n(x) = A \sin\left(\frac{2\pi x}{2L/n}\right) = A \sin\left(\frac{n\pi x}{L}\right) \tag{41.12}$$

---

[5]If the wave function were not continuous at a point, the derivative of the wave function at that point would be infinite. This result leads to difficulties in the Schrödinger equation, for which the wave function is a solution as discussed in Section 41.3.

[6]We shall show this result explicitly in Section 41.3.

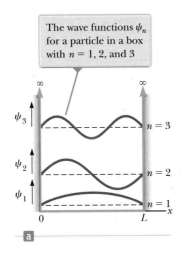

The wave functions ψ_n for a particle in a box with $n = 1, 2,$ and 3

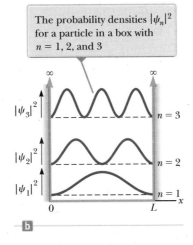

The probability densities $|\psi_n|^2$ for a particle in a box with $n = 1, 2,$ and 3

Figure 41.4 The first three allowed states for a particle confined to a one-dimensional box. The states are shown superimposed on the potential energy function of Figure 41.3b. The wave functions and probability densities are plotted vertically from separate axes that are offset vertically for clarity. The positions of these axes on the potential energy function suggest the relative energies of the states.

Normalizing this wave function shows that $A = \sqrt{2/L}$. (See Problem 18 in Enhanced WebAssign.) Therefore, the normalized wave function for the particle in a box is

$$\psi_n(x) = \sqrt{\frac{2}{L}} \sin\left(\frac{n\pi x}{L}\right) \tag{41.13}$$

◀ **Normalized wave function for a particle in a box**

Figures 41.4a and b are graphical representations of ψ_n versus x and $|\psi_n|^2$ versus x for $n = 1, 2,$ and 3 for the particle in a box.[7] Although a general wave function ψ can have positive and negative values, $|\psi|^2$ is always positive. Because $|\psi|^2$ represents a probability density, a negative value for $|\psi|^2$ would be meaningless.

Further inspection of Figure 41.4b shows that $|\psi|^2$ is zero at the boundaries, satisfying our boundary conditions. In addition, $|\psi|^2$ is zero at other points, depending on the value of n. For $n = 2$, $|\psi_2|^2 = 0$ at $x = L/2$; for $n = 3$, $|\psi_3|^2 = 0$ at $x = L/3$ and at $x = 2L/3$. The number of zero points increases by one each time the quantum number increases by one.

Because the wavelengths of the particle are restricted by the condition $\lambda = 2L/n$, the magnitude of the momentum of the particle is also restricted to specific values, which can be found from the expression for the de Broglie wavelength, Equation 40.17:

$$p = \frac{h}{\lambda} = \frac{h}{2L/n} = \frac{nh}{2L}$$

We have chosen the potential energy of the system to be zero when the particle is inside the box. Therefore, the energy of the system is simply the kinetic energy of the particle and the allowed values are given by

$$E_n = \tfrac{1}{2}mu^2 = \frac{p^2}{2m} = \frac{(nh/2L)^2}{2m}$$

$$E_n = \left(\frac{h^2}{8mL^2}\right)n^2 \qquad n = 1, 2, 3, \ldots \tag{41.14}$$

◀ **Quantized energies for a particle in a box**

This expression shows that the energy of the particle is quantized. The lowest allowed energy corresponds to the **ground state,** which is the lowest energy state for any system. For the particle in a box, the ground state corresponds to $n = 1$, for which $E_1 = h^2/8mL^2$. Because $E_n = n^2E_1$, the **excited states** corresponding to $n = 2, 3, 4, \ldots$ have energies given by $4E_1, 9E_1, 16E_1, \ldots$.

Pitfall Prevention 41.2
Reminder: Energy Belongs to a System We often refer to the energy of a particle in commonly used language. As in Pitfall Prevention 41.1, we are actually describing the energy of the *system* of the particle and whatever environment is establishing the impenetrable walls. For the particle in a box, the only type of energy is kinetic energy belonging to the particle, which is the origin of the common description.

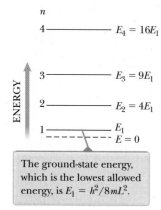

The ground-state energy, which is the lowest allowed energy, is $E_1 = h^2/8mL^2$.

Figure 41.5 Energy-level diagram for a particle confined to a one-dimensional box of length L.

Figure 41.5 is an energy-level diagram describing the energy values of the allowed states. Because the lowest energy of the particle in a box is not zero, then, according to quantum mechanics, the particle can never be at rest! The smallest energy it can have, corresponding to $n = 1$, is called the **ground-state energy.** This result contradicts the classical viewpoint, in which $E = 0$ is an acceptable state, as are *all* positive values of E.

Quick Quiz 41.2 Consider an electron, a proton, and an alpha particle (a helium nucleus), each trapped separately in identical boxes. **(i)** Which particle corresponds to the highest ground-state energy? (a) the electron (b) the proton (c) the alpha particle (d) The ground-state energy is the same in all three cases. **(ii)** Which particle has the longest wavelength when the system is in the ground state? (a) the electron (b) the proton (c) the alpha particle (d) All three particles have the same wavelength.

Quick Quiz 41.3 A particle is in a box of length L. Suddenly, the length of the box is increased to $2L$. What happens to the energy levels shown in Figure 41.5? **(a)** nothing; they are unaffected. **(b)** They move farther apart. **(c)** They move closer together.

Example 41.2   **Microscopic and Macroscopic Particles in Boxes**

(A) An electron is confined between two impenetrable walls 0.200 nm apart. Determine the energy levels for the states $n = 1, 2$, and 3.

SOLUTION

Conceptualize In Figure 41.3a, imagine that the particle is an electron and the walls are very close together.

Categorize We evaluate the energy levels using an equation developed in this section, so we categorize this example as a substitution problem.

Use Equation 41.14 for the $n = 1$ state:

$$E_1 = \frac{h^2}{8m_e L^2}(1)^2 = \frac{(6.63 \times 10^{-34} \, \text{J} \cdot \text{s})^2}{8(9.11 \times 10^{-31} \, \text{kg})(2.00 \times 10^{-10} \, \text{m})^2}$$
$$= 1.51 \times 10^{-18} \, \text{J} = \boxed{9.42 \, \text{eV}}$$

Using $E_n = n^2 E_1$, find the energies of the $n = 2$ and $n = 3$ states:

$$E_2 = (2)^2 E_1 = 4(9.42 \, \text{eV}) = \boxed{37.7 \, \text{eV}}$$
$$E_3 = (3)^2 E_1 = 9(9.42 \, \text{eV}) = \boxed{84.8 \, \text{eV}}$$

(B) Find the speed of the electron in the $n = 1$ state.

SOLUTION

Solve the classical expression for kinetic energy for the particle speed:

$$K = \tfrac{1}{2}m_e u^2 \rightarrow u = \sqrt{\frac{2K}{m_e}}$$

Recognize that the kinetic energy of the particle is equal to the system energy and substitute E_n for K:

$$(1) \quad u = \sqrt{\frac{2E_n}{m_e}}$$

Substitute numerical values from part (A):

$$u = \sqrt{\frac{2(1.51 \times 10^{-18} \, \text{J})}{9.11 \times 10^{-31} \, \text{kg}}} = \boxed{1.82 \times 10^6 \, \text{m/s}}$$

Simply placing the electron in the box results in a *minimum* speed of the electron equal to 0.6% of the speed of light!

(C) A 0.500-kg baseball is confined between two rigid walls of a stadium that can be modeled as a box of length 100 m. Calculate the minimum speed of the baseball.

▶ **41.2** continued

SOLUTION

Conceptualize In Figure 41.3a, imagine that the particle is a baseball and the walls are those of the stadium.

Categorize This part of the example is a substitution problem in which we apply a quantum approach to a macroscopic object.

Use Equation 41.14 for the $n = 1$ state:

$$E_1 = \frac{h^2}{8mL^2}(1)^2 = \frac{(6.63 \times 10^{-34}\,\text{J}\cdot\text{s})^2}{8(0.500\,\text{kg})(100\,\text{m})^2} = 1.10 \times 10^{-71}\,\text{J}$$

Use Equation (1) to find the speed:

$$u = \sqrt{\frac{2(1.10 \times 10^{-71}\,\text{J})}{0.500\,\text{kg}}} = \boxed{6.63 \times 10^{-36}\,\text{m/s}}$$

This speed is so small that the object can be considered to be at rest, which is what one would expect for the minimum speed of a macroscopic object.

WHAT IF? What if a sharp line drive is hit so that the baseball is moving with a speed of 150 m/s? What is the quantum number of the state in which the baseball now resides?

Answer We expect the quantum number to be very large because the baseball is a macroscopic object.

Evaluate the kinetic energy of the baseball: $\frac{1}{2}mu^2 = \frac{1}{2}(0.500\,\text{kg})(150\,\text{m/s})^2 = 5.62 \times 10^3\,\text{J}$

From Equation 41.14, calculate the quantum number n:

$$n = \sqrt{\frac{8mL^2E_n}{h^2}} = \sqrt{\frac{8(0.500\,\text{kg})(100\,\text{m})^2(5.62 \times 10^3\,\text{J})}{(6.63 \times 10^{-34}\,\text{J}\cdot\text{s})^2}} = 2.26 \times 10^{37}$$

This result is a tremendously large quantum number. As the baseball pushes air out of the way, hits the ground, and rolls to a stop, it moves through more than 10^{37} quantum states. These states are so close together in energy that we cannot observe the transitions from one state to the next. Rather, we see what appears to be a smooth variation in the speed of the ball. The quantum nature of the universe is simply not evident in the motion of macroscopic objects.

Example 41.3 **The Expectation Values for the Particle in a Box**

A particle of mass m is confined to a one-dimensional box between $x = 0$ and $x = L$. Find the expectation value of the position x of the particle in the state characterized by quantum number n.

SOLUTION

Conceptualize Figure 41.4b shows that the probability for the particle to be at a given location varies with position within the box. Can you predict what the expectation value of x will be from the symmetry of the wave functions?

Categorize The statement of the example categorizes the problem for us: we focus on a quantum particle in a box and on the calculation of its expectation value of x.

Analyze In Equation 41.8, the integration from $-\infty$ to ∞ reduces to the limits 0 to L because $\psi = 0$ everywhere except in the box.

Substitute Equation 41.13 into Equation 41.8 to find the expectation value for x:

$$\langle x \rangle = \int_{-\infty}^{\infty} \psi_n^{*} x \psi_n \, dx = \int_0^L x \left[\sqrt{\frac{2}{L}} \sin\left(\frac{n\pi x}{L}\right) \right]^2 dx$$

$$= \frac{2}{L} \int_0^L x \sin^2\left(\frac{n\pi x}{L}\right) dx$$

continued

▶ **41.3** continued

Evaluate the integral by consulting an integral table or by mathematical integration:[8]

$$\langle x \rangle = \frac{2}{L} \left[\frac{x^2}{4} - \frac{x \sin\left(2\dfrac{n\pi x}{L}\right)}{4\dfrac{n\pi}{L}} - \frac{\cos\left(2\dfrac{n\pi x}{L}\right)}{8\left(\dfrac{n\pi}{L}\right)^2} \right]_0^L$$

$$= \frac{2}{L} \left[\frac{L^2}{4} \right] = \boxed{\frac{L}{2}}$$

Finalize This result shows that the expectation value of x is at the center of the box for all values of n, which you would expect from the symmetry of the square of the wave functions (the probability density) about the center (Fig. 41.4b).

The $n = 2$ wave function in Figure 41.4b has a value of zero at the midpoint of the box. Can the expectation value of the particle be at a position at which the particle has zero probability of existing? Remember that the expectation value is the *average* position. Therefore, the particle is as likely to be found to the right of the midpoint as to the left, so its average position is at the midpoint even though its probability of being there is zero. As an analogy, consider a group of students for whom the average final examination score is 50%. There is no requirement that some student achieve a score of exactly 50% for the average of all students to be 50%.

Boundary Conditions on Particles in General

The discussion of the particle in a box is very similar to the discussion in Chapter 18 of standing waves on strings:

- Because the ends of the string must be nodes, the wave functions for allowed waves must be zero at the boundaries of the string. Because the particle in a box cannot exist outside the box, the allowed wave functions for the particle must be zero at the boundaries.
- The boundary conditions on the string waves lead to quantized wavelengths and frequencies of the waves. The boundary conditions on the wave function for the particle in a box lead to quantized wavelengths and frequencies of the particle.

In quantum mechanics, it is very common for particles to be subject to boundary conditions. We therefore introduce a new analysis model, the **quantum particle under boundary conditions.** In many ways, this model is similar to the waves under boundary conditions model studied in Section 18.3. In fact, the allowed wavelengths for the wave function of a particle in a box (Eq. 41.11) are identical in form to the allowed wavelengths for mechanical waves on a string fixed at both ends (Eq. 18.4).

The quantum particle under boundary conditions model *differs* in some ways from the waves under boundary conditions model:

- In most cases of quantum particles, the wave function is *not* a simple sinusoidal function like the wave function for waves on strings. Furthermore, the wave function for a quantum particle may be a complex function.
- For a quantum particle, frequency is related to energy through $E = hf$, so the quantized frequencies lead to quantized energies.
- There may be no stationary "nodes" associated with the wave function of a quantum particle under boundary conditions. Systems more complicated than the particle in a box have more complicated wave functions, and some boundary conditions may not lead to zeroes of the wave function at fixed points.

---

[8]To integrate this function, first replace $\sin^2 (n\pi x/L)$ with $\frac{1}{2}(1 - \cos 2n\pi x/L)$ (refer to Table B.3 in Appendix B), which allows $\langle x \rangle$ to be expressed as two integrals. The second integral can then be evaluated by partial integration (Section B.7 in Appendix B).

In general,

> an interaction of a quantum particle with its environment represents one or more boundary conditions, and, if the interaction restricts the particle to a finite region of space, results in quantization of the energy of the system.

Boundary conditions on quantum wave functions are related to the coordinates describing the problem. For the particle in a box, the wave function must be zero at two values of x. In the case of a three-dimensional system such as the hydrogen atom we shall discuss in Chapter 42, the problem is best presented in *spherical coordinates*. These coordinates, an extension of the plane polar coordinates introduced in Section 3.1, consist of a radial coordinate r and two angular coordinates. The generation of the wave function and application of the boundary conditions for the hydrogen atom are beyond the scope of this book. We shall, however, examine the behavior of some of the hydrogen-atom wave functions in Chapter 42.

Boundary conditions on wave functions that exist for all values of x require that the wave function approach zero as $x \to \infty$ (so that the wave function can be normalized) and remain finite as $x \to 0$. One boundary condition on any angular parts of wave functions is that adding 2π radians to the angle must return the wave function to the same value because an addition of 2π results in the same angular position.

Analysis Model **Quantum Particle Under Boundary Conditions**

Imagine a particle described by quantum physics that is subject to one or more boundary conditions. If the particle is restricted to a finite region of space by the boundary conditions, the energy of the system is quantized. Associated with each quantized energy is a quantum state characterized by a wave function and a quantum number.

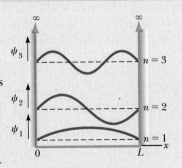

Examples:

- an electron in a quantum dot cannot escape, quantizing the energies of the electron (Section 41.4)
- an electron in a hydrogen atom is restricted to stay near the nucleus of the atom, quantizing the energies of the atom (Chapter 42)
- two atoms are bound to form a diatomic molecule, quantizing the energies of vibration and rotation of the molecule (Chapter 43)
- a proton is trapped in a nucleus, quantizing its energy levels (Chapter 44)

41.3 The Schrödinger Equation

In Section 34.3, we discussed a linear wave equation for electromagnetic radiation that follows from Maxwell's equations. The waves associated with particles also satisfy a wave equation. The wave equation for material particles is different from that associated with photons because material particles have a nonzero rest energy. The appropriate wave equation was developed by Schrödinger in 1926. In analyzing the behavior of a quantum system, the approach is to determine a solution to this equation and then apply the appropriate boundary conditions to the solution. This process yields the allowed wave functions and energy levels of the system under consideration. Proper manipulation of the wave function then enables one to calculate all measurable features of the system.

The Schrödinger equation as it applies to a particle of mass m confined to moving along the x axis and interacting with its environment through a potential energy function $U(x)$ is

$$-\frac{\hbar^2}{2m}\frac{d^2\psi}{dx^2} + U\psi = E\psi$$

(41.15) ◀ Time–independent Schrödinger equation

Erwin Schrödinger
Austrian Theoretical Physicist
(1887–1961)
Schrödinger is best known as one of
the creators of quantum mechanics. His
approach to quantum mechanics was
demonstrated to be mathematically
equivalent to the more abstract matrix
mechanics developed by Heisenberg.
Schrödinger also produced important papers in the fields of statistical
mechanics, color vision, and general
relativity.

where E is a constant equal to the total energy of the system (the particle and its environment). Because this equation is independent of time, it is commonly referred to as the **time-independent Schrödinger equation.** (We shall not discuss the time-dependent Schrödinger equation in this book.)

The Schrödinger equation is consistent with the principle of conservation of mechanical energy for an isolated system with no nonconservative forces acting. Problem 44 in Enhanced WebAssign shows, both for a free particle and a particle in a box, that the first term in the Schrödinger equation reduces to the kinetic energy of the particle multiplied by the wave function. Therefore, Equation 41.15 indicates that the total energy of the system is the sum of the kinetic energy and the potential energy and that the total energy is a constant: $K + U = E =$ constant.

In principle, if the potential energy function U for a system is known, one can solve Equation 41.15 and obtain the wave functions and energies for the allowed states of the system. In addition, in many cases, the wave function ψ must satisfy boundary conditions. Therefore, once we have a preliminary solution to the Schrödinger equation, we impose the following conditions to find the exact solution and the allowed energies:

- ψ must be normalizable. That is, Equation 41.7 must be satisfied.
- ψ must go to 0 as $x \to \pm\infty$ and remain finite as $x \to 0$.
- ψ must be continuous in x and be single-valued everywhere; solutions to Equation 41.15 in different regions must join smoothly at the boundaries between the regions.
- $d\psi/dx$ must be finite, continuous, and single-valued everywhere for finite values of U. If $d\psi/dx$ were not continuous, we would not be able to evaluate the second derivative $d^2\psi/dx^2$ in Equation 41.15 at the point of discontinuity.

The task of solving the Schrödinger equation may be very difficult, depending on the form of the potential energy function. As it turns out, the Schrödinger equation is extremely successful in explaining the behavior of atomic and nuclear systems, whereas classical physics fails to explain this behavior. Furthermore, when quantum mechanics is applied to macroscopic objects, the results agree with classical physics.

The Particle in a Box Revisited

To see how the quantum particle under boundary conditions model is applied to a problem, let's return to our particle in a one-dimensional box of length L (see Fig. 41.3) and analyze it with the Schrödinger equation. Figure 41.3b is the potential-energy diagram that describes this problem. Potential-energy diagrams are a useful representation for understanding and solving problems with the Schrödinger equation.

Because of the shape of the curve in Figure 41.3b, the particle in a box is sometimes said to be in a **square well,**[9] where a **well** is an upward-facing region of the curve in a potential-energy diagram. (A downward-facing region is called a *barrier,* which we investigate in Section 41.5.) Figure 41.3b shows an infinite square well.

In the region $0 < x < L$, where $U = 0$, we can express the Schrödinger equation in the form

$$\frac{d^2\psi}{dx^2} = -\frac{2mE}{\hbar^2}\psi = -k^2\psi \tag{41.16}$$

where

$$k = \frac{\sqrt{2mE}}{\hbar}$$

Pitfall Prevention 41.3
Potential Wells A potential well
such as that in Figure 41.3b is
a graphical representation of
energy, not a pictorial representation, so you would not see this
shape if you were able to observe
the situation. A particle moves
only horizontally at a fixed vertical
position in a potential-energy diagram, representing the conserved
energy of the system of the particle and its environment.

---

[9]It is called a square well even if it has a rectangular shape in a potential-energy diagram.

The solution to Equation 41.16 is a function ψ whose second derivative is the negative of the same function multiplied by a constant k^2. Both the sine and cosine functions satisfy this requirement. Therefore, the most general solution to the equation is a linear combination of both solutions:

$$\psi(x) = A \sin kx + B \cos kx$$

where A and B are constants that are determined by the boundary and normalization conditions.

The first boundary condition on the wave function is that $\psi(0) = 0$:

$$\psi(0) = A \sin 0 + B \cos 0 = 0 + B = 0$$

which means that $B = 0$. Therefore, our solution reduces to

$$\psi(x) = A \sin kx$$

The second boundary condition, $\psi(L) = 0$, when applied to the reduced solution gives

$$\psi(L) = A \sin kL = 0$$

This equation could be satisfied by setting $A = 0$, but that would mean that $\psi = 0$ everywhere, which is not a valid wave function. The boundary condition is also satisfied if kL is an integral multiple of π, that is, if $kL = n\pi$, where n is an integer. Substituting $k = \sqrt{2mE}/\hbar$ into this expression gives

$$kL = \frac{\sqrt{2mE}}{\hbar} L = n\pi$$

Each value of the integer n corresponds to a quantized energy that we call E_n. Solving for the allowed energies E_n gives

$$E_n = \left(\frac{h^2}{8mL^2}\right) n^2 \tag{41.17}$$

which are identical to the allowed energies in Equation 41.14.

Substituting the values of k in the wave function, the allowed wave functions $\psi_n(x)$ are given by

$$\psi_n(x) = A \sin\left(\frac{n\pi x}{L}\right) \tag{41.18}$$

which is the wave function (Eq. 41.12) used in our initial discussion of the particle in a box.

41.4 A Particle in a Well of Finite Height

Now consider a particle in a *finite* potential well, that is, a system having a potential energy that is zero when the particle is in the region $0 < x < L$ and a finite value U when the particle is outside this region as in Figure 41.6. Classically, if the total energy E of the system is less than U, the particle is permanently bound in the potential well. If the particle were outside the well, its kinetic energy would have to be negative, which is an impossibility. According to quantum mechanics, however, a finite probability exists that the particle can be found outside the well even if $E < U$. That is, the wave function ψ is generally nonzero outside the well—regions I and III in Figure 41.6—so the probability density $|\psi|^2$ is also nonzero in these regions. Although this notion may be uncomfortable to accept, the uncertainty principle indicates that the energy of the system is uncertain. This uncertainty allows the particle to be outside the well as long as the apparent violation of conservation of energy does not exist in any measurable way.

In region II, where $U = 0$, the allowed wave functions are again sinusoidal because they represent solutions of Equation 41.16. The boundary conditions, however,

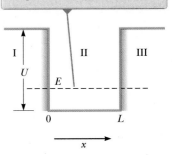

If the total energy E of the particle–well system is less than U, the particle is trapped in the well.

Figure 41.6 Potential-energy diagram of a well of finite height U and length L.

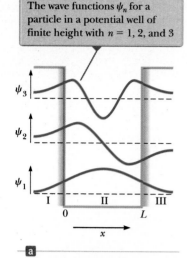

The wave functions ψ_n for a particle in a potential well of finite height with $n = 1, 2,$ and 3

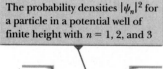

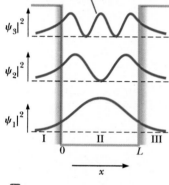

The probability densities $|\psi_n|^2$ for a particle in a potential well of finite height with $n = 1, 2,$ and 3

Figure 41.7 The first three allowed states for a particle in a potential well of finite height. The states are shown superimposed on the potential energy function of Figure 41.6. The wave functions and probability densities are plotted vertically from separate axes that are offset vertically for clarity. The positions of these axes on the potential energy function suggest the relative energies of the states.

no longer require that ψ be zero at the ends of the well, as was the case with the infinite square well.

The Schrödinger equation for regions I and III may be written

$$\frac{d^2\psi}{dx^2} = \frac{2m(U - E)}{\hbar^2}\psi \qquad (41.19)$$

Because $U > E$, the coefficient of ψ on the right-hand side is necessarily positive. Therefore, we can express Equation 41.19 as

$$\frac{d^2\psi}{dx^2} = C^2\psi \qquad (41.20)$$

where $C^2 = 2m(U - E)/\hbar^2$ is a positive constant in regions I and III. As you can verify by substitution, the general solution of Equation 41.20 is

$$\psi = Ae^{Cx} + Be^{-Cx} \qquad (41.21)$$

where A and B are constants.

We can use this general solution as a starting point for determining the appropriate solution for regions I and III. The solution must remain finite as $x \to \pm\infty$. Therefore, in region I, where $x < 0$, the function ψ cannot contain the term Be^{-Cx}. This requirement is handled by taking $B = 0$ in this region to avoid an infinite value for ψ for large negative values of x. Likewise, in region III, where $x > L$, the function ψ cannot contain the term Ae^{Cx}. This requirement is handled by taking $A = 0$ in this region to avoid an infinite value for ψ for large positive x values. Hence, the solutions in regions I and III are

$$\psi_I = Ae^{Cx} \qquad \text{for } x < 0$$

$$\psi_{III} = Be^{-Cx} \qquad \text{for } x > L$$

In region II, the wave function is sinusoidal and has the general form

$$\psi_{II}(x) = F\sin kx + G\cos kx$$

where F and G are constants.

These results show that the wave functions outside the potential well (where classical physics forbids the presence of the particle) decay exponentially with distance. At large negative x values, ψ_I approaches zero; at large positive x values, ψ_{III} approaches zero. These functions, together with the sinusoidal solution in region II, are shown in Figure 41.7a for the first three energy states. In evaluating the complete wave function, we impose the following boundary conditions:

$$\psi_I = \psi_{II} \quad \text{and} \quad \frac{d\psi_I}{dx} = \frac{d\psi_{II}}{dx} \quad \text{at } x = 0$$

$$\psi_{II} = \psi_{III} \quad \text{and} \quad \frac{d\psi_{II}}{dx} = \frac{d\psi_{III}}{dx} \quad \text{at } x = L$$

These four boundary conditions and the normalization condition (Eq. 41.7) are sufficient to determine the four constants A, B, F, and G and the allowed values of the energy E. Figure 41.7b plots the probability densities for these states. In each case, the wave functions inside and outside the potential well join smoothly at the boundaries.

The notion of trapping particles in potential wells is used in the burgeoning field of **nanotechnology,** which refers to the design and application of devices having dimensions ranging from 1 to 100 nm. The fabrication of these devices often involves manipulating single atoms or small groups of atoms to form very tiny structures or mechanisms.

One area of nanotechnology of interest to researchers is the **quantum dot,** a small region that is grown in a silicon crystal and acts as a potential well. This region can trap electrons into states with quantized energies. The wave functions for a particle in a quantum dot look similar to those in Figure 41.7a if L is on the order of nanometers. The storage of binary information using quantum dots is an active field of research. A simple binary scheme would involve associating a one with a quantum dot containing an electron and a zero with an empty dot. Other schemes involve cells of multiple dots such that arrangements of electrons among the dots correspond to ones and zeroes. Several research laboratories are studying the properties and potential applications of quantum dots. Information should be forthcoming from these laboratories at a steady rate in the next few years.

41.5 Tunneling Through a Potential Energy Barrier

Consider the potential energy function shown in Figure 41.8. In this situation, the potential energy has a constant value of U in the region of width L and is zero in all other regions.[10] A potential energy function of this shape is called a **square barrier,** and U is called the **barrier height.** A very interesting and peculiar phenomenon occurs when a moving particle encounters such a barrier of finite height and width. Suppose a particle of energy $E < U$ is incident on the barrier from the left (Fig. 41.8). Classically, the particle is reflected by the barrier. If the particle were located in region II, its kinetic energy would be negative, which is not classically allowed. Consequently, region II and therefore region III are both classically *forbidden* to the particle incident from the left. According to quantum mechanics, however, all regions are accessible to the particle, regardless of its energy. (Although all regions are accessible, the probability of the particle being in a classically forbidden region is very low.) According to the uncertainty principle, the particle could be within the barrier as long as the time interval during which it is in the barrier is short and consistent with Equation 40.26. If the barrier is relatively narrow, this short time interval can allow the particle to pass through the barrier.

Let's approach this situation using a mathematical representation. The Schrödinger equation has valid solutions in all three regions. The solutions in regions I and III are sinusoidal like Equation 41.18. In region II, the solution is exponential like Equation 41.21. Applying the boundary conditions that the wave functions in the three regions and their derivatives must join smoothly at the boundaries, a full solution, such as the one represented by the curve in Figure 41.8, can be found. Because the probability of locating the particle is proportional to $|\psi|^2$, the probability of finding the particle beyond the barrier in region III is nonzero. This result is in complete disagreement with classical physics. The movement of the particle to the far side of the barrier is called **tunneling** or **barrier penetration.**

The probability of tunneling can be described with a **transmission coefficient** T and a **reflection coefficient** R. The transmission coefficient represents the probability that the particle penetrates to the other side of the barrier, and the reflection coefficient is the probability that the particle is reflected by the barrier. Because the incident particle is either reflected or transmitted, we require that $T + R = 1$. An approximate expression for the transmission coefficient that is obtained in the case of $T \ll 1$ (a very wide barrier or a very high barrier, that is, $U \gg E$) is

$$T \approx e^{-2CL} \tag{41.22}$$

where

$$C = \frac{\sqrt{2m(U - E)}}{\hbar} \tag{41.23}$$

This quantum model of barrier penetration and specifically Equation 41.22 show that T can be nonzero. That the phenomenon of tunneling is observed experimentally provides further confidence in the principles of quantum physics.

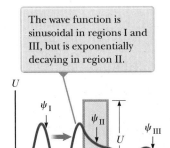

The wave function is sinusoidal in regions I and III, but is exponentially decaying in region II.

Figure 41.8 Wave function ψ for a particle incident from the left on a barrier of height U and width L. The wave function is plotted vertically from an axis positioned at the energy of the particle.

Pitfall Prevention 41.4
"Height" on an Energy Diagram The word *height* (as in *barrier height*) refers to an energy in discussions of barriers in potential-energy diagrams. For example, we might say the height of the barrier is 10 eV. On the other hand, the barrier *width* refers to the traditional usage of such a word and is an actual physical length measurement between the locations of the two vertical sides of the barrier.

---

[10] It is common in physics to refer to L as the *length* of a well but the *width* of a barrier.

⒬uick Quiz 41.4 Which of the following changes would increase the probability of transmission of a particle through a potential barrier? (You may choose more than one answer.) **(a)** decreasing the width of the barrier **(b)** increasing the width of the barrier **(c)** decreasing the height of the barrier **(d)** increasing the height of the barrier **(e)** decreasing the kinetic energy of the incident particle **(f)** increasing the kinetic energy of the incident particle

Example 41.4   Transmission Coefficient for an Electron

A 30-eV electron is incident on a square barrier of height 40 eV.

(A) What is the probability that the electron tunnels through the barrier if its width is 1.0 nm?

SOLUTION

Conceptualize Because the particle energy is smaller than the height of the potential barrier, we expect the electron to reflect from the barrier with a probability of 100% according to classical physics. Because of the tunneling phenomenon, however, there is a finite probability that the particle can appear on the other side of the barrier.

Categorize We evaluate the probability using an equation developed in this section, so we categorize this example as a substitution problem.

Evaluate the quantity $U - E$ that appears in Equation 41.23:

$$U - E = 40 \text{ eV} - 30 \text{ eV} = 10 \text{ eV} \left(\frac{1.6 \times 10^{-19} \text{ J}}{1 \text{ eV}} \right) = 1.6 \times 10^{-18} \text{ J}$$

Evaluate the quantity $2CL$ using Equation 41.23:

$$(1) \quad 2CL = 2 \frac{\sqrt{2(9.11 \times 10^{-31} \text{ kg})(1.6 \times 10^{-18} \text{ J})}}{1.055 \times 10^{-34} \text{ J} \cdot \text{s}} (1.0 \times 10^{-9} \text{ m}) = 32.4$$

From Equation 41.22, find the probability of tunneling through the barrier:

$$T \approx e^{-2CL} = e^{-32.4} = \boxed{8.5 \times 10^{-15}}$$

(B) What is the probability that the electron tunnels through the barrier if its width is 0.10 nm?

SOLUTION

In this case, the width L in Equation (1) is one-tenth as large, so evaluate the new value of $2CL$:

$$2CL = (0.1)(32.4) = 3.24$$

From Equation 41.22, find the new probability of tunneling through the barrier:

$$T \approx e^{-2CL} = e^{-3.24} = \boxed{0.039}$$

In part (A), the electron has approximately 1 chance in 10^{14} of tunneling through the barrier. In part (B), however, the electron has a much higher probability (3.9%) of penetrating the barrier. Therefore, reducing the width of the barrier by only one order of magnitude increases the probability of tunneling by about 12 orders of magnitude!

41.6 Applications of Tunneling

As we have seen, tunneling is a quantum phenomenon, a manifestation of the wave nature of matter. Many examples exist (on the atomic and nuclear scales) for which tunneling is very important.

Alpha Decay

One form of radioactive decay is the emission of alpha particles (the nuclei of helium atoms) by unstable, heavy nuclei (Chapter 44). To escape from the nucleus, an alpha particle must penetrate a barrier whose height is several times larger than the energy of the nucleus–alpha particle system as shown in Figure 41.9. The barrier results from a combination of the attractive nuclear force (discussed in Chapter 44) and the Coulomb repulsion (discussed in Chapter 23) between the alpha particle and the rest of the nucleus. Occasionally, an alpha particle tunnels through the barrier, which explains the basic mechanism for this type of decay and the large variations in the mean lifetimes of various radioactive nuclei.

Figure 41.8 shows the wave function of a particle tunneling through a barrier in one dimension. A similar wave function having spherical symmetry describes the barrier penetration of an alpha particle leaving a radioactive nucleus. The wave function exists both inside and outside the nucleus, and its amplitude is constant in time. In this way, the wave function correctly describes the small but constant probability that the nucleus will decay. The moment of decay cannot be predicted. In general, quantum mechanics implies that the future is indeterminate. This feature is in contrast to classical mechanics, from which the trajectory of an object can be calculated to arbitrarily high precision from precise knowledge of its initial position and velocity and of the forces exerted on it. Do not think that the future is undetermined simply because we have incomplete information about the present. The wave function contains all the information about the state of a system. Sometimes precise predictions can be made, such as the energy of a bound system, but sometimes only probabilities can be calculated about the future. The fundamental laws of nature are probabilistic. Therefore, it appears that Einstein's famous statement about quantum mechanics, "God does not roll dice," was wrong.

A radiation detector can be used to show that a nucleus decays by emitting a particle at a particular moment and in a particular direction. To point out the contrast between this experimental result and the wave function describing it, Schrödinger imagined a box containing a cat, a radioactive sample, a radiation counter, and a vial of poison. When a nucleus in the sample decays, the counter triggers the administration of lethal poison to the cat. Quantum mechanics correctly predicts the probability of finding the cat dead when the box is opened. Before the box is opened, does the cat have a wave function describing it as fractionally dead, with some chance of being alive?

This question is under continuing investigation, never with actual cats but sometimes with interference experiments building upon the experiment described in Section 40.7. Does the act of measurement change the system from a probabilistic to a definite state? When a particle emitted by a radioactive nucleus is detected at one particular location, does the wave function describing the particle drop instantaneously to zero everywhere else in the Universe? (Einstein called such a state change a "spooky action at a distance.") Is there a fundamental difference between a quantum system and a macroscopic system? The answers to these questions are unknown.

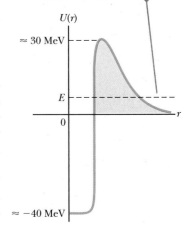

The alpha particle can tunnel through the barrier and escape from the nucleus even though its energy is lower than the height of the well.

Figure 41.9 The potential well for an alpha particle in a nucleus.

Nuclear Fusion

The basic reaction that powers the Sun and, indirectly, almost everything else in the solar system is fusion, which we shall study in Chapter 45. In one step of the process that occurs at the core of the Sun, protons must approach one another to within such a small distance that they fuse and form a deuterium nucleus. (See Section 45.4.) According to classical physics, these protons cannot overcome and penetrate the barrier caused by their mutual electrical repulsion. Quantum mechanically, however, the protons are able to tunnel through the barrier and fuse together.

Scanning Tunneling Microscopes

The scanning tunneling microscope (STM) enables scientists to obtain highly detailed images of surfaces at resolutions comparable to the size of a *single atom*. Figure 41.10 (page 1014), showing the surface of a piece of graphite, demonstrates what STMs can do. What makes this image so remarkable is that its resolution is

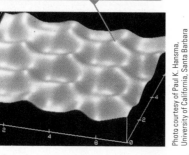

The contours seen here represent the ring-like arrangement of individual carbon atoms on the crystal surface.

Figure 41.10 The surface of graphite as "viewed" with a scanning tunneling microscope. This type of microscope enables scientists to see details with a lateral resolution of about 0.2 nm and a vertical resolution of 0.001 nm.

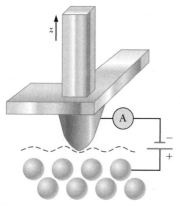

Figure 41.11 Schematic view of a scanning tunneling microscope. A scan of the tip over the sample can reveal surface contours down to the atomic level. An STM image is composed of a series of scans displaced laterally from one another. (Based on a drawing from P. K. Hansma, V. B. Elings, O. Marti, and C. Bracker, *Science* **242**:209, 1988. © 1988 by the AAAS.)

approximately 0.2 nm. For an optical microscope, the resolution is limited by the wavelength of the light used to make the image. Therefore, an optical microscope has a resolution no better than 200 nm, about half the wavelength of visible light, and so could never show the detail displayed in Figure 41.10.

Scanning tunneling microscopes achieve such high resolution by using the basic idea shown in Figure 41.11. An electrically conducting probe with a very sharp tip is brought near the surface to be studied. The empty space between tip and surface represents the "barrier" we have been discussing, and the tip and surface are the two walls of the "potential well." Because electrons obey quantum rules rather than Newtonian rules, they can "tunnel" across the barrier of empty space. If a voltage is applied between surface and tip, electrons in the atoms of the surface material can tunnel preferentially from surface to tip to produce a tunneling current. In this way, the tip samples the distribution of electrons immediately above the surface.

In the empty space between tip and surface, the electron wave function falls off exponentially (see region II in Fig. 41.8 and Example 41.4). For tip-to-surface distances $z > 1$ nm (that is, beyond a few atomic diameters), essentially no tunneling takes place. This exponential behavior causes the current of electrons tunneling from surface to tip to depend very strongly on z. By monitoring the tunneling current as the tip is scanned over the surface, scientists obtain a sensitive measure of the topography of the electron distribution on the surface. The result of this scan is used to make images like that in Figure 41.10. In this way, the STM can measure the height of surface features to within 0.001 nm, approximately 1/100 of an atomic diameter!

You can appreciate the sensitivity of STMs by examining Figure 41.10. Of the six carbon atoms in each ring, three appear lower than the other three. In fact, all six atoms are at the same height, but all have slightly different electron distributions. The three atoms that appear lower are bonded to other carbon atoms directly beneath them in the underlying atomic layer; as a result, their electron distributions, which are responsible for the bonding, extend downward beneath the surface. The atoms in the surface layer that appear higher do not lie directly over subsurface atoms and hence are not bonded to any underlying atoms. For these higher-appearing atoms, the electron distribution extends upward into the space above the surface. Because STMs map the topography of the electron distribution, this extra electron density makes these atoms appear higher in Figure 41.10.

The STM has one serious limitation: Its operation depends on the electrical conductivity of the sample and the tip. Unfortunately, most materials are not electrically conductive at their surfaces. Even metals, which are usually excellent electrical conductors, are covered with nonconductive oxides. A newer microscope, the atomic force microscope, or AFM, overcomes this limitation.

Resonant Tunneling Devices

Let's expand on the quantum-dot discussion in Section 41.4 by exploring the **resonant tunneling device.** Figure 41.12a shows the physical construction of such a device. The island of gallium arsenide in the center is a quantum dot located between two barriers formed from the thin extensions of aluminum arsenide. Figure 41.12b shows both the potential barriers encountered by electrons incident from the left and the quantized energy levels in the quantum dot. This situation differs from the one shown in Figure 41.8 in that there are quantized energy levels on the right of the first barrier. In Figure 41.8, an electron that tunnels through the barrier is considered a free particle and can have any energy. In contrast, the second barrier in Figure 41.12b imposes boundary conditions on the particle and quantizes its energy in the quantum dot. In Figure 41.12b, as the electron with the energy shown encounters the first barrier, it has no matching energy levels available on the right side of the barrier, which greatly reduces the probability of tunneling.

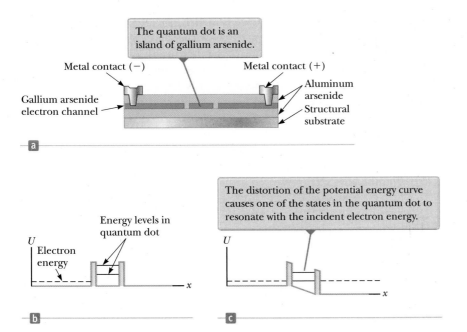

Figure 41.12 (a) The physical structure of a resonant tunneling device. (b) A potential-energy diagram showing the double barrier representing the walls of the quantum dot. (c) A voltage is applied across the device.

Figure 41.12c shows the effect of applying a voltage: the potential decreases with position as we move to the right across the device. The deformation of the potential barrier results in an energy level in the quantum dot coinciding with the energy of the incident electrons. This "resonance" of energies gives the device its name. When the voltage is applied, the probability of tunneling increases tremendously and the device carries current. In this manner, the device can be used as a very fast switch on a nanotechnological scale.

Resonant Tunneling Transistors

Figure 41.13a shows the addition of a gate electrode at the top of the resonant tunneling device over the quantum dot. This electrode turns the device into a **resonant**

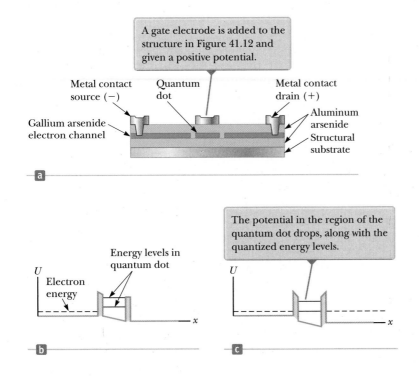

Figure 41.13 (a) A resonant tunneling transistor. (b) A potential-energy diagram showing the double barrier representing the walls of the quantum dot. (c) A voltage is applied to the gate electrode.

tunneling transistor. The basic function of a transistor is amplification, converting a small varying voltage into a large varying voltage. Figure 41.13b, representing the potential-energy diagram for the tunneling transistor, has a slope at the bottom of the quantum dot due to the differing voltages at the source and drain electrodes. In this configuration, there is no resonance between the electron energies outside the quantum dot and the quantized energies within the dot. By applying a small voltage to the gate electrode as in Figure 41.13c, the quantized energies can be brought into resonance with the electron energy outside the well and resonant tunneling occurs. The resulting current causes a voltage across an external resistor that is much larger than that of the gate voltage; hence, the device amplifies the input signal to the gate electrode.

41.7 The Simple Harmonic Oscillator

Consider a particle that is subject to a linear restoring force $F = -kx$, where k is a constant and x is the position of the particle relative to equilibrium ($x = 0$). The classical description of such a situation is provided by the particle in simple harmonic motion analysis model, which was discussed in Chapter 15. The potential energy of the system is, from Equation 15.20,

$$U = \tfrac{1}{2}kx^2 = \tfrac{1}{2}m\omega^2 x^2$$

where the angular frequency of vibration is $\omega = \sqrt{k/m}$. Classically, if the particle is displaced from its equilibrium position and released, it oscillates between the points $x = -A$ and $x = A$, where A is the amplitude of the motion. Furthermore, its total energy E is, from Equation 15.21,

$$E = K + U = \tfrac{1}{2}kA^2 = \tfrac{1}{2}m\omega^2 A^2$$

In the classical model, any value of E is allowed, including $E = 0$, which is the total energy when the particle is at rest at $x = 0$.

Let's investigate how the simple harmonic oscillator is treated from a quantum point of view. The Schrödinger equation for this problem is obtained by substituting $U = \tfrac{1}{2}m\omega^2 x^2$ into Equation 41.15:

$$-\frac{\hbar^2}{2m}\frac{d^2\psi}{dx^2} + \tfrac{1}{2}m\omega^2 x^2\psi = E\psi \qquad \textbf{(41.24)}$$

The mathematical technique for solving this equation is beyond the level of this book; nonetheless, it is instructive to guess at a solution. We take as our guess the following wave function:

$$\psi = Be^{-Cx^2} \qquad \textbf{(41.25)}$$

Substituting this function into Equation 41.24 shows that it is a satisfactory solution to the Schrödinger equation, provided that

$$C = \frac{m\omega}{2\hbar} \quad \text{and} \quad E = \tfrac{1}{2}\hbar\omega$$

It turns out that the solution we have guessed corresponds to the ground state of the system, which has an energy $\tfrac{1}{2}\hbar\omega$. Because $C = m\omega/2\hbar$, it follows from Equation 41.25 that the wave function for this state is

◀ **Wave function for the ground state of a simple harmonic oscillator**

$$\psi = Be^{-(m\omega/2\hbar)x^2} \qquad \textbf{(41.26)}$$

where B is a constant to be determined from the normalization condition. This result is but one solution to Equation 41.24. The remaining solutions that describe the excited states are more complicated, but all solutions include the exponential factor e^{-Cx^2}.

The energy levels of a harmonic oscillator are quantized as we would expect because the oscillating particle is bound to stay near $x = 0$. The energy of a state having an arbitrary quantum number n is

$$E_n = \left(n + \tfrac{1}{2}\right)\hbar\omega \quad n = 0, 1, 2, \ldots \tag{41.27}$$

The state $n = 0$ corresponds to the ground state, whose energy is $E_0 = \tfrac{1}{2}\hbar\omega$; the state $n = 1$ corresponds to the first excited state, whose energy is $E_1 = \tfrac{3}{2}\hbar\omega$; and so on. The energy-level diagram for this system is shown in Figure 41.14. The separations between adjacent levels are equal and given by

$$\Delta E = \hbar\omega \tag{41.28}$$

Notice that the energy levels for the harmonic oscillator in Figure 41.14 are equally spaced, just as Planck proposed for the oscillators in the walls of the cavity that was used in the model for blackbody radiation in Section 40.1. Planck's Equation 40.4 for the energy levels of the oscillators differs from Equation 41.27 only in the term $\tfrac{1}{2}$ added to n. This additional term does not affect the energy emitted in a transition, given by Equation 40.5, which is equivalent to Equation 41.28. That Planck generated these concepts without the benefit of the Schrödinger equation is testimony to his genius.

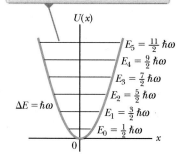

The levels are equally spaced, with separation $\hbar\omega$. The ground-state energy is $E_0 = \tfrac{1}{2}\hbar\omega$.

Figure 41.14 Energy-level diagram for a simple harmonic oscillator, superimposed on the potential energy function.

Example 41.5 | Molar Specific Heat of Hydrogen Gas

In Figure 21.6 (Section 21.3), which shows the molar specific heat of hydrogen as a function of temperature, vibration does not contribute to the molar specific heat at room temperature. Explain why, modeling the hydrogen molecule as a simple harmonic oscillator. The effective spring constant for the bond in the hydrogen molecule is 573 N/m.

SOLUTION

Conceptualize Imagine the only mode of vibration available to a diatomic molecule. This mode (shown in Fig. 21.5c) consists of the two atoms always moving in opposite directions with equal speeds.

Categorize We categorize this example as a quantum harmonic oscillator problem, with the molecule modeled as a two-particle system.

Analyze The motion of the particles relative to the center of mass can be analyzed by considering the oscillation of a single particle with reduced mass μ. (See Problem 40 in Enhanced WebAssign.)

Use the result of Problem 40 in Enhanced WebAssign to evaluate the reduced mass of the hydrogen molecule, in which the masses of the two particles are the same:

$$\mu = \frac{m_1 m_2}{m_1 + m_2} = \frac{m^2}{2m} = \tfrac{1}{2}m$$

Using Equation 41.28, calculate the energy necessary to excite the molecule from its ground vibrational state to its first excited vibrational state:

$$\Delta E = \hbar\omega = \hbar\sqrt{\frac{k}{\mu}} = \hbar\sqrt{\frac{k}{\frac{1}{2}m}} = \hbar\sqrt{\frac{2k}{m}}$$

Substitute numerical values, noting that m is the mass of a hydrogen atom:

$$\Delta E = (1.055 \times 10^{-34}\,\text{J}\cdot\text{s})\sqrt{\frac{2(573\,\text{N/m})}{1.67 \times 10^{-27}\,\text{kg}}} = 8.74 \times 10^{-20}\,\text{J}$$

Set this energy equal to $\tfrac{3}{2}k_B T$ from Equation 21.19 and find the temperature at which the average molecular translational kinetic energy is equal to that required to excite the first vibrational state of the molecule:

$$\tfrac{3}{2}k_B T = \Delta E$$

$$T = \tfrac{2}{3}\left(\frac{\Delta E}{k_B}\right) = \tfrac{2}{3}\left(\frac{8.74 \times 10^{-20}\,\text{J}}{1.38 \times 10^{-23}\,\text{J/K}}\right) = 4.22 \times 10^3\,\text{K}$$

Finalize The temperature of the gas must be more than 4 000 K for the translational kinetic energy to be comparable to the energy required to excite the first vibrational state. This excitation energy must come from collisions between

continued

▶ **41.5** continued

molecules, so if the molecules do not have sufficient translational kinetic energy, they cannot be excited to the first vibrational state and vibration does not contribute to the molar specific heat. Hence, the curve in Figure 21.6 does not rise to a value corresponding to the contribution of vibration until the hydrogen gas has been raised to thousands of kelvins.

Figure 21.6 shows that rotational energy levels must be more closely spaced in energy than vibrational levels because they are excited at a lower temperature than the vibrational levels. The translational energy levels are those of a particle in a three-dimensional box, where the box is the container holding the gas. These levels are given by an expression similar to Equation 41.14. Because the box is macroscopic in size, L is very large and the energy levels are very close together. In fact, they are so close together that translational energy levels are excited at the temperature at which liquid hydrogen becomes a gas shown in Figure 21.6.

Summary

Definitions

The **wave function** Ψ for a system is a mathematical function that can be written as a product of a space function ψ for one particle of the system and a complex time function:

$$\Psi(\vec{r}_1, \vec{r}_2, \vec{r}_3, \ldots, \vec{r}_j, \ldots, t) = \psi(\vec{r}_j)e^{-i\omega t} \quad \textbf{(41.2)}$$

where ω $(= 2\pi f)$ is the angular frequency of the wave function and $i = \sqrt{-1}$. The wave function contains within it all the information that can be known about the particle.

The measured position x of a particle, averaged over many trials, is called the **expectation value** of x and is defined by

$$\langle x \rangle \equiv \int_{-\infty}^{\infty} \psi^* x \psi \, dx \quad \textbf{(41.8)}$$

Concepts and Principles

In quantum mechanics, a particle in a system can be represented by a wave function $\psi(x, y, z)$. The probability per unit volume (or probability density) that a particle will be found at a point is $|\psi|^2 = \psi^*\psi$, where ψ^* is the complex conjugate of ψ. If the particle is confined to moving along the x axis, the probability that it is located in an interval dx is $|\psi|^2 \, dx$. Furthermore, the sum of all these probabilities over all values of x must be 1:

$$\int_{-\infty}^{\infty} |\psi|^2 \, dx = 1 \quad \textbf{(41.7)}$$

This expression is called the **normalization condition.**

If a particle of mass m is confined to moving in a one-dimensional box of length L whose walls are impenetrable, then ψ must be zero at the walls and outside the box. The wave functions for this system are given by

$$\psi(x) = A \sin\left(\frac{n\pi x}{L}\right) \quad n = 1, 2, 3, \ldots \quad \textbf{(41.12)}$$

where A is the maximum value of ψ. The allowed states of a particle in a box have quantized energies given by

$$E_n = \left(\frac{h^2}{8mL^2}\right)n^2 \quad n = 1, 2, 3, \ldots \quad \textbf{(41.14)}$$

The wave function for a system must satisfy the **Schrödinger equation.** The time-independent Schrödinger equation for a particle confined to moving along the x axis is

$$-\frac{\hbar^2}{2m}\frac{d^2\psi}{dx^2} + U\psi = E\psi \quad \textbf{(41.15)}$$

where U is the potential energy of the system and E is the total energy.

▮ **Quantum Particle Under Boundary Conditions.** An interaction of a quantum particle with its environment represents one or more boundary conditions. If the interaction restricts the particle to a finite region of space, the energy of the system is quantized. All wave functions must satisfy the following four boundary conditions: (1) $\psi(x)$ must remain finite as x approaches 0, (2) $\psi(x)$ must approach zero as x approaches $\pm\infty$, (3) $\psi(x)$ must be continuous for all values of x, and (4) $d\psi/dx$ must be continuous for all finite values of $U(x)$. If the solution to Equation 41.15 is piecewise, conditions (3) and (4) must be applied at the boundaries between regions of x in which Equation 41.15 has been solved.

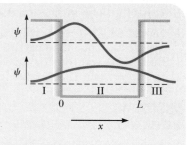

1. A beam of quantum particles with kinetic energy 2.00 eV is reflected from a potential barrier of small width and original height 3.00 eV. How does the fraction of the particles that are reflected change as the barrier height is reduced to 2.01 eV? (a) It increases. (b) It decreases. (c) It stays constant at zero. (d) It stays constant at 1. (e) It stays constant with some other value.

2. A quantum particle of mass m_1 is in a square well with infinitely high walls and length 3 nm. Rank the situations (a) through (e) according to the particle's energy from highest to lowest, noting any cases of equality. (a) The particle of mass m_1 is in the ground state of the well. (b) The same particle is in the $n = 2$ excited state of the same well. (c) A particle with mass $2m_1$ is in the ground state of the same well. (d) A particle of mass m_1 in the ground state of the same well, and the uncertainty principle has become inoperative; that is, Planck's constant has been reduced to zero. (e) A particle of mass m_1 is in the ground state of a well of length 6 nm.

3. Is each one of the following statements (a) through (e) true or false for an electron? (a) It is a quantum particle, behaving in some experiments like a classical particle and in some experiments like a classical wave. (b) Its rest energy is zero. (c) It carries energy in its motion. (d) It carries momentum in its motion. (e) Its motion is described by a wave function that has a wavelength and satisfies a wave equation.

4. Is each one of the following statements (a) through (e) true or false for a photon? (a) It is a quantum particle, behaving in some experiments like a classical particle and in some experiments like a classical wave. (b) Its rest energy is zero. (c) It carries energy in its motion. (d) It carries momentum in its motion. (e) Its motion is described by a wave function that has a wavelength and satisfies a wave equation.

5. A particle in a rigid box of length L is in the first excited state for which $n = 2$ (Fig. OQ41.5). Where is the particle most likely to be found? (a) At the center of the box. (b) At either end of the box. (c) All points in the box are equally likely. (d) One-fourth of the way from either end of the box. (e) None of those answers is correct.

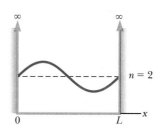

Figure OQ41.5

6. Two square wells have the same length. Well 1 has walls of finite height, and well 2 has walls of infinite height. Both wells contain identical quantum particles, one in each well. **(i)** Is the wavelength of the ground-state wave function (a) greater for well 1, (b) greater for well 2, or (c) equal for both wells? **(ii)** Is the magnitude of the ground-state momentum (a) greater for well 1, (b) greater for well 2, or (c) equal for both wells? **(iii)** Is the ground-state energy of the particle (a) greater for well 1, (b) greater for well 2, or (c) equal for both wells?

7. The probability of finding a certain quantum particle in the section of the x axis between $x = 4$ nm and $x = 7$ nm is 48%. The particle's wave function $\psi(x)$ is constant over this range. What numerical value can be attributed to $\psi(x)$, in units of $nm^{-1/2}$? (a) 0.48 (b) 0.16 (c) 0.12 (d) 0.69 (e) 0.40

8. Suppose a tunneling current in an electronic device goes through a potential-energy barrier. The tunneling current is small because the width of the barrier is large and the barrier is high. To increase the current most effectively, what should you do? (a) Reduce the width of the barrier. (b) Reduce the height of the barrier. (c) Either choice (a) or choice (b) is equally effective. (d) Neither choice (a) nor choice (b) increases the current.

9. Unlike the idealized diagram of Figure 41.11, a typical tip used for a scanning tunneling microscope is rather jagged on the atomic scale, with several irregularly spaced points. For such a tip, does most of the tunneling current occur between the sample and (a) all the points of the tip equally, (b) the most centrally located point, (c) the point closest to the sample, or (d) the point farthest from the sample?

10. Figure OQ41.10 represents the wave function for a hypothetical quantum particle in a given region. From the choices *a* through *e*, at what value of *x* is the particle most likely to be found?

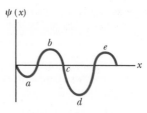

Figure OQ41.10

Conceptual Questions

1. denotes answer available in *Student Solutions Manual/Study Guide*

1. Richard Feynman said, "A philosopher once said that 'it is necessary for the very existence of science that the same conditions always produce the same results.' Well, they don't!" In view of what has been discussed in this chapter, present an argument showing that the philosopher's statement is false. How might the statement be reworded to make it true?

2. Discuss the relationship between ground-state energy and the uncertainty principle.

3. For a quantum particle in a box, the probability density at certain points is zero as seen in Figure CQ41.3. Does this value imply that the particle cannot move across these points? Explain.

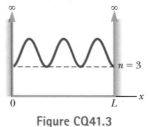

Figure CQ41.3

4. Why are the following wave functions not physically possible for all values of *x*? (a) $\psi(x) = Ae^x$ (b) $\psi(x) = A \tan x$

5. What is the significance of the wave function ψ?

6. In quantum mechanics, it is possible for the energy *E* of a particle to be less than the potential energy, but classically this condition is not possible. Explain.

7. Consider the wave functions in Figure CQ41.7. Which of them are not physically significant in the interval shown? For those that are not, state why they fail to qualify.

8. How is the Schrödinger equation useful in describing quantum phenomena?

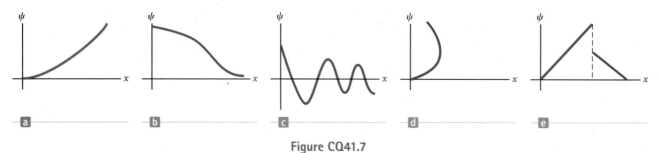

Figure CQ41.7

Problems available in ENHANCED WebAssign Access end-of-chapter problems online at www.webassign.net

List of Enhanced Problems

Problem Number	Targeted Feedback in Enhanced WebAssign	Analysis Model Tutorial in Enhanced WebAssign	Master It in Enhanced WebAssign	Watch It in Enhanced WebAssign
41.1			✓	
41.9		✓		
41.12	✓			✓
41.13	✓			✓
41.23			✓	
41.31	✓		✓	
41.32	✓			✓
41.33	✓			✓
41.34	✓			✓
41.35			✓	
41.37	✓			✓
41.43	✓			✓
41.49	✓			✓

Atomic Physics

This street in the Ginza district in Tokyo displays many signs formed from neon lamps of varying bright colors. The light from these lamps has its origin in transitions between quantized energy states in the atoms contained in the lamps. In this chapter, we investigate those transitions. *(© Ken Straiton/Corbis)*

In Chapter 41, we introduced some basic concepts and techniques used in quantum mechanics along with their applications to various one-dimensional systems. In this chapter, we apply quantum mechanics to atomic systems. A large portion of the chapter is focused on the application of quantum mechanics to the study of the hydrogen atom. Understanding the hydrogen atom, the simplest atomic system, is important for several reasons:

- The hydrogen atom is the only atomic system that can be solved exactly.
- Much of what was learned in the 20th century about the hydrogen atom, with its single electron, can be extended to such single-electron ions as He^+ and Li^{2+}.
- The hydrogen atom is an ideal system for performing precise tests of theory against experiment and for improving our overall understanding of atomic structure.

- The quantum numbers that are used to characterize the allowed states of hydrogen can also be used to investigate more complex atoms, and such a description enables us to understand the periodic table of the elements. This understanding is one of the greatest triumphs of quantum mechanics.
- The basic ideas about atomic structure must be well understood before we attempt to deal with the complexities of molecular structures and the electronic structure of solids.

The full mathematical solution of the Schrödinger equation applied to the hydrogen atom gives a complete and beautiful description of the atom's properties. Because the mathematical procedures involved are beyond the scope of this text, however, many details are omitted. The solutions for some states of hydrogen are discussed, together with the quantum numbers used to characterize various allowed states. We also discuss the physical significance of the quantum numbers and the effect of a magnetic field on certain quantum states.

A new physical idea, the *exclusion principle*, is presented in this chapter. This principle is extremely important for understanding the properties of multielectron atoms and the arrangement of elements in the periodic table.

Finally, we apply our knowledge of atomic structure to describe the mechanisms involved in the production of x-rays and in the operation of a laser.

42.1 Atomic Spectra of Gases

As pointed out in Section 40.1, all objects emit thermal radiation characterized by a *continuous* distribution of wavelengths. In sharp contrast to this continuous-distribution spectrum is the *discrete* **line spectrum** observed when a low-pressure gas undergoes an electric discharge. (Electric discharge occurs when the gas is subject to a potential difference that creates an electric field greater than the dielectric strength of the gas.) Observation and analysis of these spectral lines is called **emission spectroscopy.**

When the light from a gas discharge is examined using a spectrometer (see Fig. 38.15), it is found to consist of a few bright lines of color on a generally dark background. This discrete line spectrum contrasts sharply with the continuous rainbow of colors seen when a glowing solid is viewed through the same instrument. Figure 42.1a (page 1024) shows that the wavelengths contained in a given line spectrum are characteristic of the element emitting the light. The simplest line spectrum is that for atomic hydrogen, and we describe this spectrum in detail. Because no two elements have the same line spectrum, this phenomenon represents a practical and sensitive technique for identifying the elements present in unknown samples.

Another form of spectroscopy very useful in analyzing substances is **absorption spectroscopy.** An absorption spectrum is obtained by passing white light from a continuous source through a gas or a dilute solution of the element being analyzed. The absorption spectrum consists of a series of dark lines superimposed on the continuous spectrum of the light source as shown in Figure 42.1b for atomic hydrogen.

The absorption spectrum of an element has many practical applications. For example, the continuous spectrum of radiation emitted by the Sun must pass through the cooler gases of the solar atmosphere. The various absorption lines observed in the solar spectrum have been used to identify elements in the solar atmosphere. In early studies of the solar spectrum, experimenters found some lines that did not correspond to any known element. A new element had been discovered!

Pitfall Prevention 42.1

Why Lines? The phrase "spectral lines" is often used when discussing the radiation from atoms. Lines are seen because the light passes through a long and very narrow slit before being separated by wavelength. You will see many references to these "lines" in both physics and chemistry.

Figure 42.1 (a) Emission line spectra for hydrogen, mercury, and neon. (b) The absorption spectrum for hydrogen. Notice that the dark absorption lines occur at the same wavelengths as the hydrogen emission lines in (a). (K. W. Whitten, R. E. Davis, M. L. Peck, and G. G. Stanley, *General Chemistry*, 7th ed., Belmont, CA, Brooks/Cole, 2004.)

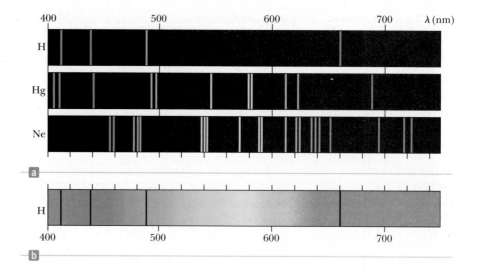

The new element was named helium, after the Greek word for Sun, *helios*. Helium was subsequently isolated from subterranean gas on the Earth.

Using this technique, scientists have examined the light from stars other than our Sun and have never detected elements other than those present on the Earth. Absorption spectroscopy has also been useful in analyzing heavy-metal contamination of the food chain. For example, the first determination of high levels of mercury in tuna was made with the use of atomic absorption spectroscopy.

The discrete emissions of light from gas discharges are used in "neon" signs such as those in the opening photograph of this chapter. Neon, the first gas used in these types of signs and the gas after which these signs are named, emits strongly in the red region. As a result, a glass tube filled with neon gas emits bright red light when an applied voltage causes a continuous discharge. Early signs used different gases to provide different colors, although the brightness of these signs was generally very low. Many present-day "neon" signs contain mercury vapor, which emits strongly in the ultraviolet range of the electromagnetic spectrum. The inside of a present-day sign's glass tube is coated with a material that emits a particular color when it absorbs ultraviolet radiation from the mercury. The color of the light from the tube results from the particular material chosen. A household fluorescent light operates in the same manner, with a white-emitting material coating the inside of the glass tube.

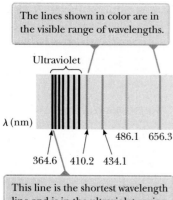

The lines shown in color are in the visible range of wavelengths.

This line is the shortest wavelength line and is in the ultraviolet region of the electromagnetic spectrum.

Figure 42.2 The Balmer series of spectral lines for atomic hydrogen, with several lines marked with the wavelength in nanometers. (The horizontal wavelength axis is not to scale.)

From 1860 to 1885, scientists accumulated a great deal of data on atomic emissions using spectroscopic measurements. In 1885, a Swiss schoolteacher, Johann Jacob Balmer (1825–1898), found an empirical equation that correctly predicted the wavelengths of four visible emission lines of hydrogen: H_α (red), H_β (blue-green), H_γ (blue-violet), and H_δ (violet). Figure 42.2 shows these and other lines (in the ultraviolet) in the emission spectrum of hydrogen. The four visible lines occur at the wavelengths 656.3 nm, 486.1 nm, 434.1 nm, and 410.2 nm. The complete set of lines is called the **Balmer series.** The wavelengths of these lines can be described by the following equation, which is a modification made by Johannes Rydberg (1854–1919) of Balmer's original equation:

Balmer series ▶

$$\frac{1}{\lambda} = R_{\mathrm{H}}\left(\frac{1}{2^2} - \frac{1}{n^2}\right) \quad n = 3, 4, 5, \ldots \tag{42.1}$$

where R_{H} is a constant now called the **Rydberg constant** with a value of $1.097\ 373\ 2 \times 10^7\ \mathrm{m^{-1}}$. The integer values of n from 3 to 6 give the four visible lines from 656.3 nm (red) down to 410.2 nm (violet). Equation 42.1 also describes the ultraviolet spectral lines in the Balmer series if n is carried out beyond $n = 6$. The **series limit** is the shortest wavelength in the series and corresponds to $n \rightarrow \infty$, with a wavelength of 364.6 nm as in Figure 42.2. The measured spectral lines agree with the empirical equation, Equation 42.1, to within 0.1%.

Other lines in the spectrum of hydrogen were found following Balmer's discovery. These spectra are called the Lyman, Paschen, and Brackett series after their discoverers. The wavelengths of the lines in these series can be calculated through the use of the following empirical equations:

$$\frac{1}{\lambda} = R_H\left(1 - \frac{1}{n^2}\right) \quad n = 2, 3, 4, \ldots \qquad (42.2)$$ ◀ Lyman series

$$\frac{1}{\lambda} = R_H\left(\frac{1}{3^2} - \frac{1}{n^2}\right) \quad n = 4, 5, 6, \ldots \qquad (42.3)$$ ◀ Paschen series

$$\frac{1}{\lambda} = R_H\left(\frac{1}{4^2} - \frac{1}{n^2}\right) \quad n = 5, 6, 7, \ldots \qquad (42.4)$$ ◀ Brackett series

No theoretical basis existed for these equations; they simply worked. The same constant R_H appears in each equation, and all equations involve small integers. In Section 42.3, we shall discuss the remarkable achievement of a theory for the hydrogen atom that provided an explanation for these equations.

42.2 Early Models of the Atom

The model of the atom in the days of Newton was a tiny, hard, indestructible sphere. Although this model provided a good basis for the kinetic theory of gases (Chapter 21), new models had to be devised when experiments revealed the electrical nature of atoms. In 1897, J. J. Thomson established the charge-to-mass ratio for electrons. (See Fig. 29.15 in Section 29.3.) The following year, he suggested a model that describes the atom as a region in which positive charge is spread out in space with electrons embedded throughout the region, much like the seeds in a watermelon or raisins in thick pudding (Fig. 42.3). The atom as a whole would then be electrically neutral.

In 1911, Ernest Rutherford (1871–1937) and his students Hans Geiger and Ernest Marsden performed a critical experiment that showed that Thomson's model could not be correct. In this experiment, a beam of positively charged alpha particles (helium nuclei) was projected into a thin metallic foil such as the target in Figure 42.4a (page 1026). Most of the particles passed through the foil as if it were empty space, but some of the results of the experiment were astounding. Many of the particles deflected from their original direction of travel were scattered through *large* angles. Some particles were even deflected backward, completely reversing their direction of travel! When Geiger informed Rutherford that some alpha particles were scattered backward, Rutherford wrote, "It was quite the most incredible event that has ever happened to me in my life. It was almost as incredible as if you fired a 15-inch [artillery] shell at a piece of tissue paper and it came back and hit you."

Such large deflections were not expected on the basis of Thomson's model. According to that model, the positive charge of an atom in the foil is spread out over such a great volume (the entire atom) that there is no concentration of positive charge strong enough to cause any large-angle deflections of the positively charged alpha particles. Furthermore, the electrons are so much less massive than the alpha particles that they would not cause large-angle scattering either. Rutherford explained his astonishing results by developing a new atomic model, one that assumed the positive charge in the atom was concentrated in a region that was small relative to the size of the atom. He called this concentration of positive charge the **nucleus** of the atom. Any electrons belonging to the atom were assumed to be in the relatively large volume outside the nucleus. To explain why these electrons were not pulled into the nucleus by the attractive electric force, Rutherford modeled them as moving in orbits around the nucleus in the same manner as the planets orbit the Sun (Fig. 42.4b). For this reason, this model is often referred to as the planetary model of the atom.

Two basic difficulties exist with Rutherford's planetary model. As we saw in Section 42.1, an atom emits (and absorbs) certain characteristic frequencies of

Stock Montage, Inc.

Joseph John Thomson
English physicist (1856–1940)
The recipient of a Nobel Prize in Physics in 1906, Thomson is usually considered the discoverer of the electron. He opened up the field of subatomic particle physics with his extensive work on the deflection of cathode rays (electrons) in an electric field.

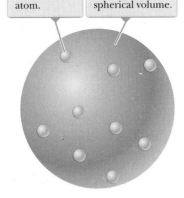

The electrons are small negative charges at various locations within the atom.

The positive charge of the atom is distributed continuously in a spherical volume.

Figure 42.3 Thomson's model of the atom.

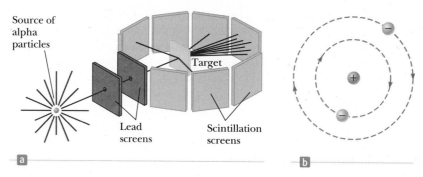

Figure 42.4 (a) Rutherford's technique for observing the scattering of alpha particles from a thin foil target. The source is a naturally occurring radioactive substance, such as radium. (b) Rutherford's planetary model of the atom.

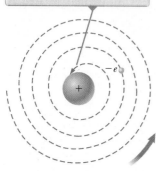

Because the accelerating electron radiates energy, the size of the orbit decreases until the electron falls into the nucleus.

Figure 42.5 The classical model of the nuclear atom predicts that the atom decays.

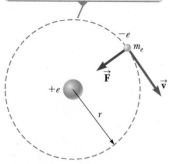

The orbiting electron is allowed to be only in specific orbits of discrete radii.

Figure 42.6 Diagram representing Bohr's model of the hydrogen atom.

electromagnetic radiation and no others, but the Rutherford model cannot explain this phenomenon. A second difficulty is that Rutherford's electrons are described by the particle in uniform circular motion model; they have a centripetal acceleration. According to Maxwell's theory of electromagnetism, centripetally accelerated charges revolving with frequency f should radiate electromagnetic waves of frequency f. Unfortunately, this classical model leads to a prediction of self-destruction when applied to the atom. Identifying the electron and the proton as a nonisolated system for energy, Equation 8.2 becomes $\Delta K + \Delta U = T_{ER}$, where K is the kinetic energy of the electron, U is the electric potential energy of the electron–nucleus system, and T_{ER} represents the outgoing electromagnetic radiation. As energy leaves the system, the radius of the electron's orbit steadily decreases (Fig. 42.5). The system is an isolated system for angular momentum because there is no torque on the system. Therefore, as the electron moves closer to the nucleus, the angular speed of the electron will increase, just like the spinning skater in Figure 11.10 in Section 11.4. This process leads to an ever-increasing frequency of emitted radiation and an ultimate collapse of the atom as the electron plunges into the nucleus.

42.3 Bohr's Model of the Hydrogen Atom

Given the situation described at the end of Section 42.2, the stage was set for Niels Bohr in 1913 when he presented a new model of the hydrogen atom that circumvented the difficulties of Rutherford's planetary model. Bohr applied Planck's ideas of quantized energy levels (Section 40.1) to Rutherford's orbiting atomic electrons. Bohr's theory was historically important to the development of quantum physics, and it appeared to explain the spectral line series described by Equations 42.1 through 42.4. Although Bohr's model is now considered obsolete and has been completely replaced by a probabilistic quantum-mechanical theory, we can use the Bohr model to develop the notions of energy quantization and angular momentum quantization as applied to atomic-sized systems.

Bohr combined ideas from Planck's original quantum theory, Einstein's concept of the photon, Rutherford's planetary model of the atom, and Newtonian mechanics to arrive at a semiclassical structural model based on some revolutionary ideas. The structural model of the Bohr theory as it applies to the hydrogen atom has the following properties:

1. *Physical components:*
   The electron moves in circular orbits around the proton under the influence of the electric force of attraction as shown in Figure 42.6.

2. *Behavior of the components:*

(a) Only certain electron orbits are stable. When in one of these **stationary states,** as Bohr called them, the electron does not emit energy in the form of radiation, even though it is accelerating. Hence, the total energy of the atom remains constant and classical mechanics can be used to describe the electron's motion. Bohr's model claims that the centripetally accelerated electron does not continuously emit radiation, losing energy and eventually spiraling into the nucleus, as predicted by classical physics in the form of Rutherford's planetary model.

(b) The atom emits radiation when the electron makes a transition from a more energetic initial stationary state to a lower-energy stationary state. This transition cannot be visualized or treated classically. In particular, the frequency f of the photon emitted in the transition is related to the change in the atom's energy and is not equal to the frequency of the electron's orbital motion. The frequency of the emitted radiation is found from the energy-conservation expression

$$E_i - E_f = hf \tag{42.5}$$

where E_i is the energy of the initial state, E_f is the energy of the final state, and $E_i > E_f$. In addition, energy of an incident photon can be absorbed by the atom, but only if the photon has an energy that exactly matches the difference in energy between an allowed state of the atom and a higher-energy state. Upon absorption, the photon disappears and the atom makes a transition to the higher-energy state.

(c) The size of an allowed electron orbit is determined by a condition imposed on the electron's orbital angular momentum: the allowed orbits are those for which the electron's orbital angular momentum about the nucleus is quantized and equal to an integral multiple of $\hbar = h/2\pi$,

$$m_e vr = n\hbar \quad n = 1, 2, 3, \ldots \tag{42.6}$$

where m_e is the electron mass, v is the electron's speed in its orbit, and r is the orbital radius.

Niels Bohr
Danish Physicist (1885–1962)
Bohr was an active participant in the early development of quantum mechanics and provided much of its philosophical framework. During the 1920s and 1930s, he headed the Institute for Advanced Studies in Copenhagen. The institute was a magnet for many of the world's best physicists and provided a forum for the exchange of ideas. Bohr was awarded the 1922 Nobel Prize in Physics for his investigation of the structure of atoms and the radiation emanating from them. When Bohr visited the United States in 1939 to attend a scientific conference, he brought news that the fission of uranium had been observed by Hahn and Strassman in Berlin. The results were the foundations of the nuclear weapon developed in the United States during World War II.

These postulates are a mixture of established principles and completely new and untested ideas at the time. Property 1, from classical mechanics, treats the electron in orbit around the nucleus in the same way we treat a planet in a circular orbit around a star, using the particle in uniform circular motion analysis model. Property 2(a) was a radical new idea in 1913 that was completely at odds with the understanding of electromagnetism at the time. Property 2(b) represents the principle of conservation of energy as described by the nonisolated system model for energy. Property 2(c) is another new idea that had no basis in classical physics.

Property 2(b) implies qualitatively the existence of a characteristic discrete emission line spectrum *and also* a corresponding absorption line spectrum of the kind shown in Figure 42.1 for hydrogen. Using these postulates, let's calculate the allowed energy levels and find quantitative values of the emission wavelengths of the hydrogen atom.

The electric potential energy of the system shown in Figure 42.6 is given by Equation 25.13, $U = k_e q_1 q_2/r = -k_e e^2/r$, where k_e is the Coulomb constant and the negative sign arises from the charge $-e$ on the electron. Therefore, the *total* energy of the atom, which consists of the electron's kinetic energy and the system's potential energy, is

$$E = K + U = \tfrac{1}{2} m_e v^2 - k_e \frac{e^2}{r} \tag{42.7}$$

The electron is modeled as a particle in uniform circular motion, so the electric force $k_e e^2 / r^2$ exerted on the electron must equal the product of its mass and its centripetal acceleration ($a_c = v^2/r$):

$$\frac{k_e e^2}{r^2} = \frac{m_e v^2}{r}$$

$$v^2 = \frac{k_e e^2}{m_e r} \tag{42.8}$$

From Equation 42.8, we find that the kinetic energy of the electron is

$$K = \tfrac{1}{2} m_e v^2 = \frac{k_e e^2}{2r}$$

Substituting this value of K into Equation 42.7 gives the following expression for the total energy of the atom:[1]

$$E = -\frac{k_e e^2}{2r} \tag{42.9}$$

Because the total energy is *negative*, which indicates a bound electron–proton system, energy in the amount of $k_e e^2/2r$ must be added to the atom to remove the electron and make the total energy of the system zero.

We can obtain an expression for r, the radius of the allowed orbits, by solving Equation 42.6 for v^2 and equating it to Equation 42.8:

$$v^2 = \frac{n^2 \hbar^2}{m_e^2 r^2} = \frac{k_e e^2}{m_e r}$$

$$r_n = \frac{n^2 \hbar^2}{m_e k_e e^2} \quad n = 1, 2, 3, \ldots \tag{42.10}$$

Equation 42.10 shows that the radii of the allowed orbits have discrete values: they are quantized. The result is based on the *assumption* that the electron can exist only in certain allowed orbits determined by the integer n (Bohr's Property 2(c)).

The orbit with the smallest radius, called the **Bohr radius** a_0, corresponds to $n = 1$ and has the value

$$a_0 = \frac{\hbar^2}{m_e k_e e^2} = 0.052\,9 \text{ nm} \tag{42.11}$$

Substituting Equation 42.11 into Equation 42.10 gives a general expression for the radius of any orbit in the hydrogen atom:

$$r_n = n^2 a_0 = n^2 (0.052\,9 \text{ nm}) \quad n = 1, 2, 3, \ldots \tag{42.12}$$

Bohr's theory predicts a value for the radius of a hydrogen atom on the right order of magnitude, based on experimental measurements. This result was a striking triumph for Bohr's theory. The first three Bohr orbits are shown to scale in Figure 42.7.

The quantization of orbit radii leads to energy quantization. Substituting $r_n = n^2 a_0$ into Equation 42.9 gives

$$E_n = -\frac{k_e e^2}{2a_0}\left(\frac{1}{n^2}\right) \quad n = 1, 2, 3, \ldots \tag{42.13}$$

Inserting numerical values into this expression, we find that

$$E_n = -\frac{13.606 \text{ eV}}{n^2} \quad n = 1, 2, 3, \ldots \tag{42.14}$$

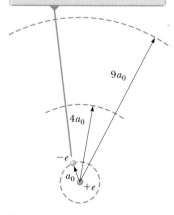

The electron is shown in the lowest-energy orbit, but it could be in any of the allowed orbits.

$9a_0$

$4a_0$

a_0

$-e$

$+e$

Figure 42.7 The first three circular orbits predicted by the Bohr model of the hydrogen atom.

Bohr radius ▶

Radii of Bohr orbits ▶
in hydrogen

[1]Compare Equation 42.9 with its gravitational counterpart, Equation 13.19.

Only energies satisfying this equation are permitted. The lowest allowed energy level, the ground state, has $n = 1$ and energy $E_1 = -13.606$ eV. The next energy level, the first excited state, has $n = 2$ and energy $E_2 = E_1/2^2 = -3.401$ eV. Figure 42.8 is an energy-level diagram showing the energies of these discrete energy states and the corresponding quantum numbers n. The uppermost level corresponds to $n = \infty$ (or $r = \infty$) and $E = 0$.

Notice how the allowed energies of the hydrogen atom differ from those of the particle in a box. The particle-in-a-box energies (Eq. 41.14) increase as n^2, so they become farther apart in energy as n increases. On the other hand, the energies of the hydrogen atom (Eq. 42.14) are inversely proportional to n^2, so their separation in energy becomes smaller as n increases. The separation between energy levels approaches zero as n approaches infinity and the energy approaches zero.

Zero energy represents the boundary between a bound system of an electron and a proton and an unbound system. If the energy of the atom is raised from that of the ground state to any energy larger than zero, the atom is **ionized.** The minimum energy required to ionize the atom in its ground state is called the **ionization energy.** As can be seen from Figure 42.8, the ionization energy for hydrogen in the ground state, based on Bohr's calculation, is 13.6 eV. This finding constituted another major achievement for the Bohr theory because the ionization energy for hydrogen had already been measured to be 13.6 eV.

Equations 42.5 and 42.13 can be used to calculate the frequency of the photon emitted when the electron makes a transition from an outer orbit to an inner orbit:

$$f = \frac{E_i - E_f}{h} = \frac{k_e e^2}{2a_0 h}\left(\frac{1}{n_f^2} - \frac{1}{n_i^2}\right) \tag{42.15}$$

Because the quantity measured experimentally is wavelength, it is convenient to use $c = f\lambda$ to express Equation 42.15 in terms of wavelength:

$$\frac{1}{\lambda} = \frac{f}{c} = \frac{k_e e^2}{2a_0 hc}\left(\frac{1}{n_f^2} - \frac{1}{n_i^2}\right) \tag{42.16}$$

Remarkably, this expression, which is purely theoretical, is *identical* to the general form of the empirical relationships discovered by Balmer and Rydberg and given by Equations 42.1 to 42.4:

$$\frac{1}{\lambda} = R_H\left(\frac{1}{n_f^2} - \frac{1}{n_i^2}\right) \tag{42.17}$$

provided the constant $k_e e^2/2a_0 hc$ is equal to the experimentally determined Rydberg constant. Soon after Bohr demonstrated that these two quantities agree to within approximately 1%, this work was recognized as the crowning achievement of his new quantum theory of the hydrogen atom. Furthermore, Bohr showed that all the spectral series for hydrogen have a natural interpretation in his theory. The different series correspond to transitions to different final states characterized by the quantum number n_f. Figure 42.8 shows the origin of these spectral series as transitions between energy levels.

Bohr extended his model for hydrogen to other elements in which all but one electron had been removed. These systems have the same structure as the hydrogen atom except that the nuclear charge is larger. Ionized elements such as He^+, Li^{2+}, and Be^{3+} were suspected to exist in hot stellar atmospheres, where atomic collisions frequently have enough energy to completely remove one or more atomic electrons. Bohr showed that many mysterious lines observed in the spectra of the Sun and several other stars could not be due to hydrogen but were correctly predicted by his theory if attributed to singly ionized helium. In general, the number of protons in the nucleus of an atom is called the **atomic number** of the element

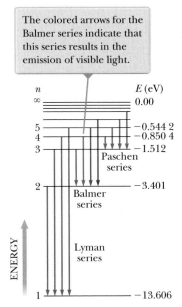

The colored arrows for the Balmer series indicate that this series results in the emission of visible light.

Figure 42.8 An energy-level diagram for the hydrogen atom. Quantum numbers are given on the left, and energies (in electron volts) are given on the right. Vertical arrows represent the four lowest-energy transitions for each of the spectral series shown.

and is given the symbol Z. To describe a single electron orbiting a fixed nucleus of charge $+Ze$, Bohr's theory gives

$$r_n = (n^2) \frac{a_0}{Z} \tag{42.18}$$

$$E_n = -\frac{k_e e^2}{2a_0}\left(\frac{Z^2}{n^2}\right) \quad n = 1, 2, 3, \ldots \tag{42.19}$$

Although the Bohr theory was triumphant in its agreement with some experimental results on the hydrogen atom, it suffered from some difficulties. One of the first indications that the Bohr theory needed to be modified arose when improved spectroscopic techniques were used to examine the spectral lines of hydrogen. It was found that many of the lines in the Balmer and other series were not single lines at all. Instead, each was a group of lines spaced very close together. An additional difficulty arose when it was observed that in some situations certain single spectral lines were split into three closely spaced lines when the atoms were placed in a strong magnetic field. Efforts to explain these and other deviations from the Bohr model led to modifications in the theory and ultimately to a replacement theory that will be discussed in Section 42.4.

Bohr's Correspondence Principle

In our study of relativity, we found that Newtonian mechanics is a special case of relativistic mechanics and is usable only for speeds much less than c. Similarly,

> quantum physics agrees with classical physics when the difference between quantized levels becomes vanishingly small.

This principle, first set forth by Bohr, is called the **correspondence principle.**[2]

For example, consider an electron orbiting the hydrogen atom with $n > 10\,000$. For such large values of n, the energy differences between adjacent levels approach zero; therefore, the levels are nearly continuous. Consequently, the classical model is reasonably accurate in describing the system for large values of n. According to the classical picture, the frequency of the light emitted by the atom is equal to the frequency of revolution of the electron in its orbit about the nucleus. Calculations show that for $n > 10\,000$, this frequency is different from that predicted by quantum mechanics by less than 0.015%.

Quick Quiz 42.1 A hydrogen atom is in its ground state. Incident on the atom is a photon having an energy of 10.5 eV. What is the result? **(a)** The atom is excited to a higher allowed state. **(b)** The atom is ionized. **(c)** The photon passes by the atom without interaction.

Quick Quiz 42.2 A hydrogen atom makes a transition from the $n = 3$ level to the $n = 2$ level. It then makes a transition from the $n = 2$ level to the $n = 1$ level. Which transition results in emission of the longer-wavelength photon? **(a)** the first transition **(b)** the second transition **(c)** neither transition because the wavelengths are the same for both

Example 42.1   Electronic Transitions in Hydrogen

(A) The electron in a hydrogen atom makes a transition from the $n = 2$ energy level to the ground level ($n = 1$). Find the wavelength and frequency of the emitted photon.

---

[2]In reality, the correspondence principle is the starting point for Bohr's property 2(c) on angular momentum quantization. To see how property 2(c) arises from the correspondence principle, see J. W. Jewett Jr., *Physics Begins with Another M . . . Mysteries, Magic, Myth, and Modern Physics* (Boston: Allyn & Bacon, 1996), pp. 353–356.

▶ **42.1** continued

SOLUTION

Conceptualize Imagine the electron in a circular orbit about the nucleus as in the Bohr model in Figure 42.6. When the electron makes a transition to a lower stationary state, it emits a photon with a given frequency and drops to a circular orbit of smaller radius.

Categorize We evaluate the results using equations developed in this section, so we categorize this example as a substitution problem.

Use Equation 42.17 to obtain λ, with $n_i = 2$ and $n_f = 1$:

$$\frac{1}{\lambda} = R_{\mathrm{H}}\left(\frac{1}{1^2} - \frac{1}{2^2}\right) = \frac{3R_{\mathrm{H}}}{4}$$

$$\lambda = \frac{4}{3R_{\mathrm{H}}} = \frac{4}{3(1.097 \times 10^7 \, \mathrm{m}^{-1})} = 1.22 \times 10^{-7} \, \mathrm{m} = \boxed{122 \, \mathrm{nm}}$$

Use Equation 34.20 to find the frequency of the photon:

$$f = \frac{c}{\lambda} = \frac{3.00 \times 10^8 \, \mathrm{m/s}}{1.22 \times 10^{-7} \, \mathrm{m}} = \boxed{2.47 \times 10^{15} \, \mathrm{Hz}}$$

(B) In interstellar space, highly excited hydrogen atoms called Rydberg atoms have been observed. Find the wavelength to which radio astronomers must tune to detect signals from electrons dropping from the $n = 273$ level to the $n = 272$ level.

SOLUTION

Use Equation 42.17, this time with $n_i = 273$ and $n_f = 272$:

$$\frac{1}{\lambda} = R_{\mathrm{H}}\left(\frac{1}{n_f^2} - \frac{1}{n_i^2}\right) = R_{\mathrm{H}}\left(\frac{1}{(272)^2} - \frac{1}{(273)^2}\right) = 9.88 \times 10^{-8} \, R_{\mathrm{H}}$$

Solve for λ:

$$\lambda = \frac{1}{9.88 \times 10^{-8} R_{\mathrm{H}}} = \frac{1}{(9.88 \times 10^{-8})(1.097 \times 10^7 \, \mathrm{m}^{-1})} = \boxed{0.922 \, \mathrm{m}}$$

(C) What is the radius of the electron orbit for a Rydberg atom for which $n = 273$?

SOLUTION

Use Equation 42.12 to find the radius of the orbit:

$$r_{273} = (273)^2 \, (0.052 \, 9 \, \mathrm{nm}) = \boxed{3.94 \, \mu\mathrm{m}}$$

This radius is large enough that the atom is on the verge of becoming macroscopic!

(D) How fast is the electron moving in a Rydberg atom for which $n = 273$?

SOLUTION

Solve Equation 42.8 for the electron's speed:

$$v = \sqrt{\frac{k_e e^2}{m_e r}} = \sqrt{\frac{(8.99 \times 10^9 \, \mathrm{N \cdot m^2/C^2})(1.60 \times 10^{-19} \, \mathrm{C})^2}{(9.11 \times 10^{-31} \, \mathrm{kg})(3.94 \times 10^{-6} \, \mathrm{m})}}$$

$$= \boxed{8.01 \times 10^3 \, \mathrm{m/s}}$$

WHAT IF? What if radiation from the Rydberg atom in part (B) is treated classically? What is the wavelength of radiation emitted by the atom in the $n = 273$ level?

Answer Classically, the frequency of the emitted radiation is that of the rotation of the electron around the nucleus.

Calculate this frequency using the period defined in Equation 4.15:

$$f = \frac{1}{T} = \frac{v}{2\pi r}$$

Substitute the radius and speed from parts (C) and (D):

$$f = \frac{v}{2\pi r} = \frac{8.02 \times 10^3 \, \mathrm{m/s}}{2\pi(3.94 \times 10^{-6} \, \mathrm{m})} = 3.24 \times 10^8 \, \mathrm{Hz}$$

Find the wavelength of the radiation from Equation 34.20:

$$\lambda = \frac{c}{f} = \frac{3.00 \times 10^8 \, \mathrm{m/s}}{3.24 \times 10^8 \, \mathrm{Hz}} = 0.927 \, \mathrm{m}$$

This value is about 0.5% different from the wavelength calculated in part (B). As indicated in the discussion of Bohr's correspondence principle, this difference becomes even smaller for higher values of n.

42.4 The Quantum Model of the Hydrogen Atom

In the preceding section, we described how the Bohr model views the electron as a particle orbiting the nucleus in nonradiating, quantized energy levels. This model combines both classical and quantum concepts. Although the model demonstrates excellent agreement with some experimental results, it cannot explain others. These difficulties are removed when a full quantum model involving the Schrödinger equation is used to describe the hydrogen atom.

The formal procedure for solving the problem of the hydrogen atom is to substitute the appropriate potential energy function into the Schrödinger equation, find solutions to the equation, and apply boundary conditions as we did for the particle in a box in Chapter 41. The potential energy function for the hydrogen atom is that due to the electrical interaction between the electron and the proton (see Section 25.3):

$$U(r) = -k_e \frac{e^2}{r} \tag{42.20}$$

where k_e is the Coulomb constant and r is the radial distance from the proton (situated at $r = 0$) to the electron.

The mathematics for the hydrogen atom is more complicated than that for the particle in a box for two primary reasons: (1) the atom is three-dimensional, and (2) U is not constant, but rather depends on the radial coordinate r. If the time-independent Schrödinger equation (Eq. 41.15) is extended to three-dimensional rectangular coordinates, the result is

$$-\frac{\hbar^2}{2m} \left(\frac{\partial^2 \psi}{\partial x^2} + \frac{\partial^2 \psi}{\partial y^2} + \frac{\partial^2 \psi}{\partial z^2} \right) + U\psi = E\psi$$

It is easier to solve this equation for the hydrogen atom if rectangular coordinates are converted to spherical polar coordinates, an extension of the plane polar coordinates introduced in Section 3.1. In spherical polar coordinates, a point in space is represented by the three variables r, θ, and ϕ, where r is the radial distance from the origin, $r = \sqrt{x^2 + y^2 + z^2}$. With the point represented at the end of a position vector $\vec{r}$ as shown in Figure 42.9, the angular coordinate θ specifies its angular position relative to the z axis. Once that position vector is projected onto the xy plane, the angular coordinate ϕ specifies the projection's (and therefore the point's) angular position relative to the x axis.

The conversion of the three-dimensional time-independent Schrödinger equation for $\psi(x, y, z)$ to the equivalent form for $\psi(r, \theta, \phi)$ is straightforward but very tedious, so we omit the details.[3] In Chapter 41, we separated the time dependence from the space dependence in the general wave function Ψ. In this case of the hydrogen atom, the three space variables in $\psi(r, \theta, \phi)$ can be similarly separated by writing the wave function as a product of functions of each single variable:

$$\psi(r, \theta, \phi) = R(r)f(\theta)g(\phi)$$

In this way, Schrödinger's equation, which is a three-dimensional partial differential equation, can be transformed into three separate ordinary differential equations: one for $R(r)$, one for $f(\theta)$, and one for $g(\phi)$. Each of these functions is subject to boundary conditions. For example, $R(r)$ must remain finite as $r \rightarrow 0$ and $r \rightarrow \infty$; furthermore, $g(\phi)$ must have the same value as $g(\phi + 2\pi)$.

The potential energy function given in Equation 42.20 depends *only* on the radial coordinate r and not on either of the angular coordinates; therefore, it appears only in the equation for $R(r)$. As a result, the equations for θ and ϕ are independent of the particular system and their solutions are valid for *any* system exhibiting rotation.

When the full set of boundary conditions is applied to all three functions, three different quantum numbers are found for each allowed state of the hydrogen atom,

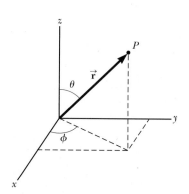

Figure 42.9 A point *P* in space is located by means of a position vector $\vec{r}$. In Cartesian coordinates, the components of this vector are *x*, *y*, and *z*. In spherical polar coordinates, the point is described by *r*, the distance from the origin; θ, the angle between $\vec{r}$ and the *z* axis; and ϕ, the angle between the *x* axis and a projection of $\vec{r}$ onto the *xy* plane.

[3]Descriptions of the solutions to the Schrödinger equation for the hydrogen atom are available in modern physics textbooks such as R. A. Serway, C. Moses, and C. A. Moyer, *Modern Physics*, 3rd ed. (Belmont, CA: Brooks/Cole, 2005).

one for each of the separate differential equations. These quantum numbers are restricted to integer values and correspond to the three independent degrees of freedom (three space dimensions).

The first quantum number, associated with the radial function $R(r)$ of the full wave function, is called the **principal quantum number** and is assigned the symbol n. The differential equation for $R(r)$ leads to functions giving the probability of finding the electron at a certain radial distance from the nucleus. In Section 42.5, we will describe two of these radial wave functions. From the boundary conditions, the energies of the allowed states for the hydrogen atom are found to be related to n as follows:

$$E_n = -\left(\frac{k_e e^2}{2a_0}\right)\frac{1}{n^2} = -\frac{13.606 \text{ eV}}{n^2} \quad n = 1, 2, 3, \ldots \qquad \textbf{(42.21)}$$

◀ **Allowed energies of the quantum hydrogen atom**

This result is in exact agreement with that obtained in the Bohr theory (Eqs. 42.13 and 42.14)! This agreement is *remarkable* because the Bohr theory and the full quantum theory arrive at the result from completely different starting points.

The **orbital quantum number,** symbolized ℓ, comes from the differential equation for $f(\theta)$ and is associated with the orbital angular momentum of the electron. The **orbital magnetic quantum number** m_ℓ arises from the differential equation for $g(\phi)$. Both ℓ and m_ℓ are integers. We will expand our discussion of these two quantum numbers in Section 42.6, where we also introduce a fourth (nonintegral) quantum number, resulting from a relativistic treatment of the hydrogen atom.

The application of boundary conditions on the three parts of the full wave function leads to important relationships among the three quantum numbers as well as certain restrictions on their values:

The values of n are integers that can range from 1 to ∞.

The values of ℓ are integers that can range from 0 to $n - 1$.

The values of m_ℓ are integers that can range from $-\ell$ to ℓ.

◀ **Restrictions on the values of hydrogen-atom quantum numbers**

For example, if $n = 1$, only $\ell = 0$ and $m_\ell = 0$ are permitted. If $n = 2$, then ℓ may be 0 or 1; if $\ell = 0$, then $m_\ell = 0$; but if $\ell = 1$, then m_ℓ may be 1, 0, or -1. Table 42.1 summarizes the rules for determining the allowed values of ℓ and m_ℓ for a given n.

For historical reasons, all states having the same principal quantum number are said to form a **shell.** Shells are identified by the letters K, L, M, . . . , which designate the states for which $n = 1, 2, 3, \ldots$. Likewise, all states having the same values of n and ℓ are said to form a **subshell.** The letters[4] s, p, d, f, g, h, . . . are used to designate the subshells for which $\ell = 0, 1, 2, 3, \ldots$. The state designated by $3p$, for example, has the quantum numbers $n = 3$ and $\ell = 1$; the $2s$ state has the quantum numbers $n = 2$ and $\ell = 0$. These notations are summarized in Tables 42.2 and 42.3 (page 1034).

States that violate the rules given in Table 42.1 do not exist. (They do not satisfy the boundary conditions on the wave function.) For instance, the $2d$ state, which

Pitfall Prevention 42.3

Energy Depends on n Only for Hydrogen The implication in Equation 42.21 that the energy depends only on the quantum number n is true only for the hydrogen atom. For more complicated atoms, we will use the same quantum numbers developed here for hydrogen. The energy levels for these atoms depend primarily on n, but they also depend to a lesser degree on other quantum numbers.

Pitfall Prevention 42.4

Quantum Numbers Describe a System It is common to assign the quantum numbers to an electron. Remember, however, that these quantum numbers arise from the Schrödinger equation, which involves a potential energy function for the *system* of the electron and the nucleus. Therefore, it is more *proper* to assign the quantum numbers to the atom, but it is more *popular* to assign them to an electron. We follow this latter usage because it is so common.

Table 42.1	Three Quantum Numbers for the Hydrogen Atom		
Quantum Number	Name	Allowed Values	Number of Allowed States
n	Principal quantum number	1, 2, 3, . . .	Any number
ℓ	Orbital quantum number	0, 1, 2, . . . , $n - 1$	n
m_ℓ	Orbital magnetic quantum number	$-\ell, -\ell + 1, \ldots, 0, \ldots, \ell - 1, \ell$	$2\ell + 1$

Table 42.2	Atomic Shell Notations
n	Shell Symbol
1	K
2	L
3	M
4	N
5	O
6	P

[4]The first four of these letters come from early classifications of spectral lines: sharp, principal, diffuse, and fundamental. The remaining letters are in alphabetical order.

Table 42.3 Atomic
Subshell Notations

ℓ	Subshell Symbol
0	s
1	p
2	d
3	f
4	g
5	h

would have $n = 2$ and $\ell = 2$, cannot exist because the highest allowed value of ℓ is $n - 1$, which in this case is 1. Therefore, for $n = 2$, the 2s and 2p states are allowed but 2d, 2f, ... are not. For $n = 3$, the allowed subshells are 3s, 3p, and 3d.

Quick Quiz 42.3 How many possible subshells are there for the $n = 4$ level of hydrogen? **(a)** 5 **(b)** 4 **(c)** 3 **(d)** 2 **(e)** 1

Quick Quiz 42.4 When the principal quantum number is $n = 5$, how many different values of **(a)** ℓ and **(b)** m_ℓ are possible?

Example 42.2 The $n = 2$ Level of Hydrogen

For a hydrogen atom, determine the allowed states corresponding to the principal quantum number $n = 2$ and calculate the energies of these states.

S O L U T I O N

Conceptualize Think about the atom in the $n = 2$ quantum state. There is only one such state in the Bohr theory, but our discussion of the quantum theory allows for more states because of the possible values of ℓ and m_ℓ.

Categorize We evaluate the results using rules discussed in this section, so we categorize this example as a substitution problem.

From Table 42.1, we find that when $n = 2$, ℓ can be 0 or 1. Find the possible values of m_ℓ from Table 42.1:

$$\ell = 0 \quad \rightarrow \quad m_\ell = 0$$
$$\ell = 1 \quad \rightarrow \quad m_\ell = -1, 0, \text{ or } 1$$

Hence, we have one state, designated as the 2s state, that is associated with the quantum numbers $n = 2$, $\ell = 0$, and $m_\ell = 0$, and we have three states, designated as 2p states, for which the quantum numbers are $n = 2$, $\ell = 1$, and $m_\ell = -1$; $n = 2$, $\ell = 1$, and $m_\ell = 0$; and $n = 2$, $\ell = 1$, and $m_\ell = 1$.

Find the energy for all four of these states with $n = 2$ from Equation 42.21:

$$E_2 = -\frac{13.606 \text{ eV}}{2^2} = \boxed{-3.401 \text{ eV}}$$

42.5 The Wave Functions for Hydrogen

Because the potential energy of the hydrogen atom depends only on the radial distance r between nucleus and electron, some of the allowed states for this atom can be represented by wave functions that depend only on r. For these states, $f(\theta)$ and $g(\phi)$ are constants. The simplest wave function for hydrogen is the one that describes the 1s state and is designated $\psi_{1s}(r)$:

Wave function for hydrogen ▶
in its ground state

$$\psi_{1s}(r) = \frac{1}{\sqrt{\pi a_0^{\,3}}} e^{-r/a_0} \tag{42.22}$$

where a_0 is the Bohr radius. (In Problem 26 in Enhanced WebAssign, you can verify that this function satisfies the Schrödinger equation.) Note that ψ_{1s} approaches zero as r approaches ∞ and is normalized as presented (see Eq. 41.7). Furthermore, because ψ_{1s} depends only on r, it is *spherically symmetric*. This symmetry exists for all s states.

Recall that the probability of finding a particle in any region is equal to an integral of the probability density $|\psi|^2$ for the particle over the region. The probability density for the 1s state is

$$|\psi_{1s}|^2 = \left(\frac{1}{\pi a_0^{\,3}}\right) e^{-2r/a_0} \tag{42.23}$$

Because we imagine the nucleus to be fixed in space at $r = 0$, we can assign this probability density to the question of locating the electron. According to Equation 41.3, the probability of finding the electron in a volume element dV is $|\psi|^2\, dV$. It is convenient to define the *radial probability density function* $P(r)$ as the probability per unit radial length of finding the electron in a spherical shell of radius r and thickness dr. Therefore, $P(r)\, dr$ is the probability of finding the electron in this shell. The volume dV of such an infinitesimally thin shell equals its surface area $4\pi r^2$ multiplied by the shell thickness dr (Fig. 42.10), so we can write this probability as

$$P(r)\, dr = |\psi|^2\, dV = |\psi|^2\, 4\pi r^2\, dr$$

Therefore, the radial probability density function for an s state is

$$P(r) = 4\pi r^2 |\psi|^2 \tag{42.24}$$

Substituting Equation 42.23 into Equation 42.24 gives the radial probability density function for the hydrogen atom in its ground state:

$$P_{1s}(r) = \left(\frac{4r^2}{a_0^{\,3}}\right) e^{-2r/a_0} \tag{42.25}$$

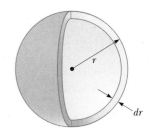

Figure 42.10 A spherical shell of radius r and infinitesimal thickness dr has a volume equal to $4\pi r^2\, dr$.

◀ **Radial probability density for the 1s state of hydrogen**

A plot of the function $P_{1s}(r)$ versus r is presented in Figure 42.11a. The peak of the curve corresponds to the most probable value of r for this particular state. We show in Example 42.3 that this peak occurs at the Bohr radius, the radial position of the electron when the hydrogen atom is in its ground state in the Bohr theory, another remarkable agreement between the Bohr theory and the quantum theory.

According to quantum mechanics, the atom has no sharply defined boundary as suggested by the Bohr theory. The probability distribution in Figure 42.11a suggests that the charge of the electron can be modeled as being extended throughout a region of space, commonly referred to as an *electron cloud*. Figure 42.11b shows the probability density of the electron in a hydrogen atom in the 1s state as a function of position in the xy plane. The darkness of the blue color corresponds to the value of the probability density. The darkest portion of the distribution appears at $r = a_0$, corresponding to the most probable value of r for the electron.

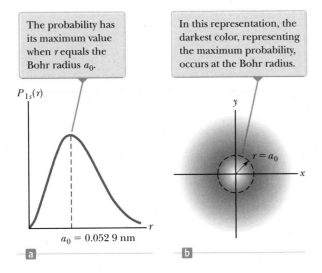

The probability has its maximum value when r equals the Bohr radius a_0.

In this representation, the darkest color, representing the maximum probability, occurs at the Bohr radius.

$P_{1s}(r)$

$a_0 = 0.052\,9$ nm

a

$r = a_0$

b

Figure 42.11 (a) The probability of finding the electron as a function of distance from the nucleus for the hydrogen atom in the 1s (ground) state. (b) The cross section in the xy plane of the spherical electronic charge distribution for the hydrogen atom in its 1s state.

| Example 42.3 | **The Ground State of Hydrogen** |

(A) Calculate the most probable value of r for an electron in the ground state of the hydrogen atom.

continued

▶ **42.3** continued

SOLUTION

Conceptualize Do not imagine the electron in orbit around the proton as in the Bohr theory of the hydrogen atom. Instead, imagine the charge of the electron spread out in space around the proton in an electron cloud with spherical symmetry.

Categorize Because the statement of the problem asks for the "most probable value of r," we categorize this example as a problem in which the quantum approach is used. (In the Bohr atom, the electron moves in an orbit with an *exact* value of r.)

Analyze The most probable value of r corresponds to the maximum in the plot of $P_{1s}(r)$ versus r. We can evaluate the most probable value of r by setting $dP_{1s}/dr = 0$ and solving for r.

Differentiate Equation 42.25 and set the result equal to zero:

$$\frac{dP_{1s}}{dr} = \frac{d}{dr}\left[\left(\frac{4r^2}{a_0^3}\right)e^{-2r/a_0}\right] = 0$$

$$e^{-2r/a_0}\frac{d}{dr}(r^2) + r^2\frac{d}{dr}(e^{-2r/a_0}) = 0$$

$$2re^{-2r/a_0} + r^2(-2/a_0)e^{-2r/a_0} = 0$$

$$(1) \quad 2r[1 - (r/a_0)]e^{-2r/a_0} = 0$$

Set the bracketed expression equal to zero and solve for r:

$$1 - \frac{r}{a_0} = 0 \quad \rightarrow \quad r = \boxed{a_0}$$

Finalize The most probable value of r is the Bohr radius! Equation (1) is also satisfied at $r = 0$ and as $r \rightarrow \infty$. These points are locations of the *minimum* probability, which is equal to zero as seen in Figure 42.11a.

(B) Calculate the probability that the electron in the ground state of hydrogen will be found outside the Bohr radius.

SOLUTION

Analyze The probability is found by integrating the radial probability density function $P_{1s}(r)$ for this state from the Bohr radius a_0 to ∞.

Set up this integral using Equation 42.25:

$$P = \int_{a_0}^{\infty} P_{1s}(r)\,dr = \frac{4}{a_0^3}\int_{a_0}^{\infty} r^2 e^{-2r/a_0}\,dr$$

Put the integral in dimensionless form by changing variables from r to $z = 2r/a_0$, noting that $z = 2$ when $r = a_0$ and that $dr = (a_0/2)\,dz$:

$$P = \frac{4}{a_0^3}\int_2^{\infty}\left(\frac{za_0}{2}\right)^2 e^{-z}\left(\frac{a_0}{2}\right)dz = \frac{1}{2}\int_2^{\infty} z^2 e^{-z}\,dz$$

Evaluate the integral using partial integration (see Appendix B.7):

$$P = -\frac{1}{2}(z^2 + 2z + 2)e^{-z}\Big|_2^{\infty}$$

Evaluate between the limits:

$$P = 0 - [-\frac{1}{2}(4 + 4 + 2)e^{-2}] = 5e^{-2} = \boxed{0.677 \text{ or } 67.7\%}$$

Finalize This probability is larger than 50%. The reason for this value is the asymmetry in the radial probability density function (Fig. 42.11a), which has more area to the right of the peak than to the left.

WHAT IF? What if you were asked for the *average* value of r for the electron in the ground state rather than the most probable value?

Answer The average value of r is the same as the expectation value for r.

Use Equation 42.25 to evaluate the average value of r:

$$r_{avg} = \langle r \rangle = \int_0^{\infty} rP(r)\,dr = \int_0^{\infty} r\left(\frac{4r^2}{a_0^3}\right)e^{-2r/a_0}\,dr$$

$$= \left(\frac{4}{a_0^3}\right)\int_0^{\infty} r^3 e^{-2r/a_0}\,dr$$

▶ **42.3** continued

Evaluate the integral with the help of the first integral listed in Table B.6 in Appendix B:

$$r_{avg} = \left(\frac{4}{a_0^3}\right)\left(\frac{3!}{(2/a_0)^4}\right) = \tfrac{3}{2}a_0$$

Again, the average value is larger than the most probable value because of the asymmetry in the wave function as seen in Figure 42.11a.

The next-simplest wave function for the hydrogen atom is the one corresponding to the 2s state ($n = 2$, $\ell = 0$). The normalized wave function for this state is

$$\psi_{2s}(r) = \frac{1}{4\sqrt{2\pi}}\left(\frac{1}{a_0}\right)^{3/2}\left(2 - \frac{r}{a_0}\right)e^{-r/2a_0} \qquad (42.26)$$

◀ **Wave function for hydrogen in the 2s state**

Again notice that ψ_{2s} depends only on r and is spherically symmetric. The energy corresponding to this state is $E_2 = -(13.606/4)$ eV $= -3.401$ eV. This energy level represents the first excited state of hydrogen. A plot of the radial probability density function for this state in comparison to the 1s state is shown in Figure 42.12. The plot for the 2s state has two peaks. In this case, the most probable value corresponds to that value of r that has the highest value of P ($\approx 5a_0$). An electron in the 2s state would be much farther from the nucleus (on the average) than an electron in the 1s state.

42.6 Physical Interpretation of the Quantum Numbers

The principal quantum number n of a particular state in the hydrogen atom determines the energy of the atom according to Equation 42.21. Now let's see what the other quantum numbers in our atomic model correspond to physically.

The Orbital Quantum Number ℓ

We begin this discussion by returning briefly to the Bohr model of the atom. If the electron moves in a circle of radius r, the magnitude of its angular momentum relative to the center of the circle is $L = m_e vr$. The direction of $\vec{L}$ is perpendicular to the plane of the circle and is given by a right-hand rule. According to classical physics, the magnitude L of the orbital angular momentum can have any value. The Bohr model of hydrogen, however, postulates that the magnitude of the angular momentum of the electron is restricted to multiples of $\hbar$; that is, $L = n\hbar$. This model must be modified because it predicts (incorrectly) that the ground state of hydrogen has one unit of angular momentum. Furthermore, if L is taken to be zero in the Bohr model, the electron must be pictured as a particle oscillating along a straight line through the nucleus, which is a physically unacceptable situation.

These difficulties are resolved with the quantum-mechanical model of the atom, although we must give up the convenient mental representation of an electron orbiting in a well-defined circular path. Despite the absence of this representation, the atom does indeed possess an angular momentum and it is still called orbital angular momentum. According to quantum mechanics, an atom in a state whose principal quantum number is n can take on the following *discrete* values of the magnitude of the orbital angular momentum:[5]

$$L = \sqrt{\ell(\ell + 1)}\,\hbar \quad \ell = 0, 1, 2, \ldots, n - 1 \qquad (42.27)$$

◀ **Allowed values of L**

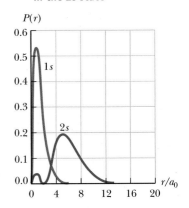

$P(r)$

Figure 42.12 The radial probability density function versus r/a_0 for the 1s and 2s states of the hydrogen atom.

[5]Equation 42.27 is a direct result of the mathematical solution of the Schrödinger equation and the application of angular boundary conditions. This development, however, is beyond the scope of this book.

Given these allowed values of ℓ, we see that $L = 0$ (corresponding to $\ell = 0$) is an acceptable value of the magnitude of the angular momentum. That L can be zero in this model serves to point out the inherent difficulties in any attempt to describe results based on quantum mechanics in terms of a purely particle-like (classical) model. In the quantum-mechanical interpretation, the electron cloud for the $L = 0$ state is spherically symmetric and has no fundamental rotation axis.

The Orbital Magnetic Quantum Number m_ℓ

Because angular momentum is a vector, its direction must be specified. Recall from Chapter 29 that a current loop has a corresponding magnetic moment $\vec{\boldsymbol{\mu}} = I\vec{\mathbf{A}}$ (Eq. 29.15), where I is the current in the loop and $\vec{\mathbf{A}}$ is a vector perpendicular to the loop whose magnitude is the area of the loop. Such a moment placed in a magnetic field $\vec{\mathbf{B}}$ interacts with the field. Suppose a weak magnetic field applied along the z axis defines a direction in space. According to classical physics, the energy of the loop–field system depends on the direction of the magnetic moment of the loop with respect to the magnetic field as described by Equation 29.18, $U_B = -\vec{\boldsymbol{\mu}} \cdot \vec{\mathbf{B}}$. Any energy between $-\mu B$ and $+\mu B$ is allowed by classical physics.

In the Bohr theory, the circulating electron represents a current loop. In the quantum-mechanical approach to the hydrogen atom, we abandon the circular orbit viewpoint of the Bohr theory, but the atom still possesses an orbital angular momentum. Therefore, there is some sense of rotation of the electron around the nucleus and a magnetic moment is present due to this angular momentum.

As mentioned in Section 42.3, spectral lines from some atoms are observed to split into groups of three closely spaced lines when the atoms are placed in a magnetic field. Suppose the hydrogen atom is located in a magnetic field. According to quantum mechanics, there are *discrete* directions allowed for the magnetic moment vector $\vec{\boldsymbol{\mu}}$ with respect to the magnetic field vector $\vec{\mathbf{B}}$. This situation is very different from that in classical physics, in which all directions are allowed.

Because the magnetic moment $\vec{\boldsymbol{\mu}}$ of the atom can be related[6] to the angular momentum vector $\vec{\mathbf{L}}$, the discrete directions of $\vec{\boldsymbol{\mu}}$ translate to the direction of $\vec{\mathbf{L}}$ being quantized. This quantization means that L_z (the projection of $\vec{\mathbf{L}}$ along the z axis) can have only discrete values. The orbital magnetic quantum number m_ℓ specifies the allowed values of the z component of the orbital angular momentum according to the expression[7]

Allowed values of L_z ▶

$$L_z = m_\ell \hbar \qquad\qquad \textbf{(42.28)}$$

The quantization of the possible orientations of $\vec{\mathbf{L}}$ with respect to an external magnetic field is often referred to as **space quantization.**

Let's look at the possible magnitudes and orientations of $\vec{\mathbf{L}}$ for a given value of ℓ. Recall that m_ℓ can have values ranging from $-\ell$ to ℓ. If $\ell = 0$, then $L = 0$; the only allowed value of m_ℓ is $m_\ell = 0$ and $L_z = 0$. If $\ell = 1$, then $L = \sqrt{2}\hbar$ from Equation 42.27. The possible values of m_ℓ are -1, 0, and 1, so Equation 42.28 tells us that L_z may be $-\hbar$, 0, or $\hbar$. If $\ell = 2$, the magnitude of the orbital angular momentum is $\sqrt{6}\hbar$. The value of m_ℓ can be -2, -1, 0, 1, or 2, corresponding to L_z values of $-2\hbar$, $-\hbar$, 0, $\hbar$, or $2\hbar$, and so on.

Figure 42.13a shows a **vector model** that describes space quantization for the case $\ell = 2$. Notice that $\vec{\mathbf{L}}$ can never be aligned parallel or antiparallel to $\vec{\mathbf{B}}$ because the maximum value of L_z is $\ell\hbar$, which is less than the magnitude of the angular momentum $L = \sqrt{\ell(\ell + 1)}\hbar$. The angular momentum vector $\vec{\mathbf{L}}$ is allowed to be perpendicular to $\vec{\mathbf{B}}$, which corresponds to the case of $L_z = 0$ and $\ell = 0$.

---

[6]See Equation 30.22 for this relationship as derived from a classical viewpoint. Quantum mechanics arrives at the same result.

[7]As with Equation 42.27, the relationship expressed in Equation 42.28 arises from the solution to the Schrödinger equation and application of boundary conditions.

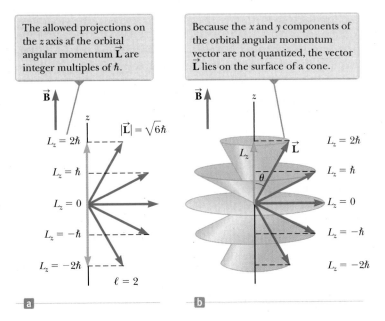

Figure 42.13 A vector model for $\ell = 2$.

The allowed projections on the z axis of the orbital angular momentum $\vec{\mathbf{L}}$ are integer multiples of $\hbar$.

Because the x and y components of the orbital angular momentum vector are not quantized, the vector $\vec{\mathbf{L}}$ lies on the surface of a cone.

The vector $\vec{\mathbf{L}}$ does not point in one specific direction. If $\vec{\mathbf{L}}$ were known exactly, all three components L_x, L_y, and L_z would be specified, which is inconsistent with an angular momentum version of the uncertainty principle. How can the magnitude and z component of a vector be specified, but the vector not be completely specified? To answer, imagine that L_x and L_y are completely unspecified so that $\vec{\mathbf{L}}$ lies anywhere on the surface of a cone that makes an angle θ with the z axis as shown in Figure 42.13b. From the figure, we see that θ is also quantized and that its values are specified through the relationship

$$\cos \theta = \frac{L_z}{L} = \frac{m_\ell}{\sqrt{\ell(\ell + 1)}} \qquad (42.29)$$

◀ **Allowed directions of the orbital angular momentum vector**

If the atom is placed in a magnetic field, the energy $U_B = -\vec{\boldsymbol{\mu}} \cdot \vec{\mathbf{B}}$ is additional energy for the atom–field system beyond that described in Equation 42.21. Because the directions of $\vec{\boldsymbol{\mu}}$ are quantized, there are discrete total energies for the system corresponding to different values of m_ℓ. Figure 42.14a shows a transition between two atomic levels in the absence of a magnetic field. In Figure 42.14b, a magnetic

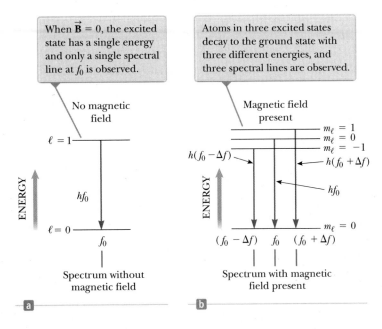

When $\vec{\mathbf{B}} = 0$, the excited state has a single energy and only a single spectral line at f_0 is observed.

Atoms in three excited states decay to the ground state with three different energies, and three spectral lines are observed.

Figure 42.14 The Zeeman effect. (a) Energy levels for the ground and first excited states of a hydrogen atom. (b) When the atom is immersed in a magnetic field $\vec{\mathbf{B}}$, the state with $\ell = 1$ splits into three states, giving rise to emission lines at f_0, $f_0 + \Delta f$, and $f_0 - \Delta f$, where Δf is the frequency shift of the emission caused by the magnetic field.

field is applied and the upper level, with $\ell = 1$, splits into three levels corresponding to the different directions of $\vec{\mu}$. There are now three possible transitions from the $\ell = 1$ subshell to the $\ell = 0$ subshell. Therefore, in a collection of atoms, there are atoms in all three states and the single spectral line in Figure 42.14a splits into three spectral lines. This phenomenon is called the *Zeeman effect*.

The Zeeman effect can be used to measure extraterrestrial magnetic fields. For example, the splitting of spectral lines in light from hydrogen atoms in the surface of the Sun can be used to calculate the magnitude of the magnetic field at that location. The Zeeman effect is one of many phenomena that cannot be explained with the Bohr model but are successfully explained by the quantum model of the atom.

Example 42.4   Space Quantization for Hydrogen

Consider the hydrogen atom in the $\ell = 3$ state. Calculate the magnitude of $\vec{L}$, the allowed values of L_z, and the corresponding angles θ that $\vec{L}$ makes with the z axis.

SOLUTION

Conceptualize Consider Figure 42.13a, which is a vector model for $\ell = 2$. Draw such a vector model for $\ell = 3$ to help with this problem.

Categorize We evaluate results using equations developed in this section, so we categorize this example as a substitution problem.

Calculate the magnitude of the orbital angular momentum using Equation 42.27:

$$L = \sqrt{\ell(\ell + 1)}\hbar = \sqrt{3(3 + 1)}\hbar = \boxed{2\sqrt{3}\,\hbar}$$

Calculate the allowed values of L_z using Equation 42.28 with $m_\ell = -3, -2, -1, 0, 1, 2,$ and 3:

$$L_z = \boxed{-3\hbar, -2\hbar, -\hbar, 0, \hbar, 2\hbar, 3\hbar}$$

Calculate the allowed values of $\cos \theta$ using Equation 42.29:

$$\cos \theta = \frac{\pm 3}{2\sqrt{3}} = \pm 0.866 \qquad \cos \theta = \frac{\pm 2}{2\sqrt{3}} = \pm 0.577$$

$$\cos \theta = \frac{\pm 1}{2\sqrt{3}} = \pm 0.289 \qquad \cos \theta = \frac{0}{2\sqrt{3}} = 0$$

Find the angles corresponding to these values of $\cos \theta$:

$$\theta = \boxed{30.0°, 54.7°, 73.2°, 90.0°, 107°, 125°, 150°}$$

WHAT IF? What if the value of ℓ is an arbitrary integer? For an arbitrary value of ℓ, how many values of m_ℓ are allowed?

Answer For a given value of ℓ, the values of m_ℓ range from $-\ell$ to $+\ell$ in steps of 1. Therefore, there are 2ℓ nonzero values of m_ℓ (specifically, $\pm 1, \pm 2, \ldots, \pm\ell$). In addition, one more value of $m_\ell = 0$ is possible, for a total of $2\ell + 1$ values of m_ℓ. This result is critical in understanding the results of the Stern–Gerlach experiment described below with regard to spin.

Wolfgang Pauli and Niels Bohr watch a spinning top. The spin of the electron is analogous to the spin of the top but is different in many ways.

The Spin Magnetic Quantum Number m_s

The three quantum numbers n, ℓ, and m_ℓ discussed so far are generated by applying boundary conditions to solutions of the Schrödinger equation, and we can assign a physical interpretation to each quantum number. Let's now consider **electron spin,** which does *not* come from the Schrödinger equation.

In Example 42.2, we found four quantum states corresponding to $n = 2$. In reality, however, eight such states occur. The additional four states can be explained by requiring a fourth quantum number for each state, the **spin magnetic quantum number m_s.**

The need for this new quantum number arises because of an unusual feature observed in the spectra of certain gases, such as sodium vapor. Close examination of one prominent line in the emission spectrum of sodium reveals that the line is, in fact, two closely spaced lines called a *doublet*.[8] The wavelengths of these lines occur in the yellow region of the electromagnetic spectrum at 589.0 nm and 589.6 nm. In 1925, when this doublet was first observed, it could not be explained with the existing atomic theory. To resolve this dilemma, Samuel Goudsmit (1902–1978) and George Uhlenbeck (1900–1988), following a suggestion made by Austrian physicist Wolfgang Pauli, proposed the spin quantum number.

To describe this new quantum number, it is convenient (but technically incorrect) to imagine the electron spinning about its axis as it orbits the nucleus as described in Section 30.6. As illustrated in Figure 42.15, only two directions exist for the electron spin. If the direction of spin is as shown in Figure 42.15a, the electron is said to have *spin up*. If the direction of spin is as shown in Figure 42.15b, the electron is said to have *spin down*. In the presence of a magnetic field, the energy associated with the electron is slightly different for the two spin directions. This energy difference accounts for the sodium doublet.

The classical description of electron spin—as resulting from a spinning electron—is incorrect. More recent theory indicates that the electron is a point particle, without spatial extent. Therefore, the electron is not modeled as a rigid object and cannot be considered to be spinning. Despite this conceptual difficulty, all experimental evidence supports the idea that an electron does have some intrinsic angular momentum that can be described by the spin magnetic quantum number. Paul Dirac (1902–1984) showed that this fourth quantum number originates in the relativistic properties of the electron.

In 1921, Otto Stern (1888–1969) and Walter Gerlach (1889–1979) performed an experiment that demonstrated space quantization. Their results, however, were not in quantitative agreement with the atomic theory that existed at that time. In their experiment, a beam of silver atoms sent through a nonuniform magnetic field was split into two discrete components (Fig. 42.16). Stern and Gerlach repeated the experiment using other atoms, and in each case the beam split into two or more components. The classical argument is as follows. If the z direction is chosen to be the direction of the maximum nonuniformity of $\vec{B}$, the net magnetic force on the atoms is along the z axis and is proportional to the component of the magnetic moment $\vec{\mu}$ of the atom in the z direction. Classically, $\vec{\mu}$ can have any orientation, so the deflected beam should be spread out continuously. According to quantum mechanics, however, the deflected beam has an integral number of discrete components and the number of components determines the number of possible values of μ_z. Therefore, because the Stern–Gerlach experiment showed split beams, space quantization was at least qualitatively verified.

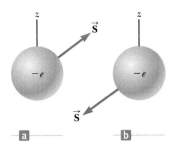

Figure 42.15 The spin of an electron can be either (a) up or (b) down relative to a specified z axis. As in the case of orbital angular momentum, the x and y components of the spin angular momentum vector are not quantized.

Pitfall Prevention 42.5
The Electron Is Not Spinning Although the concept of a spinning electron is conceptually useful, it should not be taken literally. The spin of the Earth is a mechanical rotation. On the other hand, electron spin is a purely quantum effect that gives the electron an angular momentum as if it were physically spinning.

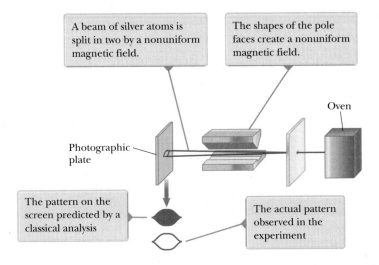

A beam of silver atoms is split in two by a nonuniform magnetic field.

The shapes of the pole faces create a nonuniform magnetic field.

Oven

Photographic plate

The pattern on the screen predicted by a classical analysis

The actual pattern observed in the experiment

Figure 42.16 The technique used by Stern and Gerlach to verify space quantization.

[8]This phenomenon is a Zeeman effect for spin and is identical in nature to the Zeeman effect for orbital angular momentum discussed before Example 42.4 except that no external magnetic field is required. The magnetic field for this Zeeman effect is internal to the atom and arises from the relative motion of the electron and the nucleus.

For the moment, let's assume the magnetic moment of the atom is due to the orbital angular momentum. Because μ_z is proportional to m_ℓ, the number of possible values of μ_z is $2\ell + 1$ as found in the What If? section of Example 42.4. Furthermore, because ℓ is an integer, the number of values of μ_z is always odd. This prediction is not consistent with Stern and Gerlach's observation of two components (an *even* number) in the deflected beam of silver atoms. Hence, either quantum mechanics is incorrect or the model is in need of refinement.

In 1927, T. E. Phipps and J. B. Taylor repeated the Stern–Gerlach experiment using a beam of hydrogen atoms. Their experiment was important because it involved an atom containing a single electron in its ground state, for which the quantum theory makes reliable predictions. Recall that $\ell = 0$ for hydrogen in its ground state, so $m_\ell = 0$. Therefore, we would not expect the beam to be deflected by the magnetic field at all because the magnetic moment $\vec{\mu}$ of the atom is zero. The beam in the Phipps–Taylor experiment, however, was again split into two components! On the basis of that result, we must conclude that something other than the electron's orbital motion is contributing to the atomic magnetic moment.

As we learned earlier, Goudsmit and Uhlenbeck had proposed that the electron has an intrinsic angular momentum, spin, apart from its orbital angular momentum. In other words, the total angular momentum of the electron in a particular electronic state contains both an orbital contribution $\vec{L}$ and a spin contribution $\vec{S}$. The Phipps–Taylor result confirmed the hypothesis of Goudsmit and Uhlenbeck.

In 1929, Dirac used the relativistic form of the total energy of a system to solve the relativistic wave equation for the electron in a potential well. His analysis confirmed the fundamental nature of electron spin. (Spin, like mass and charge, is an *intrinsic* property of a particle, independent of its surroundings.) Furthermore, the analysis showed that electron spin[9] can be described by a single quantum number s, whose value can be only $s = \frac{1}{2}$. The spin angular momentum of the electron *never changes*. This notion contradicts classical laws, which dictate that a rotating charge slows down in the presence of an applied magnetic field because of the Faraday emf that accompanies the changing field (Chapter 31). Furthermore, if the electron is viewed as a spinning ball of charge subject to classical laws, parts of the electron near its surface would be rotating with speeds exceeding the speed of light. Therefore, the classical picture must not be pressed too far; ultimately, spin of an electron is a quantum entity defying any simple classical description.

Because spin is a form of angular momentum, it must follow the same quantum rules as orbital angular momentum. In accordance with Equation 42.27, the magnitude of the **spin angular momentum $\vec{S}$** for the electron is

Magnitude of the spin angular momentum of an electron ▶

$$S = \sqrt{s(s + 1)}\,\hbar = \frac{\sqrt{3}}{2}\,\hbar \qquad \text{(42.30)}$$

Like orbital angular momentum $\vec{L}$, spin angular momentum $\vec{S}$ exhibits space quantization as described in Figure 42.17. The spin vector $\vec{S}$ can have two orientations relative to a z axis, specified by the **spin magnetic quantum number** $m_s = \pm\frac{1}{2}$. Similar to Equation 42.28 for orbital angular momentum, the z component of spin angular momentum is

Allowed values of S_z ▶

$$S_z = m_s\hbar = \pm\tfrac{1}{2}\hbar \qquad \text{(42.31)}$$

The two values $\pm\hbar/2$ for S_z correspond to the two possible orientations for $\vec{S}$ shown in Figure 42.17. The value $m_s = +\frac{1}{2}$ refers to the spin-up case, and $m_s = -\frac{1}{2}$ refers to the spin-down case. Notice that Equations 42.30 and 42.31 do not allow the spin vector to lie along the z axis. The actual direction of $\vec{S}$ is at a relatively large angle with respect to the z axis as shown in Figures 42.15 and 42.17.

---

[9]Scientists often use the word *spin* when referring to the spin angular momentum quantum number. For example, it is common to say, "The electron has a spin of one half."

As discussed in the What If? feature of Example 42.4, there are $2\ell + 1$ possible values of m_ℓ for orbital angular momentum. Similarly, for spin angular momentum, there are $2s + 1$ values of m_s. For a spin of $s = \frac{1}{2}$, the number of values of m_s is $2s + 1 = 2$. These two possibilities for m_s lead to the splitting of the beams into two components in the Stern–Gerlach and Phipps–Taylor experiments.

The spin magnetic moment $\vec{\mu}_{\text{spin}}$ of the electron is related to its spin angular momentum $\vec{S}$ by the expression

$$\vec{\mu}_{\text{spin}} = -\frac{e}{m_e}\vec{S} \qquad (42.32)$$

where e is the electronic charge and m_e is the mass of the electron. Because $S_z = \pm\frac{1}{2}\hbar$, the z component of the spin magnetic moment can have the values

$$\vec{\mu}_{\text{spin},z} = \pm\frac{e\hbar}{2m_e} \qquad (42.33)$$

As we learned in Section 30.6, the quantity $e\hbar/2m_e$ is the Bohr magneton $\mu_B = 9.27 \times 10^{-24}$ J/T. The ratio of magnetic moment to angular momentum is twice as great for spin angular momentum (Eq. 42.32) as it is for orbital angular momentum (Eq. 30.22). The factor of 2 is explained in a relativistic treatment first carried out by Dirac.

Today, physicists explain the Stern–Gerlach and Phipps–Taylor experiments as follows. The observed magnetic moments for both silver and hydrogen are due to spin angular momentum only, with no contribution from orbital angular momentum. In the Phipps–Taylor experiment, the single electron in the hydrogen atom has its electron spin quantized in the magnetic field in such a way that the z component of spin angular momentum is either $\frac{1}{2}\hbar$ or $-\frac{1}{2}\hbar$, corresponding to $m_s = \pm\frac{1}{2}$. Electrons with spin $+\frac{1}{2}$ are deflected downward, and those with spin $-\frac{1}{2}$ are deflected upward. In the Stern–Gerlach experiment, 46 of a silver atom's 47 electrons are in filled subshells with paired spins. Therefore, these 46 electrons have a net zero contribution to both orbital and spin angular momentum for the atom. The angular momentum of the atom is due to only the 47th electron. This electron lies in the $5s$ subshell, so there is no contribution from orbital angular momentum. As a result, the silver atoms have angular momentum due to just the spin of one electron and behave in the same way in a nonuniform magnetic field as the hydrogen atoms in the Phipps–Taylor experiment.

The Stern–Gerlach experiment provided two important results. First, it verified the concept of space quantization. Second, it showed that spin angular momentum exists, even though this property was not recognized until four years after the experiments were performed.

As mentioned earlier, there are eight quantum states corresponding to $n = 2$ in the hydrogen atom, not four as found in Example 42.2. Each of the four states in Example 42.2 is actually two states because of the two possible values of m_s. Table 42.4 shows the quantum numbers corresponding to these eight states.

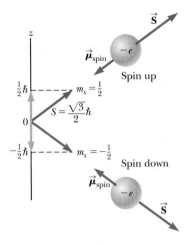

Figure 42.17 Spin angular momentum $\vec{S}$ exhibits space quantization. This figure shows the two allowed orientations of the spin angular momentum vector $\vec{S}$ and the spin magnetic moment $\vec{\mu}_{\text{spin}}$ for a spin-$\frac{1}{2}$ particle, such as the electron.

Table 42.4				**Quantum Numbers for the $n = 2$ State of Hydrogen**		
n	ℓ	m_ℓ	m_s	Subshell	Shell	Number of States in Subshell
2	0	0	$\frac{1}{2}$	$2s$	L	2
2	0	0	$-\frac{1}{2}$			
2	1	1	$\frac{1}{2}$	$2p$	L	6
2	1	1	$-\frac{1}{2}$			
2	1	0	$\frac{1}{2}$			
2	1	0	$-\frac{1}{2}$			
2	1	-1	$\frac{1}{2}$			
2	1	-1	$-\frac{1}{2}$			

42.7 The Exclusion Principle and the Periodic Table

We have found that the state of a hydrogen atom is specified by four quantum numbers: n, ℓ, m_ℓ, and m_s. As it turns out, the number of states available to other atoms may also be predicted by this same set of quantum numbers. In fact, these four quantum numbers can be used to describe all the electronic states of an atom, regardless of the number of electrons in its structure.

For our discussion of atoms with many electrons, it is often easiest to assign the quantum numbers to the electrons in the atom as opposed to the entire atom. An obvious question that arises here is, "How many electrons can be in a particular quantum state?" Pauli answered this important question in 1925, in a statement known as the **exclusion principle:**

> No two electrons can ever be in the same quantum state; therefore, no two electrons in the same atom can have the same set of quantum numbers.

If this principle were not valid, an atom could radiate energy until every electron in the atom is in the lowest possible energy state and therefore the chemical behavior of the elements would be grossly modified. Nature as we know it would not exist.

In reality, we can view the electronic structure of complex atoms as a succession of filled levels increasing in energy. As a general rule, the order of filling of an atom's subshells is as follows. Once a subshell is filled, the next electron goes into the lowest-energy vacant subshell. We can understand this behavior by recognizing that if the atom were not in the lowest energy state available to it, it would radiate energy until it reached this state. This tendency of a quantum system to achieve the lowest energy state is consistent with the second law of thermodynamics discussed in Chapter 22. The entropy of the Universe is increased by the system emitting photons, so that energy is spread out over a larger volume of space.

Before we discuss the electronic configuration of various elements, it is convenient to define an *orbital* as the atomic state characterized by the quantum numbers n, ℓ, and m_ℓ. The exclusion principle tells us that only two electrons can be present in any orbital. One of these electrons has a spin magnetic quantum number $m_s = +\frac{1}{2}$, and the other has $m_s = -\frac{1}{2}$. Because each orbital is limited to two electrons, the number of electrons that can occupy the various shells is also limited.

Table 42.5 shows the allowed quantum states for an atom up to $n = 3$. The arrows pointing upward indicate an electron described by $m_s = +\frac{1}{2}$, and those pointing downward indicate that $m_s = -\frac{1}{2}$. The $n = 1$ shell can accommodate only two electrons because $m_\ell = 0$ means that only one orbital is allowed. (The three quantum numbers describing this orbital are $n = 1$, $\ell = 0$, and $m_\ell = 0$.) The $n = 2$ shell has two subshells, one for $\ell = 0$ and one for $\ell = 1$. The $\ell = 0$ subshell is limited to two electrons because $m_\ell = 0$. The $\ell = 1$ subshell has three allowed orbitals, corresponding to $m_\ell = 1$, 0, and -1. Because each orbital can accommodate two electrons, the $\ell = 1$ subshell can hold six electrons. Therefore, the $n = 2$ shell can contain eight electrons as shown in Table 42.4. The $n = 3$ shell has three subshells ($\ell = 0, 1, 2$) and nine orbitals, accommodating up to 18 electrons. In general, each shell can accommodate up to $2n^2$ electrons.

Wolfgang Pauli
Austrian Theoretical Physicist
(1900–1958)
An extremely talented theoretician who made important contributions in many areas of modern physics, Pauli gained public recognition at the age of 21 with a masterful review article on relativity that is still considered one of the finest and most comprehensive introductions to the subject. His other major contributions were the discovery of the exclusion principle, the explanation of the connection between particle spin and statistics, theories of relativistic quantum electrodynamics, the neutrino hypothesis, and the hypothesis of nuclear spin.

Pitfall Prevention 42.6

The Exclusion Principle Is More General A more general form of the exclusion principle, discussed in Chapter 46, states that no two *fermions* can be in the same quantum state. Fermions are particles with half-integral spin ($\frac{1}{2}$, $\frac{3}{2}$, $\frac{5}{2}$, and so on).

Table 42.5   Allowed Quantum States for an Atom Up to $n = 3$

Shell	n	1	2			3								
Subshell	ℓ	0	0	1		0	1			2				
Orbital	m_ℓ	0	0	1	0	−1	0	1	0	−1	2	1	0	−1	−2
	m_s	↑↓	↑↓	↑↓	↑↓	↑↓	↑↓	↑↓	↑↓	↑↓	↑↓	↑↓	↑↓	↑↓	↑↓

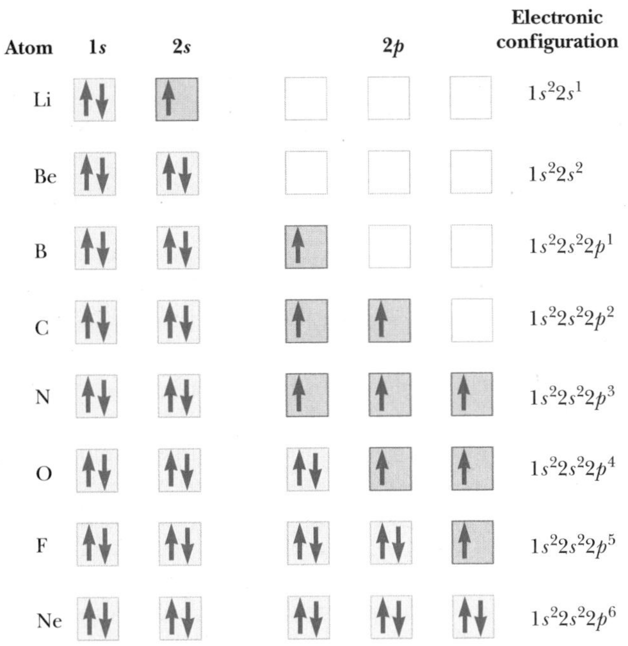

Figure 42.18 The filling of electronic states must obey both the exclusion principle and Hund's rule (page 1046).

The exclusion principle can be illustrated by examining the electronic arrangement in a few of the lighter atoms. The atomic number Z of any element is the number of protons in the nucleus of an atom of that element. A neutral atom of that element has Z electrons. Hydrogen ($Z = 1$) has only one electron, which, in the ground state of the atom, can be described by either of two sets of quantum numbers n, ℓ, m_ℓ, m_s: 1, 0, 0, $\frac{1}{2}$ or 1, 0, 0, $-\frac{1}{2}$. This electronic configuration is often written $1s^1$. The notation $1s$ refers to a state for which $n = 1$ and $\ell = 0$, and the superscript indicates that one electron is present in the s subshell.

Helium ($Z = 2$) has two electrons. In the ground state, their quantum numbers are 1, 0, 0, $\frac{1}{2}$ and 1, 0, 0, $-\frac{1}{2}$. No other possible combinations of quantum numbers exist for this level, and we say that the K shell is filled. This electronic configuration is written $1s^2$.

Lithium ($Z = 3$) has three electrons. In the ground state, two of them are in the $1s$ subshell. The third is in the $2s$ subshell because this subshell is slightly lower in energy than the $2p$ subshell.[10] Hence, the electronic configuration for lithium is $1s^2 2s^1$.

The electronic configurations of lithium and the next several elements are provided in Figure 42.18. The electronic configuration of beryllium ($Z = 4$), with its four electrons, is $1s^2 2s^2$, and boron ($Z = 5$) has a configuration of $1s^2 2s^2 2p^1$. The $2p$ electron in boron may be described by any of the six equally probable sets of quantum numbers listed in Table 42.4. In Figure 42.18, we show this electron in the leftmost $2p$ box with spin up, but it is equally likely to be in any $2p$ box with spin either up or down.

Carbon ($Z = 6$) has six electrons, giving rise to a question concerning how to assign the two $2p$ electrons. Do they go into the same orbital with paired spins ($\uparrow \downarrow$), or do they occupy different orbitals with unpaired spins ($\uparrow \uparrow$)? Experimental data show that the most stable configuration (that is, the one with the lowest energy) is the latter, in which the spins are unpaired. Hence, the two $2p$ electrons in carbon and the three $2p$ electrons in nitrogen ($Z = 7$) have unpaired spins as

---

[10]To a first approximation, energy depends only on the quantum number n, as we have discussed. Because of the effect of the electronic charge shielding the nuclear charge, however, energy depends on ℓ also in multielectron atoms. We shall discuss these shielding effects in Section 42.8.

Figure 42.18 shows. The general rule that governs such situations, called **Hund's rule,** states that

> when an atom has orbitals of equal energy, the order in which they are filled by electrons is such that a maximum number of electrons have unpaired spins.

Some exceptions to this rule occur in elements having subshells that are close to being filled or half-filled.

In 1871, long before quantum mechanics was developed, the Russian chemist Dmitri Mendeleev (1834–1907) made an early attempt at finding some order among the chemical elements. He was trying to organize the elements for the table of contents of a book he was writing. He arranged the atoms in a table similar to that shown in Figure 42.19, according to their atomic masses and chemical similarities. The first table Mendeleev proposed contained many blank spaces, and he boldly stated that the gaps were there only because the elements had not yet been discovered. By noting the columns in which some missing elements should be located, he was able to make rough predictions about their chemical properties. Within 20 years of this announcement, most of these elements were indeed discovered.

The elements in the **periodic table** (Fig. 42.19) are arranged so that all those in a column have similar chemical properties. For example, consider the elements in the last column, which are all gases at room temperature: He (helium), Ne (neon), Ar (argon), Kr (krypton), Xe (xenon), and Rn (radon). The outstanding characteristic of all these elements is that they do not normally take part in chemical reactions; that is, they do not readily join with other atoms to form molecules. They are therefore called *inert gases* or *noble gases.*

Group I	Group II				Transition elements						Group III	Group IV	Group V	Group VI	Group VII	Group 0
H 1 $1s^1$															H 1 $1s^1$	He 2 $1s^2$
Li 3 $2s^1$	Be 4 $2s^2$										B 5 $2p^1$	C 6 $2p^2$	N 7 $2p^3$	O 8 $2p^4$	F 9 $2p^5$	Ne 10 $2p^6$
Na 11 $3s^1$	Mg 12 $3s^2$										Al 13 $3p^1$	Si 14 $3p^2$	P 15 $3p^3$	S 16 $3p^4$	Cl 17 $3p^5$	Ar 18 $3p^6$
K 19 $4s^1$	Ca 20 $4s^2$	Sc 21 $3d^14s^2$	Ti 22 $3d^24s^2$	V 23 $3d^34s^2$	Cr 24 $3d^54s^1$	Mn 25 $3d^54s^2$	Fe 26 $3d^64s^2$	Co 27 $3d^74s^2$	Ni 28 $3d^84s^2$	Cu 29 $3d^{10}4s^1$	Zn 30 $3d^{10}4s^2$	Ga 31 $4p^1$	Ge 32 $4p^2$	As 33 $4p^3$	Se 34 $4p^4$	Br 35 $4p^5$	Kr 36 $4p^6$
Rb 37 $5s^1$	Sr 38 $5s^2$	Y 39 $4d^15s^2$	Zr 40 $4d^25s^2$	Nb 41 $4d^45s^1$	Mo 42 $4d^55s^1$	Tc 43 $4d^55s^2$	Ru 44 $4d^75s^1$	Rh 45 $4d^85s^1$	Pd 46 $4d^{10}$	Ag 47 $4d^{10}5s^1$	Cd 48 $4d^{10}5s^2$	In 49 $5p^1$	Sn 50 $5p^2$	Sb 51 $5p^3$	Te 52 $5p^4$	I 53 $5p^5$	Xe 54 $5p^6$
Cs 55 $6s^1$	Ba 56 $6s^2$	57–71*	Hf 72 $5d^26s^2$	Ta 73 $5d^36s^2$	W 74 $5d^46s^2$	Re 75 $5d^56s^2$	Os 76 $5d^66s^2$	Ir 77 $5d^76s^2$	Pt 78 $5d^96s^1$	Au 79 $5d^{10}6s^1$	Hg 80 $5d^{10}6s^2$	Tl 81 $6p^1$	Pb 82 $6p^2$	Bi 83 $6p^3$	Po 84 $6p^4$	At 85 $6p^5$	Rn 86 $6p^6$
Fr 87 $7s^1$	Ra 88 $7s^2$	89–103**	Rf 104 $6d^27s^2$	Db 105 $6d^37s^2$	Sg 106 $6d^47s^2$	Bh 107 $6d^57s^2$	Hs 108 $6d^67s^2$	Mt 109 $6d^77s^2$	Ds 110 $6d^97s^1$	Rg 111	Cn 112	113	Fl 114	115	Lv 116	117	118

*Lanthanide series	La 57 $5d^16s^2$	Ce 58 $5d^14f^16s^2$	Pr 59 $4f^36s^2$	Nd 60 $4f^46s^2$	Pm 61 $4f^56s^2$	Sm 62 $4f^66s^2$	Eu 63 $4f^76s^2$	Gd 64 $5d^14f^76s^2$	Tb 65 $5d^14f^86s^2$	Dy 66 $4f^{10}6s^2$	Ho 67 $4f^{11}6s^2$	Er 68 $4f^{12}6s^2$	Tm 69 $4f^{13}6s^2$	Yb 70 $4f^{14}6s^2$	Lu 71 $5d^14f^{14}6s^2$
**Actinide series	Ac 89 $6d^17s^2$	Th 90 $6d^27s^2$	Pa 91 $5f^26d^17s^2$	U 92 $5f^36d^17s^2$	Np 93 $5f^46d^17s^2$	Pu 94 $5f^67s^2$	Am 95 $5f^77s^2$	Cm 96 $5f^76d^17s^2$	Bk 97 $5f^86d^17s^2$	Cf 98 $5f^{10}7s^2$	Es 99 $5f^{11}7s^2$	Fm 100 $5f^{12}7s^2$	Md 101 $5f^{13}7s^2$	No 102 $5f^{14}7s^2$	Lr 103 $5f^{14}6d^17s^2$

Figure 42.19 The periodic table of the elements is an organized tabular representation of the elements that shows their periodic chemical behavior. Elements in a given column have similar chemical behavior. This table shows the chemical symbol for the element, the atomic number, and the electron configuration. A more complete periodic table is available in Appendix C.

We can partially understand this behavior by looking at the electronic configurations in Figure 42.19. The chemical behavior of an element depends on the outermost shell that contains electrons. The electronic configuration for helium is $1s^2$, and the $n = 1$ shell (which is the outermost shell because it is the only shell) is filled. Also, the energy of the atom in this configuration is considerably lower than the energy for the configuration in which an electron is in the next available level, the $2s$ subshell. Next, look at the electronic configuration for neon, $1s^2 2s^2 2p^6$. Again, the outermost shell ($n = 2$ in this case) is filled and a wide gap in energy occurs between the filled $2p$ subshell and the next available one, the $3s$ subshell. Argon has the configuration $1s^2 2s^2 2p^6 3s^2 3p^6$. Here, it is only the $3p$ subshell that is filled, but again a wide gap in energy occurs between the filled $3p$ subshell and the next available one, the $3d$ subshell. This pattern continues through all the noble gases. Krypton has a filled $4p$ subshell, xenon a filled $5p$ subshell, and radon a filled $6p$ subshell.

The column to the left of the noble gases in the periodic table consists of a group of elements called the *halogens:* fluorine, chlorine, bromine, iodine, and astatine. At room temperature, fluorine and chlorine are gases, bromine is a liquid, and iodine and astatine are solids. In each of these atoms, the outer subshell is one electron short of being filled. As a result, the halogens are chemically very active, readily accepting an electron from another atom to form a closed shell. The halogens tend to form strong ionic bonds with atoms at the other side of the periodic table. (We shall discuss ionic bonds in Chapter 43.) In a halogen lightbulb, bromine or iodine atoms combine with tungsten atoms evaporated from the filament and return them to the filament, resulting in a longer-lasting lightbulb. In addition, the filament can be operated at a higher temperature than in ordinary lightbulbs, giving a brighter and whiter light.

At the left side of the periodic table, the Group I elements consist of hydrogen and the *alkali metals:* lithium, sodium, potassium, rubidium, cesium, and francium. Each of these atoms contains one electron in a subshell outside of a closed subshell. Therefore, these elements easily form positive ions because the lone electron is bound with a relatively low energy and is easily removed. Therefore, the alkali metal atoms are chemically active and form very strong bonds with halogen atoms. For example, table salt, NaCl, is a combination of an alkali metal and a halogen. Because the outer electron is weakly bound, pure alkali metals tend to be good electrical conductors. Because of their high chemical activity, however, they are not generally found in nature in pure form.

It is interesting to plot ionization energy versus atomic number Z as in Figure 42.20. Notice the pattern of $\Delta Z = 2, 8, 8, 18, 18, 32$ for the various peaks. This pattern follows from the exclusion principle and helps explain why the elements repeat their chemical properties in groups. For example, the peaks at $Z = 2, 10, 18,$

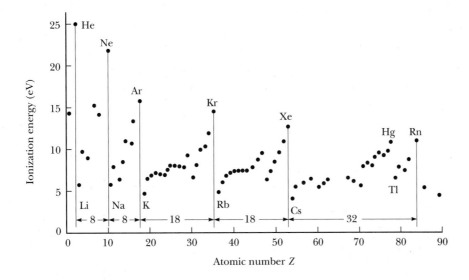

Figure 42.20 Ionization energy of the elements versus atomic number.

Allowed transitions are those that obey the selection rule $\Delta\ell = \pm 1$.

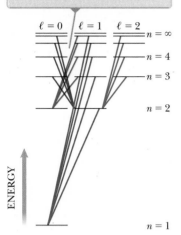

Figure 42.21 Some allowed electronic transitions for hydrogen, represented by the colored lines.

▶ Selection rules for allowed atomic transitions

The peaks represent *characteristic x-rays*. Their appearance depends on the target material.

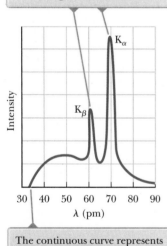

The continuous curve represents *bremsstrahlung*. The shortest wavelength depends on the accelerating voltage.

Figure 42.22 The x-ray spectrum of a metal target. The data shown were obtained when 37-keV electrons bombarded a molybdenum target.

and 36 correspond to the noble gases helium, neon, argon, and krypton, respectively, which, as we have mentioned, all have filled outermost shells. These elements have relatively high ionization energies and similar chemical behavior.

42.8 More on Atomic Spectra: Visible and X-Ray

In Section 42.1, we discussed the observation and early interpretation of visible spectral lines from gases. These spectral lines have their origin in transitions between quantized atomic states. We shall investigate these transitions more deeply in these final three sections of this chapter.

A modified energy-level diagram for hydrogen is shown in Figure 42.21. In this diagram, the allowed values of ℓ for each shell are separated horizontally. Figure 42.21 shows only those states up to $\ell = 2$; the shells from $n = 4$ upward would have more sets of states to the right, which are not shown. Transitions for which ℓ does not change are very unlikely to occur and are called *forbidden transitions*. (Such transitions actually can occur, but their probability is very low relative to the probability of "allowed" transitions.) The various diagonal lines represent allowed transitions between stationary states. Whenever an atom makes a transition from a higher energy state to a lower one, a photon of light is emitted. The frequency of this photon is $f = \Delta E/h$, where ΔE is the energy difference between the two states and h is Planck's constant. The **selection rules** for the *allowed transitions* are

$$\Delta\ell = \pm 1 \quad \text{and} \quad \Delta m_\ell = 0, \pm 1 \tag{42.34}$$

Figure 42.21 shows that the orbital angular momentum of an atom *changes* when it makes a transition to a lower energy state. Therefore, the atom alone is a *non-isolated* system for angular momentum. If we consider the atom–photon system, however, it must be an *isolated* system for angular momentum because nothing else is interacting with this system. The photon involved in the process must carry angular momentum away from the atom when the transition occurs. In fact, the photon has an angular momentum equivalent to that of a particle having a spin of 1. We have now determined over several chapters that a photon has energy, linear momentum, and angular momentum, and each of these is conserved in atomic processes.

Recall from Equation 42.19 that the allowed energies for one-electron atoms and ions, such as hydrogen and He^+, are

$$E_n = -\frac{k_e e^2}{2a_0}\left(\frac{Z^2}{n^2}\right) = -\frac{(13.6 \text{ eV})Z^2}{n^2} \tag{42.35}$$

This equation was developed from the Bohr theory, but it serves as a good first approximation in quantum theory as well. For multielectron atoms, the positive nuclear charge Ze is largely shielded by the negative charge of the inner-shell electrons. Therefore, the outer electrons interact with a net charge that is smaller than the nuclear charge. The expression for the allowed energies for multielectron atoms has the same form as Equation 42.35 with Z replaced by an effective atomic number Z_{eff}:

$$E_n = -\frac{(13.6 \text{ eV})Z_{eff}^2}{n^2} \tag{42.36}$$

where Z_{eff} depends on n and ℓ.

X-Ray Spectra

X-rays are emitted when high-energy electrons or any other charged particles bombard a metal target. The x-ray spectrum typically consists of a broad continuous band containing a series of sharp lines as shown in Figure 42.22. In Section 34.6, we mentioned that an accelerated electric charge emits electromagnetic radiation. The x-rays in Figure 42.22 are the result of the slowing down of high-energy electrons as they strike the target. It may take several interactions with the atoms of the target before the electron gives up all its kinetic energy. The amount of kinetic energy given up in any interaction can vary from zero up to the entire kinetic energy of the electron. Therefore, the wavelength of radiation from these interactions lies in a continuous range from some minimum value up to infinity. It is this general slowing down of the electrons that provides the continuous curve in Figure 42.22, which shows the cutoff of x-rays below a minimum wavelength value that depends on the kinetic energy of the incoming electrons. X-ray radiation with its origin in the slowing down of electrons is called **bremsstrahlung,** the German word for "braking radiation."

Extremely high-energy bremsstrahlung can be used for the treatment of cancerous tissues. Figure 42.23 shows a machine that uses a linear accelerator to accelerate electrons up to 18 MeV and smash them into a tungsten target. The result is a beam of photons, up to a maximum energy of 18 MeV, which is actually in the gamma-ray range in Figure 34.13. This radiation is directed at the tumor in the patient.

The discrete lines in Figure 42.22, called **characteristic x-rays** and discovered in 1908, have a different origin. Their origin remained unexplained until the details of atomic structure were understood. The first step in the production of characteristic x-rays occurs when a bombarding electron collides with a target atom. The electron must have sufficient energy to remove an inner-shell electron from the atom. The vacancy created in the shell is filled when an electron in a higher level drops down into the level containing the vacancy. The existence of characteristic lines in an x-ray spectrum is further direct evidence of the quantization of energy in atomic systems.

The time interval for atomic transitions to happen is very short, less than 10^{-9} s. This transition is accompanied by the emission of a photon whose energy equals the difference in energy between the two levels. Typically, the energy of such transitions is greater than 1 000 eV and the emitted x-ray photons have wavelengths in the range of 0.01 nm to 1 nm.

Let's assume the incoming electron has dislodged an atomic electron from the innermost shell, the K shell. If the vacancy is filled by an electron dropping from the next higher shell—the L shell—the photon emitted has an energy corresponding to the K_α characteristic x-ray line on the curve of Figure 42.22. In this notation, K refers to the final level of the electron and the subscript α, as the *first* letter of the Greek alphabet, refers to the initial level as the *first* one above the final level. Figure 42.24 shows this transition as well as others discussed below. If the vacancy in the K shell is filled by an electron dropping from the M shell, the K_β line in Figure 42.22 is produced.

Other characteristic x-ray lines are formed when electrons drop from upper levels to vacancies other than those in the K shell. For example, L lines are produced when vacancies in the L shell are filled by electrons dropping from higher shells. An L_α line is produced as an electron drops from the M shell to the L shell, and an L_β line is produced by a transition from the N shell to the L shell.

Although multielectron atoms cannot be analyzed exactly with either the Bohr model or the Schrödinger equation, we can apply Gauss's law from Chapter 24 to make some surprisingly accurate estimates of expected x-ray energies and wavelengths. Consider an atom of atomic number Z in which one of the two electrons in the K shell has been ejected. Imagine drawing a gaussian sphere immediately inside the most probable radius of the L electrons. The electric field at the position of the L electrons is a combination of the fields created by the nucleus, the single K electron, the other L electrons, and the outer electrons. The wave functions of the outer electrons are such that the electrons have a very high probability of being farther from the nucleus than the L electrons are. Therefore, the outer

Figure 42.23 Bremsstrahlung is created by this machine and used to treat cancer in a patient.

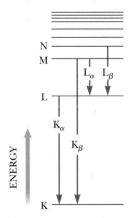

Figure 42.24 Transitions between higher and lower atomic energy levels that give rise to x-ray photons from heavy atoms when they are bombarded with high-energy electrons.

electrons are much more likely to be outside the gaussian surface than inside and, on average, do not contribute significantly to the electric field at the position of the L electrons. The effective charge inside the gaussian surface is the positive nuclear charge and one negative charge due to the single K electron. Ignoring the interactions between L electrons, a single L electron behaves as if it experiences an electric field due to a charge $(Z - 1)e$ enclosed by the gaussian surface. The nuclear charge is shielded by the electron in the K shell such that Z_{eff} in Equation 42.36 is $Z - 1$. For higher-level shells, the nuclear charge is shielded by electrons in all the inner shells.

We can now use Equation 42.36 to estimate the energy associated with an electron in the L shell:

$$E_{\text{L}} = -(Z - 1)^2 \frac{13.6 \text{ eV}}{2^2}$$

After the atom makes the transition, there are two electrons in the K shell. We can approximate the energy associated with one of these electrons as that of a one-electron atom. (In reality, the nuclear charge is reduced somewhat by the negative charge of the other electron, but let's ignore this effect.) Therefore,

$$E_{\text{K}} \approx -Z^2(13.6 \text{ eV}) \tag{42.37}$$

As Example 42.5 shows, the energy of the atom with an electron in an M shell can be estimated in a similar fashion. Taking the energy difference between the initial and final levels, we can then calculate the energy and wavelength of the emitted photon.

In 1914, Henry G. J. Moseley (1887–1915) plotted $\sqrt{1/\lambda}$ versus the Z values for a number of elements where λ is the wavelength of the K_α line of each element. He found that the plot is a straight line as in Figure 42.25, which is consistent with rough calculations of the energy levels given by Equation 42.37. From this plot, Moseley determined the Z values of elements that had not yet been discovered and produced a periodic table in excellent agreement with the known chemical properties of the elements. Until that experiment, atomic numbers had been merely placeholders for the elements that appeared in the periodic table, the elements being ordered according to mass.

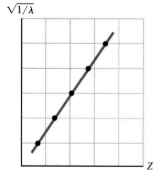

$\sqrt{1/\lambda}$

Z

Figure 42.25 A Moseley plot of $\sqrt{1/\lambda}$ versus Z, where λ is the wavelength of the K_α x-ray line of the element of atomic number Z.

Quick Quiz 42.5 In an x-ray tube, as you increase the energy of the electrons striking the metal target, do the wavelengths of the characteristic x-rays (a) increase, (b) decrease, or (c) remain constant?

Quick Quiz 42.6 True or False: It is possible for an x-ray spectrum to show the continuous spectrum of x-rays without the presence of the characteristic x-rays.

Example 42.5   **Estimating the Energy of an X-Ray**

Estimate the energy of the characteristic x-ray emitted from a tungsten target when an electron drops from an M shell ($n = 3$ state) to a vacancy in the K shell ($n = 1$ state). The atomic number for tungsten is $Z = 74$.

SOLUTION

Conceptualize Imagine an accelerated electron striking a tungsten atom and ejecting an electron from the K shell ($n = 1$). Subsequently, an electron in the M shell ($n = 3$) drops down to fill the vacancy and the energy difference between the states is emitted as an x-ray photon.

Categorize We estimate the results using equations developed in this section, so we categorize this example as a substitution problem.

Use Equation 42.37 and $Z = 74$ for tungsten to estimate the energy associated with the electron in the K shell:

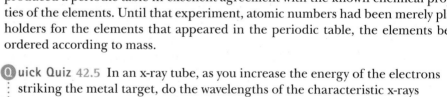

$E_{\text{K}} \approx -(74)^2(13.6 \text{ eV}) = -7.4 \times 10^4 \text{ eV}$

▶ **42.5** continued

Use Equation 42.36 and that nine electrons shield the nuclear charge (eight electrons in the $n = 2$ state and one electron in the $n = 1$ state) to estimate the energy of the M shell:

$$E_M \approx -\frac{(13.6 \text{ eV})(74 - 9)^2}{(3)^2} \approx -6.4 \times 10^3 \text{ eV}$$

Find the energy of the emitted x-ray photon:

$$hf = E_M - E_K \approx -6.4 \times 10^3 \text{ eV} - (-7.4 \times 10^4 \text{ eV})$$
$$\approx 6.8 \times 10^4 \text{ eV} = \boxed{68 \text{ keV}}$$

Consultation of x-ray tables shows that the M–K transition energies in tungsten vary from 66.9 keV to 67.7 keV, where the range of energies is due to slightly different energy values for states of different ℓ. Therefore, our estimate differs from the midpoint of this experimentally measured range by approximately 1%.

42.9 Spontaneous and Stimulated Transitions

We have seen that an atom absorbs and emits electromagnetic radiation only at frequencies that correspond to the energy differences between allowed states. Let's now examine more details of these processes. Consider an atom having the allowed energy levels labeled E_1, E_2, E_3,.... When radiation is incident on the atom, only those photons whose energy hf matches the energy separation ΔE between two energy levels can be absorbed by the atom as represented in Figure 42.26. This process is called **stimulated absorption** because the photon stimulates the atom to make the upward transition. At ordinary temperatures, most of the atoms in a sample are in the ground state. If a vessel containing many atoms of a gaseous element is illuminated with radiation of all possible photon frequencies (that is, a continuous spectrum), only those photons having energy $E_2 - E_1$, $E_3 - E_1$, $E_4 - E_1$, and so on are absorbed by the atoms. As a result of this absorption, some of the atoms are raised to excited states.

Once an atom is in an excited state, the excited atom can make a transition back to a lower energy level, emitting a photon in the process as in Figure 42.27. This process is known as **spontaneous emission** because it happens naturally, without requiring an event to trigger the transition. Typically, an atom remains in an excited state for only about 10^{-8} s.

In addition to spontaneous emission, **stimulated emission** occurs. Suppose an atom is in an excited state E_2 as in Figure 42.28 (page 1052). If the excited state is a *metastable state*—that is, if its lifetime is much longer than the typical 10^{-8} s lifetime of

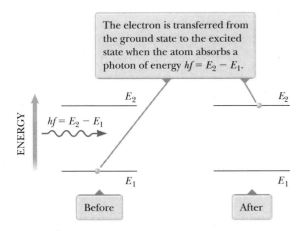

Figure 42.26 Stimulated absorption of a photon.

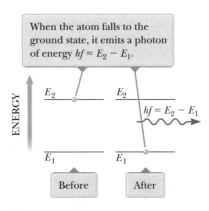

Figure 42.27 Spontaneous emission of a photon by an atom that is initially in the excited state E_2.

Figure 42.28 Stimulated emission of a photon by an incoming photon of energy $hf = E_2 - E_1$. Initially, the atom is in the excited state.

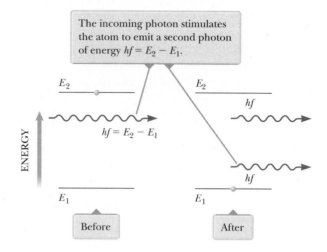

The incoming photon stimulates the atom to emit a second photon of energy $hf = E_2 - E_1$.

excited states—the time interval until spontaneous emission occurs is relatively long. Let's imagine that during that interval a photon of energy $hf = E_2 - E_1$ is incident on the atom. One possibility is that the photon energy is sufficient for the photon to ionize the atom. Another possibility is that the interaction between the incoming photon and the atom causes the atom to return to the ground state[11] and thereby emit a second photon with energy $hf = E_2 - E_1$. In this process, the incident photon is not absorbed; therefore, after the stimulated emission, two photons with identical energy exist: the incident photon and the emitted photon. The two are in phase and travel in the same direction, which is an important consideration in lasers, discussed next.

42.10 Lasers

In this section, we explore the nature of laser light and a variety of applications of lasers in our technological society. The primary properties of laser light that make it useful in these technological applications are the following:

- Laser light is coherent. The individual rays of light in a laser beam maintain a fixed phase relationship with one another.
- Laser light is monochromatic. Light in a laser beam has a very narrow range of wavelengths.
- Laser light has a small angle of divergence. The beam spreads out very little, even over large distances.

To understand the origin of these properties, let's combine our knowledge of atomic energy levels from this chapter with some special requirements for the atoms that emit laser light.

We have described how an incident photon can cause atomic energy transitions either upward (stimulated absorption) or downward (stimulated emission). The two processes are equally probable. When light is incident on a collection of atoms, a net absorption of energy usually occurs because when the system is in thermal equilibrium, many more atoms are in the ground state than in excited states. If the situation can be inverted so that more atoms are in an excited state than in the ground state, however, a net emission of photons can result. Such a condition is called **population inversion.**

Population inversion is, in fact, the fundamental principle involved in the operation of a **laser** (an acronym for *light amplification by stimulated emission of radia-*

---

[11]This phenomenon is fundamentally due to *resonance*. The incoming photon has a frequency and drives the system of the atom at that frequency. Because the driving frequency matches that associated with a transition between states—one of the natural frequencies of the atom—there is a large response: the atom makes the transition.

tion). The full name indicates one of the requirements for laser light: to achieve laser action, the process of stimulated emission must occur.

Suppose an atom is in the excited state E_2 as in Figure 42.28 and a photon with energy $hf = E_2 - E_1$ is incident on it. As described in Section 42.9, the incoming photon can stimulate the excited atom to return to the ground state and thereby emit a second photon having the same energy hf and traveling in the same direction. The incident photon is not absorbed, so after the stimulated emission, there are two identical photons: the incident photon and the emitted photon. The emitted photon is in phase with the incident photon. These photons can stimulate other atoms to emit photons in a chain of similar processes. The many photons produced in this fashion are the source of the intense, coherent light in a laser.

For the stimulated emission to result in laser light, there must be a buildup of photons in the system. The following three conditions must be satisfied to achieve this buildup:

- The system must be in a state of population inversion: there must be more atoms in an excited state than in the ground state. That must be true because the number of photons emitted must be greater than the number absorbed.
- The excited state of the system must be a *metastable state,* meaning that its lifetime must be long compared with the usually short lifetimes of excited states, which are typically 10^{-8} s. In this case, the population inversion can be established and stimulated emission is likely to occur before spontaneous emission.
- The emitted photons must be confined in the system long enough to enable them to stimulate further emission from other excited atoms. That is achieved by using reflecting mirrors at the ends of the system. One end is made totally reflecting, and the other is partially reflecting. A fraction of the light intensity passes through the partially reflecting end, forming the beam of laser light (Fig. 42.29).

One device that exhibits stimulated emission of radiation is the helium–neon gas laser. Figure 42.30 is an energy-level diagram for the neon atom in this system. The mixture of helium and neon is confined to a glass tube that is sealed at the ends by mirrors. A voltage applied across the tube causes electrons to sweep through the tube, colliding with the atoms of the gases and raising them into excited states. Neon atoms are excited to state E_3^* through this process (the asterisk indicates a metastable state) and also as a result of collisions with excited helium atoms. Stimulated emission occurs, causing neon atoms to make transitions to state E_2. Neighboring excited atoms are also stimulated. The result is the production of coherent light at a wavelength of 632.8 nm.

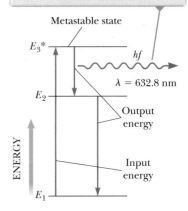

The atom emits 632.8-nm photons through stimulated emission in the transition $E_3^* - E_2$. That is the source of coherent light in the laser.

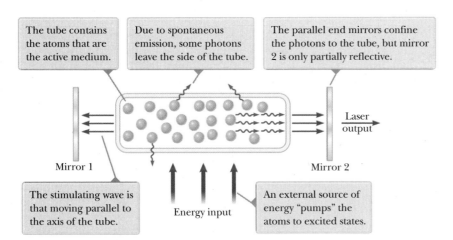

The tube contains the atoms that are the active medium.

Due to spontaneous emission, some photons leave the side of the tube.

The parallel end mirrors confine the photons to the tube, but mirror 2 is only partially reflective.

Laser output

Mirror 1

Mirror 2

The stimulating wave is that moving parallel to the axis of the tube.

Energy input

An external source of energy "pumps" the atoms to excited states.

Figure 42.29 Schematic diagram of a laser design.

Figure 42.30 Energy-level diagram for a neon atom in a helium–neon laser.

Figure 42.31 This robot carrying laser scissors, which can cut up to 50 layers of fabric at a time, is one of the many applications of laser technology.

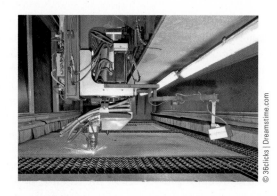

© 36clicks | Dreamstime.com

Applications

Since the development of the first laser in 1960, tremendous growth has occurred in laser technology. Lasers that cover wavelengths in the infrared, visible, and ultraviolet regions are now available. *Laser diodes* are used as laser pointers, and in surveying and construction rangefinders, fiber optic communication, DVD and Blu-ray players, and bar code readers. *Carbon dioxide lasers* are used in industry for welding and cutting, such as the process shown to cut fabric in Figure 42.31. *Excimer lasers* are used in Lasik eye surgery. A variety of other types of lasers exist and are used in various applications. These applications are possible because of the unique characteristics of laser light. In addition to being highly monochromatic, laser light is also highly directional and can be sharply focused to produce regions of extremely intense light energy (with energy densities 10^{12} times the density in the flame of a typical cutting torch).

Lasers are used in precision long-range distance measurement (range finding). In recent years, it has become important in astronomy and geophysics to measure as precisely as possible the distances from various points on the surface of the Earth to a point on the Moon's surface. To facilitate these measurements, the *Apollo* astronauts set up a 0.5-m square of reflector prisms on the Moon, which enables laser pulses directed from an Earth-based station to be retroreflected to the same station (see Fig. 35.8a). Using the known speed of light and the measured round-trip travel time of a laser pulse, the Earth–Moon distance can be determined to a precision of better than 10 cm.

Because various laser wavelengths can be absorbed in specific biological tissues, lasers have a number of medical applications. For example, certain laser procedures have greatly reduced blindness in patients with glaucoma and diabetes. Glaucoma is a widespread eye condition characterized by a high fluid pressure in the eye, a condition that can lead to destruction of the optic nerve. A simple laser operation (iridectomy) can "burn" open a tiny hole in a clogged membrane, relieving the destructive pressure. A serious side effect of diabetes is neovascularization, the proliferation of weak blood vessels, which often leak blood. When neovascularization occurs in the retina, vision deteriorates (diabetic retinopathy) and finally is destroyed. Today, it is possible to direct the green light from an argon ion laser through the clear eye lens and eye fluid, focus on the retina edges, and photocoagulate the leaky vessels. Even people who have only minor vision defects such as nearsightedness are benefiting from the use of lasers to reshape the cornea, changing its focal length and reducing the need for eyeglasses.

Laser surgery is now an everyday occurrence at hospitals and medical clinics around the world. Infrared light at 10 μm from a carbon dioxide laser can cut through muscle tissue, primarily by vaporizing the water contained in cellular material. Laser power of approximately 100 W is required in this technique. The advantage of the "laser knife" over conventional methods is that laser radiation cuts tissue and coagulates blood at the same time, leading to a substantial reduction in blood loss. In addition, the technique virtually eliminates cell migration, an important consideration when tumors are being removed.

A laser beam can be trapped in fine optical fiber light guides (endoscopes) by means of total internal reflection. An endoscope can be introduced through natural orifices, conducted around internal organs, and directed to specific interior body locations, eliminating the need for invasive surgery. For example, bleeding in the gastrointestinal tract can be optically cauterized by endoscopes inserted through the patient's mouth.

In biological and medical research, it is often important to isolate and collect unusual cells for study and growth. A laser cell separator exploits the tagging of specific cells with fluorescent dyes. All cells are then dropped from a tiny charged nozzle and laser-scanned for the dye tag. If triggered by the correct light-emitting tag, a small voltage applied to parallel plates deflects the falling electrically charged cell into a collection beaker.

An exciting area of research and technological applications arose in the 1990s with the development of *laser trapping* of atoms. One scheme, called *optical molasses* and developed by Steven Chu of Stanford University and his colleagues, involves focusing six laser beams onto a small region in which atoms are to be trapped. Each pair of lasers is along one of the x, y, and z axes and emits light in opposite directions (Fig. 42.32). The frequency of the laser light is tuned to be slightly below the absorption frequency of the subject atom. Imagine that an atom has been placed into the trap region and moves along the positive x axis toward the laser that is emitting light toward it (the rightmost laser on the x axis in Fig. 42.32). Because the atom is moving, the light from the laser appears Doppler-shifted upward in frequency in the reference frame of the atom. Therefore, a match between the Doppler-shifted laser frequency and the absorption frequency of the atom exists and the atom absorbs photons.[12] The momentum carried by these photons results in the atom being pushed back to the center of the trap. By incorporating six lasers, the atoms are pushed back into the trap regardless of which way they move along any axis.

In 1986, Chu developed *optical tweezers*, a device that uses a single tightly focused laser beam to trap and manipulate small particles. In combination with microscopes, optical tweezers have opened up many new possibilities for biologists. Optical tweezers have been used to manipulate live bacteria without damage, move chromosomes within a cell nucleus, and measure the elastic properties of a single DNA molecule. Chu shared the 1997 Nobel Prize in Physics with two of his colleagues for the development of the techniques of optical trapping.

An extension of laser trapping, *laser cooling*, is possible because the normal high speeds of the atoms are reduced when they are restricted to the region of the trap. As a result, the temperature of the collection of atoms can be reduced to a few microkelvins. The technique of laser cooling allows scientists to study the behavior of atoms at extremely low temperatures (Fig. 42.33).

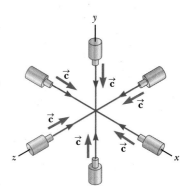

Figure 42.32 An optical trap for atoms is formed at the intersection point of six counterpropagating laser beams along mutually perpendicular axes.

The orange dot is the sample of trapped sodium atoms.

Courtesy of National Institute of Standards and Technology, U.S. Dept. of Commerce

Figure 42.33 A staff member of the National Institute of Standards and Technology views a sample of trapped sodium atoms cooled to a temperature of less than 1 mK.

Summary

Concepts and Principles

The wavelengths of spectral lines from hydrogen, called the **Balmer series,** can be described by the equation

$$\frac{1}{\lambda} = R_H \left(\frac{1}{2^2} - \frac{1}{n^2} \right) \quad n = 3, 4, 5, \ldots \tag{42.1}$$

where R_H is the **Rydberg constant.** The spectral lines corresponding to values of n from 3 to 6 are in the visible range of the electromagnetic spectrum. Values of n higher than 6 correspond to spectral lines in the ultraviolet region of the spectrum.

continued

---

[12]The laser light traveling in the same direction as the atom is Doppler-shifted further downward in frequency, so there is no absorption. Therefore, the atom is not pushed out of the trap by the diametrically opposed laser.

The Bohr model of the atom is successful in describing some details of the spectra of atomic hydrogen and hydrogen-like ions. One basic assumption of the model is that the electron can exist only in discrete orbits such that the angular momentum of the electron is an integral multiple of $h/2\pi = \hbar$. When we assume circular orbits and a simple Coulomb attraction between electron and proton, the energies of the quantum states for hydrogen are calculated to be

$$E_n = -\frac{k_e e^2}{2a_0}\left(\frac{1}{n^2}\right) \quad n = 1, 2, 3, \ldots \quad \textbf{(42.13)}$$

where n is an integer called the **quantum number,** k_e is the Coulomb constant, e is the electronic charge, and $a_0 = 0.052\ 9$ nm is the **Bohr radius.**

If the electron in a hydrogen atom makes a transition from an orbit whose quantum number is n_i to one whose quantum number is n_f, where $n_f < n_i$, a photon is emitted by the atom. The frequency of this photon is

$$f = \frac{k_e e^2}{2a_0 h}\left(\frac{1}{n_f^2} - \frac{1}{n_i^2}\right) \quad \textbf{(42.15)}$$

Quantum mechanics can be applied to the hydrogen atom by the use of the potential energy function $U(r) = -k_e e^2/r$ in the Schrödinger equation. The solution to this equation yields wave functions for allowed states and allowed energies:

$$E_n = -\left(\frac{k_e e^2}{2a_0}\right)\frac{1}{n^2} = -\frac{13.606\text{ eV}}{n^2} \quad n = 1, 2, 3, \ldots \quad \textbf{(42.21)}$$

where n is the **principal quantum number.** The allowed wave functions depend on three quantum numbers: n, ℓ, and m_ℓ, where ℓ is the **orbital quantum number** and m_ℓ is the **orbital magnetic quantum number.** The restrictions on the quantum numbers are

$$n = 1, 2, 3, \ldots$$
$$\ell = 0, 1, 2, \ldots, n-1$$
$$m_\ell = -\ell, -\ell+1, \ldots \ell-1, \ell$$

All states having the same principal quantum number n form a **shell,** identified by the letters K, L, M, $\ldots$ (corresponding to $n = 1, 2, 3, \ldots$). All states having the same values of n and ℓ form a **subshell,** designated by the letters $s, p, d, f, \ldots$ (corresponding to $\ell = 0, 1, 2, 3, \ldots$).

An atom in a state characterized by a specific value of n can have the following values of L, the magnitude of the atom's orbital angular momentum $\vec{\mathbf{L}}$:

$$L = \sqrt{\ell(\ell+1)}\,\hbar$$
$$\ell = 0, 1, 2, \ldots, n-1 \quad \textbf{(42.27)}$$

The allowed values of the projection of $\vec{\mathbf{L}}$ along the z axis are

$$L_z = m_\ell \hbar \quad \textbf{(42.28)}$$

Only discrete values of L_z are allowed as determined by the restrictions on m_ℓ. This quantization of L_z is referred to as **space quantization.**

The electron has an intrinsic angular momentum called the **spin angular momentum.** Electron spin can be described by a single quantum number $s = \frac{1}{2}$. To describe a quantum state completely, it is necessary to include a fourth quantum number m_s, called the **spin magnetic quantum number.** This quantum number can have only two values, $\pm\frac{1}{2}$. The magnitude of the spin angular momentum is

$$S = \frac{\sqrt{3}}{2}\hbar \quad \textbf{(42.30)}$$

and the z component of $\vec{\mathbf{S}}$ is

$$S_z = m_s \hbar = \pm\tfrac{1}{2}\hbar \quad \textbf{(42.31)}$$

That is, the spin angular momentum is also quantized in space, as specified by the spin magnetic quantum number $m_s = \pm\frac{1}{2}$.

The **exclusion principle** states that **no two electrons in an atom can be in the same quantum state.** In other words, no two electrons can have the same set of quantum numbers n, ℓ, m_ℓ, and m_s. Using this principle, the electronic configurations of the elements can be determined. This principle serves as a basis for understanding atomic structure and the chemical properties of the elements.

The magnetic moment $\vec{\boldsymbol{\mu}}_{\text{spin}}$ associated with the spin angular momentum of an electron is

$$\vec{\boldsymbol{\mu}}_{\text{spin}} = -\frac{e}{m_e}\vec{\mathbf{S}} \quad \textbf{(42.32)}$$

The z component of $\vec{\boldsymbol{\mu}}_{\text{spin}}$ can have the values

$$\mu_{\text{spin},z} = \pm\frac{e\hbar}{2m_e} \quad \textbf{(42.33)}$$

The x-ray spectrum of a metal target consists of a set of sharp characteristic lines superimposed on a broad continuous spectrum. **Bremsstrahlung** is x-radiation with its origin in the slowing down of high-energy electrons as they encounter the target. **Characteristic x-rays** are emitted by atoms when an electron undergoes a transition from an outer shell to a vacancy in an inner shell.

Atomic transitions can be described with three processes: **stimulated absorption,** in which an incoming photon raises the atom to a higher energy state; **spontaneous emission,** in which the atom makes a transition to a lower energy state, emitting a photon; and **stimulated emission,** in which an incident photon causes an excited atom to make a downward transition, emitting a photon identical to the incident one.

Objective Questions   1. denotes answer available in *Student Solutions Manual/Study Guide*

1. **(i)** What is the principal quantum number of the initial state of an atom as it emits an M_β line in an x-ray spectrum? (a) 1 (b) 2 (c) 3 (d) 4 (e) 5 **(ii)** What is the principal quantum number of the final state for this transition? Choose from the same possibilities as in part (i).

2. If an electron in an atom has the quantum numbers $n = 3$, $\ell = 2$, $m_\ell = 1$, and $m_s = \frac{1}{2}$, what state is it in? (a) $3s$ (b) $3p$ (c) $3d$ (d) $4d$ (e) $3f$

3. An electron in the $n = 5$ energy level of hydrogen undergoes a transition to the $n = 3$ energy level. What is the wavelength of the photon the atom emits in this process? (a) 2.28×10^{-6} m (b) 8.20×10^{-7} m (c) 3.64×10^{-7} m (d) 1.28×10^{-6} m (e) 5.92×10^{-5} m

4. Consider the $n = 3$ energy level in a hydrogen atom. How many electrons can be placed in this level? (a) 1 (b) 2 (c) 8 (d) 9 (e) 18

5. Which of the following is *not* one of the basic assumptions of the Bohr model of hydrogen? (a) Only certain electron orbits are stable and allowed. (b) The electron moves in circular orbits about the proton under the influence of the Coulomb force. (c) The charge on the electron is quantized. (d) Radiation is emitted by the atom when the electron moves from a higher energy state to a lower energy state. (e) The angular momentum associated with the electron's orbital motion is quantized.

6. Let $-E$ represent the energy of a hydrogen atom. **(i)** What is the kinetic energy of the electron? (a) $2E$ (b) E (c) 0 (d) $-E$ (e) $-2E$ **(ii)** What is the potential energy of the atom? Choose from the same possibilities (a) through (e).

7. The periodic table is based on which of the following principles? (a) The uncertainty principle. (b) All electrons in an atom must have the same set of quantum numbers. (c) Energy is conserved in all interactions. (d) All electrons in an atom are in orbitals having the same energy. (e) No two electrons in an atom can have the same set of quantum numbers.

8. (a) Can a hydrogen atom in the ground state absorb a photon of energy less than 13.6 eV? (b) Can this atom absorb a photon of energy greater than 13.6 eV?

9. Which of the following electronic configurations are *not* allowed for an atom? Choose all correct answers. (a) $2s^2 2p^6$ (b) $3s^2 3p^7$ (c) $3d^7 4s^2$ (d) $3d^{10} 4s^2 4p^6$ (e) $1s^2 2s^2 2d^1$

10. What can be concluded about a hydrogen atom with its electron in the d state? (a) The atom is ionized. (b) The orbital quantum number is $\ell = 1$. (c) The principal quantum number is $n = 2$. (d) The atom is in its ground state. (e) The orbital angular momentum of the atom is not zero.

11. **(i)** Rank the following transitions for a hydrogen atom from the transition with the greatest gain in energy to that with the greatest loss, showing any cases of equality. (a) $n_i = 2$; $n_f = 5$ (b) $n_i = 5$; $n_f = 3$ (c) $n_i = 7$; $n_f = 4$ (d) $n_i = 4$; $n_f = 7$ **(ii)** Rank the same transitions as in part (i) according to the wavelength of the photon absorbed or emitted by an otherwise isolated atom from greatest wavelength to smallest.

12. When an atom emits a photon, what happens? (a) One of its electrons leaves the atom. (b) The atom moves to a state of higher energy. (c) The atom moves to a state of lower energy. (d) One of its electrons collides with another particle. (e) None of those events occur.

13. (a) In the hydrogen atom, can the quantum number n increase without limit? (b) Can the frequency of possible discrete lines in the spectrum of hydrogen increase without limit? (c) Can the wavelength of possible discrete lines in the spectrum of hydrogen increase without limit?

14. Consider the quantum numbers (a) n, (b) ℓ, (c) m_ℓ, and (d) m_s. **(i)** Which of these quantum numbers are fractional as opposed to being integers? **(ii)** Which can sometimes attain negative values? **(iii)** Which can be zero?

15. When an electron collides with an atom, it can transfer all or some of its energy to the atom. A hydrogen atom is in its ground state. Incident on the atom are several electrons, each having a kinetic energy of 10.5 eV. What is the result? (a) The atom can be excited to a higher allowed state. (b) The atom is ionized. (c) The electrons pass by the atom without interaction.

Conceptual Questions

1. denotes answer available in *Student Solutions Manual/Study Guide*

1. Why is stimulated emission so important in the operation of a laser?

2. An energy of about 21 eV is required to excite an electron in a helium atom from the 1s state to the 2s state. The same transition for the He⁺ ion requires approximately twice as much energy. Explain.

3. Why are three quantum numbers needed to describe the state of a one-electron atom (ignoring spin)?

4. Compare the Bohr theory and the Schrödinger treatment of the hydrogen atom, specifically commenting on their treatment of total energy and orbital angular momentum of the atom.

5. Could the Stern–Gerlach experiment be performed with ions rather than neutral atoms? Explain.

6. Why is a *nonuniform* magnetic field used in the Stern–Gerlach experiment?

7. Discuss some consequences of the exclusion principle.

8. (a) According to Bohr's model of the hydrogen atom, what is the uncertainty in the radial coordinate of the electron? (b) What is the uncertainty in the radial component of the velocity of the electron? (c) In what way does the model violate the uncertainty principle?

9. Why do lithium, potassium, and sodium exhibit similar chemical properties?

10. It is easy to understand how two electrons (one spin up, one spin down) fill the $n = 1$ or K shell for a helium atom. How is it possible that eight more electrons are allowed in the $n = 2$ shell, filling the K and L shells for a neon atom?

11. Suppose the electron in the hydrogen atom obeyed classical mechanics rather than quantum mechanics. Why should a gas of such hypothetical atoms emit a continuous spectrum rather than the observed line spectrum?

12. Does the intensity of light from a laser fall off as $1/r^2$? Explain.

Problems available in Access end-of-chapter problems online at www.webassign.net

Section 42.1 Atomic Spectra of Gases
Problems 1–5

Section 42.2 Early Models of the Atom
Problems 6–7

Section 42.3 Bohr's Model of the Hydrogen Atom
Problems 8–21

Section 42.4 The Quantum Model of the Hydrogen Atom
Problems 22–24

Section 42.5 The Wave Functions for Hydrogen
Problems 25–29

Section 42.6 Physical Interpretation of the Quantum Numbers
Problems 30–39

Section 42.7 The Exclusion Principle and the Periodic Table
Problems 40–47

Section 42.8 More on Atomic Spectra: Visible and X-Ray
Problems 48–57

Section 42.9 Spontaneous and Stimulated Transitions
Section 42.10 Lasers
Problems 58–64

Additional Problems
Problems 65–88

Challenge Problems
Problems 88–91

Solutions to the following Problems are available in the *Student Solutions Manual/Study Guide*:
42.5, 42.6, 42.18, 42.21, 42.22, 42.24, 42.26, 42.34, 42.39, 42.44, 42.55, 42.61, 42.67, 42.69, 42.70, 42.81, 42.87, 42.89, and 42.91

List of Enhanced Problems

Problem Number	Targeted Feedback in Enhanced WebAssign	Analysis Model Tutorial in Enhanced WebAssign	Master It in Enhanced WebAssign	Watch It in Enhanced WebAssign
42.10			✓	
42.14		✓		
42.16	✓			✓
42.18	✓		✓	✓
42.36				✓
42.39	✓		✓	
42.54			✓	
42.55	✓		✓	
42.61	✓		✓	
42.63			✓	✓
42.64		✓		
42.68	✓			✓
42.69	✓	✓	✓	
42.79		✓	✓	

Molecules and Solids

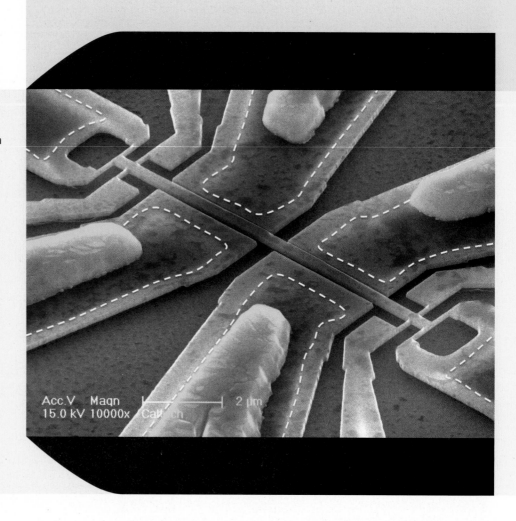

Acc.V   Magn   k   2 µm
15.0 kV 10000x Caltech

The photograph shows a *NEMS resonator*, where NEMS is an acronym for *nanoelectromechanical system*. The device employs a semiconductor bridge vibrating in a standing wave like the strings in Chapter 18. When a single molecule or other particle adheres to the bridge, the resonance frequencies of the normal modes shift in a measurable way. Scientists can determine the mass of the particle from the shifts in the frequencies. The new device shows promise in allowing the masses of molecules and many biological particles to be measured with great accuracy. *(Caltech/Scott Kelberg and Michael Roukes)*

The most random atomic arrangement, that of a gas, was well understood in the 1800s as discussed in our study of kinetic theory in Chapter 21. In a crystalline solid, the atoms are not randomly arranged; rather, they form a regular array. The symmetry of the arrangement of atoms both stimulated and allowed rapid progress in the field of solid-state physics in the 20th century. Recently, our understanding of liquids and amorphous solids has advanced. (In an amorphous solid such as glass or paraffin, the atoms do not form a regular array.) The recent interest in the physics of low-cost amorphous materials has been driven by their use in such devices as solar cells, memory elements, and fiber-optic waveguides. With the addition of liquids, amorphous solids, and some more exotic forms of matter, such as Bose-Einstein condensates, solid-state physics expanded in the middle of the 20th century to become known as *condensed matter physics.*

We begin this chapter by studying the aggregates of atoms known as molecules. We describe the bonding mechanisms in molecules, the various modes of molecular excitation, and the radiation emitted or absorbed by molecules. Next, we show how molecules combine to form solids. Then, by examining their energy-level structure, we explain the differences between insulating, conducting, semiconducting, and superconducting materials. The chapter also includes discussions of semiconducting junctions and several semiconductor devices.

43.1 Molecular Bonds

The bonding mechanisms in a molecule are fundamentally due to electric forces between atoms (or ions). Because the electric force is conservative, the forces between atoms in the system of a molecule are related to a potential energy function. A stable molecule is expected at a configuration for which the potential energy function for the molecule has its minimum value. (See Section 7.9.)

A potential energy function that can be used to model a molecule should account for two known features of molecular bonding:

1. The force between atoms is repulsive at very small separation distances. When two atoms are brought close to each other, some of their electron shells overlap, resulting in repulsion between the shells. This repulsion is partly electrostatic in origin and partly the result of the exclusion principle. Because all electrons must obey the exclusion principle, some electrons in the overlapping shells are forced into higher energy states and the system energy increases as if a repulsive force existed between the atoms.
2. At somewhat larger separations, the force between atoms is attractive. If that were not true, the atoms in a molecule would not be bound together.

Taking into account these two features, the potential energy for a system of two atoms can be represented by an expression of the form

$$U(r) = -\frac{A}{r^n} + \frac{B}{r^m} \qquad \textbf{(43.1)}$$

where r is the internuclear separation distance between the two atoms and n and m are small integers. The parameter A is associated with the attractive force and B with the repulsive force. Example 7.9 gives one common model for such a potential energy function, the Lennard–Jones potential.

Potential energy versus internuclear separation distance for a two-atom system is graphed in Figure 43.1. At large separation distances between the two atoms, the slope of the curve is positive, corresponding to a net attractive force. At the equilibrium separation distance, the attractive and repulsive forces just balance. At this point, the potential energy has its minimum value and the slope of the curve is zero.

A complete description of the bonding mechanisms in molecules is highly complex because bonding involves the mutual interactions of many particles. In this section, we discuss only some simplified models.

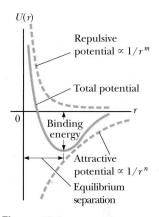

Figure 43.1 Total potential energy as a function of internuclear separation distance for a system of two atoms.

Ionic Bonding

When two atoms combine in such a way that one or more outer electrons are transferred from one atom to the other, the bond formed is called an **ionic bond.** Ionic bonds are fundamentally caused by the Coulomb attraction between oppositely charged ions.

A familiar example of an ionically bonded solid is sodium chloride, NaCl, which is common table salt. Sodium, which has the electronic configuration $1s^2 2s^2 2p^6 3s^1$, is ionized relatively easily, giving up its $3s$ electron to form a Na^+ ion. The energy required to ionize the atom to form Na^+ is 5.1 eV. Chlorine, which has the electronic configuration $1s^2 2s^2 2p^5$, is one electron short of the filled-shell structure of argon. If we compare the energy of a system of a free electron and a Cl atom with one in which the electron joins the atom to make the Cl^- ion, we find that the energy of the ion is lower. When the electron makes a transition from the $E = 0$ state to the negative energy state associated with the available shell in the atom, energy is released. This amount of energy is called the **electron affinity** of the atom. For chlorine, the electron affinity is 3.6 eV. Therefore, the energy required to form Na^+ and Cl^- from isolated atoms is $5.1 - 3.6 = 1.5$ eV. It costs 5.1 eV to remove

Figure 43.2 Total energy versus internuclear separation distance for Na$^+$ and Cl$^-$ ions.

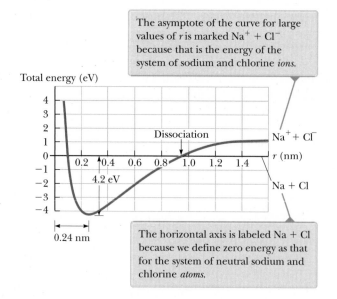

The asymptote of the curve for large values of r is marked Na$^+$ + Cl$^-$ because that is the energy of the system of sodium and chlorine *ions*.

The horizontal axis is labeled Na + Cl because we define zero energy as that for the system of neutral sodium and chlorine *atoms*.

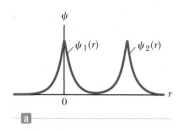

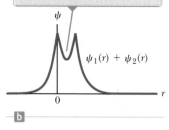

The probability amplitude for an electron to be between the atoms is high.

Figure 43.3 Ground-state wave functions $\psi_1(r)$ and $\psi_2(r)$ for two atoms making a covalent bond. (a) The atoms are far apart, and their wave functions overlap minimally. (b) The atoms are close together, forming a composite wave function $\psi_1(r) + \psi_2(r)$ for the system.

the electron from the Na atom, but 3.6 eV of it is gained back when that electron is allowed to join with the Cl atom.

Now imagine that these two charged ions interact with one another to form a NaCl "molecule."[1] The total energy of the NaCl molecule versus internuclear separation distance is graphed in Figure 43.2. At very large separation distances, the energy of the system of ions is 1.5 eV as calculated above. The total energy has a minimum value of -4.2 eV at the equilibrium separation distance, which is approximately 0.24 nm. Hence, the energy required to break the Na$^+$–Cl$^-$ bond and form neutral sodium and chlorine atoms, called the **dissociation energy**, is 4.2 eV. The energy of the molecule is lower than that of the system of two neutral atoms. Consequently, it is **energetically favorable** for the molecule to form: if a lower energy state of a system exists, the system tends to emit energy to achieve this lower energy state. The system of neutral sodium and chlorine atoms can reduce its total energy by transferring energy out of the system (by electromagnetic radiation, for example) and forming the NaCl molecule.

Covalent Bonding

A **covalent bond** between two atoms is one in which electrons supplied by either one or both atoms are shared by the two atoms. Many diatomic molecules—such as H$_2$, F$_2$, and CO—owe their stability to covalent bonds. The bond between two hydrogen atoms can be described by using atomic wave functions. The ground-state wave function for a hydrogen atom (Chapter 42) is

$$\psi_{1s}(r) = \frac{1}{\sqrt{\pi a_0^3}} e^{-r/a_0}$$

This wave function is graphed in Figure 43.3a for two hydrogen atoms that are far apart. There is very little overlap of the wave functions $\psi_1(r)$ for atom 1, located at $r = 0$, and $\psi_2(r)$ for atom 2, located some distance away. Suppose now the two atoms are brought close together. As that happens, their wave functions overlap and form the compound wave function $\psi_1(r) + \psi_2(r)$ shown in Figure 43.3b. Notice that the probability amplitude is larger between the atoms than it is on either side of the combination of atoms. As a result, the probability is higher that the electrons associated with the atoms will be located between the atoms than on the outer regions

---

[1]NaCl does not tend to form as an isolated molecule at room temperature. In the solid state, NaCl forms a crystalline array of ions as described in Section 43.3. In the liquid state or in solution with water, the Na$^+$ and Cl$^-$ ions dissociate and are free to move relative to each other.

of the system. Consequently, the average position of negative charge in the system is halfway between the atoms. This scenario can be modeled as if there were a fixed negative charge between the atoms, exerting attractive Coulomb forces on both nuclei. Therefore, there is an overall attractive force between the atoms, resulting in a covalent bond.

Because of the exclusion principle, the two electrons in the ground state of H_2 must have antiparallel spins. Also because of the exclusion principle, if a third H atom is brought near the H_2 molecule, the third electron would have to occupy a higher energy level, which is not an energetically favorable situation. For this reason, the H_3 molecule is not stable and does not form.

Van der Waals Bonding

Ionic and covalent bonds occur between atoms to form molecules or ionic solids, so they can be described as bonds *within* molecules. Two additional types of bonds, van der Waals bonds and hydrogen bonds, can occur *between* molecules.

You might think that two neutral molecules would not interact by means of the electric force because they each have zero net charge. They are attracted to each other, however, by weak electrostatic forces called **van der Waals forces.** Likewise, atoms that do not form ionic or covalent bonds are attracted to each other by van der Waals forces. Noble gas atoms, for example, because of their filled shell structure, do not generally form molecules or bond to each other to form a liquid. Because of van der Waals forces, however, at sufficiently low temperatures at which thermal excitations are negligible, noble gases first condense to liquids and then solidify. (The exception is helium, which does not solidify at atmospheric pressure.)

The van der Waals force results from the following situation. While being electrically neutral, a molecule has a charge distribution with positive and negative centers at different positions in the molecule. As a result, the molecule may act as an electric dipole. (See Section 23.4.) Because of the dipole electric fields, two molecules can interact such that there is an attractive force between them.

There are three types of van der Waals forces. The first type, called the *dipole–dipole force,* is an interaction between two molecules each having a permanent electric dipole moment. For example, polar molecules such as HCl have permanent electric dipole moments and attract other polar molecules.

The second type, the *dipole–induced dipole force,* results when a polar molecule having a permanent electric dipole moment induces a dipole moment in a nonpolar molecule. In this case, the electric field of the polar molecule creates the dipole moment in the nonpolar molecule, which then results in an attractive force between the molecules.

The third type is called the *dispersion force,* an attractive force that occurs between two nonpolar molecules. In this case, although the average dipole moment of a nonpolar molecule is zero, the average of the square of the dipole moment is nonzero because of charge fluctuations. Two nonpolar molecules near each other tend to have dipole moments that are correlated in time so as to produce an attractive van der Waals force.

Hydrogen Bonding

Because hydrogen has only one electron, it is expected to form a covalent bond with only one other atom within a molecule. A hydrogen atom in a given molecule can also form a second type of bond between molecules called a **hydrogen bond.** Let's use the water molecule H_2O as an example. In the two covalent bonds in this molecule, the electrons from the hydrogen atoms are more likely to be found near the oxygen atom than near the hydrogen atoms, leaving essentially bare protons at the positions of the hydrogen atoms. This unshielded positive charge can be attracted to the negative end of another polar molecule. Because the proton is unshielded by electrons, the negative end of the other molecule can come very close to the proton to form a bond strong enough to form a solid crystalline structure, such as

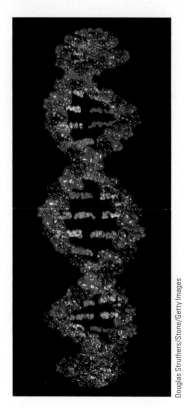

Figure 43.4 DNA molecules are held together by hydrogen bonds.

Total energy of a molecule ▶

Moment of inertia for ▶
a diatomic molecule

that of ordinary ice. The bonds within a water molecule are covalent, but the bonds between water molecules in ice are hydrogen bonds.

The hydrogen bond is relatively weak compared with other chemical bonds and can be broken with an input energy of approximately 0.1 eV. Because of this weakness, ice melts at the low temperature of 0°C. Even though this bond is weak, however, hydrogen bonding is a critical mechanism responsible for the linking of biological molecules and polymers. For example, in the case of the DNA (deoxyribonucleic acid) molecule, which has a double-helix structure (Fig. 43.4), hydrogen bonds form by the sharing of a proton between two atoms and create linkages between the turns of the helix.

Quick Quiz 43.1 For each of the following atoms or molecules, identify the most likely type of bonding that occurs between the atoms or between the molecules. Choose from the following list: ionic, covalent, van der Waals, hydrogen. **(a)** atoms of krypton **(b)** potassium and chlorine atoms **(c)** hydrogen fluoride (HF) molecules **(d)** chlorine and oxygen atoms in a hypochlorite ion (ClO⁻)

43.2 Energy States and Spectra of Molecules

Consider an individual molecule in the gaseous phase of a substance. The energy E of the molecule can be divided into four categories: (1) electronic energy, due to the interactions between the molecule's electrons and nuclei; (2) translational energy, due to the motion of the molecule's center of mass through space; (3) rotational energy, due to the rotation of the molecule about its center of mass; and (4) vibrational energy, due to the vibration of the molecule's constituent atoms:

$$E = E_{el} + E_{trans} + E_{rot} + E_{vib}$$

We explored the roles of translational, rotational, and vibrational energy of molecules in determining the molar specific heats of gases in Sections 21.2 and 21.3. The translational energy is important in kinetic theory, but it is unrelated to internal structure of the molecule, so this molecular energy is unimportant in interpreting molecular spectra. The electronic energy of a molecule is very complex because it involves the interaction of many charged particles, but various techniques have been developed to approximate its values. Although the electronic energies can be studied, significant information about a molecule can be determined by analyzing its quantized rotational and vibrational energy states. Transitions between these states give spectral lines in the microwave and infrared regions of the electromagnetic spectrum, respectively.

Rotational Motion of Molecules

Let's consider the rotation of a molecule around its center of mass, confining our discussion to the diatomic molecule (Fig. 43.5a) but noting that the same ideas can be extended to polyatomic molecules. A diatomic molecule aligned along a y axis has only two rotational degrees of freedom, corresponding to rotations about the x and z axes passing through the molecule's center of mass. We discussed the rotation of such a molecule and its contribution to the specific heat of a gas in Section 21.3. If ω is the angular frequency of rotation about one of these axes, the rotational kinetic energy of the molecule about that axis can be expressed with Equation 10.24:

$$E_{rot} = \tfrac{1}{2}I\omega^2 \tag{43.2}$$

In this equation, I is the moment of inertia of the molecule about its center of mass, given by

$$I = \left(\frac{m_1 m_2}{m_1 + m_2}\right)r^2 = \mu r^2 \tag{43.3}$$

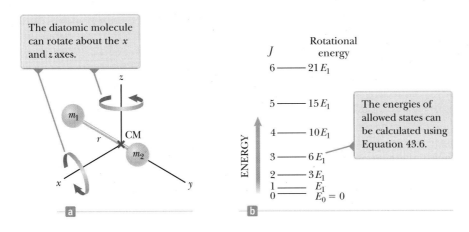

where m_1 and m_2 are the masses of the atoms that form the molecule, r is the atomic separation, and μ is the **reduced mass** of the molecule (see Example 41.5 and Problem 40 in Chapter 41 in Enhanced WebAssign):

$$\mu = \frac{m_1 m_2}{m_1 + m_2} \tag{43.4}$$

◀ **Reduced mass of a diatomic molecule**

The magnitude of the molecule's angular momentum about its center of mass is given by Equation 11.14, $L = I\omega$, which classically can have any value. Quantum mechanics, however, restricts the molecule to certain quantized rotational frequencies such that the angular momentum of the molecule has the values[2]

$$L = \sqrt{J(J+1)}\,\hbar \qquad J = 0, 1, 2, \ldots \tag{43.5}$$

◀ **Allowed values of rotational angular momentum**

where J is an integer called the **rotational quantum number.** Combining Equations 43.5 and 43.2, we obtain an expression for the allowed values of the rotational kinetic energy of the molecule:

$$E_{\text{rot}} = \tfrac{1}{2}I\omega^2 = \frac{1}{2I}(I\omega)^2 = \frac{L^2}{2I} = \frac{(\sqrt{J(J+1)}\,\hbar)^2}{2I}$$

$$E_{\text{rot}} = E_J = \frac{\hbar^2}{2I}J(J+1) \qquad J = 0, 1, 2, \ldots \tag{43.6}$$

◀ **Allowed values of rotational energy**

The allowed rotational energies of a diatomic molecule are plotted in Figure 43.5b. As the quantum number J goes up, the states become farther apart as displayed earlier for rotational energy levels in Figure 21.7.

For most molecules, transitions between adjacent rotational energy levels result in radiation that lies in the microwave range of frequencies ($f \sim 10^{11}$ Hz). When a molecule absorbs a microwave photon, the molecule jumps from a lower rotational energy level to a higher one. The allowed rotational transitions of linear molecules are regulated by the selection rule $\Delta J = \pm 1$. Given this selection rule, all absorption lines in the spectrum of a linear molecule correspond to energy separations equal to $E_J - E_{J-1}$, where $J = 1, 2, 3, \ldots$. From Equation 43.6, we see that the energies of the absorbed photons are given by

$$E_{\text{photon}} = \Delta E_{\text{rot}} = E_J - E_{J-1} = \frac{\hbar^2}{2I}[J(J+1) - (J-1)J]$$

$$E_{\text{photon}} = \frac{\hbar^2}{I}J = \frac{h^2}{4\pi^2 I}J \qquad J = 1, 2, 3, \ldots \tag{43.7}$$

◀ **Energy of a photon absorbed in a transition between adjacent rotational levels**

[2]Equation 43.5 is similar to Equation 42.27 for orbital angular momentum in an atom. The relationship between the magnitude of the angular momentum of a system and the associated quantum number is the same as it is in these equations for *any* system that exhibits rotation as long as the potential energy function for the system is spherically symmetric.

where J is the rotational quantum number of the higher energy state. Because $E_{photon} = hf$, where f is the frequency of the absorbed photon, we see that the allowed frequency for the transition $J = 0$ to $J = 1$ is $f_1 = h/4\pi^2 I$. The frequency corresponding to the $J = 1$ to $J = 2$ transition is $2f_1$, and so on. These predictions are in excellent agreement with the observed frequencies.

Quick Quiz 43.2 A gas of identical diatomic molecules absorbs electromagnetic radiation over a wide range of frequencies. Molecule 1 is in the $J = 0$ rotation state and makes a transition to the $J = 1$ state. Molecule 2 is in the $J = 2$ state and makes a transition to the $J = 3$ state. Is the ratio of the frequency of the photon that excited molecule 2 to that of the photon that excited molecule 1 equal to **(a)** 1, **(b)** 2, **(c)** 3, **(d)** 4, or **(e)** impossible to determine?

Example 43.1   Rotation of the CO Molecule

The $J = 0$ to $J = 1$ rotational transition of the CO molecule occurs at a frequency of 1.15×10^{11} Hz.

(A) Use this information to calculate the moment of inertia of the molecule.

SOLUTION

Conceptualize Imagine that the two atoms in Figure 43.5a are carbon and oxygen. The center of mass of the molecule is not midway between the atoms because of the difference in masses of the C and O atoms.

Categorize The statement of the problem tells us to categorize this example as one involving a quantum-mechanical treatment and to restrict our investigation to the rotational motion of a diatomic molecule.

Analyze Use Equation 43.7 to find the energy of a photon that excites the molecule from the $J = 0$ to the $J = 1$ rotational level:

$$E_{photon} = \frac{h^2}{4\pi^2 I}(1) = \frac{h^2}{4\pi^2 I}$$

Equate this energy to $E = hf$ for the absorbed photon and solve for I:

$$\frac{h^2}{4\pi^2 I} = hf \rightarrow I = \frac{h}{4\pi^2 f}$$

Substitute the frequency given in the problem statement:

$$I = \frac{6.626 \times 10^{-34} \text{ J} \cdot \text{s}}{4\pi^2 (1.15 \times 10^{11} \text{ s}^{-1})} = \boxed{1.46 \times 10^{-46} \text{ kg} \cdot \text{m}^2}$$

(B) Calculate the bond length of the molecule.

SOLUTION

Find the reduced mass μ of the CO molecule:

$$\mu = \frac{m_1 m_2}{m_1 + m_2} = \frac{(12 \text{ u})(16 \text{ u})}{12 \text{ u} + 16 \text{ u}} = 6.86 \text{ u}$$

$$= (6.86 \text{ u})\left(\frac{1.66 \times 10^{-27} \text{ kg}}{1 \text{ u}}\right) = 1.14 \times 10^{-26} \text{ kg}$$

Solve Equation 43.3 for r and substitute for the reduced mass and the moment of inertia from part (A):

$$r = \sqrt{\frac{I}{\mu}} = \sqrt{\frac{1.46 \times 10^{-46} \text{ kg} \cdot \text{m}^2}{1.14 \times 10^{-26} \text{ kg}}}$$

$$= 1.13 \times 10^{-10} \text{ m} = \boxed{0.113 \text{ nm}}$$

Finalize The moment of inertia of the molecule and the separation distance between the atoms are both very small, as expected for a microscopic system.

WHAT IF? What if another photon of frequency 1.15×10^{11} Hz is incident on the CO molecule while that molecule is in the $J = 1$ state? What happens?

Answer Because the rotational quantum states are not equally spaced in energy, the $J = 1$ to $J = 2$ transition does not have the same energy as the $J = 0$ to $J = 1$ transition. Therefore, the molecule will *not* be excited to the $J = 2$ state. Two

▶ **43.1** continued

possibilities exist. The photon could pass by the molecule with no interaction, or the photon could induce a stimulated emission, similar to that for atoms and discussed in Section 42.9. In this case, the molecule makes a transition back to the $J = 0$ state and the original photon and a second identical photon leave the scene of the interaction.

Vibrational Motion of Molecules

If we consider a molecule to be a flexible structure in which the atoms are bonded together by "effective springs" as shown in Figure 43.6a, we can apply the particle in simple harmonic motion analysis model to the molecule as long as the atoms in the molecule are not too far from their equilibrium positions. Recall from Section 15.3 that the potential energy function for a simple harmonic oscillator is parabolic, varying as the square of the position of the particle relative to the equilibrium position. (See Eq. 15.20 and Fig. 15.9b.) Figure 43.6b shows a plot of potential energy versus atomic separation for a diatomic molecule, where r_0 is the equilibrium atomic separation. For separations close to r_0, the shape of the potential energy curve closely resembles the parabolic shape of the potential energy function in the particle in simple harmonic motion model.

According to classical mechanics, the frequency of vibration for the system shown in Figure 43.6a is given by Equation 15.14:

$$f = \frac{1}{2\pi}\sqrt{\frac{k}{\mu}} \qquad \textbf{(43.8)}$$

where k is the effective spring constant and μ is the reduced mass given by Equation 43.4. In Section 21.3, we studied the contribution of a molecule's vibration to the specific heats of gases.

Quantum mechanics predicts that a molecule vibrates in quantized states as described in Section 41.7. The vibrational motion and quantized vibrational energy can be altered if the molecule acquires energy of the proper value to cause a transition between quantized vibrational states. As discussed in Section 41.7, the allowed vibrational energies are

$$E_{\text{vib}} = \left(v + \tfrac{1}{2}\right)hf \qquad v = 0, 1, 2, \ldots \qquad \textbf{(43.9)}$$

where v is an integer called the **vibrational quantum number.** (We used n in Section 41.7 for a general harmonic oscillator, but v is often used for the quantum number when discussing molecular vibrations.) If the system is in the lowest vibrational state, for which $v = 0$, its ground-state energy is $\tfrac{1}{2}hf$. In the first excited vibrational state, $v = 1$ and the energy is $\tfrac{3}{2}hf$, and so on.

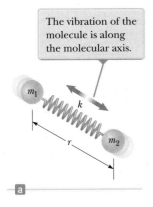

The vibration of the molecule is along the molecular axis.

a

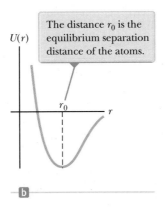

$U(r)$

The distance r_0 is the equilibrium separation distance of the atoms.

r_0

r

b

Figure 43.6 (a) Effective-spring model of a diatomic molecule. (b) Plot of the potential energy of a diatomic molecule versus atomic separation distance. Compare with Figure 15.11a.

Figure 43.7 Allowed vibrational energies of a diatomic molecule, where f is the frequency of vibration of the molecule, given by Equation 43.8.

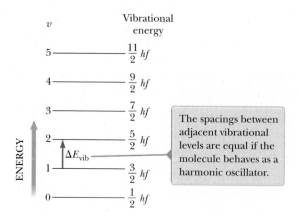

The spacings between adjacent vibrational levels are equal if the molecule behaves as a harmonic oscillator.

Substituting Equation 43.8 into Equation 43.9 gives the following expression for the allowed vibrational energies:

Allowed values of vibrational energy ▶

$$E_{vib} = \left(v + \tfrac{1}{2}\right)\frac{h}{2\pi}\sqrt{\frac{k}{\mu}} \qquad v = 0, 1, 2, \ldots \qquad \textbf{(43.10)}$$

The selection rule for the allowed vibrational transitions is $\Delta v = \pm 1$. Transitions between vibrational levels are caused by absorption of photons in the infrared region of the spectrum. The energy of an absorbed photon is equal to the energy difference between any two successive vibrational levels. Therefore, the photon energy is given by

$$E_{photon} = \Delta E_{vib} = \frac{h}{2\pi}\sqrt{\frac{k}{\mu}} \qquad \textbf{(43.11)}$$

The vibrational energies of a diatomic molecule are plotted in Figure 43.7. At ordinary temperatures, most molecules have vibrational energies corresponding to the $v = 0$ state because the spacing between vibrational states is much greater than $k_B T$, where k_B is Boltzmann's constant and T is the temperature.

Quick Quiz 43.3 A gas of identical diatomic molecules absorbs electromagnetic radiation over a wide range of frequencies. Molecule 1, initially in the $v = 0$ vibrational state, makes a transition to the $v = 1$ state. Molecule 2, initially in the $v = 2$ state, makes a transition to the $v = 3$ state. What is the ratio of the frequency of the photon that excited molecule 2 to that of the photon that excited molecule 1? **(a)** 1 **(b)** 2 **(c)** 3 **(d)** 4 **(e)** impossible to determine

---

Example 43.2 **Vibration of the CO Molecule** AM

The frequency of the photon that causes the $v = 0$ to $v = 1$ transition in the CO molecule is 6.42×10^{13} Hz. We ignore any changes in the rotational energy for this example.

(A) Calculate the force constant k for this molecule.

SOLUTION

Conceptualize Imagine that the two atoms in Figure 43.6a are carbon and oxygen. As the molecule vibrates, a given point on the imaginary spring is at rest. This point is not midway between the atoms because of the difference in masses of the C and O atoms.

Categorize The statement of the problem tells us to categorize this example as one involving a quantum-mechanical treatment and to restrict our investigation to the vibrational motion of a diatomic molecule. The molecule is analyzed with portions of the *particle in simple harmonic motion* analysis model.

▶ **43.2** continued

Analyze Set Equation 43.11 equal to the photon energy hf and solve for the force constant:

$$\frac{h}{2\pi}\sqrt{\frac{k}{\mu}} = hf \ \rightarrow \ k = 4\pi^2\mu f^2$$

Substitute the frequency given in the problem statement and the reduced mass from Example 43.1:

$$k = 4\pi^2(1.14 \times 10^{-26} \text{ kg})(6.42 \times 10^{13} \text{ s}^{-1})^2 = \boxed{1.85 \times 10^3 \text{ N/m}}$$

(B) What is the classical amplitude A of vibration for this molecule in the $v = 0$ vibrational state?

SOLUTION

Equate the maximum elastic potential energy $\frac{1}{2}kA^2$ in the molecule (Eq. 15.21) to the vibrational energy given by Equation 43.10 with $v = 0$ and solve for A:

$$\tfrac{1}{2}kA^2 = \frac{h}{4\pi}\sqrt{\frac{k}{\mu}} \ \rightarrow \ A = \sqrt{\frac{h}{2\pi}}\left(\frac{1}{\mu k}\right)^{1/4}$$

Substitute the value for k from part (A) and the value for μ:

$$A = \sqrt{\frac{6.626 \times 10^{-34} \text{ J} \cdot \text{s}}{2\pi}}\left[\frac{1}{(1.14 \times 10^{-26} \text{ kg})(1.85 \times 10^3 \text{ N/m})}\right]^{1/4}$$

$$= 4.79 \times 10^{-12} \text{ m} = \boxed{0.004 \ 79 \text{ nm}}$$

Finalize Comparing this result with the bond length of 0.113 nm we calculated in Example 43.1 shows that the classical amplitude of vibration is approximately 4% of the bond length.

Molecular Spectra

In general, a molecule vibrates and rotates simultaneously. To a first approximation, these motions are independent of each other, so the total energy of the molecule for these motions is the sum of Equations 43.6 and 43.9:

$$E = (v + \tfrac{1}{2})hf + \frac{\hbar^2}{2I}J(J + 1) \tag{43.12}$$

The energy levels of any molecule can be calculated from this expression, and each level is indexed by the two quantum numbers v and J. From these calculations, an energy-level diagram like the one shown in Figure 43.8a (page 1070) can be constructed. For each allowed value of the vibrational quantum number v, there is a complete set of rotational levels corresponding to $J = 0, 1, 2, \ldots$. The energy separation between successive rotational levels is much smaller than the separation between successive vibrational levels. As noted earlier, most molecules at ordinary temperatures are in the $v = 0$ vibrational state; these molecules can be in various rotational states as Figure 43.8a shows.

When a molecule absorbs a photon with the appropriate energy, the vibrational quantum number v increases by one unit while the rotational quantum number J either increases or decreases by one unit as can be seen in Figure 43.8. Therefore, the molecular absorption spectrum in Figure 43.8b consists of two groups of lines: one group to the right of center and satisfying the selection rules $\Delta J = +1$ and $\Delta v = +1$, and the other group to the left of center and satisfying the selection rules $\Delta J = -1$ and $\Delta v = +1$.

The energies of the absorbed photons can be calculated from Equation 43.12:

$$E_{\text{photon}} = \Delta E = hf + \frac{\hbar^2}{I}(J + 1) \quad J = 0, 1, 2, \ldots \quad (\Delta J = +1) \tag{43.13}$$

$$E_{\text{photon}} = \Delta E = hf - \frac{\hbar^2}{I}J \quad J = 1, 2, 3, \ldots \quad (\Delta J = -1) \tag{43.14}$$

Figure 43.8 (a) Absorptive transitions between the $v = 0$ and $v = 1$ vibrational states of a diatomic molecule. Compare the energy levels in this figure with those in Figure 21.7. (b) Expected lines in the absorption spectrum of a molecule. These same lines appear in the emission spectrum.

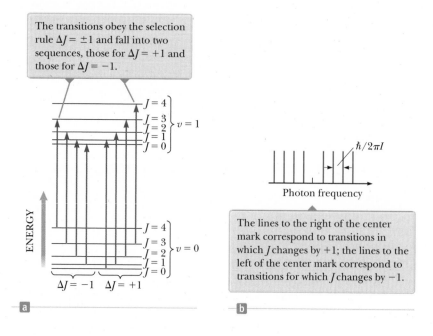

The transitions obey the selection rule $\Delta J = \pm 1$ and fall into two sequences, those for $\Delta J = +1$ and those for $\Delta J = -1$.

The lines to the right of the center mark correspond to transitions in which J changes by $+1$; the lines to the left of the center mark correspond to transitions for which J changes by -1.

where J is the rotational quantum number of the *initial* state. Equation 43.13 generates the series of equally spaced lines *higher* than the frequency f, whereas Equation 43.14 generates the series *lower* than this frequency. Adjacent lines are separated in frequency by the fundamental unit $\hbar/2\pi I$. Figure 43.8b shows the expected frequencies in the absorption spectrum of the molecule; these same frequencies appear in the emission spectrum.

The experimental absorption spectrum of the HCl molecule shown in Figure 43.9 follows this pattern very well and reinforces our model. One peculiarity is apparent, however: each line is split into a doublet. This doubling occurs because two chlorine isotopes (Cl-35 and Cl-37; see Section 44.1) were present in the sample used to obtain this spectrum. Because the isotopes have different masses, the two HCl molecules have different values of I.

The intensity of the spectral lines in Figure 43.9 follows an interesting pattern, rising first as one moves away from the central gap (located at about 8.65×10^{13} Hz, corresponding to the forbidden $J = 0$ to $J = 0$ transition) and then falling. This intensity is determined by a product of two functions of J. The first function corresponds to the number of available states for a given value of J. This function is $2J + 1$, corresponding to the number of values of m_J, the molecular rotation analog to m_ℓ for atomic states. For example, the $J = 2$ state has five substates with five values of m_J ($m_J = -2, -1, 0, 1, 2$), whereas the $J = 1$ state has only three substates ($m_J = -1, 0, 1$). Therefore, on average and without regard for the second function described below, five-thirds as many molecules make the transition from the $J = 2$ state as from the $J = 1$ state.

The second function determining the envelope of the intensity of the spectral lines is the Boltzmann factor, introduced in Section 21.5. The number of molecules in an excited rotational state is given by

$$n = n_0 e^{-\hbar^2 J(J+1)/(2Ik_BT)}$$

where n_0 is the number of molecules in the $J = 0$ state.

Multiplying these factors together indicates that the intensity of spectral lines should be described by a function of J as follows:

▶ Intensity variation in the vibration–rotation spectrum of a molecule

$$I \propto (2J + 1)e^{-\hbar^2 J(J+1)/(2Ik_BT)} \tag{43.15}$$

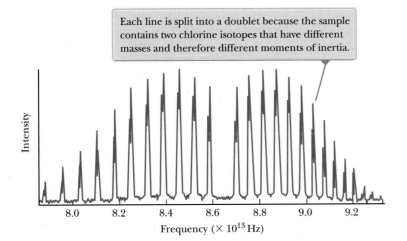

Figure 43.9 Experimental absorption spectrum of the HCl molecule.

The factor $(2J + 1)$ increases with J while the exponential second factor decreases. The product of the two factors gives a behavior that closely describes the envelope of the spectral lines in Figure 43.9.

The excitation of rotational and vibrational energy levels is an important consideration in current models of global warming. Most of the absorption lines for CO_2 are in the infrared portion of the spectrum. Therefore, visible light from the Sun is not absorbed by atmospheric CO_2 but instead strikes the Earth's surface, warming it. In turn, the surface of the Earth, being at a much lower temperature than the Sun, emits thermal radiation that peaks in the infrared portion of the electromagnetic spectrum (Section 40.1). This infrared radiation is absorbed by the CO_2 molecules in the air instead of radiating out into space. Atmospheric CO_2 acts like a one-way valve for energy from the Sun and is responsible, along with some other atmospheric molecules, for raising the temperature of the Earth's surface above its value in the absence of an atmosphere. This phenomenon is commonly called the "greenhouse effect." The burning of fossil fuels in today's industrialized society adds more CO_2 to the atmosphere. This addition of CO_2 increases the absorption of infrared radiation, raising the Earth's temperature further. In turn, this increase in temperature causes substantial climatic changes.

As seen in Figure 43.10, the amount of carbon dioxide in the atmosphere has been steadily increasing since the middle of the 20th century. This graph shows hard data that indicate that the atmosphere is undergoing a distinct change, although not all scientists agree on the interpretation of what that change means in terms of global temperatures.

The Intergovernmental Panel on Climate Change (IPCC) is a scientific body that assesses the available information related to global warming and associated effects

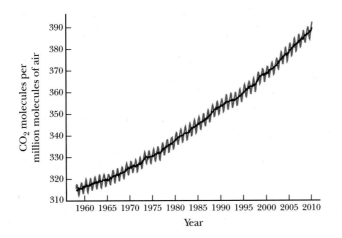

Figure 43.10 The concentration of atmospheric carbon dioxide in parts per million (ppm) of dry air as a function of time. These data were recorded at the Mauna Loa Observatory in Hawaii. The yearly variations (red-brown curve) coincide with growing seasons because vegetation absorbs carbon dioxide from the air. The steady increase in the average concentration (black curve) is of concern to scientists.

related to climate change. It was originally established in 1988 by two United Nations organizations, the World Meteorological Organization and the United Nations Environment Programme. The IPCC has published four assessment reports on climate change, the most recent in 2007, and a fifth report is scheduled to be released in 2014. The 2007 report concludes that there is a probability of greater than 90% that the increased global temperature measured by scientists is due to the placement of greenhouse gases such as carbon dioxide in the atmosphere by humans. The report also predicts a global temperature increase between 1°C and 6°C in the 21st century, a sea level rise from 18 cm to 59 cm, and very high probabilities of weather extremes, including heat waves, droughts, cyclones, and heavy rainfall.

In addition to its scientific aspects, global warming is a social issue with many facets. These facets encompass international politics and economics, because global warming is a worldwide problem. Changing our policies requires real costs to solve the problem. Global warming also has technological aspects, and new methods of manufacturing, transportation, and energy supply must be designed to slow down or reverse the increase in temperature.

Conceptual Example 43.3 **Comparing Figures 43.8 and 43.9**

In Figure 43.8a, the transitions indicated correspond to spectral lines that are equally spaced as shown in Figure 43.8b. The actual spectrum in Figure 43.9, however, shows lines that move closer together as the frequency increases. Why does the spacing of the actual spectral lines differ from the diagram in Figure 43.8?

SOLUTION

In Figure 43.8, we modeled the rotating diatomic molecule as a rigid object (Chapter 10). In reality, however, as the molecule rotates faster and faster, the effective spring in Figure 43.6a stretches and provides the increased force associated with the larger centripetal acceleration of each atom. As the molecule stretches along its length, its moment of inertia I increases. Therefore, the rotational part of the energy expression in Equation 43.12 has an extra dependence on J in the moment of inertia I. Because the increasing moment of inertia is in the denominator, as J increases, the energies do not increase as rapidly with J as indicated in Equation 43.12. With each higher energy level being lower than indicated by Equation 43.12, the energy associated with a transition to that level is smaller, as is the frequency of the absorbed photon, destroying the even spacing of the spectral lines and giving the spacing that decreases with increasing frequency seen in Figure 43.9.

43.3 Bonding in Solids

A crystalline solid consists of a large number of atoms arranged in a regular array, forming a periodic structure. The ions in the NaCl crystal are ionically bonded, as already noted, and the carbon atoms in diamond form covalent bonds with one another. The metallic bond described at the end of this section is responsible for the cohesion of copper, silver, sodium, and other solid metals.

Ionic Solids

Many crystals are formed by ionic bonding, in which the dominant interaction between ions is the Coulomb force. Consider a portion of the NaCl crystal shown in Figure 43.11a. The red spheres are sodium ions, and the blue spheres are chlorine ions. As shown in Figure 43.11b, each Na^+ ion has six nearest-neighbor Cl^- ions. Similarly, in Figure 43.11c, we see that each Cl^- ion has six nearest-neighbor Na^+ ions. Each Na^+ ion is attracted to its six Cl^- neighbors. The corresponding potential energy is $-6k_e e^2/r$, where k_e is the Coulomb constant and r is the separation distance between each Na^+ and Cl^-. In addition, there are 12 next-nearest-neighbor

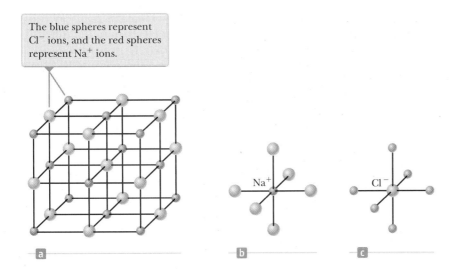

The blue spheres represent Cl⁻ ions, and the red spheres represent Na⁺ ions.

a   b   c

Figure 43.11 (a) Crystalline structure of NaCl. (b) Each positive sodium ion is surrounded by six negative chlorine ions. (c) Each chlorine ion is surrounded by six sodium ions.

Na⁺ ions at a distance of $\sqrt{2}r$ from the Na⁺ ion, and these 12 positive ions exert weaker repulsive forces on the central Na⁺. Furthermore, beyond these 12 Na⁺ ions are more Cl⁻ ions that exert an attractive force, and so on. The net effect of all these interactions is a resultant negative electric potential energy

$$U_{attractive} = -\alpha k_e \frac{e^2}{r} \tag{43.16}$$

where α is a dimensionless number known as the **Madelung constant.** The value of α depends only on the particular crystalline structure of the solid. For example, $\alpha = 1.747\,6$ for the NaCl structure. When the constituent ions of a crystal are brought close together, a repulsive force exists because of electrostatic forces and the exclusion principle as discussed in Section 43.1. The potential energy term B/r^m in Equation 43.1 accounts for this repulsive force. We do not include neighbors other than nearest neighbors here because the repulsive forces occur only for ions that are very close together. (Electron shells must overlap for exclusion-principle effects to become important.) Therefore, we can express the total potential energy of the crystal as

$$U_{total} = -\alpha k_e \frac{e^2}{r} + \frac{B}{r^m} \tag{43.17}$$

where m in this expression is some small integer.

A plot of total potential energy versus ion separation distance is shown in Figure 43.12. The potential energy has its minimum value U_0 at the equilibrium separation, when $r = r_0$. It is left as a problem (Problem 59 in Enhanced WebAssign) to show that

$$U_0 = -\alpha k_e \frac{e^2}{r_0}\left(1 - \frac{1}{m}\right) \tag{43.18}$$

This minimum energy U_0 is called the **ionic cohesive energy** of the solid, and its absolute value represents the energy required to separate the solid into a collection of isolated positive and negative ions. Its value for NaCl is −7.84 eV per ion pair.

To calculate the **atomic cohesive energy,** which is the binding energy relative to the energy of the neutral atoms, 5.14 eV must be added to the ionic cohesive energy value to account for the transition from Na⁺ to Na and 3.62 eV must be subtracted to account for the conversion of Cl⁻ to Cl. Therefore, the atomic cohesive energy of NaCl is

$$-7.84 \text{ eV} + 5.14 \text{ eV} - 3.62 \text{ eV} = -6.32 \text{ eV}$$

In other words, 6.32 eV of energy per ion pair is needed to separate the solid into isolated neutral atoms of Na and Cl.

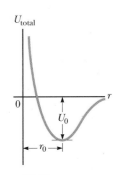

Figure 43.12 Total potential energy versus ion separation distance for an ionic solid, where U_0 is the ionic cohesive energy and r_0 is the equilibrium separation distance between ions.

Ionic crystals form relatively stable, hard crystals. They are poor electrical conductors because they contain no free electrons; each electron in the solid is bound tightly to one of the ions, so it is not sufficiently mobile to carry current. Ionic crystals have high melting points; for example, the melting point of NaCl is 801°C. Ionic crystals are transparent to visible radiation because the shells formed by the electrons in ionic solids are so tightly bound that visible radiation does not possess sufficient energy to promote electrons to the next allowed shell. Infrared radiation is absorbed strongly because the vibrations of the ions have natural resonant frequencies in the low-energy infrared region.

Covalent Solids

Solid carbon, in the form of diamond, is a crystal whose atoms are covalently bonded. Because atomic carbon has the electronic configuration $1s^2 2s^2 2p^2$, it is four electrons short of filling its $n = 2$ shell, which can accommodate eight electrons. Because of this electron structure, two carbon atoms have a strong attraction for each other, with a cohesive energy of 7.37 eV. In the diamond structure, each carbon atom is covalently bonded to four other carbon atoms located at four corners of a cube as shown in Figure 43.13a.

The crystalline structure of diamond is shown in Figure 43.13b. Notice that each carbon atom forms covalent bonds with four nearest-neighbor atoms. The basic structure of diamond is called tetrahedral (each carbon atom is at the center of a regular tetrahedron), and the angle between the bonds is 109.5°. Other crystals such as silicon and germanium have the same structure.

Carbon is interesting in that it can form several different types of structures. In addition to the diamond structure, it forms graphite, with completely different properties. In this form, the carbon atoms form flat layers with hexagonal arrays of atoms. A very weak interaction between the layers allows the layers to be removed easily under friction, as occurs in the graphite used in pencil lead.

Carbon atoms can also form a large hollow structure; in this case, the compound is called **buckminsterfullerene** after the famous architect R. Buckminster Fuller, who invented the geodesic dome. The unique shape of this molecule (Fig. 43.14) provides a "cage" to hold other atoms or molecules. Related structures, called "buckytubes" because of their long, narrow cylindrical arrangements of carbon atoms, may provide the basis for extremely strong, yet lightweight, materials.

A current area of active research is in the properties and applications of **graphene.** Graphene consists of a monolayer of carbon atoms, with the atoms arranged in hexagons so that the monolayer looks like chicken wire. Graphite flakes that are shed from a pencil while writing contain small fragments of graphene. Pioneers in graphene research include Andre Geim (b. 1958) and Konstantin Novoselov (b. 1974) of the University of Manchester, who received the Nobel Prize in Physics in 2010 for their experiments. Graphene has interesting electronic, thermal, and optical properties that are currently under investigation. Its mechanical properties include a breaking strength 200 times that of steel. Potential applications under

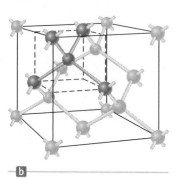

Figure 43.13 (a) Each carbon atom in a diamond crystal is covalently bonded to four other carbon atoms so that a tetrahedral structure is formed. (b) The crystal structure of diamond, showing the tetrahedral bond arrangement.

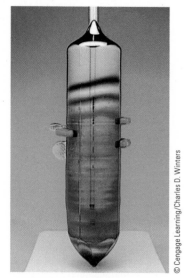

A cylinder of nearly pure crystalline silicon (Si), approximately 25 cm long. Such crystals are cut into wafers and processed to make various semiconductor devices.

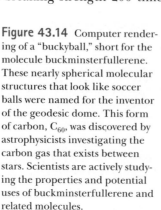

Figure 43.14 Computer rendering of a "buckyball," short for the molecule buckminsterfullerene. These nearly spherical molecular structures that look like soccer balls were named for the inventor of the geodesic dome. This form of carbon, C_{60}, was discovered by astrophysicists investigating the carbon gas that exists between stars. Scientists are actively studying the properties and potential uses of buckminsterfullerene and related molecules.

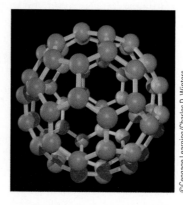

Table 43.1	**Atomic Cohesive Energies of Some Covalent Solids**
Solid	Cohesive Energy (eV per ion pair)
C (diamond)	7.37
Si	4.63
Ge	3.85
InAs	5.70
SiC	6.15
ZnS	6.32
CuCl	9.24

study include graphene nanoribbons, quantum dots, transistors, optical modulators, and integrated circuits.

The atomic cohesive energies of some covalent solids are given in Table 43.1. The large energies account for the hardness of covalent solids. Diamond is particularly hard and has an extremely high melting point (about 4 000 K). Covalently bonded solids usually have high bond energies and high melting points, and are good electrical insulators.

Metallic Solids

Metallic bonds are generally weaker than ionic or covalent bonds. The outer electrons in the atoms of a metal are relatively free to move throughout the material, and the number of such mobile electrons in a metal is large. The metallic structure can be viewed as a "sea" or a "gas" of nearly free electrons surrounding a lattice of positive ions (Fig. 43.15). The bonding mechanism in a metal is the attractive force between the entire collection of positive ions and the electron gas. Metals have a cohesive energy in the range of 1 to 3 eV per atom, which is less than the cohesive energies of ionic or covalent solids.

Light interacts strongly with the free electrons in metals. Hence, visible light is absorbed and re-emitted quite close to the surface of a metal, which accounts for the shiny nature of metal surfaces. In addition to the high electrical conductivity of metals produced by the free electrons, the nondirectional nature of the metallic bond allows many different types of metal atoms to be dissolved in a host metal in varying amounts. The resulting *solid solutions,* or *alloys* (steel, bronze, brass, etc.), may be designed to have particular properties, such as tensile strength, ductility, electrical and thermal conductivity, and resistance to corrosion.

Because the bonding in metals is between all the electrons and all the positive ions, metals tend to bend when stressed. This bending is in contrast to nonmetallic solids, which tend to fracture when stressed. Fracturing results because bonding in nonmetallic solids is primarily with nearest-neighbor ions or atoms. When the distortion causes sufficient stress between some set of nearest neighbors, fracture occurs.

The blue area represents the electron gas, and the red spheres represent the positive metal ions.

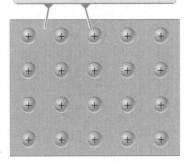

Figure 43.15 Highly schematic diagram of a metal.

43.4 Free-Electron Theory of Metals

In Section 27.3, we described a classical free-electron theory of electrical conduction in metals, a structural model that led to Ohm's law. According to this theory, a metal is modeled as a classical gas of conduction electrons moving through a fixed lattice of ions. Although this theory predicts the correct functional form of Ohm's law, it does not predict the correct values of electrical and thermal conductivities.

A quantum-based free-electron theory of metals remedies the shortcomings of the classical model by taking into account the wave nature of the electrons. In this model, based on the quantum particle under boundary conditions analysis model, the outer-shell electrons are free to move through the metal but are trapped within a three-dimensional box formed by the metal surfaces. Therefore, each electron is represented as a particle in a box. As discussed in Section 41.2, particles in a box are restricted to quantized energy levels.

Statistical physics can be applied to a collection of particles in an effort to relate microscopic properties to macroscopic properties as we saw with kinetic theory of gases in Chapter 21. In the case of electrons, it is necessary to use *quantum statistics*, with the requirement that each state of the system can be occupied by only two electrons (one with spin up and the other with spin down) as a consequence of the exclusion principle. The probability that a particular state having energy E is occupied by one of the electrons in a solid is

Fermi–Dirac distribution ▶
function

$$f(E) = \frac{1}{e^{(E-E_F)/k_B T} + 1} \qquad (43.19)$$

where $f(E)$ is called the **Fermi–Dirac distribution function** and E_F is called the **Fermi energy.** A plot of $f(E)$ versus E at $T = 0$ K is shown in Figure 43.16a. Notice that $f(E) = 1$ for $E < E_F$ and $f(E) = 0$ for $E > E_F$. That is, at 0 K, all states having energies less than the Fermi energy are occupied and all states having energies greater than the Fermi energy are vacant. A plot of $f(E)$ versus E at some temperature $T > 0$ K is shown in Figure 43.16b. This curve shows that as T increases, the distribution rounds off slightly. Because of thermal excitation, states near and below E_F lose population and states near and above E_F gain population. The Fermi energy E_F also depends on temperature, but the dependence is weak in metals.

Let's now follow up on our discussion of the particle in a box in Chapter 41 to generalize the results to a three-dimensional box. Recall that if a particle of mass m is confined to move in a one-dimensional box of length L, the allowed states have quantized energy levels given by Equation 41.14:

$$E_n = \left(\frac{h^2}{8mL^2}\right)n^2 = \left(\frac{\hbar^2 \pi^2}{2mL^2}\right)n^2 \quad n = 1, 2, 3, \ldots$$

Now imagine a piece of metal in the shape of a solid cube of sides L and volume L^3 and focus on one electron that is free to move anywhere in this volume. Therefore, the electron is modeled as a particle in a three-dimensional box. In this model, we require that $\psi(x, y, z) = 0$ at the boundaries of the metal. It can be shown (see Problem 37 in Enhanced WebAssign) that the energy for such an electron is

$$E = \frac{\hbar^2 \pi^2}{2m_e L^2}(n_x^2 + n_y^2 + n_z^2) \qquad (43.20)$$

where m_e is the mass of the electron and n_x, n_y, and n_z are quantum numbers. As we expect, the energies are quantized, and each allowed value of the energy is characterized by this set of three quantum numbers (one for each degree of freedom) and the spin quantum number m_s. For example, the ground state, corresponding to

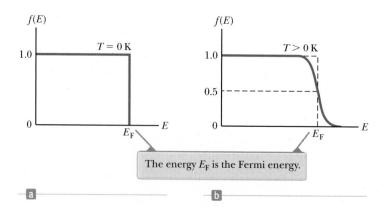

Figure 43.16 Plot of the Fermi–Dirac distribution function $f(E)$ versus energy at (a) $T = 0$ K and (b) $T > 0$ K.

The energy E_F is the Fermi energy.

$n_x = n_y = n_z = 1$, has an energy equal to $3\hbar^2\pi^2/2m_eL^2$ and can be occupied by two electrons, corresponding to spin up and spin down.

Because of the macroscopic size L of the box, the energy levels for the electrons are very close together. As a result, we can treat the quantum numbers as continuous variables. Under this assumption, the number of allowed states per unit volume that have energies between E and $E + dE$ is

$$g(E)\,dE = \frac{8\sqrt{2}\,\pi m_e^{3/2}}{h^3}E^{1/2}\,dE \qquad (43.21)$$

(See Example 43.5.) The function $g(E)$ is called the **density-of-states function.**

If a metal is in thermal equilibrium, the number of electrons per unit volume $N(E)\,dE$ that have energy between E and $E + dE$ is equal to the product of the number of allowed states per unit volume and the probability that a state is occupied; that is, $N(E)\,dE = g(E)f(E)\,dE$:

$$N(E)\,dE = \left(\frac{8\sqrt{2}\,\pi m_e^{3/2}}{h^3}E^{1/2}\right)\left(\frac{1}{e^{(E-E_F)/k_BT} + 1}\right)dE \qquad (43.22)$$

Plots of $N(E)$ versus E for two temperatures are given in Figure 43.17.

If n_e is the total number of electrons per unit volume, we require that

$$n_e = \int_0^\infty N(E)\,dE = \frac{8\sqrt{2}\,\pi m_e^{3/2}}{h^3}\int_0^\infty \frac{E^{1/2}\,dE}{e^{(E-E_F)/k_BT} + 1} \qquad (43.23)$$

We can use this condition to calculate the Fermi energy. At $T = 0$ K, the Fermi–Dirac distribution function $f(E) = 1$ for $E < E_F$ and $f(E) = 0$ for $E > E_F$. Therefore, at $T = 0$ K, Equation 43.23 becomes

$$n_e = \frac{8\sqrt{2}\,\pi m_e^{3/2}}{h^3}\int_0^{E_F} E^{1/2}\,dE = \tfrac{2}{3}\frac{8\sqrt{2}\,\pi m_e^{3/2}}{h^3}E_F^{3/2} \qquad (43.24)$$

Solving for the Fermi energy at 0 K gives

$$E_F(0) = \frac{h^2}{2m_e}\left(\frac{3n_e}{8\pi}\right)^{2/3} \qquad (43.25)$$

◀ **Fermi energy at $T = 0$ K**

The Fermi energies for metals are in the range of a few electron volts. Representative values for various metals are given in Table 43.2. It is left as a problem (Problem 39 in Enhanced WebAssign) to show that the average energy of a free electron in a metal at 0 K is

$$E_{avg} = \tfrac{3}{5}E_F \qquad (43.26)$$

In summary, we can consider a metal to be a system comprising a very large number of energy levels available to the free electrons. These electrons fill the levels in accordance with the Pauli exclusion principle, beginning with $E = 0$ and ending with E_F. At $T = 0$ K, all levels below the Fermi energy are filled and all levels above the Fermi energy are empty. At 300 K, a small fraction of the free electrons are excited above the Fermi energy.

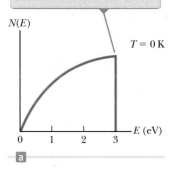

To provide a sense of scale, imagine that the Fermi energy E_F of the metal is 3 eV.

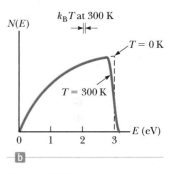

Figure 43.17 Plot of the electron distribution function versus energy in a metal at (a) $T = 0$ K and (b) $T = 300$ K.

Table 43.2	Calculated Values of the Fermi Energy for Metals at 300 K Based on the Free-Electron Theory	
Metal	Electron Concentration (m^{-3})	Fermi Energy (eV)
Li	4.70×10^{28}	4.72
Na	2.65×10^{28}	3.23
K	1.40×10^{28}	2.12
Cu	8.46×10^{28}	7.05
Ag	5.85×10^{28}	5.48
Au	5.90×10^{28}	5.53

Example 43.4 The Fermi Energy of Gold

Each atom of gold (Au) contributes one free electron to the metal. Compute the Fermi energy for gold.

SOLUTION

Conceptualize Imagine electrons filling available levels at $T = 0$ K in gold until the solid is neutral. The highest energy filled is the Fermi energy.

Categorize We evaluate the result using a result from this section, so we categorize this example as a substitution problem.

Substitute the concentration of free electrons in gold from Table 43.2 into Equation 43.25 to calculate the Fermi energy at 0 K:

$$E_F(0) = \frac{(6.626 \times 10^{-34} \text{ J} \cdot \text{s})^2}{2(9.11 \times 10^{-31} \text{ kg})} \left[\frac{3(5.90 \times 10^{28} \text{ m}^{-3})}{8\pi}\right]^{2/3}$$

$$= 8.85 \times 10^{-19} \text{ J} = \boxed{5.53 \text{ eV}}$$

Example 43.5 Deriving Equation 43.21

Based on the allowed states of a particle in a three-dimensional box, derive Equation 43.21.

SOLUTION

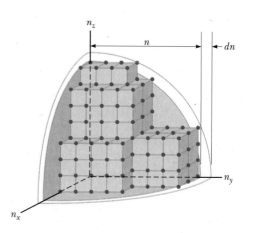

Figure 43.18 The dots representing the allowed states are located at integer values of n_x, n_y, and n_z and are therefore at the corners of cubes with sides of "length" 1.

Conceptualize Imagine a particle confined to a three-dimensional box, subject to boundary conditions in three dimensions. Imagine also a three-dimensional *quantum number space* whose axes represent n_x, n_y, and n_z. The allowed states in this space can be represented as dots located at integral values of the three quantum numbers as in Figure 43.18. This space is not traditional space in which a location is specified by coordinates x, y, and z; rather, it is a space in which allowed states can be specified by coordinates representing the quantum numbers. The number of allowed states having energies between E and $E + dE$ corresponds to the number of dots in the spherical shell of radius n and thickness dn.

Categorize We categorize this problem as that of a quantum system in which the energies of the particle are quantized. Furthermore, we can base the solution to the problem on our understanding of the particle in a one-dimensional box.

Analyze As noted previously, the allowed states of the particle in a three-dimensional box are described by three quantum numbers n_x, n_y, and n_z. For a macroscopic sample of metal, the number of allowed values of these quantum numbers is tremendous, so on a macroscopic scale, the allowed states in the number space can be modeled as continuous.

Defining $E_0 = \hbar^2\pi^2/2m_e L^2$ and $n = (E/E_0)^{1/2}$, rewrite Equation 43.20:

$$(1) \quad n_x^2 + n_y^2 + n_z^2 = \frac{2m_e L^2}{\hbar^2\pi^2}E = \frac{E}{E_0} = n^2$$

In the quantum number space, Equation (1) is the equation of a sphere of radius n. Therefore, the number of allowed states having energies between E and $E + dE$ is equal to the number of points in a spherical shell of radius n and thickness dn.

Find the "volume" of this shell, which represents the total number of states $G(E)\ dE$:

$$(2) \quad G(E)\ dE = \tfrac{1}{8}(4\pi n^2\ dn) = \tfrac{1}{2}\pi n^2\ dn$$

We have taken one-eighth of the total volume because we are restricted to the octant of a three-dimensional space in which all three quantum numbers are positive.

Replace n in Equation (2) with its equivalent in terms of E using the relation $n^2 = E/E_0$ from Equation (1):

$$G(E)\ dE = \tfrac{1}{2}\pi\left(\frac{E}{E_0}\right)d\left[\left(\frac{E}{E_0}\right)^{1/2}\right] = \tfrac{1}{2}\pi\frac{E}{(E_0)^{3/2}}d[(E)^{1/2}]$$

▶ **43.5** continued

Evaluate the differential:

$$G(E)\ dE = \tfrac{1}{2}\pi\left[\frac{E}{(E_0)^{3/2}}\right](\tfrac{1}{2}E^{-1/2}\ dE) = \tfrac{1}{4}\pi E_0^{-3/2}E^{1/2}\ dE$$

Substitute for E_0 from its definition above:

$$G(E)\ dE = \tfrac{1}{4}\pi\left(\frac{\hbar^2\pi^2}{2m_eL^2}\right)^{-3/2}E^{1/2}\ dE$$

$$= \frac{\sqrt{2}}{2}\ \frac{m_e^{3/2}L^3}{\hbar^3\pi^2}\ E^{1/2}\ dE$$

Letting $g(E)$ represent the number of states per unit volume, where L^3 is the volume V of the cubical box in normal space, find $g(E) = G(E)/V$:

$$g(E)\ dE = \frac{G(E)}{V}\ dE = \frac{\sqrt{2}}{2}\ \frac{m_e^{3/2}}{\hbar^3\pi^2}\ E^{1/2}\ dE$$

Substitute $\hbar = h/2\pi$:

$$g(E)\ dE = \frac{4\sqrt{2}\ \pi m_e^{3/2}}{h^3}\ E^{1/2}\ dE$$

Multiply by 2 for the two possible spin states in each particle-in-a-box state:

$$g(E)\ dE = \frac{8\sqrt{2}\ \pi m_e^{3/2}}{h^3}\ E^{1/2}\ dE$$

....

Finalize This result is Equation 43.21, which is what we set out to derive.

43.5 Band Theory of Solids

In Section 43.4, the electrons in a metal were modeled as particles free to move around inside a three-dimensional box and we ignored the influence of the parent atoms. In this section, we make the model more sophisticated by incorporating the contribution of the parent atoms that form the crystal.

Recall from Section 41.1 that the probability density $|\psi|^2$ for a system is physically significant, but the probability amplitude ψ is not. Let's consider as an example an atom that has a single s electron outside of a closed shell. Both of the following wave functions are valid for such an atom with atomic number Z:

$$\psi_s^+(r) = +Af(r)e^{-Zr/na_0} \qquad \psi_s^-(r) = -Af(r)e^{-Zr/na_0}$$

where A is the normalization constant and $f(r)$ is a function[3] of r that varies with the value of n. Choosing either of these wave functions leads to the same value of $|\psi|^2$, so both choices are equivalent. A difference arises, however, when two atoms are combined.

If two identical atoms are very far apart, they do not interact and their electronic energy levels can be considered to be those of isolated atoms. Suppose the two atoms are sodium, each having a lone 3s electron that is in a well-defined quantum state. As the two sodium atoms are brought closer together, their wave functions begin to overlap as we discussed for covalent bonding in Section 43.1. The properties of the combined system differ depending on whether the two atoms are combined with wave functions $\psi_s^+(r)$ as in Figure 43.19a or whether they are combined with one having wave function $\psi_s^+(r)$ and the other $\psi_s^-(r)$ as in Figure 43.19b. The choice of two atoms with wave function $\psi_s^-(r)$ is physically equivalent to that with two positive wave functions, so we do not consider it separately. When two wave functions $\psi_s^+(r)$ are combined, the result is a composite wave function in which the probability amplitudes add between the atoms. If $\psi_s^+(r)$ combines with $\psi_s^-(r)$,

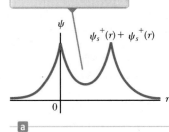

The probability of an electron being between the atoms is nonzero.

$\psi_s^+(r) + \psi_s^+(r)$

a

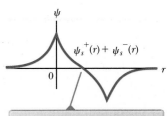

$\psi_s^+(r) + \psi_s^-(r)$

b

The probability of an electron being between the atoms is generally lower than in **a** and zero at the midpoint.

Figure 43.19 The wave functions of two atoms combine to form a composite wave function for the two-atom system when the atoms are close together. (a) Two atoms with wave functions $\psi_s^+(r)$ combine. (b) Two atoms with wave functions $\psi_s^+(r)$ and $\psi_s^-(r)$ combine.

---

[3]The functions $f(r)$ are called *Laguerre polynomials*. They can be found in the quantum treatment of the hydrogen atom in modern physics textbooks.

Figure 43.20 Energies of the 1s and 2s levels in sodium as a function of the separation distance r between atoms.

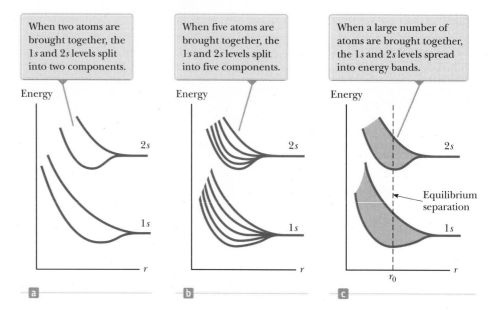

When two atoms are brought together, the 1s and 2s levels split into two components.

When five atoms are brought together, the 1s and 2s levels split into five components.

When a large number of atoms are brought together, the 1s and 2s levels spread into energy bands.

Equilibrium separation

a   b   c

however, the wave functions between the nuclei subtract. Therefore, the composite probability amplitudes for the two possibilities are different. These two possible combinations of wave functions represent two possible states of the two-atom system. We interpret these curves as representing the probability amplitude of finding an electron. The positive–positive curve shows some probability of finding the electron at the midpoint between the atoms. The positive–negative function shows no such probability. A state with a high probability of an electron *between* two positive nuclei must have a different energy than a state with a high probability of the electron being elsewhere! Therefore, the states are *split* into two energy levels due to the two ways of combining the wave functions. The energy difference is relatively small, so the two states are close together on an energy scale.

Figure 43.20a shows this splitting effect as a function of separation distance. For large separations r, the electron clouds do not overlap and there is no splitting. As the atoms are brought closer so that r decreases, the electron clouds overlap and we need to consider the system of two atoms.

When a large number of atoms are brought together to form a solid, a similar phenomenon occurs. The individual wave functions can be brought together in various combinations of $\psi_s^+(r)$ and $\psi_s^-(r)$, each possible combination corresponding to a different energy. As the atoms are brought close together, the various isolated-atom energy levels split into multiple energy levels for the composite system. This splitting in levels for five atoms in close proximity is shown in Figure 43.20b. In this case, there are five energy levels corresponding to five different combinations of isolated-atom wave functions.

As the number of atoms grows, the number of combinations of wave functions grows, as does the number of possible energies. If we extend this argument to the large number of atoms found in solids (on the order of 10^{23} atoms per cubic centimeter), we obtain a huge number of levels of varying energy so closely spaced that they may be regarded as a continuous **band** of energy levels as shown in Figure 43.20c. In the case of sodium, it is customary to refer to the continuous distributions of allowed energy levels as *s* bands because the bands originate from the *s* levels of the individual sodium atoms.

Each energy level in the atom can spread into a band when the atoms are combined into a solid. Figure 43.21 shows the allowed energy bands of sodium at a fixed separation distance between the atoms. Notice that energy gaps, corresponding to *forbidden energies*, occur between the allowed bands. In addition, some bands exhibit sufficient spreading in energy that there is an overlap between bands arising from different quantum states (3s and 3p).

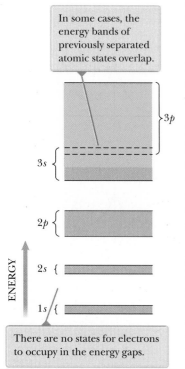

In some cases, the energy bands of previously separated atomic states overlap.

3p

3s

2p

2s

1s

ENERGY

There are no states for electrons to occupy in the energy gaps.

Figure 43.21 Energy bands of a sodium crystal. Blue represents energy bands occupied by the sodium electrons when the atom is in its ground state. Gold represents energy bands that are empty.

As indicated by the blue-shaded areas in Figure 43.21, the $1s$, $2s$, and $2p$ bands of sodium are each full of electrons because the $1s$, $2s$, and $2p$ states of each atom are full. An energy level in which the orbital angular momentum is ℓ can hold $2(2\ell + 1)$ electrons. The factor 2 arises from the two possible electron spin orientations, and the factor $2\ell + 1$ corresponds to the number of possible orientations of the orbital angular momentum. The capacity of each band for a system of N atoms is $2(2\ell + 1)N$ electrons. Therefore, the $1s$ and $2s$ bands each contain $2N$ electrons ($\ell = 0$), and the $2p$ band contains $6N$ electrons ($\ell = 1$). Because sodium has only one $3s$ electron and there are a total of N atoms in the solid, the $3s$ band contains only N electrons and is partially full as indicated by the blue coloring in Figure 43.21. The $3p$ band, which is the higher region of the overlapping bands, is completely empty (all gold in the figure).

Band theory allows us to build simple models to understand the behavior of conductors, insulators, and semiconductors as well as that of semiconductor devices, as we shall discuss in the following sections.

43.6 Electrical Conduction in Metals, Insulators, and Semiconductors

Good electrical conductors contain a high density of free charge carriers, and the density of free charge carriers in insulators is nearly zero. Semiconductors, first introduced in Section 23.2, are a class of technologically important materials in which charge-carrier densities are intermediate between those of insulators and those of conductors. In this section, we discuss the mechanisms of conduction in these three classes of materials in terms of a model based on energy bands.

Metals

If a material is to be a good electrical conductor, the charge carriers in the material must be free to move in response to an applied electric field. Let's consider the electrons in a metal as the charge carriers. The motion of the electrons in response to an electric field represents an increase in energy of the system (the metal lattice and the free electrons) corresponding to the additional kinetic energy of the moving electrons. The system is described by the nonisolated system model for energy. Equation 8.2 becomes $W = \Delta K$, where the work is done on the electrons by the electric field. Therefore, when an electric field is applied to a conductor, electrons must move upward to an available higher energy state on an energy-level diagram to represent the additional kinetic energy.

Figure 43.22 shows a half-filled band in a metal at $T = 0$ K, where the blue region represents levels filled with electrons. Because electrons obey Fermi–Dirac statistics, all levels below the Fermi energy are filled with electrons and all levels above the Fermi energy are empty. The Fermi energy lies in the band at the highest filled state. At temperatures slightly greater than 0 K, some electrons are thermally excited to levels above E_F, but overall there is little change from the 0 K case. If a potential difference is applied to the metal, however, electrons having energies near the Fermi energy require only a small amount of additional energy from the applied electric field to reach nearby empty energy states above the Fermi energy. Therefore, electrons in a metal experiencing only a weak applied electric field are free to move because many empty levels are available close to the occupied energy levels. The model of metals based on band theory demonstrates that metals are excellent electrical conductors.

Insulators

Now consider the two outermost energy bands of a material in which the lower band is filled with electrons and the higher band is empty at 0 K (Fig. 43.23). The

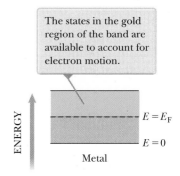

The states in the gold region of the band are available to account for electron motion.

$E = E_F$

$E = 0$

Metal

Figure 43.22 Half-filled band of a metal, an electrical conductor. At $T = 0$ K, the Fermi energy lies in the middle of the band.

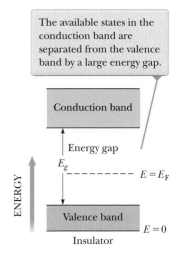

The available states in the conduction band are separated from the valence band by a large energy gap.

Conduction band

Energy gap E_g

$E = E_F$

Valence band

$E = 0$

Insulator

Figure 43.23 An electrical insulator at $T = 0$ K has a filled valence band and an empty conduction band. The Fermi level lies somewhere between these bands in the region known as the energy gap.

lower, filled band is called the **valence band,** and the upper, empty band is the **conduction band.** (The conduction band is the one that is partially filled in a metal.) It is common to refer to the energy separation between the valence and conduction bands as the **energy gap** E_g of the material. The Fermi energy lies somewhere in the energy gap[4] as shown in Figure 43.23.

Suppose a material has a relatively large energy gap of, for example, approximately 5 eV. At 300 K (room temperature), $k_BT = 0.025$ eV, which is much smaller than the energy gap. At such temperatures, the Fermi–Dirac distribution predicts that very few electrons are thermally excited into the conduction band. There are no available states that lie close in energy above the valence band and into which electrons can move upward to account for the extra kinetic energy associated with motion through the material in response to an electric field. Consequently, the electrons do not move; the material is an insulator. Although an insulator has many vacant states in its conduction band that can accept electrons, these states are separated from the filled states by a large energy gap. Only a few electrons occupy these states, so the overall electrical conductivity of insulators is very small.

Table 43.3	Energy-Gap Values for Some Semiconductors	
	E_g (eV)	
Crystal	0 K	300 K
Si	1.17	1.14
Ge	0.74	0.67
InP	1.42	1.34
GaP	2.32	2.26
GaAs	1.52	1.42
CdS	2.58	2.42
CdTe	1.61	1.56
ZnO	3.44	3.2
ZnS	3.91	3.6

Semiconductors

Semiconductors have the same type of band structure as an insulator, but the energy gap is much smaller, on the order of 1 eV. Table 43.3 shows the energy gaps for some representative materials. The band structure of a semiconductor is shown in Figure 43.24. Because the Fermi level is located near the middle of the gap for a semiconductor and E_g is small, appreciable numbers of electrons are thermally excited from the valence band to the conduction band. Because of the many empty levels above the thermally filled levels in the conduction band, a small applied potential difference can easily raise the electrons in the conduction band into available energy states, resulting in a moderate current.

At $T = 0$ K, all electrons in these materials are in the valence band and no energy is available to excite them across the energy gap. Therefore, semiconductors are poor conductors at very low temperatures. Because the thermal excitation of electrons across the narrow gap is more probable at higher temperatures, the conductivity of semiconductors increases rapidly with temperature, contrasting sharply with the conductivity of metals, which decreases slowly with increasing temperature.

Charge carriers in a semiconductor can be negative, positive, or both. When an electron moves from the valence band into the conduction band, it leaves behind a vacant site, called a **hole,** in the otherwise filled valence band. This hole (electron-deficient site) acts as a charge carrier in the sense that a free electron from a nearby site can transfer into the hole. Whenever an electron does so, it creates a new hole at the site it abandoned. Therefore, the net effect can be viewed as the hole migrating through the material in the direction opposite the direction of electron movement. The hole behaves as if it were a particle with a positive charge $+e$.

A pure semiconductor crystal containing only one element or one compound is called an **intrinsic semiconductor.** In these semiconductors, there are equal numbers of conduction electrons and holes. Such combinations of charges are called **electron–hole pairs.** In the presence of an external electric field, the holes move in the direction of the field and the conduction electrons move in the direction opposite the field (Fig. 43.25). Because the electrons and holes have opposite signs, both motions correspond to a current in the same direction.

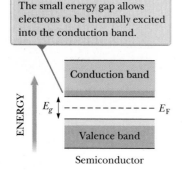

The small energy gap allows electrons to be thermally excited into the conduction band.

Figure 43.24 Band structure of a semiconductor at ordinary temperatures ($T \approx 300$ K). The energy gap is much smaller than in an insulator.

---

[4]We defined the Fermi energy as the energy of the highest filled state at $T = 0$, which might suggest that the Fermi energy should be at the top of the valence band in Figure 43.23. A more sophisticated general treatment of the Fermi energy, however, shows that it is located at that energy at which the probability of occupation is one-half (see Fig. 43.16b). According to this definition, the Fermi energy lies in the energy gap between the bands.

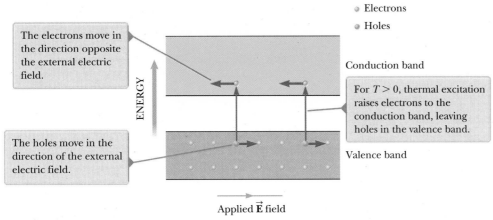

Figure 43.25 Movement of charges (holes and electrons) in an intrinsic semiconductor.

The electrons move in the direction opposite the external electric field.

The holes move in the direction of the external electric field.

● Electrons
● Holes

Conduction band

For $T > 0$, thermal excitation raises electrons to the conduction band, leaving holes in the valence band.

Valence band

Applied $\vec{\mathbf{E}}$ field

Quick Quiz 43.4 Consider the data on three materials given in the table.

Material	Conduction Band	E_g
A	Empty	1.2 eV
B	Half full	1.2 eV
C	Empty	8.0 eV

Identify each material as a conductor, an insulator, or a semiconductor.

Doped Semiconductors

When impurities are added to a semiconductor, both the band structure of the semiconductor and its resistivity are modified. The process of adding impurities, called **doping,** is important in controlling the conductivity of semiconductors. For example, when an atom containing five outer-shell electrons, such as arsenic, is added to a Group IV semiconductor, four of the electrons form covalent bonds with atoms of the semiconductor and one is left over (Fig. 43.26a). This extra electron is nearly free of its parent atom and can be modeled as having an energy level that lies in the energy gap, immediately below the conduction band (Fig. 43.26b). Such a pentavalent atom in effect donates an electron to the structure and hence is referred to as a **donor atom.** Because the spacing between the energy level of the electron of the donor atom and the bottom of the conduction band is very

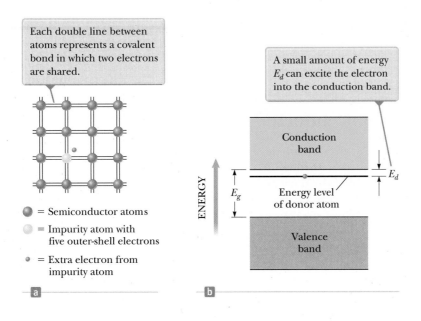

Each double line between atoms represents a covalent bond in which two electrons are shared.

A small amount of energy E_d can excite the electron into the conduction band.

Conduction band

Energy level of donor atom

Valence band

● = Semiconductor atoms

○ = Impurity atom with five outer-shell electrons

● = Extra electron from impurity atom

a   **b**

Figure 43.26 (a) Two-dimensional representation of a semiconductor consisting of Group IV atoms (gray) and an impurity atom (yellow) that has five outer-shell electrons. (b) Energy-band diagram for a semiconductor in which the nearly free electron of the impurity atom lies in the energy gap, immediately below the bottom of the conduction band.

Figure 43.27 (a) Two-dimensional representation of a semiconductor consisting of Group IV atoms (gray) and an impurity atom (yellow) having three outer-shell electrons. (b) Energy-band diagram for a semiconductor in which the energy level associated with the trivalent impurity atom lies in the energy gap, immediately above the top of the valence band.

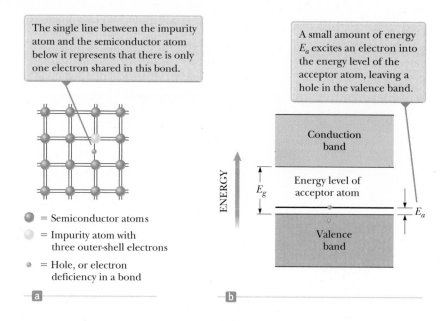

The single line between the impurity atom and the semiconductor atom below it represents that there is only one electron shared in this bond.

A small amount of energy E_a excites an electron into the energy level of the acceptor atom, leaving a hole in the valence band.

● = Semiconductor atoms

○ = Impurity atom with three outer-shell electrons

○ = Hole, or electron deficiency in a bond

small (typically, approximately 0.05 eV), only a small amount of thermal excitation is needed to cause this electron to move into the conduction band. (Recall that the average energy of an electron at room temperature is approximately $k_BT \approx$ 0.025 eV.) Semiconductors doped with donor atoms are called **n-type semiconductors** because the majority of charge carriers are electrons, which are **n**egatively charged.

If a Group IV semiconductor is doped with atoms containing three outer-shell electrons, such as indium and aluminum, the three electrons form covalent bonds with neighboring semiconductor atoms, leaving an electron deficiency—a hole—where the fourth bond would be if an impurity-atom electron were available to form it (Fig. 43.27a). This situation can be modeled by placing an energy level in the energy gap, immediately above the valence band, as in Figure 43.27b. An electron from the valence band has enough energy at room temperature to fill this impurity level, leaving behind a hole in the valence band. This hole can carry current in the presence of an electric field. Because a trivalent atom accepts an electron from the valence band, such impurities are referred to as **acceptor atoms.** A semiconductor doped with trivalent (acceptor) impurities is known as a **p-type semiconductor** because the majority of charge carriers are **p**ositively charged holes.

When conduction in a semiconductor is the result of acceptor or donor impurities, the material is called an **extrinsic semiconductor.** The typical range of doping densities for extrinsic semiconductors is 10^{13} to 10^{19} cm^{-3}, whereas the electron density in a typical semiconductor is roughly 10^{21} cm^{-3}.

43.7 Semiconductor Devices

The electronics of the first half of the 20th century was based on vacuum tubes, in which electrons pass through empty space between a cathode and an anode. We have seen vacuum tube devices in Figure 29.6 (the television picture tube), Figure 29.10 (circular electron beam), Figure 29.15a (Thomson's apparatus for measuring e/m_e for the electron), and Figure 40.9 (photoelectric effect apparatus).

The transistor was invented in 1948, leading to a shift away from vacuum tubes and toward semiconductors as the basis of electronic devices. This phase of electronics has been under way for several decades. As discussed in Chapter 41, there may be a new phase of electronics in the near future using nanotechnological devices employing quantum dots and other nanoscale structures.

In this section, we discuss electronic devices based on semiconductors, which are still in wide use and will be for many years to come.

The Junction Diode

A fundamental unit of a semiconductor device is formed when a *p*-type semiconductor is joined to an *n*-type semiconductor to form a ***p–n* junction.** A **junction diode** is a device that is based on a single *p–n* junction. The role of a diode of any type is to pass current in one direction but not the other. Therefore, it acts as a one-way valve for current.

The *p–n* junction shown in Figure 43.28a consists of three distinct regions: a *p* region, an *n* region, and a small area that extends several micrometers to either side of the interface, called a *depletion region.*

The depletion region may be visualized as arising when the two halves of the junction are brought together. The mobile *n*-side donor electrons nearest the junction (deep-blue area in Fig. 43.28a) diffuse to the *p* side and fill holes located there, leaving behind immobile positive ions. While this process occurs, we can model the holes that are being filled as diffusing to the *n* side, leaving behind a region (brown area in Fig. 43.28a) of fixed negative ions.

Because the two sides of the depletion region each carry a net charge, an internal electric field on the order of 10^4 to 10^6 V/cm exists in the depletion region (see Fig. 43.28b). This field produces an electric force on any remaining mobile charge carriers that sweeps them out of the depletion region, so named because it is a region depleted of mobile charge carriers. This internal electric field creates an internal potential difference ΔV_0 that prevents further diffusion of holes and electrons across the junction and thereby ensures zero current in the junction when no potential difference is applied.

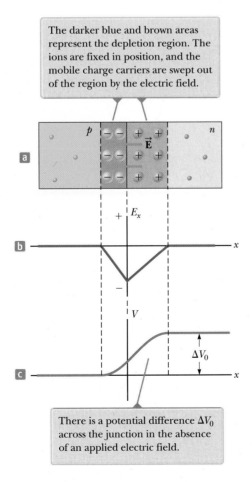

The darker blue and brown areas represent the depletion region. The ions are fixed in position, and the mobile charge carriers are swept out of the region by the electric field.

There is a potential difference ΔV_0 across the junction in the absence of an applied electric field.

Figure 43.28 (a) Physical arrangement of a *p–n* junction. (b) Component E_x of the internal electric field versus x for the *p–n* junction. (c) Internal electric potential difference ΔV versus x for the *p–n* junction.

Figure 43.29 (a) A p–n junction under forward bias. The middle diagram shows the potentials applied at the ends of the junction. Below that is a circuit diagram showing a battery with an adjustable voltage. The upper diagram shows how the potential varies across the junction. The dashed line shows the potential difference across the unbiased junction. (b) When the battery is reversed and the p–n junction is under reverse bias, the current is very small. (c) The characteristic curve for a real p–n junction.

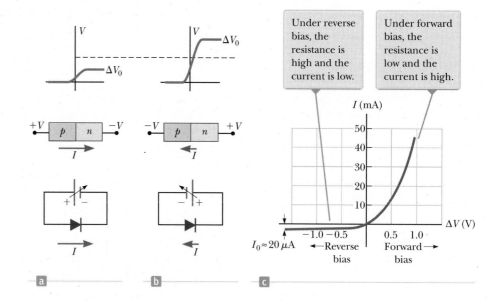

The operation of the junction as a diode is easiest to understand in terms of the potential difference graph shown in Figure 43.28c. If a voltage ΔV is applied to the junction such that the p side is connected to the positive terminal of a voltage source as shown in Figure 43.29a, the internal potential difference ΔV_0 across the junction decreases as shown at the top of the figure; the decrease results in a current that increases exponentially with increasing forward voltage, or *forward bias*. For *reverse bias* (where the n side of the junction is connected to the positive terminal of a voltage source), the internal potential difference ΔV_0 increases with increasing reverse bias as in Figure 43.29b; the increase results in a very small reverse current that quickly reaches a saturation value I_0. The current–voltage relationship for an ideal diode is

$$I = I_0 \left(e^{e\,\Delta V / k_{\mathrm{B}} T} - 1 \right) \tag{43.27}$$

where the first e is the base of the natural logarithm, the second e represents the magnitude of the electron charge, k_{B} is Boltzmann's constant, and T is the absolute temperature. Figure 43.29c shows an I–ΔV plot characteristic of a real p–n junction, demonstrating the diode behavior.

Light-Emitting and Light-Absorbing Diodes

Light-emitting diodes (LEDs) and semiconductor lasers are common examples of devices that depend on the behavior of semiconductors. LEDs are used in LCD television displays, household lighting, flashlights, and camera flash units. Semiconductor lasers are often used for pointers in presentations and in playback equipment for digitally recorded information.

Light emission and absorption in semiconductors is similar to light emission and absorption by gaseous atoms except that in the discussion of semiconductors we must incorporate the concept of energy bands rather than the discrete energy levels in single atoms. As shown in Figure 43.30a, an electron excited electrically into the conduction band can easily recombine with a hole (especially if the electron is injected into a p region). As this recombination takes place, a photon of energy E_g is emitted. With proper design of the semiconductor and the associated plastic envelope or mirrors, the light from a large number of these transitions serves as the source of an LED or a semiconductor laser.

Conversely, an electron in the valence band may absorb an incoming photon of light and be promoted to the conduction band, leaving a hole behind (Fig. 43.30b). This absorbed energy can be used to operate an electrical circuit.

One device that operates on this principle is the **photovoltaic solar cell,** which appears in many handheld calculators. An early large-scale application of arrays of

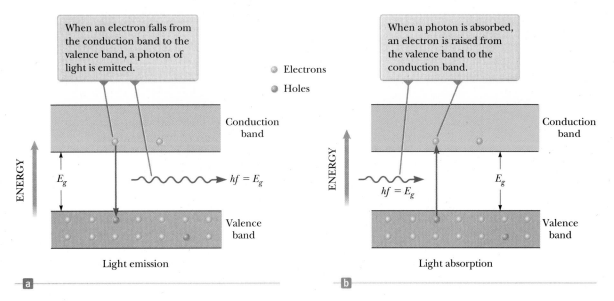

When an electron falls from the conduction band to the valence band, a photon of light is emitted.

○ Electrons
● Holes

When a photon is absorbed, an electron is raised from the valence band to the conduction band.

Conduction band

E_g

$hf = E_g$

Valence band

Light emission

a

ENERGY

Conduction band

$hf = E_g$

E_g

Valence band

Light absorption

b

ENERGY

Figure 43.30 (a) Light emission from a semiconductor. (b) Light absorption by a semiconductor.

photovoltaic cells is the energy supply for orbiting spacecraft. The solar panels of the Hubble Space Telescope can be seen in the chapter-opening photograph for Chapter 38 on page 906.

During the early years of the current century, application of photovoltaics for ground-based generation of electricity has been one of the world's fastest-growing energy technologies. At the time of this printing, the global generation of energy by means of photovoltaics is over 70 GW. A homeowner can install arrays of photovoltaic panels on the roof of his or her house and generate enough energy to operate the home as well as feed excess energy back into the electrical grid. Several photovoltaic power plants have recently been completed, including the Agua Caliente Solar Project in Arizona (200 MW completed in 2012, and 397 MW projected at completion), the Golmud Solar Park in China (200 MW), and the Charanka Solar Park in India (214 MW completed in 2012, and 500 MW projected at completion), the latter of which will be one location in the Gujarat Solar Park, a collection of several sites that is hoped to eventually supply close to 1 GW of power.

Example 43.6   Where's the Remote?

Estimate the band gap of the semiconductor in the infrared LED of a typical television remote control.

SOLUTION

Conceptualize Imagine electrons in Figure 43.30a falling from the conduction band to the valence band, emitting infrared photons in the process.

Categorize We use concepts discussed in this section, so we categorize this example as a substitution problem.

In Chapter 34, we learned that the wavelength of infrared light ranges from 700 nm to 1 mm. Let's pick a number that is easy to work with, such as 1 000 nm (which is not a bad estimate because remote controls typically operate in the range of 880 to 950 nm.)

Estimate the energy hf of the photons from the remote control:

$$E = hf = \frac{hc}{\lambda} = \frac{1\ 240\ \text{eV} \cdot \text{nm}}{1\ 000\ \text{nm}} = \boxed{1.2\ \text{eV}}$$

This value corresponds to an energy gap E_g of approximately 1.2 eV in the LED's semiconductor.

The Transistor

The invention of the transistor by John Bardeen (1908–1991), Walter Brattain (1902–1987), and William Shockley (1910–1989) in 1948 totally revolutionized the world of electronics. For this work, these three men shared the Nobel Prize in Physics in 1956. By 1960, the transistor had replaced the vacuum tube in many electronic applications. The advent of the transistor created a multitrillion-dollar industry that produces such popular devices as personal computers, wireless keyboards, smartphones, electronic book readers, and computer tablets.

A **junction transistor** consists of a semiconducting material in which a very narrow n region is sandwiched between two p regions or a p region is sandwiched between two n regions. In either case, the transistor is formed from two p–n junctions. These types of transistors were used widely in the early days of semiconductor electronics.

During the 1960s, the electronics industry converted many electronic applications from the junction transistor to the **field-effect transistor,** which is much easier to manufacture and just as effective. Figure 43.31a shows the structure of a very common device, the **MOSFET,** or **metal-oxide-semiconductor field-effect transistor.** You are likely using millions of MOSFET devices when you are working on your computer.

There are three metal connections (the M in MOSFET) to the transistor: the *source, drain,* and *gate.* The source and drain are connected to n-type semiconductor regions (the S in MOSFET) at either end of the structure. These regions are connected by a narrow channel of additional n-type material, the n channel. The source and drain regions and the n channel are embedded in a p-type substrate material, which forms a depletion region, as in the junction diode, along the bottom of the n channel. (Depletion regions also exist at the junctions underneath the source and drain regions, but we will ignore them because the operation of the device depends primarily on the behavior in the channel.)

The gate is separated from the n channel by a layer of insulating silicon dioxide (the O in MOSFET, for oxide). Therefore, it does not make electrical contact with the rest of the semiconducting material.

Imagine that a voltage source ΔV_{SD} is applied across the source and drain as shown in Figure 43.31b. In this situation, electrons flow through the upper region of the n channel. Electrons cannot flow through the depletion region in the lower part of the n channel because this region is depleted of charge carriers. Now a second voltage ΔV_{SG} is applied across the source and gate as in Figure 43.31c. The positive potential on the gate electrode results in an electric field below the gate that is directed downward in the n channel (the field in "field-effect"). This electric field exerts upward forces on electrons in the region below the gate, causing them to move into the n channel. Consequently, the depletion region becomes smaller,

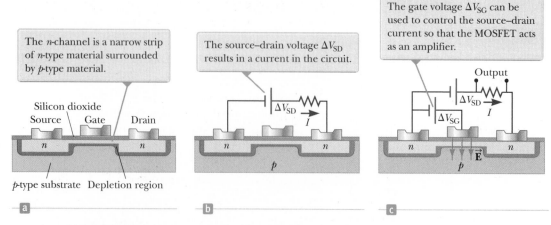

Figure 43.31 (a) The structure of a metal-oxide-semiconductor field-effect transistor (MOSFET). (b) A source–drain voltage is applied. (c) A gate voltage is applied.

widening the area through which there is current between the top of the *n* channel and the depletion region. As the area becomes wider, the current increases.

If a varying voltage, such as that generated from music stored in the memory of a smartphone, is applied to the gate, the area through which the source–drain current exists varies in size according to the varying gate voltage. A small variation in gate voltage results in a large variation in current and a correspondingly large voltage across the resistor in Figure 43.31c. Therefore, the MOSFET acts as a voltage amplifier. A circuit consisting of a chain of such transistors can result in a very small initial signal from a microphone being amplified enough to drive powerful speakers at an outdoor concert.

The Integrated Circuit

Invented independently by Jack Kilby (1923–2005, Nobel Prize in Physics, 2000) at Texas Instruments in late 1958 and by Robert Noyce (1927–1990) at Fairchild Camera and Instrument in early 1959, the integrated circuit has been justly called "the most remarkable technology ever to hit mankind." Kilby's first device is shown in Figure 43.32. Integrated circuits have indeed started a "second industrial revolution" and are found at the heart of computers, watches, cameras, automobiles, aircraft, robots, space vehicles, and all sorts of communication and switching networks.

Figure 43.32 Jack Kilby's first integrated circuit, tested on September 12, 1958.

In simplest terms, an **integrated circuit** is a collection of interconnected transistors, diodes, resistors, and capacitors fabricated on a single piece of silicon known as a *chip*. Contemporary electronic devices often contain many integrated circuits (Fig. 43.33). State-of-the-art chips easily contain several million components within a 1-cm^2 area, and the number of components per square inch has increased steadily since the integrated circuit was invented. The dramatic advances in chip technology can be seen by looking at microchips manufactured by Intel. The 4004 chip, introduced in 1971, contained 2 300 transistors. This number increased to 3.2 million 24 years later in 1995 with the Pentium processor. Sixteen years later, the Core i7 Sandy Bridge processor introduced in November 2011 contained 2 270 million transistors.

Integrated circuits were invented partly to solve the interconnection problem spawned by the transistor. In the era of vacuum tubes, power and size considerations of individual components set modest limits on the number of components that could be interconnected in a given circuit. With the advent of the tiny, low-power, highly reliable transistor, design limits on the number of components disappeared and were replaced by the problem of wiring together hundreds of thousands of components. The magnitude of this problem can be appreciated when we consider that second-generation computers (consisting of discrete transistors rather than integrated circuits) contained several hundred thousand components requiring more than a million joints that had to be hand-soldered and tested.

In addition to solving the interconnection problem, integrated circuits possess the advantages of miniaturization and fast response, two attributes critical for high-speed computers. Because the response time of a circuit depends on the time

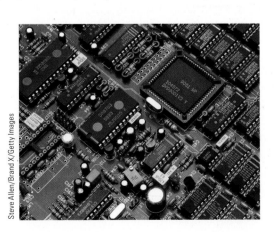

Figure 43.33 Integrated circuits are prevalent in many electronic devices. All the flat circuit elements with black-topped surfaces in this photograph are integrated circuits.

interval required for electrical signals traveling at the speed of light to pass from one component to another, miniaturization and close packing of components result in fast response times.

43.8 Superconductivity

We learned in Section 27.5 that there is a class of metals and compounds known as **superconductors** whose electrical resistance decreases to virtually zero below a certain temperature T_c called the *critical temperature* (Table 27.3). Let's now look at these amazing materials in greater detail, using what we know about the properties of solids to help us understand the behavior of superconductors.

Let's start by examining the Meissner effect, introduced in Section 30.6 as the exclusion of magnetic flux from the interior of superconductors. The Meissner effect is illustrated in Figure 43.34 for a superconducting material in the shape of a long cylinder. Notice that the magnetic field penetrates the cylinder when its temperature is greater than T_c (Fig. 43.34a). As the temperature is lowered to below T_c, however, the field lines are spontaneously expelled from the interior of the superconductor (Fig. 43.34b). Therefore, a superconductor is more than a perfect conductor (resistivity $\rho = 0$); it is also a perfect diamagnet ($\vec{B} = 0$). The property that $\vec{B} = 0$ in the interior of a superconductor is as fundamental as the property of zero resistance. If the magnitude of the applied magnetic field exceeds a critical value B_c, defined as the value of B that destroys a material's superconducting properties, the field again penetrates the sample.

Because a superconductor is a perfect diamagnet, it repels a permanent magnet. In fact, one can perform a demonstration of the Meissner effect by floating a small permanent magnet above a superconductor and achieving magnetic levitation as seen in Figure 30.27 in Section 30.6.

Recall from our study of electricity that a good conductor expels static electric fields by moving charges to its surface. In effect, the surface charges produce an electric field that exactly cancels the externally applied field inside the conductor. In a similar manner, a superconductor expels magnetic fields by forming surface currents. To see why that happens, consider again the superconductor shown in Figure 43.34. Let's assume the sample is initially at a temperature $T > T_c$ as illustrated in Figure 43.34a so that the magnetic field penetrates the cylinder. As the cylinder is cooled to a temperature $T < T_c$, the field is expelled as shown in Figure 43.34b. Surface currents induced on the superconductor's surface produce a magnetic field that exactly cancels the externally applied field inside the superconductor. As you would expect, the surface currents disappear when the external magnetic field is removed.

A successful theory for superconductivity in metals was published in 1957 by John Bardeen, L. N. Cooper (b. 1930), and J. R. Schrieffer (b. 1931); it is generally called BCS theory, based on the first letters of their last names. This theory led to a Nobel Prize in Physics for the three scientists in 1972. In this theory, two electrons can interact via distortions in the array of lattice ions so that there is a net attractive force between the electrons.[5] As a result, the two electrons are bound into an entity called a *Cooper pair*, which behaves like a particle with integral spin. Particles with integral spin are called *bosons*. (As noted in Pitfall Prevention 42.6, *fermions* make up another class of particles, those with half-integral spin.) An important feature of bosons is that they do not obey the Pauli exclusion principle. Consequently, at very low temperatures, it is possible for all bosons in a collection of such particles to be in the lowest quantum state. The entire collection of Cooper pairs in the metal is described by a single wave function. Above the energy level associated with this wave function is an energy gap equal to the binding energy of a Cooper pair. Under

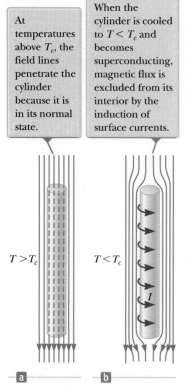

At temperatures above T_c, the field lines penetrate the cylinder because it is in its normal state.

When the cylinder is cooled to $T < T_c$ and becomes superconducting, magnetic flux is excluded from its interior by the induction of surface currents.

$T > T_c$ $T < T_c$

a   b

Figure 43.34 A superconductor in the form of a long cylinder in the presence of an external magnetic field.

---

[5]A highly simplified explanation of this attraction between electrons is as follows. The attractive Coulomb force between one electron and the surrounding positively charged lattice ions causes the ions to move inward slightly toward the electron. As a result, there is a higher concentration of positive charge in this region than elsewhere in the lattice. A second electron is attracted to the higher concentration of positive charge.

the action of an applied electric field, the Cooper pairs experience an electric force and move through the metal. A random scattering event of a Cooper pair from a lattice ion would represent resistance to the electric current. Such a collision would change the energy of the Cooper pair because some energy would be transferred to the lattice ion. There are no available energy levels below that of the Cooper pair (it is already in the lowest state), however, and none available above because of the energy gap. As a result, collisions do not occur and there is no resistance to the movement of Cooper pairs.

An important development in physics that elicited much excitement in the scientific community was the discovery of high-temperature copper oxide-based superconductors. The excitement began with a 1986 publication by J. Georg Bednorz (b. 1950) and K. Alex Müller (b. 1927), scientists at the IBM Zurich Research Laboratory in Switzerland. In their seminal paper,[6] Bednorz and Müller reported strong evidence for superconductivity at 30 K in an oxide of barium, lanthanum, and copper. They were awarded the Nobel Prize in Physics in 1987 for their remarkable discovery. Shortly thereafter, a new family of compounds was open for investigation and research activity in the field of superconductivity proceeded vigorously. In early 1987, groups at the University of Alabama at Huntsville and the University of Houston announced superconductivity at approximately 92 K in an oxide of yttrium, barium, and copper ($YBa_2Cu_3O_7$). Later that year, teams of scientists from Japan and the United States reported superconductivity at 105 K in an oxide of bismuth, strontium, calcium, and copper. Superconductivity at temperatures as high as 150 K have been reported in an oxide containing mercury. In 2006, Japanese scientists discovered superconductivity for the first time in iron-based materials, beginning with LaFePO, with a critical temperature of 4 K. The highest critical temperature that has been reported so far in the iron-based materials is 55 K, a milestone held by fluorine-doped SmFeAsO. These newly discovered materials have rejuvenated the field of high-T_c superconductivity. Today, one cannot rule out the possibility of room-temperature superconductivity, and the mechanisms responsible for the behavior of high-temperature superconductors are still under investigation. The search for novel superconducting materials continues both for scientific reasons and because practical applications become more probable and widespread as the critical temperature is raised.

Although BCS theory was very successful in explaining superconductivity in metals, there is currently no widely accepted theory for high-temperature superconductivity. It remains an area of active research.

Summary

Concepts and Principles

Two or more atoms combine to form molecules because of a net attractive force between the atoms. The mechanisms responsible for molecular bonding can be classified as follows:

- **Ionic bonds** form primarily because of the Coulomb attraction between oppositely charged ions. Sodium chloride (NaCl) is one example.
- **Covalent bonds** form when the constituent atoms of a molecule share electrons. For example, the two electrons of the H_2 molecule are equally shared between the two nuclei.
- **Van der Waals bonds** are weak electrostatic bonds between molecules or between atoms that do not form ionic or covalent bonds. These bonds are responsible for the condensation of noble gas atoms and nonpolar molecules into the liquid phase.
- **Hydrogen bonds** form between the center of positive charge in a polar molecule that includes one or more hydrogen atoms and the center of negative charge in another polar molecule.

continued

---

[6] J. G. Bednorz and K. A. Müller, *Z. Phys. B* **64**:189, 1986.

The allowed values of the rotational energy of a diatomic molecule are

$$E_{rot} = E_J = \frac{\hbar^2}{2I} J(J+1) \quad J = 0, 1, 2, \ldots \quad \textbf{(43.6)}$$

where I is the moment of inertia of the molecule and J is an integer called the **rotational quantum number.** The selection rule for transitions between rotational states is $\Delta J = \pm 1$.

The allowed values of the vibrational energy of a diatomic molecule are

$$E_{vib} = \left(v + \tfrac{1}{2}\right)\frac{h}{2\pi}\sqrt{\frac{k}{\mu}} \quad v = 0, 1, 2, \ldots \quad \textbf{(43.10)}$$

where v is the **vibrational quantum number,** k is the force constant of the "effective spring" bonding the molecule, and μ is the **reduced mass** of the molecule. The selection rule for allowed vibrational transitions is $\Delta v = \pm 1$, and the energy difference between any two adjacent levels is the same, regardless of which two levels are involved.

Bonding mechanisms in solids can be classified in a manner similar to the schemes for molecules. For example, the Na^+ and Cl^- ions in NaCl form **ionic bonds,** whereas the carbon atoms in diamond form **covalent bonds.** The **metallic bond** is characterized by a net attractive force between positive ion cores and the mobile free electrons of a metal.

In the **free-electron theory of metals,** the free electrons fill the quantized levels in accordance with the Pauli exclusion principle. The number of states per unit volume available to the conduction electrons having energies between E and $E + dE$ is

$$N(E)\, dE = \left(\frac{8\sqrt{2}\,\pi m_e^{3/2}}{h^3} E^{1/2}\right)\left(\frac{1}{e^{(E-E_F)/k_B T} + 1}\right) dE \quad \textbf{(43.22)}$$

where E_F is the **Fermi energy.** At $T = 0$ K, all levels below E_F are filled, all levels above E_F are empty, and

$$E_F(0) = \frac{h^2}{2m_e}\left(\frac{3n_e}{8\pi}\right)^{2/3} \quad \textbf{(43.25)}$$

where n_e is the total number of conduction electrons per unit volume. Only those electrons having energies near E_F can contribute to the electrical conductivity of the metal.

In a crystalline solid, the energy levels of the system form a set of **bands.** Electrons occupy the lowest energy states, with no more than one electron per state. Energy gaps are present between the bands of allowed states.

A **semiconductor** is a material having an energy gap of approximately 1 eV and a valence band that is filled at $T = 0$ K. Because of the small energy gap, a significant number of electrons can be thermally excited from the valence band into the conduction band. The band structures and electrical properties of a Group IV semiconductor can be modified by the addition of either donor atoms containing five outer-shell electrons or acceptor atoms containing three outer-shell electrons. A semiconductor **doped** with donor impurity atoms is called an **n-type semiconductor,** and one doped with acceptor impurity atoms is called a **p-type semiconductor.**

Objective Questions   1. denotes answer available in *Student Solutions Manual/Study Guide*

1. Is each one of the following statements true or false for a superconductor below its critical temperature? (a) It can carry infinite current. (b) It must carry some nonzero current. (c) Its interior electric field must be zero. (d) Its internal magnetic field must be zero. (e) No internal energy appears when it carries electric current.

2. An infrared absorption spectrum of a molecule is shown in Figure OQ43.2. Notice that the highest peak on either side of the gap is the third peak from the gap. After this spectrum is taken, the temperature of the sample of molecules is raised to a much higher value. Compared with Figure OQ43.2, in this new spectrum is the highest absorption peak (a) at the same frequency, (b) farther from the gap, or (c) closer to the gap?

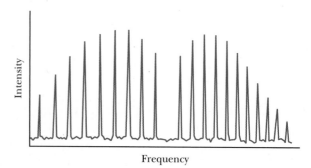

Figure OQ43.2

3. What kind of bonding likely holds the atoms together in the following solids (i), (ii), and (iii)? Choose your

answers from these possibilities: (a) ionic bonding, (b) covalent bonding, and (c) metallic bonding. **(i)** The solid is opaque, shiny, flexible, and a good electric conductor. **(ii)** The crystal is transparent, brittle, and soluble in water. It is a poor conductor of electricity. **(iii)** The crystal is opaque, brittle, very hard, and a good electric insulator.

4. The Fermi energy for silver is 5.48 eV. In a piece of solid silver, free-electron energy levels are measured near 2 eV and near 6 eV. **(i)** Near which of these energies are the energy levels closer together? (a) 2 eV (b) 6 eV (c) The spacing is the same. **(ii)** Near which of these energies are more electrons occupying energy levels? (a) 2 eV (b) 6 eV (c) The number of electrons is the same.

5. As discussed in Chapter 27, the conductivity of metals decreases with increasing temperature due to electron collisions with vibrating atoms. In contrast, the conductivity of semiconductors increases with increasing temperature. What property of a semiconductor is responsible for this behavior? (a) Atomic vibrations decrease as temperature increases. (b) The number of conduction electrons and the number of holes increase steeply with increasing temperature. (c) The energy gap decreases with increasing temperature. (d) Electrons do not collide with atoms in a semiconductor.

6. (i) Should you expect an *n*-type doped semiconductor to have (a) higher, (b) lower, or (c) the same conductivity as an intrinsic (pure) semiconductor? **(ii)** Should you expect a *p*-type doped semiconductor to have (a) higher, (b) lower, or (c) the same conductivity as an intrinsic (pure) semiconductor?

7. Consider a typical material composed of covalently bonded diatomic molecules. Rank the following energies from the largest in magnitude to the smallest in magnitude. (a) the latent heat of fusion per molecule (b) the molecular binding energy (c) the energy of the first excited state of molecular rotation (d) the energy of the first excited state of molecular vibration

Conceptual Questions

1. denotes answer available in *Student Solutions Manual/Study Guide*

Conceptual Questions 5 and 6 in Chapter 27 can be assigned with this chapter.

1. The energies of photons of visible light range between the approximate values 1.8 eV and 3.1 eV. Explain why silicon, with an energy gap of 1.14 eV at room temperature (see Table 43.3), appears opaque, whereas diamond, with an energy gap of 5.47 eV, appears transparent.

2. Discuss the three major forms of excitation of a molecule (other than translational motion) and the relative energies associated with these three forms.

3. How can the analysis of the rotational spectrum of a molecule lead to an estimate of the size of that molecule?

4. Pentavalent atoms such as arsenic are donor atoms in a semiconductor such as silicon, whereas trivalent atoms such as indium are acceptors. Inspect the periodic table in Appendix C and determine what other elements might make good donors or acceptors.

5. When a photon is absorbed by a semiconductor, an electron–hole pair is created. Give a physical explanation of this statement using the energy-band model as the basis for your description.

6. (a) Discuss the differences in the band structures of metals, insulators, and semiconductors. (b) How does the band-structure model enable you to understand the electrical properties of these materials better?

7. (a) What essential assumptions are made in the free-electron theory of metals? (b) How does the energy-band model differ from the free-electron theory in describing the properties of metals?

8. How do the vibrational and rotational levels of heavy hydrogen (D_2) molecules compare with those of H_2 molecules?

9. Discuss models for the different types of bonds that form stable molecules.

10. Discuss the differences between crystalline solids, amorphous solids, and gases.

Problems available in Access end-of-chapter problems online at **www.webassign.net**

Section 43.1 Molecular Bond
Problems 1–6

Section 43.2 Energy States and Spectra of Molecules
Problems 7–23

Section 43.3 Bonding in Solids
Problems 24–26

Section 43.4 Free-Electron Theory of Metals
Section 43.5 Band Theory of Solids
Problems 27–39

Section 43.6 Electrical Conduction in Metals, Insulators, and Semiconductors
Problems 40–46

Section 43.7 Semiconductor Devices
Problems 47–50

Section 43.8 Superconductivity
Problems 51–53

Additional Problems
Problems 54–62

Challenge Problems

Problems 63–64

Solutions to the following Problems are available in the *Student Solutions Manual/Study Guide:*

43.2, 43.9, 43.13, 43.26, 43.27, 43.33, 43.38, 43.39, 43.40, 43.50, 43.53, 43.59, and 43.61

List of Enhanced Problems

Problem Number	Targeted Feedback in Enhanced WebAssign	Analysis Model Tutorial in Enhanced WebAssign	Master It in Enhanced WebAssign	Watch It in Enhanced WebAssign
43.2	✓			✓
43.9			✓	
43.13		✓		
43.21				✓
43.32	✓			✓
43.33			✓	
43.38			✓	
43.49				✓
43.50			✓	
43.63				✓

Ötzi the Iceman, a Copper Age man, was discovered by German tourists in the Italian Alps in 1991 when a glacier melted enough to expose his remains. Analysis of his corpse has exposed his last meal, illnesses he suffered, and places he lived. Radioactivity was used to determine that he lived in about 3300 BC. Ötzi can be seen today in the Südtiroler Archäologiemuseum (South Tyrol Museum of Archaeology) in Bolzano, Italy. *(©Vienna Report Agency/Sygma/Corbis)*

The year 1896 marks the birth of nuclear physics when French physicist Antoine-Henri Becquerel (1852–1908) discovered radioactivity in uranium compounds. This discovery prompted scientists to investigate the details of radioactivity and, ultimately, the structure of the nucleus. Pioneering work by Ernest Rutherford showed that the radiation emitted from radioactive substances is of three types—alpha, beta, and gamma rays—classified according to the nature of their electric charge and their ability to penetrate matter and ionize air. Later experiments showed that alpha rays are helium nuclei, beta rays are electrons, and gamma rays are high-energy photons.

In 1911, Rutherford, Hans Geiger, and Ernest Marsden performed the alpha-particle scattering experiments described in Section 42.2. These experiments established that the nucleus of an atom can be modeled as a point mass and point charge and that most of the atomic mass is contained in the nucleus. Subsequent studies revealed the presence of a new type of force, the short-range nuclear force, which is predominant at particle separation distances less than approximately 10^{-14} m and is zero for large distances.

In this chapter, we discuss the properties and structure of the atomic nucleus. We start by describing the basic properties of nuclei, followed by a discussion of nuclear forces and binding energy, nuclear models, and the phenomenon of radioactivity. Finally, we explore the various processes by which nuclei decay and the ways that nuclei can react with each other.

44.1 Some Properties of Nuclei

All nuclei are composed of two types of particles: protons and neutrons. The only exception is the ordinary hydrogen nucleus, which is a single proton. We describe the atomic nucleus by the number of protons and neutrons it contains, using the following quantities:

- the **atomic number** Z, which equals the number of protons in the nucleus (sometimes called the *charge number*)
- the **neutron number** N, which equals the number of neutrons in the nucleus
- the **mass number** $A = Z + N$, which equals the number of **nucleons** (neutrons plus protons) in the nucleus

A **nuclide** is a specific combination of atomic number and mass number that represents a nucleus. In representing nuclides, it is convenient to use the symbol $^A_Z X$ to convey the numbers of protons and neutrons, where X represents the chemical symbol of the element. For example, $^{56}_{26}Fe$ (iron) has mass number 56 and atomic number 26; therefore, it contains 26 protons and 30 neutrons. When no confusion is likely to arise, we omit the subscript Z because the chemical symbol can always be used to determine Z. Therefore, $^{56}_{26}Fe$ is the same as ^{56}Fe and can also be expressed as "iron-56" or "Fe-56."

The nuclei of all atoms of a particular element contain the same number of protons but often contain different numbers of neutrons. Nuclei related in this way are called **isotopes.** The isotopes of an element have the same Z value but different N and A values.

The natural abundance of isotopes can differ substantially. For example $^{11}_6C$, $^{12}_6C$, $^{13}_6C$, and $^{14}_6C$ are four isotopes of carbon. The natural abundance of the $^{12}_6C$ isotope is approximately 98.9%, whereas that of the $^{13}_6C$ isotope is only about 1.1%. Some isotopes, such as $^{11}_6C$ and $^{14}_6C$, do not occur naturally but can be produced by nuclear reactions in the laboratory or by cosmic rays.

Even the simplest element, hydrogen, has isotopes: ^1_1H, the ordinary hydrogen nucleus; ^2_1H, deuterium; and ^3_1H, tritium.

Quick Quiz 44.1 For each part of this Quick Quiz, choose from the following answers: (a) protons (b) neutrons (c) nucleons. **(i)** The three nuclei ^{12}C, ^{13}N, and ^{14}O have the same number of what type of particle? **(ii)** The three nuclei ^{12}N, ^{13}N, and ^{14}N have the same number of what type of particle? **(iii)** The three nuclei ^{14}C, ^{14}N, and ^{14}O have the same number of what type of particle?

Charge and Mass

The proton carries a single positive charge e, equal in magnitude to the charge $-e$ on the electron ($e = 1.6 \times 10^{-19}$ C). The neutron is electrically neutral as its name implies. Because the neutron has no charge, it was difficult to detect with early experimental apparatus and techniques. Today, neutrons are easily detected with devices such as plastic scintillators.

Nuclear masses can be measured with great precision using a mass spectrometer (see Section 29.3) and by the analysis of nuclear reactions. The proton is approximately 1 836 times as massive as the electron, and the masses of the proton and the neutron are almost equal. The **atomic mass unit** u is defined in such a way that the mass of one atom of the isotope ^{12}C is exactly 12 u, where 1 u is equal to $1.660\ 539 \times 10^{-27}$ kg. According to this definition, the proton and neutron each have a mass of approximately 1 u and the electron has a mass that is only a small fraction of this value. The masses of these particles and others important to the phenomena discussed in this chapter are given in Table 44.1 (page 1098).

Pitfall Prevention 44.1

Mass Number Is Not Atomic Mass
The mass number A should not be confused with the atomic mass. Mass number is an integer specific to an isotope and has no units; it is simply a count of the number of nucleons. Atomic mass has units and is generally not an integer because it is an average of the masses of a given element's naturally occurring isotopes.

Table 44.1	Masses of Selected Particles in Various Units		
		Mass	
Particle	kg	u	MeV/c^2
Proton	$1.672\,62 \times 10^{-27}$	1.007 276	938.27
Neutron	$1.674\,93 \times 10^{-27}$	1.008 665	939.57
Electron (β particle)	$9.109\,38 \times 10^{-31}$	$5.485\,79 \times 10^{-4}$	0.510 999
$^{1}_{1}$H atom	$1.673\,53 \times 10^{-27}$	1.007 825	938.783
$^{4}_{2}$He nucleus (α particle)	$6.644\,66 \times 10^{-27}$	4.001 506	3 727.38
$^{4}_{2}$He atom	$6.646\,48 \times 10^{-27}$	4.002 603	3 728.40
$^{12}_{6}$C atom	$1.992\,65 \times 10^{-27}$	12.000 000	11 177.9

You might wonder how six protons and six neutrons, each having a mass larger than 1 u, can be combined with six electrons to form a carbon-12 atom having a mass of exactly 12 u. The bound system of ^{12}C has a lower rest energy (Section 39.8) than that of six separate protons and six separate neutrons. According to Equation 39.24, $E_R = mc^2$, this lower rest energy corresponds to a smaller mass for the bound system. The difference in mass accounts for the binding energy when the particles are combined to form the nucleus. We shall discuss this point in more detail in Section 44.2.

It is often convenient to express the atomic mass unit in terms of its *rest-energy equivalent*. For one atomic mass unit,

$$E_R = mc^2 = (1.660\,539 \times 10^{-27}\ \text{kg})(2.997\,92 \times 10^{8}\ \text{m/s})^2 = 931.494\ \text{MeV}$$

where we have used the conversion 1 eV = $1.602\,176 \times 10^{-19}$ J.

Based on the rest-energy expression in Equation 39.24, nuclear physicists often express mass in terms of the unit MeV/c^2.

Example 44.1   **The Atomic Mass Unit**

Use Avogadro's number to show that 1 u = 1.66×10^{-27} kg.

SOLUTION

Conceptualize From the definition of the mole given in Section 19.5, we know that exactly 12 g (= 1 mol) of ^{12}C contains Avogadro's number of atoms.

Categorize We evaluate the atomic mass unit that was introduced in this section, so we categorize this example as a substitution problem.

Find the mass m of one ^{12}C atom:

$$m = \frac{0.012\ \text{kg}}{6.02 \times 10^{23}\ \text{atoms}} = 1.99 \times 10^{-26}\ \text{kg}$$

Because one atom of ^{12}C is defined to have a mass of 12.0 u, divide by 12.0 to find the mass equivalent to 1 u:

$$1\ \text{u} = \frac{1.99 \times 10^{-26}\ \text{kg}}{12.0} = 1.66 \times 10^{-27}\ \text{kg}$$

The Size and Structure of Nuclei

In Rutherford's scattering experiments, positively charged nuclei of helium atoms (alpha particles) were directed at a thin piece of metallic foil. As the alpha particles moved through the foil, they often passed near a metal nucleus. Because of the positive charge on both the incident particles and the nuclei, the particles were deflected from their straight-line paths by the Coulomb repulsive force.

Rutherford used the isolated system (energy) analysis model to find an expression for the separation distance d at which an alpha particle approaching a nucleus head-on is turned around by Coulomb repulsion. In such a head-on collision, the mechanical energy of the nucleus–alpha particle system is conserved. The initial kinetic energy of the incoming particle is transformed completely to electric potential energy of the system when the alpha particle stops momentarily at the point of closest approach (the final configuration of the system) before moving back along the same path (Fig. 44.1). Applying Equation 8.2, the conservation of energy principle, to the system gives

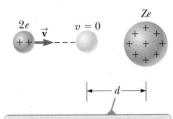

$$\Delta K + \Delta U = 0$$

$$\left(0 - \tfrac{1}{2}mv^2\right) + \left(k_e \frac{q_1 q_2}{d} - 0\right) = 0$$

where m is the mass of the alpha particle and v is its initial speed. Solving for d gives

$$d = 2k_e \frac{q_1 q_2}{mv^2} = 2k_e \frac{(2e)(Ze)}{mv^2} = 4k_e \frac{Ze^2}{mv^2}$$

Because of the Coulomb repulsion between the charges of the same sign, the alpha particle approaches to a distance d from the nucleus, called the distance of closest approach.

Figure 44.1 An alpha particle on a head-on collision course with a nucleus of charge Ze.

where Z is the atomic number of the target nucleus. From this expression, Rutherford found that the alpha particles approached nuclei to within 3.2×10^{-14} m when the foil was made of gold. Therefore, the radius of the gold nucleus must be less than this value. From the results of his scattering experiments, Rutherford concluded that the positive charge in an atom is concentrated in a small sphere, which he called the nucleus, whose radius is no greater than approximately 10^{-14} m.

Because such small lengths are common in nuclear physics, an often-used convenient length unit is the femtometer (fm), which is sometimes called the **fermi** and is defined as

$$1 \text{ fm} \equiv 10^{-15} \text{ m}$$

In the early 1920s, it was known that the nucleus of an atom contains Z protons and has a mass nearly equivalent to that of A protons, where on average $A \approx 2Z$ for lighter nuclei ($Z \leq 20$) and $A > 2Z$ for heavier nuclei. To account for the nuclear mass, Rutherford proposed that each nucleus must also contain $A - Z$ neutral particles that he called neutrons. In 1932, British physicist James Chadwick (1891–1974) discovered the neutron, and he was awarded the Nobel Prize in Physics in 1935 for this important work.

Since the time of Rutherford's scattering experiments, a multitude of other experiments have shown that most nuclei are approximately spherical and have an average radius given by

$$r = aA^{1/3} \tag{44.1}$$

◀ **Nuclear radius**

where a is a constant equal to 1.2×10^{-15} m and A is the mass number. Because the volume of a sphere is proportional to the cube of its radius, it follows from Equation 44.1 that the volume of a nucleus (assumed to be spherical) is directly proportional to A, the total number of nucleons. This proportionality suggests that *all nuclei have nearly the same density*. When nucleons combine to form a nucleus, they combine as though they were tightly packed spheres (Fig. 44.2). This fact has led to an analogy between the nucleus and a drop of liquid, in which the density of the drop is independent of its size. We shall discuss the liquid-drop model of the nucleus in Section 44.3.

Figure 44.2 A nucleus can be modeled as a cluster of tightly packed spheres, where each sphere is a nucleon.

Example 44.2 The Volume and Density of a Nucleus

Consider a nucleus of mass number A.

(A) Find an approximate expression for the mass of the nucleus.

continued

▶ **44.2** continued

SOLUTION

Conceptualize Imagine the nucleus to be a collection of protons and neutrons as shown in Figure 44.2. The mass number A counts *both* protons and neutrons.

Categorize Let's assume A is large enough that we can imagine the nucleus to be spherical.

Analyze The mass of the proton is approximately equal to that of the neutron. Therefore, if the mass of one of these particles is m, the mass of the nucleus is approximately Am.

(B) Find an expression for the volume of this nucleus in terms of A.

SOLUTION

Assume the nucleus is spherical and use Equation 44.1:   (1) $V_{\text{nucleus}} = \frac{4}{3}\pi r^3 = \frac{4}{3}\pi a^3 A$

(C) Find a numerical value for the density of this nucleus.

SOLUTION

Use Equation 1.1 and substitute Equation (1):

$$\rho = \frac{m_{\text{nucleus}}}{V_{\text{nucleus}}} = \frac{Am}{\frac{4}{3}\pi a^3 A} = \frac{3m}{4\pi a^3}$$

Substitute numerical values:

$$\rho = \frac{3(1.67 \times 10^{-27} \text{ kg})}{4\pi(1.2 \times 10^{-15} \text{ m})^3} = 2.3 \times 10^{17} \text{ kg/m}^3$$

Finalize The nuclear density is approximately 2.3×10^{14} times the density of water ($\rho_{\text{water}} = 1.0 \times 10^3 \text{ kg/m}^3$).

WHAT IF?   What if the Earth could be compressed until it had this density? How large would it be?

Answer Because this density is so large, we predict that an Earth of this density would be very small.

Use Equation 1.1 and the mass of the Earth to find the volume of the compressed Earth:

$$V = \frac{M_E}{\rho} = \frac{5.97 \times 10^{24} \text{ kg}}{2.3 \times 10^{17} \text{ kg/m}^3} = 2.6 \times 10^7 \text{ m}^3$$

From this volume, find the radius:

$$V = \frac{4}{3}\pi r^3 \rightarrow r = \left(\frac{3V}{4\pi}\right)^{1/3} = \left[\frac{3(2.6 \times 10^7 \text{ m}^3)}{4\pi}\right]^{1/3}$$

$$r = 1.8 \times 10^2 \text{ m}$$

An Earth of this radius is indeed a small Earth!

Nuclear Stability

You might expect that the very large repulsive Coulomb forces between the close-packed protons in a nucleus should cause the nucleus to fly apart. Because that does not happen, there must be a counteracting attractive force. The **nuclear force** is a very short range (about 2 fm) attractive force that acts between all nuclear particles. The protons attract each other by means of the nuclear force, and, at the same time, they repel each other through the Coulomb force. The nuclear force also acts between pairs of neutrons and between neutrons and protons. The nuclear force dominates the Coulomb repulsive force within the nucleus (at short ranges), so stable nuclei can exist.

The nuclear force is independent of charge. In other words, the forces associated with the proton–proton, proton–neutron, and neutron–neutron interactions are the same, apart from the additional repulsive Coulomb force for the proton–proton interaction.

Evidence for the limited range of nuclear forces comes from scattering experiments and from studies of nuclear binding energies. The short range of the nuclear force is shown in the neutron–proton (n–p) potential energy plot of Figure 44.3a obtained by scattering neutrons from a target containing hydrogen. The depth of the n–p potential energy well is 40 to 50 MeV, and there is a strong repulsive component that prevents the nucleons from approaching much closer than 0.4 fm.

The nuclear force does not affect electrons, enabling energetic electrons to serve as point-like probes of nuclei. The charge independence of the nuclear force also means that the main difference between the n–p and p–p interactions is that the p–p potential energy consists of a *superposition* of nuclear and Coulomb interactions as shown in Figure 44.3b. At distances less than 2 fm, both p–p and n–p potential energies are nearly identical, but for distances of 2 fm or greater, the p–p potential has a positive energy barrier with a maximum at 4 fm.

The existence of the nuclear force results in approximately 270 stable nuclei; hundreds of other nuclei have been observed, but they are unstable. A plot of neutron number N versus atomic number Z for a number of stable nuclei is given in Figure 44.4. The stable nuclei are represented by the black dots, which lie in a narrow range called the *line of stability*. Notice that the light stable nuclei contain an equal number of protons and neutrons; that is, $N = Z$. Also notice that in heavy stable nuclei, the number of neutrons exceeds the number of protons: above $Z = 20$, the line of stability deviates upward from the line representing $N = Z$. This deviation can be understood by recognizing that as the number of protons increases, the strength of the Coulomb force increases, which tends to break the nucleus apart. As a result, more neutrons are needed to keep the nucleus stable because neutrons experience only the attractive nuclear force. Eventually, the repulsive Coulomb forces between protons cannot be compensated by the addition of more neutrons. This point occurs at $Z = 83$, meaning that elements that contain more than 83 protons do not have stable nuclei.

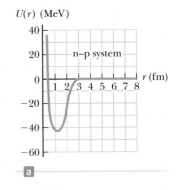

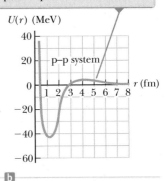

The difference in the two curves is due to the large Coulomb repulsion in the case of the proton–proton interaction.

Figure 44.3 (a) Potential energy versus separation distance for a neutron–proton system. (b) Potential energy versus separation distance for a proton–proton system. To display the difference in the curves on this scale, the height of the peak for the proton–proton curve has been exaggerated by a factor of 10.

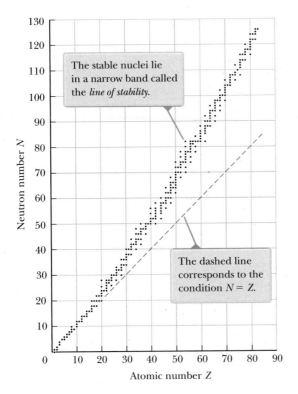

Figure 44.4 Neutron number N versus atomic number Z for stable nuclei (black dots).

44.2 Nuclear Binding Energy

As mentioned in the discussion of ^{12}C in Section 44.1, the total mass of a nucleus is less than the sum of the masses of its individual nucleons. Therefore, the rest energy of the bound system (the nucleus) is less than the combined rest energy of the separated nucleons. This difference in energy is called the **binding energy** of the nucleus and can be interpreted as the energy that must be added to a nucleus to break it apart into its components. Therefore, to separate a nucleus into protons and neutrons, energy must be delivered to the system.

Conservation of energy and the Einstein mass–energy equivalence relationship show that the binding energy E_b in MeV of any nucleus is

◀ Binding energy of a nucleus

$$E_b = [ZM(\text{H}) + Nm_n - M(^A_Z\text{X})] \times 931.494 \text{ MeV/u} \tag{44.2}$$

where $M(\text{H})$ is the atomic mass of the neutral hydrogen atom, m_n is the mass of the neutron, $M(^A_Z\text{X})$ represents the atomic mass of an atom of the isotope ^A_ZX, and the masses are all in atomic mass units. The mass of the Z electrons included in $M(\text{H})$ cancels with the mass of the Z electrons included in the term $M(^A_Z\text{X})$ within a small difference associated with the atomic binding energy of the electrons. Because atomic binding energies are typically several electron volts and nuclear binding energies are several million electron volts, this difference is negligible.

A plot of binding energy per nucleon E_b/A as a function of mass number A for various stable nuclei is shown in Figure 44.5. Notice that the binding energy in Figure 44.5 peaks in the vicinity of $A = 60$. That is, nuclei having mass numbers either greater or less than 60 are not as strongly bound as those near the middle of the periodic table. The decrease in binding energy per nucleon for $A > 60$ implies that energy is released when a heavy nucleus splits, or *fissions*, into two lighter nuclei. Energy is released in fission because the nucleons in each product nucleus are more tightly bound to one another than are the nucleons in the original nucleus. The important process of fission and a second important process of *fusion*, in which energy is released as light nuclei combine, shall be considered in detail in Chapter 45.

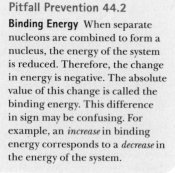

Pitfall Prevention 44.2

Binding Energy When separate nucleons are combined to form a nucleus, the energy of the system is reduced. Therefore, the change in energy is negative. The absolute value of this change is called the binding energy. This difference in sign may be confusing. For example, an *increase* in binding energy corresponds to a *decrease* in the energy of the system.

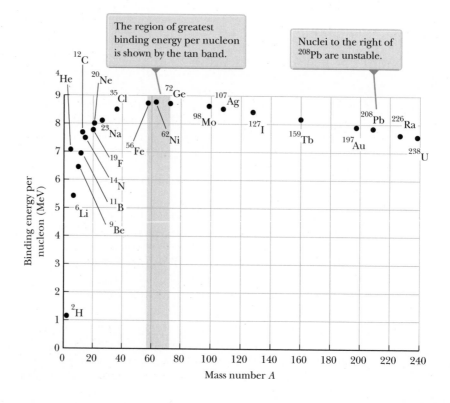

Figure 44.5 Binding energy per nucleon versus mass number for nuclides that lie along the line of stability in Figure 44.4. Some representative nuclides appear as black dots with labels.

Another important feature of Figure 44.5 is that the binding energy per nucleon is approximately constant at around 8 MeV per nucleon for all nuclei with $A > 50$. For these nuclei, the nuclear forces are said to be *saturated,* meaning that in the closely packed structure shown in Figure 44.2, a particular nucleon can form attractive bonds with only a limited number of other nucleons.

Figure 44.5 provides insight into fundamental questions about the origin of the chemical elements. In the early life of the Universe, the only elements that existed were hydrogen and helium. Clouds of cosmic gas coalesced under gravitational forces to form stars. As a star ages, it produces heavier elements from the lighter elements contained within it, beginning by fusing hydrogen atoms to form helium. This process continues as the star becomes older, generating atoms having larger and larger atomic numbers, up to the tan band shown in Figure 44.5.

The nucleus $^{63}_{28}\text{Ni}$ has the largest binding energy per nucleon of 8.794 5 MeV. It takes additional energy to create elements with mass numbers larger than 63 because of their lower binding energies per nucleon. This energy comes from the supernova explosion that occurs at the end of some large stars' lives. Therefore, all the heavy atoms in your body were produced from the explosions of ancient stars. You are literally made of stardust!

44.3 Nuclear Models

The details of the nuclear force are still an area of active research. Several nuclear models have been proposed that are useful in understanding general features of nuclear experimental data and the mechanisms responsible for binding energy. Two such models, the liquid-drop model and the shell model, are discussed below.

The Liquid-Drop Model

In 1936, Bohr proposed treating nucleons like molecules in a drop of liquid. In this **liquid-drop model,** the nucleons interact strongly with one another and undergo frequent collisions as they jiggle around within the nucleus. This jiggling motion is analogous to the thermally agitated motion of molecules in a drop of liquid.

Four major effects influence the binding energy of the nucleus in the liquid-drop model:

- **The volume effect.** Figure 44.5 shows that for $A > 50$, the binding energy per nucleon is approximately constant, which indicates that the nuclear force on a given nucleon is due only to a few nearest neighbors and not to all the other nucleons in the nucleus. On average, then, the binding energy associated with the nuclear force for each nucleon is the same in all nuclei: that associated with an interaction with a few neighbors. This property indicates that the total binding energy of the nucleus is proportional to A and therefore proportional to the nuclear volume. The contribution to the binding energy of the entire nucleus is $C_1 A$, where C_1 is an adjustable constant that can be determined by fitting the prediction of the model to experimental results.

- **The surface effect.** Because nucleons on the surface of the drop have fewer neighbors than those in the interior, surface nucleons reduce the binding energy by an amount proportional to their number. Because the number of surface nucleons is proportional to the surface area $4\pi r^2$ of the nucleus (modeled as a sphere) and because $r^2 \propto A^{2/3}$ (Eq. 44.1), the surface term can be expressed as $-C_2 A^{2/3}$, where C_2 is a second adjustable constant.

- **The Coulomb repulsion effect.** Each proton repels every other proton in the nucleus. The corresponding potential energy per pair of interacting protons is $k_e e^2/r$, where k_e is the Coulomb constant. The total electric potential energy is equivalent to the work required to assemble Z protons, initially infinitely far apart, into a sphere of volume V. This energy is proportional to the number of proton pairs $Z(Z-1)/2$ and inversely proportional to the nuclear radius. Consequently, the reduction in binding energy that results from the Coulomb effect is $-C_3 Z(Z-1)/A^{1/3}$, where C_3 is yet another adjustable constant.

- **The symmetry effect.** Another effect that lowers the binding energy is related to the symmetry of the nucleus in terms of values of N and Z. For small values of A, stable nuclei tend to have $N \approx Z$. Any large asymmetry between N and Z for light nuclei reduces the binding energy and makes the nucleus less stable. For larger A, the value of N for stable nuclei is naturally larger than Z. This effect can be described by a binding-energy term of the form $-C_4(N-Z)^2/A$, where C_4 is another adjustable constant.[1] For small A, any large asymmetry between values of N and Z makes this term relatively large and reduces the binding energy. For large A, this term is small and has little effect on the overall binding energy.

Adding these contributions gives the following expression for the total binding energy:

$$E_b = C_1 A - C_2 A^{2/3} - C_3 \frac{Z(Z-1)}{A^{1/3}} - C_4 \frac{(N-Z)^2}{A} \tag{44.3}$$

This equation, often referred to as the **semiempirical binding-energy formula,** contains four constants that are adjusted to fit the theoretical expression to experimental data. For nuclei having $A \geq 15$, the constants have the values

$$C_1 = 15.7 \text{ MeV} \qquad C_2 = 17.8 \text{ MeV}$$

$$C_3 = 0.71 \text{ MeV} \qquad C_4 = 23.6 \text{ MeV}$$

Equation 44.3, together with these constants, fits the known nuclear mass values very well as shown by the theoretical curve and sample experimental values in Figure 44.6. The liquid-drop model does not, however, account for some finer details of nuclear structure, such as stability rules and angular momentum. Equation 44.3 is a *theoretical* equation for the binding energy, based on the liquid-drop model, whereas binding energies calculated from Equation 44.2 are *experimental* values based on mass measurements.

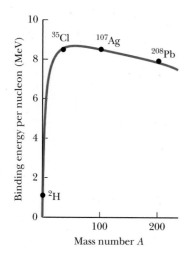

Figure 44.6 The binding-energy curve plotted by using the semiempirical binding-energy formula (red-brown). For comparison to the theoretical curve, experimental values for four sample nuclei are shown.

Example 44.3   Applying the Semiempirical Binding–Energy Formula

The nucleus ^{64}Zn has a tabulated binding energy of 559.09 MeV. Use the semiempirical binding-energy formula to generate a theoretical estimate of the binding energy for this nucleus.

SOLUTION

Conceptualize Imagine bringing the separate protons and neutrons together to form a ^{64}Zn nucleus. The rest energy of the nucleus is smaller than the rest energy of the individual particles. The difference in rest energy is the binding energy.

Categorize From the text of the problem, we know to apply the liquid-drop model. This example is a substitution problem.

For the ^{64}Zn nucleus, $Z = 30$, $N = 34$, and $A = 64$. Evaluate the four terms of the semiempirical binding-energy formula:

$C_1 A = (15.7 \text{ MeV})(64) = 1\,005 \text{ MeV}$

$C_2 A^{2/3} = (17.8 \text{ MeV})(64)^{2/3} = 285 \text{ MeV}$

$C_3 \dfrac{Z(Z-1)}{A^{1/3}} = (0.71 \text{ MeV}) \dfrac{(30)(29)}{(64)^{1/3}} = 154 \text{ MeV}$

$C_4 \dfrac{(N-Z)^2}{A} = (23.6 \text{ MeV}) \dfrac{(34-30)^2}{64} = 5.90 \text{ MeV}$

---

[1]The liquid-drop model *describes* that heavy nuclei have $N > Z$. The shell model, as we shall see shortly, *explains* why that is true with a physical argument.

▶ **44.3** continued

Substitute these values into Equation 44.3: $E_b = 1\,005\text{ MeV} - 285\text{ MeV} - 154\text{ MeV} - 5.90\text{ MeV} = \boxed{560\text{ MeV}}$

This value differs from the tabulated value by less than 0.2%. Notice how the sizes of the terms decrease from the first to the fourth term. The fourth term is particularly small for this nucleus, which does not have an excessive number of neutrons.

The Shell Model

The liquid-drop model describes the general behavior of nuclear binding energies relatively well. When the binding energies are studied more closely, however, we find the following features:

- Most stable nuclei have an even value of A. Furthermore, only eight stable nuclei have odd values for both Z and N.
- Figure 44.7 shows a graph of the difference between the binding energy per nucleon calculated by Equation 44.3 and the measured binding energy. There is evidence for regularly spaced peaks in the data that are not described by the semiempirical binding-energy formula. The peaks occur at values of N or Z that have become known as **magic numbers:**

$$Z \text{ or } N = 2, 8, 20, 28, 50, 82 \qquad (44.4)$$

◀ **Magic numbers**

- High-precision studies of nuclear radii show deviations from the simple expression in Equation 44.1. Graphs of experimental data show peaks in the curve of radius versus N at values of N equal to the magic numbers.
- A group of *isotones* is a collection of nuclei having the same value of N and varying values of Z. When the number of stable isotones is graphed as function of N, there are peaks in the graph, again at the magic numbers in Equation 44.4.
- Several other nuclear measurements show anomalous behavior at the magic numbers.[2]

These peaks in graphs of experimental data are reminiscent of the peaks in Figure 42.20 for the ionization energy of atoms, which arose because of the shell structure of the atom. The **shell model** of the nucleus, also called the **independent-particle model,** was developed independently by two German scientists: Maria Goeppert-Mayer in 1949 and Hans Jensen (1907–1973) in 1950. Goeppert-Mayer and Jensen

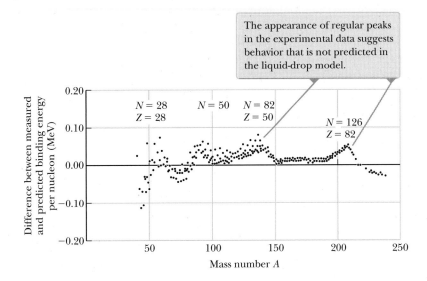

Figure 44.7 The difference between measured binding energies and those calculated from the liquid-drop model as a function of A. (Adapted from R. A. Dunlap, *The Physics of Nuclei and Particles,* Brooks/Cole, Belmont, CA, 2004.)

[2]For further details, see chapter 5 of R. A. Dunlap, *The Physics of Nuclei and Particles,* Brooks/Cole, Belmont, CA, 2004.

Maria Goeppert-Mayer
German Scientist (1906–1972)
Goeppert-Mayer was born and edu-
cated in Germany. She is best known
for her development of the shell model
(independent-particle model) of the
nucleus, published in 1950. A similar
model was simultaneously developed
by Hans Jensen, another German
scientist. Goeppert-Mayer and Jensen
were awarded the Nobel Prize in Physics in 1963 for their extraordinary
work in understanding the structure
of the nucleus.

The energy levels for the
protons are slightly higher
than those for the neutrons
because of the electric
potential energy associated
with the system of protons.

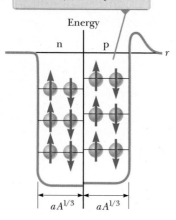

Figure 44.8 A square potential
well containing 12 nucleons. The
red spheres represent protons,
and the gray spheres represent
neutrons.

shared the 1963 Nobel Prize in Physics for their work. In this model, each nucleon
is assumed to exist in a shell, similar to an atomic shell for an electron. The nucleons exist in quantized energy states, and there are few collisions between nucleons.
Obviously, the assumptions of this model differ greatly from those made in the
liquid-drop model.

The quantized states occupied by the nucleons can be described by a set of quantum numbers. Because both the proton and the neutron have spin $\frac{1}{2}$, the exclusion
principle can be applied to describe the allowed states (as it was for electrons in
Chapter 42). That is, each state can contain only two protons (or two neutrons)
having *opposite* spins (Fig. 44.8). The proton states differ from those of the neutrons
because the two species move in different potential wells. The proton energy levels
are farther apart than the neutron levels because the protons experience a superposition of the Coulomb force and the nuclear force, whereas the neutrons experience only the nuclear force.

One factor influencing the observed characteristics of nuclear ground states is
nuclear spin–orbit effects. The atomic spin–orbit interaction between the spin of an
electron and its orbital motion in an atom gives rise to the sodium doublet discussed in Section 42.6 and is magnetic in origin. In contrast, the nuclear spin–
orbit effect for nucleons is due to the nuclear force. It is much stronger than in the
atomic case, and it has opposite sign. When these effects are taken into account,
the shell model is able to account for the observed magic numbers.

The shell model helps us understand why nuclei containing an even number of
protons and neutrons are more stable than other nuclei. (There are 160 stable even–
even isotopes.) Any particular state is filled when it contains two protons (or two
neutrons) having opposite spins. An extra proton or neutron can be added to the
nucleus only at the expense of increasing the energy of the nucleus. This increase
in energy leads to a nucleus that is less stable than the original nucleus. A careful
inspection of the stable nuclei shows that the majority have a special stability when
their nucleons combine in pairs, which results in a total angular momentum of zero.

The shell model also helps us understand why nuclei tend to have more neutrons
than protons. As in Figure 44.8, the proton energy levels are higher than those for
neutrons due to the extra energy associated with Coulomb repulsion. This effect
becomes more pronounced as Z increases. Consequently, as Z increases and higher
states are filled, a proton level for a given quantum number will be much higher
in energy than the neutron level for the same quantum number. In fact, it will be
even higher in energy than neutron levels for higher quantum numbers. Hence, it
is more energetically favorable for the nucleus to form with neutrons in the lower
energy levels rather than protons in the higher energy levels, so the number of neutrons is greater than the number of protons.

More sophisticated models of the nucleus have been and continue to be developed. For example, the *collective model* combines features of the liquid-drop and
shell models. The development of theoretical models of the nucleus continues to be
an active area of research.

44.4 Radioactivity

In 1896, Becquerel accidentally discovered that uranyl potassium sulfate crystals
emit an invisible radiation that can darken a photographic plate even though the
plate is covered to exclude light. After a series of experiments, he concluded that
the radiation emitted by the crystals was of a new type, one that requires no external stimulation and was so penetrating that it could darken protected photographic
plates and ionize gases. This process of spontaneous emission of radiation by uranium was soon to be called **radioactivity.**

Subsequent experiments by other scientists showed that other substances were
more powerfully radioactive. The most significant early investigations of this type
were conducted by Marie and Pierre Curie (1859–1906). After several years of care-

ful and laborious chemical separation processes on tons of pitchblende, a radioactive ore, the Curies reported the discovery of two previously unknown elements, both radioactive, named polonium and radium. Additional experiments, including Rutherford's famous work on alpha-particle scattering, suggested that radioactivity is the result of the *decay,* or disintegration, of unstable nuclei.

Three types of radioactive decay occur in radioactive substances: alpha (α) decay, in which the emitted particles are ^{4}He nuclei; beta (β) decay, in which the emitted particles are either electrons or positrons; and gamma (γ) decay, in which the emitted particles are high-energy photons. A **positron** is a particle like the electron in all respects except that the positron has a charge of $+e$. (The positron is the *antiparticle* of the electron; see Section 46.2.) The symbol e^- is used to designate an electron, and e^+ designates a positron.

We can distinguish among these three forms of radiation by using the scheme described in Figure 44.9. The radiation from radioactive samples that emit all three types of particles is directed into a region in which there is a magnetic field. Following the particle in a field (magnetic) analysis model, the radiation beam splits into three components, two bending in opposite directions and the third experiencing no change in direction. This simple observation shows that the radiation of the undeflected beam carries no charge (the gamma ray), the component deflected upward corresponds to positively charged particles (alpha particles), and the component deflected downward corresponds to negatively charged particles (e^-). If the beam includes a positron (e^+), it is deflected upward like the alpha particle, but it follows a different trajectory due to its smaller mass.

The three types of radiation have quite different penetrating powers. Alpha particles barely penetrate a sheet of paper, beta particles can penetrate a few millimeters of aluminum, and gamma rays can penetrate several centimeters of lead.

The decay process is probabilistic in nature and can be described with statistical calculations for a radioactive substance of macroscopic size containing a large number of radioactive nuclei. For such large numbers, the rate at which a particular decay process occurs in a sample is proportional to the number of radioactive nuclei present (that is, the number of nuclei that have not yet decayed). If N is the number of undecayed radioactive nuclei present at some instant, the rate of change of N with time is

$$\frac{dN}{dt} = -\lambda N \tag{44.5}$$

where λ, called the **decay constant,** is the probability of decay per nucleus per second. The negative sign indicates that dN/dt is negative; that is, N decreases in time.

Equation 44.5 can be written in the form

$$\frac{dN}{N} = -\lambda \, dt$$

Marie Curie
Polish Scientist (1867–1934)
In 1903, Marie Curie shared the Nobel Prize in Physics with her husband, Pierre, and with Becquerel for their studies of radioactive substances. In 1911, she was awarded a Nobel Prize in Chemistry for the discovery of radium and polonium.

Pitfall Prevention 44.3

Rays or Particles? Early in the history of nuclear physics, the term *radiation* was used to describe the emanations from radioactive nuclei. We now know that alpha radiation and beta radiation involve the emission of particles with nonzero rest energy. Even though they are not examples of electromagnetic radiation, the use of the term *radiation* for all three types of emission is deeply entrenched in our language and in the physics community.

Pitfall Prevention 44.4

Notation Warning In Section 44.1, we introduced the symbol N as an integer representing the number of neutrons in a nucleus. In this discussion, the symbol N represents the number of undecayed nuclei in a radioactive sample remaining after some time interval. As you read further, be sure to consider the context to determine the appropriate meaning for the symbol N.

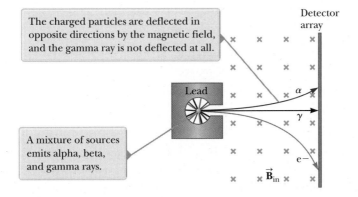

The charged particles are deflected in opposite directions by the magnetic field, and the gamma ray is not deflected at all.

A mixture of sources emits alpha, beta, and gamma rays.

Lead

Detector array

α

γ

e^-

$\vec{B}_{in}$

Figure 44.9 The radiation from radioactive sources can be separated into three components by using a magnetic field to deflect the charged particles. The detector array at the right records the events.

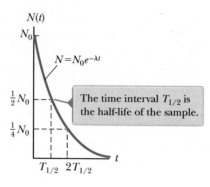

which, upon integration, gives

◀ **Exponential behavior of the number of undecayed nuclei**

$$N = N_0 e^{-\lambda t} \qquad (44.6)$$

where the constant N_0 represents the number of undecayed radioactive nuclei at $t = 0$. Equation 44.6 shows that the number of undecayed radioactive nuclei in a sample decreases exponentially with time. The plot of N versus t shown in Figure 44.10 illustrates the exponential nature of the decay. The curve is similar to that for the time variation of electric charge on a discharging capacitor in an RC circuit, as studied in Section 28.4.

The **decay rate** R, which is the number of decays per second, can be obtained by combining Equations 44.5 and 44.6:

◀ **Exponential behavior of the decay rate**

$$R = \left| \frac{dN}{dt} \right| = \lambda N = \lambda N_0 e^{-\lambda t} = R_0 e^{-\lambda t} \qquad (44.7)$$

where $R_0 = \lambda N_0$ is the decay rate at $t = 0$. The decay rate R of a sample is often referred to as its **activity**. Note that both N and R decrease exponentially with time.

Another parameter useful in characterizing nuclear decay is the **half-life** $T_{1/2}$:

> The **half-life** of a radioactive substance is the time interval during which half of a given number of radioactive nuclei decay.

Pitfall Prevention 44.5

Half-life It is *not* true that all the original nuclei have decayed after two half-lives! In one half-life, half of the original nuclei will decay. In the second half-life, half of those remaining will decay, leaving $\frac{1}{4}$ of the original number.

To find an expression for the half-life, we first set $N = N_0/2$ and $t = T_{1/2}$ in Equation 44.6 to give

$$\frac{N_0}{2} = N_0 e^{-\lambda T_{1/2}}$$

Canceling the N_0 factors and then taking the reciprocal of both sides, we obtain $e^{\lambda T_{1/2}} = 2$. Taking the natural logarithm of both sides gives

◀ **Half-life**

$$T_{1/2} = \frac{\ln 2}{\lambda} = \frac{0.693}{\lambda} \qquad (44.8)$$

After a time interval equal to one half-life, there are $N_0/2$ radioactive nuclei remaining (by definition); after two half-lives, half of these remaining nuclei have decayed and $N_0/4$ radioactive nuclei are left; after three half-lives, $N_0/8$ are left; and so on. In general, after n half-lives, the number of undecayed radioactive nuclei remaining is

$$N = N_0 \left(\tfrac{1}{2}\right)^n \qquad (44.9)$$

where n can be an integer or a noninteger.

A frequently used unit of activity is the **curie** (Ci), defined as

◀ **The curie**

$$1 \text{ Ci} \equiv 3.7 \times 10^{10} \text{ decays/s}$$

This value was originally selected because it is the approximate activity of 1 g of radium. The SI unit of activity is the **becquerel** (Bq):

◀ **The becquerel**

$$1 \text{ Bq} \equiv 1 \text{ decay/s}$$

Therefore, $1 \text{ Ci} = 3.7 \times 10^{10}$ Bq. The curie is a rather large unit, and the more frequently used activity units are the millicurie and the microcurie.

Quick Quiz 44.2 On your birthday, you measure the activity of a sample of ^{210}Bi, which has a half-life of 5.01 days. The activity you measure is 1.000 μCi. What is the activity of this sample on your next birthday? **(a)** 1.000 μCi **(b)** 0 **(c)** $\sim 0.2\ \mu$Ci **(d)** $\sim 0.01\ \mu$Ci **(e)** $\sim 10^{-22}\ \mu$Ci

---

Example 44.4 | How Many Nuclei Are Left?

The isotope carbon-14, $^{14}_{6}$C, is radioactive and has a half-life of 5 730 years. If you start with a sample of 1 000 carbon-14 nuclei, how many nuclei will still be undecayed in 25 000 years?

SOLUTION

Conceptualize The time interval of 25 000 years is much longer than the half-life, so only a small fraction of the originally undecayed nuclei will remain.

Categorize The text of the problem allows us to categorize this example as a substitution problem involving radioactive decay.

Analyze Divide the time interval by the half-life to determine the number of half-lives:

$$n = \frac{25\ 000\ \text{yr}}{5\ 730\ \text{yr}} = 4.363$$

Determine how many undecayed nuclei are left after this many half-lives using Equation 44.9:

$$N = N_0\left(\tfrac{1}{2}\right)^n = 1\ 000\left(\tfrac{1}{2}\right)^{4.363} = \boxed{49}$$

Finalize As we have mentioned, radioactive decay is a probabilistic process and accurate statistical predictions are possible only with a very large number of atoms. The original sample in this example contains only 1 000 nuclei, which is certainly not a very large number. Therefore, if you counted the number of undecayed nuclei remaining after 25 000 years, it might not be exactly 49.

Example 44.5 | The Activity of Carbon

At time $t = 0$, a radioactive sample contains 3.50 μg of pure $^{11}_{6}$C, which has a half-life of 20.4 min.

(A) Determine the number N_0 of nuclei in the sample at $t = 0$.

SOLUTION

Conceptualize The half-life is relatively short, so the number of undecayed nuclei drops rapidly. The molar mass of $^{11}_{6}$C is approximately 11.0 g/mol.

Categorize We evaluate results using equations developed in this section, so we categorize this example as a substitution problem.

Find the number of moles in 3.50 μg of pure $^{11}_{6}$C:

$$n = \frac{3.50 \times 10^{-6}\ \text{g}}{11.0\ \text{g/mol}} = 3.18 \times 10^{-7}\ \text{mol}$$

Find the number of undecayed nuclei in this amount of pure $^{11}_{6}$C:

$$N_0 = (3.18 \times 10^{-7}\ \text{mol})(6.02 \times 10^{23}\ \text{nuclei/mol}) = \boxed{1.92 \times 10^{17}\ \text{nuclei}}$$

(B) What is the activity of the sample initially and after 8.00 h?

SOLUTION

Find the initial activity of the sample using Equations 44.7 and 44.8:

$$R_0 = \lambda N_0 = \frac{0.693}{T_{1/2}} N_0 = \frac{0.693}{20.4\ \text{min}}\left(\frac{1\ \text{min}}{60\ \text{s}}\right)(1.92 \times 10^{17})$$

$$= (5.66 \times 10^{-4}\ \text{s}^{-1})(1.92 \times 10^{17}) = \boxed{1.09 \times 10^{14}\ \text{Bq}}$$

continued

▶ **44.5** continued

Use Equation 44.7 to find the activity at $t = 8.00$ h $= 2.88 \times 10^4$ s:

$$R = R_0 e^{-\lambda t} = (1.09 \times 10^{14} \text{ Bq}) e^{-(5.66 \times 10^{-4} \text{ s}^{-1})(2.88 \times 10^4 \text{ s})} = \boxed{8.96 \times 10^6 \text{ Bq}}$$

Example 44.6   A Radioactive Isotope of Iodine

A sample of the isotope ^{131}I, which has a half-life of 8.04 days, has an activity of 5.0 mCi at the time of shipment. Upon receipt of the sample at a medical laboratory, the activity is 2.1 mCi. How much time has elapsed between the two measurements?

SOLUTION

Conceptualize The sample is continuously decaying as it is in transit. The decrease in the activity is 58% during the time interval between shipment and receipt, so we expect the elapsed time to be greater than the half-life of 8.04 d.

Categorize The stated activity corresponds to many decays per second, so N is large and we can categorize this problem as one in which we can use our statistical analysis of radioactivity.

Analyze Solve Equation 44.7 for the ratio of the final activity to the initial activity:

$$\frac{R}{R_0} = e^{-\lambda t}$$

Take the natural logarithm of both sides:

$$\ln\left(\frac{R}{R_0}\right) = -\lambda t$$

Solve for the time t:

$$(1) \quad t = -\frac{1}{\lambda} \ln\left(\frac{R}{R_0}\right)$$

Use Equation 44.8 to substitute for λ:

$$t = -\frac{T_{1/2}}{\ln 2} \ln\left(\frac{R}{R_0}\right)$$

Substitute numerical values:

$$t = -\frac{8.04 \text{ d}}{0.693} \ln\left(\frac{2.1 \text{ mCi}}{5.0 \text{ mCi}}\right) = \boxed{10 \text{ d}}$$

Finalize This result is indeed greater than the half-life, as expected. This example demonstrates the difficulty in shipping radioactive samples with short half-lives. If the shipment is delayed by several days, only a small fraction of the sample might remain upon receipt. This difficulty can be addressed by shipping a combination of isotopes in which the desired isotope is the product of a decay occurring within the sample. It is possible for the desired isotope to be in *equilibrium*, in which case it is created at the same rate as it decays. Therefore, the amount of the desired isotope remains constant during the shipping process and subsequent storage. When needed, the desired isotope can be separated from the rest of the sample; its decay from the initial activity begins at this point rather than upon shipment.

44.5 The Decay Processes

As we stated in Section 44.4, a radioactive nucleus spontaneously decays by one of three processes: alpha decay, beta decay, or gamma decay. Figure 44.11 shows a close-up view of a portion of Figure 44.4 from $Z = 65$ to $Z = 80$. The black circles are the stable nuclei seen in Figure 44.4. In addition, unstable nuclei above and below the line of stability for each value of Z are shown. Above the line of stability, the blue circles show unstable nuclei that are neutron-rich and undergo a beta decay process in which an electron is emitted. Below the black circles are red circles corresponding to proton-rich unstable nuclei that primarily undergo a beta-decay process in which a positron is emitted or a competing process called electron capture. Beta decay and electron capture are described in more detail below. Further below the line of stability (with a few exceptions) are tan circles that represent very proton-rich nuclei for which the primary decay mechanism is alpha decay, which we discuss first.

Alpha Decay

A nucleus emitting an alpha particle ($^4_2$He) loses two protons and two neutrons. Therefore, the atomic number Z decreases by 2, the mass number A decreases by 4, and the neutron number decreases by 2. The decay can be written

$$^A_Z X \rightarrow {}^{A-4}_{Z-2}Y + {}^4_2He \tag{44.10}$$

where X is called the **parent nucleus** and Y the **daughter nucleus.** As a general rule in any decay expression such as this one, (1) the sum of the mass numbers A must be the same on both sides of the decay and (2) the sum of the atomic numbers Z must be the same on both sides of the decay. As examples, ^{238}U and ^{226}Ra are both alpha emitters and decay according to the schemes

$$^{238}_{92}U \rightarrow {}^{234}_{90}Th + {}^4_2He \tag{44.11}$$

$$^{226}_{88}Ra \rightarrow {}^{222}_{86}Rn + {}^4_2He \tag{44.12}$$

The decay of ^{226}Ra is shown in Figure 44.12.

When the nucleus of one element changes into the nucleus of another as happens in alpha decay, the process is called **spontaneous decay.** In any spontaneous decay, relativistic energy and momentum of the parent nucleus as an isolated system must be conserved. The final components of the system are the daughter nucleus and the alpha particle. If we call M_X the mass of the parent nucleus, M_Y the mass of the daughter nucleus, and M_α the mass of the alpha particle, we can define the **disintegration energy** Q of the system as

$$Q = (M_X - M_Y - M_\alpha)c^2 \tag{44.13}$$

The energy Q is in joules when the masses are in kilograms and c is the speed of light, 3.00×10^8 m/s. When the masses are expressed in atomic mass units u, however, Q can be calculated in MeV using the expression

$$Q = (M_X - M_Y - M_\alpha) \times 931.494 \text{ MeV/u} \tag{44.14}$$

Table 44.2 (page 1112) contains information on selected isotopes, including masses of neutral atoms that can be used in Equation 44.14 and similar equations.

The disintegration energy Q is the amount of rest energy transformed and appears in the form of kinetic energy in the daughter nucleus and the alpha particle and is sometimes referred to as the Q value of the nuclear decay. Consider the case of the ^{226}Ra decay described in Figure 44.12. If the parent nucleus is at rest before the decay, the total kinetic energy of the products is 4.87 MeV. (See Example 44.7.) Most of this kinetic energy is associated with the alpha particle because this particle is much less massive than the daughter nucleus ^{222}Rn. That is, because the system is also isolated in terms of momentum, the lighter alpha particle recoils with a much higher speed than does the daughter nucleus. Generally, less massive particles carry off most of the energy in nuclear decays.

Experimental observations of alpha-particle energies show a number of discrete energies rather than a single energy because the daughter nucleus may be left in an

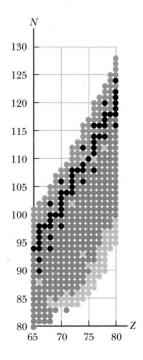

Figure 44.11 A close-up view of the line of stability in Figure 44.4 from $Z = 65$ to $Z = 80$. The black dots represent stable nuclei as in Figure 44.4. The other colored dots represent unstable isotopes above and below the line of stability, with the color of the dot indicating the primary means of decay.

- ● Beta (electron)
- ● Stable
- ● Beta (positron) or electron capture
- ● Alpha

Pitfall Prevention 44.6

Another Q We have seen the symbol Q before, but this use is a brand-new meaning for this symbol: the disintegration energy. In this context, it is not heat, charge, or quality factor for a resonance, for which we have used Q before.

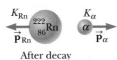

Figure 44.12 The alpha decay of radium-226. The radium nucleus is initially at rest. After the decay, the radon nucleus has kinetic energy K_{Rn} and momentum $\vec{\mathbf{p}}_{Rn}$ and the alpha particle has kinetic energy K_α and momentum $\vec{\mathbf{p}}_\alpha$.

Table 44.2 **Chemical and Nuclear Information for Selected Isotopes**

Atomic Number Z	Element	Chemical Symbol	Mass Number A (* means radioactive)	Mass of Neutral Atom (u)	Percent Abundance	Half-life, if Radioactive $T_{1/2}$
−1	electron	e^-	0	0.000 549		
0	neutron	n	1*	1.008 665		614 s
1	hydrogen	^{1}H = p	1	1.007 825	99.988 5	
	[deuterium	^{2}H = D]	2	2.014 102	0.011 5	
	[tritium	^{3}H = T]	3*	3.016 049		12.33 yr
2	helium	He	3	3.016 029	0.000 137	
	[alpha particle	$\alpha = {}^4$He]	4	4.002 603	99.999 863	
			6*	6.018 889		0.81 s
3	lithium	Li	6	6.015 123	7.5	
			7	7.016 005	92.5	
4	beryllium	Be	7*	7.016 930		53.3 d
			8*	8.005 305		10^{-17} s
			9	9.012 182	100	
5	boron	B	10	10.012 937	19.9	
			11	11.009 305	80.1	
6	carbon	C	11*	11.011 434		20.4 min
			12	12.000 000	98.93	
			13	13.003 355	1.07	
			14*	14.003 242		5 730 yr
7	nitrogen	N	13*	13.005 739		9.96 min
			14	14.003 074	99.632	
			15	15.000 109	0.368	
8	oxygen	O	14*	14.008 596		70.6 s
			15*	15.003 066		122 s
			16	15.994 915	99.757	
			17	16.999 132	0.038	
			18	17.999 161	0.205	
9	fluorine	F	18*	18.000 938		109.8 min
			19	18.998 403	100	
10	neon	Ne	20	19.992 440	90.48	
11	sodium	Na	23	22.989 769	100	
12	magnesium	Mg	23*	22.994 124		11.3 s
			24	23.985 042	78.99	
13	aluminum	Al	27	26.981 539	100	
14	silicon	Si	27*	26.986 705		4.2 s
15	phosphorus	P	30*	29.978 314		2.50 min
			31	30.973 762	100	
			32*	31.973 907		14.26 d
16	sulfur	S	32	31.972 071	94.93	
19	potassium	K	39	38.963 707	93.258 1	
			40*	39.963 998	0.011 7	1.28×10^9 yr
20	calcium	Ca	40	39.962 591	96.941	
			42	41.958 618	0.647	
			43	42.958 767	0.135	
25	manganese	Mn	55	54.938 045	100	
26	iron	Fe	56	55.934 938	91.754	
			57	56.935 394	2.119	

continued

Table 44.2 **Chemical and Nuclear Information for Selected Isotopes** (*continued*)

Atomic Number Z	Element	Chemical Symbol	Mass Number A (* means radioactive)	Mass of Neutral Atom (u)	Percent Abundance	Half-life, if Radioactive $T_{1/2}$
27	cobalt	Co	57*	56.936 291		272 d
			59	58.933 195	100	
			60*	59.933 817		5.27 yr
28	nickel	Ni	58	57.935 343	68.076 9	
			60	59.930 786	26.223 1	
29	copper	Cu	63	62.929 598	69.17	
			64*	63.929 764		12.7 h
			65	64.927 789	30.83	
30	zinc	Zn	64	63.929 142	48.63	
37	rubidium	Rb	87*	86.909 181	27.83	
38	strontium	Sr	87	86.908 877	7.00	
			88	87.905 612	82.58	
			90*	89.907 738		29.1 yr
41	niobium	Nb	93	92.906 378	100	
42	molybdenum	Mo	94	93.905 088	9.25	
44	ruthenium	Ru	98	97.905 287	1.87	
54	xenon	Xe	136*	135.907 219		2.4×10^{21} yr
55	cesium	Cs	137*	136.907 090		30 yr
56	barium	Ba	137	136.905 827	11.232	
58	cerium	Ce	140	139.905 439	88.450	
59	praseodymium	Pr	141	140.907 653	100	
60	neodymium	Nd	144*	143.910 087	23.8	2.3×10^{15} yr
61	promethium	Pm	145*	144.912 749		17.7 yr
79	gold	Au	197	196.966 569	100	
80	mercury	Hg	198	197.966 769	9.97	
			202	201.970 643	29.86	
82	lead	Pb	206	205.974 465	24.1	
			207	206.975 897	22.1	
			208	207.976 652	52.4	
			214*	213.999 805		26.8 min
83	bismuth	Bi	209	208.980 399	100	
84	polonium	Po	210*	209.982 874		138.38 d
			216*	216.001 915		0.145 s
			218*	218.008 973		3.10 min
86	radon	Rn	220*	220.011 394		55.6 s
			222*	222.017 578		3.823 d
88	radium	Ra	226*	226.025 410		1 600 yr
90	thorium	Th	232*	232.038 055	100	1.40×10^{10} yr
			234*	234.043 601		24.1 d
92	uranium	U	234*	234.040 952		2.45×10^5 yr
			235*	235.043 930	0.720 0	7.04×10^8 yr
			236*	236.045 568		2.34×10^7 yr
			238*	238.050 788	99.274 5	4.47×10^9 yr
93	neptunium	Np	236*	236.046 570		1.15×10^5 yr
			237*	237.048 173		2.14×10^6 yr
94	plutonium	Pu	239*	239.052 163		24 120 yr

Source: G. Audi, A. H. Wapstra, and C. Thibault, "The AME2003 Atomic Mass Evaluation," *Nuclear Physics A* **729**:337–676, 2003.

excited quantum state after the decay. As a result, not all the disintegration energy is available as kinetic energy of the alpha particle and daughter nucleus. The emission of an alpha particle is followed by one or more gamma-ray photons (discussed shortly) as the excited nucleus decays to the ground state. The observed discrete alpha-particle energies represent evidence of the quantized nature of the nucleus and allow a determination of the energies of the quantum states.

If one assumes ^{238}U (or any other alpha emitter) decays by emitting either a proton or a neutron, the mass of the decay products would exceed that of the parent nucleus, corresponding to a negative Q value. A negative Q value indicates that such a proposed decay does not occur spontaneously.

Quick Quiz 44.3 Which of the following is the correct daughter nucleus associated with the alpha decay of $^{157}_{72}$Hf? **(a)** $^{153}_{72}$Hf **(b)** $^{153}_{70}$Yb **(c)** $^{157}_{70}$Yb

---

Example 44.7 **The Energy Liberated When Radium Decays**   AM

The ^{226}Ra nucleus undergoes alpha decay according to Equation 44.12.

(A) Calculate the Q value for this process. From Table 44.2, the masses are 226.025 410 u for ^{226}Ra, 222.017 578 u for ^{222}Rn, and 4.002 603 u for $^4_2$He.

SOLUTION

Conceptualize Study Figure 44.12 to understand the process of alpha decay in this nucleus.

Categorize The parent nucleus is an *isolated system* that decays into an alpha particle and a daughter nucleus. The system is isolated in terms of both *energy* and *momentum*.

Analyze Evaluate Q using Equation 44.14:
$$Q = (M_X - M_Y - M_\alpha) \times 931.494 \text{ MeV/u}$$
$$= (226.025\ 410\ \text{u} - 222.017\ 578\ \text{u} - 4.002\ 603\ \text{u}) \times 931.494 \text{ MeV/u}$$
$$= (0.005\ 229\ \text{u}) \times 931.494 \text{ MeV/u} = \boxed{4.87 \text{ MeV}}$$

(B) What is the kinetic energy of the alpha particle after the decay?

Analyze The value of 4.87 MeV is the disintegration energy for the decay. It includes the kinetic energy of both the alpha particle and the daughter nucleus after the decay. Therefore, the kinetic energy of the alpha particle would be *less* than 4.87 MeV.

Set up a conservation of momentum equation, noting that the initial momentum of the system is zero:

$$(1)\quad 0 = M_Y v_Y - M_\alpha v_\alpha$$

Set the disintegration energy equal to the sum of the kinetic energies of the alpha particle and the daughter nucleus (assuming the daughter nucleus is left in the ground state):

$$(2)\quad Q = \tfrac{1}{2} M_\alpha v_\alpha^2 + \tfrac{1}{2} M_Y v_Y^2$$

Solve Equation (1) for v_Y and substitute into Equation (2):

$$Q = \tfrac{1}{2} M_\alpha v_\alpha^2 + \tfrac{1}{2} M_Y \left(\frac{M_\alpha v_\alpha}{M_Y}\right)^2 = \tfrac{1}{2} M_\alpha v_\alpha^2 \left(1 + \frac{M_\alpha}{M_Y}\right)$$
$$Q = K_\alpha \left(\frac{M_Y + M_\alpha}{M_Y}\right)$$

Solve for the kinetic energy of the alpha particle:

$$K_\alpha = Q\left(\frac{M_Y}{M_Y + M_\alpha}\right)$$

Evaluate this kinetic energy for the specific decay of ^{226}Ra that we are exploring in this example:

$$K_\alpha = (4.87 \text{ MeV})\left(\frac{222}{222 + 4}\right) = \boxed{4.78 \text{ MeV}}$$

Finalize The kinetic energy of the alpha particle is indeed less than the disintegration energy, but notice that the alpha particle carries away *most* of the energy available in the decay.

To understand the mechanism of alpha decay, let's model the parent nucleus as a system consisting of (1) the alpha particle, already formed as an entity within the nucleus, and (2) the daughter nucleus that will result when the alpha particle is emitted. Figure 44.13 shows a plot of potential energy versus separation distance r between the alpha particle and the daughter nucleus, where the distance marked R is the range of the nuclear force. The curve represents the combined effects of (1) the repulsive Coulomb force, which gives the positive part of the curve for $r > R$, and (2) the attractive nuclear force, which causes the curve to be negative for $r < R$. As shown in Example 44.7, a typical disintegration energy Q is approximately 5 MeV, which is the approximate kinetic energy of the alpha particle, represented by the lower dashed line in Figure 44.13.

According to classical physics, the alpha particle is trapped in a potential well. How, then, does it ever escape from the nucleus? The answer to this question was first provided by George Gamow (1904–1968) in 1928 and independently by R. W. Gurney (1898–1953) and E. U. Condon (1902–1974) in 1929, using quantum mechanics. In the view of quantum mechanics, there is always some probability that a particle can tunnel through a barrier (Section 41.5). That is exactly how we can describe alpha decay: the alpha particle tunnels through the barrier in Figure 44.13, escaping the nucleus. Furthermore, this model agrees with the observation that higher-energy alpha particles come from nuclei with shorter half-lives. For higher-energy alpha particles in Figure 44.13, the barrier is narrower and the probability is higher that tunneling occurs. The higher probability translates to a shorter half-life.

As an example, consider the decays of ^{238}U and ^{226}Ra in Equations 44.11 and 44.12, along with the corresponding half-lives and alpha-particle energies:

$$^{238}\text{U:} \quad T_{1/2} = 4.47 \times 10^9 \text{ yr} \qquad K_\alpha = 4.20 \text{ MeV}$$

$$^{226}\text{Ra:} \quad T_{1/2} = 1.60 \times 10^3 \text{ yr} \qquad K_\alpha = 4.78 \text{ MeV}$$

Notice that a relatively small difference in alpha-particle energy is associated with a tremendous difference of six orders of magnitude in the half-life. The origin of this effect can be understood as follows. Figure 44.13 shows that the curve below an alpha-particle energy of 5 MeV has a slope with a relatively small magnitude. Therefore, a small difference in energy on the vertical axis has a relatively large effect on the width of the potential barrier. Second, recall Equation 41.22, which describes the exponential dependence of the probability of transmission on the barrier width. These two factors combine to give the very sensitive relationship between half-life and alpha-particle energy that the data above suggest.

A life-saving application of alpha decay is the household smoke detector, shown in Figure 44.14. The detector consists of an ionization chamber, a sensitive current detector, and an alarm. A weak radioactive source (usually $^{241}_{95}$Am) ionizes the air in the chamber of the detector, creating charged particles. A voltage is maintained between the plates inside the chamber, setting up a small but detectable current in the external circuit due to the ions acting as charge carriers between the plates. As long as the current is maintained, the alarm is deactivated. If smoke drifts into the chamber, however, the ions become attached to the smoke particles. These heavier particles do not drift as readily as do the lighter ions, which causes a decrease in the detector current. The external circuit senses this decrease in current and sets off the alarm.

Beta Decay

When a radioactive nucleus undergoes beta decay, the daughter nucleus contains the same number of nucleons as the parent nucleus but the atomic number is changed by 1, which means that the number of protons changes:

$$^{A}_{Z}\text{X} \rightarrow \ ^{A}_{Z+1}\text{Y} + e^{-} \quad \text{(incomplete expression)} \tag{44.15}$$

$$^{A}_{Z}\text{X} \rightarrow \ ^{A}_{Z-1}\text{Y} + e^{+} \quad \text{(incomplete expression)} \tag{44.16}$$

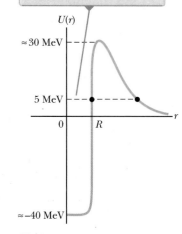

Classically, the 5-MeV energy of the alpha particle is not sufficiently large to overcome the energy barrier, so the particle should not be able to escape from the nucleus.

Figure 44.13 Potential energy versus separation distance for a system consisting of an alpha particle and a daughter nucleus. The alpha particle escapes by tunneling through the barrier.

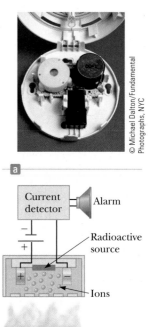

© Michael Dalton/Fundamental Photographs, NYC

Figure 44.14 (a) A smoke detector uses alpha decay to determine whether smoke is in the air. The alpha source is in the black cylinder at the right. (b) Smoke entering the chamber reduces the detected current, causing the alarm to sound.

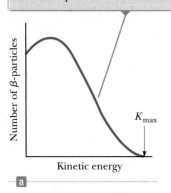

The observed energies of beta particles are continuous, having all values up to a maximum value.

K_{max}

Number of β-particles

Kinetic energy

a

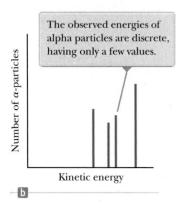

The observed energies of alpha particles are discrete, having only a few values.

Number of α-particles

Kinetic energy

b

Figure 44.15 (a) Distribution of beta-particle energies in a typical beta decay. (b) Distribution of alpha-particle energies in a typical alpha decay.

where, as mentioned in Section 44.4, e⁻ designates an electron and e⁺ designates a positron, with *beta particle* being the general term referring to either. *Beta decay is not described completely by these expressions.* We shall give reasons for this statement shortly.

As with alpha decay, the nucleon number and total charge are both conserved in beta decays. Because A does not change but Z does, we conclude that in beta decay, either a neutron changes to a proton (Eq. 44.15) or a proton changes to a neutron (Eq. 44.16). Note that the electron or positron emitted in these decays is not present beforehand in the nucleus; it is created in the process of the decay from the rest energy of the decaying nucleus. Two typical beta-decay processes are

$$^{14}_{6}\text{C} \rightarrow {}^{14}_{7}\text{N} + \text{e}^- \quad \text{(incomplete expression)} \qquad \textbf{(44.17)}$$

$$^{12}_{7}\text{N} \rightarrow {}^{12}_{6}\text{C} + \text{e}^+ \quad \text{(incomplete expression)} \qquad \textbf{(44.18)}$$

Let's consider the energy of the system undergoing beta decay before and after the decay. As with alpha decay, energy of the isolated system must be conserved. Experimentally, it is found that beta particles from a single type of nucleus are emitted over a continuous range of energies (Fig. 44.15a), as opposed to alpha decay, in which the alpha particles are emitted with discrete energies (Fig. 44.15b). The kinetic energy of the system after the decay is equal to the decrease in rest energy of the system, that is, the Q value. Because all decaying nuclei in the sample have the same initial mass, however, *the Q value must be the same for each decay.* So, why do the emitted particles have the range of kinetic energies shown in Figure 44.15a? The isolated system model and the law of conservation of energy seem to be violated! It becomes worse: further analysis of the decay processes described by Equations 44.15 and 44.16 shows that the laws of conservation of angular momentum (spin) and linear momentum are also violated!

After a great deal of experimental and theoretical study, Pauli in 1930 proposed that a third particle must be present in the decay products to carry away the "missing" energy and momentum. Fermi later named this particle the **neutrino** (little neutral one) because it had to be electrically neutral and have little or no mass. Although it eluded detection for many years, the neutrino (symbol ν, Greek nu) was finally detected experimentally in 1956 by Frederick Reines (1918–1998), who received the Nobel Prize in Physics for this work in 1995. The neutrino has the following properties:

▶ **Properties of the neutrino**

- It has zero electric charge.
- Its mass is either zero (in which case it travels at the speed of light) or very small; much recent persuasive experimental evidence suggests that the neutrino mass is not zero. Current experiments place the upper bound of the mass of the neutrino at approximately 7 eV/c^2.
- It has a spin of $\frac{1}{2}$, which allows the law of conservation of angular momentum to be satisfied in beta decay.
- It interacts very weakly with matter and is therefore very difficult to detect.

We can now write the beta-decay processes (Eqs. 44.15 and 44.16) in their correct and complete form:

▶ **Beta decay processes**

$$^{A}_{Z}\text{X} \rightarrow {}^{A}_{Z+1}\text{Y} + \text{e}^- + \bar{\nu} \quad \text{(complete expression)} \qquad \textbf{(44.19)}$$

$$^{A}_{Z}\text{X} \rightarrow {}^{A}_{Z-1}\text{Y} + \text{e}^+ + \nu \quad \text{(complete expression)} \qquad \textbf{(44.20)}$$

as well as those for carbon-14 and nitrogen-12 (Eqs. 44.17 and 44.18):

$$^{14}_{6}\text{C} \rightarrow {}^{14}_{7}\text{N} + \text{e}^- + \bar{\nu} \quad \text{(complete expression)} \qquad \textbf{(44.21)}$$

$$^{12}_{7}\text{N} \rightarrow {}^{12}_{6}\text{C} + \text{e}^+ + \nu \quad \text{(complete expression)} \qquad \textbf{(44.22)}$$

where the symbol $\bar{\nu}$ represents the **antineutrino,** the antiparticle to the neutrino. We shall discuss antiparticles further in Chapter 46. For now, it suffices to say that a neutrino is emitted in positron decay and an antineutrino is emitted in electron

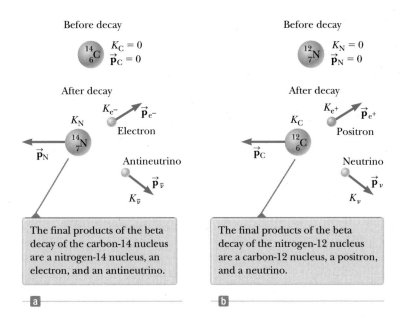

Figure 44.16 (a) The beta decay of carbon-14. (b) The beta decay of nitrogen-12.

Before decay

$K_C = 0$
$\vec{\mathbf{p}}_C = 0$

After decay

K_{e^-} $\vec{\mathbf{p}}_{e^-}$
K_N Electron
$\vec{\mathbf{p}}_N$

Antineutrino
$\vec{\mathbf{p}}_{\bar{\nu}}$
$K_{\bar{\nu}}$

The final products of the beta decay of the carbon-14 nucleus are a nitrogen-14 nucleus, an electron, and an antineutrino.

a

Before decay

$K_N = 0$
$\vec{\mathbf{p}}_N = 0$

After decay

K_{e^+} $\vec{\mathbf{p}}_{e^+}$
K_C Positron
$\vec{\mathbf{p}}_C$

Neutrino
$\vec{\mathbf{p}}_\nu$
K_ν

The final products of the beta decay of the nitrogen-12 nucleus are a carbon-12 nucleus, a positron, and a neutrino.

b

decay. As with alpha decay, the decays listed above are analyzed by applying conservation laws, but relativistic expressions must be used for beta particles because their kinetic energy is large (typically 1 MeV) compared with their rest energy of 0.511 MeV. Figure 44.16 shows a pictorial representation of the decays described by Equations 44.21 and 44.22.

In Equation 44.19, the number of protons has increased by one and the number of neutrons has decreased by one. We can write the fundamental process of e^- decay in terms of a neutron changing into a proton as follows:

$$\text{n} \rightarrow p + e^- + \bar{\nu} \qquad \textbf{(44.23)}$$

The electron and the antineutrino are ejected from the nucleus, with the net result that there is one more proton and one fewer neutron, consistent with the changes in Z and $A - Z$. A similar process occurs in e^+ decay, with a proton changing into a neutron, a positron, and a neutrino. This latter process can only occur within the nucleus, with the result that the nuclear mass decreases. It cannot occur for an isolated proton because its mass is less than that of the neutron.

A process that competes with e^+ decay is **electron capture,** which occurs when a parent nucleus captures one of its own orbital electrons and emits a neutrino. The final product after decay is a nucleus whose charge is $Z - 1$:

$$_Z^A\text{X} + _{-1}^0\text{e} \rightarrow _{Z-1}^A\text{Y} + \nu \qquad \textbf{(44.24)}$$

◀ **Electron capture**

In most cases, it is a K-shell electron that is captured and the process is therefore referred to as **K capture.** One example is the capture of an electron by $_4^7$Be:

$$_4^7\text{Be} + _{-1}^0\text{e} \rightarrow _3^7\text{Li} + \nu$$

Because the neutrino is very difficult to detect, electron capture is usually observed by the x-rays given off as higher-shell electrons cascade downward to fill the vacancy created in the K shell.

Finally, we specify Q values for the beta-decay processes. The Q values for e^- decay and electron capture are given by $Q = (M_X - M_Y)c^2$, where M_X and M_Y are the masses of neutral atoms. In e^- decay, the parent nucleus experiences an increase in atomic number and, for the atom to become neutral, an electron must be absorbed by the atom. If the neutral parent atom and an electron (which will eventually combine with the daughter to form a neutral atom) is the initial system and the final system is the neutral daughter atom and the beta-ejected electron, the system contains a free electron both before and after the decay. Therefore, in subtracting the initial and final masses of the system, this electron mass cancels.

Pitfall Prevention 44.7

Mass Number of the Electron An alternative notation for an electron, as we see in Equation 44.24, is the symbol $_{-1}^0$e, which does not imply that the electron has zero rest energy. The mass of the electron is so much smaller than that of the lightest nucleon, however, that we approximate it as zero in the context of nuclear decays and reactions.

The Q values for e^+ decay are given by $Q = (M_X - M_Y - 2m_e)c^2$. The extra term $-2m_e c^2$ in this expression is necessary because the atomic number of the parent decreases by one when the daughter is formed. After it is formed by the decay, the daughter atom sheds one electron to form a neutral atom. Therefore, the final products are the daughter atom, the shed electron, and the ejected positron.

These relationships are useful in determining whether or not a process is energetically possible. For example, the Q value for proposed e^+ decay for a particular parent nucleus may turn out to be negative. In that case, this decay does not occur. The Q value for electron capture for this parent nucleus, however, may be a positive number, so electron capture can occur even though e^+ decay is not possible. Such is the case for the decay of ^7_4Be shown above.

Quick Quiz 44.4 Which of the following is the correct daughter nucleus associated with the beta decay of $^{184}_{72}\text{Hf}$? **(a)** $^{183}_{72}\text{Hf}$ **(b)** $^{183}_{73}\text{Ta}$ **(c)** $^{184}_{73}\text{Ta}$

Carbon Dating

The beta decay of ^{14}C (Eq. 44.21) is commonly used to date organic samples. Cosmic rays in the upper atmosphere cause nuclear reactions (Section 44.7) that create ^{14}C. The ratio of ^{14}C to ^{12}C in the carbon dioxide molecules of our atmosphere has a constant value of approximately $r_0 = 1.3 \times 10^{-12}$. The carbon atoms in all living organisms have this same $^{14}\text{C}/^{12}\text{C}$ ratio r_0 because the organisms continuously exchange carbon dioxide with their surroundings. When an organism dies, however, it no longer absorbs ^{14}C from the atmosphere, and so the $^{14}\text{C}/^{12}\text{C}$ ratio decreases as the ^{14}C decays with a half-life of 5 730 yr. It is therefore possible to measure the age of a material by measuring its ^{14}C activity. Using this technique, scientists have been able to identify samples of wood, charcoal, bone, and shell as having lived from 1 000 to 25 000 years ago. This knowledge has helped us reconstruct the history of living organisms—including humans—during this time span.

A particularly interesting example is the dating of the Dead Sea Scrolls. This group of manuscripts was discovered by a shepherd in 1947. Translation showed them to be religious documents, including most of the books of the Old Testament. Because of their historical and religious significance, scholars wanted to know their age. Carbon dating applied to the material in which they were wrapped established their age at approximately 1 950 yr.

Conceptual Example 44.8 **The Age of Iceman**

In 1991, German tourists discovered the well-preserved remains of a man, now called "Ötzi the Iceman," trapped in a glacier in the Italian Alps. (See the photograph at the opening of this chapter.) Radioactive dating with ^{14}C revealed that this person was alive approximately 5 300 years ago. Why did scientists date a sample of Ötzi using ^{14}C rather than ^{11}C, which is a beta emitter having a half-life of 20.4 min?

SOLUTION

Because ^{14}C has a half-life of 5 730 yr, the fraction of ^{14}C nuclei remaining after thousands of years is high enough to allow accurate measurements of changes in the sample's activity. Because ^{11}C has a very short half-life, it is not useful; its activity decreases to a vanishingly small value over the age of the sample, making it impossible to detect.

An isotope used to date a sample must be present in a known amount in the sample when it is formed. As a general rule, the isotope chosen to date a sample should also have a half-life that is on the same order of magnitude as the age of the sample. If the half-life is much less than the age of the sample, there won't be enough activity left to measure because almost all the original radioactive nuclei will have decayed. If the half-life is much greater than the age of the sample, the amount of decay that has taken place since the sample died will be too small to measure. For example, if you have a specimen estimated

▶ **44.8** continued

to have died 50 years ago, neither ^{14}C (5 730 yr) nor ^{11}C (20 min) is suitable. If you know your sample contains hydrogen, however, you can measure the activity of ^{3}H (tritium), a beta emitter that has a half-life of 12.3 yr.

Example 44.9   **Radioactive Dating**

A piece of charcoal containing 25.0 g of carbon is found in some ruins of an ancient city. The sample shows a ^{14}C activity R of 250 decays/min. How long has the tree from which this charcoal came been dead?

SOLUTION

Conceptualize Because the charcoal was found in ancient ruins, we expect the current activity to be smaller than the initial activity. If we can determine the initial activity, we can find out how long the wood has been dead.

Categorize The text of the question helps us categorize this example as a carbon dating problem.

Analyze Solve Equation 44.7 for t:

$$(1) \quad t = -\frac{1}{\lambda} \ln\left(\frac{R}{R_0}\right)$$

Evaluate the ratio R/R_0 using Equation 44.7, the initial value of the ^{14}C/^{12}C ratio r_0, the number of moles n of carbon, and Avogadro's number N_A:

$$\frac{R}{R_0} = \frac{R}{\lambda N_0(^{14}\text{C})} = \frac{R}{\lambda r_0 N_0(^{12}\text{C})} = \frac{R}{\lambda r_0 n N_A}$$

Replace the number of moles in terms of the molar mass M of carbon and the mass m of the sample and substitute for the decay constant λ:

$$\frac{R}{R_0} = \frac{R}{(\ln 2/T_{1/2})r_0(m/M)N_A} = \frac{RMT_{1/2}}{r_0 m N_A \ln 2}$$

Substitute numerical values:

$$\frac{R}{R_0} = \frac{(250 \text{ min}^{-1})(12.0 \text{ g/mol})(5\,730 \text{ yr})}{(1.3 \times 10^{-12})(25.0 \text{ g})(6.022 \times 10^{23} \text{ mol}^{-1}) \ln 2}\left(\frac{3.156 \times 10^7 \text{ s}}{1 \text{ yr}}\right)\left(\frac{1 \text{ min}}{60 \text{ s}}\right)$$

$$= 0.667$$

Substitute this ratio into Equation (1) and substitute for the decay constant λ:

$$t = -\frac{1}{\lambda} \ln\left(\frac{R}{R_0}\right) = -\frac{T_{1/2}}{\ln 2} \ln\left(\frac{R}{R_0}\right)$$

$$= -\frac{5\,730 \text{ yr}}{\ln 2} \ln(0.667) = \boxed{3.4 \times 10^3 \text{ yr}}$$

Finalize Note that the time interval found here is on the same order of magnitude as the half-life, so ^{14}C is a valid isotope to use for this sample, as discussed in Conceptual Example 44.8.

Gamma Decay

Very often, a nucleus that undergoes radioactive decay is left in an excited energy state. The nucleus can then undergo a second decay to a lower-energy state, perhaps to the ground state, by emitting a high-energy photon:

$$^A_Z\text{X}^* \rightarrow {}^A_Z\text{X} + \gamma \qquad\qquad (44.25) \quad \blacktriangleleft \text{ Gamma decay}$$

where X* indicates a nucleus in an excited state. The typical half-life of an excited nuclear state is 10^{-10} s. Photons emitted in such a de-excitation process are called gamma rays. Such photons have very high energy (1 MeV to 1 GeV) relative to the energy of visible light (approximately 1 eV). Recall from Section 42.3 that the energy of a photon emitted or absorbed by an atom equals the difference in energy

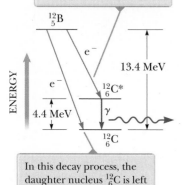

In this decay process, the daughter nucleus is in an excited state, denoted by $^{12}_{6}C^*$, and the beta decay is followed by a gamma decay.

In this decay process, the daughter nucleus $^{12}_{6}C$ is left in the ground state.

Figure 44.17 An energy-level diagram showing the initial nuclear state of a ^{12}B nucleus and two possible lower-energy states of the ^{12}C nucleus.

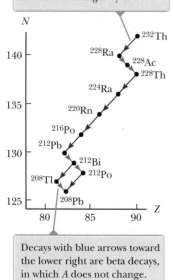

Decays with violet arrows toward the lower left are alpha decays, in which A changes by 4.

Decays with blue arrows toward the lower right are beta decays, in which A does not change.

Figure 44.18 Successive decays for the ^{232}Th series.

between the two electronic states involved in the transition. Similarly, a gamma-ray photon has an energy hf that equals the energy difference ΔE between two nuclear energy levels. When a nucleus decays by emitting a gamma ray, the only change in the nucleus is that it ends up in a lower-energy state. There are no changes in Z, N, or A.

A nucleus may reach an excited state as the result of a violent collision with another particle. More common, however, is for a nucleus to be in an excited state after it has undergone alpha or beta decay. The following sequence of events represents a typical situation in which gamma decay occurs:

$$^{12}_{5}B \rightarrow {}^{12}_{6}C^* + e^- + \bar{\nu} \qquad (44.26)$$

$$^{12}_{6}C^* \rightarrow {}^{12}_{6}C + \gamma \qquad (44.27)$$

Figure 44.17 shows the decay scheme for ^{12}B, which undergoes beta decay to either of two levels of ^{12}C. It can either (1) decay directly to the ground state of ^{12}C by emitting a 13.4-MeV electron or (2) undergo beta decay to an excited state of $^{12}C^*$ followed by gamma decay to the ground state. The latter process results in the emission of a 9.0-MeV electron and a 4.4-MeV photon.

The various pathways by which a radioactive nucleus can undergo decay are summarized in Table 44.3.

44.6 Natural Radioactivity

Radioactive nuclei are generally classified into two groups: (1) unstable nuclei found in nature, which give rise to **natural radioactivity,** and (2) unstable nuclei produced in the laboratory through nuclear reactions, which exhibit **artificial radioactivity.**

As Table 44.4 shows, there are three series of naturally occurring radioactive nuclei. Each series starts with a specific long-lived radioactive isotope whose half-life exceeds that of any of its unstable descendants. The three natural series begin with the isotopes ^{238}U, ^{235}U, and ^{232}Th, and the corresponding stable end products are three isotopes of lead: ^{206}Pb, ^{207}Pb, and ^{208}Pb. The fourth series in Table 44.4 begins with ^{237}Np and has as its stable end product ^{209}Bi. The element ^{237}Np is a *transuranic* element (one having an atomic number greater than that of uranium) not found in nature. This element has a half-life of "only" 2.14×10^6 years.

Figure 44.18 shows the successive decays for the ^{232}Th series. First, ^{232}Th undergoes alpha decay to ^{228}Ra. Next, ^{228}Ra undergoes two successive beta decays to ^{228}Th. The series continues and finally branches when it reaches ^{212}Bi. At this point, there are two decay possibilities. The sequence shown in Figure 44.18 is characterized by a mass-number decrease of either 4 (for alpha decays) or 0 (for beta or gamma decays). The two uranium series are more complex than the ^{232}Th series. In addition, several naturally occurring radioactive isotopes, such as ^{14}C and ^{40}K, are not part of any decay series.

Because of these radioactive series, our environment is constantly replenished with radioactive elements that would otherwise have disappeared long ago. For example, because our solar system is approximately 5×10^9 years old, the supply of

Table 44.3	**Various Decay Pathways**
Alpha decay	$^{A}_{Z}X \rightarrow {}^{A-4}_{Z-2}Y + {}^{4}_{2}He$
Beta decay (e^-)	$^{A}_{Z}X \rightarrow {}^{A}_{Z+1}Y + e^- + \bar{\nu}$
Beta decay (e^+)	$^{A}_{Z}X \rightarrow {}^{A}_{Z-1}Y + e^+ + \nu$
Electron capture	$^{A}_{Z}X + e^- \rightarrow {}^{A}_{Z-1}Y + \nu$
Gamma decay	$^{A}_{Z}X^* \rightarrow {}^{A}_{Z}X + \gamma$

Table 44.4   The Four Radioactive Series

Series		Starting Isotope	Half-life (years)	Stable End Product
Uranium	⎫	$^{238}_{92}U$	4.47×10^9	$^{206}_{82}Pb$
Actinium	⎬ Natural	$^{235}_{92}U$	7.04×10^8	$^{207}_{82}Pb$
Thorium	⎭	$^{232}_{90}Th$	1.41×10^{10}	$^{208}_{82}Pb$
Neptunium		$^{237}_{93}Np$	2.14×10^6	$^{209}_{83}Bi$

^{226}Ra (whose half-life is only 1 600 years) would have been depleted by radioactive decay long ago if it were not for the radioactive series starting with ^{238}U.

44.7 Nuclear Reactions

We have studied radioactivity, which is a spontaneous process in which the structure of a nucleus changes. It is also possible to stimulate changes in the structure of nuclei by bombarding them with energetic particles. Such collisions, which change the identity of the target nuclei, are called **nuclear reactions.** Rutherford was the first to observe them, in 1919, using naturally occurring radioactive sources for the bombarding particles. Since then, a wide variety of nuclear reactions has been observed following the development of charged-particle accelerators in the 1930s. With today's advanced technology in particle accelerators and particle detectors, the Large Hadron Collider (see Section 46.10) in Europe can achieve particle energies of 14 000 GeV = 14 TeV. These high-energy particles are used to create new particles whose properties are helping to solve the mysteries of the nucleus.

Consider a reaction in which a target nucleus X is bombarded by a particle a, resulting in a daughter nucleus Y and an outgoing particle b:

$$a + X \rightarrow Y + b \tag{44.28}$$

◀ Nuclear reaction

Sometimes this reaction is written in the more compact form

$$X(a, b)Y$$

In Section 44.5, the Q value, or disintegration energy, of a radioactive decay was defined as the rest energy transformed to kinetic energy as a result of the decay process. Likewise, we define the **reaction energy** Q associated with a nuclear reaction as *the difference between the initial and final rest energies resulting from the reaction:*

$$Q = (M_a + M_X - M_Y - M_b)c^2 \tag{44.29}$$

◀ Reaction energy Q

As an example, consider the reaction $^7\text{Li}(p, \alpha)^4\text{He}$. The notation p indicates a proton, which is a hydrogen nucleus. Therefore, we can write this reaction in the expanded form

$$^1_1\text{H} + {}^7_3\text{Li} \rightarrow {}^4_2\text{He} + {}^4_2\text{He}$$

The Q value for this reaction is 17.3 MeV. A reaction such as this one, for which Q is positive, is called **exothermic.** A reaction for which Q is negative is called **endothermic.** To satisfy conservation of momentum for the isolated system, an endothermic reaction does not occur unless the bombarding particle has a kinetic energy greater than Q. (See Problem 74 in Enhanced WebAssign.) The minimum energy necessary for such a reaction to occur is called the **threshold energy.**

If particles a and b in a nuclear reaction are identical so that X and Y are also necessarily identical, the reaction is called a **scattering event.** If the kinetic energy of the system (a and X) before the event is the same as that of the system (b and Y) after the event, it is classified as *elastic scattering.* If the kinetic energy of the system after the event is less than that before the event, the reaction is described as *inelastic scattering.* In this case, the target nucleus has been raised to an excited state by the event, which accounts for the difference in energy. The final system now consists of b and an excited nucleus Y*, and eventually it will become b, Y, and γ, where γ is the gamma-ray photon that is emitted when the system returns to the ground state. This elastic and inelastic terminology is identical to that used in describing collisions between macroscopic objects as discussed in Section 9.4.

In addition to energy and momentum, the total charge and total number of nucleons must be conserved in any nuclear reaction. For example, consider the reaction ^{19}F(p, α)^{16}O, which has a Q value of 8.11 MeV. We can show this reaction more completely as

$$^{1}_{1}\text{H} + {^{19}_{9}\text{F}} \rightarrow {^{16}_{8}\text{O}} + {^{4}_{2}\text{He}} \tag{44.30}$$

The total number of nucleons before the reaction $(1 + 19 = 20)$ is equal to the total number after the reaction $(16 + 4 = 20)$. Furthermore, the total charge is the same before $(1 + 9)$ and after $(8 + 2)$ the reaction.

44.8 Nuclear Magnetic Resonance and Magnetic Resonance Imaging

In this section, we describe an important application of nuclear physics in medicine called *magnetic resonance imaging*. To understand this application, we first discuss the spin angular momentum of the nucleus. This discussion has parallels with the discussion of spin for atomic electrons.

In Chapter 42, we discussed that the electron has an intrinsic angular momentum, called spin. Nuclei also have spin because their component particles—neutrons and protons—each have spin $\frac{1}{2}$ as well as orbital angular momentum within the nucleus. All types of angular momentum obey the quantum rules that were outlined for orbital and spin angular momentum in Chapter 42. In particular, two quantum numbers associated with the angular momentum determine the allowed values of the magnitude of the angular momentum vector and its direction in space. The magnitude of the nuclear angular momentum is $\sqrt{I(I + 1)}\,\hbar$, where I is called the **nuclear spin quantum number** and may be an integer or a half-integer, depending on how the individual proton and neutron spins combine. The quantum number I is the analog to ℓ for the electron in an atom as discussed in Section 42.6. Furthermore, there is a quantum number m_I that is the analog to m_ℓ, in that the allowed projections of the nuclear spin angular momentum vector on the z axis are $m_I \hbar$. The values of m_I range from $-I$ to $+I$ in steps of 1. (In fact, for *any* type of spin with a quantum number S, there is a quantum number m_S that ranges in value from $-S$ to $+S$ in steps of 1.) Therefore, the maximum value of the z component of the spin angular momentum vector is $I\hbar$. Figure 44.19 is a vector model (see Section 42.6) illustrating the possible orientations of the nuclear spin vector and its projections along the z axis for the case in which $I = \frac{3}{2}$.

Nuclear spin has an associated nuclear magnetic moment, similar to that of the electron. The spin magnetic moment of a nucleus is measured in terms of the **nuclear magneton** μ_n, a unit of moment defined as

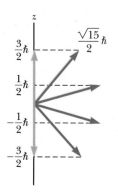

Figure 44.19 A vector model showing possible orientations of the nuclear spin angular momentum vector and its projections along the z axis for the case $I = \frac{3}{2}$.

▶ Nuclear magneton

$$\mu_n \equiv \frac{e\hbar}{2m_p} = 5.05 \times 10^{-27}\,\text{J/T} \tag{44.31}$$

where m_p is the mass of the proton. This definition is analogous to that of the Bohr magneton μ_B, which corresponds to the spin magnetic moment of a free electron (see Section 42.6). Note that μ_n is smaller than μ_B $(= 9.274 \times 10^{-24}\,\text{J/T})$ by a factor of 1 836 because of the large difference between the proton mass and the electron mass.

The magnetic moment of a free proton is $2.792\,8\mu_n$. Unfortunately, there is no general theory of nuclear magnetism that explains this value. The neutron also has a magnetic moment, which has a value of $-1.913\,5\mu_n$. The negative sign indicates that this moment is opposite the spin angular momentum of the neutron. The existence of a magnetic moment for the neutron is surprising in view of the neutron being uncharged. That suggests that the neutron is not a fundamental particle but rather has an underlying structure consisting of charged constituents. We shall explore this structure in Chapter 46.

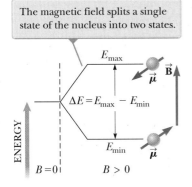

The magnetic field splits a single state of the nucleus into two states.

Figure 44.20 A nucleus with spin $\frac{1}{2}$ is placed in a magnetic field.

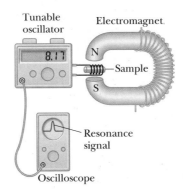

Resonance signal

Figure 44.21 Experimental arrangement for nuclear magnetic resonance. The radio-frequency magnetic field created by the coil surrounding the sample and provided by the variable-frequency oscillator is perpendicular to the constant magnetic field created by the electromagnet. When the nuclei in the sample meet the resonance condition, the nuclei absorb energy from the radio-frequency field of the coil; this absorption changes the characteristics of the circuit in which the coil is included. Most modern NMR spectrometers use superconducting magnets at fixed field strengths and operate at frequencies of approximately 200 MHz.

The potential energy associated with a magnetic dipole moment $\vec{\mu}$ in an external magnetic field $\vec{B}$ is given by $-\vec{\mu} \cdot \vec{B}$ (Eq. 29.18). When the magnetic moment $\vec{\mu}$ is lined up with the field as closely as quantum physics allows, the potential energy of the dipole–field system has its minimum value E_{min}. When $\vec{\mu}$ is as antiparallel to the field as possible, the potential energy has its maximum value E_{max}. In general, there are other energy states between these values corresponding to the quantized directions of the magnetic moment with respect to the field. For a nucleus with spin $\frac{1}{2}$, there are only two allowed states, with energies E_{min} and E_{max}. These two energy states are shown in Figure 44.20.

It is possible to observe transitions between these two spin states using a technique called **NMR,** for **nuclear magnetic resonance.** A constant magnetic field ($\vec{B}$ in Fig. 44.20) is introduced to define a z axis and split the energies of the spin states. A second, weaker, oscillating magnetic field is then applied perpendicular to $\vec{B}$, creating a cloud of radio-frequency photons around the sample. When the frequency of the oscillating field is adjusted so that the photon energy matches the energy difference between the spin states, there is a net absorption of photons by the nuclei that can be detected electronically.

Figure 44.21 is a simplified diagram of the apparatus used in nuclear magnetic resonance. The energy absorbed by the nuclei is supplied by the tunable oscillator producing the oscillating magnetic field. Nuclear magnetic resonance and a related technique called *electron spin resonance* are extremely important methods for studying nuclear and atomic systems and the ways in which these systems interact with their surroundings.

A widely used medical diagnostic technique called **MRI,** for **magnetic resonance imaging,** is based on nuclear magnetic resonance. Because nearly two-thirds of the atoms in the human body are hydrogen (which gives a strong NMR signal), MRI works exceptionally well for viewing internal tissues. The patient is placed inside a large solenoid that supplies a magnetic field that is constant in time but whose magnitude varies spatially across the body. Because of the variation in the field, hydrogen atoms in different parts of the body have different energy splittings between spin states, so the resonance signal can be used to provide information about the positions of the protons. A computer is used to analyze the position information to provide data for constructing a final image. Contrast in the final image among different types of tissues is created by computer analysis of the time intervals for the nuclei to return to the lower-energy spin state between pulses of radio-frequency photons. Contrast can be enhanced with the use of contrast agents such as gadolinium compounds or iron oxide nanoparticles taken orally or injected intravenously. An MRI scan showing incredible detail in internal body structure is shown in Figure 44.22.

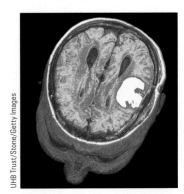

UHB Trust/Stone/Getty Images

Figure 44.22 A color-enhanced MRI scan of a human brain, showing a tumor in white.

The main advantage of MRI over other imaging techniques is that it causes minimal cellular damage. The photons associated with the radio-frequency signals used in MRI have energies of only about 10^{-7} eV. Because molecular bond strengths are much larger (approximately 1 eV), the radio-frequency radiation causes little cellular damage. In comparison, x-rays have energies ranging from 10^4 to 10^6 eV and can cause considerable cellular damage. Therefore, despite some individuals' fears of the word *nuclear* associated with MRI, the radio-frequency radiation involved is overwhelmingly safer than the x-rays that these individuals might accept more readily. A disadvantage of MRI is that the equipment required to conduct the procedure is very expensive, so MRI images are costly.

The magnetic field produced by the solenoid is sufficient to lift a car, and the radio signal is about the same magnitude as that from a small commercial broadcasting station. Although MRI is inherently safe in normal use, the strong magnetic field of the solenoid requires diligent care to ensure that no ferromagnetic materials are located in the room near the MRI apparatus. Several accidents have occurred, such as a 2000 incident in which a gun pulled from a police officer's hand discharged upon striking the machine.

Summary

Definitions

A nucleus is represented by the symbol $^{A}_{Z}X$, where A is the **mass number** (the total number of nucleons) and Z is the **atomic number** (the total number of protons). The total number of neutrons in a nucleus is the **neutron number** N, where $A = N + Z$. Nuclei having the same Z value but different A and N values are **isotopes** of each other.

The magnetic moment of a nucleus is measured in terms of the **nuclear magneton** μ_n, where

$$\mu_n \equiv \frac{e\hbar}{2m_p} = 5.05 \times 10^{-27}\,\text{J/T} \quad \textbf{(44.31)}$$

Concepts and Principles

Assuming nuclei are spherical, their radius is given by

$$r = aA^{1/3} \quad \textbf{(44.1)}$$

where $a = 1.2$ fm.

Nuclei are stable because of the **nuclear force** between nucleons. This short-range force dominates the Coulomb repulsive force at distances of less than about 2 fm and is independent of charge. Light stable nuclei have equal numbers of protons and neutrons. Heavy stable nuclei have more neutrons than protons. The most stable nuclei have Z and N values that are both even.

The difference between the sum of the masses of a group of separate nucleons and the mass of the compound nucleus containing these nucleons, when multiplied by c^2, gives the **binding energy** E_b of the nucleus. The binding energy of a nucleus can be calculated in MeV using the expression

$$E_b = [ZM(\text{H}) + Nm_n - M(^{A}_{Z}X)] \times 931.494 \text{ MeV/u}$$

$$\textbf{(44.2)}$$

where $M(\text{H})$ is the atomic mass of the neutral hydrogen atom, $M(^{A}_{Z}X)$ represents the atomic mass of an atom of the isotope $^{A}_{Z}X$, and m_n is the mass of the neutron.

The **liquid-drop model** of nuclear structure treats the nucleons as molecules in a drop of liquid. The four main contributions influencing binding energy are the volume effect, the surface effect, the Coulomb repulsion effect, and the symmetry effect. Summing such contributions results in the **semiempirical binding-energy formula**:

$$E_b = C_1 A - C_2 A^{2/3} - C_3 \frac{Z(Z-1)}{A^{1/3}} - C_4 \frac{(N-Z)^2}{A} \quad \textbf{(44.3)}$$

The **shell model,** or **independent-particle model,** assumes each nucleon exists in a shell and can only have discrete energy values. The stability of certain nuclei can be explained with this model.

A radioactive substance decays by **alpha decay, beta decay,** or **gamma decay.** An alpha particle is the ^{4}He nucleus, a beta particle is either an electron (e$^-$) or a positron (e$^+$), and a gamma particle is a high-energy photon.

If a radioactive material contains N_0 radioactive nuclei at $t = 0$, the number N of nuclei remaining after a time t has elapsed is

$$N = N_0 e^{-\lambda t} \tag{44.6}$$

where λ is the **decay constant,** a number equal to the probability per second that a nucleus will decay. The **decay rate,** or **activity,** of a radioactive substance is

$$R = \left| \frac{dN}{dt} \right| = R_0 e^{-\lambda t} \tag{44.7}$$

where $R_0 = \lambda N_0$ is the activity at $t = 0$. The **half-life** $T_{1/2}$ is the time interval required for half of a given number of radioactive nuclei to decay, where

$$T_{1/2} = \frac{0.693}{\lambda} \tag{44.8}$$

In alpha decay, a helium nucleus is ejected from the parent nucleus with a discrete set of kinetic energies. A nucleus undergoing beta decay emits either an electron (e$^-$) and an antineutrino ($\bar{\nu}$) or a positron (e$^+$) and a neutrino (ν). The electron or positron is ejected with a continuous range of energies. In **electron capture,** the nucleus of an atom absorbs one of its own electrons and emits a neutrino. In gamma decay, a nucleus in an excited state decays to its ground state and emits a gamma ray.

Nuclear reactions can occur when a target nucleus X is bombarded by a particle a, resulting in a daughter nucleus Y and an outgoing particle b:

$$a + X \rightarrow Y + b \tag{44.28}$$

The mass–energy conversion in such a reaction, called the **reaction energy** Q, is

$$Q = (M_a + M_X - M_Y - M_b)c^2 \tag{44.29}$$

Objective Questions

1. denotes answer available in *Student Solutions Manual/Study Guide*

1. In nuclear magnetic resonance, suppose we increase the value of the constant magnetic field. As a result, the frequency of the photons that are absorbed in a particular transition changes. How is the frequency of the photons absorbed related to the magnetic field? (a) The frequency is proportional to the square of the magnetic field. (b) The frequency is directly proportional to the magnetic field. (c) The frequency is independent of the magnetic field. (d) The frequency is inversely proportional to the magnetic field. (e) The frequency is proportional to the reciprocal of the square of the magnetic field.

2. When the $^{95}_{36}$Kr nucleus undergoes beta decay by emitting an electron and an antineutrino, does the daughter nucleus (Rb) contain (a) 58 neutrons and 37 protons, (b) 58 protons and 37 neutrons, (c) 54 neutrons and 41 protons, or (d) 55 neutrons and 40 protons?

3. When $^{32}_{15}$P decays to $^{32}_{16}$S, which of the following particles is emitted? (a) a proton (b) an alpha particle (c) an electron (d) a gamma ray (e) an antineutrino

4. The half-life of radium-224 is about 3.6 days. What approximate fraction of a sample remains undecayed after two weeks? (a) $\frac{1}{2}$ (b) $\frac{1}{4}$ (c) $\frac{1}{8}$ (d) $\frac{1}{16}$ (e) $\frac{1}{32}$

5. Two samples of the same radioactive nuclide are prepared. Sample G has twice the initial activity of sample H. (i) How does the half-life of G compare with the half-life of H? (a) It is two times larger. (b) It is the same. (c) It is half as large. (ii) After each has passed through five half-lives, how do their activities compare? (a) G has more than twice the activity of H. (b) G has twice the activity of H. (c) G and H have the same activity. (d) G has lower activity than H.

6. If a radioactive nuclide ^A_ZX decays by emitting a gamma ray, what happens? (a) The resulting nuclide has a different Z value. (b) The resulting nuclide has the same A and Z values. (c) The resulting nuclide has a different A value. (d) Both A and Z decrease by one. (e) None of those statements is correct.

7. Does a nucleus designated as $^{40}_{18}$X contain (a) 20 neutrons and 20 protons, (b) 22 protons and 18 neutrons, (c) 18 protons and 22 neutrons, (d) 18 protons and 40 neutrons, or (e) 40 protons and 18 neutrons?

8. When $^{144}_{60}$Nd decays to $^{140}_{58}$Ce, identify the particle that is released. (a) a proton (b) an alpha particle (c) an electron (d) a neutron (e) a neutrino

9. What is the Q value for the reaction ^{9}Be $+ \alpha \rightarrow {}^{12}$C $+$ n? (a) 8.4 MeV (b) 7.3 MeV (c) 6.2 MeV (d) 5.7 MeV (e) 4.2 MeV

10. (i) To predict the behavior of a nucleus in a fission reaction, which model would be more appropriate,

(a) the liquid-drop model or (b) the shell model? **(ii)** Which model would be more successful in predicting the magnetic moment of a given nucleus? Choose from the same answers as in part (i). **(iii)** Which could better explain the gamma-ray spectrum of an excited nucleus? Choose from the same answers as in part (i).

11. A free neutron has a half-life of 614 s. It undergoes beta decay by emitting an electron. Can a free proton undergo a similar decay? (a) yes, the same decay (b) yes, but by emitting a positron (c) yes, but with a very different half-life (d) no

12. Which of the following quantities represents the reaction energy of a nuclear reaction? (a) (final mass − initial mass)$/c^2$ (b) (initial mass − final mass)$/c^2$ (c) (final mass − initial mass)c^2 (d) (initial mass − final mass)c^2 (e) none of those quantities

13. In the decay $^{234}_{90}\text{Th} \rightarrow {}^{A}_{Z}\text{Ra} + {}^{4}_{2}\text{He}$, identify the mass number and the atomic number of the Ra nucleus: (a) $A = 230$, $Z = 92$ (b) $A = 238$, $Z = 88$ (c) $A = 230$, $Z = 88$ (d) $A = 234$, $Z = 88$ (e) $A = 238$, $Z = 86$

Conceptual Questions

1. denotes answer available in *Student Solutions Manual/Study Guide*

1. If a nucleus such as ^{226}Ra initially at rest undergoes alpha decay, which has more kinetic energy after the decay, the alpha particle or the daughter nucleus? Explain your answer.

2. "If no more people were to be born, the law of population growth would strongly resemble the radioactive decay law." Discuss this statement.

3. A student claims that a heavy form of hydrogen decays by alpha emission. How do you respond?

4. In beta decay, the energy of the electron or positron emitted from the nucleus lies somewhere in a relatively large range of possibilities. In alpha decay, however, the alpha-particle energy can only have discrete values. Explain this difference.

5. Can carbon-14 dating be used to measure the age of a rock? Explain.

6. In positron decay, a proton in the nucleus becomes a neutron and its positive charge is carried away by the positron. A neutron, though, has a larger rest energy than a proton. How is that possible?

7. (a) How many values of I_z are possible for $I = \frac{5}{2}$? (b) For $I = 3$?

8. Why do nearly all the naturally occurring isotopes lie above the $N = Z$ line in Figure 44.4?

9. Why are very heavy nuclei unstable?

10. Explain why nuclei that are well off the line of stability in Figure 44.4 tend to be unstable.

11. Consider two heavy nuclei X and Y having similar mass numbers. If X has the higher binding energy, which nucleus tends to be more unstable? Explain your answer.

12. What fraction of a radioactive sample has decayed after two half-lives have elapsed?

13. Figure CQ44.13 shows a watch from the early 20th century. The numbers and the hands of the watch are painted with a paint that contains a small amount of natural radium $^{226}_{88}$Ra mixed with a phosphorescent material. The decay of the radium causes the phosphorescent material to glow continuously. The radioactive nuclide $^{226}_{88}$Ra has a half-life of approximately 1.60×10^3 years. Being that the solar system is approximately 5 billion years old, why was this isotope still available in the 20th century for use on this watch?

© Richard Megna/Fundamental Photographs

Figure CQ44.13

14. Can a nucleus emit alpha particles that have different energies? Explain.

15. In Rutherford's experiment, assume an alpha particle is headed directly toward the nucleus of an atom. Why doesn't the alpha particle make physical contact with the nucleus?

16. Suppose it could be shown that the cosmic-ray intensity at the Earth's surface was much greater 10 000 years ago. How would this difference affect what we accept as valid carbon-dated values of the age of ancient samples of once-living matter? Explain your answer.

17. Compare and contrast the properties of a photon and a neutrino.

Section 44.1 Some Properties of Nuclei
Problems 1–14

Section 44.2 Nuclear Binding Energy
Problems 15–21

Section 44.3 Nuclear Models
Problems 22–24

Section 44.4 Radioactivity
Problems 25–34

Section 44.5 The Decay Processes
Problems 35–42

Section 44.6 Natural Radioactivity
Problems 43–46

Section 44.7 Nuclear Reactions
Problems 47–51

Section 44.8 Nuclear Magnetic Resonance and Magnetic Resonance Imaging
Problems 52–53

Additional Problems
Problems 54–76

Challenge Problems
Problems 77–78

Solutions to the following Problems are available in the *Student Solutions Manual/Study Guide:*
44.3, 44.7, 44.17, 44.19, 44.22, 44.26, 44.27, 44.29, 44.39, 44.41, 44.45, 44.50, 44.51, 44.52, and 44.69

List of Enhanced Problems

Problem Number	Targeted Feedback in Enhanced WebAssign	Analysis Model Tutorial in Enhanced WebAssign	Master It in Enhanced WebAssign	Watch It in Enhanced WebAssign
44.3	✓		✓	
44.9		✓		
44.11			✓	
44.26	✓		✓	
44.29			✓	
44.42				✓
44.46	✓			✓
44.47	✓			✓
44.54			✓	
44.55			✓	
44.69		✓	✓	

CHAPTER
45

Applications of Nuclear Physics

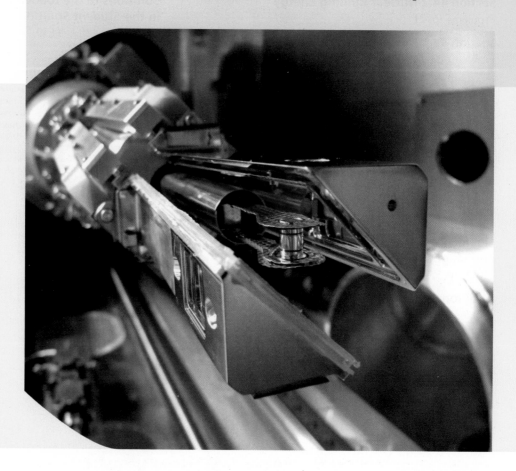

In this chapter, we study both nuclear fission and nuclear fusion. The structure above is the target assembly for the inertial confinement procedure for initiating fusion by laser at the National Ignition Facility in Livermore, California. The triangle-shaped shrouds protect the fuel pellets and then open a few seconds before very powerful lasers bombard the target. *(Courtesy of Lawrence Livermore National Library)*

In this chapter, we study two means for deriving energy from nuclear reactions: fission, in which a large nucleus splits into two smaller nuclei, and fusion, in which two small nuclei fuse to form a larger one. In both cases, the released energy can be used either constructively (as in electric power plants) or destructively (as in nuclear weapons). We also examine the ways in which radiation interacts with matter and discuss the structure of fission and fusion reactors. The chapter concludes with a discussion of some industrial and biological applications of radiation.

45.1 Interactions Involving Neutrons

Nuclear fission is the process that occurs in present-day nuclear reactors and ultimately results in energy supplied to a community by electrical transmission. Nuclear fusion is an area of active research, but it has not yet been commercially developed for the supply of energy. We will discuss fission first and then explore fusion in Section 45.4.

To understand nuclear fission and the physics of nuclear reactors, we must first understand how neutrons interact with nuclei. Because of their charge neutrality, neutrons are not subject to Coulomb forces and as a result do not interact electrically with electrons or the nucleus. Therefore, neutrons can easily penetrate deep into an atom and collide with the nucleus.

A fast neutron (energy greater than approximately 1 MeV) traveling through matter undergoes many collisions with nuclei, giving up some of its kinetic energy in each collision. For fast neutrons in some materials, elastic collisions dominate. Materials for which that occurs are called **moderators** because they slow down (or moderate) the originally energetic neutrons very effectively. Moderator nuclei should be of low mass so that a large amount of kinetic energy is transferred to them when struck by neutrons. For this reason, materials that are abundant in hydrogen, such as paraffin and water, are good moderators for neutrons.

Eventually, most neutrons bombarding a moderator become **thermal neutrons,** which means they have given up so much of their energy that they are in thermal equilibrium with the moderator material. Their average kinetic energy at room temperature is, from Equation 21.19,

$$K_{\text{avg}} = \tfrac{3}{2}k_{\text{B}}T \approx \tfrac{3}{2}(1.38 \times 10^{-23}\,\text{J/K})(300\,\text{K}) = 6.21 \times 10^{-21}\,\text{J} \approx 0.04\,\text{eV}$$

which corresponds to a neutron root-mean-square speed of approximately 2 800 m/s. Thermal neutrons have a distribution of speeds, just as the molecules in a container of gas do (see Chapter 21). High-energy neutrons, those with energy of several MeV, *thermalize* (that is, their average energy reaches K_{avg}) in less than 1 ms when they are incident on a moderator.

Once the neutrons have thermalized and the energy of a particular neutron is sufficiently low, there is a high probability the neutron will be captured by a nucleus, an event that is accompanied by the emission of a gamma ray. This **neutron capture** reaction can be written

$$^{1}_{0}\text{n} + {}^{A}_{Z}\text{X} \;\rightarrow\; {}^{A+1}_{Z}\text{X}^{*} \;\rightarrow\; {}^{A+1}_{Z}\text{X} + \gamma \qquad\qquad \textbf{(45.1)}$$

◀ **Neutron capture reaction**

Once the neutron is captured, the nucleus $^{A+1}_{Z}\text{X}^{*}$ is in an excited state for a very short time before it undergoes gamma decay. The product nucleus $^{A+1}_{Z}\text{X}$ is usually radioactive and decays by beta emission.

The neutron-capture rate for neutrons passing through any sample depends on the type of atoms in the sample and on the energy of the incident neutrons. The interaction of neutrons with matter increases with decreasing neutron energy because a slow neutron spends a larger time interval in the vicinity of target nuclei.

45.2 Nuclear Fission

As mentioned in Section 44.2, nuclear **fission** occurs when a heavy nucleus, such as ^{235}U, splits into two smaller nuclei. Fission is initiated when a heavy nucleus captures a thermal neutron as described by the first step in Equation 45.1. The absorption of the neutron creates a nucleus that is unstable and can change to a lower-energy configuration by splitting into two smaller nuclei. In such a reaction, the combined mass of the daughter nuclei is less than the mass of the parent nucleus, and the difference in mass is called the **mass defect.** Multiplying the mass defect by c^2 gives the numerical value of the released energy. This energy is in the form of kinetic energy associated with the motion of the neutrons and the daughter nuclei after the fission event. Energy is released because the binding energy per nucleon of the daughter nuclei is approximately 1 MeV greater than that of the parent nucleus (see Fig. 44.5).

Nuclear fission was first observed in 1938 by Otto Hahn (1879–1968) and Fritz Strassmann (1902–1980) following some basic studies by Fermi. After bombarding uranium with neutrons, Hahn and Strassmann discovered among the reaction products two medium-mass elements, barium and lanthanum. Shortly thereafter, Lise Meitner (1878–1968) and her nephew Otto Frisch (1904–1979) explained what had happened. After absorbing a neutron, the uranium nucleus had split into two

Pitfall Prevention 45.1
Binding Energy Reminder Remember from Chapter 44 that binding energy is the absolute value of the system energy and is related to the system mass. Therefore, when considering Figure 44.5, imagine flipping it upside down for a graph representing system mass. In a fission reaction, the system mass decreases. This decrease in mass appears in the system as kinetic energy of the fission products.

Before the event, a slow neutron approaches a ^{235}U nucleus.

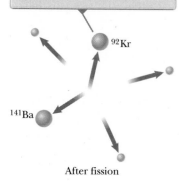

^{235}U

Before fission

After the event, there are two lighter nuclei and three neutrons.

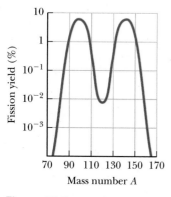

^{92}Kr

^{141}Ba

After fission

Figure 45.1 A nuclear fission event.

nearly equal fragments plus several neutrons. Such an occurrence was of considerable interest to physicists attempting to understand the nucleus, but it was to have even more far-reaching consequences. Measurements showed that approximately 200 MeV of energy was released in each fission event, and this fact was to affect the course of history in World War II.

The fission of ^{235}U by thermal neutrons can be represented by the reaction

$$^1_0n + {}^{235}_{92}U \rightarrow {}^{236}_{92}U^* \rightarrow X + Y + \text{neutrons} \qquad (45.2)$$

where ^{236}U* is an intermediate excited state that lasts for approximately 10^{-12} s before splitting into medium-mass nuclei X and Y, which are called **fission fragments.** In any fission reaction, there are many combinations of X and Y that satisfy the requirements of conservation of energy and charge. In the case of uranium, for example, approximately 90 daughter nuclei can be formed.

Fission also results in the production of several neutrons, typically two or three. On average, approximately 2.5 neutrons are released per event. A typical fission reaction for uranium is

$$^1_0n + {}^{235}_{92}U \rightarrow {}^{141}_{56}Ba + {}^{92}_{36}Kr + 3({}^1_0n) \qquad (45.3)$$

Figure 45.1 shows a pictorial representation of the fission event in Equation 45.3.

Figure 45.2 is a graph of the distribution of fission products versus mass number A. The most probable products have mass numbers $A \approx 95$ and $A \approx 140$. Suppose these products are $^{95}_{39}$Y (with 56 neutrons) and $^{140}_{53}$I (with 87 neutrons). If these nuclei are located on the graph of Figure 44.4, it is seen that both are well above the line of stability. Because these fragments are very unstable owing to their unusually high number of neutrons, they almost instantaneously release two or three neutrons.

Let's estimate the disintegration energy Q released in a typical fission process. From Figure 44.5, we see that the binding energy per nucleon is approximately 7.2 MeV for heavy nuclei ($A \approx 240$) and approximately 8.2 MeV for nuclei of intermediate mass. The amount of energy released is 8.2 MeV − 7.2 MeV = 1 MeV per nucleon. Because there are a total of 235 nucleons in $^{235}_{92}$U, the energy released per fission event is approximately 235 MeV, a large amount of energy relative to the amount released in chemical processes. For example, the energy released in the combustion of one molecule of octane used in gasoline engines is about one-millionth of the energy released in a single fission event!

Quick **Quiz** 45.1 When a nucleus undergoes fission, the two daughter nuclei are generally radioactive. By which process are they most likely to decay? **(a)** alpha decay **(b)** beta decay (e^-) **(c)** beta decay (e^+)

Quick **Quiz** 45.2 Which of the following are possible fission reactions?

(a) $^1_0n + {}^{235}_{92}U \rightarrow {}^{140}_{54}Xe + {}^{94}_{38}Sr + 2({}^1_0n)$

(b) $^1_0n + {}^{235}_{92}U \rightarrow {}^{132}_{50}Sn + {}^{101}_{42}Mo + 3({}^1_0n)$

(c) $^1_0n + {}^{239}_{94}Pu \rightarrow {}^{137}_{53}I + {}^{97}_{41}Nb + 3({}^1_0n)$

Figure 45.2 Distribution of fission products versus mass number for the fission of ^{235}U bombarded with thermal neutrons. Notice that the vertical axis is logarithmic.

Fission yield (%) axis: 10, 1, 10^{-1}, 10^{-2}, 10^{-3}

Mass number A axis: 70, 90, 110, 130, 150, 170

| Example 45.1 | The Energy Released in the Fission of ^{235}U |

Calculate the energy released when 1.00 kg of ^{235}U fissions, taking the disintegration energy per event to be $Q = 208$ MeV.

SOLUTION

Conceptualize Imagine a nucleus of ^{235}U absorbing a neutron and then splitting into two smaller nuclei and several neutrons as in Figure 45.1.

▶ **45.1** continued

Categorize The problem statement tells us to categorize this example as one involving an energy analysis of nuclear fission.

Analyze Because $A = 235$ for uranium, one mole of this isotope has a mass of $M = 235$ g.

Find the number of nuclei in our sample in terms of the number of moles n and Avogadro's number, and then in terms of the sample mass m and the molar mass M of ^{235}U:

$$N = nN_A = \frac{m}{M}N_A$$

Find the total energy released when all nuclei undergo fission:

$$E = NQ = \frac{m}{M}N_AQ = \frac{1.00 \times 10^3 \text{ g}}{235 \text{ g/mol}}(6.02 \times 10^{23} \text{ mol}^{-1})(208 \text{ MeV})$$

$$= \boxed{5.33 \times 10^{26} \text{ MeV}}$$

Finalize Convert this energy to kWh:

$$E = (5.33 \times 10^{26} \text{ MeV})\left(\frac{1.60 \times 10^{-13} \text{ J}}{1 \text{ MeV}}\right)\left(\frac{1 \text{ kWh}}{3.60 \times 10^6 \text{ J}}\right) = 2.37 \times 10^7 \text{ kWh}$$

which, if released slowly, is enough energy to keep a 100-W lightbulb operating for 30 000 years! If the available fission energy in 1 kg of ^{235}U were suddenly released, it would be equivalent to detonating about 20 000 tons of TNT.

45.3 Nuclear Reactors

In Section 45.2, we learned that when ^{235}U fissions, one incoming neutron results in an average of 2.5 neutrons emitted per event. These neutrons can trigger other nuclei to fission. Because more neutrons are produced by the event than are absorbed, there is the possibility of an ever-building chain reaction (Fig. 45.3). Experience shows that if the chain reaction is not controlled (that is, if it does not

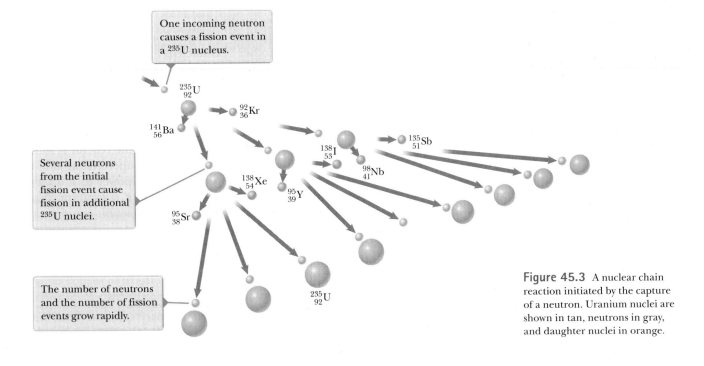

One incoming neutron causes a fission event in a $^{235}_{92}$U nucleus.

Several neutrons from the initial fission event cause fission in additional ^{235}U nuclei.

The number of neutrons and the number of fission events grow rapidly.

$^{235}_{92}$U $^{92}_{36}$Kr $^{141}_{56}$Ba $^{138}_{54}$Xe $^{95}_{38}$Sr $^{95}_{39}$Y $^{138}_{53}$I $^{98}_{41}$Nb $^{135}_{51}$Sb $^{235}_{92}$U

Figure 45.3 A nuclear chain reaction initiated by the capture of a neutron. Uranium nuclei are shown in tan, neutrons in gray, and daughter nuclei in orange.

can be varied and any power level within the design range of the reactor can be achieved.

Quick Quiz 45.3 To reduce the value of the reproduction constant K, do you **(a)** push the control rods deeper into the core or **(b)** pull the control rods farther out of the core?

Safety and Waste Disposal

The 1986 accident at the Chernobyl reactor in Ukraine and the 2011 nuclear disaster caused by the earthquake and tsunami in Japan rightfully focused attention on reactor safety. Unfortunately, at Chernobyl the activity of the materials released immediately after the accident totaled approximately 1.2×10^{19} Bq and resulted in the evacuation of 135 000 people. Thirty individuals died during the accident or shortly thereafter, and data from the Ukraine Radiological Institute suggest that more than 2 500 deaths could be attributed to the Chernobyl accident. In the period 1986–1997, there was a tenfold increase in the number of children contracting thyroid cancer from the ingestion of radioactive iodine in milk from cows that ate contaminated grass. One conclusion of an international conference studying the Ukraine accident was that the main causes of the Chernobyl accident were the coincidence of severe deficiencies in the reactor physical design and a violation of safety procedures. Most of these deficiencies have since been addressed at plants of similar design in Russia and neighboring countries of the former Soviet Union.

The March 2011 accident in Japan was caused by an unfortunate combination of a massive earthquake and subsequent tsunami. The most hard-hit power plant, Fukushima I, shut down automatically after the earthquake. Shutting down a nuclear power plant, however, is not an instantaneous process. Cooling water must continue to be circulated to carry the energy generated by beta decay of the fission by-products out of the reactor core. Unfortunately, the water from the tsunami broke the connection to the power grid, leaving the plant without outside electrical support for circulating the water. While the plant had emergency generators to take over in such a situation, the tsunami inundated the generator rooms, making the generators inoperable. Three of the six reactors at Fukushima experienced meltdown, and there were several explosions. Significant radiation was released into the environment. At the time of this printing, all 54 of Japan's nuclear power plants have been taken offline, and the Japanese public has expressed strong reluctance to continue with nuclear power.

Commercial reactors achieve safety through careful design and rigid operating protocol, and only when these variables are compromised do reactors pose a danger. Radiation exposure and the potential health risks associated with such exposure are controlled by three layers of containment. The fuel and radioactive fission products are contained inside the reactor vessel. Should this vessel rupture, the reactor building acts as a second containment structure to prevent radioactive material from contaminating the environment. Finally, the reactor facilities must be in a remote location to protect the general public from exposure should radiation escape the reactor building.

A continuing concern about nuclear fission reactors is the safe disposal of radioactive material when the reactor core is replaced. This waste material contains long-lived, highly radioactive isotopes and must be stored over long time intervals in such a way that there is no chance of environmental contamination. At present, sealing radioactive wastes in waterproof containers and burying them in deep geologic repositories seems to be the most promising solution.

Transport of reactor fuel and reactor wastes poses additional safety risks. Accidents during transport of nuclear fuel could expose the public to harmful levels of radiation. The U.S. Department of Energy requires stringent crash tests of all containers used to transport nuclear materials. Container manufacturers must demonstrate that their containers will not rupture even in high-speed collisions.

Despite these risks, there are advantages to the use of nuclear power to be weighed against the risks. For example, nuclear power plants do not produce air pollution and greenhouse gases as do fossil fuel plants, and the supply of uranium on the Earth is predicted to last longer than the supply of fossil fuels. For each source of energy—whether nuclear, hydroelectric, fossil fuel, wind, solar, or other—the risks must be weighed against the benefits and the availability of the energy source.

45.4 Nuclear Fusion

In Chapter 44, we found that the binding energy for light nuclei ($A < 20$) is much smaller than the binding energy for heavier nuclei, which suggests a process that is the reverse of fission. As mentioned in Section 39.8, when two light nuclei combine to form a heavier nucleus, the process is called nuclear **fusion.** Because the mass of the final nucleus is less than the combined masses of the original nuclei, there is a loss of mass accompanied by a release of energy.

Two examples of such energy-liberating fusion reactions are as follows:

$$^1_1H + {}^1_1H \rightarrow {}^2_1H + e^+ + \nu$$

$$^1_1H + {}^2_1H \rightarrow {}^3_2He + \gamma$$

These reactions occur in the core of a star and are responsible for the outpouring of energy from the star. The second reaction is followed by either hydrogen–helium fusion or helium–helium fusion:

$$^1_1H + {}^3_2He \rightarrow {}^4_2He + e^+ + \nu$$

$$^3_2He + {}^3_2He \rightarrow {}^4_2He + {}^1_1H + {}^1_1H$$

These fusion reactions are the basic reactions in the **proton–proton cycle,** believed to be one of the basic cycles by which energy is generated in the Sun and other stars that contain an abundance of hydrogen. Most of the energy production takes place in the Sun's interior, where the temperature is approximately 1.5×10^7 K. Because such high temperatures are required to drive these reactions, they are called **thermonuclear fusion reactions.** All the reactions in the proton–proton cycle are exothermic. An overview of the cycle is that four protons combine to generate an alpha particle, positrons, gamma rays, and neutrinos.

Quick Quiz 45.4 In the core of a star, hydrogen nuclei combine in fusion reactions. Once the hydrogen has been exhausted, fusion of helium nuclei can occur. If the star is sufficiently massive, fusion of heavier and heavier nuclei can occur once the helium is used up. Consider a fusion reaction involving two nuclei with the same value of A. For this reaction to be exothermic, which of the following values of A are impossible? **(a)** 12 **(b)** 20 **(c)** 28 **(d)** 64

> **Pitfall Prevention 45.2**
> **Fission and Fusion** The words *fission* and *fusion* sound similar, but they correspond to different processes. Consider the binding-energy graph in Figure 44.5. There are two directions from which you can approach the peak of the graph so that energy is released: combining two light nuclei, or fusion, and separating a heavy nucleus into two lighter nuclei, or fission.

Example 45.2   **Energy Released in Fusion**

Find the total energy released in the fusion reactions in the proton–proton cycle.

SOLUTION

Conceptualize The net nuclear result of the proton–proton cycle is to fuse four protons to form an alpha particle. Study the reactions above for the proton–proton cycle to be sure you understand how four protons become an alpha particle.

Categorize We use concepts discussed in this section, so we categorize this example as a substitution problem.

continued

▶ **45.2** continued

Find the initial mass of the system using the atomic mass of hydrogen from Table 44.2:

$$4(1.007\ 825\ \text{u}) = 4.031\ 300\ \text{u}$$

Find the change in mass of the system as this value minus the mass of a ⁴He atom:

$$4.031\ 300\ \text{u} - 4.002\ 603\ \text{u} = 0.028\ 697\ \text{u}$$

Convert this mass change into energy units:

$$E = 0.028\ 697\ \text{u} \times 931.494\ \text{MeV/u} = \boxed{26.7\ \text{MeV}}$$

This energy is shared among the alpha particle and other particles such as positrons, gamma rays, and neutrinos.

Terrestrial Fusion Reactions

The enormous amount of energy released in fusion reactions suggests the possibility of harnessing this energy for useful purposes. A great deal of effort is currently under way to develop a sustained and controllable thermonuclear reactor, a fusion power reactor. Controlled fusion is often called the ultimate energy source because of the availability of its fuel source: water. For example, if deuterium were used as the fuel, 0.12 g of it could be extracted from 1 gal of water at a cost of about four cents. This amount of deuterium would release approximately 10^{10} J if all nuclei underwent fusion. By comparison, 1 gal of gasoline releases approximately 10^8 J upon burning and costs far more than four cents.

An additional advantage of fusion reactors is that comparatively few radioactive by-products are formed. For the proton–proton cycle, for instance, the end product is safe, nonradioactive helium. Unfortunately, a thermonuclear reactor that can deliver a net power output spread over a reasonable time interval is not yet a reality, and many difficulties must be resolved before a successful device is constructed.

The Sun's energy is based in part on a set of reactions in which hydrogen is converted to helium. The proton–proton interaction is not suitable for use in a fusion reactor, however, because the event requires very high temperatures and densities. The process works in the Sun only because of the extremely high density of protons in the Sun's interior.

The reactions that appear most promising for a fusion power reactor involve deuterium (^2_1H) and tritium (^3_1H):

$$^2_1\text{H} + {}^2_1\text{H} \rightarrow {}^3_2\text{He} + {}^1_0\text{n} \quad Q = 3.27\ \text{MeV}$$

$$^2_1\text{H} + {}^2_1\text{H} \rightarrow {}^3_1\text{H} + {}^1_1\text{H} \quad Q = 4.03\ \text{MeV} \tag{45.4}$$

$$^2_1\text{H} + {}^3_1\text{H} \rightarrow {}^4_2\text{He} + {}^1_0\text{n} \quad Q = 17.59\ \text{MeV}$$

As noted earlier, deuterium is available in almost unlimited quantities from our lakes and oceans and is very inexpensive to extract. Tritium, however, is radioactive ($T_{1/2} = 12.3$ yr) and undergoes beta decay to ³He. For this reason, tritium does not occur naturally to any great extent and must be artificially produced.

One major problem in obtaining energy from nuclear fusion is that the Coulomb repulsive force between two nuclei, which carry positive charges, must be overcome before they can fuse. Figure 45.7 is a graph of potential energy as a function of the separation distance between two deuterons (deuterium nuclei, each having charge $+e$). The potential energy is positive in the region $r > R$, where the Coulomb repulsive force dominates ($R \approx 1$ fm), and negative in the region $r < R$, where the nuclear force dominates. The fundamental problem then is to give the two nuclei enough kinetic energy to overcome this repulsive force. This requirement can be accomplished by raising the fuel to extremely high temperatures (to approximately 10^8 K). At these high temperatures, the atoms are ionized and the system consists of a collection of electrons and nuclei, commonly referred to as a *plasma*.

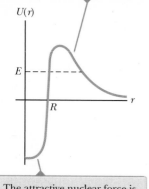

The Coulomb repulsive force is dominant for large separation distances between the deuterons.

The attractive nuclear force is dominant when the deuterons are close together.

Figure 45.7 Potential energy as a function of separation distance between two deuterons. R is on the order of 1 fm. If we neglect tunneling, the two deuterons require an energy E greater than the height of the barrier to undergo fusion.

Example 45.3	The Fusion of Two Deuterons

For the nuclear force to overcome the repulsive Coulomb force, the separation distance between two deuterons must be approximately 1.0×10^{-14} m.

(A) Calculate the height of the potential barrier due to the repulsive force.

SOLUTION

Conceptualize Imagine moving two deuterons toward each other. As they move closer together, the Coulomb repulsion force becomes stronger. Work must be done on the system to push against this force, and this work appears in the system of two deuterons as electric potential energy.

Categorize We categorize this problem as one involving the electric potential energy of a system of two charged particles.

Analyze Evaluate the potential energy associated with two charges separated by a distance r (Eq. 25.13) for two deuterons:

$$U = k_e \frac{q_1 q_2}{r} = k_e \frac{(+e)^2}{r} = (8.99 \times 10^9 \ \text{N} \cdot \text{m}^2/\text{C}^2) \frac{(1.60 \times 10^{-19} \ \text{C})^2}{1.0 \times 10^{-14} \ \text{m}}$$

$$= 2.3 \times 10^{-14} \ \text{J} = \boxed{0.14 \ \text{MeV}}$$

(B) Estimate the temperature required for a deuteron to overcome the potential barrier, assuming an energy of $\frac{3}{2} k_B T$ per deuteron (where k_B is Boltzmann's constant).

SOLUTION

Because the total Coulomb energy of the pair is 0.14 MeV, the Coulomb energy per deuteron is equal to 0.07 MeV = 1.1×10^{-14} J.

Set this energy equal to the average energy per deuteron:

$$\tfrac{3}{2} k_B T = 1.1 \times 10^{-14} \ \text{J}$$

Solve for T:

$$T = \frac{2(1.1 \times 10^{-14} \ \text{J})}{3(1.38 \times 10^{-23} \ \text{J/K})} = \boxed{5.6 \times 10^8 \ \text{K}}$$

(C) Find the energy released in the deuterium–deuterium reaction

$$^2_1\text{H} + ^2_1\text{H} \ \rightarrow \ ^3_1\text{H} + ^1_1\text{H}$$

SOLUTION

The mass of a single deuterium atom is equal to 2.014 102 u. Therefore, the total mass of the system before the reaction is 4.028 204 u.

Find the sum of the masses after the reaction:

$$3.016\ 049 \ \text{u} + 1.007\ 825 \ \text{u} = 4.023\ 874 \ \text{u}$$

Find the change in mass and convert to energy units:

$$4.028\ 204 \ \text{u} - 4.023\ 874 \ \text{u} = 0.004\ 33 \ \text{u}$$

$$= 0.004\ 33 \ \text{u} \times 931.494 \ \text{MeV/u} = \boxed{4.03 \ \text{MeV}}$$

Finalize The calculated temperature in part (B) is too high because the particles in the plasma have a Maxwellian speed distribution (Section 21.5) and therefore some fusion reactions are caused by particles in the high-energy tail of this distribution. Furthermore, even those particles that do not have enough energy to overcome the barrier have some probability of tunneling through (Section 41.5). When these effects are taken into account, a temperature of "only" 4×10^8 K appears adequate to fuse two deuterons in a plasma. In part (C), notice that the energy value is consistent with that already given in Equation 45.4.

WHAT IF? Suppose the tritium resulting from the reaction in part (C) reacts with another deuterium in the reaction

$$^2_1\text{H} + ^3_1\text{H} \ \rightarrow \ ^4_2\text{He} + ^1_0\text{n}$$

How much energy is released in the sequence of two reactions?

continued

▶ **45.3** continued

Answer The overall effect of the sequence of two reactions is that three deuterium nuclei have combined to form a helium nucleus, a hydrogen nucleus, and a neutron. The initial mass is $3(2.014\ 102\ \text{u}) = 6.042\ 306$ u. After the reaction, the sum of the masses is $4.002\ 603$ u $+ 1.007\ 825$ u $+ 1.008\ 665 = 6.019\ 093$ u. The excess mass is equal to $0.023\ 213$ u, equivalent to an energy of 21.6 MeV. Notice that this value is the sum of the Q values for the second and third reactions in Equation 45.4.

The temperature at which the power generation rate in any fusion reaction exceeds the loss rate is called the **critical ignition temperature** T_{ignit}. This temperature for the deuterium–deuterium (D–D) reaction is 4×10^8 K. From the relationship $E \approx \frac{3}{2}k_B T$, the ignition temperature is equivalent to approximately 52 keV. The critical ignition temperature for the deuterium–tritium (D–T) reaction is approximately 4.5×10^7 K, or only 6 keV. A plot of the power P_{gen} generated by fusion versus temperature for the two reactions is shown in Figure 45.8. The straight green line represents the power P_{lost} lost via the radiation mechanism known as bremsstrahlung (Section 42.8). In this principal mechanism of energy loss, radiation (primarily x-rays) is emitted as the result of electron–ion collisions within the plasma. The intersections of the P_{lost} line with the P_{gen} curves give the critical ignition temperatures.

In addition to the high-temperature requirements, two other critical parameters determine whether or not a thermonuclear reactor is successful: the **ion density** n and **confinement time** τ, which is the time interval during which energy injected into the plasma remains within the plasma. British physicist J. D. Lawson (1923–2008) showed that both the ion density and confinement time must be large enough to ensure that more fusion energy is released than the amount required to raise the temperature of the plasma. For a given value of n, the probability of fusion between two particles increases as τ increases. For a given value of τ, the collision rate between nuclei increases as n increases. The product $n\tau$ is referred to as the **Lawson number** of a reaction. A graph of the value of $n\tau$ necessary to achieve a net energy output for the D–T and D–D reactions at different temperatures is shown in Figure 45.9. In particular, **Lawson's criterion** states that a net energy output is possible for values of $n\tau$ that meet the following conditions:

$$n\tau \geq 10^{14}\ \text{s/cm}^3 \quad \text{(D–T)} \tag{45.5}$$

$$n\tau \geq 10^{16}\ \text{s/cm}^3 \quad \text{(D–D)}$$

These values represent the minima of the curves in Figure 45.9.

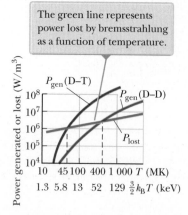

Figure 45.8 Power generated versus temperature for deuterium–deuterium (D–D) and deuterium–tritium (D–T) fusion. When the generation rate exceeds the loss rate, ignition takes place.

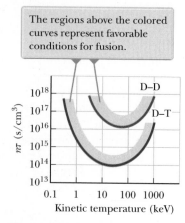

Figure 45.9 The Lawson number $n\tau$ at which net energy output is possible versus temperature for the D–T and D–D fusion reactions.

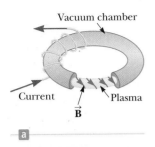

Vacuum chamber

Current   Plasma

$\vec{B}$

a

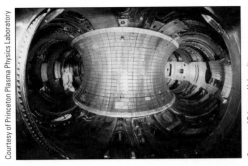

b

c

Figure 45.10 (a) Diagram of a tokamak used in the magnetic confinement scheme. (b) Interior view of the closed Tokamak Fusion Test Reactor (TFTR) vacuum vessel at the Princeton Plasma Physics Laboratory. (c) The National Spherical Torus Experiment (NSTX) that began operation in March 1999.

Lawson's criterion was arrived at by comparing the energy required to raise the temperature of a given plasma with the energy generated by the fusion process.[2] The energy E_{in} required to raise the temperature of the plasma is proportional to the ion density n, which we can express as $E_{in} = C_1 n$, where C_1 is some constant. The energy generated by the fusion process is proportional to $n^2 \tau$, or $E_{gen} = C_2 n^2 \tau$. This dependence may be understood by realizing that the fusion energy released is proportional to both the rate at which interacting ions collide ($\propto n^2$) and the confinement time τ. Net energy is produced when $E_{gen} > E_{in}$. When the constants C_1 and C_2 are calculated for different reactions, the condition that $E_{gen} \geq E_{in}$ leads to Lawson's criterion.

Current efforts are aimed at meeting Lawson's criterion at temperatures exceeding T_{ignit}. Although the minimum required plasma densities have been achieved, the problem of confinement time is more difficult. The two basic techniques under investigation for solving this problem are *magnetic confinement* and *inertial confinement*.

Magnetic Confinement

Many fusion-related plasma experiments use **magnetic confinement** to contain the plasma. A toroidal device called a **tokamak,** first developed in Russia, is shown in Figure 45.10a. A combination of two magnetic fields is used to confine and stabilize the plasma: (1) a strong toroidal field produced by the current in the toroidal windings surrounding a doughnut-shaped vacuum chamber and (2) a weaker "poloidal" field produced by the toroidal current. In addition to confining the plasma, the toroidal current is used to raise its temperature. The resultant helical magnetic field lines spiral around the plasma and keep it from touching the walls of the vacuum chamber. (If the plasma touches the walls, its temperature is reduced and heavy impurities sputtered from the walls "poison" it, leading to large power losses.)

One major breakthrough in magnetic confinement in the 1980s was in the area of auxiliary energy input to reach ignition temperatures. Experiments have shown

---

[2]Lawson's criterion neglects the energy needed to set up the strong magnetic field used to confine the hot plasma in a magnetic confinement approach. This energy is expected to be about 20 times greater than the energy required to raise the temperature of the plasma. It is therefore necessary either to have a magnetic energy recovery system or to use superconducting magnets.

that injecting a beam of energetic neutral particles into the plasma is a very efficient method of raising it to ignition temperatures. Radio-frequency energy input will probably be needed for reactor-size plasmas.

When it was in operation from 1982 to 1997, the Tokamak Fusion Test Reactor (TFTR, Fig. 45.10b) at Princeton University reported central ion temperatures of 510 million degrees Celsius, more than 30 times greater than the temperature at the center of the Sun. The $n\tau$ values in the TFTR for the D–T reaction were well above 10^{13} s/cm^3 and close to the value required by Lawson's criterion. In 1991, reaction rates of 6×10^{17} D–T fusions per second were reached in the Joint European Torus (JET) tokamak at Abington, England.

One of the new generation of fusion experiments is the National Spherical Torus Experiment (NSTX) at the Princeton Plasma Physics Laboratory and shown in Figure 45.10c. This reactor was brought on line in February 1999 and has been running fusion experiments since then. Rather than the doughnut-shaped plasma of a tokamak, the NSTX produces a spherical plasma that has a hole through its center. The major advantage of the spherical configuration is its ability to confine the plasma at a higher pressure in a given magnetic field. This approach could lead to development of smaller, more economical fusion reactors.

An international collaborative effort involving the United States, the European Union, Japan, China, South Korea, India, and Russia is currently under way to build a fusion reactor called ITER. This acronym stands for International Thermonuclear Experimental Reactor, although recently the emphasis has shifted to interpreting "iter" in terms of its Latin meaning, "the way." One reason proposed for this change is to avoid public misunderstanding and negative connotations toward the word *thermonuclear*. This facility will address the remaining technological and scientific issues concerning the feasibility of fusion power. The design is completed, and Cadarache, France, was chosen in June 2005 as the reactor site. Construction began in 2007 and will require about 10 years, with fusion operation projected to begin in 2019. If the planned device works as expected, the Lawson number for ITER will be about six times greater than the current record holder, the JT-60U tokamak in Japan. ITER is expected to produce ten times as much output power as input power, and the energy content of the alpha particles inside the reactor will be so intense that they will sustain the fusion reaction, allowing the auxiliary energy sources to be turned off once the reaction is initiated.

Example 45.4   **Inside a Fusion Reactor**

In 1998, the JT-60U tokamak in Japan operated with a D–T plasma density of 4.8×10^{13} cm^{-3} at a temperature (in energy units) of 24.1 keV. It confined this plasma inside a magnetic field for 1.1 s.

(A) Do these data meet Lawson's criterion?

SOLUTION

Conceptualize With the help of the third of Equations 45.4, imagine many such reactions occurring in a plasma of high temperature and high density.

Categorize We use the concept of the Lawson number discussed in this section, so we categorize this example as a substitution problem.

Evaluate the Lawson number for the JT-60U: $n\tau = (4.8 \times 10^{13}\ \text{cm}^{-3})(1.1\ \text{s}) = 5.3 \times 10^{13}\ \text{s/cm}^3$

This value is close to meeting Lawson's criterion of 10^{14} s/cm^3 for a D–T plasma given in Equation 45.5. In fact, scientists recorded a power gain of 1.25, indicating that the reactor operated slightly past the break-even point and produced more energy than it required to maintain the plasma.

(B) How does the plasma density compare with the density of atoms in an ideal gas when the gas is under standard conditions ($T = 0°C$ and $P = 1$ atm)?

▶ **45.4** continued

SOLUTION

Find the density of atoms in a sample of ideal gas by evaluating N_A/V_{mol}, where N_A is Avogadro's number and V_{mol} is the molar volume of an ideal gas under standard conditions, 2.24×10^{-2} m³/mol:

$$\frac{N_A}{V_{mol}} = \frac{6.02 \times 10^{23} \text{ atoms/mol}}{2.24 \times 10^{-2} \text{ m}^3/\text{mol}} = 2.7 \times 10^{25} \text{ atoms/m}^3$$
$$= 2.7 \times 10^{19} \text{ atoms/cm}^3$$

This value is more than 500 000 times greater than the plasma density in the reactor.

Inertial Confinement

The second technique for confining a plasma, called **inertial confinement,** makes use of a D–T target that has a very high particle density. In this scheme, the confinement time is very short (typically 10^{-11} to 10^{-9} s), and, because of their own inertia, the particles do not have a chance to move appreciably from their initial positions. Therefore, Lawson's criterion can be satisfied by combining a high particle density with a short confinement time.

Laser fusion is the most common form of inertial confinement. A small D–T pellet, approximately 1 mm in diameter, is struck simultaneously by several focused, high-intensity laser beams, resulting in a large pulse of input energy that causes the surface of the fuel pellet to evaporate (Fig. 45.11). The escaping particles exert a third-law reaction force on the core of the pellet, resulting in a strong, inwardly moving compressive shock wave. This shock wave increases the pressure and density of the core and produces a corresponding increase in temperature. When the temperature of the core reaches ignition temperature, fusion reactions occur.

One of the leading laser fusion laboratories in the United States is the Omega facility at the University of Rochester in New York. This facility focuses 24 laser beams on the target. Currently under operation at the Lawrence Livermore National Laboratory in Livermore, California, is the National Ignition Facility. The research apparatus there includes 192 laser beams that can be focused on a deuterium–tritium pellet. Construction was completed in early 2009, and a test firing of the lasers in March 2012 broke the record for lasers, delivering 1.87 MJ to a target. This energy is delivered in such a short time interval that the power is immense: 500 trillion watts, more than 1 000 times the power used in the United States at any moment.

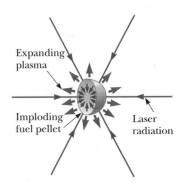

Figure 45.11 In inertial confinement, a D–T fuel pellet fuses when struck by several high-intensity laser beams simultaneously.

Fusion Reactor Design

In the D–T fusion reaction

$$^2_1\text{H} + ^3_1\text{H} \rightarrow ^4_2\text{He} + ^1_0\text{n} \qquad Q = 17.59 \text{ MeV}$$

the alpha particle carries 20% of the energy and the neutron carries 80%, or approximately 14 MeV. A diagram of the deuterium–tritium fusion reaction is shown in Figure 45.12. Because the alpha particles are charged, they are primarily absorbed by the plasma, causing the plasma's temperature to increase. In contrast, the 14-MeV neutrons, being electrically neutral, pass through the plasma and are absorbed by a surrounding blanket material, where their large kinetic energy is extracted and used to generate electric power.

One scheme is to use molten lithium metal as the neutron-absorbing material and to circulate the lithium in a closed heat-exchange loop, thereby producing steam and driving turbines as in a conventional power plant. Figure 45.13 (page 1142) shows a diagram of such a reactor. It is estimated that a blanket of lithium approximately 1 m thick will capture nearly 100% of the neutrons from the fusion of a small D–T pellet.

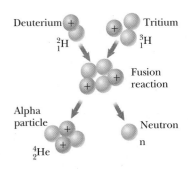

Figure 45.12 Deuterium–tritium fusion. Eighty percent of the energy released is in the 14-MeV neutron.

Figure 45.13 Diagram of a fusion reactor.

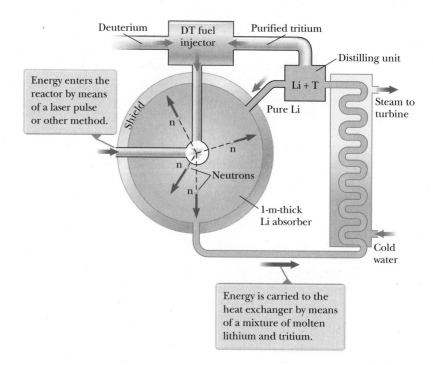

The capture of neutrons by lithium is described by the reaction

$$^{1}_{0}n + ^{6}_{3}Li \rightarrow ^{3}_{1}H + ^{4}_{2}He$$

where the kinetic energies of the charged tritium $^{3}_{1}H$ and alpha particle are transformed to internal energy in the molten lithium. An extra advantage of using lithium as the energy-transfer medium is that the tritium produced can be separated from the lithium and returned as fuel to the reactor.

Advantages and Problems of Fusion

If fusion power can ever be harnessed, it will offer several advantages over fission-generated power: (1) low cost and abundance of fuel (deuterium), (2) impossibility of runaway accidents, and (3) decreased radiation hazard. Some of the anticipated problems and disadvantages include (1) scarcity of lithium, (2) limited supply of helium, which is needed for cooling the superconducting magnets used to produce strong confining fields, and (3) structural damage and induced radioactivity caused by neutron bombardment. If such problems and the engineering design factors can be resolved, nuclear fusion may become a feasible source of energy in the twenty-first century.

45.5 Radiation Damage

In Chapter 34, we learned that electromagnetic radiation is all around us in the form of radio waves, microwaves, light waves, and so on. In this section, we describe forms of radiation that can cause severe damage as they pass through matter, such as radiation resulting from radioactive processes and radiation in the form of energetic particles such as neutrons and protons.

The degree and type of damage depend on several factors, including the type and energy of the radiation and the properties of the matter. The metals used in nuclear reactor structures can be severely weakened by high fluxes of energetic neutrons because these high fluxes often lead to metal fatigue. The damage in such situations is in the form of atomic displacements, often resulting in major alterations in the properties of the material.

Radiation damage in biological organisms is primarily due to ionization effects in cells. A cell's normal operation may be disrupted when highly reactive ions are formed as the result of ionizing radiation. For example, hydrogen and the hydroxyl radical OH^- produced from water molecules can induce chemical reactions that may break bonds in proteins and other vital molecules. Furthermore, the ionizing radiation may affect vital molecules directly by removing electrons from their structure. Large doses of radiation are especially dangerous because damage to a great number of molecules in a cell may cause the cell to die. Although the death of a single cell is usually not a problem, the death of many cells may result in irreversible damage to the organism. Cells that divide rapidly, such as those of the digestive tract, reproductive organs, and hair follicles, are especially susceptible. In addition, cells that survive the radiation may become defective. These defective cells can produce more defective cells and can lead to cancer.

In biological systems, it is common to separate radiation damage into two categories: somatic damage and genetic damage. *Somatic damage* is that associated with any body cell except the reproductive cells. Somatic damage can lead to cancer or can seriously alter the characteristics of specific organisms. *Genetic damage* affects only reproductive cells. Damage to the genes in reproductive cells can lead to defective offspring. It is important to be aware of the effect of diagnostic treatments, such as x-rays and other forms of radiation exposure, and to balance the significant benefits of treatment with the damaging effects.

Damage caused by radiation also depends on the radiation's penetrating power. Alpha particles cause extensive damage, but penetrate only to a shallow depth in a material due to the strong interaction with other charged particles. Neutrons do not interact via the electric force and hence penetrate deeper, causing significant damage. Gamma rays are high-energy photons that can cause severe damage, but often pass through matter without interaction.

Several units have been used historically to quantify the amount, or dose, of any radiation that interacts with a substance.

> The **roentgen** (R) is that amount of ionizing radiation that produces an electric charge of 3.33×10^{-10} C in 1 cm^3 of air under standard conditions.

Equivalently, the roentgen is that amount of radiation that increases the energy of 1 kg of air by 8.76×10^{-3} J.

For most applications, the roentgen has been replaced by the rad (an acronym for *radiation absorbed dose*):

> One **rad** is that amount of radiation that increases the energy of 1 kg of absorbing material by 1×10^{-2} J.

Although the rad is a perfectly good physical unit, it is not the best unit for measuring the degree of biological damage produced by radiation because damage depends not only on the dose but also on the type of the radiation. For example, a given dose of alpha particles causes about ten times more biological damage than an equal dose of x-rays. The **RBE** (relative biological effectiveness) factor for a given type of radiation is **the number of rads of x-radiation or gamma radiation that produces the same biological damage as 1 rad of the radiation being used.** The RBE factors for different types of radiation are given in Table 45.1 (page 1144). The values are only approximate because they vary with particle energy and with the form of the damage. The RBE factor should be considered only a first-approximation guide to the actual effects of radiation.

Finally, the **rem** (radiation equivalent in man) is the product of the dose in rad and the RBE factor:

$$\text{Dose in rem} \equiv \text{dose in rad} \times \text{RBE} \qquad \textbf{(45.6)} \qquad \blacktriangleleft \text{ Radiation dose in rem}$$

Table 45.1 **RBE Factors for Several Types of Radiation**

Radiation	RBE Factor
X-rays and gamma rays	1.0
Beta particles	1.0–1.7
Alpha particles	10–20
Thermal neutrons	4–5
Fast neutrons and protons	10
Heavy ions	20

Note: RBE = relative biological effectiveness.

Table 45.2 **Units for Radiation Dosage**

Quantity	SI Unit	Symbol	Relations to Other SI Units	Older Unit	Conversion
Absorbed dose	gray	Gy	= 1 J/kg	rad	1 Gy = 100 rad
Dose equivalent	sievert	Sv	= 1 J/kg	rem	1 Sv = 100 rem

According to this definition, 1 rem of any two types of radiation produces the same amount of biological damage. Table 45.1 shows that a dose of 1 rad of fast neutrons represents an effective dose of 10 rem, but 1 rad of gamma radiation is equivalent to a dose of only 1 rem.

This discussion has focused on measurements of radiation dosage in units such as rads and rems because these units are still widely used. They have, however, been formally replaced with new SI units. The rad has been replaced with the *gray* (Gy), equal to 100 rad, and the rem has been replaced with the *sievert* (Sv), equal to 100 rem. Table 45.2 summarizes the older and the current SI units of radiation dosage.

Low-level radiation from natural sources such as cosmic rays and radioactive rocks and soil delivers to each of us a dose of approximately 2.4 mSv/yr. This radiation, called *background radiation,* varies with geography, with the main factors being altitude (exposure to cosmic rays) and geology (radon gas released by some rock formations, deposits of naturally radioactive minerals).

The upper limit of radiation dose rate recommended by the U.S. government (apart from background radiation) is approximately 5 mSv/yr. Many occupations involve much higher radiation exposures, so an upper limit of 50 mSv/yr has been set for combined whole-body exposure. Higher upper limits are permissible for certain parts of the body, such as the hands and the forearms. A dose of 4 to 5 Sv results in a mortality rate of approximately 50% (which means that half the people exposed to this radiation level die). The most dangerous form of exposure for most people is either ingestion or inhalation of radioactive isotopes, especially isotopes of those elements the body retains and concentrates, such as ^{90}Sr.

45.6 Uses of Radiation

Nuclear physics applications are extremely widespread in manufacturing, medicine, and biology. In this section, we present a few of these applications and the underlying theories supporting them.

Tracing

Radioactive tracers are used to track chemicals participating in various reactions. One of the most valuable uses of radioactive tracers is in medicine. For example, iodine, a nutrient needed by the human body, is obtained largely through the

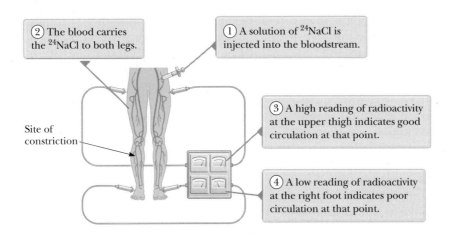

Figure 45.14 A tracer technique for determining the condition of the human circulatory system.

② The blood carries the ^{24}NaCl to both legs.

① A solution of ^{24}NaCl is injected into the bloodstream.

③ A high reading of radioactivity at the upper thigh indicates good circulation at that point.

Site of constriction

④ A low reading of radioactivity at the right foot indicates poor circulation at that point.

intake of iodized salt and seafood. To evaluate the performance of the thyroid, the patient drinks a very small amount of radioactive sodium iodide containing ^{131}I, an artificially produced isotope of iodine (the natural, nonradioactive isotope is ^{127}I). The amount of iodine in the thyroid gland is determined as a function of time by measuring the radiation intensity at the neck area. How much of the isotope ^{131}I remains in the thyroid is a measure of how well that gland is functioning.

A second medical application is indicated in Figure 45.14. A solution containing radioactive sodium is injected into a vein in the leg, and the time at which the radioisotope arrives at another part of the body is detected with a radiation counter. The elapsed time is a good indication of the presence or absence of constrictions in the circulatory system.

Tracers are also useful in agricultural research. Suppose the best method of fertilizing a plant is to be determined. A certain element in a fertilizer, such as nitrogen, can be *tagged* (identified) with one of its radioactive isotopes. The fertilizer is then sprayed on one group of plants, sprinkled on the ground for a second group, and raked into the soil for a third. A Geiger counter is then used to track the nitrogen through each of the three groups.

Tracing techniques are as wide ranging as human ingenuity can devise. Today, applications range from checking how teeth absorb fluoride to monitoring how cleansers contaminate food-processing equipment to studying deterioration inside an automobile engine. In this last case, a radioactive material is used in the manufacture of the car's piston rings and the oil is checked for radioactivity to determine the amount of wear on the rings.

Materials Analysis

For centuries, a standard method of identifying the elements in a sample of material has been chemical analysis, which involves determining how the material reacts with various chemicals. A second method is spectral analysis, which works because each element, when excited, emits its own characteristic set of electromagnetic wavelengths. These methods are now supplemented by a third technique, **neutron activation analysis.** A disadvantage of both chemical and spectral methods is that a fairly large sample of the material must be destroyed for the analysis. In addition, extremely small quantities of an element may go undetected by either method. Neutron activation analysis has an advantage over chemical analysis and spectral analysis in both respects.

When a material is irradiated with neutrons, nuclei in the material absorb the neutrons and are changed to different isotopes, most of which are radioactive. For example, ^{65}Cu absorbs a neutron to become ^{66}Cu, which undergoes beta decay:

$$^{1}_{0}n + {}^{65}_{29}Cu \rightarrow {}^{66}_{29}Cu \rightarrow {}^{66}_{30}Zn + e^- + \bar{\nu}$$

The presence of the copper can be deduced because it is known that ^{66}Cu has a half-life of 5.1 min and decays with the emission of beta particles having a maximum energy of 2.63 MeV. Also emitted in the decay of ^{66}Cu is a 1.04-MeV gamma ray. By examining the radiation emitted by a substance after it has been exposed to neutron irradiation, one can detect extremely small amounts of an element in that substance.

Neutron activation analysis is used routinely in a number of industries. In commercial aviation, for example, it is used to check airline luggage for hidden explosives. One nonroutine use is of historical interest. Napoleon died on the island of St. Helena in 1821, supposedly of natural causes. Over the years, suspicion has existed that his death was not all that natural. After his death, his head was shaved and locks of his hair were sold as souvenirs. In 1961, the amount of arsenic in a sample of this hair was measured by neutron activation analysis, and an unusually large quantity of arsenic was found. (Activation analysis is so sensitive that very small pieces of a single hair could be analyzed.) Results showed that the arsenic was fed to him irregularly. In fact, the arsenic concentration pattern corresponded to the fluctuations in the severity of Napoleon's illness as determined from historical records.

Art historians use neutron activation analysis to detect forgeries. The pigments used in paints have changed throughout history, and old and new pigments react differently to neutron activation. The method can even reveal hidden works of art behind existing paintings because an older, hidden layer of paint reacts differently than the surface layer to neutron activation.

Radiation Therapy

Radiation causes much damage to rapidly dividing cells. Therefore, it is useful in cancer treatment because tumor cells divide extremely rapidly. Several mechanisms can be used to deliver radiation to a tumor. In Section 42.8, we discussed the use of high-energy x-rays in the treatment of cancerous tissue. Other treatment protocols include the use of narrow beams of radiation from a radioactive source. As an example, Figure 45.15 shows a machine that uses ^{60}Co as a source. The ^{60}Co isotope emits gamma rays with photon energies higher than 1 MeV.

In other situations, a technique called *brachytherapy* is used. In this treatment plan, thin radioactive needles called *seeds* are implanted in the cancerous tissue. The energy emitted from the seeds is delivered directly to the tumor, reducing the exposure of surrounding tissue to radiation damage. In the case of prostate cancer, the active isotopes used in brachytherapy include ^{125}I and ^{103}Pd.

Food Preservation

Radiation is finding increasing use as a means of preserving food because exposure to high levels of radiation can destroy or incapacitate bacteria and mold spores (Fig. 45.16). Techniques include exposing foods to gamma rays, high-energy electron beams, and x-rays. Food preserved by such exposure can be placed in a sealed container (to keep out new spoiling agents) and stored for long periods of time. There is little or no evidence of adverse effect on the taste or nutritional value of food

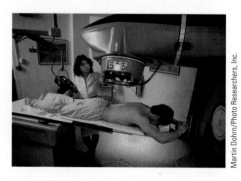

Figure 45.15 This large machine is being set to deliver a dose of radiation from ^{60}Co in an effort to destroy a cancerous tumor. Cancer cells are especially susceptible to this type of therapy because they tend to divide more often than cells of healthy tissue nearby.

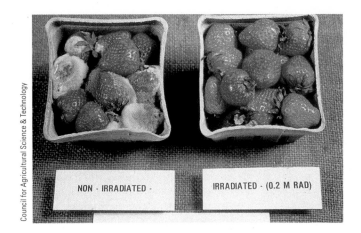

Figure 45.16 The strawberries on the left are untreated and have become moldy. The unspoiled strawberries on the right have been irradiated. The radiation has killed or incapacitated the mold spores that have spoiled the strawberries on the left.

NON - IRRADIATED -

IRRADIATED - (0.2 M RAD)

from irradiation. The safety of irradiated foods has been endorsed by the World Health Organization, the Centers for Disease Control and Prevention, the U.S. Department of Agriculture, and the Food and Drug Administration. Irradiation of food is presently permitted in more than 50 countries. Some estimates place the amount of irradiated food in the world as high as 500 000 metric tons each year.

Summary

Concepts and Principles

The probability that neutrons are captured as they move through matter generally increases with decreasing neutron energy. A **thermal neutron** is a slow-moving neutron that has a high probability of being captured by a nucleus in a **neutron capture event**:

$$\,_0^1 n + \,_Z^A X \;\rightarrow\; \,_Z^{A+1} X^* \;\rightarrow\; \,_Z^{A+1} X + \gamma \qquad \text{(45.1)}$$

where $\,_Z^{A+1} X^*$ is an excited intermediate nucleus that rapidly emits a photon.

Nuclear fission occurs when a very heavy nucleus, such as ^{235}U, splits into two smaller **fission fragments.** Thermal neutrons can create fission in ^{235}U:

$$\,_0^1 n + \,_{92}^{235} U \;\rightarrow\; \,_{92}^{236} U^* \rightarrow X + Y + \text{neutrons} \qquad \text{(45.2)}$$

where $^{236}U^*$ is an intermediate excited state and X and Y are the fission fragments. On average, 2.5 neutrons are released per fission event. The fragments then undergo a series of beta and gamma decays to various stable isotopes. The energy released per fission event is approximately 200 MeV.

The **reproduction constant** K is the average number of neutrons released from each fission event that cause another event. In a fission reactor, it is necessary to maintain $K \approx 1$. The value of K is affected by such factors as reactor geometry, mean neutron energy, and probability of neutron capture.

In **nuclear fusion,** two light nuclei fuse to form a heavier nucleus and release energy. The major obstacle in obtaining useful energy from fusion is the large Coulomb repulsive force between the charged nuclei at small separation distances. The temperature required to produce fusion is on the order of 10^8 K, and at this temperature, all matter occurs as a plasma.

In a fusion reactor, the plasma temperature must reach the **critical ignition temperature,** the temperature at which the power generated by the fusion reactions exceeds the power lost in the system. The most promising fusion reaction is the D–T reaction, which has a critical ignition temperature of approximately 4.5×10^7 K. Two critical parameters in fusion reactor design are **ion density** n and **confinement time** τ, the time interval during which the interacting particles must be maintained at $T > T_{\text{ignit}}$. **Lawson's criterion** states that for the D–T reaction, $n\tau \geq 10^{14}$ s/cm³.

Objective Questions

1. In a certain fission reaction, a ^{235}U nucleus captures a neutron. This process results in the creation of the products ^{137}I and ^{96}Y along with how many neutrons? (a) 1 (b) 2 (c) 3 (d) 4 (e) 5

2. Which particle is most likely to be captured by a ^{235}U nucleus and cause it to undergo fission? (a) an energetic proton (b) an energetic neutron (c) a slow-moving alpha particle (d) a slow-moving neutron (e) a fast-moving electron

3. In the first nuclear weapon test carried out in New Mexico, the energy released was equivalent to approximately 17 kilotons of TNT. Estimate the mass decrease in the nuclear fuel representing the energy converted from rest energy into other forms in this event. *Note:* One ton of TNT has the energy equivalent of 4.2×10^9 J. (a) 1 μg (b) 1 mg (c) 1 g (d) 1 kg (e) 20 kg

4. Working with radioactive materials at a laboratory over one year, (a) Tom received 1 rem of alpha radiation, (b) Karen received 1 rad of fast neutrons, (c) Paul received 1 rad of thermal neutrons as a whole-body dose, and (d) Ingrid received 1 rad of thermal neutrons to her hands only. Rank these four doses according to the likely amount of biological damage from the greatest to the least, noting any cases of equality.

5. If the moderator were suddenly removed from a nuclear reactor in an electric generating station, what is the most likely consequence? (a) The reactor would go supercritical, and a runaway reaction would occur. (b) The nuclear reaction would proceed in the same way, but the reactor would overheat. (c) The reactor would become subcritical, and the reaction would die out. (d) No change would occur in the reactor's operation.

6. You may use Figure 44.5 to answer this question. Three nuclear reactions take place, each involving 108 nucleons: (1) eighteen ^6Li nuclei fuse in pairs to form nine ^{12}C nuclei, (2) four nuclei each with 27 nucleons fuse in pairs to form two nuclei with 54 nucleons, and (3) one nucleus with 108 nucleons fissions to form two nuclei with 54 nucleons. Rank these three reactions from the largest positive Q value (representing energy output) to the largest negative value (representing energy input). Also include $Q = 0$ in your ranking to make clear which of the reactions put out energy and which absorb energy. Note any cases of equality in your ranking.

7. A device called a *bubble chamber* uses a liquid (usually liquid hydrogen) maintained near its boiling point. Ions produced by incoming charged particles from nuclear decays leave bubble tracks, which can be photographed. Figure OQ45.7 shows particle tracks in a bubble chamber immersed in a magnetic field. The tracks are generally spirals rather than sections of circles. What is the primary reason for this shape? (a) The magnetic field is not perpendicular to the velocity of the particles. (b) The magnetic field is not uniform in space. (c) The forces on the particles increase with time. (d) The speeds of the particles decrease with time.

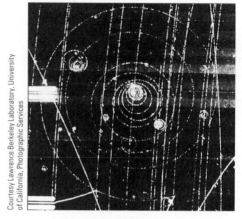

Courtesy Lawrence Berkeley Laboratory, University of California, Photographic Services

Figure OQ45.7

8. If an alpha particle and an electron have the same kinetic energy, which undergoes the greater deflection when passed through a magnetic field? (a) The alpha particle does. (b) The electron does. (c) They undergo the same deflection. (d) Neither is deflected.

9. Which of the following fuel conditions is *not* necessary to operate a self-sustained controlled fusion reactor? (a) The fuel must be at a sufficiently high temperature. (b) The fuel must be radioactive. (c) The fuel must be at a sufficiently high density. (d) The fuel must be confined for a sufficiently long period of time. (e) Conditions (a) through (d) are all necessary.

Conceptual Questions

1. What factors make a terrestrial fusion reaction difficult to achieve?

2. Lawson's criterion states that the product of ion density and confinement time must exceed a certain number before a break-even fusion reaction can occur. Why should these two parameters determine the outcome?

3. Why would a fusion reactor produce less radioactive waste than a fission reactor?

4. Discuss the advantages and disadvantages of fission reactors from the point of view of safety, pollution, and resources. Make a comparison with power generated from the burning of fossil fuels.

5. Discuss the similarities and differences between fusion and fission.

6. If a nucleus captures a slow-moving neutron, the product is left in a highly excited state, with an energy approximately 8 MeV above the ground state. Explain the source of the excitation energy.

7. Discuss the advantages and disadvantages of fusion power from the viewpoint of safety, pollution, and resources.

8. A scintillation crystal can be a detector of radiation when combined with a photomultiplier tube (Section 40.2). The scintillator is usually a solid or liquid material whose atoms are easily excited by radiation. The excited atoms then emit photons when they return to their ground state. The design of the radiation detector in Figure CQ45.8 might suggest that any number of dynodes may be used to amplify a weak signal. What factors do you suppose would limit the amplification in this device?

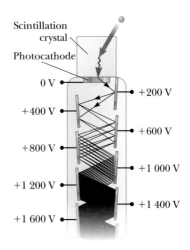

Figure CQ45.8

9. Why is water a better shield against neutrons than lead or steel?

Problems available in **ENHANCED** **WebAssign**   Access end-of-chapter problems online at **www.webassign.net**

Section 45.1 Interactions Involving Neutrons
Section 45.2 Nuclear Fission
Problems 1–11

Section 45.3 Nuclear Reactors
Problems 12–20

Section 45.4 Nuclear Fusion
Problems 21–31

Section 45.5 Radiation Damage
Problems 32–42

Section 45.6 Uses of Radiation
Problems 43–45

Additional Problems
Problems 46–70

Challenge Problems
Problems 71–73

Solutions to the following Problems are available in the *Student Solutions Manual/Study Guide:*
45.5, 45.11, 45.17, 45.29, 45.30, 45.37, 45.39, 45.43, 45.51, 45.56, 45.58, 45.59, 45.63, 45.64, and 45.67

List of Enhanced Problems

Problem Number	Targeted Feedback in Enhanced WebAssign	Analysis Model Tutorial in Enhanced WebAssign	Master It in Enhanced WebAssign	Watch It in Enhanced WebAssign
45.1			✓	
45.11	✓	✓	✓	
45.12			✓	
45.19			✓	
45.29			✓	
45.36			✓	
45.39	✓			✓
45.43			✓	
45.51			✓	
45.55	✓			✓

Particle Physics and Cosmology

The word *atom* comes from the Greek *atomos*, which means "indivisible." The early Greeks believed that atoms were the indivisible constituents of matter; that is, they regarded them as elementary particles. After 1932, physicists viewed all matter as consisting of three constituent particles: electrons, protons, and neutrons. Beginning in the 1940s, many "new" particles were discovered in experiments involving high-energy collisions between known particles. The new particles are characteristically very unstable and have very short half-lives, ranging between 10^{-6} s and 10^{-23} s. So far, more than 300 of these particles have been catalogued.

Until the 1960s, physicists were bewildered by the great number and variety of subatomic particles that were being discovered. They wondered whether the particles had no systematic relationship connecting them or whether a pattern was emerging that would provide a better understanding of the elaborate structure in the subatomic world. For example, that the neutron has a magnetic moment despite having zero electric charge (Section 44.8) suggests an underlying structure to the neutron. The periodic table explains how more than 100 elements can be formed from three types of particles (electrons, protons, and

One of the most intense areas of current research is the hunt for the Higgs boson, discussed in Section 46.10. The photo shows an event recorded at the Large Hadron Collider in July 2012 that shows particles consistent with the creation of a Higgs boson. The data is not entirely conclusive, however, and the hunt continues. *(CERN)*

neutrons), which suggests there is, perhaps, a means of forming more than 300 subatomic particles from a small number of basic building blocks.

Recall Figure 1.2, which illustrated the various levels of structure in matter. We studied the atomic structure of matter in Chapter 42. In Chapter 44, we investigated the substructure of the atom by describing the structure of the nucleus. As mentioned in Section 1.2, the protons and neutrons in the nucleus, and a host of other exotic particles, are now known to be composed of six different varieties of particles called *quarks*. In this concluding chapter, we examine the current theory of elementary particles, in which all matter is constructed from only two families of particles, quarks and leptons. We also discuss how clarifications of such models might help scientists understand the birth and evolution of the Universe.

46.1 The Fundamental Forces in Nature

As noted in Section 5.1, all natural phenomena can be described by four fundamental forces acting between particles. In order of decreasing strength, they are the nuclear force, the electromagnetic force, the weak force, and the gravitational force.

The nuclear force discussed in Chapter 44 is an attractive force between nucleons. It has a very short range and is negligible for separation distances between nucleons greater than approximately 10^{-15} m (about the size of the nucleus). The electromagnetic force, which binds atoms and molecules together to form ordinary matter, has a strength of approximately 10^{-2} times that of the nuclear force. This long-range force decreases in magnitude as the inverse square of the separation between interacting particles. The weak force is a short-range force that tends to produce instability in certain nuclei. It is responsible for decay processes, and its strength is only about 10^{-5} times that of the nuclear force. Finally, the gravitational force is a long-range force that has a strength of only about 10^{-39} times that of the nuclear force. Although this familiar interaction is the force that holds the planets, stars, and galaxies together, its effect on elementary particles is negligible.

In Section 13.3, we discussed the difficulty early scientists had with the notion of the gravitational force acting at a distance, with no physical contact between the interacting objects. To resolve this difficulty, the concept of the gravitational field was introduced. Similarly, in Chapter 23, we introduced the electric field to describe the electric force acting between charged objects, and we followed that with a discussion of the magnetic field in Chapter 29. For each of these types of fields, we developed a particle in a field analysis model. In modern physics, the nature of the interaction between particles is carried a step further. These interactions are described in terms of the exchange of entities called **field particles** or **exchange particles**. Field particles are also called **gauge bosons**.[1] The interacting particles continuously emit and absorb field particles. The emission of a field particle by one particle and its absorption by another manifests as a force between the two interacting particles. In the case of the electromagnetic interaction, for instance, the field particles are photons. In the language of modern physics, the electromagnetic force is said to be *mediated* by photons, and photons are the field particles of the electromagnetic field. Likewise, the nuclear force is mediated by field particles called *gluons*. The weak force is mediated by field particles called *W* and *Z bosons*, and the gravitational force is proposed to be mediated by field particles called *gravitons*. These interactions, their ranges, and their relative strengths are summarized in Table 46.1.

---

[1] The word *bosons* suggests that the field particles have integral spin as discussed in Section 43.8. The word *gauge* comes from *gauge theory*, which is a sophisticated mathematical analysis that is beyond the scope of this book.

Table 46.1	**Particle Interactions**			
Interactions	**Relative Strength**	**Range of Force**	**Mediating Field Particle**	**Mass of Field Particle (GeV/c^2)**
Nuclear	1	Short ($\approx$ 1 fm)	Gluon	0
Electromagnetic	10^{-2}	∞	Photon	0
Weak	10^{-5}	Short ($\approx 10^{-3}$ fm)	$W^{\pm}$, Z^0 bosons	80.4, 80.4, 91.2
Gravitational	10^{-39}	∞	Graviton	0

Paul Adrien Maurice Dirac
British Physicist (1902–1984)
Dirac was instrumental in the understanding of antimatter and the unification of quantum mechanics and relativity. He made many contributions to the development of quantum physics and cosmology. In 1933, Dirac won a Nobel Prize in Physics.

46.2 Positrons and Other Antiparticles

In the 1920s, Paul Dirac developed a relativistic quantum-mechanical description of the electron that successfully explained the origin of the electron's spin and its magnetic moment. His theory had one major problem, however: its relativistic wave equation required solutions corresponding to negative energy states, and if negative energy states existed, an electron in a state of positive energy would be expected to make a rapid transition to one of these states, emitting a photon in the process.

Dirac circumvented this difficulty by postulating that all negative energy states are filled. The electrons occupying these negative energy states are collectively called the *Dirac sea*. Electrons in the Dirac sea (the blue area in Fig. 46.1) are not directly observable because the Pauli exclusion principle does not allow them to react to external forces; there are no available states to which an electron can make a transition in response to an external force. Therefore, an electron in such a state acts as an isolated system unless an interaction with the environment is strong enough to excite the electron to a positive energy state. Such an excitation causes one of the negative energy states to be vacant as in Figure 46.1, leaving a hole in the sea of filled states. This process is described by the nonisolated system model: as energy enters the system by some transfer mechanism, the system energy increases and the electron is excited to a higher energy level. *The hole can react to external forces and is observable.* The hole reacts in a way similar to that of the electron except that it has a positive charge: it is the *antiparticle* to the electron.

This theory strongly suggested that *an antiparticle exists for every particle*, not only for fermions such as electrons but also for bosons. It has subsequently been verified that practically every known elementary particle has a distinct antiparticle. Among the exceptions are the photon and the neutral pion (π^0; see Section 46.3). Following the construction of high-energy accelerators in the 1950s, many other antiparticles were revealed. They included the antiproton, discovered by Emilio Segré (1905–1989) and Owen Chamberlain (1920–2006) in 1955, and the antineutron, discovered shortly thereafter. The antiparticle for a charged particle has the same mass as the particle but opposite charge.[2] For example, the electron's antiparticle (the *positron* mentioned in Section 44.4) has a rest energy of 0.511 MeV and a positive charge of $+1.60 \times 10^{-19}$ C.

Carl Anderson (1905–1991) observed the positron experimentally in 1932 and was awarded a Nobel Prize in Physics in 1936 for this achievement. Anderson discovered the positron while examining tracks created in a cloud chamber by electron-like particles of positive charge. (These early experiments used cosmic rays—mostly energetic protons passing through interstellar space—to initiate high-energy reactions on the order of several GeV.) To discriminate between positive and negative charges, Anderson placed the cloud chamber in a magnetic field,

An electron can make a transition out of the Dirac sea only if it is provided with energy equal to or larger than $2m_e c^2$.

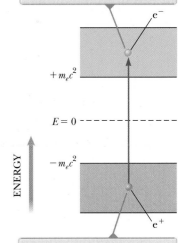

An upward transition of an electron leaves a vacancy in the Dirac sea, which can behave as a particle identical to the electron except for its positive charge.

Figure 46.1 Dirac's model for the existence of antielectrons (positrons). The minimum energy for an electron to exist in the gold band is its rest energy $m_e c^2$. The blue band of negative energies is filled with electrons.

---

[2]Antiparticles for uncharged particles, such as the neutron, are a little more difficult to describe. One basic process that can detect the existence of an antiparticle is pair annihilation. For example, a neutron and an antineutron can annihilate to form two gamma rays. Because the photon and the neutral pion do not have distinct antiparticles, pair annihilation is not observed with either of these particles.

Courtesy Lawrence Berkeley Laboratory, University of California, Photographic Services

Figure 46.2 (a) Bubble-chamber tracks of electron–positron pairs produced by 300-MeV gamma rays striking a lead sheet from the left. (b) The pertinent pair-production events. The positrons deflect upward and the electrons downward in an applied magnetic field.

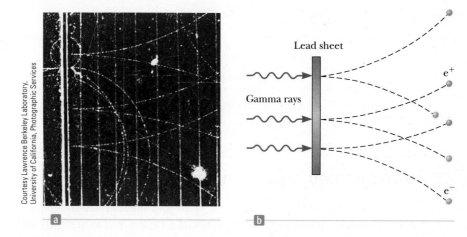

a

b

Lead sheet

Gamma rays

e^+

e^-

Pitfall Prevention 46.1

Antiparticles An antiparticle is not identified solely on the basis of opposite charge; even neutral particles have antiparticles, which are defined in terms of other properties, such as spin.

---

causing moving charges to follow curved paths. He noted that some of the electron-like tracks deflected in a direction corresponding to a positively charged particle.

Since Anderson's discovery, positrons have been observed in a number of experiments. A common source of positrons is **pair production.** In this process, a gamma-ray photon with sufficiently high energy interacts with a nucleus and an electron–positron pair is created from the photon. (The presence of the nucleus allows the principle of conservation of momentum to be satisfied.) Because the total rest energy of the electron–positron pair is $2m_ec^2 = 1.02$ MeV (where m_e is the mass of the electron), the photon must have at least this much energy to create an electron–positron pair. The energy of a photon is converted to rest energy of the electron and positron in accordance with Einstein's relationship $E_R = mc^2$. If the gamma-ray photon has energy in excess of the rest energy of the electron–positron pair, the excess appears as kinetic energy of the two particles. Figure 46.2 shows early observations of tracks of electron–positron pairs in a bubble chamber created by 300-MeV gamma rays striking a lead sheet.

Quick Quiz 46.1 Given the identification of the particles in Figure 46.2b, is the direction of the external magnetic field in Figure 46.2a **(a)** into the page, **(b)** out of the page, or **(c)** impossible to determine?

The reverse process can also occur. Under the proper conditions, an electron and a positron can annihilate each other to produce two gamma-ray photons that have a combined energy of at least 1.02 MeV:

$$e^- + e^+ \rightarrow 2\gamma$$

Because the initial momentum of the electron–positron system is approximately zero, the two gamma rays travel in opposite directions after the annihilation, satisfying the principle of conservation of momentum for the isolated system.

Electron–positron annihilation is used in the medical diagnostic technique called *positron-emission tomography* (PET). The patient is injected with a glucose solution containing a radioactive substance that decays by positron emission, and the material is carried throughout the body by the blood. A positron emitted during a decay event in one of the radioactive nuclei in the glucose solution annihilates with an electron in the surrounding tissue, resulting in two gamma-ray photons emitted in opposite directions. A gamma detector surrounding the patient pinpoints the source of the photons and, with the assistance of a computer, displays an image of the sites at which the glucose accumulates. (Glucose metabolizes rapidly in cancerous tumors and accumulates at those sites, providing a strong signal for a PET detector system.) The images from a PET scan can indicate a wide variety of disorders in the brain, including Alzheimer's disease (Fig. 46.3). In addition, because glucose metabolizes more rapidly in active areas

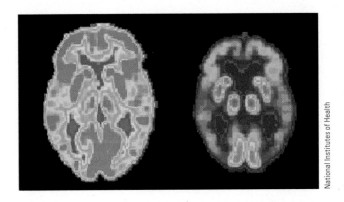

Figure 46.3 PET scans of the brain of a healthy older person *(left)* and that of a patient suffering from Alzheimer's disease *(right)*. Lighter regions contain higher concentrations of radioactive glucose, indicating higher metabolism rates and therefore increased brain activity.

of the brain, a PET scan can indicate areas of the brain involved in the activities in which the patient is engaging at the time of the scan, such as language use, music, and vision.

46.3 Mesons and the Beginning of Particle Physics

Physicists in the mid-1930s had a fairly simple view of the structure of matter. The building blocks were the proton, the electron, and the neutron. Three other particles were either known or postulated at the time: the photon, the neutrino, and the positron. Together these six particles were considered the fundamental constituents of matter. With this simple picture, however, no one was able to answer the following important question: the protons in any nucleus should strongly repel one another due to their charges of the same sign, so what is the nature of the force that holds the nucleus together? Scientists recognized that this mysterious force must be much stronger than anything encountered in nature up to that time. This force is the nuclear force discussed in Section 44.1 and examined in historical perspective in the following paragraphs.

The first theory to explain the nature of the nuclear force was proposed in 1935 by Japanese physicist Hideki Yukawa, an effort that earned him a Nobel Prize in Physics in 1949. To understand Yukawa's theory, recall the introduction of field particles in Section 46.1, which stated that each fundamental force is mediated by a field particle exchanged between the interacting particles. Yukawa used this idea to explain the nuclear force, proposing the existence of a new particle whose exchange between nucleons in the nucleus causes the nuclear force. He established that the range of the force is inversely proportional to the mass of this particle and predicted the mass to be approximately 200 times the mass of the electron. (Yukawa's predicted particle is *not* the gluon mentioned in Section 46.1, which is massless and is today considered to be the field particle for the nuclear force.) Because the new particle would have a mass between that of the electron and that of the proton, it was called a **meson** (from the Greek *meso*, "middle").

In efforts to substantiate Yukawa's predictions, physicists began experimental searches for the meson by studying cosmic rays entering the Earth's atmosphere. In 1937, Carl Anderson and his collaborators discovered a particle of mass 106 MeV/c^2, approximately 207 times the mass of the electron. This particle was thought to be Yukawa's meson. Subsequent experiments, however, showed that the particle interacted very weakly with matter and hence could not be the field particle for the nuclear force. That puzzling situation inspired several theoreticians to propose two mesons having slightly different masses equal to approximately 200 times that of the electron, one having been discovered by Anderson and the other, still undiscovered, predicted by Yukawa. This idea was confirmed in 1947 with the discovery of the **pi meson** (π), or simply **pion**. The particle discovered by Anderson in 1937, the one initially thought to be Yukawa's meson, is not really a

Hideki Yukawa
Japanese Physicist (1907–1981)
Yukawa was awarded the Nobel Prize in Physics in 1949 for predicting the existence of mesons. This photograph of him at work was taken in 1950 in his office at Columbia University. Yukawa came to Columbia in 1949 after spending the early part of his career in Japan.

Figure 46.4 Feynman diagram representing a photon mediating the electromagnetic force between two electrons.

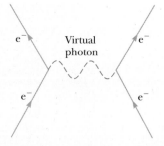

meson. (We shall discuss the characteristics of mesons in Section 46.4.) Instead, it takes part in the weak and electromagnetic interactions only and is now called the **muon** (μ).

The pion comes in three varieties, corresponding to three charge states: π^+, π^-, and π^0. The π^+ and π^- particles (π^- is the antiparticle of π^+) each have a mass of 139.6 MeV/c^2, and the π^0 mass is 135.0 MeV/c^2. Two muons exist: μ^- and its antiparticle μ^+.

Pions and muons are very unstable particles. For example, the π^-, which has a mean lifetime of 2.6×10^{-8} s, decays to a muon and an antineutrino.[3] The muon, which has a mean lifetime of 2.2 μs, then decays to an electron, a neutrino, and an antineutrino:

$$\pi^- \rightarrow \mu^- + \bar{\nu}$$
$$\mu^- \rightarrow e^- + \nu + \bar{\nu} \tag{46.1}$$

For chargeless particles (as well as some charged particles, such as the proton), a bar over the symbol indicates an antiparticle, as for the neutrino in beta decay (see Section 44.5). Other antiparticles, such as e^+ and μ^+, use a different notation.

The interaction between two particles can be represented in a simple diagram called a **Feynman diagram,** developed by American physicist Richard P. Feynman. Figure 46.4 is such a diagram for the electromagnetic interaction between two electrons. A Feynman diagram is a qualitative graph of time on the vertical axis versus space on the horizontal axis. It is qualitative in the sense that the actual values of time and space are not important, but the overall appearance of the graph provides a pictorial representation of the process.

In the simple case of the electron–electron interaction in Figure 46.4, a photon (the field particle) mediates the electromagnetic force between the electrons. Notice that the entire interaction is represented in the diagram as occurring at a single point in time. Therefore, the paths of the electrons appear to undergo a discontinuous change in direction at the moment of interaction. The electron paths shown in Figure 46.4 are different from the *actual* paths, which would be curved due to the continuous exchange of large numbers of field particles.

In the electron–electron interaction, the photon, which transfers energy and momentum from one electron to the other, is called a *virtual photon* because it vanishes during the interaction without having been detected. In Chapter 40, we discussed that a photon has energy $E = hf$, where f is its frequency. Consequently, for a system of two electrons initially at rest, the system has energy $2m_ec^2$ before a virtual photon is released and energy $2m_ec^2 + hf$ after the virtual photon is released (plus any kinetic energy of the electron resulting from the emission of the photon). Is that a violation of the law of conservation of energy for an isolated system? No; this process does *not* violate the law of conservation of energy because the virtual

Richard Feynman
American Physicist (1918–1988)
Inspired by Dirac, Feynman developed quantum electrodynamics, the theory of the interaction of light and matter on a relativistic and quantum basis. In 1965, Feynman won the Nobel Prize in Physics. The prize was shared by Feynman, Julian Schwinger, and Sin Itiro Tomonaga. Early in Feynman's career, he was a leading member of the team developing the first nuclear weapon in the Manhattan Project. Toward the end of his career, he worked on the commission investigating the 1986 *Challenger* tragedy and demonstrated the effects of cold temperatures on the rubber O-rings used in the space shuttle.

---

[3]The antineutrino is another zero-charge particle for which the identification of the antiparticle is more difficult than that for a charged particle. Although the details are beyond the scope of this book, the neutrino and antineutrino can be differentiated by means of the relationship between the linear momentum and the spin angular momentum of the particles.

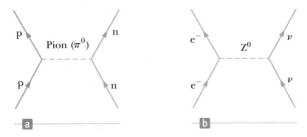

Figure 46.5 (a) Feynman diagram representing a proton and a neutron interacting via the nuclear force with a neutral pion mediating the force. (This model is *not* the current model for nucleon interaction.) (b) Feynman diagram for an electron and a neutrino interacting via the weak force, with a Z^0 boson mediating the force.

photon has a very short lifetime Δt that makes the uncertainty in the energy $\Delta E \approx \hbar/2 \, \Delta t$ of the system greater than the photon energy. Therefore, within the constraints of the uncertainty principle, the energy of the system is conserved.

Now consider a pion exchange between a proton and a neutron according to Yukawa's model (Fig. 46.5a). The energy ΔE_R needed to create a pion of mass m_π is given by Einstein's equation $\Delta E_R = m_\pi c^2$. As with the photon in Figure 46.4, the very existence of the pion would appear to violate the law of conservation of energy if the particle existed for a time interval greater than $\Delta t \approx \hbar/2 \, \Delta E_R$ (from the uncertainty principle), where Δt is the time interval required for the pion to transfer from one nucleon to the other. Therefore,

$$\Delta t \approx \frac{\hbar}{2 \, \Delta E_R} = \frac{\hbar}{2 m_\pi c^2}$$

and the rest energy of the pion is

$$m_\pi c^2 = \frac{\hbar}{2 \, \Delta t} \qquad (46.2)$$

Because the pion cannot travel faster than the speed of light, the maximum distance d it can travel in a time interval Δt is $c \, \Delta t$. Therefore, using Equation 46.2 and $d = c \, \Delta t$, we find

$$m_\pi c^2 = \frac{\hbar c}{2d} \qquad (46.3)$$

From Table 46.1, we know that the range of the nuclear force is on the order of 10^{-15} fm. Using this value for d in Equation 46.3, we estimate the rest energy of the pion to be

$$m_p c^2 \approx \frac{(1.055 \times 10^{-34} \, \text{J} \cdot \text{s})(3.00 \times 10^8 \, \text{m/s})}{2(1 \times 10^{-15} \, \text{m})}$$

$$= 1.6 \times 10^{-11} \, \text{J} \approx 100 \, \text{MeV}$$

which corresponds to a mass of 100 MeV/c^2 (approximately 200 times the mass of the electron). This value is in reasonable agreement with the observed pion mass.

The concept just described is quite revolutionary. In effect, it says that a system of two nucleons can change into two nucleons plus a pion as long as it returns to its original state in a very short time interval. (Remember that this description is the older historical model, which assumes the pion is the field particle for the nuclear force; the gluon is the actual field particle in current models.) Physicists often say that a nucleon undergoes *fluctuations* as it emits and absorbs field particles. These fluctuations are a consequence of a combination of quantum mechanics (through the uncertainty principle) and special relativity (through Einstein's energy–mass relationship $E_R = mc^2$).

In this section, we discussed the field particles that were originally proposed to mediate the nuclear force (pions) and those that mediate the electromagnetic force (photons). The graviton, the field particle for the gravitational force, has yet to be observed. In 1983, $W^\pm$ and Z^0 particles, which mediate the weak force, were discovered by Italian physicist Carlo Rubbia (b. 1934) and his associates, using a proton–antiproton collider. Rubbia and Simon van der Meer (1925–2011), both at CERN,[4] shared the 1984 Nobel Prize in Physics for the discovery of the $W^\pm$ and Z^0 particles and the development of the proton–antiproton collider. Figure 46.5b shows a Feynman diagram for a weak interaction mediated by a Z^0 boson.

46.4 Classification of Particles

All particles other than field particles can be classified into two broad categories, *hadrons* and *leptons*. The criterion for separating these particles into categories is whether or not they interact via the strong force. The nuclear force between nucleons in a nucleus is a particular manifestation of the strong force, but we will use the term *strong force* to refer to any interaction between particles made up of quarks. (For more detail on quarks and the strong force, see Section 46.8.) Table 46.2 provides a summary of the properties of hadrons and leptons.

Table 46.2   **Some Particles and Their Properties**

Category	Particle Name	Symbol	Anti-particle	Mass (MeV/c^2)	B	L_e	L_μ	L_τ	S	Lifetime(s)	Spin
Leptons	Electron	e^-	e^+	0.511	0	+1	0	0	0	Stable	$\frac{1}{2}$
	Electron–neutrino	ν_e	$\bar{\nu}_e$	$< 2\ eV/c^2$	0	+1	0	0	0	Stable	$\frac{1}{2}$
	Muon	μ^-	μ^+	105.7	0	0	+1	0	0	2.20×10^{-6}	$\frac{1}{2}$
	Muon–neutrino	ν_μ	$\bar{\nu}_\mu$	< 0.17	0	0	+1	0	0	Stable	$\frac{1}{2}$
	Tau	τ^-	τ^+	1 784	0	0	0	+1	0	$< 4 \times 10^{-13}$	$\frac{1}{2}$
	Tau–neutrino	ν_τ	$\bar{\nu}_\tau$	< 18	0	0	0	+1	0	Stable	$\frac{1}{2}$
Hadrons											
Mesons	Pion	π^+	π^-	139.6	0	0	0	0	0	2.60×10^{-8}	0
		π^0	Self	135.0	0	0	0	0	0	0.83×10^{-16}	0
	Kaon	K^+	K^-	493.7	0	0	0	0	+1	1.24×10^{-8}	0
		K_S^0	$\bar{K}_S^0$	497.7	0	0	0	0	+1	0.89×10^{-10}	0
		K_L^0	$\bar{K}_L^0$	497.7	0	0	0	0	+1	5.2×10^{-8}	0
	Eta	η	Self	548.8	0	0	0	0	0	$< 10^{-18}$	0
		η'	Self	958	0	0	0	0	0	2.2×10^{-21}	0
Baryons	Proton	p	$\bar{p}$	938.3	+1	0	0	0	0	Stable	$\frac{1}{2}$
	Neutron	n	$\bar{n}$	939.6	+1	0	0	0	0	614	$\frac{1}{2}$
	Lambda	Λ^0	$\bar{\Lambda}^0$	1 115.6	+1	0	0	0	−1	2.6×10^{-10}	$\frac{1}{2}$
	Sigma	Σ^+	$\bar{\Sigma}^-$	1 189.4	+1	0	0	0	−1	0.80×10^{-10}	$\frac{1}{2}$
		Σ^0	$\bar{\Sigma}^0$	1 192.5	+1	0	0	0	−1	6×10^{-20}	$\frac{1}{2}$
		Σ^-	$\bar{\Sigma}^+$	1 197.3	+1	0	0	0	−1	1.5×10^{-10}	$\frac{1}{2}$
	Delta	Δ^{++}	$\bar{\Delta}^{--}$	1 230	+1	0	0	0	0	6×10^{-24}	$\frac{3}{2}$
		Δ^+	$\bar{\Delta}^-$	1 231	+1	0	0	0	0	6×10^{-24}	$\frac{3}{2}$
		Δ^0	$\bar{\Delta}^0$	1 232	+1	0	0	0	0	6×10^{-24}	$\frac{3}{2}$
		Δ^-	$\bar{\Delta}^+$	1 234	+1	0	0	0	0	6×10^{-24}	$\frac{3}{2}$
	Xi	Ξ^0	$\bar{\Xi}^0$	1 315	+1	0	0	0	−2	2.9×10^{-10}	$\frac{1}{2}$
		Ξ^-	$\bar{\Xi}^+$	1 321	+1	0	0	0	−2	1.64×10^{-10}	$\frac{1}{2}$
	Omega	Ω^-	Ω^+	1 672	+1	0	0	0	−3	0.82×10^{-10}	$\frac{3}{2}$

---

[4]CERN was originally the Conseil Européen pour la Recherche Nucléaire; the name has been altered to the European Organization for Nuclear Research, and the laboratory operated by CERN is called the European Laboratory for Particle Physics. The CERN acronym has been retained and is commonly used to refer to both the organization and the laboratory.

Hadrons

Particles that interact through the strong force (as well as through the other fundamental forces) are called **hadrons.** The two classes of hadrons, *mesons* and *baryons,* are distinguished by their masses and spins.

Mesons all have zero or integer spin (0 or 1). As indicated in Section 46.3, the name comes from the expectation that Yukawa's proposed meson mass would lie between the masses of the electron and the proton. Several meson masses do lie in this range, although mesons having masses greater than that of the proton have been found to exist.

All mesons decay finally into electrons, positrons, neutrinos, and photons. The pions are the lightest known mesons and have masses of approximately 1.4×10^2 MeV/c^2, and all three pions—π^+, π^-, and π^0—have a spin of 0. (This spin-0 characteristic indicates that the particle discovered by Anderson in 1937, the muon, is not a meson. The muon has spin $\frac{1}{2}$ and belongs in the *lepton* classification, described below.)

Baryons, the second class of hadrons, have masses equal to or greater than the proton mass (the name *baryon* means "heavy" in Greek), and their spin is always a half-integer value ($\frac{1}{2}$, $\frac{3}{2}$, . . .). Protons and neutrons are baryons, as are many other particles. With the exception of the proton, all baryons decay in such a way that the end products include a proton. For example, the baryon called the Ξ^0 hyperon (Greek letter xi) decays to the Λ^0 baryon (Greek letter lambda) in approximately 10^{-10} s. The Λ^0 then decays to a proton and a π^- in approximately 3×10^{-10} s.

Today it is believed that hadrons are not elementary particles but instead are composed of more elementary units called quarks, per Section 46.8.

Leptons

Leptons (from the Greek *leptos,* meaning "small" or "light") are particles that do not interact by means of the strong force. All leptons have spin $\frac{1}{2}$. Unlike hadrons, which have size and structure, leptons appear to be truly elementary, meaning that they have no structure and are point-like.

Quite unlike the case with hadrons, the number of known leptons is small. Currently, scientists believe that only six leptons exist: the electron, the muon, the tau, and a neutrino associated with each: e^-, μ^-, τ^-, ν_e, ν_μ, and ν_τ. The tau lepton, discovered in 1975, has a mass about twice that of the proton. Direct experimental evidence for the neutrino associated with the tau was announced by the Fermi National Accelerator Laboratory (Fermilab) in July 2000. Each of the six leptons has an antiparticle.

Current studies indicate that neutrinos have a small but nonzero mass. If they do have mass, they cannot travel at the speed of light. In addition, because so many neutrinos exist, their combined mass may be sufficient to cause all the matter in the Universe to eventually collapse into a single point, which might then explode and create a completely new Universe! We shall discuss this possibility in more detail in Section 46.11.

46.5 Conservation Laws

The laws of conservation of energy, linear momentum, angular momentum, and electric charge for an isolated system provide us with a set of rules that all processes must follow. In Chapter 44, we learned that conservation laws are important for understanding why certain radioactive decays and nuclear reactions occur and others do not. In the study of elementary particles, a number of additional conservation laws are important. Although the two described here have no theoretical foundation, they are supported by abundant empirical evidence.

Baryon Number

Experimental results show that whenever a baryon is created in a decay or nuclear reaction, an antibaryon is also created. This scheme can be quantified by assigning every particle a quantum number, the **baryon number,** as follows: $B = +1$ for all baryons, $B = -1$ for all antibaryons, and $B = 0$ for all other particles. (See Table 46.2.) The **law of conservation of baryon number** states that

Conservation of baryon ▶
number

> whenever a nuclear reaction or decay occurs, the sum of the baryon numbers before the process must equal the sum of the baryon numbers after the process.

If baryon number is conserved, the proton must be absolutely stable. For example, a decay of the proton to a positron and a neutral pion would satisfy conservation of energy, momentum, and electric charge. Such a decay has never been observed, however. The law of conservation of baryon number would be consistent with the absence of this decay because the proposed decay would involve the loss of a baryon. Based on experimental observations as pointed out in Example 46.2, all we can say at present is that protons have a half-life of at least 10^{33} years (the estimated age of the Universe is only 10^{10} years). Some recent theories, however, predict that the proton is unstable. According to this theory, baryon number is not absolutely conserved.

Quick Quiz 46.2 Consider the decays **(i)** $n \rightarrow \pi^+ + \pi^- + \mu^+ + \mu^-$ and **(ii)** $n \rightarrow p + \pi^-$. From the following choices, which conservation laws are violated by each decay? (a) energy (b) electric charge (c) baryon number (d) angular momentum (e) no conservation laws

---

Example 46.1 **Checking Baryon Numbers**

Use the law of conservation of baryon number to determine whether each of the following reactions can occur:

(A) $p + n \rightarrow p + p + n + \bar{p}$

SOLUTION

Conceptualize The mass on the right is larger than the mass on the left. Therefore, one might be tempted to claim that the reaction violates energy conservation. The reaction can indeed occur, however, if the initial particles have sufficient kinetic energy to allow for the increase in rest energy of the system.

Categorize We use a conservation law developed in this section, so we categorize this example as a substitution problem.

Evaluate the total baryon number for the left side of the reaction: $\qquad 1 + 1 = 2$

Evaluate the total baryon number for the right side of the reaction: $\qquad 1 + 1 + 1 + (-1) = 2$

Therefore, baryon number is conserved and the reaction can occur.

(B) $p + n \rightarrow p + p + \bar{p}$

SOLUTION

Evaluate the total baryon number for the left side of the reaction: $\qquad 1 + 1 = 2$

Evaluate the total baryon number for the right side of the reaction: $\qquad 1 + 1 + (-1) = 1$

Because baryon number is not conserved, the reaction cannot occur.

Example 46.2   Detecting Proton Decay

Measurements taken at two neutrino detection facilities, the Irvine–Michigan–Brookhaven detector (Fig. 46.6) and the Super Kamiokande in Japan, indicate that the half-life of protons is at least 10^{33} yr.

(A) Estimate how long we would have to watch, on average, to see a proton in a glass of water decay.

SOLUTION

Conceptualize Imagine the number of protons in a glass of water. Although this number is huge, the probability of a single proton undergoing decay is small, so we would expect to wait for a long time interval before observing a decay.

Figure 46.6 (Example 46.2) A diver swims through ultrapure water in the Irvine–Michigan–Brookhaven neutrino detector. This detector holds almost 7 000 metric tons of water and is lined with over 2 000 photomultiplier tubes, many of which are visible in the photograph.

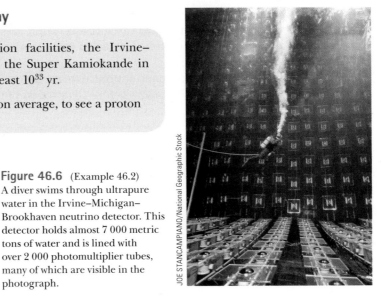

JOE STANCAMPIANO/National Geographic Stock

Categorize Because a half-life is provided in the problem, we categorize this problem as one in which we can apply our statistical analysis techniques from Section 44.4.

Analyze Let's estimate that a drinking glass contains a number of moles n of water, with a mass of $m = 250$ g and a molar mass $M = 18$ g/mol.

Find the number of molecules of water in the glass:

$$N_{molecules} = nN_A = \frac{m}{M}N_A$$

Each water molecule contains one proton in each of its two hydrogen atoms plus eight protons in its oxygen atom, for a total of ten protons. Therefore, there are $N = 10N_{molecules}$ protons in the glass of water.

Find the activity of the protons from Equation 44.7:

$$(1) \quad R = \lambda N = \frac{\ln 2}{T_{1/2}}\left(10\,\frac{m}{M}N_A\right) = \frac{\ln 2}{10^{33}\,\text{yr}}(10)\left(\frac{250\,\text{g}}{18\,\text{g/mol}}\right)(6.02 \times 10^{23}\,\text{mol}^{-1})$$

$$= 5.8 \times 10^{-8}\,\text{yr}^{-1}$$

Finalize The decay constant represents the probability that *one* proton decays in one year. The probability that *any* proton in our glass of water decays in the one-year interval is given by Equation (1). Therefore, we must watch our glass of water for $1/R \approx$ 17 million years! That indeed is a long time interval, as expected.

(B) The Super Kamiokande neutrino facility contains 50 000 metric tons of water. Estimate the average time interval between detected proton decays in this much water if the half-life of a proton is 10^{33} yr.

SOLUTION

Analyze The proton decay rate R in a sample of water is proportional to the number N of protons. Set up a ratio of the decay rate in the Super Kamiokande facility to that in a glass of water:

$$\frac{R_{Kamiokande}}{R_{glass}} = \frac{N_{Kamiokande}}{N_{glass}} \rightarrow R_{Kamiokande} = \frac{N_{Kamiokande}}{N_{glass}}R_{glass}$$

The number of protons is proportional to the mass of the sample, so express the decay rate in terms of mass:

$$R_{Kamiokande} = \frac{m_{Kamiokande}}{m_{glass}}R_{glass}$$

Substitute numerical values:

$$R_{Kamiokande} = \left(\frac{50\,000\,\text{metric tons}}{0.250\,\text{kg}}\right)\left(\frac{1\,000\,\text{kg}}{1\,\text{metric ton}}\right)(5.8 \times 10^{-8}\,\text{yr}^{-1}) \approx 12\,\text{yr}^{-1}$$

Finalize The average time interval between decays is about one-twelfth of a year, or approximately one month. That is much shorter than the time interval in part (A) due to the tremendous amount of water in the detector facility. Despite this rosy prediction of one proton decay per month, a proton decay has never been observed. This suggests that the half-life of the proton may be larger than 10^{33} years or that proton decay simply does not occur.

Lepton Number

There are three conservation laws involving lepton numbers, one for each variety of lepton. The **law of conservation of electron lepton number** states that

Conservation of electron ▶
lepton number

whenever a nuclear reaction or decay occurs, the sum of the electron lepton numbers before the process must equal the sum of the electron lepton numbers after the process.

The electron and the electron neutrino are assigned an electron lepton number $L_e = +1$, and the antileptons e^+ and $\bar{\nu}_e$ are assigned an electron lepton number $L_e = -1$. All other particles have $L_e = 0$. For example, consider the decay of the neutron:

$$n \rightarrow p + e^- + \bar{\nu}_e$$

Before the decay, the electron lepton number is $L_e = 0$; after the decay, it is $0 + 1 + (-1) = 0$. Therefore, electron lepton number is conserved. (Baryon number must also be conserved, of course, and it is: before the decay, $B = +1$, and after the decay, $B = +1 + 0 + 0 = +1$.)

Similarly, when a decay involves muons, the muon lepton number L_μ is conserved. The μ^- and the ν_μ are assigned a muon lepton number $L_\mu = +1$, and the antimuons μ^+ and $\bar{\nu}_\mu$ are assigned a muon lepton number $L_\mu = -1$. All other particles have $L_\mu = 0$.

Finally, tau lepton number L_τ is conserved with similar assignments made for the tau lepton, its neutrino, and their two antiparticles.

Quick Quiz 46.3 Consider the following decay: $\pi^0 \rightarrow \mu^- + e^+ + \nu_\mu$. What conservation laws are violated by this decay? **(a)** energy **(b)** angular momentum **(c)** electric charge **(d)** baryon number **(e)** electron lepton number **(f)** muon lepton number **(g)** tau lepton number **(h)** no conservation laws

Quick Quiz 46.4 Suppose a claim is made that the decay of the neutron is given by $n \rightarrow p + e^-$. What conservation laws are violated by this decay? **(a)** energy **(b)** angular momentum **(c)** electric charge **(d)** baryon number **(e)** electron lepton number **(f)** muon lepton number **(g)** tau lepton number **(h)** no conservation laws

---

Example 46.3   Checking Lepton Numbers

Use the law of conservation of lepton numbers to determine whether each of the following decay schemes (A) and (B) can occur:

(A) $\mu^- \rightarrow e^- + \bar{\nu}_e + \nu_\mu$

SOLUTION

Conceptualize Because this decay involves a muon and an electron, L_μ and L_e must each be conserved separately if the decay is to occur.

Categorize We use a conservation law developed in this section, so we categorize this example as a substitution problem.

Evaluate the lepton numbers before the decay: $L_\mu = +1 \qquad L_e = 0$

Evaluate the total lepton numbers after the decay: $L_\mu = 0 + 0 + 1 = +1 \qquad L_e = +1 + (-1) + 0 = 0$

Therefore, both numbers are conserved and on this basis the decay is possible.

(B) $\pi^+ \rightarrow \mu^+ + \nu_\mu + \nu_e$

▶ **46.3** continued

SOLUTION

Evaluate the lepton numbers before the decay: $\qquad L_\mu = 0 \qquad L_e = 0$

Evaluate the total lepton numbers after the decay: $\qquad L_\mu = -1 + 1 + 0 = 0 \qquad L_e = 0 + 0 + 1 = 1$

Therefore, the decay is not possible because electron lepton number is not conserved.

46.6 Strange Particles and Strangeness

Many particles discovered in the 1950s were produced by the interaction of pions with protons and neutrons in the atmosphere. A group of these—the kaon (K), lambda (Λ), and sigma (Σ) particles—exhibited unusual properties both as they were created and as they decayed; hence, they were called *strange particles*.

One unusual property of strange particles is that they are always produced in pairs. For example, when a pion collides with a proton, a highly probable result is the production of two neutral strange particles (Fig. 46.7):

$$\pi^- + p \;\rightarrow\; K^0 + \Lambda^0$$

The reaction $\pi^- + p \rightarrow K^0 + n$, where only one final particle is strange, never occurs, however, even though no previously known conservation laws would be violated and even though the energy of the pion is sufficient to initiate the reaction.

The second peculiar feature of strange particles is that although they are produced in reactions involving the strong interaction at a high rate, they do not decay into particles that interact via the strong force at a high rate. Instead, they decay very slowly, which is characteristic of the weak interaction. Their half-lives are in

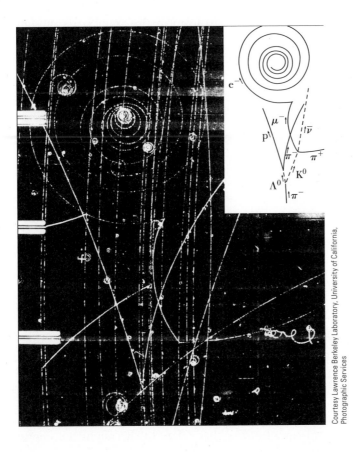

Courtesy Lawrence Berkeley Laboratory, University of California, Photographic Services

Figure 46.7 This bubble-chamber photograph shows many events, and the inset is a drawing of identified tracks. The strange particles Λ^0 and K^0 are formed at the bottom as a π^- particle interacts with a proton in the reaction $\pi^- + p \rightarrow K^0 + \Lambda^0$. (Notice that the neutral particles leave no tracks, as indicated by the dashed lines in the inset.) The Λ^0 then decays in the reaction $\Lambda^0 \rightarrow \pi^- + p$ and the K^0 in the reaction $K^0 \rightarrow \pi^+ + \mu^- + \bar{\nu}_\mu$.

the range 10^{-10} s to 10^{-8} s, whereas most other particles that interact via the strong force have much shorter lifetimes on the order of 10^{-23} s.

To explain these unusual properties of strange particles, a new quantum number S, called **strangeness,** was introduced, together with a conservation law. The strangeness numbers for some particles are given in Table 46.2. The production of strange particles in pairs is handled mathematically by assigning $S = +1$ to one of the particles, $S = -1$ to the other, and $S = 0$ to all nonstrange particles. The **law of conservation of strangeness** states that

Conservation of strangeness ▶

> in a nuclear reaction or decay that occurs via the strong force, strangeness is conserved; that is, the sum of the strangeness numbers before the process must equal the sum of the strangeness numbers after the process. In processes that occur via the weak interaction, strangeness may not be conserved.

The low decay rate of strange particles can be explained by assuming the strong and electromagnetic interactions obey the law of conservation of strangeness but the weak interaction does not. Because the decay of a strange particle involves the loss of one strange particle, it violates strangeness conservation and hence proceeds slowly via the weak interaction.

Example 46.4　**Is Strangeness Conserved?**

(A) Use the law of strangeness conservation to determine whether the reaction $\pi^0 + n \rightarrow K^+ + \Sigma^-$ occurs.

SOLUTION

Conceptualize We recognize that there are strange particles appearing in this reaction, so we see that we will need to investigate conservation of strangeness.

Categorize We use a conservation law developed in this section, so we categorize this example as a substitution problem.

Evaluate the strangeness for the left side of the reaction using Table 46.2:

$$S = 0 + 0 = 0$$

Evaluate the strangeness for the right side of the reaction:

$$S = +1 - 1 = 0$$

Therefore, strangeness is conserved and the reaction is allowed.

(B) Show that the reaction $\pi^- + p \rightarrow \pi^- + \Sigma^+$ does not conserve strangeness.

SOLUTION

Evaluate the strangeness for the left side of the reaction:

$$S = 0 + 0 = 0$$

Evaluate the strangeness for the right side of the reaction:

$$S = 0 + (-1) = -1$$

Therefore, strangeness is not conserved.

46.7 Finding Patterns in the Particles

One tool scientists use is the detection of patterns in data, patterns that contribute to our understanding of nature. For example, Table 21.2 shows a pattern of molar specific heats of gases that allows us to understand the differences among monatomic, diatomic, and polyatomic gases. Figure 42.20 shows a pattern of peaks in the ionization energy of atoms that relate to the quantized energy levels in the

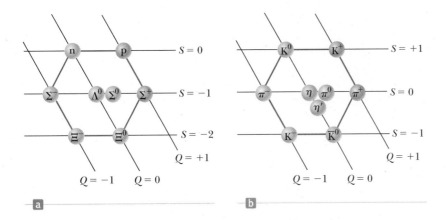

Figure 46.8 (a) The hexagonal eightfold-way pattern for the eight spin-$\frac{1}{2}$ baryons. This strangeness-versus-charge plot uses a sloping axis for charge number Q and a horizontal axis for strangeness S. (b) The eightfold-way pattern for the nine spin-zero mesons.

atoms. Figure 44.7 shows a pattern of peaks in the binding energy that suggest a shell structure within the nucleus. One of the best examples of this tool's use is the development of the periodic table, which provides a fundamental understanding of the chemical behavior of the elements. As mentioned in the introduction, the periodic table explains how more than 100 elements can be formed from three particles, the electron, the proton, and the neutron. The table of nuclides, part of which is shown in Table 44.2, contains hundreds of nuclides, but all can be built from protons and neutrons.

The number of particles observed by particle physicists is in the hundreds. Is it possible that a small number of entities exist from which all these particles can be built? Taking a hint from the success of the periodic table and the table of nuclides, let explore the historical search for patterns among the particles.

Many classification schemes have been proposed for grouping particles into families. Consider, for instance, the baryons listed in Table 46.2 that have spins of $\frac{1}{2}$: p, n, Λ^0, Σ^+, Σ^0, Σ^-, Ξ^0, and Ξ^-. If we plot strangeness versus charge for these baryons using a sloping coordinate system as in Figure 46.8a, a fascinating pattern is observed: six of the baryons form a hexagon, and the remaining two are at the hexagon's center.

As a second example, consider the following nine spin-zero mesons listed in Table 46.2: π^+, π^0, π^-, K^+, K^0, K^-, η, η', and the antiparticle $\overline{K}^0$. Figure 46.8b is a plot of strangeness versus charge for this family. Again, a hexagonal pattern emerges. In this case, each particle on the perimeter of the hexagon lies opposite its antiparticle and the remaining three (which form their own antiparticles) are at the center of the hexagon. These and related symmetric patterns were developed independently in 1961 by Murray Gell-Mann and Yuval Ne'eman (1925–2006). Gell-Mann called the patterns the **eightfold way,** after the eightfold path to nirvana in Buddhism.

Groups of baryons and mesons can be displayed in many other symmetric patterns within the framework of the eightfold way. For example, the family of spin-$\frac{3}{2}$ baryons known in 1961 contains nine particles arranged in a pattern like that of the pins in a bowling alley as in Figure 46.9. (The particles Σ^{*+}, Σ^{*0}, Σ^{*-}, Ξ^{*0},

Murray Gell-Mann
American Physicist (b. 1929)
In 1969, Murray Gell-Mann was awarded the Nobel Prize in Physics for his theoretical studies dealing with subatomic particles.

© Linh Hassel/Age Fotostock

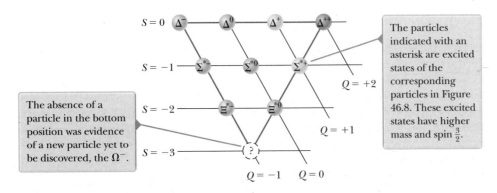

The particles indicated with an asterisk are excited states of the corresponding particles in Figure 46.8. These excited states have higher mass and spin $\frac{3}{2}$.

The absence of a particle in the bottom position was evidence of a new particle yet to be discovered, the Ω^-.

Figure 46.9 The pattern for the higher-mass, spin-$\frac{3}{2}$ baryons known at the time the pattern was proposed.

Figure 46.10 Discovery of the Ω^- particle. The photograph on the left shows the original bubble-chamber tracks. The drawing on the right isolates the tracks of the important events.

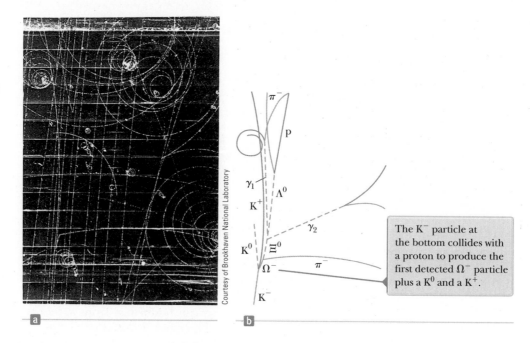

Courtesy of Brookhaven National Laboratory

The K⁻ particle at the bottom collides with a proton to produce the first detected Ω^- particle plus a K^0 and a K^+.

and Ξ^{*-} are excited states of the particles Σ^+, Σ^0, Σ^-, Ξ^0, and Ξ^-. In these higher-energy states, the spins of the three quarks—see Section 46.8—making up the particle are aligned so that the total spin of the particle is $\frac{3}{2}$.) When this pattern was proposed, an empty spot occurred in it (at the bottom position), corresponding to a particle that had never been observed. Gell-Mann predicted that the missing particle, which he called the omega minus (Ω^-), should have spin $\frac{3}{2}$, charge -1, strangeness -3, and rest energy of approximately 1 680 MeV. Shortly thereafter, in 1964, scientists at the Brookhaven National Laboratory found the missing particle through careful analyses of bubble-chamber photographs (Fig. 46.10) and confirmed all its predicted properties.

The prediction of the missing particle in the eightfold way has much in common with the prediction of missing elements in the periodic table. Whenever a vacancy occurs in an organized pattern of information, experimentalists have a guide for their investigations.

46.8   Quarks

As mentioned earlier, leptons appear to be truly elementary particles because there are only a few types of them, and experiments indicate that they have no measurable size or internal structure. Hadrons, on the other hand, are complex particles having size and structure. The existence of the strangeness–charge patterns of the eightfold way suggests that hadrons have substructure. Furthermore, hundreds of types of hadrons exist and many decay into other hadrons.

The Original Quark Model

In 1963, Gell-Mann and George Zweig (b. 1937) independently proposed a model for the substructure of hadrons. According to their model, all hadrons are composed of two or three elementary constituents called **quarks.** (Gell-Mann borrowed the word *quark* from the passage "Three quarks for Muster Mark" in James Joyce's *Finnegans Wake.* In Zweig's model, he called the constituents "aces.") The model has three types of quarks, designated by the symbols u, d, and s, that are given the arbitrary names **up, down,** and **strange.** The various types of quarks are called **flavors.** Figure 46.11 is a pictorial representation of the quark compositions of several hadrons.

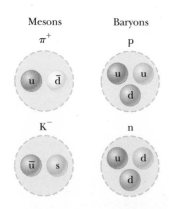

Figure 46.11 Quark composition of two mesons and two baryons.

Table 46.3 **Properties of Quarks and Antiquarks**

Quarks

Name	Symbol	Spin	Charge	Baryon Number	Strangeness	Charm	Bottomness	Topness
Up	u	$\frac{1}{2}$	$+\frac{2}{3}e$	$\frac{1}{3}$	0	0	0	0
Down	d	$\frac{1}{2}$	$-\frac{1}{3}e$	$\frac{1}{3}$	0	0	0	0
Strange	s	$\frac{1}{2}$	$-\frac{1}{3}e$	$\frac{1}{3}$	−1	0	0	0
Charmed	c	$\frac{1}{2}$	$+\frac{2}{3}e$	$\frac{1}{3}$	0	+1	0	0
Bottom	b	$\frac{1}{2}$	$-\frac{1}{3}e$	$\frac{1}{3}$	0	0	+1	0
Top	t	$\frac{1}{2}$	$+\frac{2}{3}e$	$\frac{1}{3}$	0	0	0	+1

Antiquarks

Name	Symbol	Spin	Charge	Baryon Number	Strangeness	Charm	Bottomness	Topness
Anti-up	$\bar{u}$	$\frac{1}{2}$	$-\frac{2}{3}e$	$-\frac{1}{3}$	0	0	0	0
Anti-down	$\bar{d}$	$\frac{1}{2}$	$+\frac{1}{3}e$	$-\frac{1}{3}$	0	0	0	0
Anti-strange	$\bar{s}$	$\frac{1}{2}$	$+\frac{1}{3}e$	$-\frac{1}{3}$	+1	0	0	0
Anti-charmed	$\bar{c}$	$\frac{1}{2}$	$-\frac{2}{3}e$	$-\frac{1}{3}$	0	−1	0	0
Anti-bottom	$\bar{b}$	$\frac{1}{2}$	$+\frac{1}{3}e$	$-\frac{1}{3}$	0	0	−1	0
Anti-top	$\bar{t}$	$\frac{1}{2}$	$-\frac{2}{3}e$	$-\frac{1}{3}$	0	0	0	−1

An unusual property of quarks is that they carry a fractional electric charge. The u, d, and s quarks have charges of $+2e/3$, $-e/3$, and $-e/3$, respectively, where e is the elementary charge 1.60×10^{-19} C. These and other properties of quarks and antiquarks are given in Table 46.3. Quarks have spin $\frac{1}{2}$, which means that all quarks are fermions, defined as any particle having half-integral spin, as pointed out in Section 43.8. As Table 46.3 shows, associated with each quark is an antiquark of opposite charge, baryon number, and strangeness.

The compositions of all hadrons known when Gell-Mann and Zweig presented their model can be completely specified by three simple rules:

- A meson consists of one quark and one antiquark, giving it a baryon number of 0, as required.
- A baryon consists of three quarks.
- An antibaryon consists of three antiquarks.

The theory put forth by Gell-Mann and Zweig is referred to as the *original quark model*.

Quick Quiz 46.5 Using a coordinate system like that in Figure 46.8, draw an eightfold-way diagram for the three quarks in the original quark model.

Charm and Other Developments

Although the original quark model was highly successful in classifying particles into families, some discrepancies occurred between its predictions and certain experimental decay rates. Consequently, several physicists proposed a fourth quark flavor in 1967. They argued that if four types of leptons exist (as was thought at the time), there should also be four flavors of quarks because of an underlying symmetry in nature. The fourth quark, designated c, was assigned a property called **charm.** A *charmed* quark has charge $+2e/3$, just as the up quark does, but its charm distinguishes it from the other three quarks. This introduces a new quantum number C, representing charm. The new quark has charm $C = +1$, its antiquark has charm of $C = -1$, and all other quarks have $C = 0$. Charm, like strangeness, is conserved in strong and electromagnetic interactions but not in weak interactions.

Table 46.4   Quark Composition of Mesons

		$\bar{b}$		$\bar{c}$		$\bar{s}$		$\bar{d}$		$\bar{u}$	
						Antiquarks					
	b	Y	$(b\bar{b})$	B_c^-	$(\bar{c}b)$	$\bar{B}_s^0$	$(\bar{s}b)$	$\bar{B}_d^0$	$(\bar{d}b)$	B^-	$(\bar{u}b)$
	c	B_c^+	$(\bar{b}c)$	J/Ψ	$(\bar{c}c)$	D_s^+	$(\bar{s}c)$	D^+	$(\bar{d}c)$	D^0	$(\bar{u}c)$
Quarks	**s**	B_s^0	$(\bar{b}s)$	D_s^-	$(\bar{c}s)$	η, η'	$(\bar{s}s)$	$\bar{K}^0$	$(\bar{d}s)$	K^-	$(\bar{u}s)$
	d	B_d^0	$(\bar{b}d)$	D^-	$(\bar{c}d)$	K^0	$(\bar{s}d)$	π^0, η, η'	$(\bar{d}d)$	π^-	$(\bar{u}d)$
	u	B^+	$(\bar{b}u)$	$\bar{D}^0$	$(\bar{c}u)$	K^+	$(\bar{s}u)$	π^+	$(\bar{d}u)$	π^0, η, η'	$(\bar{u}u)$

Note: The top quark does not form mesons because it decays too quickly.

Table 46.5   Quark Composition of Several Baryons

Particle	Quark Composition
p	uud
n	udd
Λ^0	uds
Σ^+	uus
Σ^0	uds
Σ^-	dds
Δ^{++}	uuu
Δ^+	uud
Δ^0	udd
Δ^-	ddd
Ξ^0	uss
Ξ^-	dss
Ω^-	sss

Note: Some baryons have the same quark composition, such as the p and the Δ^+ and the n and the Δ^0. In these cases, the Δ particles are considered to be excited states of the proton and neutron.

Evidence that the charmed quark exists began to accumulate in 1974, when a heavy meson called the J/Ψ particle (or simply Ψ, Greek letter psi) was discovered independently by two groups, one led by Burton Richter (b. 1931) at the Stanford Linear Accelerator (SLAC), and the other led by Samuel Ting (b. 1936) at the Brookhaven National Laboratory. In 1976, Richter and Ting were awarded the Nobel Prize in Physics for this work. The J/Ψ particle does not fit into the three-quark model; instead, it has properties of a combination of the proposed charmed quark and its antiquark ($c\bar{c}$). It is much more massive than the other known mesons ($\sim 3\ 100$ MeV/c^2), and its lifetime is much longer than the lifetimes of particles that interact via the strong force. Soon, related mesons were discovered, corresponding to such quark combinations as $\bar{c}d$ and $c\bar{d}$, all of which have great masses and long lifetimes. The existence of these new mesons provided firm evidence for the fourth quark flavor.

In 1975, researchers at Stanford University reported strong evidence for the tau (τ) lepton, mass 1 784 MeV/c^2. This fifth type of lepton led physicists to propose that more flavors of quarks might exist, on the basis of symmetry arguments similar to those leading to the proposal of the charmed quark. These proposals led to more elaborate quark models and the prediction of two new quarks, **top** (t) and **bottom** (b). (Some physicists prefer *truth* and *beauty*.) To distinguish these quarks from the others, quantum numbers called *topness* and *bottomness* (with allowed values $+1$, 0, -1) were assigned to all quarks and antiquarks (see Table 46.3). In 1977, researchers at the Fermi National Laboratory, under the direction of Leon Lederman (b. 1922), reported the discovery of a very massive new meson Y (Greek letter upsilon), whose composition is considered to be $b\bar{b}$, providing evidence for the bottom quark. In March 1995, researchers at Fermilab announced the discovery of the top quark (supposedly the last of the quarks to be found), which has a mass of 173 GeV/c^2.

Table 46.4 lists the quark compositions of mesons formed from the up, down, strange, charmed, and bottom quarks. Table 46.5 shows the quark combinations for the baryons listed in Table 46.2. Notice that only two flavors of quarks, u and d, are contained in all hadrons encountered in ordinary matter (protons and neutrons).

Will the discoveries of elementary particles ever end? How many "building blocks" of matter actually exist? At present, physicists believe that the elementary particles in nature are six quarks and six leptons, together with their antiparticles, and the four field particles listed in Table 46.1. Table 46.6 lists the rest energies and charges of the quarks and leptons.

Despite extensive experimental effort, no isolated quark has ever been observed. Physicists now believe that at ordinary temperatures, quarks are permanently confined inside ordinary particles because of an exceptionally strong force that prevents them from escaping, called (appropriately) the **strong force**[5] (which we

---

[5]As a reminder, the original meaning of the term *strong force* was the short-range attractive force between nucleons, which we have called the *nuclear force*. The nuclear force between nucleons is a secondary effect of the strong force between quarks.

Table 46.6	The Elementary Particles and Their Rest Energies and Charges	
Particle	**Approximate Rest Energy**	**Charge**
Quarks		
u	2.4 MeV	$+\frac{2}{3}e$
d	4.8 MeV	$-\frac{1}{3}e$
s	104 MeV	$-\frac{1}{3}e$
c	1.27 GeV	$+\frac{2}{3}e$
b	4.2 GeV	$-\frac{1}{3}e$
t	173 GeV	$+\frac{2}{3}e$
Leptons		
e^-	511 keV	$-e$
μ^-	105.7 MeV	$-e$
τ^-	1.78 GeV	$-e$
ν_e	< 2 eV	0
ν_μ	< 0.17 MeV	0
ν_τ	< 18 MeV	0

introduced at the beginning of Section 46.4 and will discuss further in Section 46.10). This force increases with separation distance, similar to the force exerted by a stretched spring. Current efforts are under way to form a **quark–gluon plasma,** a state of matter in which the quarks are freed from neutrons and protons. In 2000, scientists at CERN announced evidence for a quark–gluon plasma formed by colliding lead nuclei. In 2005, experiments at the Relativistic Heavy Ion Collider (RHIC) at Brookhaven suggested the creation of a quark–gluon plasma. Neither laboratory has provided definitive data to verify the existence of a quark–gluon plasma. Experiments continue, and the ALICE project (A Large Ion Collider Experiment) at the Large Hadron Collider at CERN has joined the search.

Quick Quiz 46.6 Doubly charged baryons, such as the Δ^{++}, are known to exist. True or False: Doubly charged mesons also exist.

46.9 Multicolored Quarks

Shortly after the concept of quarks was proposed, scientists recognized that certain particles had quark compositions that violated the exclusion principle. In Section 42.7, we applied the exclusion principle to electrons in atoms. The principle is more general, however, and applies to all particles with half-integral spin ($\frac{1}{2}$, $\frac{3}{2}$, etc.), which are collectively called fermions. Because all quarks are fermions having spin $\frac{1}{2}$, they are expected to follow the exclusion principle. One example of a particle that appears to violate the exclusion principle is the Ω^- (sss) baryon, which contains three strange quarks having parallel spins, giving it a total spin of $\frac{3}{2}$. All three quarks have the same spin quantum number, in violation of the exclusion principle. Other examples of baryons made up of identical quarks having parallel spins are the Δ^{++} (uuu) and the Δ^- (ddd).

To resolve this problem, it was suggested that quarks possess an additional property called **color charge.** This property is similar in many respects to electric charge except that it occurs in six varieties rather than two. The colors assigned to quarks are red, green, and blue, and antiquarks have the colors antired, antigreen, and antiblue. Therefore, the colors red, green, and blue serve as the "quantum numbers" for the color of the quark. To satisfy the exclusion principle, the three quarks in any baryon must all have different colors. Look again at the quarks in the baryons in Figure 46.11 and notice the colors. The three colors "neutralize" to white.

Pitfall Prevention 46.3
Color Charge Is Not Really Color The description of color for a quark has nothing to do with visual sensation from light. It is simply a convenient name for a property that is analogous to electric charge.

Figure 46.12 (a) A green quark is attracted to an antigreen quark. This forms a meson whose quark structure is (qq̄). (b) Three quarks of different colors attract one another to form a baryon.

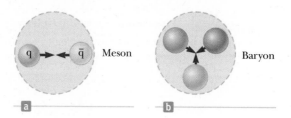

A quark and an antiquark in a meson must be of a color and the corresponding anticolor and will consequently neutralize to white, similar to the way electric charges + and − neutralize to zero net charge. (See the mesons in Fig. 46.11.) The apparent violation of the exclusion principle in the Ω^- baryon is removed because the three quarks in the particle have different colors.

The new property of color increases the number of quarks by a factor of 3 because each of the six quarks comes in three colors. Although the concept of color in the quark model was originally conceived to satisfy the exclusion principle, it also provided a better theory for explaining certain experimental results. For example, the modified theory correctly predicts the lifetime of the π^0 meson.

The theory of how quarks interact with each other is called **quantum chromo-dynamics,** or QCD, to parallel the name *quantum electrodynamics* (the theory of the electrical interaction between light and matter). In QCD, each quark is said to carry a color charge, in analogy to electric charge. The strong force between quarks is often called the **color force.** Therefore, the terms *strong force* and *color force* are used interchangeably.

In Section 46.1, we stated that the nuclear interaction between hadrons is mediated by massless field particles called **gluons.** As mentioned earlier, the nuclear force is actually a secondary effect of the strong force between quarks. The gluons are the mediators of the strong force. When a quark emits or absorbs a gluon, the quark's color may change. For example, a blue quark that emits a gluon may become a red quark and a red quark that absorbs this gluon becomes a blue quark.

The color force between quarks is analogous to the electric force between charges: particles with the same color repel, and those with opposite colors attract. Therefore, two green quarks repel each other, but a green quark is attracted to an antigreen quark. The attraction between quarks of opposite color to form a meson (qq̄) is indicated in Figure 46.12a. Differently colored quarks also attract one another, although with less intensity than the oppositely colored quark and antiquark. For example, a cluster of red, blue, and green quarks all attract one another to form a baryon as in Figure 46.12b. Therefore, every baryon contains three quarks of three different colors.

Although the nuclear force between two colorless hadrons is negligible at large separations, the net strong force between their constituent quarks is not exactly zero at small separations. This residual strong force is the nuclear force that binds protons and neutrons to form nuclei. It is similar to the force between two electric dipoles. Each dipole is electrically neutral. An electric field surrounds the dipoles, however, because of the separation of the positive and negative charges (see Section 23.6). As a result, an electric interaction occurs between the dipoles that is weaker than the force between single charges. In Section 43.1, we explored how this interaction results in the Van der Waals force between neutral molecules.

According to QCD, a more basic explanation of the nuclear force can be given in terms of quarks and gluons. Figure 46.13a shows the nuclear interaction between a neutron and a proton by means of Yukawa's pion, in this case a π^-. This drawing differs from Figure 46.5a, in which the field particle is a π^0; there is no transfer of charge from one nucleon to the other in Figure 46.5a. In Figure 46.13a, the charged pion carries charge from one nucleon to the other, so the nucleons change identities, with the proton becoming a neutron and the neutron becoming a proton.

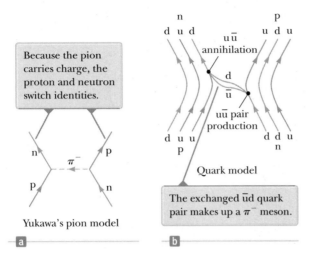

Because the pion carries charge, the proton and neutron switch identities.

n   p

π⁻

p   n

Yukawa's pion model

a

n   p
d u d   u d u
     u̅u
     annihilation
        d
        u̅
  uu̅ pair
  production
d u u   d d u
p   n

Quark model

The exchanged u̅d quark pair makes up a π⁻ meson.

b

Figure 46.13 (a) A nuclear interaction between a proton and a neutron explained in terms of Yukawa's pion-exchange model. (b) The same interaction, explained in terms of quarks and gluons.

Let's look at the same interaction from the viewpoint of the quark model, shown in Figure 46.13b. In this Feynman diagram, the proton and neutron are represented by their quark constituents. Each quark in the neutron and proton is continuously emitting and absorbing gluons. The energy of a gluon can result in the creation of quark–antiquark pairs. This process is similar to the creation of electron–positron pairs in pair production, which we investigated in Section 46.2. When the neutron and proton approach to within 1 fm of each other, these gluons and quarks can be exchanged between the two nucleons, and such exchanges produce the nuclear force. Figure 46.13b depicts one possibility for the process shown in Figure 46.13a. A down quark in the neutron on the right emits a gluon. The energy of the gluon is then transformed to create a uu̅ pair. The u quark stays within the nucleon (which has now changed to a proton), and the recoiling d quark and the u̅ antiquark are transmitted to the proton on the left side of the diagram. Here the u̅ annihilates a u quark within the proton and the d is captured. The net effect is to change a u quark to a d quark, and the proton on the left has changed to a neutron.

As the d quark and u̅ antiquark in Figure 46.13b transfer between the nucleons, the d and u̅ exchange gluons with each other and can be considered to be bound to each other by means of the strong force. Looking back at Table 46.4, we see that this combination is a π⁻, or Yukawa's field particle! Therefore, the quark model of interactions between nucleons is consistent with the pion-exchange model.

46.10 The Standard Model

Scientists now believe there are three classifications of truly elementary particles: leptons, quarks, and field particles. These three types of particles are further classified as either fermions or bosons. Quarks and leptons have spin $\frac{1}{2}$ and hence are fermions, whereas the field particles have integral spin of 1 or higher and are bosons.

Recall from Section 46.1 that the weak force is believed to be mediated by the W⁺, W⁻, and Z⁰ bosons. These particles are said to have *weak charge*, just as quarks have color charge. Therefore, each elementary particle can have mass, electric charge, color charge, and weak charge. Of course, one or more of these could be zero.

In 1979, Sheldon Glashow (b. 1932), Abdus Salam (1926–1996), and Steven Weinberg (b. 1933) won the Nobel Prize in Physics for developing a theory that unifies the electromagnetic and weak interactions. This **electroweak theory** postulates that the weak and electromagnetic interactions have the same strength when the particles involved have very high energies. The two interactions are viewed as different manifestations of a single unifying electroweak interaction. The theory makes many concrete predictions, but perhaps the most spectacular is the prediction of

Figure 46.14 The Standard Model of particle physics.

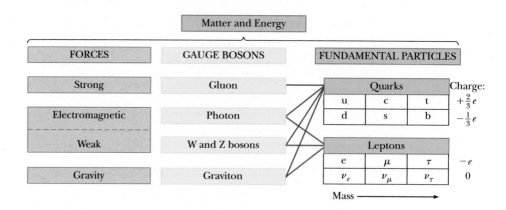

the masses of the W and Z particles at approximately 82 GeV/c^2 and 93 GeV/c^2, respectively. These predictions are close to the masses in Table 46.1 determined by experiment.

The combination of the electroweak theory and QCD for the strong interaction is referred to in high-energy physics as the **Standard Model.** Although the details of the Standard Model are complex, its essential ingredients can be summarized with the help of Fig. 46.14. (Although the Standard Model does not include the gravitational force at present, we include gravity in Fig. 46.14 because physicists hope to eventually incorporate this force into a unified theory.) This diagram shows that quarks participate in all the fundamental forces and that leptons participate in all except the strong force.

The Standard Model does not answer all questions. A major question still unanswered is why, of the two mediators of the electroweak interaction, the photon has no mass but the W and Z bosons do. Because of this mass difference, the electromagnetic and weak forces are quite distinct at low energies but become similar at very high energies, when the rest energy is negligible relative to the total energy. The behavior as one goes from high to low energies is called *symmetry breaking* because the forces are similar, or symmetric, at high energies but are very different at low energies. The nonzero rest energies of the W and Z bosons raise the question of the origin of particle masses. To resolve this problem, a hypothetical particle called the **Higgs boson,** which provides a mechanism for breaking the electroweak symmetry, has been proposed. The Standard Model modified to include the Higgs boson provides a logically consistent explanation of the massive nature of the W and Z bosons. In July 2012, announcements from the ATLAS (A Toroidal LHC Apparatus) and CMS (Compact Muon Solenoid) experiments at the Large Hadron Collider (LHC) at CERN claimed the discovery of a new particle having properties consistent with that of a Higgs boson. The mass of the particle is 125–127 GeV, within the range of predictions made from theoretical considerations using the Standard Model.

Because of the limited energy available in conventional accelerators using fixed targets, it is necessary to employ colliding-beam accelerators called **colliders.** The concept of colliders is straightforward. Particles that have equal masses and equal kinetic energies, traveling in opposite directions in an accelerator ring, collide head-on to produce the required reaction and form new particles. Because the total momentum of the interacting particles is zero, all their kinetic energy is available for the reaction.

Several colliders provided important data for understanding the Standard Model in the latter part of the 20th century and the first decade of the 21st century: the Large Electron–Positron (LEP) Collider and the Super Proton Synchrotron at CERN, the Stanford Linear Collider, and the Tevatron at the Fermi National Laboratory in Illinois. The Relativistic Heavy Ion Collider at Brookhaven National Laboratory is the sole remaining collider in operation in the United States. The Large Hadron Collider at CERN, which began collision operations in March 2010, has

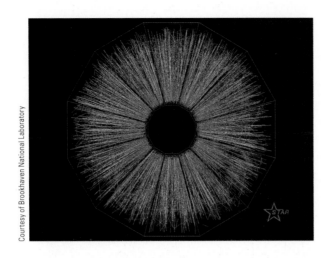

Figure 46.15 A shower of particle tracks from a head-on collision of gold nuclei, each moving with energy 100 GeV. This collision occurred at the Relativistic Heavy Ion Collider (RHIC) at Brookhaven National Laboratory and was recorded with the STAR (Solenoidal Tracker at RHIC) detector. The tracks represent many fundamental particles arising from the energy of the collision.

Courtesy of Brookhaven National Laboratory

taken the lead in particle studies due to its extremely high energy capabilities. The expected upper limit for the LHC is a center-of-mass energy of 14 TeV. (See page 680 for a photo of a magnet used by the LHC.)

In addition to increasing energies in modern accelerators, detection techniques have become increasingly sophisticated. We saw simple bubble-chamber photographs earlier in this chapter that required hours of analysis by hand. Figure 46.15 shows a complex set of tracks from a collision of gold nuclei.

46.11 The Cosmic Connection

In this section, we describe one of the most fascinating theories in all science—the Big Bang theory of the creation of the Universe—and the experimental evidence that supports it. This theory of cosmology states that the Universe had a beginning and furthermore that the beginning was so cataclysmic that it is impossible to look back beyond it. According to this theory, the Universe erupted from an infinitely dense singularity about 14 billion years ago. The first few moments after the Big Bang saw such extremely high energy that it is believed that all four interactions of physics were unified and all matter was contained in a quark–gluon plasma.

The evolution of the four fundamental forces from the Big Bang to the present is shown in Figure 46.16 (page 1174). During the first 10^{-43} s (the ultrahot epoch, $T \sim 10^{32}$ K), it is presumed the strong, electroweak, and gravitational forces were joined to form a completely unified force. In the first 10^{-35} s following the Big Bang (the hot epoch, $T \sim 10^{29}$ K), symmetry breaking occurred for gravity while the strong and electroweak forces remained unified. It was a period when particle energies were so great ($> 10^{16}$ GeV) that very massive particles as well as quarks, leptons, and their antiparticles existed. Then, after 10^{-35} s, the Universe rapidly expanded and cooled (the warm epoch, $T \sim 10^{29}$ to 10^{15} K) and the strong and electroweak forces parted company. As the Universe continued to cool, the electroweak force split into the weak force and the electromagnetic force approximately 10^{-10} s after the Big Bang.

After a few minutes, protons and neutrons condensed out of the plasma. For half an hour, the Universe underwent thermonuclear fusion, exploding as a hydrogen bomb and producing most of the helium nuclei that now exist. The Universe continued to expand, and its temperature dropped. Until about 700 000 years after the Big Bang, the Universe was dominated by radiation. Energetic radiation prevented matter from forming single hydrogen atoms because photons would instantly ionize any atoms that happened to form. Photons experienced continuous Compton scattering from the vast numbers of free electrons, resulting in a Universe that was opaque to radiation. By the time the Universe was about 700 000 years old, it had

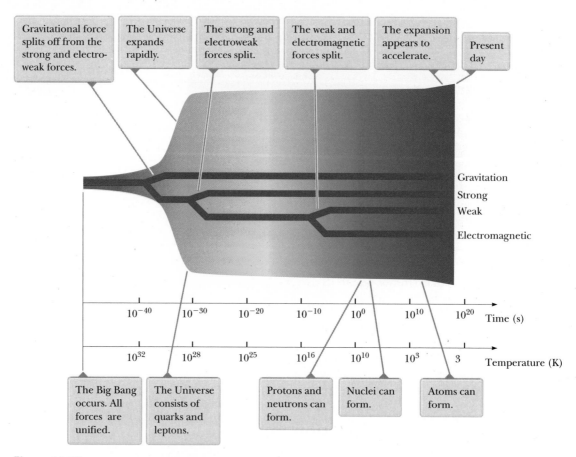

| Gravitational force splits off from the strong and electro-weak forces. | The Universe expands rapidly. | The strong and electroweak forces split. | The weak and electromagnetic forces split. | The expansion appears to accelerate. | Present day |

Gravitation
Strong
Weak
Electromagnetic

| The Big Bang occurs. All forces are unified. | The Universe consists of quarks and leptons. | Protons and neutrons can form. | Nuclei can form. | Atoms can form. |

Figure 46.16 A brief history of the Universe from the Big Bang to the present. The four forces became distinguishable during the first nanosecond. Following that, all the quarks combined to form particles that interact via the nuclear force. The leptons, however, remained separate and to this day exist as individual, observable particles.

expanded and cooled to approximately 3 000 K and protons could bind to electrons to form neutral hydrogen atoms. Because of the quantized energies of the atoms, far more wavelengths of radiation were not absorbed by atoms than were absorbed, and the Universe suddenly became transparent to photons. Radiation no longer dominated the Universe, and clumps of neutral matter steadily grew: first atoms, then molecules, gas clouds, stars, and finally galaxies.

Observation of Radiation from the Primordial Fireball

In 1965, Arno A. Penzias (b. 1933) and Robert W. Wilson (b. 1936) of Bell Laboratories were testing a sensitive microwave receiver and made an amazing discovery. A pesky signal producing a faint background hiss was interfering with their satellite communications experiments. The microwave horn that served as their receiving antenna is shown in Figure 46.17. Evicting a flock of pigeons from the 20-ft horn and cooling the microwave detector both failed to remove the signal.

The intensity of the detected signal remained unchanged as the antenna was pointed in different directions. That the radiation had equal strengths in all directions suggested that the entire Universe was the source of this radiation. Ultimately, it became clear that they were detecting microwave background radiation (at a wavelength of 7.35 cm), which represented the leftover "glow" from the Big Bang. Through a casual conversation, Penzias and Wilson discovered that a group at Princeton University had predicted the residual radiation from the Big Bang and were planning an experiment to attempt to confirm the theory. The excitement in the scientific community was high when Penzias and Wilson announced that they had already observed an excess microwave background compatible with a 3-K

AT&T Bell Laboratories

Figure 46.17 Robert W. Wilson (*left*) and Arno A. Penzias with the Bell Telephone Laboratories horn-reflector antenna.

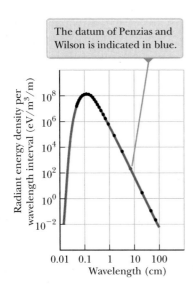

The datum of Penzias and Wilson is indicated in blue.

Figure 46.18 Theoretical black-body (brown curve) and measured radiation spectra (black points) of the Big Bang. Most of the data were collected from the COsmic Background Explorer, or COBE, satellite.

blackbody source, which was consistent with the predicted temperature of the Universe at this time after the Big Bang.

Because Penzias and Wilson made their measurements at a single wavelength, they did not completely confirm the radiation as 3-K blackbody radiation. Subsequent experiments by other groups added intensity data at different wavelengths as shown in Figure 46.18. The results confirm that the radiation is that of a black body at 2.7 K. This figure is perhaps the most clear-cut evidence for the Big Bang theory. The 1978 Nobel Prize in Physics was awarded to Penzias and Wilson for this most important discovery.

In the years following Penzias and Wilson's discovery, other researchers made measurements at different wavelengths. In 1989, the COBE (COsmic Background Explorer) satellite was launched by NASA and added critical measurements at wavelengths below 0.1 cm. The results of these measurements led to a Nobel Prize in Physics for the principal investigators in 2006. Several data points from COBE are shown in Figure 46.18. The Wilkinson Microwave Anisotropy Probe, launched in June 2001, exhibits data that allow observation of temperature differences in the cosmos in the microkelvin range. Ongoing observations are also being made from Earth-based facilities, associated with projects such as QUaD, Qubic, and the South Pole Telescope. In addition, the Planck satellite was launched in May 2009 by the European Space Agency. This space-based observatory has been measuring the cosmic background radiation with higher sensitivity than the Wilkinson probe. The series of measurements taken since 1965 are consistent with thermal radiation associated with a temperature of 2.7 K. The whole story of the cosmic temperature is a remarkable example of science at work: building a model, making a prediction, taking measurements, and testing the measurements against the predictions.

Other Evidence for an Expanding Universe

The Big Bang theory of cosmology predicts that the Universe is expanding. Most of the key discoveries supporting the theory of an expanding Universe were made in the 20th century. Vesto Melvin Slipher (1875–1969), an American astronomer, reported in 1912 that most galaxies are receding from the Earth at speeds up to several million miles per hour. Slipher was one of the first scientists to use Doppler shifts (see Section 17.4) in spectral lines to measure galaxy velocities.

In the late 1920s, Edwin P. Hubble (1889–1953) made the bold assertion that the whole Universe is expanding. From 1928 to 1936, until they reached the limits of the 100-inch telescope, Hubble and Milton Humason (1891–1972) worked at Mount Wilson in California to prove this assertion. The results of that work and of its continuation with the use of a 200-inch telescope in the 1940s showed that the speeds at which galaxies are receding from the Earth increase in direct proportion to their distance R from us. This linear relationship, known as **Hubble's law,** may be written

Hubble's law ▶

$$v = HR \qquad (46.4)$$

where H, called the **Hubble constant,** has the approximate value

$$H \approx 22 \times 10^{-3} \text{ m/(s} \cdot \text{ly)}$$

Example 46.5 Recession of a Quasar

A quasar is an object that appears similar to a star and is very distant from the Earth. Its speed can be determined from Doppler-shift measurements in the light it emits. A certain quasar recedes from the Earth at a speed of $0.55c$. How far away is it?

SOLUTION

Conceptualize A common mental representation for the Hubble law is that of raisin bread cooking in an oven. Imagine yourself at the center of the loaf of bread. As the entire loaf of bread expands upon heating, raisins near you move slowly with respect to you. Raisins far away from you on the edge of the loaf move at a higher speed.

Categorize We use a concept developed in this section, so we categorize this example as a substitution problem.

Find the distance through Hubble's law:

$$R = \frac{v}{H} = \frac{(0.55)(3.00 \times 10^8 \text{ m/s})}{22 \times 10^{-3} \text{ m/(s} \cdot \text{ly)}} = \boxed{7.5 \times 10^9 \text{ ly}}$$

WHAT IF? Suppose the quasar has moved at this speed ever since the Big Bang. With this assumption, estimate the age of the Universe.

Answer Let's approximate the distance from the Earth to the quasar as the distance the quasar has moved from the singularity since the Big Bang. We can then find the time interval from the *particle under constant speed* model: $\Delta t = d/v = R/v = 1/H \approx 14$ billion years, which is in approximate agreement with other calculations.

Will the Universe Expand Forever?

In the 1950s and 1960s, Allan R. Sandage (1926–2010) used the 200-inch telescope at Mount Palomar to measure the speeds of galaxies at distances of up to 6 billion light-years away from the Earth. These measurements showed that these very distant galaxies were moving approximately 10 000 km/s faster than Hubble's law predicted. According to this result, the Universe must have been expanding more rapidly 1 billion years ago, and consequently we conclude from these data that the expansion rate is slowing.[6] Today, astronomers and physicists are trying to determine the rate of expansion. If the average mass density of the Universe is less than some critical value ρ_c, the galaxies will slow in their outward rush but still escape to infinity. If the average density exceeds the critical value, the expansion will eventually stop and contraction will begin, possibly leading to a superdense state followed by another expansion. In this scenario, we have an oscillating Universe.

Example 46.6 The Critical Density of the Universe

(A) Starting from energy conservation, derive an expression for the critical mass density of the Universe ρ_c in terms of the Hubble constant H and the universal gravitational constant G.

---

[6]The data at large distances have large observational uncertainties and may be systematically in error from effects such as abnormal brightness in the most distant visible clusters.

▶ **46.6** continued

SOLUTION

Conceptualize Figure 46.19 shows a large section of the Universe, contained within a sphere of radius R. The total mass in this volume is M. A galaxy of mass $m \ll M$ that has a speed v at a distance R from the center of the sphere escapes to infinity (at which its speed approaches zero) if the sum of its kinetic energy and the gravitational potential energy of the system is zero.

Figure 46.19 (Example 46.6) The galaxy marked with mass m is escaping from a large cluster of galaxies contained within a spherical volume of radius R. Only the mass within R slows the escaping galaxy.

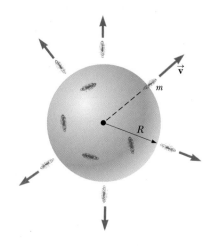

Categorize The Universe may be infinite in spatial extent, but Gauss's law for gravitation (an analog to Gauss's law for electric fields in Chapter 24) implies that only the mass M inside the sphere contributes to the gravitational potential energy of the galaxy–sphere system. Therefore, we categorize this problem as one in which we apply Gauss's law for gravitation. We model the sphere in Figure 46.19 and the escaping galaxy as an *isolated system* for *energy*.

Analyze Write the appropriate reduction of Equation 8.2, assuming that the galaxy leaves the spherical volume while moving at the escape speed:

$$\Delta K + \Delta U = 0$$

$$(0 - \tfrac{1}{2}mv^2) + \left[0 - \left(-\frac{GmM}{R} \right) \right] = 0$$

Substitute for the mass M contained within the sphere the product of the critical density and the volume of the sphere:

$$\tfrac{1}{2}mv^2 = \frac{Gm(\tfrac{4}{3}\pi R^3 \rho_c)}{R}$$

Solve for the critical density:

$$\rho_c = \frac{3v^2}{8\pi G R^2}$$

From Hubble's law, substitute for the ratio $v/R = H$:

$$(1) \quad \rho_c = \frac{3}{8\pi G}\left(\frac{v}{R}\right)^2 = \frac{3H^2}{8\pi G}$$

(B) Estimate a numerical value for the critical density in grams per cubic centimeter.

SOLUTION

In Equation (1), substitute numerical values for H and G:

$$\rho_c = \frac{3H^2}{8\pi G} = \frac{3[22 \times 10^{-3}\ \text{m}/(\text{s} \cdot \text{ly})]^2}{8\pi (6.67 \times 10^{-11}\ \text{N} \cdot \text{m}^2/\text{kg}^2)} = 8.7 \times 10^5\ \text{kg}/\text{m} \cdot (\text{ly})^2$$

Reconcile the units by converting light-years to meters:

$$\rho_c = 8.7 \times 10^5\ \text{kg}/\text{m} \cdot (\text{ly})^2 \left(\frac{1\ \text{ly}}{9.46 \times 10^{15}\ \text{m}}\right)^2$$

$$= 9.7 \times 10^{-27}\ \text{kg}/\text{m}^3 = \boxed{9.7 \times 10^{-30}\ \text{g}/\text{cm}^3}$$

Finalize Because the mass of a hydrogen atom is 1.67×10^{-24} g, this value of ρ_c corresponds to 6×10^{-6} hydrogen atoms per cubic centimeter or 6 atoms per cubic meter.

Missing Mass in the Universe?

The luminous matter in galaxies averages out to a Universe density of about 5×10^{-33} g/cm³. The radiation in the Universe has a mass equivalent of approximately 2% of the luminous matter. The total mass of all nonluminous matter (such as interstellar gas and black holes) may be estimated from the speeds of galaxies orbiting each other in a cluster. The higher the galaxy speeds, the more mass in the cluster. Measurements on the Coma cluster of galaxies indicate, surprisingly, that the amount of nonluminous matter is 20 to 30 times the amount of luminous matter present in stars and luminous gas clouds. Yet even this large, invisible component of *dark matter* (see Section 13.6), if extrapolated to the Universe as a whole, leaves the observed mass density a factor of 10 less than ρ_c calculated in Example 46.6. The deficit, called *missing mass*, has been the subject of intense theoretical and experimental work, with exotic particles such as axions, photinos, and superstring particles suggested as candidates for the missing mass. Some researchers have made the more mundane proposal that the missing mass is present in neutrinos. In fact, neutrinos are so abundant that a tiny neutrino rest energy on the order of only 20 eV would furnish the missing mass and "close" the Universe. Current experiments designed to measure the rest energy of the neutrino will have an effect on predictions for the future of the Universe.

Mysterious Energy in the Universe?

A surprising twist in the story of the Universe arose in 1998 with the observation of a class of supernovae that have a fixed absolute brightness. By combining the apparent brightness and the redshift of light from these explosions, their distance and speed of recession from the Earth can be determined. These observations led to the conclusion that the expansion of the Universe is not slowing down, but is accelerating! Observations by other groups also led to the same interpretation.

To explain this acceleration, physicists have proposed *dark energy*, which is energy possessed by the vacuum of space. In the early life of the Universe, gravity dominated over the dark energy. As the Universe expanded and the gravitational force between galaxies became smaller because of the great distances between them, the dark energy became more important. The dark energy results in an effective repulsive force that causes the expansion rate to increase.[7]

Although there is some degree of certainty about the beginning of the Universe, we are uncertain about how the story will end. Will the Universe keep on expanding forever, or will it someday collapse and then expand again, perhaps in an endless series of oscillations? Results and answers to these questions remain inconclusive, and the exciting controversy continues.

46.12 Problems and Perspectives

While particle physicists have been exploring the realm of the very small, cosmologists have been exploring cosmic history back to the first microsecond of the Big Bang. Observation of the events that occur when two particles collide in an accelerator is essential for reconstructing the early moments in cosmic history. For this reason, perhaps the key to understanding the early Universe is to first understand the world of elementary particles. Cosmologists and physicists now find that they have many common goals and are joining hands in an attempt to understand the physical world at its most fundamental level.

Our understanding of physics at short distances is far from complete. Particle physics is faced with many questions. Why does so little antimatter exist in the Universe? Is it possible to unify the strong and electroweak theories in a logical and consistent manner? Why do quarks and leptons form three similar but distinct families? Are muons the same as electrons apart from their difference in mass, or do they have other subtle differences that have not been detected? Why are some particles charged and others neutral? Why do quarks carry a fractional charge? What determines the masses of the elementary constituents of matter? Can isolated quarks exist? Why do electrons and protons have *exactly* the same magnitude of

---

[7]For an overview of dark energy, see S. Perlmutter, "Supernovae, Dark Energy, and the Accelerating Universe," *Physics Today* **56**(4): 53–60, April 2003.

charge when one is a truly fundamental particle and the other is built from smaller particles?

An important and obvious question that remains is whether leptons and quarks have an underlying structure. If they do, we can envision an infinite number of deeper structure levels. If leptons and quarks are indeed the ultimate constituents of matter, however, scientists hope to construct a final theory of the structure of matter, just as Einstein dreamed of doing. This theory, whimsically called the Theory of Everything, is a combination of the Standard Model and a quantum theory of gravity.

String Theory: A New Perspective

Let's briefly discuss one current effort at answering some of these questions by proposing a new perspective on particles. While reading this book, you may recall starting off with the *particle* model in Chapter 2 and doing quite a bit of physics with it. In Chapter 16, we introduced the *wave* model, and there was more physics to be investigated via the properties of waves. We used a *wave* model for light in Chapter 35; in Chapter 40, however, we saw the need to return to the *particle* model for light. Furthermore, we found that material particles had wave-like characteristics. The quantum particle model discussed in Chapter 40 allowed us to build particles out of waves, suggesting that a *wave* is the fundamental entity. In the current Chapter 46, however, we introduced elementary *particles* as the fundamental entities. It seems as if we cannot make up our mind! In this final section, we discuss a current research effort to build particles out of waves and vibrations on strings!

String theory is an effort to unify the four fundamental forces by modeling all particles as various quantized vibrational modes of a single entity, an incredibly small string. The typical length of such a string is on the order of 10^{-35} m, called the **Planck length.** We have seen quantized modes before in the frequencies of vibrating guitar strings in Chapter 18 and the quantized energy levels of atoms in Chapter 42. In string theory, each quantized mode of vibration of the string corresponds to a different elementary particle in the Standard Model.

One complicating factor in string theory is that it requires space–time to have ten dimensions. Despite the theoretical and conceptual difficulties in dealing with ten dimensions, string theory holds promise in incorporating gravity with the other forces. Four of the ten dimensions—three space dimensions and one time dimension—are visible to us. The other six are said to be *compactified;* that is, the six dimensions are curled up so tightly that they are not visible in the macroscopic world.

As an analogy, consider a soda straw. You can build a soda straw by cutting a rectangular piece of paper (Fig. 46.20a), which clearly has two dimensions, and rolling it into a small tube (Fig. 46.20b). From far away, the soda straw looks like a one-dimensional straight line. The second dimension has been curled up and is not visible. String theory claims that six space–time dimensions are curled up in an analogous way, with the curling being on the size of the Planck length and impossible to see from our viewpoint.

Another complicating factor with string theory is that it is difficult for string theorists to guide experimentalists as to what to look for in an experiment. The

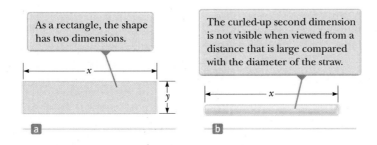

Figure 46.20 (a) A piece of paper is cut into a rectangular shape. (b) The paper is rolled up into a soda straw.

Planck length is so small that direct experimentation on strings is impossible. Until the theory has been further developed, string theorists are restricted to applying the theory to known results and testing for consistency.

One of the predictions of string theory, called **supersymmetry,** or SUSY, suggests that every elementary particle has a superpartner that has not yet been observed. It is believed that supersymmetry is a broken symmetry (like the broken electroweak symmetry at low energies) and the masses of the superpartners are above our current capabilities of detection by accelerators. Some theorists claim that the mass of superpartners is the missing mass discussed in Section 46.11. Keeping with the whimsical trend in naming particles and their properties, superpartners are given names such as the *squark* (the superpartner to a quark), the *selectron* (electron), and the *gluino* (gluon).

Other theorists are working on **M-theory,** which is an eleven-dimensional theory based on membranes rather than strings. In a way reminiscent of the correspondence principle, M-theory is claimed to reduce to string theory if one compactifies from eleven dimensions to ten dimensions.

The questions listed at the beginning of this section go on and on. Because of the rapid advances and new discoveries in the field of particle physics, many of these questions may be resolved in the next decade and other new questions may emerge.

Summary

Concepts and Principles

Before quark theory was developed, the four fundamental forces in nature were identified as nuclear, electromagnetic, weak, and gravitational. All the interactions in which these forces take part are mediated by **field particles.** The electromagnetic interaction is mediated by photons; the weak interaction is mediated by the $W^{\pm}$ and Z^0 bosons; the gravitational interaction is mediated by gravitons; and the nuclear interaction is mediated by gluons.

A charged particle and its **antiparticle** have the same mass but opposite charge, and other properties will have opposite values, such as lepton number and baryon number. It is possible to produce particle–antiparticle pairs in nuclear reactions if the available energy is greater than $2mc^2$, where m is the mass of the particle (or antiparticle).

Particles other than field particles are classified as hadrons or leptons. **Hadrons** interact via all four fundamental forces. They have size and structure and are not elementary particles. There are two types, **baryons** and **mesons.** Baryons, which generally are the most massive particles, have nonzero **baryon number** and a spin of $\frac{1}{2}$ or $\frac{3}{2}$. Mesons have baryon number zero and either zero or integral spin.

Leptons have no structure or size and are considered truly elementary. They interact only via the weak, gravitational, and electromagnetic forces. Six types of leptons exist: the electron e^-, the muon μ^-, and the tau τ^-, and their neutrinos ν_e, ν_μ, and ν_τ.

In all reactions and decays, quantities such as energy, linear momentum, angular momentum, electric charge, baryon number, and lepton number are strictly conserved. Certain particles have properties called **strangeness** and **charm.** These unusual properties are conserved in all decays and nuclear reactions except those that occur via the weak force.

Theorists in elementary particle physics have postulated that all hadrons are composed of smaller units known as **quarks,** and experimental evidence agrees with this model. Quarks have fractional electric charge and come in six **flavors:** up (u), down (d), strange (s), charmed (c), top (t), and bottom (b). Each baryon contains three quarks, and each meson contains one quark and one antiquark.

According to the theory of **quantum chromodynamics,** quarks have a property called **color;** the force between quarks is referred to as the **strong force** or the **color force.** The strong force is now considered to be a fundamental force. The nuclear force, which was originally considered to be fundamental, is now understood to be a secondary effect of the strong force due to gluon exchanges between hadrons.

The electromagnetic and weak forces are now considered to be manifestations of a single force called the **electroweak force.** The combination of quantum chromodynamics and the electroweak theory is called the **Standard Model.**

The background microwave radiation discovered by Penzias and Wilson strongly suggests that the Universe started with a Big Bang about 14 billion years ago. The background radiation is equivalent to that of a black body at 3 K. Various astronomical measurements strongly suggest that the Universe is expanding. According to **Hubble's law,** distant galaxies are receding from the Earth at a speed $v = HR$, where H is the **Hubble constant,** $H \approx 22 \times 10^{-3}$ m/(s · ly), and R is the distance from the Earth to the galaxy.

Objective Questions

1. denotes answer available in *Student Solutions Manual/Study Guide*

1. What interactions affect protons in an atomic nucleus? More than one answer may be correct. (a) the nuclear interaction (b) the weak interaction (c) the electromagnetic interaction (d) the gravitational interaction

2. In one experiment, two balls of clay of the same mass travel with the same speed v toward each other. They collide head-on and come to rest. In a second experiment, two clay balls of the same mass are again used. One ball hangs at rest, suspended from the ceiling by a thread. The second ball is fired toward the first at speed v, to collide, stick to the first ball, and continue to move forward. Is the kinetic energy that is transformed into internal energy in the first experiment (a) one-fourth as much as in the second experiment, (b) one-half as much as in the second experiment, (c) the same as in the second experiment, (d) twice as much as in the second experiment, or (e) four times as much as in the second experiment?

3. The Ω^- particle is a baryon with spin $\frac{3}{2}$. Does the Ω^- particle have (a) three possible spin states in a magnetic field, (b) four possible spin states, (c) three times the charge of a spin $-\frac{1}{2}$ particle, or (d) three times the mass of a spin $-\frac{1}{2}$ particle, or (e) are none of those choices correct?

4. Which of the following field particles mediates the strong force? (a) photon (b) gluon (c) graviton (d) $W^\pm$ and Z bosons (e) none of those field particles

5. An isolated stationary muon decays into an electron, an electron antineutrino, and a muon neutrino. Is the total kinetic energy of these three particles (a) zero, (b) small, or (c) large compared to their rest energies, or (d) none of those choices are possible?

6. Define the average density of the solar system ρ_{SS} as the total mass of the Sun, planets, satellites, rings, asteroids, icy outliers, and comets, divided by the volume of a sphere around the Sun large enough to contain all these objects. The sphere extends about halfway to the nearest star, with a radius of approximately 2×10^{16} m, about two light-years. How does this average density of the solar system compare with the critical density ρ_c required for the Universe to stop its Hubble's-law expansion? (a) ρ_{SS} is much greater than ρ_c. (b) ρ_{SS} is approximately or precisely equal to ρ_c. (c) ρ_{SS} is much less than ρ_c. (d) It is impossible to determine.

7. When an electron and a positron meet at low speed in empty space, they annihilate each other to produce two 0.511-MeV gamma rays. What law would be violated if they produced one gamma ray with an energy of 1.02 MeV? (a) conservation of energy (b) conservation of momentum (c) conservation of charge (d) conservation of baryon number (e) conservation of electron lepton number

8. Place the following events into the correct sequence from the earliest in the history of the Universe to the latest. (a) Neutral atoms form. (b) Protons and neutrons are no longer annihilated as fast as they form. (c) The Universe is a quark–gluon soup. (d) The Universe is like the core of a normal star today, forming helium by nuclear fusion. (e) The Universe is like the surface of a hot star today, consisting of a plasma of ionized atoms. (f) Polyatomic molecules form. (g) Solid materials form.

Conceptual Questions

1. denotes answer available in *Student Solutions Manual/Study Guide*

1. The W and Z bosons were first produced at CERN in 1983 by causing a beam of protons and a beam of antiprotons to meet at high energy. Why was this discovery important?

2. What are the differences between hadrons and leptons?

3. Neutral atoms did not exist until hundreds of thousands of years after the Big Bang. Why?

4. Describe the properties of baryons and mesons and the important differences between them.

5. The Ξ^0 particle decays by the weak interaction according to the decay mode $\Xi^0 \rightarrow \Lambda^0 + \pi^0$. Would you expect this decay to be fast or slow? Explain.

6. In the theory of quantum chromodynamics, quarks come in three colors. How would you justify the statement that "all baryons and mesons are colorless"?

7. An antibaryon interacts with a meson. Can a baryon be produced in such an interaction? Explain.

8. Describe the essential features of the Standard Model of particle physics.

9. How many quarks are in each of the following: (a) a baryon, (b) an antibaryon, (c) a meson, (d) an antimeson? (e) How do you explain that baryons have half-integral spins, whereas mesons have spins of 0 or 1?

10. Are the laws of conservation of baryon number, lepton number, and strangeness based on fundamental properties of nature (as are the laws of conservation of momentum and energy, for example)? Explain.

11. Name the four fundamental interactions and the field particle that mediates each.

12. How did Edwin Hubble determine in 1928 that the Universe is expanding?

13. Kaons all decay into final states that contain no protons or neutrons. What is the baryon number for kaons?

Problems available in Access end-of-chapter problems online at **www.webassign.net**

Section 46.1 The Fundamental Forces in Nature
Section 46.2 Positrons and Other Antiparticles
Problems 1–5

Section 46.3 Mesons and the Beginning of Particle Physics
Problems 6–7

Section 46.4 Classification of Particles
Section 46.5 Conservation Laws
Problems 8–19

Section 46.6 Strange Particles and Strangeness
Problems 20–27

Section 46.7 Finding Patterns in the Particles
Section 46.8 Quarks
Section 46.9 Multicolored Quarks
Section 46.10 The Standard Model
Problems 28–36

Section 46.11 The Cosmic Connection
Problems 37–49

Section 46.12 Problems and Perspectives
Problem 50

Additional Problems
Problems 51–66

Challenge Problems
Problems 67–73

Solutions to the following Problems are available in the *Student Solutions Manual/Study Guide:*
46.5, 46.6, 46.9, 46.13, 46.15, 46.19, 46.25, 46.27, 46.32, 46.45, 46.53, 46.56, 46.64, 46.67, and 46.71

List of Enhanced Problems

Problem Number	Targeted Feedback in Enhanced WebAssign	Analysis Model Tutorial in Enhanced WebAssign	Master It in Enhanced WebAssign	Watch It in Enhanced WebAssign
46.5	✓		✓	
46.27	✓		✓	
46.34			✓	
46.45			✓	
46.47	✓			✓
46.56	✓		✓	
46.59		✓		
46.69	✓			✓

Tables

Table A.1 Conversion Factors

Length

	m	cm	km	in.	ft	mi
1 meter	1	10^2	10^{-3}	39.37	3.281	6.214×10^{-4}
1 centimeter	10^{-2}	1	10^{-5}	0.393 7	3.281×10^{-2}	6.214×10^{-6}
1 kilometer	10^3	10^5	1	3.937×10^4	3.281×10^3	0.621 4
1 inch	2.540×10^{-2}	2.540	2.540×10^{-5}	1	8.333×10^{-2}	1.578×10^{-5}
1 foot	0.304 8	30.48	3.048×10^{-4}	12	1	1.894×10^{-4}
1 mile	1 609	1.609×10^5	1.609	6.336×10^4	5 280	1

Mass

	kg	g	slug	u
1 kilogram	1	10^3	6.852×10^{-2}	6.024×10^{26}
1 gram	10^{-3}	1	6.852×10^{-5}	6.024×10^{23}
1 slug	14.59	1.459×10^4	1	8.789×10^{27}
1 atomic mass unit	1.660×10^{-27}	1.660×10^{-24}	1.137×10^{-28}	1

Note: 1 metric ton = 1 000 kg.

Time

	s	min	h	day	yr
1 second	1	1.667×10^{-2}	2.778×10^{-4}	1.157×10^{-5}	3.169×10^{-8}
1 minute	60	1	1.667×10^{-2}	6.994×10^{-4}	1.901×10^{-6}
1 hour	3 600	60	1	4.167×10^{-2}	1.141×10^{-4}
1 day	8.640×10^4	1 440	24	1	2.738×10^{-5}
1 year	3.156×10^7	5.259×10^5	8.766×10^3	365.2	1

Speed

	m/s	cm/s	ft/s	mi/h
1 meter per second	1	10^2	3.281	2.237
1 centimeter per second	10^{-2}	1	3.281×10^{-2}	2.237×10^{-2}
1 foot per second	0.304 8	30.48	1	0.681 8
1 mile per hour	0.447 0	44.70	1.467	1

Note: 1 mi/min = 60 mi/h = 88 ft/s.

Force

	N	lb
1 newton	1	0.224 8
1 pound	4.448	1

(Continued)

Table A.1   Conversion Factors (continued)

Energy, Energy Transfer

	J	ft · lb	eV
1 joule	1	0.737 6	6.242×10^{18}
1 foot-pound	1.356	1	8.464×10^{18}
1 electron volt	1.602×10^{-19}	1.182×10^{-19}	1
1 calorie	4.186	3.087	2.613×10^{19}
1 British thermal unit	1.055×10^3	7.779×10^2	6.585×10^{21}
1 kilowatt-hour	3.600×10^6	2.655×10^6	2.247×10^{25}

	cal	Btu	kWh
1 joule	0.238 9	9.481×10^{-4}	2.778×10^{-7}
1 foot-pound	0.323 9	1.285×10^{-3}	3.766×10^{-7}
1 electron volt	3.827×10^{-20}	1.519×10^{-22}	4.450×10^{-26}
1 calorie	1	3.968×10^{-3}	1.163×10^{-6}
1 British thermal unit	2.520×10^2	1	2.930×10^{-4}
1 kilowatt-hour	8.601×10^5	3.413×10^2	1

Pressure

	Pa	atm
1 pascal	1	9.869×10^{-6}
1 atmosphere	1.013×10^5	1
1 centimeter mercury[a]	1.333×10^3	1.316×10^{-2}
1 pound per square inch	6.895×10^3	6.805×10^{-2}
1 pound per square foot	47.88	4.725×10^{-4}

	cm Hg	lb/in.²	lb/ft²
1 pascal	7.501×10^{-4}	1.450×10^{-4}	2.089×10^{-2}
1 atmosphere	76	14.70	2.116×10^3
1 centimeter mercury[a]	1	0.194 3	27.85
1 pound per square inch	5.171	1	144
1 pound per square foot	3.591×10^{-2}	6.944×10^{-3}	1

[a] At 0°C and at a location where the free-fall acceleration has its "standard" value, 9.806 65 m/s².

Table A.2   Symbols, Dimensions, and Units of Physical Quantities

Quantity	Common Symbol	Unit[a]	Dimensions[b]	Unit in Terms of Base SI Units
Acceleration	$\vec{a}$	m/s²	L/T^2	m/s²
Amount of substance	n	MOLE		mol
Angle	θ, ϕ	radian (rad)	1	
Angular acceleration	$\vec{\alpha}$	rad/s²	T^{-2}	s^{-2}
Angular frequency	ω	rad/s	T^{-1}	s^{-1}
Angular momentum	$\vec{L}$	kg · m²/s	ML^2/T	kg · m²/s
Angular velocity	$\vec{\omega}$	rad/s	T^{-1}	s^{-1}
Area	A	m²	L^2	m²
Atomic number	Z			
Capacitance	C	farad (F)	Q^2T^2/ML^2	$A^2 \cdot s^4/kg \cdot m^2$
Charge	q, Q, e	coulomb (C)	Q	A · s

(Continued)

Table A.2 Symbols, Dimensions, and Units of Physical Quantities (continued)

Quantity	Common Symbol	Unit[a]	Dimensions[b]	Unit in Terms of Base SI Units
Charge density				
Line	λ	C/m	Q/L	$A \cdot s/m$
Surface	σ	C/m^2	Q/L^2	$A \cdot s/m^2$
Volume	ρ	C/m^3	Q/L^3	$A \cdot s/m^3$
Conductivity	σ	$1/\Omega \cdot m$	Q^2T/ML^3	$A^2 \cdot s^3/kg \cdot m^3$
Current	I	AMPERE	Q/T	A
Current density	J	A/m^2	Q/TL^2	A/m^2
Density	ρ	kg/m^3	M/L^3	kg/m^3
Dielectric constant	κ			
Electric dipole moment	$\vec{\mathbf{p}}$	$C \cdot m$	QL	$A \cdot s \cdot m$
Electric field	$\vec{\mathbf{E}}$	V/m	ML/QT^2	$kg \cdot m/A \cdot s^3$
Electric flux	Φ_E	$V \cdot m$	ML^3/QT^2	$kg \cdot m^3/A \cdot s^3$
Electromotive force	$\boldsymbol{\varepsilon}$	volt (V)	ML^2/QT^2	$kg \cdot m^2/A \cdot s^3$
Energy	E, U, K	joule (J)	ML^2/T^2	$kg \cdot m^2/s^2$
Entropy	S	J/K	ML^2/T^2K	$kg \cdot m^2/s^2 \cdot K$
Force	$\vec{\mathbf{F}}$	newton (N)	ML/T^2	$kg \cdot m/s^2$
Frequency	f	hertz (Hz)	T^{-1}	s^{-1}
Heat	Q	joule (J)	ML^2/T^2	$kg \cdot m^2/s^2$
Inductance	L	henry (H)	ML^2/Q^2	$kg \cdot m^2/A^2 \cdot s^2$
Length	ℓ, L	METER	L	m
Displacement	$\Delta x, \Delta \vec{\mathbf{r}}$			
Distance	d, h			
Position	$x, y, z, \vec{\mathbf{r}}$			
Magnetic dipole moment	$\vec{\boldsymbol{\mu}}$	$N \cdot m/T$	QL^2/T	$A \cdot m^2$
Magnetic field	$\vec{\mathbf{B}}$	tesla (T) (= Wb/m^2)	M/QT	$kg/A \cdot s^2$
Magnetic flux	Φ_B	weber (Wb)	ML^2/QT	$kg \cdot m^2/A \cdot s^2$
Mass	m, M	KILOGRAM	M	kg
Molar specific heat	C	$J/mol \cdot K$		$kg \cdot m^2/s^2 \cdot mol \cdot K$
Moment of inertia	I	$kg \cdot m^2$	ML^2	$kg \cdot m^2$
Momentum	$\vec{\mathbf{p}}$	$kg \cdot m/s$	ML/T	$kg \cdot m/s$
Period	T	s	T	s
Permeability of free space	μ_0	N/A^2 (= H/m)	ML/Q^2	$kg \cdot m/A^2 \cdot s^2$
Permittivity of free space	ϵ_0	$C^2/N \cdot m^2$ (= F/m)	Q^2T^2/ML^3	$A^2 \cdot s^4/kg \cdot m^3$
Potential	V	volt (V)(= J/C)	ML^2/QT^2	$kg \cdot m^2/A \cdot s^3$
Power	P	watt (W)(= J/s)	ML^2/T^3	$kg \cdot m^2/s^3$
Pressure	P	pascal (Pa)(= N/m^2)	M/LT^2	$kg/m \cdot s^2$
Resistance	R	ohm (Ω)(= V/A)	ML^2/Q^2T	$kg \cdot m^2/A^2 \cdot s^3$
Specific heat	c	$J/kg \cdot K$	L^2/T^2K	$m^2/s^2 \cdot K$
Speed	v	m/s	L/T	m/s
Temperature	T	KELVIN	K	K
Time	t	SECOND	T	s
Torque	$\vec{\boldsymbol{\tau}}$	$N \cdot m$	ML^2/T^2	$kg \cdot m^2/s^2$
Velocity	$\vec{\mathbf{v}}$	m/s	L/T	m/s
Volume	V	m^3	L^3	m^3
Wavelength	λ	m	L	m
Work	W	joule (J)(= $N \cdot m$)	ML^2/T^2	$kg \cdot m^2/s^2$

[a]The base SI units are given in uppercase letters.

[b]The symbols M, L, T, K, and Q denote mass, length, time, temperature, and charge, respectively.

Mathematics Review

This appendix in mathematics is intended as a brief review of operations and methods. Early in this course, you should be totally familiar with basic algebraic techniques, analytic geometry, and trigonometry. The sections on differential and integral calculus are more detailed and are intended for students who have difficulty applying calculus concepts to physical situations.

B.1 Scientific Notation

Many quantities used by scientists often have very large or very small values. The speed of light, for example, is about 300 000 000 m/s, and the ink required to make the dot over an i in this textbook has a mass of about 0.000 000 001 kg. Obviously, it is very cumbersome to read, write, and keep track of such numbers. We avoid this problem by using a method incorporating powers of the number 10:

$$10^0 = 1$$
$$10^1 = 10$$
$$10^2 = 10 \times 10 = 100$$
$$10^3 = 10 \times 10 \times 10 = 1\,000$$
$$10^4 = 10 \times 10 \times 10 \times 10 = 10\,000$$
$$10^5 = 10 \times 10 \times 10 \times 10 \times 10 = 100\,000$$

and so on. The number of zeros corresponds to the power to which ten is raised, called the **exponent** of ten. For example, the speed of light, 300 000 000 m/s, can be expressed as 3.00×10^8 m/s.

In this method, some representative numbers smaller than unity are the following:

$$10^{-1} = \frac{1}{10} = 0.1$$

$$10^{-2} = \frac{1}{10 \times 10} = 0.01$$

$$10^{-3} = \frac{1}{10 \times 10 \times 10} = 0.001$$

$$10^{-4} = \frac{1}{10 \times 10 \times 10 \times 10} = 0.000\,1$$

$$10^{-5} = \frac{1}{10 \times 10 \times 10 \times 10 \times 10} = 0.000\,01$$

In these cases, the number of places the decimal point is to the left of the digit 1 equals the value of the (negative) exponent. Numbers expressed as some power of ten multiplied by another number between one and ten are said to be in **scientific notation.** For example, the scientific notation for 5 943 000 000 is 5.943×10^9 and that for 0.000 083 2 is 8.32×10^{-5}.

When numbers expressed in scientific notation are being multiplied, the following general rule is very useful:

$$10^n \times 10^m = 10^{n+m} \tag{B.1}$$

where n and m can be *any* numbers (not necessarily integers). For example, $10^2 \times 10^5 = 10^7$. The rule also applies if one of the exponents is negative: $10^3 \times 10^{-8} = 10^{-5}$.

When dividing numbers expressed in scientific notation, note that

$$\frac{10^n}{10^m} = 10^n \times 10^{-m} = 10^{n-m} \tag{B.2}$$

Exercises

With help from the preceding rules, verify the answers to the following equations:

1. $86\ 400 = 8.64 \times 10^4$
2. $9\ 816\ 762.5 = 9.816\ 762\ 5 \times 10^6$
3. $0.000\ 000\ 039\ 8 = 3.98 \times 10^{-8}$
4. $(4.0 \times 10^8)(9.0 \times 10^9) = 3.6 \times 10^{18}$
5. $(3.0 \times 10^7)(6.0 \times 10^{-12}) = 1.8 \times 10^{-4}$
6. $\dfrac{75 \times 10^{-11}}{5.0 \times 10^{-3}} = 1.5 \times 10^{-7}$
7. $\dfrac{(3 \times 10^6)(8 \times 10^{-2})}{(2 \times 10^{17})(6 \times 10^5)} = 2 \times 10^{-18}$

B.2 Algebra

Some Basic Rules

When algebraic operations are performed, the laws of arithmetic apply. Symbols such as x, y, and z are usually used to represent unspecified quantities, called the **unknowns.**

First, consider the equation

$$8x = 32$$

If we wish to solve for x, we can divide (or multiply) each side of the equation by the same factor without destroying the equality. In this case, if we divide both sides by 8, we have

$$\frac{8x}{8} = \frac{32}{8}$$

$$x = 4$$

Next consider the equation

$$x + 2 = 8$$

In this type of expression, we can add or subtract the same quantity from each side. If we subtract 2 from each side, we have

$$x + 2 - 2 = 8 - 2$$

$$x = 6$$

In general, if $x + a = b$, then $x = b - a$.

Now consider the equation

$$\frac{x}{5} = 9$$

If we multiply each side by 5, we are left with x on the left by itself and 45 on the right:

$$\left(\frac{x}{5}\right)(5) = 9 \times 5$$

$$x = 45$$

In all cases, *whatever operation is performed on the left side of the equality must also be performed on the right side.*

The following rules for multiplying, dividing, adding, and subtracting fractions should be recalled, where a, b, c, and d are four numbers:

	Rule	Example
Multiplying	$\left(\dfrac{a}{b}\right)\left(\dfrac{c}{d}\right) = \dfrac{ac}{bd}$	$\left(\dfrac{2}{3}\right)\left(\dfrac{4}{5}\right) = \dfrac{8}{15}$
Dividing	$\dfrac{(a/b)}{(c/d)} = \dfrac{ad}{bc}$	$\dfrac{2/3}{4/5} = \dfrac{(2)(5)}{(4)(3)} = \dfrac{10}{12}$
Adding	$\dfrac{a}{b} \pm \dfrac{c}{d} = \dfrac{ad \pm bc}{bd}$	$\dfrac{2}{3} - \dfrac{4}{5} = \dfrac{(2)(5) - (4)(3)}{(3)(5)} = -\dfrac{2}{15}$

Exercises

In the following exercises, solve for x.

Answers

1. $a = \dfrac{1}{1 + x}$ $x = \dfrac{1 - a}{a}$

2. $3x - 5 = 13$ $x = 6$

3. $ax - 5 = bx + 2$ $x = \dfrac{7}{a - b}$

4. $\dfrac{5}{2x + 6} = \dfrac{3}{4x + 8}$ $x = -\dfrac{11}{7}$

Powers

When powers of a given quantity x are multiplied, the following rule applies:

$$x^n x^m = x^{n+m} \tag{B.3}$$

For example, $x^2 x^4 = x^{2+4} = x^6$.

When dividing the powers of a given quantity, the rule is

$$\frac{x^n}{x^m} = x^{n-m} \tag{B.4}$$

For example, $x^8/x^2 = x^{8-2} = x^6$.

A power that is a fraction, such as $\frac{1}{3}$, corresponds to a root as follows:

$$x^{1/n} = \sqrt[n]{x} \tag{B.5}$$

For example, $4^{1/3} = \sqrt[3]{4} = 1.587\,4$. (A scientific calculator is useful for such calculations.)

Finally, any quantity x^n raised to the mth power is

$$(x^n)^m = x^{nm} \tag{B.6}$$

Table B.1 summarizes the rules of exponents.

Table B.1   Rules of Exponents

$x^0 = 1$
$x^1 = x$
$x^n x^m = x^{n+m}$
$x^n/x^m = x^{n-m}$
$x^{1/n} = \sqrt[n]{x}$
$(x^n)^m = x^{nm}$

Exercises

Verify the following equations:

1. $3^2 \times 3^3 = 243$
2. $x^5 x^{-8} = x^{-3}$

3. $x^{10}/x^{-5} = x^{15}$

4. $5^{1/3} = 1.709\ 976$ (Use your calculator.)

5. $60^{1/4} = 2.783\ 158$ (Use your calculator.)

6. $(x^4)^3 = x^{12}$

Factoring

Some useful formulas for factoring an equation are the following:

$$ax + ay + az = a(x + y + z)$$ common factor

$$a^2 + 2ab + b^2 = (a + b)^2$$ perfect square

$$a^2 - b^2 = (a + b)(a - b)$$ differences of squares

Quadratic Equations

The general form of a quadratic equation is

$$ax^2 + bx + c = 0 \tag{B.7}$$

where x is the unknown quantity and a, b, and c are numerical factors referred to as **coefficients** of the equation. This equation has two roots, given by

$$x = \frac{-b \pm \sqrt{b^2 - 4ac}}{2a} \tag{B.8}$$

If $b^2 \geq 4ac$, the roots are real.

Example B.1

The equation $x^2 + 5x + 4 = 0$ has the following roots corresponding to the two signs of the square-root term:

$$x = \frac{-5 \pm \sqrt{5^2 - (4)(1)(4)}}{2(1)} = \frac{-5 \pm \sqrt{9}}{2} = \frac{-5 \pm 3}{2}$$

$$x_+ = \frac{-5 + 3}{2} = -1 \qquad x_- = \frac{-5 - 3}{2} = -4$$

where x_+ refers to the root corresponding to the positive sign and x_- refers to the root corresponding to the negative sign.

Exercises

Solve the following quadratic equations:

Answers

1. $x^2 + 2x - 3 = 0$	$x_+ = 1$	$x_- = -3$
2. $2x^2 - 5x + 2 = 0$	$x_+ = 2$	$x_- = \frac{1}{2}$
3. $2x^2 - 4x - 9 = 0$	$x_+ = 1 + \sqrt{22}/2$	$x_- = 1 - \sqrt{22}/2$

Linear Equations

A linear equation has the general form

$$y = mx + b \tag{B.9}$$

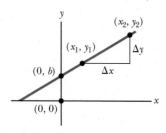

Figure B.1 A straight line graphed on an xy coordinate system. The slope of the line is the ratio of Δy to Δx.

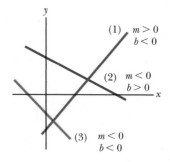

Figure B.2 The brown line has a positive slope and a negative y-intercept. The blue line has a negative slope and a positive y-intercept. The green line has a negative slope and a negative y-intercept.

where m and b are constants. This equation is referred to as linear because the graph of y versus x is a straight line as shown in Figure B.1. The constant b, called the **y-intercept,** represents the value of y at which the straight line intersects the y axis. The constant m is equal to the **slope** of the straight line. If any two points on the straight line are specified by the coordinates (x_1, y_1) and (x_2, y_2) as in Figure B.1, the slope of the straight line can be expressed as

$$\text{Slope} = \frac{y_2 - y_1}{x_2 - x_1} = \frac{\Delta y}{\Delta x} \tag{B.10}$$

Note that m and b can have either positive or negative values. If $m > 0$, the straight line has a *positive* slope as in Figure B.1. If $m < 0$, the straight line has a *negative* slope. In Figure B.1, both m and b are positive. Three other possible situations are shown in Figure B.2.

Exercises

1. Draw graphs of the following straight lines: (a) $y = 5x + 3$ (b) $y = -2x + 4$ (c) $y = -3x - 6$

2. Find the slopes of the straight lines described in Exercise 1.

Answers (a) 5 (b) -2 (c) -3

3. Find the slopes of the straight lines that pass through the following sets of points: (a) $(0, -4)$ and $(4, 2)$ (b) $(0, 0)$ and $(2, -5)$ (c) $(-5, 2)$ and $(4, -2)$

Answers (a) $\frac{3}{2}$ (b) $-\frac{5}{2}$ (c) $-\frac{4}{9}$

Solving Simultaneous Linear Equations

Consider the equation $3x + 5y = 15$, which has two unknowns, x and y. Such an equation does not have a unique solution. For example, $(x = 0, y = 3)$, $(x = 5, y = 0)$, and $(x = 2, y = \frac{9}{5})$ are all solutions to this equation.

If a problem has two unknowns, a unique solution is possible only if we have *two* pieces of information. In most common cases, those two pieces of information are equations. In general, if a problem has n unknowns, its solution requires n equations. To solve two simultaneous equations involving two unknowns, x and y, we solve one of the equations for x in terms of y and substitute this expression into the other equation.

In some cases, the two pieces of information may be (1) one equation and (2) a condition on the solutions. For example, suppose we have the equation $m = 3n$ and the condition that m and n must be the smallest positive nonzero integers possible. Then, the single equation does not allow a unique solution, but the addition of the condition gives us that $n = 1$ and $m = 3$.

Example B.2

Solve the two simultaneous equations

$$(1)\quad 5x + y = -8$$

$$(2)\quad 2x - 2y = 4$$

SOLUTION

From Equation (2), $x = y + 2$. Substitution of this equation into Equation (1) gives

$$5(y + 2) + y = -8$$

$$6y = -18$$

▶ **B.2** continued

$$y = \boxed{-3}$$
$$x = y + 2 = \boxed{-1}$$

Alternative Solution Multiply each term in Equation (1) by the factor 2 and add the result to Equation (2):

$$10x + 2y = -16$$
$$\underline{2x - 2y = 4}$$
$$12x \qquad = -12$$
$$x = \boxed{-1}$$
$$y = x - 2 = \boxed{-3}$$

   Two linear equations containing two unknowns can also be solved by a graphical method. If the straight lines corresponding to the two equations are plotted in a conventional coordinate system, the intersection of the two lines represents the solution. For example, consider the two equations

$$x - y = 2$$
$$x - 2y = -1$$

These equations are plotted in Figure B.3. The intersection of the two lines has the coordinates $x = 5$ and $y = 3$, which represents the solution to the equations. You should check this solution by the analytical technique discussed earlier.

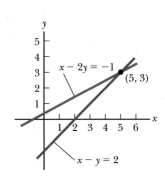

Figure B.3 A graphical solution for two linear equations.

Exercises

Solve the following pairs of simultaneous equations involving two unknowns:

	Answers
1. $x + y = 8$	$x = 5, y = 3$
$x - y = 2$	
2. $98 - T = 10a$	$T = 65, a = 3.27$
$T - 49 = 5a$	
3. $6x + 2y = 6$	$x = 2, y = -3$
$8x - 4y = 28$	

Logarithms

Suppose a quantity x is expressed as a power of some quantity a:

$$x = a^y \qquad \text{(B.11)}$$

The number a is called the **base** number. The **logarithm** of x with respect to the base a is equal to the exponent to which the base must be raised to satisfy the expression $x = a^y$:

$$y = \log_a x \qquad \text{(B.12)}$$

Conversely, the **antilogarithm** of y is the number x:

$$x = \text{antilog}_a y \qquad \text{(B.13)}$$

   In practice, the two bases most often used are base 10, called the *common* logarithm base, and base $e = 2.718\ 282$, called Euler's constant or the *natural* logarithm base. When common logarithms are used,

$$y = \log_{10} x \quad (\text{or } x = 10^y) \qquad \text{(B.14)}$$

When natural logarithms are used,

$$y = \ln x \quad (\text{or } x = e^y) \tag{B.15}$$

For example, $\log_{10} 52 = 1.716$, so $\text{antilog}_{10}\ 1.716 = 10^{1.716} = 52$. Likewise, $\ln 52 = 3.951$, so $\text{antiln}\ 3.951 = e^{3.951} = 52$.

In general, note you can convert between base 10 and base e with the equality

$$\ln x = (2.302\ 585) \log_{10} x \tag{B.16}$$

Finally, some useful properties of logarithms are the following:

$$\left.\begin{array}{l} \log(ab) = \log a + \log b \\ \log(a/b) = \log a - \log b \\ \log(a^n) = n \log a \end{array}\right\} \text{any base}$$

$$\ln e = 1$$

$$\ln e^a = a$$

$$\ln\left(\frac{1}{a}\right) = -\ln a$$

B.3 Geometry

The **distance** d between two points having coordinates (x_1, y_1) and (x_2, y_2) is

$$d = \sqrt{(x_2 - x_1)^2 + (y_2 - y_1)^2} \tag{B.17}$$

Two angles are equal if their sides are perpendicular, right side to right side and left side to left side. For example, the two angles marked θ in Figure B.4 are the same because of the perpendicularity of the sides of the angles. To distinguish the left and right sides of an angle, imagine standing at the angle's apex and facing into the angle.

Radian measure: The arc length s of a circular arc (Fig. B.5) is proportional to the radius r for a fixed value of θ (in radians):

$$s = r\theta$$
$$\theta = \frac{s}{r} \tag{B.18}$$

Table B.2 gives the **areas** and **volumes** for several geometric shapes used throughout this text.

The equation of a **straight line** (Fig. B.6) is

$$y = mx + b \tag{B.19}$$

where b is the y-intercept and m is the slope of the line.

The equation of a **circle** of radius R centered at the origin is

$$x^2 + y^2 = R^2 \tag{B.20}$$

The equation of an **ellipse** having the origin at its center (Fig. B.7) is

$$\frac{x^2}{a^2} + \frac{y^2}{b^2} = 1 \tag{B.21}$$

where a is the length of the semimajor axis (the longer one) and b is the length of the semiminor axis (the shorter one).

The equation of a **parabola** the vertex of which is at $y = b$ (Fig. B.8) is

$$y = ax^2 + b \tag{B.22}$$

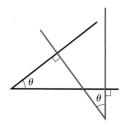

Figure B.4 The angles are equal because their sides are perpendicular.

Figure B.5 The angle θ in radians is the ratio of the arc length s to the radius r of the circle.

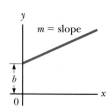

Figure B.6 A straight line with a slope of m and a y-intercept of b.

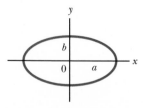

Figure B.7 An ellipse with semimajor axis a and semiminor axis b.

Table B.2   **Useful Information for Geometry**

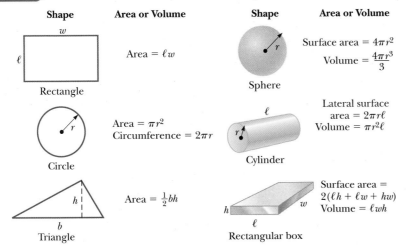

Shape	Area or Volume	Shape	Area or Volume
Rectangle	Area $= \ell w$	Sphere	Surface area $= 4\pi r^2$ Volume $= \frac{4\pi r^3}{3}$
Circle	Area $= \pi r^2$ Circumference $= 2\pi r$	Cylinder	Lateral surface area $= 2\pi r \ell$ Volume $= \pi r^2 \ell$
Triangle	Area $= \frac{1}{2} bh$	Rectangular box	Surface area $=$ $2(\ell h + \ell w + hw)$ Volume $= \ell wh$

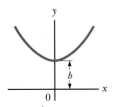

Figure B.8 A parabola with its vertex at $y = b$.

The equation of a **rectangular hyperbola** (Fig. B.9) is

$$xy = \text{constant} \qquad \text{(B.23)}$$

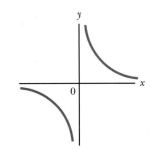

Figure B.9 A hyperbola.

B.4 Trigonometry

That portion of mathematics based on the special properties of the right triangle is called trigonometry. By definition, a right triangle is a triangle containing a 90° angle. Consider the right triangle shown in Figure B.10, where side a is opposite the angle θ, side b is adjacent to the angle θ, and side c is the hypotenuse of the triangle. The three basic trigonometric functions defined by such a triangle are the sine (sin), cosine (cos), and tangent (tan). In terms of the angle θ, these functions are defined as follows:

$$\sin \theta = \frac{\text{side opposite } \theta}{\text{hypotenuse}} = \frac{a}{c} \qquad \text{(B.24)}$$

$$\cos \theta = \frac{\text{side adjacent to } \theta}{\text{hypotenuse}} = \frac{b}{c} \qquad \text{(B.25)}$$

$$\tan \theta = \frac{\text{side opposite } \theta}{\text{side adjacent to } \theta} = \frac{a}{b} \qquad \text{(B.26)}$$

a = opposite side
b = adjacent side
c = hypotenuse

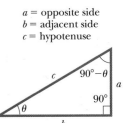

Figure B.10 A right triangle, used to define the basic functions of trigonometry.

The Pythagorean theorem provides the following relationship among the sides of a right triangle:

$$c^2 = a^2 + b^2 \qquad \text{(B.27)}$$

From the preceding definitions and the Pythagorean theorem, it follows that

$$\sin^2 \theta + \cos^2 \theta = 1$$

$$\tan \theta = \frac{\sin \theta}{\cos \theta}$$

The cosecant, secant, and cotangent functions are defined by

$$\csc \theta = \frac{1}{\sin \theta} \quad \sec \theta = \frac{1}{\cos \theta} \quad \cot \theta = \frac{1}{\tan \theta}$$

Table B.3   **Some Trigonometric Identities**

$\sin^2 \theta + \cos^2 \theta = 1$	$\csc^2 \theta = 1 + \cot^2 \theta$
$\sec^2 \theta = 1 + \tan^2 \theta$	$\sin^2 \dfrac{\theta}{2} = \frac{1}{2}(1 - \cos \theta)$
$\sin 2\theta = 2 \sin \theta \cos \theta$	$\cos^2 \dfrac{\theta}{2} = \frac{1}{2}(1 + \cos \theta)$
$\cos 2\theta = \cos^2 \theta - \sin^2 \theta$	$1 - \cos \theta = 2 \sin^2 \dfrac{\theta}{2}$
$\tan 2\theta = \dfrac{2 \tan \theta}{1 - \tan^2 \theta}$	$\tan \dfrac{\theta}{2} = \sqrt{\dfrac{1 - \cos \theta}{1 + \cos \theta}}$

$\sin (A \pm B) = \sin A \cos B \pm \cos A \sin B$

$\cos (A \pm B) = \cos A \cos B \mp \sin A \sin B$

$\sin A \pm \sin B = 2 \sin \left[\frac{1}{2}(A \pm B)\right] \cos \left[\frac{1}{2}(A \mp B)\right]$

$\cos A + \cos B = 2 \cos \left[\frac{1}{2}(A + B)\right] \cos \left[\frac{1}{2}(A - B)\right]$

$\cos A - \cos B = 2 \sin \left[\frac{1}{2}(A + B)\right] \sin \left[\frac{1}{2}(B - A)\right]$

The following relationships are derived directly from the right triangle shown in Figure B.10:

$$\sin \theta = \cos (90° - \theta)$$

$$\cos \theta = \sin (90° - \theta)$$

$$\cot \theta = \tan (90° - \theta)$$

Some properties of trigonometric functions are the following:

$$\sin (-\theta) = -\sin \theta$$

$$\cos (-\theta) = \cos \theta$$

$$\tan (-\theta) = -\tan \theta$$

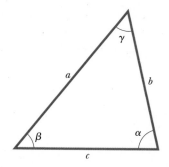

Figure B.11 An arbitrary, non-right triangle.

The following relationships apply to *any* triangle as shown in Figure B.11:

$$\alpha + \beta + \gamma = 180°$$

Law of cosines $\begin{cases} a^2 = b^2 + c^2 - 2bc \cos \alpha \\ b^2 = a^2 + c^2 - 2ac \cos \beta \\ c^2 = a^2 + b^2 - 2ab \cos \gamma \end{cases}$

Law of sines $\quad \dfrac{a}{\sin \alpha} = \dfrac{b}{\sin \beta} = \dfrac{c}{\sin \gamma}$

Table B.3 lists a number of useful trigonometric identities.

Example B.3

Consider the right triangle in Figure B.12 in which $a = 2.00$, $b = 5.00$, and c is unknown. From the Pythagorean theorem, we have

$$c^2 = a^2 + b^2 = 2.00^2 + 5.00^2 = 4.00 + 25.0 = 29.0$$

$$c = \sqrt{29.0} = 5.39$$

To find the angle θ, note that

$$\tan \theta = \frac{a}{b} = \frac{2.00}{5.00} = 0.400$$

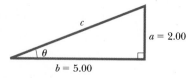

Figure B.12 (Example B.3)

▶ **B.3** c o n t i n u e d

Using a calculator, we find that

$$\theta = \tan^{-1}(0.400) = \boxed{21.8°}$$

where $\tan^{-1}(0.400)$ is the notation for "angle whose tangent is 0.400," sometimes written as arctan (0.400).

Exercises

1. In Figure B.13, identify (a) the side opposite θ (b) the side adjacent to ϕ and then find (c) $\cos\theta$, (d) $\sin\phi$, and (e) $\tan\phi$.

Answers   (a) 3   (b) 3   (c) $\frac{4}{5}$ (d) $\frac{4}{5}$ (e) $\frac{4}{3}$

2. In a certain right triangle, the two sides that are perpendicular to each other are 5.00 m and 7.00 m long. What is the length of the third side?

Answer   8.60 m

3. A right triangle has a hypotenuse of length 3.0 m, and one of its angles is 30°. (a) What is the length of the side opposite the 30° angle? (b) What is the side adjacent to the 30° angle?

Answers   (a) 1.5 m   (b) 2.6 m

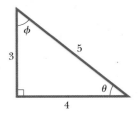

Figure B.13 (Exercise 1)

B.5 Series Expansions

$$(a + b)^n = a^n + \frac{n}{1!}a^{n-1}b + \frac{n(n-1)}{2!}a^{n-2}b^2 + \cdots$$

$$(1 + x)^n = 1 + nx + \frac{n(n-1)}{2!}x^2 + \cdots$$

$$e^x = 1 + x + \frac{x^2}{2!} + \frac{x^3}{3!} + \cdots$$

$$\ln(1 \pm x) = \pm x - \tfrac{1}{2}x^2 \pm \tfrac{1}{3}x^3 - \cdots$$

$$\left.\begin{array}{l}\sin x = x - \dfrac{x^3}{3!} + \dfrac{x^5}{5!} - \cdots \\[2mm] \cos x = 1 - \dfrac{x^2}{2!} + \dfrac{x^4}{4!} - \cdots \\[2mm] \tan x = x + \dfrac{x^3}{3} + \dfrac{2x^5}{15} + \cdots \quad |x| < \dfrac{\pi}{2}\end{array}\right\} x \text{ in radians}$$

For $x \ll 1$, the following approximations can be used:[1]

$$(1 + x)^n \approx 1 + nx \qquad \sin x \approx x$$
$$e^x \approx 1 + x \qquad\quad \cos x \approx 1$$
$$\ln(1 \pm x) \approx \pm x \qquad \tan x \approx x$$

B.6 Differential Calculus

In various branches of science, it is sometimes necessary to use the basic tools of calculus, invented by Newton, to describe physical phenomena. The use of calculus is fundamental in the treatment of various problems in Newtonian mechanics, electricity, and magnetism. In this section, we simply state some basic properties and "rules of thumb" that should be a useful review to the student.

---

[1]The approximations for the functions $\sin x$, $\cos x$, and $\tan x$ are for $x \le 0.1$ rad.

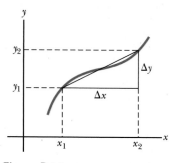

Figure B.14 The lengths Δx and Δy are used to define the derivative of this function at a point.

First, a **function** must be specified that relates one variable to another (e.g., a coordinate as a function of time). Suppose one of the variables is called y (the dependent variable), and the other x (the independent variable). We might have a function relationship such as

$$y(x) = ax^3 + bx^2 + cx + d$$

If a, b, c, and d are specified constants, y can be calculated for any value of x. We usually deal with continuous functions, that is, those for which y varies "smoothly" with x.

The **derivative** of y with respect to x is defined as the limit as Δx approaches zero of the slopes of chords drawn between two points on the y versus x curve. Mathematically, we write this definition as

$$\frac{dy}{dx} = \lim_{\Delta x \to 0} \frac{\Delta y}{\Delta x} = \lim_{\Delta x \to 0} \frac{y(x + \Delta x) - y(x)}{\Delta x} \qquad \text{(B.28)}$$

where Δy and Δx are defined as $\Delta x = x_2 - x_1$ and $\Delta y = y_2 - y_1$ (Fig. B.14). Note that dy/dx *does not* mean dy divided by dx, but rather is simply a notation of the limiting process of the derivative as defined by Equation B.28.

A useful expression to remember when $y(x) = ax^n$, where a is a *constant* and n is *any* positive or negative number (integer or fraction), is

$$\frac{dy}{dx} = nax^{n-1} \qquad \text{(B.29)}$$

If $y(x)$ is a polynomial or algebraic function of x, we apply Equation B.29 to *each* term in the polynomial and take $d[\text{constant}]/dx = 0$. In Examples B.4 through B.7, we evaluate the derivatives of several functions.

Special Properties of the Derivative

A. Derivative of the product of two functions If a function $f(x)$ is given by the product of two functions—say, $g(x)$ and $h(x)$—the derivative of $f(x)$ is defined as

$$\frac{d}{dx}f(x) = \frac{d}{dx}[g(x)h(x)] = g\frac{dh}{dx} + h\frac{dg}{dx} \qquad \text{(B.30)}$$

B. Derivative of the sum of two functions If a function $f(x)$ is equal to the sum of two functions, the derivative of the sum is equal to the sum of the derivatives:

$$\frac{d}{dx}f(x) = \frac{d}{dx}[g(x) + h(x)] = \frac{dg}{dx} + \frac{dh}{dx} \qquad \text{(B.31)}$$

C. Chain rule of differential calculus If $y = f(x)$ and $x = g(z)$, then dy/dz can be written as the product of two derivatives:

$$\frac{dy}{dz} = \frac{dy}{dx}\frac{dx}{dz} \qquad \text{(B.32)}$$

D. The second derivative The second derivative of y with respect to x is defined as the derivative of the function dy/dx (the derivative of the derivative). It is usually written as

$$\frac{d^2y}{dx^2} = \frac{d}{dx}\left(\frac{dy}{dx}\right) \qquad \text{(B.33)}$$

Some of the more commonly used derivatives of functions are listed in Table B.4.

Table B.4   **Derivative for Several Functions**

$$\frac{d}{dx}(a) = 0$$

$$\frac{d}{dx}(ax^n) = nax^{n-1}$$

$$\frac{d}{dx}(e^{ax}) = ae^{ax}$$

$$\frac{d}{dx}(\sin ax) = a\cos ax$$

$$\frac{d}{dx}(\cos ax) = -a\sin ax$$

$$\frac{d}{dx}(\tan ax) = a\sec^2 ax$$

$$\frac{d}{dx}(\cot ax) = -a\csc^2 ax$$

$$\frac{d}{dx}(\sec x) = \tan x \sec x$$

$$\frac{d}{dx}(\csc x) = -\cot x \csc x$$

$$\frac{d}{dx}(\ln ax) = \frac{1}{x}$$

$$\frac{d}{dx}(\sin^{-1} ax) = \frac{a}{\sqrt{1 - a^2x^2}}$$

$$\frac{d}{dx}(\cos^{-1} ax) = \frac{-a}{\sqrt{1 - a^2x^2}}$$

$$\frac{d}{dx}(\tan^{-1} ax) = \frac{a}{1 + a^2x^2}$$

Note: The symbols a and n represent constants.

Example B.4

Suppose $y(x)$ (that is, y as a function of x) is given by

$$y(x) = ax^3 + bx + c$$

where a and b are constants. It follows that

$$y(x + \Delta x) = a(x + \Delta x)^3 + b(x + \Delta x) + c$$
$$= a(x^3 + 3x^2 \Delta x + 3x \Delta x^2 + \Delta x^3) + b(x + \Delta x) + c$$

so

$$\Delta y = y(x + \Delta x) - y(x) = a(3x^2 \Delta x + 3x \Delta x^2 + \Delta x^3) + b\Delta x$$

Substituting this into Equation B.28 gives

$$\frac{dy}{dx} = \lim_{\Delta x \to 0} \frac{\Delta y}{\Delta x} = \lim_{\Delta x \to 0} \left[3ax^2 + 3ax \Delta x + a \Delta x^2\right] + b$$

$$\frac{dy}{dx} = 3ax^2 + b$$

Example B.5

Find the derivative of

$$y(x) = 8x^5 + 4x^3 + 2x + 7$$

SOLUTION

Applying Equation B.29 to each term independently and remembering that d/dx (constant) $= 0$, we have

$$\frac{dy}{dx} = 8(5)x^4 + 4(3)x^2 + 2(1)x^0 + 0$$

$$\frac{dy}{dx} = 40x^4 + 12x^2 + 2$$

Example B.6

Find the derivative of $y(x) = x^3/(x + 1)^2$ with respect to x.

SOLUTION

We can rewrite this function as $y(x) = x^3(x + 1)^{-2}$ and apply Equation B.30:

$$\frac{dy}{dx} = (x + 1)^{-2} \frac{d}{dx}(x^3) + x^3 \frac{d}{dx}(x + 1)^{-2}$$

$$= (x + 1)^{-2} 3x^2 + x^3 (-2)(x + 1)^{-3}$$

$$\frac{dy}{dx} = \frac{3x^2}{(x + 1)^2} - \frac{2x^3}{(x + 1)^3} = \frac{x^2(x + 3)}{(x + 1)^3}$$

Example B.7

A useful formula that follows from Equation B.30 is the derivative of the quotient of two functions. Show that

$$\frac{d}{dx}\left[\frac{g(x)}{h(x)}\right] = \frac{h\dfrac{dg}{dx} - g\dfrac{dh}{dx}}{h^2}$$

SOLUTION

We can write the quotient as gh^{-1} and then apply Equations B.29 and B.30:

$$\frac{d}{dx}\left(\frac{g}{h}\right) = \frac{d}{dx}(gh^{-1}) = g\frac{d}{dx}(h^{-1}) + h^{-1}\frac{d}{dx}(g)$$

$$= -gh^{-2}\frac{dh}{dx} + h^{-1}\frac{dg}{dx}$$

$$= \frac{h\dfrac{dg}{dx} - g\dfrac{dh}{dx}}{h^2}$$

B.7 Integral Calculus

We think of integration as the inverse of differentiation. As an example, consider the expression

$$f(x) = \frac{dy}{dx} = 3ax^2 + b \tag{B.34}$$

which was the result of differentiating the function

$$y(x) = ax^3 + bx + c$$

in Example B.4. We can write Equation B.34 as $dy = f(x)\,dx = (3ax^2 + b)\,dx$ and obtain $y(x)$ by "summing" over all values of x. Mathematically, we write this inverse operation as

$$y(x) = \int f(x)\,dx$$

For the function $f(x)$ given by Equation B.34, we have

$$y(x) = \int (3ax^2 + b)\,dx = ax^3 + bx + c$$

where c is a constant of the integration. This type of integral is called an *indefinite integral* because its value depends on the choice of c.

A general **indefinite integral** $I(x)$ is defined as

$$I(x) = \int f(x)\,dx \tag{B.35}$$

where $f(x)$ is called the *integrand* and $f(x) = dI(x)/dx$.

For a *general continuous* function $f(x)$, the integral can be interpreted geometrically as the area under the curve bounded by $f(x)$ and the x axis, between two specified values of x, say, x_1 and x_2, as in Figure B.15.

The area of the blue element in Figure B.15 is approximately $f(x_i)\,\Delta x_i$. If we sum all these area elements between x_1 and x_2 and take the limit of this sum as $\Delta x_i \to 0$,

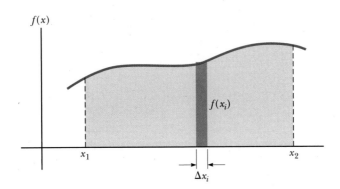

Figure B.15 The definite integral of a function is the area under the curve of the function between the limits x_1 and x_2.

we obtain the *true* area under the curve bounded by $f(x)$ and the x axis, between the limits x_1 and x_2:

$$\text{Area} = \lim_{\Delta x_i \to 0} \sum_i f(x_i)\Delta x_i = \int_{x_1}^{x_2} f(x)\, dx \tag{B.36}$$

Integrals of the type defined by Equation B.36 are called **definite integrals.**

One common integral that arises in practical situations has the form

$$\int x^n\, dx = \frac{x^{n+1}}{n+1} + c \quad (n \neq -1) \tag{B.37}$$

This result is obvious, being that differentiation of the right-hand side with respect to x gives $f(x) = x^n$ directly. If the limits of the integration are known, this integral becomes a *definite integral* and is written

$$\int_{x_1}^{x_2} x^n\, dx = \frac{x^{n+1}}{n+1}\Bigg|_{x_1}^{x_2} = \frac{x_2^{\,n+1} - x_1^{\,n+1}}{n+1} \quad (n \neq -1) \tag{B.38}$$

Examples

1. $\displaystyle\int_0^a x^2\, dx = \frac{x^3}{3}\Bigg]_0^a = \frac{a^3}{3}$

2. $\displaystyle\int_0^b x^{3/2}\, dx = \frac{x^{5/2}}{5/2}\Bigg]_0^b = \tfrac{2}{5}b^{5/2}$

3. $\displaystyle\int_3^5 x\, dx = \frac{x^2}{2}\Bigg]_3^5 = \frac{5^2 - 3^2}{2} = 8$

Partial Integration

Sometimes it is useful to apply the method of *partial integration* (also called "integrating by parts") to evaluate certain integrals. This method uses the property

$$\int u\, dv = uv - \int v\, du \tag{B.39}$$

where u and v are *carefully* chosen so as to reduce a complex integral to a simpler one. In many cases, several reductions have to be made. Consider the function

$$I(x) = \int x^2 e^x\, dx$$

which can be evaluated by integrating by parts twice. First, if we choose $u = x^2$, $v = e^x$, we obtain

$$\int x^2 e^x\, dx = \int x^2\, d(e^x) = x^2 e^x - 2\int e^x x\, dx + c_1$$

Now, in the second term, choose $u = x$, $v = e^x$, which gives

$$\int x^2 e^x \, dx = x^2 e^x - 2x e^x + 2\int e^x \, dx + c_1$$

or

$$\int x^2 e^x \, dx = x^2 e^x - 2xe^x + 2e^x + c_2$$

The Perfect Differential

Another useful method to remember is that of the *perfect differential*, in which we look for a change of variable such that the differential of the function is the differential of the independent variable appearing in the integrand. For example, consider the integral

$$I(x) = \int \cos^2 x \sin x \, dx$$

This integral becomes easy to evaluate if we rewrite the differential as $d(\cos x) = -\sin x \, dx$. The integral then becomes

$$\int \cos^2 x \sin x \, dx = -\int \cos^2 x \, d(\cos x)$$

If we now change variables, letting $y = \cos x$, we obtain

$$\int \cos^2 x \sin x \, dx = -\int y^2 \, dy = -\frac{y^3}{3} + c = -\frac{\cos^3 x}{3} + c$$

Table B.5 lists some useful indefinite integrals. Table B.6 gives Gauss's probability integral and other definite integrals. A more complete list can be found in various handbooks, such as *The Handbook of Chemistry and Physics* (Boca Raton, FL: CRC Press, published annually).

Table B.5   Some Indefinite Integrals (An arbitrary constant should be added to each of these integrals.)

$$\int x^n \, dx = \frac{x^{n+1}}{n+1} \text{ (provided } n \neq 1)$$

$$\int \ln ax \, dx = (x \ln ax) - x$$

$$\int \frac{dx}{x} = \int x^{-1} \, dx = \ln x$$

$$\int xe^{ax} \, dx = \frac{e^{ax}}{a^2}(ax - 1)$$

$$\int \frac{dx}{a + bx} = \frac{1}{b}\ln(a + bx)$$

$$\int \frac{dx}{a + be^{cx}} = \frac{x}{a} - \frac{1}{ac}\ln(a + be^{cx})$$

$$\int \frac{x \, dx}{a + bx} = \frac{x}{b} - \frac{a}{b^2}\ln(a + bx)$$

$$\int \sin ax \, dx = -\frac{1}{a}\cos ax$$

$$\int \frac{dx}{x(x + a)} = -\frac{1}{a}\ln\frac{x + a}{x}$$

$$\int \cos ax \, dx = \frac{1}{a}\sin ax$$

$$\int \frac{dx}{(a + bx)^2} = -\frac{1}{b(a + bx)}$$

$$\int \tan ax \, dx = -\frac{1}{a}\ln(\cos ax) = \frac{1}{a}\ln(\sec ax)$$

$$\int \frac{dx}{a^2 + x^2} = \frac{1}{a}\tan^{-1}\frac{x}{a}$$

$$\int \cot ax \, dx = \frac{1}{a}\ln(\sin ax)$$

$$\int \frac{dx}{a^2 - x^2} = \frac{1}{2a}\ln\frac{a + x}{a - x}\,(a^2 - x^2 > 0)$$

$$\int \sec ax \, dx = \frac{1}{a}\ln(\sec ax + \tan ax) = \frac{1}{a}\ln\left[\tan\left(\frac{ax}{2} + \frac{\pi}{4}\right)\right]$$

$$\int \frac{dx}{x^2 - a^2} = \frac{1}{2a}\ln\frac{x - a}{x + a}\,(x^2 - a^2 > 0)$$

$$\int \csc ax \, dx = \frac{1}{a}\ln(\csc ax - \cot ax) = \frac{1}{a}\ln\left(\tan\frac{ax}{2}\right)$$

(*Continued*)

| Table B.5 | **Some Indefinite Integrals** *(continued)* |

$$\int \frac{x\,dx}{a^2 \pm x^2} = \pm\tfrac{1}{2}\ln(a^2 \pm x^2)$$

$$\int \frac{dx}{\sqrt{a^2 - x^2}} = \sin^{-1}\frac{x}{a} = -\cos^{-1}\frac{x}{a}\;(a^2 - x^2 > 0)$$

$$\int \frac{dx}{\sqrt{x^2 \pm a^2}} = \ln(x + \sqrt{x^2 \pm a^2})$$

$$\int \frac{x\,dx}{\sqrt{a^2 - x^2}} = -\sqrt{a^2 - x^2}$$

$$\int \frac{x\,dx}{\sqrt{x^2 \pm a^2}} = \sqrt{x^2 \pm a^2}$$

$$\int \sqrt{a^2 - x^2}\,dx = \tfrac{1}{2}\left(x\sqrt{a^2 - x^2} + a^2\sin^{-1}\frac{x}{|a|}\right)$$

$$\int x\sqrt{a^2 - x^2}\,dx = -\tfrac{1}{3}(a^2 - x^2)^{3/2}$$

$$\int \sqrt{x^2 \pm a^2}\,dx = \tfrac{1}{2}x\sqrt{x^2 \pm a^2} \pm a^2\ln(x + \sqrt{x^2 \pm a^2})$$

$$\int x(\sqrt{x^2 \pm a^2})\,dx = \tfrac{1}{3}(x^2 \pm a^2)^{3/2}$$

$$\int e^{ax}\,dx = \frac{1}{a}e^{ax}$$

$$\int \sin^2 ax\,dx = \frac{x}{2} - \frac{\sin 2ax}{4a}$$

$$\int \cos^2 ax\,dx = \frac{x}{2} + \frac{\sin 2ax}{4a}$$

$$\int \frac{dx}{\sin^2 ax} = -\frac{1}{a}\cot ax$$

$$\int \frac{dx}{\cos^2 ax} = \frac{1}{a}\tan ax$$

$$\int \tan^2 ax\,dx = \frac{1}{a}(\tan ax) - x$$

$$\int \cot^2 ax\,dx = -\frac{1}{a}(\cot ax) - x$$

$$\int \sin^{-1} ax\,dx = x(\sin^{-1} ax) + \frac{\sqrt{1 - a^2 x^2}}{a}$$

$$\int \cos^{-1} ax\,dx = x(\cos^{-1} ax) - \frac{\sqrt{1 - a^2 x^2}}{a}$$

$$\int \frac{dx}{(x^2 + a^2)^{3/2}} = \frac{x}{a^2\sqrt{x^2 + a^2}}$$

$$\int \frac{x\,dx}{(x^2 + a^2)^{3/2}} = -\frac{1}{\sqrt{x^2 + a^2}}$$

| Table B.6 | **Gauss's Probability Integral and Other Definite Integrals** |

$$\int_0^\infty x^n e^{-ax}\,dx = \frac{n!}{a^{n+1}}$$

$$I_0 = \int_0^\infty e^{-ax^2}\,dx = \frac{1}{2}\sqrt{\frac{\pi}{a}} \quad \text{(Gauss's probability integral)}$$

$$I_1 = \int_0^\infty xe^{-ax^2}\,dx = \frac{1}{2a}$$

$$I_2 = \int_0^\infty x^2 e^{-ax^2}\,dx = -\frac{dI_0}{da} = \frac{1}{4}\sqrt{\frac{\pi}{a^3}}$$

$$I_3 = \int_0^\infty x^3 e^{-ax^2}\,dx = -\frac{dI_1}{da} = \frac{1}{2a^2}$$

$$I_4 = \int_0^\infty x^4 e^{-ax^2}\,dx = \frac{d^2 I_0}{da^2} = \frac{3}{8}\sqrt{\frac{\pi}{a^5}}$$

$$I_5 = \int_0^\infty x^5 e^{-ax^2}\,dx = \frac{d^2 I_1}{da^2} = \frac{1}{a^3}$$

$$\vdots$$

$$I_{2n} = (-1)^n \frac{d^n}{da^n} I_0$$

$$I_{2n+1} = (-1)^n \frac{d^n}{da^n} I_1$$

B.8 Propagation of Uncertainty

In laboratory experiments, a common activity is to take measurements that act as raw data. These measurements are of several types—length, time interval, temperature, voltage, and so on—and are taken by a variety of instruments. Regardless of the measurement and the quality of the instrumentation, **there is always uncertainty associated with a physical measurement.** This uncertainty is a combination of that associated with the instrument and that related to the system being measured. An example of the former is the inability to exactly determine the position of a length measurement between the lines on a meterstick. An example of uncertainty related to the system being measured is the variation of temperature within a sample of water so that a single temperature for the sample is difficult to determine.

Uncertainties can be expressed in two ways. **Absolute uncertainty** refers to an uncertainty expressed in the same units as the measurement. Therefore, the length of a computer disk label might be expressed as (5.5 ± 0.1) cm. The uncertainty of ± 0.1 cm by itself is not descriptive enough for some purposes, however. This uncertainty is large if the measurement is 1.0 cm, but it is small if the measurement is 100 m. To give a more descriptive account of the uncertainty, **fractional uncertainty** or **percent uncertainty** is used. In this type of description, the uncertainty is divided by the actual measurement. Therefore, the length of the computer disk label could be expressed as

$$\ell = 5.5 \text{ cm} \pm \frac{0.1 \text{ cm}}{5.5 \text{ cm}} = 5.5 \text{ cm} \pm 0.018 \quad \text{(fractional uncertainty)}$$

or as

$$\ell = 5.5 \text{ cm} \pm 1.8\% \quad \text{(percent uncertainty)}$$

When combining measurements in a calculation, the percent uncertainty in the final result is generally larger than the uncertainty in the individual measurements. This is called **propagation of uncertainty** and is one of the challenges of experimental physics.

Some simple rules can provide a reasonable estimate of the uncertainty in a calculated result:

Multiplication and division: When measurements with uncertainties are multiplied or divided, add the *percent uncertainties* to obtain the percent uncertainty in the result.

Example: The Area of a Rectangular Plate

$$A = \ell w = (5.5 \text{ cm} \pm 1.8\%) \times (6.4 \text{ cm} \pm 1.6\%) = 35 \text{ cm}^2 \pm 3.4\%$$

$$= (35 \pm 1) \text{ cm}^2$$

Addition and subtraction: When measurements with uncertainties are added or subtracted, add the *absolute uncertainties* to obtain the absolute uncertainty in the result.

Example: A Change in Temperature

$$\Delta T = T_2 - T_1 = (99.2 \pm 1.5)°\text{C} - (27.6 \pm 1.5)°\text{C} = (71.6 \pm 3.0)°\text{C}$$

$$= 71.6°\text{C} \pm 4.2\%$$

Powers: If a measurement is taken to a power, the percent uncertainty is multiplied by that power to obtain the percent uncertainty in the result.

Example: The Volume of a Sphere

$$V = \tfrac{4}{3}\pi r^3 = \tfrac{4}{3}\pi (6.20 \text{ cm} \pm 2.0\%)^3 = 998 \text{ cm}^3 \pm 6.0\%$$

$$= (998 \pm 60) \text{ cm}^3$$

For complicated calculations, many uncertainties are added together, which can cause the uncertainty in the final result to be undesirably large. Experiments should be designed such that calculations are as simple as possible.

Notice that uncertainties in a calculation always add. As a result, an experiment involving a subtraction should be avoided if possible, especially if the measurements being subtracted are close together. The result of such a calculation is a small difference in the measurements and uncertainties that add together. It is possible that the uncertainty in the result could be larger than the result itself!

Periodic Table of the Elements

Group I	Group II				Transition elements				
H 1 1.007 9 $1s$									
Li 3 6.941 $2s^1$	**Be** 4 9.0122 $2s^2$								
Na 11 22.990 $3s^1$	**Mg** 12 24.305 $3s^2$								

Symbol — **Ca** 20 — Atomic number
Atomic mass† — 40.078
$4s^2$ — Electron configuration

K 19 39.098 $4s^1$	**Ca** 20 40.078 $4s^2$	**Sc** 21 44.956 $3d^14s^2$	**Ti** 22 47.867 $3d^24s^2$	**V** 23 50.942 $3d^34s^2$	**Cr** 24 51.996 $3d^54s^1$	**Mn** 25 54.938 $3d^54s^2$	**Fe** 26 55.845 $3d^64s^2$	**Co** 27 58.933 $3d^74s^2$	
Rb 37 85.468 $5s^1$	**Sr** 38 87.62 $5s^2$	**Y** 39 88.906 $4d^15s^2$	**Zr** 40 91.224 $4d^25s^2$	**Nb** 41 92.906 $4d^45s^1$	**Mo** 42 95.94 $4d^55s^1$	**Tc** 43 (98) $4d^55s^2$	**Ru** 44 101.07 $4d^75s^1$	**Rh** 45 102.91 $4d^85s^1$	
Cs 55 132.91 $6s^1$	**Ba** 56 137.33 $6s^2$	57–71*	**Hf** 72 178.49 $5d^26s^2$	**Ta** 73 180.95 $5d^36s^2$	**W** 74 183.84 $5d^46s^2$	**Re** 75 186.21 $5d^56s^2$	**Os** 76 190.23 $5d^66s^2$	**Ir** 77 192.2 $5d^76s^2$	
Fr 87 (223) $7s^1$	**Ra** 88 (226) $7s^2$	89–103**	**Rf** 104 (261) $6d^27s^2$	**Db** 105 (262) $6d^37s^2$	**Sg** 106 (266)	**Bh** 107 (264)	**Hs** 108 (277)	**Mt** 109 (268)	

*Lanthanide series

La 57 138.91 $5d^16s^2$	**Ce** 58 140.12 $5d^14f^16s^2$	**Pr** 59 140.91 $4f^36s^2$	**Nd** 60 144.24 $4f^46s^2$	**Pm** 61 (145) $4f^56s^2$	**Sm** 62 150.36 $4f^66s^2$
Ac 89 (227) $6d^17s^2$	**Th** 90 232.04 $6d^27s^2$	**Pa** 91 231.04 $5f^26d^17s^2$	**U** 92 238.03 $5f^36d^17s^2$	**Np** 93 (237) $5f^46d^17s^2$	**Pu** 94 (244) $5f^67s^2$

**Actinide series

Note: Atomic mass values given are averaged over isotopes in the percentages in which they exist in nature.
†For an unstable element, mass number of the most stable known isotope is given in parentheses.

		Group III	Group IV	Group V	Group VI	Group VII	Group 0
						H 1 1.007 9 $1s^1$	**He** 2 4.002 6 $1s^2$
		B 5 10.811 $2p^1$	**C** 6 12.011 $2p^2$	**N** 7 14.007 $2p^3$	**O** 8 15.999 $2p^4$	**F** 9 18.998 $2p^5$	**Ne** 10 20.180 $2p^6$
		Al 13 26.982 $3p^1$	**Si** 14 28.086 $3p^2$	**P** 15 30.974 $3p^3$	**S** 16 32.066 $3p^4$	**Cl** 17 35.453 $3p^5$	**Ar** 18 39.948 $3p^6$

Ni 28 58.693 $3d^84s^2$	**Cu** 29 63.546 $3d^{10}4s^1$	**Zn** 30 65.41 $3d^{10}4s^2$	**Ga** 31 69.723 $4p^1$	**Ge** 32 72.64 $4p^2$	**As** 33 74.922 $4p^3$	**Se** 34 78.96 $4p^4$	**Br** 35 79.904 $4p^5$	**Kr** 36 83.80 $4p^6$
Pd 46 106.42 $4d^{10}$	**Ag** 47 107.87 $4d^{10}5s^1$	**Cd** 48 112.41 $4d^{10}5s^2$	**In** 49 114.82 $5p^1$	**Sn** 50 118.71 $5p^2$	**Sb** 51 121.76 $5p^3$	**Te** 52 127.60 $5p^4$	**I** 53 126.90 $5p^5$	**Xe** 54 131.29 $5p^6$
Pt 78 195.08 $5d^96s^1$	**Au** 79 196.97 $5d^{10}6s^1$	**Hg** 80 200.59 $5d^{10}6s^2$	**Tl** 81 204.38 $6p^1$	**Pb** 82 207.2 $6p^2$	**Bi** 83 208.98 $6p^3$	**Po** 84 (209) $6p^4$	**At** 85 (210) $6p^5$	**Rn** 86 (222) $6p^6$
Ds 110 (271)	**Rg** 111 (272)	**Cn** 112 (285)	113†† (284)	**Fl** 114 (289)	115†† (288)	**Lv** 116 (293)	117†† (294)	118†† (294)

Eu 63 151.96 $4f^76s^2$	**Gd** 64 157.25 $4f^75d^16s^2$	**Tb** 65 158.93 $4f^85d^16s^2$	**Dy** 66 162.50 $4f^{10}6s^2$	**Ho** 67 164.93 $4f^{11}6s^2$	**Er** 68 167.26 $4f^{12}6s^2$	**Tm** 69 168.93 $4f^{13}6s^2$	**Yb** 70 173.04 $4f^{14}6s^2$	**Lu** 71 174.97 $4f^{14}5d^16s^2$
Am 95 (243) $5f^77s^2$	**Cm** 96 (247) $5f^76d^17s^2$	**Bk** 97 (247) $5f^86d^17s^2$	**Cf** 98 (251) $5f^{10}7s^2$	**Es** 99 (252) $5f^{11}7s^2$	**Fm** 100 (257) $5f^{12}7s^2$	**Md** 101 (258) $5f^{13}7s^2$	**No** 102 (259) $5f^{14}7s^2$	**Lr** 103 (262) $5f^{14}6d^17s^2$

†† Elements 113, 115, 117, and 118 have not yet been officially named. Only small numbers of atoms of these elements have been observed.
Note: For a description of the atomic data, visit *physics.nist.gov/PhysRefData/Elements/per_text.html*.

Table D.1   SI Units

	SI Base Unit	
Base Quantity	Name	Symbol
Length	meter	m
Mass	kilogram	kg
Time	second	s
Electric current	ampere	A
Temperature	kelvin	K
Amount of substance	mole	mol
Luminous intensity	candela	cd

Table D.2   Some Derived SI Units

Other Quantity	Name	Symbol	Expression in Terms of Base Units	Expression in Terms of SI Units
Plane angle	radian	rad	m/m	
Frequency	hertz	Hz	s^{-1}	
Force	newton	N	$kg \cdot m/s^2$	J/m
Pressure	pascal	Pa	$kg/m \cdot s^2$	N/m^2
Energy	joule	J	$kg \cdot m^2/s^2$	$N \cdot m$
Power	watt	W	$kg \cdot m^2/s^3$	J/s
Electric charge	coulomb	C	$A \cdot s$	
Electric potential	volt	V	$kg \cdot m^2/A \cdot s^3$	W/A
Capacitance	farad	F	$A^2 \cdot s^4/kg \cdot m^2$	C/V
Electric resistance	ohm	Ω	$kg \cdot m^2/A^2 \cdot s^3$	V/A
Magnetic flux	weber	Wb	$kg \cdot m^2/A \cdot s^2$	$V \cdot s$
Magnetic field	tesla	T	$kg/A \cdot s^2$	
Inductance	henry	H	$kg \cdot m^2/A^2 \cdot s^2$	$T \cdot m^2/A$

Answers to Quick Quizzes

Chapter 1

1. (a)
2. False
3. (b)

Chapter 2

1. (c)
2. (b)
3. False. Your graph should look something like the one shown below. This v_x–t graph shows that the maximum speed is about 5.0 m/s, which is 18 km/h (= 11 mi/h), so the driver was not speeding.

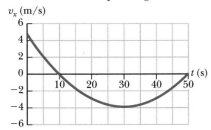

4. (b)
5. (c)
6. (a)–(e), (b)–(d), (c)–(f)
7. (i) (e)   (ii) (d)

Chapter 3

1. vectors: (b), (c); scalars: (a), (d), (e)
2. (c)
3. (b) and (c)
4. (b)
5. (c)

Chapter 4

1. (a)
2. (i) (b)   (ii) (a)
3. 15°, 30°, 45°, 60°, 75°
4. (i) (d)   (ii) (b)
5. (i) (b)   (ii) (d)

Chapter 5

1. (d)
2. (a)
3. (d)
4. (b)
5. (i) (c)   (ii) (a)
6. (b)

7. (b) Pulling up on the rope decreases the normal force, which, in turn, decreases the force of kinetic friction.

Chapter 6

1. (i) (a)   (ii) (b)
2. (i) Because the speed is constant, the only direction the force can have is that of the centripetal acceleration. The force is larger at Ⓒ than at Ⓐ because the radius at Ⓒ is smaller. There is no force at Ⓑ because the wire is straight.   (ii) In addition to the forces in the centripetal direction in part (a), there are now tangential forces to provide the tangential acceleration. The tangential force is the same at all three points because the tangential acceleration is constant.

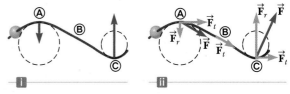

3. (c)
4. (a)

Chapter 7

1. (a)
2. (c), (a), (d), (b)
3. (d)
4. (a)
5. (b)
6. (c)
7. (i) (c)   (ii) (a)
8. (d)

Chapter 8

1. (a) For the television set, energy enters by electrical transmission (through the power cord). Energy leaves by heat (from hot surfaces into the air), mechanical waves (sound from the speaker), and electromagnetic radiation (from the screen).   (b) For the gasoline-powered lawn mower, energy enters by matter transfer (gasoline). Energy leaves by work (on the blades of grass), mechanical waves (sound), and heat (from hot surfaces into the air).   (c) For the hand-cranked pencil sharpener, energy enters by work (from your hand turning the crank). Energy leaves by work (done on the pencil), mechanical waves (sound), and heat due to the temperature increase from friction.
2. (i) (b)   (ii) (b)   (iii) (a)

A-25

3. (a)
4. $v_1 = v_2 = v_3$
5. (c)

Chapter 9

1. (d)
2. (b), (c), (a)
3. (i) (c), (e)   (ii) (b), (d)
4. (a) All three are the same. (b) dashboard, seat belt, air bag
5. (a)
6. (b)
7. (b)
8. (i) (a)   (ii) (b)

Chapter 10

1. (i) (c)   (ii) (b)
2. (b)
3. (i) (b)   (ii) (a)
4. (i) (b)   (ii) (a)
5. (b)
6. (a)
7. (b)

Chapter 11

1. (d)
2. (i) (a)   (ii) (c)
3. (b)
4. (a)

Chapter 12

1. (a)
2. (b)
3. (b)
4. (i) (b)   (ii) (a)   (iii) (c)

Chapter 13

1. (e)
2. (c)
3. (a)
4. (a) Perihelion (b) Aphelion (c) Perihelion (d) All points

Chapter 14

1. (a)
2. (a)
3. (c)
4. (b) or (c)
5. (a)

Chapter 15

1. (d)
2. (f)
3. (a)
4. (b)
5. (c)
6. (i) (a)   (ii) (a)

Chapter 16

1. (i) (b)   (ii) (a)
2. (i) (c)   (ii) (b)   (iii) (d)
3. (c)
4. (f) and (h)
5. (d)

Chapter 17

1. (c)
2. (b)
3. (b)
4. (e)
5. (e)
6. (b)

Chapter 18

1. (c)
2. (i) (a)   (ii) (d)
3. (d)
4. (b)
5. (c)

Chapter 19

1. (c)
2. (c)
3. (c)
4. (c)
5. (a)
6. (b)

Chapter 20

1. (i) iron, glass, water (ii) water, glass, iron
2. The figure below shows a graphical representation of the internal energy of the system as a function of energy added. Notice that this graph looks quite different from Figure 20.3 in that it doesn't have the flat portions during the phase changes. Regardless of how the temperature is varying in Figure 20.3, the internal energy of the system simply increases linearly with energy input; the line in the graph below has a slope of 1.

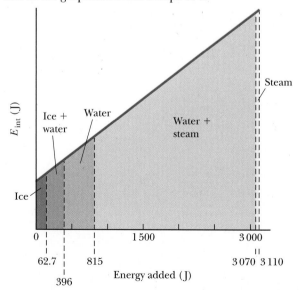

3.

Situation	System	Q	W	ΔE_{int}
(a) Rapidly pumping up a bicycle tire	Air in the pump	0	+	+
(b) Pan of room-temperature water sitting on a hot stove	Water in the pan	+	0	+
(c) Air quickly leaking out of a balloon	Air originally in the balloon	0	−	−

4. Path A is isovolumetric, path B is adiabatic, path C is isothermal, and path D is isobaric.

5. (b)

Chapter 21

1. (i) (b)   (ii) (a)
2. (i) (a)   (ii) (c)
3. (d)
4. (c)

Chapter 22

1. (i) (c)   (ii) (b)
2. (d)
3. C, B, A
4. (a) one   (b) six
5. (a)
6. false (The adiabatic process must be *reversible* for the entropy change to be equal to zero.)

Chapter 23

1. (a), (c), (e)
2. (e)
3. (b)
4. (a)
5. A, B, C

Chapter 24

1. (e)
2. (b) and (d)
3. (a)

Chapter 25

1. (i) (b)   (ii) (a)
2. Ⓑ to Ⓒ, Ⓒ to Ⓓ, Ⓐ to Ⓑ, Ⓓ to Ⓔ
3. (i) (c)   (ii) (a)
4. (i) (a)   (ii) (a)

Chapter 26

1. (d)
2. (a)
3. (a)
4. (b)
5. (a)

Chapter 27

1. (a) > (b) = (c) > (d)
2. (b)
3. (b)
4. (a)
5. $I_a = I_b > I_c = I_d > I_e = I_f$

Chapter 28

1. (a)
2. (b)
3. (a)
4. (i) (b)   (ii) (a)   (iii) (a)   (iv) (b)
5. (i) (c)   (ii) (d)

Chapter 29

1. (e)
2. (i) (b)   (ii) (a)
3. (c)
4. (i) (c), (b), (a)   (ii) (a) = (b) = (c)

Chapter 30

1. $B > C > A$
2. (a)
3. $c > a > d > b$
4. $a = c = d > b = 0$
5. (c)

Chapter 31

1. (c)
2. (c)
3. (b)
4. (a)
5. (b)

Chapter 32

1. (c), (f)
2. (i) (b)   (ii) (a)
3. (a), (d)
4. (a)
5. (i) (b)   (ii) (c)

Chapter 33

1. (i) (c)   (ii) (b)
2. (b)
3. (a)
4. (b)
5. (a) $X_L < X_C$ (b) $X_L = X_C$ (c) $X_L > X_C$
6. (c)
7. (c)

Chapter 34

1. (i) (b)   (ii) (c)
2. (c)
3. (c)
4. (b)
5. (a)
6. (c)
7. (a)

Chapter 35

1. (d)
2. Beams ② and ④ are reflected; beams ③ and ⑤ are refracted.
3. (c)
4. (c)
5. (i) (b)   (ii) (b)

Chapter 36

1. false
2. (b)
3. (b)
4. (d)
5. (a)
6. (b)
7. (a)
8. (c)

Chapter 37

1. (c)
2. The graph is shown here. The width of the primary maxima is slightly narrower than the $N = 5$ primary width but wider than the $N = 10$ primary width. Because $N = 6$, the secondary maxima are $\frac{1}{36}$ as intense as the primary maxima.

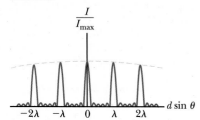

3. (a)

Chapter 38

1. (a)
2. (i)
3. (b)
4. (a)
5. (c)
6. (b)
7. (c)

Chapter 39

1. (c)
2. (d)
3. (d)
4. (a)
5. (a)
6. (c)
7. (d)
8. (i) (c)   (ii) (a)
9. (a) $m_3 > m_2 = m_1$ (b) $K_3 = K_2 > K_1$ (c) $u_2 > u_3 = u_1$

Chapter 40

1. (b)
2. Sodium light, microwaves, FM radio, AM radio.
3. (c)
4. The classical expectation (which did not match the experiment) yields a graph like the following drawing:

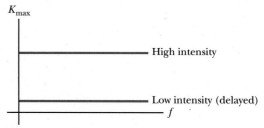

5. (d)
6. (c)
7. (b)
8. (a)

Chapter 41

1. (d)
2. (i) (a)   (ii) (d)
3. (c)
4. (a), (c), (f)

Chapter 42

1. (c)
2. (a)
3. (b)
4. (a) five   (b) nine
5. (c)
6. true

Chapter 43

1. (a) van der Waals   (b) ionic   (c) hydrogen   (d) covalent
2. (c)
3. (a)
4. A: semiconductor; B: conductor; C: insulator

Chapter 44

1. (i) (b)   (ii) (a)   (iii) (c)
2. (e)
3. (b)
4. (c)

Chapter 45

1. (b)
2. (a), (b)
3. (a)
4. (d)

Chapter 46

1. (a)
2. (i) (c), (d)   (ii) (a)
3. (b), (e), (f)
4. (b), (e)
5.

$S = 0$ d   u

$S = -1$ s

$Q = -\frac{1}{3}$ $Q = +\frac{2}{3}$

6. false

Index

Some Physical Constants

Quantity	Symbol	Value[a]
Atomic mass unit	u	$1.660\ 538\ 782\ (83) \times 10^{-27}$ kg
		$931.494\ 028\ (23)$ MeV/c^2
Avogadro's number	N_A	$6.022\ 141\ 79\ (30) \times 10^{23}$ particles/mol
Bohr magneton	$\mu_B = \dfrac{e\hbar}{2m_e}$	$9.274\ 009\ 15\ (23) \times 10^{-24}$ J/T
Bohr radius	$a_0 = \dfrac{\hbar^2}{m_e e^2 k_e}$	$5.291\ 772\ 085\ 9\ (36) \times 10^{-11}$ m
Boltzmann's constant	$k_B = \dfrac{R}{N_A}$	$1.380\ 650\ 4\ (24) \times 10^{-23}$ J/K
Compton wavelength	$\lambda_C = \dfrac{h}{m_e c}$	$2.426\ 310\ 217\ 5\ (33) \times 10^{-12}$ m
Coulomb constant	$k_e = \dfrac{1}{4\pi\epsilon_0}$	$8.987\ 551\ 788\ldots \times 10^9$ N·m^2/C^2 (exact)
Deuteron mass	m_d	$3.343\ 583\ 20\ (17) \times 10^{-27}$ kg
		$2.013\ 553\ 212\ 724\ (78)$ u
Electron mass	m_e	$9.109\ 382\ 15\ (45) \times 10^{-31}$ kg
		$5.485\ 799\ 094\ 3\ (23) \times 10^{-4}$ u
		$0.510\ 998\ 910\ (13)$ MeV/c^2
Electron volt	eV	$1.602\ 176\ 487\ (40) \times 10^{-19}$ J
Elementary charge	e	$1.602\ 176\ 487\ (40) \times 10^{-19}$ C
Gas constant	R	$8.314\ 472\ (15)$ J/mol·K
Gravitational constant	G	$6.674\ 28\ (67) \times 10^{-11}$ N·m^2/kg^2
Neutron mass	m_n	$1.674\ 927\ 211\ (84) \times 10^{-27}$ kg
		$1.008\ 664\ 915\ 97\ (43)$ u
		$939.565\ 346\ (23)$ MeV/c^2
Nuclear magneton	$\mu_n = \dfrac{e\hbar}{2m_p}$	$5.050\ 783\ 24\ (13) \times 10^{-27}$ J/T
Permeability of free space	μ_0	$4\pi \times 10^{-7}$ T·m/A (exact)
Permittivity of free space	$\epsilon_0 = \dfrac{1}{\mu_0 c^2}$	$8.854\ 187\ 817\ldots \times 10^{-12}$ C^2/N·m^2 (exact)
Planck's constant	h	$6.626\ 068\ 96\ (33) \times 10^{-34}$ J·s
	$\hbar = \dfrac{h}{2\pi}$	$1.054\ 571\ 628\ (53) \times 10^{-34}$ J·s
Proton mass	m_p	$1.672\ 621\ 637\ (83) \times 10^{-27}$ kg
		$1.007\ 276\ 466\ 77\ (10)$ u
		$938.272\ 013\ (23)$ MeV/c^2
Rydberg constant	R_H	$1.097\ 373\ 156\ 852\ 7\ (73) \times 10^7$ m^{-1}
Speed of light in vacuum	c	$2.997\ 924\ 58 \times 10^8$ m/s (exact)

Note: These constants are the values recommended in 2006 by CODATA, based on a least-squares adjustment of data from different measurements. For a more complete list, see P. J. Mohr, B. N. Taylor, and D. B. Newell, "CODATA Recommended Values of the Fundamental Physical Constants: 2006." *Rev. Mod. Phys.* **80:**2, 633–730, 2008.

[a]The numbers in parentheses for the values represent the uncertainties of the last two digits.

Solar System Data

Body	Mass (kg)	Mean Radius (m)	Period (s)	Mean Distance from the Sun (m)
Mercury	3.30×10^{23}	2.44×10^6	7.60×10^6	5.79×10^{10}
Venus	4.87×10^{24}	6.05×10^6	1.94×10^7	1.08×10^{11}
Earth	5.97×10^{24}	6.37×10^6	3.156×10^7	1.496×10^{11}
Mars	6.42×10^{23}	3.39×10^6	5.94×10^7	2.28×10^{11}
Jupiter	1.90×10^{27}	6.99×10^7	3.74×10^8	7.78×10^{11}
Saturn	5.68×10^{26}	5.82×10^7	9.29×10^8	1.43×10^{12}
Uranus	8.68×10^{25}	2.54×10^7	2.65×10^9	2.87×10^{12}
Neptune	1.02×10^{26}	2.46×10^7	5.18×10^9	4.50×10^{12}
Pluto[a]	1.25×10^{22}	1.20×10^6	7.82×10^9	5.91×10^{12}
Moon	7.35×10^{22}	1.74×10^6	—	—
Sun	1.989×10^{30}	6.96×10^8	—	—

[a]In August 2006, the International Astronomical Union adopted a definition of a planet that separates Pluto from the other eight planets. Pluto is now defined as a "dwarf planet" (like the asteroid Ceres).

Physical Data Often Used

Average Earth–Moon distance	3.84×10^8 m
Average Earth–Sun distance	1.496×10^{11} m
Average radius of the Earth	6.37×10^6 m
Density of air (20°C and 1 atm)	1.20 kg/m^3
Density of air (0°C and 1 atm)	1.29 kg/m^3
Density of water (20°C and 1 atm)	1.00×10^3 kg/m^3
Free-fall acceleration	9.80 m/s^2
Mass of the Earth	5.97×10^{24} kg
Mass of the Moon	7.35×10^{22} kg
Mass of the Sun	1.99×10^{30} kg
Standard atmospheric pressure	1.013×10^5 Pa

Note: These values are the ones used in the text.

Some Prefixes for Powers of Ten

Power	Prefix	Abbreviation	Power	Prefix	Abbreviation
10^{-24}	yocto	y	10^1	deka	da
10^{-21}	zepto	z	10^2	hecto	h
10^{-18}	atto	a	10^3	kilo	k
10^{-15}	femto	f	10^6	mega	M
10^{-12}	pico	p	10^9	giga	G
10^{-9}	nano	n	10^{12}	tera	T
10^{-6}	micro	μ	10^{15}	peta	P
10^{-3}	milli	m	10^{18}	exa	E
10^{-2}	centi	c	10^{21}	zetta	Z
10^{-1}	deci	d	10^{24}	yotta	Y

Standard Abbreviations and Symbols for Units

Symbol	Unit	Symbol	Unit
A	ampere	K	kelvin
u	atomic mass unit	kg	kilogram
atm	atmosphere	kmol	kilomole
Btu	British thermal unit	L	liter
C	coulomb	lb	pound
°C	degree Celsius	ly	light-year
cal	calorie	m	meter
d	day	min	minute
eV	electron volt	mol	mole
°F	degree Fahrenheit	N	newton
F	farad	Pa	pascal
ft	foot	rad	radian
G	gauss	rev	revolution
g	gram	s	second
H	henry	T	tesla
h	hour	V	volt
hp	horsepower	W	watt
Hz	hertz	Wb	weber
in.	inch	yr	year
J	joule	Ω	ohm

Mathematical Symbols Used in the Text and Their Meaning

Symbol	Meaning
$=$	is equal to
$\equiv$	is defined as
$\neq$	is not equal to
$\propto$	is proportional to
$\sim$	is on the order of
$>$	is greater than
$<$	is less than
$\gg (\ll)$	is much greater (less) than
$\approx$	is approximately equal to
Δx	the change in x
$\displaystyle\sum_{i=1}^{N} x_i$	the sum of all quantities x_i from $i = 1$ to $i = N$
$\lvert x \rvert$	the absolute value of x (always a nonnegative quantity)
$\Delta x \rightarrow 0$	Δx approaches zero
$\dfrac{dx}{dt}$	the derivative of x with respect to t
$\dfrac{\partial x}{\partial t}$	the partial derivative of x with respect to t
$\displaystyle\int$	integral